Granddad's Wonderful Book of Chemistry

By Kurt Saxon

CONTENTS

Library of Congress Catalogue Number 87-072414
ISBN 0-9602258-0-6
Seventh Printing, May, 1991
Printed in U. S. A.

8 1/2 X 11, 480 Pages - $18.00

Money Making & Saving; Hundreds of ideas for Cottage Industry, Trades, Hobbies. Cut Food Bills; Never Hunger. Burglar-Proof Your Home. Surviving a Nuclear Winter. Make Soy Milk, Sprouts, Tofu. Toy Making for Fun and Profit. Making Imitation Gold & Silver. Make a 1 or 2 Horsepower Windmill from Scrap. Foot-Powered Lathe. Make See-Through Mirrors. Black Powder; How to Make the Best. Set Up a Home Foundry. Build Concrete Furniture.

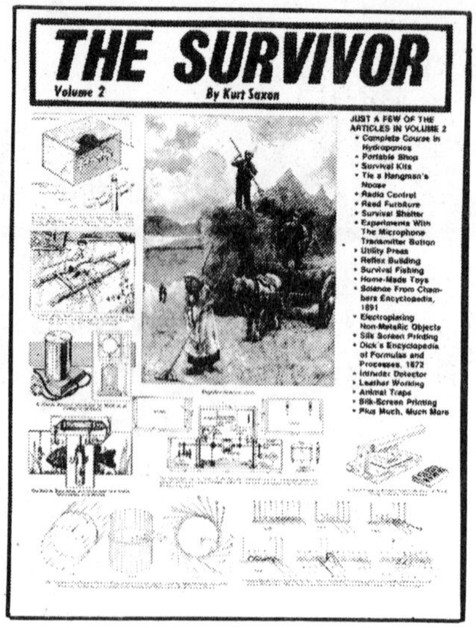

8½ x 11, 464 Pages — $18.00

More trades, hobbies, crafts, 19th century science. This volume also completes the total reprinting of Dick's 1872 Encyclopedia of Formulas and Processes. Full instructions in the trades of Dying, Soap-Making, Liquors, Perfumery, Cosmetics, Cements, Ceramic Glazes, Paints, Metalurgy, Electroplating, Etc. Also in this volume: THE TANKS ARE COMING — complete course in soilless growing, inside or outside. GUIDE TO RADIO CONTROL.

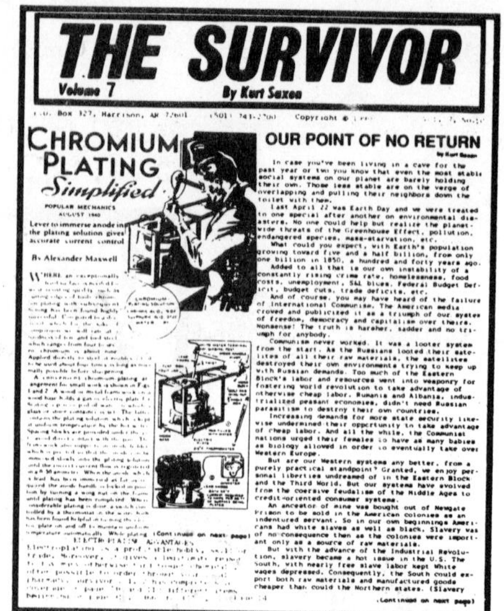

THE SURVIVOR, 12 ISSUES - $10.00

THE SURVIVOR is the first publication preparing its readers to survive the total collapse of our economy and social system. THE SURVIVOR is not an alarmist publication, nor does it give false hope to those who believe they can ride out the coming calamities by "wise investing." Instead, THE SURVIVOR shows you how our great grandparents provided for their own needs and those of their neighbors. With THE SURVIVOR you can, without buying commercial "survival" food and products, protect your home, feed your family and be of such value to your neighbors that they will defend you, rather than loot you. New subscribers get all available back issues with any book order. Also, subscribers get 10% off on all Atlan books, tapes and products.

8 1/2 X 11, 477 Pages - $18.00

(All other editions hopelessly obsolete.) The ultimate armed civilian. This first of a set of three contains the original PMJB. Revised, Corrected, Enlarged; plus Full Auto Conversion of the AR-15, Mark 1, Mini-14, and Sten Gun. FIREWORKS & EXPLOSIVES LIVE GRANDDAD USED TO MAKE, EXPLOSIVES, MATCHES & FIREWORKS. PYROTECHNY-Weingart. AMERICAN PYROTECHNIST. WE SHALL FIGHT IN THE STREETS. ARSON BY ELECTRONICS. U.S. MARINES & ARMY HAND-TO-HAND COMBAT. Plus, making Potassium Cyanide, Fire Grenades & The Ultimate Booby Trap.

8½ x 11, 484 Pages — $18.00

Five Classics in one volume! THE CHEMISTRY OF POWDER & EXPLOSIVES, Tenny Davis, 1943. AMERICAN JIU-JITSU, 1942. CHEMICALS IN WAR, 1937. THE POOR MAN'S ARMORER. THE ARMORER, long out of print, but finally back with a whole arsenal of improvised weaponry not in any other books. Homemade Bazooka, Silencers, Booby Traps, Bolas, Concealed Weaponry, Spear Gun, Smoke/Gas Grenades, Mines, Caltrops, Ingenious Zip Gun, Do-it-yourself Machine Pistol, Knife-Throwing Techniques, plus, ARMY IMPROVISED MUNITIONS HAND-BOOK.

8½ x 11, 412 Pages — $18.00

THE WEAPONEER: Poison Gas Grenades, Wallet Pistol, Killer Darts, Poisons, Bombs, Explosives, Stenches. Includes Ricin, the deadliest organic poison known and its simple manufacture. THE GUNSMITH'S MANUAL, 1883. Repair or reconstruct 19th Century guns. Hand forge iron & steel for gun parts & knives. Case hardening, tempering & annealing. SILENCERS FROM THE HOME WORKSHOP, Bill Holmes. KNIGHT'S MECHANICAL DICTIONARY. Military ordnance from 1884 back. BOOBY TRAPS 1965. Make your territory impregnable.

OUTSIDE U.S. SEND BANK OR POSTAL MONEY ORDERS MADE OUT IN U.S. FUNDS

Item	Amount	Order From	
The Survivor — Volume 1 .... **$18**		Visa/Mastercard Accepted	Name
The Survivor — Volume 2 .... **$18**		$2.00 Postage Any Size Order	
Subscription **$10**		Order From:	Address
The Poor Man's James Bond **$18**		**ATLAN "R" FORMULARIES**	
The Poor Man's James Bond, Vol. 2 ..... **$18**		P.O. Box 327	City State
The Poor Man's James Bond, Vol. 3 ..... **$18**		Harrison, AR 72601	
Arkansans Add 6% Sales Tax		(501) 743-2700	Zip **Date**
Total Enclosed			

DICTIONARY OF OLD-FASHIONED TERMS AND SYNONYMS

The numbers following many of the terms indicate that the compound is described in the Dick's Formulary section beginning on page 221. If the compound is not in that section it is in another book of Atlan Formularies in its representative section.

A

Abelmosk, abelmusk: A bushy herb (Abelmoschus moschatus), native to tropical Asia and the East Indies, whose musky seeds are used in perfumery and to flavor coffee.
Ablution: Act of washing or cleansing.
Abortifacient: A drug used to cause abortion.
Absinthe: Wormwood. A green alcoholic liquor derived from wormwood. Large and small absinthe (797) probably refers to grades of wormwood chips.
Absolute alcohol: Pure, with no water. See 1441.
Acescent: To turn sour.
Acetaldehyde: 4308 CH_3CHO Aldehyde.
Acetate: An ester or salt of acetic acid.
Acetate of alumina: $Al_2(C_2H_3O_2)_3$ Aluminum acetate.
Acetate of ammonia, ammonium: See ammonium acetate.
Acetate of amyl: See amyl acetate.
Acetate of baryta: See barium acetate.
Acetate of cobalt: 4253 $Co(C_2H_3O_2)_2$ Cobalt acetate.
Acetate of copper: See cupric acetate.
Acetate of ethyl: See ethyl acetate.
Acetate of iron: See ferrous acetate.
Acetate of lead: See lead acetate.
Acetate of lime: 4244 See calcium acetate.
Acetate of manganese: See manganous acetate.
Acetate of morphia: 4267 $C_{17}H_{19}O_3N{+}H_2O$ Morphine.
Acetate of oxide of ethyl: See ethyl acetate.
Acetate of potash, potassa, potassium: See potassium acetate.
Acetate of protoxide of manganese: See manganous acetate.
Acetate of soda: See sodium acetate.
Acetate of zinc: See zinc acetate.
Acetic acid: 3889-99 $HC_2H_3O_2$ Acetulic acid; hydrated acetic acid; hydric acetate; hydrogen acetate; pyroligneous acid. When free from water, even though it is still a liquid, it is called : acetic hydrate; acid of vinegar; glacial acetic acid (3891-2); monohydrated acetic acid; radical vinegar.
Acetic ether: See ethyl acetate.
Acetic hydrate: See acetic acid.
Acetification: To change into vinegar or acetic acid.
Acetimetry, acetometry: 69, Method of determining amount of acetic acid present in a substance.
Acetous fermentation: Runaway fermentation which results when fermenting cider or wine turns to vinegar.
Acetulic acid: See acetic acid.
Acetum: Vinegar.
Acicular: Shaped like needles.
Acid carbonate of ammonia: See ammonium bicarbonate.
Acidimeter, acidimetry: 78, 82 Acid measurements.
Acid of vinegar: See acetic acid.
Acid of potassium carbonate: See potassium bicarbonate.
Acid of potassium sulphate: See potassium bisulphate.
Acid of tartrate of potassa: See potassium bitartrate.
Acidulate: To make acidulous or to sour.
Acidulated water: Water with acid in it.
Acidulous: Sour.
Aconite: From the Aconitum napellus. Monk's hood; wolf bane. Roots yield a famous alkaloidal poison and sedative.
Adamantine spar: See aluminum oxide.
Admixture: Something added and mixed in.
Aerated soda, water: 4431 Water charged with carbonic acid gas.
Affinity: Has attraction to and mixes well with.
Affusion: Flooding with water or other liquid.
Agar agar: A gelatinous extract of seaweed, used for bacterial cultures, laxtives, etc. Dried in sheets, it is called seaweed isinglass.
Agaric: Any gill fungus, as the common mushroom.
Agglutination: Clumping together without mixing.
Ague: Prolonged fever and/or chills.
Albuminized paper: 3178 Paper treated with albumin.
Albuminous: Like or containing albumin or egg white.
Alcohol solution of potassa, potassium: KOH, A solution of alcohol and potassium hydroxide.
Alcoholate: A salt in which alcohol appears to replace the water of crystallization, as is the case of certain chlorides, nitrates, etc., as chloral alcoholate, $CCl_3CHO.C_2H_6O$. Some of them may be formed by simple solution and crystallization of the salt in alcohol.
Alcoholmetry (See 53), alcoholometry: The method of finding the amount of pure alcohol in alcoholic liquids.
Aldehyde: See acetaldehyde.
Alembic: A still.
Aleppo galls: Hard, brittle, spherical bodies, resembling hickory nuts produced on the twigs of Quercus infectoria by Cynips tinctoria.
Alizarin, alizarine: $C_{14}H_6O_2(OH)_2$ A dye. Formerly prepared from madder and now produced artificially from anthracine, and forming, when pure, a reddish-yellow powder or orange-red crystals.
Alkali: Any base or hydroxide, as soda, potash, etc., that is soluble in water and can neutralize acid. Turns red litmus paper blue.
Alkalimeter: See 82 An instrument to show the strength of alkalies, or the quantity of alkali in a mixture.
Alkaline antimoniate: A general term for any salt of antimonic acid.
Alkanet: A European plant (Alkanna tinctoria); also its root. Also a red dyestuff prepared from alkanet root and used to color tinctures, pomades, beverages, etc.
Alloxan: A crystalline oxidation product, $C_4H_4N_2O_5$. A substance formed by the action of nitric acid on uric acid.
Alloy: A metal that is a mixture of two or more metals, or of a metal and something else.
Alteratives: 5161 Medicines which alter the patient's condition.
Althaea paper: 4428 A test paper from hollyhock flowers.
Alum: 4256 $KAl(SO_4)_2.12H_2O$, Potassium aluminum sulphate.
Alumina: See aluminum oxide.
Alumine: See aluminum oxide.
Aluminized charcoal: 4314 A decolorizing agent.
Aluminous clay: Clay containing aluminum oxide. Kaolin is an example.
Aluminum chloride: Al_2Cl_6 Chloralum (impure); chloride of aluminum; sesquichloride of aluminum.
Aluminum hydrate: 4257 $Al_2(HO)_6$ Aluminum hydroxide; hydrated alumina.
Aluminum oxide: Al_2O_3 Adamantine spar; alumina; alumine; corundum; emery.
Aluminum sulphate: 4259 $Al_2(SO_4)_3.18H_2O$ Alunogen; (cake or concentrated alum, erroneous) feather alum; hair salt; neutral sulphate of alumina; sesquisulphate of alumina.
Alum plumb: A dying mordant using alum added to a decoction of logwood. Plumb probably refers to "a sediment that settles out like lead".

Alunogen: See aluminum sulphate.

Amalgam: Any alloy of mercury with another metal or metals.

Amandine: 1117-8 A cosmetic skin softener.

Amber: A yellow or brownish-yellow translucent fossil resin.

Ambergris: A grayish, waxy substance, secreted by sperm whales and found floating in tropical seas. Used in making perfumes.

Amber seed: The seed of the abelmosk, somewhat resembling millet and having a musky flavor. Musk seed.

Ambretta, ambrette: An extract of amber seed (abelmosk), used as a fixative in the manufacture of perfumes.

Amenorrhea: Any drug used to promote menstrual discharge.

American isinglass: Isinglass made from local fish. See isinglass.

American pearlash: Locally refined potassium carbonate.

Amianthus: Asbestos.

Ammonia: 3981, 4067 NH_3 Ammonia gas; ammoniacal gas; anhydrous ammonia; volatile air; volatile alkali; terhydride of nitrogen.

Ammoniac: The aromatic gum resin of the ammoniac plant. It is sold in the form of yellowish tears or lumps, and has a bittersweet, somewhat nauseous and acrid taste. Used as an expectorant and in the formulation of certain plasters.

Ammoniacal gas: See ammonia.

Ammoniacal liquor: Impure ammonia water obtained as a by-product in the distillation of coal, tar, bones, etc.

Ammoniacal lye: 1716 A mild bleaching solution formed of five parts of river water and one part stale urine.

Ammoniacal secret salt of glauber: See ammonium sulphate.

Ammoniacal sulphate of copper: See ammonio-sulphate of copper.

Ammoniac chloride muriate of ammonia: See ammonium chloride.

Ammonia gas: See ammonia.

Ammonia normal carbonate: See ammomium carbonate.

Ammonia, solution of: Ammonia; ammonia water; ammonium hydrate; aqua ammonia; liquor of ammonia; spirits of hartshorn; water of ammonia.

Ammoniated copper: A solution of cupric ions in ammonia.

Ammoniated tincture: Any tincture with ammonia added.

Ammonia water: See ammonia, solution of.

Ammonic: Read as ammonium. Ammonic chloride=ammonium chloride, etc.

Ammonio-chloride: A double salt of a chloride dissolved in an aqueous solution of ammonium chloride.

Ammonio-chloride of mercury: See mercuric ammonium chloride.

Ammonio-chloride of zinc: 4110.

Ammonio-citrate of iron: See ammonio-ferric citrate.

Ammonio-ferric alum: See ammonio-ferric sulphate.

Ammonio-ferric citrate: 4162 $(NH_4)_3Fe(C_6H_5O_7)_2$ Ammonio-citrate of iron; citrate of iron and ammonium; ferric and ammonium citrate.

Ammonio-ferric sulphate: 4738 $Fe_2(NH_4)_2(SO_4)_4.23H_2O$ Ammonio-ferric alum; sulphate of iron and ammonium.

Ammonio-hydric carbonate: See ammonium bicarbonate.

Ammonio-nitrate of copper: 4067.

Ammonio-nitrate of silver: $Ag(NH_3)_2{+}NO_3$.

Ammonio-pyrophosphate of iron: 4737 An elixer.

Ammonio-sulphate of copper: 4090 $(N_2H_8Cu)SO_4$ Ammoniacal sulphate of copper.

Ammonio-sulphate of iron: See ammonio-ferric sulphate.

Ammonium acetate: 4218 $(NH_4)C_2H_3O_2$ Acetate of ammonium; spirit of minderus.

Ammonium and hydrogen carbonate: See ammonium bicarbonate.

Ammonium bicarbonate: 4221 $H(NH_4)CO_3$ Acid carbonate of ammonium; ammonio-hydric carbonate; ammonium and hydrogen carbonate; bicarbonate of ammonia; mono-ammoniac carbonate.

Ammonium bichromate: $(NH_4)_2Cr_2O_7$ Bichromate of ammonia.

Ammonium carbonate: 4219 $(NH_4)_2CO_3$ Ammonia normal carbonate; carbonate of oxide of ammonium; di-ammonic carbonate; neutral carbonate of ammonium; sal volatile; volatile salt.

Ammonium chloride: 4222 NH_4Cl Ammoniac chloride muriate of ammonia; ammonic chloride; hydrochlorate of ammonia; muriate of ammonia; sal ammoniac; salmiac.

Ammonium hydrate: See ammonia, solution of.

Ammonium iodide: 4225 NH_4I Hydroiodate of ammonia.

Ammonium nitrate: $(NH_4)NO_3$ Nitrate of ammonia; nitrate of oxide of ammonia; nitrous ammoniacal salt.

Ammonium oxalate: $(NH_4)_2C_2O_4$ Ammonic oxalate; oxalate of ammonia.

Ammonium sesquicarbonate: 4220 $(NH_4)_4H_2(CO_3)_3.H_2O$ Half-acid carbonate of ammonia; preston salts; smelling salts; tetra ammonio-dihydric carbonate; volatile spirits of hartshorn.

Ammonium sulphate: 4223 $(NH_4)_2SO_4$ Ammoniacal secret salt of glauber; glauber's secret sal ammoniac; glauber's secret salt; sulphate of ammonia; sulphate of ammonia; sulphate of oxide of ammonia.

Ammonium sulphide: 1203, 4228 $(NH_4)_2S$ Ammonic sulphide; hydrosulphuret of ammonia; sulphuret of ammonia.

Ammoniuret: An ammoniate; a compound with ammonia or ammonium.

Ammoniuret of gold: 2132, 3725 Fulminate of gold.

Amorphous sulphur: 4350 Sulphur melted for molding.

Amygdaline: Pertaining to or resembling an almond or almonds.

Amyl acetate: 4302 $CH_3CO_2C_5H_{11}$ Banana oil; pearl oil.

Amyl valeriate: 1469, 4303 $C_4H_9COOC_5H_{11}$ Apple oil; valerianate of amyl.

Angelica: Angelica archangelica, whose root stalks are candied and whose roots and fruit furnish the angelica oil used as a flavoring for liqueurs and as a perfume.

Anglesite: See lead sulphate.

Angostura: An aromatic, bitter bark used as a tonic and fever reducer taken from a South American tree (Cusparia angostura) of the family rutaceae.

Anhydrous: Without water. Having no water of crystallization.

Anhydrous acetic acid: 3894.

Anhydrous alcohol: See absolute alcohol.

Anhydrous alumina: Al_2O_3 Aluminum oxide with no water in its composition.

Anhydrous ammonia: See ammonia.

Anhydrous bromide of sodium: $NaBr$ Anhydrous sodium bromide.

Anhydrous carbonate of soda: Na_2CO_3 Anhydrous sodium carbonate.

Anhydrous protoxide of tin: 4119 SnO Stannous oxide.

Anhydrous salts: Compounds of acids and bases without water of crystallization.

Anhydrous sesquioxide of iron: 4153 Fe_2O_3.

Aniline: An oily, poisonous, basic liquid $C_6H_5NH_2$. Colorless when pure. Obtainable by the distillation of indigo, coal tar, etc.

Animal black: Bone charcoal.

Anime: Any of various resins or oleoresins.

Annatto, annotto, arnotto: A red or yellowish-red dyestuff prepared from the pulp surrounding the seeds of the annatto tree. Used for coloring oils, butter, etc.

Anneal: See 2. To heat (glass, metals, etc.) and then to cool slowly to prevent brittleness. To temper.

Anodyne: 5130 Anything that relieves or lessens pain.

Anthelmentic: Expelling or destroying intestinal worms. An anthelmentic remedy; vermifuge.

Anthos seed: The seed of the herb, rosemary.

Antichlor: $Na_2S_2O_3.5H_2O$ Sodium thiosulphate.

Antimonial wine: Wine of antimony. A solution of tartar emetic in sherry wine.

Antimoniate of potash: Calcined antimony.

Antimonic acid: 3339 Acid left from the action of nitric acid on antimony.

Antimonious chloride: See antimony chloride.

Antimonious oxide: See antimony trioxide.

Antimonious sulphide: See antimony sulphide, black.

Antimony bloom: See antimony trioxide.

Antimony chloride: 4131 $SbCl_3$ Antimonious chloride; butter of antimony; caustic antimony; chloride of antimony; sesquicloride of antimony; terchloride of antimony; trichloride of antimony.

Antimony glance: See antimony sulphide, black.

Antimony oxide: 4127 Sb_2O_3 Flowers of antimony; white oxide of antimony.

Antimony pentasulphide: 4133 Sb_2S_5 Golden sulphuret of antimony; kermes mineral; oxysulphuret of antimony; pentasulphuret of antimony; precipitated sulphide of antimony; schlippe's salt; sulphurated antimony.

Antimony sulphide, black: 4132 Sb_2S_3 Antimonius sulphide; antimony glance; antimony trisulphide; crude antimony; gray antimony; sesquisulphuret of antimony; stibnite antimonite; sulphide of antimony; sulphuret of antimony; tersulphuret of antimony.

Antimony tartrate: 4129 $KSbOC_4H_4O_6.H_2O$ Emetic tartar; potasso-tartrate of antimony; tartarated antimony; tartar emetic; tartarized antimony.

Antimony trioxide: 4130 Sb_2O_3 Antimonious oxide; hypantimonious acid; oxide of antimony; protoxide of antimony; teroxide of antimony; valentinite.

Antimony trisulphide: See antimony sulphide, black.

Antipyretic: A medicine for relieving fever.

Aperient: Laxative.

Apple oil: See amyl valerianate.

Aqua ammonia: $NH_4OH{+}H_2O$ Ammonium hydroxide.

Aqua fortis: See nitric acid.

Aqua regia: See nitrohydrochloric acid.

Aqueous: Of or like water; watery.

Aqueous infusion: The liquid extract that results when a substance is soaked in water.

Archil: A blue, red or violet dyestuff obtained from several species of lichen such as Roccella tinctoria, R. fuciformis, and Lecanora tartarea, by action of air and ammonia. Cudbear.

Areca nut: Betel nut. The charcoal of this nut was favored as an ingredient of toothpaste.

Areometer: See 64. A hydrometer. A floating instrument for determining specific gravity, especially of liquids, and thence the strength of alcoholic liquors, saline solutions, etc.

Argal, argol: 4197 A grayish or reddish crystalline crust in wine casks; crude tartar; unprocessed potassium bitartrate.

Argand lamp: See 13, fig. 1. A lamp with a tubular wick which admits a current of air inside as well as outside of the flame.

Argentic: Read as silver. Argentic chloride=silver chloride, etc.

Argentiferous: Producing or containing silver, as ore.

Argentine flowers: See antimony oxide.

Argentite: See silver sulphide.

Armenian bole: A soft, clayey, bright-red earth found in Armenia, Tuscany, etc. Used as a coloring material and, formerly, in medicine.

Arnica: The rhizome and roots of the herb Arnica montana, used for the stimulant and local irritant effect, especially in the form of the tincture, as an embrocation for bruises, swellings, sprains, etc; hence, the tincture.

Aromatic: A plant, drug or medicine characterized by a fragrant smell, and

usually by a warm, pungent taste, as ginger, cinnamon, spices, etc.
Aromatic spirits, distilled: 941 An essence with a little water added.
Arrack: A strong Oriental liquor made from rice or molasses.
Arsenamine: See arsenic hydride.
Arsenate, arseniate: 3938 H_3AsO_4 A salt or ester of arsenic acid.
Arsenate of copper: See copper arsenate.
Arsenate of potash: See potassium arsenate.
Arsenate of silver: See silver arsenate.
Arsenate of soda: See sodium arsenate.
Arsenic: 3935 As_2O_3 Arsenic trioxide; arsenious acid; arsenious anhydride; arsenolite; white arsenic.
Arsenic acid: 3938-9 $H_3AsO_4 . 1/2H_2O$ Any of three acids derived from arsenic pentoxide, As_2O_5.
Arsenic bisulphide: See arsenic disulphide.
Arsenic disulphide: As_2S_2 Arsenic bisulphide; realgar; red sulphide of arsenic; red sulphuret of arsenic; sulphide of arsenic; sulphuret of arsenic.
Arsenic hydride: AsH_3 Arsenamine; arseneted hydrogen; arseniureted hydrogen; hydride of arsenic.
Arsenic trichloride: $AsCl_3$ Arsenious chloride; chloride of arsenic; fuming liquor of arsenic; sesquichloride of arsenic; terchloride of arsenic.
Arsenic trioxide: See arsenic.
Arsenic trisulphide: 4356 As_2S_3 Arsenious sulphide; king's yellow; orpiment; sesquisulphide of arsenic; tersulphide of arsenic; tersulphuret of arsenic; yellow sulphuret of arsenic.
Arseneted hydrogen: See Arsenic hydride.
Arsenious acid: See arsenic.
Arsenious anhydride: See arsenic.
Arsenious chloride: See arsenic trichloride.
Arsenious sulphide: See arsenic trisulphide.
Arsenite of copper: See copper arsenite.
Arseniureted hydrogen: See arsenic hydride.
Arsenolite: See arsenic.
Artificial oil of bitter almonds: See nitro-benzene.
Asafetida, asafoetida: A bad-smelling gum resin obtained from various Asiatic plants of the carrot family and used as an antispasmodic.
Asarabacca: Any plant of the genus Asarum.
Asbestos: $Mg_3Si_2O_7 . 2H_2O$ Magnesium silicate.
Asbolite: See cobalt oxide.
Aseptin: 1616 A preservative for milk, meat, etc.
Asphalt: A brown or black tarlike variety of bitumen obtained by evaporating petroleum.
Asphaltum: Asphalt.
Asafoetida: See asafetida.
Astringent: 5555 Any substance which contracts body tissue and blood vessels, checking the flow of blood. Styptic.
Atrophia: See atropine.
Atropine: $C_{17}H_{23}NO_3$ A poisonous white crystalline alkaloid extracted from the belladonna and other plants of the family Solanaceae. It is used to relieve spasms, to diminish secretions, to relieve pain and to dilate the pupil of the eye.
Attar: A perfume made from the petals of flowers.
Auric chloride: 4075 $AuCl_3$ Perchloride of gold; terchloride of gold.
Aurin, aurine: $C_{19}H_{14}O_3$ A poisonous red synthetic dye derived from triphenylmethane; called also rosalic acid and pararosalic acid.
Aurochloride of sodium: $Na(AuCl_4) . 4H_2O$.
Aurous chloride: 4075 AuCl Gold chloride; gold monochloride; protochloride of gold.
Aurous oxide: Au_2O gold monoxide; gold oxide.
Avoirdupois: 5935.
Azotic acid: See nitric acid.
Azurite: $2CuCO_3 . Cu(OH)_2$ Blue basic carbonate of copper; a natural ore of copper. It has been used as a pigment but is not very permanent. Stone blue.

B

Badigeon: 2158 A cement used by artists to cover holes and defects.
Bain marie: See 5. A water bath or double boiler.
Baize: A coarse woolen cloth used to cover pool tables, etc.
Baldwin's phosphorous: 4334 A luminous substance.
Ball licorice: Pure licorice prepared in the form of balls.
Balm: An aromatic gum resin obtained from certain trees and plants and used as medicine. Balsam. Any fragrant oil or ointment.
Balm of Gilead: An aromatic ointment prepared from a small Asiatic and African tree of the myrrh family. Balm of Mecca.
Balm of Mecca: See balm of Gilead.
Balsam: 5090 to 5116 See balm.
Balsam of copaiba: See copaiba.
Balsam of Peru: A reddish-brown syrupy balsam, obtained from a tropical American tree (Myroxylon pereirae), used as a digestive tonic and expectorant and in the treatment of ulcers, etc.
Balsam of sulphur: 5114 A boiled mixture of sulphur and olive oil used as a medicine and as an ingredient in a metallic glaze.
Balsam of tolu: A fragrant reddish-brown or yellowish-brown semisolid or solid balsam, obtained from a South American tree (Myroxylon balsamum). It is used as a stimulating expectorant, an antiseptic and a flavoring for cough syrups.
Banana oil: $CH_3CO_2C_5H_{11}$ Amyl acetate. Used as flavoring, solvent, etc.
Banca, banka tin: Tin of high purity from an East Indian island.
Bandoline: A sticky, perfumed hair dressing.
Baric: Read as barium. Baric acetate=barium acetate, etc.
Barilla: An impure sodium carbonate made from the ashes of Salola soda and other salt water plants. Formerly much used in making soap, glass, etc.
Barite: See barium sulphate.
Barium acetate: 4223 $Ba(C_2H_3O_2)$ Acetate of baryta.
Barium chloride: 4234 $BaCl_2 . 2H_2O$ Baric chloride; chloride of barium.
Barium dioxide: See barium peroxide.
Barium monoxide: See barium oxide.
Barium monosulphide: See barium sulphide.
Barium nitrate: 4230 $Ba(NO_3)_2$ Nitrate of baryta.
Barium oxide: 3985 BaO Barium monoxide; baryta; caustic baryta; oxide of barium; protoxide of barium.
Barium peroxide: 4236 BaO_2 Barium dioxide; binoxide of barium; deutoxide of barium; hyperoxide of barium.
Barium sulphate: 4231 $BaSO_4$ Barite; heavy spar; sulphate of baryta.
Barium sulphide: 4237 BaS Baric sulphide; barium monosulphide; sulphuret of baryta.
Bark: See tanbark.
Barm: Yeast formed on brewing liquors.
Barwood: A hard, red dyewood of tropical Africa, from a tree of the genus Lingoum, but supposed by some to be from Baphia nitida. Camwood.
Baryta: See barium oxide.
Barytes: See barium sulphate.
Base: A substance which forms a salt when it reacts with an acid.
Basic lead citrate: See tribasic lead acetate.
Basic mercuric sulphate: See mercurous sulphate.
Basic nitrate of bismuth: See bismuth subnitrate.
Basilicon: 4964 An ointment.
Bath: See 3. In the general nature of a double boiler with water, sand, oil, etc. being in the bottom part and the substance to be bathed in the top part.
Bath brick: An unbaked brick of silaceous material used to scour and polish metals.
Bath of chalk: See chalk bath.
Batting: Vigorous churning.
Baum nerval: 5113 A brand of nerve tonic.
Baum's hydrometer: See 61, 63 (page 131) See Hydrometer.
Bay: An evergreen tree with glossy, leathery leaves. Laurel tree. Myrtle wood.
Bay salt: A very coarse grained variety of common salt, originally obtained from sea water.
Beef's gall: Gall from cattle. See gall.
Beladonna: A poisonous plant with reddish bell-shaped flowers and black berries; deadly nightshade. A drug, also called atropine, obtained from this plant and used to dilate the pupil of the eye, stimulate the heart, relieve spasms, etc.
Ben: The seed of any species of Moringa. The oil expressed from it, called oil of ben, is used for extracting perfumes and for lubricating delicate machinery.
Bengal catechu: See catechu.
Benjamin: See benzoic acid.
Benzene: 4321 C_6H_6 A volatile and inflamable hydrocarbon, formed by distilling benzoic acid with lime, and by the action of heat on various organic substances. Also called benzoin; hydride of phenyl; phenyl hydride.
Benzoate: A salt or ester of benzoic acid.
Benzoated: Mixed or treated with benzoic acid or gum benzoin.
Benzoated tincture: A tincture of gum benzoin.
Benzoate of ether, ethyl: See ethyl benzoate.
Benzoate of soda: $NaC_7H_5O_2$ Sodium benzoate.
Benzoatic fat: 1516 A method of preserving animal fats with benzoin.
Benzoic acid: 3942-3 $C_6H_5CO_2H$ An acid, crystallizing in the form of light, white, satiny flakes, which occurs in benzoin and other resins, in cranberries and (combined) in the urine of herbivorous animals. Commercially it is prepared from toluene. It is used in medicine as an antiseptic, in dye and perfume manufacture, and as a preservative of foods, etc. Also called flowers of benjamin or benzoin; hydrate of benzoyl; salt of benzoin.
Benzoic ether: See ethyl benzoate.
Benzoin: A balsamic resin obtained from Styrax benzoin and from other species of Styrax. It appears commercially as hard, brittle tears or masses having a fragrant odor and slightly aromatic taste, and is used in making benzoic acid, as a stimulant and expectorant, and as a perfume and incense. Called also gum benzoin; benjamin, gum benjamin; asa dulcis.
Benzol, benzole: See benzine.
Bergamot: A kind of orange (Citrus bergamia) having a pear-shaped fruit whose rind yields an essential oil much used in perfumery. Also the essence or perfume made from bergamot fruit.
Bibulous paper: Readily absorbing fluids, as blotting paper.
Bicarbonate of ammonia: See ammonium bicarbonate.
Bicarbonate of potassa: See potassium bicarbonate.
Bicarbonate of soda: See sodium bicarbonate.
Bicarburet of hydrogen: 4321 $CH_2:CH_2$ Ethylene.
Bichloride of copper: See cuprous chloride.

Bichloride of gold: See aurous chloride.
Bichloride of mercury: See mercuric chloride.
Bichloride of platinum: See platinous chloride.
Bichloride of tin: See stannis chloride.
Bichromate of ammonia: See ammonium bichromate.
Bichromate of potash, potassa: See potassium dichromate.
Bicyanide of mercury: See mercuric cyanide.
Bigarade: The bitter orange; Citrus aurantium.
Bihydrate of lime: See Calcium sulphate.
Bile: The bitter, greenish fluid secreted by the liver and found in the gall bladder.
Billet: A wooden club.
Binacetate of ammonia: Probably ammonium acetate.
Binacetate of copper: See copper acetate.
Binoxalate of potass: See potassium oxalate.
Binoxide of barium: See barium peroxide.
Binoxide of hydrogen: See hydrogen peroxide.
Binoxide of manganese: See manganese dioxide.
Binoxide of mercury: See mercuric oxide.
Binoxide of tin: See stannic oxide.
Birck's alkalimeter: See 82. See alkalimeter.
Bird pepper: A pepper (Capsicum baccatum) having very small, oblong, red fruits, which are among the most pungent of all red peppers.
Biscuit ware: Unglazed pottery or porçelain.
Bismuth nitrate: 4134-5 $Bi(NO_3)_3.5H_2O$ Neutral nitrate of bismuth; nitrate of bismuth; ternitrate of bismuth.
Bismuthous oxide: See bismuth trioxide.
Bismuth subnitrate: 4135 $BiONO_3$ Basic nitrate of bismuth; flake white; magistry of bismuth; pearl white; trisnitrate of bismuth.
Bismuth trioxide: 4136 Bi_2O_3 Bismouthous oxide; flowers of bismuth; oxide of bismuth; protoxide of bismuth; teroxide of bismuth.
Bistre: 2692 A brown color prepared from the root of beech-wood.
Bisulphate of potash, potassa: See potassium bisulphate.
Bisulphate of soda: See sodium bisulphate.
Bisulphide: Same as disulphide, usually.
Bisulphide of carbon: See carbon disulphide.
Bisulphide of mercury: See mercuric sulphide.
Bisulphite of potash: See potassium hydrogen sulphite.
Bisulphite of sodium: See sodium hydrogen sulphite.
Bisulphuret of carbon: See carbon disulphide.
Bisulphuret of iron: See iron bi or disulphide.
Bisulphuret of tin: SnS_2 Stannic sulphide.
Bitartrate of potassa: See potassium bitartrate.
Bitter almond: A variety (Amygdalus communus amara) of the common almond. Also, its extremely bitter seed.
Bitter apple, cucumber, gourd: See colocynth.
Bittern: The bitter mother liquor that remains in saltworks after the salt has crystallized out. From it are obtained magnesium chloride and sulphate, sodium sulphate and bromides and iodides.
Bitumen: Any of several substances obtained as asphaltic residue in the distillation of coal tar, petroleum, etc., or occuring as natural asphalt.
Black antimony: The mineral stibnite, especially when prepared as a black powder, as for medicinal use.
Black antimony sulphide: See antimony sulphide, black.
Black cohosh: The bugbane (Cimicifuga racemosa); black snake root.
Black jack: See zinc sulphide.
Black lead: See graphite.
Black liquor: See protoacetate of iron.
Black oxide of cobalt: See cobalt oxide.
Black oxide of copper: See cupric oxide.
Black oxide of manganese: See manganous peroxide.
Black oxide of mercury: See mercurous oxide.
Black pitch: Asphalt.
Black precipitate: See mercurous oxide.
Black resin: Common resin - dark.
Black sulphur: See sulphur vivum.
Black sulphuret of antimony: See antimony sulphide, black.
Black sulphuret of mercury: See mercuric sulphide.
Bladder: 6239 Animal bladders used as containers and as pliable covers for pots, etc., as we use plastic wrap.
Blanch: To remove vegetable skins by scalding.
Bland medium: Neutral; neither caustic nor acid.
Blanquette: A crude soda made from the ashes of marine plants.
Blast lamp: A lamp provided with some arrangement for intensifying combustion by means of a blast.
Bleaching liquor: 104 A solution of calcium hypochlorite.
Bleaching powder: See calcium hypochlorite.
Blistered steel: Crude steel formed from wrought iron by cementation (see 3274) - so called from its blistered surface.
Block tin: Solid tin as opposed to tin plate.
Blond: A silk bobbin lace originally the color of raw silk or blond hair, but now usually bleached white or dyed black.
Blood cement: 2172 Bullock's blood thickened with calcium oxide.
Blood heat: 98.6° F.
Blood root: A North American plant of the poppy family, (Sanguinaria canadensis). It has acrid emetic properties and the rootstock is used as a stimulant expectorant.
Bloom sugar: See 1368. A stage in preparing sugar for making candies.
Blowing of flowers: To bloom.
Blue, bluing: A blue liquid or powder, usually of indigo, used in rinsing white fabrics to prevent yellowing.
Blue copperas: See cupric sulphate.
Blue galls: See aleppo galls.
Blue stone: See cupric sulphate.
Blue verditer: 2688 Azurite blue.
Blue vitriol: See cupric sulphate.
Bodkin: A dagger.
Boiled oil: 2725 An oil whose drying properties have been increased by heating, usually with dryers, as lead and manganese oxides or salts. Boiled linseed oil is used in paints, varnishes, printing inks, linoleum, etc.
Bolted: Any substance sifted through a cloth or sieve.
Bone ash: See calcium phosphate.
Bone dust: Bone meal.
Bone fat: The fatty matter in bones, extracted by means of heat or solvents, and used chiefly in candles and cheap soap. It is drab to dark-brown and usually contains free fatty acids, lime soaps, calcium phosphate, sand, etc.
Bone gelatine: 4367 Gelatine taken directly from bones.
Boneset: Comfrey.
Bone-spirit: An ammoniacal liquid obtained along with bone oil in the dry distillation of bones.
Bone-spirit salt: The residue left from evaporating bone-spirit.
Book muslin: Bookbinder's muslin.
Boracic acid: H_3BO_3 Boric acid.
Borate of soda, sodium: See sodium borate.
Borax: See sodium borate.
Boss: A ball of cotton wool enclosed in fine silk on which polishing material is put for smoothing surfaces.
Bougie: 6369 A long slender instrument, introduced into the urethra, esophagus, or rectum, to overcome any narrowing of those canals.
Boullay's filter: See 41. One type of filter.
Bouquet: French for an aromatic extract.
Boxwood: The very close-grained, heavy, tough, hardwood of the box (Buxus). It is white or light-yellow in color, and its hardness and fine grain make it valuable in wood engraving and in the manufacture of musical instruments, rules, tool handles, carvings, etc.
Bran bath: See 189. A process of drying newly dyed material.
Brandy marc: Brandy distilled from the dregs, refuse and pulp left after making wine.
Brazil, brazilwood: The wood of various tropical American species of Caesalpinia or related genera. It is used in cabinetwork, but more especially, being a soluble redwood, in making red and purple dyes.
Bright lead shavings: New lead shavings; unoxidized.
Brimstone: 4349 Sulphur.
Britannia, Britannia metal: 3348, 3417 A silver-white alloy of tin, antimony, bismuth and copper. Largely used for tableware.
British oil: 5362 A rubifacient liniment.
Bromate of potash: $KBrO_3$ Potassium bromate.
Bromide of ammonium: 4227 NH_4Br Ammonium bromide.
Bromide of cadmium: 4263 $CdBr_2$ Cadmium bromide.
Bromide of potassium: 4198 KBr Potassium bromide.
Bromide of soda: 4214 NaBr Sodium bromide.
Brown blaze: In zinc processing, the color of the blaze which indicates cadmium and arsenic. See 3310.
Brown oxide of lead: PbO_2 Lead dioxide.
Brown oxide of manganese: MnO_2 Manganese dioxide.
Brown red: See ferric oxide.
Brown sulphur: 4350 Sulphur melted for molding.
Brucia, brucine: 4006 $C_{23}H_{26}N_2O_4$ A poisonous alkaloid found with strychnine in the seeds of species of Strychnos. A victim appears to be having a heart attack.
Buchu: The dried leaves of Barosma betulina or B. crenulata. Used as a diuretic and diaphoretic.
Buckbean: A plant (Menyanthes trifoliata) growing in bogs in America, having clusters of white or purplish flowers and intensely bitter leaves, sometimes used in medicine.
Buckram: A coarse cloth of linen or hemp, stiffened with sizing or glue, used in hats or garments to keep them in shape.
Bugbane: See cimifuga.
Bullock's gall: Gall from bulls. See gall.
Bung down: To hold immersed as with holding pickles or kraut immersed in liquid.
Burgundy pitch: A yellowish-brown, hard, viscous resin, prepared from the exudation of the Norway spruce (Picea abies); also, resin from any of various pines, sold under this name. Used in medicinal plasters.
Burned selenite: Selenite roasted to a powder. See selenite.
Burnet: The plant, Poterium sanguisorba, the common burnet, often occuring as a weed in fields of sain foin, which it resembles.
Burning fluid: Probably a mixture of wood alcohol and turpentine.
Burnt alum: Alum which has been dried at 200°C. and powdered. It is a caustic used to remove dead tissues. Also called dried alum.
Burnt borax: Calcined sodium borate.
Burnt ivory: Bones burned to a powder.

Burnt soot: Probably lampblack.
Burnt sugar: Caramel.
Butt: A cask holding two hogsheads, or 129.7 gallons.
Butter of antimony: See antimony chloride.
Butter of cacao: Cocoa butter.
Butter of nutmeg: A soft yellow or brownish fat of nutmeg odor obtained from nutmegs. Used in pomades and medicines.
Butter of tin: See stannic chloride.
Butter of zinc: See zinc chloride.
Butyrate of amyl: $C_3H_7.COOC_5H_{11}$ Amyl butyrate.
Butyrate of ethyl: 4293 $C_3H_7.COO.C_2H_5$ Ethyl butyrate.
Butyrate of magnesia: $Mg(C_3H_7CO_2)_2$ Magnesium butyrate.
Butyric acid: 3966 $C_3H_7CO_2H$.
Butyric ether: See ethyl butyrate.
Butyrine: 4260 $(C_3H_7.COO)_3C_3H_5$.

C

Cadmium iodide: 4262 CdI_2 Hydroiodate of cadmium.
Cadmium sulphide: CdS Cadmium yellow; greenockite; sulphide of cadmium.
Cadmium yellow: See cadmium sulphide.
Cajeput, cajuput: An East Indian tree (Melaleuca leucadendron); called also paperbark and white tree. It yields a pungent greenish oil known as oil of cajuput, used as a stimulant, antispasmodic, and anodyne, but chiefly as a local application in skin diseases.
Cake alum: See aluminum sulphate.
Calabria licorice: Licorice from Calabria, Italy.
Calamine: 5761 H_2ZnSiO_5 Zinc silicate.
Calcic: Read as calcium. Calcic sulphate=calcium sulphate, etc.
Calcic hydrate: See calcium hydroxide.
Calcinated alum: Alum burned to a powder.
Calcination: 3849 To burn or roast to a powder.
Calcine, calcined: See calcination.
Calcined antimony: See diaphoretic antimony.
Calcined magnesia: See magnesium oxide.
Calcined oyster shells: See calcium oxide.
Calcined plaster: See calcium sulphate.
Calcined sulphate of iron: Calcined ferric sulphate.
Calcite: See calcium carbonate.
Calc spar: See calcium carbonate.
Calcium carbonate: $CaCO_3$ Calcite; calc spar; chalk; creta praecipitata; limestone; marble; whitening; whiting.
Calcium chloride: 4247 $CaCl_2$ Fixed ammoniacal salt; muriate of lime.
Calcium hydroxide: 3994 $Ca(OH)_2$ Calcic hydrate; hydrated lime; lime water; slaked lime.
Calcium hypochlorite: 4245 $CaOCl_2$ Bleaching liquor; bleaching powder; chloride of lime; chlorinated lime; oxymuriate of lime.
Calcium light: Limelight. A light source produced by directing an oxyhydrogen flame on a cylinder of lime.
Calcium monosulphide: See calcium sulphide.
Calcium monoxide: See calcium oxide.
Calcium nitrate: 2223 $Ca(NO_3)_2$ Lime saltpeter; nitrate of lime.
Calcium oxide: CaO Calcic oxide; calcined oyster shells; calcium monoxide; quicklime; unslaked lime.
Calcium phosphate: 4635 $Ca_3(PO_4)_2$ Bone ash; phosphate of lime; tricalcic phosphate.
Calcium sulphate: $CaSO_4.2H_2O$ Bihydrate of lime; calcic sulphate; gypsum; plaster of Paris; selenite; sulphate of lime.
Calcium sulphide: CaS Calcium monosulphide; sulphide of calcium; sulphuret of calcium.
Calisaya: Yellow cinchona bark from Cinchona calisaya, C. ledgeriana, or a hybrid of either of these with other species of Cinchona; yellow bark.
Calomel: See mercurous chloride.
Caliche: See sodium nitrate.
Calomba, calombo, colombo, columbo: The root of an African plant (Jateorhiza columba), family Menispermacal. It contains the bitter principle columbin and is used as a tonic.
Camel's hair brush, or pencil: A small brush used by painters, made usually of hair from the tails of squirrels.
Camomile: The dried flower heads of Anthemis nobilis and Matricaria chamomilla, used as aromatic bitters.
Campeachy logwood: Logwood from Campechy, Mexico.
Camphene: 4317 Purified spirits of turpentine.
Camphorated acetic acid: 3895. CH_3COOH.
Camphorated spirits: 4862 (CH_3OH) Two parts camphor mixed with one part alcohol.
Camwood: See barwood.
Canada balsam: The turpentine yielded by the balsam fir.
Canella, cannella: A genus of trees of the type of the canella-bark family (Canellaceae). The only species is C. winteranda of southern Florida and the West Indies, which is called the cinnamon bark, whitewood and wild cinnamon.
Canker: In plants a lesion of the bark, also sometimes involving the wood of a tree or shrub or, more rarely, of an herbaceous stem, caused either by bacteria, fungi, or other agencies.
Cantharidal collodion: 4742.
Cantharides: A preparation of dried blister beetles or Spanish fly (Cantharis vesicatoria), used externally as a rubefacient and vesicatory, and internally as a stimulant to the genitourinary mucous membranes.
Canton's phosphorous: 4335.
Caotchouc: Rubber; India rubber; pure rubber.
Caotchoucine: A mixture of hydrocarbons (including isoprene, dipentene, etc.) obtained by the dry distillation of pure rubber; called also rubber oil and caoutchouc oil. It is used as a rust preventive, as a denaturant of alcohol, etc.
Cape aloe: A much-branched South African plant (Aloe ferox) with reddish, prickly, succulent leaves.
Capillaire: A syrup prepared from the maidenhair fern.
Capital: The top part of a still. See G of the diagram in 13.
Capsicum: The dried ripe fruit of Capsicum fructescens, a hot red pepper, containing capsaicin, and used as a gastric and intestinal stimulant and and as a rubefacient.
Capsule: An evaporating dish.
Caramel: 694 Sugar burned for coloring and/or a flavor.
Carbolic acid: See phenol.
Carbolic acid paper: 1614 Phenol impregnated paper for wrapping meats to preserve them.
Carbonate of ammonia: See ammonium carbonate.
Carbonate of baryta: 4233 $BaCO_3$ Barium carbonate.
Carbonate of cobalt: 4252 $CoCO_3$ Cobalt carbonate.
Carbonate of copper: See copper carbonate.
Carbonate of iron: See ferrous carbonate.
Carbonate of lead: See lead carbonate.
Carbonate of lime: See calcium carbonate.
Carbonate of lithia: See lithium carbonate.
Carbonate of magnesia: 4240 $MgCO_3$ Magnesium carbonate.
Carbonate of oxide of ammonia: See ammonium carbonate.
Carbonate of potash, potassa: See potassium carbonate.
Carbonate of soda: See sodium carbonate.
Carbonate of strontia: See strontium carbonate.
Carbonate of zinc: See zinc carbonate.
Carbonated alkali: An alkali metallic carbonate such as calcium carbonate or sodium carbonate.
Carbon bisulphide: See carbon disulphide.
Carbon disulphide: 4309-11 CS_2 Bisulphide of carbon; bisulphuret of carbon; carbon bisulphide; carbon sulphide; carburet of sulphur.
Carbonic acid gas: 3913-4 CO_2 Carbon dioxide.
Carbonic gas: Carbon dioxide.
Carbonic oxide: 4064-5 CO Carbon monoxide; gaseous oxide of carbon; protoxide of carbon.
Carbon monoxide: See carbonic oxide.
Carbon sulphide: See carbon disulphide.
Carboy: A large glass bottle enclosed in a box, or formerly in wickerwork, now chiefly for carrying corrosive liquids.
Carburet: A carbide; also to roast to blackness.
Carburet of iron: See graphite.
Carburet of sulphur: See carbon disulphide.
Cardamom, cardamon, cardamum: The aromatic capsular fruit of an East Indian herb (Eletaria cardamomum). The seeds are used as a condiment and in medicine as an adjuvant to other aromatics, stimulants and stomachics.
Carmine: A red color made from cochineal.
Carminitive: A medicine to relieve flatulence.
Carron oil: 5513 A lotion of equal parts of linseed oil and limewater, applied to burns and scalds.
Cartridge paper: A thick, stout paper for making cartridges; a rough tinted paper used for covering walls; an inferior kind of drawing paper.
Cascarilla: The aromatic bark of a Bahamian shrub (Croton eluteria). It is used used for making incense and as a digestive tonic.
Case harden: 3297 The process of giving a steel surface to iron.
Casein: A protein that is one of the chief constituents of milk and the basis of cheese.
Cassia: The bark (cassia bark) of certain tropical evergreen trees, used like cinnamon.
Cassiterite: See stannic oxide.
Castor bottle: Bottles used for holding castor oil.
Catarrh: Inflamation of a mucous membrane of the nose or throat with an increased flow of mucus; head cold.
Catechu: An extract of the heartwood of either of two East Indian acacias, (Acacia catechu and A. catechu sundra).
Cathartic: A strong laxative.
Caustic: Something that will burn or destroy living tissue by chemical action; corrosive.
Caustic antimony: See antimony chloride.
Caustic baryta: See barium oxide.
Caustic lye: Redundant. All lye is caustic.
Caustic potash, potassa: See potassium hydroxide.
Caustic soda: See sodium hydroxide.
Cedrat, cedrate: The citron.

Celsius' thermometer: See 85. A centigrade thermometer.
Cementation: A process which consists in surrounding a solid body with a powder of other substances, and heating the whole to a degree not sufficient to cause fusion, the physical properties of the body being changed by chemical combination with the powder; thus, iron becomes steel by cementation with charcoal and green glass becomes porcelain by cementation with sand.
Centaurium: A genus of herbs of the gentian family.
Centigrade: See 85, 86. A thermometer which measures the freezing point at zero and the boiling point at 100°.
Cerate: An ointment prepared for external application, consisting essentially of wax (sometimes resin or spermaceti) mixed with oil, lard and medicinal ingredients.
Ceruleo sulphate of ammonia: See ammonium sulphate.
Ceruleo sulphate of potassa: See potassium sulphate.
Cerussite: See lead carbonate.
Cevadilla: See sabadilla.
Ceylon cinnamon: Ceylonese Cinnamomum zeylanicum.
Chalcontrite: See cupric sulphate.
Chalk: See calcium carbonate.
Chalk bath: A hot water and brine solution of calcium carbonate in which newly dyed material is worked. See 189.
Chalybeate: Anything impregnated with salts of iron; having a taste due to iron. A chalybeate water, liquid or medicine.
Chamber lye: Urine.
Chamomile: See camomile.
Charged: Filled; impregnated.
Checkerberry: See oil of wintergreen.
Chemic: A solution of indigo sulphate whose acid has been neutralized with sodium carbonate.
Chili saltpeter: See sodium nitrate.
Chimogene: The lightest liquid known in 1872. Made from petroleum; extremely volatile and combustable; actually, a liquified gas.
China grass: The stiff, dried, hand-cleaned but not degummed fiber of ramie.
Chinchona: See cinchona.
Chio turpentine: Most likely a brand name.
Chirata, chirayta, chiretta: The dried plant, Swertia chirata, of northern India. It is used as a bitter tonic.
Chloralum: See aluminum chloride.
Chlorate of barium, baryta: $Ba(ClO_3)_2$ Barium chlorate.
Chlorate of potash, potassa: See potassium chlorate.
Chlorate of soda: $NaClO_3$ Sodium chlorate.
Chlorate of thallium: $TlClO_3$ Thallium chlorate.
Chlorethane: See ethyl chloride.
Chlorhydric acid: See hydrochloric acid.
Chlorhydric ether: See ethyl chloride.
Chloric ether: See ethylene chloride.
Chloride of aluminum: See aluminum chloride.
Chloride of ammonium: See ammonium chloride.
Chloride of antimony: See antimony chloride.
Chloride of arsenic: See arsenic trichloride.
Chloride of barium: See barium chloride.
Chloride of benzoyle: $C_6H_5CO.Cl$ Benzoyle chloride.
Chloride of calcium: See calcium chloride.
Chloride of cobalt: See cobalt chloride.
Chloride of copper: See copper chloride.
Chloride of ethyl: See ethyl chloride.
Chloride of gold: See gold chloride.
Chloride of iron: See ferric chloride.
Chloride of lime: See calcium hypochlorite.
Chloride of magnesium: See magnesium chloride.
Chloride of manganese: See manganese chloride.
Chloride of mercury: See mercurous chloride.
Chloride of nickel: See nickel chloride.
Chloride of platinum: See platinum chloride.
Chloride of potash, potassa, potassium: See potassium chloride.
Chloride of silver: See silver chloride.
Chloride of soda, sodium: See sodium chloride.
Chloride of tin: See stannous chloride.
Chloride of zinc: See zinc chloride.
Chlorinated lime: See calcium hypochlorite.
Chlorine water: Common bleach or chlorine in solution.
Chlorodyne: 5200 A patent pain-killer with a chloroform base.
Chloroform: 4271-5 $CHCl_3$ An easily-made anasthetic.
Chlorophyl: The green coloring matter of plants.
Choke damp: Carbonic acid gas.
Cholagogue: Anything promoting the discharge of bile from the system.
Chromate of baryta, barytes: $BaCrO_4$ Barium chromate.
Chromate of lead: See lead chromate.
Chromate of potash, potassa: See potassium chromate.
Chrome alum: $CrK(SO_4)_2.12H_2O$ Chromium acid sulphate. Used in tanning and as a mordant in dying.
Chrome green: See chromous oxide.
Chrome orange, red: See lead dichromate.
Chrome yellow: See lead chromate.
Chromic acid: 3945-6 H_2CrO_4.
Chromous oxide: 2701 CrO Chrome green; green oxide of chromium; monoxide of chromium; protoxide of chromium.
Cicatrize: To induce the formation of a scar.
Cimifuga: 4592 A small genus of herbs of the crowfoot family (Ranunculaceae), the bugbanes. The roots of Cimifuga racemosa, the black cohosh, used in rheumatism and dysmenorrhea.
Cinchona: The dried bark of any of several species of Chinchona (especially C. ledgeriana and C. succirubra) or their hybrids, containing the alkaloids quinine, cinchonine, quinidine, cinchonidine, etc. It is used as a cure in malaria, as a reducer in other fevers and as a bitter tonic and stomachic. Called also Peruvian bark ; jesuit's bark.
Cinnabar: See mercuric sulphide.
Citrate of bismuth: $BiC_6H_5O_7$ Bismuth citrate.
Citrate of iron: See ferric citrate.
Citrate of magnesia: 4805 $Mg_3(C_6H_5O_7)_2$ Magnesium citrate.
Citrate of sesquioxide of iron: See ferric citrate.
Citrine, citron: Pertaining to citrus fruits.
Civet: A substance found in a pouch near the sexual organs of the true civet cats. It is used as an ingredient in perfumes.
Clap: In washing, stretching the wet material, relaxing it, then bringing it out again sharply.
Claret: A red wine.
Clarified: Made clear.
Coagulum: A coagulated mass or substance; a clot or curd.
Coal tar creosote: See phenol.
Coaly: Containing carbon.
Cobalt black: See cobalt oxide.
Cobalt chloride: 4251 $CoCl_2$ Chloride of cobalt; cobalt dichloride; cobaltous chloride; hydrochloride of cobalt; muriate of cobalt.
Cobaltic oxide: See cobalt peroxide.
Cobalt monoxide: See cobalt protoxide.
Cobalt nitrate: 4250 $Co(NO_3)_2$ Cobaltous nitrate; nitrate of cobalt.
Cobaltous nitrate: See cobalt nitrate.
Cobaltous oxide: See cobalt oxide.
Cobalt oxide: 4250 CoO Asbolite; black oxide of cobalt; cobalt black; cobalt monoxide; cobaltous oxide; gray oxide of cobalt; oxide of cobalt; protoxide of cobalt.
Cobalt peroxide: 4250 Co_2O_3 Cobaltic oxide; cobalt sesquioxide.
Cobalt sesquioxide: See cobalt peroxide.
Cocculus indicus: The berry of a woody vine (Anamirta cocculous) of the East Indies. It is very poisonous and yields picrotoxin.
Cochineal: A dyestuff consisting of the dried bodies of females of a scale insect (Dactylopius coccus).
Cocoa butter: The yellowish-white fat of the cacao seed.
Coddle: To cook slowly and gently, as eggs or fruit, in water just below the boiling point.
Cohobation: To repeat the distillation of something. To pour the liquor back upon the matter remaining in the vessel or upon fresh material.
Colchicum: The dried root or dried seeds of Colchicum autmnale, which contain the alkaloid colchicine and possess an emetic, diuretic, and cathartic action. It is used chiefly for gout and reumatism.
Colcothar: See ferric oxide.
Cold drawn linseed oil: Prepared without the use of heat. Raw, unboiled.
Collodion: 4743-4 A viscous solution of pyroxylin in a mixture of alcohol and ether; also, sometimes, a similar solution of pyroxylin in some other solvent, as acetone, used as a coating for wounds, for photographic films, small balloons, membranes, etc.
Colocynth: A Mediteranian and African herbacious vine (Citrullus colocynthis) allied to the watermelon; also, its fruit, from which is prepared a powerful cathartic. Called also bitter apple, bitter cucumber and bitter gourd.
Colombo, columbo: See calomba.
Colza: Rapeseed.
Compound: A combination of different elements.
Compound ethers: See ester.
Concentrated alum: See aluminum sulphate.
Condiment: A seasoning or relish for food.
Congelation: To go from a liquid to a solid through cooling, as water to ice or molten metal to solid.
Congreve matches: An early friction match.
Conium: The dried, full-grown, but unripe fruit of the poison hemlock (Conium maculatum), containing the alkaloids conine and methyl conine and used as a sedative. Also, one of the deadliest poisons known.
Conserve of roses, etc.: A medicinal confection made of undried vegetable substances mixed with finely powdered refined sugar so as to form a uniform mass.
Coomb: Four bushels.
Copaiba: An oleoresin obtained from several South American species of copaifera as a viscid, transparent, pale-yellow or brown liquid of aromatic odor. It is a stimulant and diuretic and is much used in affections of the mucous membranes.
Copal: A hard resin obtained from various tropical trees and used in varnishes.
Copper acetate: 4088-9 $Cu(C_2H_3O_2)_2$ Binacetate of copper; crystallized verdigris; crystals of venus; neutral acetate of copper; normal acetate of copper; subacetate of copper.
Copper arsenite: 2711 $Cu(AsO_2)_2$ Arsenite of copper; mineral green; Paris green; Scheele's green.
Copperas: See ferrous sulphate.

Copper carbonate: $CuCO_3$ Dibasic carbonate of copper; dicarbonate of copper; green copper carbonate; malachite.
Copper dust: Finely powdered copper.
Copper ferrocyanide: 4098 $CuFeCy_6$ Ferrocyanide of copper; prussiate of copper.
Copper foundings: Copper melted and cast.
Copper pyrites: $CuFeS_2$ Chalcopyrite; copper-iron sulphide.
Copper scales: The black scaly coating of oxide on the surface of heated copper.
Copper sulphate: See cupric sulphate.
Coralline: A poisonous yellow dye, the sodium salt of aurin (rosolic acid).
Cordial: An aromatized and sweetened spirit, used as a beverage or as a restorative; a liqueur.
Corm, corms: Plant bulbs such as tulip bulbs, etc.
Corrosive: Any agent which destroys matter; a caustic.
Corrosive chloride of mercury: See mercuric chloride.
Corrosive sublimate: See mercuric chloride.
Corundum: See aluminum oxide.
Cosmetic mercury: See mercuric ammonium chloride.
Cotyledons: The first leaf, or one of the first pair or whorl of leaves at the end of a stem, such as cloves.
Cowhage: A tropical woody vine (Mucuna pruriens) having crooked pods covered with barbed, brittle hairs which cause intolerable itching. Also, these hairs, mixed with honey or other vehicles, used to expel worms. In the U.S. the trumpet creeper.
Crackled sugar: In candy making, the state that the melted sugar that hangs to a stick dipped into it, and dipped directly into cold water, is not dissolved off, but turns hard and snaps.
Crape, crepe: A kind of thin worsted stuff, formerly used for the gowns of clergymen.
Crawly root: Coral root, corallorhiza odontorhiza.
Cream of tartar: See potassium bitartrate.
Creosote oil: That part of the wood-tar distillate from which creosote is obtained by purification.
Creta praecipitata: See calcium carbonate.
Crocus, crocus martis, crocus of mars: See ferric oxide.
Crocus of antimony: 4128 A brownish-yellow product, mainly sodium or potassium thioantimonite obtained as a slag in refining antimony.
Crocus powder: See ferric oxide.
Croton oil: A viscid, acrid, yellow to brown fixed oil, obtained from the seeds of Croton tiglium, a small East Indian tree. It is a drastic cathartic, a vesicant and a pustulant.
Crude antimony: See antimony sulphide, black.
Crude native sulphur: See sulphur.
Crude tartar: Unprocessed potassium bitartrate.
Crutch: A long stick, having the form of an inverted T, used for stirring soap.
Crystallized soda: $NaHCO_3$ Crystalline sodium carbonate.
Crystallized verdigris: See copper acetate.
Crystals of tartar: See potassium bitartrate.
Crystals of tin: See Stannous chloride.
Crystals of venus: See copper acetate.
Cubeb, cubebs: The dried, unripe nearly full-grown fruit of Piper cubeba, which is crushed and smoked in cigarettes for asthma and from which there is prepared an oleoresin used in the treatment of gonorrhea and nasal catarrh.
Cubic nitre: See sodium nitrate.
Cucurbit: The main body of a still. See 13 A. A vessel or flask for distillation, used with, or forming part of an alembic; matrass.
Cudbear: See archil.
Culcifuge: Mosquito and flea repellant.
Cullender: Collander.
Cullet: Broken or refuse glass, a certain amount of which is necessary in the batch.
Culm: 3314 A kind of coal found in Wales.
Cupel: A small, shallow, porous cup, especially of bone ash used in assaying to separate precious metals from lead, etc.
Cupellation: 3191 The refinement of gold or silver, etc., in the cupel by exposure to a high temperature in a draft or blast of air, by which the metal is oxidized and sinks into the porous cupel.
Cupric chloride: 4097 $CuCl_2$ Chloride of copper; copper chloride; neutral chloride of copper.
Cupric ferrocyanide: $Cu_2Fe(CN)_6.7H_2O$ Ferrocyanide of copper; prussiate of copper.
Cupric nitrate: 4091 $Cu(NO_3)_2.3H_2O$ Nitrate of copper.
Cupric oxide: 4092 CuO Black oxide of copper; deutoxide of copper; melaconite; oxide of copper; protoxide of copper; tenorite.
Cupric sulphate: 4096 $CuSO_4.5H_2O$ Blue copperas; blue stone; blue vitriol; chalcontrite; copper sulphate; Roman vitriol.
Cuprite: See cuprous oxide.
Cuprous chloride: $CuCl$ Bichloride of copper; dichloride of copper; subchloride of copper.
Cuprous oxide: 4094 Cu_2O Cuprite; dioxide of copper; red copper ore; red oxide of copper; suboxide of copper.
Curcuma: See tumeric.
Curd: In soap making, the granular mass of soap which separates from the lye and rises when salt is added to the boiled liquid.
Curd soap: A soap making process during which the soap rises to the top in granular curds, leaving behind an impure, alkaline liquor.
Currier: One who curries and dresses leather after it is tanned.
Cwt.: See 5935 Hundredweight.
Cyamelid, cyamelide: $C_3H_3N_3O_3$ A white, amorphous compound formed by the polymerization of cyanic acid.
Cyanate of ammonia: 4323 Heated urea.
Cyanate of potassa: $KCNO$ Potassium cyanate.
Cyanate of silver: $AgCNO$ Silver cyanate.
Cyanide of mercury: See mercuric cyanide.
Cyanide of potash, potassa: See potassium cyanide.
Cyanide of silver: See silver cyanide.
Cyanide of zinc: See zinc cyanide.
Cyanogen: $(CN)_2$ A colorless, very poisonous gas, prepared by heating a mixture of 1 part mercuric cyanide and 1 part mercuric chloride.
Cyanuret of potassium: See potassium cyanide.
Cyanuret of zinc: See zinc cyanide.
Cyanuric acid: $C_3N_3(OH)_3$ A white, crystalline, almost tasteless acid.

D

Damar, gum: A resin derived from various evergreen trees of the genus agathis, in Australia, New Zealand and the West Indies. Used mostly for making a colorless varnish.
Damping off: A diseased condition of seedlings or cuttings caused by certain parasitic fungi which invade the plant tissues near the ground, producing rotting, usually with moist lesions on the stem.
Dash wheel: A washing machine consisting of a revolving drum divided into compartments.
Decantation: See 10.
Decoction: See 34.
Defecation: See 3845.
Deflagration: See 11.
Deleterious: Hurtful or destructive; noxious.
Deliquescent: Having a strong attraction to water.
Delphine, delphinine: A poisonous crystalline alkaloid obtained from various larkspurs (genus Delphinium) and used in an ointment to relieve neuralgia. Also used to kill lice.
Denatured: Made unfit for drinking.
Deodorized alcohol: Alcohol distilled free from odorous substances.
Dephlogisticated: Any substance having had all its oxygen removed.
Depilatory: An agent used to remove hair or wool.
Depuration: The act of purifying or freeing fluids from miscellaneous matter. This is done by decantation, when the unwanted matter is deposited on the bottom of the vessel; or by skimming, fining or clarifying.
Desiccation: See 12.
Despumation: To skim; to clarify by removing scum or impurities.
Desqumation: To peel off in the form of scales. To scale off.
Deutoxide of barium: See barium peroxide.
Deutoxide of copper: See cupric oxide.
Deutoxide of hydrogen: See hydrogen peroxide.
Deutoxide of manganese: See manganese dioxide.
Deutoxide of mercury: See mercuric oxide.
Dextrin, dextrine: See 4345.
Diacetate of lead: See tribasic lead acetate.
Diachylon: 5043 A plaster made out of litharge and either olive oil or olive oil and lard, hence consisting essentially of lead oleate mixed with small amounts of glycerin and oleic acid. It is used for skinned surfaces and wounds and as an adhesive.
Di-ammonic carbonate: See ammonium carbonate.
Diaphoretic: See sudorific.
Diaphoretic antimony: Calcined antimony.
Diaphram: A dividing membrane, or thin partition.
Diatomaceous earth: See infusorial earth.
Dibasic carbonate of copper: See copper carbonate.
Dicarbonate of copper: See copper carbonate.
Dichloride of copper: See cuprous chloride.
Dichloride of manganese: See manganese chloride.
Dichloride of platinum: See platinous chloride.
Dichloride of tin: See stannous chloride.
Dichromate of lead: See lead dichromate.
Diffused: Spread out; not concentrated.
Digestion: See 40.
Digestive salt of silvius: See potassium acetate.
Dihydric sulphide: See hydrogen sulphide.
Dilute percolate: Diluted alcohol added to an already percolated substance in order to remove the last of its essence.
Dilute spirits: Alcohol and water.
Dilute sulphuric acid: See 3857.
Dioxide of copper: See cupric oxide.
Dioxide of mercury: See mercurous oxide.
Dioxide of tin: See stannic oxide.
Dipotassic carbonate: See potassium carbonate.
Dipotassic sulphate: See potassium sulphate.
Discutient: A medicine which will scatter or drive away tumors.
Disodic orthophosphate: See sodium phosphate.

Displacement: See 41.
Distillate: The product of distillation.
Distilled zinc: Zinc vaporized and collected after cooling. Usually obtained from ores roasted in a retort furnace. See zinc oxide.
Disulphate of quinia: See quinine sulphate.
Disulphide: Same as bisulphide.
Diuretic: Tending to increase the secretion and discharge of urine.
Diuretic salt: See potassium acetate.
Double muriate: ?
Double oil of vitriol: (303) ?
Double water of roses: (1131) ?
Drachm, dram: See 5952-55. 1/16th av. ounce.
Dragon's blood: The resin exuding from the fruit of a Malayan rattan palm (Calamus draco) or from other species of calamus. This is the true dragon's blood of commerce, used for coloring varnish and other substances.
Draught, draft: Act of drinking or inhaling; a drink.
Draw: To extract the essence or strength of; as by infusion; as to draw tea.
Dredged with flour: To sprinkle, as a roast, with flour, etc.; also, to sift or sprinkle with flour.
Drop lake: Lac resin in the form of drops.
Dross ivory black: 2719 Burned and pulverized shavings of ivory.
Dross of lead: The oxidized scum which forms on the surface of melting lead.
Dry acetic acid: Acetic acid which contains no water. See anhydrous acetic acid, 3894.
Dryers: Substances such as varnish, litharge, etc., which when put in house paints, etc., cause them to dry faster.
Dry ice: CO_2 Solid carbon dioxide.
Drying oils: Oils, such as linseed, which dry quickly in air.
Dryopteris: A wide-spread genus of medium-sized ferns of the polypody family (Polypodiacae), the shield ferns. In medicine this fern is a worm killer.
Ductile: Capable of being drawn out or hammered thin; such as metals easily worked and capable of being drawn out into wire or thread.
Dutch foil, gold, leaf, metal: Tombac rolled or beaten into thin sheets.
Dutch liquid: $CH_2Cl.CH_2Cl$ Ethylene dichloride.
Dutch rushes: Tough, springy rushes used for scouring.
Dyer's acetate of iron: See protoacetate of iron.
Dyer's spirits: A solution made by dissolving tin in nitro-muriatic acid.

E

Ebullition: An act or process of boiling or bubbling up.
Edulcoration: 3847 To free from acids, salts or other soluble substances, by washing; to purify.
Effervescence: The act of bubbling and hissing as in fermenting liquors, carbonated water, etc.
Efflorescent: To change on the surface, or throughout, to a whitish, mealy or crystalline powder from the loss of water of crystallization on exposure to the air.
Effluvia: Anything flowing out.
Egg powder: Baking powder with 1/2 dram of tumeric to the pound.
Elaic acid: See oleic acid.
Elecampane: A large, coarse European herb (Inula helenium) with yellow rayed flowers, naturalized in the United States. The root was formerly used in pulmonary diseases.
Elective: Tending to combine with, or act upon, one substance rather than another; as elective affinity or attraction.
Electuary: A medicine composed of powders, or other ingredients, mixed with honey or syrup so as to form a pasty mass.
Elemi: A fragrant oleoresin obtained from various tropical trees, used in making varnishes, and medicinally in ointments and plasters. Commonly from an Arabian tree (Boswellia freereana) sometimes called African elemi.
Elixers: 4697 Medicines made of drugs in alcoholic solutions.
Elutriation: See 14.
Embrocation: A lotion to moisten and rub a diseased part of the body.
Emery: See aluminum oxide.
Emetic: A medicine or other substance that causes vomiting.
Emetic tartar: See antimony tartrate.
Emollient: A medicine that has a softening or soothing effect on surface tissues.
Empyreuma: The peculiar smell of the products of organic substances burned in close vessels.
Empyreumatic oil: Any of certain oils obtained by distilling organic substances at high temperatures.
Emulsion: See 43. A preparation of an oily substance held in suspension in a watery liquid.
English oil of lavender: Oil of lavender made in England.
English soda: Sodium hydroxide from England with a strength of 80 to 83 degrees.
Epispastic: Causing a blister or producing a serious discharge by exciting inflammation. A blistering agent; vesicant.
Epsom salts: See magnesium sulphate.
Equivalent: One part being the same weight, etc., as another.
Ergot: A fungus growth on the grains of rye and other cereal plants.
Erysipelas: An acute infectious disease of the skin or mucous membranes caused by several kinds of streptoccus and characterized by local inflammation and fever.
Escharotic: Caustic.
Espaliers: Trellises on which fruit trees and bushes are trained.
Essence: The term loosely applied to a preparation of almost any kind, that is supposed to contain in a high degree, the essential or distinctive principle or quality of some substance.
Essence of mirbane: See nitro benzene.
Essential oil of lemons: See potassium binoxalate.
Essential oils: See 46 and 1464.
Essential salt: A salt obtained by crystallizing a plant juice or animal liquid.
Essential salt of lemons: See potassium binoxalate.
Ester: An organic compound, comparable to an inorganic salt, formed by the reaction of an acid and an alcohol.
Ether: See ethyl oxide.
Ethereal: Having ether as a main ingredient.
Ethin, ethine: HC:CH Acetylene.
Ethiops: Any of various preparations of a black or very dark color.
Ethiops mineral: See mercurous sulphide.
Ethule, ethyl, ethyle: C_2H_5 A univalent hydrocarbon radical, whatever that means.
Ethyl acetate: 4291 $C_2H_5C_2H_3O_2$ Acetate of oxide of ethyl; acetic ether; pyroligneous ether.
Ethyl benzene: 6409 $C_6H_5.C_2H_5$ Phenylethane; xylene; xylol.
Ethyl benzoate: 4294 $C_2H_5C_7H_5O_2$ Benzoate of ether, ethyl; benzoic ether.
Ethyl butyrate: 4293 $C_2H_5C_4H_7O_2$ Burtic ether; pineapple oil.
Ethyl chloride: 4290 C_2H_5Cl Chlorethane; chlorhydric ether; chloride of ether; light hydrochloric ether; muriatic ether.
Ethylene chloride: 4297 $CH_2Cl.CH_2Cl$ Chloric ether; Glycol dichloride.
Ethylic ether: See ethyl oxide.
Ethyl iodide: C_2H_5I Iodethane hydriotic ether.
Ethyl nitrate: 4287 $C_2H_5NO_3$ Nitric ether.
Ethyl nitrite: 4288 $C_2H_5NO_3$ Hyponitrous ether: hyponitrite of ether; nitrite of ether; nitrite of oxide of ethyl; nitrous ether.
Ethyl oxide: 4280 $C_4H_{10}O$ Ether; ethylic ether; hydrate of ether; sulphuric ether.
Excipient: An inert substance used in preparing remedies (as pills) to give them a suitable form or consistency.
Excoriation: Having the skin stripped or worn off, flayed, abraded.
Expressing: See 45. Pressing out.
Extract: See 44.
Extractive: Something meant to be extracted.
Extract of lead: See lead diacetate.
Extract of Saturn: See lead diacetate.

F

Factitious: Imitation; artificial.
Faeces: Sediment after infusion or distillation; dregs; refuse.
Fahrenheit: See 6, 85, 86.
Faints: The weak and impure spirit which comes over last in the distillation of whiskey or other liquor and is caught seperately.
Farina: Starch; specifically potato starch.
Fat oil: Fatty or fixed oils with all stearine removed and which remain liquid at room temperature.
Fatty acids: CnH_2nO_2 Any one of the series of saturated acids, as acetic acid, formic acid, etc. - so called because some of the members, as stearic and palmitic acids, occur (in their glycerine esters) in the natural fats, and are fat-like substances.
Feather alum: See aluminum sulphate.
Feathered sugar: The state of boiled sugar in which, on drawing out, it forms floss, as with cotton candy.
Feathering: To pour molten metal from some height into cold water.
Feather-shot: Copper granulated by being poured molten into cold water.
Febrifuge: A medicine which reduces or cures a fever.
Febrifuge salt: See potassium chloride.
Febrifuge salt of silvius: See potassium chloride.
Fecula: Starch.
Feculence: muddiness, foulness; sediment; dregs; feces.
Feints: See faints.
Felon: A painful inflammation of the structures of a finger, usually of the last joint; whitlow.
Ferric and ammonium citrate: See ammonio-ferric citrate.
Ferric and potassium tartrate: $KFeOC_4H_4O_6$ Ferric, ferro tartrate of potassium; potassio-ferric tartrate; tartarated iron; tartrate of iron and potassium; tartrate of potassa and iron.
Ferric bi or disulphide: See iron bi or disulphide.
Ferric chloride: 4165-6 $Fe_2Cl_6.12H_2$ Chloride of iron; perchloride of iron; permuriate of iron; sesquichloride of iron.
Ferric citrate: 4160 $Fe_2(C_6H_5O_7)_2$ Citrate of iron; citrate of sesquioxide of iron; prussiate of iron.
Ferric ferrocyanide: 4167 $Fe_4(Fe(CN)_6)_3$ Berlin Blue Ferrocyanide of iron; ferrocyanuret of iron; ferroprussiate of iron; percyanide of iron; Prussian blue; sesquiferrocyanide of iron.

Ferric nitrate: 4171 $Fe_2(NO_3)_3$ Nitrate of iron: nitrate of sesquioxide of iron; protonitrate of iron.
Ferric orthophosphate: See ferric phosphate.
Ferric oxide: 4154 Fe_2O_3 Brown red; colcothar; crocus; hematite; Indian red; iron oxide; iron peroxide; jeweler's red; red bole; red ochre; red oxide of iron; rouge; sesquioxide of iron; specular iron ore; venetian red.
Ferric phosphate: 4631 $Fe_2H_3(PO_4)_3$ Ferric orthophosphate; phosphate of iron.
Ferric sulphate: $Fe_2(SO_4)_3$ Sesquisalt of iron; sesquisulphate of iron.
Ferric tartrate of potassium: See ferric and potassium tartrate.
Ferricyanide of potassium: See potassium ferricyanide.
Ferridcyanide of iron: 2674 $Fe(Fe(CN)_6)$ Ferric ferricyanide; Turnbull's Prussian blue.
Ferridcyanide of lead: $Pb_3(Fe(CN)_6)$ Lead ferricyanide.
Ferridcyanide of potassium: See potassium ferricyanide.
Ferridcyanogen: $H_3(Fe(CN)_6)$ Ferricyanic acid.
Ferrochyazic acid: 3956 $H_4(Fe(CN)_6)$ Hydroferrocyanic acid.
Ferrocyanic acid: $H_4Fe(CN)_6$ A colorless, crystalline acid obtained by treating ferrocyanides with acids. Paris blue.
Ferrocyanide of copper: See copper ferrocyanide.
Ferrocyanide of iron: See ferric ferrocyanide.
Ferrocyanide of potassium: See potassium ferrocyanide.
Ferrocyanuret of potassium: See potassium ferrocyanide.
Ferroprussiate of iron: See ferric ferrocyanide.
Ferroprussiate of potassa: See potassium ferrocyanide.
Ferrosulphate of potassa: $KFe(SO_4)_2.12H_2O$ Potassium ferrosulphate.
Ferrotartrate of potassa: See ferric and potassium tartrate.
Ferrous acetate: 4159 $Fe(C_2H_3O_2)_2$ Acetate of iron.
Ferrous carbonate: 4163 $FeCO_3$ Carbonate of iron; protocarbonate of iron; siderite; spathic iron ore; subcarbonate of iron.
Ferrous chloride: 4165 $FeCl_2$ Hydrated chloride of iron; muriate of iron; protochloride of iron.
Ferrous oxide: 4151 FeO Monoxide of iron; protoxide of iron.
Ferrous sulphate: 4146 $FeSO_4.7H_2O$ Copperas; green copperas; green vitriol; iron sulphate; iron vitriol; protosulphate of iron; salt of colcothar; salt of iron; salt of steel; shoemaker's black; sulphate of iron; tincture of iron.
Ferrous sulphide: 4053, 4147 FeS Iron sulphide; monosulphide of iron; protosulphide of iron; sulphide of iron; sulphuret of iron.
Ferruginous marl: Iron bearing clay.
Filbert: Hazelnut.
Filtering powders: See 3839.
Fine, fining: In brewing, the addition of a solution of isinglass to beer to effect rapid clarification.
Fining: In glass making, the process of effecting complete solution of the batch and producing clarity by keeping the materials at a high heat in the melting pot.
Fining metal: The conversion of pig iron into wrought iron in a hearth or charcoal fire.
First lye: See 519 in Soap.
Fix: In dying or photography, to make permanent.
Fixed ammoniacal salt: See calcium chloride.
Fixed oil: A non-volatile oil - applied especially to the liquid fats found in many animals (in the fatty tissue) and plants (usually in the seeds), as distinguished from volatile or essential oil.
Fixidity, fixity: The state of being fixed, or stable.
Flake white: See lead carbonate. The same term is also applied to bismuth subnitrate.
Flatting: A mode of painting in which the paint leaves the work without a gloss.
Flatting mill: A machine with rollers which flattens metals, glass, foil, etc.
Flavin, flavine: A dye prepared from quercitron bark.
Fleabane: See pennyroyal.
Flint glass: A heavy, brilliant glass containing lead.
Flocculent: Of the appearance of wool, wooly; flocky.
Florence flask: A round or pear-shaped flask with a long neck.
Florentine orris: A European iris (Iris florintina) having large white flowers with lavender tinged falls (the three drooping petals). Its rhizome yields the perfume known as orris.
Florentine receiver: A reciever having a tube or spout attached near the base and rising nearly to the level of the top. It is used in distilling oils in steam, for separating the condensed water from the upper oily layer.
Flour-emery: Finely ground emery or corundum.
Flour of mustard: Finely ground mustard.
Flowers of antimony: See antimony oxide.
Flowers of bismuth: See bismuth trioxide.
Flowers of sal ammoniac: Ammonium carbonate condensed from vapors.
Flowers of sulphur: 4354 In distilling sulphur from impurities, the first vapors are condensed as a yellow powder called flowers of sulphur; sublimed sulphur.
Flowers of zinc: See zinc oxide.
Fluid extract: See 4571.
Fluoric acid: See hydrofluoric acid.
Fluoride of ammonia: NH_4F Ammonium fluoride.
Fluorspar: CaF_2 Natural calcium fluoride.
Fluosilicic acid: H_2SiF_6 An unstable acid known only in aqueous solution, being formed by the action of silicon fluoride, SiF_4, on water.
Flux: Any substance or mixture used to promote fusion, especially the fusion of metals or minerals. Common metallurgical fluxes are silica and silicates (acidic), lime and limestone (basic), and fluorite (neutral). Also any substance applied to surfaces to be joined by soldering or welding, just prior to or during the operation, to clean and free them from oxide, thus promoting their union.
Foeted, fetid: Having a rotten smell.
Fomentation: See 5155. The application of warm, moist substances, as wet cloths, to the body for easing pain. Also the lotion thus applied.
Foreign asphaltum, or foreign anything: An obscure classification, seldom important. Use a local brand of the product.
Formalin: HCOH Formaldehyde.
Formiate of ether: 4295 C_2H_5OCHO Formic ether.
Founding: The art of melting and casting metal.
Frangible: Breakable; brittle; fragile.
French berries: Buckthorn berries.
French chalk: A soft, white, granular variety of steatite used for drawing lines on cloth, in tailoring, dry cleaning, etc.
French nut oil: Oil from either the English walnut or the European chestnut.
French sand: Sand from French beaches. Neither practical nor necessary.
French spirits: Alcohol from France.
Fretting in: The partial production of a second fermentation, for the purpose of mellowing down the flavor of foreign ingredients (chiefly brandy) added to wine.
Friable: Easily crumbled, pulverized, or reduced to powder.
Frit: A compound of alkalies, boric acid and lime, with silica and sometimes lead oxide, partly or wholly fused and used as a basis for certain glazes.
Fuchin, fuchine: 2553 A dye producing bluish red. Also called magenta.
Fuccus crispus: Irish moss.
Fuller's earth: A highly absorbent, claylike substance used to remove grease from woolen cloth in fulling; to purify oil, etc.
Fully blown: Mature.
Fully proof: Neither over nor under proof. See proof spirit, 1436.
Fulminates: 2131-5 Explosive powders made by combining fulminic acid with with certain metals.
Fuming liquor of arsenic: See arsenic trichloride.
Fused: Melted together and united.
Fused nitrate of strontia: Melted strontium nitrate.
Fusel oil: An oily, acrid, poisonous liquid occuring in insufficiently distilled alcoholic products.
Fusible: Capable of being fused, melted, liquified.
Fusible metal: Any easily fusible metal or alloy.
Fusion: See 19.
Fustic: The wood of a common tropical American tree (Chlorophora tinctoria), which yields a light-yellow dye.

G

Galanga, galingale: The pungent, aromatic rhizome of various plants related to the true ginger, as Kaempferia galanga and species of Alpinia (A. galanga and A. officinarum). It is used medicinally and to some extent in perfumes.
Galena: PbS Lead sulphite. Used as a flux for earthenware glazes.
Galipot: 4331 The crude turpentine oleoresin formed as an exudation upon the bark of the cluster pine (Pinus pinaster) in southern Europe.
Gall: A bitter, greenish fluid secreted by the liver and stored in the gall bladder; bile.
Gallate of iron: $Fe(C_6H_2(OH)_3CO)_2$ Ferrous gallate.
Gallic acid: 3905-10 $C_6H_2(OH)_3CO_2H$ Trihydroxybenzoic acid.
Galline: See pyrogallic acid.
Galling: Fixing or mordanting dyed goods with boiled gallnuts.
Gallipoli oil: An inferior kind of olive oil brought from Gallipoli.
Gallipot: A small vessel used by druggists to hold medicines.
Gallnut: A nut-like gall or tumor on oak leaves, high in tannic acid.
Gamboge: An orange-red gum resin, becoming bright yellow when powdered. It is gotten from various trees of the genus Garcinia, the best quality (G. hanburyi) in Asia. It is used as a yellow pigment and in medicine as a cathartic.
Garancine: Madder-red.
Gaseous ammonia: See ammonia.
Gaseous oxide of carbon: See carbonic oxide.
Gas lime: Slaked lime that has been used in purifying gas and hence contains calcium carbonate, calcium sulphide, etc.
Gas tar: Coal tar obtained as a by-product in the manufacture of illuminating gas.
Gay Lussac's pouret: See 82, figure 1.
Gay Lussac's pyrophorus: 4340. See 4340.
Gelatine: See 4365.
Gelatine wafers: See 4370.
Gelsemium: The root of the yellow jasmine, used as a spasmodic, diaphoretic and neuralgic.
Gem salt: Rock salt crystals.
Gendar's hydrometer: See 58.
Gentian: The rhizome and roots of the yellow gentian (Gentiana lutea), used as a tonic and stomachic.

Gerb: A firework throwing a shower of sparks.
Gerlach's table: See 627.
German camomile: Camomile extracted from Matricaria chamomilla.
German fuming acid: A reference to fuming sulphuric acid made in Germany.
German soda: Sodium hydroxide at a strength of 95%. See 82.
Gill: See 5966-69. 1/4 pint.
Ginger-grass: An East Indian grass (Cymbopogon flexuosus) which yields ginger-grass oil.
Glacial acetic acid: See acetic acid.
Glaire: Egg white; to apply egg white or white of egg.
Glass of antimony: 2370 Sb_4O_6 Gray oxide of antimony; vitrified antimony; vitrified oxide of antimony.
Glass of borax: 2377 Calcined borax.
Glass paper: 1933 Sandpaper made with powdered glass.
Glauberite: See sodium sulphate.
Glauber's salt: See sodium sulphate.
Glauber's secret sal ammoniac: See ammonium sulphate.
Glauber's secret salt: See ammonium sulphate.
Gleet: A persistent transparent mucous discharge from the urethra; formerly, any morbid discharge.
Glue-water: Water boiled with a bit of gelatine. Used for sizing.
Glutinous: Rich in, or having a high content of gluten.
Glycerol: $HOCH_2.CHOH$ Glycerin.
Glycol dichloride: See ethylene chloride.
Glyconine: 5491 Egg white dissolved in glycerin.
Goebel's pyrophorus: 4344. See 4340.
Gold chloride: See aurous chloride.
Golden sulphuret of antimony: See antimony pentasulphide.
Gold monochloride: See aurous chloride.
Gold oxide: See aurous oxide.
Gold purple: See purple of cassius.
Gold size: 3580 Any adhesive used for attaching gold leaf to surfaces.
Gold solution: A solution of aurous chloride.
Goslarite: See zinc sulphate.
Goulard's acetate of lead: See tribasic lead acetate.
Goulard's extract: See tribasic lead acetate.
Goulard's lotion: See tribasic lead acetate.
Goulard's water: See tribasic lead acetate.
Grafting paper: 1881 Paper covered with grafting wax.
Grain: To granulate or powder.
Grain black: Blacking used on grain of leather. Probably 2968-9.
Grain gold: 3194 Granulated gold.
Grain musk: Raw musk in the form of a coarse powder.
Grain silver: 3217 Silver dust.
Grains of paradise: The pungent seeds of a West African plant (Amomum
Grain tin: 3316 Granulated tin.
Gram, gramme: 6027, 6031 1/28th of an av. ounce.
Granulation: See 20.
Graphite: 4164 Fe_3C Black lead; carburet of iron; iron carbide; plumbago.
Gravel: A deposit of small, calculous concretions in the kidneys and urinary bladder; the disease causing it.
Gray antimony: See antimony sulphide, black.
Gray oxide of antimony: See antimony glass.
Gray oxide of cobalt: See cobalt oxide.
Gray oxide of mercury: See mercurous oxide.
Green copperas: See ferrous sulphate.
Green copper carbonate: See copper carbonate.
Greenockite: See cadmium sulphide.
Green oxide of chromium: See chromous oxide.
Green oxide of copper: See verdigris.
Green sulphate of iron: See ferrous sulphate.
Green vitriol: See ferrous sulphate.
Guaiac: The tonka bean.
Guaiacum chips: Chips from the guaiacum officinale. See lignum vitae.
Guano: The excrement of sea fowl or bats. Rich in phosphates.
Guarana: A dried paste made from the seeds of Paullinia cupana, used in making an astringent drink for treatment of diarrhea. Besides tannin, it contains up to 4.5% of caffeine.
Guarancine: A derivitive of gurana.
Gum ammoniac: See ammoniac.
Gum arabic: A gum, obtained from several species of Acacia, composed chiefly of the calcium, magnesium and potassium salts of arabic acid. The best gum is nearly white or colorless and translucent and is obtained from Acacia senegal and A. arabica. Used in making adhesives, inks, candy and in pharmacy.
Gum benzoin: See benzoin.
Gum damar: See damar.
Gum kino: See kino.
Gum lac: An inferior kind of lac produced by a scale insect, Gascardia madascariensis.
Gum mastic: See mastic.
Gum myrr: See myrrh.
Gum sandarac: See sandarac.
Gum senegal: Senegal gum.
Gum tragacanth: See tragacanth.
Gun cotton: See 2141. An explosive.
Gutta percha: A substance resembling rubber but containing more resin, from the latex of several Maylasian trees of the genera Payena and Palaquium. It is nearly white to brown, hard and rather elastic, softens on heating and can be vulcanized. It is used as an electric insulator and in temporary fillings in teeth.
Gyle: A brewing; the beer produced at one brewing.
Gypsum: See calcium sulphate.

H

Hair bag: A felt bag used for pressing fluid from dregs.
Hair cloth: Felt. Used to filter large amounts of liquid.
Hair pencil: A brush or pencil made of fine hair for painting.
Hair salt: See aluminum sulphate.
Hair sieve: A sieve or strainer with a felt bottom.
Half-acid carbonate of ammonia: See ammonium sesquicarbonate.
Half refined tartar: See potassium bitartrate.
Halite: See sodium chloride.
Hank: A coil of yarn; skein.
Hard soap: Soap, the base of which is sodium hydroxide.
Hare's pyrophorus: 4342. See 4340.
Haricot: The ripe seeds or the unripe pod, of the common string bean (Phaseolus vulgaris), or other species of the same genus.
Hartshorn: Ammonia in water solution. Also ammonium carbonate, used in smelling salts. So-called because formerly obtained from deer's antlers.
Hatter's card: A flat board studded with thousands of wire teeth, used for arranging mats of hair for pressing into felt.
Hay saffron: A deep orange-colored substance consisting of the aromatic stigmas of saffron (Crocus sativus). It is used to color and flavor foods, and was formerly much used as a dyestuff, and as a stimulant antispasmodic emmenagogue in medicine.
Head: The top part of the still which directs the distillate to the cooler.
Heavy spar: See barium sulphate.
Hellebore: The dried rhizome and roots of any medicinal herb of the genus Helleborus, especially white hellebore, containing a number of alkaloids, as veratrine, etc., which acts as a sedative and a depressant. Used externally as a parasiticide and insecticide; pokeroot; pokeweed.
Hematite: An important ore of iron, the sesquioxide of iron, so-called because of the red color of its powder. See ferric oxide.
Hematitic: Resembling hematite; red iron ore.
Hemlock: Conium; the dried, full grown, but unripe fruit of the poison hemlock (Conium maculatum), containing the alkaloids conine and methyl conine and used as a narcotic and sedative.
Hemp tow: The coarse and broken part of hemp separated and ready for spinning.
Henbane: A herb of the Hyoscyamus niger. It is a deadly poison, especially to fowl. An extract of the leaves is used in medicine, having properties similar to those of belladonna.
Henna: A paste made by mixing powdered henna leaves with catechu, used for tinting the hair red.
Hepar: Any of several sulphur compounds, liver-colored, made by fusing sulphur or a sulphide with an alkali metal compound or metallic oxide; as hepar antimonii (impure antimony and potassium sulphide: hepar calcis (calcium sulphide); hepar sulphurus or liver of sulphur (impure potassium sulphide).
Hepar of sulphur: See potassium sulphide.
Herpes: A virus disease of the skin, characthertzed by the eruption of small blisters on the skin or mucous membranes.
Hessian crucible: A cheap, brittle and fragile, but very refractory crucible, composed of the finest fire clay and sand, and commonly used for a single heating. It is used for assaying, etc.
Highly rectified: Distilled more than once.
Hive syrup: The compound syrup of squill, employed as an emetic in croup and as an expectorant.
Hogshead: A large cask or barrel holding from 54 to 140 gallons.
Holy thistle: An annual thistlelike herb (Cnicus benedictus).
Homberg's pyrophorus: 4341. See 4340.
Homogenous: A compound so well mixed that no different chemicals can be discerned.
Honduras cochineal: Cochineal imported from Honduras.
Horn quicksilver: See mercuric chloride.
Horn silver: See silver chloride.
Hot chamber: Oven.
Huiles: French for oils.
Hydrastin: $C_{21}H_{21}O_6N$ A bitter, crystalline, non-poisonous alkaloid found in the rootstock of the goldenseal and used as a tonic and febrifuge.
Hydrate: A compound formed by the union of water with some other substance and represented as actually containing water, even though a crystal.
Hydrated: A hydrate.
Hydrated acetic acid: See acetic acid.
Hydrated antimonic acid: H_3SbO_4 Orthantimonic acid.
Hydrated benzoyl: See benzoic acid.
Hydrated chloride of iron: See ferrous chloride.
Hydrated oxide of phenyle: See phenol.

Hydrated oxide of potash, potassa: See potassium hydroxide.
Hydrated peroxide of iron: See ferric oxide.
Hydrated persulphuret of iron: See iron bi or disulphide.
Hydrated phosphorous acid: H_3PO_2 Phosphorous acid.
Hydrated protosulphuret of iron: See ferrous sulphide.
Hydrated sesquioxide of iron: See ferric oxide.
Hydrated sulphuric ether: See ethyl oxide.
Hydrate of alumina: See aluminum hydrate.
Hydrate of chloral: See 4276-8. $CCl_3.CH(OH)_2$ Chloral hydrate.
Hydrate of copper: $Cu(H_2O)_2Cl_2$ Diaquo copper II dichloride.
Hydrate of ether: See ethyl oxide.
Hydrate of lime: See calcium hydroxide.
Hydrate of phenyle: See phenol.
Hydrate of potash, potassa: See potassium hydroxide.
Hydrate of quinine: $C_{20}H_{24}N_2O_2.3H_2O$ Quinine hydrate.
Hydrate of soda: See sodium hydroxide.
Hydraulic lime: A limestone containing some silica and aluminum oxide, and yielding a quicklime that will set or harden under water.
Hydric acetate: See acetic acid.
Hydric chloride: See hydrochloric acid.
Hydric sulphide: See hydrogen sulphide.
Hydride of arsenic: See arsenic hydride.
Hydride of phenyl: See benzine.
Hydriodate of ammonia: See ammonium iodide.
Hydriodate of barytes: Ba_2I Barium iodide.
Hydriodate of cadmium: See cadmium iodide.
Hydriodate of iron: Fe_2I_6 Ferric iodide.
Hydriodate of quinine: $C_{20}H_{24}N_2O_2.3HI$ Quinine iodide.
Hydriodic, hydriotic acid: See hydrogen iodide.
Hydrobromic acid: HBr Hydrogen bromide.
Hydrocarbon: A compound containing only hydrogen and carbon.
Hydrochlorate of ammonia: See ammonium chloride.
Hydrochlorate of aniline: $C_6H_5NH_2HCl$ Aniline hydrochloride.
Hydrochlorate of copper: CuCl Copper chloride.
Hydrochlorate of magnesia: See magnesium chloride.
Hydrochloric acid: 3882-4 HCl Chlorhydric acid; hydric chloride; hydrogen chloride; marine acid; muriatic acid; salt acid; spirit of salts.
Hydrochloric ether: See ethyl chloride.
Hydrochloride of cobalt: See Cobalt chloride.
Hydrochloride of soda: See sodium chloride.
Hydrocyanate of silver: See silver cyanide.
Hydroferricyanic acid: 3954 $H_3Fe(CN)_6$ Hydroferridcyanic acid.
Hydrofluoric acid: 3960 HF Fluoric acid; hydtogen fluoride.
Hydrogen acetate: See acetic acid.
Hydrogen and sodium bicarbonate: See sodium bicarbonate.
Hydrogen binoxide: See hydrogen peroxide.
Hydrogen chloride: See hydrochloric acid.
Hydrogen fluoride: See hydrofluoric acid.
Hydrogen iodide: 3970 HI Hydriotic acid; iohydric acid.
Hydrogen monosulphide: See hydrogen sulphide.
Hydrogen peroxide: H_2O_2 Deutoxide of hydrogen; hydrogen binoxide; hydroxl; oxygenated water; peroxide.
Hydrogen phosphide: See phosphorous hydride.
Hydrogen potassium carbonate: See potassium bicarbonate.
Hydrogen sulphide: 4052 H_2S Dihydric sulphide; hydric sulphide; hydrogen monosulphide; hydrosulphuric acid; sulphphydric acid; sulphuretted hydrogen.
Hydroiodate of ammonium: See ammonium iodide.
Hydroiodate of cadmium: See cadmium iodide.
Hydrometer: See 53.
Hydrophosphorous acid: ?
Hydrosodic carbonate: See sodium bicarbonate.
Hydrosodic phosphate: See sodium phosphate.
Hydrosulphurate: Read as sulphide. Hydrosulphurate of potassium=potassium sulphide.
Hydrosulphuric acid: See hydrogen sulphide.
Hydrosulphydrate of potassium: See potassium sulphide.
Hydroxyl: See hydrogen peroxide.
Hyoscyamus: The leaves of the henbane, containing the alkaloids hyoscyamine and scopolamine, and used as an antispasmodic and sedative.
Hypantimonious acid: See antimony trioxide.
Hypernic: See Brazilwood.
Hyperoxide of barium: See barium peroxide.
Hypochlorite of lime: See calcium hypochlorite.
Hypochlorite of zinc: $Zn(ClO)_2$ Zinc hypochlorite.
Hyponitrite of ether: See ethyl nitrite.
Hyponitrite of oxide of ethyl: See ethyl nitrite.
Hyponitrous acid: $H_2N_2O_2$ An explosive prepared by reducing nitrates or nitrites.
Hyponitrous ether: See ethyl nitrite.
Hypophosphite of iron: 4642 $Fe(H_2PO_2)_3$ Ferric hypophosphite.
Hypophosphite of lime: $Ca(H_2PO_2)_2$ Calcium hypophosphite.
Hypophosphite of potash, potassa: KH_2PO_2 Potassium hypophosphite.
Hypophosphite of soda: $NaH_2PO_2.H_2O$ Sodium hypophosphite.
Hyposulphate of soda: $Na_2S_2O_3.5H_2O$ Sodium hyposulphate; sodium thiosulphate.
Hyposulphite of lead: $PbS_2O_3.H_2O$ Lead hydrosulphite; lead thiosulphite
Hyposulphite of lime: 4248
Hyposulphite of soda: 4211 $NaHSO_2$ Sodium hyposulphite.

I

Igneous fusion: Fusion by heat alone, unassisted by solution in the water of crystallization.
Ignition: See 3850.
Impalpable: Not perceptable to the touch.
Imperial: See 6032-6. Old British weights and measures.
Impregnate: To cause one compound to absorb another.
Indelible: That which cannot be removed, washed away, blotted out.
India, indian ink: A black pigment prepared by mixing lamp black or ivory black with a glutinous binder.
Indian meal: Ground Indian corn or maize; multi-colored corn.
India red: See ferric oxide.
India, Indian rhubarb: A stout herb (Peltiphyllum peltatum), family saxifragaceae.
India rubber: Natural rubber; crude rubber obtained from latex.
Indigo: A blue dye obtained from several plants, especially the species of Indigofera (I. tinctoria, I. anil), etc. and woad (Isatis tinctoria).
Indigo carmine: A blue dye, the sodium or potassium salt of indigotindisulphonic acid (indigo extract), used as a microscopic stain, for coloring foods and for dying.
Indurated: Hardened.
Inert: An inactive substance with no importance to a given mixture but used only as a filler.
Infusible white precipitate: See mercuric ammonium chloride.
Infusion: See 37.
Infusorial earth: Composed of fossilized infusoria. See kieselguhr.
Ingress: To go in; to enter.
Innoxious: Free from hurtful qualities or effects; harmless.
Inodorous alcohol: Having no scent other than its own.
Insoluble: Incapable of being dissolved.
In vacuo: In a vacuum.
Iodate of soda: $NaIO_3$ Sodium iodate.
Iodic acid: 3968-9 HIO_3.
Iodide of ammonium: See ammonium iodide.
Iodide of barium: Ba_2I Barium iodide.
Iodide of cadmium: See cadmium iodide.
Iodide of ethyl: See ethyl iodide.
Iodide of iron: 4161 Fe_2I_6 Ferric iodide.
Iodide of lead: 4103 PbI_2 Lead iodide.
Iodide of potassium: See potassium iodide.
Iodide of sodium: 4216 NaI Sodium iodide.
Iodide of zinc: ZnI_2 Zinc Iodide.
Iodurretted: Iodized; any element which has been combined with iodine.
Iohydric acid: See hydrogen iodide.
Ipecac, ipecacuanha: The dried rhizome and roots of Cephaelis ipecacuanha. It contains the alkaloids emetine and cephaeline, and is used as an expectorant and emetic and as a specific in amoebic dysentary.
Iron bi or disulphide: 4148 FeS_2 Bisulphuret of iron; ferric bi or disulphide; iron bi or disulphide; iron pyrites; marcasite.
Iron carbide: See graphite.
Iron dust: Fine particles of iron made by filing or rasping.
Iron liquor: 118 A black liquid solution of crude ferrous acetate used as a mordant by dyers and calico printers. See protoacetate of iron.
Iron mold, mould: A spot on cloth, etc., stained by rusty iron or by ink.
Iron oxide: See ferric oxide.
Iron peroxide: See ferric oxide.
Iron pin: An iron pipe or bar used like a rolling pin for flattening and as a weight.
Iron pyrites: See iron bi or disulphide.
Iron sand: Iron filings as used in fireworks.
Iron soap: See 1562. A waterproofing agent.
Iron sulphate: See ferrous sulphate.
Iron sulphide: See ferrous sulphide.
Iron vitriol: See ferrous sulphate.
Isinglass: A semitransparent, whitish, and very pure form of gelatine made from the air bladders of fish. Used for making jellies, glue, as a clarifier, etc. Also mica in thin sheets. Also agar agar or kanten.
Ivory nut: The nutlike seed of a South American palm (Phytelephas macrocarpa). It is as large as a hen's egg and contains a very hard endosperm, which under the name of vegetable ivory, is used for turning and carving, as for buttons, etc.

J

Jalap: The purgative tuberous root of a mexican plant (Exogonium jalapa) or a powdered drug obtained from it.

Jamaica ginger: An alcoholic extract of ginger used as a flavoring essence.
Jamaica rum: Rum made in Jamaica.
Japan: To cover with a coat of Japan, or of some other hard, brilliant varnish, in the manner of the Japanese; to laquer. Also, to give a high, glossy black to, supplemented by heating.
Japanese cement: 2159 Glue made from powdered rice.
Japanese wax: A yellowish fat obtained from the berries of certain species of sumac, chiefly in Japan, and used similarly to beeswax.
Javell's bleaching liquor: 4787 A solution of potassium chloride.
Javelle water: 6414 $KOCl+H_2O$ Originally potassium hypochlorite solution, now usually sodium hypochlorite solution $NaOCl+H_2O$.
Jelly bag: A bag, as of cheesecloth or flannel, through which the material for jelly is strained.
Jesuit's bark: See cinchona.
Jeweller's red: See ferric oxide.
Jeweller's rouge: See ferric oxide.
Jordan almonds: Almonds imported from Malaga and used mainly in confections.

K

Kalsomine: 2794 Calcimine; whitewash.
Kanten: See agar agar.
Kermes mineral: See antimony pentasulphide.
Kieselguhr: Loose or porous diatomite; diatomacious earth.
Kilo: Kilogram or kilometer.
Kinching: Putting a noose or twist in a hank to be dyed.
King's yellow: See arsenic trisulphide.
Kino: A dark red or blackish tanniferous product similar to catechu, obtained from various tropical trees. It is commonly used in medicine as an astringent and hemostatic, but less often than catechu in tanning and dying.
Kyanize: 1681 Wood preserving treatment by corrosive sublimate.

L

Labarraque's solution: See javelle water.
Lac: A resinous substance secreted by a scale insect (Tachardia lacca) which lives on the twigs of various trees, especially certain species of the genus Ficus.
Lac dye: A scarlet dye obtained from crude lac.
Lac resin: A purplish-red pigment prepared from lac by precipitation of the coloring matter with a metallic compound. Also called shellac.
Lac sulphur: See precipitated sulphur.
Lactate of iron: $Fe(CH_3CH(OH)COO)$ Ferrous lactate.
Lactate of lime: 3959 $Ca(C_3H_5O_3)_2.5H_2O$ Calcium lactate.
Lactin: Lactose.
Lactose: $C_{12}H_{22}O_{11}$ A sugar present in milk and separable from the whey by evaporation as hard crystals containing a molecule of water; milk sugar.
Lactucarium: 4666 The dried, milky juice of a prickly lettuce (Lactuca virosa) resembling opium in physical properties and formerly used (without adaquate basis) as a mild narcotic.
Lactuca virosa: See lactucarium.
Lake: Lac resin.
Laminated: Divided into thin plates or layers; pressed between such.
Langour: Enfeebling disease; sorrowful estate.
Latent tube:?
Laudanum: Formerly, any of various preparations of opium: now a tincture of opium.
Lavender water: A perfume consisting primarily of an alcoholic solution of oil of lavender.
Lawn sieve: A fine sieve, often of lawn or silk, used in straining potter's slip, cooking, etc.
Leach: To percolate or run a liquid through something, as ashes.
Lead acetate: 4101 $Pb(C_2H_3O_2)_2.3H_2O$ Lead diacetate; plumbic acetate; salt of satum; sugar of lead.
Lead carbonate: 2693 $PbSO_4$ Cerussite; flake white; white lead.
Lead chromate: 4105 $PbCrO_4$ Leipsig yellow; lemon yellow; Paris yellow.
Lead diacetate: 4775 See lead acetate.
Lead dichromate: $PbCr_2O_7$ Chrome orange; chrome red.
Lead glass: Glass containing lead. See flint glass.
Lead monoxide: See lead oxide.
Lead oxide: 4106 PbO Lead monoxide; lead protoxide; litharge; massicot; scale litharge; yellow oxide of lead; yellow protoxide of lead.
Lead oxide, red: Pb_3O_4 Minium; plumbate of oxide of lead; red lead; triplumbic tetroxide.
Lead protoxide: See lead oxide.
Lead subacetate: See tribasic lead acetate.
Lead sulphate: $PbSO_4$ Anglesite; sulphate of lead.
Lead wash, water: A dilute solution of basic lead acetate used in medicine as an external application, especially in inflamatory conditions arising from sprains, bruises, etc.
Leer, lehr: An oven in which glass is annealed.
Lees: Dregs.
Leipsig yellow: Lead chromate.
Lemon grass: A tropical grass (Cymbopogon citratus) which yields lemon grass oil.
Lemon thyme: A lemon-scented variety (Thymus serpylum vulgarus) of the wild thyme.
Lemon yellow: See lead chromate.
Lenitive: Having the quality of easing pain.
Leopard's bane: See arnica.
Leucorrhoea: A discharge from the vagina.
Levigate: To make smooth by grinding to a fine powder while in a moist condition; sometimes, merely to stir with water so as to separate the finer particles.
Lewery's white precipitate: See mercuric ammonium chloride.
Ley: Lie.
Liard: A lubricating compound.
Libavius' fuming liquor: See stannic chloride.
Liebig's condenser: A condenser consisting of two tubes, one inside the other, with space within for circulation of water.
Life root: The ragwort Senecio aureus.
Light hydrochloric ether: See ethyl chloride.
Lignum vitae: A tropical American tree (Guaiacum officinale) found especially in the West Indies, also, its very hard, heavy wood.
Lima wood: A soluble red wood derived from Caesalpinia tinctoria of Equador and Peru, used in dying.
Lime: See calcium oxide.
Lime pyrolignite: $Ca(C_2H_3O_2)_2.H_2O$ Calcium acetate.
Lime saltpeter: See calcium nitrate.
Lime soaps: Calcium matter which forms as a curd when soap is used in hard water . An example is a bathtub ring.
Limestone: See calcium carbonate.
Limpid: Clear; transparent.
Liniment: A medicated liquid to be rubbed on the skin to soothe sore or inflamed areas.
Liquation: See 21.
Liquid carmine: A solution of carmine in ammonia.
Liquid glass: 2816 K_2SiO_3 Potassium silicate.
Liquid silex: Sodium or potassium silicate.
Liquifaction: See 22.
Liquor: Any liquid substance, as water, milk, blood, sap, juice, etc.
Liquor of ammonia: See ammonia, solution of.
Liquor of potash, potassa: A solution of potassium hydroxide and water.
Litharge: See lead oxide.
Lithia: 3983 Li_2O Lithium oxide.
Lithic: Read as lithium. Lithic carbonate=lithium carbonate, etc.
Lithium aluminum silicate: $LiAl(Si_2O_5)_2$ Petalite; spodumene.
Lithium carbonate: 4238 Li_2CO_3 Carbonate of lithia; lithic carbonate.
Litmus: 4417-18 A dyestuff obtained from certain lichens. Paper soaked in litmus solution is used to test acidity/alkalinity.
Litre: 2 1/10 pints.
Little pan: A process for making soap. See 583.
Liver of antimony: 4128. See hepar.
Liver of sulphur: See potassium sulphide.
Lixivation: See 23.
Lixivial salts: The salts obtained by evaporating a lixivium.
Lixivium: Any solution obtained by lixivation, especially lye.
Lobelia: The leaves and tops of Lobelia inflata, commonly known as Indian tobacco or emetic weed. Used as an antispasmodic and expectorant.
Logwood: The very hard, brown or brownish-red wood of a central American tree (Haematoxylon campechianum). It contains the coloring principal haematoxylin used in staining cells for microscopy.
Lorme's tables: 622-626. Tables showing strengths of lye.
Lotion: 4817 A liquid medicinal preparation for washing.
Lucifer match: The friction match whose tip usually contains phosphorous sulphide mixed with other combustibles and with oxidizing material, as potassium chlorate, potassium nitrate or red lead.
Lunar caustic: See silver nitrate.
Lupulin: The fine, yellow, resinous powder on the strobiles of hops. It is used as a sedative.
Lute: A composition employed to secure the joints of chemical vessels, or as a covering to protect them from the violence of the fire.
Lycopodium: A fine yellowish, highly inflammable powder composed of the spores of various species of Lycopodium, especially of L. clavatum. It is used in fireworks and in medicine as an absorbent in excoriations of the skin.
Lye: A strong, alkaline solution obtained by leaching wood ashes. Commercial lye is sodium hydroxide crystals.
Lynn sand: (2413) ?

M

Macerate: See 39.

Madder red: The color imparted by madder.
Magistry of bismuth: See bismuth subnitrate.
Magnate of baryta: 4229 $Ba(MnO_4)$ Barium manganate.
Magnesia: See magnesium oxide.
Magnesian limestone: Dolomite.
Magnesite: 4240 $MgCO_3$ Magnesium carbonate.
Magnesium chloride: 4243 $MgCl_2$ Hydrochlorate of magnesia; magnesium hydrochlorate; muriate of magnesia.
Magnesium hydrochlorate: See magnesium chloride.
Magnesium oxide: 3990 MgO Calcined magnesia; magnesia; oxide of magnesium; perclastite.
Magnesium sulphate: 4241 $MgSO_4.7H_2O$ Bitter purging salt; cathartic salt; epsomite; epsom salts; magnesic sulphate; sal amer; sal anglicum; salt of canal; sulphate of magnesia.
Maiden hair: Any fern of the genus Adiantum.
Malachite: See copper carbonate.
Malates: See 3967. The salts of malic acid.
Male fern: See dryopteris.
Maleic, malic acid: 3967 $C_2H_2(CO_2H)_2$.
Malleable: Capable of being extended or shaped by beating with a hammer, or by the pressure of rollers. Most metals are malleable.
Mallow, malva: Any plant of the genus Malva, especially the common wild mallow.
Manganese borate: 2735 MnB_4O_7.
Manganese chloride: $MnCl_2$ Dichloride of manganese; manganous chloride; muriate of manganese; protochloride of manganese.
Manganese dioxide: 4173, 4254 MnO_2 Black oxide of manganese; binoxide of manganese; deutoxide of manganese; manganic peroxide; manganous oxide; oxide of manganese; permanganic oxide; peroxide of manganese; pyrolusite.
Manganese peroxide: See manganese dioxide.
Manganic acid: 3941 H_2MnO_4 Permanganic acid.
Manganic peroxide: See manganese dioxide.
Manganous acetate: $Mn(C_2H_3O_2)_2$ Acetate of protoxide of manganese.
Manganous chloride: See manganese chloride.
Manganous oxide: See manganese dioxide.
Manganous phosphate: 4633 $MnH.PO_4.6H_2O$ Phosphate of manganese; phosphate of protoxide of manganese.
Manganous sulphate: $MnSO_4.4H_2O$ Sulphate of manganese; sulphate of protoxide of manganese.
Manna: The sweetish exudate of the European flowering ash Flaxinus ornus and of several related species, obtained in the form of flakes, fragments or as a viscid mass. Its chief constituent is mannitol. Manna is used medicinally as a gentle laxative, demulcent, and expectorant.
Mannite, mannitol: $C_6H_8(OH)_6$ A white crystalline compound obtained from manna.
Marble: See calcium carbonate.
Marc: The refuse matter remaining after pressing seeds, fruits, especially grapes, sugar beets, sugar cane, etc.
Marcasite: See iron bi or disulphide.
Marine acid: See hydrochloric acid.
Marine glue: 229? A waterproof glue.
Marine salt: See sodium chloride.
Marl: An earth composed of clay and calcium carbonate and used as a fertilizer.
Marsh gas: CH_4 Methane.
Massicot: See lead oxide.
Mastic, mastich: A resin exuding from the mastic tree Pistacia lintiscus, obtained by incision. It is used as an astringent and an aromatic, also as an ingredient in varnishes and laquers.
Matico: The leaves of a common tropical American shrub (Piper angustifolium) used as a styptic and for venereal disease.
Matrass: See cucurbit.
Mayence: See strass.
Meal powder: 2140 Gunpowder made without water.
Mecca balsam: See balm of giliad.
Meconic acid: $C_5H(OH)(CO_2H)_2O_2$ A bitter white, crystalline acid found in opium.
Meershaum: $H_4Mg_2Si_3O_{10}$ A fine, white, claylike mineral, soft and light enough when in dry masses to float in water; sepiolite. It is a hydrous magnesium silicate.
Melaconite: See cupric oxide.
Melilot: Any plant of the genus melilotus; sweet clover.
Melissa: A mint. Melissa officinalis is the commonly cultivated aromatic.
Menstru, menstruum: Any substance which dissolves a solid body; a solvent.
Mercurammonium chloride: See mercuric ammonium chloride.
Mercurial pneumatic trough: 4031 A pneumatic trough filled with mercury for collecting gasses which would be absorbed by water.
Mercuric ammonium chloride: 4140-42 $HgNH_2Cl$ Ammonio-chloride of mercury; cosmetic mercury; infusible white precipitate; Lewry's white precipitate; mercurammonium chloride; white precipitate.
Mercuric chloride: 4139 $HgCl_2$ Bichloride of mercury; corrosive chloride of mercury; corrosive sublimate; horn quicksilver; perchloride of mercury.
Mercuric cyanide: $HgCy_2$ Bicyanide of mercury; cyanide of mercury; prussiate of mercury.
Mercuric nitrate: 4144 $Hg(NO_3)_2.2H_2O$ Nitrate of mercury; pernitrate of mercury; protonitrate of mercury.
Mercuric oxide: 4141 HgO Binoxide of mercury; deutoxide of mercury; monoxide of mercury; oxide of mercury; peroxide of mercury; red mercuric oxide; red oxide of mercury; red precipitate; yellow mercuric oxide; yellow oxide of mercury.
Mercuric sulphide: 2681 HgS Bisulphide of mercury; cinnabar; protosulphide of mercury; red mercuric sulphide; red sulphide of mercury; sulphide of mercury; vermillion.
Mercurous chloride: 4138 $HgCl$ Calomel; chloride of mercury; mercury chloride; protochloride of mercury; subchloride of mercury; submuriate of mercury.
Mercurous oxide: 4143 Hg_2O Black oxide of mercury; Black precipitate; dioxide of mercury; gray oxide of mercury; protoxide of mercury; suboxide of mercury.
Mercurous sulphate: Hg_2SO_4 Basic mercuric sulphate; protosulphate of mercury; subsulphate of mercury; sulphate of the suboxide of mercury; turpeth mineral; yellow subsulphate of mercury.
Mercurous sulphide: Hg_2S Ethiop's mineral.
Mercury chloride: See mercurous chloride.
Metallic antimony: See antimony.
Metallic arsenic: As Pure arsenic.
Metallic bath: A bath of metallic elements in which objects are immersed for coating; as in galvanizing.
Metallic gauze: A fine metallic screen.
Metallic oxide: An oxide of a metal.
Metallic potassium: K Pure potassium.
Metallic soap powder: A bronzing material made of soap, copper and iron sulphates.
Metallic sulphuret: A sulphide of metal.
Methanol: CH_3OH Methyl alcohol; methylic alcohol; wood alcohol.
Methylated ether: A mixture of wood alcohol and ether.
Methylated spirits: Grain alcohol denatured with wood alcohol.
Mezeron bark: The bark from a small European shrub Daphne mezereum.
Miasma: Poisonous vapor formerly supposed to arise from decomposing animal or vegetable matter, swamps, etc.
Mildew: Discoloration caused by parasitic fungi on cloth, leather, etc.
Milk of lime: Slaked lime mixed to a milky consistency with water.
Milk of magnesia: $Mg(OH)_2$ Magnesium hydroxide in water.
Milk of sulphur: See precipitated sulphur.
Millefleurs: See 1065. A perfume made from several different flowers.
Mine: A piece comprising various small fireworks which are scattered into the air with a loud report.
Mineral acid: Any inorganic acid.
Mineral blue: Azurite ground as a pigment.
Mineral green: See copper arsenite.
Minim: The smallest liquid measure or one drop.
Minium: See lead oxide, red.
Miscible: Capable of being mixed; mixable.
Mitcham: Oil of lavender.
Moire: A watered, clouded or frosted appearance on textile fabrics or metallic surfaces.
Monesia: An astringent vegetable extract derived from the bark (monesia bark) of the South American tree Pradosia lactescens.
Monks hood: See aconite.
Monoammoniac carbonate: See ammonium bicarbonate.
Monocarbonate of soda: See sodium carbonate.
Monochloride of silver: See silver chloride.
Monochromate of potash, potassa: See potassium chromate.
Monohydrated acetic acid: See acetic acid.
Monopotassic carbonate: See potassium bicarbonate.
Monopotassic sulphate: See potassium bisulphate.
Monosodic carbonate: See sodium bicarbonate.
Monosulphide of iron: See ferrous sulphide.
Monoxide of chromium: See chromous oxide.
Monoxide of mercury: See mercuric oxide.
Monoxide of nickel: See nickelous oxide.
Monoxide of tin: See stannous oxide.
Mordant: In dying, any substance which, by combining with a dyestuff to form an insoluble compound or lake, serves to produce a fixed color in a textile fiber, in leather, etc.
Morphia: $C_{17}H_{19}O_3N{+}H_2O$ Morphine.
Mother barm: Yeast collected from natural brewing, which is then used to generate other brews.
Mother liquid, liquor: The residual solution, often impure or complex, which remains after the substances readily, or regularly crystallizing, have been removed. Also called mother water.
Mother water: See mother liquid.
Mother wort: A bitter, old world mint (Leonorus cardiaca) or any other species of Leonorus.
Moulder's damp sand: Sand dampened for casting metals.
Mucilaginous: Like mucilage; moist and sticky; viscid.
Mucilaginous fermentation: See 16.
Muffle: A plate or compartment in a furnace which keeps fire off the objects being heated.
Muller: A pestle; any object used to mash substances to a paste or powder.
Murexide: 4224 $C_8N_6H_8O_6$ Purpuriate of ammonium muriate.

Muriate: A chloride.
Muriate of ammonia: See ammonium chloride.
Muriate of arsenic: See arsenic trichloride.
Muriate of copper: See cupric chloride.
Muriate of iron: See ferric chloride.
Muriate of lead: 4102 $PbCl_2$ Lead chloride.
Muriate of lime: See calcium chloride.
Muriate of magnesia: See magnesium chloride.
Muriate of manganese: See manganous chloride.
Muriate of morphia: $C_{17}H_{19}O_3N.HCl$ Morphine hydrochloride.
Muriate of nickel: See nickel chloride.
Muriate of soda: See sodium chloride.
Muriate of tin: See stannous chloride.
Muriate of zinc: See zinc chloride.
Muriatic acid: See hydrochloric acid.
Muriatic ether: See ethyl chloride.
Musk: A substance with a strong, penetrating odor, obtained from a musk deer. It was used as the basis for numerous perfumes.
Musk mallow: A European mallow (Malva moschata) with faintly musk-scented foliage.
Musk seed: See amber seed.
Muslin: A fine cotton cloth of plain weave.
Must: Any juice in process of fermenting.
Myrrh: A fragrant, bitter-tasting gum resin exuded from several shrubs of Arabia and East Africa, used in making incense, perfumes, etc.

N

Naples asphaltum: Asphalt made in Naples. Difference unknown and probably unimportant.
Naples soap: Unknown brand of soap from Naples.
Natron: See sodium carbonate.
Neatsfoot oil: A pale yellow fixed oil made by boiling the feet and shin bones of cattle, used chiefly as a leather dressing and fine lubricant.
Neroli: A yellowish essential oil obtained from orange flowers, chiefly by distillation and used in cologne and other perfumes, in liqueurs, etc.
Nerval: A salve for the sinews.
Nervine balsam: See baume nerval.
Neutral: Neither acid nor alkali.
Neutral carbonate of ammonium: See ammonium carbonate.
Neutral chloride of copper: See cupric chloride.
Neutral chromate of potash: See potassium chromate.
Neutral copper acetate: See copper acetate.
Neutralize: To cause to be neither acid nor alkali.
Neutral nitrate of bismuth: See bismuth nitrate.
Neutral sulphate of alumina: See aluminum sulphate.
Neutral sulphate of quinine: $(C_{20}H_{24}O_2N_2)_2H_2O_4.7H_2O$ Quinine sulphate.
Nicaragua wood: See Brazilwood.
Nickel chloride: 4174 $NiCl_2$ Muriate of nickel; nickelous chloride.
Nickelic oxide: See nickel peroxide.
Nickelous chloride: See nickel chloride
Nickelous oxide: 4175 NiO Monoxide of nickel; protoxide of nickel.
Nickel peroxide: 4176 Ni_2O_3 Nickelic oxide; sesquioxide of nickel.
Niter: See potassium nitrate.
Nitrate of ammonium: See ammonium nitrate.
Nitrate of baryta: See barium nitrate.
Nitrate of bismuth: See bismuth nitrate.
Nitrate of cobalt: See cobalt nitrate.
Nitrate of copper: See cupric nitrate.
Nitrate of iron: See ferric nitrate.
Nitrate of lead: 4107 $Pb(NO_3)_2$ Lead nitrate.
Nitrate of lime: See calcium nitrate.
Nitrate of mercury: See mercuric nitrate.
Nitrate of oxide of ammonium: See ammonium nitrate.
Nitrate of potash, potassa: See potassium nitrate.
Nitrate of sesquioxide of iron: See ferric nitrate.
Nitrate of silver: See silver nitrate.
Nitrate of soda: See sodium nitrate.
Nitrate of strichnia: $C_{21}H_{22}O_2N_2.HNO_3$ Strichnine nitrate.
Nitrate of strontia: $Sr(NO_3)_2$ Strontium nitrate.
Nitrate of urea: 4323-4 $CONH_2NO_3$ Urea nitrate.
Nitre: See potassium nitrate.
Nitric acid: 3873 HNO_3 Aqua fortis; azotic acid.
Nitric and hydrochloric acid: See nitrohydrochloric acid.
Nitric ether: See ethyl nitrate.
Nitrite of oxide of ethyl: See ethyl nitrite.
Nitro-benzene: 4322 $C_6H_5NO_2$ Artificial oil of bitter almonds; essence of mirbane; nitro-benzole; oil of mirbane.
Nitrohydrochloric acid: 3879 HNO_3+HCl Aqua regia; nitric and hydrochloric acid; nitro-muriatic acid.
Nitro-muriatic acid: See nitrohydrochloric acid.
Nitroprusside: 4217 Pertaining to or designating a red, crystalline, unstable acid, $H_2Fe(NO)(CN)_5$, obtained in the form of its salts by action of nitric acid on ferrocyanides, by action of sodium nitrate and acetic acid on ferrocyanides, etc.
Nitrous acid: 3877 HNO_2 Fuming nitric acid; red fuming nitric acid.
Nitrous ammoniacal salt: See ammonium nitrate.
Nitrous ether: See ethyl nitrite.
Noisettes: Hazel nuts.
Nordhausen acid: 3858
Normal copper acetate: See copper acetate.
Normal potassium carbonate: See potassium carbonate.
Normal potassium sulphate: See potassium sulphate.
Norwegian tar: A brand of tar from Norway probably little different from ours.
Nostrum: A patent medicine.
Noyau: A kernel such as cherry or peach seeds.
Nut gall: Any nut-like gall, especially one of those produced by the oak.
Nut oil: Oil obtained from nuts.
Nux vomica: The poisonous seed of the Asiatic tree strichnos nux-vomica. It contains several alkaloids, chiefly strychnine and brucine.

O

Oakum: Loose fiber obtained by untwisting and picking old hemp ropes. It was used especially for calking the seams of ships.
Ochre: An earthy and often impure ore of iron, usually red (hematite) or yellow (limonite), extensively used as a pigment in making paints, etc.
Octadecanoic acid: See stearic acid.
Odorine: C_6H_7N Impure picoline. An intensely smelly oil distilled from bone fat.
Oenanthate of oxide of ethyl: See oenanthic ether.
Oenanthic ether: 4296 $C_{14}H_{26}O_2$ Oenanthate of oxide of ethyl; pelargonic ether.
Officinal: Standardized drugs and medicines stocked by pharmacies.
Oil cloth: Cloth treated with oil or paint and used for garments, table and shelf coverings, etc.
Oil cognac: See oil of cognac.
Oil of ben: See ben.
Oil of cassia: Oil extracted crom cinnamon bark.
Oil of cloves: Oil extracted from cloves.
Oil of mirbane: See nitro-benzene.
Oil of rue: See rue.
Oil of salt: See bittern.
Oil of stone: 5362 A rubefacient liniment.
Oil of sulphur: Oil obtained from expressed marc of olives by extraction with carbon disulphide. Used in soap making.
Oil of turpentine: 4317 Pyrogenous oil of turpentine; spirits of turpentine.
Oil of vitriol: Sulphuric acid.
Oil of wintergreen: See wintergreen.
Oil silk: Oiled silk used to cover burns.
Oil size: 3571, 3580 A preparation used in gilding.
Oil soap: Soft soap made of olive oil and potassium hydroxide. See 605.
Oil varnish: A linseed oil based varnish.
Ointment: 4931 A semisolid medicinal preparation for anointing or smearing on the skin; an unguent.
Oleaginous: Containing or producing oil.
Oleic: Pertaining to, derived from, or contained in oil.
Oleic acid: $HC_{18}H_{34}O_2$ Elaic acid.
Oleine: The liquid part of any fat, distinguished from the stearin, or solid portion.
Oleoresin: A liquid or similiquid preparation extracted by means of ether, consisting essentially of fixed or volatile oil holding resin in solution.
Oleoresin capsicum: 4580 Any hot red pepper will do.
Olibanum: The fragrant gum resin, frankincence.
Olivine: A cosmetic skin softener with an olive oil base.
Opal pictures: Lantern slides for projecting pictures on a screen.
Opaque: Not transparent.
Opodeldoc: 4869-70 A soap-based liniment.
Orange apples: The fruit of Maclura pomifera, the Osage-orange hedge.
Orange shellac: Purified lac resin.
Orangetis: (816) ?
Orchil: See archil.
Origanum: See sweet marjoram.
Ormolu: 3425 An alloy resembling gold.
Oroide: An alloy, chiefly of copper and zinc or tin, resembling gold in color and brilliancy, and used in making cheap jewelry.
Orpiment: See arsenic trisulphide.
Orris: The fragrant root stock of any of several European irises, especially the Florentine iris. The powdered root is used pulverized in perfumery and medicine, and as an ingredient of sachet and tooth powders.
Osier: Any of various willows whose pliable twigs are used for furniture, basketry, etc.
Otto: See attar.
Oxalate: A salt of oxalic acid.
Oxalate of ammonia: See ammonium oxalate.
Oxalate of aniline: $(C_6H_5NH)_2C_2O_4$ Aniline oxalate.

Oxalate of iron: 4158 FeC_2O_4 Ferrous oxalate.
Oxalate of lime: $CaC_2O_4.H_2O$ Calcium oxalate.
Oxalate of nickel: 4178 NiC_2O_4 Nickelous oxalate.
Oxalate of potassa: See potassium oxalate.
Oxalate of strontia: $SrC_2O_4.H_2O$ Strontium oxalate.
Oxalate of tin: SnC_2O_4 Stannous oxalate.
Ox gall: The fresh gall of the domestic ox. Purified and dried, it is used in medicine.
Oxide of antimony: See antimony trioxide.
Oxide of barium: See barium oxide.
Oxide of bismuth: See bismuth trioxide.
Oxide of chrome: See chromous oxide.
Oxide of cobalt: See cobalt oxide.
Oxide of copper: See cupric oxide.
Oxide of ethyl: See ethyl oxide.
Oxide of gold: See aurous oxide.
Oxide of lead: See lead oxide.
Oxide of magnesium: See magnesium oxide.
Oxide of manganese: See manganese dioxide.
Oxide of mercury: See mercuric oxide.
Oxide of nickel: See nickelous oxide.
Oxide of silver: See silver oxide.
Oxide of tin: See stannous oxide.
Oxide of zinc: See zinc oxide.
Oxychloride of phosphorous: See phosphorous oxychloride.
Oxygenated water: See hydrogen peroxide.
Oxymel: 4690 A mixture of honey and dilute acetic acid, used as excipient.
Oxymuriate of lime: See calcium hypochlorite.
Oxymuriate of mercury: See mercuric chloride.
Oxymuriatic acid: $HClO$ Hypochlorous acid.
Oxysulphuret of antimony: See antimony pentasulphide.
Ozone ether: 4284 A disinfecting agent.
Ozonometer: 4429 A test paper for registering atmospheric ozone.

P

Paeonia: Peony
Palatine: A bluish-red dye.
Pallisandre: Brazilian rosewood.
Palmitic acid: $CH_3(CH_2)_{14}CO_2H$ A white crystalline fatty acid occurring free in palm oil, as the myricyl ester in beeswax.
Palm oil: A solid yellow or reddish fat or butter obtained from the flesh of the fruit of several species of palms, especially the oil palm and used in making soap and candles, etc.
Palsy: An illness in which one becomes paralytic, tottering, shaky.
Papier mache: Shredded, chewed or otherwise pulverized paper, mixed with paste, glue, etc., to be molded and shaped into various articles.
Papier saxe: See 3130. A brand of photographic paper.
Pappy: Like pap; soft; succulent.
Papyrine: 1968 Artificial parchment.
Paramalic acid: 3967 $C_2H_2(CO_2H)_2$ Fumaric acid.
Parchment: Sheep or goat skin prepared for writing on.
Paragoric: A medicine that eases pain. Usually refers to tincture of opium.
Pareira brava: The root of a South American vine (Chondrodendron tomentosum) widely used in medicine as a diuretic.
Paris blue: See ferrocyanic acid.
Paris green: See copper arsenite.
Paris white: A fine grade of calcium carbonate.
Parting: Separation of the constituents of alloys, especially the separation of gold from silver, as in refining.
Partridge berry: An American trailing plant (Mitchella repens) with an edible but insipid scarlet berry; squaw vine.
Paste: Imitation gems.
Paste blacking: A mixture of ivory black or lampblack and a little oil.
Pasteboard: A stiff materialmade by pasting several sheets of paper one upon another.
Pastilles: Cones of incense used for fumigating or refreshing rooms.
Patchouli: An East Indian shrubbery mint (Pogostemon patchouli) yielding a fragrant essential oil.
Patent medicines: 5171 Packaged medicines put up for immediate use and protected by letters of patent. The package is labeled with the name of the medicine and carries directions for its use.
Peachwood: See Brazilwood.
Pearlash, pearl ashes: See potassium carbonate.
Pearlash lye: See potassium carbonate.
Pearl oil: 4302 $C_5H_{11}C_2H_3O_2$ Amyl acetate.
Pearl starch: Fine-grained starch.
Pearl white: See bismuth subnitrate.
Peck: See 5970. 1/4 bushel.
Pegging: In wine making the occasional removing of the plug to examine the batch. Also called spiling.
Pelargonic ether: See oenanthic ether.
Pelargoniums: Geraniums.
Pellicle: A film formed on top of an evaporating liquid.
Pellitory: The root of Anacyclus pyrethrum, used to cause salivation and as a constituant of certain dentifrices.
Pennyroyal: A mint (Hedeoma pulegioides). It yields the commercial oil of pennyroyal used as a culcifuge; fleabane.
Penta-sulphuret of antimony: See antimony pentasulphide.
Pepsin, pepsine: An extract of pepsin from the stomachs ofcalves, etc., used as a medicine in aiding digestion.
Perchlorate of potassa: 4185 $KClO_4$ Potassium perchlorate.
Perchloride: A chloride containing a relatively high proportion of chlorine.
Perchloride of gold: See auric chloride.
Perchloride of iron: See ferric chloride.
Perchloride of manganese: $MnCl_4$ Manganese perchloride.
Perchloride of mercury: See mercuric chloride.
Perchloride of potassa: 4185 Formula actually indicates potassium perchlorate $KClO_4$.
Perchloride of tin: See stannic chloride.
Percolation: See 41, 4572.
Percyanide of iron: See ferric ferrocyanide.
Periclastite: See magnesium oxide.
Periodide of phosphorous: PIO_6 Phosphorous periodide.
Perlate salt: See sodium phosphate.
Permanganate of baryta: $Ba(MnO_4)_2$ Barium permanganate.
Permanganate of manganese: See manganese peroxide.
Permanganate of potash, potassa: See potassium permanganate.
Permanganic acid: See manganic acid.
Permanganic oxide: See manganese dioxide.
Permeable: Allowing gas or liquid to pass through. Such as blotting paper.
Permuriate of iron: See ferrous chloride.
Permuriate of tin: See stannic chloride.
Pernitrate of iron: 4172 $Fe(NO_3)_2$ Ferrous nitrate.
Pernitrate of mercury: See mercuric nitrate.
Peroxide: When used alone it usually means hydrogen peroxide.
Peroxide of barium: See barium peroxide.
Peroxide of cobalt: See cobalt peroxide.
Peroxide of copper: 4095 $CuO_2.H_2O$ Copper peroxide.
Peroxide of hydrogen: See hydrogen peroxide.
Peroxide of iron: See ferric oxide.
Peroxide of lead: See lead oxide.
Peroxide of manganese: See manganese dioxide.
Peroxide of mercury: See mercuric oxide.
Peroxide of nickel: See nickel peroxide.
Peroxide of tin: See stannic oxide.
Peroxydizement: To oxidize to the utmost, or so as to form a peroxide.
Persalt: A salt containing a relatively large proportion of the acidic element or group; as, the persalts of iron.
Persalt of iron: $Fe_2(SO_4)_3$ Ferric sulphate.
Persian berry: The fruit of any species of buckthorn (Rhamnus).
Persulphate of iron: FeS_2O_5 Ferrous sulphate.
Peruvian bark: See cinchona.
Peruvian niter: See sodium nitrate.
Petalite: See lithium aluminum silicate.
Petite-grain: Oil of orange leaf.
Pharmacopoeial strength: Any product made according to the U.S. Pharmacopoeia, adopted as the standard of the country in the food and drug act of 1906. Earlier references are according to the British Pharmacopoeia.
Phenic acid: See phenol.
Phenol: 3916-7 C_6H_6O Carbolic acid; coal tar creosote; phenic acid; phenol alcohol; six carbon phenol.
Phenol alcohol: See phenol.
Phenol hydride: See benzene.
Phenol paper: 1936 Paper treated with phenol and used to wrap meats.
Phenylethane: See ethyl benzene.
Phial: A vial; especially a small glass medicine bottle.
Phosphate of baryta: $Ba_2(PO_4)_3$ Barium phosphate.
Phosphate of iron: See ferric phosphate.
Phosphate of lime: See calcium phosphate.
Phosphate of manganese: See manganous phosphate.
Phosphate of protoxide of manganese: See manganous phosphate.
Phosphate of soda: See sodium phosphate.
Phosphine: See phosphorous hydride.
Phosphoric oxychloride: See phosphorous oxychloride.
Phosphoric monoxychloride: See phosphorous oxychloride.
Phosphorous hydride: 4055 PH_3 Hydrogen phosphide; phosphine; phosphorous trihydride; phosphureted hydrogen.
Phosphorous oxychloride: PCl_3O Phosphoric monoxychloride; phosphoric oxychloride.
Phosphorous trihydride: See phosphorous hydride.
Phosphureted hydrogen: See phosphorous hydride.
Pimpernell, pimpinella root: The dried rhizome and roots of the burnet saxifrage, used as a diaphoretic and diuretic.
Pineapple oil: See ethyl butyrate.
Pink root: Any of several herbs of the genus Spigelia, especially S. marilandica, the American wormroot. Used as an anthelmintic.

Pipe, pipe cask: Two hogsheads or 129.7 gallons.
Pipe clay: Highly plastic and fairly pure clay of a grayish-white color, used in making pipes, etc.
Pipkin: A small earthen pot, usually one having a horizontal handle.
Pipsissewa: Any evergreen herb of the genus Chimaphila, especially C. corymbosa. Its astringent leaves have been used as a tonic and diuretic.
Pitch: A black or dark-colored viscous substance obtained as a residue in distilling coal tar, wood tar, etc.
Plaster: 5040 An external application of a consistency harder than ointment, prepared for use by spreading it on linen, etc., to heal or soothe.
Plaster of Paris: See calcium sulphate.
Platin: Any of various imitation platinums.
Platina: See platinum.
Platina mohr: See platinum black.
Platinous chloride: 3320, 4084 $PtCl_2$ Bichloride of platinum; dichloride of platinum; platinum chloride.
Platinum black: 3338 Finely divided platinum; platina mohr.
Platinum chloride: See platinous chloride.
Pleurisy root: The butterfly weed Asclepias tuberosa; also its root, used as a diaphoretic and expectorant especially in pleurisy.
Plumb: A mordant in dying, using tin and acids or ammonium.
Plumbago: See graphite.
Plumbate of oxide of lead: See lead oxide, red.
Plumbate of soda: Red lead oxide, dissolved in sodium hydroxide.
Plumbic acetate: See lead acetate.
Podophyllin, podophyllum: The rhizome and rootlet of the may apple (Podophyllum peltatum), used as a cholagogue and cathartic.
Poke root, weed: See hellebore.
Polisher's putty: See stannic oxide.
Polypody: Any fern of the genus Polypodium. P. vulgaris is common in North America. Also called male polypody.
Pomace: The substance of apples or other fruit, crushed by grinding, as in making cider.
Pony glass: A small liqueur glass or undersized beer glass. Shot glass.
Porcelain: A fine, white, translucent, hard earthenware; china.
Porous: Full of pores, hense capable of absorbing moisture.
Porphyrization: Grinding a substance to a fine powder on porphry stone.
Porphyry: A slab or block of porphyry used for triturating drugs, etc.
Porter: 856 A weak stout, containing about 4% alcohol.
Port fire: 2102 A fuse for firing cannon.
Portugal: Possibly the Portugal laurel, A European evergreen shrub (Laurocerasus lusitanica) with handsome foliage and white flowers.
Potash, potashes: 4181 Potassium carbonate, especially that obtained by leaching wood ashes, evaporating the lye, and calcining the residue. This yields crude potash, colored and very impure. Purified potash, a white solid, is often called pearlash or pearl ashes.

Potash of commerce: See potassium carbonate.
Potash solution of sulphur: Probably sulphur dissolved in a solution of potassium carbonate.
Potash water: Potassium hydroxide in solution.
Potassa: 3974.
Potassa hydrate: See potassium hydrate.
Potassic: Read as potassium. Potassic sulphate=potassium sulphate, etc.
Potassio-ferric tartrate: See ferric and potassium tartrate.
Potassio-tartrate of antimony: See antimony tartrate.
Potassio-tartrate of soda: See sodium and potassium tartrate.
Potassium acetate: 4180 $K(C_2H_3O_2)$ Acetate of potash; digestive salt of silvius; diuretic salt; potassic acetate.
Potassium acid oxalate: $KH_3(C_2O_4)$ Potassium quadroxalate; potassium trihydrogen oxalate; quadroxalate of potassa.
Potassium bicarbonate: 4183 $KHCO_3$ Acid potassium carbonate; bicarbonate of potassa; hydrogen potassium carbonate; monopotassic carbonate.
Potassium bichromate: 4187 $K_2Cr_2O_7$ Bichromate of potassium; potassic acid chromate; potassic dichrome; potassium dichromate; red chromate of potash.
Potassium binoxalate: KHC_2O_4 Essential salt of lemons; potassium hydrogen oxalate; salt of sorrel.
Potassium bisulphate: $KHSO_4$ Acid potassium sulphate; bisulphate of potash; monopotassic sulphate; potassium hydrogen sulphate.
Potassium bitartrate: 4197 $KHC_4H_4O_6$ Acid tartrate of potassa; argal; argol; cream of tartar; crystals of tartar; half refined tartar; supertartrate of potassa.
Potassium carbonate: 4181 K_2CO_3 Carbonate of potassa; dipotassic carbonate; normal potassium carbonate; pearlash; potash; potash of commerce; potassium neutral carbonate; salt of tartar; salt of wormwood; subcarbonate of potassa.
Potassium chlorate: 4184 $KClO_3$ Chlorate of potash; potassic chlorate.
Potassium chloride: 4199 KCl Chloride of potash; chloride of potassa; chloride of potassium; febrifuge salt; febrifuge salt of sylvius; sylvite.
Potassium chromate: 4186 K_2CrO_4 Chromate of potassa; monochromate of potassa; neutral chromate of potassa; potassic chromate; salt of chrome; yellow chromate of potassa.
Potassium cyanide: 4202 KCy Cyanide of potash; cyanuret of potassium; potassic cyanide.
Potassium dichromate: 4187 $K_2Cr_2O_7$ Bichromate of potassium; potassic acid chromate; potassic dichrome; red chromate of potash.
Potassium ferricyanide: 4200 $K_3Fe(CN)_6$ Ferridcyanide of potassium; potassium ferricyanuret; red prussiate of potash.
Potassium ferricyanuret: See potassium ferricyanide.
Potassium ferrocyanide: 4201 $K_4Fe(CN)_6$ Ferrocyanuret of potassium; ferroferroprussiate of potassa; potassic ferrocyanide; prussiate of potash; yellow prussiate of potash.
Potassium hydrate: See potassium hydroxide.
Potassium hydrogen oxalate: See potassium binoxalate.
Potassium hydrogen sulphate: See potassium bisulphate.
Potassium hydroxide: 101, 3976 KOH Caustic potash, potassa; hydrated oxide oxide of potassa; hydrate of potassa; potassa hydrate; potassic hydrate; potassium hydroxide potassa; stock lye.
Potassium hydroxide potassa: See potassium hydroxide.
Potassium iodide: 4203 KI Potassic iodide.
Potassium neutral carbonate: See potassium carbonate.
Potassium nitrate: 4194 KNO_3 Niter; nitrate of potassa; nitre; saltpeter.
Potassium oxalate: $K_2C_2O_4$ Neutral oxalate of potassa; salt of lemon.
Potassium permanganate: 3941, 4190 $KMnO_4$ Permanganate of potash.
Potassium quadroxalate: See potassium acid oxalate.
Potassium sulphate: K_2SO_4 Dipotassic sulphate; normal potassium sulphate; potassic sulphate; sal polychrest; sulphate of potassa; vitriolated tartar.
Potassium sulphide: 4204 K_2S Hepar of sulphur; liver of sulphur; sulphuret of potassium.
Potassium sulphocyanate: See potassium sulphocyanide.
Potassium sulphocyanide: 4205 KCNS Potassic sulphocyanide; potassium sulphocyanate; sulphocyanuret of potassium.
Potassium tartrate: 4196 $K_2C_4H_4O_6$ Neutral tartrate; neutral tartrate of potassium; soluble tartrate; tartrate of potassa; vegetable salt.
Potassium trihydrogen oxalate: See potassium acid oxalate.
Potter's clay, earth: A clay fairly free from iron, used in making pottery.
Poultice: 5018 A hot, soft, moist mass, as of mustard, applied to a sore part of the body..
Pounce: 1954 A fine powder, as of cuttlefish bone, formerly used to prevent ink from blotting. Also powdered charcoal, etc., sprinkled over a stencil to make a design, as on cloth.
Pouret: A vessel used for pouring by the drop. See 82, figures 1-4.
Precipitated chalk: 1291 An ingredient in tooth paste.
Precipitated sulphide of antimony: See antimony pentasulphide.
Precipitated sulphur: 4351 A milk of sulphur, used in treating skin affections and as a mild laxative.
Precipitation: See 24.
Preston salts: See ammonium sesquicarbonate.
Price's glycerine: Probably a brand name so try plain glycerin.
Pricked paper: Paper with fine perforations for transferring patterns with chalk, lampblack, etc.
Pricked wine: Decaying, discolored or spoiled wine.
Prince Rupert's drop: 2373 A specially made drop of glass which shatters when its tail is snapped.
Privet berries: The fruit of the Ligustum vulgare.
Proof spirit: See 1436.
Protoacetate of iron: 118 Black liquor; dyer's acetate of iron; iron liquor; pyrolignite of iron; tar iron.
Protocarbonate of iron: See ferrous carbonate.
Protochloride of gold: See aurous chloride.
Protochloride of iron: See ferrous chloride.
Protochloride of manganese: $MnCl_2$ Manganous chloride.
Protochloride of mercury: See mercurous chloride.
Protochloride of tin: See stannous chloride.
Protonitrate of iron: See ferrous nitrate.
Protonitrate of mercury: See mercurous nitrate.
Protosalts: Any salt corresponding to a peroxide.
Protosulphate of iron: See ferrous sulphate.
Protosulphate of mercury: See mercurous sulphate.
Protosulphide of iron: See ferrous sulphide.
Protosulphide of mercury: See mercuric sulphide.
Protosulphuret of iron: See ferrous sulphide.
Protoxide of antimony: See antimony trioxide.
Protoxide of barium: See barium oxide.
Protoxide of bismuth: See bismuth trioxide.
Protoxide of carbon: See carbonic oxide.
Protoxide of chromium: See chromous oxide.
Protoxide of copper: See cuprous oxide.
Protoxide of iron: See ferrous oxide.
Protoxide of mercury: See mercurous oxide.
Protoxide of nickel: See nickelous oxide.
Protoxide of silver: See silver oxide.
Protoxide of tin: See stannous oxide.
Proud flesh: An exuberant growth of granulation tissue in a wound or ulcer.
Prussian alkali: See 84. Sodium hydroxide at a strength of 95 degrees.
Prussian blue: See ferric ferrocyanide.
Prussiate cake: Ferric ferrocyanide or ferrous ferricyanide pressed into cakes and used as blueing to whiten clothes.
Prussiate of copper: See copper ferrocyanide.
Prussiate of iron: See ferric citrate.
Prussiate of mercury: See mercuric cyanide.
Prussiate of potash, potassa: See potassium ferrocyanide.

Pulverulent: Something easily pulverized to a fine powder.
Pumice stone: A light, porous, volcanic rock.
Purified salt: Any salt with all its impurities removed.
Purple of cassius: 2720 Gold purple; a purple pigment.
Purple precipitate of cassius: Same as purple of cassius.
Purpurate of ammonia: $C_8H_8N_6O_6.H_2O$ Murexide of ammonia.
Pustulant: Producing pustules.
Pustule: A small circumscribed elevation of the skin with an inflamed base, containing pus.
Putrescible: A substance, usually nitrogenous, which is liable to undergo bacterial decomposition when in contact with air and moisture at ordinary temperatures.
Putrid fermentation: Putrefaction; rotting.
Pyro: See pyrogallic acid.
Pyroborate: See sodium borate.
Pyrogallic acid: 3909-10 $HC_6H_5O_3$ galline; pyro; pyrogallol.
Pyrogallol: See pyrogallic acid.
Pyrogenous oil of turpentine: See oil of turpentine.
Pyroligneous acid: See acetic acid.
Pyroligneous ether: See ethyl acetate.
Pyrolignite of iron: See protoacetate of iron.
Pyrolusite: See manganese dioxide.
Pyrophorous: 4340 Any of several substances, or mixtures, which ignite spontaneously on exposure to air, as a carbonized mixture of alum and sugar, or finely divided lead or iron.
Pyrophosphate of soda: See sodium pyrophosphate.
Pyroxilic spirit: Wood naptha.
Pyroxylin: 2141 $C_{12}H_{14}(ONO_2)_6O_4$ Cellulose nitrate; guncotton.

Q

Quadrisilicate: See sodium silicate.
Quadroxalate of potassa: See potassium acid oxalate.
Quartation: The alloying with silver of a button rich in gold, to reduce the gold to such a proportion (usually 1/4 or less) that the acid used in parting may act as desired.
Quassia: A bitter drug extracted from quassia wood.
Queen's root: The dried root of the queenroot, Stillingia sylvatica, used as an emetic and cathartic.
Quercitron bark: Bark of the black oak Quercus velutrina, used in tanning and dying.
Quicklime: See calcium oxide.
Quick match: 2060 A thread or wick of cotton impregnated with an inflammable mixture (usually a paste of gunpowder and starch). It is used in carrying fire from one part to another in fireworks and in lighting flares.
Quicksilver: Mercury.
Quinine sulphate: 4265 $(C_{20}H_{24}O_2N_2)_2H_2SO_4+7H_2O$ Disulphate of quinia; sulphate of quinine.
Quintal: 112 lbs. English; 100 lbs. U.S.

R

Radical vinegar: See acetic acid.
Rakasiri: 5115 Oil of rosemary dissolved in common gin.
Ramie: See China grass.
Rancid: Spoiled.
Raw milk: Unpasturized milk.
Raw oil: Unboiled linseed oil.
Raw sugar: Unrefined sugar.
Reagent: 4372 Any substance which, because it takes part in certain reactions, is used in detecting, examining, or measuring other substances, in preparing material, etc.
Realgar: See arsenic disulphide.
Reaumur's thermometer: See 85.
Recent: Something made recently and so is at full potency.
Rectified: 3848 Any substance distilled more than once for added purity.
Rectified spirits: 1435 Alcohol distilled until it has only 16% water.
Red argol: See argal.
Red arsenic: See arsenic disulphide.
Red bark: A red variety of cinchona bark obtained from Cinchona succiruba and its hybrids.
Red bole: See ferric oxide.
Red chromate of potash: See potassium dichromate.
Red copper ore: See cuprous oxide.
Red coral: A Mediteranian coral, Corallium nobile. Also known as C. rubrum.
Red lake: See lac resin.
Red lead: See lead oxide, red.
Red liquor: 100 A solution consisting chiefly of aluminum acetate, used as a mordant, especially in dying red.
Red mercuric oxide: See mercuric oxide.
Red mercuric sulphide: See mercuric sulphide.
Red ochre: See ferric oxide.
Red oxide of copper: See cuprous oxide.
Red oxide of iron: See ferric oxide.
Red oxide of mercury: See mercuric oxide.
Red Peruvian bark: See cinchona.
Red precipitate: See mercuric oxide.
Red prussiate: See potassium ferricyanide.
Red sanders, saunders: Red sandalwood.
Red spirits: 108 A tin spirit used in dying red.
Red sulphide of arsenic: See arsenic disulphide.
Red sulphide of mercury: See mercuric sulphide.
Red sulphuret of arsenic: See arsenic disulphide.
Red sulphuret of mercury: See mercuric sulphide.
Red tartar: Potassium bitartrate deposited during the fermentation of red wine.

Reduction: See 26.
Refractory clay: A heat resistant clay high in silica and aluminum oxide.
Reguline, regulus: The more or less impure button, globule or mass of metal formed beneath the slag in melting ores.
Regulus of antimony: Sb Antimony.
Rennet: The lining membrane of the stomach of the unweaned calf or other animal, especially the fourth stomach of ruminants. Used in curdling milk for making cheese.
Reprise: The second part of the operation of parting gold.
Residuum: Dregs; precipitates, etc.
Resin: A solid or semisolid organic substance exuded from various plants and trees, as the pines.
Resolution: Act of separating a compound into its elements or component parts.
Resolvent: That which has the power to disperse inflamatory or other lesions; a discutient; anything which aids the absorbtion of effused products.
Reverbatory furnace: A furnace or kiln in which the flame or heat is reflected from the inside surfaces onto the material being treated.
Rhamnus: See vervain.
Rhatany root: The dried root of either of two American shrubs, Krameria triandra and K. argentea, used as an astringent.
Rhizome: A root, usually one which sends out shoots.
Rhodium wood: The fragrant wood of the root and stem of either of the herbs Convolvulus scoparius and C. virgatus.
Rhombic phosphate of soda: See sodium phosphate.
Rice spirit: Alcohol from rice fermentation.
Rice water: A drink, chiefly for invalids, made by boiling a small quantity of rice in water.
Riddle: A sieve with coarse meshes, usually of wire.
Roche alum: 4256 A variety of ordinary alum.
Rochelle salt: See sodium potassium tartrate.
Roll annotto: Annatto put up in rolls. See annatto.
Roll sulphur: 4353 Sulphur distilled and condensed as a liquid which is cast into sticks or rolls.
Roman alum: Ordinary alum crystallized in cubes.
Roman vitriol: See cupric sulphate.
Ropey: A cloudy, oily look to wine needing tannin. Also capable of being drawn into a thread; viscous; tenacious; glutinous.
Rosaniline: 2553 The base of many aniline dyes.
Rose geranium: Geranium oil.
Rose pink: 2685 Whiting dyed pink as a pigment, dentifrice, etc.
Rose water: A watery solution of the essence of the rose made by distilling the fresh flowers with water.
Rosolic acid: $C_{20}H_{16}O_3$ A crystalline acid obtained by heating a mixture of phenol and cresylic acid with sulphuric acid arsenic acid; by the action of nitrous acid on rosaniline, etc.
Rot steep: To steep (cloth) in an alkiline lye in order to remove the oil, grease and dressing.
Rotten stone: A friable siliceous stone, the residue of a siliceous limestone whose calcerous matter has been removed by the solvent action of water. Also called tripoli.
Rouge: See ferric oxide.
Rubefacient: A substance for external application, producing redness of the skin.
Rubine: Any of certain red dyes.
Rue: A European, strong scented perennial woody herb (Ruta graveolens) with a bitter taste.
Runnings: That which runs off in distilling.
Russia isinglass: Isinglass made from the bladders of the huso or large sturgeon.

S

Sabadilla: A Mexican plant (Skoinolon officinale), family Melanthacreae; also its seeds, used as a source of veratrine and in the preparation of an insecticide, especially for head lice.
Sacharify: To convert starch into sugar.

Sacharine fermentation: See 16. That by which starch and gum are converted into sugar.
Saccharometer: Any device for measuring sugar in solution.
Saffron: A species of crocus (Crocus sativus) with purple flowers, and the drug, flavor and dyestuff it yields.
Sago: A dry, granulated or powdered starch from the pith of the sago palm, used as food and for stiffening textiles.
Saint John's bread: Carob.
Sal ammonia, ammoniac: See ammonium chloride.
Sal enixum: The refuse from the making of nitric acid.
Saleratus: See sodium bicarbonate.
Saleratus water: A solution of sodium bicarbonate.
Salicin, salicine: 4021 $C_{13}H_{18}O_7$ A bitter white crystalline glucoside, found in the bark and leaves of several species of willow (Salix) and poplar. On hydrolysis (by the enzyme emulsion or otherwise) it yields dextrose and saligenin. It is used in medicine as an antipyretic, antirheumatic, and tonic.
Salicylate of methyl: $C_6H_4(OH)CO_2CH_3$ The methyl salt of salicylic acid.
Salifiable: Capable of being formed into a salt.
Saligenin: $C_6H_4(OH)CH_2OH$ A white crystalline compound obtained by the hydrolysis of salacin.
Saline: Containing a salt.
Saline draughts: Sour drinks such as lemonade.
Salmiac, salmiak: See ammonium chloride.
Sal polychrest: See potassium sulphate.
Sal soda: Crystalline sodium carbonate.
Salt: Usually means sodium chloride.
Salt acid: See hydrochloric acid.
Salt of amber: See succinic acid.
Salt of antimony: A double salt of antimony fluoride and ammonium sulphate used as a mordant.
Salt of barilla: See sodium carbonate.
Salt of benzoin: See benzoic acid.
Salt of chrome: See potassium chromate.
Salt of colcothar: See ferrous sulphate.
Salt of iron: See ferrous sulphate.
Salt of lemon: See potassium binoxalate.
Salt of satum: See lead acetate.
Salt of seignette: See sodium potassium tartrate.
Salt of soda: See sodium carbonate.
Salt of sorrel: See potassium binoxalate.
Salt of steel: See ferrous sulphate.
Salt of tartar: See potassium carbonate.
Salt of vitriol: See zinc sulphate.
Salt of wormwood: See potassium carbonate.
Saltpeter, saltpetre: See potassium nitrate.
Saltpeter paper: See touch paper.
Sal volatile: See ammonium carbonate.
Sandal: Sandalwood.
Sandarac: A brittle, faintly aromatic, translucent resin obtained from the sandarac tree, usually in the form of small pale-yellow grains or tears. Used in varnish and as incense.
Sand bath: See 4. A pan of hot sand in which vessels to be heated are partly immersed.
Santal: Denotes the sandalwood tree. Check context for which part of tree.
Santaline: Sandalwood oil.
Santonate of soda: 4649 $C_{15}H_{19}O_4N_2$ Sodium santonate.
Santonica: An anthelmintic drug consisting of the unexpanded dried heads of the wormwood, Artemisia pauciflora; levant wormseed.
Santonin, santonine: $C_{15}H_{18}O_3$ A colorless, crystalline, slightly bitter compound, occuring in santonica and in other species of Artimisia, and used as an anthelmintic. In overdose it produces poisoning and disturbance of vision.
Sap green: A dull-green lake prepared from buckthorn berries.
Sapolio: A trade name for an old brand of English complexion soap.
Saponaceous: Resembling soap; having the qualities of soap.
Saponification: Act, process or result of soap making; saponify.
Sapphire: The color sapphire blue.
Sarsaparilla: Any of various species of Smilax, as S. officinalis, S. papyracea and S. medica; the dried, cordlike roots of any of these, used in the form of a decoction, infusion, fluid extract or syrup as a mild tonic and alterative.
Sassafras: The dried bark of the root of Sassafras varifolium, used as a diaphoretic, a flavoring agent, and aromatic bitters. It also yields an aromatic, volatile oil used in perfumes.
Saturation: See 27.
Savin, savine: A Eurasian evergreen (Juniperus sabina). Its bitter, acrid tops are sometimes used in medicine for gout, amenorrhea, as an abortifacient, etc.
Savonettes: Small balls of perfumed soap, usually for guests.
Savory: An aromatic European mint (Satureia hortensis) much used in cooking. Also called summer savory.
Scabies: Itch caused by mites; mange.
Scald: A sudden covering of a substance with boiling water.
Scale litharge: Newly formed litharge before being ground.
Scammony: The dried root of Convolvulus scammonia; the resin obtained as an exudation from the living root or prepared by extracting the dried root with alcohol and precipitating with water. It is a drastic cathartic.
Scheel's green: See copper arsenite.
Schiff's table 628 A table of soda solutions.
Schlippe's salt: See antimony pentasulphide.
Schuster's alkalimeter: See 82.
Scio, or chio, turpentine: Turpentine obtained from Pinus pistalia terebinthus.
Scorbutic: One having scurvy; an antiscorbutic.
Scoria. scoriae: The refuse from the smelting of metals, reduction of ores, etc.; dross; slag, etc.
Scratch brush: One made from fine brass wire.
Scrofula: Tuberculosis of the lymphatic glands, especially of the neck, with enlargement of the glands; king's evil.
Scruple: An apothecaries' weight equal to 1/3 dram.
Scurfy: A skin condition producing flakey scales.
Scurvy grass: Any of several cresses, especially Cochlearia officinalis, found in arctic regions. It is a remedy for scurvy.
Sebacic ether: Sebic or sebacic acid; obtained from the distillation of oleic acid.
Seed lac: A resinous substance secreted by a scale insect (Tachardia lacca). When crushed and washed it becomes granular and is known as seed lac.
Seignette's salt: See sodium and potassium tartrate.
Selenite: See calcium sulphate.
Senega, seneka: The dried root of the Polygala senega, containing an irritating saponin, senegin and used as an expectorant.
Senna: The dried leaflets of certain species of Cassia, especially Cassia acutifolia, used in medicine as a purgative.
Sepia bone: Cuttlefish bone.
Serpentaria: The dried rhizome and roots of the birthwort Aristolochia serpentaria and of the Texas snakeroot (A. reticulata) used for aromatic bitters.
Sesquicarbonate of ammonia: See ammonium sesquicarbonate.
Sesquicarbonate of soda: See sodium sesquicarbonate.
Sesquichloride of aluminum: See aluminum chloride.
Sesquichloride of antimony: See antimony chloride.
Sesquichloride of arsenic: See arsenic trichloride.
Sesquichloride of iron: See ferric chloride.
Sesquiferrocyanide of iron: See ferric ferrocyanide.
Sesquioxide of chromium: Cr_2O_3 Chromic oxide.
Sesquioxide of iron: See ferric oxide.
Sesquioxide of manganese: Mn_2O_3 Manganic oxide.
Sesquioxide of nickel: See nickel peroxide.
Sesquioxide of tin: 4120 Sn_2O_3
Sesquiphosphate of iron: See ferric phosphate.
Sesquisalt of iron: See ferric sulphate.
Sesquisulphate of alumina: See aluminum sulphate.
Sesquisulphate of iron: See ferric sulphate.
Sesquisulphide of arsenic: See arsenic trisulphide.
Sesquisulphuret of antimony: See antimony sulphide, black.
Sharp sand: A sand in which the grains are angular rather than rounded.
Shave grass: The scouring rush, Equisetum.
Sheepfoot jelly: The gelatinous substance obtained by boiling the bird's foot trefoil, Lotus corniculatus.
Shellbark: Shagbark; shellbark hickory.
Shell lime: Lime gotten by calcining shells.
Shoemaker's black: See ferrous sulphate.
Siderite: See ferrous carbonate.
Sienna: An earthy substance, brownish-yellow (raw sienna) and orange-red or reddish-brown when burnt (burnt sienna), much used as a pigment. It owes its colors to oxides of iron and, usually, of manganese.
Silex: SiO_2 Silica, especially in the form of quartz.
Silica jell, jelly: 2819 A form of colloidal silica, like coarse sand in appearance, but possessing many fine pores and therefore extremely absorbent.
Silicate: To combine with silica or silicates; to coat or impregnate with silica or silicates.
Silicate of potassa: 2816 K_2SiO_3 Potassium silicate.
Silicate of soda: See sodium silicate.
Silicious sand: Sand with a high silica content.
Silicic acid: In old formulas, silica, unless elaborated on.
Silicum: silica.
Silver chloride: 3214, 4083 AgCl Argentic chloride; chloride of silver; monochloride of silver.
Silver cyanide: 3697 AgCn Argentic cyanide; hydrocyanate of silver.
Silver glance: See silver sulphide.
Silver monoxide: See silver oxide.
Silver nitrate: 4077 $AgNO_3$ Argentic nitrate.
Silver oxide: 4078 Ag_2O Argentic oxide; protoxide of silver; silver monoxide.
Silver sand: A hard, heavy, silver-colored sand used by lithographers for rubbing stones to a level surface.
Silver sulphide: 4082 Ag_2S Argentite; silver glance; sulphuret of silver; vitreous silver.
Simple syrup: Syrup made up of only syrup and water.
Six carbon phenol: See phenol.
Size: Any substance used to fill the surface pores of cloth, paper, etc.
Slake: To cause (lime) to heat and crumble by treatment with water; to hydrate; also to alter (lime) by exposure to air.
Slaked lime: Lime which has been treated with water.
Small beer: Beer with a very low alcoholic content.

Smalt, smalts: 2687 A blue pigment.
Smartweed: The water pepper (Persicaria hydropiper); also any of several species of persicaria having acrid juice.
Smelling salts: See ammonium sesquicarbonate.
Smithsonite: See zinc carbonate.
Soaper salts: Impure potassium hydroxide.
Soap lees: The impure alkaline liquor left over from soap making.
Soap lye: Same as soap lees.
Soap varnish: 3824 A mixture of soap, linseed oil and litharge for bronzing.
Socotrine aloes: Aloes from the island of Socotra off Cape Guardaful, Africa.
Soda: See sodium carbonate.
Soda ash: See sodium carbonate.
Soda crystals: See sodium carbonate.
Soda glass: See sodium silicate.
Sodic: Read as sodium. Sodic carbonate=sodium carbonate, etc.
Sodic carbonate: See sodium carbonate.
Sodic hydrate: See sodium hydroxide.
Sodium acetate: 4206 $Na(C_2H_3O_2).3H_2O$ Acetate of soda.
Sodium acid sulphate: See sodium bisulphate.
Sodium and potassium tartrate: See sodium potassium tartrate.
Sodium biborate: See sodium borate.
Sodium bicarbonate: 4209 $NaHCO_3$ Hydrogen and sodium bicarbonate; hydrosodic carbonate; monosodic carbonate; sodium hydrocarbonate.
Sodium bisulphate: $NaHSO_4$ Bisulphate of soda; sodium acid sulphate.
Sodium borate: $Na_2B_4O_7.10H_2O$ Borate of sodium; borax; pyroborate; sodium biborate; sodium tetraborate.
Sodium carbonate: 4208 $Na_2CO_3.10H_2O$ Carbonate of soda; monocarbonate of soda; natron; sal soda; salt of barilla; salt of soda; soda; soda ash; soda crystals; sodic carbonate; subcarbonate of soda; washing soda.
Sodium chloride: 4215 NaCl Common salt; halite; marine salt; muriate of soda.
Sodium hydrocarbonate: See sodium bicarbonate.
Sodium hydroxide: 102, 3979 NaOH Caustic soda; hydrate of soda; sodic hydrate.
Sodium nitrate: $NaNO_3$ Caliche; chilian saltpeter; cubic niter; Peruvian saltpeter.
Sodium phosphate: 4210 $Na_2HPO_4.12H_2O$ Disodic orthophosphate; hydrosodic phosphate; perlate salt; phosphate of soda; rhombic phosphate of soda; tasteless salt; tribasic phosphate of soda.
Sodium potassium tartrate: 4213 $KNaC_4H_4O_6.4H_2O$ Potassio-tartrate of soda; rochelle salts; salt of seignette; sodium and potassium tartrate; tartarated soda; tartarized soda; tartrate of potassium and soda.
Sodium pyrophosphate: $Na_4P_2O_7+10H_2O$ Calcined sodium phosphate; normal sodium pyrophosphate.
Sodium silicate: 2816 $Na_2Si_4O_2$ Quadrisilicate; soda glass; soluble glass; tetra silicate; water glass.
Sodium sulphate: 4207 $Na_2SO_4.10H_2O$ Glauberite; glauber's salt; sulphate of soda; thenardite miribilite; wonderful salt.
Sodium sulphite: Na_2SO_3 Sodic sulphite; sulphite of soda.
Sodium tetraborate: See sodium borate.
Soft soap: Soap with potassium hydroxide for a base.
Solid bitumen: Asphalt.
Solomon's seal: An herb, Polygonatum, used to flavor beer, etc.
Soluble glass: See sodium silicate.
Soluble indigo: A blue pigment, ammonium sulphate and/or potassium sulphate.
Solution of ammonia: See ammonia, solution of.
Solution of tin: Tin dissolved in acids for dying. See 107-13.
Sonorous cake stearine: Stearine so pure and hard it rings.
Sour: See 105. A sulphuric acid and water solution used for dying.
Spanish brown: Earth of dark reddish-brown (due to the presence of iron oxides), used as a pigment.
Spanish licorice: The common licorice.
Spanish white: Whiting. Also bismuth subnitrate.
Spathic iron ore: See ferrous carbonate.
Specular iron ore: See ferric oxide.
Sperm oil: Oil from the sperm whale.
Spermaceti: A white, waxy solid which separates from the oil obtained from the sperm whale, the dolphin and allied cetaceans.
Sphalerite: See zinc sulphide.
Spile: A plug for a wine vat.
Spirit, spirits: See 17. Any liquid produced by distillation. Usually alcoholic; also, various solutions used in dying, such as tin spirits, etc.
Spirit of minderous: See ammonium acetate.
Spirit of salts: See hydrochloric acid.
Spirit of tar: (2417)?
Spirit of wine: Alcohol; rectified spirit.
Spirit lamp: A lamp which burns alcohol.
Spiritous: Refined; pure.
Spirits of hartshorn: See ammonia, solution of.
Spirits of niter: 70% absolute nitric acid, HNO_3, and 30% water.
Spirit varnish: 2903 Varnish with an alcoholic base.
Spirit tin: 107 Tin digested in acids and used as mordants.
Spirit turpentine: See oil of turpentine.
Spodumene: See lithium aluminum sulphate.
Spongiopilene: A kind of cloth interwoven with small pieces of sponge and rendered waterproof on one side. When moistened with hot water it was used as a poultice.
Spongy platinum: 3336-7, 4086.
Spontaneous: Self-acting with no external cause.
Sprat: A small herring.
Squaw vine: See partridgeberry.
Squill: The cut and dried fleshy inner scales of the white variety of the bulb Urginea scilla, or the younger bulbs of U. indica. It is used as an expectorant, cardiac stimulant and diuretic.
Squirrel corn: A North American herb (Dicentra canadensis); turkey corn.
Stannate of potassa: $K_2SnO_3.3H_2O$ Potassium stannate.
Stannate of soda: $Na_2SnO_3.3H_2O$ Sodium stannate.
Stannic acid: 4121 H_2SnO_3.
Stannic chloride: 4124 $SnCl_4$ Bichloride of tin; butter of tin; chloride of tin; Libavius' fuming liquor; perchloride of tin; permuriate of tin; tetrachloride of tin.
Stannic oxide: 4121-2 SnO_2 Binoxide of tin; cassiterite; dioxide of tin; peroxide of tin; tin putty.
Stannous chloride: 4123 $SnCl_2$ Dichloride of tin; muriate of tin; protochloride of tin; tin crystals.
Stannous chloride, hydrated: $SnCl_2.2H_2O$ Tin salt.
Stannous oxide: 4119 SnO Monoxide of tin; oxide of tin; protoxide of tin.
Star anise: The dried fruit of the illicium, used as a spice.
Starch gum: Dextrine.
Stavesacre: A Eurasian larkspur (Delphinium staphisagria). Its seeds contain delphinine, which is violently emetic and cathartic and which is used as a fish poison. A tincture or ointment prepared from the seeds is used to kill lice on the head or body.
Steam bath: To subject a substance to steam as in a kitchen steamer.
Steam glue: (2287) ?
Steam jacket: An outer casing between which and an inner cylinder (or other body) can be passed a current of steam, air, etc.
Stearic acid: 4325 $CH_3(CH_2)_{16}CO_2$ A white, crystalline fatty acid obtained by saponification of tallow or other hard fats containing stearin; octadecanoic acid. It melts to an oily liquid at 69.3° C. (156.7° F.). It is used chiefly for making candles, usually with the addition of parrafin. Commercial stearic acid (stearine) is commonly a mixture of stearic and palmitic acids.
Stearine: See stearic acid.
Stearopten: 1472 The part of an essential oil separated as a solid on cooling or long standing.
Steep: See 39. Macerate.
Stereo-chromic painting: A process of mural painting in which the pigment is fixed by a series of reactions between the lime, fluosilicic acid, and sodium silicate.
Stibnite antimonite: See antimony sulphide, black.
Stillingia: 4587 The dried root of the queensroot Stillingia silvatica, used as an emetic and cathartic.
Stock lye: See potassium hydroxide.
Stone blue: See azurite.
Stone paper: 1934 Sandpaper made with powdered pumice stone.
Stone ware: A coarse potter's ware, glazed and fired to a density.
Stoneware head: One used for distilling corrosive substances.
Stopcock: A cock or valve for stopping or regulating the flow through a pipe, etc.
Storax: A resin derived from various trees of the genus styrax, especially Styrax officinalis. It was formerly used as incense.
Stramonium: 4499 The dried leaves of the jimson weed, used in medicine similarly to belladonna, especially in asthma. It contains the alkaloids atropine, hyoscyamine and scopolomine.
Strass: 2352, 2419 A brilliant lead glass used in making artificial gems.
Stratum: Layers.
Stream tin: Cassiterite accumulated in alluvial deposits.
Strike: To precipitate (a dye) by a mordant. Also, to cause (a dye) to be absorbed by an inert base.
Strong muriatic acid: At least 35% hydrochloric acid in water.
Strontanite: See strontium carbonate.
Strontium carbonate: $SrCO_3$ Carbonate of strontia; strontanite.
Strichnia, strichnine: 4005 $C_{21}H_{22}N_2O_2$ A very poisonous alkaloid obtained from various plants of the genus Strychnos, as from the seeds of the St. Ignatius' bean (S. ignatii) and from nux vomica. It is used in medicine (usually as a sulphate) as a tonic and stimulant for the central nervous system.
Styptic: See astringent.
Styrax calamita: A small shrub or tree of Asia Minor yielding a resin similar to storax and used in medicine.
Styrol: See 4315.
Subacetate of copper: See copper acetate.
Subacetate of lead: See tribasic lead acetate.
Subcarbonate of iron: See ferrous carbonate.
Subcarbonate of potash: See potassium carbonate.
Subcarbonate of soda: See sodium carbonate.
Subchloride of copper: See cuprous chloride.
Subchloride of mercury: See mercurous chloride.
Sublimation, sublime: See 30.
Sublimed sulphur: See flowers of sulphur.

Submuriate of mercury: See mercurous chloride.
Subnitrate of bismuth: See bismuth subnitrate.
Suboxide of copper: See cuprous oxide.
Suboxide of mercury: See mercurous oxide.
Subsalt: Chemically, an oxysalt, as bismuth oxychloride or bismuth oxynitrate, formed from hydrolysis of bismuth salts in water.
Subsidence: To sink or fall to the bottom; to settle.
Substantive colors: See 93. Dyes which color without a mordant.
Subsulphate of mercury: See mercurous sulphate.
Succinic acid: 4306 $(CH_2CO_2H)_2$ Salt of amber.
Sudorific: Causing or inducing sweat, diaphoretic.
Suet: The hard fat about the kidneys and loins in beef and mutton, which, when melted and freed from the membranes, forms tallow.
Sugar of lead: See lead acetate.
Sugar of milk: See lactose.
Sugar of resin: 4313 A substance used to waterproof gunpowder, matches, etc.
Sulphantimoniate of potassa: K_3SbS_4 Potassium thioantimoniate.
Sulphate: To treat or impregnate with sulphuric acid or a sulphate; to convert into a sulphate.
Sulphate of alumina: See aluminum sulphate.
Sulphate of ammonia: See ammonium sulphate.
Sulphate of atropia: $(C_{17}H_{23}O_3N)_2H_2SO_4$ Atropine sulphate.
Sulphate of baryta: See barium sulphate.
Sulphate of chromium: $CrSO_4.7H_2O$ Chromium sulphate.
Sulphate of cobalt: $Co_2(SO_4)_3$ Cobalt sulphate.
Sulphate of copper: See cupric sulphate.
Sulphate of indigo: $C_{16}H_8N_2O_2.SO_4$ Indigo sulphate.
Sulphate of iron: See ferrous sulphate.
Sulphate of lead: See lead sulphate.
Sulphate of lime: See calcium sulphate.
Sulphate of lithia: 4239 Li_2SO_4 Lithium sulphate.
Sulphate of magnesia: See magnesium sulphate.
Sulphate of manganese: See manganous sulphate.
Sulphate of nickel: 4177 $NiSO_4.7H_2O$ Nickelous sulphate.
Sulphate of oxide of ammonia: See ammonium sulphate.
Sulphate of peroxide of iron: (2676)?
Sulphate of potassa: See potassium sulphate.
Sulphate of protoxide of manganese: See manganous sulphate.
Sulphate of quinine: See quinine sulphate.
Sulphate of rosaniline: Rosaniline sulphate.
Sulphate of silver: 4081 Ag_2SO_4 Silver sulphate.
Sulphate of soda: See sodium sulphate.
Sulphate of strychnia: $(C_{21}H_{22}O_2N_2)_2.H_2SO_4+5H_2O$ Strychnine sulphate.
Sulphate of the suboxide of mercury: See mercurous sulphate.
Sulphate of zinc: See zinc sulphate.
Sulphide of ammonia: See ammonium sulphide.
Sulphide of antimony: See antimony sulphide, black.
Sulphide of arsenic: See arsenic disulphide.
Sulphide of barium: BaS Barium sulphide.
Sulphide of cadmium: See cadmium sulphide.
Sulphide of calcium: See calcium sulphide.
Sulphide of carbon: See carbon disulphide.
Sulphide of copper: CuS Cupric sulphide.
Sulphide of hydrogen: See hydrogen sulphide.
Sulphide of iron: See ferrous sulphide.
Sulphide of magnesia: 4247 MgS Magnesium sulphide.
Sulphide of mercury: See mercuric sulphide.
Sulphide of potassium: K_2S Potassium sulphide.
Sulphide of silver: See silver sulphide.
Sulphide of soda: Na_2S Sodium sulphide.
Sulphite: 3864 The salts formed by the combination of sulphurous acid with a base.
Sulphite of lime: $CaSO_3.2H_2O$ Calcium sulphite.
Sulphite of potassa: $K_2SO_3.2H_2O$ Potassium sulphite.
Sulphite of soda: See sodium sulphite.
Sulpho-carbolate of zinc: $(C_6H_4SO_3H)_2Zn$ Zinc sulphophenolate.
Sulpho-carbonic acid: See carbon disulphide.
Sulphocyanide of ammonium: 4226 NH_4CNS Ammonium sulphocyanide.
Sulphocyanide of potassium: See potassium sulphocyanide.
Sulphocyanuret of potassium: See potassium sulphocyanide.
Sulphydric acid: See hydrogen sulphide.
Sulphurated antimony: See antimony pentasulphide.
Sulphuration: The process in which materials are bleached by the exposure to the fumes of burning sulphur.
Sulphuret: Read as sulphide. Sulphuret of potassium=potassium sulphide, etc.
Sulphureted hydrogen: See hydrogen sulphide.
Sulphureted sulphide of calcium: CaS_2 Calcium disulphide.
Sulphureted water: 4462 Water impregnated with hydrogen sulphide.
Sulphuric acid: 3854-60 H_2SO_4 Oil of vitriol. See page 221.
Sulphuric ether: See ethyl oxide.
Sulphurous acid: 1718, 3864-5-6-8, 4066 H_2SO_3 Sulphurous anhydride.
Sulphurous anhydride: See sulphurous acid.
Sulphur vivum: 4355 S Black sulphur; crude sulphur; horse brimstone.
Sumac, sumach: A material used in tanning and dying, consisting of the dried and powdered leaves and flower clusters of various species of sumac. Sumac contains from 10 to 30% tannin.
Summer savory: See savory.
Superadded: A substance added by choice, over and above what is desired.
Superincumbent: Lying or resting on so as to exert pressure.
Supernatant: Floating on the surface, as oil on water. Also, the liquid left over after precipitation.
Supersaturate: To add to beyond saturation.
Supertartrate of potassium: See potassium bitartrate.
Sweet flag: The aromatic root of the Acorus calamus, used as a carminative and tonic in dyspepsia and colic.
Sweet marjoram: An aromatic European herb (Marjorana hortensis), closely related to the wild marjoram (Oreganum vulgare).
Sweetmouth glue: Glue which is to be wet with the tongue, as stamps, envelopes, slightly sweetened so it won't taste bad.
Sweet oil: Any mild, edible oil, as olive oil.
Sweet spirits of niter: 4289 An alcoholic solution of nitrous ether.
Sweet yeast: (730)?
Swilled: Swirled.
Sylvite: See potassium chloride.
Syrup of ether: 4653 A calming medicine made of ethyl oxide and syrup.
Syrup of lemons: Lemon juice extract, etc., mixed with syrup.
Syrup of saffron: Saffron mixed with syrup.
Syrup of violets: Violet scent mixed with syrup.

T

Tallow: The fat of animals like the ox and sheep, extracted from membranous and fibrous matter by melting. It is solid, white and almost tasteless when pure, and is used in soap, candles, oleomargarine, etc.
Tanbark: Any bark rich in tannin, bruised or cut into small pieces, and used in tanning.
Tan liquor: Tanning liquor.
Tannate of iron: 4170 $Fe(C_{13}H_9O_7CO_2)_2$ Ferrous tannate.
Tannate of lime: 2330 Boiler encrustations broken down by tannin.
Tannate of quinia: Quinine impregnated with tannin.
Tannic acid: 3911-12 $C_{14}H_{10}O_9$ Tannin.
Tannin: An amorphous, strongly astringent substance obtained in the form of brownish-white shining scales from gallnuts (of which it constitutes 50% or more), sumac, valonia and other plant products.
Tansy: Any plant of the genus Tanacetum. The common tansy (T. vulgare) has a strong aromatic odor and a very bitter taste. It was formerly much used in cookery and in medicine as a bitter tonic.
Tapioca paper: 3157 A paper used for copying photographs by artificial light.
Tar: A thick, dark-brown or black, viscous liquid obtained by the distillation of wood, coal and other organic materials, and having a varied composition according to the temperature and material employed in obtaining it.
Taraxacum: The dried rhizome and roots of the dandelion Taraxacum officinale, used as a bitter and laxative.
Tar iron: See protoacetate of iron.
Tartar: A substance existing in the juice of grapes and deposited in winecasks as a pale or dark reddish crust or sediment.
Tartarated antimony: See antimony tartrate.
Tartarated iron: See ferric and potassium tartrate.
Tartarated soda: See sodium potassium tartrate.
Tartar emetic: See antimony tartrate.
Tartaric acid: 3929-30 $C_4H_6O_6$ A colorless crystalline acid found in fruit juices, etc. and obtained from tartar. Used in cookery, medicine, etc.
Tatrarized antimony: See antimony tartrate.
Tartrate of antimony: See antimony tartrate.
Tartrate of iron and potassium: See ferric and potassium tartrate.
Tartrate of lead: $PbC_4H_4O_6$ Lead tartrate.
Tartrate of lime: $CaC_4H_4O_6.4H_2O$ Calcium tartrate.
Tartrate of potassa: See potassium tartrate.
Tartrate of potassa and iron: See ferric and potassium tartrate.
Tartrate of potassium and soda: See sodium potassium tartrate.
Tartrate of silver: $Ag_2C_4H_4O_6$ Silver tartrate.
Tasteless alcohol: That which is pure and free from foreign matter.
Tasteless salt: See sodium phosphate.
Tears: Drop-size bits.
Teel: Sesame oil.
Tenacity: The resistance to being pulled asunder by the force of tension.
Tenorite: See cupric oxide.
Terchloride of antimony: See antimony chloride.
Terchloride of arsenic: See arsenic trichloride.
Terchloride of gold: See auric chloride.
Terhydride of nitrogen: See ammonia.
Ternitrate of bismuth: See bismuth nitrate.
Teroxide of antimony: See antimony trioxide.
Teroxide of bismuth: See bismuth trioxide.
Terpin, terpine: 4312 $C_{10}H_{18}(OH)_2$ Terpin.
Terpinol, terpinole: See terpin.
Terra merita: Tumeric.

Tersulphate of iron: 4168 $Fe_2(SO_4)_3$ Ferric sulphate.
Tersulphide of antimony: See antimony sulphide, black.
Tersulphide of arsenic: See arsenic trisulphide.
Tersulphite of arsenic: $As_2(SO_3)_3$ Arsenic trisulphite.
Tersulphuret of antimony: See antimony sulphide, black.
Tersulphuret of arsenic: See arsenic trisulphide.
Test papers: 4408 Papers, which, when dipped into solutions, show acidity or alkalinity.
Tetra-ammonio dihydric carbonate: See ammonia sesquicarbonate.
Tetrachloride of tin: See stannic chloride.
Tetrasilicate: See sodium silicate.
Tetter: Any of various vesicular skin diseases, as eczema, herpes and ringworm.
Theriac: Any antidote to poison.
Thenardite miribilite: See sodium sulphate.
Thrip: A small insect which feeds on plant juices.
Tin ashes: (2394)?
Tincal: Crude native borax.
Tin crystals: See stannous chloride.
Tincture: A solution of medical substance in alcohol. Also a dye product.
Tincture of galls: Tannin extracted from galls.
Tincture of iodine: 4491 Iodine dissolved in alcohol.
Tincture of iron: See ferrous sulphate.
Tincture of litmus: Litmus used in solution instead of dried in paper.
Tinned copper: Tin plated copper.
Tin powder: 3317-8 Finely powdered tin.
Tin putty: See stannic oxide.
Tin salt: See stannous chloride.
Tin spirits: 107-13 Tin dissolved by acids and used as mordants.
Tin stone: See stannic oxide.
Tolu: See balsam of tolu.
Toluidine: 2552-3 A derivitive of toluene used in making dyes.
Tombac: 3442-3 An alloy consisting mainly of copper and zinc, used for cheap jewelry, gilding, etc. See Dutch foil.
Tone: 3137-8 In photography, achieving the proper shade of the image.
Tonics: 5117 Medicines that increase the tone of the muscular fiber and impart vigor to the system.
Tonka bean: The seed of a tropical South American tree (Dipteryx odorata). It has a pleasant odor, due to the presence of coumarin. It is used to make coumarin, in perfumes and as a flavoring. Tonquin.
Tonquin: See tonka bean.
Tormentilla: A Eurasian herb (Potentilla tormentilla), the root of which is powerfully astringent, and is used sometimes in medicine and in tanning and dying.
Touch needle: A small bar of gold, either pure, or alloyed with silver in a known proportion, for trying the fineness of a gold or silver article by comparing the streaks made by the article and the bar on a touchstone.
Touch paper; 2059 Paper impregnated with potassium nitrate. It burns steadily without flame and is used for igniting fireworks. Saltpeter paper.
Touchstone: 3190 A stone or piece of black pottery used for assaying.
Tow: The coarse and broken part of flax, hemp or jute, separated and ready for spinning.
Trace chain: Chain used to fasten a horse's collar to the whippletree.
Tragacanth: A gum obtained from Astragalus gummifer. It comes in thin, white, translucent pieces, or flakes. The gum swells up in water and is used in the arts and as an emulsifying agent and excipient in pharmacy.
Train oil: Oil from the whale or other marine animal.
Tralles: See 54-59. A hydrometer.
Tranquil fusion: The melting of a compound without any chemical change.
Translucent: Partially transparent.
Transudation: Sweating through.
Tribasic lead acetate: $Pb(C_2H_3O_2)_2.2PbO$ Basic lead citrate; diacetate of lead; Goulard's acetate of lead; Goulard's lotion; Goulard's extract; Goulard's water; subacetate of lead.
Tribasic phosphate of soda: See sodium phosphate.
Tricalcic phosphate: See calcium phosphate.
Trichloride of antimony: See antimony chloride.
Trihydroxybenzoic acid: See gallic acid.
Trinidad asphaltum: A natural asphalt found in a pitch lake in Trinidad.
Triplumbic tetroxide: See lead oxide, red.
Tripoli: A very friable, soft silica, including diatomacious earth, found in Tripoli.
Trisnitrate of bismuth: See bismuth subnitrate.
Trituration: See 31.
Troy weight: See 5942.
True asphaltum: Asphalt.
Truncated: Cut off or cut short.
Tubulated: See 13, figure 1.
Tubulure: A short tubular opening as at the top of a retort.
Tubus: Tubule. A small pipe or fistular body; a little tube.
Tum: To card wool or other fiber as a preliminary to finer carding.
Tun: A large cask holding 4 hogsheads or 252 gallons.
Tungstate of lime: $CaWO_4$ Calcium tungstate.
Tungstate of soda: 4212 $Na_2WO_4.2H_2O$ Sodium tungstate.
Tungstic acid: 4212 H_2WO_4.
Tungstic glue: 2281 An artificial hard rubber.
Turbid: Having the lees or sediment disturbed; muddy; unclear.
Turkey corn: See squirrel corn.
Turkey rhubarb: Chinese rhubarb formerly imported through Turkey.
Turmuric: The starch from the root of the Curcuma longa.
Turmuric paper: Paper impregnated with turmuric, used as a test for alkaline substances, which turn it from yellow to brown, and for boric acid, which turns it red-brown.
Turnbull's blue: $Fe_3(Fe(CN)_6)_2$ Ferrous ferricyanide.
Turner: Lathe.
Turnings: Fine slivers, usually metal, from a lathe.
Turpeth's mineral: See mercurous sulphate.
Tutty: 4113 Impure zinc oxide.
Twaddell's hydrometer: See 68.

U

Ultramarine: A pigment obtained by powdering the product from roasting a mixture of kaolin, soda ash, sulphur and charcoal.
Umber: A brown earth, highly valued by artists as a permanent pigment and used either in the raw state or calcined.
Unctious oils: Oils used as unguents, for anointing.
Unexceptionable: Perfect.
Unguent: A lubricant or salve for sores, burns; an ointment.
Unicorn root: The dried rhizome and roots of Aletris farinosa.
Unkilned malt: Malt which has had no heat treatment.
Unsized: Not treated with any sizing, especially of paper.
Unvolatilized: Not evaporated; in liquid form.
Urea: 4323 $CO(NH_2)_2$ The chief solid constituent of the urine of man and other animals.
Ure's test: See 76,77. Tests for the strength of acetic acid.
Uric acid: $C_5H_4N_4O_3$ A white, odorless and tasteless, nearly insoluble diacid, present in small quantity in the urine of man and most animals.

V

Valentenite: See antimony trioxide.
Valerian: A drug consisting of the dried rhizome and roots of the common valerian, Valeriana officinalis, used as a carminative, especially in nervous affections.
Valerianate of amyl: See amyl valeriate.
Valerianate of amyloxide: 1469 $C_{10}H_{20}O_2$ Amyl isovaleriate.
Valerianate of ethyl: 4300 $C_4H_9.COO.C_2H_5$ Ethyl valeriate.
Valerianate of potassa: 4305 $C_4H_9.COOK$ Potassium valeriate.
Valerianate of soda: 1469 $C_4H_9.COONa$ Sodium valeriate.
Valerianic acid: 4305 $CH_3.(CH_2)_3.COOH$ Valeric acid.
Vegetable acid: An acid occuring plants as citric acid; an organic acid.
Vegetable black: A fine variety of lampblack made by the combustion of vegetable oils.
Vegetable salt: See potassium tartrate.
Vegetal: Vegetable; from plants.
Venetian, Venice glass: A glass made at Murano, near Venice.
Venetian red: See ferric oxide.
Venetian turpentine: 4318 Turpentine from the Larix decidua.
Veratria, veratrine: A mixture of alkaloids (chiefly cevadine and veratridine) obtained as a white or grayish powder from sabadilla seeds and used as a counterirritant in neuralgia and arthritis. It is an intense local irritant and a powerful muscle and nerve poison.
Verbena: Lippia citriodora, etc.; Indian lemon grass.
Verdigris: 4088 $2Cu(C_2H_3O_2)_2.CuO$ Basic copper acetate; verditer.
Verditer: See verdigris.
Verjuice: The sour juice of crabapples, of green or unripe grapes, apples, etc.; also an acid liquor made from such juice.
Vermifuge: See anthelmentic.
Vermillion: See mercuric sulphide.
Vervain: Any plant of the genus Verbena.
Vesicant: A blistering application or plaster; a vesicatory; an epispastic.
Vesicle: A small and more or less circular, elevation of the outer layer of the skin, containing a clear watery fluid.
Vinaigrette: A small bottle holding smelling salts.
Vinegar of squill: A solution of the active principal of squill in vinegar or dilute acetic acid, usually prepared by maceration.
Vinous: Of or pertaining to wine.
Vinous fermentation: See 16.
Virgin wax: Bee's wax, pure and unadulterated.
Viscid: Sticking and adhering, and having a ropy or glutinous consistency; sticky, as tar, gums, etc.
Viscous fermentation: See 16.
Vitivert: The cuscuta plant of the family Cuscutaceae.
Vitrified antimony: See glass of antimony.
Vitrified lead: ?
Vitrified oxide of antimony: See glass of antimony.
Vitrify: To change into glass or a glassy substance by heat and fusion.

Vitriol: A sulphate of any of various metals, as copper (blue vitriol), iron (green vitriol), etc.
Vitriolated tartar: See potassium sulphate.
Vitrious silver: See silver sulphide.
Volatile: Easily passing away by evaporation; readily vaporizable, as volatile oil or liquids.
Volatile air: See ammonia.
Volatile alkali: See ammonia.
Volatile salt: See ammonium carbonate.
Volatile spirits of hartshorn: See ammonium sesquicarbonate.
Volatilize: To render or become volatile; to exhale or evaporate; to cause to pass off in vapor.
Vulcanize: Heat treatment to harden natural rubber.

W

Wahoo root: A shrub (Evonymus atropurpureus). The dried bark of the root is used as a cathartic.
Wall saltpeter: Calcium nitrate, found as an efflorescence on walls exposed to decaying nitrogenous matter or constant dampness.
Walnut: Usually means the black walnut (Juglans nigra).
Walnut peel, rind: The rind of the black walnut used as a brown dye.
Warp: The threads which extend lengthwise in fabric as opposed to the woof, which threads go over and under the warp threads.
Wash: The alcoholic mixture resulting from fermentation, before distilling.
Washing: See 32.
Washing soda: See sodium carbonate.
Wash leather: Split sheepskin dressed with oil, in imitation of chamois, and used for cleaning, dusting, washing, etc.
Waste soda ashes: Ashed of marine plants lixivated and leached of their salts.
Watch glass: A watch lens used as an evaporator.
Water bath: See 5.
Water glass: See sodium silicate.
Water of ammonia: See ammonia, solution of.
Water of crystallization: Water which is regarded as present (chemically combined) in many crystallized substances, and which is expelled from them by heat, usually with the loss by the substance of its crystalline properties.
Water plate: A plate heated by hot water in a receptacle beneath.
Wedgewood ware mortar: Since wedgewood is a decorating style with pottery, it is supposed that any ceramic mortar would do.
Weld: The black oak Quercus velutina; also its bark, used in tanning and dying.
Whale oil foot: The sediment produced in refining whale oil.
Whey: The watery part of milk, separated from the thicker, coagulating part, or curd.
White agaric: A common white fungus (Fomes, syn. polyphorous, officinalis), used for its medicinal properties; purging agaric.
White arsenic: See arsenic.
White copperas: $Fe_2(SO_4)_3.9H_2O$ Hydrous ferric sulphate. The term white copperas is also used for zinc sulphate.
White glass: Clear, uncolored glass.
White glycerine: Probably either chemically pure or a brand name.
White gold: See platinum.
White leadL See lead carbonate.
White liquor: An emulsion of potassium carbonate and fat oil in dying. See 189.
White merino: Wool from the merino sheep.
Whitening: See calcium carbonate.
White precipitate: See mercuric ammonium chloride.
White size: 1951 Sodium silicate: Called white because it is for white paper.
White tartar: Potassium bitartrate deposited during the fermentation of white wine.
White vitriol: See zinc sulphate.
White wax: Beeswax rendered white by bleaching.
Whiting: See calcium carbonate.
Whitlow: See felon.
Wine glass: Two fluid ounces. See 5963.
Wine of antimony: See antimonial wine.
Wine pipes: See pipe.
Wintergreen: In North America, a low evergreen herb (Gaultheria procumbens) whose aromatic leaves yield oil of wintergreen.
Winter savory: An aromatic mint (Satureia montana).
Winter strained lard oil: Lard oil gathered in winter. Sopposedly more potent.
Witness: Part of a process in assaying gold. See 3191.
Wolf bane: See aconite.
Wolff's apparatus: (4227)?
Wonderful salt: See sodium sulphate.
Wood naptha: A distillation product of wood, containing chiefly methanol (wood alcohol), with some acetone, methyl acetate, etc.; wood spirit.
Wood spirit: See wood naptha.
Woof: See warp.
Worm: A spiral condensing tube used in distilling.
Worm seed: The fruit of the Mexican tea (Chenopodium ambrosioides) used as an anthelmintic. Also the drug santonica, called specifically Levant wormseed.
Worm tub: The vessel containing cool water in which the worm is fixed.
Worsted: A smooth-surfaced yarn spun from long-stapled pure wool, combed so that its fibers lie parallel to each other.
Wort: An infusion of malt which ferments and forms beer.
White oxide of antimony: See antimony oxide.

X

Xylene: See ethyl benzene:
Xylol: See ethyl benzene.

Y

Yarrow: 4588 A strong-scented common herb (Achillea millefolium), used as an alterative, astringent and diuretic. Also put on wounds and some people even smoked it.
Yellow bark: See calisaya.
Yellow chromate of potassa: See potassium chromate.
Yellow dock: A common herb (Rumex crispus). Its roots are stomachic, tonic and purgative.
Yellow jasmine, jessamine: Gelsemium sempervirens. Its roots are used as a spasmodic, diaphoretic and neuralgic.
Yellow mercuric oxide: See mercuric oxide.
Yellow oxide of lead: See lead oxide.
Yellow oxide of mercury: See mercuric oxide.
Yellow protoxide of lead: See lead oxide.
Yellow prussiate of potash: See potassium ferrocyanide.
Yellow sanders: A tropical tree (Ximinia americana) with yellow wood.
Yellow subsulphate of mercury: See mercurous sulphate.
Yellow sulphuret of arsenic: See arsenic trisulphide.

Z

Zaffre: Crude cobalt oxide calcined with siliceous sand.
Zedoary: A fragrant East Indian drug of a warm, bitter aromatic taste, formerly used in medicine as a stimulant. It is the rhizome of different species of curcuma, especially Curcuma zedoaria.
Zeiodite: 2213 A hard substance made from sulphur and glue, used for making water-tight, air-tight cells for galvanic batteries.
Zest: The aromatic oil yielded by any citrus peel.
Zinc blende: See zinc sulphide.
Zinc carbonate: 4112 $ZnCO_3$ Carbonate of zinc; smithsonite.
Zinc chloride: 4109-11 $ZnCl_2$ Chloride of zinc; butter of zinc; muriate of zinc.
Zinc cyanide: 3753, 4115 $ZnCy_6$ Cyanide of zinc; cyanureret of zinc.
Zinc gray: Zinc dust used as a pigment.
Zincite: See zinc oxide.
Zinc oxide: 4117 ZnO Distilled zinc; flowers of zinc; peroxide of zinc; red zinc ore; tutty; zincite; zinc white.
Zinc sulphate: 4114 $ZnSO_4.7H_2O$ Goslarite; salt of vitriol; white copperas; white vitriol; zinc vitriol.
Zinc sulphide: ZnS Black jack; sphalerite; zinc blende.
Zinc vitriol: See zinc sulphate.
Zinc white: See zinc oxide.

CHEMICAL MAGIC

With Full Explanations, including "Patter," for giving an Entertainment

BY

V. E. JOHNSON, M.A.

LONDON

C. ARTHUR PEARSON, LIMITED

HENRIETTA STREET, W.C.

1920

CHEMICAL MAGIC - GLOSSARY:

For most of the unfamiliar terms, check the DICTIONARY OF OLD-FASHIONED TERMS on page two through twenty-two.

Those terms which were not in that section are here.

Acidulous tartrite of potash: Cream of tartar.
Black resin or colophony: Common resin - dark.
Chloride of platinum: Platinous chloride.
Copper salt: Use copper sulphate.
Gum water: Mucilage thinned with water.
Heavy paraffin oil: Kerosene.
Lamp cotton: Lamp wick or the cotton from it.
Lithium salt: Lithium crystals; may be crushed.
Loaf sugar: Pure white sugar.
Nitrate of lithia: Lithium nitrate.
Nitrate of tin: Stannic nitrate.
Oil of almonds: Oill pressed from almonds.
Phosphuret of lime: Calcium phosphate.
Rock alum: Alum in crystalline fragments.
Sodium hydrate: Sodium hydroxide, lye.
Spirits of wine: Brandy.
Sulphocyanide of mercury: Mercuric sulphocyanide.
Sulphur dioxide gas: See 4052, 4053, page 52.
Sulphurin: Sulphur.
Vesta: Common match.
Yellow soap: Common lye soap as opposed to a detergent-based soap.

Contents

CHEMICAL MAGIC

CHAPTER I.

INTRODUCTORY.

Chemistry has been called, on many occasions, a fascinating science, and with much truth, for it certainly affords more recreation than any other. From times immemorable it has never failed to be a veritable storehouse of mystery and wonder; for ages it has been used by conjurers and magicians for the production of effects which have not only astonished and perplexed our forefathers, but our contemporaries as well.

Without a doubt no other science lends itself so readily to the production of magical effects. Nevertheless, although it is quite true that magicians for the last 2,000 years and more *have* used it for the producing of some of their effects, its use has always been strictly limited; and this in spite of the fact that fresh discoveries are ever being made in chemistry which are both startling and even unexpected, not only to those who have no knowledge of the subject, but even to those who have.

The true reason probably is that such " tricks " have too often been undertaken by people of a totally unscientific character, devoid of all such training, and presented in a totally unscientific and therefore in reality " careless " manner. Under such conditions failure is a certainty.

2

Now the reader need not be in the least alarmed by the above statement or think that a certain knowledge of chemistry or a certain amount of scientific training is necessary before he or she can carry out successfully the experiments, etc., related in the following pages.

This is not so at all—all that is necessary *is to be scientific in your method of carrying them out.* Haphazard methods are worse than useless, and in the end can only lead to failure, and no failure is worse than a chemical illusion or experiment carried out in a careless, slovenly manner.

PRECAUTIONS AND GENERAL HINTS.

First of all, good pure chemicals *must* be used, this is the first condition of success. All such chemicals *must* be kept in *well-corked* (or better still glass-stoppered) bottles; tins and cardboard boxes should not be used. Chemicals cannot be depended on, under such circumstances, for the delicate tests that they have to fulfil.

As already stated, keep all bottles well and tightly corked and clearly labelled—they must be kept in a dry place—this may also be considered as essential.

Any solutions or acids which at any time become discoloured must be straightway poured down the sink.

For all solutions, etc., distilled water only should be used; in fact, it is well and pays best in the end to use it always unless otherwise stated.

Whenever possible use clear glass jugs and decanters—the magicians of old used opaque vessels, and their illusions thereby lost much of their effect—such would not do at the present day.

3

Everything must be kept scrupulously clean and washed and carefully dried again before using; if you cannot use distilled water, for final rinsing, use water which has been boiled and allowed to cool.

The actual apparatus required is neither expensive nor difficult to make, and will be dealt with further in Chapter IX. But in studying and preparing the experiments a fairly good pair of scales are essential—such a pair as are often used by amateur photographers or those shown in Fig 1, in which the scale

FIG.1.

pans are of glass or horn rather than metal, are to be preferred. It is a great advantage if such scales are mounted on a stand with an upright (no matter how rough) from which they can be hung. See that the baseboard is horizontal, *i.e.* the upright vertical when measuring. In addition to a pair of scales a graduated measure (preferably of glass) is essential. It is a great

4

advantage though not essential to use the metric system of weights and measures.

When commencing experimenting use only very small quantities of chemicals and small test-tubes as glasses—a considerable saving is expected thereby—and you can at any time increase sizes and amounts by the use of your balance and graduated measure.

Carefully note the effect produced by varying the relative amounts of the chemicals used in any experiment; and having found out *by actual experiment with certain given chemicals* the exact relative proportions and amounts to produce the best effect, make a careful note of the same and keep for future reference.

Remember that in Chemical Magic all the effects produced must be sudden and startling; the less time given to your audience to think the less chance have they of discovering your secrets.

For example, if water is to be turned into wine, *i.e.* a clear solution to become a coloured one, the colouring must be instantaneous—in the twinkling of an eye; conversely, if wine is to be turned into water, then the bleaching or decolourising agent must act in a similarly rapid manner.

The reader will not fail to note that in the following pages the exact amount of the various ingredients to be used is very often not stated. This varies with the nature and purity of the chemicals, and also with the process of their manufacture and the care with which they may have been kept.

On purchasing a fresh supply or repeating any experiment after more than a few weeks, it is advis-
5
able to make a preliminary test or you may experience a surprise as well as your audience.

If the above "Hints" are carefully carried out, *success is certain*; if they are not, you may be successful—but your success will partake more of the nature of a "fluke" and you have only to go on long enough to come a most "howling cropper"—as more than one professional conjurer has done in exactly similar circumstances.

Remember too that such a failure can only be a most biting frost—almost certainly admitting of no explanation or quibbling whatever, and it will be entirely your own fault because the remedy was there before you all the time and you refused to take it.

Above all things never present any experiment to your friends or in public until you have thoroughly mastered it in private *in every detail.*

The mere fact of doing the experiments once is useless, try it half a dozen times at least under slightly varying conditions, paying the most careful attention to all small points. This not only saves endless trouble and disappointment, but it leaves your mind free.

Moreover, the actual working of the experiment should be purely mechanical—with the mind left free for the necessary "patter," *i.e.* the art of *mis*direction, without which any magical entertainment would lose about nine-tenths of its interest and charm.
6

CHAPTER II.

WINE AND WATER AND OTHER TRANSMUTATIONS OF LIQUID COLOURS.

EXPERIMENT I.

WINE AND WATER (NON-POISONOUS).

This has ever been a favourite trick amongst magicians—the following method has this advantage, that the contents of the glasses can be drunk or at any rate sipped without any harm.

Apparatus required.—A clear glass jug and two clear glass tumblers or large wine glasses, the latter with partly hollow stems *if possible.*

Chemicals required.—Phenolphthalein ; carbonate of soda (caustic soda is better), and tartaric acid.

Phenolphthalein, which produces a violent colour when combined with an alkali, must be used sparingly, and then the result will be red. Mix a teaspoonful or less of phenolphthalein in a small decanter of water (distilled for preference).

In one glass place a little of a fairly strong solution of carbonate of soda or a smaller amount of caustic soda solution, in the other tumbler a little concentrated solution of tartaric acid.

Call the first glass *A* and the second *B*. The clear liquid in the decanter becomes instantaneously coloured when it is poured into *A*, and this coloured solution is decolourised when it, in turn, is poured into *B*.
7

It should be understood that when the phenolphthalein solution comes in contact with an acid it is colourless, but in an alkaline solution it turns red.

EXPERIMENT II. (2)

TO SEPARATE TWO WINES THAT HAVE BEEN MIXED.

Apparatus required.—Three decanters and two tumblers.

Chemicals.—Sulphocyanide of potassium, sulphocyanide of mercury, bichloride of mercury. As these ingredients are very poisonous they must on no account be tasted or be left standing about.

In decanter *A* is a solution of sulphocyanide of potassium (port) ; in decanter *B*, a similar solution of sulphocyanide of mercury, turned to sherry by means of the bichloride of mercury.

These are mixed in the third decanter *C*. In the first glass is a little bichloride of mercury to turn the mixture sherry coloured, and in the second glass a little dab of red dye to turn it port coloured.

EXPERIMENT III. (3)

TO TURN MILK TO WATER AND *Vice Versâ.*

To make a fluid resembling milk, mix together concentrated solutions of calcium chloride and sodium carbonate; to clear it use hydrochloric acid.

Another method—sulphurous acid and acetate of lead; clear with strong nitric acid. The following
8
is also very good but requires care—and common salt is useless.

Nitrate of silver and a saturated solution of *pure* chloride of sodium ; cleared by ammonia. This must on no account be shaken or it will curdle straight away.

EXPERIMENT IV. (4)

TO TURN MILK INTO WINE.

Clear the milk as per last experiment and add a little dye, *i.e.* clear it in one glass and pour it into another containing a dab of dye. To turn wine to milk, clear the wine and add a milk producer.

EXPERIMENT V. (5)

WATER TO INK; INK TO CLARET; CLARET TO WATER.

Apparatus.—A decanter capable of holding about a pint, filled with *distilled* water; seven wine or other small glasses of same size and similar in appearance.

Chemicals.—Tannin, perchloride of iron, oxalic acid, ammonia, sulphuric acid. Dissolve in the distilled water in the decanter about as much tannin as will lie on a sixpence. Of the glasses one and three are unprepared; numbers two and four contain two drops of perchloride of iron; number five about ten drops of a saturated solution of oxalic acid; number six about the same quantity of strong liquid ammonia; and number seven a small teaspoonful of sulphuric acid.

On pouring from the decanter into the glasses,

9

number one gives clear water; number two, ink; number three clear water; number four, ink once more.

The contents of these four glasses are then returned to the decanter, the whole then appearing to be ink.

The first four glasses are then again filled from the decanter, appearing of course as ink. A wave of the magic wand over the fifth glass changes the ink (when poured from the decanter) into clear water; a wave in the reverse direction over the next glass changes the ink to claret (this glass it being remembered is the sixth).

Again are the contents of all the glasses dealt with poured back into the decanter, which now appears as claret.

The six glasses are again filled, all claret coloured—another kind of wave over the seventh glass gives once again clear water.

To produce this excellent chemical trick in a proper manner—the glasses must not be arranged in a row—but apparently in a more or less haphazard fashion, so that the latter changes are not apparently made in any particular glass.

N.B.—The contents must not be sipped or tasted.

EXPERIMENT VI. (6)

HOW TO STIR A TUMBLER OF CLEAR WATER FIRST TO YELLOW—THEN BY GRADATIONS TO SCARLET AND FINALLY BACK TO CLEAR WATER AGAIN.

Apparatus required.—Specially constructed tumbler—ordinary glass rod, rubber tubing, rubber ball, etc.

10

Chemicals.—Bichloride of mercury, iodide of potassium.

This is a particularly effective chemical trick if well prepared and presented with plenty of palaver or "patter."

The old method of presenting the experiment was by means of a faked rod or stick. The present one allows the glass rod to be passed round for examination.

A small rather thin tumbler—*i.e.* thin in the bottom should be selected, and a small hole drilled through the centre of the bottom, about one-eighth of an inch in diameter, this can be done at a china shop or wherever china is riveted.

A small rubber ball (see Fig 2), about an inch in diameter, and fitted with a piece of rubber or glass tubing and a small piece of cycle valve tubing in order to make it watertight when the tube passes through the hole in the tumbler. The tube and ball in one should be obtainable at any good chemist's.

Cut or shorten the tube of the rubber ball with a fine saw, so that it does not project through the bottom of the tumbler more than is necessary. It is necessary that there be a hole in the table top or box on which the tumbler stands in order to allow the rubber ball to pass through, and a little practice is necessary in picking up the glass so as to hide the ball; but a few trials soon show you how to do this.

The glass can be inverted so long as the ball is not squeezed. When picking up the glass, it is obvious that it is the back of the hand that is turned to the audience, the ball being in the palm of the hand.

11

The tumbler should be grasped close to the bottom by the thumb and first finger (or first and second fingers), with the rubber ball resting against the ball of the thumb.

The ball is squeezed by pressing gently on it with the second or third finger; any such movement can be sufficiently masked by the simultaneous stirring of the liquid in the tumbler. See Fig 3.

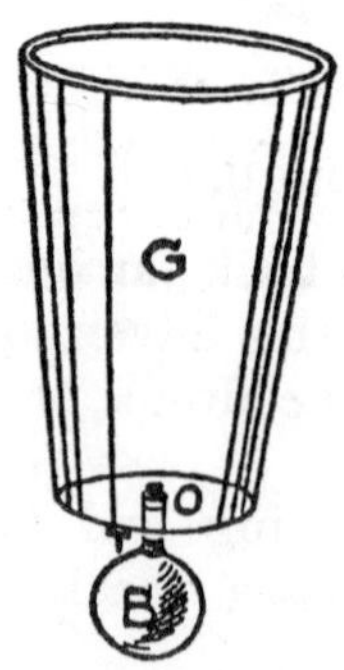

FIG. 2.

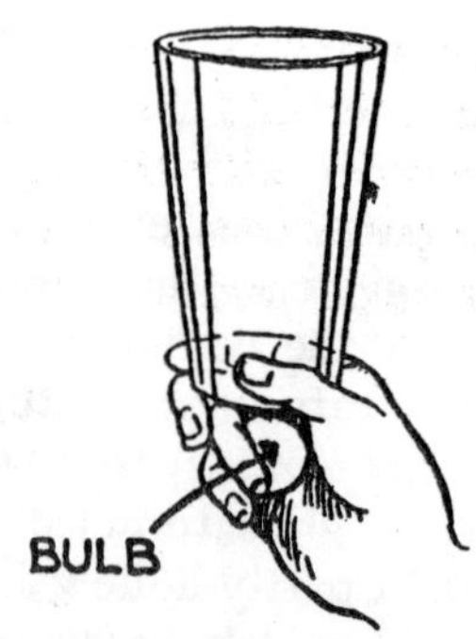

FIG. 3.

As more liquid is pressed up from the ball into the tumbler, the latter naturally gets fuller, and to avoid any apparent increase a *small* quantity should be poured out from time to time into another (opaque) vessel, on the pretext of showing that the liquid is really changing colour.

Place a rather weak solution of bichloride of mercury in the decanter; this is very poisonous and must on no account be left standing about. When a strong solution of iodide of potassium is mixed with it, little

12

by little, it gradually turns yellow, then brownish-red, then red, then scarlet, and finally, as still more is added, colourless once more.

The rubber ball should be filled in the following manner: Pour the iodide of potassium into the tumblers in small quantities and squeeze the ball and then allow it to expand, repeat the process until full.

The tumbler *X* can then be carefully washed, dried and inverted without any of the solution coming out of the ball, provided the latter is not squeezed in any way.

When exhibiting the experiment, pour some of the solution from the decanter into the glass, then squeeze the ball gently and slowly and stir at the same time with the glass rod.

EXPERIMENT VII. (7)

CHAMELEON OR RAINBOW WATER.

The magical idea contained in this experiment is that a chameleon has been dissolved in the water in the decanter and that when poured out into different glasses various colours appear, or it may be suggested that the decanter contains a piece broken off a rainbow.

If the latter idea be adopted then the aniline dyes used must correspond with the colours of the rainbow, viz. red, orange-yellow, green, blue, violet, indigo.

Apparatus.—Clear glass decanter, half a dozen glasses; seven in the case of the rainbow.

Chemicals.—Six little packets of aniline dyes and a little glycerine.

Mix first of all a little water and glycerine in equal

13

proportions. Procure half a dozen little pots, such as chemists use for cold cream, etc., and mix each dye powder with the glycerine and water until it reaches a pasty consistency; place each dye paste in a separate pot and label the same.

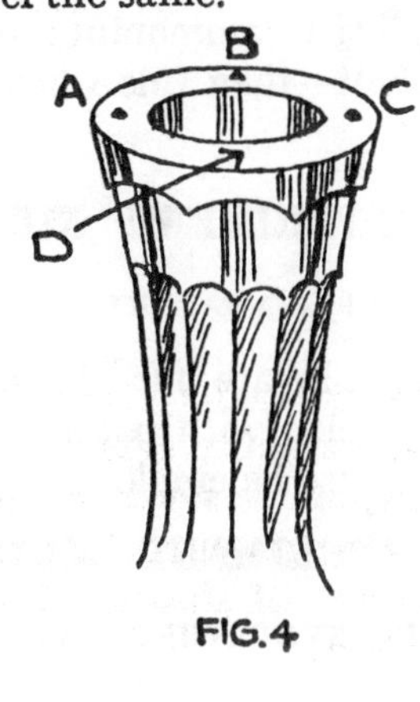

FIG. 4

The decanter and some of the glasses are prepared as follows: Calling the glasses 1, 2, 3, 4, 5, 6, the inside bottoms of glasses 1, 2, 3 have each a tiny dab of dye paste, of whatever colours preferred—in the case of the rainbow, however, they would be red, orange and yellow.

The lip of the decanter (see Fig 4), has three dye dabs on it as shown at *a*, *b* and *c*, or, if preferred, they can be placed slightly lower down the neck. In filling glasses 1, 2, and 3 the decanter must be so held that the clear water in the decanter shall *not* run over any of the three dye dabs, this is quite easy, *i.e.* it must run over the part marked *d*, Fig 4. Three

14

glasses are thus filled; having got thus far you pretend you overhear a remark about the glasses having been prepared beforehand.

The remaining three glasses are passed round for examination, and the rest of the colours (which *can* be produced in any order demanded) are produced by holding the decanter in the proper position for the water to run over the correct dye.

Now, to finish the trick properly, one ought to be able to pour the contents of all the glasses back into the decanter and have clear water once more.

This can be done in the following manner: instead of one—*two* exactly similar decanters must be used.

Call these 1 and 2; decanter one contains only clear water, decanter two contains a strong solution of chlorine water (see Chapter IX.)—this decanter can be substituted for No. 1 whilst the performer searches for a funnel with which to pour the contents of the glasses back again.

If the colours do not change instantly it does not much matter, it is naturally more difficult to make a broken and disintegrated thing whole than to do the opposite.

In a room strong chlorine water can be smelt, hence it must be killed with more powerful odour of a pleasant kind. Some dyes bleach more easily than others, and careful experiments are necessary to ensure producing the best effect.

EXPERIMENT VIII. (8)

THE OBEDIENT COLOURS.

Apparatus.—A decanter and four glass tumblers.

Chemicals.—Four aniline dyes not easily bleached,

15

and four other colours (easy to bleach), *i.e.* vegetable dyes or sensitive indicators like phenolphthalein, litmus solution, chlorophyll (the green colouring matter of plants).

Let the tumblers, not more than two-thirds full, contain (say), liquids of the following colours, yellow, red, blue, green. These colours can be changed as desired as follows:—

The change being brought about by the decanter containing a not too strong solution of chlorine water with its lip prepared as in the last experiment.

The chlorine water being sufficiently strong to bleach the coloured liquid in the tumbler but not the dye dabs over which it passes.

EXPERIMENT IX. (9)

TO RENDER A GREEN FLUID BLUE AND THEN RED BY BLOWING OR BREATHING INTO IT.

Apparatus.—A glass tumbler and a piece of glass tubing.

Chemicals.—Tincture of cabbage, liquid ammonia. Take some fresh leaves from a red cabbage, cut them into small pieces and pour over them distilled water which has been boiled, let the whole macerate for several hours. Decant the clean fluid and (preferably) mix with it one-eighth of its own bulk of spirits of wine. This is tincture of cabbage ready for use; it will not keep any length of time. Only a small quantity, not more than two-thirds of a small tumbler full, need be prepared.

Take this and mix with it a very minute portion of liquid ammonia—just sufficient to turn it slightly green; any other alkali besides ammonia can be used.

If the air from the lungs be breathed or blown through this—making use of the piece of glass tubing for the purpose—the green tincture will at first become blue again, and if the action be continued it will become reddened.

The carbonic acid gas from the lungs is the cause of these changes.

Note on "Sensitive Indicators."

Phenolphthalein dissolved in an acid or neutral medium is colourless—in an alkaline solution it turns red.

Litmus is red in an acid, violet in a neutral, and blue in an alkaline solution.

Methyl Orange is yellow in an alkaline solution and red in an acid.

Cochineal is violet in an alkaline solution and yellow-red in an acid.

Sodium and potassium carbonate are the best as alkaline solutions for our purpose, and tartaric acid about the best to render a liquid acid, since it is not poisonous. Other commonly used acids can of course be employed.

As to the tincture of cabbage referred to in Experiment IX., an alum solution turns this green; a potash purple; and hydrochloric acid crimson.

CHAPTER III.

DISAPPEARING AND REAPPEARING WRITING, ETC., OR SYMPATHETIC INKS

In the following experiments the idea is to show a person a piece of paper or other material apparently blank and then cause characters to appear on it at will. In other words, to write a letter, execute a drawing or picture, with some chemical which will leave no visible trace whatsoever, and then develop up the characters or design by means of a suitable reagent whether chemical or physical. In carrying out such experiments the use of a clean quill pen is greatly to be preferred.

EXPERIMENT X. (10)

GREEN SYMPATHETIC INK.

Write or draw on paper with a solution of cobalt chloride; breath upon it (if necessary) and the writing disappears to reappear when heated. Such writing can be made to appear and disappear by alternately warming and cooling, if care can be taken not to expose the paper to a greater degree of heat than is necessary to render the writing legible.

If a drawing be made of the trunks and branches of trees in the ordinary manner and a tracing of the leaves with this sympathetic ink, then such a tree appears leafless until it is heated, when it suddenly

becomes covered with a beautiful foliage. If the heating be secret (by electricity say), we have here all that is necessary for the production of an excellent magical effect.

EXPERIMENT XI. (11)

BLUE SYMPATHETIC INK.

For this the easiest method is to use potassium ferrocyanide and sponge or spray it with a dilute solution of iron sulphate; or use copper sulphate and sponge or spray with ammonium hydrate. If we use acetate of cobalt—then this when heated gives a very similar result.

EXPERIMENT XII. (12)

BROWN SYMPATHETIC INK.

Employ copper sulphate for the invisible writing or drawing and develop with potassium ferrocyanide, or heat or milk can also be employed.

EXPERIMENT XIII. (13)

RED SYMPATHETIC INK.

For a red use chloride of mercury and develop with iodide of potassium.

EXPERIMENT XIV. (14)

YELLOW SYMPATHETIC INK.

Use chloride of mercury and develop with sodium hydrate or caustic soda, or use lead acetate and develop with potassium chromate [19] or dissolve equal parts of copper sulphate and ammonium chloride in water and develop by heat. This image, as in Experiment X, etc., disappears on cooling.

EXPERIMENT XV. (15)

BLACK SYMPATHETIC INK.

For this colour use a solution of iron sulphate and develop with pyrogallic, gallic or tannic acid, or a very dilute solution of sulphuric acid and develop with heat.

EXPERIMENT XVI. (16)

A ROSE-COLOURED SYMPATHETIC INK.

Dissolve a small quantity of oxide of cobalt and a little potassium nitrate in acetic acid—develop with heat. The image or writing disappears on cooling as before.

EXPERIMENT XVII. (17)

TO RENDER AN INVISIBLE WRITING VISIBLE, BY IMMERSING THE PAPER IN WATER.

If the characters be formed with a solution of nitrate of bismuth, they will, when dry, be invisible; but, on wetting the paper with a little water, the writing will appear in dense white marks.

EXPERIMENT XVIII. (18)

TO RENDER AN INVISIBLE WRITING OR DRAWING VISIBLE BY EXPOSURE TO LIGHT.

If the writing or drawing be made with a solution of silver nitrate, sufficiently dilute, so as not to injure [20] the paper, the characters or drawing, when dry, will be invisible and will remain so if the paper be closely folded up or be in any other way kept from the light; but if the paper be exposed to the sun's rays or to a strong light, the characters speedily become visible, at first assuming a brown colour and finally turning black.

The same result can be obtained if the characters be exposed to the fire.

EXPERIMENT XIX. (19)

A GREEN SYMPATHETIC INK, WHICH BECOMES BLUE WHEN HELD OVER A CERTAIN LIQUID AND GREEN AGAIN ON EXPOSURE TO AIR.

Write on a paper with a solution of copper sulphate, if the solution be rather a strong one the characters will be visible in green; if the solution be weak they will be invisible; but if the paper be held over a saucer or similar vessel containing some ammonia, the writing at once assumes a blue colour, becoming visible if previously invisible. This colour again departs on warming the paper by the fire or even by leaving it exposed to the air for a certain time.

EXPERIMENT XX. (20)

ONION JUICE AS A SYMPATHETIC INK.

Write or draw with the juice of an onion—the writing or drawing is invisible until warmed.

[21]

EXPERIMENT XXI. (21)

HOW TO PRODUCE MAGICAL PICTURES ON GLASS.

This is a very effective experiment indeed. It consists in showing a sheet or square of glass on which nothing can be seen, not even on the closest inspection; but when the glass is breathed upon so as to coat its surface with a fine layer of moisture, a picture (see Fig 5)

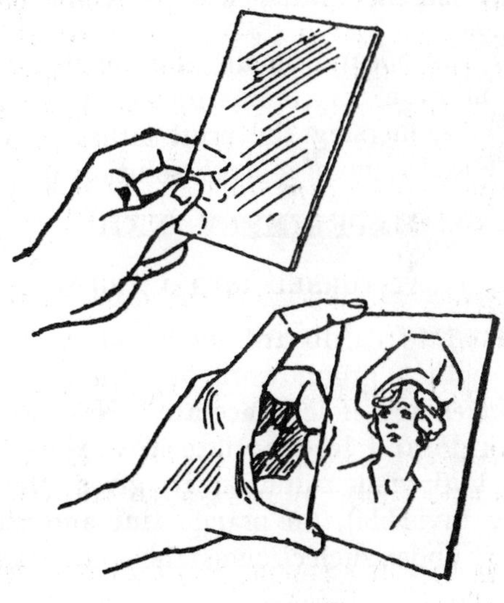

FIG. 5.

at once comes into view, disappearing again as soon as the moisture leaves the glass.

To produce a picture of this kind, you must first of all procure a piece of good plate glass free from blemishes of any kind. Procure also a small quantity of hydro- [22] fluoric acid—this will be supplied and must be kept in a rubber bottle, and great care must be employed in using it—on no account must it be allowed to come in contact with the fingers, and when finished with should either be poured away down the sink or kept under lock and key. The drawing, etc., is made on the glass by means of this liquid and a quill pen, which must have been very thoroughly cleaned beforehand, in fact a new pen is essential.

In about five minutes, or eight or nine at the most, the glass must be well washed under a tap and carefully dried with a cloth. It is now ready for use, and it is only necessary to breathe on it for the picture to appear.

Some little experimenting will, in all probability, be necessary to know exactly how long to leave the acid on, which again will depend on the particular strength of the acid you are using. Too long an exposure to the action of the acid would leave a *visible* picture which is just exactly what is not wanted, and too short a one—an imperfect picture.

The experiment is really one in glass etching.

EXPERIMENT XXII. (22)

THE INVISIBLE MADE VISIBLE.

Take any ordinary photographic negative—and take a print from it on chloride of silver paper (not self-toning), and without toning fix the print in a ten per cent. solution of hypo. Wash thoroughly. Next float the print on a five per cent. solution of chloride of mercury. The picture will rapidly fade

23

away and disappear. Again thoroughly wash the print and allow to dry. The image can be made to reappear by immersing the print either in a five per cent. solution of hypo or a *weak* solution of ammonia.

EXPERIMENT XXIII. (23)

A CAT TURNED INTO A TIGER.

Make a white cardboard model of a tabby cat, and treat with a groundwork of antimony chloride, cross by markings of lead acetate. Now when antimony chloride and lead acetate are exposed to the fumes of hydrogen sulphide, it gives the former (previously invisible), an orange tint and the latter a black one: under such circumstances our cat becomes a tiger.

For the production of a magical effect the model of the cat must be put under a glass shade or jar—such as a 3 lb. jam jar—which must be placed upside down on a suitable pedestal or stand.

The hydrogen sulphide fumes can be generated as follows: place a little pulverised iron sulphide in a piece of filtering or blotting paper behind the cat. A little dilute sulphuric acid must be put on the other side of the baseboard, which should be slightly lower on this side than on the other. The jar and baseboard can then be picked up and slightly tilted so that the sulphuric acid comes in contact with the iron sulphide—thereby generating the necessary gas. Various other methods of generating the gas could obviously be made use of—these I leave to the reader's ingenuity.

24

EXPERIMENT XXIV. (24)

ANOTHER ZOOLOGICAL TRANSFORMATION.

In this case our model is a leopard, and the experiment is to show that—chemically at any rate—the leopard *can* change his spots.

Our leopard in this case will be a somewhat special brand—since he will have yellow spots on a light or white ground.

The spots are painted with a mixture of potassium chromate and hydrochloric acid. The gas used in this case is sulphur dioxide, which will be changed from yellow to green. The reader should experiment for himself in transformations of this nature, because the subject is obviously one that can be extended; for instance, flowers, figures, etc., can be drawn upon silk with a solution of silver nitrate, when they are invisible, but if moistened and exposed to the action of hydrogen gas, then they become visible. Various chlorides, viz. chlorides of gold, tin, etc., can be similarly dealt with.

EXPERIMENT XXV. (25)

THE BLUSHING PORTRAIT.

Take a plain print or drawing of a young lady and paint the face, neck, etc., with a solution of equal parts of water and methylated spirits to which a few grains of phenolphthalein have been added. When such a portrait is subjected to ammonia fumes—it will be found to "blush" most beautifully, especially if it be slightly damp. And when withdrawn from the ammonia influence it will resume its natural colour once more.

25

EXPERIMENT XXVI. (26)

THE COLOUR CHANGING PICTURE

The sketch or picture must be drawn with a solution of copper bromide. Anything which should in the resultant picture become green such as trees, grass, etc., must be painted with a solution of cobalt chloride.

The sky and any water effects must be painted with a solution of cobalt acetate.

The sketch must then be allowed to dry. When such a picture is heated, it will appear in its natural colour.

EXPERIMENT XXVII. (27)

THE DREAR AND FRUITFUL CORNFIELDS.

Draw a landscape and finish it with sympathetic inks as described in Experiment X, cornfields being painted in and finished with a dilute solution of nitrate of copper. The whole will have a very drear and bleak appearance until warmed, when it will instantly assume a cheerful and lively appearance, as if by magic.

EXPERIMENT XXXI. (31)

THE INVISIBLE MADE VISIBLE BY WETTING.

Write or draw with a weak solution of alum in lemon juice. The characters or drawing will be invisible until wetted with water, which will render them of a greyish colour, and quite transparent. On a letter, written with a weak solution of rock-alum alone, being dried, and having a small quantity of water poured over it, the writing will appear of a whiter colour than the paper.

26

EXPERIMENT XXXII. (32)

A VIOLET SYMPATHETIC INK.

Write with rice water, when dry the writing will be invisible; but if washed over with a solution of iodine, the writing will appear of a violet colour. This particular method of secret writing was practised with great success in a certain war in India. The first letter of this kind being received from Jellalabad concealed in a quill. On opening it a small paper was unfolded on which appeared only the single word "iodine." The liquid was applied, and an important despatch from Sir Robert Sale at once became visible.

EXPERIMENT XXXIII. (33)

FLUORESCENT WRITING.

Write or draw on white a letter or design, using a solution of sulphate of quinine as the ink; when dry this will be quite invisible. But if the same be exposed in the dark to the light of a vacuum or Geissler tube, the design or writing will appear in a blue violet.

In exactly a similar manner if calcium tungstate were used and the other side of the paper turned to an X-ray tube in action (the tube itself being wrapped up in black paper), the writing or design will appear in white. Remember that side of the paper or cardboard on which the design is drawn must be turned *towards* the reader's eyes and not towards the tube. There are other chemicals which do this also, but most of them are expensive.

27

CHAPTER IV.

SPONTANEOUS COMBUSTION—QUICK BLEACHING AGENTS—MAGICAL USES OF THE SAME AND ALSO OF CERTAIN GASES, SMOKES AND VAPOURS. (34) — (41)

EXPERIMENTS XXXIV—XLI.

SPONTANEOUS COMBUSTIONS.

(i.) Mix a little chlorate of potash with twice its quantity of loaf sugar, reduced to a powder; place this mixture on a plate, dip a piece of wire in sulphuric acid, and let a single drop fall from its end on the mixture, it will immediately burst into flame. The mixing should be done on a plate or piece of paper with a paper knife.

(ii.) Put a little iodine in a saucer, and drop on it a piece of phosphorus; these two substances have a strong liking for one another, and will very shortly combine with the evolution of flame, producing a mixture of the two iodides of phosphorus, which will be left in the saucer as a compound of a reddish colour.

(iii.) Bruise and slightly moisten with water a few crystals of nitrate of copper, add a few drops of water, then roll up quickly and tightly in a piece of tinfoil. In about a minute the tinfoil will begin to smoke and soon after take fire (water slightly if it does not), and

28

explode with a slight crackling noise. No more water must be added than will make it into a liquid paste, a few filaments of tow assists the experiment.

(v.) Put a little *dry* pulverized charcoal into a *warm* tea cup, and pour over it a little strong nitric acid. Combustion will follow as in the last experiment. The cup should be placed in a soup plate or dish containing a little water.

(vi.) Pour a tablespoonful of turpentine into an old cup, and place it in the open air; then put about half that quantity of nitric acid, mixed with a few drops of sulphuric, into a test tube *fastened to the end of a long stick;* pour this upon the turpentine, and it will immediately burst into flames and burn with much heat and light, and a prodigious amount of smoke.

(vii.) Take a metal button, and rub it briskly for a short time to make it warm or even hot, then touch a piece of phosphorus with it, the phosphorus will immediately burst into flame and burn.

(viii.) To six grains of chlorate of potash, reduced to a fine powder, add three of finely pulverized charcoal; mix them on a piece of paper with the least possible friction.

Add to this mixture two grains of sulphur. When this mixture is forcibly rubbed with a pestle in a mortar, it inflames with a rapid flash, like gunpowder. The hand must be covered with a glove when per-

29

forming this experiment, or better still, the pestle fixed to the end of a stick not less than three feet long.

EXPERIMENT XLII. (42)

THE SPONTANEOUS COMBUSTION OF THREE METALS.

In a *perfectly dry* ladle place a small piece of potassium, together with a like piece of sodium—mix them by rubbing them together with the point of a knife. The mixture so far will take place quietly, but if this alloy be brought into contact with a little mercury—this compound when shaken or agitated instantly bursts into flame and burns vividly.

The ladle must be *quite* dry—because the least dampness would or might fire the potassium.

EXPERIMENT XLIII. (43)

A SAFE MAGICAL EXPLOSIVE.

Take a few flakes of iodine—*but no more;* grind them carefully to a powder in a pestle and mortar. Then place them in a *clean* glass test tube and pour about a teaspoonful of *strong* liquid ammonia on them. Allow this mixture to stand for at least half an hour, and then carefully filter through white blotting paper. The dark brown substance left behind must be carefully divided into minute portions whilst still wet and then partially dried, but conveyed whilst still damp to the article on which it is proposed to use it.

It is a powerful fulminate and explodes of its own

accord if dried in too warm a place. Even a pellet the size of a pin's head will produce quite an audible explosion, but if used in minute quantities there is absolutely no danger.

EXPERIMENT XLIV. (44)

ANOTHER FULMINATING MIXTURE.

Take three parts by weight of nitre, two of *dry* carbonate of potash, and one of sulphur. Dry thoroughly and then grind them *separately* to a fine powder; then mix them carefully with a paper knife or bone spatula.

Take a pinch of this powder *but no more,* place in an iron dish or ladle, apply a gentle heat. It will explode with a deafening report. On no account place the face near the substance when it is being heated. In its solid or powdered state this mixture keeps indefinitely—but combustion immediately follows heating if that heating be sufficient to cause fusion.

EXPERIMENT XLV. (45)

HOW TO MAKE FLASH PAPER.

Any thin *unsized* paper can be used. Procure a suitable flat dish just large enough to hold the sheet, say a quarter size, which is a handy size to work with. Fill the dish three parts full with a solution (by measure) of four parts of sulphuric acid and five parts of nitric acid. The two acids should be well mixed by means of a glass rod. Take one of the sheets of

paper and immerse it in the solution for about ten minutes. Remove it by the aid of the glass rod and place it in a pan or dish full of warm distilled water. It must remain there for one hour, then pour off the water and refill the pan afresh—this must be repeated several times, or better still place the sheet or sheets in running warm water until every trace of the acid is removed. If this is not done thoroughly, not only will the paper be liable to spontaneous combustion, but will explode violently if struck. After a thorough washing, the papers must be dried slowly by a gentle heat. If, after drying, a small pellet of the paper be lighted at one point and thrown into the air, a flame of intense light is produced. So perfect is the combustion that no perceptible ash is left. Colour effects may be obtained if, after drying, the sheets be taken and immersed for five minutes in a warm saturated solution of a lithium salt (for crimson), copper chlorate (for blue), barium chlorate (for green), and potassium nitrate (for violet); strontium chlorate also gives a crimson but not so good as the lithium salt. They must afterwards be hung up to dry.

EXPERIMENT XLVI. (46)

THE BLEACHING POWER OF CHLORINE.

We have already had some example of the bleaching action of chlorine water on coloured solutions in Chapter I.

In the following instance the gas and not the aqueous solution is used. This experiment can be given a magical effect by introducing the gas *secretly*

when the colour will appear to be *shaken out of the liquid.*

In a tall glass jar, or bottle, put a solution of indigo, not much more than half-filling the jar. Invert a bottle or flask of chlorine gas over the mouth of the jar, the gas being about 2.5 times as heavy as air will pour out like water. Close the mouth of the bottle or jar with a cork or bung and shake. All trace of colour will almost immediately disappear.

EXPERIMENT XLVII. (47)

FLOWER BLEACHING.

Natural flowers, more especially roses, dahlias and violets, are soon bleached when thrown into a jar full of sulphur dioxide gas. The colouring matter is not, however, destroyed as when chlorine is used, and may even be restored by exposing the flowers to the air for some time.

The most successful method, however, is to immerse them in an alkaline solution. Ammonia, by itself, is not very successful, but when mixed with ether, the restoring action is very rapid.

As well as flowers, silk, straw, wool, etc., may be bleached by the action of sulphur dioxide.

Both this and chlorine gas are extremely evil smelling agents and must on no account be inhaled. When used in their gaseous state, be careful to remember that any article which it is required to bleach must be *damp*. Unless this is the case the experiment will not succeed.

When used as gases in magic, means must be taken to overcome the unpleasant smell referred to.

EXPERIMENT XLVIIA. (47A)

TO REMOVE AND AFTERWARDS RESTORE THE COLOUR OF A RIBBON.

Dip a coloured ribbon (rose coloured is very suitable), into nitric acid, diluted with eight to ten parts of water, and as soon as the colour disappears, which it will do in a short time, take out the ribbon and put it into a very weak alkaline solution ; when the alkaline solution will quickly neutralize the acid, and the colour will re-appear.

EXPERIMENT XLVIII. (48)

THE FADED ROSE RESTORED.

Take a red rose that is quite faded, and put a little sulphur on a chafing-dish of hot coals ; then hold the rose over the fumes of sulphur, and it will quickly become white ; in this state dip it into water, then put it in a box or drawer for three or four hours, and when taken out it will be quite red again.

EXPERIMENT XLIX. (49)

THE MAGIC DYE.

Dissolve indigo in a diluted solution of sulphuric acid, and add to it an equal amount of a solution of potassium carbonate. If a piece of white cloth be

dipped in this mixture, it will be changed to blue ; a piece of yellow cloth to green, and a piece of red to purple ; and blue litmus paper will be turned red.

EXPERIMENT L. (50)

TO MAKE TWO INVISIBLE VAPOURS VISIBLE.

Take an ordinary glass tumbler, in the bottom of which are a few drops of ammonium hydrate (commonly called ammonia, although ammonia, strictly speaking, is a gas), and place over it a plate or saucer, on the *bottom* of which is a little hydrochloric acid, then the tumbler becomes filled with smoke. As a trick this is generally shown by the performer throwing a handkerchief over the plate and tumbler, then walking to the other side of the stage and puffing smoke from a cigarette. After a time the handkerchief is removed and the smoke apparently is found in the tumbler.

Another method is for the performer to produce two clay pipes, both empty, and to all appearance unprepared ; but on putting one of the pipes into his mouth and holding the other bowl inverted over it, smoke at once begins to pour forth from between them. Care must be taken to *exhale* (blow out), and not *inhale* (breathe in) the fumes.

EXPERIMENT LI. (51)

AN EXPERIMENT IN WEIGHING NOTHING.

The gas known as carbon dioxide is one and a half times as heavy as ordinary air ; it can therefore be

poured from one open vessel (such as a jam jar or cocoa tin) into another.

This naturally suggests an experiment in weighing nothing ; the apparatus required being a pair of fairly sensitive scales and two tins or jam jars.

In one scale pan is placed one jar or tin, and weights in the other scale pan, to balance it. The other jar (jars are better than tins), is full of carbon dioxide. When this jar is inverted over the balanced jar, as if you were pouring something from one jar into the other (which you are really doing, only it is invisible), the balanced jar will no longer balance but go down, and as nothing has been seen to pass from one jar to another, nothing has apparently been weighed, not even the jar now for it no longer balances.

EXPERIMENT LII. (52)

WHAT PUTS THE CANDLES OUT, OR THE INVISIBLE EXTINGUISHER.

If a bottle or jar full of carbon dioxide be inverted over a row of lighted candles, the jar slowly sweeping over them from end to end, they will all be put out in succession. A more effective way to perform this trick is as follows; partly fill a bladder—such as a football bladder—and have this concealed beneath your coat, under your left arm, have a tube running from it down the right sleeve into the palm of the hand, the end being closed by a clip. When using, remove the clip and squeeze the bladder between the arm and side, at the same time passing the hand over the candles.

36

EXPERIMENT LIII. (53)

HOW TO PRODUCE FULLY INFLATED BALLOONS FROM A HAT.

As the balloons or bladders used in this experiment are of special manufacture, they should be procured from a reliable conjuring depôt. They are of two different sizes, of which the smaller is much the weaker, it being made from a much weaker material; the larger is not only extra strong but has a wider neck.

The agents used for inflating the balloons are soda bicarbonate and tartaric acid; about a quarter of a pound of each is enough to inflate half a dozen balloons. The tartaric acid should be emptied into a clean jam jar, just covered with water and allowed to stand. With the aid of a syringe pump this solution into one of the small balloons until it is as large as a small egg. Next tie up the mouth tight, and insert it through the neck of one of the larger balloons. In the latter place a tablespoonful of bicarbonate of soda and tie up this neck as well. To inflate the large balloon, you merely squash the little one inside the big one from the outside and the action of the tartaric acid on the bicarbonate does the rest.

As many as two dozen balloons or more have been produced from a tall hat in this manner. To be perfect, the balloon should float in the air—to do this with such a small balloon hydrogen gas would be necessary, coal gas might just do it but it has only half the lifting power of hydrogen.

37

CHAPTER V.

CHEMICAL PYROTECHNICS—FIRE PRODUCTION—FIRE AND WATER—FIRE-EATING.

EXPERIMENT LIV. (54)

PHARAOH'S SERPENTS.

This remarkable chemical illusion has probably caused more wonder than any other similar kind of experiment.

Many have assumed that the conjurations of the Ancient Egyptians in the time of Moses were produced by some chemical preparation of like nature. Pharaoh's serpents (see Fig 6), are composed of sulphocyanide of mercury, which is generally rolled up in a cone of tinfoil; and which being lighted at its apex, there issues forth slowly a serpent-like coil, which

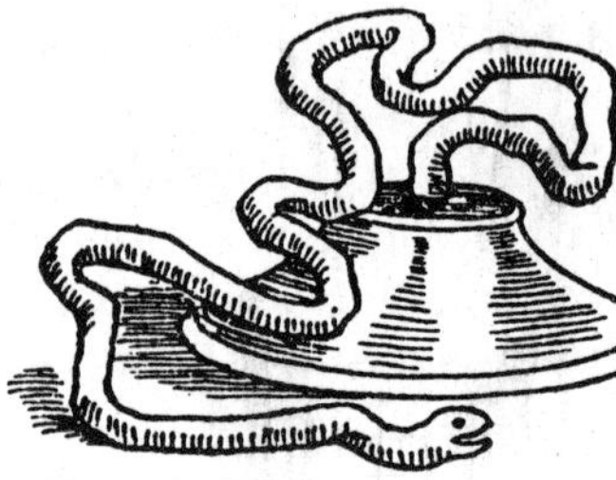

FIG.6.

38

twists and turns in all directions and continues to issue from the cone in a marvellously curious manner, until all the composition is consumed. The serpent, snake or viper *can* be handled without breaking, but it is very brittle and also very light. Internally it has a black metallic appearance; but externally it is of a dirty white colour, not unlike that of a real snake.

But the most strange-looking chemical serpents are made when the sulphocyanide of mercury is mixed up into a paste with a little gum water, and either roughly shaped into a cone or made long like your finger. These, when placed in the fender or on a plate, burn with the most astonishing fierceness, producing a snake of a light or dark brown colour, many yards long. The fumes being poisonous must not be inhaled, and the "egg" should be lighted in the fender, and then the noxious gases can escape up the chimney.

Another recipe for the above is: make the sulphocyanide of mercury into a paste by adding a little mucilage of tragacanth, with about five per cent. nitrate of potash. Since not only are the fumes given off during combustion injurious, but the sulphocyanide of mercury is a deadly poison, and the residue poisonous as well, these serpents can only be made under proper supervision. A non-poisonous, although it must be confessed a much less effective, "serpent" can be made as follows: take bichromate of potash, two parts; nitrate of potash, one part; powdered sugar, one part. Powder these in a pestle and mortar separately, and then thoroughly mix. Make small

39

paper cones and press the mixture into them. Remember always to light these eggs or cones at the top.

EXPERIMENT LV. (55)

JAPANESE MATCHES.

Take five parts by weight of lamp black, eleven of sulphur, and twenty-five of meal powder, and thoroughly mix. Add a weak solution of gum water, sufficient only to make the mixture into the consistence of a thick paste.

Roll or flatten this out and then cut it up into cubes of about one-quarter inch sides. Place on a warm dish and allow them to dry very slowly. The more slowly they dry the better, and a week is not too long for the process. When perfectly dry, fix the little cubes on straw-like fibres similar to those used in carpet brooms, or pieces of very fine split bamboo.

To fire them hold them stem downwards and light in a candle, etc. They first of all blaze up, then form a molten mass, from which very curious coruscations soon appear.

EXPERIMENT LVI. (56)

BURNING WATER.

Apparatus required.—A brass or metal bowl, or fairly deep tin; a glass decanter; a tumbler.

Chemicals.—Ether (pure); potassium (the metal). In the decanter is clear water, some of which is poured into the tumbler and drunk. On pouring the same

water into the bowl or pan, it at once catches fire and burns--produces flames a foot or so high. The bowl or pan contains a little ether—a few teaspoonfuls, on which is placed a piece of potassium no larger than a pea. If the ether be pure no chemical action takes place between the ether and the potassium, but as soon as the water is poured in the potassium at once sets free the hydrogen from the water to such an extent that the hydrogen is ignited, and in its turn fires the ether—naturally it is the ether floating on the top of the water that burns and not the water.

EXPERIMENT LVII. (57)

HOW TO RENDER BODIES LUMINOUS IN THE DARK.

If a four or six ounce flask or bottle containing a few ounces of liquid phosphorus be unstopped or uncorked in darkness, the empty space in the flask above the liquid emits a sufficient light to show the time if a watch or clock face be held near it. When the flask is again corked, the light vanishes, but re-appears instantly on re-opening it. In cold weather, it is necessary to hold the flask or bottle in the hand, before the cork is removed. Without this precaution no light will be emitted.

Liquid phosphorus may likewise be used for forming luminous writings or drawings. It may be smeared on the face or hands, or any warm object to render it luminous, without any risk; but when rubbing it on the face be careful to shut the eyes. The appearance in the dark is simply hideously frightful, all the

parts of the face appear to be covered with a luminous lambent flame of a bluish-white colour, whilst the eyes and mouth are shown as vacuous cavities.

HOW TO PREPARE LIQUID PHOSPHORUS.

The best method is as follows: heat very gently, for two hours, one part of phosphorus with six of oil of almonds. This oil charged with phosphorus must be kept in a well-corked bottle; the bottle also must be kept in the dark, because light decomposes the solution of phosphorus.

In dealing with phosphorus the young experimenter must be careful, it should not be used in pieces larger than a moderately large shot or small pea; it should be *cut under* water with a sharp penknife, holding the stick with a pair of forceps. It should be dried on blotting-paper on a metal tray and not touched with the fingers, as the warmth of the hand, if it be dry, may cause it to ignite spontaneously, and the result is a nasty burn, because once ignited it is not easily extinguished.

EXPERIMENT LVIII. (58)

LUMINOUS WRITING—ANOTHER METHOD.

Fix, with a pair of forceps, a small piece of solid phosphorus in a quill pen and write with it (*i.e.* the phosphorus), upon a piece of paper. If the paper be taken into a dark room, the writing will be plainly visible. This luminosity arises from the *slow* combustion of the phosphorus.

EXPERIMENT LIX. (59)

TO RENDER THE SURFACE OF WATER SELF-LUMINOUS.

Wet a piece of lump sugar with phosphorized ether and throw it into a basin of water; the surface of the water will become luminous in the dark, and by gently blowing upon it, phosphorescent undulations will be formed, which illuminate the air above the water for a considerable distance above the fluid.

In winter the water must be rendered blood warm. If the phosphorized ether be applied to the hand or warm objects (which may safely be done), it renders them self-luminous or phosphorescent in the dark.

HOW TO PREPARE PHOSPHORIZED ETHER.

Allow sulphuric ether to stand, for several weeks, over a considerable quantity of phosphorus in a well-stopped bottle or flask.

EXPERIMENT LX. (60)

TO LIGHT A CANDLE WITH A GLASS OF WATER.

Take a little piece of phosphorus the size of a pin's head, and with a little piece of tallow stick it on the edge of a drinking glass. Then take a candle, one which has been burning for a minute or two, blow it out and apply it to the glass so that the hot wick touches the phosphorus, it will at once ignite the phosphorus and be relighted, the glass must be tipped at the same time, so that water nearly but not quite

reaches the phosphorus, so that it looks as if it is the water which is relighting it.

EXPERIMENT LXI. (61)

GREEN FIRE UNDER WATER.

Put into a glass tumbler two ounces of water, and add, first, a piece or two of phosphorus about the size of a pea, then thirty or forty grains of chlorate of potash; next pour upon these by means of a funnel with a long neck reaching to the bottom of the glass, five or six drachms of sulphuric acid. As soon as the sulphuric acid comes in contact with the ingredients already put in, flashes of fire begin to dart from under the surface of the fluid.

When this takes place drop into the mixture a few pieces of phosphuret of lime; this will at once illumine the bottom of the vessel and cause a stream of fire of an emerald green colour to pass through the fluid.

EXPERIMENT LXII. (62)

WILL-O'-THE WISPS.

Take a glass tumbler three parts full of water, and drop into it two or three lumps of phosphuret of lime (phosphide of calcium), decomposition will take place and phosphuretted hydrogen be produced, bubbles of which will rise to the surface and take fire immediately they burst through the surface or skin of the water, terminating in beautiful rings of smoke.

Another method is to place some phosphorus in a glass retort, with a fairly strong solution of potash and some ether and heat gently—the end of the retort must dip under water—much larger quantities of the gas can be given off in this manner. Coal gas and caustic soda can be used instead of ether and potash.

The ether is used for the purpose of preventing the combustion of the first gas bubbles formed in the retort, which by their explosion would in all probability break the retort.

Another way to generate this gas is as follows: Add gradually, in the form of a very fine stream, one ounce of sulphuric acid to about six ounces of water, contained in an earthenware basin; throw in about an ounce of granulated zinc, and a small piece or two of phosphorus, phosphuretted hydrogen gas is again produced, and taking fire (as before) the moment that it comes in contact with the air; in a short time the whole surface will become luminous and continue so as long as the gas is generated, which may be seen darting from the bottom through the liquid with great rapidity.

It is better to use coal gas and caustic-soda instead of potash and ether, but see that the coal gas *completely* displaces the air in the flask or retort, in other words pass enough in.

It is best of all to use carbon dioxide instead of coal gas—a lighted taper plunged into this gas goes out at once, thus showing when the flask is full.

An eight-ounce flask and a rather large delivery tube with its end well under water is better than an ordinary retort. First, put in a few pieces of caustic

potash, then some small pieces of phosphorus and a little water. Fill the upper part of flask with carbon dioxide, test, cork up and fit delivery tube, etc.

Phosphuretted hydrogen or phosphine has an unpleasant odour, like putrid fish, is poisonous, and must not be inhaled.

The gas, if left standing in the light, loses its will-o'-the-wisp properties.

EXPERIMENT LXIII. (63)

FIRE BURNING UNDER WATER.

The following two mixtures are said to burn under water—the writer has not tried them personally.

Take strontium nitrate forty parts, flower of sulphur thirteen parts, chlorate of potash five parts, sulphuret of antimony four parts. Powder these ingredients *separately* with a pestle and mortar and then mix carefully and thoroughly. Next take a strong paper cylinder or case about two to two and a half inches long, well closed and varnished at one end so as to be quite watertight (if not varnished soak in melted candle wax). Pack the case tightly with the mixture, light and put into the water mouth downwards—when it will continue to burn—the best effect being produced if a tall glass jar be used.

Another mixture similarly treated which will do the same thing is the following:—

Take one part nitrate of potash, three parts of gunpowder and the same quantity of sulphur. Remember always to grind all ingredients in *any* experi-

ment *separately: if sulphur and chlorate of potash* (for instance), *be rubbed together in a mortar they are extremely liable to explode.* The first composition given above should be freshly made—it must not be kept as it is liable to spontaneous combustion.

EXPERIMENT LXIV. (64)

TO DRAW A DESIGN ON A PIECE OF PAPER BY SETTING FIRE TO IT.

Take a sheet of paper and draw upon it a design with a concentrated solution of potassium nitrate and allow it to dry. If the glowing end of a piece of string be applied to any part of the design, the curious effect is produced of its burning itself out without injury to the other parts of the paper.

A simple design should be chosen without much detail. A written word or short message for magical use is as good as anything.

EXPERIMENT LXV. (65)

TO DRAW A DESIGN ON A PIECE OF PAPER, TO BURN THE PAPER AND HAVE THE DESIGN LEFT.

Take half a pint of asbestos paint and add a *few* drops of nitric acid. Paint the design or writing with this. The best way is by means of a stencil and stencil brush in the case of a design, but this is not essential.

After painting, a thorough drying must be permitted; on setting light to the paper obviously the

47

part treated will not burn since the painting has rendered it fireproof.

Another incombustible solution is alum and white of egg—but the former is to be preferred.

EXPERIMENT LXVI. (66)

EATING BURNING COALS.

The following is in all probability the method adopted by professional performers: a good charcoal fire is made and just before the experiment is performed three or four pieces of soft pine are thrown on. When burnt this is not to be distinguished from the charcoal save by sticking a fork into it. The fork easily enters the pine wood, but not the charcoal. The charcoal would burn anyone who touched it, the pine wood does not; the great point being to close the mouth immediately the piece of pinewood is put in so as to exclude all air. Similarly raisins can be plucked out of burning brandy or alcohol by means of a pickle fork and eaten straight away provided the same precaution is adopted.

EXPERIMENT LXVII. (67)

CHINESE FIRE EATING.

Take some pieces of *soft thick string* and soak them for twelve hours in a saturated solution of potassium nitrate. Then take them out, dry them thoroughly, and cut them up into one inch lengths.

Light one of these and place it in a small ball of tow.

48

This can safely be placed in the mouth, and clouds of smoke and even sparks sent out. Once in the mouth it is only necessary to blow in order to do this. If the ball gets too hot, simply close the mouth and breathe through the nose. Never in any of these experiments *inhale the breath through the mouth* or you will be certain to cough and thereby absolutely spoil the effect.

EXPERIMENT LXVIII. (68)

HOW TO EAT LIGHTED WAX VESTAS.

This is quite an easy and safe experiment if the following directions are carefully carried out.

Strike the vesta as usual and allow it to burn for a short time, next pick off the glowing head of the vesta with the finger and thumb—quite easy if done quickly.

The lighted vesta is then *boldly* placed in the mouth, which is *at once* tightly shut so as to put out the flame; the breath at the same time being *exhaled* (breathed out), through the nose.

The mouth should be moist at the time and the vesta can be heard to go out with a faint hiss. Place the vesta well in the mouth, not on the lips. Do the experiment boldly and you will never burn yourself—if not done boldly it is best not done at all.

It need scarcely be added, perhaps, that the vesta should not be swallowed.

EXPERIMENT LXIX. (69)

EATING A LIGHTED CANDLE FOR DESERT.

This is a genuine trick in a way, but far less awful than it sounds. The candle is not a wax one—but

49

fashioned from a turnip or, if preferred, an apple; its wick is made from a piece of sweet almond or Brazil nut—all of which can of course be eaten.

If carried out as above it would only make a small piece of candle—if stuck on to the genuine article and bitten off and the genuine article passed round, the trick is much enhanced.

EXPERIMENT LXX. (70)

LICKING A RED HOT POKER.

An Italian chemist named Sementini found a strong solution of alum applied to the skin rendered that part (temporarily), very unsusceptible to fire, in other words a red hot iron could be applied for a short time with impunity. Soap impregnated with alum and rubbed on the hand was found to be still better. He then discovered that a layer of powdered sugar covered with soap was sufficient to render his tongue insensible to heat.

Licking a red hot poker was a very common performance at country fairs many years ago—such tricks are, however, best left to the professional.

EXPERIMENT LXXI. (71)

SALAMANDER'S HAIR.

According to tradition the Salamander is an amphibious animal, which is incombustible and puts out fire. Soak some stoutish thread in a saturated solution of salt and water ; allow it to dry thoroughly and

50

again soak it and dry it. This must be done several times. As a result the usual appearance of the thread is unaltered.

Suspend a piece of it from a suitable support and hang on it a very light ring.

Experiment alone can show how heavy a ring can be used.

Set fire to the thread, which will burn from end to end, but not break, unless the ring is too heavy. The thread is burnt but not the salt, a little tube of this substance being left of sufficient strength to hold up a very light ring.

The experiment will not succeed in a draught. The following is a great improvement on the above and makes a really good trick if properly carried out :

Take four pieces of thread and fasten them to the corners of a piece of muslin about three inches by three. Prepare both thread and muslin as before.

Hang up the muslin by the four threads attached to its corners ; and place in it a smallish blown egg.

Set fire to the hammock, etc., the result will be the same as with the ring

If you can " palm " and substitute the " blown " egg for a similar " full " one passed round for examination the trick is greatly improved.

EXPERIMENT LXXII. (72)

SELF-LUMINOUS STEAM.

If a little phosphorus be placed in a copper kettle full of water and the water boiled in the dark—and every outlet be carefully closed save the spout (even

51

this is best partly closed so as to project the steam in the form of a small jet), then this jet of steam will be quite luminous, owing to the minute particles of phosphorus carried out mechanically with the steam. The effect produced is a very pretty one indeed.

EXPERIMENT LXXIII. (73)

TO SET FIRE TO A NEWSPAPER BY BREATHING ON IT.

Apparatus.—An ordinary newspaper unprepared; a plate.

Chemicals.—Some perchlorate of potash and sugar ; sulphuric acid.

On one side of the plate is a little of the chlorate of potash and sugar (chlorate, one part; powdered loaf sugar, two parts), and on the other side of the plate a little sulphuric acid. The plate must rest on the table or stand so that the side on which is the sulphuric acid is a little lower than the other.

The newspaper more or less crumpled up, is put lightly on the plate, and the whole lifted off the plate still in the same slightly inclined position. One of the audience (or two) is invited to come forward and blow on the paper. When the performer thinks the time ripe the plate is slightly tilted over to the other side, the sulphuric acid and chlorate and sugar take fire, igniting the newspaper and the trick is done.

EXPERIMENT LXXIV. (74)

HOW TO LIGHT A CIGARETTE WITH A PIECE OF ICE.

Apparatus.—A cigarette—absolutely dry.

Chemicals.—Small piece of potassium; a piece of ice.

52

In the end of the cigarette is concealed a small piece of potassium, a *tiny* piece only projecting—when this touches a piece of moist ice it is fired and the cigarette lighted.

Certain features are, however, essential for success—the tobacco in the cigarette must be absolutely dry, it is best dried in an oven. The piece of potassium should partake of the nature of a stick—as thick as a darning needle and about a quarter of an inch long.

The exact amount must be derived from experiment. If the tobacco at the end holding the potassium is just *moistened* with a drop of naphtha—failure is practically impossible. The writer has, however, never found it necessary to do this.

Try the trick on your fishmonger and note its effect.

EXPERIMENT LXXV. (75)

HOW TO MELT A COIN IN A NUTSHELL.

Mix very carefully and thoroughly three parts of dry nitre, one part of sulphur, and one of very fine sawdust.

If some of this powder be pressed down into a walnut shell, and a threepenny bit or farthing, be put in this, and the nutshell afterwards filled and heaped up with more of the powder pressed down hard, and the powder be set on fire, the coin will be found melted in the mass when the combustion has been completed, whilst the nutshell will only be blackened.

53(76) -(80)

EXPERIMENTS LXXVI—LXXX.

COLOURED FIRES.

Dissolve crystals of any copper salt in spirits of wine, light the solution and it will burn with a beautiful *emerald green flame*, more especially if this be done in an iron ladle or spoon which is held over some source of heat such as a Bunsen burner at the same time. Pieces of sponge soaked in spirit and salts hung up by wires and lighted have often been used in theatrical effects; strips of flannel saturated with it and wrapped round tridents, swords, etc., produce, when lighted, the flaming swords and fireforks brandished by the mimic demons in melodrama.

Crimson Fire.—Put into a gallipot one ounce of spirits of salts, and add to it as much powdered nitrate of strontia as will make a thick paste. Next put a gridiron over a slow fire and set on it the gallipot. Allow it to remain there until the contents are very nearly dry, taking care to avoid the fumes which are poisonous. When the mixture has cooled add about four ounces of pyroxylic spirit and pour the whole into a bottle for use.

When required, wind some lamp-cotton on a two inch nail and drive this into an imitation torch and pour on to it as much liquid as the cotton wick will absorb—without allowing it actually to drop off; light it, and a very beautiful crimson flame will result.

Other Coloured Flames.—The following salts, if finely powdered, mixed with spirits of wine and set fire to will communicate to the flame their peculiar

54

colours, more especially if the burning spirit and salt be heated as well at the same time: chloride of sodium (common salt), yellow; nitrate of potash, pale violet; nitrate of lime, brick red; nitrate of lithia, red; borax, green; chloride of calcium, orange; chloride of lithium, purple. Another method is to mix the powdered salt with powdered sulphur, place the mixture in an iron shovel and ignite the same.

Silver Fire.—Put upon a piece of burning charcoal a morsel of the dried crystals of nitrate of silver, it will at once throw out beautiful sparks, whilst the surface of the charcoal will be coated with silver.

Bengal Lights.—Take nitre, one pound; sulphur, five ounces; sulphide of antimony, two and a half ounces; mix well together on a plate or large dish with a paper knife and light in small quantities with a match or fusee. If well mixed it should burn with a brilliant white light. If magnesium powder be added, the light is greatly increased.

55

CHAPTER VI.

HEAT AND COLD.

EXPERIMENT LXXXI. (81)

HOT AND COLD: WHICH IS WHICH?

Take four basins, put cold water into one, lukewarm water into the second and third, and hot water into the fourth. Invite a person to come forward and place one of his hands in the cold water and the other into the hot, let them remain there for, say, a quarter of a minute, then ask him to take them out, and plunge them in the two basins of lukewarm water, one hand in each, and tell you which water is the hotter of the two, or which water is hot and which is cold.

To the hand which has been in the hot water the water will appear cold and *vice versâ*.

(82) - (87)

EXPERIMENTS LXXXII—LXXXVII.

FREEZING MIXTURES.

Take eleven drachms of nitrate of ammonia, ten of nitrate of potash, and sixteen of sulphate of soda, reduce each *separately* to a fine powder, and mix them gradually in a glass or better in a thin metal tin—such as a cocoa tin—with five ounces of water—the size of the tin should be just sufficient to hold the mixture; as the salts dissolve great cold will be produced as can be seen by immersing a thermometer in the liquid, which should sink below freezing point.

56

A little water, about half an ounce, in a small test-tube, when immersed in the mixture during its solution should be frozen in about ten minutes. The salts used should be fresh and not damp; a good deal depends on this.

Dissolve 500 grammes of potassium sulphocyanide in 400 grammes of cold water—this should give a temperature of —20°.

Crystallized calcium chloride four parts, and snow three parts (or ice), when mixed produce a temperature from 0° to —45°.

Mix together nitrate of lime and snow. To produce the greatest effect, equal parts of the salt finely powdered and newly fallen snow must be *quickly* mixed together. This mixture has been used to freeze mercury.

Set a quart pot on a stool, on which a little water has been previously thrown, before the fire, put a handful of snow into the pot and also a handful of common salt. Hold the pot with one hand and with a short stick stir the contents with the other ; in a few minutes the pot will be frozen so hard to the stool that it can only be pulled off with difficulty.

Fill a common thermometer with cold water and hang it upon the air by a piece of string. If the thermometer tube be continually sprayed with ether, the water will ere long become ice ; the quick evaporation of the ether carrying away the heat from the water.

(88) —(89)

EXPERIMENTS LXXXVIII—LXXXIX.

TWO HEAT EXPERIMENTS.

Put in an earthenware cup or pot a small quantity,

57

say two ounces, of water and pour on to this in the form of a thin stream about three ounces of sulphuric acid, the temperature of the mixture will be raised many degrees above that of boiling water.

N.B.—*On no account must water be poured on to sulphuric acid.*

Dissolve a little lime in nitric acid, then pour some of the liquid into a glass and add a few drops of sulphuric acid, the whole will become nearly a solid mass, and, at the same time, give out a strong heat.

EXPERIMENT XC. (90)

A CHANGE OF TEMPERATURE.

In a narrow tall glass jar, or a large long test-tube or glass flask, place as many lumps of baryta as will nearly fill it.

The lumps must be large and quite dry, with plenty of air spaces between them.

Pour cold water over the baryta lumps. The heat developed will, in all probability, be such as to make the water boil and ere long some of the pieces of baryta will become red hot.

EXPERIMENT XCI. (91)

TO CAUSE WATER TO BOIL BY THE APPLICATION OF COLD, AND TO CEASE TO BOIL BY THE APPLICATION OF HEAT.

Half fill a large glass chemical flask with water and place it over a Bunsen burner or other suitable source

58

of heat. Let the water boil briskly for a few minutes, then cork up the flask quickly, using either a good fitting rubber stopper or a cork which has been greased, or an ordinary cork and a slip of moist bladder tied over the same.

The *essential* point is to cork up the flask so as to exclude all air.

On removing the flask from the source of heat (which should be turned out or down when the cork is pushed into the flask), the water will keep on boiling for a short time, and when this boiling ceases—if the flask be inverted and placed in the ring of a retort stand and cold water be squeezed over it from a sponge, the water will once again begin to boil furiously, at once ceasing to boil if hot water be applied to the flask. Provided there is no breakage on the part of the flask, boiling and non-boiling by the alternate applications of cold and heat may be kept up for some time. The water need not be very hot nor intensely cold, but careful experiment is necessary.

EXPERIMENT XCII. (92)

A CURIOUS EXPERIMENT IN THE INFLUENCE OF COLOUR ONLY, ON THE POWER OF BODIES, IN ABSORBING AND REFLECTING HEAT.

On a winter's day, when the ground is covered with snow, take four pieces of woollen cloth of the same size, and if possible thickness, but of different colours, viz. black, blue, brown and white, and lay them, at the same time, on the surface of the snow, in the

59

immediate neighbourhood of one another, where the sun's rays can fall upon them ; in a few hours they will be found to have sunk into the snow to various depths.

EXPERIMENT XCIII. (93)

AN EASY METHOD OF BREAKING GLASS IN ANY DIRECTION.

Dip a piece of worsted thread into turpentine, wrap it round the glass in the direction in which you require it to be broken, and then set fire to the thread ; or apply a piece of thick red hot copper wire round the glass, and if it does not crack as required at once, throw cold water on it whilst the wire remains hot. By the above means glass that is broken irregularly may often be refashioned and again rendered useful for chemical and other purposes.

EXPERIMENT XCIV. (94)

A METAL WHICH MELTS IN BOILING WATER.

Melt (by weight) four parts of bismuth, two and a half of lead, and one and a half of tin, in an iron pot or ladle over the fire, to form one mass. If a piece of this metal be put into water, it melts when the water begins to boil and remains melted as long as the water is kept boiling.

A spoon formed of this alloy, when used for stirring water which is boiling briskly, melts in the hand of the stirrer.

The lead should be melted first, then the tin, and bismuth added in tiny pieces.

60

An alloy known as Wood's Fusible Alloy, which melts at a much lower temperature even, is composed of: tin, two parts; lead, two parts; bismuth, seven parts; cadimum, one to two parts. This melts at about 68° centigrade, boiling point 100°.

The ingredients should be ground to a fine powder before mixing.

61

CHAPTER VII.

SOME MISCELLANEOUS EXPERIMENTS.

(95) (99)

EXPERIMENTS XCV—XCIX.

CHEMICAL VEGETATION.

Lead Trees.—In a three pound glass jam jar, nearly filled with filtered soft or rain water, put one to two ounces of acetate of lead, previously well powdered, if necessary; shake the mixture and allow it to stand undisturbed for two or three days; decant the clear fluid from the insoluble residue (if any), reject the latter, and, after having rinsed the jar with water, return into it the clear solution; next fit the top of jar with a bung or piece of wood from which can be suspended thin strips of zinc, so twisted and shaped as to resemble as much as possible the branches of trees. On no account shake the jar—stand it on one side for several hours, when a beautiful metallic tree-like growth will be found to have formed.

Tin Trees.—Put three drachms of nitrate of tin in three ounces of water contained in a small bottle; add ten to twelve drops of nitric acid and shake well till fully dissolved. Suspend zinc as before and a similar result will follow.

Silver Trees.—This is the name given to a curious and beautiful precipitation of silver, by means of mercury, in an arborescent form. Lemery was the

62

first to observe this, and the following is his original method: Dissolve one part of silver in dilute nitric acid. Dilute the solution with twenty parts of distilled water and add to it two parts of mercury. The time required for the formation of the silver tree takes about forty days. The following is another and much quicker acting method:

Dissolve four drachms of nitrate of silver in half a pint of distilled water, set the containing vessel on one side for a time where it will not be disturbed, then pour in two drachms of mercury (quicksilver).

Gold Trees.—These can be produced as in the previous experiment by using a solution of chloride of gold and mercury.

Camphor Trees.—Dissolve some camphor in warm spirits of wine until the spirit will dissolve no more. Pour some of this solution on to a clean cold sheet of glass, the camphor will instantly crystallize into beautiful tree-like forms.

EXPERIMENT C. (100)

HOW TO PASS WATER THROUGH A SOLID GLASS JAR.

Pour some water into a plate, light a piece of loosely crumpled paper and throw it into a glass jar; then turn the glass upside-down (with the burning paper in it) in the water in the plate, the burning paper will soon go out—but the water will rise up in the glass until it very likely becomes half full. The air has become rarified inside the jar—and the pressure outside is greater than inside, hence the water rises in the jar until equilibrium is established.

63

The previous paragraph will explain the following:

Take a plate, *i.e.* a deep plate, like a soup plate, and nearly fill it with water; then produce an empty bottle—rather long and narrow for preference—into which you assure the company you can pour water through the solid glass bottom, and not through the mouth as ordinary people do. Having declared that it is necessary for the bottle to be perfectly dry, when the experiment is performed (for the purpose of opening the pores of the glass, if you should be asked the reason), take the opportunity of heating it at the fire so that the bottle save the neck by which you hold it is quite hot, too hot to hold if possible. The bottle can be held to the fire by a stick thrust in the neck if preferred.

As soon as it is quite hot, place it quickly, upside down, with its mouth in the water in the plate; then pour a teaspoonful of water on the bottom, just as if you were pouring it through the glass; every time you repeat this experiment, the bottle will become fuller and fuller of water, and as this takes place whenever water is poured on the bottom of the bottle, it presents (if skilfully done) every appearance of having passed through the solid glass.

After the experiment—pour water into the bottle in the usual way—and have it passed round for examination.

A better way is to have the bottle brought in by an assistant, hot and ready for action—the audience supposing it an ordinary *cold* bottle; and cold or cool it will be by the time you have done with it, whence it can be passed round as before.

64

EXPERIMENT CI. (101)

SINGULAR INSTANTANEOUS CRYSTALLIZATION.

Make a concentrated solution of sulphate of soda or glauber's salt, by adding portions of it gradually in boiling water, till this fluid will dissolve no more (an ounce and a half of boiling water will dissolve about two ounces of salt). Having done this, pour the solution, whilst boiling hot, in a glass jar or bottle previously warmed, and cork it tightly up, as all air must be excluded—a good rubber stopper or greased cork should be used. This being done, set the bottle by in a quiet place, without shaking.

This solution will cool to the temperature of the air and remain perfectly fluid, but the moment the cork is withdrawn and atmospheric air gets in, it will begin to crystallize on its upper surface in fine satin-like crystals, which will shoot downwards in a few seconds, like a dense white cloud, and so much heat will be evolved, as to make the bottle very sensibly warm to the hand.

When crystallization is completely accomplished, the whole mass is usually so solidified that not a drop falls out. If the crystallization does not ensue immediately on drawing out the cork, the slightest agitation or the dropping in of a minute crystal of the salt will almost invariably start the required action.

EXPERIMENT CII. (102)

THE ALUM BASKET.

Make a small basket, about the size of the hand, of iron wire or split willow; then take some lamp

65

cotton, untwist it and wind it round every portion, of the basket. Then mix alum in the proportion of one pound of alum to a quart of water and boil it until the alum is all dissolved. Pour the solution into a deep pan, and in the liquid suspend the basket, so that no part of it touches the containing vessel. Let the whole remain perfectly at rest for twenty-four hours; when on taking out the basket, the alum will be found crystallized all over the cotton frame. In like manner, a cinder; a piece of coke; the sprig of a plant, or any other object, suspended in the solution by means of a thread, will become covered with beautiful crystals. If powdered turmeric be added to the hot solution the crystals will be of a bright yellow; if litmus be used instead they will be of the brightest red; logwood imparts to them a purple colour, and common writing ink a black tint; or if sulphate of copper be used instead of alum, the crystals will be a fine blue. But the coloured alum—crystals are much more brittle than those of alum alone and the colours fade. The best way to preserve them is to place them under a glass shade with a saucer containing a little water, thus keeping the air within the glass saturated with moisture; under such circumstances the crystals retain both their texture and colour.

EXPERIMENT CIII. (103)

CHEMICAL FROST OR SNOW.

Arrange in a tasteful manner in a bottomless glass jar, vessel or bell jar, some sprigs, etc., of holly or other evergreen, or imitation paper flowers. Place the

66

glass vessel on a metal or non-inflammable metal stand, with a hole in the centre. In this place a small porcelain evaporating dish containing a little benzoic acid, heat it from below with a methylated spirit lamp, improvised out of an old twopenny ink bottle and a piece of cotton wool. A very elegant and beautiful imitation of hoar frost or snow will thereby be produced on the sprigs, etc., contained in the vessel.

The stopper or cork in the top of the jar, should be left open at first, and closed when the spirit lamp is withdrawn.

EXPERIMENT CIIIA. (103A)

THE ROTATORY MOTION OF CAMPHOR UPON WATER, OR THE ELIXIR VITÆ.

Fill a saucer, or broad basin, with water, and let fall upon it camphor reduced to the fineness of coarse sand. The floating particles of camphor will instantly begin to move, and acquire a progressive rotatory motion, which continues for some minutes and finally subsides.

If the water be touched by a substance which is the least bit greasy, all the particles of camphor dart back briskly, and are, as if by a stroke of magic, instantly deprived of their motion and vivacity.

EXPERIMENT CIV. (104)

THE TRANSMUTATION OF METALS.

Immerse a piece of polished iron into a solution of sulphate of copper, the latter metal will become

67

precipitated in a metallic form on the surface of the iron.

Immerse a piece of silver in the solution and no such effect is produced; but if the two metals, the silver and the iron, are brought into contact, and then immersed together in the solution, *both* metals will speedily receive a coating of copper.

EXPERIMENT CV. (105)

TO TURN STEEL INTO GOLD.

The following is a very simple way of gilding steel: mix together some sulphuric ether and nitrate of gold. If into the ether thus charged with gold a *piece of highly-polished steel* be immersed—a test tube and large needle are ample for the experiment—the ether speedily evaporates, leaving a real coat of gold upon the steel.

By applying a fine brush or a pen, all kind of figures may be delineated on highly-polished steel instruments, such as razors, scissors, penknives, etc.

The moment the steel is gilded it should be washed in water; and the adhesion of the gold to the steel is much improved by burnishing.

EXPERIMENT CVI. (106)

TO TURN IVORY INTO SILVER.

The following is a very easy method indeed of silvering ivory. Take a piece of polished ivory, and immerse it in a solution of nitrate of silver until it has

68

acquired a bright yellow colour; this must be done in a subdued light; then take it out of the solution, immerse it in a tumbler of distilled water and expose it, in the distilled water, to the direct rays of the sun.

After the ivory has thus been exposed for about two or three days to the action of the sun's light, it appears black; but on rubbing it, the black surface will be changed into a bright metallic one resembling silver.

Although this coating of revived metal is extremely thin, yet if the ivory be well impregnated with the nitrate of silver, the solution penetrates to a considerable depth and as fast as the silver wears off from the surface of the ivory, the oxide below it becoming uncovered and exposed to the sun's rays, a new coating of revived metal will be formed to replace it, and the surface of the ivory will not lose its metallic appearance.*

* This most interesting experiment I have taken *verbatim* from a book more than 100 years old.—Ed.

EXPERIMENT CVII. (107)

THE MAGICAL SCENT SPRAY.

Six white feathers and an ordinary scent spray or atomizer are shown; when sprayed upon, each feather becomes any colour desired. The feathers have previously been carefully dusted with six different aniline dye powders and then shaken, so that to all appearance, all the dye powder is shaken off. If alcohol or eau-de-cologne be sprayed over them, the

69

feathers at once become coloured. Each feather must bear a private mark so that the performer knows which is which.

EXPERIMENT CVIII. (108)

A MAGICAL DRAWING.

Various salts—such as the salts of chlorides of platinum, gold, iridium, etc., are strongly affected by mercurial vapours.

Suppose a design (see Fig. 7) drawn on a sheet of

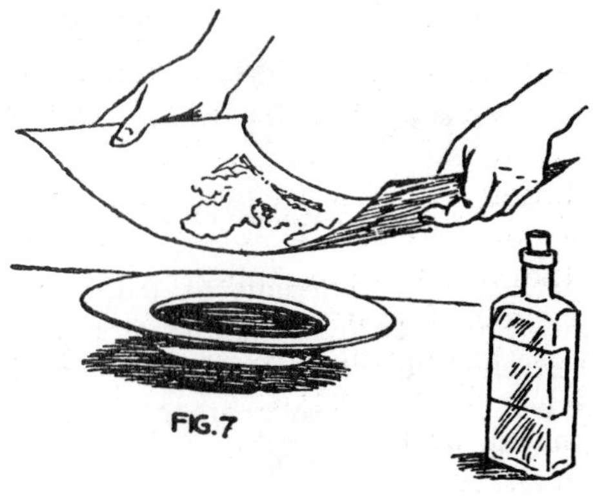

FIG. 7

white paper with platinum chloride, such a drawing is absolutely invisible. If, however, the drawing be held over some mercury contained in a dish, the design will at once become visible in dark tints. The experiment shows how very diffuse the vapours of mercury are in their action on certain salts.

70

EXPERIMENT CIX. (109)

CHLORINE GAS AS A SUPPORTER OF COMBUSTION.

If a piece of paper be moistened with turpentine and plunged into a jar full of dry chlorine gas, it will take fire and burn. A piece of phosphorus does the same thing, igniting more quickly even than in oxygen gas. Dutch metal and powdered antimony will do the same thing.

EXPERIMENT CX. (110)

SOAP BUBBLES FLOATING ON CARBON DIOXIDE.

If a large and deep glass bowl—some twelve to fifteen inches in diameter and eight to ten inches in depth—be filled with carbon dioxide and soap bubbles be blown and detached over the bowl of gas, they will not descend to the bottom of the bowl but remain floating about on the top of the invisible gas in a most mysterious and fascinating manner.

In order to improve the bubbles and make them last longer, glycerine as well as soap should be added to the water.

Use common yellow soap, distilled water, adding glycerine in the proportion of one part to four of water.

EXPERIMENT CXI. (111)

RESIN GRAPES.

Dip the bowl of a tobacco pipe into melted resin; blow through the pipe, when bubbles will be formed

71

of a brilliant silvery hue, and a variety of colours. This is the way in which Italians imitate bunches of grapes; the bubbles being fastened together and dusted with powder—blue, to resemble bloom.

EXPERIMENT CXII. (112)

HOW TO PRODUCE A LIQUID FROM TWO SOLIDS.

Mix together equal portions of sulphate of soda and acetate of lead, both in the form of a fine powder, let them be well rubbed together in a mortar, when the two solids will be found so to react on one another that a fluid will be produced.

EXPERIMENT CXIII. (113)

TO SEPARATE OIL AND WATER.

The reader is doubtless aware that oil is lighter than water and floats upon its surface. Half fill a vessel of any convenient description with water, and then pour some oil on the top, the oil may be easily separated from the water in the following manner: place in it the end of a cotton wick, the other end of the wick being carried over the top of the vessel into another vessel preferably at a lower level; the oil, obedient to the laws of capillary attraction, will rise gradually in the cotton wick and fall, drop by drop, from the other end of it into the vessel placed ready to receive it.

72

EXPERIMENT CXIV. (114)

HOW TO MAKE A LUMP OF SUGAR FLOAT ON WATER.

The lump of sugar must first be immersed in collodion. When taken out the collodion solution quickly evaporates and leaves behind a transparent layer, very like gold-beater's skin, possessing considerable tenacity. The lump of sugar must be put in a current of air for a day or two, in order that the ether may entirely evaporate. In appearance it will then be exactly like any other lump of sugar, and it may be placed in a sugar basin with other similar looking lumps. On dropping it into a glass of water, it first sinks to the bottom, but after a few seconds, it rises to the top, where it remains. The prepared lump of course does really dissolve in the water—the collodionized shell alone remaining. To the eye the illusion is perfect, but not of course to the touch, and it must on no account be handled.

It may be shown as a trick in many ways, and if accompanied by some sleight of hand, can be made a first-class illusion.

First allow one or two of the audience to drop in lumps of sugar taken from the basin which has no prepared lumps. Then, by a little piece of sleight of hand, the prepared lump is introduced and dropped in.

EXPERIMENT CXV. (115)

CHAMELEON LIQUIDS.

Expose to a red heat in a portable furnace and in a covered crucible, one part of finely-powdered oxide

73

of manganese, with three parts of nitre for one hour, thus forming an alcalized oxide of manganese; put into a wine glass a few grains of this substance, and put a similar quantity into another glass.

If cold water is now poured into one glass, a green solution is obtained, which very rapidly changes to purple and gradually becomes red.

And if the other glass be filled with warm water, a violet coloured solution is produced which quickly changes to crimson.

The intensity of the change of colour is influenced by the quantity of oxide in the water. With about ten grains and one pint of cold water, the solution is a beautiful green colour, and changes in a few seconds to a deep purple, which lastly becomes red; and if upon a like quantity of the oxide four ounces of water be poured, the solution is of a deep green colour, and on adding to it more water it acquires a rose-red colour, allowing a yellowish precipitate to fall. A few drops of nitric acid added to the green or to the purple solution, change it instantly to a vivid red.*

* Another way of preparing the powder is to mix together and expose to a strong heat, in an open crucible, for rather more than a quarter of an hour, three parts of nitrate of potash and one of manganese dioxide, both in a finely powdered state. When the liquid changes slowly, the following range of colours can easily be traced—green, blue, violet, indigo, purple, red. Experiment with varied proportions of the two ingredients.

EXPERIMENT CXVI. (116)

CHAMELEON POWDER.

Take some mercuric oxide (an orange-coloured

74

powder), and put it on a white-enamelled plate or dish; on heating the plate, the powder changes first to a bright red, purple, etc., eventually becoming black. On cooling it returns to its original colour. If the heat be secretly applied—quite a pretty little trick can be worked up with respect to the same.

EXPERIMENT CXVII. (117)

THE TWISTING OF ROPES AND RAISING OF HEAVY BURDENS BY CAPILLARY ATTRACTION.

If we hang up a rope some ten to twelve feet long, in a vertical position, and fasten to its lower extremity a heavy weight to stretch it as much as possible, the rope, when wetted, will be shortened, and if the length be such as to permit the weight to just touch the ground when dry, then when wet the weight will be raised from the ground. The following ancedote respecting the famous obelisk erected by Pope Sixtus V. before St. Peter's at Rome is full of interest. The cavalier Fontana, who had undertaken to raise this monument, was, it is said, on the point of failing, just when the column was about to be placed on its pedestal. It was suspended in mid air ; and the ropes had stretched a little so that the base of the obelisk could not reach the summit of the pedestal, a man cried out, " Wet the ropes." This advice was followed, and the column, as of its own accord, gradually rose to the required height, to be placed upright on the pedestal prepared for it.

(118)-(119) EXPERIMENTS CXVIII—CXIX.

THE PRECIPITATION OF COPPER AND SILVER IN A CRYSTALLINE AND METALLIC FORM.

Allow a piece of phosphorus to remain immersed for about twelve hours in a solution of sulphate of copper, it will become enveloped in a coat of extremely brilliant and crystalline metallic copper, impervious to air.

Substitute a solution of nitrate of silver for the copper sulphate and the silver will be precipitated on the phosphorus in the form of fine dendritic crystals.

EXPERIMENT CXX. (120)

HOW TO SILVER COPPER, BRASS, ETC.

First, clean the article intended to be silvered by means of a little dilute nitric acid, or by scouring it with a mixture of common salt and alum.

When it is perfectly bright, moisten a little of the powder, known as *silvering powder*, with water, and rub it for some time on the perfectly clean surface of copper or brass, which will become covered with a coat of metallic silver. It may afterwards be polished with soft leather.

How to make the Silvering Powder.—Dissolve silver in nitric acid, and put pieces of copper into the solution ; this will throw down the silver in the state of a metallic powder.

Take fifteen or twenty grains of this powder and

mix with it two drachms of acidulous tartrite of potash, the same quantity of common salt, and half a drachm of alum.

Another Method.—Precipitite silver powder as before. To half an ounce of this add common salt and nitrate of ammonia, of each two ounces and one drachm of corrosive sublimate ; rub them together and make them into a paste with water. With this copper utensils of every kind that have previously been boiled with acidulous tartrite of potash and alum can be silvered by being first rubbed, and afterwards made red hot and burnished.

EXPERIMENT CXXI. (121)

THE SINKING AND FLOATING EGG.

If a glass jar—preferably tall—be filled with dilute hydrochloric acid and an egg be placed in it, it will first of all sink. It will, however, be found after a short time to rise to the surface and slowly rotate. Notice the formation of bubbles of carbonic acid gas on the shell, it is to their formation that the rotation is due.

EXPERIMENT CXXII. (122)

A VERY EFFICIENT AND BEAUTIFUL PARLOUR FIREWORK (KNOWN AS) SPUR ROWELS.

Mix together nitre, lamp black and sulphurin, the proportion of 8 : 3 : 4.

The sulphur and nitre must be powdered separately

and again ground when mixed, and then the charcoal carefully rubbed in, till the entire mixture has a uniform dark grey tint. Test from time to time by running a little of the mixture into a little paper case and setting it on fire. If the stars come out in clusters and spread well without being mixed up with other or duller sparks, it may be taken that the whole is well mixed. If not mix until they do. Do not carry the pounding process too far or the sparks will be too small. The sparks thrown into the air have the shape of the rowel of a spur, hence the name. The chief difficulty consists in bringing about a complete mixture of the charcoal. The sparks will not burn the hand when falling on it, nor set fire to anything owing to the perfect combustion.

EXPERIMENT CXXIII. (123)

THE BLACK AND WHITE STATUE.

If a figure or image be constructed of white paper or cardboard and be moistened with a solution of acetate of lead, then when such a statue is exposed to the fumes of sulphuretted hydrogen gas it turns black.

Sulphuretted hydrogen (an evil smelling and

poisonous gas) can be prepared by the action of dilute sulphuric or hydrochloric acid on iron sulphide, placed in a flask fitted with a thistle funnel and delivery tube. It should be collected over warm and not cold water, owing to the great solubility of the gas in cold water.

78

EXPERIMENT CXXIV. (124)

BURNING COPPER WIRE.

Raise some sulphur to boiling point in a test tube or similar vessel. Insert a coil of *fine* copper wire; vivid combustion will follow.

EXPERIMENT CXXV. (125)

BEAUTIFUL CRYSTALS OF BISMUTH.

Having melted some of the metal bismuth in an iron ladle, remove, by means of a knife or otherwise, the dross or scum that floats on the top, then allow it to rest until a crust of hard metal is formed on the surface. As soon as that is done make a hole in the edge of this crust and pour off the metal that still remains in the centre of the mass; when it is quite cold, remove the upper crust by means of a file, and if the experiment has been properly carried out, the inside of the mass will be found covered with a group of beautiful crystals of bismuth, sufficiently curious and pleasing to form an interesting mantel-piece ornament.

EXPERIMENT CXXVI. (126)

TO PLUNGE YOUR HAND INTO WATER AND YET NOT WET IT.

This is a very taking trick if properly carried out. You drop some bright object into a clear glass dish

79

nearly full of water. You then announce to the company that you are about to pick the object out of the water with your bare fingers without wetting them. You then sprinkle the surface of the water with lycopodium powder. The powder forms around your fingers, etc., like a proper glove, and the water has no more action on your fingers than it has on the feathers on a duck's back. A better method is to rub the hand well with the powder.

EXPERIMENT CXXVII. (127)

TO PLUNGE A PIECE OF WHITE PAPER INTO AN INKPOT FULL OF INK WITHOUT BLACKENING IT.

All that you require are a large inkstand with a wide mouth and some well-powdered black resin or colophony.

You invite one of the audience to dip a sheet of white paper (rolled up into the form of a cylinder, into the inkstand which is about half full of ink. It should be well dipped in so that plenty of ink is brought out when it is withdrawn. On the excuse of filling up the inkpot, you pretend to pour some more ink into the pot, but in reality you pour in rosin powder and water. When the surface of the ink has a layer of this fluid above it, the roll of paper can be plunged in without any fear of actual contact with the ink. When taking out the paper it should be shaken slightly so as to prevent any of the rosin powder sticking to it.

80

EXPERIMENT CXXVIII. (128)

THE DISAPPEARING INK THAT LEAVES NO TRACE.

Whereas inks exist which under the action of light entirely disappear—so far as our ordinary vision are concerned—many such, however, leave a trace on an ordinary photographic plate, and others can be revivified by suitable chemical reagents.

In fact, in nearly all cases it is possible to revive the characters.

In the following case, however, this cannot be done. The "ink" is made as follows: dilute some starch in water to the consistency of a cream and add a few drops of tincture of iodine. Dip the pen in the ink thus made and write on ordinary paper. The writing is plainly visible, and dries very quickly. To obliterate it completely, it is only necessary to rub it with a handkerchief or even with the hand—and it will disappear as easily and completely as chalk from a blackboard.

EXPERIMENT CXXIX. (129)

CURIOUS ACTION OF ICE IN AMMONIA GAS.

Place or drop a piece of ice into a vessel full of ammonia gas, so violent a combination between the two at once takes place as almost immediately to melt the ice.

(130) (131)
EXPERIMENTS CXXX—CXXXI.

THE COLOUR CHANGING FOUNTAINS.

Fit a flask with a cork and short tube. The tube

81

must be drawn out to a point at one end and should project an inch or so into the flask. Now fill the flask, by air displacement, with ammonia gas, and when full insert the cork, etc., and dip the outer end of the tube into a dish full of water reddened with a little litmus. Immediately the ammonia gas begins to dissolve in the water and a fountain is produced in the flask, due to the partial vacuum in the flask, the red colour of the fountain jet is changed to blue—a very magical effect to the uninitiated.

If the flask be filled with hydrochloric acid gas and the water in the dish be the ordinary (diluted) blue litmus solution, the fountain will change from blue to red.

Red litmus solution as used above is made by adding a drop or two of acid to the ordinary blue solution. In neither case should the colour be too deep.

EXPERIMENT CXXXII. (132)

HOW TO MAKE ONE BAD SMELL KILL ANOTHER BAD SMELL.

Both chlorine and sulphuretted hydrogen gases are extremely evil-smelling ones, especially the latter which partakes of the odour of rotten eggs and cabbage water combined. Place a little sulphuretted hydrogen solution, *i.e.* water charged with the gas, in a dish; the smell of rotten eggs is at once perceptible. Now add gradually some chlorine water, the solution will become milky, but all smell will vanish.

82

If more chlorine water be added the smell of chlorine will become noticeable, and this excess of chlorine diffusing into the air around will destroy the sulphuretted hydrogen odour which before filled the room.

EXPERIMENT CXXXIII. (133)

ANOTHER SPONTANEOUS COMBUSTION EXPERIMENT.

If a piece of phosphorus be placed on a piece of blotting paper and dried and it be covered with lamp-black or soot, the phosphorus will melt and ignite of its own accord.

EXPERIMENT CXXXIV. (134)

A GHOST PAINT EXPERIMENT.

Half fill a test tube with sweet oil, add a small piece of phosphorus, warm the tube and contents by immersing the tube in hot water for a few minutes. The phosphorus will become dissolved in the oil to a certain extent. The hands, face, or any surface may be rubbed over with the mixture and will present quite a luminous appearance in a dark room. The very small quantity of phosphorus dissolved makes the experiment a perfectly safe one.

EXPERIMENT CXXXV. (135)

HOW TO MAKE "SOLUBLE GLASS."

Melt together in a crucible fifteen parts of sand, ten parts of carbonate of potash and one part of charcoal.

83

Or mix together one part of powdered silica, five to six times the amount of carbonate of soda and an equal quantity of carbonate of potash, put in a crucible as before and heat to redness in a fire. When the effervescence has ceased, remove the crucible from the fire, allow it to cool, chip out the product and boil it in water; a solution of sodium silicate or soluble glass is the result. In the first given mixture, after heating wash away all excess of alkali with cold water; after this is done the powdered glass should be boiled in water in the proportion of one of the former to five of the latter, when it gradually dissolves. This solution may be evaporated to a thick pasty fluid, which looks like a jelly when cool, and on exposure to the air in thin films changes to a transparent and brittle, but not hard glass.

EXPERIMENT CXXXVI. (136)

A GOOD "GREEN FIRE" RECIPE.

Mix very carefully in *small* quantities forty-five parts of barium nitrate, fifteen of sulphur, ten of potassium chlorate, two and a half of lamp-black and one half of antimony sulphide, having previously reduced the constituents to a fine powder *separately*. The product—an explosive mixture—burns with a brilliant green flame.

EXPERIMENT CXXXVII. (137)

HOW TO MAKE "GERMAN SILVER."

Mix together one part of nickel, one part of zinc and

84

two parts of copper, and fuse in a crucible or iron ladle. The alloy produced is known as German Silver.

EXPERIMENT CXXXVIII. (138)

BURNING THE AIR.

Fill a large glass jam jar with coal gas and put it, mouth downwards, on the ring of a retort stand, the mouth, *pro tem.*, being covered with a greased glass plate.

Fill a football bladder with air and attach it by means of a piece of rubber tubing to a piece of glass tubing, with the end drawn out to a point in order to make it resemble a gas jet. Press lightly on the bladder, if necessary, to produce a steady stream of air. Light the coal gas at the mouth of the jar and steadily introduce the jet of air, which will be ignited at the mouth of the jar by means of the flame from the burning coal gas, and as it is pushed up into the interior of the jar will continue to burn, thereby proving that ordinary air will burn in an atmosphere of coal gas, just as coal gas burns in air.

Great care must be taken that coal gas only and not coal gas mixed with the air is in the jar or there may be an explosion.

EXPERIMENT CXXXIX. (139)

A STRANGE ACTING ALLOY.

If small pieces of the metal sodium be added to mercury and warmed under a layer of heavy paraffin oil, a very curious alloy indeed is produced. It is

85

liquid when it contains 1.5 per cent. of sodium, and solid when it contains more. If to this alloy there be added a concentrated solution of ammonium chloride, a remarkable phenomenon occurs, the amalgam swelling up enormously whilst still retaining its metallic appearance.

EXPERIMENT CXL. (140)

A MAGICAL PICTURE.

On a wooden frame stretch tightly a piece of unbleached muslin; make separate solutions of the four following chemicals in warm water, sulphate of iron, nitrate of bismuth, sulphate of copper, prussiate of potash.

Use the first three for painting your "invisible" picture, the fourth sprayed on is the developer. Use a separate brush for each colour, avoid much detail, allow one colour to dry thoroughly before applying the next. When dry all are invisible; when sprayed on with the prussiate of potash, the sulphate of iron comes up blue, the nitrate of bismuth yellow, and the sulphate of copper brown.

86

CHAPTER VIII.

HOW TO GIVE AN ENTERTAINMENT IN CHEMICAL MAGIC.

It is one thing to perform a chemical experiment in a laboratory or even in private; it is a totally different thing to present a series of such experiments as a magical entertainment in public or before one's friends. Success or non-success in the latter case depends far more on the *manner* in which they are presented than on the actual experiments themselves. In magic a trick or experiment (for they are in a sense the same thing) is just what the presenter or performer makes it—nothing more, nothing less.

It is quite possible for anyone to make himself quite proficient in a series of experiments, learn perfectly by heart a suggested "line" of patter—and yet when presented the whole affair might fall as flat as ditchwater. Never be content to learn and get up an experiment *exactly* on the lines suggested—that is only to see it through the mind of the writer; always add some variation of your own, you thus make it *your* experiment and *your* knowledge.

WAYS OF PRESENTING AN ENTERTAINMENT.

An entertainment in Chemical Magic may naturally be presented in a great variety of ways.

(i.) As part of an entertainment.
(ii.) As the entertainment itself.

87

The first method is undoubtedly the best, especially if you have an acquaintance with some of the ordinary branches of magic as now practised in entertainment form. Three or four tricks in chemical magic interspersed with others add greatly to any magical show.

Their selection in this case we leave entirely to the personal taste of the reader, adding only one word of warning: if you are not in any way familiar with scientific experiments, make absolutely certain that you know them inside out before presenting them in public—more than one well-known professional conjurer has come a most howling cropper in such circumstances. The magic kettle—that boils on ice—is a case in point. One very important thing remains: the introduction of chemical experiments into a magical entertainment should appear *natural, i.e.* should follow as a natural consequence from something that has gone before. Appropriate "patter" can bridge most streams of this character, so perhaps it is hardly necessary to add that—since magic is the art of *mis*-direction there is not the least reason for introducing the word "Chemistry" in connection with such experiments, but very much the reverse.

Passing on to consider No. 2 in which the entertainment is to be, practically speaking, confined to Chemical Magic, the first and most important item is not to make it long. Half an hour at first—and later perhaps three quarters—will be as long as you will be able to command the full attention of your audience—especially if juvenile in part or whole. Make it an invariable rule, if possible, to cut your "show" short when signs of inattention or the least weariness

88

becomes apparent on the part of your audience. Far better to give too short an entertainment than one too long. Some amateurs fail lamentably in this respect. Never give your audience *all* they want—always let them go away wanting a bit more.

Above all, never give an experiment away *intentionally*—such is an unpardonable crime in magic; moreover, you will sink ninety per cent. in the eyes of your audience as soon as you have done it. If something goes wrong, as it does at times with the most proficient, pass it off—with all the "cheek" (in a quiet way) that you are capable of and always in a pleasant manner, and proceed to get on with the next. Never make excuses. If you can make the entertainment tell a story—so much the better—as an example of this I quote: "King Ko Ko," by Professor Hoffman; if not, let one experiment follow naturally from another, or one naturally lead up to the next; let the apparatus be absolutely of the most ordinary kind, ordinary tumblers and glasses, ordinary decanters, jam jars, etc., etc., and not "chemical" in appearance. As to what you will call the entertainment or what name you will give it, must be left entirely to the reader's taste and inclinations. If you wish to give it a scientific or pseudo-scientific character—then *Alchemy* naturally suggests itself.

In this case you could well adopt the Astrologer's dress and some of the curious looking apparatus of the middle ages.

Again your entertainment might be given as the outcome of some parchments found in an old moorish tower—an ancient Egyptian tomb, etc., etc. The

89

great point to remember is to have everything appropriate, or at any rate sufficiently so to pass muster. Here—as elsewhere—you must, if you wish to be really successful, make the entertainment your own.

One point which, of course, is always paramount in any question of this kind is the matter of £. s. d.; fortunately here, however, the band is a very elastic one. Quite a good entertainment can be given for a comparatively small sum, even in these expensive days—as well as for a very large one; according (1) to the nature and scale on which the experiments are performed, and (2) more especially on the character of the accessories—such as dress, furniture, scenery, etc., with which they are accompanied. Let us pass on now to consider the presentation of a cheap and yet quite effective entertainment—unaccompanied by any out of the way accessories in either apparatus, dress, scenery, or furniture. The subject might bear a score or so of different titles. I have called it simply:

"SOME EXPERIMENTS IN MODERN MAGIC, OR THINGS ARE NOT WHAT THEY SEEM."

The selected experiments are numbers: 1, 7, 17 or 31, 21, 50, 51 or 52, 60, 66 or 67, 68 or 69, 71, 73, 74, 91, 100, 105, and one other experiment not so far related: *i.e.* 16 experiments in all, which should be more than ample to commence with; in fact, it could well be cut down to a round dozen for a start. Later, the number can be increased with, however, 19 or 20 as the absolute limit.

90

We propose taking the experiments in the following order: 74; 50; 73; 1; 7, original; 105; 17 or 31; 91; 71; 51; 60; 52; 69; 21.

There is first of all the question of an assistant to be considered; this is, practically speaking, almost a necessity, and not only an assistant, but a capable one as well; the capability mainly consisting in doing what he or she has been told—neither more nor less. As to the question of sex—the writer prefers a lady assistant; a lady giving the show might again prefer a gentleman assistant; it is purely one of choice and convenience. However it be arranged, it is all important that the one be the performer and the other the assistant only. If the performer gets *everything* in readiness as he or she should do, then the assistant need know nothing further than clearly to understand his or her direction. All this may seem making a good deal of fuss about minor details, only, unfortunately or fortunately as the case may be, it is just such careful attention to minor details that makes for success.

Faraday, than whom no greater experimentalist ever lived, was always most careful to see, before any lecture, that the stoppers of any bottles (which might be required) were loose, *i.e.* not too tight fitting, in order that there should be no unnecessary delay.

Two or at most three small tables of the usual drawing-room type arranged fairly well apart, with a larger table somewhat in the rear—partly curtained off—should be all that is required in the way of furniture. Be careful to see that you can move

91

freely between the tables—to upset one—and it is quite easily done—is a fiasco not easily overcome.

I have already spoken about the use of *ordinary and non-chemical looking apparatus*; and jam jars have also been mentioned. But you must not consider yourself, however, under any obligation to use such, save on the score of expense.

It is idle to try to pretend that handsome tables—suitably draped in some cases—handsome glass decanters, glass bowls, etc., do not add to the effect; undoubtedly in nine cases out of ten they do. Your chief aim, however, should be one of *appropriateness all round*—because such, in the end, always produces the best effect.

Say you are using as suggested three small tables, one right, one centre, one left, and that you work them right to left.

Then when you are working the centre and left, your assistant can clear the right, preparing it for the next experiment belonging to that table, and so on with the centre and the left.

In every case such a rule need not be followed, but *you must have a rule or order of doing things and you must not get in one another's way.*

If you are working without an assistant, and some will, I know, prefer this—then, if possible, four tables are better than three, and everything else wanted must be ready on the larger table behind—where ample space ought also to be left to put everything when finished with, so far as such is necessary to conclude the entertainment.

On table on right (say) is all you want (save the

92

cigarette case which is of course in your pocket) for experiments 1 and 2 and 3; on table in centre, all you require for 4 and 5, and on table on left the necessary apparatus for numbers 6 and 7, etc.

The remainder ready on the most convenient part of the larger table in the rear.

Coming forward you begin in something after the following style:

"Ladies and gentlemen, I have undertaken, in my case, the somewhat onerous task of trying both to interest and amuse you this evening by showing you some of the latest achievements in modern magic, a mighty subject, so vast indeed that I shall have to confine myself almost entirely to one particular branch of the subject. As to what that particular branch is—eh! well—eh! of course I need not waste or take up any of your time telling you what particular branch that is, because you will *see* for yourselves—or, if you don't see, you will—well—eh! you have other organs, and you will no doubt make use of them all to their fullest extent—if for no other reason than to catch me tripping.

"Of all intangible things—scarcely anything is more intangible than 'smoke'—many of you have no doubt tried to catch a handful of it and failed. People sometimes—more often now than formerly—speak of their money 'disappearing like smoke,' and smoke like money which has disappeared never comes back. Permit me, however, to show you an experiment which proves all this to be wrong, or in other words the exception which proves the rule.

"First of all, it is necessary to procure some 'smoke.'

93

Now of all smoke—tobacco smoke, being the most agreeable and even to an anti-smoker the least disagreeable, is the particular kind that I propose to make use of. [*Performer here brings out his cigarette case and takes out* (*prepared*) *cigarette and proceeds to feel in his pockets for matches.*]

"Dear me—how annoying, I have forgotten my matches. Can any gentleman—but, eh! I was quite forgetting, *pro tem.*, my magical powers—force of habit—you know—it is years now since I carried such things.

"You wouldn't think much of a magician who wanted 'matches' to light his cigarettes with, would you? Permit me—I see my assistant has kindly provided me with a piece of ice—to cool my fevered brow, I presume, later, should it be necessary. See, I take this piece of ice, quite unprepared, imported direct from the refrigerating machine, two blocks away—I utter the magic word: 'pentadekyl para to lylketone,' and on presenting it to my cigarette it lights at once, as you all can see.

"Now on this table I have an ordinary tumbler, a ditto saucer, both as you can see quite unprepared, an ordinary handkerchief, fresh from the laundry and quite unprepared, and, marvellous, it is without a single hole. I close up the tumbler by placing the saucer over the top, and, further, to isolate it I throw this handkerchief over it as well. I now retire to the end of the room—puffing as you can see beautiful wreaths of curling smoke into the air—which quickly disappear—to reappear anon, as you will see—where I will—in this case in the covered up tumbler on yonder table. [*Returning slowly to table—this must*

94

be carefully timed beforehand.] See, I withdraw the handkerchief, the glass as you can see is now full of smoke—I withdraw the saucer, it issues forth—eh! I beg your pardon, sir—the smoke issuing from the glass is not the same in colour as that which I puffed into the air?

"My dear sir—was the smoke issuing from the end of my cigarette—the same as that I respired—good word that—from my lungs?

"No—exactly then, how can you expect there to be no difference—after it has disappeared—passed into the covered tumbler and once again become concentrated and visible?

"I am, I assure you, ladies and gentlemen, always most willing to oblige—but really between you and me and the onion some people *are* particular.

"Talking of onions reminds me that some very peculiar traditions exist with respect to 'onions.' In the south of Europe there is a curious superstition that if an ordinary steel magnet be rubbed with an onion it loses its magnetic properties; many people even now seem still to hold that popular fallacy—that eating a dozen or so raw onions would have some slight effect on your breath. Some even go further and put the adjecture 'unpleasant' in. I have even heard another adjective used—viz. 'strong.'

"It is most curious how one thing leads on to another—do you know I once knew of a man who had—who could set fire to a newspaper by simply breathing on it; I don't mind confessing I have done it myself before now—in fact, I don't mind showing you the experiment if you would like to see it.

95

"I have here part of an ordinary newspaper which I crumple up, so as to make it occupy less space, and place it on this plate. I pick up the plate with the newspaper on it and by——but I think it would be more interesting if someone from the audience actually performed the experiment. Do you; sir, mind coming forward and you? I prefer two; I will first of all try you one at a time.

"Would you mind standing here on the left and you on the right—not too close, please. Now [*holding plate aloft*] if you [*referring to the person on the left*] will please blow gently—we will see what happens, nothing so far—will you [*referring to the person on the right*] please try—just a little harder, no—no sign as yet—now will you please both blow*—ah! [*tilting plate slightly*]. Now—yes—here she goes—congratulations gentlemen on your combined breaths—I mean your success. One moment ere you retire. Permit me to offer you a little refreshment as a slight appreciation of your assistance.

* See that the paper is not blown off the plate—a thumb or finger over a part of it is advisable; make sure also the paper hides the chemicals.

"As you see we have here two decanters, one of wine the other of water—you, sir—you would prefer wine, and you, sir, water—good—allow me"—[*pours out water from the decanter into one glass—in the other glass it changes to wine, the next remarks depend on what the two members of the audience do, they may taste it—they may not—they may express approval—they may not—probably they will taste it and not care for it*]. "You find it good—but a little of a very good thing goes a

96

long, long way—so you prefer to have the major portion—good. Let us pass on to our next—whilst on the subject of 'drinks' and liquids I wonder whether any amongst you have heard of 'rainbow water.' What is rainbow water, it is nothing more nor less than water which has had some pieces of broken-off rainbow dissolved in it, and which when poured out into separate glasses will show all the colours of the rainbow in turn. It is not so well known as it might be; you won't find it mentioned in 'What's What' or eyen in the 'Child's Guide to Knowledge.' No chemist keeps it, though he won't admit it, and always tries to palm something else off on you as being the genuine article.

"To tell you the strict truth, I believe I'm the only person in the world who happens to have a sample just now. It was given me by an aviator friend of mine, who one day whilst travelling at the rate of about 750 miles an hour—more or less—rather less in fact—went bang through a rainbow before it could get out of his way, and some of the knocked off bits fell in his machine and he passed them on to me—knowing that I was interested in things of that sort.

"They are here in this decanter [*showing it*]. You see the water is quite clear—I have here a number of glasses [*indicating them*], commencing at the left hand end—you will note I pour out in turn red, orange, yellow, green—I beg your pardon—the glasses have been chemically prepared before hand—madam [*setting down decanter*], allow me [*picking up three remaining glasses*] to pass you the remaining glasses—examine them for yourself—[*after they have been*

97

returned the remaining three colours, blue, violet, indigo, are poured out by means of the coloured dabs on the decanter].

"The fact that my aviator friend was unable to charge through the rainbow without breaking some of it, led to a very interesting discussion at the Hocus Pocus club the other night on 'The Penetrability of Matter'—or can two things be in the same place at the same time. I beg your pardon—it doesn't matter—now that's just where you make the mistake—as I will proceed to demonstrate. What's that? The answer is as plain as the nose on your face—Just so! Just so! I know what you would say—all matter occupies a certain space, and whilst in that space nothing else can occupy it. I beg pardon. You can drive a nail into a piece of wood. I'm afraid not, madam—I'm afraid not—not in the sense you mean it. You thrust aside the wood—but the wood is not where the nail is—or the nail where the wood is. Two things cannot then be in the same space at the same time. *I* cannot be where *you* are—nor can *you* be where *I* am.

"So the scientists tell us.

"I will now proceed to demonstrate the very opposite. I have here (*indicating apparatus on table on the right*) a large glass bowl nearly full of water, and on the water floats a cork. Here are two large glass jars—I see my assistant has placed them upside down, but that is of no consequence. I pick up this one (*picking up the one containing ordinary air and thrusting it down into the large glass bowl in such a manner as to carry down the cork inside the jar*).

98

"Note carefully the cork does not rise up inside the jar—in other words the jar is not full of water. Why? Because the jar is full of air—and before the water can come in the air must come out. *Disponendo airum et aqua intruenda est,* or displace the air and the water will enter.

"Now I have here a precisely similar jar and will perform precisely the same experiment with it [*per-*

former does so, having first removed the other glass jar the water rushes up and fills the jar.]

" Now, lest any of you should think that this jar is not an ordinary one—*i.e.* has some trickery in connection with it—a hole in the bottom for instance—I will remove it and invert it and pour water into it—as you can see—not a drop runs out.

" It was impossible for the jar to have been a vacuum with nothing in it, for the end was open, whilst I transferred it to the top of the water. I beg pardon—but the jars were upside down—certainly—since they were about to be placed in the water in an inverted position—that was the most suitable position—very thoughtful of my assistant to save me the trouble of having to turn them over. Saves time and trouble. Anything else—I thank you—and can only say in concluding this little experiment—who can now deny the truth of my assertion, that two things can be in the same place at the same time—Hang—[*exclamation caused by performer dropping—or knocking over—one of jars—or anything else—on to the floor and breaking it*]. I beg pardon, ladies and gentlemen, but even a magician cannot overcome the law of gravity—which ultimately brings everything and everybody to the ground.

99

" Talking of gravitation—which as some of you know was first discovered by Sir Isaac Newton, when an apple hit him in the eye, whilst sitting in his garden under a gooseberry bush—naturally brings us to the Einstein theory of the same—according to which light is bent off the straight when passing near the Sun. Well, after all, there is nothing new in this, we know of plenty of other things besides light that get off the straight and on the bend when passing near the 'Pig and Whistle' to say nothing of the 'Rising Sun.' According to this strange theory, if we understand aright—you have only to travel with the speed of electricity and light, some mere trifle over 280,000 miles a second, and there is no such thing as time. In other words, you would grow no older—and if you were old and could only go a bit faster you would begin to get young again. A sort of *Elixir Vitae*, as it were, depending on your velocity. In fact, if you could only get up enough speed, you could travel round the world and come back to the place where you were—before you were—where you were.

" Perhaps, after this, I need hardly add that Einstein is a fourth-dimensional man; in fact, we are each a sort of line or cylinder in this world of four dimensions. You see, if you take a surface and move it through space you trace out a solid in space, and if you take something solid and move it through space you trace a what—why, something having four dimensions of course—from which it at once follows, that one solid must be capable of passing through another solid. Now magicians have been doing this for centuries—making things pass through hats and

100

tabletops and what nots; but people always maintained—although, of course, they couldn't ever really prove it—that it was all a fake—that they really went through trap-doors, revolving shutters, and all sorts of silly things like that. Some of them, too, didn't fail to point out that the performer very frequently, if not almost invariably, covered them up—just before the psychological moment arrived. Now I have here an ordinary soup plate nearly full of water—and I have also a tallish narrow glass jar—eh!—*i.e.* I should have but my assistant seems to have forgotten it—excuse me one moment while I get it [retires, immediately returning with hot jar*.] Now look carefully because *I* do not propose at any time to cover anything up—I place it inverted in this soup plate of water—I take this little spoon—since I wish the experiment to work slowly in order that you may have a full opportunity of really seeing what happens, and I pour a teaspoonful through the bottom of the glass and willing with all my will-power I will it to pass through—now another—note carefully the water is really going through—look, you can see it slowly rising in the jar as I continue the operation—so often as I pour water on the top so often does it rise in the jar.

* How hot the jar really must be made in order to produce a sufficient effect, must be found out by careful experiment; and if too hot to be carried in the hand, can be brought back held in a handkerchief, a suitable excuse—as to the undesirability of any contact between the fingers and glass, etc., etc.—being made. This is a point best left to the reader.

" What's that—some is running down the side, quite so—I was a little too quick—as you can well under-

101

stand the experiment is no easy one. And now to show you that the jar is not faked or prepared in any way, I remove it from the water and holding it right way up you see no water runs out. The jar being, save at my wish or command, quite watertight. *Apparatus removed.*

" Whilst speaking to you just now about this four-dimensional world of ours I mentioned the subject of *Elixir Vitae*, one of the most earnestly sought after things of all ages—the rejunevator of youth—claimed by many—achieved as yet by none. Closely connected with this was the transmutation of metals—the turning of the baser or less valuable metals into gold. I have here a small piece of steel in the shape of a darning needle, and in this very small narrow glass [*a test tube can be used*] a little, a very little, of a most precious liquid discovered by an ancestor of mine, seventy-nine times removed. Although most carefully preserved, this is all that now remains. I place this darning needle in it and there we will leave it for a time—in order that the magic influence of my ancestor's will may have full sway.

"It is also necessary that I say in silence a certain secret incantation over the same—during which I must ask for absolute stillness—that stillness which can be felt—as the ghost said when he trod on a tin tack and said—well, never mind, it was a naughty ghost and so was the word it said. See—I remove the one time steel needle from the solution and it is now no longer steel—but GOLD. Thunders of applause —I don't think—this trick never gets applauded—due to jealously of course—who would be a 'profi-

102

teer' I mean 'prophet-here' provided he could only be one somewhere else.

"Let me, however, try once again to re-establish our mutual bond of 'sympathy'—by means of a little experiment with what is known as 'sympathetic inks,' or secret writing by means of which two people can correspond, apparently sending to one another merely blank sheets of paper—the writing on which is absolutely invisible and can be only rendered visible by one in the secret; will anyone who may happen to have a pen or pencil and a piece of paper please write down first a cipher, next prefix fifty, to the right place five, and to the whole add one-fifth of eight. The result will give you a most important factor in human happiness. You have it—the answer—you give it up—well let us see what we learn from this piece of paper—which at present you can see is quite plain—I have here a tumbler of water—I drink some —to show you it is merely water—I immerse this slip of paper and on taking it out—you see in dense white characters—the word L O V E; which, I think you will admit, is the correct solution. L stands for fifty, the cipher for the letter O, V for five, whilst E the first letter in eight—is most assuredly the fifth part. Sympathetic Ink—Sympathy and Love—most appropriate, isn't it?

"Talking of 'appropriate' defined as 'to take and make one's own,' you may know the story of the man who had never read any of Shakespeare's plays, but who went to see one acted.

"He expressed himself delighted with the same except that: 'It was so awfully full of quotations

103

you know,' which was certainly getting hold of the wrong end of the broom. Think how often those celebrated Shakespearean lines:

"*To Boil or not to Boil: That is the question whether 'tis hotter to leave upon the hub, or* . . . has been cribbed and quoted.

"Now I put it to you quite candidly—as man to man —why should anything boil—why should it cease to boil when it has once boiled—why should it not keep on boiling? Yes! Yes! I know what you would tell me—a thing boils when it is heated and ceases to boil when it cools. Does it really? That is the general opinion—all right—we all live and learn, don't we? I'll now proceed to show you that you are all irretrievably, irrevocably, irresponsibly and irreversibly wrong. [*Performer brings forward from the rear—flask of water on retort stand with spirit lamp underneath nearly on the boil.*] I have here, as you can see, a flask about half full of water nearly on the boil, which my assistant has got ready for me so that no time may be wasted. As you can now see, the water is beginning to boil and in a few seconds will be boiling vigorously. This now being the case, I cork up the flask tightly so that no air shall get in or out. To make sure it is tightly corked. I invert it, and, as you can see, this is really the case. The water has now ceased to boil. In order to make it boil again, you would naturally apply more heat to it, instead of that I will squeeze out of this sponge some *cold* water over it, and as you can plainly perceive the water again begins to boil furiously. I have here in this jug some *hot* water. I pour some over the flask and the boiling

104

at once ceases. I again try the sponge and the boiling begins again. Hot water once more and once more it ceases. Well! Well! We live and learn as the 250 years' old parrot at the Zoo said on seeing the first aeroplane, and these *parrots*—I beg your pardon—*paradoxes* of nature will *crop* up, hue and cry them down as we may. Everyone thinks he knows as much as everybody else—until he finds some other fellow who knows a bit more.

"I wonder how many of you have heard of the Salamander—that amphibious animal which dies in water and cannot live on land. Amongst the Arabs, and even in later times, it was believed that the Salamander was born in fire and lived upon it, being thus mixed up with the Phoenix—the legendary bird not the Fire Insurance Co. Whereas if we consult older writers, such as Aristotle for instance—speaking of the Salamander he cites it as: 'that which, when it walks through fire extinguishes it'—this as a proof of its incombustibility. Ælian says that when those who work with forges find their bellows fail to quicken the flames, they look for the Salamander and put things right by killing it. The asbestos of Marco Polo was described as being made of the hairs of Salamanders, and in Bacon Salamander's wool or hair also occurs as a name for asbestos. Francis I. chose it as his emblem with the motto '*J'y vis et je l'éteins.*'

"We see, therefore, that the Salamander is neither born in fire nor does it eat it, but that we have a great concensus of opinion as to its incombustibility especially of its 'hairs.' Of this, I now propose to give you ocular demonstration.

105

" I have here a small net (in appearance just like muslin) made of Salamander hair, to the four corners four similar hairs are attached as shown, the latter four hairs being fixed to a ring attached to the upright in this little stand.

" I have here a small egg which I place in the little Salamander net or hammock, it holds it up as you would expect. (*Removing egg and passing it round*). Please see that the egg is a real heavy one and not a fake . . . (*taking back egg and palming it for a similar blown egg afterwards placed in hammock*). You will observe the suspension threads and hammock are quite flexible, *i.e.* they are not made of wire or any fake of that sort—I now place the egg back in the hammock and proceed to set fire to both hammock and threads, they burn as you can see and are apparently consumed—but that this is not really the case, can be seen from the fact that the egg does not fall. Who is there who can now doubt the fire resisting, the Salamandrinic properties, of the Salamander ?*

* Best covered over before removed—to hide probable collapse in so doing.

"*Ex nihilo nihil fit*—a silly old motto which every conjurer is of course constantly refuting. Fancy any magician worthy of the name admitting, that out of nothing he could only produce nothing—well, he'd soon cease to be a magician. The production of something out of nothing or of nothing effecting something is just what he glories in.

" A real magician delights in ' nothings ' ; they are like bread to ordinary mortals—his very ' staff ' of life.

" Permit me to show you a little experiment in

106

' nothing '—I beg pardon—oh dear no, nothing of the kind—I said an *experiment* not a quibble or catch. I have here an ordinary pair of rather accurate—*i.e.* sensitive scales—not capable of weighing an elephant it is true—but capable of weighing any small object with considerable accuracy. Will one or two of you please come forward and examine them to see there is no trickery about them (*often this has been done*) ? I now place this small glass jar on one scale pan and proceed to balance it accurately with weights put in the other scale pan.

" I have here another glass jar, which as you can see is like the other jar, full of nothing save air. I bring the mouth of this jar over the other and tip it up just as if I were pouring something from one jar into the other. See, the jar no longer balances but goes down, just as if something had really passed from one jar into the other. And we have now weighed what ? Why, nothing—not even the jar since it no longer balances. Could you possibly imagine a more nothingless experiment ? You could imagine nothing ? No, I thought not—I mean it—most unwise to let your imagination run away with you as you never can possibly tell where it will lead you. I have here half a dozen candles, like skittles, all in a row—at present they are not lighted—now I want you all to imagine them lighted—imagine it please, very strongly—a little stronger all of you please—Ah ! that's better, and now as you can see quite clearly the candles *are* all lighted.

" Behold what the concentrated essence of all your imaginations has done.

107

" I beg your pardon. The candles are *not* lighted. Oh ! Dear ! Now you've gone and spoilt everything. If you'd only believed the candles really lighted—I mean if your concentrated imaginations had only been sufficiently powerful to have produced the necessary ' toddy-idic ' force adequate to . . . but what's the use of talking (*aside*). How the deuce am I to get these candles lighted and invisibly extinguished now ? (*Passing to another table—performer lifts up a glass of water as if to drink—still aside.*) Ah ! I have it :

" Ladies and gentlemen, I regret very much to say that your strange lack of imaginative power has rendered the carrying out of the next two experiments I was about to show you an almost impossible feat. Nevertheless I will attempt them in the only way left me. Should I succeed, I need scarcely add that the greater honour will be mine to do—and yours to see* (*bows*). I have here a glass of water. With your permission I will drink a little as I am feeling rather thirsty (*performer does so*). See—watch closely, I bring the rim of the glass near the wick of this candle (*an end one*) ; look—I tip up the glass—and presto—sacracissimo—as soon as the water and wick meet the candle is ignited. With this candle I will now proceed to light the other five—just in the good old ordinary way.

* Note. N.B. In the experiment as here shown, a tiny piece of potassium must be used instead of phosphorus, since the candle has not been previously lighted. Experiment carefully beforehand. The potassium is best concealed in the candle wick.

Having now got my candles ignited—by the way I suppose you are all prepared to believe the candles

108

are now really lighted—are you sure it is not your imagination—how many candles are there lighted—six—you're sure they are lighted and that the number is six (*performer brings his hand rapidly over one candle which almost immediately goes out*). What's that, there are only five candles alight now—I beg your pardon—what, only three now (*performer has extinguished two more*). Three—still three—no, two now—still two—what, only one now. You're quite sure there is one still alight.

"Well, well—some people *can* imagine things—and

I thought you had no imagination. Ladies and gentlemen—I apologise most humbly. (*Assistant removes unlighted candles rapidly.*) It almost looks as if our last experiment was going to be a successful one after all. Before showing it to you—I am sure you will not mind my partaking of a little nourishment—its exhausting work this you know—as you are now about to perceive, I am somewhat 'Esquimauxian' in my tastes (*picking up remaining candlestick and candle**)—that is to say I'm Og! Gog! and Magog! on candles. I simply love 'em (*performer proceeds to bite off end of lighted candle—chew it and swallow it with much gusto*).

* The top part of this candle is prepared *à la* experiment 69—and it is this piece that is bitten off and eaten.

"Ah! that's better—and now like a giant refreshed let us tackle our last experiment.

"There is a very old saying:—

"'*What wonders may be brought to pass*
By the optician's magic-glass.'

109

"Now I have here such a magic piece of glass; as you can see it is to all appearance just a piece of ordinary glass—you can see through it (*holding it up*) as usual—and to prove to you that there cannot be anything on it, I will wash it and dry it on this towel—observe, there is no deception—I wash both sides and dry them. (*Coming forward.*) Will some one please give it a final rub over? (*After this has been done.*) Now, what would you like to see appear on this piece of glass—this magic mirror as it were—since it is about to mirror your thoughts or wishes?"

. . . .

(*What follows here must of necessity be left to the performer—it depends on what picture has been etched on the glass. If the performer cannot, by skilful suggestion, force the requisite wish from the audience, i.e. from some one, he will always have a choice. He must employ a confederate and follow out his or her suggestion or wish.*

If he has two glasses prepared with different pictures and can secretly substitute the second for the first and produce a second and different picture, then the effect is much heightened and the performance brought to a very satisfactory conclusion.)

110

CHAPTER IX.

PREPARING AND ARRANGING THE NECESSARY APPARATUS—GAS GENERATION—LABORATORY DIRECTIONS—HINTS AND CAUTIONS.

With regard to the preparation and arrangement of the necessary apparatus and as to certain precautions that should be taken in dealing with certain chemicals, something has already been said. In this chapter some further hints and precautions and also directions are added, accompanied with the necessary illustrations for the making of certain gases such as chlorine, carbon dioxide, etc., used in some of the foregoing experiments.

First then with regard to certain further

HINTS AND CAUTIONS.

Always keep phosphorus in a bottle with sufficient water to well cover it; if the bottle gets upset—refill it at once.

Always keep potassium in a bottle under oil; never place a piece in the mouth.

Never pour out ether, or bring an uncorked bottle of the same near a naked light. Always put the stopper or cork back in the bottle *immediately* after use; do this with all other chemicals as well.

111

Always put the same cork back into the same bottle—don't think it doesn't matter because it does. Always put all bottles back into their proper places. Always wash and clean everything thoroughly after use, especially the pestle and mortar—the latter should be well scrubbed, lest traces be left of a substance which, when ground with some other, will produce an explosion.

Never grind two substances together in a mortar unless expressly told to do so.

Always carefully pour away down the sink all solutions, especially if poisonous, as soon as done with.

Never use concentrated acids when told to use diluted ones.

When powders or solid substances are to be mixed always powder them well first, and then stir them together till the mixture is as uniform as possible, they cannot be too well mixed.

When using a Bunsen burner, see that the holes at the bottom are open; see also that the gas is lighted only at the top of the burner and not at the bottom. The lighting at the bottom shows that too much air is entering; turn out, regulate, and light again.

Always apply heat gently at first, to avoid cracking flask; when containing powders and not liquids, use a piece of wire gauze.

Always carefully dry any article that will rust. Don't throw sand or any solid substances down drains.

Cork borers are useful. Always bore through from the small end of the cork turning and pushing at the same time.

By first slightly greasing the glass tubing and then

112
the cork, a perfectly airtight joint can be made without melting parafin or sealing wax round it, which is liable to crack under the action of heat.

When generating gas in a flask or jar do not insert the tube taking off the gas—the delivery tube—more than a quarter of an inch below the cork. Glass tubing can be easily bent in an ordinary gas burner; hold it the long way of the flame, and by rolling it over and over in your fingers, heat it equally all round. When soft its own weight will practically bend it into a curve. To round off sharpened edges, make them rust red in a Bunsen burner and allow them to cool.

To cut glass tubing, make a notch in the tubing with a small triangular file and then snap—pulling and bending at the same time. When collecting gases, two or sometimes three pound glass jam jars are very convenient, for preliminary experiments at any rate. It is better, but not essential, that their tops be ground. This can be done as follows: place on a flat surface a sheet of emery paper of medium coarseness. It is best glued to a flat piece of wood, some one and a half to two inches thick and not less than eight inches square. Hold the jar firmly and grind with a circular movement.

The joint between the jar top and the glass plate or cover can be rendered air tight by means of a little grease.

Always protect all corks from the action of chlorine by dipping them in hot melted parafin wax.

If rubber tubing be used in the generation of chlorine gas, it *must* be *well* washed in *running* water afterwards. When mixing a solid and a liquid in a flask

113
which is afterwards to be heated, see that no dry powder is left at the bottom of the flask or it may crack. On no account hold the face near a piece of ignited potassium when it is moving about on the surface of water, it often finishes with a slight explosion and a particle might lodge in the eye.

If any acid be spilt on the hand, place it *at once* under a copious stream of water from the tap. If spilt on the table or floor scatter whitening on it. Never put anything away dirty, a piece of apparatus that would take fifteen seconds to wash well directly after use may take as many minutes to clean a week later.

Wash glass vessels first under the tap, and finally in distilled water when it is necessary for them to be free from all traces of salts. Invert them and allow them to drain and dry; the *insides* need not be wiped.

HOW TO PREPARE CHLORINE GAS.

In the illustration, Fig 8, we see an ordinary chemical glass flask fitted with a (paraffin-waxed) cork, through which a hole has been bored; a longish piece of bent glass tubing; a piece of wire gauze; a retort stand and two iron rings—one to hold the neck of the flask and the other the piece of wire gauze; a Bunsen burner; a tall glass jar, and a piece of cardboard, pierced with a central hole for the glass tube, to cover the top of the jar—a three pound jam jar answers quite well. The cork in the flask and the glass tube must fit tightly.

Instead of using a retort stand and two rings, a tripod (circular top) and a piece of wire gauze on

114
which the flask stands can be used. It is advantageous but not necessary to use a thistle funnel (see any text book on chemistry).

The chemicals used are black oxide of manganese, and strong hydrochloric acid.

Put about a tablespoonful of the manganese oxide in the flask, about one fourth this quantity of water and enough acid to cover the manganese; see that none of the manganese clings to the bottom of the

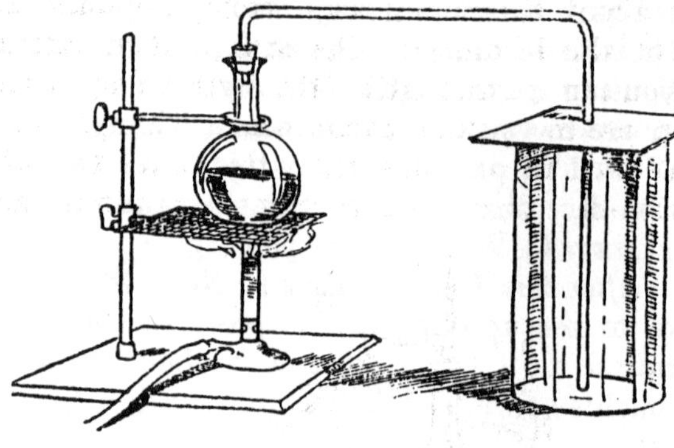

FIG. 8

flask, but that the whole is well wetted. The wire gauze is a safeguard against the cracking of the flask, not an essential.

Fit on the cork and glass tube and arrange the apparatus as shown. The right or long limb of the glass tubing should descend to one-eight of an inch of the bottom of the jar. If the end is irregular it can well rest on the bottom. On no account heat the contents too strongly or they will boil over into the collecting jar which is just what you do not want.

115
THE PREPARATION OF CHLORINE WATER.

As this is simply prepared by letting the gas pass through water, all that is necessary is to fill the collecting jar with water and allow the gas to bubble through the water for some time.

Instead of using one collecting jar or bottle it is better to use several connected, as shown in the illustration, Fig. 9, as by this means much less gas is lost and several bottles of chlorine water are obtained differing in strength, which is a decided advantage.

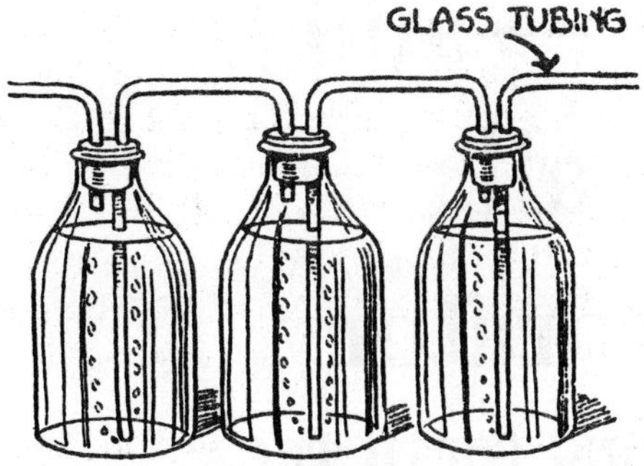

FIG. 9.

All such bottles must be kept well corked and all corks previously soaked in hot wax, or if glass stoppers be used these too must be greased. The preparation of both chlorine gas and chlorine water are best carried out in the open air.

HOW TO PREPARE SULPHUROUS ACID.

Take about half an ounce of copper filings, place them in an eight ounce flask fitted with a thistle

116

funnel and delivery tube (see Fig 10), leading into a jar of water.

Pour about three ounces of strong sulphuric acid down the thistle funnel. The flask is then carefully heated—using a sand bath—the heat being lowered as soon as the action commences. The gas given off is allowed to pass into the water, until the latter is saturated. This then becomes sulphurous acid.

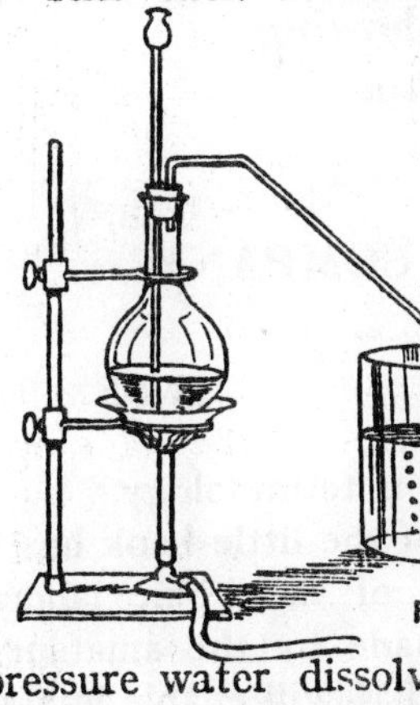

FIG. 10.

At normal pressure water dissolves eighty times its volume of the gas. It cannot therefore be collected over water—but since the gas is twice as heavy as air the gas may be conveniently collected by air displacement, like chlorine gas, see above.

HOW TO PREPARE CARBON DIOXIDE.

This is one of the simplest and easiest of all gases to produce, it being at once brought about by the action of an acid on a carbonate or bicarbonate.

117

In the laboratory it is usually prepared as follows: some lumps or pieces of marble are placed in a flask and hydrochloric acid poured down a thistle funnel (sufficient must be poured in to cover the bottom of the thistle funnel or the gas will escape up the thistle funnel). Since the gas is half as heavy again as ordinary air, the gas can be collected like chlorine (see above). It is a colourless gas, without smell

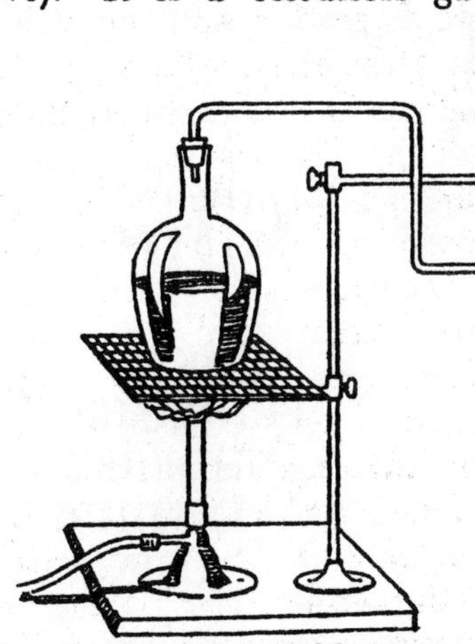

FIG. 11

and with a very faint acid taste, and but slightly soluble in water. It is best to allow it to bubble through water before collecting it in order to remove all traces of hydrochloric acid.

HOW TO PREPARE AMMONIA GAS.

Arrange the apparatus as shown in Fig 11, the inverted flask for collecting the ammonia gas, resting on the large ring of the retort stand.

118

In the flask are placed two parts of powdered quicklime and one of sal ammoniac, the flask is heated and the gas collected as shown. Both the ingredients *must* be finely powdered and quite dry. To insure that the lime is really "quick," a new "lime" such as is used in magic lanterns should be used.

FIG 12

The ones sold in hermetically sealed glass tubes are the only ones you can be really sure of.

Another method is as follows:

Put some ammonia liquid, *i.e.* ammonium hydrate—commonly called ammonia in a flask and boil as shown in Fig. 12, allowing the gas that is given off to pass over quicklime or solid caustic potash in order to dry it.

How to Make—and Use A Small Chemical Laboratory

A BOOK FOR BEGINNERS SETTING FORTH THE FUNDAMENTALS OF CHEMICAL SCIENCE IN UNDERSTANDABLE TERMS. INCLUDES INSTRUCTIONS FOR BUILDING AND USING A SMALL HOME LABORATORY.

By

RAYMOND FRANCIS YATES

Illustrated with specially made engravings, showing construction and use of chemical apparatus.

NEW YORK
THE NORMAN W. HENLEY PUBLISHING COMPANY
2 WEST 45th STREET
1926

INDEX PAGE 88

PREFACE

This little book has been prepared for those who wish to become acquainted with the great fundamentals of chemistry. The author resolved at the beginning not to write the treatment in the prescribed "text book" way, and it was therefore decided to explain the Electron Theory as an introduction to the study of inorganic chemistry.

The Electron Theory has done much to make chemistry a more vitally interesting subject. It is a new philosophy that has brought the science of chemistry to a more understandable basis. It offers explanations to many phenomena that up until a few years ago were absolutely mysterious. Today, chemists speak in terms of the Electron Theory, and nobody desirous of entering into the study of chemistry should neglect to become familiar with its outstanding features at least. These, the author has attempted to outline in common terminology.

The second part of the little book has been devoted to the construction of the home laboratory. Such things as can be made by the amateur chemist are described in a way that will enable him to build them with a minimum of trouble.

The third portion of the book has been given over to chemical experiments. Experiments have been chosen that will give the reader a broader understanding of chemical science in general. These experiments also cover the various phases of chemistry.

This book will only serve as an introduction to the study of chemistry. The author has endeavored to explain the fundamentals in an understandable way so that the student, upon completing this treatment, will be able to start the study of chemistry from a more elaborate book without becoming confused and discouraged, as many do.

RAYMOND FRANCIS YATES.

TABLE OF CONTENTS

PART I

INTRODUCTION TO THE STUDY OF CHEMISTRY

PART II

EQUIPPING A SMALL CHEMICAL LABORATORY

PART III

CHEMICAL EXPERIMENTS

How to Make and Use a Small Chemical Laboratory

PART I

INTRODUCTION TO THE STUDY OF CHEMISTRY

Elements and Compounds—Chemical Symbols—Molecules—Atoms—Electrons—How Electrons Form Atoms—Explanation of Chemical Activity—Chemical Formulas—Speed of Chemical Reactions—Physical and Chemical Changes—The Law of Definite Proportions—Different Kinds of Chemical Reaction—Classification of Chemical Compounds—The Law of Combining Weights—Solutions—Acids and Bases—Neutral Substances—Organic and Inorganic Chemistry—Electrochemistry—Experiments in Electrochemistry.

Many people are inclined to regard chemistry as one of the more occult sciences. Of course, it must be admitted that the science of chemistry has many ramifications, but the study and practice of ordinary inorganic chemistry is quite within the range of any person with enough ambition to master the essentials.

ELEMENTS AND COMPOUNDS

The study of chemistry is generally started by an understanding of the words "element" and "compound." Scientists have found that this world of ours, as great as it is, is merely composed of eighty odd elements or basic substances. There is a great difference between an element and a compound. If

a chemist takes a quantity of water to analyze, he finds that he is able to break the water up into two gases, i.e., hydrogen and oxygen. These two gases, when separated, resist any further attempts to be "divided." That is, they are basic substances—they are elements. A compound, then, is a combination of elements. Water is a compound because it is composed of oxygen and hydrogen. Iron is an element because, we might say, it is composed of nothing but iron—the chemist is not able to break up the iron into any other constituents.

It was stated before that there are but eighty different elements in the whole world. However, these eighty odd elements combine in different proportions to form many thousand compounds. The physical and chemical properties and characteristics of the compounds formed are entirely different from those of the individual elements which go to make up the compound. For instance, water is a liquid, but its constituents are the gases hydrogen and oxygen.

Some of the elements are gaseous, some solid and a few liquid. The list of chemical elements—elements of the whole world—follows:

	Symbol	Combining or atomic weight		Symbol	Combining or atomic weight
Aluminum	Al	27.1	Columbium	Cb	93.5
Antimony	Sb	120.2	Copper	Cu	63.57
Argon	A	39.88	Dysprosium	Dy	162.5
Arsenic	As	74.96	Erbium	Er	167.7
Barium	Ba	137.37	Europium	Eu	152.0
Bismuth	Bi	208.0	Fluorine	F	19.0
Boron	B	11.0	Gadolinium	Gd	157.3
Bromine	Br	79.92	Gallium	Ga	69.9
Cadmium	Cd	112.40	Germanium	Ge	72.6
Calcium	Ca	40.07	Glucinum	Gl	9.1
Carbon	C	12.00	Gold	Au	197.2
Cerium	Ce	140.25	Helium	He	3.99
Cesium	Cs	132.81	Hydrogen	H	1.008
Chlorine	Cl	35.46	Indium	In	114.8
Chromium	Cr	52.0	Iodine	I	126.92
Cobalt	Co	58.97	Iridium	Ir	193.1

9

	Symbol	Combining or atomic weight
Iron	Fe	55.84
Krypton	Kr	82.92
Lanthanum	La	139.0
Lead	Pb	207.10
Litecium	Lu	174.0
Lithium	Li	6.94
Magnesium	Mg	24.32
Manganese	Mn	54.93
Mercury	Hg	200.6
Molybdenum	Mo	96.0
Neodymium	Nd	144.3
Neon	Ne	20.2
Nickel	Ni	58.68
Niton (radium emanation)	Nt	222.4
Nitrogen	N	14.01
Osmium	Os	190.9
Oxygen	O	16.00
Palladium	Pd	106.7
Phosphorus	P	31.04
Platinum	Pt	195.2
Potassium	K	39.10
Praseodymium	Pr	140.6
Radium	Ra	226.4
Rhodium	Rh	102.9
Rubidium	Rb	85.45

	Symbol	Combining or atomic weight
Ruthenium	Ru	101.7
Samarium	Sa	150.4
Scandium	Sc	44.1
Selenium	Se	79.2
Silicon	Si	28.3
Silver	Ag	107.88
Sodium	Na	23.00
Strontium	Sr	87.63
Sulphur	S	32.07
Tantalum	Ta	181.5
Tellurium	Te	127.5
Terbium	Tb	159.2
Thallium	Tl	204.0
Thorium	Th	232.4
Thulium	Tm	168.5
Tin	Sn	119.0
Tintanium	Ti	48.1
Tungsten.	W	184.0
Uranium	U	238.5
Vanadium	V	51.0
Xenon	Xe	130.2
Ytterbium	Yb	172.0
Yttrium	Yt	89.0
Zinc	Zn	65.37
Zirconium	Zr	90.6

CHEMICAL SYMBOLS

After each of the elements in the above list will be found a letter or symbol which chemists use in place of writing the entire word out. In place of writing out the word hydrogen, merely the first letter, H, is used. This may be called the shorthand method of chemical expression and the symbol for all of the common elements should be memorized by the amateur chemist. Upon referring to the list it will be seen that the symbol for some of the elements is composed of two letters, and that these two letters in no way correspond to the letters in the word. Sodium is represented by the letters Na, and mercury by Hg. Many of the elements start with the same letter, and if this letter was used for each element, it

10

would lead to confusion. Therefore, the first two letters of the Latin name are used. Cu, which represents copper, is taken from the Latin **cuper**. In some cases only the first letter is employed.

MOLECULES

Before going farther with the study of chemistry it will be well to understand the terms molecule, atom and electron. The molecule is composed of two or more atoms, usually of different compounds. Water, which is composed of hydrogen and oxygen, has a molecule made up of two atoms of hydrogen and one atom of oxygen. Molecules, however, are not always associated with compounds. There are a few elements that have a molecule which is made up of two or more of their own atoms.

Molecules are extremely small. In fact, they are far beyond the range of the most powerful optical instruments of magnification. Molecules are generally measured in the millionths of an inch. Some molecules are composed of two atoms and some of many hundred atoms. Of course, those containing many hundred atoms are much larger than those with a smaller number.

ATOMS

If molecules are made up of atoms and molecules are measured in the millionth of an inch, we can readily appreciate the fact that the atom must be a much smaller particle. This is quite true. The reader will understand that compounds do not have atoms in the sense that an element has atoms. True, the ultimate constituent of a compound is the atom, but owing to the fact that the atoms are grouped together

11

to form molecules, chemists are inclined to regard a compound as being made up of molecules rather than atoms. This will be made more clear later.

We are now aware of the fact that a molecule is composed of two or more atoms and that the atom is that infinitesimal particle that goes to make up elements. Thus, we have the atom of sodium, the atom of potassium, the atom of copper, etc. We cannot say, however, that we have the atom of water because water is made of two elements in combination. In place of atom, the term molecule is used.

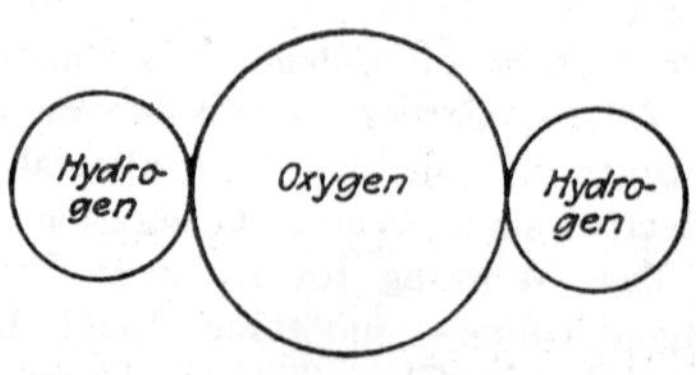

Fig. 1.—A Diagrammatic Representation of a Molecule of Water.

A molecule of water is pictured diagrammatically in Fig. 1. It will be seen that the atom of oxygen is really larger than the atom of hydrogen. It will also be seen that there are two atoms of hydrogen to one atom of oxygen. Upon referring to the list of elements, it will be seen that the atomic weight of oxygen is greater than the atomic weight of hydrogen. Owing to the fact that atoms are all made up of the same particles—electrons, which we will consider later—it will be understood that the elements with the heaviest atomic weights must be the largest. The chemical expression for water is written H_2O. This means that there are two atoms of hydrogen and one atom of oxygen. The figure two after the H signifies the number of hydrogen atoms in the compound. The subject of

12

the chemical expression of formulas will be taken up more in detail later.

ELECTRONS

At this point we are ready to consider the last and ultimate constituent of matter—the electron is ushered in. The electron is really a comparatively recent discovery of science. It is responsible for the Electron Theory which would seem to reduce the phenomenon of the whole world to a purely electrical problem.

HOW ELECTRONS FORM ATOMS

The electron is absolutely the smallest particle. It goes to make up atoms. Each atom has an identity. We have atoms of copper, iron, etc. All electrons are the same. Just as atoms in various numbers form molecules, electrons in various numbers form atoms. However, we cannot draw a distinct parallel between the two classes.

The electron is really a unit of negative electricity. It has a mass of 1/1700ths that of an atom of hydrogen, which, as a glance at the list of chemical elements will show, is the smallest and lightest atom. It may confuse many readers to think somewhat vaguely of an electron being a "unit of negative electricity." However, this is about the best explanation that can be given. In a strict sense, the electron cannot be considered as matter. The single electron we must accept as a unit of mysterious "something" which we have called negative electricity. Atoms are composed of various numbers of electrons. Each atom has a specified number of electrons. Thus, each element has a certain identity.

Are we to picture the atom as being composed of

13

a specified number of electrons arranged in a haphazard manner? Not so. In fact, we know that "like charges of electricity repel one another and dislike

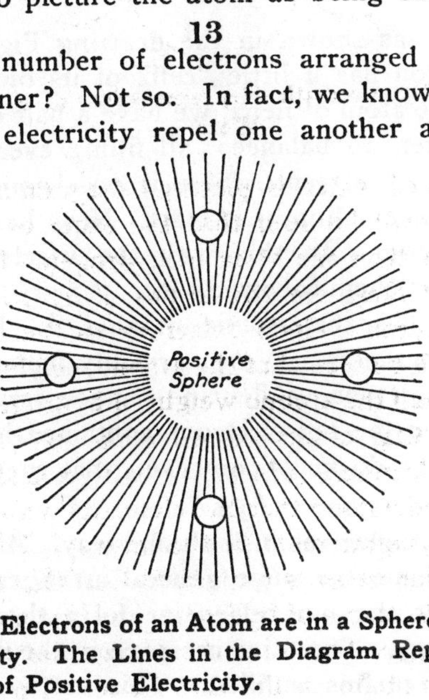

Fig. 2.—The Electrons of an Atom are in a Sphere of Positive Electricity. The Lines in the Diagram Represent This Sphere of Positive Electricity.

charges attract one another." With this fact in mind, then, how can we think of our hypothetical atom as being made up of electrons when these electrons are negative units of electricity and when negative charges repel one another? If this was the case, our hypothetical atom of electricity would soon dissipate.

The electron theory describes an atom as a particle composed of a definite number of electrons or negatively charged units revolving about a sphere of positive electricity. Of course, this "sphere of positive electricity" is just as much a mystery to us as the electron itself. Nevertheless, it furnishes us with a reasonable hypothesis. If dislike charges of electricity attract one another, then the sphere of positive electricity which forms the nucleus of the atom will cause

14

the electrons, which are units of negative electricity, to remain in proximity to the nucleus.

A very crude diagram which is intended to represent at least the arrangement of the electrons in an atom is shown in Fig. 2. It must be remembered that every atom of oxygen, for instance, has exactly the same number of electrons. This is true of all the atoms of the different elements. From this we see that each chemical element is different because it contains a certain number of electrons in its atom. All matter must be regarded as being composed of electrons; these electrons are grouped into atoms and the atoms into molecules.

From the above information, we assume, then, that each atom is in an electrically neutral condition. If the positive charge is strong enough to bring about an electrical equilibrium, this would be the case. Let us, for the moment, assume that this is the case. We will picture a group of electrons held about the nucleus of an atom with neither the positive electricity of the nucleus nor the negative electricity of the electrons predominating. We will say that we have an electrical equilibrium. What if one electron was in some way to escape from the atom? The equalization of charges or the equilibrium would then be destroyed and the positive charge would predominate. Again, we can imagine one hypothetical atom coming in contact with another atom, the second atom losing an electron which is immediately picked up by the first atom. Investigation has proven that this interchange of electrons actually does take place.

Let us review what we have said concerning the electron theory a little more thoughtfully. The importance of the theory to present-day science justifies this.

If we were to arrange all of the elements in the

15

list shown on pages 8 and 9 according to their atomic weight, starting with hydrogen as one because it is the lightest, and giving the second place to the next heaviest, and so on, part of the list would be as follows: Hydrogen, 1; Helium, 2; Lithium, 3; Glucinum, 4; Boron, 5; Carbon, 6; Nitrogen, 7; Oxygen, 8; Fluorine, 9; Neon, 10; Sodium, 11; Magnesium, 12; Aluminum, 13; Silicon, 14; Phosphorus, 15; Sulphur, 16; Chlorine, 17; Argon, 18; Potassium, 19; Calcium, 20; Scandium, 21; Titanium, 22; Vanadium, 23; Chromium, 24; Manganese, 25; Iron, 26; Cobalt, 27; Nickel, 28; Copper, 29; Zinc, 30, etc., etc. Here the elements are arranged according to their atomic weight, so if copper, for instance, occupies the twenty-ninth place, we will call 29 the atomic number of copper. Thus the atomic number of carbon will be 6, that of Fluorine, 9, etc. We must remember not to confuse this *atomic number* with *atomic weight,* although it is related to it in a certain respect.

EXPLANATION OF CHEMICAL ACTIVITY

A young English scientist recently promulgated the theory that the atomic number corresponds with the electro-positive charges that form the nucleus of any atom. In other words, if the atomic number corresponds to the number of units of positive electricity that go to make up the atomic nuclei, there must be one electron or negative unit. Thus, the hydrogen atom contains one electron, the helium atom two electrons, lithium three electrons and so on through the entire list.

By themselves, we remember, electrons would repel each other, but in the presence of positive charges, they show a disposition to arrange themselves in definite groups or configurations. If hydrogen has

16

but one electron, how does it conform to any particular systematic arrangement? This does not necessarily interfere with the hypothesis we have in mind. We can picture the hydrogen atom as being made up of one charge of positive electricity and one electron. The lone electron which goes to make up the hydrogen atom is ever trying to form a pair and therefore the atoms group into molecules and hydrogen is one of the elements that has a molecule.

Let us consider the second element or helium, which has an atomic number of two. We understand that

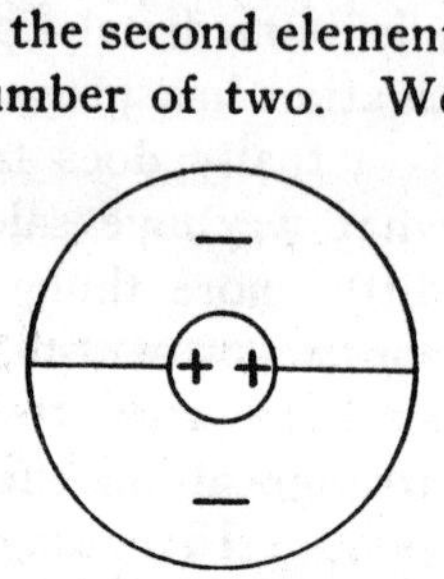

Fig. 3.—How the Electrons and Charges in an Atom of Helium are Arranged.

this element has two positive charges to form its nucleus and therefore two electrons to counterbalance these charges. The atom of helium can be pictured as shown in Fig. 3. The + marks represent the positive charges and the — marks the electrons. Here we have the ideal atom, with the electrons ideally arranged and electrically balanced. As a result, helium is an extremely inactive element. Chemists call it inert. It shows absolutely no disposition to enter into chemical relationship with any other element. Why should it? It is electrically balanced—it neither wants to gain nor lose electrons.

Let us pass on to the gas neon. Neon, although 9 places from helium in the list of atomic numbers, is

17

next to it in point of stability or chemical inactivity. It is another of the inert elements. Neon has ten electrons in its atom. We remarked some time ago that the electrons tended to arrange themselves systematically about the positive charges of a nucleus. Here we have 10 electrons. Let us see how they would arrange themselves, in the light of our theory. The first two electrons arrange themselves about the nucleus the same way as the two electrons in the atom of the helium atom. The remainder, in the words of the theory, show a disposition to group themselves

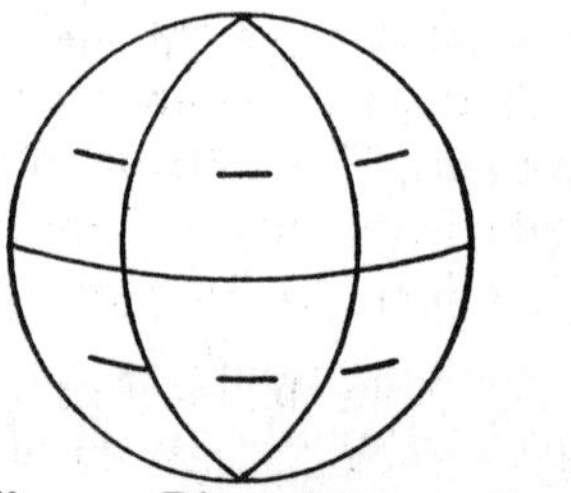

Fig. 4.—Electrons Show a Disposition to Group Themselves in Octets as Shown.

into octets, as shown in the drawing Fig. 4. Thus, each electron has a little "cell" of its own, we may say. In the atom of neon, we have a pair of electrons and an octet, all balanced, all filling every available space, and, as a result, there is no chemical activity to neon. Like helium it is inert. Neon has an ideally arranged atom. In this way the Electron Theory accounts for chemical inactivity.

Fluorine has the atomic number 9. Let us see what happens now. We can readily understand that the nine electrons of fluorine cannot arrange themselves as did the ten electrons contained in the atom of neon. We do know, however, that they will tend

18

to arrange themselves in the same way. **All electrons tend to assume the same general arrangement.** We assume that the nine electrons form themselves in the same way. There is one space in the octet, however, that remains unfilled. This empty space accounts for the extreme chemical activity of fluorine, as we shall soon see. Fluorine has two electrons about its nucleus just as neon does. The other seven electrons are in the outer shell or octet, leaving one space unoccupied—one electron is lacking to make the octet complete. In this condition, the atom of fluoriné is not favored with a stable electrical condition. It needs one more electron to make it stable. The cumulative effect of its seven outside electrons to form an octet is so great as to give the element fluorine an intense chemical activity. In fact, fluorine is a very active element. We might say that its atom has an intense appetite for one more electron to complete its octet. We can also understand that the atom of fluorine is electro-positive in nature as it is constantly in need of one more electron to bring about an electrical equilibrium.

There are many other elements like fluorine. In fact, every element aside from the inert gases, helium, neon, krypton, xenon and argon does not have its outer shell satisfied or occupied, and it is always trying to make pairs or octets to complete its arrangement and to bring about an electrical equilibrium. **This is the basis of all chemical combination and activity.**

What arrangement do the electrons make in atoms with more than ten? The electrons over ten form another octet or shell over the first one. If there are enough electrons, the third and even the fourth octet or shell is formed. The outside shell or octet, of all the atoms, aside from those of the inert gases, is incomplete. Therefore, all of the elements are chem-

19

ically active aside from the inert gases. We must look upon chemical action as merely an interchange of electrons.

CHEMICAL FORMULAS

Having covered the cause of chemical reaction, we are now in a position to understand chemical formula and expression. We now know that water is expressed chemically H_2O. This signifies that there are two atoms of hydrogen and one of oxygen in the water molecule. The reaction between oxygen and hydrogen can be expressed in this way:

$$2H + O = H_2O.$$

This means that hydrogen plus oxygen equals water. It must be understood here that all chemical elements do not react when brought together. Some elements are very inert or chemically inactive. In other words, they have very stable atoms in perfect electrical equilibrium. Thus, the gas helium is extremely inert. In fact, it is not possible to make it combine with any known element. We must also remember that a chemical reaction can take place between compounds as well as between elements. When two elements or compounds react with one another, it is said to be due to "chemical affinity." However, we now know that this chemical affinity is really electrical in nature.

Let us see what would happen if we were to drop a small piece of the element sodium (Na) into water (H_2O). If this is done, a very energetic reaction to hydrochloric acid (HCl). The result can be represented by the formula:

$$Na + H_2O = NaOH + H.$$

If we study this carefully we can account for all of the elements that entered into the reaction. We can

20

also see that we have made or caused to be made a new compound. The NaOH is called sodium hydroxide (the word hydroxide is produced by a combination of hydrogen and oxygen). The resulting molecule which makes up the mass of the sodium hydroxide is composed of one atom of sodium, one of hydrogen and one of oxygen. We must remember, however, that we had two atoms of hydrogen when we started out. What became of the second one? The + H on the end of the chemical expression signifies that the second atom of hydrogen escaped in a free condition. If we were to place a test tube over the sodium immediately it was placed in the water, we could collect the hydrogen as it left the surface of the water.

The amateur chemist must learn that when two chemical compounds are placed together and a reaction takes place, each one of the elements that entered into the reaction must be accounted for. Let us place two more substances together and see what happens. This time we will add a small piece of iron (Fe) to hydrochloric acid (HCl). The result can be represented in this way:

$$Fe + HCl = FeCl + H.$$

The result of this chemical reaction is iron chloride and free hydrogen. We can see from this that the iron must have had a greater chemical affinity for the gas, chlorine, than the hydrogen had. Hence, the metal and the hydrogen merely changed places.

SPEED OF CHEMICAL REACTIONS

At this point it will be well to mention that all chemical reactions do not take place at the same rate of speed. When the sodium was placed in water,

21

the reaction took place very rapidly. In fact, if the piece of sodium had been large enough, the reaction would have been so violent and the heat generated so great, that the hydrogen escaping would have been ignited and an explosion would have resulted. All reactions, however, are not so violent as this one.

Some substances take weeks and even months to undergo a tangible reaction. On the other hand many reactions take place in a remarkably short space of time. The explosion of dynamite takes place in 1/24,000th of a second.

All chemical reactions are not so simple as those illustrated above. The more elements present in any chemical reaction the more complex it will be. The following reaction is more complicated than those mentioned hitherto:

$$2NaOH + H_2SO_4 = 2H_2O + Na_2SO_4$$

This is the action that results when sodium hydroxide is placed in contact with sulphuric acid. The result is water and sodium sulphate.

PHYSICAL AND CHEMICAL CHANGES

When water is frozen, is the change a chemical or a physical one? This is a question that must be decided before we go farther. The change of liquid water into ice is entirely physical. That is, no fundamental change has taken place in the molecular arrangement of the water; the change is not a permanent one that gives the water a different chemical nature. A chemical change always produces an entirely new substance. For instance, when coal burns, a white ash is left and there is nothing left that resembles the original material. This is a chemical change.

Chemical action is always accompanied either by

22

the liberation or absorption of heat. When aluminum burns in oxygen, forming aluminum oxide, a terrific heat is produced by the reaction. In other cases, heat is absorbed by a reaction. However, in most cases heat is liberated. The more rapid a chemical reaction takes place, the higher the temperature produced will be. In the oxidation of aluminum, as carried out in the Thermit process of welding, a temperature of nearly 5000 degrees Fahr. is produced. In slow chemical reactions that extend over a great length of time, heat is liberated but not perceptibly. In some cases, there is such a slight rise in temperature that it is almost immeasurable. In every case, a specific amount of heat is liberated, and this depends entirely upon the amount of the substances or elements that enter into the reaction.

THE LAW OF DEFINITE PROPORTIONS

When two chemical elements react to form a compound, they always do so in definite proportions. For instance, when copper burns in oxygen, forming copper oxide, there is always the same proportion by weight of oxygen and copper. The substance, if analyzed, will be found to contain 79.96 per cent copper and the remainder oxygen. If there is any excess of either one of the two elements present during the reaction, it will remain unchanged. This law holds true throughout the entire science of chemistry, and it is called the Law of Definite Proportions. It must be stated as follows: A pure compound always contains the same elements in constant ratio by weight.

23

DIFFERENT KINDS OF CHEMICAL REACTION

We are now ready to learn of the three different kinds of chemical action. If the red oxide of mercury is heated in a test tube over a Bunsen flame, oxygen will be given off and pure metallic mercury will be left in the tube. The application of heat brings about a separation of the elements oxygen and mercury. In other words a decomposition has taken place and this chemical action is therefore called one of decomposition. We will define a chemical action of decomposition, then, as one in which a compound is divided or broken up into its constituents.

If iron filings are placed in a test tube with powdered sulphur and heated, the mass will gradually change to a black appearing substance called iron sulphide. This chemical action is said to be one of synthesis or combination—two elements combined to form a new compound.

If a little sodium hydroxide (NaOH) is placed in a test tube containing a solution of copper sulphate ($CuSO_4$) a white precipitate will fall to the bottom of the tube. After this action has taken place, there is present in the tube copper hydroxide and sodium sulphate. The sodium changed place with the copper and the copper changed place with the sodium. Such a chemical action is said to be one of exchange or substitution.

CLASSIFICATION OF CHEMICAL COMPOUNDS

When iron reacts with the gas chlorine, a compound known as iron chloride is produced. If copper was to react with the gas in place of iron, we would have

24

copper chloride. In fact, most of the metals will react with chlorine to form what is known as a "chloride" of the particular metal which enters into the reaction. If the metals react with sulphur, sulphates or sulphides result. If they react with water, hydroxides result. If sodium reacts with the gas nitrogen, a compound known as sodium nitrate is the product of the reaction. Thus, we also have a number of compounds known as nitrates and nitrides. When the word nitrate, sulphate, or chloride is used, we know, then, that the compound mentioned will contain either nitrogen, sulphur or chlorine. Many of the metals combine with the gas, carbon dioxide. The resulting compound is called a carbonate. Thus, we have potassium carbonate, sodium carbonate, calcium carbonate, etc. When other elements combine with phosphorus, phosphates are produced. We have sodium phosphate, calcium phosphate and potassium phosphate. The reader, however, should not confuse himself by thinking that all the elements combine with phosphorus, nitrogen or chlorine to form phosphates, nitrates and chlorides.

THE LAW OF COMBINING WEIGHTS

The law of combining weights is a very important part of chemistry. The best we can do in the limited space we have at our disposal is to give the outstanding features of it. Upon referring to the list of elements which appear on pages 8 and 9 we will find the atomic or combining weight written after each element. Thus, after nitrogen we find 14, after oxygen 16, etc. This number, we must remember, represents the weight of each element. By it we are able to ascertain, in a comparative way, the weights of the different elements. We know that hydrogen, which

is 1.008, must be much lighter than lead, which is given as 207.10. Now, if lead combines with any other element, the weight of the resulting compound will be 207.10 plus the combining or atomic weight of the element which entered into the reaction with the lead. When oxygen and hydrogen combine to form water, we know the formula to be $2H + O = H_2O$. The subscript 2 means that two combining weights (or we may say atoms) of hydrogen unite with one combining weight of oxygen. We know that the combining weight of hydrogen is 1.008 and that of oxygen 16. Knowing this, then, we can easily figure the molecular weight of water. We say molecular weight because it is the sum of the weights of the atoms contained in the molecule. It is figured in this way: $1.0008 \times 2 = 2.016 + 16 = 18.016$ molecular weight of water. When hydrogen combines with the gas chlorine to form hydrogen chloride, the molecular weight of the resulting compound will be 35.46 (chlorine) + 1.008 = 36.468 molecular weight of hydrogen chloride. Hydrogen chloride is expressed chemically in this way, HCl. Here we see that only one combining weight of hydrogen enters into combination with one combining weight of chlorine.

SOLUTIONS

At this point, we will consider solution. When ordinary table salt is dissolved in water, we say we have a solution. Solutions play a very important part in chemistry. When a substance is capable of passing into solution, we say it is "soluble." Different substances have different degrees of solubility. Some are so backward in this respect that we call them insoluble. Others are extremely soluble. Not only are solids capable of forming solutions, but some gases

also. Hydrogen chloride (which is really called hydrochloric acid gas) is very soluble in water and when in solution it is called hydrochloric acid.

Temperature plays an important part in the phenomena of solutions. In general it can be said that the higher the temperature of water is the greater its dissolving powers will be. The illustration in Fig.

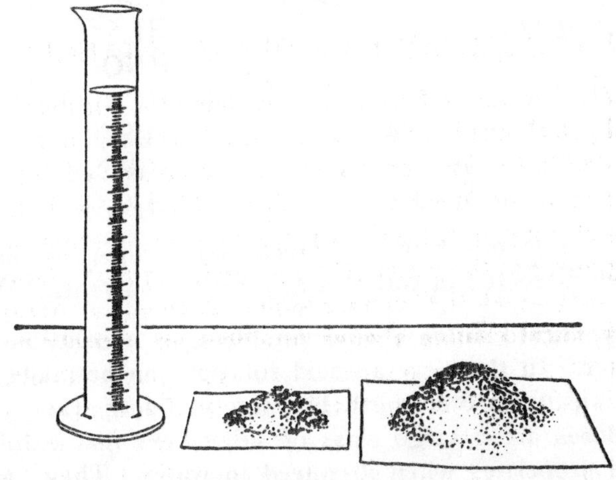

Fig. 5.—The Illustration Shows the Difference in the Amount of Sodium Nitrate Dissolved in 100 c.c. of Hot and Cold Water.

5 shows the difference in the amount of sodium nitrate dissolved in a given volume of hot and cold water. The higher the temperature of the water the greater the amount of sodium nitrate it will be capable of taking into solution. We may say that for any given temperature a given volume of water will be capable of dissolving a specific amount of any soluble substance.

What would happen if we were to dissolve a salt

in hot water and then permit the water to cool? Upon cooling, a certain portion of the salt would recrystallize and precipitate to the bottom of the vessel.

ACIDS AND BASES

We now come to a consideration of acids and bases. A substance is either acid, base or neutral. Some substances are strongly base, others only slightly so.

Acids have many properties in common, by means of which they can be identified. They all taste sour, redden litmus paper and contain hydrogen as one of their essential constituents. Of course, this does not mean that all compounds containing hydrogen are acid. However, there is no acid that does not contain hydrogen. Another general characteristic of acids is that they liberate hydrogen when they come in contact with magnesium, or zinc. In fact, most of the acids liberate hydrogen upon contact with most any metal. The following shows the reaction caused by bringing zinc in contact with hydrochloric acid:

$$Zn + HCl = ZnCl + H.$$

The following is a short list of the more common acids:

Hydrochloric acid	HCl
Hydrobromic acid	HBr
Hydriodic acid	HI
Nitric acid	HNO_3
Acetic acid	$HC_2H_3O_2$
Sulphuric acid	H_2SO_4
Phosphoric acid	H_3PO_4
Oxalic acid	$H_2C_2O_4$

Acids manifest their characteristic properties only when dissolved in water or a few other solvents. We

28

may regard an acid solution, then, as a solution in water. In this way an acid solution can be made of practically any strength by merely adding water.

Bases are a large class of substances that exhibit like properties when dissolved in water. They have a distinct alkaline taste, a soapy feel and they all turn red litmus paper blue. They all contain hydrogen and oxygen in what has become known as the hydroxyl group, which is represented by OH. Bases, like acids, do not exhibit their characteristic properties unless dissolved in water. The common properties of the bases are attributed to the OH or hydroxyl group. Every compound that contains the hydroxyl group is not necessarily a base. The identity is brought about by the action of the substance when dissolved in water.

NEUTRAL SUBSTANCES

A substance that is neither an acid nor a base is said to be neutral. It must be understood that we cannot refer to the elements as being either acid or neutral. Elements are all neutral; they exhibit neither the properties of bases nor acids. From the foregoing we find that the chemist divides all substances up into three classes—acids, bases or neutral.

It would be quite natural to think that when a base is added to an acid that a neutral solution would result. This is exactly what happens. This can be proved by adding hydrochloric acid to a solution of sodium hydroxide in the proper proportion. If this is done, we will find that the resulting solution will neither turn blue litmus paper red nor red litmus paper blue. It has neither an alkaline nor sour taste nor a soapy feeling. It will not evolve hydrogen when it comes in contact with either magnesium or zinc. In fact, the solution exhibits none of the properties com-

29

mon to acids or bases. If the solution is tasted it will be found to have a distinct salty taste. An outline of the reaction which takes place will account for this:

$$NaOH + HCl = H_2O + NaCl.$$

The resulting product of the reaction is sodium chloride ($NaCl$), or common table salt. The salt can be obtained by completely evaporating the water by the application of heat. The salt will be left in the receptacle in its usual crystalline form.

All the bases and acids neutralize each other when they come in contact. Of course, the resulting reaction is not the same as the one outlined above. When potassium hydroxide (KOH) and nitric acid (HNO_3) are brought together the following reaction takes place:

$$KOH + HNO_3 = H_2O + KNO_3.$$

The potassium nitrate (KNO_3) is a salt of the metal which formed part of the hydroxide. This is the case in all reactions of this nature; a salt of the metal is always formed.

A list of the common bases is given below:

Sodium hydroxide	$NaOH$
Ammonium hydroxide	NH_4OH
Potassium hydroxide	KOH
Barium hydroxide	$Ba(OH)_2$
Calcium hydroxide	$Ca(OH)_2$
Strontium hydroxide	$Sr(OH)_2$

These bases are largely concerned with the general study of chemistry and the student reader is urged to memorize their formula as far as possible.

30

ORGANIC AND INORGANIC CHEMISTRY

The reader has probably often heard the terms organic and inorganic chemistry. To avoid confusion it might be well to make the distinction clear at this point. Organic chemistry is the chemistry of the vegetable and mineral kingdoms. It is concerned largely with the elements carbon, nitrogen, hydrogen, and carbon which unite in an unending variety of combinations to form many thousand different compounds. The following is a typical organic formula:

C_2H_5OH (ethyl or grain alcohol).

The following is also an organic reaction which shows

what takes place when an acid is brought in contact with ethyl alcohol:

$$CH_3COOH + C_2H_5OH + H_2O + CH_3COOC_2H_5.$$

It will be very easy to distinguish this reaction from those that we have been considering throughout this book. The study of organic chemistry is quite out of the scope of this book, which is prepared to give the beginner an insight into organic chemistry.

ELECTROCHEMISTRY

Many years ago, early chemical investigators discovered a remarkable connection or relationship between chemistry and electricity. With the outline of the electron theory fresh in our minds this statement should not be amazing. The science which deals with electricity and chemistry has been called electrochemistry from a combination of the two words. It will be within the province of this book to briefly consider this particular phase of chemistry, which is very important.

31

We now come to a consideration of an ion, and having in mind the meaning of an atom, we need only add an electrical charge to the atom and we have an ion. But, in doing this, we must keep in mind that there are two different kinds of electrical charges—negative and positive. That like charges of electricity repel one another and dislike charges attract, is also an important factor. We mean by this that two negative charges would repel each other, but a negative and a positive charge would attract each other. If these charges were distributed upon two bodies, the bodies would have a very noticeable tendency to attract or repel each other, depending upon the nature of the charges that were on their surfaces. We will conclude this explanation of an ion by assuming that it is an ordinary atom of matter carrying with it either a positive or negative electrical charge. If the charge is negative, we say it is a negative ion, and if it is positive, we say it is a positive ion.

The average reader will be surprised to learn that chemically pure water is a non-conductor of the electric current—that is, it conducts to such a very small extent that it is called a non-conductor. The fact may be proved by a very simple experiment, providing chemically pure water is at hand. The water is placed in a vessel and made to form a part of an electric circuit in which a current is flowing. If a sensitive indicating instrument is placed in the circuit, we find that the water will not permit a current to pass through it. Let us drop a pinch of table salt or sodium chloride into the water and see what happens. It will do us no good to look into the water, as the process that is taking place there is far beyond our sight, but let us look at the indicating instrument. Upon the addition of the sodium chloride to the water, we notice that the instrument gives a marked indica-

32

tion, proving beyond all possible doubt that a current is passing through the water. Something very wonderful has certainly happened to the water. It behooves us to learn just what has taken place.

The chemical philosophy of Svante Arrhenius, whose hypothesis of electrolysis we set out to learn, tells us that the molecule of the table salt or sodium chloride that we dropped into the water has broken up into ions. We mean by this that the atom of sodium and the atom of chlorine in the molecule of sodium chloride have dissociated. Not only this, but that they have suddenly acquired an electrical charge and have become ions. The theory tells us that the sodium ion is positively charged and that the chlorine ion is negatively charged. After being told that the electrodes that lead the current into the water from the battery are electrically charged, we immediately decide that something very interesting is about to take place. The electrode that leads the current into the water is called the cathode and is negatively charged. Remembering that we have tiny charged atoms in the water, and in accordance with the preceding explanation concerning the attraction of differently charged bodies, we naturally expect the anode and the cathode to attract the ions in the water. This is just what happens. The sodium ions being positively charged, are attracted by the cathode or negative electrode, and the chlorine ions, being negatively charged, are attracted by the anode or positive electrode. Thus we can picture an ionic migration taking place within the water, the tiny particles with the positive charge rushing one way to the cathode and the negatively charged particles rushing the other way to the anode. We have forgotten something. In assuming that the water contains free atoms of sodium, we have forgotten that these same atoms are very sensitive to

33

water and that a great chemical affinity exists between water and sodium. Indeed, if metallic sodium in sufficient quantity is thrown into water, it will react with explosive violence. Why do our atoms or ions of sodium not react with the water in the vessel? Is it because of their electric charge? Arrhenius tells us that it is. He claimed that the electric charge carried by the sodium ion protects it from the water and prevents a chemical reaction from taking place. Keeping this fact in mind, we will follow one of these little ions of sodium through the water until it suddenly bumps into the negatively charged cathode. If our eyes were powerful enough to see, we would find that the ion of sodium immediately reacts with the water upon reaching the cathode. Is this a hitch in our theory or is it in accordance with what should take place? It is just what should take place. Upon reaching the negatively charged cathode, the tiny positive charge carried by the sodium ion is neutralized and the ion is therefore restored to a normal atom of metallic sodium. Being robbed of its positive charge, it is no longer protected from the water and immediately reacts with it to form a substance called sodium hydroxide. The same thing happens at the anode. The negatively charged ion of chlorine is neutralized upon reaching the anode, and having no chemical affinity for the water, it rushes to the surface and escapes into the atmosphere.

Having proved that sodium chloride will change water from a non-conductor, we will see if other substances will cause the same result. We will start with cane sugar. No matter how much sugar is dissolved in the chemically pure water the conductivity will not be increased in the least, as will be noted by watching the indicating instrument. We decide, then, that all substances that dissolve in water do

not act like salt. In the words of our theory, they do not ionize. A substance that ionizes or causes water to become a conductor of the electric current when it is dissolved in it is termed an electrolyte. A substance that does not affect water in this way, such as sugar, is called a non-electrolyte. There are many different substances that are electrolytes and many that are non-electrolytes. The following table contains names of a few of the most prominent ones of each class:

Non-electrolytes	Electrolytes
Cane sugar	Sodium nitrate
Methyl alcohol	Sodium chloride
Ethyl alcohol	Sulphuric acid
Benzene	Nitric acid
Ether	Hydrochloric acid
Chloroform	Potassium hydroxide
Acetone	Sodium hydroxide
Glycerine	Acetic acid
	Oxalic acid
	Silver nitrate
	Ammonium hydroxide
	Potassium sulphate
	Potassium nitrate

We could add many more substances to this list before it would be complete. In fact, every substance is either an electrolyte or non-electrolyte.

While we have only considered the electrolytic decomposition of sodium chloride in the foregoing paragraphs, we must not confuse ourselves by thinking that this is the only substance that will act in this manner. The electrolysis of sodium chloride presented a simple case and was merely taken as an example. It will be understood that the products of

electrolysis will depend entirely upon the chemical composition of the electrolyte and nothing else. Chemically, sodium chloride is a very simple compound, being composed of the metal sodium and the gas chlorine. All electrolytes do not have this very simple formula, however, and we are interested to know just how a more complex compound of matter would act, or, to be more explicit, just how it would ionize when dissolved. Glancing at our list of electrolytes, we find sulphuric acid among those named, and we will see just how this common acid will perform when dissolved in water. It is represented by the chemical symbol H_2SO_4, but this need not confuse us, as it merely indicates that it is composed of hydrogen (H), sulphur (S) and oxygen (O). Will each atom of the respective elements that form sulphuric acid act as ions or will they group themselves? The theory of Arrhenius tells us they should group themselves, the positive ions in one group and the negative ions in another group. We find this to be true with sulphuric acid, as the hydrogen atoms forms the positive ions while the atoms of sulphur and oxygen remain together and form compound negative ions. Before placing the sulphuric acid in water we knew the atom of hydrogen would form the positive ion and the combined atoms of sulphur and oxygen the negative ion, but just what foretold this action? The Arrhenius theory contains a general rule that tells us how the atoms of the different elements will act when they are ionized—that is, whether they will be positive or negative. The rule is a very simple one and can be easily remembered. It states that the atoms of all the metals and hydrogen will form positive ions when placed in solution. If this is the case, the atoms of all the elements outside of these will form negative ions. We will see, then, that sulphuric acid substan-

tiated this rule. The hydrogen atom or ion was positive and the atoms or ions of the sulphur and oxygen were negative.

A very beautiful experiment showing this profound positive nature of the metals can be performed. A dilute solution of sulphuric acid is placed in a small tube equipped with two electrodes. A tiny globule of mercury is then placed in the tube and a direct current is sent through the solution. The little globule of mercury will travel rapidly to the negative pole, thereby proving that it is of a positive nature. In this way it acts just as a ion of mercury would in solution.

Having in mind the general characteristics of electrolytes, we will place a problem before ourselves and see if we can solve it. Again referring to our list of electrolytes, we find sodium hydroxide. This is recognized chemically by the formula NaOH. The Na stands for sodium and the OH for oxygen and hydrogen respectively. Remembering our rule that all the metals and hydrogen are electro-positive, we see that we have two positive elements in this compound: sodium and hydrogen. We would naturally expect the atoms of sodium and hydrogen to unite and form a complex positive ion like the atoms of oxygen and sulphur united to form a complex negative ion when sulphuric acid was electrolyzed. We are indeed surprised to learn that this does not hold true with sodium and hydrogen, and yet it does not necessarily disprove our theory. This apparent hitch is overcome by assuming that the negative charge on the oxygen ion is stronger than the positive charge on the hydrogen ion and it therefore insists upon holding the hydrogen ion in close relationship to it. The sodium hydroxide then ionizes according to the equation $NaOH = Na^{+} + OH^{-}$ where the Na represents sodium

37

and the O and H oxygen and hydrogen respectively. The ion OH^{-} is called the hydroxyl ion.

EXPERIMENTS IN ELECTROCHEMISTRY

We are now prepared, in the light of the theory set forth in the preceding paragraphs, to make a few experiments in electrochemistry, for the subject of electrolysis is nothing but a study of electrochemistry. We will start our experiments with a reproduction of Sir Humphry Davy's experiment, in which he isolated the element sodium from one of its compounds. The arrangement of the apparatus for the experiment is

38

shown in Fig. 6. It is, of course, understood that this experiment must be carried out with direct current. The space between the block of sodium hydroxide dissolves to form the electrolyte. The mercury merely acts as an electrode. As fast as the metallic sodium is set free at the negative electrode or mercury it forms a sodium-mercury amalgam. When the sodium becomes amalgamated with the mercury it is protected from the water in the electrolyte, as we must not forget that sodium is chemically very sensitive to water. After the current has passed for a few minutes, we find that the mercury becomes hard and stiff owing to the sodium amalgamating with it. It will be necessary to keep the surface between the mercury and the sodium hydroxide well moistened during the experiment. The sodium may be separated from the mercury by a process of distillation. We

39

can obtain potassium from its hydroxide in the same manner.

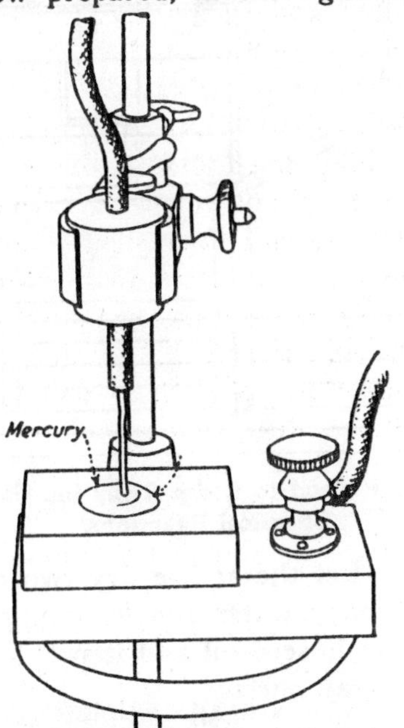

Fig. 6.—A Reproduction of Sir Humphry Davy's Experiment of Producing Metallic Sodium.

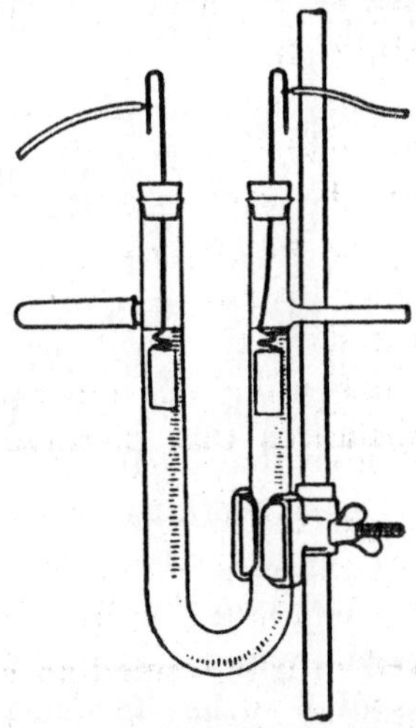

Fig. 7.—Apparatus for the Electrolysis of Zinc Chloride to Obtain Metallic Zinc.

In our next experiment we will produce more metallic zinc by a process of electrolysis. The apparatus used is shown in Fig. 7. It consists merely of a U-tube with two electrodes and a solution of zinc chloride. If the current is permitted to pass through this solution for a short time, we notice a beautiful formation of metallic zinc at the negative electrode, while the anode product will be the gas chlorine.

40

PART II

THE CHEMICAL LABORATORY

Method of Obtaining Heat—Bottles—A Chemical Balance—Test Tube Racks—A Ring Stand—A Hydrogen Sulphide Generator — Necessary Apparatus — Test Tubes — Evaporating Dishes—Fire Clay Crucibles—Filter Paper—Graduated Cylinder—Chemical Thermometers—Pipette—Red and Blue Litmus Paper — Beakers — Pinch Cocks — Thistle Tubes — Flasks — Clamps—Chemicals for the Laboratory.

The chemical laboratory should be located in a light, airy place and should occupy a space at least 15 ft. x 15 ft. Running water should also be available, although this is not absolutely essential. Illuminating gas should also be available for heating purposes, but there are substitutes for this in the event gas is unobtainable.

Benches and shelves should be placed on at least two sides of the room, so that plenty of space will be provided to work with and to store apparatus and chemicals. The shelves should be placed back of the benches, as shown in Fig. 8. Both the shelves and the benches can be made from ¾-in. yellow pine. The bench legs can be formed by 2 x 4's. The benches do not need to be wider than 3 ft. and the shelves 8 inches. This leaves a working space of 2 feet, 4 inches on the bench. A tier of four shelves is generally high enough. The first two shelves can have a space of one foot between them for large bottles, and the next two a space of 8 inches for small bottles and apparatus.

Part of one of the benches should be covered with
41
some acid-resisting material, such as slate or marble This space can then be used when working with corrosive acids to prevent the boards in the workbench

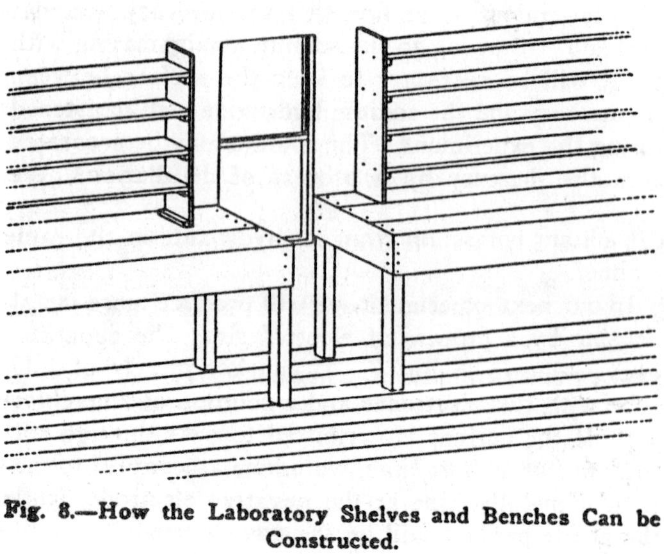

Fig. 8.—How the Laboratory Shelves and Benches Can be Constructed.

from becoming unsightly. If the amateur chemist desires he can paint all of the bench surface over with an acid-proof preparation. The author is indebted to Mr. Nomas Tallman, a contributor to Everyday Engineering Magazine, for the following receipt:

"A fire and chemical proof paint, which leaves a nice looking finish is made by boiling 790 parts of copper sulphate, 790 parts of potassium chromate and 1000 parts water in the same container.

When the solution is boiling, apply one coat to the article to be finished. Allow the solution to dry one day, then boil the solution again and apply the second coat allowing this to dry two days. Heat 100 parts of aniline oil, 100 parts water in another container to a
42
high boiling point and apply one coat to the article, and let dry for one day.

Make a new solution of aniline oil, hydrochloric acid and water, like the previous one and apply in the

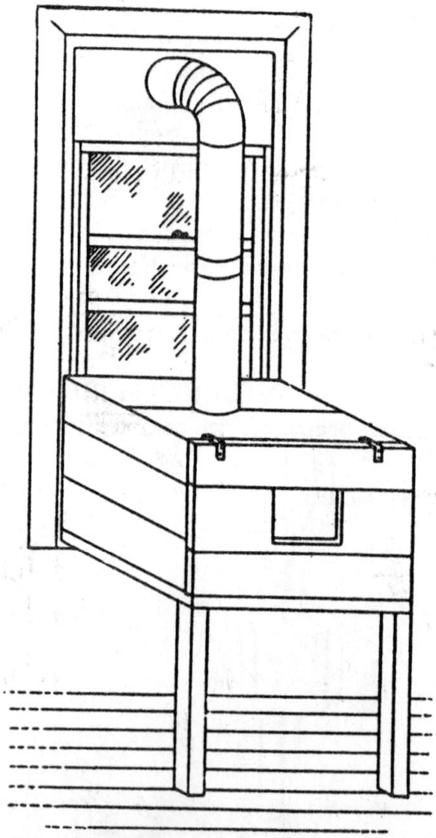

Fig. 9.—A Simple Hood to Carry Away the Gases Caused by Chemical Reactions.

same manner. Let the article dry two days. Wash the article in soapy water and let it dry thoroughly, then rub it with linseed oil and it will leave a highly polished dark gray surface.

43

The parts used are taken by weight, and the solutions may be applied with a paint brush. Care should be taken not to mix the two solutions before applying, and a new brush should be used for the second solution. The brushes should be washed after each coating."

Provision must be made to dispose of the obnoxious smelling and dangerous gaseous products given off during certain chemical reactions. In large laboratories these reactions generally take place under what is known as a hood which leads the gases to the outer atmosphere. An improvised hood for home use can be very easily made. Such a hood is shown in Fig. 9. The box should be large enough to accommodate different combinations of the chemical apparatus found in the laboratory. A box three feet square should prove to be a very suitable size. This can be made of galvanized iron or wood. In fact, wood is preferable. A hole should be cut in the top of the box to accommodate a piece of stove pipe. The pipe should lead to the outer atmosphere. A good method of doing this is shown in the figure. A piece of wood is cut just large enough to fit in the window. The pipe passes through this. A door of some kind should be arranged on the hood so that trouble will not be had in setting apparatus inside the box. The door should also fit well to prevent gases from escaping. The door should also have a glass window in order to see how any reaction is progressing. A small electric light placed inside the hood will make this possible.

METHOD OF OBTAINING HEAT

We now come to a consideration of the heat. If gas is available, the heating problem is a simple one. Two Bunsen burners should be obtained. Bunsen

44

burners can be purchased from any chemical supply house and they are so inexpensive that it would really not pay one to bother constructing them. Several feet of rubber hose should be bought with the burners so that they can be moved to any part of the bench.

In the event illuminating gas is not available the next best substitute seems to be alcohol. A good alcohol burner is shown in Fig. 10. A large wide

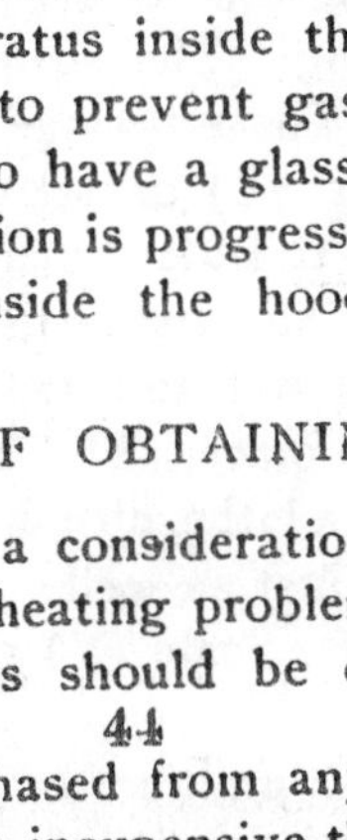

Fig. 10.—A Simple Alcohol Burner for the Laboratory.

mouthed bottle, a rubber cork, a brass tube and a wick are all that are needed for its construction. The parts are assembled as shown. The ordinary little alcohol lamps, although suitable for many purposes about the laboratory, will not be found to produce sufficient heat for all purposes. The temperature they produce is quite high enough for ordinary purposes, but the volume of heat is insufficient.

BOTTLES

A word about bottles. While it contributes much to the general appearance of the laboratory to have bottles of uniform size and shape, such bottles are

45

very expensive. Often times it is possible to accumulate many bottles of the same size and shape in which some patent medicine or other substance is sold. A number of small bottles should also be collected in which the more costly chemicals are kept. Each bottle should be labeled properly with both the formula and the name of the substance in it. This is an example:

H_2SO_4

Sulphuric Acid.

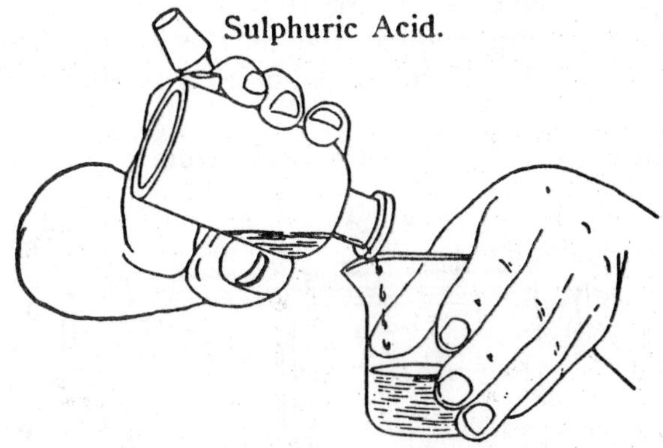

Fig. 11.—When Pouring a Chemical from a Bottle, the Stopper Should be Held Between the Fingers as Shown.

Great care should be taken that no mistakes are made in putting the wrong chemical in any bottle. This would probably result disastrously in some cases.

It may be well to mention at this point how to pour a liquid from a bottle. First, the liquid should be poured from the side of the bottle opposite the label. If a drop of the liquid runs down the side of the bottle, it will not destroy the label. Fig. 11 shows the proper method of pouring liquid from a bottle.

46

A CHEMICAL BALANCE

A good chemical balance on which different amounts of chemicals can be determined accurately is badly needed. Mr. Dana S. Greenlaw, a contributor to Everyday Engineering Magazine, describes the construction of a home-made balance which the author believes is hard to improve upon. Mr. Greenlaw's description follows:

"For serious chemical work the experimenter should have a small balance in his laboratory. Such a balance, however, is very costly and, therefore, many experimenters find it necessary to work without one.

The balance described in this article is very inexpensive and easily constructed. The materials necessary are not uncommon and all of the parts can be made with little labor.

Fig. 12 illustrates the completed balance. The handle (Q) raises the beam lift (H) and thereby not only removes the weight from the bearing, but also prevents the beam from swinging. The lever (O) permits the stop (K) to drop and allows the pans to swing. The bearings are made from an old hack saw blade.

The beam of the scale will be considered first. This is made of either brass or aluminum and is ⅛ inch thick, ½ inch wide and 10 inches long. In the center of the beam a ⅛ inch slot is sawed. A ½ inch hack saw blade is then ground to a knife edge. A piece 1½ inches long is soldered into the slot. (See A.) This forms the main bearing of the beam. Exactly ¼ inch from each end of the beam a small V-shaped groove (D) is made.

The base of the balance is cut from a piece of ½ inch hard wood. It is 7½ inches wide by 13 inches long. In the center of the base the upright (J) is

47

placed. This upright is ½ inch thick, 1 inch wide and 8 inches long. At the top of the upright or standard the slot for the beam lift (H) to slide in is made. This slot is ⅜ inch wide and 1¾ inches long. The bearings (BB) are screws, the slots of which are filed out to a V shape.

The beam lift (H) is made of ⅜ inch stock. It is

48

one inch wide and 6½ inches long. 2½ inches from each end it begins to taper. At the end it is ⅜ inches wide. The brass rods (II) pass through the base and are soldered to the brass strip (X). The cam (Y) causes these rods to be either raised or lowered. This cam is belted or soldered to the rod (W) which is attached to the handle (Q). The rod (W) is held in place by the brass strips (UP) which are screwed to the base.

The pieces (KK) prevent the pans from swinging. The nuts adjust them so that they will come to the proper height. These bolts are 3/16 inch in diameter and 1½ inches long. The springs (RR) hold these bolts in position. These springs should be made very weak. The handle (Z) causes the rod (S) to turn. The hooks on this rod lift the spring and thus allows the bolts to drop. The writer used a rib from an umbrella to make the rod and handle. The guide (Y) holds the handle in place. The legs of the balance are turned to shape on a lathe, although square ones would serve the purpose just as well. The pointer (L) is a fine knitting needle. Scale (Y) shows when the beam is in balance.

The pans are of aluminum or nickled brass. They are attached to the beam by the two knitting needles (FF). These needles are soldered to the bearing (E) which is made of a hack-saw blade. These bearings swing in the groove previously produced in the beam.

The more carefully this balance is made the more accurate and sensitive it will be and the smaller the quantity it will register. The base may be made large enough to enclose the balance in a glass case. If such a case is used the front should have a door which opens. If a case is used a small beaker of sulphuric acid or calcium chloride should be kept in it to prevent moisture from attacking the balance."

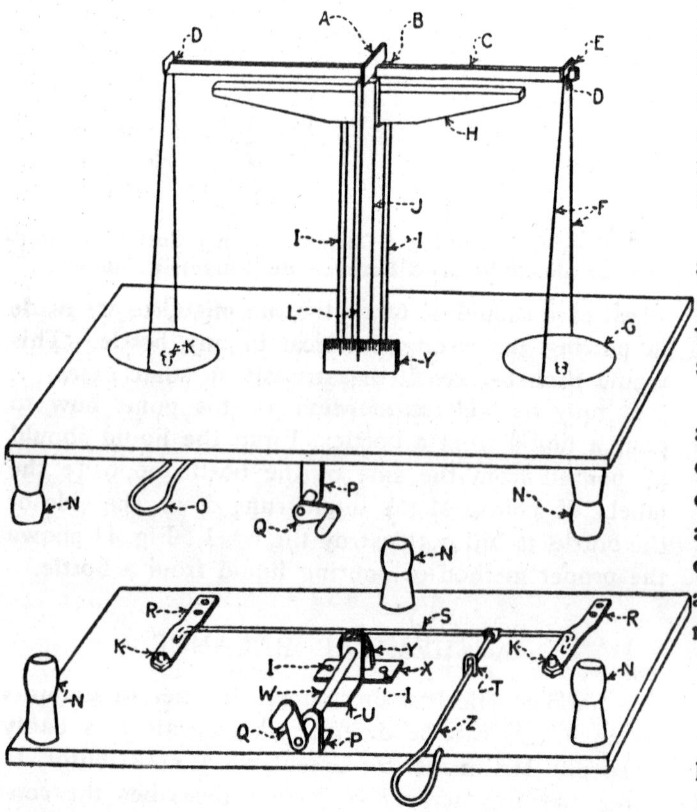

Fig. 12.—Constructional Details of a Small Chemical Balance.

49

TEST TUBE RACKS

Several small test tube racks should be made for the laboratory. These are very simple to construct and no description should be necessary as the drawing (Fig. 13) gives all the necessary details.

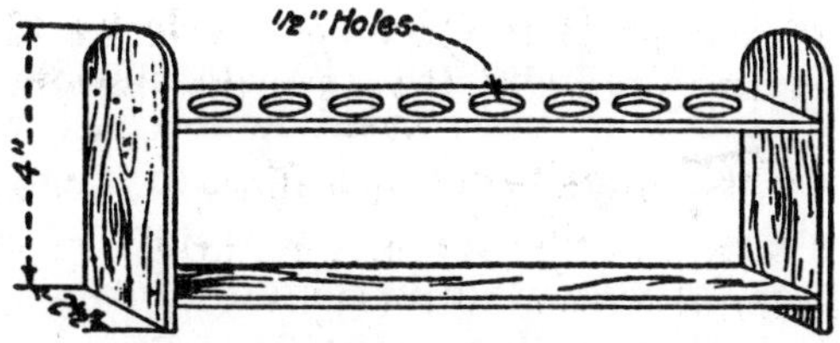

Fig. 13.—Dimensions of a Small Test Tube Holder.

A RING STAND

A ring stand can be very easily made. The complete details are given in Fig. 14. The base of the

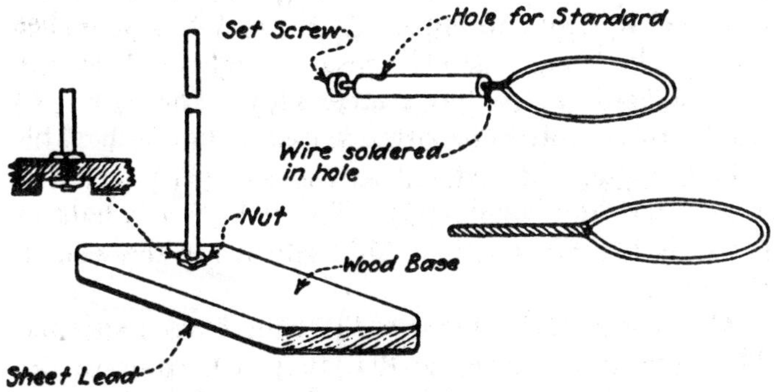

Fig. 14.—Details of a Home-made Laboratory Ring Stand.

stand should be cut to shape from wood and covered with sheet lead to give it sufficient weight so that the stand will not tip over easily when in use. The top

of the base should be covered with some acid-proof preparation. The standard is fastened to the base with two nuts as illustrated. The ring is very easily produced. The wire used should be very heavy so as to be able to hold its shape. The method of holding it in the brass piece which slips over the standard is clearly shown. About three of these rings should be made for different sized dishes and crucibles. A small piece of copper gauze is often used to place over the ring when very small dishes are being heated. Ring-stands are used very much about the chemical laboratory not only as a convenient holder for dishes being heated, but also for general apparatus set-ups.

A HYDROGEN SULPHIDE GENERATOR

A hydrogen sulphide generator is quite necessary for the small laboratory as this gas is used in many different ways. Hydrogen sulphide is generated when hydrochloric acid comes in contact with iron sulphide. These two compounds are used for the production of

the gas in what is known as a Kipp generator. The construction of a simple Kipp generator is shown clearly in Fig. 15. The bottle in which the students lamp chimney is placed should of necessity be one

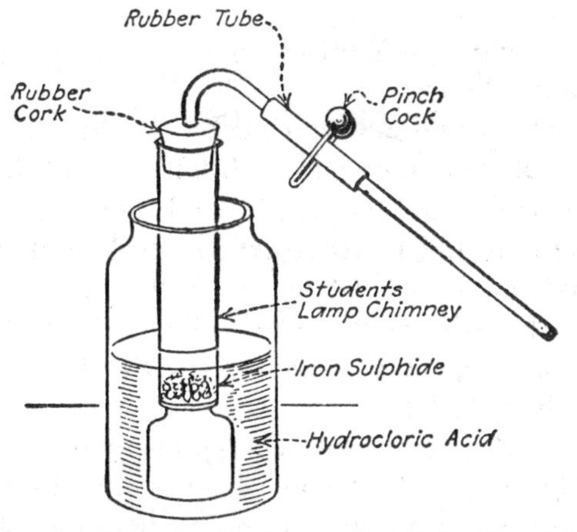

Fig. 15.—A Banks Hydrogen-Sulphide Generator.

with a very wide mouth. A large button is placed in the narrow part of the chimney. Chunks of iron sulphide are placed in the upper portion of the chimney. The remainder of the apparatus is very clear. The operation of the device is simple in the extreme. The larger bottle is half-filled with hydrochloric acid. The chimney is then placed in the bottle **with the pinch-cock closed.** The air is compressed in the chimney, which prevents the hydrochloric acid from reaching the iron sulphide. When the pinch-cock is opened, this air pressure is released and the hydrochloric acid succeeds in coming in contact with the iron sulphide. The following reaction occurs:

$$FeS + 2HCl = H_2S + FeCl_2$$

The hydrogen sulphide is forced out of the rubber tube. When enough of the gas has been produced, the pinch-cock is closed and the pressure of the H_2S increases until the hydrochloric acid is forced from the chimney.

NECESSARY APPARATUS

While the amateur chemist can make a great number of the necessary articles that should be in his laboratory, there are certain things which are quite beyond his facilities and ability. Fortunately, none of these articles are expensive. While some of the things in the following list could be made, they are so cheap that the author believes more time and trouble would be wasted on them than they are worth:

1 Test-tube holder.
1 Small evaporating dish.
1 Large evaporating dish.
1 Small fire-clay crucible.
1 Box of filter papers.
1 Graduated cylinder.
1 pound assorted glass tubing.

1 Thermometer, 200 degrees C.
1 Pipette.
2 Funnels (large and small).
1 Box of blue litmus paper.
1 Box of red litmus paper.
6 Beakers from 25 c.c. to 200 c.c.
4 Pinch-cocks.
1 Test-tube brush.
1 Water-bottle.
6 feet rubber tubing.
12 Large test-tubes.
12 Small test-tubes.
1 U-tube.
1 Thistle tube.
2 Erlenmeyer flasks.
1 Clamp for ring-stand.
2 Flasks.
Mortar and pestle.

The reader will find in this list many articles about which he knows little or nothing.

There will be a few articles in the above list with which the amateur chemist is not familiar. For this reason a description will be given of each one.

53

TEST TUBES

A test tube is used to heat liquids and solids over a Bunsen burner. During the heating the top

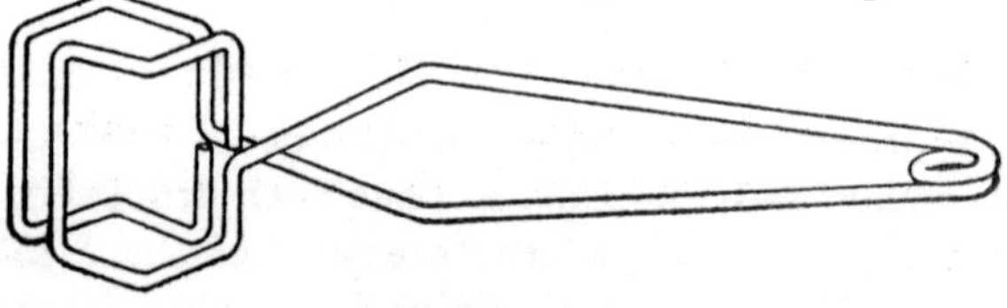

Fig. 16.—A Wire Clip Used for Holding Hot Test Tubes.

of the tube becomes so hot that it is impossible to hold it by hand. A little wire test tube holder as shown in Fig. 16 is used.

EVAPORATING DISHES

Evaporating dishes are used for a multitude of purposes. A small evaporating dish is shown in Fig. 17. The principal use of the evaporating dish is to evaporate liquids. For instance, if we had a solution of

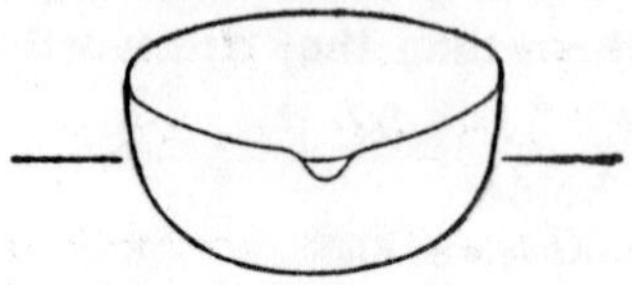

Fig. 17.—An Evaporating Dish Which is Generally Used for Boiling Purposes.

sodium chloride in water and we wished to obtain the sodium chloride in the original crystalline condition, we would place it in the evaporating dish and boil the water off, leaving the dry sodium chloride in the bot-

54

tom of the evaporating dish. Evaporating dishes are made of glass and possess considerable heat-resisting qualities. They are also impervious to anything but extremely sudden changes in temperature. Evaporating dishes are obtainable in various sizes.

FIRE CLAY CRUCIBLES

A small fire clay crucible of about 25 c.c. capacity is a useful little article to have about the laboratory and it is capable of standing much higher temperatures than ordinary glassware, and for this reason it can be used where ordinary chemical glassware would be unsuitable.

FILTER PAPER

Filter paper has an extremely important use. It is used for the filtration of any solution or liquid that has

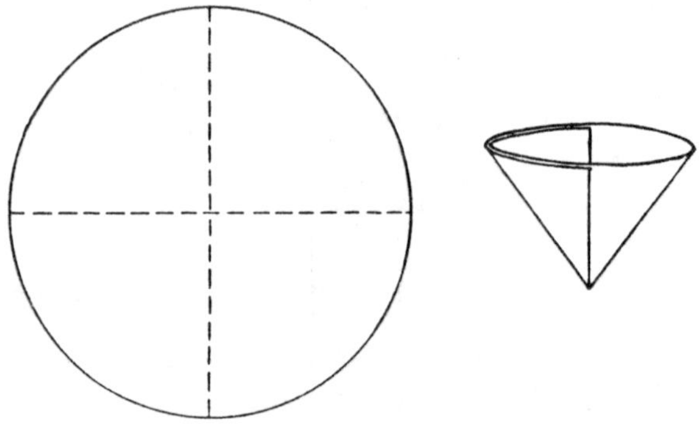

Fig. 18A—How Filtering Paper is Folded.

solid matter in it. The liquid is allowed to pass freely while the solid matter is arrested. Filter paper must

55

be chemically pure, and for this reason it is very carefully prepared for use. It comes in various sizes for use with different sized funnels. The use of filter paper will be very clearly seen by referring to Fig. 18A.

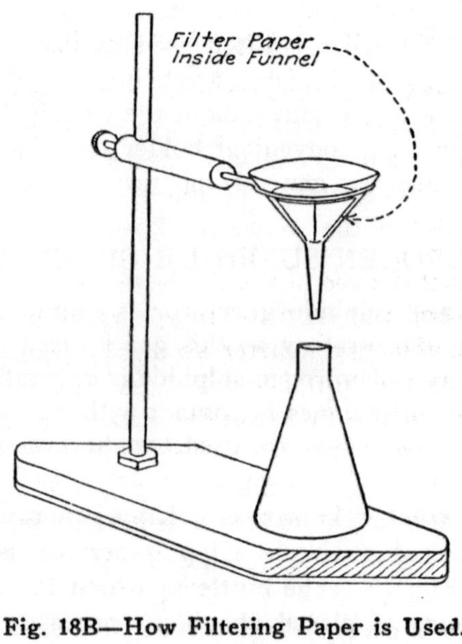

Fig. 18B—How Filtering Paper is Used.

The method of holding filter paper in place in the funnel is shown in Fig. 18B. The size of the filter paper would depend largely upon the size of the funnel that it is to be used with.

GRADUATED CYLINDER

A graduated cylinder is shown in Fig. 19. This is a much needed article about the laboratory. Each mark of the cylinder represents a cubic centimeter

56

and cylinders are obtainable with different capacities. 100 centimeters is a very good size for the amateur chemist and it should be included in his equipment.

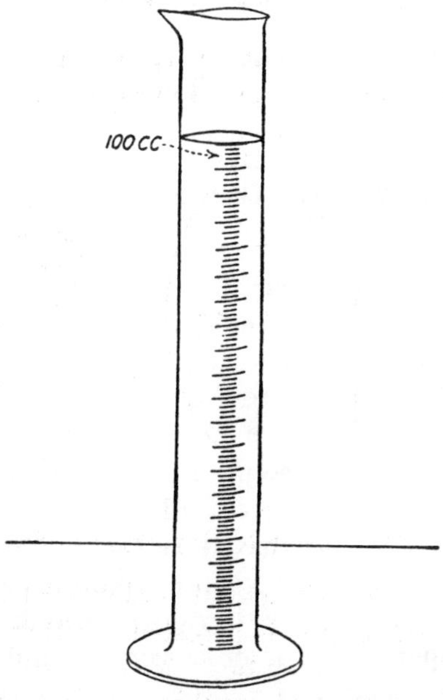

Fig. 19.—A 100 c.c. Graduated Cylinder Used in Measuring Liquids.

The cylinder is used largely for measuring off different solutions and liquids.

CHEMICAL THERMOMETERS

The chemical thermometer used in the small laboratory should be graduated in Centigrade degrees as the Centigrade system is considered more scientific than the Fahrenheit system, and therefore it finds more universal use in the world of science. A word of

57

caution about the use of the thermometer. If we have a liquid to boil and wish to bring it up to 100 degrees Centigrade, the thermometer should be placed in the liquid before the heat is applied so that the rise in temperature of the thermometer will be gradual. If we were to permit the liquid to reach 100 degrees and then place the thermometer in it, it would break immediately, due to the sudden change in temperature.

PIPETTE

A pipette is a very simple device and one is shown in Fig. 20. This is used for conveying liquid from one receptacle to another. One end of the pipette is

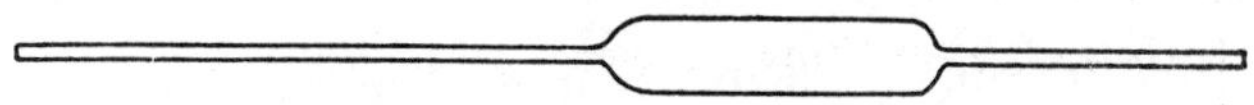

Fig. 20.—A Pipette, the Use of Which is Described in the Text.

placed in the mouth and the other end in the liquid to be taken into it. By sucking gently on the end of the pipette the liquid will rise into it. If a particularly poisonous liquid is being handled, care should be taken that none of it is drawn into the mouth.

RED AND BLUE LITMUS PAPER

Red and blue litmus paper is used for testing alkaline and acid solutions. For instance, if we had a solution and wished to know whether it was a base or acid, we would take a piece of red or blue litmus paper and dip it in it. If the red paper turned blue it would indicate a base. If it remained red we would not know whether or not it was neutral or alkaline. Therefore, we would try it with blue litmus paper. If the blue litmus paper would turn red we would then under-

58

stand that it was an acid solution. If neither the red nor blue litmus paper changed in color we would understand that the solution being tested was neutral.

BEAKERS

A beaker is shown in Fig. 21. At least 6 of these should be in the student's laboratory. Beakers are used for various purposes. Reactions are permitted

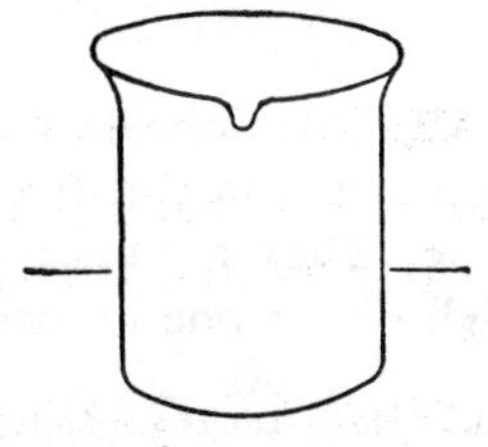

Fig. 21.—A Small Beaker Which Finds a Multitude of Uses in the Chemical Laboratory.

to take place in them, precipitates are formed in them and they are also used for heating purposes. A beaker is a very convenient receptacle to handle and liquids are very easily poured from them.

PINCH-COCKS

Pinch-cocks are used in connection with rubber tubing to either cut off a supply of liquid or gas. The pinch-cock merely presses the tube together and holds it tightly so that nothing can pass. Owing to the nature of the rubber perfectly gas tight and liquid stoppage is made.

A test tube brush is shown in Fig. 22. This is used

for the purpose of cleaning out dirty test tubes. In ordinary cases water will do very nicely for cleaning out the tubes, but if a particularly clean receptacle is

59

desired for an extremely accurate chemical determination, the tube should first be washed out with water and this should be followed with sulphuric acid. Sul-

Fig. 22.—A Small Brush Used in Cleaning Test Tubes.

phuric acid is then carefully washed out with water and the final washing is made with ethyl alcohol, after which the tube is allowed to dry.

Thistle tubes are extremely simple and one of their

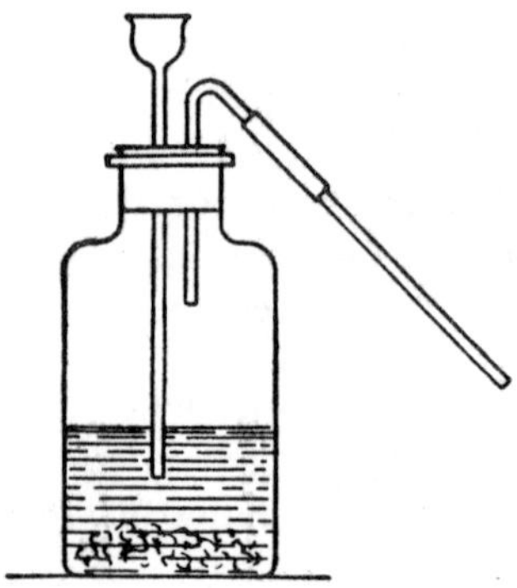

Fig. 23.—Showing the Use of a Thistle Tube.

uses is shown in Fig. 23. Thistle tubes cost but a few cents each and one should be in the laboratory of every chemical student.

FLASKS

An Erlenmeyer flask has a wide variety of uses and it is advisable to include two of these in the general chemical equipment. They are often used with a rubber cork into which either one or two glass tubes are

60

placed. The flask when arranged in this way would become an important part of many different apparatus

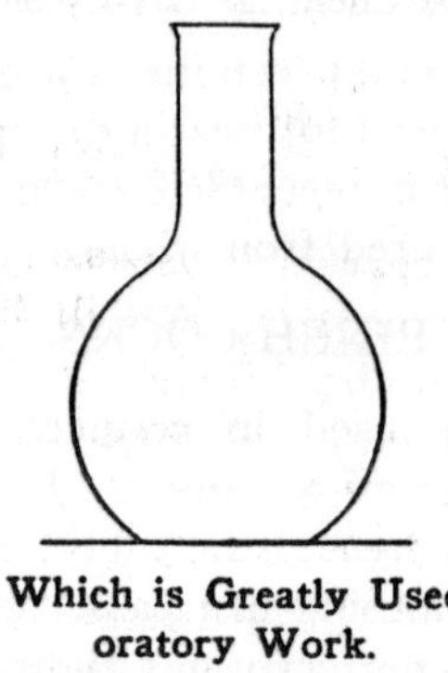

Fig. 24.—A Flask Which is Greatly Used in Chemical Laboratory Work.

set-ups. Flasks similar to those shown in Fig. 24 are very important.

WASH BOTTLE

One use to which such a flask can be put is that of forming a wash bottle. A simple wash bottle can easily be made by the student and one is shown in

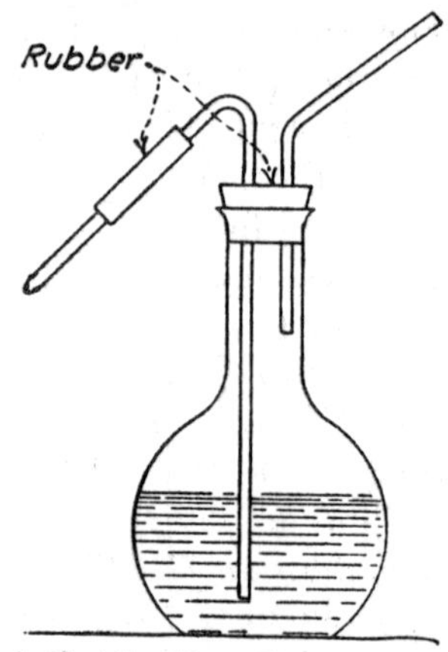

Fig. 25.—A Wash Bottle That Can be Assembled by the Home Chemist.

61

Fig. 25. It will be seen that by blowing on the one tube the pressure is developed on the surface of the water and the water is forced out of the other tube. Here is an instance where the bottle can be used. We will assume that we are filtering solid matter from the solution. If the solution is passed through the filter paper the solid matter will be left about the sides of the paper. Being that it is the solid matter and not the solution that we wish to obtain it will be necessary to wash the solid matter down into the bottom of the filter paper. This is done very carefully with the wash bottle.

CLAMPS

A clamp for use in connection with the ring stand that was described in a previous part of this book is shown in Fig. 26. These clamps have a very wide use

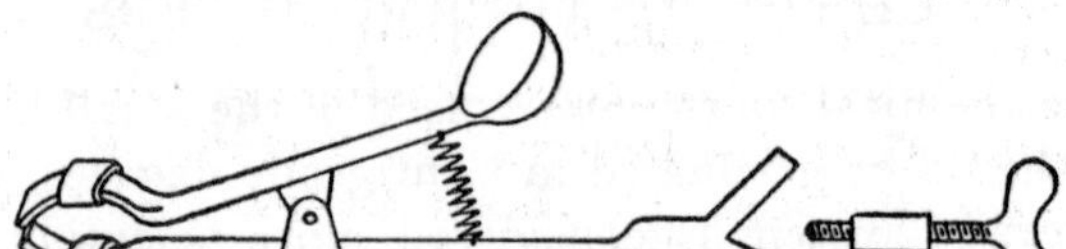

Fig. 26.—A Clamp to be Used in Connection With the Ring Stand.

and they come in extremely handy for many different purposes. For instance, they can be used for gripping test tubes, flasks and glass tubing.

CHEMICALS FOR THE LABORATORY

We now come to a consideration of the chemicals with which the small laboratory should be equipped. Many of these compounds can be obtained at the corner drug store, but it will be necessary to purchase the majority from some chemical supply house.

62

Copper sulphate ($CuSO_4$), ¼ lb.
Sodium nitrate ($NaNO_3$), ¼ lb.
Sodium carbonate (Na_2CO_3), ¼ lb.
Potassium carbonate (K_2CO_3), ¼ lb.
Potassium nitrate (KNO_3), ¼ lb.
Sulphuric acid (H_2SO_4), ½ pt.
Hydrochloric acid (HCl), ½ pt.
Nitric acid (HNO_3), ½ pt.
Sulphur (S), ½ lb.
C.p. zinc (Zn), ¼ lb.
Copper filings (Cu), 2 oz.
Iron filings (Fe), 2 oz.
Ammonia (NH_3), ½ pt.
Ammonium chloride (NH_4Cl), ¼ lb.
Sodium chloride (table salt) ($NaCl$), ¼ lb.
Potassium chlorate ($KClO_3$), ¼ lb.
Manganese dioxide, 2 oz. MnO_2.
Calcium chloride ($CaCl_2$), ¼ lb.
Sodium hydroxide ($NaOH$), 2 oz.
Potassium hydroxide (KOH), 2 oz.
Magnesium ribbon, 2 oz. Mn.
C.p. aluminum (Al), 3 oz.
Charcoal, C ½ bb.
Ethyl alcohol ($C_6H_{12}O_6$), ½ pt.
Mercury, Hq., 6 oz.
C.p. tin (Sn), 3 oz.
C.p. lead (Pb), 6 oz.
Mercuric oxide (HgO), 4 oz.
Ferric oxide, ¼ lb.

63

PART III

CHEMICAL EXPERIMENTS

Preparation of Oxygen — Breaking Glass Tubing — Catalytic Agents—Experiments with Oxygen—Production of Chlorine —Experiments with Chlorine—Neutralization—Distillation of Water — Practical Distillation — Thermit — A Small Electric Furnace—Furnace Connections—Experiments in Catalysis—Negative Catalysis — Detecting the Presence of Carbon Dioxide — Production of Calcium Carbonate — Electrolytic Production of Lead Carbonate—Production of Nitric Acid —Colloidal State of Matter — Crystalloidal and Colloidal Solution—Electrolysis of Water—Generation of Hydrogen in Quantities—Glass Working for Amateur Chemists—Cutting and Breaking—Drilling—Smoothing Rough Edges—Bending Glass Tubing—Cutting Glass Tubing—Drawing Glass Jets—Sealing a Platinum Wire in Glass—Joining Tubing and Making Tees — Spreading the Ends of Tubing — Glass Spiral Tubes—Miscellaneous Information.

Now that we have our chemical laboratory partly equipped, we will conduct a few experiments that will not only broaden our knowledge of chemistry but will assist us in understanding the manipulation of chemical apparatus. During these experiments it will be necessary to add to the equipment of the laboratory by making several new adjuncts. These will be described with the experiment with which they are used.

PREPARATION OF OXYGEN

The first experiment will be the preparation of oxygen. We understand that there is a great abundance of oxygen in the air, but here it forms a physical

64

combination with the other atmospheric gases, notably among them nitrogen.

Oxygen can be produced in a free, undiluted state by the decomposition of potassium chlorate ($KClO_3$). The decomposition is brought about merely by breaking it up into potassium chloride and oxygen. The arrangement of the apparatus necessary to carry out the

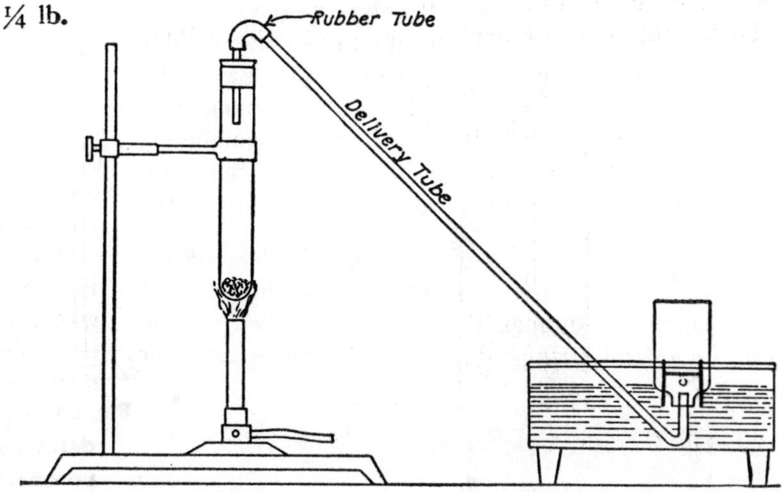

Fig. 27.—Apparatus Necessary for the Generation of Oxygen.

experiment is shown clearly in Fig. 27. The pneumatic trough is a simple device and is used for collecting gases. The collecting receptacle is so arranged in the pneumatic trough that the water keeps the receptacle continuously sealed so that no gas can escape. The gas bubbles through the water into the mouth of the receptacle. The pneumatic trough can be made easily by the amateur chemist. The dimensioned drawing which forms Fig. 28 will give all the necessary details.

65

BREAKING GLASS TUBING

In setting up the apparatus necessary for the experiment of producing oxygen, it will be necessary to break some glass tubing. There is a certain way to do this. First, the glass is nicked with a three-cornered file at the point where it is to be broken. It is then gripped in the hands as shown in Fig. 29. With the thumbs placed in this manner the tube is broken by a quick motion. On small tuning under ⅜ inch diameter, a nice clean break will be produced in this way.

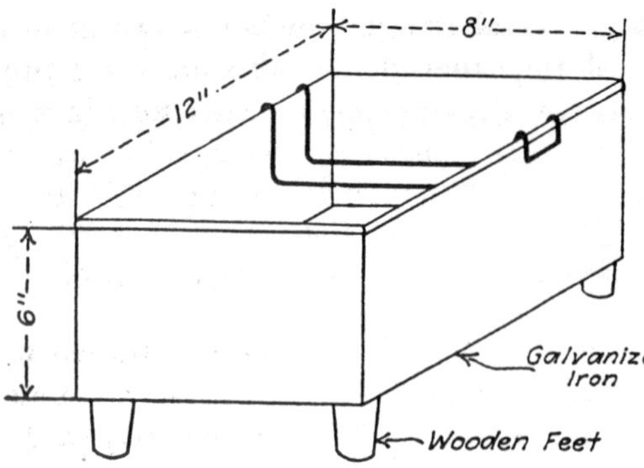

Fig. 28.—Dimensions of a Small Pneumatic Trough.

CATALYTIC AGENTS

We are now ready to continue with the experiment. While potassium chlorate will evolve oxygen at a temperature well above 300° Fahr., it can be made to give it off at a much lower temperature (200° Fahr.) by mixing it with one-third of its weight of manganese dioxide. After the potassium chlorate is completely decomposed, we are surprised to find that the manganese dioxide is left quite unchanged. Was it the

66

mere presence of the compound that accelerated the decomposition of potassium chlorate? This is true. The manganese dioxide in this case is called a catalyzer or a catalytic agent. We may define a catalytic

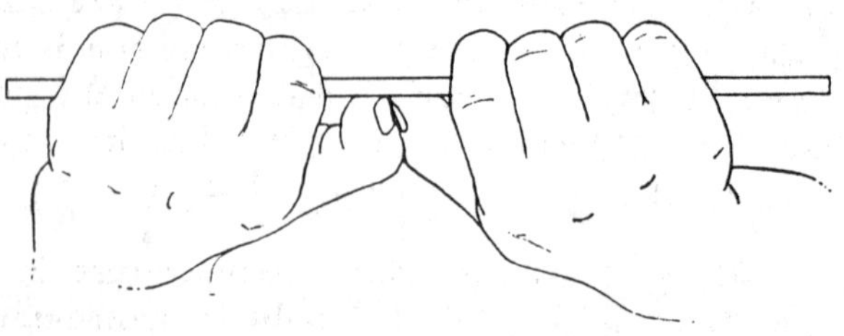

Fig. 29.—How Small Glass Tubing is Held When it is to be Broken.

agent, then, as a substance, which by its mere presence, causes a chemical reaction to take place. The presence of platinum or mercury in hydrogen peroxide (H_2O_2) causes it to decompose into water and oxygen.

EXPERIMENTS WITH OXYGEN

Having produced the oxygen, many interesting experiments can be performed by its aid. Oxygen is an extremely active element. It combines with all but a few of the elements to form oxides. This process is called oxidation. If we heat a piece of soft iron wire to redness and introduce it in the bottle of oxygen, it will become intensely hot and burn (oxidize) with a shower of tiny sparks. If a match is lit and the flame extinguished, it will be reestablished as the match is introduced into the bottle immediately after the flame is extinguished. Many other similar interesting experiments can be performed with oxygen. The reader

67

will become familiar with these when he extends his chemical studies beyond the confines of this book.

PRODUCTION OF CHLORINE

The next experiment will be that of producing the gas chloride. A word of caution will be advisable here. Chlorine is a very poisonous gas and the experimenter should not inhale any if possible. Of course, a little will escape into the air and it will be

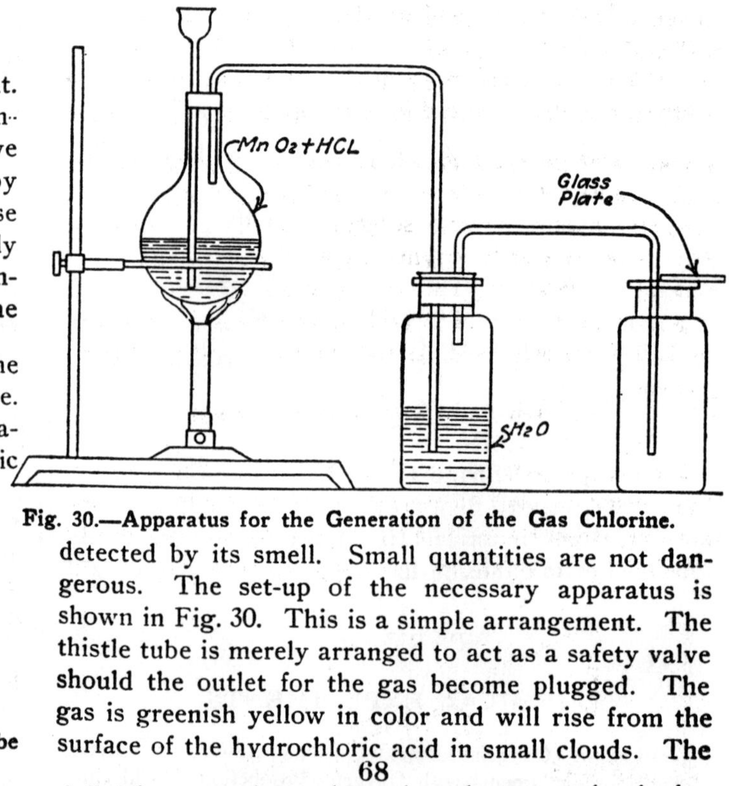

Fig. 30.—Apparatus for the Generation of the Gas Chlorine.

detected by its smell. Small quantities are not dangerous. The set-up of the necessary apparatus is shown in Fig. 30. This is a simple arrangement. The thistle tube is merely arranged to act as a safety valve should the outlet for the gas become plugged. The gas is greenish yellow in color and will rise from the surface of the hydrochloric acid in small clouds. The

68

following reaction takes place between the hydrochloric acid and manganese dioxide:

$$MnO_2 4HCl = Cl_2 + MnCl_2 + 2H_2O$$

There are really three products of this reaction: water, chlorine and manganese chloride.

EXPERIMENTS WITH CHLORINE

Chlorine is a notable bleaching agent. In fact, it finds a great commercial use for this purpose. The bleaching properties of chlorine can be very easily demonstrated by placing a piece of wet calico in the jar containing the free gas. If the cloth is left there long enough, it will be bleached almost white when it is taken out.

Chlorine is somewhat soluble in water. If some of the gas is dissolved in water and exposed to sunlight, the following reaction takes place, the sunlight acting as a "catalytic" agent:

$$2H_2O + 2Cl_2 = 4HCl + O_2$$

Here we find the production of hydrochloric acid and free oxygen.

NEUTRALIZATION

Next, we will try a little experiment in neutralization. We will take a small quantity of dilute acid (any kind) in a test tube. After the acid, we will pour in a like amount of an alkaline solution. Sodium hydroxide (NaOH) will do nicely for this purpose. After this, we will test the solution with litmus paper to see whether it is acid or alkaline. If it is acid, we will add a little more of the NaOH. If it is alkaline, we will add a little more acid. The acid or alkaline solution should be added in very small quantities until

a point will be reached where the red litmus will remain red and the blue litmus will remain blue. This indicates that a neutral solution has been produced. This solution can be changed either way by the addition of a slight amount of either an acid or a base. If a half drop of acid is added, the solution will immediately become acid, as will be indicated by litmus paper.

DISTILLATION OF WATER

The next experiment will be that of distilling water. The author decided to treat this not so much as an interesting experiment, but to convey to the reader the principles of distillation and condensation. If we had

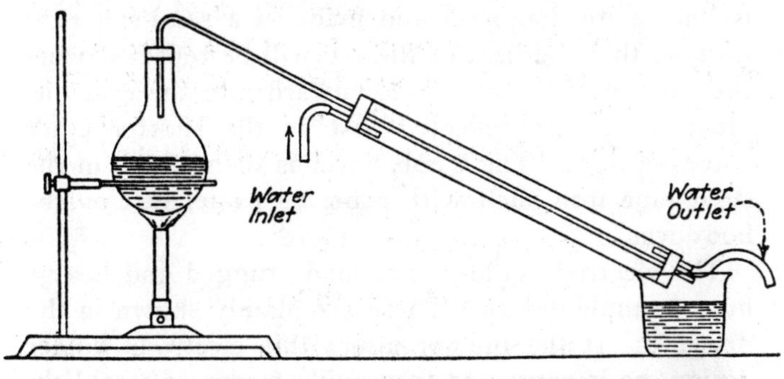

Fig. 31.—Showing Apparatus Used for Distilling Water in the Laboratory.

a vapor rising off a boiling solution which we wished to obtain in the liquid state, we would proceed in the same way that we are proceeding with the water.

The apparatus necessary for the distillation of water is shown in Fig. 31. The condenser is a simple device and is in need of very little description. It consists of a large glass tube with a rubber cork in each end. Two holes are in each of the rubber corks to permit

the two glass tubes shown to pass. Running water is made to circulate through the condenser by means of the tubes A and B.

When the water reaches the boiling point in the receptacle, its vapor passes off into the glass tube that passes through the condenser. This tube is kept cool and the water vapor or steam condenses to water when it enters. The condensed water runs into the receptacle.

FRACTIONAL DISTILLATION

If we had a mixed solution and we desired to separate the various constituents, it could be done by a process of distillation. All substances have a different vapor pressure or temperature of evaporation, we may say. If we had a solution of alcohol and water which we wanted to separate, we would heat the liquid slightly to evaporate the alcohol, and then increase the temperature to evaporate the water. This process is called fractional evaporation.

THERMIT

A simple experiment in high temperature can be made with a little iron oxide and powdered aluminum. The aluminum can be the kind used in aluminum paint. This is very flaky and will be very suitable for the purpose. A small carbon or graphite crucible will also be needed to conduct the experiment properly. The aluminum and iron oxide are placed as shown. (Fig. 32.) The aluminum and iron oxide are mixed in equal proportions. The magnesium ribbon is used to start the reaction, which is based upon the great "affinity" of aluminum for oxygen. The iron oxide loses its oxygen to the aluminum. The reaction takes

place with great violence and a temperature in the neighborhood of 3000° C. is produced momentarily. When the reaction is completed, a small pillet of

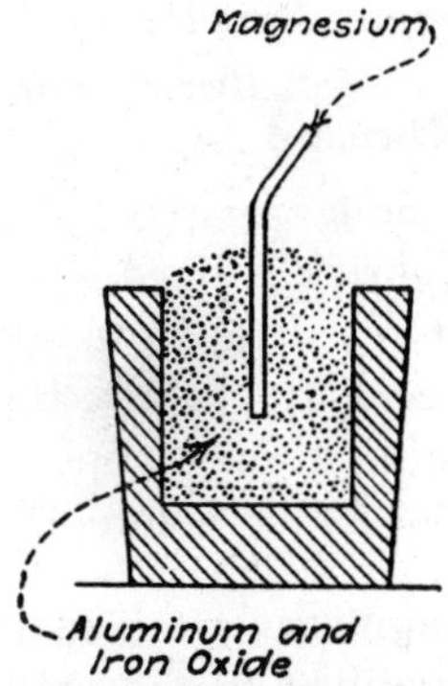

Fig. 32.—How Thermit is Ignited.

molten iron will be left in the bottom of the crucible. Many of the metallic oxides can be reduced in this way to obtain the pure metal.

A SMALL ELECTRIC FURNACE

A small electric furnace will make a valuable addition to the experimenter's laboratory. It will provide him with means of conducting many interesting experiments that would otherwise be quite impossible. The ensuing description is of a small laboratory furnace.

The body and crucible of the furnace (see Fig. 33) will be considered first. The body is made from half a standard brick of some good heat-insulating material of considerable refractory endurance. There are several good brands of refractory bricks on the market, such as Sil-O-Cel, Nonpareil, Osceola, etc. A hole 1½

72

inches deep is chiseled in the center of the brick. A groove is then cut through the center of the brick to accommodate the electrodes, as shown in the drawing. As the electrodes are ½ inch in diameter, this groove

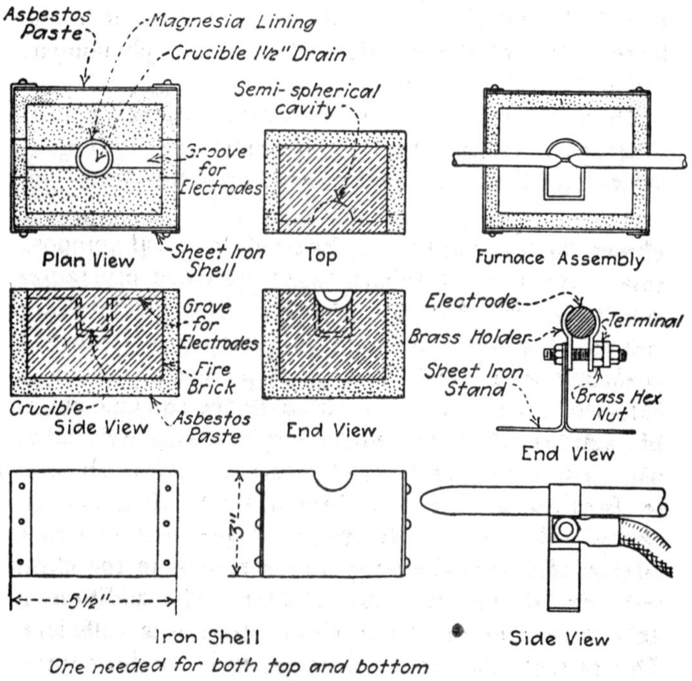

Fig. 33.—Details of a Small Electric Furnace for Use in the Chemical Laboratory.

should be just a trifle larger than ½ inch. The cavity in the center of the brick is lined with magnesia which may be obtained at any drug store. Being in the powdered form, it is necessary to mix it with some bonding material. While it may appear peculiar to the builder, either molasses or common syrup is used as

73

the bond. The magnesia is mixed into a thick, consistent paste and applied to the walls and bottom of the cavity or crucible of the furnace. The whole brick is then placed in a hot oven and allowed to remain until the magnesia lining becomes baked. To carry this baking further and to completely remove all the volatile organic constituents of the bond, an arc is established directly over the cavity to thoroughly heat the magnesia. Silicate of soda may also be used successfully as a bonding substance.

An outer casing of stove-pipe iron is now made for the furnace. This is made ½ inch larger than the refractory brick so that the brick can be placed in the casing and the intervening space between the two filled with some good heat-insulating material. Heat insulation is a prime factor in designing laboratory furnaces. The heat produced in a furnace escapes very readily, and if high temperatures are to be produced the furnace must be properly insulated. The thermal resistivity of contacting surfaces is very high, therefore a small furnace should have several different layers of refractory or heat-insulating substances in its body. The great thermal resistance of contacting surfaces is the reason it takes an onion so long to boil. The heat-insulating substance used between the shell or casing and the brick of the furnace need not have a high refractory power. Ordinary asbestos paste will do very nicely. If the little furnace is not provided with an outer covering of a heat-insulating substance, it will operate at a very low point of efficiency, as the surface rediation will account for a dissipation of heat as fast as it can be produced beyond a certain temperature of the arc.

A cover of high refractory power is now made. The builder should take great care in making this cover. It should fit as accurately as possible in order to pre-

74

vent dissipation of heat from the crucible. The other half of the brick used in making the body of the furnace may be employed for the cover. The cavity and two grooves are cut in the brick as shown. The cavity is lined with magnesia and made in a semi-spherical form so that the heat of the arc will be reflected trom the surface of the cavity downward into the crucible. Heat waves are reflected just as the shorter ether waves of light. The whole cover is then placed in the stove-pipe iron shell with asbestos paste, just as the body was.

The electrode holders are made rugged and heavy, but of simple design. These are plainly shown in the drawing. If the builder alters this electrode holder design, he is cautioned to provide means of establishing a firm, low resistance contact between the electrode and the feed cable. With a low pressure and high current strength, a poor contact will heat badly and cause a lowering of working efficiency. The electrodes should be of carbon ½ inch in diameter. They are adjusted by pushing them in or out of the furnace. The advantages of an electrode adjusting device do not justify the trouble of its construction. The electrodes and terminals rest on the sheet iron standard as shown. The furnace sits on a slab of ½-inch slate.

FURNACE CONNECTIONS

A suitable transformer for use with a furnace of this

type should have a core that measures 10 inches long, 8 inches wide made up of strips of iron 2½ inches wide. The thickness of the core should be 2¾ inches. The secondary of the transformer consists of fifty turns of doubled No. 10 magnet wire. The primary consists of 150 turns of No. 14 magnet wire. The con-

75

nections for the furnace and transformer are shown in Fig. 34. If the transformer is not used, the furnace

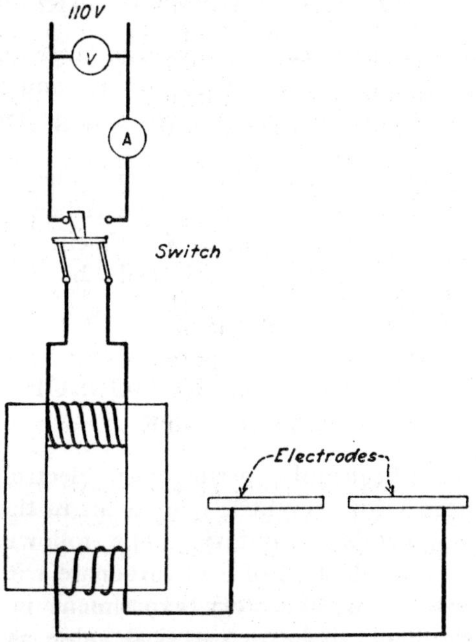

Fig. 34.—How the Electric Furnace is Connected With a Suitable Transformer.

can be run in connection with a water rheostat as shown in Fig. 35.

EXPERIMENTS IN CATALYSIS

A few experiments in catalysis will give the reader a broader understanding of this wonderful process. Probably one of the most beautiful experiments in catalysis is that of the decomposition of hydrogen peroxide (H_2O_2) by the catalytic influence of mer-

76

cury. If a little clean mercury is placed in a test tube and a small amount of a 10 per cent solution of hydrogen peroxide poured over it, the hydrogen peroxide will at once begin to decompose into water and oxygen. The decomposition, however, is not continuous, but periodic. If we watch the tube closely, we will see a little gaseous cloud burst from the surface at regular intervals—vibratory decomposition taking place with chronological accuracy! If our eyes were only powerful enough to see; if our intelligence were only great enough to comprehend! The mercury is seemingly impotent, yet what mysterious power it possesses. But mercury is not the only substance whose "presence" will cause hydrogen peroxide to break up into water and oxygen. Finely divided platinum or pyrolusite will bring about the same change.

Such an ordinary substance as water is a wonderful catalyst for certain reactions. Thus, carbon monoxide and oxygen will react with explosive violence if ignited in the presence of water vapor. If the water vapor is completely removed from these gases, however, they absolutely refuse to combine.

It is amazing to learn what minute quantities of catalytic agents are capable of effecting certain chemical reactions. It requires about .0045 gram of platinum to cause the union of twelve parts of oxygen and hydrogen. But this is a comparatively large amount when it is considered that .000,000,000,1 of ¼ of an ounce of copper sulphate will effect a certain reaction.

In the foregoing paragraphs we briefly considered chemical catalysis. We will now deviate a little and learn a few of the wonders of a similar phenomena called "physical catalysis."

There are two chemical derivatives of the element chromium, called chromic chloride and chromous chloride. The "ic" on chromic and the "ous" on

77

chromous each signifies a different chemical composition. Chromic chloride, a beautiful violet crystalline substance, is what the chemist calls insoluble; it refuses to dissolve in water no matter how much it is shaken or agitated. If a small amount of chromous chloride is placed in the water with the chromic chloride, solution of the chromic chloride immediately takes place and a beautiful violet solution results. Just as in chemical catalysis, the chromous chloride suffers no change whatsoever. Its powerful presence causes a release of potential energy which results in the chromic chloride passing into solution. The addition of .000025 of a gram of chromous chloride is sufficient. This is truly the magic of nature in its highest form.

The catalytic power of a catalyst is inexhaustible. It may be used over and over again, yet it remains just as potent as ever. Some catalysts exert their cryptic influence in more than one reaction. This is notably the case with platinum, which is a veritable chemical busybody. The oxides of iron are also powerful catalysts and bring about many chemical partnerships that would otherwise be impossible. That great "king of chemicals," sulphuric acid, is manufactured by a process in which catalysis plays the leading rôle. It would be quite impossible in this limited space to mention all the chemical processes of today in which catalysis is paramount. Catalysts are also busy in the various processes of nature. Fermentation is brought about by the catalytic influence of bacteria. In the human body, many catalytic agents are exerting their persuasive influences, and recent investigations seem to prove that the very problem of life itself is wrapped up in the mysteries of catalysis.

78

NEGATIVE CATALYSTS

There are certain substances which act in a manner directly opposite to that of catalysts; their mere presence, instead of producing chemical action, prevents it. Such substances are called "negative catalysts" and their potency in retarding chemical union is far greater than that of a catalyst in accelerating it. Prussic acid, carbon disulphide, phosphorus and sulphuretted hydrogen are powerful negative catalysts even in the most minute quantities. In considering these few negative catalysts, we are reminded that they are very poisonous to the human system and we ask ourselves: "Is it because they retard the important processes of catalysis that are constantly going on within the body?" Present-day scientists are inclined to answer in the affirmative.

There is a colloidal substance in the blood called enzyme. Enzyme is a potent catalyst and gives a wonderful exhibit of its ability in this respect when we place hydrogen peroxide upon an open wound in our flesh. A characteristic effervescence immediately takes place and the hydrogen peroxide is chemically decomposed into water and oxygen, by the wonderful influence of this enzyme in the blood. The enzyme is not changed in the least.

DETECTING THE PRESENCE OF CARBON DIOXIDE

The presence of carbon dioxide can easily be determined by allowing the gas to pass through a solution of lime water. When carbon dioxide passes through such a solution a white, curdy precipitate is produced which forms in clouds. This precipitate is calcium carbonate. A very simple experiment which will illustrate

79

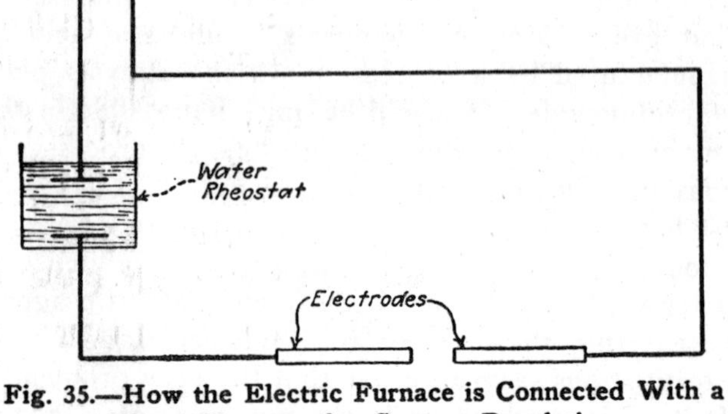

Fig. 35.—How the Electric Furnace is Connected With a Water Rheostat for Current Regulation.

this simple reaction between carbon dioxide and lime water can be made by bubbling our breath through the lime water placed in a test tube. It is a generally known fact that the exhaled breath from the human system consists of a very large percentage of carbon dioxide. The breath is blown through a small tube, the opposite end of which is immersed in the lime water.

PRODUCTION OF CALCIUM CARBONATE

Calcium carbonate can be formed in large quantities by setting up a carbon dioxide generator and allowing the gas generated to pass through a large receptacle

80

containing the solution of lime water. The setting up of a carbon dioxide generator is described in one of the later experiments in this book. Calcium carbonate produced by the reaction can easily be obtained in the solid state by evaporating the lime water solution to dryness.

ELECTROLYTIC PRODUCTION OF LEAD CARBONATE

Some very beautiful experiments in electrolysis can be produced with very little equipment in the experimental engineer's laboratory. The following lines describe the production of lead carbonate with simple apparatus. As a laboratory experiment in electrochemistry that will be of practical value as well as illustrate some of the most interesting actions of electrolysis, it is recommended to the experimenter.

Two lead electrodes, a glass vessel to hold them, and a carbon dioxide generator with a delivery tube is all the apparatus required for this beautiful experiment. The apparatus is set up as shown in the drawing, Fig. 36. A dilute solution of sulphuric acid acting on a few pieces of marble (calcium carbonate) in the gas generator will produce a sufficient quantity of carbon dioxide (CO_2) for the experiment. The electrolyte in the cell is a solution consisting of 15 grams of sodium chlorate and 5 grams of sodium carbonate. The current from a small storage battery will be sufficient to produce the electrolysis. After the electric circuit is closed, start the CO_2 generator, which delivers the gas to the negative electrode. As the CO_2 comes in contact with the cathode, a thick cloud of white lead or lead carbonate is produced and slowly precipitates to the bottom of the cell. While lead carbonate is the product obtained at the anode, sodium

81

carbonate is reproduced at the cathode. As a practical laboratory or lecture experiment in electrochemistry, this one is hard to beat. For best results, the electrolyte should be constantly agitated with a glass stirring rod while the experiment is being conducted.

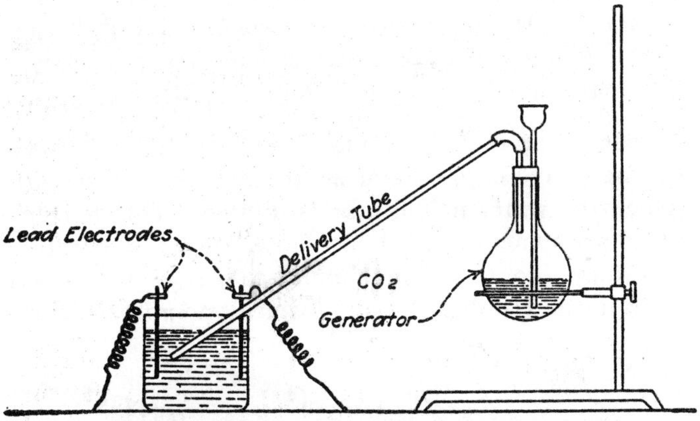

Fig. 36.—The Apparatus for the Experiment of Producing Lead Carbonate in the Laboratory.

PRODUCTION OF NITRIC ACID

The next experiment will be that of producing nitric acid. It will be understood that these experiments are outlined to give the amateur chemist a broader insight into general chemical processes rather than merely for the value of the chemicals produced. Nitric acid, of course, is very cheap and can be purchased in considerable quantities for a small amount. However, the following experiment will be interesting as well as valuable as a matter of experience.

A retort similar to that shown in Fig. 37 is used. This is an old-fashioned retort, but there will be many

82

cases where it can be employed successfully. First, we will place some solid sodium sulphate in the retort from the opening at the top. After this, sulphuric acid is poured into the retort over the sodium sulphate. A

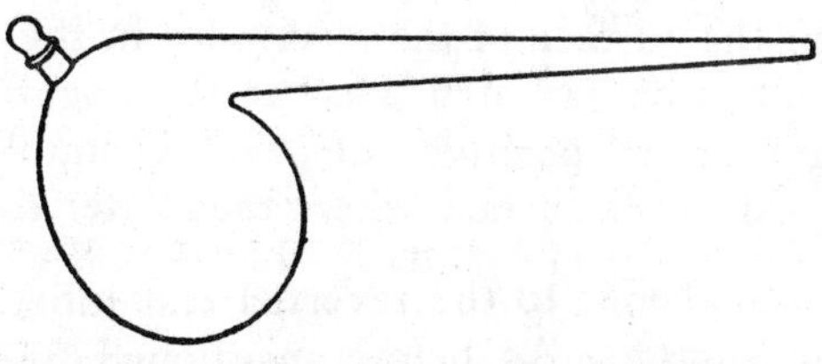

Fig. 37.—A Retort Which Finds Many Uses in the Chemical Laboratory.

test tube is placed over the free end of the retort as shown in Fig. 38 and the test tube is in turn immersed in a receptacle of cold water. With the general application of heat under the retort the following reaction will take place:

$$Na_2SO_4 + H_2SO_4 = HNO_3 + Na_2SO_4$$

The nitric acid passes off in a vaporous state and upon reaching the cool atmosphere of the test tube it condenses to a liquid form. It will be seen that the product of the reaction, aside from nitric acid, is sodium sulphate. This can be obtained in the solid state by evaporating the solution left in the retort to dryness.

ALLOTROPIC FORMS OF MATTER

At this point we will consider allotropic forms of matter. Carbon exists in three different states, i.e., amorphous carbon, graphite and diamond. In all of these three substances we have what the chemist recognizes as carbon with different physical proper-

83

ties. Thus, graphite is soft and unctious, while the diamond is the hardest substance known. Graphite and the diamond are the allotropic forms of carbon. There are other elements that have allotropic forms. Phosphorus is one of these.

COLLOIDAL STATE OF MATTER

The student of chemistry should not overlook that which is known as the colloidal state of matter. In the forerunning part of this book we learned something about solutions. Substances like sugar, salt, etc., are called crystalloids because they are able to pass into solution and may be again restored to their original condition by evaporating the liquid in which they are dissolved.

A colloid will not dissolve in this manner. In fact, colloids as they are generally understood, are not capable of passing into solution. The colloidal condition is a phenomena which usually associates itself with what are known as insoluble substances. If we would break the graphite tip off of our lead pencil and try to dissolve it in water, we would find that it absolutely refused to pass into solution, no matter how strongly we agitated it with every means at our disposal. If we place a little tannin in the water in which the tip of the graphite pencil was placed and removed the solution to a mortar, we would be able to make the graphite pass into a colloidal state by crushing it with a pestle. This would be a long and tedious process, however, as the graphite must be agitated considerably with the pestle before a colloidal preparation is formed. Fifteen minutes of agitation would be sufficient to bring about a "solution" of colloidal graphite. After fifteen minutes of work the dark appearing liquid in the mortar should be poured off and diluted

84

with about three times the amount of water. If the graphite passes into a colloidal condition, the resulting solution will be very dark and entirely opaque. We will find that this colloidal solution will be able to pass through filter paper as easily as a solution of sugar.

CRYSTALLOIDAL AND COLLOIDAL SOLUTION

At this point we must differentiate between a crystalloidal solution and a colloidal solution. Physicists believe that sugar, when dissolved in water, is reduced to a true molecular condition. This, however, is not true of colloids. The particles in the graphite solution which we just considered have been found

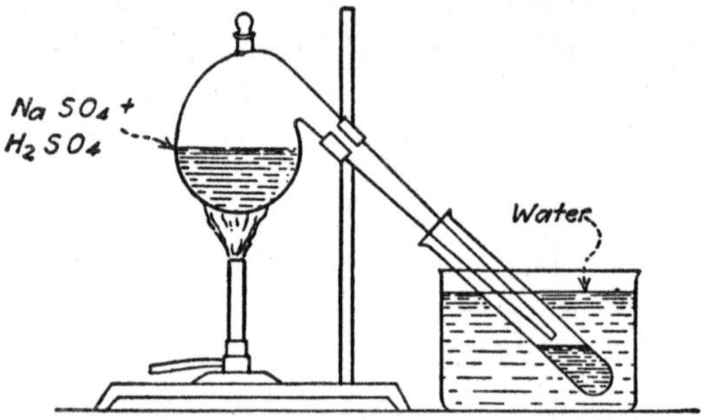

Fig. 38.—Producing Nitric Acid With Simple Apparatus as a Laboratory Experiment.

to measure about 1/250,000 of an inch. It has been further proven that these tiny particles are held in suspension by electrical charges. It is quite reasonable to suppose that we would be able to precipitate these particles by relieving them of their electrical charges.

85

In fact, this is very easily done by putting a few drops of acid or alkaline solution in the receptacle containing the colloidal particles of graphite in suspension. Upon the addition of the acid or alkaline solution the colloidal particles of graphite immediately form larger aggregations which precipitate to the bottom of the receptacle.

Colloidal solutions of most all the metals can be obtained by forming an arc under water using an electrode of the metal of which the colloidal preparation is desired. For instance, if we desire to prepare a colloidal solution of silver we would draw an arc between two silver wires under water. It will be found difficult to maintain this subaqueous arc for any length of time. However, every time the wires are touched a small cloud of colloidal silver will be produced in the water. The water used should be as pure as possible as the colloidal particles are extremely sensitive to foreign matter.

ELECTROLYSIS OF WATER

A beautiful experiment in electrolysis is that of decomposing water into its constituent elements, hydrogen and oxygen. There is probably no experiment in the realm of electrochemistry that will do more to benefit the young scientist than this. It is an experiment that can be very easily done with simple apparatus. The set-up for the apparatus is shown in Fig. 39. A U-tube is provided with two corks and two electrodes. The electrodes should be made up of small squares of sheet platinum soldered to wires that lead to the outside of the tube. However, if platinum is not available for this experiment, and it is not usually accessible to the amateur experimenter, carbon will make a very good substitute. Electrodes made of

86

copper, brass or iron would be readily attacked both by the active oxygen and by the sulphuric acid with which the water in the tube is diluted. The delivery tubes are generally made as shown in the illustration and they should not be immersed below the liquid level in the U-tube. An improvised pneumatic trough

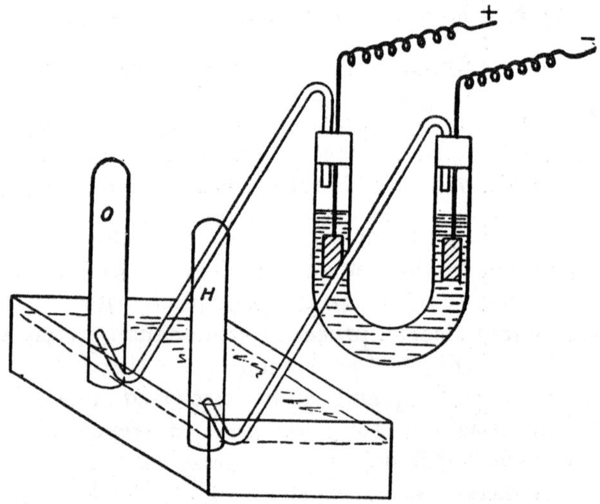

Fig. 39.—Apparatus for the Electrolysis of Water.

is used to collect the gas. Two inverted test tubes are placed over the ends of the delivery tubes as illustrated. The wires leading from the electrodes in the U-tube should be connected to a small storage battery or several dry cells. Upon the completion of the electric circuit a very energetic action will be noticed on the surface of the electrodes in the U-tube. Small bubbles of gas, hydrogen at the negative pole and oxygen at the positive pole, will accumulate and pass upward to the surface where they enter the deliv-

87

ery tube and thence to the inverted test tube and the pneumatic trough. As before mentioned, the water used in the U-tube should be made slightly acid by

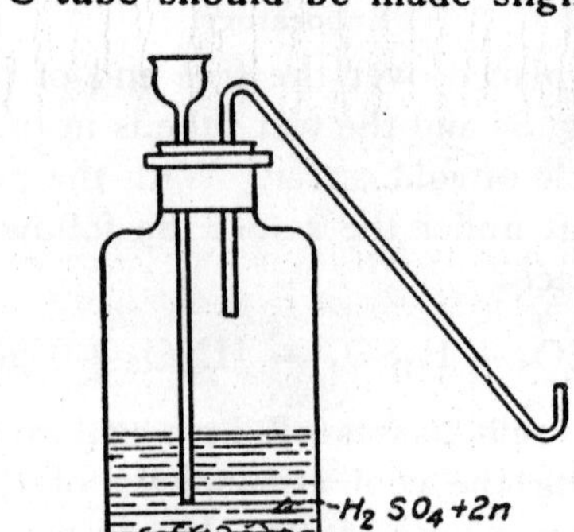

Fig. 40.—A Generator for Producing Hydrogen.

the addition of sulphuric acid. This increases the

conductivity of the water, but does not act in any other way.

GENERATION OF HYDROGEN IN QUANTITIES

Hydrogen can be produced more abundantly by allowing sulphuric acid to act upon one of the metals, preferably zinc. As mentioned in a previous part of this book, sulphuric acid, upon coming in contact with zinc, results in the production of zinc sulphide and hydrogen, according to the following formula:

$$H_2SO_4 + Zn = ZnSO_4 + 2H$$

The apparatus necessary for the production of hydrogen in quantities is shown in Fig. 40. It consists merely of a large mouthed bottle with a delivery tube and thistle tube placed in the cork.

88

Hydrogen can be collected by merely placing a test tube over the end of the delivery tube. It is understood that hydrogen is the lightest substance known

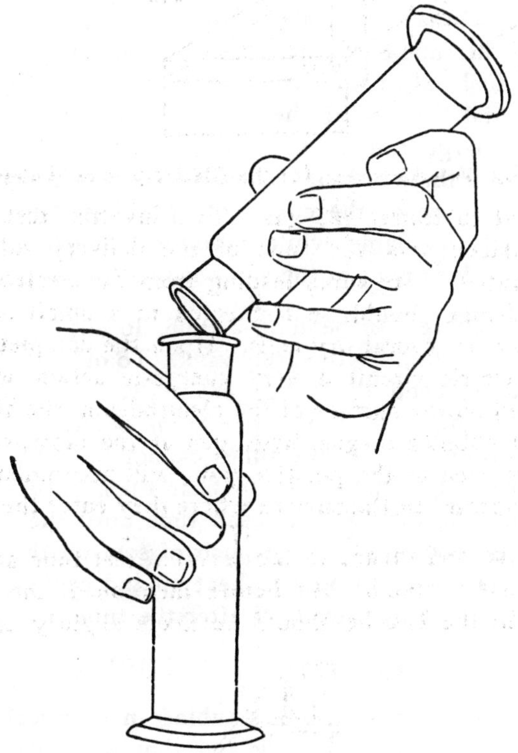

Fig. 41.—Hydrogen is Much Lighter Than Air and it Must Therefore be Poured *upward*.

and, therefore, it will have a tendency to rise in the tube and gradually displace the air. Hydrogen can be poured from one tube or bottle to another, as shown in Fig. 41. It will be seen that hydrogen is

89

not poured like water, but just the opposite. The bottle into which the hydrogen is being poured is inverted.

After the test tube has been held over the delivery tube of the hydrogen generator, a match is brought to the mouth of the tube. The hydrogen will explode with a sharp report. This is owing to the fact that air or oxygen is mixed with it as it comes from the generator. If the generator is used for some time this oxygen content will be reduced and almost pure hydrogen will be available at the delivery tube. When air or oxygen is mixed with hydrogen it forms a very explosive combination. The hydrogen oxidizes or burns and, of course, the product is water. After the generator is delivering pure hydrogen a match should be introduced into the test tube containing the gas. The match will be extinguished, but the hydrogen at the entrance of the tube will be ignited and burn quietly with an almost invisible flame.

GLASS WORKING FOR AMATEUR CHEMISTS

The amateur chemist must be able to work with glass. A knowledge of the manipulation of glass in the setting-up of various combinations of apparatus in the performance of certain experiments is quite necessary.

Glass is a very difficult substance with which to work and it requires considerable care, patience and experience to handle it successfully. The experimental engineer often finds it necessary to drill, etch, cut or cement glass and, unless he is provided with proper instructions, the result of his labor is very apt to be discouraging. The various operations of glass working are described in the following paragraphs.

90

CUTTING AND BREAKING

The best and most certain method of cutting glass plates is by means of the ordinary wheel cutter and straight edge. Very little pressure should be used and the glass should be broken gently after the scratch is made. The glass should be laid on a perfectly smooth surface while it is being scratched as it is very apt to break if there are any irregularities underneath it when the pressure is applied. If a glass cutter is not at hand, the sheets may be cut with a red hot iron in the following manner: File a notch in the edge of the glass, and, starting at the notch, draw the iron slowly over the surface along the line it is desired to cut. If the iron is moved with the proper speed, the glass will separate at the point of contact. Irregular pieces can be fashioned in this way. Another method, which is probably more curious than practical, employs a pair of heavy scissors. The sheet of glass is held under water in a horizontal position and cut in the usual way. Successful cutting in this manner depends somewhat upon the grade and nature of the glass being cut. The scissors should be moved very slowly and with care. Otherwise the glass will split in various directions.

It is a very difficult matter to cut off glass bottles with an ordinary cutter, and some other means must be used. One of the best and most simple methods of accomplishing this is to fill the bottle with kerosene to the height it is desired to cut it. A red hot iron is then plunged into the oil and the bottle will break evenly around the surface line of the oil, due to unequal expansion. Of all the methods of cutting off bottles, this is the most practical.

Another method of cutting a bottle or jar off with some chance of success employs a string soaked with

91

alcohol which is wound around the receptacle at a point where the cut is to be made. The string is ignited and allowed to burn out. Immediately after this, the bottle is plunged into cold water. It should then crack where the string was wound around.

DRILLING

The drilling of glass presents a very difficult problem. The softer grades may be drilled with ordinary metal drills lubricated with camphorated turpentine or a dilute solution of sulphuric acid. The drill should bear very lightly upon the glass and the lubricant should be applied very freely. A five per cent solution of commercial sulphuric acid will be found to give the best results.

The harder grades of glass may be drilled with metal drills that have been super-tempered by bringing them to a red heat and plunging them into a solution of salt water that has been previously boiled. A three-cornered file can be broken off and used to drill glass when it is placed in a chuck. The end of the file should be ground to a point on a grinding wheel.

Another rather uncertain method of making a hole in glass is to build a wall of putty where the hole is desired and fill the cup thus formed with alcohol and ignite it. After the alcohol is burned out, cold water should be splashed on the glass and the circle will drop out.

SMOOTHING ROUGH EDGES

Glass may be filed with an ordinary file using turpentine as a lubricant, but the process is slow and tedious. The only real practical method is to employ

92

a Carborundum wheel traveling at high speed. The wheel should be of very fine grit, otherwise it chips off small pieces of glass and produces a ragged edge. The proper speed of the wheel depends somewhat upon its size. Three to five-inch wheels should travel at least 4000 r.p.m. for satisfactory results. A very good bevel can be ground on glass plates in this manner by employing a wheel of extra fine grit. In the case of using Carborundum, the grit should be about FF. The beveled edge of the glass can be polished with fine abrasive powder.

The rough edges of thin glass can be easily smoothed by "fire polishing." The edge of the glass is held in the flame of a Bunsen burner until it starts to soften. This produces a surface like the original one. In doing this, remove the glass from the flame very gradually, as it will break if cooled too suddenly.

BENDING GLASS TUBING

This is a seemingly simple operation, yet it requires no little skill to do it properly. A "fishtail" gas burner should be used to obtain the best results, although an ordinary Bunsen burner can be used. If the latter, the air supply at the base should be shut off, in order to secure a luminous flame. Place the tube in the yellow part of the flame and revolve it with the fingers so that the heat will be properly distributed. When the tube becomes heated to that point where it starts to bend of its own weight, remove it from the flame and bend it at the desired radius. If a bend with a large radius is wished, the tube should be heated for three or four inches of its length (depending upon the size of the curve) by running it back and forth in the flame while revolving it between the fingers. It is bent in the same manner as a tube with

93

a small curve. In bending a tube, do not do so with a jerk as the tube will surely kink. To obtain a graceful curve, bend the tube gently yet with sufficient speed to accomplish the operation before the glass becomes chilled. Large tubes cannot be bent in this manner. It is necessary to heat them successively, bending the tube a little each time until the desired curve is produced. The deposit of soot on the tubes, which causes them both to heat up and cool off more quickly, is easily wiped off after the tube has cooled.

CUTTING GLASS TUBING

Methods of cutting glass tubing in chemical work must of necessity be quick and simple. Small tubes may be cut by filing a notch with a three-corner file and breaking the tube at the notch by placing the thumbs at each side of it and snapping the tube off with a quick movement. Larger tubes require a scratch to be filed completely around the outside before they will break. When breaking large tubes in this manner, they should be wrapped in a towel as the hands are very apt to be lacerated, if the tubes should happen to splinter. It will be found that the foregoing methods cannot be applied to tubing with a diameter over one inch. In cutting large tubes, two strips of wet blotting paper should be placed around the tube each side of the line that has been previously filed. The line is then followed around with either a small pointed flame or a red-hot iron.

DRAWING GLASS JETS

Glass jets are often used in chemical work and may be easily made as follows: Heat the tube in the same manner as when bending and, after taking it out of

the flame, pull it out gently until the diameter reaches about 1/8 inch. Then allow it to cool and cut it at the desired point. After it is cut, the ends may be "fire polished" in a hot flame.

SEALING A PLATINUM WIRE IN GLASS

The only difficult thing about this operation is obtaining the platinum. The tube or glass bulb is first drawn out to a jet and the wire is then inserted. The point of the jet is revolved in the flame until the glass melts. It is then allowed to cool very slowly. It is, of course, understood that wire made of other metals cannot be sealed into glass owing to the unequal co-efficient of expansion of the two materials.

JOINING TUBING AND MAKING TEES

The operation of making a "Tee" joint requires no little patience, and the experimenter should not become discouraged if his first attempt is unsuccessful. A seven or eight inch length of the tubing is first closed at one end by heating it in a flame, rotating it until the end is completely fused together. A pointed flame is then directed on the spot where it is desired to make the joint. When this spot becomes thoroughly heated, the open end of the tube is blown through, which forces the glass to part and make a hole where the wall was heated with the pointed flame. This hole should be of about the same diameter as the tube which is to form the other part of the tube which is to be joined to the hole. During this operation, the end of the first tube should be kept closed with the finger and, when both the tubes are hot, the end of the second tube is carefully brought in contact with the edges of the hole. After the tubes

become slightly welded together in this position, they are taken out of the flame. The end of the first tube is then blown into, after which the joint is again placed in the flame and heated until the glass is thoroughly fused together. The "Tee" is then blown into until the point of junction entirely disappears. The superfluous tubing is then cut off.

Tubes are joined together in much the same manner. If it is desired to join a large tube to a small one, the diameter of the large tube is first reduced by drawing the end into a jet. The ends of the tubes are then heated and joined. The junction is fashioned by blowing into the end of the tube.

SPREADING THE ENDS OF TUBING

Many times use is found for a small glass tube with a large opening in one end. The end of a small glass tube can be spread in the following manner. Take a piece of iron wire and smear it over with a thin film of vaseline at the end. Heat the end of the wire and the tube in the flame and when the glass becomes sufficiently soft, insert the hot iron wire in the end and rotate the tube slowly. Take care that the tube does not become too hot.

GLASS SPIRAL TUBES

Spirals of glass tubing are often used in chemical experiments. The operation in making these spirals is not so difficult as it may seem at first thought. It will be necessary to make a winding jig. In winding the hot tubing on the form or jig it will be necessary to have an assistant turn the windlass. In starting the spiral, the tube is first heated and bent at right angles. It is then fastened to the pipe by means of a few turns

of wire. The blow-pipe is then lighted and the flame directed on the glass tube. As the glass becomes heated to the proper point, the windlass is slowly turned. It should not be turned either too fast or too slow, but just at the right speed. If the tube is heated too much, it will flatten out as it is wound on the pipe. If it is not heated enough, it will crack. It is advisable to use tubing with a thick wall in making glass spirals, as it is not so liable to collapse.

INDEX

The Medical Student's Manual of CHEMISTRY

BY

R. A. WITTHAUS, A.M., M.D.

PROFESSOR OF CHEMISTRY AND PHYSICS IN THE UNIVERSITY OF THE CITY OF NEW YORK; PROFESSOR OF CHEMISTRY AND TOXICOLOGY IN THE UNIVERSITY OF BUFFALO; PROFESSOR OF CHEMISTRY AND TOXICOLOGY IN THE UNIVERSITY OF VERMONT; MEMBER OF THE CHEMICAL SOCIETIES OF PARIS AND BERLIN; MEMBER OF THE AMERICAN CHEMICAL SOCIETY; FELLOW OF THE AMERICAN ACADEMY OF MEDICINE, ETC.

INDEX PAGE 217

PREFACE TO THE PRESENT EDITION.

THE arrangement and classification adopted in the first edition have been continued, with the following modifications:

That portion treating of chemical physics has been contracted, until it contains nothing not absolutely essential to an understanding of what follows. This has been done in the belief that the sciences of chemistry and physics have each assumed a degree of individual importance in their applications to medical science that they should be treated of as distinct subjects.

The chemistry of the metals has been made to follow that of the nonmetals, in spite of the illogical character of such an arrangement, because it is believed that the student should be given as full a drilling in the simpler branches of the subject as possible before he is called upon to face the more complex chemistry of the carbon compounds.

The formulæ of the acids and salts have been changed from the continental method SO_4H_2, NO_3K, etc., adopted in the first edition, to that more generally followed in this country and in England, H_2SO_4, KNO_3, etc.; the latter method being more in consonançe with our system of nomenclature.

That portion of the work treating of the chemistry of the carbon compounds has been much extended, and in great part rewritten. The prominence given to this portion of the subject the author believes to be justified, notwithstanding its intricacy and the consequent difficulty of teaching it to medical students, by reason of the intimate connection of organic chemistry with physiology and pharmacy, and the rapidly increasing use of complex organic products as medicines.

R. A. W.

NEW YORK,
August 21, 1887.

PREFACE TO THE FIRST EDITION.

IN venturing to add another to the already long list of chemical text-books, the author trusts that he may find some apology in this, that the work is intended solely for the use of a class of students whose needs in the study of this science are peculiar.

While the main foundations of chemical science, the philosophy of chemistry, must be taught to and studied by all classes of students alike, the subsequent development of the study in its details must be moulded to suit the purposes to which the student will subsequently put his knowledge. And particularly in the case of medical students, in our present defective methods of medical teaching, should the subject be confined as closely as may be to the general truths of chemistry and its applications to medical science.

In the preparation of this Manual the author has striven to produce a work which should contain as much as possible of those portions of special chemistry which are of direct interest to the medical practitioner, and at the same time to exclude so far as possible, without detriment to a proper understanding of the subject, those portions which are of purely technological interest. The descriptions of processes of manufacture are therefor made very brief, while chemical physiology and the chemistry of hygiene, therapeutics, and toxicology

have been dwelt upon.

The work has been divided into three parts. In the first part the principles of chemical science are treated of, as well as so much of chemical physics as is absolutely requisite to a proper understanding of that which follows. A more extended study of physics is purposely avoided, that subject being, in the opinion of the author, rather within the domain of physiology than of chemistry.

The second part treats of special chemistry, and in this certain departures from the methods usually followed in chemical text-books are to be noted. The elements are classed, not in metals and metal-loids, a classification as arbitrary as unscientific, but into classes and groups according to their *chemical* characters.

In the text the formula of a substance is used in most instances in place of its name, *after it has been described*, with a view to giving the student that familiarity with the notation which can only be obtained by continued use.

As the distinction between inorganic and organic chemistry is merely one of convenience, the consideration of the carbon compounds is made to follow in its logical place after that of the element carbon.

In the third part those operations and manipulations which will be of utility to the student and physician are briefly described; not with the expectation that these directions can take the place of actual experience in the laboratory, but merely as an outline sketch in aid thereto.

Although the Manual puts forth no claim as a work upon analytical chemistry, we have endeavored to bring that branch of the subject rather into the foreground so far as it is applicable to medical chemistry. The qualitative characters of each element are given under the appropriate heading, and in the third part, systematic schemes for the examination of calculi and of simple chemical compounds are given. Quantitative methods of interest to the physician are also described in their appropriate places. In this connection the author would not be understood as saying that the methods recommended are in all instances the best known, but simply that they are the best adapted to the limited facilities of the physician.

The author would have preferred to omit all mention of Troy and Apothecaries' weight, but in deference to the opinions of those venerable practitioners who have survived their student days by a half century, those weights have been introduced in brackets after the metric, as the value of degrees Fahrenheit have been made to follow those Centigrade.

R. A. W.

BUFFALO, N, Y.,
September 16, 1883.

TABLE OF CONTENTS.

THE MEDICAL STUDENT'S

MANUAL OF CHEMISTRY

PART I.

INTRODUCTION.

The simplest definition of chemistry is a modification of that given by Webster: *That branch of science which treats of the composition of substances, their changes in composition, and the laws governing such changes.*

If a bar of soft iron be heated sufficiently it becomes luminous; if caused to vibrate it emits sound; if introduced within a coil of wire through which a galvanic current is passing, it becomes magnetic and attracts other iron brought near it. Under all these circumstances the iron is still iron, and so soon as the heat, vibration, or galvanic current ceases, it will be found with its original characters unchanged; it has suffered no change in *composition.* If now the iron be heated in an atmosphere of oxygen gas it burns and is converted into a substance which, although it contains iron, has neither the appearance nor the properties of that metal. The iron and a part of the oxygen have disappeared and have been converted into a new substance, differing from either; there has been change in *composition,* there has been *chemical action.* Changes wrought in matter by physical forces, such as light, heat and electricity are temporary, and last only so long as the force is in activity; except in the case of changes in the state of aggregation, as when a substance is pulverized or fashioned into given shape. Changes in chemical composition are permanent, lasting until some other change is brought about by another manifestation of chemical action.

However distinct chemical may thus be from physical forces, it is none the less united with them in that grand correlation whose existence was first announced by Grove, in 1842. As, from chemical action, manifestations of every variety of physical force may be obtained: light, heat, and mechanical force from the oxidation of carbon; and electrical force from the action of zinc upon sulphuric acid—so does chemical action have its origin, in many instances, in the physical forces. Luminous rays bring about the chemical decomposition of the salts of silver, and the chemical union of chlorine and hydrogen; by electrical action a decomposition of many compounds into their constituents is instituted, while instances are

2

abundant of reactions, combinations, and decompositions which require a certain elevation of temperature for their production. While, therefor, chemistry in the strictest sense of the term, deals only with those actions which are attended by a change of composition in the material acted upon, yet chemical actions are so frequently, nay universally, affected by existing physical conditions, that the chemist is obliged to give his attention to the science of physics, in so far, at least, as it has a bearing upon chemical reactions, to *chemical physics*—a branch of the subject which has afforded very important evidence in support of theoretical views originating from purely chemical reactions.

General Properties of Matter.

Indestructibility.—The result of chemical action is change in the composition of the substance acted upon, a change accompanied by corresponding alterations in its properties. Although we may cause matter to assume a variety of different forms and render it, for the time being, invisible, yet in none of these changes is there the smallest particle of matter destroyed. When carbon is burned in an atmosphere of oxygen, it disappears, and, so far as we can learn by the senses of sight or touch, is lost; but the result of the burning is an invisible gas, whose weight is equal to that of the carbon which has disappeared, plus the weight of the oxygen required to burn it.

Weight.—All bodies attract each other with a force which is in direct proportion to the amount of matter which they contain. The force of this attraction exerted upon surrounding bodies by the earth becomes sensible as weight, when the motion of the attracted body toward the centre of gravity of the earth is prevented.

In chemical operations we have to deal with three kinds of weight: *absolute, apparent,* and *specific.*

The Absolute Weight of a body is its weight in vacuo. It is determined by placing the entire weighing apparatus under the receiver of an air-pump.

The Apparent Weight, or Relative Weight, of a body is that which we usually determine with our balances, and is, if the volume of the body weighed be greater than that of the counterpoising weights, less than its true weight. Every substance in a liquid or gaseous medium suffers a loss of apparent weight equal to that of the volume of the medium so displaced. For this reason the apparent weight of some substances may be a minus quantity; thus, if the air contained in a vessel suspended from one arm of a poised balance be replaced by hydrogen, that arm of the balance to which the vessel is attached will rise, indicating a diminution in weight. (See Weighing; Part III.)

The Specific Weight or Specific Gravity of a substance is the weight of a given volume of that substance, as compared with the weight of an equal bulk of some substance, accepted as a standard of comparison, under like conditions of temperature and pressure. The sp. gr. of solids and liquids are referred to water; those of gases to air or to hydrogen. Thus the sp. gr. of sulphuric acid being 1.8, it is, volume for volume, one and eight-tenths times as heavy as water. As, by reason of their different rates of expansion by heat, solids and liquids do not have the same sp. gr. at all temperatures, that at which the observation is made should always be noted, or some standard temperature adopted. The standard tempera-

3

ture adopted by some continental writers and in the U. S. P. is 15° (59° F.); other standard temperatures are 4° (39.2° F.), the point of greatest density of water, used by most continental writers, and 15.6° (60° F.), used in Great Britain and to some extent in this country.

The determination of the specific weight of a substance is frequently of great service. Sometimes it affords a rapid means of distinguishing between two substances similar in appearance; sometimes in determining the quantity of an ingredient in a mixture of two liquids, as alcohol and water; and frequently in determining approximately the quantity of solid matter in solution in a liquid. It is the last object which we have in view in determining the sp. gr. of the urine.

An aqueous solution of a solid has a higher sp. gr. than pure water, the increase in sp. gr. following a regular but different rate of increase with each solid. In a simple solution—one of common salt in water, for instance—the proportion of solid in solution can be determined from the sp. gr. In complex solutions, such as the urine, the sp. gr. does not indicate the proportion of solid in solution with accuracy. In the absence of sugar and albumen, a determination of the sp. gr. of urine affords an indication of the amount of solids sufficiently accurate for usual clinical purposes. Moreover, as urea is much in excess over other urinary solids, the oscillations in the sp. gr. of the urine, if the quantity passed in twenty-four hours be considered, and in the absence of albumen and sugar, indicate the variations in the elimination of urea, and consequently the activity of disassimilation of nitrogenous material.

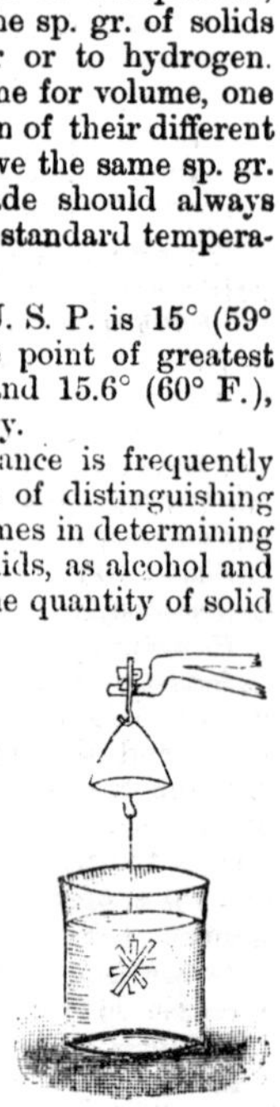

Fig. 1.

To determine the sp. gr. of substances, different methods are adopted, according as the substance is in the solid, liquid, or gaseous state; is in mass or in powder; or is soluble or insoluble in water.

Solids.—*The substance is heavier than water, insoluble in that liquid, and not in powder.*—It is attached by a fine silk fibre or platinum wire to a hook arranged on one arm of the balance, and weighed. A beaker full of pure water is then so placed that the body is immersed in it (Fig. 1.), and a second weighing made. By dividing the weight in air by the loss in water, the sp. gr. (water = 1.00) is obtained. Example:

A piece of lead weighs in air	82.0
A piece of lead weighs in water	74.9
Loss in water	7.1

$$\frac{82.0}{7.1} = 11.55 = \text{sp. gr. of lead.}$$

The substance is in powder, insoluble in water.—The specific gravity bottle (Fig. 2) filled with water, and the powder previously weighed and in a separate vessel, are weighed together. The water is poured out of the bottle, into which the powder is introduced with enough water to fill the bottle completely: the weight of the bottle and its contents is now determined. The weight of the powder alone, divided by the loss between the first and second weighings, is the specific gravity. Example:

Weight of iron filings used	6.562
Weight of iron filings and sp. gr. bottle filled with water	148.327
Weight of sp. gr. bottle containing iron filings and filled with water	147.470
Water displaced by iron	0.857

$$\frac{6.562}{0.857} = 7.65 = \text{sp. gr. of iron.}$$

4

The substance is lighter than water.—A sufficient bulk of some heavy substance, whose sp. gr. is known, is attached to it and the same method followed, the loss of weight of the heavy substance being subtracted from the total loss. Example:

A fragment of wo d weighs	4.?946
A fragment of lea weighs	10.6193
Wood with lead attached weighs	15.0139
Wood with lead attached weighs in water	5.9295
Loss of weight of combination	9.0844
Loss of weight of lead in water	0.7903
Loss of weight of wood	8.2941

$$\frac{4.3946}{8.2941} = 0.529 = \text{sp. gr. of wood.}$$

The substance is soluble in or decomposable by water.—Its specific gravity, referred to some liquid not capable of acting on it, is determined, using that liquid as water is used in the case of insoluble substances. The sp. gr. so obtained, multiplied by that of the liquid used, is the sp. gr. sought. Example:

A piece of potassium weighs	2.576
A sp. gr. bottle full of naphtha, sp. gr. 0.758, weighs	22.784
	25.360
The bottle with potassium and naphtha weighs	23.103
Loss	2.257

$$\frac{2.576}{2.257} = 1.141 \times 0.758 = 0.865 = \text{sp. gr. of potassium.}$$

LIQUIDS.—The sp. gr. of liquids is determined by the specific gravity bottle, sometimes called *picnometer*, or by the *spindle* or *hydrometer*.

By the bottle.—This method is the more accurate, and, if a balance be at hand, is easily conducted. A bottle of thin glass (Fig. 2) is so made as to contain a given volume of water, say 100 c.c., at 15° C., and its weight is determined once for all. To use the picnometer, it is filled with the liquid to be examined and weighed. The weight obtained, minus that of the bottle, is the sp. gr. sought if the bottle contain 1000 c.c.; $\frac{1}{10}$ if 100 c.c., etc. Example: Having a bottle whose weight is 35.35, and which contains 100 c.c.; filled with urine it weighs 137.91, the sp. gr. of the urine is 137.91−35.35=102.56×10=1025.6——Water=1000.

By the spindle.—The method by the hydrometer is based upon the fact that a solid will sink in a liquid whose sp. gr. is greater than its own, until it has displaced a volume of the liquid whose weight is equal to its own; and all forms of hydrometers are simply contrivances to measure the volume of liquid which they displace when immersed. The hydrometer most used by physicians is the urinometer (Fig. 3); it should not be chosen too small, as the larger the bulb, and the thinner and longer the stem, the more accurate are its indications. The most convenient method of using the instrument is as follows: The cylinder, which should have a foot and rim, but no pouring lip, is filled to within an inch of the top; the spindle is then floated and the cylinder completely filled with the liquid under examination (Fig. 3). The reading is then taken at the highest point *a*, where the surface of the liquid comes in contact with the spindle.*

* The advantages of the method described over that usually followed are: Greater facility in reading, less liability to error, the possibility of taking the reading in opaque liquids, and the fact that readings are made upward, not downward. The spindles require to be specially graduated, and are made by Baudin, of Paris, and Eimer & Amend, of New York.

5

In all determinations of sp. gr. the liquid examined should have the temperature for which the instrument is graduated, as all liquids expand with heat and contract when cooled, and consequently the result obtained will be too low if the urine or other liquid be at a temperature above that at which the instrument is intended to be used, and too high if below that temperature. An accurate correction may be made for temperature in simple solutions; in a complex fluid like the urine, however, this can only be done roughly by allowing 1° of sp. gr. for each 3° C. (5.4° Fahr.) of variation in temperature.

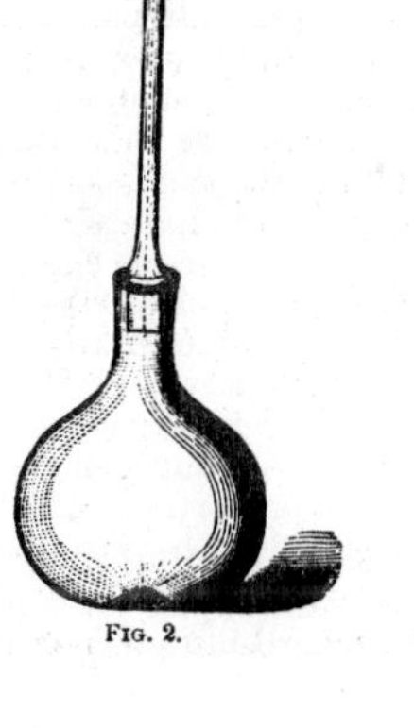

FIG. 2.

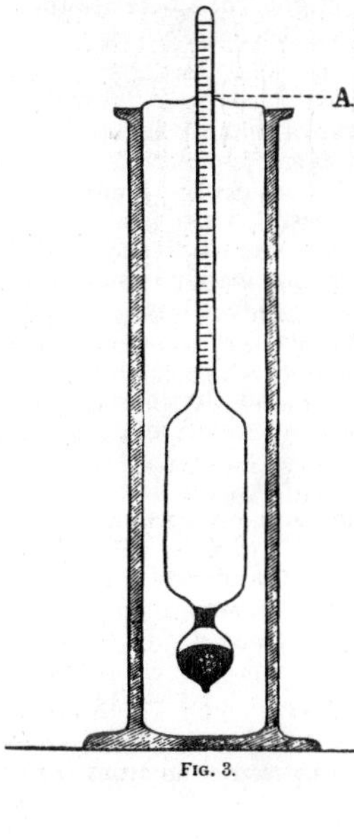

FIG. 3.

GASES AND VAPORS.—The specific gravities of gases and vapors are of great importance in theoretical chemistry, as from them we can determine molecular weights, in obedience to the law of Avogadro (p. 14).

Gases.—The specific gravities of gases are obtained as follows: A glass flask of about 300 c.c. capacity, having a neck 20 centimetres long and 6 millimetres in diameter, and fitted with a glass stopcock, is filled with mercury; reversed over mercury; and filled with the gas to just below the stopcock. The stopcock is now closed; the temperature, *t*; the barometric pressure, H; and the height of the mercurial column in the neck above that in the trough, *h*, are determined, and the flask weighed. Let P be the weight found, and V the capacity of the flask, determined once for all, then

$$\frac{V(H-h)}{760\,(1+0.00366\,t)} = V_0 = \text{the volume of the gas at } 0° \text{ and } 760 \text{ mm.}$$

The flask is then brought under the receiver of an air-pump, the glass stopcock being open, and the air alternately exhausted and allowed to enter until the gas in the flask is replaced by air. The temperature *t′*, the barometric pressure H′, and the weight of the flask filled with air P′, are now determined. From these results the weight, K, of the gas occupying the volume V_0 is obtained by the formula:

$$K = P - P' + \frac{V\,H'}{760\,(1+0.00366\,t')} \times 0.001293$$

The sp. gr. referred to air is found by the formula:

$$\frac{K}{V_0 \times 0.001293}$$

and that referred to hydrogen by the formula:

$$\frac{K}{V_0 \times 0.001293 \times 0.06927}$$

6

Vapors.—The specific gravity of vapors is best determined by Meyer's method, as follows: A small, light glass vessel (Fig. 4) is filled completely with the solid or liquid whose vapor density is to be determined, and weighed; from this weight that of the vessel is subtracted; the difference being the weight of the substance P. The small vessel and contents are now introduced into the large branch of the apparatus (Fig. 5), whose weight is then determined. The apparatus is now filled with mercury, the capillary opening at the top of the larger branch is closed by the blow-pipe, and the whole again weighed. The apparatus is suspended by a metallic wire near the bottom of a long tube closed at the bottom, and containing about 50 c.c. of some liquid whose boiling-point is constant and higher than that of the substance experimented on. When the liquid has been heated to active boiling, and the mercury ceases to escape from the small tube, the barometric pressure and the temperature of the air are observed. After the apparatus is cooled, the tube (Fig. 5), with its contents, is weighed, and the difference in the level of mercury which existed in the two branches during the heating determined by breaking the capillary point, tilting the apparatus until the smaller branch is completely filled, marking the level of mercury in the larger branch, and afterward measuring the distance from that point to the opening.

FIG. 4.

By the above process the following factors are determined:

P = weight of substance;
T = boiling-point of external liquid;
t = temperature of air;
H = barometric pressure reduced to 0°;
h = difference in level of mercury in two branches of tube;
h′ = tension of vapor of mercury at T;
a = weight of mercury used;
q = weight of mercury required to fill the tube Fig. 4;
r = weight of mercury remaining in the apparatus after heating.

From these the specific gravity, air = 1, is obtained by the equation:

$$D = \frac{P\,760\,(1+0.00367\,T)\,13.59}{(H+h+h')\,0.0012932\,[(a+q)\{1+0.0000303\,(T-t)\}-r\{1+0.00018\,(T-t)\}][1+0.00018\,t]}$$

The sp. gr. in terms of air = 1 may be reduced to sp. gr. referred to hydrogen = 2, by dividing by 0.06927.

States of Matter.—Matter exists in one of three states; solid, liquid, and gaseous. In the solid form the particles of matter are comparatively close together, and are separated with more difficulty than are those of liquid or gaseous matter; or, in other words, the cohesion of solid matter is greater than that of the other two forms. In the liquid the particles are less firmly bound together and are capable of freer motion about one another. In the gas the mutual attraction of the particles disappears entirely, and their distance from each other depends upon the pressure to which the gas is subjected.

The term *fluid* applies to both liquids and gases, the former being designated as *incompressible*, from the very slight degree to which their volume can be reduced by pressure. The gases are designated as *compressible fluids*, from the fact that their volume can be reduced by pressure to an extent limited only by their passage into the liquid form.

It is highly probable that all substances, which are not decomposed when heated, are capable of existing in the three forms of solid, liquid, and gas. There are, however, some substances which are only known in two forms—as alcohol; or in a single form—as carbon; probably because we are as yet unable to produce artificially a temperature sufficiently low to solidify the one, or sufficiently high to liquefy or volatilize the other. Since the liquefaction of the so-called permanent gases the distinction between gases and *vapors* is only one of degree and of convenience.

The passage of a substance from one form to another is always attended by the absorption or liberation of a definite amount of heat. In passing from the solid to the gaseous form, a body absorbs a definite amount of heat with each change of form. If a given quantity of ice at a temperature below the freezing-point of water be heated, its temperature gradually rises until the thermometer marks 0° C., at which point it remains stationary until

FIG. 5.

7

the last particle of ice has disappeared. At that time another rise of the thermometer begins, and continues until 100° C. is reached (at 760 mm. of barometric pressure), when the water boils, and the thermometer remains stationary until the last particle of water has been converted into steam ; after which, if the application of heat be continued, the thermometer again rises. During these two periods of stationary thermometer, heat is taken up by the substance, but is not indicated by the thermometer or by the sense. Not being sensible, it is said to be *latent*, a term which is liable to mislead, as conveying the idea that heat is stored up in the substance *as heat ;* such is not the case. During the period of stationary thermometer the heat is not sensible as heat, for the reason that it is being used up in the *work* required to effect that separation of the particles of matter which constitutes its passage from solid to liquid or from liquid to gas.

The amount of heat required to bring about the passage of a given weight of a given substance from the denser to the rarer form is always the same, and the *temperature* indicated by the thermometer during this passage is always the same for that substance, unless in either case a modification be caused by a variation in pressure. The degree of temperature indicated by the thermometer while a substance is passing from the solid to the liquid state is called its *fusing-point ;* that indicated during its passage from the liquid to the gaseous form, its *boiling-point.*

The absorption of heat by a volatilizing liquid is utilized in the arts and in medicine for the production of cold (which is simply the absence of heat), in the manufacture of artificial ice, and in the production of local anæsthesia by the ether-spray. The removal of heat from the body in this way, by the evaporation of perspiration from the surface, is an important factor in the maintenance of the body temperature at a point consistent with life.

When a substance passes from a rarer to a denser form it gives out—liberates—an amount of heat equal to that which it absorbed in its passage in the opposite direction. It is for this reason that, while we apply heat to convert a liquid into a vapor, we apply cold to reduce a gas to a liquid. As a rule, the thermometrical indication is the same in whichever direction the change of form occurs ; some substances, however, solidify at a temperature slightly different from that at which they fuse.

Most solids, when heated, are first converted into liquids, and these into gases ; there are, however, some exceptions to this rule. Most vapors when condensed pass into the liquid form, and this in turn into the solid ; some substances, however, are condensed from the form of vapor directly to that of solid, in which case they are said to *sublime.*

Divisibility.—All substances are capable of being separated, with greater or less facility, by mechanical means into minute particles. With suitable apparatus, gold may be divided into fragments, visible by the aid of the microscope, whose weight would be $\frac{1}{5000000000000}$ of a grain ; and it is probable that when a solid is dissolved in a liquid a still greater subdivision is attained.

Although we have no direct experimental evidence of the existence of a limit to this divisibility, we are warranted in believing that matter is not infinitely divisible. A strong argument in favor of this view being that, after physical subdivision has reached the limit of its power with regard to compound substances, these may be further divided into dissimilar bodies by chemical means.

The limit of mechanical subdivision is the *molecule* of the physicist, the smallest quantity of matter with which he has to deal.

8

Elements.

If we examine the various substances existing upon and in our earth, we find that many of them can be so decomposed as to yield two or more other substances, distinct in their properties from the substance from whose decomposition they resulted, and from each other. If, for example, sugar be treated with sulphuric acid it blackens, and a mass of charcoal separates. Upon further examination we find that water has also been produced. From this water we may obtain two gases, differing from each other widely in their properties. Sugar is therefor made up of carbon and the two gases, hydrogen and oxygen ; but it has the properties of sugar, and not those of either of its constituent parts. There is no method known by which carbon, hydrogen, and oxygen can be split up, as sugar is, into other dissimilar substances.

An element or simple substance is a substance which cannot by any known means be split up into other dissimilar bodies.

The number of well-characterized elements at present known is sixty-six. During a few years past the discovery of other elements not included in the above number, *decipium, philippium, davyium, norwegium,* and *neptunium,* has been announced.

Laws Governing the Combination of Elements.

The alchemists, Arabian and European, contented themselves in accumulating a store of knowledge of isolated phenomena, without, as far as we know, attemping, in any serious way, to group them in such a manner as to learn the laws governing their occurrence. It was not until the latter part of the last century, 1777, that Wenzel, of Dresden, implied, if he did not distinctly enunciate, what is known as the law of reciprocal proportions. A few years later, Richter, of Berlin, confirming the work of Wenzel, added to it the law of definite proportions, usually called Dalton's first law. Finally, as the result of his investigations from 1804 to 1808, Dalton added the law of multiple proportions, and, reviewing the work of his predecessors, enunciated the results clearly and distinctly.

Considering these laws, not in the order of their discovery, but in that of their natural sequence, we have :

THE LAW OF DEFINITE PROPORTIONS.—*The relative weights of elementary substances in a compound are definite and invariable.* If, for example, we analyze water, we find that it is composed of eight parts by weight of oxygen for each part by weight of hydrogen, and that this proportion exists in every instance, whatever the source of the water. If, instead of decomposing, or *analyzing* water, we start from its elements, and by *synthesis,* cause them to unite to form water, we find that, if the mixture be made in the proportion of eight oxygen to one hydrogen by weight, the entire quantity of each gas will be consumed in the formation of water. But if an excess of either have been added to the mixture, that excess will remain after the combination.

Compounds are substances made up of two or more elements united with each other in definite proportions. Compounds exhibit properties of their own, which differ from those of the constituent elements to such a degree that the properties of a compound can never be deduced from a knowledge of those of the constituent elements. Common salt, for instance, is com-

9

posed of 39.32 per cent. of the light, bluish-white metal, sodium, and 60.68 per cent of the greenish-yellow, suffocating gas, chlorine.

A mixture is composed of two or more substances, elements or compounds, mingled in any proportion. The characters of a mixture may be predicated from a knowledge of the properties of its constituents. Thus sugar and water may be mixed in any proportion and the mixture will have the sweetness of the sugar, and will be liquid or solid according as the liquid or solid ingredient predominates in quantity.

THE LAW OF MULTIPLE PROPORTIONS.—*When two elements unite with each other to form more than one compound, the resulting compounds contain simple multiple proportions of one element as compared with a constant quantity of the other.*

Oxygen and nitrogen, for example, unite with each other to form no less than five compounds. Upon analysis we find that in these the two elements bear to each other the following relations by weight :

In the first, 14 parts of nitrogen to 8 of oxygen.
In the second, 14 parts of nitrogen to $8 \times 2 = 16$ of oxygen.
In the third, 14 parts of nitrogen to $8 \times 3 = 24$ of oxygen.
In the fourth, 14 parts of nitrogen to $8 \times 4 = 32$ of oxygen.
In the fifth, 14 parts of nitrogen to $8 \times 5 = 40$ of oxygen.

THE LAW OF RECIPROCAL PROPORTIONS.—*The ponderable quantities in which substances unite with the same substance express the relation, or a simple multiple thereof, in which they unite with each other.* Or, as Wenzel stated it, "the weights b, b', b'' of several bases which neutralize the same weight a of an acid are the same which will neutralize a constant weight a' of another acid ; and the weights a, a', a'' of different acids which neutralize the same weight b of a base are the same which will neutralize a constant weight of another base b'."

The Atomic Theory.

The laws of Wenzel, Richter, and Dalton, given above, are simply generalized statements of certain groups of facts, and, as such, not only admit of no doubt, but are the foundations upon which chemistry as an exact science is based. Dalton, seeking an explanation of the reason of being of these facts, was led to adopt the view, held by the Greek philosopher Democritus, that matter was not infinitely divisible. He retained the name *atom* (ἄτομος = indivisible), given by Democritus to the ultimate particles of which matter was supposed by him to be composed ; but rendered the idea more precise by ascribing to these atoms real magnitude and a definite weight, and by considering elementary substances as made up of atoms of the same kind, and compounds as consisting of atoms of different kinds.

This hypothesis, the first step toward the atomic theory as entertained to-day, afforded a clear explanation of the numerical results stated in the three laws. If hydrogen and oxygen always unite together in the proportion of one of the former to eight of the latter, it is because, said Dalton, the compound consists of an atom of hydrogen, weighing 1, and an atom of oxygen, weighing 8. If, again, in the compounds of nitrogen and oxygen, we have the two elements uniting in the proportions 14 : 8——14 : 8×2——14 : 8×3——14 : 8×4——14 : 8×5, it is because they are

10

severally composed of an atom of nitrogen weighing 14, united to 1, 2, 3,

4, or 5 atoms of oxygen, each weighing 8. Further, that compounds do not exist in which any fraction of 8 oxygen enters, because 8 is the weight of the indivisible atom of oxygen.

One of the chief advantages of Dalton's hypothesis is in the introduction of this precise and simple relation between the quantities of the constituents of a compound. Chemists before Dalton's day, in expressing the results of their analyses, did not progress beyond statements of the percentage composition. Expressing the composition of four of the carbon compounds in percentages, we have:

	Carbon.	Hydrogen.	Oxygen.	
Marsh gas.............	75.0	25.0	....	=100
Olefiant gas	85.7	14.3	....	=100
Carbonic oxide	42.9	....	57.1	=100
Carbonic acid...........	27.3	....	72.7	=100

These figures convey nothing beyond the mere centesimal composition of the substances which they express. The cardinal point of Dalton's discovery lies in his translation of them into the simple relations:

	Carbon.	Hydrogen.	Oxygen.
Marsh gas........................	6	2	..
Olefiant gas......................	6	1	..
Carbonic oxide...................	6	..	8
Carbonic acid.....................	6	..	16

Dalton's hypothesis of the existence of atoms as definite quantities did not, however, meet with general acceptance. Davy, Wollaston, and others considered the quantities in which Dalton had found the elements to unite with each other, as mere *proportional numbers* or *equivalents*, as they expressed it, nor is it probable that Dalton's views would have received any further recognition until such time as they might have been exhumed from some musty tome, had their publication not been closely followed by that of the results of the labors of Humboldt and of Gay Lussac, concerning the *volumes* in which gases unite with each other.

In the form of what are known as Gay Lussac's laws, these results are:

First.—There exists a simple relation between the volumes of gases which combine with each other.

Second.—There exists a simple relation between the sum of the volumes of the constituent gases, and the volume of the gas formed by their union. For example:

1 volume chlorine unites with 1 volume hydrogen to form 2 volumes hydrochloric acid.
1 volume oxygen unites with 2 volumes hydrogen to form 2 volumes vapor of water.
1 volume nitrogen unites with 3 volumes hydrogen to form 2 volumes ammonia.
1 volume oxygen unites with 1 volume nitrogen to form 2 volumes nitric oxide.
1 volume oxygen unites with 2 volumes nitrogen to form 2 volumes nitrous oxide.

Berzelius, basing his views upon these results of Gay Lussac, modified the hypothesis of Dalton and established a distinction between the *equivalents* and *atoms*. The composition of water he expressed, in the notation which he was then introducing, as being H_2O, and not HO as Dalton's hypothesis called for. As, however, Berzelius still considered the atom of oxygen as weighing 8, he was obliged also to consider the atoms of hydrogen and of certain other elements as double atoms—a fatal defect in his system, which led to its overthrow and the re-establishment of the formula HO for water.

11

It was reserved to Gerhardt to clearly establish the distinction between atom and molecule; to observe the bearing of the discoveries of Avogadro and Ampère upon chemical philosophy; and thus to establish the atomic theory as entertained at present.

As a result of his investigations in the domain of organic chemistry, Gerhardt found that, if Dalton's equivalents be adhered to, whenever carbonic acid or water is liberated by the decomposition of an organic substance, it is invariably in double equivalents, never in single ones; always $2CO_2$ or 2HO or some multiple thereof, never CO_2 or HO. He further found that if the equivalents C=6, H=1, and O=8 be retained, the formulæ became such that the equivalents of carbon are always divisible by two. In fact, he found the same objections to apply to the notation then in use that had been urged against that of Berzelius.

In 1811, Avogadro, from purely physical researches, had been enabled to state the law which is now known by his name, to the effect that *equal volumes of all gases, under like conditions of temperature and pressure, contain equal numbers of molecules.*

In the hands of Gerhardt this law, in connection with those of Gay Lussac, became the foundation of what is sometimes called the "new chemistry." Bearing in mind Avogadro's law, we may translate the first three combinations given in the table on p. 10 into the following:

1 molecule chlorine unites with 1 molecule hydrogen to form 2 molecules hydrochloric acid.
1 molecule oxygen unites with 2 molecules hydrogen to form 2 molecules vapor of water.
1 molecule nitrogen unites with 3 molecules hydrogen to form 2 molecules ammonia.

But the ponderable quantities in which these combinations take place are:

35.5 chlorine to............................ 1 hydrogen.
16 oxygen to................................ 2 hydrogen.
14 nitrogen to.............................. 3 hydrogen.

And as single molecules of hydrogen, oxygen, and nitrogen are in these combinations subdivided to form 2 molecules of hydrochloric acid, water, and ammonia, it follows that these molecules must each contain two equal quantities of hydrogen, oxygen, and nitrogen, less in size than the molecules themselves. And, further, as in these instances each molecule contains two of these smaller quantities, or *atoms*, the relation between the weights of the molecules must be also the relation between the weights of the atoms, and we may therefor express the combinations thus:

1 atom chlorine weighing 35.5 unites with 1 atom hydrogen weighing 1;
1 atom oxygen weighing 16 unites with 2 atoms hydrogen weighing 2;
1 atom nitrogen weighing 14 unites with 3 atoms hydrogen weighing 3;

and consequently, if the atom of hydrogen weighs 1, that of chlorine weighs 35.5, that of oxygen 16, and that of nitrogen 14.

Atomic and Molecular Weights.

Atomic Weight.—The distinction between molecules and atoms may be expressed by the following definitions:

A molecule is the smallest quantity of any substance that can exist in the free state.

12

An atom is the smallest quantity of an elementary substance that can enter into a chemical reaction.

The molecule is always made up of atoms, upon whose nature, number, and arrangement with regard to each other, the properties of the substance depend. In an elementary substance the atoms composing the molecules are the same in kind, and usually two in number. In compound substances they are dissimilar and vary in quantity from two in a simple compound, like hydrochloric acid, to hundreds or thousands in more complex substances. *The word atom can only be used in speaking of an elementary body, and that only while it is passing through a reaction. The term molecule applies indifferently to elements and compounds.*

The atoms have definite relative weights; and upon an exact determination of these weights depends the entire science of quantitative analytical chemistry. They have been determined by repeated and careful analyses of perfectly pure compounds of the elements, and *express the weight of one atom of the element as compared with the weight of one atom of hydrogen, that being the lightest element known.* It is also the weight of a volume of the element, in the form of gas, which would occupy the same volume, under like pressure and temperature, as an amount of hydrogen weighing one. What the *absolute weight* of an atom of any element may be we do not know, nor would the knowledge be of any service did we possess it.

The following table contains a list of the elements at present known, with their atomic weights:

ELEMENTS.

NAME.	A. Symbol.	B. Atomic weight.	NAME.	A. Symbol.	B. Atomic weight.
Aluminium.........	Al.	27.02	Hydrogen	H.	1
Antimony	Sb.	120	Indium	In.	113.4
Arsenic	As.	74.9	Iodine	I.	126.85
Barium	Ba.	136.8	Iridium	Ir.	192.7
Bismuth	Bi.	206.5	Iron	Fe.	55.9
Boron..............	Bo.	11	Lanthanium	La.	138.5
Bromine	Br.	79 952	Lead.	Pb.	206 92
Cadmium	Cd.	111.8	Lithium............	Li.	7
Cæsium	Cs.	132.6	Magnesium.........	Mg.	24
Calcium............	Ca.	40	Manganese	Mn.	54
Carbon.............	C.	11.974	Mercury............	Hg.	199 7
Cerium.............	Ce.	141	Molybdenum	Mo.	95.5
Chlorine	Cl.	35 457	Nickel	Ni.	58
Chromium..........	Cr.	52.4	Niobium	Nb.	94
Cobalt..............	Co.	58.9	Nitrogen	N.	14.044
Copper	Cu.	63.2	Osmium...	Os.	198.5
Didymium	D.	144.78	Oxygen	O.	16
Erbium	E.	165.9	Palladium..........	Pd.	105.7
Fluorine	Fl.	19	Phosphorus.........	P.	31
Gallium............	Ga.	68.8	Platinum..........	Pt.	194.4
Glucinum	Gl.	9	Potassium..........	K.	39.137
Gold...............	Au.	196.2	Rhodium...........	Rh.	104.1

13

ELEMENTS.—*Continued.*

NAME.	A. Symbol.	B. Atomic weight.	NAME.	A. Symbol.	B. Atomic weight.
Rubidium	Rb.	85.3	Thallium	Tl.	203.7
Ruthenium	Ru.	104.2	Thorium	Th.	233
Scandium	Sc.	44	Tin	Sn.	117.7
Selenium	Se.	78.8	Titanium	Ti.	49.85
Silicon	Si.	28	Tungsten	W.	183.6
Silver	Ag.	107.675	Uranium	U.	238.5
Sodium	Na.	22.998	Vanadium	V.	51.3
Strontium	Sr.	87.4	Ytterbium	Yb.	172.7
Sulphur	S.	31.984	Yttrium	Y.	89.8
Tantalum	Ta.	182	Zinc	Zn.	64.9
Tellurium	Te.	128	Zirconium	Zr.	89.6

In some cases the results of analyses are such as would agree with two values as the atomic weight of an element equally well. In this case we can decide which is the correct value by the law of Dulong and Petit. These observers found that while the atomic weights of the elements vary greatly from each other, the specific heats (see p. 33) differ from each other in an opposite manner, and to such an extent that the product obtained by multiplying the two together does not vary much from 6.4. This product is known as the *atomic heat.* When by analysis it is not possible to determine which of two numbers is the correct atomic weight of an element, that one is selected which, when multiplied by the specific heat, gives a result most nearly approaching 6.4.

The atomic heats of boron, carbon, silicon, sulphur, and phosphorus are subject to great variations, as is shown in the following table:

		Specific heat.	Atomic heat.			Specific heat.	Atomic heat.
BORON.				SILICON.			
Crystallized	at − 39.6°	0.1915	2.11	Crystallized	at − 39.8°	0.1360	3.81
Crystallized	at + 76.7°	0.2737	3.01	Crystallized	at + 128.7°	0.1964	5.50
Crystallized	at + 233.2°	0.3663	3.99	Crystallized	at + 232.4°	0.2029	5.68
Amorphous		0.255	2.81	Fused	at + 100°	0.175	4.90
				SULPHUR.			
				Orthorhombic	at + 45°	0.163	5.22
CARBON.				Orthorhombic	at + 99°	0.1776	5.68
Diamond	at − 50.5°	0.0625	0.76	Liquid	at + 150°	0.234	7.49
Diamond	at + 140°	0.2218	2.66	Recently fused	at + 98°	0.20259	6.48
Diamond	at + 985°	0.4589	5.51	PHOSPHORUS.			
Graphite	at − 50.3°	0.1138	1.37	Yellow	at − 78°	0.174	5.39
Graphite	at + 138.5°	0.2542	3.05	Yellow	at + 36°	0.202	6.26
Graphite	at + 977.9°	0.4670	5.60	Liquid	at + 100°	0.212	6.57
Wood charcoal		0.2415	2.90	Amorphous	at + 98°	0.170	5.27

It will be observed that, as the temperature of the *solid* element is increased, the atomic heat more nearly approaches 6.4. It will further be noticed that those elements with which the perturbations occur are those which are capable of existing in two or more allotropic forms (see p. 31). As in the passage of an element from one allotropic condition to another, absorption or liberation of heat always takes place, as the result of "interior work;" it is probable that these perturbations are due to a constant tendency of the element to pass from one allotropic condition to another.

14

The atomic heats of those elementary gases which have only been liquified by enormous cold and pressure are tolerably constant at about 2.4.

Molecular Weight.—*The molecular weight of a substance is the weight of its molecule as compared with the weight of an atom of hydrogen.* It is also, obviously, the sum of the weights of all the atoms making up the molecule.

A very ready means of determining the molecular weight of any substance which we can convert into a gas is based upon Avogadro's law. The sp. gr. of a gas is the weight of a given volume as compared with that of an equal volume of hydrogen. But these equal volumes contain equal numbers of molecules (p. 11), and therefor, in determining the sp. gr. of a gas, we obtain the weight of its molecule as compared with that of a molecule of hydrogen; and, as the molecule contains two atoms of hydrogen, while one atom of hydrogen is the unit of comparison, it follows that *the specific gravity of a gas, multiplied by two, is its molecular weight.*

For example, the gas acetylene and the liquid benzene each contain 92.31 per cent. of carbon, and 7.69 per cent. of hydrogen; which is equivalent to 24 parts, or two atoms of carbon; and 2 parts, or two atoms of hydrogen. The sp. gr. of acetylene, referred to hydrogen = 2, is 13; its molecular weight is, therefor, 26, and its molecule contains two atoms of carbon and two atoms of hydrogen. The sp. gr. of vapor of benzene is 39; its molecular weight is, therefor, 78, and its molecule contains six atoms of carbon and six atoms of hydrogen.

The vapor densities of comparatively few elements are known:

	Vapor density.	Atomic weight.	Molecular weight.		Vapor density.	Atomic weight.	Molecular weight.
Hydrogen	1	1	2	Iodine	127	127	254
Oxygen	16	16	32	Phosphorus	62	31	124
Sulphur	32	32	64	Arsenic	150	75	300
Selenium	82	79	164	Nitrogen	14	14	28
Tellurium	130	128	260	Potassium	39	39	78
Chlorine	35.5	35.5	71	Cadmium	56	112	112
Bromine	80	80	160	Mercury	100	200	200

The atomic weight being, in most of the above instances, equal to the vapor density, and to half the molecular weight, it may be inferred that *the molecules of these elements consist of two atoms.* Noticeable discrepancies exist in the case of four elements. The molecular weights of phosphorus and arsenic, as obtained from their vapor densities, are not double but four times as great as their atomic weights. The molecules of phosphorus and arsenic are, therefor, supposed to contain four atoms. Those of cadmium and mercury contain but one atom.

Valence or Atomicity

It is known that the atoms of different elements possess different powers of combining with and of replacing atoms of hydrogen. Thus:

One atom of chlorine combines with one atom of hydrogen,
One atom of oxygen combines with two atoms of hydrogen,
One atom of nitrogen combines with three atoms of hydrogen,
One atom of carbon combines with four atoms of hydrogen.

The valence, atomicity, or equivalence of an element is the saturating power of one of its atoms as compared with that of one atom of hydrogen.

Elements may be classified according to their valence into—

15

Univalent elements or monads	Cl′
Bivalent elements or dyads	O″
Trivalent elements or triads	B‴
Quadrivalent elements or tetrads	C^iv
Quinquivalent elements or pentads	P^v
Sexvalent elements or hexads	W^vi

Elements of even valence, *i.e.*, those which are bivalent, quadrivalent, or sexvalent, are sometimes called *artiads;* those of uneven valence being designated as *perissads.*

In notation the valence is indicated, as above, by signs placed to the right and above the symbol of the element.

But the valence of the elements is not fixed and invariable. Thus, while chlorine and iodine each combine with hydrogen, atom for atom, and in those compounds are consequently univalent, they unite with each other to form two compounds—one containing one atom of iodine and one of chlorine, the other containing one atom of iodine and three of chlorine. Chlorine being univalent, iodine is obviously trivalent in the second of these compounds. Again, phosphorus forms two chlorides, one containing three, the other five atoms of chlorine to one of phosphorus.

In view of these facts, we must consider, either: 1, that the valence of an element is that which it exhibits in its most saturated compounds, as phosphorus in the pentachloride, and that the lower compounds are nonsaturated and have free valences; or 2, that the valence is variable. The first supposition depends too much upon the chances of discovery of compounds in which the element has a higher valence than that which might be considered as the maximum to-day. The second supposition—notwithstanding the fact that, if we admit the possibility of two distinct valences, we must also admit the possibility of others—is certainly the more tenable and the more natural. *In speaking, therefor, of the valence of an element, we must not consider it as an absolute quality of its atoms, but simply as their combining power in the particular class of compounds under consideration.* Indeed, compounds are known in whose molecules the atoms of one element exhibit two distinct valences; thus, ammonium cyanate contains two atoms of nitrogen: one in the ammonium group is quinquivalent, one in the acid radical is trivalent.

When an element exhibits different valences, these differ from each other by two. Thus, phosphorus is trivalent or quinquivalent; platinum is bivalent or quadrivalent.

Symbols—Formulæ—Equations.

SYMBOLS.—These are conventional abbreviations of the names of the elements, whose purpose it is to introduce simplicity and exactness into descriptions of chemical actions. They consist of the initial letter of the Latin name of the element, to which is usually added one of the other letters. If there be more than two elements whose names begin with the same letter, the single-letter symbol is reserved for the commonest element. Thus, we have nine elements whose names begin with C; of these the commonest is Carbon, whose symbol is C; the others have double-letter symbols, as Chlorine, Cl; Cobalt, Co; Copper, Cu (Cuprum), etc.

These symbols do not indicate simply an indeterminate quantity, but one atom of the corresponding element.

16

When more than one atom is spoken of, the number of atoms which it is desired to indicate is written either before the symbol or, in small figures, after and below it; thus, H indicates one atom of hydrogen; 2Cl, two atoms of chlorine; C_4, four atoms of carbon, etc.

FORMULÆ.—What the symbol is to the element, the formula is to the compound; by it the number and kind of atoms of which the molecule of a substance is made up are indicated. The simplest kind of formulæ are what are known as *empirical formulæ*, which indicate only the kind and number of atoms which form the compound. Thus, HCl indicates a molecule composed of one atom of hydrogen united with one atom of chlorine; $5H_2O$, five molecules, each composed of two atoms of hydrogen and one atom of oxygen, the number of molecules being indicated by the proper numeral placed before the formula, in which place it applies to all the symbols following it. Sometimes it is desired that a numeral shall apply to a part of the symbols only, in which case they are enclosed in parentheses; thus, $Al_2(SO_4)_3$ means twice Al and 3 times SO_4.

For other varieties of formulæ, see p. 23.

EQUATIONS are combinations of formulæ and algebraic signs so arranged as to indicate a chemical reaction and its results. The signs used are the plus and equality signs; the former being equivalent to "and," and the second meaning "have reacted upon each other and have produced." The substances entering into the reaction are placed before the equality sign, and the products of the reaction after it; thus, the equation

$$2KHO + H_2SO_4 = K_2SO_4 + 2H_2O$$

means, when translated into ordinary language: two molecules of potash, each composed of one atom of potassium, one atom of hydrogen, and one atom of oxygen, *and* one molecule of sulphuric acid, composed of one atom of sulphur, four atoms of oxygen, and two atoms of hydrogen, *have reacted upon each other and have produced* one molecule of potassium sulphate, composed of one atom of sulphur, four atoms of oxygen, and two atoms of potassium, *and* two molecules of water, each composed of two atoms of hydrogen and one atom of oxygen.

As no material is ever lost or created in a reaction, the number of each kind of atom occurring before the equality sign in an equation must always be the same as that occurring after it.

Electrolysis.

When a galvanic current of sufficient power is made to pass through a compound liquid, or a solution of a compound capable of conducting the current, a decomposition of the compound almost invariably ensues.

The terminals by which the current is conducted into the liquid are known as the *poles* or *electrodes*, and for this purpose are best made of sheets of platinum. The *pole connected with the copper, carbon, or platinum end of the battery is known as the positive pole; that connected with the zinc end as the negative pole.* The decomposition by the voltaic current is known as *electrolysis*, and the liquid subjected to decomposition is called an *electrolyte.*

When compounds are subjected to electrolysis the constituent elements are not discharged throughout the mass, although the decomposition occurs at all points between the electrodes. In compounds made up of

17

two elements only, *binary compounds*, one element is given off at each of the poles, entirely unmixed with the other, and always from the same pole. Thus, if hydrochloric acid be subjected to electrolysis, pure hydrogen is given off at the negative pole and pure chlorine at the positive pole.

In the case of compounds containing more than two elements, a similar decomposition occurs; one element being liberated at one pole and the remaining group of elements separating at the other. This primary decomposition is frequently modified as to its final products by intercurrent chemical reactions; indeed, the group of elements liberated at one pole is rarely capable of separate existence. When, for instance, a solution of potassium sulphate is subjected to electrolysis the liquid in the arm of the tube connected with the positive pole becomes acid in reaction, and gives off oxygen; at the same time the liquid on the negative side becomes alkaline, and gives off a volume of hydrogen double that of the oxygen liberated. In the first place, the potassium sulphate molecule is decomposed into potassium and the group SO_4:

$$K_2SO_4 = SO_4 + K_2.$$

The potassium liberated at the negative pole immediately decomposes the surrounding water, forming potash and liberating hydrogen; and the group SO_4 liberated at the positive pole immediately reacts with water to form sulphuric acid and liberate oxygen:

$$K_2 + 2H_2O = 2KHO + H_2 \text{ and } SO_4 + H_2O = H_2SO_4 + O.$$

In the electrolysis of chemical compounds the different elements and groups of elements, such as SO_4 in the example given above, known as *residues* or *radicals*, seem to be possessed of definite electrical characters, and are given off at one or the other pole in preference. Those which are given off at the *positive* or platinum pole are supposed to be negatively electrified, and are therefor known as *electro-negative or acidulous elements or residues;* those given off at the *negative* pole, being positively electrified, are known as *electro-positive or basylous elements or residues.* The following are the electrical characters of the principal elements and residues:

ELECTRO-NEGATIVE OR ACIDULOUS.		ELECTRO-POSITIVE OR BASYLOUS.	
Oxygen,	Molybdenum,	Hydrogen,	Nickel,
Sulphur,	Tungsten,	Potassium,	Cobalt,
Nitrogen,	Boron,	Sodium,	Cerium,
Chlorine,	Carbon,	Lithium,	Lead,
Iodine,	Antimony,	Barium,	Tin,
Fluorine,	Tellurium,	Strontium,	Bismuth,
Phosphorus,	Niobium,	Calcium,	Uranium,
Selenium,	Titanium,	Magnesium,	Copper,
Arsenic,	Silicon,	Glucinium,	Silver,
Chromium,	Osmium,	Yttrium,	Mercury,
		Aluminium,	Palladium,
Residues of acids remaining after the removal of a number of hydrogen atoms equal to the basicity of the acid.		Zirconium,	Platinum,
		Manganese,	Rhodium,
		Zinc,	Iridium,
		Cadmium,	Gold,
		Iron,	Alcoholic radicals.

18

Acids, Bases, and Salts.

An acid is a compound of an electro-negative element or residue with hydrogen; which hydrogen it can part with in exchange for an electro-positive element without formation of a base. An acid may also be defined as a *compound body which evolves water by its action upon pure caustic potash or soda.*

No substance which does not contain hydrogen can, therefor, be called an acid.

The basicity of an acid is the number of replaceable hydrogen atoms contained in its molecule.

A monobasic acid is one containing a single replaceable atom of hydrogen, as nitric acid, HNO_3; *a dibasic acid* is one containing two such replaceable atoms, as sulphuric acid, H_2SO_4; *a tribasic acid* is one containing three replaceable hydrogen atoms, as phosphoric acid, H_3PO_4. *Polybasic acids* are such as contain more than one atom of replaceable hydrogen.

Hydracids are acids containing no oxygen; *oxacids* or *oxyacids* contain both hydrogen and oxygen.

The term *base* is regarded by many authors as applicable to any compound body capable of neutralizing an acid; it is, however, more consistent with modern views to limit the application of the name to such *compound substances as are capable of entering into double decomposition with acids to form salts and water.* They may be considered as one or more molecules of water in which one-half of the hydrogen has been replaced by an electro-positive element or radical; or as compounds of such elements or radicals with one or more groups, OH. Being thus considered as derivable from water, they are also known as *basic hydrates.* They have the general formula, $M_N(OH)_n$. They are *monatomic, diatomic, triatomic*, etc., according as they contain one, two, three, etc., groups *oxhydryl* (OH).

A double decomposition is a reaction in which both of the reacting compounds are decomposed to form two new compounds.

Sulphobases, or hydrosulphides, are compounds in all respects resembling the bases, except that in them the oxygen of the base is replaced by sulphur.

Salts are substances formed by the substitution of basylous radicals or elements for a part or all of the replaceable hydrogen of an acid. They are always formed, therefor, when bases and acids enter into double decomposition. They are not, as was formerly supposed, formed by the union of a metallic with a non-metallic oxide, but, as stated above, by the substitution of one or more atoms of an element or radical for the hydrogen of the acid. Thus, the compound formed by the action of sulphuric acid upon quicklime is not SO_3CaO, but $CaSO_4$, formed by the interchange of atoms:

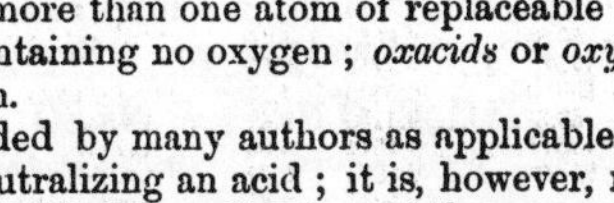

and not

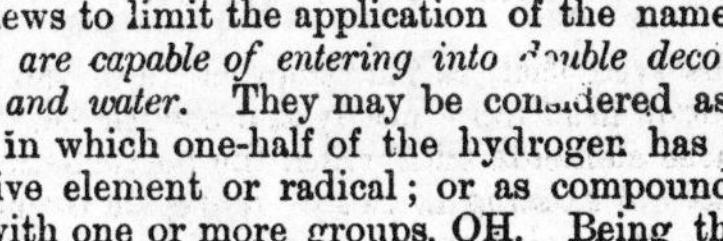

it is, therefor, *calcium sulphate*, and not *sulphate of lime.*

19

The term *salt*, as used at present, applies to the compound formed by the substitution of another element for the hydrogen of *any* acid ; and indeed, as used by some authors, to the acids themselves, which are considered as salts of hydrogen. It is probable, however, that eventually the name will be limited to such compounds as correspond to acids whose molecules contain more than two elements. Indeed, from the earliest times of modern chemistry a distinction has been observed between the *haloid salts, i.e.*, those the molecules of whose corresponding acids consisted of hydrogen united with one other element, on the one hand; and the salts of the oxacids, *i.e.*, those into whose composition oxygen entered, on the other hand. This distinction, however, has gradually fallen into the background, for the reason that the methods and conditions of formation of the two kinds of salts are usually the same *when the basylous element belongs to that class usually designated as metallic.*

There are, however, important differences between the two classes of compounds. There exist compounds of all of the elements corresponding to the hydracids, binary compounds of chlorine, bromine, iodine, and sulphur. There is, on the other hand, a large class of elements which are incapable of forming salts corresponding to the oxacids ; no salt of an oxacid with any one of the elements usually classed as metalloids (excepting hydrogen) has been obtained.

Haloid salts may be formed by direct union of their constituent elements ; oxysalts are never so produced.

Action of Acids and Bases on Salts, and of Salts on each other.

If an acid be added to a solution of a salt whose acid it nearly equals in chemical activity, the salts of both acids and the free acids themselves will probably exist in the solution, provided both acids and salts are soluble. Thus :

$$2H_2SO_4 + 3KNO_3 = K_2SO_4 + KNO_3 + H_2SO_4 + 2HNO_3$$

Sulphuric acid. Potassium nitrate. Potassium sulphate. Potassium nitrate. Sulphuric acid. Nitric acid.

If an acid be added to a solution of a salt whose acid it greatly exceeds in activity, the salt is decomposed, with formation of the salt of the stronger acid and liberation of the weaker acid ; both acids and salts being soluble :

$$H_2SO_4 + 2C_2H_3O_2Na = Na_2SO_4 + 2C_2H_3O_2H$$

Sulphuric acid. Sodium acetate. Sodium sulphate. Acetic acid.

If to a solution of a salt whose acid is insoluble in the solvent used, an acid be added capable of forming a soluble salt with the basylous element, such soluble salt is formed and the acid is deposited :

$$H_2SO_4 + 2C_{18}H_{35}O_2Na = Na_2SO_4 + 2C_{18}H_{35}O_2H$$

Sulphuric acid. Sodium stearate. Sodium sulphate. Stearic acid.

If to a salt whose acid is volatile at the existing temperature, an acid capable of forming with the basylous element a salt fixed at the same temperature be added, the fixed salt is formed and the volatile acid expelled. Thus, with the application of heat :

$$H_2SO_4 + 2NaNO_3 = Na_2SO_4 + 2HNO_3$$

Sulphuric acid. Sodium nitrate. Sodium sulphate. Nitric acid.

20

If to a solution of a salt an acid be added which is capable of forming an insoluble salt with the base, such insoluble salt is formed and precipitated :

$$H_2SO_4 + Ba(NO_3)_2 = BaSO_4 + 2HNO_3$$

Sulphuric acid. Barium nitrate. Barium sulphate. Nitric acid.

If to a solution of a salt whose basylous element is insoluble a soluble base is added, capable of forming a soluble salt with the acid, such soluble salt is formed, with precipitation of the insoluble base :

$$CuSO_4 + 2KHO = K_2SO_4 + CuH_2O_2$$

Cupric sulphate. Potassium hydrate. Potassium sulphate. Cupric hydrate.

If a base be added to a solution of a salt with whose acid it is capable of forming an insoluble salt, such insoluble salt is formed and precipitated, and the base of the original salt, if insoluble, is also precipitated :

$$BaH_2O_2 + K_2SO_4 = BaSO_4 + 2KHO$$

Barium hydrate. Potassium sulphate. Barium sulphate. Potassium hydrate.

$$BaH_2O_2 + Ag_2SO_4 = BaSO_4 + 2AgHO$$

Barium hydrate. Silver sulphate. Barium sulphate. Silver hydrate.

When solutions of two salts, the acids of both of which form soluble salts with both bases, are mixed, the resultant liquid contains the four salts :

$$3K_2SO_4 + 3NaNO_3 = 2K_2SO_4 + Na_2SO_4 + 2KNO_3 + NaNO_3$$

Potassium sulphate. Sodium nitrate. Potassium sulphate. Sodium sulphate. Potassium nitrate. Sodium nitrate.

or in some other proportion.

If solutions of two salts, the acid of one of which is capable of uniting with the base of the other to form an insoluble salt, are mixed, such insoluble salt is precipitated :

$$Ba(NO_3)_2 + Na_2SO_4 = BaSO_4 + 2NaNO_3$$

Barium nitrate. Sodium sulphate. Barium sulphate. Sodium nitrate.

Nomenclature.

The names of the elements are mostly of Greek derivation, and have their origin in some prominent property of the substance ; thus, *phosphorus*, φῶς, light, and φέρειν, to bear. Some are of Latin origin, as *silicon*, from *silex*, flint ; some of Gothic origin, as *iron*, from *iarn ;* and others are derived from modern languages, as *potassium*, from *pot-ash*. Very little system has been followed in naming the elements, beyond applying the termination *ium* to the metals, and *ine* or *on* to the metalloids ; and even to this rule we find such exceptions as a metal called *manganese* and a metalloid called *sulphur*.

The names of *compound substances* were formerly chosen upon the same system, or rather lack of system, as those of the elements. So long as the number of compounds with which the chemist had to deal remained small, the use of these fanciful appellations, conveying no more to the mind than perhaps some unimportant quality of the substances to which they applied, gave rise to comparatively little inconvenience. In these

21

later days, however, when the number of compounds has risen high in the thousands, some systematic method has become absolutely necessary.

The principle at the base of the system of nomenclature at present used is that the name shall itself convey, as far as possible, the composition and character of the substance.

Compounds consisting of two elements, or of an element and a radical only, *binary compounds*, are designated by compound names made up of the name of the more electro-positive, followed by that of the more electro-negative, in which the termination *ide* has been substituted for the terminations *ine, on, ogen, ygen, orus, ium*, and *ur*. For example : the compound of potassium and chlorine is called potassium chlor*ide*, that of potassium and oxygen, potassium ox*ide*, that of potassium and phosphorus, potassium phosph*ide*.

In a few instances the older name of a compound is used in preference to the one which it should have under the above rule, for the reason that the substance is one which is typical of a number of other substances, and therefor deserving of exceptional prominence ; such are *ammonia*, NH_3 ; *water*, H_2O.

When, as frequently happens, two elements unite with each to form more than one compound, these are usually distinguished from each other by prefixing to the last word of the name the Greek numeral corresponding to the number of atoms of the element designated by that word, as compared with a *fixed* number of atoms of the other element.

Thus, in the series of compounds of nitrogen and oxygen, most of which contain two atoms of nitrogen, N_2 is the standard of comparison, and consequently the names are as follows :

N_2O = Nitrogen *mon*oxide.
NO ($=N_2O_2$) = Nitrogen *di*oxide.
N_2O_3 = Nitrogen *tri*oxide.
N_2O ($=N_2O_4$) = Nitrogen *tetr*oxide.
N_2O_5 = Nitrogen *pent*oxide.

Another method of distinguishing two compounds of the same two elements consists in terminating the first word in *ous*, in that compound which contains the less proportionate quantity of the more electro-negative element, and in *ic* in that containing the greater proportion ; thus :

SO_2 = Sulphur*ous* oxide.
SO_3 = Sulphur*ic* oxide.

Hg_2Cl_2 (2Hg : 2Cl) = Mercur*ous* chloride.
$HgCl_2$ (2Hg : 4Cl) = Mercur*ic* chloride.

This method, although used to a certain extent in speaking of compounds composed of two elements of Class II. (see p. 27), is used chiefly in speaking of binary compounds of elements of different classes.

In naming the *oxacids* the word *acid* is used, preceded by the name of the electro-negative element other than oxygen, to which a prefix or suffix is added to indicate the degree of oxidation. If there be only two, the least oxidized is designated by the suffix *ous*, and the more oxidized by the suffix *ic*, thus :

HNO_2 = Nitr*ous* acid.
HNO_3 = Nitr*ic* acid.

22

If there be more than two acids, formed in regular series, the least oxidized is designated by the prefix *hypo* and the suffix *ous* ; the next by the suffix *ous* ; the next by the suffix *ic* ; and the most highly oxidized by the prefix *per* and the suffix *ic* ; thus :

$HClO$ = *Hypo*chlor*ous* acid.
$HClO_2$ = Chlor*ous* acid.
$HClO_3$ = Chlor*ic* acid.
$HClO_4$ = *Per*chlor*ic* acid.

Certain elements, such as sulphur and phosphorus, exist in acids which are derived from those formed in the regular way, and which are specially designated.

The names of the oxysalts are derived from those of the acids by dropping the word *acid*, changing the termination of the other word from *ous* into *ite*, or from *ic* into *ate*, and prefixing the name of the electro-positive element or radical ; thus :

H_2SO_3 Sulphur*ous* acid.	K_2SO_3 Potassium sulph*ite*.
H_2SO_4 Sulphur*ic* acid.	K_2SO_4 Potassium sulph*ate*.
$HClO$ Hypochlor*ous* acid.	$KClO$ Potassium hypochlor*ite*.

Acids whose molecules contain more than one atom of replaceable hydrogen are capable of forming more than one salt with electro-negative elements, or radicals, whose valence is less than their basicity. Ordinary phosphoric acid, for instance, contains in each molecule three atoms of basic hydrogen, and consequently is capable of forming three salts by the replacement of one, two, or three of its hydrogen atoms by one, two, or three atoms of a univalent element ; to distinguish these the Greek prefixes *mono*, *di*, and *tri* are used, thus :

H_2KPO_4 = *Mono*potassic phosphate.
HK_2PO_4 = *Di*potassic phosphate.
K_3PO_4 = *Tri*potassic phosphate.

The first is also called *dihydro*potassic phosphate, and the second, *hydrodi*potassic phosphate.

In the older works, salts in which the hydrogen has not been entirely displaced are sometimes called *bi*salts (bicarbonates), or *acid* salts ; those in which the hydrogen has been entirely displaced being designated as *neutral* salts.

Some elements, such as mercury, copper, and iron, form two distinct series of salts ; these are distinguished, in the same way as the acids, by the use of the suffix *ous* in the names of those containing the less proportion of the electro-negative group, and the suffix *ic* in those containing the greater proportion, *e.g.* :

$(Cu_2)_2SO_4$ .... $(1SO_4 : 4Cu)$ = Cupr*ous* sulphate.
Cu_2SO_4 .... $(2SO_4 : 4Cu)$ = Cupr*ic* sulphate.
$FeSO_4$ .... $(2SO_4 : 2Fe)$ = Ferr*ous* sulphate.
$(Fe_2)(SO_4)_3$............ $(3SO_4 : 2Fe)$ = Ferr*ic* sulphate.

The names, *basic* salts, *subs*alts, and *oxy*salts have been applied indifferently to salts, such as the lead subacetates, which are compounds containing the normal acetate and the hydrate or oxide of lead ; and to salts such as the so-called bismuth subnitrate, which is a nitrate, not of bismuth, but of the univalent radical $(Bi'''O'')'$.

By *double salts* are meant such as are formed by the substitution of different elements or radicals for two or more atoms of replaceable hydrogen of the acid, such as ammonio-magnesian phosphate, $PO_4Mg''(NH_4)'$.

Radicals.

A radical, or compound radical, is a non-saturatea group of atoms which behaves like an atom of an element. Such radicals are capable of passing from one compound into another, and are sometimes, although rarely, capable of separate existence. Marsh gas has the composition CH_4 ; by acting upon it in suitable ways we can cause the atom of carbon, accompanied by three of the hydrogen atoms, to pass into a variety of other compounds, such as : $(CH_3)Cl$; $(CH_3)OH$; $(CH_3)_2O$; $C_2H_3O_2(CH_3)$. Marsh gas, therefor, consists of the radical (CH_3) combined with an atom of hydrogen : $(CH_3)'H$.

It is especially among the compounds of carbon that the existence of radicals comes into prominent notice ; they, however, occur in inorganic substances also ; thus the nitric acid molecule consists of the radical NO_2, combined with the group OH.

Like the elements, the radicals possess different valences, depending always upon the number of unsatisfied valences which they contain. Thus the radical (CH_3) is univalent, because three of the four valences of the carbon atom are satisfied by atoms of hydrogen, leaving one free valence ; the radical (PO) of phosphoric acid is trivalent, because two of the five valences of the phosphorus atom are satisfied by the two valences of the bivalent oxygen atom, leaving three free valences.

In notation the radicals are usually enclosed in brackets, as above, to indicate their nature. The names of radicals terminate in *yl* or in *gen* ; thus : (CH_3) = methyl ; (CN) = cyanogen.

The terms *radical* and *residue*, although sometimes used as synonyms, are not such in speaking of electrical decompositions (see p. 17). Thus the *radical* of sulphuric acid is SO_2 ; but when sulphuric acid is electrolyzed it is decomposed into hydrogen and the *residue* SO_4.

Constitution. Typical and Graphic Formulæ.

The composition of a compound is the number and kind of atoms contained in its molecule ; and is shown by its empirical formula.

The constitution of a compound is the number and kind of atoms and their relations to each other, within its molecule ; and is shown by its typical or graphic formula.

The characters of a compound depend not only upon the kind and number of its atoms, but also upon the manner in which they are attached to each other, upon their constitution. There are, for instance, two substances, each having the empirical formula $C_2H_4O_2$, one of which is a strong acid, the other a neutral ether. As the molecule of each contains the same number and kind of atoms, the differences in their properties

must be due to differences in the manner in which the atoms are linked together.

In the system of *typical formulæ* all substances are considered as being so constituted that their rational formulæ may be referred to one of three classes or types, or to a combination of two of these types. These three classes, being named after the most common substance occurring in each, are expressed thus :

The hydrogen type.	The water type.	The ammonia type.
$\left.\begin{matrix}H\\H\end{matrix}\right\}$	$\left.\begin{matrix}H\\H\end{matrix}\right\}O$	$\left.\begin{matrix}H\\H\\H\end{matrix}\right\}N$
$\left.\begin{matrix}H_2\\H_2\end{matrix}\right\}$ etc.,	$\left.\begin{matrix}H_2\\H_2\end{matrix}\right\}O_2$ etc.,	$\left.\begin{matrix}H_2\\H_2\\H_2\end{matrix}\right\}N_2$ etc.,

it being considered that the formula of any substance of known constitution can be indicated by substituting the proper element, or radical, for one or more of the atoms of the type, thus :

Hydrochloric acid.	Alcohol.	Ethylamine.	Calcium chloride.	Sulphuric acid.	Urea.
$\left.\begin{matrix}Cl\\H\end{matrix}\right\}$	$\left.\begin{matrix}(C_2H_5)'\\H\end{matrix}\right\}O$	$\left.\begin{matrix}(C_2H_5)'\\H\\H\end{matrix}\right\}N$	$\left.\begin{matrix}Cl_2\\Ca\end{matrix}\right\}$	$\left.\begin{matrix}(SO_2)''\\H_2\end{matrix}\right\}O_2$	$\left.\begin{matrix}(CO)''\\H_2\\H_2\end{matrix}\right\}N_2$

Typical formulæ are of great service in the classification of compound substances, as well as to indicate, to a certain degree, their nature and the method of the reactions into which they enter. Thus in the case of the two substances mentioned above as both having the composition $C_2H_4O_2$, we find on examination that one contains the group $(CH_3)'$, while the other contains the group $(C_2H_3O)'$, united to one atom of replaceable hydrogen. The difference in their constitution at once becomes apparent in their typical formulæ, $\left.\begin{matrix}(CHO)'\\(CH_3)'\end{matrix}\right\}O$ and $\left.\begin{matrix}(C_2H_3O)'\\H\end{matrix}\right\}O$, indicating differences in their properties, which we find upon experiment to exist. The first substance is neutral in reaction and possesses no acid properties ; it closely resembles a salt of an acid having the formula $\left.\begin{matrix}(CHO)\\H\end{matrix}\right\}O$. The second substance, on the other hand, has a strongly acid reaction, and markedly acid properties, as indicated by the oxidized radical and the extra-radical hydrogen. It is capable of forming salts by the substitution of an atom of a univalent, basylous element for its single replaceable atom of hydrogen : $\left.\begin{matrix}(C_2H_3O)'\\Na\end{matrix}\right\}O$.

Although typical formulæ have been, and still are, of great service, many cases arise, especially in treating of the more complex organic substances, in which they do not sufficiently indicate the relations between the atoms which constitute the molecule, and thus fail to convey a proper idea of the nature of the substance. Considering, for example, the ordinary lactic acid, we find its composition to be $C_3H_6O_3$, which, expressed typically, would be $\left.\begin{matrix}(C_3H_4O)''\\H_2\end{matrix}\right\}O_2$, a constitution supported by the fact that the radical $(C_3H_4O)''$ may be obtained in other compounds, as $\left.\begin{matrix}(C_3H_4O)'' \\ Cl_2\end{matrix}\right\}$. This constitution, however, cannot be the true one, because in the first place, lactic acid is not dibasic, but monobasic; and, in the second place, there is another acid, called paralactic acid, having an identical composition, yet differing in its products of decomposition. These differences in the properties of the two acids must be due to a different *arrangement* of atoms in their molecules, a view which is supported by the sources from which they are obtained and the nature of their products of decomposition.

To express the constitution of such bodies, *graphic formulæ* are used, in which the position of each atom in relation to the others is set forth. The constitution of the two lactic acids would be expressed by graphic formulæ in this way:

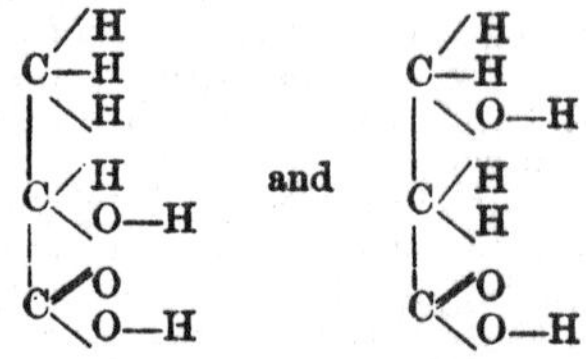

or,

Ordinary lactic acid. Paralactic acid.

It must be understood that these graphic formulæ are simply intended to show the relative attachments of the atoms, and are in nowise intended to convey the idea that the molecule is spread out upon a flat surface with the atoms arranged as indicated in the diagram.

Great care and much labor are required in the construction of these graphic formulæ, the positions of the atoms being determined by a close study of the methods of formation, and of the products of decomposition of the substance under consideration. Naturally, in a matter of this nature, there is always room for differences of opinion—indeed, the entire atomic theory is open to question, as is the theory of gravitation itself. But, whatever may be advanced, two facts cannot be denied: first, that chemistry owes its advancement within the past half-century to the atomic theory, which to-day is more in consonance with observed facts than any substitute which can be offered; second, that without the use of graphic formulæ it is impossible to offer any adequate explanation of the reactions which we observe in dealing with the more complex organic substances.

In chemistry, as in other sciences, a sharp distinction must always be made between facts and theory: the former, once observed, are immutable additions to our knowledge; the latter are of their nature subject to change with our increasing knowledge of facts. We have every reason for believing, however, that the supports upon which the atomic theory

rests are such that, although it may be modified in its details, its essential features will remain unaltered.

Classification of the Elements.

Berzelius was the first to divide all the elements into two great classes, to which he gave the names *metals* and *metalloids*. The metals, being such substances as are opaque, possess what is known as metallic lustre, are good conductors of heat and electricity, and are electro-positive; the *metalloids*, on the other hand, such as are gaseous, or, if solid, do not possess metallic lustre, have a comparatively low power of conducting heat and electricity, and are electro-negative.

This division, based purely upon physical properties, which, in many cases, are ill-defined, has become insufficient. Several elements formerly classed under the above rules with the metals, resemble the metalloids in their chemical characters much more closely than they do any of the metals; indeed, by the characters mentioned above, it is impossible to draw any line of demarcation which shall separate the elements distinctly into two groups.

The classification of the elements should be such that each group shall contain elements whose *chemical* properties are similar—the *physical* properties being considered only in so far as they are intimately connected with the chemical (see p. 13). The arrangement of elements into groups is not equally easy in all cases; some groups, as the chlorine group, are sharply defined, while the members of others differ from each other more widely in their properties. The positions of most of the more recently discovered elements are still uncertain, owing to the imperfect state of our knowledge of their properties.

The method of classification which we will adopt, and which we believe to be more natural than any hitherto suggested, is based upon the chemical properties of the oxides and upon the valence of the elements. We abandon the division into metals and metalloids, and substitute for it a division into four great classes, according to the nature of the oxides and the existence or non-existence of oxysalts. In the first of these classes hydrogen and oxygen are placed together, for the reason that, although they differ from each other in many of their properties, they together form the basis of our classification, and may, for this and other reasons, be regarded as *typical elements*. They both play important parts in the formation of acids, and neither would find a suitable place in either of the other classes. Our primary division would then be as follows:

Class I.—*Typical elements.*

Class II.—*Elements whose oxides unite with water to form acids, never to form bases. Which do not form oxysalts.*

This class contains all the so-called metalloids except hydrogen and oxygen.

Class III.—*Elements whose oxides unite with water, some to form bases, others to form acids. Which form oxysalts.*

Class IV.—*Elements whose oxides unite with water to form bases; never to form acids. Which form oxysalts.*

In this class are included the more strongly electro-positive metals.

Within the classes a further subdivision is made into groups, each group containing those elements within the class which have equal va-

lences, which form corresponding compounds, and whose *chemical characters* are otherwise similar.

For the sake of convenience the term *metal* is retained to apply to the members of Classes III. and IV.; the term *non-metal* being used for those belonging to Class II.

Class I.

Group I.—Hydrogen.
Group II.—Oxygen.

Class II.

Group I.—Fluorine, chlorine, bromine, iodine.
Group II.—Sulphur, selenium, tellurium.
Group III.—Nitrogen, phosphorus, arsenic, antimony.
Group IV.—Boron.
Group V.—Carbon, silicon.
Group VI.—Vanadium, niobium, tantalium.
Group VII.—Molybdenum, tungsten, osmium (?).

Class III.

Group I.—Gold.
Group II.—Chromium, manganese, iron.
Group III.—Glucinium, aluminium, scandium, gallium, indium.
Group IV.—Uranium.
Group V.—Lead.
Group VI.—Bismuth.
Group VII.—Titanium, zirconium, tin.
Group VIII.—Palladium, platinum.
Group IX.—Rhodium, ruthenium, iridium.

Class IV.

Group I.—Lithium, sodium, potassium, rubidium, cæsium, silver.
Group II.—Thallium.
Group III.—Calcium, strontium, barium.
Group IV.—Magnesium, zinc, cadmium.
Group V.—Nickel, cobalt.
Group VI.—Copper, mercury.
Group VII.—Yttrium, cerium, ytterbium, lanthanium, didymium, erbium.
Group VIII.—Thorium.

Physical Characters of Chemical Interest.

Crystallization.—Solid substances exist in two forms, *amorphous* and *crystalline*. In the former they assume no definite shape; they conduct heat equally well in all directions; they break irregularly; and, if transparent, allow light to pass through them equally well in all directions. A solid in the crystalline form has a definite geometrical shape; conducts heat more readily in some directions than in others; when

28

broken, separates in certain directions, called *planes of cleavage,* more readily than in others; and modifies the course of luminous rays passing through it differently when they pass in certain directions than when they pass in others.

Crystals are formed in one of four ways: 1.) An amorphous substance,

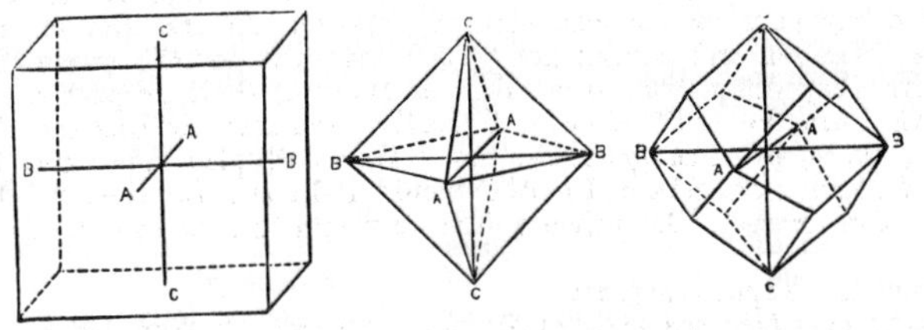

FIG. 7.

by slow and gradual modification, may assume the crystalline form; as vitreous arsenic trioxide (*q. v.*) passes to the crystalline variety. 2.) A fused solid, on cooling, crystallizes; as bismuth. 3.) When a solid is sublimed it is usually condensed in the form of crystals. Such is the case with arsenic trioxide. 4.) The usual method of obtaining crystals is by the evaporation of a solution of the substance. If the evaporation be slow and the solution at rest, the crystals are large and well-defined. If the crystals separate by the sudden cooling of a hot solution, especially if it be agitated during the cooling, they are small.

Most crystals may be divided by imaginary planes into equal, symmetrical halves; such planes are called *planes of symmetry.* Thus in the crystals in Fig. 7 the planes *ab ab, ac ac,* and *bc bc* are planes of symmetry.

When a plane of symmetry contains two or more equivalent linear directions passing through the centre, it is called the *principal plane of symmetry;* as in Fig. 8 the plane *ab ab,* containing the equal linear directions *aa* and *bb.*

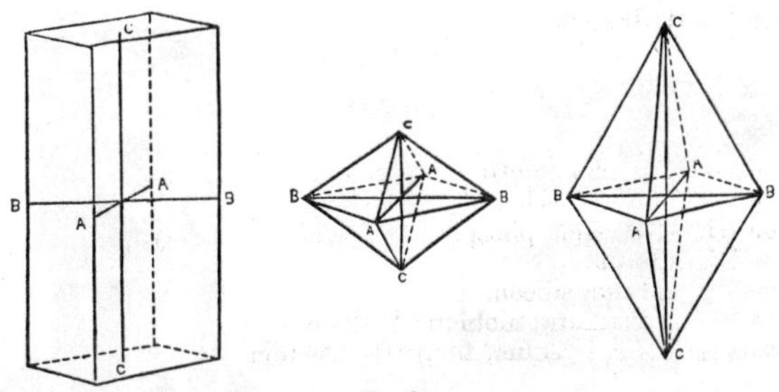

FIG. 8.

Any normal erected upon a plane of symmetry, and prolonged in both directions until it meets opposite parts of the exterior of the crystal, at equal distances from the plane, is called an *axis of symmetry.*

The axis normal to the principal plane is the *principal axis.* Thus in Fig. 8, *aa, bb,* and *cc* are axes of symmetry, and *cc* is the principal axis.

29

Upon the relations of these imaginary planes and axes a classification of all crystalline forms into six *systems* has been based.

I. THE CUBIC, REGULAR, OR MONOMETRIC SYSTEM.—The crystals of this system have three equal axes, *aa, bb, cc,* Fig. 7, crossing each other at right

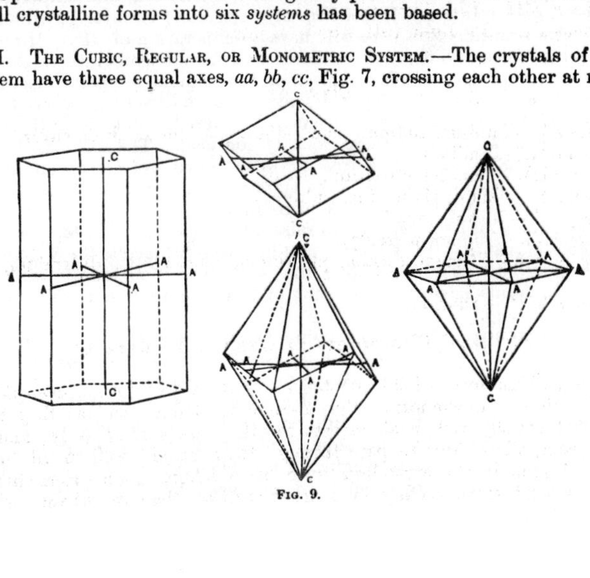

FIG. 9.

angles. The simple forms are the *cube;* and its derivatives, the *octahedron, tetrahedron,* and *rhombic dodecahedron.* The crystals of this system expand equally in all directions when heated, and are not doubly refracting.

II. THE RIGHT SQUARE PRISMATIC, PYRAMIDAL, QUADRATIC, TETRAGONAL, OR DIMETRIC SYSTEM contains those crystals having three axes placed at right angles to each other—two as *aa* and *bb,* Fig. 8, being equal to each other and the third, *cc,* either longer or shorter. The simple forms are the *right square prism* and the *right square based octahedron.* The crystals of this system expand equally only in two directions when heated; they refract light doubly in all directions except through one axis of single refraction.

III. THE RHOMBOHEDRAL OR HEXAGONAL SYSTEM includes crystals having four axes, three of which *aa, aa, aa,* Fig. 9, are of equal length and cross each other at 60° in the same plane; to which plane the fourth axis, *cc,* longer or shorter than the others, is at right angles. The simple forms are the *regular six-sided prism,* the *regular dodecahedron,* the *rhombohedron,* and the *scalenohedron.* These crystals expand equally in two directions when heated, and refract light singly through the principal axis, but in other directions refract it doubly.

IV. THE RHOMBIC, RIGHT PRISMATIC, OR TRIMETRIC SYSTEM.—The axes of crystals of this system are three in number, all at right angles to each other, and all of unequal length. Fig. 8 represents crystals of this system, supposing *aa, bb,* and *cc* to be unequal to each other. The simple forms are the *right rhombic octahedron,* the *right rhombic prism,* the *right rectangular octahedron,* and the *right rectangular prism.* The crystals of this system, like those of the two following, have no true principal plane or axis.

V. THE OBLIQUE, MONOSYMMETRIC, OR MONOCLINIC SYSTEM.—The crystals of this system have three axes, two of which, *aa,* and *cc,* Fig. 10, are at

30

right angles; the third, *bb,* is perpendicular to one and oblique to the other; they may be equal or all unequal in length. The simple forms are *the oblique rectangular and oblique rhombic prism and octahedron.*

VI. THE DOUBLY OBLIQUE, ASYMMETRIC, TRICLINIC, OR ANORTHIC SYSTEM contains crystals having three axes of unequal length, crossing each other

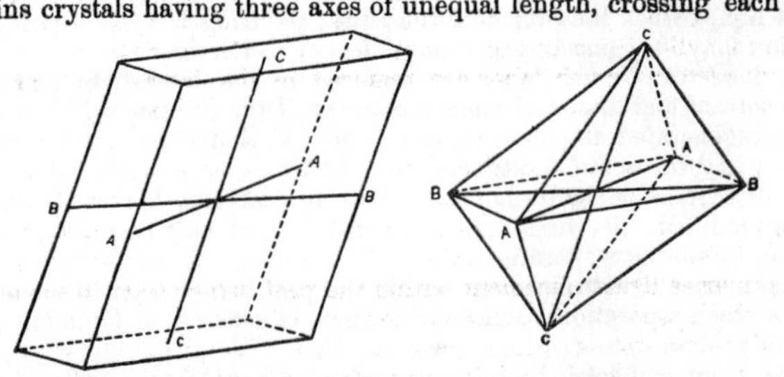

FIG. 10.

at angles not right angles; Fig. 10, *aa, bb,* and *cc* being unequal and the angles between them other than 90°.

The crystals of the fourth, fifth, and sixth systems, when heated, expand equally in the directions of their three axes; they refract light doubly except in two axes.

SECONDARY FORMS.—The crystals occurring in nature or produced artificially have some one of the forms mentioned above, or some modification

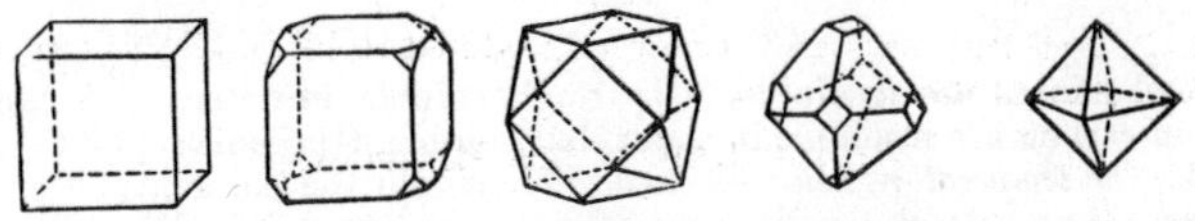

FIG. 11.

of those forms. These modifications or *secondary forms* may be produced by symmetrically removing the angles or edges, or both angles and edges, of the primary forms; thus, by progressively removing the angles of the cube, the secondary forms shown in Fig. 11 are produced.

It sometimes happens in the formation of a derivative form that alternate faces are excessively developed, producing at length entire obliteration of the others, as shown in Fig. 12. Such crystals are said to be

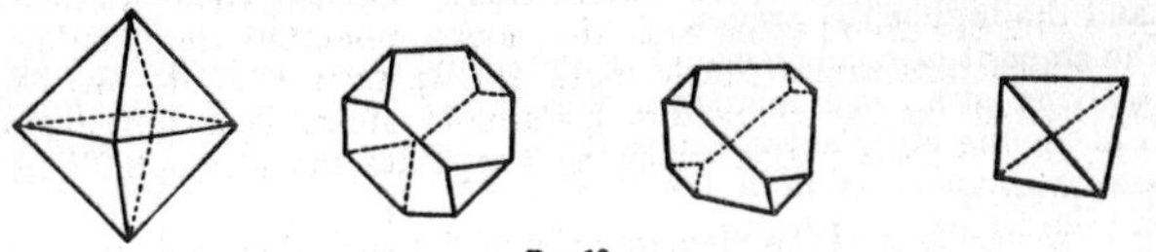

FIG. 12.

hemihedral; they can be developed only in a system having a principal axis.

ISOMORPHISM.—In many instances two or more substances crystallize in forms identical with each other, and, in most cases, such substances resemble each other in their chemical constitution; they are said to be

31

isomorphous. This identity of crystalline form does not depend so much upon the nature of the elements themselves, as upon the structure of the molecule. The protoxide and peroxide of iron do not crystallize in the same form, nor can they be substituted for each other in reactions without radically altering the properties of the resultant compound. On the other hand, all that class of salts known as alums are isomorphous ; not only are their crystals identical in shape, but a crystal of one alum, placed in a saturated solution of another, grows by regular deposition of the second upon its surface. Other alums may be subsequently added to the crystal, a section of which will then exhibit the various salts, layer upon layer.

DIMORPHISM.—Although most substances crystallize, if at all, in one simple form or in some of its modifications, a few bodies are capable of assuming two crystalline forms belonging to different systems ; such are said to be *dimorphous.* Thus, sulphur, as obtained by the evaporation of its solution in carbon disulphide, forms octahedra belonging to the fourth system ; when obtained by cooling melted sulphur, the crystals are oblique prisms, belonging to the fifth system. Occasional instances of *trimorphism,* of the formation of crystals belonging to three different systems by the same substance, are also known.

ALLOTROPY.—Dimorphism apart, a few substances are known to exist in more than one solid form. These varieties of the same substance exhibit different physical properties, while their chemical qualities are the same in kind. Such modifications are said to be *allotropic.* One or more allotropic modifications of a substance are usually *crystalline,* the other or others *amorphous* or *vitreous.* Sulphur, for example, exists not only in two dimorphous varieties of crystals, but also in a third, *allotropic* form, in which it is flexible, amorphous, and transparent. Carbon exists in three allotropic forms : two crystalline, the diamond and graphite ; the third amorphous.

In passing from one allotropic modification to another, a substance absorbs or gives out heat.

Solution.—A solid, liquid, or gas is said to *dissolve,* or form a *solution* with a liquid when the two substances unite to form a homogeneous liquid. Solution may be a purely physical process or a chemical combination.

In *simple* or *physical solution* there is no modification of the properties of the solvent and dissolved substance, beyond the liquefaction ; the latter can be regenerated in its primitive form by simple evaporation of the former ; and the act of solution is attended by a *diminution of temperature.* In *chemical solution* the properties of both solvent and dissolved are more or less modified ; the dissolved substance can not be obtained from the solution by simple evaporation of the solvent, unless the compound formed be decomposable, with formation of the original substance, at the temperature of the evaporation. The act of chemical solution is attended by an *elevation of temperature.*

The amount of solid, liquid, or gas which a liquid is capable of dissolving by simple solution depends upon the following conditions :

1. *The nature of the solvent and substance to be dissolved.*—No rule can be given which will apply in a general way to the solvent power of liquids or the solubility of substances. Water is of all liquids the best solvent of most substances ; in it some substances are so readily soluble that they absorb a sufficiency from the atmosphere to form a solution ; as calcium chloride. Such substances are said to be *deliquescent.* Other substances are insoluble in water in any proportion ; as barium sulphate. Elemen-

32

tary substances are insoluble, or sparingly soluble, in water. Substances rich in carbon are insoluble in water, but soluble in organic liquids.

2. *The temperature* has a marked influence on the solubility of a substance. As a rule, water dissolves a greater quantity of a solid substance as the temperature is increased. This increase in solubility is different in the case of different soluble substances ; thus the increase in solubility of the chlorides of barium and of potassium is directly in proportion to the increase of temperature ; the solubility of sodium chloride is almost imperceptibly increased by elevation of temperature ; the solubility of sodium sulphate increases rapidly up to 33° (91°.4 F.), above which temperature it again diminishes.

The solubility of gases in water is the greater the lower the temperature and the higher the pressure.

The amount of a substance that a given quantity of solvent is capable of dissolving at a given temperature is fixed. A solution containing as much of the dissolved substance as it is capable of dissolving is said to be *saturated ;* if made at high temperatures it is said to be a *hot saturated,* and if at ordinary temperatures a *cold saturated solution.*

If a hot saturated solution of a salt be cooled, the solid is in most instances separated by crystallization. If in the case of certain substances, such as sodium sulphate, however, the solution be allowed to cool while undisturbed, no crystallization occurs, and the solution at the lower temperature contains a greater quantity of the solid than it could dissolve at that temperature. Such a solution is said to be *supersaturated.* The contact of particles of solid material with the surface of a supersaturated solution induces immediate crystallization, attended with elevation of temperature.

3. *The presence of other substances already dissolved.*—If to a saturated solution of potassium nitrate, sodium chloride be added, a further quantity of potassium nitrate may be dissolved. In this case there is double decomposition between the two salts, and the solution contains, besides them, potassium chloride and sodium nitrate.

4. *The presence of a second solvent.*—If two solvents, *a* and *b,* incapable of mixing with each other, be brought in contact with a substance which both are capable of dissolving ; neither *a* nor *b* take up the whole of the substance to the exclusion of the other, however greatly the solvent power or bulk of the one may exceed that of the other. The relative quantities taken up by each solvent is in a constant ratio.

Diffusion of Liquids—Dialysis.—If a liquid be carefully floated upon the surface of a second liquid, of greater density, with which it is capable of mixing, two distinct layers will at first be formed. Even at perfect rest, mixture will begin immediately, and progress slowly until the two liquids have *diffused* into each other to form a single liquid whose density is the same throughout.

Substances differ from each other in the rapidity with which they diffuse. Substances capable of crystallization, *crystalloids,* are much more diffusible than those which are incapable of crystallization—*colloids.*

If, in place of bringing two solutions in contact with each other, they be separated by a solid or semi-solid, moist, colloid layer, diffusion takes place in the same way through the interposed layer. Advantage is taken of this fact to separate crystalloids from colloids by the process of *dialysis.* The mixed solutions of crystalloid and colloid are brought into the inner vessel of a dialyser, Fig. 13, whose bottom consists of a layer of moist parchment paper, while the outer vessel is filled with pure water. Water

33

passes into the inner vessel, and the crystalloid passes into the water in the outer vessel. By frequently changing the water in the outer vessels, solutions of the albuminoids or of ferric hydrate, etc., almost entirely free from crystalloids, may be obtained.

FIG. 13.

Specific Heat.—Equal volumes of different substances at the same temperature contain different amounts of heat. If two equal volumes of the same liquid of different temperatures be mixed together, the resulting mixture has a temperature which is the mean between the temperatures of the original volumes. If one litre of water at 4° be mixed with a litre at 38°, the resulting two litres will have a temperature of 21°. Mixtures of equal volumes of different substances, at different temperatures, do not have a temperature which is the mean of the original temperatures of its constituents. A litre of water at 4°, mixed with a litre of mercury at 38°, forms a mixture whose temperature is 27°. Mercury and water, therefor, differ from each other in their capacity for heat. The same difference exists in a more marked degree between equal *weights* of dissimilar bodies ; if a pound of water at 4° be agitated with a pound of mercury at 70°, both liquids will have a temperature of 67°.

The amount of heat required to raise a kilo of water 1° in temperature is a definite quantity. The *specific heat* of any substance is the amount of heat required to raise one kilo of that substance 1° in temperature, expressed in terms having the amount of heat required to raise a kilo of water 1° as unity.

Spectroscopy.—Light in passing through a prism is not only *refracted* into a different course, but is also decomposed or *dispersed* into different colors, which make up a *spectrum.* A spectrum is one of three kinds : 1.) *Continuous,* consisting of a continuous band of colors : red, orange, yellow, green, blue, indigo, and violet. Such spectra are produced by light from white-hot solids and liquids, from gas-light, candle-light, lime-light, and electric light. 2.) *Bright-line spectra,* composed of bright lines upon a dark ground, are produced by glowing vapors and gases. 3.) *Absorption spectra* consist of continuous spectra, crossed by dark lines or bands, and are produced by light passing through a solid, liquid, or gas, capable of absorbing certain rays. Examples of bright-line and absorption spectra are shown in Fig. 14.

34

The spectrum of sun-light belongs to the third class. It is not continuous, but is crossed by a great number of dark lines, known as Fraun-

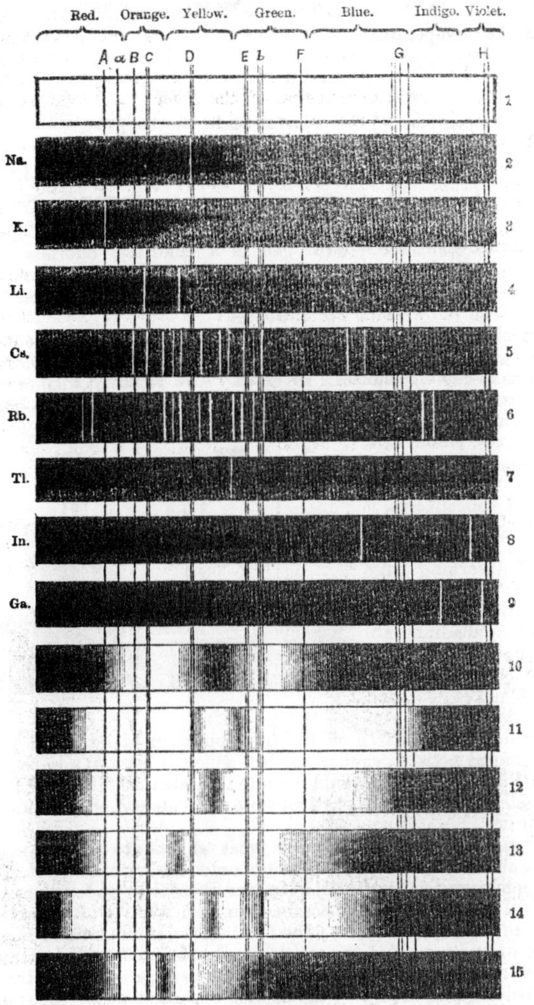

FIG. 14.

hofer's lines, the most distinct of which are designated by letters (No. 1, Fig. 14).

The spectroscope consists of four essential parts: 1st, the slit, *a*, Fig. 15; a linear opening between two accurately straight and parallel knife-

35

edges. 2d, the collimating lens, *b*; a biconvex lens in whose principal focus the slit is placed, and whose object it is to render the rays from the

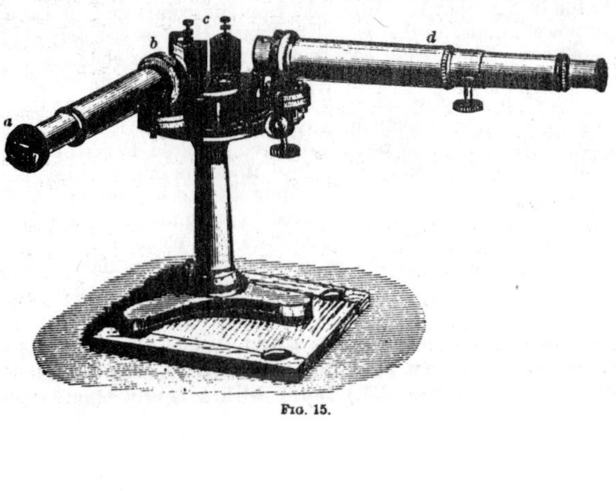

FIG. 15.

slit parallel before they enter the prism. 3d, the prism, or prisms, *c*, of dense glass, usually of 60°, and so placed that its refracting edge is parallel to the slit. 4th, an observing telescope, *d*, so arranged as to receive the rays as they emerge from the prisms. Besides these parts spectroscopes are usually fitted with some arbitrary graduation, which serves to fix the location of lines observed. In direct vision spectroscopes a compound prism is used, so made up of prisms of different kinds of glass that the emerging ray is nearly in the same straight line as the entering ray.

As the spectra produced by different substances are characterized by the positions of the lines or bands, some means of fixing their location is required. The usual method consists in determining their relation to the principal Fraunhofer lines. As, however, the relative positions of these lines vary with the nature of the substance of which the prism is made, although their position with regard to the colors of the spectrum is fixed, no two of the arbitrary scales used will give the same reading.

The most satisfactory method of stating the positions of lines and bands is in wave-lengths. The lengths of the waves of rays of different degrees of refrangibility have been carefully determined, the unit of measurement being the tenth-metre, of which 10^{10} make a metre. The wave-lengths, $= \lambda$, of the principal Fraunhofer lines, are:

A.......	7604.00	D.......	5892.12	G.......	4307.25
a.......	7185.00	E.......	5269.13	H_1.......	3968.01
B.......	6867.00	*b*.......	5172.00	H_2.......	3933.00
C.......	6562.01	F.......	4860.72		

The scale of wave-lengths can easily be used with any spectroscope having an arbitrary scale, with the aid of a curve constructed by interpolation. To construct such a curve, paper is used which is ruled into square inches and tenths. The ordinates are marked with a scale of wave-lengths, and the abscisses with the arbitrary scale of the instrument. The position of each principal Fraunhofer line is then carefully determined in terms of the arbitrary scale, and marked upon the paper with a × at the

36

point where the line of its wave-length and that of its position in the arbitrary scale cross each other. Through these × a curve is then drawn as regularly as possible. In noting the position of an absorption-band, the position of its centre in the arbitrary scale is observed, and its value in wave-lengths obtained from the curve, which, of course, can only be used with the scale and prism for which it has been made.

Polarimetry.—A ray of light passing from one medium into another of different density, at an angle other than 90° to the plane of separation of the two media, is deflected from its course, or *refracted*. Certain substances have the power, not only of deflecting a ray falling upon them in certain directions, but also of dividing it into two rays, which are peculiarly modified. The splitting of the ray is termed *double refraction*, and the altered rays are said to be *polarized*. When a ray of such polarized light meets a mirror held at a certain angle, or a crystal of Iceland spar peculiarly cut (a Nichol's prism), also at a certain angle, it is extinguished. The crystal which produces the polarization is called the *polarizer*, and that which produces the extinction the *analyzer*.

If, when the polarizer and analyzer are so adjusted as to extinguish a ray passing through the former, certain substances are brought between them, light again passes through the analyzer; and in order again to produce extinction, the analyzer must be rotated upon the axis of the ray to the right or to the left. Substances capable of thus influencing polarized light are said to be *optically active*. If, to produce extinction, the analyzer is turned in the direction of the hands of a watch, the substance is said to be *dextrogyrous*; if in the opposite direction, *lævogyrous*.

The distance through which the analyzer must be turned depends upon the peculiar power of the optically active substance, the length of the column interposed, the concentration if in solution, and the wave-length of the original ray of light. The *specific rotary power* of a substance is the rotation produced, in degrees and tenths, by one gram of the substance, dissolved in one cubic centimetre of a non-active solvent, and examined in a column one decimetre long. The specific rotary power is determined by dissolving a known weight of the substance in a given volume of solvent, and observing the angle of rotation produced by a column of given length. Then let $p =$ weight in grams of the substance contained in 1 c.c. of solution; l the length of the column in decimetres; a the angle of rotation observed; and $[a]$ the specific rotary power sought, we have

$$[a] = \frac{a}{pl}.$$

In most instruments monochromatic light, corresponding to the D line of the solar spectrum, is used, and the specific rotary power for that ray is expressed by the sign $[a]_D$. The fact that the rotation is right-handed is expressed by the sign +, and that it is left-handed by the sign —.

It will be seen from the above formula that, knowing the value of $[a]_D$ for any given substance, we can determine the weight of that substance in a solution by the formula

$$p = \frac{a}{[a]_D \times l}.$$

The polarimeter or sarcharometer is simply a peculiarly constructed polariscope used to determine the value of *a*.

PART II.

SPECIAL CHEMISTRY.

CLASS I.

TYPICAL ELEMENTS.

HYDROGEN—OXYGEN.

ALTHOUGH, in a strict sense, hydrogen is regarded by most chemists as the one and only type-element—that whose atom is the unit of atomic and molecular weights—the important part which oxygen plays in the formation of those compounds whose nature forms the basis of our classification, its acid-forming power in organic compounds, and the differences existing between its properties and those of the elements of the sulphur group, with which it is usually classed, warrant us in separating it from the other elements and elevating it to the position it here occupies.

HYDROGEN.

Symbol = **H**—*Univalent—Atomic weight* = 1—*Molecular weight* = 2—*Sp. gr.* = 0.06926 A*—*One litre weighs* 0.0896 *gram* †—100 *cubic inches weigh* 2.1496 *grains* ‡—1 *gram measures* 11.19 *litres* †—1 *grain measures* 46.73 *cubic inches* ‡—*Name derived from* ὕδωρ = *water, and* γεννάω = *I produce—Discovered by Cavendish, in* 1766.

* Air = 1. When the sp. gr. is referred to H = 1, A is replaced by H.
† At 0° C. and 760 mm. barometric pressure.
‡ At 60° F. and 30 inches bar. pressure.

OCCURRENCE.—Occurs free in volcanic gases, in fire-damp, occluded in meteorites, in the gases exhaled from the lungs, and in those of the stomach and intestine. In combination in water, hydrogen sulphide, ammoniacal compounds, and in many organic substances.

PREPARATION.—(1.) By electrolysis of water, H is given off at the negative pole. Utilized when pure H is required.

38

(2.) By the disassociation of water at very high temperatures.

(3.) By the decomposition of water by certain metals. The alkaline metals decompose water at the ordinary temperature:

$$\underset{\text{Sodium.}}{Na_2} + \underset{\text{Water.}}{2H_2O} = \underset{\text{Sodium hydrate.}}{2NaHO} + \underset{\text{Hydrogen.}}{H_2}$$

What part the water plays in the reaction is still a subject of discussion; it is probable that its action is rather physical than chemical. Chemically pure zinc, or zinc whose surface has been coated with an alloy of zinc and mercury, does not decompose the acid unless it forms part of a galvanic battery whose circuit is closed. The zincs of galvanic batteries are therefor coated with the alloy mentioned — are amalgamated — to prevent waste of zinc and acid.

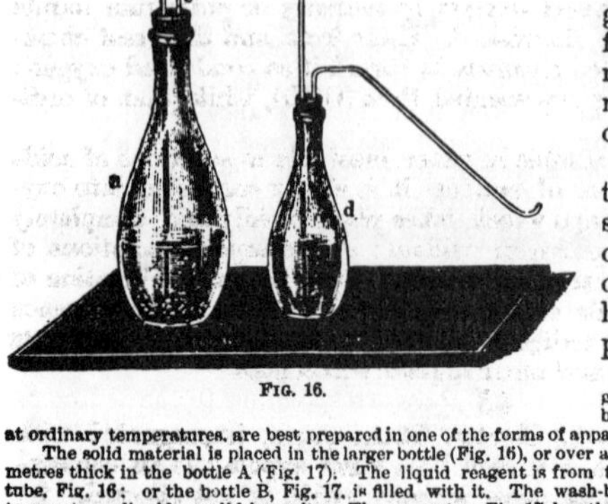

FIG. 16.

This method is resorted to for obtaining H; the gas so obtained is, however, contaminated with small quantities of other gases, hydrogen phosphide, sulphide, and arsenide.

Hydrogen, carbon dioxide, hydrogen sulphide, and other gases produced by the action of a liquid upon a solid at ordinary temperatures, are best prepared in one of the forms of apparatus shown in Figs. 16 and 17.

The solid material is placed in the larger bottle (Fig. 16), or over a layer of broken glass about five centimetres thick in the bottle A (Fig. 17). The liquid reagent is from time to time introduced by the funnel tube, Fig. 16; or the bottle B, Fig. 17, is filled with it. The wash-bottles are partially filled with water to arrest any liquid or solid impurity. The apparatus, Fig. 17, has the advantage of being always ready for use; when the stopcock is open the gas escapes, when it is closed the internal pressure depresses the level of the liquid in A into the layer of broken glass, and the action is arrested.

Some other metals, such as iron and copper, effect the decomposition only at high temperatures:

$$\underset{\text{Iron.}}{3Fe_2} + \underset{\text{Water.}}{8H_2O} = \underset{\text{Triferric tetroxide.}}{2Fe_3O_4} + \underset{\text{Hydrogen.}}{8H_2}$$

(4.) By decomposition of water, passed over red hot coke:

$$\underset{\text{Carbon.}}{C} + \underset{\text{Water.}}{2H_2O} = \underset{\text{Carbon dioxide.}}{CO_2} + \underset{\text{Hydrogen.}}{2H_2}$$

or at a higher temperature:

$$\underset{\text{Carbon.}}{2C} + \underset{\text{Water.}}{2H_2O} = \underset{\text{Carbon monoxide.}}{2CO} + \underset{\text{Hydrogen.}}{2H_2}$$

(5.) By decomposition of mineral acids, in the presence of water, by zinc and certain other metals:

$$\underset{\text{Zinc.}}{Zn} + \underset{\text{Sulphuric acid.}}{H_2SO_4} + \underset{\text{Water.}}{xH_2O} = \underset{\text{Zinc sulphate.}}{ZnSO_4} + \underset{\text{Hydrogen.}}{H_2} + \underset{\text{Water.}}{xH_2O}$$

PROPERTIES.—*Physical.*—Hydrogen is a colorless, odorless, tasteless gas; 14.47 times lighter than air, being the lightest substance known. The weight of a litre, 0.0896 gram, is called a *crith* (κριθή = barleycorn).

39

It is almost insoluble in water and alcohol. In obedience to the law: *The diffusibility of two gases varies inversely as the square roots of their densities*, it is the most rapidly diffusible of gases. The rapidity with which this diffusion takes place renders the use of hydrogen, which has been kept for even a short time in gas-bags or gasometers, dangerous. At −140° (−229° F.), under a pressure of 650 atmospheres, it forms a steel-blue liquid.

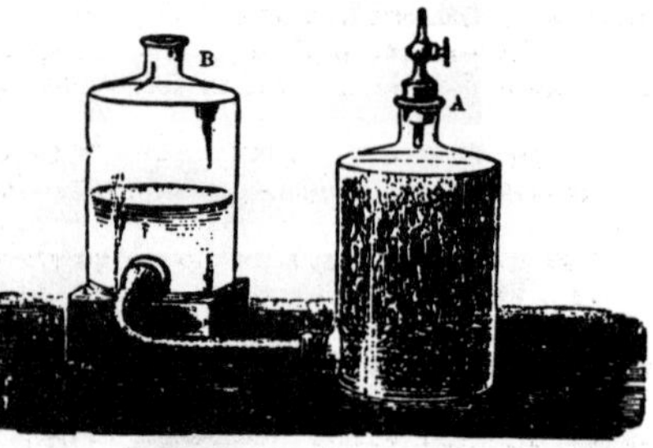

FIG. 17.

Certain metals have the power of absorbing large quantities of hydrogen, which is then said to be *occluded*. Palladium absorbs 376 volumes at the ordinary temperature; 643 vols. at 90° (194° F.), and 526 vols. at 245° (473° F.). The occluded gas is driven off by the application of heat, and possesses great chemical activity, similar to that which it has when in the nascent state. This latter quality would seem to indicate that the gas is contained in the metal, not in a mere physical state of condensation, but in chemical combination.

Chemical.—Hydrogen exhibits no great tendency to combine with other elements at ordinary temperatures; the only one with which it combines under such circumstances is chlorine, and then only under the influence of light. It does not support combustion, but, when ignited, burns with a pale blue and very hot flame; the result of the combination being water. Mixtures of hydrogen and oxygen (in the proportion of 2H to O) explode violently on the approach of flame or by the passage of the electric spark, the explosion being caused by the sudden expansion of the vapor of water formed, under the influence of the heat of the reaction. Hydrogen also unites with oxygen when brought in contact with spongy platinum.

Many compounds containing oxygen give up that element when heated in an atmosphere of hydrogen:

$$\underset{\text{Cupric oxide.}}{CuO} + \underset{\text{Hydrogen.}}{H_2} = \underset{\text{Copper.}}{Cu} + \underset{\text{Water.}}{H_2O}.$$

The removal of oxygen from a compound is called a reduction or deoxidation.

At the instant that H is liberated from its compounds it has a deoxidizing power similar to that which ordinary H possesses only at elevated temperatures. The greater energy of H, and of other elements as well, in this *nascent state*, may be thus explained: free H exists in the form of molecules, each one of which is composed of two atoms. At the instant of its liberation from a compound, on the other hand, it is in the form of individual atoms, and that portion of force required to split up the molecule into atoms, necessary when free H enters into reaction, is not required when the gas is in the nascent state, and consequently a less addition of force in the shape of heat is required to bring about the reaction.

In its physical and chemical properties, this element more closely resembles those usually ranked as metals than it does those forming the class of metalloids, among which it is usually placed; its conducting power,

40

its appearance in the liquid form, as well as its relation to the acids, which may be considered as salts of H, tend to separate it from the metalloids.

ANALYTICAL CHARACTERS.—(1.) Burns with a faintly blue flame, which deposits water on a cold surface brought in contact with it; (2.) Mixed with oxygen, explodes on contact with flame, producing water.

OXYGEN.

Symbol = O—*Bivalent*—*Atomic weight* = 16; *molecular weight* = 32—*sp. gr.* = 1.10563 A (*calculated* = 1.1088); 15.95 H; *sp. gr. of liquid* = 0.9787—*One litre weighs* 1.4300 *grams* = 16 *criths*—100 *cubic inches weigh* 34.27 *grains*—*Name derived from* ὀξύς = *acid, and* γεννάω = *I produce*—*Discovered by Mayow in* 1674; *re-discovered by Priestley in* 1774.

OCCURRENCE.—Oxygen is the most abundant of the elements. It exists free in atmospheric air; in combination in a great number of substances, mineral, vegetable, and animal.

PREPARATION.—(1.) By heating certain oxides:

$$\underset{\text{Mercuric oxide.}}{2HgO} = \underset{\text{Mercury.}}{2Hg} + \underset{\text{Oxygen.}}{O_2}$$

This was the method used by Priestley: 100 grams of mercuric oxide produce 5.16 litres of oxygen:

$$\underset{\text{Manganese dioxide.}}{3MnO_2} = \underset{\text{Trimanganic tetroxide.}}{Mn_3O_4} + \underset{\text{Oxygen.}}{O_2}$$

The black oxide of manganese is heated to redness in an iron or clay retort (Scheele, 1775); and 100 grams yield 8.51 litres of oxygen.

(2.) By the electrolysis of water, acidulated with sulphuric acid, O is given off at the positive pole.

(3.) By the action of sulphuric acid upon certain compounds rich in O: manganese dioxide, potassium dichromate, and plumbic peroxide:

$$\underset{\text{Manganese dioxide.}}{2MnO_2} + \underset{\text{Sulphuric acid.}}{2H_2SO_4} = \underset{\text{Manganous sulphate.}}{2MnSO_4} + \underset{\text{Water.}}{2H_2O} + \underset{\text{Oxygen.}}{O_2}$$

100 grams of manganese dioxide produce 12.82 litres of O.

(4.) By decomposing H_2SO_4 at a red heat, $2H_2SO_4 = 2SO_2 + 2H_2O + O_2$.

(5.) By the decomposition by heat of certain salts rich in O: alkaline permanganates, nitrates, and chlorates.

The best method, and that usually adopted, is by heating a mixture of potassium chlorate and manganese dioxide in equal parts, moderately at first and more strongly toward the end of the reaction. At the end of the operation the manganese dioxide remains, apparently unaltered, and it is probable that during the action it goes through a series of oscillating oxidations and deoxidations, which take place at a lower temperature than that required for the decomposition of the chlorate alone.

The chlorate gives up all its O (27.26 litres from 100 grains of the salt), according to the equation:

$$\underset{\text{Potassium chlorate.}}{2KClO_3} = \underset{\text{Potassium chloride.}}{2KCl} + \underset{\text{Oxygen.}}{3O_2}$$

41

The operation may be conducted in the apparatus shown in Fig. 18, or, on a large scale, with a copper or iron retort.

PROPERTIES.—*Physical.*—Oxygen is a colorless, odorless, tasteless gas, very sparingly soluble in water, somewhat more soluble in absolute alcohol. It liquefies at − 140° (229° F.) under a pressure of 300 atmospheres.

Chemical.—Oxygen is characterized, chemically, by the strong tendency which it exhibits to enter into combination with other elements, only one of which is known, *i.e.*, fluorine, that does not form an oxygenated compound. With most elements it unites directly, especially at elevated temperatures. In many instances this union is attended by the appearance of light, and always by the extrication of heat. The luminous union of O with another element constitutes the familiar phenomenon of *combustion*, and is the principal source from which we obtain so-called artificial heat and light. A body is said to be *combustible* when it is capable of so energetically combining with the oxygen of the air as to liberate light as well as heat. Gases are said to be *supporters of combustion*, when combustible substances will unite with them, or with some of their constituents, the union being attended with the appearance of heat and light. The distinction between combustible substances and supporters of combustion is, however, one of mere convenience; the action taking place between the two substances, one is as much a party to it as the other. A jet of air burns in an atmosphere of coal-gas as readily as a jet of coal-gas burns in air.

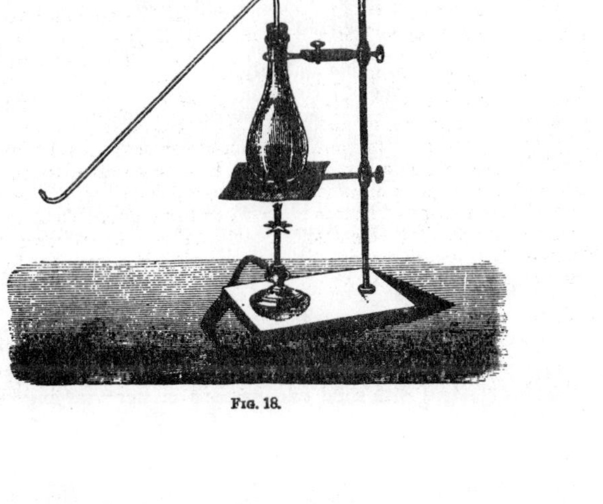

FIG. 18.

The compounds of oxygen—the oxides—are divisible into three groups:

1. *Anhydrides*—oxides capable of combining with water to form *acids*. Thus *sulphuric anhydride*, SO_3, unites with water to form *sulphuric acid*, H_2SO_4.

The term "anhydride" is not limited in application to *binary* compounds, but applies to any substance capable of combining with water to form an acid. Thus the compound $C_4H_6O_3$ is known as acetic anhydride,

42

because it combines with water to form acetic acid: $C_4H_6O_3 + H_2O = 2C_2H_4O_2$. (See compounds of arsenic and sulphur, p. 88.)

2. *Basic oxides* are such as combine with water to form bases. Thus, calcium oxide, CaO, unites with water to form calcium hydrate, CaH_2O_2.

3. *Saline, neutral, or indifferent oxides* are such as are neither acid nor basic in character. In some instances they are essentially neutral, as in the case of the protoxide of hydrogen or water. In other cases they are formed by the union of two other oxides, one basic, the other acid in quality, such as the red oxide of lead, Pb_3O_4, formed by the union of a molecule of the acidulous peroxide, PbO_2, with two of the basic protoxide, PbO. It is to oxides of this character that the term "saline" properly applies.

The process of respiration is very similar to combustion, and as oxygen gas is the best supporter of combustion, so, in the diluted form in which it exists in atmospheric air, it is not only the best, but the only supporter of animal respiration. (See carbon dioxide.)

ANALYTICAL CHARACTERS.—1.) A glowing match-stick bursts into flame in free oxygen. 2.) Free O when mixed with nitrogen dioxide produces a brown gas.

OZONE.—ALLOTROPIC OXYGEN.

Air through which discharges of static electricity have been passed, and oxygen obtained by decomposition of water (if electrodes of gold or platinum be used), have a peculiar odor, somewhat resembling that of sulphur, which is due to the conversion of a part of the oxygen into ozone.

Ozone has not been obtained free from oxygen; indeed, the highest degree of concentration which has been reached does not exceed ten per cent. of ozone. Thus diluted, ozone is produced: 1.) By the decomposition of water by the battery. 2.) By the slow oxidation of phosphorus in damp air. 3.) By the action of concentrated sulphuric acid upon barium dioxide. 4.) By the passage of silent electric discharges through air or oxygen.

In the preparation of ozonized oxygen the best results are obtained by passing a slow current of oxygen through an apparatus made entirely of glass and platinum, cooled by a current of cold water, and traversed by the invisible discharge of an induction coil.

Under the most favorable conditions hitherto attained, the nearest approach to pure ozone has been ten parts in one hundred, the remainder being unaltered oxygen.

When oxygen is ozonized it contracts slightly in volume, and when the ozone is removed from ozonized oxygen by mercury or potassium iodide the volume of the gas is not diminished. These facts, and the great chemical activity of ozone, have led chemists to regard it as condensed oxygen; the molecule of ozone being represented thus (OOO), while that of ordinary oxygen is (OO).

Ozone is very sparingly soluble in water, insoluble in solutions of acids and alkalies. In the presence of moisture it is slowly converted into oxygen at 100° (212° F.), a change which takes place rapidly and completely at 237° (459° F.). It is a powerful oxidant; it decomposes solutions of potassium iodide with formation of potassium hydrate and liberation of iodine; it oxidizes all metals except gold and platinum, in the presence of moisture; it decolorizes indigo and other organic pigments, and acts rapidly upon rubber, cork, and other organic substances.

43

ANALYTICAL CHARACTERS.—1.) Neutral litmus paper, impregnated with solution of potassium iodide, is turned blue when exposed to air containing ozone. The same litmus paper without iodide is not affected. 2.) Manganous sulphate solution is turned brown by ozone. 3.) Solutions of thallous salts are colored yellow or brown by ozone.

When inhaled, air containing 0.07 gram of ozone per litre causes intense coryza and hæmoptysis. It is probable that ozone is by no means as constant a constituent of the atmosphere as was formerly supposed. (See Hydrogen dioxide.)

Compounds of Hydrogen and Oxygen.

Two are known—hydrogen oxide or water, H_2O; hydrogen peroxide or oxygenated water, H_2O_2.

Water.

H_2O—*Molecular weight* = 18—*Sp. gr.* = 1—*Vapor density* = 0.6218 A —*Composition discovered by Priestley in* 1780.

Occurrence.—In unorganized nature H_2O exists in the gaseous form in atmospheric air and in volcanic gases; in the liquid form very abundantly; and as a solid in snow, ice, and hail.

As *water of crystallization* it exists in definite proportion in certain crystals, to the maintenance of whose shape it is necessary.

In the organized world H_2O forms a constituent part of every tissue and fluid.

Formation.—Water is formed: 1. By union, brought about by elevation of temperature, of one vol. O with two vols. H.

2. By burning H or substances containing it in air or O.

3. By heating organic substances containing H to redness with cupric oxide, or with other substances capable of yielding O. This method of formation is utilized to determine the amount of H contained in organic substances.

4. When an acid and a hydrate react upon each other to form a salt:

$$\underset{\text{Sulphuric acid.}}{H_2SO_4} + \underset{\text{Potassium hydrate.}}{2KHO} = \underset{\text{Potassium sulphate.}}{K_2SO_4} + \underset{\text{Water.}}{2H_2O}$$

5. When a metallic oxide is reduced by hydrogen:

$$\underset{\text{Cupric oxide.}}{CuO} + \underset{\text{Hydrogen.}}{H_2} = \underset{\text{Copper.}}{Cu} + \underset{\text{Water.}}{H_2O}$$

6. In the reduction and oxidation of many organic substances.

Pure H_2O is not found in nature. When required pure it is separated from suspended matters by filtration, and from dissolved substances by distillation.

Properties.—*Physical.*—With a barometric pressure of 760 mm. H_2O is solid below 0° (32° F.); liquid between 0° (32° F.) and 100° (212° F.); and gaseous above 100° (212° F.). When H_2O is enclosed in capillary tubes, or is at complete rest, it may be cooled to −15° (5° F.) without solidifying. If at this temperature it be agitated, it solidifies instantly and the temperature suddenly rises to 0° (32 F.). The melting-point of ice is lowered 0.0075° (0.0135° F.) for each additional atmosphere of pressure.

44

The boiling-point is subject to greater variations than the freezing-point. It is the lower as the pressure is diminished, and the higher as it is increased. Advantage is taken of the reduced boiling-point of solutions in vacuo for the separation of substances, such as cane sugar, which are injured at the temperature of boiling H_2O. On the other hand, the increased temperature that may be imparted to liquid H_2O under pressure is utilized in many processes, in the laboratory and in the arts, for effecting solutions and chemical actions which do not take place at lower temperatures. The boiling-point of H_2O holding solid matter in solution is higher than that of pure H_2O, the degree of increase depending upon the amount and nature of the substance dissolved. On the other hand, mixtures of H_2O with liquids of lower boiling-point boil at temperatures less than 100° (212° F.). Although the conversion of water into water-gas takes place most actively at 100° (212° F.), water and ice evaporate at all temperatures.

Water is the best solvent we have, and acts in some instances as a simple solvent, in others as a chemical solvent.

When a solid absorbs sufficient water from the air to form a solution it is said to deliquesce.

Gases are more soluble in cold than in hot liquids. Hydrogen forms an exception to this rule, being equally soluble in all temperatures.

The solubility of a gas in water varies directly as the pressure.

In most cases solids are more soluble in hot than in cold liquids.

When a liquid contains as much of a dissolved substance as it is capable of holding at the existing temperature, it is said to be saturated.

Solutions of certain salts, saturated at high temperatures, may be cooled without depositing any of the salt; they are then *super-saturated*, and contain more of the dissolved substance than they could take up at the lower temperature.

A saturated solution of one substance in H_2O is often capable of dissolving considerable quantities of another substance, and of then becoming capable of taking up a further quantity of the first substance.

The power of H_2O to dissolve gases increases with increased pressure.

Fats, resins, and, in general, organic substances containing a large number of carbon atoms, are insoluble in H_2O.

Vapor of water is colorless, transparent, and invisible. Sp. gr. 0.6234 A or 9 H. A litre of vapor of water weighs 0.8064. The latent heat of vaporization of water is 536.5; that is, as much heat is required to vaporize 1 kilo. of water at 100° as would suffice to raise 536.5 kilos. of water 1° in temperature. In passing from the liquid to the gaseous state, water expands 1,696 times in volume.

Chemical.—Water may be shown to consist of 1 vol. O and 2 vols. H, or 8 by weight of O and 1 by weight of H, either by analysis or synthesis.

Analysis is the reducing of a compound to its constituent elements.

Synthesis is the formation of a compound from its elements. A partial synthesis is one in which a complex compound is produced from a simpler one, but not from the elements.

Water may be resolved into its constituent gases: 1st. By electrolysis of acidulated water; H being given off at the negative and O at the positive pole. 2d. By passing vapor of H_2O through a platinum tube heated to a whiteness, or through a porcelain tube heated to about 1,100°. 3d. By the action of the alkaline metals. Hydrogen is given off, and the metallic hydrate remains in solution in an excess of H_2O. 4th. By passing vapor of H_2O over red-hot iron. Oxide of iron remains and H is given off.

45

Water combines with oxides to form new compounds, some of which are acids and others bases, known as *hydrates*.

A hydrate is a compound formed by the replacement of part of the hydrogen of water by another element or radical.

The hydrates of the electro-negative elements and radicals are *acids*; most of those of the electro-positive elements and radicals are *basic hydrates*.

A compound capable of combining with water to form an acid is called an anhydride.

Certain substances, in assuming the crystalline form, combine with a definite proportion of water, which is known as *water of crystallization*, and whose presence, although necessary to the maintenance of certain physical characters, such as color and crystalline form, does not modify their chemical reactions. In many instances a portion of the water of crystallization may be driven off at a comparatively low temperature, while a much higher temperature is required to expel the remainder. This latter is known as *water of constitution*.

The symbol Aq (Latin, *aqua*) is frequently used to designate the water of crystallization, the water of constitution being indicated by H_2O. Thus $MgSO_4\,H_2O + 6\,Aq$ represents magnesium sulphate with one molecule of water of constitution and six molecules of water of crystallization. We consider it preferable, however, as the distinction between water of crystallization and water of constitution is only one of degree and not of kind, to use the symbol Aq to designate the sum of the two; thus, $MgSO_4 + 7\,Aq$.

Crystals which lose their water of crystallization on exposure to air are said to *effloresce*; those which do not are said to be *permanent*.

Water decomposes the chlorides of the second class of elements (those of carbon only at high temperatures and under pressure); while the chlorides of the elements of the third and fourth classes are either insoluble, or soluble without decomposition.

Natural Waters.—Water, as it occurs in nature, always contains solid and gaseous matter in solution, and frequently solids in suspension.

Natural waters may be classified, according to the nature and quantity of foreign matters which they contain, into *potable* and *unpotable* waters. To the first class belong rain-water, snow- and ice-water, spring-water (fresh), river-water, lake-water, and well-water. To the second class belong stagnant waters, sea-water, and the waters of mineral springs.

Rain-water is usually the purest of natural waters, so far as dissolved solids are concerned, containing very small quantities of the chlorides, sulphates, and nitrates of sodium and ammonium. Owing to the large surface exposed during condensation, rain-water contains relatively large quantities of dissolved gases—oxygen, nitrogen, and carbon dioxide; and sometimes hydrogen sulphide and sulphur dioxide. The absence of carbonates and the presence of nitrates and oxygen render rain-water particularly prone to dissolve lead when in contact with that metal. In summer, rain-water is liable to become charged with vegetable organic matter suspended in the atmosphere.

Ice-water contains very small quantities of dissolved solids or gases, which, during freezing, remain in great part in the unfrozen water. Suspended impurities are imprisoned in the ice and liberated when this is melted.

Melted snow contains about the same proportion of fixed solid matter as rain-water, but a less proportion of ammoniacal salts and of gases.

Spring-water is rain-water which, having percolated through a portion

46

of the earth's crust (in which it may also have been subjected to pressure), has become charged with solid and gaseous matter; varying in kind and quantity according to the nature of the strata through which it has percolated, the duration of contact, and the pressure to which it was subject during such contact.

Spring-waters from igneous rocks and from the older sedimentary formations are fresh and sweet, and any spring-water may be considered such whose temperature is less than 20° (68° F.), and which does not contain more than 0.4 gram of solid matter to the litre (28 grains per gal.); provided that a large proportion of the solid matter does not consist of salts having a medicinal action, and that sulphurous gases and sulphides are absent.

Artesian wells are artificial springs, produced by boring in a low-lying district, until a pervious layer between two impervious strata is reached; the outcrop of the system being in an adjacent elevated region.

River-water is a mixture of rain-water, spring-water, and the drainage water of the district through which the river flows, to which snow-water, ice-water, or sea-water is sometimes added. The water of a river flowing rapidly through a granitic region is, unless polluted by man, bright, fresh, and highly aërated; that of a stream flowing sluggishly through rich alluvial land is unaërated, and rich in dissolved and suspended solids.

The amount of dissolved solids in river water increases with the distance from its source. The chief sources of pollution of river-water are by the discharge into them of the sewage of towns and cities, or of the waste products of factories.

Lake-water is an accumulation of river- and rain-water. As the waters of lakes are kept in constant agitation both by the wind and by the current, they become to a great extent purified from organic contamination.

Well-water may be very good or very bad. If the well be simply a reservoir dug over a spring, and removed from sources of contamination, it has all the characters of fresh spring-water. If, on the other hand, it be simply a hole dug in the earth, the water which it contains is the surface water which has percolated through the thin layer of earth corresponding to the depth of the well, and is consequently warm, unaërated and charged with organic impurity.

Wells dug near dwellings are very liable to become charged with the worst of contaminations, animal excreta, by their filtration through the soil, either by reasŏn of the fracture of the house-drain or otherwise.

Impurities in Potable Waters.—A water to be fit for drinking purposes should be cool, limpid, and odorless. It should have an agreeable taste, neither flat, salty, nor sweetish, and it should dissolve soap readily, without formation of any flocculent precipitate.

Although it is safe to condemn a water which does not possess the above characters, it is by no means safe to regard all waters which do possess them as beyond suspicion. To determine whether a water is potable it must be more carefully examined as to the following constituents:

Total solids.—The amount of solid material dissolved in potable waters varies from 5 to 40 in 100,000; and a water containing more than the latter quantity (28 grains per gall.) is to be condemned on that account alone.

To determine the quantity of total solids 25 c.c. of the filtered water are evaporated to dryness in a previously weighed platinum dish, over the water-bath. The dish with the contained dry residue is cooled in a desiccator and again weighed. The increase in weight, multiplied by 4,000, gives the total solids in parts per 100,000.

Hardness.—The greater part of the solid matter dissolved in natural waters consists of the salts of calcium, accompanied by less quantities of the salts of magnesium. The calcium salt is usually the carbonate or the sulphate; sometimes the chloride, phosphate, or nitrate.

A water containing an excess of calcareous salt is said to be *hard*, and one not so charged is said to be *soft*. If the hardness be due to the presence of the carbonate it is *temporary*, if due to the sulphate it is *permanent*. Calcium carbonate is almost insoluble in pure water, but in the presence of free carbonic acid the more soluble bicarbonate is dissolved, but on the water being boiled, it is decomposed with precipitation of the carbonate if the quantity exceed 0.5 gram per litre. As calcium sulphate is held in solution by virtue of its own, albeit sparing, solubility, it is not deposited when the water is boiled.

An accurate determination of the quantity of calcium and magnesium salts in water is rarely required; it is, however, frequently desirable to determine their quantity approximately, the result being the *degree of hardness*.

For this purpose a solution of soap of known strength is required. This is made by dissolving 10 grams of air-dried, white Castile soap, cut into thin shavings, in a litre of dilute alcohol (sp. gr. 0.949). To determine whether this solution contains the proper amount of soap, 10 c.c. of a solution, made by dissolving 1.11 grams of pure, recently fused calcium chloride in a litre of water, are diluted with 60 c.c. of water and the soap solution added until a persistent lather is produced on agitation. If 11 c.c. of soap solution have been used it has the proper strength; if a greater or less quantity have been added it must be concentrated or diluted in proportion to the excess or deficiency. The soap solution must not be filtered, but, if turbid, must be shaken before using.

To determine the hardness, 70 c.c. of the water to be tested are placed in a glass-stoppered bottle of 250 c.c. capacity, and the soap solution gradually added from a burette. After each addition of soap solution the bottle is shaken, and allowed to lie upon its side five minutes. This is continued until at the end of five minutes a lather remains upon the surface of the liquid in the bottle. At this time the hardness is indicated by the number of c.c. of soap solution added, minus one. If more than 16 c.c. of soap solution are added the liquid in the bottle must be diluted with 70 c.c. of distilled water.

A good drinking-water should not have a hardness of more than fifteen.

Chlorides.—The presence of the chlorides of the alkaline metals, in quantities not sufficient to be detectable by the taste, is of no importance *per se*; but in connection with the presence of organic impurity, a determination of the amount of chlorine affords a ready method of indicating the probable source of the organic contamination. As vegetable organic matter brings with it but small quantities of chlorides, while animal contaminations are rich in those compounds, the presence of a large amount of chlorine serves to indicate that organic impurity is of animal origin. Indeed, when time presses, as during an epidemic, it is best to rely upon determinations of chlorine, and condemn all waters containing more than 1.5 in 100,000 (one grain per gallon) of that element.

For the determination of chlorine two solutions are required: a solution of silver nitrate containing 4.79 grams per litre; a strong solution of potassium chromate. One hundred c.c. of the water are placed in a beaker with enough of the chromate solution to communicate a distinct yellow color. If the reaction be acid it is rendered neutral or *faintly* alkaline by the addition of sodium carbonate solution. The silver solution is now allowed to flow in from a burette, drop by drop, during constant agitation, until a *faint* reddish tinge persists. At this time the burette reading is taken; each c.c. of silver solution added represents 0.01 of chlorine per litre.

Organic Matter.—The most serious of the probable contaminations of drinking-water is that by organic matters containing nitrogen. When these are present in even moderate quantity, and when, at the same time, the proportion of chlorine is greater than usual, the water has been contaminated by animal excreta and contains, under suitable conditions, the causes of disease, be they germs or poisons.

Of the methods suggested for the determination of the amount of organic matter in natural waters there is unfortunately none which is easy

of application and at the same time reliable. That which yields the best results is Wanklyn's process:

The following solutions are required; *a*. Made by dissolving 200 grams of potassium hydrate and 8 grams of potassium permanganate in a litre of water. The solution is boiled down to about 725 c.c., cooled, and brought to its original bulk by the addition of boiled distilled water. *b*. *Nessler's reagent.* 35 grams of potassium iodide and 13 grams of mercuric chloride are dissolved in 800 c.c. of water by the aid of heat and agitation. A cold, saturated solution of mercuric chloride is then added, drop by drop, until the red precipitate which is formed is no longer redissolved on agitation; 160 grams of potassium hydrate are then dissolved in the liquid, to which a slight excess of mercuric chloride solution is finally added, and the bulk of the whole made up to a litre with water. The solution is allowed to stand, decanted, and preserved in completely filled, well-stoppered bottles. *c*. *Standard solutions of ammonia.* The stronger of these is made by dissolving 3.15 grams of ammonium chloride in a litre of water. The weaker, by mixing one volume of the stronger with 99 volumes of water. The latter contains 0.00001 gram of ammonia in each c.c., and is the one used in the determinations, the stronger solution serving only for its convenient preparation. *d*. *A saturated solution of sodium carbonate.* *e*. *Distilled water.* The middle third of the distillate, 100 c.c. of which must not be perceptibly colored in ten minutes by the addition of 2 c.c. of Nessler's reagent.

The testing of a water is conducted as follows: Half a litre of the water to be tested (before taking the sample, the demijohn or other vessel containing the water must be thoroughly shaken) is introduced, by a funnel, into a tubulated retort capable of holding one litre. If the water be acid, 10 c.c. of the solution of sodium carbonate *d* are added. Having connected the retort with a Liebig's condenser, the joint being made tight by a packing of moistened filter-paper, the water is made to boil as soon as possible by applying the flame of a Bunsen burner brought close to the bottom of the naked retort. The first 50 c.c. of distillate are collected in a cylindrical vessel of clear glass, about an inch in diameter. The following 150 c.c. are collected and thrown away, after which the fire is withdrawn. While these are passing over, the first 50 c.c. are Nesslerized (*vide infra*), and the result, plus one-third as much again, is the amount of free ammonia contained in the half-litre of water.

When 200 c.c. have distilled over, all the free ammonia has been removed, and it now remains to decompose the organic material, and determine the amount of ammonia formed. To effect this, 50 c.c. of the permanganate solution *a* are added through the funnel to the contents of the retort, which is shaken, stoppered, and again heated. The distillate is now collected in separate portions of 50 c.c. each, in glass cylinders, until 3 such portions have been collected. These are then separately Nesslerized as follows: 2 c.c. of the Nessler reagent are added to the sample of 50 c.c. of distillate; if ammonia be present, a yellow or brown color will be produced, dark in proportion to the quantity of ammonia present. Into another cylinder a given quantity of the standard solution of ammonia *c* is allowed to flow from a burette; enough water is added to make the bulk up to 50 c.c., and then 2 c.c. of Nessler reagent. This cylinder, and that containing the 50 c.c. of Nesslerized distillate, are then placed side by side upon a sheet of white paper and their color examined. If the shade of color in the two cylinders be *exactly* the same, the 50 c.c. of distillate contain the same amount of ammonia as the quantity of standard solution of ammonia used. If the colors be different in intensity, another comparison-cylinder must be arranged, using more or less of the standard solution, as the first comparison-cylinder was lighter or darker than the distillate. When the proper similarity of shades has been attained, the number of cubic centimetres of the standard solution used is determined by the reading on the burette. This process, which, with a little practice, is neither difficult nor tedious, is to be repeated with the first 50 c.c. of distillate and with the three portions of 50 c.c. each, distilled after the addition of the permanganate solution. If, for example, it required 1 c.c. of standard solution in Nesslerizing the first 50 c.c., and for the others 3.5 c.c., 1.5 c.c., and 0.2 c.c., the following is the result and the usual method of recording it:

Free ammonia	.01	Albuminoid ammonia.........	.035
Correction............	.003		.015
			.002
	.013		.052
Free ammonia per litre....	.026 milligr.	Albuminoid ammonia per litre....	.104 milligr.

If a water yield no albuminoid ammonia it is organically pure, even if it contains much free ammonia and chlorides; if it contains from .02 to .05 milligrams per litre, it is still quite pure; when the albuminoid ammonia reaches 0.1 milligr. per litre the water is to be looked upon with suspicion; and it is to be condemned when the proportion reaches 0.15. When free ammonia is also present in considerable quantity, a water yielding 0.05 of albuminoid ammonia is to be looked upon with suspicion.

Poisonous Metals.—Those most liable to occur in drinking waters are iron, copper, and lead, and of these the last is the most important.

The power possessed by a water of dissolving lead varies materially with the nature of the substances which it holds in solution. The presence of nitrates is favorable to the solution of lead, an influence which is, however, much diminished by the simultaneous presence of other salts. A water highly charged with oxygen dissolves lead readily, especially if the metallic surface be so exposed to the action of the water as to be alternately acted upon by it and by the air. On the other hand, waters containing carbonates or free carbonic acid may be left in contact with lead with comparative impunity, owing to the formation of a protective coating of the insoluble carbonate of lead on the surface of the metal. This does not apply, however, to water charged with a large excess of carbon dioxide under pressure. Of all natural waters that most liable to contamination with lead is rain-water; it contains ammonium nitrate with very small quantities of other salts; and it is highly aërated, but contains no carbonates and comparatively small quantities of carbon dioxide. Obviously, therefor, rain-water should neither be collected from a leaden roof, nor stored in leaden tanks, nor drank after having been long in contact with lead pipes. As a rule, the purer the water the more liable it is to dissolve lead when brought in contact with that metal, especially if the contact occur when the water is at a high temperature, or when it lasts for a long period.

To determine the power of water for dissolving lead, take two tumblers of the water to be tested; in one place a piece of lead, whose surface has been scraped bright, and allow them to stand twenty-four hours. At the end of that time remove the lead and pass sulphuretted hydrogen through the water in both tumblers; if the one which contained the metal become perceptibly darker than the other, the water has a power of dissolving lead such as to render its contact with surfaces of that metal dangerous if prolonged beyond a short time.

To test for the presence of poisonous metals, solution of ammonium sulphydrate is added to the water contained in a porcelain capsule. If a dark color be produced, which is not discharged on addition of hydrochloric acid, the water is contaminated with lead or copper.

For quantitative determinations solutions containing known quantities of the poisonous metals are used: for iron 4.96 grams of ferrous sulphate in a litre of water; for copper 3.93 grams of cupric sulphate to the litre; and for lead 1.66 gram of lead acetate to the litre. One c.c. of each solution contains 0.001 gram of the metal. To use the solutions 100 c.c. of the water to be tested and 100 c.c. of pure water are placed in two porcelain capsules, to each of which some ammonium sulphydrate is then added. The appropriate standard solution is then allowed to flow into the capsule containing the pure water, until the shade of color produced is the same as that of the liquid in the other capsule. The burette reading at this time gives the number of centigrams of the metal in a litre of water.

Suspended solids.—Most natural waters deposit, on standing, more or less solid, insoluble material. These substances have been either suspended mechanically in the water, which deposits them when it remains at rest, or they have been in solution, and are deposited by becoming insoluble as the water is deprived of carbon dioxide by exposure to air and by relief from pressure.

The suspended particles should be collected by subsidence in a conical glass, and should be examined microscopically for low forms of animal and vegetable life. The quantity of suspended solids is determined by passing a litre of the turbid water through a dried and weighed filter, which with the collected deposit, is again dried and weighed. The difference between the two weights is the weight of suspended matter in a litre of the water.

Purification of water.—The artificial means of rendering a more or less contaminated water fit for use are of five kinds: 1. Distillation; 2. Subsidence; 3. Filtration; 4. Precipitation; 5. Boiling.

The method of *distillation* is used in the laboratory when a very pure water is desired, and also at sea upon steamships, and even on sailing vessels upon occasion. Distilled water is, however, too pure for continued use, being hard of digestion, and flat to the taste from the absence of gases and of solid matter in solution. When circumstances oblige the use of such water, it should be agitated with air, and should be charged with inorganic matter to the extent of about 0.03 gram each of calcic bicarbonate and sodium chloride to the litre.

Purification by *subsidence* is adopted only as an adjunct to precipitation and filtration, and for the separation of the heavier particles of suspended matter.

50

The ideal process of *filtration* consists in the separation of all particles of suspended matter, without any alteration of such substances as are held in solution. In the filtration of potable waters on a large scale, however, the more minute particles of suspended matters are only partially separated, while, on the other hand, an important change in the dissolved materials takes place, at least in certain kinds of filters, in the oxidation of organic matters, whether in solution or in suspension. In the filtration of large quantities of water it is passed through sand or charcoal, or through both substances arranged in alternate layers. Filtration through charcoal is much more effective than that through sand, owing to the much greater activity of the oxidation of nitrogenized organic matter in the former case.

Precipitation processes are only adapted to hard waters, and are designed to separate the excess of calcium salt, and at the same time a considerable quantity of organic matter, which is mechanically carried down with the precipitate. The method usually followed consists in the addition of lime (in the form of lime-water), in just sufficient quantity to neutralize the excess of carbon dioxide present in the water. The added lime, together with the calcium salt naturally present in the water, is then precipitated, except that small portion of calcium carbonate which the water, freed from carbon dioxide, is capable of dissolving. To determine when sufficient lime-water has been added, take a sample from time to time during the addition, and test it with solution of silver nitrate until a brown precipitate is formed. At this point cease the addition of lime-water and mix the limed water with further portions of the hard water, until a sample, treated with silver-nitrate solution, gives a yellowish in place of a brown color.

The purification of water by *boiling* can only be carried on upon a small scale; it is, however, of great value for the softening of temporarily hard waters, and for the destruction of organized impurities, for which latter purpose it should never be neglected during outbreaks of cholera and typhoid, if, indeed, water be drank at all at such times.

MINERAL WATERS.—Under this head are classed all waters which are of therapeutic or industrial value, by reason of the quantity or nature of the dissolved solids which they contain; or which have a temperature greater than 20° (68° Fah.).

The composition of mineral waters varies greatly, according to the nature of the strata or veins through which the water passes, and to the conditions of pressure and previous composition under which it is in contact with these deposits.

The substances almost universally present in mineral waters are: oxygen, nitrogen, carbon dioxide; sodium carbonate, bicarbonate, sulphate and chloride; calcium carbonate and bicarbonate. Of substances occasionally present the most important are: sulphydric acid; sulphides of sodium, iron and magnesium; bromides and iodides of sodium and magnesium; calcium and magnesium chlorides; carbonate, bicarbonate, sulphate, peroxide, and crenate of iron; silicates of sodium, calcium, magnesium, and iron; aluminium salts; salts of lithium, cæsium, and rubidium; free sulphuric, silicic, arsenic, and boric acids; and ammoniacal salts.

Although a sharply defined classification of mineral waters is not possible, one which is useful, if not accurate, may be made, based upon the predominance of some constituent, or constituents, which impart to the water a well-defined therapeutic value. A classification which has been generally adopted is into five classes:

I. *Acidulous waters;* whose value depends upon dissolved carbonic acid.

51

They contain but small quantities of solids, principally the bicarbonates of sodium and calcium and sodium chloride.

II. *Alkaline waters;* which contain notable quantities of the carbonates or bicarbonates of sodium, potassium, lithium, and calcium, sufficient to communicate to them an alkaline reaction, and frequently a soapy taste; either naturally or after expulsion of carbon dioxide by boiling.

III. *Chalybeate waters;* which contain salts of iron in greater proportion than 40 milligrams per litre (2.8 grains per gall.). They contain ferrous bicarbonate, sulphate, crenate, and apocrenate, calcium carbonate, sulphates of potassium, sodium, calcium, magnesium, and aluminium, notable quantities of sodium chloride, and frequently small amounts of arsenic. They have the taste of iron and are usually clear as they emerge from the earth. Those containing ferrous bicarbonate deposit a sediment on standing, by loss of carbon dioxide and formation of ferrous carbonate.

IV. *Saline waters;* which contain neutral salts in considerable quantity. The nature of the salts which they contain is so diverse that the group may well be subdivided:

α. Chlorine waters; which contain large quantities of sodium chloride, accompanied by less amounts of the chlorides of potassium, calcium, and magnesium. Some are so rich in sodium chloride that they are not of service as therapeutic agents, but are evaporated to yield a more or less pure salt. Any natural water containing more than 3 grams per litre (210 grains per gall.) of sodium chloride belongs to this class, provided it do not contain substances more active in their medicinal action in such proportion as to warrant its classification elsewhere. Waters containing more than 15 grams per litre (1,050 grains per gall.) are too concentrated for internal administration.

β. Sulphate waters are actively purgative from the presence of considerable proportions of the sulphates of sodium, calcium, and magnesium. Some contain large quantities of sodium sulphate, with mere traces of the calcium and magnesium salts, while in others the proportion of the sulphates of magnesium and calcium is as high as 30 grams per litre (2,100 grains per gall.), to 20 grams per litre (1,400 grains per gall.) of sodium sulphate. They vary much in concentration; from 5 grams (350 grains per gall.) of total solids to the litre in some, to near 60 grams per litre (4,200 grains per gall.) in others. They have a salty, bitter taste, and vary much in temperature.

γ. Bromine and iodine waters are such as contain the bromides or iodides of potassium, sodium, or magnesium in sufficient quantity to communicate to them the medicinal properties of those salts.

V. *Sulphurous waters;* which hold hydrogen sulphide or metallic sulphides in solution. They have a disagreeable odor and are usually warm. They contain 0.2 to 4 grams of solids per litre (14–280 grains per gall.).

PHYSIOLOGICAL.—Water is taken into the body both as a liquid and as a constituent of every article of food; the amount ingested by a healthy adult being 2.25 to 2.75 litres (2⅓ to 3 quarts) per diem. The greater the elimination and the drier the nature of the food the greater is the amount of H_2O taken in the liquid form.

Water is a constituent of every tissue and fluid of the body, varying from 0.2 per cent. in the enamel of the teeth to 99.5 per cent. in the perspiration and saliva. It constitutes about 60 per cent. of the weight of the body.

The consistency of the various parts does not depend entirely upon the relative proportion of solids and H_2O, but is influenced by the nature of

52

the solids. The blood, although liquid in the ordinary sense of the term, contains a less proportional amount of H_2O than does the tissue of the kidneys, and about the same proportion as the tissue of the heart. Although the bile and mucus are not as fluid as the blood, they contain a larger proportion of H_2O to solids than does that liquid.

Water is discharged by the kidneys, intestine, skin, and pulmonary surfaces. The quantity discharged is greater than that ingested; the excess being formed in the body by the oxidation of the H of its organic constituents.

Hydrogen Dioxide.

Hydrogen peroxide--Oxygenated water.

H_2O_2—*Molecular weight* = 34—*Sp. gr.* = 1.455—*Discovered by Thénard in* 1818.

This substance may be obtained in a state of purity by accurately following the process of Thénard. It may also be obtained, mixed with a large quantity of H_2O, by passing a rapid current of carbon dioxide through H_2O holding hydrate of barium dioxide in suspension—$BaO_3H_2 + CO_2 = BaCO_3 + H_2O_2$. It is also formed in small quantity during the slow oxidation of many elements and compounds, such as P, Pb, Zn, Cd, Al, alcohol, ether, and the essences.

The pure substance is a colorless, syrupy liquid, which, when poured into H_2O, sinks under it before mixing. It has a disagreeable, metallic taste, somewhat resembling that of tartar emetic. When taken into the mouth it produces a tingling sensation, increases the flow of saliva, and bleaches the tissues with which it comes in contact. It is still liquid at $-30°$ ($-22°$ F.). It is very unstable, and, even in darkness and at ordinary temperatures, is gradually decomposed. At 20° (68° F.) the decomposition takes place more quickly, and at 100° (212° F.) rapidly and with effervescence. The dilute substance, however, is comparatively stable, and may be boiled and even distilled without suffering decomposition.

Hydrogen peroxide acts both as a reducing and an oxidizing agent. Arsenic, sulphides, and sulphur dioxide are oxidized by it at the expense of half its oxygen. When it is brought in contact with silver oxide both substances are violently decomposed, water and elementary silver remaining. By certain substances, such as gold, platinum, and charcoal in a state of fine division, fibrin, or manganese dioxide, it is decomposed with evolution of oxygen; the decomposing agent remaining unchanged.

The pure substance, when decomposed, yields 475 times its volume of oxygen; the dilute 15 to 20 volumes.

In dilute solution it is used as a bleaching agent and in the renovation of old oil-paintings.

ANALYTICAL CHARACTERS.—1. To a solution of starch a few drops of cadmium iodide solution are added, then a small quantity of the fluid to be tested, and, finally, a drop of a solution of ferrous sulphate. A blue color is produced in the presence of hydrogen peroxide, even if the solution contain only 0.05 milligram per litre.

2. Add freshly prepared tincture of guaiacum and a few drops of a cold infusion of malt. A blue color—1 in 2,000,000.

53

3. Add the liquid to be tested to mixed solutions of ferric chloride and potassium ferricyanide (which should have no blue tinge). A blue color —1 in 10,000,000.

4. Add to 6 c.c. of the liquid sulphuric acid, iodide of zinc, starch-paste, 2 drops of a two per cent. solution of cupric sulphate, and a little one-half per cent. solution of ferrous sulphate, in the order named. A blue color.

ATMOSPHERIC HYDROGEN DIOXIDE.—Atmospheric air constantly contains small quantities of hydrogen dioxide, which is also present in rain-water and in hail, and in less proportion in snow and hoar-frost. The amount present in rain-water varies from 0.008 to 0.499 milligram per litre (0.000454 to 0.028 grain per U. S. gallon), according to the direction of the wind and the season of the year. It is more abundant with equatorial than with polar winds, and more abundant in summer than in winter.

54

CLASS II.—ACIDULOUS ELEMENTS.

ELEMENTS ALL OF WHOSE HYDRATES ARE ACIDS, AND WHICH DO NOT FORM SALTS WITH THE OXACIDS.

I. CHLORINE GROUP.

FLUORINE. CHLORINE. BROMINE. IODINE.

The elements of this group are univalent. With hydrogen they form acid compounds, composed of one volume of the element in the gaseous state with one volume of hydrogen. Their hydrates are monobasic acids when they exist (fluorine forms no hydrate). The first two are gases, the third liquid, the fourth solid at ordinary temperatures. They are known as the *halogens*. The relations of their compounds to each other are shown in the following table:

Hydro-ic acid.	Monoxide.	Trioxide.	Tetroxide.	Hypo-ous acid.	-ous acid.	-ic acid.	Per-ic acid.
HF,	——	——	——	——	——	——	——
HCl,	Cl_2O	Cl_2O_3	Cl_2O_4	HClO	$HClO_2$	$HClO_3$	$HClO_4$
HBr	——	——	——	HBrO	——	$HBrO_3$	$HBrO_4$
HI	——	——	I_2O_4	HIO	HIO_2	HIO_3	HIO_4

FLUORINE.

Symbol = **F**—*Atomic weight* = 19—*Discovered by Sir H. Davy in* 1812.

Although many attempts have been made to isolate this element, it has probably never been obtained in the free state, unless the colorless gas obtained by G. J. and Th. Knox, by the decomposition of mercury fluoride and of hydrofluoric acid in vessels of fluor-spar was the element. Fluorine forms compounds with all the other elements except oxygen.

Hydrogen Fluoride.

Hydrofluoric acid = **HF**—*Molecular weight* = 20.

Hydrofluoric acid is obtained by the action of an excess of sulphuric acid upon fluor-spar, with the aid of gentle heat: $CaFl_2 + H_2SO_4 = CaSO_4 + 2HF$. If a solution be desired, the operation is conducted in a platinum or lead retort, whose beak is connected with a U-shaped receiver of the same metal, which is cooled and contains a small quantity of water.

The aqueous acid is a colorless liquid, highly acid and corrosive, and having a penetrating odor. Great care must be exercised that neither the solution nor the gas come in contact with the skin, as they produce painful ulcers which heal with difficulty, and also constitutional symptoms which may last for days. When the acid has accidentally come in contact with the skin the part should be washed with dilute solution of potash, and the vesicle which forms should be opened.

55

Both the gaseous acid and its solution remove the silica from glass, a property utilized in etching upon that substance, the parts upon which no action is desired being protected by a coating of wax.

The presence of fluorine in a compound is detected by reducing the substance to powder, moistening it with sulphuric acid in a platinum crucible, over which is placed a slip of glass prepared as above; at the end of half an hour the wax is removed from the glass, which will be found to be etched if the substance examined contained a fluoride.

CHLORINE.

Symbol = Cl—*Atomic weight* = 35.5—*Molecular weight* = 71—*Sp. gr.* = 2.4502 *A—One litre weighs* 3.17 *grams*—100 *cubic inches weigh* 76.3 *grains—Name derived from* χλωρός = *yellowish-green—Discovered by Scheele in* 1774.

OCCURRENCE.—Only in combination, most abundantly in sodium chloride.

PREPARATION.—(1.) By heating together manganese dioxide and hydrochloric acid (Scheele). The reaction takes place in two stages: manganic chloride is first formed according to the equation: $MnO_2 + 4HCl = MnCl_4 + 2H_2O$; and is subsequently decomposed into manganous chloride and chlorine: $MnCl_4 = MnCl_2 + Cl_2$.

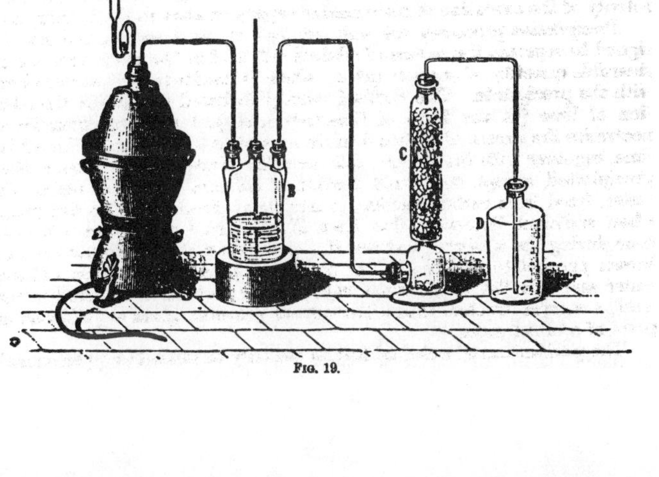

FIG. 19.

This and similar operations are usually conducted in an apparatus such as that shown in Fig. 19. The earthenware vessel A (which on a small scale may be replaced by a glass flask) is two-thirds filled with lumps of manganese dioxide of the size of hazel-nuts, and adjusted in the water-bath; hydrochloric acid is poured in through the safety-tube and the bath heated. The disengaged gas is caused to bubble through the small quantity of water in B, is then dried by passage over the fragments of calcium chloride in C, and is finally collected by displacement of air in the vessel D.

When the vessel A has become half filled with liquid it is best to decant the solution of manganous chloride, wash the remaining oxide with water and begin anew. A kilo. of oxide yields 257.5 litres of Cl.

(2.) By the action of manganese dioxide upon hydrochloric acid in the presence of sulphuric acid, manganous sulphate being also formed: MnO_2 +

56

$2HCl + H_2SO_4 = MnSO_4 + 2H_2O + Cl_2$. The same quantity of chlorine is obtained as in (1), with the use of half the amount of hydrochloric acid.

(3.) By heating a mixture of one part each of manganese dioxide and sodium chloride, with three parts of sulphuric acid. Hydrochloric acid and sodium sulphate are first formed: $H_2SO_4 + 2NaCl = Na_2SO_4 + 2HCl$; and the acid is immediately decomposed by either of the reactions indicated in (1) and (2), according as sulphuric acid is or is not present in excess.

(4.) By the action of potassium dichromate upon hydrochloric acid; potassium and chromic chlorides being also formed: $K_2Cr_2O_7 + 14HCl = 2KCl + Cr_2Cl_6 + 7H_2O + 3Cl_2$. Two parts of powdered dichromate are heated with 17 parts of acid of sp. gr., 1.16; 100 grams of the salt yielding 22.5 litres of Cl.

(5.) When a slow evolution of Cl, extending over a considerable period of time, is desired, as for ordinary disinfection, moistened chloride of lime is exposed to the air, the calcium hypochlorite being decomposed by the atmospheric carbon dioxide. If a more rapid evolution of gas be desired, the chloride of lime is moistened with dilute hydrochloric acid in place of with water.

Properties.—*Physical.*—A greenish yellow gas, at the ordinary temperature and pressure; it has a penetrating odor, and is, even when highly diluted, very irritating to the respiratory passages. Being soluble in H_2O to the extent of one volume to three volumes of the solvent, it must be collected by displacement of air, as shown in Fig. 19. A saturated aqueous solution of Cl is known to chemists as *chlorine water*, and in pharmacy as *aqua chlori* (*U. S.*), *Liquor chlori* (*Br.*); it should bleach, but not redden, litmus paper. Under a pressure of 6 atmospheres at 0° (32° F.), or $8\frac{1}{2}$ atmospheres at 12° (53°.6 F.), Cl becomes an oily, yellow liquid, of sp. gr. 1.33; and boiling at − 33.6° (− 28°.5 F.).

Chemical.—Chlorine exhibits a great tendency to combine with other elements, with all of which, except F, O, N, and C, it unites directly, frequently with evolution of light as well as heat, and sometimes with an explosion. With H it combines slowly, to form hydrochloric acid, under the influence of diffuse daylight, and violently in direct sunlight or in highly actinic artificial lights. A candle burns in Cl with a faint flame and thick smoke, its H combining with the Cl, while carbon becomes free.

At a red heat Cl decomposes H_2O rapidly, with formation of hydrochloric acid. The same change takes place slowly under the influence of sunlight, hence chlorine water should be kept in the dark or in bottles of yellow glass.

In the presence of H_2O, chlorine is an active bleaching and disinfecting agent. It acts as an indirect oxidant, decomposing H_2O, the nascent O from which then attacks the coloring or odorous principle.

Chlorine is readily fixed by many organic substances, either by addition or substitution. In the first instance, as when Cl and olefiant gas unite to form ethylene chloride, the organic substance simply takes up one or more atoms of chlorine: $C_2H_4 + Cl_2 = C_2H_4Cl_2$. In the second instance, as when Cl acts upon marsh gas to produce methyl chloride: $CH_4 + Cl_2 = CH_3Cl + HCl$, each substituted atom of Cl displaces an atom of H, which combines with another Cl atom to form hydrochloric acid.

Hydrate of chlorine, $Cl5H_2O$, is a yellowish green, crystalline substance, formed when Cl is passed through chlorine water cooled to 0° (32° F.). It is decomposed at 10° (50° F.).

Analytical Characters.—See p. 62.

57

Hydrogen Chloride.

Hydrochloric Acid.—Muriatic Acid.—Acidum Hydrochloricum (*U. S.*; *Br.*)—HCl—*Molecular weight* = 36.5—*Sp. gr.*, 1.259 *A—A litre weighs* 1.6293 *gram.*

Occurrence.—In volcanic gases and in the gastric juice of the mammalia.

Preparation.—(1.) By the direct union of its constituent elements.

(2.) By the action of sulphuric acid upon a chloride, a sulphate being at the same time formed: $H_2SO_4 + 2NaCl = Na_2SO_4 + 2HCl$.

This is the reaction by which the HCl used in the arts is produced, either as a separate industry or as an incidental product in Leblanc's process for obtaining sodium carbonate (q. v.).

(3.) Hydrochloric acid is also formed in a great number of reactions, as when Cl is substituted in an organic compound.

Properties.—*Physical.*—A colorless gas, acid in reaction and taste, having a sharp, penetrating odor, and producing great irritation when inhaled. It becomes liquid under a pressure of 40 atmospheres at 4° (39° F.). It is very soluble in H_2O, one volume of which dissolves 480 volumes of the gas at 0° (32° F.).

Chemical.—Hydrochloric acid is neither combustible nor a supporter of combustion, although certain elements, such as K and Na, burn in it. It forms white clouds on contact with moist air.

Solution of Hydrochloric Acid.—It is in the form of aqueous solution that this acid is usually employed in the arts and in pharmacy. It is, when pure, a colorless liquid (yellow when impure), acid in taste and reaction, whose sp. gr. and boiling-point vary with the degree of concentration. When heated, it evolves HCl, if it contain more than 20 per cent. of that gas, and H_2O if it contain less. A solution containing 20 per cent. boils at 111° (232° F.), is of sp. gr. 1.099, has the composition $HCl + 8H_2O$, and distils unchanged.

Commercial muriatic acid is a yellow liquid; sp. gr. about 1.16; contains 32 per cent. HCl; and contains iron, sodium chloride, and arsenical compounds.

Acidum hydrochloricum is a colorless liquid, containing small quantities of impurities. It contains 31.9 per cent. HCl and its sp. gr. is 1.16 (U. S.; Br.). The dilute acid is the above diluted with water. Sp. gr. 1.049 = 10 per cent. HCl (U. S.); sp. gr. 1.052 = 10.5 per cent. HCl (Br.).

C. P. (*chemically pure*) *acid* is usually the same as the strong pharmaceutical acid and far from pure (see below).

Hydrochloric acid is classed, along with nitric and sulphuric acids, as one of the three *strong mineral acids*. It is decomposed by many elements, with formation of a chloride and liberation of hydrogen: $2HCl + Zn = ZnCl_2 + H_2$. With oxides and hydrates of elements of the third and fourth classes it enters into double decomposition, forming H_2O and a chloride: $CaO + 2HCl = CaCl_2 + H_2O$ or $CaH_2O_2 + 2HCl = CaCl_2 + 2H_2O$. Most of the metallic chlorides are soluble in H_2O, those of Ag, Pb, and Hg (ous) being exceptions. The chlorides of the non-metals are decomposed on contact with H_2O.

Oxidizing agents decompose HCl with liberation of Cl. A mixture of hydrochloric and nitric acids in the proportion of three molecules of the former to one of the latter, is the *acidum nitrohydrochloricum* (*U. S.*; *Br.*),

58

or *aqua regia*. The latter name alludes to its power of dissolving gold, by combination of the nascent Cl which it liberates with that metal to form the soluble auric chloride.

Impurities.—A chemically pure solution of this acid is exceedingly rare. The impurities usually present are: *Sulphurous acid*—hydrogen sulphide is given off when the acid is poured upon zinc; *Sulphuric acid*—a white precipitate is formed with barium chloride; *Chlorine* colors the acid yellow; *Lead* gives a black color when the acid is treated with hydrogen sulphide; *Iron*—the acid gives a red color with ammonium sulphocyanate; *Arsenic*—the method of testing by hydrogen sulphide is not sufficient. If the acid is to be used for toxicological analysis, a litre, diluted with half as much H_2O, and to which a small quantity of potassium chlorate has been added, is evaporated over the water-bath to 400 c.c.; 25 c.c. of sulphuric acid are then added, and the evaporation continued until the liquid measures about 100 c.c. This is introduced into a Marsh apparatus and must produce *no* mirror during an hour.

Analytical Characters.—See p. 62.

Toxicology.—*Poisons and corrosives.—A poison is any substance which, after absorption into the blood, produces death or serious bodily harm.*

A corrosive is a substance capable of producing death by its chemical action upon a tissue with which it comes in direct contact, without absorption by the blood.

Under the above definitions the strong mineral acids act as corrosives rather than as poisons. They produce their injurious results by destroying the tissues with which they come in contact, and will cause death as surely by destroying a large surface of skin as when they are taken into the stomach.

The object of the treatment in corrosion by the mineral acids is to neutralize the acid and convert it into a harmless salt. For this purpose the best agent is magnesia (magnesia usta) suspended in a small quantity of water; or, if this be not at hand, a strong solution of soap. Chalk and the carbonates and bicarbonates of sodium and potassium should *not* be given, as they generate large volumes of gas. The scrapings of a plastered wall, or oil, are entirely useless. The stomach-pump, or any attempt at the introduction of a tube into the œsophagus, is not to be thought of.

Compounds of Chlorine and Oxygen.

Three compounds of chlorine and oxygen have been isolated, two being anhydrides. They are all very unstable, and prone to sudden and violent decomposition.

Chlorine Monoxide—Cl_2O—87— = *hypochlorous anhydride* or *oxide*, is formed as a blood-red liquid by the action, below 20° (68° F.), of dry Cl

upon precipitated mercuric oxide : $HgO + 2Cl_2 = HgCl_2 + Cl_2O$.

On contact with H_2O it forms *hypochlorous acid,* HClO, which, owing to its instability, is not used industrially, although the hypochlorites of Ca, K, and Na are.

CHLORINE TRIOXIDE = *chlorous anhydride* or *oxide,* Cl_2O_3—119—is a yellowish-green gas, formed by the action of dilute nitric acid upon potassium chlorate in the presence of arsenic trioxide. At 50° (122° F.) it explodes. It is a strong bleaching agent; is very irritating when inhaled, and readily soluble in H_2O, the solution probably containing *chlorous acid,* $HClO_2$.

59

CHLORINE TETROXIDE = *chlorine peroxide,* Cl_2O_4—135—is a violently explosive body, produced by the action of sulphuric acid upon potassium chlorate. Below − 20° (− 4° F.), it is an orange-colored liquid ; above that temperature, a yellow gas. There is no corresponding hydrate known ; and if it be brought in contact with an alkaline hydrate, a mixture of chlorate and chlorite is formed.

Besides the above, two oxacids of Cl are known, the anhydrides corresponding to which have not been isolated.

CHLORIC ACID—$HClO_3$ — 84.5—obtained in aqueous solution as a strongly acid, yellowish, syrupy liquid, by decomposing its barium salt by the proper quantity of sulphuric acid.

PERCHLORIC ACID—$HClO_4$—100.5—is the most stable of the series. It is obtained by boiling potassium chlorate with hydrofluosilicic acid, decanting the cold fluid, evaporating until white fumes appear, decanting from time to time, and finally distilling. It is a colorless, oily liquid ; sp. gr. 1.782 ; which explodes on contact with organic substances or charcoal.

BROMINE.

Symbol = **Br**—*Atomic weight* = 80—*Molecular weight* = 160—*Sp. gr. of liquid* = 3.1872 at 0° ; *of vapor* = 5.52 *A*—*Freezing-point* = − 24°.5 (−12°.1 *F.*)—*Boiling-point* = 63° (145°.4 *F.*)—*Name derived from* βρῶμος = *a stench.*—*Discovered by Balard in* 1826—*Bromum* (*U. S. ; Br.*).

OCCURRENCE.—Only in combination, most abundantly with Na and Mg in sea-water and the waters of mineral springs.

PREPARATION.—It is obtained from the mother liquors left by the evaporation of sea-water and of that of certain mineral springs, and from seaweed. These are mixed with sulphuric acid and manganese dioxide and heated, when the bromides are decomposed by the Cl produced, and Br distils.

PROPERTIES.—*Physical.*—A dark reddish brown liquid, volatile at all temperatures above − 24°.5 (− 12°.1 F.); giving off brown-red vapors which produce great irritation when inhaled. Soluble in water to the extent of 3.2 parts per 100 at 15° (59° F.) ; more soluble in alcohol, carbon disulphide, chloroform, and ether.

Chemical.—The chemical characters of Br are similar to those of Cl, but less active. With H_2O it forms a crystalline hydrate at 0° (32° F.): $Br.5H_2O$. Its aqueous solution is decomposed by exposure to light, with formation of hydrobromic acid.

It is highly poisonous.

ANALYTICAL CHARACTERS.—See p. 62.

Hydrogen Bromide.

Hydrobromic acid—*Acidum hydrobromicum dil.* (*U. S.*) = **HBr**—*Molecular weight* = 81—*Sp. gr.* = 2.71 *A*—*A litre weighs* 3.63 *grams*—*Liquefies at* − 69° (−92°.2 *F.*)—*Solidifies at* − 73° (−99°.4 *F.*).

PREPARATION.—This substance cannot be obtained from a bromide as HCl is obtained from a chloride. It is produced, along with phosphorous acid, by the action of H_2O, upon phosphorus tribromide : $PBr_3 + 3H_2O = H_3PO_3 + 3HBr$; or by the action of Br upon paraffine.

60

PROPERTIES.—A colorless gas ; produces white fumes with moist air ; acid in taste and reaction, and readily soluble in H_2O, with which it forms a hydrate, $HBr2H_2O$.

Its chemical properties are similar to those of the corresponding Cl compound.

ANALYTICAL.—See page 62.

Oxacids of Bromine.

No oxides of bromine are known, although three oxacids exist, either in the free state or as salts :

HYPOBROMOUS ACID—**HBrO**—97—is obtained, in aqueous solution, by the action of Br upon mercuric oxide, silver oxide, or silver nitrate. When Br is added to concentrated solution of potassium hydrate, no hypobromite is formed, but a mixture of bromate and bromide, having no decolorizing action. With sodium hydrate, however, sodium hypobromite is formed in solution ; and such a solution, freshly prepared, is used in Knop's process for determining urea (q. v.).

BROMIC ACID—$HBrO_3$—129—has only been obtained in aqueous solution or in combination. It is formed by decomposing barium bromate with an equivalent quantity of sulphuric acid : $Ba(BrO_3)_2 + H_2SO_4 = 2HBrO_3 + BaSO_4$. In combination it is produced, along with the bromide, by the action of Br on caustic potassa : $3Br_2 + 6KHO = KBrO_3 + 5KBr + 3H_2O$.

PERBROMIC ACID—$HBrO_4$—145—is obtained on a comparatively stable, oily liquid, by the decomposition of perchloric acid by Br, and concentrating over the water-bath.

It is noticeable in this connection that, while HCl and the chlorides are more stable than the corresponding Br compounds, the oxygen compounds of Br are more permanent than those of Cl.

IODINE.

Symbol = I—*Atomic weight* = 127—*Molecular weight* = 254—*Sp. gr. of solid* = 4.948 ; *of vapor* = 8.716 *A*—*Fuses at* 113°.6 (236°.5 *F.*)—*Boils at* 175° (347° *F.*)—*Name derived from* ἰώδης = *violet*—*Discovered by Courtois in* 1811—*Iodum* (*U. S. ; Br.*).

OCCURRENCE.—In combination with Na, K, Ca, and Mg, in sea-water, the waters of mineral springs, marine plants and animals ; cod-liver oil contains about 37 parts in 100,000.

PREPARATION.—It is obtained from the ashes of sea-weed, called *kelp* or *varech.* These are extracted with H_2O, and the solution evaporated to small bulk. The mother liquor, separated from the other salts which crystallize out, contains the iodides, which are decomposed by Cl, aided by heat, and the liberated iodine condensed.

PROPERTIES.—*Physical.*—Blue-gray, crystalline scales, having a metallic lustre. Volatile at all temperatures, the vapor having a violet color and a peculiar odor. It is sparingly soluble in H_2O, which, however, dissolves larger quantities on standing over an excess of iodine, by reason of the formation of hydriodic acid. The presence of certain salts, notably potassium iodide, increase the solvent power of H_2O for iodine. The *Liq. Iodi*

61

Comp. (*U. S*), (*Liq. Iodi, Br.*) is solution of potassium iodide containing free iodine. Very soluble in alcohol ; *Tinct. iodi* (*U. S. ; Br*) ; in ether chloroform, benzol, and carbon disulphide. With the three last-named solvents it forms violet solutions, with the others brown solutions.

Chemical.—In its chemical characters I resembles Cl and Br, but is less active. It decomposes H_2O slowly and is a weak bleaching and oxidizing agent. It decomposes hydrogen sulphide with formation of hydriodic acid and liberation of sulphur. It does not combine directly with oxygen, but does with ozone. Potassium hydrate solution dissolves it, with formation of potassium iodide and some hypoiodite. Nitric acid oxidizes it to iodic acid. With ammonium hydrate solution it forms the explosive nitrogen iodide.

IMPURITIES.—*Non-volatile substances* remain when the I is volatilized. *Water* separates as a distinct layer when I is dissolved in carbon disulphide. *Cyanogen iodide* appears in white, acicular crystals among the crystals of sublimed I when half an ounce of the substance is heated over the water-bath for twenty minutes, in a porcelain capsule, covered with a flat-bottomed flask filled with cold water. The last named is the most serious impurity as it is actively poisonous.

TOXICOLOGY.—Taken internally, iodine acts both as a local irritant and as a true poison. It is discharged as an alkaline iodide by the urine and perspiration, and when taken in large quantity it appears in the fæces.

The poison should be removed as rapidly as possible by the use of the stomach-pump and of emetics. Farinaceous substances may also be given.

ANALYTICAL CHARACTERS.—See below.

Hydrogen Iodide.

Hydriodic acid—**HI**—*Molecular weight* = 128—*Sp. gr.* **4.443 A.**

PREPARATION.—By the decomposition of phosphorous triiodide by water : $PI_3 + 3H_2O = H_3PO_3 + 3HI$. Or, in solution, by passing hydrogen sulphide through water holding iodine in suspension : $H_2S + I_2 = 2HI + S$.

PROPERTIES.—A colorless gas, forming white fumes on contact with air, and of strong acid reaction. Under the influence of cold and pressure it forms a yellow liquid, which solidifies at −55° (−67° F.). Water dissolves it to the extent of 425 volumes for each volume of the solvent at 10° (50° F.).

It is partly decomposed into its elements by heat. Mixed with O it is decomposed, even in the dark, with formation of H_2O and liberation of I. Under the influence of sunlight the gas is slowly decomposed, although its solutions are not so affected, if they be free from air. Chlorine and bromine decompose it, with liberation of iodine. With many metals it forms iodides. It yields up its H readily and is used in organic chemistry as a source of that element in the nascent state.

ANALYTICAL CHARACTERS.—CHLORINE, BROMINE, AND IODINE, AND THEIR BI-

NARY COMPOUNDS.—*Chlorine.*—(1.) Color.

(2.) Odor.

(3.) **Is dissolved by solutions of the alkaline hydrates, to which it communicates bleaching powers.**

(4.) With silver nitrate solution it gives a white ppt., soluble in NH_4HO, insoluble in HNO_3.

***Bromine.*—(1.) Color of liquid and vapor.**

62

(2.) Chloroform or carbon disulphide, when shaken with solution of Br, assume a yellow or brown color.

(3.) Colors starch paste yellow.

Iodine.—(1.) Color of vapor.

(2.) Dissolves in chloroform and carbon disulphide with a violet color.

(3.) Colors starch-paste deep violet-blue, the color disappearing on heating and returning on cooling.

Chlorides.—(1.) With $AgNO_3$, a white ppt., insoluble in HNO_3, readily soluble in NH_4HO.

(2.) With $Hg_2(NO_3)_2$, a white ppt., which turns black with NH_4HO.

Bromides.—(1.) With $AgNO_3$, a yellowish-white ppt., insoluble in HNO_3, sparingly soluble in NH_4HO.

(2.) With chlorine water, a yellow color, and when shaken with chloroform the latter is colored yellow; or colors starch-paste yellow.

Iodides.—(1.) With $AgNO_3$, a yellowish-white ppt., insoluble in HNO_3, almost insoluble in NH_4HO.

(2.) With fuming HNO_3, a yellow color, and when shaken with chloroform the latter is colored violet; or colors starch-paste dark blue.

(3.) With $PdCl_2$, a dark brown ppt.

Oxacids of Iodine.

The best known of these are the highest two of the series—iodic and periodic acids.

IODIC ACID—**HIO_3**—176—is formed as an iodate, whenever I is dissolved in a solution of an alkaline hydrate: $I_6 + 6KHO = KIO_3 + 5KI + 3H_2O$; as the free acid, by the action of strong oxidizing agents, such as nitric acid or chloric acid, upon I; or by passing Cl for some time through H_2O holding I in suspension.

Iodic acid appears in white crystals, decomposable at 170° (338° F.), and quite soluble in H_2O, the solution having an acid reaction, and a bitter, astringent taste.

It is an energetic oxidizing agent, yielding up its O readily, with separation of elementary I or of HI. It is used as a test for the presence of morphine (*q. v.*).

PERIODIC ACID—**HIO_4**—192—is formed by the action of Cl upon an alkaline solution of sodium iodate. The sodium salt thus obtained is dissolved in nitric acid, treated with silver nitrate, and the resulting silver periodate decomposed with H_2O. From the solution the acid is obtained in colorless crystals, fusible at 130° (266° F.), very soluble in water, and readily decomposable by heat.

II. SULPHUR GROUP.

SULPHUR—SELENIUM—TELLURIUM.

The elements of this group are bivalent. With hydrogen they form compounds composed of one volume of the element, in the form of vapor, with two volumes of hydrogen—the combination being attended with a condensation in volume of one-third. Their hydrates are dibasic acids

63

They are all solid at ordinary temperatures. The relation of their compounds to each other are shown in the following table:

Hydro-ic acid.	Dioxide.	Trioxide.	Hypo-ous acid.	-ous acid.	-ic acid.
H_2S	SO_2	SO_3	H_2SO_2	H_2SO_3	H_2SO_4
H_2Se	SeO_2	SeO_3	——	H_2SeO_3	H_2SeO_4
H_2Te	TeO_2	TeO_3	——	H_2TeO_3	H_2TeO_4

SULPHUR.

Symbol = **S**—*Atomic weight* = 32—*Molecular weight* = 64—*Sp. gr. of vapor* = 2.22 A—*Fuses at* 114° (237.2° F.)—*Boils at* 447.3° (837° F.).

OCCURRENCE.—Free in crystalline powder, large crystals, or amorphous in volcanic regions. In combination in sulphides and sulphates, and in albuminoid substances.

PREPARATION.—By purification of the native sulphur, or decomposition of *pyrites*, natural sulphides of iron.

Crude sulphur is the product of a first distillation. A second distillation in more perfectly constructed apparatus yields *refined sulphur*. During the first part of the distillation, while the air of the condensing chamber is still cool, the vapor of S is suddenly condensed into a fine, crystalline powder, which is *flowers of sulphur, sulphur sublimatum* (*U. S.*). Later, when the temperature of the condensing chamber is above 114°, the liquid S collects at the bottom, whence it is drawn off and cast into sticks of *roll sulphur*.

PROPERTIES.—*Physical.*—Sulphur is usually yellow in color; at low temperatures, and in minute subdivision, as in the precipitated *milk of sulphur, sulphur præcipitatum* (*U. S.*), it is almost or quite colorless. Its taste and odor are faint but characteristic. At 114° (237°.2 F.) it fuses to a thin yellow liquid, which at 150°–160° (302°–320° F.) becomes thick and brown; at 330°–340° (626°–642°.2 F.) it again becomes thin and light in color; finally it boils, giving off brownish-yellow vapor at a temperature variously stated between 440° (824° F) and 448° (838°.4 F.). If heated to about 400° (752° F.) and suddenly cooled it is converted into *plastic sulphur*, which may be moulded into any desired form. It is insoluble in water, sparingly soluble in anilin, phenol, benzol, benzine, and chloroform; readily soluble in protochloride of sulphur and carbon disulphide. It is dimorphous; when fused sulphur crystallizes it does so in oblique rhombic prisms; its solution in carbon disulphide deposits it on evaporation in rhombic octahedra. The prismatic variety is of sp. gr. 1.95 and fuses at 120° (248° F.); the sp. gr. of the octahedral is 2.05, and its fusing point 114°.5 (238° F.). The prismatic crystals by exposure to air become opaque, by reason of a gradual conversion into octahedra.

Chemical.—**Sulphur unites readily with other elements, especially at high temperatures. Heated in air or O, it burns with a blue flame to sulphur dioxide, SO_2. In H it burns with formation of hydrogen sulphide, H_2S. The compounds of S are similar in constitution, and to some extent in chemical properties, to those of O. In many organic substances S may replace O, as in sulphocyanic acid, CNSH, corresponding to cyanic acid, CNOH.**

USES.—Sulphur is used principally in the manufacture of gunpowder; also to some extent in making sulphuric acid, sulphur dioxide, and matches, and for the prevention of fungoid and parasitic growths.

64

Hydrogen Sulphide.

Sulphydric acid—Hydrosulphuric acid—Sulphuretted hydrogen.

H_2S—*Molecular weight* = 34—*Sp. gr.* = 1.19 A.

OCCURRENCE.—In volcanic gases; as a product of the decomposition of organic substances containing S; in solution in the waters of some mineral springs; and occasionally in small quantity in the gases of the intestine.

PREPARATION.—(1.) By direct union of the elements; either by burning S in H, or by passing H through molten S.

(2.) By the action of nascent H upon sulphuric acid if the mixture become heated. (See Marsh test for arsenic.)

(3.) By the action of HCl upon antimony trisulphide: $Sb_2S_3 + 6HCl = 2SbCl_3 + 3H_2S$.

(4.) By the action of dilute sulphuric acid upon ferrous sulphide: $FeS + H_2SO_4 = FeSO_4 + H_2S$.

(5.) By the action of HCl upon calcium sulphide: $CaS + 2HCl = CaCl_2 + H_2S$.

The gas is usually obtained in the laboratory by reaction (4), either in an apparatus such as that shown in Fig. 17 (p. 41) or in one of the forms of apparatus shown in Figs. 20, 21. The sulphide is put into the bulb *b*, Fig. 20, through the opening *e*, or into the bottle *b*, Fig. 21. The dilute acid with which the uppermost and lowest bulbs, Fig. 20, are filled comes in contact with the sulphide when the stopcock is opened, or in the apparatus, Fig. 21, is poured through the funnel tube *c*. *a* is a wash-bottle partly filled with water.

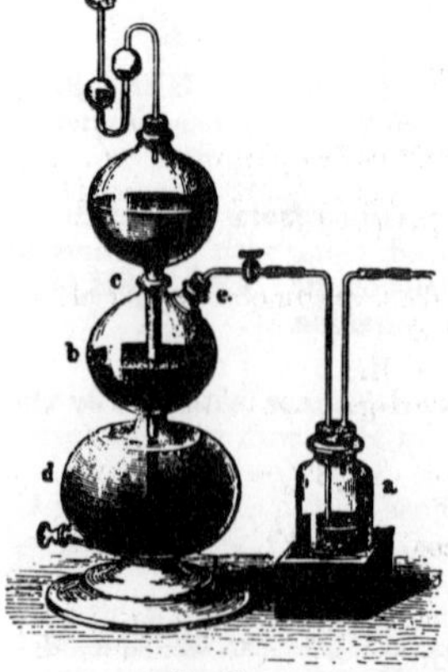

FIG. 20.

FIG. 21.

As ferrous sulphide is liable to contain arsenic, and as hydrogen sulphide generated from it may be contaminated with hydrogen arsenide, the gas, when required for toxicological analysis should always be obtained by reaction (5) in the apparatus, Fig. 20.

PROPERTIES.—*Physical.*—A colorless gas, having the odor of rotten eggs and a disgusting taste; soluble in H_2O to the extent of 3.23 parts to 1 at 15° (59° F.); soluble in alcohol. Under 17 atmospheres pressure, or at − 74° (− 101°.2 F.) at the ordinary pressure, it liquefies; at − 85.5° (− 122° F.) it forms white crystals.

65

Chemical.—Burns in air with formation of sulphur dioxide and water: $2H_2S + 3O_2 = 2SO_2 + 2H_2O$. If the supply of oxygen be deficient, H_2O is formed and sulphur liberated: $2H_2S + O_2 = 2H_2O + S_2$. Mixtures of H_2S and air or O explode on contact with flame. Solutions of the gas when exposed to air become oxidized with deposition of S. Such solutions should be made with boiled H_2O and kept in bottles which are completely filled and well corked. Oxidizing agents, Cl, Br, and I remove its H with deposition of S. Hydrogen sulphide and sulphur dioxide mutually decompose each other into water, pentathionic acid and sulphur: $4SO_2 + 3H_2S = 2H_2O + H_2S_5O_6 + S_2$.

When the gas is passed through a solution of an alkaline hydrate its S displaces the O of the hydrate to form a sulphydrate: $H_2S + KHO = H_2O + KHS$. With solutions of metallic salts H_2S usually relinquishes its S to the metal: $CuSO_4 + H_2S = CuS + H_2SO_4$, a property which renders it of great value in analytical chemistry.

Analytical Characters — *Hydrogen sulphide.*—(1.) Blackens paper moistened with lead acetate solution. (2.) Has an odor of rotten eggs.

Sulphides.—(1.) Heated in the oxidizing flame of the blowpipe, give a blue flame and odor of SO_2.

(2.) With a mineral acid give off H_2S (except sulphides of Hg, Au, and Pt).

Physiological.—Hydrogen sulphide is produced in the intestine by the decomposition of albuminous substances or of taurochloric acid; it also

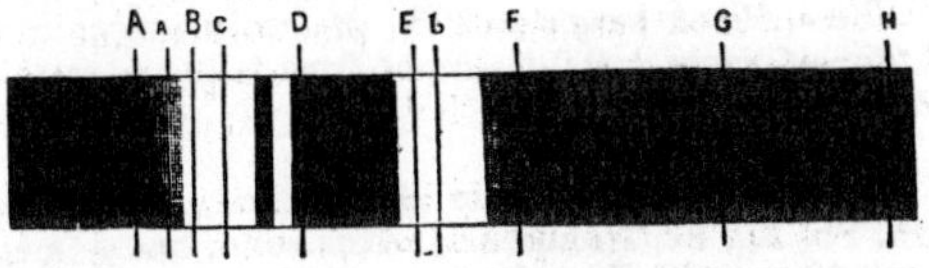

Fig. 22.

occurs sometimes in abscesses, and in the urine in tuberculosis, variola, and cancer of the bladder. It may also reach the bladder by diffusion from the rectum.

Toxicology.—An animal dies almost immediately in an atmosphere of pure H_2S, and the diluted gas is still rapidly fatal. An atmosphere containing one per cent. may be fatal to man, although individuals habituated to its presence can exist in an atmosphere containing three per cent. Even when highly diluted it produces a condition of low fever, and care is to be taken that the air of laboratories in which it is used shall not become contaminated with it. Its toxic powers are due primarily, if not entirely, to its power of reducing and combining with the blood-coloring matter.

The form in which hydrogen sulphide generally produces deleterious effects is as a constituent of the gases emanating from sewers, privies, burial vaults, etc. These give rise to either slow poisoning, as when sewer gases are admitted to sleeping and other apartments by defective plumbing, or to sudden poisoning, as when a person enters a vault or other locality containing the noxious atmosphere.

The treatment should consist in promoting the inhalation of pure air, artificial respiration, cold affusions, and the administration of stimulants.

After death the blood is found to be dark in color, and gives the spectrum shown in Fig. 22, due to sulphæmoglobin.

66

Sulphur Dioxide.

Sulphurous oxide, anhydride or acid—Acidum sulphurosum (U. S.; Br.)—**SO_2**—*Molecular weight* = 64—*Sp. gr. of gas* = 2.213; *of liquid* = 1.45—*Boils at* − 10° (14° *F.*); *solidifies at* − 75° (− 103° *F.*).

Occurrence.—In volcanic gases and in solution in some mineral waters.

Preparation.—(1.) By burning S in air or O.

(2.) By roasting iron pyrites in a current of air.

(3.) During the combustion of coal or coal-gas containing S or its compounds.

(4.) By heating sulphuric acid with copper: $2H_2SO_4 + Cu = CuSO_4 + 2H_2O + SO_2$.

(5.) By heating sulphuric acid with charcoal: $2H_2SO_4 + C = 2SO_2 + CO_2 + 2H_2O$.

When the gas is to be used as a disinfectant it is usually obtained by reaction (1); in sulphuric acid factories (2) is used; (3) indicates the method in which atmospheric SO_2 is chiefly produced; in the laboratory (4) is used; (5) is the process directed by the U. S. and Br. Pharmacopæias.

Properties.—*Physical.*—A colorless, suffocating gas, having a disagreeable and persistent taste. Very soluble in H_2O, which at 15° (59° F.) dissolves about 40 times its volume (see below); also soluble in alcohol. At − 10° (14° F.) it forms a colorless, mobile, transparent liquid, by whose rapid evaporation a cold of − 65° (− 85° F.) is obtained.

Chemical.—Sulphur dioxide is neither combustible nor a supporter of combustion. Heated with H it is decomposed: $SO_2 + 2H_2 = S + 2H_2O$. With nascent hydrogen H_2S is formed: $SO_2 + 3H_2 = H_2S + 2H_2O$.

Water not only dissolves the gas but combines with it to form the true sulphurous acid, H_2SO_3. With solutions of metallic hydrates it forms metallic sulphites: $SO_2 + KHO = KHSO_3$ or $SO_2 + 2KHO = K_2SO_3 + H_2O$. A hydrate having the composition H_2SO_3, $8H_2O$ has been obtained as a crystalline solid, fusible at +4° (39°.2 F.).

Sulphur dioxide and sulphurous acid solution are powerful reducing agents, being themselves oxidized to sulphuric acid: $SO_2 + H_2O + O = H_2SO_4$ or $H_2SO_3 + O = H_2SO_4$. It reduces nitric acid with formation of sulphuric acid and nitrogen tetroxide: $SO_2 + 2HNO_3 = H_2SO_4 + 2NO_2$. It decolorizes organic pigments, without, however, destroying the pigment, whose color may be restored by an alkali or a stronger acid. It destroys H_2S, acting in this instance, not as a reducing, but as an oxidizing agent: $4SO_2 + 3H_2S = 2H_2O + H_2S_5O_6 + S_2$. With Cl it combines directly under the influence of sunlight to form *sulphuryl chloride* $(SO_2)''Cl_2$. Sulphurous acid is dibasic.

Analytical Characters.—(1.) Odor of burning sulphur.

(2.) Paper moistened with starch-paste and iodic acid solution turns blue in air containing 1 in 3,000 of SO_2.

Sulphites.—(1.) With HCl give off SO_2.

(2.) With Zn and HCl give off H_2S.

(3.) With $AgNO_3$ a white ppt., soluble in excess of sulphite. When the mixture is boiled elementary Ag is deposited.

(4.) With $Ba(NO_3)_2$ a white ppt., soluble in HCl. Solution of Cl added to this solution forms a white ppt., insoluble in acids.

67

Sulphur Trioxide.

Sulphuric oxide or anhydride—**SO_3**—*Molecular weight* = 80—*Sp. gr.* 1.95—*Fuses at* 18.3° (65° F.)—*Boils at* 46° (114.8° F.).

Preparation.—(1.) By union of SO_2 and O at 250°–300° (482°–572° F.) or in presence of spongy platinum.

(2.) By heating sulphuric acid in presence of phosphoric anhydride: $H_2SO_4 + P_2O_5 = SO_3 + 2HPO_3$.

(3.) By heating dry sodium pyrosulphate: $Na_2S_2O_7 = Na_2SO_4 + SO_3$.

(4.) By heating pyrosulphuric acid below 100° (212° F.) in a retort fitted with a receiver, cooled by ice and salt: $H_2S_2O_7 = H_2SO_4 + SO_3$.

Properties.—White, silky, odorless crystals which give off white fumes in damp air. It unites with H_2O with a hissing sound and elevation of temperature to form sulphuric acid. When dry it does not redden litmus.

Oxacids of Sulphur.

H_2SO_2 Hydrosulphurous acid.	$H_2S_2O_7$ Pyrosulphuric acid.
H_2SO_3 Sulphurous acid.	$H_2S_2O_6$ Dithionic acid.
H_2SO_4 Sulphuric acid.	$H_2S_3O_6$ Trithionic acid.
	$H_2S_4O_6$ Tetrathionic acid.
$H_2S_2O_3$ Hyposulphurous acid.	$H_2S_5O_6$ Pentathionic acid.

Hydrosulphurous Acid—H_2SO_2—66.

Is an unstable body only known in solution, obtained by the action of zinc upon solution of sulphurous acid. It is a powerful bleaching and deoxidizing agent.

Sulphuric Acid.

Oil of Vitriol—Acidum sulphuricum (U. S.; Br.)—**H_2SO_4**—98.

Preparation.—(1.) By the union of sulphur trioxide and water: $SO_3 + H_2O = H_2SO_4$.

(2.) By the oxidation of SO_2 or of S in the presence of water: $2SO_2 + 2H_2O + O_2 = 2H_2SO_4$; or $S_2 + 2H_2O + 3O_2 = 2H_2SO_4$.

The manufacture of H_2SO_4 may be said to be the basis of all chemical industry, as there are but few processes in chemical technology into some part of which it does not enter. The method followed at present, the result of gradual improvement, may be divided into two stages: 1st, the formation of a dilute acid; 2d, the concentration of this product.

The first part is carried on in immense chambers of timber, lined with lead, and furnishes an acid having a sp. gr. of 1.55, and containing 65 per cent. of true sulphuric acid, H_2SO_4. Into these chambers SO_2, obtained by burning sulphur or by roasting pyrites, is driven along with a large excess of air. In the chambers it comes in contact with nitric acid, at the expense of which it is oxidized to H_2SO_4, while nitrogen tetroxide (red fumes) is formed: $SO_2 + 2HNO_3 = H_2SO_4 + 2NO_2$. Were this the only reaction, the disposal of the red fumes would present a serious difficulty

68

and the amount of nitric acid consumed would be very great. A second reaction occurs between the red fumes and H_2O, which is injected in the form of steam, by which nitric acid and nitrogen dioxide are produced: $3NO_2 + H_2O = 2HNO_3 + NO$. The nitrogen dioxide in turn combines with O to produce the tetroxide, which then regenerates a further quantity of nitric acid, and so on. This series of reactions is made to go on continuously, the nitric acid being constantly regenerated, and acting merely as a carrier of O from the air to the SO_2, in such manner that the sum of the reactions may be represented by the equation: $2SO_2 + 2H_2O + O_2 = 2H_2SO_4$.

The acid is allowed to collect in the chambers until it has the sp. gr. 1.55, when it is drawn off. This *chamber acid*, although used in a few industrial processes, is not yet strong enough for most purposes. It is concentrated, first by evaporation in shallow leaden pans until its sp. gr. reaches 1.746; at this point it begins to act upon the lead, and is transferred to platinum stills, where the concentration is completed.

VARIETIES.—Sulphuric acid is met with in several conditions of concentration and purity:

(1.) The *commercial oil of vitriol*, largely used in manufacturing processes, is a more or less deeply colored, oily liquid, varying in sp. gr. from 1.833 to 1.842, and in concentration from 93 per cent. to 99½ per cent. of true H_2SO_4.

(2.) *C. P. acid = Acidum sulphuricum, U. S.; Br.*, of sp. gr. 1.84, colorless and comparatively pure (see below).

(3.) *Glacial sulphuric acid* is a hydrate of the composition H_2SO_4,H_2O, sometimes called *bihydrated sulphuric acid*, which crystallizes in rhombic prisms, fusible at + 8°.5 (47°. 3 F.) when an acid of sp. gr. 1.788 is cooled to that temperature.

(4.) *Ac. sulph. dil.* (*U. S.; Br.*) is a dilute acid of sp. gr. 1.069 and containing between 9 and 10 per cent. H_2SO_4 (U. S.), or of sp. gr. 1.094, containing between 12 and 13 per cent. H_2SO_4 (Br.).

PROPERTIES.—*Physical.*—A colorless, heavy, oily liquid; sp. gr. 1.842 at 12° (53°.6 F.); crystallizes at 10°.5 (50°.9 F.); boils at 338° (640°.4 F.). It is odorless, intensely acid in taste and reaction, and highly corrosive. It is non-volatile at ordinary temperatures. Mixtures of the acid with H_2O have a lower boiling-point and lower sp. gr. as the proportion of H_2O increases.

Chemical.—At a red heat vapor of H_2SO_4 is partly dissociated into SO_3 and H_2O; or, in the presence of platinum, into SO_2,H_2O and O. When heated with S, C, P, Hg, Cu, or Ag, it is reduced, with formation of SO_2.

Sulphuric acid has a great tendency to absorb H_2O, the union being attended with elevation of temperature, increase of bulk and diminution of sp. gr. of the acid, and contraction of volume of the mixture. Three parts, by weight, of acid of sp. gr. 1.842, when mixed with one part of H_2O produce an elevation of temperature to 130° (266° F.), and the resulting mixture occupies a volume ⅛ less than the sum of the volumes of the constituents. Strong H_2SO_4 is a good desiccator of air or gases. It should not be left exposed in uncovered vessels lest, by increase of volume, it overflow. When it is to be diluted with H_2O, the acid should be added to the H_2O in a vessel of thin glass, to avoid the projection of particles or the rupture of the vessel. It is by virtue of its affinity for H_2O that H_2SO_4 chars or dehydrates organic substances. Sulphuric acid is a powerful dibasic acid.

IMPURITIES.—The commercial acid is so impure that it is only fit for

69

manufacturing and the coarsest chemical uses. The so-called C. P. acid may further contain: *Lead;* becomes cloudy when mixed with 10 times its volume of H_2O, if the quantity of Pb be sufficient; the dilute acid gives a black color with H_2S. *Salts;* leave a fixed residue when the acid is evaporated. *Sulphur dioxide;* gives off H_2S when the acid, diluted with an equal volume of H_2O, comes in contact with Zn. *Carbon;* communicates a brown color to the acid. *Arsenic;* is very frequently present. When the acid is to be used for toxicological analysis, the test by H_2S is not sufficient; the acid, diluted with an equal volume of H_2O, is to be introduced into a Marsh apparatus, in which no visible stain should be produced during an hour. *Oxides of nitrogen;* are almost invariably present; they communicate a pink or red color to pure brucine.

ANALYTICAL CHARACTERS.—(1.) Barium chloride (or nitrate); a white ppt., insoluble in acids. The ppt., dried and heated with charcoal, forms BaS, which, with HCl, gives off H_2S.

(2.) Plumbic acetate forms a white ppt., insoluble in dilute acids.

(3.) Calcium chloride forms a white ppt., either immediately or on dilution with two volumes of alcohol, insoluble in dilute HCl or HNO_3.

TOXICOLOGY.—Sulphuric acid is an active corrosive and may be, if taken in sufficient quantity in a highly diluted state, a true poison. The concentrated acid causes death, either within a few hours by corrosion and perforation of the walls of the stomach and œsophagus, or, after many weeks, by starvation due to destruction of the gastric mucous membrane and closure of the pyloric orifice of the stomach.

The treatment is the same as that for corrosion by HCl. (See p. 58.)

Pyrosulphuric Acid.

Fuming sulphuric acid—Nordhausen oil of vitriol—Disulphuric hydrate—$\mathbf{H_2S_2O_7}$—*Molecular weight* = 178—*Sp. gr.* = 1.9—*Boils at* 52.°2 (126° *F.*).

PREPARATION.—By distilling dry ferrous sulphate; and purification of the product by repeated crystallizations and fusions, until a substance fusing at 35° (95° F.) is obtained.

PROPERTIES.—The commercial Nordhausen acid, which is a mixture of $H_2S_2O_7$ with excess of SO_3, or of H_2SO_4, is a brown, oily liquid, which boils below 100° (212° F.) giving off SO_3; and is solid or liquid according to the temperature.

SELENIUM.

Symbol = **Se**—*Atomic weight* = 79.5—*Molecular weight* = 159—*Sp. gr. of solid* = 4.788; *of vapor* = 5.68*A*—*Name from* σελήνη = *moon*—*Discovered by Berzelius in* 1817.

A rare element, occurring in combination with Cu, Fe, Ag and Hg and accompanying S. It is capable of existing in three allotropic forms. Its compounds are similar in constitution to those of S.

70

TELLURIUM.

Symbol = **Te**—*Atomic weight* = 128—*Molecular weight* = 256—*Sp. gr. of solid* = 6.25; *of vapor* = 9.0*A*—*Name from* tellus = *earth*—*Discovered in* 1782 *by Müller.*

One of the least common of the elements, it occurs free and combined with Bi, Pb, Ag, Sb, Ni and Au. It is solid, has a metallic lustre, fuses at about 500° (932° F.). Its compounds are similar to those of Se and S.

III. NITROGEN GROUP.

NITROGEN—PHOSPHORUS—ARSENIC—ANTIMONY.

The elements of this group are either trivalent or quinquivalent. With hydrogen they form non-acid compounds composed of one volume of the element in the gaseous state with three volumes of hydrogen, the union being attended with a condensation of volume of one-half. Their hydrates are acids containing one, two, three, or four atoms of replaceable hydrogen.

Bismuth, frequently classed in this group, is excluded, owing to the existence of the nitrate $Bi(NO_3)_3$. The relations existing between the compounds of the elements of this group are shown in the following table:

NH_3,	N_2O,	NO,	N_2O_3,	NO_2,	N_2O_5,	—	—	—	—	HNO_3
PH_3,	—	—	P_2O_3,	—	P_2O_5,	H_3PO_2,	H_3PO_3,	H_3PO_4,	$H_4P_2O_7$,	HPO_3
AsH_3,	—	—	As_2O_3,	—	As_2O_5,	—	H_3AsO_3,	H_3AsO_4,	$H_4As_2O_7$,	$HAsO_3$
SbH_3,	—	—	Sb_2O_3,	—	Sb_2O_5,	—	—	H_3SbO_4,	$H_4Sb_2O_7$,	$HSbO_3$
Hyd-ride.	Mon-oxide.	Di-oxide.	Tri-oxide.	Tetr-oxide.	Pent-oxide.	Hypo-ous acid.	-ous acid.	Ortho-ic acid.	Pyro-ic acid.	Meta-ic acid.

NITROGEN.

Azote—Symbol = **N**—*Atomic weight* = 14—*Molecular weight* = 28—*Sp. gr.* = 0.9701—*One litre weighs* 1.254 *grams*—*Name from* νίτρον = *nitre*, γένεσις = *source*; *or from* ἀ, *privative* ζωή = *life*—*Discovered by Mayow in* 1669.

OCCURRENCE.—Free in atmospheric air and in volcanic gases. In combination in the nitrates, in ammoniacal compounds and in a great number of animal and vegetable substances.

PREPARATION.—(1.) By removal of O from atmospheric air, or by burning P in air, or by passing air slowly over red-hot copper. It is contaminated with CO_2, H_2O, etc.

(2.) By passing Cl through excess of ammonium hydrate solution. If ammonia be not maintained in excess, the Cl reacts with the ammonium chloride formed, to produce the explosive nitrogen chloride.

(3.) By heating ammonium nitrite: or a mixture of ammonium chloride and potassium nitrite.

PROPERTIES.—A colorless, odorless, tasteless, non-combustible gas; not a supporter of combustion; very sparingly soluble in water.

It is very slow to enter into combination, and most of its compounds

71

are very prone to decomposition, which may occur explosively or slowly. Nitrogen combines directly with O under the influence of electric discharges; and with H under like conditions and indirectly during the decomposition of nitrogenized organic substances.

Nitrogen is not poisonous, but is incapable of supporting respiration.

Atmospheric Air.

The alchemists considered air as an element until Mayow in 1669 demonstrated its complex nature. It was not, however, until 1770 that Priestley repeated the work of Mayow; and that the compound nature of air and the characters of its constituents were made generally known by the labors (1770–1781) of Priestley, Rutherford, Lavoisier and Cavendish. The older chemists used the terms *gas* and *air* as synonymous.

Composition.—Air is not a chemical compound, but a mechanical mixture of O and N with smaller quantities of other gases. Leaving out of consideration about 0.4 to 0.5 per cent. of other gases, air consists of 20.93 O and 79.07 N, by volume; or 23 O and 77 N, by weight; proportions which vary but very slightly at different times and places; the extremes of the proportion of O found having been 20.908 and 20.999.

That air is not a compound is shown by the fact that the proportion of its constituents does not represent a relation between their atomic weights or between any multiples thereof; as well as by the solubility of air in water. Were it a compound it would have a definite degree of solubility of its own, and the dissolved gas would have the same composition as when free. But each of its constituents dissolves in H_2O according to its own solubility and air dissolved in H_2O at 13° (55°.4 F.) consists of N and O, not in the proportion given above, but in the proportion 65.27 to 34.72.

Besides these two main constituents, air contains about 4–5 thousandths of its bulk of other substances: vapor of water, carbon dioxide, ammoniacal compounds, hydrocarbons, ozone, oxides of nitrogen, and solid particles held in suspension.

Vapor of water.—Atmospheric moisture is either visible, as in fogs and clouds, when it is in the form of a finely divided liquid; or invisible, as vapor of water. The amount of H_2O which a given volume of air can hold without precipitation varies according to the temperature. It happens rarely that air is as highly charged with moisture as it is capable of being for the existing temperature. The difference between the amount of water which the air is capable of holding at the existing temperature and that which it actually does hold is its *fraction of saturation*, or *hygrometric state.* Ordinarily air contains from 66 to 70 per cent. of its possible amount of moisture; if the quantity be less than this the air is too dry and causes a parched sensation and the sense of "stuffiness" so common in furnace-heated houses; if it be greater, evaporation from the skin is impeded and the air is oppressive if warm.

The actual amount of moisture in air is determined by passing a known volume through tubes filled with calcium chloride; whose increase in weight represents the amount of H_2O in the volume of air used. The fraction of saturation is determined by instruments called *hygrometers, hygroscopes* or *psychrometers.*

***Carbon dioxide.*—The quantity of carbon dioxide in free air varies from 3 to 6 parts in 10,000 by volume. (See Carbon dioxide.)**

***Ammoniacal compounds.*—Carbonate, nitrate, and nitrite of ammonium**

72

occur in small quantity (0.1 to 6.0 parts per million of NH_3) in air, as products of the decomposition of nitrogenized organic substances. They are absorbed and assimilated by plants.

Nitric and nitrous acids, usually in combination with ammonium, are produced either by the oxidation of combustible substances containing N, or by direct union of N and H_2O during discharges of atmospheric electricity. Rain-water falling during thunder-showers has been found to contain as much as 3.71 per million of HNO_3.

Sulphuric and sulphurous acids occur in combination with NH_4 in the air over cities and manufacturing districts, where they are produced by the oxidation of S existing in coal and coal-gas.

Hydrocarbons have been detected in the air of cities and of swampy places, in small quantities.

Solid particles of the most diverse nature are always present in air and become visible in a beam of sunlight. Chloride of sodium is almost always present, always in the neighborhood of salt water. Air contains myriads of germs of vegetable organisms, mould, etc., which are propagated by the transportation of these germs by air-currents. Whether or no certain diseases are thus propagated by germs or poisons; and whether low forms of organized beings can or cannot make their appearance without the introduction of germs are questions, both sides of which are supported by active partisans, and concerning which little or nothing is known with certainty.

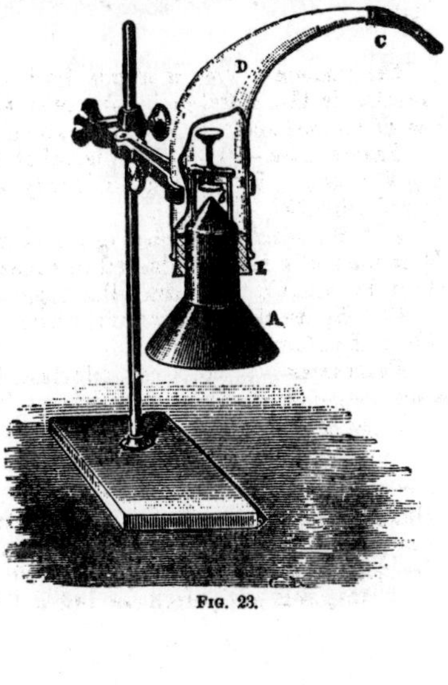

Fig. 23.

The continued inhalation of air containing large quantities of solid particles in suspension may cause severe pulmonary disorder by mere mechanical irritation, and apart from any poisonous quality in the substance; such is the case with the air of carpeted ball-rooms, and of the workshops of certain trades, furniture-polishers, metal-filers, etc.

Atmospheric dust is best collected by an instrument such as is shown in Fig. 23. A disk of thin glass is fastened upon the plate *b*, over the small opening in A, and its lower surface moistened with a mixture of equal parts of water and glycerin, the opening C is connected with an aspirator. After one or more cubic metres of air have been drawn through the apparatus, the thin glass is detached and the deposit examined microscopically.

Ammonia.

Hydrogen nitride—Volatile alkali—**NH_3**—*Molecular weight* = 17—*Sp. gr.* = 0.589*A*—*Liquefies at* − 40° (− 40° *F.*)—*Boils at* − 33°.7 (− 28°.7 *F.*)—*Solidifies at* −75° (−103° *F.*)—*A litre weighs* 0.7655 *grams.*

Preparations.—(1.) By union of nascent H with N.

(2.) By decomposition of organic matter containing N, either spontaneously or by destructive distillation.

73

(3.) By heating a mixture of dry slacked lime with ammonium chloride: $2NH_4Cl + CaH_2O_2 = CaCl_2 + 2H_2O + 2NH_3$.

(4.) By heating solution of ammonium hydrate: $NH_4HO = NH_3 + H_2O$.

Properties.—*Physical.*—A colorless gas, having a pungent odor and an acrid taste. It is very soluble in H_2O, 1 volume of which at 0° (32° F.) dissolves 1050 vols. NH_3 and at 15° (59° F.), 727 vols. NH_3. Alcohol and ether also dissolve it readily. Liquid ammonia is a colorless, mobile fluid, used in ice machines for producing artificial cold, the liquid absorbing a great amount of heat in volatilizing.

Chemical.—At a red heat ammonia is decomposed into a mixture of N and H, occupying double the volume of the original gas. It is similarly decomposed by the prolonged passage through it of discharges of electricity. It is not readily combustible, yet it burns in an atmosphere of O with a yellowish flame. Mixtures of NH_3 with O, nitrogen monoxide, or nitrogen dioxide, explode on contact with flame. Water dissolves ammonia with elevation of temperature and probably with formation of ammonium hydrate, NH_4HO (q. v.). It combines directly with acids to produce ammonium salts, without separation of hydrogen. (See Ammonium.)

Nitrogen Monoxide.

Nitrous oxide—Laughing gas—Nitrogen protoxide—N_2O—*Molecular weight* = 44—*Sp. gr.* = 1.527*A*—*Fuses at* −100° (−148° *F.*)—*Boils at* −87° (−124° *F.*)—*Discovered in* 1776 *by Priestley.*

Preparation.—By heating ammonium nitrate: $(NH_4)NO_3 = N_2O + 2H_2O$. To obtain a pure product there should be no ammonium chloride present (as an impurity of the nitrate), and the heat should be applied gradually and not allowed to exceed 250° (482° F.), and the gas formed should be passed through wash-bottles containing sodium hydrate and ferrous sulphate.

Properties.—*Physical.*—A colorless, odorless gas, having a sweetish taste; soluble in H_2O, more so in alcohol. Under a pressure of 30 atmospheres, at 0° (32° F.), it forms a colorless, mobile liquid, which, when dissolved in carbon disulphide and evaporated *in vacuo*, produces a cold of − 140° (− 220° F.)

Chemical.—It is decomposed by a red heat and by the continuous passage of electric sparks. It is not combustible, but is, after oxygen, the best supporter of combustion known.

Physiological.—Although, owing to the readiness with which N_2O is decomposed into its constituent elements, and the nature and relative proportions of these elements, it is capable of maintaining respiration longer than any gas except oxygen or air; an animal will live for a short time only in an atmosphere of pure nitrous oxide. When inhaled, diluted with air, it produces the effects first observed by Davy in 1799: first an exhilaration of spirits, frequently accompanied by laughter, and a tendency to muscular activity, the patient sometimes becoming aggressive; afterward there is complete anæsthesia and loss of consciousness. **It has been much used, by dentists especially, as an anæsthetic in operations of short duration, and in one or two instances anæsthesia has been maintained by its use for nearly an hour.**

A solution in water under pressure, containing five volumes of the gas, is sometimes used for internal administration.

74

Nitrogen Dioxide.

***Nitric oxide*—NO—*Molecular weight* = 30—*Sp. gr.* = 1.039*A*—*Discovered by Hales in* 1772.**

Preparation.—By the action of copper on moderately diluted nitric acid in the cold: $3Cu + 8HNO_3 = 3Cu(NO_3)_2 + 4H_2O + 2NO$; the gas being collected after displacement of air from the apparatus.

PROPERTIES.—A colorless gas, whose odor and taste are unknown; very sparingly soluble in H_2O; more soluble in alcohol.

It combines with O when mixed with that gas or with air, to form the reddish-brown nitrogen tetroxide. It is absorbed by solution of ferrous sulphate, to which it communicates a dark brown or black color. It is neither combustible nor a good supporter of combustion, although ignited C and P continue to burn in it, and the alkaline metals, when heated in it, combine with its O with incandescence.

Nitrogen Trioxide.

Nitrous anhydride.—N_2O_3—76.

Has not been obtained in a condition of purity. A mixture of 95 per cent. of N_2O_3 with 5 per cent. of N_2O may, however, be obtained by decomposing liquefied nitrogen tetroxide with a small quantity of H_2O at a low temperature: $4NO_2 + H_2O = 2HNO_3 + N_2O_3$. This is a dark indigo-blue liquid, which, boiling at about 0° (32° F.), is partly decomposed.

Nitrogen Tetroxide.

Nitrogen peroxide—Hyponitric acid—Nitrous fumes—NO_2—*Molecular weight* = 46—*Sp. gr.* = 1.58*A* (*at* 154° *C.*)—*Boils at* 22° (71°.6 *F.*)—*Solidifies at* 9° (15°.8 *F.*).

PREPARATION.—(1.) By mixing one volume O with two volumes NO; both dry and ice-cold.

(2.) By heating perfectly dry lead nitrate, O being also produced: $2Pb(NO_3)_2 = 2PbO + 4NO_2 + O_2$.

(3.) By dropping strong nitric acid upon a red-hot platinum surface.

PROPERTIES.—When pure and dry it is an orange-yellow liquid at the ordinary temperature; the color being darker the higher the temperature. The red fumes which are produced when nitric acid is decomposed by starch or by a metal consist of NO_2 mixed with N_2O_3. It dissolves in nitric acid, forming a dark yellow liquid, which is blue or green if N_2O_3 be also present. With SO_2 it combines to form a solid, crystalline compound, which is sometimes produced in the manufacture of H_2SO_4, and known as *lead chamber crystals*. A small quantity of H_2O decomposes it into HNO_3 and N_2O_3, which latter colors it green or blue; a larger quantity of H_2O decomposes it into HNO_3 and NO. By bases it is transformed into a mixture of nitrite and nitrate: $2NO_2 + 2KHO = KNO_2 + KNO_3 + H_2O$.

It is an energetic oxydant, for which it is largely used. With certain organic substances it does not behave as an oxydant, but becomes

75

substituted as an univalent radical; thus with benzol it forms nitro-benzol: $C_6H_5(NO_2)$.

TOXICOLOGY.—The brown fumes given off during many processes, in which nitric acid is decomposed, are dangerous to life. All such operations, when carried on on a small scale, as in the laboratory, should be conducted under a hood or some other arrangement, by which the fumes are carried into the open air. When, in industrial processes, the volume of gas formed becomes such as to be a nuisance when discharged into the air, it should be utilized in the manufacture of H_2SO_4 or absorbed by H_2O or an alkaline solution.

An atmosphere contaminated with brown fumes is more dangerous than one containing Cl, as the presence of the latter is more immediately annoying. At first there is only coughing, and it is only two to four hours later that a difficulty in breathing is felt, death occurring in ten to fifteen hours. At the autopsy the lungs are found to be extensively disorganized and filled with black fluid.

Even air containing small quantities of brown fumes, if breathed for a long time, produces chronic disease of the respiratory organs. To prevent such accidents, thorough ventilation in locations where brown fumes are liable to be formed is imperative. In cases of spilling nitric acid, safety is to be sought in retreat from the apartment until the fumes have been replaced by pure air from without.

Nitrogen Pentoxide.

Nitric anhydride—N_2O_5—*Molecular weight* = 108—*Fuses at* 30° (86° *F.*)—*Boils at* 47° (116°.6 *F.*).

PREPARATION.—(1.) By decomposing dry silver nitrate with dry Cl in an apparatus entirely of glass: $4AgNO_3 + 2Cl_2 = 4AgCl + 2N_2O_5 + O_2$.

(2.) By removing water from fuming nitric acid with phosphorus pentoxide: $6HNO_3 + P_2O_5 = 2H_3PO_4 + 3N_2O_5$.

PROPERTIES.—Prismatic crystals at temperatures above 30° (86° F.). It is very unstable, being decomposed by a heat of 50° (122° F.); on contact with H_2O, with which it forms nitric acid; and even spontaneously. Most substances which combine readily with O, remove that element from N_2O_5.

Nitrogen Acids.

Three are known, either free or in combination, corresponding to the three oxides containing uneven numbers of O atoms:

$$N_2O + H_2O = 2HNO\text{—Hyponitrous acid.}$$
$$N_2O_3 + H_2O = 2HNO_2\text{—Nitrous acid.}$$
$$N_2O_5 + H_2O = 2HNO_3\text{—Nitric acid.}$$

Hyponitrous acid—HNO—31—Known only in combination. **Silver hyponitrite is formed by reduction of sodium nitrate by nascent H and decomposition with silver nitrate.**

Nitrous acid—HNO_2—47—has not been isolated, although its salts, the nitrites, are well-defined compounds: $M'NO_2$ or $M''(NO_2)_2$.

76

Nitric Acid.

Aquafortis—Hydrogen nitrate—Acidum nitricum—U. S.; Br.—HNO_3—63.

PREPARATION.—(1.) By the direct union of its constituent elements under the influence of electric discharges.

(2.) By the decomposition of an alkaline nitrate by strong H_2SO_4. With moderate heat a portion of the acid is liberated: $2NaNO_3 + H_2SO_4 = NaHSO_4 + NaNO_3 + HNO_3$, and at a higher temperature the remainder is given off: $NaNO_3 + NaHSO_4 = Na_2SO_4 + HNO_3$. This is the reaction used in the manufacture of HNO_3.

VARIETIES.—*Commercial*—a yellowish liquid, very impure, and of two degrees of concentration: *single aquafortis*; sp. gr. about 1.25 = 39% HNO_3; and *double aquafortis*; sp. gr. about 1.4 = 64% HNO_3.

Fuming—a reddish yellow liquid, more or less free from impurities; charged with oxides of nitrogen. Sp. gr. about 1.5. Used as an oxidizing agent.

C. P.—a colorless liquid, sp. gr. 1.522, which should respond favorably to the tests given below.

Acidum nitricum, U. S.; Br.—a colorless acid, of sp. gr. 1.42 = 70% HNO_3.

Acidum nitricum dilutum, U. S.; Br.—the last mentioned, diluted with H_2O to sp. gr. 1.059 = 10% HNO_3 (*U. S.*), or to sp. gr. 1.101 = 17.44% HNO_3 (*Br.*).

PROPERTIES.—*Physical.*—The pure acid is a colorless liquid; sp. gr. 1.522; boils at 86° (186°.8 F.); solidifies at—40° (—40° F.); gives off white fumes in damp air; and has a strong acid taste and reaction. The sp. gr. and boiling-point of dilute acids vary with the concentration. If a strong acid be distilled, the boiling-point gradually rises from 86° (186°.8 F.) until it reaches 123° (253°.4 F.), when it remains constant, the sp. gr. of distilled and distillate being 1.42 = 70% HNO_3. If a weak acid be taken originally the boiling-point rises until it becomes stationary at the same point.

Chemical.—When exposed to air and light, or when heated to redness, HNO_3 is decomposed into NO_2; H_2O and O. Nitric acid is a valuable oxydant; it converts I, P, S, C, B, and Si or their lower oxides into their highest oxides; it oxidizes and destroys most organic substances, although with some it forms products of substitution. Most of the metals dissolve in HNO_3 as nitrates, a portion of the acid being at the same time decomposed into NO and H_2O: $4HNO_3 + 3Ag = 3AgNO_3 + NO + 2H_2O$. The so-called "noble metals," gold and platinum, are not dissolved by either HNO_3 or HCl, but dissolve as chlorides in a mixture of the two acids, called *aqua regia*. In this mixture the two acids mutually decompose each other according to the equations: $HNO_3 + 3HCl = 2H_2O + NOCl + Cl_2$ and $2HNO_3 + 6HCl = 4H_2O + 2NOCl_2 + Cl_2$ with formation of *nitrosyl chloride*, NOCl, and *bichloride* $NOCl_2$; and nascent Cl; the last named combining with the metal. When HNO_3 is decomposed by zinc or iron or in the porous cup of a Grove battery, N_2O_4 and NO_2 are formed and dissolve in the acid, which is colored dark yellow, blue or green. An acid so charged is known as *nitroso-nitric acid*. Nitric acid is monobasic.

IMPURITIES.—*Oxides of Nitrogen* render the acid yellow, and decolorize

77

potassium permanganate when added to the dilute acid. *Sulphuric acid* produces cloudiness when $BaCl_2$ is added to the acid, diluted with two volumes of H_2O. *Chlorine, iodine* cause a white ppt. with $AgNO_3$. *Iron* gives a red color when the diluted acid is treated with ammonium sulphocyanate. *Salts*, leave a fixed residue when the acid is evaporated to dryness on platinum.

ANALYTICAL CHARACTERS.—(1.) Add an equal volume of concentrated H_2SO_4, cool, and float on the surface of the mixture a solution of $FeSO_4$. The lower layer becomes gradually colored brown, black or purple, beginning at the top.

(2.) Boil in a test-tube a small quantity of HCl containing enough sulphindigotic acid to communicate a blue color, add the suspected solution and boil again ; the color is discharged.

(3.) If acid neutralize with KHO, evaporate to dryness, add to the residue a few drops of H_2SO_4 and a crystal of brucine (or some sulphanilic acid) ; a red color is produced.

(4.) Add H_2SO_4 and Cu to the suspected liquid and boil, brown fumes appear (best visible by looking into the mouth of the test-tube).

The above appearances are caused by free nitric acid, but if the tests be conducted as directed a nitrate, if present, is decomposed with liberation of the acid.

All neutral nitrates are soluble in H_2O ; some so-called basic salts are insoluble, as bismuthyl nitrate : $(BiO)NO_3$.

Toxicology.—Although most of the nitrates are poisonous when taken internally in sufficiently large doses, their action seems to be due rather to the metal than to the acid radical. Nitric acid itself is one of the most powerful of corrosives.

Any animal tissue with which the concentrated acid comes in contact is immediately disintegrated ; a yellow stain, afterward turning to dirty brownish, or, if the action be prolonged, an eschar, is formed. When taken internally its action is the same as upon the skin, but, owing to the more immediately important function of the parts, is followed by more serious results (unless a large cutaneous surface be destroyed).

The symptoms following its ingestion are the same as those produced by the other mineral acids, except that all parts with which the acid has come in contact, including vomited shreds of mucous membrane, are colored yellow. The treatment is the same as that indicated when SO_4H_2 or HCl have been taken ; i.e. neutralization of the corrosive by magnesia or an alkali.

Compounds of Nitrogen with the Halogens.

Nitrogen chloride—NCl_3—120.5—is formed by the action of excess of Cl upon NH_3 or an ammoniacal compound. It is an oily, light yellow liquid ; sp. gr. 1.653 ; has been distilled at 71° (159°.8 F.). When heated to 96° (204°.8 F.), when subjected to concussion, or when brought in contact with phosphorus, alkalies or greasy matters it is decomposed, with a violent explosion, into one volume N and three volumes Cl.

Nitrogen bromide—NBr_3—254—has been obtained as a reddish brown, syrupy liquid, very volatile, and resembling the chloride in its properties, by the action of potassium bromide upon nitrogen chloride.

Nitrogen iodide—NI_3—395—When iodine is brought in contact with ammonium hydrate solution, a dark brown or black powder, highly

78

explosive when dried, is formed. This substance varies in composition according to the conditions under which the action occurs ; sometimes the iodide alone is formed ; under other circumstances it is mixed with compounds containing N, I and H.

PHOSPHORUS.

Symbol = **P**—*Atomic weight* = 31—*Molecular weight* = 124—*Sp. gr. of vapor* = 4.2904 *A—Name from* φῶς = *light*, φέρω = *I bear—Discovered by Brandt in* 1669—*Phosphorus* (*U. S.; Br.*).

Occurrence.—Only in combination ; in the mineral and vegetable worlds as phosphates of Ca, Mg, Al, Pb, K, Na. In the animal kingdom as phosphates of Ca, Mg, K and Na, and in organic combination.

Preparation.—From bone-ash, in which it occurs as tricalcic phosphate. Three parts of bone-ash are digested with 2 parts of strong H_2SO_4, diluted with 20 volumes H_2O, when insoluble calcic sulphate and the soluble monocalcic phosphate, or "superphosphate," are formed : $Ca_3(PO_4)_2 + 2H_2SO_4 = H_4Ca(PO_4)_2 + 2CaSO_4$. The solution of superphosphate is filtered off and evaporated, the residue is mixed with about one-fourth its weight of powdered charcoal and sand, and the mixture heated, first to redness, finally to a white heat, in earthenware retorts, whose beaks dip under water in suitable receivers. During the first part of the heating the monocalcic phosphate is converted into metaphosphate : $CaH_4(PO_4)_2 = Ca(PO_3)_2 + 2H_2O$; which is in turn reduced by the charcoal with formation of carbon monoxide and liberation of phosphorus, while the calcium is combined as silicate : $2Ca(PO_3)_2 + 2SiO_2 + 5C_2 = 2CaSiO_3 + 10CO + P_4$.

The crude product is purified by fusion, first under a solution of bleaching powder, next under ammoniacal water, and finally under water containing a small quantity of sulphuric acid and potassium dichromate. It is then strained through leather and cast into sticks under warm water.

Properties.—*Physical.*—Phosphorus is capable of existing in four allotropic forms :

(1.) *Ordinary, or yellow variety,* in which it usually occurs in commerce. This is a yellowish, translucid solid of the consistency of wax. Below 0° (32° F.) it is brittle ; it fuses at 44°.3 (111°.7 F.) ; and boils at 290° (554° F.) in an atmosphere not capable of acting upon it chemically. Its vapor is colorless ; sp. gr. = 4.5A—65 H at 1040° (1940° F.). It volatilizes below its boiling-point, and water boiled upon it gives off steam charged with its vapor. Exposed to air it gives off white fumes and produces ozone. It is luminous in the dark. It is insoluble in water ; sparingly soluble in alcohol and ether ; soluble in carbon disulphide, and in the fixed and volatile oils. It crystallizes on evaporation of its solutions in octahedræ or dodecahedræ. Sp. gr. 1.83 at 10° (50° F.).

(2.) *White phosphorus* is formed as a white, opaque pellicle upon the surface of the ordinary variety when this is exposed to light under aërated water. Sp. gr. 1.515 at 15° (59° F.). When fused it reproduces ordinary phosphorus without loss of weight.

(3.) *Black variety* is formed when ordinary phosphorus is heated to 70° (158° F.) and suddenly cooled.

(4.) *Red variety* is produced from the ordinary by maintaining it at from 240° (464° F.) to 280° (536° F.) for two or three days in an atmosphere

79

of carbon dioxide ; and, after cooling, washing out the unaltered yellow phosphorus with carbon disulphide. It is also formed upon the surface of the yellow variety when it is exposed to direct sunlight.

It is a reddish, odorless, tasteless solid, which does not fume in air, nor dissolve in the solvents of the yellow variety. Sp. gr. 2.1. Heated to 500° (932° F.) with lead, in the absence of air, it dissolves in the molten metal, from which it separates on cooling in violet black, rhombohedral crystals, of sp. gr. 2.34. If prepared at 250° (482° F.) it fuses below that temperature, and at 260° (500° F.) is transformed into the yellow variety which distils. The crystalline product does not fuse. It is not luminous at ordinary temperatures.

Chemical.—The most prominent property of P is the readiness with which it combines with O. The yellow variety ignites and burns with a bright flame if heated in air to 60° (140° F.), or if exposed in a finely divided state to air at the ordinary temperature ; with formation of P_2O_3 ; P_2O_5 ; H_3PO_3 or H_3PO_4 according as O is present in excess or not and according as the air is dry or moist. The temperature of ignition of yellow P is so low that it must be preserved under boiled water. By directing a current of O upon it, P may be burned under H_2O, heated above 45° (113° F.). The red variety combines with O much less readily and may be kept in contact with air without danger.

The luminous appearance of yellow P is said to be due to the formation of ozone. It does not occur in pure O at the ordinary temperature, nor in air under pressure, nor in the absence of moisture, nor in the presence of minute quantities of carbon disulphide, oil of turpentine, alcohol, ether, naphtha, and many gases.

Yellow phosphorus burns in Cl with formation of PCl_3 or PCl_5 according as P or Cl is present in excess. Both yellow and red varieties combine directly with Cl, Br, and I.

Phosphorus is not acted on by HCl or cold H_2SO_4. Hot H_2SO_4 oxidizes it with formation of phosphorous acid and sulphur dioxide : $P_4 + 6H_2SO_4 = 4H_3PO_3 + 6SO_2$. Nitric acid oxidizes it violently to phosphoric acid and nitrogen di- and tetr-oxides : $12HNO_3 + P_4 = 4H_3PO_4 + 8NO_2 + 4NO$.

Phosphorus is a reducing agent. When immersed in cupric sulphate solution it becomes covered with a coating of metallic copper. In silver nitrate solution it produces a black deposit of silver phosphide.

Toxicology.—The red variety differs from the other allotropic forms of phosphorus in not being poisonous, probably owing to its insolubility, and in being little liable to cause injury by burning.

The burns produced by yellow phosphorus are more serious than a like destruction of cutaneous surface by other substances. A burning fragment of P adheres tenaciously to the skin, into which it burrows. One of the products of the combustion is metaphosphoric acid (q. v.) which, being absorbed, gives rise to true poisoning. Burns by P should be washed immediately with dilute javelle water, liq. sodæ chlorinatæ, or solution of chloride of lime. Yellow P should never be allowed to come in contact with the skin, except it be under cold water.

Yellow P is one of the most insidious of poisons. It is taken or administered usually as "ratsbane" or match-heads. The former is frequently starch paste charged with phosphorus ; the latter, in the ordinary sulphur match, a mixture of potassium chlorate, very fine sand, phosphorus, and a coloring matter. The symptoms in acute phosphorus-poisoning appear with greater or less rapidity, according to the dose, and the presence or absence in the stomach of substances which favor its

80

absorption. Their appearance may be delayed for days, but as a rule they appear within a few hours. A disagreeable garlicky taste in the mouth, and heat in the stomach are first observed, the latter gradually developing into a burning pain, accompanied by vomiting of dark-colored matter, which, when shaken in the dark, is phosphorescent ; low temperature and dilatation of the pupils. In some cases death follows at this point suddenly, without the appearance of any further marked symptoms ; usually, however, the patient rallies, seems to be doing well, until suddenly jaundice makes its appearance, accompanied by retention of urine, and frequently delirium, followed by coma and death.

There is no known chemical antidote to phosphorus; the treatment is, therefor, limited to the removal of the unabsorbed portions of the poison by the action of an emetic, zinc sulphate or apomorphia, as expeditiously

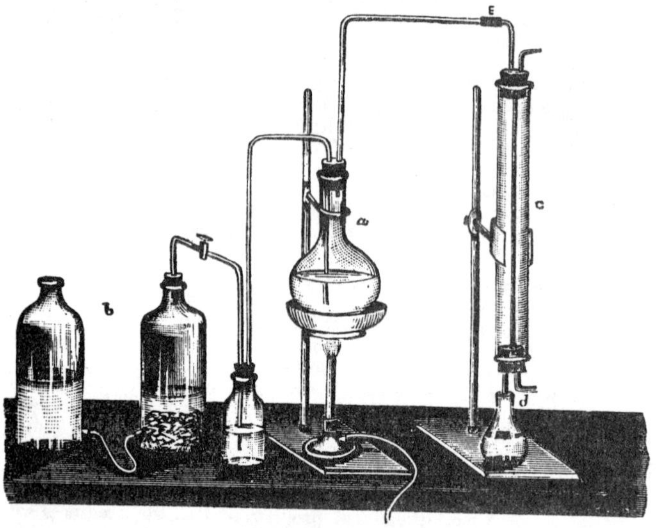

Fig. 24.

as possible, and the administration of oil of turpentine—the older the oil the better—as a physiological antidote. The use of fixed oils or fats is to be avoided, as they favor the absorption of the poison by their solvent action. The prognosis is very unfavorable.

As commercial phosphorus is usually contaminated with arsenic, the effects of the latter substance may also appear in poisoning by the former.

Analysis.—When, after a death supposed to be caused by phosphorus, chemical evidence of the existence of the poison in the body, etc., is desired, the investigation must be made as soon after death as possible, for the reason that the element is rapidly oxidized, and the detection of the higher stages of oxidation of phosphorus is of no value as evidence of the administration of the element, because they are normal constituents of the body and of the food.

The detection of elementary phosphorus in a systematic toxicological

81

analysis is connected with that of prussic acid, alcohol, ether, chloroform, and other volatile poisons. The substances under examination are diluted with H_2O, acidulated with H_2SO_4, and heated over a sand-bath in the flask *a* (Fig. 24). This flask is connected with a CO_2 generator, *b*, whose stop-cock is closed, and with a Liebig's condenser, *c*, which is in darkness (the operation is best conducted in a dark room) and so placed as to deliver the distillate into the flask *d*. The odor of the distillate is noted. In the presence of P it is usually alliaceous. The condenser is also observed. If at the point of greatest condensation a faint, luminous ring be observed (in the absence of *all* reflections), it is proof positive of the presence of unoxidized phosphorus; the absence, however, of that poison is not to be inferred from the absence of the luminous ring (see above). If this fail to appear when one-third the fluid contents of the flask *a* have distilled over, the condenser is disconnected at *e*, and in its place the absorbing apparatus,

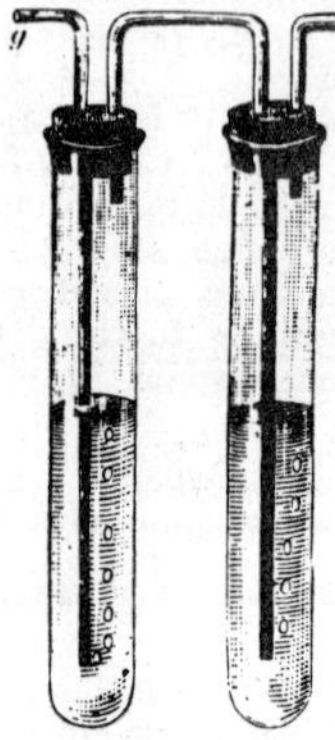

Fig. 25.

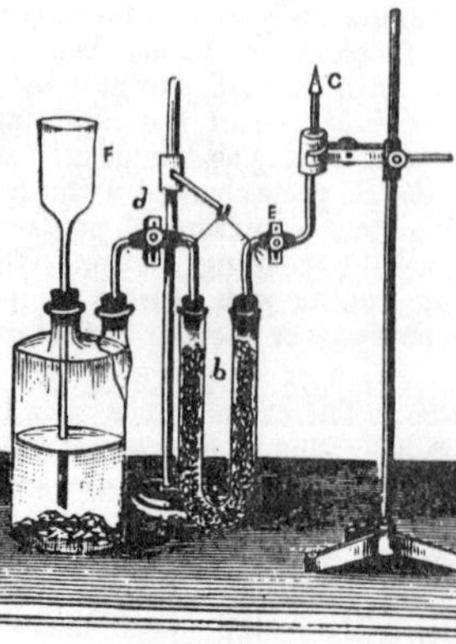

Fig. 26.

Fig. 25, partly filled with a neutral solution of silver nitrate, is adjusted by a rubber tube attached at *g*, and a slow and constant stream of CO_2 is caused to traverse the apparatus from *b*. If during continuation of the distillation no black deposit is formed in the silver solution, the absence of P may be inferred. If a black deposit be formed, it must be further examined to determine if it be silver phosphide. For this purpose the apparatus shown in Fig. 26 is used. In the bottle *a* hydrogen is generated from *pure* Zn and H_2SO_4, the gas passing through the drying-tube *b*, filled with fragments of $CaCl_2$, and out through the platinum tip at *c*; *d* and *e* are pinch-cocks. When the apparatus is filled with hydrogen, *d* is closed until the funnel-tube *f* is three-quarters filled with the liquid from *a*; then *e* is closed and *d* opened, and the black silver deposit, which has been collected on a filter and washed, is thrown into *f*; *e* is then slightly opened and the escaping gas ignited at *c*, the size of the flame being regulated by *e*. If the deposit contain P the flame will have a green color; and when examined with the spectroscope will give the spectrum of bright bands shown in Fig. 27.

Chronic phosphorus poisoning, or *Lucifer disease*, occurs among operatives engaged in the dipping, drying, and packing of phosphorus matches.

82

Those engaged in the manufacture of phosphorus itself are not so affected. Sickly women and children are most subject to it. The cause of the disease has been ascribed to the presence of arsenic, and to the formation of oxides of phosphorus and ozone. The progress of the disorder is slow, and its culminating manifestation is the destruction of one or both maxillæ by necrosis.

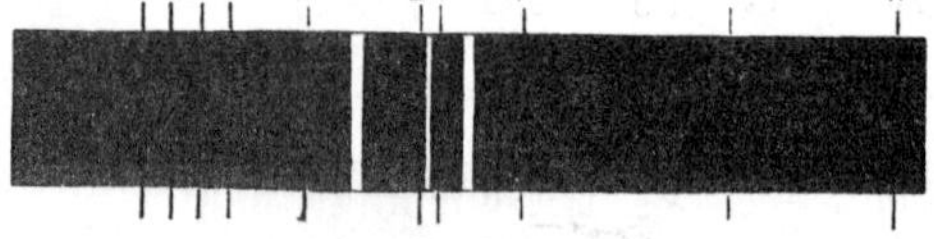

Fig. 27.

The frequency of the disease may be in some degree diminished by thorough ventilation of the shops, by frequent washing of the face and mouth with a weak solution of sodium carbonate, and by exposing oil of turpentine in saucers in the workshops. None of these methods, however, effect a perfect prevention, which can only be attained by the substitution of the red variety of phosphorus for the yellow in this industry.

Hydrogen Phosphides.

Gaseous hydrogen phosphide—*Phosphonia, Phosphamine*—**PH_3**—34—A colorless gas, having a strong alliaceous odor, which is obtained pure by decomposing phosphonium iodide, PH_4I, with H_2O. Mixed with H and vapor of P_2H_4 it is produced as a spontaneously inflammable gas, by the action of hot, concentrated solution of potassium hydrate on P, or by decomposition of calcium phosphide by H_2O. It is highly poisonous. After death the blood is found to be of a dark violet color, and to have, in a great measure, lost its power of absorbing oxygen.

Liquid hydrogen phosphide—**P_2H_4**—66—is the substance whose vapor communicates to PH_3 its property of igniting on contact with air. It is separated by passing the spontaneously inflammable PH_3 through a bulb tube surrounded by a freezing mixture.

It is a colorless, heavy liquid, which is decomposed by exposure to sunlight or to a temperature of 30° (86° F.).

Solid hydrogen phosphide—**P_4H_2**—126—is a yellow solid, formed when P_2H_4 is decomposed by sunlight. It is not phosphorescent and only ignites at 160° (320° F.).

Oxides of Phosphorus.

Two are known: P_2O_3 and P_2O_5.

Phosphorus trioxide—*Phosphorous anhydride*—**P_2O_3**—110—is formed when P is burned in a very limited supply of perfectly dry air or O. It is a white, flocculent solid, which, on exposure to air, ignites by the heat developed by its union with water to form phosphorous acid.

Phosphorus pentoxide—*Phosphoric anhydride*—**P_2O_5**—142—is formed when P is burned in an excess of dry O. It is a white, flocculent solid, which has almost as great a tendency to combine with H_2O as has P_2O_5. It absorbs moisture rapidly, deliquescing to a highly acid liquid, containing, not orthophosphoric, but metaphosphoric acid. It is used as a drying agent.

83

Phosphorus Acids.

Acid	Formula	Acid	Formula
Hypophosphorous acid	H_3PO_2.	Pyrophosphoric acid	$H_4P_2O_7$.
Phosphorous acid	H_3PO_3.	Metaphosphoric acid	HPO_3.
Phosphoric acid	H_3PO_4.		

Their basicities are as follows: H_3PO_2 is monobasic; H_3PO_3 is dibasic;

H_3PO_4 is tribasic; $H_4P_2O_7$ is tetrabasic, and HPO_3 is monobasic.

Hypophosphorous acid—H_3PO_2—66—is a crystalline solid, or, more usually, a strongly acid, colorless syrup. It is oxidized by air to a mixture of H_3PO_3 and H_3PO_4.

Phosphorous acid—H_3PO_3—82—is formed by decomposition of phosphorous trichloride by water: $PCl_3 + 3H_2O = H_3PO_3 + 3HCl$. It is a highly acid syrup, is decomposed by heat, and is a strong reducing agent.

Orthophosphoric acid—*Common, or tribasic, phosphoric acid—Acidum phosphoricum, U. S.; Br.*—H_3PO_4—98—does not occur free in nature, but is widely disseminated in combination in the phosphates, in the three kingdoms of nature.

It is prepared: (1) By converting bone phosphate, $Ca_3(PO_4)_2$, into the corresponding lead or barium salt, $Pb_3(PO_4)_2$ or $Ba_3(PO_4)_2$, and decomposing the former by H_2S, or the latter by H_2SO_4. (2) By oxidizing P by dilute HNO_3 aided by heat. The operation should be conducted with caution and heat gradually applied by the sand-bath. It is best to use red phosphorus. This is the process directed by the U. S. and Br. Pharm.

The concentrated acid is a colorless, transparent, syrupy liquid; still containing H_2O, which it gives off on exposure over H_2SO_4, leaving the pure acid in transparent, deliquescent, prismatic crystals. It is decomposed by heat to form, first, pyrophosphoric acid, then metaphosphoric acid. It is tribasic.

If made from arsenical phosphorus, and commercial phosphorus is usually arsenical, it is contaminated with arsenic acid, whose presence may be recognized by Marsh's test (q. v.). The acid should not respond to the indigo and ferrous sulphate tests for HNO_3.

Analytical Characters of the Orthophosphates.—(1) With ammoniacal solution of silver nitrate, a yellow precipitate. (2) With solution of ammonium molybdate in HNO_3, a yellow precipitate. (3) With magnesia mixture,* a white crystalline precipitate, soluble in acids, insoluble in ammonium hydrate.

* Made by dissolving 11 pts. crystallized magnesium chloride and 28 pts. ammonium chloride in 130 pts. water, adding 70 pts. dilute ammonium hydrate and filtering after two days.

Pyrophosphoric acid—$H_4P_2O_7$—178.—When orthophosphoric acid (or hydro-disodic phosphate) is maintained at 213° (415°.4 F.), two of its molecules unite, with the loss of the elements of a molecule of water: $2H_3PO_4 = P_2O_7H_4 + H_2O$, to form pyrophosphoric acid.

Metaphosphoric acid—*Glacial phosphoric acid*—HPO_3—80—is formed by heating H_3PO_4 or $H_4P_2O_7$ to near redness: $H_3PO_4 = HPO_3 + H_2O$; or $H_4P_2O_7 = 2HPO_3 + H_2O$. It is usually obtained from bone phosphate; this is first converted into ammonium phosphate, which is then subjected to a red heat.

84

It is a white, glassy, transparent solid, odorless, and acid in taste and reaction. Slowly deliquescent in air, it is very soluble in H_2O, although the solution takes place slowly, and is accompanied by a peculiar crackling sound. In constitution and basicity it resembles HNO_3.

Action of the Phosphates on the Economy.—The salts of orthophosphoric acid are important constituents of animal tissues, and give rise, when taken internally in reasonable doses, to no untoward symptoms. The acid itself may act deleteriously by virtue of its acid reaction. Meta- and pyro-phosphoric acids, even when taken in the form of neutral salts, have a distinct action (the pyro being the more active) upon the motor ganglia of the heart, producing diminution of the blood-pressure, and, in comparatively small doses, death from cessation of the heart's action.

Compounds of Phosphorus with the Halogens.

Phosphorus trichloride—PCl_3—137.5—is obtained by heating P in a limited supply of Cl. It is a colorless liquid; sp. gr. 1.61; has an irritating odor; fumes in air; boils at 76° (169° F.). Water decomposes it with formation of H_3PO_3 and HCl.

Phosphorus pentachloride—PCl_5—208.5—is formed when P is burnt in excess of Cl. It is a light yellow, crystalline solid: gives off irritating fumes; and is decomposed by H_2O.

Phosphorus oxychloride—$POCl_3$—153.5—is formed by the action of a limited quantity of H_2O on the pentachloride: $PCl_5 + H_2O = POCl_3 + 2HCl$. It is a colorless liquid; sp. gr. 1.7; boils at 110° (230° F.); and solidifies at −10° (+14° F.).

With **bromine** P forms compounds similar in composition and properties to the chlorine compounds. With **iodine** it forms two compounds, PI_2 and PI_3. With **fluorine** it forms two compounds, PF_3 and PF_5, the former liquid, the second gaseous.

ARSENIC.

Symbol = **As**—*Atomic weight* = 75—*Molecular weight* = 300—*Sp. gr. of solid* = 5.75; *of vapor* = 10.6*A at* 860° (1580° *F.*)—*Name from* ἀρσενικόν = *orpiment.*

Occurrence.—Free in small quantity; in combination as arsenides of Fe, Co, and Ni, but most abundantly in the sulphides, orpiment and realgar, and in arsenical iron pyrites or mispickel.

Preparation.—(1.) By heating mispickel in clay cylinders which communicate with sheet iron condensing tubes.

(2.) By heating a mixture of arsenic trioxide and charcoal; and purifying the product by resublimation.

Properties.—*Physical.*—A brittle, steel gray solid, having a metallic lustre. At the ordinary pressure, and without contact of air, it volatilizes without fusion at 180° (256° F.); under strong pressure it fuses at a dull red heat. Its vapor is yellowish, and has the odor of garlic. It is insoluble in H_2O and in other liquids unless chemically altered.

Chemical.—Heated in air it is converted into the trioxide and ignites somewhat below a red heat. In O it burns with a brilliant, bluish white light. In dry air it is not altered, but in the presence of moisture its surface becomes tarnished by oxidation. In H_2O it is slowly oxidized, a

85

portion of the oxide dissolving in the water. It combines readily with Cl, Br, I, and S, and with most of the metals. With H it only combines when that element is in the nascent state. Warm, concentrated H_2SO_4 is decomposed by As with formation of SO_2; As_4O_6 and H_2O. Nitric acid is readily decomposed, giving up its O to the formation of arsenic acid. With hot HCl, arsenic trichloride is formed. When fused with potassium hydrate, arsenic is oxidized, H is given off, and a mixture of potassium arsenite and arsenide remains, which by greater heat is converted into arsenic, which volatilizes, and potassium arsenate, which remains.

Compounds of Arsenic and Hydrogen.

Two are known: the solid As_2H (?), and the gaseous, AsH_3.

Hydrogen arsenide—*Arseniuretted or arsenetted hydrogen = Arsenia—Arsenamine*—**AsH**—*Molecular weight* = 78—*Sp. gr.* = 2.695*A*—*Liquefies at* − 40° (−40° F.).

Formation.—(1.) By the action of H_2O upon an alloy obtained by fusing together native sulphide of antimony, 2 pts.; cream of tartar, 2 pts.; and arsenic trioxide, 1 pt.

(2.) By the action of dilute HCl or H_2SO_4 upon the arsenides of Zn and Sn.

(3.) Whenever a reducible compound of arsenic is in presence of nascent hydrogen. (See Marsh test.)

(4.) By the action of H_2O upon the arsenides of the alkaline metals.

(5.) By the combined action of air, moisture and organic matter upon arsenical pigments.

Properties.—*Physical.*—A colorless gas; having a strong alliaceous odor; soluble in 5 vols. of H_2O, free from air.

Chemical.—It is neutral in reaction. In contact with air and moisture its H is slowly removed by oxidation, and elementary As deposited. It is also decomposed into its elements by the passage through it of luminous electric discharges; and when subjected to a red heat. It is not acted on by dry O at ordinary temperatures, but a mixture of the two gases containing 3 vols. O and 2 vols. AsH_3 explodes when heated, forming As_4O_6 and H_2O; if the proportion of O be less, elementary As is deposited.

The gas burns with a greenish flame, from which a white cloud of arsenic trioxide arises. A cold surface held above the flame becomes coated with a white, crystalline deposit of the oxide. If the flame be cooled by the introduction of a cold surface into it the H alone is oxidized and elementary As is deposited. Chlorine decomposes the gas explosively with formation of HCl and arsenic trioxide. Bromine and iodine behave similarly, but with less violence.

All oxidizing agents decompose it readily; H_2O and arsenic trioxide being formed by the less active oxidants, and H_2O and arsenic acid by the more active. Solid potassium hydrate decomposes the gas partially and becomes coated with a dark deposit which seems to be elementary arsenic. Solutions of the alkaline hydrates absorb and decompose it; H is given off and an alkaline arsenite remains in the solution. Many metals, when heated in AsH_3, decompose it with formation of a metallic arsenide and liberation of hydrogen. Solution of silver nitrate is reduced by it; elementary silver is deposited, and the solution contains arsenic trioxide.

Although H_2S and AsH_3 decompose each other to a great extent, with formation of arsenic trisulphide, the latter gas is capable of existing, to

86

some extent at least, in presence of the former. Hence in making H_2S for use in toxicological analysis materials free from As must be used.

Compounds of Arsenic and Oxygen.

Two are known: As_4O_6 and As_2O_5.

Probably the gray substance formed by the action of moist air on elementary arsenic is a lower oxide.

Arsenic trioxide—*Arsenious anhydride—White arsenic—Arsenic*—

Arsenious acid—Acidum arseniosum, U. S.; Br.—**As_2O_3**—198.

PREPARATION.—(1.) By roasting the native sulphides of arsenic in a current of air.

(2.) By burning arsenic in air or oxygen.

PROPERTIES.—*Physical.*—It occurs in two distinct forms: *crystallized* or "*powdered*," and *vitreous*. When freshly fused, it appears in colorless or faintly yellow, transparent, vitreous masses, having no visible crystalline structure. Shortly, however, these masses become opaque upon the surface, and present the appearance of porcelain; this change, which is due to the substance assuming the crystalline form, slowly progresses toward the centre of the mass, which, however, remains vitreous for a long time. The change is attended by the slow liberation of heat, and, if it be made to take place more rapidly, a faint light is visible in obscurity. When arsenic trioxide is sublimed, if the vapors be condensed upon a cool surface, it is deposited in the form of brilliant octahedral crystals, which are larger and more perfect the nearer the temperature of the condensing surface is to 180° (356° F.). The crystalline variety may be converted into the vitreous by keeping it for some time at a temperature near its point of volatilization.

The taste of arsenic trioxide is at first faintly sweet, afterward acrid, metallic, and nauseating. It is odorless; in aqueous solution (see below) it has a faintly acid reaction. The sp. gr. of the vitreous variety is 3.785; that of the crystalline, 3.689.

Its solubility in water varies with the temperature, the method of making the solution, the presence of foreign substances and the nature of the oxide:

	Transparent form.	Opaque form.	Fresh crystalline oxide.
1,000 parts of cold distilled water, after standing 24 hours, dissolved.......	1.74 parts.	1.16 parts.	2.0 parts.
1,000 parts of boiling water poured on the oxide, and allowed to stand for 24 hours, dissolved	10.12 parts.	5.4 parts.	15.0 parts.
1,000 parts of water boiled for one hour, the quantity being kept uniform by the addition of boiling water from time to time, and filtered immediately, dissolved..................	64.5 parts.	76.5 parts.	87.0 parts

The vitreous variety is more soluble than the crystalline, but by prolonged boiling the crystalline is converted into the vitreous, or, at all events, the solubility of the two forms becomes the same. The solution of the crystallized oxide in cold H_2O is always very slow (the vitreous oxide dissolves more rapidly), and continues for a long time. If white arsenic be thrown

87

upon cold H_2O, only a portion of it sinks, the remainder floating upon the surface, notwithstanding its high specific gravity. This is due to a repulsion of the H_2O from the surfaces of the crystals, which also accounts, to some extent at least, for its slow solution. Even after several days, cold H_2O does not dissolve all the oxide with which it is in contact. If one part of oxide be digested with 80 parts of H_2O, at ordinary temperatures for several days, the resulting solution contains $\frac{1}{90}$; with 160 parts H_2O, $\frac{1}{180}$; with 240 parts, $\frac{1}{280}$; with 1,000 parts H_2O, $\frac{1}{1200}$; and even when 16,000 or 100,000 parts of H_2O are used, a portion of the oxide remains undissolved. Arsenious oxide which had remained in contact with cold H_2O in closed vessels for eighteen years, dissolved to the extent of 1 part in 54 of H_2O, or 18.5 parts in 1,000, which may be given as the maximum solubility of the crystallized oxide in cold water. The power of H_2O of holding the acid in solution, once it is dissolved, is not the same as its power of dissolving it. If a concentrated solution be made by boiling H_2O upon the oxide and filtering hot, the filtrate may be evaporated down to one-half its original bulk without depositing any of the acid, of which this concentrated fluid now contains as much as one part in six of H_2O, or 166.6 parts per 1,000. If a hot solution of the acid be allowed to cool, the solution will contain 62.5 parts per 1,000 at 16° (60°.8 F.), and 50 parts per 1,000 at 7° (44°.6 F.)

The solubility of the oxide in alcohol varies with the strength of the spirit and the nature of the oxide, the vitreous variety being more soluble in strong than in weak alcohol, while the contrary is the case with the crystalline, as is shown in the following table:

1,000 parts dissolve	Alcohol at 56 per cent.	Alcohol at 79 per cent.	Alcohol at 86 per cent.	Absolute alcohol.
Crystallized oxide { At 15° (59° F.)................	16.80	14.30	7.15	0.25
Crystallized oxide { At the boiling-point..........	48.95	45.51	31.97	34.02
Vitreous oxide at 15° (59° F.)	5.04	5.40	.....	10.60

The presence of the mineral acids and alkalies, ammonia and ammoniacal salts, alkaline carbonates, tartaric acid, and the tartrates, increases the solubility of arsenic trioxide in water. It is less soluble in fluids containing fats or extractive or other organic matters (the various liquid articles of food), than it is in pure water.

In chemico-legal cases, in which the question of the solubility of arsenic is likely to arise, it must not be forgotten that the quantity of As_2O_3 which a person may unconsciously take in a given quantity of fluid is not limited, under certain circumstances, to that which the fluid is capable of *dissolving*; a much greater quantity than this may be taken while in suspension in the liquid, especially if it be mucilaginous.

CHEMICAL.—Its solutions are acid in reaction and probably contain the true arsenious acid, H_3AsO_3; they are neutralized by bases with formation of *arsenites*. Solutions of sodium or potassium hydrate dissolve it with formation of the corresponding arsenite. It is readily reduced, with separation of As, when heated with hydrogen, carbon, or potassium cyanide, and at lower temperatures by more active reducing agents. Oxidizing agents, such as HNO_3, the hydrates of chlorine, chromic acid, convert it into arsenic pentoxide or arsenic acid. Its solution, acidulated with HCl and boiled in presence of copper, deposits on the metal a gray film composed of an alloy of Cu and As.

Arsenic pentoxide—*Arsenic anhydride*—**As_2O_5**—230—is obtained by heating arsenic acid to redness. It is a white, amorphous solid, which, when exposed to the air, slowly absorbs moisture. It is fusible at a dull

88

red heat, and at a slightly higher temperature decomposes to As_2O_3 and O. It dissolves slowly in H_2O, forming arsenic acid, H_3AsO_4.

Arsenic Acids.

Arsenious acid........	H_3AsO_3.	Pyroarsenic acid........	$H_4As_2O_7$.
Arsenic acid	H_3AsO_4.	Metarsenic acid	$HAsO_3$.

Arsenious acid—H_3AsO_3—126—is supposed to exist in aqueous solutions of the trioxide, although it has not been separated. Corresponding to it are important salts, called *arsenites*, which have the general formulæ HM'_2AsO_3, $HM''AsO_3$, $H_4M''(AsO_3)_2$.

Orthoarsenic acid—*Arsenic acid*—**H_3AsO_4**—142—is obtained by oxidizing As_2O_3 with HNO_3 in the presence of H_2O: $As_2O_3 + 2H_2O + 2HNO_3 = 2H_3AsO_4 + N_2O_3$. A similar oxidation is also effected by Cl, aqua regia and other oxidants.

A syrupy, colorless, strongly acid solution is thus obtained, which, at 15° (59° F.) becomes semi-solid from the formation of transparent crystals containing 1 Aq. These crystals, which are very soluble and deliquescent, lose their Aq. at 100° (212° F.) and form a white, pasty mass composed of minute white, anhydrous needles. At higher temperatures it is converted into $H_4As_2O_7$, $HAsO_3$, and As_2O_5.

In presence of nascent H it is decomposed into H_2O and AsH_3. It is reducible to H_3AsO_3 by SO_2. If H_2S be passed through solutions of arsenic acid or of an arsenate, the first portions of the gas reduce the arsenical compound to the lower state of oxidation, while S separates; afterward the arsenious acid is decomposed, with formation of arsenic trisulphide.

Like phosphoric acid, arsenic acid is tribasic; and the arsenates resemble the phosphates in composition, and in many of their chemical and physical properties.

Pyroarsenic acid—$H_4As_2O_7$—266.—Arsenic acid when heated to 160° (320° F.) is converted into compact masses of pyroarsenic acid: $2H_3AsO_4 = H_4As_2O_7 + H_2O$. It is very prone to revert to orthoarsenic acid by taking up water.

Metarsenic acid—$HAsO_3$—124.—At 200°–206° (392°–403° F.) $As_2O_7H_4$ gradually loses H_2O to form metarsenic acid: $H_4As_2O_7 = 2HAsO_3 + H_2O$. It forms white, pearly crystals, which dissolve readily in H_2O with regeneration of H_3AsO_4. It is monobasic.

Compounds of Arsenic and Sulphur.

Arsenic disulphide—*Red sulphide of arsenic—Realgar—Red orpiment—Ruby sulphur—Sandarach*—**As_2S_2**—214—occurs in nature in translucent, ruby red crystals. It is also prepared by heating a mixture of As_2O_3 and S; as so obtained it appears in brick-red masses.

It is fusible, insoluble in H_2O, but soluble in solutions of the alkaline sulphides and in boiling solution of potassium hydrate.

89

Arsenic trisulphide—*Orpiment—Auripigmentum—Yellow sulphide of arsenic—King's yellow*—**As_2S_3**—246—occurs in nature in brilliant golden yellow flakes. Obtained by passing H_2S through an acid solution of As_2O_3; or by heating a mixture of As and S, or of As_2O_3 and S in equivalent proportions.

When formed by precipitation it is a lemon yellow powder, or in orange yellow, crystalline masses when prepared by sublimation. Almost insoluble in cold H_2O, but sufficiently soluble in hot H_2O to communicate to it a distinct yellow color; by continued boiling with H_2O it is decomposed into H_2S and As_2O_3. Insoluble in dilute HCl; but readily soluble in solutions of the alkaline hydrates, carbonates, and sulphides. It volatilizes when heated.

Nitric acid oxidizes it, forming H_3AsO_4 and H_2SO_4. A mixture of HCl and potassium chlorate has the same effect. It corresponds in constitution to As_2O_3, and, like it, may be regarded as an anhydride, for, although sulpharsenious acid, H_3AsS_3, has not been separated, the sulpharsenites, pyro- and meta-sulpharsenites are well-characterized compounds.

Arsenic pentasulphide—As_2S_5—310—is said to have been formed by fusing a mixture of As_2S_3 and S in proper proportions, as a yellow, fusible solid, capable of sublimation in absence of air. There exist well-defined sulpharsenates, pyro- and meta-sulpharsenates.

Compounds of Arsenic with the Halogens.

Arsenic trifluoride—AsF_3—132.—A colorless, fuming liquid, boiling at 63° (145° F.), obtained by distilling a mixture of As_2O_3, H_2SO_4 and fluorspar. It attacks glass.

Arsenic trichloride—$AsCl_3$—181.5.—Obtained by distilling a mixture of As_2O_3, H_2SO_4 and NaCl, using a well-cooled receiver.

It is a colorless liquid, boils at 134° (273° F.), fumes when exposed to the air, and volatilizes readily at temperatures below its boiling-point. Its formation must be avoided in processes for the chemico-legal detection of arsenic, lest it be volatilized and lost. It is formed by the action of HCl, even when comparatively dilute, upon As_2O_3 at the temperature of the water-bath; but, if potassium chlorate be added, the trioxide is oxidized to arsenic acid, and the formation of the chloride thus prevented. Arsenic trioxide, when fused with sodium nitrate, is converted into sodium arsenate, which is not volatile; if, however, small quantities of chlorides be present, $As Cl_3$ is formed. It is highly poisonous.

Arsenic tribromide—$As Br_3$—315.—Obtained by adding powdered As to Br, and distilling the product at 220° (428° F.). A solid, colorless, crystalline body, fuses at 20°–25° (68°–77° F.), boils at 220° (428° F.), and is decomposed by H_2O.

Arsenic triiodide—*Arsenii iodidum, U.S.*—**AsI_3**—456.—Formed by adding As to a solution of I in carbon disulphide; or by fusing together As and I in proper proportions. A brick-red solid, fusible and volatile. Soluble in a large quantity of H_2O. Decomposed by a small quantity of H_2O into HI, As_2O_3, H_2O and a residue of $As I_3$.

Action of Arsenical Compounds upon the Animal Economy.

The poisonous nature of many of the arsenical compounds has been known from remote antiquity, and it is probable that more murders have

90

been committed by their use than by that of all other toxic substances combined. Even at the present time—notwithstanding the fact that, suspicion once aroused, the detection of arsenic in the dead body is certain and comparatively easy—criminal arsenical poisoning is still quite common, especially in rural districts.

The poison is usually taken by the mouth, but it has also been introduced by other channels; the skin, either uninjured or abraded; the rectum, vagina or male urethra. The forms in which it has been taken are: (1.) *Elementary arsenic,* which is not poisonous *so long as it remains such;* in contact with water, or with the saliva, however, it is converted into an oxide, which is then dissolved, and, being capable of absorption, produces the characteristic effects of the arsenical compounds. Fly-paper is coated with a paste containing As, a portion of which has been oxidized by the action of air and moisture. (2.) *Hydrogen arsenide,* the most actively poisonous of the inorganic compounds of arsenic, has been the cause of several accidental deaths, among others, that of the chemist Gehlen, who died in consequence of having inhaled a few bubbles of the gas while experimenting upon it. In other cases death has followed the inhalation of hydrogen, made from zinc, or sulphuric acid contaminated with arsenic. (3.) *Arsenic trioxide* is the compound most frequently used by criminals. It has been given by every channel of entrance to the circulation; in some instances concealed with great art, in others merely held in suspension by stirring in a transparent fluid given to an intoxicated person. If the poison have been given in quantity, and undissolved, it may be found in the stomach after death in the form of eight-sided crystals, more or less worn by the action of the solvents with which it has come in contact.

The lethal dose is variable, death having occurred from two and one-half grains, and recovery having followed the taking of a dose of two ounces. It is more active when taken fasting than when taken on a full stomach, in which latter case all, or nearly all, the poison is frequently expelled by vomiting, before there has been time for the absorption of more than a small quantity. (4.) *Potassium arsenite,* the active substance in "Fowler's solution," although largely used by the laity in malarial districts as an ague-cure, has, so far as the records show, produced but one case of fatal poisoning. (5.) *Sodium arsenite* is sometimes used to clean metal vessels, a practice whose natural results are exemplified in the death of an individual who drank beer from a pewter mug so cleaned; and in the serious illness of 340 children in an English institution, in which this material had been used for cleaning the water-boiler. (6.) *Arsenic acid and arsenates.*—The acid itself has, so far as we know, been directly fatal to no one. The cases of death and illness, however, which have been put to the account of the red anilin dyes, are not due to them directly, but to arsenical residues remaining in them as the result of defective processes of manufacture. (7.) *Sulphides of arsenic.*—Poisoning by these is generally due to the use of orpiment, introduced into articles of food as a coloring matter, by a combination of fraud and stupidity, in mistake for turmeric. (8.) *The arsenical greens.*—Scheele's green or cupric arsenite, and Schweinfurth green or cupric aceto-metarsenite (the latter commonly known in the United States as Paris green, a name applied in Europe to one of the aniline pigments). These substances, although rarely administerd with murderous intent, have been the cause of death in a great number of cases. Among suicides in the lower orders of the population in large cities, Paris green has been the favorite.

The arsenical pigments may also produce disastrous results by "acci-

91

dent;' by being incorporated in ornamental pieces of confectionery; by being used in the dyeing of textile fabrics, from which they may be easily rubbed off; and by being used in the manufacture of wall-paper. Many instances of chronic or subacute arsenical poisoning have resulted from inhabiting rooms hung with paper whose whites, reds, or greens were produced by arsenical pigments. From such paper the poison is disseminated in the atmosphere of the room in two ways: either as an impalpable powder, mechanically detached from the paper and floating in the air, or, as Fleck has shown, by their decomposition, and the consequent diffusion of volatile arsenical compounds in the air.

The *treatment* in acute arsenical poisoning is the same, whatever may be the form in which the poison has been taken, if it have been taken by the mouth. The first indication is the removal of any unabsorbed poison from the alimentary canal. If vomiting have not occurred from the effects of the toxic, it should be induced by the administration of zinc sulphate, or by mechanical means. The stomach-pump should not be used unless the case is seen soon after the taking of the poison. When the stomach has been emptied, the chemical antidote is to be administered, with a view to the transformation in the stomach of any remaining arsenical compound into the insoluble, and, therefor, innocuous ferrous arsenate. From recent experiences, it would seem that the preparation known as "dialyzed iron" is very efficacious; failing this, ferric hydrate must be prepared extemporaneously, as when dry or not recently prepared it has no longer the power of combining with the arsenical compound. To prepare this substance a solution of ferric sulphate, *Liq. ferri tersulphatis* (U. S.) = *Liq. ferri persulphatis* (Br.), is diluted with three volumes of water and treated with aqua ammoniæ in slight excess. The precipitate formed is collected upon a muslin filter and washed with water until the washings are nearly tasteless. The contents of the filter—*Ferri oxidum hydratum* (U. S.), *Ferri peroxidum humidum* (Br.) is to be given moist in repeated doses of one to two teaspoonfuls, until an amount of the hydrate equal to 20 times the weight of white arsenic taken has been administered.

Precautions to be taken by the Physician in cases of suspected Poisoning.

It will rarely happen that in a case of suspected homicidal poisoning by arsenic, or by other poisons, the physician in charge will be willing or competent to conduct the chemical analysis upon which probably the conviction or acquittal of the accused will mainly depend. Upon his knowledge and care, however, the success or futility of the chemist's labors depend in a great measure.

It is, as a rule, the physician who first suspects foul play; and, while it is undoubtedly his duty to avoid any public manifestation of his suspicion, it is just as certainly his duty toward his patient and toward the community to satisfy himself as to the truth or falsity of his suspicion by the application of a simple test to the excreta of the patient during life, the result of which may enable him to prevent a crime, or, failing that, take the first step toward the punishment of the criminal.

In a case in which, from the symptoms, the physician suspects poisoning by any substance, he should himself test the urine or fæces, or both, and govern his treatment and his actions toward the patient, and those

92

surrounding the patient, by the results of his examination. Should the case terminate fatally, he should at once communicate his suspicions to the prosecuting officer, and require a post-mortem investigation, which should, if at all possible, be conducted in the presence of the chemist who is to conduct the analysis; for, be the physician as skilled as he may be, there are odors and appearances, observable in many cases at the opening of the body, full of meaning to the toxicological chemist, which are ephemeral, and whose bearing upon the case is not readily recognized by those not thoroughly experienced.

Cases frequently arise in which it is impossible to bring the chemist upon the ground in time for the autopsy; in such cases the physician should remember that that portion of the poison remaining in the alimentary tract (we are speaking of true poisons) is but the residue of the dose in excess of that which has been necessary to produce death; and, if the processes of elimination have been active, there may remain no trace of the poison in the alimentary canal, while it still may be detectable in deeper-seated organs. Moreover, the finding of poison in the stomach alone would not, at the present time, be sufficient to procure conviction of the criminal, who might raise the question as to whether the poison was not injected by some malicious person into that viscus after death.

For these reasons it is not sufficient to send the stomach alone for analysis; the chemist should also receive the entire intestinal canal, at least one-half the liver, the spleen, one or both kidneys, a piece of muscular tissue, the brain, and any urine that may remain in the bladder. The intestinal canal should be removed and sent to the chemist *without having been opened*, and with ligatures enclosing the contents at the two ends of the stomach and at the lower end of the intestine. The brain and alimentary canal are to be placed in separate jars, and the other viscera in another jar together; the urine in a vial by itself. All of these vessels are to be new and clean, and are to be closed by new corks, or by glass stoppers, or covers (not zinc screw-caps), which are then coated with paraffine (not sealing-wax), and so fastened with strings and seals that it is impossible to open the vessels without cutting the strings or breaking the seals. If the physician fail to observe these precautions, he has probably made the breach in the evidence through which the criminal will escape, and has at the outset defeated the aim of the analysis.

Analytical Characters of the Arsenical Compounds.

Arsenious Compounds.—(1.) H_2S, a yellow color in neutral or alkaline liquids; a yellow ppt. in acid liquids. The ppt. dissolves in solutions of the alkaline hydrates, carbonates, and sulphydrates; but is scarcely affected by HCl. Hot HNO_3 decomposes it.

(2.) $AgNO_3$, in the presence of a little NH_4HO, gives a yellow ppt. This test is best applied by placing the neutral arsenical solution in a porcelain capsule, adding neutral solution of $AgNO_3$, and blowing upon it over the stopper of the NH_4HO bottle, moistened with that reagent.

(3.) $CuSO_4$ under the same conditions as in (2) gives a yellowish green ppt.

(4.) A small quantity of solid As_2O_3 is placed in the point *a* of the tube, Fig. 28; above it, at *b*, a splinter of recently ignited charcoal; *b* is first heated to redness, then *a*; the vapor of As_2O_3 passing over the hot charcoal is reduced, and elementary As is deposited at *e* in a metallic ring.

93

The tube is then cut between *b* and *c*, the larger piece held with *d* uppermost, and heated at *c*; the deposit is volatilized, the odor of garlic is observed, and bright, octahedral crystals appear in the cool part of the tube.

(5.) *Reinsch test.*—The suspected liquid is acidulated with one-sixth its bulk of HCl; strips of electrotype copper are immersed in the liquid, which is boiled. In the presence of an arsenious compound a gray or bluish deposit is formed upon the Cu. A similar deposit is produced by other substances (Bi, Sb, Hg). To complete the test the Cu is removed, washed, and dried between folds of filter-paper, without removing the deposit. The copper, with its adherent film, is rolled into a cylinder, and introduced into a dry piece of Bohemian tubing, about ¼-inch in diameter and six inches long, which is held at the angle shown in Fig. 29 and heated at the point containing the copper. If the deposit consists of arsenic a white deposit is formed at *a*, which contains brilliant specks, and when examined with a magnifier is found to consist of minute octahedral crystals, Fig. 30.

FIG. 28.

The advantages of this test are: it may be applied in the presence of organic matter, to the urine for instance; it is easily conducted; and its positive results are not misleading, *if the test be carried to completion.* These advantages render it the most suitable method for the physician to use, *during the life of the patient.* It should not be used *after death* by the physician, as by it copper is introduced into the substances under examination, which may subsequently interfere seriously with the analysis. The purity of the Cu and HCl must be proved by a blank testing before use. Reinsch's test is not as delicate as Marsh's, and it does not react when the arsenic is in the higher stage of oxidation, nor in presence of oxidizing agents.

(6.) *Marsh's test* is based upon the formation of AsH_3 when a reducible compound of arsenic is in presence of nascent H; and the subsequent decomposition of the arsenical gas by heat, with separation of elementary arsenic.

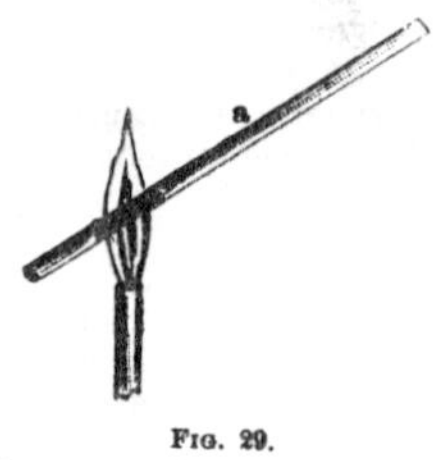

FIG. 29.

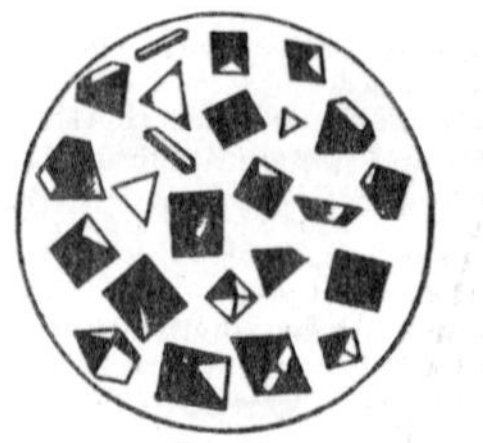
FIG. 30.

The apparatus used (Fig. 31) consists of a glass flask *a*, of about 150cc.

94

(5 fl ℥), through the cork of which pass a funnel-tube *c*, and a right angle bulb-tube *b*. The latter is connected with a tube *d*, filled with fragments of calcium chloride; which in turn connects with the Bohemian glass tube *gg*, whose middle third is bent into a spiral (Fig. 32). The other end of

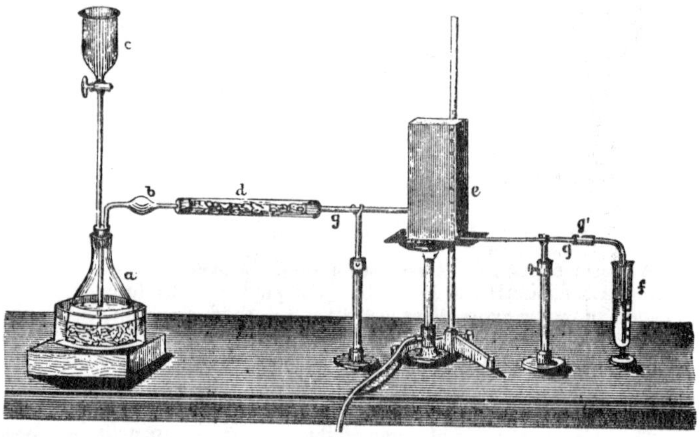

FIG. 31.

gg is bent downward, and dips into a solution of silver nitrate in the test-tube *f*. The coiled portion of *gg*, which is to be strongly heated by a large Bunsen burner, is supported by a coarse wire gauze and enclosed in a sheet-iron chimney *e*.

The flask *a* is first charged with about 25 grams (6½ ℥) of pure granulated zinc, which has been in contact with a diluted solution of platinic chloride for half an hour and then washed. The apparatus is then connected in such a manner that all joints are gas-tight, and the funnel-tube *c* about half filled with H_2SO_4, diluted with an equal bulk of H_2O, and cooled. By opening the stopcock the acid is brought in contact with the zinc in small quantities, in such a manner that during the entire testing bubbles of gas pass through *f*, at the rate of 60-80 per minute. After fifteen minutes the Bunsen burner is lighted and the heating continued, during evolution of gas from zinc and H_2SO_4, for an hour. At the end of that time, if no stain have formed in *g* beyond *e*, then zinc and acid may be considered pure and the suspected solution, prepared as described on page 98, introduced slowly through the funnel-tube.

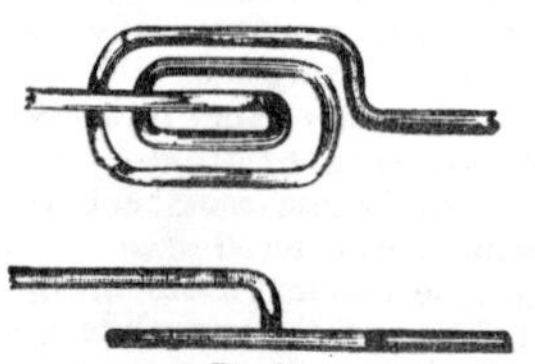
FIG. 32.

If arsenic be present in the substance examined, a hair-brown or gray deposit is formed in the cool part of *g* beyond *e*; at the same time the contents of *f* are darkened.

To distinguish the stains produced by arsenical compounds from the similar ones produced by antimony the following differences are noted:

95

The Arsenical Stain.	*The Antimonial Stain.*
First.—Is farther removed from the heated portion of the tube, and, if small in quantity, is double—the first hair-brown, the second steel-gray.	*First.*—Is quite near the heated portion of the tube.
Second.—Volatilizes readily when heated in an atmosphere of hydrogen, being deposited farther along in the tube. The escaping gas has the odor of garlic.	*Second.*—Requires a much higher temperature for its volatilization; fuses before volatilizing. Escaping gas has no alliaceous odor.
Third.—When cautiously heated in a current of oxygen, brilliant white octahedral crystals of arsenic trioxide are deposited farther along in the tube.	*Third.*—No crystals formed by heating in oxygen.
Fourth.—Instantly soluble in solution of sodium hypochlorite.	*Fourth.*—Insoluble in solution of sodium hypochlorite.
Fifth.—Slowly dissolved by solution of	*Fifth.*—Dissolves quickly in solution of

ammonium sulphydrate; more rapidly when warmed.

***Sixth.*—The solution obtained in 5 leaves, on evaporation over the water-bath, a bright yellow residue.**

***Seventh.*—The residue obtained in 6 is soluble in aqua ammoniæ, but insoluble in hydrochloric acid.**

***Eighth.*—Is soluble in warm nitric acid; the solution on evaporation yields a white residue, which turns brick-red when moistened with silver nitrate solution.**

***Ninth.*—Is not dissolved by a solution of stannous chloride.**

ammonium sulphydrate.

Sixth.—The solution obtained in 5 leaves, on evaporation over the water-bath, an orange-red residue.

Seventh.—The residue obtained in 6 is insoluble in aqua ammoniæ, but soluble in hydrochloric acid.

Eighth.—Is soluble in warm nitric acid; the solution on evaporation yields a white residue, which is not colored when moistened with silver nitrate solution.

Ninth.—Dissolves slowly in solution of stannous chloride.

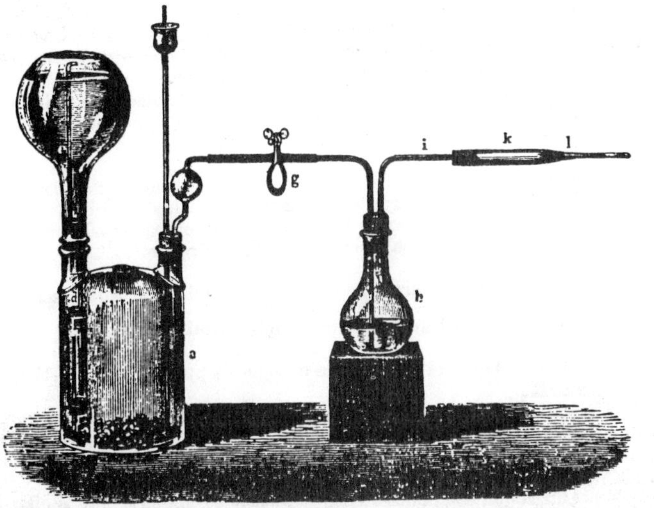

Fig. 33.

If, however, the process described on p. 98 have been followed, there can be no antimony in the liquid which would contain arsenic, if present.

96

The silver solution in *f* is tested for arsenious acid by floating upon its surface a layer of diluted NH_4HO solution, which, in the presence of arsenic, produces a yellow (not brown) band at the point of junction of the two liquids.

In place of bending the tube *gg'* downward, it may be bent upward and drawn out at *g'*. If the escaping gas be then ignited, the heating of the coil being discontinued, a white deposit of As_2O_3 may be collected on a glass surface held *above* the flame; or a brown deposit of elementary As upon a cold, porcelain surface held *in* the flame.

(7.) *Frezenius' and von Babo's test.*—The sulphide obtained in (1) is dried and mixed with 12 parts of a *dry* mixture of 3 pts. sodium carbonate and 1 pt. potassium cyanide, and the mixture brought into the tube, Fig. 33 at *k*. The apparatus is then connected as in the figure and filled with CO_2 which is allowed to pass through it in a slow current from *a*. The tube is then heated to redness at *k*, when, if arsenic be present, a gray deposit is formed at *l*; which has the characters of the arsenical stain indicated on p. 97.

Arsenic Compounds.—(1.) H_2S does not form a ppt. in neutral or alkaline solutions. In acid solutions it first reduces the arsenic to an arsenious compound, which is then decomposed with precipitation of the yellow As_2S_3.

(2.) $AgNO_3$, under the same conditions as with the arsenious compounds, produces a brick-red ppt. of silver arsenate.

(3.) $CuSO_4$ under like circumstances produces a bluish-green ppt.

Arsenic compounds behave like arsenious compounds with the tests 4, 6 and 7 for the latter.

Method of Analysis for Mineral Poisons.

In cases of suspected poisoning a systematic course of analysis is to be followed by which the presence or absence of *all* the more usual poisons can be determined.

In the search for mineral poisons (see alkaloids, p. 252) the first step is the destruction of organic matter. To this end the material to be examined, if liquid, is thinned with H_2O: and if solid is divided into small pieces and suspended in H_2O. About $^1/_{10}$ the volume of concentrated HCl and a small quantity of potassium chlorate are added and the mixture heated over a water-bath in a porcelain capsule. Potassium chlorate in small quantities, and, if necessary, HCl, are added from time to time, while the mixture is occasionally stirred and lumps of solid matter crushed with a flattened glass rod, until the mass has a uniform light yellow color. If the liquid smell strongly of Cl, CO_2 is passed through it. When the odor of Cl has disappeared, the liquid is filtered and the residue washed with hot water. If a deposit form on cooling the liquid is again filtered. The clear filtrate and washings, if strongly acid, are *partially* neutralized with sodium carbonate and treated with H_2S; the gas being passed slowly through the liquid for about half-an-hour at a time, at intervals of 4–6 hours, during 3 days; the vessel being well corked during the interval. The precipitate formed, which may contain Sn, As, Sb, Hg, Pb, Bi or Cu, is collected on a filter and washed with H_2O containing a small quantity of H_2S, until the washings fail to give the *faintest* cloudiness when boiled, acidulated with HNO_3 and treated with silver nitrate.

Solution of ammonium sulphydrate is added to the precipitate on the filter, which is then washed with water. The solution passing through may contain As, Sb, Sn and Cu; the residue on the filter (A) may contain Hg, Pb, Bi and Cu. The solution is evaporated over the water-bath to dryness, and the residue moistened with fuming HNO_3, dried, moistened with H_2O and dried several times, and then, after neutralization with caustic soda, fused with a mixture of sodium carbonate and nitrate, until it is colorless, or contains only a black, granular deposit, the heat being slowly increased. The cooled residue of fusion is dissolved in a small quantity of warm H_2O, and CO_2 is passed through the solution, whether it be clear or cloudy. The solution, if not perfectly clear, is filtered. Any deposit retained by the filter (B) may contain Sn, Sb or Cu. The filtrate is strongly acidulated with H_2SO_4, and slowly evaporated and heated, with addition of more H_2SO_4 if necessary, until abundant white fumes are given off. The cooled residue, which may contain As, is dissolved in H_2O and introduced into the Marsh apparatus when cold.

The residue B, if black, is dissolved in hot NO_3H and the solution tested for Cu. If it be white, it is ignited with the filter in a porcelain crucible; fused with potassium cyanide; and washed with H_2O. The residue is extracted with warm HCl and the solution tested for Sn. If any residue remain it is extracted with HCl to which a few drops of HNO_3 have been added, and the solution tested for Sb.

The residue A, after washing, is boiled with HNO_3, diluted with H_2O and filtered. The filtrate is tested for Cu, Bi and Pb. The residue, if any, is tested for Hg and Pb.

97

ANTIMONY.

Symbol = **Sb** (*Latin, stibium*)—*Atomic weight* = 120—*Molecular weight* = 240 (?)—*Sp. gr.* = 6.175—*Fuses at* 450° (842° *F.*).

Occurrence.—Free in small quantity; principally in the trisulphide, Sb_2S_3.

Preparation.—The native sulphide (black or crude antimony) is roasted and then reduced by heating with charcoal. The commercial antimony so obtained may be purified by fusing a mixture of antimony, 16 pts.; native sulphide of antimony, 1 pt.; and dry sodium carbonate, 2 pts. After cooling, the button is powdered and fused with 1½ pts. sodium carbonate and 1% ferrous sulphide. The antimony is again separated, powdered, and fused with sodium carbonate and a small quantity of sodium nitrate. Each fusion is maintained for an hour.

Properties.—*Physical.*—A bluish-gray, brittle solid, having a metallic lustre; readily crystallizable; tasteless and odorless; volatilizes at a red heat, and may be distilled in an atmosphere of H.

Chemical.—Is not altered by dry or moist air at ordinary temperatures. When sufficiently heated in air it burns with formation of Sb_2O_3, as a white, crystalline solid. It also combines directly with Cl, Br, I, S, and many metallic elements. It combines with H under the same circumstances as does As. Cold, dilute H_2SO_4 does not affect it; the hot, concentrated acid forms with it antimonyl sulphate, $(SbO)_2SO_4$ and SO_2. Hot HCl dissolves it when finely divided, with evolution of H. It is readily oxidized by HNO_3, with formation of H_3SbO_4 or Sb_2O_4. Aqua regia dissolves it as $SbCl_3$ or $SbCl_5$. Solutions of the alkaline hydrates do not act on it.

The element itself does not form salts with the oxides. There are, however, compounds, formed by the substitution of the group *antimonyl* (SbO), for the basic hydrogen of those acids. (See tartar emetic.)

Hydrogen Antimonide.

Antimoniuretted hydrogen—Stibamine—Stibonia—$\mathbf{SbH_3}$—123.—It has not been obtained in a condition of purity, but is produced, mixed with H, when a reducible compound of Sb is in presence of nascent H.

It is a colorless, odorless, combustible gas, subject to the same decompositions as AsH_3; from which it differs in being by no means as poisonous, and in its action upon silver nitrate solution. The arsenical gas acts upon the silver salt according to the equation: $6AgNO_3 + AsH_3 + 3H_2O = 6HNO_3 + H_3AsO_3 + 3Ag_2$, and the precipitate formed is elementary silver, while H_3AsO_3 remains in the solution. In the case of SbH_3 the reaction is $3AgNO_3 + SbH_3 = 3HNO_3 + SbAg_3$, all of the Sb being precipitated in the black silver antimonide.

Compounds of Antimony and Oxygen.

Antimony trioxide—*Antimonous anhydride—Oxide of antimony—Antimonii oxidum* (*U. S.; Br.*)—$\mathbf{Sb_2O_3}$—**288—occurs in nature; and is prepared artificially by decomposing the oxychloride; or by heating Sb in air.**

98

It is an amorphous, insoluble, tasteless, odorless powder; white at ordinary temperatures, but yellow when heated. It fuses readily, and may be distilled in absence of oxygen. Heated in air it burns like tinder and is converted into Sb_2O_4.

It is reduced with separation of Sb when heated with charcoal or in H. It is also readily oxidized by HNO_3 or potassium permanganate. It dissolves in HCl as $SbCl_3$; in Nordhausen sulphuric acid, from which solution brilliant crystalline plates of antimonyl pyrosulphate, $(SbO)_2S_2O_7$, separate; and in solutions of tartaric acid and hydropotassic tartrate (see tartar emetic). Boiling solutions of alkaline hydrates convert it into antimonic acid.

Antimony pentoxide—*Antimonic anhydride*—$\mathbf{Sb_2O_5}$—320—is obtained by heating metantimonic acid to dull redness. It is an amorphous, tasteless, odorless, pale lemon-yellow colored solid; very sparingly soluble in water and in acids. At a red heat it is decomposed into Sb_2O_4 and O.

Antimony antimoniate—*Intermediate oxide — Diantimonic tetroxide*—$\mathbf{Sb_2O_4}$—304—occurs in nature, and is formed when the oxides or hydrates of Sb are strongly heated, or when the lower stages of oxidation or the sulphides are oxidized by HNO_3, or by fusion with sodium nitrate. It is insoluble in H_2O; but is decomposed by HCl, hydropotassic tartrate and potash.

Antimony Acids.

The normal antimonous acid, H_3SbO_3, corresponding to H_3PO_3, is unknown; but the series of antimonic acids: ortho—H_3SbO_4, pyro—$H_4Sb_2O_7$, and meta—$HSbO_3$, is complete, either in the form of salts or in that of the free acids. There also exists, in its sodium salt, a derivative of the lacking antimonous acid: *metantimonous acid*, $HSbO_2$.

The compound sometimes used in medicine under the name *washed diaphoretic antimony* is potassium metantimonate, united with an excess of the pentoxide: $2KSbO_3$, Sb_2O_5. The hydropotassic pyroantimonate, $K_2H_2Sb_2O_7,6Aq$ is a valuable reagent for the sodium compounds. It is obtained by calcining a mixture of one part of antimony with four parts of potassium nitrate and fusing the product with its own weight of potassium carbonate.

Chlorides of Antimony.

Antimony trichloride—*Protochloride* or *butter of antimony*—**$SbCl_3$**—226.5—is obtained by passing dry Cl over an excess of Sb_2S_3; by dissolving Sb_2S_3 in HCl; or by distilling mixtures, either of Sb_2S_3 and mercuric chloride, or of Sb and mercuric chloride, or of antimonyl pyrosulphate and sodium chloride.

At low temperatures it is a solid, crystalline body; at the ordinary temperature a yellow, semi-solid mass, resembling butter; at 73°.2 (164° F.) it fuses to a yellow, oily liquid, which boils at 223° (433°.4 F.). Obtained by solution of Sb_2S_3 in HCl of the usual strength it forms a dark yellow solution, which, when concentrated to sp. gr. 1.47, constitutes the *Liq. Antimonii chloridi* (*Br.*).

It absorbs moisture from air and is soluble in a small quantity of H_2O; with a larger quantity it is decomposed with precipitation of a white

99

powder, *powder of Algaroth*, whose composition is SbOCl if cold H_2O be used, and $Sb_4O_5Cl_2$ if the H_2O be boiling. In H_2O containing 15 per cent. or more HCl, $SbCl_3$ is soluble without decomposition.

Antimony pentachloride—$SbCl_5$—297.5—is formed by the action of Cl in excess upon Sb or $SbCl_3$ and purified by distillation in a current of Cl.

It is a fuming, colorless liquid, which solidifies at −20° (−4° F.), the solid fusing at −6° (21°.2 F.). It absorbs moisture from air. With a small quantity of H_2O, and by evaporation over H_2SO_4, it forms a hydrate, $SbCl_5 4H_2O$, which appears in transparent, deliquescent crystals. With more H_2O a crystalline oxychloride, $SbOCl_3$, is formed; and with a still greater quantity, a white precipitate of orthoantimonic acid, H_3SbO_4.

Sulphides of Antimony.

Antimony trisulphide—*Sesquisulphide of antimony—Black antimony—Antimonii sulphidum* (*U. S.*)—*Antimonium nigrum* (*Br.*)—**Sb_2S_3**—336—is the chief ore of antimony; and is formed when H_2S is passed through a solution of tartar emetic.

The native sulphide is a steel-gray, crystalline solid; the artificial product an orange-red or brownish-red, amorphous powder. The *crude antimony* of commerce is in conical loaves, prepared by simple fusion of the native sulphide. It is soft, fusible, readily pulverized, and has a bright metallic lustre.

Heated in air it is decomposed into SO_2 and a brown, vitreous, more or less transparent mass, composed of varying proportions of oxide and oxysulphides, known as *crocus*, or *liver*, or *glass of antimony*. Sb_2S_3 is an anhydride, corresponding to which are salts known as *sulphantimonites*, having the general formula M'_2HSbS_3. If an excess of Sb_2S_3 be boiled with a solution of potash or soda, a liquid is obtained which contains an alkaline sulphantimonite and an excess of Sb_2S_3. If this solution be filtered and decomposed by an acid while still hot, an orange-colored, amorphous precipitate is produced, which is the *antimonium sulphuratum* (*U. S.; Br.*) and consists of a mixture in varying proportions of Sb_2S_3 and Sb_2O_3. If, however, the solution be allowed to cool, a brown, voluminous, amorphous precipitate separates, which consists of antimony trisulphide and trioxide, potassium or sodium sulphide, and alkaline sulphantimonite in varying proportions; and is known as *Kermes mineral.* If now the solution from which the Kermes has been separated, be decomposed with H_2SO_4, a reddish-yellow substance separates, which is the *golden sulphuret of antimony*, and consists of a mixture of Sb_2S_3 and Sb_2S_5. The precipitate obtained when H_2S acts upon a solution of an antimonial compound is, according to circumstances, Sb_2S_3 or Sb_2S_5, mixed with free S. By the action of HCl on Sb_2S_5, H_2S is produced.

Antimony pentasulphide.—Sb_2S_5—400—is obtained by decomposing an alkaline sulphantimonate by an acid. It is a dark orange-red, amorphous powder, readily soluble in solutions of the alkalies and alkaline sulphides, with which it forms *sulphantimonates.*

An oxysulphide, $Sb_6S_6O_3$, is obtained by the action of a solution of sodium hyposulphite upon Sb_2Cl_3 or tartar emetic. It is a fine, red powder used as a pigment, and called *antimony cinnabar* or *antimony vermilion.*

100

Action of Antimony Compounds on the Economy.

The compounds of antimony are poisonous, and act with greater or less energy as they are more or less soluble. The compound which is most frequently the cause of antimonial poisoning is tartar emetic (q. v.), which has caused death in a dose of half a grain, although recovery has followed the ingestion of half an ounce in several instances. Indeed, the chances of recovery seem to be better with large than with small doses, probably owing to the more rapid and complete removal of the poison by vomiting with large doses. Antimonials have been sometimes criminally administered in small and repeated doses, the victim dying of exhaustion. In such a case an examination of the urine will reveal the cause of the trouble.

If vomiting have not occurred in cases of acute antimonial poisoning it should be provoked by warm water, or the stomach should be evacuated by the pump. Tannin in some form (decoction of oak bark, cinchona, nutgalls, tea) should then be given with a view to rendering any remaining poison insoluble.

Medicinal antimonials are very liable to contamination with arsenic.

Analytical characters of Antimonial Compounds.

(1.) With H_2S in acid solution, an orange-red ppt., soluble in NH_4HS and in hot HCl.

(2.) A strip of bright copper suspended in a boiling solution of an Sb compound, acidulated with HCl, is coated with a blue-gray deposit. This deposit when dried (on the copper) and heated in a tube open at both ends yields a white, *amorphous* sublimate (see No. 5, p. 95).

(3.) Antimonial compounds yield a deposit by Marsh's test, similar to that obtained with arsenical compounds, but differing in the particulars given above (see No. 6, p. 97).

If, in cases of suspected poisoning, the examination have been conducted as directed on p. 98 any Sb present is separated during the fusion with sodium nitrate and carbonate, and the subsequent solution and filtration, so completely that As and Sb cannot be mistaken for one another.

IV.—BORON GROUP.

BORON

Symbol = **B**—*Atomic weight* = 11—*Molecular weight* = 22 (?)—*Isolated by Davy in* 1807.

Boron constitutes a group by itself; it is trivalent in all of its compounds; it forms but one oxide, which is the anhydride of a tribasic acid; and it forms no compound with H.

It is separable in two allotropic modifications. *Amorphous boron* is prepared by decomposition of the oxide, by heating with metallic potassium or sodium. It is a greenish-brown powder; sparingly soluble in H_2O; infusible and capable of direct union with Cl, Br, O, S and N.

Crystallized boron is produced when the oxide, chloride or fluoride is re-

101

duced by Al. It crystallizes in quadratic prisms; more or less transparent, and varying in color from a faint yellow to deep garnet-red; very hard; sp. gr. 2.68. It burns when strongly heated in O, and readily in Cl; it also combines with N, which it is capable of removing from NH_3 at a high temperature.

Boron trioxide.

Boric or *boracic anhydride*—**B_2O_3**—70—is obtained by heating boric acid to redness in a platinum vessel. It is a transparent, glass-like mass, used in blowpipe analysis under the name *vitreous boric acid.*

Boric Acids.

Orthoboric acid—*Boric* or *boracic acid—acidum boricum* (*U. S.*)—**H_3BO_3**—62—occurs in nature; and is prepared by slowly decomposing a boiling, concentrated solution of borax with an excess of H_2SO_4, and allowing the acid to crystallize.

It forms brilliant crystalline plates, unctuous to the touch; odorless; slightly bitter; soluble in 25 parts H_2O at 10° (50° F.); soluble in alcohol. Its solution reddens litmus but turns turmeric paper brown. When its aqueous solution is distilled a portion of the acid passes over.

If H_3BO_3 be heated for some time at 80° (176° F.), it loses H_2O and is converted into *metaboric acid*, HBO_2. If maintained at 100° (212° F.) for several days it loses a further quantity of H_2O and is converted into *pyroboric acid*, $H_2B_4O_7$, whose sodium salt is borax.

V.—CARBON GROUP.

Carbon—Silicon.

The elements of this group are bivalent or quadrivalent. The saturated oxide of each is the anhydride of a dibasic acid. They are both combustible, and each occurs in three allotropic forms.

CARBON.

***Symbol* = C—*Atomic weight* = 12—*Molecular weight* = 24 (?).**

Occurrence.—Free in its three allotropic forms: The *diamond* in octahedral crystals; in alluvial sand, clay, sandstone and conglomerate; *graphite*, in amorphous or imperfectly crystalline forms; *amorphous*, in the different varieties of anthracite and bituminous coal, jet, etc. In combination it is very widely distributed in the so-called organic substances.

Properties.—*Diamond.*—The crystals of diamond, which is almost pure carbon, are usually colorless or yellowish, but may be blue, green, pink, brown or black. It is the hardest substance known, and the one which refracts light the most strongly; its index of refraction is 2.47 to 2.75. It is very brittle; a bad conductor of heat and of electricity; sp. gr. 3.50 to 3.55. When very strongly heated in vacuo it swells up and is converted into a black mass resembling coke.

102

Graphite is a form of carbon almost as pure as the diamond, capable of crystallizing in hexagonal plates; sp. gr. 2.2; dark gray in color; opaque; soft enough to be scratched by the nail; and a good conductor of electricity. It is also known as *black lead* or *plumbago*. It has been obtained artificially by allowing molten cast-iron, containing an excess of carbon, to cool slowly, and dissolving the iron in HCl.

Amorphous carbon is met with in a great variety of forms, natural and artificial, in all of which it is black; sp. gr. 1.6–2.0; more or less porous; and a conductor of electricity.

Anthracite coal is hard and dense; it does not flame when burning; is difficult to kindle, but gives great heat with a suitable draught. It contains 80–90 per cent. of carbon. *Bituminous* coal differs from anthracite in that, when burning, it gives off gases which produce a flame. Some varieties are quite soft, while others, such as *jet*, are hard enough to assume a high polish. It is usually compact in texture, and very frequently contains impressions of leaves and other parts of plants. It contains about 75 per cent. of carbon.

Charcoal, carbo ligni, U. S., is obtained by burning woody fibre with an insufficient supply of air. It is brittle and sonorous; has the form of the wood from which it was obtained, and retains all the mineral matter present in the woody tissue. Its sp. gr. is about 1.57. It has the power of condensing within its pores odorous substances and large quantities of gases; 90 volumes of ammonia, 55 of hydrogen sulphide, 9.25 of oxygen. This property is taken advantage of in a variety of ways. Its power of absorbing odorous bodies renders it valuable as a disinfecting and filtering agent, and in the prevention of putrefaction and fermentation of certain liquids. The efficacy of charcoal as a filtering material is due also, in a great measure, to the oxidizing action of the oxygen condensed in its pores; indeed, if charcoal be boiled with dilute HCl, dried, and heated to redness, the oxidizing action of the oxygen, which it thus condenses, is very energetic.

Lamp-black is obtained by incomplete combustion of some resinous or tarry substance, or natural gas, the smoke or soot from which is directed into suitable condensing-chambers. It is a light, amorphous powder, and contains a notable quantity of oily and tarry material, from which it may be freed by heating in a covered vessel. It is used in the manufacture of printer's ink.

Coke is the substance remaining in gas-retorts after the distillation of bituminous coal in the manufacture of illuminating gas. It is a hard, grayish substance, usually very porous, dense, and sonorous. When iron retorts are used, a portion of the gaseous products are decomposed by contact with the hot iron surface, upon which there is then deposited a layer of very hard, compact, grayish carbon, which is a good conductor of electricity, and furnishes the best material for making the carbons of galvanic batteries and the points for the electric light. It does not form when gas is made in clay retorts.

Animal charcoal is obtained by calcining animal matters in closed vessels. If prepared from bones it is known as _bone-black, carbo animalis, U. S._; if from ivory, _ivory black_; the latter is used as a pigment, the former as a decolorizing agent. Bones yield about 60 per cent. of bone-black, which contains, besides carbon, nitrogen and the phosphates and other mineral substances of the bones. It possesses in a remarkable degree the power of absorbing coloring matters. When its decolorizing power is lost by saturation with pigmentary bodies, it may be restored, although not

103

completely, by calcination. For certain purposes purified animal charcoal, *i.e.*, freed from mineral matter, *carbo animalis purificatus, U. S.*, is required, and is obtained by extracting the commercial article with HCl and washing it thoroughly; its decolorizing power is diminished by this treatment. Animal charcoal has the power of removing from a solution certain crystalline substances, notably the alkaloids, and a method has been suggested for separating these bodies from organic mixtures by its use.

All forms of carbon are insoluble in any known liquid.

Chemical.—All forms of C combine with O at high temperatures with light and heat. The product of the union is carbon dioxide if the supply of air or O be sufficient; if O be present in limited quantity carbon monoxide is formed. The affinity of C for O renders it a valuable reducing agent. Many metallic oxides are reduced when heated with C, and steam is decomposed when passed over red-hot C: $H_2O + C = CO + H_2$. At elevated temperatures C also combines directly with S, to form carbon disulphide. With H carbon also combines directly under the influence of the voltaic arc.

For Compounds of Carbon see page 168.

SILICON.

Symbol = **Si**—*Atomic weight* = 28—*Molecular weight* = 56 (?)—*Discovered by Davy* 1807—*Name from* silex = *flint.*

Also known as *silicium*; occurs in three allotropic forms: *Amorphous silicon*, formed when silicon chloride is passed over heated K or Na, is a dark brown powder, heavier than water. When heated in air it burns with a bright flame to the dioxide. It dissolves in potash and in hydrofluoric acid, but is not attacked by other acids. *Graphitoid silicon* is obtained by fusing potassium fluosilicate with aluminium. It forms hexagonal plates, of sp. gr. 2.49, which do not burn when heated to whiteness in O, but may be oxidized at that temperature by a mixture of potassium chlorate and nitrate. It dissolves slowly in alkaline solutions, but not in acids. *Crystallized silicon*, corresponding to the diamond, forms crystalline needles, which are only attacked by a mixture of nitric and hydrofluoric acids.

Silicon, although closely related to C, exists in nature in but few compounds; it has been caused to form artificial combinations, however, which indicate its possible capacity to exist in substances corresponding to those C compounds vulgarly known as organic, *e.g., silicichloroform* and *silicibromoform*, $SiHCl_3$ and $SiHBr_3$.

Hydrogen silicide—SiH_4—32—is obtained as a colorless, insoluble, spontaneously inflammable gas, by passing the current of a galvanic battery of twelve cells through a solution of common salt, using a plate of aluminium, alloyed with silicon, as the positive electrode.

Silicon chloride—$SiCl_4$—170—a colorless, volatile liquid, having an irritating odor; sp. gr. 1.52; boils at 59° (138°.2 F.); formed when Si is heated to redness in Cl.

Silicic oxide—*Silicic anhydride—Silex*—**SiO_2**—60—is the most important of the compounds of silicon. It exists in nature in the different varieties of quartz, and in the rocks and sands containing that mineral, in agate, carnelian, flint, etc. Its purest native form is *rock crystal*; its hy-

104

drates occur in the opal, and in solution in natural waters. When crystallized it is fusible with difficulty; when heated to redness with the alkaline carbonates it forms *silicates*, which solidify to glass-like masses on cooling. It unites with H_2O to form a number of acid hydrates. The normal hydrate, H_4SiO_4, has not been isolated, although it probably exists in the solution obtained by adding an excess of HCl to a solution of sodium silicate. A gelatinous hydrate, soluble in water and in acids and alkalies, is obtained by adding a small quantity of HCl to a concentrated solution of sodium silicate.

Hydrofluosilicic acid—H_2SiF_6—144—is obtained in solution by passing the gas disengaged by gently heating a mixture of equal parts of fluorspar and pounded glass, and 6 pts. H_2SO_4, through water; the disengagement tube being protected from moisture by a layer of mercury. It is used in analysis as a test for K and Na.

VI. VANADIUM GROUP.

Vanadium—Niobium—Tantalum.

The elements of this group resemble those of the N group, but are more distinctly quadrivalent; particularly Nb and Ta.

Vanadium—V—51.3—a brilliant, crystalline metal; sp. gr. = 5.5; which forms a series of oxides similar to those of N. No salts of V are known, but salts of vanadyl (VO) are numerous, and are used in the manufacture of aniline black.

Niobium—Nb—94—a bright, steel-gray metal; sp. gr. 7.06; which burns in air to Nb_2O_5 and in Cl to $NbCl_5$; not attacked by acids.

Tantalum—Ta—182—closely resembles Nb in its chemical characters.

VII. MOLYBDENUM GROUP.

MOLYBDENUM—TUNGSTEN—OSMIUM.

The position of this group is doubtful; and it is probable that the lower oxides will be found to be basic in character: in which case the group should be transferred to the third class.

Molybdenum—Mo—95.5—a brittle white metal. The oxide MoO_3, ***molybdic anhydride***, combines with H_2O to form a number of acids; the ammonium salt of one of which is used as a reagent for PO_4H_3: with which it forms a conjugate acid, ***phosphomolybdic acid***, used as a reagent for the alkaloids.

Tungsten—*Wolfram*—**W**—183.6—a hard, brittle metal: sp. gr. 17.4. The oxide, WO_3, ***tungstic anhydride***, is a yellow powder, forming with H_2O several acid hydrates; one of which, ***metatungstic acid***, is used as a test for the alkaloids, as are also the conjugate *silicotungstic* and ***phosphotungstic acids***. Tissues impregnated with *sodium tungstate* are rendered uninflammable.

Osmium—Os—198.5—occurs in combination with Ir in Pt ores: combustible and readily oxidized to OsO_4. This oxide, known as *osmic acid*, forms colorless crystals, soluble in H_2O, which give off intensely irritating vapors. It is used as a staining agent by histologists, and also in dental practice.

105

CLASS III.—AMPHOTERIC ELEMENTS.

ELEMENTS WHOSE OXIDES UNITE WITH WATER, SOME TO FORM BASES, OTHERS TO FORM ACIDS. WHICH FORM OXYSALTS.

I. GOLD GROUP.

GOLD.

Symbol = **Au (AURUM)**—*Atomic weight* = 196.2—*Molecular weight* = 392.4 (?)—*sp. gr.* = 19.258-19.367—*Fuses at* 1200° (2192° *F.*).

This, the only member of the group, forms two series of compounds; in one, AuCl, it is univalent; in the other, $AuCl_3$, trivalent. Its hydrate, *auric acid*, $Au(OH)_3$, corresponds to the oxide Au_2O_3. Its oxysalts are unstable.

It is yellow or red by reflected light, green by transmitted light, reddish-purple when finely divided; not very tenacious; softer than silver; very malleable and ductile. It is not acted on by H_2O or air at any temperature, nor by any single acid. It combines directly with Cl, Br, I, P, Sb, As, and Hg. It dissolves in nitromuriatic acid as auric chloride. It is oxidized by alkalies in fusion on contact with air.

Auric chloride—*Gold trichloride*—$AuCl_3$—302.7—obtained by dissolving Au in aqua regia, evaporating at 100° (212° F.), and purifying by crystallization from H_2O. Deliquescent yellow prisms, very soluble in H_2O, alcohol and ether; readily decomposed with separation of Au, by contact with P, or with reducing agents. Its solution, treated with the chlorides of tin, deposits a purple double stannate of Sn and Au, called "purple of cassius." With alkaline chlorides it forms double chlorides, *chloraurates* (*auri et sodii chloridum, U.S.*).

Analytical Characters.

(1.) With H_2S, from neutral or acid solution, a blackish-brown ppt. in the cold; insoluble in HNO_3 and HCl; soluble in aqua regia and in yellow NH_4HS.

(2.) With stannous chloride and a little chlorine water, a purple-red ppt., insoluble in HCl.

(3.) With ferrous sulphate a brown deposit, which assumes the lustre of gold when dried and burnished.

II. IRON GROUP.

CHROMIUM—MANGANESE—IRON.

The elements of this group form two series of compounds: in one they are bivalent, as in $Fe''Cl_2$ or $Mn''SO_4$, while in the other they are quad-

106

rivalent; but when quadrivalent the atoms do not enter into combination singly, but grouped two together to form a hexavalent unit $\left[\begin{matrix}Fe\equiv \\ | \\ Fe\equiv\end{matrix}\right]'''$, as in $(Fe_2)^{vi}Cl_6$, $(Cr_2)^{vi}O_3$. They form several oxides; of which the oxide MO_3 is an anhydride, corresponding to which are acids and salts; most of the other oxides are basic.

CHROMIUM.

Symbol = **Cr**—*Atomic weight* = 52.4—*Molecular weight* = 104.8 (?)—*Sp. gr.* = 6.8—*Discovered by Vauquelin*, 1797—*Name from* χρῶμα = color.

The element is separated with difficulty by reduction of its oxide by charcoal, or of its chloride by sodium. It is a hard, crystalline, almost infusible metal. Combines with O only at a red heat; it is not attacked by acids, except HCl; is readily attacked by alkalies.

Chromic Oxide—*Sesquioxide, or green oxide of chromium*—Cr_2O_3—152.8—obtained, amorphous, by calcining a mixture of potassium dichromate and starch, or, crystallized, by heating neutral potassium chromate to redness in Cl.

It is green; insoluble in H_2O, acids, and alkalies; fusible with difficulty, and not decomposed by heat; not reduced by H. At a red heat in air, it combines with alkaline hydrates and nitrates to form chromates. It forms two series of salts, the terms of one of which are green, those of the other violet. The alkaline hydrates separate a bluish green hydrate from solutions of the green salts, and a bluish violet hydrate from those of the violet salts.

Chromium green, or *emerald green*, is a green hydrate, formed by decomposing a double borate of chromium and potassium by H_2O. It is used in the arts as a substitute for the arsenical greens, and is non-poisonous.

Chromic Anhydride—*Acidum chromicum* (*U. S.*)—CrO_3—100.4—is formed by decomposing a solution of potassium dichromate by excess of H_2SO_4 and crystallizing.

It crystallizes in deliquescent crimson prisms, very soluble in H_2O and in dilute alcohol. It is a powerful oxidant, capable of igniting strong alcohol.

The true chromic acid has not been isolated, but salts are known which correspond to three acid hydrates: H_2CrO_4 = *chromic acid;* $H_2Cr_2O_7$ = *dichromic acid;* and $H_2Cr_3O_{10}$ = *trichromic acid.*

Chlorides.—Two chlorides and one oxychloride of chromium are known. *Chromous chloride*, $CrCl_2$, is a white solid, soluble with a blue color in H_2O. *Chromic chloride*, $(Cr_2)Cl_6$, forms large, red crystals, insoluble in H_2O when pure.

Sulphates.—A *violet* sulphate crystallizes in octahedra, $(Cr)_2(SO_4)_3$ + 15 Aq, and is very soluble in H_2O; at 100° it is converted into a *green* salt, $(Cr)_2(SO_4)_3$ + 5 Aq, soluble in alcohol; which at higher temperatures is converted into the red, insoluble, anhydrous salt. Chromic sulphate forms double sulphates, containing 24 Aq, with the alkaline sulphates. (See Alums.)

107

Analytical Characters.

CHROMOUS SALTS.—(1.) Potash, a brown ppt.

(2.) Ammonium hydrate, greenish white ppt.

(3.) Alkaline sulphides, black ppt.

(4.) Sodium phosphate, blue ppt.

CHROMIC SALTS.—(1.) Potash, green ppt.; an excess of precipitant forms a green solution, from which Cr_2O_3 separates on boiling.

(2.) Ammonium hydrate, greenish-gray ppt.

(3.) Ammonium sulphydrate, greenish ppt.

CHROMATES.—(1.) H_2S in acid solution, brownish color, changing to green.

(2.) Ammonium sulphydrate, greenish ppt.

(3.) Barium chloride, yellowish ppt.

(4.) Silver nitrate, brownish-red ppt., soluble in HNO_3 or NH_4HO.

(5.) Lead acetate, yellow ppt., soluble in potash, insoluble in acetic acid.

Action on the Economy.

Chromic anhydride oxidizes organic substances, and is used as a caustic. The *chromates*, especially potassium dichromate (*q. v.*), are irritants, and have a distinctly poisonous action as well. Workmen handling the dichromate are liable to a form of chronic poisoning.

In acute chromium-poisoning, emetics, and subsequently magnesium carbonate in milk, are to be given.

MANGANESE.

Symbol = **Mn**—*Atomic weight* = 54—*Molecular weight* = 108 (?)—*Sp. gr.* = 7.138-7.206.

A hard, grayish, brittle metal; fusible with difficulty; obtained by reduction of its oxides by C at a white heat. It is not readily oxidized by cold, dry air; but is superficially oxidized when heated. It decomposes H_2O, liberating H; and dissolves in dilute acids.

Compounds of Manganese.

Oxides.—Manganese forms six oxides or compounds representing them: *Manganous oxide*, MnO; *manganoso-manganic oxide*, Mn_3O_4; *manganic oxide*, Mn_2O_3; *permanganic oxide*, MnO_2, and *permanganic anhydride*, Mn_2O_7, are known free. *Manganic anhydride*, MnO_3, has not been isolated. MnO and Mn_2O_3 are basic; Mn_3O_4 and MnO_2 are indifferent oxides; and MnO_3 and Mn_2O_7 are anhydrides, corresponding to the *manganates* and *permanganates*.

PERMANGANIC OXIDE.—*Manganese dioxide, or black oxide—Mangani oxidum nigrum* (*U. S.*)—*Manganesii ox. nig.* (*Br.*)—**MnO_2**—86—exists in nature as *pyrolusite*, the principal ore of manganese, in steel gray or brownish-black, imperfectly crystalline masses.

108

At a red heat it loses 12 per cent. of O : $3MnO_2 = Mn_3O_4 + O_2$; and at a white heat a further quantity of O is given off : $2Mn_3O_4 = 6MnO + O_2$. Heated with H_2SO_4 it gives off O and forms manganous sulphate : $2MnO_2 + 2H_2SO_4 = 2MnSO_4 + 2H_2O + O$. With HCl it yields manganous chloride, H_2O and Cl : $MnO_2 + 4HCl = MnCl_2 + 2H_2O + Cl_2$. It is not acted on by HNO_3.

Chlorides.—Two chlorides of Mn are known : *manganous chloride*, $MnCl_2$, a pink, deliquescent, soluble salt, occurring, mixed with ferric chloride, in the waste liquid of the preparation of Cl ; and *manganic chloride*, Mn_2Cl_6.

Salts of Manganese.

Manganese forms two series of salts : *Manganous salts*, containing Mn'' ; and *manganic salts*, containing $(Mn_2)^{vi}$; the former are colorless or pink, and soluble in water ; the latter are unstable.

MANGANOUS SULPHATE.—*Mangani sulphas* (*U. S.*)—**$MnSO_4 + nAq$**—150 +n18—is formed by the action of H_2SO_4 on MnO_2. Below 6° (42°.8 F.) it crystallizes with 7 Aq, and is isomorphous with ferrous sulphate ; between 7°–20° (44°.6–68° F.) it forms crystals with 5 Aq, and is isomorphous with cupric sulphate ; between 20°–30° (68°–86° F.) it crystallizes with 4 Aq. It is rose-colored, darker as the proportion of Aq increases, soluble in H_2O, insoluble in alcohol. With the alkaline sulphates it forms double salts, with 6 Aq.

Analytical Characters.

MANGANOUS.—(1.) Potash, white ppt., turning brown.
(2.) Alkaline carbonates, white ppts.
(3.) Ammonium sulphydrate, flesh-colored ppt., soluble in acids, sparingly soluble in excess of precipitant.
(4.) Potassium ferrocyanide, faintly reddish white ppt., in neutral solution ; soluble in HCl.
(5.) Potassium cyanide, rose-colored ppt., forming brown solution with excess.

MANGANIC.—(1.) H_2S, ppt. of sulphur.
(2.) Ammonium sulphydrate, flesh-colored ppt.
(3.) Potassium ferrocyanide, greenish ppt.
(4.) Potassium ferricyanide, brown ppt.
(5.) Potassium cyanide, light brown ppt.

MANGANATES—are green salts, whose solutions are only stable in presence of excess of alkali, and turn brown when diluted and acidulated.

PERMANGANATES—form red solutions, which are decolorized by SO_2, other reducing agents, and many organic substances.

IRON.

Symbol = **Fe** (**FERRUM**)—*Atomic weight* = 55.9—*Molecular weight* = 111.8 ?—*Sp. gr.* = 7.25–7.9 *Fuses at* 1600° (1912° F.)—*Name from the Saxon*, iren.

OCCURRENCE.—Free in small quantity only in platinum ores and meteorites. As Fe_2O_3 in *red hæmatite* and *specular iron ;* as hydrates of Fe_2O_3 in *brown hæmatite* and *oölitic iron ;* as Fe_3O_4 in *magnetic iron ;* as

109

$FeCO_3$ in *spathic iron, clay ironstone* and *bog ore ;* and as FeS_2 in *pyrites*. It is also a constituent of most soils and clays, exists in many mineral waters, and in the red blood pigment of animals.

PREPARATION.—In working the ores, reduction is first effected in a *blast-furnace*, into which alternate layers of ore, coal, and limestone are fed from the top, while air is forced in from below. In the lower part of the furnace CO_2 is produced at the expense of the coal ; higher up it is reduced by the incandescent fuel to CO, which at a still higher point reduces the ore ; the fused metal so liberated collects at the lowest point under a layer of *slag*, and is drawn off to be cast as *pig iron*. This product is then purified by burning out impurities, in the process known as *puddling*.

Pure iron is prepared by reduction of ferrous chloride or of ferric oxide by H at a temperature approaching redness.

VARIETIES.—*Cast iron* is a brittle, white or gray, crystalline metal, consisting of Fe 89–90% ; C 1–4.5% ; and Si, P, S, and Mn. As *pig iron*, it is the product of the blast-furnace.

Wrought or *bar iron* is a fibrous, tough metal, freed in part from the impurities of cast iron by *refining* and *puddling*.

Steel is Fe combined with a quantity of C less than that existing in cast, and greater than that in bar iron. It is prepared by *cementation ;* which consists in causing bar iron to combine with C ; or by the *Bessemer method ;* which, as now used, consists in burning the C out of molten cast iron, to which the proper proportion of C is then added in the shape of *spiegel eisen*, an iron rich in Mn and C.

The purest forms of commercial iron are those used in piano-strings, the teeth of carding machines, and electro-magnets ; known as *soft iron*.

Reduced iron—Ferrum reductum (*U. S.*)—*Fer. redactum* (*Br.*)—is Fe, more or less mixed with Fe_2O_3 and Fe_3O_4, obtained by heating Fe_2O_3 in H.

PROPERTIES.—*Physical.*—Pure iron is silver-white ; quite soft ; crystallizes in cubes or octahedra. *Wrought iron* is gray, hard ; very tenacious ; fibrous ; quite malleable and ductile ; capable of being welded ; highly magnetic but only temporarily so. *Steel* is gray ; very hard and brittle if *tempered*, soft and tenacious if not ; permanently magnetic.

Chemical.—Iron is not altered by dry air at the ordinary temperature. At a red heat it is oxidized. In damp air it is converted into a hydrate ; *iron rust. Tinplate* is sheet iron, coated with tin ; *galvanized iron* is coated with zinc, to preserve it from the action of damp air.

Iron unites directly with Cl, Br, I, S, N, P, As, and Sb. It dissolves in HCl as ferrous chloride, while H is liberated. Heated with strong H_2SO_4, it gives off SO_2 ; with dilute H_2SO_4, H is given off and ferrous sulphate formed. Dilute HNO_3 dissolves Fe, but the concentrated acid renders it *passive*, when it is not dissolved by either concentrated or dilute HNO_3, until the passive condition is destroyed by contact with Pt, Ag or Cu, or by heating to 40° (104° F.).

Compounds of Iron.

Oxides.—Three oxides of iron exist free : FeO ; Fe_2O_3 ; Fe_3O_4.

FERROUS OXIDE—*Protoxide of iron*—**FeO**—71.9—is formed by heating Fe_2O_3 in CO or CO_2.

FERRIC OXIDE—*Sesquioxide or peroxide of iron—Colcothar—Jeweller's rouge—Venetian red*—**Fe_2O_3**—159.8—occurs in nature (see above); and is formed when ferrous sulphate is strongly heated, as in the manufacture of

110

pyrosulphuric acid. It is a reddish, amorphous solid, is a weak base, and is decomposed at a white heat into O and Fe_3O_4.

MAGNETIC OXIDE—*Black oxide—Ferri oxidum magneticum* (*Br.*)—**Fe_3O_4**—231.7—is the *natural loadstone*, and is formed by the action of air or steam upon iron at high temperatures. It is probably a compound of ferrous and ferric oxides (FeO, Fe_2O_3), as acids produce with it mixtures of ferrous and ferric salts.

Hydrates.—FERROUS.—When a solution of a ferrous salt is decomposed by an alkaline hydrate, a greenish-white hydrate, FeH_2O_2, is deposited ; which rapidly absorbs O from the air, with formation of ferric hydrate.

FERRIC.—When an alkali is added to a solution of a ferric salt, a brown, gelatinous precipitate is formed, which is the *normal ferric hydrate*, $(Fe)_2H_6O_6$ = *Ferri peroxidum hydratum* (*U. S.*) ; *Fer. perox. humidum* (*Br.*). It is not formed in the presence of fixed organic acids, or of sugar in sufficient quantity. If preserved under H_2O it is partly oxidized, forming an oxyhydrate which is incapable of forming ferrous arsenate with As_2O_3.

If the hydrate $(Fe_2)H_6O_6$ be dried at 100° (212° F.), it loses $2H_2O$, and is converted into $(Fe_2)O_2, H_2O_2$, which is the *Ferri peroxidum hydratum* (*Br.*).

If the normal hydrate be dried *in vacuo* it is converted into $(Fe_2)_2H_6O_9$, and this, when boiled for some hours with H_2O, is converted into the *colloid*, or *modified hydrate* $(Fe_2)H_2O_4$ (?), which is brick-red in color ; almost insoluble in HNO_3 and HCl ; gives no Prussian blue reaction, and forms a turbid solution with acetic acid. If recently precipitated ferric hydrate be dissolved in solution of ferric chloride or acetate, and subjected to dialysis, almost all the acid passes out, leaving in the dialyzer a dark-red solution, which probably contains this colloid hydrate, and which is instantly coagulated by a trace of H_2SO_4, by alkalies, many salts, and by heat (*dialyzed iron*).

FERRIC ACID—**$H_2Fe_2O_4$**.—Neither the free acid nor the oxide, FeO_3, are known in the free state ; the *ferrates*, however, of Na, K, Ba, Sr, and Ca are known.

Sulphides.—FERROUS SULPHIDE—*Protosulphide of iron*—**FeS**—87.9—is formed :

(1) By heating a mixture of finely divided Fe and S to redness ; (2) by pressing roll sulphur on white hot iron ; (3) in a hydrated condition, FeS, H_2O, by treating a solution of a ferrous salt with an alkaline sulphydrate.

The dry sulphide is a brownish, brittle, magnetic solid ; insoluble in H_2O, soluble in acids with evolution of H_2S. The hydrate is a black powder, which absorbs O from the air, turning yellow, by formation of Fe_2O_3, and liberation of S. It occurs in the fæces of persons taking chalybeate waters or preparations of iron.

FERRIC SULPHIDE—*Sesquisulphide*—**Fe_2S_3**—207.8—occurs in nature in *copper pyrites* and is formed when the disulphide is heated to redness.

FERRIC DISULPHIDE—**FeS_2**—119.9—occurs in the *white* and *yellow Martial pyrites* used in the manufacture of H_2SO_4. When heated in air it is decomposed into SO_2 and *magnetic pyrites :* $3FeS_2 + 2O_2 = Fe_3S_4 + 2SO_2$.

Chlorides.—FERROUS CHLORIDE—*Protochloride*—**$FeCl_2$**—129.9—is produced : (1) by passing dry HCl over red hot Fe ; (2) by heating ferric chloride in H ; (3) as a hydrate, $FeCl_2$, $4H_2O$, by dissolving Fe in HCl.

The anhydrous compound is a yellow, crystalline, volatile, and very soluble solid; the hydrated is in greenish, oblique rhombic prisms, deliques-

111

cent and very soluble in H_2O and alcohol. When heated in air it is converted into ferric chloride and an oxychloride.

FERRIC CHLORIDE—*Sesquichloride—Perchloride—Ferri chloridum* (*U. S.*) —**Fe_2Cl_6**—324.8—is produced in the anhydrous form by heating Fe in Cl. As a hydrate, $Fe_2Cl_6,4H_2O$ or $Fe_2Cl,6H_2O$; it is formed: (1) by solution of the anhydrous compound; (2) by dissolving Fe in aqua regia; (3) by dissolving ferric hydrate in HCl; (4) by the action of Cl or of HNO_3 on solution of ferrous chloride. It is by the last method that the pharmaceutical product is obtained.

The anhydrous compound forms reddish-violet, crystalline plates, very deliquescent. The hydrates form yellow, nodular, imperfectly crystalline masses, or rhombic plates, very soluble in H_2O, soluble in alcohol and ether. In solution it is converted into $FeCl_2$ by reducing agents. The *Liq. ferri chloridi* (*U. S.*) = *Liq. fer. perchloridi* (*Br.*) is an aqueous solution of this compound, containing excess of acid. The *Tinct. fer. chlor.* (*U. S.*) and *Tinct. fer. perchl.* (*Br.*) are the solution diluted with alcohol; and contain ethyl chloride and ferrous chloride.

Bromides.—FERROUS BROMIDE—**$FeBr_2$**—215.9—is formed by the action of Br on excess of Fe in presence of H_2O.

FERRIC BROMIDE—**Fe_2Br_6**—591.8—is prepared by the action of excess of Br on Fe.

Iodides.—FERROUS IODIDE—*Ferri iodidum* (*U. S.*; *Br.*)—**FeI_2**—309.9—is obtained, with $4H_2O$, by the action of I upon excess of Fe in the presence of warm H_2O. When anhydrous, it is a white powder; hydrated, it is in green crystals. In air it is rapidly decomposed, more slowly in the presence of sugar.

FERRIC IODIDE—**Fe_2I_6**—873.8—is formed by the action of excess of I on Fe.

Salts of Iron.

Sulphates.—FERROUS SULPHATE—*Protosulphate—Green vitriol—Copperas—Ferri sulphas* (*U. S.*; *Br.*)—**$FeSO_4$ + 7 Aq**—151.9 + 126—is formed: (1) by oxidation of the sulphide, Fe_2S_3, formed in the manufacture of H_2SO_4; (2) by dissolving Fe in dilute H_2SO_4.

It forms green, efflorescent, oblique rhombic prisms, quite soluble in H_2O, insoluble in alcohol. It loses 6 Aq at 100° (212° F.) (*Ferr. sulph. exsiccatus U. S.*); and the last Aq at about 300° (572° F.). At a red heat it is decomposed into Fe_2O_3, SO_2 and SO_3. By exposure to air it is gradually converted into a basic ferric sulphate, $(Fe_2)(SO_4)_3,5Fe_2O_3$.

FERRIC SULPHATES are quite numerous, and are formed by oxidation of ferrous sulphate under different conditions. The normal sulphate, $(Fe_2)(SO_4)_3$, is formed by treating solution of $FeSO_4$ with HNO_3, and evaporating, after addition of one molecule of H_2SO_4 for each two molecules of $FeSO_4$. The *Liq. fer. tersulphatis* (*U. S.*) contains this salt. It is a yellowish-white, amorphous solid.

Of the many *basic ferric sulphates*, the only one of medical interest is *Monsel's salt*, $5(Fe_2)(SO_4)_3 + 4Fe_2O_3$, which exists in the *Liq. ferri subsulphatis* (*U. S.*) and *Liq. fer. persulphatis* (*Br.*). Its solution is decolorized, and forms a white deposit with excess of H_2SO_4.

Nitrates.—FERROUS NITRATE—**$Fe(NO_3)_2$**—179.9—a greenish, unstable salt, formed by double decomposition between barium nitrate and ferrous sulphate; or by the action of HNO_3 on FeS.

FERRIC NITRATES.—The *normal nitrate*—**$(Fe_2)(NO_3)_6$**—483.8—is ob-

112

tained in solution by dissolving Fe in HNO_3 of sp. gr. 1.115; or by dissolving ferric hydrate in HNO_3. It therefor exists in the *Liq. ferri nitratis* (*U. S.*). It crystallizes in rhombic prisms with 18 Aq, or in cubes with 12 Aq.

Several basic nitrates are known, all of which are uncrystallizable, and by their presence (as when Fe is dissolved in HNO_3 to saturation) prevent the crystallization of the normal salt.

Phosphates.—TRIFERROUS PHOSPHATE—**$Fe_3(PO_4)_2$**—357.7.—A white precipitate, formed by adding disodic phosphate to a solution of a ferrous salt, in presence of sodium acetate. By exposure to air it turns blue; a part being converted into ferric phosphate. The *ferri phosphas* (*Br.*), is such a mixture of the two salts. It is insoluble in H_2O; sparingly soluble in H_2O containing carbonic or acetic acid.

It is probably this phosphate, capable of turning blue, which sometimes occurs in the lungs in phthisis, in blue pus, and in long-buried bones.

FERRIC PHOSPHATE—**$(Fe_2)(PO_4)_2$**—301.8—is produced by the action of an alkaline phosphate on ferric chloride. It is soluble in HCl, HNO_3, citric and tartaric acids, insoluble in phosphoric acid and in solution of hydrosodic phosphate. The *ferri phosphas* (*U. S.*) is a compound, or mixture of this salt with disodic citrate, which is soluble in water.

There exist quite a number of basic ferric phosphates.

FERRIC PYROPHOSPHATE—**$(Fe_2)_2(P_2O_7)_3$**—745.6—is precipitated by decomposition of a solution of a ferric compound by sodium pyrophosphate; an excess of the Na salt dissolves the precipitate when warmed, and, on evaporation, leaves scales of a double salt, $(Fe_2)_2(P_2O_7)_3$, $Na_8(P_2O_7)_2$ + 20 Aq.

The *ferri pyrophosphas* (*U. S.*) is probably a mixture, or compound (?) of ferric pyrophosphate, trisodic citrate, and ferric citrate.

Acetates.—FERROUS ACETATE—**$Fe(C_2H_3O_2)_2$**—173.9—is formed, by decomposition of ferrous sulphate by calcium acetate, in soluble, silky needles.

FERRIC ACETATES.—The *normal salt*, $(Fe_2)(C_2H_3O_2)_6$, is obtained by adding slight excess of ferric sulphate to lead acetate, and decanting after twenty-four hours. It is dark red, uncrystallizable, very soluble in alcohol and in H_2O. If its solution be heated it darkens suddenly, gives off acetic acid, and contains a basic acetate; when boiled it loses all its acetic acid and deposits ferric hydrate; when heated in closed vessels to 100° (212° F.), and treated with a trace of mineral acid, it deposits the modified ferric hydrate.

Ferrous Carbonate—$FeCO_3$—115.9—occurs as an ore of iron, and is obtained in a hydrated form by adding an alkaline carbonate to a ferrous salt. It is a greenish, amorphous powder, which, on exposure to air, turns red by formation of ferric hydrate; a change which is retarded by the presence of sugar, hence the addition of that substance in the *ferri carbonas saccharatus* (*U. S.*; *Br.*). It is insoluble in pure H_2O, but soluble in H_2O containing carbolic acid, probably as *ferrous bicarbonate*, $H_2Fe(CO_3)_2$, in which form it occurs in chalybeate waters.

Ferrous Lactate—*Ferri lactas* (*U. S.*)—**$Fe(C_3H_5O_3)_2$ + 3 Aq**—233.9 + 54—is formed when iron filings are dissolved in lactic acid. It crystallizes in greenish-yellow needles; soluble in H_2O; insoluble in alcohol; permanent in air when dry.

Ferrous Oxalate—*Ferri oxalas* (*U. S.*)—**FeC_2O_4 + Aq**—143.9 + 36—is a yellow, crystalline-powder; sparingly soluble in H_2O formed by dissolving iron filings in solution of oxalic acid.

113

Tartrates.—FERROUS TARTRATE—**$FeC_4H_4O_6$ + 2 Aq**—203.9 + 36.—A white, crystalline powder; formed by dissolving Fe in hot concentrated solution of tartaric acid.

FERRIC TARTRATE—**$Fe_2(C_4H_4O_6)_3$ + 3 Aq**—555.8 + 54.—A dirty yellow, amorphous mass, obtained by dissolving recently precipitated ferric hydrate in tartaric acid solution, and evaporating below 50° (122° F.).

A number of double tartrates, containing the group $(Fe_2O_2)''$ are also known. Such are: *Ferrico-ammonic tartrate* = *ferri et ammonii tartras* (*U. S.*), $(C_4H_4O_6)_2(Fe_2O_2)$, (NH_4) + 4 Aq, and *Ferrico-potassic tartrate* = *ferri et potassii tartras* (*U. S.*), $(C_4H_4O_6)_2(Fe_2O_2)K_2$. They are prepared by dissolving recently precipitated ferric hydrate in hot solutions of the hydro-alkaline tartrate. They only react with ferro- and sulphocyanides after addition of a mineral acid.

Citrates.—FERRIC CITRATE—*Ferri citras* (*U. S.*)—**$(Fe_2)(C_6H_5O_7)_2$ + 6 Aq**—489.8 + 108—is in garnet-colored scales, obtained by dissolving ferric hydrate in solution of citric acid and evaporating the solution at about 60° (140° F.). It loses 3 Aq at 120° (248° F.), and the remainder at 150° (302° F.). If a small quantity of ammonia be added before the evaporation, the product consists of the *modified citrate* = *ferri et ammonii citras* (*U. S.*), which only reacts with potassium ferrocyanide after addition of HCl.

The various citrates of iron and alkaloids are not definite compounds.

Ferric Ferrocyanide—*Prussian blue*—**$(Fe_2)_2(FeC_6N_6)_3$ + 18 Aq**—859.3 + 324—is a dark blue precipitate, formed when potassium ferrocyanide is added to a ferric salt. It is insoluble in H_2O, alcohol and dilute acids; soluble in oxalic acid solution (blue ink). Alkalies turn it brown.

Ferrous Ferricyanide—*Turnbull's blue*—**$Fe_3(Fe_2C_{12}N_{12})$ + *n* Aq**—591.5 + *n* 18—is a dark-blue substance produced by the action of potassium ferricyanide on ferrous salts. Heated in air it is converted into Prussian blue and ferric oxide.

Analytical Characters.

FERROUS.—Are acid; colorless when anhydrous; pale green when hydrated; oxidized by air to basic ferric compounds.

(1.) Potash: greenish-white ppt.; insoluble in excess; changing to green or brown in air.

(2.) Ammonium hydrate: greenish ppt.; soluble in excess; not formed in presence of ammoniacal salts.

(3.) Ammonium sulphydrate: black ppt.; insoluble in excess; soluble in acids.

(4.) Potassium ferrocyanide (in absence of ferric salts): white ppt.; turning blue in air.

(5.) Potassium ferricyanide: blue ppt.; soluble in KHO; insoluble in HCl.

FERRIC—Are acid, and yellow or brown.

(1.) Potash or ammonium hydrate: voluminous, red-brown ppt.; insoluble in excess.

(2.) Hydrogen sulphide: in acid solution; milky ppt. of sulphur; ferric reduced to ferrous compound.

(3.) Ammonium sulphydrate : black ppt ; insoluble in excess ; soluble in acids.

114

(4.) Potassium ferrocyanide : dark-blue ppt. ; insoluble in HCl ; soluble in KHO.

(5.) Potassium sulphocyanate : dark red color ; prevented by tartaric or citric acid ; discharged by mercuric chloride.

(6.) Tannin : blue-black color.

III. ALUMINIUM GROUP.

Glucinium—Aluminium—Scandium—Gallium—Indium.

This group is placed in the third class by virtue of the existence of the aluminates, and of the relations between the compounds of these elements and some of those of the previous group. They form, however, but one series of compounds, corresponding to the ferric, containing the group $(M_2)^{vi}$. No acids or salts of the members of the group, other than aluminium, are known ; yet their resemblances in other points are such as to forbid their separation.

GLUCINIUM.

Symbol = **Gl** or **Be** (*Beryllium*)—*Atomic weight* = 9—*Sp. gr.* = 2.1.

A rare element occurring in the emerald and beryl. The metal resembles aluminium and its compounds resemble those of Al, and, in some respects, those of Mg. Its soluble salts are sweet in taste (γλυκύς = sweet).

ALUMINIUM.

Symbol = **Al**—*Atomic weight* = 27—*Molecular weight* = 55 (?)—*sp. gr.* = 2.56–2.67—*Fuses at about* 700° (1292° F.)—*Name from* alumen = *alum* —*Discovered by Wöhler*, 1827.

Occurrence.—Exceedingly abundant in the clays as *silicate*.

Preparation.—(1.) By decomposing vapor of aluminium chloride by Na or K (Wöhler).

(2.) Aluminium hydrate, mixed with sodium chloride and charcoal, is heated in Cl, by which a double chloride of Na and Al (2NaCl, Al_2Cl_6) is formed. This is then heated with Na, when Al and NaCl are produced. (The industrial process.)

Properties.—*Physical.*—A bluish-white metal ; hard ; quite malleable and ductile when annealed from time to time ; slightly magnetic ; a good conductor of electricity ; non-volatile ; very light, and exceedingly sonorous.

Chemical.—It is not affected by air or O, except at very high temperatures, and then only superficially ; if, however, it contain Si, it burns readily in air, forming aluminium silicate. It does not decompose H_2O at a red heat ; but in contact with Cu, Pt, or I it does so at 100° (212° F.). It combines directly with Bo, Si, Cl, Br, and I. It is attacked by HCl, gaseous or in solution, with evolution of H, and formation of Al_2Cl_6. It dissolves in alkaline solutions, with formation of aluminates, and liberation of H. It alloys with Cu to form a golden yellow metal (aluminium bronze).

115

Compounds of Aluminium.

Aluminium Oxide—*Alumina*—$\mathbf{Al_2O_3}$—102—occurs in nature, nearly pure, as *corundum, emery, ruby, sapphire,* and *topaz ;* and is formed artificially by calcining the hydrate, or ammonia alum, at a red heat.

It is a light, white, odorless, tasteless powder ; fuses with difficulty ; and, on cooling, solidifies in very hard crystals. Unless it have been heated to bright redness, it combines with H_2O, with elevation of temperature. It is almost insoluble in acids and alkalies. H_2SO_4, diluted with an equal bulk of H_2O, dissolves it slowly as $(Al_2)(SO_4)_3$. Fused potash and soda combine with it to form aluminates. It is not reduced by charcoal.

Aluminium Hydrate—*Aluminii hydras* (*U. S.*)—$\mathbf{Al_2H_6O_6}$—156—is formed when a solution of an aluminium salt is decomposed by an alkali, or alkaline carbonate. It constitutes a gelatinous mass, which, when dried, leaves an amorphous, translucid mass ; and when pulverized a white, tasteless, amorphous powder. When the liquid in which it is formed contains coloring matters, these are carried down with it, and the dried deposits are used as pigments, called *lakes*.

When freshly precipitated, it is insoluble in H_2O ; soluble in acids and solutions of the fixed alkalies. When dried at a temperature above 50° (122° F.), or after 24 hours contact with the mother liquor, its solubility is greatly diminished. With acids it forms salts of aluminium ; and with alkalies, aluminates of the alkaline metal. Heated to near redness it is decomposed into Al_2O_3 and H_2O. A soluble modification is obtained by dialysing a solution of $Al_2H_6O_6$ in Al_2Cl_6, or by heating a dilute solution of aluminium acetate for 240 hours.

Aluminates are for the most part crystalline, soluble compounds, obtained by the action of metallic oxides or hydrates upon alumina. *Potassium aluminate*, $K_2Al_2O_4 + 3$ Aq—is formed by dissolving recently precipitated aluminium hydrate in potash solution. It forms white crystals ; very soluble in H_2O, insoluble in alcohol ; caustic and alkaline. By a large quantity of H_2O it is decomposed into aluminium hydrate and a more alkaline salt, $K_6Al_4O_9$.

Sodium Aluminate.—The aluminate $Na_2Al_2O_4$ is not known. That having the composition $Na_6Al_4O_9$ is prepared by heating to redness a mixture of 1 pt. sodium carbonate and 2 pts. of a native ferruginous aluminium hydrate (beauxite). It is insoluble in H_2O, and is decomposed by carbonic acid, with precipitation of aluminium hydrate.

Aluminium Chloride—$\mathbf{Al_2Cl_6}$—267—is prepared by passing Cl over a mixture of Al_2O_3 and C, heated to redness ; or by heating clay in a mixture of gaseous HCl and vapor of CS_2.

It crystallizes in colorless, hexagonal prisms ; fusible ; volatile ; deliquescent ; very soluble in H_2O and in alcohol. From a hot, concentrated solution, it separates in prisms with 12 Aq.

The disinfectant called *chloralum* is a solution of impure Al_2Cl_6.

Salts of Aluminium.

Aluminium Sulphate—*Aluminii sulphas* (*U. S.*)—$\mathbf{(Al_2)(SO_4)_3}$ + **18 Aq**—342 + 324—is obtained by dissolving $Al_2H_6O_6$ in H_2SO_4 ; or (industrially) by heating clay with H_2SO_4.

It crystallizes with difficulty in thin, flexible plates ; soluble in H_2O ; very sparingly soluble in alcohol. Heated, it fuses in its Aq, which it

116

gradually loses up to 200° (392° F.), when a white, amorphous powder, $(Al_2)(SO_4)_3$, remains ; this is decomposed at a red heat, leaving a residue of pure alumina.

Alums—are double sulphates of the alkaline metals, and the higher sulphates of this or the preceding group. When crystallized, they have the general formula : $(M_2)^{vi}(SO_4)_3$, $R'_2SO_4 + 24$ Aq, in which (M) may be (Fe_2), (Mn_2), (Cr_2), (Al_2), or (Ga_2) ; and R_2 may be K_2, Na_2, Rb_2, Cs_2, Tl_2, or $(NH_4)_2$. They are isomorphous with each other.

Alumen (*U. S.*)—$Al_2(SO_4)_3$, $K_2SO_4 + 24$ Aq—516 + 432—is manufactured from "alum shale," and is formed when solutions of the sulphates of Al and K are mixed in suitable proportion.

It crystallizes in large, transparent, regular octahedra ; has a sweetish, astringent taste, and is readily soluble in H_2O. Heated, it fuses in its Aq at 92° (197°.6 F.) ; and gradually loses 45.5 per cent. of its weight of H_2O as the temperature rises to near redness. The product, known as *burnt alum = alumen exsiccatum* (*U. S.*), is $(Al)_2(SO_4)_3$, K_2SO_4, and is slowly, but completely soluble in 20–30 pts. H_2O. At a bright red heat, SO_2 and O are given off and Al_2O_3 and potassium sulphate remain ; at a higher temperature, potassium aluminate is formed. Its solutions are acid in reaction ; dissolve Zn and Fe with evolution of H ; and deposit $Al_2H_6O_6$ when treated with ammonium hydrate.

Alumen (*Br.*)—$Al_2(SO_4)_3$,$(NH_4)_2SO_4 + 24$ Aq—474 + 432—is the compound now usually met with as *alum*, both in this country and in England. It differs from potash alum in being more soluble in H_2O between 20°–30° (68°–86° F.), and less soluble at other temperatures ; and in the action of heat upon it. At 92° (197°.6 F.) it fuses in its Aq ; at 205° (401° F.) it loses its ammonium sulphate, leaving a white, hygroscopic substance, very slowly and incompletely soluble in H_2O. More strongly heated, it leaves alumina.

Silicates—are very abundant in the different varieties of *clay, feldspar, albite, labradorite, mica,* etc. The clays are hydrated aluminium silicates, more or less contaminated with alkaline and earthy salts and iron, to which last certain clays owe their color. The purest is *kaolin*, or porcelain clay, a white or grayish powder. They are largely used in the manufacture of the different varieties of bricks, terra cotta, pottery, and porcelain. *Porcelain* is made from the purer clays, mixed with sand and feldspar ; the former to prevent shrinkage, the latter to bring the mixture into partial fusion, and to render the product translucent. The fashioned articles are subjected to a first baking ; the porous, baked clay is then coated with a *glaze*, usually composed of oxide of lead, sand, and salt. During a second baking, the glaze fuses and coats the article with a hard, impermeable layer. The coarser articles of pottery are glazed by throwing sodium chloride into the fire ; the salt is volatilized, and, on contact with the hot aluminium silicate, deposits a coating of the fusible sodium silicate, which hardens on cooling.

Analytical Characters.

(1.) Potash, or soda ; white ppt. ; soluble in excess.

(2.) Ammonium hydrate ; white ppt. ; almost insoluble in excess, especially in presence of ammoniacal salts.

(3.) Sodium phosphate ; white ppt. ; readily soluble in KHO and NaHO, but not in NH_4HO ; soluble in mineral acids, but not in acetic acid.

117

(4.) Blowpipe—on charcoal does not fuse, and moistened with cobalt nitrate solution turns dark sky-blue.

SCANDIUM.

Symbol = Sc—*Atomic weight* = 44.9—*Discovered by Nilson* (1879)—*Name from* Scandia.

Occurs in minute traces in gadolinite and euxenite. It forms an oxide, Sc_2O_3; a light, white, infusible powder; sp. gr. 3.8; resembling alumina.

GALLIUM.

Symbol = Ga—*Atomic weight* = 68.8—*Sp. gr.* = 5.9—*Fuses at* 36° (86° *F.*)—*Name from* Gallia—*Discovered by Lecoq de Boisbaudran* (1876).

Occurs in very small quantity in certain zinc blendes. It is a hard, white metal; soluble in hot NO_3H, in HCl and in KHO solution. In chemical characters it closely resembles Al; forms an oxide Ga_2O_3, and a series of alums.

The discovery of Sc and Ga afford most flattering verifications of predictions based upon purely theoretical considerations.

It has been observed that there exist numerical relations between the atomic weights of the elements, which, in groups of allied elements, differ from each other by (approximately) some multiple of eight. Upon this variation Mendelejeff has based what is known as the *Periodic Law*, to the effect that: "*The properties of elements, the constitution of their compounds, and the properties of the latter, are periodic functions of the atomic weights of the elements.*"

In accordance with this law the elements may be thus arranged:

Series.	Group I.	Group II.	Group III.	Group IV.	Group V.	Group VI.	Group V.	Group VI.
	....	....	....	RH_4	RH_3	RH_2	RH	(R_2H)
	R_2O	RO	R_2O_3	RO_2	R_2O_5	RO_3	R_2O_7	(RO_4)
1	H=1	....	....	....	....	....	....	....
2	Li=7	Be=9	B=11	C=12	N=14	O=16	F=19	....
3	Na=23	Mg=24	Al=27	Si=28	P=31	S=32	Cl=35	Cu=63 Fe=56
4	K=39	Ca=40	Sc=44	Ti=48	V=51	Cr=52	Mn=55	Co=59 Ni=59
5	(Cu=63)	Zn=65	Ga=68	?=72	As=75	Se=78	Br=80	Ru=104 Rh=104
6	Rb=85	Sr=87	Yt=88	Zr (?)=90	Nb=94	Mo=96	?=100	Pd=106 Ag=108
7	(Ag=108)	Cd=112	In=113	Sn=118	Sb=120	Fe=125	I=127	Os=195 Ir=197
8	Cs=133	Ba=137	D=138 (?)	Ce=140	....	....	....	Pb=198 Au=196
9	....	....	....	....	....	....	....	....
10	....	....	E=178 (?)	L=180 (?)	Ta=182	W=184	?=190	....
11	(Au=196)	Hg=200	Tl=204	Pb=207	Bi=208	....	....	....
12	....	....	....	Th=231	....	U=250	....	....

The atomic weights and chemical characters, which were announced by Mendelejeff in 1870 as those of the undiscovered elements which would occupy the positions 4 and 5 in group III., have been since found to be those of Sc and Ga.

INDIUM.

Symbol = In—*Atomic weight* = 113.4—*Sp. gr.* = 7.42—*Fuses at* 176° (348°.8 F)—*Discovered by Reich and Richter in* 1863.

A soft, silver-white, ductile metal, which occurs in small quantity in certain zinc blendes. It is characterized spectroscopically by two principal lines—λ = 4511 and 4101.

118

IV. URANIUM GROUP.

URANIUM.

Symbol = Ur—*Atomic weight* = 238.5—*Sp. gr.* = 18.4—*Discovered by Klaproth* (1789).

This element is usually classed with Fe and Cr, or with Ni and Co. It does not, however, form compounds resembling the ferric; it forms a series of well-defined *uranates*, and a series of compounds of the radical *uranyl* (UO)′. Standard solutions of its acetate or nitrate are used for the quantitative determination of PO_4H_3.

V. LEAD GROUP.

LEAD.

Symbol = **Pb (PLUMBUM)**—*Atomic weight* = 206.9—*Molecular weight* = 413.8 (?)—*Sp. gr.* = 11.445—*Fuses at* 325° (617° F.)—*Name from* lœd = heavy (*Saxon*).

Lead is usually classed with Cd, Bi, or Cu and Hg. It differs, however, from Bi in being bivalent or quadrivalent, but not trivalent, and in forming no compounds resembling those of bismuthyl (BiO); from Cd, in the nature of its O compounds; and from Cu and Hg in forming no compounds similar to the mercurous and cuprous salts. Indeed, the nature of the Pb compounds is such that the element is best classed in a group by itself, which finds a place in this class by virtue of the existence of potassium plumbate.

Occurrence.—Its most abundant ore is *galena*, PbS. It also occurs in *white lead ore*, $PbCO_3$, in *anglesite*, $PbSO_4$, and in *horn lead*, $PbCl_2$.

Preparation.—Galena is first roasted with a little lime. The mixture of PbO, PbS, and $PbSO_4$, so obtained, is strongly heated in a reverberatory furnace, when SO_2 is driven off. The impure *work lead* so formed is purified by fusion in air and removal of the film of oxides of Sn and Sb. If the ore be rich in Ag, that metal is extracted by taking advantage of the greater fusibility of an alloy of Pb and Ag, than of Pb alone; and subsequent oxidation of the remaining Pb.

Properties.—*Physical.*—It is a grayish white metal; brilliant upon freshly cut surfaces; very soft and pliable; not very malleable or ductile; crystallizes in octahedra; a poor conductor of electricity; a better conductor of heat. When expanded by heat it does not, on cooling, return to its original volume.

Chemical.—When exposed to air it is oxidized, more readily and completely at high temperatures. The action of H_2O on Pb varies with the conditions: Pure unaërated H_2O has no action upon it. By the combined action of air and moisture Pb is oxidized, and the oxide dissolved in the H_2O, leaving a metallic surface for the continuance of the action. The solvent action of H_2O upon Pb is increased, owing to the formation of basic salts, by the presence of nitrogenized organic substances, nitrates, nitrites, and chlorides. On the other hand, carbonates, sulphates, and phosphates, by their tendency to form insoluble coatings, diminish the corroding action of H_2O. Carbonic acid in small quantity, especially in presence of carbonates, tends to preserve Pb from solution, while H_2O highly charged with it (soda water) dissolves the metal readily. Lead is dissolved, as the nitrate, by HNO_3. H_2SO_4 when cold and moderately con-

119

centrated, does not affect it; but, when heated, dissolves it the more readily as the acid is more concentrated. It is attacked by HCl of sp. gr. 1.12, especially if heated. Acetic acid dissolves it as acetate, or in the presence of CO_2, converts it into white lead.

Compounds of Lead.

Oxides.—Lead Monoxide—*Protoxide*—*Massicot*—*Litharge*—*Plumbi oxidum* (*U. S.; Br.*)—**PbO**—222.9—is prepared by heating Pb or its carbonate or nitrate in air. If the product have been fused, it is *litharge*; if not, *massicot*. It forms copper-colored, mica-like plates, or a yellow powder; or crystallizes from its solution in soda or potash in white, rhombic dodecahedra, or in rose-colored cubes. It fuses near a red heat, and volatilizes at a white heat; sp. gr. 9.277–9.5. It is sparingly soluble in H_2O, forming an alkaline solution.

Heated in air to 300° (572° F.) it is oxidized to *minium*. It is readily reduced by H or C. With Cl it forms $PbCl_2$ and O. It is a strong base; decomposes alkaline salts, with liberation of the alkali. It dissolves in HNO_3 and in hot acetic acid, as nitrate or acetate. When ground up with oils it saponifies the glycerin ethers, the Pb combining with the fatty acids to form Pb soaps, one of which, *lead oleate*, is the *emplastrum plumbi* (*U. S.; Br.*). It also combines with the alkalies and earths to form *plumbites*. *Calcium plumbite*, $CaPb_2O_3$, is a crystalline salt, formed by heating PbO with milk of lime, and used in solution as a hair-dye.

Plumboso-plumbic Oxide—*Red oxide*—*Minium*—*Red lead*—$\mathbf{Pb_3O_4}$—684.7—is prepared by heating massicot to 300° (572° F.) in air. It ordinarily has the composition Pb_3O_4, and has been considered as composed of PbO_2, 2PbO; or as a basic lead salt of plumbic acid, PbO_3Pb, PbO. An orange-colored variety is formed when lead carbonate is heated to 300° (572° F.).

It is a bright red powder, sp. gr. 8.62. It is converted into PbO when strongly heated, or by the action of reducing agents. HNO_3 changes its color to brown, dissolving PbO and leaving PbO_2. It is decomposed by HCl, with formation of $PbCl_2$, H_2O and Cl.

Lead Dioxide—*Peroxide*, or *puce oxide*, or *brown oxide*, or *binoxide of lead*—*Plumbic anhydride*—$\mathbf{PbO_2}$—238.9—is prepared, either by dissolving the PbO out of red lead by dilute HNO_3, or by passing a current of Cl through H_2O holding lead carbonate in suspension.

It is a dark, reddish-brown, amorphous powder; sp. gr. 8.903–9.190; insoluble in H_2O. Heated, it loses half its O and is converted into PbO. It is a valuable oxidant. It absorbs SO_2 to form $PbSO_4$. It combines with alkalies to form *plumbates*, M_2PbO_3.

Plumbic Acid—$\mathbf{PbO_3H_2}$—256.9—forms crystalline plates, at the + electrode, when alkaline solutions of the Pb salts are decomposed by a weak current.

Lead Sulphide—*Galena*—**PbS**—238.9—exists in nature. It is also formed by direct union of Pb and S; by heating PbO with S or vapor of CS_2; or by decomposing a solution of a Pb salt by H_2S or an alkaline sulphide.

The native sulphide is bluish-gray, and has a metallic lustre; sp. gr. 7.58; that formed by precipitation is a black powder; sp. gr. 6.924. It fuses at a red heat and is partly sublimed, partly converted into a subsulphate. Heated in air it is converted into $PbSO_4$, PbO and SO_2. Heated

120

in H it is reduced. Hot HNO_3 oxidizes it to $PbSO_4$. Hot HCl converts it into $PbCl_2$. Boiling H_2SO_4 converts it into $PbSO_4$ and SO_2.

Lead Chloride—$\mathbf{PbCl_2}$—277.9—is formed by the action of Cl upon Pb at a red heat; by the action of boiling HCl upon Pb; and by double decomposition between a lead-salt and a chloride.

It crystallizes in plates, or hexagonal needles; sparingly soluble in cold H_2O, less soluble in H_2O containing HCl; more soluble in hot H_2O, and in concentrated HCl.

Several oxychlorides are known. *Cassel, Paris, Verona*, or *Turner's yellow* is $PbCl_2$,7PbO.

Lead Iodide—*Plumbi iodidum* (*U. S.; Br.*)—PbI_2—460.9—is deposited as a bright yellow powder, when a solution of potassium iodide is added to a solution of a Pb salt. Fused in air it is converted into an oxyiodide. Light and moisture decompose it, with liberation of I. It is almost insoluble in H_2O, soluble in solutions of ammonium chloride, sodium hyposulphite, alkaline iodides, and potash.

Salts of Lead.

Nitrates.—LEAD NITRATE—*Plumbi nitras*—(*U. S.; Br.*)—**$Pb(NO_3)_2$**—330.9—is formed by solution of Pb or of its oxides in excess of HNO_3. It forms anhydrous crystals; soluble in H_2O. Heated, it is decomposed into PbO; O and NO_2.

Besides the neutral nitrate, basic lead nitrates are known, which seem to indicate the existence of nitrogen acids similar to those of phosphorus; $Pb_3(NO_4)_2$—*orthonitrate*; and $Pb_2N_2O_7$—*pyronitrate.*

Lead Sulphate—$PbSO_4$—302.9—is formed by the action of hot, concentrated H_2SO_4 on Pb; or by double decomposition between a sulphate and a Pb salt in solution. It is a white powder; almost insoluble in H_2O; soluble in concentrated H_2SO_4 from which it is deposited by dilution.

Lead Chromate—*Chrome yellow*—**$PbCrO_4$**—323.3—is formed by decomposing $Pb(NO_3)_2$ with potassium chromate. It is a yellow, amorphous powder; insoluble in H_2O; soluble in alkalies.

Acetates.—NEUTRAL LEAD ACETATE—*Salt of Saturn—Sugar of Lead—Plumbi acetas* (*U. S.; Br.*)—**$Pb(C_2H_3O_2)_2$ + 3 Aq**—324.9 + 54—is formed by dissolving PbO in acetic acid; or by exposing Pb in contact with acetic acid to air.

It crystallizes in large, oblique rhombic prisms, sweetish, with a metallic after-taste; soluble in H_2O and alcohol; its solutions being acid. In air it effloresces, and is superficially converted into carbonate. It fuses at 75°.5 (167°.9 F.); loses Aq, and a part of its acid at 100° (212° F.), forming the sesquibasic acetate; at 280° (536° F.) it enters into true fusion, and, at a slightly higher temperature, is decomposed into CO_2; Pb, and acetone. Its aqueous solution dissolves PbO, with formation of basic acetates.

SEXBASIC LEAD ACETATE—**$Pb(C_2H_3O_2)OH$, 2PbO**—728.7—is the main constituent of *Goulard's extract = Liq. plumbi subacetatis* (*U. S.; Br.*), and is formed by boiling a solution of the neutral acetate with Pb in fine powder. The solution becomes milky on addition of ordinary H_2O from formation of the sulphate and carbonate.

Lead Carbonate—$PbCO_3$—266.9—occurs in nature as *cerusite*; and is formed, as a white, insoluble powder, when a solution of a Pb com-

pound is decomposed by an alkaline carbonate, or by passing CO_2 through a solution containing Pb.

The *plumbi carbonas* (*U. S.; Br.*), or *white lead* or *ceruse*, is a basic carbonate, $(PbCO_3)_2$, PbH_2O_2—774.7—mixed with varying proportions of other basic carbonates. It is usually prepared by the action of CO_2 on a solution of the subacetate, prepared by the action of acetic acid on Pb and PbO. It is a heavy, white powder; insoluble in H_2O, except in the presence of CO_2; soluble in acids with effervescence; and decomposed by heat into CO_2 and PbO.

Analytical Characters.

(1.) Hydrogen sulphide, in acid solution: a black ppt.; insoluble in alkaline sulphides, and in cold, dilute acids.

(2.) Ammonium sulphydrate: black ppt.; insoluble in excess.

(3.) Hydrochloric acid: white ppt.; in not too dilute solution; soluble in boiling H_2O.

(4.) Ammonium hydrate: white ppt.; insoluble in excess.

(5.) Potash; white ppt.: soluble in excess, especially when heated.

(6.) Sulphuric acid: white ppt.; insoluble in weak acids, soluble in solution of ammonium tartrate.

(7.) Potassium iodide: yellow ppt.; sparingly soluble in boiling H_2O; soluble in large excess.

(8.) Potassium chromate: yellow ppt.; soluble in KHO solution.

(9.) Iron or zinc separate the element from solutions of its salts.

Action on the Economy.

All the soluble compounds of Pb, and those which, although not soluble, are readily convertible into soluble compounds by H_2O, air, or the digestive fluids, are actively poisonous. Some are also injurious by their local action upon tissues with which they come in contact; such are the acetate, and, in less degree, the nitrate.

The *chronic* form of lead intoxication, *painter's colic, etc.*, is purely poisonous, and is produced by the continued absorption of minute quantities of Pb, either by the skin, lungs, or stomach. The acute form presents symptoms referable to the local as well as to the poisonous action of the Pb salt, and is usually caused by the ingestion of a single dose of the acetate or carbonate.

Metallic Pb, although probably not poisonous of itself, causes chronic lead-poisoning by the readiness with which it is converted into compounds capable of absorption. The sources of poisoning by metallic Pb are: the contamination of drinking water which has been in contact with the metal (see p. 48); the use of articles of food or of chewing tobacco which has been packed in tin-foil containing an excess of Pb; the drinking of beer or other beverages which have been in contact with pewter; or the handling of the metal and its alloys.

Almost all the compounds of Pb may produce painter's colic. The carbonate, in painters, artists, manufacturers of white lead, and in persons sleeping in newly painted rooms; the oxides, in the manufactures of glass, pottery, sealing-wax, and litharge, and by the use of lead-glazed pottery;

by other compounds, by the inhalation of the dust of cloth factories, and by the use of lead hair-dyes.

Acute lead-poisoning is by no means of as common occurrence as the chronic form, and usually terminates in recovery. It is caused by the ingestion of a single large dose of the acetate, subacetate, carbonate, or of red lead. In such cases the administration of magnesium sulphate is indicated; it enters into double decomposition with the Pb salt to form the insoluble $PbSO_4$.

Lead once absorbed is eliminated very slowly, it becoming fixed by combination with the albuminoids, a form of combination which is rendered soluble by potassium iodide. The channels of elimination are by the perspiration, urine, and bile.

In the analysis for mineral poisons (see p. 96), the major part of the Pb is precipitated as PbS in the treatment by H_2S. The PbS remains upon the filter after extraction with ammonium sulphydrate; it is treated with warm HCl, which decolorizes it by transforming the sulphide into chloride. The $PbCl_2$ thus formed is dissolved in hot H_2O, from which it crystallizes on cooling. The solution still contains $PbCl_2$ in sufficient quantity to respond to the tests for the metal.

Although Pb is not a normal constituent of the body, the every-day methods by which it may be introduced into the economy, and the slowness of its elimination are such as to render the greatest caution necessary in drawing conclusions from the detection of Pb in the body after death.

VI. BISMUTH GROUP.

BISMUTH.

Symbol = **Bi**—*Atomic weight* = 207.5—*Molecular weight* = 420 (?)—*Sp. gr.* = 9.677–9.935—*Fuses at* 268° (514°.4 *F.*).

This element is usually classed with Sb; by some writers among the metals, by others in the phosphorus group. We are led to class Bi in our third class, and in a group alone, because: (1) while the so-called salts of Sb are not salts of the element, but of the radical (SbO)′, *antimonyl*, Bi enters into saline combination, not only in the radical *bismuthyl* (BiO)′, but also as an element; (2) while the compounds of the elements of the N group in which those elements are quinquivalent are, as a rule, more stable than those in which they are trivalent, Bi is trivalent in all its known compounds except one, which is very unstable, in which it is quinquivalent; (3) the hydrates of the N group are strongly acid, and their corresponding salts are stable and well defined; but those hydrates of Bi which are acid are but feebly so, and the bismuthates are unstable; (4) no compound of Bi and H is known.

OCCURRENCE.—Occurs principally free, also as Bi_2O_3 and Bi_2S_3.

PROPERTIES.—Crystallizes in brilliant, metallic rhombohedra; hard and brittle.

It is only superficially oxidized in cold air. Heated to redness in air, it becomes coated with a yellow film of oxide. In H_2O containing CO_2 it forms a crystalline subcarbonate. It combines directly with Cl, Br, and I. It dissolves in hot H_2SO_4 as sulphate, and in HNO_3 as nitrate.

It is usually contaminated with As, from which it is best purified by

heating to redness a mixture of powdered bismuth, potassium carbonate, soap, and charcoal, under a layer of charcoal. After an hour the mass is cooled; the button is separated and fused until its surface begins to be coated with a yellowish-brown oxide.

Compounds of Bismuth.

Oxides.—Four oxides are known: Bi_2O_2; Bi_2O_3; Bi_2O_4; and Bi_2O_5.

BISMUTH TRIOXIDE—*Bismuthous oxide — Protoxide*—**Bi_2O_3**—468—is formed by heating Bi, or its nitrate, carbonate, or hydrate. It is a pale yellow, insoluble powder; sp. gr. 8.2; fuses at a red heat; soluble in HCl, HNO_3 and H_2SO_4 and in fused potash.

Hydrates.—Bismuth forms at least four hydrates.

BISMUTHOUS HYDRATE—**BiH_3O_3**—261—is formed as a white precipitate when potash or ammonium hydrate is added to a cold solution of a Bi salt. When dried, it loses H_2O and is converted into *bismuthyl hydrate*

(BiO)HO.

Bismuthic Acid—$(BiO_2)HO$—259—is deposited as a red powder when Cl is passed through a boiling solution of potash, holding bismuthous hydrate in suspension.

Pyrobismuthic Acid—$H_4Bi_2O_7$—536—is a dark brown powder, precipitated from solution of bismuth nitrate by potassium cyanide.

Bismuth Trichloride — *Bismuthous chloride* — $BiCl_3$ — 316.5 — is formed by heating Bi in Cl; by distilling a mixture of Bi and mercuric chloride; or by distilling a solution of Bi in aqua regia. It is a fusible, volatile, deliquescent solid; soluble in dilute HCl. On contact with H_2O it is decomposed with formation of *bismuthyl chloride* (BiO)Cl, or *pearl white.*

Salts of Bismuth.

Bismuth Nitrate—$Bi(NO_3)_3 + 5Aq$—396 + 90—obtained by dissolving Bi in HNO_3. It crystallizes in large, colorless prisms; at 150° (302° F.), or by contact with H_2O, it is converted into bismuthyl nitrate; at 260° (500° F.) into Bi_2O_3.

Salts of Bismuthyl.

Bismuthyl Nitrate — *Trisnitrate* or *subnitrate of bismuth* — *Flake white*—*Bismuthi subnitras* (*U. S.*; *Br.*)—$(BiO)NO_3H_2O$—306—is formed by decomposing a solution of $Bi(NO_3)_3$ with a large quantity of H_2O. It is a white, heavy, faintly acid powder; soluble to a slight extent in H_2O when freshly precipitated, the solution depositing it again on standing. It is decomposed by pure H_2O, but not by H_2O containing $\frac{1}{500}$ ammonium nitrate. It usually contains 1 Aq, which it loses at 100° (212° F.).

Bismuth subnitrate, as well as the subcarbonate, is liable to contamination with arsenic, which accompanies bismuth in its ores. The method for separating this dangerous impurity, directed by the British Pharmacopœia, is more perfect than that usually followed in this country. The metal is first purified by fusion with potassium nitrate, which dissolves any arsenic present in the form of sodium arsenite, and the purified metal

is then converted into nitrate by solution in HNO_3, and this in turn into subnitrate by decomposition with a large volume of H_2O.

The maximum amount of arsenic which has been found in commercial bismuth subnitrate is one-tenth of one per cent.

To detect the presence of arsenic, the subnitrate (or subcarbonate) is boiled for half an hour with an equal weight of *pure* sodium carbonate, dissolved in ten times its weight of H_2O. The solution is filtered; the filtrate evaporated to dryness; the residue strongly heated; and, after cooling, cautiously decomposed with strong H_2SO_4. The mass is then gradually heated, during stirring, until dense white fumes are given off. The cooled residue is finally treated with water and the liquid introduced into a Marsh apparatus. (See page 93.)

Bismuthyl Subcarbonate—*Bismuthi subcarbonas* (*U. S.*)—*Bismuthi carbonas* (*Br.*)—$(BiO)_2CO_3H_2O$—530—is a white or yellowish, amorphous powder, formed when a solution of an alkaline carbonate is added to a solution of $Bi(NO_3)_3$. It is odorless and tasteless, and insoluble in H_2O and in alcohol.

When heated to 100° (212° F.), it loses H_2O and is converted into $(BiO)_2CO_3$. At a higher temperature it is further decomposed into Bi_2O_3 and CO_2.

Analytical Characters.

(1.) Water: white ppt., even in presence of tartaric acid, but not of HNO_3, HCl, or H_2SO_4.

(2.) Hydrogen sulphide: black ppt.; insoluble in dilute acids and in alkaline sulphides.

(3.) Ammonium sulphydrate: black ppt.; insoluble in excess.

(4.) Potash, soda, or ammonia: white ppt.; insoluble in excess, and in tartaric acid; turns yellow when the liquid is boiled.

(5.) Potassium ferrocyanide: yellowish ppt.; insoluble in HCl.

(6.) Potassium ferricyanide: yellowish ppt.; soluble in HCl.

(7.) Infusion of galls: orange ppt.

(8.) Potassium iodide: brown ppt.; soluble in excess.

(9.) Reacts with Reinsch's test (*q. v.*), but gives no sublimate in the glass tube.

Action on the Economy.

Although the medicinal compounds of bismuth probably are poisonous, if taken in sufficient quantity, the ill effects ascribed to them are in most, if not all cases, referable to contamination with arsenic. Symptoms of arsenical poisoning have not only been frequently observed when the subnitrate has been taken internally, but also when it has been used as a cosmetic.

When preparations of bismuth are administered, the alvine discharges contain bismuth sulphide as a dark brown powder.

VII. TIN GROUP.

Titanium. Zirconium. Tin.

Ti and Sn are bivalent in one series of compounds, $SnCl_2$, and quadrivalent in another, $SnCl_4$. Zr, so far as known, is always quadrivalent. Each of these elements forms an acid (or salts corresponding to one) of the composition H_2MO_3, and a series of oxysalts of the composition $M^{iv}(NO_3)_4$.

TITANIUM.

Symbol = **Ti**—*Atomic weight* = 48—*sp. gr.* = 5.3.

Occurs in clays and iron ores and as TiO_2 in several minerals. ***Titanic anhydride***, TiO_2, is a white, insoluble, infusible powder, used in the manufacture of artificial teeth; dissolves in fused KHO as potassium titanate. Titanium combines readily with N, which it absorbs from air when heated. When NH_3 is passed over red-hot TiO_2, it is decomposed with formation of the violet nitride, TiN_2. Another compound of Ti and N forms hard, copper-colored, cubical crystals.

ZIRCONIUM.

Symbol = **Zr**—*Atomic weight* = 89.6—*sp. gr.* = 4.15.

Occurs in zircon and hyacinth. Its oxide, *zirconia*, ZrO_2, is a white powder, insoluble in KHO. Being infusible and not altered by exposure to air, it is used in pencils to replace lime in the calcium light.

TIN.

Symbol = **Sn** **(STANNUM)** — *Atomic weight* = 117.7 — *Molecular weight* = 235.4 (?)—*sp. gr.* = 7.285–7.293—*Fuses at* 228° (442°.4 F.).

Occurrence.—As *tinstone* (SnO_2) and in *stream tin.*

Preparation.—The commercial metal is prepared by roasting the ore, extracting with H_2O, reducing the residue by heating with charcoal, and refining.

Pure tin is obtained by dissolving the metal in HCl; filtering; evaporating; dissolving the residue in H_2O; decomposing with ammonium carbonate; and reducing the oxide with charcoal.

Properties.—A soft, malleable, bluish-white metal; but slightly tenacious; emits a peculiar sound, the *tin-cry*, when bent. A good conductor of heat and electricity. Air affects it but little, except when it is heated; more rapidly if Sn be alloyed with Pb. It oxidizes slowly in H_2O, more rapidly in the presence of sodium chloride. Its presence with Pb accelerates the action of H_2O upon the latter. It dissolves in HCl as $SnCl_2$. In presence of a small quantity of H_2O, HNO_3 converts it into metastannic acid. Alkaline solutions dissolve it as metastannates. It combines directly with Cl, Br, I, S, P, and As.

Tin plates are thin sheets of Fe, coated with Sn. *Tin foil* consists of thin laminæ of Sn, frequently alloyed with Pb. Copper and iron vessels are *tinned*, after brightening, by contact with molten Sn. Pewter, bronze, bell metal, gun metal, britannia metal, speculum metal, type metal, solder, and fusible metal contain Sn.

Compounds of Tin.

Oxides.— Stannous Oxide—*Protoxide*—**SnO** — 133.7 — obtained by heating the hydrate or oxalate without contact of air. It is a white, amorphous powder, soluble in acids and in hot concentrated solution of potash. It absorbs O readily.

Stannic Oxide—*Binoxide of tin*—SnO_2—149.7—occurs native as *tinstone* or *cassiterite*, and is formed when Sn or SnO is heated in air.

Hydrates.—Stannous Hydrate—SnH_2O_2—151.7—is a white precipitate formed by alkaline hydrates and carbonates in solutions of $SnCl_2$.

Stannic Acid—H_2SnO_3—167.7—is formed by the action of alkaline hydrates on solutions of $SnCl_4$. It dissolves in solutions of the alkaline hydrates, forming *stannates.*

Metastannic Acid—$H_2Sn_5O_{11}$—766.5—is a white, insoluble powder, formed by acting on Sn with HNO_3.

Chlorides.—Stannous Chloride—*Protochloride*—*Tin crystals*—$SnCl_2 + 2Aq$—188.7 + 36—is obtained by dissolving Sn in HCl. It crystallizes in colorless prisms; soluble in a small quantity of H_2O; decomposed by a large quantity, unless in the presence of free HCl, with formation of an oxychloride. Loses its Aq at 100° (212° F.). In air it is transformed into stannic chloride and oxychloride. Oxidizing and chlorinating agents convert it into $SnCl_4$. It is a strong reducing agent.

Stannic Chloride—*Bichloride*—*Liquid of Libavius*—$SnCl_4$—259.7—by acting on Sn or $SnCl_2$ with Cl, or by heating Sn in aqua regia. It is a fuming, yellowish liquid; sp. gr. 2.28; boils at 120° (248° F.).

Analytical Characters.

Stannous.—(1.) Potash or soda: white ppt.; soluble in excess; the solution deposits Sn when boiled.

(2.) Ammonium hydrate: white ppt.; insoluble in excess; turns olive-brown when the liquid is boiled.

(3.) Hydrogen sulphide: dark brown ppt.; soluble in KHO, alkaline sulphides, and hot H_2O.

(4.) Mercuric chloride: white ppt.; turning gray and black.

(5.) Auric chloride: purple or brown ppt., in presence of small quantity of HNO_3.

(6.) Zinc: deposit of Sn.

STANNIC.—(1.) Potash or ammonia: white ppt.; soluble in excess.

(2.) Hydrogen sulphide: yellow ppt.; soluble in alkalies, alkaline sulphides, and hot HCl.

(3.) Sodium hyposulphite: yellow ppt. when heated.

VIII. PLATINUM GROUP.

PALLADIUM. PLATINUM.

IX. RHODIUM GROUP.

RHODIUM. RUTHENIUM. IRIDIUM.

The elements of these two groups, together with osmium, are usually classed as "metals of the platinum ores." They all form hydrates (or salts representing them) having acid properties. Osmium has been removed because the relations existing between its compounds and those of molybdenum and tungsten are much closer than those which they exhibit to the compounds of these groups. The separation of the remaining

127

platinum metals into two groups is based upon resemblances in the composition of their compounds, as shown in the following table:

Chlorides.					*Oxides.*				
$PdCl_2$	$PtCl_2$	$RhCl_2$	$RuCl_2$	?	PdO	PtO	RhO	RuO	IrO
$PdCl_4$	$PtCl_4$	——	$RuCl_4$	$IrCl_4$	——	——	Rh_2O_3	Ru_2O_3	Ir_2O_3
——	——	Rh_2Cl_6	Ru_2Cl_6	Ir_2Cl_6	PdO_2	PtO_2	RhO_2	RuO_2	IrO_2
					——	——	RhO_3	RuO_3	IrO_3
					——	——	——	RuO_4	——

PLATINUM.

Symbol = **Pt**—*Atomic weight* = 194.4—*Molecular weight* = 388.8 (?) —*sp. gr.* = 21.1–21.5.

OCCURRENCE.—Free and alloyed with Os, Ir, Pd, Rh, Ru, Fe, Pb, Au, Ag, and Cu.

PROPERTIES.—The compact metal has a silvery lustre; softens at a white heat; may be welded; fuses with difficulty; highly malleable, ductile and tenacious. *Spongy platinum* is a grayish, porous mass, formed by heating the double chloride of Pt and NH_4. *Platinum black* is a black powder, formed by dissolving Pt Cl_2 in solution of potash and heating with alcohol. Both platinum black and platinum sponge are capable of condensing large quantities of gas, and act as indirect oxidants.

Platinum is not oxidized by air or O; it combines directly with Cl, P, As, Si, S, and C; is not attacked by acids, except aqua regia, in which it dissolves as $PtCl_4$. It forms fusible alloys when heated with metals or reducible metallic oxides. It is attacked by mixtures liberating Cl, and by contact with heated phosphates, silicates, hydrates, nitrates, or carbonates of the alkaline metals.

PLATINIC CHLORIDE—*Tetrachloride* or *perchloride of platinum*—$PtCl_4$—336.4—is obtained by dissolving Pt in aqua regia, and evaporating. It crystallizes in very soluble, deliquescent, yellow needles. Its solution is used as a test for compounds of NH_4 and K.

PALLADIUM.

Symbol = **Pd**—*Atomic weight* = 105.7—*Molecular weight* = 211.4 (?)—*sp. gr.* = 11.5.

A white metal resembling Pt in appearance, but usually exhibiting a reddish reflection. It is harder, much lighter, and more readily fusible than Pt. It dissolves in HNO_3 as $Pd(NO_3)_2$. It possesses the property of *occluding* gases, notably hydrogen, in a much more marked degree than any other metal. One volume of palladium condenses 640 volumes of hydrogen at 100° (212° F.).

RHODIUM.

Symbol = **Rh**—*Atomic weight* = 104.1—*Molecular weight* = 208.2 (?)—*sp. gr.* = 11.4.

A hard, malleable, white metal, insoluble in aqua regia. Its compounds are mostly red, whence its name, from ῥόδον, a rose.

128

RUTHENIUM.

Symbol = **Ru**—*Atomic weight* = 104.2—*sp. gr.* = 11.4.

A hard, brittle, very difficultly fusible metal, not dissolved by aqua regia, occurring in small quantity in platinum ores.

IRIDIUM.

Symbol = **Ir**—*Atomic weight* = 192.7—*sp. gr.* = 22.3.

A hard, brittle metal which occurs in nature in platinum ores, and alloyed with osmium. It is not attacked by aqua regia. It is used to make an alloy with platinum which is less fusible, more rigid, harder, denser, and less readily attacked chemically than pure platinum.

129

CLASS IV.—BASYLOUS ELEMENTS.

ELEMENTS WHOSE OXIDES UNITE WITH WATER TO FORM BASES; NEVER TO FORM ACIDS. WHICH FORM OXYSALTS.

I. SODIUM GROUP.

Alkaline Metals.

LITHIUM—SODIUM—POTASSIUM. RUBIDIUM—CÆSIUM—SILVER.

Each of the elements of this group forms a single chloride, M'Cl, and one or more oxides, the most stable of which has the composition M'_2O; they are, therefor, univalent. Their hydrates, M'HO, are more or less alkaline and have markedly basic characters. Silver resembles the other members of the group in chemical properties, although it does not in physical characters.

LITHIUM.

Symbol = **Li**—*Atomic weight* = 7—*Molecular weight* = 14 (?)—*Sp. gr.* = 0.589—*Fuses at* 180° (356° F.)—*Discovered by Arfvedson in* 1817—*Name from* λίθειος = stony.

OCCURRENCE.—Widely distributed in small quantity; in many minerals and mineral waters; in the ash of tobacco and other plants; in the milk and blood.

PROPERTIES.—A silver-white, ductile, volatile metal; the lightest of the solid elements; burns in air with a crimson flame; decomposes H_2O at ordinary temperatures, without igniting.

Lithium Oxide—Li_2O—30—is a white solid, formed by burning Li in dry O. It dissolves slowly in H_2O to form *lithium hydrate*—LiHO.

Lithium Chloride—LiCl—42.5—crystallizes in deliquescent, regular octahedra; very soluble in H_2O and in alcohol.

Lithium Bromide—*Lithii bromidum* (*U. S.*)—**LiBr**—87—is formed by decomposing lithium sulphate with potassium bromide; or by saturating a solution of HBr with lithium carbonate. It crystallizes in very deliquescent, soluble needles.

Lithium Carbonate—*Lithii carbonas* (*U. S.*; *Br.*)—Li_2CO_3—74—is a white, sparingly soluble, alkaline, amorphous powder. With uric acid it forms lithium urate (*q. v.*).

Analytical Characters.

(1.) Ammonium carbonate: white ppt. in concentrated solutions; not in dilute solutions or in presence of ammoniacal salts.

(2.) Sodium phosphate: white ppt. in neutral or alkaline solution; soluble in acids and in solutions of ammoniacal salts.

130

(3.) It colors the Bunsen flame red; and exhibits a spectrum of two lines—λ = 6705 and 6102 (Fig. 14, No. 4).

SODIUM.

Symbol = **Na** (**NATRIUM**)—*Atomic weight* = 23—*Molecular weight*, = 46 (?)—*Sp. gr.* = 0.972—*Fuses at* 95°.6 (204°.1 F.)—*Discovered by Davy*, 1807.

OCCURRENCE.—As chloride very abundantly and widely distributed; also as carbonate, nitrate, sulphate, borate, etc.

PREPARATION.—By heating a mixture of dry sodium carbonate, chalk, and charcoal to whiteness in iron retorts, connected with suitable condensers in which the distilled metal collects under a layer of coal naphtha.

PROPERTIES.—A silver-white metal, rapidly tarnished and coated with a yellow film in air. Waxy at ordinary temperatures; volatile at a white heat.

In air it is gradually oxidized from the surface, but may be kept in closed vessels without the protection of a layer of naphtha. It decomposes H_2O, sometimes explosively. Burns with a yellow flame. Combines directly with Cl, Br, I, S, P, As, Pb, and Sn.

Compounds of Sodium.

Oxides.—Two oxides are known: *Sodium monoxide*—Na_2O—a grayish-white mass; formed when Na is burnt in dry air, or by the action of Na on NaHO. *Sodium dioxide*—Na_2O_2—a white solid, formed when Na is heated in dry air to 200° (392° F.).

Sodium Hydrate—*Caustic soda—Soda (U. S.)—Soda caustica (Br.)*—**NaHO**—40—is formed: (1) when H_2O is decomposed by Na; (2) by decomposing sodic carbonate by calcium hydrate: $Na_2CO_3 + CaH_2O_2 = CO_3Ca + 2NaHO$ (soda by lime); (3) in the same manner as in (2), using barium hydrate in place of lime (soda by baryta). It frequently contains considerable quantities of As.

It is an opaque, white, fibrous, brittle solid; fusible below redness; sp. gr. 2.00; very soluble in H_2O, forming strongly alkaline and caustic solutions (*soda lye* and *liq. sodæ*). When exposed to air, solid or in solution, it absorbs H_2O and CO_2 and is converted into carbonate. Its solutions attack glass.

Sodium Chloride—*Common salt—Sea salt—Table salt—Sodii chloridum (U. S., Br.)*—**NaCl**—58.5—occurs very abundantly in nature, deposited in the solid form as *rock salt*; in solution in all natural waters, especially in sea and mineral spring waters; in suspension in the atmosphere; and as a constituent of almost all animal and vegetable tissues and fluids. It is formed in an infinite variety of chemical reactions. It is obtained from rock salt, or from the waters of the sea or of saline springs; and is the source from which all the Na compounds are usually obtained, directly or indirectly.

It crystallizes in anhydrous, white cubes or octahedra; sp. gr. 2.078; fuses at a red heat, and crystallizes on cooling: sensibly volatile at a white heat; quite soluble in H_2O, the solubility varying but slightly with the

131

temperature. Dilute solutions yield pure ice on freezing. It is precipitated from concentrated solutions by HCl. It is insoluble in absolute alcohol; sparingly soluble in dilute spirit. It is decomposed by H_2SO_4 with formation of HCl and sodium sulphate: $2NaCl + H_2SO_4 = 2HCl + Na_2SO_4$.

Physiological.—Sodium chloride exists in every animal tissue and fluid, and is present in the latter, especially the blood, in tolerably constant proportion. It is introduced with the food, either as a constituent of the alimentary substances, or as a condiment. In the body it serves to aid the phenomena of osmosis and to maintain the solution of the albuminoids. It is probable, also, that it is decomposed in the gastric mucous membrane with formation of free hydrochloric acid.

It is discharged from the economy by all the channels of elimination, notably by the urine, when the supply by the food is maintained. If, however, the food contain no salt, it disappears from the urine before it is exhausted from the blood.

The amount of Cl (mainly in the form of NaCl) voided by a normal male adult in 24 hours is about 10 grams (154 grains), corresponding to 16.5 grams (255 grains) of NaCl. When normal or excessive doses are taken, the amount eliminated by the urine is less than that taken in; when small quantities are taken, the elimination is at first in excess of the supply. The hourly elimination increases up to the seventh hour, when it again diminishes. The amount of NaCl passed in the urine is less than the normal in acute, febrile diseases; in intermittent fever it is diminished during the paroxysms, but not during the intervals. In diabetes it is much increased, sometimes to 29 grams (448 grains) per diem.

Quantitative determination of chlorides in urine.—The process is based upon the formation of the insoluble silver chloride, and upon the formation of the brown silver chromate in neutral liquids, in the absence of soluble chlorides. The solutions required are: (1) *A solution of silver nitrate* of known strength, made by dissolving 29.075 grams of pure, fused silver nitrate (see p. 143) in a litre of water; (2) *a solution of neutral potassium chromate.*

To conduct the determination, 5–10 c.c. of the urine are placed in a platinum basin, 2 grams of sodium nitrate (free from chloride) are added; the whole is evaporated to dryness over the water-bath, and the residue heated gradually until a colorless, fused mass remains. This, on cooling, is dissolved in H_2O, the solution placed in a small beaker, treated with pure, dilute HNO_3 to faintly acid reaction, and neutralized with calcium carbonate. Two or three drops of the chromate solution are added, and then the silver solution from a burette, during constant stirring of the liquid in the beaker, until a faint reddish tinge remains permanent. Each c.c. of the silver solution used represents 10 milligrams NaCl (or 6.065 milligrams Cl) in the amount of urine used.

Example.—5 c.c. urine used; 6 c.c. silver solution added; 1,200 c.c. urine passed in 24 hours: $\therefore \frac{0.01 \times 6}{5} \times 1{,}200 = 14.4$ grams NaCl in 24 hours.

If the urine contain iodides or bromides, they must be removed by acidulating the solution of the residue of incineration with SO_4H_2, removing the iodine or bromine by shaking with carbon disulphide, neutralizing the aqueous solution with calcium carbonate and proceeding as above.

Sodium Bromide—*Sodii bromidum (U. S.)*—**NaBr**—103—is formed by dissolving Br in solution of NaHO to saturation; evaporating; calcining

132

at dull redness; redissolving; filtering; and crystallizing. It crystallizes in anhydrous cubes; quite soluble in H_2O, soluble in alcohol.

Sodium Iodide—*Sodii iodidum (U. S.)*—**NaI**—150—is prepared by heating together H_2O, Fe, and I in fine powder; filtering; adding an equivalent quantity of sodium sulphate and some slacked lime; boiling; decanting and evaporating. Crystallizes in anhydrous cubes; very soluble in H_2O; soluble in alcohol.

Salts of Sodium.

Sodium Nitrate—*Cubic* or *Chili saltpetre—Sodii nitras (U. S.)—Sodæ nitras (Br.)*—**$NaNO_3$**—85—occurs in natural deposits in Chili and Peru. It crystallizes in anhydrous, deliquescent rhombohedra; cooling and somewhat bitter in taste; fuses at 310° (590° F.); very soluble in H_2O. Heated with H_2SO_4 it is decomposed, yielding HNO_3 and hydrosodic sulphate: $H_2SO_4 + NaNO_3 = HNaSO_4 + HNO_3$.

Sulphates.—Hydrosodic Sulphate—*Acid sodium sulphate—Bisulphate*—**$HNaSO_4$**—120—crystallizes in long, four-sided prisms; is unstable and decomposed by air, H_2O or alcohol, into H_2SO_4 and Na_2SO_4. Heated to dull redness it is converted into *sodium pyrosulphate*, $Na_2S_2O_7$, corresponding to Nordhausen sulphuric acid.

Sodic Sulphate—*Neutral sodium sulphate—Glauber's salt—Sodii sulphas (U. S.)—Sodæ sulphas (Br.)*—**Na_2SO_4 + n Aq**—142 + n 18—occurs in nature in solid deposits and in solution in natural waters. It is obtained principally as a step in the manufacture of the carbonate by the action of H_2SO_4 on NaCl.

It crystallizes with 7 Aq, from saturated or supersaturated solutions at 5° (41° F.); or, more usually, with 10 Aq. As usually met with it is in large, colorless, oblique rhombic prisms with 10 Aq; which effloresce in air and gradually lose all their Aq. It fuses at 33° (91°.4 F.) in its Aq, which it gradually loses. If fused at 33° (91°.4 F.) and allowed to cool, it remains liquid in *supersaturated* solution, from which it is deposited, the entire mass becoming solid, on contact with a small particle of solid matter. It dissolves in HCl with considerable diminution of temperature.

Physiological.—The neutral sulphates of Na and K seem to exist in small quantity in all animal tissues and fluids, with the exception of milk, bile, and gastric juice; certainly in the blood and urine. They are partially introduced with the food, and partly formed as a result of the metamorphosis of those constituents of the tissues which contain S in organic combination.

The principal elimination of the sulphates is by the urine. All the sulphuric acid in the urine is not in simple combination with the alkaline metals; a considerable amount exists in the form of the alkaline salts of conjugate, monobasic ether acids, which on decomposition yield an aromatic organic compound. The amount of H_2SO_4 discharged by the urine in 24 hours, in the form of alkaline sulphates, is from 2.5 to 3.5 grams (38.5–54 grains); that eliminated in the salts of conjugate acids, 0.617 to 0.094 gram (9.5–1.5 grains).

Sodium Sulphite—*Sodii sulphis (U. S.)*—**Na_2SO_3 + 7 Aq**—126 + 126—is formed by passing SO_2 over crystallized Na_2CO_3. It crystallizes in efflorescent, oblique prisms; quite soluble in H_2O, forming an alkaline solution. It acts as a reducing agent.

133

Sodium Hyposulphite—*Sodii hyposulphis (U. S.)*—**$Na_2S_2O_3$ + 5 Aq**—158 + 90—is obtained by dissolving S in hot concentrated solution of Na_2SO_3, and crystallizing.

It forms large, colorless, efflorescent prisms; fuses at 45° (113° F.); very soluble in H_2O; insoluble in alcohol. Its solutions precipitate alumina from solutions of Al salts, without precipitating Fe or Mn; they dissolve many compounds insoluble in H_2O; cuprous hydrate, iodides of Pb, Ag and Hg, sulphates of Ca and Pb. It acts as a disinfectant and antiseptic.

Silicates.—Quite a number of silicates of Na are known. If silica and Na_2CO_3 be fused together, the residue extracted with H_2O, and the solution evaporated, a transparent, glass-like mass, soluble in warm water, remains; this is *soluble glass* or *water glass*. Exposed to air in contact with stone, it becomes insoluble, and forms an impermeable coating.

Phosphates.—Trisodic Phosphate—*Basic sodium phosphate*—**Na_3PO_4 + 12 Aq**—164 + 216—is obtained by adding NaHO to disodic phosphate solution and crystallizing. It forms six-sided prisms; quite soluble in H_2O. Its solution is alkaline, and, on exposure to air, absorbs CO_2 with formation of HNa_2PO_4 and Na_2CO_3.

Disodic Phosphate—*Hydro-disodic phosphate—Neutral sodium phosphate—Phosphate of soda—Sodii phosphas (U. S.)—Sodæ phosphas (Br.)*—**HNa_2PO_4 + 12 Aq**—142 + 216—is obtained by converting tricalcic phosphate into monocalcic phosphate and decomposing that salt with sodium carbonate: $Ca(PO_4H_2)_2 + 2Na_2CO_3 = CaCO_3 + H_2O + CO_2 + 2HNa_2PO_4$.

Below 30° (86° F.) it crystallizes in oblique rhombic prisms with 12 Aq; at 33° (91°.4 F.) it crystallizes with 7 Aq. The salt with 12 Aq effloresces in air and parts with 5 Aq; and is very soluble in H_2O. The salt with 7 Aq is not efflorescent and less soluble in H_2O. Its solutions are faintly alkaline.

MONOSODIC PHOSPHATE—*Acid sodium phosphate*—**H_2NaPO_4 + Aq**—120 + 18—crystallizes in rhombic prisms; forming acid solutions. At 100° (212° F.) it loses Aq; at 200° (392° F.) it is converted into *acid pyrophosphate*, $Na_2H_2P_2O_7$; and at 204° (399°.2 F.) into the *metaphosphate*, $NaPO_3$.

PHYSIOLOGICAL.—All the sodium phosphates exist, accompanied by the corresponding K salts, in the animal economy. The disodic and dipotassic phosphates are the most abundant, and of these two the former. They exist in every tissue and fluid of the body, and are more abundant in the fluids of the carnivora than in those of the herbivora. In the blood, in which the Na salt predominates in the plasma, and the K salt in the corpuscles, they serve to maintain an alkaline reaction. With strictly vegetable diet the proportion of phosphates in the blood diminishes, and that of the carbonates (the predominating salts in the blood of the herbivora) increases.

The monosodic and monopotassic phosphates exist in the urine, the former predominating, and to their presence the acid reaction of that fluid is largely due. They are produced by decomposition of the neutral salts by uric acid. The urine of the herbivora, whose blood is poor in phosphates, is alkaline in reaction.

The greater part of the phosphates in the body are introduced with the food. A portion is formed in the economy by the oxidation of phosphorized organic substances, the lecithins.

Disodic Tetraborate—*Sodium pyroborate—Borate of sodium—Borax—Tincal—Sodii boras* (*U. S.*)—*Borax* (*Br.*)—**$Na_2B_4O_7$ + 10 Aq**—202 + 180—is prepared by boiling boracic acid with Na_2CO_3 and crystallizing. It

134

crystallizes in hexagonal prisms with 10 Aq; permanent in moist air, but efflorescent in dry air; or in regular octahedra with 5 Aq, permanent in dry air. Either form, when heated, fuses in its Aq, swells considerably; at a red heat becomes anhydrous; and, on cooling, leaves a transparent, glass-like mass. When fused it is capable of dissolving many metallic oxides, forming variously colored masses, hence its use as a flux and in blow-pipe analysis.

Sodium Hypochlorite—NaClO—74.5—only known in solution—*Liq. sodæ chloratæ* (*U. S.; Br.*) or *Labarraque's solution*—obtained by decomposing a solution of chloride of lime by Na_2CO_3. It is a valuable source of Cl, and is used as a bleaching and disinfecting agent.

Sodium Manganate—Na_2MnO_4 + 10 Aq—164 + 180—faintly colored crystals, forming a green solution with H_2O—*Condy's green disinfectant.*

Sodium Permanganate—$Na_2Mn_2O_8$—282—prepared in the same way as the K salt (*q. v.*), which it resembles in its properties It enters into the composition of *Condy's fluid*, and of "*chlorozone*," which contains $Na_2Mn_2O_8$ and NaClO.

Sodium Acetate—*Sodii acetas* (*U. S.*)—*Sodæ acetas* (*Br.*)—**$NaC_2H_3O_2$ + 3 Aq**—82 + 54—crystallizes in large, colorless prisms; acid and bitter in taste; quite soluble in H_2O; soluble in alcohol; loses its Aq in dry air, and absorbs it again from moist air. Heated with soda lime, it yields marsh gas. The anhydrous salt, heated with H_2SO_4, yields glacial acetic acid.

Carbonates.—Three are known: Na_2CO_3; $HNaCO_3$, and $H_2Na_4(CO_3)_3$.

SODIC CARBONATE—*Neutral carbonate—Soda—Sal soda—Washing soda—Soda crystals—Sodii carbonas* (*U. S.*)—*Sodæ carbonas* (*Br.*)—**Na_2CO_3 + 10 Aq**—106 + 180—industrially the most important of the Na compounds, is manufactured by Leblanc's or Solvay's processes; or from *cryolite*, a native fluoride of Na and Al.

Leblanc's process, in its present form, consists of three distinct processes: (1.) The conversion of NaCl into the sulphate by decomposition by H_2SO_4. (2.) The conversion of the sulphate into carbonate by heating a mixture of the sulphate with calcium carbonate and charcoal. The product of this reaction, known as *black ball soda*, is a mixture of sodium carbonate with charcoal and calcium sulphide and oxide. (3.) The purification of the product obtained in (2). The ball black is broken up, disintegrated by steam, and lixiviated. The solution on evaporation yields the *soda salt* or *soda* of commerce.

Of late years Leblanc's process has been in great part replaced by Solvay's method, or *ammonia process*, which is more economical and yields a purer product. In this process sodium chloride and ammonium bicarbonate react upon each other, with production of the sparingly soluble sodium bicarbonate and the very soluble ammonium chloride. The sodium bicarbonate is then simply collected, dried, and heated, when it is decomposed into Na_2CO_3, H_2O, and CO_2.

The anhydrous carbonate, *Sodii carbonas exsiccatus* (*U. S.*), Na_2CO_3, is formed as a white powder by calcining the crystals. It fuses at dull redness and gives off a little CO_2. It combines with and dissolves in H_2O with elevation of temperature.

The crystalline sodium carbonate, Na_2CO_3 + 10 Aq, forms large rhombic crystals, which effloresce rapidly in dry air; fuse in their Aq at 34° (93°.2 F.); are soluble in H_2O, most abundantly at 38° (100°.4 F.). The solutions are alkaline in reaction.

135

HYDROSODIC CARBONATE—*Monosodic carbonate—Bicarbonate of soda—Acid carbonate of soda—Vichy salt—Sodii bicarbonas*—(*U. S.*)—*Sodæ bicarbonas* (*Br.*)—$NaHCO_3$—84—exists in solution in many mineral waters. It is obtained by the action of CO_2 upon the disodic salt in the presence of H_2O.

It crystallizes in rectangular prisms, anhydrous and permanent in dry air; in damp air it gives off CO_2 and is converted into the sesquicarbonate, $Na_4H_2(CO_3)_3$. When heated, it gives off CO_2 and H_2O, and leaves the disodic carbonate; quite soluble in water; above 70° (158° F.) the solution gives off CO_2. The solutions are alkaline.

PHYSIOLOGICAL.—The fact that the carbonates of Na and K are almost invariably found in the ash of animal tissues and fluids, is no evidence of their existence there in life, as the carbonates are produced by the incineration of the Na and K salts of organic acids. There is, however, excellent indirect proof of the existence of the alkaline carbonates in the blood, especially of the herbivora, in the urine of the herbivora at all times, and in that of the carnivora and omnivora when food rich in the salts of the organic acids, with alkaline metals, is taken. The carbonates in the blood are both the mono- and disodic and potassic; and the carbonic acid in the plasma is held partially in simple solution, and partly in combination in the monometallic carbonates.

Analytical Characters.

(1.) Hydrofluosilicic acid: gelatinous ppt., if not too dilute.

(2.) Potassium pyroantimonate: in neutral solution and in absence of metals, other than K and Li: a white flocculent ppt.; becoming crystalline on standing.

(3.) Periodic acid in excess: white ppt., in not too dilute solutions.

(4.) Colors the Bunsen flame yellow, and shows a brilliant double line at $\lambda = 5895$ and 5889 (Fig. 14, No. 2).

POTASSIUM.

Symbol = **K** (**KALIUM**)—*Atomic weight* = 39—*Molecular weight* = 78 (?)—*Sp. gr.* = 0.865—*Fuses at* 62°.5 (144°.5 F.)—*Discovered by* Davy, 1807—*Names from* pot ash, *and* Kali = ashes (Arabic).

It is prepared by a process similar to that followed in obtaining Na; is a silver-white metal; brittle at 0° (32° F.); waxy at 15° (59° F.); fuses at 62°.5 (144°.5 F.); distils in green vapors at a red heat, condensing in cubic crystals.

It is the only metal which oxidizes at low temperatures in dry air, in which it is rapidly coated with a white layer of oxide or hydrate, and frequently ignites, burning with a violet flame; it must, therefor, be kept under naphtha. It decomposes H_2O or ice with great energy, the heat of the reaction igniting the liberated H. It combines with Cl with incandescence, and also unites directly with S, P, As, Sb, and Sn. Heated in CO_2 it is oxidized and liberates C.

136

Compounds of Potassium.

Oxides.—Three are known: K_2O; K_2O_2; and K_2O_4.

Potassium Hydrate—*Potash—Potassa—Common caustic—Potassa* (*U. S.*)—*Potassa caustica* (*Br.*)—**KHO**—56—is obtained by a process similar to that used in manufacturing NaHO. It is purified by solution in alcohol, evaporation and fusion in a silver basin and casting in silver moulds—*potash by alcohol*; it is then free from KCl and K_2SO_4, but contains small quantities of K_2CO_3 and frequently As.

It is usually met with in cylindrical sticks, hard, white, opaque, and brittle. The KHO by alcohol has a bluish tinge and a smoother surface than the common; sp. gr. 2.1; fuses at dull redness; is freely soluble in H_2O, forming a strongly alkaline and caustic liquid; less soluble in alcohol. In air, solid or in solution, it absorbs H_2O and CO_2, and is converted into K_2CO_3. Its solutions dissolve Cl, Br, I, S, and P. It decomposes the ammoniacal salts with liberation of NH_3; and the salts of many of the metals, with formation of a K salt and a metallic hydrate. It dissolves the albuminoids; and, when heated, decomposes them with formation of leucin, tyrosin, etc. It oxidizes the carbohydrates with formation of potassium oxalate and carbonate.

Sulphides.—Five are known: K_2S, K_2S_2, K_2S_3, K_2S_4, and K_2S_5; also a sulphydrate: **KHS**.

POTASSIUM MONOSULPHIDE—**K_2S**—110—is formed by the action of KHO on KHS.

POTASSIUM DISULPHIDE—**K_2S_2**—142—is an orange-colored solid, formed by exposing an alcoholic solution of KHS to the air.

POTASSIUM TRISULPHIDE—**K_2S_3**—174—a brownish-yellow mass, obtained by fusing together K_2CO_3 and S in the proportion: $4K_2CO_3 + 10S = SO_4$

$K_2 + 3K_2S_3 + 4CO_2$.

POTASSIUM PENTASULPHIDE—K_2S_5—238—is formed, as a brown mass, when K_2CO_3 and S are fused together in the proportion : $4K_2CO_3 + 16S = 4CO_2 + 3K_2S_5 + K_2SO_4$.

Liver of sulphur—hepar sulphuris—potassii sulphuratum (*U. S.*; *Br.*)—is a mixture of K_2S_3 and K_2S_5.

POTASSIUM SULPHYDRATE—**KHS**—72—is formed by saturating a solution of KHO with H_2S.

Potassium Chloride—*Sal digestivum Sylvii*—**KCl**—74.5—exists in nature, either pure or mixed with other chlorides; principally as *carnallite*, KCl, $MgCl_2$ + 6 Aq. It crystallizes in anhydrous, permanent cubes, soluble in H_2O.

Potassium Bromide—*Potassii bromidum* (*U. S.*; *Br.*)—**KBr**—119—is formed either by decomposing ferrous bromide by K_2CO_3, or by dissolving Br in solution of KHO. In the latter case the *bromate* formed is converted into KBr by calcining the product. It crystallizes in anhydrous cubes or tables ; has a sharp, salty taste ; very soluble in H_2O, sparingly so in alcohol. It is decomposed by Cl with liberation of Br.

Potassium Iodide—*Potassii iodidum* (*U. S.*; *Br.*)—**KI**—166—is obtained by saturating KHO solution with I, evaporating, and calcining the resulting mixture of iodide and iodate with charcoal. It frequently contains iodate and carbonate. It crystallizes in cubes, transparent if pure ; permanent in air ; anhydrous ; soluble in H_2O and in alcohol. It is decomposed by Cl, HNO_3 and HNO_2, with liberation of I. It combines with other iodides to form double iodides.

137

Salts of Potassium.

Potassium Nitrate—*Nitre—Saltpetre—Potassii nitras* (*U. S.*)—*Potassæ nitras* (*Br.*)—**KNO_3**—101—occurs in nature and is produced artificially as a result of the decomposition of nitrogenized organic substances. It is usually obtained by decomposing native $NaNO_3$ by boiling solution of K_2CO_3 or KCl.

It crystallizes in six-sided, rhombic prisms, grooved upon the surface ; soluble in H_2O with depression of temperature ; more soluble in H_2O containing NaCl ; very sparingly soluble in alcohol ; fuses at 350° (662° F.) without decomposition ; gives off O and is converted into nitrite below redness ; more strongly heated, it is decomposed into N, O, and a mixture of K oxides. It is a valuable oxidant at high temperatures ; heated with charcoal it deflagrates.

Gunpowder is an intimate mixture of KNO_3 with S and C, in such proportion that the KNO_3 yields all the O required for the combustion of the S and C.

Potassium Chlorate—*Potassii chloras* (*U. S.*)—*Potassæ chloras* (*Br.*)—**$KClO_3$**—122.5—is prepared : (1) by passing Cl through a solution of KHO ; (2) by passing Cl over a mixture of milk of lime and KCl, heated to 60° (140° F.). It crystallizes in transparent, anhydrous plates ; soluble in H_2O ; sparingly soluble in weak alcohol.

It fuses at 400° (752° F.). If further heated it is decomposed into KCl and perchlorate, and at a still higher temperature the perchlorate is decomposed into KCl and O : $2KClO_3 = KClO_4 + KCl + O_2$ and $KClO_4 = KCl + 2O_2$. It is a valuable source of O, and a more active oxidant than KNO_3. When mixed with readily oxidizable substances, C, S, P, sugar, tannin, resins, etc., the mixtures explode when subjected to shock. With strong H_2SO_4 it gives off Cl_2O_4, an explosive yellow gas. It is decomposed by HNO_3 with formation of KNO_3, $KClO_4$, and liberation of Cl and O. Heated with HCl it gives off a mixture of Cl and Cl_2O_4, the latter acting as an energetic oxidant in solutions in which it is generated.

Potassium Hypochlorite—**KClO**—90.5—is formed in solution by imperfect saturation of a cooled solution of KHO with hypochlorous acid. An impure solution is used in bleaching : *Javelle water*.

Sulphates.—POTASSIC SULPHATE—*Dipotassic sulphate—Potassii sulphas* (*U. S.*)—*Potassæ sulphas* (*Br.*)—**K_2SO_4**—174—occurs native ; in the ash of many plants ; and in solution in mineral waters. It crystallizes in right rhombic prisms ; hard ; permanent in air ; salt and bitter in taste ; soluble in H_2O.

HYDROPOTASSIC SULPHATE—*Monopotassic sulphate—Acid sulphate*—**$KHSO_4$**—136—is formed as a by-product in the manufacture of HNO_3. When heated it loses H_2O, and is converted into the *pyrosulphate*, $K_2S_2O_7$, which, at a higher temperature, is decomposed into K_2SO_4 and SO_3.

Potassic Sulphite—*Dipotassic sulphite—Potassii sulphis* (*U. S.*)—**K_2SO_3**—158—is formed by saturating solution of K_2CO_3 with SO_2, and evaporating over H_2SO_4. It crystallizes in oblique rhombohedra ; soluble in H_2O. Its solution absorbs O from air, with formation of K_2SO_4.

Potassium Dichromate—*Bichromate of potash—Potassii bichromas* (*U. S.*)—*Potassæ bichromas* (*Br.*)—**$K_2Cr_2O_7$**—294.8—is formed by heating a mixture of *chrome iron ore* with KNO_3, or K_2CO_3 in air ; extracting with H_2O ; neutralizing with dilute H_2SO_4 ; and evaporating. It forms large, reddish-orange colored prismatic crystals ; soluble in H_2O ; fuses below

redness, and at a higher temperature is decomposed into O, potassium chromate, and sesquioxide of chromium. Heated with HCl, it gives off Cl.

Potassium Permanganate—*Potassii permanganas* (*U. S.*)—*Potassæ permanganas* (*Br.*)—**$K_2Mn_2O_8$**—314—is obtained by fusing a mixture of manganese dioxide, KHO, and $KClO_3$, and evaporating the solution to crystallization ; K_2MnO_4 and KCl are first formed ; on boiling with H_2O the manganate is decomposed into $K_2Mn_2O_8$ and KHO and MnO_2.

It crystallizes in dark prisms, almost black, with greenish reflections, which yield a red powder when broken. Soluble in H_2O, communicating to it a red color, even in very dilute solution. It is a most valuable oxidizing agent. With organic matter its solution is turned to green by the formation of the manganate, or deposits the brown sesquioxide of manganese, according to the nature of the organic substance ; in some instances the reaction takes place best in the cold, in others under the influence of heat ; in some better in acid solutions, in others in alkaline solutions. Mineral reducing agents act more rapidly. Its oxidizing powers render its solutions valuable as disinfectants.

Potassium Acetate—*Potassii acetas* (*U. S.*)—*Potassæ acetas* (*Br.*)—**$KC_2H_3O_2$**—110—exists in the sap of plants ; and it is by its calcination that the major part of the carbonate of wood ashes is formed. It is prepared by neutralizing acetic acid with K_2CO_3 or $KHCO_3$.

It forms crystalline needles, deliquescent, and very soluble in H_2O ; less soluble in alcohol. Its solutions are faintly alkaline.

Carbonates—POTASSIC CARBONATE—*Salt of tartar—Pearl ash—Potasii carbonas* (*U. S.*)—*Potassæ carbonas* (*Br.*)—**K_2CO_3**—138—exists in mineral waters and in the animal economy. It is prepared industrially in an impure form, known as *potash* or *pearlash*, from wood ashes, from the molasses of beet-sugar, and from the native Stassfurth chloride. It is obtained pure by decomposing the monopotassic salt, purified by several recrystallizations, by heat or by calcining a potassium salt of an organic acid. Thus cream of tartar, mixed with nitre and heated to redness, yields a black mixture of C and K_2CO_3, called *black flux* ; on extracting which with H_2O, a pure carbonate, known as *salt of tartar*, is dissolved.

Anhydrous, it is a white, granular, deliquescent, very soluble powder. At low temperatures it crystallizes with 2 Aq. Its solution is alkaline.

HYDROPOTASSIC CARBONATE—*Monopotassic carbonate—Bicarbonate—Potassii bicarbonas* (*U. S.*)—*Potassæ bicarbonas* (*Br.*)—**$HKCO_3$**—100—is obtained by dissolving K_2CO_3 in H_2O and saturating the solution with CO_2. It crystallizes in oblique rhombic prisms, much less soluble than the carbonate. In solution it is gradually converted into the dipotassic salt when heated, when brought into a vacuum, or when treated with an inert gas. The solutions are alkaline in reaction and in taste, but are not caustic.

The substance used in baking, under the name *salæratus*, is this or the corresponding Na salt. Its extensive use in some parts of the country is undoubtedly in great measure the cause of the prevalence of dyspepsia. When used alone in baking it "raises" the bread by decomposition into carbon dioxide and dipotassic (or disodic) carbonate, the latter producing disturbances of digestion by its strong alkaline reaction.

Hydropotassic Oxalate—*Monopotassic oxalate—Binoxalate of Potash*—**HKC_2O_4**—128—forms transparent, soluble, acid needles. It occurs, along with the *quadroxalate*, HKC_2O_4, $H_2C_2O_4$ + 2 Aq, in *salt of lemon* or *salt of sorrel*, used in straw bleaching and for the removal of ink-stains, etc. It

139

closely resembles Epsom salt in appearance, and has been fatally mistaken for it.

Tartrates.—POTASSIC TARTRATE—*Dipotassic tartrate—Soluble tartar—Neutral tartrate of potash—Potassii tartras* (*U. S.*)—*Potassæ tartras* (*Br.*)—**$K_2C_4H_4O_6$**—226—is prepared by neutralizing the hydropotassic salt with potassium carbonate. It forms a white, crystalline powder, very soluble in H_2O, the solution being dextrogyrous, $[a]_D = +28°.48$; soluble in alcohol. Acids, even acetic, decompose its solution with precipitation of the monopotassic salt.

HYDROPOTASSIC TARTRATE—*Monopotassic tartrate—Cream of tartar—Potassii bitartras* (*U. S.*)—*Potassæ bitartras* (*Br.*)—**$HKC_4H_4O_6$**—188.—During the fermentation of grape-juice, as the proportion of alcohol increases, crystalline crusts collect in the cask. These constitute the *crude tartar* or *argol* of commerce, which is composed, in great part, of monopotassic tartrate. The crude product is purified by repeated crystallization from boiling H_2O ; digesting the purified tartar with HCl at 20° (68° F.) ; washing with cold H_2O, and crystallizing from hot H_2O.

It crystallizes in hard, opaque (translucent when pure), rhombic prisms, which have an acidulous taste, and are very sparingly soluble in H_2O, still less soluble in alcohol. Its solution is acid, and dissolves many metallic oxides with formation of double tartrates. When boiled with antimony trioxide, it forms tartar emetic.

It is used in the household, combined with monosodic carbonate, in baking, the two substances reacting upon each other to form Rochelle salt, with liberation of carbon dioxide.

BAKING-POWDERS are now largely used as substitutes for yeast in the manufacture of bread. Their action is based upon the decomposition of $HNaCO_3$ by some salt having an acid reaction, or by a weak acid. In addition to the bicarbonate and flour, or corn starch (added to render the bulk convenient to handle and to diminish the rapidity of the reaction), they contain cream of tartar, tartaric acid, alum, hydrochloric acid, or acid phosphates. Sometimes ammonium sesquicarbonate is used, in whole or in part, in place of sodium carbonate.

The reactions by which the CO_2 is liberated are:

$$1.\ \underset{\text{Hydropotassic tartrate.}}{HKC_4H_4O_6} + \underset{\text{Hydrosodic carbonate.}}{NaHCO_3} = \underset{\text{Sodium potassium tartrate.}}{NaKC_4H_4O_6} + \underset{\text{Water.}}{H_2O} + \underset{\text{Carbon dioxide.}}{CO_2}.$$

$$2.\ \underset{\text{Tartaric acid.}}{H_2C_4H_4O_6} + \underset{\text{Hydrosodic carbonate.}}{2NaHCO_3} = \underset{\text{Disodic tartrate.}}{Na_2C_4H_4O_6} + \underset{\text{Water.}}{2H_2O} + \underset{\text{Carbon dioxide.}}{2CO_2}.$$

$$3.\ \underset{\text{Aluminium potassium alum.}}{Al_2(SO_4)_3, K_2SO_4} + \underset{\text{Hydrosodic carbonate.}}{6NaHCO_3} = \underset{\text{Potassic sulphate.}}{K_2SO_4} + \underset{\text{Sodic sulphate.}}{3Na_2SO_4} + \underset{\text{Aluminium hydrate.}}{Al_2H_6O_6} + \underset{\text{Carbon dioxide.}}{6CO_2}.$$

$$4.\ \underset{\text{Aluminium ammonium alum.}}{Al_2(SO_4)_3, (NH_4)_2SO_4} + \underset{\text{Hydrosodic carbonate.}}{6NaHCO_3} = \underset{\text{Ammonic sulphate.}}{(NH_4)_2SO_4} + \underset{\text{Sodic sulphate.}}{3Na_2SO_4} + \underset{\text{Aluminium hydrate.}}{Al_2H_6O_6} + \underset{\text{Carbon dioxide.}}{6CO_2}.$$

$$5.\ \underset{\text{Aluminium sulphate.}}{Al_2(SO_4)_3} + \underset{\text{Hydrosodic carbonate.}}{6NaHCO_3} = \underset{\text{Sodic sulphate.}}{3Na_2SO_4} + \underset{\text{Aluminium hydrate.}}{Al_2H_6O_6} + \underset{\text{Carbon dioxide.}}{6CO_2}.$$

140

$$6.\ \underset{\text{Hydrochloric acid.}}{HCl} + \underset{\text{Hydrosodic carbonate.}}{NaHCO_3} = \underset{\text{Sodium chloride.}}{NaCl} + \underset{\text{Water.}}{H_2O} + \underset{\text{Carbon dioxide.}}{CO_2}.$$

$$7.\ \underset{\text{Monosodic phosphate.}}{NaH_2PO_4} + \underset{\text{Hydrosodic carbonate.}}{NaHCO_3} = \underset{\text{Disodic phosphate.}}{Na_2HPO_4} + \underset{\text{Water.}}{H_2O} + \underset{\text{Carbon dioxide.}}{CO_2}.$$

$$8.\ \underset{\text{Aluminium sulphate.}}{2Al_2(SO_4)_3} + \underset{\text{Ammonium sesquicarbonate.}}{3[2([NH_4]_2CO_3)CO_2]} + \underset{\text{Water.}}{6H_2O} = \underset{\text{Ammonium sulphate.}}{6(NH_4)_2SO_4} + \underset{\text{Aluminium hydrate.}}{2Al_2H_6O_6} + \underset{\text{Carbon dioxide.}}{9CO_2}.$$

No. 1 is the reaction which takes place when cream of tartar and soda, or a baking-powder composed of those substances, are used in baking. The solid product of the reaction is Rochelle salt. No. 2 is that which occurs between tartaric acid and soda, and is but seldom utilized. No. 3 is that between burnt potassium alum and soda. It is not utilized at present, as the ammonium alum is more economical. Nos. 4 and 5 are those which occur in alum baking-powders, the burnt ammonia alum being anhydrous ammonium aluminium sulphate, or aluminium sulphate, according to the degree of heat used in its manufacture. The solid residues of the reaction are sodic sulphate and aluminium hydrate. No. 6 is a reaction very little used, owing to the inconvenience of handling a liquid, to the too rapid action of the substances upon each other, and to the danger of introducing arsenic with the acid. No. 7 is used to a certain extent, and has the advantage that the solid residue of the reaction is a normal constituent of the body. No. 8 is occasionally utilized as an adjunct to No. 5.

In our opinion, while yeast is to be preferred to any baking-powder, an alum-powder is in no way more liable to produce disturbances of digestion than one compounded of cream of tartar and soda. Referring to Equation 5, above, and taking the amount of powder generally used, 35 grains per pound of bread, it will be seen that that amount of powder, containing 9.26 grains of aluminum sulphate, when neutralized during baking, produces 11.5 grains of Glauber's salt, 4.24 grains of aluminium hydrate, and 7.12 grains of carbon dioxide. On the other hand, a cream of tartar powder to produce, according to reaction above, the same quantity, 7.12 grains, of carbon dioxide, forms at the same time 33.98 grains of Rochelle salt. Assuming that one to two pounds is the average amount of bread consumed by an adult in twenty-four hours, there can be but little choice between taking on the one hand 4.24–8.48 grains of alumina and 11.5–23.0 grains of Glauber's salt; and on the other hand, 33.98–67.96 grains of Rochelle salt. Indeed, there is more danger to be apprehended from the tendency of repeated small doses of Rochelle salt to render the urine alkaline and thus favor the formation of phosphatic calculi, than from any supposed deleterious action of alumina, whose local action, even in considerable doses, is that of a very mild astringent, and whose absorption is very doubtful.

Sodium Potassium Tartrate—*Rochelle salt—Sel de seignette—Potassii et sodii tartras* (*U. S.*)—*Soda tartarata* (*Br.*)—$NaKC_4H_4O_6 + 4Aq$—210 + 72—is prepared by saturating hydropotassic tartrate with sodium carbonate. It crystallizes in large, transparent prisms, which effloresce superficially in dry air and attract moisture in damp air. It fuses at 70–80° (158°–176° F.), and loses 3 Aq at 100° (212° F.). It is soluble in H_2O, the solutions being dextrogyrous, $[a]_D = +29°.67$.

141

Potassium Antimonyl Tartrate—*Tartarated antimony—Tartar emetic—Antimonii et potassii tartras* (*U. S.*)—*Antimonium tartaratum* (*Br.*)—$(SbO)'KC_4H_4O_6$—323—is prepared by boiling a mixture of 3 pts. Sb_2O_3 and 4 pts. $HKC_4H_4O_6$ in H_2O for an hour, filtering, and allowing to crystallize; when required pure, it must be made from pure materials.

It crystallizes in transparent, soluble, right rhombic octahedra, which turn white in air. Its solutions are acid in reaction, have a nauseating, metallic taste, are lævogyrous, $[a]_D = +156°.2$, and are precipitated by alcohol. The crystals contain ½ Aq, which they lose entirely at 100° (212° F.), and partially by exposure to air. It is decomposed by the alkalies, alkaline earths, and alkaline carbonates, with precipitation of Sb_2O_3. The precipitate is redissolved by excess of soda or potash, or by tartaric acid. HCl, H_2SO_4 and HNO_3 precipitate corresponding antimonyl compounds from solutions of tartar emetic. It converts mercuric into mercurous chloride. It forms double tartrates with the tartrates of the alkaloids.

Potassium Cyanide—*Potassii cyanidum* (*U. S.*)—**KCN**—65—is obtained by heating a mixture of potassium ferrocyanide and dry K_2CO_3 as long as effervescence continues; decanting and crystallizing.

It is usually met with in dull, white, amorphous masses; odorless when dry, it has the odor of hydrocyanic acid when moist. It is deliquescent, and very soluble in H_2O; almost insoluble in alcohol. Its solution is acrid, and bitter in taste, with an after-taste of hydrocyanic acid. It is very readily oxidized to the cyanate, a property which renders it valuable as a reducing agent. Solutions of KCN dissolve I, AgCl, the cyanides of Ag and Au, and many metallic oxides.

It is actively poisonous, and produces its effects by decomposition and liberation of hydrocyanic acid (*q. v.*).

Potassium Ferrocyanide—*Yellow prussiate of potash—Potassii ferrocyanidum* (*U. S.*)—*Potassæ prussias flava* (*Br.*)—$K_4[Fe(CN)_6] + 3\,Aq$—367.9 + 54.—This salt, the source of the other cyanogen compounds, is manufactured by adding organic matter (blood, bones, hoofs, leather, etc.) and iron to K_2CO_3 in fusion; or by other processes in which the N is obtained from the residues of the purification of coal-gas, from atmospheric air, or from ammoniacal compounds.

It forms soft, flexible, lemon-yellow crystals, permanent in air at ordinary temperatures. They begin to lose Aq at 60° (140° F.), and become anhydrous at 100° (212° F.). Soluble in H_2O; insoluble in alcohol, which precipitates it from its aqueous solution. When calcined with KHO or K_2CO_3, potassium cyanide and cyanate are formed, and Fe is precipitated. Heated with dilute H_2SO_4, it yields an insoluble white or blue salt, potassium sulphate, and hydrocyanic acid. Its solutions form with those of many of the metallic salts insoluble ferrocyanides; those of Zn, Pb, and Ag are white, cupric ferrocyanide is mahogany-colored, ferrous ferrocyanide is bluish-white, ferric ferrocyanide (Prussian blue) is dark blue. Blue ink is a solution of Prussian blue in a solution of oxalic acid.

Potassium Ferricyanide—*Red prussiate of potash*—$K_6Fe_2(CN)_{12}$—657.8—is prepared by acting upon the ferrocyanide with chlorine; or, better, by heating the white residue of the action of H_2SO_4 upon potassium ferrocyanide, in the preparation of hydrocyanic acid, with a mixture of 1 vol. HNO_3 and 20 vols. H_2O; the *blue* product is digested with H_2O and potassium ferrocyanide, the solution filtered and evaporated.

It forms red, oblique, rhombic prisms, almost insoluble in alcohol. With solutions of ferrous salts it gives a dark blue precipitate, *Turnbull's blue*.

142

Analytical Characters.

(1.) Platinic chloride, in presence of HCl: yellow ppt.; crystalline if slowly formed; sparingly soluble in H_2O, much less so in alcohol.

(2.) Tartaric acid, in not too dilute solution: white ppt.; soluble in alkalies and in concentrated acids.

(3.) Hydrofluosilicic acid: translucent, gelatinous ppt.; forms slowly; soluble in strong alkalies.

(4.) Perchloric acid: white ppt.; sparingly soluble in H_2O; insoluble in alcohol.

(5.) Phosphomolybdic acid: white ppt.; forms slowly.

(6.) Colors the Bunsen flame violet (the color is only observable through blue glass in presence of Na), and exhibits a spectrum of two bright lines: $\lambda = 7860$ and 4045 (Fig. 14, No 3).

Action of the Sodium and Potassium Compounds on the Economy.

The hydrates of Na and of K, and in a less degree the carbonates, disintegrate animal tissues, dead or living, with which they come in contact, and, by virtue of this action, act as powerful caustics upon a living tissue. Upon the skin they produce a soapy feeling and in the mouth a soapy taste. Like the acids, they cause death, either immediately, by corrosion or perforation of the stomach ; or secondarily after weeks or months, by closure of one or both openings of the stomach, due to thickening, consequent upon inflammation.

The treatment consists in the neutralization of the alkali by an acid, dilute vinegar. Neutral oils and milk are of service, more by reason of their emollient action than for any power they have to neutralize the alkali by the formation of a soap at the temperature of the body.

The other compounds of Na, if the acid be not poisonous, are without deleterious action, unless taken in excessive quantity. Common salt has produced paralysis and death in a dose of half a pound. The neutral salts of K, on the contrary, are by no means without true poisonous action when taken internally, or injected subcutaneously in sufficient quantities ; causing dyspnœa, convulsions, arrest of the heart's action, and death. In the adult human subject, death has followed the ingestion of doses of ℥ ss. – ℥ j. of the nitrate, in several instances ; doses of ʒ ij. – ℥ ij. of the sulphate have also proved fatal.

Cæsium—*Symbol* = **Cs**—*Atomic weight* = 132.6 ; and **Rubidium**—*Symbol* = **Rb**—*Atomic weight* = 85.3 —are two rare elements, discovered in 1860 by Kirchoff and Bunsen while examining spectroscopically the ash of a spring water. They exist in very small quantity in *lepidolite*. They combine with O and decompose H_2O even more energetically than does K, forming strongly alkaline hydrates.

SILVER.

Symbol = **Ag (ARGENTUM)**—*Atomic weight* = 107.9 — *Molecular weight* = 216 (?)—*Sp. gr.* = **10.4–10.54**—*Fuses at* **1,000° (1,832° F.).**

Although silver is usually classed with the "noble metals," it differs from Au and Pt widely in its chemical characters, in which it more closely resembles the alkaline metals.

143

When pure Ag is required, coin silver is dissolved in HNO_3 and the diluted solution precipitated with HCl. The silver chloride is washed until the washings no longer precipitate with silver nitrate ; and reduced either (1) by suspending it in dilute H_2SO_4 in a platinum basin, with a bar of pure Zn, and washing thoroughly after complete reduction ; or (2) by mixing it with chalk and charcoal (AgCl, 100 parts ; C, 5 parts ; $CaCO_3$, 70 parts) and gradually introducing the mixture into a red-hot crucible.

Silver is a white metal ; very malleable and ductile ; the best known conductor of heat and electricity. It is not acted on by pure air, but is blackened in air containing a trace of H_2S. It combines directly with Cl, Br, I, S, P, and As. Hot H_2SO_4 dissolves it as sulphate, and HNO_3 as nitrate. The caustic alkalies do not affect it. It alloys readily with many metals ; its alloy with Cu is harder than the pure metal.

Oxides.—Three oxides of silver are known : Ag_4O, Ag_2O, and Ag_2O_2.

SILVER MONOXIDE—*Protoxide—Argenti oxidum*—(*U. S.* ; *Br.*)—**Ag_2O**—231.8—formed by precipitating a solution of silver nitrate with potash. It is a brownish powder ; faintly alkaline and very slightly soluble in H_2O ; strongly basic. It readily gives up its oxygen. On contact with ammonium hydrate it forms a fulminating powder.

Chloride—AgCl—143.4—formed when HCl or a chloride is added to a solution containing silver. It is white ; turns violet and black in sunlight ; volatilizes at 260° (500° F.) ; sparingly soluble in HCl ; soluble in solutions of the alkaline chlorides, hyposulphides, and cyanides, and in ammonium hydrate.

Bromide—AgBr ; and **Iodide**—AgI—are yellowish precipitates, formed by decomposing silver nitrate with potassium bromide and iodide.

Argentic Nitrate—*Argenti nitras* (*U. S.* ; *Br.*)—**$AgNO_3$**—169.9—is prepared by dissolving Ag in HNO_3, evaporating, fusing, and recrystallizing. It crystallizes in anhydrous, right rhombic plates ; soluble in H_2O. The solutions are colorless and neutral. In the presence of organic matter it turns black in sunlight.

The salt, fused and cast into cylindrical moulds, constitutes *lunar caustic, lapis infernalis* ; *argenti nitras fusa* (*U. S.*). If, during fusion, the temperature be raised too high, it is converted into nitrite, O, and Ag ; and if sufficiently heated leaves pure Ag.

Dry Cl and I decompose it, with liberation of anhydrous HNO_3. It absorbs NH_3 to form a white solid, $AgNO_3,3NH_3$, which gives up its NH_3 when heated. Its solution is decomposed very slowly by H, with deposition of Ag.

Argentic Cyanide—*Argenti cyanidum* (*U. S.*)—**AgCN**—133.9—is prepared by passing HCN through a solution of $AgNO_3$. It is a white, tasteless powder ; gradually turns brown in daylight ; insoluble in dilute acids ; soluble in ammonium hydrate, and in solutions of ammoniacal salts, cyanides, or hyposulphites. The strong mineral acids decompose it with liberation of HCN.

Analytical Characters.

(1.) Hydrochloric acid : white, flocculent ppt. ; soluble in NH_4HO ; insoluble in HNO_3.

(2.) Potash or soda : brown ppt. ; insoluble in excess ; soluble in NH_4HO.

(3.) Ammonium hydrate, from neutral solutions : brown ppt. ; soluble in excess.

144

(4.) Hydrogen sulphide or ammonium sulphydrate : black ppt. ; insoluble in NH_4HS.

(5.) Potassium bromide : yellowish-white ppt. ; insoluble in acids, if not in great excess ; soluble in NH_4HO.

(6.) Potassium iodide : same as KBr, but the ppt. is less soluble in NH_4HO.

Action on the Economy.

Silver nitrate acts both locally as a corrosive, and systemically as a true poison. Its local action is due to its decomposition by contact with organic substances, resulting in the separation of elementary Ag, whose deposition causes a black stain, and liberation of free HNO_3, which acts as a caustic. When absorbed, it causes nervous symptoms, referable to its poisonous action. The blue coloration of the skin, observed in those to whom it is administered for some time, is due to the reduction of the metal under the combined influence of light and organic matter ; especially of the latter, as the darkening is observed, although it is less intense, in internal organs.

In acute poisoning by silver nitrate, sodium chloride or white of egg should be given ; and, if the case be seen before the symptoms of corrosion are far advanced, emetics.

AMMONIUM COMPOUNDS.

The ammonium theory.—**Although the radical *ammonium*, NH_4, has probably never been isolated, its existence in the ammoniacal compounds is almost universally admitted. The ammonium hypothesis is based upon the following facts : (1) the close resemblance of the ammoniacal salts to those of K and Na ; (2) when ammonia gas and an acid gas come together, they unite, *without liberation of hydrogen*, to form an ammoniacal salt ; (3) the diatomic anhydrides unite directly with dry ammonia with formation of the ammonium salt of an amido acid :**

$$\underset{\text{Sulphur trioxide.}}{SO_3} + \underset{\text{Ammonia.}}{2NH_3} = \underset{\text{Ammonium sulphamate.}}{SO_2(NH_2)(NH_4)}$$

(4) when solutions of the ammoniacal salts are subjected to electrolysis, a mixture having the composition $NH_3 + H$ is given off at the negative pole ; (5) amalgam of sodium, in contact with a concentrated solution of ammonium chloride, increases much in volume, and is converted into a light, soft mass, having the lustre of mercury. This *ammonium amalgam* is decomposed gradually, giving off ammonia and hydrogen in the proportion $NH_3 + H$; (6) if the gases $NH_3 + H$, given off by decomposition of the amalgam, exist there in simple solution, the liberated H would have the ordinary properties of that element ; if, on the other hand, they exist in combination, the H would exhibit the more energetic affinities of an element in the nascent state. The hydrogen so liberated is in the nascent state.

Compounds of Ammonium.

Ammonium Hydrate—*Caustic ammonia*—**NH_4HO**—35—has never been isolated, probably owing to its tendency to decomposition : $NH_4HO = NH_3 + H_2O$. It is considered as existing in the so-called aqueous solutions

145

of ammonia. These are colorless liquids ; of less sp. gr. than H_2O ; strongly alkaline ; and having the taste and odor of ammonia, which gas they give off on exposure to air, and more rapidly when heated. They are neutralized by acids, with elevation of temperature and formation of ammoniacal salts. The *Aqua ammoniæ* (*U. S.*) and *Liq. Ammoniæ* (*Br.*) are such solutions.

Sulphides.—Four are known : $(NH_4)_2S$; $(NH_4)_2S_2$; $(NH_4)_2S_4$; and $(NH_4)_2S_5$; as well as a sulphydrate $(NH_4)HS$.

AMMONIUM SULPHYDRATE—**NH_4HS**—51—is formed in solution by saturating a solution of NH_4HO with H_2S ; or anhydrous by mixing equal volumes of dry NH_3 and dry H_2S.

The anhydrous compound is a colorless, transparent, volatile and soluble solid ; capable of sublimation without decomposition. The solution when freshly prepared is colorless, but soon becomes yellow from oxidation and formation of ammonium disulphide and hyposulphite, and finally deposits sulphur.

The sulphides and hydrosulphide of ammonium are also formed during the decomposition of albuminoids, and exist in the gases formed in burial vaults, sewers, etc.

Ammonium Chloride—*Sal ammoniac—Ammonii chloridum* (*U. S.* ; *Br.*)—**NH_4Cl**—53.5—is obtained from the ammoniacal water of gas-works. It is a translucid, fibrous, elastic solid ; salty in taste, neutral in reaction ; volatile without fusion or decomposition ; soluble in H_2O. Its solution is neutral, but loses NH_3 and becomes acid when boiled.

Ammonium chloride exists in small quantity in the gastric juice of the sheep and dog ; also in the perspiration, urine, saliva, and tears.

Ammonium Bromide—*Ammonii bromidum* (*U. S.*)—**$(NH_4)Br$**—98

—is formed either by combining NH_3 and HBr ; by decomposing ferrous bromide with NH_4HO ; or by double decomposition between KBr and $SO_4(NH_4)_2$. It is a white, granular powder, or crystallizes in large prisms, which turn yellow on exposure to air ; quite soluble in H_2O ; volatile without decomposition.

Ammonium Iodide—*Ammonii iodidum* (*U. S.*)—**NH_4I** — 145 — is formed by union of equal volumes of NH_3 and HI ; or by double decomposition of KI and $(NH_4)_2SO_4$. It crystallizes in deliquescent, soluble cubes.

Salts of Ammonium.

Ammonium Nitrate—*Ammonii nitras* (*U. S.*)—**$(NH_4)NO_3$**—80—is prepared by neutralizing HNO_3 with ammonium hydrate or carbonate. It crystallizes in flexible, anhydrous, six-sided prisms ; very soluble in H_2O with considerable diminution of temperature ; fuses at 150° (302° F.), and decomposes at 210° (410° F.), with formation of nitrous oxide : $(NH_4)NO_3 = N_2O + 2\,H_2O$. If the heat be suddenly applied or allowed to surpass 250° (482° F.), NH_3, NO, and N_2O are formed. When fused it is an active oxidant.

Sulphates—Ammonic Sulphate—*Diammonic sulphate—Ammonii sulphas* (*U. S.*)—**$(NH_4)_2SO_4$**—132—is obtained by collecting the distillate from a mixture of ammoniacal gas liquor and lime in H_2SO_4. It forms anhydrous, soluble, rhombic crystals ; fuses at 140° (284° F.), and is decomposed at 200° (392° F.) into NH_3 and $H\,(NH_4)SO_4$.

Hydroammonic Sulphate—*Mono-ammonic sulphate—Bisulphate of ammonia*—**$H(NH)_4SO_4$**—115—is formed by the action of H_2SO_4 on $(NH_4)_2SO_4$. It crystallizes in right rhombic prisms, soluble in H_2O and alcohol.

146

Ammonium Acetate—**$(NH_4)C_2H_3O_2$**—77—is formed by saturating acetic acid with NH_3, or with ammonium carbonate. It is a white, odorless, very soluble solid ; fuses at 86° (186°.8 F.), and gives off NH_3 ; then acetic acid, and finally acetamide. *Liq. ammonii acetatis = Spirit of Mindererus* is an aqueous solution of this salt.

Carbonates.—Ammonic Carbonate—*Diammonic carbonate—Neutral ammonium carbonate*—**$(NH_4)_2CO_3$+Aq**—96 + 18—has been obtained as a white crystalline solid. In air it is rapidly decomposed into NH_3 and $H(NH_4)CO_3$.

Hydroammonic Carbonate — *Monoammonic carbonate — Acid carbonate of ammonia*—**$H(NH_4)CO_3$**—79—is prepared by saturating a solution of NH_4HO or ammonium sesquicarbonate with CO_2. It crystallizes in large, rhombic prisms ; quite soluble in H_2O. At 60° (140° F.) it is decomposed into NH_3 and CO_2.

Ammonium Sesquicarbonate—*Sal volatile—Preston salts—Ammonii carbonas* (*U. S.*)—*Ammoniæ carbonas* (*Br.*)—**$(NH_4)_4H_2(CO_3)_3$**—254—is prepared by heating a mixture of NH_4Cl and chalk, and condensing the product. It crystallizes in rhombic prisms ; has an ammoniacal odor and an alkaline reaction ; soluble in H_2O. By exposure to air or by heating its solution it is decomposed into H_2O, NH_3, and $H(NH_4)CO_3$.

Analytical Characters.

(1.) Entirely volatile at high temperatures.

(2.) Heated with KHO, the ammoniacal compounds give off NH_3, recognizable : (*a*) by changing moist red litmus to blue ; (*b*) by its odor ; (*c*) by forming a white cloud on contact with a glass rod moistened with HCl.

(3.) With platinic chloride : a yellow, crystalline ppt.

(4.) With hydro-sodic tartrate, in moderately concentrated and neutral solution : a white, crystalline ppt.

Action on the Economy.

Solutions of the hydrate and carbonate act upon animal tissues in the same way as the corresponding Na and K compounds. They, moreover, disengage NH_3, which causes intense dyspnœa, irritation of the air passages, and suffocation.

The treatment indicated is the neutralization of the alkali by a dilute acid. Usually the vapor of acetic acid or of dilute HCl must be administered by inhalation.

II. THALLIUM GROUP.

THALLIUM.

Symbol = **Tl**—*Atomic weight* = 203.7 — *sp. gr.* = 11.8–11.9—*Fuses at* 294° (561° F.)—*Discovered by Crookes* (1861).

A rare element, first obtained from the deposits in flues of sulphuric acid factories, in which pyrites from the Hartz were used. It resembles Pb in appearance and in physical properties, but differs entirely from that element in its chemical characters. It resembles Au in being univalent and trivalent, but differs from it, and resembles the alkaline metals in being readily oxidized, in forming alums, and in forming no acid hydrate. It differs from the alkaline metals in the thallic compounds, which contain Tl'''. It is characterized spectroscopically by a bright green line—$\lambda = 5349$.

147

III. CALCIUM GROUP.

Metals of the Alkaline Earths.

Calcium—Strontium—Barium.

The members of this group are bivalent in all their compounds ; each forms two oxides : MO and MO_2 ; each forms a hydrate having well-marked basic characters.

CALCIUM.

Symbol = **Ca**—*Atomic weight* = 40—*Molecular weight* = 80 (?)—*Sp. gr.* = 1.984—*Discovered by Davy in* 1808—*Name from* calx = lime.

Occurs only in combination, as limestone, marble, chalk ($CaCO_3$) ; gypsum, selenite, alabaster ($CaSO_4$), and many other minerals. In bones, egg-shells, oyster-shells, etc., as $Ca_3(PO_4)_2$ and $CaCO_3$, and in many vegetable structures.

The element is a hard, yellow, very ductile, and malleable metal ; fusible at a red heat ; not sensibly volatile. In dry air it is not altered, but is converted into CaH_2O_2 in damp air ; decomposes H_2O ; burns when heated in air.

Compounds of Calcium.

Calcium Monoxide—*Quick lime—Lime—Calx* (*U. S.* ; *Br.*)—**CaO**—56—is prepared by heating a native carbonate (limestone) ; or, when required pure, by heating a carbonate prepared by precipitation.

It occurs in white or grayish, amorphous masses ; odorless ; alkaline ; caustic ; almost infusible ; sp. gr. 2.3. With H_2O it gives off great heat and is converted into the hydrate (*slacking*). In air it becomes *air-slacked*, falling into a white powder, having the composition $CaCO_3$, CaH_2O_2.

Calcium Hydrate—*Slacked lime—Calcis hydras* (*Br.*)—**CaH_2O_2**—74—is formed by the action of H_2O on CaO. If the quantity of H_2O used be one-third that of the oxide, the hydrate remains as a dry, white, odorless powder ; alkaline in taste and reaction ; more soluble in cold than in hot H_2O. If the quantity of H_2O be greater a creamy or milky liquid remains, *cream* or *milk of lime* ; a solution holding an excess in suspension. With a sufficient quantity of H_2O the hydrate is dissolved to a clear solution, which is *lime water—Liquor calcis* (*U. S.* ; *Br.*). The solubility of CaH_2O_2 is diminished by the presence of alkalies, and is increased by sugar or mannite : *Liq. calc. saccharatus* (*Br.*). Solutions of CaH_2O_2 absorb CO_2 with formation of a white deposit of $CaCO_3$.

Calcium Chloride—*Calcii chloridum* (*U. S.* ; *Br.*)—**$CaCl_2$**—111—is obtained by dissolving marble in HCl : $CaCO_3 + 2HCl = CaCl_2 + H_2O + CO_2$. It is bitter ; deliquescent ; very soluble in H_2O ; crystallizes with 6 Aq, which it loses when fused, leaving a white, amorphous mass ; used as a drying agent.

Chloride of Lime—*Bleaching powder—Calx chlorata* (*U. S.* ; *Br.*)—is a mixture composed chiefly of $CaCl_2$ and calcium hypochlorite $Ca(ClO)_2$; prepared by passing Cl over CaH_2O_2, maintained in excess. It is a grayish

148

white powder ; bitter and acrid in taste ; soluble in cold H_2O ; decomposed by boiling H_2O, and by the weakest acids with liberation of Cl. It is decomposed by CO_2 with formation of $CaCO_3$, and liberation of hypochlorous acid, if it be moist ; or of Cl, if it be dry. A valuable disinfectant.

Salts of Calcium.

Calcium Sulphate—**$CaSO_4$**—136—occurs in nature as *anhydrite* ; and with 2 Aq in *gypsum, alabaster, selenite* ; and in solution in natural waters. *Terra alba* is ground gypsum. It crystallizes with 2 Aq in right rhombic prisms ; sparingly soluble in H_2O, more soluble in H_2O containing free acid or chlorides. When the hydrated salt (gypsum) is heated to 80° (176° F.), or more rapidly between 120°–130° (248°–266° F.), it loses its Aq and is converted into a white, opaque mass ; which, when ground, is *plaster-of-Paris*.

The *setting* of plaster when mixed with H_2O, is due to the conversion of the anhydrous into the crystalline, hydrated salt. The ordinary plastering should never be used in hospitals, as, by reason of its irregularities and porosity it soon becomes saturated with the transferrers of septic disease, be they germs or poisons, and cannot be thoroughly purified by disinfectants. Plaster surfaces may, however, be rendered dense and be highly polished, so as to be smooth and impermeable, by adding glue and alum, or an alkaline silicate to the water used in mixing.

Phosphates.—Three are known : $Ca_3(PO_4)_2$; $Ca_2(HPO_4)_2$, and $Ca(H_2PO_4)_2$.

Tricalcic Phosphate—*Tribasic* or *neutral phosphate—Bone phosphate—Calcii phosphas præcipitatus* (*U. S.*)—*Calcis phosphas* (*Br.*)—**$Ca_3(PO_4)_2$**—310—occurs in nature in soils, guano, *coprolites*, phosphorite, in all plants, and in every animal tissue and fluid. It is obtained by dissolving bone-ash in HCl, filtering, and precipitating with NH_4HO ; or by double decomposition between $CaCl_2$ and an alkaline phosphate. When freshly precipitated it is gelatinous ; when dry, a light, white, amorphous powder ; almost insoluble in pure H_2O; soluble to a slight extent in H_2O containing ammoniacal salts, or NaCl or $NaNO_3$; readily soluble in dilute acids, even in H_2O charged with carbonic acid. It is decomposed by H_2SO_4 into $CaSO_4$ and $Ca(H_2PO_4)_2$. *Bone-ash* is an impure form of $Ca_3(PO_4)_2$, obtained by calcining bones, and used in the manufacture of P and of superphosphate.

DICALCIC PHOSPHATE—$\mathbf{Ca_2(HPO_4)_2 + 2\,Aq}$—272 + 36—is a crystalline, insoluble salt; formed by double decomposition between $CaCl_2$ and HNa_2PO_4 in acid solution.

MONOCALCIC PHOSPHATE—*Acid calcium phosphate—Superphosphate of lime*—$\mathbf{Ca(H_2PO_4)_2}$—234—exists in brain tissue and in those animal liquids whose reaction is acid. It is also formed when $Ca_3(PO_4)_2$ is dissolved in an acid, and is manufactured, for use as a manure, by decomposing bone-ash with H_2SO_4. It crystallizes in pearly plates; very soluble in H_2O. Its solutions are acid.

PHYSIOLOGICAL.—All three calcium phosphates, accompanied by the corresponding Mg salts, exist in the animal economy. The *tricalcic* salt occurs in all the solids of the body and in all fluids not having an acid reaction, being held in solution in the latter by the presence of chlorides. In the fluids it is present in very small quantity, except in the milk, in which it is comparatively abundant; 2.5 to 3.95 parts per 1,000 in human milk, and 1.8 to 3.87 parts per 1,000 in cow's milk; constituting about 70 per cent

149

of the ash. The *bones* contain about 35 parts of organic matter, combined with 65 parts of mineral material. The average of human bone-ash is: $Ca_3(PO_4)_2$—83.89; $CaCO_3$—13.03; Ca, combined with Cl, Fl, and organic acids—0.35; Fl—0.23; Cl—0.18. The average quantity of $Ca_3(PO_4)_2$ in male adult bones is 57 per cent.; that of $CaCO_3$, 10 per cent.; and that of $Mg_3(PO_4)_2$, 1.3 per cent. In pathological conditions the composition of bone is modified as shown in the following table:

ANALYSES OF BONES.

In 100 parts.	Healthy male, aged 40; femur.	Osteomalacia, male, aged 40; femur.	Osteomalacia, male, aged 60; femur.	Osteomalacia, child; vertebra.	Rachitis; femur.	Rachitis; humerus.	Caries; femur.	Caries, female, aged, 40; vertebra.	Necrosis.
Collagen	......	48.83	32.54	75.22	72.20	}81.12}	35.69	41.42	19.58
Fats		29.18	4.15	6.12	7.20		3.00	8.36	1.22
Tricalcic phosphate	56.9	17.56	53.25	12.56	14.78	15.60	}51.53	44.05	72.63
Calcium fluoride	......	......	......	......	1.00	......			
Calcium carbonate	10.2	3.04	7.49	3.20	3.00	2.66	5.44	3.45	4.03
Trimagnesic phosphate	1.3	0.23	1.22	0.92	0.80	*	3.43	1.02	1.93
Other salts	......	0.37	1.35	1.98	1.02	0.62	0.91	1.70	0.61
Organic matter	35.8	78.01	36.69	81.24	79.40	81.12	38.69	49.78	20.80
Ash	64.2	21.20	63.31	19.66	20.60	18.88	61.31	50.22	79.20
* Included in tricalcic phosphate.	Frémy.	Lehmann.	Von Bibra.	Marchand.	Marchand.	Ragsky.	Becquerel and Rodier.	Becquerel and Rodier.	Von Bibra.

The *teeth* consist largely of $Ca_3(PO_4)_2$; the dentine of human molars containing 66.72 per cent., and the enamel 89.82 per cent.

From the *urine*, tricalcic phosphate is frequently deposited, either in the form of an amorphous, granular sediment, or as calculi. The *dicalcic* salt occurs occasionally in urinary sediments, in the form of needle-shaped crystals arranged in rosettes, and also in urinary calculi. The *monocalcic* salt is always present in acid urine, constituting, with the corresponding magnesium salt, the *earthy phosphates.* The total elimination of H_3PO_4 by the urine is about 2.75 grams (42.5 grains) in 24 hours; of which two-thirds are in combination with Na and K; and one-third with Ca and Mg. The hourly elimination follows about the same variation as that of the chlorides. The total elimination is greater with animal than with vegetable food; is diminished during pregnancy; and is above the normal during excessive mental work. The elimination of earthy phosphates is greatly increased in osteomalacia, often so far that they are in excess of the alkaline phosphates.

So long as the urine is *acid*, it contains the soluble acid phosphates; when the reaction becomes alkaline, or even on loss of CO_2 by exposure to air, the acid phosphate is converted into the insoluble $Ca_3(PO_4)_2$. Alkaline urines are for this reason almost always turbid, and become clear on the addition of acid. It is in such urine that phosphatic calculi are invariably formed, usually about a nucleus of uric acid or of a foreign body. If the alkalinity be due to the formation of ammonia, the trimagnesic phosphate is not formed, but ammonio-magnesian phosphate (*q. v.*).

Quantitative determination of phosphates in urine.—A process for determining the quantity of phosphates in urine is based upon the formation of the insoluble uranium phosphate, and upon the production of a

150

brown color when a solution of a uranium salt is brought in contact with a solution of potassium ferrocyanide. Four solutions are required: (1) *a standard solution of disodic phosphate,* made by dissolving 10.085 grams of crystallized, non-effloresced HNa_2PO_4 in H_2O, and diluting to a litre; (2) *an acid solution of sodium acetate,* made by dissolving 100 grams sodium acetate in H_2O, adding 100 c.c. glacial acetic acid, and diluting with H_2O to a litre; (3) *a strong solution of potassium ferrocyanide*; (4) *a standard solution of uranium acetate,* made by dissolving 20.3 grams of yellow uranic oxide in glacial acetic acid, and diluting with H_2O to nearly a litre. Solution 1 serves to determine the true strength of this solution, as follows: 50 c.c. of Solution 1 are placed in a beaker, 5 c.c. of Solution 2 are added, the mixture heated on a water-bath, and the uranium solution gradually added from a burette until a drop from the beaker produces a brown color when brought in contact with a drop of the ferrocyanide solution. At this point the reading of the burette, which indicates the number of c.c. of the uranium solution, corresponding to 0.1—P_2O_5, is taken. A quantity of H_2O, determined by calculation from the result thus obtained, is then added to the remaining uranium solution, such as to render each c.c. equivalent to 0.005 gram P_2O_5.

To determine the *total phosphates* in a urine: 50 c.c. are placed in a beaker, 5 c.c. sodium acetate solution are added; the mixture is heated on the water-bath, and the uranium solution delivered from a burette until a drop, removed from the beaker and brought in contact with a drop of ferrocyanide solution, produces a brown tinge. The burette reading, multiplied by 0.005, gives the amount of P_2O_5 in 50 c.c urine; and this, multiplied by $\frac{1}{50}$ the amount of urine passed in 24 hours, gives the daily elimination.

To determine the *earthy phosphates,* a sample of 100 c.c. urine is rendered alkaline with NH_4HO and set aside for 12 hours; the precipitate is then collected upon a filter, washed with ammoniacal water, brought into a beaker, dissolved in a small quantity of acetic acid; the solution diluted to 50 c.c. with H_2O, treated with 5 c.c. sodium acetate solution, and the amount of P_2O_5 determined as above.

Calcium Carbonate—$\mathbf{CaCO_3}$—100—the most abundant of the natural compounds of Ca, exists as *limestone, calcspar, chalk, marble, Iceland spar,* and *arragonite*; and forms the basis of corals, shells of crustacea and of molluscs, etc.

The *precipitated chalk—Calcii carbonas præcipitata* (*U. S.*; *Br.*)—is prepared by precipitating a solution of $CaCl_2$ with one of Na_2CO_3. *Prepared chalk—Creta præparata* (*U. S.*; *Br.*)—is native chalk, purified by grinding with H_2O, diluting, allowing the coarser particles to subside, decanting the still turbid liquid, collecting, and drying the finer particles; a process known as *elutriation.*

It is a white powder, almost insoluble in pure H_2O; much more soluble in H_2O containing carbonic acid, the solution being regarded as containing *hydrocalcic carbonate* $H_2Ca(CO_3)_2$. At a red heat it yields CO_2 and CaO. It is decomposed by acids with liberation of CO_2.

PHYSIOLOGICAL.—Calcium carbonate is much more abundant in the lower than in the higher forms of animal life. It occurs in the egg-shells of birds, in the bones and teeth of all animals; in solution in the saliva and urine of the herbivora, and deposited in the crystalline form, as *otoliths,* in the internal ear of man. It is deposited pathologically in calcifications, in parotid calculi, and occasionally in human urinary calculi and sediments.

Calcium Oxalate—*Oxalate of lime*—$\mathbf{CaC_2O_4}$—128—exists in the

151

sap of many plants, and is formed as a white, crystalline precipitate, by double decomposition between a Ca salt and an alkaline oxalate. It is insoluble in H_2O, acetic acid, or NH_4HO; soluble in the mineral acids and in solution of H_2NaPO_4.

PHYSIOLOGICAL.—Calcium oxalate is taken into the body in vegetable food, and is formed in the economy, where its production is intimately connected with that of uric acid.

It occurs in the urine, in which it is increased in quantity when large amounts of vegetable food are taken; when sparkling wines or beers are indulged in; and when the carbonates of the alkalies, lime-water and lemon-juice, are administered. It is deposited as a urinary sediment in the form of small, brilliant octahedra, having the appearance of the backs of square letter-envelopes; or in dumb-bells. It is usually deposited from acid urine, and accompanied by crystals of uric acid. Sometimes, however, it occurs in urines undergoing alkaline fermentation, in which case it is accompanied by crystals of ammonio-magnesian phosphate.

The renal or vesical calculi of calcium oxalate, known as *mulberry calculi,* are dark brown or gray, very hard, occasionally smooth, generally tuberculated, soluble in HCl without effervescence; and when ignited, they blacken, turn white, and leave an alkaline residue.

Analytical Characters.

(1.) Ammonium sulphydrate: nothing, unless the Ca salt be the phosphate, oxalate or fluoride, when it forms a white ppt.

(2.) Alkaline carbonates: white ppt.; not prevented by the presence of ammoniacal salts.

(3.) Ammonium oxalate: white ppt.; insoluble in acetic acid; soluble in HCl, or HNO_3.

(4.) Sulphuric acid: white ppt., from solutions which are not too dilute; very sparingly soluble in H_2O; insoluble in alcohol; soluble in sodium hyposulphite solution.

(5.) Sodium tungstate: dense white ppt., even from dilute solutions.

(6.) Colors the flame of the Bunsen burner reddish-yellow, and exhibits a spectrum of a number of bright bands, the most prominent of which are: $\lambda = 6265, 6202, 6181, 6044, 5982, 5933, 5543$, and 5517.

STRONTIUM.

Symbol = **Sr**—*Atomic weight* = 87.4—*Sp. gr.* = 2.54.

An element, not as abundant as Ba, occurring principally in the minerals *strontianite* (CO_3Sr) and *celestine* (SO_4Sr). Its compounds resemble those of Ca and Ba. Its nitrate is used in making red fire.

Analytical characters.—(1.) Behaves like Ba with alkaline carbonates and PO_4Na_2H. (2.) Calcium sulphate: a white ppt. which forms slowly; accelerated by addition of alcohol. (3.) The Sr compounds color the Bunsen flame red, or, as observed through blue glass, purple or rose color. The Sr flame gives a spectrum of many bands, of which the most prominent are: $\lambda = 6694, 6664, 6059, 6031, 4607$.

BARIUM.

Symbol = **Ba**—*Atomic weight* = 136.8—*Molecular weight* = 273.6 (?)—*Sp. gr.* = 4.0—*Discovered by Davy*, 1808—*Name from* βαρὺς = heavy.

Occurs only in combination, principally as *heavy spar* ($BaSO_4$) and *witherite* ($BaCO_3$). It is a pale yellow, malleable metal, quickly oxidized in air, and decomposing H_2O at ordinary temperatures.

152

Compounds of Barium.

Oxides.—Barium Monoxide—**BaO**—152.8—is prepared by calcining the nitrate. It is a grayish-white or white, amorphous, caustic solid. In air it absorbs moisture and CO_2, and combines with H_2O as does CaO.

Barium Dioxide—**BaO_2**—168.8—is prepared by heating the monoxide in O. It is a grayish-white, amorphous solid. Heated in air it is decomposed: $BaO_2 = BaO + O$. Aqueous acids dissolve it with formation of a barytic salt and H_2O_2.

Barium Monohydrate—*Caustic baryta*—**BaH_2O_2**—170.8—is prepared by the action of H_2O on BaO. It is a white, amorphous solid, soluble in H_2O. Its aqueous solution, *baryta water*, is alkaline, and absorbs CO_2 with formation of a white deposit of $BaCO_3$.

Barium Chloride—$BaCl_2$ + Aq—207.8 + 36—is obtained by treating BaS or $BaCO_3$ with HCl. It crystallizes in prismatic plates, permanent in air, soluble in H_2O.

Salts of Barium.

Barium Nitrate—$Ba(NO_3)_2$—260.8—is prepared by neutralizing HNO_3 with $BaCO_3$. It forms octahedral crystals, soluble in H_2O.

Barium Sulphate—$BaSO_4$—232.8—occurs in nature as *heavy spar* and is formed as an amorphous, white powder, insoluble in acids, by double decomposition between a Ba salt and a sulphate in solution. It is insoluble in H_2O and in acids. It is used as a pigment, *permanent white*.

Barium Carbonate—$BaCO_3$—196.8—occurs in nature as *witherite*, and is formed by double decomposition between a Ba salt and a carbonate in alkaline solution. It is a heavy, amorphous, white powder, insoluble in H_2O, soluble with effervescence in acids.

Analytical Characters.

(1.) Alkaline carbonates: white ppt., in alkaline solution.

(2.) Sulphuric acid, or calcium sulphate: white ppt.; insoluble in acids.

(3.) Sodium phosphate: white ppt.; soluble in HNO_3.

(4.) Colors the Bunsen flame greenish-yellow, and exhibits a spectrum of several lines, the most prominent of which are: $\lambda = 6108, 6044, 5881, 5536$.

Action on the Economy.

The oxides and hydrate act as corrosives by virtue of their alkalinity, and also as true poisons. All soluble compounds of Ba, and those which are readily converted into soluble compounds in the stomach, are actively poisonous. Soluble sulphates, followed by emetics, are indicated as antidotes.

IV. MAGNESIUM GROUP.

Magnesium—Zinc—Cadmium.

Each of these elements forms a single oxide—a corresponding basic hydrate, and a series of salts in which its atoms are bivalent.

153

MAGNESIUM.

Symbol = **Mg**—*Atomic weight* = 24—*Molecular weight* = 48 (?)—*Sp. gr.* = 1.75—*Fuses at* 1000° (1832° F.)—*Discovered by Davy*, 1808.

Occurs as carbonate in *dolomite* or *magnesian limestone*, and as silicate in *mica, asbestos, soapstone, meerschaum, talc*, and in other minerals. It also accompanies Ca in the forms in which it is found in the animal and vegetable worlds.

It is prepared by heating its chloride with Na. It is a hard, light, malleable, ductile, white metal. It burns with great brilliancy when heated in air (magnesium light), but may be distilled in H. It decomposes vapor of H_2O when heated; reduces CO_2 with the aid of heat, and combines directly with Cl, S, P, As, and N. It dissolves in dilute acids, but is not affected by alkaline solutions.

Compounds of Magnesium.

Magnesium Oxide—*Calcined magnesia—magnesia* (*U. S.; Br.*)—**MgO**—40—is obtained by calcining the carbonate, hydrate, or nitrate. It is a light, bulky, tasteless, odorless, amorphous, white powder; alkaline in reaction; almost insoluble in H_2O; readily soluble without effervescence in acids.

Magnesium Hydrate—MgH_2O_2—58—occurs in nature, and is formed when a solution of a Mg salt is precipitated with excess of NaHO in absence of ammoniacal salts. It is a heavy, white powder, insoluble in H_2O; absorbs CO_2.

Magnesium Chloride—$MgCl_2$—95—is formed when MgO or $MgCO_3$ is dissolved in HCl. It is an exceedingly deliquescent, soluble substance, which is decomposed into HCl and MgO when its aqueous solutions are evaporated to dryness.

Salts of Magnesium.

Magnesium Sulphate—*Epsom salt—Sedlitz salt—Magnesii sulphas* (*U. S.*)—*Magnesiæ sulphas* (*Br.*)—**$MgSO_4$ + 7 Aq**—120 + 126—exists in solution in sea-water and in the waters of many mineral springs, especially those known as *bitter waters*. It is formed by the action of H_2SO_4 on Mg CO_3. It crystallizes in right rhombic prisms; bitter; slightly effervescent, and quite soluble in H_2O. Heated, it fuses and gradually loses 6 Aq up to 132° (269°.6 F.); the last Aq it loses at 210° (410° F.).

Phosphates.—Resemble those of Ca in their constitution and properties, and accompany them in the situations in which they occur in the animal body, but in much smaller quantity.

Magnesium also forms double phosphates, constituted by the substitution of one atom of the bivalent metal for two of the atoms of basic hydrogen, of a molecule of phosphoric acid and of an atom of an alkaline metal, or of an ammonium group, for the remaining basic hydrogen.

Ammonio-Magnesian Phosphate—*Triple phosphate*—**$Mg(NH_4)PO_4$ + 6 Aq**—137 + 108—is produced when an alkaline phosphate and NH_4HO are

154

added to a solution containing Mg. When heated it is converted into magnesium pyrophosphate $Mg_2P_2O_7$, in which form H_3PO_4 and Mg are usually weighed in quantitative analysis.

In the urine, alkaline phosphates and magnesium salts are always present, and consequently when, by decomposition of urea, the urine becomes alkaline, the conditions for the formation of this compound are fulfilled; and being practically insoluble, especially in the presence of excess of phosphates and of ammonia, it is deposited in crystals, usually tabular, sometimes feathery and stellate in form. When it is formed in the bladder, in the presence of some body to serve as a nucleus, the crystallization takes place upon the nucleus and a *fusible calculus* is produced.

Carbonates.—Magnesium Carbonate—*Neutral carbonate*—**$MgCO_3$**—84—exists native in *magnesite*, and, combined with $CaCO_3$, in *dolomite*. It cannot be formed, like other carbonates, by decomposing a Mg salt with an alkaline carbonate, but may be obtained by passing CO_2 through H_2O holding tetramagnesic tricarbonate in suspension.

Trimagnesic Dicarbonate—**$(MgCO_3)_2MgH_2O_2$ + 2 Aq**—226 + 36—is formed in small crystals when a solution of $MgSO_4$ is precipitated with excess of Na_2CO_3 and the mixture boiled.

Tetramagnesic Tricarbonate—*Magnesia alba—Magnesii carbonas* (*U. S.*)—*Magnesiæ carbonas* (*Br.*)—3**$(MgCO_3)MgH_2O_2$**+3 **Aq**—310+54—occurs in commerce in light, white cubes, composed of a powder which is amorphous or partly crystalline. It is prepared by precipitating a solution of $MgSO_4$ with one of Na_2CO_3; if the precipitation occur in cold dilute solutions (*Magnesiæ carbonas lævis, Br.*), very little CO_2 is given off; a light, bulky precipitate falls, and the solution contains magnesium, probably in the form of the bicarbonate $Mg(HCO_3)_2$; this solution, on standing, deposits crystals of the carbonate, $MgCO_3$ + 3 Aq. If hot concentrated solutions be used and the liquid then boiled upon the precipitate, CO_2 is given off, and a denser, heavier precipitate is formed, which varies in composition according to the length of time during which the boiling is continued, and to the presence or absence of excess of sodium carbonate. The pharmaceutical product frequently contains $4(MgCO_3),MgH_2O_2 + 4H_2O$, or even $2(MgCO_3),MgH_2O_2 + 2H_2O$. All of these compounds are very sparingly soluble in H_2O, but much more soluble in H_2O containing ammoniacal salts.

Analytical Characters.

(1.) Ammonium hydrate: voluminous, white ppt. from neutral solutions.

(2.) Potash or soda: voluminous, white ppt. from warm solutions; prevented by the presence of NH_4 salts and of certain organic substances.

(3.) Ammonium carbonate: slight ppt. from hot solutions; prevented by the presence of NH_4 salts.

(4.) Sodium or potassium carbonate: white ppt., best from hot solution; prevented by the presence of NH_4 compounds.

(5.) Disodic phosphate: white ppt. in hot, not too dilute solutions.

(6.) Oxalic acid: nothing alone, but in presence of NH_4HO a white ppt.; not formed in presence of NH_4Cl or salts of NH_4.

155

ZINC.

Symbol = **Zn**—*Atomic weight* = 64.9—*Molecular weight* = 64.9—*Sp. gr.* = 6.862—7.215—*Fuses at* 415° (779° *F.*)—*Distils at* 1040° (1904° *F.*).

Occurs principally in *calamine* ($ZnCO_3$); and *blende* (ZnS); also as oxide and silicate; never free. It is separated from its ores by calcining, roasting, and distillation.

It is a bluish-white metal; crystalline, granular, or fibrous; quite malleable and ductile when pure. The commercial metal is usually brittle. At 130°–150° (266°–302° F.) it is pliable, and becomes brittle again above 200°–210° (392°–410° F.).

At 500° (932° F.) it burns in air with a greenish-white flame, and gives off snowy white flakes of the oxide (*lana philosophica; nil album; pompholix*). In moist air it becomes coated with a film of hydrocarbonate. It decomposes steam when heated.

Pure H_2SO_4 and pure Zn do not react together in the cold; if the acid be diluted, however, it dissolves the Zn with evolution of H and formation of $ZnSO_4$, in the presence of a trace of Pt or Cu. The commercial metal dissolves readily in dilute H_2SO_4, with evolution of H and formation of $ZnSO_4$, the action being accelerated in presence of Pt, Cu, or As. Zinc surfaces thoroughly coated with a layer of an amalgam of Hg and Zn are only attacked by H_2SO_4 if they form part of closed galvanic circuit; hence the zincs of galvanic batteries are protected by *amalgamation*. Zinc also decomposes HNO_3, HCl, and acetic acid.

When required for toxicological analysis, zinc must be perfectly free from As and sometimes from P. It is better to test samples until a pure one is found than to attempt the purification of a contaminated metal.

Zinc surfaces are readily attacked by weak organic acids; vessels of *galvanized iron* or *sheet zinc* should therefor never be used to contain articles of food or medicines.

Compounds of Zinc.

Zinc Oxide—*Zinci oxidum* (*U. S.; Br.*)—**ZnO**—80.9—is prepared either by calcining the precipitated carbonate, or by burning Zn in a current of air. An impure oxide, known as *tutty*, is deposited in the flues of zinc furnaces and in those in which brass is fused. When obtained by calcination of the carbonate, it forms a soft, white, tasteless, and odorless powder; when produced by burning the metal, it occurs in light, voluminous, white masses. It is neither fusible, volatile, nor decomposable by heat, and is completely insoluble in neutral solvents. It dissolves in dilute acids, with formation of the corresponding salts.

It is used in the arts as a white pigment in place of lead carbonate, and is not darkened by H_2S.

Zinc Hydrate—ZnH_2O_2—98.9—is not formed by union of ZnO and H_2O; but is produced when a solution of a Zn salt is treated with KHO. Freshly prepared, it is very soluble in alkalies and in solutions of NH_4 salts.

Zinc Chloride—*Butter of zinc*—*Zinci chloridum* (*U. S.; Br.*)—$ZnCl_2$ **+ Aq**—135.9 + 18—is obtained by dissolving Zn in HCl; or by heating Zn in Cl. It is a soft, white, very deliquescent, fusible, volatile mass;

156

very soluble in H_2O, somewhat less so in alcohol. Its solution has a burning metallic taste; destroys vegetable tissues; dissolves silk; and exerts a strong dehydrating action upon organic substances in general.

In dilute solution it is used as a disinfectant and antiseptic (*Burnett's fluid*), as a preservative of wood and as an embalming injection.

Salts of Zinc.

Zinc Sulphate—*White vitriol*—*Zinci sulphas* (*U. S.; Br.*)—$ZnSO_4$ $+nAq$—160.9$+n$18—is formed when Zn, ZnO, ZnS, or $ZnCO_3$ is dissolved in diluted H_2SO_4. It crystallizes below 30° (86° F.) with 7 Aq; at 30° (86° F.) with 6 Aq; between 40°–50° (104°–122° F.) with 5 Aq; at 0° (32° F.) from concentrated acid solution with 4 Aq; from a boiling solution it is precipitated by concentrated H_2SO_4 with 2 Aq; from a saturated solution at 100° (212° F.) with 1 Aq; and anhydrous when the salt with 1 Aq is heated to 238° (460° F.).

The salt usually met with is that with 7 Aq, which is in large, colorless, four-sided prisms; efflorescent; very soluble in H_2O; sparingly soluble in weak alcohol. Its solutions have a strong, styptic taste; coagulate albumin when added in moderate quantity, the coagulum dissolving in an excess; and form insoluble precipitates with the tannins.

Carbonates.—Zinc Carbonate—$ZnCO_3$—124.9—occurs in nature as *calamine*. If an alkaline carbonate be added to a solution of a Zn salt, the neutral carbonate, as in the case of Mg, is not formed, but an oxycarbonate, $nZnCO_3$, $nZnH_2O_2$ [*Zinci carbonas* (*U. S.; Br.*)], whose composition varies with the conditions under which it is formed.

Analytical Characters.

(1.) Hydrate of K, Na or NH_4: white ppt., soluble in excess.

(2.) Carbonate of K or Na: white ppt., in absence of NH_4 salts.

(3.) Hydrogen sulphide, in neutral solution: white ppt. In presence of an excess of a mineral acid, the formation of this ppt. is prevented unless sodium acetate be also present.

(4.) Ammonium sulphydrate: white ppt., insoluble in excess, in KHO, NH_4HO, or acetic acid; soluble in dilute mineral acids.

(5.) Ammonium carbonate: white ppt., soluble in excess.

(6.) Disodic phosphate, in absence of NH_4 salts: white ppt., soluble in acids or alkalies.

(7.) Potassium ferrocyanide: white ppt., insoluble in HCl.

Action on the Economy.

All the compounds of Zn which are soluble in the digestive fluids behave as true poisons; and solutions of the chloride (in common use by tinsmiths, and in disinfecting fluids) have also well-marked corrosive properties. When Zn compounds are taken, it is almost invariably by mistake for other substances: the sulphate for Epsom salt, and solutions of the chloride for various liquids, gin, fluid magnesia, vinegar, etc.

Metallic zinc is dissolved by solutions containing NaCl, or organic acids, for which reason articles of food kept in vessels of galvanized iron

157

become contaminated with zinc compounds, and, if eaten, produce more or less intense symptoms of intoxication. For the same reason materials intended for analysis, in cases of supposed poisoning, *should never be packed in jars closed by zinc caps.*

CADMIUM.

Symbol = **Cd**—*Atomic weight* = 111.8—*Molecular weight* = 111.8—*Sp. gr.* =8.604—*Fuses at* 227°.8 (442° F.)—*Boils at* 860° (1580° F.).

A white metal, malleable and ductile at low temperature, brittle when heated; which accompanies Zn in certain of its ores. It resembles zinc in its physical as well as its chemical characters. It is used in certain fusible alloys, and its iodide is used in photography.

Analytical Characters.—Hydrogen sulphide: bright yellow ppt.; insoluble in NH_4HS and in dilute acids and alkalies, soluble in boiling HNO_3 or HCl.

V. NICKEL GROUP.

Nickel—Cobalt.

These two elements bear some resemblance chemically to those of the Fe group; from which they differ in forming, so far as known, no compounds similar to the ferrates, chromates, and manganates. They form compounds corresponding to Fe_2O_3, but those corresponding to the ferric series are either wanting or exceedingly unstable.

NICKEL.

Symbol = **Ni**—*Atomic weight* = 58—*Sp. gr.* = 8.637.

Occurs in combination with S, and with S and As.

It is a white metal, hard, slightly magnetic, not tarnished in air. *German silver* is an alloy of Ni, Cu, and Zn. Its salts are green.

Analytical Characters.

(1.) Ammonium sulphydrate: black ppt.; insoluble in excess.

(2.) Potash or soda: apple-green ppt., in absence of tartaric acid; insoluble in excess.

(3.) Ammonium hydrate: apple-green ppt.; soluble in excess, forming a violet solution which deposits the apple-green hydrate when heated with KHO.

COBALT.

Symbol = **Co**—*Atomic weight* = 58.9—*Sp. gr.* = 8.5–8.7.

Occurs in combination with As and S. Its salts are red when hydrated, and usually blue when anhydrous. Its phosphate is used as a blue pigment.

158

Analytical Characters.

(1.) Ammonium sulphydrate: black ppt.; insoluble in excess.

(2.) Potash: blue ppt.; turns red, slowly in the cold, quickly when heated; not formed in the cold in presence of NH_4 salts.

(3.) Ammonium hydrate: blue ppt.; turns red in absence of air, green in its presence.

VI. COPPER GROUP.

Copper—Mercury.

Each of these elements forms two series of compounds: one contains compounds of the bivalent group $\left(\begin{matrix}Cu\\|\\Cu\end{matrix}\right)''$ or $(Hg_2)''$ which are designated by the termination *ous;* the other contains compounds of single, bivalent atoms Cu'' or Hg'', which are designated by the termination *ic.*

COPPER.

Symbol = **Cu (CUPRUM)**—*Atomic weight* = 63.1—*Molecular weight* = 127 (?)—*Sp. gr.* = 8.914–8.952—*Fuses at* 1091° (1996° *F.*).

Occurrence.—It is found free in crystals or amorphous masses, sometimes of great size; also a sulphide, *copper pyrites;* oxide, *ruby ore* and *black oxide;* and basic carbonate, *malachite.*

Properties.—*Physical.*—A yellowish-red metal; dark brown when finely divided; very malleable, ductile, and tenacious; a good conductor of heat and electricity; has a peculiar, metallic taste and a characteristic odor.

Chemical.—It is unaltered in dry air at the ordinary temperature; but when heated to redness is oxidized to CuO. In damp air it becomes coated with a brownish film of oxide; a green film of basic carbonate; or, in salt air, a green film of basic chloride. Hot H_2SO_4 dissolves it with formation of $CuSO_4$ and SO_2; it is dissolved by HNO_3 with formation of $Cu(NO_3)_2$ and NO; and by HCl with liberation of H. Weak acids form with it soluble salts in presence of air and moisture. It is dissolved by NH_4HO, in presence of air, with formation of a blue solution. It combines directly with Cl, frequently with light.

Compounds of Copper.

Oxides.—Cuprous Oxide—*Suboxide or red oxide of copper*—$(Cu_2)O$—142.4—is formed by calcining a mixture of $(Cu_2)Cl_2$ and Na_2CO_3; or a mixture of CuO and Cu. It is a red or yellow powder; permanent in air; sp. gr. 5.749–6.093; fuses at a red heat; easily reduced by C or H. Heated in air it is converted into CuO.

Cupric Oxide—*Binoxide or black oxide of copper*—**CuO**—79.2—is prepared by heating Cu to dull redness in air; or by calcining $Cu(NO_3)_2$; or by prolonged boiling of the liquid over a precipitate produced by heating

159

a solution of a cupric salt, in presence of glucose, with KHO. By the last method it is sometimes produced in Trommer's test for sugar, when an excessive quantity of $CuSO_4$ has been used.

It is a black, or dark reddish-brown, amorphous solid; readily reduced by C, H, Na, or K at comparatively low temperatures. When heated with organic substances it gives up its O, converting the C into CO_2 and the H into H_2O: $C_2H_6O + 6CuO = 6Cu + 2CO_2 + 3H_2O$; a property which renders it valuable in organic analysis, as by heating a known weight of organic substance with CuO and weighing the amount of CO_2 and H_2O produced, the percentage of C and H may be obtained. It dissolves in acids with formation of salts.

Hydrates.—Cuprous Hydrate—$(Cu)_2H_2O_2$ (?)—160.4 (?)—is formed as a yellow or red powder when mixed solutions of $CuSO_4$ and KHO are heated in presence of glucose. By boiling the solution it is rapidly dehydrated with formation of $(Cu_2)O$.

Cupric Hydrate—CuH_2O_2—97.2—is formed by the action of KHO upon solution of $CuSO_4$, in absence of reducing agents and in the cold. It is a bluish, amorphous powder; very unstable, and readily dehydrated, with formation of CuO.

Sulphides.—Cuprous Sulphide—*Subsulphide or protosulphide of copper*—Cu_2S—158.4—occurs in nature as *copper glance* or *chalcosine,* and in many double sulphides, *pyrites.*

Cupric Sulphide—CuS—95.2—is formed by the action of H_2S or of NH_4HS on solutions of cupric salts. It is almost black when moist, greenish-brown when dry. Hot HNO_3 oxidizes it to $CuSO_4$; hot HCl converts it into $CuCl_2$, with separation of S, and formation of H_2S. It is sparingly soluble in NH_4HS, its solubility being increased by the presence of organic matter.

Chlorides.—Cuprous Chloride—*Subchloride or protochloride*—$(Cu_2)Cl_2$—197.4—is prepared by heating Cu with one of the chlorides of Hg; by dissolving $(Cu_2)O$ in HCl, without contact of air; or by the action of reducing agents on solutions of $CuCl_2$. It is a heavy, white powder; turns violet and blue by exposure to light; soluble in HCl; insoluble in H_2O. It forms a crystallizable compound with CO; and its solution in HCl is used in analysis to absorb that gas.

Cupric Chloride—*Chloride or deutochloride*—$CuCl_2$—134.2—is formed by dissolving Cu in aqua regia; if the Cu be in excess, it reduces $CuCl_2$ to $(Cu_2)Cl_2$. It crystallizes in bluish-green, rhombic prisms with 2 Aq; deliquescent; very soluble in H_2O and in alcohol.

Salts of Copper.

Cupric Nitrate—$Cu(NO_3)_2$—187.2—is formed by dissolving Cu, CuO, or $CuCO_3$ in HNO_3. It crystallizes at 20°–25° (68°–77° F.) with 3 Aq; below 20° (68° F.) with 6 Aq, forming blue, deliquescent needles. Strongly heated, it is converted into CuO.

Cupric Sulphate—*Blue vitriol—Blue stone—Cupri sulphas* (*U. S.; Br.*)—$CuSO_4 + 5$ **Aq**—159.2 + 90—is prepared: (1) by roasting CuS; (2) from the water of copper mines; (3) by exposing Cu, moistened with dilute H_2SO_4, to air; (4) by heating Cu with H_2SO_4.

As ordinarily crystallized, it is in fine, blue, oblique prisms; soluble in H_2O; insoluble in alcohol; efflorescent in dry air at 15° (59° F.), losing 2 Aq. At 100° (212° F.) it still retains 1 Aq, which it loses at 230° (446°

160

F.), leaving a white, amorphous powder of the anhydrous salt, which, on taking up H_2O, resumes its blue color. Its solutions are blue, acid, styptic, and metallic in taste.

When NH_4HO is added to a solution of $CuSO_4$, a bluish-white precipitate falls, which redissolves in excess of the alkali, to form a deep blue solution; strong alcohol floated over the surface of this solution separates long, right rhombic prisms, having the composition $CuSO_4,4NH_3 + H_2O$, which are very soluble in H_2O. This solution constitutes *ammonio-sulphate of copper* or *aqua sapphirina.*

Arsenite—*Scheele's green—Mineral green*—is a mixture of cupric arsenite and hydrate; prepared by adding potassium arsenite to solution of $CuSO_4$. It is a grass-green powder, insoluble in H_2O; soluble in NH_4HO, or in acids. Exceedingly poisonous.

Schweinfurt Green—*Mitis green* or *Paris green*—is the most frequently used, and the most dangerous of the cupro-arsenical pigments. It is prepared by adding a thin paste of neutral cupric acetate with H_2O to a boiling solution of arsenious acid, and continuing the boiling during a further addition of acetic acid. It is an insoluble, green, crystalline powder, having the composition $(C_2H_3O_2)_2Cu + 3(As_2O_4Cu)$. It is decomposed by prolonged boiling in H_2O, by aqueous solutions of the alkalies, and by the mineral acids.

Carbonates.—The existence of cuprous carbonate is doubtful. *Cupric carbonate*—$CuCO_3$—exists in nature, but has not been obtained artificially. *Dicupric carbonate*—$CuCO_3,CuH_2O_2$—exists in nature as *malachite.* When a solution of a cupric salt is decomposed by an alkaline carbonate, a bluish precipitate, having the composition $CuCO_3,CuH_2O_2 + H_2O$, is formed, which, on drying, loses H_2O, and becomes green; it is used as a pigment under the name *mineral green.* *Tricupric carbonate—Sesquicarbonate of copper*—$2(CuCO_3),CuH_2O_2$—exists in nature as a blue mineral called *azurite* or *mountain blue,* and is prepared by a secret process for use as a pigment known as *blue ash.*

Acetates.—Cupric Acetate—*Diacetate—Crystals of Venus—Cupri acetas* (*U. S.*)—$Cu(C_2H_3O_2)_2$ + **Aq**—181.2 + 18—is formed when CuO or verdigris is dissolved in acetic acid; or by decomposition of a solution of $CuSO_4$ by $Pb(C_2H_3O_2)_2$. It crystallizes in large, bluish-green prisms, which lose their Aq at 140° (284° F.). At 240°–260° (464°–500° F.) they are decomposed with liberation of glacial acetic acid.

Basic Acetates—*Verdigris*—is a substance prepared by exposing to air piles composed of alternate layers of grape-skins and plates of copper, and removing the bluish-green coating from the copper. It is a mixture, in varying proportions, of three different substances: $(C_2H_3O_2)_2CuH_2O_2$ + 5 Aq; $[(C_2H_3O_2)_2Cu]_2,CuH_2O_2$ + 5 Aq; and $(C_2H_3O_2)_2Cu,2(CuH_2O_2)$.

Analytical Characters.

Cuprous—are very unstable and readily converted into cupric compounds.

(1.) Potash: white ppt.; turning brownish.

(2.) Ammonium hydrate, in absence of air: a colorless liquid; turns blue in air.

Cupric—are white when anhydrous; when soluble in H_2O they form blue or green, acid solutions.

(1.) Hydrogen sulphide: black ppt.; insoluble in KHS or NaHS; spar-

161

ingly soluble in NH_4HS; soluble in hot concentrated HNO_3 and in KCN.

(2.) Alkaline sulphydrates: same as H_2S.

(3.) Potash or soda: pale blue ppt.; insoluble in excess. If the solution be heated over the ppt., the latter contracts and turns black.

(4.) Ammonium hydrate, in small quantity: pale blue ppt.; in larger quantity, deep blue solution.

(5.) Potassium or sodium carbonate: greenish-blue ppt.; insoluble in excess; turning black when the liquid is boiled.

(6.) Ammonium carbonate: pale blue ppt.; soluble with deep blue color in excess.

(7.) Potassium cyanide: greenish-yellow ppt.; soluble in excess.

(8.) Potassium ferrocyanide: chestnut-brown ppt.; insoluble in weak acids; decolorized by KHO.

(9.) Iron is coated with metallic Cu.

Action on the Economy.

The opinion, until recently universal among toxicologists, that all the compounds of copper are poisonous, has been much modified by recent researches. Certain of the copper compounds, such as the sulphate, having a tendency to combine with albuminoid and other animal substances, produce symptoms of irritation by their direct local action, when brought in contact with the gastric or intestinal mucous membrane. One of the characteristic symptoms of such irritation is the vomiting of a greenish matter, which develops a blue color upon the addition of NH_4HO.

Cases are not wanting in which severe illness, and even death, has followed the use of food which has been in contact with imperfectly tinned copper vessels; cases in which nervous and other symptoms referable to a truly poisonous action have occurred. As, however, it has also been shown that non-irritant, *pure* copper compounds may be taken in considerable doses with impunity, it appears at least probable that the poisonous action attributed to copper is due to other substances. The tin and solder used in the manufacture of copper utensils contain lead, and in some cases of so-called copper-poisoning, the symptoms have been such as are as consistent with lead-poisoning as with copper-poisoning. Copper is also notoriously liable to contamination with arsenic, and it is by no means improbable that compounds of that element are the active poisonous agents in some cases of supposed copper-intoxication. Nor is it improbable that articles of food allowed to remain exposed to air in copper vessels should undergo those peculiar changes which result in the formation of poisonous substances, such as the sausage- or cheese-poisons, or the ptomaines.

The treatment, when irritant copper compounds have been taken, should consist in the administration of white of egg or of milk, with whose albuminoids an inert compound is formed by the copper salt. If vomiting do not occur spontaneously, it should be induced by the usual methods.

The detection of copper in the viscera after death is not without interest, especially if arsenic have been found, in which case its discovery or non-discovery enables us to differentiate between poisoning by the arsenical greens and that by other arsenical compounds. The detection of mere traces of copper is of no significance, because, although copper is

162

not a physiological constituent of the body, it is almost invariably present, having been taken with the food.

Pickles and canned vegetables are sometimes intentionally *greened* by the addition of copper; this fraud is readily detected by inserting a large needle into the pickle or other vegetable; if copper be present the steel will be found to be coated with copper after half an hour's contact.

MERCURY.

Symbol = **Hg (HYDRARGYRUM)**—*Atomic weight* = 199.7 — *Molecular weight* = 199.7—*Sp. gr. of liquid* = 13.596; *of vapor* = 6.97—*Fuses at* −38°.8 (−37°.9 F.)—*Boils at* 350° (662° F.).

Occurrence.—Chiefly as *cinnabar* (HgS); also in small quantity free and as chloride.

Preparation.—The commercial product is usually obtained by simple distillation in a current of air: $HgS + O_2 = Hg + SO_2$. If required pure, it must be freed from other metals by distillation, and agitation of the redistilled product with mercurous nitrate solution, solution of Fe_2Cl_6, or dilute HNO_3.

Properties.—*Physical.*—A bright metallic liquid; volatile at all temperatures. Crystallizes in octahedra of sp. gr. 14.0. When pure it rolls over a smooth surface in round drops; the formation of tear-shaped drops indicates the presence of impurities.

Chemical.—If pure it is not altered by air at the ordinary temperature, but if contaminated with foreign metals its surface becomes dimmed. Heated in air it is oxidized superficially to HgO. It does not decompose H_2O. It combines directly with Cl, Br, I and S. It alloys readily with most metals to form *amalgams*. It amalgamates with Fe and Pt only with difficulty. Hot concentrated H_2SO_4 dissolves it with evolution of SO_2 and formation of $HgSO_4$. It dissolves in cold HNO_3 with formation of a nitrate.

Elementary mercury is insoluble in H_2O, and probably in the digestive liquids. It enters, however, into the formation of three medicinal agents: *hydrargyrum cum creta* (*U. S.; Br.*); *massa hydrargyri* (*U. S.*) = *pilula hydrargyri* (*Br.*); and *unguentum hydrargyri* (*U. S.; Br.*), all of which owe their efficacy, not to the metal itself, but to a certain proportion of oxide produced during their manufacture. The fact that blue mass is more active than mercury with chalk is due to the greater proportion of oxide contained in the former. It is also probable that absorption of vapor of Hg by cutaneous surfaces is attended by its conversion into $HgCl_2$.

Compounds of Mercury.

Oxides.—Mercurous Oxide—*Protoxide or black oxide of mercury*—**$(Hg_2)O$**—415.4—is obtained by adding a solution of $(Hg_2)(NO_3)_2$ to an excess of solution of KHO. It is a brownish-black, tasteless powder; very prone to decomposition into HgO and Hg. It is converted into $(Hg_2)Cl_2$ by HCl; and by other acids into the corresponding mercurous salts.

It is formed by the action of CaH_2O_2 on mercurous compounds, and exists in *black wash.*

163

Mercuric Oxide—*Red, or binoxide of mercury—Hydrargyri oxidum flavum* (*U. S.; Br.*)—*Hydrargyri oxidum rubrum* (*U. S.; Br.*)—**HgO**—215.7—is prepared by two methods: (1) by calcining $Hg(NO_3)_2$ as long as brown fumes are given off (*Hydr. oxid. rubr.*); or, (2) by precipitating a solution of a mercuric salt by excess of KHO (*Hydr. oxid. flavum*). The products obtained, although the same in composition, differ in physical characters and in the activity of their chemical actions. That obtained by (1) is red and crystalline; that obtained by (2) is yellow and amorphous. The latter is much the more active in its chemical and medicinal actions.

It is very sparingly soluble in H_2O, the solution having an alkaline reaction and a metallic taste. It exists both in solution and in suspension in *yellow wash*, prepared by the action of CaH_2O_2 on a mercuric compound.

Exposed to light and air it turns black, more rapidly in presence of organic matter, giving off O and liberating Hg: $HgO = Hg + O$. It decomposes the chlorides of many metallic elements in solution, with formation of a metallic oxide and mercuric oxychlorides. It combines with alkaline chlorides to form soluble double chlorides, called *chloromercurates* or *chlorhydrargyrates;* and forms similar compounds with alkaline iodides and bromides.

Sulphides.—Mercurous Sulphide—**$(Hg_2)S$**—431.4—a very unstable compound, formed by the action of H_2S on mercurous salts.

Mercuric Sulphide—*Red sulphide of mercury—Cinnabar—Vermilion—Hydrargyri sulphidum rubrum* (*U. S.*)—**HgS**—231.7—exists in nature in amorphous red masses, or in red crystals, and is the chief ore of Hg. If Hg and S be ground up together in the cold, or if a solution of a mercuric salt be completely decomposed by H_2S, a black sulphide is obtained, which is the *Æthiops mineralis* of the older pharmacists.

A red sulphide is obtained for use as a pigment (*vermilion*), by agitating for some hours at 60° (140° F.) a mixture of Hg, S, KHO, and H_2O. It is a fine, red powder, which turns brown, and finally black, when heated. Heated in air, it burns to SO_2 and Hg. It is decomposed by strong H_2SO_4, but not by HNO_3 or HCl.

Chlorides.—Mercurous Chloride—*Protochloride or mild chloride of mercury—Calomel—Hydrargyri chloridum mite* (*U. S.*)—*Hydrargyri subchloridum* (*Br.*)—**$(Hg_2)Cl_2$**—470.4—is now principally obtained by mutual decomposition of NaCl and $(Hg_2)SO_4$. Mercuric sulphate is first obtained by heating together 2 pts. Hg and 3 pts. H_2SO_4; the product is then caused to combine with a quantity of Hg equal to that first used, to form $(Hg_2)SO_4$; which is then mixed with dry NaCl, and the mixture heated in glass vessels, connected with condensing chambers; $2NaCl + (Hg_2)SO_4 = Na_2SO_4 + (Hg_2)Cl_2$.

In practice, varying quantities of $HgCl_2$ are also formed, and must be removed from the product by washing with boiled, distilled H_2O until the washings no longer precipitate with NH_4HO. The presence of $HgCl_2$ in calomel may be detected by the formation of a black stain upon a bright iron surface, immersed in the calomel, moistened with alcohol; or by the production of a black color by H_2S in H_2O which has been in contact with and filtered from calomel so contaminated.

Calomel is also formed in a number of other reactions: (1) by the action of Cl upon excess of Hg; (2) by the action of Hg upon Fe_2Cl_6; (3) by the action of HCl, or of a chloride, upon $(Hg_2)O$, or upon a mercurous salt; (4) by the action of reducing agents, including Hg, upon $HgCl_2$.

Calomel crystallizes in nature, and when sublimed, in quadratic prisms. When precipitated it is deposited as a heavy, amorphous, white powder,

164

faintly yellowish, and producing a yellowish mark when rubbed upon a dark surface. It sublimes, without fusing, between 420° and 500° (788°–932° F.), is insoluble in cold H_2O and in alcohol; soluble in boiling H_2O to the extent of 1 part in 12,000; when boiled with H_2O for some time, it suffers partial decomposition, Hg is deposited and $HgCl_2$ dissolves.

Although Hg_2Cl_2 is insoluble in H_2O, in dilute HCl, and in pepsin solution, it is dissolved at the body temperature in an aqueous solution of pepsin acidulated with HCl.

When exposed to light, calomel becomes yellow, then gray, owing to partial decomposition, with liberation of Hg and formation of $HgCl_2$: $(Hg_2)Cl_2 = Hg + HgCl_2$. It is converted into $HgCl_2$ by Cl or aqua regia: $(Hg_2)Cl_2 + Cl_2 = 2HgCl_2$. In the presence of H_2O, I converts it into a mixture of $HgCl_2$ and HgI_2: $(Hg_2)Cl_2 + I_2 = HgCl_2 + HgI_2$. It is also converted into $HgCl_2$ by HCl and by alkaline chlorides: $(Hg_2)Cl_2 = HgCl_2 + Hg$. This change occurs in the stomach when calomel is taken internally, and that to such an extent when large quantities of NaCl is taken with the food, that calomel cannot be used in naval practice as it may be with patients who do not subsist upon salt provisions. It is converted by KI into $(Hg_2)I_2$: $(Hg_2)Cl_2 + 2KI = 2KCl + (Hg_2)I_2$; which is then decomposed by excess of KI into Hg and HgI_2, the latter dissolving: $(Hg)_2I_2 = Hg + HgI_2$. Solutions of the sulphates of Na, K, and NH_4 dissolve notable quantities of $(Hg_2)Cl_2$. The hydrates and carbonates of K and Na decompose it with formation of $(Hg_2)O$: $(Hg_2)Cl_2 + Na_2CO_3 = (Hg_2)O + CO_2 + 2NaCl$; and the $(Hg_2)O$ so formed is decomposed into HgO and Hg. If alkaline chlorides be also present, they react upon the HgO so produced, with formation of $HgCl_2$.

MERCURIC CHLORIDE—*Perchloride or bichloride of mercury—Corrosive sublimate—Hydrargyri chloridum corrosivum* (*U. S.*)—*Hydrargyri perchloridum* (*Br.*)—**$HgCl_2$**—270.7—is prepared by heating a mixture of 5 pts. dry $HgSO_4$ with 5 pts. dry NaCl, and 1 pt. MnO_2 in a glass vessel communicating with a condensing chamber.

It crystallizes by sublimation in octahedra, and by evaporation of its solutions in flattened, right rhombic prisms; fuses at 265° (509° F.), and boils at about 295° (563° F.); soluble in H_2O and in alcohol; very soluble in hot HCl, the solution gelatinizing on cooling. Its solutions have a disagreeable, acid, styptic taste, and are highly poisonous.

It is easily reduced to $(Hg_2)Cl_2$ and Hg, and its aqueous solutions are so decomposed when exposed to light; a change which is retarded by the presence of NaCl. Heated with Hg it is converted into $(Hg_2)Cl_2$. When dry $HgCl_2$ or its solution is heated with Zn, Cd, Ni, Fe, Pb, Cu, or Bi, those elements remove part of all of its Cl, with separation of $(Hg_2)Cl_2$ or Hg. Its solution is decomposed by H_2S with separation of a yellow sulphochloride, which, with an excess of the gas, is converted into black HgS. It is soluble without decomposition in H_2SO_4, HNO_3, and HCl. It is decomposed by KHO or NaHO, with separation of a brown oxychloride if the alkaline hydrate be in limited quantity; or of the orange-colored HgO if it be in excess. A similar decomposition is effected by CaH_2O_2 and MgH_2O_2; which does not, however, take place in presence of an alkaline chloride, or of certain organic matters, such as sugar and gum. Many organic substances decompose it into $(Hg_2)Cl_2$ and Hg, especially under the influence of sunlight. Albumen forms with it a white precipitate, which is insoluble in H_2O, but soluble in an excess of fluid albumen and in solutions of alkaline chlorides. It readily combines with metallic chlorides, to form soluble double chlorides, called *chloromercurates* or *chlor-*

165

hydrargyrates. One of these, obtained in flattened, rhombic prisms, by the cooling of a boiling solution of $HgCl_2$ and NH_4Cl, has the composition $HgCl_2$, $2(NH_4Cl) + Aq$, and was formerly known as *sal alembroth* or *sal sapientiæ*.

MERCURAMMONIUM CHLORIDE — *Mercury chloramidide — Infusible white precipitate—Ammoniated mercury—Hydrargyrum ammoniatum* (*U. S.; Br.*) —**NH_2HgCl**—251.1—is prepared by adding a slight excess of NH_4HO to a solution of $HgCl_2$. It is a white powder, insoluble in alcohol, ether, and cold H_2O; decomposed by hot H_2O with separation of a heavy, yellow powder. It is entirely volatile without fusion. The *fusible white precipitate* is formed in small crystals when a solution containing equal parts of $HgCl_2$ and NH_4Cl is decomposed by Na_2CO_3. It is *mercurdiammonium chloride*, NH_2HgCl,NH_4Cl.

Iodides.—MERCUROUS IODIDE—*Protoiodide* or *yellow iodide—Hydrargyri iodidum viride* (*U. S.; Br.*)—**Hg_2I_2**—653.4—is prepared by grinding together 200 pts. Hg and 127 pts. I with a little alcohol until a green paste is formed. It is a greenish-yellow, amorphous powder, insoluble in H_2O and in alcohol. When heated it turns brown and volatilizes completely. When exposed to light, or even after a time in the dark, it is decomposed into HgI_2 and Hg. The same decomposition is brought about instantly by KI; more slowly by solutions of alkaline chlorides and by HCl when heated. NH_4HO dissolves it with separation of a gray precipitate.

MERCURIC IODIDE—*Biniodide* or *red iodide—Hydrargyri iodidum rubrum* (*U. S.; Br.*)—**HgI_2**—453.7—is obtained by double decomposition between $HgCl_2$ and KI, care being had to avoid too great an excess of the alkaline iodide, that the soluble potassium iodhydrargyrate may not be formed.

It is sparingly soluble in H_2O; but forms colorless solutions with alcohol. It dissolves readily in many dilute acids and in solutions of ammoniacal salts, alkaline chlorides, and mercuric salts; and in solutions of alkaline iodides. Iron and copper convert it into $(Hg_2)I_2$, then into Hg. The hydrates of K and Na decompose it into oxide or oxyiodide, and combine with another portion to form iodhydrargyrates, which dissolve. NH_4HO separates from its solution a brown powder, and forms a yellow solution which deposits white flocks.

Cyanides.—MERCURIC CYANIDE—*Hydrargyri cyanidum* (*U. S.*)—**$Hg(CN)_2$**—251.7—is best prepared by heating together, for a quarter of an hour, potassium ferrocyanide, 1 pt.; $HgSO_4$, 2 pts.; and H_2O, 8 pts. It crystallizes in quadrangular prisms; soluble in 8 pts. of cold H_2O, much less soluble in alcohol; highly poisonous. When heated dry it blackens, and is decomposed into $(CN)_2$ and Hg; if heated in presence of H_2O it yields HCN, Hg, CO_2, and NH_3. Hot concentrated H_2SO_4, and HCl, HBr, HI, and H_2S in the cold, decompose it with liberation of HCN. It is not decomposed by alkalies.

Salts of Mercury.

Nitrates.—There exist, besides the normal nitrates: $(Hg_2)(NO_3)_2$ and $Hg(NO_3)_2$, three basic mercurous nitrates, three basic mercuric nitrates, and a mercuroso-mercuric nitrate.

MERCUROUS NITRATE—**$(Hg_2)(NO_3)_2$ + 2 Aq**—523.4 + 36—is formed when excess of Hg is digested with HNO_3, diluted with ½ vol. H_2O; until short, prismatic crystals separate.

166

It effloresces in air; fuses at 70° (158° F.); dissolves in a small quantity of hot H_2O, but with a larger quantity is decomposed with separation of the yellow, basic *trimercuric nitrate*, $Hg(NO_3)_2$, $2HgO + Aq$.

DIMERCUROUS NITRATE.—**$(Hg_2)(NO_3)_2,Hg_2O$ + Aq**—938.8 + 18—is formed by acting upon the preceding salt with cold H_2O until it turns lemon-yellow; or by extracting with cold H_2O the residue of evaporation of the product obtained by acting upon excess of Hg with concentrated HNO_3.

TRIMERCUROUS NITRATE—**$(Hg_2)_2(NO_3)_2$, Hg_2O + 3 Aq**—1462.2 + 54—is obtained in large, rhombic prisms, when excess of Hg is boiled with HNO_3, diluted with 5 pts. H_2O, for 5–6 hours, the loss by evaporation being made up from time to time.

MURCURIC NITRATE—**$Hg(NO_3)_2$**—323.7—is formed when Hg or HgO is dissolved in excess of HNO_3, and the solution evaporated at a gentle heat. A syrupy liquid is obtained, which, over quick-lime, deposits large, deliquescent crystals, having the composition $2[Hg(NO_3)_2] + Aq$, while there remains an uncrystallizable liquid, $Hg(NO_3)_2 + 2\ Aq$.

This salt is soluble in H_2O, and exists in the *Liq. hydrargyri nitratis* (*U. S.*), *Liq. hydrargyri nitratis acidus* (*Br.*); in the volumetric standard solution used in *Liebig's process* for urea; and probably in *citrine ointment* = *Ung. hydrar. nitratis* (*U. S.; Br.*).

DIMERCURIC NITRATE—**$Hg(NO_3)_2$, HgO + Aq**—539.4—is formed when HgO is dissolved to saturation in hot HNO_3, diluted with 1 vol. H_2O; and crystallizes on cooling. It is decomposed by H_2O into *trimercuric nitrate*, $Hg(NO_3)_2$, 2HgO, and $Hg(NO_3)_2$.

HEXAMERCURIC NITRATE—**$Hg(NO_3)_2$, 5HgO**—1402.2—is formed as a red powder, by the action of H_2O on trimercuric nitrate.

Sulphates.—MERCUROUS SULPHATE—**$(Hg_2)SO_4$**—495.4—is a white, crystalline powder, formed by gently heating together 2 pts. Hg and 3 pts. H_2SO_4, and causing the product to combine with 2 pts. Hg. Heated with NaCl it forms $(Hg_2)Cl_2$.

MERCURIC SULPHATE—*Hydrargyri sulphas* (*Br.*)—**$HgSO_4$**—295.7—is obtained by heating together Hg and H_2SO_4; or Hg, H_2SO_4, and HNO_3. It is a white, crystalline, anhydrous powder, which on contact with H_2O is decomposed with formation of *trimercuric sulphate*, $HgSO_4$, 2HgO; a yellow, insoluble powder known as *turpeth mineral* = *Hydrargyri subsulphas flavus* (*U. S.*).

Analytical Characters.

MERCUROUS.—(1.) Hydrochloric acid: white ppt.; insoluble in H_2O and in acids; turns black with NH_4HO; when boiled with HCl, deposits Hg, while $HgCl_2$ dissolves.

(2.) Hydrogen sulphide: black ppt.; insoluble in alkaline sulphydrates, in dilute acids, and in KCN; partly soluble in boiling HNO_3.

(3.) Potash: black ppt.; insoluble in excess.

(4.) Potassium iodide: greenish ppt.; converted by excess into Hg which is deposited, and HgI_2, which dissolves.

MERCURIC.—(1.) Hydrogen sulphide: black ppt. If the reagent be slowly added, the ppt. is first white, then orange, finally black.

(2.) Ammonium sulphydrate: black ppt.; insoluble in excess, except in the presence of organic matter.

(3.) Potash or soda: yellow ppt.; insoluble in excess.

167

(4.) Ammonium hydrate: white ppt.; soluble in great excess and in solutions of NH_4 salts.

(5.) Potassium carbonate: red ppt.

(6.) Potassium iodide: yellow ppt., rapidly turning to salmon color, then to red; easily soluble in excess of KI, or in great excess of mercuric salt.

(7.) Stannous chloride, in small quantity: white ppt.; in larger quantity gray ppt.; and when boiled, deposit of globules of Hg.

Action on the Economy.

Mercury, in the metallic form, is without action upon the animal economy so long as it remains such; on contact, however, with alkaline chlorides it is converted into a soluble double chloride, and this the more readily the greater the degree of subdivision of the metal. The mercurials insoluble in dilute HCl are also inert until they are converted into soluble compounds.

Mercuric chloride, a substance into which many other compounds of Hg are converted when taken into the stomach or applied to the skin, not only has a distinctly corrosive action, by virtue of its tendency to unite with albuminoids, but when absorbed it produces well-marked poisonous effects, somewhat similar to those of arsenical poisoning; indeed, owing to its corrosive action and to its greater solubility, and more rapid absorption, it is a more dangerous poison than As_2O_3. In poisoning by $HgCl_2$, the symptoms begin sooner after the ingestion of the poison than in arsenical poisoning, and those phenomena referable to the local action of the toxic are more intense.

The treatment should consist in the administration of white of egg, not in too great quantity, and the removal of the compound formed, by emesis, before it has had time to redissolve in the alkaline chlorides contained in the stomach.

Absorbed Hg tends to remain in the system in combination with albuminoids, from which it may be set free, or, more properly, brought into soluble combination, at a period quite removed from the date of last administration, by the exhibition of alkaline iodides.

Mercury is eliminated principally by the saliva and urine, in which it may be readily detected. The fluid is faintly acidulated with HCl, and in it is immersed a short bar of Zn, around which a spiral of dentist's gold-foil is wound in such a way as to expose alternate surfaces of Zn and Au. After 24 hours, if the saliva or urine contain Hg, the Au will be whitened by amalgamation; and, if dried and heated in the closed end of a small glass tube, will give off Hg, which condenses in globules, visible with the aid of a magnifier, in the cold part of the tube.

168

COMPOUNDS OF CARBON.

Organic Substances.

In the seventeenth and eighteenth centuries, chemists had observed that there might be extracted from animal and vegetable bodies substances which differed much in their properties from those which could be obtained from the mineral world; substances which burned without leaving a residue, and many of which were subject to the peculiar changes wrought by the processes of fermentation and putrefaction. It was not until the beginning of the present century, however, that chemistry was divided into the two sections of *inorganic* and *organic*.

In the latter class were included all such substances as existed only in the organized bodies of animals and vegetables, and which seemed to be of a different essence from that of mineral bodies, as chemists had been unable to produce any of these organic substances by artificial means. Later in the history of the science it was found that these bodies were all made up of a very few elements, and that they all contained carbon. Gmelin at this time proposed to consider as organic substances all such as contained more than one atom of C, his object in thus limiting the minimum number of atoms of C being that substances containing one atom of C, such as carbonic acid and marsh-gas, were formed in the mineral kingdom, and consequently, according to then existing views, could not be considered as *organic*. Illogical as such a distinction is, we find it still adhered to in text-books of very recent date.

The notion that organic substances could only be formed by some mysterious agency, manifested only in organized beings, was finally exploded by the labors of Wöhler and Kolbe. The former obtained urea from ammonium cyanate; while the latter, at a subsequent period, formed acetic acid, using in its preparation only such unmistakably mineral substances as coal, sulphur, aqua regia, and water.

During the half-century following Wöhler's first synthesis, chemists have succeeded not only in making from mineral materials many of the substances previously only formed in the laboratory of nature, but have also produced a vast number of carbon compounds which were previously unknown, and which, so far as we know, have no existence in nature. At the present time, therefor, *we must consider as an organic substance any compound containing carbon, whatever may be its origin and whatever its properties.* Indeed, the name *organic* is retained merely as a matter of convenience, and not in any way as indicating the origin of these compounds. Although, owing to the great number of the carbon compounds, it is still convenient to treat of them as forming a section by themselves, their relations with the compounds of other elements is frequently very close; indeed, within the past few years, compounds of silicon have been obtained, which indicate the possibility that that element is capable of forming series of compounds as interesting in numbers and variety as those of carbon.

Nevertheless, there are certain peculiarities exhibited by C in its compounds, which are not possessed to a like extent by any other element,

169

and which render the study of organic substances peculiarly interesting and profitable.

In the study of the compounds of the other elements, we have to deal with a small number of substances, relatively speaking, formed by the union with each other of a large number of elements. With the organic substances the reverse is the case; for, although compounds have been formed which contain C along with each of the other elements, the great majority of the organic substances are made up of C, combined with a very few other elements; H, O and N occurring in them most frequently.

It is chiefly in the study of the carbon compounds that we have to deal with *radicals* (see p. 23). Among mineral substances there are many whose molecules consist simply of a combination of two atoms; among organic substances there is none which does not contain a radical: indeed, organic chemistry has been defined as "the chemistry of compound radicals."

The atoms of carbon possess in a higher degree than those of any other element the power of uniting with each other, and in so doing of interchanging valences. Were it not for this property of the C atoms, we could have but one saturated compound of carbon and hydrogen, CH_4, or, expressed graphically:

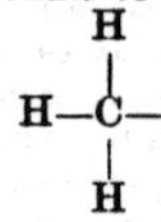

There exist, however, a great number of such compounds, which differ from each other by one atom of C and two atoms of H. In these substances the atoms of C may be considered as linked together in a continuous chain, their free valences being satisfied by H atoms; thus:

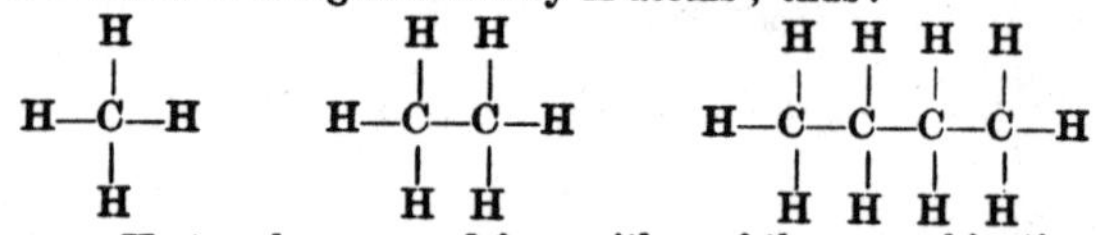

If now one H atom be removed from either of these combinations, we have a group possessing one free valence, and consequently univalent. The decompositions of these substances show that they contain such radicals, and that their typical formulæ are:

$$\left.\begin{matrix}CH_3\\H\end{matrix}\right\}; \qquad \left.\begin{matrix}C_2H_5\\H\end{matrix}\right\}; \qquad \left.\begin{matrix}C_4H_9\\H\end{matrix}\right\}.$$

Homologous Series.

It will be observed that these formulæ differ from each other by CH_2, or some multiple of CH_2, more or less. In examining numbers of organic substances, which are closely related to each other in their properties, we find that we can arrange the great majority of them in series, each term of which differs from the one below it by CH_2; such a series is called an

170

homologous series. It will be readily understood that such an arrangement in series vastly facilitates the remembering of the composition of organic bodies. In the following table, for example, are given the saturated hydrocarbons and their more immediate derivatives. At the head of each vertical column is an algebraic formula, which is the *general formula* of the entire series below it; *n* being equal to the numerical position in the series.

HOMOLOGOUS SERIES.

Saturated hydrocarbons, C_nH_{2n+2}.	Alcohols, $C_nH_{2n+2}O$.	Aldehydes $C_nH_{2n}O$.	Acids, $C_nH_{2n}O_2$.	Ketones, $C_nH_{2n}O$.
CH_4	CH_4O	......	CO_2H_2	......
C_2H_6	C_2H_6O	C_2H_4O	$C_2O_2H_4$	......
C_3H_8	C_3H_8O	C_3H_6O	$C_3O_2H_6$	C_3H_6O
C_4H_{10}	$C_4H_{10}O$	C_4H_8O	$C_4O_2H_8$	C_4H_8O
C_5H_{12}	$C_5H_{12}O$	$C_5H_{10}O$	$C_5O_2H_{10}$	$C_5H_{10}O$
C_6H_{14}	$C_6H_{14}O$	$C_6H_{12}O$	$C_6O_2H_{12}$	.....
C_7H_{16}	$C_7H_{16}O$	$C_7H_{14}O$	$C_7O_2H_{14}$	.....
C_8H_{18}	$C_8H_{18}O$	$C_8H_{16}O$	$C_8O_2H_{16}$	.....
C_9H_{20}	$C_9H_{20}O$	....	$C_9O_2H_{18}$	.....
$C_{10}H_{22}$	$C_{10}H_{22}O$	....	$C_{10}O_2H_{20}$	.....
$C_{11}H_{24}$	....	....	....	.....
$C_{12}H_{26}$	....	....	$C_{12}O_2H_{24}$	.....
$C_{13}H_{28}$	....	....	....	.....
$C_{14}H_{30}$	....	....	$C_{14}O_2H_{28}$	.....

But the arrangement in homologous series does more for us than this. The properties of substances in the same series vary in regular gradation according to their position in the series ; thus, in the series of alcohols in the above table, the boiling-points of the first six are, 66.5.°, 78.4°, 96.7°, 111.7°, 132.2°, 153.9° ; from which it will be seen that the boiling-point of any one of them can be determined, with a maximum error of 3°, by taking the mean of those of its neighbors above and below. In this way we may prophecy, to some extent, the properties of a wanting member in a series before its discovery. The terms of any homologous series must all have the same *constitution*, *i.e.*, their constituent atoms must be similarly arranged within the molecule.

Isomerism—Metamerism—Polymerism.

Two substances are said to be isomeric, or to be isomeres of each other, when they have the same centesimal composition. If, for instance, we analyze acetic acid and methyl formiate, we find that each body consists of C, O and H, in the following proportions :

Carbon	40	$24 = 12 \times 2$
Oxygen	53.33	$32 = 16 \times 2$
Hydrogen	6.67	$4 = 1 \times 4$
	100.00	60

171

This similarity of centesimal composition may occur in two ways : the two substances may each contain in a molecule the same numbers of each kind of atom ; or one may contain in each molecule the same kind of atoms as the other, but in a higher multiple. In the above instance, for example, each substance may have the composition $C_2H_4O_2$; or one may have that formula and the other, $C_6H_{12}O_6$, or $C_2H_4O_2 \times 3$. In the former case the substances are said to be *metameric*, in the latter *polymeric*. Whether two substances are metameric or polymeric can only be determined by ascertaining the weights of their molecules, which is usually accomplished by determining the sp. gr. of their vapors (see p. 14).

The sp. gr. of the vapor of acetic acid is the same as that of methyl formiate, and, consequently, each substance is made up of molecules, each containing $C_2H_4O_2$. But the two substances differ from each other greatly in their properties, and their differences are at once indicated by their typical or graphic formulæ :

$$\left.\begin{matrix}(C_2H_3O)\\ H\end{matrix}\right\}O \quad \text{and} \quad \left.\begin{matrix}(CHO)'\\ (CH_3)'\end{matrix}\right\}O;$$

or graphically :

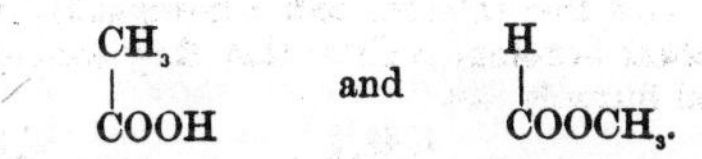

Classification of Organic Substances.

As the compounds of the other elements may be divided into classes, such as acids, bases, salts, etc., according to their chemical functions, the compounds of carbon also arrange themselves into certain well-defined groups, called by the French chemists *functions*—a term which it would be well to introduce into our own nomenclature. The properties of the functions of organic substances do not depend, like those of other compounds, upon the *kind* of atoms of which they are composed, but rather upon the arrangement of the atoms within the molecule ; and in this point we find the most prominent distinction between organic and mineral substances. Arsenic, for instance, is poisonous in whatever form of chemical combination it may be, provided only that it can be rendered soluble, and therefor capable of absorption. Carbon, oxygen, and hydrogen, on the other hand, combine with each other to form substances having the most diverse action upon the economy—the fats and sugars, ordinary articles of food, on the one hand, and substances having such marked toxic powers as ether and oxalic acid, on the other—the differences between the properties of the two substances depending entirely upon the numbers and positions in the molecule of the same kind of atoms.

172

SATURATED HYDROCARBONS AND THEIR DERIVATIVES.

FIRST SERIES OF HYDROCARBONS.

SERIES C_nH_{2n+2}.

A hydrocarbon is a compound of carbon and hydrogen only. It is saturated when all the valences of all the constituent atoms are satisfied.

The hydrocarbons of this series at present known are the following :

Name.	Formula.	Specific gravity of liquid.	Boiling-point. Centigrade.	Name.	Formula.	Specific gravity of liquid.	Boiling-point. Centigrade.
Methyl hydride...	CH_3H	......	......	Nonyl hydride....	$C_9H_{19}H$	0.741 at 18°	136°–138°
Ethyl hydride....	C_2H_5H	......	......	Decyl hydride.....	$C_{10}H_{21}H$	0.757 at 18°	158°–162°
Propyl hydride...	C_3H_7H	......	......	Undecyl hydride..	$C_{11}H_{23}H$	0.766 at 18°	180°–182°
Butyl hydride....	C_4H_9H	0.600 at 0°	0°	Dodecyl hydride..	$C_{12}H_{25}H$	0.778 at 18°	198°–200°
Amyl hydride.....	$C_5H_{11}H$	0.628 at 18°	30°	Tridecyl hydride..	$C_{13}H_{27}H$	0.796 at 18°	218° 220°
Hexyl hydride....	$C_6H_{13}H$	0.669 at 18°	68°	Tetradecyl hydride	$C_{14}H_{29}H$	0.809 at 18°	236°–240°
Heptyl hydride....	$C_7H_{15}H$	0.690 at 18°	92°–94°	Pentadecyl hydride	$C_{15}H_{31}H$	0.825 at 18°	258°–262°
Octyl hydride	$C_8H_{17}H$	0.726 at 18°	116°–118°	Hexadecyl hydride	$C_{16}H_{33}H$	......	about 280°

They form an homologous series whose general formula is C_nH_{2n+2}, and are known as *paraffines* from their stability (*parum* = little, *affinis* = affinity). Their constitution is expressed typically by the formula $\left.\begin{matrix}C_nH_{2n+1}\\ H\end{matrix}\right\}$, and the *radicals* C_nH_{2n+1}, of which they are the *hydrides*, are designated as the radicals of the monoatomic alcohols.

Corresponding to the higher terms of the series (those above the third) there are one or more isomeres, which may be arranged in four classes. (1.) The *normal* or regularly formed series, in which each C atom is linked to two other C atoms. (2.) Those in which one C atom is linked to three others. (3.) Those in which two C atoms are each linked to three others. (4.) Those in which one C atom is linked to four others. The constitution of these series is explained by the graphic formulæ :

(1.) (2.) (3.) (4.)

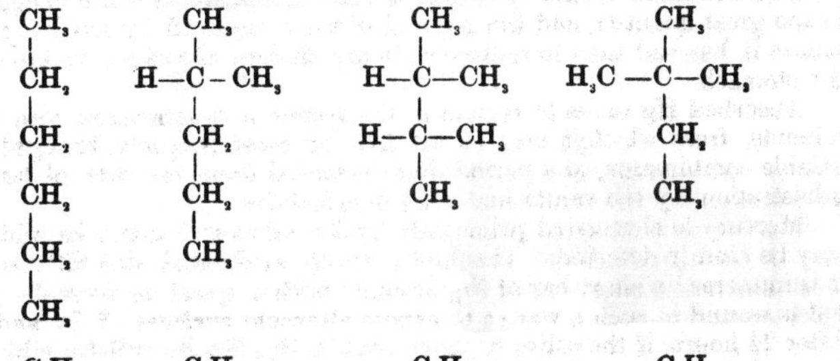

C_6H_{14} C_6H_{14} C_6H_{14} C_6H_{14}

As all of these compounds are saturated they are incapable of being modified by *addition*, i.e. by the simple insertion of other atoms into the

173

molecule ; they may, however, be modified by *substitution*, i.e. by the removal of one or more of their atoms and the substitution therefor of an atom or atoms of different kind.

Methyl hydride—*Methane*—*Marsh-gas*—*Light carburetted hydrogen*—*Fire-damp*—CH_4—16—is given off in swamps as a product of decomposition of vegetable matter, in coal mines, and in the gases issuing from the earth in the vicinity of petroleum deposits. Coal-gas contains it in the proportion of 36–50 per cent. It may be prepared by strongly heating a mixture of sodium acetate with sodium hydrate and quicklime.

It is a colorless, odorless, tasteless gas ; very sparingly soluble in H_2O ; sp. gr. 0.559A. At high temperatures it is decomposed into C and H. It burns in air with a pale yellow flame. Mixed with air or O it explodes violently on contact with flame, producing water and carbon dioxide ; the latter constituting the *after-damp* of miners. It is not affected by Cl in the dark, but under the influence of diffuse daylight one or more of the H atoms are displaced by an equivalent quantity of Cl. In direct sunlight the substitution is accompanied by an explosion.

Petroleum.—Crude petroleum differs in composition and in physical properties in the products of different wells, even in the same section of country. It varies in color from a faintly yellowish tinge to a dark brown, nearly black, with greenish reflections. The lighter-colored varieties are limpid, and the more highly colored of the consistency of thin syrup. The sp. gr. varies from 0.74 to 0.92. Crude petroleums contain all the hydrocarbons mentioned in the list on p. 172 (the first of the series, being found in the gases accompanying petroleum, is also held in solution by the oil under the pressure it supports in natural pockets), besides hydrocarbons of the olefine series, and of the benzol series.

The crude oil is highly inflammable, usually highly colored, and is prepared for its multitudinous uses in the arts by the processes of distillation and refining. The distillation is usually so conducted as to divide the product into four parts:

Naphtha..............	Sp. gr. 0.700—12–15%	*Burning oil*..........	Sp. gr. 0.788—60%
Benzine...............	Sp. gr. 0.730— 9–12%	*Residuum and loss*...............	18–19%

The *naphtha*, or *petroleum ether*, is further separated by distillation into other products : *Rhigoline*, a highly inflammable liquid ; sp. gr. about 0.60, which boils at about 21° (70° F.). It is used to produce cold by its rapid evaporation, but its low boiling-point and inflammability render its use dangerous. *Gasoline;* sp. gr. about 0.63–0.61; boils at about 76° (170° F.).

Benzine or *benzoline*, sp. gr. about 0.73; boils at about 148° (298° F.), and is largely used in the arts as a solvent. It must not be confounded with *benzol* or *benzene*, C_6H_6 (q. v.).

The most important product of petroleum is that portion which distils above 183° (361° F.) and which constitutes *kerosene*, and other oils used for burning in lamps. An oil to be safely used for burning in lamps should not "flash," or give off inflammable vapor, below 60° (140° F.); and should not burn at temperatures below 65°.5 (150° F.).

From the residue remaining after the separation of the kerosene, a variety of other products are obtained. *Lubricating oils*, of too high boiling-point for use in lamps. *Paraffine*, a white, crystalline solid, fusible at 45°–65° (113°–149° F.), which is used in the arts for a variety of purposes formerly served by wax, such as the manufacture of candles. In the laboratory it is very useful for coating the glass stoppers of bottles, and for

174

other purposes, as it is not affected by acids or by alkalies. It is odorless, tasteless, insoluble in H_2O and in cold alcohol; soluble in boiling alcohol and in ether, fatty and volatile oils, and mineral oils. It is also obtained by the distillation of certain varieties of coal, and is found in nature in *fossil wax* or *ozocerite*.

The products known as *vaseline*, *petrolatum* (*U. S.*), *cosmoline*, etc., which are now so largely used in pharmacy and perfumery, are mixtures of paraffine and the heavier petroleum oils. Like petroleum itself, its various commercial derivatives are not definite compounds, but mixtures of the hydrocarbons of this series.

Haloid Derivatives of the Paraffines.

By the action of Br upon the paraffines, or by the action of HCl, HBr or HI upon the corresponding hydrates, compounds are obtained in which one of the H atoms of the hydrocarbon has been replaced by an atom of Cl, Br or I: $C_2H_6 + Br_2 = C_2H_5Br + HBr$, or $C_2H_5OH + HCl = C_2H_5Cl + H_2O$. These compounds may be considered as the chlorides, bromides or iodides of the alcoholic radicals; and are known as *haloid ethers*.

When Cl is allowed to act upon CH_4, it replaces a further number of H atoms until finally carbon tetrachloride, CCl_4, is produced. Considering marsh gas as *methyl hydride*, CH_3,H, the first product of substitution is *methyl chloride*, CH_3,Cl; the second *monochlormethyl chloride*, CH_2Cl, Cl; the third *dichlormethyl chloride*, or chloroform, $CHCl_2Cl$; and the fourth *carbon tetrachloride*, CCl_4.

Similar derivatives are formed with Br and I and with the other hydrocarbons of the series.

Methyl chloride—CH_3Cl—50.5—is a colorless gas, slightly soluble in H_2O, and having a sweetish taste and odor. It is obtained by distilling together H_2SO_4, sodium chloride and methyl achohol. It may be condensed to a liquid which boils at −22° (−7°.6 F.). It burns with a greenish flame. Heated with potassium hydrate it is converted into methyl alcohol.

Monochlormethyl chloride—*Methene chloride—Dichloromethane—Methylene chloride—Chloromethyl*—CH_2Cl,Cl—85—is obtained by the action of Cl upon CH_3Cl; or by shaking an alcoholic solution of chloroform with powdered zinc and a little ammonium hydrate. In either case the product must be purified.

It is a colorless, oily liquid, boils at 40°–42° (104°–107°.6 F.); sp. gr. 1.36; its odor is similar to that of chloroform; it is very slightly soluble in H_2O; and is not inflammable. Like most of the chlorinated derivatives of this series, it is possessed of anæsthetic powers. Its use as an anæsthetic is attended with the same (if not greater) danger as that of chloroform.

Dichlormethyl chloride—*Methenyl chloride—Formyl chloride—Trichloromethane—Chloroform—Chloroformum* (*U. S., Br.*)—$CHCl_2,Cl$—120.5—is obtained by heating in a capacious still, 35–40 litres (9–11 gall.) of H_2O, adding 5 kilos (11 lbs.) of recently slacked lime and 10 kilos (22 lbs.) of chloride of lime; 2.5 kilos (2½ qts.) of alcohol are then added and the temperature quickly raised until the product begins to distil, when the fire is withdrawn, heat being again applied toward the end of the reaction. The crude chloroform so obtained is purified, first by agitation

175

with H_2SO then by mixing with alcohol and recently ignited potassium carbonate, and distilling the mixture.

It is a colorless, volatile liquid, having a strong, agreeable, ethereal odor, and a sweet taste; sp. gr. 1.497; very sparingly soluble in H_2O; miscible with alcohol and ether in all proportions; boils at 60°.8 (141°.4 F.). It is a good solvent for many substances insoluble in H_2O, such as phosphorus, iodine, fats, resins, caoutchouc, gutta-percha and the alkaloids.

It ignites with difficulty, but burns from a wick with a smoky, red flame, bordered with green. It is not acted on by H_2SO_4, except after long contact, when HCl is given off. In direct sunlight Cl converts it into CCl_4 and HCl. The alkalies in aqueous solution do not act upon it, but when heated with them in alcoholic solution it is decomposed with formation of chloride and formiate of the alkaline metal. When perfectly pure it is not altered by exposure to light; but if it contain compounds of N, even in very minute quantity, it is gradually decomposed by solar action into HCl, Cl and other substances.

Impurities.—*Alcohol*, if present in large amount, lowers the sp. gr. of the chloroform, and causes it to fall through H_2O in opaque, pearly drops. If present in small amount it produces a green color with ferrous dinitrosulphide (obtained by acting on ferrous chloride with a mixture of potassium nitrate and ammonium hydrosulphide). *Aldehyde* produces a brown color when $CHCl_3$ containing it is heated with liquor potassæ. *Hydrochloric acid* reddens blue litmus, and causes a white precipitate in an aqueous solution of silver nitrate shaken with chloroform. *Methyl and empyreumatic compounds* are the most dangerous of the impurities of chloroform. Their absence is recognized by the following characters: (1.) When the chloroform is shaken with an equal volume of colorless H_2SO_4, and allowed to stand 24 hours; the upper (chloroform) layer should be perfectly colorless, and the lower (acid) layer colorless or faintly yellow. (2.) When a small quantity is allowed to evaporate spontaneously, the last portions should have no pungent odor, and the remaining film of moisture should have no taste or odor other than those of chloroform.

Analytical Characters.—(1.) Add a little alcoholic solution of potash and 2–3 drops of aniline and warm; a disagreeeble odor, resembling that of witch-hazel, is produced.

(2.) Vapor of $CHCl_3$, when passed through a red-hot tube, is decomposed with formation of HCl and Cl, the former of which is recognized by the production of a white ppt., soluble in ammonium hydrate, in an acid solution of silver nitrate. This test does not afford reliable results when the substance tested contains a free acid and chlorides.

(3.) Dissolve about 0.01 Gm. of β naphthol in a small quantity of KHO solution, warm, and add the suspected liquid; a blue color is produced.

Toxicology.—The action of chloroform varies as it is taken by the stomach or by inhalation. In the former case, owing to its insolubility, but little is absorbed, and the principal action is the local irritation of the mucous surfaces. Recovery has followed a dose of four ounces, and death has been caused by one drachm, taken into the stomach. Chloroform vapor acts much more energetically, and seems to owe its potency for evil to its paralyzing influence upon the nerve-centres, notably upon those of the heart. While persons suffering from heart disease are particularly susceptible to the paralyzing effect of chloroform vapor, there are many cases recorded of death from the inhalation of small quantities, properly diluted, in which no heart lesion was found upon a post-mortem examina-

176

tion. Chloroform is apparently not altered in the system, and is eliminated with the expired air.

No chemical antidote to chloroform is known. When it has been swallowed, the stomach-pump and emetics are indicated; when taken by inhalation, a free circulation of air should be established about the face; artificial respiration and the application of the induced current to the sides of the neck should be resorted to.

The nature of the poison is usually revealed at the autopsy by its peculiar odor, which is most noticeable on opening the cranial and thoracic cavities. In a toxicological analysis, chloroform is to be sought for especially in the lungs and blood. These are placed in a flask; if acid, neutralized with sodium carbonate; and subjected to distillation at the temperature of the water-bath. The vapors are passed through a tube of difficultly fusible glass; at first the tube is heated to redness for about an inch of its length, and test No. 2 applied to the issuing gas. The tube is then allowed to cool, and the distillate collected in a pointed tube, from the point of which any $CHCl_3$ is removed by a pipette and tested according to Nos. 1 and 3 above.

Carbon tetrachloride—*Chlorocarbon*—CCl_4—154—is formed by the prolonged action, in sunlight, of Cl upon CH_3Cl or $CHCl_3$; or more rapidly, by passing Cl, charged with vapor of carbon disulphide, through a red-hot tube, and purifying the product.

It is a colorless, oily liquid, insoluble in H_2O; soluble in alcohol and in ether; sp. gr. 1.56; boils at 78° (172°.4 F.). Its vapor is decomposed at a red heat into a mixture of the *dichloride*, C_2Cl_4, *trichloride*, C_2Cl_6, and free Cl.

Methyl bromide—CH_3Br—95.—A colorless liquid; sp. gr. 1.664; boils at 13° (55.4° F.); formed by the combined action of P and Br on methyl hydrate.

Dibromomethyl bromide—*Methenyl bromide—Formyl bromide—Bromoform*—$CHBr_2$, Br—253—is prepared by gradually adding Br to a cold solution of potassium hydrate in methyl alcohol, until the liquid begins to be colored; and rectifying over calcium chloride.

A colorless, aromatic, sweet liquid; sp. gr. 2.13; boils at 150°–152° (302°–306° F.); solidifies at −9° (15°.8 F.); sparingly soluble in H_2O; soluble in alcohol and ether. Boiled with alcoholic potash it is decomposed in the same way as is $CHCl_3$.

Its physiological action is similar to that of $CHCl_3$. It occurs as an impurity of commercial Br, accompanied by *carbon tetrabromide*, CBr_4.

Methyl iodide—CH_3I—142—a colorless liquid, sp. gr. 2.237 ; boils at 45° (113° F.) ; burns with difficulty, producing violet vapor of iodine. It is prepared by a process similar to that for obtaining the bromide ; and is used in the aniline industry.

Diiodomethyl iodide—*Methenyl iodide—Formyl iodide—Iodoform—Iodoformum, U. S.*—CHI_2I—394.—Formed, like chloroform and bromoform, by the combined action of potash and the halogen upon alcohol ; it is also produced by the action of I upon a great number of organic substances, and is usually prepared by heating a mixture of alkaline carbonate, H_2O, I and ethylic alcohol, and purifying the product by recrystallization from alcohol.

Iodoform is a solid, crystallizing in yellow, hexagonal plates, which melt at 115°–120° (239°–248° F.). It may be sublimed, a portion being decomposed. It is insoluble in water, acids, and alkaline solutions : soluble in alcohol, ether, carbon disulphide, and the fatty and essential oils :

the solutions, when exposed to the light, undergo decomposition and assume a violet-red color. It has a sweet taste and a peculiar, penetrating odor, resembling, when the vapor is largely diluted with air, that of saffron. When heated with potash, a portion is decomposed into formiate and iodide, while another portion is carried off unaltered with the aqueous vapor. It contains 96.7% of its weight of iodine.

Ethyl chloride—*Hydrochloric or muriatic ether*—C_2H_5Cl—64.5.—A colorless, white, ethereal liquid ; boils at 11° (51°.8 F.) ; obtained by passing gaseous HCl through ethylic alcohol to saturation and distilling over the water-bath.

Ethyl bromide — *Hydrobromic ether* — C_2H_5Br — 109.—A colorless, ethereal liquid ; boils at 40°.7 (105°.3 F.) ; obtained by the combined action of P and Br on ethylic alcohol.

Ethyl iodide—*Hydriodic ether*—C_2H_5I—156—is prepared by placing absolute alcohol and P in a vessel surrounded by a freezing mixture and gradually adding I ; when the action has ceased, the liquid is decanted, distilled over the water-bath, and the distillate washed and rectified.

It is a colorless liquid ; boils at 72°.2 (162° F.) ; has a powerful, ethereal odor ; burns with difficulty. It is largely used in the aniline industry.

MONOATOMIC ALCOHOLS.

Series $C_nH_{2n+2}O$.

The following is a list of the terms of the primary series which have been studied, and their prominent physical properties.

Name.	Empirical formula.	Typical formula.	Fusing-point.	Boiling-point.	Specific gravity.
Methyl hydrate..........	CH_4O	$\left.\begin{matrix}CH_3\\H\end{matrix}\right\}O$	..	66°.5	0.814
Ethyl hydrate...........	C_2H_6O	$\left.\begin{matrix}C_2H_5\\H\end{matrix}\right\}O$	..	78°.3	0.8095
Propyl hydrate..........	C_3H_8O	$\left.\begin{matrix}C_3H_7\\H\end{matrix}\right\}O$	..	96°.7	0.820
Butyl hydrate...........	$C_4H_{10}O$	$\left.\begin{matrix}C_4H_9\\H\end{matrix}\right\}O$	..	114°.7	0.817
Amyl hydrate............	$C_5H_{12}O$	$\left.\begin{matrix}C_5H_{11}\\H\end{matrix}\right\}O$	−20°	132°	....
Hexyl hydrate...........	$C_6H_{14}O$	$\left.\begin{matrix}C_6H_{13}\\H\end{matrix}\right\}O$	..	150°	0.820
Heptyl hydrate..........	$C_7H_{16}O$	$\left.\begin{matrix}C_7H_{15}\\H\end{matrix}\right\}O$	..	168°	....
Octyl hydrate...........	$C_8H_{18}O$	$\left.\begin{matrix}C_8H_{17}\\H\end{matrix}\right\}O$	..	186°	....
Nonyl hydrate...........	$C_9H_{20}O$	$\left.\begin{matrix}C_9H_{19}\\H\end{matrix}\right\}O$	..	204°	....
Decyl hydrate...........	$C_{10}H_{22}O$	$\left.\begin{matrix}C_{10}H_{21}\\H\end{matrix}\right\}O$	..	....	...
Cetyl hydrate...........	$C_{16}H_{34}O$	$\left.\begin{matrix}C_{16}H_{33}\\H\end{matrix}\right\}O$	49°	....	....
Ceryl hydrate...........	$C_{27}H_{56}O$	$\left.\begin{matrix}C_{27}H_{55}\\H\end{matrix}\right\}O$	79°	....	....
Myricyl hydrate.........	$C_{30}H_{62}O$	$\left.\begin{matrix}C_{30}H_{61}\\H\end{matrix}\right\}O$	85°	....	....

The name *alcohol*, formerly applied only to the substance now popularly so called, has gradually come to be used to designate a large class of important bodies, of which vinic alcohol is the representative. These substances are mainly characterized by their power of entering into double decomposition with acids, to form neutral compounds, called *compound ethers*, water being at the same time formed, at the expense of both alcohol and acid. *They are the hydrates of hydrocarbon radicals, and as such resemble the metallic hydrates, while the compound ethers are the counter parts of the metallic salts :*

$$\left.\begin{matrix}(C_2H_5)\\H\end{matrix}\right\}O+\left.\begin{matrix}(C_2H_3O)\\H\end{matrix}\right\}O=\left.\begin{matrix}(C_2H_3O)\\(C_2H_5)\end{matrix}\right\}O+\left.\begin{matrix}H\\H\end{matrix}\right\}O$$

Ethyl hydrate. Acetic acid. Ethyl acetate. Water.

$$\left.\begin{matrix}K\\H\end{matrix}\right\}O+\left.\begin{matrix}(C_2H_3O)\\H\end{matrix}\right\}O=\left.\begin{matrix}(C_2H_3O)\\K\end{matrix}\right\}O+\left.\begin{matrix}H\\H\end{matrix}\right\}O$$

Potassium hydrate. Acetic acid. Potassium acetate. Water.

As the metallic hydrates may be considered as formed by the union of one atom of the metallic element with a number of groups OH′, corresponding to its valence, so the alcohols are formed by union of an unoxidized radical with a number of groups OH′, equal to or less than the number of free valences of the radical. When the alcohol contains one OH, it is designated as *monoatomic ;* when two, *diatomic ;* when three, *triatomic*, etc.

The simplest alcohols are those of this series derivable from the saturated hydrocarbons, and having the general formula $C_nH_{2n+2}O$, or $C_nH_{2n+1}OH$. They may be formed synthetically : (1.) By acting upon the corresponding iodide with potassium hydrate : $C_2H_5I + KHO = KI + C_2H_5OH$. (2.) From the alcohol next below it in the series, by direct addition of CH_2, only, however, by a succession of five reactions. (3.) By the action of H_2SO_4 and H_2O upon the corresponding hydrocarbon of the series C_nH_{2n}.

The saturated monoatomic alcohols are, however, not limited to one corresponding to each alcoholic radical. There exist—corresponding to the higher alcohols—a number of substances having the same centesimal composition and the same alcoholic properties, but differing in their physical characters and in their products of decomposition and oxidation. These isomeres have been the subject of much careful study of late years. It has been found that the molecules of methyl, ethyl, and other higher alcohols are made up of the group $(CH_2OH)'$ united to H or to C_nH_{2n+1}, thus :

CH_2OH	CH_2OH	CH_2OH
|	|	|
H	CH_3	C_2H_5
Methyl alcohol.	Ethyl alcohol.	Propyl alcohol.

and all monoatomic alcohols containing this group, CH_2OH, have been designated as *primary alcohols*. Isomeric with these are other bodies,

which, in place of the group $(CH_2OH)'$, contain the group $(CHOH)''$, and are distinguished as *secondary alcohols*. Thus we have :

Primary propyl alcohol.	Secondary propyl alcohol.
$(CH_2OH)'$	CH_3
	|
|	$(CHOH)''$
C_2H_5	|
	CH_3
C_3H_8O	C_3H_8O

And further, other isomeric substances are known which contain the group $(COH)'''$, and which are called *tertiary alcohols*, thus :

Primary amylic alcohol.	Secondary amylic alcohol.	Tertiary amylic alcohol.
$(CH_2OH)'$	C_2H_5	CH_3
|	|	|
C_4H_9	$(CHOH)''$	$(C_2H_5)—(COH)'''$
	|	|
	C_2H_5	CH_3
$C_5H_{12}O$	$C_5H_{12}O$	$C_5H_{12}O$

The alcohols of these three classes are distinguished from each other principally by their products of oxidation. The primary alcohols yield by oxidation, first an aldehyde and then an acid, each containing the same number of C atoms as the alcohol, and formed, the aldehyde by the removal of H_2 from the group (CH_2OH), and the acid by the substitution of O for H_2 in the same group, thus:

CH_2OH	COH	COOH
|	|	|
CH_3	CH_3	CH_3
Ethyl alcohol.	Ethyl aldehyde.	Acetic acid.

In the case of the secondary alcohols, the first product of oxidation is a *ketone*, containing the same number of C atoms as the alcohol, and formed by the substitution of O for HOH in the distinguishing group :

CH_3
|
CHOH
|
CH_3

Secondary propyl alcohol.

CH_3
|
CO
|
CH_3

Propyl ketone or acetone.

The tertiary alcohols yield by oxidation ketones or acids, whose molecules contain a less number of C atoms than the alcohol from which they are derived.

But the complication does not end here; isomeres exist corresponding to the higher alcohols, which are themselves primary alcohols, and contain the group $(CH_2OH)'$. Thus there exist no less than seven distinct sub-

180

stances, all having the centesimal composition of amyl alcohol, $C_5H_{12}O$, and the properties of alcohols; and theoretical considerations point to the probable existence of an eighth. Of these eight substances, four are primary, three secondary alcohols, and the remaining one a tertiary alcohol. As each of these bodies contains the group of atoms characteristic of the class of alcohol to which it belongs, it is obvious that the differences observed in their properties are due to differences in the arrangement of the other atoms of the molecule. Experimental evidence, which it would require too much space to discuss in this place, has led chemists to ascribe the following formulæ of constitution to these isomeres.

Primary amylic alcohols:

$CH_3—CH_2—CH_2—CH_2—CH_2,OH$

Normal amylic alcohol.

$CH_3 \backslash$
$CH_3 /$ CH—CH_2—CH_2,OH

Active amylic alcohol of fermentation.

$CH_3 \backslash$
$CH_3—CH_2 /$ CH—CH_2,OH

Inactive amylic alcohol of fermentation.

$CH_3 \backslash$
CH_3—C—CH_2,OH
$CH_3 /$

Unknown.

Secondary amylic alcohols:

$CH_3—CH_2 \backslash$
$CH_3—CH_2 /$ CH,OH

Diethyl carbinol.

$CH_3 \backslash$
$CH_3—CH_2—CH_2 /$ CH,OH

Methyl-propyl carbinol.

$CH_3 \backslash$ $CH_3 \backslash$
$CH_3 /$ CH / CH,OH

Methyl-isopropyl carbinol.

Tertiary amylic alcohol:

$CH_3 \backslash$
CH_3—C,OH
$CH_3—CH_2 /$

Methyl hydrate—*Carbinol—Pyroxylic spirit—Wood spirit*—**CH_3HO—32**—may be formed from marsh-gas, CH_3H, by first converting it into the iodide and acting upon this with potassium hydrate: $CH_3I + KHO = KI + CH_3HO$. It is usually obtained by the destructive distillation of wood. The *crude wood vinegar* so produced is a mixture of acetic acid and methyl alcohol with a variety of other products. The crude vinegar, separated from tarry products, is redistilled; the first tenth of the distil-

181

late is treated with quicklime and again distilled; the distillate treated with dilute H_2SO_4; decanted and again distilled. The product, still quite impure, is the *wood alcohol, wood naphtha*, or *pyroxylic spirit* of commerce. The pure hydrate can only be obtained by decomposing a crystalline compound, such as methyl oxalate, and rectifying the product until the boiling-point is constant at 66°.5 (151°.7 F.).

Pure methyl alcohol is a colorless liquid, having an ethereal and alcoholic odor, and a sharp, burning taste; sp. gr. 0.814 at 0°; boils at 66°.5 (151°.7 F.); burns with a pale flame, giving less heat than that of ethylic alcohol; mixes with water, alcohol, and ether in all proportions; is a good solvent of resinous substances, and also dissolves sulphur, phosphorus, potash, and soda.

Methyl hydrate is not affected by exposure to air under ordinary circumstances, but in the presence of platinum-black it is oxidized, with formation of the corresponding aldehyde and acid, formic acid. Hot HNO_3 decomposes it with formation of nitrous fumes, formic acid and methyl nitrate. It is acted upon by H_2SO_4 in the same way as ethyl alcohol. The organic acids form methyl ethers with it. With HCl under the influence of a galvanic current, it forms an oily substance having the composition C_2H_3ClO.

Methylated spirit is ethyl alcohol containing sufficient wood spirit to render it unfit for the manufacture of ardent spirits, by reason of the disgusting odor and taste which crude wood alcohol owes to certain empyreumatic products which it contains. Spirits so treated are not subject to the heavy duties imposed upon ordinary alcohol, and are, therefor, largely used in the arts and for the preservation of anatomical preparations. It contains one-ninth of its bulk of wood naphtha.

Ethyl hydrate—*Ethylic alcohol—Methyl carbinol—Vinic alcohol—Alcohol—Spirits of wine*—**C_2H_5HO—46.**

PREPARATION.—Industrially alcohol and alcoholic liquids are obtained from substances rich in starch or glucose.

The manufacture of alcohol consists of three distinct processes: 1st, the conversion of starch into sugar; 2d, the fermentation of the saccharine liquid; 3d, the separation, by distillation, of the alcohol formed by fermentation. The raw materials for the first process are *malt* and some substance (grain, potatoes, rice, corn, etc.) containing starch. Malt is barley which has been allowed to germinate, and, at the proper stage of germination, roasted. During this growth there is developed in the barley a peculiar nitrogenous principle called *diastase*. The starchy material is mixed with a suitable quantity of malt and water, and the mass maintained at a temperature of 65°-70° (149°-158° F.) for two to three hours, during which the diastase rapidly converts the starch into *dextrin*, and this in turn into *glucose*.

The saccharine fluid, or *wort*, obtained in the first process, is drawn off, cooled, and *yeast* is added. As a result of the growth of the yeast-plant, a complicated series of chemical changes take place, the principal one of which is the splitting up of the glucose into carbon dioxide and alcohol: $C_6H_{12}O_6 = 2C_2H_5OH + 2CO_2$. There are formed at the same time small quantities of glycerin, succinic acid, and propyl, butyl, and amyl alcohols.

An aqueous fluid is thus obtained which contains 3-15 per cent. of alcohol; this is then separated by the third process, that of distillation and rectification. The apparatus used for this purpose has been so far perfected that by a single distillation an alcohol of 90-95 per cent. can be obtained.

182

In some cases alcohol is prepared from fluids rich in glucose, such as grape-juice, molasses, syrup, etc.; in such cases the first process becomes unnecessary.

Commercial alcohol always contains H_2O, and when pure or *absolute* alcohol is required, the commercial product must be mixed with some hygroscopic solid substance, such as quicklime, from which it is distilled after having remained in contact twenty-four hours.

FERMENTATION.—This term, derived from *fervere* = to boil, was originally applied to alcoholic fermentation, by reason of the bubbling of the saccharine liquid caused by the escape of CO_2; subsequently it came to be applied to all decompositions similarly attended by the escape of gas.

At present it is used by many authors to apply to a number of heterogeneous processes; and some writers distinguish between "true" and "false" fermentation. It is best, we believe, to limit the application of the term to those decompositions designated as *true fermentations*.

Fermentation is a decomposition of an organic substance, produced by the processes of nutrition of a low form of animal or vegetable life.

The true ferments are therefor all *organized beings*, such as *torula cerevisiæ*, producing alcoholic fermentation; *penicillium glaucum*, producing lactic acid fermentation; and *mycoderma aceti*, producing acetic acid fermentation.

The false fermentations are not produced by an organized body, but by a soluble, unorganized, nitrogenous substance, whose method of action is as yet imperfectly understood. They may be, therefor, designated by the term *cryptolysis*. Diastase, pepsin and trypsin are *cryptolytes*.

PROPERTIES.—Alcohol is a thin, colorless, transparent liquid, having a spirituous odor, and a sharp, burning taste; sp. gr. 0.8095 at 0°, 0.7939 at 15° (59° F.); it boils at 78°.5 (173°.3 F.), and has not been solidified; at temperatures below −90° (−130° F.) it is viscous. It mixes with water in all proportions, the union being attended by elevation in temperature and contraction in volume (after cooling to the original temperature). It also attracts moisture from the air to such a degree that absolute alcohol only remains such for a very short time after its preparation. It is to this power of attracting H_2O that alcohol owes its preservative power for animal substances. It is a very useful solvent, dissolving a number of gases, most of the mineral and organic acids and alkalies, most of the chlorides and carbonates, some of the nitrates, all the sulphates, essences, and resins. Alcoholic solutions of fixed medicinal substances are called *tinctures*; those of volatile principles, *spirits*.

The action of oxygen upon alcohol varies according to the conditions. Under the influence of energetic oxidants, such as chromic acid, or, when alcohol is burned in the air, the oxidation is rapid and complete, and is attended by the extrication of much heat, and the formation of carbon dioxide and water: $C_2H_6O + 3O_2 = 2CO_2 + 3H_2O$. Mixtures of air and vapor of alcohol explode upon contact with flame. If a less active oxidant be used, such as platinum-black, or by the action of atmospheric oxygen at low temperatures, a simple oxidation of the alcoholic radical takes place, with formation of acetic acid $\left.\begin{matrix}C_2H_5\\H\end{matrix}\right\}O + O_2 = \left.\begin{matrix}C_2H_3O\\H\end{matrix}\right\}O + H_2O$, a reaction which is utilized in the manufacture of acetic acid and vinegar. If the oxidation be still further limited, *aldehyde* is formed: $2C_2H_6O + O_2 = 2C_2H_4O + 2H_2O$. If vapor of alcohol be passed through a tube filled with platinum sponge and heated to redness, or if a coil of heated platinum wire be introduced into an atmosphere of alcohol vapor, the products of

183

oxidation are quite numerous: among them are water, ethylene, aldehyde, acetylene, carbon monoxide, and acetal. Heated platinum wire introduced into vapor of alcohol continues to glow by the heat resulting from the oxidation, a fact which has been utilized in the thermocautery.

Chlorine and bromine act energetically upon alcohol, producing a number of chlorinated and brominated derivatives, the final products being *chloral* and *bromal* (q. v.). If the action of Cl be moderated, aldehyde and HCl are first produced. Iodine acts quite slowly in the cold, but old solutions of I in alcohol (tr. iodine) are found to contain HI, ethyl iodide, and other imperfectly studied products. In the presence of an alkali, I acts upon alcohol to produce iodoform. Potassium and sodium dissolve in alcohol with evolution of H; upon cooling, a white solid crystallizes, which is the double oxide of ethyl and the alkaline metal. Nitric acid, aided by a gentle heat, acts violently upon alcohol, producing nitrous ether, brown fumes, and products of oxidation. For the action of other acids upon alcohol see the corresponding ethers. The hydrates of the alkaline metals dissolve in alcohol, but react upon it slowly; the solution turns brown and contains an acetate. If alcohol be gently heated with HNO_3 and nitrate of silver or of mercury, a gray precipitate falls, which is silver or mercury fulminate.

Varieties.—It occurs in different degrees of concentration: *absolute alcohol* is pure alcohol, C_2H_6O. It is not purchasable and must be made as required; the so-called absolute alcohol of the shops is rarely stronger that 98 per cent. *Alcohol* (*U. S.*), sp. gr. 0.820, contains 94 per cent. by volume, and *spiritus rectificatus* (*Br.*), sp. gr. 0.838, contains 84 per cent. This is the ordinary rectified spirit used in the arts. *Alcohol dilutum* (*U. S.*) =*Spiritus tenuior* (*Br.*), sp. gr. 0.920, used in the preparation of tinctures, contains 53 per cent. It is of about the same strength as the *proof spirit* of commerce.

Analytical Characters.—(1.) Heated with a small quantity of solution of potassium dichromate and H_2SO_4, the liquid assumes an emerald-green color, and if the quantity of C_2H_6O be not very small, the peculiar fruity odor of aldehyde is developed.

(2.) Warmed and treated with a few drops of potash solution and a small quantity of iodine, an alcoholic liquid deposits a yellow, crystalline ppt. of iodoform, either immediately or after a time.

(3.) If HNO_3 be added to a liquid containing C_2H_6O, nitrous ether, recognizable by its odor, is given off. If a solution of mercurous nitrate with excess of HNO_3 be then added, and the mixture heated, a further evolution of nitrous ether occurs, and a yellow-gray deposit of fulminating mercury is formed, which may be collected, washed, dried and exploded.

(4.) If an alcoholic liquid be heated for a few moments with H_2SO_4 diluted with H_2O and distilled, the distillate, on treatment with H_2SO_4 and potassium permanganate, and afterward with sodium hyposulphite, yields aldehyde, which may be recognized by the production of a violet color with a dilute solution of fuchsin.

None of the above reactions, *taken singly*, is characteristic of alcohol.

Action on the Economy.—In a concentrated form, alcohol exerts a dehydrating action upon animal tissues with which it comes in contact; causing coagulation of the albuminoid constituents. When diluted, ethylic alcohol may be a food, a medicine, or a poison, according to the dose and the condition of the person taking it. When taken in excessive doses, or in large doses for a long time, it produces symptoms and lesions characteristic of pure alcoholism, acute or chronic, modified or

184

aggravated by those produced by other substances, such as amyl alcohol, which accompany it in the alcoholic fluids used as beverages. Taken in moderate quantities, with food, it aids digestion and produces a sense of comfort and exhilaration. As a medicine it is the most valuable of stimulants.

Much has been written concerning the value of alcohol as a food. If it have any value as such, it is as a producer of heat and force by its oxidation in the body; experiments have failed to show that more than a small percentage (16 per cent. in 24 hrs.) of medium doses of alcohol ingested are eliminated by all channels; the remainder, therefor, disappears in the body, as the idea that it can there "accumulate" is entirely untenable. That some part should be eliminated unchanged is to be expected from the rapid diffusion and the high volatility of alcohol.

On the other hand, if alcohol be oxidized in the body, we should expect, in the absence of violent muscular exercise, an increase in temperature, and the appearance in the excreta of some product of oxidation of alcohol: aldehyde, acetic acid, carbon dioxide, or water, while the elimination of nitrogenous excreta, urea, etc., would remain unaltered or be diminished. While there is no doubt that excessive doses of alcohol produce a diminution of body temperature, the experimental evidence concerning the action in this direction of moderate doses is conflicting and incomplete. Of the products of oxidation, aldehyde has not been detected in the excreta, and acetic acid only in the intestinal canal. The elimination of carbonic acid, as such, does not seem to be increased, although positive information upon this point is wanting. If acetic acid be produced, this would form an acetate, which in turn would be oxidized to a carbonate, and eliminated as such by the urine. The elimination of water under the influence of large doses of alcohol is greater than at other times: but whether this water is produced by the oxidation of the hydrogen of the alcohol, or is removed from the tissues by its dehydrating action, is an open question.

While physiological experiment yields only uncertain evidence, the experience of arctic travellers and others shows that the use of alcohol tends to diminish rather than increase the capacity to withstand cold. The experience of athletes and of military commanders is that intense and prolonged muscular exertion can be best performed without the use of alcohol. The experience of most literary men is that long-continued mental activity is more difficult with than without alcohol.

In cases of acute poisoning by alcohol, the stomach-pump and catheter should be used as early as possible. A plentiful supply of air, the cold douche, and strong coffee are indicated.

Alcoholic Beverages.—The variety of beverages in whose preparation alcoholic fermentation plays an important part is very great, and the products differ from each other materially in their composition and in their physiological action. They may be divided into four classes, the classification being based upon the sources from which they are obtained and upon the method of their preparation.

I.—Those prepared by the fermentation of malted grain—*beers, ales, and porters*.

II.—Those prepared by the fermentation of grape juice—*wines*.

III.—Those prepared by the fermentation of the juices of fruits other than the grape—*cider, fruit-wines*.

IV.—Those prepared by the distillation of some fermented saccharine liquid—*ardent spirits*.

185

Beer, ale, and porter are aqueous infusions or decoctions of malted grain, fermented and flavored with hops; they contain, therefor, the soluble constituents of the grain employed; dextrin and glucose, produced during the malting; alcohol and carbon dioxide, produced during the fermentation; and the soluble constituents of the flavoring material. The alcoholic strength of malt liquors varies from 1.5 to 9 per cent. Weiss beer contains 1.5-1.9 per cent.; lager, 4.1-4.5 per cent.; bock beer, 3.88-5.23 per cent.; London porter, 5.4-6.9 per cent.; Burton ale, 5.9 per cent.; Scotch ale, 8.5-9 per cent. Malt liquors all contain a considerable quantity of nitrogenous material (0.4-1 per cent. N), and succinic, lactic, and acetic acids. The amount of inorganic material, in which the phosphates of potassium, sodium, and magnesium predominate largely, varies from 0.2 to 0.3 per cent. The sp. gr. is from 1.014 to 1.033.

The adulterations of malt liquors are numerous and varied. Sodium carbonate is added with the double purpose of neutralizing an excess of acetic acid and increasing the foam. The most serious adulteration consists in the introduction of bitter principles other than hops, and notably of strychnine, cocculus indicus (picrotoxin), and picric acid.

Wines are produced by the fermentation of grape-juice: in the case of red wines the *marc*, or mass of skins, seed and stems, is allowed to remain in contact with the *must*, or fermenting juice, until, by production of alcohol, the liquid dissolves a portion of the coloring-matter of the skins. A certain proportion of tannin is also dissolved, whose presence is necessary to prevent *stringiness*. Sweet wines are produced from must rich in glucose and by arresting the fermentation before that sugar has been completely decomposed. Dry wines are obtained by more complete fermentation of must less rich in glucose. Tartaric acid is the predominating acid in grape-juice, and as the proportion of alcohol increases during fermentation the acid potassium tartrate is deposited.

Most wines of good quality improve in flavor with age, and this improvement is greatly hastened by the process of *pasteuring*, which consists in warming the wine to a temperature of 60° C. (140° F.), without contact of air.

Light wines are those whose percentage of alcohol is less than 12 per cent. In this class are included the clarets, Sauternes, Rhine, and Moselle wines; champagnes, Burgundies, the American wines (except some varieties of California wine) Australian, Greek, Hungarian, and Italian wines.

The champagnes and some Moselle wines are sparkling, a quality which is communicated to them by bottling them before the fermentation is completed, thus retaining the carbon dioxide, which is dissolved by virtue of the pressure which it exerts. When properly prepared they are agreeable to the palate, and assist the digestion; when new, however, they are liable to communicate their fermentation to the contents of the stomach and thus seriously disturb digestion.

Of the still wines, the most widely used are the *clarets, Vinum rubrum* (U. S.), or red Bordeaux wines, and the *hocks, Vinum album* (U. S.), or white Rhine, Moselle and American wines. The former are of low alcoholic strength, mildly astringent, and contain but a small quantity of nitrogenous material, qualities which render them particularly adapted to table use and as mild stimulants. The Rhine wines are thinner and more acid, and generally of lower alcoholic strength than the clarets. The Burgundy and Rhone wines are celebrated for their high flavor and body; they are not strongly alcoholic, but contain a large quantity of nitrogenous material, to which they are indebted for their notoriety as developers of

186

gout. Our native American wines, particularly those of the Ohio Valley and of California, are yearly improving in flavor and quality; they more closely resemble the Rhine wines and Sauternes than other European wines.

Heavy wines are those whose alcoholic strength is greater than 12 per cent., usually 14 to 17 per cent.; they include the sherries, ports, Madeiras, Marsala, and some California wines, and are all the products of warm climates. Sherry is an amber-colored wine, grown in the south of Spain, *Vinum Xericum* (Br.). Marsala closely resembles sherry in appearance, and is frequently substituted for it. Port is a rich, dark red wine, grown in Portugal.

The adulteration of wine by the addition of *foreign* substances is confined almost entirely to their artificial coloration, which is produced by the most various substances, indigo, logwood, fuchsine, etc. The addition of natural constituents of wines, obtained from other sources, and the mixing of different grades of wine are, however, extensively practised. Water and alcohol are the chief substances so added; an excess of the former may be detected by the taste, and the low sp. gr. after expulsion of the alcohol. Most wines intended for export are *fortified* by the addition of alcohol; when the alcoholic spirit used is free from amyl alcohol, and is added in moderate quantities, there can be no serious objection to the practice, especially when applied to certain wines which, without such treatment, do not bear transportation. The mixing of fine grades of wine with those of a poorer quality is extensively practised, particularly with champagnes, clarets, and Burgundies, and is perfectly legitimate. The same cannot be said, however, of the manufacture of factitious wine, either entirely from materials not produced from the grape, or by converting white into red wines, or by mixing wines with coloring matters, alcohol, etc., to produce imitations of wines of a different class, an industry which flourishes extensively in Normandy, at Bingen on the Rhine, and at Hamburg. The wines so produced are usually heavy wines, port and sherry so-called.

Cider is the fermented juice of the apple, prepared very much in the same way as wine is from grape-juice, and containing 3.5 to 7.5 per cent. of alcohol. It is very prone to acetous fermentation, which renders it sour and not only unpalatable, but liable to produce colic and diarrhœa with those not hardened to its use.

Spirits are alcoholic beverages, prepared by fermentation and distillation. They differ from beers and wines in containing a greater proportion of alcohol, and in not containing any of the non-volatile constituents of the grains or fruits from which they are prepared. Besides alcohol and water they contain acetic, butyric, valerianic and œnanthic ethers, to which they owe their flavor; sometimes tannin and coloring matter derived from the cask; amylic alcohol remaining after imperfect purification; sugar intentionally added; and caramel. It is to the last-named substance that all dark spirits owe their color; although, after long keeping in wood a naturally colorless spirit assumes a straw color.

The varieties of spirituous beverages in common use are:

Brandy, spiritus vini gallici (U. S., Br.), obtained by the distillation of wine, and manufactured in France and in California and Ohio. It is of sp. gr. 0.929 to 0.934, is dark or light in color, according to the quantity of burnt sugar added, and contains about 1.2 per cent. of solid matter. *American whiskey, spiritus frumenti* (U. S.), prepared from wheat, rye, barley, or Indian corn; has a sp. gr. of 0.922 to 0.937 and contains 0.1 to

187

0.3 per cent. of solids. *Scotch and Irish whiskies*, colorless spirits distilled from fermented grains: sp. gr. 0.915 to 0.920, having a peculiar smoky flavor produced by drying the malted grain by a peat fire. *Gin*, also distilled from malted grain, sp. gr. 0.930 to 0.944, flavored with juniper, and sometimes fraudulently with turpentine. *Rum*, a spirit distilled from molasses, and varying in color and flavor from the dark *Jamacia rum* to the colorless *St. Croix rum*. The former is of sp. gr. 0.914 to 0.926, and contains one per cent. of solid matter.

Liqueurs are spirits sweetened and flavored with vegetable aromatics, and frequently colored; *anisette* is flavored with aniseed; *absinthe*, with wormwood; *curaçoa*, with orange-peel; *kirschwasser*, with cherries, the stones being cracked and the spirits distilled from the bruised fermented fruit; *kümmel*, with cummin and caraway seeds; *maraschino*, with cherries; *noyeau*, with peach and apricot kernels.

Propyl hydrate—*Ethyl carbinol—Primary propyl alcohol*—C_3H_7OH—60—is produced, along with ethylic alcohol, during fermentation, and obtained by fractional distillation of marc brandy, from *cognac oil, huile de marc* (not to be confounded with oil of wine), an oily matter, possessing the flavor of inferior brandy, which separates from marc brandy, distilled at high temperatures; and from the residues of manufacture of alcohol from beet-root, grain, molasses, etc. It is a colorless liquid, has a hot alcoholic taste, and a fruity odor; boils at 96.7° (206°.1 F.); and is miscible with water. It has not been put to any use in the arts. Its intoxicating and poisonous actions are greater than those of ethyl alcohol. It exists in small quantity in cider.

Butyl alcohols—C_4H_9OH—74.—Of the four butyl alcohols theoretically possible three are known to exist:

Primary normal butyl alcohol—Butyl alcohol of fermentation—Propyl carbinol—$CH_3—CH_2—CH_2—CH_2OH$—is formed in small quantities during alcoholic fermentation, and may be obtained by repeated fractional distillation from the oily liquid left in the rectification of vinic alcohol. It is a colorless liquid; boils at 114°.7 (238°.5 F.). It is more actively poisonous than ethyl or methyl alcohol.

Secondary butyl alcohol; ethyl-methyl carbinol—$\begin{matrix}CH_3—CH_2\\CH_3\end{matrix}\!>\!CHOH$— a liquid which boils at 99° (210°.2 F.).

Tertiary butyl alcohol; trimethyl carbinol, $\begin{matrix}CH_3\\CH_3\\CH_3\end{matrix}\!—COH$—a crystalline solid, which fuses at 20°–25° (68°–77° F.), and boils at 82° (179°.6 F.).

Amylic alcohols—$C_5H_{11}OH$—88.—Of the eight amyl alcohols theoretically possible (see p. 180) seven have been obtained. The substance usually known as **amylic alcohol**, *potato spirit, fusel oil, alcohol amylicum* (Br.), is a mixture in varying proportions of the two primary alcohols; $\begin{matrix}CH_3\\CH_3\end{matrix}\!>\!CH—CH_2—CH_2OH$ and $\begin{matrix}CH_3—CH_2\\CH_3\end{matrix}\!>\!CH—CH_2OH$; the former differing from the latter in that it deviates the plane of polarization to the left ($[a]_D = -4°.36'$); in its boiling-point being 2° (3°.6 F.) lower, and in the greater solubility of the amyl-sulphate of barium obtained from it.

It is formed during alcoholic fermentation of glucose in greater abundance than any of the alcohols other than the ethylic. Owing to its high boiling-point, it is in great part retained in the oily material which collects in the still during the rectification of alcohol and spirits; a portion, however, passes over and is removed by subsequent treatment (see below). It

188

is obtained from the last milky products of rectification of alcoholic fluids made from grain or potatoes; these are shaken with H_2O to remove ethyl alcohol, the supernatant oily fluid is decanted, dried by contact with fused calcium chloride, and distilled; that portion which passes over between 128° and 132° (262°.4–269°.6 F.) being collected.

It is a colorless, oily liquid, has an acrid taste and a peculiar odor, at first not unpleasant, afterward nauseating and provocative of severe headache; it boils at 132° (269°.6 F.) and crystallizes at −20° (4° F.); sp. gr. 0.8184 at 15° (5° F.); it mixes with alcohol and ether, but not with water. It burns difficultly with a pale blue flame.

When exposed to air it oxidizes very slowly; quite rapidly, however, in contact with platinum-black, forming valerianic acid. The same acid, along with other substances, is produced by the action of the more powerful oxidants upon amyl alcohol. Chlorine attacks it energetically, forming amyl chloride, HCl, and other chlorinated derivatives. Sulphuric acid dissolves in amyl alcohol, with formation of amyl-sulphuric acid, $SO_4(C_5H_{11})H$, corresponding to ethyl-sulphuric acid. It also forms similar acids with phosphoric, oxalic, citric, and tartaric acids. Its ethers, when dissolved in ethyl alcohol, have the taste and odor of various fruits, and are used in the preparation of artificial fruit-essences. Amyl alcohol is also used in analysis as a solvent, particularly for certain alkaloids, and in pharmacy for the artificial production of valerianic acid and the valerianates.

Its vapor, when inhaled, produces severe headache, a sense of suffocation, giddiness, and, in large doses, death. The liquid, taken internally, especially when in alcoholic solution, is much more actively poisonous than ethylic alcohol. Even in very dilute solution it produces the rapid intoxication, and severe headache and vertigo, which are prominent effects of inferior whiskey.

To free spirits of amyl alcohol, to *defuselate* them, advantage is usually taken of the absorbent power of freshly burnt wood charcoal, which is either placed in the still or made into a filter, through which the spirit is passed after distilliation, or, preferably, the vapor from the still is made to pass through a layer of charcoal before condensation. Spirits properly freed of fusel oil give off no irritating or foul fumes, when hot; they are not colored red when mixed with three parts C_2H_6O and one part strong H_2SO_4; they are not colored red or black by ammoniacal silver nitrate solution; when 150 parts of the spirit mixed with 1 part potash, dissolved in a little H_2O, are evaporated down to 15 parts, and mixed with an equal volume of dilute H_2SO_4, no offensive odor should be given off.

Cetyl hydrate—*Cetylic alcohol—Ethal*—$C_{16}H_{33}OH$—242—is obtained by the saponification of spermaceti (its palmitic ether). It is a white crystalline solid; fusible at 49° (120°.2 F.); insoluble in H_2O; soluble in alcohol and ether; tasteless and odorless.

Ceryl hydrate—$C_{27}H_{55}OH$—396—and **Myricyl hydrate**—$C_{30}H_{61}OH$—438—are obtained as white, crystalline solids: the former from China wax; the latter from beeswax, by saponification.

SIMPLE ETHERS.

Oxides of Alcoholic Radicals of the Series C_nH_{2n+1}.

The term *ether* was originally applied to any volatile liquid obtained by the action of an acid upon an alcohol.

The simple ethers are the oxides of the alcoholic radicals. They bear the

189

same relation to the alcohols that the oxides of the basylous elements bear to their hydrates:

$\left.\begin{matrix}C_2H_5\\C_2H_5\end{matrix}\right\}O$	$\left.\begin{matrix}K\\K\end{matrix}\right\}O$	$\left.\begin{matrix}C_2H_5\\H\end{matrix}\right\}O$	$\left.\begin{matrix}K\\H\end{matrix}\right\}O$
Ethyl oxide (ethylic ether).	Potassium oxide.	Ethyl hydrate (alcohol.)	Potassium hydrate.

When the two alcoholic radicals are the same, as in the above instance, the ether is designated as *simple;* when the radicals are different, as in methyl-ethyl oxide, $\left.\begin{matrix}CH_3\\C_2H_5\end{matrix}\right\}O$, they are called *mixed ethers.*

Methyl oxide—$\left.\begin{matrix}CH_3\\CH_3\end{matrix}\right\}O$—46—isomeric with ethyl alcohol, is obtained by the action of H_2SO_4 and boric acid upon methyl alcohol, or by the action of silver oxide on methyl iodide. It is a colorless gas; has an ethereal odor; burns with a pale flame; liquefies at −36° (−32°.8 F.); and boils at −21° (−5°.8 F.); is soluble in H_2O, H_2SO_4 and ethyl alcohol.

Ethyl oxide—*Ethylic ether—Ether—Sulphuric ether—Æther fortior* (*U. S.*)—*Æther purus* (*Br.*)—$\left.\begin{matrix}C_2H_5\\C_2H_5\end{matrix}\right\}O$—74.

Preparation.—A mixture is made of 5 pts. of alcohol, 90 %, and 9 pts. of concentrated H_2SO_4, in a vessel surrounded by cold H_2O. This mixture is introduced into a retort, over which is conveniently arranged a vessel from which a slow stream of alcohol can be made to enter the retort. Heat is applied by a sand-bath, and the addition of alcohol and the heat are so regulated that the temperature does not rise above 140° (284° F.). The retort is connected with a well-cooled condenser, and the process continued until the temperature in the retort rises above the point indicated. It is important that the tube by which the alcohol is introduced be drawn out to a small opening, and dip well down below the surface of the liquid. The distillate thus obtained contains, besides ether, alcohol, water, and gases resulting from the decomposition of the alcohol and H_2SO_4, notably SO_2. It is subjected to a first purification by shaking with H_2O containing potash or lime, decanting the supernatant ether and redistilling. The product of this process is "washed ether," or *æther* (U. S.). It is still contaminated with water and alcohol, and when desired pure, as for producing anæsthesia and for processes of analysis, it is subjected to a second purification. It is again shaken with H_2O, decanted after separation, shaken with recently fused calcium chloride and newly burnt lime, with which it is left in contact 24 hours, and from which it is then distilled.

It was known at an early day that a small quantity of H_2SO_4 is capable of converting a large quantity of alcohol into ether, and that at the end of the process the H_2SO_4 remains in the retort unaltered, except by secondary reactions. A metaphysical explanation of the process was found in the assertion that the acid acted by its mere presence, by *catalysis*, as it was said; in other words, it acted because it acted, a very ready but a very feminine method of explaining what is not understood, which is still invoked by some authors as a covering for our ignorance of the rationale of certain chemico-physiological phenomena. It was only in 1850 that Alex. Williamson, by a series of ingenious experiments, determined the true nature of the process. In the conversion of alcohol into ether, an intermediate substance, sulphovinic acid, is alternately formed at the expense of the alcohol, and destroyed with formation of ether and regen-

190

eration of H_2SO_4. At first H_2SO_4 and alcohol act upon each other, molecule for molecule, to form H_2O and sulphovinic acid: $\left.\begin{matrix}C_2H_5\\H\end{matrix}\right\}O + \left.\begin{matrix}SO_2\\H_2\end{matrix}\right\}O_2 = \left.\begin{matrix}H\\H\end{matrix}\right\}O + \left.\begin{matrix}SO_2\\C_2H_5\\H\end{matrix}\right\}O_2$. The new acid, as soon as formed, reacts with a second molecule of alcohol, with regeneration of H_2SO_4 and formation of ether: $\left.\begin{matrix}SO_2\\C_2H_5\\H\end{matrix}\right\}O_2 + \left.\begin{matrix}C_2H_5\\H\end{matrix}\right\}O = \left.\begin{matrix}SO_2\\H_2\end{matrix}\right\}O_2 + \left.\begin{matrix}C_2H_5\\C_2H_5\end{matrix}\right\}O.$

Theoretically, therefor, a given quantity of H_2SO_4 could convert an unlimited amount of alcohol into ether. Such would also be the case in practice, were it not that the acid gradually becomes too dilute, by admixture with the H_2O formed during the reaction, and at the same time is decomposed by secondary reactions, into which it enters with impurities in the alcohol; causes which in practice limit the amount of ether produced to about four to five times the bulk of acid used.

Properties.—*Physical.*—Ether is a colorless, limpid, mobile, highly refracting liquid; it has a sharp, burning taste, and a peculiar, tenacious odor, characterized as ethereal. Sp. gr. 0.723 at 12°.5 (54°.5 F.); it boils at 34°.5 (94°.1 F.), and crystallizes at −31° (−23°.8 F.). Its tension of vapor is very great, especially at high temperatures; it should, therefor, be stored in strong bottles, and should be kept in situations protected from elevations of temperature. It is exceedingly volatile, and, when allowed to evaporate freely, absorbs a great amount of heat, of which property advantage is taken to produce local anæsthesia, the part being benumbed by the cold produced by the rapid evaporation of ether sprayed upon the surface. Water dissolves one-ninth its weight of ether. Ethylic and methylic alcohols are miscible with it in all proportions. Ether is an excellent solvent of many substances not soluble in water and alcohol, while, on the other hand, it does not dissolve many substances soluble in those fluids. The resins and fats are readily soluble in ether; the salts of the alkaloids and many vegetable coloring matters are soluble in alcohol and water, but insoluble in ether, while the free alkaloids are for the most part soluble in ether, but insoluble, or very sparingly soluble, in water.

Chemical.—Ether, whether in the form of vapor or of liquid, is highly inflammable; and burns with a luminous flame. The vapor forms with air a violently explosive mixture. It is denser than air, through which it falls and diffuses itself to a great distance; great caution is therefor required in handling ether in a locality in which there is a light or fire, especially if the fire be near the floor.

Pure ether is neutral in reaction, but, on exposure to air or O, especially in the light, it becomes acid from the formation of a small quantity of acetic acid. H_2SO_4 mixes with ether with elevation of temperature and formation of sulphovinic acid; sulphuric anhydride forms ethyl sulphate. HNO_3, aided by heat, oxidizes ether to carbon dioxide and acetic and oxalic acids. Ether, saturated with HCl and distilled, yields ethyl chloride. Cl, in the presence of H_2O, oxidizes ether, with formation of aldehyde, acetic acid, and chloral. In the absence of H_2O, however, a series of products of substitution are produced, in which 2, 4 and 10 atoms of H are replaced by a corresponding number of atoms of Cl. These substances in turn, by substitution of alcoholic radicals, or of atoms of elements, for atoms of Cl, give rise to other derivatives.

191

Action on the Economy.—Ether is largely used in medicine for producing anæsthesia, either locally by diminution of temperature due to its rapid evaporation, or generally by inhalation. When taken in overdose it causes death, although it is by no means as liable to give rise to fatal accidents as is chloroform. Patients suffering from an overdose may, in the vast majority of cases, be resuscitated by artificial respiration and the induced current, one pole to be applied to the nape of the neck, and the other carried across the body just below the anterior attachments of the diaphragm.

In cases of death from ether the odor is generally well marked in the clothing and surroundings, and especially on opening the thoracic cavity. In the analysis it is sought for in the blood and lungs at the same time as chloroform (q. v.).

MONOBASIC ACIDS. Series $C_nH_{2n}O_2$.

As the higher terms of this series are obtained from the fats, and the lower terms are volatile liquids, these acids are sometimes designated as the *volatile fatty acids.* The known terms are:

Name.	Formula.	Fusing-point.	Boiling-point.	Name.	Formula.	Fusing-point.	Boiling-point.
Formic acid	CHO_2H	1°	100°	Lauric acid	$C_{12}H_{23}O_2H$	43.5°	....
Acetic acid	$C_2H_3O_2H$	17°	119	Myristic acid	$C_{14}H_{27}O_2H$	53.8°	....
Propionic acid	$C_3H_5O_2H$	....	140	Palmitic acid	$C_{16}H_{31}O_2H$	62°	....
Butyric acid	$C_4H_7O_2H$	−20°	160	Margaric acid	$C_{17}H_{33}O_2H$	60°	....
Valerianic acid	$C_5H_9O_2H$	....	175	Stearic acid	$C_{18}H_{35}O_2H$	69°	....
Caproic acid	$C_6H_{11}O_2H$	9°	198	Arachic acid	$C_{20}H_{39}O_2H$	75°	....
Œnanthylic acid	$C_7H_{13}O_2H$	...	212	Benic acid	$C_{22}H_{43}O_2H$	76°	....
Caprylic acid	$C_8H_{15}O_2H$	14°	236	Hyænic acid	$C_{25}H_{49}O_2H$	77°	...
Pelargonic acid	$C_9H_{17}O_2H$	18°	260	Cerotic acid	$C_{27}H_{53}O_2H$	78°	....
Capric acid	$C_{10}H_{19}O_2H$	27°	....	Melissic acid	$C_{30}H_{59}O_2H$	88°	....

Although formed in a variety of ways, these acids may be considered as being derived from the primary monoatomic alcohols, by the substitution of O for H_2 in the group CH_2OH:

$$CH_3—CH_2—CH_2—CH_2—CH_2,OH$$

Normal amylic alcohol.

$$CH_3—CH_2—CH_2—CH_2—CO,OH$$

Normal valerianic acid.

Considered typically, the substitution of O for H_2 occurs in the radical: $\left.\begin{matrix}C_5H_{11}\\H\end{matrix}\right\}O$—$\left.\begin{matrix}C_5H_9O\\H\end{matrix}\right\}O$, and communicates to the radical electro-negative or acid qualities.

Formic acid - HCO,OH—46—occurs in the acid secretion of red ants, in the stinging hairs of certain insects, in the blood, urine, bile, perspiration, and muscular fluid of man, in the stinging-nettle, and in the leaves of trees of the pine family. It is produced in a number of reactions; by the oxidation of many organic substances: sugar, starch, fibrin, gelatin, albumin, etc.; by the action of potash upon chloroform and kindred

192

bodies; by the action of mineral acids in hydrocyanic acid; during the fermentation of diabetic urine; by the direct union of carbon monoxide and water; by the decomposition of oxalic acid under the influence of glycerin at about 100° (212° F.).

It is a colorless liquid, having an acid taste and a penetrating odor; it acts as a vesicant; it boils at 100° (212° F.), and, when pure, crystallizes at 0° (32° F.). It is miscible with H_2O in all proportions.

The mineral acids decompose it into H_2O and carbon monoxide. Oxidizing agents convert it into H_2O and carbon dioxide. Alkaline hydrates decompose it with formation of a carbonate and liberation of H. It acts as a reducing agent with the salts of the noble metals.

Acetic acid—*Acetyl hydrate—Hydrogen acetate—Pyroligneous acid—Acidum aceticum (U. S.; Br.)*—**CH_3CO,OH—60.**

Formation.—(1.) By the oxidation of alcohol:

$$CH_3CH_2,OH + O_2 = CH_3CO,OH + H_2O.$$

(2.) By the dry distillation of wood.

(3.) By the decomposition of natural acetates by mineral acids.

(4.) By the action of potash in fusion on sugar, starch, oxalic, tartaric, citric acids, etc.

(5.) By the decomposition of gelatin, fibrin, casein, etc., by H_2SO_4 and manganese dioxide.

(6.) By the action of carbon dioxide upon sodium methyl: $CO_2 + NaCH_3 = C_2H_3O_2Na$; and decomposition of the sodium acetate so produced.

The acetic acid used in the arts and in pharmacy is prepared by the destructive distillation of wood. The products of the distillation, which vary with the nature of the wood used, are numerous. Charcoal remains in the retort, while the distilled product consists of an acid, watery liquid; a tarry material; and gaseous products. The gases are carbon dioxide, carbon monoxide, and hydrocarbons; they are sometimes used for illuminating purposes, but are usually directed into the furnace, where they serve as fuel. The tar is a mixture of empyreumatic oils, hydrocarbons, phenol, oxyphenol, acetic acid, ammonium acetate, etc.

The acid water is very complex, and contains, besides acetic acid, formic, propionic, butyric, valerianic, and oxyphenic acids, acetone, naphthalene, benzene, toluene, cumene, creasote, methyl alcohol, and methyl acetate, etc. Partially freed from tar by decantation, it still contains about 20 per cent. of tarry and oily material, and about 4 per cent. of acetic acid; this is the *crude pyroligneous acid* of commerce.

The crude product is subjected to a first purification by distillation; the first portions are collected separately and yield methyl alcohol (q. v.); the remainder of the distillate is the *distilled pyroligneous acid*, used to a limited extent as an antiseptic, but principally for the manufacture of acetic acid and the acetates. It can only be freed from the impurities which it still contains by chemical means; to this end slacked lime and chalk are added, at a gentle heat, to neutralization; the liquid is boiled and allowed to settle twenty-four hours; the clear liquid, which is a solution of calcium acetate, is decanted and evaporated; the calcium salt is converted into sodium acetate, which is then purified by calcination at a temperature below 330° (626° F.), dissolved, filtered, and recrystallized; the salt is then decomposed by a proper quantity of H_2SO_4, and the liberated acetic acid separated by distillation.

The product so obtained is a solution of acetic acid in water, contain-

193

ing 36 per cent. of true acetic acid, and being of sp. gr. 1.047, U. S. (the acid of the Br. Ph. is weaker—33 per cent. $C_2H_4O_2$, and sp. gr. 1.044).

Pure acetic acid, known as *glacial acetic acid, acidum aceticum glaciale* (U. S.), is obtained by decomposition of a pure dry acetate by heat.

Properties.—Acetic acid is a colorless liquid. Below 17° (62°.6 F.), when pure, it is a crystalline solid. It boils at 119° (246°.2 F.); sp. gr. 1.0801 at 0° (32° F.); its odor is penetrating and acid; in contact with the skin it destroys the epidermis and causes vesication; it mixes with H_2O in all proportions, the mixtures being less in volume than the sum of the volumes of the constituents. The sp. gr. of the mixtures gradually increase up to that containing 23 per cent. of H_2O, after which they again diminish, and all the mixtures containing more than 43 per cent. of acid are of higher sp. gr. than the acid itself.

Vapor of acetic acid burns with a pale, blue flame; and is decomposed at a red-heat. It only decomposes calcic carbonate in the presence of H_2O. Hot H_2SO_4 decomposes and blackens it, SO_2 and CO_2 being given off. Under ordinary circumstances Cl acts upon it slowly, more actively under the influence of sunlight, to produce *monochloracetic acid*, CH_2ClCO,OH; *dichloracetic acid*, $CHCl_2CO,OH$; and *trichloracetic acid*, CCl_3CO,OH. The last named is an odorless acid, strongly vesicant, crystalline solid; fuses at 46° (114°.8 F.) and boils at 195°–200° (383°–392° F.).

Analytical Characters.—(1.) Warmed with SO_4H_2 it blackens.

(2.) With silver nitrate a white crystalline ppt., partly dissolved by heat; no reduction of Ag on boiling.

(3.) Heated with H_2SO_4 and C_2H_6O, acetic ether, recognizable by its odor, is given off.

(4.) When an acetate is calcined with a small quantity of As_2O_3 the foul odor cacodyl oxide is developed.

(5.) Neutral solution of ferric chloride produces in neutral solutions of acetates a deep red color, which turns yellow on addition of free acid.

Vinegar is an acid liquid owing its acidity to acetic acid, and holding certain fixed and volatile substances in solution. It is obtained from some liquid containing 10 per cent. or less of alcohol, which is converted into acetic acid by the transferring of atmospheric oxygen to the alcohol during the process of nutrition of a peculiar vegetable ferment, known as *mycoderma aceti*, or, popularly, as *mother of vinegar*. Vinegar is now manufactured principally by one of two processes—the German method and that of Pasteur. In the former, the alcoholic fluid, which must also contain albuminous matter, is allowed to trickle slowly through barrels containing beech-wood shavings, supported by a perforated false bottom. By a suitable arrangement of holes and tubes, an ascending current of air is made to pass through the barrel. The acetic ferment clings to the shavings, and under its influence acetification takes place rapidly, owing to the large surface exposed to the air. In Pasteur's process, the ferment is sown upon the surface of the alcoholic liquid, contained in large, shallow, covered vats, from which the vinegar is drawn off after acetification has been completed; the mother is collected, washed, and used in a subsequent operation.

The liquids from which vinegar is made are wine, cider, and beer, to which dilute alcohol is frequently added; the most esteemed being that obtained from white wine. *Wine vinegar* has a pleasant, acid taste and odor; it consists of water, acetic acid (about 5 per cent.), potassium bitartrate, alcohol, acetic ether, glucose, malic acid, mineral salts present in wine, a fermentescible, nitrogenized substance, coloring matter, etc. Sp.

194

gr. 1.020 to 1.025. When evaporated, it yields from 1.7 to 2.4 per cent. of solid residue.

Vinegars made from alcoholic liquids other than wine contain no potassium bitartrate, contain less acetic acid, and have not the aromatic odor of wine vinegar. *Cider vinegar* is of sp. gr. 2.0; is yellowish, has an odor of apples, and yields 1.5 per cent. of extract on evaporation. *Beer vinegar* is of sp. gr. 3.2; has a bitterish flavor, and an odor of sour beer; it leaves 6 per cent. of extract on evaporation.

The principal *adulterations* of vinegar are: *sulphuric acid*, which produces a black or brown color when a few drops of the vinegar and some fragments of cane-sugar are evaporated over the water-bath to dryness. *Water*, an excess of which is indicated by a low power of saturation of the vinegar, in the absence of mineral acids. Two parts of good wine vinegar neutralize 10 parts of sodium carbonate; the same quantity of cider vinegar, 3.5 parts; and of beer vinegar, 2.5 parts of carbonate. *Pyroligneous acid* may be detected by the creosote-like odor and taste. *Pepper, capsicum*, and other acrid substances, are often added to communicate fictitious strength. In vinegar so adulterated an acrid odor is perceptible after neutralization of the acid with sodium carbonate. *Copper, zinc, lead, and tin* frequently occur in vinegar which has been in contact with those elements, either during the process of manufacture or subsequently.

Distilled vinegar is prepared by distilling vinegar in glass vessels; it contains none of the fixed ingredients of vinegar, but its volatile constituents (acetic acid, water, alcohol, acetic ether, odorous principles, etc.), and a small quantity of aldehyde.

When dry acetate of copper is distilled, a blue, strongly acid liquid passes over; this, upon rectification, yields a colorless, mobile liquid, which boils at 56° (132°.8 F.), has a peculiar odor, and is a mixture of acetic acid, water, and acetone, known as *radical vinegar*.

Toxicology.—When taken internally, acetic acid and vinegar (the latter in doses of 4–5 fl. ℥) act as irritants and corrosives, causing in some instances perforation of the stomach, and death in 6–15 hours. Milk of magnesia should be given as an antidote, with the view to neutralizing the acid.

Propionic acid—C_2H_5CO,OH—74—is formed in many decompositions of organic substances: By the action of caustic potassa upon sugar, starch, gum, and ethyl cyanide; during fermentation, vinous or acetic; in the distillation of wood; during the putrefaction of peas, beans, etc.; by the oxidation of normal propylic alcohol, etc. It is best prepared by heating ethyl cyanide with potash until the odor of the ether has disappeared; the acid is then liberated from its potassium compound by H_2SO_4 and purified.

It is a colorless liquid, sp. gr. 0.996, does not solidify at −21° (−5°.8 F.), boils at 140° (284° F.), mixes with water and alcohol in all proportions, resembles acetic acid in odor and taste. Its salts are soluble and crystallizable.

Butyric acid—C_3H_7CO,OH—88—has been found in the milk, perspiration, muscular fluid, the juices of the spleen and of other glands, the urine, contents of the stomach and large intestine, fæces, and guano; in certain fruits, in yeast, in the products of decomposition of many vegetable substances; and in natural waters; in fresh butter in small quantity. more abundantly in that which is rancid.

It is formed by the action of H_2SO_4 and manganese dioxide, aided by heat, upon cheese, starch, gelatin, etc.; during the combustion of tobacco

195

(as ammonium butyrate; by the action of HNO_3 upon oleic acid; during the putrefaction of fibrin and other albuminoids; during a peculiar fermentation of glucose and starchy material in the presence of casein or gluten. This fermentation, known as the butyric, takes place in two stages; at first the glucose is converted into lactic acid: $C_6H_{12}O_6 = 2(C_3H_6O_3)$; and this in turn is decomposed into butyric acid, carbon dioxide, and hydrogen: $2C_3H_6O_3 = C_4H_8O_2 + 2CO_2 + 2H_2$.

Butyric acid is obtained from the animal charcoal which has been used in the purification of glycerin, in which it exists as calcium butyrate. It is also formed by subjecting to fermentation a mixture composed of glucose, water, chalk, and cheese or gluten. The calcium butyrate is decomposed by H_2SO_4, and the butyric acid separated by distillation.

Butyric acid is a colorless, mobile liquid, having a disagreeable, persistent odor of rancid butter, and a sharp, acid taste; soluble in water, alcohol, ether, and methyl alcohol; boils at 164° (327°.2 F.), distilling unchanged; solidifies in a mixture of solid carbon dioxide and ether; sp. gr. 0.974 at 15° (59° F.); a good solvent of fats.

It is not acted upon by H_2SO_4 in the cold, and only slightly under the influence of heat. Nitric acid dissolves it unaltered in the cold, but on the application of heat, oxidizes it to succinic acid. Dry Cl under the influence of sunlight, and Br under the influence of heat and pressure, form products of substitution with butyric acid. It readily forms ethers and salts.

Butyric acid is formed in the intestine, by the process of fermentation mentioned above, at the expense of those portions of the carbohydrate elements of food which escape absorption, and is discharged with the fæces as ammonium butyrate.

Isobutyric acid, an isomere of butyric acid, which boils at 152° (305°.6 F.), has also been found in human fæces. It corresponds to isobutyl alcohol.

Valerianic acids—C_4H_9CO,OH—102.—Corresponding to the four primary amylic alcohols, there are four amylic or valerianic acids:

I. $CH_3—CH_2—CH_2—CH_2—CO,OH.$ II. $\begin{matrix}CH_3\\CH_3\end{matrix}\rangle CH—CH_2—CO,OH.$

III. $\begin{matrix}CH_3\\CH_3—CH_2\end{matrix}\rangle CH—CO,OH.$ IV. $\begin{matrix}CH_3\diagdown\\CH_3—C—CO,OH.\\CH_3\diagup\end{matrix}$

I. **Normal valerianic acid**—*Butylformic acid—Propylacetic acid*—is obtained by the oxidation of normal amylic alcohol. It is an oily liquid, boils at 185° (365° F.), and has an odor resembling that of butyric acid.

II., III. **Ordinary valerianic acid**—*Delphinic acid—Phocenic acid—Isovaleric acid—Isopropyl acetic acid—Isobutylformic acid—Acidum valerianicum* (Br.).—This acid exists in the oil of the porpoise, and in valerian root and in angelica root. It is formed during putrid fermentation or oxidation of albuminoid substances. It occurs in the urine and fæces in typhus, variola, and acute atrophy of the liver. It is also formed in a variety of chemical reactions and notably by the oxidation of amylic alcohol.

It is prepared either by distilling water from valerian root, or, more economically, by mixing rectified amylic alcohol with H_2SO_4, adding when cold, a solution of potassium dichromate, and distilling after the reaction has become moderated: the distillate is neutralized with sodium carbonate; and the acid is obtained from the sodium valerianate so produced, by decomposition by H_2SO_4 and rectification.

196

The properties and nature of the acid differ according to those of the amyl alcohol from which it is obtained. The active alcohol yields the acid, $\begin{matrix}CH_3\\CH_3\end{matrix}\rangle CH—CH_2—CO,OH$, which is itself optically active; which forms an uncrystallizable and exceedingly soluble barium salt, and whose boiling-point is 173°.5 (344°.3 F.). The inactive alcohol yields by oxidation the acid, $\begin{matrix}CH_3\\CH_3—CH_2\end{matrix}\rangle CH—CO,OH$, which is optically inactive; whose barium salt is readily crystallizable and soluble in water to the extent of 48 parts in 100; and whose boiling-point is 174°.5 (346° F.).

The acid obtained from valerian root is identical with the acid obtained by the oxidation of *optically inactive* amylic alcohol. The artificial product, being obtained from the commercial mixture of active and inactive alcohols, is a mixture in different proportions of the two acids mentioned above.

The ordinary valerianic acid is an oily, colorless liquid, having a penetrating odor, and a sharp, acrid taste. It solidifies at −16° (3°.2 F.); boils at 173°–175° (343°.4–347° F.); sp. gr. 0.9343–0.9465 at 20° (68° F.); burns with a white, smoky flame. It dissolves in 30 parts of water, and in alcohol and ether in all proportions. It dissolves phosphorus, camphor, and certain resins.

IV. *Trimethyl acetic acid—Pivalic acid*—is a crystalline solid, which fuses at 35.5° (96° F.) and boils at 163°.7 (326°.7 F.); sparingly soluble in H_2O; obtained by the action of cyanide of mercury upon tertiary butyl iodide.

Caproic acids—*Hexylic acids*—$C_5H_{11}CO,OH$—116.—There probably exist quite a number of isomeres having the composition indicated above, some of which have been prepared from butter, cocoa-oil, and cheese, and by decomposition of amyl cyanide, or of hexyl alcohol.

The acid obtained from butter, in which it exists as a glyceric ether, is a colorless, oily liquid, boils at 205° (401° F.); sp. gr. 0.931 at 15° (59° F.); has an odor of perspiration and a sharp, acid taste; is very sparingly soluble in water, but soluble in alcohol.

Œnanthylic acid—*Heptylic acid*—$C_6H_{13}CO,OH$—130—exists in spirits distilled from rice and maize, and is formed by the action of HNO_3 on fatty substances, especially castor oil. It is a colorless oil; sp. gr. 0.9167; boils at 212° (413°.6 F.).

Caprylic acid—*Octylic acid*—$C_7H_{15}CO,OH$—144—accompanies caproic acid in butter, cocoa-oil, etc. It is a solid; fuses at 15° (59° F.); boils at 236° (457° F.); almost insoluble in H_2O.

Pelargonic acid—*Nonylic acid*—$C_8H_{17}CO,OH$—158.—A colorless oil, solid below 10° (50° F.); boils at 260° (500° F.); exists in oil of geranium, and is formed by the action of HNO_3 on oil of rue.

Capric acid—*Decylic acid*—$C_9H_{19}CO,OH$—172—exists in butter, cocoa-oil, etc., associated with caproic and caprylic acids in their glyceric ethers, and in the residues of distillation of Scotch whiskey, as amyl caprate. It is a white, crystalline solid; melts at 27°.5 (81°.5 F.); boils at 273° (523°.4 F.).

Lauric acid—*Laurostearic acid*—$C_{11}H_{23}CO,OH$—200—is a solid, fusible at 43°.5 (110°.3 F.) obtained from laurel berries, cocoa-butter, and other vegetable fats.

Myristic acid—$C_{13}H_{27}CO,OH$—228.—A crystalline solid, fusible at 54° (129°.2 F.); existing in many vegetable oils, cow's butter, and spermaceti.

Palmitic acid—*Ethalic acid*—$C_{15}H_{31}CO,OH$—256—exists in palm-oil, in combination when the oil is fresh, and free when the oil is old; it also

197

enters into the composition of nearly all animal and vegetable fats. It is obtained from the fats, palm-oil, etc., by saponification with caustic potassa and subsequent decomposition of the soap by a strong acid. It is also formed by the action of caustic potash in fusion upon cetyl alcohol (ethal), and by the action of the same reagent upon oleic acid.

Palmitic acid is a white, crystalline solid; odorless; tasteless; lighter than H_2O, in which it is insoluble; quite soluble in alcohol and in ether; fuses at 62° (143°.6 F.); distils unchanged with vapor of water.

Margaric acid—$C_{17}H_{33}CO,OH$—270—formerly supposed to exist as a glyceride in all fats, solid and liquid. What had been taken for margaric acid was a mixture of 90 per cent. of palmitic and 10 per cent. of stearic acid. It is obtained by the action of potassium hydrate upon cetyl cyanide, as a white, crystalline body; fusible at 59°.9 (140° F.).

Stearic acid—$C_{17}H_{35}CO,OH$—284—exists as a glyceride in all solid fats, and in many oils, and also free to a limited extent.

To obtain it pure, the fat is saponified with an alkali, and the soap decomposed by HCl; the mixture of fatty acids is dissolved in a large quantity of alcohol, and the boiling solution *partly* precipitated by the addition of a concentrated solution of barium acetate. The precipitate is collected, washed, and decomposed by HCl; the stearic acid which separates is washed and recrystallized from alcohol. The process is repeated until the product fuses at 70° (158° F.).

Pure stearic acid is a colorless, odorless, tasteless solid; fusible at 70° (158° F.); unctuous to the touch; insoluble in H_2O; very soluble in alcohol and in ether. The alkaline stearates are soluble in H_2O; those of Ca, Ba, and Pb are insoluble.

Stearic and palmitic acids exist free in the intestine during the digestion of fats, a portion of which is decomposed by the action of the pancreatic secretion into fatty acids and glycerin. The same decomposition also occurs in the presence of putrefying albuminoid substances.

Arachic acid—$C_{19}H_{39}CO,OH$—312—exists as a glyceride in peanut-oil (now largely used as a substitute for olive-oil), in oil of ben, and in small quantity in butter. It is a crystalline solid, which melts at 75° (167° F.).

COMPOUND ETHERS.

As the alcohols resemble the mineral bases, and the organic acids resemble those of mineral origin, so the compound ethers are similar in constitution to the salts, being *formed by the double decomposition of an alcohol with an acid, mineral or organic*, as a salt is formed by double decomposition of an acid and a mineral base, the radical playing the part of an atom of corresponding valence.

$$\left.\begin{matrix}K'\\H\end{matrix}\right\}O + \left.\begin{matrix}(NO_2)\\H\end{matrix}\right\}O = \left.\begin{matrix}H\\H\end{matrix}\right\}O + \left.\begin{matrix}(NO_2)\\K'\end{matrix}\right\}O$$

Potassium hydrate. — Nitric acid. — Water. — Potassium nitrate.

$$\left.\begin{matrix}(C_2H_5)'\\H\end{matrix}\right\}O + \left.\begin{matrix}(NO_2)\\H\end{matrix}\right\}O = \left.\begin{matrix}H\\H\end{matrix}\right\}O + \left.\begin{matrix}(NO_2)\\(C_2H_5)'\end{matrix}\right\}O$$

Ethyl hydrate (alcohol). — Nitric acid. — Water. — Ethyl nitrate (nitric ether).

198

Methyl nitrate—$\left.\begin{matrix}NO_2\\CH_3\end{matrix}\right\}O$—77.—A colorless liquid; sp. gr. 1.182 at 22° (71°.6 F.); boils at 66° (150°.8 F.); gives off vapor which detonates at 150° (302° F.). Prepared by the action of potassium nitrate and H_2SO_4 on methyl alcohol.

Methyl nitrite—$\left.\begin{matrix}NO\\CH_3\end{matrix}\right\}O$—61—obtained by heating methyl alcohol with HNO_3 and Cu. Below—12° (10°.4 F.) it is a yellowish liquid; above that temperature a gas.

Ethyl nitrate—*Nitric ether*—$\left.\begin{matrix}NO_2\\C_2H_5\end{matrix}\right\}O$—91.—A colorless liquid; has a sweet taste and bitter after-taste; sp. gr. 1.112 at 17° (62°.6 F.); boils at 85° (185° F.); gives off explosive vapors. Prepared by distilling a mixture of HNO_3 and C_2H_6O in the presence of urea.

Ethyl nitrite—*Nitrous ether*—$\left.\begin{matrix}NO\\C_2H_5\end{matrix}\right\}O$—75—is best prepared by directing the nitrous fumes, produced by the action of starch on HNO_3 under the influence of heat, into alcohol, contained in a retort connected with a well-cooled receiver.

It is a yellowish liquid; has an apple-like odor, and a sharp, sweetish taste; sp. gr. 0.947; boils at 18° (64°.4 F.); gives off inflammable vapor; very sparingly soluble in H_2O; readily soluble in alcohol and ether.

Warm H_2O decomposes it into C_2H_6O; HNO_3 and NO. Alkalies decompose it into malate and nitrate of the alkaline element. It is energetically attacked by H_2SO_4, H_2S and the alkaline sulphides. It is liable to spontaneous decomposition, especially in the presence of H_2O, into NO and malic acid.

Its vapor rapidly produces anæsthesia; it is, however, used only in alcoholic solution: *Spiritus ætheris nitrosi* (U. S., Br.), which also contains aldehyde. Owing to the presence of the last-named substance, and to the presence of H_2O, the spirit is very liable to become acid, either from the formation of acetic acid by the oxidation of the aldehyde, or from the decomposition of the ether under the influence of H_2O, a change which renders it unfit for use in many of the prescriptions in which it is frequently used, especially in that with potassium iodide, from which it liberates iodine. The presence of free acid may be detected by effervescence when the spirit is shaken with hydrosodic carbonate. Its acidity may be corrected by shaking with potassium carbonate, and decanting, provided it does not contain H_2O.

Ethyl sulphates.—These are two in number: $(C_2H_5)HSO_4$—*Ethyl-sulphuric, or sulphovinic acid*; $(C_2H_5)_2SO_4$—*Ethyl-sulphate—Sulphuric ether.*

Ethyl-sulphuric Acid—$\left.\begin{matrix}SO_2\\(C_2H_5)\\H\end{matrix}\right\}O_2$—126—is formed as an intermediate product in the manufacture of ethylic ether (q. v.).

Pure ethyl-sulphuric acid is a colorless, syrupy, highly acid liquid; sp. gr, 1.316; soluble in water and alcohol in all proportions, insoluble in ether.

It decomposes slowly at ordinary temperatures, more rapidly when heated. When heated alone or with alcohol, it yields ether and H_2SO_4. When heated with H_2O, it yields alcohol and H_2SO_4. It forms crystalline salts, known as *sulphovinates*; one of which, *sodium sulphovinate*, $(C_2H_5)NaSO_4$, has been used in medicine. It is a white, deliquescent solid,

199

either crystalline with 1Aq., or granular and anhydrous; soluble in H_2O. Its solution should give no precipitate with barium chloride.

Ethyl Sulphate—$\left.\begin{matrix}SO_2\\(C_2H_5)_2\end{matrix}\right\}O_2$—154—the true sulphuric ether, is obtained by passing vapor of SO_3 into *pure* ethylic ether, thoroughly cooled.

It is a colorless, oily liquid; has a sharp, burning taste, and the odor of peppermint; sp. gr. 1.120; it cannot be distilled without decomposition; in contact with H_2O it is decomposed with formation of sulphovinic acid.

By the action of an excess of H_2SO_4 upon alcohol; by the dry distillation of the sulphovinates; and in the last stages of manufacture of ether, a yellowish, oily liquid, having a penetrating odor and a sharp, bitter taste, is formed; this is *sweet* or *heavy oil of wine*, and its ethereal solution is *Oleum æthereum* (U. S.). It seems to be a mixture of ethyl-sulphate with hydrocarbons of the series C_nH_{2n}. On contact with H_2O or an alkaline solution, it is decomposed, sulphovinic acid is formed, and there separates a colorless oil, of sp. gr. 0.917, boiling at 280° (536° F.), which is *light oil of wine*. This oil is polymeric with ethylene, and is probably *cetene*, $C_{16}H_{32}$; it is sometimes called *etherine* or *etherol*.

Ethyl acetate—*Acetic ether*—*Æther aceticus* (*U. S.*)—$\left.\begin{matrix}C_2H_3O\\C_2H_5\end{matrix}\right\}O$—88—is obtained by distilling a mixture of sodium acetate, alcohol and H_2SO_4; or by passing carbon dioxide through an alcoholic solution of potassium acetate.

It is a colorless liquid, has an agreeable, ethereal odor; boils at 74° (165°.2 F.); sp. gr. 0.89 at 15° (59° F.); soluble in 6 pts. water, and in all proportions in methyl and ethyl alcohols and in ether; a good solvent of essences, resins, cantharidine, morphine, gun-cotton, and, in general, of substances soluble in ether; burns with a yellowish white flame. Chlorine acts energetically upon it, producing products of substitution, varying according to the intensity of the light from $C_4H_6Cl_2O_2$ to $C_4Cl_8O_2$.

Amyl nitrate—$\left.\begin{matrix}NO_2\\C_5H_{11}\end{matrix}\right\}O$—133—obtained by distilling a mixture of HNO_3 and amylic alcohol in the presence of a small quantity of urea. It is a colorless, oily liquid; sp. gr. 0.994 at 10° (50° F.); boils at 148° (298°.4 F.) with partial decomposition.

Amyl nitrite—*Amyl nitris* (*U. S.*)—$\left.\begin{matrix}NO\\C_5H_{11}\end{matrix}\right\}O$—117—prepared by directing the nitrous fumes, evolved by the action of NO_3H upon starch, into amyl alcohol contained in a retort heated over a water-bath; purifying the distillate by washing with an alkaline solution; and rectifying.

It is a slightly yellowish liquid; sp. gr. 0.877; boils at 95° (203° F.); its vapor explodes when heated to 260° (500° F.); insoluble in water; soluble in alcohol in all proportions; vapor orange-colored. Alcoholic solution of potash decomposes it slowly, with formation of potassium nitrite and oxides of ethyl and amyl. When dropped upon fused potash, it ignites and yields potassium valerianate.

Amyl nitrite is frequently impure; its boiling-point should not vary more than two or three degrees from that given above.

Cetyl palmitate—*Cetine*—$\left.\begin{matrix}C_{16}H_{31}O\\C_{16}H_{33}\end{matrix}\right\}O$—480—is the chief constituent of *spermaceti* = *cetaceum* (U. S., Br.). This is the concrete portion, obtained by expression and crystallization from alcohol, of the oil contained in the cranial sinuses of the sperm-whale. It forms white, crystal-

200

line plates; fusible at 49° (120°.2 F.); slightly unctuous to the touch; tasteless, and almost odorless; insoluble in water; soluble in alcohol and ether; burns with a bright flame. Besides cetine, it contains ethers not only of palmitic, but also of stearic, myristic, and laurostearic acids; and of the alcohols: *lethal*, $C_{12}H_{26}O$; *methal*, $C_{14}H_{30}O$; *ethal*, $C_{16}H_{34}O$; and *stethal*, $C_{18}H_{38}O$.

Melissyl palmitate — *Melissin*—$\left.\begin{matrix}C_{16}H_{31}O\\C_{30}H_{61}\end{matrix}\right\}O$—676.—Beeswax consists mainly of two substances; *cerotic acid*, $C_{27}H_{53}O,OH$, which is soluble in boiling alcohol, and melissyl palmitate, insoluble in that liquid, united with minute quantities of substances which communicate to the wax its color and odor. Yellow wax melts at 62°-63° (143°.6-145°.4 F.); after bleaching, which is brought about by exposure to light, air, and moisture, it does not fuse below 66° (150°.8 F.). *China wax*, a white substance resembling spermaceti, is a vegetable product, consisting chiefly of *ceryl cerotate*, $C_{27}H_{53}O_2(C_{27}H_{55})$.

ALDEHYDES.

Series $C_nH_{2n}O$.

Formic aldehyde	CH_2O.	Valerianic aldehyde	$C_5H_{10}O$.
Acetic aldehyde	C_2H_4O.	Caproic aldehyde	$C_6H_{12}O$.
Propionic aldehyde	C_3H_6O.	Œnanthylic aldehyde	$C_7H_{14}O$.
Butyric aldehyde	C_4H_8O.	Caprylic aldehyde	$C_8H_{16}O$.
Isobutyric aldehyde	C_4H_8O.	Palmitic aldehyde	$C_{16}H_{32}O$.

It will be remembered that the monobasic acids are obtained from the alcohols by oxidation of the radical:

$$\left.\begin{matrix}(C_2H_5)\\H\end{matrix}\right\}O \qquad \left.\begin{matrix}(C_2H_3O)'\\H\end{matrix}\right\}O$$

Ethyl alcohol. Acetic acid.

These oxidized radicals are capable of forming compounds similar in constitution to those of the non-oxidized radicals. There are chlorides, bromides, and iodides; their hydrates are the *acids,* $\left.\begin{matrix}(C_2H_3O)\\H\end{matrix}\right\}O$ = acetic acid; their oxides are known as *anhydrides,* $\left.\begin{matrix}(C_2H_3O)\\(C_2H_3O)\end{matrix}\right\}O$ = acetic anhydride; and their hydrides are the *aldehydes* $\left.\begin{matrix}(C_2H_3O)\\H\end{matrix}\right\}$ = acetic aldehyde. The name aldehyde is a corruption of *alcohol dehydrogenatum,* from the method of their formation, by the removal of hydrogen from alcohol.

The aldehydes all contain the group of atoms (COH)′, and their constitution may be thus graphically indicated:

$$\begin{matrix}COH\\|\\CH_3\end{matrix} \qquad \begin{matrix}COH\\|\\CH_2\\|\\CH_3\end{matrix}$$

Acetic aldehyde. Propionic aldehyde.

They are capable, by fixing H_2, of regenerating the alcohol; and, by fixing O, of forming the corresponding acid:

$$\begin{matrix}COH\\|\\CH_3\end{matrix} \qquad \begin{matrix}CH_2OH\\|\\CH_3\end{matrix} \qquad \begin{matrix}CO,OH\\|\\CH_3\end{matrix}$$

Acetic aldehyde. Ethylic alcohol. Acetic acid.

Acetic aldehyde—*Acetyl hydride*—$\left.\begin{matrix}C_2H_3O\\H\end{matrix}\right\}$—44—is formed in all reactions in which alcohol is deprived of H without introduction of O. It is prepared by distilling from a capacious retort, connected with a well-cooled condenser, a mixture of H_2SO_4, 6 pts.; H_2O, 4 pts.; alcohol, 4 pts.; and powdered manganese dioxide, 6 pts. The product is redistilled from calcium chloride below 50° (122° F.). The second distillate is mixed with two volumes of ether, cooled by a freezing mixture, and saturated with dry NH_3; there separate crystals of *ammonium acetylide,* C_2H_3O, NH_4, which are washed with ether, dried, and decomposed in a distilling apparatus, over the water-bath, with the proper quantity of dilute H_2SO_4; the distillate is finally dried over calcium chloride and rectified below 35° (95° F.).

Aldehyde is a colorless, mobile liquid; has a strong, suffocating odor; sp. gr. 0.790 at 18° (64°.4 F.); boils at 21° (69°.8 F.); soluble in all proportions in water, alcohol, and ether. If perfectly pure, it may be kept unchanged; but if an excess of acid have been used in its preparation, it gradually decomposes. When heated to 100° (212° F.), it is decomposed into water and crotonic aldehyde.

In the presence of nascent H, aldehyde takes up H_2 and regenerates alcohol. Cl converts it into acetyl chloride, C_2H_3O, Cl, and other products. Oxidizing agents quickly convert it into acetic acid. At the ordinary temperature H_2SO_4; HCl; and SO_2 convert it into a solid substance called *paraldehyde,* $C_6H_{12}O_3$ (?), which fuses at 10.5° (50°.9 F.); boils at 124° (255°.2 F.), and is more soluble in cold than in warm water. When heated with potassium hydrate, aldehyde becomes brown, a brown resin separates, and the solution contains potassium formiate and acetate. If a watery solution of aldehyde be treated, first with NH_3 and then with H_2S, a solid, crystalline base, *thialdine,* $C_6H_{13}NS_2$, separates. It also forms crystalline compounds with the alkaline bisulphites. It decomposes solutions of silver nitrate, separating the silver in the metallic form, and under conditions which cause it to adhere strongly to glass.

Vapor of aldehyde, when inhaled in a concentrated form, produces asphyxia, even in comparatively small quantity; when diluted with air it is said to act as an anæsthetic. When taken internally it causes sudden and deep intoxication, and it is to its presence that the first products of the distillation of spirits of inferior quality owe in a great measure their rapid, deleterious action.

Trichloraldehyde—*Trichloracetyl hydride*—*Chloral*—$\left.\begin{matrix}C_2Cl_3O\\H\end{matrix}\right\}$—147.5—is one of the final products of the action of Cl upon alcohol, and is obtained by passing dry Cl through absolute alcohol to saturation; applying heat toward the end of the reaction, which requires several hours for its completion. The liquid separates into two layers; the lower is removed and shaken with an equal volume of concentrated H_2SO_4 and again allowed to separate into two layers; the upper is decanted; again mixed with H_2SO_4, from which it is distilled; the distillate is treated with quicklime,

from which it is again distilled, that portion which passes over between 94° and 99° (201°.2–210°.2 F.) being collected. It sometimes happens that chloral in contact with H_2SO_4 is converted into a modification, insoluble in H_2O, known as *metachloral;* when this occurs it is washed with H_2O, dried and heated to 180° (356° F.), when it is converted into the soluble variety, which distils over.

Chloral is a colorless liquid, unctuous to the touch; has a penetrating odor and an acrid, caustic taste; sp. gr. 1.502 at 18° (64°.4 F.); boils at 94.4° (201°.9 F.); very soluble in water, alcohol, and ether; dissolves Cl, Br, I, S and P. Its vapor is highly irritating. It distils without alteration.

Although chloral has not been obtained by the direct substitution of Cl for H in aldehyde, its reactions show it to be an aldehyde; it forms crystalline compounds with the bisulphites; it reduces solutions of silver nitrate in the presence of NH_3; NH_3 and H_2S form with it a compound similar to thialdine; with nascent H it regenerates aldehyde; oxidizing agents convert it into trichloracetic acid. Alkaline solutions decompose it with formation of chloroform and a formiate.

With a small quantity of H_2O chloral forms a solid, crystalline hydrate, heat being at the same time liberated. This hydrate has the composition C_2HCl_3O, H_2O, and its constitution, as well as that of chloral itself, is indicated by the formulæ:

$$\begin{matrix}CH_3\\|\\CHO\end{matrix} \qquad \begin{matrix}CCl_3\\|\\CHO\end{matrix} \qquad \begin{matrix}CCl_3\\|\\CH(OH)_2\end{matrix}$$

Aldehyde. Trichloraldehyde (chloral). Chloral hydrate.

Chloral hydrate—*Chloral* (*U. S.*)—is a white, crystalline solid; fuses at 57° (134°.6 F.); boils at 98° (208°.4 F.), at which temperature it suffers partial decomposition into chloral and H_2O; volatilizes slowly at ordinary temperatures; is very soluble in H_2O; neutral in reaction; has an ethereal odor, and a sharp, pungent taste. Concentrated H_2SO_4 decomposes it with formation of chloral and *chloralide.* HNO_3 converts it into trichloracetic acid. When pure it gives no precipitate with silver nitrate solution, and is not browned by contact with concentrated H_2SO_4.

Chloral also combines with alcohol, with elevation of temperature, to form a solid, crystalline body—*chloral alcoholate:* $CCl_3—CH\langle\begin{matrix}OH\\O—C_2H_5\end{matrix}$.

Action of Chloral Hydrate upon the Economy.—Although it was the ready decomposition of chloral into a formiate and chloroform which first suggested its use as a hypnotic to Liebreich, and although this decomposition was at one time believed to occur in the body under the influence of the alkaline reaction of the blood, more recent investigations have shown that the formation of chloroform from chloral in the blood is, to say the least, highly improbable, and that chloral has, in common with many other chlorinated derivatives of this series, the property of acting directly upon the nerve-centres.

Neither the urine nor the expired air contain chloroform when chloral is taken internally; when taken in large doses, chloral appears in the urine. The fact that the action of chloral is prolonged for a longer period than that of the other chlorinated derivatives of the fatty series is probably due, in a great measure, to its less volatility and less rapid elimination.

When taken in overdose, chloral acts as a poison, and its use as such

is rapidly increasing as acquaintance with its powers becomes more widely disseminated.

No chemical antidote is known. The treatment should be directed to the removal of any chloral remaining in the stomach by the stomach-pump, and to the maintenance or restoration of respiration.

In fatal cases of poisoning by chloral that substance may be detected in the blood, urine, and contents of the stomach by the following method: the liquid is rendered strongly alkaline with potassium hydrate; placed in a flask, which is warmed to 50°–60° (122°–140° F.), and through which a slow current of air, heated to the same temperature, is made to pass; the air, after bubbling through the liquid, is tested for chloroform by the methods described on p. 113. If affirmative results are obtained in this testing, it remains to determine whether the chloroform detected existed in the fluid tested in its own form, or resulted from the decomposition of chloral; to this end a fresh portion of the suspected liquid is *rendered acid* and tested as before. A negative result is obtained in the second testing when chloral is present.

Bromal—$\left.\begin{matrix}C_2Br_3O\\H\end{matrix}\right\}$—281.—A colorless, oily, pungent liquid; sp. gr. 3.34; boils at 172° (341°.6 F.); neutral; soluble in H_2O, alcohol, and ether. It combines with H_2O to form *bromal hydrate,* $CBr_3,CH(OH)_2$; large transparent crystals; soluble in H_2O; decomposed by alkalies into bromoform and a formiate. Produces anæsthesia without sleep; very poisonous.

KETONES OR ACETONES.

Series $C_nH_{2n}O$.

Although the aldehydes are not acid in reaction, and are not usually regarded as acids, there exist substances known as ketones or acetones, which may be regarded as formed by the substitution of an alcoholic radical for the H of the group COH. These substances all contain the group of atoms (CO)″, and their constitution may be represented graphically thus:

$$\begin{array}{c} CH_3 \\ | \\ CO \\ | \\ CH_3 \\ \text{Dimethyl ketone} \\ \text{(acetone).} \end{array} \qquad \begin{array}{c} CH_3 \\ | \\ CO \\ | \\ CH_2 \\ | \\ CH_3 \\ \text{Methyl-ethyl ketone.} \end{array}$$

the first being a *symmetrical ketone* and the latter a *non-symmetrical*. The ketones are isomeric with the aldehydes, from which they are distinguished: 1st, by the action of H, which produces a primary alcohol with an aldehyde, and a secondary alcohol with a ketone:

$$\begin{array}{c} COH \\ | \\ CH_2 \\ | \\ CH_3 \\ \text{Propionic aldehyde.} \end{array} + H_2 = \begin{array}{c} CH_2OH \\ | \\ CH_2 \\ | \\ CH_3 \\ \text{Propyl alcohol.} \end{array}$$

204

$$\begin{array}{c} CH_3 \\ | \\ CO \\ | \\ CH_3 \\ \text{Acetone.} \end{array} + H_2 = \begin{array}{c} CH_3 \\ | \\ CH,OH \\ | \\ CH_3 \\ \text{Isopropyl alcohol.} \end{array}$$

2d, by the action of O, which unites directly with an aldehyde to produce the corresponding acid, while it causes the disruption of the molecule of the ketone, with formation of two acids:

$$\begin{array}{c} COH \\ | \\ CH_2 \\ | \\ CH_3 \\ \text{Propionic aldehyde.} \end{array} + O = \begin{array}{c} CO,OH \\ | \\ CH_2 \\ | \\ CH_3 \\ \text{Propionic acid.} \end{array}$$

$$\begin{array}{c} CH_3 \\ | \\ CO \\ | \\ CH_3 \\ \text{Acetone.} \end{array} + O_2 = \begin{array}{c} CO,OH \\ | \\ H \\ \text{Formic acid.} \end{array} + \begin{array}{c} CO,OH \\ | \\ CH_3 \\ \text{Acetic acid.} \end{array}$$

Dimethyl ketone—*Acetone*—*Acetylmethylide*—*Pyroacetic ether* or *spirit*—$CO\langle^{CH_3}_{CH_3}$—58—is formed as one of the products of the dry distillation of the acetates; by the decomposition of the vapor of acetic acid at a red heat; by the dry distillation of sugar, tartaric acid, etc.; and in a number of other reactions. It is obtained by distilling dry calcium acetate in an earthenware retort at a dull red heat; the distillate, collected in a well-cooled receiver, is freed from H_2O by digestion with fused calcium chloride, and rectified; those portions being collected which pass over at 60° (140° F.). It is also formed in large quantity in the preparation of aniline.

It is a limpid, colorless liquid; sp. gr. 0.7921 at 18° (64°.4 F.); boils at 56° (132°.8 F.); soluble in H_2O, alcohol, and ether; has a peculiar, ethereal odor, and a burning taste; is a good solvent of resins, fats, camphor, gun-cotton; readily inflammable. It forms crystalline compounds with the alkaline bisulphites. Cl and Br, in the presence of alkalies, convert it into chloroform or bromoform; Cl alone produces with acetone a number of chlorinated products of substitution. Certain oxidizing agents transform it into a mixture of formic and acetic acids; others into oxalic acid.

Acetone has been found to exist in the blood and urine in certain pathological conditions, and notably in diabetes; the peculiar odor exhaled by diabetics is produced by this substance, which has also been considered by some authors as being the cause of the respiratory derangements and coma which frequently occur in the last stages of the disease.

That acetone exists in the blood in such cases is certain; it is not certain, however, that its presence produces the condition designated as *acetonæmia*. It can hardly be doubted that the acetone thus existing in the blood is indirectly formed from diabetic sugar, and it is probable also that a complex acid, known as *ethyldiacetic*, $C_6H_9O_3H$, is formed as an intermediate product.

205

MONAMINES.

The *monamines* are substances which may be considered as being derived from one molecule of NH_3 by the substitution of one, two, or three alcoholic radicals for one, two, or three H atoms. They are designated as *primary, secondary*, and *tertiary*, according as they contain one, two, or three alcoholic radicals:

$$\begin{array}{c} H \\ | \\ N{-}H \\ | \\ H \\ NH_3 \\ \text{Ammonia.} \end{array} \quad \begin{array}{c} H \\ | \\ N{-}CH_2{-}CH_3 \\ | \\ H \\ (C_2H_5)H_2N \\ \text{Ethylamine} \\ \text{(primary).} \end{array} \quad \begin{array}{c} H \\ | \\ N{-}CH_2{-}CH_3 \\ | \\ CH_2{-}CH_3 \\ (C_2H_5)_2HN \\ \text{Diethylamine} \\ \text{(secondary).} \end{array} \quad \begin{array}{c} CH_2{-}CH_3 \\ | \\ N{-}CH_2{-}CH_3 \\ | \\ CH_2{-}CH_3 \\ (C_2H_5)_3N \\ \text{Triethylamine} \\ \text{(tertiary).} \end{array}$$

They are also known as *compound ammonias*, and resemble ammonia in their chemical properties; uniting with acids, without elimination of H_2O, to form salts resembling those of ammonium. They also combine with H_2O to form *quaternary ammonium hydrates*, similar in constitution to ammonium hydrate. The alkalinity and solubility in H_2O of the primary monamines are greater than those of the secondary, and those of the secondary greater than those of the tertiary. Their chlorides form sparingly soluble compounds with platinic chloride.

The *primary* monamines are formed by the action of potassium hydrate upon the corresponding cyanic ether:

$$\underset{\text{Ethyl cyanate.}}{CNO,C_2H_5} + \underset{\text{Potash.}}{2KHO} = \underset{\text{Ethylamine.}}{NH_2,C_2H_5} + \underset{\text{Potassium carbonate.}}{CO_3K_2};$$

or by heating together an alcoholic solution of ammonia and an ether:

$$\underset{\text{Ethyl iodide.}}{C_2H_5I} + \underset{\text{Ammonia.}}{NH_3} = \underset{\text{Hydriodic acid.}}{HI} + \underset{\text{Ethylamine.}}{NH_2,C_2H_5};$$

or by the action of nascent H upon the cyanides of the alcoholic radicals:

$$\underset{\text{Methyl cyanide.}}{CN,CH_3} + \underset{\text{Hydrogen.}}{2H_2} = \underset{\text{Ethylamine.}}{NH_2,C_2H_5}.$$

The *secondary* monamines are formed by the action of the iodides or bromides of the alcoholic radicals upon the primary monamines.

The *tertiary* monamines are produced by the distillation of the hydrates or iodides of the quaternary ammoniums, or by the action of the iodides of the alcoholic radicals upon the secondary monamines.

It is obvious from the above described properties of these substances that they are true alkaloids (see p. 331), among which also belong the diamines and triamines (see p. 249).

Methylamine—*Methylia*—$\left.\begin{array}{c}CH_3\\H_2\end{array}\right\}$ N—31—is a colorless gas; has a fishy, ammoniacal odor; inflammable; is the most soluble gas known, one

206

volume of H_2O dissolving 1,154 volumes of methylia at 12°.5 (54°.5 F.); the solution is strongly alkaline and caustic.

The aqueous solution possesses the odor of the gas, and is highly caustic and alkaline. It neutralizes the acids with formation of methylammonium salts (*e.g.*, CH_3H_3N,NO_3 = methylammonium nitrate), which are for the most part crystallizable and very soluble in H_2O. Its chloraurate crystallizes in beautiful golden-yellow needles, soluble in water, alcohol, and ether. Its chloroplatinate crystallizes in golden-yellow scales, soluble in water, insoluble in alcohol.

See trimethylamine, below.

Dimethylamine—*Dimethylia*—$\left.\begin{array}{c}(CH_3)_2\\H\end{array}\right\}$ N—45—is a liquid below 8° (46°.4 F.); has an ammoniacal odor, and is quite soluble in H_2O. It constitutes about 50 per cent. of the commercial trimethylamine, which also contains 5 to 10 per cent. of trimethylamine, the remainder being a mixture of monomethylamine, isobutylamine, and propylamine. Its chloroplatinate forms long needles.

See trimethylamine, below.

Trimethylamine—*Trimethylia*—$(CH_3)_3N$—59—is formed by the action of methyl iodide upon NH_3, and as a product of decomposition of many organic substances, it being one of the products of the action of potash on many vegetable substances, alkaloids, etc. It also occurs naturally in cod-liver oil, ergot, chenopodium, yeast, guano, human urine, the blood of the calf, and many flowers.

It is an oily liquid, having a disagreeable odor of fish; boils at 9° (48°.2 F.); alkaline; soluble in H_2O, alcohol, and ether; inflammable. It combines with acids to form salts of trimethyl ammonium, which are crystallizable.

Trimethylamine has long been known to exist in the pickle in which herrings have been preserved. More recently it has been found to be an important product of putrefactive changes in fish, starch-paste, brain-tissue, muscular tissue, and other albuminoid substances, being accompanied by lesser quantities of monomethylamine, dimethylamine, ethylamine, and diethylamine, as well as by other peculiar alkaloidal bodies. It has also been observed accompanying more active alkaloids in blood-serum, etc., which have served for the culture of various bacilli. (See choline and neurine, below, and ptomaïnes, p. 343.)

Its chloroplatinate crystallizes in octahedra, insoluble in alcohol.

The commercial trimethylamine, obtained by the dry distillation of distillery-waste, contains but $\frac{6}{10}$ per cent. of the substance whose name it bears. (See dimethylamine, above.) It has frequently been mistaken by writers upon materia medica for its isomere *propylamine*, $\left.\begin{matrix}(C_3H_7)\\H_2\end{matrix}\right\}N$, which differs from it in odor and in boiling at 50° (122° F.). Its chloride, under the names *chloride of propylamia, of secalia, of secalin*, has been used in the treatment of gout and of rheumatism.

Tetramethyl ammonium hydrate—$(CH_3)_4$**N,OH**—91.—This substance, whose constitution is similar to that of ammonium hydrate, is obtained by decomposing the corresponding iodide, $(CH_3)_4NI$, formed by the action of methyl iodide upon trimethylamine. It is a crystalline solid; deliquescent; very soluble in H_2O; caustic; not volatile without decomposition. It attracts carbon dioxide from the air, and combines with acids to form crystallizable salts.

The iodide is said to exert an action upon the economy similar to that of curare.

207

Choline—*Trimethyloxethylammonium hydrate*—$\left.\begin{matrix}(CH_3)_3\\(CH_2,CH_2—OH)\end{matrix}\right\}$ **N,OH** $= C_5H_{15}NO_2$—121—is a quaternary monammonium hydrate, containing three methyl groups and one ethylene hydroxide (oxethyl) group. It does not occur in the normal body in the free state, but is widely disseminated as a component part of an exceedingly important class of substances, the lecithins (see p. 273). It is also obtained from bile and from the yolk of eggs. It is one of the earliest products of cadaveric putrefaction, resulting, in all probability, from decomposition of the lecithins.

Choline has been obtained synthetically by the action of a concentrated solution of trimethylamine upon ethylene oxide, or upon ethylene chlorhydrin. When heated, it splits up into glycol and trimethylamine.

It appears as a thick syrup, soluble in H_2O and in alcohol, and strongly alkaline in reaction. Even in dilute aqueous solution it prevents the coagulation of albumin and redissolves coagulated albumin and fibrin. It is a strong base; attracts carbon dioxide from the air; forms with HCl a salt, soluble in alcohol, which crystallizes in plates and needles, very much resembling in appearance those of cholesterin. Its chloroplatinate is purified with difficulty; its chloraurate readily. Solutions of its chloride differ in their behavior with alkaloidal reagents from those of neurine in forming no precipitate with tannic acid, and in forming a voluminous white precipitate with phosphomolybdic acid, which becomes crystalline on standing.

Administered hypodermically to animals it causes muscarine-like intoxication, although much less intense in its action than either that alkaloid or neurine.

Amanitine—*Trimethyloxethylideneammonium hydrate*—$\left.\begin{matrix}(CH_3)_3\\(CH_3—CHOH)\end{matrix}\right\}$ **N,OH** $= C_5H_{15}NO_2$—121—is an isomere of *choline*, existing along with muscarine (see below) in *Agaricus muscarius*. By oxidation with HNO_3 it yields muscarine.

Muscarine—$\left.\begin{matrix}(CH_3)_3\\(C_2H_5O_2)\end{matrix}\right\}$ **N,OH** $= C_5H_{15}NO_3$—is a substituted tetramethylammonium hydrate closely related to choline and amanitine, from the former of which it may be obtained by oxidation.

It occurs in nature in *Agaricus muscarius*, and is produced during putrefactive decomposition of albuminoid substances. Its formation under such circumstances is of great importance, not only by reason of its actively poisonous qualities, but for the reason that, with the exception of the amines above mentioned, it is the only alkaloid formed during putrefaction which is known to be a product of the vegetable world as well.

The free alkaloid occurs in very deliquescent, irregular crystals, or, if not perfectly dry, a colorless, odorless, and tasteless, but strongly alkaline syrup; readily soluble in all proportions in water and in alcohol; very sparingly soluble in chloroform; insoluble in ether. It is a more powerful base than ammonium hydrate, forming an alkaline carbonate and neutral salts with other acids. When decomposed it yields trimethylamine. Its platinochloride crystallizes in octahedra. Its chloride forms colorless, brilliant, deliquescent needles.

When administered to animals, muscarine causes increased secretion of saliva and tears; vomiting; evacuation of fæces, at first solid, later liquid; contraction of the pupils, almost to the extent of closure; diminution of the rapidity of the pulse; interference with respiration and locomotion; gradual sinking of the heart's action and respiration; and death. Atropine

208

prevents the action of muscarine, and diminishes its intensity when already established.

Neurine—*Trimethylvinylammonium hydrate*, $\left.\begin{matrix}(CH_3)_3\\(C_2H_3)\end{matrix}\right\}$ **NOH** $= C_5H_{13}NO$, is a substance nearly related to choline, and long confounded with it, supposed by Liebreich to exist in the brain. The same body is one of the alkaloids produced by the putrefaction of muscular tissues, and is endowed with poisonous qualities, resembling, but less intense than, those of muscarine.

Another cadaveric alkaloid, related to neurine and produced under similar conditions, is a diamine; *neuridine*, $C_5H_{14}N_2$.

MONAMIDES.

These bodies differ from the amines in containing *oxygenated*, or *acid* radicals, in place of alcoholic radicals. Like the amines, they are divisible into *primary*, *secondary*, and *tertiary*. They are the nitrides of the acid radicals, as the amines are the nitrides of the alcoholic radicals.

The monamides may also be regarded as the acids in which the OH of the group COOH has been replaced by (NH_2):

$$\begin{matrix}CH_3\\|\\COOH\end{matrix}\qquad\qquad\begin{matrix}CH_3\\|\\CONH_2\end{matrix}$$

Acetic acid. Acetamide.

The *primary* monamides, containing radicals of the acids of the acetic series, are formed: (1.) By the action of heat upon an ammoniacal salt:

$$\left.\begin{matrix}(C_2H_3O)'\\NH_4\end{matrix}\right\}O=\left.\begin{matrix}H\\H\end{matrix}\right\}O+\left.\begin{matrix}(C_2H_3O)'\\H_2\end{matrix}\right\}N$$

Ammonium acetate. Water. Acetamide.

(2.) By the action of a compound ether upon ammonia:

$$\left.\begin{matrix}(C_2H_3O)'\\(C_2H_5)'\end{matrix}\right\}O+\left.\begin{matrix}H\\H\\H\end{matrix}\right\}N=\left.\begin{matrix}(C_2H_3O)'\\H_2\end{matrix}\right\}N+\left.\begin{matrix}C_2H_5\\H\end{matrix}\right\}O$$

Ethyl acetate. Ammonia. Acetamide. Alcohol.

(3.) By the action of the chloride of an acid radical upon dry NH_3:

$$\left.\begin{matrix}(C_2H_3O)'\\Cl\end{matrix}\right\}+2\left(\left.\begin{matrix}H\\H\\H\end{matrix}\right\}N\right)=\left.\begin{matrix}NH_4\\Cl\end{matrix}\right\}+\left.\begin{matrix}(C_2H_3O)'\\H_2\end{matrix}\right\}N$$

Acetyl chloride. Ammonia. Ammonium chloride. Acetamide.

The *secondary* monamides of the same class are obtained: (1.) By the action of the chlorides of acid radicals upon the primary amides:

$$\left.\begin{matrix}(C_2H_3O)'\\H_2\end{matrix}\right\}N+\left.\begin{matrix}(C_2H_3O)'\\Cl\end{matrix}\right\}=\left.\begin{matrix}(C_2H_3O)_2\\H\end{matrix}\right\}N+\left.\begin{matrix}H\\Cl\end{matrix}\right\}$$

Acetamide. Acetyl chloride. Diacetamide. Hydrochloric acid.

209

(2.) By the action of HCl upon the primary monamides at high temperatures:

$$2\left(\left.\begin{matrix}(C_2H_3O)'\\H_2\end{matrix}\right\}N\right)+\left.\begin{matrix}H\\Cl\end{matrix}\right\}=\left.\begin{matrix}(C_2H_3O)_2\\H\end{matrix}\right\}N+\left.\begin{matrix}NH_4\\Cl\end{matrix}\right\}$$

Acetamide. Hydrochloric acid. Diacetamide. Ammonium chloride.

The *tertiary* monamides of this series of radicals have been but imperfectly studied; some of them have been obtained by the action of the chlorides of acid radicals upon metallic derivatives of the secondary amides.

The *primary* monamides containing radicals of the fatty acids are solid, crystallizable, neutral in reaction, volatile without decomposition, mostly soluble in alcohol and ether, and mostly capable of uniting with acids to form compounds similar in constitution to the ammoniacal salts. They are capable of uniting with H_2O to form the ammonical salt of the corresponding acid, and with the alkaline hydrates to form the metallic salt of the corresponding acid, and ammonia. The *secondary* monamides,

containing two radicals of the fatty series, are acid in reaction, and their remaining atom of extra-radical H may be replaced by an electro-positive atom.

Acetamide—$\left.\begin{matrix}(C_2H_3O)'\\H_2\end{matrix}\right\}N$**—59—is obtained by heating, under pressure, a mixture of ethyl acetate and aqua ammoniæ, and purifying by distillation. It is a solid, crystalline substance, very soluble in H_2O, alcohol, and ether; fuses at 78° (172°.4 F.); boils at 221° (429°.8 F.); has a sweetish, cooling taste, and an odor of mice. Boiling potassium hydrate solution decomposes it into potassium acetate and ammonia. Phosphoric anhydride deprives it of H_2O, and forms with it** *acetonitrile* **or** *methyl cyanide.*

AMIDO-ACIDS OF THE FATTY SERIES.

These compounds, also known as *glycocols*, are of mixed function, acid and basic, obtained by the substitution of the univalent group $(NH_2)'$ for an atom of radical H of an acid:

$$\begin{array}{c}CH_3\\|\\COOH\end{array}\qquad\qquad\begin{array}{c}CH_2(NH_2)\\|\\COOH\end{array}$$

Acetic acid. Amido-acetic acid (glycocol).

Some of them, and many of their derivatives, exist in animal bodies. Corresponding to them are many isomeres belonging to other series.

Amido-acetic acid—*Glycocol—Sugar of gelatin—Glycolamic acid—Glycine*—$\begin{array}{l}CH_2NH_2\\|\\COOH\end{array}$**—75—was first obtained by the action of H_2SO_4 upon gelatin. It is best prepared by acting upon glue with caustic potassa, NH_3 being liberated; H_2SO_4 is then added, and the crystals of potassium sulphate separated; the liquid is evaporated, the residue dissolved in alcohol, from which solution the glycocol is allowed to crystallize.**

210

It may also be obtained synthetically by a method which indicates its constitution—by the action of ammonia upon chloracetic acid:

$$\begin{array}{l}CH_2Cl\\|\\COOH\end{array} + \begin{array}{l}H\diagdown\\H{-}N\\H\diagup\end{array} = \begin{array}{l}CH_2NH_2\\|\\COOH\end{array} + \begin{array}{l}H\\Cl\end{array}$$

Chloracetic acid. Ammonia. Amido-acetic acid. Hydrochloric acid.

It may be obtained from ox-bile, in which it exists as the salt of a conjugate acid; from uric acid by the action of hydriodic acid; and by the union of formic aldehyde, hydrocyanic acid, and water. It is isomeric with *glycolamide.*

It has been found to exist free in animal nature only in the muscle of the scallop, and, when taken internally, its constituents are eliminated as urea. In combination it exists in the gelatinoids, and with cholic acid as sodium glycocholate (*q. v.*) in the bile. It is one of the products of decomposition of glycocholic acid, hyoglycocholic acid, and hippuric acid by dilute acids and by alkalies, and of the decomposition of tissues containing gelatinoids.

It appears as large, colorless, transparent crystals; has a sweet taste; melts at 170° (338° F.); decomposes at higher temperatures; sparingly soluble in cold H_2O; much more soluble in warm H_2O; insoluble in absolute alcohol and in ether; acid in reaction.

It combines with acids to form crystalline compounds, which are decomposed at the temperature of boiling water; hot H_2SO_4 carbonizes it; HNO_3 converts it into glycolic acid (*q. v.*); with HCl it forms a chloride; heated under pressure with benzoic acid it forms hippuric acid. Its acid function is more marked; it expels carbonic and acetic acids from calcium carbonate and plumbic acetate. The presence of a small quantity of glycocol prevents the precipitation of cupric hydrate from cupric sulphate solution by potassium hydrate; the solution becomes dark blue, does not yield cuprous hydrate on boiling, and precipitates crystalline needles of copper glycolamate on the addition of alcohol to the cold solution. With ferric chloride it gives an intense red solution, whose color is discharged by acids, and reappears on neutralization. With phenol and sodium hypochlorite it gives a blue color, as does ammonia. By oxidation with potassium permanganate in alkaline solution it yields carbon dioxide, oxalic, carbonic, and oxamic acids, and water. It also forms crystalline compounds with many salts and ethers. Methyl amido-acetate is isomeric with sarcosine:

$$\begin{array}{l}CH_2NH_2\\|\\COOH\end{array}\qquad\begin{array}{l}CH_2NH_2\\|\\COOCH_3\end{array}\qquad\begin{array}{l}CH_2NH(CH_3)\\|\\COOH\end{array}$$

Glycocol (amido-acetic acid). Methyl amido-acetate. Sarcosine (methyl-glycocol).

Methyl-glycocol—*Sarcosine*—$\begin{array}{l}CH_2[NH(CH_3)]\\|\\COOH\end{array}$—89—isomeric with alanine and with lactamide (*q. v.*), does not exist as such in animal nature, but has been obtained from creatine (*q. v.*) by the action of barium hydrate:

$$\underset{\text{Creatine.}}{C_4H_9N_3O_2} + \underset{\text{Water.}}{H_2O} = \underset{\text{Sarcosine.}}{C_3H_7NO_2} + \underset{\text{Urea.}}{CON_2H_4};$$

urea being formed at the same time, and decomposed by the further action of the barium hydrate into NH_3 and barium carbonate.

211

Its constitution is indicated by its synthetic formation from chloracetic acid and methylamine:

$$\begin{array}{l}CH_2Cl\\|\\COOH\end{array} + \begin{array}{l}CH_3\diagdown\\H{-}N\\H\diagup\end{array} = \begin{array}{l}CH_2[NH(CH_3)]\\|\\COOH\end{array} + \begin{array}{l}H\\Cl\end{array}$$

Chloracetic acid. Methylamine. Sarcosine. Hydrochloric acid.

It crystallizes in colorless, transparent prisms; very soluble in water; sparingly soluble in alcohol and ether. Its aqueous solution is not acid, and has a sweetish taste; it unites with acids to form crystalline salts, but does not form metallic salts. It is capable of combining with cyanamide to form creatine.

Betaine—*Trimethylglycocol—Oxyneurine—Oxycholine*—$\begin{array}{l}CH_2{-}CO\\|\qquad|\\(CH_3)_3N{-}O\end{array}$ $= C_5H_{11}NO_2$**—117—was first obtained from the juice of the sugar-beet; afterward it was obtained by oxidation of choline; and is also** produced synthetically, either by acting upon trimethylamine with monochloracetic acid, as glycocol is obtained by the action of the same acid upon ordinary ammonia; or by acting upon glycocol itself with methyl iodide.

Betaïne crystallizes in large, brilliant crystals, containing one molecule of water of crystallization. At the ordinary temperature they are deliquescent, but at 100° (212° F.) effloresce, and lose their Aq. It is very soluble in water and in alcohol. It is decomposed by heat, with evolution of trimethylamine. It forms crystalline salts. Its chloraurate is crystalline and very sparingly soluble in cold water.

The method of its synthesis and the composition of its chloride indicate it to be related to tetramethylammonium hydrate, but when its chloride is decomposed by silver oxide, it is not with substitution of OH for Cl, but with separation of Cl + H_2O.

Betaïne is the type of a number of similar compounds derivable from the amido acids by substitution of various hydrocarbon radicals.

Amidopropionic Acid—*Alanine*—$\begin{array}{l}CH_2{-}CH_2(NH_2)\\|\\COOH\end{array}$—89.—Isomeric with sarcosine and with lactamide; does not exist, as far as is known at present, in nature. It is obtained by the action of alcoholic ammonia upon bromopropionic acid:

$$\begin{array}{l}CH_2Br\\|\\CH_2\\|\\COOH\end{array} + 2\left(\begin{array}{l}H\diagdown\\H{-}N\\H\diagup\end{array}\right) = \begin{array}{l}CH_2(NH_2)\\|\\CH_2\\|\\COOH\end{array} + BrNH_4$$

Bromopropionic acid. Ammonia. Amidopropionic acid. Ammonium bromide.

It may also be prepared by starting from lactic acid, from which it differs by containing NH_2 in place of OH.

It crystallizes in large, oblique, rhombic prisms; very soluble in H_2O; sparingly soluble in alcohol; insoluble in ether. Its aqueous solution is neutral and sweet. Nitrous acid converts it into lactic acid, N, and H_2O. It dissolves in acids without neutralizing them, but yet, in certain cases, with the formation of crystalline compounds. Its Ba, Pb, Cu, and Ag salts are soluble and crystalline.

212

Amidobutyric Acid—*Butalanine*—$C_4H_9NO_2$—and **Amidovalerianic acid**—$C_5H_{11}NO_2$—are only of theoretic interest at present. The latter has been found in the tissue of the pancreas and among the products of the action of pancreatic juice upon albumin. They are among the products of the decomposition of albumin by caustic baryta.

Amidocaproic Acid—*Leucine*—$\begin{array}{l}CH_3{-}C_3H_6{-}CH_2(NH_2)\\|\\COOH\end{array} = C_6H_{13}NO_2$—131—exists widely distributed in animal nature; it has been obtained from the normal spleen, pancreas, salivary, lymphatic, thymus, and thyroid glands, lungs, and liver. Pathologically, its quantity in the liver is much increased in diseases of that organ, and in typhus and variola; in the bile in typhus; in the blood in leucocythæmia, and in yellow atrophy of the liver; in the urine in yellow atrophy of the liver, in typhus, and in variola; in choleraic discharges from the intestine; in pus; in the fluids of dropsy; and of atheromatous cysts. In these situations it is usually accompanied by tyrosine (*q. v.*). It is much more abundant in the tissues of the lower forms of animal life, and has also been found in vegetable tissues.

It is formed by the decomposition of nitrogenized animal and vegetable substances, by heating with strong alkalies or dilute acids; by the decomposition of elastic tissues it is formed with a small quantity of tyrosine; by that of gelatinoid materials, leucine and glycine are obtained; by that of albuminoids, leucine and a small, but variable, quantity of tyrosine are formed; and that of epidermic tissues yields leucine and tyrosine. It is also one of the products of the putrefaction of animal and vegetable albuminoids, and of the action of pancreatic juice upon fibrin. It has also been formed synthetically by the action of NH_3 upon bromocaproic acid, in the same way that alanine is formed from bromopropionic acid (see above).

It may be obtained by a variety of methods, the most advantageous of which consists in boiling 1 pt. horn-shavings with 4 pts. H_2SO_4 and 12 pts. H_2O, for 36 hours, renewing the H_2O as it evaporates; the acid liquid is saturated with milk of lime and boiled again for 24 hours; it is then filtered through linen, a slight excess of H_2SO_4 is added, and the liquid again filtered and evaporated; tyrosine first crystallizes out and is separated, after which leucine separates in crystals, which are purified by recrystallization from a small quantity of H_2O, the crystals first formed being rejected. The leucine so obtained is further purified by solution in hot H_2O; digestion with lead hydrate; filtration; treatment with H_2S; filtration; treatment with animal charcoal; filtration and crystallization.

Leucine crystallines from alcohol in soft, pearly plates, lighter than H_2O, and somewhat resembling cholesterin; sometimes in round masses composed of closely grouped needles radiating from a centre. It is sparingly soluble in cold H_2O; readily in warm H_2O; almost insoluble in cold alcohol and ether; soluble in boiling alcohol, which deposits it on cooling; it is odorless and tasteless, and its solutions are neutral. Its solubility in H_2O is increased by the presence of acetic acid or of potassium acetate. It sublimes at 170° (338° F.) without decomposition; if suddenly heated above 180° (356° F.), it is decomposed into amylamine and carbon dioxide.

When heated to 140° (284° F.), with hydriodic acid under pressure, it is decomposed into caproic acid and ammonia. Nitrous acid converts it into *leucic acid*, $C_6H_{12}O_3$, H_2O and N. It unites with acids to form soluble, crystalline salts. It also dissolves readily in solutions of alkaline hydrates, forming crystalline compounds with the metallic elements.

The formation of leucine in the body is one of the steps of the trans-

213

formation of at least some part of the albuminoids into urea. That leucine is formed at the expense of the albuminoids by some fermentation-like process, there can be no doubt. As it is only discharged in the urine in certain exceptional pathological conditions, and as at the same time the elimination of urea is greatly diminished, it seems highly probable that under normal conditions the N of leucine finally makes its exit from the body as urea, notwithstanding the fact that chemists have hitherto been unable to obtain urea from leucine artificially. As to the nature of the changes by which leucine is converted into urea in the body, we are as yet in the dark. When leucine and tyrosine appear in the urine, that fluid is poor in urea and usually contains biliary coloring matters; the substitution of leucine for urea may be so extensive that the urine contains no urea, and contains leucine in such quantity that it crystallizes out spontaneously.

Analytical Characters.—The presence of leucine and tyrosine in the urine may be detected as follows: the freshly collected urine is treated with basic lead acetate, filtered, the filtrate treated with H_2S, filtered from the precipitated lead sulphide, and the filtrate evaporated over the water-bath; leucine and tyrosine crystallize; they may be separated by extraction of the residue with hot alcohol, which dissolves the leucine and leaves the tyrosine. The leucine left by evaporation of the alcoholic solution may be recognized by its crystalline form and by the following characters: (1) a small portion is moistened on platinum foil with HNO_3, which is then cautiously evaporated; a colorless residue remains, which, when warmed with caustic soda solution, turns yellow or brown, and by further concentration is converted into oily drops, which do not adhere to the platinum (Scherer's test); (2) a portion of the residue is heated in a dry test-tube; it melts into oily drops, and the odor of amylamine (odor of ammonia combined with that of fusel oil) is observed; (3) if a boiling mixture of leucine and solution of neutral lead acetate be carefully neutralized with ammonia, brilliant crystals of a compound of leucine and lead oxide separate; (4) leucine carefully heated in a glass tube, open at both ends, to 170° (338° F.), sublimes without fusing, and condenses in flocculent shreds, resembling those of sublimed zinc oxide. If heated beyond 180° (356° F.), the decomposition mentioned in 2d occurs.

Tyrosine—$C_9H_{11}NO_3$—145—is a substance which does not belong to this series, and is probably an amido-acid of the aromatic series; nevertheless, as its constitution is still undetermined, and as it is almost universally found to accompany leucine in animal tissues and in the products of their decomposition, it may be considered in this place.

The methods of its formation and preparation are given under leucine. It crystallizes from its watery and ammoniacal solutions in silky needles, arranged in stellate bundles; very sparingly soluble in cold H_2O; almost insoluble in alcohol; more soluble in hot H_2O. When heated, it turns brown and yields an oily matter having the odor of phenol; when heated in small quantities to 270° (518° F.), it is decomposed into carbon dioxide and a white solid, having the composition $C_8H_{11}NO$, which sublimes. It combines with both acids and bases.

It has been found in animal nature in the same situations as leucine. When taken into the stomach it is not altered in the economy, but is eliminated in the urine and fæces.

Analytical Characters.—(1) its crystalline form; (2) when heated it gives off an odor of phenol; (3) when moistened with HNO_3 and carefully evaporated, a deep yellow residue remains, which turns darker with

214

NaHO; (4) with concentrated H_2SO_4 and slightly warmed, it dissolves with a transient red color—the solution, filtered and neutralized with $CaCO_3$, gives a violet color with Fe_2Cl_6 solution; (5) when boiled with acid nitrate of mercury solution, a pink color, and later, a red precipitate.

Biliary Acids.—The bile of most animals contains the sodium salts of two amido-acids of complex constitution. These acids may be decomposed into a non-nitrogenized acid (cholic acid), and either an amido-acid (glycocol), or an amido-sulphurous acid (taurine). The following biliary acids have been described:

Glycocholic acid—$C_{26}H_{43}NO_6$—465—(sometimes designated as *acide cholique*, *cholsäure*, *cholic acid*, by French and German writers). It exists as its sodium salt in the bile of the herbivora, and in much smaller proportion in that of the carnivora; it exists in small quantity in human blood and urine in icterus; in human bile its quantity varies with the diet.

It is best obtained from ox-bile; this is evaporated to one-fourth of its original volume, the residue is ground up with animal charcoal, and dried at 100° (212° F.); the dry mass, while still hot, is broken up and introduced into a flask, in which it is digested with absolute alcohol, with repeated agitation, for some days; the colorless, filtered alcoholic solution is partially evaporated, but not to the extent of becoming syrupy, then mixed with an excess of anhydrous ether, which, if the reagents were free from H_2O, causes the immediate separation of a crystalline precipitate of the mixed biliary salts. If the alcohol or ether used contain H_2O, the precipitate is at first resinous and only becomes crystalline after standing, or does not become crystalline if the proportion of H_2O be too great. The crystalline deposit is collected upon a filter, washed with ether and dissolved in a small quantity of H_2O; to the aqueous solution a small quantity of ether is added, and then enough dilute H_2SO_4 to render the mixture permanently cloudy; the glycocholic acid gradually crystallizes out, and may be further purified by solution in alcohol, and precipitation with a great excess of ether.

Glycocholic acid forms brilliant, colorless, transparent needles, which are sparingly soluble in cold H_2O, readily soluble in warm H_2O and in alcohol, almost insoluble in ether. The watery solution is acid in reaction, and tastes at first sweet, afterward intensely bitter. Its alcoholic solution exerts a right-handed polarization $[a]_D = +29°$; when evaporated it leaves the acid in a resinous form.

When heated with potash, baryta, or dilute H_2SO_4 or HCl, it is decomposed into *cholic acid* and glycocol:

$$\underset{\text{Glycocholic acid.}}{C_{26}H_{43}NO_6} + \underset{\text{Water.}}{H_2O} = \underset{\text{Cholic acid.}}{C_{24}H_{40}O_5} + \underset{\text{Glycocol.}}{C_2H_5NO_2}.$$

Glycocholic acid dissolves unchanged in cold concentrated H_2SO_4, and is precipitated on dilution of the solution with H_2O; if the mixture be warmed the bile acid is decomposed, and there separate oily drops of *cholonic acid*, $C_{26}H_{41}NO_5$, differing from glycocholic acid by $-H_2O$. When allowed to remain long in contact with concentrated H_2SO_4, glycocholic acid is converted into a colorless, resinous mass, which slowly forms a saffron-yellow solution with the mineral acid, which turns flame-red when warmed, and which, on dilution, deposits a flocculent material which is colorless, greenish, or brownish, according to the temperature at which it

215

is formed. Glycocholic acid, altered by contact with concentrated H_2SO_4, absorbs O when exposed to the air, and turns red, then blue, and finally brown after a few days.

Sodium Glycocholate, $C_{26}H_{42}NO_6Na$, exists in the bile; it crystallizes in stellate needles, very soluble in H_2O, less so in absolute alcohol, and insoluble in ether; its acoholic solution exerts right-handed polarization $[a]_D = +25°.7$.

Lead Glycocholate, $(C_{26}H_{42}NO_6)_2Pb$ (?), is formed as a white, flocculent precipitate, when solution of lead subacetate is added to a solution of a glycocholate or of glycocholic acid; with the *neutral* acetate the precipitation does not occur in the presence of an excess of acetic acid. It is soluble in alcohol, and in an excess of lead acetate solution.

The glycocholates of the alkaline earths are soluble in H_2O. Glycocholic acid and the glycocholates react with Pettenkofer's test (see below).

Glycocholic acid forms compounds with the alkaloids, some of which are crystalline, others amorphous; they are for the most part very sparingly soluble in H_2O, but readily soluble in solutions of the biliary salts and in bile.

Taurocholic acid—$C_{26}H_{45}NO_7S$—515—(*choleic acid* of Strecker)—exists as its sodium salt in the bile of man and of the carnivora, and in much less abundance in that of the herbivora; in the bile of the dog it seems to be unaccompanied by any other biliary acid. It may be obtained from dog's bile by a modification of the method described under glycocholic acid; the watery solution is not treated with H_2SO_4, as in the preparation of that acid, but with solution of basic lead acetate and ammonia. The precipitate so formed is extracted with boiling alcohol, the solution filtered hot and treated with H_2S; the clear liquid, filtered from the precipitated lead sulphide, is evaporated to a small bulk and treated with a large excess of ether; the acid is precipitated in the resinous form, but, after standing for a varying period, assumes the crystalline form.

When carefully prepared it forms silky, crystalline needles, which, when exposed to the air, deliquesce rapidly, and which, even under absolute ether, are gradually converted into a transparent, amorphous, resinous mass. It is soluble in H_2O and alcohol; insoluble in ether; its aqueous solution is very bitter; in alcoholic solution it deviates the plane of polarization to the right, $[a]_D = +24°.5$; its solutions are acid in reaction.

Taurocholic acid is very readily decomposed by heating with barium hydrate, with dilute acids, and even by evaporation of its solution, into cholic acid and *taurine*:

$$\underset{\text{Taurocholic acid.}}{C_{26}H_{45}NO_7S} + \underset{\text{Water.}}{H_2O} = \underset{\text{Cholic acid.}}{C_{24}H_{40}O_5} + \underset{\text{Taurine.}}{C_2H_7NO_3S}$$

The same decomposition occurs in the presence of putrefying material and in the intestine. Taurocholic acid has not been found to accompany glycocholic in the urine of icteric patients.

The taurocholates are neutral in reaction; those of the alkaline metals are soluble in alcohol and in water; and by long contact with ether they assume the crystalline form. They may be separated from the glycocholates in watery solution, either: (1) by dilute H_2SO_4 in the presence of a small quantity of ether, which precipitates glycocholic acid alone; or (2) by adding *neutral* lead acetate to the solution of the mixed salts (which must be neutral in reaction) lead glycocholate is precipitated and separated by

216

filtration; to the mother liquor basic lead acetate and ammonia are added, when lead taurocholate is precipitated. The acids are obtained from the hot alcoholic solutions of the Pb salts by decomposition with H_2S, filtration, concentration, and precipitation by ether.

Solutions of the taurocholates, like those of the glycocholates, have the power of dissolving cholesterin and of emulsifying the fats; they also form with the salts of the alkaloids compounds which are insoluble in H_2O, but soluble in an excess of the biliary salt. The taurocholate of morphine is crystallizable. They react with Pettenkofer's test.

Hyoglycocholic acid, $C_{27}H_{43}NO_5$, and **Hyotaurocholic acid**, $C_{27}H_{45}NO_6S$, (?) are conjugate acids of *hyocholic acid*, $C_{25}H_{40}O_4$, and glycocol and taurine, which exist in the bile of the pig. **Chenotaurocholic acid**, a conjugate acid of taurine and *chenocholic acid*, $C_{27}H_{44}O_4$, is obtained from the bile of the goose.

Cholic acid—$C_{24}H_{40}O_5$—408—(*cholalic acid* of Strecker), is a product af decomposition of glyco- and taurocholic acids, obtained as indicated above. It also occurs, as the result of a similar decomposition, in the intestines and fæces of both herbivora and carnivora. It forms large, clear, deliquescent crystals; sparingly soluble in H_2O, readily soluble in alcohol and ether; intensely bitter in taste, with a sweetish aftertaste; in alcoholic solution it is dextrogyric $[a]_D = +35°$. The alkaline cholates are crystallizable and readily soluble in H_2O, the others difficultly soluble. Cholic acid and the cholates respond to Pettenkofer's test.

By boiling with acids or by continued heating to 200° (392° F.), cholic acid loses the elements of H_2O, and is transformed into *dyslysin*, $C_{24}H_{36}O_3$, a neutral, resinous material, insoluble in H_2O and alcohol, sparingly soluble in ether.

THE PETTENKOFER REACTION.—All of the biliary acids, and the cholic acid and dyslysin obtained by their decomposition, have the property of forming a yellow solution with concentrated H_2SO_4, the color of which rapidly increases in intensity, and which exhibits a green fluorescence. Their watery solutions also, when treated with a small quantity of cane-sugar and with concentrated H_2SO_4, so added that the mixture acquires a temperature of 70° (158° F.) but does not become heated much beyond that point, develop a beautiful cherry-red color, which gradually changes to dark reddish purple. Although this reaction is observed in the presence of very small quantities of the biliary acids, it loses its value, unless applied as directed below, from the fact that many other substances give the same reaction, either with H_2SO_4 alone, or in the presence of cane-sugar. Among these substances are many which exist naturally in animal fluids, or which may be introduced with the food or as medicines; such are cholesterin, the albuminoids, lecithin, oleic acid, cerebrin, phenol, turpentine, tannic acid, salicylic acid, morphine, codeine, many oils and fats, cod-liver oil, etc. It has been suggested that a distinction could be made between the color produced by the Pettenkofer test with the biliary acids and those produced by the same test with other substances, by spectroscopic observation; the test with biliary acids in watery solution exhibiting a single dark and broad absorption-band (Fig. 34, No. 2); the same test in alcoholic solution shows two bands (No. 1); but while this spectrum differs from those observed in the purple solutions obtained with many other substances, such as albumin (No. 3); it does not differ sufficiently from that obtained with the morphine salts (No. 4) to render it a safe method for controlling the test.

The following method of applying Pettenkofer's test to the urine and

217

other fluids removes, we believe, every source of error. The urine, etc., is first evaporated to dryness at the temperature of the water-bath, a small quantity of coarse animal charcoal having been added; the residue is extracted with absolute alcohol, the alcoholic liquid filtered, partially evaporated, and treated with ten times its bulk of absolute ether; after standing an hour or two, any precipitate which may have formed is collected upon a small filter, washed with ether, and dissolved in a small quantity of H_2O; this aqueous solution is placed in a test-tube, a drop or two of a strong aqueous solution of cane-sugar (sugar, 1; water, 4), and then pure concentrated H_2SO_4 are added; the addition of the acid being so regulated, and the test-tube dipped from time to time in cold water, that the temperature shall be from 60°–75° (140°–167° F.). In the presence of biliary acids the mixture usually becomes turbid at first, and then turns cherry-red and finally purple, the intensity of the color varying with the amount of biliary acid present.

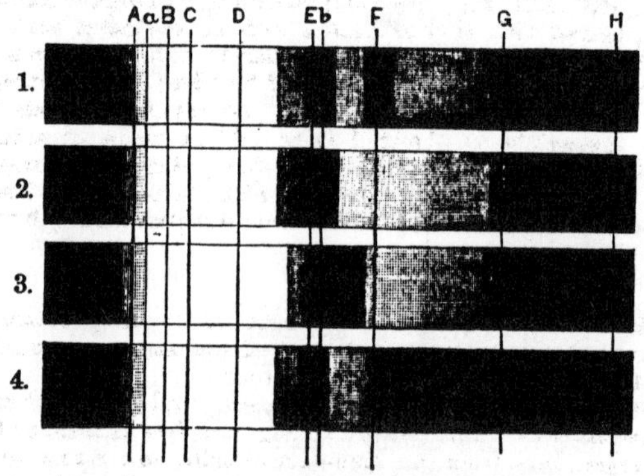

FIG. 34.

PHYSIOLOGICAL CHEMISTRY OF THE BILIARY ACIDS.—These substances do not normally pre-exist in the blood, and are consequently formed in the liver, and they are not reabsorbed from the intestine unchanged. Solutions of the biliary salts, injected into the circulation in small quantity, cause a diminution in the frequency of the pulse and of the respiratory movements, a lowering of the temperature and arterial tension, and disintegration of the blood-corpuscles. In large doses (2–4 grams [30–60 grains] for a dog) they produce the same effects to a more marked degree; epileptiform convulsions, black and bloody urine, and death more or less rapidly. These effects do not follow the injection of the products of decomposition of the biliary acids, except cholic acid, and in that case the symptoms are much less marked. Nor are the biliary acids discharged unaltered with the fæces; they are decomposed in the intestine. The extract, suitably purified, of the contents of the upper part of the small intestine, gives a well-marked reaction with Pettenkofer's test; while similar extracts of the contents of the lower part of the large intestine, or of the fæces, fail to give the reaction, and consequently are free from glyco- or taurocholic, cholic acid, or dyslysin; the fæces, moreover, do not contain either taurine or glycocol. During the processes, at

218

present but imperfectly understood, which take place in the intestine, the bile-acids are undoubtedly decomposed into cholic acid and taurine or glycocol, which are subsequently reabsorbed, either as such, or after having been subjected to further decomposition; and as a consequence of their decomposition they probably have some influence upon intestinal digestion.

The biliary salts are precipitated from their aqueous solution, or from bile, by fresh gastric juice from the same animal; but they are not so

precipitated if the gastric juice contain peptone. The proportion of biliary salts in human bile seems to vary considerably, as shown by the following analyses :

	I.	II.	III.	IV.	V.	VI.	VII.	VIII.	IX.
Mucin	2.66	2.98	2.21	1.45	....	2.48	1.29	....	1.29
Cholesterin	0.16	0.26 }	4.73	3.09	{ 0.25	0.25	0.34	....	0.35
Fats	0.32	0.92 }			{ 0.04	0.05	0.36	....	0.73
Taurocholate of sodium, }	7.22	9.14	10.79	5.65	{	0.75	1.93	1.57	0.87
Glycocholate of sodium }					{ 4.48	2.09	0.44	4.90	3.03
Scaps	....	....	....	....	0.64	0.82	1.63	1.46	1.39
Mineral salts	0.65	0.77	1.08	0.63	3.86	0.46?	1.46?	....	....
Water	86.00	85.92	82.27	89.81	....	90.88	91.08	....	....
Total solids	14.00	14.08	17.73	10.19	....	9.12	8.92	....	....

I. Frerichs : Bile from man, æt. 18, killed by a fall. II. Frerichs : Male, æt. 22, died of a wound. III. Gorup-Besanez : Male, æt. 49, decapitated IV. Gorup-Besanez : Female, æt. 29, decapitated. V. Jacobsen : Male, biliary fistula. VI., VII. Trifanowski : Males. VIII. Socolof : Mean of six analyses of human bile. IX. Hoppe-Seyler : Mean of five analyses of bile from subjects with healthy livers.

Pathologically, the biliary acids may be detected in the blood and urine in icterus and acute atrophy of the liver, although by no means as frequently as the biliary coloring matters.

Creatine—$C_4H_9N_3O_2$+Aq—131+18—is another complex amido-acid, which occurs as a normal constituent of the juices of muscular tissue, voluntary and involuntary, of brain, blood, and amniotic fluid.

It is best obtained from the flesh of the fowl, which contains 0.32 per cent., or from beef-heart, which contains 0.14 per cent., by hashing, warming with alcohol and expressing strongly ; the alcohol is distilled off, the residual liquid precipitated with lead acetate, filtered, treated with H_2S, again filtered, the filtrate evaporated to a syrup, from which the creatine crystallizes. It is soluble in boiling H_2O and in alcohol, insoluble in ether ; crystallizes in brilliant, oblique, rhombic prisms ; neutral, tasteless, loses aq. at 100° (212° F.) ; fuses and decomposes at higher temperatures. When long heated with H_2O or treated with concentrated acids, it loses H_2O, and is converted into creatinine. Baryta water decomposes it into sarcosine and urea. It is not precipitated by silver nitrate, except when it is in excess and in presence of a small quantity of potassium hydrate ; the white precipitate so obtained is soluble in excess of potash, from which a jelly separates which turns black, slowly at ordinary temperatures, rapidly at 100° (212° F.). A white precipitate, which turns black when heated, is also formed when a solution of creatine is similarly treated with mercuric chloride and potash.

219

Creatinine—$C_4H_7N_3O$—113—a product of the dehydration of creatine, is a normal and constant constituent of the urine and amniotic fluid, and also exists in the blood and muscular tissue.

It crystallizes in oblique, rhombic prisms, soluble in H_2O and in hot alcohol ; insoluble in ether. It is a strong base, has an alkaline taste and reaction ; expels NH_3 from the ammoniacal salts, and forms well-defined salts, among which is the double chloride of zinc and creatinine ($(C_4H_7N_3O)_2ZnCl_2$), obtained in very sparingly soluble, oblique prismatic crystals, when alcoholic solutions of creatinine and zinc chloride are mixed.

The quantity of creatinine eliminated is slightly greater than that of uric acid, 0.6–1.3 gram (9.25–20 grains) in 24 hours ; it is not increased by muscular exercise, but is diminished in progressive muscular atrophy. It is obtained from the urine by precipitation with zinc chloride.

Xantnine—*Xanthic oxide—Urous acid*—$C_5H_4N_4O_2$—152—occurs in a rare form of urinary calculus ; in the pancreas, spleen, liver, thymus, and brain of mammals and fishes ; and in human urine after the use of sulphur baths or inunctions.

It is an amorphous, yellowish-white powder ; very slightly soluble in cold H_2O. If dissolved in HNO_3 and the solution evaporated, xanthine leaves a yellowish residue, which turns reddish-yellow on the addition of potash solution, and violet-red when heated.

Xanthine calculi vary in size from that of a pea to that of a pigeon's egg. They are rather hard, brownish-yellow, smooth, shining, and made up of well-defined, concentric layers. Their broken surfaces assume a waxy polish when rubbed.

Hypoxanthine—*Sarcine*—$C_5H_4N_4O$—136—occurs in the spleen, muscular tissue, thymus, suprarenal capsules and brain of mammals ; in the liver in acute yellow atrophy ; and in the blood and urine in leucocythæmia. It may be obtained from the mother liquor of the preparation of creatine (*q. v.*).

It forms nodular masses ; soluble in 300 parts of cold, and 78 parts of boiling H_2O. It is produced from uric acid or from xanthine by the action of sodium amalgam, and when oxidized by HNO_3 it yields xanthine.

Guanine—$C_5H_5N_5O$—151—occurs in guano, in the excrements of the lower animals, and in the pancreas, lungs, and liver of certain mammalians. It is a white or yellowish, amorphous, odorless and tasteless solid ; almost insoluble in H_2O, alcohol and ether ; readily soluble in acids and alkalies, with which it forms compounds.

Carnine—$C_7H_8N_4O_3 + H_2O$—196 + 18—is obtained from Liebig's meat extract in chalky, microscopic crystals, readily soluble in warm H_2O. It forms compounds with acids and alkalies, similar to those of hypoxanthine.

COMPOUNDS OF THE ALCOHOLIC RADICALS WITH OTHER ELEMENTS.

The organic substances hitherto considered are composed of seven elements only : C, H, O, N, Cl, Br and I ; but compounds of C containing every known element have been observed to exist in nature, or have been produced artificially. Of these quite a number may be considered as containing the radicals of the series C_nH_{2n+1}, which exist in the monoatomic alcohols. These bodies are almost exclusively the products of the labora-

220

tory, and resemble in constitution some of the compounds already considered.

Sulphides.—The compounds of the alcoholic radicals with S are the same in constitution as those with O, S taking the place of O :

$$\left.\begin{matrix}C_2H_5\\H\end{matrix}\right\}O'' \qquad \left.\begin{matrix}C_2H_5\\C_2H_5\end{matrix}\right\}O'' \qquad \left.\begin{matrix}C_2H_5\\H\end{matrix}\right\}S'' \qquad \left.\begin{matrix}C_2H_5\\C_2H_5\end{matrix}\right\}S''$$

Ethyl hydrate (alcohol). Ethyl oxide (ether). Ethyl sulphydrate (mercaptan). Ethyl sulphide.

Ethyl Sulphydrate, usually known as *mercaptan*, from its tendency to unite with mercury (*corpus mercurium captans*), is formed in a variety of reactions. It is best prepared by treating alcohol with H_2SO_4, as in the preparation of sulphovinic acid (*q. v.*) ; mixing the crude product with excess of potash ; separating from the crystals of potassium sulphate ; saturating with H_2S ; and distilling.

It is a mobile, colorless liquid ; sp. gr. 0.8325 ; has an intensely disagreeable odor, combined of those of garlic and H_2S ; boils at 36°.2 (97°.2 F.) ; ignites readily and burns with a blue flame ; may be readily frozen by the cold produced by its own evaporation ; neutral in reaction ; sparingly soluble in H_2O, soluble in all proportions in alcohol and ether ; dissolves I, S and P.

Potassium and sodium act with mercaptan as with alcohol, replacing the extra-radical hydrogen. In its behavior toward the oxides it more closely resembles the acids than the alcohols, being capable even of entering into double decomposition to form salts, called *sulphethylates* or *mercaptides*. Its action with mercuric oxide is characteristic, forming a white, crystalline sulphide of ethyl and mercury :

$$2\left(\left.\begin{matrix}C_2H_5\\H\end{matrix}\right\}S\right) + Hg''O = \left.\begin{matrix}(C_2H_5)_2\\Hg''\end{matrix}\right\}S_2 + H_2O$$

Ethyl sulphydrate. Mercuric oxide. Ethyl-mercuric sulphide. Water.

Ethyl Sulphide, a colorless liquid ; having a penetrating, disagreeable odor of garlic ; boiling at 73° (163°.4 F.) ; insoluble in H_2O, soluble in alcohol ; inflammable ; obtained by the action of ethyl chloride upon potassium sulphide.

Phosphines, arsines, and **stibines** are compounds resembling the amines in constitution, in which the N is replaced by P, As, or Sb. Like the amines, they may be primary, secondary, or tertiary :

$$\left.\begin{matrix}C_2H_5\\H\\H\end{matrix}\right\}N \qquad \left.\begin{matrix}C_2H_5\\H\\H\end{matrix}\right\}P \qquad \left.\begin{matrix}C_2H_5\\C_2H_5\\H\end{matrix}\right\}As \qquad \left.\begin{matrix}C_2H_5\\C_2H_5\\C_2H_5\end{matrix}\right\}Sb.$$

Ethylamine (primary). Ethylphospine (primary). Diethyl-arsine (secondary). Triethyl-stibine (tertiary).

There also exist compounds containing P, As, or Sb, which are similar in constitution to the hydrates and salts of ammonium, and of the compound ammoniums :

$$NH_4I \qquad N(CH_3)_4I \qquad As(CH_3)_4I$$

Ammonium iodide. Tetramethyl ammonium iodide. Tetramethyl arsenium iodide.

221

Most of these compounds, which are very numerous, are as yet only of theoretic interest. One of them, however, is deserving of notice here :

Dimethyl Arsine, $\left.\begin{matrix}CH_3\\CH_3\\H\end{matrix}\right\}As$—106—which may be considered as being the hydride of the radical [$As(CH_3)_2$], does not exist as such ; there is however, a liquid known as the *fuming liquor of Cadet*, or *alkarsin*, which is obtained by distilling a mixture of potassium acetate and arsenic trioxide. This liquid contains the oxide of the above radical, and a substance which ignites on contact with air, and which consists of the same radical united to itself $2[As(CH_3)_2]$. This radical, called *cacodyle* (κακός = evil), is capable of entering into a great number of other combinations. Cacodyle and its compounds are all exceedingly poisonous, especially the cyanide, an ethereal liquid, very volatile, the presence of whose vapor in inspired air, even in minute traces, produces symptoms referable both to arsenic and to hydrocyanic acid.

Organo-metallic substances are compounds of the alcoholic radicals with metals. They are very numerous, usually obtained by the action of the iodide of the alcoholic radical upon the metallic element, in an atmosphere of H. They are substances which, although they have been put to no uses in the arts or in medicine, have been of great service in chemical research. As typical of this class of substances we may mention:

ZINC-ETHYL—$\left.\begin{matrix}C_2H_5\\C_2H_5\end{matrix}\right\}$ Zn—123—obtained by heating at 130° (266° F.) in a sealed tube a mixture of perfectly dry zinc amalgam with ethyl iodide; the contents of the tube are then distilled in an atmosphere of coal-gas, or H, and the distillate collected in a receiver, in which it can be sealed by fusion of the glass without contact with air.

It is a colorless, transparent, highly refracting liquid; sp. gr. 1.182; boils at 118° (244°.4 F.). On contact with air it ignites and burns with a luminous flame, bordered with green, and gives off dense clouds of zinc oxide, a property which renders it very dangerous to handle. On contact with H_2O it is immediately decomposed into zinc hydrate and ethyl hydride. It is chiefly useful as an agent by which the radical ethyl can be introduced into organic molecules.

ALLYLIC SERIES.

The compounds heretofore considered may be derived more or less directly from the saturated hydrocarbons; in the derivatives, as in the hydrocarbons, the valences of the C atoms are all satisfied, and that in the simplest and most complete manner, two neighboring C atoms always exchanging a *single* valence. There exist, however, other compounds, containing less H in proportion to C than those already considered, and yet resembling them in being monoatomic. These compounds have usually been considered as *non-saturated*, because all the *possible* valences are not satisfied, and the substances are therefor capable of forming products of *addition*, while the *saturated* compounds can only form products of *substitution*.

In this sense the substances composing this series are non-saturated, but they are not so in the sense that they contain C or other atoms whose valences are not satisfied. The following formulæ indicate the constitu-

222

tion of the substances of this series, and their relation to those of the previous one. It will be observed that in the allyl compounds two neighboring C atoms exchange *two* valences:

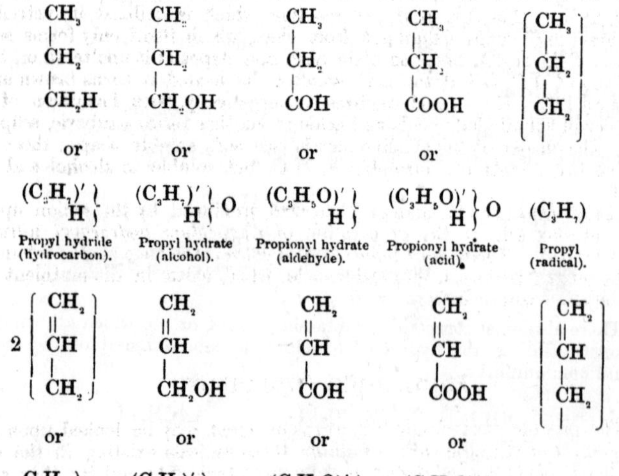

Diallyl—$\left.\begin{matrix}C_3H_5\\C_3H_5\end{matrix}\right\}$—82—formerly known as *allyl*, is obtained by the action of sodium upon allyl iodide, and is not, as its empirical formula would seem to indicate, a superior homologue of acetylene and allylene (*q. v.*).

It is a colorless liquid, having a peculiar odor, somewhat resembling that of horseradish; boils at 59° (138°.2 F.); sp. gr. 0.684 at 14° (57°.2 F.).

Allyl hydrate—*Allylic alcohol*—$\left.\begin{matrix}C_3H_5\\H\end{matrix}\right\}$ O—58—may be obtained by the action of sodium upon dichlorhydrine in ethereal solution; or by heating four parts of glycerin with one part of crystallized oxalic acid.

Allylic alcohol is a colorless, mobile liquid; solidifies at −54° (−65°.2 F.); boils at 97° (206°.6 F.); sp. gr. 0.8507 at 25° (77° F.); soluble in H_2O; has an odor resembling the combined odors of alcohol and essence of mustard; burns with a luminous flame.

Allyl alcohol is isomeric with propylic aldehyde and with acetone. Being an unsaturated compound, it is capable of forming products of addition with Cl, Br and I, etc., which are isomeric or identical with products of substitution obtained by the action of the same elements upon glycerin. Oxidizing agents convert it first into acrolein, acrylic aldehyde, C_3H_4O, and finally into acrylic acid. It does not combine readily with H, but in the presence of nascent H combination takes place slowly, with formation of propylic alcohol.

223

Allyl oxide—*Allylic ether*—$\left.\begin{matrix}C_3H_5\\C_3H_5\end{matrix}\right\}$ O—98—exists in small quantities in crude essence of garlic. It is obtained as a colorless liquid, having an alliaceous odor; insoluble in H_2O; boiling at 82° (179°.6 F.), by a number of reactions, but best by the action of allyl iodide upon sodium allyl oxide.

Allyl sulphide—*essence of garlic*—$\left.\begin{matrix}C_3H_5\\C_3H_5\end{matrix}\right\}$ S—114—is obtained by the action of an alcoholic solution of potassium sulphide upon allyl iodide; also as a constituent of the volatile oil of garlic, by macerating garlic, or other related vegetables, in water, and distilling. Crude essence of garlic is thus obtained as a heavy, fetid, brown oil; this is purified by redistillation below 140° (284° F.); contact with potassium and subsequent redistillation from calcium chloride.

It is a colorless, transparent oil; lighter than H_2O, sparingly soluble in H_2O, very soluble in alcohol and ether; boils at 140° (280° F.); has an intense odor of garlic. It does not exist naturally in the plant, but is formed during the process of extraction by the action of H_2O, probably in a manner similar to that in which essence of mustard is formed under similar circumstances. It is to the formation of allyl sulphide, which is highly volatile, that garlic owes the odor which it emits.

Allyl sulphocyanate—*Essential oil of mustard*—*Oleum sinapis volatile* (*U. S.*)—$\left.\begin{matrix}CN\\C_3H_5\end{matrix}\right\}$ S—99.—If the seeds of white or black mustard be strongly expressed, a bland, neutral oil is obtained, which resembles rapeseed and colza oils in its physical properties, and in being composed of the glycerides of stearic, oleic, and erucic acids. The cake remaining after the expression of this oil from *black mustard*, or the black-mustard seeds themselves, pulverized and moistened with H_2O, gives off a strong, pungent odor. If the H_2O be now distilled, a volatile oil passes over with it, which is the crude essential oil of mustard.

In practice the powdered cake of black-mustard seeds, from which the fixed oil has been expressed, is digested with H_2O for 24 hours, after which the H_2O is distilled as long as any oily matter passes over; the oil is collected, dried by contact with calcium chloride, and redistilled. Essence of mustard may also be obtained synthetically by the action of allyl bromide or iodide upon potassium sulphocyanate, or by the action of allyl iodide upon silver sulphocyanate.

This essence does not exist preformed in the mustard, but results from the decomposition of a peculiar constituent of the seeds, *potassium myronate*, determined by cryptolytic action set up by another constituent, *myrosine*, in the presence of H_2O.

Potassium myronate exists only in appreciable quantity in the black variety of mustard, from which it may be obtained in the shape of short prismatic crystals, transparent, odorless, bitter; very soluble in H_2O, sparingly so in alcohol.

Myrosine is a nitrogenized cryptolite, existing in the white as well as in the black mustard, and in other seeds. It may be obtained from white-mustard seeds, in an impure form, by extraction with cold H_2O, filtering and evaporating the solution at a temperature below 40° (104° F.); the syrupy fluid so obtained is precipitated with alcohol, the precipitate washed with alcohol, redissolved in H_2O, and the solution evaporated below 40° (104° F.) to dryness.

At temperatures above 40° (104° F.) myrosine becomes coagulated and

224

incapable of decomposing potassium myronate, a change which is also produced by contact with acetic acid. As the rubefacient and vesicant actions of mustard when moistened with H_2O, are due to the production of allyl sulphocyanate, neither vinegar, acetic acid, nor heat greater than 40°

(104° F.) should be used in the preparation of mustard cataplasms.

Pure allyl sulphocyanate is a transparent, colorless oil; sp. gr. 1.015 at 20° (68° F.); boils at 143° (289°.4 F.); has a penetrating, pungent odor, sparingly soluble in H_2O, very soluble in alcohol and ether. When exposed to the light it gradually turns brownish yellow and deposits a resinoid material. When applied to the skin it produces rubefaction, quickly followed by vesication.

ACIDS AND ALDEHYDES OF THE ACRYLIC SERIES.

These substances bear the same relation to the alcohols of the allyl series that the volatile fatty acids and the corresponding aldehydes bear to the ethylic series of alcohols. The following terms of the series have been obtained:

Acids. $C_nH_{2n-2}O_2$,		*Aldehydes.* $C_nH_{2n-2}O$.	
Acrylic acid	$C_3O_2H_4$	Acrolein	C_3OH_4
Crotonic	$C_4O_2H_6$	Crotonic Aldehyde	C_4OH_6
Angelic	$C_5O_2H_8$		
Pyroterebic	$C_6O_2H_{10}$		
Oleic	$C_{18}O_2H_{34}$		

The acids of this series differ from those containing the same number of C atoms in the formic series, by containing two atoms of H less; they are readily converted into acids of the formic series by the action of potassium hydrate in fusion.

Acrylic acid—$\left.\begin{matrix}C_3H_3O\\H\end{matrix}\right\}O$—72—is obtained by oxydation of acrolein by silver oxide, and is formed in a number of other reactions. It is a colorless, highly acid liquid; has a penetrating odor; solidifies at 7° (44°.6 F.); boils at 140° (284° F.). Nascent H unites with it to form propionic acid. It forms crystalline salts and ethers.

Acrylic aldehyde—*Allylic aldehyde—Acrolein*—$\left.\begin{matrix}C_3H_3O\\H\end{matrix}\right\}$—56.— When the fats and fixed oils are decomposed by heat, a disagreeable, irritating odor is produced, which is due to the formation of acrolein by the dehydration of the glycerin contained in the fatty material. Acrolein may be obtained by heating glycerin with strong H_2SO_4, or with hydropotassic sulphate. Glycerin is the alcohol (hydrate) of a radical having the same composition as allyl, but so differing from it in constitution as to be trivalent in place of univalent.

$$\underset{\text{Glycerin.}}{(C_3H_5)'''(OH)_3} = \underset{\text{Water.}}{2H_2O} + \underset{\text{Acrolein.}}{(C_3H_3O)'H}$$

Acrolein is a colorless, limpid liquid; lighter than H_2O; boils at 52°.4 (126°.3 F.); sparingly soluble in H_2O, more soluble in alcohol; very volatile; its vapor is very pungent and irritating. When freshly prepared it is neutral in reaction, but on contact with air it rapidly becomes acid by oxidation. For the same reason it does not keep well, even in closed ves-

225

sels; on standing it deposits a flocculent material, which has been called *disocryl*, while at the same time formic, acetic, and acrylic acids are formed. Oxydizing agents convert it into acrylic acid, or, if they be energetic, into a mixture of formic and acetic acids. The caustic alkalies produce from it resinoid substances similar to those formed from acetic aldehyde. With NH_3 it forms a crystalline, odorless compound, which behaves as a base.

Acrolein is formed whenever glycerin, or any substance containing it or its compounds with the fatty acids, is heated to a temperature sufficient to effect its decomposition; for this reason, and because of the irritating action of the acrolein, the heavy petroleum-oils are preferable to those of vegetable or animal origin for the lubricating of machinery operated in enclosed places.

Crotonic acid—$\left.\begin{matrix}C_4H_5O\\H\end{matrix}\right\}O$—86—was first obtained from croton-oil, *oleum tiglii* (*U. S.*), *oleum crotonis* (*Br.*), in which it exists in combination with glycerin, and accompanied by the glycerin ethers of several other fatty acids; it is, however, neither the vesicant nor the purgative principle of the oil. It may be obtained by saponification of croton-oil, or, better, by the action of potassium hydrate upon allyl cyanide.

It is an oily liquid; solidifies at −5° (23° F.); acrid in taste; gives off highly irritating vapors at temperatures slightly above 0° (32° F.). When taken internally it acts as an irritant poison.

An acid obtained by oxidation of crotonic aldehyde is probably an isomere, as it is in the form of crystals at ordinary temperatures, and only fuses at 73° (163°.4 F.).

Crotonic aldehyde—$\left.\begin{matrix}C_4H_5O\\H\end{matrix}\right\}$—70.—If aldehyde, H_2O, and HCl, be mixed together at a low temperature, and the mixture exposed to diffused daylight for some days, an oily liquid is formed, which, after purification, has the composition $C_4H_8O_2$. This substance, known as *aldol*, when exposed to heat, is decomposed into water and crotonic aldehyde: $C_4H_8O_2 = H_2O + C_4H_6O$.

Crotonic aldehyde is a colorless liquid; boils at 105° (221° F.); gives off highly irritating vapors. It bears the same relation to croton chloral that aldehyde does to chloral.

Croton chloral—*Trichlorocroton aldehyde*—$\left.\begin{matrix}C_4H_2Cl_3O\\H\end{matrix}\right\}$—173.5—a substance which has been used as an anæsthetic whose action is particularly directed to the sensory nerves distributed to the head and face. It is prepared by directing a current of Cl through acetic aldehyde, as ordinary chloral is obtained by the action of Cl upon ethylic alcohol. The first action is to convert ethylic aldehyde into crotonic aldehyde by condensation and elimination of H_2O; in the second stage of the reaction the substitution of three atoms of Cl for an equal number of atoms of H in the croton aldehyde thus formed takes place.

Angelic acid—$\left.\begin{matrix}C_5H_7O\\H\end{matrix}\right\}O$—100—exists in angelica root, in the flowers of chamomile, *Anthemis* (U. S.), and in croton-oil.

It crystallizes in colorless prisms, which fuse at 45°.5 (113°.9 F.); boils at 185° (365° F.); has an aromatic odor and an acid, pungent taste; sparingly soluble in cold H_2O; readily soluble in hot H_2O, alcohol, and ether. By the action of heat it is converted into its isomere, *methylcrotonic acid*, $\left.\begin{matrix}C_4H_4(CH_3)O\\H\end{matrix}\right\}O$.

226

Oleic acid—*Acidum oleicum* (*U. S.*)—$\left.\begin{matrix}C_{18}H_{33}O\\H\end{matrix}\right\}O$—246—exists as its glyceric ether, *olein*, in most, if not in all the fats and in all fixed oils. It is obtained in an impure form on a large scale as a by-product in the manufacture of candles. This product is, however, very impure; to purify it, it is first cooled to 0° (32° F.), the liquid portion collected; cooled to −10° (14° F.), expressed, and the solid portion collected; this is melted and treated with half its weight of massicot; the lead oleate so obtained is dissolved out by ether; the decanted ethereal solution is shaken with HCl, the ethereal layer decanted and evaporated, when it leaves oleic acid, contaminated with a small quantity of oxyoleic acid, from which it can be purified only by a tedious process.

Pure oleic acid is a white, pearly, crystalline solid, which fuses to a colorless liquid at 14° (57°.2 F.); it is odorless and tasteless; soluble in alcohol, ether, and cold H_2SO_4; insoluble in H_2O; sp. gr. 0.808 at 19° (66°.2 F.). Neutral in reaction. It can be distilled in vacuo without decomposition, but when heated in contact with air, it is decomposed with formation of hydrocarbons, volatile fatty acids, and sebacic acid. It dissolves the fatty acids readily, forming mixtures whose consistency varies with the proportions of liquid and solid acid which they contain. The solid acid is but little altered by exposure to air, but when liquid it absorbs O rapidly, becomes yellow, rancid, acid in reaction, and incapable of solidifying when cooled; these changes take place the more rapidly the higher the temperature.

Cl and Br attack oleic acid with formation of products of substitution. If oleic acid be heated with an excess of caustic potassa to 200° (392° F.), it is decomposed into palmitic and acetic acids; $C_{18}H_{34}O_2 + 2KHO = C_{16}H_{31}O_2K + C_2H_3O_2K + H_2$; a reaction which is utilized industrially to obtain hard soaps, palmitates, from olein, which itself only forms soft soaps. Cold H_2SO_4 dissolves oleic acid, and deposits it unaltered on the addition of H_2O, but if the acid solution be heated it turns brown and gives off SO_2. Nitric acid oxidizes it energetically, with formation of a number of volatile fatty acids and acids of another series—suberic, adipic, etc. The oleates of the alkaline metals are soft, soluble soaps; those of the earthy metals are insoluble in H_2O, but soluble in alcohol and in ether.

Elaidic acid is an isomere of oleic acid, produced by the action upon it of nitrous acid in the preparation of *Unguentum hydrargyri nitratis* (*U. S.*; *Br.*). The nitrous fumes formed convert the oleic acid, contained in the oil and lard used, into elaidic acid, which exists in the ointment in combination with mercury.

POLYATOMIC COMPOUNDS.

The organic compounds hitherto considered may be looked upon as compounds of *univalent* carbon radicals, these radicals existing in the alcohols and acids in combination with an atom each of O and H; they are called *monoatomic* because they contain a single atom of H capable of being replaced by an alcoholic radical. There exist other C compounds, in which the radicals, containing a less number of H atoms as compared with the number of C atoms, have a valence greater than one; these radicals form acids, alcohols, etc., in which the number of atoms of replaceable H is greater than one, and which are designated as polyatomic.

227

HYDROCARBONS.

1st Series. C_nH_{2n+2}	2d Series. C_nH_{2n}	3d Series. C_nH_{2n-2}	4th Series. C_nH_{2n-4}	5th Series. C_nH_{2n-6}	6th Series. C_nH_{2n-8}	7th Series. C_nH_{2n-10}	8th Series. C_nH_{2n-12}	9th Series. C_nH_{2n-14}	10th Series. C_nH_{2n-16}	11th Series. C_nH_{2n-18}
CH_4 Methane.	......	......	......	......	......	......	......	......	......	......
C_2H_6 Ethane.	C_2H_4 Ethene.	C_2H_2 Acetylene.	......	......	......	......	......	......	......	......
C_3H_8 Propane.	C_3H_6 Propene.	C_3H_4 Allylene.	......	......	......	......	......	......	......	......
C_4H_{10} Butane.	C_4H_8 Butene.	C_4H_6 Crotonylene.	C_4H_4	......	......	......	......	......	......	......
C_5H_{12} Pentane.	C_5H_{10} Pentene.	C_5H_8 Valerylene.	C_5H_6 Valylene.	......	......	......	......	......	......	......
C_6H_{14} Hexane.	C_6H_{12} Hexene.	C_6H_{10} Hexylene.	C_6H_8	C_6H_6 Benzene.	......	......	......	......	......	......
C_7H_{16} Heptane.	C_7H_{14} Heptene.	C_7H_{12} Œnanthylidene	C_7H_{10}	C_7H_8 Toluene.	......	......	......	......	......	......
C_8H_{18} Octane.	C_8H_{16} Octene.	C_8H_{14} Caprylidene.	C_8H_{12}	C_8H_{10} Xylene.	C_8H_8 Cinnamene.	......	......	......	......	......
C_9H_{20} Nonane.	C_9H_{18} Nonene.	C_9H_{16}	C_9H_{14}	C_9H_{12} Cumene.	C_9H_{10}	......	......	......	......	......
$C_{10}H_{22}$ Decane.	$C_{10}H_{20}$ Decene.	$C_{10}H_{18}$ Decenylene.	$C_{10}H_{16}$ Terebenthene.	$C_{10}H_{14}$ Cymene.	$C_{10}H_{12}$	$C_{10}H_{10}$ Naphthydrene.	$C_{10}H_8$ Naphthalene.	......	......	......
$C_{11}H_{24}$ Undecane.	$C_{11}H_{22}$ Undecene.	$C_{11}H_{20}$	$C_{11}H_{18}$	$C_{11}H_{16}$ Laurene.	$C_{11}H_{14}$	$C_{11}H_{12}$	$C_{11}H_{10}$	......	......	......
$C_{12}H_{26}$ Dodecane.	$C_{12}H_{24}$ Dodecene.	$C_{12}H_{22}$	$C_{12}H_{20}$	$C_{12}H_{18}$	$C_{12}H_{16}$	$C_{12}H_{14}$	$C_{12}H_{12}$	$C_{12}H_{10}$ Acenaphthalene.	......	......
$C_{13}H_{28}$ Tridecane.	$C_{13}H_{26}$ Tridecene.	$C_{13}H_{24}$	$C_{13}H_{22}$	$C_{13}H_{20}$	$C_{13}H_{18}$	$C_{13}H_{16}$	$C_{13}H_{14}$	$C_{13}H_{12}$	$C_{13}H_{10}$ Fluorene.	......
$C_{14}H_{30}$ Tetradecane.	$C_{14}H_{28}$ Tetradecene.	$C_{14}H_{26}$	$C_{14}H_{24}$	$C_{14}H_{22}$	$C_{14}H_{20}$	$C_{14}H_{18}$	$C_{14}H_{16}$	$C_{14}H_{14}$	$C_{14}H_{12}$ Stilbene.	$C_{14}H_{10}$ Anthracene.

228

NON-SATURATED HYDROCARBONS.

Besides the compounds of C and H described on pp. 172 *et seq.*, in which all the valences of the C atoms are satisfied, either by the attachment of H atoms, or by the interchange of a *single* valence between neighboring C atoms, there exist many others in which the proportion of H to C is less. These compounds are *non-saturated*, in this, that they are capable of uniting directly with atoms of other elements, or with radicals, to form products of addition, while the composition of the *saturated* hydrocarbons can only be modified by *substitution*; they are not, however, to be considered as containing any unsatisfied valence.

These hydrocarbons are very numerous, and may be arranged in homologous series, as shown in the table on page 227, each succeeding series containing a less amount of H in proportion to the C:

SECOND SERIES OF HYDROCARBONS—OLEFINES.

Series C_nH_{2n}.

The terms of this series contain two H atoms less than the corresponding terms of the first series; they differ in constitution in this, that, while in the first series a single valence is exchanged between each two neighboring C atoms, in the second series two valences are exchanged between two of the C atoms:

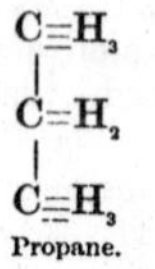

Propane.

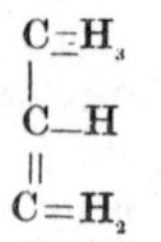

Propylene.

They are designated as *olefines*; or, to distinguish them from the terms of the first series, by the terminations *ylene* or *ene*, thus the second is called *ethylene* or *ethene*. They behave as bivalent radicals.

Ethene—*Ethylene—Olefiant gas—Elayl—Heavy carburetted hydrogen*—$CH_2 \| CH_2$—28—is formed by the dry distillation of fats, resins, wood, and coal, and is one of the most important constituents of illuminating gas. It is also obtained by the dehydration of alcohol or ether.

It has been obtained synthetically: (1) by passing a mixture of H_2S and carbon monoxide over iron or copper heated to redness; (2) by heating acetylene in the presence of H, or by the action of nascent H upon copper acetylide; (3) by the action of H upon the chloride C_2Cl_4, obtained by the action of Cl upon carbon disulphide. It is prepared in the laboratory by the dehydration of alcohol: a mixture of 4 pts. H_2SO_4 and 1 pt. alcohol is placed in a flask containing enough sand to form a thin paste, and gradually heated to about 170° (338° F.); the gas, which is given off in abundance, is purified by causing it to pass through wash-bottles containing H_2O, an alkaline solution, and concentrated H_2SO_4.

229

Pure ethylene is a colorless gas; tasteless; has a faint odor resembling that of salt water, or an ethereal odor when impure; irrespirable; sparingly soluble in H_2O, more soluble in alcohol. It burns with a luminous, white flame, and forms explosive mixtures with air and oxygen. When heated for some time at a dull red heat it is converted into acetylene, ethyl and methyl hydrides, a tarry product, and carbon.

Ethylene readily enters into combination. It unites with H to form ethyl hydride, C_2H_6. With O it unites explosively on the approach of a flame, with formation of carbon dioxide and H_2O. Oxidizing agents, such as potassium permanganate in alkaline solution, convert it into oxalic acid and H_2O. A mixture of Cl and ethene, in the proportion of two volumes of the former to one of the latter, unite with an explosion on contact with flame, the union being attended with a copious deposition of C and the formation of HCl. Chlorine and ethene, mixed in equal volumes and exposed to diffused daylight, unite slowly, with formation of an oily liquid; *ethene chloride*, $C_2H_4Cl_2$=*Dutch liquid*, to whose formation ethene owes the name *olefiant gas*. By suitable means ethene may also be made to yield chlorinated products of substitution, the highest of which is *carbon dichloride*, C_2Cl_4. Br and I also form products of addition and of substitution with ethene. By union with (OH), it forms glycol (*q. v.*). It slowly dissolves in ordinary H_2SO_4, with formation of sulphovinic acid; with fuming H_2SO_4 it combines with elevation of temperature and formation of ethionic anhydride.

When inhaled, diluted with air, ethene produces effects somewhat similar to those of nitrous oxide.

Pentene—*Amylene* or *valerene*—C_5H_{10}—70—**a colorless, mobile liquid, boiling at 39° (102°.2 F.); obtained by heating alcohol with a concentrated solution of zinc chloride. Its use as an anaesthetic has been suggested.**

Ethene chloride—*Bichloride of ethylene—Dutch liquid*—$CH_2Cl | CH_2Cl$—99—**is obtained by passing a current of ethene through a retort in which Cl is being generated, and connected with a cooled receiver. The distillate is washed with a solution of caustic potassa, afterward with H_2O, and is finally rectified.**

It is a colorless, oily liquid, which boils at 82.5° (180°.5 F.); has a sweetish taste and an ethereal odor. It is isomeric with the chloride of mo-

nochlorinated ethyl, $\begin{array}{l} C_2H_4Cl \\ | \\ Cl \end{array}$, which boils at 64° (147°.2 F.). It is capable of fixing other atoms of Cl by substitution for H, and thus forming a series of chlorinated derivatives, the highest of which is C_2Cl_6.

DIATOMIC ALCOHOLS.

SERIES $C_nH_{2n+2}O_2$.

These substances are usually designated as *glycols*. They are the hydrates of the hydrocarbons of the series C_nH_{2n}, and consist of those hydrocarbons, playing the part of bivalent radicals, united with two groups OH ; their general typical formula is then $\left.\begin{array}{r}(C_nH_{2n})'' \\ H_2\end{array}\right\} O_2$. We have seen (p. 178) that the primary monoatomic alcohols contain the group of

230

atoms (CH_2OH), united with $n(C_nH_{2n+1})$; the *primary* glycols are similarly constructed, and consist of twice the group (CH_2OH), united in the higher terms to $n(CH_2)$. The constitution of the glycols and their relations to the monoatomic alcohols are indicated by the following formulæ:

CH_2OH | CH_2 | CH_3	CH_2OH | CH_2 | CH_2OH
Primary propyl alcohol.	Primary propyl glycol.

As the monoatomic alcohols are such by containing in their molecules a group (OH), closely attached to an electro-positive group, and capable of removal and replacement by an electro-negative group or atom, so the glycols are *diatomic* by the fact that they contain two such groups (OH). As the monoatomic alcohols are therefor only capable of forming a single ether with a monobasic acid, the glycols are capable of forming two such ethers:

$CH_2\ (C_2H_3O_2)'$ | CH_3	$CH_2\ (C_2H_3O_2)'$ | CH_2OH	$CH_2\ (C_2H_3O_2)'$ | $CH_2\ (C_2H_3O_2)'$
Ethyl acetate	Monoacetic glycol.	Diacetic glycol.

Ethene glycol—*Ethylene glycol* or *Alcohol* or *Hydrate*—$\begin{array}{l} CH_2OH \\ | \\ CH_2OH \end{array}$—62.—This, the best known of the glycols, is prepared by the action of dry silver acetate upon ethylene bromide. The ether so obtained is purified by redistillation, and decomposed by heating for some time with barium hydrate.

It is a colorless, slightly viscous liquid ; odorless ; faintly sweet ; sp. gr. 1.125 at 0° (32° F.) ; boils at 197° (386°.6 F.) ; sparingly soluble in ether ; very soluble in water and in alcohol.

It is not oxidized by simple exposure to air, but on contact with platinum black it is oxidized to glycolic acid ; more energetic oxidants transform it into oxalic acid. Chlorine acts slowly upon glycol in the cold ; more rapidly under the influence of heat, producing chlorinated and other derivatives. By the action of dry HCl upon cooled glycol, a product is formed, intermediate between it and ethylene chloride, a neutral compound—*ethene chlorhydrate* or *ethene chlorhydrin*, $\begin{array}{l} CH_2OH \\ | \\ CH_2Cl \end{array}$, which boils at 130° (266° F.).

Ethene oxide—*Ethylene oxide*—$(C_2H_4)''O$—44.—This substance, isomeric with aldehyde, is obtained by the action of potassium hydrate upon ethene chlorhydrate.

It is a transparent, volatile liquid ; boils at 13°.5 (54°.3 F.) ; gives off inflammable vapors ; mixes with H_2O in all proportions. It is capable of uniting directly with H_2O to form glycol ; and with HCl gas to regenerate ethene chlorhydrate.

Taurine—$SO_3C_2H_7N$—125—is isomeric with a derivative of glycol, *isethionamide*. It is obtained from ox-bile by boiling with dilute HCl ;

231

decanting and concentrating the liquid ; separating from the sodium chloride which crystallizes ; evaporating further, and precipitating with alcohol. The deposit is purified by recrystallization from alcohol.

It crystallizes in large, transparent, oblique, rhombic prisms, permanent in air, soluble in H_2O, almost insoluble in absolute alcohol and ether.

Taurine has acid properties and forms salts ; it is not attacked by H_2SO_4, HNO_3, or nitromuriatic acid, but is oxidized by nitrous acid, with formation of H_2O, N, and isethionic acid.

It exists in the animal economy, in the bile in taurocholic acid (*q. v.*) ; and has also been detected in the intestine and fæces, muscle, blood, liver, kidneys, and lungs. The *pneumic acid*, described as existing in the lung, is taurine. When taken internally, it is eliminated by the urine, not in its own form, but as *taurocarbamic* or *isethionuric acid*, $C_3H_8N_2SO_4$.

ACIDS DERIVED FROM THE GLYCOLS.

As the acids of the acetic series are obtained from the primary monoatomic alcohols by the substitution of O for H_2 in the characterizing group CH_2OH:

CH_3 | CH_2,OH	CH_3 | CO,OH
Ethyl alcohol.	Acetic acid.

so the diatomic alcohols may, by oxidation, be made to yield acids, formed by the same substitution of O for H_2. But the glycols differ from the monoatomic alcohols in containing two groups CH_2OH, and they consequently yield two acids, as the substitution occurs in one or both of the alcoholic groups:

CH_2,OH | CH_2,OH	CH_2,OH | CO,OH	CO,OH | CO,OH
Ethene glycol.	Glycolic acid.	Oxalic acid.

A study of these two acids shows them to be possessed of peculiar differences of function. Each of them contains two groups (OH), whose hydrogen is capable of replacement by an acid or alcoholic radical:

CH_2,OC_2H_5 | $COOH$	CH_2,OH | CO,OC_2H_5	$CH_2OC_2H_5$ | CO,OC_2H_5	CO,OH | CO,OC_2H_5	CO,OC_2H_5 | CO,OC_2H_5
Ethylglycolic acid.	Ethyl glycolate.	Ethyl ethylglycolate.	Ethyloxalic acid.	Ethyl oxalate.

They are, therefor, both said to be *diatomic*. The ability, however, of the two acids to form salts is not the same, for while oxalic acid is capable of forming two salts of univalent metals, and a salt of a bivalent metal with a single molecule of the acid ; glycolic acid only forms a single salt of an univalent metal, and two of its molecules are required to form a salt of a bivalent metal ; in other words, glycolic acid is monobasic while oxalic acid is dibasic. It is only that H atom which is contained in the electro-negative group COOH, which is replaceable as acid hydrogen, while that of

232

the electro-positive group CH_2OH is only replaceable, as is the corresponding hydrogen of an alcohol.

In general terms, therefor, the *atomicity* of an organic acid may be greater th'n its *basicity*, the former representing the number of H atoms contained in its molecule, which are capable of being displaced by alcoholic radicals, while the latter represents the number of H atoms replaceable by electro-positive elements or radicals, with formation of salts or of ethers.

There may, therefor, be obtained from the glycols, by more or less complete oxidation, two series of acids; those of the first are diatomic and monobasic ; those of the second diatomic and dibasic.

DIATOMIC AND MONOBASIC ACIDS.

SERIES $C_nH_{2n}O_3$.

The acids of this series at present known are:

(Carbonic acid)............ CO_3H_2	Butylactic acid .... $C_4O_3H_8$	Leucic acid .... $C_6O_3H_{12}$
Glycolic acid .... $C_2O_3H_4$	Oxyvaleric acid .... $C_5O_3H_{10}$	(?) Œnanthic acid...... $C_{14}O_3H_{28}$
Ethyleno-lactic acid...... $C_3O_3H_6$		

The first-named of these acids, although not capable, so far as yet known, of existing in the free state, is widely represented in nature in the shape of its salts, the carbonates. Its position in this series is an anomaly, and at first sight a contradiction, as it is certainly not a monobasic, but a distinctly dibasic acid, or, more properly speaking, would be such were it obtained in a state of purity. It is, however, in this position, as the inferior homologue of glycolic acid, that carbonic acid is most naturally placed, and the dibasic nature of the latter acid does not present any valid objection to such a position, for if we consider one term of a series as derivable from its superior homologue by the subtraction of CH_2, and if we bear in mind that the basic nature of the hydrogen atom in a group OH depends upon its close union with the group CO (or with

some other electro-negative group), it will become evident that the inferior homologue of glycolic acid must contain two groups OH united to one CO, and must, therefor, be dibasic:

$$\begin{array}{l}CH_2OH\\ |\\ CO{,}OH\end{array} - CH_2 = \begin{array}{l}OH\\ |\\ CO{,}OH\end{array} \text{ or } CO\Big\langle\begin{array}{l}OH\\ OH\end{array}$$

Glycolic acid. Carbonic acid.

The other acids of the series are formed: (1.) By the partial oxidation of the corresponding glycol:

$$\begin{array}{l}CH_2OH\\ |\\ CH_2OH\end{array} + O_2 = \begin{array}{l}CH_2OH\\ |\\ CO{,}OH\end{array} + \begin{array}{l}H\\ H\end{array}\Big\rangle O$$

Glycol. Glycolic acid. Water.

233

(2.) By the combined action of water and silver oxide upon the monochlor-acid of the acetic series, or by heating the alkaline salt of such an acid with water or potassium hydrate:

$$\begin{array}{l}CH_2Cl\\ |\\ COOK\end{array} + \begin{array}{l}H\\ H\end{array}\Big\rangle O = \begin{array}{l}CH_2OH\\ |\\ CO{,}OH\end{array} + KCl$$

Potassium monochloracetate. Water. Glycolic acid. Potassium chloride.

(3.) By reducing the corresponding acid of the oxalic series by nascent hydrogen:

$$\begin{array}{l}COOH\\ |\\ COOH\end{array} + 2H_2 = \begin{array}{l}CH_2OH\\ |\\ COOH\end{array} + \begin{array}{l}H\\ H\end{array}\Big\rangle O$$

Oxalic acid. Glycolic acid. Water.

Carbonic acid—$CO\Big\langle\begin{array}{l}OH\\ OH\end{array}$—62.—Although this acid has not been isolated, it probably exists in aqueous solutions of CO_2, which have an acid reaction, while dry CO_2 is neutral. Its salts, the carbonates, are well characterized.

Oxides of Carbon.

Carbon monoxide—*Carbonous oxide—Carbonic oxide*—**CO—28.**

FORMATION.—(1.) By burning C with a limited supply of air.

(2.) By passing dry carbon dioxide over red-hot charcoal.

(3.) By heating oxalic acid with H_2SO_4 : $C_2O_4H_2 = H_2O + CO + CO_2$; and passing the gas through sodic hydrate to separate CO_2.

(4.) By heating potassium ferrocyanide with H_2SO_4.

PROPERTIES.—A colorless, tasteless gas; sp. gr. 0.9678A; very sparingly soluble in H_2O and in alcohol.

It burns in air with a blue flame and formation of carbon dioxide; it forms explosive mixtures with air and oxygen; it is oxidized to carbon dioxide by cold chromic acid. It is a valuable reducing agent, and is used for the reduction of metallic oxides at a red heat. Ammoniacal solutions of the cuprous salts absorb it readily. Being non-saturated, it unites readily with O to form CO_2, and with Cl to form $COCl_2$, the latter a colorless, suffocating gas, known as *phosgene*, or *carbonyl chloride*.

TOXICOLOGY.—Carbon monoxide is an exceedingly poisonous gas, and is the chief toxic constituent of the gases given off from blast-furnaces, from defective flues, and open coal or charcoal fires, and of illuminating gas. An atmosphere containing but a small proportion of this gas produces asphyxia and death, even if the quantity of oxygen present be equal to or even greater than that normally existing in the atmosphere; 0.5 per cent. of CO in air is sufficient to kill a small bird in a few moments, and one per cent. proves fatal to small mammals.

Poisoning by CO may occur in several ways. By inhalation of the gases discharged from blast-furnaces and from copper-furnaces, the former containing 25 to 32 per cent., and the latter 13 to 19 per cent. of CO. By the fumes given off from charcoal burned in a confined space, which consist of a mixture of the two oxides of carbon, the dioxide predominating

234

largely, especially when the combustion is most active. The following is the composition of an atmosphere produced by burning charcoal in a confined space, and which proved rapidly fatal to a dog: oxygen, 19.19; nitrogen, 76.62; carbon dioxide, 4.61; carbon monoxide, 0.54; marsh-gas, 0.04. Obviously the deleterious effects of charcoal-fumes are more rapidly fatal in proportion as the combustion is imperfect and the room small and ill-ventilated.

A fruitful source of CO poisoning, sometimes fatal, but more frequently producing languor, headache, and debility, is to be found in the stoves, furnaces, etc., used in heating our dwellings and other buildings, especially when the fuel is anthracite coal. This fuel produces in its combustion, when the air-supply is not abundant, considerable quantities of CO, to which a further addition may be made by a reduction of the dioxide, also formed, in passing over red-hot iron; this poisonous gas may find its way into the rooms either through cracks or other defects in the stoves, flues, or pipes; by occasional downward currents of air passing over fires in open fireplaces, or, much more frequently, by direct passage through the heated metal. Experiment has shown that metals, notably cast-iron, are quite pervious to gases when heated to redness; when, therefor, a stove or the fire-box of a hot-air furnace becomes red-hot, a portion of the gases, formed by the combustion of the fuel, passes through the pores of the metal to contaminate the air without, and gives rise to CO poisoning to a degree depending upon the degree of imperfection of the ventilation, the nature of the fuel, and the amount of air supplied to it. The precautions required to avoid this form of what may be called chronic CO poisoning, and which is by no means uncommon, are: (1) To have the stoves or furnaces lined with fire-clay, which tends to prevent their overheating and to diminish their perviousness to gases; (2) to avoid heating to redness; (3) to furnish an abundant supply of air to the fuel; (4) to secure proper ventilation; and (5), in the case of hot-air furnaces, to obtain, by an abundant supply of external air to the air-chamber, a large supply of moderately heated air rather than a small quantity of very hot air.

Of late years cases of fatal poisoning by coal-gas are of very frequent occurrence, caused either by accidental inhalation, by inexperienced persons blowing out the gas, or by suicides. The most actively poisonous ingredient of coal-gas is CO, which exists in the ordinary illuminating gas in the proportion of 4 to 7.5 per cent., and in water-gas, made by decomposing superheated steam by passage over red-hot coke, and subsequent charging with vapor of hydrocarbons, in the large proportion of 30-35 per cent.

The method in which CO produces its fatal effects is by forming with the blood-coloring matter a compound which is more stable than oxyhæmoglobin, and thus causing asphyxia by destroying the power of the blood-corpuscles of carrying O from the air to the tissues. This compound of CO and hæmoglobin is quite stable, and hence the symptoms of this form of poisoning are very persistent, lasting until the place of the coloring-matter thus rendered useless is supplied by new-formation. The prognosis is very unfavorable when the amount of the gas inhaled has been at all considerable; the treatment usually followed, *i.e.*, artificial respiration, and inhalation of O, failing to restore the altered coloring-matter. There would seem to be no form of poisoning in which transfusion of blood is more directly indicated than in that by CO.

Detection after death.—The blood of those asphyxiated by CO is persistently bright red in color. When suitably diluted and examined with

235

the spectroscope, it presents an absorption spectrum (Fig. 35) of two bands similar to that of oxyhæmoglobin (Fig. 14, No. 11) but in which the two bands are more equal and somewhat nearer the violet end of the spectrum. Owing to the greater stability of the CO compound, its spectrum may be readily distinguished from that of the O compound by the addition of a reducing agent (an ammoniacal solution of ferrous tartrate), which changes the spectrum of oxyhæmoglobin to the single-band spectrum of hæmoglobin (Fig. 14, No. 12), while that of the CO compound remains unaltered, or only fades partially.

If a solution of caustic soda of sp. gr. 1.3 be added to normal blood, a black, slimy mass is formed, which, when spread upon a white plate, has a greenish-brown color; the same reagent added to blood altered by CO forms a firmly clotted mass, which in thin layers upon a white surface is bright red in color.

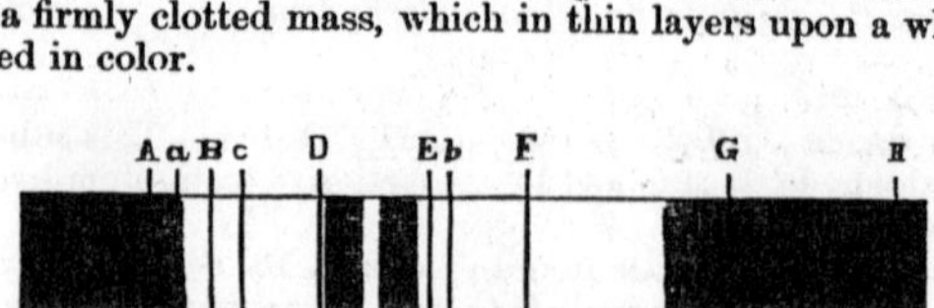

FIG. 35.

For the method of detecting and determining CO in gaseous mixtures, see p. 243.

Carbon dioxide—*Carbonic anhydride—Carbonic acid gas*—**CO_2—44.**

PREPARATION.—(1.) By burning C in air or O.

(2.) By decomposing a carbonate (marble = $CaCO_3$) by a mineral acid (HCl diluted with an equal volume of H_2O).

PROPERTIES.—At ordinary temperatures and pressures it is a colorless, **suffocating gas; has an acidulous taste; sp. gr. 1.529A; soluble in an equal volume of H_2O at the ordinary pressure; much more soluble as the pressure increases. *Soda water* is a solution of carbonic acid in H_2O under increased pressure. When compressed to the extent of 38 atmospheres** at 0° (32° F.); 50 atm. at 15° (59° F.); or 73 atm. at 30° (86° F.) it forms a transparent, mobile liquid, by whose evaporation, when the pressure is relieved, sufficient cold is produced to solidify a portion into a snow-like mass, which by spontaneous evaporation in air, produces a temperature of −90° (−130° F.).

Carbon dioxide neither burns nor does it support combustion. When heated to 1,300° (2,370° F.), it is decomposed into CO and O. A similar decomposition is brought about by the passage through it of electric sparks. When heated with H it yields CO and H_2O. When K, Na or Mg is heated in an atmosphere of CO_2, the gas is decomposed with formation of a carbonate and separation of carbon. When caused to pass through solutions of the hydrates of Na, K, Ca, or Ba, it is absorbed, with formation of the carbonates of those elements, which, in the case of the last two, are deposited as white precipitates. Solution of potash is frequently used in analysis to absorb CO_2, and lime and baryta water as tests for its presence. The hydrates mentioned also absorb CO_2 from moist air.

Atmospheric Carbon Dioxide.—Carbon dioxide is a constant constituent of atmospheric air in small and varying quantities; the mean amount in free country air being about 4 in 10.000. The variations in amount under different conditions is shown in the following table:

236

Amount of Carbon Dioxide in Air.

Collected at	Parts in 10,000.	Determined by
Paris	3.190	Boussingault and Lewy.
Andilly—twenty miles from Paris	2.989	Boussingault and Lewy.
Paris—Day	3.9	Boussingault.
Night	4.2	Boussingault.
Ocean—Day	5.42	Lewy.
Night	3.346	Lewy.
Geneva	4.68	Saussure.
Meadow—three-fourths mile from Geneva:		
Dry months	4.79 to 5.18	Saussure.
After long rains	3.57 to 4.56	Saussure.
December, damp and cloudy	3.85 to 4.25	Saussure.
January, frost	4.57	Saussure.
January, thaw	4.27	Saussure.
Lake Geneva	4.39	Saussure.
Arctic regions	4.83 to 6.41	Moss.
Gosport barracks	6.45	Chaumont.
Anglesey barracks	14.04	Chaumont.
Hilsey Hospital	4.72	Chaumont.
Portsmouth Hospital	9.76	Chaumont.
Cell in Pentonville Prison	9.89	Chaumont.
Cell in Chatham Prison	16.91	Chaumont.
Boys' school—69 cubic feet per head	31.0	Roscoe.
Room—51 cubic feet per head	52.8	Weaver.
Girls' school—150 cubic feet per head	72.3	Pettenkofer.
Greenhouse—Jardin des Plantes	1.0	
Theatre—Parquet	23.0	
Near ceiling	43.0	
Lead mine—Lamps burn	80.0	F. Leblanc.
Lamps extinguished	390.0	F. Leblanc.
Grotto del Cane	7,360.0	F. Leblanc.

It will be observed that on land the amount is greater by night than by day, while the reverse is the case at sea; on land the green parts of plants absorb CO_2 during the hours of sunlight, but not during those of darkness. The increase in the amount in air over large bodies of water during the daytime is due to the less solubility of CO_2 in the surface-water when heated by the sun's rays. The absence of vegetation accounts for the large quantity of CO_2 in the air of the polar regions, and the same cause, aided by an increased production, for its excess in the air of cities over that of the country.

The sources of atmospheric CO_2 are:

(1.) *The respiration of animals.*—The air expired from the lungs of animals contains a quantity of CO_2, varying with the age, sex, food, and muscular development and activity, while, at the same time, a much smaller quantity is discharged by the skin and in solution in the urine.

In females the increase of elimination follows the same rule as with males until puberty, when it ceases, and the amount exhaled remains about the same until the menopause, when the elimination of CO_2 suddenly increases to nearly the same as that occurring in males of the same age, and subsequently gradually declines with advancing age. During pregnancy the elimination of CO_2 is temporarily increased. In both sexes and at all ages the exhalation of CO_2 is greater during muscular activity than when the individual is at rest, and greater in those whose muscular development is more perfect. An adult man discharges 20.77 litres = three-fourths cubic foot, of CO_2 per hour, or 498.88 litres = 18 cubic feet, per diem.

237

The following table, from the experiments of Andral and Gavarret, indicates the quantity of CO_2 eliminated by males of various ages:

Elimination of Carbon Dioxide.

Age.	Mean weight.		Carbon eliminated, in grams.		Carbon dioxide eliminated, in grams		Oxygen absorbed, in grams.		Carbon dioxide eliminated, in litres.		Oxygen absorbed, in litres.	
	In kilos.	In lbs.	In 1 hour.	In 24 hours.	In 1 hour.	In 24 hours.	In 1 hour.	In 24 hours.	In 1 hour.	In 24 hours.	In 1 hour.	In 24 hours.
8 years	22.26	49.07	5.0	120.8	18.3	442.9	15.613	374.70	9.30	225.16	8.68	207.22
15 years	46.41	102.32	8.7	208.8	31.9	765.6	27.166	651.98	16.21	389.22	18.91	453.89
16 years	53.39	117.70	10.8	259.2	39.6	950.4	33.723	809.36	20.13	483.17	23.48	563.42
18 to 20 years	60.88	134.22	11.4	273.6	41.8	1003.2	35.599	854.32	21.25	510.01	24.78	594.79
20 to 24 years	66.90	147.49	12.2	292.8	44.7	1073.6	38.094	914.28	22.72	545.81	26.52	636.47
40 to 60 years	67.15	148.04	10.1	242.4	37.0	888.8	31.537	756.89	18.81	451.85	21.95	526.92
60 to 80 years	63.35	139.66	9.2	220.8	33.7	809.6	28.727	689.45	17.13	411.59	20.00	479.98

The expired air under ordinary conditions contains about 4.5 per cent. by volume of CO_2, the proportion being greater the slower the respiration.

(2.) *Combustion.*—The greater part of the atmospheric CO_2 is a product of the oxidation of C in some form as a source of light and heat. In the following table are given the amounts of CO_2 produced, and of air consumed, by different kinds of fuel and illuminating materials; by comparing them with the quantities of the same gases produced and consumed by an adult man it will be seen that, in equal times, an ordinary gas-burner produces nearly six times as much CO_2, and consumes nearly ten times as much air as a man. The amount of air consumed by fuel is, for practical purposes, greater than that given in the table, as the oxidation is never complete, the air in the chimney frequently containing ten per cent. of oxygen by volume (see below).

Combustion of Fuel.

Fuel.	Average amount burned in one hour.	Average percentage of		Carbon dioxide produced by				Air deoxidized by				Heat units.	Light in standard candles, 100.
		Carbon.	Hydrogen.	One volume in volumes.	One part by weight in parts by weight.	one hour. In kilos.	one hour. In litres.	One volume in volumes.	One kilo in cubic metres.	In one hour. In kilos.	In one hour. In litres.		
Hydrogen	......	....	100.0	..	...	....	....	2.39	26.89	...	...	34462	..
Carbon to CO_2	......	100.0	..	..	3.65	....	...	...	9.83	....	....	8080	..
Carbon to CO	......	100.0	....	..	....	...	....	...	4.93	....	....	2474	..
Carbon monoxide	......	42.86	....	1.0	1.57	....	....	2.39	0.44	....	....	2403	..
Marsh-gas	......	75.0	25.09	1.0	2.75	....	....	9.55	13.45	....	....	13063	..
Ethene	......	85.72	14.28	2.0	3.14	....	....	14.33	12.67	....	....	11857	..
Coal-gas	140 litres	40.0	55.0	0.80	1.67	0.221	112	7.14	11.04	1.293	1000	11000	..
Crude petroleum	......	84.0	13.0	..	3.08	....	....	...	12.07	....	....	11775	..
Kerosene	15 gr.	87.0	13.0	..	3.17	0.048	25	...	12.12	0.235	182	11055	180
Wax	10 gr.	79.2	13.2	..	2.89	0.029	15	...	11.24	0.146	113	10496	100
Stearic acid	10 gr.	76.05	12.68	..	2.9	0.029	15	...	8.69	0.112	86.9	9716	84
Colza-oil	42 gr.	70.43	10.5	..	2.81	0.118	60	...	8.28	0.450	348	...	159
Wood (dry pine)	......	39.10	4.90	..	1.43	....	....	...	5.16	....	....	3600	..
Wood charcoal	......	85.0	...	..	3.10	....	....	...	8.36	....	....	7640	..
Peat	......	45.0	1.5	..	1.64	....	....	...	4.82	....	....	3000	..
Coke	......	87.0	....	..	3.17	....	....	...	8.55	....	....	...	..
Anthracite	......	90.0	2.5	..	3.29	....	....	...	9.22	....	....	6000	..
Alcohol	......	52.17	13.04	..	1.90	....	....	...	8.64	....	..	7188	..
Adult man	10 gr. C.	....	....	..	....	0.037	19	...	...	0.134	104	...	..

238

(3.) *Fermentation.*—Most fermentations, including putrefactive changes, are attended by the liberation of CO_2; thus, alcoholic fermentation takes place according to the equation:

$$\underset{180}{C_6H_{12}O_6} = \underset{92}{2C_2H_6O} + \underset{44}{2CO_2},$$

and consequently discharges into the air 44 parts by weight of CO_2 for every 92 parts of alcohol formed, or 191.5 litres of gas for every litre of absolute alcohol obtained.

(4.) *Tellural sources.*—Volcanoes in activity discharge enormous quantities of CO_2, and, in volcanic countries, the same gas is thrown out abundantly through fissures in the earth. All waters, sweet and mineral, hold this gas in solution, and those which have become charged with it under pressure in the earth's crust, upon being relieved of the pressure when they reach the surface, discharge the excess into the air.

(5.) *Manufacturing processes.*—Large quantities of CO_2 are added to the air in the vicinity of lime- and brick-kilns, cement-works, etc.

(6.) *In mines*, after explosions of "fire-damp." These explosions are caused by the sudden union of the C and H of CH_4 with the O of the air, and are consequently attended by the formation of large volumes of CO_2, known to miners as *after-damp.*

Constancy of the amount of atmospheric carbon dioxide.—It has been roughly estimated by Poggendorff that 2,500,000,000,000 cubic metres of CO_2 are annually discharged into our atmosphere, and that this quantity represents one eighty-sixth of the total amount at present existing therein. This being the case, with the present production, the percentage of atmospheric CO_2 would be doubled in eighty-six years; no such increase has, however, been observed, and the average percentage found by Angus

Smith, in 1872, is about the same as that observed by Boussingault in 1840, *i.e.*, four parts in ten thousand. The CO_2 discharged into the air is, therefor, removed from it about as fast as it is produced. This removal is effected in two ways: (1) by the formation of deposits of earthy carbonates by animal organisms, corals, mollusks, etc.; (2) principally by the process of nutrition of vegetables, which absorb CO_2 both by their roots and leaves, and in the latter, under the influence of the sun's rays, decompose it, retaining the C, which passes into more complex molecules; and discharging a volume of O about equal to that of the CO_2 absorbed.

Air contaminated with excess of carbon dioxide, and its effects upon the organism.—When, from any of the above sources, the air of a given locality has received sufficient CO_2 to raise the proportion above 7 in 10,000 by volume, it is to be considered as contaminated; the seriousness of the contamination depending not only upon the amount of the increase, but also upon the source of the CO_2. If the gas be derived from fermentation, or from tellural or manufacturing sources, it is simply added to the otherwise unaltered air, and the absolute amount of oxygen present remains the same; when, however, it is produced in a confined space by the processes of combustion and respiration, the composition of the air is much more seriously modified, as not only is there addition of a deleterious gas, but a simultaneous removal of an equal volume of O; hence the importance of providing, by suitable ventilation, for the supply of new air from without to habitations and other places where human beings are collected within doors, especially where the illumination is artificial.

Although an adult man deoxidizes a little over 100 litres of air in an hour, a calculation of the quantity which he would require in a given time

239

cannot be based exclusively upon that quantity, as the deoxidation cannot be carried to completeness; indeed, when the proportion of CO_2 in air exceeds five per cent., it becomes incapable of supporting life, while a much smaller quantity, one per cent., is provocative of severe discomfort, to say the least.

In calculating the quantity of air which should be supplied to a given enclosed space, most authors have agreed to adopt as a basis that the percentage of CO_2 should not be allowed to exceed 0.6 volume per 1,000; of which 0.4 is normally present in air, and 0.2 the product of respiration or combustion. Taking the amount of CO_2 eliminated by an adult at 19 litres (=0.7 cubic foot) per hour, a man will have brought the air of an air-tight space of 100 cubic metres (=3,500 cubic feet) up to the permissible maximum of impurity in an hour. The following table is given by Parkes to indicate the contamination of air by the respiration of an adult in an hour, and the supply of external air required to restore the proper equilibrium:

Amount of cubic space (breathing-space) for one man in cubic feet.	Ratio per 1,000 of CO_2 from respiration at the end of one hour, if there have been no change of air.	Amount of air necessary to dilute to standard of 0.2, or including initial CO_2 of 0.6 per 1,000 volumes during the first hour.	Amount necessary to dilute to the given standard every hour after the first.
100	6.00	2,900	3,000
200	3.00	2,800	3,000
300	2.00	2,700	3,000
400	1.50	2,600	3,000
500	1.20	2,500	3,000
600	1.00	2,400	3,000
700	0.85	2,300	3,000
800	0.75	2,200	3,000
900	0.66	2,100	3,000
1,000	0.60	2,000	3,000

Practically, owing to the imperfect closing of doors and windows, and to ventilation by chimneys, inhabited spaces are never hermetically closed, and a less quantity of air-supply than that indicated in the table may usually be considered as sufficient.

A sleeping-room occupied by a single person should have a cubic space of 30 to 50 cubic metres (=1,050 to 1,800 cubic feet), conditions which are fulfilled in rooms measuring 10 × 13 × 8 feet, and 13 × 15.6 × 9 feet.

In calculating the space of dormitories to be occupied by several healthy people, the smallest air-space that should, under any circumstances, be allowed, is 12 cubic metres (=420 cubic feet) for each person. To determine the number of individuals that may sleep in a room, multiply its length, width, and height together, and divide the product by 420 if the measurement be in feet, or by 12 if it be in metres. Thus, a dormitory 40 feet long, 20 feet wide, and 10 feet high, is fitted for the accommodation of 19 persons at most; for 40 × 20 × 10 = 8,000 and $\frac{8000}{420}$ = 19.05.

As a rule, in places where many persons are congregated, it is necessary to resort to some scheme of ventilation by which a sufficient supply

240

of fresh air shall be introduced and the vitiated air removed, the quantity to be supplied varying according to circumstances. Experiment has shown that, in order to keep the air pure to the senses, the quantity of air which must be supplied per head and per hour in temperate climates are as shown in the table:

Situation.	Cubic metres.	Cubic feet.	Situation.	Cubic metres.	Cubic feet.
Barracks (daytime)......	35	1,236	Hospital wards (surgical)	170	6,004
Barracks (night-time)....	70	2,472	Contagious and lying-in..	170	6,004
Workshops (mechanical)	70	2,472	Mines, metalliferous.....	150	5,297
School-rooms............	35	1,236	Mines, coal..............	170	6,004
Hospital wards..........	85	3,002			

The amounts given are the smallest permissible, and should be exceeded wherever practicable.

Lights.—The amount of air to be supplied to each individual, given in the last section, are, with the exception of those furnished in mines, based upon the supposition that coal-gas is not used as a means of artificial illumination, or that the burners are so arranged with reference to the ventilating-flues that the products of combustion pass out immediately. Each cubic foot of illuminating-gas consumes in its combustion a quantity of O equal to that contained in 7.14 cubic feet of air, and produces 0.8 cubic feet of CO_2, besides a large quantity of watery vapor, and less amounts of H_2SO_4, SO_2, and sometimes CO; and an ordinary gas-burner consumes about three feet per hour. It is obvious, therefor, that a much larger quantity of pure air must be furnished to maintain the atmosphere of an apartment at the standard of 0.6 per 1,000 of CO_2, when the vitiation is produced by the combustion of gas, than when it is the result of the respiration of a human being, and that to such an extent that a single three-foot burner requires a supply of air which would be sufficient for six human beings. As a basis for computation, it may be considered that, for each cubic foot of gas consumed, 1,800 cubic feet of air should be furnished by ventilation.

The contamination of air by gas-lights becomes a question of serious importance in our dwellings upon occasions of social gatherings, and in theatres and other places of public resort which are used during the hours of darkness. The average size of a parlor in a city dwelling is 15 × 25 × 15 feet; it therefor contains 4,875 cubic feet, and its atmosphere would, if it were hermetically closed, be brought to the standard of maximum allowable contamination by the respiration of four adults in an hour, allowing 1,200 cubic feet per head, per hour. If such an apartment be illuminated, upon the occasion of an evening party at which fifty adults are present for four hours, by ten three-feet gas-burners, the amounts of air which should be supplied by ventilation are as follows in cubic feet:

	If the products of combustion of gas be discharged into the room.		If the products of combustion of the gas be carried off.	
	Per hour.	For four hours.	Per hour.	For four hours.
For fifty persons........	60,000	240,000	60,000	240,000
For ten gas-burners......	54,000	216,000	......	....
Totals..................	114,000	456,000	60,000	240,000

241

In the first instance, in which the products of the combustion of gas are discharged into the apartment, an adequate ventilation can only be secured by a complete change of the air every 2.6 minutes, which can only be attained by the use of mechanical contrivances, and with the production of draughts; in the second instance, in which it is presumed that the gas-burners are so situated, with reference to a ventilating-shaft or shafts, that the products of combustion are immediately carried off, not only is the period in which a complete change of air is required extended to 4.8 minutes, but the heat of the burners, causing an uptake current in the ventilator, favors the exit of the vitiated air, and the consequent entrance of external air to take its place.

In theatres the contamination of the air by the burning of gas should be entirely eliminated by placing the burners either under the dome ventilator, or in boxes which open to the air of the house only below the level of the burner, and which are in communication with a ventilating-shaft. Even under these conditions it is necessary, to ensure perfect ventilation, to resort to some mechanical contrivance to remove the air vitiated by respiration and to supply its place by fresh air from without, which may be previously warmed or cooled according to the season, and which, in cities, should be filtered.

When artificial illumination is obtained from lamps or candles, or from gas in small quantity and for a short time, the contamination of the air is sufficiently compensated by the ventilation through imperfect closing of the windows. A room without a window should never be used for human habitation.

One important advantage of the electric light, if it ever become practicable, will be that it consumes no O and produces no CO_2.

Although, by the combustion of fuel, O is consumed and CO_2 produced, heating arrangements only become a source of vitiation of air under the circumstances detailed above (see p. 234); indeed, in the majority of cases, if properly arranged, they are the means of ventilation, either by aspirating the vitiated air of the apartment, or by the introduction of air from without.

Action on the economy.—An animal introduced into an atmosphere of pure CO_2 dies almost instantly, and without entrance of the gas into the lungs, death resulting from spasm of the glottis, and consequent apnœa.

When diluted with air, the action of CO_2 varies according to its proportion, and according to the proportion of O present.

First.—When the proportion of O is not diminished, the poisonous action of CO_2 is not as manifest, in equal quantities, as when the air is poorer in oxygen. An animal will die rapidly in an atmosphere composed of 21 per cent. O, 59 per cent. N, and 20 per cent. CO_2 by volume; but will live for several hours in an atmosphere whose composition is 40 per cent. O, 37 per cent. N, 23 per cent. CO_2. If CO_2 be added to normal air, of course the relative quantity of O is slightly diminished, while its absolute quantity remains the same; this is the condition of affairs existing in nature when the gas is discharged into the air; under these circumstances an addition of 10–15 per cent. of CO_2 renders an air rapidly poisonous, and one of 5–8 per cent. will cause the death of small animals more slowly. Even a less proportion than this may become fatal to an individual not habituated.

In the higher states of dilution, CO_2 produces immediate loss of muscular power, and death without a struggle; when more dilute, a sense of irritation of the larynx, drowsiness, pain in the head, giddiness, gradual loss of muscular power, and death in coma.

242

Second.—If the CO_2 present in air be produced by respiration or combustion, the proportion of O is at the same time diminished, and much smaller absolute and relative amounts of the poisonous gas will produce the effects mentioned above; thus, an atmosphere containing in volumes 19.75 per cent. O, 74.25 per cent. N, 6 per cent. CO_2, is much more rapidly fatal than one composed of 21 per cent. O, 59 per cent. N, 20 per cent. CO_2. With a corresponding reduction of O, 5 per cent. of CO_2 renders an air sufficiently poisonous to destroy life; 2 per cent. produces severe suffering; 1 per cent. causes great discomfort, while 0.1 per cent., or even less, is recognized by a sense of closeness.

The *treatment* in all cases of poisoning by CO_2 consists in the inhalation of pure air (to which a small excess of O may be added), aided, if necessary, by artificial respiration, the cold douche, galvanism, and friction.

When it chances that an individual entering an atmosphere containing an excess of CO_2, or other noxious gas, is seen to fall insensible, it is simply multiplying the number of victims, for others to follow, unprotected, with a view to effecting a rescue. Probably the most readily obtainable protection is a towel saturated with lime-water, and so held over the mouth and nostrils that the inspired air passes through it, and also through two or three layers of dry towelling interposed between the moistened part and the skin.

Detection of carbon dioxide and analysis of confined air.—Carbon dioxide, or air containing it, causes a white precipitate when caused to bubble through lime or baryta water; normal air contains enough of the gas to form a scum upon the surface of these solutions when exposed to it.

It was at one time supposed that air in which a candle continued to burn was also capable of maintaining respiration. This is, however, by no means necessarily true; a candle introduced into an atmosphere in which the normal proportion of O is contained, burns readily in the presence of 8 per cent. of CO_2; is perceptibly dulled by 10 per cent.; is usually extinguished with 13 per cent.; always extinguished with 16 per cent. Its extinction is caused by a less proportion of CO_2, 4 per cent., if the quantity of O be at the same time diminished. Moreover, a contaminated atmosphere may not contain enough CO_2 to extinguish, or perceptibly dim the flame of a candle, and at the same time contain enough of the monoxide to render it fatally poisonous if inhaled.

The presence of CO_2 in a gaseous mixture is determined by its absorption by a solution of potash; its quantity either by measuring the diminution in bulk of the gas or by noting the increase in weight of an alkaline solution. To determine the proportions of the various gases present in air the apparatus shown in Fig. 36 is used. A is an aspirator of known capacity, filled with water at the beginning of the operation. It connects by a flexible tube from its upper part with an absorbing apparatus consisting of *a*, a U-shaped tube containing fragments of pumice stone, moistened with H_2SO_4; by the increase in weight of this tube the weight of watery vapor in the volume of air drawn through by the aspirator is determined; *b*, a Liebig's bulb filled with a solution of potash; *c*, a U-tube filled with fragments of pumice moistened with H_2SO_4; *b* and *c* are weighed together and their increase in weight is the weight of CO_2 in the volume of air operated on. Every gram of increase in weight represents 0.50607 litre, or 31.60356 cubic inches; *d* is a tube of difficultly fusible glass, filled with black oxide of copper and heated to redness; *e* is a U-tube filled with pumice moistened with H_2SO_4; its increase in weight represents H_2O obtained from decomposition of CH_4. Every gram of increase in weight of

243

e represents 0.444 gram, or 0.621 litre, or 38.781 cubic inches of marsh gas; *f* and *g* are similar to *b* and *c*, and their increase in weight represents CO_2 formed by oxidation of CO and CH_4 in *d*. From this the amount of CO_2 is thus calculated: First, 2.75 grams are deducted from the increase of weight of *f* and *g* for each gram of CH_4 formed by *e*; of the remainder, every gram represents 0.6364 gram, or 0.5085 litre, or 31.755 cubic inches of CO. The air is drawn through the apparatus by opening the stopcock of A to such an extent that about 30 bubbles a minute pass through *b*.

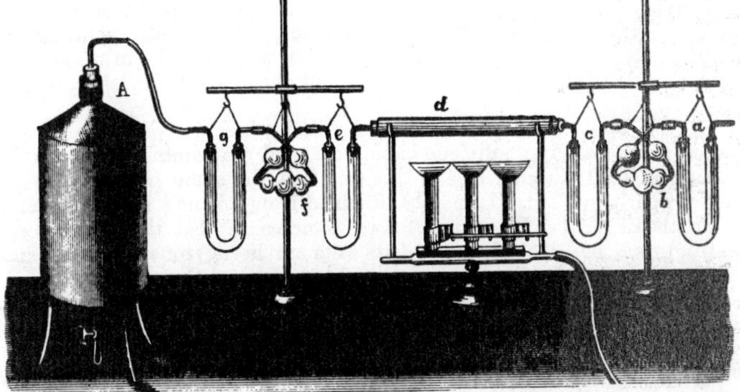

Fig. 36.

Carbon disulphide—*Bisulphide of carbon*—*Carbonei bisulphidum* (*U. S.*)—CS_2—76—is formed by passing vapor of S over C heated to redness, and is partly purified by rectification.

It is a colorless liquid; when pure it has a peculiar, but not disagreeable odor, the nauseating odor of the commercial product being due to the presence of another sulphurated body; boils at 47° (116°.6 F.); sp. gr. 1.293; very volatile; its rapid evaporation in vacuo produces a cold of −60° (−76° F.); it does not mix with H_2O; it refracts light strongly.

It is highly inflammable, and burns with a bluish flame, giving off CO_2 and SO_2; its vapor forms highly explosive mixtures with air, which detonate on contact with a glass rod heated to 250° (482° F.). Its vapor forms a mixture with nitrogen dioxide, which, when ignited, burns with a brilliant flame, rich in actinic rays.

There also exists a substance intermediate in composition between CO_2 and CS_2, known as *carbon oxysulphide*, CSO, which is an inflammable, colorless gas, obtained by decomposing potassium sulphocyanate with dilute H_2SO_4.

Toxicology.—Cases of acute poisoning by CS_2 have hitherto only been observed in animals; its action is very similar to that of chloroform.

Workmen engaged in the manufacture of CS_2 and in the vulcanization of rubber, as well as others exposed to the vapor of the disulphide, are subject to a form of chronic poisoning which may be divided into two stages. The first, or stage of excitation, is marked by headache, vertigo, a disagreeable taste, cramps in the legs; the patient talks, laughs, sings, and weeps immoderately, and sometimes becomes violently delirious. In the

244

second stage the patient becomes sad and sleepy, sensibility diminishes, sometimes to the extent of complete anæsthesia, especially of the lower extremities, the headache becomes more intense, the appetite is greatly impaired, and there is general weakness of the limbs, which terminates in paralysis.

The only remedy which has been suggested is thorough ventilation of the workshops, and abandonment of the trade at the first appearance of the symptoms.

Glycollic acid—$\begin{matrix} CH_2OH \\ | \\ COOH \end{matrix}$—76—is formed by the oxidation of glycol, by the action of nitrous acid on glycocol, and by the action of potash on monochloracetic acid.

It forms deliquescent, acicular crystals; very soluble in water; soluble in alcohol and ether; has a strongly acid taste and reaction; fuses at 78° (172°.4 F.); is decomposed at 150° (302° F.); at an intermediate temperature it loses H_2O, forming *glycollide*, or *glycollic anhydride*, $C_2H_2O_2$.

Lactic acids—$C_3H_6O_3$—90.—There are probably three, certainly two acids having this composition. Two of these would seem, from their products of decomposition, to be of similar constitution, while the molecular composition of the third is distinct; the two of similar constitution are sometimes designated as *ethylidene lactic acids*, because of their containing the group of atoms CH_3, while the third is designated as *ethyleno-lactic acid*, as it contains the group CH_2; the constitution is expressed by the formulæ:

$$\begin{array}{ccc} CH_3 & \qquad & CH_2OH \\ | & & | \\ CH,OH & & \cdot CH_2 \\ | & & | \\ COOH & & COCH \\ \text{Ethylidene lactic acid.} & & \text{Ethyleno-lactic acid.} \end{array}$$

Obviously it is the ethylene acid which is the superior homologue of glycollic acid.

Ethyleno-lactic Acid.—Muscular tissue contains a mixture of this and optically active ethylidene lactic acid, which has been known as *sarcolactic acid.*

Ethyleno-lactic acid may be obtained from muscular tissue or from Liebig's extract of meat. It is optically inactive, as are also solutions of its salts; its zinc salt contains 2 Aq, and is very soluble in water and quite soluble in alcohol. When oxidized by chromic acid it yields malonic acid.

Of the two ethylidene lactic acids, that which is *optically active* is the one accompanying ethylene lactic acid, and predominating over it in amount, in dead muscle; it is to this acid that the name *paralactic acid* is most properly applied. It may be obtained from Liebig's meat extract.

Paralactic acid differs from its two isomeres in that its solutions are dextrogyrous, and the solutions of its salts are lævogyrous. The specific rotary power of the acid is $[a]_D = +3^\circ.5$; that of the zinc salt $[a]_D = -7.6^\circ$; and of the calcium salt $[a]_D = -3^\circ.8$. Its products of decomposition are the same as those of ordinary lactic acid.

Ordinary Lactic Acid—*Lactic acid of fermentation—Optically inactive ethylidene lactic acid—Acidum lacticum* (*U. S.*)—exists in nature, widely distributed in the vegetable kingdom, and as the product of a fermenta-

245

tion which is designated as the lactic, in milk, sour-krout, fermented beet-juice, and rice, and in the liquid refuse of starch factories and tanneries.

Lactic acid is obtained as a product of the fermentation of certain sugars, milk-sugar and grape-sugar; as a result of the processes of nutrition of a minute vegetable, the lactic ferment, in which the sugar is converted into its polymere: $C_6H_{12}O_6 = 2C_3H_6O_3$. It is usually produced by allowing a mixture of cane-sugar, tartaric acid, water, rotten cheese, skim milk and chalk to ferment for 10 days at 35° (95° F.). The calcium lactate produced is separated, purified and decomposed with an equivalent quantity of H_2SO_4.

It has also been obtained synthetically by oxidation of the propylglycol of Wurtz, which is a secondary glycol, a synthesis which indicates its constitution:

$$\begin{array}{ccccccc} CH_3 & & & & CH_3 & & \\ | & & & & | & & \\ CHOH & + & O_2 & = & CHOH & + & H_2O \\ | & & & & | & & \\ CH_2OH & & & & COOH & & \\ \text{Propylglycol.} & & \text{Oxygen.} & & \text{Lactic acid} & & \text{Water.} \end{array}$$

It is a colorless, syrupy liquid; sp. gr. 1.215 at 20° (68° F.); does not solidify at −24° (−11°.2 F.); soluble in water, alcohol, and ether; is not capable of distillation without decomposition; when heated to 130° (266° F.) it loses water and is converted into *dilactic acid*, $C_6H_{10}O_5$, and, when heated to 250° (482° F.), into *lactide*, $C_3H_4O_2$. It is a good solvent of tricalcic phosphate.

Oxidizing agents convert this acid into formic and acetic acids, without the formation of any malonic acid.

Physiological.—The three lactic acids occur in animal nature, either free or in combination. Free lactic acid of fermentation occurs in the contents of the small intestine, and, when vegetable food has been taken, in the stomach; it is not, however, the acid to which the *normal, unmixed* gastric juice owes its acidity. Its salts have been found to exist in the contents of the stomach and those of the intestines, chyle, bile, parenchymatous fluid of spleen, liver, thymus, thyroid, pancreas, lungs, and brain; urine. Pathologically in the blood in leucocythæmia, pyæmia, puerperal fever, and after excessive muscular effort; in the fluids of ovarian cysts and transudations. In the urine it is abundant in phosphorus-poisoning, in acute atrophy of the liver, and in rachitis and osteo malachia.

Muscular tissue, after death or continued contractions, contains the mixture of acids known to the older authors as sarcolactic acid. Normal, quiescent muscle is neutral in reaction; but, when rigor mortis appears, or if the muscle be tetanized, its reaction becomes acid from the liberation of sarcolactic acid. Whether these acids are formed *de novo* during the contraction of the muscle, or whether they are produced by the decomposition of lactates existing in the quiescent muscle, is still undetermined; certain it is, however, that a given quantity of muscle has, when separated from the circulation, a fixed maximum of acid-producing capacity, which is greater in a muscle that has been tetanized during the interval between its removal and the establishment of rigor, than in one which has been at rest.

There exist no grounds upon which to base the supposition that, in rheumatic fever, lactic acid is present in the blood.

246

DIATOMIC AND DIBASIC ACIDS.

Series $C_nH_{2n-2}O_4$.

Oxalic acid	$C_2O_4H_2$	Adipic acid	$C_6O_4H_{10}$	Azelaic acid	$C_9O_4H_{16}$
Malonic acid	$C_3O_4H_4$	Pimelic acid	$C_7O_4H_{12}$	Sebacic acid	$C_{10}O_4H_{18}$
Succinic acid	$C_4O_4H_6$	Suberic acid	$C_8O_4H_{14}$	Roccellic acid	$C_{17}O_4H_{32}$
Deoxyglutanic acid	$C_5O_4H_8$				

They are derived from the primary glycols by complete oxidation; they are diatomic and dibasic, and contain two groups, CO, OH. They form two series of salts with the univalent metals, and two series of ethers, one of which contains neutral, and the other acid ethers. They may be obtained from the corresponding glycols, or from acids of the preceding series, by oxidation.

Oxalic acid—$\begin{array}{l} COOH \\ | \\ COOH \end{array}$—90—$C_2O_4H_2$,2Aq—126—does not occur free in nature, but in the oxalates of K, Na, Ca, Mg, and Fe in the juices of many plants, sorrel, rhubarb, cinchona, oak, etc.; as a native ferrous oxalate; and in small quantity in human urine. It is prepared artificially by oxidizing sugar or starch by HNO_3, or by the action of an alkaline hydrate in fusion upon sawdust. The soluble alkaline oxalate obtained by the latter method is converted into the insoluble Ca or Pb salt, which is washed and decomposed by an equivalent quantity of H_2SO_4 or H_2S; and the liberated acid purified by recrystallization.

Oxalic acid is also formed by the oxidation of many organic substances—alcohol, glycol, sugar, etc.; by the action of potassa in fusion upon the alkaline formiates; and by the action of K or Na upon CO_2.

It crystallizes in transparent prisms, containing 2Aq, which effloresce on exposure to air, and lose their Aq slowly but completely at 100° (212° F.), or in a dry vacuum. It fuses at 98° (208°.4 F.) in its Aq; at 110°–132° (230°–269°.6 F.) it sublimes in the anhydrous form, while a portion is decomposed; above 160° (320° F.) the decomposition is more extensive; H_2O, CO_2, CO, and formic acid are produced, while a portion of the acid is sublimed unchanged. It dissolves in 15.5 parts of water at 10° (50° F.); the presence of HNO_3 increases its solubility. It is quite soluble in alcohol. It has a sharp taste and an acid reaction in solution.

Oxalic acid is readily oxidized; in watery solution it is converted into CO_2 and H_2O, slowly by simple exposure to air, more rapidly in the presence of platinum black or of the salts of platinum and gold; under the influence of sunlight; or when heated with HNO_3, manganese dioxide, chromic acid, Br, Cl, or hypochlorous acid. Its oxidation, when it is triturated dry with pure oxide of lead, is sufficiently active to heat the mass to redness. H_2SO_4, H_3PO_4, and other dehydrating agents decompose it into H_2O, CO, and CO_2.

Analytical Characters.—(1.) In neutral or alkaline solution a white ppt. with a solution of a Ca salt.

(2.) Silver nitrate, a white ppt., soluble in HNO_3 and in NH_4HO. The ppt. does not darken when the fluid is boiled, but, when dried and heated on platinum foil, it explodes.

(3.) Lead acetate, in solutions not too dilute, a white ppt., soluble in HNO_3, insoluble in acetic acid.

247

Toxicology.—Although certain oxalates are constant constituents of vegetable food and of the human body, the acid itself, as well as hydropotassic oxalate, is a violent poison when taken internally, acting both locally as a corrosive upon the tissues with which it comes in contact, and as a true poison, the predominance of either action depending upon the concentration of the solution. Dilute solutions may produce death without pain or vomiting, and after symptoms resembling those of narcotic poisoning. Death has followed a dose of ʒj. of the solid acid, and recovery a dose of ℥j. in solution. When death occurs, it may be almost instantaneously, usually within half an hour; sometimes after weeks or months, from secondary causes.

The treatment, which must be as expeditious as possible, consists in the administration, *first*, of lime or magnesia, or a salt of Ca or Mg suspended or dissolved in a *small* quantity of H_2O or mucilaginous fluid; afterward, if vomiting have not occurred spontaneously, and if the symptoms of corrosion have not been severe, an emetic may be given. In the treatment of this form of poisoning several points of negative caution are to be observed. As in all cases in which a corrosive has been taken internally, the use of the stomach-pump is to be avoided. The alkaline carbonates are of no value in cases of oxalic acid poisoning, as the oxalates which they form are soluble, and almost as poisonous as the acid itself. The ingestion of water, or the administration of warm water as an emetic, is contraindicated when the poison has been taken in the solid form (or where doubt exists as to what form it was taken in), as they dissolve, and thus favor the absorption of the poison.

Analysis.--In fatal cases of poisoning by oxalic acid the contents of the stomach are sometimes strongly acid in reaction ; more usually, owing to the administration of antidotes, neutral, or even alkaline. In a systematic analysis the poison is to be sought for in the residue of the portion examined for prussic acid and phosphorus ; or, if the examination for those substances be omitted, in the residue or final alkaline fluid of the process for alkaloids (see p. 332 *et seq.*). If oxalic acid alone is to be sought for, the contents of the stomach, or other substances *if acid*, are extracted with water, the liquid filtered, the filtrate evaporated, the residue extracted with alcohol, the alcoholic fluid evaporated, the residue redissolved in water (solution No. 1). The portion undissolved by alcohol is extracted with alcohol acidulated with hydrochloric acid, the solution evaporated after filtration, the residue dissolved in water (solution No. 2). Solution No. 1 contains any oxalic acid which may have existed free in the substances examined ; No. 2 that which existed in the form of *soluble* oxalates. If lime or magnesia have been administered as an antidote, the substances must be boiled for an hour or two with potassium carbonate (not the hydrate), filtered, and the filtrate treated as above. In the solutions so obtained, oxalic acid is characterized by the tests given above. The urine is also to be examined microscopically for crystals of calcium oxalate. The stomach may contain small quantities of oxalates as normal constituents of certain foods.

Malonic acid— $\begin{array}{l} CH_2—COOH \\ | \\ COOH \end{array}$ —104—is a product of oxidation of ethyleno-lactic acid, and is identical with the *nicotic acid* of tobacco. It forms prismatic crystals, very soluble in H_2O, alcohol, and ether ; which fuse at 140° (284° F.), and are decomposed at 150° (302° F.).

248

Succinic acid— $\begin{array}{l} CH_2—COOH \\ | \\ CH_2—COOH \end{array}$ —118—exists in amber, coal, fossil wood, and in small quantity in animal and vegetable tissues. Its presence has been detected in the normal urine after the use of fruits and of asparagus, in the parenchymatous fluids of the spleen, thyroid, and thymus, and in the fluids of hydrocele and of hydatid cysts. It is also formed in small quantity during alcoholic fermentation ; as a product of oxidation of many fats and fatty acids ; and by synthesis from ethylene cyanide.

It may be obtained by dry distillation of amber, or, preferably, by the fermentation of malic acid.

It crystallizes in large prisms or hexagonal plates, which are colorless, odorless, permanent in air, acid in taste, soluble in water, sparingly so in ether and in cold alcohol. It fuses at 180° (356° F.), and distils with partial decomposition at 235° (455° F.). It withstands the action of oxidizing agents ; reducing agents convert it into the corresponding acid of the fatty series, butyric acid ; with Br it forms products of substitution ; H_2SO_4 is without action upon it ; phosphoric anhydride removes H_2O and converts it into *succinic anhydride*, $C_4H_4O_3$.

COMPOUND ETHERS OF THE ACIDS OF THE SERIES

$C_nH_{2n}O_3$ AND $C_nH_{2n-2}O_4$.

The members of both of these series contain two atoms of H replaceable by alcoholic radicals. In those of the series $C_nH_{2n}O_3$ (with the exception of carbonic acid), being monobasic, although diatomatic, it is not immaterial which H is so replaced. If it be that of the group CH_2OH, the resulting compound is a monobasic acid, in which the H of the group COOH may be replaced by another alcoholic radical to form a neutral ether of the new acid ; if, on the other hand, the H of the group COOH be first replaced, a neutral compound ether is formed. In the members of the series $C_nH_{2n-2}O_4$, which are dibasic, the substitution of an alcoholic radical for the H of either group COOH produces a monobasic acid, in which the H of the other COOH may be replaced by another radical to form a neutral ether. The following formulæ indicate the differences in the nature of these compounds :

$\begin{array}{l} CH_2OH \\ \vert \\ COOH \end{array}$	$\begin{array}{l} CH_2OC_2H_5 \\ \vert \\ COOH \end{array}$	$\begin{array}{l} CH_2OH \\ \vert \\ COOC_2H_5 \end{array}$	$\begin{array}{l} CH_2OC_2H_5 \\ \vert \\ COOC_2H_5 \end{array}$
Glycolic acid.	Ethylglycolic acid.	Ethyl glycolate.	Ethyl ethylglycolate.

$\begin{array}{l} COOH \\ \vert \\ COOH \end{array}$	$\begin{array}{l} COOC_2H_5 \\ \vert \\ COOH \end{array}$	$\begin{array}{l} COOC_2H_5 \\ \vert \\ COOC_2H_5 \end{array}$
Oxalic acid.	Ethyloxalic acid.	Ethyl oxalate.

249

ALDEHYDES AND ANHYDRIDES OF THE SERIES

$C_nH_{2n}O_3$ AND $C_nH_{2n-2}O_4$.

In treating of the monoatomic compounds, it was stated that substances existed corresponding to the fatty acids, known as aldehydes and anhydrides, the former differing from the acids in that they contained the group COH instead of COOH ; the latter being the oxides of the acid radicals. Similar compounds exist corresponding to the acids of these two series.

The aldehydes corresponding to the series $C_nH_{2n}O_3$ contain the group COH in place of the group COOH, and as they also contain the group CH_2OH, they are possessed of the double function of primary alcohol and aldehyde. Those of the series $C_nH_{2n-2}O_4$ form two series ; in one of which only one of the groups COOH is deoxidized to COH ; in the other, both. Those of the first series, still containing a group COOH, are monobasic acids as well as aldehydes :

$\begin{array}{l} CH_2OH \\ \vert \\ COOH \end{array}$	$\begin{array}{l} CH_2OH \\ \vert \\ COH \end{array}$	$\begin{array}{l} COOH \\ \vert \\ COOH \end{array}$	$\begin{array}{l} COOH \\ \vert \\ COH \end{array}$	$\begin{array}{l} COH \\ \vert \\ COH \end{array}$
Glycolic acid.	Glycolic aldehyde.	Oxalic acid.	Glyoxalic acid.	Glyoxol.

While the anhydrides of the fatty series may be considered as derived from the acids by the subtraction of H_2O from two molecules of the acid ; those of both the series of acids under consideration are derived from a single molecule of the acid by the subtraction of H_2O :

$\begin{array}{l} CH_3 \\ \vert \\ COOH \end{array}$	$\begin{array}{l} CH_2OH \\ \vert \\ COOH \end{array}$	$\begin{array}{l} CH_2—COOH \\ \vert \\ CH_2—COOH \end{array}$
Acetic acid.	Glycolic acid.	Succinic acid.
$\left.\begin{array}{l} CH_3—CO \\ CH_3—CO \end{array}\right\rangle O$	$\left.\begin{array}{l} CH_2 \\ \vert \\ CO \end{array}\right\rangle O$	$\left.\begin{array}{l} CH_2—CO \\ \vert \\ CH_2—CO \end{array}\right\rangle O$
Acetic anhydride.	Glycolic anhydride.	Succinic anhydride.

AMINES OF THE GLYCOLS.

ETHYLENIC COMPOUND AMMONIAS.

These substances are derived from a double molecule of NH_3, or of ammonium hydrate, by the substitution of the diatomic radicals of the glycols (hydrocarbons of the series C_nH_{2n}) for an equivalent number of H atoms. They are distinguished from the corresponding compounds of the radicals of the monoatomic alcohols, the *monamines*, by the designation of *diamines*.

When it is considered that in the formation of these substances double

250

H atoms can be replaced by diatomic radicals to form primary, secondary, and tertiary amines :

$\left.\begin{array}{r} H_2 \\ H_2 \\ H_2 \end{array}\right\} N_2$	$\left.\begin{array}{r} (C_2H_4)'' \\ H_2 \\ H_2 \end{array}\right\} N_2$	$\left.\begin{array}{r} (C_2H_4)'' \\ (C_2H_4)'' \\ H_2 \end{array}\right\} N_2$	$\left.\begin{array}{r} (C_2H_4)'' \\ (C_2H_4)'' \\ (C_2H_4)'' \end{array}\right\} N_2$;
Double ammonia molecule.	Ethylene amine. *Primary.*	Diethylene amine. *Secondary.*	Triethylene amine. *Tertiary.*

that others exist in which two univalent radicals replace a bivalent radical ; others, again, in which H atoms have been replaced by groups OH ; and finally, that similar compounds of P, As and Sb exist, it is not astonishing that the study of the vast number of substances, the possibility of whose existence is thus indicated, is still in its infancy.

Some recently discovered alkaloids, produced during putrefaction (see Ptomaïnes, p. 343), are diamines ; and there is strong probability that further investigation will show some of the vegetable alkaloids, whose constitution is as yet unknown, to belong in this class.

AMIDES OF THE ACIDS OF THE SERIES

$C_nH_{2n}O_3$ AND $C_nH_{2n-2}O_4$.

This class of substances, formed by the substitution of radicals of the acids for H atoms in NH_3 molecules, contains some substances of the greatest medical interest. The radicals of the acids of the series $C_nH_{2n}O_2$, except carbonic acid, being univalent, form amides similar in constitution to those of the acids of the series $C_nH_{2n}O_2$ (p. 208).

In the case of the dibasic acids no less than three series of amides are known to exist; thus we have, corresponding to oxalic acid:

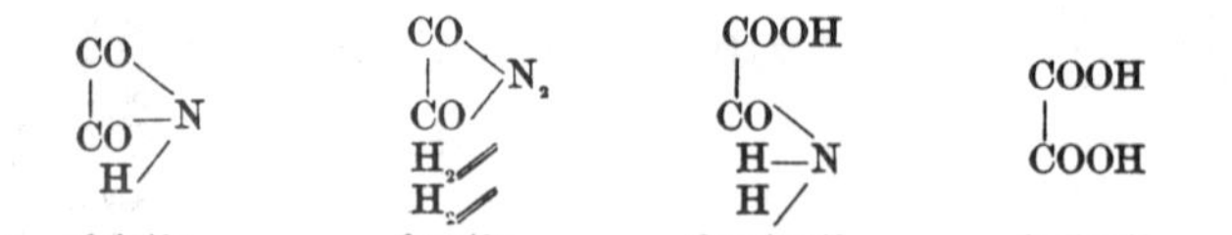

Oximide. *Secondary monamide.* — Oxamide. *Primary diamide.* — Oxamic acid. *Primary monamide.* — Oxalic acid *Acid.*

In the first of these, two H atoms of a single NH_3 molecule are replaced by the bivalent radical of the acid; these are distinguished as *imides.* Those of the second series are normally formed *diamides.* In the third series, the univalent remainder, left by the removal of OH from the acid, replaces an atom of H in one molecule of NH_3, and the resulting compound, still containing a group COOH, has the functions of a monobasic acid.

Amides of Carbonic Acid.

Carbimide—$\left.\begin{matrix}(CO)''\\ H\end{matrix}\right\} N$—43.—Although cyanic acid (q.v.) **has frequently** been regarded as the imide of carbonic acid, there are many **reasons,**

251

drawn from the methods of formation and properties of cyanic acid, which lead us to assign to it the constitution $\begin{matrix}CN\\ |\\ OH\end{matrix}$, rather than that given above, and to consider it as an isomere of the hitherto undiscovered carbimide.

Carbamide—Urea—$\left.\begin{matrix}(CO)''\\ H_2\\ H_2\end{matrix}\right\} N_2$—60.

Occurrence.—Urea does not occur in the vegetable world. It exists principally in the urine of the mammalia; also in smaller quantity in the excrements of birds, fishes, and some reptiles; in the mammalian blood, chyle, lymph, liver, spleen, lungs, brain, vitreous and aqueous humors, saliva, perspiration, bile, milk, amniotic and allantoic fluids, muscular tissue, and in serous fluids (see below).

Formation.—(1.) As a product of the decomposition of uric acid, usually by oxidation:

$$\underset{\text{Uric acid.}}{C_5H_4N_4O_3} + \underset{\text{Water.}}{H_2O} + \underset{\text{Oxygen.}}{O} = \underset{\text{Urea.}}{CON_2H_4} + \underset{\text{Alloxan.}}{C_4H_2N_2O_4}$$

(2.) By the oxidation of oxamide.

(3.) By the action of caustic potassa upon creatin:

$$\underset{\text{Creatin.}}{C_4H_9N_3O_2} + \underset{\text{Water.}}{H_2O} = \underset{\text{Urea.}}{CON_2H_4} + \underset{\text{Sarcosine.}}{C_3H_7NO_2}$$

(4.) By the limited oxidation of albuminoid substances, by potassium permanganate, and during the processes of nutrition.

(5.) By the action of carbon oxychloride on dry ammonia.

(6.) By the action of ammonium hydrate on ethyl carbonate at 180° (356° F.).

(7.) By heating ammonium carbonate in sealed tubes to 130° (266° F.).

(8.) By the slow evaporation of an aqueous solution of hydrocyanic acid.

(9.) By the molecular transformation of its isomeride, ammonium cyanate.

$$\underset{\text{Ammonium cyanate.}}{\begin{matrix}CN\\ |\\ O\,(NH_4)\end{matrix}} = \underset{\text{Urea.}}{\left.\begin{matrix}(CO)\\ H_2\\ H_2\end{matrix}\right\} N_2}$$

Preparation.—(1.) *From the urine.*—Fresh urine is evaporated to the consistency of a syrup over the water-bath; the residue is cooled and mixed with an equal volume of colorless HNO_3 of sp. gr. 1.42; the crystals are washed with a small quantity of cold H_2O, and dissolved in hot H_2O; the solution is decolorized, so far as possible, without boiling, with animal charcoal, filtered, and neutralized with potassium carbonate; the liquid is then concentrated over the water-bath, and decanted from the crystals of potassium nitrate which separate; then evaporated to dryness over the water-bath, and the residue extracted with strong, hot alcohol; the alcoholic solution, on evaporation, leaves the urea more or less colored by urinary pigment.

252

(2.) *By synthesis.*—Urea is more readily obtained in a state of purity from potassium cyanate. This is dissolved in cold H_2O, and dry ammonium sulphate is added to the solution. Potassium sulphate crystallizes out and is separated by decanting the liquid, which is then evaporated over the water-bath, fresh quantities of potassium sulphate crystallizing and being separated during the first part of the evaporation; the dry residue is extracted with strong, hot alcohol; this, on evaporation, leaves the urea, which, by a second crystallization from alcohol, is obtained pure.

Properties.—*Physical.*—Urea crystallizes from its aqueous solution in long, flattened prisms, and by spontaneous evaporation of its alcoholic solution in quadratic prisms with octahedral ends. It is colorless and odorless; has a cooling, bitterish taste, resembling that of saltpetre; is neutral in reaction; soluble in one part of H_2O at 15° (59° F.), the solution being attended with diminution of temperature; soluble in five parts of cold alcohol (sp. gr. 0.816) and in one part of boiling alcohol; very sparingly soluble in ether. When its powder is mixed with that of certain salts, such as sodium sulphate, the Aq. of the salt separates, and the mass becomes soft or even liquid. When pure it is not deliquescent, but is slightly hygrometric, and when it is to be weighed it should be dried at 100° (212° F.) and cooled in a dessicator. Fuses at 130° (266° F.).

Chemical.—Heated a few degress above 130° (266° F.) urea boils, giving off ammonia and ammonium carbonate, and leaves a residue of *ammelide,* $C_6H_9N_9O_3$. When heated to 150°–170° (302°–338° F.), it is decomposed, leaving a mixture of ammelide, cyanuric acid, and biuret:

$$\underset{\text{Urea.}}{8CON_2H_4} = \underset{\text{Carbon dioxide.}}{2CO_2} + \underset{\text{Ammelide.}}{C_6H_9N_9O_3} + \underset{\text{Ammonia.}}{7NH_3} + \underset{\text{Water.}}{H_2O}$$

$$\underset{\text{Urea.}}{3CON_2H_4} = \underset{\text{Cyanuric acid.}}{C_3O_3N_3H_3} + \underset{\text{Ammonia.}}{3NH_3}$$

$$\underset{\text{Urea.}}{2CON_2H_4} = \underset{\text{Biuret.}}{C_2H_5N_3O_2} + \underset{\text{Ammonia.}}{NH_3}$$

If urea is maintained at 150°–170° (302°–338° F.) for some time, a dry, grayish mass remains, which consists principally of cyanuric acid. In this reaction, the volatile products contain urea, not that that substance is volatile, but because a portion of the cyanuric acid and ammonia unite to regenerate urea by the reverse action to that given above.

Dilute aqueous solutions of urea are not decomposed by boiling; but if the solution be concentrated, or the boiling prolonged for a long time, the urea is partially decomposed into CO_2 and NH_3. The same decomposition takes place more rapidly and completely when a solution of urea is heated under pressure to 140° (284° F.). A pure aqueous solution of urea is not altered by exposure to filtered air. If urine be allowed to stand, putrefactive changes take place under the influence of a peculiar, organized ferment, or of a diastase-like body which is a constituent of normal urine.

Chlorine decomposes urea with production of CO_2, N, and HCl. Solutions of the alkaline hypochlorites and hypobromites effect a similar decomposition in the presence of an excess of alkali, according to the equation:

$$\underset{\text{Urea.}}{CON_2H_4} + \underset{\text{Sodium hypochlorite.}}{3NaClO} = \underset{\text{Carbon dioxide.}}{CO_2} + \underset{\text{Water.}}{2H_2O} + \underset{\text{Nitrogen.}}{N_2} + \underset{\text{Sodium chloride.}}{3NaCl}$$

253

Upon this decomposition are based the quantitative processes of Knop, Hüfner, Yvon, Davy, Leconte, etc.

Nitrous acid, or HNO_3 charged with nitrous vapors, decomposes urea according to the equation:

$$\underset{\text{Urea.}}{CON_2H_4} + \underset{\text{Nitrogen trioxide.}}{N_2O_3} = \underset{\text{Carbon dioxide.}}{CO_2} + \underset{\text{Nitrogen.}}{N_4} + \underset{\text{Water.}}{2H_2O} \quad (1)$$

or the equation:

$$\underset{\text{Urea.}}{2CON_2H_4} + \underset{\text{Nitrogen trioxide.}}{N_2O_3} = \underset{\text{Ammonium carbonate.}}{CO_3(NH_4)_2} + \underset{\text{Nitrogen.}}{N_4} + \underset{\text{Carbon dioxide.}}{CO_2} \quad (2)$$

If the mixture be made in the cold, of one molecule of nitrogen trioxide to two molecules of urea, the decomposition is that indicated by Equation 2. If, on the other hand, the trioxide be gradually added to the previously warmed urea solution in the same proportion, half the urea is decomposed while the remainder remains unaltered, and, upon the addition of a further and sufficient quantity of the trioxide, all the urea is decomposed according to Equation 1. Upon this reaction are based the processes of Gréhant, Boymond, Draper, etc.

When heated with mineral acids or alkalies, urea is decomposed with formation of CO_2 and NH_3; if the decomposing agent be an acid, CO_2 is given off, and an ammoniacal salt remains; if an alkali, a carbonate of the alkaline metal remains, and NH_3 is given off. Upon this decomposition are based the processes of Heintz and Ragsky, Bunsen, etc.

Urea forms definite compounds, not only with acids, but also with certain oxides and salts. Of the compounds which it forms with *acids*, the most important are those with nitric and oxalic acids.

Urea nitrate—CON_2H_4,HNO_3—is formed as a white, crystalline mass when a concentrated solution of urea is treated, in the cold, with HNO_3. It is much less soluble in H_2O than is urea, especially in the presence of an excess of HNO_3. It decomposes the carbonates with liberation of urea. If a solution of urea nitrate be evaporated over the water-bath, it is decomposed, bubbles of gas being given off beyond a certain degree of concentration, and large crystals of urea, covered with smaller ones of urea nitrate, separate.

Urea oxalate—$2CON_2H_4,H_2C_2O_4$—separates as a fine, crystalline powder from mixed aqueous solutions of urea and oxalic acid of sufficient concentration. It is acid in taste and reaction, less soluble in cold H_2O than the nitrate, and less soluble in the presence of an excess of oxalic acid than in pure H_2O. Its solution may be evaporated at the temperature of the water-bath without suffering decomposition.

Of the compounds of urea with *oxides*, the most interesting are those with mercuric oxide, three in number:

a. $CON_2H_4,2HgO$ is formed by gradually adding mercuric oxide to a solution of urea, heated to near its boiling-point; the filtered liquid, on standing twenty-four hours, deposits crystalline crusts of the above composition.

β. $CON_2H_4,3HgO$ is formed as a gelatinous precipitate when mercuric chloride solution is added to a solution of urea containing potassium hydrate.

254

γ. $CON_2H_4,4HgO$ is formed as a white, amorphous precipitate when a dilute solution of mercuric nitrate is gradually added to a dilute alkaline solution of urea, and the excess of acid neutralized from time to time. A yellow tinge in the precipitate indicates the formation of mercuric subnitrate after the urea has been all precipitated (Liebig's process).

Of the compounds of urea with *salts*, that with sodium chloride is the only one of importance:

$CON_2H_4,NaCl,H_2O$.—It is obtained in prismatic crystals when solutions of equal molecules of urea and sodium chloride are evaporated together. It is deliquescent and very soluble in water. Its solution, when mixed with solution of oxalic acid, only forms urea oxalate after long standing, or on evaporation.

Physiology.—Urea is a constant constituent of normal mammalian blood and urine, and is the chief product of the oxidation of albuminoid substances which occur in the body; the bulk of the N assimilated from the food ultimately making its exit from the body in the form of urea in the urine.

The determinations of the amount of urea in the blood and fluids other than the urine are, owing to imperfections in the processes of analysis, not as accurate as could be desired, the error being generally a minus one. Some of the more prominent are given in the following table:

Quantity of Urea in Parts per 1,000 in Animal Fluids other than Urine.

Normal blood—dog	0.24–0.53	Munk.
Normal blood—human	0.2 -0.4	Gamgee.
Normal blood—human	0.16	Pickard.
Normal blood—human	0.14–0.18	Gautier.
Normal blood—human placental	0.28–0.62	Picard.
Normal blood—human fœtal	0.27	Picard.
Blood of dog before nephrotomy	0.26–0.88	Gréhant.
Blood of dog, three hours after nephrotomy	0.45–0.93	Gréhant.
Blood of dog, twenty-seven hours after nephrotomy	2.06–2.76	Gréhant.
Human blood in cholera	2.4	Voit.
Human blood in cholera	3.6	Chalvet.
Human blood in Bright's	15.0	Bright & Babington.
Lymph—dog	0.16	Wurtz.
Lymph—cow	0.19	Wurtz.
Chyle—cow	0.19	Wurtz.
Milk	0.13	Picard.
Saliva	0.35	Picard.
Bile	0.30	Picard.
Fluid of ascites	0.15	Picard.
Perspiration	0.38	Funke.
Perspiration	0.88	Picard.

The quantity of urea contained in human urine under various circumstances of health and disease has been the subject of a great number of investigations, and a determination of the amount voided in a given case is frequently of great importance to the physician, as indicating the amount of disassimilation of nitrogenous material occurring in the body at the time. Under normal conditions, the quantity of urea voided in twenty-four hours is subject to considerable variations, as is shown in the subjoined table:

255

Amount of Urea in Human Urine—Normal.

	Parts per 1,000.	Grams in total urine of 24 hours.	
Urine of sp. gr. 1009.2	9.88	......	Millon.
Urine of sp. gr. 1011.6	11.39	......	Millon.
Urine of sp. gr. 1019.0	18.58	......	Boymond.
Urine of sp. gr. 1026.0	25.80	......	Millon.
Urine of sp. gr. 1027.7	29.70	......	Millon.
Urine of sp. gr. 1028.0	27.08	......	Boymond.
Urine of sp. gr 1029.0	31.77	......	Millon.
Urine of adult male (average)	30.0	......	Berzelius.
Urine of adult male (average)	....	28.052	Lecanu.
Urine of adult male (average)	25–32	22–35	Neubauer.
Urine of adult male (average)	....	32–43	Kerner.
Urine of adult male (average)	23.3	35	Vogel.
Urine of adult male, animal food	....	51–92	Franque.
Urine of adult male, mixed food	....	36–38	Franque.
Urine of adult male, vegetable food	....	24–28	Franque.
Urine of adult male, non-nitrogenized food	....	16	Franque.
Urine of old men, 84–86 years	....	8.11	Lecanu.
Urine of adult female (average)	....	19.116	Lecanu.
Urine of pregnant female	....	30–38	Quinquand.
Urine of female, 24 hours after delivery	....	20–22	Quinquand.
Urine of infant, first day	....	0.03–0.04	Quinquand.
Urine of infant, fifth day	....	0.12–0.15	Quinquand.
Urine of infant, eighth day	....	0.2 –0.28	Quinquand.
Urine of infant, fifteenth day	....	0.3 –0.04	Quinquand.
Urine of child four years old	.. .	4.505	Lecanu.
Urine of child eight years old	....	13.471	Lecanu.
Urine of boy eighteen months old	....	8–12	Harley.
Urine of girl eighteen months old	....	6–9	Harley.

The variations are produced by:

(1.) *Age.*—In new-born children the elimination of urea is insignificant. By growing children the amount voided is absolutely less than that discharged by adults, but, relatively to their weight, considerably greater; thus, Harley gives the following amounts of urea in grams for each pound of body-weight in twenty-four hours: boy, eighteen months, 0.4; girl, eighteen months, 0.35; man, twenty-seven years, 0.25; woman, twenty-seven years, 0.20. During adult life the mean elimination of urea remains stationary, unless modified by other causes than age. In old age the amount sinks to below the absolute quantity discharged by growing children.

(2.) *Sex.*—At all periods of life females eliminate less urea than males. The proportion given by Beigel differs slightly from that of Harley, viz.: one kilo of male, 0.35 grams urea in twenty-four hours; one kilo of female, 0.25 grams. During pregnancy females discharge more urea than males; very shortly after delivery the amount sinks to the normal, below which it passes during lactation.

(3.) *Food.*—The quantity of urea eliminated is in direct proportion to the amount of N contained in the food. The ingestion of large quantities of watery drinks increases the amount, and a contrary effect is produced by tea, coffee, and alcohol. With insufficient food the excretion of urea is diminished, although not arrested, even in extreme starvation.

(4.) *Exercise.*—The question whether the elimination of urea is increased during violent muscular exercise is one which has been the subject of many observations and of much discussion. An examination of the

256

various results shows that, while the excretion of urea is slightly greater during violent exercise than during periods of rest, that increase is so insignificant in comparison to the work done, and, in some instances, to the loss of body-weight, as to render the assumption that muscular force is the result of the oxidation of the nitrogenized constituents of muscle improbable. (See Gamgee: "Physiological Chemistry," i., pp. 385–401, for a full review of the subject.)

The percentage of urea in the urine of the same individual, is not the same at different times of the day. The minimum hourly elimination is in the morning hours; an increase begins immediately after the principal meal, and reaches its height in about six hours, when a diminution sets in and progresses to the time of the next meal. Gorup-Besanez gives a curve representing the hourly variations in the elimination of urea, which, reduced to figures, gives the following:

Hour.	Urea in grams.	Hour.	Urea in grams.	Hour.	Urea in grams.
8–9 A.M.	1.5	4– 5 P.M.	2.6	12–1 A.M.	1.9
9–10 A.M.	1.5	5– 6 P.M.	3.1	1–2 A.M.	1.9
10–11 A.M.	1.4	6– 7 P.M.	2.8	2–3 A.M.	1.9
11 A.M.–12 M.	1.3	7– 8 P.M.	2.5	3–4 A.M.	1.8
12 M.–1 P.M.	1.8	8– 9 P.M.	2.3	4–5 A.M.	1.6
1–2 P.M.	1.9	9–10 P.M.	2.0	5–6 A.M.	1.6
2–3 P.M.	2.1	10–11 P.M.	2.0	6–7 A.M.	1.6
3–4 P.M.	2.3	11–12 P.M.	2.3	7–8 A.M.	1.5

The total of which, however, represents a quantity above the normal.

The absolute amount of urea eliminated in twenty-four hours is increased by the exhibition of diuretics, alkalies, colchicum, turpentine, rhubarb, alkaline silicates, and compounds of antimony, arsenic, and phosphorus. It is diminished by digitalis, caffein, potassium iodide, and lead acetate; not sensibly affected by quinine.

Pathologically the quantity of urea voided may be either increased or diminished: an increase above the normal indicating an increased oxidation of nitrogenous material, or the retention of the urea formed within the body; and a diminution a deficient oxidation of the same class of substances, or, as is frequently the case, a diminution in the supply of nitrogen to the body from loss of appetite or power of assimilation.

In acute febrile diseases both the relative and absolute amounts of urea eliminated augments, with some oscillations, until the fever is at its height; there is, however, no constant relation between the amount of urea eliminated and the body temperature. During the period of defervescence, the amount of urea eliminated in twenty-four hours is diminished below the normal; during convalescence it again slowly increases. If the malady terminate in death the diminution of urea is continuous to the end. In intermittent fever the amount of urea discharged is increased on the day of the fever and diminished during the interval. In cholera, during the algid stage, the elimination of urea by the kidneys is almost completely arrested, while the quantity in the blood is greatly increased. When the secretion of urine is again established, the excretion of urea is greatly increased (60–80 grams = 926–1235 grains a day), and the abundant perspiration is also rich in urea. In cardiac diseases, attended

257

with respiratory difficulty, but without albuminuria, the elimination of urea is diminished and that of uric acid increased. In nephritis, attended with albuminuria, the elimination of urea at first remains normal; later it diminishes, and the urea, accumulating in the blood, gives rise to *uræmic poisoning.* The quantity of urea in the urine is also diminished in all diseases attended with dropsical effusions; but is increased when the dropsical fluid is reabsorbed. In true diabetes the amount of urea *in the urine of twenty-four hours* is greater than normal. In chronic diseases the elimination of urea is below the normal, owing to imperfect oxidation.

Analytical Characters.—To detect the presence of urea in a fluid, it is mixed with three to four volumes of alcohol, and filtered after having stood several hours in the cold; the filtrate is evaporated on the water-bath, and the residue extracted with strong alcohol; the filtered alcoholic fluid is evaporated, and the residue tested as follows:

(1.) A small portion is heated in a dry test-tube to about 160° (320° F.), until the odor of ammonia is no longer observed; the residue is treated with a few drops of caustic potassa solution and one drop of cupric sulphate solution. If urea be present, the biuret resulting from its decomposition by heat causes the solution of the cupric oxide with a reddish-violet color.

(2.) A portion of the residue is dissolved in a drop or two of H_2O, and an equal quantity of *colorless* concentrated HNO_3 added; if urea be present in sufficient quantity there appear white, shining, hexagonal or rhombic, crystalline plates or six-sided prisms of urea nitrate.

(3.) A portion dissolved in water, as in (2), is treated with a solution of oxalic acid; rhombic plates of urea oxalate crystallize.

Determination of Quantity of Urea in Urine.—It must not be forgotten that, in all quantitative determinations of constituents of the urine, the question to be solved is not how much of that constituent is contained in a given quantity of urine, but how much of that substance the patient is discharging in a given time, usually twenty-four hours. Quantitative determinations are, therefor, in most cases, barren of useful results, unless the quantity of urine passed by the patient in twenty-four hours is known; and, in view of diurnal variations in elimination, unless the urine examined be a sample taken from the *mixed urine* of twenty-four hours.

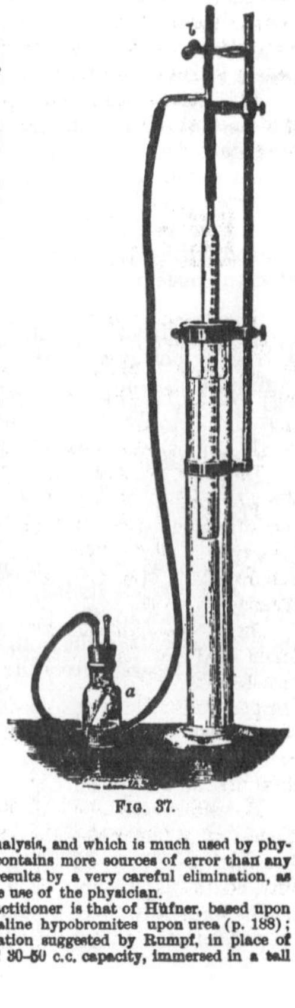

Fig. 37.

The process giving the most accurate results is that of Bunsen, in which the urea is decomposed into CO_2 and NH_3, the former of which is weighed as barium carbonate. Unfortunately, this process requires an expenditure of time and a degree of skill in manipulation, which render its application possible only in a well-appointed laboratory.

A process which is described in most text-books upon urinary analysis, and which is much used by physicians, is that of Liebig. As this method is one, however, which contains more sources of error than any other, and as it can only b made to yield approximately correct results by a very careful elimination, as far as possible, of those defects, it is not one which is adapted to the use of the physician.

Probably the most satisfactory process in the hands of the practitioner is that of Hüfner, based upon the reaction, to which attention was first called by Knop, of the alkaline hypobromites upon urea (p. 188); using, however, Dietrich's apparatus, or the more simple modification suggested by Rumpf, in place of that of Hüfner. The apparatus (Fig. 37) consists of a burette of 30–50 c.c. capacity, immersed in a tall

258

Table of the Weight of One Cubic Centimetre of Nitrogen.

Temperature in Centigrade.	720	722	724	726	728	730	732	734	736	738	740	742	744
10°	1.1338	1.1370	1.1402	1.1434	1.1466	1.1498	1.1529	1.1561	1.1593	1.1625	1.1657	1.1689	1.1721
11°	1.1288	1.1320	1.1352	1.1384	1.1415	1.1447	1.1479	1.1511	1.1542	1.1574	1.1606	1.1638	1.1670
12°	1.1237	1.1269	1.1301	1.1333	1.1364	1.1396	1.1428	1.1459	1.1491	1.1523	1.1554	1.1586	1.1618
13°	1.1187	1.1219	1.1251	1.1282	1.1314	1.1345	1.1377	1.1409	1.1440	1.1472	1.1503	1.1535	1.1566
14°	1.1136	1.1168	1.1200	1.1231	1.1263	1.1294	1.1326	1.1357	1.1389	1.1420	1.1452	1.1483	1.1515
15°	1.1085	1.1117	1.1149	1.1180	1.1211	1.1243	1.1274	1.1305	1.1337	1.1368	1.1399	1.1431	1.1462
16°	1.1034	1.1066	1.1097	1.1128	1.1160	1.1191	1.1222	1.1253	1.1285	1.1316	1.1347	1.1378	1.1409
17°	1.0983	1.1014	1.1045	1.1076	1.1107	1.1138	1.1170	1.1201	1.1232	1.1263	1.1294	1.1325	1.1356
18°	1.0930	1.0961	1.0992	1.1023	1.1054	1.1085	1.1117	1.1148	1.1179	1.1209	1.1241	1.1272	1.1303
19°	1.0877	1.0908	1.0939	1.0970	1.1001	1.1032	1.1063	1.1094	1.1125	1.1156	1.1187	1.1218	1.1248
20°	1.0825	1.0855	1.0886	1.0917	1.0948	1.0979	1.1009	1.1040	1.1071	1.1102	1.1133	1.1164	1.1194
21°	1.0771	1.0802	1.0832	1.0863	1.0894	1.0924	1.0955	1.0986	1.1017	1.1047	1.1078	1.1109	1.1139
22°	1.0717	1.0747	1.0778	1.0808	1.0839	1.0870	1.0900	1.0931	1.0961	1.0992	1.1025	1.1053	1.1084
23°	1.0662	1.0692	1.0723	1.0753	1.0784	1.0814	1.0845	1.0875	1.0906	1.0936	1.0967	1.0997	1.1028
24°	1.0606	1.0636	1.0667	1.0697	1.0728	1.0758	1.0789	1.0819	1.0849	1.0880	1.0910	1.0940	1.0971
25°	1.0550	1.0580	1.0610	1.0641	1.0671	1.0701	1.0732	1.0762	1.0792	1.0823	1.0853	1.0883	1.0913
	720	722	724	726	728	730	732	734	736	738	740	742	744

746	748	750	752	754	756	758	760	762	764	766	768	770	Temperature in Centigrade.
1.1753	1.1785	1.1817	1.1848	1.1880	1.1912	1.1944	1.1976	1.2008	1.2040	1.2072	1.2104	1.2136	10°
1.1701	1.1733	1.1765	1.1717	1.1829	1.1860	1.1892	1.1924	1.1956	1.1988	1.2019	1.2051	1.2083	11°
1.1649	1.1681	1.1713	1.1744	1.1776	1.1808	1.1839	1.1871	1.1903	1.1934	1.1966	1.1998	1.2029	12°
1.1598	1.1630	1.1661	1.1693	1.1724	1.1756	1.1787	1.1819	1.1851	1.1882	1.1914	1.1945	1.1977	13°
1.1546	1.1577	1.1609	1.1640	1.1672	1.1703	1.1735	1.1766	1.1798	1.1829	1.1861	1.1892	1.1923	14°
1.1493	1.1525	1.1556	1.1587	1.1619	1.1650	1.1681	1.1713	1.1744	1.1775	1.1807	1.1838	1.1869	15°
1.1441	1.1472	1.1503	1.1534	1.1566	1.1597	1.1628	1.1659	1.1691	1.1722	1.1753	1.1784	1.1816	16°
1.1397	1.1419	1.1450	1.1481	1.1512	1.1543	1.1574	1.1605	1.1636	1.1667	1.1699	1.1730	1.1761	17°
1.1334	1.1365	1.1396	1.1427	1.1458	1.1489	1.1520	1.1551	1.1582	1.1613	1.1644	1.1675	1.1706	18°
1.1279	1.1310	1.1341	1.1372	1.1403	1.1434	1.1465	1.1496	1.1527	1.1558	1.1589	1.1620	1.1650	19°
1.1225	1.1256	1.1287	1.1318	1.1348	1.1379	1.1410	1.1441	1.1472	1.1502	1.1533	1.1564	1.1595	20°
1.1170	1.1201	1.1231	1.1262	1.1293	1.1324	1.1354	1.1385	1.1416	1.1446	1.1477	1.1508	1.1539	21°
1.1115	1.1145	1.1176	1.1206	1.1237	1.1268	1.1298	1.1329	1.1359	1.1390	1.1421	1.1451	1.1482	22°
1.1058	1.1089	1.1119	1.1150	1.1180	1.1211	1.1241	1.1272	1.1302	1.1333	1.1363	1.1394	1.1424	23°
1.1001	1.1032	1.1062	1.1092	1.1123	1.1153	1.1184	1.1214	1.1244	1.1275	1.1305	1.1336	1.1366	24°
1.0944	1.0974	1.1004	1.1035	1.1065	1.1095	1.1126	1.1156	1.1186	1.1216	1.1247	1.1277	1.1307	25°
746	748	750	752	754	756	758	760	762	764	766	768	770	

Barometric pressure in millimetres.

glass cylinder filled with water, and supported in such a way as to admit of being raised or lowered at pleasure. The upper end of the burette communicates with the evolution bottle *a*, which has a capacity of 75 c.c., by means of a rubber tube.

The reagent required is made as follows: 27 c.c. of a solution of caustic soda, made by dissolving 100 grams NaHO in 250 c.c. H_2O, are brought into a glass-stoppered bottle, 2.5 c.c. bromine are added, the mixture shaken, and diluted with water to 150 c.c. The caustic soda solution may be kept in a glass-stoppered bottle, whose stopper is well paraffined, but the mixture must be made up as required.

To conduct a determination, about 20 c.c. of the hypobromite solution are placed in the bottle *a*; 5 c.c. of the urine to be examined are placed in the short test-tube, which is then introduced into the position shown in the figure, care being had that no urine escapes. The cork with its fittings is then introduced, the pinch cock *b* opened, and closed again when the level of liquid in the burette is the same as that in the cylinder. The decomposing vessel *q* is then inclined so that the urine and hypobromite solution mix; the decomposition begins at once, and the evolved N passes into the burette, which is raised from time to time, so as to keep the external and internal levels of water about equal; the CO_2 formed is retained by the soda solution. In about an hour (the decomposition is usually complete in fifteen minutes, but it is well to wait an hour) the height is so adjusted that the inner and outer levels of water are exactly even, and the graduation is read, while the standing of the barometer and thermometer are noted at the same time.

In calculating the percentage of urea from the volume of N obtained, it is essential that a correction should be made for differences of temperature and pressure, without which the result from an ordinary sample of urine may be vitiated by an error of ten per cent. If, however, the temperature and barometric pressure have been noted, the correction is readily made by the use of the preceding table, computed by Dietrich.

In the square of the table in which the horizontal line of the observed temperature crosses the vertical line of the observed barometric pressure will be found the weight, in milligrams, of a c.c. of N; this, multiplied by the observed volume of N, gives the weight of N produced by the decomposition of the urea contained in 5 c.c. urine. But as 60 parts urea yield 28 parts N, the weight of N, multiplied by 2.14, gives the weight of urea in milligrams in 5 c.c. urine. This quantity, multiplied by twice the amount of urine in 24 hours, and divided by 10,000, gives the amount of urea eliminated in 24 hours in grams. If the result be desired in grains the amount in grams is multiplied by 15.434.

Example.—5 c.c. urine decomposed; barometer = 736 mm; thermometer = 10°; burette reading before decomposition = 64.2; same after decomposition = 32.6; c.c. N collected = 31.6. From the table 1 c.c. N. at 10° and 736 mm. BP. weighs 1.1593. The patient passes 1500 c.c. urine in 24 hours:

$$31.6 \times 1.1593 = 36.6339 = \text{milligr. N in 5 c.c. urine,}$$

$$36.6339 \times 2.14 = 78.3965 = \text{milligr. urea in 5 c.c. urine.}$$

$$\frac{78.3965 \times 3000}{10.000} = 23.519 = \text{grams urea in 24 hours.}$$

$$23.519 \times 15.434 = 362.99 = \text{grains urea in 24 hours.}$$

In using this process it is well to have the urea solution as near the strength of one per cent. as possible; therefor, if the urine be concentrated, it should be diluted. Even when carefully conducted, the process is not strictly accurate; creatinin and uric acid are also decomposed with liberation of N, thus causing a slight plus error; on the other hand, a minus error is caused by the fact, that in the decomposition of urea by the hypobromite, the theoretical result is never obtained within about eight per cent. in urine. These errors may be rectified to a great extent by multiplying the result by 1.044.

A process which does not yield as accurate results as the preceding, but which is more easy of application, is that of Fowler, based upon the loss of sp. gr. of the urine after the decomposition of its urea by hypochlorite. To apply this method the sp. gr. of the urine is carefully determined, as well as that of the liq. sodæ chlorinatæ (Squibb's). One volume of the urine is then mixed with exactly seven volumes of the

259

liq. sod. chlor., and, after the first violence of the reaction has subsided, the mixture is shaken from time to time during an hour, when the decomposition is complete; the sp. gr. of the mixture is then determined. As the reaction begins instantaneously when the urine and reagent are mixed, the sp. gr. of the mixture must be calculated by adding together once the sp. gr. of the urine and seven times the sp. gr. of the liq. sod. chlor., and dividing the sum by eight. From the quotient so obtained the sp. gr. of the mixture after decomposition is subtracted; every degree of loss in sp. gr. indicates 0.7791 gram of urea in 100 c.c. of urine. The sp. gr. determinations must all be made at the same temperature; and that of the mixture only when the evolution of gas has ceased entirely.

Finally, when it is only desired to determine whether the urea is greatly in excess or much below the normal, advantage may be taken of the formation of crystals of urea nitrate. Two samples of the urine are taken, one of 5 c.c. and one of 10 c.c. ; the latter is evaporated, at a low temperature, to the bulk of the former, and cooled ; to both one-third volume of colorless HNO_3 is added. If crystals do not form within a few moments in the concentrated sample, the quantity of urea is below the normal ; if they do in the unconcentrated sample, it is in excess. In using this very rough method, regard must be had to the quantity of urine passed in 24 hours ; the above applies to the normal amount of 1200 c.c. ; if the quantity be greater or less, the urine must be concentrated or diluted in proportion. This method cannot be used if the urine is albuminous.

Compound Ureas.

These compounds, which are exceedingly numerous, may be considered as formed by the substitution of one or more alcoholic or acid radicals for one or more of the remaining H atoms of urea.

Those containing alcoholic radicals may be obtained, as urea is obtained from ammonium cyanate, from the cyanate of the corresponding compound ammonium ; or by the action of NH_3, or of the compound ammonias, upon the cyanic ethers.

Those containing acid radicals have received the distinctive name of *ureids* ; some of them are derivatives of uric acid, which is itself probably

260

an ureid. We will limit our consideration of these bodies to uric acid and the ureids obtained from and related to it.

Uric acid—*Lithic acid*—$C_5H_2N_4O_3H_2$—168.—Occurrence.—So far as yet known, uric acid is exclusively an animal product. It exists in the urine of man and of the carnivora, and in that of the herbivora when, during early life or starvation, they are for the time being carnivora ; as a constituent of urinary calculi ; and very abundantly in the excrement of serpents, tortoises, birds, molluscs, and insects, also in guano. It is present in very small quantity in the blood of man, more abundantly in that of gouty patients and in that of birds. The so-called "chalk-stones" deposited in the joints of gouty patients are composed of sodium urate. It also occurs in the spleen, lungs, liver, pancreas, brain, and muscular fluid.

Preparation.—Although uric acid may be obtained from calculi, urine, and guano, the source from which it is most readily obtained in a state of purity is the solid urine of large serpents, which is composed almost entirely of uric acid and the acid urates of sodium, potassium, and ammonium. This is dried, powdered, and dissolved in a solution of potassium hydrate, containing one part of potash to 20 of water ; the solution is boiled until all odor of NH_3 has disappeared. Through the filtered solution CO_2 is passed, through a wide tube, until the precipitate, which was at first gelatinous, has become granular and sinks to the bottom ; the acid potassium urate so formed is collected on a filter, and washed with cold H_2O until the wash-water becomes turbid when added to the first filtrate ; the deposit is now dissolved in hot dilute caustic potassa solution, and the solution filtered hot into HCl, diluted with an equal volume of H_2O. The precipitated uric acid is washed and dried.

Properties—*Physical.*—Uric acid, when pure, crystallizes in small, white, rhombic, rectangular or hexagonal plates, or in rectangular prisms, or in dendritic crystals of a hydrate, $C_5H_4N_4O_3,2H_2O$. As crystallized from urine it is more or less colored with urinary pigments, and forms rectangular or rhombic plates, usually with the angles rounded so as to form lozenges, which are arranged in bundles, daggers, crosses, or dendritic groups, sometimes of considerable size. It is almost insoluble in H_2O, requiring for its solution 1900 parts of boiling H_2O and 15,000 parts of cold H_2O ; insoluble in alcohol and ether ; its aqueous solution is acid to test-paper ; cold HCl dissolves it more readily than H_2O, and on evaporation deposits it in rectangular plates. It is tasteless and odorless.

Chemical.—When heated, it is decomposed without fusion or sublimation. Its constitution is unknown. Heated in Cl it yields cyanuric acid and HCl. When Cl is passed for some time through H_2O holding uric acid in suspension, alloxan, parabanic and oxalic acids, and ammonium cyanate are formed. Similar decomposition is produced by Br and I. It is simply dissolved by HCl. It is dissolved by H_2SO_4 ; from a hot solution in which a deliquescent, crystalline compound, $C_5H_4N_4O_3$, $4H_2SO_4$ is deposited ; it is partly decomposed by H_2SO_4 at 140° (284° F.). It dissolves in cold HNO_3 with effervescence and formation of alloxan, alloxantine, and urea ; with hot HNO_3 parabanic acid is produced. Solutions of the alkalies dissolve uric acid with formation of neutral urates. Uric acid is dibasic.

Ammonium urates.—The neutral salt, $C_5H_2N_4O_3(NH_4)_2$, is unknown. The acid salt, $C_5H_3N_4O_3(NH_4)$, exists as a constituent of the urine of the lower animals, and occurs, accompanying other urates and free uric acid, in urinary sediments and calculi. Sediments of this salt are rust-yellow or pink in color, amorphous, or composed of globular masses, set with projecting points, or elongated dumb-bells, and are formed in alkaline urine. It is very sparingly soluble in H_2O ; soluble in warm HCl, from which solution crystalline plates of uric acid are deposited.

261

Potassium urates.—The neutral salt, $C_5H_2N_4O_3K_2$, is obtained when a solution of potassium hydrate, free from carbonate, is saturated with uric acid ; the solution on concentration deposits the salt in fine needles. It is soluble in 44 parts of cold H_2O and in 35 parts of boiling H_2O. It is alkaline in taste, and absorbs CO_2 from the air.

The acid salt, $C_5H_3N_4O_3K$, is formed as a granular (at first gelatinous) precipitate when a solution of the neutral salt is treated with CO_2. It dissolves in 800 parts of cold H_2O and in 80 parts of boiling H_2O. The occurrence of potassium urates in urinary sediments and calculi is very exceptional.

Sodium urates.—The neutral salt, $C_5H_2N_4O_3Na_2$, is formed under similar conditions as the corresponding potassium salt. It forms nodular masses, soluble in 77 parts of cold H_2O and in 75 of boiling H_2O ; it absorbs CO_2 from the air.

The acid salt, $C_5H_3N_4O_3Na$, is formed when the neutral salt is treated with CO_2. It is soluble in 1200 parts of cold H_2O and in 125 parts of boiling H_2O. It occurs in urinary sediments and calculi, very rarely crystallized. The arthritic calculi of gouty patients are almost exclusively composed of this salt, frequently beautifully crystallized.

Calcium urates.—The neutral salt, $C_5H_2N_4O_3Ca$, is obtained by dropping a solution of neutral potassium urate into a boiling solution of calcium chloride until the precipitate is no longer redissolved, and then boiling for an hour. A granular powder, soluble in 1500 parts of cold H_2O and in 1440 parts of boiling H_2O.

The acid salt, $(C_5H_3N_4O_3)_2Ca$, is obtained by decomposing a boiling solution of acid potassium urate with calcium chloride solution. It crystallizes in needles, soluble in 603 parts of cold H_2O and in 276 parts of boiling H_2O. It occurs occasionally in urinary sediments and calculi, and in "chalk-stones."

Lithium urates.—The acid salt, $C_5H_3N_4O_3Li$, is formed by dissolving uric acid in a warm solution of lithium carbonate. It crystallizes in needles, which dissolve in 60 parts of H_2O at 50° (122° F.) and do not separate when the solution is cooled. It is with a view to the formation of this, the most soluble of the acid urates, that the compounds of lithium are given to patients suffering with the uric acid diathesis.

Physiology.—Uric acid exists in the economy chiefly in combination as its sodium salts ; it is occasionally found free, and from the probable method of its formation it is difficult to understand how all the uric acid in the economy should not have existed there free, at least at the instant of its formation. It can scarcely be doubted that uric acid is one of the products of the oxidation of the albuminoid substances—an oxidation intermediate in the production of urea ; and that consequently diseases in which there is an excessive formation of uric acid, such as gout, have their origin in defective oxidation.

In human urine the quantity of uric acid varies with the nature of the food in the same manner as does urea, and in about the same proportion :

	Urea.	Uric acid.	Proportion of uric acid to urea.
Animal food	71.5	1.25	57.2
Mixed food	37.0	0.76	48.7
Vegetable food	26.0	0.50	52.0
Non-nitrogenised food	16 0	0 34	47.0

262

The mean elimination of uric acid in the urine is from one-thirty-fifth to one-sixtieth of that of urea, or about 0.5 to 1.0 gram (7.7–15.4 grains) in 24 hours. With a strictly vegetable diet the elimination of 24 hours may fall to 0.3 gram (4.6 grains), and with a surfeit of animal food it may rise to 1.5 gram (23 grains). The hourly elimination is increased after meals, and diminished by fasting and by muscular and mental activity.

Deposits of free uric acid occur in acid, concentrated urines. In gout the proportion of uric acid in the urine is diminished, although, owing to the small quantity of urine passed, it may be relatively great ; during the paroxysms the quantity of uric acid is increased, both relatively and absolutely. The proportion of uric acid in the blood is invariably increased in gout.

Analytical Characters.—Uric acid may be recognized by its crystalline form and by the *murexid test.* To apply this test the substance is moistened with HNO_3, which is evaporated nearly to dryness at a low temperature ; the cooled residue is then moistened with ammonium hydrate. If uric acid be present, a yellow residue—sometimes pink or red when the uric acid was abundant—remains after the evaporation of the HNO_3, and this, on the addition of the alkali, assumes a rich purplish-red color.

To detect uric acid in the blood, about two drachms of the serum are placed in a flat glass dish and faintly acidulated with acetic acid ; a very fine fibril of linen thread is placed in the liquid, which is set aside and allowed to evaporate to the consistency of a jelly ; the fibril is then removed and examined microscopically. If the blood contain uric acid in abnormal proportion, the thread will have attached to it crystals of uric acid.

Quantitative Determination.—The best method for the determination of the quantity of uric acid in urine is the following: 250 c.c. of the filtered urine are acidulated with 10 c.c. of HCl, and the mixture set aside for 24 hours in a cool place. A small filter is washed, first with dilute HCl and then with H_2O, dried at 100° (212° F.), and weighed. At the end of 24 hours this filter is moistened in a funnel, and the crystals of uric acid collected upon it (those which adhere to the walls of the precipitating vessel are best separated by a small section of rubber tubing passed over the end of a glass rod, and used as a brush). No H_2O is to be used in this part of the process, the filtered urine being passed through a second time, if this be required, to bring all the crystals upon the filter. The deposit on the filter is now washed with 35 c.c. of pure H_2O, added in small portions at a time; the filter and its contents are then dried and weighed. The difference between this weight and that of the dry filter alone is the weight of uric acid in 250 c.c. of urine. If from any cause more than 35 c.c of wash-water have been used, $0^{mgr.}.043$ must be added to this weight for every c.c. of extra wash-water.

If the urine contain albumen, this must first be separated by adding two or three drops of acetic acid, heating to near 100° (212° F.), until the coagulum becomes flocculent, and filtering.

Ureids derived from Uric Acid.—These substances are quite numerous, and are divisible into ureids, diureids, triureids, and uræmic acids, according as they are formed by substitution in one, two, or three molecules of urea, and according as the acid radical substituted does or does not retain a group COOH. Some of these substances require a brief mention:

Oxalylurea—*Parabanic acid*—$\left.\begin{matrix}(CO)''\\(C_2O_2)''\\H_2\end{matrix}\right\}N_2$—114—is urea in which two atoms of H have been replaced by the bivalent radical $(C_2O_2)''$, of

oxalic acid. It is obtained by oxidizing uric acid or alloxan by hot HNO_3.

Allantoin—$C_4H_6N_4O_3$—130—occurs in the allantoic fluid of the cow; in the urine of sucking calves, in that of dogs and cats when fed on meat, in that of children during the first eight days of life, in that of adults after the ingestion of tannin, and in that of pregnant women. It is produced artificially by oxidizing uric acid, suspended in boiling H_2O, with lead dioxide.

It crystallizes in small, tasteless, neutral, colorless prisms; sparingly soluble in cold H_2O, readily soluble in warm H_2O. Heated with alkalies it yields oxalic acid and NH_3; and with dilute acids, *allanturic acid*, $C_3H_4N_2O_3$.

Allantoïn has been obtained synthetically by heating together glyoxylic acid and urea.

Mesoxalylurea—*Alloxan*—$C_4H_2N_2O_4$—142—is a product of the limited oxidation of uric acid. It has been found in the intestinal mucus in a case of diarrhœa. It forms colorless crystals, readily soluble in H_2O. It gradually turns red in air, and stains the skin red.

Oxaluric acid—$C_3H_4N_2O_4$—132—occurs in its ammonium salt, as a normal constituent, in small quantity, in human urine. It may be obtained by heating oxalylurea with calcium carbonate.

It is a white, sparingly soluble powder, which is converted into urea and oxalic acid when boiled with water or alkalies. Its ammonium salt crystallizes in white, glistening, sparingly soluble needles. Its ready conversion into urea and oxalic acid and its formation from oxalylurea, itself a product of oxidation of uric acid, render it probable that oxaluric acid is one of the many intermediate products of the oxidation of the nitrogenous constituents of the body.

Dialuric acid—*Oxybarbituric acid*—$C_4H_4N_2O_4$—a dibasic acid, produced by reduction of alloxan.

Alloxantine—$C_8H_4N_4O_7$—is a substance crystallizing in small, brilliant, very sparingly soluble prisms, produced by the action of reducing agents upon alloxan, whose action is less powerful than that required to convert alloxan into dialuric acid.

Murexide—*Ammonium purpurate*—$C_8H_4(NH_4)N_5O_6$—is produced by oxidation of uric acid, of alloxan, and of a number of other derivatives of uric acid with subsequent contact of ammonium hydrate. It is supposed to be the ammonium salt of a hypothetical and non-isolated acid. The ammonium salt is of a brilliant, but evanescent purple color. (See Murexid test for uric acid, p. 262.)

Hydurilic acid—$C_8H_6N_4O_6$—is produced as a yellowish, crystalline, sparingly soluble powder by heating together glycerin and dialuric acid. It is a strong dibasic acid.

Violuric acid—$C_4H_3N_3O_4$—is produced, along with alloxan, by the action of nitric acid upon hydurilic acid. It forms small, readily soluble, octahedral crystals. It is a strong monobasic acid, whose salts are brilliantly colored.

TRIATOMIC ALCOHOLS.

Series $C_nH_{2n+2}O_3$.

There is as yet only one alcohol known containing a trivalent radical. This is *glycerin*, whose relation to the monoatomic and diatomic alcohols is shown by the following formulæ:

CH_3	CH_3	CH_2OH	CH_2OH
|	|	|	|
CH_2	CH_2	CH_2	$CHOH$
|	|	|	|
CH_3	CH_2OH	CH_2OH	CH_2OH
Propane.	Propyl alcohol.	Propyl glycol.	Glycerin.

Glycerin—*Glycerinum* (*U. S.*)—$C_3H_5(OH)_3$—92—was first obtained as a secondary product in the manufacture of lead plaster; it is now produced as a by-product in the manufacture of soaps and of stearin candles. It exists free in palm-oil and in other vegetable oils; it is produced in small quantity during alcoholic fermentation, and is consequently present in wine and beer. It is much more widely disseminated in its ethers, the neutral fats, in the animal and vegetable kingdoms.

It has been obtained by partial synthesis, by heating for some time a mixture of allyl tribromide, silver acetate and acetic acid, and saponifying the triacetin so obtained.

The glycerin obtained by the process now generally followed—the decomposition of the neutral fats and the distillation of the product in a current of superheated steam—is free from the impurities which contaminated the product of the older processes. The only impurity likely to be present is water, which may be recognized by the low sp. gr.

Glycerin is a colorless, odorless, syrupy liquid, has a sweetish taste; sp. gr. 1.26 at 15° (59° F.). Although it cannot usually be caused to crystallize by the application of the most intense cold, it does so sometimes under imperfectly understood conditions, forming small, white needles of sp. gr. 1.268, and fusible between 7° and 8° (44°.6–46°.4 F.). It is soluble in all proportions in water and alcohol, insoluble in ether and in chloroform. The sp. gr. of mixtures of glycerin and water increase with the proportion of glycerin. It is a good solvent for a number of mineral and organic substances (*glycerites* and *glyceroles*). It is not volatile at ordinary temperatures. When heated, a portion distils unaltered at 275°–280° (527°–536° F.), but the greater part is decomposed into acrolein, acetic acid, carbon dioxide, and combustible gases. It may be distilled unchanged in a current of superheated steam between 285° and 315° (545°–599° F.).

Concentrated glycerin, when heated to 150° (302° F.) ignites and burns without odor and without leaving a residue, and with a pale-blue flame. It may also be burnt from a short wick.

Glycerin is readily oxidized, yielding different products with different degrees of oxidation. Platinum black oxidizes it, with formation, finally, of H_2O and CO_2; oxidized by manganese dioxide and H_2SO_4, it yields CO_2 and formic acid. If a layer of glycerin, diluted with an equal volume of H_2O be floated on the surface of HNO_3 of sp. gr. 1.5, a mixture of several acids, is formed: oxalic, $2C_2O_4H_2$; glyceric, $C_3H_6O_4$; formic, CH_2O_2; glycollic, $C_2H_4O_3$; glyoxylic, $C_2H_4O_4$; and tartaric, $C_4H_6O_6$. When glycerin is heated with potassium hydrate, a mixture of potassium acetate and formiate is

produced. When glycerin, diluted with 20 volumes of H_2O, is heated with Br; CO_2, bromoform, glyceric acid, and HBr are produced. Phosphoric anhydride removes the elements of H_2O from glycerin, with formation of acroleïn (see p. 224). A similar action is effected by heating with H_2SO_4, or with potassium hydrosulphate. Heated with oxalic acid, glycerin yields CO_2 and formic acid.

The presence of glycerin in a liquid may be detected as follows: Add NaHO to feebly alkaline reaction, and dip into it a loop of Pt wire holding a borax bead; then heat the bead in the blow-pipe flame, which is colored green if the liquid contain $\frac{1}{100}$ of glycerin.

The glycerin used for medicinal purposes should respond to the following tests: (1) its sp. gr. should not vary much from that given above; (2) it should not rotate polarized light; (3) it should not turn brown when heated with sodium hydrate; (4) it should not be colored by H_2S; (5) when dissolved in its own weight of alcohol, containing one per cent. of H_2SO_4, the solution should be clear; (6) when mixed with an equal volume H_2SO_4, of sp. gr. 1.83, it should form a limpid, brownish mixture, but should not give off gas.

ACIDS DERIVABLE FROM THE GLYCERINS.

Two series of acids are derivable from the glycerins by substitution of O for H_2 in the group CH_2OH:

CH_2OH	CH_2OH	$COOH$
|	|	|
$CHOH$	$CHOH$	$CHOH$
|	|	|
CH_2OH	$COOH$	$COOH$
Glycerin.	Glyceric acid.	Tartronic acid.

The terms of both series are triatomic; those of the glyceric series are monobasic, and those of the tartronic series are dibasic (see p. 231).

Malic acid—$C_4H_6O_5$—134—is the second term of the tartronic series, and is therefor dibasic. It exists in the vegetable kingdom; either free or combined with K, Na, Ca, Mg, or organic bases; principally in fruits, such as apples, cherries, etc.; accompanied by citrates and tartrates.

It crystallizes in brilliant, prismatic needles; odorless; acid in taste; fusible at 100° (212° F.); loses H_2O at 140° (284° F.); deliquescent; very soluble in H_2O and in alcohol. Heated to 175°–180° (347°–356° F.), it is decomposed into H_2O and *maleic acid*, $C_4H_4O_4$. The malates are oxidized to carbonates in the body.

ETHERS OF GLYCERIN.

GLYCERIDES.

As glycerin is a triatomic alcohol, it contains three oxhydryl groups which may be removed, combining with H from an acid to form H_2O, and leaving a univalent, bivalent, or trivalent remainder, which may replace the H of monobasic acids to form three series of ethers. As, further, the

· 266

OH groups differ from each other in that two of them are contained in the primary group CH_2OH, the other in the secondary group CHOH, there exist two isomeres of each mono- and di-glyceride.

Glycerin.	Monoacetin.	Diacetin.	Triacetin.
CH_2OH	CH_2—O—C_2H_3O	CH_2—O—C_2H_3O	CH_2—O—C_2H_3O
CHOH	CHOH	CH—O—C_2H_3O	CH—O—C_2H_3O
CH_2OH	CH_2OH	CH_2OH	CH_2—O—C_2H_3O

Of the many substances of this class, only a few, principally those entering into the composition of the neutral fats, require consideration here.

Tributyrin—$C_3H_5\ (O,C_4H_7O)_3$—302—exists in butter. It may also be obtained by heating glycerin with butyric acid and H_2SO_4. It is a pungent liquid, very prone to decomposition, with liberation of butyric acid.

Trivalerin—$C_3H_5\ (O,C_5H_9O)_3$—344—exists in the oil of some maritime mammalia, and is identical with the *phocenine* of Chevreul.

Tricaproin — $C_3H_5\ (O,C_6H_{11}O)_3$ — 386 — **Tricaprylin** $C_3H_5\ (O,C_8H_{15}O)_3$—470—and **Tricaprin**—$C_3H_5\ (O,C_{10}H_{19}O)_3$—554—exist in small quantities in milk, butter, and cocoa-butter.

Tripalmitin—$C_3H_5\ (O,C_{16}H_{31}O)_3$—806—exists in most animal and vegetable fats, notably in palm-oil; it may also be obtained by heating glycerin with 8 to 10 times its weight of palmitic acid for 8 hours at 250° (482° F.). It forms crystalline plates, very sparingly soluble in alcohol, even when boiling; very soluble in ether. It fuses at 50° (122° F.) and solidifies again at 46° (114°.8 F.).

Trimargarin—$C_3H_5\ (O,C_{17}H_{33}O)_3$—848—has probably been obtained artificially as a crystalline solid, fusible at 60° (140° F.), solidifiable at 52° (125°.6 F.). The substance formerly described under this name as a constituent of animal fats is a mixture of tripalmitin and tristearin.

Tristearin—$C_3H_5\ (O,C_{18}H_{35}O)_3$—890—is the most abundant constituent of the solid fatty substances. It is prepared in large quantities as an industrial product in the manufacture of stearin candles, etc., but is obtained in a state of purity only with great difficulty.

In as pure a form as readily obtainable, it forms a hard, brittle, crystalline mass; fusible at 68° (154°.4 F.), solidifiable at 61° (141°.8 F.); soluble in boiling alcohol, almost insoluble in cold alcohol, readily soluble in ether.

Triolein—$C_3H_5\ (O,C_{18}H_{33}O)_3$—884—exists in varying quantity in all fats, and is the predominant constituent of those which are liquid at ordinary temperatures; it may be obtained from animal fats by boiling with alcohol, filtering the solution, decanting after twenty-four hours' standing; freezing at 0° (32° F.), and expressing.

It is a colorless, odorless, tasteless oil; soluble in alcohol and ether, insoluble in water; sp. gr. 0.92.

Trinitro-glycerin—*Nitro-glycerin* — $C_3H_5\ (ONO_2)_3$—227—used as an explosive, both pure and mixed with other substances, in dynamite, giant powder, etc., is obtained by the combined action of H_2SO_4 and HNO_3 upon glycerin. Fuming HNO_3 is mixed with twice its weight of H_2SO_4 in a cooled earthen vessel; 33 parts by weight of the mixed acids are placed in a porcelain vessel, and 5 parts of glycerin, of 31° Beaumé, are gradually added with constant stirring, while the vessel is kept well cooled; after five minutes the whole is thrown into 5–6 volumes of cold

267

water; the nitro-glycerin separates as a heavy oil which is washed with cold water.

Nitro-glycerin is an odorless, yellowish oil; has a sweetish taste; sp. gr. 1.6; insoluble in water, soluble in alcohol and ether; not volatile; crystallizes in prismatic needles when kept for some time at 0° (32° F.); fuses again at 8° (46°.4 F.).

When pure nitro-glycerin is exposed to the air at 30° (86° F.) for some time, it decomposes, without explosion and with production of glyceric and oxalic acids. When heated to 100° (212° F.) it volatilizes without decomposition; at 185° (365° F.) it boils, giving off nitrous fumes; at 217° (422°.6 F.) it explodes violently; if quickly heated to 257° (494°.6 F.) it assumes the spheroidal form, and volatilizes without explosion. Upon the approach of flame at low temperatures it ignites and burns with slight decrepitations. When subjected to shock, it is suddenly decomposed into CO_2; N; vapor of H_2O, and O, the decomposition being attended with a violent explosion.

In order to render this explosive less dangerous to handle, it is now usually mixed with some inert substance, usually diatomaceous earth, in which form it is known as *dynamite*, etc.

When taken internally, nitro-glycerin is an active poison, producing effects somewhat similar to those of strychnine; in drop-doses, diluted, it causes violent headache, fever, intestinal pain, and nervous symptoms. It has been latterly used as a therapeutic agent, and has been used by the homœopaths under the name of *glonoin*.

NEUTRAL OILS AND FATS.

These are mixtures in varying proportions of tripalmitin, tristearin, and triolein, with small quantities of other glycerides, coloring and odorous principles, which are obtained from animal and vegetable bodies. The oils are fluid at ordinary temperatures, the solid glycerides being in solution in an excess of the liquid triolein. The fats, owing to a less proportion of the liquid glyceride, are solid or semi-solid at the ordinary temperature of the air. Members of both classes are fluid at sufficiently high temperatures, and solidify when exposed to a sufficiently low temperature. They are, when pure, nearly tasteless and odorless, unctuous to the touch, insoluble in and not miscible with H_2O, upon which they float; combustible, burning with a luminous flame; when rubbed upon paper they render it translucent. When heated with the caustic alkalies or in a current of superheated steam, they are *saponified*, *i.e.*, decomposed into glycerin and a fatty acid. If the saponification be produced by an alkali, the fatty acid combines with the alkaline metal to form a *soap* (*q. v.*).

Most of the fats and many of the oils, when exposed to the air, absorb O, are decomposed with liberation of volatile fatty acids, and acquire an acid taste and odor, and an acid reaction. A fat which has undergone these changes is said to have become *rancid*. Many of the vegetable oils are, however, not prone to this decomposition. Some of them, by oxidation on contact with the air, become thick, hard and dry, forming a kind of varnish over surfaces upon which they are spread; these are designated as *drying* or *siccative* oils. Others, although they become more dense on exposure to air, become neither dry nor gummy; these are known as *non-drying, greasy*, or *lubricating* oils.

Under ordinary conditions, oils and melted fats do not mix with

268

water, and, if shaken with that fluid, form a *temporary* milky mixture, which, on standing for a short time, separates into two distinct layers, the oil floating on the water. In the presence, however, of small quantities of certain substances, such as albumen, pancreatin (*q. v.*), ptyalin, etc., the milky mixture obtained by shaking together oil and water does not separate into distinct layers on standing; such a mixture, in which the fat is held in a *permanent* state of suspension in small globules in a watery fluid, is called an *emulsion*. Perfect emulsions may be easily obtained by agitating an oil containing a trace of free oleic acid with a very dilute solution of sodium carbonate and borax.

Fixed oils.—These substances are designated as "fixed," to distinguish them from other vegetable products having an oily appearance, but which differ from the true oils in their chemical composition and in their physical properties, especially in that they are volatile without decomposition, and are obtained by distillation, while the fixed oils are obtained by expression, with or without the aid of a gentle heat.

Palm oil is a reddish-yellow solid at ordinary temperatures, has a bland taste and an aromatic odor. It saponifies readily, and is usually acid and contains free glycerin from spontaneous decomposition.

Rape seed and *colza oils*, produced from various species of *Brassica*, are yellow, limpid oils having a strong odor and disagreeable taste.

Croton-oil—*Oleum tiglii* (*U. S.*)—*Oleum crotonis* (*Br.*)—varies much in color and activity, according to its source; that which is obtained from the East is yellowish, liquid, transparent, and much less active than that prepared in Europe from the imported seeds, which is darker, less fluid, caustic in taste, and wholly soluble in absolute alcohol. Croton-oil contains, beside the glycerides of oleic, crotonic and fatty acids, about four per cent. of a peculiar principle called *crotonol*, to which the oil owes its vesicating properties; it also contains an alkaloid-like substance, also existing in castor-oil, called *ricinine*. None of these bodies, however, are possessed of the drastic powers of the oil itself.

Peanut-oil—Ground-nut oil—an almost colorless oil, very much resembling olive-oil, in place of which it is frequently used for culinary purposes, intentionally or otherwise. It is readily saponifiable, yielding two peculiar acids, *arachaïc* and *hypogaïc* (see Olive-oil).

Cotton-seed oil—Oleum gossypii seminis (*U. S.*)—a pale yellow, bland oil, also resembling olive-oil, for which it is frequently substituted.

Almond-oil—Oleum amygdalæ expressum (*U. S.*)—*Oleum amygdalæ* (*Br.*)—a light yellow oil, very soluble in ether, soluble in alcohol; nearly inodorous; has a bland, sweetish taste. The pure oil has no odor of bitter almonds.

Olive-oil—Oleum olivæ (*U. S., Br.*).—A well-known oil of a yellow or greenish-yellow color, almost odorless, and of a bland and sweetish taste. The finest grades have a yellow tinge and a faint taste of the fruit; they are prepared by cold pressure; they are less subject to rancidity than the lower grades. Olive-oil is very frequently adulterated, chiefly with poppy-oil, sesame oil, cotton-seed oil and peanut-oil; the presence of the first is detected by Pontet's reagent (made by dissolving 6 parts Hg in 7.5 parts of HNO_3 of 36° in the cold), which converts pure olive-oil into a solid mass, while an oil adulterated with a drying oil remains semi-solid. A contamination with oil of sesame is indicated by the production of a green color, with a mixture of HNO_3 and H_2SO_4. Peanut-oil, an exceedingly common adulterant in this country, is recognized by the following method: ten grams of the oil are saponified; the soap is decomposed with HCl; the

269

liberated fatty acids dissolved in 50 c.c. of strong alcohol; the solution precipitated with lead acetate; the precipitate washed with ether; the residue decomposed with hot dilute HCl; the oily layer separated and extracted with strong alcohol; the alcoholic fluid, on evaporation, yields crystals of *arachaïc acid*, if the oil contains peanut-oil.

Cocoa-butter—Oleum theobromæ (*U. S., Br.*)—is, at ordinary temperatures, a whitish or yellowish solid of the consistency of tallow, and having an odor of chocolate and a pleasant taste; it does not easily become rancid. The most reliable test of its purity is its fusing-point, which should not be much below 33° (91°.4 F.).

Linseed oil—Flaxseed oil—Oleum lini (*U. S., Br.*)—is a dark, yellowish-brown oil of disagreeable odor and taste. In it oleic acid is, at least partially, replaced by *linoleic acid*, whose presence causes the oil, on exposure to air, to absorb oxygen and become thick and finally solid. This drying power is increased by boiling the oil with litharge (boiled oil).

Castor-oil—Oleum ricini (*U. S., Br.*)—is usually obtained by expression of the seeds, although in some countries it is prepared by decoction or by extraction with alcohol. It is a thick, viscid, yellowish oil, has a faint odor and a nauseous taste. It is more soluble in alcohol than any other fixed vegetable oil, and is also very soluble in ether. It saponifies very readily. Ammonia separates from it a crystalline solid, fusible at 66° (158°.8 F.) *ricinolamide*. Hot HNO_3 attacks it energetically, and finally converts it into *suberic acid*.

Whale-oil—Train-oil—obtained by *trying out* the fat or blubber of the "right whale" and of other species of *balænæ*. It is of sp. gr. 0.924 at 15° (59° F.); brownish in color; becomes solid at about 0°; has a very nauseous taste and odor. It is colored yellow by H_2SO_4; and is blackened by Cl.

Neat's-foot oil—is obtained by the action of boiling H_2O upon the feet of neat cattle, horses, and sheep, deprived of the flesh and hoofs. It is straw-yellow or reddish-yellow, odorless, not disagreeable in taste, not prone to rancidity, does not solidify at quite low temperatures; sp. gr. at 15° (59° F.) = 0.916. It is bleached, not colored, by chlorine.

Lard-oil—Oleum adipis (*U. S.*)—obtained in large quantities in the United States as a by-product in the manufacture of candles, etc., from pig's fat. A light yellow oil, used principally as a lubricant; is not colored by H_2SO_4, but is colored brown by a mixture of H_2SO_4 and HNO_3.

Tallow-oil—obtained by expression with a gentle heat from the fat of the ox and sheep. Sp. gr. 0.9003; light yellow in color. Colored brown by H_2SO_4. Formerly this oil, under the trade name of "oleic acid," was simply a by-product in the manufacture of stearin candles; of late years, however, it is specially prepared for the manufacture of oleo-margarine.

Cod-liver oil—Oleum morrhuæ (*U. S., Br.*)—is obtained from the livers of cod-fish, either by extraction with water heated to about 80° (176° F.), or by hanging the livers in the sun and collecting the oil which drips from them. There are three commercial varieties of this oil: *a. Brown.*—Dark brown, with greenish reflections; has a disagreeable, irritating taste; faintly acid; does not solidify at −13° (8°.6 F.). *b. Pale brown.*—Of the color of Malaga wine; has a peculiar odor and a fishy, irritating taste; strongly acid. *c. Pale.*—Golden yellow; deposits a white fat at −13° (8°.6 F.); has a fresh odor, slightly fishy, and a not unpleasant taste, without after-taste.

Pure cod-liver oil, with a drop of H_2SO_4, gives a bluish-violet aureole, which gradually changes to crimson, and later to brown. A drop of fuming HNO_3 dropped into the oil is surrounded by a pink aureole if the oil

270

be pure; if largely adulterated with other fish-oils, the pink color is not observed and the oil becomes slightly cloudy. Fresh cod-liver oil is not colored by rosaniline. If a third of the oil be distilled, the distillate becomes solid; while if it be contaminated with vegetable oils, the distillate becomes liquid.

Cod-liver oil contains, besides the glycerides of oleic, palmitic and stearic acids, those of butyric and acetic acids; certain biliary principles (to whose presence the sulphuric acid reaction given above is probably due), a phosphorized fat of undetermined composition; small quantities of bromine and iodine, probably in the form of organic compounds; a peculiar fatty acid called *gadinic acid*, which solidifies at 60° (140° F.); and a brown substance called *gaduin* or *gadinine*.

To which, if to any, of these substances cod-liver oil owes its value as a therapeutic agent is still unknown, although many theories have been advanced. Certain it is, however, that one of the chief values of this oil is as a food in a readily assimilable form.

Solid Animal Fats.—The glycerides of stearic, palmitic, and oleic acids exist, in health, in nearly all parts of the body; in the fluids in solution or in suspension, in the form of minute oil-globules; incorporated in the solid or semi-solid tissues, or deposited in collections in certain locations, as under the skin, enclosed in cells of connective tissue, in which the mixture of the three glycerides is in such proportion that the contents of the cells are fluid at the temperature of the body.

The total amount of fat in the body of a healthy adult is from 2.5 to 5 per cent. of the body-weight, although it may vary considerably from that proportion in conditions not, strictly speaking, pathological. The approximate quantities of fat in 100 parts of the various tissues and fluids, in health, are the following:

Urine	?	Blood	0.4	Cortex of brain	5.5
Perspiration	0.001	Cartilage	1.3	Brain	8.0
Vitreous humor	0.002	Bone	1.4	Hen's egg	11.6
Saliva	0.02	Bile	1.4	White matter of brain	20.0
Lymph	0.05	Crystalline lens	2.0	Nerve-tissue	22.1
Synovial fluid	0.06	Liver	2.4	Spinal cord	23.6
Amniotic fluid	0.2	Muscle	3.3	Fat-tissue	82.7
Chyle	0.3	Hair	4.2	Marrow	96.0
Mucus	0.4	Milk	4.3		

The amount of fat, under normal conditions, is usually greater in women and children than in men; generally greater in middle than in old age, although in some individuals the reverse is the case; greater in the inhabitants of cold climates than in those of hot countries.

In wasting from disease and from starvation the fats are rapidly absorbed, and are again as rapidly deposited when the normal condition of affairs is restored.

Besides, as a result of the tendency to corpulence, which in some individuals amounts to a pathological condition, fats may accumulate in certain tissues as a result of morbid changes. This accumulation may be due either to degeneration or to infiltration. In the former case, as when muscular tissue degenerates in consequence of long disuse, the natural tissue disappears and is replaced by fat; in the latter case, as in fatty infiltration of the heart, oil-globules are deposited between the natural morphological elements, whose change, however, may subsequently take place by true fatty degeneration, due to pressure. Fatty degeneration of the liver and of other organs occurs also in phthisis, chronic heart, and lung affections,

271

as a result of overfeeding, from the abuse of alcoholic stimulants, and from the action of certain poisons, especially of phosphorus. Tumors composed of adipose tissue occur and are known as "lipomata."

The greater part of the fat of the body enters it as such with the food. Not unimportant quantities are, however, formed in the body, and that from the albuminoid as well as from the starchy and saccharine constituents of the food. By what steps this transformation takes place is still uncertain, although there is abundant evidence that it does occur.

Those fats taken in with the food are unaltered by the digestive fluids, except in that they are freed from their enclosing membranes in the stomach, until they reach the duodenum; here, under the influence of the pancreatic juice, the major part is converted into a fine emulsion, in which form it is absorbed by the lacteals. A smaller portion is saponified, and the products of the saponification, free fatty acids, soaps, and glycerin, subsequently absorbed by lacteals and blood-vessels.

The service of the fats in the economy is undoubtedly as a producer of heat and force by its oxidation; and by its low power of conducting heat, and the position in which it is deposited under the skin, as a retainer of heat produced in the body. The fats are not discharged from the system in health, except the excess contained in the food over that which the absorbents are capable of taking up, which passes out with the fæces; a small quantity distributed over the surface in the perspiration and sebaceous secretion (which can hardly be said to be eliminated); and a mere trace in the urine.

Butter.—The fat of milk, separated and made to agglomerate by agitation, and more or less salted to insure its keeping. It consists of the glycerides of stearic, palmitic, oleic, butyric, capric, caprylic, and caproic acids, with a small amount of coloring matter, more or less water and salt, and caseine. Good, natural butter contains 80–90 per cent. of fat, 6–10 per cent. of water, 2–5 per cent. of curd, and 2–5 per cent. of salt; fuses at from 32°.8 to 34°.9 (91°–94°.8 F.).

Butter is adulterated with excess of water and salt, starch, animal fats other than those of butter, and artificial coloring matters.

Excess of salt and water are usually worked in together, the former up to 14 per cent. and the latter to 15 per cent. To determine the presence of an excess of water, about 4 grams (60 grains) of the butter, taken from the middle of the lump, are weighed in a porcelain capsule, in which it is heated over the water-bath, as long as it loses weight; it is then weighed again; the loss of weight is that of the quantity of water in the original weight of butter, less that of the capsule. The proportion of salt is determined by incinerating a weighed quantity of butter and determining the chlorine in the ash by the nitrate of silver method (see Sodium chloride). Roughly, the weight of the ash may be taken as salt. Starch is detected by spreading out a thin layer of butter, adding solution of iodine, and examining under the microscope for purple spots.

The detection of foreign fats in butter, formerly a most unsatisfactory problem to the analyst, has now become one which may be answered with great certainty. All of the chemical processes used are based upon a peculiar difference in the composition of butter-fat from other animal and vegetable fats and oils. When butter-fat is saponified, it yields from 5 to 8 per cent. of butyric acid and its near homologues, which are soluble in H_2O, and may be distilled without suffering decomposition, and from 85.5 to 87.5 of stearic, palmitic, and oleic acids, which are neither soluble in water nor capable of being distilled. The other fats and oils, when

272

saponified, yield mere traces of the volatile or soluble fatty acids, and much larger quantities (95.3 to 95.7 per cent.) of insoluble acids. These variations are utilized directly in some processes, such as those of Hehner and Reichert, in which the percentage of fixed and volatile acids are directly determined. In other processes, such as that of Koettstorfer, advantage is taken of the different neutralizing power of the two groups of acids. Thus, as butyric acid, $C_4H_8O_2$, and stearic acid, $C_{18}H_{36}O_2$, are each capable of neutralizing KHO, molecule for molecule, it follows that their neutralizing power is in proportion to their molecular weights, and that 56 parts KHO will require for neutralization 88 parts of butyric acid, or 284 parts of stearic acid. For descriptions of processes the student is referred to Allen: "Commercial Organic Analysis," 2d ed., II.

Methods for detecting admixture of foreign fats by physical means are unreliable. One of the best, which may be of service for preliminary testing, is that of Angell and Hehner. A pear-shaped bulb of thin glass is made of such size as to displace 1 c.c. water, is weighted with mercury until it weighs 3.4 grams (52.5 grains), and the pointed end closed by fusion. The butter to be tested is fused in a beaker over the water-bath, and when quite fluid is poured out into a test-tube, about ¾ inch diameter and 6 inches long, which is kept moderately warm and upright until the fat has separated in a clear layer above the water, and then immersed in water at 15° (59° F.) until the fat has solidified. The test-tube is then arranged as shown in Fig. 38, the bulb being laid upon the surface of the fat. The water in the beaker is now heated until the globular part of the bulb has just sunk below the surface of the fat, at which time the height of the thermometer is noted; this is the "sinking-point."

The sinking-point of pure butter is 34°.3 to 36°.3 (93°.7–97°.3 F.), that of oleomargarine is lower, that of butter adulterated with other fats is higher.

FIG. 38.

"Oleomargarine" is a product made in imitation of butter, which it resembles very closely in color, taste, odor, and general appearance. Under the original patent, it is made from beef-fat, which is hashed, steamed, and subjected to pressure at a carefully regulated temperature. Under this treatment it is separated into two fatty products, one a white solid "stearine," the other a faintly yellow oil, "oleo-oil." This oil is then mixed with milk, the mixture colored and churned. The subsequent treatment of the product is the same as that of butter. "Butterine," "suine," etc., are products made, by a modification of the above process, from beef or mutton tallow, lard, and cotton-seed oil.

Butter is frequently, and oleomargarine is always, colored with some foreign pigment, "butter color," which is usually a preparation of annoto.

Soaps—are the metallic salts of stearic, palmitic, and oleic acids; those of K, Na, and NH_4 are soluble, those of the other metals insoluble; those of Na are hard, those of K soft.

Soap is made from almost any oil or fat, the best from olive-oil, or peanut, or palm-oil, and lard. The first step in the process of manufacture is the *saponification* of the fat, which consists in the decomposition of the glyceric ethers into glycerin and the fatty acids, and the combination of the latter with an alkaline metal. It is usually effected by gradually adding fluid fat to a weak boiling solution of caustic soda, or potassa, to

273

saturation. From this weak solution the soap is separated by "salting," which consists in adding, during constant agitation, a solution of caustic alkali, heavily charged with common salt, until the soap separates in grumous masses, which float upon the surface and are separated. Finally the soap is pressed to separate adhering water, fused, and cast into moulds.

White castile soap—Sapo (*U. S.*) *Sapo durus* (*Br.*)—is a Na soap made from olive-oil; strongly alkaline, hard, not greasy, very soluble; contains 21 per cent. H_2O. *Sapo mollis* (*Br.*) is a K soap made from olive oil, and contains an excess of alkali and glycerin. *Yellow soap* is made from tallow or other animal fat, and contains about ⅓ its weight of rosin. *Emplastrum plumbi* (*U. S.; Br.*) is a lead soap, prepared by saponifying olive oil with litharge.

The soaps are decomposed by weak acids, with liberation of the fatty acid; by compounds of the alkaline earths, with formation of an insoluble soap; and in the same way by most of the metallic salts.

LECITHINS—NERVE TISSUE.

Lecithin—is a substance first obtained from the yolk of hen's eggs, and subsequently found to exist in brain-tissue, particularly the gray substance, nerve tissue, semen, blood-corpuscles, blood-serum, milk, bile, and other animal tissues and fluids.

As obtained from brain-tissue lecithin is a colorless or faintly yellowish, imperfectly crystalline solid, or sometimes of a waxy consistency. It is very hygroscopic. It does not dissolve in H_2O, in which, however, it swells up and forms a mass like starch-paste. It dissolves in alcohol or ether, very sparingly in the cold, but readily under the influence of heat. It dissolves in chloroform and in benzol. Lecithin is very prone to decomposition, particularly at slightly elevated temperatures. Its chloride combines with $PtCl_4$ to form an insoluble yellowish chloroplatinate.

When an alcoholic solution of lecithin is brought into contact with hot solution of barium hydrate it yields *barium glycerophosphate*, *barium stearate*, and *choline* (see p. 207). This decomposition indicates the constitution of lecithin and its relations to the fats. Glycerophosphoric acid is orthophosphoric acid in which an atom of hydrogen has been replaced by the univalent remainder $CH_2OH—CHOH—CH_2$—left by the removal of OH from glycerin.

$$\begin{array}{l} \quad\;\; /OH \\ O{=}P{-}OH \\ \quad\;\; \backslash O{-}CH_2{-}CHOH{-}CH_2OH. \end{array}$$

In lecithin the remaining oxhydryl groups of the glycerin remainder are removed by union with the basic hydrogen of two molecules of stearic acid, and one of the two remaining basic hydrogen atoms of the phosphoric acid is displayed by choline. It is obvious that the number of lecithins is not limited to one, but that many may exist, and probably do, into whose composition any one, or any combination of two, of the acids of the same series as stearic acid may enter.

$$\begin{array}{l} \qquad\qquad /(CH_3)_3 \\ \quad\;\; /O{-}N{-\!-}CH_2{-}CH_2{-}OH \\ O{=}P{-}O{-}H \\ \quad\;\; \backslash O{-}CH_2{-}CH(C_{18}H_{35}O_2){-}CH_2(C_{18}H_{35}O_2). \end{array}$$

Distearyl-lecithin.

274

Nerve-tissue, which is exceedingly complex in its chemical composition, and whose chemistry is still in a most rudimentary condition, seems to contain similar constituents in its different parts, which differ, however materially in their quantitative composition.

The following substances have been obtained from cerebral tissue:

Mineral Substances.

Water.
Phosphates of Na, K, Ca, Mg.
Ferric oxide.
Silicic oxide.
Traces of sulphates, chlorides, and fluorides.

Albuminoids.

Substance related to myosin.
Soluble albuminoid, coagulable at 75° (167° F.).
Casein (?).

Products of Decomposition.

Glycerophosphoric acid.
Oleophosphoric acid.
Volatile fatty acids.
Lactates.
Hypoxanthine.
Xanthine.
Creatine.

Organic Substances.

Elastin.	Lecithin.
Neurokeratin.	Fats (?).
Neucleïn.	Inosite.
Cerebrin.	Cholesterin.

The composition of white and gray matter differs quantitatively, as shown below:

	Gray matter.	White matter.
Albuminoids	55.37	24.72
Lecithin	17.24	9.90
Cholesterin and fats	18.68	51.91
Cerebrin	0.53	9.55
Extractive matters, insoluble in ether	6.71	3.34
Salts	1.46	0.57

Cerebrin is a substance deposited in the crystalline form from hot ethero-alcoholic extracts of brain-tissue. It is white, very light, odorless, and tasteless; insoluble in water or in cold alcohol or ether. Its solutions are neutral. It does not contain phosphorus.

The substance known as *protagon*, described by Liebreich as having been obtained from brain-tissue, would seem to exist there notably in the white substance of Schwann. It appears to be a compound formed by the union of lecithin with cerebrin.

Neurokeratin is a substance, occurring principally in the gray matter, which is insoluble in all solvents, and is not acted upon by digestive liquids.

275

THIRD SERIES OF HYDROCARBONS.

SERIES C_nH_{2n-2}.

The terms of this series at present known are:

Acetylene	C_2H_2	Crotonylene	C_4H_6	Rutylene	$C_{10}H_{18}$
Allylene	C_3H_4	Valerylene	C_5H_8	Benylene	$C_{15}H_{28}$

Acetylene—*Ethene*—C_2H_2—26—exists in coal gas and is formed in the decomposition, by heat or otherwise, of many organic substances. It is best prepared by passing a slow current of coal gas through a narrow tube, traversed by induction sparks; directing the gas through a solution of cuprous chloride; and collecting and decomposing the precipitate by HCl. It may be obtained by direct synthesis from H and C, by producing the electric arc between carbon points in a glass globe filled with hydrogen.

It is a colorless gas, rather soluble in H_2O; has a peculiar, disagreeable odor; such as is observed when a Bunsen burner burns within the tube. It forms explosive mixtures with O. It unites with N, under the influence of the electric discharge, to form hydrocyanic acid. Mixed with Cl, it detonates violently in diffuse daylight, without the aid of heat. It may be made to unite with itself to form its polymeres benzene, C_6H_6, styrolene, C_8H_8 and naphthydrene, $C_{10}H_{10}$.

Its presence may be detected by the formation in an ammoniacal solution of cuprous chloride of a blood red precipitate, which is explosive when dry. It is probable that explosions which sometimes occur in brass or copper pipes, through which illuminating gas is conducted, are due to the formation of this compound.

Illuminating gas—is now manufactured by a variety of processes, almost every company using some modification of the method, or of the nature and proportion of the materials; thus we have gas made from wood, from coal, from fats, from petroleum, and by the decomposition of H_2O and subsequent charging of the gas with the vapor of naphtha. The typical process is that in which the gas is produced by heating bituminous coal to bright redness in retorts. As it issues from the retorts the gas is charged with substances volatile only at high temperatures; these are deposited in the condensers or coolers, and form coal- or gas-tar. From the condensers the gas passes through what are known as "scrubbers" and "lime-purifiers," in which it is deprived of ammonical compounds and other impurities. As it comes from the condensers, coal-gas contains:

* Acetylene.	* Styrolene.	† Hydrogen.	† Hydrogen sulphide.
* Ethylene.	* Naphthalene.	† Carbon monoxide.	† Carbon disulphide.
* Marsh-gas.	* Acenaphthalene.	† Carbon dioxide.	† Sulphuretted hydrocarbons.
* Butylene.	* Fluorene.	† Ammonia.	† Nitrogen.
* Propylene.	* Propyl hydride.	† Cyanogen.	† Aqueous vapor.
* Benzene.	* Butyl hydride.	† Sulphocyanogen.	

In passing through the purifiers the gas is freed of the impurities to a greater or less extent, and, as usually delivered to consumers, contains:

* Marsh-gas.	* Ethylene.	† Nitrogen.	† Carbon monoxide.
* Acetylene.	‡ Hydrogen.	† Aqueous vapor.	† Carbon dioxide.
	* Vapors of Hydrocarbons.		

* Illuminating constituents. † Impurities. ‡ Diluent.

276

TETRATOMIC ALCOHOLS.

SERIES $C_nH_{2n} + {}_2O_4$.

Very few of these compounds have yet been obtained. They may be regarded as the hydrates of the hydrocarbons C_nH_{2n-2}; as the glycols are the hydrates of the ethylene series.

Erythrite—*Phycite*—

```
CH2OH
|
CHOH
|          =C4H6(OH)4—122—
CHOH
|
CH2OH
```

$=C_4H_6(OH)_4$—122—is a product of decomposition of *erythrine*, $C_{20}H_{22}O_{10}$, which exists in the lichens of the genus *rocella*. It crystallizes in large, brilliant prisms; very soluble in H_2O and in hot alcohol, almost insoluble in ether; sweetish in taste; its solutions neither affect polarized light, nor reduce Fehling's solution, nor are capable of fermentation. Its watery solution, like that of sugar, is capable of dissolving a considerable quantity of lime, and from this solution alcohol precipitates a definite compound of erythrite and calcium. By oxidation with platinum black it yields *erythroglucic acid*, $C_4H_8O_5$. With fuming HNO_3 it forms a tetranitro compound, which explodes under the hammer.

ACIDS DERIVABLE FROM ERYTHRITE.

Theoretically erythrite should, by simple oxidation, yield two acids; one of the series $C_nH_{2n}O_3$, and another of the series $C_nH_{2n-2}O_4$. Although both of these acids are known, only the first, *erythroglucic acid*, has been obtained by oxidation of erythrite:

```
CH2OH        COOH         COOH
|            |            |
CHOH         CHOH         CHOH
|            |            |
CHOH         CHOH         CHOH
|            |            |
CH2OH        CH2OH        COOH
Erythrite.   Erythroglucic acid.   Tartaric acid.
```

Tartaric acids—*Acidum tartaricum* (*U. S., Br.*)—$C_4H_6O_6$—150.—There exist four acids having the composition $C_4H_6O_6$, which differ from each other only in their physical properties, and are very readily converted into one another; they are designated as: 1st, *Right*; 2d, *Left*; 3d, *Inactive tartaric acid*; 4th, *Racemic acid*. *Right* or *Dextrotartaric acid* crystallizes in large, oblique, rhombic prisms, having hemihedral facettes. Solutions of the acid and its salts are dextrogyrous.

Lævotartaric acid crystallizes in the same form as dextrotartaric acid, only the hemihedral facettes are on the opposite sides, so that crystals of

277

the two acids, when held facing each other, appear like the reflections one of the other. Its solution and those of its salts are lævogyrous to the same degree that corresponding solutions of dextrotartaric acid are dextrogyrous. *Racemic acid* is a compound of the two preceding; it forms crystals having no hemihedral facettes, and its solutions are without action on polarized light. It is readily separated into its components. *Inactive tartaric acid*, although resembling racemic acid in its crystalline form and inactivity with respect to polarized light, differs essentially from that acid in that it cannot be decomposed into right and left acids, and in the method of its production.

The tartaric acid which exists in nature is the dextrotartaric; it occurs, both free and in combination, in the sap of the vine and in many other vegetable juices and fruits. Although this is probably the only tartaric acid existing in nature, all four varieties may and do occur in the commercial acid, being formed during the process of manufacture.

Tartaric acid is obtained in the arts from hydropotassic tartrate, or cream of tartar (*q. v.*). This salt is dissolved in H_2O and the solution boiled with chalk until its reaction is neutral; calcic and potassic tartrates are formed. The insoluble calcic salt is separated and the potassic salt decomposed by treating the solution with calcic chloride. The united deposits of calcium-tartrate are suspended in H_2O, decomposed with the proper quantity of H_2SO_4, the solution separated from the deposit of calcium sulphate, and evaporated to crystallization.

The ordinary tartaric acid crystallizes in large prisms; very soluble in H_2O and alcohol; acid in taste and reaction. It fuses at 170° (338° F.); at 180° (356° F.) it loses H_2O, and is gradually converted into an anhydride; at 200°–210° (392°–410° F.) it is decomposed with formation of *pyruvic acid*, $C_3H_4O_3$, and *pyrotartaric acid*, $C_5H_8O_4$; at higher temperatures CO_2, CO, H_2O, hydrocarbons and charcoal are produced. If kept in fusion some time, two molecules unite, with loss of H_2O, to form *tartralic* or *ditartaric acid*, $C_8H_{10}O_{11}$.

Tartaric acid is attacked by oxidizing agents with formation of CO_2, H_2O, and, in some instances, formic and oxalic acids. Certain reducing agents convert it into malic and succinic acids. With fuming HNO_3 it forms a dinitro-compound, which is very unstable, and which, when decomposed below 36° (96°.8 F.), yields tartaric acid. It forms a precipitate with lime-water, soluble in an excess of H_2O; in not too dilute solution it forms a precipitate with potassium sulphate solution; it does not precipitate with the salts of Ca. When heated with a solution of auric chloride it precipitates the gold in the metallic form. As its formula indicates (see above), tartaric acid is tetratomic and dibasic. It has a great tendency to the formation of double salts, such as tartar emetic (*q. v.*).

When taken into the economy, as it constantly is in the form of tartrates, the greater part is oxidized to carbonic acid (carbonates); but, if taken in sufficient quantity, a portion is excreted unchanged in the urine and perspiration. The free acid is poisonous in large doses.

Citric acid—*Acidum citricum* (*U. S.*, *Br.*)—$C_6H_8O_7$ + Aq—192 + 18—is best considered in this place, although its constitution is different from that of tartaric acid. It exists in the juices of many fruits—lemon, strawberry, etc.

It is obtained from lemon-juice, which is filtered, boiled, and saturated with chalk. The insoluble calcium citrate is separated and decomposed with H_2SO_4, the solution filtered, and evaporated to crystallization.

It crystallizes in large, right rhombic prisms, which lose their aq. at

278

100° (212° F.); very soluble in water, less soluble in alcohol, sparingly soluble in ether; heated to 100° (212° F.) it fuses; at 175° (347° F.) it is decomposed, with loss of H_2O and formation of *aconitic acid*, $C_6H_6O_6$; at a higher temperature CO_2 is given off, and *itaconic acid*, $C_5H_6O_4$, and *citraconic acid*, $C_5H_6O_4$, are formed.

Concentrated H_2SO_4 decomposes it with evolution of CO; oxidizing agents convert it into formic acid and CO_2, or into acetone and CO_2, or into oxalic and acetic acids and CO_2. It is tetratomic and tribasic. In the body its salts are oxidized to carbonates.

FOURTH SERIES OF HYDROCARBONS.

SERIES C_nH_{2n-4}.

But one of the lower terms of this series is known; this is *valylene*, C_5H_6, obtained by the action of an alcoholic solution of potash on valerylene dibromide. It is a liquid, boiling at 45° (113° F.).

Among the higher terms of the series are many substances of industrial and medical importance.

Terebenthene—$C_{10}H_{16}$—136—is the type of a great number of isomeric substances existing in the *volatile oils* or *essences*. It is the chief constituent of oil of turpentine.

To obtain it in a state of purity, oil of turpentine is mixed with an alkaline carbonate, and distilled in vacuo over a water-bath, or by fractional distillation of the crude oil, those portions being collected which pass over at about 156° (312°.8 F.).

Pure terebenthene is a colorless, mobile liquid; has the peculiar odor of turpentine; boils at about 156° (312°.8 F.); burns with a smoky, luminous flame. Obtained from the turpentine of *pinus maritima*, it is lævogyrous, purified by distillation in vacuo, $[a]_D = -42°.36$, by fractional distillation, $[a]_D = -40°.32$; that obtained from *pinus australis* is dextrogyrous, $[a]_D = +18°.9$; specific gravity at 0° (32° F.) = 0.8767.

It absorbs oxygen rapidly from the air, whether pure or in the commercial essence, becoming thick and finally gummy. Oxidizing agents, such as HNO_3, attack it energetically, causing it to ignite and burn suddenly, with separation of a large volume of carbon. HCl unites with it to form a number of compounds, as do also HI and HBr—all the compounds having the odor of camphor. When mixed with HNO_3, diluted with alcohol, and exposed to the air, it forms a crystalline pseudo-glycol, *terpine*. Cl, Br and I form compounds of substitution or of addition.

TURPENTINE—*Terebenthina* (*U. S.*)—is the name given to the concrete juice of various species of trees of the genera *Pinus*, *Abies*, and *Larix*, which consist of terebenthene, its isomeres, and resinous and other substances. The product differs in composition and properties according to the kind of tree from which it is produced.

White turpentine—Common American turpentine—obtained from *Pinus palustris* and *P. tæda*; is yellowish-white, semi-fluid at summer temperature, hard and solid when cooled; on exposure to air it becomes dry, hard, and brittle. It is usually subjected to distillation near the place of its collection, by which process it is separated into the volatile *oil*; or *essence of turpentine* (*q. v.*), and *rosin*, or *colophony* (*q. v.*). *European turpentine—Bordeaux turpentine*—obtained from *P. sylvestris* and *P. maritima*. *Canada turpentine—Canada balsam—Balsam of fir—Terebenthina canadensis* (*U. S.*)

279

—is from *abies balsamea*. It is a tenacious semi-solid, of the consistency of honey when fresh, colorless or yellowish, sticky, bitter in taste, and having a balsamic odor; when long exposed to the air, or when heated over the water-bath, its volatile constituents are lost, and it is converted into a hard brittle mass. *Venice turpentine*—produced from *larix Europæa*—is a thick, viscid liquid, yellowish or greenish in color; soluble in alcohol; does not concrete as readily as other turpentines. *Chian turpentine*, the product of *pistachia terebinthus*, is a thick, greenish-yellow liquid.

ESSENCE OF TURPENTINE—*Oil of turpentine—Spirits of turpentine—Oleum terebinthinæ* (*U. S.*, *Br.*)—is the volatile product of the distillation of turpentine. It is not identical with terebenthene, although that substance is its main constituent; it contains also hydrocarbons isomeric with turpentine and substances containing oxygen, which either pre-exist in the turpentine, or, more usually, result from the method of preparing the oil. When recently distilled, it is a colorless, limpid, neutral liquid; sp. gr. 0.86; usually lævogyrous, sometimes dextrogyrous. When exposed to the air it rapidly becomes yellow and viscid. The action of reagents upon it is practically the same as upon terebenthene.

The number of isomerides existing in oil of turpentine is very great; some are optically active, others inactive; they also vary in their sp. gr., fusing- or boiling-points, and capacity for absorbing oxygen.

Isomeres of Terebenthene.—There exist a great number of bodies, the products of distillation of vegetable substances, which are known as *essences*, *essential oils*, *volatile oils* or *distilled oils*. They resemble each other in being odorous, oily, sparingly soluble in water, more or less soluble in alcohol and ether; colorless or yellowish, inflammable, and prone to become resinous on exposure to air. They are not simple chemical compounds, but mixtures, and in many of them the principal ingredient is a hydrocarbon, isomeric with terebenthene, and consequently having the composition $_nC_{10}H_{16}$. Some contain hydrocarbons, others aldehydes, acetones, phenols, and ethers.

Of the numerous other hydrocarbons closely related to terebenthene, but two require further consideration as being the principal constituents of *caoutchouc* and *gutta-percha*.

Caoutchouc—*India-rubber*—is a peculiar substance existing in suspension in the milky juice of quite a number of trees growing in warm climates. It is, when pure, a mixture of two hydrocarbons—*caoutchene*, $C_{10}H_{16}$, and *isoprene*, C_5H_8.

The commercial article is yellowish-brown; sp. gr. 0.919 to 0.942; soft, flexible; almost impermeable, but still capable of acting as a dialyzing membrane when used in sufficiently thin layers. It is insoluble in H_2O and alcohol, both of which, however, it absorbs by long immersion, the former to the extent of 25 per cent., and the latter of 20 per cent., of its own weight; it is soluble in ether, petroleum, fatty and essential oils; its best solvent is carbon disulphide, either alone, or, better, mixed with 5 parts of absolute alcohol.

It is not acted upon by dilute mineral acids, but is attacked by concentrated HNO_3 and H_2SO_4, and especially by a mixture of the two. Alkalies tend to render it tougher, although a solution of soda of 40° B. renders it soft after an immersion of a few hours. Cl attacks it after a time, depriving it of its elasticity, and rendering it hard and brittle. When heated it becomes viscous at 145° (293° F.), and fuses at 170°–180° (347°–356° F.) to a thick liquid, which, on cooling, remains sticky and only regains its primitive character after a long time. On contact with flame it ignites,

280

burning with a reddish, smoky flame, which is extinguished with difficulty.

The most valuable property of india-rubber, apart from its elasticity, is that which it possesses of entering into combination with S to form what is known as *vulcanized rubber*, which is produced by heating together the normal caoutchouc and S to 130°–150° (266°–302° F.). Ordinary vulcanized rubber differs materially from the natural gum in its properties; its elasticity and flexibility are much increased; it does not harden when exposed to cold; it only fuses at 200° (392° F.); finally, it resists the action of reagents, of solvents, and of the atmosphere much better than does the natural gum.

Frequently rubber tubing is too heavily charged with sulphur for certain chemical uses, in which case it may be desulphurized by boiling with dilute caustic soda solution.

Hard rubber, *vulcanite*, *or ebonite*, is a hard, tough variety of vulcanized rubber, susceptible of a good polish, and a non-conductor of electricity. It contains 20 to 35 per cent. of S (the ordinary vulcanized rubber contains 7 to 10 per cent.).

Gutta-percha—is the concrete juice of *isonandra gutta*. It is a tough, inelastic, brownish substance, having an odor similar to that of caoutchouc; at ordinary temperatures it is rather hard, but when warmed it becomes soft and may be moulded, or even cast, so as to assume any form, which it retains on cooling; it may be welded at slightly elevated temperatures, is a good insulating and waterproofing material, and is tough and pliable. It is insoluble in water, alkaline solutions, dilute acids, including hydrofluoric, and in fatty oils; it is soluble in benzene, oil of turpentine, essential oils, chloroform, and especially in carbon disulphide. A solution in chloroform is known as *traumaticine* or *Liq. gutta perchœ* (*U. S.*), and is used to obtain, by its evaporation, a thin film of gutta-percha over parts which it is desired to protect from the air. It is attacked by HNO_3 and H_2SO_4.

When exposed to air and light, it is gradually changed from the surface inward, assuming a sharp, acid odor, becoming hard and cracked, and even a conductor of electricity.

Gutta-percha is a more complex substance than caoutchouc, and seems to be made up of three substances: *Gutta*, $C_{20}H_{32}$, 75–82 per cent., a white, tough substance, fusing at 150° (302° F.), soluble in the ordinary solvents of gutta-percha, but insoluble in alcohol and ether. *Albane*, $C_{20}H_{32}O_2$, 14–19 per cent., a white, crystalline resin, heavier than water, fusible at 160° (320° F.); soluble in benzene, essence of turpentine, carbon disulphide, ether, chloroform, and hot absolute alcohol; not attacked by HCl. *Fluviale*, 4–6 per cent., $C_{20}H_{32}O$, a yellowish resin, slightly heavier than water, hard and brittle at 0° (32° F.), soft at 50° (122° F.), liquid at 100° (212° F.); soluble in the solvents of gutta-percha.

Camphors and Resins.—Most of the essential oils yield on distillation two products of different boiling-points; one of these is a hydrocarbon, in most instances of the terebenthene series, liquid at ordinary temperatures, and sometimes known as an *eleoptene*. The other, of higher boiling-point, and solid at ordinary temperatures, designated a *stearoptene*, is an oxidized product, and either exists as such in the vegetable exudation, or is produced under subsequent treatment. The *camphors* are probably aldehydes or alcohols corresponding to hydrocarbons related to terebenthene, although their constitution is still uncertain.

Common camphor—*Japan camphor—Laurel camphor—Campholic aldehyde—Camphora* (*U. S., Br.*)—$C_{10}H_{16}O$—152.—Three modifications are

281

known, which seem to differ from each other only in their action upon polarized light: (1.) *Dextro camphor = camphore officinarum;* obtained from *laurus camphora*—$[a]_D = +47°.4$. (2.) *Laevo camphor;* obtained from *matricaria postlanium*—$[a]_D = -47°.4$. (3.) *Inactive camphor*, obtained from the essential oils of rosemary, sage, lavender, and origanum.

The first is the ordinary camphor of the shops. It is a white, translucent, crystalline solid; sp. gr. 0.986–0.996, hot and bitter in taste; aromatic; sparingly soluble in H_2O; quite soluble in ether, acetic acid, methylic and ethylic alcohols and the oils; fuses at 175° (347° F.); boils at 204° (399°.2 F.); sublimes at all temperatures.

It ignites readily and burns with a luminous flame. Cold HNO_3 dissolves it, and from the solution H_2O precipitates it unchanged. Boiling HNO_3, or potassium permanganate, oxidizes it to *dextro camphoric acid*, $C_{10}H_{16}O_4$. Concentrated H_2SO_4 forms with it a black solution, from which H_2O precipitates an oily material called *camphene*. Distilled with P_2O_5, it yields *cymene*, $C_{10}H_{14}$. Alkaline solutions, by long heating under pressure, convert it into *camphic acid*, $C_{10}H_{16}O_2$, and *borneol*. Cl attacks it with difficulty. Br unites with it to form an unstable compound, which forms ruby-red crystals having the composition $C_{10}H_{14}OBr_2$. These crystals, when heated to 80°–90° (176°–194° F.), fuse and give off HBr, there remaining an amber-colored liquid, which solidifies on cooling and yields, by recrystallization from boiling alcohol, long, hard, rectangular crystals of *monobromo camphor—camphora monobromata* (*U. S.*)—$C_{10}H_{15}OBr$. When vapor of camphor is passed over a mixture of fused potash and lime, heated to 300°–400° (572°–752° F.), it unites directly with the potash to form the K salt of *campholic acid*, $C_{10}H_{18}O_2$.

Borneol—*Borneo camphor—Camphol—Camphyl alcohol*—$C_{10}H_{18}O$—154—is usually obtained from *dryobalanops camphora*, although it may be obtained from other plants, and even artificially by the hydrogenation of laurel camphor. The product from these different sources is the same chemically, so far as we can determine, but varies, like the modifications of camphor, in its action on polarized light.

It forms small, white, transparent, friable crystals; has an odor which recalls at the same time those of laurel camphor and of pepper; has a hot taste; is insoluble in water, readily soluble in alcohol, ether, and acetic acid; fuses at 198° (388°.4 F.), boils at 212° (413°.6 F.).

It is a true alcohol, and enters into double decomposition with acids to form ethers. When heated with P_2O_5, it yields a hydrocarbon, *borneene*, $C_{10}H_{16}$. Oxidized by HNO_3, it is converted into laurel camphor.

Menthol—*Menthyl alcohol*—$C_{10}H_{20}O$—156—exists in essential oil of peppermint. It crystallizes in colorless prisms; fusible at 36° (96°.8 F.); sparingly soluble in water; readily soluble in alcohol, ether, carbon disulphide, and in acids. Corresponding to it are a series of *menthyl* ethers.

Eucalyptol—$C_{12}H_{20}O$—180—is contained in the leaves of *eucalyptus globulus;* it is liquid at ordinary temperatures, and boils at 175° (347° F.); by distillation with phosphoric anhydride it yields *eucalyptene*, $C_{12}H_{18}$.

Terpine—*Terebenthene bihydrate*—$C_{10}H_{16},2H_2O + Aq$—172 + 18—is sometimes spontaneously deposited from oil of turpentine containing water; it may be obtained by frequently agitating for a month or more a mixture of oil of turpentine, alcohol, and ordinary nitric acid. It forms fine, large, rhombic prisms; sp. gr. 1.0994; sparingly soluble in cold water; soluble in hot water, alcohol, and ether; fusible at 103° (217°.4 F.).

Terpinol—$(C_{10}H_{16})_2H_2O$—290—is formed when terpine in solution in warm water is treated with a very small quantity of H_2SO_4 or HCl, and

282

distilled. It is a colorless liquid; has an odor of hyacinth; sp. gr. 0.852; boils at 168° (334°.4 F.), at which temperature it suffers partial decomposition. It appears to possess the function of an ether.

Resins—are generally the products of oxidation of the hydrocarbons allied to terebenthene; are amorphous (rarely crystalline); insoluble in water; soluble in alcohol, ether, and essences. Many of them contain acids.

They may be divided into several groups, according to the nature of their constituents: (1.) *Balsams*, which are usually soft or liquid, and are distinguished by containing free *cinnamic* or *benzoic acid* (*q. v.*). The principal members of this group are *benzoin, liquidambar, Peru balsam, styrax*, and *balsam tolu*. (2.) *Oleo-resins* consist of a true resin mixed with an oil, and usually with an oxidized product other than cinnamic or benzoic acid. The principal members of this group are *Burgundy* and *Canada pitch, Mecca balsam*, and the resins of *capsicum, copaiva, cubebs, elemi, labdanum*, and *lupulin*. (3.) *Gum-resins* are mixtures of true resins and gums. Many of them are possessed of medicinal qualities; *aloes, ammoniac, asafœtida, bdellium, euphorbium, galbanum, gamboge, guaiac, myrrh, olibanum, opoponax*, and *scammony*. (4.) *True resins* are hard substances obtainable from the members of the three previous classes, and containing neither essences, gums, nor aromatic acids. Such are *colophony* or *rosin, copal, dammar, dragon's blood, jalap, lac, mastic*, and *sandarac*. (5.) *Fossil resins*, such as *amber, asphalt*, and *ozocerite*.

CARBOHYDRATES.

These substances are composed of C, H and O; they all contain C_6, or some multiple thereof; and the H and O which they contain is always in the proportion of H_2 to O. Their constitution is still unknown; probably some are aldehydes, others alcohols and others ethers. Most of them are constituents of animal or vegetable organisms, and have not been obtained by complete synthesis.

They are divisible into three groups, the members of each of which are isomeric with each other:

I. GLUCOSES.	II. SACCHAROSES.	III. AMYLOSES.
$_n(C_6H_{12}O_6)$	$_n(C_{12}H_{22}O_{11})$	$_n(C_6H_{10}O_5)$
+Glucose. (Dextrose.)	+Saccharose.	+Starch.
−Lævulose.	+Lactose.	+Glycogen.
Mannitose.	+Maltose.	+Dextrin.
+Galactose.	+Melitose.	−Inulin.
Inosite.	+Melezitose.	Tunicin.
−Sorbin.	+Trehalose.	Cellulose.
−Eucalin.	+Mycose.	Gums.
	Synanthrose.	
	+Parasaccharose.	

Glucoses, $C_6H_{12}O_6$—180.

Glucose—*Grape-sugar—Dextrose—Liver-sugar—Diabetic sugar.*—The substance from which this group takes its name exists in all sweet and acidulous fruits; in many vegetable juices; in honey; in the animal economy in the contents of the intestines, in the liver, bile, thymus, heart, lungs,

283

blood, and in small quantity in the urine. Pathologically it is found in the saliva, perspiration, fæces, and largely increased in the blood and urine in diabetes mellitus (see below). It may also be obtained by decomposition of certain vegetable substances called *glucosides* (*q. v.*).

It is prepared artificially by heating starch or cellulose for 24 to 36 hours with a dilute mineral acid (H_2SO_4). Glucose obtained by this method is liable to contamination with traces of arsenic, which it receives from the H_2SO_4. Starch is also converted into glucose by the influence of *diastase*, formed during the germination of grain.

Glucose crystallizes with difficulty from its aqueous solution, in white, opaque, spheroidal masses containing 1 aq.; from alcohol in fine, transparent, anhydrous prisms; at about 60° (140° F.) in dry air the hydrated variety loses H_2O. It is soluble in all proportions in hot H_2O; very soluble in cold H_2O; soluble in alcohol. It is less sweet and less soluble than cane sugar. Its solutions are dextrogyrous: $[a]_D = +52°.85$.

At 170° (338° F.) it loses H_2O and is converted into *glucosan*, $C_6H_{10}O_5$. Hot dilute mineral acids convert it into a brown substance, *ulmic acid*, and, in the presence of air, formic acid. It dissolves in concentrated H_2SO_4, without coloration, forming *sulphoglucic acid*. Cold concentrated HNO_3 converts it into *nitro-glucose;* hot dilute HNO_3 oxidizes it to a mixture of oxalic and oxysaccharic acids. With organic acids it forms ethers. Its solutions dissolve potash, soda, lime, baryta, and the oxides of Pb and Cu, with which it forms compounds. When its solutions are heated with an alkali they assume a yellow or brown color, and give off a molasses-like odor, from the formation of *glucic* and *melassic* acids. Glucose in alkaline solution exerts a strong reducing action, which is favored by heat; Ag, Bi, and Hg are precipitated from their salts; and cupric are reduced to cuprous compounds with separation of cuprous oxide. In the presence of yeast, at suitable temperatures, glucose undergoes alcoholic fermentation.

Physiological.—The greater part of the glucose in the economy in health is introduced with the food, either in its own form or as other carbohydrates, which by digestion are converted into glucose; a certain quantity is also produced in the liver at the expense of glycogen, a formation which continues for some time after death. In some forms of diabetes the production of glucose in the liver is undoubtedly greatly increased. The quantity of sugar normally existing in the blood varies from 0.81 to 1.231 part per thousand; in diabetes it rises as high as 5.8 parts per thousand.

Under normal conditions, and with food not too rich in starch and saccharine materials, the quantity of sugar eliminated as such is exceedingly small—so small indeed that some observers have contested the fact of any being eliminated in health. It is oxidized in the body, and the ultimate products of such oxidation eliminated as CO_2 and H_2O. Whether or no intermediate products are formed, is still uncertain; the probability, however, is that there are. The oxidation of sugar is impeded in diabetes. Where this oxidation, or any of its steps, occurs, is at present a matter of conjecture merely; if, as is usually believed, glucose disappears to a marked extent in the passage of the blood through the lungs, the fact is a strong support of the view that its transformation into CO_2 and H_2O does not occur as a simple oxidation, as the notion that sugar or any other substance is "burned" in the lung, beyond the small amount required by the nutrition of the organ itself, is scarcely tenable at the present day.

So long as the quantity of glucose in the blood remains at or below

284

the normal percentage, it is not eliminated in the urine in quantities appreciable by the tests usually employed; when, however, the amount of glucose in the blood surpasses this limit from any cause, the urine becomes saccharine, and that to an extent proportional to the increase of glucose in the circulating fluids. The causes which may bring about such an increase are numerous and varied; many of them are entirely consistent with health, and the mere presence of increased quantities of sugar in the urine is no proof, taken by itself, of the existence of diabetes.

Sugar is detectable by the ordinary tests in the urine under the following circumstances:

Physiologically.—(1.) In the urine of pregnant women and during lactation. It appears in the latter stages of gestation and does not disappear entirely until the suppression of the lacteal secretion.

(2.) In small quantities in sucking children from eight days to two and one-half months.

(3.) In the urine of old persons (seventy to eighty years).

(4.) In those whose food contains a large amount of starchy or saccharine material. To this cause is due the apparent prevalence of diabetes in certain localities, as in districts where the different varieties of sugar are produced.

Pathologically.—(1.) In abnormally stout persons, especially in old persons and in women at the period of the menopause. The quantity does not exceed 8 to 12 grams per 1,000 c.c. (3.5–5.5 grains per ounce), and disappears when starchy and saccharine food is withheld. This form of glycosuria is liable to develop into true diabetes when it appears in young persons.

(2.) In diseases attended with interference of the respiratory processes —lung diseases, etc.

(3.) In diseases where there is interference with the hepatic circulation —hepatic congestion, compression of the portal vein by biliary calculi, cirrhosis, atrophy, fatty degeneration, etc.

(4.) In many cerebral and cerebro-spinal disturbances—general paresis, dementia, epilepsy; by puncture of the fourth ventricle.

(5.) In intermittent and typhus fevers.

(6.) By the action of many poisons—carbon monoxide, arsenic, chloroform, curari; by injection into an artery of ether, ammonia, phosphoric acid, sodium chloride, amyl nitrite, glycogen.

(7.) In true diabetes the elimination of sugar in the urine is constant, unless arrested by suitable regulation of diet, and not temporary, as in the conditions previously mentioned. The quantity of urine is increased, sometimes enormously, and it is of high sp. gr. The elimination of urea is increased absolutely, although the quantity in 1,000 c.c. may be less than that normally existing in that bulk of urine. The quantity of sugar in diabetic urine is sometimes enormous; an elimination of 200 grams (6.4 ounces) in twenty-four hours is by no means uncommon; instances in which the amount has reached 400 to 600 grams (12.9–19.3 ounces) are recorded, and one case in which no less than 1,376 grams (45 ounces) were discharged in one day. The elimination is not the same at all hours of the day; during the night less sugar is voided than during the day; the hourly elimination increases after meals, reaching its maximum in 4 hours, after which it diminishes to reach the minimum in 6 to 7 hours, when it may disappear entirely; this variation is more pronounced the more copious the meal. It is obvious from the above, that, in order that quantitative determinations of sugar in urine shall be of clinical value, it is necessary

285

that the determination be made in a sample taken from the mixed urine of twenty-four hours.

The relation existing between the quantity of sugar in the blood and its elimination by the urine in diabetes is well shown by the following results of Pavy, which also show the beneficial effects of restricting the diet:

	Urine.				Blood.
	Quantity in 24 hours.	Specific gravity.	Sugar excreted in 24 hours.	Sugar in 1,000 parts.	Sugar in 1,000 parts.
Case I. Mixed diet............	6608 c.c.	1040	751.6 grams.	109.91	5.763
Case II. Mixed diet............	6474 c.c.	1041	633.0 grams.	94.08	5.545
Case II. Restricted diet........	3407 c.c.	1031	245.2 grams.	61.34	2.625
Case III. Mixed diet...........	5878 c.c.	1036	567.7 grams.	93.39	4.970
Case III. Restricted diet.......	2470 c.c.	1033	115.8 grams.	45.49	2.789
Case IV. Partly restricted diet..	1704 c.c	1036	21.81 grams.	48.11	1.848
Case IV. Partly restricted diet, 3½ months later....	852 c c	1034	14.40 grams.	31.75	1.543

Analytical Characters.—A saccharine urine is usually abundant in quantity, pale in color, of high sp. gr., covered with a persistent froth on being shaken, and exhales a peculiar odor; when evaporated it leaves a sticky residue. The presence of glucose in urine is indicated by the following tests:

If the urine be albuminous, it is indispensable that the albumen be separated before any of the tests for sugar are applied; this is done by adding one or two drops of acetic acid, or, if the urine be alkaline, just enough acetic acid to turn the reaction to acid, *and no more,* heating over the water-bath until the albumen has separated in flocks, and filtering.

(1.) When examined by the polarimeter (see p. 36) it deviates the plane of polarization to the right.

(2.) When mixed with an equal volume of liquor potassæ and heated, it turns yellow, and, if sugar be abundant, brown. A molasses-like odor is observable on adding HNO_3 (Moore's test).

(3.) The urine, rendered faintly blue with indigo solution and faintly alkaline with sodium carbonate, and heated to boiling without agitation, turns violet and then yellow if sugar be present; on agitation the blue color is restored (Mulder-Neubauer test).

(4.) About 1 c.c. of the urine, diluted with twice its bulk of water, is treated with two or three drops of cupric sulphate solution and about 1 c c. of caustic potassa solution; if sugar be present the bluish precipitate is dissolved on agitation, forming a blue solution; the clear blue fluid, when heated to near boiling, deposits a yellow, orange, or red precipitate of cuprous oxide if sugar be present (Trommer's test). In the application of this test an excess of cupric sulphate is to be avoided, lest the color be masked by the formation of the black cupric oxide. Sometimes no precipitate is formed, but the liquid changes in color from blue to yellow; this occurs in the presence of small quantities of cupric salt and large quantities of sugar, the cuprous oxide being held in solution by the excess of glucose; in this case the test is to be repeated, using a sample of urine more diluted with water. In some instances, also, the reaction is interfered with by excess of normal constituents of the urine, uric acid, creatinine, coloring matter, etc., and instead of a bright precipitate, a muddy deposit is formed; when this occurs the urine is heated with ani-

286

mal charcoal and filtered; the filtrate evaporated to dryness; the residue extracted with alcohol; the alcoholic extract evaporated; the residue redissolved in water, and tested as described above.

(5.) Four or five c.c. of Fehling's solution (see p. 287) are heated in a test-tube to boiling; it should remain unaltered. The urine is then added guttatim; if it contain sugar, the mixture turns green, and a yellow or red precipitate of cuprous oxide is formed, usually darker in color than that obtained by Trommer's test. The absence of glucose is not to be inferred until a bulk of urine equal to that of the Fehling's solution used has been added, and the mixture boiled from time to time without the formation of a precipitate. This test is the most convenient and the most reliable for clinical purposes.

(6.) A few c.c. of the urine are mixed in a test-tube with an equal volume of solution of sodium carbonate (1 pt. crystal. carbonate and 3 pts. water), a few granules of bismuth subnitrate are added, and the mixture boiled for some time (until it begins to "bump," if necessary). If sugar be present, the bismuth powder turns brown or black by reduction to elementary bismuth (Boettger's test). No other normal constituent of the urine reacts with this test; a fallacy is, however, possible from the presence of some compound, which, by giving up sulphur, may cause the formation of the black bismuth sulphide; to guard against this, when an affirmative result has been obtained, another sample of urine is rendered alkaline and boiled with pulverized litharge; the powder should not turn black.

(7.) A solution of sugar, mixed with good yeast and kept at 25° (77° F.) is decomposed into CO_2 and alcohol. To apply the fermentation-test to urine, take three test-tubes, A, B, and C, place in each some washed (or compressed) yeast, fill A completely with the urine to be tested, and place it in an inverted position, the mouth below the surface of some of the same urine in another vessel (the entrance of air being prevented, during the inversion, by closing the opening of the tube with the finger, or a cork on the end of a wire, until it has been brought below the surface of the urine). Fill B completely with some urine to which glucose has been added, and C with distilled water, and invert them in the same way as A; B in saccharine urine, and C in distilled water. Leave all three tubes in a place where the temperature is about 25° (77° F.) for twelve hours, and then examine them. If gas have collected in B over the surface of the liquid, and none in A, the urine is free from sugar; if gas have collected in both A and B, and not in C, the urine contains sugar; if no gas have collected in B, the yeast is worthless, and if any gas be found in C, the yeast itself has given off CO_2. In the last two cases the process must be repeated with a new sample of yeast.

QUANTITATIVE DETERMINATION OF GLUCOSE.—(1.) *By the polarimeter.*—The filtered urine is observed by the polariscope (see p. 38) and the mean of half a dozen readings taken as the angle of deviation; from this the percentage of sugar is determined by the formula $p = \frac{a}{52.85 \times l}$ in which p = the weight, in grams, of glucose in 1 c.c. of urine; a = the angle of deviation; l = the length of the tube in decimeters. The same formula may be used for other substances by substituting for 52.85 the value of $[a]_D$ for that substance. If the urine contain albumen, it must be removed *before determining the value of a.*

(2.) *By specific gravity: Robert's method.*—The sp. gr. of the urine is carefully determined at 25° (77° F.); yeast is then added, and the mixture

287

kept at 25° (77° F.) until fermentation is complete; the sp. gr. is again observed, and will be found to be lower than before. Each degree of diminution represents 0.2196 gram of sugar in 100 c.c. (1 grain per ounce) of urine.

(3.) *By Fehling's solution.*—Of the many formulæ for Fehling's solutions, the one to which we give the preference is that of Dr. Piffard. Two solutions are required:

I. Cupric sulphate (pure, crystals)	51.98 grams.
Water	500.0 c.c.
II. Rochelle salt (pure, crystals)	259.9 grams.
Sodic hydrate solution, sp. gr. 1.12	1000.0 c.c.

When required for use, one volume of No. I. is mixed with two volumes of No. II. The copper contained in 20 c.c. of this mixture is precipitated as cuprous oxide by 0.1 gram glucose.

To use the solution, 20 c.c. of the mixed solutions are placed in a flask of 250-300 c.c. capacity, 40 c.c. of distilled water are added, the whole thoroughly mixed and heated to boiling. On the other hand, the urine to be tested is diluted with four times its volume of water if poor in sugar, and with nine times its volume if highly saccharine (the degree of dilution required is, with a little practice, determined by the appearance of the deposit obtained in the qualitative testing); the water and urine are thoroughly mixed and a burette filled with the mixture. A few drops of aqua ammoniæ are added to the Fehling's solution and the diluted urine added, in small portions toward the end, until the blue color is entirely discharged—the contents of the flask being made to boil briskly between each addition from the burette. When the liquid in the flask shows no blue color, when looked through with a white background, the reading of the burette is taken; this reading, divided by five if the urine was diluted with four volumes of water, or by ten if with nine volumes, gives the number of c.c. of urine containing 0.1 gram of glucose; and consequently the elimination of glucose in twenty-four hours, in decigrams, is obtained by dividing the number of c.c. of urine in twenty-four hours by the result obtained above.

Example.—20 c.c. Fehling's solution used, and urine diluted with four volumes of water.

Reading of burette: 36.5 c.c. $\frac{36.5}{5} = 7.3$ c.c. urine contain 0.1 gram glucose. Patient is passing 2,436 c.c. urine in twenty-four hours. $\frac{2,436}{7.3} = 333.6$ decigr. = 33.36 grams glucose in twenty-four hours.

The accuracy of the determination may be controlled by filtering off some of the fluid from the flask at the end of the reaction; a portion of the filtrate is acidulated with acetic acid and treated with potassium ferrocyanide solution; if it turn reddish brown the reduction has not been complete, and the result is affected with a plus error. To another portion of the filtrate a few drops of cupric sulphate solution are added and the mixture boiled; if any precipitation of cuprous oxide be observed, an excess of urine has been added, and the result obtained is less than the true one.

This method, when carefully conducted with accurately prepared and

288

undeteriorated solutions, is the best adapted to clinical uses. The copper solution should be kept in the dark, in a well-closed bottle, and the stopper and neck of the No. II. bottle should be well coated with paraffin.

(4.) *Gravimetric method.*—When more accurate results than are obtainable by Fehling's volumetric process are desired, recourse must be had to a determination of the weight of cuprous oxide obtained by reduction. A small quantity of freshly prepared Fehling's solution is heated to boiling in a small flask; to it is gradually added, with the precautions observed in the volumetric method, a known volume of urine, such that at the end of the reduction there shall remain an excess of unreduced copper salt. The flask is now completely filled with boiling H_2O, corked, and allowed to cool. The alkaline fluid is separated as rapidly as possible from the precipitated oxide, by decantation and filtration through a small double filter, and the precipitate and flask repeatedly washed with hot H_2O until the washings are no longer alkaline; a small portion of the precipitate remains adhering to the walls of the flask. The filter and its contents are dried and burned in a weighed porcelain crucible; when this has cooled, the flask is rinsed out with a small quantity of HNO_3; this is added to the contents of the crucible, evaporated over the water-bath, the crucible slowly heated to redness, cooled, and weighed; the difference between this last weight and that of the crucible + that of the filter-ash, is the weight of cupric oxide, of which 220 parts = 100 parts of glucose.

Lævulose—*Uncrystallizable sugar*—forms the uncrystallizable portion of the sugar of fruits and of honey, in which it is associated with glucose; it is also produced artificially by the prolonged action of boiling water upon *inulin;* and as one of the constituents of inverted sugar.

Lævulose is not capable of crystallization, but may be obtained as a thick syrup; very soluble in water, insoluble in absolute alcohol; it is sweeter but less readily fermentable than glucose, which it equals in the readiness with which it reduces cupro-potassic solutions. Its prominent physical property, and that to which it owes its name, is its strong left-handed polarization, $[a]_D = -106°$ at 15° (59° F.). At 170° (338° F.) it is converted into the solid, amorphous *lævulosan*, $C_6H_{10}O_5$.

Mannitose—is obtained by the oxidation of mannite. It is a yellow, uncrystallizable sugar, having many of the characters of glucose, but optically inactive.

Galactose—sometimes improperly called lactose—is formed by the action of dilute acids upon lactose (milk sugar) as glucose is formed from saccharose. It differs from glucose in crystallizing more readily, in being very sparingly soluble in cold alcohol, in its action upon polarized light, $[a]_D = +83°.33$, and in being oxidized to mucic acid by HNO_3.

Inosite—*Muscle-sugar*—exists in the liquid of muscular tissue, in the lungs, kidneys, liver, spleen, brain, and blood; pathologically in the urine in Bright's, diabetes, and after the use of drastics in uræmia, and in the contents of hydatid cysts; also in the seeds and leaves of certain plants. What the source and function of inosite in the animal economy may be is still a matter of conjecture.

It forms long, colorless, monoclinic crystals, containing 2 Aq., usually arranged in groups having a cauliflower-like appearance. It effloresces in dry air; has a distinctly-sweet taste; is easily soluble in water, difficultly in alcohol; insoluble in absolute alcohol and in ether; it is without action upon polarized light.

The position of inosite in this series is based entirely upon its chemical composition, as it does not possess the other characteristics of the

289

group. It does not enter directly into alcoholic fermentation, although upon contact with putrefying animal matters it produces lactic and butyric acids; when boiled with barium or potassium hydrate, it is not even colored; in the presence of inosite, potash precipitates with cupric sulphate solution, the precipitate being redissolved in an excess of potash; but no reduction takes place upon boiling the blue solution.

The presence of inosite is indicated by the following reactions: *Scherer's.*—Treated with HNO_3, the solution evaporated to near dryness, and the residue moistened with ammonium hydrate and calcium chloride, and again evaporated; a rose-pink color is produced. Succeeds only with nearly pure inosite. *Gallois'.*—Mercuric nitrate produces, in solutions of inosite, a yellow precipitate, which, on cautious heating, turns red; the color disappears on cooling, and reappears on heating.

Saccharoses, $C_{12}H_{22}O_{11}$—342.

Saccharose—*Cane-sugar—Beet-sugar—Saccharum* (*U. S.*)—The most

important member of the group, exists in many roots, fruits, and grasses, and is produced from the sugar-cane, *saccharum officinarum*, sorghum, *sorghum saccharatum*, beet, *beta vulgaris*, and sugar-maple, *acer saccharinum*.

For the extraction of sugar the expressed juice is heated in large pans to about 100° (212° F.); milk of lime is added, which causes the precipitation of albumen, wax, calcic phosphate, etc.; the clear liquid is drawn off, and "delimed" by passing a current of CO_2 through it; the clear liquid is again drawn off and evaporated, during agitation, to the crystallizing-point; the product is drained, leaving what is termed *raw* or *muscovado sugar*, while the liquor which drains off is *molasses*. The sugar so obtained is purified by the process of "refining," which consists essentially in adding to the raw sugar, in solution, albumen in some form, which is then coagulated, filtering first through canvas, afterward through animal charcoal; the clear liquid is evaporated in "vacuum-pans," at a temperature not exceeding 72° (161°.6 F.), to the crystallizing-point. The product is allowed to crystallize in earthen moulds; a saturated solution of pure sugar is poured upon the crystalline mass in order to displace the uncrystallizable sugar which still remains; and the loaf is finally dried in an oven. The liquid displaced as above is what is known as *sugar-house syrup*.

Pure sugar should be entirely soluble in water; the solution should not turn brown when warmed with dilute potassium hydrate solution; should not reduce Fehling's solution, and should give no precipitate with ammonium oxalate.

***Beet-sugar* is the same as cane-sugar, except that, as usually met with in commerce, it is lighter, bulk for bulk. *Sugar-candy*, or *rock-candy*, is cane-sugar allowed to crystallize slowly from a concentrated solution without agitation. *Maple-sugar* is a partially refined, but not decolorized variety of cane-sugar.**

Saccharose crystallizes in small, white, monoclinic prisms; or, as sugar-candy, in large, yellowish, transparent crystals; sp. gr. 1.606. It is very soluble in water, dissolving in about one-third its weight of cold water, and more abundantly in hot water. It is insoluble in absolute alcohol or ether, and its solubility in water is progressively diminished by the addi-

290

tion of alcohol. Aqueous solutions of cane-sugar are dextrogyrous, $[a]_D = +73°.8$.

When saccharose is heated to 160° (320° F.) it fuses, and the liquid, on cooling, solidifies to a yellow, transparent, amorphous mass, known as *barley-sugar;* at a slightly higher temperature, it is decomposed into glucose and lævulosan; at a still higher temperature, H_2O is given off, and the glucose already formed is converted into glucosan; at 210° (410° F.) the evolution of H_2O is more abundant, and there remains a brown material known as *caramel*, or *burnt sugar;* a tasteless substance, insoluble in strong alcohol, but soluble in H_2O or aqueous alcohol, and used to communicate color to spirits; finally, at higher temperatures, methyl hydride and the two oxides of carbon are given off; a brown oil, acetone, acetic acid, and aldehyde distil over; and a carbonaceous residue remains.

If saccharose be boiled for some time with H_2O, it is converted into *inverted sugar*, which is a mixture of glucose and lævulose: $C_{12}H_{22}O_{11} + H_2O = C_6H_{12}O_6 + C_6H_{12}O_6$. With a solution of saccharose the polarization is dextrogyrous, but, after invertion, it becomes lævogyrous, because the left-handed action of the molecule of lævulose produced, $[a]_D = -106,°$ is only partly neutralized by the right-handed action of the glucose, $[a]_D = +52°.85$. This inversion of cane-sugar is utilized in the testing of samples of sugar. On the other hand, it is to avoid its occurrence, and the consequent loss of sugar, that the vacuum-pan is used in refining—its object being to remove the H_2O at a low temperature.

Those acids which are not oxidizing agents act upon saccharose in three ways, according to circumstances: (1) if tartaric and other organic acids be heated for some time with saccharose to 100°—120° (212°-248° F.), compounds known as *saccharides*, and having the constitution of ethers, are formed; (2) heated with mineral acids, even dilute, and less rapidly with some organic acids, saccharose is quickly converted into inverted sugar; (3) concentrated acids decompose cane-sugar entirely, more rapidly when heated than in the cold; with HCl, formic acid and a brown, flocculent material (ulmic acid?) are formed; with H_2SO_4, SO_2 and H_2O are formed, and a voluminous mass of charcoal remains. Oxalic acid, aided by heat, produces CO_2, formic acid, and a brown substance (humine?).

Oxidizing agents act energetically upon cane-sugar, which is a good reducing agent. With potassium chlorate, sugar forms a mixture which detonates when subjected to shock, and which deflagrates when moistened with H_2SO_4. Dilute HNO_3, when heated with saccharose, oxidizes it to saccharic and oxalic acids. Concentrated HNO_3, alone or mixed with H_2SO_4, converts it into the explosive *nitro-saccharose*. Potassium permanganate, in acid solution, oxidizes it completely to CO_2 and H_2O.

Cane-sugar reduces the compounds of Ag, Hg and Au, when heated with their solutions; it does not reduce the cupro-potassic solutions in the cold, but effects their reduction when heated with them to an extent proportional to the amount of excess of alkali present.

When moderately heated with liquor potassæ, cane-sugar does not turn brown, as does glucose; but by long ebullition it is decomposed by the alkalies much less readily than glucose, with formation of acids of the fatty series and oxalic acid.

With the bases, saccharose forms definite compounds called *sucrates* (improperly saccharates, a name belonging to the salts of saccharic acid). With Ca it forms five compounds. Hydrate of calcium dissolves readily in solutions of sugar, with formation of a Ca compound, soluble in H_2O,

291

containing an excess of sugar. A solution containing 100 parts of sugar in 600 parts of H_2O dissolves 32 parts of calcic oxide. These solutions have an alkaline taste; are decomposed, with formation of a gelatinous precipitate, when heated, and, with deposition of calcium carbonate and regeneration of saccharose, when treated with CO_2. Quantities of calcium sucrates are frequently introduced into sugars to increase their weight—an adulteration the less readily detected, as the sucrate dissolves with the sugar. Calcium sucrates exist in the *liq. calcis saccharatus* (*Br.*).

Yeast causes fermentation of solutions of cane-sugar, but only after its conversion into glucose. Fermentation is also caused by exposing a solution of sugar containing ammonium phosphate to the air.

During the process of digestion, probably in the small intestine, cane-sugar is converted into glucose.

Lactose—*Milk-sugar*—*Lactine*—*Saccharum lactis* (*U. S.*, *Br.*)—has hitherto been found only in the milk of the mammalia. It may be obtained from skim-milk by coagulating the casein with a small quantity of H_2SO_4, filtering, evaporating, redissolving, decolorizing with animal charcoal, and recrystallizing.

It forms prismatic crystals; sp. gr. 1.53; hard, transparent, faintly sweet, soluble in 6 parts of cold and in 2.5 parts of boiling H_2O; soluble in acetic acid; insoluble in alcohol and in ether; its solutions are dextrogyrous, $[a]_D = +59°.3$. The crystals, dried at 100° (212° F.), contain 1 Aq., which they lose at 150° (302° F.).

Lactose is not altered by contact with air. Heated with dilute mineral or with strong organic acids, it is converted into galactose. HNO_3 oxidizes it to mucic and oxalic acids. A mixture of HNO_3 and H_2SO_4 converts it into an explosive nitro-compound. With organic acids it forms ethers. With soda, potash, and lime it forms compounds similar to those of saccharose, from which lactose may be recovered by neutralization, unless they have been heated to 100° (212° F.), at which temperature they are decomposed. It reduces Fehling's solution, and reacts with Trommer's test.

In the presence of yeast, lactose is capable of alcoholic fermentation, which takes place slowly, and, as it appears, without previous transformation of the lactose into either glucose or galactose. On contact with putrefying albuminoids it enters into lactic fermentation.

The average proportion of lactose in different milks is as follows: Cow, 5.5 per cent.; mare, 5.5; ass, 5.8; human, 5.3; sheep, 4.2; goat, 4.0. When taken internally, it is converted into galactose by the pancreatic secretion; when injected into the blood, it does not appear in the urine, which, however, contains glucose.

Maltose—A sugar closely resembling glucose in many of its properties, is formed along with dextrine during the conversion of starch into sugar by the action of diastase and of the cryptolytes of the saliva and pancreatic juice. It crystallizes as does glucose, but differs from that sugar in being less soluble in alcohol and in exerting a dextrogyratory power three times as great.

Amyloses, $_n(C_6H_{10}O_5)$—$n162$.

Starch—*Amylum* (*U. S.*)—the most important member of the group, exists in the roots, stems, and seeds of all plants. It is prepared from rice, wheat, potatoes, maniot, beans, sago, arrow-root, etc. The comminuted vegetable tissue is steeped for a considerable time in H_2O ren-

292

dered faintly alkaline with soda; the softened mass is then rubbed on a sieve under a current of water, which washes out the starch granules; the washings are allowed to deposit the starch, which, after washing by decantation, is dried at a low temperature.

Starch is a white powder, having a peculiar slippery feel, or it appears in short columnar masses. The granules of starch differ in size and appearance according to the kind of plant from which they have been obtained. They are rounded or egg-shaped masses, having at the centre or toward one end a spot, called the *hilum*, around which are a series of concentric lines more or less well marked. Differences in size, shape, and markings of starch granules are shown in Fig. 39.

Starch is not altered by exposure to air, except that it absorbs moisture. Commercial starch contains 18 per cent. of H_2O, of which it loses 8 per cent. in vacuo, and the remaining 10 per cent. at 145° (293° F.). It is insoluble in alcohol, ether and cold water. If 15 to 20 parts of H_2O be gradually heated with 1 part of starch, the granules swell at about 55° (131° F.), and at 80° (176° F.) they have reached 30 times their original dimensions; their structure is no longer distinguishable, and they form a translucent, gelatinous mass, commonly known as starch

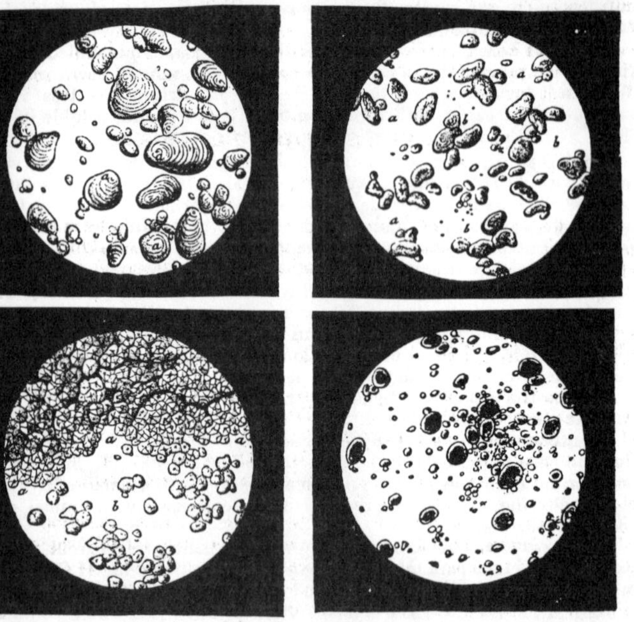

Fig. 39.

293

paste. In this state the starch is said to be hydrated, and, if boiled with much H_2O, and the liquid filtered, a solution of starch passes through, which is opalescent from the suspension in it of undissolved particles. Cold dilute solutions of the alkalies produce the same effects on starch as does hot water. Hydrated starch is dextrogyrous, $[a]_D = + 216°$. Dry heat causes the granules of starch to swell and burst; at 200° (392° F.) it is converted into *dextrin;* at 230° (446° F.) it forms a brownish-yellow, fused mass, composed principally of *pyrodextrin.* Hydrated starch is converted into dextrin by heating with H_2O at 160° (320° F.), and, if the action be prolonged, the new product is changed to glucose.

The amount of starch contained in food vegetables varies from about 5 per cent. in turnips to 89 per cent. in rice, as will be observed in the following table:

COMPOSITION OF VEGETABLE FOODS.

	Nitrogen-ized matter.	Starch.	Dextrin, etc.	Cellu-lose.	Fat.	Mineral matter.	Carbo-hydrate.	Water.	Vegetable fibre, etc.	Authority.
Wheat, hard......	22.75	58.62	9.50	3.50	2.61	3.02	....	....	..	Payen.
Wheat, hard.......	19.50	65.07	7.60	3.0	2.12	2.71	....	....	..	Payen.
Wheat, hard.......	20.0	63.80	8.0	3.10	2.25	2.85	....	....	..	Payen.
Wheat, semi-hard..	15.25	70.05	7.0	3.0	1.95	2.75	....	....	..	Payen.
Wheat, soft	12.65	76.51	6.05	2.80	1.87	2.12	....	....	..	Payen.
Rye	12.50	64.65	14.90	3.10	2.25	2.60	....	....	..	Payen.
Barley.............	12.96	66.43	10.0	4.75	2.76	3.10	....	....	..	Payen.
Oats...............	14.39	60.59	9.25	7.06	5.50	3.25	....	....	..	Payen.
Maize..............	12.50	67.55	4.0	5.90	8.80	1.25	....	....	..	Payen.
Rice...............	7.55	88.65	1.0	1.10	0.80	0.90	....	....	..	Payen.
Flour..............	14.45	....	....	....	1.25	1.60	68.48	14.22	..	Payen.
Flour..............	10.80	....	....	....	2.0	1.70	70.50	15.0	..	Letheby.
Bread..............	8.10	....	....	....	1.60	2.30	51.00	37.0	..	Letheby.
Oatmeal............	12.60	....	....	....	5.60	3.0	63.80	15.0	..	Letheby.
Buckwheat..........	13.10	64.90	....	3.50	3.0	2.50	....	13.0	..	Payen.
Quinoa seeds.......	22.86	56.80	....	....	5.74	5.05	....	....	9.53	Voelcker.
Quinoa flour.......	19.0	60.0	....	....	5.0	..	....	16.0	..	Voelcker.
Horse-bean.........	30.80	48.30	....	3.0	1.90	3.50	....	12.50	..	Payen.
Broad bean.........	29.65	55.85	....	1.05	2.0	3.65	....	8.40	..	Payen.
White bean.........	25.50	55.70	....	2.09	2.80	3.20	....	9.90	..	Payen.
Peas, dried........	23.80	58.70	....	3.50	2.10	2.10	....	8.30	..	Payen.
Lentils............	25.20	56.0	....	2.40	2.60	2.30	....	11.50	..	Payen.
Potato.............	2.50	20.0	1.09	1.04	0.11	1.26	....	74.0	..	Payen.
Potato	2.10	18.80	3.20	...	0.20	0.70	....	75.0	..	Letheby.
Sweet potato.......	1.50	16.05	10.20	0.45	0.30	2.60	....	67.50	1.10	Payen.
Carrots............	1.30	8.40	6.10	....	0.20	1.0	....	83.0	..	Letheby.
Parsnip............	1.10	9.60	5.80	....	0.50	1.0	....	82.0	..	Letheby.
Turnip.............	1.20	5.10	2.10	....	..	0.60	....	91.0	..	Letheby.

If starch be ground up with dilute H_2SO_4, after about half an hour the mixture gives only a violet color with I (see below); if now the acid be neutralized with chalk and the filtered liquid evaporated, it yields a white, granular product, which differs from starch in being soluble in H_2O, especially at 50° (122° F.), and in having a lower rotary power, $[a]_D = + 211°$. If the action be prolonged, the value of $[a]_D$ continues to sink until it reaches $+73.7°$, when the product consists of a mixture of dextrin and glucose. Concentrated HNO_3 dissolves starch in the cold, forming a nitro-product called *xylodin* or *pyroxam,* which is insoluble in H_2O, soluble in a mixture of alcohol and ether; explosive. HCl and oxalic acid convert starch into glucose. When starch is heated under pressure to 120° (248° F.) with stearic or acetic acid, compounds are formed which seem to be ethers, and to indicate that starch is the hydrate of a trivalent, oxygenated radical, $(C_6H_7O_2)'''$. Potash and soda in dilute solution convert starch into the soluble modification mentioned above.

294

A dilute solution of I produces a more or less intense blue-violet color with starch, either dry, hydrated, or in solution, the color disappearing on the application of heat, and returning on cooling. If to a solution of starch, blued by I, a solution of a neutral salt be added, there separates a blue, flocculent deposit of the so-called *iodide of starch.* Iodine renders starch soluble in water, and a soluble iodized starch, *Amylum iodatum* (*U. S.*), is obtained by triturating together 19 pts. starch, 2 pts. water, and 1 pt. iodine, and drying below 40° (104° F.).

Starch has not been found in the animal economy outside of the alimentary canal, in which, as a prerequisite to its absorption, it must be converted into dextrin and glucose. This change is partially effected by the action of the saliva; more rapidly with hydrated than with dry starch, and more rapidly with the saliva of some animals than that of others; those of man and of the rabbit acting much more quickly than those of the horse and dog. A great part of the starch taken with the food passes into the small intestine unchanged; here, under the influence of a pancreatic cryptolyte, a further transformation into glucose, and of a portion into lactic and butyric acids, takes place.

During the germination of grain, as in the process of malting, a peculiar, nitrogenized substance is produced, which is known as *diastase.* Under the influence of this body the starch is more or less completely converted into glucose, in very much the same way as the conversion occurs in the body.

This "diastatic" action, whether produced by vegetable or animal processes, does not take place by a simple conversion of starch into glucose, by some such single reaction as that expressed by $C_6H_{10}O_5 + H_2O = C_6H_{12}O_6$, but by successive stages in which "soluble starch" is first produced, then several bodies called *dextrines,* then maltose, and finally glucose. (See Dextrin, p. 295.)

Glycogen occurs in the liver, the placenta, white blood-corpuscles, pus-cells, young cartilage-cells, in many embryonic tissues, and in muscular tissue. During the activity of muscles the amount of glycogen which they contain is diminished, and that of sugar increased.

Pure glycogen is a snow-white, floury powder; amorphous, tasteless, and odorless; soluble in H_2O, insoluble in alcohol and ether. In H_2O it swells up at first, and forms an opalescent solution, which becomes clear on the addition of potash. Its solutions are dextrogyrous to about three times the extent of those of glucose.

Dilute acids, ptyalin, pancreatin, extract of liver-tissue, blood, diastase, and albuminoids convert glycogen into a sugar having all the properties of glucose. Cold HNO_3 converts it into xyloidin; on boiling, into oxalic acid. Its solutions dissolve cupric hydrate, which is, however, not reduced on boiling. Iodine colors glycogen wine-red.

Concerning the method of formation of glycogen in the economy, but little is known with certainty; there is little room for doubting, however, that while the bulk of the glycogen found in the liver results from modification of the carbohydrates, it may be and is produced from the albuminoids as well. The ultimate fate of glycogen is undoubtedly its transformation into sugar under the influence of the many substances existing in the body capable of provoking that change. This transformation is continuous in the liver during life, and is accomplished through the same series of intermediary changes into dextrins and maltose as in the case of the conversion of starch into sugar, except that possibly the structure of the dextrins may be different.

295

Dextrin—*British gum*—a substance resembling gum arabic in appearance and in many properties, is obtained by one of three methods: (1) by subjecting starch to a dry heat of 175° (347° F.); (2) by heating starch with dilute H_2SO_4 to 90° (194° F.) until a drop of the liquid gives only a wine-red color; neutralizing with chalk, filtering, concentrating, precipitating with alcohol; (3) by the action of diastase (infusion of malt) upon hydrated starch. As soon as the starch is dissolved the liquid must be rapidly heated to boiling to prevent saccharification.

Commercial dextrin is a colorless, or yellowish, amorphous powder, soluble in H_2O in all proportions, forming mucilaginous liquids. When obtained by evaporation of its solution, it forms masses resembling gum arabic in appearance. Its solutions are dextrogyrous, and reduce cupropotassic solutions under the influence of heat, to amounts varying with the method of formation of the sample. It is colored wine-red by iodine. It is extensively used in the manufacture of mucilage.

Recent investigations have shown that by the action of diastase upon starch, four dextrins are produced: 1st, *Erythrodextrin,* which is colored red by iodine, and which is easily attacked by diastase; 2d, *Achroödextrin* α, not colored by iodine; partially converted into sugar by diastase;

rotary power $[a]_D = +210°$; reducing power (glucose = 100) = 12; 3d, *Achroödextrin β*, not colored by iodine, nor decomposable in 24 hours by diastase; rotary power + 190°; reducing power = 12; 4th, *Achroödextrin γ*, not colored by iodine, nor decomposed by diastase; slowly converted into glucose by dilute H_2SO_4; rotary power = + 150°; reducing power = 28.

An explanation of this series of transformations has been suggested in the supposition that the molecule of starch consists of $50(C_{12}H_{20}O_{10})$; that this is first converted into soluble starch $10(C_{12}H_{20}O_{10})$, and that this is then converted into the different forms of dextrin by a series of hydrations attended by simultaneous formation of maltose, of which the final result might be represented by the equation:

$$\underset{\text{Soluble starch.}}{10(C_{12}H_{20}O_{10})} + \underset{\text{Water.}}{8(H_2O)} = \underset{\text{Achroödextrin.}}{2(C_{12}H_{20}O_{10})} + \underset{\text{Maltose.}}{8(C_{12}H_{22}O_{11})}$$

Cellulose—*Cellulin*—*Lignin*—forms the basis of all vegetable tissues; it exists, almost pure, in the pith of elder and of other plants, in the purer, unsized papers, in cotton, and in the silky appendages of certain seeds. Cotton, freed from extraneous matter by boiling with potash, and afterward with dilute HCl, yields pure cellulose.

It is a white material, having the shape of the vegetable structure from which it was obtained; insoluble in the usual neutral solvents, but soluble in the deep-blue liquid obtained by dissolving copper in ammonia in contact with air.

Vegetable parchment, or *parchment paper*, is a tough material, possessing all the valuable properties of parchment, made by immersing unsized paper for an instant in moderately strong H_2SO_4, washing thoroughly, and drying.

Nitro-cellulose. By the action of HNO_3 upon cellulose (cotton) three different products of substitution may be obtained: *mononitro-cellulose*, soluble in acetic acid, insoluble in a mixture of ether and alcohol; *dinitro-cellulose*, insoluble in acetic acid, soluble in a mixture of ether and alcohol; *trinitro-cellulose*, soluble in both the above solvents. *Gun-cotton* or *pyroxy-*

296

lin is composed of varying proportions of these three derivatives. When gun-cotton is required as an explosive agent, the process is so managed that the product shall contain the greatest possible proportion of trinitro-cellulose, the most readily inflammable of the three. When required for the preparation of collodion, for use in medicine or in photography, dinitro-cellulose is the most valuable. To obtain this, a mixture is made of equal weights of HNO_3 and H_2SO_4 (of each about 5 times the weight of the cotton to be treated); in this the cotton is immersed and well stirred for about three minutes, after which it is well stirred in a large vessel of water, washed with fresh portions of water until the washings are no longer precipitated by barium chloride, and dried. *Collodion* is a solution of dinitro-cellulose in a mixture of three volumes of ether and one volume of alcohol.

Celluloid is gun-cotton and camphor compacted under pressure.

Gums—are substances of unknown constitution, existing in plants; amorphous; soluble in water, insoluble in alcohol; converted into glucose by boiling with dilute H_2SO_4.

Lichenin is obtained from various lichens by extraction with boiling water, forming a jelly on cooling; it is oxidized to oxalic acid by HNO_3; is colored yellow by iodine; and is precipitated from its solutions by alcohol.

Arabin is the soluble portion of gum arabic and gum senegal—*Acacia* (*U. S.*). To separate it, gum arabic is dissolved in water acidulated with HCl, and precipitated by alcohol. It is a white, amorphous, tasteless substance, which is not colored by iodine; is oxidized by HNO_3 to mucic and saccharic acids; is converted by H_2SO_4 into a non-fermentable sugar, *arabinose*; and has the composition, $C_{12}H_{20}O_{10} + 1$ Aq.

Bassorin constitutes the greater part of gum tragacanth; it is insoluble in water, but swells up to a jelly in that fluid.

Cerasin is an insoluble gum exuded by cherry- and plum-trees; water acts upon it as upon bassorin.

AROMATIC SUBSTANCES.

The name of *aromatic substances* was first given to a class of bodies related to benzoic acid, and including a number of products possessed of aromatic odors. At present the meaning of the term has been extended to include a great number of bodies belonging to, or derivable from, the hydrocarbons of the fifth and higher series, all of which may, in fact, be considered as products of addition or of substitution, or both, derivable from benzene, C_6H_6.

A few of these substances, such as benzoic acid, have long been known, and occur in nature in quantities sufficient to readily supply all present demands. Others, such as salicylic acid, although existing in nature, are found in small amount, and are now manufactured artificially by processes which could only have been devised after a knowledge of their constitution was obtained. By far the greater number of aromatic compounds at present known have no existence in nature, and are obtained as products of the laboratory or of manufacturing industries. Among these are many substances for which valuable uses have already been found in the arts and in medicine—*e.g.*, the aniline, anthracene, and naphthalene dyes, carbolic and cresylic acids—while hardly a day passes without a suggestion of the practical utility of some substance formerly known only as a "chemical curiosity."

297

FIFTH SERIES OF HYDROCARBONS.

SERIES C_nH_{2n-6}.

The hydrocarbons of this series are the starting-points from which the major part of that numerous and important class of substances usually classed as *aromatic* are obtainable or derivable. Those of the series at present known are:

Benzene	C_6H_6	boils at 80°.4 (176°. 7 F.)	Cumene	C_9H_{12}	boils at 151°.4 (304°. 5 F.)
Toluene	C_7H_8	boils at 110°.3 (230°. 5 F.)	Cymene	$C_{10}H_{14}$	boils at 175°.0 (347°. 0 F.)
Xylene	C_8H_{10}	boils at 142°.0 (287°. 6 F.)	Laurene	$C_{11}H_{16}$	boils at 188°.0 (370°. 4 F.)

Benzene—*Benzol*—*phenyl hydride*—C_6H_6—78—(not to be confounded with the commercial *benzine*, a mixture of hydrocarbons of the series C_nH_{2n+2}, obtained from petroleum) does not exist in nature, but is produced in a number of reactions. It is obtained by one or two methods, according as it is required chemically pure or mixed with other substances.

To obtain it pure, recourse must be had to the decomposition of one of its derivatives, benzoic acid; this substance is intimately mixed with 3 pts. slacked lime, and the mixture heated to dull redness in an earthenware retort, connected with a well-cooled receiver; the upper layer of distilled liquid is separated, shaken with potassium hydrate solution, again separated, dried by contact with fused calcium chloride, and redistilled over the water-bath.

For use in the arts, and for most chemical purposes, benzene is obtained from coal- or gas-tar, an exceedingly complex mixture, containing some forty or fifty substances, among which are:

Hydrocarbons.		Acids.	Bases.	
Benzene,	Acenaphthalene.	Carbolic.	Pyridine.	Iridoline.
Toluene.	Fluorene.	Cresylic.	Aniline.	Cryptidine.
Xylene.	Anthracene.	Phlorylic.	Picoline.	Acridine.
Cumene.	Retene.	Rosolic.	Lutidine.	Coridine.
Cymene.	Chrysene.	Oxyphenic.	Collidine.	Rubidine.
Naphthalene.	Pyrene.		Leucoline.	Viridine.

By a primary distillation of coal-tar the most volatile constituents, including benzene, are separated as *light oil*; this is washed, first with H_2SO_4, and then with caustic soda, and afterward redistilled; that portion being collected which passes between 80° and 85° (176°–185° F.). This is the commercial benzene, a product still contaminated with the higher homologues of the same series, from which it is almost impossible to separate it, but whose presence is rather advantageous than otherwise to the principal use to which benzol is put—the manufacture of aniline dyes.

Benzene is a colorless, mobile liquid, having, when pure, an agreeable odor; sp. gr. 0.86 at 15° (59° F.); crystallizing at +4°.5 (40°.1 F.); boiling at 80°.5 (176°.9 F.); very sparingly soluble in water, soluble in alcohol, ether, and acetone. It dissolves I, S, P, resins, caoutchouc, gutta-percha, and almost all the alkaloids. It is inflammable, and burns with a luminous, smoky flame.

Benzene unites with Cl or Br to form products of addition, or of substitution; the corresponding iodine compounds can only be obtained by indirect methods. Sulphuric acid combines with benzene to form a neu-

298

tral substance, *sulpho-benzide*, when the anhydrous acid is used, and *phenyl-sulphurous acid* with the ordinary H_2SO_4.

If fuming HNO_3 of sp. gr. 1.52 be slowly added to benzene, a reddish liquid is formed; from which, on the addition of H_2O a reddish-yellow oil separates, and is purified by washing with H_2O and with sodium carbonate solution, drying and rectifying. This oily material is *mononitro-benzene* (see p. 313). If benzol be boiled with fuming HNO_3, or if it be dropped into a mixture of HNO_3 and H_2SO_4, so long as the fluids mix, a crystalline product, *dinitro-benzene*, is formed.

The constitution of benzene, the nucleus of the aromatic compounds, differs in character from that of the hydrocarbons of the series hitherto considered, and is of importance in connection with the formation of its numerous derivatives. Writing the molecular formulæ of the sixth of each of the first three series (the constitution of those of the terebenthene series is still doubtful) we have:

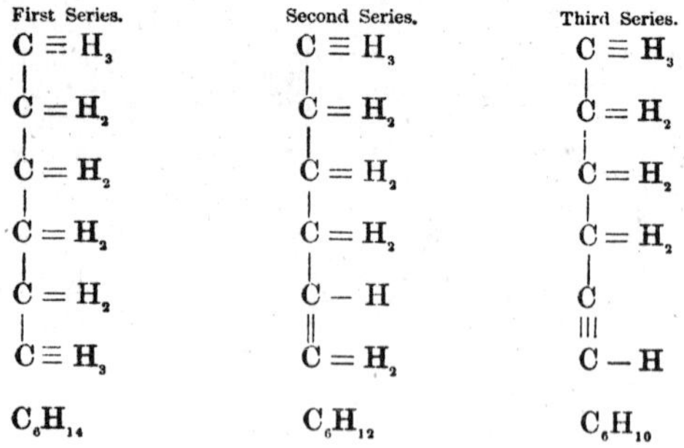

It will be observed that in each of these the chain of C atoms is an open one, and that the series differ in this, that in the first each of the C atoms exchanges with its neighbor a single valence; in the second two neighboring C atoms exchange two valences between them; and that in the third there is an exchange of three valences between two neighboring C atoms. And, further, that in terms above the second in the first two series, and the third in the third series, superior homologues may be considered as formed by interpolation of CH_2 in the chain of the one next below.

In the case of benzene the C atoms are arranged, not in an open, but a closed chain, and exchange with each other alternately one and two valences, and consequently the molecular formula of benzol is:

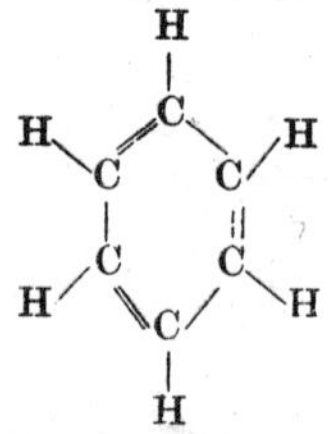

The superior homologues of benzene are derived from it by the substitution of CH_3 for H, and all the derivatives of benzol are formed by

299

such substitution of a group or groups for an atom or atoms of H, in such a way that they all contain one or more groups of six atoms of C arranged as above:

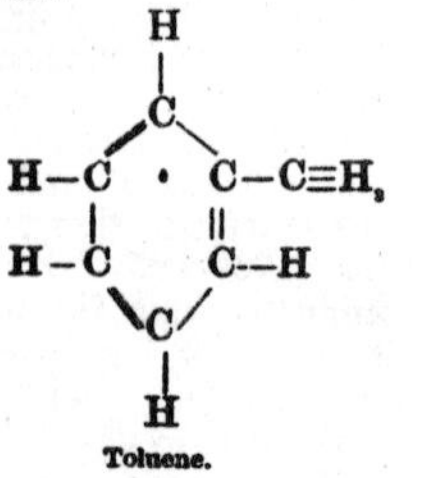

Toluene.

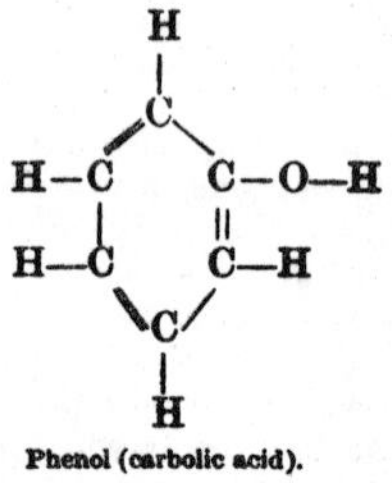

Phenol (carbolic acid).

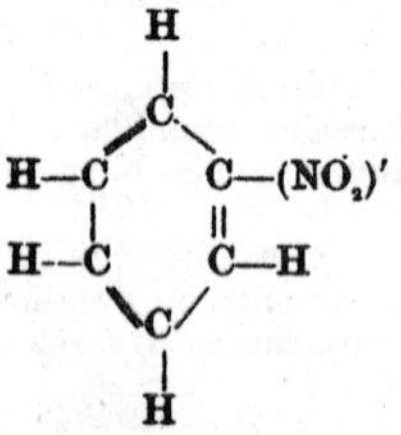

Nitro-benzene.

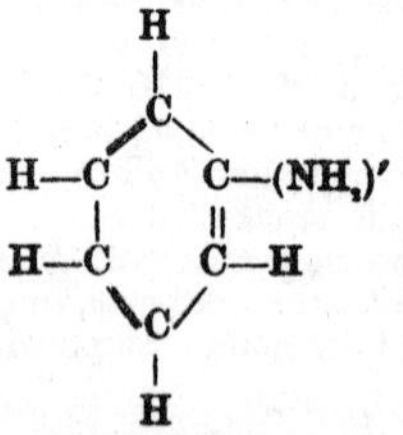

Amido-benzene (aniline).

The superior homologues of benzene include many isomeres. As they are derivable from benzene by substitution of a hydrocarbon radical or radicals CnH_2n+_1 for one or more atoms of hydrogen, the following isomeres may exist:

$C_6H_4(CH_3)_2$ = Dimethylbenzene } C_8H_{10}
$C_6H_5(C_2H_5)$ = Ethylbenzene
$C_6H_3(CH_3)_3$ = Trimethylbenzene }
$C_6H_5(C_3H_7)$ = Propylbenzene } = C_9H_{12}
$C_6H_4(CH_3)(C_2H_5)$ = Methylethylbenzene }
$C_6H_2(CH_3)_4$ = Tetramethylbenzene }
$C_6H_4(C_2H_5)_2$ = Diethylbenzene }
$C_6H_5(C_4H_9)$ = Butylbenzene } = $C_{10}H_{14}$
$C_6H_3(CH_3)_2(C_2H_5)$ = Dimethylethylbenzene }
$C_6H_4(CH_3)(C_3H_7)$ = Methylpropylbenzene }

The number of isomeres of the higher terms of the series is further increased by the occurrence of increasing numbers of isomeres in the radicals themselves in C H, and all higher terms. (See graphic formula, p. 172.)

In these hydrocarbons and in other derivatives of benzene the six atoms of carbon belonging to benzene constitute what is known as the *benzene nucleus, benzene ring,* or the *principal chain;* while the substituted groups are designated as the *lateral chains.*

300

Toluene—*Toluol—Methyl-benzene*—C_6H_5,CH_3—92— exists in the products of distillation of wood, coal, etc., and as one of the constituents of commercial benzene. It has been formed synthetically by acting upon a mixture of monobromo-benzene and methyl iodide with sodium.

It is a colorless liquid, having a peculiar odor, differing somewhat from that of benzene; boils at 110°.3 (230°.5 F.); does not solidify at −20° (−4° F.); sp. gr. 0.872 at 15° (59° F.); almost insoluble in water, soluble in alcohol, ether, carbon disulphide. It burns with a bright, but very smoky flame. It yields a number of derivatives similar to those of benzene, among which may be mentioned *nitro-toluene* and *toluidine,* the homologues of nitro-benzene and aniline, which accompany those substances in the commercial products; *cresylol,* the superior homologue of carbolic acid, and *benzylic alcohol.*

Xylene—*Xylol—Dimethyl-benzene*—$C_6H_4(CH_3)_2$—106—accompanies its inferior homologues in coal-tar. When pure it is a liquid of an aromatic odor; sp. gr. 0.865 at 20° (68° F.); boils at 142° (287°.6 F.); insoluble in water, soluble in ether, benzene, etc., sparingly soluble in alcohol.

There are three isomeric substances having this composition, and differing in the position in which the substituted CH_3 groups are placed. Each of these corresponds to a series of derivatives parallel to those of benzene.

Cumene—*Cumol—Propyl-benzene*—$C_6H_5(C_3H_7)$—120—**is obtained by distilling a mixture of cuminic acid and lime, as benzene is prepared from benzoic acid. It is a limpid liquid, having a strong aromatic odor; boils at 151°.4 (304°.5 F.); insoluble in H_2O, very soluble in alcohol and ether.**

There are several isomeres of this substance, among which are *pseudo-cumene,* or *trimethyl-benzene,* C_6H_3 $(CH_3)_3$, and *mesitylene,* or *methyl-ethyl-benzene,* C_6H_4 $(CH_3)(C_2H_5)$; each corresponding to a series of derivatives.

Cymene—*Cymol.*—**There are many isomeres, of which one exists ready formed in essence of cumin, and in hemlock. It is a colorless, oily liquid; has an odor of lemon; sp. gr. 0.857 at 16° (60°.8 F.); boils at 175° (347° F.); insoluble in water, but readily soluble in alcohol, ether, and essential oils.**

HALOID DERIVATIVES.

By the substitution of atoms of Cl, Br, or I for the hydrogen of the principal and lateral chains of the hydrocarbons, products are obtained which include numerous and peculiar cases of isomery.

In the case of benzene itself there exist products of substitution containing 1, 2, 3, 4, 5, and 6 atoms of Cl, Br or I, or combinations of two or three of those elements. In the case of the unisubstituted derivatives, C_6H_5Cl, C_6H_5Br, and C_6H_5I, but one of each exists. Of the bisubstituted, trisubstituted, and quadrisubstituted derivatives three of each are known.

From the existence of but one unisubstituted derivative it is obvious that it is immaterial in which of the CH groups this substitution occurs, and hence these six groups are equal to each other in value. The existence of isomeres of the higher products of substitution depends upon differences in the *relative positions* of the substituted atoms to each other, their *orientation* as it is sometimes called, and not to their absolute positions.

If we represent the molecule of benzene by a hexagon, leaving out the

301

C and H symbols for the sake of brevity, we may start at any angle and number the angles corresponding to each C and H from one to six:

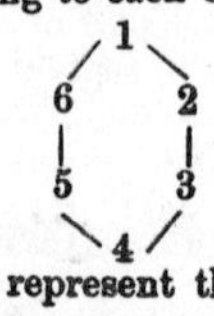

In such a hexagon we may represent the formulæ of the three bisubstituted Br derivates thus:

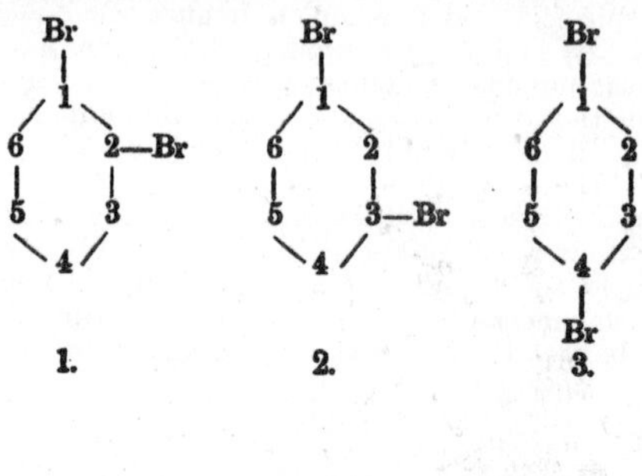

1. 2. 3.

In No. 1 the positions of the substituted atoms are *consecutive*, and as the *absolute* positions in the molecule have no influence, it follows that 2—3; 3—4; 4—5; 5—6; 6—1, all are the same as 1—2. In number 2 the positions are *unsymmetrical*, or separated from each other by a single H atom; and 2—4; 3—5; 4—6, and 5—1 are equal to 1—3. In number 3 the positions are *symmetrical*, or separated from each other by two H atoms; and 2—5; 3—6; 5—2, and 6—3, are the same as 1—4. From this it appears that but three bisubstituted Br products of benzene can exist.

The three series of bi- and tri-substituted derivatives of benzene, whether the substitution be of a halogen or of any univalent element or radical, are designated by the prefixes *ortho*, *meta*, and *para*. Thus, in the figure above:

No. 1 = 1—2 = Orthobibromo-benzene.
No. 2 = 1—3 = Metabibromo-benzene.
No. 3 = 1—4 = Parabibromo-benzene.

The distinction between the three groups is best made by the relations between the bi- and tri-substituted derivatives. The consecutive or *ortho* bisubstituted derivatives can produce by further substitution two tri-derivatives; the unsymmetrical, or *meta*, can produce three trisubstituted derivatives; and the symmetrical, or *para*, can produce but one trisubstituted derivative.

In expressing the constitution of substituted derivatives it is customary either to use the prefixes ortho, para, and meta, as explained above, or to designate the substance by the numerical positions of the substituted atoms or radicals, as in the following notices of the chlorine derivatives of benzene:

***Monochloro-benzene*—C_6H_5Cl—liquid; boils at 132° (269°.6 F.); sp. gr. 1.128 at 0°; obtained by the action of Cl upon C_6H_6 in the cold, in the presence of a little I.**

***Orthodichloro-benzene*—1—2—liquid; boils at 179° (354°.2); sp. gr. 1.328 at 0°; obtained by the action of Cl on C_6H_6.**

***Metadichloro-benzene*—1—3—liquid; boils at 172° (341°.6 F.); sp. gr. 1.307 at 0°; obtainable indirectly.**

302

Paradichloro-benzene—1—4—crystalline; fuses at 56°.4 (133°.5 F.); boils at 173° (343°.4 F.); is the principal product of the action of Cl on C_6H_6 in presence of I.

Trichloro-benzene—1—2—4—crystals; fuses at 17° (62°.6 F.); boils at 213° (415°.4 F.).

Trichloro-benzene—1—3—5—crystals; fuses at 63°.4 (146°.1 F.); boils at 208° (406°.4 F.).

Tetrachloro-benzene—1—2—3—5—crystals; fuses at 50° (122° F.); boils at 246° (474°.8 F.).

Tetrachloro-benzene—1—2—4—5—crystals; fuses at 137° (278°.6 F.); boils between 243°–246° (469°.4–474°.8 F.).

PHENOLS.

The hydrocarbons of the benzene series, unlike those previously considered, form two distinct kinds of hydrates, differing from each other materially in their properties. The terms of one of these series exhibit all the functions of the alcohols, and are known as *aromatic alcohols*. The terms of the other series differ in function from any substance thus far considered, and are known as *phenols*. The difference between them and the aromatic alcohols is due to the fact that in the phenols the OH is directly attached to a C atom, while in the alcohols it forms part of the group of atoms CH_2OH, characteristic of the alcohols:

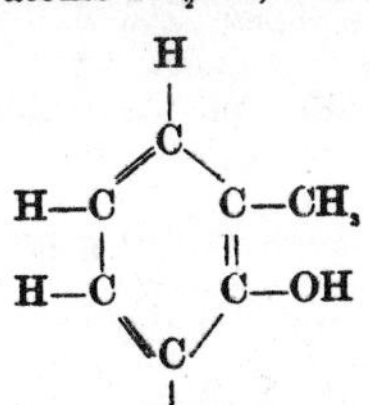

Benzylic Phenol.

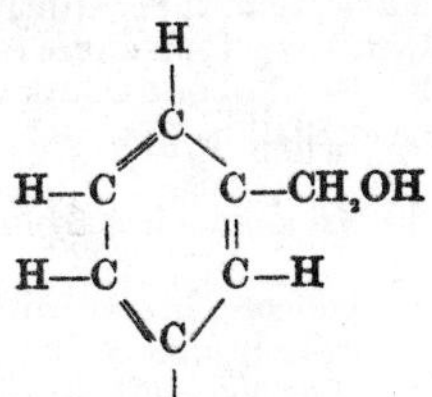

Benzylic alcohol.

The phenols differ from the alcohols in not furnishing by oxidation corresponding aldehydes and acids; in not dividing into water and hydrocarbon under the influence of dehydrating agents; in not reacting with acids to form ethers; in combining to form directly products of substitution with Cl and Br; and in forming with metallic elements compounds more stable than similar compounds of the true alcohols. In short, the phenols appear to have, besides an alcoholic function, more or less of the function of acids.

Phenol—*Phenyl hydrate—Phenic acid—Carbolic acid—Acidum carbolicum (U. S., Br.)*—C_6H_5OH—94—exists in considerable quantity in coal- and wood-tar, and in small quantity in castoreum, and possibly in urine.

It is formed: (1) by fusing sodium phenylsulphide with an excess of alkali; (2) by heating phenyl iodide with potassium hydrate to 320° (608° F.); (3) by heating together salicylic acid and quicklime; (4) by total synthesis from acetylene; (5) by dry distillation of benzoin.

The source from which it is obtained is that portion of the product of distillation of coal-tar which passes over between 150° and 200° (302°–392° F.). This is treated with a saturated solution of potash, containing undissolved alkali; a solid phenate is formed, which is dissolved in hot H_2O; the liquid is allowed to separate into two layers, the lower of which

303

is drawn off and neutralized with HCl; the phenol rises to the surface, is separated, washed with water, dried over calcium chloride, redistilled, crystallized at −10° (14° F.), and the crystals drained.

Pure phenol crystallizes in long, colorless, prismatic needles, fusible at 35° (95° F.), boiling at 187° (368°.6 F.). It has a peculiar, well-known odor, and an acrid, burning taste; very sparingly soluble in water, readily soluble in alcohol and in ether; sp. gr. 1.065 at 18° (64°.4 F.); neutral in reaction. On contact with the skin or with mucous surfaces, it produces a white stain; it coagulates albuminoids, and is a powerful antiseptic.

It may be distilled without decomposition. It absorbs H_2O from damp air to form a hydrate, which crystallizes in six-sided prisms, fusible at 16° (60°.8 F.). Its vapor is reduced to benzene when heated with Zn. It combines with H_2SO_4 to form *phenylsulphuric acids*. It forms *trinitrophenic acid* (*q. v.*) with HNO_3 of 36° B. When heated with H_2SO_4 and oxalic acid it forms *rosolic acid* or *corallin*, which is a mixture from which the pigments *aurin*, *peonin*, *azulin*, and *phenicin* are obtained.

Analytical Characters.—(1.) Its peculiar odor.

(2.) Mix with one-quarter volume of NH_4HO; add two drops sodium hypochlorite solution, and warm; a blue or green color. Add HCl to acid reaction; turns red.

(3.) Add two drops of liquid to a little HCl, add one drop HNO_3; a purple red color.

(4.) Boil with HNO_3 as long as red fumes are given off. Neutralize with KHO; a yellow, crystalline precipitate.

(5.) With $FeSO_4$ solution; a lilac color.

(6.) Float the liquid on H_2SO_4, add powdered KNO_3; violet color.

(7.) With excess of Br water; a yellowish-white precipitate.

Toxicology.—When taken internally, phenol is an active poison, and one whose use by suicides has become quite common. When it has been taken the mouth is whitened by its caustic action, and there is a marked odor of carbolic acid in the breath. It is eliminated by the urine, partly unchanged, and partly in the form of colored derivatives, which color the urine greenish, brownish, or even black. The treatment consists in the administration of albumen (white of egg) and of emetics.

To detect phenol in the urine, that liquor must not be distilled with H_2SO_4, as sometimes recommended, as it contains normally substances which by such treatment yield carbolic acid. The best method consists in adding an excess of bromine water to about 500 c.c. (1 pint) of the urine; on standing some hours, a yellowish precipitate collects at the bottom of the vessel; this is removed, washed, and treated with sodium amalgam, when the characteristic odor of phenol is developed. From other parts of the body, phenol may be recovered by acidulating with tartaric acid; distilling; extracting the distillate by shaking with ether; evaporating the ethereal solution; extracting the residue with a small quantity of water, and applying to this solution the tests described above.

Cresylol—*Cresol—Cresylic acid—Benzylic phenol—Cresylic phenol*—$C_6H_4(CH_3)OH$—108—accompanies phenol in coal- and wood-tars, from which it may be obtained by fractional distillation; it is more readily obtained pure from toluene.

When pure it is a crystalline solid, fusible at 34°.5 (94°.1 F.); as usually met with, however, it is a liquid, which does not solidify at −18° (−0°.4 F.), and boils at 203° (397°.4 F.); it has an odor of creasote. Its properties, decompositions, and products resemble those of phenol.

Creasote—*Creasotum (U. S.)*—is a complex mixture, containing phenol, cresylol, *creasol*, $C_8H_{10}O_2$, and other substances, obtained from wood-tar,

304

and formerly extensively used as an antiseptic. It is an oily liquid, colorless when freshly prepared, but becoming brownish on exposure to light; it has a burning taste and a strong, peculiar odor; it boils at 203° (397°.4 F.), and does not solidify at −27° (−16°.6 F.).

Crude phenol is often substituted for creasote; the two substances may be distinguished by the following characters:

Phenol.	Creasote.
Soluble in commercial glycerin.	Insoluble in commercial glycerin.
Precipitates nitro-cellulose from collodion.	Does not precipitate collodion.
Gives a brown color with ferric chloride and alcohol	Gives a green color with ferric chloride and alcohol.
Gives a violet color with ferric chloride and ammonium hydrate.	Gives a green color, passing to brown, with ferric chloride and ammonium hydrate.

Xenols—*Xylenols*—$C_6H_3(CH_3)_2OH$—122.—Theoretically there are six possible xenols derivable from corresponding xylenes; of these, four have been thus far obtained by the general methods of obtaining the phenols. None is of practical interest.

Thymol—*Cymylic phenol*—$C_6H(CH_3)_4OH$—150—exists, accompanying cymene and thymene, $C_{10}H_{16}$, in essence of thyme, from which it is obtained. The essence contains about one-half its weight of thymol, which is separated by agitation with a concentrated solution of caustic soda; separation of the alkaline liquid, which is diluted and neutralized with HCl; thymol separates and is purified by rectification at 230° (446° F.).

It crystallizes in large, transparent, rhombohedral tables; has a peppery taste and an agreeable, aromatic odor; it fuses at 44° (111°.2 F.), and boils at 230° (446° F.); is sparingly soluble in water, very soluble in alcohol and ether; with the alkalies it forms definite compounds, which are very soluble in water. Its reactions are very similar to those of phenol.

Thymol is an excellent disinfecting and antiseptic agent, and one of the best of embalming materials; possessing the advantage over phenol of having itself a pleasant odor.

SUBSTITUTED PHENOLS.

We have seen above (p. 301) how three bi- and tri-substituted derivatives are derivable from benzene. Phenol is a unisubstituted derivative of the same substance and hence still contains five H atoms which may be replaced by other elements or radicals. So long as but one other univalent atom or radical is introduced, the number of possible derivatives remains the same as if but one kind of atom or radical were introduced, as the reversal of the order Cl Br or Br Cl cannot influence the nature of the compound. But when the number of substituted atoms, differing in kind, is increased beyond two, or the valence of one or more of them exceeds one, the number of possible isomeres is progressively increased. Thus, while there are but three tribromo-benzenes:

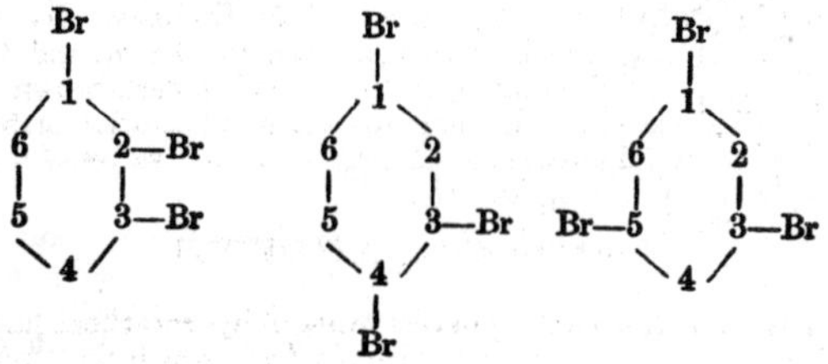

305

there are six chlorobromo-benzenes:

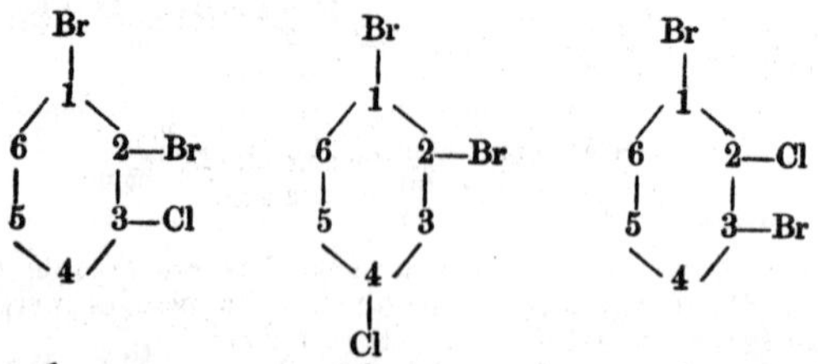

1. 2. 3.

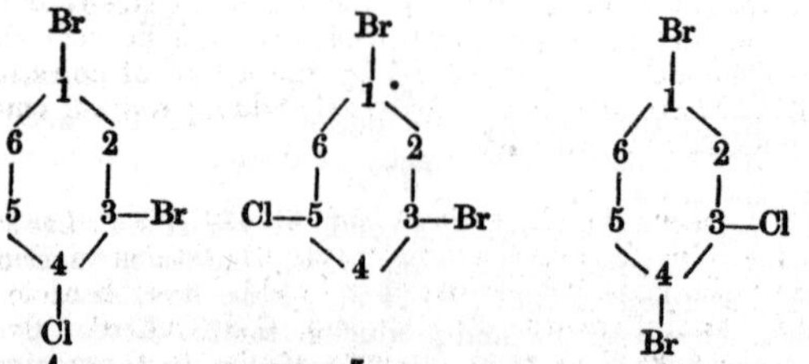

4. 5. 6.

of which 1 and 2 are derivable from orthobibromobenzene (see p. 301); 3, 4, and 5 from metabibromobenzene, and 6 from parabibromobenzene. If, in place of two elements or radicals, we have three, the number of trisubstituted derivatives is increased to ten.

In the place of Cl and Br in the above examples any univalent atom or radical may be substituted, thus giving rise to a great number of derivatives. Certain of such substituted radicals determine the function of the original unisubstituted derivative of benzene and of all of its polysubstituted derivatives. Thus the group (OH) is characteristic of the phenols; (CH_3) or ($C_nH_{2n}+1$) of the superior homologues of benzene; (CH_2OH) of the alcohols; (COOH) of the acids; (NO_2) of the nitro-derivatives; (NH_2) of the amines, etc.

The naming of such polysubstituted derivatives presents many difficulties. Adherence to the principle that the name of a compound shall indicate its constitution, involves the construction of names which are frequently of unwieldy length. It is usual to consider the characterizing group as occupying the position 1 in the hexagon, and to prefix the term *ortho* to the name of that radical or atom occupying one of the ortho-positions 2 and 6 with relation to the characterizing group; *meta* to that occupying one of the meta-positions 3 and 5; and *para* to that occupying the para-position 4.

Thus the substance having the constitution indicated by the formula 1 (see next page) is designated by the name *orthonitroparabromo-phenol.* But even this is not always sufficiently definite, for to each of the substances 2 and 3 (see next page), although differing in characters, the name *orthonitrometabromo-phenol* applies. It has been suggested, to avoid this difficulty, that the prefix *allortho* be used to designate the second or-

306

tho-position 6, and the prefix *allometa* to designate the second meta-position 5.

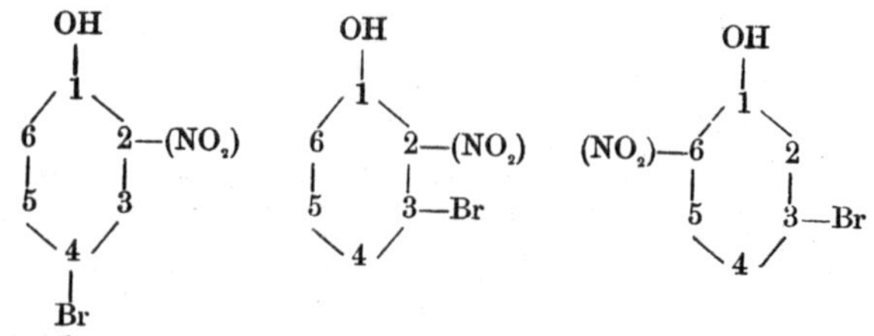

1. 2. 3.

The name of No. 3 would thus become *metabromoallorthonitro-phenol.*

When formulæ are used, all confusion may be readily avoided, even in the most complex substances, by the use of the numeral corresponding to the position in the benzene chain, enclosed in brackets. Thus, the formulæ of 2 and 3 above may be written:

$$C_6H_3(OH)(NO_2)_{(2)}Br_{(3)}; \text{ and}$$
$$C_6H_3(OH)Br_{(3)}NO_2)_{(6)}.$$

Nitro-phenols—*Mononitro-phenols*—**$C_6H_4(NO_2)OH$**—(1—2), (1—3), and (1—4) are formed by the action of HNO_3 on C_6H_5OH. The ortho compound (1—2) crystallizes in large yellow needles, sparingly soluble, and capable of distillation with steam. The meta and para compounds are both colorless, non-volatile, crystalline bodies. Two *dinitro-phenols,* $C_6H_3OH(NO_2)_{2\,(2-4)}$ and $C_6H_3OH(NO_2)_{2(2-6)}$ are obtained by the action of strong nitric acid on phenol, or on ortho- or para-mononitro-phenol. They are both solid, crystalline substances, converted by further nitration into picric acid.

Trinitro-phenols—**$C_6H_2(NO_2)_3OH$.** Two are known. (1.) *Picric acid—Carbazotic acid—Trinitro-phenic acid*—(NO_2) in 2—4—6. It is formed by nitrification of phenol, or of 1—2—4 or 1—2—6 dinitro-phenols, and also by the action of HNO_3 on indigo, silk, wool, resins, etc. It crystallizes in brilliant, yellow, rectangular plates, or in six-sided prisms; it is odorless, and has an intensely bitter taste, whence its name (from πικρὸς = bitter); it is acid in reaction; sparingly soluble in water, very soluble in alcohol, ether, and benzene; it fuses at 122°.5 (252°.5 F.), and may, if heated with caution, be sublimed unchanged; but, if heated suddenly or in quantity, it explodes with violence. It behaves as a monobasic acid, forming salts, which are for the most part soluble, yellow, crystalline, and decomposed with explosion when heated.

Picric acid is valuable as a dye-stuff, coloring silk and wool yellow; as a staining medium in histological investigations; and as a reagent for the alkaloids, with many of which it forms crystalline precipitates. It is also sometimes fraudulently added to beer and to other food articles, to communicate to them either a bitter taste or a yellow color.

Analytical Characters.—(1.) Its intensely bitter taste.

(2.) Its alcoholic solution, when shaken with a potassium salt, gives a yellow crystalline ppt.

307

(3.) An ammoniacal solution of cupric sulphate gives a green, crystalline ppt.

(4.) Glucose, heated with a dilute alkaline solution of picric acid, communicates to it a blood-red color.

(5.) Warmed with an alkaline solution of potassium cyanide, an intense red color is produced (the same effect is produced by ammonium sulphydrate).

(6.) Unbleached wool, immersed in boiling solution of picric acid, is dyed yellow.

Nos. 1, 3, 5, and 6 are quite delicate.

When taken internally in overdose, it acts as a poison; it may be separated from animal fluids or from beer by evaporation to a syrup, extracting with 95 per cent. alcohol, acidulated with H_2SO_4; filtering; evaporating; and applying the tests to a solution of the residue.

DIATOMIC PHENOLS.

Diatomic phenols are derived from the benzene series of hydrocarbons by the substitution of two (OH) groups for two atoms of hydrogen. In obedience to the laws of substitution already discussed, three such compounds exist, corresponding to each hydrocarbon. Thus, in the case of benzene :

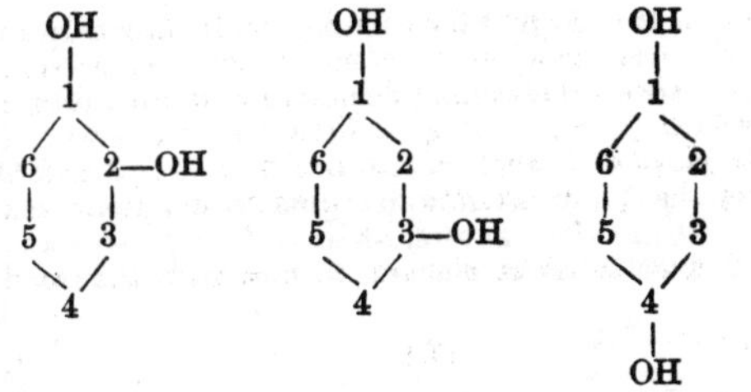

Ortho. 1—2 Pyrocatechin. Meta. 1—3 Resorcin. Para. 1—4 Hydroquinone.

Pyrocatechin—*Oxyphenic acid*—*Orthodioxy-benzene*—$C_6H_4(OH)_2$—1—2—is obtained from catechin or from morintannic acid by dry distillation ; also by the action of KHO on orthochlor- or orthoiodo-phenol, or by decomposing its methyl ether, *guaiacol*, by HI at 200° (392° F.). It crystallizes in short, square prisms ; fuses at 104° (219°.2 F.), and boils at 245°.5 (473°.9 F.). Readily soluble in water, alcohol, and ether. Its aqueous solution gives a dark-green color with Fe_2Cl_6 solution, changing to violet on addition of NH_4HO, $NaHCO_3$, or tartaric acid.

Resorcin—*Metadioxy-benzene*—$C_6H_4(OH)_2$—1—3—is obtained by the action of fused KHO on parachlor- or iodo-phenol. It is usually prepared by dry distillation of extract of Brazil wood.

It forms short, thick, colorless and odorless, rhombic prisms. Fuses at 104° (219°.2 F.), and boils at 271° (519°.8 F.). It is very soluble in water, alcohol, and ether. Its aqueous solution is neutral in reaction, and intensely sweet. With Fe_2Cl_6 its solutions assume a dark-violet color, which is discharged by NH_4HO. Its ammoniacal solution, by exposure to air, assumes a pink color, changing to brown and, on evaporation, green

308

and dark blue. Heated with phthalic anhydride at 195° (383° F.) it yields *fluorescein* (see page 309). It dissolves in fuming H_2SO_4, forming an orange-red solution, which becomes darker and then changes to greenish-black and then pure blue, and to purple on being warmed.

Resorcin has been recently used in medical practice.

Hydroquinone—*Paradioxy-benzene*—$C_6H_4(OH)_2$—1—4—is formed by fusing paraiodo-phenol with KHO at 180° (356° F.), by dry distillation of oxysalicylic acid, or of quinic acid, and by the action of reducing agents on *quinone*. It forms colorless, rhombic prisms, which fuse at 169° (336°.2 F.). Readily soluble in water, alcohol, or ether. Its aqueous solution is turned red-brown by NH_4HO. Oxidizing agents convert it into quinone.

Quinone—$C_6H_4(OO)''$—is the representative of a number of similar compounds, derivable from the aromatic hydrocarbons. It is produced by the oxidizing action of $MnO_2 + H_2SO_4$, or of dilute chromic acid, upon quite a number of para-benzene derivatives ; but best by the limited oxidation of quinic acid.

It crystallizes in yellow prisms ; fuses at 116° (240°.8 F.) ; sublimes at ordinary temperatures ; is sparingly soluble in cold, but readily soluble in hot water and in alcohol or ether. It gives off a peculiar pungent odor and stimulates the lachrymal secretion. Reducing agents convert it into hydroquinone.

There is no similar substance known corresponding either to pyrocatechin or to resorcin.

Orsin—*Dimetadioxy-toluene*—$C_6H_3(CH_3)_{(1)}(OH)_{(3)}(OH)_{(5)}$—exists in nature in those lichens which are used as sources of archil and litmus (*Rocella tinctoria*, etc.). It crystallizes in six-sided prisms ; is sweet ; readily soluble in water, alcohol, or ether ; fuses at 58° (136°.4 F.). Its aqueous solution is colored violet-blue by Fe_2Cl_6. It unites with NH_3 to form a compound which absorbs O from the air and is converted into *orcein*, $C_7H_7NO_3$; a dark red or purple body, which is the chief constituent of the dye-stuff known as archil, cudbear, French purple, and litmus.

TRIATOMIC PHENOLS.

The only compounds of this class at present known with certainty are two isomeric triatomic phenols, which owe the differences in properties existing between them to a different placing of the OH groups. They are *phloroglucin* and *pyrogallol*.

Phloroglucin—$C_6H_3(OH)_3$—126—is obtained by the action of potash upon phloretin, quercitrin, maclurin (see Glucosides), catechin, kino, etc. It crystallizes in rhombic prisms, containing 2 Aq ; is very sweet ; very soluble in water, alcohol, and ether.

Pyrogallol—*Pyrogallic acid*—$C_6H_3(OH)_3$—126—is formed when gallic acid (*q. v.*) is heated to 200° (392° F.). It crystallizes in white needles ; neutral in reaction ; very soluble in water ; very bitter ; fuses at 115° (239° F.) ; boils at 210° (410° F.) ; poisonous. Its most valuable property is that of absorbing oxygen, for which purpose it is used in the laboratory in the form of a solution of potassium pyrogallate.

309

PHENOL DYES.

Aurin—$C_{19}H_{14}O_3$ and **Rosolic acid**—$C_{20}H_{16}O_3$—are substances existing in the dye obtained by the action of oxalic acid upon phenol in presence of H_2SO_4, known as *coralline* or *pæonine*, which communicates to silk or wool a fine yellow-red color.

Aurin crystallizes in fine, red needles from its solution in HCl. It is insoluble in H_2O, but soluble in HCl, alcohol, and glacial acetic acid. It forms a colorless compound with potassium bisulphite.

Phthaleins.—These substances are produced by heating the phenols with phthalic anhydride, $C_8H_4O_3$, water being at the same time eliminated.

Their constitution is that of a benzene nucleus, two of whose H atoms have been replaced by two acetone groups (CO), whose remaining valences attaches them to two phenol groups by exchange with an atom of hydrogen.

Thus **Phenol-phthalein**, the simplest of the group, has the constitution

$$C_6H_4\begin{cases}CO-C_6H_4(OH)\\CO-C_6H_4(OH)\end{cases}$$

Phenol-phthalein is a yellow, crystalline powder, insoluble in water, but soluble in alcohol. Its alcoholic solution, perfectly colorless if neutral, assumes a brilliant magenta-red in the presence of an alkali. This property renders phenol-phthalein very valuable as an *indicator* of reaction.

Resorcin-phthalein—*Fluorescëin*—$C_{20}H_{12}O_5$—bears the same relation to resorcin that phenol-phthalein does to phenol, and is obtained from resorcin by a corresponding method. It is a dark-brown crystalline powder, which dissolves in ammonia to form a red solution, exhibiting the most brilliant green fluorescence. A tetrabromo-derivative of fluorescëin is used as a dye under the name *eosin*.

AROMATIC ALCOHOLS.

The alcohols corresponding to this series of hydrocarbons have the same composition as the corresponding phenols, from which they differ in constitution and in having the functions of true alcohols. They yield on oxidation, first an aldehyde and then an acid, and they contain the characterizing group of the primary alcohols, CH_2OH ; once if the alcohol be monoatomic, twice if diatomic, etc. Thus :

C_6H_5,CH_2OH = Benzylic alcohol.
C_6H_5,COH = Benzoic aldehyde.
$C_6H_5,COOH$ = Benzoic acid.

As they contain the benzene nucleus they are capable of yielding isomeric products of further substitution, ortho, para, or meta, according to the position of the substituted atom or radical.

Benzylic alcohol—*Benzoic alcohol*—*Benzyl hydrate*—$C_6H_5(CH_2OH)$—108—does not exist in nature, and is of interest chiefly as corresponding to two important compounds, benzoic acid and benzoic aldehyde (oil of bitter almonds). It is obtained by the action of potassium hydrate upon oil of bitter almonds, or by slowly adding sodium amalgam to a boiling solution of benzoic acid.

310

It is a colorless liquid ; boils at 206°.5 (403°.7 F.) ; has an aromatic odor ; is insoluble in water, soluble in all proportions in alcohol, ether, and carbon disulphide. By oxidation it yields, first, benzoic aldehyde, $C_6H_5(COH)$; and afterward, benzoic acid, $C_6H_5(COOH)$. By the same means it may be made to yield products similar to those obtained from the alcohols of the saturated hydrocarbons.

ALPHENOLS.

These substances are intermediate in function between the alcohols and the phenols, and contain both substituted groups (OH) and CH_2OH.

Saligenin, $C_6H_4\begin{cases}CH_2OH\\OH\end{cases}$—124—is obtained from salicin (*q. v.*) in large, tabular crystals ; quite soluble in alcohol, water, and ether. Oxidizing agents convert it into salicylic aldehyde, which by further oxidation yields salicylic acid. It is also formed by the action of nascent hydrogen on salicylic aldehyde.

ALDEHYDES.

Benzoic aldehyde—*Benzoyl hydride*—$C_6H_5(COH)$—106—is the main constituent of oil of bitter almonds, although it does not exist in the almonds (see p. 329) ; it is formed, along with hydrocyanic acid and glucose, by the action of water upon amygdalin. It is also formed by a number of general methods of producing aldehydes : by the dehydration of benzylic alcohol; by the dry distillation of a mixture in molecular proportions of calcium benzoate and formiate; by the action of nascent hydrogen upon benzoyl cyanide, etc.

It is obtained from bitter almonds. The crude oil contains, besides benzoic aldehyde, hydrocyanic and benzoic acids and cyanobenzoyl; to purify it, it is treated with three to four times its volume of a concentrated solution of sodium bisulphite; the crystalline mass is expressed, dissolved in a small quantity of water, and decomposed with a concentrated solution of sodium carbonate—the treatment being repeated, if necessary.

It is a colorless oil, having an acrid taste and the odor of bitter almonds; sp. gr. 1.043; boils at 179°.4 (354°.9 F.); soluble in 30 parts of water, and in all proportions in alcohol and ether. Oxidizing agents convert it into benzoic acid, a change which occurs by mere exposure to air. Nascent hydrogen converts it into benzylic alcohol. With Cl and Br it forms benzoyl chloride or bromide. H_2SO_4 dissolves it when heated, forming a purple-red color, which turns black if more strongly heated.

When *perfectly pure*, benzoic aldehyde exerts no deleterious action when taken internally; owing, however, to the difficulty of completely removing the hydrocyanic acid, the substances usually sold as *oil of bitter almonds, ratafia*, and *almond flavor*, are almost always poisonous, if taken in sufficient quantity. They may contain as much as 10–15 per cent. of hydrocyanic acid, although said to be "purified." The presence of the poisonous substances may be detected by the tests given on page 327.

Salicylic aldehyde—*Salicyl hydride—Salicylol—Salicylous acid*—**$C_6H_4(OH)COH$**—122—exists in the flowers of *spiræa ulmaria*, and is the

311

principal ingredient of the essential oil of that plant. It is best obtained by oxidizing salicin (*q. v.*).

It is a colorless oil; turns red on exposure to air; has an agreeable, aromatic odor, and a sharp, burning taste; sp. gr. 1.173 at 13°.5 (56°.3 F.); boils at 196°.5 (385°.7 F.); soluble in water, more so in alcohol and ether.

It is, as we should suspect from its origin, a substance of mixed function, possessing the characteristic properties of aldehyde and phenol. It produces a great number of derivatives, some of which have the characters of salts and ethers.

Methyl-protocatechuic aldehyde—*Vanillin*—**$C_6H_3(OH)(OCH_3)COH$**—is the odoriferous principle of vanilla. It is produced artificially by oxidation of *coniferin*, $C_{16}H_{22}O_8$, a glucoside occurring in coniferous plants. It crystallizes in needles, fuses at 80° (176° F.); is sparingly soluble in water, readily soluble in alcohol or ether. It has a pungent taste, and a faint odor of vanilla, the latter more marked when the substance is heated. On exposure to air it becomes partially oxidized to *vanillic acid*, $C_8H_8O_4$.

ACIDS CORRESPONDING TO THE AROMATIC HYDRATES.

The acids, possibly derivable from benzene by the substitution of (COOH), or of (COOH) and (OH), for atoms of hydrogen, would form, were they all known, a great number of series; there are, however, comparatively few of them which have been as yet obtained, although the number of acid series known is greater than that of corresponding alcohols. Each series of mono- and diatomic alcohols furnishes a corresponding series of acids; thus:

$C_6H_5-CH_2OH$	$C_6H_4\left\langle\begin{matrix}CH_2OH\\CH_2OH\end{matrix}\right.$	$C_6H_4\left\langle\begin{matrix}CH_2OH\\OH.\end{matrix}\right.$
Benzoic alcohol.	Toluyl glycol.	Saligenin.
C_6H_5-COOH	$C_6H_4\left\langle\begin{matrix}COOH\\COOH\end{matrix}\right.$	$C_6H_4\left\langle\begin{matrix}COOH\\OH\end{matrix}\right.$
Benzoic acid.	Terephthalic acid.	Salicylic acid.

There are still a number of other series of acids possibly derivable directly from benzene, without speaking of substituted acids of more complex nature; of these, however, the majority are wanting.

By the progressive substitution of groups (COOH) for atoms of hydrogen in benzene, we may obtain six series of acids, five of which have been isolated:

$C_6H_5(COOH)$ $-C_nH_{2n-8}O_2$ ....Benzoic series.
$C_6H_4(COOH)_2-C_nH_{2n-10}O_4$ ....Phthalic series.
$C_6H_3(COOH)_3-C_nH_{2n-12}O_6$ ....Trimellitic series.
$C_6H_2(COOH)_4-C_nH_{2n-14}O_8$ ....Prehnitic series.
$C_6H(COOH)_5$ $-C_nH_{2n-16}O_{10}$............Wanting.
$C_6(COOH)_6$ $-C_nH_{2n-18}O_{12}$............Mellitic series.

The alphenols, containing a single group (OH), are at present represented by a single series:

$$C_6H_4(OH)(COOH)—C_nH_{2n-8}O_3—\text{Salicylic series.}$$

312

Corresponding to the unknown alphenols, containing a greater number of (OH) groups, there are at present but two series of acids known:

$$C_6H_3(OH)_2(COOH)—C_nH_{2n-8}O_4—\text{Veratric series,}$$

and

$$C_6H_2(OH)_3(COOH)—C_nH_{2n-8}O_5—\text{Gallic series.}$$

In each of these series the basicity is, as usual, equal to the number of groups (COOH).

Benzoic acid—*Acidum benzoicum* (*U. S.*)—**$C_6H_5(COOH)$**—122—exists ready formed in benzoin, tolu balsam, castoreum, and several resins. It does not exist in animal nature, so far as is at present known; in those situations in which it has been found, it has resulted from decomposition of hippuric acid (*q. v.*), or has been introduced from without. When taken in moderate doses, it does not pass out in its own form, but is converted into hippuric acid; in excessive doses a portion is eliminated unchanged in the urine. It is obtained from benzoin, or from the urine of herbivorous animals; and is formed in a variety of reactions.

It crystallizes in white, transparent plates; odorless; acid; fuses at 122° (251°.6 F.); sublimes at 145° (293° F.); boils at 240° (464° F.); sparingly soluble in cold water; soluble in hot water, alcohol, and ether. Dilute HNO_3 does not attack it. It dissolves in ordinary H_2SO_4, and is precipitated unchanged by H_2O. Its salts are all soluble.

Hippuric acid—*Benzyl-glycocol—Benzyl-amido-acetic acid*—**$C_9H_9NO_3$**—179—is a constant constituent of the urine of the herbivora, and of human urine to the extent of 0.29–2.84 grams (4.5–43.8 grains) in 24 hours. It is more abundant with a purely vegetable diet, after the administration of benzoic acid, and in diabetes mellitus and chorea.

It crystallizes in transparent, colorless, odorless, bitter prisms; sparingly soluble in water; fuses at 130° (266° F.). It dissolves unchanged in HCl; but on boiling the solution it is decomposed into benzoic acid and glycocol. The same decomposition is effected by dilute H_2SO_4, HNO_3, and oxalic acid, and by a ferment developed in putrefying urine. Oxidizing agents convert it into benzoic acid, benzamide, and CO_2.

The characters of hippuric acid are: (1) when heated in a dry tube it fuses and gives off a sublimate of benzoic acid and an odor of hydrocyanic acid; (2) it gives a brown ppt. with ferric chloride; (3) when heated with lime it gives off benzine and ammonia.

Salicylic Acid—*Oxybenzoic acid—Acidum salicylicum* (*U. S.*)—**$C_6H_4(OH)COOH$**—138—was first obtained from essence of *spiræa*, which consists largely of salicylic aldehyde, and subsequently from oil of wintergreen (*gaultheria*), which contains methyl salicylate; and also from *salicin*, a glucoside yielding salicylic aldehyde. It is now obtained from phenol. This is fused, and, while a current of dry CO_2 is passed through it, small portions of Na are added; the sodium salicylate thus formed is dissolved in H_2O and decomposed with HCl, when the liberated salicylic acid is precipitated.

It crystallizes in fine white needles; very sparingly soluble in cold water, quite soluble in hot water, alcohol, and ether; it fuses at 158° (316°.4 F.), and may be distilled with but slight decomposition, if it be pure. Cl and Br form with it products of substitution. Fuming HNO_3 forms with it a nitro-derivative and, if the action be prolonged, converts it into picric acid. With ferric chloride, its aqueous solution assumes a fine violet color.

313

Salicylic acid and its salts (it is monobasic, although diatomic) are extensively used in medicine, both externally as antiseptics and internally in the treatment of rheumatism, etc. It is not without caustic properties, and hence, when taken internally, it should be largely diluted.

Gallic acid—*Acidum gallicum* (*U. S.*)—**$C_6H_2(OH)_3COOH$**—170—exists in nature in certain leaves, seeds, and fruits. It is best obtained from gall-nuts, which contain its glucoside, gallotannic acid (*q. v.*). It can be obtained from salicylic acid.

It crystallizes in long silky needles with 1 Aq; odorless; acidulous in taste; sparingly soluble in cold water, very soluble in hot water and in alcohol; its solutions are acid. When heated to 210°–215° (410°–419° F.) it yields CO_2 and pyrogallol (*q. v.*). Its solution does not precipitate gelatin, nor the salts of the alkaloids, as does tannin. It forms four series of salts.

NITRO-DERIVATIVES OF BENZENE.

By substitution of the univalent radical (NO_2) for the hydrogen of benzene a series of substitution products are obtainable, corresponding to the series of haloid derivatives, phenols, etc. (see pp. 301, 304, 305).

Nitro-benzol—*Nitro-benzene—Mono-nitro-benzene—Essence of Mirbane*—**$C_6H_5(NO_2)$**—123—is obtained by the moderated action of fuming HNO_3, or of a mixture of HNO_3 and H_2SO_4, on benzene.

It is a yellow, sweet liquid, with an odor of bitter almonds; sp. gr. 1.209 at 15° (59° F.); boils at 213° (415°.4 F.); almost insoluble in water; very soluble in alcohol and ether. Concentrated H_2SO_4 dissolves, and, when boiling, decomposes it. Boiled with fuming HNO_3, it is converted into *binitro-benzol*. It is converted into *aniline* by reducing agents.

It has been used in perfumery as *artificial essence of bitter almonds*; but as inhalation of its vapor, even largely diluted with air, causes headache,

drowsiness, difficulty of respiration, cardiac irregularity, loss of muscular power, convulsions, and coma, its use for that purpose is to be condemned. Taken internally it is an active poison.

Nitro-benzol may be distinguished from oil of bitter almonds (benzoic aldehyde) by H_2SO_4, which does not color the former; and by the action of acetic acid and iron filings, which convert nitro-benzol into aniline, whose presence is detected by the reactions for that substance (*q. v.*).

AMIDO-DERIVATIVES OF BENZENE.

These substances are derivable from benzene and its homologues by the substitution of one or more univalent groups (NH_2) (amidogen) for atoms of hydrogen. They may also be considered as *phenylamines*, produced by the substitution of the univalent radical *phenyl* (C_6H_5), or its homologues, derivable from the benzene nucleus, for the hydrogen of ammonia. They all are strongly basic in character.

Aniline — *Amido-benzene — Amido-benzol — Phenylamine — Kyanol — Cristalline* — $\left.\begin{matrix}C_6H_5\\H_2\end{matrix}\right\}$ N—93—**exists in small quantity in coal-tar and is one of the products of the destructive distillation of indigo. It is prepared by the reduction of nitro-benzene by hydrogen: $C_6H_5(NO_2) + 3H_2 = C_6H_5(NH_2) + 2H_2O$; the hydrogen being liberated in the nascent state in contact with nitro-benzol by the action of iron filings on acetic acid.**

314

Pure aniline is a colorless liquid; has a peculiar, aromatic odor, and an acrid, burning taste; sp. gr. 1.02 at 16° (60°.8 F.); boils at 184°.8 (364°.6 F.); crystallizes at −8° (17°.6 F.); soluble in 31 pts. of cold water, soluble in all proportions in alcohol, ether, carbon disulphide, etc.; when exposed to air, it turns brown, the color of the commercial "oil," and, finally, resinifies; it is neutral in reaction. Oxidizing agents convert it into blue, violet, red, green, or black derivatives. Cl, Br, and I act upon it violently to produce products of substitution. Concentrated H_2SO_4 converts it according to the conditions, into *sulphanilic* or *disulphanilic acid*. With acids it unites, after the manner of the ammonia, without liberation of H_2O or H to form salts, most of which are crystallizable, soluble in water, and colorless, although by exposure to air, especially if moist, they turn red.

Analytical Characters.—(1.) With a nitrate and H_2SO_4, a red color.

(2.) Cold H_2SO_4 does not color it alone; on addition of potassium dichromate, a fine blue color is produced, which, *on dilution with water*, passes to violet, and, if not diluted, to black.

(3.) With calcium hypochlorite, a violet color.

(4.) Heated with cupric chlorate, a black color.

(5.) Heated with mercuric chloride, a deep crimson color.

Toxicology.—Aniline itself, when taken in the liquid form or by inhalation, is an active poison, producing symptoms similar to those caused by nitro-benzol (*q. v.*). Its salts, *if pure*, seem to have but slight deleterious action.

DERIVATIVES OF ANILINE.

By the substitution of other radicals or elements for the remaining hydrogen atoms of the benzene nucleus, or for the hydrogen atoms of the amidogen group, NH_2, a great number of derivatives, including many isomeres, are produced.

In all of these derivatives the group (NH_2) is considered as occupying the position 1.

Chloranilines.—**Three** *monochloranilines* **are known, of which two, ortho- (1—2) and meta- (1—3), are liquid. The other, para- (1—4), is solid and crystalline.**

Four *dichloranilines*, **1—2—4, 1—2—5, 1—3—5, and 1—3—4, are known, all solid and crystalline.**

Two *trichloranilines*, **1—2—4—6 and 1—2—4—5 are known, both solid and crystalline.**

The corresponding *bromanilines* **are also known; also a** *tetrabromaniline*, **1—2—3—4—6, and a pentabromaniline, $C_6(NH_2)Br_5$.**

Of the possible *iodanilines*, **but four have been described:** *Metamonoiodaniline* **(1—3);** *paramoniodaniline* **(1—4); the** *diiodaniline* **(1—2—4); and the** *triiodaniline* **(1—2—4—6).**

Nitranilines.—The three isomeres, ortho-, meta-, and para- *Mononitranilines*, $C_6H_4(NH_2)(NO_2)$ are formed by imperfect reduction of the dinitro-benzenes.

Two *dinitranilines*, $C_6H_3(NH_2)(NO_2)_2$ (1—2—4) and (1—2—6), are known.

A single *trinitraniline*, $C_6H_2(NH_2)(NO_2)_3$ (1—2—4—6), has been obtained by the action of alcoholic ammonia upon the ethylic or methylic ether of picric acid. It is also called *picramide*.

Anilides.—These are compounds in which one of the H atoms of the amidogen group has been replaced by an acid radical. Or they may also

315

be considered as amides, whose remaining hydrogen has been more or less replaced by phenyl, C_6H_5.

Acetanilide—$C_6H_5(NH.C_2H_3O)$ = *Phenyl-acetamide*—is obtained either by heating together aniline and glacial acetic acid for several hours, or, better, by the action of acetyl chloride on aniline. It forms colorless, shining, crystalline scales; fuses at 112°.5 (234°.5 F.), and volatilizes unchanged at 295° (563° F.). It is sparingly soluble in cold water, soluble in hot water and in alcohol.

It has been recently introduced into medical practice as an antiperiodic, under the name *antifebrine*.

ANILINE DYES.

It was observed at an early period that when crude aniline was acted upon by oxidizing agents a brilliant red color was produced. Efforts to isolate this color, beginning in 1856, have led, not only to success in the end desired, but also to the discovery of a great number of substances, many of which are valuable as dye-stuffs communicating not only brilliant colors, but the greatest variety of shades and colors. Among the substances commercially classified as aniline dyes are many pigments which do not properly belong here, being derivatives of phenol, naphthalene, anthracene, etc.

Of the true aniline dyes the most important, and that from which most of the others are industrially derived, is *fuchsine*, also called *magenta*, *aniline red*, *roseine*, *azaleine*, *etc.*

Although fuchsine is obtainable by a great variety of methods, those industrially used are limited to modifications of two: the oxidation of commercial aniline by arsenic acid, or by a mixture of nitro-benzene, hydrochloric acid, and iron filings; and the purification of the product, after combination with an acid, by repeated recrystallizations.

The commercial fuchsine, which varies much in quality, is a hard, more or less crystalline substance of a brilliant green color, sparingly soluble in cold water, readily soluble in hot water and in alcohol, the solutions having a brilliant red color.

The commercial fuchsines are salts, usually the chloride or acetate, of a base which is itself colorless, called *rosaniline*, whose constitution has been but recently determined, having the empirical formula $C_{20}H_{19}N_3H_2O$.

Rosaniline is one of a series of homologous substances the first term of which is *pararosaniline*, $C_{19}H_{19}N_3O$—whose molecule:

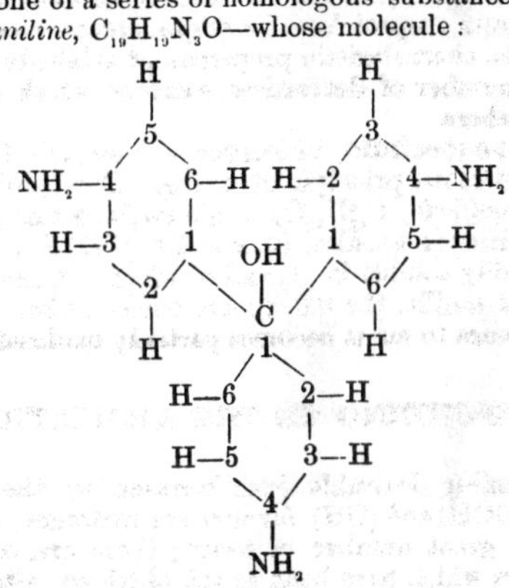

316

consists of three benzene nuclei, united by a group (COH), the para H atom of each of the benzene nuclei being replaced by a group (NH_2). The remaining H atoms of the benzene nuclei may be replaced, either by CH_3 to produce the higher homologues, or by other atoms or radicals.

Neither pure aniline nor pure toluidine will produce a red color by the action of oxidizing agents, the formation of a rosaniline requiring a combination of the two.

The rosanilines are powerful triacid bases, forming salts which are all colored, and from which the colorless bases may be separated by decomposing concentrated solutions of their salts with concentrated KHO solution.

Hoffman's violet is *triethyl-rosaniline chloride*, produced by heating together ethyl iodide and rosaniline.

Lyons blue = *triphenyl-rosaniline chloride*, obtained by heating rosaniline with an excess of aniline; *gas green*, obtained by heating rosaniline chloride with aldehyde and sulphuric acid; *Paris violet*, obtained by the oxidation of methyl aniline.

Mauvein is a base whose sulphate, obtained by mixing cold dilute solutions of potassium dichromate and aniline sulphate, is a fine, purple dye. A blue dye is also obtained by heating mauvein with aniline.

Aniline black is obtained by acting on aniline with a mixture of cupric sulphide or a vanadium salt and potassium chlorate.

Saffronin is a base derived from commercial oils, rich in the superior homologues of aniline (toluidines). Its hydrochlorate is used in place of safflower.

AZO AND DIAZO DERIVATIVES.

The azo compounds are derivable from the aromatic hydrocarbons by loss of two H atoms from two molecules of the hydrocarbon, and union of the remainders through the intermediary of a group (—N = N—)″. They are formed by the action of certain reducing agents upon the nitro-derivatives, and may be considered as intermediary products in the reduction of the nitro-derivatives to amines. Thus in the case of benzene:

$$\underset{\text{Nitrobenzene.}}{[C_6H_5(NO_2)]_2} + H_6 = 3H_2O + \underset{\text{Azoxybenzene.}}{\left.\begin{matrix}C_6H_5-N\\C_6H_5-N\end{matrix}\right>O}$$

$$\underset{\text{Azoxybenzene.}}{\left.\begin{matrix}C_6H_5-N\\C_6H_5\;\;N\end{matrix}\right>O} + H_2 = H_2O + \underset{\text{Azobenzene.}}{\left.\begin{matrix}C_6H_5-N\\C_6H_5-N\end{matrix}\right>}$$

$$\underset{\text{Azobenzene.}}{\left.\begin{matrix}C_6H_5-N\\C_6H_5-N\end{matrix}\right>} + H_2 = \underset{\text{Hydrazobenzene.}}{\left.\begin{matrix}C_6H_5NH\\C_6H_5NH\end{matrix}\right>}$$

$$\underset{\text{Hydrazobenzene.}}{\left.\begin{matrix}C_6H_5NH\\C_6H_5NH\end{matrix}\right>} + H_2 = \underset{\text{Aniline.}}{2[C_6H_5(NH_2)]}$$

The diazo compounds consist of an univalent remainder of an aromatic hydrocarbon, united by the group (—N = N—) with a haloid atom, or an acid residue: $C_6H_5—N = N—Br$ = Diazobenzene bromide.

317

HYDRAZINES.

The hydrazines are theoretically derivable from the group $H_2N—NH_2{}'$, diamidogen, by the substitution of acid, alcoholic, or phenylic radicals for one or more of the hydrogen atoms.

Phenyl hydrazine—$C_6H_5—HN—NH_2$—is obtained by the action of zinc-dust and acetic acid on diazo-amidobenzene. It is a yellow oil, sparingly soluble in water, soluble in alcohol and in ether; possessed of strong reducing power, and acting as a monacid base to form crystallized salts.

PYRIDINE BASES.

These interesting substances, closely related to the vegetable alkaloids as well as to some of the alkaloids produced during putrefactive decomposition of animal matters, were first discovered in 1851, as constituents of *oil of Dippel* = *oleum animale* = *oleum cornu cervi* = *bone-oil*, an oil produced during the dry distillation of bones, horns, etc., and as a by-product in the manufacture of ammoniacal compounds from those sources. They also occur in coal-tar, naphtha, and in commercial ammonia, methylic spirit, and fusel oil.

The pyridine bases at present known are:

	Formula.	Boiling-point.	Sp. Gr. at 22°.
Pyridine	C_5H_5N	115°	0.924
Picoline	C_6H_7N	134°	0.933
Lutidine	C_7H_9N	154°	0.945
Collidine	$C_8H_{11}N$	170°	0.953
Parvoline	$C_9H_{13}N$	188°	0.966
Coridine	$C_{10}H_{15}N$	211°	0.974
Rubidine	$C_{11}H_{17}N$	230°	1.017
Viridine	$C_{12}H_{19}N$	251°	1.024

It will be observed that these compounds are metameric with the anilines, from which they differ in constitution, as shown by the structural formulæ of picoline and aniline:

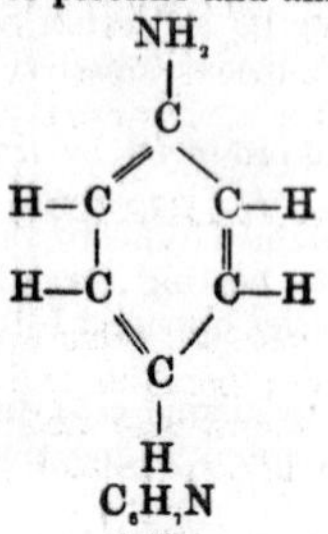

C_6H_7N
Aniline.

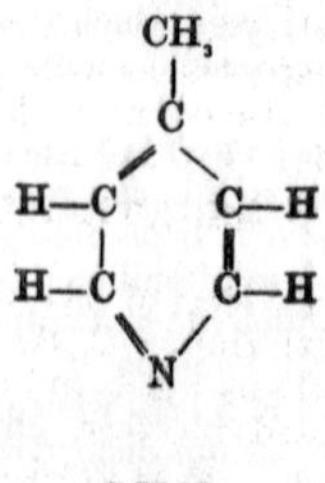

C_6H_7N
Picoline.

318

They are all liquid at the ordinary temperature, behave as tertiary monamines, react with several of the general reagents of the alkaloids, and form chloroplatinates which are decomposed by boiling water.

Pyridine—$C\left<\begin{matrix}CH=CH\\CH-CH\end{matrix}\right>N$—is obtained from oil of Dippel, and is also obtainable synthetically from *piperidine*, $CH_2\left<\begin{matrix}CH_2-CH_2\\CH_2-CH_2\end{matrix}\right>N-H$, which is itself a derivative of *piperine*, $C_{17}H_{19}O_3N$, a constituent of black and white pepper.

It is a colorless, mobile liquid, having a peculiar, very penetrating odor. It boils at 115° (239° F.). It mixes with water in all proportions. It is strongly alkaline, and combines with acids as does NH_3. Like all the bases of this series, it is very stable, and withstands the action of such oxidizing agents as fuming HNO_3 and chromic acid. It forms crystalline salts.

Parvoline, $C_9H_{13}N$; *Collidine*, $C_8H_{11}N$; and *Hydrocollidine*, $C_8H_{13}N$—have been noted as products of putrefactive decomposition of albuminoids.

Pyrrol—$HN\left<\begin{matrix}CH=CH\\CH=CH\end{matrix}\right>$—is a weak base accompanying the pyridine bases in oil of Dippel, and also obtainable from other sources. It is a colorless, oily liquid, whose odor resembles that of chloroform.

By acting upon pyrrol with an ethereal solution of iodine, a quadrisubstituted derivative, tetriodopyrrol, C_4HI_4N, is obtained as a brown powder, which has been used under the name *Iodol* as a substitute for iodoform in surgical practice.

CHINOLINE BASES.

The bases of this group at present known are:

Chinoline	C_9H_7N	Pentahiroline	$C_{13}H_{15}N$
Lepidine	$C_{10}H_9N$	Isoline	$C_{14}H_{17}N$
Cryptidine	$C_{11}H_{11}N$	Ettidine	$C_{15}H_{19}N$
Tetrahiroline	$C_{12}H_{13}N$	Validine	$C_{16}H_{21}N$

whose constitution and relations to the pyridine bases are shown by the formulæ:

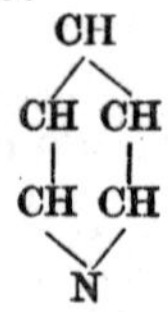

Pyridine.
C_5H_5N.

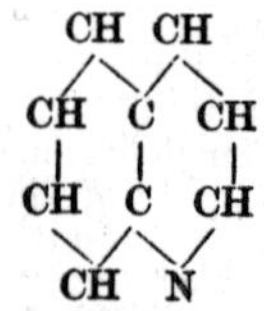

Chinoline.
C_9H_7N.

They are obtained by the destructive distillation of the cinchonine, quinine, and other natural alkaloids, to which they are closely related.

Chinoline—C_9H_7N—is a mobile liquid; boils at 238° (460°.4 F.); becomes rapidly colored on contact with air. It has an intensely bitter and acrid taste, and an odor somewhat like that of bitter almonds. It is sparingly soluble in water, readily soluble in alcohol and ether.

319

Chinoline is the nucleus of a vast number of products of substitution, among which are four substances which have recently assumed medical importance:

Thalline = *Tetrahydroparachinanisol*—$C_{10}H_{13}NO$—is a derivative of the paramethyl ether of chinoline. It is met with in the form of sulphate and tartrate in the shape of crystalline powders. The odor of the sulphate is similar to that of anisol (methyl phenol); that of the tartrate to that of coumarin. The taste of both is bitter, acrid, and salty. Both salts are readily soluble in H_2O, the sulphate the more readily. Solutions of thalline salts assume, even when very dilute, a magnificent emerald-green color with Fe_2Cl_6 solution.

Ethylthalline—$C_{12}H_{17}NO$—is a derivative of thalline, whose chloride is hygroscopic; readily forming solutions which are acid in reaction, bitter in taste, and assume a red-brown color with Fe_2Cl_6.

Antipyrine = *Dimethyloxychinizine*—$C_{11}H_{12}N_2O$—is obtained by heating methyloxychinizine with methyl iodide and methyl alcohol in sealed vessels at 100° (212° F.); the first-named substance having been previously obtained by the action of acetylacetic ether upon phenyl hydrazine.

It constitutes a voluminous, reddish, crystalline powder; readily soluble in water, ether, alcohol, and chloroform.

Its solution with Fe_2Cl_6 is colored deep red-brown, the color being discharged by H_2SO_4. Nitrous acid colors dilute solutions of antipyrine a bright green, which persists for several days at the ordinary temperature. If the mixture be heated, and a drop of fuming HNO_3 added, the color changes to light red, then blood red, and the liquid deposits a purple oil on continued heating. Addition of a drop of fuming HNO_3 to a cold concentrated solution of antipyrine produces precipitation of small green crystals.

Kairine—*Methyloxychinoline hydride*—$C_{10}H_{13}NO$—is more nearly derived from chinoline than the substances previously mentioned. Its chloride is a crystalline, nearly white, easily soluble powder, whose taste is at once bitter, aromatic, and salty.

Thalline, ethylthalline, antipyrine, and kairine are possessed of antiperiodic and antipyretic properties.

INDIGO GROUP.

In this group are included a number of substances, derivable from indigo-blue, which are evidently closely related to the benzene group, as is shown by the number of benzene derivatives which are obtained by their decomposition, yet whose constitution is not yet definitely established.

Indigotin—*Indigo-blue*—$C_{16}H_{10}N_2O_2$—constitutes the greater part of the commercial *indigo.* It does not exist preformed in the plants from which it is obtained, whose juice is naturally colorless, but is produced by decomposition of a glucoside contained in them (see Indican, p. 320).

Indigotin may be obtained by the action of phosphorus trichloride on *isatin;* or, in a nearly pure form, by cautiously subliming commercial indigo. It forms purple-red, somewhat metallic, orthorhombic prisms or plates, odorless, tasteless, neutral, insoluble in water, ether, or dilute acids or alkalies. By dry distillation it yields aniline and other products.

320

By moderate heating with dilute HNO_3 it gives off gas and is converted into isatin.

Indigo Sulphonic Acids.—When indigo is heated for some time with fuming H_2SO_4 it dissolves. If the solution be diluted with H_2O, a blue powder, soluble in H_2O, but insoluble in dilute acids, is precipitated. This is *Indigo-monosulphonic* or *phœnicin-sulphonic acid*—$C_{16}H_9N_2O_2SO_3H$.

The filtrate from the last-mentioned precipitate contains *Indigo-disulphonic, sulphindylic,* or *sulphindigotic acid*—$C_{16}H_8N_2O_2(SO_3H)_2$—whose K and Na salts constitute soluble pastes known in the arts as *soluble indigo,* or *indigocarmine.*

Isatin—$C_8H_5NO_2$—obtained by oxidation of indigo-blue, forms shining, transparent, red-brown prisms. It is odorless, sparingly soluble in water, readily soluble in alcohol.

Dioxindol—*Hydrindic acid*—$C_8H_7NO_2$—is formed by the action of Na on isatin suspended in H_2O. It forms yellow prisms, soluble in H_2O, and combines with both bases and acids.

Oxindol—C_8H_7NO—is obtained from dioxindol by reduction with Na amalgam in acid solution. It crystallizes in easily soluble, colorless needles, and combines with acids and bases.

Indol—C_8H_7N—is produced by distilling oxindol over zinc-dust, or by heating orthonitrocinnamic acid with KHO and Fe filings.

It crystallizes in large, shining, colorless plates, having the odor of naphthylamine. It is a weak base, forming salts with acids, which are, however, decomposed by boiling water. Its aqueous solution, acidulated with HCl, is colored rose-red by KNO_2. It is converted into aniline by fused KHO.

It is one of the products of putrefaction of albuminoid substances, and is formed during the action of the pancreatic secretion upon albuminoids. It is partly eliminated with the fæces and partly reabsorbed.

In the intestine and fæces indol is invariably accompanied by **Skatol,** C_9H_9N, its superior homologue, which may also be obtained by the action of Sn and HCl on indigo. It crystallizes in brilliant plates, and is less soluble than indigo. The product obtained from indigo has a penetrating but not disagreeable odor, while that obtained from putrid albumin and from fæcal or intestinal matter has a disgusting odor, probably due to the presence of foreign substances.

Indican—$C_{26}H_{31}N_{17}$—is a glucoside existing in the different varieties of indigo-producing plants, and also in the urine and blood of man and the herbivora.

It is a yellow or light-brown syrup, which cannot be dried without decomposition, bitter and disagreeable to the taste, acid in reaction, and soluble in water, alcohol, and ether.

It is very prone to decomposition. Even slight heating decomposes it into leucin, *indicanin,* $C_{20}H_{23}NO_{12}$, and *indiglucin,* $C_6H_{10}O_6$. A characteristic decomposition is that when heated in acid solution, or under the influence of certain ferments (?) it is decomposed into indigo-blue and indiglucin, the latter a glucose:

$$\underset{\text{Indican.}}{2C_{26}H_{31}NO_{17}} + \underset{\text{Water.}}{4H_2O} = \underset{\text{Indigotin.}}{C_{16}H_{10}N_2O_2} + \underset{\text{Indiglucin.}}{6C_6H_{10}O_6}$$

321

SIXTH SERIES OF HYDROCARBONS.

SERIES C_nH_{2n-8}.

This series has at present but two representatives, derivable from benzene by the substitution of one lateral chain for an atom of hydrogen.

Cinnamene—*Styrolene—Cinnamol—Styrol—Liquid essence of styrax*—C_8H_8—104—exists ready formed in essential oil of styrax; it is also formed by decomposition of cinnamic acid (*q. v.*), or, synthetically, by the action of a red heat upon pure acetylene, a mixture of acetylene and benzene, or a mixture of benzene and ethylene. It is a colorless liquid, has a penetrating odor, recalling those of benzene and naphthalene, and a peppery taste; boils at 143° (289°.4 F.); soluble in all proportions in alcohol and water; neutral in reaction.

ALCOHOLS.

SERIES $C_nH_{2n-8}O$.

There are but two alcohols of this series known:

Cinnyl alcohol............ $C_9H_{10}O$ | Cholesterin.............. $C_{26}H_{44}O$

Cholesteric alcohol—*Cholesterin*—$C_{26}H_{43}OH$—372—is an alcohol, although usually classed by physiologists among the fats, because it is greasy to the touch and soluble in ether.

It occurs in the animal economy, normally in the bile, blood (especially that coming from the brain), nerve-tissue, brain, spleen, sebum, contents of the intestines, meconium, and fæces; pathologically in biliary calculi, in the urine in diabetes and icterus, in the fluids of ascites, hydrocele, etc., in tubercular and cancerous deposits, in cataracts, in atheromatous degenerations, and sometimes, in masses of considerable size, in certain cerebral tumors. It also exists in the vegetable world in peas, beans, olive-oil, wheat, etc. It has not been obtained by synthesis. It is best obtained from biliary calculi, the lighter-colored varieties of which consist almost entirely of this substance. The calculi are pulverized, extracted with boiling ether, the solution filtered hot, the ether distilled off, the residue dissolved in boiling alcohol, and the solution allowed to cool; the crystals which separate are heated for some time with alcohol containing a little potash; on cooling, crystals form, which are finally washed with alcohol so long as the washings are colored or alkaline, and recrystalized from ether.

Cholesterin crystallizes with or without Aq.; from benzol, petroleum, chloroform or anhydrous ether, it separates in delicate, colorless, silky needles, having the composition $C_{26}H_{44}O$; from hot alcohol, or a mixture of alcohol, and ether, it crystallizes in rhombic plates, usually with one obtuse angle wanting, having the composition $C_{26}H_{44}O + 1$ Aq.; these crystals, transparent at first, become opaque on exposure to air, from loss of aq. It is insoluble in water, in alkalies and dilute acids, difficultly soluble

322

in cold alcohol, readily soluble in hot alcohol, ether, benzol, acetic acid, glycerin, and solutions of the biliary acids. It is odorless and tasteless. When anhydrous it fuses at 145° (293° F.) and solidifies at 137° (278°.6 F.); sp. gr. 1.046. It is lævogyrous, $[\alpha]_D = 31°.6$ in any solvent.

It combines readily with the volatile fatty acids. From its solution in glacial acetic acid a compound having the composition $C_{26}H_{44}O,C_2H_4O_2$ separates in fine curved crystals, which are decomposed on contact with water or alcohol; when heated with acids under pressure, it forms true ethers. Hot HNO_3 oxidizes it to *cholesteric acid,* $C_8H_{10}O_5$, which is also produced by the oxidation of biliary acids; a fact which indicates the probable existence of some relation between the methods of formation of cholesterin and of the biliary acids in the economy.

ANALYTICAL CHARACTERS.—(1.) Moistened with HNO_3, and evaporated to dryness, a yellow residue remains, which turns brick-red on addition of NH_4HO.

(2.) It is colored violet when a mixture of 2 vols. H_2SO_4 (or HCl) and 1 vol. ferric chloride solution is evaporated upon it.

(3.) When ground up with H_2SO_4 and chloroform added, a blue-red or violet color is produced, which changes to green on exposure to air.

SEVENTH SERIES OF HYDROCARBONS.

SERIES C_nH_{2n-10}.

The only representative of this series at present known is

Naphthydrene—*Naphthylene hydride*—$C_{10}H_{10}$—130—obtained by heating naphthalene with potassium, and decomposing the product with water. It also occurs in heavy petroleum. It is a colorless liquid; boils at 205° (401° F.), and has a strong, disagreeable odor.

EIGHTH SERIES OF HYDROCARBONS.

SERIES C_nH_{2n-12}.

The only term of this series is

Naphthalene—$C_{10}H_8$—128—occurring in coal-tar. It has been formed by a synthesis which indicates its constitution; benzene and ethylene, when heated together, unite to form, first, cinnamene and afterward naphthalene. It is constituted by the fusion of two benzol groups by two C atoms, thus:

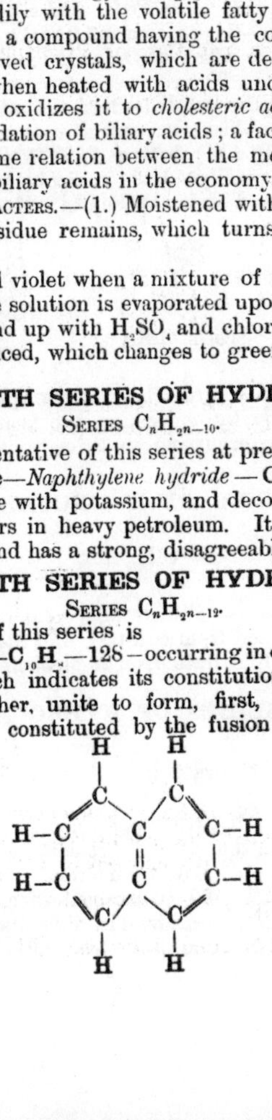

323

It crystallizes in large, brilliant plates; has a burning taste and a faint aromatic odor; fuses at 80° (176° F.) and boils at 217° (422°.6 F.), subliming, however, at lower temperatures; burns with a bright, smoky flame; insoluble in water, soluble in alcohol, ether, and essences. It forms substitution compounds with Cl, Br, I, HNO_3, and H_2SO_4.

SUBSTITUTION DERIVATIVES OF NAPHTHALENE.

By the replacement of the hydrogen atoms of naphthalene by other atoms or by radicals, substitution products are obtained somewhat in the same manner as in the case of benzene (see pp. 301, 304, 305). In the case of naphthalene, however, the number of isomeres is much greater than with benzene. In the structural formula of naphthalene the positions

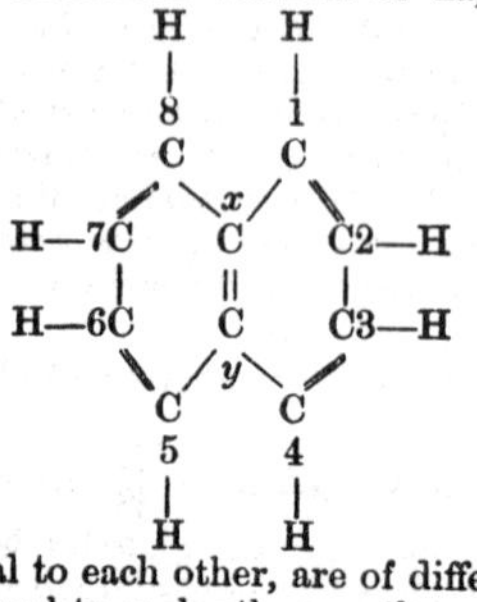

1, 4, 5, 8, although equal to each other, are of different value from the positions 2, 3, 6, 7, also equal to each other, as they are differently disposed with regard to the carbon atoms *x* and *y*. There exist, therefor, two possible unisubstituted derivatives of naphthalene for a single such derivative of benzene, etc. If the substituted group occupy the approximate positions 1, 4, 5, or 8, it is called an α-derivative; if it occupy the remote positions 2, 3, 6, or 7, it is a β-derivative.

Of the numerous derivatives of naphthalene, the only ones of present medical interest are those corresponding to the monophenols:

Naphthols—$C_{10}H_7OH$—of which there are two:

α-Naphthol has been obtained by heating phenyl-isocrotonic acid; also by boiling an aqueous solution of diazonaphthalene nitrate with nitrous acid, or by fusing α-naphthalene-sulphonic acid with KHO.

It crystallizes in colorless prisms; fuses at 94° (201°.2 F.); boils at 280° (536° F.); is nearly insoluble in water, but soluble in alcohol and in ether, and gives a transient violet color with Fe_2Cl_6 and a hypochlorite.

β-Naphthol = Isonaphthol—is prepared industrially by fusion of β-naphthalene sulphonate of sodium with NaHO, for the manufacture of a yellow dye-stuff: *Campobello yellow*. The commercial product is in reddish-gray, friable, light masses. The pure substance forms colorless, silky, crystalline plates, having a faint, phenol-like odor, and an evanescent, sharp, burning taste. It fuses at 123° (253°.4 F.), boils at 286° (514°.8 F.), and is sparingly soluble in water, but readily soluble in alcohol and ether. Its aqueous solutions are not colored violet by Fe_2Cl_6. The pure substance is a valuable antiseptic.

324

NINTH SERIES OF HYDROCARBONS.

SERIES C_nH_{2n-14}.

Is represented by a single hydrocarbon: **Acenaphthalene**—$C_{12}H_{10}$—154—produced synthetically by continuing the heating of naphthalene with ethylene, the reaction occurring in three steps. It also exists in coal-tar.

TENTH SERIES OF HYDROCARBONS.

SERIES C_nH_{2n-16}.

Is represented by two terms: **Fluorene**, a solid, crystalline body, boiling at 305° (581° F.), obtained from coal-tar; and **Stilbene**, obtained by the action of ammonium sulphydrate upon an alcoholic solution of benzoic aldehyde.

ELEVENTH SERIES OF HYDROCARBONS.

SERIES C_nH_{2n-18}.

Anthracene—$C_{14}H_{10}$—178—exists as a constituent of coal-tar, and is obtained by expression from the substance remaining in the still after the distillation of naphthalene, etc. The commercial product thus obtained is a yellowish mass containing 50–80 per cent. of anthracene, the purification of which is a matter of considerable difficulty. It has also been obtained synthetically, by the action of the heat on benzyl-toluene, and in other ways.

When pure, anthracene crystallizes in rhombic tables having a bluish fluorescence; fusible at 210° (410° F.) and boiling above 360° (680° F.); its best solvents are benzene and carbon disulphide, in which, however, it is only sparingly soluble.

The constitution of anthracene is that of two benzene nuclei united through two of their C atoms:

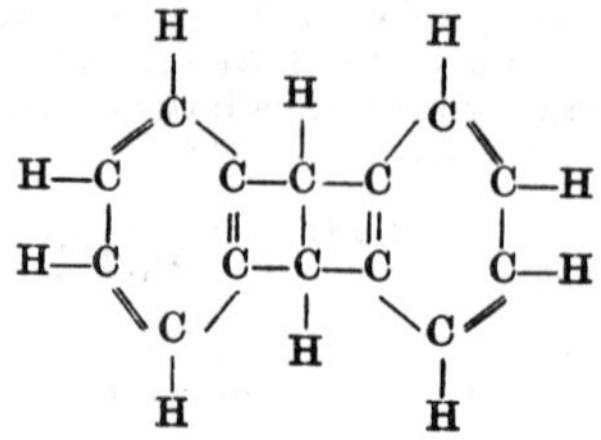

Oxidizing agents convert anthracene into anthraquinone. Reducing agents decompose it into three hydrocarbons, $C_{14}H_{20}$, C_7H_{16}, and an oily

325

hydrocarbon boiling above 360° (648° F.). Br and Cl attack it violently, I more slowly, forming products of addition.

An isomere, *Phenanthrene*, $C_{14}H_{10}$, which boils between 320° and 350° (608°–662° F.), accompanies anthracene in the crude product.

DERIVATIVES OF ANTHRACENE.

As may be inferred from the complex molecule of anthracene, the number of possible derivatives of substitution and of addition, including many isomeres, is very great.

Our knowledge of these derivatives is as yet fragmentary, and but few of those known are of present medical interest.

Anthraquinone—$C_6H_4\langle{}^{CO}_{CO}\rangle C_6H_4$—is formed by oxidation of anthracene. It forms yellow needles, which fuse at 273° (523°.4 F.). It is not easily oxidized, but is converted into anthracene by sufficiently active reducing agents.

Dioxyanthraquinone—*Alizarin*—$C_6H_4\langle{}^{CO}_{CO}\rangle C_6H_2\langle{}^{OH}_{OH}$—is the red pigment of the madder root (*Rubia tinctoria*). Artificial alizarin has now almost completely displaced the natural product in dyeing. It is obtained by the action of fused KHO on many anthracene derivatives, the one generally used being *anthraquinone-disulphonic acid*, $C_{14}H_6O_2(SO_3H)_2$.

Methylanthracene—$C_{14}H_9,CH_3$—is obtainable by synthesis, and also by heating chrysophanic acid, emodin, or eloïn with zinc-dust.

Chrysophanic Acid—*Parietic Acid—Rheic Acid—Rhein*, $C_{15}H_{10}O_4$—is a derivative of methyl anthracene, which exists in the lichens *Parmelia parietina* and *Squamaria elegans*, in senna, and in rhubarb, and obtainable to the extent of 80 per cent. from *Goa powder = Chrysarobin*, $C_{30}H_{26}O_7$.

Chrysophanic acid crystallizes in golden, orange-yellow, interlaced needles. It is almost tasteless and odorless; fuses at 162° (291°.6 F.); almost insoluble in cold water, sparingly soluble in hot water, alcohol, and ether, readily soluble in benzene. It forms a red solution with H_2SO_4, from which it is deposited, unchanged by water. It also forms red solutions with alkalies. Reducing agents convert it into methylanthracene.

Trioxymethylanthraquinone—*Emodin*—$C_{14}H_4(CH_3)(OH)_3O_2$—occurs in the bark of *Rhamnus frangula*, and accompanies chrysophanic acid in rhubarb. It crystallizes in long, orange-red prisms which fuse at 250° (482° F.), and yield methylanthracene when heated with zinc-dust.

HIGHER SERIES OF HYDROCARBONS.

The twelfth series is not at present represented. Of the thirteenth series, one hydrocarbon, *pyrene*, $C_{16}H_{10}$, is known; and one of the fourteenth series, *chrysene*, $C_{18}H_{12}$—both obtained from coal-tar.

Pyrene crystallizes in plates; fuses at 142° (287°.6 F.). It forms a compound with picric acid, which crystallizes in red needles.

Chrysene crystallizes in bright-yellow, glistening scales; is sparingly soluble in alcohol, and forms a compound with picric acid which crystallizes in brown needles.

326

CYANOGEN COMPOUNDS.

The substances which we have so far considered are all derivable, more or less directly, from the various hydrocarbons, and may be considered, upon the theory of types, as produced by the substitution of radicals composed of C and H, C and O, or C, H and O, for atoms of H of the three typical substances H_2, H_2O and H_3N.

The substances of this class are typically considered as containing the radical (CN)′, which is known as *cyanogen*, and has the same power of passing unchanged from compound to compound, as do methyl and ethyl.

Dicyanogen—$(CN)_2$—52—is prepared by heating mercuric cyanide. It is a colorless gas; has a pronounced odor of bitter almonds; sp. gr. 1.8064 A.; burns in air with a purple flame, giving off N and CO_2. It is quite soluble in H_2O, the solution turning brown in air.

With H_2O alone, or with H_2O and NH_3, dicyanogen enters into combinations which indicate the relations existing between the cyanogen compounds and those previously considered:

$$\underset{\text{Dicyanogen.}}{(CN)_2} + \underset{\text{Water.}}{4H_2O} = \underset{\text{Ammonium oxalate.}}{C_2O_4(NH_4)_2}$$

$$\underset{\text{Dicyanogen.}}{(CN)} + \underset{\text{Water.}}{H_2O} = \underset{\text{Cyanic acid.}}{CNOH} + \underset{\text{Hydrocyanic acid.}}{CNH}$$

$$\underset{\text{Cyanic acid.}}{CNOH} + \underset{\text{Water.}}{H_2O} = \underset{\text{Ammonia.}}{NH_3} + \underset{\text{Carbon dioxide.}}{CO_2}$$

$$\underset{\text{Cyanic acid.}}{CNOH} + \underset{\text{Ammonia.}}{NH_3} = \underset{\text{Urea.}}{CON_2H_4}$$

It has a very deleterious action upon both animal and vegetable life, even when largely diluted with air.

Hydrogen cyanide—*Cyanogen hydride—Hydrocyanic acid—Prussic acid*—$\left.\begin{matrix}CN\\H\end{matrix}\right\}$—27—exists ready formed in the juice of *cassava*, and is formed by the action of H_2O upon bitter almonds, cherry-laurel leaves, etc. It is also formed in a great number of reactions: by the passage of the electric discharge through a mixture of acetylene and N; by the action of chloroform on NH_3; by the distillation of, or the action of HNO_3 upon, many organic substances; by the decomposition of cyanides.

It is always prepared by the decomposition of a cyanide. Its preparation in the pure form is an operation attended with the most serious danger, and should only be attempted by those well trained in chemical manipulation. For medical uses a very dilute acid is required; the *acid. hydrocyanicum dil.* (*U. S., Br.*) contains, if freshly and properly prepared, two per cent. of anhydrous acid; that of the French Codex is much stronger—ten per cent.

The pure acid is a colorless, mobile liquid, has a penetrating and characteristic odor; sp. gr. 0.7058 at 7° (44°.6 F.); crystallizes at −15° (5° F.); boils at 26°.5 (79°.7 F.); is rapidly decomposed by exposure to light. The dilute acid of the U. S. P. is a colorless liquid, having the odor of the acid; faintly acid, the reddened litmus returning to blue on exposure to air; sp. gr. 0.997; 10 grams of the acid should be accurately neutralized by 1.27 gram of silver nitrate. The dilute acid deteriorates

327

on exposure to light, although more slowly than the concentrated; a trace of phosphoric acid added to the solution retards the decomposition.

Most strong acids decompose HCN. The alkalies enter into double decomposition with it to form *cyanides*. It is decomposed by Cl and Br, with formation of cyanogen chloride or bromide. Nascent H converts it into methylamine.

Analytical Characters.—(1.) With silver nitrate a dense, white ppt.; which is not dissolved on addition of HNO_3 to the liquid, but dissolves when separated and heated with concentrated HNO_3; soluble in solutions of alkaline cyanides or hyposulphites.

(2.) Treated with NH_4HS, evaporated to dryness, and ferric chloride added to the residue; a blood-red color.

(3.) With potash and then a mixture of ferrous and ferric sulphates: a greenish ppt., which is partly dissolved with a deep blue color by HCl.

(4.) Heated with a dilute solution of picric acid and then cooled: a deep red color.

(5.) Moisten a piece of filter paper with a freshly prepared alcoholic solution of guaiac; dip the paper into a very dilute solution of $CuSO_4$, and, after drying, into the liquid to be tested. In the presence of HCN it assumes a deep blue color.

Toxicology.—Hydrocyanic acid is a violent poison, whether it be inhaled as vapor or swallowed, either in the form of dilute acid, of soluble cyanide, or of the pharmaceutical preparations containing it, such as oil of bitter almonds and cherry-laurel water; its action being more rapid when taken by inhalation or in aqueous solution than in other forms. When the medicinal acid is taken in poisonous dose, its lethal effect may seem to be produced instantaneously; nevertheless, several respiratory efforts usually are made after the victim seems to be dead, and instances are not wanting in which there was time for considerable voluntary motion between the time of the ingestion of the poison and unconsciousness. In the great majority of cases the patient is either dead or fully under the influence of the poison on the arrival of the physician, who should, however, not neglect to apply the proper remedies if the faintest spark of life remain. Chemical antidotes are, owing to the rapidity of action of the poison, of no avail, although possibly chlorine, recommended as an antidote by many, may have a chemical action on that portion of the acid already absorbed. The treatment indicated is directed to the maintenance of respiration; cold douche, galvanism, artificial respiration, until elimination has removed the poison. If the patient survive an hour after taking the poison, the prognosis becomes very favorable; in the first stages it is exceedingly unfavorable, unless the quantity taken has been very small.

In cases of death from hydrocyanic acid a marked odor of the poison is almost always observed in the apartment and upon opening the body, even several days after death. In cases of suicide or accident, the vessel from which the poison has been taken will usually be found in close proximity to the body, although the absence of such vessel is *not* proof that the case is one of homicide.

Notwithstanding the volatility and instability of the poison, its presence has been detected two months after death, although the chances of separating it are certainly the better the sooner after death the analysis is made. The search for hydrocyanic acid is combined with that for phosphorus; the part of the distillate containing the more volatile products is examined by the tests given above; it is best, when the presence of free hydrocyanic acid is suspected, to distil at first without acidulating.

328

In such cases the stomach should never be opened until immediately before the analysis.

Cyanic acid—*Cyanogen hydrate*—$\left.\begin{matrix}CN\\H\end{matrix}\right\rangle O$—43—does not exist in nature; it is obtained by calcining the cyanides in presence of an oxidizing agent; or by the action of dicyanogen upon solutions of the alkalies or alkaline carbonates; or by the distillation of cyanuric acid.

It is a colorless liquid; has a strong odor, resembling that of formic acid; its vapor is irritating to the eyes, and it produces vesication when applied to the skin; it is soluble in water. When free it is readily changed by exposure to air into cyamelide.

Sulphocyanic acid—*Cyanogen sulphydrate*—$\left.\begin{matrix}CN\\H\end{matrix}\right\rangle S$—59—bears the same relation to cyanic acid that CS_2 does to CO_2. It is obtained by the decomposition of its salts, which are obtained by boiling a solution of the cyanide with S; by the action of dicyanogen upon the metallic sulphide; and in several other ways.

The free acid is a colorless liquid; crystallizes at −12°.5 (9°.5 F.); boils at 102°.5 (216°.5 F.); acid in reaction. The prominent reaction of the acid and of its salts is the production of a deep red color with the ferric salts; the color being discharged by solution of mercuric chloride, but not by HCl.

Sulphocyanic acid exists in human saliva in combination, probably, with sodium. The free acid is actively poisonous and its salts were formerly supposed to be so also; it is probable, however, that much of the deleterious action of the potassium salt—that usually experimented with—is due as much to the metal as to the acid.

Metallocyanides.—The radical cyanogen, besides combining with metallic elements to form true cyanides, in which the radical (CN) enters as a univalent atom, is capable of combining with certain metals (notably those of the iron and platinum groups) to form complex radicals. These combining with H, form acids, and with basic elements form salts in which the analytical reactions of the metallic element entering into the radical are completely masked. Of these *metallocyanides* the best known are those in which iron enters into the radical. As iron is capable of forming two series of compounds, in one of which the single atom Fe″ enters in its bivalent capacity, and in the other of which the hexavalent double atom $(Fe_2)^{vi}$ is contained; so, uniting with cyanogen, iron forms two ferrocyanogen radicals: $[(CN)'_6Fe'']^{iv}$, *ferrocyanogen*, and $[(CN)'_{12}(Fe_2)^{vi}]^{vi}$ *ferricyanogen;* each of which unites with hydrogen to form an acid, corresponding to which are numerous salts: $(C_6N_6Fe)H_4$, *hydroferrocyanic acid*, tetrabasic; and $(C_{12}N_{12}Fe_2)H_6$, *hydroferricyanic acid*, hexabasic (see potassium and iron salts).

COMPOUNDS OF UNKNOWN CONSTITUTION.

GLUCOSIDES.

Under this head are classed a number of substances, some of them important medicinal agents, which are the products of vegetable or animal nature. Their characteristic property is that, under the influence of a dilute mineral acid, they yield glucose, phloroglucin or mannite, together with some other substance. Under the supposition that glucose and its

329

congeners are alcohols, it is quite probable that the glucosides are their corresponding ethers.

Amygdalin, $C_{20}H_{27}NO_{11}$—457—exists in cherry-laurel and in bitter almonds, but not in sweet almonds. Its characteristic reaction is that, in the presence of *emulsin*, which exists in sweet as well as in bitter

almonds, and of water, it is decomposed into glucose, benzoic aldehyde, and hydrocyanic acid. The same reaction is brought about by boiling with dilute H_2SO_4 or HCl. Bitter almonds contain about 2 per cent. of amygdalin.

Digitalin.—The pharmaceutical products sold under the above name, and obtained from *digitalis*, are mixtures in varying proportions of several glucosides. *Digitonin*, $C_{31}H_{52}O_{17}$, an amorphous, yellowish substance, very soluble in aqueous alcohol. *Digitalin*, $C_5H_8O_2$, the principal constituent of the French digitalin, is a colorless, very bitter, crystalline solid, insoluble in water, soluble in alcohol. *Digitaleïn*, a white, intensely bitter, amorphous solid, very soluble in water, soluble in alcohol. *Digitoxin*, $C_{21}H_{32}O_7$, a colorless crystalline solid, insoluble in water, sparingly soluble in alcohol. It is not a glucoside, and is converted into *toxiresin* by dilute acids.

The *abstractum digitalis* (*U. S.*) probably contains all the above, the extraction of the first being more complete with weak alcohol, that of the others with strong alcohol.

Glycyrrhizin.—A non-crystallizable, yellowish, pulverulent principle, obtained from liquorice; soluble with difficulty in cold water, soluble in hot water, alcohol, and ether; bitter-sweet in taste. By long boiling with dilute acids it is decomposed into glucose and *glycyrrhetin*, $C_{18}H_{26}O_4$.

Jalapin—$C_{34}H_{56}O_{16}$—720—is the active principle of scammony, and exists also to a limited extent in jalap (see below). It is an insipid, colorless, amorphous substance, which is decomposed by dilute acids into glucose and *jalapinol*. The active ingredient of jalap is not, as the name would imply, jalapin, but a resinous substance called *convolvulin*, which is insoluble in ether, odorless, and insipid. It is not attacked by dilute H_2SO_4, although the concentrated acid dissolves it with a carmine-red color, slowly turning to brown; in alcoholic solution it is decomposed by gaseous HCl into glucose and *convolvulinic acid*.

Quinovin—*Quinovatic acid.*—A bitter principle, possessed of acid functions, obtained from the false bark, known as *cinchona nova*; it is a glucoside, being decomposed by dilute acids into a sugar resembling mannitan and *quinovic acid*.

Salicin—*Salicinum* (*U. S.*)—$C_{13}H_{18}O_7$—286—occurs in the bark of the willow (*salix*). It is a white, crystalline substance; insoluble in ether, soluble in water and in alcohol; very bitter, its solutions are dextrogyrous, $[a]_D = +55°.8$. Dilute acids decompose it into glucose and *saligenin* (*q. v.*). Concentrated H_2SO_4 colors it red, the color being discharged on the addition of water. When taken into the economy it is converted into salicylic aldehyde and acid, which are eliminated in the urine.

Santonin—*Santonic acid*—$C_{15}H_{18}O_3$—*Santoninum* (*U. S., Br.*)—246.—A glucoside having distinct acid properties; obtained from various species of *Artemisia*. It crystallizes in colorless, rectangular prisms, which turn yellow on exposure to light; odorless and tasteless; insoluble in cold water, sparingly soluble in hot water, alcohol, and ether; its solutions are faintly acid in reaction. Santonin, in solution, gives a chamois-colored precipitate with the ferric salts, and a white precipitate with silver, zinc, and mercurous salts; no precipitate with mercuric salts.

Patients taking santonin pass urine having the appearance of that containing bile, which, when treated with potash, turns cherry-red or

330

crimson, the color being discharged by an acid, and regenerated on neutralization.

Solanin.—A glucoside, having basic properties, existing in different plants of the genus *Solanum*. It crystallizes in fine, white, silky needles; having an acrid, bitter taste; insoluble in water, and but sparingly soluble in ether and in alcohol. By the action of hot dilute acids it is decomposed into glucose and a basic substance, *solanidin*. When not heated, solanin combines with acids to form uncrystallizable salts. Cold concentrated H_2SO_4 colors it orange-yellow, and finally forms with it a brown solution; HNO_3 dissolves it, the solution being at first colorless, afterward rose-pink.

Tannins—*Tannic acid*—$C_{14}H_{10}O_9$—322.—Quite a number of different substances of vegetable origin, principally derived from barks, leaves, and seeds. They are amorphous, soluble in water, astringent, capable of precipitating albumen, and of forming imputrescible compounds with the gelatinoids. They are, with one possible exception, glucosides.

Gallo-tannic Acid—*Acidum tannicum* (*U. S., Br.*)—is the best known of the tannins, and is obtained from nut-galls, *galla* (*U. S., Br.*), which are excrescences produced upon oak-trees by the puncture of minute insects. It appears as a yellowish, amorphous, odorless, friable mass; has an astringent taste; very soluble in water, less so in alcohol, almost insoluble in ether; its solutions are acid in reaction, and on contact with animal tissues give up the dissolved tannin, which becomes fixed by the tissue to form a tough, insoluble, and non-putrescible material (leather).

A freshly prepared solution of pure gallo-tannic acid gives a dark blue precipitate with ferric salts, but not with ferrous salts. If, however, the solution have been exposed to the air, it is altered by oxidation, and gives, with ferrous salts, a black color (in whose production gallic acid probably plays an important part), which is the coloring material of ordinary writing-ink.

Caffetannic Acid—exists in saline combination in coffee and in Paraguay tea. It colors the ferric salts green, and does not affect the ferrous salts, except in the presence of ammonia: it precipitates the salts of quinine and of cinchonine, but does not precipitate tartar emetic or gelatin. It is a glucoside, being decomposed by suitable means into *caffeic acid* and mannitan.

Cachoutannic Acid—obtained from catechu, is soluble in water, alcohol, and ether. Its solutions precipitate gelatin, but not tartar emetic; they color the ferric salts grayish green.

Morintannic Acid—*Maclurin*—a yellow, crystalline substance, obtained from fustic; more soluble in alcohol than in water. Its solutions precipitate green with ferroso-ferric solutions; yellow with lead acetate; brown with tartar emetic; yellowish-brown with cupric sulphate. It is decomposable into phoroglucin and *protocatechuic acid*.

Quercitannic acid is the active tanning principle of oak-bark; it differs from gallo-tannic acid in not being capable of conversion into gallic acid, and in not furnishing pyrogallol on dry distillation. It forms a violet-black precipitate with ferric salts. The tannin existing in black tea seems to be quercitannic acid.

Quinotannic acid, a tannin existing in cinchona barks, probably in combination with the alkaloids. It is a light yellow substance; soluble in water, alcohol, and ether; its taste is astringent, but not bitter. Dilute H_2SO_4 decomposes it, at a boiling temperature, into glucose and a red substance—*quinova red*.

331

ALKALOIDS.

***The alkaloids are organic, nitrogenized substances, alkaline in reaction, and capable of combining with acids to form salts in the same way as does ammonia.* They are also known as *vegetable* or *organic bases* or *alkalies*, and are probably amines of complex constitution. The similarity between the relation of the free alkaloids to their salts and that of ammonia to the ammoniacal salts is shown in the following equations:**

$$\underset{\text{Ammonia.}}{2NH_3} + \underset{\text{Sulphuric acid.}}{H_2SO_4} = \underset{\text{Ammonium sulphate.}}{(NH_4)_2SO_4}$$

$$\underset{\text{Morphia.}}{2C_{17}H_{19}NO_3} + \underset{\text{Sulphuric acid.}}{H_2SO_4} = \underset{\text{Morphonium sulphate.}}{(C_{17}H_{20}NO_3)_2SO_4}$$

Classification.—The natural alkaloids are temporarily arranged in two groups:

(1.) Those which are liquid and volatile, and consist of C, H and N. The synthesis of one of their number shows that they are true amines.

(2.) Those which are solid, crystalline, volatile with difficulty, if at all, and consist of C, H, N and O. No representative of this class has yet been obtained by synthesis.

General Physical Characters.—As a rule they are insoluble, or nearly so, in water; more soluble in alcohol, chloroform, petroleum-ether, and benzol. Their salts are, for the most part, soluble in water and insoluble or sparingly soluble in petroleum-ether, benzol, ether, chloroform, and amyl alcohol. All exert a rotary action on polarized light:

Quinine	$[a] = -126°.7$	Codeine	$[a] = -118°.2$
Quinidine	$[a] = +250°.75$	Narceine	$[a] = -6°.7$
Cinchonine	$[a] = +190°.4$	Strychnine	$[a] = -132°.07$
Cinchonidine	$[a] = -144°.61$	Brucine	$[a] = -61°.27$
Morphine	$[a] = -88°.4$	Nicotine	$[a] = -93°.5$
Narcotine	$[a] = -103°.5$		

Generally, combination with an acid diminishes their rotary power; with quinine the reverse is the case. Free narcotine is lævogyrous; its salts are dextrogyrous. They are all bitter in taste.

General Chemical Reactions.—*Potash, soda, ammonia, lime, baryta, and magnesia* **precipitate the alkaloids from solutions of their salts.**

Phosphomolybdic acid forms a precipitate which is bright yellow, with aniline, morphine, veratrine, aconitine, emetine, atropine, hyoscyamine, theine, theobromine, coniïne, and nicotine; brownish-yellow with narcotine, codeine, and piperine; yellowish-white with quinine, cinchonine and strychnine; yolk-yellow with brucine.

The reagent is prepared as follows: Ammonium molybdate is dissolved in H_2O, the solution filtered, and a quantity of hydrodisodic phosphate 1/5 in weight of the molybdate used is added, and then NO_3H to strong acid reaction. The mixture is warmed; set aside for a day; the yellow ppt. collected on a filter; washed with H_2O acidulated with NO_3H; and while still moist transferred to a porcelain capsule, to which the liquid obtained by exhausting the remainder on the filter with NH_4HO is added. The fluid is warmed and gradually treated with pulverized sodium carbonate until a colorless solution is obtained. This is evaporated to dryness: a small quantity of sodium nitrate is added, and the whole gradually heated to quiet fusion and until all NH_3 is expelled. The residue is dissolved in warm H_2O (1 to 10), acidulated with NO_3H, and decanted.

To use the reagent, a drop of the suspected liquid is placed on a glass plate with a black background, and near it a drop of the reagent; and the two drops are made to mix slowly by a pointed glass rod.

332

***Potassium iodhydrargyrate* gives a yellowish precipitate with alkaloidal solutions which are acid, neutral, or faintly alkaline in reaction.**

The reagent is obtained by dissolving 13.546 grams of mercuric chloride and 49.8 grams of potassium iodide in a litre of water.

The solution may be used for quantitative determinations. The reagent is added from a burette to the solution of alkaloid until a drop, filtered from the solution which is being tested, and placed upon a black surface, gives no precipitate with a drop of the reagent. Each c.c. of reagent used indicates the presence in the volume of liquid tested of the following quantities of alkaloids, in grams:

Aconitine	0.0267	Brucine	0.0233	Nicotine	0.00405
Atropine	0.0145	Veratrine	0.0269	Quinine	0.0108
Narcotine	0.0213	Morphine	0.0200	Cinchonine	0.0102
Strychnine	0.0167	Coniïne	0.00416	Quinidine	0.0120

Of course, the process can be used only in a solution containing a single alkaloid.

Separation of Alkaloids from Organic Mixtures and from Each Other.—One of the most difficult of the toxicologist's tasks is the separation from a mixture of organic material (contents of stomach, viscera) of an alkaloid in such a state of purity as to render its identification perfect. The difficulty is the greater if the amount present be small, as is usually the case; and if the search be not confined to a single alkaloid, as frequently occurs. Some of these substances, as strychnine, are detectable with much greater facility and certainty than others.

Of the processes hitherto suggested, the best is that of Dragendorff, devised for the detection of any alkaloid or poisonous organic principle present in the substances examined. It is very exhaustive, and well adapted to cases frequently arising in chemico-legal practice; but, on the other hand, is too intricate to be serviceable to the general practitioner.

An abridgment of this process may be of use to detect the presence of the more commonly used alkaloids in a mixture of organic material. *The physician should, however, bear in mind that, in cases liable to give rise to legal proceedings, these may become seriously complicated by the analysis of any parts of the body, dejecta, or suspected articles of food, etc., by any process open to attack by the most searching cross-examination.*

The substances to be examined are reduced to a fine state of subdivision, and are digested for an hour or more in water acidulated with SO_4H_2, at a temperature of 40° to 50° (104°–122° F.); this is repeated three times, the liquid being filtered and the solid material expressed. The united extracts are evaporated at the temperature of the water-bath to a *thin* syrup; this is mixed with three or four volumes of alcohol, the mixture kept at about 35° (95° F.) for 24 hours, cooled well and filtered; the residue being washed with seventy per cent. alcohol. The alcohol is distilled from the filtrate, and the watery residue diluted with H_2O and filtered.

The filtrate so obtained contains the sulphates of the alkaloids, and from it the alkaloids themselves are separated by the following steps:

A. The acid watery liquid is shaken with freshly rectified petroleum ether (which should boil at about 65°–70° (149°–158° F.), and should be used with caution, as it is very inflammable): after several agitations the ether layer is allowed to separate and is removed; this treatment is repeated so long as the ether dissolves anything. The residue obtained by the evaporation of the ether—*Residue* I.—is mostly composed of coloring matters, etc., which it is desirable to remove.

B. The same treatment of the watery liquid is repeated with benzene, which on evaporation yields *Residue* II., which is, if crystalline, to be tested for cantharidin, santonin, and digitalin (*q. v.*); if amorphous, for elaterin and colchicin.

C. The acid, aqueous fluid is then treated in the same way with chloroform to obtain *Residue* III., which is examined for cinchonine, digitalin, and picrotoxin by the proper tests.

D. The watery fluid, after one more shaking with petroleum ether and removal of the ethereal layer, is rendered alkaline with ammonium hydrate and shaken with petroleum ether at 40° (104° F.), the ethereal layer being removed as quickly as possible while still warm; this is repeated two or three times, and repeated with *cold* petroleum ether, which is removed after a time. The warm ethereal layers yield *Residue* IV. *a*; the cold ones *Residue* IV. *b*. The former is tested for strychnine, quinine, brucine, veratrine; the latter for coniïne and nicotine.

E. The alkaline, watery fluid is shaken with benzene, which, on evaporation, yields *Residue* V., which may contain strychnine, brucine, quinine, cinchonine, atropine, hyoscyamine, physostigmine, aconitine, codeine, thebaine, and narceine.

F. A similar treatment with chloroform yields *Residue* VI., which may contain a trace of morphine.

G. The alkaline liquid is then shaken with amyl alcohol, which is separated and evaporated; *Residue* VII. is tested for morphine, solanin, and salicin.

H. Finally, the watery liquid is itself evaporated with pounded glass, the residue extracted with chloroform, and *Residue* VIII., left by the evaporation of the chloroform, tested for curarine.

333

Volatile Alkaloids.

Coniïne—*Conicine*—*Cicutine*—$C_8H_{15}N$—125—is obtained from *Conium maculatum*, in which it is accompanied by two other alkaloids, *methylconiïne*, $C_8H_{14}N(CH_3)$, and *conhydrine*, $C_8H_{17}NO$—the former a volatile liquid, the second a crystalline solid.

Coniïne is a colorless, oily liquid; has an acid taste and a disagreeable penetrating odor; sp. gr. 0.878; can be distilled when protected from air; boils at 212° (413°.6 F.); exposed to air it resinifies; it is very sparingly soluble in water, but is more soluble in cold than in hot water; soluble in all proportions in alcohol, soluble in six volumes of ether, very soluble in fixed and volatile oils.

The vapor which it gives off at ordinary temperatures forms a white cloud when it comes in contact with a glass rod moistened with HCl, as does NH_3. It forms salts which crystallize with difficulty. Cl and Br combine with it to form crystallizable compounds; I in alcoholic solution forms a brown precipitate in alcoholic solutions of coniïne, which is soluble without color in an excess. Oxidizing agents attack it with production of butyric acid (see below). The iodides of ethyl and methyl combine with it to form iodides of ethyl and methyl-conium. It has been obtained synthetically by first allowing butyric aldehyde and an alcoholic solution of ammonia to remain some months in contact at 30° (86° F.), when dibutyraldine is formed.

$$\underset{\text{Butyric aldehyde.}}{2(C_4H_8O)} + \underset{\text{Ammonia.}}{NH_3} = \underset{\text{Dibutyraldine.}}{C_8H_{17}NO} + \underset{\text{Water.}}{H_2O}$$

The dibutyraldine thus obtained is then heated under pressure to 150°–180° (302°–356° F.), when it loses water:

$$\underset{\text{Dibutyraldine.}}{C_8H_{17}NO} = \underset{\text{Coniïne.}}{C_8H_{15}N} + \underset{\text{Water.}}{H_2O}$$

A synthesis which, in connection with the decompositions of coniïne, shows its rational formula to be $\left.\begin{matrix}(C_4H_7)'\\(C_4H_7)'\\H\end{matrix}\right\}N$.

Analytical Characters.—(1.) With dry HCl gas it turns reddish purple, and then dark blue.

(2.) Aqueous HCl of sp. gr. 1.12 evaporated from coniïne leaves a green-blue, crystalline mass.

(3.) With iodic acid a white ppt. from alcoholic solutions.

(4.) With H_2SO_4 and evaporation of the acid: a red color, changing to green, and an odor of butyric acid.

Nicotine—$C_{10}H_{14}N_2$—162—exists in tobacco in the proportion of 2–8 per cent.

It is a colorless, oily liquid, which turns brown on exposure to light and air; has a burning, caustic taste and a disagreeable, penetrating odor; it distils at 250° (392° F.); it burns with a luminous flame; sp. gr. 1.027 at 15° (59° F.); it is very soluble in water, alcohol, the fatty oils, and ether; the last-named fluid removes it from its aqueous solution when the two are shaken together; it absorbs water rapidly from moist air. Its salts are deliquescent and crystallize with difficulty.

334

Analytical Characters.—(1.) Its ethereal solution, added to an ethereal solution of iodine, separates a reddish-brown, resinoid oil, which gradually becomes crystalline.

(2.) With HCl, a violet color.

(3.) With HNO_3, an orange color.

Both nicotine and coniïne are actively poisonous, producing death by asphyxia, sometimes as rapidly as prussic acid.

Fixed Alkaloids.

These are much more numerous than those which are volatile, and form the active principles of a great number of poisonous plants. As we are yet in the dark as to the constitution of these bodies, the classification which we adopt is the temporary one, based upon the botanic characters of the plants from which they are derived.

Opium Alkaloids.

Opium is the inspissated juice of the capsules of the poppy. It is of exceedingly complex composition, and contains, besides a neutral body called *meconin* (probably a polyatomic alcohol, $C_{10}H_{10}O_4$), a peculiar acid, *meconic acid* (*q. v.*), lactic acid, gum, albumen, wax, and a volatile matter—no less than eighteen different alkaloids, one or two of which, however, are probably formed during the process of extraction, and do not pre-exist in opium.

The following is a list of the constituents of opium, those marked * being of medical interest:

Name.	Formula.	Percent. in Smyrna opium.	Percent. in Constantinople opium.	Name.	Formula.	Percent. in Smyrna opium.	Percent. in Constantinople opium.
*Meconic acid	$C_7H_4O_7$	4.70	4.38	Laudanine	$C_{20}H_{25}NO_4$	....	....
Lactic acid	$C_3H_6O_3$	1.25	....	Papaverine	$C_{21}H_{21}NO_4$	1.00	....
Meconine	$C_{10}H_{10}O_4$	0.08	0.30	Opianine	$C_{21}H_{21}NO_7$	....	....
*Morphine	$C_{17}H_{19}NO_3$	10.30	4.50	Meconidine	$C_{21}H_{23}NO_4$	....	....
Pseudomorphine	$C_{17}H_{19}NO_4$	....	....	Cryptopine	$C_{21}H_{23}NO_5$	....	....
Hydrocotarnine	$C_{12}H_{15}NO_3$	....	....	Laudanosine	$C_{21}H_{27}NO_4$	....	....
*Codeine	$C_{18}H_{21}NO_3$	0.25	1.52	*Narcotine	$C_{22}H_{23}NO_7$	1.30	3.47
*Thebaine	$C_{19}H_{21}NO_3$	0.15	....	Lanthopine	$C_{23}H_{25}NO_4$	....	....
Protopine	$C_{20}H_{19}NO_5$	....	....	*Narceine	$C_{23}H_{29}NO_9$	0.71	0.42
Rhœadine	$C_{20}H_{21}NO_6$	....	....	Porphyroxine	....	....	....
Codamine	$C_{20}H_{25}NO_4$	....	....				

Morphine—*Morphina* (*U. S.*)—$C_{17}H_{19}NO_3 + Aq$—285 + 18—crystallizes in colorless prisms; odorless, but very bitter; it fuses at 120° (248° F.), losing its Aq. More strongly heated, it swells up, becomes carbonized, and finally burns. It is soluble in 1,000 pts. of cold water, in 100 pts. of boiling water; in 20 pts. of alcohol of 0.82, and in 13 pts. of boiling alcohol of the same strength; in 390 pts of cold amyl alcohol, much more soluble in the same liquid warm; almost insoluble in aqueous ether;

335

rather more soluble in alcoholic ether; almost insoluble in benzene; soluble in 60 pts. of chloroform. All the solvents dissolve morphine more readily and more copiously when it is freshly precipitated from solutions of its salts than when it has assumed the crystalline form.

Morphine combines with acids to form crystallizable salts, of which the chloride, sulphate, and acetate are used in medicine. If morphine be heated for some hours with excess of HCl, under pressure, to 150° (302° F.), it loses water, and is converted into a new base—*apomorphine*, $C_{17}H_{17}NO_2$.

By the action of H_2SO_4 on morphine at 100°, two amorphous, basic products of condensation, *trimorphine* and *tetramorphine*, are produced.

By heating together acetic anhydride and morphine, three modifications, α, β, γ of *acetyl-morphine*, $C_{17}H_{18}(C_2H_3O)NO_3$, are formed. Similarly substituted *butyryl-*, *benzoyl-*, *succinyl-*, *camphoryl-*, *methyl-*, and *ethyl-morphine* are also known.

Although the synthesis of morphine has not yet been accomplished, enough is known of its constitution to indicate that it contains the phenolic group (OH), and that it is a derivative of phenanthrene (see p. 325).

The salts of morphine are crystalline. The *acetate—Morphinæ Acetas, U. S.—Morphiæ Acetas, Br.*—is a white, crystalline powder, soluble in 12 parts of water, which decomposes on exposure to air, with loss of acetic acid. The *chloride—Morphinæ Hydrochloras, U. S.*—is less soluble, but more permanent than the acetate. The *sulphate—Morphinæ Sulphas, U. S.—Morphiæ Sulphas, Br.*—is the form in which morphine is the most frequently used in medicine. It is a very light, crystalline, feathery powder; odorless, bitter, and neutral in reaction. It dissolves in 24 parts of water. Its solutions deposit morphine as a white precipitate on addition of an alkali. The crystals contain 5 Aq., which they lose at 130° (266° F.).

ANALYTICAL CHARACTERS.—(1.) It is colored red, changing to yellow, by HNO_3.

(2.) Cold concentrated H_2SO_4 dissolves it, forming a colorless solution, which after 24 hours turns pink on addition of a trace of HNO_3; and the fluid when warmed, cooled, and diluted with H_2O, turns deep mahogany-brown on the addition of a splinter of potassium dichromate.

(3.) A mixture of morphine and cane-sugar (1 to 4) added to concentrated H_2SO_4 gives a dark red color, which is intensified by a drop of bromine-water.

(4.) If iodic acid solution and a drop of chloroform be added to morphine, free iodine is liberated, which colors the chloroform violet. If now dilute NH_4HO be floated on the surface of the liquid, a dark brownish zone is formed.

(5.) A neutral solution of a morphine salt gives a blue color with neutral solution of ferric chloride.

(6.) A solution of molybdic acid in H_2SO_4 (Fröhde's reagent) gives with morphine a violet color, changing to blue, dirty green, and faint pink. Water discharges the color.

(7.) Solution of morphine acetate produces a gray ppt. when warmed with ammoniacal silver nitrate solution; and the filtrate turns red or pink with HNO

(8.) Auric chloride gives a yellow ppt., turning violet-blue, with solutions of morphine salts.

336

(9.) Add solution of Fe_2Cl_6 (2–16) to solution of potassium ferri-cyanide (the mixture must not assume a blue color), add morphine solution —a deep blue color.

(10.) Heat morphine with concentrated H_2SO_4 to 200° (392° F.) until green-black; add a drop of the liquid cautiously to water; the solution turns blue. Shake a portion with ether; the ether turns purple. Shake another portion with chloroform; the chloroform turns blue.

(11.) Warm the solid alkaloid with concentrated H_2SO_4; add cautiously a few drops of alcoholic solution of KHO (30 %); a yellow color is produced, changing to dirty red, then steel blue, and sky blue, and, with a further quantity of KHO solution, cherry red.

Codeine—*Codeina* (*U. S.*)—$C_{18}H_{21}NO_3 + Aq$—299 + 18—crystallizes in large rhombic prisms, or from ether, without Aq., in octahedra; bitter; soluble in 80 pts. cold water; 17 pts. boiling water; very soluble in alcohol, ether, chloroform, benzene; almost insoluble in petroleum ether.

ANALYTICAL CHARACTERS.—(1.) Cold concentrated H_2SO_4 forms with it a colorless solution, which turns blue after some days, or when warmed.

(2.) Fröhde's reagent dissolves it with a dirty green color, which after a time turns blue.

(3.) Chlorine water forms with it a colorless solution which turns yellowish red with NH_4HO.

Narceine—$C_{23}H_{29}NO_9 + 2Aq$—463 + 36—crystallizes in bitter, prismatic needles; sparingly soluble in water, alcohol, and amyl alcohol; insoluble in ether, benzol, and petroleum ether.

ANALYTICAL CHARACTERS.—(1.) Concentrated H_2SO_4 dissolves it with a gray brown color, which changes to red, slowly at ordinary temperatures, rapidly when heated.

(2.) Fröhde's reagent colors it dark olive green, passing to red after a time, or when heated.

(3.) Iodine solution colors it blue violet, like starch.

Narcotine—$C_{22}H_{23}NO_7$—413—crystallizes in transparent prisms, almost insoluble in water and in petroleum ether; soluble in alcohol, ether, benzol, and chloroform. Its salts are mostly uncrystallizable, unstable, and readily soluble in water and alcohol.

ANALYTICAL CHARACTERS.—(1.) Concentrated H_2SO_4 forms with it a solution, at first colorless, in a few moments yellow, and after a day or two, red.

(2.) Its solution in dilute H_2SO_4, if gradually evaporated until the acid volatilizes, turns orange red, bluish violet, and reddish violet.

(3.) Fröhde's reagent dissolves it with a greenish color, passing to cherry red.

Thebaine—*Paramorphine*—$C_{19}H_{21}NO_3$—311—crystallizes in white plates; tasteless when pure; insoluble in water; soluble in alcohol, ether, and benzol.

ANALYTICAL CHARACTERS.—(1.) With concentrated H_2SO_4 an immediate bright red color, turning to yellowish red.

(2.) Its solution in chlorine water turns reddish-brown with NH_4HO.

(3.) With Fröhde's reagent same as 1.

Meconic acid—$C_7H_4O_7 + 3Aq$—200 + 54—is a tribasic acid, peculiar to opium, in which it exists in combination with a part, at least, of the alkaloids. It crystallizes in small prismatic needles; acid and astringent in taste; loses its Aq at 120° (248° F.); quite soluble in water; soluble in alcohol; sparingly soluble in ether.

ANALYTICAL CHARACTERS.—With ferric chloride, a blood-red color, which

337

is not discharged by dilute acids or by mercuric chloride; but is discharged by stannous chloride and by the alkaline hypochlorites.

Apomorphine—$C_{17}H_{17}NO_2$—is used hypodermically as an emetic in the shape of the chloride, *apomorphinæ hydrochloras, U. S.* It is obtained by sealing morphine with an excess of strong HCl in a thick glass tube, and heating the whole to 140° (252° F.) for two to three hours. It is obtained also by the same process from codeine. The free alkaloid is a white, amorphous solid, difficultly soluble in water. The chloride forms colorless, shining crystals, which have a tendency to assume a green color on exposure to light and air. It is odorless, bitter, and neutral; soluble in 6.8 parts of cold water.

TOXICOLOGY OF OPIUM AND ITS DERIVATIVES.—Opium, its preparations and the alkaloids obtained from it are all active poisons. They produce drowsiness, stupor, slow and stertorous respiration, contraction of the pupil; small and irregular pulse, coma, and death. The symptoms set in from 10 minutes to 3 hours, sometimes immediately, sometimes only after 18 hours. Death has occurred in from 45 minutes to 3 days, usually in 5–18 hours. After 24 hours the prognosis is favorable. Death has been caused in an adult by one-half grain of acetate of morphia, while 30 grains a day have been taken by those accustomed to its use without ill effects.

The alkaloids of opium have not the same action. In soporific action, beginning with the most powerful, they rank thus: Narceine, morphine, codeine; in tetanizing action: thebaine, papaverine, narcotine, codeine, morphine; in toxic action: thebaine, codeine, papaverine, narceine, morphine, narcotine.

The treatment should consist in the removal of unabsorbed poison from the stomach by emesis and the stomach-pump, and washing out of the stomach after injection into it of powdered charcoal in suspension, or tea or coffee infusion. Cold affusions should be used and the patient kept awake.

After death the reactions for meconic acid and narcotine permit of distinguishing whether the poisoning was by opium or its preparations, or by morphine.

Cinchona Alkaloids.

Although by no means so complex as opium, cinchona bark contains a great number of substances: *quinine, cinchonine, quinidine, cinchonidine, aricine; quinic, quinotannic, and quinovic acids; cinchona red*, etc. Of these the most important are quinine and cinchonine.

Quinine—*Quinina* (*U. S.*)—$C_{20}H_{24}N_2O_2 + n\,Aq$—324 + n18—exists in the bark of a variety of trees of the genera *Cinchona* and *China*, indigenous in the mountainous regions of the north of South America, which vary considerably in their richness in this alkaloid, and consequently in value; the best samples of calisaya bark contain from 30 to 32 parts per 1,000 of the sulphate; the poorer grades 4 to 20 parts per 1,000; inferior grades of bark contain from mere traces to 6 parts per 1,000.

It is known in three different states of hydration, with 1, 2, and 3 Aq. and anhydrous. The anhydrous form is an amorphous, resinous substance, obtained by evaporation of solutions in anhydrous alcohol or ether. The first hydrate is obtained in crystals by exposing to air recently precipitated and well-washed quinine. The second by precipitating by ammonia a solution of quinine sulphate, in which H has been previously liberated by the action of Zn upon H_2SO_4; it is a greenish, resinous

338

body, which loses H_2O at 150° (302° F.). The third, that to which the following remarks apply, is formed by precipitating solution of quinine salts with ammonia.

It crystallizes in hexagonal prisms; very bitter; fuses at 57° (134°.6 F.); loses Aq. at 100° (212° F.) and the remainder at 125° (257° F.); becomes colored, swells up, and, finally, burns with a smoky flame. It does not sublime. It dissolves in 2,200 pts. of cold H_2O, in 760 of hot H_2O; very soluble in alcohol and chloroform; soluble in amyl alcohol, benzene, fatty and essential oils, and ether. Its alcoholic solution is powerfully lævogyrous, $[a]_D = -270°.7$ at 18° (64°.4 F.), which is diminished by increase of temperature, but increased by the presence of acids.

Analytical Characters.—(1.) Dilute H_2SO_4 dissolves quinine in colorless but fluorescent solution (see below).

(2.) Solutions of quinine salts turn green when treated with Cl and then with NH_3.

(3.) Cl passed through H_2O holding quinine in suspension forms a red solution.

(4.) Solution of quinine treated with Cl water and then with fragments of potassium ferrocyanide becomes pink, passing to red.

Sulphate—*Disulphate—Quininæ sulphas* (*U.S.*)—*Quiniæ sulphas* (*Br.*)—$SO_4(C_{20}H_{25}N_2O_2)_2 + 7Aq$—746 + 126—crystallizes in prismatic needles; very light; intensely bitter; phosphorescent at 100° (212° F.); fuses readily; loses its Aq. at 120° (248° F.), turns red, and finally carbonizes; effloresces in air, losing 6 Aq.; soluble in 740 pts. H_2O at 13° (55°.4 F.), in 30 pts. boiling H_2O, and 60 pts. alcohol. Its solution with alcoholic solution of I deposits brilliant green crystals of *iodoquinine sulphate*.

Hydrosulphate—*Quininæ bisulphas* (*U. S.*)—$SO_4H(C_{20}H_{25}N_2O_2) + 7Aq$—422 + 126—is formed when the sulphate is dissolved in excess of dilute H_2SO_4. It crystallizes in long, silky needles, or in short, rectangular prisms; soluble in 10 pts. H_2O at 15° (59° F.). Its solutions exhibit a marked fluorescence, being colorless, but showing a fine pale blue color when illuminated by a bright light against a dark background.

Impurities.—Quinine sulphate should respond to the following tests:

(1.) When 1 gram (15.4 grains) is shaken in a test tube with 15 c.c. (4 fl ʒ) of ether, and 2 c.c. (32 ♏.) of NH_4HO; the liquids should separate into two clear layers, without any milky zone between them (cinchonine).

(2.) Dissolved in hot H_2O, the solution precipitated with an alkaline oxalate, the filtrate should not ppt. with NH_4HO (quinidine).

(3.) It should dissolve completely in dilute H_2SO_4 (fats, resins).

(4.) It should dissolve completely in boiling, dilute alcohol (gum, starch, salts).

(5.) It should not blacken with H_2SO_4 (cane-sugar).

(6.) It should not turn red or yellow with H_2SO_4 (salicin and phlorizin).

(7.) It should leave no residue when burnt on platinum foil (mineral substances).

By the action of alkaline hydrates upon quinine, formic acid, chinoline (see p. 318), and pyridine bases (see p. 317) are produced.

Concentrated HCl at 140°–150° (284°–302° F.) decomposes quinine, with separation of methyl chloride and formation of *apoquinine*, $C_{19}H_{22}N_2O_2$, an amorphous base.

Oxidizing agents produce from quinine oxalic acid and acids related to pyridine, notably *pyridindicarbonic* or *cinchomeronic* acid, $C_5H_3N(COOH)_2$, which are also formed by oxidation of cinchonine.

339

Although cinchonine (see below) differs from quinine in composition by + O, and although the decompositions of the two bases show them both to be related to the chinoline and pyridine bases, attempts to convert cinchonine into quinine have resulted only in the formation of other products, among which is an isomere of quinine, *oxycinchonine*.

Methylquinine, $C_{20}H_{24}N_2O_2CH_3$, is a base which has a curare-like action.

Cinchonine—*Cinchonina* (*U. S.*)—$C_{19}H_{22}N_2O$—294—occurs in Peruvian bark in from 2 to 30 pts. per 1,000. It crystallizes without Aq. in colorless prisms; fuses at 150° (302° F.); soluble in 3,810 pts. H_2O at 10° (50° F.), in 2,500 pts. boiling H_2O; in 140 pts. alcohol and in 40 pts. chloroform. The salts of cinchonine resemble those of quinine in composition; are quite soluble in H_2O and alcohol; are not fluorescent; permanent in air; phosphorescent at 100° (212° F.).

Quinidine and Quinicine—are bases isomeric with quinine; the former occurring in cinchona bark, and distinguishable from quinine by its strong dextrorotary power; the second a product of the action of heat on quinine, not existing in cinchona.

Cinchonidine—a base, isomeric with cinchonine, occurring in certain varieties of bark; lævogyrous. At 130° (266° F.) H_2SO_4 converts it into another isomere, *cinchonicine*.

Caffeine—*Theine—Guaranine—Caffeina* (*U. S.*)—$C_8H_{10}N_4O_2$ + Aq—194 + 18—exists in coffee, tea, Paraguay tea, and other plants. It crystallizes in long, silky needles; faintly bitter; soluble in 75 pts. H_2O at 15° (59° F.); less soluble in alcohol and ether. Hot fuming HNO_3 converts it into a yellow liquid, which after evaporation, turns purple with NH_4HO.

Alkaloids of the Loganiaceæ.

Strychnine—*Strychnina* (*U. S.*)—$C_{21}H_{22}N_2O_2$—334—exists in the seeds and bark of different varieties of *strychnos*.

It crystallizes on slow evaporation of its solutions in orthorhombic prisms, by rapid evaporation as a crystalline powder; very sparingly soluble in H_2O and in strong alcohol; soluble in 5 pts. chloroform. Its aqueous solution is intensely bitter, the taste being perceptible in a solution containing 1 pt. in 600,000.

It is a powerful base; neutralizes and dissolves in concentrated H_2SO_4 without coloration; and precipitates many metallic oxides from solutions of their salts. Its salts are mostly crystallizable, soluble in H_2O and alcohol, and intensely bitter. The *acetate* is the most soluble. The *neutral sulphate* crystallizes, with 7 Aq., in rectangular prisms. The iodides of methyl and ethyl react with strychnine to produce the iodides of *methyl* or *ethylstrychnium*, white crystalline basic substances, producing an action on the economy similar to that of curare. When acted on by H_2SO_4 and potassium chlorate, with proper precautions, *strychnic* or *igasuric acid* is formed.

Analytical Characters.—(1.) Dissolves in concentrated H_2SO_4 without color. The solution deposits strychnine when diluted with H_2O, or when neutralized with magnesia or an alkali.

(2.) If a fragment of potassium dichromate (or other substance capable of yielding nascent O) is drawn through a solution of strychnine in H_2SO_4, it is followed by a streak of color; at first blue (very transitory and frequently not observed), then a brilliant violet, which slowly passes to rose-pink and finally to yellow. Reacts with $\frac{1}{50000}$ grain of strychnine.

340

(3.) A dilute solution of potassium dichromate forms a yellow, crystalline ppt. in strychnine solutions; which, when washed and heated with concentrated H_2SO_4 gives the play of colors indicated in 2.

(4.) If a solution of strychnine be evaporated on a bit of platinum foil, the residue moistened with concentrated H_2SO_4, the foil connected with the + pole of a single Grove cell, and a platinum wire from the − pole brought in contact with the surface of the acid, a violet color appears upon the surface of the foil.

(5.) Strychnine and its salts are intensely bitter.

(6.) A solution of strychnine introduced under the skin of the back of a frog causes difficulty of respiration and tetanic spasms, which are aggravated by the slightest irritation, and twitching of the muscles during the intervals between the convulsions. With a small frog, whose surface has been dried before injection of the solution, $\frac{1}{10000}$ grain of acetate of strychnine will produce tetanic spasms in 10 minutes and death in 2 hours.

(7.) Solid strychnine, moistened with a solution of iodic acid in H_2SO_4, produces a yellow color, changing to brick-red and then to violet-red.

(8.) Moderately concentrated HNO_3 colors strychnine yellow in the cold. A pink or red color indicates the presence of brucine.

Toxicology.—Strychnine is one of the most active and most frequently used of poisons. It produces a sense of suffocation, thirst, *tetanic spasms*, usually opisthotonos, sometimes emprosthotonos, occasionally vomiting, contraction of the pupils during the spasms, and death, either by asphyxia during a paroxysm or by exhaustion during a remission. The symptoms appear in from a few minutes to an hour after taking the poison, usually in about 20 minutes; and death in from 5 minutes to 6 hours, usually within 2 hours. Death has been caused by ¼ grain, and recovery has followed the taking of 20 grains.

The treatment should consist of the removal of the unabsorbed poison by the stomach-pump, injecting charcoal, and pumping it out after about 5 minutes; under the influence of chloroform if necessary. Chloral hydrate should be given.

Strychnine is one of the most stable of the alkaloids, and may remain for a long time in contact with putrefying organic matter without suffering decomposition.

Brucine—$C_{23}H_{26}N_2O_4$ + 4 Aq—394 + 72—accompanies strychnine. It forms oblique rhomboidal prisms, which lose their Aq. in dry air. Sparingly soluble in H_2O; readily soluble in alcohol, chloroform, and amyl alcohol; intensely bitter. It is a powerful base and most of its salts are soluble and crystalline. Its action on the economy is similar to that of strychnine but much less energetic.

Analytical Characters.—(1.) Concentrated HNO_3 colors it bright red, soon passing to yellow; stannous chloride, or colorless NH_4HS, change the red color to violet.

(2.) Chlorine water, or Cl, color brucine bright red, changed to yellowish brown by NH_4HO.

Alkaloids of the Solanaceæ.

Solanine—$C_{43}H_{71}NO_{16}$—857—obtained from many species of *Solanum*; crystallizes in small, white, bitter, sparingly soluble prisms. Concentrated H_2SO_4 colors it orange red, passing to violet and then to brown. It is colored yellow by concentrated HCl. It dissolves in concentrated

341

HNO_3, the solution being at first colorless, but after a time becomes purple.

Atropine—*Daturine*—*Atropina, U. S.*—*Atropia,* Br.—$C_{17}H_{23}NO_3$—289—occurs in *atropa belladonna* and in *datura stramonium.* It forms colorless, silky needles, which are sparingly soluble in cold water, more readily soluble in hot water, very soluble in chloroform. It is odorless, but has a disagreeable, persistent bitter taste. It is distinctly alkaline, and neutralizes acids with formation of salts. One of these, the sulphate—*Atropinæ Sulphas, U. S.*—is a white, crystalline powder, readily soluble in water, which is the form in which atropine is usually administered.

Toxicology.—It is actively poisonous, producing drowsiness, dryness of the mouth and throat, dilatation of the pupils, loss of speech, diplopia, dizziness, delirium, coma.

The treatment should consist in the administration of emetics and the use of the stomach-pump.

Analytical Characters.—(1.) If a fragment of potassium dichromate be dissolved in a few drops of H_2SO_4, the mixture warmed, a fragment of atropine and a drop or two of H_2O added, and the mixture stirred, an odor of orange-blossoms is developed.

(2.) A solution of atropine dropped upon the eye of a cat produces dilatation of the pupil.

(3.) The dry alkaloid (or salt) is moistened with fuming HNO_3 and the mixture dried on the water-bath. When cold it is moistened with an alcoholic solution of KHO—a violet color which changes to red.

When atropine is heated with concentrated HCl to 120°–130° (248°–266° F.) for several hours, or when it is warmed with baryta-water to 58° (136°.4 F.) it is decomposed into a base related to the pyridines: **Tropine**—C_7H_{11}—**OH, NCH_3**—and, at first, **tropic acid**—$C_9H_{10}O_3$—but later **atropic acid**—CH_2—$C(C_6H_5)COOH$. Tropine is also produced by a similar decomposition of hyoscyamine.

Hyoscyamine—$C_{15}H_{23}NO_3$—occurs, along with another base, *hyoscine,* isomeric with atropine, in *hyoscyamus niger.* It crystallizes, when pure, in odorless, white, silky needles, whose taste is very sharp and disagreeable, and which are very sparingly soluble in water. As most commonly met with, it forms a yellowish, soft, hygroscopic mass which gives off a peculiar, tobacco-like odor. It neutralizes acids. Its sulphate—*Hyoscyaminæ Sulphas, U. S.*—forms yellowish crystals, very soluble in water, hygroscopic, and neutral in reaction.

Alkaloids from other Sources.

Ergotine—$C_{50}H_{52}N_2O_3$—and **Ecboline**—are two brown, amorphous, faintly bitter, and alkaline alkaloids obtained from ergot. They are readily soluble in water and form amorphous salts. The medicinal preparations known as ergotine are not the pure alkaloid.

Colchicine—$C_{17}H_{19}NO_5$—occurs in all portions of *colchicum autumnale* and other members of the same genus. It is a yellowish-white, gummy, amorphous substance, having a faintly aromatic odor and a persistently bitter taste. It is slowly but completely soluble in water, forming faintly acid solutions. It forms salts which are, however, very unstable.

342

Concentrated HNO_3, or, preferably, a mixture of H_2SO_4 and $NaNO_2$ colors colchicine blue-violet. If the solution be then diluted with H_2O, it becomes yellow, and on addition of NaHO solution, brick-red.

Veratrine—*Veratrina, U. S.*—$C_{32}H_{52}N_2O_8$—occurs in *veratrum officinalis = asagræa officinalis,* accompanied by **Sabadilline**—$C_{20}H_{26}N_2O_5$—**Jervine**—$C_{30}H_{46}N_2O_3$—and other alkaloids. The substance to which the name *Veratrina, U. S.,* applies is not the pure alkaloid, but a mixture of those occurring in the plant.

Concentrated H_2SO_4 dissolves veratrine, forming a yellow solution turning orange in a few moments, and then, in about half an hour, bright carmine red. Concentrated HCl forms a colorless solution with veratrine, which turns dark red when cautiously heated.

Piperine—$C_{17}H_{19}NO_3$—occurs in black and white pepper. It crystallizes in colorless, transparent prisms; almost tasteless when pure; very sparingly soluble in water. It is a very weak base.

If piperine be heated with alcoholic KHO, it is decomposed into *piperidine*—$C_5H_{11}N$—and *piperic acid*—$C_{12}H_{10}O_4$. If piperidine be treated with silver oxide, pyridine (see p. 317) is formed.

Berberine—*Xanthopicrite*—$C_{20}H_{17}NO_4$—occurs in *berberis vulgaris, cocculus palmatus,* and many other plants. It crystallizes in fine yellow needles or prisms; bitter in taste and neutral in reaction. It is difficultly soluble in cold water, readily soluble in alcohol and in boiling water. It forms well-defined, crystalline, yellow salts.

Aconitine—$C_{26}H_{35}NO_7(OH)_2O(CO,C_6H_5)$—is an alkaloid obtained from *aconitum napellus* and other species of *aconitum.* It is a colorless and odorless powder, possessed of an intensely bitter taste, and sharp, burning after-taste. It is strongly alkaline; almost insoluble in water, readily soluble in alcohol, ether, chloroform, or benzene. It neutralizes acids completely, with formation of well-defined, crystalline salts.

Aconite contains, besides aconitine, three other alkaloids, if not a greater number: *Napelline, acolyctine,* and *lycoctonine.* These three alkaloids, notably the first named, along with small quantities of aconitine, constitute the English or Morson's "aconitine," which is probably made from *aconitum ferox.* Probably, also, all commercial samples of aconitine are mixtures of aconitine and nepalline with lesser quantities of the other alkaloids and *aconine* and *pseudaconine.*

If aconitine be heated in sealed tubes with H_2O to 140°–150° (284°–302° F.) for several hours, it is decomposed into benzoic acid and *aconine,* $C_{26}H_{35}NO_7(OH)_4$.

A Japanese variety of aconite contains a peculiar alkaloid: *Japaconitine,* $C_{66}H_{88}N_2O_2$.

Analytical Characters.—(1.) Concentrated H_2SO_4 dissolves aconitine, forming a light, yellow-brown solution, which slowly turns darker, and changes to light yellow on addition of HNO_3. (2.) If aconitine be dissolved in aqueous phosphoric acid, and the solution very gradually evaporated, a violet color is produced.

Toxicology.—Aconite and aconitine have been the agents used in quite a number of homicidal poisonings.

The symptoms usually manifest themselves within a few minutes; sometimes are delayed for an hour. There is numbness and tingling, first of the mouth and fauces, later becoming general. There is a sense of dryness and of constriction in the throat. Persistent vomiting usually occurs, but is absent in some cases. There is diminished sensibility, with numbness,

343

great muscular feebleness, giddiness, loss of speech, irregularity and failure of the heart's action. Death may result from shock if a large dose of the alkaloid be taken, but more usually it is by syncope.

The treatment should be directed to the removal of unabsorbed poison by the stomach-pump, and washing out of the stomach with infusion of tea holding powdered charcoal in suspension. Stimulants should be freely administered.

Pilocarpine—$C_{11}H_{16}N_2O_2$—is the principal alkaloid of *jaborandi.* It forms a colorless, amorphous mass, readily soluble in water, alcohol, ether, and chloroform. It readily forms salts. Its chloride—*Pilocarpinæ hydrochloras, U. S.*—occurs in white, deliquescent, odorless crystals.

Cocaine—$C_{17}H_{21}O_4$—is an alkaloid obtained from the leaves of *erythroxylon coca.* It crystallizes in large, six-sided prisms. Its taste is at first bitter, producing paralysis of the sense of taste subsequently. It is strongly alkaline. Its chloride, extensively used for the production of local anæsthesia, crystallizes in well-formed prismatic needles, readily soluble in water.

When heated with concentrated HCl, it is decomposed into benzoic acid, methyl alcohol, and a new base, *ecgonin,* $C_9H_{15}NO_3$.

Physostigmine—*Eserine*—$C_{15}H_{21}N_3O_2$—is an alkaloid existing in the Calabar bean, *physostogma venenosum.* It is a colorless, amorphous solid, odorless and tasteless, alkaline and difficultly soluble in water. It neutralizes acids completely, with formation of tasteless salts. Its salicylate—*Physostigminæ Salicylas, U. S.*—forms short, colorless, prismatic crystals, sparingly soluble in water.

Concentrated H_2SO_4 forms a yellow solution with physostigmine or its salts, which soon turns olive-green. Concentrated HNO_3 forms with it a yellow solution. If a solution of the alkaloid in H_2SO_4 be neutralized with NH_4HO, and the mixture warmed, it is gradually colored red, reddish yellow, green, and blue.

Curarine—$C_{36}H_{35}N$ (?)—is an alkaloid obtainable from the South American arrow-poison, *curare,* or *woorara.* It crystallizes in four-sided, colorless prisms, which are hygroscopic, faintly alkaline, and intensely bitter.

Curarine dissolves in H_2SO_4, forming a pale violet solution, which slowly changes to red. If a crystal of potassium dichromate be drawn through the H_2SO_4 solution, it is followed by a violent coloration, which differs from the similar color obtained with strychnine under similar circumstances, in being more permanent, and in the absence of the following pink and yellow tints.

Emetine—$C_{28}H_{40}N_2O_5$—an alkaloid existing in *ipecacuanha,* which crystallizes in colorless needles or tabular crystals, slightly bitter and acrid; odorless, and sparingly soluble in water.

It dissolves in concentrated H_2SO_4, forming a green solution, which gradually changes to yellow. With Fröhde's reagent it gives a red color, which soon changes to yellowish-green and then to green.

Ptomaïnes.

This name, derived from πτῶμα = that which is fallen—*i.e.,* a corpse—was first suggested by Selmi to apply to a class of substances, first distinctly recognized by him, which are produced from albuminoid substances under

344

the influence of putrefactive decomposition, and which are distinctly alkaloidal in character.

The ptomaïnes are possessed of all of the distinguishing characters of the vegetable alkaloids. They are alkaline in reaction, and combine with acids to form salts. Some are liquid, others are solid and crystalline. Some are actively poisonous, others are practically inert. They behave toward the general reagents for alkaloids in much the same way as do the vegetable alkaloids.

Although the names *ptomaïnes* and *cadaveric alkaloids* are applied to alkaloids of animal origin, it is certain that such alkaloids may be and are produced during life in the animal economy.

It was feared that, as alkaloidal substances in many respects resembling those of vegetable origin are produced in the animal body, not only after death, but during life, grave doubts would be cast upon the results of analyses made to detect the presence of poisonous vegetable alkaloids in the cadaver in cases of suspected poisoning. Such fears were by no means groundless, as there is abundant evidence that ptomaïnes have been mistaken for vegetable alkaloids in chemico-legal analyses. The ptomaïnes, however, as well as the vegetable alkaloids, may be positively identified by a careful analysis based upon the use, not of a single reaction, but of *all known reactions* for the alkaloid in question. Therefor, it is possible to positively predicate the existence or non-existence of a given vegetable alkaloid in a cadaver, but it can only be done after a thorough and conscientious examination by all physiological and chemical reactions.

The ptomaïnes have of recent years assumed great importance to the physician by reason of their bearing upon the etiology of disease, and sufficient experimental evidence has already been obtained to warrant the belief that the method of action of many of the known pathogenic bacteria is by their production of alkaloidal poisons (see below).

One of the first of the putrid alkaloids to be formed in cadaveric matter is *choline* (see pp. 207, 273), which undoubtedly has its origin in the decomposition of the lecithins.

Neuridine—$C_5H_{14}N_2$ (?)—is a diamine, related to *neurine* (see p. 208), which is formed during the early stages of cadaveric putrefaction. It is gelatinous, readily soluble in water, insoluble in alcohol and ether, and very prone to decomposition, yielding dimethylamine and trimethylamine. It forms a chloride which crystallizes in long, transparent needles, very soluble in water. It is non-poisonous.

Cadaverine—$C_5H_{14}N_2$—identical with pentamethylendiamine, NH_2—$(CH_2)_5$—NH_2, is formed at a somewhat later stage of cadaveric putrefaction, along with putrescine and saprine (see below).

Its chloride is crystalline, hygroscopic, very soluble in water, insoluble in strong alcohol and ether. Like most of the ptomaïnes and several of the vegetable alkaloids, it gives a distinct blue color with ferric chloride and potassium ferricyanide. It is non-poisonous.

Putrescine—$C_4H_{12}N_2$—and **Saprine—$C_5H_{14}N_2$**—are two non-poisonous diamines produced along with cadaverine. They are both liquid, and each forms a crystalline chloride.

Mydaleïne is a putrid alkaloid, of undetermined composition, forming a difficultly crystallizable, hygroscopic chloride, which is actively poisonous. Five milligrammes administered hypodermically to a cat causes death after profuse diarrhœa and secretion of saliva, violent convulsions, and paralysis beginning with the extremities and extending to the muscles of respiration.

345

Neurine (see p. 208) is produced during the later stages of putrefaction. It is actively poisonous, and produces symptoms similar to those caused by muscarine. Atropine is a powerful antidote to its action.

Mydine—$C_8H_{11}NO$—is a base produced after continued putrefaction at comparatively low temperatures. It is a powerful base, and a strong reducing agent, and has an ammoniacal odor. It is non-poisonous.

Mydatoxine—$C_6H_{13}NO_2$—is a strongly alkaline syrup, which produces, when administered to animals, violent clonic spasms, followed by paralysis and death.

Other ptomaïnes produced during putrefaction of meat, fish, etc., are **methylguanidine, $C_2H_7N_3$**—poisonous; **muscarine, $C_5H_{15}NO_3$**—poisonous; and **gadinine, $C_7H_{17}NO_2$**—non-poisonous.

An alkaloid, many of whose chemical reactions have been determined, although its composition is unknown, has been obtained from the internal organs, and dejecta of cholera victims, as well as from cultures of the comma bacillus. This alkaloid, when administered to animals, causes symptoms of poisoning and death.

From the cultures of the Koch-Eberth typhus bacillus an alkaloid has been isolated—**Typhotoxine—$C_7H_{17}NO_2$**—which, when administered to animals, causes paralysis, copious diarrhœa, and death.

Tetanine—$C_{13}H_{30}N_2O_4$—is an alkaloid obtained from cultures of a bacillus originating from a wound which had been the cause of death by tetanus. It forms a deliquescent chloride, and a very soluble chloroplatinate. The free base or its chloride, when injected into mice or guinea-pigs, causes clonic or tonic convulsions of the greatest intensity, which terminate in death.

Mytilitoxine—$C_6H_{15}NO_2$—is an alkaloid obtained from poisonous mussels, which, when administered to animals in small amount, causes the same symptoms as are produced by the mussels.

ALBUMINOIDS AND GELATINOIDS.

Protein Bodies.

The substances of this class are never absent in living vegetable or animal cells, to whose "life" they are indispensable. They are as yet the products exclusively of the organized world.

Physical Characters.—They are almost all uncrystallizable and incapable of dialysis. Some are soluble in water, others only in water containing traces of other substances, others are insoluble. Their solutions are all lævogyrous. Some are separated as solids from their solutions, in a permanently modified form, by heat and by certain reagents; a change called *coagulation*. When once coagulated they cannot be redissolved. The temperature at which coagulation by heat occurs varies with different albuminoids, and is of value in distinguishing them from one another.

Composition.—They consist of C, N, H, O, and usually a small quantity of S, and form highly complex molecules whose exact composition is uncertain. Of their *constitution* nothing is definitely known, although there is probability that they are highly complex amides, related to the ureids, and formed by the combination of glycollamine, leucine, tyrosine, etc., with radicals of the acetic and benzoic series.

346

General Reactions.—They all respond to the following tests;

(1.) A purple red color when warmed to 70° (158° F.) with Millon's reagent. The reagent is made by dissolving, by the aid of heat, 1 pt. Hg in 2 pts. HNO_3 of sp. gr. 1.42; diluting with 2 vols. H_2O, and decanting after 24 hours.

(2.) A yellow color with HNO_3; changing to orange with NH_4HO (Xanthoproteic reaction).

(3.) A purple color with Pettenkofer's test (*q. v.*).

(4.) With a drop or two of cupric sulphate solution and liquor potassæ a violet color.

(5.) A solution of an albuminoid in excess of glacial acetic acid is colored violet and rendered faintly fluorescent by concentrated H_2SO_4.

(6.) With potassium ferrocyanide, in solutions strongly acid with acetic acid, a white ppt.

Decompositions.—Dilute acids decompose them into two substances: one insoluble, amorphous, yellowish, called *hemiprotein*; the other soluble in water, insoluble in alcohol, faintly acid, called *hemialbumin*. A prolonged boiling with moderately concentrated H_2SO_4 decomposes them, forming well-defined substances—glycocol, leucine, tyrosine; aspartic and glutamic acids. Alkalies dissolve them more or less readily; on boiling the solution, part of the sulphur is converted into sulphide and hyposulphite. Their alkaline solutions, when neutralized by acids, deposit Mulder's *proteine*. Concentrated alkalies decompose them into amido-acids. By fusion with alkalies, alkaline cyanides are also produced. When they are heated with caustic baryta and water at 100° (212° F.) carbonate, sulphate, oxalate, and phosphate of barium are deposited, and CO_2 and NH_3 are given off in the same proportions as when urea is similarly treated; when the temperature is raised, under pressure, finally to 200° (392° F.), a crystalline mass is formed which contains oxalic and acetic acids, a number of amido-acids, aspartic and glutamic acids, and a substance resembling dextrin. Heated with H_2O, under pressure, they are partly dissolved and partly decomposed. A mixture of H_2SO_4 and manganese dioxide, or potassium dichromate, produces from the albuminoids, aldehydes, and acids of the fatty and benzoic series, hydrocyanic acid, and cyanides. When heated under pressure with Br and H_2O they yield CO_2, oxalic and aspartic acids, amido acids, and bromine derivatives of the fatty and benzoic series. Potassium permanganate produces from them urea, CO_2, NH_3 and H_2O.

Putrefaction—*is a decomposition of dead albuminoid and gelatinous matter, attended by the evolution of fetid gas, and by the appearance of low forms of organized beings (bacteria).*

That it may occur there must have been contact with air, and there must be presence of moisture and a temperature between 5°–90° (41°–194° F.). It is attended by the breaking down and liquefaction of the material if it be solid; or its clouding and the formation of a scum upon the surface if it be liquid. The products of putrefaction vary with the conditions under which it occurs, the most prominent are: N, H, hydrocarbons, H_2S, NH_3, CO_2, certain ill-defined phosphorized and sulphurated bodies, acids of the acetic and lactic series, amido acids, and alkaloidal substances.

Under certain imperfectly defined conditions, buried animal matter is converted into a substance resembling tallow, and called *adipocere*, which consists chiefly of palmitate, stearate, and oleate of ammonium, phosphate and carbonate of calcium, and an undetermined nitrogenous substance.

Putrefaction may be prevented by: (1) exclusion of air; (2) removal

347

of water; (3) maintaining the temperature below 5° (41° F.); (4) the action of antiseptics.

Antiseptics are substances which prevent or restrain putrefaction.

Deodorizers, or air purifiers, are substances which destroy the odorous products of putrefaction.

Disinfectants are substances which restrain infectious diseases by destroying their specific poisons.

Certain substances are antiseptic, deodorant, and disinfectant; such are: chlorine, bromine, iodine, the hypochlorites, and sulphur dioxide; others lack one of the powers, as the mineral acids and the non-volatile "disinfectants," which are antiseptic and disinfectant, but not deodorant. Still others exert but one of the powers, as water and air, which may be mechanical deodorants, but neither disinfectants nor antiseptics.

There occurs a decomposition of vegetable tissues under the influence of warmth and moisture, which is known as *eremacausis*, differing from putrefaction in that the substances decomposed are the carbohydrate instead of the azotized constituents, and in the products of the decomposition, there being no fetid gases evolved (except there be simultaneous putrefaction), and the final product is a brownish material (humus or ulmin).

CLASSIFICATION.—In the present unsatisfactory state of our knowledge of the chemical constitution of these substances, we can only adopt a temporary classification, based upon their physical and physiological characters.

A. ALBUMINOIDS:

I. *Soluble in pure water; coagulated by heat.*—The true *albumins* of the white of egg, serum, and vegetable albumin.

II. *Insoluble in pure water; soluble in water without alteration in presence of neutral salts, alkalies and acids; and capable of precipitation unchanged from these solutions.*

1. *Globulins.*—Vitellin, myosin, paraglobulin, fibrinogen.
2. *Animal caseins.*—Milk casein, serum casein.
3. *Vegetable caseins.*—Gluten casein, legumin, conglutin.
4. *First terms of decomposition of the albuminoids by acids, alkalies, and cryptolytes.*—Albuminates (so called), acid albumin, syntonin, hemiprotein, peptone.

III. *Insoluble in water and only soluble after decomposition. Cannot be separated without alteration from their solutions in acids and alkalies.*—Gluten fibrin, gliadin, mucedin.

IV. *Coagulated.*—Coagulated albumin and fibrin.

V. *Amyloid matter.*—Lardacein.

B. GELATINOIDS:

I. *Collagenes.*—Collagen, elastin, ossein and its derivatives, chondrigen? chondrin? gelatin, keratin.

II. *Mucilaginous bodies.*—Mucin, paralbumin, colloidin.

ALBUMINOIDS.

I.—**Egg albumin** exists in solution, imprisoned in a network of delicate membranes, in the white of egg. It is obtained in an impure condition by cutting the whites of eggs with scissors, expressing through linen, diluting with an equal volume of water, filtering and concentrating the filtrate at a temperature below 40° (104° F.); mineral salts, which adhere to it tenaciously, are separated by dialysis. It seems to be a mixture of two different substances, one of which coagulates at 63° (145°.4 F.), and has the rotary power $[a]_o = -43°$; the other coagulates at 74° (165°.2 F.),

348

and has the value of $[a]_D = -26°$.

Its solutions are not precipitated by a small quantity of HCl, but an excess of that acid produces a deposit which is difficultly soluble in HCl, H_2O, and salt solution. Its characteristic reaction is that it is coagulated by agitation with ether.

Serum-albumin exists in blood-serum, chyle, lymph, pericardial fluid, the fluids of cysts and of transudations, in milk and, pathologically, in the urine. It is best obtained from blood-serum, after removal of paraglobulin (*q. v.*), by a tedious process, and only then in a state of doubtful purity. It is less abundant in the blood of some animals than paraglobulin, but more abundant in that of man.

Solutions of serum-albumin are lævogyrous $[a]_o = -56°$; they are not precipitated by CO_2, by acetic or orthophosphoric acid, by ether or by magnesium sulphate. They are precipitated by mineral acids, tannic acid, metaphosphoric acid, and most metallic salts. When heated they become opalescent at 60° (140° F.), and coagulate in the flocculent form at 72°–75° (161°.6–167° F.).

DETECTION AND DETERMINATION OF ALBUMIN IN URINE.—If the urine be not perfectly clear it is filtered, if this do not render it *perfectly* transparent, it is treated with a few drops of magnesia mixture (p. 85 note), and again filtered. The filtrate, if alkaline, is rendered *just* acid by adding acetic acid guttatim (nitric acid should not be used, and the acidulation of alkaline urine is imperative). The urine is now heated to near boiling, and if a cloudiness or precipitate be formed. HNO_3 is added slowly to the extent of about 10 drops. If heat produce a cloudiness, which clears up completely on addition of HNO_3, it is due to an excess of earthy phosphates. If a cloudiness produced by heat do not clear up (it may increase) on addition of HNO_3, it is due to albumin.

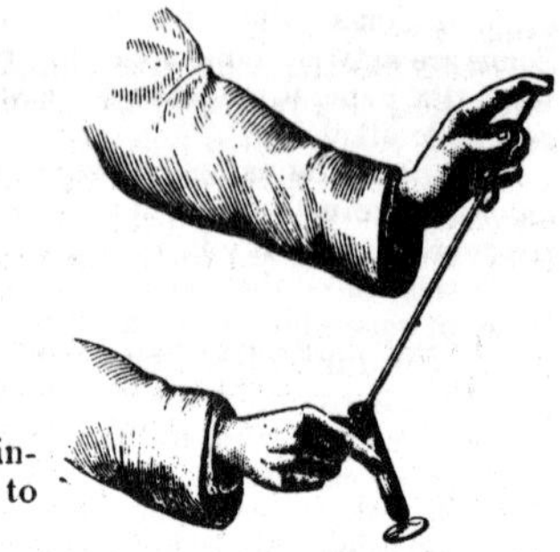

FIG. 40.

Small quantities of albumin may sometimes be better detected by Heller's test: A layer of HNO_3 is placed in a test-tube, which is then held at an angle and the urine allowed to flow *slowly* upon its surface (Fig. 40) so as to form a distinct layer, with the minimum of mixing of the two liquids; the test-tube is then brought to the vertical slowly, and the point of junction of the two liquids examined against a dark background. If albumin be present a white, opaque band, whose upper and lower borders are *sharply defined*, will be seen at the line of junction of the two liquids. When urates are present in excess, a white band will be observed, but its position will be rather above the line of junction, and its upper border will not be sharply defined, but gradually diminish in density from below upward. In non-albuminous urines there is usually a darkening, but never an opacity at the line of junction.

Quantity.—The only method of determining the quantity of albumin in urine, with an approach to accuracy, is gravimetric: 20–50 c.c. (5.4–13.5 fl ʒ) of the filtered urine (according as the qualitative testing shows albumin to be present in large or small quantity) are slowly heated over the water-bath, and, as the boiling temperature is approached, 3–4 drops of acetic acid are added. After the urine has boiled for a few moments. it is thrown upon a filter. The coagulum is washed with boiling H_2O, then with H_2O acidulated with NO_3H, then with alcohol and finally with ether. By these washings impurities are removed, and the albumin is caused to contract firmly, so that it can be easily detached and transferred to a weighed watch-glass: upon this it is dried at 115° (239° F.) and the whole weighed. The difference between the last weight and that of the watch-glass, is the weight of dry albumin in the volume of urine used.

349

Vegetable albumin—exists in solution in all vegetable juices, and forms the most valuable constituent of those vegetables which are used as food. It is coagulated from its solutions at 61°–63° (141°.8–145°.4 F.), and by nearly all acids.

II.—**Vitelin** exists in the yolk of egg and in the crystalline lens. It is soluble in dilute solution of sodium chloride, from which it is precipitated by excess of H_2O; by heating to 75°–80° (167°–176° F.); and by alcohol. It is not precipitated by solid sodium chloride. It dissolves in weak alkaline solutions without alteration and in very dilute HCl (1–1000), by which it is quickly converted into syntonin.

Myosin—is one of the principal constituents of the muscular fibre in rigor mortis. It is a faintly yellow, opalescent, distinctly alkaline liquid, which, when dropped into distilled H_2O, deposits the myosin in globular masses, while the H_2O assumes an acid reaction. It is insoluble in H_2O, easily soluble in dilute salt solution, from which it is precipitated by the addition of solid sodium chloride, or by a heat of 55°–60° (131°–140° F.). Very dilute HCl dissolves and converts it into syntonin.

Paraglobulin.—This substance has been described by various authors under the names: *plasmine* (Denis), *serum casein* (Panum), *serum globuline, fibrino-plastic matter* (Schmidt), *serin* (Denis). It exists in blood-serum, in pericardial fluid, hydrocele fluid, lymph and chyle, from which it is obtained by diluting with 10–15 volumes of ice-cold H_2O, treatment of the solution with strong current of CO_2, and washing the collected deposit with H_2O as long as a portion of the filtrate precipitates with acetic acid and potassium ferrocyanide, or with silver nitrate. It is a granular substance, which gradually becomes more compact; insoluble in H_2O, sparingly soluble in H_2O containing CO_2; soluble in dilute alkalies, in lime-water, in solutions of neutral alkaline salts, in dilute acids. Its solution in very dilute alkaline fluids is perfectly neutral and is not coagulated by heat, except after *faint* acidulation with acetic or mineral acids; it is precipitated by a large volume of alcohol; its solutions are also precipitated incompletely by dissolving sodium chloride in them to saturation, and completely by similar solution of magnesium sulphate; this last method of precipitation is used for the separation of paraglobulin from serum-albumin (see Fibrin).

Fibrinogen—after the separation of paraglobulin from blood-plasma, as described above, if the liquid be still further diluted and again treated with CO_2, a substance is obtained which, although closely resembling paraglobulin in many characters, is distinct from it, and, unlike paraglobulin, it cannot be obtained from the serum separated from coagulated blood.

Paraglobulin and fibrinogen are both soluble in a solution of sodium chloride containing 5–8 per cent. of the salt; when the degree of concentration of the salt solution is raised to 12–16 per cent., the fibrinogen is precipitated, while the paraglobulin remains in solution and is only precipitated, and then incompletely, when the percentage of salt surpasses twenty (see Fibrin).

Milk casein—the most abundant of the albuminoids of the milk of mammalia, closely resembles alkali albuminates, with which it is probably identical, as the main point of distinction has been found to be without significance. Unlike pure alkali albuminates, casein is coagulated from its solution by rennet (the product of the fourth stomach of the calf) at 40° (104° F.); but it has been found that alkali albuminate is also so coagu-

350

lated when milk-sugar and fat are added to the solution.

MILK.—The secretion of the mammary gland is water holding in solution casein, albumin, lactose, and salts; and fat in suspension. *Cream* consists of the greater part of the fat, with a small proportion of the other constituents of the milk. *Skim milk* is milk from which the cream has been removed. *Buttermilk* is cream from which the greater part of the fat has been removed, and consequently is of about the same composition as skim milk.

The composition of milk differs in animals of different species:

	Human.	Cow.	Goat.	Sheep.	Ass.	Mare.	Cream.	Condensed milk.
Water....	88.35	84.28	86.85	83.30	89.01	90.45	45.99	25.68
Solids....	11.65	15.72	13.52	16.60	10.99	9.55	54.01	74.32
Casein ... } Albumin.. }	3.15	3.57 / 0.78	2.53 / 1.26	5.73	3.57	2.53	6.33	16.83
Fat	3.87	6.47	4.34	6.05	1.85	1.31	43.97	10.27
Lactose ..	4.37	4.34	3.78	3.96	5.05	5.43	3.28	44.33*
Salts.....	0.26	0.63	0.65	0.68		0.29	0.42	2.80

* Including 28.98 parts of cane-sugar.

The composition of cows' milk varies considerably according to the age, condition, breed and food of the cow; to the time and frequency of milking; and to whether the sample examined is from the first, middle, or last part of each milking.

Cows' milk is very frequently adulterated, both by the removal of the cream and the addition of water. For ordinary purposes, the purity of the milk may be determined by observing the sp. gr. and the percentage of cream by the lactometer and creamometer, *neither of which, used alone, affords indications which can be relied upon*. The sp. gr. should be observed at the temperature for which the instrument is made, as in a complex fluid such as milk no valid correction for temperature is practical; it ranges in pure milk from 1027 to 1034, it being generally the lower in milk which has been watered, and in such as is very rich in cream, and the higher the less cream is present. The average sp. gr. is 1030; the average percentage of cream 13.

The percentage of cream is determined by the creamometer: a glass tube about a foot long and half an inch in diameter, the upper fifth (excluding about an inch from the top) being graduated into hundredths of the whole, the 0 being at the top. To use it, it is simply filled to the 0 with the milk to be tested, set aside for twenty hours and the point of separation between milk and cream read off. It should be above eight per cent.

This method of determining the purity of milk, although sufficient for ordinary purposes, should not be considered as affording evidence upon which to base legal proceedings; in such cases nothing short of a chemical determination of the percentage of fats, and of solids not fat, should be accepted as evidence of the impurity of milk.

Serum-casein is a substance obtained from blood-serum diluted with 10 volumes of H_2O, freed from paraglobulin by CO_2, and from albumin by acetic acid and heat. It is insoluble in salt solutions, slowly soluble in a one per cent. solution of sodium hydrate. Such a solution is partially pre-

351

cipitated by CO_2, almost completely by acetic acid, and completely by heating with excess of powdered sodium chloride; incompletely soluble in dilute HCl.

Gluten-casein.—That portion of crude gluten (a soft, elastic, grayish, material best obtained from flour) which is insoluble in alcohol, hot or cold; **Legumin**—a sparingly soluble albuminoid obtained from peas, beans, etc.; and **Conglutin**—a substance closely related to legumin and to *gliadin*, but differing from them in some characters, obtained from almonds, are three vegetable albuminoids resembling casein.

They are insoluble in pure water, readily soluble in dilute alkaline solutions, from which they are precipitated by acids and by rennet.

Alkali albuminates—*proteins* of Hoppe Seyler—are formed when an albuminoid is dissolved in concentrated solutions of potassium and sodium hydrates; it is very probable that they are identical with serum and milk-casein.

Acid albumins—are substances obtained by precipitating solutions of albuminoids by the simultaneous addition of an acid and a large quantity of a neutral salt; they vary exceedingly in composition and properties.

Syntonin—*Parapeptone*—is extracted from contractile tissues. The same substance is formed by the action of dilute acids upon the albuminoids, and as the first product of the action of the gastric juice, or of mixtures of pepsin and dilute acid upon albuminoids. It resembles serum casein closely, the only divergence in their properties being that syntonin is much more readily soluble in a 0.1 per cent. solution of HCl, and in faintly alkaline liquids.

Peptone—*Albuminose*—is the product of the action of the gastric and pancreatic juices upon albuminoids during the process of digestion. It is soluble in H_2O, insoluble in alcohol and in ether. Its watery solution is neutral, not precipitable by acids or alkalies, or by heat when faintly acid. Alcohol precipitates it in white, casein-like flocks, which, if slowly heated to 90° (194° F.) while still moist form a transparent, yellowish liquid, and, on cooling, an opaque, yellowish, glassy mass. It has a greater power than other albuminoids of combining with acids and bases.

The most important character of peptone, in which it differs from other albuminoids, is that it is readily dialysable. Its presence in the blood has not been demonstrated, and it is probable that immediately upon its entrance into the circulation it is converted into albuminoids resembling, yet differing from, those from which it was derived.

Peptone is produced by the action of many chemical reagents upon albuminoids; and also as one of the first products of putrefaction. When produced by putrefaction, or by artificial digestion, it is accompanied by *peptotoxine*, a crystallizable and actively poisonous alkaloidal substance.

It has been claimed that the gastric digestion of different albuminoids produces, not a single substance, but a distinct peptone for each albuminoid. If such be the case, and the present state of our knowledge does not permit of a definite answer to the question, these bodies are very closely related.

Peptone responds to the general reactions for the albuminoids (see p. 346), from which it may be distinguished by the *biuret reaction*. If a mere trace of $CuSO_4$ solution be added to a solution of peptone and then KHO or NaHO solution, a purple or reddish violet color is produced. A similar appearance is produced with acid albumins.

352

IV.—Coagulated albumins—are obtained, as described above, from the soluble varieties by the action of acids, heat, alcohol, etc. They are insoluble in water, alcohol, solutions of neutral salts; difficultly soluble in dilute alkaline solutions. In acetic acid they swell up and dissolve slowly; from this solution they are precipitated by concentrated salt solution. Concentrated HCl dissolves them with formation of syntonin. By the action of gastric juice, natural or artificial, they are converted first into syntonin, then into peptone.

Fibrin—is obtained when blood is allowed to coagulate or is whipped with a bundle of twigs. When pure it is at first a gelatinous mass, which contracts to a white, stringy, tenacious material, made up of numerous minute fibrils; when dried it is hard, brittle, and hygroscopic. It is insoluble in water, alcohol, ether; in dilute acid it swells up and dissolves slowly and incompletely. When heated with water to 72° (161°.6 F.), or by contact with alcohol, it is contracted, and is no longer soluble in dilute acids, but soluble in dilute alkalies. In solutions of many neutral salts of 6–10 per cent., it swells up and is partially dissolved; from this solution it separates on the addition of water, or upon the application of heat to 73° (163°.4 F.), or by acetic acid or alcohol. Moist fibrin has the property of decomposing oxygenated water with copious evolution of oxygen.

Fibrin does not exist as such in the blood, and the method of its formation and of the clotting of blood has been the subject of much experiment and argument; nor can the question be said to be definitely set at rest. In e light of the researches of Denis, Schmidt, and especially of Hammarsten, it may be considered as almost proven that fibrin is formed from fibrinogen under favorable circumstances, and by a transformation which is not yet understood. Whether paraglobulin plays any part directly in the formation of fibrin or not, is still an open question.

V.—Amyloid—is a pathological product, occurring in fine grains, resembling starch-granules in appearance, in the membranes of the brain and cord, in waxy and lardaceous liver, and in the walls of the blood-vessels. Its composition is that of the albuminoids, from which it differs in being colored red by iodine; violet or blue by iodine and H_2SO_4. Soluble in HCl with formation of syntonin; and in alkalies. It is not attacked by the gastric juice, and is not as prone to putrefaction as the other albuminoids.

GELATINOIDS.

I.—Collagen.—Bony tissue is made up mainly of tricalcic phosphate, combined with an organic material called *ossein*, which is a mixture of collagen, elastin, and an albuminoid existing in the bone-cells. Collagen also exists in all substances which, when treated with H_2O, under the influence of heat and pressure, yield gelatin. It is insoluble in cold H_2O, but by prolonged boiling is converted into gelatin, which dissolves. It is dissolved by alkalies.

Gelatin—obtained as above, from ossein, exists in the commercial product of that name, and in a less pure form in glue. When pure it is an amorphous, translucent, yellowish, tasteless substance, which swells up in cold H_2O, without dissolving, and forms, with boiling H_2O, a thick, sticky solution, which on cooling becomes, according to its concentration, a hard glassy mass or a soft jelly—the latter even when the solution is very dilute. It is insoluble in alcohol and ether, but soluble, on warming, in glycerin;

353

the solution in the last-named liquid forms, on cooling, a jelly which has recently been applied to various contrivances for copying writing. A film of gelatin impregnated with potassium dichromate becomes hard and insoluble on exposure to sunlight.

Chondrin is the name given to a substance obtained from cartilaginous tissue and supposed to be distinct from gelatin. It is probably a mixture of gelatin and mucin.

Elastin—is obtained from elastic tissues by successive treatment with boiling alcohol, ether, water, concentrated acetic acid, dilute potash solution and water. It is fibrous, yellowish; swells up in water and becomes elastic; soluble with a brown color in concentrated potash solution. It contains no S, and on boiling with H_2SO_4 yields glycol.

Keratin—is the organic basis of horny tissues, hair, nails, feathers, whalebone, epithelium, tortoise-shell, etc. It is probably not a distinct chemical compound, but a mixture of several closely related bodies.

II.—**Mucin**—is a substance containing no S and existing in the different varieties of mucus, in certain pathological fluids, in the bodies of molluscs, in the saliva, bile, connective tissues, etc. Its solutions, like the fluids in which it occurs, are viscid. It is precipitated by acetic acid and by HNO_3, but is dissolved by an excess of the latter; it dissolves readily in alkaline solutions, and swells up in H_2O, with which it forms a false solution. It is not coagulated by heat.

ANIMAL CRYPTOLYTES.

Soluble Animal Ferments.

Under this head are classed substances somewhat resembling the albuminoids, of unknown composition, occurring in animal fluids, and having the power of effecting changes in other organic substances, the method of whose action is undetermined. (See p. 182.)

Ptyalin—is a substance occurring in saliva, and having the power of converting starch into dextrin and a sugar resembling glucose (ptyalose), in liquids having an alkaline, neutral, or faintly acid reaction.

Pepsin—is the cryptolyte of the gastric juice. Attempts to separate it without admixture of other substances have hitherto proved fruitless; nevertheless, mixtures containing it and exhibiting its characteristic properties more or less actively have been obtained by various methods. The most simple consists in macerating the finely divided mucous membrane of the stomach in alcohol for 48 hours, and afterward extracting it with glycerin; this forms a solution of pepsin, which is quite active and resists putrefaction well, and from which a substance containing the pepsin is precipitated by a mixture of alcohol and ether.

If pepsin be required in the solid form, it is best obtained by Brücke's method. The mucous membrane of the stomach of the pig is cleaned and detached from the muscular coat by scraping; the pulp so obtained is digested with dilute phosphoric acid at 38° (100°.4 F.), until the greater part of it is dissolved; the filtered solution is neutralized with lime-water; the precipitate is collected, washed with H_2O, and dissolved in dilute HCl; to this solution a saturated solution of cholesterin, in a mixture of 4 pts. alcohol and 1 pt. ether, is gradually added; the deposit so formed is repeatedly shaken with the liquid, collected on a filter, washed with H_2O and then with dilute acetic acid, until all HCl is removed; it is then

354

treated with ether and H_2O: the former dissolves cholesterin and is poured off, the latter the pepsin; after several shakings with ether the aqueous liquor is evaporated at 38° (100°.4 F.), when it leaves the pepsin as an amorphous, grayish-white substance; almost insoluble in pure H_2O, readily soluble in acidulated H_2O; probably forming a compound with the acid, which possesses the property of converting albuminoids into peptone.

The so-called *pepsina porci* is either the calcium precipitate obtained as described in the first part of the above method: or, more commonly, the mucous membrane of the stomach of the pig, scraped off, dried, and mixed with rice-starch or milk sugar.

Pancreatin.—Under this name, substances obtained from the pancreatic secretion, and from extracts of the organ itself, have been described, and to some extent used therapeutically. They do not, however, contain all the cryptolytes of the pancreatic juice, and in many instances are inert albuminoids. The actions of the pancreatic juice are: (1) it rapidly converts starch, raw or hydrated, into sugar; (2) in alkaline solution—its natural reaction—it converts albuminoids into peptone; (3) it emulsifies neutral fats; (4) it decomposes fats, with absorption of H_2O and liberation of glycerin and fatty acids.

The pancreatic secretion probably contains a number of cryptolytes—certainly two. The one of these to which it owes its peptone-forming power has been obtained in a condition of comparative purity by Kühne, and called by him *trypsin*; in aqueous solution it digests fibrin almost immediately, but it exerts no action upon starch.

The diastatic (sugar-forming) cryptolyte of the pancreatic juice has not been separated, although a glycerin extract of the finely divided pancreatic tissue contains it, along with trypsin.

ANIMAL COLORING MATTERS.

Hæmoglobin and its Derivatives—*Hæmato-crystallin.*—The coloring matter of the blood is a highly complex substance, resembling the albuminoids in many of its properties, but differing from them in being crystallizable and in containing iron.

Hæmoglobin exists in the red-blood corpuscles in two conditions of oxidation; in the form in which it exists in arterial blood it is loosely combined with a certain quantity of oxygen, and is known as *oxyhæmoglobin.* The mean of many nearly concording analyses shows its composition to be $C_{600}H_{960}N_{154}FeS_3O_{179}$. When obtained from the blood of man and from that of many of the lower animals, it crystallizes in beautiful red prisms or rhombic plates; that from the blood of the squirrel in hexagonal plates; and that from the guinea-pig in tetrahedra. The crystals are always doubly refracting. It may be dried in vacuo at 0° (32° F.); if thoroughly dried below 0° (32° F.), it may be heated to 100° (212° F.) without decomposition, but the presence of a trace of moisture causes its decomposition at a much lower temperature. Its solubility in water varies with the species of animal from whose blood it was obtained; thus, that from the guinea-pig is but sparingly soluble, while that from the pig is very soluble. It is also dissolved unchanged by very weak alkaline solutions, but is decomposed by acids or salts having an acid reaction.

Hæmoglobin, or *reduced hæmoglobin*, is formed from oxyhæmoglobin in

355

the economy during the passage of arterial into venous blood; and by the action of reducing agents, or by boiling its solution at 40° (104° F.) in the vacuum of the mercury pump.

Oxyhæmoglobin is of a much brighter color than the reduced, and has a different absorption spectrum. The spectrum of oxyhæmoglobin varies with the concentration. In concentrated solutions the light is entirely absorbed, in more dilute solutions the spectrum 10, Fig. 14, is observed, and in still further dilutions 11, Fig. 14; in which the band at D is narrower, darker, and more sharply defined than the other. In highly diluted solution the band at D is alone visible. The spectrum of hæmoglobin consists of a single band much broader and fainter than either of the oxyhæmoglobin bands (12, Fig. 14).

Hæmoglobin, in contact with O or air, is immediately converted into oxyhæmoglobin. With CO it forms a compound resembling oxyhæmoglobin in the color of its solution, but in which the CO cannot be replaced by O; for which reason hæmoglobin, once combined with CO, becomes permanently unfit to fulfil its function in respiration (see p. 234).

When a solution of oxyhæmoglobin is boiled, it becomes turbid, and a dirty, brownish-red coagulum is deposited; the hæmoglobin has been decomposed into an albuminoid (or mixture of albuminoids), called by Preyer *globin*, and *hæmatin.* The latter, at one time supposed to be the blood-coloring matter, is a blue-black substance, having a metallic lustre and incapable of crystallization; it is insoluble in water, alcohol, ether, and dilute acids; soluble in alkaline solutions. It has the composition $C_{68}H_{70}N_8Fe_2O_{10}$. Its alkaline solutions exhibit the spectrum 13, Fig. 14. Although itself uncrystallizable, hæmatin combines with HCl to form a compound which crystallizes in rhombic prisms, and which is identical with the earliest known crystalline blood-pigment, *hæmin*, or Teichmann's crystals.

When reduced hæmoglobin is decomposed as above, in the absence of oxygen, hæmatin is not produced, but a substance identical with that called *reduced hæmatin*, and called by Hoppe-Seyler *hæmocromogen*; whose spectrum is shown in 14, Fig. 14.

If a solution of hæmoglobin be exposed for some time to air it changes in color from red to brownish, and assumes an acid reaction; it then exhibits the spectrum 15, Fig. 14, due to the production of *methæmoglobin*, probably a stage in the conversion of hæmoglobin into hæmatin and globin.

Biliary pigments.—There are certainly four, and probably more, pigmentary bodies obtainable from the bile and from biliary calculi, some of which consist in great part of them.

Bilirubin—$C_{32}H_{36}N_4O_6$—is, when amorphous, an orange-yellow powder, and when crystalline, in red rhombic prisms. It is sparingly soluble in H_2O, alcohol, and ether; readily soluble in hot chloroform, carbon disulphide, benzene, and in alkaline solutions. When treated with HNO_3 containing nitrous acid, or with a mixture of concentrated HNO_3 and H_2SO_4, it turns first green, then blue, then violet, then red, and finally yellow. This reaction, known as Gmelin's, is very delicate, and is used for the detection of bile-pigments in icteric urine and in other fluids.

Biliverdin—$C_{32}H_{36}N_4O_8$—is a green powder, insoluble in H_2O, ether, and chloroform; soluble in alcohol and in alkaline solutions. It exists in green biles, but its presence in yellow biles or biliary calculi is doubtful. It responds to Gmelin's test. In alkaline solution it is changed after a time into biliprasin.

Bilifuscin—$C_{16}H_{20}N_2O_4$—obtained in small quantity from human gall-stones, is an almost black substance, sparingly soluble in H_2O, ether, and

356

chloroform; readily soluble in alcohol and in dilute alkaline solutions Its existence in the bile is doubtful.

Biliprasin—$C_{16}H_{22}N_2O_6$(?)—exists in human gall-stones, in ox-gall, and in icteric urine. It is a black, shining substance, insoluble in H_2O, ether, and chloroform; soluble in alcohol and in alkaline solutions.

Urobilin—*Hydrobilirubin*—$C_{32}H_{40}N_4O_7$.—Under the name urobilin, Jaffé described a substance which he obtained from dark, febrile urine, and which he regarded as the normal coloring matter of that fluid; subsequently he obtained it from dog's bile and from human bile, from gall-stones and from fæces. *Stercobilin*, from the fæces, is identical with urobilin.

Urinary pigments.—Our knowledge of the nature of the substances to which the normal urinary secretion owes its color is exceedingly unsatisfactory. Jaffé in his discovery of urobilin shed but a transient light upon the question, as that substance exists in but a small percentage of normal urines, although they certainly contain a substance readily convertible into it. Besides the substance convertible into urobilin, and sometimes urobilin itself, human and mammalian urines contain at least one other pigmentary body, *uroxanthin*, or *indigogen*. This substance was formerly considered as identical with *indican*, a glucoside existing in plants of the genus *Isatis*, which, when decomposed, yields, among other substances, indigo-blue. Uroxanthin, however, differs from indican in that the former is not decomposed by boiling with alkalies, and does not yield any glucose-like substance on decomposition; the latter is almost immediately decomposed by boiling alkaline solutions, and, under the influence of acids and of certain ferments, yields, besides indigo-blue, *indiglucin*, a sweet, non-fermentable substance, which reduces Fehling's solution.

Uroxanthin is a normal constituent of human urine, but is much increased in the first stage of cholera, in cases of cancer of the liver, Addison's disease, and intestinal obstruction. It has also been detected in the perspiration.

In examining the color of urine it should be rendered strongly acid with HNO_3 or HCl, and allowed to stand six hours to liberate combined pigment, and then examined by transmitted light in a beaker three inches in diameter.

Melanin is the black pigment of the choroid, melanotic tumors, and skin of the negro; and occurs pathologically in the urine and deposited in the air-passages.

PART III.

LABORATORY TECHNICS.

Chemistry is essentially a science of experiment; and not only is a knowledge of its truths much more rapidly and easily acquired by the student through the actual performance of experiment, than by any amount of reading or attendance upon illustrated lectures; but it is even doubtful whether a thorough knowledge of the facts and theories of the science can be obtained in any other way than by personal observation.

A description of the various manipulations of the general chemical laboratory would fill volumes. A short account of the more prominent of those required in a study of rudimentary chemistry, and in those processes of analysis which are likely to be of service to the physician will, we believe, not be out of place in a work of this nature.

GENERAL RULES.

"Cleanliness," said John Wesley, "is next to godliness." The chemist, whatever his supply of godliness, must be thoroughly imbued with the spirit of cleanliness; not so much as regards himself, for he who fears to soil his fingers is not of the material whereof chemists are made, but as regards the vessels and reagents which are his tools. Any substance foreign to the matter under examination and the reagents used, whatever be its nature, is *dirt* to the chemist.

Glass vessels should always be cleaned as soon as possible after using, as foreign substances are much more readily removed then than after they have dried upon the glass. Usually rinsing with clear water, and friction with a probang or bottle brush is sufficient; greasy and resinous substances may be removed with KHO solution; and other adherent deposits usually with HCl or HNO_3; the alkali or acid being removed by clear water. After washing, the vessels are drained upon a clean surface, and are not to be put away unless *perfectly bright*.

Order and system are imperative, especially if several operations are conducted at the same time. If there be "a place for everything, and everything in its place," much time will be spared. If a process be of such a nature that it requires a number of vessels, each vessel should be numbered with a small gum label, and the notes of the operation should indicate the stage of the process in each vessel.

The habit of taking full and systematic notes of experiments and analyses in a book kept especially for the purpose, is one which the student cannot contract too early. He will be surprised, in looking over and

358

comparing his notes, at the amount of information he will have collected in a short time; much of which, had the memory been trusted to, would have been lost.

REAGENTS.

The stock of reagents required varies, of course, with the nature of the work to be done; from the small number required in urinary analysis, to the array on the shelves of a fully appointed analytical laboratory.

The liquid reagents and solutions should always be kept in glass-stoppered bottles (the 4½ ℥ bottles, with labels blown in the glass, serve very well). The solid reagents may be kept in cork-stoppered or, preferably, glass-stoppered bottles. The ordinary glass stoppers should never be laid upon the table, lest they take up particles of foreign matter and contaminate the contents of the bottle; but should be held between the third and little fingers of the left hand.

The reagents required for ordinary urinary analysis are:

Nitric acid,	Potassium hydrate,	Cupric sulphate,
Sulphuric acid,	Ammonium hydrate,	Fehling's solution,
Acetic acid,		Test papers.

Those required for ordinary qualitative analysis are:

Hydrochloric acid,	Ammonium carbonate,	Barium chloride,
Nitric acid,	Ammonium oxalate,	Calcium sulphate,
Sulphuric acid,	Sodium carbonate,	Magnesium sulphate,
Acetic acid,	Hydro-disodic phosphate,	Cupric sulphate,
Hydrogen sulphide,	Potassium ferrocyanide,	Argentic nitrate,
Ammonium sulphide,	Potassium ferricyanide,	Mercuric chloride,
Ammonium hydrate,	Potassium sulphocyanate,	Plumbic acetate,
Potassium hydrate,	Potassium carbonate,	Ferric chloride,
Ammonium chloride,	Potassium chromate,	Platinic chloride.

The chemicals must be C. P. (= chemically pure); and the solutions *must* be made with distilled H_2O. It is well to put corresponding numbers on each bottle and stopper to prevent their becoming mixed in cleaning.

GLASS TUBING.

The tubing used in making all usual connections and apparatus is the soft German or American tubing. When the tube is to be strongly heated, Bohemian tubing must be used. The fashioning of tubing of the diameter generally used for gas connections is a simple matter.

Cutting into desired lengths is accomplished by making a scratch with a triangular file at the desired point; holding the tube as shown in Fig. 41; and partly drawing, and partly bending it.

Fig. 41.

Larger glass surfaces may be cut in any required direction, by first making a deep scratch with the file; starting the break by bringing in contact with scratched spot a piece of red-hot glass tubing; and leading the break in the desired direction by applying a heated piece of ¼-inch iron wire, as shown in Fig. 42. Cut ends of tubing should always be rendered smooth by heating them to incipient fusion.

359

Bending is done by heating the tube at the desired point in an ordinary gas flame (not a blow-pipe flame), without rotating it, until softened; removing from the flame and bending *toward* that surface which was nearest the orifice of the gas jet.

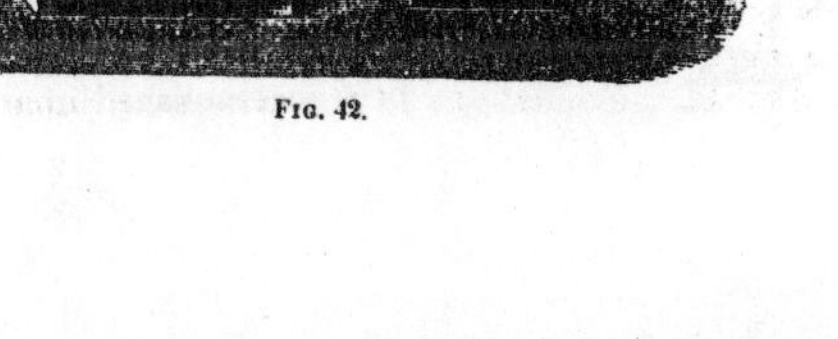

Fig. 42.

Closing.—For this and other operations with glass tubing, the glass-blower's flame, obtained with a burner (Fig. 43) which permits of the injection of air into the gas flame, is required. To make a test-tube a piece of tubing of the length of two test-tubes is drawn out at the middle (see below). The small end of each piece is then heated and the superfluous glass removed by a warm glass rod, which is brought into contact for an instant and then drawn away. The closed end is then heated during rotation until soft, and rendered hemispherical by gently blowing into the open end. The open end is then heated and while hot formed into a lip by a circular motion with a hot iron wire.

Drawing out consists in heating the tube at the point desired, during rotation, and drawing it apart after removal from the flame.

Joining.—Two pieces of tubing of different diameters may be joined end for end if they be of the same kind of glass. The ends of each are closed, heated, and blown out into thin bulbs. The bulb is then broken off, the ends heated, pressed firmly together, and re-heated during alternate pressure and drawing apart, and gentle blowing into one end while the other is closed, until an even joint is obtained.

Stirring rods are made by cutting glass rods to the required length and rounding the ends by fusion.

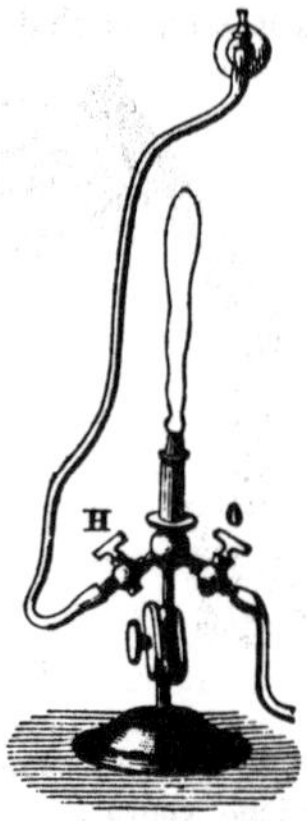

FIG. 43.

COLLECTION OF GASES.

Gases are collected over the *pneumatic trough*, by *displacement* of air; or over the *mercurial trough*.

In the pneumatic trough (Fig. 44) gases are collected over water in bell jars filled with that liquid. This method of collection can only be used for insoluble or sparingly soluble gases; and if heat have been used in the generation of the gas the disengagement tube must be removed from the water *before* the heat is discontinued, to avoid an explosion.

360

Soluble gases are collected over mercury or by upward or downward displacement of air, according as they are without action on Hg, or heavier or lighter than air.

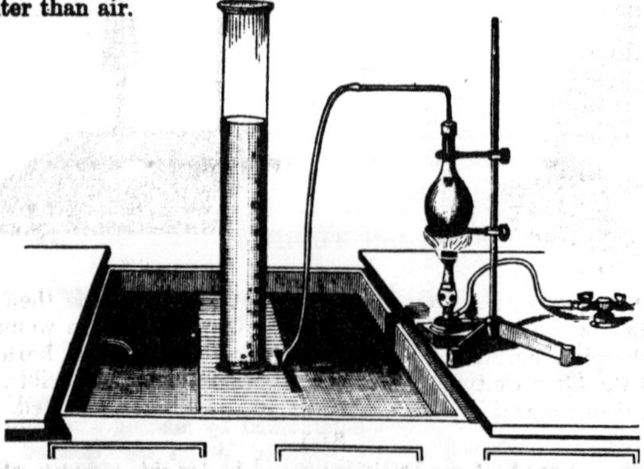

FIG. 44.

SOLUTION.

As the particles of liquids can be brought into closer contact than those of solids, reactions are usually facilitated by bringing the reagents into solution or into fusion.

FIG. 45.

At a given temperature solution of a solid is more rapid the greater the surface exposed to the solvent, *i.e.*, the greater the degree of subdivision.

Ordinary salts are ground to powder in Wedgwood or glass mortars. Very hard substances are first coarsely powdered in steel mortars and then finely ground in agate mortars. Soft substances are best subdivided either by hashing, as in the case of muscular tissue, or by forcing through the meshes of a fine sieve, as in the case of white of egg, brain tissue, etc.

When only certain constituents of the substance are to be dissolved, *percolation* may be resorted to. The substance to be extracted is packed in a percolator in such a manner that the extracting liquid filters through it slowly.

When the solvent is a volatile liquid—ether, chloroform, carbon disulphide—extraction is best accomplished in an apparatus such as that shown in Fig. 45, in which the liquid is boiled in A; the vapor passing through *a*, *b*, is liquefied in the condenser and flows back over the substance in B. The extract collects in A.

361

PRECIPITATION—DECANTATION—FILTRATION—WASHING.

When the conversion of an ingredient of a solution into an insoluble compound, and its separation from the liquid are desired, both the liquid and the reagent should be in clear solution, and the latter should be added to the former, which has been warmed. The vessel is then set in a warm place until the precipitate has subsided, a few drops of the precipitant are added to the clear liquid, and if no cloudiness be produced the precipitation is complete. Precipitation should be effected in Erlenmeyer flasks (Fig. 46) or in precipitating jars (Fig. 47) that the precipitate may not collect on the sides, and may be readily detached by the wash-bottle.

FIG. 46.

FIG. 47.

Precipitates are separated from the liquid in which they have been formed by *decantation* or *filtration*.

FIG. 48.

DECANTATION consists in allowing the precipitate to subside and pouring off the supernatant liquid; it should always be employed as a preliminary to filtration, and is sometimes used exclusively, when the precipitate is washed by repeatedly pouring on clear water and decanting it until it no longer contains any solid matter.

362

In pouring liquid from one vessel to another it should be guided by a glass rod, as shown in Fig. 48; the outer surface of the lip of the pouring vessel having been slightly greased.

FILTRATION is resorted to more frequently than decantation. Filters are made from muslin, paper, asbestos, or glass wool.

FIG. 49.

FIG. 50.

Muslin filters are only used for coarse filtration.

Paper filters are the most frequently used. For coarse work the ordinary gray or German white paper is used; but for analytic work a paper which leaves but a small amount of ash is required; the best now in the market is Schleicher & Schüll's Nos. 597 and 589. The filter should be taken of such size that when folded it will be smaller than the funnel in which it is to rest. It is folded across one diameter, and again over the radius at right angles to the first diameter; one of the four layers of paper, then seen at the circular portion of the filter, is separated from the other three, in such a way as to form a cone. The filter so formed is brought into the funnel, and, while held in position by a finger-nail over one of the folds, is wetted with water from the wash-bottle. After the paper has been brought in contact with the funnel by a glass rod, the liquid to be filtered is introduced, care being had not to overflow the filter, and to allow any supernatant liquid in the precipitating jar to pass through, before bringing the precipitate itself upon the filter. Funnels used for filtering should have an angle of 60°, and a long stem, the point of which is ground off at an acute angle.

Asbestos and glass wool plugs loosely introduced into the stem of a funnel, are used in filtering such liquids as would destroy paper.

For filtrations which take place slowly the *filter-pump* is now extensively used. It is simply an appliance for exhausting the air in the stem of the funnel, and thus taking advantage of atmospheric pressure. A simple and effective form of pump is that shown in Fig. 49, in which the water (under 10 feet or more of pressure) enters at *a* and aspirates the air from *b* through *c*. When the pump is used a small cone of platinum

363

must be placed at the apex of the funnel to support the point of the *t*, **which would otherwise be ruptured.**

When the precipitate has been collected upon the filter, it must be *washed* **until free from extraneous matter. This is effected by blowing into the tube** *a* **of the wash-bottle, Fig. 50, while the end of the tube** *b* **is held so as to deliver a** *gentle* **stream into the filter; care being had that the precipitate is not lost by spurting, overflowing, or** *creeping* **up the sides of the funnel. The completeness of the washing is** *not to be guessed at* **but is to be judged by adding reagents, suitable to the case, to portions of the filtrate until they fail to cause a cloudiness.**

EVAPORATION—DRYING—IGNITION.

Evaporations are usually conducted on the sand- or water-bath. The sand-bath is simply a flat, iron vessel, filled with sand and heated. By its use the heat is more evenly distributed than with the naked flame.

The water-bath, usually of the form shown at *a* Fig. 51, is used where the temperature is to be kept below 100° (212° F.). It should *always* be used in evaporating liquids containing organic matter, and care should be had that it does not become dry.

Fig. 51.

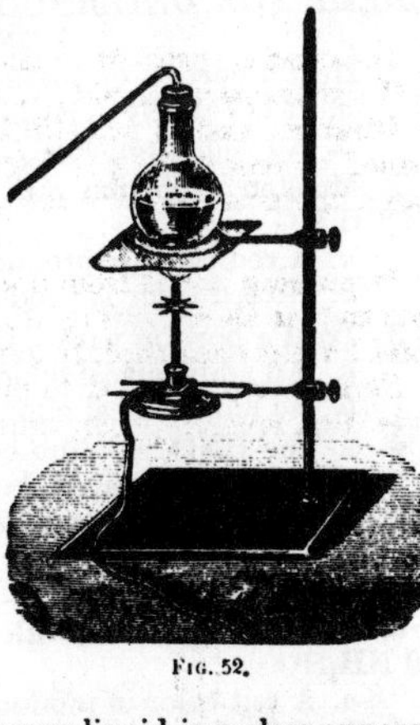

Fig. 52.

In cases where it is desired to boil an aqueous liquid in a glass or porcelain vessel; this is supported on a piece of wire gauze and a Bunsen burner or spirit lamp brought under it (Fig. 52). A piece of sheet-iron may be substituted for the wire gauze, with flat-bottomed vessels. The outside of the heated vessel must be *dry*.

In heating liquids in test-tubes, *the mouth of the tube must be held away from the person.* It is best held by a piece of thick paper bent around the upper end of the tube (Fig. 53). The tube should be heated near, not at its bottom.

In no case should flame, or the sand of the sand-bath, come in contact with a glass vessel above the level of the liquid within.

Drying is always necessary as a preliminary to weighing, whether the

364

substance is hygroscopic or not. It is usually effected in water ovens (Fig. 54), if a temperature of 100° (212° F.) be sufficient; or in air ovens, somewhat similarly constructed, if a higher temperature be desired. As a substance can never be accurately weighed while it is warm, it is removed from the oven and placed in the desiccator (Fig. 55), over $H\,SO_4$ or $CaCl_2$, until it has cooled.

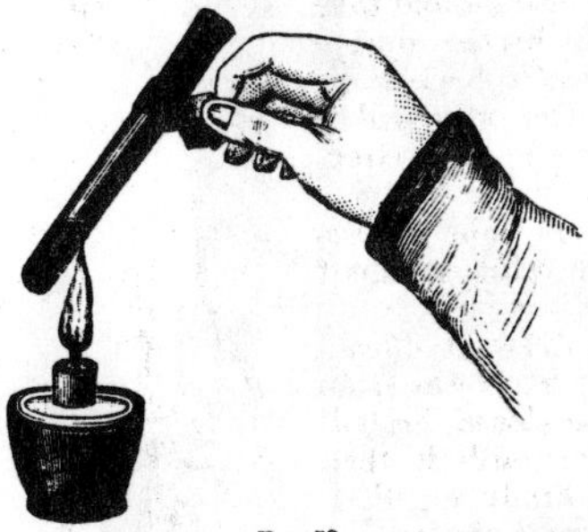

Fig. 53.

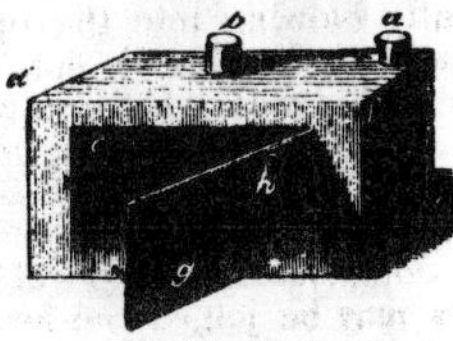

Fig. 54.

In cases where the substance would be injured by elevation of temperature, it is dried by allowing it to remain in the desiccator until it ceases to lose weight.

Ignition has for its object the removal of organic matter by burning, and is conducted in platinum or porcelain crucibles. If a filter and precipitate are to be ignited, they are first well dried; as much as possible of the precipitate is detached and brought into the crucible, placed upon a

Fig. 55.

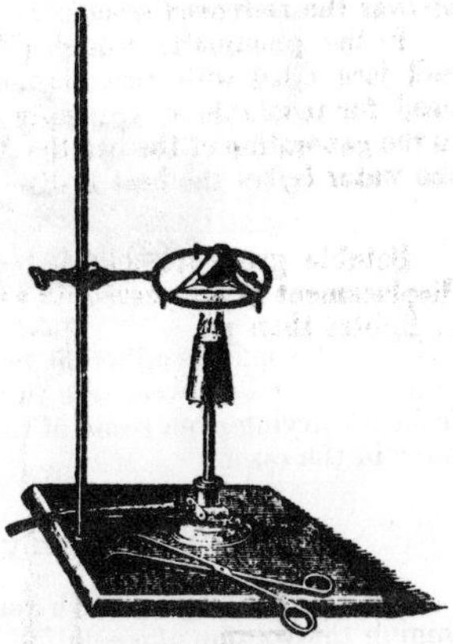

Fig. 56.

sheet of white paper; the filter, with adherent precipitate, is then rolled into a thin cone, around which a piece of platinum wire is wound; by means of the platinum wire the filter is held in the flame and burnt; the remains of the filter are then added to the contents of the crucible, which is supported in the position shown in Fig. 56, in which it is heated, at first

365

moderately, and the heat gradually increased to bright redness, at which it is maintained until no carbon remains. Before weighing, the crucible is to be cooled in the desiccator.

In igniting it must not be forgotten that mineral substances may be modified or lost. Carbon at high temperature deoxidizes easily reducible substances; alkaline chlorides are partly volatilized; mineral bases combined with organic acids are converted into carbonates. In every instance only that amount of heat which is required is to be applied. In some cases it is well to accelerate the oxidation by the addition of ammonium nitrate.

WEIGHING.

The balance, Fig. 57, should always be kept in a glass case, containing a vessel with $CaCl_2$, and in a situation protected from the fumes of the laboratory. The weights should be kept in a box by or in the balance case, which is to be closed when not in use.

In weighing observe the following rules:

(1.) See that the balance is in adjustment before using, especially if more than one person use it. (2.) Always put the substance to be weighed in the same pan, usually the left hand one, and the weights in the other. (3.) Never bring any chemical in contact with the pans, but have a pair of large watch-glasses of equal weight, one in either pan. Pieces of paper will *not* **serve the purpose. (4.) Never add to or remove from either pan a weight of more than 0.5 gram without putting the balance out of action. (5.) Never weigh anything warm. (6.) In weighing a substance which has been dried do not consider the weight correct until two successive weigh**

FIG 57.

ings, with an intervening drying of a half hour, give identical results. (7.)

366

In adding the weights, do so in regular order from above downward. (8.) In counting the weights, reckon the amount first by the empty holes in the box, and then tally in replacing the weights. (9.) Substances liable to absorb moisture from the air are to be weighed in closed vessels. Thus,

FIG. 58.

when a filter and its adherent precipitate are to be weighed together, they must be placed between the two watch-glasses (Fig. 58) as soon as taken from the drying oven; one of the watch-glasses being used to support the filter in the oven.

MEASURING—VOLUMETRIC ANALYSIS.

The principle upon which volumetric analysis is based is that by determining the volume of a solution of known strength, required to accurately neutralize another solution of unknown strength, the amount of active substance in the latter may be calculated.

If, for example, we have a solution of silver nitrate which contains 170 grams to the litre, and we find that 12 c.c. of this solution precipitate all the chlorine from 10 c.c. of a solution of NaCl, it follows that the NaCl solution contains 70.20 grams of that substance per litre, because:

$$\underset{170}{NO_3Ag} + \underset{58.5}{NaCl} = \underset{85}{NO_3Na} + \underset{143.5}{AgCl}$$

and therefor each c.c. of the NO_3Ag solution will accurately precipitate 0.0585 grm. NaCl; but as it has required 12 c.c. of the NO_3Ag solution to neutralize 10 c.c. of the NaCl solution, the latter contains $0.0585 \times 12 = 0.702$ grm. NaCl or 1,000 contain $0.702 \times 100 = 70.20$ grms. NaCl.

It is obvious, therefor, that the value of volumetric methods depends, among other things, greatly upon the accuracy of the *standard solutions*, as the solutions of known strength are called, and upon the accuracy of the measurements of volume.

FIG. 59.

A standard solution containing in a litre of liquid a number of grams of the active substance, equal to its molecular weight, is a *normal* solution; one containing $\frac{1}{10}$ that amount is a *decinormal solution*.

An *indicator* is a substance which, by some characteristic reaction (*end reaction*), which will occur only when the substance to be determined has been completely removed, indicates the point when a proper volume of the standard solution has been added.

The apparatus required for volumetric analysis consists of:

(1.) A *litre-flask* (Fig. 59); a flask of such size that, when filled to the mark on the neck, at the temperature for which it has been graduated, it contains exactly 1,000 c.c. of water.

367

(2.) A *burette*, which is a glass tube graduated into cubic centimetres, and having a stopcock or pinchcock at its lower extremity.

(3.) A series of *pipettes* (Fig. 60), which are glass tubes, having bulbs

FIG. 60.

blown upon them of such size that when they are filled to a mark on the tube above the bulb, they contain a given number of cubic centimetres.

(4.) Small beakers; stirring rods; bottles for standard solutions.

In making a standard solution the object to be attained is to have a solution, one litre of which shall contain a known quantity of the active material. If then in the formula for the normal solution of silver nitrate:

Silver nitrate	**170 grams.**
Distilled water	**1,000 c.c.**

FIG. 61.

we weigh out the NO_3Ag on the one hand, and measure the H_2O on the other, and mix the two, we will have, not what is desired, a solution containing 170 grms. NO_3Ag in 1,000 c.c. H_2O, but a solution of 170 grms. NO_3Ag in 1,000 + x c.c. H_2O, in which x = the volume occupied by the NO_3Ag. Therefor, in making standard solutions, weigh out the active substances; introduce them into the litre-flask; and then fill that to the mark with H_2O. Too much caution cannot be used in having pure chemicals and making accurate weighings in preparing volumetric solutions; indeed the great disadvantage of the use of these methods by physicians is that the solutions which they use are carelessly prepared and, consequently, the time which they spend in obtaining inaccurate, but seemingly accurate results is worse than thrown away.

To use a volumetric solution it is poured into the burette, whose stopcock has been closed, until above the o mark; the stopcock is then slightly opened so as to expel *all* air from the delivery tube. The float (Fig. 61) is now introduced from above, and touched with a glass rod to free it from adhering air-bubbles; and the solution allowed to flow out from below until the mark on the float is opposite the o of the burette. All is now ready for use; a given quantity of the solution to be analyzed is measured into a pipette and placed in a beaker, a few drops of the indicator solution are added, and the standard solution allowed to flow in until the end reaction is reached. The reading of the burette is then taken and the calculation made.

368

SCHEME FOR DETERMINING THE COMPOSITION OF CALCULI.

1. Heat a portion on platinum foil:
- ***a.*** **It is entirely volatile** **2**
- ***b.*** **A residue remains** **5**

2. Moisten a portion with HNO_3; evaporate to dryness at low heat; add NH_4HO:
- ***a.*** **A red color is produced** **3**
- ***b.*** **No red color is produced** **4**

3. Treat a portion with KHO, without heating:
- ***a.*** **An ammoniacal odor is observed** ...... ***Ammonium urate.***
- ***b.*** **No ammoniacal odor** .... ***Uric acid.***

4. *a.* The HNO_3 solution becomes yellow when evaporated; the yellow residue becomes reddish-yellow on addition of KHO, and, on heating with KHO, violet red..*Xanthin.*
- ***b.*** **The HNO_3 solution becomes dark brown on evaporation** .... ***Cystin.***

5. Moisten a portion with HNO_3; evaporate to dryness at low heat; add NH_4HO:
- ***a.*** **A red color is produced** **6**
- ***b.*** **No red color is produced** **9**

6. Heat before the blow-pipe on platinum foil:
- ***a.*** **Fuses** **7**
- ***b.*** **Does not fuse** **8**

7. Bring into blue flame on platinum wire:
- ***a.*** **Colors flame yellow** .... ***Sodium urate.***
- ***b.*** **Colors flame violet** .... ***Potassium urate.***

8. The residue from 6:
- ***a.*** **Dissolves in dil. HCl with effervescence; the solution forms a white ppt. with ammonium oxalate.... *Calcium urate.***
- ***b.*** **Dissolves with slight effervescence in dil. H_2SO_4; the solution, neutralized with NH_4HO, gives a white ppt. with**

HNa_2PO_4........................*Magnesium urate.*

9. Heat before the blow-pipe on platinum foil:
a. It fuses.................*Ammonio-magnesian phosphate.*
b. It does not fuse.................................... 10

10. The residue from 9, when moistened with H_2O, is:
a. Alkaline 11
b. Not alkaline ....*Tricalcic phosphate.*

11. The original substance dissolves in HCl:
a. With effervescence ....*Calcium carbonate.*
b. Without effervescence..................*Calcium oxalate.*

NOTE.—A fresh portion of the powdered calculus is to be taken for each operation except where otherwise stated.

369

SCHEME FOR DETERMINING THE COMPOSITION OF AN INORGANIC COMPOUND. SOLUBLE IN WATER OR IN ACIDS.

DETERMINATION OF BASES.

1. Acidulate with HCl:
a. No ppt. is formed.. 5
b. A white ppt. is formed.................................. 2

2. Add HCl drop by drop to complete precipitation, collect on filter, wash:
a. Filtrate... 5
b. Precipitate.. 3

3. Treat ppt. on filter with boiling H_2O, test filtrate with H_2S:
a. H_2S produces a black or brown color ....*Lead.*
b. H_2S does not cause darkening 4

4. Treat ppt. on filter with NH_4HO:
a. Ppt. turns gray or black ....*Mercury(ous).*
b. Filtrate gives white ppt. with HNO_3 ....*Silver.*

5. Pass H_2S through clear, acid liquid:
a. No ppt. is formed.. 18
b. A ppt. is formed... 6

6. Treat with H_2S, with occasional warming, to complete precipitation; collect ppt. on filter; wash with H_2O containing trace of H_2S:
a. Filtrate... 18
b. Precipitate 7

7. Treat a portion of ppt. with NH_4HS, warmed in test-tube:
a. Ppt. is dissolved... 8
b. A residue remains undissolved......................... 13

8. Dry the remainder of ppt. from 6, mix it with equal parts of Na_2CO_3 and $NaNO_3$, and throw mixture in small portions into red-hot porcelain crucible; when cold dissolve residue in H_2O; filter:
a. Filtrate... 9
b. Residue 10

9. Add to the filtrate NH_4HO,$MgSO_4$ and NH_4Cl, and rub inside of test-tube with glass rod:
a. A white, crystalline ppt. forms immediately or after a time..*Arsenic.*
b. No ppt. forms ....*Absence of As.*

10. The residue is:
a. White 11
b. Brown or black.. 12

11. Heat a portion of the residue in a platinum capsule with HCl, place a small piece of Zn in liquid:
a. The platinum surface turns black ....*Antimony.*
b. The HCl liquid, removed by decantation, gives a white ppt. with excess of $HgCl_2$ sol.............................*Tin.*

370

12. The original solution:
a. Gives a brown ppt. with $FeSO_4$ sol..................*Gold.*
b. Does not give a brown ppt. with $FeSO_4$ sol., but gives a yellow ppt. with KCl sol......................*Platinum.*

13. Wash undissolved residue and boil with dil. HNO_3 in porcelain capsule, filter:
a. Filtrate 14
b. Residue (if any) 17

14. Add dil. H_2SO_4 to a portion of filtrate, warm, and let stand some time:
a. A ppt. forms. Mix whole of filtrate with H_2SO_4 dil., evap. over water-bath, extract residue with H_2O, filter, and treat filtrate according to 15 ....*Lead.*
b. No ppt. forms 15

15. Add NH_4HO to remainder of filtrate (or to filtrate from 14 *a*):
a. A ppt. is formed. Filter and test filtrate according to 16..*Bismuth.*
b. No ppt. is formed.. 16

16. Add SO_2 and CNSK to the liquid, evaporate, dissolve residue in H_2O, add H_2S to solution:
a. The solution 15 *b.* was blue ....*Copper.*
b. The treatment 16 produced a yellow ppt.......*Cadmium.*

17. Is black, dissolves in aqua regia, and the solution gives a gray ppt. with $SnCl_2$ ....*Mercury(ic).*

18. Boil portion of liquid to expel H_2S, add a few drops HNO_3, boil, add NH_4HO just to alkaline reaction, add NH_4HS:
a. Neither NH_4HO nor NH_4HS caused ppt 31
b. NH_4HS caused ppt., NH_4HO did not................. 20
c. NH_4HO caused ppt.. 19

19. The original liquid is:
a. Neutral.. 20
b. Alkaline or acid 28

20. Add to remainder of liquid 5 *a.* or 6 *a.* NH_4Cl, NH_4HO just to alkaline reaction, and excess NH_4HS, warm, filter, wash:
a. Filtrate... 31
b. Deposit... 21

21. The deposit is:
a. White.. 22
b. Colored 25

22. Dissolve deposit in small quantity HCl, boil, concentrate to small bulk, add NaHO, boil some time:
a. A ppt. forms, which afterward dissolves............ 23
b. A ppt. forms, which does not redissolve 24

23. The solution 22 *a.* is divided into two parts:
a. Treated with a *small* quantity of H_2S gives a white ppt.*Zinc.*
b. Treated with HCl to acid reaction, and then with slight excess NH_4HO, gives, when heated, a white ppt. insoluble in NH_4Cl...........................*Aluminium.*

24. Dilute, filter; test filtrate for Zn and Al as in 23. Dissolve ppt. in HCl, evaporate to small bulk, dilute, neutralize nearly with Na_2CO_3, add $BaCO_3$, filter, after standing:

371

a. Filtrate, treated with H_2SO_4 and again filtered, gives solution which, when made alkaline with NaHO, gives white ppt. with H_2S ....*Zinc.*
b. Residue (if any), heated with Na_2CO_3 in outer blow-pipe flame, gives bead which is green when hot and bluish-green and opaque when cold ....*Manganese.*

25. The deposit is:
a. Completely dissolved in dil. HCl.......................... 26
b. Not dissolved in dil. HCl 27

26. Boil to expel H_2S, add HNO_3, boil, filter. Concentrate, add excess NaHO sol., boil, filter from residue *b.*:
a. Filtrate. Test for Zn and Al as in 23.
b. Divide residue into 3 parts:
aa. Dissolved in HCl dil. gives red color with CNSK..*Iron.*
bb. Fused with CO_3Na_2 and $KClO_3$ forms yellow mass, which forms yellow sol. in H_2O........*Chromium.*
cc. Treated as in 24 *b.* gives same results.....*Manganese.*

27. Filter, wash, examine filtrate according to 26. Heat portion of residue with borax on platinum wire in blow-pipe flame:
a. A transparent blue bead is obtained.............*Cobalt.*
b. A bead is obtained, which is yellow when hot, nearly colorless when cold...............................*Nickel.*

28. Add to remainder of liquid 5 *a.* or 6 *a.*, NH_4Cl, NH_4HO just to alkaline reaction, and NH_4HS, warm, filter:
a. A residue remains 29
b. No residue remains....................................... 30

29. Treat filtrate as in 30. Examine residue for Ni and Co as in 27.

30. Boil to expel H_2S, divide into 2 parts:
a. Add dil. H_2SO_4. If a ppt. form, filter, wash, fuse ppt. with Na_2CO_3, wash, dissolve in HCl, and test sol. for Ca, Ba, and Sr, according to 32.
b. Heat with HNO_3, test small portion for Fe with CNSK, add Fe_2Cl_6, evaporate, add H_2O,Na_2CO_3 to near neutralization, and $BaCO_3$; stir, let stand until liquid is colorless. Separate ppt. *aa.* from filtrate *bb.*:
aa. Boil ppt. with NaHO sol, filter; test filtrate for Al by 23 *b.* and residue for Cr by 26 *bb.*
bb. Mix filtrate with few drops HCl, boil, add NH_4HO and NH_4HS. If a ppt. form, test for Mn and Zn, as in 24. If no ppt. form, mix sol. with excess H_2SO_4, boil, filter, add excess NH_4HO and $(NH_4)_2C_2O_4$, filter, add

HNa_2PO_4 to filtrate, a white ppt. . . . . *Magnesium.*

31. Add to a small portion of the liquid NH_4Cl, $(NH_4)_2CO_3$ and NH_4HO, warm:

a. A ppt. forms 32
b. No ppt. forms 36

32. Treat the whole of liquid with NH_4Cl, $(NH_4)_2CO_3$ and NH_4HO as in 31, filter:

a. Filtrate 36
b. Precipitate 33

372

33. Wash, dissolve in small quantity dil. HCl, evaporate over water-bath, dissolve in a little H_2O, add $CaSO_4$ to a small portion of liquid:

a. A ppt. forms 34
b. No ppt. forms 35

34. Add H_2SiF_6 to another portion of solution 33:

a. A ppt. is formed. A portion of the original solid colors the Bunsen flame green . . . . *Barium.*
b. No ppt. formed. A portion of the original solid colors the Bunsen flame red . . . . *Strontium.*

35. Mix another portion of liquid 33 with $(NH_4)_2C_2O_4$, a white ppt. . . . . *Calcium.*

36. Add HNa_2PO_4 sol. to a small portion of liquid, rub inner surface of test-tube with glass rod:

a. A white, cystalline ppt. . . . . *Magnesium.*
b. No ppt 37

37. Evaporate, ignite, dissolve in small quantity H_2O, divide solution into two parts:

a. Forms yellow, crystalline ppt. with $PtCl_4$; colors flame violet (observe through blue glass) . . . . *Potassium.*
b. Produces crystalline ppt. with potassium pyroantimonate; colors flame yellow . . . . *Sodium.*

38. Triturate original substance with CaH_2O_2 and H_2O; it develops an odor of ammonia . . . . *Ammonium.*

Determination of Mineral Acids.

After determination of bases, bear in mind what acids can possibly form soluble salts with the bases found (see Table I, p. 354), and limit the search to those. Examine separate portions of the original solution according to 1, 3, 4, 8, 10, 12, and 13.

1. Add HCl:

a. Effervesces 2
b. A gelatinous ppt. is formed . . . . *Silicate.*

2. The gas given off in 1 *a.* has:

a. No odor, and forms a white ppt. when passed through lime-water . . . . *Carbonate.*
b. An odor of rotten eggs, and blackens paper moistened with $Pb(C_2H_3O_2)_2$ . . . . *Sulphide.*

3. In testing for bases As was found; add sol. $AgNO_3$ and NH_4HO:

a. A yellow ppt. . . . . *Arsenite.*
b. A brick-red ppt. . . . . *Arsenate.*

4. Add $Ba(NO_3)_2$, and, if acid, add NH_4HO to faint alkaline reaction:

a. No ppt. formed 8
b. A ppt. is formed 5

5. Add HNO_3 to acid reaction to a portion of 4 *b.*:

a. The ppt. does not redissolve completely; filter; examine filtrate by *b.* . . . . *Sulphate.*
b. The ppt. redissolves 6

373

6. Treat another portion of 4 *b.* or 5 *a.* with acetic acid:

a. It dissolves completely . . . . *Phosphate.*
b. It does not dissolve completely 7

7. Filter:

a. Filtrate (in absence of As) gives white ppt. with NH_4HO, NH_4Cl, and $MgSO_4$ . . . . *Phosphate.*
b. Ppt. dissolves in dil. HCl; sol. gives ppt. with $CaCl_2$ in neutral solution . . . . *Oxalate.*

8. Acidulated with HNO_3; add sol. $AgNO_3$:

a. A ppt. is formed 9
b. No ppt. is formed 10

9. Filter; treat ppt. with HNO_3:

a. It dissolves completely 10
b. It does not dissolve completely 12

10. The solid substance:

a. Produces a yellow color with H_2SO_4 . . . . *Chlorate.*
b. Does not produce a yellow color with H_2SO_4 11

11. Divide liquid 9 *a.* into 4 parts:

a. Gives white ppt. with NH_4HO, NH_4Cl, and $MgSO_4$ . . . . *Phosphate.*
b. Acidulated slightly with HCl, turns turmeric paper red . . . . *Borate.*
c. Acidulate with HCl, evaporate to dryness, add HCl, an insoluble residue remains . . . . *Silicate.*
d. A portion of original substance, moistened with H_2SO_4, gives off gas which corrodes glass . . . . *Fluoride.*

12. The original liquid gives:

a. A blue color with a drop of chlorine water and starch paste . . . . *Iodide.*
b. A blue ppt. with sol. $FeSO_4 + Fe_2(SO_4)_3$ . . . . *Cyanide.*
c. Is colored yellow or brown by chlorine water, but does not react as in 12 *a.* . . . . *Bromide.*
d. Ppt. 8 *a.* is readily soluble in NH_4HO . . . . *Chloride.*

13. Heat the dry salt with Cu and H_2SO_4 and conduct the gas through sol. $Fe_2(SO_4)_3$, which it turns brown . . . . *Nitrate.*

374

TABLE I.—SOLUBILITIES. Frezenius.

W or w = soluble in H_2O. A or a = insoluble in H_2O; soluble in HCl, NO_3H, or aqua regia. I or i = insoluble in H_2O and acids. W-A = sparingly soluble in H_2O, but soluble in acids. W-I = sparingly soluble in H_2O and acids. A-I = insoluble in H_2O, sparingly soluble in acids. Capitals indicate common substances.

	Aluminium.	Ammonium.	Antimony.	Barium.	Bismuth.	Cadmium.	Calcium.	Chromium.	Cobalt.	Copper.	Ferrous.	Ferric.	Lead.	Magnesium.	Manganese.	Mercurous.	Mercuric.	Nickel.	Potassium.	Silver.	Sodium.	Strontium.	Stannous.	Stannic.	Zinc.	
Acetate	W	W	..	W	w	w	a	W	w	W	w	W	W	w	w	w-a	w	w	W	w	W	w	w	w	W	Acetate.
Arsenate	a	w	a	a	a	a	a	a	a	a	a	a	a	a	a	a	a	a	W	a	W	a	a	a	..	Arsenate.
Arsenite	..	w	a	a	..	..	a	..	a	A	a	a	a	a	a	a	a	a	w	a	w	a	a	..	..	Arsenite.
Benzoate	W	w	..	W	..	w	W	..	..	a	W	a	a	w	w	a	w-a	..	w	w-a	w	..	..	..	..	Benzoate.
Borate	a	w	..	a	a	w-a	a	a	a	a	a	a	a	w-a	a	..	..	a	W	a	W	a	a	..	a	Borate.
Bromide	W	W	w-a	w	w-a	W	W	w-i	w	w	W	W	w-i	w	w	a-i	w	w	W	a	W	w	..	..	w	Bromide.
Carbonate	a	W	..	A	A	a	A	a	A	A	A	a	A	A	A	a	a	A	W	a	W	A	..	..	A	Carbonate.
Chlorate	W	w	..	W	w	w	w	w	w	w	w	W	w	w	w	w	w	w	W	w	w	w	w	..	w	Chlorate.
Chloride	W	W[2]	W-A[6]	W	W-A[10]	W	W	W-I	W	W	W	W	W-I	W	W	A-I	W[16]	W	W[20]	I	W	W	W	W	W	Chloride.
Chromate	..	w	a	a	a	a	w-a	a	a	w	..	w	A-I	w	w	a	w-a	a	W	a	w	w-a	a	..	w	Chromate.
Citrate	W	w	..	a	..	a	w-a	w	w	w	w	W	a	w	a	a	w-a	w	w	a	W	a	..	..	w-a	Citrate.
Cyanide	..	w	..	w-a	..	a	w	a	a-i	a	a-i	..	a	w	a	..	W	a-i	W	i	w	W	..	..	a	Cyanide.
Ferricyanide	..	w	..	..	..	..	w	..	i	..	I	W	w-a	w	i	..	..	i	W	i	w	..	..	..	a	Ferricyanide.
Ferrocyanide	..	w	..	w-a	..	..	w	..	i	i	i	I	a	w	a	..	..	i	W	i	w	w	..	..	a-i	Ferrocyanide.
Fluoride	w	W	w	a-i	w	w-a	A	w	w-a	a	w-a	U	a	a-i	a	..	w-a	w-a	w	w	w	a-i	w	w	w-a	Fluoride.
Formate	w	w	..	w	w	w	w	w	w	w	w	w	w-a	w	w	W	w	w	w	w	w	w	w	..	w	Formate.
Hydrate	A	W	A	W	a	a	W-A	A	A	a	a	A	a	A	a	..	..	a	W	..	W	w	a	a	a	Hydrate.
Iodide	w	W	w-a	w	a	W	w	w	w	w	W	w	W-A	w	w	A	A	w	W	i	w	w	w	w	w	Iodide.
Malate	w	w	..	w-a	..	..	w-a	..	..	..	..	w	w-a	w	w	a	w-a	..	w	w-a	w	w	w	w	w	Malate.
Nitrate	w	W	..	W	W[11]	w	w	W	W	W	W	W	W	w	w	W	W	W	W	W	W	W	..	..	w	Nitrate.
Oxalate	a	W	a	a	a	a	A	w-a	A	a	a	a	a	a	w-a	a	a	a	W	a	W	a	a	w	a	Oxalate.
Oxide	A-I	..	a[7]	W	a	a	W-A	A-I	A	A	a	A	A	A	A[15]	A	A	A	W	a	W	W	a	A-I	A	Oxide.
Phosphate	a	W[3]	w-a	w-a	a	a	W-A	a	a	a	a	a	a	a[3]	a	a	a	a	w	a	W	a	a	a	a	Phosphate.
Silicate	A-I	..	..	a	..	a	a	a	a	a	a	a	a	a	a	..	..	a	W	..	W	a	..	..	a	Silicate.
Succinate	w-a	w	..	w-a	..	w	w-a	..	w-a	w-a	w	..	a	w	w	a	w-a	w	w	a	w	w-a	..	a	w-a	Succinate.
Sulphate	W[1]	W[4]	a	A	w	W	W-I	W-A[12]	W[13]	W	W	W	A-I	W	W	w-a	W[17]	W	W[12]	W-A	W	I	w	..	W	Sulphate.
Sulphide	a	W	A[8]	W	a	A	W-A	a-i	a	A	A	A	A	a	a	a	A[18]	A[19]	W	a[21]	W	w	a[22]	A[22]	A[23]	Sulphide.
Tartrate	w	w[5]	a[9]	a	a	w-a	a	w	w	w	w-a	W[14]	a	w-a	w-a	w-a	a	a	W	a	w	a	a	..	a	Tartrate.

[1] $(SO_4)_4(Al_2)(NH_4)_2$ = W; $(SO_4)_4(Al_2)K_2$ = W. [2] $As(NH_4)Cl_4$ = W; $Pt(NH_4)Cl_3$ = W-I. [3] $PO_4HNa(NH_4)$ = W; $PO_4Mg(NH_4)$ = A. [4] $(SO_4)_2Fe(NH_4)_2$ = W; $(SO_4)_2Cu(NH_4)_2$ = W. [5] $C_4H_4O_6K(NH_4)$ = W. [6] SbOCl = A. [7] Sb_2O_3 = soluble in HCl, not in NO_3H. [8] Sb_2S_3 = Sol. in hot HCl, slightly in NO_3H. [9] $C_4H_4O_6K(SbO)$ = W. [10] BiOCl = A. [11] $NO_3(BiO)$ = A. [12] $(SO_4)_4(Cr_2)K_2$ = W. [13] CoS = easily sol. in NO_3H, very slowly in HCl. [14] $(C_4H_4O_6)_4(Fe_2)K_2$ = W [15] MnO_2 = sol. in HCl; insol. in NO_3H. [16] Mercurammonium chloride = A. [17] Basic sulphate = A. [18] HgS = insol. in HCl and in NO_3H, sol. in aq. regia. [19] See 13. [20] $PtKCl_3$ = W-A. [21] Only soluble in NO_3H. [22] Sn sulphides = sol. in hot. HCl; oxidized, not dissolved by NO_3H. Sublimed $SnCl_4$ only sol. in aq. regia. [23] Hardly sol. in NO_3H, difficultly in HCl.
Au_2S = insol. in HCl and in NO_3H, sol. in aq. regia. $AuBr_3$, $AuCl_3$, and $Au(CN)_3$ = w; AuI_3 = a. PtS_2 = insol. in HCl, slightly sol. in hot NO_3H; sol. in aq. regia. $PtBr_4$, $PtCl_4$, $Pt(CN)_4$, $(NO^3)_4Pt$, $(C_2O_4)_2Pt$, $(SO_4)_2Pt$ = w; PtO_2 = a; PtI_4 = i.

375

TABLE II.—WEIGHTS AND MEASURES.

Measures of Length.

1 millimetre	= 0.001 metre	=	0.0394 inch.	
1 centimetre	= 0.01 "	=	0.3937 "	
1 decimetre	= 0.1 "	=	3.9371 inches.	
1 METRE		=	39.3708 "	
1 decametre	= 10 metres	=	32.8089 feet.	
1 hectometre	= 100 "	=	328.089 "	
1 kilometre	= 1000 "	=	0.6214 mile.	

Inch.		Millimetres.	Inches.		Centimetres.	Inches.		Centimetres.
1/64	=	0.3819	2	=	5.08	9	=	22.86
1/32	=	0.7638	3	=	7.62	10	=	25.40
1/16	=	1.5875	4	=	10.16	11	=	27.94
1/8	=	3.175	5	=	12.70	12	=	30.48
1/4	=	6.35	6	=	15.24	18	=	45.72
1/2	=	12.7	7	=	17.78	24	=	60.96
1	=	25.4	8	=	20.32	36	=	91.44

Measures of Capacity.

1 millilitre	= 1 c.c.	= 0.001 litre	=	0.0021 U. S. pint.	
1 centilitre	= 10 "	= 0.01 "	=	0.0211 "	"
1 decilitre	= 100 "	= 0.1 "	=	0.2113 "	"
1 LITRE	= 1000 "		=	1.0567 "	quart.
1 decalitre		= 10 litres	=	2.6418 "	galls.
1 hectolitre		= 100 "	=	26.418 "	"
1 kilolitre		= 1000 "	=	264.18 "	"

♏.		c.c.	♏.		c.c.	♏.		c.c.	Fl ʒ.		c.c.	Fl ℥.		c.c.
1	=	0.06	21	=	1.29	41	=	2.52	1	=	3.70	11	=	325.25
2	=	0.12	22	=	1.36	42	=	2.58	2	=	7.39	12	=	354.82
3	=	0.19	23	=	1.42	43	=	2.66	3	=	11.09	13	=	384.40
4	=	0.25	24	=	1.48	44	=	2.72	4	=	14.79	14	=	413.97
5	=	0.31	25	=	1.54	45	=	2.77	5	=	18.48	15	=	443.54
6	=	0.37	26	=	1.60	46	=	2.84	6	=	22.18	16	=	473.11
7	=	0.43	27	=	1.66	47	=	2.90	7	=	25.88			
8	=	0.49	28	=	1.73	48	=	2.96	8	=	29.57	O.		Litres.
9	=	0.55	29	=	1.79	49	=	3.02	Fl ℥.			1	=	0.47
10	=	0.62	30	=	1.85	50	=	3.08	1	=	29.57	2	=	0.95
11	=	0.68	31	=	1.91	51	=	3.14	2	=	59.14	3	=	1.42
12	=	0.74	32	=	1.98	52	=	3.20	3	=	88.67	4	=	1.89
13	=	0.80	33	=	2.04	53	=	3.26	4	=	118.24	5	=	2.36
14	=	0.86	34	=	2.10	54	=	3.32	5	=	147.81	6	=	2.84
15	=	0.92	35	=	2.16	55	=	3.39	6	=	177.39	7	=	3.31
16	=	0.99	36	=	2.22	56	=	3.46	7	=	206.96	8	=	3.79
17	=	1.05	37	=	2.28	57	=	3.52	8	=	236.53	9	=	4.26
18	=	1.11	38	=	2.34	58	=	3.58	9	=	266.10	10	=	4.73
19	=	1.17	39	=	2.40	59	=	3.64	10	=	295.68	11	=	5.20
20	=	1.23	40	=	2.46	60	=	3.70				12	=	5.67

376

Weights.

1 milligram	= 0.001 gram	=	0.015 grain Troy.	
1 centigram	= 0.01 "	=	0.154 "	"
1 decigram	= 0.1 "	=	1.543 "	"
1 GRAM		=	15.432 grains	"
1 decagram	= 10 grams	=	154.324 "	"
1 hectogram	= 100 "	=	0.268 lb.	"
1 kilogram	= 1000 "	=	2.679 lbs.	"

Grains.		Grams.	Grains.		Grams.	Grains.		Grams.	℥		Grams.
1/64	=	0.001	21	=	1.361	47	=	3.046	1	=	31.103
1/32	=	0.002	22	=	1.426	48	=	3.110	2	=	62.207
1/16	=	0.004	23	=	1.458	49	=	3.175	3	=	93.310
1/8	=	0.008	24	=	1.555	50	=	3.240	4	=	124.414
1/4	=	0.016	25	=	1.620	51	=	3.305	5	=	155.517
1/2	=	0.032	26	=	1.685	52	=	3.370	6	=	186.621
1	=	0.065	27	=	1.749	53	=	3.434	7	=	217.724
2	=	0.130	28	=	1.814	54	=	3.499	8	=	248.828
3	=	0.194	29	=	1.869	55	=	3.564	9	=	279.931
4	=	0.259	30	=	1.944	56	=	3.629	10	=	311.035
5	=	0.324	31	=	2.009	57	=	3.694	11	=	342.138
6	=	0.389	32	=	2.074	58	=	3.758	12	=	373.250
7	=	0.454	33	=	2.139	59	=	3.823			
8	=	0.518	34	=	2.204	60	=	3.888	lbs.		Kilos.
9	=	0.583	35	=	2.268				1	=	0.373
10	=	0.648	36	=	2.332	ʒ			2	=	0.747
11	=	0.713	37	=	2.397	1	=	3.888	3	=	1.120
12	=	0.778	38	=	2.462	2	=	7.776	4	=	1.493
13	=	0.842	39	=	2.527	3	=	11.664	5	=	1.866
14	=	0.907	40	=	2.592	4	=	15.552	6	=	2.240
15	=	0.972	41	=	2.657	5	=	19.440	7	=	2.613
16	=	1.037	42	=	2.722	6	=	23.328	8	=	2.986
17	=	1.102	43	=	2.787	7	=	27.216	9	=	3.359
18	=	1.166	44	=	2.852	8	=	31.103	10	=	3.733
19	=	1.231	45	=	2.916						
20	=	1.296	46	=	2.980						

INDEX.

DICK'S ENCYCLOPEDIA

OF

PRACTICAL RECEIPTS AND PROCESSES.

1872

INDEX PAGE 355

Manipulations. Under this heading will be found a brief description of the various methods of chemical manipulation, constantly employed in this work. This is deemed especially necessary, as many, if not all, of the processes described, depend greatly on careful and skillful manipulation in the preparation as well as in the combination of the necessary ingredients. *(See No. 3830.)*

2. Annealing. The process by which glass is rendered less frangible, and metals, which have become brittle, again rendered tough and malleable. *Glass* vessels, and other articles of glass, are annealed by being placed in an oven or apartment near the furnaces at which they are formed, called the *"leer,"* where they are allowed to cool slowly, the process being prolonged according to their bulk. *Steel, iron, and other metals,* are annealed by heating them and allowing them to cool slowly on the hearth of the furnace, or any other suitable place, unexposed to the cold.

3. Bath. In cases where an equable heat has to be sustained at, or not to exceed, a certain fixed degree, it is evident that an open fire or flame would be too variable for the purpose. To obviate this difficulty, the vessel to be heated is immersed or imbedded, to a convenient depth, in another vessel containing water, oil, saline solution, sand, metal, etc., as circumstances require, to which the heat is applied and whose temperature can be regulated, if necessary, by the use of a thermometer. Steam is also applied to this purpose; but, of course, requires special apparatus. The baths most commonly used are the water bath and the sand bath.

4. Sand Bath. An iron or copper vessel should be employed for this purpose. Sufficient sea or river sand, previously washed clean and dried, must be put in to cover the bottom completely. The vessel to be acted on is then introduced, and the intervening space around it filled up to the desired height with sand, and the whole placed over a furnace. The object of the sand is to cut off direct communication with the fire and produce a gradual and equable heat.

5. Water Bath, or Bain-Marie. This arrangement is used where the heat required is not over 212° Fah., and consists of one vessel within another, secured so that they cannot come in contact at any point below the level of the water which has been introduced to fill up the space between them. A double glue-pot is a water bath.

As the temperature of water cannot be increased, in an open vessel, above its boiling point, 212°, a vessel immersed in it can never be heated above that point; and, by keeping the water boiling, this degree can be steadily sustained. Where other degrees of heat are requisite, the following table, showing the boiling points of different substances and saturated solutions, will serve as a guide. A still higher degree of heat may be reached by using, with appropriate vessels, metals whose melting point is known. *(See Index for Melting Point of Metals.)*

6. Table exhibiting in degrees of Fahrenheit the Boiling Heat of different liquids.

Liquid	° F.
Ether	96°
do sp. grav.: .7365 at 48°	100
Carburet of Sulphur	113
Alcohol, sp. gr. .813	173⅛
Nitric Acid, sp. gr. 1.42	247
Water	212
Ammonia	140
Muriatic Acid, sp. gr. 1.094	232
Rectified Petroleum	306
Oil of Turpentine	316
Sulphuric Acid, sp. gr. 1.848	600
do do do 1.810	473
do do do 1.780	435
do do do 1.700	374
do do do 1.650	350
do do do 1.520	290
do do do 1.408	260
do do do 1.300	240
Phosphorus	554
Linseed Oil	640
Whale Oil	630
Mercury	662

7. Table showing the Boiling Heat of various Saturated Solutions.

Saturated solution of	° F.
Muriate of Lime	285°
Acetate of Soda	256
Nitrate of Soda	246
Rochelle Salt	240
Nitre	238
Muriate of Ammonia	236
Tartrate of Potash	234
Sea Salt	224⅛
Muriate of Soda	224
Sulphate of Magnesia	222
Borax	222
Phosphate of Soda	222
Carbonate of Soda	220
Alum	220
Chlorate of Potash	218
Sulphate of Copper	216
Acetate of Lead	215⅜
Glauber Salt	213⅛

8. Concentration. The volatilization or evaporation of part of a liquid in order to increase the strength of the remainder. The operation can only be performed on solutions of substances of greater fixity than the menstrua or liquids in which they are dissolved. Many of the liquid acids, solutions of the alkalis, etc., are concentrated by distilling off their water.

9. Crystallization. Crystals are symmetrical forms assumed by certain bodies in solidifying from a liquid or gaseous state: and as the same substances, under similar circumstances, always assume the same crystalline shape, their crystals afford a means of distinguishing substances otherwise similar in appearance; as for instance oxalic acid and Epsom salts. Sulphur, anhydrous salts, lead, tin, and other fusible substances which are unalterable by heat are crystallized by *fusion.* They are to be melted at the lowest possible temperature, and allowed to cool very gradually. As soon as a crust forms on the surface (which then becomes furrowed) it must be pierced with a rod, and the fluid portion decanted, and the crystals will be found coating the interior of the vessel. Volatile solids, such as iodine, camphor, etc., when heated so as to produce *Sublimation* (*see No.* 30), yield vapors which, in cooling, take the form of crystals.

Soluble substances are crystallized by the evaporation of a saturated solution of the substance. The solution should be made and, if necessary, clarified and filtered at boiling point, in which state more of the substance is held in solution than when cool; this excess is deposited in crystalline form as the solution cools or evaporates. The crystals thus obtained are strained from the remaining liquid, or *mother water,* and dried.

If strings be suspended in the hot solution, crystals will form upon them during cooling or evaporation; in this manner rock-candy, blue vitriol (sulphate of copper), alum, etc., are crystallized. Crystallization is also sometimes the result of chemical reaction; silver, for instance, precipitated from its solutions by zinc, forms a crystalline deposit.

10. Decantation. The operation of pouring off the clear portion of a liquid from its sediment. This is performed either by gently inclining the vessel, or by means of a syphon. When a liquid is set aside to settle for future decantation by the first method, it is best to use a bell shaped vessel, or one provided with a lip, for convenience in pouring; as in decanting from a full vessel whose side is straight, the liquid is very apt to flow down the outside of the vessel. This can, however, be obviated by holding a glass rod or stick, previously wetted in the liquid, nearly upright, with one end resting in or suspended over the receptacle into which the liquid is to be decanted; the liquid is poured gently down the upper side of the stick, keeping the rim of the vessel in contact with it. The liquid will be more strongly attracted by the wet stick, than by the dry surface of the outside of the vessel. *(See illustration.)*

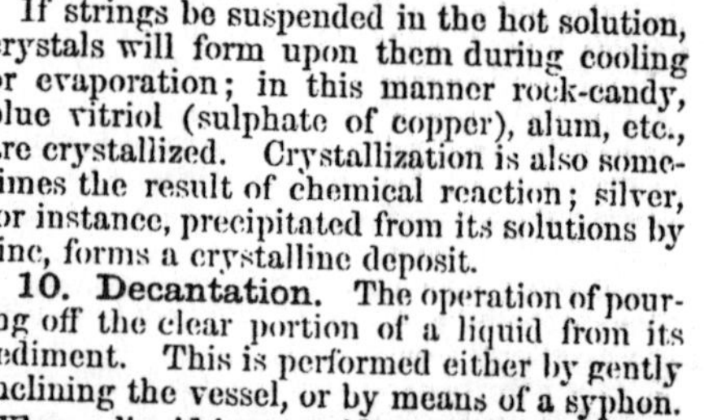

If this method of decanting is inconvenient, or, from the nature of the vessel, impossible, a syphon must be used. This is a tube of glass or metal, bent at an angle of about 30°, with one leg or end longer than the other. A piece of india-rubber tubing makes an excellent and easily adjusted syphon for decanting liquids which will not affect that material. The syphon must be first filled and then the shorter leg inserted in the liquid, care being taken to keep its extremity always below the surface, and the liquid will flow continuously out of the longer leg as long as there is any left in the vessel. For decanting caustic liquids, acids, &c., syphons of different kinds are provided, constructed especially for the purpose.

11. Deflagration. The sudden combustion of any substance, for the purpose of producing some change in its composition, by the joint action of heat and oxygen. The process is commonly performed by projecting into a red hot crucible, in small portions at a time, a mixture of about equal parts of nitre and of the body to be oxidized.

12. Desiccation. The evaporation or drying off of the aqueous portion of solid bodies. Plants and chemical preparations are deprived of their humidity by exposure to the sun, a current of dry air, an atmosphere rendered artificially dry by sulphuric acid, or by the direct application of heat by means of a water-bath, a sand-bath, or a common fire. Planks and timber are now seasoned, on the large scale, in this way, by which a condition may be attained in 2 or 3 days, which, on the old system, took as many years to produce.

13. Distillation. Distillation consists in vaporizing a liquid in one vessel, and conducting the vapor into another vessel, where it is condensed and collected. The process is used for separating a liquid from solid substances with which it may be mixed; for impregnating a liquid with the volatile principles of plants, as in the preparation of Eau de Cologne and other aromatic spirits, and for separating a more volatile liquid from one less so, as alcohol from water.

For example, as alcohol is transformed into vapor at the temperature of 176°, while water remains, at this temperature, in a liquid state, it is only necessary to heat the mixed liquids to 176°, when the alcohol rises in vapor, and the water is left behind. The vessel in which the liquids are heated is closed by an air-tight cover, and from this cover a pipe is led and coiled through a cask of cold water; as the alcoholic vapor enters this cold pipe it is condensed to the liquid form. This process of evaporating and condensing a liquid is called distillation; the apparatus is called a still or retort, and the coiled pipe is the "worm of the still," or the condenser.

On the small scale distillation is performed in the simplest way by means of the common glass retort (*a*,) and the receiver (*b*,) as in

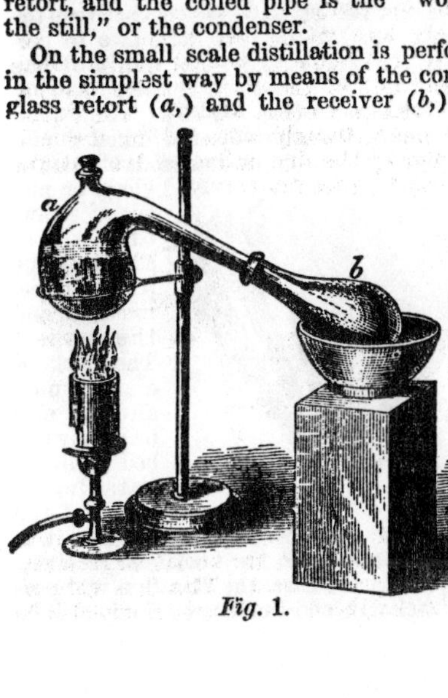

Fig. 1.

Fig. 1. The retort may be either simple, as in Fig. 2, or tubulated as in Fig. 1, and sometimes the receiver has a tubulure to allow the escape of gas or expanded air, as in Fig. 3. The great advantages of the glass retort are that it admits of constant observation of the materials within, that it is acted upon or injured by but few substances, and may be cleaned generally with facility. Its great disadvantage is its brittleness.

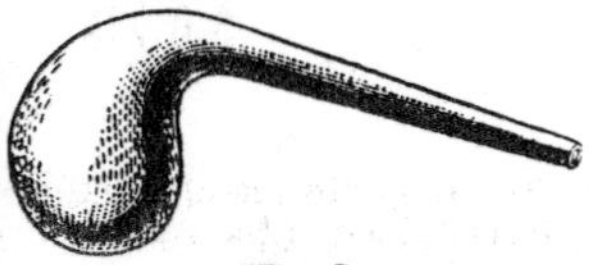

Fig. 2.

The tubulated retort is more liable to crack than the plain one, on account of the necessarily greater thickness of the glass in the neighborhood of the tubulature; nevertheless it is very convenient on account of the facility which it offers for the introduction of the materials.

When the common glass retort and receiver are used for the distillation of liquids, care should be taken not to apply the luting until the atmospheric air is expelled (*see Lute*), unless the receiver has a tubulure for its escape. The operator should aim at keeping the body of the retort hot, and the neck and receiver cool. A hood of pasteboard will facilitate the former; and the latter will be accomplished by keeping the neck and receiver wrapped in wet cloths, on which a stream of cold water is kept running. This may be conveniently done by means of a syphon, made by dipping one end of a strip of cotton in a vessel of water, and allowing the other end to hang down upon the cloths, bound loosely around the receiver and the neck of the retort. Retorts are heated in a water or sand bath, placed over the naked fire, or they may be held by a circle of metal, in which case the retort may be heated by the argand gas flame, as in Fig. 1, or by live coals. Where it is to be subjected to a heat sufficient to soften the glass, the bulb may be previously coated with a mixture of clay and sand, and dried. (*See Nos.* 1695 *and following.*)

Fig. 3

Even on the small scale it is sometimes necessary to employ distillatory apparatus constructed of other materials besides glass.

The still in general use (*see page* 12) may be considered as composed of three or four parts:

I. The cucurbit or body of the still, A. This portion of the apparatus receives the direct action of the fire, and contains the liquid to be distilled when the process is to be conducted by a naked fire. It is in the form of a truncated reversed cone, A, mounted on a rounded portion, *a a*, which rests on the furnace, X X, and terminated at the top by a collar of somewhat smaller diameter than the lower part.

C is a hole by which the liquid is introduced into the body of the apparatus; *d d* are the handles.

II. The water-bath, B, a cylindrical vessel of tin or tinned copper, which is placed in the cucurbit, A, closing it lightly by means of the collar, *m*, which rests on the collar, *b b*. This vessel is used only when the mixture to be distilled is not exposed to the direct heat of the fire; in this case the cucurbit, A, fulfills the office of a water-bath, and the vessel, B, takes the place of the cucurbit.

When, instead of distilling by the naked fire, the water-bath is employed, water only is put into the cucurbit, in which the vessel, B, is placed containing the liquid to be distilled.

III. The head of the capital, G. This part may be placed either on the cucurbit, when distilling by naked fire, or on the vessel, B, if used, care having been taken to make both openings of the same size; it is very nearly the shape of the upper part of a retort, and is furnished with a large pipe by which the vapor is to be carried off to the worm or cooler.

n. A hole which, during the operation, is kept closed by a screw top, *c*, and its use is to introduce fresh liquid into the cucurbit without having to disconnect the apparatus.

IV. The cooler or worm, D. This is a long tin pipe, bent in the form of a screw, and enclosed in a copper or wooden vessel full of cold water. The upper part of the pipe, which is often enlarged in a globular form, receives from the beak of the capital the vapors arising from the cucurbit; the lower portion is open below, so that the condensed liquid flows into a vessel placed underneath.

All the joints of the apparatus are to be luted with bands of paper soaked in paste; the joint of the cucurbit, when used as a water-bath, must not be tight, in order to allow of the escape of the steam from the boiling water. (*See Lute.*)

g g. Tin rests for supporting and fixing the worm in the vessel.

h. A vertical pipe fixed to the side of the vessel, open at both ends and terminated at the top by a funnel.

This pipe serves to renew the water in the cooler; cold water is poured in at the top which flows to the bottom of the vessel, and being of a lower specific gravity than the hot water, forces it out at the escape pipe, *i*.

k. A tap, by which all the water in the worm tub can be discharged.

f. A connecting pipe inserted between the beak of the capital and the collar of the still is of precisely the same height as the collar, *m*, of the cucurbit, B, and is only used in distilling by the water-bath; when a naked fire is used this pipe is unnecessary, as the beak will reach down to the collar of the still without it.

In distilling perfumes and cordials, the object is to extract or separate the odorous and aromatic principle from the roots, flowers, seed, or spices used to impart the characteristic odor and taste to the liquor, and it is usual to macerate such ingredients in strong alcohol several days before distillation. Great care should be taken that *the heat should, in all cases, be as gentle and uniform as possible.*

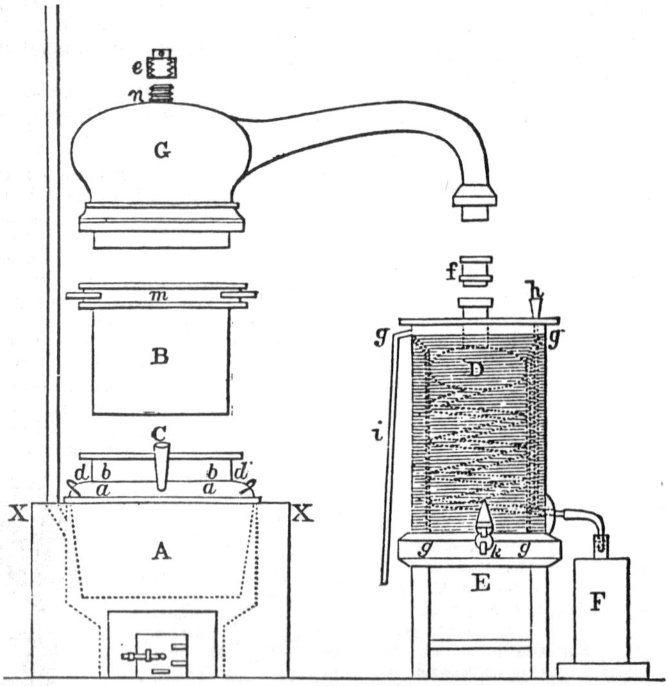

Remember that accidents may be effectually prevented by distilling spirits in a water-bath, which, if sufficiently large, will perform the operation with all the dispatch requisite for the most extensive business.

14. Elutriation. In chemistry, the operation of washing insoluble powders with water, to separate them from foreign matter, or the coarser portion. It is usually performed by grinding or triturating the mass with a little water, until reduced to a very fine powder, and this paste is suddenly diffused through a large quantity of water in a deep vessel, from which, after the subsidence of the grosser portion, the liquid is poured into another vessel, and allowed to deposit the fine powder it still holds in suspension. When this has taken place, the clear supernatant liquor is decanted, and the sediment drained and dried. The coarse sediment deposited in the first vessel is now submitted to a fresh grinding and diffusion through water, and the entire operation is repeated, until the whole of the pulverizable portion is washed over. The proper length of time for the liquid to remain in the first vessel, depends solely on the density of the powder, and the degree of fineness required in the product; heavy powders subsiding almost immediately, while light ones often take several minutes to deposit their coarser portion. Sometimes three or more vessels are employed, and the muddy liquor, after remaining a short time in the first, is poured into the next one, and this, in a short time longer, into the third, and so on, until the last vessel is filled, by which means, powders of different degrees of fineness are obtained; that deposited in the last vessel being in the minutest state of division.

15. Evaporation. The conversion of a fluid into vapor by means of heat, diminished atmospheric pressure, or exposure to a dry atmosphere. The process of evaporation is resorted to;—1. For the vapor as a source of heat or power, as in steam boilers, &c.;—2. To separate volatile fluids from other bodies which are either fixed or less volatile;—3. To recover solid bodies from their solutions;—4. To concentrate or strengthen a solution by expelling a portion of the liquid;—5. To purify liquids by expelling any volatile matters which they may contain. As evaporation is, under ordinary circumstances, confined to the surface of the liquid, wide shallow vessels are the best for the purpose; the process is greatly facilitated by exposing the surface to a current of dry air, especially if the air be heated. On a small scale, shallow capsules of glass, wedgwood ware, porcelain or metal, are commonly employed, and are exposed to heat by placing them over a lamp, open fire, or in a water or sand-bath. (*See No.* 44.)

16. Fermentation. Chemists distinguish fermentation into five kinds, viz:

The *saccharine fermentation*, by which starch and gum are converted into *sugar*.

The *alcoholic* or *vinous fermentation*, by which sugar is converted into *alcohol*.

The *viscous* or *mucilaginous fermentation*, which converts sugar into *slime* or *mucilage*, instead of alcohol.

The *acetous fermentation*, by which alcohol is converted into *vinegar*.

The *putrid fermentation*, or *putrefaction*, which is exhibited in its most marked form in the putrefaction of animal substances.

17. Filtration. The word *filtration* is absolutely synonymous with *straining;* but, in the language of the laboratory, the former is usually applied to the operation of rendering liquids transparent, or nearly so, by passing them through *fine* media, as filtering paper, for instance; the latter to the mere separation of the grosser portion, by running them through *coarse* media, as flannel, horse-hair cloth, etc., through which they flow with considerable rapidity. *Filtration* is distinguished from *clarification*, by the former removing the solid matter, or cause of opacity or foulness, by mere mechanical means, whereas the latter consists in the clearing of a liquid by depuration, or the subsidence of the suspended substances or fæces, arising from their gravity being naturally greater than the fluid with which they are mixed, or being rendered so by heat or the addition of some foreign substance. (*See Fining.*)

The apparatus, vessels, or media, employed for filtration, are called *filters*, and are commonly distinguished from *strainers* by the superior fineness of their pores, as above noticed.

Both strainers and filters act on the same principles as the common sieve on powders; they all, in like manner, retain or hold back the coarser matter, but permit the liquid or smaller and more attenuated particles to pass through. The term *medium* has been applied to the substance through the pores of which the liquid percolates.

The forms of filters, and the substances of which they are composed, are various, and depend upon the nature of the liquids for which they are intended. On the small scale, funnels of tin, zinc, copper, wedgwood ware, earthenware, glass, or porcelain, are commonly employed as the containing vessels. The filtering medium may be any substance of a sufficiently spongy or porous nature to allow of the free percolation of the liquid, and whose pores are, at the same time, sufficiently fine to render it limpid or transparent. Unsized paper, flannel, linen, muslin, cotton-wood, felt, sand, coarsely-powdered charcoal, porous stone or earthenware, and numerous other substances of a similar kind are employed for this purpose.

Filters of unsized paper are well suited for all liquids that are not of a corrosive or viscid nature, and are universally employed for filtering small quantities of liquids in the laboratory. A piece of the paper is taken, of a size proportionate to the quantity of the substance to be filtered, and is first doubled from corner to corner into a triangle (*see Fig.* 1, *below*), which is again doubled into a smaller triangle, and the angular portion of the margin being rounded off with a pair of scissors, constitutes a paper cone, which is placed on a funnel and nearly filled with the liquid. A piece of paper so cut, when laid flat upon a table, should be nearly circular. Another method of forming a paper filter, preferred by some persons, is to double the paper once, as above described, and then to fold it in a similar way to a fan, observing so to open it (*see Fig.* 2) and lay it on the funnel that a sufficient interval be left between the two to permit of the free percolation of the liquid. (*See Fig.* 3.)

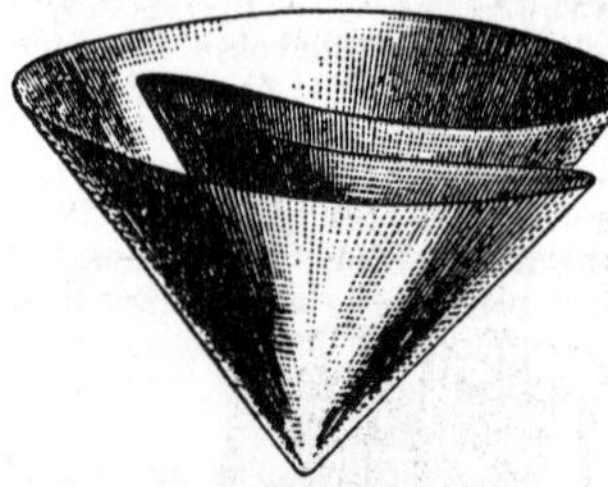

Fig. 1.

To promote the same object, a funnel should be deeply ribbed inside, or small rods of wood or glass, or pieces of straw, or quills, should be placed between it and the paper. The neck of a funnel should also be deeply ribbed or fluted outside, to permit of the free outward passage of the air when it is placed in a narrow-mouthed bottle or receiver. Unless this is the case, the filtration will proceed but slowly, and the filtered liquid will be

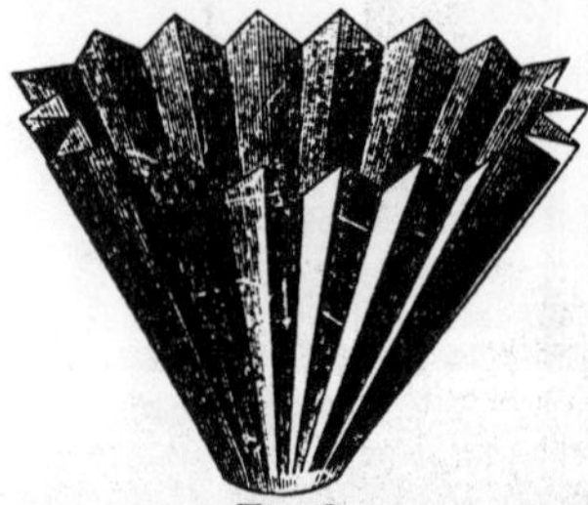

Fig. 2.

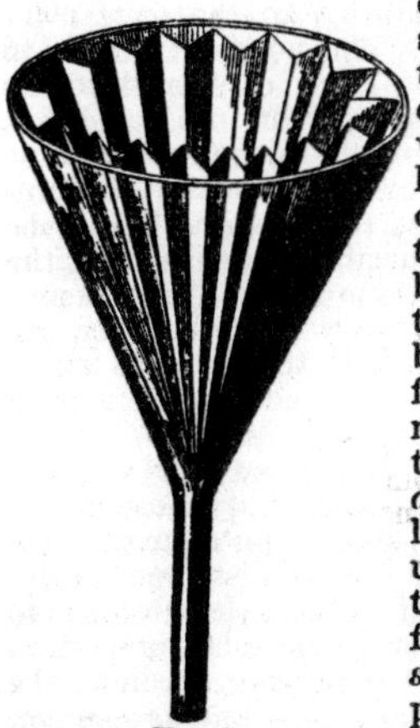

Fig. 3.

driven up the outside of the neck of the funnel by the confined air, and will be continually hissing and flowing over the mouth of the vessel. The breadth of a funnel, to filter well, should be about three-fourths of its height, reckoning from the throat or neck. If *deeper*, the paper is liable to be continually ruptured from the pressure of the fluid; and when *shallower*, filtration proceeds slowly and an unnecessarily large surface of the liquid is exposed to evaporation. To lessen this as much as possible, the upper edge of the glass is frequently ground perfectly smooth, and a piece of smooth plate-glass is laid thereon. When paper filters are of large dimensions, or for aqueous fluids that soften the texture of the paper, or for collecting heavy powders or metallic precipitates, it is usual to support them on linen or muslin to prevent their breaking. This is best done by folding the cloth up with the paper and cutting the filter out of the two, in the same way as would be done with doubled paper, observing so to place it in the funnel that the paper and muslin may remain close together, especially towards the bottom.

The filtration of small quantities of liquids, as in chemical experiments, may often be conveniently performed by merely placing the paper on the circular top of a recipient; or on a ring of glass or earthenware laid on the top of any suitable vessel. A filter of this kind, that will hold one fluid ounce, will filter many ounces of some liquids in an hour.

Good filtering paper should contain no soluble matter, and should not give more than one two hundred and fiftieth to one two hundred and thirtieth of its weight of ashes. The soluble matter may be removed by washing it, first with *very dilute* muriatic acid, and secondly with distilled water.

For filtering a larger quantity of a liquid than can be conveniently managed with a funnel, and also for substances that are either too viscid or too much loaded with feculence to allow them to pass freely through paper, conical bags made of flannel, felt, twilled cotton cloth or *Canton* flannel, linen, or muslin, and suspended to iron hooks by rings or tapes, are commonly employed. (*See Fig.* 4.) The first two of the above substances are preferable for saccharine, mucilaginous, and acidulous liquids; the third for oily ones; and the remainder for tinctures, weak alkaline lyes, and similar solutions. These bags have the disadvantage of sucking up a considerable quantity of the fluid poured into them, and are therefore objectionable, except for large quantities, or when continued in actual use as filters for some time. On the large scale, a number of them are usually worked together, and are generally enclosed in cases to prevent evaporation, and to exclude dirt from the filtered liquor that trickles down their outsides.

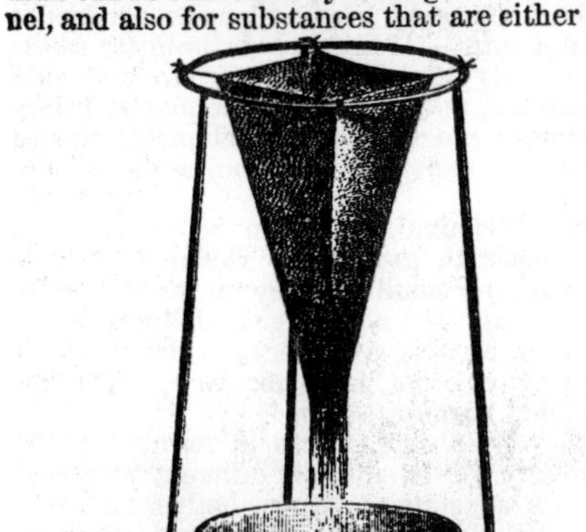

Fig. 4.

A very simple mode of filtering aqueous fluids, which are not injured by exposure to the air, is to draw them off from one vessel to another, by means of a number of threads of loosely twisted cotton or worsted arranged in the form of a syphon. The little cotton rope at once performs the operations of decantation and filtration. This method is often convenient for sucking off the water from small quantities of precipitates.

When pulverulent substances, as *sand*, *coarsely-powdered charcoal*, etc., are employed as the media for filtration, vessels of

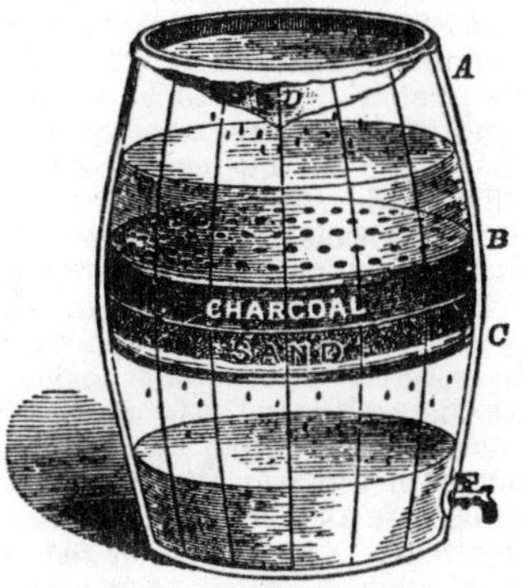

Fig. 5.

wood, or stoneware, are employed to contain them and the supernatant liquid. In these cases, the filtering medium is usually arranged as a shelf or diaphragm, and divides the vessel into two compartments; the upper one being intended to contain the liquid, and the under one to receive the same when filtered. Such an apparatus is set in operation by merely filling the upper chamber, and may, at any time, be readily cleaned out by reversing it and passing clean water through it in an opposite direction. The following is a filter of this description, and very simple in its arrangement. (*See Fig.* 5.) *A* is a common cask, *B* and *C* are false bottoms, fitting in perfectly air tight, but perforated with one-fourth inch holes. *C* should be covered with canvas, and above that a sheet of cotton wadding; above the wadding is a bed of *perfectly clean* sand, 3 inches deep. The sand should be covered over with flannel, and above the flannel should be a bed of *granulated* animal charcoal (*sifted* and *fanned from the dust*), 4 inches in depth. After having done this, fit in the false bottom, *B*, and cover it with a piece of cotton cloth. *D* is a bag made of Canton flannel to prevent the liquor being filtered from coming with too much force upon the false bottom. By substituting cotton wadding instead of the charcoal in the above filter, a fine filter for brandy and other liquors may be obtained.

A filter which possesses the advantages of being easily and cheaply cleaned when dirty, and which very thoroughly purifies brandy or water with great rapidity, may be formed by placing a stratum of sponge between *two perforated metallic plates*, united by a central screw, and arranged in such a manner as to permit of the sponge being *compressed* to any required degree. Brandy or water, under gentle pressure, flows with great rapidity through the pores of compressed sponge.

It is often of great advantage to render a filter self-acting, or to construct it in such a way that it may feed itself, so that it may continue full and at work without the constant attention of the operator. On the *small scale*, this may be readily effected by an arrangement as represented in Fig. 6; and on the *large scale* by placing the vessel containing the unfiltered liquid on a higher level than the filter, and by having the end of the supply-pipe fitted with a ballcock, to keep the liquid in the filter constantly at the same height. (*See No.* 3840.)

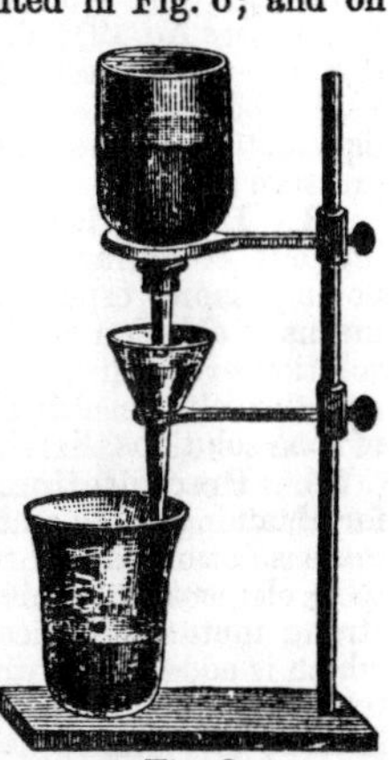

Fig. 6.

The rapidity of filtration depends upon the *porosity of the filtering medium — the extent of filtering surface—the relative viscidity or limpidness of the filtering liquid, and the porosity and fineness of the substances it holds in suspension.* The most efficient filter is produced when the first two are so graduated to the latter, that the liquid filters rapidly and is rendered perfectly transparent. (*See No.* 3838.) (*Cooley.*)

Tinctures and dilute spirits are usually filtered through bibulous paper placed on a funnel, or through thin and fine cotton bags. In general, tinctures clarify themselves by the subsidence of the suspended matter, when allowed to repose for a few days. Hence it is the bottoms alone that require filtering; the supernatant clear portion need only be run through a small hair sieve, a piece of tow or cotton placed in the throat of a funnel, or some other coarse medium, to remove any floating substances, as pieces of straw, &c. *Spirits* largely loaded with essential oil, as those of aniseed, &c., run rapidly through paper or muslin, but usually require the addition of a spoonful or two of magnesia before they will flow *quite* clear. When possible, tinctures, spirits, and all similar volatile fluids, are better cleared by subsidence or clarification than by filtration, as, in the latter way, part is lost by evaporation. (*See Nos.* 3834, &c.)

18. Gun-cotton as a Filter. Gun-cotton, carefully prepared, is scarcely acted on by the most energetic chemical agents at ordinary temperatures. It may therefore be used as a filter for solutions containing strong acids, alkalies, etc.

19. Fusion. *Aqueous* fusion is the dissolving of crystalline compounds in their own water of crystallization, by the application of heat. *Igneous* fusion is a term applied to the liquefaction of bodies by heat alone. The containing vessels used for igneous fusion should be of a material capable of sustaining the requisite degree of heat without either melting or cracking. Crucibles made of very refractory clay are used for high temperatures, metallic or earthenware vessels for lower degrees of heat.

20. Granulation. The reduction of metals into grains, drops, or coarse powder. This is done by pouring them, in the melted state, into water. The same effect is obtained by violently agitating the molten metal until cool, in a wooden box, well chalked inside. (*See No.* 25.) In many cases the metal is allowed to run through the holes of a kind of colander or sieve to produce minute division; if the drops are allowed to fall from a sufficient height, they will become spherical; in this way lead shot is made.

21. Liquation. The process of sweating out, by heat, the more fusible metals of an alloy.

22. Liquefaction. The conversion of a solid into the liquid state, either by heat—*fusion*, (*see No.* 19); absorption of water from the air—*deliquescence;* or the action of a fluid body—*solution.* (*See No.* 29.) The liquefaction of gases and vapors is effected by pressure and cold.

23. Lixiviation. The process of dissolving out or extracting the saline matter of bodies, more especially of ashes, &c., by means of ablution or digestion in water. The solution so obtained is called a lye or lixivium, and the salts resulting from the evaporation of such solutions, lixivial salts.

24. Precipitation. This is the method for obtaining solid matter, by mixing two or more solutions of substances containing certain elementary equivalents which have a strong mutual chemical affinity. That fluid which is added to another to produce precipitation is called the *precipitant.* If a solution is to be precipitated, it is best, unless otherwise directed, to first heat it by means of a sand bath. (*See No.* 4.) A tall bell-shaped glass with a mouth is the best for precipitating. The precipitant is to be added gradually, stirring the mixture continually with a glass rod, until precipitation ceases. The liquid should then be allowed to settle until clear. In order to ascertain whether there is any matter left in the liquid unprecipitated, let one drop of the precipitant fall into the mixture; if any signs of precipitation ensue, more must be added; if the mixture remains unchanged and clear, the operation is complete. The liquid may then be carefully decanted and the precipitated matter, which is called a *precipitate*, filtered and dried. When the precipitate is the chief object of the process, it is usually necessary to wash it after filtration. This operation requires but little attention when the precipitate is insoluble in water; but when it is *in some degree* soluble in that liquid, great attention is required to prevent the loss which might result from the use of too much water. Precipitates soluble in water, but insoluble in alcohol, are frequently, on a small scale, washed with spirit more or less concentrated. (*See No.* 14.)

25. Pulverization. The reduction of any substance to dust or powder is generally performed by means of a pestle and mortar, or, on a larger scale, by stamping, grinding or milling. A few soft substances, as carbonate of magnesia, carbonate of lead, &c., may be pulverized by simply rubbing through a fine sieve; while many hard and gritty, and some soft substances, such as chalk, antimony, &c., are pulverized on a large scale by elutriation. (*See No.* 14.) Others will only yield to a rasp or file. Whichever method is adopted, the substance to be pulverized must be very dry, and may even require artificial drying or desiccation. (*See No.* 12.) On the other hand, a few substances, as rice, sago, nux vomica, &c., are often soaked in water, or steamed, before being pulverized. In some cases, some other substance or *intermedium* is introduced to aid in the operation; thus, sugar is used in pulverizing civet, musk, nutmeg, and vanilla; absorbing the moisture which could not otherwise be readily got rid of. The addition of a very small quantity of alcohol renders the powdering of camphor easy. Gold leaf is pulverized by mixing with sulphate of potassa, and then removing the potassa by washing with water. (*See also No.* 2517.) Fusible metals are reduced by melting and rubbing in a mortar until cold, or by agitating when melted in a box covered inside with chalk or whiting. Glass, quartz and silicated stones require to be heated red hot and then thrown into cold water, to make them sufficiently friable for pulverization. When powdering very dusty or costly articles in a mortar, it should be covered with a loose skin of leather, fastened firmly round the top of the mortar and the pestle, to prevent loss of the dust, and possible injury to the operator's lungs. When a substance is required to be reduced to an impalpable powder, a slab and muller are used; this process is termed *porphyrization.*

26. Reduction. This term is applied to a process by which the oxygen is withdrawn from a metallic oxide, leaving the base in its original state. This is effected by heating the oxide with carbon or hydrogen; or by exposing it to the action of some other body which has a powerful affinity for oxygen. A portion of the metallic oxide to be reduced, is mixed with finely powdered charcoal and exposed in a crucible to the heat of a furnace. The metallic residue, which remains after reduction by this means, is usually mixed with coal dust. This is prevented by lining the crucible with charcoal dust made into a dough with clay and water, leaving a space in the middle to receive the metallic oxide, *not mixed* with charcoal, as in the former instance; the crucible must be covered, and then heated. The reduction in this way is slower, but the metal will be pure and free from coal dust.

When hydrogen is employed for reduction, the metallic oxide is heated to redness in a glass or porcelain tube, and subjected to a current of hydrogen gas, which absorbs the oxygen, and leaves the metal pure. Other agents are sometimes used for reducing, as tallow, oil, resin, sugar, and starch; but carbon and hydrogen are the agents generally employed.

27. Saturation. A liquid is said to be saturated with some other substance when it ceases to dissolve any more of it. An acid is saturated with an alkali when sufficient of the alkali has been added to completely neutralize the acid, and vice versa.

28. Sifting. This is a means employed to obtain uniformity of fineness in a pulverized substance; and is also of use in mixing different substances powdered to the same degree of fineness. The sieves used for this purpose are furnished with cloths of various materials and different degrees of fineness; consisting of brass wire, horse hair, buckram, book muslin, gauze, or raw silk; this last constituting a bolting cloth for sifting impalpable powders. These are stretched over a wooden cylinder in the same manner as the head of a drum. During the process of pulverizing, the use of the sieve is necessary from time to time to separate the finer powder from the coarser particles, which have to be returned after each sifting, to the mortar for further trituration. The powder is made to pass through the meshes of the sieve by gently agitating it between the hands; a rough jarring motion will force through some of the coarser particles, and destroy the uniformity of the powder. A sieve should be fitted with a drum head, top and bottom, the upper one to confine the dust of the substance being sifted, and the lower one to catch the sifted powder as it falls through the sieve. An arrangement of this kind is called a drum or box sieve.

29. Solution. Under the head of solutions, are properly included only those liquids which consist of water or an aqueous menstruum, in which has been dissolved an appropriate quantity of any soluble substance to impart to the liquor its peculiar properties. When spirit is the dissolving medium, the liquid receives the name of *alcoholic solution, spirit*, or *tincture*, while substances dissolved in water form *aqueous solutions.* In cases where a substance is dissolved in an acid or alkaline solution, whose acid or alkali is afterwards neutralized by means of an alkali (to counteract the acid), or an acid (to destroy alkali), the solution is then termed a *neutral solution.* A saturated solution is a solution made according to No. 27.

Professor Youmans, in the "Hand Book of Household Science," says: "Solids should be crushed or pulverized, to expose the largest surface to the action of the solvent liquid. Substances which in the lump would remain for days undissolved, when reduced to powder are liquefied in a short time. When a solid, as common salt or alum, is placed in a vessel of water to dissolve, it rests at the bottom. The water surrounding it becomes saturated, and being heavier, remains also at the bottom, so that the solution proceeds very slowly. By stirring, the action is hastened, but this takes up much time. The best plan is to suspend the salt in a colander, basket, or coarse bag, at the surface of the liquid. As the particles of water take up the particles of salt, they become heavier and sink; other particles take their places, dissolve more of the salt, and sink in turn, so that the action of a constant current of liquid is kept up on the suspended crystals, and always at that portion most capable of dissolving them."

30. Sublimation. The process by which volatile solid substances are reduced to the state of vapor by heat, and again condensed in solid form. It differs from ordinary distillation only in being confined to dry solid substances, and in the heat employed being, in general, much greater. Calomel, corrosive sublimate, and sal ammoniac, are thus prepared.

31. Trituration. The reduction of a solid body to powder by rubbing. This is effected on a small scale by means of a pestle and mortar; and on a large scale by grinding in a mill, or with a muller or a slab made of porphyry or other hard substance; this latter is termed *porphyrization.*

32. Washing. This is resorted to in chemistry for two widely different purposes. When a substance contains both soluble and insoluble matter, the soluble portion can be separated from the insoluble by washing; this is called *Lixiviation.* (*See No.* 23.)

When it is desired to cleanse or remove impurities from an insoluble powder, this is also effected by washing. (*See Nos.* 14 *and* 3841.)

Preparations.

The following methods of preparing decoctions, extracts, tinctures, &c., are from the best practical sources. Other directions for making extracts, essences, attars, &c., for the special purposes of Perfumery, &c., will be found under their respective headings.

34. To Prepare Decoctions. Decoctions are solutions of the properties of vegetables obtained by boiling, which is presumed to be a more effective method of extracting their properties than mere infusion.

For making decoctions, the substances should be well bruised, or reduced to a *very coarse* powder, or, if fresh and soft, they should be sliced small. In the former case, any *very fine* powder or *adhering dust* should be removed with a sieve, as its presence would tend to make the product thick and disagreeable, and also more troublesome to strain. The vessel in which the boiling is conducted should be furnished with an accurately fitting cover, the better to exclude the air, and the heat should be so regulated that the fluid may be kept *"simmering,"* or only *gently boiling*, as *violent* boiling is not only quite unnecessary, but absolutely injurious. In every case the liquor should be strained while *hot*, but *not* boiling, and the best method of doing this is to employ a fine hair sieve, or a coarse flannel bag. In general it is found, that as decoctions cool, a sediment is formed, in consequence of the boiling water dissolving a larger portion of vegetable matter than it can retain in solution when cold. This deposit for the most part consists of the active principles of the solution, and should be mingled with the clear liquid by agitation, when the decoction is used. It will thus be seen that the common practice of leaving the filtration until the liquid has become cold, and also of rejecting the sediment, is injudicious, and should be scrupulously avoided; as, however much decoctions so prepared may please the eye, they are not only *inferior in strength*, but, in many cases, *nearly inert*. It may be further remarked, that long boiling is in no case necessary, and should be avoided, especially in decoctions prepared from aromatic vegetables, or those abounding in extractive. The colleges, in such cases, direct the ingredients "to be boiled for a *short time*," or "for 10 minutes," or they limit the time of boiling by stating the quantity that must be volatilized, as—"boil to a *pint*, and strain," the latter method being generally employed for those substances that do not suffer by lengthened boiling.

Distilled water, or perfectly clean *rain* water, should alone be used for decoctions. Spring and river water, from their containing lime, have less solvent powers.

Decoctions of all vegetables not exerting a *very powerful* action on the human system may be made by boiling 1 ounce of the vegetable matter in 1 pint of water for 10 or 15 minutes. The ordinary dose of such a decoction is the same as that of a similar infusion. (*See No.* 37.)

When the medicinal properties of vegetables are volatile, or are injured by a *strong* heat, infusion should be had recourse to, in preference to boiling; but when a solution of the fixed constituents is alone sought, decoction is preferable. In preparing compound decoctions, those ingredients should be boiled first which least readily impart their active principles, and those which most readily impart them should be added afterwards. In many cases it will be proper simply to infuse the more aromatic substances in the hot decoction of the other ingredients, by which means their volatile principles will be preserved.

35. To Prepare Tinctures. Tinctures are solutions of vegetable and animal drugs, and sometimes of mineral substances, in spirituous liquids. The spirit most commonly employed is proof-spirit; sometimes rectified spirit is used, and occasionally ether. Ammonia is sometimes conjoined with the spirit, in which case the solution is termed an ammoniated tincture. Rectified spirit is alcohol, with 16 per cent. of water, and its specific gravity is .838. Proof-spirit is composed of 5 parts of rectified spirit mixed with 3 parts of water, the resulting compound containing 47.5 per cent. of water, specific gravity .920. The choice between proof and rectified spirit depends on their respective solvent powers over the active principles of the drugs employed.

Tinctures are usually prepared by reducing the solid ingredients to small fragments, coarse powder, or fine powder, macerating them for 7 days or upwards in proof or rectified spirit, straining the solution through linen or muslin, or paper, and finally expressing the residuum strongly, to obtain what fluid is still retained in the mass. They are also prepared by the method of displacement. (*See No.* 41.) All tinctures should be prepared in close glass or stoneware vessels, and be shaken frequently during the process of maceration. Tinctures are better clarified by repose than by filtration, as in the latter case a considerable portion is retained by the filtering medium, and lost by evaporation. In ordinary cases, it will be sufficient to allow the tincture to settle for a few days, and then to pour off the clear supernatant portion through a funnel loosely choked with a piece of sponge or tow, to keep back any floating fragments of straw or other light substances; after which the remaining foul portion of the liquid may be filtered through paper. When it is absolutely necessary to filter a tincture, and the quantity is large, conical bags should be employed. The filtration should be conducted as rapidly as possible, for the double purpose of lessening the amount lost by evaporation, and the action of the air on the fluid. Tinctures long exposed to the air frequently lose their transparency within a few days after their filtration, owing to the oxidation and precipitation of some portion of the matter previously held in solution. Resinous and oily tinctures, as those of myrrh, tolu, and lavender, may be usually restored to their former brightness by the addition of a quantity of spirit, equal to that which they have lost by evaporation; but many tinctures resist this mode of treatment, and require refiltering. Ethereal tinctures are best prepared by percolation, and should be both made and kept in stoppered bottles.

When both the substances are fluid, as in the case of certain balsams, the spirituous solution is made by merely mixing the two together in suitable proportions. For instance —Tincture or essence of Tolu consists of 3 drachms balsam of Tolu and 1 quart of alcohol.

The tinctures of the drug-stores are usually very uncertain and inferior preparations. Not only is their manufacture carelessly conducted, without reference to the respective characters of their ingredients, but the ingredients themselves are often deficient in strength and quantity.

We will now proceed to explain the various methods by which good tinctures are obtained.

36. To obtain Tinctures by Infusion, Maceration, and Digestion. In order to extract the soluble principles of substances which cannot be advantageously distilled, infusion is often resorted to. This consists in submitting them for a greater or less period of time to the action of a liquid, with or without the aid of heat.

This is known by the name of infusion, digestion, or maceration, terms all signifying the same process with different modifications in the way of conducting it.

37. Infusion. When the principles to be extracted are soluble in water, and at the same time but slightly volatile, boiling water is poured on the substance of which the infusion is required, the vessel is carefully covered, and the whole allowed to remain untouched for some minutes or even some hours, according to the greater or less penetrability of the substance, and the required strength of the infusion; the result is an INFUSION, properly so called.

If an infusion is required of dried leaves or flowers, they are first moistened with a little boiling water, and a time allowed for them to swell and soften before adding the rest of the water. Infusions made by adding all the water at once, as is still frequently practiced, are deficient both in flavor and perfume. The infusion of tea is an every-day illustration of this; as all who can make a good cup of tea know how necessary it is to first *draw* the tea with a small portion of water; and yet, strange to say, this principle is utterly neglected in the case of coffee, where its application is just as effective. (*See French Coffee.*)

Infusions of all vegetables that do not exert a *very powerful* action on the human frame, may be made by pouring 1 pint of boiling water on 1 ounce of the vegetable matter and allowing it to macerate for from ½ to 1 hour. *The ordinary dose* of such infusions is 1 to 2 ounces three or four times a day.

Infusions, like decoctions, are liable to undergo spontaneous decomposition by keeping, especially in warm weather, when a few hours are often sufficient for their passage into a state of active fermentation; they should therefore be prepared for use daily, as beyond 24 hours they cannot be depended on.

Infusions should be made in vessels which cannot be attacked by any of the substances with which they are in contact, and closed sufficiently tight to prevent the loss of the most volatile principles.

The tin cucurbit, with cover, is in these two respects best adapted for infusions in water.

38. Concentrated Infusions. These are now very generally met with in trade, and are made of 8 times the pharmacopœial strength. They are mostly prepared by employing 8 times the usual quantity of ingredients, and only three-fourths of the proper quantity of water, and adding to the strained liquor, when cold, sufficient spirit of wine to bring the liquid up to the proper strength (about one-third of the weight of the strained infusion). A still better plan is to treat 8 times the usual quantity of the ingredients with a mixture of rectified spirits 1 part and cold water 3 parts; in the usual way for making tinctures, either by *maceration* for 7 to 14 days, or by *percolation*. Concentrated infusions made in this way keep well, and deposit scarcely any sediment. Many houses, that are remarkable for the brilliancy and beauty of these preparations, employ one-third spirit of wine and two-thirds water as the menstruum. It may, however, be taken as a general rule, that for vegetable substances that abound in woody fibre, and contain but little extractive matter soluble in water (as quassia for instance), one-sixth to one-fifth part of spirit is sufficient for their preservation; while for those abounding in mucilage or fecula, or that readily soften and become pulpy and glutinous in *weak* spirit (as rhubarb), one-fifth to one-third is required. By macerating in the infusion as much bruised mustard seed as can be added without flavoring the liquor, along with a little bruised cloves, most vegetable infusions may be preserved without either fermenting or becoming mouldy with very little spirit (one-ninth or one-tenth).

39. Maceration. When an infusion is made without the aid of heat it is termed maceration. This takes a much longer time than an infusion, properly so called; it rarely requires less than 7 days, sometimes several weeks. Those substances to which heat would be injurious, or which are easily soluble, are treated in this way. In many distillations this method is made use of to soften the substances before putting into the still; and to facilitate the extraction of their odorous principle.

Tinctures, when prepared by maceration, should be frequently shaken during the process, which should be conducted in glass vessels well stopped.

40. Digestion is a prolonged infusion which is usually conducted at a medium temperature between that required for an *infusion*, properly so called, and that of a *maceration*. Its object is usually to impregnate alcohol with the principles of a substance which would be but slowly extracted without the aid of a certain amount of heat, such as that of the sun or of hot ashes.

Mixing together two or more liquors and allowing them to stand for some days, is also called digestion.

Maceration and digestion are usually performed in vessels of stoneware or glass, which are placed on the sand-bath, in cases where a regular and uniform heat is required.

Whatever may be the form or nature of the vessels employed, care must be taken not to fill them full, also to cover those which are to be placed on the sand-bath with a damp piece of parchment tightly tied round the top, with many pin holes pricked in it. If this latter precaution be neglected, the increased volume produced by the heat and also the expansion of the air may burst it. Moreover, the process is never so well conducted in a vessel that is too full.

41. To obtain Tinctures by Displacement or Percolation. The kind of filtration commonly called the *process of displacement*, for extracting the essence from roots, herbs, seeds, barks, &c., is effected in the following manner: It is first necessary that the articles to be acted upon should be ground in a drug mill to the condition of a coarse powder; then weigh each powder by itself, and mix them together in the proportions demanded by the recipe, and moisten the mass thoroughly with alcohol, allowing it to *macerate* for 12 hours in a vessel well covered. Next is required a hollow instrument of cylindrical form, having one end shaped liked a funnel, so that it can be inserted in the neck of a glass bottle, and having inside, near the lower end, a partition pierced with numerous small holes, like the strainer of a French coffee-pot, which is a simple coffee percolator; in the absence of such a partition, soft cotton, or any insoluble substance, may be substituted, and being placed in the inside at the lower end of the instrument, will answer as well as the strainer. This instrument is called a percolator. Boullay's filter or percolator is usually employed. Macerate the ingredients to be acted upon, for the time named—introduce them into the percolator, and slightly press them upon the partition. Any portion of the liquid used in the maceration, not absorbed by the powder, should be poured upon the mass in the instrument, and allowed to percolate. Now gradually pour into the percolator sufficient of the alcohol, or other liquid to be filtered, to drive before it, or *displace*, the liquid contained in the mass; the portion introduced must in like manner be *displaced* by another portion; and so on, till the required quantity of filtered liquor is obtained. This extract is called a tincture. In case the liquor which first passes through should be thick and turbid, again introduce it into the instrument, being very careful not to have the powder too coarse or loosely pressed, or it will permit the liquid to pass too quickly; and on the other hand it should not be too fine and compact, or it may offer an unnecessary resistance. Should the liquor flow too rapidly, return it to the instrument, and close it beneath for a time, and thus permit the finer parts of the powder to subside, and cause a slower percolation.

The method of percolation is now preferred by all who have made sufficient trial of it to apply it correctly.

The first portion of liquid obtained by the method of displacement is always in a state of high concentration. In general it is a simple solution of the soluble ingredients of the crude drug in the fluid employed. But sometimes the solvent, if compound, is resolved into its compound parts, and the fluid which passes through it at any given time is only *one* of these, holding in solution only the most soluble parts of the drug.

Thus, if diluted alcohol be poured over powder of myrrh, in the cylinder of the percolator, the fluid which first drops into the receiver is a solution of an oily consistence chiefly composed of resin and volatile oil dissolved in alcohol. In like manner when the powder of gall-nuts is treated in the same way by hydrated sulphuric ether, two layers of fluid are obtained, one of which is a highly concentrated solution of tannin in the water of the ether, and the other a weak solution of the same principle in pure ether. In all cases, therefore, in which it is not otherwise directed, it is absolutely necessary to agitate the several portions of the liquid obtained by percolation together, in order to insure a product of uniform strength, or activity.

To illustrate the operation of displacement, and describe an excellent percolator for making perfume tinctures, we will suppose that benzoin is under treatment. The apparatus made wholly of glass, having been arranged as shown by the engraving, and a plug of raw cotton dropped loosely at *b*, the benzoin in coarse powder is then poured into the tube portion, *A*, until it reaches the line, *c*. Alcohol (95 per cent.) is next added, until it rises to the line, *d*. As soon as the first portion sinks into the benzoin, a fresh addition must be made; and thus the succeeding relays go on displacing those which preceded them without mingling with them. Each stratum becomes more and more charged with soluble matter as it descends; and when it reaches the bottom of the mass, under the pressure of the superincumbent liquor, it runs out saturated. When, by successive additions of fresh alcohol, the benzoin under treatment has become exhausted, the

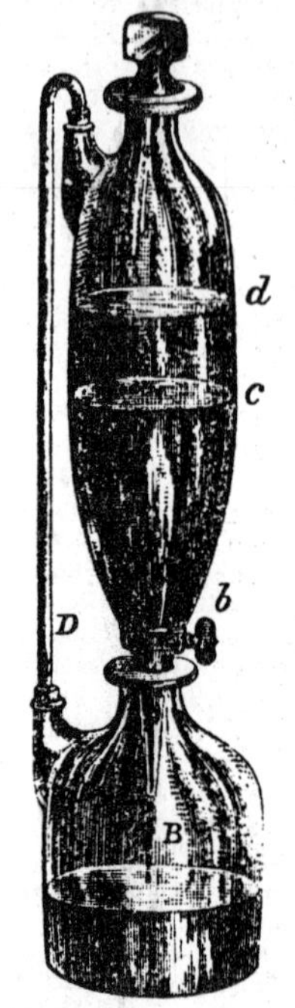

liquid passes through the mass, and falls into the receiver, *B*, as tasteless and colorless as when first poured in. This indicates the completion of the process.

As atmospheric pressure is an important element in the operation, it will not answer to shut it off by closing the top of the displacer, without making some compensating arrangement; and, therefore, a communication between the upper and lower vessels is established by means of a latent-tube arrangement, *D*. In this manner the apparatus is kept close, and the evaporation of alcohol prevented, while the pressure produced is distributed throughout the apparatus, and rendered uniform. As the runnings are clear, filtration is rarely necessary. The quantity of alcohol thus consumed need not be more than sufficient to exhaust the material; and the resulting tincture must therefore be diluted to the proper strength. For perfumes, deodorized alcohol must always be used.

The method of displacement has the advantage of expedition, economy, and yielding products possessing uniformity of strength; but it requires considerable experience to adapt it to all substances. The art rests in properly packing the ingredients in the cylinder, some substances requiring considerable pressure to be used, while others, when even lightly packed, scarcely permit the fluid to pass through them. An excellent plan applicable to all substances, but especially those of a glutinous or mucilaginous nature, is to mix the powder with an equal bulk of well-washed sand, before rubbing it up with the menstruum. The coarseness of the powder must also be attended to. Substances that readily become soft and pappy when wetted by the menstruum, should not be used so fine as those that are more woody and fibrous. The method of displacement answers well for the preparation of all tinctures that are not of a resinous nature, and for most infusions of woody and fibrous substances, as roots, woods, barks, leaves, seeds, insects, &c. It is especially adapted for the preparation of concentrated infusions and essences, as they may thus be obtained of any required strength, without loss, or requiring concentration by heat, which is so destructive to their virtues.

When ordinary tinctures are made in *large quantities*, displacement is never likely to supersede maceration, on account of any practical advantages it may possess. If the prescribed directions be duly attended to, the process of maceration is unexceptionable. The process is more simple than the other; the mode of operating more uniform; it is, in fact, always the same; it requires less of skill and dexterity in conducting it; it requires less constant attention during its progress, which, in operating on large quantities, is a consideration; and finally, the apparatus required is less complicated. When, however, only small quantities of tincture are to be made at a time, and kept in stock, the adoption of the process of displacement will often be found convenient and advantageous. It offers the means of making a tincture in two or three hours, which, by the other process, would require as many weeks. (*See No.* 4572.)

42. Proportion of Ingredients used for making Tinctures. The following are the proportions usually employed for the most important perfume tinctures:

Tincture.		Troy.	Alcohol.
Vanilla	Vanilla bean, rasped	1 lb.	8 pts.
Musk	Grain musk	2 drachms.	8 pts.
Frangipani	Powder a la frangipani	1 lb.	6 pts.
Rhodium	Rhodium-wood, rasped.	1 lb.	2 qts.
Civet	Civet, orris-root	½ oz	2 qts.
Tonquin	Tonka bean.	1 lb.	8 pts.
Orris	Orris-root	7 pts	8 pts.
Alkanet—red col.	Alkanet	½ oz	1 qt.
Turmeric—yellow.	Turmeric	½ oz	1 qt.

43. To Prepare Emulsions. These are milky liquids, formed by the mechanical admixture of oil, balsam, or resin, with water, by means of some other substance that possesses the property of combining with both. There are numerous preparations of the kind in pharmacy and medicine, which, in the later pharmacopœias, have received the name of "mixtures." There are also several emulsions employed as cosmetics, either alone, or as vehicles for other ingredients. The common name of emulsions is "*milk*," but the term is often incorrectly extended to opaque white liquids of an entirely distinct character.

The successful *preparation* of *emulsions* is a matter requiring some little skill and care. In some instances, as with the almond, the two substances necessary to produce a perfect emulsion are presented by nature, ready to our hand, in the same vegetable production; nothing more is necessary than to reduce it with the pestle, and triturate it with water, gradually added. In other cases, and which are far the more numerous, we have to operate on oily or resinous ingredients in their common form. These we are enabled to suspend in water, or mechanically combine with it, by the intervention of *thick mucilage, almonds*, or *yolk of egg*. It is found that 1 drachm (60 grs.) of the *first*—made with equal parts of good *gum-arabic* and *water* (powdered gum is sometimes used instead of mucilage)—1 ounce of the *second*, (usually about 26 in number), and one of the *last*, will form 2 drachms of oil or resinous matter into an emulsion with about 1 fluid ounce of water, gradually added; and such an emulsion, if properly made, will then, in most instances, bear further dilution with water. (The *yolk* of an ordinary-sized hen's egg is referred to. It should be remembered, that emulsions formed with yolk of egg will not keep long, owing to the putrescible nature of the latter.) Of these, mucilage is the medium most commonly employed. According to Montgomery, for conversion into permanent emulsions, "*oils* require about three-fourths their weight; *balsams* and *spermaceti*, equal parts; *resins*, twice their weight; and *musk* and *ambergris* 5 times its weight." In some cases instead of the above substances, a little *liquor of potassa* is employed, when a saponaceous emulsion is formed, which differs considerably in its properties from an emulsion of the same ingredients produced by means of a bland medium.

In making an emulsion, the *gum*, or other medium employed, should be first put into the mortar, and rendered thoroughly homogeneous with the pestle. If *almonds* are used, they should be treated as noticed under "almond-paste" (*see No.* 1123), a few drops of water being added to prevent "oiling," and to reduce them to a smooth, soft paste. The oil or resinous matter may then be gradually added and rubbed in, carefully observing not to add it more quickly than it can be subdued by the pestle; and if, during this part of the manipulation, the mixture should begin to exhibit a "breaking" or "curdling" appearance at the edges, a few drops of water must be immediately incorporated with it, before adding the remainder of the oil. If this be not done, the emulsive mixture in the mortar will, in general, suddenly lose its tenacious consistence, and the process will fail. After the whole of the oil, balsam, or resinous matter is thoroughly incorporated, the water or other aqueous vehicle intended to form the bulk of the emulsion, should be added gradually and with care, each portion being perfectly blended with the liquid mass in the mortar, by patient trituration, before adding the next. If any alcoholic liquid is employed, it should be added at the very end of the process, and then only very gradually, as otherwise it will cause the separation of the ingredients.

It must be observed that soluble salts, spirit, acids, and astringents, are, as a rule, incompatible with the emulsive form. If saline matter must be introduced, it should only be added in a very minute quantity, and in the state of solution, to the ready-formed emulsion; and in this case emulsion of almonds is the most suitable vehicle. (*See No.* 1125.) Spirits and acids act by precipitating the mucilaginous matter, or yolk. Even the addition of a very little lemon juice, or of a portion of slightly acescent syrup, will often entirely destroy an emulsion. This inevitably occurs with emulsions made with liquor of potassa, or other alkaline medium, owing to the absolute incompatibility of acids and alkalies in the same liquid.

It is found that volatile oils are more readily made into emulsions if mixed with an equal volume of some simple fixed oil, as that of the almond or olive, before proceeding to operate on them.

All emulsions should be well shaken before use. (*Cooley.*)

44. To Prepare Extracts. The process of obtaining an extract of a substance involves two distinct operations: First, the production of a solution of the soluble portion of the substance operated on; and next, the reduction of this solution to a proper consistence by evaporation. The substance is first, where practicable, reduced to coarse powder by bruising, or sliced with a knife, so that every portion may be fully exposed to the action of the solvent. Refractory substances are first softened by the solvent and then sliced. Other substances whose nature does not require reducing, are used without preparation.

Different fluids are used for solvents, as best adapted to the solubility of the substance under treatment. Some bodies, such as fresh vegetables, yield their juice by expression alone. In the preparation of *aqueous* extracts, the ingredients are treated with rain or distilled water, until all the soluble matter that is desired to obtain from them is dissolved. This is effected by either *maceration, percolation, infusion*, or *decoction*, as circumstances require: the solution thus obtained is poured off and the remaining soluble matter either pressed or washed out, and added to the solution; it is next allowed time to settle, then decanted, and strained or filtered; and if this fails to render the liquid clear, it is clarified by white of egg, and filtered; Canton flannel, first soaked in water, being generally employed for this purpose. When water acidulated with acetic acid is employed, vegetable substances are usually macerated in it in the cold, or the dilute acid is sprinkled over the bruised plant, if fresh, and the juice expressed by strong pressure.

When the principles to be extracted are insoluble, or only slightly soluble, in water, alcohol is employed, either in the form of rectified spirit, proof spirit, or diluted. These produce *alcoholic* or *spirituous* extracts; and are generally obtained by either *maceration* or *digestion*.

Ether is well adapted for obtaining extracts from bodies whose principles consist of volatile oils or resin, on account of its strong affinity for those substances. Such are termed *ethereal* extracts. In nearly all cases, filtration is necessary to insure a pure extract.

The means usually employed for evaporating an *aqueous* solution, are rapid boiling over a fire until the extract is thick enough to offer some risk of burning, and the evaporation finished over a water bath or in shallow vessels at a moderate heat, the further escape of vapor being promoted by continuous stirring with a wooden spoon or stick. It is not always advisable to heat a solution to the boiling point, but if boiling is resorted to, it cannot be done too rapidly, as the heat cannot rise above its boiling point, and rapid ebullition hastens evaporation. The fluid must never be stirred while ebullition is going on.

Two fundamental rules are:—to conduct evaporation at as low a temperature as is consistent with other objects; and,—to exclude atmospheric air; or, at least, to expose the liquid to its action for as short a time as possible, as most solutions lose more or less of their active principles by heat and exposure. Solutions which will not bear boiling without loss of strength are evaporated in a vacuum, either in a closed still, or under the receiver of an air pump, in which a vessel is placed containing strong sulphuric acid; this has a powerful affinity for water and absorbs its vapor as quickly as it comes in contact with it.

A good plan for evaporation, though slow, is to place the liquid in a broad shallow vessel, exposed in a stove or drying room to a temperature of about 100° Fahr., allowing free access for the air. The extracts thus evaporated are said to be lighter in color and more transparent than by most other ways.

The method for evaporating an *alcoholic* or an *ethereal* solution is substantially the same as that pursued with an aqueous solution; except that, as a matter of economy, the vapor may be led off and condensed again.

A *good extract* should be free from grit, and wholly soluble in 20 parts of the solvent used for making the extract, forming a nearly clear solution; it should be of a proper consistence and of uniform texture and color, smooth and glossy in appearance; this latter can only be arrived at by assiduous and laborious stirring as the extract thickens; and may be promoted by adding 3 or 4 per cent. each of olive oil and gum arabic, with 1 or 2 per cent. of spirit of wine. Extracts should be put into pots as soon as made, securely tied down with bladder, and kept in a dry place. Any tendency to become mouldy may be prevented by adding, the last thing before removing from the evaporating pan, a few drops of oil of cloves, or a still less quantity of creosote, dissolved in a little alcohol; or by moistening with oil of cloves or creosote, the inside of the bladder used for covering the pots.

45. To obtain Vegetable Juices by Expression. The juices of plants are obtained by bruising the fresh leaves in a marble mortar, or in a mill, and expressing the juice which, after *defecation* for some hours in a cool situation, is either filtered through paper, or strained after coagulating its albuninous matter by heat. Some plants require the addition of $\frac{1}{8}$ its quantity of water before pressing. The expression of the juice of lemons, oranges, quinces, &c., is facilitated by previously mixing the pulp with clean chopped straw. Mulberries, &c., after being crushed between the hands, are left 3 or 4 days to undergo a slight fermentation, before pressing. A very powerful screw press is required for this purpose. The PRESERVATION of the juices of the narcotic plants, and some other vegetables, has lately assumed considerable interest, from these preparations having been proposed as substitutes for the common tinctures. It appears that the juice of young plants just coming into flower, yields only $\frac{2}{3}$ the amount of extract which may be obtained from the same quantity of juice expressed from the matured plant, or when the flowers are fully blown; and the strength of the product is also inferior. The leaves alone should be preferably employed, and should be exclusively of the second year's growth, when the plants are biennials.

Bruise the leaves in a marble mortar (on the large scale, in a mill), and submit them to the action of a powerful press; allow the juice to remain for 24 hours in a cold place, then decant the clear portion, add $\frac{1}{4}$ part by measure of spirit (90 per cent.), agitate, and in 24 hours again decant the clear, and filter it through paper. Keeps well under ordinary circumstances.

The method directed by the *Paris Codex* is as follows: to the fresh leaves, bruised in a marble mortar, is added an equal weight of rectified spirit, and after *maceration* for 15 days, the whole is pressed, and the resulting tincture filtered.

The commencing dose of the narcotic juices is about 5 drops. In the above manner are prepared the *preserved juices of aconite, belladonna, colchicum* (corms), *hemlock, henbane, foxglove, lactuca virosa, taraxacum,* &c.

46. To Extract Essential Oil from Wood, Barks, Roots, Herbs etc. Take balm, mint, sage, or any other herb, &c., put into a bottle, and pour upon it a spoonful of ether; keep in a cool place a few hours, and then fill the bottle with cold water; the essential oil will swim on the surface, and may be easily separated.

Specific Gravity

Specific Gravity is the density of the matter of which any body is composed, compared with the density of another body, assumed as the standard, or 1.000. This standard is pure distilled water for liquids and solids, and atmospheric air for gaseous bodies and vapors. In the United States and England the specific gravity, unless when otherwise expressed, is always taken at 60° F.; but in France at 32°, or the temperature of melting ice. In most cases, however, it is sufficient merely to note the temperature, and to apply a correction, depending on the known density of water or air, at the different degrees of the thermometric scale.

The above plan has been adopted, because the weight of an equal *bulk* of different substances varies greatly. Thus, as gold is 19 and silver 10 times heavier than water, those numbers, 19 and 10, are said to represent the specific gravity of gold and silver. The heaviest of all known substances is the very hard metal used for making points to the so-called diamond gold pens. It is called iridium; its specific gravity is 23. Next comes platinum, 21; gold, 19; mercury, 13.5; lead, 11.3; silver, 10; copper, 8; iron, 7; zinc, 6; different kinds of stones, from 4 to 1; aluminum, 2.5. Flax and all woody fibres have a specific gravity of 1.4, and are thus heavier than water, but wood will float or sink according to the number of its pores into which the water does not penetrate. So ebony and many kinds of hard wood sink, pine and all kinds of soft wood float. Cork is the lightest wood, its specific gravity being only 0.24, less than one-quarter that of water. Alcohol is about three-quarters the weight of water, and as the strength of liquor depends on the amount of alcohol it contains, this strength is simply found out by its specific gravity indicated by the more or less floating of a little instrument called a hydrometer, the weaker liquid being little lighter than water has the strongest buoyant power; solutions of different salts, sugar, etc., being heavier than water, have a stronger buoyant power; vessels therefore will draw less water in the sea than in fresh water, and it is more difficult to swim in the latter than in the sea. The lightest of all liquids has a specific gravity of 0.6; it is called chimogene, and is made from petroleum; it is exceedingly volatile and combustible; in fact, it is a liquefied gas. Carbonic acid gas or choke damp is about 500 times lighter than water; common air, 800; street gas about 2,000, and pure hydrogen, the lightest of all substances, 12,000 times. The heaviest substance has thus 23×12,000 or more than a quarter of a million times more weight than an equal bulk of the lightest; and the substance of which comets consist, has by astronomers been proved to be even several thousand times lighter than hydrogen gas.

48. To find the Specific Gravity of a Substance heavier or lighter than Water. In order to ascertain the specific gravity of a body heavier than water, the following method is adopted. First weigh it in air, then weigh it immersed in water. The difference between these two weights will be its *loss* of weight in water, or, in other words, the weight of the water displaced. Then divide the weight in air by its loss in water, and the result is the specific gravity.

Thus, suppose a substance weighs,
12 pounds in air,
and 10 pounds in water.
—

Its *loss* is 2 pounds in water.

Divide 12 (weight in air) by 2 (loss in water), and the result is its specific gravity, 6.—That is, the substance is, *bulk for bulk*, 6 times as heavy as water.

If the substance will not sink in water, then weight must be added to make it *just sink* below the surface. This extra weight, *added to* the weight in air, show its *loss* in water. Thus, if a substance weighs 8 pounds in air, but requires 2 pounds to be added to submerge it in water, its *loss* of weight in water is 2 added to 8=10 pounds.

Proceeding as before, we divide its weight in air, 8, by its loss in water, 10 and we have it specific gravity $\frac{8}{10}$=.8.

49. To find the Specific Gravity of a Liquid or a Gas. Weigh it in a specific gravity bottle, glass flask, or other vessel of known capacity; and dividing that weight by the weight of the same bulk of water, the quotient is, as before, the specific gravity.

50. To find the Specific Gravity of a Solid Body Soluble in Water. Take its specific gravity in regard to some liquid which does not dissolve it, and multiply by the specific gravity of the liquid. Thus, a piece of sugar, whose weight is 400 grains, is found to lose 217.5 grains if weighed when immersed in oil of turpentine; this would make its specific gravity, as compared with oil of turpentine, $\frac{400}{217.5}$=1.84. The specific gravity of the turpentine is .87; then, 1.84×.87=1.6, the real specific gravity of the sugar.

51. To find the Specific Gravity of a Body in Powder Insoluble in Water. Introduce it into a bottle whose capacity is known; fill the bottle with pure water at 60°. It will hold as much less water as is equal to the bulk of the powder, and the weight of the powder in air divided by this difference will give the specific gravity. Thus, supposing the bottle to hold 1000 grains of water, 100 grains of emery are introduced, and the bottle filled up with water. If no water were displaced the two should weigh 1100 grains; they really weigh 1070; the difference, 30 grains, is the weight of water displaced; 100÷30=3.333, specific gravity of the emery.

52. To Determine the Weight of a Body from its Specific Gravity. A cubic foot of water weighs 1000 ounces; hence, to determine the weight of a given bulk of any body the specific gravity of which is known, multiply the cubic content in feet by 1000, and this by the specific gravity, and the product will be the weight in ounces avoirdupois.

Alcoholmetry

Alcoholmetry. The percentage of absolute alcohol in any spirituous liquid may be given either by *volume* or weight, but as liquors are sold by measure, not weight, it is generally preferred to know the percentage by volume. The per cent. of weight remains the same in all temperatures, but the per cent. by volume varies with the temperature or heat of the liquid. Many instruments have been introduced to determine the quantity of absolute alcohol contained in any spirituous liquors, and these are known as *hydrometers*, or *alcoholmeters.* Hydrometers made by different inventors have come into use in different countries; thus the hydrometer made by Tralles has been adopted by the governments of the United States and Prussia; that made by Gay Lussac has been legally sanctioned in France and Sweden; while that invented by Sikes has been approved and made the excise standard in Great Britain.

54. Tralles' Hydrometer. Tralles' hydrometer is the instrument used by our government to ascertain the strength of *imported* liquors, and is made of glass. Tralles has adopted as the standard of comparison pure or absolute alcohol in volume at the temperature of 60° Fah., the strength of which he expresses by a scale divided into 100 degrees or parts, each of which represents $\frac{1}{100}$ part of alcohol. When floated in any spirituous liquor at a temperature of 60° Fah., it immediately indicates the strength. For instance, if in a brandy at that temperature it sinks to 65, it shows that 65 parts of the liquor is absolute alcohol, and 35 parts water; should it sink to 90, it indicates that the liquor is 90 parts or per cent. strong, and so on.

An increase of heat causes liquids to expand in volume, and a decrease produces contraction; therefore spirits over the normal temperature of 60° Fah. appear stronger than they really are, and below 60° they are really stronger than they appear to be.

It is therefore evident that the degrees of percentage of this hydrometer are only correct when the spirit under trial *has* the normal temperature of 60° Fah. When the temperature varies from 60°, the percentage can only be ascertained by a long and tedious calculation. To avoid this Mr. Tralles has constructed a simple table by which the real percentage of alcohol is found in liquids of different temperatures from the results exhibited by the instrument. (*See No.* 55.) The horizontal line at the top shows the various temperatures given by the thermometer; the column of figures under 60° shows the *true* percentage of strength at the normal or standard temperature of 60°; the figures under the other degrees of temperature show the *observed* or *apparent* degrees of strength as indicated by the hydrometers.

As an example of the simple manner by which this table may be used, we will suppose that the temperature of the spirits to be tested is at 75°, Fah., and that the hydrometer sinks to 53° on the scale; this would be the *observed* or apparent degree or percentage of strength. Now to find the *real* percentage of strength at 60°, we turn to the table and find the upright or vertical column of figures headed 75°, we then run down the figures until we arrive at 53.0; having ascertained this, we then trace the horizontal line to the *left* or right to the outside column headed 60°, and at the point when the horizontal line running from 53.0 meets the column headed 60°, will be found the number 50. We thus ascertain that a spirit at 75° having an *observed* strength of 53 has only a *real* percentage of 50 at the normal or established temperature of 60°.

Suppose that another sample of brandy, instead of being at 75° is at 50°, and the instrument still sinks to 53. In the same way we select the column headed 50°, and run down the figures until we find 53.0, then by tracing the horizontal line until we arrive at the outside column headed 60° (either the first or last column), we find the number 55, which is the true percentage of the brandy at 60° Fah.

Again, if an alcoholic liquid at a temperature of 30° be found to contain 23.5 per cent. by volume, by reference to the table 30 will be found to express its actual strength at 60° Fah.

We might multiply examples, but the above are sufficient to show the manner by which the table may be worked.

55. Table to find the true percentage of Absolute Alcohol *by volume in a liquid at* 60° *from the observed percentage indicated by a Glass Hydrometer at any other temperature.*

60°	30°	35°	40°	45°	50°	55°	65°	70°	75°	80°	85°	60°
0	—0.2	—0.4	—0.4	—0.5	—0.4	—0.2	+0.2	+0.6	+1.0	+1.4	+1.9	0
5	+4.6	+4.5	+4.5	+4.5	+4.6	+4.8	5.3	5.8	6.2	6.7	7.3	5
10	9.1	9.0	9.1	9.2	9.3	9.7	10.4	11.0	11.6	12.3	13.0	10
15	13.0	13.1	13.3	13.5	13.9	14.5	15.6	16.3	17.1	18.0	19.0	15
20	16.5	16.9	17.4	17·8	18.5	19.2	20.8	21.8	22.8	23.8	24.9	20
25	19.9	20.6	21.4	22.2	23.0	24.1	25.9	27.0	28.2	29.4	30.5	25
30	23.5	24.5	25.7	26.6	27.7	28.8	31.1	32.2	33.4	34.5	35.7	30
35	28.0	29.2	30.4	31.6	32.7	33.8	36.2	37.3	38.4	39.5	40.6	35
40	33.0	34.2	35.4	36.7	37.8	39.0	41.1	42.2	43.3	44.3	45.4	40
45	38.4	39.6	40.7	41.8	42.9	43.9	46.1	47.1	48.2	49.2	50.3	45
50	43.7	44.7	45.8	46.9	47.9	49.0	51.0	52.0	53.0	54.0	55.1	50
55	49.0	50.0	51.0	52.0	53.0	54.0	54.9	56.9	57.9	58.9	59.9	55
60	54.2	55.2	56.2	57.1	58.1	59.0	60.9	61.9	62.9	63.8	64.9	60
65	59.4	60.3	61.2	62.2	63.1	64.0	65.9	66.8	67.7	68.6	69.6	65
70	64.6	65.5	66.4	67.3	68.2	69.1	70.8	71.7	72.6	73.5	74.5	70
75	69.8	70.7	71.5	72.4	73.3	74.2	75.8	76.7	77.6	78.4	79.3	75
80	75.0	75.8	76.6	77.5	78.4	79.2	80.8	81.7	82.4	83.2	84.1	80
85	80.3	81.1	81.8	82·6	83.5	84.3	85.7	86.5	87.3	88.0	88.8	85
90	85.6	86.4	87.1	87.9	88.6	89.3	90.7	91.4	92.0	92.7	93.4	90

The following table gives the *richness* or the per cent. of alcohol by volume, in reference to the volume of the liquid at the temperature when tested; it therefore requires that the liquor should be tested exactly at the same temperature at which it was measured.

56. Table to find the true percentage of Absolute Alcohol *in a liquid of any temperature from the observed percentage indicated by the Glass Hydrometer at the same temperature.*

True per ct. of Alcohol by Volume.	Observed per cent. indicated by the Glass Hydrometer.										
	30°	35°	40°	45°	50°	55°	65°	70°	75°	80°	85°
0	—0.2	—0.4	—0.4	—0.5	—0.4	—0.2	+0.2	+0.6	+1.0	+1.4	+1.9
5	+4.6	+4.5	+4.5	+4.5	+4.6	+4.8	5.3	5.8	6.2	6.7	7.3
10	9.1	9.0	9.1	9.2	9.3	9.7	10.4	11.0	11.6	12.3	13.0
15	13.0	13.1	13.3	13.6	14.1	14.5	15.6	16.3	17.1	18.0	19.0
20	16.5	16.9	17.4	17.9	18.5	19.2	20.8	21.8	22.9	23.9	25.0
25	19.8	20.5	21.3	22.2	23.0	24.1	25.9	27.1	28.3	29.5	30.7
30	23.3	24.3	25.5	26.5	27.6	28.8	31.2	32.3	33.5	34.6	35.9
35	27.7	28.9	30.2	31.4	32.6	33.8	36.3	37.5	38.6	39.7	40.9
40	32.5	33.8	35.1	36.5	37.7	38.9	41.2	42.4	43.5	44.6	45.8
45	37.8	39.1	40.3	41.5	42.7	43.8	46.2	47.3	48.5	49.6	50.8
50	43.1	44.2	45.4	46.6	47.7	48.9	51.1	52.2	53.4	54.5	55.6
55	48.3	49.4	50.5	51.6	52.8	53.9	56.1	57.2	58.3	59.4	60.5
60	53.4	54.5	55.6	56.7	57.8	58.9	61.1	62.2	63.3	64.4	65.5
65	58.4	59.5	60.6	61.7	62.8	63.9	66.0	67.1	68.2	69.3	70.4
70	63.5	64.6	65.7	66.8	67.9	69.0	71.0	72.1	73.2	74.3	75.4
75	68.6	69.7	70.7	71.8	72·9	74.0	76.0	77.1	78.2	79.2	80.3
80	73.7	74.8	75.8	76.9	78.0	79.0	81.0	82.1	83.1	84.1	85.2
85	78.8	79.8	80.9	81.9	83.0	84.0	86.0	87.0	88.0	89.0	90.0
90	84.0	85.1	86.1	87.1	88.1	89.1	91.0	91.9	92.8	93.7	94.6

Thus, if the Hydrometer indicated 59.4 per cent. in a liquid at 80° Fah., the table in No. 57 would give its true percentage (richness) to 55 per cent.; that is, 100 volumes of the liquid at 80° contains 55 volumes of alcohol. Tralles' Hydrometer gives the *per cent.* by volume only. If it be desired to know the *per cent.* by *weight*, it may be ascertained from the *percentage* in volume of the liquid at 60° Fah. by table in No. 57.

57. Table of Comparison *between the per cent. of Alcohol by volume at* 60° *(Tralles') and per cent. by weight.*

Per Cent. by Vol	Per Cent. by Weight.	Per Cent. by Vol	Per Cent. by Weight.	Per Cent. by Weight.	Per Cent. by Volume.	Per Cent. by Weight.	Per Cent. by Volume.
0	0.	55	47.29	0	0.	55	63.97
5	4.00	60	52.20	5	6.25	60	68.97
10	8.05	65	57.25	10	12.42	65	73.79
15	12.15	70	62.51	15	18.52	70	78.40
20	16.28	75	67.93	20	24.57	75	82.80
25	20.46	80	73.59	25	30.55	80	86.97
30	24.69	85	79.50	30	36.45	85	90.88
35	28.99	90	85.75	35	42.25	90	94.46
40	33.39	95	92.46	40	47.92	95	97.61
45	37.90	100	100.00	45	53.43	100	100.00
50	42.52			50	58,79		

58. Gendar's Hydrometer. Annexed we give a comparative view of the scales of Tralles and Gendar, the former used by the revenue officers of the United States for *imported liquors*, and indicating the per cent. by volume of alcohol in spirituous liquors, and the latter used throughout the whole country for *domestic liquors*, determining the per cent. above and below *proof.*

This is inserted for the convenient comparison of the American standards. Tables of other areometers reduced to specific gravity will be found in Nos. 6155, &c.

The first column of the table exhibits the specific gravities at 60° Fah., for *mixtures* of pure alcohol and water;—taking water at the temperature of its greatest density, about 39.5° Fah., as 1.0000, and, therefore, having at 60° Fah. a specific gravity of 0.9991. Of the above mixtures, each 100 gallons or measures contain the number of gallons or measures of alcohol indicated in the second column (Tralles' hydrometer scale) if measured at 60° Fah.

In the Tralles' hydrometer scale there is no reference to proof of any denomination; and in that of Gendar's there is but one proof, marked P. on the hydrometer; the others, such as 2d, 3d and 4th proofs, were, at all times, incorrect and deceptive. The National Tax Law, of August 1st, 1862, says that "the term proof shall be construed, and is hereby declared to mean that proof of a liquor which corresponds to 50 degrees of Tralles' hydrometer at the temperature of 60 degrees Fah." Proof spirit is, therefore, by law, of the alcoholic strength of 50 per cent. by volume, having a specific gravity of 0.9335, or a mixture of equal quantities of absolute alcohol at the specific gravity of 0.793, and distilled water at 60° Fah. In other words, proof spirit is one-half pure water and half absolute alcohol.

To ascertain what strength any liquor above proof by the Gendar hydrometer would be by the Tralles hydrometer, add 100 to the given proof if above proof, or deduct, if below proof, from 100 on the Gendar scale, and divide by 2. Say a liquor is at 40 above P. on the Gendar scale, you then add 100, making 140, and divide by 2, which will show 70 on the Tralles' scale. If below P. deduct the proof from 100 and divide the remainder by 2. Say a liquor is 35 below P., consequently you have a remainder of 65, and divide the 65 by 2, which will show $32\frac{1}{2}$ on Tralles' scale. Having ascertained the degree of strength of any liquor by volume on Tralles' scale above 50 (which is proof), multiply the degrees by 2, and cut off the two right hand figures, and it will show the degrees above proof on Gendar's scale. Thus we will suppose the spirit to show 70 per cent. strength on Tralles' scale. We therefore multiply 70 by 2, which gives us 140; we now cut off the two right hand figures, thus, 1.40, and find that the liquor is 40 above proof on Gendar's scale.

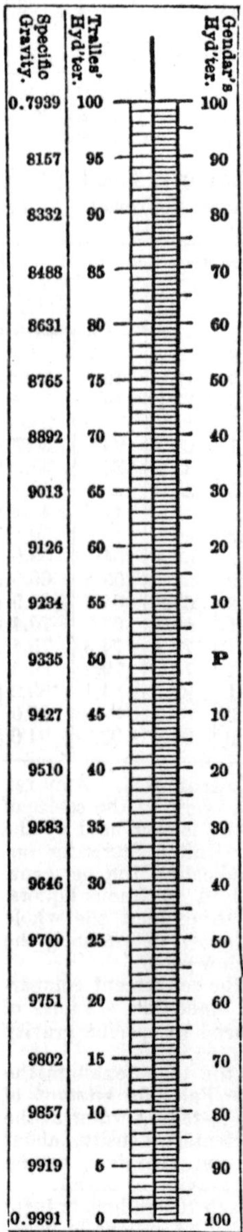

Or suppose a liquor is below 50 on Tralles' scale (which is below proof). To find what degree it is below on Gendar's scale, multiply the degrees on Tralles' scale by 2 and add a number sufficient to make 100; the number required to be added will show the degree below proof.

59. Tralles' Table of Percentage of Alcohol. When the temperature of the spirit is 60° Fah., the first column of the table on page 26 gives at *once* the percentage of alcohol by *measure; when the temperature is below* 60° an addition must be made of 1 measure per cent. for every 5 degrees of the thermometer; and when above 60° a like quantity must be deducted. This correction will amount to the fraction $\frac{1}{5}$ or the decimal 2 for every single degree, and is very easily made. *If the specific gravity* sought cannot be found exactly in the table, the *difference* between it and the next greater specific gravity in the table must be taken, which will give the numerator of a fraction, having for its denominator the number found in the *third* column against the next greater number just employed. This fraction, added to the percentage of alcohol in the first column of the table against the said specific gravity, will give the true percentage sought. Thus, if the specific gravity of a spirituous liquor is .9605, what is its alcoholic content? Here .9605 is not in the table, but the next greater number is .9609; the former must therefore be deducted from the latter, and the difference (4) put as the numerator of the fraction, having for its denominator the number (13) in the column of differences against .9609. The fraction $\frac{4}{13}$ so found, added to the percentage against .9609 in the first column, gives $33\frac{4}{13}$ as the true percentage of alcohol in the given sample

Tralles' Table exhibiting the percentage, by volume, of Alcohol, corresponding to any given specific gravity.

Alcohol in 100 Measures of Spirit.	Specific Gravity at 60° F.	Difference of Specific Gravity.	Alcohol in 100 Measures of Spirit.	Specific Gravity at 60° F.	Difference of Specific Gravity.	Alcohol in 100 Measures of Spirit.	Specific Gravity at 60° F.	Difference of Specific Gravity.
Pure water	.9919	00	34	.9596	13	68	.8941	24
1	.9976	15	35	.9583	13	69	.8917	24
2	.9961	15	36	.9570	13	70	.8892	25
3	.9947	14	37	.9556	14	71	.8867	25
4	.9933	14	38	.9541	15	72	.8842	25
5	.9919	14	39	.9526	15	73	.8817	25
6	.9906	13	40	.9510	16	74	.8791	26
7	.9893	13	41	.9494	16	75	.8765	26
8	.9881	12	42	.9478	16	76	.8739	26
9	.9869	12	43	.9461	17	77	.8712	27
10	.9857	12	44	.9444	17	78	.8685	27
11	.9845	12	45	.9427	17	79	.8658	27
12	.9834	11	46	.9409	18	80	.8631	27
13	.9823	11	47	.9391	18	81	.8603	28
14	.9812	11	48	.9373	18	82	.8575	28
15	.9802	10	49	.9354	19	83	.8547	28
16	.9791	11	50	.9335	19	84	.8518	29
17	.9781	10	51	.9315	20	85	.8488	30
18	.9771	10	52	.9295	20	86	.8458	30
19	.9761	10	53	.9275	20	87	.8428	30
20	.9751	10	54	.9254	21	88	.8397	31
21	.9741	10	55	.9234	20	89	.8365	32
22	.9731	10	56	.9213	21	90	.8332	33
23	.9720	11	57	.9192	21	91	.8299	33
24	.9710	10	58	.9170	22	92	.8265	34
25	.9700	10	59	.9148	22	93	.8230	35
26	.9689	11	60	.9126	22	94	.8194	36
27	.9679	10	61	.9104	22	95	.8157	37
28	.9668	11	62	.9082	22	96	.8118	39
29	.9657	11	63	.9059	23	97	.8077	41
30	.9646	11	64	.9036	23	98	.8034	43
31	.9634	12	65	.9013	23	99	.7988	46
32	.9622	12	66	.8989	24	Pure Alcohol	.7939	49
33	.9609	13	67	.8965	24			

60. Table for reducing the strength of Alcohol. The following Table given by Booth, shows the quantity of water that must be added to alcohol of a given strength, in order to produce an alcohol of inferior strength.

The upper horizontal column contains the percentage of strength of the stronger alcohol to be diluted; the vertical columns below, denote the volumes of water which must be added to 100 volumes of it, in order to produce a spirit of the strength indicated in the left hand column.

Desired strength in per cent.	90	85	80	75	70	65	60	55	50
85	6.56								
80	13.79	6.83							
75	21.89	14.48	7.20						
70	31.05	23.14	15.35	7.64					
65	41.53	33.03	24.66	16.37	8.15				
60	53.65	44.48	35.44	26.47	17.58	8.56			
55	67.87	57.90	48.07	38.32	28.63	19.02	9.47		
50	84.71	73.90	63.04	52.43	41.73	31.25	20.47	10.35	
45	105.34	93.30	81.38	69.54	57.78	46.09	34.46	22.90	11.41
40	130.80	117.34	104.01	90.76	77.58	64.48	51.43	38.46	25.[illegible]5
35	163.28	148.01	132.88	117.82	102.84	87.93	73.08	58.31	43.59
30	206.22	188.57	171.05	153.61	136.04	118.94	101.71	84.54	67.45
25	266.12	245·15	224.30	203.53	182.83	162.21	141.65	121.16	100.73
20	355.80	329.84	304.01	278.26	252.58	226.98	201.43	175.96	150.55
15	505.27	471.	436.85	402.81	368.83	334.91	301.07	267.29	233.64
10	804.54	753.65	702.89	752.21	601.60	551.06	500.59	450.19	399.85

Illustration. If we have alcohol of 70 per cent. strength, and desire to reduce its strength to 40 per cent.—we look for 40 in the left-hand column, and the figures on a line with it in the column headed 70, we find to be 77.58. This shows that we must add 77.58, or a trifle over 77½ gallons of water to 100 gallons of our 70 per cent. alcohol, to produce a spirit of 40 per cent. strength.

61. Baumé's Hydrometer for Liquids Lighter than Water. In Baumé's hydrometer for liquids lighter than water, the instrument is poised, so that the 0 of the scale is at the bottom of the stem, when it is floating in a solution of 1 ounce common salt in 9 ounces water, and the depth to which it sinks in distilled water shows the 10th degree; the space between these fixed points being equally divided.

62. Table showing the Specific Gravity *corresponding with the several degrees of Baumé's Hydrometer for liquids lighter than water.*

Degrees Baumé	Specific Gravity.	Degrees Baumé	Specific Gravity.
60°	.745	34°	.859
59	.749	33	.864
58	.753	32	.869
57	.757	31	.874
56	.760	30	.880
55	.764	29	.885
54	.768	28	.890
53	.773	27	.896
52	.777	26	.901
51	.781	25	.907
50	.785	24	.913
49	.789	23	.918
48	.794	22	.924
47	.798	21	.930
46	.802	20	.936
45	.807	19	.942
44	.811	18	.948
43	.816	17	.954
42	.820	16	.960
41	.825	15	.967
40	.830	14	.973
39	.834	13	.980
38	.839	12	.986
37	.844	11	.993
36	.849	10	1.000
35	.854		

63. Baumé's Hydrometer for Liquids Heavier than Water. In the hydrometer for liquids heavier than water, the position of the fixed points is reversed; for the 0 is at the top of the stem, and denotes the level to which the hydrometer sinks in distilled water: the 10th degree is lower down, and shows the level to which it sinks in the saline solution, and the graduation is continued downwards.

64. Baumé's Areometer, or Saccharometer for Liquids Heavier than Water. This instrument is generally in use in this country and in France, when it is necessary to ascertain the strength or density of a liquid heavier than water. In England, Twaddel's hydrometer is mostly employed for the purpose. Baumé's instrument is principally used by confectioners to test the density of syrup; also by brewers and distillers to discover the quantity of saccharine matter in wort; and by soap manufacturers and dyers to prove the strength of their lyes and dyeing materials. This variety of Baumé's hydrometer is usually called a saccharometer, and when plunged in pure water at 58° Fahr., marks 0 upon its scale; in a solution containing 15 per cent. of common salt and 85 of water by weight, it marks 15°; so that each degree on the scale is meant to indicate a density corresponding to the percentage of the salt.

The temperature at which Baumé's hydrometer was originally adjusted was 54½° Fahr.; it is now commonly adjusted to 58° or 60° Fahr.; hence arise the discrepancies observable in the published tables of the "correspondence between degrees of Baumé's and real specific gravities."

65. Table showing the Specific Gravity *corresponding with the several degrees of Baumé's Hydrometer for liquids heavier than water.*

Degrees of Baumé.	Specific Gravity.	Degrees of Baumé.	Specific Gravity.
0	1000	39	1372
1	1007	40	1384
2	1014	41	1398
3	1022	42	1412
4	1029	43	1426
5	1036	44	1440
6	1044	45	1454
7	1052	46	1470
8	1060	47	1485
9	1067	48	1501
10	1075	49	1516
11	1083	50	1532
12	1091	51	1549
13	1100	52	1566
14	1108	53	1583
15	1116	54	1601
16	1125	55	1618
17	1134	56	1637
18	1143	57	1656
19	1152	58	1676
20	1161	59	1695
21	1171	60	1715
22	1180	61	1736
23	1190	62	1758
24	1199	63	1779
25	1210	64	1801
26	1221	65	1823
27	1231	66	1847
28	1242	67	1872
29	1252	68	1897
30	1261	69	1921
31	1275	70	1946
32	1286	71	1974
33	1298	72	2002
34	1309	73	2031
35	1321	74	2059
36	1334	75	2087
37	1346	76	2116
38	1359		

66. To Convert Degrees Baumé into Specific Gravity. I. For liquids heavier than water.—Subtract the degree of Baumé from 145, and divide into 145; the quotient is the specific gravity.

II. For liquids lighter than water.—Add the degree of Baumé to 130, and divide it into 140; the quotient is the specific gravity.

67. To Convert Specific Gravity into Degrees Baumé. I. For liquids heavier than water.—Divide the specific gravity into 145, and subtract from 145; the remainder is the degree of Baumé.

II. For liquids lighter than water.—Divide the specific gravity into 140 and subtract 130 from the quotient; the remainder will be the degree of Baumé.

68. Twaddell's Hydrometer. This Hydrometer is much used in the bleaching and dyeing establishments in Scotland, and some parts of England. According to this scale 0 is equal to 1000, or the specific gravity of distilled water, and every additional 5 degrees of specific gravity adds 1 degree to Twaddell's scale. So that, in order to find the specific gravity corresponding to any degree of Twaddell's scale, multiply the degree by 5 and add 1000; thus, if this hydrometer shows 30°, 30 multiplied by 5 gives 150, and 1000 added makes 1150, the specific gravity. To find the degree of Twaddell corresponding to any specific gravity, deduct 1000 from the specific gravity, and divide the remainder by 5; the quotient will be the corresponding degree of Twaddell.

Thus, if it be required to find the degree of Twaddell corresponding to 1150 specific gravity, deduct 1000 from 1150, and divide the remainder, 150, by 5, and the quotient, 30, gives the degrees of Twaddell required. In this way the corresponding degrees of Twaddell and Baumé can easily be found. Thus, 31 degrees of Baumé are equivalent to a specific gravity of 1275; and this, according to the above rule, will give 55 degrees Twaddell. By reversing this process, Twaddell can as readily be reduced to Baumé.

Acetimetry.

Acetimetry. The art of determining the strength of acetic acid and vinegar. Several methods are employed for the purpose, based on—the quantity of acid required for saturation;—the specific gravity after the liquid has been neutralized with hydrate of lime;—and the simple specific gravity. In all these methods, account should be taken of any mineral acid which may have been added, as is common with vinegars, to impart artificial strength.

70. To find the Comparative Weights of Dry and Glacial Acetic Acid. As both *dry* and *glacial* (or *hydrated*) acetic acid are referred to in many places, in speaking of strengths, it may be convenient to know that 51 parts of *dry* acetic acid are equal to 60 parts of *glacial*. (*See No.* 81.) Hence the weight of *glacial* acid multiplied by .8512, gives the weight of *dry* acid; and the weight of *dry* acid, multiplied by 1.1748 gives a very close approximation to the weight of *glacial* acid.

71. Precautions in Testing Acids. It is essential to success, in testing acetic or other acids by saturation, to hit the *exact* point of neutralization. It will be found greatly to simplify matters to tint with litmus (*see No.* 78) either the sample under examination, or the test liquid; but when litmus is used, it is advisable to apply a gentle heat to the test tube when saturation appears *nearly* reached; the heat will expel from the liquor the free carbonic acid, which itself has the property of reddening litmus. A glass or wooden rod should be used for stirring, and the test liquid added drop by drop.

72. To find the strength of Acetic Acid by its Saturating Power. Dissolve 196¼ grains pure crystallized bicarbonate of potassa in a little water; add to the solution sufficient water to make up exactly 1000 minims, or the 100 divisions of an *acidimeter*, a graduated glass tube of 100 divisions, each division representing 10 minims. (*See illustrations, No.* 82.) A solution is thus formed, which, when added by degrees to 100 minims of the acetic acid or vinegar under examination, until the latter is exactly saturated, indicates the exact amount of acid present in the sample. Each minim of the alkaline solution thus employed represents 1 per cent. of dry acetic acid. The test liquid must be added a drop at a time to avoid the risk of loss by excessive effervescence.

73. To find the strength of strong Acetic Acid. If *strong* acetic acid be under inspection, it will be found convenient, previously to testing it, to dilute it with from 2 to 8 times its weight of distilled water, according to its degree of concentration. Dilute acid and vinegar require no further dilution.

Instead of 196½ grains *crystallized* bicarbonate of potassa, may be used either 135 grains *dry* (*see No.* 12) carbonate of potassa, 281 grains crystallized carbonate of soda, or 104 grains *dry* carbonate of soda. (*See No.* 80.)

By using 98¼ grains (half the quantity) of the bicarbonate of potassa, we obtain a still more delicate test liquid; as each minim used for saturating a sample of acid will represent only ½ of 1 per cent. of dry acid.

74. To find the strength of Acetic Acid by Saturation without an Acidimeter. The foregoing method can also be applied to test by weight, instead of by an acidimeter; 1000 grains of the test liquid are used in testing 100 grains of acid. Every grain of the test liquid necessary to produce saturation indicates $\frac{1}{10}$ grain of dry acid, and every ten grains are equal to 1 per cent. Schuster's alkalimeter is a convenient instrument for this process. (*See No.* 82.) 1000 grains of the test liquid are introduced into the alkalimeter, and the whole weighed; the weight of the bottle and solution, after using such portion of its contents as is required for testing, deducted from the previous weight of the whole, gives the exact quantity in grains of the solution consumed; this, divided by 10, gives the percentage of acid in the sample tested. This method admits of great accuracy.

75. Practical test of the strength of Acetic Acid. A less accurate, but more convenient method for practical purposes, is as follows:—To 100 or 1000 parts (or grains) of a sample under inspection, add cautiously from a weighed quantity of powdered pure *dry* bicarbonate of potassa, sufficient to produce exact neutralization; carefully re-weigh the bicarbonate unconsumed. Double the loss in grains will indicate the percentage of acid in the liquid tested.

76. Ure's Test of the strength of Acetic Acid. Ure's test gives very accurate results, if the ammonia employed is of the proper specific gravity. To 100 grains of a sample, very slightly reddened with neutral (blue) tincture of litmus, add liquor of ammonia of specific gravity .992 from an acidimeter (*see No.* 82) until perfect neutralization is effected, indicated by the original blue color of the litmus being restored. The number of acidimetric divisions of ammonia expended, multiplied by 51 (for *dry*) or by 60 (for *glacial*) and the product divided by 100, will give, respectively, the percentage of *dry* or *glacial* acid in the sample. Thus:—if a sample of vinegar takes 10 acidimetric divisions of ammonia to neutralize it, then 10 multiplied by 51, and divided by 100, gives 5.10, equivalent to $5\frac{1}{10}$ per cent. of *dry* acid:—or, 10 multiplied by 60 and divided by 100, gives 6 per cent. of *glacial* or *hydrated* acid in the sample.

77. Ure's Test, by Grains, of the strength of Acetic Acid. The same strength of ammonia is to be used in the acidimeter as in the preceding test, and the number of grain-measures of ammonia employed for a multiplier instead of acidimetric divisions. The only difference is, that the product in each case must be divided by 1000 instead of 100, to give the percentage of acid.

Acidimetry.

Acidimetry. The estimation of the quantity of an acid contained in any given sample.

The methods used are founded chiefly on the capacity of acids to saturate or neutralize alkaline bases; and, in some of the liquid acids, on specific gravity.

The accuracy of the tests, when saturation is resorted to, depends greatly on the *exact* point of neutralization, as already remarked under the head of *Acetimetry*. The proper point is arrived at when the liquid, after being slightly heated, ceases to redden litmus, or does not alter the color of turmeric paper (*see Test Papers*); if it turns the latter brown, too much test-liquid has been added, and the operation becomes useless. A good method is to tint either the acid sample or the test-liquid with a few drops of litmus (*see No.* 71), when the reddish shade will gradually deepen to purple as the point of saturation is approached, and the blue color be restored as soon as that point is reached.

79. To test the strength of an Acid by Saturation. Place in a test tube 100 grains of the acid to be examined; if the acid be liquid, dilute it—if solid, dissolve it—in 6 or 8 times its weight of distilled water. Then *exactly* neutralize it with an alkali added drop by drop. The known quantity of alkali consumed for this purpose represents an equivalent quantity of the actual acid contained in the test tube. The common practice is to dissolve 1 equivalent (*see No.* 80) of an alkaline test in water, and to make up the solution to 1000 grains (100 acidimetric divisions). The equivalent value of the test-liquid is then 100; hence, the quantity of the sample tested will bear the same proportion to the equivalent number (*see No.* 81) of the acid under examination, that the acidimetric divisions of the test-liquid consumed, bear to the percentage of acid sought. For example: Suppose 100 grains of a sample of sulphuric acid require 60 acidimetric divisions (600 grains) of the test-liquid to neutralize them; what is the percentage of the acid? The equivalent of dry sulphuric acid is 40 (*see No.* 81); therefore by the rule of proportion, since 100 : 40 : : 60 : 24, the sample contains 24 per cent. of dry sulphuric acid.

In this method the choice of the re-agent must depend on the operator. Some prefer the ammonia test (*see No.* 76), which is very convenient and easily applied; others give a preference to bicarbonates or carbonates of potassa or soda. Whichever be adopted, it must be perfectly pure. A test solution, once carefully prepared of the proper strength, may be kept unharmed for any length of time in a stoppered bottle, and will be always ready for application.

80. Table of Equivalents of Alkalis.

	GRAINS.
Pure ammonia	17
Dry carbonate of soda	53
Crystallized carbonate of soda	143
Crystallized bicarbonate of soda	84
Dry carbonate of potassa	69
Crystallized carbonate of potassa	87
Crystallized bicarbonate of potassa	100
Pure or caustic soda	31
Pure or caustic potash	47
Sesquicarbonate of soda	85
Neutral carbonate of ammonia	43¼
Sesquicarbonate of ammonia	59
Bicarbonate of ammonia	79

1000 grain measures of pure water of ammonia of specific gravity .992, contain 17 grains or 1 equivalent of pure gaseous ammonia.

It is understood that all crystals must be perfectly free from attached water, but not in the least effloresced.

81. Table of Equivalents of Acids. This table is based on the foregoing table of alkalis; so that, for instance, 1 equivalent (17 grains) of pure ammonia will exactly neutralize 1 equivalent (22 grains) dry carbonic acid, &c.

	GRAINS.
Anhydrous acetic acid	51
Hydrated or crystallized acetic acid	60
Dry benzoic acid	113
Crystallized benzoic acid	122
Dry boracic acid	35
Crystallized boracic acid	62
Dry carbonic acid	22
Dry citric acid	58
Crystallized citric acid	67
Dry hydrochloric acid	36½
Liquid hydrochloric acid (sp.gr.1.16)	109
Dry malic acid	58
Dry nitric acid	54
Liquid nitric acid (sp. gr. 1.5)	67
" " " (sp. gr.1.42)	90
Dry oxalic acid	36
Crystallized oxalic acid	63
Dry sulphuric acid	40
Liquid sulphuric acid (sp. gr. 1.845	49
Dry tartaric acid	66
Crystallized tartaric acid	75

82. Acidimeter. An acidimeter is a glass tube, graduated with 100 divisions, each division representing 10 grains of distilled water, termed grain measures. The acidimeter is used for testing acids and alkalis, and is usually furnished with a lip for convenience in pouring by drops. Where great delicacy is required in pouring or dropping, various appliances are resorted to, by which the outward flow can be instantly arrested, merely by placing the finger or thumb on an orifice arranged for the ingress of air. In the illustrations, *c* denotes the place of egress for

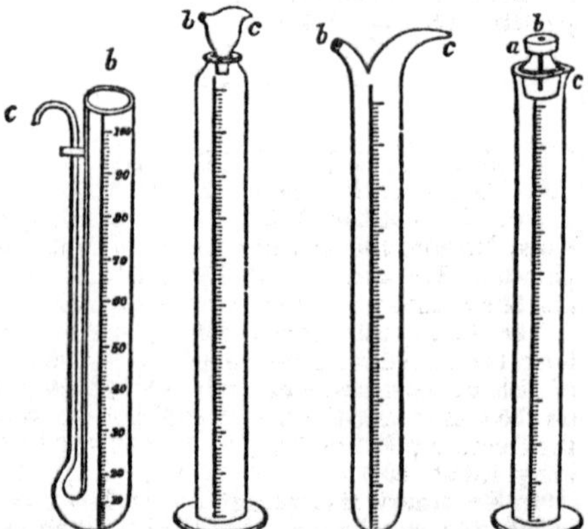

Fig. 1. *Fig.* 2. *Fig.* 3. *Fig.* 4.

the contained liquid; *b*, the orifice for the ingress of air, to be stopped by the finger or thumb; in Fig. 2, both orifices are in a hollow movable stopper; in Fig. 4, the air-hole only is in the stopper, *a*.

Fig. 1 represents Gay Lussac's Pouret.

Fig. 2, Normandy's modification of Schuster's Alkalimeter.

Fig. 3, Birck's Alkalimeter.

Fig. 4 is a simple acidimeter, with a stopper fitted to it, having a groove to correspond with the lip, and a vent-hole drilled through it to admit the air.

These modifications of the simple acidimeter are employed to allow of the test-liquid being added a single drop at a time, which is absolutely necessary during the first part of the process, to prevent undue effervescence, and consequent danger of loss of the liquid; and in the latter part it is equally indispensable in order to attain exact saturation. They dispense with the use of a separate *pipette*, being, in fact, acidimeters and pipettes combined.

Alkalimetry.

Alkalimetry. The method of estimating the strength of alkalis. The processes used are the same as in *acidimetry;* only that the unknown quantity sought is an alkali, and the test applied is an acid. The test acid is 1 equivalent (40 grains, *see No.* 81) of sulphuric acid sp. gr. 1.032 at 60° Fahr. inserted in an acidimeter (*see No.* 82) and made up with distilled water to 100 acidimetric divisions.

84. To find the strength of an Alkali. Place 100 grains of the alkali in a tube, and agitate it with about ½ ounce hot water. When settled, pour off the clear into a vessel for trial. Repeat this process until nothing soluble remains in the test tube, shown by the last washing *not* affecting the color of turmeric paper. Care must be taken not to waste the smallest portion of the liquid, as it would render the results inaccurate.

Next, *exactly* neutralize the alkaline solution by adding sufficient of the test acid drop by drop. If the saturation is complete, it will neither turn litmus paper red, nor turmeric paper brown. (*See No.* 78.) The weight of alkali tested, bears the same relation to its *equivalent weight* (*see No.* 80), that the acidimetric divisions of acid used, do to the percentage of alkali sought.

Thus:—If we test 100 grains of potash and find it requires 35 acidimetric divisions of test acid to saturate it, we refer to table No. 80, and find that the equivalent of pure potash is 47 grains. Then 100 : 47 : : 35 : 16.45.

That is, the sample of potash under examination contains nearly 16½ per cent. of *pure* potash. (*See No.* 587.)

The Thermometer.

The Thermometer. In Fahrenheit's Thermometer, which is universally employed in this country and Great Britain, the freezing point of water is placed at 32°, and the boiling point at 212° and the number of intervening degrees is 180.

The Centigrade thermometer, which has long been used in Sweden under the name of Celsius' thermometer, and is now employed on the continent of Europe generally, marks the freezing point at *Zero* or 0°, and the boiling point at 100°.

In Reaumur's thermometer, used in France before the revolution, the freezing point is *Zero,* and the boiling point 80°.

Degrees below zero are distinguished by prefixing the minus sign, thus —; so that —17° Fahr. represent a temperature of 17° lower than zero, equivalent to 49 degrees below freezing point.

86. To Convert degrees of Centigrade into degrees of Fahrenheit. Multiply the degrees of Centigrade by 9, and divide the result by 5:—then add 32.

Thus: to find the degrees of Fahrenheit equivalent to 30 degrees of Centigrade.

	30 degrees Centigrade.
Multiplied by	9
Divided by 5)270	
	54
Add	32
Answer,	86 degrees Fahrenheit.

87. To reduce degrees of Fahrenheit to the corresponding degrees of Centigrade. Reverse the above process—First deduct 32 from the degrees of Fahrenheit, then multiply the difference by 5, and lastly divide the result by 9.

Thus,	86 degrees Fahrenheit.
Deduct	32
	54
Multiplied by	5
Divided by 9)270	
Answer,	30 degrees Centigrade.

88. To Reduce degrees of Reaumur to the corresponding degrees of Fahrenheit. Multiply the degrees of Reaumur by 9, divide the result by 4, and then add 32.

Thus,	24° Reaumur.
Multiplied by	9
Divided by 4)216	
	54
Add	32
Answer,	86° Fahrenheit.

89. To reduce degrees of Fahrenheit to corresponding degrees of Reaumur. Reverse the above process.

90. To reduce degrees of Reaumur to Centigrade. Add to the degrees of Reaumur their one-fourth part.

Thus,	40° Reaumur.
Add one-fourth,	10
Answer	50° Centigrade.

91. To reduce degrees of Centigrade to Reaumur. Deduct one-fifth part.

Thus,	50° centigrade
Deduct one-fifth	10
Answer,	40° Reaumur.

92. Table of corresponding degrees of Fahrenheit, Reaumur and the Centigrade.

	Fahrenheit.	Reaumur.	Centigrade.
Boiling.	212	80	100
	203	76	95
	194	72	90
	185	68	85
	176	64	80
	167	60	75
	158	56	70
	149	52	65
	140	48	60
	131	44	55
	122	40	50
	113	36	45
	104	32	40
	95	28	35
	86	24	30
	77	20	25
	68	16	20
	59	12	15
	50	8	10
	41	4	5
Freezing.	32	0	0
	23	— 4	— 5
	14	— 8	—10
	5	—12	—15
	— 4	—16	—20
	—13	—20	—25
	—22	—24	—30
	—31	—28	—35
	—40	—32	—40

All intermediate degrees can be obtained by the preceding rules.

Bitters.

Bitters. Bitters are considered as tonic and stomachic, and to improve the appetite when taken in moderation. The best time is early in the morning, or an hour before meals. An excessive use of bitters tends to weaken the stomach. They should not be taken for a longer period than a fortnight at one time, allowing a similar period to elapse before again having recourse to them.

815. To Make French Cognac Bitters. Take 1½ pounds each red Peruvian bark, calisaya bark, bitter orange peel, and sweet orange peel; 2 ounces calamus root; 4 ounces cardamom seeds; 1½ ounces each cinnamon, cloves, and nutmegs; 4 ounces caraway seed, and 3 pounds wild cherry bark. Pound all these ingredients to a coarse powder and steep for 15 days in 45 gallons proof spirit (or 60 gallons spirit 25 below proof), stirring occasionally. Then rack it off, and mix sufficient caramel (*see No.* 694) to make it a dark red; add 15 pounds white sugar dissolved in 15 gallons water; let the whole settle, then filter. If the bitters are required to be of an amber color, omit the wild cherry bark and the caramel coloring.

816. To Make Angostura Bitters. Take 4 ounces gentian root; 10 ounces each calisaya bark, Canada snake-root, Virginia snake-root, liquorice root, yellow bark, allspice, dandelion root, and Angostura bark; 6 ounces cardamom seeds; 4 ounces each balsam of tolu, orangetis, Turkey rhubarb, and galanga; 1 pound orange peel; 1 pound alkanet root; 1½ ounces caraway seed; 1½ ounces cinnamon; ½ ounce cloves; 2 ounces each nutmegs, coriander seed, catechu, and wormwood; 1 ounce mace; 1¼ pounds red saunders, and 8 ounces curcuma. Pound these ingredients and steep them as in the last receipt, in 50 gallons spirit; and, before filtering, add 30 pounds honey.

817. Amazon Bitters. Take 90 gallons plain proof spirit; 3¼ pounds red Peruvian bark; 3¼ pounds calisaya bark; 1⅛ pounds calamus root; 4¾ pounds orange peel; 3½ ounces cinnamon; 3½ ounces cloves; 3½ ounces nutmeg; 2 ounces cassia buds; 6½ pounds red saunders. First mash all the ingredients, put them in the spirit, and let them infuse 14 days, being careful to stir the mixture well twice every day. Then rack off and color with 11 pints brandy coloring, to get a dark red tint. Stir ¼ hour. Dissolve 30 pounds white sugar in 30 gallons water; add, and again stir ½ hour. Let the mixture rest 4 or 5 days, and when bright, bottle. If the red saunders is not used, the color will be a bright amber. This is the finest bitters in the market. Compounded according to the above directions, the dealer will obtain 120 gallons 25 below proof.

818. Boker's Bitters. Take 1½ ounces quassia; 1½ ounces calamus; 1½ ounces catechu (powdered); 1 ounce cardamom; 2 ounces dried orange peel. Macerate the above 10 days in ½ gallon strong whiskey, and then filter and add 2 gallons water. Color with mallow or malva flowers.

819. Stoughton Bitters. To 12 pounds dry orange peel, 3 pounds Virginia snake-root, 1 pound American saffron, 16 pounds gentian root, add 1 pound red saunders. Grind all the above ingredients to a coarse powder, and macerate for 10 days in 20 gallons 65 per cent. alcohol, then filter.

820. Stoughton Bitters. (Another Receipt.) 2 pounds ginsing; 2 pounds gentian root; 1½ pounds dry orange peel; ½ pound Virginia snake-root; 1 ounce quassia;

¼ pound cloves; 3 ounces red saunders wood; 3 gallons alcohol 95 per cent.; 3 gallons soft water. Grind all the ingredients to coarse powder, infuse 10 days, and filter.

821. Wild Cherry Bitters. Take of wild cherry bark, 4 pounds; squaw vine (Partridge berry), 1 pound; Juniper berries, 8 ounces. Pour boiling water over the above and let it stand for 24 hours; strain, and pour again boiling water on the ingredients; let it macerate for 12 hours, then express and filter through paper, so that the whole will make 5 gallons, to which add of sugar, 3½ pounds; molasses, 1½ gallons; tincture of peach kernels, 6 ounces; tincture of prickly ash berries, 3 ounces; alcohol, 2 quarts.

822. To Make Peruvian Bitters. Take 8 ounces red Peruvian bark; 8 ounces orange peel; 1½ drachms each cinnamon, cloves, and nutmeg; and 75 cayenne pepper seeds. Infuse them, well bruised, in 8 gallons proof spirits, for 15 to 20 days, stirring every day. Draw off and filter.

823. Brandy Bitters. Grind to coarse powder 3 pounds gentian root, 2 pounds dry orange peel, 1 pound cardamom seeds, 2 ounces cinnamon, 2 ounces cochineal. Infuse 10 days in 1 gallon brandy, 8 gallons water, and filter.

824. Nonpareil Bitters. Grind to coarse powder 2 ounces Peruvian bark, ½ ounce sweet orange peel, ½ ounce bitter orange peel, 25 grains cinnamon, 25 grains cloves, 25 grains nutmeg, 15 cayenne seeds. Infuse ten days in 2 gallons 65 per cent. alcohol, then filter.

825. Spanish Bitters. Grind to coarse powder 5 ounces polypody, 6 ounces calamus root, 8 ounces orris root, 2½ ounces coriander seed, 1 ounce centaurium, 3 ounces orange peel, 2 ounces German camomile flowers; then macerate with 4¾ gallons 95 per cent. alcohol and add 5¼ gallons water and 1½ ounces of sugar. Filter and color brown.

826. Aromatic Bitters. Macerate 2¾ pounds ground dried small orange apples, ½ pound ground dried orange peel, 2 ounces ground dried calamus root, 2 ounces ground dried pimpinella root, 1 ounce ground dried cut hops, for 14 days, with 10 gallons of spirit at 45 per cent.; press, and add 2½ pints brown sugar syrup. Filter. Color dark brown.

827. Stomach Bitters. Grind to a coarse powder ½ pound cardamom seeds, ⅛ pound nutmegs, ¼ pound grains of Paradise, ½ pound cinnamon, ¼ pound cloves, ¼ pound ginger, ¼ pound galanga, ¼ pound orange peel, ⅛ pound lemon peel; then macerate with 4¾ gallons 95 per cent. alcohol, and add a syrup made of 4½ gallons water and 12 pounds sugar; then filter.

828. Hamburg Bitters. Grind to a coarse powder 2 ounces agaric, 5 ounces cinnamon, 4 ounces cassia buds, ½ ounce grains of Paradise, 3 ounces quassia wood, ¾ ounce cardamom seeds, 3 ounces gentian root, 3 ounces orange apples dried, 1½ ounces orange peel; macerate with 4¼ gallons 95 per cent. alcohol, mixed with 5¾ gallons water; add 2¾ ounces acetic ether. Color brown.

829. Bitters made with Essences. 40 gallons proof spirit, 1 drachm oil of anise, 1 drachm oil of caraway, ½ drachm oil of cloves, 1 drachm oil of lemon, 1 drachm oil of oranges, 1 drachm oil of cinnamon, ½ drachm oil of bitter almonds, 1 gallon sugar syrup. Cut the oils in 95 per cent. alcohol, and mix. Color with brandy coloring.

830. Bitter Filter. A fine bitter filter may be made according to fig. 5, No. 17.

831. Orange Bitters. Macerate 6 pounds orange peel for 24 hours with 1 gallon water, cut the yellow part of the peel from off the white, and chop it fine; macerate with 4¾ gallons 95 per cent. alcohol for two weeks, or displace (*see No.* 41); then add a syrup made of 4¼ gallons water and 16 pounds sugar. Filter through Canton flannel.

Aromatic Vinegar—Vinaigre Aromatique.

This is a compound of strong acetic acid with certain powerful essential oils. To produce the finer qualities of aromatic vinegar, glacial acetic acid must alone be employed. Aromatic vinegar is used as a pungent and refreshing nasal stimulant in languor, faintness, nervous headaches, dimness of sight, &c. For this purpose it is generally dropped on a small piece of sponge placed in a stoppered bottle, or a vinaigrette, which is only smelt at. It forms a useful caustic for warts and corns. As it is highly corrosive, it should be kept from contact with the skin and clothes. (*Cooley.*)

1084. Fine Aromatic Vinegar. Take of glacial acetic acid, 1 pound avoirdupois; rectified spirit, 2 Imperial fluid ounces; camphor (pure, crushed small), 2½ ounces; oil of cloves (finest), 1½ drachms; oil of rosemary, 1 drachm; oil of bergamot, oil of cinnamon, oil of lavender, oil of pimento, neroli (or essence de petit-grain), of each, ½ drachm; mix (in a stoppered bottle), and agitate until the whole of the camphor is dissolved. Very fine, and highly esteemed.

1085. Aromatic Vinegar. Take of camphor, 1 ounce avoirdupois; oil of cloves, 1 drachm; oil of cedrat, and lavender (Mitcham), of each 40 grains; oil of bergamot and thyme, of each 20 grains; oil of cinnamon, 10 grains; glacial acetic acid, ½ pound; mix as before. Very fine.

1086. Henry's Aromatic Vinegar. This resembles the preceding, except in being strongly scented with the oils of cloves, lavender, rosemary, and calamus aromaticus only.

1087. Vinaigre Aromatique. Take of camphor, 1 ounce avoirdupois; oil of cloves, 15 grains; oil of cinnamon, 10 grains; oil of lavender (English), 5 or 6 grains; glacial acetic acid, ½ pint. As the last. It is improved by doubling the quantities of the essential oils.

1088. Acetic Perfumes. The stronger aromatic or perfumed vinegars fall under this class of preparations; as do also various esprits and eaux (alcoöliques) to which a marked acetic odor has been given by the addition of concentrated acetic acid. The latter may be conveniently prepared by simply adding 1 to 1¼ fluid ounces of glacial acetic acid to each ½ pint of scented spirit. For acetic eau de Cologne and other like perfumes, 1¼ to 2 ounces of acid, per pint, is generally sufficient.

Smelling Salts.

Sesquicarbonate of ammonia commonly passes under this name, and, with the addition of a few drops of essential oil, is frequently employed to fill smelling bottles. Its pungency, however, is neither so great nor so durable as that of the true or neutral carbonate of ammonia. The latter salt continues unchanged in composition, and preserves its pungency as long as a particle of it remains unvolatilized. The portion only which flies off suffers decomposition as it volatilizes, separating into gaseous ammonia and carbonic acid. The pungency of the sesquicarbonate, on the other hand, depends solely on its gradual decomposition, in the solid state, into carbonate of ammonia, which flies off under exposure to the air; and into bicarbonate of ammonia, which is much less volatile and only slightly pungent, and which remains behind; the weight of the latter being far greater than one-half the weight of the original salt. Carbonate of ammonia, and not the sesquicarbonate, should, therefore, be alone used in filling smelling bottles, if a strong, agreeable, and durable pungency be desired. It is employed, either directly or indirectly, by the makers of all the more esteemed smelling salts of the day; and their predecessors did the same, even long before the chemistry of the two salts, and the rationale of the properties which cause a preference for the one, were known. (*Cooley.*)

1090. Fine Smelling Salts. Take of carbonate of ammonia (crushed small), 1 pound avoirdupois; oil of lavender (Mitcham), oil of bergamot, of each 1 Imperial fluid ounce; oil of cloves, 2 fluid drachms; oil of cassia, 1 fluid drachm. Rub them thoroughly together, sublime at a very gentle heat into a well-cooled receiver, and at once put the product into a well-stoppered bottle, or bottles. The sublimation may be omitted, but the quality of the product suffers. This is varied in some samples, by substituting 1 ounce of oil of lemon, or a little of the oils of rosemary and sweet flag (calamus aromaticus), for the oils of cloves and cassia; or by adding (after sublimation) a dash (2 or 3 drops per bottle) of essence of musk or essence royale.

1091. Smelling Salts. As before, but taking as perfume, oil of bergamot, 2 fluid ounces; oil of verbena, ½ fluid ounce; attar of roses, 1 to 2 drachms. It is varied as in the last.

1092. Smelling Salts. Same as No. 1090, but using oil of bergamot and lemon, of each, ¾ fluid ounce; essence de petit-grain, 3 fluid drachms; oil of cloves and cassia, of each, 1 fluid drachm; varied, as before, at will.

1093. Inexhaustible Smelling Salts. Take 1 pint liquid ammonia, 1 drachm attar of rosemary, 1 drachm attar of lavender, ½ drachm attar of bergamot, and ½ drachm attar of cloves. Mix together by agitation in a very strong, well-stoppered bottle. To prepare a smelling-bottle of this mixture, fill a stopper-bottle with pieces of sponge, previously well beaten, washed and dried; pour into the bottle as much of the mixture as the sponge will absorb, but not sufficient for a drop to escape if the bottle be inverted.

1094. Aromatic Spirit of Ammonia. Take of carbonate of ammonia, 8 ounces avoirdupois; strong liquor of ammonia (.882) 4 Imperial fluid ounces; volatile oil of nutmeg, 4 fluid drachms; oil of lemon, 6 fluid drachms; rectified spirit, 6 pints; water, 3 pints; mix, and distill 7 pints. Specific gravity .870. This is now the only authorized formula. The product is excellent, and very agreeable in use. (*Br. Ph.*)

1095. Ammoniated Perfumes. These are prepared by either adding strong liquor of ammonia to the liquid perfumes (eaux, esprits, &c.,) in sufficient quantity to impart to them a pungent ammoniacal odor, or by adding to the articles, before distillation, the ingredients that, by their mutual reaction, produce ammonia. In the former case, ½ to 1½ fluid ounces of liquor of ammonia (.880-.882), per pint, will be required, according to the nature of the preparation and the degree of pungency desired; and in general, when much essential oil is present, a spirit of higher strength than usual should be employed for the esprit, to compensate for its subsequent dilution by the ammonia. In the other case, 4 to 5 drachms of sal ammoniac, and 7 to 8 drachms of carbonate of potash for each pint of the product intended to be drawn over, are mixed with the cold ingredients just before distillation. For this use the liquor of ammonia must be perfectly free from tarry or empyreumatic matter, and have a purely ammoniacal odor.

1096. Ammoniated Eau de Cologne; Ammoniacal Cologne Water. As a perfume, this is best prepared by either of the methods noticed under ammoniated perfumes. It is now very extensively employed as a substitute for spirit of sal volatile. When intended for use in this way, a more agreeable and effective article may be produced by adding 1 ounce of carbonate (sesquicarbonate) of ammonia, and ½ fluid ounce of the strong liquor of ammonia to each pint of the product, or intended product, which will then have about the strength of the officinal spirit of sal volatile (spiritus ammoniæ aromaticus) of the British Ph. That of the stores has usually only little more than half this strength.

1097. Eau de Lavande Ammoniacale. To each Imperial pint of eau de lavande (*see No.* 989), add of liquor of ammonia (.880-.882), ½ to 1 fluid ounce.

1098. Ammoniacal Lavender Water. Take of oil of lavender (English) 1 fluid ounce; spirit of ammonia (caustic) 1½ pints; mix. The product is the officinal preparation of the French. Used as a stimulating pungent scent, in fainting, headaches, &c.

Alcohol.

Alcohol is a light, transparent, colorless, volatile, inflammable fluid; mixes in all proportions with water, with evolution of heat and condensation of the mixture, but some hours elapse before the union is complete. It dissolves resins, essential oils (*see No.* 940), camphor, bitumen, soaps, sugar, the alkaloids, wax, spermaceti, and various other substances. Boils at 172°, and in a vacuum at 56° Fahr.; curdles milk; coagulates albumen, and separates both starch and gum from their mucilages; uncongealable by cold; powerfully antiseptic to animal or vegetable substances immersed in it; with acids it forms ethers. Its evaporation, like that of ether, produces intense cold. By undergoing the acetic fermentation it is converted into vinegar. Dilute alcohol may be procured by the ordinary process of distillation, from *all* fermented liquors; when drawn from wine, as in France, it is called brandy; when from rice, as in the East Indies, it is called arrack or *toddy;* when from grain or malt, as in the United States or Great Britain, it is called whiskey, and when from molasses or the juice of the sugar-cane, as in the West Indies, it is called rum.

Whiskey is the spirit from which alcohol is usually obtained in this country.

By distilling a hundred gallons of whiskey, between 50 and 60 gallons of alcohol are received in the condenser of a specific gravity of 0.835. By a second distillation, taking care to collect only the first portions, and cautiously managing the heat so as not to allow it to rise to the temperature of boiling water, alcohol may be obtained of a specific gravity of 0.825, which is the lightest spirit that can be received by ordinary distillation. At this stage it contains 11 per cent. of water and some small portions of fusel oil.

The best alcohol is that manufactured under Attwood's patent process, in which manganic acid is used to destroy the fusel oil and other foreign substances. This alcohol withstands the tests of nitrate of silver and sulphuric acid remarkably well. (*See No.* 1444.)

The high wine, or rectified spirit, distilled and rectified in the United States, and often sold as French pure spirit, is free from all deleterious substances, and nearly scentless. Its strength is usually from 84 to 95 per cent. (*See Nos.* 53, *&c.*)

1436. Proof Spirit contains 52¼ per cent. by volume of pure alcohol; has a specific gravity of .920 at 60° Fahr.; and is no more than a mixture of 49 parts by weight pure alcohol with 51 parts water. This is the strength of the proof spirit usually employed by perfumers, and for medicinal purposes; but by law (*see No.* 58), proof spirit is equal parts by volume of absolute alcohol and distilled water, having a specific gravity of .933.

1437. Dilute Alcohol. Alcohol dilutum (*U. S. Ph.*) consists of equal measures of officinal alcohol and water; it contains 39 per cent. by weight, or 46.33 per cent. by volume, of pure or absolute alcohol, and has a specific gravity of .941, equal to 19° of Baumé's light hydrometer.

1438. Alcohol. Officinal alcohol (*U. S. Ph.*) contains 85 per cent. by weight, or 89 per cent. by volume, of pure alcohol; its specific gravity is .835, or 38.45° Baumé.

1439. Stronger Alcohol. Alcohol fortius (*U. S. Ph.*) has 92 per cent. by weight, or 94.65 per cent. by volume, of pure alcohol; and a specific gravity of .817, or about 42° Baumé.

1440. Amylic Alcohol. A peculiar oily, nearly colorless acrid liquid, known also as *Fusel oil,* obtained by distilling fermented grain or potatoes, by continuing the process after the ordinary spirit has ceased to come over. Its specific gravity is .818, and its boiling point 268° to 272° Fahr. (*U. S. Ph.*)

1441. Absolute Alcohol. To procure absolute or anhydrous alcohol, take the bladder of an ox or calf, soak it for some time in water, then inflate it and carefully free it from the attached fat and vessels; this must be done on both sides. After it is again inflated and dried, smear over the outer surface *twice,* and the inner surface *four* times, with a solution of isinglass. Then nearly fill it with the spirit to be concentrated, leaving only a small space vacant; it is then to be securely fastened, and suspended in a warm situation, at a temperature of about 122° Fahr., over a sand bath, or in the neighborhood of an oven or fire. In six to twelve hours, if the heat be properly conducted, the spirit will be concentrated, and in a little time longer may be rendered nearly free from water (anhydrous) or of the strength of 97 or 98 per cent.

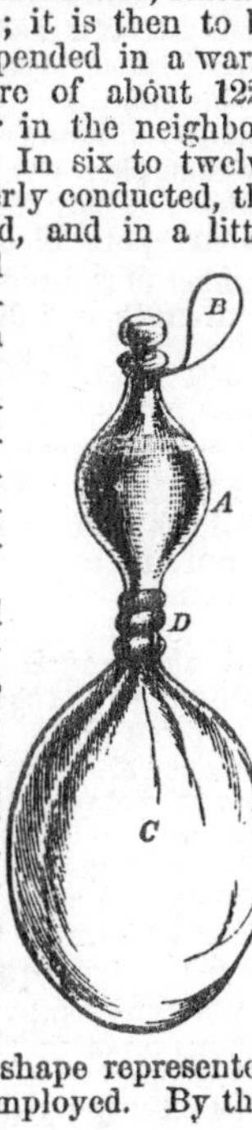

This alcohol will be sufficiently pure for all the common purposes of the manufacturers, and is an excellent spirit for making varnishes, &c.

The same bladder will serve more than one hundred times; and in fact a *common* bladder, thoroughly cleansed from fat, and washed and dried, may be used without any further preparation. The bladder should be kept very nearly full, or else a portion of the spirit will escape through the empty part. To prevent this accident, a bottle with a double neck, of the shape represented in the engraving, may be employed. By this means the bladder may be kept always full.

A, A bottle with two necks, the upper furnished with a ground-glass stopper.

B, Loop of cord to hang up the apparatus.

C, Bladder containing spirit, filled by means of the bottle, *A*.

D, Neck of bladder accurately secured to the lower neck of the bottle, *A*.

After the first or second time of using the bladder, it gives alcohol sufficiently pure for most experimental purposes. Before hanging the apparatus up, it is better to enclose and suspend it in a coarse netting, which will prevent any accident arising from the strain on the neck of the bladder. Should weaker spirit than that directed in the preceding formula be used, to procure alcohol by either method, it must be previously concentrated, or the operation repeated a second time.

Absolute alcohol is used to dissolve resins by the varnish maker; essential oils, by the perfumer; pyroxyline (gum cotton), by the photographer; and by the pharmaceutist to prepare tinctures and for many other purposes.

1442. Chemical Method of Procuring Absolute Alcohol. Take 1 gallon of the alcohol of commerce; throw 1 pound freshly made chloride of calcium into the alcohol, and, as soon as it is dissolved, distill off 7 pints and 5 fluid ounces. Or, take of rectified spirit 1 imperial pint; lime, 18 ounces; break the lime into small fragments, mix with the alcohol in a retort properly connected, and expose the mixture to a gentle heat until the lime begins to slake; then withdraw the heat until the slaking is finished. Now raise the heat gently and distill off 17 fluid ounces. Alcohol thus obtained will have a density, when the operation is carefully managed, of 0.796.

1443. To Increase the Strength of Common Alcohol. Take a pint of common spirits, and put it into a bottle which it will only fill about ¾ full. Add to it ½ ounce pearlash or salt of tartar, powdered as much as it can be without occasioning any great loss of its heat. Shake the mixture frequently for about half an hour, before which time a considerable sediment, like phlegm, will be separated from the spirits, and will appear along with the undissolved pearlash or salt at the bottom of the bottle. Then pour the spirit off into another bottle, being careful to bring none of the sediment or salt along with it. For this purpose an instrument called a *separating funnel* is well adapted. To the quantity just poured off add ½ ounce pearlash, powdered and heated as before, and repeat the same treatment. Continue to do this as often as necessary, till little or no sediment forms; when this is the case, 1 ounce of alum, powdered and made hot, but not burned, must be put into the spirits, and suffered to remain some hours, the bottle being frequently shaken during the time; after which the spirit, when poured off, will be found free from all impurities, and equal to the best rectified spirits of wine.

1444. To Test the Purity of Alcohol. The presence of water may be detected by its specific gravity. Fusel oil may be detected by adding a little of a solution of nitrate of silver to the alcohol. Dissolve 10 grains nitrate of silver in 1 ounce of pure distilled water. Then take half a tumblerful of the suspected liquor and drop into it 25 drops of the above solution; and if the liquid should contain any grain oil, it will assume the form of a black powder and float on the surface. The action of this test is not always immediate, for it is sometimes necessary to wait from 1 to 30 hours when testing a sample of alcohol which has been well rectified, before any evidence of the oil or powder can be perceived floating on the liquid, and even then it is necessary to expose the glass to a strong light before the powder can be discovered.

For detecting fusel oil in alcohol, Mr. E. N. Kent finds pure sulphuric acid the best test. Half fill a test tube with the spirit to be tested, then fill up slowly with pure concentrated sulphuric acid. Pure spirit remains colorless; impure spirit becomes colored in proportion to the amount of fusel oil present. 1 per cent. of wood spirit (wood naphtha) in alcohol, will cause it to turn yellow or brown with the addition of caustic potassa. Pure alcohol is neutral to test paper; should be colorless; will evaporate entirely by heat; retains its transparency when combined with water or ether; tastes and smells vinous.

1445. To Free Alcohol from Fusel Oil. This may be effected by digesting the alcohol with charcoal. By Schaeffer's method the alcohol is filtered through alternate layers of sand, wood-charcoal, boiled wheat, and broken oyster shells; this removes all other impurities as well. The fusel oil can be extracted from small quantities of alcohol, by adding a few drops of olive oil to the spirit, agitating thoroughly in a bottle, and, after settling, decanting. The olive oil dissolves and retains the fusel oil.

1446. To Deodorise Whiskey or Alcohol and free it from Fusel Oil. To the barrel of liquor add about a gallon (or more) of water saturated with chlorine; stir up thoroughly, and let it rest for 12 hours. Then saturate with chalk; add another gallon of water, and distill.

1447. To Filter Alcohol. The following method of filtering alcohol, or its solutions, is said to be very satisfactory, and is used extensively in North Germany, where it constitutes one of the secrets of the trade. Clean, unsized paper (Swedish filtering paper is the best), is torn into shreds and stirred into the liquid to be clarified. The whole is then strained through a flannel bag, when the resulting liquid will be found to possess the utmost clearness and limpidity. A filter may also be made by spreading thin paper pulp evenly upon stretched flannel or woolen cloth. When dry, the cloth so coated will be found to give better results than the felts, etc., commonly employed as filters. (*See Nos.* 714 *and* 811.)

1448. To Test the Strength of Alcohol. Alcohol dissolves chloroform, so that when a mixture of alcohol and water is shaken up with chloroform, the alcohol and chloroform unite, leaving the water separate. On this fact Basile Rakowitsch, of the Imperial Russian Navy, has founded his invention. The instrument he uses is a graduated glass tube into which a measured quantity of chloroform is poured, and to this is added a given quantity of the liquid to be tested; these are well mixed together and then left to subside; the chloroform takes up the alcohol and leaves the water, which, being lighter than the chloroform, will float on the top; and the quantity of water that has been mixed with the spirit will be at once seen.

1449. Arithmetical Rules for the Treatment of Alcohol. The following excellent rules, derived from various sources, contain, and will yield to the manufacturer, much information of a very useful character.

1450. To Ascertain the Cost of any Quantity of Alcohol at any Degree or Percentage of Strength Above or Below Proof. Alcohol is always bought and sold at so much above or below proof. To ascertain the price of a quantity of alcohol, add the percentage over proof, or deduct the percentage under proof, and multiply by the price per gallon. Thus: what will 40 gallons of alcohol, 25 per cent. over proof, cost at 28 cents proof? We first find 25 per cent. of 40, which is 10; we then add that number to 40, the number of gallons, and we get 50; we then multiply 50 by 28, the price per gallon proof, and get $14.00, or 35 cents per gallon. Again, what will 40 gallons alcohol, 25 per cent under proof, cost, at 28 cents per gallon proof? Again, we find that 25 per cent. of 40 is 10; we then deduct 10 from 40, this leaves us 30; by multiplying 30 by 28 we get $8.40, or 21 cents per gallon.

1451. To Ascertain How Much Water Should be Added to Spirits, to Reduce it from a Given Degree of Strength to a Lower Degree or Percentage of Strength. The manufacturer may sometimes find it necessary to reduce or increase the strength of spirit, according as circumstances may require. To accomplish this, we give the following rules, which will be found useful to the dealer: multiply the number of gallons by the actual degree of strength of the spirit, and divide the amount by the degree of strength sought to be obtained, and from the answer subtract 100; the amount thus obtained will show the quantity of water to be added to the spirit in order to reduce it to the degree sought. For example: suppose you have 100 gallons of spirit at 80° by Tralles' hydrometer, and wish to reduce it to 50° or proof. Multiply 100 by 80, and divide the amount by 50, then from the answer subtract 100; this will show that 60 gallons of water must be added to the spirit in order to reduce it to 50° Tralles', or proof.

Thus, 100 gallons
Multiplied by 80
—
Divided by 50)8000(160
Deduct 100
—
60

1452. To Ascertain the Quantity of Pure or Absolute Alcohol in any Given Amount of Liquor. The quantity of alcohol contained in any amount of liquor is readily ascertained after testing the strength with Tralles' hydrometer at 60° Fahr., by simply multiplying the figures expressing the quantity of liquor, by the ascertained strength; for example: a barrel of brandy containing 32 gallons, 60° strong at 60° Fah., contains 19⅕ gallons pure alcohol. Rule.—**Multiply the number of gallons by the ascertained degrees of strength, and divide by 100.** Thus:

32 gallons,
60° Tralles' at 60° Fahr.
—
19.20, or 19⅕ gallons pure alcohol.

1453. To Ascertain the Number of Gallons at any Required Number Below Proof, in any Given Number of Proof Gallons. Multiply the given number of proof gallons by 100, and then divide the product thus obtained by a number found by deducting the required number of degrees below proof from 100. The quotient will be the answer. For example: How many gallons, 25 below proof, are there in 35 gallons proof?

100 35 gallons proof,
25 B. P. 100
— —
75)3500(46⅔ gallons 25 below proof.

We thus see by the above example that 35 gallons proof spirit is equal to 46⅔ gallons 25 below proof.

1454. To Increase the Strength of a Spirit from any Degree to a Higher given Degree, or Percentage. To increase the degree of strength of a spirit, multiply the number of gallons by the actual degree of strength of the spirit, and divide by the degree of strength sought to be obtained. For example: suppose you have 100 gallons of spirit at proof, or 50° by Tralles' hydrometer, and wish to increase its strength to 80°. Multiply 100 gallons by 50 and divide by 80; the answer will give you the number of gallons of spirit, 62½, to be added to the 100 gallons in spirit in order to increase its volume to 80° by Tralles' hydrometer.

Thus, 100
50
—
80)5000
—
62.4, or 62½.

1455. To Reduce Spirit a Given Number Above Proof to a Required Number Below Proof, by the Addition of Water. Multiply the number of gallons of spirit by the sum of the given degree above proof and the required degree below proof, and divide the product by a number to be found by subtracting the required proof from 100. The quotient will give the number of gallons of water to be added.

Suppose you want to reduce 40 gallons spirit 20 above proof to 10 below proof, how much water must be added to accomplish the result?

100 40 gallons.
Required proof, 10 30
— —
90)1,200(13⅓ gals. water.

It will thus be seen that, to reduce 40 gallons spirit 20 above proof to 10 below proof, it will be necessary to add 13⅓ gallons of water, making 53⅓ gallons in all.

1456. To Reduce High Proof Spirit to a Required Lower Proof, by the Addition of Water. First multiply the number of gallons by a number expressing the difference in degrees of strength between the given proof of the spirit to be reduced and the required degree, or proof, to which it is to be reduced. Divide the product thus ascertained by a number to be found by adding the required proof to 100.

Suppose you desire to reduce 72 gallons spirit at 30 above proof to 10 above proof, how much water must you add?

30, given strength.
10, required strength.
—
20, difference.

Required strength, 10 72, No. of gals.
100 20, difference.
— —
110)1,440($13\frac{1}{11}$ gals.

Thus it will be seen that, to reduce 72 gallons spirit at 30 above proof to 10 above proof, it is necessary to add $13\frac{1}{11}$ gallons of water, making about 85 gallons in all.

1457. To Reduce Spirit of a Given Number Above Proof to a Required Number Below Proof, by the Substitution of Water for Spirit. Deduct the number below proof from 100, and multiply the number of gallons by the remainder. Then add the number which the given liquor is above proof to 100, and divide the above product by the number thus obtained. The quotient, deducted from the original number of high proof gallons, will give the answer required. All small fractions may be rejected.

Suppose you want to reduce a cask of 40 gallons spirit at 20 above proof to 10 below proof.

100
10

Multiply 90
by 40

To 100 add 20=120)3,600(30
Original number of gallons, 40
Deduct quotient, 30

Answer, 10 gallons.

Thus it will be seen that 10 gallons should be removed, and their place supplied with water, in order to make the mixture equal to 10 degrees below proof.

1458. To Reduce Spirit of a Given Number Above Proof to Proof Spirit, by the Substitution of Water for Spirit. Multiply the number of gallons by 100, then add the number which the spirit is above proof to 100, and divide the above product by the number thus obtained; subtract the quotient from the number expressing the original quantity of spirit, and the answer will give the number of gallons to be removed from the spirit and replaced with water, in order to reduce the high proof spirit down to proof.

Suppose you want to reduce a cask of 24 gallons of spirit 20 above proof to proof spirit.

Above proof, 20 24
100 100

120)2,400(20
Original quantity 24
20

Answer, 4

It will be seen by the above example that 4 gallons have to be taken from the spirit and the same quantity of water added, to reduce it to proof.

1459. To Raise Spirit of a Given Number Under Proof to a Required Strength Above Proof, by the Substitution of High Proof Spirit. Multiply the number of gallons by the number expressing the difference in degrees of strength between the high proof spirit to be added and the required degree to which it is to be raised. Divide the product thus found by a number to be obtained by adding the given number below proof to the number the high spirit is above proof; then subtract the quotient from the original number of gallons, and the remainder will show the quantity of low spirit to be removed and its place supplied by the addition of the same quantity of high proof spirit.

Suppose you desire to raise a cask of 40 gallons at 10 below proof to 15 above proof, by means of spirit 40 above proof:

40 40 A. P. 40 number of gals.
15 10 B. B. 25 multiplied by diff.

Diff. 25 50)1000(20
40 gals. original quantity to be raised.
20 deduct quotient.

20 answer.

The above example shows that 20 gallons should be taken from the low proof spirit, and the same quantity of spirit added at 40 above proof, to raise it to 15 above proof.

1460. To Raise Spirit of a Given Number Below Proof to Proof Spirit, by the Substitution of High Proof Spirit. Multiply the number of gallons by the number which the high proof spirit is above proof, divide the product by a number to be found by adding the given number the spirit is below proof to the number the high spirit is above proof; subtract the quotient from the original number of gallons, and the remainder will show the quantity of low proof spirit to be removed, and its place to be supplied by the addition of high proof spirit.

Suppose you desire to raise a cask of 40 gallons at 5 below proof, to proof, by means of spirit 35 degrees above proof.

35 A. P. 40 number of gallons.
5 B. P. 35 above proof.

40)1400(35 quotient.
40 gallons,
35 quotient,

5 answer.

It will thus be seen that 5 gallons should be taken from the low proof spirit, and the same quantity of spirit added at 35 above proof, in order to raise it to proof strength.

1461. To Raise Spirit of a Given Number Above Proof to a Still Higher Degree of Strength, by the Addition of High Proof Spirit. First multiply the number of gallons by a number expressing the difference in degrees of strength between the given proof of the spirit to be raised, and the required degree to which it is to be raised. Divide the product thus ascertained, by a number to be found by subtracting the difference in degrees between the spirit to be raised and the high proof spirit employed to raise it. The quotient will show the number of gallons of a higher proof which must be added.

Suppose you desire to raise a cask of 35 gallons spirit 15 above proof to 20 above proof, by the addition of spirit 30 above proof.

20 required proof,
15 given proof,

5 difference.

From 30 35 number of gallons.
Subtract 15 5 multiplied by difference.

15)175(11⅔ answer.

1462. To Reduce Low Proof Spirit to a Still Lower Proof, by the Addition of Water. First multiply the number of gallons by the difference in degrees of strength between the given proof of the spirit to be reduced, and the required proof to which it is to be reduced. Divide the product by a number ascertained by subtracting the given proof from 100, and the quotient will give the number of gallons of water to be added.

Suppose you want to reduce 40 gallons spirit 10 below proof, to 15 below proof.

Required proof 15
Given proof 10

Difference 5

100 40 gallons
10 given proof 5 difference

90)200(2⅔ gals. water

1463. To Raise a Low Proof Spirit to a Higher Required Proof by the Addition of High Proof Spirit. Multiply the number of gallons by a number expressing the difference in degrees of strength between the given proof of the spirit to be raised, and the required proof to which it is to be raised. Divide the product thus ascertained by the sum of the given proof, and the high proof spirit to be added, and the quotient will give the answer.

Suppose you desire to raise 40 gallons spirit 15 below proof to 10 below proof with spirit 10 above proof.

Given proof 15
Required proof 10

Difference 5

Given proof 15 40 gallons
High proof 10 5 difference

25)200(8 gals. answer.

Essential Oils; Volatile Oils.

The essential or volatile oils are an extensive and important class of bodies derived from the vegetable kingdom, and found in almost every part of the larger number of the plants which produce them, except the cotyledons of the seeds, which, in general, form the exclusive repository of the fixed oils. It is the volatile oils which confer upon flowers, leaves, fruit, seeds, roots, barks, and woods, their peculiar and characteristic odors; but among these they are not equally distributed in the same individual, and are often altogether absent from some of them. To them we are indebted for our most delightful perfumes, and our choicest aromatics and spices. All of them, when perfectly pure, are colorless; though, before rectification, nearly the whole of them have a pale yellow tint, and some of them are brown, blue, or green. They mix in all proportions with the fixed oils, dissolve freely in both alcohol and ether, and are sparingly soluble in water, forming perfumed or medicated waters. (*See Nos. 1080, &c.*) Their boiling point usually ranges between 310° and 325° Fahr., and is always considerably higher than water. They resist saponification and (excepting oil of cloves) do not combine with the salifiable bases. Their density fluctuates a little on either side of water. The lightest oil is that of citrons (specific gravity 0.847), and the heaviest, that of sassafras (specific gravity 1.096). When cooled sufficiently they all soldify. The common temperature of the atmosphere is sufficient for this with some of them, as the oils of roses and aniseed; whilst others require to be cooled below the freezing point of water before they assume the solid form. By exposure to the air they rapidly absorb oxygen, and become partially converted into resin. This is the cause of the deposit that usually forms in them (especially in the expressed oil of orange) when kept in an imperfectly stopped bottle. (*Cooley.*)

1465. To Obtain Essential Oils. All essential oils which are more or less volatile can be obtained from substances by distilling the articles along with an equal weight (some use a larger proportion) of water; but some substances that give out their oil with difficulty, are first soaked for 24 hours in twice their weight of water, to each gallon of which 1 pound of common salt has been added, by which its boiling point is raised, and consequently the oil comes over more easily. In such cases a quick fire is used, and when one half the water has come over, it is returned into the still, and this is repeated until the distilled water ceases to come over mixed with oil. The heat of steam or a salt water

Fig. 1.

bath should be preferably employed; but if a naked fire be used, the still should be deep and narrow, by which means the bottom will be more perfectly covered when the quantity of water becomes small, and burning prevented. When the distilled water is to be repeatedly poured back on the ingredients, a very convenient plan is to so arrange the apparatus that, after the water has separated from the oil, it shall flow back again into the still, by which much time and trouble will be saved. The separation of the oil and water is effected by allowing the mixed liquids to drop into a Florentine receiver (*see Fig.* 1), when the oil is lighter than water, by which means the latter accumulates at *a*, and the water flows over by the spout, *b*. The essential oil is obtained in this manner from the following: Anise, caraway, wormseed, cubebs, fennel, pennyroyal, juniper, lavender, lemon, cinnamon, peppermint, spearmint, horsemint, origanum, pimento, rosemary, savine, sassafras, valerian, &c. The empyreumatic oil of tobacco is obtained by introducing the dry leaves in coarse powder into a green glass retort, heating it in a sand-bath to a dull red heat. Separate the oily liquid from the watery portion as it comes over, and keep for use. (*See No.* 46.) The same receiver may be employed for oils heavier than water, by reversing the arrangement; but a glass separator (*see Fig.* 2) will be found more convenient.

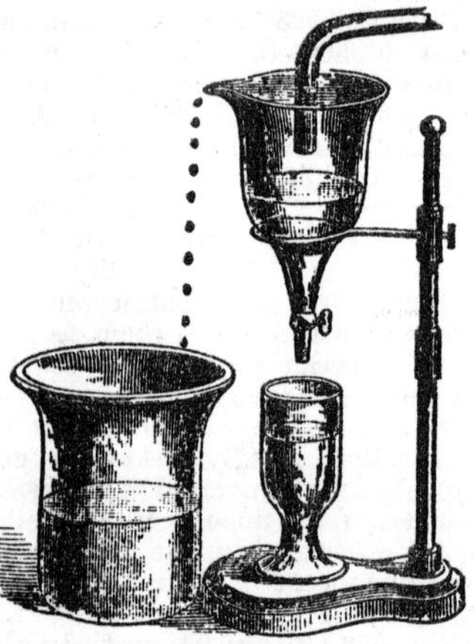

Fig. 2.

In this case the oil accumulates at the bottom of the vessel, and may be drawn off by the cock. The oil of cloves and other heavy essential oils are obtained by macerating 5 pounds coarsely powdered material for 48 hours in 10 pounds water containing 1 pound salt; and distilling until the product is no longer milky. After the oil has deposited, the remaining water is again distilled, and this repeated until all the oil has been extracted from the water. After 10 days, the oil is cleaned and clarified by filtering. The essential oil of cloves, cinnamon, rhodium-wood, sandal, calamus, aloes, &c., are thus obtained. That of bitter almonds and of mustard are obtained by making a thin paste of the material with water; and, after 24 hours' maceration, distilling by steam-bath. The essential oils of lemons, oranges, and some other fruits, are chiefly obtained by submitting the yellow rind to powerful pressure; but in this way they are not so white, nor do they keep so well as when distilled. Volatile oils should be kept in well-closed and nearly full bottles, in the dark, and opened as seldom as possible, as by age and frequent exposure they become resinous. The process of distillation should be done as rapidly as possible, and the light oils collected soon after its separation from the water.

1466. Special Directions for Distilling Essential Oils. Substances yielding volatile oils are generally distilled with water, the proportion of which varies with each article, but under all circumstances must be sufficient to prevent the substance from burning before the whole of the oil has passed over. To prevent the risk of burning, it has been recommended to suspend the substance to be distilled in a basket, or a bag of wire-work, in the water, so as not to touch the bottom or sides of the alembic; or to place the substance on a perforated shelf in the upper part of the alembic above the surface of the water. Some substances, such as mustard, bitter almonds, &c., which are mixed to a paste with water, are distilled by the action of a current of steam heated to the necessary degree and admitted into the bottom of the alembic. An excess above what is necessary acts injuriously by holding some of the oil in solution after the mixed vapors are condensed; on the other hand, if too small a quantity be employed, besides the danger of burning, the whole of the oil will not be distilled. Dried plants require more water than the fresh and succulent.

The form of the alembic has an influence over the quantity of water distilled, which depends more upon the extent of surface than the amount of liquid; by employing a high and narrow vessel the disadvantage of an excess of water is much obviated.

The temperature should be equable, and regulated so as not to exceed the required degree of heat; and, as some oils are more volatile than others, an appropriate temperature must be obtained and sustained; the use of a higher temperature than is necessary being injurious. Any degree of heat can be steadily applied by the use of a bath, either of water or of some solution (weaker or stronger as required) of which the boiling point is known. (*See No.* 7.)

The more volatile oils pass freely with the steam into the neck of the receiver, but some that are less volatile are apt to condense in the head, and return into the body of the still; for these a still should be employed with a large and low head, having a rim or gutter inside, in which the oil may be received as it condenses, and thence led into the neck of the condensing tube (*see No.* 1077), which is better straight than coiled, for convenience in cleaning, as the alembic and all its appurtenances must be perfectly clean before distilling each kind of essential oil.

Certain flowers, such as orange flowers and roses, yield little or no oil when dry, and must be preserved fresh, either with salt, or by means of glycerine, to keep them in condition for distilling their oils. (*See No.* 1349.)

The most of the aromatic herbs are usually distilled while fresh, although it is thought by some that they yield a larger product when moderately dried. Dried substances require, previous to distillation, to be thoroughly macerated with water; and to facilitate this end, should be prepared by slicing, rasping, bruising, or other appropriate means. Sometimes the proportion of oil in the substance employed is so small that it is wholly dissolved in the water distilled, even though the smallest necessary quantity of water has been employed in the alembic. In this case the distilled solution must be redistilled several times with fresh quantities of the substance, until more oil passes over than the water will dissolve. This process is called *cohobation.*

1467. Millon's Method of Obtaining Essential Oils. The flowers are placed in a percolating apparatus (*see No.* 41) and then ether or sulphide of carbon is poured over them. After leaving the flowers in contact for 15 minutes the liquid is drawn off and a fresh supply added and drawn off in a similar manner. This completely dissolves all the essential oil of the flowers, leaving them quite scentless. The liquid is next distilled, and the ether or sulphide of carbon, being volatile at a much lower temperature than the fragrant principle, is drawn over alone, and leaves a residue containing all the perfume of the flower. This residue, more or less solid, is exposed to the heat of the sun until it loses the unpleasant smell of the solvent used. No degree of natural heat is capable of altering the perfume or turning it rancid. The product has a much finer odor than essential oil prepared by any other system.

1468. Cognac Oil. Oil of cognac is prepared by dissolving the fusel oil of brandy marc in strong rectified spirit, and then adding a sufficient quantity of concentrated sulphuric acid to form a sulphate; alcohol and excess of acid are removed by washing the newly formed compound with water. To 100 pounds marc add $\frac{1}{2}$ pound sulphuric acid; the oil is generally formed towards the end of the distillation, and is found floating in blackish drops on the surface of the distillate. According to a distinguished French chemist, this oil is a compound of potato oil and œnanthic ether.

1469. Oil of Apple. Mix cautiously 1 part fusel oil, 3 parts sulphuric acid, and 2 parts water. Dissolve $2\frac{1}{2}$ parts bichromate of potash in $4\frac{1}{2}$ parts water, introduce this into a large tubulated retort, and gradually add the former liquid, so that the boiling continues very slowly. The distillate, which is principally valerianic acid, is saturated with carbonate of soda, and evaporated to dryness. Take of the valerianate of soda, thus formed, $1\frac{1}{2}$ parts; fusel oil, 1 part; sulphuric acid, 1 part; mix cautiously, heat by a water-bath, and mix with water; the impure valerianate of amyloxide will separate. It is washed several times with water, then with a solution of carbonate of soda, and finally with water. This is dissolved in from 6 to 8 parts of water.

1470. Oil of Jargonelle Pear. This is made from the heavy fusel oil which comes over last in distillation. To purify the fusel oil, wash it with soda and water, and distill between 254° and 284° Fahr. Of this take 1 pound; glacial acetic acid, 1 pound; sulphuric acid, $\frac{1}{2}$ pound. Digest for some hours at 254°. The ether separates upon the addition of water, and is purified by washing with soda and water. Mixed with $\frac{1}{10}$ part acetic ether, and 7 parts of deodorized alcohol, it gives the essence of pears.

1471. Oil of Quince—Pelargonic Ether—is made from oil of rue by treating it with double its volume of dilute nitric acid, heating the mixture until it begins to boil. After some time two layers are seen. The lower one is separated with a pipette, and freed from nitric acid by evaporation in a chloride of zinc bath; it is then filtered, mixed with deodorized alcohol, and digested at a gentle heat until the fruity odor is noticed. This ether seems identical with the ethereal oil of wine, which gives the bouquet. It is sometimes sold as oil of cognac.

1472. To Restore the Fragrance of Oil of Lemon. There are several oils that, by absorption of oxygen from the air, will become camphorated, grow turbid, deposit a residue, generally called stearopten, and lose more or less of their flavor, instead of which they acquire the odor of turpentine. Those oils that are free from oxygen are chiefly subject to these changes, and it is therefore necessary to keep them in full bottles, well stoppered, and in a cool place. When they have deteriorated in the way indicated, they may be improved, but can never be restored to their original quality. Many means have been proposed for this purpose, but the one now generally employed in France is to shake the oil with warm water several times, letting it settle, and drawing it off by means of a syphon; it may lastly be filtered either through paper or linen.

1473. To Keep Oil of Lemon Fragrant. To every pound of oil, 1 ounce alcohol is to be added and well mixed; then 1 ounce water is put with it, which again withdraws the alcohol from the oil, and collects at the bottom of the bottle as dilute alcohol, where it should be permitted to remain until the oil has been used, with, perhaps, an occasional shake-up when the bottle has been opened. Oil of lemon treated in this manner has been kept fresh and fragrant for over a year. *Oil of orange* may be treated in the same manner with excellent effect.

1474. To Purify Essential Oils that have Deteriorated from Age. The method most commonly pursued is by redistillation, mixing them first with water, and sometimes with alkali. There are, however, other processes that have been recommended, which are believed to be equally as efficacious, and at the same time more simple. M. Curieux proposes to submit them to the action of a solution of borax with animal black. The solution of borax is mixed with the animal charcoal to form a thin consistency; the oil is then added and agitated for a quarter of an hour. At the end of that time the borax mixture is found adhering to the sides of the bottle, while the oil flows limpid. The oil of lavender, neroli, and peppermint, M. Curieux had restored or purified in this manner. Mr. Charles Bullock, of Philadelphia, has found that permanganate of potash is admirably adapted to the purpose of the restoration of resinified essential oils. A large can of oil of lemon having become unsaleable, he agitated a solution of the potash with the oil for a length of time, then decanted, mixed with fresh water, and warmed gently, till the oil floated perfectly clear on the surface. The solution of the permanganate was in the proportion of 1 ounce of the salt to 8 ounces of water. This quantity was enough for 4 pounds of the oil.

1475. To Detect the Presence of Fatty Oil and Resins in Essential Oils. The presence of fatty oil, resin, or spermaceti, may be readily detected by placing a single drop of the suspected oil on a piece of white paper, and exposing it for a short time to heat. If the oil under examination be pure, it will entirely evaporate; but if it be adulterated with one of these substances, a greasy or translucent spot will be left on the paper. These substances also remain undissolved when the oil is agitated with three or four times its volume of strong rectified spirit.

1476. To Detect the Presence of Alcohol in Essential Oils. The presence of alcohol or rectified spirit may be detected by agitation with the oil a few small fragments of dried chloride of calcium. These will remain unaltered if the oil be pure, but will dissolve in one containing alcohol, and the resulting solution will form a distinct stratum at the bottom of the vessel. The milkiness and loss of volume, when such an oil is agitated with a little water, is another test of the presence of spirit. A more delicate test of the presence of alcohol in an essential oil than the preceding, is effected by potassium. Place 12 drops of the oil on a perfectly dry watch-glass, and put a piece of potassium, the size of an ordinary pin's head, in the middle of it. If the potassium remains unchanged for 12 or 15 minutes, no alcohol is present; but if it disappears after 5 minutes, the oil contains at least 4 per cent. of alcohol; if it disappears in less than 1 minute, it proves the presence of not less than 25 per cent. of alcohol. This species of adulteration is very common. It is a very general practice of the druggists to add strong rectified spirit to their essential oils, to render them transparent, especially in cold weather. Oil of cassia and oil of cinnamon are nearly always so treated by them.

1477. To Detect the Admixture of one Essential Oil with Another. The admixture of an inferior essential oil with another more costly, is readily detected by a connoisseur or expert, by placing a drop or two on a piece of clean blotting-paper, shaking it in the air, and smelling it occasionally. The difference of the odor at the beginning and towards the end of the evaporation will show the adulteration, especially if the adulterant be oil of turpentine. This last may also be detected by remaining undissolved when the oil is agitated with about thrice its volume of strong rectified spirit. Highly rectified oil of turpentine is very largely used to adulterate the stronger scented essential oils. Foreign oil of lavender and oil of peppermint, for example, are usually compounds of 1 ounce of the genuine oil with 9 ounces of oil of turpentine. Even American and English oil of peppermint are adulterated with ⅛ part rectified spirit, besides a considerable quantity of oil of spearmint, and often turpentine.

1478. To Detect the Adulteration of a Heavy Oil with a Light One. The adulteration of a heavy oil with a light one may be detected by agitating the suspected sample with water, when, in general, the two will separate and form distinct layers.

1479. To Test the Purity of Essential Oil of Almonds. Essential oil of almonds is very generally adulterated with cheaper oils, particularly nitrobenzole (artificial oil of bitter almonds), and in nearly every case with alcohol or rectified spirit. The pure oil, when mixed with *oil of vitriol*, turns of a clear crimson-red color, without visible decomposition:—mixed with *alcoholic solution of potassa*, crystals are eliminated:—*iodine* dissolves only partially and slowly in it, without further visible results:—*chromate of potassa* does not affect it:—*nitric acid* of the specific gravity 1.42 causes no immediate reaction, but crystals of benzoic acid begin to form in 3 or 4 days; if only 7 or 8 per cent. of alcohol be present, violent effervescence speedily commences, and colored nitrous fumes are evolved. Nitric acid of specific gravity 1.5 produces the same effects in a marked degree, even when the smallest quantity only of alcohol is present. The specific gravity of the pure oil, when recent, is never less than 1.052; and when old, never greater than 1.081; that of trade averages about 1.075. Nitrobenzole has the specific gravity 1.209, and its boiling point is 415° Fahr., or fully 100° higher than that of essential oil of almonds.

1480. To Test the Purity of Oil of Bergamot. Oil of bergamot is very frequently adulterated with rectified spirit, or with the oil of lemon, orange peel, and turpentine. These may be detected in the way previously noticed. (*See No.* 1476, *&c.*) The presence of the foreign oils, particularly the last, lessens its solubility in rectified spirit. The pure oil is freely soluble in liquor of potassa, forming a clear solution. Its specific gravity is .875 to .885.

1481. To Test the Purity of Oil of Cinnamon. The common adulterants are highly rectified spirit and oil of cassia. When pure, its specific gravity is 1.035. Oil of cassia, of which the specific gravity is 1.071 to 1.073, and when old, even 1.078 to 1.090, increases it; but before trying it, it must be tested for spirit, which has a contrary effect.

1482. To Test the Purity of Oil of Lavender. Alcohol is here also the common adulterant. The finest quality—that from the flowers, has specific gravity .877 to .905. The lightest is esteemed the best. Santaline is insoluble, or very nearly so, in the pure oil, but is freely soluble in that adulterated with alcohol. The presence of oil of turpentine, and other inferior oils, may be detected by the blotting-paper test, noticed above. (*See No.* 1475.

1483. To Test the Purity of Oil of Neroli. This is the oil of orange flowers, and is commonly adulterated with alcohol, or with the oil of orange leaf (essence de petit-grain), and generally with both. The presence of the first is easily determined (*see No.* 1476); that of the second only by comparing the odor of a drop of the suspected oil, placed on a piece of paper, with a drop of pure neroli similarly treated.

1484. To Test the Purity of Otto of Roses. Cooley says: "The common adulterants are the oils of rhodium, sandal wood, and geranium, with campher, and occasionally with spermaceti, to give the spurious article the usual crystalline appearance. Pure otto has a bland, sweet taste; if it be bitter, it contains oil of rhodium or sandal wood; if it be pungent or bite the palate, it contains either oil of geranium or camphor, and most probably both; if it imparts an unctuous sensation to the palate, or if it leaves a greasy stain on paper, it contains spermaceti. A single drop of pure otto of roses exposed for some hours under a bell-glass, in the cold, to the vapor of a few grains of iodine, remains white, and continues so on subsequent exposure to the air. A sample adulterated with foreign oil, on the contrary, becomes yellow or yellowish-brown, and continues subsequently to darken, until it becomes of a deep brown color, or even perfectly black, according to the extent of the adulteration. A single drop of pure otto placed on a watch glass with one drop of concentrated sulphuric acid (oil of vitriol), and stirred with a glass rod, retains the purity of its color and odor; but a sample adulterated with other oil becomes more or less brown, and evolves peculiar odors—that from oil of geranium being strong and disagreeable; that from oil of rhodium being increased and rendered unctuous and cubeb-like; that from camphor, characteristic and combined with acidity; that from spermaceti, unctuous and clearly perceptible." Dr. R. Baur, of Constantinople, has had the opportunity of preparing a standard otto of rose on the spot, and was also in a position such as scarcely any other chemist ever was for investigating the whole subject. He says that pure otto gives, with iodine and with iodide of potassium and starch, the same reactions as when it is mixed with geranium oil, and even those with pure geranium oil are hardly different. He further says that many attempts have been made to discover some chemical reaction which would reveal the falsification of otto with geranium oil, but hitherto mostly in vain.

1485. To Test the Purity of Oil of Cloves. Oil of cloves is frequently adulterated with inferior essential oils, but when pure it exhibits the following results: When shaken with pure *liquor of ammonia*, it coagulates, and crystallizes after fusion by a gentle heat: Treated with an alcoholic *solution of potassa*, it congeals into a crystalline mass, with total loss of its odor: A solution of *chromate of potassa* converts it into brown flakes, whilst the salt loses its yellow color.

1486. To Test the Purity of Oil of Rue. This oil is nearly always adulterated. When pure, it forms a clear solution with *rectified spirit*; *Iodine* dissolves in it slowly, without apparent reaction beyond a darkening and a slight increase in viscidity: It is unaffected by a solution of *chromate of potassa*; *Nitric acid* very slowly changes it into a greenish yellow liquid balsam.

Fixed Oils and Fats.

These are compounds of carbon, hydrogen, and oxygen (hydrocarbons), obtained from the organic kingdom, and chiefly distinguished by their insipidity, unctuosity, insolubility in water, and being lighter than that fluid. Olive oil, obtained from the vegetable, and spermaceti oil, from the animal kingdom, may be taken as types of the rest. The fixed oils are chiefly found in the fruit and seeds of plants, and in thin membranous cells in various parts of the bodies of animals. Some of these oils are solid at ordinary temperatures; as palm oil, cocoanut oil, &c.; but the majority are fluid, except when considerably cooled, when they separate into two portions: the one solid, consisting mostly of stearine, and the other liquid, consisting chiefly of oleine. Nearly all the fixed oils, when freely exposed to the air, absorb oxygen, and either gradually harden, or become rancid. The former are termed drying oils, and are used by painters; the latter are used in cookery, for machinery, lamps, &c. The fixed oils, except where otherwise directed, are obtained from the bruised or ground fruit or seed, by means of powerful pressure, in screw or hydraulic presses, and are either allowed to clarify themselves by subsidence, or are filtered. Another method is by boiling the bruised seed in water, and skimming off the oil as it rises to the surface. This is the plan adopted for castor oil in the West Indies. The specific gravities of the fixed oils range between 0.865 and 0.970. (*Cooley.*)

1488. Davidson's Process of Deodorizing Putrid Whale Oil. This cheap method of purification consists in the employment of chloride of lime, the quantity depending on the degree of putrefaction of the whale oil. In general 1 pound is sufficient for 112 pounds oil; but if it is in a state of great putrefaction, there may be 1½ or 2 pounds required. With 1 pound chloride of lime about 12 times the quantity of water must be employed. The chloride is bruised in a mortar, and the water added by degrees till it forms a soft and liquid paste, and afterwards by the addition of the remainder of the water it takes the consistency of cream. This is to be mixed with the oil and often carefully stirred. After some hours 1 pound sulphuric acid, diluted with 20 or 30 times its bulk of water, is poured on the mixture, and the whole brought to a boil over a moderate fire, and stirred continually until drops of oil run off at the end of the stirring pole. It is then left for some hours for the oil to precipitate, and the acidulated water is drawn off. A common cast-iron boiler, with sheets of lead at the bottom, is the best for the purpose, or a copper or iron vessel may be used when the quantity of acid is not too great. The chloride of lime must not be bruised in a copper or iron mortar.

1489. To Restore Rancid Oil and Fat. Rancid oil and fat may be recovered by agitating them, at a gentle heat, with fresh-burnt and coarsely-powdered charcoal (which has been thoroughly freed from dust by sifting and fanning), followed by filtration through flannel; or by simple filtration through charcoal in bags of Canton flannel, according to the common method.

1490. To Restore Rancid Fat or Oil. Another method is to thoroughly wash them with hot water, frequently renewed, or to blow steam through them, until the desired effect be produced. Air freely employed for some time, instead of steam, succeeds admirably with many oils, and its use has the advantage of not introducing moisture into the article. Another method is to boil oil or fat, for 15 to 30 minutes, with a little water and calcined magnesia.

1491. To Prevent Oils and Fats from Becoming Rancid. The tendency of oils and fats to become rancid may be prevented, or greatly retarded, by artificial means. One of the simplest methods is to dissolve about 2 per cent. of gum-benzoin (in fine powder), or about one per cent. benzoic acid, in the oil or fat, by the aid of a gentle heat. This addition renders oils, pomades, ointments, &c., peculiarly soothing to an irritable or highly sensitive skin. It should be done before the addition of the scents. When the preparations are intended for exportation to hot climates, the percentage of the gum or acid should be somewhat increased. This is the plan generally adopted by the manufacturing perfumers and druggists. In the wholesale trade, carefully rendered lard, suet, &c.; simple pomades and oils, so prepared, are now common articles of stock and sale.

1492. An Excellent Preventive of Rancidity in Oils, &c. Nitric ether, or its alcoholic solution (sweet spirits of nitre), is highly recommended as a most effective preventive of rancidity. It is said that a few drops of the ether will effect this object, and will even remove the disagreeable odor of rancidity when present. Oil so treated, after being heated to remove the alcohol, when the solution has been used, is quite bright, clear, and scentless, if it were originally so. Poplar-buds, crushed and digested at a gentle heat, in oil or fat, will also remove, or greatly retard, its tendency to become rancid. Fatty bodies in a globular state may be kept a long time without becoming rancid. This peculiar state can be imparted to fatty matters by melting them at 130° Fahr. and adding a small quantity of yolk of egg, or bile, or albuminous substances, or best, a solution of alkali (composed of 5 to 10 parts for every 100 of oil), at the same temperature. The whole is then agitated for some time to bring the fatty matter into a globular condition.

1493. To Prevent Fats and Oils from Becoming Rancid. Heat the oil or melted fat for a few minutes with powdered slippery-elm bark, in the proportion of 1 drachm of the powder to one pound of fat. The bark shrinks and gradually subsides, after which the fat is poured off. It communicates an odor like that of the hickory-nut. *Butter* thus treated has been kept unchanged for a year.

1494. To Purify Vegetable Oil for Use in Lamps. To 100 pounds oil add 25 ounces alum, dissolved in 9 pounds of boiling water. After stirring it about half an hour, add 15 ounces nitric acid, still continuing to stir it. Let it stand 48 hours, when the fine oil will swim on the surface, and then draw it off. Such oil is used all over Continental Europe, and an equal quantity yields double the light of whale and fish-oil, without its offensive odor

1495. Bancroft's Process for Refining Lubricating Oils. Mr. Bancroft's process for refining common olive oil, lard oil, &c., for lubricating purposes, is to agitate them with from 3½ to 8 per cent. caustic soda lye, of 1.2 specific gravity. If, on trial of a small quantity, the lye be found to settle clear at the bottom, enough has been added. The oil is allowed to rest for 24 hours, for the soapy matter to subside; the supernatant oil is then filtered. (*See No.* 1551.) Another plan of purifying oils is to agitate them with a strong solution of common salt.

1496. Calvert's Tests for the Purity of Oils. In the use of the following tests, the result of a series of experiments by Mr. F. G. Calvert, he recommends especial care in the preparation of the reagents used for testing, not only as regards their exact strength and purity, but also in following strictly the prescribed method of using them. carefully noting the time required for their action and effects to become apparent.

1497. Calvert's Caustic Soda Test for Oils. A solution of caustic soda, specific gravity 1.340, is useful to distinguish fish from other animal and vegetable oils, owing to the distinct red color which the fish oil assumes; the presence of 1 per cent. of fish oil will be detected by the test. Add one volume of the test to 5 volumes of the oil, well mixed, and heated to the boiling point. *Hempseed oil* acquires a brown-yellow color, and becomes so thick as to entirely lose its fluidity. *Linseed oil* assumes a much brighter yellow color, and remains fluid. *India nut oil, gallipoli oil,* and *pale rape oils,* become a solid white mass in 5 minutes, while the other oils remain fluid.

1498. Calvert's Sulphuric Acid Tests for Oils. I. Sulphuric acid of specific gravity 1.475 will detect oils adulterated with *hempseed* and *linseed oils* to the amount of 10 per cent. *Fish oil* may be detected to the amount of 1 per cent. by the red color it assumes, this being noticed more particularly when the fish oil is allowed to separate by standing. To apply the test agitate 1 volume with 5 volumes of the oil, and allow the mixture to stand for fifteen minutes.

II. For the detection of *hemp, linseed, fish, gallipoli,* and *French nut oils,* 1 volume of sulphuric acid of specific gravity 1.530, agitated with 5 volumes of oil, and the mixture allowed to stand for 5 minutes. Under this test the above mentioned oils alone assume a decided coloration.

III. Sulphuric acid of specific gravity 1.635, used similar to the preceding, and the effects noted after standing 2 minutes, affords a test under which the colorations are distinct and well marked, and will detect 10 per cent. of rapeseed oil in *olive oil,* of lard oil in *poppy oil,* of French nut oil in *olive oil,* and of fish oil in *neat's foot oil.*

A stronger acid than this carbonizes the oils and destroys the coloration.

1499. Calvert's Nitric Acid Tests for Oils. The successive application of nitric acid of specific gravity 1.330, and of a solution of caustic soda of specific gravity 1.340, can be successfully applied to detect the following very frequent cases of adulteration:

I. *Gallipoli oil* with fish oils; the former assumes no distinct color with the acid, and gives with soda a mass of fibrous consistency, while fish oils are colored red, and become mucilaginous with the alkali.

II. *Castor oil* with poppy oil; the former, if adulterated, acquires a reddish tinge, and the mass with the alkali loses much of its fibrous appearance.

III. *Rapeseed oil* with French nut oil; under the nitric acid test the former, if adulterated, assumes a reddish tinge, more or less intense, which alkali increases, and renders the semi-saponified mass more fibrous.

1500. To Test the Purity of Olive Oil. Cooley says: When pure olive oil is shaken in a phial only half filled, the bead or bubbles formed very rapidly disappear, but with the adulterated oil they remain much longer before they burst. If olive oil contains ¼ part of poppy oil, part of it remains liquid at 36° Fahr., its proper freezing temperature; and if it contains ⅓ of poppy oil, it does not solidify at all, unless cooled much below the freezing point of water. Pure olive oil well agitated for some time with $\frac{1}{12}$ of its volume of nitric solution of mercury, becomes quite solid in 3 or 4 hours, without any separation of liquid oil. (The mercurial solution is made by dissolving 1 ounce mercury in 2 fluid ounces 1½ drachms nitric acid specific gravity 1.500.) According to M. Boudet, 1 grain of hyponitrous acid (hyponitric?) mixed with 3 grains of nitric acid, will cause the perfect solidification of 200 grains of pure olive oil in 75 to 78 minutes.

1501. To Test the Purity of Castor Oil. Castor oil is frequently adulterated with rape oil; but this may be detected by its not dissolving in strong alcohol, and also by its less density. Pure castor oil is soluble in an equal weight of alcohol specific gravity 0.820.

1502. To Refine Olive Oil. Olive oil intended for huiles antiques (*see No.* 1244) and other like uses, is commonly refined by violently agitating it in glass or stoneware, with about 1½ to 2 per cent. of its weight of concentrated sulphuric acid. This renders it opaque, and causes it to assume a greenish color. After about 2 weeks' repose, it deposits much coloring matter, and is then found to have acquired greater fluidity, to have become much paler, to be more emollient and glossy as a lubricator, and to burn with greater brilliancy. The clear portion is now decanted, well washed with steam or hot water, and, after sufficient repose in a close vessel, at a temperature about 60° Fahr., again decanted, and, if necessary, filtered through Canton flannel or bibulous paper. This plan is also applied to other fixed oils, and answers well for most of the recently expressed vegetable oils.

1503. To Purify and Sweeten Castor Oil. The American Journal of Pharmacy gives the following receipt for this purpose: Take 1000 parts of the oil, 25 parts purified bone-black, 10 parts calcined magnesia. Mix them carefully in a convenient vessel of glass or tinned iron, and let it stand during 3 days, with occasional agitation, and filter through paper or felt. (*See No.* 1504).

1504. To Bleach the Vegetable Oils. According to Cooley, almond, ben, castor, colza, linseed, nut, olive, poppy, rape, teel, and other like vegetable oils, are readily bleached by exposure, in glass bottles, to the light. For this purpose, 2-quart to 4-quart pale green glass or blue glass bottles filled with the oil, and covered with white gallipots inverted over them, are suitably placed, a small distance apart, on the roofs of houses or sheds, or in any other suitable position, fully exposed to the sun during the greater portion of the day, or at all events to the south-east and south. 14 to 21 days' exposure to the sun, in clear weather, during summer, is usually sufficient to decolor castor oil and almond oil; but 4, 5, or even 6 weeks, is commonly required to render linseed oil very pale. This is the common plan adopted by the wholesale druggists to whiten their castor oil, by some of the perfumers for their almond oil and olive oil, and by the oilmen for their pale linseed oil for artists. A better plan, however, when this method is adopted, is to cork the bottles loosely air-tight, but not firmly down, when the sun has been on them two or three hours, and whilst they are still heated with it. In this way the oil suffers less from the exposure than by the loose gallipot system in common use. Almond, olive, and the other sweet oils, thus treated, are apt to lose some of their blandness, and to acquire a slight sulphurous smell, and smoky flavor, whilst castor oil loses its original blandness, and assumes the strong, nauseous flavor characteristic of the white castor oil of the stores. These qualities may be removed by agitation with a little fresh animal charcoal, dry freshly prepared alumina, or calcined magnesia, and subsequent filtration; or, what is even better, though more troublesome, by well washing the oil with hot water, and subsequent repose out of contact with the air, and subsequent decantation. (*See No.* 1503.)

1505. To Bleach Vegetable Oils. Another method pursued for bleaching oils is as follows: The oil is placed in a porcelain, stoneware, or well-tinned vessel, along with some dry filtering powder, 1 to 2 pounds to each gallon of the oil, or some dry and recently prepared hydrated alumina (¼ to ½ pound per gallon of oil; but much less is often sufficient if the article be of proper quality); and the heat of steam or boiling water being applied, is vigorously stirred, with a clean wooden or stoneware spatula, for about an hour. It is then thrown into a Canton flannel oil-bag, and filtered, in the usual manner, observing to return the runnings until they become quite white and clear. This is the way perfumers and wholesale druggists usually prepare their *white almond oil, white olive oil,* and *white oil of ben.* Formerly fresh burnt animal charcoal was chiefly used for the purpose, and is still so employed by some houses; but the other substances answer better and are more convenient. (*Cooley.*)

1506. To Bleach Vegetable Oils. The oils referred to in No. 1504, as well as all other oils and fats, may be rendered perfectly colorless by agitating them with a little chromic acid; or, what is cheaper and more convenient, with a mixed solution of bichromate of potassa and sufficient sulphuric acid to seize on the alkali of the bichromate and to liberate its chromic acid. 1 to 2 drachms of the bichromate, mixed with 3 times its weight of oil of vitriol (previously diluted with about twice its volume of water, and allowed to cool), is ordinarily sufficient, when skillfully used, to perfectly bleach 2 or 3 pints of oil. It should be added gradually to the oil, with continued violent agitation, and this should be kept up for some considerable time after the last portion is added. The mixture must be made in a vessel of glass, porcelain, stoneware, or wood, and nothing metallic must touch it. In some cases a few drops of strong nitric acid (diluted with about twice its bulk of water), if added towards the end of the agitation, will facilitate the process; or, with colza, linseed, nut, and rape oil, instead of the diluted nitric acid, a few drops of hydrochloric acid without dilution. After the final agitation, the oil must be allowed to repose at a temperature of about 60° Fahr. When it has settled, the clear portion should be decanted, thoroughly washed with hot water, again allowed to repose for some time, and then finally decanted for use. If necessary, it may lastly be filtered. (*Cooley.*)

1507. Berlandt's Method of Bleaching Fixed Oils. Shake strongly for some minutes, 300 parts of the oil with 40 parts water containing 1 part permanganate of potassa; allow the mixture to stand in a warm place for some hours, and then filter. This renders the oil colorless.

1508. Dieterich's Method of Bleaching Fixed Oils. Dissolve 2¼ pounds (avoirdupois) permanganate of potassa in 31½ quarts water, in a wooden tub having a faucet in its bottom. Stir into the mixture 52½ quarts of the oil to be bleached, and keep all well stirred for 2 days. Then add 21 quarts boiling water and 11 pounds commercial hydrochloric acid, and keep the whole stirred for 2 days longer. Draw off the acid water, and wash the oil repeatedly with boiling water until all acid is removed from it.

1509. Engelhardt's Method of Bleaching Palm Oil. Heat 1000 parts by weight palm oil in an iron vessel to about 143° Fahr., and let it stand all night, sustaining the temperature. Next day pour it off into a clean vessel and let it cool down to about 100°. Meanwhile, dissolve 15 parts bichromate of potash in 45 parts boiling water; when the solution has cooled a little, pour into it 60 parts hydrochloric acid. Add this mixture to the palm oil, stirring quickly, and in about 5 minutes it will assume a sombre green color; by continued stirring the oil gradually clarifies and becomes quite limpid. It should become quite white after washing it with warm water; but if not entirely colorless, the operation must be repeated, using ½ part bichromate of potash, and 1 part hydrochloric acid. This is a quick method, and Engelhardt claims that it produces better results than the means usually employed. (*See No.* 537.)

1510. To Bleach Cotton Seed Oil. Use 1 gallon English caustic soda, in a solution of about 40° Baumé, to about 20 gallons crude oil. The oil, previous to being mixed with the solution, must be heated to about 90° Fahr. Stir constantly while adding the cold solution. If the oil is not now sufficiently light, add more of the solution to bring it to a light yellow or straw color.

1511. Keyer's Process for Purifying Oils. The process of M. Keyer, which is applicable to all oils, has given excellent results in a manufactory of rape seed oil. Into 1000 parts by weight of oil, put a mixture of 6 parts solution of ammonia and 6 parts water, and agitate the barrel well until the alkali is perfectly mixed, which may be done in 15 minutes. The barrel is then sealed hermetically, and, after 3 days' repose, the oil is decanted and filtered. The residue is used for the manufacture of soap. Oil thus worked contains no trace of acid, and the mucilaginous impurities are destroyed or precipitated.

1512. Liebig's Method of Obtaining Non-poisonous Oil of Almonds. Agitate the crude distilled oil with binoxide of mercury in slight excess; and, after a few days' contact, rectify the oil from a little fresh binoxide of mercury. The product is quite pure, if properly managed, as the hydrocyanic acid (the poisonous principle) of the oil, unites with the binoxide to form a bicyanide of mercury.

1513. Neat's-foot or Trotter Oil. Obtained by boiling neat's-foot, tripe, etc., in water. It is a coarse animal oil, very emollient, and much used to soften leather.

1514. To Refine Neat's-foot Oil. Put

a quart of the oil with ½ pound bright lead shavings, and ½ pound quicklime pounded, into a glass bottle, let it stand in the sun and light for 2 or 3 weeks, then put the oil and lime into a saucepan with ½ pound washing soda, boil gently 15 minutes, then set in the coldest place possible till the next day, when it will be found congealed; place it into a filter of white blotting paper, place a clean glass bottle under the filter, and you will get the finest oil, suitable for the most delicate machinery. Any one requiring a little nice oil would do well to try this in preference to buying it ready done. It must be kept perfectly cold while filtering, or the soda will go through.

1515. Hirzel's Method of Preserving Animal Fats. Mix 14 pounds of recently melted fat with 5 drachms salt and 15 grains alum in fine powder; heat until a scum is formed on the surface; remove the scum, and when the clear fat is cool, wash and knead it in water, frequently changing the water, so as to remove all the salt; then evaporate the water at a heat insufficient to injure the fat.

1516. To Preserve Animal Fats for a Long Time. The following mode of benzoating all kinds of animal fats will be found the most effectual for preserving them for a long time. Make a saturated solution of gum benzoin in alcohol by simple heat, allow the liquid to settle clear, then strain and mix with equal parts of fresh castor oil. Of this mixture add 4 ounces to each gallon of fat or ointment while warm. The proportion of the solution of benzoin may be increased for pomades, as it forms, by its aromatic odor, an excellent basis for perfumes. The benzoatic fat should not be kept in tin, but in well-covered jars. Steam-rendered lard, or that treated with salt and alum, should be carefully re-melted in a water-bath, to allow all the water to settle so as to pour off the pure fat. In preparing ointment and pomades it is important that the wax should be first melted, and the oil or fat warmed before adding to the wax. This precaution, which will save much time and trouble, is often neglected by young beginners. (*See Nos.* 1253 *and* 1254.)

1517. Boillot's Process for Purifying Fats. Melt 2¼ pounds avoirdupois of the fat with 2 quarts lime-water; stir actively over the fire for 2 or 3 hours, and cool. Then press in flannel and allow it to stand a day or two to harden. By melting it with acidulated water to remove the excess of lime, a hard fat results, suitable for making candles.

1518. Hog's Lard. This is obtained, like the rest of the animal fats, from the raw lard, by chopping it fine, or rather rolling it out, to break the cells in which the fat is lodged, and then melting the fat in a water-bath, or other gentle heat, and straining it while warm. Some boil them in water; but the fats thus obtained are apt to grow rank much sooner than when melted by themselves. (*See No.* 525.)

1519. To Try out Lard. This should be done in the open air. Set a large kettle over a fire, in some sheltered place, on a still day. It will cook much quicker in large quantities. Put into the kettle while the lard is cold, a little saleratus, say 1 table-spoonful to every 20 pounds; stir almost constantly when nearly done till the scraps are brown and crisp, or until the steam ceases to rise; then there is no danger of its moulding; strain out into pans, and the first will be ready to empty into crocks when the last is strained.

1520. To Detect Water in Lard. The presence of water is very easily detected by merely melting the lard, when the water collects at the bottom of the vessel as a distinct layer. The weight and volume of lard can be greatly increased by the incorporation of water with it; and purchasers of a pound of lard will frequently find that they have paid the price of the lard for as much as 4 ounces of water. Lard is also adulterated with from 2 to 5 per cent. of *milk of lime* (slacked lime mixed to a milky consistence with water); this gives the lard a beautifully white appearance, and also allows of 25 per cent. of water being stirred into it while cooling.

1521. Benzoated Lard. Take benzoin in coarse powder, 1 ounce; fresh lard, 1 pound. Heat together for 2 or 3 hours in a water-bath, and then strain.

1522. To Bleach Lard. Lard may be bleached by applying a mixture of bichromate of potassa and muriatic acid, in minute proportions, to the fat. (*See Nos.* 1509 *and* 1523, *also No.* 537.)

1523. To Bleach and Harden Tallow. In a copper boiler, put ½ gallon water, and 100 pounds rendered tallow; melt over a slow fire, and add, while stirring, 1 pound of oil of vitriol, previously diluted with 12 of water; afterwards, ½ pound bichromate of potassa, in powder; and lastly, 13 pints water, after which the fire is suffered to go down, when the tallow will collect on the surface of the dark green liquid, from which it is separated. It is then of a fine white, slightly greenish color, and possesses a considerable degree of hardness. (*See No.* 1509.)

1524. Factitious or Imitation Spermaceti. White spermaceti, 10 parts; sonorous cake stearine, 20 parts; potato starch, 5 parts; mucilage, 1 part. Melt the first three and unite well, then let the mass cool to the consistence of dough; turn it out on an oiled marble or lead slab, and roll it into a cake; next sprinkle a little mucilage on it, double it, and roll again; repeat the process as often as required; lastly allow it to cool. If it has been properly managed, it will flake when broken up, and resemble spermaceti.

1525. Extraction of Fat from Bones. A process has been adopted abroad for extracting oil and fat from bones and other animal refuse, by digesting it in a closed and heated vessel with benzole or similar hydrocarbon. After a few hours the liquid is drawn off, the hydrocarbon separated by distillation, and the oil is left ready for use. The bones may then be used for the manufacture of gelatine. This is very similar to a method lately proposed of obtaining oil from oleaginous seeds, but in this latter case, as would probably be preferable in the former, bisulphide of carbon is the menstruum employed.

Petroleum, or Crude Coal Oil.

The name of petroleum is now applied to all the native liquid substances which have a bituminous character. It consists, therefore, of an inflammable and more or less volatile oily substance, ranging in color and appearance from a yellowish white, transparent fluid, to a brown or almost black, opaque viscid mass. The former used to be called naphtha, but this name is now given to any oil of this description, whether native, or distilled from a darker grade of petroleum. The latter is the form in which the bulk of the petroleum is found in America; and this, when exposed to the air, gradually passes into *asphaltum*, or solid bitumen.

1527. To Purify Petroleum. Tank-shaped stills of a capacity of 500 to 2500 barrels are filled with crude oil, and heat applied by furnaces beneath them, causing vapors to arise, which are carried forward through pipes immersed in water, and condensed into a liquid, which runs out at the end of the pipe. The first product is gasoline, a very light hydrocarbon, marking as high as 83° and as low as 75° of Baumé's coal oil hydrometer. The heat is then somewhat increased, and the next product obtained is called *naphtha, benzine* (*not benzole*), which marks from 75° to 63° Baumé; and, when combined, will average about 67°. The heat being allowed to increase further, produces distillate, or crude burning oil. This passes over until about 8 or 10 per cent. of the original quantity contained in the still remains, which is called residuum or tar, and may be redistilled for the purpose of obtaining paraffine and lubricating oil. *Paraffine* is a fatty material, resembling sperm in appearance. The distillate or crude burning oil is converted into ordinary kerosene by a process of purification. For this purpose it is placed in a tank, where it is violently agitated by forcing air through it, and while thus agitated, 1½ to 2 per cent. sulphuric acid is added, after which the agitation is continued 15 to 30 minutes. The oil is then allowed to settle, when the acid and impurities are drawn from the bottom. The oil is then washed, first with water and then with caustic soda, by which means the remaining impurities are removed, and any acid remaining in the oil is neutralized. It is then taken to shallow bleaching tanks, where it is exposed to light and air, and allowed to settle; it is next heated by means of a coil of steam pipe running through it, to expel all gaseous vapors which will ignite at a temperature below 110° Fahr. The oil is now called a *fire test* oil, and is ready to be barreled and sent to market.

1528. To Clarify Coal Oil. Place in a close vessel 100 pounds crude coal oil, 25 quarts water, 1 pound chloride of lime, 1 pound soda, and ½ pound oxide of manganese. The mixture is violently agitated, and allowed to rest for 24 hours, when the clear oil is decanted and distilled. The 100 pounds coal oil are to be mixed with 25 pounds resin oil; this is one of the principal points in the manipulation; it removes the gummy parts from the oil, and renders them inodorous. The distillation spoken of may terminate the process, or the oils may be distilled before they are defecated and precipitated.

1529. To Decolorize Kerosene Oil. Kerosene oil is decolorized by stirring it up with 1 or 2 per cent. of oil of vitriol, which will carbonize the coloring matter, then with some milk of lime or some other caustic alkali, settling, and redistilling. The latter appears to be indispensable.

1530. Why Kerosene or Coal Oils Explode. No oil is explosive in and of itself; it is only when the vapor arising therefrom becomes mixed in the proper proportions with air, that it will explode. There should be no inflammable vapor from any oil used for burning in lamps at ordinary temperature. A volatile oil is unfit for the purpose of illumination.

1531. To Test Kerosene or Coal Oil. Burning oil is often adulterated with heavy oil, or with benzine. The adulteration with the former is shown by dimness of the flame after having burned for some time, accompanied by a charring of the wick. The latter may be readily detected by means of a thermometer, a little warm water, and a table-spoonful of the oil. Fill the cup with warm water, the temperature of which is to be brought to 110° Fahr. Pour the oil on the water; apply flame to the floating oil by match or otherwise. If the oil is unsafe it will take fire, and its use in the lamp is dangerous, for it is liable to explode. But if the oil is safe and good it will not take fire. All persons who sell kerosene that will not stand the fire test at 110° are liable to prosecution.

1532. To Extinguish the Flame of Petroleum or Benzine. Water, unless in overwhelming quantity, will not extinguish the flame of petroleum or benzine. It may, however, be speedily smothered by a woolen cloth or carpet, or a wet muslin or linen cloth, or earth or sand being thrown over it. These act by excluding the air, without which combustion cannot be maintained.

1533. To Deodorize Benzine. Shake repeatedly with plumbate of soda (oxide of lead dissolved in caustic soda), and rectify. The following plan is said to be better: Shake repeatedly with fresh portions of metallic quicksilver; let stand for 2 days, and rectify.

1534. To Manage Kerosene Lamps. These are so much used that a few hints on their management will no doubt be acceptable. There are very few common illuminating substances that produce a light as brilliant and steady as kerosene oil, but its full brilliancy is rarely attained, through want of attention to certain requisite points in its management. By following the directions here given, the greatest amount of light will be obtained, combined with economy in the consumption of the oil. The wick, oil, lamp, and all its appurtenances, must be perfectly clean. The chimney must be not only clean, but clear and bright. The wick must be trimmed exactly square, across the wick-tube, and not over the curved top of the cupola used to spread the flame; after trimming, raise the wick, and cut off the extreme corners or points. A wick cannot be trimmed well with dull scissors; the sharper the scissors, the better the shape of the flame. These hints, simple as they appear, are greatly disregarded, and the consequence is a flame dull, yellow, and apt to smoke. The burners made with an immovable cupola, and straight, cylindrical chimneys, require especial care in trimming; the wick has to be raised above the cupola, and has therefore no support when being trimmed. A kerosene lamp, with the wick turned down, so as to make a small flame, should not be placed in a sleeping room at night. A wick made of felt is greatly superior in every way to the common cotton wicks.

1535. To Keep Kerosene Lamps from Getting Greasy. The upper part of a kerosene oil lamp, after standing for a short time, frequently gets oily, from the condensation of the vapor of the oil. This will be greatly, if not entirely prevented, by taking a piece of felt and cutting a hole in it so as to fit exactly around the socket into which the burner is screwed; trim the felt off so as to leave a rim about ½ inch wide, and place this felt ring on the socket.

1536. To Cement the Socket on a Kerosene Lamp. The socket of a kerosene lamp, into which the burner is screwed, frequently becomes loose or comes off. To fasten this, take the socket off, pick out the old cement, and wash it with hot soap and water, with a little soda, to remove all trace of grease. Empty the lamp, and wash it in the same manner, especially the lip or neck which fits into the socket. Next take a cork which fits (not too tight) into the socket; grease it slightly, and screw it into the socket (the same way the burner is screwed in), until the end of the cork is nearly level with the bottom of the socket; this will leave a circular trench to receive the cement. Take the best plaster of Paris, mix it quickly as thick as it will flow, fill the trench in the socket, reverse the lamp, and press the lip of the glass firmly into the socket until the edge of the socket fits closely to the glass. This operation must be done quickly, before the plaster has had time to set. Let the whole remain about 12 hours in a warm place before using. Then unscrew the cork and scrape off any adhering plaster. (*See No.* 2260.)

1537. To Clean Vessels Used to Contain Kerosene. Wash the vessel with thin milk of lime, which forms an emulsion with the petroleum, and removes every trace of it, and by washing a second time with milk of lime and a very small quantity of chloride of lime, and allowing the liquid to remain in it about an hour, and then using it with cold water, even the smell may be so completely removed as to render the vessel thus cleansed fit for keeping beer in. At the same time the external surface of the vessel is to be washed with a rag dipped in the same substance. If the milk of lime be used warm, instead of cold, the operation is rendered much shorter. If particles of thickened petroleum adhere to the glass after the first washing, these can be removed by washing with fine sand, or by other mechanical means.

1538. To Clean Kerosene Lamps. Wash the lamp inside and out thoroughly with hot soap and water, and a little washing soda. When clean, rinse repeatedly so as to leave no trace of soap; let it drain till dry.

Chemical Manipulations.

Some of the operations employed in the preparation and use of chemicals have already been given at the commencement of this book (*see No.* 1); but, as the work progressed, it was deemed advisable, for the sake of greater precision, to add further directions for special manipulations, and descriptions of indispensable apparatus.

3831. Separating Funnels. These are glass funnels furnished with a stop-cock, and are used for separating mixed fluids of different densities. The mixed liquid is poured in-

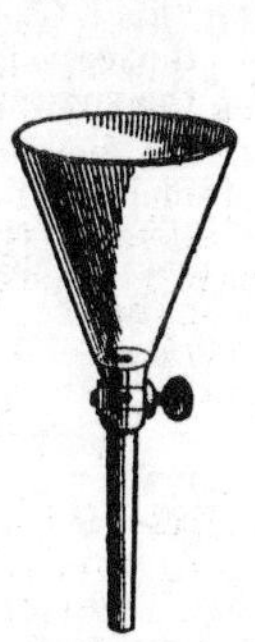

Fig. 1.

Fig. 2.

to the funnel, and, after sufficient time has been allowed for the heavier liquid to settle, it can be drawn off by opening the stop-cock, closing it immediately after the heavy liquid has passed. Fig. 1 represents a separating funnel, such as is used for ordinary purposes; but for separating a mixed liquid containing ether or other volatile fluid, a funnel, closed with a stopper similar in construction to Fig. 2, is employed to prevent evaporation while the heavier liquid is settling. For very small quantities a pipette (*see No.* 3832) is the best instrument.

3832. Pipettes. These are glass instruments used for measuring liquids in drops, and so constructed that the flow of the liquid from them is under the complete control of the operator. They may be made in any form which may be suggested to adapt them to special purposes; but pipettes for general use are usually constructed as follows: Fig. 1 is an ordinary pipette, and consists of a small cylinder of glass with an upper and lower tube, the lower end terminating in a fine orifice for the discharge of the fluid, and the upper end adapted for the finger or thumb, by which the outward flow can be instantly arrested. This is filled by the suction of the mouth. Fig. 2 is made on the same principle, having a fine orifice (*b*), and a thumb-hole (*a*), but fitted with a mouth and stopper on the upper side, for convenience of filling, or inserting a measured quantity of liquid. The lower side being flat, to allow of the instrument being laid down without risk of waste of contents.

Fig. 1.

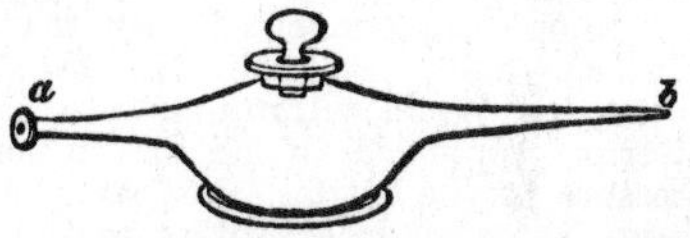

Fig. 2.

A pipette affords also a ready means of separating two liquids, too small in quantity to allow of separation by decantation or other methods usually employed. To this end, the upper or lower stratum of the mixed liquids (oil and water, for instance), may be drawn by the mouth into the pipette; or the whole may be sucked into the pipette, and the lower stratum allowed to flow out.

Graduated pipettes of various forms, especially useful in acidimetry, &c., will be found described in No. 82. These instruments are also useful, and in many cases indispensable, in conducting delicate tests.

3833. Goniometer. An instrument used for measuring the angles of crystals. The only accurate and simple instrument of this kind is the reflective goniometer invented by Dr. Wollaston.

3834. To Filter Vegetable Juices. These should be allowed to deposit their feculous matter before filtration. The supernatant liquid will often be found quite clear; when this is not the case, filtration will be necessary through coarse filtering paper. (*See No.* 17.) Some vegetable juices can be made clear simply by heating them to 180° to 200° Fahr., by which their albumen becomes coagulated.

Others admit of clarification in the same manner as syrups. (*See No.* 1357.) Many of these, again, such as hemlock, henbane, aconite, &c., are greatly injured by heat, and must be filtered or decanted after repose.

3835. To Filter Vegetable Infusions. In many instances vegetable infusions and decoctions may be clarified by defecation and decantation of the clear liquid. A convenient method of straining, when that is necessary, is by securing the corners of a square piece of flannel to a frame, which can be laid over the mouth of a pan; or by laying the flannel across the mouth of a coarse hair-sieve. Concentrated infusions and decoctions, being usually weak tinctures, may be filtered as tinctures. (*See No.* 17.) Viscid vegetable solutions may be clarified (*see No.* 1357); or may be made to filter rapidly by the addition of acetic, sulphuric, or other strong acid.

3836. To Filter Corrosive Liquids. Strong acids, &c., are filtered through powdered glass or siliceous sand, supported on pebbles in the throat of a glass funnel, or through asbestos placed in the same manner.

3837. To Filter Precipitates. When filtration is employed to separate precipitated matter from the solution in which it is suspended, the filtering medium should be such that the powder may be easily reclaimed from it with as little loss as possible. Linen or smooth bibulous paper are the best for this purpose. A camel-hair pencil should be used, if needed, in preference to a knife, to remove adhering powder from a filter, and the precipitate should be first washed down the sides of the filter by a small stream of water, so as to collect the most of it to one spot at the bottom.

The first runnings in filtration should always be returned to the filter.

3838. Bunsen's Method of Rapid Filtration. A great deal of time is frequently lost in washing precipitates, by having to wait for the liquid to pass through a filter. Bunsen's improvement consists in fixing the filtering funnel air-tight, by means of a perforated cork in the neck of a bottle which has an opening connected with the receiver of an air-pump. By exhausting the air in the bottle, the liquid will run faster through the filter in proportion to the diminution of the pressure in the bottle. Comparative experiments, some made according to the old, and others according to the new method, showing that the filtration, washing, and drying of a precipitate, which took 7 hours by the old plan, could be performed, by filtration into an exhausted bottle, in 13 minutes.

3839. Filtering Powders. In many cases a liquid will not readily become transparent by simply passing through the filter; hence has arisen the use of filtering powders, substances which rapidly choke up the pores of the media in a sufficient degree to make the fluid pass clear. These powders should not be in too fine a state of division, nor used in large quantities, as they then wholly choke up the filter, and absorb a large quantity of the liquid. For some liquids these substances are employed for the purpose of decoloring or whitening them. In such cases, it is preferable first to pass the fluid through a layer of the substance in coarse powder, from which it will run but slightly contaminated into the filters; or, if the substance be mixed with the whole body of the liquid, to pass it through some coarser medium, to remove the cruder portion, before allowing it to run into the filter. Fuller's earth, pipe clay, or potter's clay, washed, dried without heat, and reduced to coarse powder, are used to filter and bleach oils.

Fuller's earth or clay, 1 part, and 2 parts fine silicious sand, first separately washed and drained, then mixed together and dried, constitutes a filtering powder well adapted for glutinous oils.

Granulated animal charcoal, sifted and fanned free from dust, is used to filter and bleach syrups and vegetable solutions.

Carbonate of magnesia and powdered glass, or pumice stone, are used for filtering weak alcoholic solutions of essential oils, and in the preparation of perfumed waters. (*See Nos.* 976, 1029, 1080, *and* 1081.)

3840. Self-Feeding Filter. It is usually a matter of more or less importance in filtration, that the filter should be kept full. To effect this requires unremitting attention, which, when the filtration occupies a considerable time, is at least tedious. By the use of a simple apparatus, this is avoided, and filtration will continue, without any personal attention, until the operation is complete. A bottle or jar, of sufficient capacity to contain the liquid to be filtered, is placed in a convenient position, above the level of the filter (*see illustration*); through the cork, which must fit air-tight, are inserted two bent tubes; one end of the tube *b* must reach nearly to the bottom of the jar, the other end descending deep into the filter; the tube *a* terminates at one end just below the cork of the jar, the outer end being adjusted in the filter at the height which it is desired that the liquid shall be kept at in the filter. The apparatus is set in working order by sucking the liquid into the tube *b*, so as to fill it. The liquid will continue to flow until its surface in the filter rises sufficiently to reach to and close the end of the tube *a*, cutting off the ingress of air into the bottle, and thus stopping the further flow, until, by the falling of the filtrate into the vessel placed to receive it, the liquid in

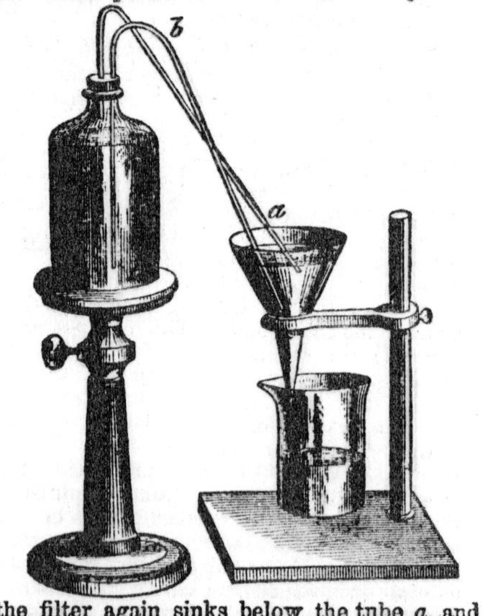

the filter again sinks below the tube *a*, and allows the flow to be resumed until again stopped as before. (*See No.* 17, *Fig.* 6.)

3841. Chemical Washing. When precipitation takes place, the deposit requires to undergo edulcoration, or cleansing from the liquid from which it was precipitated. With heavy and bulky precipitates, this is done by repeated washing, and, after the deposit has again settled, decantation of the supernatant liquid (*see No.* 3847); but when the powder is light, and separates with less facility from the liquid, the washing is better performed by a continuous stream of water passing through a filter on which the precipitate has been previously collected. The apparatus employed for a self-feeding filter (*see No.* 3840) is admirably adapted for this purpose. *Lixiviation*, or the separation of soluble matter from an insoluble powder, can be performed in the same way. (*See Nos.* 14, 23, *and* 32.)

3842. Chemical Drying. In order to deprive chemical substances of water or moisture, the simplest means is evaporation. This may be performed either by merely exposure in open shallow vessels to the natural action of a dry atmosphere, called *spontaneous evaporation*; or by the application of heat, either directly or by a water-bath, &c. (*see No.* 12); this is not always advisable or necessary, as some substances undergo change by heat, and must be dried by other means. By enclosing the substance to be dried in a box or drying-chamber in which is placed an open vessel containing strong sulphuric acid or chloride of calcium, the strong affinity for water that these substances possess, keeps the air perfectly dry, and absorbs the moisture from it as fast as the water evaporates from the material which is being dried. The water of crystalline bodies is usually driven out by exposing the crystals in a capsule or evaporating dish to heat, only just sufficient being applied to effect the purpose. Some crystals part with their water of crystallization spontaneously by exposure to the air, crumbling into powder; such crystals are called *efflorescent*, to distinguish them from those *deliquescent* crystalline bodies which spontaneously liquefy or dissolve in their own water of crystallization. Others will yield their water in an artificially dried atmosphere, as above stated; while many have sufficient affinity for water to retain it until driven off by heat, more or less intense. Crystalline substances which have been deprived of the water of crystallization, that is, have undergone desiccation, are said to be *dry*.

3843. Decarbonization. This operation is performed on cast iron, to convert it into steel or soft iron. The articles to be decarbonized are packed in finely-powdered hematite, or native oxide of iron, to which iron filings are often added, and exposed for some time to a strong red heat, by which the excess of carbon is abstracted or burnt out. The process somewhat resembles annealing or cementation.

3844. Decoloration. The blanching or loss of the natural color of any substance. Syrups, and many animal, vegetable, and saline solutions, are decolored or whitened by agitation with animal charcoal, and subsequent subsidence or filtration. Many fluids rapidly lose their natural color by exposure to light, especially the direct rays of the sun. In this way, castor, nut, poppy, and several other oils are whitened. Fish oils are partially deodorized and decolored by filtration through animal charcoal. (*See No.* 3839.) By the joint action of light, air, and moisture, cottons and linens are commonly bleached. The peculiar way in which light produces this effect has never been satisfactorily explained. The decoloration of textile fabrics and solid bodies, generally, is called bleaching.

3845. Defecation. In chemistry, the separation of a liquid from its lees, dregs, or impurities. This is usually performed by subsidence and decantation, and is commonly applied to the purification of saline solutions, on the large scale, in preference to filtration, than which it is both more expeditious and inexpensive.

3846. Neutralization. The admixture of an alkali or base with an acid in such proportions that neither shall predominate. A neutral compound neither turns turmeric paper brown, nor litmus paper red. The term *saturation* is also applied to complete neutralization (*see No.* 27); but saturation has two distinct meanings; chemically, it denotes that a given alkali has been neutralized completely by an acid, or *vice versa*. Pharmaceutically, it implies that a given solvent is charged to its utmost capacity with an active ingredient; this point is, however, so difficult to determine, that the term is scarcely ever applied accurately.

3847. Edulcoration. The affusion of water on any substance for the purpose of removing the portion soluble in that fluid. Edulcoration is usually performed by agitating or triturating the article with water, and removing the latter after subsidence, by decantation or filtration. It is the method commonly adopted to purify precipitates and other powders which are insoluble in water.

3848. Rectification. A second distillation of a fluid, for the purpose of rendering it purer. In rectifying alcohol containing water, the distillation is conducted at a temperature high enough to evaporate the alcohol and cause it to distill over into the receiver, but not high enough to boil the water, the greater part of which, therefore, remains behind in the body of the still. It is difficult to obtain an anhydrous product without employing some agent having a strong affinity for water.

3849. Calcination. The separation or expulsion, by heat, of volatile from fixed matter. By this means crystalline salts are obtained in a dry or anhydrous form, by depriving them of their water of crystallization; in this particular, the process is similar to *desiccation.* (*See No.* 12.) Calcination is also employed for the ignition of silica, &c., in order to render it more easily reducible to fragments or powder.

The operation of calcining is conducted on the small scale in platinum spoons or crucibles, and heat applied by the flame of a spirit lamp, or other appropriate means. When large quantities of matter are calcined, metal or earthenware crucibles and the heat of a furnace are employed. Charcoal is thus obtained from wood, bone-black from bones, &c.

3850. Ignition. The heating of a substance to redness. It is especially resorted to for the calcination of a substance at a high degree of heat. (*See No.* 3849.)

3851. To Bend Glass Tubes. Small glass tubes may be bent over the flame of a spirit lamp; for larger tubes, the heat of a blow-pipe flame is necessary. The tube should be heated to a dull red about an inch either way beyond the point of curvature, by revolving it in the flame; as soon as the glass begins to yield, bend the tube very gradually until curved as desired. Stopping one end of the tube, and blowing into the other while bending it, will prevent wrinkling or collapsing at the point of curvature. It requires some tact to bend a tube with an even curve and without collapsing its sides; and it is recommended by an experienced chemist to use a Bunsen burner, having the extremity flattened out so as to give a short and thin, but broad flame, something like the flame of an ordinary gas burner. The tube is placed in this flame and turned around until a good heat is given to the tube; it is then withdrawn from the flame and bent, when it does so with a perfect curve and no collapse on the sides of the tube. Of course this is only intended for the smaller tubes, but a tube of one-third of an inch and more can be thus bent very readily.

3852. To Find the Dry Weight of a Pulp or Moist Precipitate. Pulps or precipitates, such as the metallic colors, chrome yellow, white lead, &c., are of different consistence at the top from what they are at or near the bottom of the vessel in which they are contained; and the actual weight of the precipitate in the dry state can therefore not be arrived at by merely taking a sample from top or bottom, but, in most cases, only guessed at. When, however, the specific gravity of such a precipitate in its dry state is known, as well as that of the surrounding liquid, the operation of obtaining the accurate dry weight of the same while in pulp can be reduced to the simple manipulation of weighing it in a vessel. Find the weight of a vessel full of the pulp; then weigh the same vessel full of the same liquid that the pulp is moistened with, and note down the difference between the weights. Next divide this difference of weight by the difference between the specific gravities of the pulp and the liquid; lastly add this quotient to the difference of weight already noted down, and the sum will be the dry weight of the pulp.

Acids.

An acid in chemistry is any electro-negative compound, capable of combining in definite proportions with bases to form salts. Most of the liquid acids possess a sour taste, and redden litmus paper. The acids have been variously classed by different writers, as into *organic* and *inorganic; metallic* and *non-metallic; oxygen acids, hydrogen acids, and acids destitute of either of these elements;* the names being applied according to the kingdom of nature, or class of bodies to which the radical belonged, or after the element which was presumed to be the acidifying principle. Acids are in various forms; some are gaseous, as carbonic acid; some are liquid, as nitric and acetic acid; others are solid, as citric and oxalic acid; others again under peculiar conditions assume more than one of these forms. Acids, which are soluble or liquid, are corrosive, and more or less poisonous when concentrated. They change vegetable blues to red and neutralize the effects of alkalies on vegetable blues and yellows. Most of the acids are soluble in water in all proportions; they neutralize the alkalies, effervesce with the carbonates, and combine with the bases generally, forming compounds called *salts.* The methods for estimating the strength or neutralizing power of acids, as well as the strength of their solutions, will be found under ACIDIMETRY, No. 78. The names of the acids end either in *-ic* or *-ous;* the former being given to that containing the larger portion of the electro-negative element, or oxygen, and the latter to that containing the smaller quantity. As sulphur*ic* acid, an acid of sulphur, containing 3 atoms of oxygen; sulphur*ous* acid, another sulphur acid, containing only 2 atoms of oxygen. When a base forms more than 2 acid compounds with oxygen, the Greek preposition *hypo* is added to that containing the smaller portion, as hyposulphuric and hyposulphurous acids. The prepositions *per, hyper,* and the syllable *oxy* are also prefixed to the names of acids when it is intended to denote an increase of oxygen, as hypernitrous acid, perchloric acid, oxymuriatic acid, &c. The prefix *hydro* to the name of an acid denotes that the acid combination is with hydrogen, and not with oxygen. All the strong liquid acids should be kept in glass bottles, furnished with perfectly tight ground-glass stoppers; glass vessels should be used in measuring them, and they should be dispensed in stoppered vials. Fluoric acid must be kept in a bottle made of lead, silver, platinum, or pure gutta-percha, as it acts readily on glass. In the combination of acids with bases to form salts, distinctive terminations are employed to denote the kind of acid present. The name of a salt of an acid ending in *-ic,* terminates in *-ate;* thus, sulphate of soda, formed from sulphuric acid and soda. The name of a salt of an acid ending in *-ous,* terminates in *-ite;* as sulphite of lime, formed from sulphur*ous* acid and lime. The names of compounds formed by the union of non-metallic elements, and certain other bodies, with the metals or with each other, terminate in *-ide* or *-uret;* thus, sulphide or sulphuret of silver, formed of silver and sulphur. (*Cooley.*) In accordance with the scope of this work it has been found advisable to omit a number of acids, both simple and compound, of limited practical use; the selection being confined to acids of more general utility and adaptation to practical purposes.

3854. Sulphuric Acid. This is a colorless, odorless acid, and highly corrosive liquid, formed by the union of 1 equivalent of sulphur and 3 of oxygen. It is immediately colored by contact with organic matter. It attracts water so rapidly from the atmosphere, when freely exposed to it, as to absorb ⅓ its weight in 24 hours; and, under continued exposure, will absorb 6 times its weight. When 4 parts water are suddenly mixed with 1 part sulphuric acid, the temperature of the mixture rises to about 300° Fahr. Whilst 4 parts pounded ice mixed with 1 part acid, sinks the thermometer to some degrees below zero. Sulphuric acid boils and distills over at 620° Fahr., and freezes at about 20° below zero. The salts formed by the union of sulphuric acid with a base are called SULPHATES.

3855. To Obtain Commercial Sulphuric Acid. This is commonly called *oil of vitriol,* and has a specific gravity not less than 1.840, nor more than 1.845. It was first obtained by the distillation of green vitriol (sulphate of iron), but it is now made by bringing the fumes of sulphurous acid (*see No.* 3865) into contact with those evolved from a mixture of nitre and oil of vitriol, so that the former becomes oxidized at the expense of the latter. This process is conducted in a series of leaden chambers, having a little water on the floor, to absorb the acid, and so arranged as to prevent the loss of gas. As soon as the water is found to have acquired a specific gravity of 1.350 to 1.450, it is drawn off, and concentrated (*see No.* 8) in leaden boilers to a density of 1.659 to 1.700; after which it is further concentrated in green glass or platinum retorts until the specific gravity reaches 1.842 to 1.844. When cold, the clear acid is put into carboys (large globular bottles of green glass) packed securely with straw in strong wooden cases, the neck being left exposed for convenience in obtaining the acid without unpacking.

3856. Anhydrous Sulphuric Acid. Anhydrous or dry sulphuric acid is obtained by heating Nordhausen acid (*see No.* 3858) in a glass retort connected with a well-cooled receiver.

It is also prepared in the following manner: 2 parts strongest oil of vitriol are gradually added to 3 parts anhydrous phosphoric acid, contained in retort surrounded by a freezing mixture; when the compound has become brown, the retort is removed from the freezing bath and connected with a receiver which takes its place in the freezing mixture; a gentle heat is applied to the retort, when white vapors pass over and condense in the receiver under the form of beautiful silky crystals. The product equals in weight that of the phosphorus originally employed. The addition of a few drops of water to these crystals produces a dangerous explosion. They deliquesce rapidly and fume in the air; introduced into water, they hiss like red-hot iron. They melt at 66° Fahr., and boil at about 1050°, and do not redden dry litmus paper.

3857. Dilute Sulphuric Acid. The officinal strength of this acid, according to the U. S. Pharmacopœia, is thus obtained: Take 2 troy ounces sulphuric acid; add gradually to it 14 fluid ounces distilled water; filter through paper, and pass sufficient distilled water through the filter to make the diluted acid measure 1 pint. The specific gravity of this mixture is 1.082. The officinal strength of the British Pharmacopœia is somewhat greater; sufficient distilled water is added to 1350 grains sulphuric acid, so that, after it has been shaken and cooled down to 60° Fahr., it measures 1 imperial pint. The specific gravity of this is 1.094.

3858. Nordhausen Sulphuric Acid. This is also known as *fuming* sulphuric acid. It is a brown, oily liquid, which fumes in the air, is intensely corrosive, and has a specific gravity of about 1.900, and is chiefly used for dissolving indigo. It is prepared by distilling calcined sulphate of iron (green vitriol) in an earthen retort.

3859. Table Showing the Percentage of Liquid and Dry Sulphuric Acid in Dilute Acid at Different Densities.

Liquid.	Sp. Gr.	Dry.	Liquid.	Sp. Gr.	Dry.
100	1.8485	81.54	50	1.3884	40.77
99	1.8475	80.72	49	1.3788	39.95
98	1.8460	79.90	48	1.3697	39.14
97	1.8439	79.09	47	1.3612	38.32
96	1.8410	78.28	46	1.3530	37.51
95	1.8376	77.46	45	1.3440	36.69
94	1.8336	76.65	44	1.3345	35.88
93	1.8290	75.83	43	1.3255	35.06
92	1.8233	75.02	42	1.3165	34.25
91	1.8179	74.20	41	1.3080	33.43
90	1.8115	73.39	40	1.2999	32.61
89	1.8043	72.57	39	1.2913	31.80
88	1.7962	71.75	38	1.2826	30.98
87	1.7870	70.94	37	1.2740	30.17
86	1.7774	70.12	36	1.2654	29.35
85	1.7673	69.31	35	1.2572	28.54
84	1.7570	68.49	34	1.2490	27.72
83	1.7465	67.68	33	1.2409	26.91
82	1.7360	66.86	32	1.2334	26.09
81	1.7245	66.05	31	1.2260	25.28
80	1.7120	65.23	30	1.2184	24.46
79	1.6993	64.42	29	1.2108	23.65
78	1.6870	63.60	28	1.2032	22.83
77	1.6750	62.78	27	1.1956	22.01
76	1.6630	61.97	26	1.1876	21.20
75	1.6520	61.15	25	1.1792	20.38
74	1.6415	60.34	24	1.1706	19.57
73	1.6321	59.52	23	1.1626	18.75
72	1.6204	58.71	22	1.1549	17.94
71	1.6090	57.89	21	1.1480	17.12
70	1.5975	57.08	20	1.1410	16.31
69	1.5868	56.26	19	1.1330	15.49
68	1.5760	55.45	18	1.1246	14.68
67	1.5648	54.63	17	1.1165	13.86
66	1.5503	53.82	16	1.1090	13.05
65	1.5390	53.00	15	1.1019	12.23
64	1.5280	52.18	14	1.0953	11.60
63	1.5170	51.37	13	1.0887	10.41
62	1.5066	50.55	12	1.0809	9.78
61	1.4960	49.74	11	1.0743	8.97
60	1.4860	48.92	10	1.0682	8.15
59	1.4760	48.11	9	1.0614	7.34
58	1.4660	47.29	8	1.0544	6.52
57	1.4560	46.48	7	1.0477	5.71
56	1.4460	45.66	6	1.0405	4.89
55	1.4360	44.85	5	1.0336	4.08
54	1.4265	44.03	4	1.0268	3.26
53	1.4170	43.22	3	1.0206	2.446
52	1.4073	42.40	2	1.0140	1.63
51	1.3977	41.58	1	1.0074	0.8154

3860. To Purify Oil of Vitriol. Commercial sulphuric acid frequently contains *nitrous acid, arsenic,* and *saline matter.* These impurities must be removed in order to obtain the acid in any high degree of purity.

Nitrous acid is removed by adding about $1\frac{1}{2}$ grains sugar to each fluid ounce of the sulphuric acid, heated to nearly its boiling point, and continuing the heat until the dark color at first produced disappears, when it should be distilled. Another method is by adding $\frac{1}{8}$ to $\frac{1}{2}$ of 1 per cent. of sulphate of ammonia to the acid, and heating to ebullition for a few minutes. In this way the most impure acid may be rendered absolutely free from nitric acid and nitrous oxide.

Arsenic can be got rid of by adding a little sulphuret of barium, or of copper foil, to the acid, agitating the mixture well, and, after repose, decanting or distilling it.

Saline matter may be removed by simply redistilling (*rectification.*) The distillation is best conducted on the small scale, in a glass retort containing a few platinum chips, heated by a sand-bath or gas flame, rejecting the first $\frac{1}{2}$ fluid ounce that comes over.

3861. Test for Nitric Acid in Sulphuric Acid. Place in a watch glass a small portion pure and concentrated sulphuric acid at a density of 1.84; then pour, drop by drop, half the quantity of a solution of sulphate of aniline, prepared by mixing commercial aniline with diluted sulphuric acid. A glass rod is dipped in the liquid to be tested, and then stirred in the contents of the watch glass; from time to time the experimenter should blow slowly on the agitated liquid; if the liquid thus stirred contains traces of nitric acid, circular lines of a deep red are soon visible, coloring the whole liquid to a pink. On adding a very small quantity of nitric acid to the mixture, the liquid becomes of a carmine color; the addition of a single drop of very dilute nitric acid renders the liquid a deep red, and afterwards a dead red.

3862. To Remove Nitric Acid from Sulphuric Acid. Diluted sulphuric acid may be deprived of any small quantity of nitric acid it may contain, by shaking it up for a few minutes with a little powdered (freshly burned) charcoal, and afterwards filtering it. This will not answer for concentrated sulphuric acid; nitric acid is separated from it with great difficulty, and only by very protracted methods.

3863. To Decolorize Sulphuric Acid. Acid which has become brown by exposure may be decolorized by heating it gently; the carbon of the organic substances is thus converted into carbonic acid.

3864. Sulphurous Acid. This acid is used to bleach silks, woolens, *&c.*, (*see No.* 1716), and to remove vegetable stains and iron-moulds from linen. For these purposes it is prepared from sawdust, or any other refuse carbonaceous matter. The salts formed by the combination of sulphurous acid with a base are called SULPHITES. (*See Nos.* 1717 *and* 1718.)

3865. To Obtain Sulphurous Acid. In the gaseous form this acid is freely evolved by burning sulphur in air or in dry oxygen. It is also given off during the digestion of metals in hot sulphuric acid. When charcoal, wood, or cork chips, or sawdust are digested in hot sulphuric acid, a mixture of sulphurous and carbonic acids is obtained, which is used for bleaching and cleansing purposes.

3866. Pure Gaseous Sulphurous Acid. This is evolved during the action of sulphuric acid on mercury or clippings of copper. It is also obtained pure by heating in a glass retort, a mixture of 100 parts black oxide of manganese, and 12 or 14 parts sulphur. The gas evolved should be collected in a receiver over mercury.

3867. Sulphurous Acid Solution. The gas obtained according to the last method is to be passed through water, which is capable of dissolving or absorbing 30 times its bulk of the gas. To avoid waste in preparing the solution, the unabsorbed gas which escapes from the water is usually again passed through water, and the same arrangement repeated through a series of vessels of water so long as any gas escapes undissolved.

3868. Pure Sulphurous Acid. In order to prepare sulphurous acid from sulphuric acid and charcoal, it is better to employ an acid of .74 per cent., or 1.825 specific gravity. If we take a stronger acid, a part of it is entirely deoxidized to sulphur, and if weaker acid be employed, sulphuretted hydrogen is evolved. To obtain absolutely pure sulphurous acid, it is well to put sulphite of lead and coarse charcoal in the wash bottle. With these precautions, it is possible to obtain pure sulphurous acid from sulphuric acid and charcoal.

3869. Pure Liquid Sulphurous Acid. This can only be obtained by passing the pure dry gas through a glass tube surrounded by a powerful freezing mixture. The specific gravity of the pure liquid gas is 1.45; its boiling point is 14° Fahr., and causes intense cold by its evaporation.

3870. Hydrosulphuric Acid, also Called Sulphuretted Hydrogen. When sulphur acts upon paraffine at a temperature a little above the melting point of sulphur, hydrosulphuric acid gas is evolved in large quantities, and this method may be advantageously used for its generation in the laboratory. A flask, holding about a pound of the material, is fitted with a tube bent at right angles, about $\frac{1}{2}$ inch bore and 12 to 18 inches long, containing cotton wool, and to this is attached the small tube for precipitation. The production of gas may be stopped by removing the heat. Heavy paraffine oil, stearic acid, or suet, may be used as a substitute for paraffine.

3871. Nitro-Sulphuric Acid. Dissolve 1 part nitre in 9 parts sulphuric acid. This is used to separate the silver from the copper and solder of old plated goods. At about 200° Fahr. it readily dissolves silver, but scarcely acts on copper, lead, or tin, unless diluted, or assisted by a much higher temperature.

3872. Nitric Acid. There are five compounds of nitrogen and oxygen. The union of 1 equivalent of nitrogen with 1 of oxygen produces nitrous oxide, or *laughing gas;* with 2 oxygen, nitric oxide; with 3 oxygen, nitrous acid; with 4 oxygen, hyponitric acid; and with 5 equivalents of oxygen, nitric acid. Pure liquid nitric acid is colorless, highly corrosive, and possesses powerful acid properties. It is employed in assaying, to dye silk and woolens yellow, and to form various salts. In medicine, it is used as a caustic, &c. The officinal strength of nitric acid of the U. S. and British pharmacopœias has a specific gravity of 1.42, and boils at 250° Fahr. Nitric acid of less density than 1.42 parts with water and becomes stronger at lower temperatures; but acid of higher specific gravity is weakened by exposure to heat. It freezes when exposed to extreme cold. It rapidly oxidizes the metals, and unites with them and the other bases, forming salts called NITRATES. Two strengths of this acid occur in the arts, known as double and single aqua-fortis. Double aqua-fortis has usually a specific gravity of 1.36, and single, or ordinary aqua-fortis 1.22. Both are frequently sold at lower strengths. This can easily be ascertained by acidimetry. (*See No.* 78.)

3873. To Obtain Nitric Acid. The usual method adopted for obtaining this acid is to add to nitrate of potassa in coarse powder, contained in a glass retort, an equal weight of strong sulphuric acid, poured in through a funnel, so as not to wet the neck of the retort. The materials should not exceed two-thirds of the capacity of the retort. A moderate heat is at first applied, increasing as the materials begin to thicken. Red vapors will at first arise and pass over into the receiver; these will disappear in the course of the distillation, but subsequently renewed, showing that the process is completed. The pale yellow acid thus obtained may be rendered colorless, if desired, by heating it gently in a retort.

3874. To Purify Nitric Acid. The nitric acid of commerce frequently contains chlorine, muriatic and sulphuric acids, and sometimes iodine, from which it may be purified by the addition of a little nitrate of silver, as long as it produces any cloudiness, and, after repose, decanting the clear acid, and rectifying it at a heat under 212° Fahr. A perfectly colorless product may be obtained, by introducing a small portion of pure black oxide of manganese into the retort. Nitric acid may also be purified by rectification at a gentle heat, rejecting the first liquid that comes over, receiving the middle portion as genuine acid, and leaving a residuum in the retort. Another method is to agitate it with a little red-lead before rectification.

3875. Tests for Nitric Acid. It stains the skin yellow. When mixed with a little muriatic acid or sal-ammoniac, it acquires the power of dissolving gold leaf. When mixed with dilute sulphuric acid, and poured on a few fragments of zinc or iron in a tube, the evolved gas burns with a greenish white flame. Substitute alcohol for zinc in the last test. Morphia, brucia, and strychnia give it a red color, which is heightened by ammonia in excess. When placed in a tube, and a solution of protosulphate of iron cautiously added, a dark color is developed at the line of junction, which is distinctly visible when only $\frac{1}{24000}$ part of nitric acid is present. When mixed with a weak solution of sulphate of indigo, and heated, the color is destroyed.

3876. Dilute Nitric Acid. Mix 3 troy ounces nitric acid specific gravity 1.42 in a glass vessel with sufficient distilled water to make the dilute acid measure 1 pint. The specific gravity of officinal dilute nitric acid is 1.068, U. S. Dis.

3877. Fuming Nitric Acid. The red fuming nitrous or nitric acid of commerce is simply nitric acid loaded with nitrous or hyponitric acid. It may be thus prepared: Put into an iron or stoneware pot, nitre or nitrate of soda, add rather more than half its weight of strong sulphuric acid, and lute on a stoneware head. The vapor is conducted into a series of two-necked stoneware vessels, containing each $\frac{1}{6}$ of their capacity of water. The acid is usually obtained of the density of about 1.45. It is colored with nitrous acid gas, forming what is commonly, but improperly, termed nitrous acid. By gently heating the colored acid in a retort, the nitrous acid is driven off, and the acid remains nearly colorless, usually of the density of 1.38 to 1.42.

3878. Ure's Table of Percentage of Nitric Acid. This table is useful for finding the strength of dilute acids.

Specific Gravity.	Liq. Acid in 100.	Dry Acid in 100.	Specific Gravity.	Liq. Acid in 100.	Dry Acid in 100.
1.5000	100	79.700	1.2947	50	39.850
1.4980	99	78.903	1.2887	49	39.053
1.4960	98	78.106	1.2826	48	38.256
1.4940	97	77.309	1.2765	47	37.459
1.4910	96	76.512	1.2705	46	36.662
1.4880	95	75.715	1.2644	45	35.865
1.4850	94	74.918	1.2583	44	35.068
1.4820	93	74.121	1.2523	43	34.271
1.4790	92	73.324	1.2462	42	33.474
1.4760	91	72.527	1.2402	41	32.677
1.4730	90	71.730	1.2341	40	31.880
1.4700	89	70.933	1.2277	39	31.083
1.4670	88	70.136	1.2212	38	30.286
1.4640	87	69.339	1.2148	37	29.489
1.4600	86	68.542	1.2084	36	28.692
1.4570	85	67.745	1.2019	35	27.895
1.4530	84	66.948	1.1958	34	27.098
1.4500	83	66.155	1.1895	33	26.301
1.4460	82	65.354	1.1833	32	25.504
1.4424	81	64.557	1.1770	31	24.707
1.4385	80	63.760	1.1709	30	23.900
1.4346	79	62.963	1.1648	29	23.113
1.4306	78	62.166	1.1587	28	22.316
1.4269	77	61.369	1.1526	27	21.519
1.4228	76	60.572	1.1465	26	20.722
1.4189	75	59.755	1.1403	25	19.925
1.4147	74	58.978	1.1345	24	19.128
1.4107	73	58.181	1.1286	23	18.331
1.4065	72	57.384	1.1227	22	17.534
1.4023	71	56.587	1.1168	21	16.737
1.3978	70	55.790	1.1109	20	15.940
1.3945	69	54.993	1.1051	19	15.143
1.3882	68	54.196	1.0993	18	14.346
1.3833	67	53.399	1.0935	17	13.549
1.3783	66	52.602	1.0878	16	12.752
1.3732	65	51.805	1.0821	15	11.955
1.3681	64	51.068	1.0764	14	11.158
1.3630	63	50.211	1.0708	13	10.361
1.3579	62	49.414	1.0651	12	9.564
1.3529	61	48.617	1.0595	11	8.767
1.3477	60	47.820	1.0540	10	7.970
1.3427	59	47.023	1.0485	9	7.173
1.3376	58	46.226	1.0430	8	6.376
1.3323	57	45.429	1.0375	7	5.579
1.3270	56	44.632	1.0320	6	4.782
1.3216	55	43.835	1.0267	5	3.985
1.3163	54	43.038	1.0212	4	3.188
1.3110	53	42.241	1.0159	3	2.391
1.3056	52	41.444	1.0106	2	1.594
1.3001	51	40.647	1.0053	1	0.797

3879. Nitro-Muriatic Acid. Aqua regia. This is used in the arts, chiefly as a solvent for gold. By the mutual action of nitric and muriatic acids a compound of chlorine, nitrogen, and oxygen is formed. The best proportions and strength of the acids are variously stated. Colorless nitric acid must be used. Elkington employs 21 parts of nitric acid, specific gravity 1.45; 17 parts of muriatic acid 1.15 specific gravity; and 14 parts of water. This dissolves 5 parts of gold. (*See No.* 3588.) According to Cooley this acid is prepared by mixing 1 part by measure nitric acid and 2 parts hydrochloric acid. The mixture should be kept in a bottle in a cold and dark place. (*See No.* 3193.) With a base, this compound acid forms a NITRO-MURIATE.

3880. Dyer's Aqua-Fortis. Another mixture of nitric and hydrochloric acids, known as Dyer's aqua-fortis, is used by dyers, as it dissolves tin without oxidizing it. Mix 10 pounds colorless nitric acid, specific gravity 1.17, with 1 pound hydrochloric acid 1.19.

3881. Dilute Nitro-Muriatic Acid. Mix $1\frac{1}{2}$ troy ounces nitric acid, and $2\frac{1}{2}$ troy ounces muriatic acid in a pint bottle. Shake occasionally during 24 hours, and add distilled water to make up to 1 pint. Keep in a cool place, protected from the light. (*U. S. Ph.*)

3882. Muriatic or Hydrochloric Acid. Pure muriatic acid is a colorless invisible gas, having a pungent odor and an acid taste, and fuming on coming into contact with air. It is irrespirable and uninflammable. Its specific gravity is 1.2695. Under a pressure of 40 atmospheres it is liquid. Water at 40° Fahr. absorbs 480 times its volume of this gas, and acquires the specific gravity 1.2109. One cubic inch of water at 69° Fahr. absorbs 418 cubic inches, and the specific gravity becomes 1.1958. The aqueous solution of the gas constitutes the liquid form of the acid. The combinations of muriatic acid with a base are MURIATES, OR HYDROCHLORATES.

3883. To Obtain Muriatic Acid. The acid solution in water is thus obtained: Introduce 48 ounces (avoirdupois) dried chloride of sodium into a flask capable of containing an imperial gallon. Pour 44 fluid ounces sulphuric acid slowly into 32 fluid ounces water; and, when cool, add the mixture to the chloride of sodium in the flask. Connect the flask, by corks and a glass tube, with a three-necked wash-bottle, furnished with a safety tube, and containing 4 ounces water. Apply heat to the flask, conduct the disengaged gas through the wash-bottle, and thence, by means of a glass tube, into another bottle containing 50 fluid ounces distilled water, the end of the tube dipping about $\frac{1}{2}$ inch below the surface. Continue the process until the product measures 66 fluid ounces, or till the liquid has acquired a specific gravity of 1.16. The bottle must be kept cool during the process.

The muriatic acid of commerce is now chiefly obtained from the manufacturers of carbonate of soda, who procure it as a secondary product. When, however, it is directly prepared from sea-salt, an iron or stoneware boiler, set in brickwork over an open fire, furnished with a stoneware head, and connected with a series of capacious double-necked stoneware bottles, usually constitutes the distillatory and condensing apparatus.

3884. Gregory's Method of Obtaining Pure Muriatic Acid. Put into a matrass 6 parts, by weight, of purified salt, and 10 ounces oil of vitriol previously diluted with 4 of water, and cooled. Fix in the matrass a tube twice bent at right angles and having a bulb blown on the descending limb. Into a bottle surrounded with ice and water introduce distilled water equal in weight to the salt employed, and let the bent tube dip $\frac{1}{8}$ of an inch into the water. Apply a gentle heat of a sand-bath to the matrass as long as acid comes over. In about 2 hours the operation will be finished. The water is increased $\frac{3}{8}$ in bulk, and converted into hydrochloric acid of 1.14 or 1.15 specific gravity. To procure it of 1.21 specific gravity, employ part of this acid during the first half of a similar operation, and it will be speedily saturated. Phillips says a perfectly colorless acid may be obtained from the commercial sulphuric acid and common salt.

3885. Dilute Muriatic Acid. Mix 4 troy ounces muriatic acid with sufficient distilled water to make a pint. The specific gravity of the diluted acid is 1.038. (*U. S. Ph.*)

3886. Ure's Table of Percentage of Chlorine and Muriatic Acid Gas in Liquid Muriatic Acid.

Acid of 1.20 in 100.	Specific Gravity.	Chlorine	Muriatic Gas.	Acid of 1.20 in 100.	Specific Gravity.	Chlorine	Muriatic Gas.	Acid of 1.20 in 100.	Specific Gravity.	Chlorine	Muriatic Gas.
100	1.2000	39.675	40.777	66	1.1328	26.186	26.913	32	1.0637	12.697	13.049
99	1.1982	39.278	40.369	65	1.1308	25.789	26.505	31	1.0617	12.300	12.641
98	1.1964	38.882	39.961	64	1.1287	25.392	26.098	30	1.0597	11.903	12.233
97	1.1946	38.485	39.554	63	1.1267	24.996	25.690	29	1.0577	11.506	11.825
96	1.1928	38.089	39.146	62	1.1247	24.599	25.282	28	1.0557	11.109	11.418
95	1.1910	37.692	38.738	61	1.1226	24.202	24.874	27	1.0537	10.712	11.010
94	1.1893	37.296	38.330	60	1.1206	23.805	24.466	26	1.0517	10.316	10.602
93	1.1875	36.900	37.923	59	1.1185	23.408	24.058	25	1.0497	9.919	10.194
92	1.1857	36.503	37.516	58	1.1164	23.012	23.650	24	1.0477	9.522	9.786
91	1.1846	36.107	37.108	57	1.1143	22.615	23.242	23	1.0457	9.126	9.379
90	1.1822	35.707	36.700	56	1.1123	22.218	22.834	22	1.0437	8.729	8.971
89	1.1802	35.310	36.292	55	1.1102	21.822	22.426	21	1.0417	8.332	8.563
88	1.1782	34.913	35.884	54	1.1082	21.425	22.019	20	1.0397	7.935	8.155
87	1.1762	34.517	35.476	53	1.1061	21.028	21.611	19	1.0377	7.538	7.747
86	1.1741	34.121	35.068	52	1.1041	20.632	21.203	18	1.0357	7.141	7.340
85	1.1721	33.724	34.660	51	1.1020	20.235	20.796	17	1.0337	6.745	6.932
84	1.1701	33.328	34.252	50	1.1000	19.837	20.388	16	1.0318	6.348	6.524
83	1.1681	32.931	33.845	49	1.0980	19.440	19.980	15	1.0298	5.951	6.116
82	1.1661	32.535	33.437	48	1.0960	19.044	19.572	14	1.0279	5.554	5.709
81	1.1641	32.136	33.029	47	1.0939	18.647	19.165	13	1.0259	5.158	5.301
80	1.1620	31.746	32.621	46	1.0919	18.250	18.757	12	1.0239	4.762	4.893
79	1.1599	31.343	32.213	45	1.0899	17.854	18.349	11	1.0220	4.365	4.486
78	1.1578	30.946	31.805	44	1.0879	17.457	17.941	10	1.0200	3.968	4.078
77	1.1557	30.550	31.398	43	1.0859	17.060	17.534	9	1.0180	3.571	3.670
76	1.1536	30.153	30.990	42	1.0838	16.664	17.126	8	1.0160	3.174	3.262
75	1.1515	29.755	30.582	41	1.0818	16.267	16.718	7	1.0140	2.778	2.854
74	1.1494	29.361	30.174	40	1.0798	15.870	16.310	6	1.0120	2.381	2.447
73	1.1473	28.964	29.767	39	1.0778	15.474	15.902	5	1.0100	1.984	2.039
72	1.1452	28.567	29.359	38	1·0758	15.077	15.494	4	1.0080	1.588	1.631
71	1.1431	28.171	28.951	37	1.0738	14.680	15.087	3	1.0060	1.191	1.224
70	1.1410	27.772	28.544	36	1.0718	14.284	14.679	2	1.0040	0.795	0.816
69	1.1389	27.376	28.136	35	1.0697	13.887	14.271	1	1.0020	0.397	0.408
68	1.1369	26.979	27.728	34	1.0677	13.490	13.863				
67	1.1349	26.583	27.321	33	1.0657	13.094	13.456				

3887. Tests for Muriatic Acid. When a glass rod, dipped in liquor of ammonia, is held near it, it gives off white fumes. With nitrate of silver it gives a white, cloudy precipitate, insoluble in nitric acid, freely soluble in liquor of ammonia, and blackened by exposure to the light.

3888. To Purify Muriatic Acid. Commercial muriatic acid may be purified by diluting it with an equal weight of water, gently heating it in a retort, and receiving the evolved gas into a fresh quantity of pure water. Iodine and arsenic may be removed by agitating it for a few minutes with some small pieces of bright copper foil previously to rectification.

3889. Acetic Acid. This is the well-known acid principle of vinegar. It is one of the common products of fermentation, of the oxygenation of alcohol, and of the destructive distillation of wood and other vegetable matter. The officinal strength of acetic acid adopted by the U. S. Pharmacopœia has a specific gravity of 1.047. Special methods for testing the strength of acetic acid are given under Acetimetry, No 69. With bases this acid forms ACETATES.

Commercial acetic acid is principally manufactured on the large scale from acetate of soda, which yields a sufficiently strong and pure acid for commercial purposes, without the trouble of rectification. In this process, shallow copper vessels formed without rivets or solder in those parts exposed to the action of the acid, are employed for the purpose of the distillation. A coil of drawn copper pipe, heated by steam, having a pressure of 30 to 35 pounds to the inch, traverses the bottom of the apparatus. The refrigeratory consists of well cooled earthenware vessels, and the adopter or pipe connecting the still with the receivers is also of the same materials. Stills of earthenware are also frequently employed, and even worms and condensers of silver are sometimes used. The crystalline acetate of soda is placed in the still, and 35 to 36 parts of strong oil of vitriol are added to every 100 parts of the acetate of soda, and the whole stirred together with a wooden spatula. The head of the still is then luted on and the distillation commenced. This produces an acid of a specific gravity of about 1.050, and, after being agitated with a little animal charcoal, and passed through a prepared muslin filter, is ready for sale. Some manufacturers add a little acetic ether to it. By this process 4 pounds of acetic acid of the strength above mentioned is obtained for every 3 pounds of the acetate of soda employed. (*See No.* 1741.)

3890. Dilute Acetic Acid. The U. S. Pharmacopœia directs 1 pint acetic acid to be mixed with 7 pints distilled water, producing an acid of specific gravity 1.006; 100 grains of dilute acetic acid saturate 7.6 grains bicarbonate of potassa.

3891. To Obtain Pure Glacial or Hydrated Acetic Acid. Place 30 parts dry and finely powdered pure acetate of soda in a capacious retort, and pour on it 97 parts pure sulphuric acid. The heat developed by the action of the ingredients will cause one-eighth of the acetic acid to pass over. The retort may then be placed in a sand bath until the contents become quite liquid. The product, carefully rectified, yields 2 parts of pure acid containing only 20 per cent. of water. By exposing the latter portion, which comes over in a closed vessel, to a temperature below 40° Fahr., crystals of hydrated (glacial) acetic acid will be deposited. The liquid portion being then poured off, the crystals are again melted and re-crystallized by cooling. These last crystals, separated from the liquid, are perfectly pure.

3892. To Obtain Glacial or Hydrated Acetic Acid Without Distillation. The acid may also be obtained without resorting to distillation, thus: Place 100 parts powdered acetate of soda (pure commercial) in a hard-glazed stoneware or glass pan; pour 35 or 36 parts concentrated sulphuric acid gradually into the pan, so that the acid may flow under the powder, and as little heat as possible be generated by the operation. In furtherance of this necessary end, the process is best conducted in a cool apartment, and the pan kept well cooled. The whole must now be covered and allowed to stand for some hours, when crystalline grains of sulphate of soda will be found covering the inside of the vessel, and hydrated acetic acid, partly liquid and partly in crystals, in the upper portion. The temperature must then be raised just sufficiently to liquefy the crystals of acetic acid, the fluid poured off, and a very small quantity of pure acetate of lime added gradually, until it yields no trace of sulphuric acid on evaporation. After repose it may be decanted for use.

3893. To Obtain Pure Acetic Acid. Triturate together 10 parts crystallized neutral acetate of lead, and 3 parts effloresced (dry) sulphate of soda; mix together 2½ parts each of sulphuric acid and water, and, when cold, pour it on the acetate and sulphate, previously placed in a retort; then distill to dryness in a sand bath. The acid that comes over in the distillation by this process is very pure, and may be used as a test acid for chemical analyses.

3894. To Obtain Anhydrous Acetic Acid. This is acetic acid free from water, as it exists in dry acetates. Mix, in a glass retort, well-fused acetate of potassa with half its weight of chloride of benzoyle; apply a gentle heat, collect the liquid that distills over, and rectify it carefully. Hot water added to this resolves it into hydrated or glacial acetic acid.

3895. Camphorated Acetic Acid. Pulverize 1 ounce camphor in 1 fluid drachm rectified spirit, and dissolve in 10 fluid ounces strong acetic acid. This is fragrant and refreshing, and used as an embrocation in rheumatism and neuralgia, and as a fumigation in fever, &c.

3896. To Obtain Strong Acetic Acid from Vinegar. Expose the vinegar to the action of a freezing mixture, or place in the air in very cold weather; the water separates and becomes ice, and the strong acid remaining fluid may be drained from it. (*See No.* 1749.)

3897. Mohr's Table of the Specific Gravity of Acetic Acid at Various strengths. The following table, drawn up by M. Mohr, exhibits the specific gravity of acetic acid of almost every strength.

Per cent. of Glacial Acid.	Sp. Gr.	Per cent. of Glacial Acid.	Sp. Gr.	Per cent. of Glacial Acid.	Sp. Gr.
100	1.0635	67	1.069	34	1.045
99	1.0635	66	1.069	33	1.044
98	1.067	65	1.068	32	1.0424
97	1.0680	64	1.068	31	1.041
96	1.069	63	1.068	30	1.040
95	1.070	62	1.067	29	1.039
94	1.0706	61	1.067	28	1.038
93	1.0708	60	1.067	27	1.036
92	1.0716	59	1.066	26	1.035
91	1.0721	58	1.066	25	1.034
90	1.0730	57	1.065	24	1.033
89	1.0730	56	1.064	23	1.032
88	1.0730	55	1.064	22	1.031
87	1.0730	54	1.063	21	1.029
86	1.0730	53	1.063	20	1 027
85	1.0730	52	1.062	19	1.026
84	1.0730	51	1.061	18	1.025
83	1.0730	50	1.060	17	1.024
82	1.0730	49	1.059	16	1.023
81	1.0732	48	1.058	15	1.022
80	1.0735	47	1.056	14	1.020
79	1.0732	46	1.055	13	1.018
78	1.0732	45	1.055	12	1.017
77	1.073	44	1.054	11	1.016
76	1.072	43	1.053	10	1.015
75	1.072	42	1.052	9	1.013
74	1.072	41	1.0515	8	1.012
73	1.071	40	1.0513	7	1.010
72	1.071	39	1.050	6	1.008
71	1.071	38	1.049	5	1.0067
70	1.070	37	1.048	4	1.0065
69	1.070	36	1.047	3	1.004
68	1.070	35	1.046	2	1.002
				1	1.001

3898. To Concentrate Acetic Acid. Acid containing 20 per cent. of water may be deprived of a good deal of its superfluous water by standing over dry sulphate of soda. It may then be used either with or without distillation. Acetic acid of ordinary strength may be concentrated to any degree of rectification once or oftener from dry acetate of potassa or soda, rejecting the first and last portions that come over. The same acetate may be used repeatedly. The heat employed must not exceed 500° to 570° Fahr. Pure hydrated acetic acid liquefies above 62° Fahr.; at 50° to 55° it crystallizes in brilliant, colorless, transparent needles and plates; at 40° it is a crystalline solid. Free acetic acid reddens litmus paper, and may be recognized by its odor and volatility.

3899. Tests for the Purity of Acetic Acid. By heat it escapes *entirely* in vapor. Either nitrate of silver or chloride of barium being added to it, will produce no precipitate. When a thin plate of silver is digested in it, and hydrochloric acid subsequently dropped in, no precipitate is formed. Its color is unchanged by the addition of hydrosulphuric acid, or ammonia, or by ferrocyanide of potassium added after the ammonia. The presence of sulphuric acid is indicated by a white precipitate being formed on the addition of a little peroxide of lead.

3900. Oxalic Acid. This consists of colorless crystals, possessing considerable volatility, and a strong, sour taste; when exposed to a very dry atmosphere they effloresce slightly. Oxalic acid sublimes at 180° Fahr., and melts at 280°; is soluble in about nine times its weight of cold, and in its own weight of boiling water; soluble also, but in a less degree, in alcohol. It has a strong affinity for lime, and is therefore a good test for its presence, by yielding a precipitate insoluble in excess of the acid. With the bases, oxalic acid forms OXALATES.

3901. To Obtain Oxalic Acid. Liebig proposes: Nitric acid (specific gravity 1.42), 5 parts; water, 10 parts; mix, add sugar, or preferably potato starch, 1 part, and digest by a gentle heat as long as gaseous products are evolved; evaporate and crystallize, dry the crystals, redissolve in the smallest possible quantity of boiling water, and crystallize; 12 parts of potato starch yield 5 of acid. The mother water, treated with more nitric acid, and again warmed, will yield a second crop of crystals; and this should be repeated till the solution is exhausted.

Schlesinger gives the following method: Sugar 4 parts (dried at 257° Fahr.); nitric acid (specific gravity 1.38) 33 parts; the mixture, as soon as the evolution of gas ceases, is to be boiled down to one-sixth its original volume, and allowed to crystallize. The whole process may be executed in 2 hours, and yields of beautifully crystallized oxalic acid from 56 to 60 per cent. of the sugar employed.

On the large scale, the first part of the process is usually conducted in salt-glazed stoneware pipkins, about two-thirds filled and set in a water-bath; but on the small scale a glass retort or capsule may be used. The evaporation should be preferably conducted by steam. The evolved nitrous vapors are usually allowed to escape, but if conveyed into a chamber filled with cold damp air, and containing a little water, they will absorb oxygen, and be recondensed into fuming nitric acid. In England an equivalent proportion of molasses is usually substituted for sugar. Another process consists in first converting potato fecula into grape sugar with sulphuric acid, and then decomposing the sugar thus obtained by nitric acid, in the usual way. Dr. Ure recommends the use of a little sulphuric acid along with the nitric acid, which, he says, contributes to increase the product; 15 pounds of sugar yielding fully 17 pounds of crystallized oxalic acid.

3902. Dale's Process for Obtaining Oxalic Acid. At present much of the oxalic acid of commerce is obtained by heating sawdust with a mixture of 2 parts caustic soda with 1 part caustic potassa. A watery solution of the mixed alkalies is evaporated to specific gravity 1.35, and then mixed with sawdust to a paste. This is heated on iron plates to 400° Fahr., and kept at that temperature for 1 or 2 hours, with constant stirring; the heat is continued until the mass is quite dry, but not charred. It now contains 28 to 30 per cent. of oxalic acid combined with the alkalies. By washing the powder on a filter with a solution of carbonate of soda, all traces of potassa are washed out. The oxalate of soda is converted, by heated milk of lime, into oxalate of lime, and the resulting oxalate of lime is treated with sulphuric acid, leaving a solution of oxalic acid ready to be evaporated into crystals. Two pounds of sawdust yield 1 pound oxalic acid.

3903. Chemically Pure Oxalic Acid. Chemically pure oxalic acid is best prepared by precipitating a solution of binoxalate of potash with acetate of lead, washing the precipitate with water, and decomposing it, while still moist, with dilute sulphuric acid or sulphuretted hydrogen. Filter and evaporate gently, so that crystals may form as it cools.

3904. To Distinguish Oxalic Acid from Epsom Salts. Oxalic acid has occasionally been mistaken for Epsom salts, with fatal results. They may be easily distinguished. Epsom salts taste extremely *bitter* and *nauseous;* oxalic acid tastes extremely *sour.* It is safer to taste a weak solution in applying this test. Epsom salts, dissolved in water and mixed with carbonate of soda, or carbonate of potash, *turn milky*, and, after a time, a white sediment subsides; oxalic acid, mixed with carbonate of soda or carbonate of potash, *effervesces*, and the liquid, in a few seconds, becomes transparent.

3905. Gallic Acid. When pure, gallic acid forms small, feathery, and nearly colorless crystals, which have a beautiful silky lustre. Commercial gallic acid has usually a pale yellow color, soluble in both water and alcohol. Its aqueous solution decomposes by exposure to the air. It blackens the salts of iron. Dissolved in hot oil of vitriol, it forms a deep, rich, red solution, which, when thrown into water, drops the gallic acid, deprived of some of its water. Gallic acid forms GALLATES with the bases.

3906. To Obtain Gallic Acid. Mix 36 troy ounces nut-gall, in fine powder, with sufficient distilled water to make a thin paste; expose the mixture to the air in a shallow glass or porcelain vessel, in a warm place, for a month, occasionally stirring with a glass rod, and adding sufficient distilled water to preserve the original consistence. Then press out the water, boil the residue in 8 pints distilled water for a few minutes, and filter while hot through purified animal charcoal. (*See No.* 1752). Set aside to crystallize, and dry the crystals on bibulous paper. If not sufficiently free from color, dissolve the crystals in boiling distilled water, filter through a fresh portion of the charcoal, and crystallize again. (*U. S. Ph.*)

3907. To Obtain Gallic Acid from Tannin. Add a strong aqueous solution of tannic acid (tannin) to sulphuric acid, as long as a precipitate falls; collect the powder, wash, and dissolve it by the aid of heat in diluted sulphuric acid; boil for a few minutes, cool, and collect the crystals of gallic acid which will form in considerable quantity.

3908. To Distinguish Gallic Acid from Tannic Acid. Gallic acid does not affect solutions of gelatine, the protosalts of iron, or the salts of the alkaloids; but it produces a black precipitate with the sesquisalts of iron, which disappears when the liquid is heated.

3909. Pyrogallic Acid. This acid is formed in white, shining scales, inodorous, very bitter; soluble in water, alcohol, and ether; fusible at 239° Fahr., and subliming at 410°. When quite pure, it has no action on litmus paper. It is used in photography. A solution of the crude acid mixed with a little alcohol imparts a fine brown color to the hair, but stains the skin also.

3910. To Obtain Pyrogallic Acid. It may be prepared by heating gallic acid (previously dried at 212° Fahr.) in a glass retort, by means of a chloride of zinc bath, to 410°, when the pure acid sublimes, and forms in crystals on the neck of the retort, and in the receiver, which should be kept well cooled.

3911. Tannic Acid, also called *Tannin.* Pure tannic acid is solid, uncrystallizable, white, or slightly yellow; strongly astringent, but without bitterness; very soluble in water, less so in alcohol and ether, and insoluble in fixed or volatile oils. Its solution reddens litmus. With the bases tannic acid forms TANNATES.

Among the incompatibles of tannin are the alkaloids of opium, and it is altogether unavoidable that if solutions of them are brought together, a precipitate will form of tannates; also, if the preparation of opium contain saffron, as in acetum opii and Sydenham's laudanum, this will cause a further precipitation of the extractive of saffron. (*See No.* 3908.)

3912. To Obtain Tannic Acid. Expose nut-gall in fine powder to a damp atmosphere for 24 hours, then mix it with sufficient ether, previously washed with water, to form a soft paste. Set this aside, closely covered, for 6 hours; then envelope it quickly in a close canvas cloth, and obtain the liquid portion by pressing powerfully between tinned plates. Reduce the resulting cake to powder, mix it with sufficient ether shaken with $\frac{1}{16}$ its bulk of water, to form again a soft paste, and express as before. Mix the liquids, and evaporate spontaneously to a syrupy consistence; then spread it on glass or tinned plates, and dry quickly in a drying closet. Put the dry residue in a well-stopped bottle.

3913. Carbonic Acid. An acid compound, formed by the union of carbon with oxygen, sometimes called *choke-damp.* A colorless gas possessing a pungent odor and acidulous taste, rapidly absorbed by water, forming liquid carbonic acid. The agreeable pungency of ale, beer, porter, wine, &c., is in a great measure owing to the presence of carbonic acid, which they lose on exposure to the air, and then become flat and stale. Spring and well water contain carbonic acid, and water that has been boiled has an insipid taste, from its absence. Under a pressure of 36 atmospheres at 32° Fahr. it becomes fluid, and on the pressure being removed, congeals, from the cold produced by its rapid evaporation. It has been estimated that the temperature falls to 180° in this experiment. Carbonic acid gas is destructive to life, and extinguishes combustion. An atmosphere containing more than its natural quantity (about $\frac{1}{1000}$), is unfit for respiration. The air of wells, cellars, brewers' vats, &c., is frequently contaminated with this gas (choke-damp); hence the necessity of the old plan of letting down a burning candle before venturing in. If the candle will not burn, man cannot breathe there. With the bases, this acid forms CARBONATES.

3914. To Obtain Carbonic Acid. Dilute muriatic acid with 4 times its weight of water, then pour it upon fragments of marble, previously placed in a tubulated retort. Carbonic acid gas will be rapidly evolved, and may either be collected in the mercurial pneumatic trough, or applied to immediate use. When wanted perfectly dry, it must be passed over dried chloride of calcium, or through concentrated oil of vitriol. This is the most convenient way of procuring the gas on the small scale, or in the laboratory. Or: Dilute oil of vitriol with 3 or 4 times its weight of water, then pour it on whiting placed in a suitable vessel, and apply agitation. This is the plan adopted on the large scale by the soda water makers. (*See No.* 718.)

3915. Tests for Carbonic Acid. It reddens litmus paper, extinguishes the flame of a burning taper, and forms a white precipitate in aqueous solutions of lime and baryta, which is soluble in acetic acid. By the last test, a very small quantity of this gas may be easily detected in the atmosphere of rooms, &c.

3916. Carbolic Acid, also called *Phenol, Phenic acid,* and *hydrate of Phenyle.* It consists of long, colorless prismatic crystals, which melt at about 90° Fahr. into an oily liquid resembling creosote. The crystals deliquesce in moist air, forming a sort of hydrate, which boils at 370° and has a specific gravity of 1.065. Heated with ammonia, it yields aniline and water; and nitric acid converts it into picric acid. Commercial creosote consists principally of hydrated carbolic acid, but is easily distinguishable from it, as carbolic acid coagulates collodion, creosote does not. It has come into prominent notice as an efficient disinfectant.

3917. To Obtain Carbolic Acid. This is obtained from that portion of coal-tar which distills over between 300° and 400° Fahr.; this, when mixed with a hot concentrated solution of hydrate of potassa, is resolved, on the addition of water, into a light oil and a heavier alkaline liquid. By separating the latter, and neutralizing it with muriatic acid, impure carbolic acid will float on the surface in the form of a light oil. If this be distilled from dried chloride of calcium to separate the water, and the distillate be exposed to a low temperature, carbolic acid congeals in a colorless deliquescent crystalline mass, which may be separated from the liquid by pressure in bibulous paper. At 95° Fahr. the crystals melt and constitute the liquid carbolic acid. The introduction of a crystal of carbolic acid into the acid to be congealed, greatly facilitates its crystallization.

3918. Tests for the Purity of Carbolic Acid. If it becomes brown under the influence of light and air it is impure.

Put 1 fluid drachm of the liquid acid in a bottle with ½ pint warm water, and shake occasionally for half an hour; the amount of oily residue will indicate the measure of adulteration.

Mix 1 part caustic soda with 10 parts of the acid, and shake them well together. Any undissolved residue is impurity.

3919. To Remove the Odor from Carbolic Acid. It may be interesting to know of a method which will entirely remove this odor, substituting for it a delicate trace of geranium leaves, which may, perhaps, be improved upon by adding a few drops of that oil. The process, as recently published by Professor Church, consists in pouring 1 pound of the best carbolic acid of commerce (the white crystallized) into 2 gallons cold distilled water, taking care not to permit the whole of the acid to enter into solution. With a good sample, if, after shaking repeatedly at intervals, between 2 and 3 ounces of the acid remain at the bottom of the vessel used, this will be a sufficient residue to hold and contain all the impurities; with bad samples, less water must be used, and more acid. The watery solution is to be syphoned off, and filtered, if necessary, through fine filtering paper, till perfectly clear. It is then placed in a tall cylinder, and pure powdered common salt added, with constant agitation, till it no longer dissolves. On standing for a time, the greater part of the carbolic acid will be found floating as a yellow oily layer on the top of the saline liquor, and merely requires to be removed to be ready for use. As it contains 5 per cent. or more of water, it does not generally crystallize, but it may be made to do so by distilling it from a little lime. The portion collected has, at ordinary temperatures, and up to 365° Fahr., scarcely any odor save a faint one resembling that of geranium leaves. The addition of about 4 drops per fluid ounce of the French oil of geranium will still further mask the slight odor of the acid, and has an additional advantage of liquefying the pure crystallized product. The pure acid may be dissolved in 230 parts of water, and used as a gargle, or in 25 parts of water for painting the throat, or in 50 parts for the carbolic spray. By this process it becomes sufficiently deodorized for toilet purposes.

3920. Phosphoric Acid. This acid, in its pure or anhydrous state, can only be obtained by the direct combination of its elements, phosphorus and oxygen, 1 equivalent of phosphorus combining with 5 of oxygen. It consists of a white, flaky, extremely deliquescent powder, which, when fused and cooled, assumes a vitreous appearance. It is capable of assuming three separate conditions in combination with water as a base; the union of 1 equivalent of anhydrous acid with 1 equivalent of water produces ***monobasic*** or glacial phosphoric acid, called also ***metaphosphoric*** acid; 1 equivalent of anhydrous acid, with 2 of water, gives ***bibasic*** or ***pyrophosphoric*** acid; 1 of anhydrous acid with 3 of water forms *tribasic,* or commercial phosphoric acid. This last is the common form of the acid. These three forms of the acid are not pure phosphoric acid in different degrees of dilution, as they have distinguishing characteristics. Monobasic phosphoric acid coagulates albumen, and gives white gelatinous uncrystallizable precipitates with the soluble salts of baryta, lime, and silver; the ***bibasic*** does not coagulate albumen, and makes, when neutralized only, a white precipitate with nitrate of silver; the ***tribasic*** does not affect albumen, and, when neutralized, throws down a yellow precipitate (phosphate of silver) from nitrate of silver. Tribasic phosphoric acid is the usual form under which phosphoric acid combines with the bases to form PHOSPHATES.

3921. To Obtain Phosphoric Acid. This is obtained by heating nitric acid in a tubulated retort connected with a receiver; small fragments of phosphorus are dropped into the acid, singly and at intervals. As soon as the phosphorus is dissolved, the heat is increased, and the undecomposed acid distilled off. The residuum is then evaporated to a syrupy consistence, and forms the ***phosphoric acid*** of commerce.

3922. To Obtain Hydrated or Glacial Phosphoric Acid. Phosphoric acid (*see last receipt*) is gradually heated to redness in a platinum crucible, and the glacial acid obtained by evaporation. Solid hydrated or glacial phosphoric acid contains 89 per cent. of real acid, and 11 per cent. of water. It is a highly deliquescent, glassy-looking substance, very soluble in water, yielding a solution exhibiting powerful acid properties. Its concentrated solution has nearly the same properties as the solid acid; its dilute solution is not poisonous, and does not precipitate albumen. (*Cooley.*)

3923. Anhydrous Phosphoric Acid. This is evolved by burning phosphorus in a stream of dry air, or under a bell-jar, copiously supplied with dry air. The product is anhydrous phosphoric acid in snow-like flakes. These must be collected immediately, and put into a warm, dry, well-stoppered bottle. A few seconds' exposure to the air causes the anhydrous acid to deliquesce into a syrupy liquid, its attraction for water being intense. Its anhydrous state cannot be restored after deliquescence or solution.

3924. Dilute Phosphoric Acid. Mix 5 troy ounces nitric acid with ½ pint distilled water in a porcelain capsule of the capacity of 2 pints; add 6 drachms phosphorus and invert over it a glass funnel of such dimensions that its rim may rest on the inside of the capsule, near the surface of the liquid. Place the capsule on a sand-bath, and apply a moderate heat until the phosphorus is dissolved, and red vapors cease to rise. If the reaction becomes too violent, add a little distilled water; and if the red vapors cease to be evolved before the phosphorus is all dissolved, gradually add nitric acid (diluted as before) until the solution is effected. Remove the funnel, continue the heat until the excess of nitric acid is driven off, and a syrupy liquid, free from odor and weighing 2 troy ounces, remains. Mix this, when cold, with sufficient distilled water to measure 20 fluid ounces, and filter through paper.

Or: Dissolve 1 troy ounce glacial phosphoric acid in 3 fluid ounces distilled water; add 40 grains nitric acid, boil to a syrupy liquid, free from the odor of nitric acid, add distilled water to make up to 12½ fluid ounces, and filter.

3925. Tests for the Purity of Phosphoric Acid. The U. S. Pharmacopœia directs that an aqueous solution of the acid should yield no precipitate with sulphuretted hydrogen, showing the absence of metals; it should cause a white precipitate with chloride of barium, soluble in excess of acid; and, with an excess of ammonia, should cause only a slight turbidness, proving the almost total absence of earthy salts. If the presence of arsenic is denoted by the tests for that metal, it may be separated by boiling with muriatic acid, so as to convert the arsenic into a volatile chloride, which would escape with vapors of the muriatic acid.

3926. Test for the Presence of Phosphoric Acid. Hydrochloric acid is added to the solution to acid reaction, and afterwards 1 or 2 drops of a concentrated solution of sesquichloride of iron; a solution of acetate of potassa is next added in excess, when a flocculent white precipitate (sesqui-phosphate of iron) will be found if phosphoric acid was present in any form or combination in the original liquor. Arsenious acid, if present, should be removed by sulphuretted hydrogen before applying the test. (*Cooley.*)

3927. Phosphorous Acid. This is prepared by burning phosphorus under a bell-glass with a very limited supply of air. White and pulverulent. It is a powerful deoxidizing agent. With the bases it unites to form PHOSPHITES.

3928. Hypophosphoric Acid. A name erroneously given by M. Dulong to a mixture of phosphoric and phosphorous acids. (*Cooley.*)

3929. Tartaric Acid. Tartaric acid forms inodorous, sour, scarcely transparent prisms, soluble in 2 parts of water at 60°, and its own weight of boiling water. It contains about 9⅜ of combined water, fuses at 220° Fahr., boils at 260°; and, at about 400°, after losing ¼ of its water, is converted into tartralic acid. With the bases it forms salts called TARTRATES. Tartaric acid is chiefly employed in calico printing, and in medicine, as a substitute for citric acid and lemon juice, for the preparation of cooling drinks and saline draughts.

3930. To Obtain Tartaric Acid. On the small scale it is prepared as follows: Dissolve 4 pounds cream of tartar in 2 gallons boiling water; add gradually 12 ounces 7 drachms chalk; and, when the effervescence ceases, add another like portion of chalk, dissolved in 26½ fluid ounces muriatic acid, diluted with 4 pints water; collect the precipitated tartrate of lime, and well wash it with water, then boil it for 15 minutes in 8 pints 1 fluid ounce dilute sulphuric acid; next filter, evaporate to the density 1.38, and set it aside to crystallize. The crystals must be dissolved and crystallized a second and a third time.

On the large scale, the decomposition of the tartar is usually effected in a copper boiler, and that of the tartrate of lime in a leaden cistern. This part of the process is often performed by mere digestion for a few days, without the application of heat. Leaden or stoneware vessels are used as crystallizers. Good cream of tartar requires 26 per cent. of chalk, and 28.5 per cent. of dry chloride of calcium for its perfect decomposition. Dry tartrate of lime requires 75 per cent. of oil of vitriol to liberate the whole of the tartaric acid. A very slight excess of sulphuric acid may be advantageously employed. Some manufacturers bleach the colored solution of the first crystals by treating it with animal charcoal; but for this purpose the latter substance should be first purified by digesting it in muriatic acid, and afterwards by edulcorating it with water, and exposing it to a dull red heat in a covered vessel. The general management of this manufacture resembles that of citric acid. (*Cooley.*)

3931. To Detect Tartaric Acid in Citric Acid. Citric acid is sometimes adulterated with tartaric acid. This is readily detected by adding a solution of carbonate of potassa to a solution of the suspected acid; if tartaric acid be present, a crystalline precipitate of bitartate of potassa (cream of tartar) will be found. A more delicate test is to digest the suspected acid with hydrated sesquioxide of iron in a test tube, and afterwards to raise the heat slowly to the boiling point; allowing the excess of oxide to subside, decant the clear liquid, and evaporate it to a syrupy consistence. If the citric acid was pure, the liquid remains clear and of a fine red color; the presence of only 1 per cent. of tartaric acid renders it cloudy, and deposits tartrate of the sesquioxide. (*U. S. Dis.*)

3932. Citric Acid. This is an agreeable acid, cooling and antiseptic; 20 grains of citric acid are equivalent to 5 fluid drachms lemon juice. When used for making saline draughts, it is preferable to use bicarbonate of potassa as the neutralizing alkali. Their respective saturating equivalents will be found in Nos. 80 and 81. With the bases it forms CITRATES.

3933. To Prepare Citric Acid. Add 4½ ounces chalk by degrees to 4 pints lemon juice, heated, and mix; set by, that the powder may precipitate; afterwards pour off the supernatant liquor. Wash the precipitated citrate of lime frequently with warm water; then pour upon it 27¼ fluid ounces diluted sulphuric acid and 2 pints distilled water, and boil for 15 minutes; press the liquor strongly through a linen cloth, and filter it. Evaporate the filtered liquor with a gentle heat, and set it aside that crystals may form. To obtain the crystals pure, dissolve them in water a second and a third time; filter each solution, evaporate, and set it apart to crystallize. The preparation of citric acid has become an important branch of chemical manufacture, from the large consumption of this article in various operations in the arts. In conducting this process some little expertness and care are necessary to ensure success. The chalk employed should be dry, and in fine powder, and be added to the juice until it be perfectly neutralized, and the quantity consumed must be exactly noted. The precipitated citrate of lime should be well washed with water, and the sulphuric acid diluted with 6 or 8 times its weight of water, poured upon it while still warm, and thoroughly mixed with it. The agitation must be occasionally renewed for 8 or 10 hours, when the dilute citric acid must be poured off, and the residuum of sulphate of lime thoroughly washed with warm water, and the washings added to the dilute acid. The latter must then be poured off from the impurities that may have been deposited, and evaporated in a leaden boiler, over the naked fire, until it acquires a specific gravity of 1.13, when the process must be continued at a lower temperature until a pellicle appears upon the surface. This part of the process requires great attention and judgment, as, if not properly conducted, the whole batch may be carbonized and spoiled. At this point the evaporation must be stopped, and the concentrated solution emptied into warm and clean crystallizing vessels, set in a dry apartment, where the thermometer does not fall below temperate. At the end of 4 days the crystals will be ready to remove from the pans, when they must be well drained, redissolved in as little water possible, and, after being allowed to stand for a few hours to deposit impurities, again evaporated and crystallized. When the process has been well managed, the acid of the second crystallization will usually be sufficiently pure; but if this be not the case, a third, or even a fourth crystallization must be had recourse to. The mother liquors from the several pans are collected together, and, by evaporation, yield a second or third crop of crystals obtained by evaporation as before. Citric acid crystallizes with great ease, but in some cases, where all the citrate of lime has not undergone decomposition by the sulphuric acid, a little of that salt is taken up by the free citric acid, and materially obstructs the crystallization. This is best avoided by exactly apportioning the quantity of the sulphuric acid to that of the chalk used, always remembering that it requires a quantity of liquid sulphuric acid, containing exactly 40 parts of dry acid, to decompose 50 parts of carbonate of lime. Commercial sulphuric acid is usually of the specific gravity of 1.845; it will therefore take exactly 49 pounds of this acid for 50 pounds of chalk. In practice it is found that a very slight excess of sulphuric acid is better than leaving any citrate of lime undecomposed. The first crop of crystals is called "brown citric acid," and is much used by the calico printers. Sometimes a little nitric acid is added to the solution of the colored crystals, for the purpose of whitening them, but in this way a minute quantity of oxalic acid is formed. Good lemon juice yields fully 5 per cent. of lemon acid, or 2 gallons yield about 1 pound of crystals. If the imported citrate of lime be used, a given quantity must be heated to redness, and then weighed, when the percentage of lime present will be ascertained; every 28 pounds of which will require 49 pounds of sulphuric acid of 1.845 (or a corresponding quantity containing exactly 40 parts of dry acid) for its complete decomposition.

3934. Tests for the Purity of Citric Acid. When pure, it does not yield a crystalline precipitate when added in excess to a solution of carbonate of potassa; such a precipitate indicates the presence of tartaric acid. It is entirely soluble in water, and what is thrown down by acetate of lead from this solution, is entirely soluble in dilute nitric acid. No salt of potassa, except the tartrate, yields a precipitate with the aqueous solution. It is entirely decomposed by heat; added sparingly to cold lime water, it does not render it turbid, and when a few drops of a solution of citric acid are added to lime water, a clear liquid results, which, when heated, deposits a white powder, soluble in acids without effervescence.

3935. Arsenious Acid. This is the *arsenic* or *white arsenic* of commerce, imported chiefly from Germany, also manufactured in quantity in Cornwall, England. It consists of large, glassy, colorless or yellowish-white, semi-transparent cakes or porcelain-like masses, which soon become opaque on their exterior, and sometimes friable and pulverulent. The transparent arsenic is found to be more than three times as soluble in water at 55° Fahr. than the opaque. In taste it is slightly sweetish, with a slight acidity and astringency, not perceived until some minutes after being swallowed, hence its dangerous character as a poison. Crude arsenic is obtained, as a collateral product, during the smelting of cobalt ores. Pure arsenic is obtained from the crude, by a second sublimation in cast-iron vessels. The arsenic, as imported, has usually been thus purified; and, unless otherwise adulterated, is sufficiently pure for general purposes. It is sometimes kept in fine powder, and in this state is occasionally found adulterated with powdered lime or chalk; it is, therefore, better to purchase it in the lump. The salts of arsenious acid are called ARSENITES.

3936. Self-Detecting Arsenious Acid. By adding a small quantity of any of the following substances to ordinary white arsenic, the mixture changes color when mixed with liquids. This is proposed as a method of preventing mistake in the use of this poisonous article.

The addition of a small quantity of a mixture of dry calomel and quicklime to the arsenic turns *black* when mixed with a liquid.

A mixture of thoroughly dry sulphate of iron and ferrocyanide of potassium turns it *blue*.

Dry sulphate of iron and dry sulphate of soda turns *green*.

3937. Tests for the Presence of Arsenious Acid. A weak solution of ammonio-acetate of copper added to a solution of white arsenic (arsenious acid) throws down a grass green precipitate of arsenite of copper (*Scheele's green*) This precipitate, after being washed, is soluble in nitric acid, and in ammonia; is turned a brownish-red by a solution of sulphuretted hydrogen, blood-red by ferrocyanide of potassium, and yellow by nitrate of silver.

Arsenious acid in solution throws down a yellow precipitate of arsenite of silver from a solution of ammonio-nitrate of silver.

There are a number of delicate tests employed for detecting the presence of arsenic in organic matter, such as the contents of the stomach or other viscera, all more or less involving the preparation of the matter before applying the tests, and requiring the manipulation of an experienced analytical chemist. A very susceptible test, and recommended by Cooley for its simplicity, is as follows: A solution of the suspected matter is strongly acidulated with muriatic acid in the proportion of 1 part muriatic acid to from 5 to 9 parts of the solution; this is boiled in a porcelain or glass vessel containing bright and clean metallic copper in the form of sheet, gauze, or wire. In about 15 minutes, if the solution be weak, or less, if strong, presence of arsenic will be noted by the characteristic iron-gray film of arsenic deposited on the surface of the copper. The copper, having been carefully washed and dried, may be cut into small pieces and heated in a test tube over a spirit lamp, when the metallic arsenic is volatilized, and will be condensed either in metallic form or in crystals of arsenious acid. This is known as *Rensch's test*.

3938. Arsenic Acid. An acid formed by the combination of metallic arsenic with oxygen. It is sour, reddens litmus, and forms salts with the bases, which are termed ARSENIATES. By careful evaporation it may be obtained under the form of small grains, but usually has the consistence of syrup, being very deliquescent.

3939. To Obtain Arsenic Acid. Pour 6 parts of strong nitric acid on 1 part of white arsenic (arsenious acid) in a glass vessel, and distill until the solution acquires the consistence of a syrup, then transfer it into a platina crucible, and expose it for some time to a faint dull red heat, to expel the nitric acid. The addition of a little muriatic acid facilitates the process.

3940. Tests for the Presence of Arsenic Acid. Sulphuretted hydrogen gives a yellow precipitate; nitrate of silver added to the solution of an arseniate gives a precipitate of a brick red color; nitrate of lead gives a white one, and the salts of copper a bluish colored one. Pure lump sugar dissolved in an aqueous solution of arsenic acid, becomes in a few hours of a reddish color, and afterwards of a magnificent purple. For some test purposes it will be advisable to add sulphurous acid to the suspected liquor, and boil it for a short time, when the arsenic acid will be reduced to arsenious acid, in which state it will be susceptible of more delicate tests. (*See No.* 3937.)

3941. Manganesic Acid—also called *permanganic acid*—may be obtained by mixing 8 parts of binoxide of manganese with 7 parts of chlorate of potassa, both in fine powder, adding 10 parts of hydrate of potassa, dissolved in a small quantity of water, evaporating to dryness, powdering, exposing the powder to a low red heat in a platinum crucible, dissolving the mass in a large quantity of water, decanting, evaporating, and crystallizing. These crystals are *permanganate of potassa*, from which the acid may be obtained by conversion into permanganate of baryta, and by careful decomposition by dilute sulphuric acid. (*Gregory.*) It has a fine red color, bleaches, and is rapidly decomposed by organic matter. It unites with some of the bases to form PERMANGANATES.

3942. Benzoic Acid. This is also called *flowers of benzoin* or *benjamin*. It has the form of white crystalline needles of a silky lustre, possessing an agreeable odor. Benzoic acid fuses at 230° Fahr., is volatile when heated, dissolves sparingly in cold water, with less difficulty in boiling water, and very freely in alcohol. Its salts are called BENZOATES.

3943. To Obtain Benzoic Acid. Put coarsely triturated benzoin into an iron pot with a flat bottom, whose diameter is from 8 to 9 inches; the benzoin forming therein a layer of from 1 to 2 inches in depth. The open end of the pot is then to be covered with a sheet of soft and loose blotting-paper, which must be attached to the rim with paste. A cone, formed with strong and thick paper, (cartridge paper), is then to be capped over the top of the pot, including the blotting paper; and this is also to be attached with paste and string. The apparatus, thus prepared, should then be placed on the sand-bath, and exposed from 4 to 6 hours to a gentle heat. After this lapse of time, it may be removed from the sand-bath, inverted, and the string detached, when beautiful white needles, of a silky lustre, possessing the agreeable odor of benzoic acid, will be found in the paper cone.

3944. To Obtain Anhydrous Benzoic Acid. Add oxychloride of phosphorus to an excess of benzoate of soda; agitate together, and wash the mixture with boiling water. The anhydrous benzoic acid sinks like a heavy oil, and crystallizes on cooling.

3945. Chromic Acid. This consists of acicular crystals of a crimson-red color and an acid metallic taste, deliquescent, and very soluble in water, forming an orange-yellow solution. With the bases this acid forms CHROMATES. Chromate of lead forms the pigment known as *chrome-yellow*.

3946. To Obtain Chromic Acid. Take 10 measures of a saturated cold solution of bichromate of potassa, mix with it 15 measures sulphuric acid, and allow the mixture to cool. The chromic acid is deposited in crystals, which, after decanting the mother liquid, are placed on a tile to drain, covered with a bell glass.

3947. Hydrocyanic Acid. This is also called *prussic acid*, and consists of a thin, colorless, and volatile liquid, having a strong odor of peach kernels. It boils at 79° Fahr. and solidifies at 45°; its specific gravity is .7058. It constitutes one of the most deadly poisons known. Its salts are HYDROCYANATES and METALLIC CYANIDES. Prussic acid, even when dilute, is very liable to spontaneous decomposition, and this speedily occurs when it is exposed to the light. To promote its preservation, it is usual to surround the bottles containing it with thick purple paper, and to keep them inverted in an obscure situation. The addition of a very small quantity of muriatic acid renders it much less liable to change, and is generally made by manufacturers for that purpose.

3948. To Obtain Anhydrous Prussic Acid. Pure crystallized ferrocyanide of potassium, 15 parts; water and sulphuric acid, of each 9 parts; distill in a glass retort into a well-cooled receiver, containing chloride of calcium in coarse fragments, 5 parts; stop the process as soon as the chloride in the receiver is perfectly covered by the distilled fluid, and decant the acid into a bottle furnished with a good stopper. Keep it in the dark, with the bottle inverted.

3949. Dilute Prussic Acid. Mix 41 grains muriatic acid with 1 fluid ounce distilled water, add 50½ grains cyanide of silver, and shake together in a well stopped phial. When the precipitate has subsided, pour off the clear dilute acid and keep for use. (*See No.* 3947.) (*U. S. Ph.*)

3950. Tests for the Presence of Prussic Acid. It is distinguished by a strong odor of bitter almonds.

Neutralized by potash, and tested with a solution of sulphate or tincture of iron, it gives a blue precipitate, or one turning blue on the addition of dilute sulphuric or muriatic acid. This test may be applied by spreading a single drop of solution of potassa over the bottom of a white saucer or porcelain capsule, and inverting it over another vessel of the same size containing the matter under examination. After 2 to 5 minutes remove the upper capsule; add to the potassa upon it, a single drop of a solution of sulphate or tincture of iron, and expose it to the air for a few seconds. Next add 1 or 2 drops of dilute sulphuric acid, when a blue color will be developed if hydrocyanic acid is present in the matter tested.

Nitrate of silver gives a white clotty precipitate, soluble in boiling nitric acid; and which, when dried and heated in a test tube, evolves fumes of cyanogen, which burn with a violet or bluish colored flame. A watch glass, moistened with this test and inverted over matter containing hydrocyanic acid, becomes opaque and white from the formation of cyanide of silver.

Liebig's test is considered the most delicate. Moisten a watch-glass or porcelain capsule with 1 or 2 drops of yellow hydrosulphuret of ammonia; invert it over the matter as before, and after a few minutes dry it with a gentle heat. A glass rod dipped in a solution of a persalt or sesquisalt of iron, drawn over the glass, will form a blood-red streak if the smallest quantity of hydrocyanic acid is present. (*Cooley.*)

3951. Test for the Strength of Prussic Acid. For estimating the strength of the commercial acid the following plan, proposed by Dr. Ure, will be found very exact and convenient. To 100 grains, or any other convenient quantity of the acid contained in a small phial, add in succession, small quantities of the peroxide of mercury in fine powder, till it ceases to be dissolved on agitation. The weight of the red precipitate taken up being divided by 4, gives a quotient representing the quantity of real prussic acid present. By weighing out beforehand, on a piece of paper or a watch-glass, 40 or 50 grains of the peroxide, the residual weight of it shows at once the quantity expended. The operation may be always completed in five minutes, for the red precipitate dissolves as rapidly in the dilute prussic acid, with the aid of slight agitation, as sugar dissolves in water. Should the presence of muriatic acid be suspected, then the difference in the volatility of prussiate and muriate of ammonia may be had recourse to with advantage; the former exhaling at a very gentle heat, the latter requiring a subliming temperature of about 300° Fahr. After adding ammonia in slight excess to the prussic acid, if we evaporate to dryness at a heat of 212°, we may infer from the residuary sal-ammoniac the quantity of muriatic acid present. Every grain of sal-ammoniac corresponds to .6822 grains of muriatic acid.

3952. Cyanic Acid. A compound of cyanogen and oxygen only known in its hydrated state in combination with 1 equivalent of water. It combines with bases to form CYANATES. When in contact with water for a few hours it suffers decomposition, and is converted into bicarbonate of ammonia. It cannot be preserved for any length of time, as it soon passes spontaneously into a white, opaque, solid mass, to which the name of *cyamelide* has been given, which may be reconverted into cyanic acid by distillation. It reddens litmus strongly.

3953. To Obtain Cyanic Acid. Distill dry cyanuric acid, or cyamelide, in a retort, and collect the product in a well-cooled receiver. It is also formed when cyanogen is transmitted over carbonate of potassa heated to redness; a cyanate of potassa results.

Or: Pass a current of sulphuretted hydrogen gas through water in which cyanate of silver is diffused. The sulphuretted hydrogen must not be passed so long as to decompose all the cyanate of silver; for then the cyanic acid is converted into other products by the excess of the sulphuretted hydrogen.

3954. Hydroferridcyanic Acid. This is sometimes written *hydroferricyanic acid*, and is a compound of ferridcyanogen and hydrogen. With the oxides of metals this acid forms FERRIDCYANIDES; the ferridcyanide of potassium is the *red prussiate of potash* used in the arts.

3955. To Obtain Hydroferridcyanic Acid. Prepared by decomposing recently precipitated ferridcyanide of lead by sulphuretted by hydrogen, or by sulphuric acid carefully added. A yellow solution is thus obtained, which yields a deep brown powder when evaporated by heat, or yellow crystals by spontaneous evaporation.

3956. Hydroferrocyanic Acid. A peculiar compound of cyanogen, hydrogen, and iron, discovered by M. Porret, and called by him *ferrochyazic acid.* It consists of white or yellowish white crystals, soluble in water and alcohol. With metallic oxides it combines to form FERROCYANIDES or PRUSSIATES. The ferrocyanide of potassium is the *yellow prussiate of potash* of commerce.

3957. To Obtain Hydroferrocyanic Acid. It may be obtained from a concentrated boiling solution of ferrocyanide of potassium, cooled out of contact with the air, and muriatic acid added in excess. The mixture is then agitated with a little ether, which separates the acid; the latter is collected by filtration, and dried.

3958. Lactic Acid. This is a limpid, syrupy liquid, colorless or of a pale wine color, with a slight odor and very sour taste. It is found in sour milk, and some other animal fluids, and in several vegetable juices, especially in that of beet-root. It unites with bases to form LACTATES.

3959. To Obtain Lactic Acid. Ferment whey by keeping it at a temperature of 70° to 80° Fahr.; evaporate the liquor to ⅛ its bulk; decant and filter, and then saturate it with milk of lime. This converts the lactic acid into lactate of lime, which remains in solution. The liquor is filtered again and precipitated by oxalic acid, which throws down oxalate of lime and sets free the lactic acid. The liquid is again filtered, and the filtrate consists of a solution of lactic acid, containing some sugar of milk (*lactin*) and certain salts. Next concentrate the solution to a syrupy consistence, and treat it with alcohol, which dissolves the acid and precipitates all the other matter. The solution is finally filtered and the lactic acid obtained pure by distilling off the alcohol. (*U. S. Dis.*)

3960. Fluoric Acid. This is more strictly *hydrofluoric acid*, as it is a compound of hydrogen and fluorine. Its combinations with bases are called FLUORIDES or HYDROFLUORATES. The well known mineral, *fluor-spar*, is a fluoride of calcium. Fluoric acid readily dissolves glass and silica, hence it is kept in bottles of lead, silver, platinum, or pure gutta-percha. It is highly corrosive and its vapor is poisonous. It is a colorless fluid which evaporates at 59° Fahr. in dense white fumes when exposed to the air, and has a powerful affinity for water.

3961. To Obtain Fluoric Acid. The anhydrous acid is made by distilling powdered fluor-spar with twice its weight of oil of vitriol in a leaden, or better, a silver alembic, the pipe of which fits into a bottle of the same material, surrounded with ice. But as it is usually required in a diluted state, water equal in weight to the spar may be put into the receiver. Great care must be taken, as the acid, both in its gaseous and liquid form, is very destructive.

3962. Chloric Acid. This is a yellowish liquid, smelling like nitric acid; it sets fire to paper or other dry organic matter. It is a compound of chlorine and oxygen, and in combination with bases forms CHLORATES.

3963. To Obtain Chloric Acid. Dissolve chlorate of baryta in 16 times its weight of water; then add dilute sulphuric acid until all the baryta be precipitated as sulphate. The clear liquid may then be concentrated by evaporation to a thin, oily consistence.

3964. Perchloric Acid. A colorless liquid of about 1.65 specific gravity, which fumes slightly in the air, attracts moisture, and distills unchanged at about 392° Fahr. (*Cooley.*) With bases it forms PERCHLORATES.

3965. To Obtain Perchloric Acid. To finely powdered perchlorate of potassa contained in a retort, add about ⅓ its weight of strong sulphuric acid, previously diluted with an equal weight of water. At about 284° Fahr., vapors of perchloric acid pass over and condense in the receiver. (*Cooley.*) No organic matter should be used as a lute for the joints of the apparatus; if any be needed, it should be of asbestos. By distilling the concentrated liquid acid with oil of vitriol at a gentle heat, crystals of perchloric acid will be deposited on the neck of the retort and in the receiver. These crystals fuse at 113° Fahr., and are very deliquescent. (*Booth.*)

3966. Butyric Acid. A thin, colorless, oily liquid, soluble in water and alcohol; specific gravity .963; boils at 327° Fahr. It may be procured from the butyrate of magnesia by adding a little sulphuric acid in quantity not quite sufficient to decompose the whole of the butyrate used; filter and distill the clear liquor, when the product will be butyric acid, from which the water may be removed by chloride of calcium. It forms *butyrates* with some of the bases. (*See No.* 1625.)

3967. Malic Acid. Malic acid is very soluble in water, slightly deliquescent, has a pleasant acidulous taste, and, when neutralized with the bases, forms salts called *malates.* When kept fused for some time at a low heat, it is converted into *paramalic* or *fumaric acid;* and when quickly distilled, it yields *maleic acid*, while fumaric acid is left in the retort. Malic acid forms with bases, MALATES; maleic acid, MALEATES. Take the juice of the fruit of the mountain ash, immediately after it has turned red, but still unripe; heat it to the boiling point, skim, filter, nearly neutralize with ammonia, and precipitate with a solution of 1 part of acetate of lead to every 72 parts of juice; filter, and again precipitate with nitrate of lead; allow the whole to stand until it forms a mass of crystals, then well wash, dry, powder, suspend in water, and decompose by a current of sulphuretted hydrogen; again filter, neutralize with ammonia, decolor with animal charcoal, a second time precipitate with nitrate of lead, and decompose the resulting nitrate of lead by sulphuretted hydrogen; lastly, filter, evaporate and crystallize. (*Winckler.*) Mr. Everett proposes the juice of the leaf-stalks of garden rhubarb as a source of malic acid. One imperial gallon of this juice contains 11,139¼ grs. of dry malic acid. The stalks should be peeled before pressing out the juice, as the cuticle contains much color. Everett's process is as follows: Neutralize with hydrate of lime, boil, filter, precipitate with nitrate of lead, allow it to stand for a few hours, boil, cool, filter, decompose the precipitate with sulphuric acid, avoiding excess, throw down the excess of lead from the supernatant portion with sulphuretted hydrogen, evaporate, and crystallize. Malic acid is also obtained from the juice of apples and several other sorts of fruit.

3968. Iodic Acid. A compound of iodine and oxygen, forming IODATES with the bases. It is deliquescent and very soluble in water, and detonates with inflammable bodies like the nitrates and chlorates.

3969. To Obtain Iodic Acid. Dissolve iodate of soda in sulphuric acid in considerable excess, boil for 15 minutes, and set the solution aside to crystallize.

Or: Iodine, 1 part; strongest (monohydrated) nitric acid, 4 parts; mix, and apply a gentle heat until the color of the iodine disappears, then evaporate to dryness and leave the residuum in the open air at a temperature of about 59° Fahr. When, by attracting moisture, it has acquired the consistence of a syrup, put it into a place where the temperature is higher and the air drier, when, in a few days, very fine white crystals of rhomboidal shape will form.

Mr. A. Connell's method is as follows: Put 50 grains of iodine into a large, tall flask: add 1 ounce of fuming nitric acid, boil, and as the iodine sublimes and condenses on the sides of the flask, continually wash it back again with the acid. Continue this until none of the iodine remains unchanged. Then pour the whole into a shallow evaporating dish, and evaporate to dryness. Redissolve, and again evaporate till all the nitrous acid is got rid of.

3970. Hydriodic Acid. This is a compound of iodine and hydrogen. In gaseous form it is colorless, fumes in the air, and is very soluble in water. In liquid form, when strong, it is very liable to change, and should be kept in well stoppered bottles. In combination with bases it forms HYDRIODATES. The hydriodates may be easily formed by saturating the acid with the oxides or hydrates of the bases, or more economically, by acting on the bases in water, with iodine. When the hydriodates are deprived of their water, they are true IODIDES; that is, simple combinations with iodine. (*See No.* 3853.)

3971. To Obtain Hydriodic Acid. Pour a little water over some periodide of phosphorus, previously put into a small glass retort, and apply a gentle heat, when gaseous hydriodic acid will be evolved, and phosphoric acid remain behind. The gas may be either collected over mercury or passed into water, when liquid hydriodic acid will be formed.

Or: Place iodide of barium in a retort, and decompose it with sulphuric acid, when pure hydriodic acid will be evolved.

3972. Dilute Hydriodic Acid. Take 1 troy ounce iodine in fine powder. Mix 30 grains of the iodine with 5 fluid ounces distilled water in a tall glass-stoppered ½ pint bottle, and pass into the mixture hydrosulphuric acid gas until the color of the iodine entirely disappears, and a turbid liquid remains. Next, gradually add the remainder of the iodine, stirring at the same time. Again pass the gas through the liquid until it becomes colorless, and decant it into a small matrass which it must nearly fill; boil it until it ceases to give off the odor of hydrosulphuric acid, and filter through paper, passing sufficient distilled water through the filter to bring the filtered liquid to 6 fluid ounces. Keep it in a well-stoppered bottle. (*U. S. Ph.*)

Alkalies.

Substances which possess the property of neutralizing acids and combining with them in fixed proportions, forming salts, and for the most part of turning the vegetable blues to greens, and yellow turmeric paper brown. The principal alkalies are soda, potassa, and ammonia. The first has been called the mineral, the second the vegetable, and the third the volatile alkali; but this distinction is now obsolete. Soda and potassa have also been called the fixed alkalies, from their permanence in the fire. The alkalies are strictly metallic oxides. The salts of the alkalies, both alone and carbonated, are generally freely soluble in water. The methods for ascertaining the strength of alkalies and their solutions will be found under Alkalimetry, No. 83.

3974. Potassa. Pure potash (not the potash of commerce, which is an impure carbonate of potassa), is the oxide of potassium. It occurs in two forms, *anhydrous* and *hydrated* potassa. As a general distinction, the term *potash* applies to the crude commercial, and *potassa* to the more purified or chemical preparations.

3975. Anhydrous Potassa. This is a volatile, fusible, white substance, intensely corrosive, and passing into the hydrate of potassa when moistened with water. It is obtained by the combustion of potassium in hot dry air.

3976. Hydrate of Potassa. Hydrated or *caustic potassa*, when perfectly pure, is white, solid, very soluble in water and in alcohol, very deliquescent, and corrosive. To obtain it, evaporate solution of potassa rapidly in an iron vessel over the fire until ebullition ceases and the potassa melts. Pour this into suitable moulds, and, when cold, put it into stoppered bottles.

3977. Tests for Potassa. Potassa may be distinguished from the other fixed alkalies (soda and lithia), by affording, when in solution, a white crystalline precipitate (cream of tartar) with an excess of tartaric acid; and a yellow one with bichloride of platinum. The flame of burning alcohol containing potassa has a reddish tint; soda colors it yellow.

3978. Soda. This substance bears the same relation to its metallic base, *sodium*, that potassa does to potassium, but its basic and alkaline action are rather less powerful than those of potassa.

3979. To Obtain Soda. Pure soda and hydrate of soda (caustic soda) are obtained from carbonate of soda in a similar manner to the same preparations of potassa. Caustic soda is occasionally called *sodic hydrate.*

3980. Tests for Soda. The flame of burning alcohol containing soda is of a yellow color. Hydrate of soda, after it has deliquesced in the air, speedily resolidifies by the absorption of carbonic acid, forming carbonate of soda, a salt marked by being easily crystallizable, and rapidly efflorescing in dry air. In solution, soda is not precipitated by tartaric acid. With sulphuric acid it yields a salt, which by its taste (intensely bitter) and form (six-sided prisms, transparent, and extremely efflorescent) is easily recognized as sulphate of soda (Glauber's salt).

3981. Ammonia. Pure ammonia is an incondensable colorless gas, possessing great pungency and acridness, and powerful alkaline properties. Water readily absorbs about 500 times its volume of this substance, and in this state forms strong liquid ammonia, which, when much more dilute, is popularly known as spirits of hartshorn, or water of ammonia. As usually met with in the form of a semi-crystalline whitish mass, commonly called smelling salts, it is combined with carbonic acid and water, forming a sesquicarbonate of this base. According to the theory of Berzelius, ammonia should be the oxide of ammonium, a supposed but undiscovered metal. Its presence can always be detected by its pungent odor.

3982. To Obtain Ammonia. Mix unslacked lime with an equal weight of sal-ammoniac, both dry and in fine powder; introduce the mixture into a glass retort, and join the beak by a collar of India-rubber to a glass tube about 18 inches long, which must lie horizontally, and have its beak bent up ready to be placed under a glass jar, on the shelf of a mercurial pneumatic trough. Heat being applied by means of a spirit-lamp, and the air contained in the apparatus having been expelled, the gas may be collected for use. Ammonia cannot be dried by means of chloride of calcium.

3983. Lithia. This is the oxide of lithium; is caustic, alkaline, and sparingly soluble in water. One of its most remarkable properties is its power of corroding platinum. In the form of the hydrate it is white and translucent, does not deliquesce, but absorbs carbonic acid and becomes opaque. It is to be obtained from various minerals, and is also found in some mineral waters; among which is that of the Gettysburg spring. Pure lithia may be obtained by decomposing sulphate of lithia by acetate of baryta, and by expelling the acetic acid from the filtered solution by heat.

3984. Tests for Lithia. It colors the flame of alcohol containing it a carmine red. It is distinguished from potassa and soda by its phosphate and carbonate being only sparingly soluble in water; from baryta, strontia, and lime, by forming crystallizable and soluble salts with sulphuric or oxalic acid; and from magnesia, by the solution of its carbonate exhibiting an alkaline reaction.

3985. Baryta. This alkaline earth is the oxide of barium, and is found abundantly in the form of native sulphate and carbonate of baryta. With the acids it forms salts which are all more or less white; except the sulphate, they are soluble in water, or in dilute muriatic acid, and are extremely poisonous.

3986. To Obtain Pure Baryta. Ignite pure crystallized nitrate of baryta in a capacious porcelain crucible, until red vapors cease to be evolved. This forms a grayish white mass or powder, which, on the addition of water, slacks like lime, but with the evolution of more heat.

3987. To Obtain Hydrated Baryta. It may be precipitated from a solution of either nitrate or chloride of barium, by adding to it a solution of pure potassa or soda, collecting and drying the precipitate.

It is obtained in crystals, by boiling a strong solution of sulphuret of barium with successive portions of black oxide (protoxide) of copper, until it ceases to give a black precipitate with a salt of lead. The liquid, after filtration, yields crystals of hydrate of baryta on cooling.

3988. Test for Baryta. Its solutions give an immediate clear white precipitate with dilute sulphuric acid, which is insoluble in both acids and alkalies.

3989. Strontia. An alkaline earth, the oxide of a metal called strontium. It greatly resembles baryta. Hydrate of strontia is freely soluble in boiling water, and the saturated solution deposits crystals on cooling. The solution exhibits an alkaline reaction, and, like baryta, is precipitated white by sulphuric acid and the alkaline sulphates and carbonates. It is distinguished from baryta by its inferior solubility and by its soluble salts giving a red tinge to flame, while the salts of baryta impart a yellow tinge. The salts of strontia may all be prepared by dissolving the native carbonate in the respective acids. The nitrate is the only one met with in commerce, and is employed to form colored fireworks.

3990. Magnesia. An alkaline earth, the oxide of the metal magnesium, in the form of a very light, white, odorless and tasteless powder, almost insoluble in cold and boiling water. It slowly absorbs carbonic acid from the atmosphere. With the acids it forms salts, most of which may be made by the direct solution of the earth, or its hydrate or carbonate. It dissolves in hydrochloric acid without effervescence. Neither bicarbonate of potassa nor chloride of barium throws down anything from the solution. It turns turmeric paper brown when moistened.

3991. To Obtain Magnesia. Expose carbonate of magnesia in a crucible to a full red heat for 2 hours, or till the powder suspended in water does not effervesce on the addition of muriatic acid. On the large scale, covered crucibles, made of porous earthenware, are employed as the containing vessels, and the heat is applied by placing them in a sort of furnace or oven heated with coke.

3992. Test for Magnesia. Magnesia is precipitated as a bulky white hydrate, by pure alkalies; and as a bulky white carbonate by the carbonates of potassa and soda. Both the above precipitates dissolve in nitric and muriatic acid, forming salts which are very deliquescent, and soluble in alcohol.

Solutions of magnesian salts are not precipitated by the alkaline sulphates or sulphuric acid, nor, when very dilute, by oxalate of ammonia. By these tests it may be distinguished and separated from lime. These tests distinguish it from the other earths, and its insolubility in alkaline solutions marks its difference from alumina.

3993. Lime. A highly acrid, alkaline and caustic earth, less insoluble in cold than in hot water. It is the oxide of calcium. When heated to a high degree, it becomes intensely luminous, and is well known in use as the calcium light.

3994. To Obtain Lime. Lime, or *quicklime*, is obtained by exposing limestone, or chalk, which are carbonates of lime, to a red heat. *Shell-lime* is got in the same manner from the shells of the oyster and other shell-fish. When sprinkled with water, heat is generated, and the lime, combining with the water, crumbles down into a powder, which is *hydrate of lime*, or slacked lime.

3995. Tests for Lime. The alkaline carbonates, phosphates, oxalates, and sulphates, occasion white precipitates in solutions of lime. The precipitates occasioned by the first three tests are soluble in dilute nitric or muriatic acid; that by the last is insoluble in those menstrua, but soluble in solution of salt, and not reprecipitated by dilute sulphuric acid.

Oxalate of ammonia or potassa is the most delicate test of lime. If the substance under examination be a solid, dissolve it in muriatic acid, filter, evaporate to dryness, redissolve in water, and test as above. All the soluble salts of lime tinge the flame of alcohol of an orange color, but this may be confounded with the color produced by the salts of strontia.

Alkaloids.

Alkaloids. Substances of a vegetable origin, analogous to the alkaline bases, in which the medicinal activity of the plants in which they are found appear to reside. (*Cooley.*) Among the natural organic bases, or alkaloids, the following are the principal, as enumerated by Professor Fownes.

3997. Morphine or Morphia. This is the chief active principle of opium. The morphia of commerce is a white crystalline powder; but when crystallized in alcohol, forms brilliant, prismatic, transparent, and colorless crystals, which turn nitric acid red. In powder, unlike strychnine, it is fusible without decomposition, and strongly decomposes iodic acid. It is insoluble in ether, scarcely soluble in water, and freely soluble in alcohol. Potassa and ammonia precipitate morphia from the solutions of its salts.

3998. To Find the Percentage of Morphia in Opium. An excellent process for ascertaining the quality of opium is to boil an infusion of 100 grains opium with 25 grains quicklime, made into a milk with water; to filter while hot, saturate with a dilute hydrochloric acid, and to precipitate the morphia by ammonia. After expelling any excess of ammonia by heat, the precipitate is collected, dried, and weighed; the weight in grains will nearly represent the percentage of morphia in the opium.

3999. Narcotine. An alkaloid found in the insoluble portion of opium, and forms small, colorless, brilliant crystals, which give to nitric acid an orange tint.

4000. Codeine, or Codeia. Obtained from hydrochlorate of morphia, in colorless, transparent, eight-sided crystals, which do not color nitric acid red.

4001. Thebaine, or Paramorphine. This is also obtained from opium in colorless needles like those of narcotine. It does not color nitric acid red, and is much less soluble in water than codeine.

4002. Cinchonine, or Cinchonia. This is the active principle of Peruvian bark, contained in the largest quantity in the pale bark. It crystallizes in small, brilliant, transparent, four-sided prisms, insoluble in ether. *Cinchonicine* and *cinchonidine* are other varieties of this alkaloid.

4003. Quinine, or Quinia. This is also obtained from Peruvian bark, being found in largest quantity in the yellow variety of the bark. It crystallizes in small white needles. It may be distinguished from cinchonine by the form of its crystals, and its solubility in ether.

4004. Quinoidine, or *Amorphous Quinine*, is a yellow or brown resinous mass, identical in composition with quinine. *Quinicine* and *quinidine* are also varieties of quinine. (*See Nos.* 4025, *&c.*)

4005. Strychnine, or Strychnia. This is an alkaloid contained in nux vomica, and some other vegetable substances. Crystallizes in small, brilliant, eight-sided crystals, insoluble in absolute alcohol, and slightly soluble in water. It suffers decomposition on fusing, and does not decompose iodic acid; it may be thus distinguished from morphine.

4006. Brucine, or Brucia. Is obtained from the same sources as strychnine, and resembles it in many respects, but is readily soluble in all strengths of alcohol, and insoluble in water. Brucine turns nitric acid red, which becomes violet on the addition of protochloride of tin.

4007. Veratrine, or Veratria. The alkaloid principle of cevadilla seeds, and of white hellebore. When pure, it is a white powder; but as usually met with, the powder is yellowish or greenish-white, insoluble in water.

4008. Colchicine. Extracted from the seeds of the common meadow saffron; has similar properties to veratrine, but is crystalline, and soluble in water.

4009. Harmaline. A substance forming yellowish prismatic crystals, obtained from the Peganum Harmala, a plant abounding in southern Russia. By oxidation it yields *Harmine*, a fine red dye-stuff, also possessing basic properties.

4010. Theine, or Caffeine. This is an alkaloid principle extracted from tea, coffee, Paraguay tea, &c. It forms in tufts of white silky needles.

4011. Theobromine. A white crystalline powder obtained from the cacao-nuts from which chocolate is prepared. Its properties are somewhat similar to theine.

4012. Xanthine. A white powder, which may be obtained from guanine, which it resembles in its properties. It dissolves easily in ammonia or potash.

4013. Creatine. This alkaloid, called by some *kreatine*, is a crystallizable substance obtained from the juice of the muscular fibre of animals. It forms brilliant, colorless prismatic crystals. Creatine is a neutral body combining with neither acids nor alkalies. By the action of strong acids it is converted into *creatinine*, a powerful organic base, with a strong alkaline reaction, and forming crystallizable salts with acids. Creatine, treated by boiling with a solution of baryta, produces *Sarcosine*.

4014. Sarcine. This base is a constituent of the flesh of animals. It forms in delicate white microscopic needles, soluble with difficulty in cold water, easily in boiling water. It is obtained from the same source as creatine. (*See No.* 4013.)

4015. Guanine. A base obtained from guano. It is a colorless, crystalline powder, insoluble in water, alcohol, ether or ammonia. By treating guanine with muriatic acid and chlorate of potassium, *guanidine* is obtained in colorless crystals, readily soluble in water and alcohol.

Guanine, Sarcine, and Xanthine greatly resemble one another.

4016. Berberine. An alkaloid crystallizing in fine yellow needles slightly soluble in water, extracted from Barberry root.

4017. Piperine. An alcoholic extract of pepper forming colorless or yellowish crystals. Insoluble in water.

4018. Conine, or Conia. An alkaloid extract of hemlock, in the form of a volatile, oily liquid. It evolves an odor of hemlock on being moistened with a solution of potassa.

4019. Nicotine, or Nicotia. This is also a volatile, oily, acrid liquid, soluble in water, ether, alcohol, and oils. Nicotine, moistened with a solution of potassa, evolves a strong odor of tobacco.

4020. Sparteine. An alkaloid obtained from broom, also a volatile, oily liquid. Conine, nicotine, and sparteine are similar in character, being very poisonous, possessing strong alkaline reaction, and forming crystallizable salts with the acids.

4021. Salicine. A white, crystalline substance, found in the bark and leaves of several kinds of poplar and willow; but most abundantly in the white willow and the aspen. It is obtained by the careful evaporation of an infusion in cold water.

4022. To Obtain Alkaloids. Some of these substances require special processes for extracting them from the substances in which they are found, but the following methods will apply for general purposes:

When the base is insoluble in water, non-volatile, and existing in the plant in an insoluble form. Boil or macerate the bruised plant in water acidulated with muriatic acid, filter, neutralize the acid with an alkali, (ammonia, lime, or magnesia), and collect the precipitate, which must be purified by resolution in dilute acid, digestion with animal charcoal, and subsequent crystallization or precipitation by an alkali; or the first precipitate may be purified by dissolving it repeatedly in alcohol.

When the base is insoluble in water, and non-volatile, but existing in the plant in a soluble state. Boil or macerate in hot water as before; filter and precipitate by adding an alkali; purify as last.

When the base is soluble in water, and non-volatile. Make an infusion with a dilute acid (muriatic); concentrate by a gentle heat; treat the liquor with potassa and ether (conjointly); decant and evaporate.

When the base is both soluble in water and

volatile. The vegetable or its extract may be mixed with potassa and distilled; the product, neutralized with oxalic or sulphuric acid, carefully evaporated to dryness, and digested in alcohol, and this solution agitated with potassa and ether; the ethereal solution thus formed, if carefully evaporated, leaves the base nearly pure. It may be further purified by cautious distillation.

4023. Tests for Distinguishing Alkaloids. Perchloride of gold is a decisive test of certain vegetable alkalies. The following are the colors of the precipitates which it produces with the salts of the annexed alkalies dissolved in water; *quinine*, buff-colored; *cinchonine*, sulphur-yellow; *morphine*, yellow, then bluish, and lastly, violet; in this last state the gold is reduced, and the precipitate is insoluble in water, alcohol, the caustic alkalies, and sulphuric, nitric, or hydrochloric acids; *brucine*, milk, coffee, and then chocolate-brown; *strychnine*, canary-yellow; *veratrine*, slightly greenish-yellow. All these precipitates, with the exception mentioned, are very soluble in alcohol, insoluble in ether, and slightly soluble in water. Among the reactions of chloride of gold, there are two which appear to be especially important: they are those which occur with morphine and brucine; these are sufficiently marked to prevent these alkalies from being mistaken for each other, and also yield pretty good characteristics for distinguishing brucine from strychnine.

4024. Alkaloids Detected by Picric Acid. Hager has found that this acid precipitates various alkaloids from their solutions, such as brucine, strychnine, veratrine, quinine, cinchonine, and some alkaloids of opium. Morphine and atropine, however, are precipitated only from neutral and concentrated solutions, and the precipitate dissolves pretty easily in water. Glucosides, casein, and pseudo-morphine resist the action of the picric acid.

4025. Quinometry. The method of estimating the quantity of quinine in cinchona bark and the salts obtained from it. The following tests give very accurate results in examining the bark; and the salts are tested in the same way, but the result is not quite so accurate, as it includes any *quinidine* (*see No.* 4028) that may be present in the quinine; and makes, therefore, the apparent richness of the sample greater than it really is. (*Cooley.*)

4026. Test for the Strength of Quinine. Make a decoction of 100 grains of bark in 2 fluid ounces distilled water; filter, and precipitate with a sufficient quantity of a concentrated solution of carbonate of soda. Heat the fluid until the precipitate is dissolved; and when cold, dry and weigh it. It should weigh 2 grains or more, and dissolve entirely in a solution of oxalic acid. To render the result strictly accurate, the bark should be exhausted with ether, and the mixed solutions evaporated. Salts of quinine may be tested in the same manner. (*Cooley.*)

4027. Test for the Percentage of Quinine in Bark, &c. Exhaust 100 grains of bark with acidulated water; filter the solution, and render it alkaline with liquor of potassa; next agitate it with about ⅓ its volume of chloroform, and allow it to repose a short time; the chloroform, holding the alkaloid in solution, sinks to the bottom of the vessel in a distinct stratum, from which the supernatant liquid must be separated by decantation; the chloroformic solution, either at once or after being washed with a little cold water, is allowed to evaporate, and the weight of the residuum in grains gives the percentage of richness of the sample. Ether may be used instead of chloroform, but the ethereal solution will form the upper stratum instead of the lower. This test is also applicable to the salts of quinine, but with restrictions referred to in No. 4025. (*Rebourdain.*)

4028. Quinidine. An alkaloid found in quinia which has been prepared by precipitation. It is distinguished from quinine by not striking a green color when treated with chlorine followed by ammonia, as quinine does. In medicinal character its powers are comparatively feeble. It is present in nearly all the ordinary sulphate (disulphate) of quinine as sold, either through careless preparation or wilful adulteration, and is not detected by, and consequently included in, the results of the usual tests for quinine. (*See Nos.* 4025, *&c.*) *Cinchonine* is another feebler alkaloid also found in quinia.

4029. Ure's Test for the Presence of Quinidine or Cinchonine in Quinine. This test is applicable to quinine salts generally, but more especially refers to the sulphate (disulphate) of quinine. Place 10 grains of the salt into a strong test tube, furnished with a tightly-fitting cork; add 10 drops of a mixture of 1 part sulphuric acid and 5 parts water, and 15 drops water, accelerating solution by a gentle heat. When dissolved and entirely cooled, add 60 drops officinal sulphuric ether with 20 drops spirits of ammonia, close the test tube with the thumb, and shake it well; cork the tube closely and shake gently from time to time, so that the bubbles of air may readily enter the layer of ether. If the salt be free from, or contain no greater proportion than 10 per cent. of quinidine, it will be entirely dissolved; while on the surface of contact between the two strata of fluid, the mechanical impurities only will be separated. From this it appears that 10 grains of the salt may contain 1 grain of quinidine, and still a complete solution take place; but, in this case, the quinidine will shortly begin to crystallize in a layer of ether. If more than 10 per cent. of quinidine be present, there will be found an insoluble precipitate between the strata of the fluid. If this be quinidine, it will be dissolved by the addition of ether, while *cinchonine* (*see No.* 4002) will be unaffected.

Gas. A general term applied to all aëriform or permanently elastic fluids, excepting the compound of oxygen and nitrogen constituting the atmosphere, which is distinguished from the other gaseous bodies by the name of *air.* (*See No.* 4072.) Gases for chemical purposes are usually generated in a bottle of glass or other appropriate material; or, where the application of heat is necessary, in a retort. A connecting tube of convenient shape is fitted air-tight into the neck or beak of the generating vessels, through which the gas is led into receiving vessels. These are usually bottles, with accurately fitting stoppers.

4031. Pneumatic Trough. A vessel or tank nearly filled with water, provided with a shelf placed 1 or 2 inches below the surface. The receiving bottles are first immersed in and filled with the water and then placed neck downwards on the shelf, which is furnished with holes to allow of the passage of the gas into the receivers from the connecting tube, the end of which is brought immediately under one of the holes. For gases which are easily absorbed by water, mercury or some other fluid is necessary in place of the water. As the gas ascends into the receiving bottle, the water is displaced; when full, and the gas begins to escape, the bottle should be closed with a greased stopper, and removed from the trough.

4032. To Find the Weight of a Gas. Multiply the specific gravity of the gas by 309⅓ (the weight in grains of 1000 cubic inches of air), the product will be the weight of 1000 cubic inches of the gas.

4033. To Prevent the Escape of Gas from India-Rubber Tubing. India-rubber tubing is slightly permeable to gas. The amount which escapes through the walls of the tube is very small; but it may be advisable sometimes to render an escape impossible. This can be done by giving the tubing a thin coating of a varnish made by dissolving 1½ parts molasses and 2 parts gum-arabic in 7 parts of white wine and 3¼ parts strong alcohol. The molasses and gum must first be dissolved in the white wine, and the alcohol must be added very slowly, constantly stirring the mixture, or the gum will be thrown down.

4034. Oxygen. An elementary gas, colorless, tasteless, odorless, and incombustible, having a specific gravity of about 1.057. Oxygen enters largely into the composition of all nature; being a constituent part of the atmosphere (*see No.* 4072), upon which it confers the power of supporting life and combustion; and water, present more or less through the whole world, contains about 88 per cent. by weight, or 33 per cent. by volume, of oxygen; it constitutes also a portion of the majority of the mineral bodies that form the bulk of our globe. It is a powerful supporter of combustion, and its presence is essential to the existence of animal and vegetable life. Oxygen unites with certain other bodies in fixed proportions to form a class of acids distinguished as oxygen acids or *oxacids.* (*See No.* 3853.)

4035. To Obtain Oxygen Gas. Heat in a retort or flask, finely powdered chlorate of potassa, mixed with about one-fourth its weight of black oxide of manganese. The gas must be collected by attaching a tube to the flask, and passed into a receiving bottle in a pneumatic trough. (*See No.* 4031.)

Or: Take chloride of potassa, or red oxide of mercury, expose it to the heat of a spirit-lamp, in a suitable vessel, and collect the gas.

4036. Oxygen Gas from Bleaching Powder. Oxygen gas can be readily prepared by boiling bleaching powder (hypochlorite of lime) and nitrate of cobalt in a flask. Make a clear solution of the powder in water, put it into any convenient flask provided with a perforated cork and tube, and pour in a few drops of a solution of nitrate or chloride of cobalt, and set it to boil. The gas, as it is evolved, is collected in a receiving bottle. (*See No.* 4031.)

4037. To Obtain Oxygen Without Heat. According to M. Boettger, oxygen can be obtained in a very pure state by employing binoxide of barium and peroxide of lead. Take equal parts of these substances and pour on weak nitric acid; the reaction commences immediately, and the gas can be collected as usual over cold water. (*See No.* 4031.)

4038. Pure Oxygen for Inhalation. Eliot recommends for the preparation of oxygen gas, to be used in medicine, the employment of a mixture of equal parts of peroxide of barium and peroxide of lead. By pouring dilute nitric acid upon these salts, there is a violent effervescence and a copious evolution of pure oxygen gas. For greater security, the gas may be afterwards washed in water. As very little heat is necessary, the operation can be performed in any stout bottle, thus dispensing with the usual retorts. For great purity, the first portion of gas that evolves should be allowed to escape, as it contains the air which was in the apparatus.

4039. To Obtain Oxygen on the Large Scale. Nitre is exposed to a dull red heat in an iron retort or gun barrel; 1 pound of nitre thus yields about 1200 cubic inches of oxygen, slightly contaminated with nitrogen. (*Ure.*)

4040. Tests for Oxygen. It is distinguished from other gases by yielding nothing but pure water when mixed with twice its volume of hydrogen and exploded, or when a jet of hydrogen is burned in it. A recently extinguished taper, with the wick still red hot, instantly inflames when plunged into this gas. A small spiral piece of iron wire, ignited at the point and suddenly plunged into a jar of oxygen, burns with great brilliancy and rapidity.

4041. Hydrogen. A gaseous element, colorless, combustible, and the lightest of ponderable bodies, its specific gravity being only .06935. It is a constituent part (about 12 per cent. by weight, and 67 per cent. by volume) of water. According to Dumas, "it is a gaseous metal, as mercury is a liquid metal." It forms an ingredient in all bodies that possess the power of burning with flame; it burns with a pale blue flame, and, in combination with carbon, constitutes the illuminating gas in general use. In contact with spongy platinum it inflames spontaneously; and, from its extreme lightness, is the best means employed for inflating balloons. It is one of the most useful elements in the material world. Hydrogen forms, with other bodies, a class of acids called hydrogen acids or *hydracids*. (*See No.* 3853.)

4042. To Obtain Hydrogen Gas. Hydrogen gas is readily procured by pouring on fragments of zinc, in a glass bottle, or flask with a bent tube, or retort, some diluted sulphuric acid (1 measure of strong acid to 5 of water). It may be collected over water. If zinc be not at hand, fine iron wire, or the turnings or filings of iron, may be substituted for it. To procure gas of great purity, distilled zinc must be used, and it is advisable to pass the gas first through alcohol, and then through a concentrated solution of pure potassa. Care must be taken that all the air has been driven out of each vessel before any light is applied, or an explosion will ensue.

4043. Cheap Method of Obtaining Hydrogen. Take quicklime, slack it, let it cool and crumble into a dry hydrate; then mix it with charcoal, coke, or peat, and heat in a retort. The hydrate of lime (slacked lime) gives up the water that was used in slacking it, and becomes quicklime. The water is decomposed into hydrogen and carbonic acid, and these two gases can be separated by passing them through water, or the carbonic acid may be economized by employing it in the manufacture of bicarbonates. The quicklime can be again slacked and used as often as required.

4044. Hydrogen Gas for Balloons. For this purpose hydrogen may be obtained by pouring slightly diluted muriatic acid upon an equal weight of zinc, in a covered vessel having a small tap or stop-cock in the top for filling the balloons. The vessel should be made of lead, to prevent corrosion.

4045. To Estimate the Buoyant Power of Balloons. It will take about 12 cubic feet of the hydrogen gas, used for inflating balloons to balance or suspend 1 pound in the air. The rule used for balloons is as follows: The specific gravity of the gas compared with the air is .0693; 1 cubic foot of air weighs 527.04 grains, the cubic foot of gas weighs 36.93 grains; and therefore there are 527.04−36.93=490.11 grains difference between the air and gas, in one cubic foot. Multiply this difference by the number of cubic feet in the balloon, and divide by 7,000. This will give the capacity or buoyancy of the balloon, in pounds; then subtract the weight of the balloon and car.

4046. For Obtaining Hydrogen in Quantities. Place iron wire in a gun-barrel, or a porcelain tube, open at both ends, to one of which attach a retort containing water, and to the other a bent tube, connected with a pneumatic trough. The gun-barrel must now be heated to redness, and the water in the retort brought into a state of brisk ebullition, when the vapor will be decomposed, the oxygen being absorbed by the iron, and the hydrogen escaping into the gas receiver. The gas evolved may be purified, if desired, by passing it through alcohol, &c. (*See No.* 4042.)

4047. Tests For Hydrogen. Hydrogen is recognized by its combustibility; by the pale color of its flame; by producing water only when burnt in air or oxygen; by extinguishing the flame of other bodies; and by exploding when mixed with half its weight of oxygen and fired. (*Cooley.*)

4048. Carburetted Hydrogen. There are two leading gaseous compounds of carbon and hydrogen, known as carburetted hydrogen, and distinguished as light and heavy.

The light carburetted hydrogen is often abundantly disengaged in coal mines, and called *methane*, and *fire damp*. It consists of 2 equivalents of hydrogen and 1 of carbon, and burns with a yellowish flame. This gas also escapes in bubbles from the mud on the bottom of stagnant pools, combined with carbonic acid, from which it may be freed by passing through milk of lime, or a solution of caustic potassa. (*Cooley.*) It has a specific gravity of about .559. (*Fownes.*)

Heavy carburetted hydrogen is a combination of 2 equivalents of carbon and 2 of hydrogen (4 carbon and 4 hydrogen—*Booth*), and burns with a white luminous flame; it is a little lighter than air, having a specific gravity of .981. It is also called *Ethine.*

4049. To Obtain Light Carburetted Hydrogen. When 2 parts crystallized acetate of soda, 2 parts dry hydrate of potassa, and 3 parts powdered quicklime, are strongly heated in a flask or retort, this gas is abundantly evolved, and may be collected over water. (*See No.* 4031.)

4050. To Obtain Heavy Carburetted Hydrogen. Heavy carburetted hydrogen is prepared by heating in a retort 1 part of alcohol with 6 or 7 of oil of vitriol until it blackens, and conducting the mixed gases through milk of lime, which retains the sulphurous acid; and afterwards through oil of vitriol, which absorbs water, ether, and alcohol. This may also be prepared by passing the vapor of boiling alcohol through a mixture of 10 parts oil of vitriol and 3 parts water, heated to ebullition (320° to 330° Fahr.), and purifying the vapor as before.

4051. Olefiant Gas. A name given to heavy carburetted hydrogen, arising from its producing, in combination with chlorine, an oily-looking liquid. It is the presence of this gas which gives the illuminating power to *coal-gas*, which is a combination of light, heavy, and other hydrocarbons.

4052. Sulphuretted Hydrogen. A compound of hydrogen and sulphur; a colorless gas, possessing a powerful odor of rotten eggs; specific gravity 1.171; it is absorbed by water, forming liquid sulphuretted hydrogen, or hydrosulphuric acid. It is a powerful poison. Being considerably denser than air, it may be poured from its generating bottle into cavities, a scheme successfully employed by M. Thenard to destory rats in their holes, a method equally applicable to other vermin. It forms saline compounds with the alkalies, and the earths termed HYDROSULPHATES or HYDROSULPHURETS, and it precipitates metallic sulphurets from solutions of most of the metals; hence its value as a test. Air containing $\frac{1}{10000}$ of sulphuretted hydrogen will sensibly blacken a piece of white paper, moistened with a solution of acetate of lead. Sulphuretted hydrogen is the active ingredient in the sulphurous mineral waters.

4053. To Obtain Sulphuretted Hydrogen. Mix together 2 parts of iron filings with 1 of sulphur into a thin pap with water, and heat it gently in an iron vessel. Combination takes place with the evolution of heat forming *sulphuret of iron*. Cover it till cold. On this compound, contained in a glass bottle, or other suitable apparatus, pour sulphuric acid previously diluted with 7 parts of water. If more acid be afterwards required, dilute the strong acid with only 4 of water. The resulting gas is absorbed by water, and is therefore collected, in preference, over mercury. This is the plan commonly adopted in the laboratory.

To obtain it pure, mix 1 part finely powdered tersulphuret of antimony, and 5 parts strong muriatic acid, in a small glass retort or flask: apply the heat of a spirit lamp, and collect the gas over mercury. (*See No.* 4031.)

4054. Tests for Sulphuretted Hydrogen. Sulphuretted hydrogen may be recognized by the odor, and by its blackening moist carbonate of lead, and tarnishing silver, and also by its precipitating arsenious acid yellow, tartar emetic orange, and the salts of lead black.

4055. Phosphuretted Hydrogen. This is a gaseous combination of phosphorus and hydrogen; colorless, very fetid, slightly soluble in water, and burns with a white flame. It has a specific gravity of 1.24.

4056. To Obtain Phosphuretted Hydrogen. The pure gas may be evolved by gently heating hydrated phosphorus acid in a small retort, and collecting it by a pneumatic trough. (*See No* 4031.)

The spontaneously inflammable variety of this gas is made by boiling phosphorus with solution of potash in a small retort, the beak of which is kept under water: as each bubble of gas rises from the water, it inflames, and forms a ring of white smoke, which dilates as it ascends. The spontaneous inflammability of the gas, when mixed with atmospheric air or oxygen, renders caution necessary in its preparation.

4057. Nitrogen or Azote. An elementary gaseous body. Pure nitrogen is a colorless, odorless, tasteless gas, neither combustible nor capable of supporting combustion or respiration. It is neutral to test paper, does not affect lime water, and is only slightly absorbed by pure water. Liebig places its specific gravity at 0.9722, Berzelius at 0.976.

4058. Tests for Nitrogen. It is recognized by its purely negative qualities, and by its forming nitric acid when mixed with oxygen, and exposed to the electric spark; or when a jet of hydrogen is burnt in the mixed gases. (*Cooley.*)

4059. To Obtain Nitrogen. Atmospheric air may be made to yield an unlimited supply of nitrogen, by exposing it to the action of substances which combine with its oxygen. By burning a small piece of phosphorus, placed on a capsule floating on the water in a pneumatic trough, under a large bell-glass, and allowing it to stand over the water a few hours, nearly pure nitrogen is obtained, which may be further purified by agitating it with solution of pure potassa. It may be dried by passing it through concentrated oil of vitriol.

Nitrogen may be evolved by passing chlorine gas into a solution of pure ammonia, and drying, as before, through sulphuric acid.

Another plan, well recommended, is to heat bichromate of ammonia in a retort. The evolved nitrogen is deprived of all aqueous vapor by sulphuric acid as above, or by letting it stand over fused chloride of calcium.

4060. Protoxide of Nitrogen. This gas is also called *nitrous oxide*, and is largely used by inhalation, under the name of *laughing gas*, to produce insensibility to pain. It is colorless, possesses an agreeable odor, and a sweetish taste. At 45° Fahr., and under a pressure of 50 atmospheres, it is liquid. Its specific gravity is 1.5241; it supports combustion, and is absorbed by water. Its most remarkable property is its action on the system when inspired. A few deep inspirations are usually succeeded by a pleasing state of excitement, and a strong propensity to laughter and muscular exertion, which soon subside, without being followed by languor or depression. Its effects, however, vary with different constitutions.

4061. To Prepare Laughing Gas. Evaporate a solution of nitrate of ammonia until a drop of the fused mass placed on a cold plate instantly solidifies; cool, break the lump into pieces, and place it in a stoppered bottle. For use, a portion is introduced into a glass retort, and heat applied by means of a spirit lamp. As soon as the heat reaches 480° Fahr., protoxide of nitrogen is evolved, and may be collected in bladders, gas bags, a gasometer, or in the pneumatic trough over warm water. (*See No.* 4031.) Should white fumes appear within the retort after the evolution of the gas has commenced, the heat should be lowered, as, when heated to about 600°, nitrate of ammonia explodes with violence. Nitrous oxide may also be made in the same way from crystallized nitrate of ammonia, or by exposing nitric oxide for some days over iron filings, but it requires great care in its preparation.

4062. Test for Pure Laughing Gas. When pure, it is colorless, has an agreeable odor, and does not affect a solution of nitrate of silver.

4063. Carbonic Acid. An invisible acidulous gas, formed by the union of 1 equivalent of carbon with 2 of oxygen, having a specific gravity of 1.524, and highly soluble in water. Its general properties and the methods of obtaining it will be found in Nos. 3913, &c. The application of this gas to the purposes of wine-making, &c., is given in No. 718. The methods for obtaining carbonic acid gas are given in No. 3914.

4064. Carbonic Oxide. A colorless, neutral gas, formed of equal equivalents of carbon and oxygen, and has a specific gravity of .915. It burns with a pale blue flame, and is even more poisonous than carbonic acid.

4065. To Obtain Carbonic Oxide. Carbonic oxide may be obtained from carbonic acid gas by passing the latter over fragments of charcoal heated to redness in a tube of porcelain or iron.

Also, by treating binoxalate of potassa with 5 or 6 times its weight of oil of vitriol in a glass retort, at a gentle heat.

Equal weights of chalk (or carbonate of soda) and iron filings (or charcoal), strongly heated in an iron retort or gun barrel, will evolve the gas rapidly.

Whichever way the gas is evolved, it must be passed first through a caustic alkaline solution or milk of lime, to deprive it of carbonic acid, and next over dried chloride of calcium, to deprive it of moisture. It may be collected either over mercury or water, as the latter absorbs but very little.

4066. Sulphurous Acid. This is a gaseous combination of 1 equivalent of sulphur and 2 of oxygen, having a specific gravity of 1.45, and very soluble in water, which will absorb 30 times its volume of the gas. Its properties, preparation, and application to the arts. will be found in Nos. 3864, &c.

4067. Ammonia. A highly pungent gas formed by the union of 1 equivalent of nitrogen with three of hydrogen. Its specific gravity is .589. (*See Nos.* 3981, *&c.*) Double salts of ammonia are sometimes called AMMONIURETS. Thus, sulphate or nitrate of copper precipitated in solution by ammonia, and the precipitate redissolved by an excess of ammonia, may be called ammoniurets of copper, but more correctly ammonio-sulphate, or ammonio-nitrate of copper.

4068. Hydrochloric Acid. A gaseous acid formed of equal equivalents of hydrogen and chlorine. (*See Nos.* 3882, *&c.*)

4069. Chlorine. An elementary gas, of a yellowish green color, a pungent, suffocating odor, and an astringent taste. Its specific gravity is 2.47. Under a pressure of 4 atmospheres it condenses into a yellow limpid liquid. Its most remarkable properties are its power of destroying almost all animal and vegetable color, and the putrid odor of decomposing organic matter. It has a very strong attraction for metals. With bases chlorine forms CHLORIDES or CHLORURETS. (*See No.* 3853.)

4070. To Obtain Chlorine. This gas is obtained, for laboratory use, &c., by mixing together in a glass flask or retort, strong muriatic acid with half of its weight of finely-powdered peroxide of manganese. Or: Pour common muriatic acid, diluted with an equal weight of water, upon half its weight of chloride of lime. Chlorine gas is immediately evolved even in the cold, but much more rapidly on the application of a gentle heat. This gas must be collected in clean dry bottles by displacement. The tube conducting the gas must reach to the bottom of the bottle, when the chlorine, being heavier than the air, will displace the latter, without mixing with it. The bottle is known to be full by the gas overflowing the mouth, which is easily perceived by its green color. The bottle must now be closed up with an accurately fitting stopper, previously greased, and an empty one put in its place, which is subsequently treated in like manner. To free the gas entirely from muriatic acid, it may be passed through water; and to render it dry, it may be passed over dry chloride of calcium. Chlorine gas may also be collected over a saturated solution of common salt in the pneumatic trough, if the presence of moisture be no objection. (*See No.* 4031.)

4071. Tests for Chlorine. This gas is readily distinguished from other gases by its color, odor, and bleaching properties. Its aqueous solution dissolves gold leaf, and instantly blackens a piece of silver plunged into it. It rapidly destroys the color of iodide of starch, solution of indigo, litmus, and turmeric. A simple method of detecting free chlorine is to hold a rod, dipped in water of ammonia, over it, when white fumes of sal-ammoniac will be formed; this, coupled with the property of bleaching colors, may, in most cases, be taken as evidence of the presence of this substance.

4072. Air. The air or atmosphere which surrounds the earth is a mixture (not combination—*Fownes*) of 77 parts by weight (or 79.19 parts by measure) of nitrogen, and 23 parts by weight (or 20.81 by measure) of oxygen. It usually contains also a variable amount of moisture, a very small proportion of carbonic acid, a trace of ammonia, and sometimes of carburetted hydrogen; these last are found incidentally in the air, in a variable degree. It is the standard in the comparative or specific gravity of gaseous bodies. (*See No.* 47.) At 60° Fahr., and with the barometer at 30 inches, 100 cubic inches of air weigh 30.935 grains; and water (the standard of specific gravity for fluids) weighs just 816 times as much as air.

4073. Tests for Pure Air. A simple method of ascertaining the presence of impurity (carbonic acid) in the atmosphere, is to nearly fill a glass tumbler with limewater, and to place it in any convenient position, as on the mantelpiece of a room. The rapidity with which a pellicle forms on its surface, or the water becomes cloudy, corresponds to the amount of the carbonic acid present in the atmosphere that surrounds it.

A little moist carbonate of lead put on a plate or saucer, and exposed in the same way, will turn black, should any sulphuretted hydrogen be contained in the air. This is a very delicate test for that destructive gas

---

Miscellaneous Chemicals.

It is proposed, in this place, to give a concise description of the chemical compounds referred to in the various departments of this book. A complete list of chemicals would not be necessary for the scope of the work, which is a purely practical one; such information only is therefore given as is deemed necessary to render the whole thoroughly intelligible, and as complete as possible. A considerable number of them are inserted, for the sake of clearness, in connection with the process or special purpose for which their use is directed. These will be found in their proper place by reference to the INDEX.

4075. Chloride of Gold. Gold unites with chlorine under two different proportions, and are distinguished as the *protochloride* and the *terchloride* of gold.

The terchloride of gold, or *auric chloride*, consists of 3 equivalents of chlorine and 1 of gold, and is obtained by dissolving 1 part gold in 3 parts nitro-muriatic acid (aqua-regia); evaporate until vapors of chlorine begin to be disengaged, and then set the solution aside to crystallize. It forms orange-red crystalline needles, or ruby-red prismatic crystals; is deliquescent, and soluble in water, ether, and alcohol, forming a deep yellow solution. (*Cooley.*)

The protochloride of gold, or *aurous chloride*, consists of 1 equivalent each of chlorine and gold, and is obtained by evaporating the terchloride to dryness and exposing it to a heat of 450° (440° *Fownes*) Fahr., until chlorine ceases to be evolved. It forms a yellowish-white mass, insoluble in water; but it is decomposed by water, slowly when cold, and rapidly by the aid of heat, into metallic gold and terchloride of gold. (*Cooley.*)

4076. Tests for the Presence of Gold in Solutions. Protosulphate of iron gives a brown precipitate, which acquires a metallic lustre when rubbed.

Protochloride of tin (preferably containing a little perchloride) gives a violet, purple, or blackish precipitate, insoluble in muriatic acid.

Sulphuretted hydrogen and hydrosulphuret of ammonia give a black precipitate, insoluble in simple acids.

Ammonia gives a reddish-yellow precipitate (fulminating gold) with tolerably concentrated solutions, either at once, or on boiling the liquid.

Liquor of potassa gives, with neutral solutions of gold, a similar precipitate to that formed by ammonia, insoluble in excess.

4077. Fused Nitrate of Silver. Take 3 ounces refined silver, 2 fluid ounces nitric acid, and 5 fluid ounces distilled water; mix in a glass flask and apply a gentle heat until the metal is dissolved. Transfer the solution to a porcelain capsule or crucible, decanting it off a heavy black powder which appears at the bottom of the flask; evaporate the solution to dryness; raise the heat, in a dark room, until the mass liquefies, then pour it into hinged brass or iron moulds furnished with cylindrical cavities of the size of a goose-quill. Keep the product, which is *Lunar Caustic*, or *fused nitrate of silver*, in well stopped bottles, impervious to the light.

Crystallized (or crystals of) *Nitrate of Silver* is obtained by dissolving grain silver (*see No.* 3217) in nitric acid diluted with twice its weight of water; evaporating the solution until it will crystallize on cooling very slowly. (*See No.* 3215.)

4078. Oxide of Silver. Dissolve 2 parts nitrate of silver, and 1 part hydrate of potassa, each separately, in distilled water; mix the solution, and, after frequent agitation during an hour, collect and wash the precipitate, and dry it by a gentle heat in the shade. This is more strictly the *protoxide of silver*, and is in the form of a pale brown powder.

4079. To Reduce Solid Silver from its Chloride. Mix together the dry chloride of silver in $\frac{1}{3}$ its weight of powdered black resin; heat moderately in a crucible until the flame ceases to have a greenish blue color; then increase the heat suddenly until the silver fuses into a button at the bottom of the crucible. Some parties recommend an addition of a little powdered calcined borax, sprinkled on the surface before increasing the heat. (*See No.* 3214.)

4080. To Prepare Nitrate of Silver from an Alloy of Silver and Copper. Palm's method. When it is desired to prepare nitrate of silver from silver containing copper—coins for example—filter the nitric acid solution, dissolve the alloy in nitric acid, evaporate it nearly to the consistence of oil, not to dryness, and add to a part of this concentrated metallic solution, $\frac{4}{5}$ part of nitric acid free from chlorine. The silver salt precipitates in the form of crystals and the copper remains in the solution. Wash the precipitate 2 or 3 times with concentrated nitric acid, and evaporate to dryness. The more concentrated the nitric acid, the more completely is the silver salt precipitated; an acid of 1.250 specific gravity is sufficient, however, to separate completely the two metals. (*See No.* 3216.)

4081. Sulphate of Silver. Prepared by dissolving silver in sulphuric acid containing one-tenth of nitric acid; or by precipitating a solution of the nitrate by another of sulphate of soda. It dissolves in 80 parts of hot water, and falls in small needles as the solution cools. (*Cooley*). According to Fownes it dissolves in 88 parts boiling water.

4082. Sulphuret of Silver. A greyish-black substance prepared by passing sulphuretted hydrogen gas through a solution of nitrate of silver. It may also be obtained by melting sulphur and silver together.

4083. Tests for Silver in Solution. Silver is entirely soluble in diluted nitric acid. This solution, treated with an excess of muriate of soda, gives a white precipitate entirely soluble in ammonia water, and a fluid which is not affected by sulphuretted hydrogen. The nitric solution of silver also gives a white curdy precipitate (chloride of silver) with muriatic acid, soluble in ammonia and insoluble in nitric acid, and blackened by exposure to light. It gives white precipitates with solutions of the alkaline carbonates, oxalates, and prussiates. It gives yellow precipitates with the alkaline arsenites and phosphates. With the arseniates, red precipitates. With the fixed alkalies, brown precipitates. With sulphuretted hydrogen, a black powder. With phosphorus and metallic copper or zinc, a precipitate consisting of pure silver.

4084. Chloride of Platinum. The commercial chloride of platinum is the *bichloride*, formed by dissolving platinum in nitro-muriatic acid (aqua-regia), and evaporating the solution to dryness at a gentle heat. It is reddish-brown, deliquescent, and very soluble in water and in alcohol, yielding orange-colored solutions. (*Cooley.*) (*See No.* 3220.)

4085. Protochloride of Platinum. This is formed by exposing the dried and powdered bichloride (*see No.* 4084) for some time to a temperature of 450° Fahr. It is a greenish-grey, powder, insoluble in water, but soluble in muriatic acid.

4086. Ammonio-Chloride of Platinum. A solution of sal-ammoniac is added to a strong solution of bichloride of platinum (*see No* 4084), avoiding excess; the precipitate is collected on a filter, washed with a little weak alcohol, and dried at a heat not exceeding 180° Fahr. It consists of minute, transparent, yellow crystals, very feebly soluble in water, less so in dilute alcohol, and insoluble in acids. By heating to redness, it is converted into spongy platinum. (*See No.* 3336.)

4087. Tests for Solutions of Platinum. Sulphuretted hydrogen throws down from neutral and acid solutions of platinum, a blackish-brown precipitate, which is only formed after a time in the cold, but immediately on heating the liquid. Sal-ammoniac and chloride of potassium give yellow crystalline precipitates, insoluble in acids, but soluble in excess of the precipitant, upon the application of heat, and decomposable by heat, with production of spongy platinum. Ammonia and potassa also give similar precipitates in solutions previously acidulated with hydrochloric acid. (*Cooley*).

4088. Subacetate of Copper. A green or bluish-green powder, better known as *verdigris.* This may be made by spreading the marc of grapes, or pieces of cloth dipped in crude acetic acid, upon plates of copper, with exposure to the air for several weeks. (*Fownes.*)

4089. Binacetate of Copper. Verdigris, dissolved in vinegar with the aid of heat, forms dark green or blue crystals of binacetate of copper. This is the commercial *acetate of copper*.

4090. Ammonio-Sulphate of Copper. A dark blue pulverulent substance, formed by rubbing together 1 ounce sulphate of coppper and $\frac{1}{2}$ ounce sesquicarbonate of ammonia, until carbonic acid ceases to be evolved; then drying the product, wrapped in bibulous paper, in the air.

4091. Nitrate of Copper. This consists of deep blue, very deliquescent crystals, obtained by dissolving pure copper in dilute nitric acid. (*See No.* 97.)

4092. Protoxide of Copper—also known as *black oxide of copper*—may be formed by calcining metallic copper, nitrate of copper, or the hydrate, thrown down from solutions of the salts of copper by means of pure potassa. This preparation was formerly called the deutoxide of copper. It is not changed by heat, but readily gives out its oxygen when heated with combustible matter; hence its general use in organic analysis for supplying oxygen. It communicates a beautiful green color to glass and enamels.

4093. Sulphite of Copper. To a concentrated solution of bisulphite of potash add a cold solution of sulphate of copper, filter, and heat gently.

4094. Suboxide of Copper. This is the *red oxide of copper*, obtained by heating together in a covered crucible 4 parts copper filings, and 5 parts black oxide of copper. (*See No.* 4092.) Or: Mix 100 parts sulphate of copper with 57 parts carbonate of soda, (both in crystals), and fuse them at a gentle heat; cool, pulverize, add 25 parts fine copper filings, ram the mixture into a crucible, cover it over, and expose it for 20 minutes to a white heat. It consists of a superb red powder with a metallic lustre. It is used as a pigment and a bronze, and as a stain for glass and enamel, to which it gives a rich red color. Heat converts it into the black oxide. With ammonia it forms a colorless solution, which rapidly becomes blue from the action of the air. (*Cooley.*)

4095. Peroxide of Copper. An oxide formed by the action of peroxide of hydrogen water on the hydrated black oxide.

4096. Sulphate of Copper. The *blue vitriol* of commerce is obtained from the native sulphuret of copper (copper pyrites). Pure sulphate of copper is made by the direct solution of the metal, or preferably, of its oxide or carbonate, in sulphuric acid. It consists of fine blue crystals, slightly efflorescent in the air. By heat it loses its water of crystallization and crumbles into a white powder. (*See No.* 120.)

4097. Chloride of Copper. Dissolve black oxide of copper in muriatic acid; evaporate and crystallize. It forms green, deliquescent crystals, soluble in alcohol, the flame of which it turns green; exposed to a heat under 400° Fahr. it becomes anhydrous, assuming the form of a yellow powder.

4098. Ferrocyanide of Copper. Called also *Prussiate of Copper.* Precipitate a solution of a salt of copper (sulphate of copper, for instance,) with another of yellow prussiate of potash; collect the powder, wash it with water, and dry. Has a beautiful red-dish-brown color, and is sometimes used as a pigment.

4099. Tests for Copper Solutions. The solutions of copper possess a blue or green color, which they retain, even when considerably diluted with water.

With caustic potassa they give a light blue bulky precipitate, turning blackish-brown or black on boiling the liquid.

Ammonia and carbonate of ammonia produce a bluish-white precipitate, soluble in excess of ammonia, yielding a rich deep blue solution.

The carbonates of potassa give a similar precipitate to the last, but insoluble in excess of the precipitate.

Ferrocyanide of potassium gives a reddish-brown precipitate. Sulphuretted hydrogen and hydrosulphuret of ammonia give a blackish-brown or black one.

A polished rod of iron, on immersion in an acidulated solution, quickly becomes coated with metallic copper.

4100. Delicate Test for Iron and Copper. The alcohol tincture of logwood will produce a blue or bluish-black tint in water which has been run through iron or copper pipes, when neither tincture of galls, sulphocyanide, nor the ferrid and ferrocyanides of potassium show any reaction.

4101. Acetate of Lead. Acetate of lead should be completely soluble in distilled water, and when the lead is exactly precipitated with dilute sulphuric acid, or by sulphuretted hydrogen, the clear supernatant liquid should be wholly volatilized by heat without residue. Sulphuric acid poured on acetate of lead evolves acetic vapors. Acetate of lead is powerfully astringent. Take 4 pounds 2 ounces oxide of lead (litharge), acetic acid (specific gravity 1.048), and distilled water, of each 4 pints; mix the fluids, add the oxide, dissolved by a gentle heat, strain, evaporate, and crystallize. On the large scale it is usually prepared by gradually sprinkling oxide of lead into strong vinegar, heated in a copper boiler rendered negative-electric by having a large flat piece of lead soldered within it, constant stirring being employed until the acid is saturated, when the mother liquors of a former process may be added, the whole heated to the boiling point, allowed to settle till cold, decanted, evaporated to about the specific gravity 1.266 or 1.267, and then run into salt-glazed stoneware vessels to crystallize. The best proportions are, finely powdered litharge 13 parts, and acetic acid specific gravity 1.0482 to 1.0484, 23 parts. These ingredients should produce about 38½ parts of crystallized sugar of lead. A very slight excess of acid should be preserved in the liquid during the boiling and crystallization, to prevent the formation of any basic acetate, which would impede the formation of regular crystals.

4102. Chloride of Lead. This is a white crystalline powder, called also *muriate of lead.* Precipitate a solution of 19 ounces acetate of lead in 3 pints boiling distilled water, with a solution of 6 ounces chloride of sodium in 1 pint boiling water; when cold, wash and dry the precipitate. It may be obtained in brilliant colorless needle-shaped crystals, by dissolving finely powdered litharge in boiling dilute hydrochloric acid. Filter while hot, and the crystals form on cooling.

4103. Iodide of Lead. Acetate of lead, 9 ounces; water, 6 pints; dissolve; iodide of potassium (pure), 7 ounces; water, 2 pints; dissolve. Add the latter solution to the former, wash and dry the precipitate, and keep it from the light. Or: Iodide of potassium and nitate of lead, of each 1 ounce; dissolve each separately in ¾ pint of water, mix, collect the precipitate in a muslin or linen filter, and wash it with water; then boil it in 3 gallons water, soured with pyroligneous (acetic) acid, 3 fluid ounces; let the solution settle (still keeping the liquid near the boiling point), and decant the clear; as the water cools, the iodide will subside in brilliant golden yellow lamellæ, or minute crystals.

The latter is the best process, as any adhering oxide of lead is dissolved out by the acid. (*Cooley*).

4104. Chromate of Lead. To a filtered solution of acetate or nitrate of lead, add a filtered solution of chromate of potassa, as long as a precipitate forms, which is collected, washed with water, and dried. This forms *chrome-yellow.* (*See No.* 2705.)

4105. Dichromate of Lead. Boil pure carbonate of lead with chromate of potassa in excess, until it assumes a proper red color; then wash it with pure water, and dry it in the shade. This is the pigment known as *chrome-red.* (*See No.* 2706.)

4106. Litharge. Litharge is an oxide of lead prepared by scraping off the dross that forms on the surface of melted lead exposed to a current of air (dross of lead), and heating it to a full red, to melt out any undecomposed metal. The fused oxide in cooling forms a yellow or reddish semi-crystalline mass, which readily separates into scales; these, when ground, constitute the powdered litharge of commerce. Litharge is also prepared by exposing red lead to a heat sufficiently high to fuse it, and English litharge is obtained as a secondary product by liquefaction, from argentiferous lead ore. The litharge of commerce is distinguished by its color into *litharge of gold,* which is dark colored and impure, and *litharge of silver,* which is purer, and paler colored. The dark color of the former is chiefly owing to the presence of red lead. In grinding litharge, about 1 pound of olive oil is usually added to each 1 cwt., to prevent dust. Litharge is employed in pharmacy, to make plasters and several other preparations of lead; by painters as a dryer for oils, and for various other purposes in the arts.

4107. Nitrate of Lead. Litharge, 4½ ounces; diluted nitric acid, 1 pint; dissolve by a gentle heat, and set the solution aside to crystallize. Employed as external application in cutaneous affections, &c. A very weak solution is an excellent remedy for chapped hands, &c.

4108. Tests for the Presence of Lead in its Solutions. The presence of lead in solutions may be recognized by the effects produced by the following reagents:

The addition of sulphuretted hydrogen, hydrosulphuret of ammonia, or the alkaline sulphurets, to a solution containing lead, give black precipitates, insoluble in cold dilute acids, alkalies, alkaline sulphurets, and cyanide of potassium.

Caustic potassa or soda gives a white precipitate, soluble in excess of the precipitant.

Ammonia throws down a white precipitate, insoluble in excess, from all the solutions of lead salts, except that of the acetate.

Dilute sulphuric acid, in excess, also solutions of the sulphates, give a white precipitate, insoluble in dilute nitric acid, but soluble in a solution of potassa.

Chromate of potassa gives a yellow precipitate, whose soluble qualities are the same as that from sulphuric acid last mentioned.

Iodide of potassium gives a yellow precipitate, soluble in acetic acid, a solution of potassa, alcohol, and boiling water; from boiling water it is deposited in small, brilliant, golden-yellow scales, as the liquid cools. (*See also Nos.* 2694, &c.)

4109. To Prepare Chloride of Zinc. Dilute 1 pint hydrochloric acid with 1 quart water, add to it 7 ounces zinc in small pieces; when the effervescence is nearly finished, apply heat until bubbles cease to be evolved; decant the clear and evaporate to dryness. Fuse the product in a lightly covered crucible, by a red heat; pour it out on a flat, smooth stone, and, when cold, break it into small pieces, and preserve it in a well-stoppered bottle.

4110. Ammonio-Chloride of Zinc. By dissolving 68 parts chloride of zinc and 54 parts sal-ammoniac, a crystallizable salt is formed, which dissolves oxides of copper and of iron, and is useful in tinning or zincing those metals.

4111. Chloride of Zinc. Dissolve 2½ troy ounces zinc in small pieces, in sufficient muriatic acid; strain the solution, add 60 grains nitric acid, and evaporate to dryness. Dissolve the mass in 5 fluid ounces water, add 60 grains chalk, and let the mixture stand for 24 hours; then filter, and evaporate to dryness. Lastly, fuse the dry mass, pour it out on a flat stone, and, when it has congealed, break the mass in pieces and keep in a well-stoppered bottle. (*U. S. Disp.*)

4112. Precipitated Carbonate of Zinc. Take 12 troy ounces each sulphate of zinc and carbonate of soda; dissolve each separately in 4 pints water; mix the solutions and let the powder subside; pour off the liquid, wash the precipitate with hot water until the washings are nearly tasteless, and dry with a gentle heat. (*U. S. Ph.*)

4113. Tutty Powder. Impure oxide of zinc. It is a substance which collects in the chimneys of the furnaces in which the ores of zinc are smelted.

4114. To Prepare Pure Sulphate of Zinc. Mix 4 ounces laminated or granulated zinc with 3 fluid ounces oil of vitriol, and 1 pint water, in a porcelain capsule, and when gas ceases to be evolved, boil for 10 minutes, filter through muslin, and evaporate to dryness; next dissolve it in 1 pint water, agitate this solution frequently during 6 hours with ½ ounce prepared chalk, and filter it; add to the filtered solution 1 fluid drachm each commercial nitric acid and dilute sulphuric acid; evaporate the mixture until a pellicle forms on the surface, and set it aside to crystallize; dry the crystals on bibulous paper without heat, and preserve them in a bottle. The mother liquor will yield more crystals by further evaporation. This substance is also known as *white vitriol.*

4115. Cyanide of Zinc. Add a solution of cyanide of potassium to a solution of pure sulphate of zinc; wash and dry the precipitate.

4116. Flowers of Zinc. This is obtained by the rapid combustion of zinc in a deep crucible, placed sideways in a furnace, so that the flowers (oxide of zinc) may be collected as they form.

4117. Oxide of Zinc. Place carbonate of zinc in a covered clay crucible, and expose to a very low red heat, until a portion taken from the centre of the mass ceases to effervesce on being dropped into dilute sulphuric acid. This is the commercial zinc-white. (*See No.* 2696.)

4118. Tests for the Solutions of Zinc. The solutions of zinc are precipitated white by the pure alkalies and carbonate of ammonia, but are completely redissolved by excess of the precipitant. The carbonates of potassa and soda give a permanent white precipitate of carbonate of zinc. Hydrosulphuret of ammonia also gives a white precipitate, and so does sulphuretted hydrogen when the solution is quite neutral. Prussiate of potash gives a gelatinous white, or bluish-white precipitate.

4119. Protoxide of Tin. Usually termed *oxide of tin.* Precipitate a solution of protochloride of tin with carbonate of potassa, wash and dry the powder at a heat under 166° Fahr., with as little exposure to the air as possible. It is a white or greyish-white powder, soluble in acids and in the pure fixed alkalies. If it be heated in an atmosphere of carbonic acid it loses its water and changes to a dense black powder, which is anhydrous protoxide. (*Cooley.*)

4120. Sesquioxide of Tin. A grey, slimy precipitate, soluble in muriatic acid, and in ammonia, obtained by mixing fresh, moist hydrated peroxide of iron with a solution of the neutral protochloride of tin. (*Fuchs*).

4121. Binoxide or Peroxide of Tin. Obtained by adding potassa, or an alkaline carbonate, to a solution of perchloride of tin. This substance is also known as *Stannic acid;* hence, its compounds with alkalies are sometimes called STANNATES. It is soluble in acids, and in pure alkalies. If grain tin be heated in a test tube with nitric acid, the tin is converted, with evolution of yellow fumes, into a white powder, peroxide of tin. The nitric acid will convert the tin into an oxide, but it cannot combine with the oxide produced. (*Stöckhardt.*) From this it appears that *nitrate of tin* is a misnomer.

4122. Tin or Polishers' Putty. Melt tin with rather more than an equal quantity of lead, then rapidly raise the heat till the mixture is red hot; the tin will then be thrown off in dross, which should be removed as it forms. This dross is the *peroxide of tin*, or tin putty; the dross may be calcined until it becomes whitish, and then reduced to powder.

4123. Protochloride of Tin. *Muriate of tin* is obtained by distilling a mixture of chloride of mercury and tin in fine powder. It is grey, solid, resin-like, fusible, and volatile. (*Cooley.*)

4124. Perchloride of Tin. Called also *Bichloride* and *Permuriate of Tin.* The pure bichloride is obtained by heating the protochloride in chlorine gas, or by distilling a mixture of 8 parts of grain tin with 24 parts of corrosive sublimate, when a very volatile, colorless liquid comes over, which was formerly called *Libavius' fuming liquor.* A solution of the bichloride or permuriate of tin is obtained by dissolving tin in nitromuriatic acid. This solution is much used by dyers, under the name of Spirits of Tin, Dyers' Spirits, &c. (*See Nos.* 107, *&c.*) For this purpose, the acid is best made by mixing 2 parts of muriatic acid with 1 part each of nitric acid and water, all by measure. (*Liebig*). The tin should be added by degrees, one portion being allowed to dissolve before adding another; as, without this precaution, the action is apt to become violent, the temperature rise, and peroxide of tin to be deposited. (*See No.* 108.) A process which has been highly recommended, is to prepare a simple solution of the protochloride, and to convert it into the bichloride, either by the addition of nitric acid and a gentle heat, or by passing chlorine through it.

4125. Tests for the Salts of Tin. The salts of tin are characterized by the following general properties: Ferroprussiate of potash gives a white precipitate. Hydrosulphuret of potash, a brown-black with the protoxide, and a golden-yellow with the peroxide. Galls do not affect the solutions of these salts. Corrosive sublimate occasions a black precipitate with the protoxide salts; a white with the peroxide. A plate of lead frequently throws down metallic tin, or its oxide, from the saline solutions. Chloride of gold gives, with the protoxide solutions, the purple precipitate of Cassius. Chloride of platinum occasions an orange precipitate with the protoxide salts. (*Cooley.*)

4126. Ethiops of Antimony. Triturate together 3 parts sulphuret of antimony, and 2 parts black sulphuret of mercury.

4127. Flowers of Antimony. Throw powdered sulphuret of antimony, by spoonfuls, into an ignited tubulated retort that has a short and very wide neck, until as many flowers collect in the receiver as are required.

The *argentine flowers* are thus prepared: Keep metallic antimony melted in a vessel, freely exposed to the air, and furnished with a cool place for the flowers to rest upon; collect the flowers as deposited. According to Berzelius, these are sesquioxide of mercury.

4128. Liver of Antimony. Melt together 1 part sulphuret of antimony, and 2 parts dry carbonate of soda (or potash), and heat until it acquires a proper color; then cool and powder it. *Crocus of antimony* is sometimes sold for the above, but the latter is prepared by deflagrating equal parts of antimony and saltpetre (nitrate of potassa), a small portion at a time, and the fused mass, separated from the dross, reduced to fine powder. (*Cooley.*)

4129. Potassio-Tartrate of Antimony. Commercial *Tartar Emetic.* Take 2 troy ounces oxide of antimony, and 2½ troy ounces bitartrate of potassa, both in very fine powder; mix them together, and add them to 18 fluid ounces boiling distilled water in a glass vessel. Boil for 1 hour, filter while hot, and set aside to crystallize. Dry the crystals, and keep in a well-stoppered bottle. By further evaporation the mother-water will yield more crystals, which should be purified by a second crystallization. (*U. S. Ph.*)

4130. Oxide of Antimony. Insert 4 troy ounces sulphuret of antimony in very fine powder into a quart flask; add 18 troy ounces muriatic acid, and digest in a sand-bath until effervescence ceases. Then remove the bath and add 600 grains nitric acid, and when nitrous fumes cease to be given off, and the liquid has grown cold, add it to ½ pint water, and filter. Pour the filtrate gradually into 12 pints water, constantly stirring, and wash the precipitate twice by decantation, using each time spirits water; drain it through muslin, and then wash it with water until the washings cease to have an acid reaction. Add 1½ fluid ounces water of ammonia, and, after standing 2 hours, filter through wet muslin, and wash with distilled water as long as the washings form a precipitate with nitrate of silver. Then dry with a gentle heat on bibulous paper. (*U. S. Ph.*) A greyish-white powder, insoluble in water, soluble in muriatic and tartaric acids.

4131. Butter of Antimony. The liquid *chloride of antimony,* commercially known by this name, is usually made by dissolving crude or roasted black antimony in muriatic acid with the addition of a little nitric acid. It usually contains pernitrate of iron.

4132. Sulphuret of Antimony. The black sulphuret (*tersulphuret*) of antimony is prepared from commercial sulphuret of antimony or by elutriation, in the same manner as directed for prepared chalk. (*See No.* 1292.) The *commercial sulphuret* is obtained from the native gray antimony ore by fusion; this separates the sulphuret from the less fusible earthy matter; it is then run into cakes. (*Cooley.*)

Mixtures of an acidulated menstruum or syrup with a sulphuret of antimony, are apt to disengage sulphuretted hydrogen, when there is much of them, if kept in a warm room. The rule should be to prepare as small a quantity as possible, and to keep the bottle cool. (*Eymael.*)

4133. Penta-Sulphuret of Antimony. Called also *golden sulphuret of antimony.* Boil together for some hours 72 parts tersulphuret of antimony, 68 parts dry carbonate of soda, 52 parts fresh hydrate of lime, and 13 parts sulphur; filter, evaporate, and crystallize. Redissolve the crystals (*Schlippe's salt*), add dilute sulphuric acid, collect the golden floculent precipitate, wash it with cold distilled water, and dry with a gentle heat. (*Liebig.*)

4134. Nitrate of Bismuth. The *neutral nitrate* is made from 2 ounces pure bismuth broken into fragments, dissolved by heat in 6 ounces nitric acid, adding more acid, if necessary, to effect entire solution. Add to the solution half its volume of distilled water, filter through powdered glass, and crystallize by evaporation. (*Cooley.*)

4135. Subnitrate of Bismuth. This is also called *trisnitrate of bismuth, magistery of bismuth*, and *pearl white.* It is insoluble in water, but freely soluble in nitric acid. Dissolve 2 ounces bismuth in 3 fluid ounces nitric acid, previously diluted with 2 fluid ounces distilled water; then add 3 quarts cold water, and allow the white precipitate to subside. Afterwards decant the clear liquor, wash the powder, and dry it by a gentle heat. (*Br. Ph.*)

4136. Oxide of Bismuth. The *anhydrous oxide* is made by exposing the nitrate or subnitrate to gentle ignition in a crucible. This is a straw-yellow colored powder. The *hydrated oxide* is a rich-looking white powder, obtained thus: Dissolve 2 pounds bismuth in 2½ pounds nitric acid, and drop it gradually into a solution of 3 pounds carbonate of potassa in twice it weight of water, rendered caustic by previous treatment with quicklime (*see No.* 101); wash the precipitate well with cold water.

4137. Tests for the Salts of Bismuth. Tin, copper, iron, and zinc throw down bismuth from its solutions in the metallic state. If a salt of bismuth be heated with carbonate of soda by the flame of a blowpipe, a bead of the metal, surrounded by a crust of yellow oxide, is obtained. The brittleness of the bead under the hammer distinguishes it from lead. The salts of bismuth are mostly devoid of color; some are soluble, others insoluble. The soluble salts redden litmus paper; and, when the solution contains but little free acid, and is largely diluted with water, a subsalt, more or less soluble, is deposited. This property of forming subsalts is very characteristic. (*Makins.*)

4138. Chloride of Mercury. This preparation is usually known as *calomel.* Boil, by means of a sand-bath, 24 troy ounces mercury with 36 troy ounces sulphuric acid, until a dry white mass is left. Rub this, when cold, with 24 ounces mercury in an earthenware mortar until thoroughly mixed; add 18 troy ounces chloride of sodium, triturate until the globules of mercury cease to appear, and sublime the mixture. Reduce the sublimate to a very fine powder and wash it with boiling distilled water until the washings afford no precipitate with water of ammonia, and dry it. (*U. S. Ph.*)

4139. Bichloride of Mercury. The *corrosive sublimate* of the drug stores. Boil 24 troy ounces mercury in 36 troy ounces sulphuric acid, by means of a sand-bath. When cold, rub the dry white mass with 18 troy ounces chloride of sodium in an earthenware mortar; then sublime with a gentle heat. (*U. S. Ph.*)

4140. White Precipitate. This is the *ammonio-chloride of mercury,* and is prepared by dissolving, with heat, 6 ounces bichloride of mercury (corrosive sublimate) in 3 quarts distilled water; when cool, add 8 fluid ounces liquor of ammonia, frequently shaking it. Wash the precipitate with water, and dry it. It is used to make an ointment for skin diseases; also to destroy small vermin.

4141. Red Precipitate. *Red oxide* or *binoxide of mercury* is now used in medicine as an escharotic, also to induce salivation. Dissolve 4 ounces bichloride of mercury in 6 pints water; add 28 fluid ounces liquor of potassa; wash the precipitate in distilled water, and dry by a gentle heat.

4142. Chloride of Mercury and Ammonia. This is obtained by triturating together equal parts of bichloride of mercury and sal-ammoniac. This addition of sal-ammoniac renders the corrosive sublimate more soluble in water, for use in lotions and injections.

4143. Black Precipitate. *Protoxide of mercury* is obtained by agitating together 1 ounce calomel with 1 gallon lime-water; decanting the clear liquid after subsidence, and washing the sediment with distilled water, after which it is dried on bibulous paper.

4144. Protonitrate of Mercury. Mix together in a wide-bottomed glass vessel, equal parts of quicksilver and nitric acid (specific gravity 1.32); after digestion for 24 hours in a cool place, remove the crystals that have formed, wash them with a little nitric acid, drain them, and keep from the air in a stoppered bottle. (*Paris Codex.*)

4145. Tests for the Salts of Mercury. The salts of mercury are all volatilized at a dull red heat—give a white precipitate with prussiate of potash, a black one with sulphuretted hydrogen and hydrosulphurets, an orange-yellow one with gallic acid, and with a plate of polished copper, a white coat of metallic mercury.

Solutions of the protosalts of mercury yield a grey or black precipitate with alkalies, a yellowish or greenish-yellow one with iodide of potassium, a white one with muriate of soda.

Solutions of the persalts of mercury yield with caustic alkalies, yellowish or red precipitates; with alkaline carbonates, a brick-red one; with iodide of potassium, a scarlet one.

4146. Sulphate of Iron. Commercial sulphate of iron is known also as *Copperas, Green Vitriol, Shoemakers' Black,* &c. For medicinal purposes it requires some preparation: Mix 1 fluid ounce sulphuric acid with 4 pints water; add 4 pounds commercial sulphate of iron, and 1 ounce iron wire; digest with heat and occasional agitation until the sulphate is dissolved; strain while hot, and set aside so that crystals may form; evaporate the mother-liquor for more crystals, and dry the whole. (*Cooley.*)

4147. Sulphuret of Iron. Mix together 4 parts sublimed sulphur, and 7 parts iron filings. Heat in a crucible in a common fire till the mixture begins to glow; then remove the crucible from the fire, and cover it up until the reaction is at an end and the whole has become cold.

4148. Bisulphuret of Iron. This is found in large quantities in mineral form, and is known as *Iron pyrites.* It may also be obtained by projecting a mixture of 5 parts sulphur, and 4 parts iron filings, into a red-hot crucible, excluding the air as much as possible. It melts easily, and takes sharp casts, and may be colored red with vermilion.

4149. Hydrated Protosulphuret of Iron. This is a black, insoluble substance, rapidly decomposed by exposure to the air. A neutral solution of protosulphate of iron made with recently boiled or distilled water, is precipitated by adding a solution of hydrosulphuret of ammonia, or of sulphuret of potassium. Collect the precipitate on a filter, wash it as quickly as possible with recently boiled water, squeeze in a linen cloth, and preserve in its pasty state under water.

This preparation of iron is proposed by Mialhe as an antidote to the salts of arsenic, antimony, bismuth, lead, mercury, &c., and to arsenious acid, more especially to white arsenic and corrosive sublimate. On contact with the latter substance it is instantly converted into protochloride of iron and sulphuret of mercury, two comparatively inert substances.

4150. Hydrated Persulphuret of Iron. Prepared by adding, very gradually, a diluted solution of sulphuret of potassium, or of hydrosulphuret of ammonia, to a neutral solution of persulphate of iron, collecting, &c., the precipitate, in the same way as in hydrated protosulphuret of iron. Bouchardat and Sandras recommend this persulphuret as a substitute for the protosulphuret, to which, they say, it is preferable.

4151. Protoxide of Iron. Dry protoxide of iron is a black powder; in its hydrated state it is white, and when exposed to the air rapidly absorbs oxygen, assuming first a greyish-green color, and then a brownish-red, which is much brightened by exposure to a red heat, at the same time that its solubility in acids is considerably lessened. The salts of protoxide of iron have a greenish color, but yield nearly colorless solutions, except when concentrated. The white hydrate is precipitated from solutions of the protosalts of iron by the pure alkalies. (*Cooley.*)

4152. Tests for Solutions of the Salts of Protoxide of Iron. When acidulated they are not precipitated by sulphuretted hydrogen; even neutral solutions with weak acids are incompletely precipitated; whilst alkaline solutions are precipitated of a black color.

Neutral solutions are precipitated black by hydrosulphuret of ammonia.

Ammonia and potassa give a greenish-white precipitate, gradually becoming green, and then brown in the air. The presence of ammoniacal salts interferes with these tests.

Ferrocyanide of potassium (yellow prussiate of potash) gives a nearly white precipitate, becoming gradually blue in the air, and immediately so on the addition of a little weak nitric acid or chlorine water.

Ferridcyanide of potassium (red prussiate of potash, produces a rich deep blue precipitate, insoluble in muriatic acid. In highly dilute solutions the effect is only a deep bluish-green coloration.

Aurochloride of sodium gives a purple precipitate; and phosphate of soda a blue one.

Cochineal freed from fat by ether, and then digested in water (or very weak spirit), gives a solution which is colored violet by the protosalts of iron.

4153. Anhydrous Sesquioxide of Iron. A pure anhydrous sesquioxide is obtained by precipitating a solution of sesquisulphate or sesquichloride of iron with ammonia in excess, and washing, drying, and igniting the resulting hydrated peroxide.

4154. Jewelers' Rouge. The best jewelers' rouge is prepared by calcining the precipitated peroxide of iron (*see No.* 4153) until it becomes scarlet. The rust of iron contains some combined water, and is more soluble than the oxide prepared by calcination; but it is less soluble than that recently precipitated from its solution in an acid. This is also called *Colcothar, Crocus,* or *Crocus Martis.*

4155. Hydrated Sesquioxide of Iron. Take 4 ounces sulphate of iron; 3½ fluid ounces oil of vitriol; water, 1 quart; mix, dissolve, and boil, then gradually add 9 fluid drachms nitric acid; stirring well and boiling for a minute or two after each addition, until the liquor yields a yellowish-brown precipitate with ammonia, when it must be filtered and precipitated with 3½ ounces strong liquor of ammonia, rapidly added and well mixed in; collect, wash well with water, drain on a calico filter, and dry at a heat not exceeding 180° Fahr. When intended as an antidote for arsenic it should not be dried, but kept in the moist or gelatinous state. It should be kept in a well-stoppered bottle filled with recently distilled or boiled water. This preparation is also called *hydrated peroxide of iron.*

4156. Peroxide of Iron. Peroxide, or sesquioxide of iron, is a brownish-red powder, known as the red oxide or rust of iron; in its hydrated form it is very soluble in acids, but less so when anhydrous. The salts of peroxide of iron have for the most part a reddish-yellow color, and redden blue litmus paper. (*Cooley.*)

4157. Tests for the Solutions of the Salts of Peroxide of Iron. Sulphuretted hydrogen throws down a black precipitate from alkaline solutions.

Hydrosulphuret of ammonia does the same with neutral solutions; in very dilute solutions the precipitate is blackish-green; the precipitate in both cases being soluble in muriatic and acetic acids.

Ammonia and potassa produce bulky red-dish-brown precipitates insoluble in excess of the precipitant.

Ferrocyanide of potassium (yellow prussiate of potash) gives a rich blue precipitate, insoluble in muriatic acid, and readily decomposed by potassa.

Ferridcyanide of potassium (red prussiate of potash) deepens the color, but does not give a blue precipitate, as it does with the protoxide. (*See No.* 4152.)

Sulphocyanide of potassium gives an intense ruby-red color to neutral or acid solutions; this is the most sensitive test known.

Meconic acid and the meconiates also give a red color.

A tincture or infusion of galls strikes a black color; and phosphate of soda throws down a white precipitate.

4158. To Obtain Pure Oxalate of Iron. Vogel recommends the precipitation of a solution of an ordinary protosulphate of iron by oxalic acid. The filtered solutions exclude all insoluble matter, and the precipitated oxalate needs but sufficient washing and drying to obtain the oxalate of iron in a state of purity and of constant composition. This salt gently heated, with exposure to the air, takes fire, or may be kindled, and then continues to burn until the whole becomes converted into impalpable peroxide of iron. This cheap, rapid, and perfect method of obtaining a perfect oxide of iron, free from all grit and eminently fitted for all the finer polishing purposes, had led to the use of this article for polishing the finest optical glasses. By heating the product to a higher temperature, a much harder substance may be obtained, useful rather for grinding than for polishing purposes. By adding salts of alumina, chromium and other similar salts to the iron solution, we may obtain in the final result—using sufficient heat—products nearly, if not quite, equal to emery, and of extraordinary fineness.

4159. Acetate of Iron. Dissolve 20 ounces sulphate of iron in 7 ounces strong sulphuric acid, and heat in a porcelain dish nearly to boiling. Then add gradually 10 ounces strong nitric acid; and, when action ceases, while still hot, add sufficient ammonia to precipitate all the iron as sesquioxide.

Collect this on a linen cloth, and wash with water until the washings taste no longer saline. While still moist, put the sesquioxide into a bottle with sufficient strong acetic acid to dissolve it.

Twenty ounces of sulphate of iron contain 4 ounces iron; hence, if sufficient water be added to make the acetate up to 50 ounces, the solution of acetate of iron thus obtained will contain 8 per cent. of iron.

4160. Citrate of Iron. This salt is easily formed by digesting iron filings or wire with citric acid, and evaporating the solution as quickly as possible out of contact with the air. It presents the appearance of a white powder, nearly insoluble in water, and rapidly passing to a higher state of oxidation by exposure to the air. Its taste is highly metallic. It is usually administered in the form of pills, mixed with gum or syrup, to prevent premature decomposition.

4161. Iodide of Iron. Mix together 6 ounces iodine, 2 ounces iron filings, and 4½ pints water; boil in a sand-bath until the liquid turns to a pale green, filter, wash the residue with a little water, and evaporate the mixed liquors in an iron vessel, at 212°, to dryness, and immediately put the iodide into well-stoppered bottles. A great deal has been written and said about the preparation of iodide of iron, but there is in reality very little difficulty in the process. As soon as iodine and iron are mixed together under water, much heat is evolved, and if too much water be not used, the combination is soon completed, and the liquor merely requires to be evaporated to dryness, out of contact with the air, at a heat not exceeding 212°. This is most cheaply and easily performed by employing a glass flask, with a thin broad bottom and narrow mouth, by which means the evolved steam will exclude air from the vessel. The whole of the uncombined water may be known to be evaporated when vapor ceases to condense on a piece of cold glass held over the mouth of the flask; a piece of moistened starch paper occasionally applied in the same way will indicate whether free iodine be evolved; should such be the case, the heat should be immediately lessened. When the evaporation is completed, the mouth of the flask should be stopped up by laying a piece of sheet India-rubber on it, and over that a flat weight; the flask must be then removed, and, when cold, broken to pieces, the iodide weighed, and put into dry and warm stoppered wide-mouthed glass phials, which must be immediately closed, tied over with bladder and the stoppers dipped into melted wax.

4162. Ammonio-Citrate of Iron. Take 12½ ounces carbonate of soda, and 12 ounces sulphate of iron; dissolve each separately in 6 pints boiling distilled water. Mix the solutions while hot, and allow the precipitate to subside. Decant the liquor, and, after washing the precipitate frequently with water, drain it. Then add to it 6 ounces citric acid in powder, and dissolve the mixture by a gentle heat. When cool, add 9 fluid ounces liquor of ammonia of specific gravity .960. It must then be filtered, gently evaporated to the consistence of syrup, and spread very thinly on warm sheets of glass to dry, which it will rapidly do, if exposed in an atmosphere of warm dry air, and may then be easily detached from the glass, in thin scales of great brilliancy and beauty. Only a gentle heat must be employed, not exceeding that of a water-bath. This is the method of producing those beautiful transparent ruby-colored scales which are so much admired. It must be kept in well-stopped bottles.

4163. Saccharine Carbonate of Iron. A sweet-tasted greenish mass or powder. It is one of the best of the chalybeates in doses of 5 to 10 grains. When pure it should be easily soluble in hydrochloric acid with brisk effervescence. Take 4 ounces sulphate of iron, and 4¼ ounces carbonate of soda; dissolve each separately in 1 quart boiling water. Mix the solutions while hot; and, after allowing time for subsidence, collect the precipitate, wash it frequently with water, and drain. Then add 2 ounces sugar previously dissolved in 2 fluid ounces water, evaporate over a water-bath to dryness, and keep in a well-stopped bottle.

4164. Carburet of Iron. *Plumbago, or black-lead*, is the native carburet of iron. To purify it for chemical use, heat it to redness with caustic potassa in a covered crucible, then wash it well with water, boil it in nitric acid and in nitro-muriatic acid (aqua regia); again wash it in water, dry it, and expose at a white heat to a stream of dry chlorine gas. Lastly, wash it with water and again heat it to dull redness. (*Dumas.*)

4165. Chloride of Iron. The *muriate* or *protochloride* of iron is obtained by dissolving iron filings or scales in hydrochloric acid, and crystallizing by evaporation. It forms soluble green crystals, and is sometimes called *hydrated chloride of iron*. The above is not quite pure, but to obtain a pure white crystalline protochloride, transmit dry hydrochloric acid gas over iron heated to redness. This is volatile at a high temperature. (*See No.* 117.)

4166. Perchloride of Iron. The *permuriate* or *sesquichloride of iron* is obtained by dissolving sesquioxide or rust of iron in hydrochloric acid, evaporating to the consistence of syrup, and crystallizing. It forms red crystals, not quite pure. The pure perchloride is formed by passing chlorine over heated iron. This is soluble in water, alcohol, and ether, very deliquescent and corrosive, and is dissipated by a heat a little above 212° Fahr. (*Cooley.*) Perchloride of iron should not be given in mixtures containing medicated syrups or gum-arabic, since the latter, as well as all substances containing tannin, which is the case with those syrups, are incompatible with ferric salts. The proper menstruum is simple sugared water; it is also necessary to keep these mixtures from the light, on account of the chemical reduction produced by the latter. (*Eymael.*)

4167. Ferrocyanide of Iron. This is *pure Prussian blue*. Dissolve 9 troy ounces ferrocyanide of potassium in 2 pints water, and add it gradually, with stirring, to 1 pint of the solution of tersulphate of iron previously diluted with 1 pint water. Filter the mixture, and wash the precipitate on the filter with boiling water until the washings pass nearly tasteless. Lastly dry it and rub it into powder. (*U. S. Ph.*)

4168. Solution of Tersulphate of Iron. Take 2⅛ troy ounces sulphuric acid, and 1⅝ troy ounces nitric acid; mix them with ½ pint water in a large capsule, heat to the boiling point, and add 12 troy ounces sulphate of iron in coarse powder, 3 ounces at a time, stirring after each addition till effervescence ceases. Continue the heat until the solution acquires a reddish-brown color, and is free from nitrous odor. When nearly cold add water to make it up to 1½ pints. (*U. S. Ph.*)

4169. Ferridcyanide of Iron. This is better known as Turnbull's Prussian blue. (*See No.* 2674.)

4170. Tannate of Iron. Dissolve 1 part of tannin in 150 of boiling water; add 9 parts hydrated sesquioxide of iron, freshly precipitated, washed, and dried in the water-bath; evaporate gently to one half; filter, then add 1 part sugar, evaporate to dryness, and keep in a close vessel. Or: 1 part sesquioxide of iron and 2 of tannic acid evaporated to dryness with 3 parts alcohol.

4171. Nitrate of Iron. The *protonitrate of iron* is obtained by dissolving protosulphuret of iron in dilute nitric acid in the cold, and evaporating the solution in a vacuum. It forms small green crystals, very soluble, and liable to oxidation.

4172. Pernitrate of Iron. A deep red liquid formed by digesting nitric acid diluted with about half its weight of water on the sesquioxide of iron. It is also prepared from the metal. (*See No.* 116.)

4173. Oxide of Manganese. There are, according to Cooley, seven distinct compounds of oxygen and manganese, but the only one directly employed in the arts is the ***black oxide*** *(binoxide or deutoxide)* ***of manganese.*** It is a very plentiful mineral production, and is found in great abundance in many parts of Europe. The manganese of commerce is prepared by washing, to remove the earthy matter, and grinding in mills. The blackest samples are esteemed the best. It is chiefly used to supply oxygen gas, and in the manufacture of glass and chlorine; in dyeing, and to form the salts of manganese.

4174. Chloride of Nickel. Neutralize muriatic acid with oxide (protoxide) of nickel, and evaporate gently; small green crystals of chloride (*muriate*) of nickel. If these crystals are pure, they are rendered yellow and anhydrous by heat; if cobalt be present the salt retains a green tint.

4175. Protoxide of Nickel. The protoxide (*oxide*) of nickel is obtained in an anhydrous form by heating oxalate of nickel to redness in an open vessel. The *hydrated oxide* is an ash-grey powder formed by precipitating the oxalate of nickel with caustic potassa.

4176. Peroxide of Nickel. The peroxide (*sesquioxide*) is obtained by passing chlorine through water holding the hydrated oxide in suspension.

4177. Sulphate of Nickel. By neutralizing the protoxide of nickel with dilute sulphuric acid, green prismatic crystals of sulphate of nickel are obtained.

4178. Oxalate of Nickel. This is a pale bluish-green precipitate formed by adding a strong solution of oxalic acid to a like solution of sulphate of nickel.

4179. Tests for Solutions of the Salts of Nickel. Caustic alkalies give a pale-green precipitate, insoluble in excess of the precipitant, but soluble in a solution of carbonate of ammonia, yielding a greenish-blue liquid. Ferrocyanide of potassium gives a greenish-white precipitate. Sulphuretted hydrogen occasions no change in solutions of nickel containing free mineral acid; but with alkaline solutions gives a black precipitate.

4180. Acetate of Potassa. Mix together 26 fluid ounces acetic acid with 12 fluid ounces distilled water; add gradually 1 pound or more, until saturation, of carbonate of potassa; filter, and evaporate, by a sand-bath, to dryness.

4181. Carbonate of Potassa. This is also known under the name *Salt of Tartar*, and *Salt of Wormwood*. The crude carbonate is obtained by lixiviating (*see No.* 23) wood ashes, evaporating the solution to dryness, and fusing in iron pots for several hours. This constitutes the *potash* of commerce.

Another method of preparation is to transfer the product of the first evaporation to an oven or furnace so constructed that the flame is made to play over the alkaline mass, kept constantly stirred with an iron rod. The ignition is continued until the impurities are burned out, and the mass becomes of a bluish-white; this is commercial *pearlash*. The U. S. Pharmacopœia directs, for general purposes, the impure carbonate to be dissolved in water, filtered, and evaporated until it thickens, and then granulated in the manner directed for the pure carbonate.

4182. Pure Carbonate of Potassa. Put 12 troy ounces bicarbonate of potassa, in coarse powder, into a large iron crucible; heat gradually until the water of crystallization is driven off, then raise the heat to redness and maintain it at that heat for 30 minutes. When cool, dissolve it in distilled water, filter, and evaporate over a gentle fire until it thickens, then remove it from the fire and stir it constantly with an iron spatula until it granulates. (*U. S. Ph.*)

4183. Bicarbonate of Potassa. Dissolve 48 ounces carbonate of potassa in 10 pints distilled water; pass carbonic acid gas through the solution to saturation (the gas may be evolved from chalk by diluted oil of vitriol). Filter, and evaporate, that crystals may form, at a heat not exceeding 160° Fahr.; decant the clear, and dry the crystals. (*U. S. Ph.*)

4184. Chlorate of Potassa. Transmit chlorine gas through a moderately strong and warm solution of pure caustic potassa, or its carbonate, until the alkali be completely neutralized, then boil for a few minutes, gently evaporate until a pellicle forms on the surface, and set it aside, where it will cool very slowly. Crystals of the chlorate will form as the liquor cools, and must be collected, carefully washed with a little ice-cold water, and purified by re-solution and crystallization; the product is pure chlorate of potassa. The mother liquor, which contains much chloride potassium, by evaporation will yield more crystals, less pure than the former, or it may be saved for a future operation. This salt crystallizes in four and six-sided pearly scales; dissolves in 16 parts of water at 60°, and in 2½ parts at 212°. At about 450° it undergoes the igneous fusion, and on increasing the heat almost to redness, effervescence ensues, and fully 39 per cent. of pure oxygen gas is given off and the residue becomes changed into chloride of potassium. When mixed with inflammable substances, and triturated, heated, or subjected to a smart blow, it explodes with great violence. It also fulminates when thrown into strong acids. (*See No.* 2124.) (*Cooley.*)

4185. Perchlorate of Potassa. To concentrated sulphuric acid, gently warmed in an open vessel, add, in small portions at a time, an equal weight of well-dried and finely powered chloride of potassa. The *bisulphate of potassa* formed, is washed off with a little cold water, and the remaining perchloride of potassa dissolved in boiling water and crystallized.

4186. Chromate of Potassa. The *yellow chromate of potash* of commerce is only prepared on the large scale from the crude chrome ore, and is the common source of nearly all the other compounds of chromium. The ore, freed as much as possible from its impurities, is ground to powder in a mill, and mixed with ½ or ⅓ of its weight of bruised nitre, and in this state exposed to a powerful heat for several hours, on the hearth of a reverberatory furnace, during which time it is frequently stirred up with iron rods. The calcined matter is next raked out and lixiviated with hot water. A beautiful yellow-colored solution results, which is evaporated briskly over a naked fire, when the chromate of potash falls down under the form of a granular yellow salt, which is removed from time to time with a ladle, and thrown into a wooden vessel, furnished with a bottom full of holes, called the draining-box, where it is left to drain and dry. In this state it forms the commercial chromate of potash. By a second solution and crystallization, it may be obtained in larger and more regular crystals. (*Cooley.*)

4187. Bichromate of Potassa. The *red chromate of potash* is obtained from a concentrated solution of the yellow chromate, by adding sulphuric (or, still better, acetic) acid in quantity equal to half that required for the neutralization of the salt. (*See No.* 83.) The liquid is then concentrated by evaporation, and slowly cooled, so that crystals may form.

4188. Substitute for Bichromate of Potassa. One of the German scientific journals calls attention to the fact that for many purposes, such as for dyeing wool black, Glauber's salt and sulphuric acid can be economically substituted for bichromate of potassa. It gives the following recipe for dyeing 100 pounds of loose wool—namely, 6 pounds sulphate of soda, 2 pounds sulphuric acid, and 2 pounds sulphate of copper, which are to be boiled together for an hour, and colored with 40 to 50 pounds logwood, and 1 pound sulphate of copper, and finally colored black by means of a little sulphate of iron. The black thus obtained is pronounced to be beautiful, cheap, and easily spun, remaining loose and soft.

4189. Nitrite of Potassa. It is obtained mixed with a little nitre and potash by heating nitre to redness. To purify the residuum, dissolve it in boiling water, set aside for 24 hours, pour off the liquid from the deposited nitre, neutralize the free alkali with acetic acid, and add twice its volume of alcohol. In a few hours more, nitre crystallizes, and the liquid separates into two layers; the upper is alcoholic solution of acetate of potash, the lower is solution of nitrate of potash, which may be evaporated to dryness, or kept in solution. (*Beasley.*)

Or, pass nitrous acid gas, formed by acting on 1 part of starch with 10 of nitric acid, through a solution of caustic potash, specific gravity 1.38, until it becomes acid; then add a little caustic potash, so as to render it distinctly alkaline. It may then be kept in the liquid form, or evaporated to dryness. (*Corenwinder.*)

4190. Permanganate of Potassa. This consists of slender, prismatic crystals, of a dark-purple color, inodorous, and of a sweetish, astringent taste. It is a powerful disinfectant, and oxidizing agent, from the facility with which it parts with its oxygen. It has been found useful in medicine in various ways, and forms an excellent, though unstable hair dye. (*See No.* 1211.) It may be obtained by mixing 8 parts of peroxide of manganese with 7 parts chlorate of potassa, both in fine powder, adding 10 parts of hydrate of potassa, dissolved in a small quantity of water, evaporating to dryness, powdering, exposing the powder to a low red heat in a plantinum crucible, dissolving the mass in a large quantity of water, decanting, evaporating, and crystallizing. These crystals are permanganate of potassa. The PERMANGANATES or basic compounds of permanganic (manganesic) acid are all marked by their rapid decomposition when in contact with organic matter. (*Cooley.*)

4191. Tests for Permanganate of Potassa. A very dilute solution has a rose-color, free from green tinge, and is instantly decolorized by arsenite of potassa, with the formation of a brown precipitate. (*U. S. Ph.*) Dissolve 44 grains granulated sulphate of iron in 2 fluid drachms dilute sulphuric acid; the solution should completely decolorize 5 grains of the permanganate dissolved in water. (*Br. Ph.*)

4192. Hydrate of Potassa. This is also known under the name of *caustic potash.* Liquor of potassa, 1 gallon; evaporate in a clean iron vessel over the fire until the ebullition being finished, the hydrate of potassa liquefies; pour this into proper moulds. A pale greyish or bluish solid, very soluble in water and alcohol. It should be totally soluble in alcohol. Its solution should be scarcely affected by the nitrates of baryta and silver. It is chiefly used as a caustic, and in chemistry. (*Cooley.*)

4193. Potassa with Lime. Rub together, in a warm mortar, 1 ounce each of hydrate of potassa and quicklime, and keep the powder from the air in a well-stopped bottle. This is a caustic, but less manageable than either nitrate of silver (lunar caustic) or hydrate of potassa (caustic potash.)

4194. Nitrate of Potassa. Called also *nitre* and *saltpetre.* This salt is spontaneously generated in the soil, owing to the action of the atmosphere, and crystallizes upon its surface in various parts of the world, especially in the East Indies. It is also produced artificially by exposing a mixture of calcareous soil and animal matter to the atmosphere, when nitrate of lime is slowly formed, and is extracted by lixiviation. The liquid is then decomposed by adding carbonate of potash, by which carbonate of lime is precipitated and nitrate of potash remains in solution.

4195. To Purify Nitre. Nitre or saltpetre is purified for medicinal use in the following manner: Dissolve 4 pounds commercial nitre in 1 quart boiling distilled water; withdraw the heat, and stir constantly as it cools. The minute crystals, thus obtained, are to be drained, and washed in a glass or earthenware percolator, with cold distilled water, until the washings cease to give a precipitate with a solution of nitrate of silver. The contents of the percolator are then to be withdrawn and dried in an oven. (*Cooley.*)

4196. Tartrate of Potassa. Dissolve 8 ounces carbonate of potash in 2 quarts distilled water; whilst boiling hot, add gradually 1 pound, more or less, of bitartrate of potassa (cream of tartar) in fine powder, until the solution is neutralized, or ceases to change the color of either blue or reddened litmus paper. Filter through muslin, and evaporate until a pellicle forms on the surface; then set it aside to crystallize. After 12 hours, collect the crystals, dry them on bibulous paper, and keep preserved from the air.

4197. Bitartrate of Potassa. This is well known under the name of *cream of tartar,* and is found deposited as a crust on the sides of the casks and vats used for the fermentation of grape juice. The deposit from white wine is *white tartar;* that from red wine is *red tartar,* or *argol.* It is purified by boiling it in water, and crystallizing; it is then again dissolved in boiling water, and decolorized with charcoal (*see No.* 1729), and aluminous clay; the resulting clear liquid is allowed to cool slowly, forming crystals of the cream of tartar of commerce.

4198. Bromide of Potassium. Put 1 troy ounce iron filings into 1½ pints distilled water; add 2 troy ounces bromine, stirring frequently during 30 minutes; heat gently until the liquid assumes a greenish color, and add gradually 2½ troy ounces pure carbonate of potassa (previously dissolved in 1¼ pints distilled water), until it ceases to produce a precipitate; continue the heat for 30 minutes, then filter. Wash the precipitate with 1 pint boiling distilled water, and filter. Mix the filtered liquids, and crystallize by evaporation. Dry the crystals on bibulous paper and keep them in a well-stoppered bottle. (*U. S. Ph.*)

4199. Chloride of Potassium. This is obtained from the mother liquor after making chlorate of potassa (*see No.* 4184), by evaporating it to dryness, and heating it to a dull redness; it is then dissolved in water, purified by defecation and crystallized by evaporation.

4200. Ferridcyanide of Potassium. This is the *red prussiate of potash*, and is obtained from a solution of 1 part ferrocyanide of potassium in 16 parts cold water, by passing chlorine gas slowly through it, with constant agitation, until the liquid appears of a reddish green color, and ceases to give a blue precipitate, or even a blue tinge, to a solution of a sesquisalt of iron, an excess of chlorine being carefully avoided. The liquor is then evaporated till a pellicle forms on the surface, filtered while hot, and set aside to cool; the crystals are again dissolved and crystallized. (*Cooley.*)

4201. Ferrocyanide of Potassium. This *yellow prussiate of potash* is the *prussiate of potash* of commerce. It is obtained by exposing 10 parts potash or pearlash; 10 parts coke, cinders, or coal; and 5 parts iron turnings, all in coarse powder, to a full red heat in an open crucible, stirring occasionally until small jets of purple flame are no longer seen. When cool, the soluble matter is dissolved out of it, the solution filtered, evaporated, and crystallized. The crystals obtained are redissolved in hot water and cooled very slowly, forming large yellow crystals of the ferrocyanide of potassium of commerce. In order to obtain a pure article, fuse effloresced commercial prussiate of potash in a glass vessel, dissolve the fused mass in water, neutralize any excess of alkali with acetic acid, and precipitate the salt with strong alcohol; wash the precipitate with a little weak alcohol, redissolve it in water, and crystallize. (*Cooley.*)

4202. Cyanide (Cyanuret) of Potassium. Mix thoroughly 8 ounces of dry ferrocyanide of potassium and 3 ounces dry carbonate of potassa; throw the mixture into a deep red-hot earthen crucible, the heat being sustained until effervescence ceases, and the fluid portion of the mass becomes colorless; after a few minutes' rest, to allow the contents to settle, the clear portion is poured from the heavy black sediment at the bottom on a clean marble slab; and, while yet warm, broken up and placed in well-closed bottles. When pure, this salt is colorless and odorless, its crystals are cubic or octahedral, and are anhydrous. If it effervesces with acids, it contains carbonate of potassa. If it be yellow, it contains iron. (*Liebig.*)

4203. Iodide of Potassium. This important medicinal compound is obtained in various ways. The United States Pharmacopœia gives the following formula for its preparation: To 6 troy ounces potassa, dissolved in 3 pints boiling distilled water, add gradually finely powdered iodine, stirring after each addition until the solution becomes colorless, and continue the addition until the liquid remains slightly colored from excess of iodine. (This will require about 16 troy ounces of iodine.) Evaporate the solution to dryness, stirring in 2 troy ounces finely powdered charcoal towards the close of the operation, so that it may be intimately mixed with the dried salt. Rub this to powder, and heat it to dull redness in an iron crucible, maintaining that temperature for 15 minutes. After it has cooled, dissolve out the saline matter with distilled water, filter the solution, evaporate, and set it aside to crystallize. An additional quantity of crystals may be obtained from the mother water by further evaporation.

A solution of iodide of potassium keeps decidedly better when there is neither plain nor aromatic syrup or sugar in it. When gargles of honey of roses, with alum and water, have a black color, though that of honey be of the proper shade, this is owing to the presence of iron in the alum, which is by no means a rare occurrence.

4204. Sulphuret of Potassium. Rub together 1 ounce sublimed sulphur, and 2 ounces dry carbonate of potassa; heat it gradually in a covered crucible until it ceases to swell and is completely melted. Pour the liquid on to a marble slab, and, when cold, break the mass into pieces, and keep in well-stopped bottle of green glass. (*U. S. Ph.*)

4205. Sulphocyanide of Potassium. Take 3 parts cyanide of potassium, and 1 part sulphur; digest them for some time in 6 parts water, then add 3 parts more water; filter, evaporate, and crystallize. It forms long, slender, colorless prisms, which are anhydrous, deliquescent, and fusible; very soluble in water and in alcohol, and not poisonous.

4206. Acetate of Soda. This is prepared from carbonate of soda, by the same method directed for acetate of potassa (*see No.* 4180), except that the resulting solution is evaporated to a pellicle, and set aside to crystallize.

4207. Sulphate of Soda. Also called *Glauber's salt.* This is usually obtained by dissolving 2 pounds of the chloride of sodium left after the distillation of muriatic acid (*see No.* 3883) in 1 quart of boiling water; the solution is next neutralized with carbonate of lime evaporated, and crystallized. It is soluble in cold water, its solubility decreasing as the temperature of the water is raised; insoluble in alcohol, and fuses when heated.

4208. Carbonate of Soda. The carbonate of soda of commerce is either prepared by lixiviating the ashes of sea-weed, or from sulphate of soda. The ashes of marine plants have been long an article of commerce, under the names of *barilla, barilla ashes, kelp, blanquette, &c.*, but the carbonate made from them is of a very impure description. That made from the sulphate is much purer, and, when the process is well managed, merely contains a trace of sulphuric acid. The sulphate of soda is mixed with an equal weight of chalk and about half its weight of coal, each being previously ground to powder, and the mixture is exposed to a great heat in a reverberatory furnace, and during the calcination is frequently stirred with a long iron rod. The dark grey product usually contains about 22 or 23 per cent. of carbonate of soda. This is now lixiviated with tepid water, and the solution, after defecation, evaporated to dryness, mixed with a little sawdust, and roasted in a reverberatory furnace at a heat not exceeding 700° Fahr., until all the sulphur is expelled. The product now receives the name of soda-ash, or soda-salt, and contains about 50% of alkali. It may be purified by solution in water, defecation, evaporation, and crystallization; it then becomes commercial crystallized carbonate of soda, consisting of large transparent crystals, which effloresce by exposure to the air, crumbling into a white dry powder. The carbonate used in medicine is prepared from the commercial crystals by dissolving, filtering, and careful crystallization.

4209. Bicarbonate of Soda. This may be prepared from a solution of carbonate of soda treated in the same manner as for bicarbonate of potassa. (*See No.* 4183.) The U. S. Pharmacopœia directs carbonate of soda in small pieces to be enclosed in a box (having an air-tight lid, and an inner bottom perforated with holes), and thus subjected, until saturated, to a stream of carbonic acid gas previously passed through water.

Cooley recommends the following process: Mix together 1 part carbonate of soda with 2 parts dried carbonate of soda, both in powder, and surround them with an atmosphere of carbonic acid gas, under pressure. Let the action go on till no more gas is absorbed, which will generally occupy 10 to 14 hours, according to the pressure employed, then remove the salt and dry it at a heat not above 120° Fahr.

4210. Phosphate of Soda. Mix 10 pounds powdered bone ashes with 44 fluid ounces sulphuric acid; add gradually 6 pints water, and digest for 3 days, replacing the water which evaporates; then add 6 pints boiling water, strain through linen, and wash the residue on the filter with boiling water. Mix the liquors, and, after defecation, decant and evaporate to 6 pints; let the impurities again settle, and neutralize the clear fluid, heated to boiling, with a solution of carbonate of soda in slight excess; crystals will be deposited as the solution cools, and by successively evaporating, adding a little soda to the mother liquor till it is feebly alkaline, and cooling, more crystals may be obtained. Keep it in closed vessels. (*Ed. Ph.*)

4211. Hyposulphite of Soda. Mix together 1 pound dried carbonate of soda and 10 ounces flowers of sulphur, and slowly heat the powder in a porcelain dish until the sulphur melts; stir freely, to expose it to the atmosphere, until the incandescence flags, then dissolve the mass in water, and immediately boil the filtered liquid with some flowers of sulphur; lastly, carefully concentrate the solution for crystallization. (*Cooley.*)

It may also be prepard by dissolving 8 parts carbonate of soda in 16 parts water; add 1 part sublimed sulphur, and pass sulphurous acid gas, in excess, into the solution; boil the liquid in a glass matrass for a few minutes, filter, gently evaporate the filtrate to ⅓ its volume, and set it aside in a cool place to crystallize. (*Paris Codex.*)

4212. Tungstate of Soda. This is formed by dissolving tungstic acid in a concentrated solution of pure soda. *Tungstic acid* is a yellow powder obtained by digesting native tungstate of lime, finely powdered, in nitric acid. It forms TUNGSTATES with metals and bases.

4213. Potassio-Tartrate of Soda. Known in commerce as *Seignette's* or *Rochelle salt.* Dissolve 12 ounces carbonate of soda in 2 quarts boiling water; add gradually 16 ounces bitartrate of potassa in fine powder. Strain, evaporate to a pellicle or crust (*see No.* 9), and set it aside to crystallize. The mother liquor may be further evaporated for a second supply of crystals. (*Cooley.*) The U. S. Pharmacopœia adopts the same method, but directs 5 pints of boiling water to be used.

4214. Bromide of Sodium. This is now employed to a great extent instead of bromide of potassium; it is more active than the latter, is more quickly absorbed, and more regularly eliminated. To prepare it pure and in large quantities the following method is recommended: Bromide of ammonium is decomposed by an equivalent quantity of caustic or carbonate of soda, which, of course, must be free from sulphuric and hydrochloric acids. The solution yields, after evaporation, small cubes of anhydrous bromide of sodium.

4215. Chloride of Sodium. This is a muriate of soda, or common table salt, and is largely obtained by the evaporation of sea water, or from the water of salt springs. It dissolves in about $2\frac{1}{2}$ parts of water at 60° Fahr.; is insoluble in pure alcohol; fuses at a red heat; and at a higher temperature becomes volatile.

4216. Iodide of Sodium. This is obtained from soda in the same manner as iodide of potassium. (*See No.* 4203.)

4217. Nitro-Prusside of Sodium. To 213 parts of powdered ferroprussiate of potash, in a porcelain basin, add 450 parts of nitric acid of 1.42 density (or $337\frac{1}{2}$ parts at 1.50), adding all the acid at once. When dissolved, transfer to a bolt-head, and digest in a water-bath until the solution precipitates salts of protoxide of iron of a slate color. Neutralize, when cold, with a cold solution of carbonate of soda; then boil, and separate the precipitate by filtration. Evaporate the liquid again, filter, and allow the nitrates of potash and soda to crystallize out. Evaporate the liquid again, and remove the prismatic crystals of nitro-prusside as they form. They may be dissolved in water and recrystallized by cooling.

4218. Acetate of Ammonia. Mix together equal parts of sal-ammoniac and acetate of potassa, and distill; binacetate of ammonia passes over into the receiver, as an oily liquid, which, on cooling, forms a radiated crystalline mass. By passing dry ammoniacal gas into this salt, melted by a gentle heat, it is transformed into the neutral acetate, and becomes solid and inodorous.

Or: By saturating strong acetic acid with ammonia, and evaporating over sulphuric acid in vacuo, crystals of acetate of ammonia may be obtained. Very soluble both in alcohol and water, and very deliquescent.

4219. Carbonate of Ammonia. The *Neutral Carbonate* is prepared by mixing equal parts sal-ammoniac, powdered and well dried, and dried carbonate of soda, and subliming, by a gradually increased heat, from an earthen retort into a refrigerated receiver.

4220. Sesquicarbonate of Ammonia. This is the *commercial carbonate of ammonia*, and is prepared as follows; Sal-ammoniac, or pure commercial sulphate of ammonia, and chalk, equal parts, both dry and in powder. Mix and sublime from an iron pot, into a long earthen or leaden receiver, well cooled. The receiver is usually fitted with a moveable lead cover, secured by a water-joint, and has an open lead pipe in the bottom, to allow the liquid products of the distillation to drain off into a second receiver. When made of the impure sulphate of ammonia, it must be re-sublimed in iron pots, furnished with leaden heads kept cool. A little water is commonly introduced into the subliming pots, to render the product translucent. The heat is usually applied by means of a common furnace, but a steam or water bath is preferable, as the temperature required for this purpose does not exceed 200° Fahr.

4221. Bicarbonate of Ammonia. The commercial carbonate reduced to fine powder, and exposed to the air for 24 hours, becomes a bicarbonate spontaneously. It can also be obtained by passing a stream of carbonic acid gas through a solution of the sesquicarbonate until saturated, and drying the crystals which form without heat.

4222. Muriate of Ammonia. Also called *sal-ammoniac* and *hydrochlorate of ammonia*. This substance was formerly prepared in Egypt by the sublimation of the soot from camels' dung, which yields from $\frac{1}{6}$ to $\frac{1}{4}$ its weight. The sal-ammoniac of commerce is now wholly prepared at the great chemical works, and never by the small consumer, by whom it is merely occasionally refined or purified. The crude ammoniacal salt of the gas-works is placed in iron pots, lined with clay, and a leaden dome or head adapted, and heat applied until the whole has sublimed. When the crude salt is a sulphate, it is mixed with a sufficient quantity of muriate of soda before sublimation, and the sal-ammoniac is formed by the double decomposition of the ingredients. The preparation of sal-ammoniac from bone-spirit salt is nearly similar. The sal-ammoniac of commerce is found under the form of large hemispherical, cup-like cakes or masses, having a semi-crystalline texture, and varying in weight from 100 to 1000 pounds. It forms a clear and colorless solution with water, and wholly volatilizes by heat. Mixed with lime or caustic potassa, it evolves the pungent odor of ammonia; it gives a white curdy precipitate with nitrate of silver. The sal-ammoniac of commerce is generally sufficiently pure for all the purposes of the arts, but when wanted of greater purity, it may be broken into pieces and re-sublimed from an earthenware vessel into a large receiver of earthenware or glass, in which state it is known as "flowers of sal-ammoniac," from being in fine powder. Chemically pure hydrochlorate of ammonia may be prepared by adding the pure carbonate of ammonia to dilute hydrochloric acid until saturated. (*Cooley.*)

4223. Sulphate of Ammonia. The commercial sulphate is obtained by saturating with weak oil of vitriol the ammoniacal liquor of the gas-works, or bone-spirit. For medicinal purposes it is prepared by saturating dilute sulphuric acid with sesquicarbonate of ammonia in slight excess; it is then filtered, evaporated by a gentle heat, and crystallized.

4224. Murexide. This is the *purpurate of ammonia*, and consists of iridescent crystals, which reflect a beautiful green color, but transmit an equally fine reddish-purple color. It is obtained from *alloxan*, a substance formed by the action of nitric acid on uric acid.

4225. Iodide of Ammonium. Place a portion of iodine in a flask with a little water; add to it a solution of hydrosulphuret of ammonia, until the mixture loses its red color, and is turbid from the separation of sulphur; by shaking the flask, the most of the sulphur will form into a mass. Pour off the liquid, and boil it until all odor of sulphuretted hydrogen and of ammonia is lost. Then filter it, and evaporate it, constantly stirring, over a flame, until it becomes pasty, and then in a water-bath until it forms a dry salt. (*U. S. Dis.*)

4226. Sulphocyanide of Ammonium. Saturate 2 parts of common water of ammonia (specific gravity 0.950) with sulphuretted hydrogen; and add 6 parts of the same ammonia. To this mixture add 2 parts of sulphur, and the product of the distillation of 6 parts of prussiate of potash, 3 of sulphuric acid, and 18 of water. Digest till the sulphur is no longer acted on, and the liquid becomes yellow. Boil the liquid till it becomes colorless, filter, evaporate, and crystallize.

4227. Bromide of Ammonium. For the preparation of bromide of ammonium, bromine is added very gradually to diluted ammonia. The ensuing reaction produces much heat, which may cause ammonia and bromine to volatilize with the escaping nitrogen. The combination, therefore, is effected in a Wolffe's apparatus, which will condense and retain both perfectly. The evaporation of the fluid is also best done in an iron retort connected with a stoneware receiver, in which ammonia and some bromide of ammonium are condensed.

4228. Sulphuret of Ammonium. Usually called *hydrosulphuret of ammonia*. This is prepared from strong liquor of ammonia, by saturating it with sulphuretted hydrogen gas, and then adding a second portion of liquor of ammonia, equal in strength and quantity to that first used. Keep it in well-stoppered bottles. (*See No.* 1203.)

4229. Manganate of Baryta. The manganate of baryta, and of other alkalies, is formed by igniting the nitrate of the alkalies with peroxide of manganese, with excess of air, and dissolving in water. (*Booth.*)

4230. Nitrate of Baryta. It is prepared in the same manner as muriate of baryta (*see No.* 4234), substituting pure nitric acid for the muriatic acid.

4231. Sulphate of Baryta. This occurs as a native mineral, and is white, if pure. It occasionally contains iron, which may be removed by washing first with dilute sulphuric acid, and afterwards with pure water. (*See No.* 2697.)

4232. Acetate of Baryta. Dilute acetic acid neutralized with carbonate of baryta, and evaporated to form crystals.

4233. Carbonate of Baryta. A heavy white powder found in the crude state abundantly in nature, and sufficiently pure for general purposes. The pure carbonate may be precipitated from a solution of chloride of barium by the addition of any pure alkaline carbonate, washing and drying the product. (*Cooley.*)

4234. Chloride of Barium. Also called *Muriate of Baryta*. Mix gradually 10 ounces carbonate of baryta in small pieces, with $\frac{1}{2}$ pint muriatic acid diluted with 1 quart distilled water; evaporate to a pellicle or crust (*see No.* 9), and set aside to crystallize.

4235. Protoxide of Barium. This is the oxide of barium or baryta. (*See No.* 3985.)

4236. Peroxide of Barium. The peroxide or *binoxide* is prepared from pure baryta, heated to a full red heat in a porcelain tube, and exposed to a stream of pure dry oxygen gas. Instead of baryta, its nitrate may be used, but the nitrous fumes must be allowed to pass off entirely before applying the oxygen.

4237. Sulphuret of Barium. Calcine and reduce to powder 2 pounds sulphate of baryta, mix it with 4 ounces finely powdered charcoal; submit the mixture for 3 hours to a low white heat in a covered crucible. When cool, powder, and boil for 5 minutes in 5 pints water; decant the clear, and repeat the operation with 3 pints more water; unite the liquors, and crystallize by cooling.

4238. Carbonate of Lithia. Precipitate a solution of sulphate of lithia, by a strong solution of sesquicarbonate of ammonia; collect the precipitate, drain and press it, wash it with a little rectified spirit, and dry it. Dissolve in boiling water, and crystallize by slow evaporation.

4239. Sulphate of Lithia. Finely powdered petalite, 1 part; fluorspar, 2 parts; mix, add oil of vitriol, 10 parts, and heat the mixture as long as acid vapors are evolved. The residuum must be dissolved in pure water of ammonia, boiled, filtered, the solution evaporated to dryness, and the dry mass heated to redness. The matter left is pure sulphate of lithia. (*Berzelius.*)

Petalite or *Spondumene* is a mineral found in various parts of Europe, also in Massachusetts and Connecticut. (*Booth.*)

4240. Carbonate of Magnesia. There are two simple carbonates of magnesia, the heavy and the light.

The *heavy carbonate* is prepared from a saturated solution of sulphate of magnesia, 1 part by measure; water, 3 parts; heat to the boiling point, then add cold saturated solution of carbonate of soda, 1 part; boil, with constant agitation, till effervescence ceases, then add boiling water, 100 parts, agitate well, decant off the clear liquid, drain, and wash the precipitate with hot water, in a linen cloth, and finish the drying by heating it in an iron pot.

The *light carbonate* is obtained from 4 pounds sulphate of magnesia, and 4 pounds 9 ounces carbonate of soda, each separately dissolved in 2 gallons water. Mix and boil the liquors, constantly stirring for 15 minutes; after subsidence, decant the clear, wash the precipitate with boiling water, and dry it. The carbonate of magnesia of commerce is usually made up into cakes or dice, while drying, or is permitted to drain and dry in masses, which are then cut into shapes with a thin knife. It is powdered by rubbing it through a wire sieve. (*Cooley.*)

4241. Sulphate of Magnesia. This is the well-known *Epsom salts* of commerce, called after the saline springs of Epsom, in England, from the waters of which it was originally obtained. It is prepared on the large scale from *Dolomite*, or magnesian limestone. Heat the mineral with sufficient dilute sulphuric acid to convert all its carbonate into sulphate of lime, wash out all the sulphate of magnesia with hot water, and, after defecation, evaporate and crystallize.

Or, from bittern. Boil the residual liquor, or mother-water of sea-salt, for some hours, skim, and decant the clear, then concentrate by evaporation, and run the solution into wooden coolers; in 1 or 2 days ⅓ part of Epsom salts will have crystallized out. This is called *singles.* By re-solution in water, and re-crystallization, *doubles*, or Epsom salts, fit for the market, are obtained.

4242. Sulphuret of Magnesia. The *sulphide*, or sulphuret, is prepared by fusing together, in a covered crucible, 5 parts calcined magnesia and 4 parts sulphur.

4243. Chloride of Magnesium. Dissolve magnesia in muriatic acid, evaporate to dryness, add an equal weight of muriate of ammonia, project the mixture into a red hot platinum crucible, and continue the heat until tranquil fusion is attained. Pour out the fused mass on to a clean stone; and, when solid, break it into pieces, and transfer to a warm, dry bottle. (*Cooley.*)

Or: Dissolve magnesia in muriatic acid; evaporate to a specific gravity of 1.384; and put it, while hot, into a wide-mouthed flask to crystallize. (*Paris Codex.*) This chloride of magnesium is also called *hydrochlorate* or *muriate* of *magnesia.*

4244. Acetate of Lime. Neutralize acetic acid with prepared chalk (*see No.* 1292), filter the solution, evaporate by a gentle heat, and allow to crystallize.

4245. Chloride of Lime—called also *hypochlorite* and *oxymuriate of lime, bleaching powder*, and *chlorinated lime*—is seldom, if ever, made on the small scale, as it can be purchased of the large manufacturer of better quality and cheaper than it could possibly be made by the druggist. On the large scale the chlorine is generated in leaden vessels, heated by steam, and the gas, after passing through water, is conveyed by a leaden tube into an apartment built of silicious sandstone, and arranged with shelves or trays, containing fresh-slacked lime, placed one above another about an inch asunder. The process must be continued for 4 days to produce a good article of chloride of lime. During this time the lime is occasionally agitated by means of iron rakes, the handles of which pass through boxes of lime placed in the walls of the chamber, which act as valves.

4251. Chloride of Cobalt. Dissolve carbonate of cobalt in muriatic acid; the solution deposits rose-colored crystals on standing, which contain water. By evaporating the solution, anhydrous blue crystals of the chloride are obtained. (*Cooley.*)

4252. Carbonate of Cobalt. This is precipitated from a solution of nitrate of cobalt, by carbonate of potassa, producing a pale peach-colored powder, soluble in acids. (*Cooley.*)

4253. Acetate of Cobalt. The acetate is obtained by dissolving carbonate of cobalt in acetic acid. Acetate of cobalt forms a sympathetic ink. (*See No.* 2540.)

4254. Manganese. A hard, brittle, greyish-white metal, very easily oxidized, fuses with difficulty, unaffected by cold water, but dissolving freely in dilute sulphuric acid, evolving hydrogen gas. It has a specific gravity of 8.013. It is obtained by calcination in a crucible, at a strong heat, of 10 parts by weight of an oxide of manganese, made into a paste with oil, and combined with 1 part calcined borax.

4255. Peroxide of Manganese. The *black oxide* is the only oxide of manganese that is directly employed in the arts. It is a plentiful mineral production in a crude state; and is purified by grinding the native mineral or *pyrolusite* in mills, and removing the earthy matter by washing. The blackest samples are esteemed the best.

4256. Alum. The alum of commerce is a sulphate of alumina and potassa, obtained by lixiviation (*see No.* 23) from crude alum ore, or *schist.* It is obtained in large crystals, slightly efflorescent. It is applied in the arts to a great variety of purposes. When deprived of its water of crystallization by heat, it becomes *burnt* or *dried alum.* Pure *red* or *roche alum* was originally imported from Italy, where it is found in a native state. This has a reddish tinge, which extends more or less through the crystals.

4257. Hydrate of Alumina. Dissolve alum in 6 times its weight of boiling water, add a solution of carbonate of potassa, in slight excess, agitate the mixture for a few minutes, and then allow it to repose. After a time, pour the clear supernatant liquor from the precipitate or sediment, and wash the latter three or four times with tepid distilled or soft water. Next collect the precipitate on a fine calico filter, and again wash it with tepid water. When it has drained, press it between bibulous paper, and, lastly, dry it either without heat, or at a temperature not higher than 120° to 130° Fahr. The product is a soft white powder. (*Cooley.*)

4258. Acetate of Alumina. Add a solution of acetate of baryta to another of sulphate of alumina, and filter. Or, add 5 parts alum to 6 parts sugar of lead, each being first dissolved separately in hot water, and allowed to cool before mixing; decant the clear liquor. The pure acetate is made from pure hydrate of alumina, by digesting it in cold, strong acetic acid, until the latter is saturated. By spontaneous evaporation long transparent crystals form.

4259. Sulphate of Alumina. Saturate dilute sulphuric acid with hydrate of alumina; evaporate and crystallize.

4260. Butyrine. An oily fluid obtained from butter. Keep clarified butter in a porcelain vessel, at a heat of 66°, for some days; carefully collect the oily portion which separates, and agitate it with an equal weight of absolute alcohol for 24 hours, then pour off the clear and evaporate, treat the oily residuum with a little carbonate of magnesia to remove any free acid, and wash off the ***butyrate*** of ***magnesia*** thus formed with water; next heat the remaining fatty matter in alcohol, filter, and evaporate, to obtain the butyrine.

4261. Bromine. A dark reddish-colored liquid, having an odor resembling chlorine. It freezes at —4°, boils at about 135° Fahr., is very soluble in ether, less so in alcohol, and only slightly so in water. With hydrogen it forms *hydrobromic acid*, and, with the bases, compounds called BROMIDES OR HYDROBROMATES. It is obtained as follows: A current of chlorine is passed through the uncrystallizable residuum of sea-water, called ***bittern***, which then assumes an orange tint, in consequence of bromine being set free from its combinations; sulphuric ether is then agitated with it, and the mixture allowed to stand until the ethereal portion, holding the bromine in solution, floats upon the surface. By decanting, and evaporating the ether, a crude bromine may be obtained at once. To get it pure, the ethereal solution is carefully decanted, and agitated with a solution of potassa, by which means bromide of potassium and bromate of potash are formed. The whole is next evaporated to dryness, and submitted to a dull red heat; the residuum is then powdered, mixed with pure peroxide of manganese, and placed in a retort; sulphuric acid, diluted with half its weight of water, is now poured in. Red vapors immediately arise, and condense into drops of bromine, and are collected by plunging the neck of the retort to the bottom of a small receiver containing cold water. The bromine forms a stratum beneath the water, and may be collected and further purified by distillation from dry chloride of calcium. (*Cooley.*)

4262. Iodide of Cadmuim. This is prepared by mixing iodine and cadmuim filings in a moist state. This is freely soluble in water or alcohol, and may be crystallized by evaporation from ether solution, in large white transparent crystals. (*U. S. Disp.*)

4263. Bromide of Cadmium. This is made from cadmium filings and bromine, in the same manner as the iodide of cadmium from iodine. It consists of long, white, efflorescent, crystalline needles.

4264. Hydriodate of Quinine. To a concentrated solution of neutral sulphate of quinine, add, drop by drop, a concentrated solution of iodide of potassium; dry the precipitate in the shade; or, heat the liquid nearly to the boiling point, and allow it to crystallize.

4265. Sulphate of Quinine. This is the disulphate of quinia. Boil 48 troy ounces coarsely powdered yellow cinchona, in 13 pints of water containing 1⅛ troy ounces muriatic acid, and strain through muslin. Boil the residue twice successively with the same quantity of water and acid as before, and strain. Mix the decoctions, and, while the liquid is hot, gradually add 5 troy ounces finely powdered lime, previously mixed with

4246. Chloride of Calcium. Known also as *muriate of lime.* From the strong affinity this salt has for water, it is much used for drying gases and absorbing the water from ethereal and oily liquids, in organic analyses. For this purpose it is used in the dry state. In its hydrous or crystallized form, it is much used in the preparation of freezing mixtures with snow. In this case, the evaporation need only be conducted so far that the whole becomes a solid mass on removal from the fire. For both this and the last-mentioned use it is reduced to powder. It is also much used as a test for sulphuric acid, with which it produces a white precipitate insoluble in nitric acid; in the rectification of alcohol, and for forming a water-bath with a high boiling point. As a medicine, it has been given in some scrofulous and glandular diseases, and has also been used as a bath in the same cases.

4247. To Prepare Chloride of Calcium. To hydrochloric acid, diluted with an equal weight of water, add powdered chalk or white marble, in small fragments, until effervescence entirely ceases, and the liquid no longer reddens litmus paper. Filter, evaporate to one-half, and set it aside to crystallize. Then collect the crystals, dry them by pressure between bibulous paper, and keep in a stoppered bottle. The mother-liquid will yield more crystals by further evaporation.

4248. Hyposulphite of Lime. Slack 5 ounces lime with enough water to make 4 pints, boil up with 10 ounces of flowers of sulphur, and pass into the solution sulphurous acid gas (free from carbonic acid) until it has become colorless. Then filter and evaporate to crystallization, at a temperature not exceeding 140° Fahrenheit. Another way to prepare this salt is to mix 44 ounces (by weight) of a solution of fused chloride of calcium of 1.238 specific gravity, with a warm solution of 25 ounces hyposulphite of soda in 30 ounces water; evaporate to 38 ounces, and pour off, while warm, from the crystals of chloride of sodium; then allow to crystallize, and purify the crystals by re-solution.

4249. Cobalt. A metal found in ores associated with arsenic and other metals; also present in meteoric iron. It is white, brittle, and does not change in the air; has a high melting point, and is strongly magnetic. Specific gravity 8.5. (*Cooley.*)

4250. Nitrate of Cobalt. This may be obtained by dissolving metallic cobalt in nitric acid, and collecting the crystals. These crystals are ready soluble in water; of a red color; deliquescent, and melt below 212° Fahr. At a higher heat, nitrous fumes are given off, and *peroxide of cobalt* remains.

SEE PAGE 435

2 pints of water, stirring constantly until the quinia is completely precipitated. Wash the precipitate with distilled water; and, having pressed, dried, and powdered it, digest it in boiling alcohol. Pour off the liquid, and repeat the digestion several times until the alcohol is no longer rendered bitter. Mix the liquids, and distill off the alcohol until a brown viscid mass remains. Transfer it to a suitable vessel, and pour upon it 4 pints distilled water; and, having heated the mixture to the boiling point, add as much sulphuric acid as may be necessary to dissolve the quinia. Then add 1½ troy ounces animal charcoal, boil for 2 minutes, filter while hot, and set it aside to crystallize. Should the liquid before filtration be entirely neutral, acidulate it very slightly with sulphuric acid; should it, on the contrary, change the color of litmus paper to a bright red, add more charcoal. Separate the crystals from the liquid, dissolve them in boiling distilled water slightly acidulated with sulphuric acid, add a little animal charcoal, filter the solution, and set it aside to crystallize. Lastly dry the crystals on bibulous paper with a gentle heat, and keep them in a well-stopped bottle. The mother-water may be made to yield an additional quantity of sulphate of quinia by precipitating the quinia with water of ammonia, and treating the precipitate with distilled water, sulphuric acid, and animal charcoal, as before. (*U. S. Ph.*) When pure it forms light, delicate, white needles. It is entirely soluble in hot water, and more readily so when an acid is present. Precipitated by ammonia, the residuary liquid, after evaporation, should not taste of sugar. By a gentle heat it loses 8 or 10 per cent. of water. It is wholly consumed by heat. If chlorine be first added, and then ammonia, it becomes green. A solution of 10 grains in 1 fluid ounce distilled water, and 2 or 3 drops of sulphuric acid, if decomposed by a solution of ½ ounce carbonate of soda, in two waters, and heated till the precipitate shrinks and fuses, yields on cooling a solid mass, which, when dry, weighs 7.4 grains, and in powder dissolves entirely in a solution of oxalic acid.

4266. Tests for the Purity of Sulphate of Quinine. This salt is frequently adulterated with starch, magnesia, gum, sugar, &c. The first three remain undissolved when the salt is digested in spirit; the fourth is dissolved out by cold water, and the last may be detected by precipitating the quinine by liquor of potassa, and dissolving the precipitate in boiling alcohol; cinchona crystallizes out as the solution cools, but the quinine remains in the mother liquor. (*Cooley.*)

Dr. Stonelen proposes a test for the presence of salicine in sulphate of quinine. He employs three kinds of sulphuric acid—viz.: the fuming, pure concentrated acid, free from arsenic and nitric acid; ordinary concentrated sulphuric acid of commerce, containing a trace of nitric acid; and, lastly, sulphuric acid, to which, purposely, nitric acid had been added. Watch glasses having been placed on a sheet of white paper, and a drop or two of the acids above referred to (each in a separate glass) having been poured therein, a few crystals of sulphate of quinine are put on the acid; if pure, there is no coloration; but, even with 1 per cent. of salicine, the two first-named acids cause a distinct red coloration, which does not ensue with the acid containing nitric acid. This latter acid is not even colored by pure salicine.

4267. Acetate of Morphia. The acetate of morphia of commerce is usually in the form of a whitish powder, and is prepared by the mere evaporation of the solution to dryness by a gentle heat. During the process a portion of the acetic acid is dissipated, and hence this preparation is seldom perfectly soluble in water, unless it be slightly acidulated with acetic acid. It is prepared by dissolving 6 drachms morphia in 3 fluid drachms acetic acid specific gravity 1.048, diluted with 4 fluid ounces distilled water; evaporate gently, and crystallize. 100 measures of a solution of 10 grains in ½ fluid ounce water, and 5 minims of acetic acid, heated to 212°, and decomposed by a very slight excess of ammonia, yield by agitation a precipitate, which, in 24 hours, occupies 15½ measures of the liquid.

4268. Opium. The juice obtained by cutting the unripe fruit of the white poppy, and hardened by exposure to the air. It yields several alkaloids, the principal of which is morphine. The best opium comes from Smyrna, in Turkey. Sometimes the commercial article is found adulterated with various substances in order to increase its weight.

4269. To Test the Strength of Opium. Take 25 grains quicklime made into a milk with water, boil in this 100 grains opium, and filter the solution while hot; saturate the filtrate with dilute hydrochloric acid, and then precipitate the morphia by the addition of liquor of ammonia, any excess of the latter being expelled by heat. Collect the precipitate, dry, and weigh it; the weight in grains will represent the percentage of morphia in the sample of opium tested. (*Couerle.*)

4270. To Test the Purity of Opium. Macerate 100 grains opium for 24 hours in 2 fluid ounces water; filter and express the residue; then precipitate with a solution of ¼ ounce carbonate of soda in 2 fluid ounces cold water; gently heat the precipitate until it fuses, then cool and weigh it. It should weigh at least 10 grains; and, when powdered, be entirely soluble in a solution of oxalic acid.

4271. Chloroform. A thin, colorless liquid, of agreeable ethereal odor, and sweetish but slightly acrid taste. Its specific gravity (water standard) is 1.49, and the specific gravity of its vapor (air standard) is 4.2. It kindles with difficulty, burning with a greenish flame, and gives a dull, smoky-yellow color to the flame of alcohol. It occupies a prominent place among the *anæsthetics* (substances used to produce insensibility to pain by inhaling them), but has in later times been, to a certain extent, superseded by nitrous oxide. (*See No.* 4060.) Externally applied, it is refrigerant, soothing, and allays pain. It neither reddens nor bleaches litmus paper.

4272. To Obtain Chloroform. This is prepared on the large scale, by mixing, in a capacious retort or still, 4 pounds chloride of lime, 12 pounds water, and 12 fluid ounces rectified alcohol; distill cautiously as long as a dense liquid is produced, which sinks in and separates from the water with which it passes over. Separate the lower stratum of chloroform from the water, agitate it with a little sulphuric acid, and distill it by the heat of a water-bath from carbonate of baryta. (*Dumas.*)

4273. To Obtain Pure Chloroform. Place in a capacious still 3 gallons water and 30 fluid ounces rectified spirit, and raise the temperature to 100° Fahr. Add 10 pounds *chlorinated lime* (slacked lime saturated with chlorine gas), and 5 pounds slacked lime, mixing thoroughly. Apply heat, which must be withdrawn as soon as distillation has commenced, and distill 50 ounces; agitate it with ½ gallon water, and allow the crude chloroform to settle. Separate and wash the chloroform with 3 fluid ounces distilled water, repeating this operation 3 times, each time with fresh distilled water. Next agitate the chloroform for 5 minutes with an equal volume of sulphuric acid; when settled, transfer the upper stratum to a flask containing 2 ounces chloride of calcium in small pieces, and ½ ounce perfectly dry slacked lime. Agitate thoroughly, and, after an hour, distill the pure chloroform over a water-bath. Keep in a well-stoppered bottle, in a cool place. The U. S. Dispensatory has transferred this from the British Pharmacopœia, consequently avoirdupois weight and Imperial measure are adopted.

4274. To Purify Commercial Chloroform. To 102 troy ounces commercial chloroform add 17 troy ounces sulphuric acid, occasionally shaking during 24 hours. Separate the lighter liquid and mix it with 6 fluid drachms stronger alcohol. Then add 2 troy ounces carbonate of potassa, previously heated to redness, and rubbed into powder while warm. Agitate thoroughly and distill to dryness. Keep the distilled liquid in well-stopped bottles. (*U. S. Ph.*)

4275. Tests for the Purity of Chloroform. Its specific gravity should not be less than 1.490, nor more than 1.494; and should boil at 140° Fahr. When dropped into water, it sinks in transparent globules without milkiness. When mixed in a bottle with an equal bulk of sulphuric acid, it should produce no warmth; and after standing for 24 hours, neither liquid should be discolored, or, at most, a faint yellow tinge imparted to the lower or acid stratum; more discoloration than this would denote the presence of empyreumatic oily matter. When evaporated on a porcelain plate, it leaves behind a slightly aromatic odor, but free from pungency.

4276. Chloral. Chloral is an oily liquid, possessing an ethereal smell; it is soluble in alcohol, ether, and water, but its solution in the latter rapidly changes into a semisolid crystalline mass of *hydrate* of *chloral*, soluble in a larger quantity of water. Chloral boils at 202°, and has a specific gravity of 1.502.

4277. To Obtain Chloral. Place anhydrous alcohol in a tubulated retort, and pass dry chlorine gas through it, at first in the cold, but afterwards with the application of a gentle heat. As soon as the chlorine passes undecomposed through the liquor at the boiling temperature, the process is complete. On cooling, the liquid in the retort solidifies, forming a crystalline mass of hydrated chloral. This must be melted by gentle heat, and agitated with thrice its volume of oil of vitriol, when, on increasing the heat a little, an oily stratum of impure chloral will rise to the surface. This must be removed, boiled for some time, to drive off some free hydrochloric acid and alcohol, and next distilled with an equal volume of oil of vitriol; lastly, it must be rectified from finely-powdered quicklime, stopping the process as soon as the surface of the lime becomes dry. The chlorine is best introduced by a tube inserted into the tubulature of the retort, and a long tube, bent upwards, should be connected with the beak to convey away the hydrochloric acid gas extricated, and to allow the volatilized alcohol and chloral to condense and flow back into the retort.

4278. To Purify Hydrate of Chloral. There is perhaps scarcely a liquid in which chloral hydrate is insoluble at ordinary temperature; four parts of it dissolve gradually in one part of water, the solution crystallizes at 32° Fahr., but not in well-formed crystals. Alcohol and ether dissolve it to such an extent that it likewise does not crystallize well on evaporating these solvents; absolute alcohol must be excluded, because it combines with chloral. Chloroform and benzole are well adapted for recrystallization, but the first is too dear, and the last cannot be entirely removed from the crystals. The same holds good for most other liquid solvents, but uniformly satisfactory results are obtained with bisulphide of carbon; 45 parts of it dissolve at 60° to 65° Fahr., but 1 part chloral hydrate; it precipitates ethereal and alcoholic solutions of the latter. But at temperatures below the boiling of bisulphide of carbon, 4 to 5 parts of it are sufficient for dissolving 1 part chloral hydrate. If allowed to cool slowly, beautiful crystals, often an inch in length, are obtained, easily collected, and readily freed from the last traces of the solvent by exposing them in thin layers to the air. (*Flückiger.*)

4279. Sulphuric Ether—also called *oxide of ethyl*—is a colorless, transparent, very limpid fluid, having a penetrating and agreeable smell and a burning taste.

4280. To Obtain Sulphuric Ether. Put 2 pounds rectified spirit into a glass retort, and add 2 pounds sulphuric acid; place the retort on a sand-bath, and apply heat so that the liquor may boil as quickly as possible, and the ether pass into a receiver cooled by ice or water; continue to distill until a heavier fluid begins to pass over; then lower the heat, add another pound of spirit, and distill as before. Mix the distilled liquors together, pour off the supernatant portion, add 1 ounce carbonate of potassa (previously ignited), and agitate occasionally for one hour; finally, distill the ether from a large retort, and keep it in a well-stoppered bottle. This ether should have a specific gravity of .750.

It is recommended to mix only a portion of the alcohol at first with the acid, and as soon as it reaches boiling point (about 280° Fahr.), add the remainder only fast enough to replace the fluid as it distills over; also not to allow the heat to exceed 286°.

Another method is, to heat the sulphuric acid to 280°, and then introduce the alcohol in a fine stream, by means of a tube with a fine lower orifice, introduced through a cork fitted to the mouth of the retort; a thermometer being adjusted in a similar manner, so that its bulb is immersed in the contents of the retort. By this means the danger of the heat rising above 286° is obviated.

4281. Stronger Ether. *Officinal Æther Fortior.* Take 3 pints each of ether and water; shake them thoroughly together in a bottle; and, when the water has subsided, separate the ether from it, and agitate it well with 1 troy ounce each of chloride of calcium and lime, both in fine powder. After standing for 24 hours, decant the ether into a retort, with a Liebig's condenser, connected with a receiver surrounded by ice-cold water, and distill 1¼ pints stronger ether, which should be of a specific gravity not exceeding .728.

4282. To Purify Ether. Ordinary ether is purified by first agitating it with 2 or 3 times its volume of distilled water containing a few grains of carbonate of potassa, or a few drops of milk of lime; and, after decantation, again agitated with a like quantity of water only. This may be used for inhalations. The washed ether is afterwards digested on chloride of calcium to deprive it of retained moisture.

4283. Cautions About Ether. The vapor of ether is very inflammable, and when mixed with atmospheric air it forms a violently explosive mixture. The density of this vapor is 2.586, that of air being 1, hence it rapidly sinks, and frequently accumulates in the lower parts of buildings, especially cellars which are badly ventilated. Every crack, every joint in the floors of rooms, the space beneath doors, &c., offer a road for the passage of this vapor, which, though invisible, as surely runs out of every orifice, and finds its level, as a stream of water would do. The only remedy is thorough ventilation. Many serious accidents have arisen from this cause; a light carried where such vapor is present causes an explosion.

4284. Ozone Ether. By agitating ether in a flask with binoxide of barium, adding gradually perfectly pure and very dilute hydrochloric acid, occasionally cooling and subsequently allowing the ether to settle, we obtain a liquid which has been recommended as a disinfecting, bleaching, and cleansing agent, and as a test for chromic acid, which it instantly turns indigo blue. According to Boettger, this does not contain ozone, but binoxide of hydrogen, which is equivalent to it.

4285. Tests for the Purity of Ether. Pure ether should be neutral to test paper; vaporize totally when exposed to the air; when shaken in a graduated tube with half its volume of a concentrated solution of chloride of calcium, its volume should not be lessened; water should dissolve only $\frac{1}{10}$ its volume of ether, and remain transparent. Dry carbonate of potassa or tannin shaken with ether in a test-tube will become moist or form a syrupy solution, in case any water is present. The presence of alcohol is shown by shaking the ether with water, its solubility in water being the greater in direct proportion to the quantity of alcohol which it contains.

4286. To Find the Percentage of Ether in a Mixture of Ether and Alcohol. By ascertaining the specific gravity at 60° Fahr. of a mixture of ether and alcohol, the following table will give the percentage of absolute ether contained in the mixture:

TABLE OF PERCENTAGE OF ETHER.

Spec. Grav.	Per cent.	Spec. Grav.	Per cent.
0.7198	100	.7673	65
.7246	95	.7636	60
.7293	90	.7701	55
.7343	85	.7772	50
.7397	80	.7840	45
.7455	75	.7880	40
.7514	70		

4287. Nitric Ether. Take 50 parts nitric acid, specific gravity 1.375, dissolve in it 2 or 3 parts nitrate of urea, and add 50 parts alcohol. Distill until ⅞ of the whole has passed over; agitate the distillate with a little water to separate the ether, and preserve the heavier portion. It has a specific gravity of 1.112; its vapor is explosive when strongly heated, consequently great care is necessary in the distillation, to keep the heat down to the lowest working point, and to distill only small quantities at a time. (*Millon.*) (*See directions for Sulphuric Ether, in No.* 4280.)

4288. Nitrous Ether. Nitrous or hyponitrous ether has a pale yellow color, boils at 62° Fahr.; at 60° its specific gravity is .947; it is very volatile. Take starch, 1 part; nitric acid, specific gravity 1.30, 10 parts; alcohol of 85 per cent., 2 parts; water, 1 part; introduce the starch and acid into a capacious retort connected with a wide tube 2 or 3 feet long, bent at right angles, and terminating near the bottom of a two-necked bottle, containing the alcohol and water mixed together, and surrounded with a freezing mixture or very cold water. The other neck of the bottle must be connected by a wide and long glass tube, with a good refrigerator or condenser. The heat of a water-bath must be cautiously applied to the retort, when pure hyponitrous acid will be set free, and, passing into the alcohol, will form hyponitrite of oxide of ethyl (ether), which will distill in a gentle stream. The tube connecting the retort and bottle must be cooled by means of a rag or moist paper, wetted from time to time with ice-cold water; for if the tube and the alcohol be not carefully cooled, the latter becomes spontaneously hot, and boils violently, when the product is vitiated. This process is very productive and economical, and yields perfectly pure hyponitrous ether. (*Liebig.*)

4289. Sweet Spirit of Nitre. This is an alcoholic solution of nitrous ether. The mixture should have, according to the U. S. Pharmacopœia, a specific gravity of .837. It becomes acid by age.

4290. Hydrochloric Ether. This is the chloride of ethyl, and is distilled in a retort, from rectified spirit of wine saturated with dry hydrochloric acid gas. (Thénard directs equal volumes of concentrated hydrochloric acid and absolute alcohol.) The retort is connected with a Wolffe's apparatus, the first bottle of which should be two-thirds full of tepid water (70° to 75° Fahr.), and the remainder surrounded with salt and ice. To render it perfectly anhydrous, it must be digested on a few fragments of fused chloride of calcium. (*Cooley.*)

4291. Acetic Ether. This is a colorless fluid, and bears a considerable resemblance to sulphuric ether, of which it is strictly an acetate. Liebig assigns it a specific gravity of .89 at 60° Fahr., dissolving in 7 times its bulk of water; Ure gives it a specific gravity of .866 at 45°, dissolving in 8 parts water. It is decomposed by alkalies and strong acids. (*Cooley.*) It is also called *acetate of ethyl.*

4292. To Obtain Acetic Ether. Mix together 3 parts acetate of potassa (or an equivalent quantity of acetate of soda (*see No.* 80), 3 parts 85 per cent. alcohol, and 2 parts strongest oil of vitriol. Distill them in a glass retort or earthenware still, connected with a well-cooled receiver; agitate the product with a little water, to remove undecomposed alcohol, then digest it with a little chalk, to remove acidity, and afterwards with fused chloride of calcium, to absorb water. Lastly, rectify by a gentle heat. (*Fownes.*)

4293. To Prepare Butyric Ether. This is the pine-apple oil of commerce; and, largely diluted with rectified spirit, is the pine-apple essence used for flavoring. It is prepared from crude butyric acid saponified with caustic potassa, and the resulting soap distilled along with alcohol and oil of vitriol. It is sparingly soluble in water, very soluble in alcohol; boils at 230°. It is also called *butyrate of ethyl.*

4294. Benzoic Ether. A colorless oily liquid, slightly heavier than water, aromatic in taste and odor. It boils at 410° Fahr. It is prepared as follows: Take 4 parts 90 per cent. alcohol, 2 parts crystallized benzoic acid, and 1 part concentrated muriatic acid; distill them together, and, as soon as the product turns milky when mixed with water, change the receiver and collect the subsequent distillate; add water to it, decant the ether from the surface of the water, and boil it with water and a little oxide of lead (to separate the benzoic acid); lastly, free it from water by allowing it to stand over chloride of calcium. Benzoic ether is also called *benzoate of ethyl.*

4295. Formic Ether—also called *formiate of ethyl,* is a limpid, aromatic fluid, lighter than water; soluble in 10 parts of that fluid; has a specific gravity of .915, and boils at 130° Fahr. To obtain it, mix in a retort, with a well-cooled receiver, 7 parts dry formiate of soda, 10 parts oil of vitriol, and 6 parts 90 per cent. alcohol. The greater part will distill over by the heat spontaneously developed, after which the heat of a water-bath may be applied. Purify it by agitation, first with milk of lime, and afterwards with chloride of calcium. (*Cooley.*)

4296. Œnanthic Ether—named also *œnanthylate of ethyl,* and *pelargonic ether* (*see No.* 1471)—is colorless, and has a powerful intoxicating vinous odor. Its specific gravity is .862, and boils at 480° Fahr. It is obtained towards the end of the distillation of fermented liquors, especially wines, and purified by agitation with a weak solution of carbonate of potassa. (*Cooley.*) This ether has the odor of quince, and dissolved in a due proportion of alcohol, forms *quince essence.*

4297. Chloric Ether. This is synonymous with chloroform. Medicinal chloric ether consists of 1 part chloroform in 8 parts rectified spirit.

4298. Ethyl. This is a colorless, inflammable gas, of a specific gravity a little over 2 (air standard). Under a pressure of 2¼ atmospheres, at 37.5 Fahr., it assumes the form of a colorless ethereal liquid. It forms the basis of ether, which is oxide of ethyl; and of alcohol, which is the hydrated oxide of ethyl; its usefulness lays chiefly in its compounds with acids. The following are the principal ones in use, and will serve as a guide for the preparation of most of the others.

4299. Acetate of Ethyl. Heat together in a retort, 3 parts acetate of potassium, 3 parts strong alcohol, and 2 parts oil of vitriol. The distilled product is mixed with water to separate the alcohol; digested first with a little chalk, and afterwards with fused chloride of calcium; lastly, it is rectified. A fragrant, limpid liquid, having a density of .890, and boiling at 165° Fahr. (*Fownes.*)

4300. Valerianate of Ethyl. Pass dry hydrochloric acid gas through an alcoholic solution of valerianic acid. Its odor resembles butyric ether.

4301. Amyl. This is the basis of the fusel oil compounds; fusel oil being the oxide of amyl. It is a colorless, ethereal liquid, boiling at 311° Fahr. Like ethyl, its acid compounds are most used. (*See No.* 1440.)

4302. Acetate of Amyl. Mix together 1 part fusel oil and 2 parts dry acetate of potassa (potassium—*Fownes*); add 1 part concentrated sulphuric acid, and distill. Purify the distillate by washing it with a dilute solution of potassa, and again distill it from dry chloride of calcium. (*Cooley.*) Acetate of amyl, diluted with alcohol, forms the *essence of Jargonelle or Bergamot pear.*

4303. Valerianate of Amyl. Mix carefully 4 parts fusel oil with 4 parts sulphuric acid; when cold, add 5 parts valerianic acid. Warm the mixture for a few minutes in a water-bath, then mix it with a little water, which causes the ether to separate. Purify this by washing it with water, and a weak solution of carbonate of soda. An alcoholic solution of valerianate of amyl constitutes *apple essence.*

4304. Methyl. This is the basis of methylic alcohol or pyroxylic spirit, forming compounds with the acids, analogous to those of ethyl.

4305. Valerianic Acid. A volatile, fatty acid, obtained by distilling valerian root along with water, and acting on the product with caustic potassa, when valerianate of potassa is formed, and a volatile oil is separated; by evaporating to dryness, the latter is dissipated, and the dry mixture, treated with dilute sulphuric acid and distilled, yields an aqueous solution of valerianic acid. By careful redistillation it may be deprived of water. Valerianic acid may also be produced artificially, by heating fused potassa along with the oil of potato, or corn spirit, when valerianate of potassa is obtained, the acid of which is identical in all respects with that obtained from the root of valerian. (*Liebig.*) It is colorless, limpid, oleaginous; boils at 270° Fahr.; soluble in alcohol and ether, and in 30 parts of water; smells strongly of valerian; with the bases it forms salts called VALERIANATES, most of which are soluble.

4306. Succinic acid. This is obtained by mixing coarsely powdered amber with an equal weight of sand, and distilling it by a gradually increased heat; the product is purified by pressing it between bibulous paper, to remove the oil, and then subliming it. It forms colorless, inodorous crystalline scales, soluble in 5 parts cold or 2½ parts boiling water; is fusible and volatile without decomposition. (*Cooley.*)

4307. Aldehyd-Ammonia. Take sulphuric acid, 6 parts; water, 4 parts; alcohol of 80 per cent., 4 parts; peroxide of manganese in fine powder, 6 parts. Dilute the acid with the water, then carefully add the alcohol, and next the manganese; agitate and distill with a gentle heat, from a spacious retort into a receiver surrounded with ice, and connected with the former perfectly air-tight. When 6 parts have distilled, re-distill this portion from its own weight of dried chloride of calcuim until 3 parts have come over, which must be again rectified in the same manner, until 1½ parts of liquid are obtained in the receiver. This liquid must then be mixed with an equal bulk of ether, and the mixture saturated with dry ammoniacal gas; brilliant colorless prismatic crystals will form, which, after washing with ether and drying, are pure aldehyd-ammonia. It smells like turpentine; melts at 160° Fahr.; volatilizes, unchanged, at 212°; decomposed by exposure to the air; soluble in most menstrua except ether.

4308. Aldehyde. Dissolve 8 parts aldehyd-ammonia in 8 parts water; place the solution in a retort, and add 7 parts sulphuric acid, diluted with about half its weight of water; then distill as directed in last receipt. Rectify the product twice from its own weight of dried muriate of lime, at a heat not exceeding 86° Fahr. It is an ethereous liquid, boiling at 72°; neutral, inflammable, mixed with water, alcohol, and ether; decomposed by exposure to the air, into liquid acetic acid; spoils by age.

4309. Sulphuret of Carbon. A colorless, pungent, fœtid liquid, exceedingly volatile and combustible. It exceeds all substances in refractive power. In dispersive power it exceeds all fluid substances except oil of cassia. It produces intense cold by its evaporation. A spirit thermometer, having its bulb covered with cotton, if dipped into this fluid and suspended in the air, rapidly sinks from 60° to 0°, and if put into the receiver of an air-pump it will fall to —81°. Mercury may be readily frozen in this way.

4310. To Prepare Sulphuret of Carbon. Heat together in a close vessel 5 parts bisulphuret of iron, and 1 part well dried charcoal; or transmit the vapor of sulphur over fragments of charcoal heated to redness in a porcelain tube. In either case the resulting compound should be carried off as soon as formed, by means of a glass tube plunged into pounded ice, beneath which it will collect. It may be afterwards freed from adhering moisture and sulphur by distilling it at a low temperature from chloride of calcium.

4311. Bisulphide or Bisulphuret of Carbon. This is used in the arts as a solvent for India-rubber, gutta percha, &c. To procure it, Mulder recommends the following process as the most convenient. Provide an iron bottle (a quicksilver bottle answers very well), and make a second opening into it. To one opening adapt a copper tube bent twice at right angles; and to the other a straight tube dipping into the bottle. Having nearly filled the bottle with pieces of charcoal (recently heated to redness), and having screwed on the bent and straight tubes, place the bottle in a furnace, closing the mouth of the latter with a stone or clay cover in two pieces, hollowed in the centre so as to fit the upper part of the bottle, and defend it from the action of the fire. Connect the curved tube with a Wolffe's bottle half-filled with water, and placed in a freezing mixture; and when the iron bottle is sufficiently heated, introduce by the straight tube fragments of sulphur, and immediately close the mouth of the tube with a plug. The bisulphuret, as it comes over, falls to the bottom of the water. Separate it from the water, and distill over dry chloride of calcium.

4312. Terpine. Leave oil of turpentine for a long time in contact with a mixture of nitric acid and alcohol. Crystals of terpine form. By boiling an aqueous solution of terpine with a small quantity of sulphuric or other acid, terpinole is formed, and may be separated by distillation. It has the odor of hyacinths.

4313. Sugar Resin. Mix 16 parts strong sulphuric acid with 8 of the strongest nitric acid; when cooled to 70° Fahr., stir in 1 part of finely-powdered sugar. In a few seconds, when the sugar has become pasty, take it out of the acid and plunge it into cold water. Add more sugar to the acid, and proceed as before. Wash the resinous matter carefully, and dissolve it in alcohol or ether. Evaporate the solution with a gentle heat. It is very combustible. Its solution may be used to render gunpowder, lucifer matches, &c., waterproof.

4314. Aluminized Charcoal. This is recommended by Dr. Stenhouse as a cheap and very efficient decolorizing agent. Dissolve in water 54 parts of the sulphate of alumina of commerce, and mix with 92½ parts of finely powdered wood charcoal. When the charcoal is saturated, evaporate to dryness, and heat to redness in covered Hessian crucibles till the water and acid are dissipated. The charcoal contains just 7½ per cent. of anhydrous alumina.

4315. Styrol. Mix 20 parts of storax with 7 of carbonate of soda, and put them into a retort with water, and apply heat. A limpid fluid distills, which becomes, when heated to a certain point, a transparent solid.

4316. Turpentine. An oleo-resin flowing from the trunk, after removing the bark of the pitch or swamp pine. It is viscid, transparent, and of the consistence of honey.

4317. Oil of Turpentine. Oil or *spirits of turpentine* is obtained by distilling crude turpentine along with water. The remainder left in the still after distillation is *resin*. It congeals at 14°, and boils at 312° Fahr.; its specific gravity is about 870°. It is very inflammable, and becomes resinous by exposure to the air. When purified, by redistilling with 3 or 4 times its volume of water, it produces the *camphene* of commerce.

4318. Venice Turpentine. A liquid resin which exudes from the larch tree. The Venice turpentine usually met with is a factitious article composed of 2 gallons oil of turpentine added to 48 pounds melted black resin. (*Cooley.*)

4319. To Purify Turpentine. However carefully the oil of turpentine may have been distilled, it always leaves, after evaporation, a disagreeable odor, firmly adhering to the goods that have been treated with it. The same is the case with benzine and the lighter petroleum oils. This may be obviated, according to Bremer, by distillation over tannin. Articles treated with oil of turpentine that has been distilled in this way, are heated to 150° Fahr., when they lose every trace of odor. Bremer adds that this preparation is less inflammable, cheaper, and more agreeable to the workman than benzine.

4320. Benzine. This is the name given in the United States to one of the products distilled from petroleum, having a specific gravity of about .73, or 65° of Baumé's light hydrometer. (*See No.* 1527.) It has not yet been frozen, and is dangerously volatile at all temperatures. (*See No.* 346.) Benzine scarcely attacks asphaltum or pitch, and cannot (like benzole), be converted by nitric acid into nitro-benzole. It is consequently useless for the preparation of aniline. Benzine consists of about 84 per cent. carbon and 16 per cent. hydrogen. (*See No.* 440.)

4321. Benzole. In 1825, Faraday discovered a peculiar liquid which was deposited by condensation by ordinary coal-gas, and gave it the name of bicarburet of hydrogen. Some years afterwards Mitscherlich, of Berlin, obtained the same liquid from benzoic acid, and proposed for it the name of benzine. Faraday objected to this name, as too similar to the distinctive names of the alkaloids, as strychnine, morphine, &c., and decided to call it benzole. The French, however, adhered to Mitscherlich's name, and continue to call it benzine, causing considerable confusion; as benzole, from coal-tar, is a different liquid from benzine, obtained from petroleum. (*See No.* 1527.) Benzole has a specific gravity of .85, and freezes at 37° Fahr; it dissolves asphaltum or pitch rapidly, is volatile at all temperatures, but less so than benzine. Benzole can be converted by nitric acid into nitro-benzole, and, by further treatment, into aniline. (*See No.* 2552.) It contains about 92.5 per cent. of carbon, and 7.5 per cent. of hydrogen.

4322. Nitro-Benzole. A yellowish, oily fluid, insoluble in water; boils at 415° Fahr., and has a specific gravity of 1.209; known also as *essence of mirbane*. The method of preparing it is as follows: Place 10 parts fuming nitric acid in a tubulated retort capable of holding 3 times the quantity; apply heat sufficient to produce gentle ebullition. Insert a glass tube through the upper neck of the retort, and through it introduce by degrees, a drop at a time, benzole (not *benzine, see No.* 4321), so long as nitrous vapors are evolved; the liquid which passes into the receiver being poured back from time to time into the retort. When the red vapors have ceased to rise, distill off the excess of benzole, if any, from the acid. Then pour the contents remaining in the retort into 120 to 150 parts cold water, and let it stand for a few days, when the nitro-benzole will be found separated at the lower part of the vessel. Decant the upper stratum of acid, wash the nitro-benzole with water, and keep it in stoppered bottles. This substance is used as a *factitious oil of bitter almonds*, being, although poisonous, far less so than the prussic acid of which the real article consists. (*Hager.*)

4323. Urea. A crystalline, colorless, transparent substance, consisting of *cyanate of ammonia*. Fresh urine, gently evaporated to the consistence of a syrup, is to be treated with its own volume of nitric acid at 24 deg.; the mixture is to be shaken and immersed in an ice-bath to solidify the crystals of *nitrate of urea;* these are washed with ice-cold water, drained, and pressed between sheets of blotting paper. When they are thus separated from foreign matters, they are to be dissolved in water to which subcarbonate of potash is added, whereby the nitric acid is taken up, and the urea set at liberty. This new liquor is evaporated at a gentle heat, nearly to dryness; the residue is treated with pure alcohol, which only dissolves the urea; the solution is concentrated, and the urea crystallizes. (*Thénard.*)

Or: Mix 28 parts of perfectly dry ferrocyanide of potassium with 14 parts of black oxide of manganese, both pure and in fine powder; then place them on a smooth iron plate, and heat them to a dull red, over a charcoal fire. When the mass begins to burn, it must be frequently stirred; after which cool and dissolve in cold water, filter, and add 20½ parts of dry sulphate of ammonia, and decant the clear from the precipitated sulphate of potassa. Concentrate at a heat below 212°, again decant, evaporate to dryness, and digest in boiling alcohol of 80%; crystals of urea will be deposited as the solution cools. (*Liebig.*)

4324. Nitrate of Urea. This may be prepared as in last receipt from urea; or by saturating the artificial urea (Liebig's preparation) with nitric acid.

4325. Stearine. The solid portion of fats which is insoluble in cold alcohol. Pure strained mutton suet is melted in a glass flask with 7 or 8 times its weight of ether, and the solution allowed to cool; the soft pasty mass is then transferred to a cloth, and is strongly pressed, as rapidly as possible, to avoid evaporation; the solid portion is then dissolved again in ether, and the solution allowed to crystallize. The product is nearly pure.

4326. Iodine. A chemical element found both in the animal, vegetable, and mineral kingdoms, but exists in greatest abundance in sea-weed. It is principally manufactured from the mother-waters of kelp. Iodine is usually met with under the form of semi-crystalline lumps, having a metallic lustre, or friable scales, somewhat resembling gunpowder. It has a greyish-black color, a hot, acrid taste, and a disagreeable odor not much unlike that of chlorine. It fuses at 225° Fahr., volatilizes slowly at ordinary temperatures, boils at 347°, and when mixed with water rapidly rises along with its vapor at 212°. It dissolves in 7000 parts of water, and freely in alcohol and ether. It may be crystallized in large rhomboidal plates, by exposing to the air a solution of it in hydriodic acid. Iodine, like chlorine, has an extensive range of affinity; with the salifiable bases it forms compounds termed IODIDES, IODURETS, or HYDRIODATES; and it destroys vegetable colors.

4327. To Obtain Iodine. Saturate the residual liquor of the manufacture of soap from kelp (or other iodine lye) of a specific gravity of 1.374, heated to 230° Fahr., with sulphuric acid diluted with half its weight of water; cool, decant the clear, strain, and to every 12 fluid ounces add 1000 grains of black oxide of manganese, in powder; put the mixture into a glass globe, or matrass with a wide neck, over which invert another glass globe, and apply heat with a charcoal fire; iodine will sublime very copiously, and condense in the upper vessel, which, as soon as warm, should be replaced by another; and the two globes thus applied in succession as long as violet vapor arises. It may be washed out of the globes with a little cold water. A thin disc of wood, having a hole in its centre, should be placed over the shoulder of the matrass, to prevent the heat from acting on the globular receiver. On the large scale, a leaden still may be employed, and receivers of stoneware economically substituted for glass ones. The top of the leaden still is usually furnished with a moveable stopper, by which the process may be watched, and additions of manganese or sulphuric acid made, if required. The addition of the sulphuric acid should be made in a wooden or stoneware basin or trough. To render the iodine pure, it should be dried as much as possible, and then resublimed in a glass or stoneware vessel. (*Ure.*)

Or: Extract all the soluble part of kelp by water, and crystallize the soda by evaporation; to the mother-lye add oil of vitriol in excess, and boil the liquid, then strain it to separate some sulphur, and mix the filtered liquor with as much manganese as there was oil of vitriol: on applying heat, the iodine sublimes in the form of greyish-black scales, with a metallic lustre. The boiling is conducted in a leaden vessel; and a cylindrical leaden still with a very short head, and connected with 2 or 3 large globular glass receivers, is used for the subliming apparatus. Care must be taken to watch the process, and prevent the neck of the still becoming choked with condensed iodine. (*Cooley.*)

4328. To Dissolve Iodine in Cod Liver Oil. To effect this it is best to triturate the iodine with half its weight of iodide of potassium, and to add gradually the oil so as to form a uniform mixture. After standing for a few hours all the iodide will be found at the bottom of the flask, leaving the iodine in perfect solution, the oil having but little of its taste. (*Eymael.*)

4329. Tests for Iodine. Free iodine may be recognized by — The violet color of its vapor.—Striking a blue color with starch; this test is so delicate that water containing only $\frac{1}{450000}$ part of iodine acquires a perceptible blue tinge on the addition of starch.—Nitrate of silver causes a white precipitate in solutions containing iodine.—It strikes a blue color with opium and narceine. Iodine in combination, as it exists in iodic acid and the iodates, does not strike a blue color with starch, without the addition of some deoxydizing agent, as sulphurous acid or morphia; and as it exists in the iodides, not until the base is saturated with an acid (as the sulphuric or nitric), when iodine being set free, immediately reacts upon the starch. An excess of either acid or alkali destroys the action of the test. By mixing the liquid containing the iodine with the starch and sulphuric acid, and lightly pouring thereon a small quantity of aqueous chlorine, a very visible blue zone will be developed at the line of contact. (*Balard.*) Solutions containing iodates yield, with nitrate of silver, a white precipitate, soluble in ammonia; the iodides, under the same circumstances, give a pale yellowish precipitate with nitrate of silver, scarcely soluble in ammonia; a bright yellow one with acetate of lead; and a scarlet one with bichloride of mercury. The iodates deflagrate when thrown on burning coals, but the iodides do not. The iodates may also be tested as iodides, by first heating them to redness, by which they lose their oxygen, and are converted into iodides.

4330. Kelp. The alkaline ashes obtained by burning various kinds of sea-weed.

4331. Galipot. A French term for that portion of turpentine which concretes on the trunk of the tree when wounded, and is removed during the winter.

4332. Phosphorus. Phosphorus is a pale yellow, semi-transparent, and highly combustible solid; specific gravity 1.77 (water standard); melts at 108° Fahr., and unites with oxygen, forming acids, and with the metals, forming PHOSPHIDES or PHOSPHURETS. It is soluble in ether, naphtha, and the oils. From its great inflammability it can only be safely kept under water. In commerce it is always packed in tin cylinders, soldered airtight. It is a powerful corrosive poison. The specific gravity of its vapor is 4.327 (air standard).

4333. To Obtain Phosphorus. Ground bone-ash, 12 parts; water, 24 parts; mix to a pap in a large tub, and add in a slender stream (still stirring) oil of vitriol, 8 parts; work well together, adding more water if required; in 24 hours thin with water, agitate well, and, if convenient, heat the mixture in a leaden pan, and as soon as the paste has lost its granular character, transfer it into a series of tall casks; largely dilute with water, and, after settling, decant the clear portion; wash the residue well with water, mix the clear liquids, and evaporate in a copper or lead pan, till the calcareous deposit (gypsum) becomes considerable, then cool, decant the clear, and drain the sediment on a filter; evaporate the clear liquid to the consistence of honey (say to 4 parts), add 1 part of powdered charcoal, and evaporate to dryness in an iron pot, or till the bottom of the latter becomes red hot; the dry mixture, when cold, is put into earthen retorts well covered with luting and properly dried, and heat is applied sideways rather than at the bottom, by means of an air furnace. The beak of the retort is connected with a copper tube, the other end of which is made to dip about ¼ inch beneath the surface of lukewarm water placed in a trough or wide-mouthed bottle. The distilled product is purified by squeezing it through chamois leather under warm water, and is then moulded for sale by melting it under water heated to about 145° Fahr., plunging the wider end of a slightly tapering but straight glass tube into the water, sucking this up to the top of the glass, so as to warm and wet it, next immersing the end into the liquid phosphorus, and sucking it up to any desired height. The bottom of the tube being now closed with the finger, it is withdrawn, and transferred to a pan of cold water to congeal the phosphorus, which will then commonly fall out, or may be easily expelled by pressure with a piece of wire. Keep it in places where neither light nor heat has access, in phials filled with cold water which has been boiled, to expel all air, and enclose the phials in opaque cases.

4334. Baldwin's Phosphorus. Heat nitrate of lime till it melts; keep it fused for 10 minutes, and pour it into a heated iron ladle. When cool, break it into pieces, and keep it in a closely-stoppered bottle. After exposure to the sun's rays, it emits a white light in the dark.

4335. Canton's Phosphorus. Put calcined oyster shells in layers, alternately with sulphur, and heat strongly in a covered crucible for an hour. This is also luminous in the dark after exposure to the sun.

4336. Phosphorus Bottles. Put 12 grains phosphorus with ½ ounce olive oil in a 1 ounce phial; and place it, loosely corked, in a basin of hot water; as soon as the phosphorus is melted, remove the phial, cork it securely, and agitate it until nearly cold. On being uncorked it emits sufficient light in the dark to see the time by a watch, and will retain this property for some years if not too frequently employed.

4337. To Coat Phosphorus with Copper. Dr. Siewert, of Halle, suggests a method by which the sticks can be kept, even in the light, without undergoing deterioration. For this purpose, he takes advantage of the well-known property of phosphorus to reduce some metals from their solutions. The sticks of phosphorus are put into a cold saturated solution of the sulphate of copper. Presently they become coated with a deposit of metallic copper, and in this state resemble copper rods. They can now be removed to a bottle containing water, and will keep for years. When a stick is wanted for any purpose, on removing the metallic film, and scraping off a black deposit underneath it, the phosphorus will be found to have retained its translucency, as if it had been freshly cast.

4338. To Reduce Phosphorus to Powder. Melt the phosphorus in a phial containing some fresh urine, or a solution of pure urea, by the heat of hot water, and agitate until cold. Rectified spirit may be used instead of urine or urea. (*See No.* 1899.)

4339. Phosphorescent Oil. Dissolve 1 grain phosphorus in 1 ounce olive oil in a test tube by the heat of hot water, or add a larger quantity to some oil of lavender, in which it will dissolve spontaneously. Keep in a close phial.

4340. Pyrophorus. This is a term given to substances which inflame spontaneously when exposed to the air. When a small quantity of any of the powders given below is exposed to the air, it rapidly becomes hot and inflames. Their action is quicker in a damp atmosphere, or by the moisture of the breath.

4341. Homberg's Pyrophorus. Stir equal parts of alum and brown sugar (or 3 parts alum and 1 part wheat flour) in an iron ladle over the fire until dry; then put it into an earthen or coated glass phial, and keep it at a red heat so long as flame is emitted; it must then be carefully stopped up and cooled.

4342. Hare's Pyrophorus. Lampblack, 3 parts; burnt alum, 4 parts; carbonate of potash, 8 parts; as above.

4343. Gay Lussac's Pyrophorus. Sulphate of potash, 9 parts; calcined lampblack, 5 parts; as last.

4344. Goebel's Pyrophorus. Heat tartrate of lead red hot in a glass tube, and then hermetically seal it.

4345. Dextrine or Starch Gum. Heat 4 gallons water in a water-bath to between 77° and 86° Fahr.; stir in 1½ or 2 pounds finely ground malt; raise the temperature to 140°, add 10 pounds potato or other starch; mix all thoroughly, raise the heat to 158°, and keep it between that and 167° for 20 or 30 minutes. When the liquor becomes thin, instantly raise the heat to the boiling point, to prevent the formation of sugar. Strain the liquor, and evaporate it to dryness, as the dextrine will not keep long in a liquid form. Another method is to boil solution of starch with a few drops of sulphuric acid, filter the solution, and add alcohol to throw down the dextrine.

Or: Mix 500 parts potato starch with 1500 parts of cold distilled water and 8 parts of pure oxalic acid; place this mixture in a suitable vessel on a water-bath, and heat until a small sample tested with iodine solution does not produce the reaction of starch. When this is found to be the case, immediately remove the vessel from the water-bath, and neutralize the liquid with pure carbonate of lime. After having been left standing for a couple of days the liquor is filtered, and the clear filtrate evaporated upon a water-bath until the mass has become quite a paste, which is removed by a spatula, and, having been made into a thin cake, is placed upon paper and further dried in a warm place; 220 parts of pure dextrine are thus obtained. (*See No.* 2925.)

4346. Albumen. A substance which enters largely into the composition of animal bodies. It is scarcely soluble in water, but dissolves readily by adding to the water a small portion of caustic soda or potassa. White of egg is a solution of albumen.

4347. To Make Albumen. Expose the strained white of egg, or the serum of bullock's blood in a thin stratum, to a current of dry air, until it hardens into a solid transparent substance.

Or: Agitate strained white of egg with 10 or 12 times its bulk of alcohol, and collect the flocculent precipitate on a muslin filter. Dry it at a temperature not over 120° Fahr.

4348. Tests for Albumen. A solution of bichloride of mercury dropped into a fluid containing albumen, occasions a white precipitate. Tannin or tincture of galls gives a yellow, pitchy precipitate.

4349. Sulphur. Sulphur or *brimstone* is usually of a pale yellow color; melts to a clear, thin fluid, and volatile at about 232° Fahr., when it inflames spontaneously in the open air, and burns with a bluish flame. It is insoluble in water and in alcohol; soluble in turpentine and fat oils, and freely so in bisulphuret of carbon and hot liquor of potassa. With oxygen it forms sulphuric and sulphurous acids, and with the metals it combines as SULPHURETS or SULPHIDES. Its specific gravity is from 1.982 to 2.045 (water standard). The specific gravity of its vapor is 6.648 (air standard).

4350. Amorphous or Brown Sulphur. Prepared from sublimed sulphur by melting it, increasing the heat to 320° Fahr., and continuing it at that temperature for about 30 minutes, or until it becomes brown and viscid; it is then poured into water. It is now ductile like wax, may be easily moulded, and when cooled does not again become fluid below 600° Fahr.

4351. Precipitated Sulphur. Sublimed sulphur, 1 part; dry slacked lime, 2 parts; water, 25 parts; boil for 2 or 3 hours, dilute with 25 parts more water, filter, precipitate by muriatic acid, and drain; well wash, and dry the precipitate. Resembles sublimed sulphur in its general properties, but is much paler, and in a finer state of division.

4352. To Purify Precipitated Sulphur. The precipitated sulphur of the shops contains about two-thirds of its weight of sulphate of lime (plaster of Paris), owing to the substitution of sulphuric for muriatic acid in its preparation. This fraud is detected by heating a little of the suspected sample in an iron spoon or shovel, when the sulphur is volatilized, and leaves behind the sulphate of lime, which, when mixed with water and gently dried, gives the amount of the adulteration. A still simpler plan is to dissolve out the sulphur with a little hot oil of turpentine or liquor of potassa.

4353. Roll Sulphur. Crude sulphur, purified by melting and skimming it, is poured into cylindrical moulds. Common roll sulphur frequently contains from 3 to 7 per cent. of yellow arsenic.

4354. Sublimed Sulphur. Sometimes called *Flowers of Sulphur.* This is prepared by subliming sulphur in iron vessels. For medical purposes it is well washed with water and dried by a gentle heat. (*Cooley.*) An aqueous solution of pure anhydrous carbonate of soda will dissolve an appreciable quantity of flowers of sulphur, by digesting for 10 hours at 212° Fahr. (*Pole.*)

4355. Sulphur Vivum. Crude native sulphur, or *black sulphur*, is of a grey or mouse-colored powder. The same names are given to the residuum in the subliming pots, after the preparation of flowers of sulphur; it generally contains arsenic.

4356. Tersulphuret of Arsenic. The tersulphuret or *tersulphide of arsenic* is a fine golden yellow substance in lumps or powder. It is found, ready formed, in nature, or is prepared artificially by sublimation from arsenious acid and sulphur. The artificial sulphuret, *King's Yellow*, often contains 80 to 90 per cent. of white arsenic.

4357. Camphor. The camphor of commerce is a natural production. It is principally extracted from the laurel camphor tree, but it is also found in several other members of the vegetable kingdom. It is a white, semi-crystalline solid, very volatile at common temperatures, soluble in alcohol, ether, oils, and acetic acid, and slightly but sufficiently so in water to impart its characteristic smell and taste. The Chinese and Japanese extract the camphor by cutting the wood into small pieces, and boiling it with water in iron vessels—which are covered with large earthen capitals or domes—lined with rice straw. As the water boils, the camphor is volatilized along with the steam, and condenses on the straw, under the form of greyish granulations. In this state it is collected and transported to Europe, when it undergoes the process of refining into white camphor. To refine it, 100 parts of crude camphor are mixed with 2 parts each of quicklime and animal charcoal, and placed in a thin globular glass vessel sunk in a sand-bath. The heat is then cautiously applied, and the vessel gradually and carefully raised out of the sand as the sublimation goes on. When this is completed, the whole is allowed to cool. If the process be conducted too slowly, or at a heat under 375° Fahr., the product will be flaky, and consequently unsaleable, without remelting or subliming.

4358. To Pulverize Camphor. Camphor may be beaten in a mortar for some time, without being reduced to powder, but if it be first broken with the pestle, and then sprinkled with a few drops of spirit of wine, it may be readily pulverized. By adding water to an alcoholic or ethereal solution of camphor, it is precipitated under the form of an impalpable powder of exquisite whiteness, which may be collected and spontaneously dried on a filter; the addition of a minute quantity of carbonate of magnesia to the water (say 1 drachm for each 16 ounces of camphor), before mixing it with the camphor solution, will prevent the powdered camphor from hardening again after drying.

4359. Glycerine. This is a sweet, syrupy liquid, formed during the saponification of oils and fats. Its various uses will be found embodied in their respective receipts.

4360. To Obtain Commercial Glycerine. The sweet stearine liquor of the stearine manufacturers is used for this purpose. The lime contained in it is precipitated by a stream of carbonic acid gas, or by a solution of carbonate of soda, carefully avoiding adding the latter in excess; the liquor thus obtained is then boiled a little, filtered, and evaporated to a syrupy consistence. Glycerine is also obtained from the water and washings left in the manufacture of lead or litharge plaster, by mixing them together, filtering, and submitting them to the action of a stream of sulphuretted hydrogen, which precipitates the lead; the clear liquid, after settling, is decanted, filtered, and evaporated to the consistence of syrup, in a water-bath.

4361. Solvent Power of Glycerine. Klever gives the following parts of various chemicals soluble in 100 parts glycerine.

	PARTS.
Arsenious acid	20
Arsenic acid	20
Benzoic acid	10
Boracic acid	10
Oxalic acid	15
Tannic acid	50
Alum	40
Carbonate of ammonia	20
Muriate of ammonia	20
Tartrate of antimony and potassa	5.50
Atropia	3
Sulphate of atropia	33
Chloride of barium	10
Brucia	2.25
Sulphide of calcium	5
Quinine	.50
Sulphate of quinine	6.70
Tannate of quinia	.25
Acetate of copper	10
Sulphate of copper	30
Tartrate of iron and potassa	8
Lactate of iron	16
Sulphate of iron	25
Corrosive sublimate	7.50
Cyanide of mercury	27
Iodine	1.90
Morphia	.45
Acetate of morphia	20
Muriate of morphia	20
Phosphorus	.20
Acetate of lead	20
Arseniate of potassa	50
Chlorate of potassa	3.50
Bromide of potassium	25
Cyanide of potassium	32
Iodide of potassium	40
Arseniate of soda	50
Bicarbonate of soda	8
Borate of soda	60
Carbonate of soda	98
Chlorate of soda	20
Sulphur	.10
Strychnia	.25
Nitrate of strychnia	4
Sulphate of strychnia	22.50
Urea	50
Veratria	1
Chloride of zinc	50
Iodide of zinc	40
Sulphate of zinc	35

4362. To Purify Glycerine. Commercial glycerine is rendered pure by diluting it with water; it is then decolored with a little animal charcoal (*see No.* 1729), filtered, and evaporated to the consistence of a thin syrup, after which it is further evaporated in a vacuum, or over sulphuric acid, until it acquires a specific gravity of 1.265.

4363. To Purify Glycerine. Bottles sent out from wholesale and manufacturing houses, labeled, "Pure Glycerine," do not always contain what their labels declare. Some samples called pure are rich in lead, others contain chlorine, most are diluted with water, and the best is generally acid. It is necessary, therefore, to purify even the best samples by digesting them for several days with powdered chalk, allowing the latter to subside, and decanting. (*Schacht.*)

4364. Tests for the Purity of Glycerine. Pure glycerine has a neutral reaction, and on evaporation in a porcelain dish leaves only a very slight carbonaceous crust, while the impure has a much greater percentage of coaly matter. The pure article does not become brown when treated, drop by drop, with concentrated sulphuric acid, even after several hours; the impure becomes brown even when but slightly adulterated. Pure glycerine, treated with pure nitric acid and a solution of nitrate of silver, does not become cloudy, while the impure exhibits a decidedly milky appearance. Sometimes the impure article becomes blackened with the sulphide of ammonium. Oxalate of ammonia produces a black clouding; lime-water sometimes causes a milky discoloration. Pure glycerine, however, constantly remains perfectly uncolored, and clear as water, the impure becoming colored to a greater or less extent. If a few drops are rubbed between the fingers, pure glycerine causes no fatty smell; the contrary is the case with the impure, especially if a few drops of dilute sulphuric acid be introduced. (*Köller.*) (*See No.* 1151.)

4365. Gelatine. Animal jelly obtained retains the form of the bone, is washed in a by heat from the organic tissue of the bones, tendons, and ligaments, the cellular tissue, the skin, and the serous membranes in contact with water. Glue and size are coarse varieties of this substance, prepared from hoofs, hides, skins, &c.; and isinglass is a purer kind, prepared from the air-bladders and some other membranes of fish. Gelatine is insoluble in cold water, but dissolves with greater or less readiness on the application of heat, forming a tremulous and transparent jelly on cooling. It is insoluble in alcohol or ether, and is decomposed by strong alkali or acid.

4366. To Obtain Gelatine from Bones. The bones of good meat form most excellent materials for making soups and gravies, as is well known to every good cook. In France, soup is extensively made by dissolving bruised bones in a steam heat of 2 or 3 days' continuance, and also by dissolving out the earthy part by digestion in weak muriatic acid, when a lump of gelatine is obtained, which, after being well washed with water, will dissolve by boiling, and is equal to isinglass for all the purposes of making soups and jellies. Proust has recommended the following process for making bone gelatines: Crush the bones small, then boil them for 15 minutes in a kettle of water, cool, and skim the fat off, which is fit for all common purposes. The bones are then ground, and boiled in 8 to 10 times their weight of water, of which that already used must form a part, until evaporated to one-half, when a very nutritious jelly is obtained. A copper vessel should not be used, as the jelly acts upon this metal. An iron digester is the most suitable. The bones of boiled meat are nearly as productive as those of fresh meat, but roasted meat bones scarcely afford any jelly.

4367. Bone Gelatine. The bones are boiled to remove the fat, then digested in diluted muriatic acid till the earthy matter of the bone is dissolved. The gelatine, which stream of water, plunged in hot water, and again in cold, to remove all remains of acid, and sometimes put into a solution of carbonate of soda. When well washed, it is dried on open baskets or nets. By steeping the raw gelatine in cold water, dissolving it in boiling water, evaporating the jelly, and cutting it into tablets, it may be dried and preserved in that form.

4368. Nelson's Patent Gelatine. This is made from cuttings of the hides of cattle, and skins of calves. These, freed from hair, flesh, fat, &c., are washed and scoured, then macerated for 10 days in a lye of caustic soda, and afterwards placed in covered vessels at a temperature of 60° to 70° Fahr. until they become tender; then washed from the alkali, exposed to the vapor of burning sulphur until they become sensibly acid, dissolved in earthen vessels heated to 150°, strained, put into settling vessels heated to 100° or 120° for nine hours, the clear liquor drawn off and poured on the cooling slabs to the depth of ½ an inch. When cold, the jelly is cut in pieces, washed till free from acid, redissolved at 85°, poured on slabs, cut up, and dried on nets.

4369. Gelatine Wafers. Dissolve fine glue or isinglass in water, so that the solution, when cold, may be consistent. Pour it hot on a plate of glass (previously warmed with steam and slightly greased), fitted in a metallic frame whose edges are just as high as the wafers should be thick. Lay on the surface a second glass plate, also hot and greased, so as to touch every point of the gelatine while resting on the edges of the frame. By its pressure the thin cake is rendered uniform. When the glass plates have cooled, the gelatine will be solid, and may be removed. It is cut into discs of different sizes by means of proper punches.

4370. Tests for Gelatine. Gelatine dissolved in water is recognized by forming a jelly on cooling; it is precipitated by alcohol; corrosive sublimate throws down a whitish, flocculent precipitate; a solution of tannin, or an infusion of galls, gives a curdy, yellowish-white precipitate, which, on being stirred, coheres into an elastic mass, insoluble in water, and, when dry, assumes the appearance of over-tanned leather.

4371. Asbestos. A natural substance, resembling flax, capable of withstanding unchanged a considerable degree of heat; it may, therefore, be cleansed or purified by fire. It is also called *Amianthus.*

Tests or Reagents.

These are substances employed to determine the name or character of any other substance, or to detect its presence in compounds. They are used in both the solid and fluid state; generally the latter, when they are known as *liquid tests.* Their application as reagents is called *testing.* For this purpose they are commonly added drop by drop to the liquid to be tested, contained in a test-tube or test-glass. A simple way of employing them is to place a few drops or a small portion of the liquid or substance for examination on a slip of common white glass, and to add to them a drop of the test liquid. By placing the glass over a sheet of white paper, the effect will be rendered more perceptible.

A number of tests, not included here, referring to substances which hold a prominent place in some special process, have been introduced in immediate connection with the description of those substances, and will be found in the index under the head of the article to be tested.

4373. Test for Chicory in Coffee. Place a spoonful of ground coffee gently on the surface of a glass of cold water. The pure coffee will float for some time, and scarcely color the water; the chicory, if any be present, rapidly absorbs the water and sinks to the bottom, communicating a deep reddish-brown tint as it falls.

Or a spoonful of ground coffee may be placed in a small bottle of cold water, and shaken for a moment; if the sample of coffee is pure, it will rise to the surface and hardly tinge the water, whilst if the coffee is adulterated with chicory, the latter will fall to the bottom and color the water as before. A similar coloration of the water will be produced, however, if the coffee be adulterated with burnt sugar, which is the basis of the so-called "coffee essences or extracts."

4374. To Test Tea. Pure China tea is not turned black by being put into water impregnated with sulphuretted hydrogen gas, nor does it tinge spirit of hartshorn blue. The infusion is amber-colored, and is not reddened by adding a few drops of oil of vitriol to it.

4375. To Detect Copper in Liquids. Spirit of hartshorn turns them blue. Therefore tea has not been dried on copper if an infusion of it is not turned blue by this mixture. Cider, being passed through brass pots, is detected by this experiment.

4376. To Detect Watered Milk. The cheapest and easiest method of adulterating milk is by adding water, and we may readily ascertain the exact extent of adulteration by the following plan. If a glass tube, divided into 100 parts, be filled with milk and left standing for 24 hours, the cream will rise to the upper part of the tube, and occupy from 11 to 13 divisions, if the milk is genuine.

4377. To Detect Chalk in Milk. Dilute the milk with water; the chalk, if there be any, will settle to the bottom in an hour or two; put to the sediment an acid, vinegar for instance, and if effervescence take place it is chalk.

4378. To Detect Mineral Substances in Flour. The presence of a mineral adulteration of flour or meal may be readily detected. A small quantity of the suspected flour is shaken up in a glass tube with chloroform. All mineral adulterations will collect at the bottom, while the flour will float on the liquid.

4379. How to Know Good Flour. When flour is genuine or of the best kind, it holds together in a mass when squeezed by the hand, and shows the impressions of the fingers, and even of the marks of the skin, much longer than when it is bad or adulterated; and the dough made with it is very gluey, ductile, and elastic, easy to be kneaded; and may be flattened and drawn in every direction without breaking.

4380. To Detect Adulterations in Sugar. Sugar is largely adulterated. Pure cane and beet sugars may be known by their solutions bending the luminous rays in circumpolarization to the right, whereas grape and fecula sugars bend it to the left. Pure cane sugar boiled in a solution of caustic potassa remains colorless, but if starch sugar is present the liquid turns brown. A filtered solution of 33 grains cane or beet sugar in 1 ounce water, mixed with 3 grains pure caustic potassa, and then agitated with 1½ grains sulphate of copper in a close vessel, remains clear, even after the lapse of several days; but if starch sugar is present, a red precipitate is formed after some time, and if present in considerable quantity the copper will be wholly converted into oxide within 24 hours; the solution first turns blue or green, and then entirely loses its color. Of late years moist sugar has been largely adulterated with the sweet waste liquor (solution of glycerine) of the stearine manufactories; but this fraud may be detected by its inferior sweetness, and by its moist and dirty appearance.

4381. Test for Starch. The old and familiar test for starch is the blue color which free iodine produces when brought in contact with it; but this is not the only reagent by means of which we can detect the presence of starch in combination with similar bodies. Bromine is nearly as good as iodine, and tannin is said, in some instances, to be better. A solution of 50 grains tannin in ½ pint distilled water will answer for making the test. A drop of this tannin solution will cause a precipitate in extremely dilute solutions of starch; the precipitate dissolves when warmed and reappears when the solution cools; and where the starch paste is old, the reaction is said to be more sensitive than that of iodine.

4382. To Test Arrow-Root. Genuine arrow-root is odorless and tasteless, and produces a sort of crackling noise when pressed or rubbed, and emits no peculiar odor when mixed with muriatic acid. Stirred up in a mortar with double its weight of a mixture of equal parts of aqua-fortis and water, it does not become gelatinous and adhesive in less than 15 minutes.

4383. To Detect Arsenic in Colored Paper. Take a fragment of the paper and put it into a solution of ammonia. If arsenic be present the liquid will assume a bluish color. In case a further test is required, pour a little of the ammoniacal solution on crystals of nitrate of silver, and arsenic, if present, will show itself by leaving a yellow deposit on the crystals. As arsenic is used in coloring all qualities of paper, from the cheapest to the costliest, a knowledge of this test will be of service.

4384. To Detect Gum Arabic in Gum Tragacanth. Make the gum into a clear mucilage, and filter carefully; pour strong alcohol upon it, and if it retains its solubility and transparency, no gum arabic is present, but if it becomes opaque, or deposits a powder at the bottom, it contains gum arabic or some similar substance.

4385. To Test Slates. The test of a superior slate is its ability to remain unbroken, after being made red hot in a furnace and suddenly immersed in cold water while at that heat.

4386. To Test Silver or Gold. For testing gold or silver, slightly wet the metal and rub gently with lunar caustic. If genuine gold or silver the mark will be faint; but if an inferior metal it will be quite black.

4387. To Test Mushrooms. The following are said to be tests of the wholesomeness of mushrooms: Sprinkle a little salt on the spongy part or gills of the sample to be tried: if they turn yellow, they are poisonous; if black, they are wholesome.

False mushrooms have a warty cap, or else fragments of membrane adhering to the upper surface; are heavy, and emerge from a vulva or bag; they grow in tufts or clusters in woods, on the stumps of trees, &c.; whereas the true mushrooms grow in pastures.

False mushrooms have an astringent, styptic, and disagreeable taste; when cut they turn blue; they are moist on the surface, and are generally of a rose or orange color.

The gills of the true mushroom are of a pinky red, changing to a liver color; the flesh is white; the stem is white, solid, and cylindrical.

Introduce a silver spoon, or an onion, into a vessel in which mushrooms are seething; if, on taking either of them out, they assume a dark, discolored appearance, the circumstance denotes the presence of poison existing among them; if, on the other hand, the metal or onion, on being withdrawn from the liquor, wears its natural appearance, the fruit may be regarded as genuine, and of the right sort.

Rub the upper skin with a gold ring or any piece of gold: the part rubbed will turn yellow if it is a poisonous fungus.

4388. To Test the Hardness of Water. Hard water contains more or less carbonate of lime; the presence of this substance in waters is tested thus: Soap, or a solution of soap in proof spirit, mixes easily and perfectly with *pure* water, but is curdled and precipitated in water containing carbonates, chlorides, or sulphates. The degree of hardness of water depends on the amount of carbonate of lime held in it in solution, and is ascertained as follows: Dissolve 1 drachm finest white soap in 1 pint proof spirit; so adjust the strength (if not already so) that exactly 32 measures are required to be added to 100 measures of the standard solution of chloride of calcium (*see No.* 4786), before a lather can be produced. Every measure of this test solution of soap and alcohol, which is required to produce the same effect on 100 measures of a sample of hard water, represents $\frac{1}{4}$ grain of carbonate of lime or $\frac{1}{2}°$ of hardness; 2 measures equal 1° of hardness or 1 grain of carbonate per gallon, &c.

4389. To Test the Purity of Borax. Its *strength* is best ascertained by the quantity of sulphuric acid required to neutralize a given weight of borax. (*See Alkalimetry.*) The *impurities* in borax are common salt and alum, which are mixed with it to lower the value.

Common salt may be detected by a solution of the borax in hot water yielding with nitrate of silver a curdy white precipitate which is soluble in ammonia; this must be distinguished from the white pulverulent precipitate of borate of silver which will be thrown down from pure borax.

The presence of alum is determined by addition of ammonia water to a solution of the borax giving a bulky white precipitate.

4390. To Test the Purity of Musk. Musk is often largely adulterated with dried blood, the presence of which may be detected by the inferiority of the odor; by an assay for the iron contained in the blood; or by microscopic examination. The ashes left after burning pure musk are neither red nor yellow, but grey, and should not exceed 6 per cent. of the amount burned.

4391. To Test the Purity of Ambergris. From the high price of the genuine ambergris, it is very frequently adulterated. When quite pure and of the best quality it is nearly wholly soluble in hot alcohol and ether, and yields about 85 per cent. of the odorous principle (ambreine). It is also easily punctured with a heated needle, and on withdrawing it not only should the odor be immediately evolved, but the needle should come out clean, without anything adhering to it.

4392. To Test Diamonds. If you have a doubtful stone, put it into a leaden or platinum cup, with some powdered fluor-spar and a little oil of vitriol; warm the vessel over some lighted charcoal, in a fireplace, or wherever there is a strong draught to carry away the noxious vapors that will be copiously evolved. When these vapors have ceased rising let the whole cool, and then stir the mixture with a glass rod to fish out the diamond. If you find it intact it is a genuine stone; but if it is false it will be corroded by the hydrofluoric acid that has been generated around it. A small paste diamond would disappear altogether under the treatment. This test is given by Massimo Levi, an Italian chemist.

4393. Test for the Presence of Blood. Gunning has discovered in acetate of zinc a reagent that precipitates the slightest traces of the coloring matter of blood from solutions, even where the liquids are so dilute as to be colorless. Blood washed from the hands in a pail of water can readily be detected in this way. The flocculent precipitate thrown down by acetate of zinc must be washed by decantation, and finally collected on a watch-glass, and allowed to dry, when the microscope will readily reveal crystals if any blood be present. (*See No.* 6415.)

4394. Test for the Presence of a Free Acid. Dissolve chloride of silver in just sufficient ammonia to make a clear solution. If a little of the test be added to ordinary spring water, the carbonic acid present in the latter will neutralize the ammonia and precipitate the chloride. The above forms a good lecture experiment, the test being a very delicate one.

4395. Permanganate of Potassa as a Test for Organic Matter. As a test for organic matter in air and water, its accuracy has been called in question, on the ground that it does not attack all kinds of organic matter with equal facility—some, as starch, resisting its action for a long time. It must be admitted, however, that it is, at present, the only practical test that we have, and certainly shows very rapidly and clearly the presence of hurtful organic matter in water or in air. It can be applied by any one, it being only necessary to use a weak solution; the disappearance of the color indicates the presence of organic matter. In time of epidemics, such as cholera or dysentery, this test might be of much value in singling out the contaminated from the pure water. It is, perhaps, well also to recall the fact that this test forms the readiest means of purifying foul water.

4396. Trommer's Test for the Presence of Sugar in Urine. Put some of the suspected urine into a large test-tube, and add a few drops of solution of sulphate of copper, then sufficient solution of potash to render it strongly alkaline. If sugar be present, the precipitated oxide redissolves into a blue liquid, and on boiling red oxide of copper is precipitated. White merino that has been wet with a solution of bichloride of tin is said to form a ready test for sugar in urine, &c. A portion wet with the suspected liquor, and exposed to 260° to 300° of heat, becomes blackened if sugar is present.

4397. Quantitative Test for Sugar in Urine. Dissolve 400 grains pure crystallized sulphate of copper in 1600 grains of distilled water; add this gradually to a solution of 1600 grains neutral tartrate of potash in a little water mixed with 6000 or 7000 grains solution of caustic soda of 1.12 specific gravity. Add water to make up the whole 11,544 grain measures. 1000 grain measures are equivalent to 5 grains of grape sugar.

4398. Pettenkofer's Test for Bile in Urine, &c. Put a small quantity of the suspected liquid into a test-tube, and add to it, drop by drop, strong sulphuric acid till it becomes warm, taking care not to raise the temperature above 122° Fahr. Then add from 2 to 5 drops of syrup, made with 5 parts sugar to 4 of water, and shake the mixture. If the liquid contain bile, a violet coloration is observed. Acetic acid, and those substances which are converted into sugar by sulphuric acid, may be substituted for sugar.

4399. To Detect Sulphur in Coal-Gas. The presence of sulphur in coal-gas can be proved in the following simple manner: Let a platinum basin be filled with a pint of water, and the basin be heated over a spirit lamp until all the liquid has evaporated; the basin will be found to be coated on the outside, where it has been struck by the flame, with a dirty, greasy looking substance, which, on being washed off with pure distilled water, and tested, proves to be sulphuric acid. The glass chimneys used with Argand gas-burners soon become coated over internally with a white substance, which, on being washed off with distilled water, will be found to be, on testing, sulphate of ammonia. The glass panes of a room wherein gas is burned for a few evenings consecutively, will, when rubbed with the fingers of a clean hand, impart to it a substance which, on the hand being rinsed in distilled water, will yield a precipitate of sulphate of baryta with chloride of barium, and a brick-red precipitate with potassio-iodide of mercury.

4400. Test for Benzole. For distinguishing genuine benzole, or that made of coal tar, from that prepared from petroleum, Brandberg recommends us to place a small piece of pitch in a testing tube, and pour over it some of the substance to be examined. The genuine will immediately dissolve the pitch to a tar-like mass, while that derived from petroleum will scarcely be colored. (*See*

Nos. 4320 *and* 4321.)

4401. To Detect Cotton in Linen. Unravel a piece of the fabric, both warp and weft, and plunge it into a solution of aniline and fuchsine. This will dye the whole red. Take it out, wash it, and while moist dip into ammonia; the cotton threads will lose their color, while the linen will remain red. (*See No* 296, *&c.*)

4402. Hahnemann's Test for Lead in Wine. Take 1 ounce quick-lime, 1½ ounces flowers of sulphur; heat in a covered crucible for 5 or 6 minutes; take 2 drachms of this compound (which is *sulphuret of lime*), 2 drachms tartaric acid; powder, mix, and shake in a stoppered bottle with a pint of water; let it settle, pour off the clear liquid, and add 1½ ounces tartaric acid. The above test will throw down the least quantity of lead from wines, as a very sensible black precipitate.

4403. Paris Test for Lead in Wine. Expose equal parts of sulphur and powdered oyster shells to a white heat for 15 minutes, and, when cold, add an equal quantity of cream of tartar; these are to be put into a strong bottle, with common water, to boil for an hour, and the solution is afterwards to be decanted into ounce phials, adding 20 drops muriatic acid to each. Both the above tests will throw down the least quantity of lead from wines, as a very senible black precipitate. As iron might be accidentally contained in the wine, the muriatic acid is added, to prevent the precipitation of that metal. This acts in the same manner as Hahnemann's test. (*See No.* 4402.)

4404. To Distinguish Artificially Colored Wines. As the real coloring matter of wine is of difficult solubility in water free from tartaric acid, Blume proposes to make this fact of practical use in testing the purity of wine. A crumb of bread saturated in the supposed wine is placed in a plate of water; if artificially colored, the water soon partakes of the color; but if natural, a slight opalescence only will be perceptible after a quarter of an hour.

4405. To Detect Logwood in Wine. M. Lapeymere, having observed that hæmatine, the coloring principle of logwood, gives a sky-blue color in the presence of salts of copper, proposes the following test for logwood in wines: Paper is saturated with a strong solution of neutral acetate of copper, and dried. A strip of this is dipped into the suspected liquor, and, after removal, the adhering drops are made to move to and fro over the paper, which is finally to be carefully dried. If the wine contain logwood, the paper will assume a violet-blue color; but if the wine possess its natural coloring matter the paper will have a grey tint.

4406. To Detect Artificial Coloring in Wine. Use, as test liquid, a solution of potash and a solution of liquid ammonia and potash.

If the wine is colored by the coloring matter of the grape, potash changes the red color to a bottle green or brownish-green; ammonia changes the color to brownish-green or greenish-brown; a solution of alum to which some potash has been added gives a dirty grey precipitate.

If the wine is artificially colored, potash gives the following colored precipitates: Dwarf elder, mulberry, or beet root give a violet precipitate; pokeweed berries, a yellow; Indian wood, a violet red; pernambuco, a red; litmus, a violet blue; orchil or cudbear, a dirty lees color.

Or: Pour into the wine to be tested a solution of alum, and precipitate the alumina it contains, by adding potash, and the precipitates will have the same characteristic colors as above.

4407. Test for Rum. Dr. Wiederbold proposes the following method for distinguishing between true rum and the factitious liquid sold under this name: Mix a little of the rum to be tested with about a third of its bulk of sulphuric acid, and allow the mixture to stand. If the rum is genuine its peculiar odor remains after the liquid has cooled, and even after 24 hours' contact may still be distinguished. If, on the contrary, the rum is not genuine, contact with sulphuric acid promptly and entirely deprives it of all its aroma.

Test Papers.

These consist of paper which has been wetted thoroughly and uniformly with a solution of some appropriate substance, dried and cut into convenient strips, and is used to test, by its change of color, the presence of some other substance known to produce that change. This is effected by dipping a strip of the proper test paper into, or wetting it with, the liquor to be tested, and the effect noted.

4409. Brazil-wood Test Paper. Made by preparing the paper with a decoction of Brazil-wood. Alkalies turn it purple or violet; strong acids, red.

4410. Buckthorn Test Paper. From a decoction of the berries; is reddened by acids.

4411. Cherry-juice Test Paper. From the juice of cherries; has the same properties as buckthorn paper.

4412. Dahlia Test Paper. Made from an infusion of the petals of the violet dahlia (georgina purpurea); alkalies turn it green; acids, red; strong caustic alkalies turn it yellow. This is a very delicate test paper. The juice of elderberries will make a similar test paper.

4413. Indigo Test Paper. From a solution of indigo; loses its color in contact with chlorine.

4414. Iodide of Potassium Test Paper. From a solution of it in distilled water; turned blue by an acidulated solution of starch.

4415. Starch and Iodine Test Paper. Prepared by mixing starch paste with iodide of potassium; turned blue by chlorine, ozone, and the mineral acids, and by the air containing them.

4416. Lead Test Paper. From a solution of either acetate or diacetate of lead; used as a test for sulphuretted hydrogen and hydrosulphuret of ammonia, which turn it black.

4417. Blue Litmus Test Paper. Triturate 1 ounce litmus in a wedgwood-ware mortar with 3 or 4 fluid ounces boiling water; put the mixture into a flask, and add more boiling water until the liquid measures fully ¼ pint; agitate it frequently until cold, then filter it; divide the filtered fluid into 2 equal portions, stir one portion with a glass rod dipped into very dilute sulphuric acid, repeating this until the liquid begins to be very slightly tinged red, then add the other portion and mix them thoroughly. Prepare the paper with this infusion. Acids turn it red; alkalies, green. The neutral salts of most of the heavy metallic oxides redden this as well as the other blue test papers that are affected by acids.

4418. Red Litmus Paper. Treat the whole of a blue infusion, made as above, with the rod dipped in dilute sulphuric acid until it turns distinctly red. Alkalies, alkaline earths, and their sulphurets, restore its blue color; alkaline carbonates and the soluble borates produce the same effect. Red litmus paper may also be made by holding a strip of the blue litmus paper over a jar into which 2 or 3 drops of muriatic (hydrochloric) acid have been thrown.

4419. Mallow Test Paper. From an infusion of the purple flowers of the common mallow.

4420. Manganese Test Paper. From a solution of sulphate of manganese; turns black by contact with ozone.

4421. Rhubarb Test Paper. From a strong infusion of the powdered root; alkalies turn it brown, but boracic acid and its salts do not affect it.

4422. Rose Test Paper. Made with a strong infusion of the petals of the red rose; alkalies turn it a bright green.

4423. Starch Test Paper. From a cold decoction of starch; free iodine turns it blue.

4424. Sulphate of Iron Test Paper. Made with a solution of the protosulphate; as a test for hydrocyanic acid and the soluble cyanides.

4425. Turmeric Test Paper. Prepared with a decoction of 2 ounces turmeric to 1 pint water; is turned brown by alkalies, and by boracic acid and the soluble borates.

4426. Cabbage Test Paper. Make a strong infusion of red cabbage leaves, strain it, and evaporate it by a gentle heat till considerably reduced. Then dip the paper in it and dry it in the air. (This paper is of a greyish color; alkalies change it to green, acids to red. It is a very delicate test; if rendered slightly green by an alkali, carbonic acid will restore the color.)

4427. Alkanet Test Paper. The red principle of the alkanet root (Anchusa tinctoria, L.) is, as is well known, a most sensitive reagent for alkalies and acids; it is used for the preparation of test paper, and is prepared like litmus paper, by saturating unsized paper with a solution of the alkanet red. This is obtained by extracting dry alkanet root with ether; the filtered liquid is ready for use. The blue paper may be obtained from the red one by dipping it in an aqueous solution of carbonate of soda of specific gravity 1.5. A paper, answering for both alkaline and acid test, may be prepared by dividing the ethereal solution of alkanet red into two equal parts; to one is added, drop by drop, a watery solution of carbonate of soda, until the red just has changed to a distinct blue hue; then both liquids are mixed and used for the preparation of the paper. This, when dried, has to be kept in tightly closed bottles.

4428. Test Paper from Hollyhock Flowers. Some years ago Prof. Aiken, of the University of Maryland, proposed paper stained with an infusion of the petals, as a substitute for litmus paper. His althæa paper is purplish-blue when dry; acids impart a carmine hue, which is turned to bluish-green by alkalies, the neutral tint being purplish-blue; it is superior in intensity of reaction to turmeric, and quite equal to litmus, and is not affected by light, as is the case with the latter. The alkaline reaction is produced in natural or atmospheric waters; and the presence of nitrites, which change the red paper to purple, is indicated in greater dilution than with iodide starch.

4429. Ozonometer. This name has been given to paper prepared with a mixed solution of starch and iodide of potassium.

It is white, but is turned blue by ozonized air when exposed to it in a slightly moistened state. This test is sufficiently delicate to detect the presence of ozone in the atmosphere.

Factitious Mineral Waters.

These are the imitations of different celebrated springs, whose waters have more or less medicinal properties; they are prepared by adding to pure water the ingredients which the original spring water is found, by chemical analysis, to contain. Under this class are also included the ordinary aerated or carbonated waters, which are known as soda waters. The majority, whether plain or medical, are charged with carbonic acid gas by the powerful apparatus employed by manufacturers of soda waters (*see No.* 718); the gas being evolved by the action of weak sulphuric acid on marble chalk, whiting, &c. Some few obtain their carbonic acid gas by the action of an acid and an alkali introduced into the bottle, and instantly corked. The quantity of gas introduced is usually about 5 times the volume of the liquid. In making chalybeate and sulphuretted water, the water should be previously boiled, to expel all air from it.

4431. Simple Aerated Water. Carbonic acid gas water. Water charged with five or more volumes of carbonic acid gas, by means of a suitable apparatus. (*See No.* 718.)

4432. Alkaline Aerated Waters. Aerated soda and potash waters should be made by dissolving 1 drachm of the carbonated alkali in each pint of water, and charging it strongly with carbonic acid gas. The soda water usually offered for sale contains little or no soda.

4433. Aerated Magnesia Water. This is a solution of magnesia of various strengths, charged with carbonic acid gas in the same manner as other aerated waters.

4434. Murray's Fluid Magnesia may be thus made: To a boiling solution of 16 ounces sulphate of magnesia in 6 pints water, add a solution of 19 ounces crystallized carbonate of soda in the same quantity of water; boil the mixture till gas ceases to escape, stirring constantly; then set it aside to settle; pour off the liquid, and wash the precipitate on a cotton or linen cloth, with warm water, till the latter passes tasteless. Mix the precipitate, without drying it, with a gallon of water, and force carbonic acid gas into it under strong pressure, till a complete solution is effected. The *Eau Magnésienne* of the French Codex is about a third of this strength; and some fluid magnesias prepared in this country are not much stronger. Dinneford's preparation is similar to the above.

4435. Carbonated Lime-Water—Carrara Water. Lime-water (prepared from lime made by calcining Carrara marble) is supersaturated, by strong pressure, with carbonic acid, so that the carbonate of lime at first thrown down is redissolved. It contains 8 grains carbonate of lime in 10 fluid ounces water.

4436. Aerated Lithia Water. This may be conveniently made from the fresh precipitated carbonate, dissolved in carbonated water, as directed for fluid magnesia. Its antacid and antilithic properties are found useful.

4437. Baden Water. Muriate of magnesia, 2 grains; muriate of lime, 40 grains; muriate of iron, ½ grain (or 3 minims of the tincture); muriate of soda, 30 grains; sulphate of soda, 10 grains; carbonate of soda, 1 grain; water, 1 pint; carbonic acid gas, 5 volumes.

4438. Carlsbad Water. Dissolve 8 grains of muriate of lime, 1 drop of tincture of sesquichloride of iron, 50 grains of sulphate of soda, 60 grains of carbonate of soda, 8 grains of muriate of soda, in one pound of water.

4439. Carlsbad Water. Muriate of lime, 8 grains; tincture of muriate of iron, 1 drop; sulphate of soda, 50 grains; carbonate of soda, 60 grains; muriate of soda, 8 grains; carbonated water, 1 pint.

4440. Congress Water. Take common salt, 7¾ ounces; hydrate of soda, 23 grains; bicarbonate of soda, 20 grains; and calcined magnesia, 1 ounce. Add the above ingredients to 10 gallons of water, and charge with gas.

4441. Eger Water. Carbonate of soda, 5 grains; sulphate of soda, 4 scruples; muriate of soda, 10 grains; sulphate of magnesia, 3 grains; muriate of lime, 5 grains; carbonated water, 1 pint.

Or it may be made without apparatus thus: Bicarbonate of soda, 30 grains; muriate of soda, 8 grains; sulphate of magnesia, 3 grains; water, 1 pint; dissolve and add 1 scruple dry bisulphate of soda, and close the bottle immediately.

4442. Ems Water. Carbonate of soda, 2 scruples; sulphate of potash, 1 grain; sulphate of magnesia, 5 grains; muriate of soda, 10 grains; muriate of lime, 3 grains; carbonated water, 1 pint.

4443. Kissingen Water. Mix together bicarbonate of soda, 1 drachm; carbonate of lime, 8 scruples; precipitated carbonate of iron, 2 scruples; common salt, 8 ounces; muriate of ammonia, 4 grains; sulphate of soda, 8 scruples; sulphate of magnesia, 2 ounces; phosphate of soda, 13 grains; phosphate of lime, 8 scruples. Add water, ½ gallon. Let it stand half a day, filter, add carbonate of magnesia, 10 scruples, and 10 gallons water. Lastly, charge with gas by means of the usual apparatus. (*See No.* 718.)

4444. Marienbad Water. Carbonate of soda, 2 scruples; sulphate of soda, 96 grains; sulphate of magnesia, 8 grains; muriate of soda, 15 grains; muriate of lime, 10 grains; carbonated water, 1 pint.

Or, bicarbonate of soda, 50 grains; sulphate of soda, 1 drachm; muriate of soda, 15 grains; sulphate of magnesia, 10 grains; dissolve in 1 pint water, add 25 grains dry bisulphate of soda, and cork immediately.

4445. Marienbad Purging Salts. Bicarbonate of soda, 5 ounces; dried sulphate of soda, 12 ounces; dry muriate of soda, 1½ ounces; sulphate of magnesia, dried, 2 ounces; dried bisulphate of soda, 2½ ounces. Mix the salts, previously dried, separately, and keep them carefully from the air.

4446. Pullna Water. Sulphate of soda, 4 drachms; sulphate of magnesia, 4 drachms; muriate of lime, 15 grains; muriate of magnesia (dry), 1 scruple; muriate of soda, 1 scruple; bicarbonate of soda, 10 grains; water slightly carbonated, 1 pint. One of the most active of the purgative saline waters, and deserving of wider popularity.

It may be prepared without apparatus as follows: Bicarbonate of soda, 50 grains; sulphate of magnesia, 4 drachms; sulphate of soda, 3 drachms; muriate of soda, 1 scruple; dissolve in 1 pint of water; add, lastly, 2 scruples bisulphate of soda, and close the bottle immediately.

4447. Salts for Making Pullna Water. Dry bicarbonate of soda, 1 ounce; sulphate of soda, 2 ounces; sulphate of magnesia, 1½ ounces; muriate of soda, 2 drachms; tartaric acid, ⅞ ounce (or rather, bisulphate of soda, 1 ounce). All the ingredients must be previously dried.

4448. Pyrmont Water. Carbonate of lime, 12 grains; crystallized carbonate of soda, 31 grains; sulphate of soda in crystals, 7½ grains; sulphate of lime, 14 grains; sulphate of magnesia, 20 grains; sulphate of iron, 2 grains; chloride of sodium, 2 grains; chloride of magnesium, 4 grains; chloride of manganese, $\frac{1}{30}$ grain; water, 2 pints; carbonic acid, 5 volumes. Dissolve the sulphate of iron in part of the water; dissolve the other soluble salts in the remainder of the water, add the insoluble salts to the solution, and charge it with the carbonic acid. Mix the two solutions in a bottle, and cork it immediately.

4449 Seidlitz Water. This is usually imitated by strongly aerating a solution of 2 drachms sulphate of magnesia in 1 pint of water. It is also made with 4, 6, and 8 drachms of the salts to 1 pint of water, according to the strength required.

4450. Seidlitz Powders. The common Seidlitz powders do not resemble the water. A closer imitation would be made by using effloresced sulphate of magnesia instead of the potassio-tartrate of soda. A still more exact compound will be the following: Effloresced sulphate of magnesia, 2 ounces; bicarbonate of soda, ½ ounce; dry bisulphate of soda, ½ ounce; mix, and keep in a close bottle.

4451. Seidlitz Powders. Mix together thoroughly 1 troy ounce bicarbonate of soda, and 3 troy ounces Rochelle salt, both in fine powder, and divide into 12 equal parts. Divide 420 grains tartaric acid also into 12 equal parts. Put up the parts, severally, of the mixture and of the acid in separate papers, each kind of a distinctive color. (*U. S. Ph.*) The alkaline mixture is usually put up in blue, and the acid in white papers.

4452. Seidschutz Water. Sulphate of magnesia, 3 drachms; muriate of lime, nitrate of lime, bicarbonate of soda, of each 8 grains; sulphate of potash, 5 grains; aerated water, 1 pint.

4453. Seltzer or Selters Water. The seltzer water, as commonly sold, is prepared as follows: Prepare a solution of fused chloride of calcium, 1 part in 9 of water (specific gravity should be 1.088 to 1.089); a solution of calcined carbonate of soda, 1 part in 10 of water (specific gravity 1.105); a solution of chloride of magnesium, by dissolving calcined magnesia at the rate of 20 grains in dilute hydrochloric acid to make 1 fluid ounce of saturation (specific gravity 1.086); lastly, a solution of dry sulphate of soda in 10 parts water (specific gravity 1.092). These solutions are mixed with water in the following proportions: Solution of carbonate of soda, 1000 grains; solution of chloride of calcium, 200 grains; solution of chloride of magnesium, 150 grains; solution of sulphate of soda, 20 grains; added to 250 to 300 ounces (troy) of water, afterwards to be charged with carbonic acid.

4454. Seltzer Water. Muriate of lime and muriate of magnesia, of each 4 grains; dissolve these in a small quantity of water, and add it to a similar solution of 8 grains bicarbonate of soda, 20 grains muriate of soda, and 2 grains phosphate of soda; mix, and add a solution of ¼ of a grain sulphate of iron; put the mixed solution into a 20-ounce bottle, and fill up with aerated water. An imitation of seltzer water is also made by putting into a stone seltzer bottle, filled with water, 2 drachms bicarbonate of soda and 2 drachms citric acid in crystals, corking the bottle immediately.

4455. Vichy Water. Sulphate of potassa, 2 drachms; sulphate of soda, 4 scruples; phosphate of soda, 25 grains; common salt, 6 drachms; bicarbonate of soda, 5¼ ounces; carbonate of ammonia, 10 grains. Mix. Add water, ½ gallon. Let it stand half a day; filter, add 10 gallons water, and charge with gas.

4456. Vichy Water. Bicarbonate of soda, 1 drachm; muriate of soda, 2 grains; sulphate of soda, 8 grains; sulphate of magnesia, 3 grains; tincture of muriate of iron, 2 drops; aerated water, 1 pint. Dorvault directs 75 grains of bicarbonate of soda, 4 grains of chloride of sodium, ⅛ grain sulphate of iron, 10 grains sulphate of soda, and 3 grains sulphate of magnesia, to a pint of water. By adding 45 grains (or less) of citric acid, an effervescing water is obtained.

4457. Vichy Water. Soubeiran, relying on the analysis of Longchamps, imitates Vichy water by the following combination: Bicarbonate of soda, 135 grains; chloride of sodium, 2½ grains; crystallized chloride of calcium, 12 grains; sulphate of soda, 11⅓ grains; sulphate of magnesia, 3⅔ grains; tartrate of iron and potash ⅓ grain; water, 2$\frac{1}{10}$ pints; carbonic acid, 305 cubic inches (10½ pints). Dissolve the salts of soda and iron in part of the water, dissolve and add the sulphate of magnesia, and then the chloride of calcium in the remaining water. Charge now with the carbonic acid gas under pressure.

4458. Vichy Salts. Bicarbonate of soda, 1½ ounces; muriate of soda, 15 grains; effloresced sulphate of soda, 1 drachm; effloresced sulphate of magnesia, 1 scruple; dry tartarized sulphate of iron, 1 grain; dry tartaric acid, 1 ounce (or dry bisulphate of soda); mix the powders, previously dried, and keep them in a close bottle.

4459. Sea-Water. Muriate of soda, 4 ounces; sulphate of soda, 2 ounces; muriate of lime, ¼ ounce; muriate of magnesia, 1 ounce; iodide of potassium, 4 grains; bromide of potassium, 2 grains; water, 1 gallon. A common substitute for sea-water as a bath is made by dissolving 5 or 6 ounces of common salt in a gallon of water.

4460. Dry Salt to Imitate Sea-Water. The following mixture of dry salts may be kept for the immediate production of a good imitation of sea-water. Chloride of sodium (that obtained from evaporating sea-water and not recrystallized, in preference), 85 ounces; effloresced sulphate of soda, 15 ounces; dry muriate of lime, 4 ounces; dry muriate of magnesia, 16 ounces; iodide of potassium, 2 drachms; bromide of potassium, 1 grain. Mix and keep dry. Put 5 or 6 ounces to a gallon of water.

4461. Balaruc Water. Muriate of soda, 1 ounce; muriate of lime, 1 ounce; muriate of magnesia, ½ ounce; sulphate of soda, 3 drachms; bicarbonate of soda, 2 drachms; bromide of potassium, 1 grain; water, 1 gallon. Chiefly used for baths.

4462. Simple Sulphuretted Waters. Pass sulphuretted hydrogen into cold water (previously deprived of air by boiling, and cooled in a close vessel), till it ceases to be absorbed.

4463. Aix-la-Chapelle Water. Bicarbonate of soda, 12 grains; muriate of soda, 25 grains; muriate of lime, 3 grains; sulphate of soda, 8 grains; simple sulphuretted water, 2½ ounces; water slightly carbonated, 17½ ounces.

4464. Bareges Water. (*Cauterets, Bagnères de Luchon, Bonnes St. Sauveur,* may be made in the same manner.) Crystallized hydrosulphate of soda, crystallized carbonate of soda, and chloride of sodium, of each 2½ grains; water (free from air), 1 pint. A stronger solution for adding to baths is thus made: Crystallized hydrosulphate of soda, crystallized carbonate of soda, and muriate of soda, of each 2 ounces; water, 10 ounces; dissolve. To be added to a common bath at the time of using.

4465. Naples Water. Crystallized carbonate of soda, 15 grains; fluid magnesia, 1 ounce; simple sulphuretted water, 2 ounces; aerated water, 16 ounces. Introduce the sulphuretted water into the bottle last.

4466. Harrogate Water. Chloride of sodium, 100 grains; muriate of lime, 10 grains; muriate of magnesia, 6 grains; bicarbonate of soda, 2 grains; water, 18½ ounces. Dissolve and add simple sulphuretted water, 1½ ounces.

4467. Simple Chalybeate Water. Water, freed from air by boiling, 1 pint; sulphate of iron, ½ grain.

4468. Aerated Chalybeate Water. Sulphate of iron, 1 grain; carbonate of soda, 4 grains; water, deprived of air and charged with carbonic acid gas, 1 pint. Dr. Pereira recommends 10 grains each of sulphate of iron and bicarbonate of soda to be taken in a bottle of ordinary soda-water. This is equivalent to 4 grains of carbonate of iron.

4469. Brighton Chalybeate. Sulphate of iron, muriate of soda, muriate of lime, of each 2 grains; carbonate of soda, 3 grains; carbonated water, 1 pint.

4470. Bussang Water. Dissolve from ½ to ⅔ grain of sulphate of iron, 2 or 3 grains carbonate of soda, 1 grain sulphate of magnesia, and 1 of muriate of soda, in a pint of aerated water. *Forges, Provins,* and other similar waters can be imitated in the same manner.

4471. Mont d'Or Water. Bicarbonate of soda, 70 grains; sulphate of iron, ⅔ grain; muriate of soda, 12 grains; sulphate of soda, ½ grain; muriate of lime, 4 grains; muriate of magnesia, 2 grains; aerated water, 1 pint. (*See No.* 4431.)

4472. Passy Water. Sulphate of iron, 2 grains; muriate of soda, 3 grains; carbonate of soda, 4 grains; muriate of magnesia, 2 grains; aerated water, 1 pint.

4473. Pyrmont Water. Sulphate of magnesia, 20 grains; muriate of magnesia, 4 grains; muriate of soda, 2 grains; bicarbonate of soda, 16 grains; sulphate of iron, 2 grains; Carrara water, 1 pint. (*See No.* 4435.)

4474. Mialhe's Aerated Chalybeate Water. Water, 1 pint; citric acid, 1 drachm; citrate of iron, 15 grains; dissolve, and add 75 grains bicarbonate of soda.

4475. Trousseau's Martial Aerated Water. Potassio-tartrate of iron, 10 grains; artificial Seltzer water, 1 pint.

4476. Bouchardat's Gaseous Purgative. Phosphate of soda, 1½ ounces; carbonated water, 1 pint.

4477. Mialhe's Ioduretted Gaseous Water. Iodide of potassium, 15 grains; bicarbonate of soda, 75 grains; water, 1 pint; dissolve, and add sulphuric acid diluted with its weight of water, 75 grains. Cork immediately.

4478. Dupasquier's Gaseous Water of Iodide of Iron. Solution of iodide of iron (containing 10 per cent. of dry iodide), 30 grains; syrup of gum, 2½ ounces; aerated water, 17½ ounces.

Medicinal Tinctures.

These are solutions of the active principles of bodies, obtained by digesting them in alcohol more or less dilute. Ethereal tinctures are similar solutions prepared with ether. (*See Nos.* 35, *&c.*) Where percolation is resorted to in the preparation of tinctures, the directions laid down in No. 41 should be carefully followed to ensure success.

4480. Tincture of Assafetida. Macerate 4 troy ounces assafetida in 2 pints alcohol for 2 weeks, and filter through paper. (*U. S. Ph.*)

4481. Tincture of Aconite Leaf. Take 4 troy ounces recently dried aconite leaf in fine powder; moisten with 2 fluid ounces diluted alcohol; pack it firmly in a conical percolator, and gradually pour diluted alcohol sufficient to displace 2 pints of tincture. (*U. S. Ph.*)

4482. Tincture of Aconite Root. Take 12 troy ounces aconite root in fine powder; moisten with 6 fluid ounces alcohol; pack it firmly in a cylindrical precolator, and gradually pour alcohol upon it until 2 pints of tincture are obtained. (*U. S. Ph.*)

4483. Tincture of Arnica. Take 6 troy ounces of arnica flowers; mix 1½ pints alcohol and ½ pint water; moisten the arnica slightly with this mixture, and bruise it thoroughly in a mortar. Then pack it firmly in a cylindrical percolator, and pour upon it first the remainder of the mixture, and afterwards sufficient diluted alcohol to make the tincture measure 2 pints. (*U. S. Ph.*)

4484. Tincture of Belladonna. Moisten 4 troy ounces recently dried belladonna leaf, in fine powder, with 2 fluid ounces diluted alcohol; pack it firmly in a conical percolator, and gradually pour diluted alcohol upon it until 2 pints of tincture are obtained. (*U. S. Ph.*)

4485. Tincture of Hemp. Dissolve 360 grains purified extract of hemp in 1 pint alcohol, and filter through paper. (*U. S. Ph.*)

4486. Tincture of Capsicum. Moisten 1 troy ounce capsicum, in fine powder, with ½ fluid ounce diluted alcohol; pack it in a conical percolator, and gradually pour diluted alcohol upon it until 2 pints of tincture are obtained. (*U. S. Ph.*)

4487. Tincture of Cinchona. Moisten 6 troy ounces yellow cinchona, in moderately fine powder, with 2 fluid ounces diluted alcohol; pack it firmly in a glass percolator and displace, with diluted alcohol, 2 pints of tincture. (*U. S. Ph.*)

4488. Compound Tincture of Cinchona. Take 4 troy ounces red cinchona, 3 troy ounces bitter orange peel, 6 drachms serpentaria (Virginia snakeroot), 3 drachms red saunders, all in moderately fine powder; and 3 drachms saffron in moderately coarse powder. Mix the powders, moisten with 4 fluid ounces diluted alcohol, pack it firmly in a glass percolator, and displace, with diluted alcohol, 2½ pints of tincture. (*U. S. Ph.*)

4489. Tincture of Hemlock. Moisten 4 troy ounces recently dried hemlock, in fine powder, with 2 fluid ounces diluted alcohol; pack it firmly in a conical percolator, and gradually pour diluted alcohol upon it until 2 pints of tincture are obtained. (*U. S. Ph.*)

4490. Tincture of Digitalis. Moisten 4 troy ounces recently dried digitalis (fox glove), in fine powder, with 2 fluid ounces diluted alcohol; pack it firmly in a conical percolator, and displace, with diluted alcohol, 2 pints of tincture. (*U. S. Ph.*)

4491. Tincture of Iodine. Dissolve 1 ounce iodine in 1 pint alcohol. (*U. S. Ph.*) Tincture of iodine may be readily prepared by placing the iodine in a glass funnel, having previously filled the neck with broken glass, and pouring on the alcohol as it passes through. To prevent evaporation, cover the funnel with a close-fitting glass top. *Spirits of camphor* may also be speedily made in this way.

4492. Tincture of Turkey-Corn. Take 3 ounces powdered Turkey-corn root (corydalis) and make 1 pint tincture by maceration or displacement with diluted alcohol. (*Am. Dis.*)

4493. Tincture of Yellow Jasmine (Gelseminum). Cut into small pieces 8 ounces of the fresh root of yellow jasmine (gelseminum); macerate for 14 days in 2 pints diluted alcohol, express and filter. This forms a saturated tincture. (*Am. Dis.*)

4494. Universal Tincture. Bruise the following ingredients and digest for several days in 18 ounces brandy: 10 drachms aloes; 8 drachms each white agaric, rhubarb root, zedoary root, gentian root, galanga root, gum myrrh, and molasses electuary; 2 drachms saffron, and 4 ounces sugar. Express and filter.

4495. Compound Tincture of Black Pepper. This is prepared with 30 parts capsicums; 40 parts black pepper; 15 parts each grains of paradise, cinnamon, ginger, and calamus; 15 parts by measure acetate of potassa, and 60 parts alcohol.

4496. Tincture of American Hellebore. Moisten 16 troy ounces American Hellebore (veratrum viride), in moderately fine powder, with 4 fluid ounces alcohol. Pack it firmly in a cylindrical percolator, and displace, with alcohol, 2 pints of tincture. (*U. S. Ph.*)

4497. Compound Tincture of Dewberry. Take 4 ounces Dewberry (rubus trivialis) root, ½ ounce powdered Aleppo galls, 3 drachms powdered cinnamon, 10 grains powdered capsicum, 1 drachm powdered cloves, and ½ ounce gum kino. Digest for 14 days in 2 pints best brandy. Filter, and add 1 ounce tincture of opium, 1 ounce essence of peppermint, and 1 pint white sugar. Dose, 1 tea-spoonful for an adult.

4498. Tincture of Skunk-Cabbage. Take 3 ounces skunk-cabbage root in powder, and 1 pint diluted alcohol. Make a tincture by maceration, or displace 1 pint from a percolator. (*Am. Dis.*)

4499. Tincture of Stramonium. Make 1 pint of tincture from 2 ounces bruised stramonium seed and diluted alcohol. (*Am. Dis.*)

4500. Tincture of Monesia. Take 1 part extract of monesia, 6 parts alcohol, and 14 parts water. Mix and filter. (*Am. Dis.*)

4501. Tincture of St. John's Wort. Macerate for 14 days 5 ounces blossoms of St. John's wort, in 1 pint alcohol. Express and filter. (*Am. Dis.*)

4502. Compound Tincture of Kino. Take 4 drachms each powdered opium, gum kino, and cochineal; 3 drachms each camphor and cloves; and 4 drachms aromatic spirits of ammonia. Macerate in 4 pints dilute alcohol. Express and filter.

4503. Camphorated Tincture of Soap. There has been some difficulty in preparing this liniment as directed in the dispensatory, on account of its coagulating. The following formula makes a tincture which remains fluid at all temperatures. Take 4 ounces castile soap, 2 ounces camphor, ½ ounce oil of rosemary, 16 ounces water, and 20 ounces 95 per cent. alcohol.

4504. Tincture of Chloride of Iron. Introduce 3 troy ounces of iron wire, cut into pieces, into a flask of the capacity of 2 pints; pour upon it 11 troy ounces muriatic acid, and allow the mixture to stand until effervescence has ceased. Then heat it to the boiling point, decant the liquid from the undissolved iron, filter it through paper, and, having rinsed the flask with a little boiling distilled water, add this to it through the filter. Pour the filtrate into a 4-pint capsule, add 6¼ troy ounces muriatic acid; and, having heated the mixture nearly to the boiling point, add 1½ troy ounces nitric acid. When effervescence has ceased, drop in nitric acid, constantly stirring, until it no longer produces effervescence. Lastly, when the liquid is cold, add sufficient distilled water to make it measure 1 pint, and mix it with 3 pints alcohol. (*U. S. Ph.*)

4505. Tincture of Guaiac. Reduce 6 troy ounces guaiac to a moderately coarse powder, mix it with an equal bulk of dry sand, pack the mixture moderately in a conical percolator; and, having covered it with a layer of sand, gradually pour alcohol upon it until 2 pints of tincture are obtained. (*U. S. Ph.*)

4506. Tincture of Black Hellebore. Moisten 4 troy ounces black hellebore in moderately fine powder, with 1 fluid ounce diluted alcohol. Pack it in a cylindrical percolator, and gradually pour diluted alcohol upon it until 2 pints of tincture are obtained. (*U. S. Ph.*)

4507. Tincture of Mandrake (Podophyllin). Make 1 pint of tincture from 3 ounces mandrake-root in powder, with alcohol, either by maceration or percolation. (*Am. Dis.*)

4508. Tincture of Queen's Root (Stillingia). Take 3 ounces queen's root, bruised and cut into small pieces, and make 1 pint with diluted alcohol, either by maceration or displacement. (*Am. Dis.*)

4509. Tincture of Leopard's Bane (Arnica Flowers). Macerate 2 ounces arnica flowers in 1 pint dilute alcohol; or put the arnica-flowers in a percolator, and with diluted alcohol displace 1 pint. (*Am. Dis.*)

4510. Tincture of Hops. Moisten 5 troy ounces hops, in moderately coarse powder, with 2 fluid ounces diluted alcohol. Pack it very firmly in a cylindrical percolator, and displace, with diluted alcohol, 2 pints of tincture. (*U. S. Ph.*)

4511. Tincture of Henbane. Moisten 4 troy ounces henbane leaf, in fine powder, with 2 fluid ounces diluted alcohol. Pack it firmly in a conical percolator, and gradually pour diluted alcohol upon it until 2 pints of tincture are obtained. (*U. S. Ph.*)

4512. Tincture of Kino. Reduce 6 drachms kino to fine powder. Mix the powdered kino thoroughly with an equal bulk of dry sand; introduce the mixture into a conical glass percolator, and displace ½ pint of tincture, using a menstruum composed of 2 parts alcohol and 1 part water. (*U. S. Ph.*)

4513. Tincture of Lobelia. Moisten 4 troy ounces lobelia, in fine powder, with 2 fluid ounces diluted alcohol; pack it firmly in a conical percolator, and displace, with diluted alcohol, 2 pints of tincture. (*U. S. Ph.*)

4514. Tincture of Cimicifuga Racemosa (Black Cohosh, or Black Snake-Root). Black cohosh root, in fine powder, 4 troy ounces; alcohol, 1 pint. Make 1 pint of tincture by maceration or displacement. (*Am. Dis.*)

4515. Norwood's Tincture of Veratrum Viride (American Hellebore). Macerate 8 ounces of the recently dried, coarsely powdered root, in 16 ounces of alcohol for 14 days; express and filter through paper. (*Am. Dis.*)

4516. Tincture of Chiretta. Macerate 2½ ounces (avoirdupois) chiretta, cut small and bruised, in 15 Imperial fluid ounces rectified spirit, for 48 hours. Then transfer to a percolator, pouring on 5 additional fluid ounces rectified spirit; press the residuum, and filter; lastly, add rectified spirit to make up to 1 Imperial pint. (*B. Ph.*)

4517. Tincture of Ergot. Take 5 ounces (avoirdupois) ergot, and proceed in the same manner as for tincture of chiretta. (*B. Ph.*)

4518. Tincture of Blue-Flag. Macerate 3 ounces powdered blue-flag in 1 pint alcohol; or, make 1 pint by percolation. (*Am. Dis.*)

4519. Tincture of Lupulin. Pack 4 troy ounces lupulin in a narrow cylindrical percolator, and gradually pour alcohol upon it until 2 pints of tincture are obtained. (*U. S. Ph.*)

4520. Tincture of Nux Vomica. Digest with a gentle heat, 8 troy ounces finely powdered nux-vomica in 1 pint alcohol, for 24 hours in a close vessel. Then transfer the mixture to a cylindrical percolator, and gradually pour alcohol upon it until 2 pints of tincture are obtained. (*U. S. Ph.*)

4521. Tincture of Tobacco. Take a convenient quantity of the expressed juice of fresh-gathered tobacco leaves; mix it with an equal quantity of rectified spirits, and filter the mixture. This tincture, diluted with half its weight of spirits of nitric ether, is a specific for cramps or spasms of the bladder. For this purpose it is administered in doses of 10 to 20 drops, at intervals of about 2 hours.

4522. Tincture of Rhubarb. Mix together 3 troy ounces rhubarb in moderately coarse powder, and ½ troy ounce cardamom in moderately fine powder; moisten with 1 fluid ounce diluted alcohol, pack moderately in a conical percolator, and displace, with diluted alcohol, 2 pints of tincture. (*U. S. Ph.*)

4523. Tincture of Rhubarb and Senna. Reduce to a moderately coarse powder, 1 troy ounce rhubarb, 2 drachms senna, 2 drachms red saunders, 1 drachm each coriander and fennel, ½ drachm each saffron and liquorice, and 6 troy ounces raisins deprived of their seeds. Macerate for 14 days in 3 pints diluted alcohol, and filter through paper. (*U. S. Ph.*)

4524. Tincture of Bloodroot. Moisten 4 troy ounces bloodroot (sanguinaria), in moderately fine powder, with 1 fluid ounce diluted alcohol; pack it in a conical percolator, and displace, with diluted alcohol, 2 pints of tincture. (*U. S. Ph.*)

4525. Tincture of Serpentaria. Moisten 4 troy ounces serpentaria (Virginia snakeroot), in moderately fine powder, in 1 fluid ounce diluted alcohol. Pack it in a conical percolator, and gradually pour diluted alcohol upon it until 2 pints of tincture are obtained. (*U. S. Ph.*)

4526. Tincture of Valerian. This is obtained in the same manner as the tincture of serpentaria. (*See last formula.*) (*U. S. Ph.*)

4527. Camphorated Tincture of Opium. (Paregoric Elixir). This is a camphorated tincture of opium. Macerate 1 drachm each powdered opium and benzoic acid, 1 fluid drachm oil of anise, 2 ounces clarified honey, and 2 scruples camphor, in 2 pints diluted alcohol for 7 days, and filter through paper. (*U. S. Ph.*)

4528. Cummings' Quick Method of Making Paregoric. Take pulverized opium, 1 drachm; camphor gum, 2 scruples; benzoic acid, 1 drachm; oil of aniseed, 1 fluid drachm; clarified honey, 2 ounces; hot water and alcohol, 1 pint each. Dissolve the camphor and oil of aniseed in the alcohol; triturate the powdered opium in a mortar with some of the hot water for about 10 minutes, filter, and pass the remaining water through the dregs. To the fluid obtained add the alcoholic solution of oil and camphor, and dissolve finally the honey and benzoic acid in the mixture. By passing this once more through the pulverized opium, the latter will become perfectly exhausted. The addition of 10 grains of santal gives the preparation a beautiful rich tint.

4529. Tincture of Opium. (Laudanum). Macerate $2\frac{1}{2}$ ounces opium, in moderately fine powder, in 1 pint water for 3 days, with frequent agitation. Add 1 pint alcohol, and macerate for 3 days longer. Percolate, and displace 2 pints tincture by adding dilute alcohol in the percolator. (*U. S. Ph.*)

4530. Ammoniated Tincture of Opium. Digest 6 drachms benzoic acid, 6 drachms hay saffron, 4 drachms sliced opium, and 1 drachm oil of aniseed, in 1 quart spirit of ammonia for a week, and filter. Stimulant, anti-spasmodic, and anodyne. Dose, 20 to 80 drops.

4531. Squibb's Compound Tincture of Opium. This mixture is composed of tincture of opium, tincture of capsicum, spirit of camphor, each 1 fluid ounce; purified chloroform, 3 fluid drachms; and a sufficient quantity of stronger alcohol to make the whole measure 5 fluid ounces. Each fluid drachm, or tea-spoonful, contains about 100 drops, consisting of 12 minims of each of the first three ingredients, and $4\frac{1}{2}$ minims or 18 drops of chloroform. Dose, for persons over 18 years of age, a tea-spoonful; 2 to 6, ten to thirty drops; infants, one to ten drops, according to age. In time of epidemic cholera or diarrhea, when any person has two movements of the bowels more than natural within the twenty-four hours, the second one should be followed by a dose of this mixture; the dose to be repeated after every movement that follows. If the movements increase in frequency or in copiousness after the second dose of the medicine has been taken, a physician should be sent for at once, and a double dose be taken after each movement, until he arrives. Immediately after taking the first dose, the person should go to bed, and remain there for twelve hours after the diarrhea has entirely ceased.

4532. Compound Tincture of Pellitory. Take of bruised pellitory, 4 drachms; camphor, 3 drachms; oil of cloves, 2 drachms; powdered opium, 1 drachm; rectified spirit, 6 fluid ounces; digest for 8 days. The product is a most serviceable form of toothache-drops.

4533. Ethereo-alcoholic Tincture of Pellitory for Tooth and Face-ache. Take of bruised pellitory, 1 ounce; pure ether, 2 fluid ounces; strongest rectified spirit, 3 fluid ounces; digest them together in a stoppered bottle, in a cool place, for a week, with frequent agitation, then express the tincture, but avoid filtration. Some persons use equal parts of ether and spirit, but the product does not then keep so well. An excellent remedy for tooth-ache and face-ache, often giving almost immediate relief in the former case.

4534. Decoction of Balm of Gilead. For the decoction, simmer 1 ounce of the buds in a quart of soft water, down to half a pint. Take a wine-glassful or more, when the cough is troublesome.

4535. Tincture of Balm of Gilead. Infuse 2 ounces of the buds in a quart of good rum, and 4 ounces of sugar. Digest for 4 days. Take 2 or 3 tea-spoonfuls at a time. It greatly relieves cough, pains in the chest, and other pulmonary affections. The tincture and decoction form excellent remedies for cough, asthma, wheezing, &c.

4536. Tincture of Prickly-ash Berries. Macerate 8 ounces prickly-ash berries (Xanthoxylum) for 14 days in 2 pints diluted alcohol; or, displace 2 ounces of tincture by percolation. This tincture possesses all the virtues of the berries. In cholera, the dose is from $\frac{1}{2}$ to 1 fluid ounce, repeated as often as required; in ordinary cases from 1 to 4 fluid drachms, given in water. (*Am. Dis.*)

4537. Tincture of Aloes. Take 1 troy ounce socotrine aloes in fine powder, and 3 troy ounces liquorice; macerate for 14 days in $\frac{1}{2}$ pint alcohol and $1\frac{1}{2}$ pints distilled water, and filter through paper. (*U. S. Ph.*)

4538. Tincture of Aloes and Myrrh. Take 3 troy ounces each socotrine aloes and myrrh, both in moderately fine powder; 1 troy ounce saffron in moderately coarse powder; mix together, moisten with 2 fluid ounces alcohol, pack it moderately in a conical percolator, and displace, with alcohol, 2 pints of the tincture. This tincture may also be prepared by maceration for 14 days with 2 pints alcohol, and filtering through paper. (*U. S. Ph.*)

4539. Tincture of Cantharides. Moisten 1 troy ounce cantharides, in fine powder, with $\frac{1}{2}$ fluid ounce diluted alcohol; pack it in a conical percolator, and displace, with diluted alcohol, 2 pints of tincture. (*U. S. Ph.*)

4540. Tincture of Cardamom. Moisten 4 troy ounces cardamom, in fine powder, with 2 fluid ounces diluted alcohol; pack it firmly in a cylindrical percolator, and displace, with diluted alcohol, 2 pints of tincture. (*U. S. Ph.*)

4541. Tincture of Castor. Macerate 2 troy ounces bruised castor for 7 days in 2 pints alcohol; express, and filter through paper.

4542. Acetous Tincture of Valerian. Valerian root, bruised, 4 ounces; acetic acid, $1\frac{1}{2}$ ounces; diluted alcohol, $1\frac{1}{2}$ pints. Digest for 10 days in a closed vessel, and then filter. The tincture, as thus prepared, is of a beautiful red color with the predominating smell of the valerian—taste bitter and slightly astringent; may be given in doses of a dessert spoonful every 3 hours.

4543. Dover's Tincture. Pulverized ipecacuanha and opium, of each 8 grains; diluted alcohol, 1 fluid ounce. Macerate for 14 days and filter; or macerate 6 hours and displace 1 fluid ounce with diluted alcohol, 1 fluid drachm; equivalent to 10 grains Dover's powder. Used in combination with spirit of Mindererus effervescing draught, and other anti-febrile remedies in liquid form.

4544. Sweet Tincture of Red Bark (Cinchona). Red cinchona bark, in fine powder, 4 troy ounces; strong alcohol and syrup, sufficient quantity; dilute alcohol (alcohol 3 parts to 1 part water), $1\frac{1}{2}$ fluid drachms. Moisten the cinchona with the dilute alcohol, and pack in a glass funnel, in the neck of which sufficient tow (free from tar) has been placed, to act as a filter; cover the surface with a piece of perforated paper, and pour on alcohol previously mixed with an equal volume of syrup until it has reached the tow and the surface of the powder is covered; cork the neck of the funnel and allow it to macerate 48 hours; then remove the cork and continue the percolation with equal parts of alcohol and syrup, mixed, until 16 fluid ounces have been obtained.

4545. Sweet Tincture of Rhubarb. Take of rhubarb, bruised, and liquorice root, bruised, of each 2 ounces; aniseed, bruised, and sugar, of each 1 ounce; diluted alcohol, 2 pints. Macerate for 14 days, express, and filter.

4546. Aqueous Tincture of Rhubarb: Take of alkaline fluid extract of rhubarb, 3 fluid ounces. (*See No.* 4591.) Neutral carbonate of potassa, 240 grains; cinnamon water, 4 troy ounces; dissolve the carbonate in the cinnamon water; add the fluid extract, and then sufficient water to make the whole weigh 14 troy ounces. The above is an improvement on the preparation in the Prussian Ph., but are in officinal proportions, and yield a strictly officinal result.

4547. Tincture of Catechu. Take 3 troy ounces catechu, and 2 troy ounces cinnamon, both in moderately coarse powder. Mix, and moisten with 1 fluid ounce diluted alcohol; pack it into a conical glass percolator, and displace, with diluted alcohol, 2 pints of tincture. (*U. S. Ph.*)

4548. Tincture of Cinnamon. Mix 2 measures alcohol with 1 of water; moisten 3 troy ounces finely powdered cinnamon with 1 fluid ounce of the mixture; pack it in a conical percolator, and displace with the mixture 2 pints of tincture. (*U. S. Ph.*)

4549. Tincture of Colchicum. Moisten 4 troy ounces colchicum seed, in moderately fine powder, with 1 fluid ounce diluted alcohol; pack it in a cylindrical percolator, and displace, with diluted alcohol, 2 pints of the tincture. (*U. S. Ph.*)

4550. Tincture of Columbo. moisten 4 troy ounces columbo, in moderately one powder, and percolate 2 pints tincture in the same manner as the colchicum in last formula. (*U. S. Ph.*)

4551. Tincture of Cubeb. Percolate 2 pints tincture from 4 troy ounces cubeb, following the formula laid down for colchicum. (*See No.* 4549.) (*U. S. Ph.*)

4552. Tincture of Tar. Macerate 2 ounces tar in 16 ounces alcohol, until dissolved.

4553. Hamilton's Tincture of Dog-Wood. Bark of dogwood, 1 ounce; rectified spirit, 12 fluid ounces; mix, macerate for 14 days, and filter.

4554. Tincture of Colocynth. Colocynth, 8 parts; star anise, 1 part; alcohol, 96 parts. Macerate for 3 days, and filter. Dose, 15 to 20 drops.

4555. Compound Tincture of Squills and Benzoin. This is also known as *Wedel's Elixir*. Take of squills, orris root, and elecampane, each 25 drachms; liquorice root, 2 drachms; aniseed and myrrh, of each 4 scruples; saffron, 18 grains; dilute alcohol, 22 fluid ounces. Macerate for 15 days, express and filter. Dose, 40 to 60 drops, in catarrh, asthma, &c.

4556. Wood's Tincture of Kino. Kino in fine powder, $1\frac{1}{2}$ ounces; alcohol (.835), 8 fluid ounces; water, 4 fluid ounces; glycerine, 4 fluid ounces. Mix the alcohol, water, and glycerine together, and, having mixed the kino with an equal bulk of clean sand, introduce in a percolator and pour on the menstruum. This menstruum seems to thoroughly exhaust the drug of its astringent principle, and also makes a nice-looking preparation that will not deteriorate by exposure.

4557. Compound Tincture of Kino. This is made in the same way as other tinctures (*see No.* 35), with the following ingredients: 1 drachm each powdered opium, kino, and cochineal; $1\frac{1}{2}$ drachms each camphor and cloves; 1 fluid ounce aromatic spirit of ammonia, and 1 pint alcohol.

4558. Tincture of Ginger. Moisten 8 troy ounces ginger, in fine powder, with 2 fluid ounces alcohol; pack it firmly in a cylindrical percolator, and displace, with alcohol, 2 pints of tincture. (*U. S. Ph.*)

4559. Tincture of Jalap. Mix 2 measures alcohol with 1 water; moisten 6 troy ounces jalap, in fine powder, with 2 fluid ounces of the mixture; pack it moderately in a cylindrical percolator, and displace with the mixture 2 pints of tincture. (*U. S. Ph.*)

4560. Tincture of Myrrh. Take 3 troy ounces myrrh in moderately coarse powder; press it moderately into a conical percolator, and displace with alcohol 2 pints of tincture. (*U. S. Ph.*)

4561. Tincture of Nutgall. Moisten 4 troy ounces nutgall, in moderately fine powder, with 1 fluid ounce diluted alcohol; pack it in a glass percolator, and displace, with diluted alcohol, 2 pints tincture. (*U. S. Ph.*)

4562. Tincture of Quassia. Moisten 2 troy ounces quassia, in moderately fine powder, with 1 fluid ounce diluted alcohol; pack it in a percolator and displace, with diluted alcohol, 2 pints of tincture. (*U. S. Ph.*)

4563. Tincture of Rhatany. Moisten 6 troy ounces rhatany, in moderately fine powder, with 2 fluid ounces diluted alcohol; pack it in a cylindrical glass percolator, and displace, with diluted alcohol, 2 pints of tincture. (*U. S. Ph.*)

4564. Tincture of Squill. Moisten 4 troy ounces squill, in moderately coarse powder, with 1 fluid ounce diluted alcohol; pack it in a conical percolator, and displace, with diluted alcohol, 2 pints tincture. (*U. S. Ph.*)

4565. Tincture of Stramonium. Take 4 troy ounces stramonium seed, in moderately fine powder, and percolate 2 pints of tincture in the same manner as the squill in the last formula. (*U. S. Ph.*)

4566. Tincture of Tolu. Macerate 3 troy ounces balsam of tolu in 2 pints alcohol until it is dissolved; then filter. (*U. S. Ph.*)

4567. Compound Tincture of Benzoin. Macerate 3 troy ounces benzoin, ½ troy ounce socotrine aloes, both in coarse powder, and 2 troy ounces storax, for 14 days in 2 pints alcohol; filter through paper. (*U. S. Ph.*)

4568. Compound Tincture of Cardamom. Take 6 drachms cardamom, 2 drachms caraway, 5 drachms cinnamon, 1 drachm cochineal, all in moderately fine powder; mix together, and moisten with ½ fluid ounce diluted alcohol; pack it in a cylindrical percolator, and displace, with diluted alcohol, 6 fluid ounces of tincture. Lastly mix this with 2 troy ounces clarified honey, and filter through paper. (*U. S. Ph.*)

4569. Compound Tincture of Gentian. Mix together 2 troy ounces gentian, 1 troy ounce bitter orange peel, ½ troy ounce cardamom, all in fine powder; moisten with 1½ fluid ounces diluted alcohol; pack it in a conical percolator, and displace, with diluted alcohol, 2 pints of tincture. (*U. S. Ph.*)

4570. Compound Tincture of Iodine. Dissolve ½ troy ounce iodine and 1 of iodide of potassium in 1 pint alcohol. (*U. S. Ph.*)

Fluid Extracts.

This form of medicinal extracts was introduced into the United States Pharmacopœia in 1850, for the first time as a distinct class of preparations. Their distinctive character is the concentration of the active ingredients of a substance into a small bulk and in liquid form. Their advantages consist in greater convenience of administration, and in the fact that, not having been subjected to excessive evaporation, the active principles they contain are less liable to have suffered injury by heat. The main difficulty lies in their liquid form increasing the liability to undergo spontaneous decomposition; this is counteracted in some cases by the addition of sugar, in others by alcohol, and in others again by a mixture of both. Some fluid extracts have a tendency to deposit matter when combined with sugar, rendering the extract turbid or cloudy in appearance; instead of sugar, Mr. Alfred B. Taylor has proposed the use of glycerine, which, while it has the same preservative influence, possesses the property of dissolving the matter which would be deposited by the use of sugar. Fluid extracts are obtained by percolation, and the menstruum used is alcohol or alcohol and water, the proportions of each depending on the nature of the substance to be extracted. The price of alcohol has greatly increased since 1860, and a regard to economy has probably, in some cases, induced deviations in officinal preparations. This point will probably receive due consideration at the next revision of the Pharmacopœia.

4572. Grahame's Method of Percolation. Professor Grahame, of the Maryland College of Pharmacy, has proposed a modification of the displacement process which may be thus stated: Reduce the substance, by contusion, to a powder which will pass through a sieve of 40 meshes to the linear inch (if of close texture a sieve of 60 meshes is to be preferred); now add just sufficient of the menstruum to dampen the powder without wholly destroying its mobility; this usually requires about one-fourth as much menstruum as of the powder. Transfer to a glass funnel with a plug of cotton in the neck, and pack it with little or much pressure, according to its tenacity or disposition to adhere (more firmly when alcohol or ether is the menstruum than when water is to be used); if the particles of the moistened powder move freely on each other, the packing should be with as much force as a glass vessel will bear, the whole of the powder being introduced at once, and packed with a pestle or packing-stick. The whole quantity of the menstruum may now be poured on, or to the capacity of the funnel, and the process allowed to proceed to completion, without in any case repassing the first portions of the liquid. By this process, if carefully followed, very concentrated solutions are obtained. Indeed, most of the fluid extracts may be completed with little or no evaporation.

4573. Procter's Classified Formula for Making Fluid Extracts. In order to obtain as great a uniformity in the preparation of fluid extracts as the nature of the various drugs would permit, the following practical classification was drawn up by Professor William Procter, Jr., and submitted to the American Pharmaceutical Association, by whom the matter had been entrusted to him for investigation. In order to economize space, we give it in a somewhat condensed form. The paramount object in obtaining the fluid extract of a drug, is to extract, as far as possible, all the valuable ingredients; to condense them to some uniform standard strength, so that, for instance, each fluid ounce of the extract should contain the virtues of, and represent 1 ounce of the drug; and to leave the fluid in the best possible condition for retaining in solution the active principles of the drug. The process of percolation is adopted, as best adapted to effect the desired objects, and admitting a greater degree of accuracy than that of maceration. Glass funnels answer a good purpose, but cylindrical percolators may be employed. In either case, if the powder has been properly compacted, the menstruum, when added, passes very deliberately, by drops, and it will be found that the proportion of the percolate which is directed to be reserved will contain nearly all of the most valuable parts of the drug. In this way the action of the heat and air is entirely prevented on the most important part of the extracted matter, and where volatile oils are concerned this fact is particularly important. The ingredients are first reduced to a powder; and, in order to ensure the required result, different degrees of fineness are recommended, suitable to the degree of solubility and other natural peculiarities of the various drugs employed. This end is attained by sifting the powder through sieves containing a certain number of meshes to the linear inch. A sieve of 40 meshes to the inch will produce a powder designated as No. 40, &c.

A new class of oleo-resinous fluid extracts has been suggested, in which the stronger aromatics have been introduced, such as cloves, cinnamon, cardamom, &c., and which possess, for certain uses, very desirable advantages from their concentration. The number of oleo-resins has been considerably increased, on the ground that they represent their respective sources more completely and in smaller bulks than in any other form of fluid or semi-fluid extracts. On account of their superior strength, they should occupy a distinct position under the name of "*Oleo-resins,*" to distinguish them more particularly from all those preparations which go by the name of fluid extracts.

4574. Class No. 1, of Classified Fluid Extracts. The following substances are to be reduced to a powder of No. 60 degree of fineness; with the exception of *Buchu,* which should be in No. 40 powder.

Aconite Leaves.	Henbane Leaves.
Belladonna Leaves.	Matico.
Buchu.	Thorn-Apple (Stramonium).
Digitalis (Fox glove).	Valerian.

The menstruum employed is 2 pints alcohol diluted with 1 pint water. Moisten 16 troy ounces of the powdered drug evenly with 4 fluid ounces of the diluted alcohol; pack it firmly in a percolator, cover the surface of the powder with a disc of cloth (muslin, linen, lint, or any insoluble porous tissue, to prevent the disturbance of the powder); then pour on the menstruum gradually, so as to displace 3 pints; reserve the first 12 fluid ounces, and evaporate the remainder on a water-bath at 150° Fahr., to 4 fluid ounces; mix this with the reserved tincture; and, after standing 24 hours, filter through paper. The alcoholic strength of these extracts is nearly 50 per cent.

4575. Class No. 2, of Classified Fluid Extracts. The drugs included under this class should also be in at least No. 60 powder. Ipecacuanha and jalap may be reduced to dust with advantage. The fluid to be used is alcohol having a specific gravity of .835.

Aconite Root.	Jalap.
Black Snakeroot (Cimicifuga, or Black Cohosh).	May-apple Root (Podophyllum or Mandrake).
Black Hellebore.	Blood Root (Sanguinaria).
Ipecacuanha.	American Hellebore (Veratrum viride).

Moisten 16 troy ounces of the drug with 6 fluid ounces of the alcohol; displace 3 pints as directed in class 1, reserving the first ½ pint of percolate; distill the remainder until reduced to ½ pint, and, while hot, mix the distillate with the reserved tincture. After standing 24 hours, filter through paper.

4576. Class No. 3, of Classified Fluid Extracts. The substances included under this class require to be used in No. 50 powder; except columbo, No. 40; and squill, on account of its gummy nature, No. 30. The extract of colchicum deposits, by standing, a whitish sediment, which is believed to be in no wise connected with the activity of the preparation; it is recommended to allow this deposit to form before proceeding to filtration. Dilute alcohol is employed for making these extracts.

Colchicum Root.	Gentian.
Columbo.	Squill (Scilla).
Chiretta.	Seneka.
Boneset (Eupatorium).	Virginia Snake Root (Serpentaria).

Moisten 16 troy ounces of the substance with 4 fluid ounces dilute alcohol, percolate 3 pints, as in class 1, reserving the first 12 fluid ounces, evaporate the remainder to 4 fluid ounces by a water-bath at 150° Fahr.; mix with the reserved tincture; and, after 24 hours, filter.

4577. Class No. 4, of Classified Fluid Extracts. This class consists of *saccharine fluid extracts*, the sugar being introduced as a preservative agent. A decided advantage is gained by adding the sugar to the extract before the completion of the evaporation; in some cases it might be better to add the sugar previous to any evaporation. The fluid extracts of pipsissewa, bittersweet, pomegranate, pink-root, and sarsaparilla, fully represent the several drugs; and, combined with 3 times their bulk of simiple syrup, afford syrups of the ordinary strength. The menstruum used in these preparations is dilute alcohol; and the drugs are to be reduced to No. 50 powder, except galls, which should be No. 40.

Yellow Peruvian Bark (Cinchona Calisaya).	Pomegranate-root Bark (Granatum).
Pipsissewa (Chimaphila).	Blackberry Root (Rubus).
Bittersweet (Dulcamara).	Sarsaparilla.
Galls.	Pink Root (Spigelia).
Cranesbill (Geranium).	Bearberry Leaves (Uva Ursi).

Moisten 16 troy ounces of the powdered drug with ½ pint dilute alcohol; let it stand 30 minutes, then percolate as directed for class 1, until 3 pints have passed through; evaporate at a moderate heat on a water-bath to 1 pint; add 10 ounces sugar, evaporate to 1 pint, and strain while hot.

4578. Class No. 5, of Classified Fluid Extracts. The extracts obtained by this process are termed *acetic fluid extracts*. The acetic acid is introduced to control the tendency to decomposition, caused by the existence, in the drugs treated in this manner, of a salt consisting of an alkaloid and an organic acid. The fluid used is a mixture of ½ fluid ounce acetic acid and 3 pints diluted alcohol; and the drugs should be reduced to a No. 60 powder.

Ergot. Lobelia Leaves. Hemlock (Conium).

Moisten 16 troy ounces of the powder with ½ pint of the acetic mixture; pack it in a conical percolator, and displace 3 pints, reserving the first 12 fluid ounces, using dilute alcohol during the last part of the percolation. Evaporate the latter percolate to 4 fluid ounces, at a temperature not exceeding 150° Fahr.; mix this with the reserved tincture, and filter through paper.

4579. Class No. 6, of Classified Fluid Extracts. Under this division are placed *oleoresinous fluid extracts.* (*See No.* 4573.) The menstruum employed is deodorized alcohol, and the drugs are used in No. 50 powder; except canella, Ceylon cinnamon, elecampane, and orris root, used in No. 60, and myrrh in No. 30 powder.

Capsicum.	Orris Root, (Iris Florentina).
Canella.	
Cardamom.	Myrrh.
Cloves (Caryophyllum).	Pellitory Root (Pyrethrum).
Ceylon Cinnamon.	
Cubebs.	Allspice (Pimento).
Elecampane (Inula).	Prickly Ash Bark (Xanthoxylum).
Lupulin.	

The oleoresin of the above substances are to be obtained by percolation, and distilling off the alcohol.

This process of obtaining the oleoresins was modified before adoption in the U. S. Ph., by substituting ether for deodorized alcohol as the menstruum employed. The five following oleoresins are officinal preparations.

4580. Oleoresin of Capsicum. Take 12 troy ounces capsicum in fine powder, press it firmly in a cylindrical percolator, and gradually pour ether on it sufficient to displace 24 fluid ounces. Recover from this, by distillation on a water-bath, 18 fluid ounces of ether, and expose the residue in a capsule until the remaining ether has evaporated; lastly, remove, by straining, the fatty matter which separates on standing, and keep the oleoresin in a well stopped bottle. (*U. S. Ph*).

4581. Oleoresin of Cubebs. Moderately press 12 troy ounces cubebs in fine powder into a cylindrical percolator, and treat by the same process as the capsicum in the last formula. (*U. S. Ph*).

4582. Oleoresin of Lupulin. Press 12 troy ounces lupulin into a narrow cylindrical percolator, and displace with ether 30 fluid ounces; complete the process by distillation and subsequent evaporation in the same way as for capsicum. (*See No.* 4580.) (*U. S. Ph*).

4583. Oleoresin of Black Pepper. Treat 12 troy ounces black pepper in fine powder, by ethereal percolation and distillation, in the same manner as laid down in No. 4580; expose the residue after distillation in a capsule, until the remaining ether has evaporated and the deposition of piperin in crystals has ceased. Lastly, separate the oleoresin from the piperin by expression through a muslin strainer, and keep in a well-stopped bottle. (*U. S. Ph*).

4584. Oleoresin of Ginger. Take 12 troy ounces ginger in fine powder, press it firmly into a cylindrical percolator, and pour upon it 12 fluid ounces stronger ether; continue the percolation with alcohol sufficient to displace 12 fluid ounces in all. Recover from this, by distillation in a water-bath, 9 fluid ounces ether, and expose the residue in a capsule until the volatile part has evaporated. Lastly, keep the oleoresin in a well-stoppered bottle. (*U. S. Ph.*)

4585. Oleoresin of Male Fern. Pack closely 2 pounds avoirdupois, male fern, in coarse powder in a percolator; displace with 4 imperial pints ether, or until it passes colorless Let the ether evaporate on a water-bath, or recover it by distillation, and preserve the oily extract. (*Br. Ph.*) This preparation by its character decidedly belongs to the oleoresins; it has long been known and much used in Europe, under the name of *oil of fern*, in the treatment of the tapeworm. It is believed to have all the vermifugal powers of the male fern, and may be given in ½ fluid drachm doses. (*U. S. Dis.*)

4586. Fluid Extract of Rhubarb and Potassa. Grind or coarsely bruise 2 pounds avoirdupois best India rhubarb, 1 pound cassia or cinnamon, and 1 pound golden seal; macerate for 24 hours or more in 1 gallon good French brandy; express strongly, and add 1 fluid drachm oil of peppermint previously dissolved in a little 90 per cent. alcohol. Break up the compressed residue, and percolate with warm water until exhausted. Evaporate this solution to 4 pints, and, while warm (not too hot), dissolve in it 2 pounds bicarbonate of potassa, and 3 pounds refined sugar; evaporate, if necessary, to the quantity that the first macerated tincture lacks of 1½ gallons. Lastly mix the two together. It is used for the same purposes as the compound powder of rhubarb, 2 fluid drachms of the extract being equivalent to 1 drachm of the powder. (*Am. Dis.*) A simple alkaline extract of rhubarb is given in No. 4591.

4587. Fluid Extract of Stillingia. Cut fresh root of stillingia, 16 troy ounces, into small pieces; cover with alcohol, and digest for 24 hours. Then transfer to a percolator, and pack it so as to run slowly; add alcohol gradually, returning the first that passes until it runs clear. Reserve the first 12 fluid ounces percolated; then continue the percolation, with diluted alcohol, until the residuum is nearly exhausted; add 8 ounces white sugar to this dilute percolate, and evaporate by moderate heat to 4 fluid ounces. Add to this the reserved tincture, and 1 fluid drachm oil of caraway, and make 1 pint fluid extract. The dose is from 2 to 5 drops.

4588. Fluid Extract of Yarrow. Take of yarrow (the recently dried herb) in coarse powder, 8 ounces; dilute alcohol (2 parts 95 per cent. alcohol and 1 part water), a sufficient quantity. Pour over the powdered herb 4 ounces of the diluted alcohol, and work through with the hands until thoroughly moistened; allow it to stand in a covered jar for 24 hours. Pack closely in a funnel or other displacer and proceed to displace until 24 fluid ounces are obtained, which, if performed with proper care, will exhaust the herb, as tested, by tasting the droppings. The resulting liquid should be exposed in a shallow dish (in summer to a draught of air under an open window, in winter on a shelf near the top of the room), and allowed to evaporate spontaneously until it measures 16 fluid ounces; 30 or 40 grains bicarbonate of potassa in powder may then be added, which retains the extractive in solution, and clears the liquid without interfering with its properties. The evaporation of this fluid extract may be continued. if desired, with a very gentle heat (in a water-bath) until reduced to the consistence of an ordinary extract. The result in either case, fluid or solid, possesses in a marked degree the sensible and other properties of the herb, each tea-spoonful representing 30 grains of the herb.

4589. Procter's Fluid Extract of Wild Cherry Bark. Take of wild cherry bark, 24 ounces; sweet almonds, 3 ounces; and pure granulated sugar, 36 ounces. Macerate the powdered bark in 2 pints of 88 per cent. alcohol for 8 hours, introduce it into a percolator, and pour alcohol on it until 5 pints have passed, observing to regulate the passage of the liquid by a cork or stop-cock. Introduce the tincture into a capsule (or still, if the alcohol is to be regained), and evaporate it to a syrupy consistence; add ½ pint water, and again evaporate until all the alcohol is removed. Beat the almonds, without bleaching. into a smooth paste with a little of the water, and then sufficient to make the emulsion measure 1½ pints, and pour it into a quart bottle, previously containing the solution of the extract of bark; cork it securely and agitate occasionally for 24 hours, so as to give time for the decomposition of the amygdaline. The mixture is then to be quickly expressed and filtered into a bottle containing the sugar. Water should be added to the dregs and they again expressed till sufficient liquor is obtained to make the fluid extract measure 3 pints. The proportion of sugar, though less than that in syrup, is sufficient to preserve the preparation, aided by the presence of hydrocyanic acid.

4590. Parrish's Compound Fluid Extract of Buchu. Take of buchu, in coarse powder, 12 ounces; alcohol, 3 pints; water, 6 pints, or sufficient. Treat the leaves by maceration and displacement, first with a portion of the alcohol, and then with the remainder mixed with the water; evaporate the resulting liquid with a gentle heat to 3 pints, and add 2½ pounds sugar. Continue the heat till it is dissolved, and, after removing from the fire, add oil of cubebs, oil of juniper, of each 1 fluid drachm; spirit of nitric ether, 12 fluid ounces, previously mixed. Stir together.

4591. Alkaline Fluid Extract of Rhubarb. Take of fluid extract of rhubarb (by repercolation), 1 fluid ounce; neutral carbonate of potassa, 80 grains; water, 1 fluid ounce. Dissolve the carbonate in the water; to this add the fluid extract, and let the mixture repose 6 to 12 hours; then strain through muslin, and filter, if desirable. The alkaline fluid extract of rhubarb can be mixed with water in any proportion, affording a perfectly clear and transparent liquid of a deep red color. Another alkaline fluid extract of rhubarb will be found in No. 4586.

4592. Moore's Fluid Extract of Cimicifuga Racemosa (Black Cohosh or Black Snakeroot.) Take of cimicifuga, in No. 50 powder, 16 ounces, troy; alcohol 95 per cent., diluted alcohol, of each a sufficient quantity. Moisten the root with the alcohol, pack closely in the displacer, and pour on alcohol gradually until 8 fluid ounces have passed through, which reserve in a covered vessel to prevent evaporation, then proceed with dilute alcohol until the root is thoroughly exhausted. Evaporate over a water-bath until all the alcohol is driven off; set it aside to cool, that the resinous portion extracted may be deposited, which separate and add to the alcoholic portion first obtained; then proceed with the evaporation until reduced to 8 fluid ounces, and mix the two products; allow it to stand 48 hours, and then filter. (*See No.* 4575.)

4593. Compound Fluid Extract of Squills. This is alcoholic, in which 3 parts alcohol are diluted with 1 part water. Take of squills and seneka, each 16 ounces troy, reduced to a moderately coarse powder. Moisten with about 12 ounces of the liquid, and pack firmly in a conical percolator; cover the surface with a cloth and pour on of the same menstruum until 6 pints have slowly passed, reserving carefully the first 24 ounces. Evaporate the remainder in a water-bath at 150° Fahr., until reduced to 8 fluid ounces. Mix it with the reserved tincture, and, after standing, with occasional agitation, for 24 hours, filter, dropping sufficient of the menstruum on the filter to make the whole measure 2 pints. *Hive Syrup* may now be prepared from this extract by taking: compound fluid extract of squills, 4 fluid ounces; tartar emetic, 24 grains; simple syrup, 20 fluid ounces; hot water, ½ fluid ounce. Dissolve the tartar emetic in the water, and mix with the other ingredients.

4594. Procter's Fluid Extract of Hops. Take hops in coarse powder, 16 troy ounces. Mix in 4 ounces dilute alcohol; pack it in a conical percolator, cover the surface with cloth, and add dilute alcohol until 3 pints of tincture have slowly passed, carefully reserving the first 12 ounces. Evaporate the remainder of the tincture in a water-bath still to 4 fluid ounces, mix it with the reserved tincture, agitate occasionally during 24 hours, and filter, dropping sufficient dilute alcohol on the filter to make the measure of a pint.

4595. Procter's Fluid Extract of Liquorice. Take of Calabria liquorice, 8 troy ounces; and sugar in coarse powder, 10 troy ounces. Bruise the liquorice till it is reduced to pieces the size of a pea, enclose it in a gauze cloth, suspend it in a pint vessel, cover it with cold water, let it stand 12 hours (if in summer in a cool place), pour off the dense solution, renew the water, and again macerate and decant. Mix the two liquids, evaporate to 12 fluid ounces, dissolve in it the sugar, and again evaporate until the measure of 1 pint is obtained.

4596. Grahame's Fluid Extract of Burdock. Take of burdock, in No. 50 powder, 16 ounces; dilute alcohol (alcohol 9 parts, water 7 parts), a sufficient quantity. Dampen the powder with the menstruum and pack it in a suitable glass displacer: having covered the surface with a piece of muslin or perforated paper, pour on the menstruum, and continue the percolation to exhaustion, reserving 1½ ounces of the first runnings, evaporate the remainder over a water-bath until reduced to 9 fluid ounces, to which add 4 ounces sugar and dissolve. Strain, if necessary, and add the reserved portion. The dose of the extract is one tea-spoonful, representing 80 grains of the root. Burdock is one of the best vegetable alteratives, or blood depurents, and it is believed that this fluid extract might be advantageously substituted for that of sarsaparilla, as a more efficient and reliable alterative, or at least as a valuable addition to it.

4597. Fluid Extract of Chamomile. Take of fresh chamomile flowers, 1 pound; alcohol of specific gravity .871. Moisten the chamomile in coarse powder, with the alcohol, then pack in a percolator, and cover with the alcohol; digest 6 days, and draw off 12 ounces, which set aside. Continue the displacement with diluted alcohol, until it is freely exhausted of its bitterness, which evaporate in a vacuum to 4 fluid ounces. Mix and fiter. 1 drachm of this preparation represents 60 grains of chamomile flowers, which is usually given in doses of 20 grains, as a tonic, to 1 drachm, as an antiperiodic—making the dose for like cases from 20 minims to 1 fluid drachm.

4598. Fluid Extract of Seneka. The formula for making this extract will be found in No. 4576, but seneka yields its active principles so easily and entirely, that an extract of it may be obtained of standard strength without evaporation. If a convenient quantity of seneka in No. 50 powder be divided into 3 equal parts, and repercolated with 85 per cent. alcohol, an extract will be obtained, each fluid ounce of which will represent a troy ounce of the root.

4599. Fluid Extract of Ipecacuanha. Moisten 16 troy ounces ipecacuanha in fine powder with 6 fluid ounces alcohol; press it firmly into a conical percolator, and displace 3 pints of tincture, or until the ipecacuanha is exhausted. Distill the tincture over a water-bath until the residue is of a syrupy consistence. Mix with 1 fluid ounce acetic acid and 10 fluid ounces water; boil until reduced to ½ pint, and the resinous matter has separated. Filter when cold, and add water through the filter to make the filtrate up to ½ pint. Mix with ½ pint alcohol. (*U. S. Ph.*)

It is affirmed that syrup made from extract prepared according to the above formula is apt to become cloudy. It is proposed to avoid this result by dividing ipecacuanha in No. 50 powder into 3 parts, and obtaining the extract by repercolation in the same manner as the seneka in No. 4598.

4600. Fluid Extract of Sumach. Take 4 pints 76 per cent. alcohol, and 1 pound of the recently dried bark of Rhus Glabrum (sumach) in coarse powder. Moisten the powdered bark with sufficient alcohol and let it macerate for 24 hours, then percolate with the remainder of the alcohol, returning the first that passes until it runs clear. Reserve the first 4 clear fluid ounces of tincture, evaporate the remainder to 4 fluid ounces, and set aside. Then percolate the residuum nearly to exhaustion with hot water, evaporate this aqueous solution to ½ pint, then add to it 4 ounces white sugar, evaporate to 8 fluid ounces, and, while warm, mix it with the reserved 8 ounces of tincture to make 1 pint of fluid extract. (*Am. Dis.*)

4601. Fluid Extract of Scullcap. This is prepared from 1 pound of the dried leaves of scullcap (scutellaria) in precisely the same manner as directed for fluid extract of sumach in preceding receipt. (*Am. Dis.*)

4602. Fluid Extract of Life-Root is obtained from 1 pound recently dried life-root (senecio aureus) in the same manner as the sumach in No. 4600. (*Am. Dis.*)

4603. Fluid Extract of Senna and Jalap. Take 6 pints 76 per cent. alcohol. Mix together 1 pound senna and ½ pound jalap root, both in coarse powder; moisten them with some of the alcohol, and macerate for 24 hours. Transfer to a percolator and displace with the remainder of the alcohol; reserve the first 6 fluid ounces; evaporate the remainder to 6 fluid ounces and set also aside. Nearly exhaust the residuum with diluted alcohol and evaporate it to 12 fluid ounces; add 8 ounces white sugar; again evaporate to 12 fluid ounces, and, while warm, add 6 drachms carbonate of potassa, 40 minims oil of cloves dissolved in 1½ fluid drachms Hoffman's anodyne, and the 12 ounces reserved extract, making altogether 1½ pints fluid extract. (*Am. Dis.*)

4604. Fluid Extract of Blessed Thistle. Take 16 troy ounces blessed thistle (carduus benedictus) in No. 40 powder, dampen it with about 6 ounces dilute alcohol, and pack it in a suitable glass percolator; having covered the surface with a piece of muslin or a layer of clean sand (which is more convenient), displace with dilute alcohol. When 1 pint of liquid shall have passed, put it aside in a warm place for spontaneous evaporation until reduced to 10 fluid ounces. Continue the percolation with diluted alcohol until 2 more pints of liquid have passed; to these add 6 ounces sugar and reduce by evaporation over a water-bath to 6 fluid ounces, adding, while still hot, the 10 ounces of concentrated tincture; on cooling, the mixture becomes slightly turbid, but by the addition of a few drops of alcohol the resinous matter is redissolved, making a dark brown fluid extract which may be filtered if necessary.

4605. Fluid Extract of Cinchona. Take cinchona (calisaya) in powder, 8 troy ounces; simple (officinal) syrup, 4 fluid ounces; glycerine, 4 fluid ounces; alcohol, concentrated and diluted, a sufficient quantity. Moisten the cinchona with 6 fluid ounces of diluted alcohol; allow it to stand in a covered jar for three hours, and then transfer it to a cylindrical percolator. Pack it firmly, and gradually pour upon it diluted alcohol, until 12 fluid ounces of the tincture have been obtained. Set this aside, and continue the percolation with dilute alcohol, until the cinchona is thoroughly exhausted. To the last percolate add the syrup and glycerine, and evaporate by means of a water-bath to about 10 fluid ounces. To this add the reserve tincture, and continue the evaporation to 14 fluid ounces. Remove from the water-bath, and, when nearly cold, add sufficient alcohol to make the whole measure 16 fluid ounces. Each pint of the fluid extract contains nearly 2½ ounces of alcohol. (*See No.* 4577.)

4606. Fluid Extract of Pareira. Digest for 24 hours 1 pound pareira root, in coarse powder, in 1 pint boiling distilled water; then pack it in a percolator, and displace 1 gallon, or until the pareira root is exhausted. Evaporate over a water-bath to 13 fluid ounces; when cold add 3 fluid ounces rectified spirit, and filter through paper. This is the officinal formula of the British Pharmacopœia, consequently avoirdupois weight and imperial measure are to be used in preparing it. The dose consists of 1 to 2 fluid drachms.

4607. Moore's Fluid Extract of Vanilla. Take 8 troy ounces vanilla, and an equal weight of crushed loaf sugar. Slit the pods from end to end with a knife; then take them in small bundles, held tightly between the fingers, and cut them transversely into very small pieces. Of these, beat small portions at a time in an iron mortar, with a little of the sugar, until reduced to a damp powder, which must be rubbed with the hand through a No. 20 sieve; any coarse particles which will not pass through the sieve must be returned to the mortar, and, with fresh portions of vanilla and sugar, again treated as before. This is to be continued until the whole is reduced to a No. 20 powder. This is then to be mixed with 5 pints of a mixture of 3 parts alcohol and 1 part water, and the whole introduced into a 1-gallon stone jug, which must be tightly corked. The jug is then placed in a water-bath, resting upon folds of paper, and the mixture digested for 2 hours at a temperature of from 160° to 170° Fahr. The upper part of the jug must be kept cool (to prevent the undue expansion of vapor), by wrapping around it a towel or other cloth kept saturated by having cold water squeezed upon it from a sponge every 15 or 20 minutes. The jug should also be removed from the bath after each application of the water, and its contents well shaken, keeping the hand upon the cork to prevent its expulsion, and perhaps consequent loss of material. When the digestion has been completed, and the mixture has cooled, it is to be expressed through muslin. Pack the residue, previously rubbed with the hands to a uniform condition, firmly in a glass funnel prepared for percolation, and gradually pour upon it first the expressed liquid, and when this has all disappeared from the surface, continue the percolation with a mixture of 3 parts alcohol and 1 part water, until 8 pints of percolate are obtained.

Medicinal Essences.

The usual rule for making essences, is to mix 1 ounce of the essential oil with 1 quart of alcohol; although much is sold that contains only ½ ounce, and even ¼ ounce of the oil to the quart. A strong essence would consist of 1 ounce of oil to 1 pint of alcohol; from 10 to 30 drops of this would make a dose.

4609. To Color Medicinal Essences. Essence of peppermint is generally colored with tincture of turmeric; essence of cinnamon with tincture of red sandal wood; wintergreen with tincture of kino. The best way of coloring an essence is to steep for 12 hours the green leaf or other substance from which the oil is made, and then filter. The coloring is merely a matter of appearance; the essences are just as good without it.

4610. Essence of Peppermint. Oil of peppermint, 1 ounce; herb peppermint, ½ ounce; spirit of wine, 1 pint. Digest for a week, or until sufficiently colored. Palish-green, and very strong of the peppermint. Essence of peppermint is not conceived to be good by the ignorant unless it has a pale tint of green, which they presume is a proof of its being genuine. The most harmless way is to steep a little of the green peppermint in the spirit for this purpose (as above), or if this is not at hand, a little parsley will do equally as well, and in fact improve the flavor.

4611. Essence of Camphor, also called Liquor of Camphor; Concentrated Tincture of Camphor; Camphor-Drops. Dissolve 4½ drachms (avoirdupois) clear camphor, in 1 imperial pint rectified spirit. This forms the ordinary essence of camphor and the best *spirit of camphor* of the stores. Added to 15 times its bulk of pure cold water, it forms (by agitation) a transparent solution exactly resembling the camphor-julep, camphor-water, or camphor-mixture used in medicine, and which, either alone or with a little more water, forms an excellent wash for the teeth and mouth, as noticed elsewhere. (*See No.* 1335.)

Dissolve 1 avoirdupois ounce camphor in 10 ounces rectified spirit. This forms the *Concentrated Essence of Camphor* of the druggists. 10 or 12 drops added to 1 fluid ounce of pure cold water form the transparent *camphor-julep* or *camphor-water* before noticed.

4612. Essence of Coltsfoot. Balsam of tolu, 1 ounce; compound tincture of benzoin and rectified spirit of wine, of each 2 ounces; dissolve.

4613. Essence of Chamomile. Essential oil of chamomile, ½ ounce to 1 ounce; spirit of wine, 1 pint; mix. White. Or: Gentian root, sliced or bruised, 1 pound; dried orange peel, ½ pound; spirit of wine, 1 gallon; essential oil of chamomile, 5 ounces; macerate a week. Slightly colored. Some persons use ½ pound of quassia wood, instead of the gentian and orange peel. Both the above are stomachic and tonic.

4614. Essence of Spearmint. 1 ounce of essential oil to 1 pint of spirit of wine tinged green. Process, use, and dose, the same as essence of peppermint. (*See No.* 4610.)

4615. Bitter Essence. Wormwood, 4 parts; gentian root, bitter orange peel, and blessed thistle, of each 1 part; alcohol, 45 parts; digest for a week. Dose, ½ drachm to 2 drachms, combined with mixtures. Tonic and stomachic.

4616. Essence of Beef. Chop fine 1 pound lean beef, place it with ½ pint of water in a bottle which they will only half fill, and agitate violently for half an hour; then throw the whole on a sieve, and receive the liquid in a jug. Next, boil the undissolved portion in 1 pint of water for 20 minutes; strain, and mix the decoction with the cold infusion; evaporate the liquid to the consistence of thin syrup, adding spice, salt, &c., to suit the taste, and pour the essence, while boiling hot, into bottles (*see next receipt*), or jars, or (still better) tin cans, which must be closed up air-tight, and kept in a cool place. (*See No.* 1634.)

4617. To Fill Glass Bottles with Boiling Liquid. If boiling liquid be poured into cold bottles, there is a great risk of the bottle breaking, involving probably the loss of the contents. To prevent this, stand the bottles in a wide pan with sufficient cool water to reach nearly to the top of the bottles; pour sufficient water in each bottle to prevent it floating, and then let the water in the pan be brought gradually to a boil. As each bottle is to be filled, take it out of the pan, empty the water out of it, and fill it immediately.

4618. Ellis's Essence of Beef. Take lean beef, sliced thin, sufficient to fill the body of a porter bottle; cork it loosely, and place it in a pot of cold water, attaching the neck by means of a string to the handle of the pot; boil for 1½ or 2 hours, then decant the liquid and skim it. This can be seasoned and packed as in receipt No. 4616.

4619. Concentrated Essence of Ginger. Unbleached, well-bruised Jamaica ginger, 4 ounces; rectified spirit of wine, 1 pint; digest for 2 weeks, press and filter.

4620. Oxley's Concentrated Essence of Jamaica Ginger. The same as the preceding, with the addition of a very small quantity of essence of cayenne.

4621. Very Strong Concentrated Essence of Ginger. Bruised unbleached Jamaica ginger, 12 pounds; rectified spirit of wine, 2½ gallons; digest 14 days, press, strain, and reduce the essence by distillation to 1 gallon; cool and filter. This produces a most beautiful article. It is at once inexpensive and easily performed, as the spirit distilled off may be used with advantage for preparing the common tincture of ginger, and several other articles; 2 ounces of this essence are regarded as equivalent to 3 ounces of the finest ginger. A single drop swallowed will almost produce suffocation.

4622. Concentrated Essence of Ginger. Ginger and animal charcoal, both in coarse powder, equal parts; add enough rectified spirits of wine to perfectly moisten them, and after 24 hours put the mass into a percolator, return the first runnings 2 or 3 times, then change the receiver, and pour on spirit gradually as required, and at intervals, until as much essence is obtained as there was ginger employed. Quality excellent. The mass remaining in the percolator may be treated with fresh spirit until exhausted, and the tincture so obtained may be advantageously employed, instead of spirit, in making more essence with fresh ginger. The last portion of spirit in the mass may be obtained by adding a little water. (*See Percolation, No.* 41.)

4623. Concentrated Essence of Guaiacum. Guaiacum shavings, from which the dust has been sifted, 3 cwt. Exhaust the wood by boiling with water, as in preparing an extract, using as little of that fluid as is absolutely necessary; evaporate to exactly 1¼ gallons; let it stand until cold, stirring it all the time to prevent the deposit of resinous matter; put the whole into a bottle; add spirit of wine, 5 pints; agitate repeatedly for a week, then allow it to settle for 7 or 8 days, and decant the clear into another bottle. This preparation is frequently substituted for guaiacum shavings in the preparation of compound decoction of sarsaparilla. 1 pint of this essence is considered equivalent to 19 pounds of guaiacum in substance.

4624. Essence of Quinine. Take dilute sulphate of quinine, 1 drachm; rectified spirit, 1 fluid ounce; mix, add of dilute sulphuric acid (specific gravity 1.087 to 1.090), ½ fluid drachm (or less, on no account more), and agitate it thoroughly until solution is complete. A few drops added to water form an excellent wash for foul, spongy, and tender gums, loose teeth, &c.; also for weak hair.

Medicated Syrups.

Syrup is a concentrated solution of sugar in watery fluids. If made with pure water, it is termed *syrup* or *simple syrup*. Where the water contains one or more medicinal agents, it is called *medicated syrup*. Full information as to preparation, &c., will be found in Nos. 1356, &c.

4626. Syrup of Phosphate of Zinc. Phosphate of zinc, 192 grains; water, 11 fluid drachms; syrupy phosphoric acid (specific gravity 1.5), 5 fluid drachms; syrup, 10 fluid ounces. Rub the phosphate with the water, add the acid, and filter into the syrup. Each fluid drachm contains 2 grains of zinc phosphate and about 18 minims of dilute phosphoric acid. In this formula, avoirdupois weight and Imperial measure are adopted.

4627. Syrup of Phosphate of Quinine. Take of phosphate of quinia, 96 grains; water, 13½ fluid drachms; syrupy phosphoric acid (specific gravity 1.5), 2½ fluid drachms; syrup, 10 fluid drachms. Mix the acid with the water, add the quinia, and filter into the syrup. Each fluid drachm contains 1 grain of phosphate of quinine and acid equal to about 10 minims of the dilute phosphoric acid.

The same weight of quinia, prepared by precipitating an acidulated solution of the disulphate by solution of ammonia, collecting, washing, and drying at 100° Fahr., may be used, in the absence of the phosphate. In this formula avoirdupois weight and Imperial measure are intended.

4628. Syrup of Phosphate of Iron with Quinine. Take of phosphate of iron, 192 grains; phosphate of quinia, 96 grains; water, 7 fluid drachms; syrupy phosphoric acid (specific gravity 1.5), 9 fluid drachms; syrup, 10 fluid ounces. Rub the powders with the water, add the acid, and filter into the syrup. Each fluid drachm contains 2 grains of phosphate of iron and 1 grain of phosphate of quinine. In the absence of the phosphate of quinia, the same weight of quinia may be prepared as directed in No. 4627.

In this formula avoirdupois weight and Imperial measure are adopted.

4629. Easton's Syrup of Phosphate of Iron, Quinine, and Strychnine. Take of phosphate of iron, 192 grains; phosphate of quinia, or quinia prepared as directed in No. 4627, 96 grains; strychnia (in crystals), 3 grains; water, 7 fluid drachms; syrupy phosphoric acid (specific gravity 1.5), 9 fluid drachms; syrup, 10 fluid ounces. Rub the phosphate of iron with 5 drachms of the water in a glass mortar, dissolve the strychnia and quinia in the acid, previously mixed with the remaining 2 drachms of water; mix and filter into the syrup. Each fluid drachm contains 2 grains of phosphate of iron, 1 grain of phosphate of quinine, and $\frac{1}{32}$ part of a grain of strychnine.

In this formula avoirdupois weight and Imperial measure are adopted.

4630. Syrup of Phosphate of Iron and Strychnine may be prepared in the same manner as the last, omitting the phosphate of quinine.

4631. Phosphate of Iron. Dissolve 3 ounces sulphate of iron in 2 pints boiling distilled water, dissolve also 1 ounce acetate of soda and 2½ ounces phosphate of soda in another 2 pints boiling distilled water. Mix the 2 solutions, filter the precipitate through muslin, wash it with hot distilled water till the washings no longer form a precipitate with chloride of barium. Dry at a heat not exceeding 120° Fahr. (*Br. Ph*).

4632. Syrup of Phosphate of Iron. Phosphate of iron, 96 grains; water, 9 fluid drachms; syrupy phosphoric acid (specific gravity 1.5), 7 fluid drachms; syrup, 10 fluid ounces. Rub the phosphate of iron with the water in a glass mortar, add the phosphoric acid, and filter the mixture into the syrup. As thus prepared, it contains the same proportion of iron, about 2 minims less of the dilute acid (25 instead of 27), and rather more sugar than when prepared according to the British Pharmacopœia. The phosphate of iron is made by the Br. Ph. process, and dried at a temperature not exceeding 100° Fahr. The specimens found in the ordinary course of trade are not readily soluble in the acid. This want of solubility is believed to be due to the length of time they have been kept before sale, as the best results have been obtained with the phosphate only a few days old. In this formula avoirdupois weight and Imperial measure are adopted.

4633. Syrup of Phosphate of Manganese may be prepared in a similar manner with the following ingredients: Phosphate of manganese, 96 grains; water, 9 fluid drachms; syrupy phosphoric acid (specific gravity 1.5), 9 fluid drachms; syrup, 10 fluid ounces. Strength, 1 grain phosphate of manganese, and acid equal to about 25 minims of the dilute phosphoric acid in each fluid drachm. The *phosphate of manganese* is made in the same manner as the phosphate of iron, substituting sulphate of manganese for the sulphate of iron. In this formula avoirdupois weight and Imperial measure are intended.

4634. Syrup of Phosphate of Iron with Manganese. Phosphate of iron, 72 grains; phosphate of manganese, 48 grains; water, 8 fluid drachms; syrupy phosphoric acid, 8 fluid drachms; syrup, 10 fluid ounces. Rub the powders with the water, add the acid, and filter into the syrup. Each fluid drachm contains ¾ grain phosphate of iron, ½ grain phosphate of manganese, and acid equal to about 30 minims of the dilute phosphoric acid, B. Ph. Avoirdupois weight and Imperial measure are understood in the above formula.

4635. Syrup of Phosphate of Iron and Lime. Take of phosphate of iron, 96 grains; phosphate of lime, 192 grains; water, 8 fluid drachms; syrupy phosphoric acid, (specific gravity 1.5), 8 fluid drachms; syrup, 10 fluid ounces. Mix the powders with the water in a glass mortar, add the acid, and filter into the syrup. Each fluid drachm contains 1 grain of phosphate of iron, 2 grains of phosphate of lime, and an amount of acid equal to about 30 minims of the dilute phosphoric acid, B. Ph. The *phosphate of lime* is made by precipitation from solutions of chloride of calcium and phosphate of soda, and dried at 100° Fahr., and should not be kept too long before use. In this formula avoirdupois weight and Imperial measure are adopted.

4636. Durand's Syrup of Phosphate of Lime. Take of precipitated phosphate of lime, 128 grains; glacial phosphoric acid, 240 grains; sugar, in coarse powder, 7½ ounces; distilled water, 4 fluid ounces; essence of lemon, 12 drops. Mix the phosphate of lime with the water in a porcelain capsule, over a spirit or gas lamp, or in a sand-bath; add gradually the phosphoric acid until the whole of the phosphate of lime is dissolved. To this solution add sufficient water to compensate for the evaporation, then dissolve the sugar by a gentle heat, and, when perfectly cold, add the essence of lemon. The syrup of phosphate of lime, thus prepared, is colorless, transparent, of an acid taste, and contains two grains of the phosphate of lime, and nearly four grains of phosphoric acid to each tea-spoonful. When diluted it forms a phosphoric lemonade, not unpleasant to the taste. Dose, a tea-spoonful.

4637. Wiegand's Syrup of Phosphate of Lime. Dissolve 1 ounce precipitated phosphate of lime in 1 fluid ounce water by means of 4 fluid drachms muriatic acid; filter, and add 6½ fluid ounces water; then add 12 fluid ounces sugar, and strain. Dose, a tea-spoonful. This preparation is not so acid as Durand's, which is thought to be an advantage in some cases.

4638. Syrup of Rhubarb. The officinal method of preparing the fluid extract of rhubarb employed for the syrup involves much concentration by evaporation, and results in an unsightly preparation, and liable to an objectional resinous precipitation. By a modified process a fluid extract of rhubarb, equal to the officinal in strength, is first obtained by repercolating rhubarb, in moderately fine powder, with a mixture of 3 parts officinal alcohol and 1 part water. This menstruum exhausts rhubarb completely with the greatest facility. To make the syrup, take of this fluid extract, 3 fluid ounces; sugar, 28 troy ounces; water, a sufficient quantity. Add the fluid extract to 12 fluid ounces of water, filter, make up the filtrate to the measure of a pint by adding water through the filter, and dissolve in it the sugar with the aid of a gentle heat, and strain through muslin. The result is splendid. An equal product is obtained by mixing the officinal fluid extract with water, letting it repose some hours, filtering, and then completing as above.

4639. Syrup of Rhubarb and Senna. Digest for 14 days 6 ounces each bruised rhubarb root and senna leaves, and 1½ ounces cardamom seeds, in 6 pints dilute alcohol; filter, and evaporate to 3 pints. Mix 12 ounces of this with syrup made of 2 pounds sugar evaporated to 1½ pints, and mix while hot. This produces a syrup of 30° Baumé, which will not ferment.

4640. Stewart's Simple Syrup of Rhubarb. Macerate 6 ounces bruised rhubarb in 4 ounces dilute alcohol; press and filter, and evaporate to 2 pints. Mix 8 fluid ounces of this tincture with 28 fluid ounces simple syrup.

4641. Procter's Compound Syrup of Hypophosphites. Take of hypophosphite of lime, 256 grains; hyposulphite of soda, 192 grains; hyposulphite of potassa, 128 grains; hyposulphate of iron (recently precipitated), 96 grains; white sugar, 9 ounces; extract of vanilla, ½ ounce. Dissolve the salts of lime, soda, and potassa, in six ounces of water; put the iron salt in a mortar and gradually add a solution of hypophosphorus acid till it is dissolved. To this add the solution of the other salts, after it has been rendered slightly acidulous with the same acid, and then water, till the whole measures 12 fluid ounces. Dissolve in this the sugar, with heat, and flavor with the vanilla. Without flavoring, this syrup is not unpleasant.

4642. Hypophosphite of Iron. Hypophosphite of iron is obtained when 128 grains of hypophosphite of soda, dissolved in 2 ounces of water, are decomposed with a slight excess of solution of persulphate of iron, and the white precipitate well washed on a filter with water.

4643. Parrish's Compound Syrup of Hypophosphites. Take of hypophosphite of lime, 1½ ounces; hypophosphite of soda, ½ ounce; hypophosphite of potassa, ½ ounce; cane sugar, 1 pound, troy; hot water, 20 fluid ounces; orange water, 1 fluid ounce. Make a solution of the mixed salts in the hot water, filter through paper, dissolve the sugar in the solution by the aid of heat; strain, and add the orange-flower water. Dose, a teaspoonful, containing nearly five grains of the mixed salts.

4644. Compound Syrup of Phosphate

of Iron. Dissolve 10 drachms protosulphate of iron in 2 fluid ounces boiling water; also dissolve 12 drachms phosphate of soda in 4 fluid ounces boiling water; mix the solutions and wash the precipitated phosphate of iron till the washings are tasteless. Dissolve 12 drachms phosphate of lime in 4 fluid ounces boiling water with sufficient muriatic acid to make a clear solution, precipitate it with water of ammonia, and wash the precipitate. To these two precipitates add 20 drachms glacial phosphoric acid dissolved in water; when clear add 2 scruples carbonate of soda, and 1 drachm carbonate of potassa. Next add sufficient muriatic acid to dissolve the precipitate; and lastly 2 drachms powdered cochineal mixed with 3 pounds (troy) sugar; apply heat, and, when the syrup is formed, strain. It is a question whether a simple syrup of phosphate of iron is not equally efficacious with Professor Parrish's more complicated preparation given above, and known as *Parrish's Chemical Food.* Each tea-spoonful contains 1 grain phosphate of iron, 2½ grains phosphate of lime, with smaller quantities of the alkaline phosphates, all in perfect solution.

4645. Chemical Food. This is prepared by the same formula as Professor Parrish's (*see No.* 4644), omitting the cochineal and muriatic acid, and with this modification was adopted, as well as the two following receipts, by the Newark Pharmaceutical Association.

4646. Compound Syrup of Hypophosphites and Iron. Dissolve 256 grains each of the hypophosphites of soda, lime, and potassa, and 126 grains hypophosphite of iron, in 12 ounces water, by means of a water-bath. Filter, and add sufficient water to make up for the evaporation. Add 18 ounces sugar by gentle heat, to make 21 fluid ounces syrup. Each fluid ounce contains 12 grains each of the hypophosphites of soda, lime, and potassa, and 6 grains hypophosphite of iron. (*Newark P. A.*)

4647. Compound Syrup of Hypophosphites. Prepared by the same formula as the last, omitting the iron. (*Newark P. A.*)

4648. Aitken's Syrup of Iron, Quinia, and Strychnia. Dissolve 5 drachms sulphate of iron in 1 ounce of boiling water, and 1 ounce phosphate of soda in 2 ounces of the same. Mix the solutions and wash the precipitates on strainers until the washings are tasteless; dissolve 192 grains sulphate of quinia with sufficient sulphuric acid in 2 ounces of water, precipitate the clear solution by a very slight excess of water of ammonia, collect and carefully wash it. Dissolve both precipitates, and also 6 grains strychnia, in 14 ounces dilute phosphoric acid, then add 14 ounces white sugar, and dissolve the whole without heat. This syrup contains about one grain of phosphate of strychnia in each drachm. The dose might therefore be about a tea-spoonful 3 times a day. It is perfectly miscible with water, has a strongly styptic and chalybeate taste, and an after-taste of quinia. It is employed mainly as a preparative to the use of cod-liver oil, and in certain cases as a concomitant to this food substitute in scrofulous diseases, in cases of delicate children, with equal parts of the phosphatic syrup known as chemical food.

4649. Santonate of Soda. Put into a flask, 2 ounces santoninic acid, 4 fluid ounces pure caustic soda lye, and 12 fluid ounces distilled water. Heat the flask in a sand-bath or over a stove to 70° or 80° Fahr., until the santonine solution is complete; which usually requires about half an hour; then remove from the fire, and, when cold, it is conveniently evaporated.

4650. Syrup of Santonate of Soda. Boil 18 fluid ounces syrup until it marks 32° Baumé; let it cool a few minutes, then add 30 grains santonate of soda dissolved in 1 ounce distilled water. You obtain 18 fluid ounces of a transparent syrup, without a bitter taste, of 35° when cold. Each fluid ounce contains one grain of santonine. This syrup is an excellent vermifuge.

4651. Syrup of Ipecacuanha. Mix 2 fluid ounces officinal fluid extract of ipecacuanha with 30 fluid ounces syrup. (*U. S. Ph*). This syrup is said to become cloudy occasionally, and the following preparation claims to be free from this objection.

Moisten 2 troy ounces ipecacuanha with 1 fluid ounce diluted alcohol, and let it stand for 24 hours. Then transfer it to a conical percolator, and gradually pour upon it diluted alcohol until 1 pint of tincture has passed. Evaporate this by means of a water-bath to 6 fluid ounces, add 10 fluid ounces warm water, and, having rubbed it thoroughly with 45 grains carbonate of magnesia, in a mortar, filter, and add sufficient warm water through the filter to make the filtrate measure 1 pint; then add 29 troy ounces sugar, and dissolve it with the aid of a gentle heat, and, having strained the hot syrup, add sufficient warm water, through the strainer, to make it measure 2 pints when cold.

The same advantages are claimed for a syrup made in the following manner:—To 2 fluid ounces of the fluid extract made by repercolation, add 2 fluid ounces water and heat the mixture to the boiling point; then add 12 fluid ounces water, filter, and pour sufficient water through the filter to make the liquid measure 1 pint; in this dissolve 28 troy ounces sugar with the aid of heat, and strain through muslin. Both preparations will be perfectly clear, beautiful, and identical in strength and appearance, the latter possessing the natural odor and taste of ipecacuanha in an eminent degree.

4652. Compound Syrup of Squills. Take 4 troy ounces squill in No. 30 powder, and the same of seneka in No. 50 powder, mix them together, moisten with ½ pint diluted alcohol, and allow it to stand for an hour. Then transfer it to a conical percolator and pour diluted alcohol upon it until 3 pints of tincture have passed. Boil this for a few minutes, evaporate it by means of a water-bath to 1 pint, add 6 fluid ounces of boiling water, rub the liquid with 1 troy ounce carbonate of mangnesia in a mortar till thoroughly mixed, filter, and add through the filter sufficient warm water to make the filtrate measure 22 fluid ounces. Dissolve 42 troy ounces sugar in the filtered liquid, and, having heated the solution to the boiling point, strain it while hot. Then dissolve 48 grains tartrate of antimony and potassa in the solution while still hot, and add sufficient boiling water, through the strainer, to make it measure 3 pints when cold. Lastly, mix the whole thoroughly together. The above process is similar to that laid down in the U. S. Ph., except in the addition of magnesia before filtration, this being considered an improvement, as the gummy nature of the squills renders filtration unsatisfactory without it.

This syrup may also be prepared from the fluid extracts of squill and of seneka, by mixing 4 fluid ounces of each, evaporating the mixture by means of a sand-bath to a syrupy consistence; triturating this with the carbonate of magnesia, and proceeding precisely as in the above formula.

4653. Syrup of Ether. The combination of sulphuric ether with simple syrup, as usually prepared, is very unsatisfactory, whether for use alone, or mixed with other ingredients; a portion of the ether always separates and floats on the surface of the mixture, bringing with it also some impurities of the syrup. In pouring out a portion from the bottle containing it, the floating layer of ether and scum will come first, unless these be again mixed in by agitating the bottle. The following improvement is taken from the Paris Codex: Provide a bottle which has a small neck inserted in the side close to the bottom (*see illustration*); this, as well as the upper neck, should have a closely-fitting cork. The bottle must be of a size to contain 1 pint simple syrup and 1 ounce sulphuric ether. Insert these in it and shake well 3 or 4 times a day for 6 days; after which, if allowed to repose, a thin film of ether will rise and float on the surface of the syrup, separated from it by a layer of scum. The syrup, which is now saturated with ether, can be drawn through the lower neck, as required; it will be perfectly free from impurity, and no further separation of ether will take place.

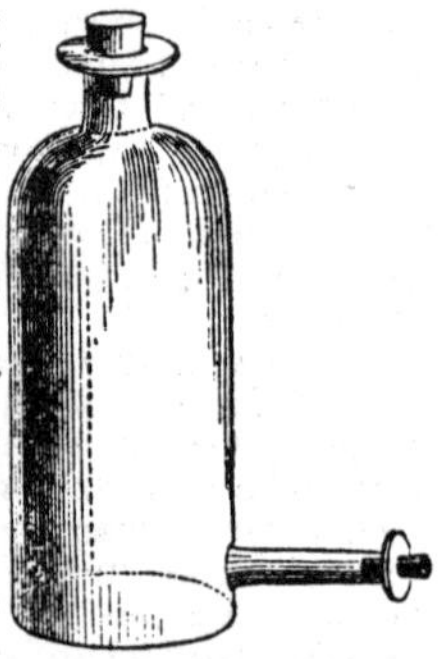

4654. Compound Syrup of Black Cohosh. Macerate 2 ounces black cohosh (black snake-root), 1 ounce seneka root, ½ ounce liquorice root, and ¼ ounce ipecacuanha root in dilute alcohol for 24 hours; then transfer to a percolator and run through two pints; evaporate the excess of alcohol by a water-bath, and convert into a syrup with sufficient quantity of sugar; lastly, treat 2 ounces wild cherry bark with half a pint of cold water, which add to the syrup previously cooled.

4655. Compound Syrup of Sarsaparilla. Reduce the following to moderately coarse powder, adopting the troy ounce throughout: 24 ounces sarsaparilla, 3 ounces guaiacum wood, 2 ounces each pale rose, senna, and liquorice root. Mix with 3 pints diluted alcohol, and allow the mixture to stand for 24 hours. Transfer to a cylindrical percolator, and displace 10 pints with diluted alcohol. Evaporate by a water-bath to 4 pints; filter, and add 96 ounces coarsely powdered sugar by the aid of heat, and strain while hot. Lastly take 5 minims each of the oils of sassafras and anise; and 3 minims oil of gaultheria; rub these oils with a small portion of the solution, and mix them thoroughly with the remainder. (*U. S. Ph.*)

4656. Scovill's Compound Syrup of Sarsaparilla. Take 8 ounces each sarsaparilla, burdock root and yellow dock; 6 ounces stillingia root (queen's root), 2 ounces turkey pea, 4 ounces false bitter-sweet, 3 ounces dandelion root, 3 ounces juniper berries, 1 ounce prickly-ash berries, 2 ounces guaiacum wood, and 9 ounces bamboo briar root. Coarsely bruise the above ingredients, and moisten them with alcohol. Let them stand 2 or 3 days, then put them in a steam displacement apparatus, and pass through the vapor of 3 pints strong alcohol. Continue the displacement with the steam of water till the strength is exhausted; set aside the 3 pints of tincture which first passed, and evaporate the remaining decoctions to 1 quart; mix this with the tincture, add 3 quarts sugar-house syrup, and, when cold, add 1¼ ounces iodide of potassium.

4657. Osborne's Syrup. This is one of the most valuable preparations that can be made for children. Simmer 11½ drachms each, rhubarb root, anise seed, and liquorice root, in 45 ounces boiling water over a slow fire till reduced to two-thirds. Then make a syrup with 4¾ troy pounds white sugar, add 2 1⁄16 drachms each manna and compound tincture of opium (paregoric), and 225 grains salt of tartar. In warm weather, add a wine-glass of French brandy.

4658. Syrup of Seneka. Evaporate 4 fluid ounces of the fluid extract (*see No.* 4598) by means of a sand or water-bath to a syrupy liquid, triturate this with ¼ ounce carbonate magnesia, and gradually add 8 fluid ounces of water, constantly stirring; filter, and add sufficient water, through the filter, to make the liquid measure 8 fluid ounces, then dissolve in it 16 troy ounces sugar, with the aid of heat, and strain through muslin while hot. The product, for its permanence and elegant appearance, cannot be surpassed. To prepare this syrup directly from a fluid extract by merely mixing that with simple syrup, would render the preparation uncommonly thin, and introduce an excessively large proportion of alcohol, which would be an unquestionable and serious objection.

4659. Compound Chloroform Syrup. This formula for an anodyne containing chloroform will remain combined and mix readily with either spirit or water. Macerate for 2 or 3 days 16 grains resin of cannabis, 2 grains capsicum, and 8 drops oil of peppermint in 4 drachms chloroform and 1½ drachms ether; filter the product. To about 1 ounce syrup add ½ drachm each of water and perchloric acid, and dissolve in this by a water-bath, 16 grains muriate of morphia; when cold add 96 minims Scheele's hydrocyanic acid, add to this the filtrate first made, and syrup sufficient to make the whole up to 4 ounces.

4660. Syrup of Chloride of Iron. Mix ½ troy ounce chloride of iron with 1 pint simple syrup. Flavoring may be added to taste. Dose, a teaspoonful, as a tonic and astringent, adapted to weak and relaxed conditions of the stomach and bowels, and to anæmic symptoms generally. *Parrish.*

The *syrup of the protochloride of iron* is prepared by dissolving 1 drachm dry protochloride of iron (ferrous chloride) in ½ troy ounce orange-flower water; mix separately 4½ troy ounces syrup of orange flower with 20 troy ounces syrup of acacia; add the iron solution to the syrup mixture. Each tablespoonful of this preparation contains about 1½ grains of the protochloride of iron. *Pharm. Soc. of Paris.*

For preparing the *syrup of perchloride of iron*, Dr. Duroy proposes combining 2 parts solution of perchloride of iron, (ferric chloride) specific gravity 30° Baumé, with 100 parts distilled water, and 25 parts syrup of orange flower. This will contain 1 per cent. of the iron salt. *Paris Codex.*

4661. Syrup of Lactate of Iron. Dissolve 1 drachm lactate of iron in 6 fluid ounces boiling water, and add 12 drachms sugar. Dose, 2 to 4 tea-spoonfuls.

4662. Syrup of Bark and Chloride of Iron. Take 1 pint of the saccharine tincture of red bark, add to this 160 minims each syrup of chloride of iron and hydrochloric acid. This contains 120 grains of red bark and 10 drops of syrup chloride iron to each fluid ounce. If it be desirable to mix in any other proportion, add one measure of hydrochloric acid for each measure of syrup of chloride of iron. This is a deep red, clear tincture, rather pleasantly bitter; if any doubt exists as to whether it has blackened, add dilute alcohol to a small quantity, until it becomes transparent enough to observe it thoroughly.

4663. Lahache's Syrup of Iodide of Potassium and Iron. Take of iodide of potassium, 308 grains; iodide of iron (in solution 1 to 3), 230 grains; orange-flower water, 462 grains; simple syrup (concentrated), 33⅓ fluid ounces; dissolve the iodide of potassium in the orange-flower water, add the other solution and incorporate the syrup. Preserve it cool and free from light.

4664. Syrup of Tannate of Iron. Citrate of iron, 2½ drachms dissolved in 1 ounce diluted acetic acid, is added to 12 ounces simple syrup, 3 ounces raspberry syrup, and 1 drachm extract of galls rubbed up with a portion of the syrup.

4665. Phillip's Syrup of Sesquichloride of Iron. Dissolve 286 grains sesquioxide of iron in 1200 grains hydrochloric acid and 2 ounces water. Filter, and add 16 ounces simple syrup. Dose, a tea-spoonful.

4666. Syrup of Lactucarium. Triturate 1 troy ounce lactucarium to powder, and heat it with 8 fluid ounces water to the boiling point; maintain the temperature for a few moments, then strain by wringing through muslin; add to the strained liquid gradually, and with constant trituration, 120 grains carbonate of magnesia; filter through paper, pouring sufficient water through the filter to make the filtrate measure 8 fluid ounces, in which dissolve 14 troy ounces sugar with heat, and strain through muslin. This makes an excellent syrup and of fine appearance.

4667. French Syrup of Balsam of Copaiba. Triturate 2½ drachms calcined magnesia with the yolk of 4 eggs; thoroughly mix with this 5½ ounces balsam copaiba, and add 10¼ ounces simple syrup. This preparation keeps well.

4668. French Syrup of Santonin. Dissolve 55½ grains santonin in a little alcohol, add it to 16 troy ounces boiling simple syrup. The strength of the syrup will be about 3 grains to the ounce.

4669. Moore's Syrup of Tar. Take of tar (strained), 1 ounce (troy); pulverized sugar (refined), 12 ounces; carbonate of magnesia, 3 ounces, rubbed to powder on a sieve; alcohol, 2 fluid ounces. Mix the alcohol with 6 fluid ounces of water, rub the tar in a mortar of sufficient capacity with 1 ounce of the sugar, and then with the carbonate of magnesia, gradually added, until the whole is reduced to a uniform, pulverulent mixture. To this gradually add, with constant trituration, which should be continued for 15 or 20 minutes, 4 fluid ounces of the mixture of alcohol and water; then strain with strong expression. Return the residue to the mortar, and again triturate, first with 1 ounce of the sugar and then with the remaining 4 fluid ounces of the mixture of alcohol and water, gradually added, as before; finally strain and strongly express, and then reduce the dregs by trituration to a smooth and uniform condition, and pack firmly in a glass funnel prepared for percolation, and adjusted to the neck of a graduated bottle containing the remainder of the sugar, and pour upon this the expressed liquid; and when it has all disappeared from the surface, continue the percolation with water until the whole measures 1 pint. Agitate occasionally, until the sugar is dissolved, and strain if necessary. Dose from a dessert to a table-spoonful. The *strained tar*, such as is usually sold in gallon cans, answers well for this purpose, but when it is not at hand the crude tar may be dissolved in a small quantity of ether, and strained, and the ether allowed to evaporate spontaneously.

4670. Syrup of Capsicum. Take of cayenne pepper in fine powder, 2 drachms; carbonate of magnesia, 1 drachm; sugar, in coarse powder, 14 ounces, troy. Rub the cayenne pepper first with the carbonate of magnesia and sugar, and then with 1 fluid ounce of alcohol, and slowly pour in water until 6 fluid ounces have been added. The whole is then to be transferred to a proper filter; and when the liquor has ceased to pass, pour on water until 9 fluid ounces of filtered liquor are obtained. To this add the remainder of the sugar, and by a gentle heat form a pint of syrup. Made in this manner syrup of capiscum is a pungent yellowish-brown syrup, each tea-spoonful of which contains nearly 2 grains of cayenne pepper

4671. Syrup of Valerianate of Ammonia. Take of valerianic acid, 2 fluid drachms; dilute alcohol, ½ fluid ounce. Saturate the valerianic acid with carbonate of ammonia, having previously mixed it with the diluted alcohol, then add the syrup sufficient to make ½ pint. Dose, a fluid drachm containing 2 grains of the valerianate.

4672. Syrup of Stillingia (Queen's Root). Take of queen's root, 3 pounds; prickly-ash berries, 1½ pounds; refined sugar, 18 pounds. Grind and mix the articles together; place the whole 4½ pounds in a convenient vessel, cover them with alcohol of 76 per cent., and macerate for three days. Then transfer the whole to a displacement apparatus, and gradually add alcohol until 5 pints of the alcoholic tincture have been obtained, which retain and set aside. Then continue the percolation with water until the liquor passes almost tasteless, add the sugar to it, and evaporate by gentle heat until 13 pints are obtained, to which add the reserved 5 pints of alcoholic tincture, and make 18 pints of syrup. It may be flavored with a sufficient quantity of the essence of sassafras if required. (*Am. Dis*).

4673. Compound Syrup of Stillingia (Queen's Root). Take queen's root and root of turkey corn, of each 2 pounds; blue flag-root, elder flowers, and pipsissewa leaves, of each 1 pound; coriander seed and prickly-ash berries of each ½ pound. Grind and mix the articles together; place the whole 8 pounds in a convenient vessel, cover them with alcohol of 76 per cent., and macerate for three days. Then convey the whole to a displacement apparatus, and gradually add alcohol until 4 pints of the alcoholic tincture have been obtained, which retain and set aside. Then continue the percolation with water, and of this second solution reserve so much as contains a sensible amount of spirit, and distill or evaporate the alcohol from it. Continue the displacement by water until the solution obtained is almost tasteless, and boil down this weaker infusion until, when added to the second solution after the evaporation of its alcohol, it will make 24 pints. To these two solutions combined, add 24 pounds of refined sugar and dissolve it by heat, carefully removing any scum which arises as it comes to the point of boiling; and if it exceeds 28 pints, evaporate to that point with constant stirring. Then remove from the fire, and, when nearly cold, add the 4 pints of reserved alcoholic tincture, and make 4 gallons of syrup, each pint of which will be equal to 4 ounces of the ingredients in medicinal virtue. (*Am. Dis.*)

4674. German Syrup of Rhubarb. Take of alkaline fluid extract of rhubarb, 3 fluid ounces (*see No.* 4591); oil of cinnamon, 3 minims; sugar, 36 troy ounces. Mix the oil of cinnamon with the fluid extract, then add sufficient water to make the whole mixture weigh 20 troy ounces; in this dissolve the sugar with the aid of heat, and strain. The above formula for syrup of rhubarb, of the Prussian pharmacopœia, is in officinal proportions, and yields a strictly official result.

4675. Alkaline Syrup Rhubarb. Take of alkaline fluid extract of rhubarb, 6 fluid ounces (*see No.* 4591); oil of cinnamon, 3 minims; sugar, 36 troy ounces. Mix the oil of cinnamon with the fluid extract; then add sufficient water to make the whole mixture weigh 20 troy ounces; in this dissolve the sugar, with the aid of heat, and strain.

4676. Syrup of Guaiac. Decidedly the most agreeable manner of administering guaiac in liquid form, so far as tried, is that of a syrup prepared as follows: Take of guaiac, 1 ounce; solution of potassa, ½ fluid ounce; sugar, 14 ounces, troy. Macerate the guaiac in the solution of potassa mixed with 2 fluid ounces of water for 2 or 3 days; then percolate with water till 8 fluid ounces of liquid are obtained, in which dissolve the sugar.

4677. Procter's Syrup of Tolu. Balsam of tolu and carbonate of magnesia, of each, ½ ounce; alcohol, 1 fluid ounce; refined sugar, 2½ pounds. Triturate the balsam of tolu and carbonate of magnesia together with 1 ounce of the sugar, gradually adding the alcohol, and then water enough to make the whole measure 12 fluid ounces. Filter, add water enough to make 1 pint of filtrate, to which add the rest of the sugar, and dissolve by a very gentle heat. If required, strain the syrup, while hot, through a damp cotton-flannel bag. This forms a beautiful, clear syrup, free from turbidness, possessing a decided taste of the balsam, with most of its medicinal virtues.

4678. Syrup of Chamomile. Take of fluid extract of chamomile, 4 ounces; syrup, 12 ounces. Mix with the syrup moderately warm, and strain through flannel. The preparation is as clear as that made from the flowers, with the convenience of being made at will. The dose is one-fourth that of the fluid extract, or from 2 to 4 drachms.

4679. Syrup of Hydrate of Chloral. Mix together 2 scruples hydrate of chloral, 1 drachm water, and 7 drachms simple syrup.

4680. Syrup of Citric Acid. Dissolve 60 grains citric acid in fine powder in sufficient warm or hot water, and add the solution to 16 fluid ounces syrup containing 30 minims spirits of lemon, shaking them all together until thoroughly mixed. Syrup made according to this formula has a better appearance, and retains its brilliance and flavor longer than that prepared according to the U. S. Pharmacopœia.

4681. Compound Syrup of Hemlock. Bruise well 2 ounces each of water hemlock (Phellandrium aquaticum) seeds, queen's-root (stillingia silvatica), and red Peruvian bark. Simmer them with 2 pints boiling water for 20 minutes; and, when cold, strain. Then evaporate to 1 pint, add 2 pounds white sugar, dissolve with a gentle heat, removing any scum that may arise, and strain the mixture while hot. Dose: 1 to 3 drachms 3 or 4 times daily.

4682. Cadet's Compound Syrup of Ipecacuanha. Mix 2 ounces each syrup of ipecacuanha and syrup of poppies, 1 ounce syrup of orange flowers, and 1½ oxymel of squill. 2 tea-spoonfuls constitute a dose in whooping-cough.

4683. Compound Syrup of Yellow-dock. Grind and mix together 2 pounds yellow-dock root (rumex), 1 pound bark of false bitter-sweet root, ½ pound American ivy bark, and ½ pound figwort. Cover them with 76 per cent. alcohol, and let them stand for 2 days. Then displace through a percolator with hot water 2 pints extract, which reserve. Continue the percolation with hot water, and reserve so much of this second solution as contains a sensible amount of spirit, distill the alcohol from it, and set it also aside. Continue the displacement with hot water until near exhaustion, and boil down this until, when mixed with the second solution, the two combined will make 12 pints. To the mixture of these two add 16 pounds refined sugar; dissolve by heat, carefully removing the scum, evaporate to 14 pints. When nearly cold add the 2 pints first reserved alcoholic tincture, making in all 2 gallons syrup. Each pint will contain the virtue of 4 ounces of the ingredients. (*Am. Dis.*)

4684. Corvisart's Syrup of Pepsine. Heat 15 parts by weight of syrup of cherries to 70° or 75° Fahr.; mix with 1 part starchy pepsine, and, after 30 minutes, filter.

4685. Goddard's Aromatic Blackberry Syrup. Make a syrup of the following ingredients: 2 pints blackberry juice, 1 pound sugar, 1 pint brandy, 6 nutmegs grated, ½ ounce bruised cinnamon, 2 drachms cloves, and 2 drachms allspice. The astringent properties of blackberry juice adapt it, particulary in combination with carminatives, to the treatment of bowel complaints.

4686. Compound Syrup of Assafœtida. The disagreeable smell and taste of assafœtida prevents to a great extent the general use of this valuable drug. Mr. Rambo, in the Journal of Pharmacy, proposes the following recipe, which unites the properties of assafœtida with those of wild cherry, and is free from above objections. Take 1 ounce assafœtida and 2 ounces carbonate of magnesia; rub these together, gradually adding 1 pint infusion of wild cherry bark, and filter. Transfer the filtrate to a bottle, and dissolve in it by agitation 24 ounces white sugar. This preparation resembles the syrup of wild cherry in appearance.

4687. Syrup of Milk. Evaporate, with constant stirring, 6 pounds of skimmed milk to 3 pounds; add 4½ pounds of sugar; dissolve with a gentle heat, and strain. It may be flavored with the addition of 1 ounce of cherry-laurel water. Milk may be preserved by first heating it, and, when cold, charging it with carbonic acid gas.

4688. Grimault's Syrup of Horseradish. Hager gives the following directions: 50 parts each of fresh scurvy-grass, buckbean, and watercress, 60 parts of horseradish, 40 of fresh orange berries, are infused with 3 parts of cinnamon in 50 parts white wine, and, after a day, expressed; 250 parts sugar are dissolved in the filtrate.

4689. Grimault's Iodinized Syrup of Horseradish. This contains 10 parts iodine, and 5 parts iodine of potassium, in 8000 parts of the above syrup of horseradish.

Oxymel.

An acidulous syrup made of honey and vinegar. The ingredients in an oxymel should preferably be of such character, and in such proportions, as to produce a mixture of the proper consistence without further evaporation.

4691. Simple Oxymel. Liquefy by heat 40 ounces (avoirdupois) clarified honey, and mix it with 5 imperial fluid ounces each acetic acid and distilled water. (*Br. Ph.*)

4692. Oxymel of Squills. Mix together 1 imperial pint vinegar of squills and 2 pounds (avoirdupois) clarified honey. Evaporate in a water-bath until it attains, when cold, a specific gravity of 1.32. (*Br. Ph.*)

4693. Clarified Honey. Melt a convenient quantity of honey by means of a water-bath, and then remove the scum. (*U. S. Ph.*)

4694. Honey of Roses. Moisten 2 troy ounces red rose, in moderately fine powder, with ½ fluid ounce diluted alcohol; pack it firmly in a conical glass percolator, and displace 6 fluid drachms with diluted alcohol. Reserve this, and percolate ½ pint more; evaporate this last by a water-bath to 10 fluid drachms, add the reserved liquid, and mix with 25 troy ounces clarified honey. (*U. S. Ph.*) Added to water, it makes an elegant astringent wash and gargle for foul and tender gums, sore mouth, sore throat, relaxed uvula, &c.

4695. Honey of Borax. Mix together 60 grains borate of soda in fine powder and 1 troy ounce clarified honey. (*U. S. Ph.*) A common application in sore gums, mouth, and lips, in thrush, salivation, &c.; also for sore nipples, excoriations, &c., a little being applied on the tip of the finger. Diluted with water it forms an excellent skin and mouth wash or lotion.

4696. Honey of Violets. Take of expressed juice of violets (clear), 1 fluid ounce; clarified honey, 2 ounces; mix without heat in a glass vessel. Used chiefly as a mouth wash, to perfume the breath, as honey of roses.

Elixirs.

A tincture with more than one base; or a compound of various medicinal substances held in solution by alcohol in some form. Under elixirs are included medicated wines, mixtures, &c.

4698. Elixir of Calisaya Bark: Reduce to a moderate powder, 8 ounces Calisaya bark; 4 ounces each orange peel, cinnamon, and coriander seed; ½ ounce each anise seed, caraway seed, and cardamoms. Percolate the above ingredients with 4 pints alcohol diluted with 12 pints water, and add 2 pints simple syrup.

4699. Ferro-phosphorated Elixir of Calisaya Bark. The percolate obtained in the last receipt, *without the syrup*, should be digested with fresh hydrated oxide of iron; this is obtained from the solution of tincture of chloride of iron (prepared according to the formula of the U. S. Pharmacopœia, *before the alcohol is added*), 8 ounces of which solution, precipitated by sufficient ammonia, furnish the requisite quantity of hydrated oxide of iron. After standing for 12 to 24 hours, with frequent shaking, test a small quantity with a few drops of tincture of iron; if it blackens with this test, more hydrated oxide must be added, until all the cincho-tannic acid is removed, which would otherwise blacken the iron salt hereafter to be added. When the oxide of iron test ceases to blacken, filter the mixture. After which add 2 pints simple syrup, and 2 ounces pyrophosphate of iron dissolved in the least possible quantity of water. Lastly, after standing for 12 hours, filter the whole. This produces a beautifully clear and pale colored ferro-phosphorate of Calisaya bark of an agreeable taste, and free from all blackness.

4700. Ferro-phosphorated Elixir of Calisaya Bark and Bismuth. This preparation is made according to the last formula, with the addition of 2 ounces citrate of bismuth, dissolved in a sufficiency of equal parts of water and liquor of ammonia at a gentle heat. The bismuth solution is added to the elixir at the same time as the pyrophosphate of iron, and the mixture filtered.

4701. Elixir of Calisaya Bark and Bismuth. This may be prepared in the same manner as the ferro-phosphorated elixir (*see No.* 4669); substituting, in the place of the pyrophosphate of iron, 2 ounces citrate of bismuth, dissolved as directed in No. 4700.

4702. Elixir of Peruvian Bark and Protoxide of Iron. Take 4 ounces Calisaya bark, 1 ounce cinnamon, 1 drachm caraway seed, and 6 ounces orange peel. Reduce them to coarse powder and percolate with 1½ pints each of alcohol and water. Next dissolve 4 ounces carbonate of iron in 4 ounces muriatic acid and 2 ounces nitric acid; dilute the solution with 8 ounces water, and filter; precipitate with sufficient liquor of ammonia, and wash the precipitate. Digest the wet precipitate with the percolated tincture for 24 hours, with occasional shaking. This must then be tested with a few drops of tincture of iron, for any cincho-tannic acid that may be left. (*See No.* 4699.) When all the acid has been removed, filter, and add 2½ pints simple syrup, and caramel to color; lastly, for every fluid ounce add 3 grains pure crystallized sulphuret of iron. This is said to be an excellent imitation of Nichol's preparation of Peruvian bark.

4703. Squibb's Liquor of Iodide of Iron. Take of iodine, 2 ounces; iron-wire, 5 drachms; sugar, 12 ounces. Make this sugar into syrup by boiling it up with 8 fluid ounces distilled water, and filtering through paper into a flask marked at the point up to which it holds 20 fluid ounces. Meanwhile shake the iodine and iron with 3 fluid ounces water in a small flask until a clear green liquid results. Add to this a small portion of the syrup, and filter the whole through a new filter into the syrup, keeping but a small portion of the solution in the filter at a time. Drain, but do not wash the filter; and, finally, add to the liquid in the bottle enough distilled water to make up 20 fluid ounces. Shake it well, and keep it in small bottles, filled and well stoppered.

4704. Physic's Bitter Wine of Iron. Take of iron filings, 3 ounces; ginger, bruised, gentian, bruised, each, 1 ounce; orange-peel, bruised, ½ ounce; strong old cider, 1 pint. Macerate in a bottle loosely corked, for 2 weeks or longer, then express and filter for use. A reaction occurs between the iron filings and the acid of the cider, resulting in the formation of malate, and perhaps some acetate of protoxide of iron, with the evolution of hydrogen gas, which swells up the ingredients, and requires that the maceration should be conducted in a bottle of twice the capacity of the ingredients. This preparation has a dark, almost black color, very bitter aromatic taste, and is a good, though not an elegant chalybeate, in the dose of a tea-spoonful.

4705. Hubbell's Wine of Iron. Take citrate (of magnetic oxide) of iron, 128 grains; precipitated extract of Calisaya bark, 256 grains. (*See next receipt.*) White wine (sherry), 1 pint; curaçoa (the best), 5¼ fluid ounces. Dissolve the precipitated extract of bark in the wine by aid of a sufficient quantity of citric acid, then add the citrate of iron, filter the solution, and add to it the curaçoa, and mix. The peculiarities of this preparation are, that it consists of iron and cinchona, and yet is free from any inky taste or appearance, is perfectly transparent, of a light brown color not very different from that of sherry wine, and a bitter, not disagreeable taste. The label claims for it the presence of citrate of the magnetic oxide of iron, as the ferruginous ingredient. The dose of this preparation is a tea-spoonful.

4706. Hubbell's Precipitated Extract of Calisaya Bark. The precipitated extract of bark employed by Mr. Hubbell is not the commercial extract, nor yet that of Wetherill, nor of Ellis, but is made by himself, by a process based on that of Mr. Herring, of London, for the manufacture of quinine. Any quantity of Calisaya bark is treated with a solution of caustic soda (2 parts to 100 of water), until it has removed the coloring matter, kinic and tannic acids, and extractive matters. The residue is washed with water, dried, and extracted with alcohol till exhausted, and the alcohol distilled off so as to obtain an extract. The extract consists almost wholly of quinia and cinchonia, and is free from tannin, and, though not soluble in wine alone, becomes so by aid of citric acid.

4707. Shinn's Bitter Wine of Iron. Take of sulphate of cinchona, 6 drachms; sulphate of quinia, 2 drachms; citrate of iron, 4 ounces; citric acid, 1 ounce; sherry wine, 4 pints; alcohol, 1 pint; orange syrup, 1 pint. Dissolve the sulphates and citric acid in 1½ pints of hot water, and the citrate of iron in ½ pint of the same; mix the solutions, and add the other ingredients.

4708. Aromatic Wine of Iron. Digest 1 ounce iron filings for 2 or 3 days in 3 fluid ounces lemon juice; add ½ ounce each bruised gentian and cinnamon, and 16 ounces Rhenish (or sherry) wine. After 24 hours decant and filter. Gentian contains no tannin, and will not blacken the iron in the solution.

4709. To Prevent Sediment in Preparations of Peruvian Bark. The formation of a sediment in this and other simple preparations of Peruvian bark may be avoided by displacing or digesting its powder first with a solution of soda which will extract the tannin, kinovin, &c.; after washing off the last traces of the alkali by means of water, the alcoholic or vinous tincture may then be prepared as usual, and will remain clear, because free from the principles extracted by the alkaline solution. The alkaloids of the bark do not dissolve in weak mineral alkalies.

4710. Cottereau's Wine of Cinchonia is made as follows: Dissolve 24 grains sulphate of cinchonia in 2 pints Madeira wine, and filter. Dose, 1 to 4 ounces.

4711. Wine of Calisaya Bark. Digest 1 part powdered Peruvian bark in 12 parts white wine for 24 hours, and filter. A similar preparation may be made of 20 parts of red wine and 1 part extract of Peruvian bark.

4712. Aromatic Mixture of Iron. Take Peruvian bark in powder, 1 ounce; columba root in coarse powder, 3 drachms; bruised cloves, 2 drachms; filings of iron, separated by a magnet, ½ ounce; digest for 3 days with occasional agitation in a covered vessel, with as much peppermint water as will give 12 ounces of a filtered product, and then add compound tincture of cardamoms, 3 fluid ounces, and tincture of orange peel, 2 fluid drachms. This mixture should be kept in a well-stoppered bottle. Properties, tonic, and valuable in various states of debility; dose from ½ to 2 fluid ounces.

4713. Procter's Rennet Wine. Take of fresh rennets (about 3), 24 troy ounces; chloride of sodium, 3 ounces; alcohol, 6 fluid ounces; white wine, 16 fluid ounces. Wash the rennets in water until perfectly clean, cut them up, and macerate them for 14 days with frequent agitation in the wine, then add the alcohol, and filter for use. Dose, 1 tea-spoonful immediately after eating.

4714. Wine of Wild Cherry Bark. Professor Parrish gives the following formula in his "Elements of Pharmacy." Alcoholic extract (from 24 ounces) of wild cherry bark, 5¼ ounces; sweet almonds, 3 ounces; water, 1 pint; and cherry wine, 2 pints. Beat the almonds with the water to a paste, rub down the extract with ½ pint of the wine, and mix the two liquids in a bottle of the capacity of 3 pints, stop it closely, and permit it to stand for 3 days, with occasional agitation; then add the remainder of the wine, allow it to stand a week, and filter. By this mode of proceeding, opportunity is afforded for the development of the hydrocyanic acid before the menstruum is made so alcoholic as to retard the reaction which favors its formation. Thus made, wine of wild cherry bark is a transparent, wine-red liquid, having an astringent bitter-almond taste and odor. The dose of this preparation as a tonic and sedative is a tea-spoonful.

4715. Ferrated Wine of Wild Cherry. Exhaust 12 ounces bruised wild cherry bark of its tonic principles with alcohol, and carefully evaporate the alcoholic tincture so as to expel the alcohol; add 6 ounces water and ½ ounce hydrated sesquioxide of iron. Macerate this with occasional agitation for 6 hours, and filter into a bottle containing an emulsion of 2 ounces sweet almonds in 6 ounces water. When reaction has ceased, filter again, and add 12 ounces white sugar, and for every ounce thus prepared, add 24 grains citrate of iron, previously dissolved in water sufficient to make the whole fluid extract measure 24 fluid ounces. The addition of iron to the bitter principle and hydrocyanic acid of the simple extract of wild cherry should render it much more efficient as a tonic, and greatly add to the value of the preparation.

4716. Ferrated Elixir of Wild Cherry. Take of fluid extract of wild cherry bark, 4 fluid ounces; curaçoa cordial, 11 fluid ounces; pyrophosphate of iron, 256 grains; boiling water, 1 fluid ounce. Mix the fluid extract with the curaçoa cordial. Dissolve the pyrophosphate of iron in the boiling water, and mix all together. Dose, a tea-spoonful 3 times daily.

4717. Elixir de Garus. Digest 2 parts by weight each of aloes and myrrh, and 1 part Spanish saffron, in 24 parts of 60 per cent. alcohol, and 2 of diluted sulphuric acid. Filter.

Or: Digest for some hours 3 parts by weight each of aloes and myrrh, and 2 parts each of nutmegs and cloves, in 576 parts rectified spirit diluted with an equal weight of water. Then add 864 parts orange-flower syrup, 192 parts orange-flower water and 2 each of cochineal and Spanish saffron. Filter. Dose of either of the above preparations, 1 tea-spoonful 3 or 4 times a day. (*Prussian Ph.*)

4718. Elixir of Pepsine. Dissolve 1 part by weight starchy pepsine in 8 parts water; filter the solution, and add 3 parts elixir of garus and 4 parts syrup of cherries. Dose, 1, 2 or 3 table-spoonfuls twice during the meals.

4719. Corvisart's Elixir of Pepsine. Saturate 1 part by weight starchy pepsine with 15 parts elixir of garus. Macerate for half an hour in a covered vessel, and filter through wetted paper. Dose, 1 table-spoonful before or during meals.

4720. Mialhe's Elixir of Pepsine. Macerate 1 part by weight of starchy pepsine, and 5 parts sugar, in 2 parts proof spirit, 9 parts white wine, and 4 parts water, until the sugar is dissolved; then filter. Dose, 1 table-spoonful before or during meals. This has an agreeable taste.

4721. French Pepsine Wine. This is

prepared by macerating starchy pepsine in 20 times its weight of white wine.

4722. Wine of Beef and Iron. Dissolve 1 ounce Liebig's extract of meat in 4 ounces water and ½ drachm bruised allspice; after standing 10 hours add 16 ounces sherry wine and 2 ounces syrup. Then dissolve 96 grains citrate of iron in 2 ounces water. Mix, filter, and add water to make the whole 24 fluid ounces. Each ounce contains 1 ounce fresh beef and 4 grains citrate of iron. Dose, 1 table-spoonful. This and the 6 following formulæ have been adopted by the Newark Pharmaceutical Association.

4723. Nutritive Wine. This is prepared in the same manner as the last receipt, omitting the citrate of iron. (*Newark P. A.*)

4724. Elixir of Pepsine, Bismuth, and Strychnia. Triturate 256 grains Hawley's pepsine with 2 ounces glycerine in 4 ounces water; dissolve 64 grains citrate of bismuth, 2 ounces orange-flower water, and add to the pepsine; then add 2 ounces deodorized alcohol, 4 ounces orange-flower water, 2 ounces syrup, and lastly 1 grain strychnia dissolved in a few drops acetic. Each fluid ounce contains: pepsine, 16 grains; citrate of bismuth, 4 grains; strychnia, $\frac{1}{16}$ grain. (*Newark P. A.*)

4725. Ferro-Phosphorated Elixir of Gentian. Take 1 drachm each coriander and mace; 1 ounce orange peel, 1 ounce gentian root. Reduce to powder and percolate with a mixture of 4 ounces deodorized alcohol, 4 ounces water, and 2 ounces orange-flower water; displace 10 ounces, dissolve in it 256 grains pyrophosphate of iron, add 6 ounces syrup, and filter. Each fluid ounce represents 16 grains pyrophosphate of iron and 30 grains gentian. (*Newark P. A.*)

4726. Wine of Pepsine. Triturate 160 grains Hawley's pepsine in 4 ounces sherry wine and 1 drachm dilute muriatic acid; pour this on a filter and pass 12 ounces more sherry wine through it. Each fluid ounce contains 10 grains pepsine. (*Newark P. A.*)

4727. Aromatic Elixir. Take 4 drachms orange peel, 2 drachms coriander seed, 2½ drachms angelica seed, and 1 drachm cochineal. Pulverize and percolate with 12 ounces deodorized alcohol and 10 ounces water. Add 5 ounces glycerine and 6 ounces syrup, to make 2 pints. This is a pleasant vehicle for administering nauseous remedies. (*Newark P. A.*)

4728. Elixir of Valerianate of Ammonia. Dissolve 96 grains valerianate of ammonia in 4 ounces water, and add it to a mixture composed of 6 drachms syrup of orange peel, 2 drachms tincture of prickly ash, and ½ ounce each of fluid extract of vanilla and compound tincture of cardamoms. Each drachm contains 2 grains valerianate of ammonia.

4729. Elixir of Taraxacum. Take of taraxacum root, 6 ounces (or fluid extract of taraxacum, 6 ounces); liquorice root, 1 ounce; simple syrup, 2½ pints. The dry ingredients must be reduced to a suitable degree of fineness for percolation. Moisten the powder with 6 ounces alcohol diluted with twice its bulk of water, then pack in a conical percolator and pour on of the alcohol and water mixture until 6½ pints are obtained, then add the syrup and mix them.

4730. Chloroform Elixir. Take 1½ ounces each chloroform, tincture of opium, tincture of camphor, and aromatic spirit of ammonia; 20 drops oil of cinnamon, and 2 ounces brandy. This is an excellent mixture for colic. Dose, ½ fluid drachm.

4731. Mynsicht's Elixir of Vitriol. This elixir is also known by the name of *acid aromatic tincture*. Take cinnamon, 2 ounces; lesser cardamoms, cloves, galanga root, and ginger, of each ½ ounce; sulphuric acid (specific gravity 1.845), 1 drachm; rectified spirit, (specific gravity .897 to .900), 2 pounds. Mix the acid and spirit, and pour them on the other ingredients reduced to a coarse powder; macerate for 8 days in a close vessel, with frequent agitation, then press it out and strain. It should be of a brownish-red color. (*Prussian Ph.*) Another formula directs as follows: Take sweet flag root, and galanga root, of each 1 ounce; ginger, cinnamon, cloves, and nutmeg, of each 3 drachms; lemon peel, 4 drachms; white sugar, 3 ounces; proof spirit, 2 pounds; dilute sulphuric acid, 3 ounces. Macerate for 6 days, then press and filter, so as to make 27 ounces. (*Austrian Ph.*)

4732. Elixir of Valerianate of Ammonia. Extract of valerian, 2 scruples; fluid extract of valerian, 2 fluid drachms; water, 7 fluid ounces. Dissolve the extract in the fluid extract and water, filter, and add valerianate of ammonia, 2 drachms; orange-flower water and simple syrup, of each ½ fluid ounce. Dose, a tea-spoonful.

4733. Goddard's Elixir of Valerianate of Ammonia. Valerianic acid (from the root), 6 fluid drachms; carbonic acid water, 8 fluid ounces; red Curaçoa cordial, 20 fluid ounces; orange-flower water, 8 fluid ounces; mucilage of gum-arabic, 2 fluid ounces. Saturate the valerianic acid with sufficient carbonate of ammonia diluted with the carbonic acid water, then add it to the flavoring ingredients and mucilage, and filter. Dose, a tea-spoonful.

4734. Moore's Elixir of Valerianate of Ammonia. Take of valerianic acid, 1 fluid ounce; distilled water, 24 fluid ounces; inodorous alcohol, 12 fluid ounces; simple syrup, 12 fluid ounces; peach water, 8 fluid ounces; saturated tincture of red saunders, 4 fluid drachms; saturated tincture of recent orange peel, 1 fluid ounce; oil of bitter almonds, 5 minims; and oil of sweet orange, 20 minims. Mix the valerianic acid and the distilled water, and a sufficient quantity of carbonate of ammonia to saturate the acid; then add the other ingredients, with a sufficient quantity of caramel to impart a brownish shade to the mixture, and filter through paper.

4735. McMunn's Elixir of Opium. The following receipt is said to have been found among the effects of the late Dr. Chilton: Take 5 pounds of Turkey opium, cut in small pieces and dried, and put it into a large strong glass jar with a wide mouth, and pour on it sulphuric ether enough to a little more than cover it; then stop the jar tight with a glass stopper, to prevent its evaporation; set it away in a cool place, and stir it daily with a stick, so that all the lumps may be broken. At the end of a week drain off the ether, and again pour on as much more, and repeat stirring it every day for a week longer, when it may be drained off as before. Then stop the jar tight, and lay it down on its side, so that all the ether that accumulates near its mouth may be drained off, and repeat doing so until the opium is all dry. Then expose it to the open air for a few days. The sulphuric ether extracts from the opium the narcotine which is its most deleterious principle, and also deprives it of its peculiar noxious odor, so that the elixir will not smell of it thereafter. Now to free the opium of the smell of the ether, and to extract its valuable medicinal principles, boil it in water, as follows: Pour into a tin boiler 4 gallons pure soft water, and when hot (but not boiling), put in the opium, when a great ebullition will take place, which is owing to the evaporation of the ether. Then let it boil 10 or 12 minutes, occasionally stirring it, so that the lumps of opium may be all broken and dissolved. Then set it away till the next day, when it should be strained through a cloth strainer, and if there be not 4 gallons of the solution, pour on the residue of opium boiling water enough to make that quantity when it is strained and clear. When in the state of watery solution, it is better to be kept in stone crocks that will hold about 2 or 3 gallons each, and in a cool place; after standing 5 or 6 days the clear solution should be carefully dipped off into a large tin can. The skimmings and dregs should be strained, and when clear put with the other. To this 4 gallons of watery solution, add 5½ gallons alcohol and stir the mixture thoroughly; then cover the can tight, so as to prevent evaporation. After standing a few days, the clear elixir may be carefully dipped off into another can, and the dregs at the bottom strained, and, when clear, poured into the other. After standing undisturbed for a few weeks it will be fit to use. It will be equivalent to laudanum, both in its strength and the size of its dose.

4736. Compound Elixir of Taraxacum. As prepared by Mr. Candidus for Dr. Cochran, of Mobile. Reduce the following ingredients to a moderately fine powder: 6 ounces taraxacum root, 4 ounces wild cherry bark, 1 ounce gentian root, 2 ounces orange peel, 1 ounce cinnamon, 1 ounce coriander seed, 2 drachms each anise, caraway and cardamom seeds, and 1 ounce liquorice root. Dilute sufficient alcohol with twice its bulk of water, and moisten the powdered ingredients with 6 ounces of it, pack in a conical percolator and displace 6¼ pints with the diluted alcohol. Add to this 2½ pints simple syrup. Dose, from ½ to 1 ounce. This elixir is an excellent vehicle for quinine, the taste of which it completely destroys.

4737. Squibb's Ammonio-Pyrophosphate of Iron. Take of pyrophosphate of soda, 4 parts by weight; solution of tersulphate of iron, 8 parts; citric acid, 2¾ parts; water of ammonia, 6¾ parts. Dissolve the *pyrophosphate of soda* (which is prepared by first drying and then calcining common phosphate of soda) in 60 parts water by means of heat; cool the solution to 50° Fahr. and filter it into a bottle of the capacity of 250 parts. Then add the solution of tersulphate of iron (*see No.* 4816), shake the mixture well, fill the bottle up with water, again agitate it, and set it aside for 24 hours to settle. Decant the clear liquid from the precipitate by means of a syphon, and repeat the washing and decantation twice. Then pour the precipitate upon a strainer, drain it for 24 hours and transfer to a tarred porcelain basin. Upon the citric acid, contained in a suitable vessel, pour the solution of ammonia, a little at a time, with constant stirring, till the crystals are dissolved and the acid accurately saturated. Then add this solution to the precipitate in the basin, and apply heat. Stir the mixture constantly till perfectly dissolved, and evaporate the solution to 24 parts; then filter through paper. Finally pour the solution upon plates, dry the salt by a moderate heat, and keep it in well-closed bottles. The yield is a little more than 7½ parts. The salt is deliquescent, in the form of pale yellowish green scales.

4738. Ammonio-Ferric Alum. This elegant styptic remedy has recently been much prescribed, especially in leucorrhœa; it is made as follows: Take of crystallized protosulphate of iron, 8 ounces; sulphuric acid, 7 fluid drachms; nitric acid, 1½ fluid ounces; sulphate of ammonia, 18 drachms. Boil the sulphate of iron in 2 pints water and add to it the sulphuric acid; when dissolved, add the nitric acid gradually, boiling for a minute or two after each addition, until the nitric acid ceases to produce a black color; boil violently, to separate deutoxide of nitrogen, and reduce the liquid to about one half, then add the sulphate of ammonia and a little sulphuric acid and set it aside to crystallize. Wash the crystals thoroughly in a little cold water to which a small portion of sulphuric acid has been added. This salt is in elegant violet-tinted crystals. Its peculiar merit consists in its marked astringency without the stimulating properties of some of this class of salts. It is easily assimilated when taken internally. Dose, 3 to 6 grains, and while it controls excessive discharges, is often useful in correcting their cause. Though called an alum, this salt contains no alumina; it is similar to the double sulphate of potassa and iron, which is called iron alum, but is more soluble.

4739. Concentrated Infusion of Roses. Rose petals or leaves, 3 pounds; boiling water, 2 gallons; infuse 2 hours, with constant agitation, then press out the liquor in a very clean tincture press, strain through flannel or a hair sieve, add diluted sulphuric acid, 24 fluid ounces, agitate well, and filter through paper supported on coarse muslin; lastly, add 6 pounds finest white sugar broken up into small lumps, but perfectly free from dust and dirt. When dissolved, put the infusion into clean, stoppered green glass bottles, and keep it from the light in a cool place. Product very superior.

Or: Take rose leaves, acid, and cold water, as last, mix, and infuse for 48 hours in a clean, covered, earthenware vessel, then press out the liquid with the hands, filter, and add the sugar as before. Product very fine, and, keeps well. In employing the first formula, care should be taken that the utensils be perfectly clean, especially the press, and earthenware glazed with lead should be avoided. The pressing should also be conducted as rapidly as possible, to avoid the color being injured by the iron, though clean iron does not readily injure infusion of roses before the addition of the acid. Should not the infusion filter quite clear through paper, the addition of the whites of 2 or 3 eggs, diluted with 2 or 3 ounces of water, followed by violent agitation of the liquid for a few minutes, and repose for 1 or 2 hours, will usually render it fine, when it may either be decanted or filtered should it require it. It will now pass rapidly through ordinary filtering paper, and at once run clear.

4740. Elixir of Vitriol. Called also *aromatic sulphuric acid.* In order that elixir of vitriol may be miscible with water without precipitation, aromatics of an oleo-resinous nature cannot be used. Add gradually 3 troy ounces sulphuric acid to ½ pint alcohol, and pour 1 fluid ounce boiling water on 2 drachms red rose leaves; when both liquids have become cool, add 1 fluid ounce fluid extract of orange-peel, and add alcohol enough to make the whole up to 18 fluid ounces. Mix and filter. Elixir of vitriol thus prepared has a pleasant aromatic odor and flavor, and the beautiful red color of the rose leaves, heightened by the presence of the acid. It is miscible with water without turbidity, and a specimen, after long keeping, has deposited but a trace of sediment.

4741. Alcoholized Sulphuric Acid. To 3 parts rectified spirits, add, very gradually, 1 part sulphuric acid. It is usually colored by letting it stand over a little cochineal. Its properties are internally refrigerant, externally caustic. As a refrigerant, it is administered in doses of ½ fluid drachm to 1 pint water.

4742. Cantharidal Collodion. Take 8 troy ounces finely powdered cantharides, press it firmly in a cylindrical percolator, and pour on it 1½ pints stronger ether. When 15 fluid ounces have passed, set the liquid aside in a close vessel, and continue percolation with stronger alcohol until ½ pint more liquid is obtained. Set this last aside to evaporate spontaneously until reduced to 1 fluid ounce; then mix it with the reserved liquid. Next add 100 grains dry collodion cotton (*see next receipt*), and agitate until dissolved. (*U. S. Ph.*)

4743. To Prepare Gun Cotton for Collodion. To 10 troy ounces nitrate of potassa, add 15½ troy ounces sulphuric acid, and stir until uniformly mixed. When cooled below 122° Fahr., add ½ troy ounce cotton, freed from impurities, stirring with a glass rod; cover the vessel closely, and, after standing 24 hours, transfer the cotton to a larger vessel, and wash it, first with cold water until the washings cease to have an acid taste, and then wash with boiling water. Press it as dry as possible with the hand, pack it tightly in a conical percolator, and pour on it stronger alcohol until the remaining water is displaced. Lastly, press it as dry as possible with the hand. The cotton thus prepared, and dried at a temperature of 212°, weighs 336 grains.

4744. To Prepare Collodion. Mix 21 fluid ounces stronger ether with 6 fluid ounces stronger alcohol in a suitable bottle, add the quantity of moist prepared cotton (as prepared in the preceding receipt), and shake occasionally until dissolved.

4745. Morphia Collodion. Collodion, 30 parts; muriate of morphia, 1 part. Applied to the affected parts in obstinate neuralgia.

4746. To Administer Hydrate of Chloral. Physicians should prescribe only the crystals, and should be very certain that they are pure. The taste of hydrate of chloral is quite unpleasant, but orange-juice completely covers it, and so does peppermint water or essence of peppermint. If taken in aqueous solution, let the patient be directed to suck the juice of an orange immediately after swallowing the dose, or mix with the solution a little peppermint water, with syrup of tolu. The following is a good formula: Take chloral hydrate, 1 drachm; peppermint water, ½ ounce; syrup tolu, ½ ounce; water, 2 ounces. Dose, from ½ ounce to 2 ounces, as may be required. The mixture should not be prepared in large quantities, nor be kept for any length of time.

4747. Improved Formula for Chalk Mixture. To obviate unpleasant and dangerous souring of chalk mixture as commonly prepared, glycerine may be substituted for the sugar, according to the following formula: Take of prepared chalk and glycerine, of each ½ ounce; pure gum acacia, 2 drachms; cinnamon water and pure water, of each 4 ounces. Rub well together until thoroughly mixed. This mixture will keep during a whole summer. The glycerine exerts a positively soothing effect upon the bowels, as well as in some degree arresting fermentation.

4748. Phosphorated Ether. Dissolve 2 grains phosphorus in ½ drachm oil of peppermint; when dissolved add sulphuric ether, ½ fluid ounce; mix well. Dose, 2 to 6 drops. This was recommended by Augustin in epilepsy, paralysis, and other like nervous affections.

4749. Compound Spirit of Ether. This preparation is known by the name of *Hoffmann's Anodyne,* and consists of ½ pint ether, 1 pint alcohol, and 6 fluid drachms ethereal oil.

4750. Moore's Extract of Black Cohosh. Moisten black cohosh root (black snake-root, or cimicifuga racemosa) in No. 50 powder, with 95 per cent. alcohol, and pack closely in a displacer; add gradually more of the alcohol until the resinous portion is exhausted; evaporate the alcoholic portion to dryness, powder the product and pass it through a fine sieve. Proceed to displace with diluted alcohol (1 part alcohol to 2 of water) until the root is perfectly exhausted, evaporate the resulting product over a water-bath until it is of nearly the required consistence of a good extract, then mix the powdered resinous portion, while the fluid is still warm, and stir constantly until cold. In this way the resin is intimately and smoothly mixed with the extractive portion; is much more readily rubbed down with aqueous solutions, and contains all the active ingredients of the root; but, however carefully prepared, change of temperature is liable to separate the resin more or less from the extract.

4751. Procter's Alcoholic Extract of Arnica. Take arnica flowers, 12 ounces, troy; alcohol, 3 pints; water, 1 pint. Mix the alcohol and water, and pour 2 pints of the mixture over the arnica, previously finely bruised. Allow it to stand for 48 hours, pack it firmly in a percolator, and pour on the remainder of the mixture until 3 pints are obtained. Evaporate this tincture in a water-bath (or still) till reduced to a soft resinous extract.

4752. Medicated Oils. These are prepared by *infusion* or *decoction.* The bruised ingredients are either simply digested in 2 to 4 times their weight of olive oil for some days, or they are gently boiled in it until they become dry and crisp, care being taken that the heat towards the end of the process is not greater than that of boiling water. As soon as either process is complete, the oil is allowed to drain from the ingredients, which may be, if necessary, submitted to the action of a press. The product is usually strained through flannel or a hair sieve while still warm, and, after standing a week or 10 days to settle, the clear portion is decanted from the dregs. Green plants are usually employed for this purpose, but in many cases the dried plants, reduced to powder, and digested for 6 or 8 hours in the oil at the heat of hot water, with frequent agitation, yield a much more valuable product. These oils are nearly all employed as external applications only.

The oil is obtained from the following, in the green state: Balsam apple, the seeds first taken out; belladonna leaves; elder flowers; fox glove leaves; garden nightshade leaves; fox glove leaves; garlic; hemlock leaves; henbane leaves; juniper berries, crushed; white lilies; poison oak leaves; roses, the petals of the flowers; fresh rue; St. John's wort flowers; fresh tobacco leaves.

Others are used dry, and reduced to powder, such as: Cantharides (Spanish flies); capsicums; dried chamomile flowers; fenugreek seeds; marsh-mallow root; mudar bark; opium; pellitory root; black pepper, &c.

Medicated Waters.

These are aqueous solutions of different substances for medicinal and other purposes. The methods of preparing them generally require special arrangements to dissolve the oils, &c., otherwise insoluble in water. (*See*

No. 1070, *&c.*)

4754. Camphor Water. Pulverize 120 grains camphor in a mortar with 40 minims alcohol; triturate it first with ½ troy ounce carbonate of magnesia, then with 2 pints distilled water, added gradually. Filter through paper. (*U. S. Ph.*)

4755. Bitter Almond Water. Rub 16 minims oil of bitter almonds with 1 drachm carbonate of magnesia, adding 2 pints water gradually. Filter through paper. (*U. S. Ph.*)

4756. Cinnamon Water. Treat ½ fluid drachm oil of cinnamon in the same manner as in the last receipt. Or, by distilling 18 troy ounces coarsely powdered cinnamon in 16 pints water, preserving only the first 8 pints of the distillate. (*U. S. Ph.*)

4757. Fennel Water. Treat ½ fluid drachm oil of fennel in the same way as last receipt. Or, by distillation from fennel in coarse powder. (*U. S. Ph.*)

4758. Peppermint Water. Same as last, using ½ fluid drachm oil of peppermint, or 18 troy ounces peppermint. (*U. S. Ph.*)

4759. Spearmint Water. Same as last, from oil of spearmint.

4760. Lime Water. Take of lime, 2 ounces; distilled water, 2 quarts. Slack the lime with a little of the water; pour on the remainder of the water and stir them together; then immediately cover the vessel and let it rest for 4 hours. Keep the solution, with the undissolved lime, in glass-stoppered bottles, and when wanted for use, pour off the clear liquor. It is an anti-acid tonic, kills worms, and frees the bowels from slimy and morbific matter. It promotes digestion; it is valuable in looseness, scrofula, diabetes, and whites. Mixed with a decoction of Peruvian bark, it wonderfully strengthens the debilitated, and those threatened with atrophy.

4761. Lobelia Water. Lobelia leaves and capsules, or powder, 1 ounce; boiling water, ½ pint; brandy, ½ pint. Infuse a week. Good for sore and inflamed eyes, erysipelas, ringworms, &c.

4762. Fever Drink. The juice of a lemon; cream of tartar, 1 tea-spoonful; water, 1 pint. Sweeten with loaf sugar. When the patient is thirsty, let him drink freely.

4763. Saline Mixture. Take fresh lemon juice, 1½ ounces; carbonate of potassa, 1 drachm; white sugar, 3 drachms; pure water, 12 ounces; essence of peppermint, 30 drops. Mix. A tea-cupful to be taken often in inflammatory fevers and sore throat.

4764. Tar Water. Take of tar, 2 pints; water, 1 gallon. Mix, by stirring them with a wooden rod for a quarter of an hour, and, after the tar has subsided, strain the liquor, and keep it in well-corked-phials. Tar-water should have the color of white wine, and an empyreumatic taste. It is frequently used as a remedy in chronic bronchitis. It acts as a stimulant, raising the pulse and increasing the discharge by the skin and kidneys. It may be drunk to the extent of a pint or two in the course of a day.

4765. Tar Water. M. Magnes Lahens suggests a method of preparing this water, which is more expeditious and convenient than the plan commonly followed. He mixes the tar with sand, previously washed and dried, throws the mixture into a percolator, and shakes the instrument gently to secure proper adjustment of the mixture. Water is then poured on, the first part of the filtrate is rejected, and the latter portion is kept for use. He uses ½ ounce tar and 26 ounces of sand to obtain 2 pints of the medicated water, which corresponds in strength with that of the Paris codex.

4766. Camphor Water. Take ¼ ounce of camphor and enclose it with a glass marble in a muslin bag; put this into a wide-mouthed bottle, such a one as is used for preserved fruit. Now fill up the bottle with water that has boiled a few minutes and has been allowed to become cold. The glass marble is used to keep the camphor from floating, which it otherwise would do. After about 3 days the water will become saturated with the camphor, and may be poured off as required. A wine-glassful is a dose. It is very useful as an anti-spasmodic in hysteric and nervous affections.

4767. Barley Water. Wash away with cold water all extraneous matter from 2 ounces pearl barley; then boil for a short time in ½ pint water, throw this away, and boil the parboiled barley in 4 pints water down to 2 pints, and strain.

4768. Distilled Water. Take 10 gallons of spring water; distill it, rejecting the first quart that comes over, and preserving the next 8 gallons of the remainder.

Solutions.

In pharmacy, a solution consists of water in which a certain fixed quantity of a soluble substance has been dissolved. (*See No.* 29).

4770. Solution of Acetate of Morphia. Mix 4 drachms acetate of morphia with 15 drops acetic acid, 1 pint distilled water, and ½ pint proof spirit. Dose, from 5 to 20 drops.

4771. Solution of Sulphate of Morphia. Dissolve 1 grain sulphate of morphia in 1 fluid ounce distilled water. Dose, 1 teaspoonful, used in the same cases as opium

4772. Compound Solution of Alum. Rub together 1 ounce each alum and sulphate of zinc; dissolve in 3 pints boiling water. If necessary, filter. This is detergent and astringent, and is used as a lotion for old ulcers, excoriations &c.; and, largely diluted with water, as an eye-wash and injection.

4773. Solution of Ammonio-Nitrate of Silver. Dissolve 44 grains pure crystallized nitrate of silver in 1 fluid ounce distilled water; add gradually ammonia water until the precipitate at first thrown down is very nearly, but not entirely, redissolved. This solution is used as a test for arsenious acid, in combination with which it forms a yellow precipitate, arsenite of silver.

4774. Solution of Chloride of Barium. Dissolve 1 drachm chloride of barium in 1 fluid ounce water, and filter the solution. Dose, 5 drops, gradually increased to 10 or 12 drops, 2 or 3 times a day, for scrofula, scirrhous affections, and worms. Is used externally, largely diluted, as a lotion in scrofulous ophthalmia; also as a test for sulphuric acid and the soluble sulphates, in contact with which it makes a heavy white precipitate, insoluble in either hydrochloric or nitric acid. It is said to detect the presence of $\frac{1}{10000}$ part of sulphuric acid.

4775. Solution of Diacetate of Lead —sometimes called *Extract of Lead.* Boil 27 ounces acetate of lead, and 16 ounces finely powdered litharge, in 3 quarts water for ½ an hour, constantly stirring; then add sufficient distilled water to make up 3 quarts. If required, filter, and keep in a closed vessel. This solution is almost the same in strength and preparation as the *solution of subacetate of lead* of the U. S. Pharmacopœia.

4776. Goulard's Water or Lotion. Mix 1½ fluid drachms diacetate of lead with 2 fluid drachms proof spirits and 1 pint distilled water. This lotion is sedative, refrigerant, and astringent. This is the *dilute solution of diacetate (or subacetate) of lead.*

4777. Donovan's Arsenic and Mercury Solution. Triturate 6 grains finely powdered pure arsenic, 16 grains pure mercury, and 50½ grains pure iodine, with ½ fluid drachm alcohol, until dry; then add gradually 8 fluid ounces water, triturating constantly; heat the mixture in a flask until it begins to boil, and, when cold and filtered, add sufficient water to make up to 8 fluid ounces 6 fluid drachms. Dose 10 to 30 drops, 2 or 3 times a day, soon after a meal, for scaly skin diseases.

4778. Standard Solution of Chloride of Calcium. Dissolve carefully 2 grains pure carbonate of lime in a little pure hydrochloric acid; evaporate the solution to dryness, and dissolve the residuum in 1 pint pure water. This forms the standard solution of 16° of hardness. 1 measure of this solution mixed with 15 of water constitutes a solution of 1° of hardness; 2 measures of it with 14 of water make a solution of 2° of hardness &c. This solution is the standard used in testing the hardness of water.

4779. Solution of Iodide of Potassium. Dissolve 10 grains iodide of potassium and 5 grains iodine in 1 pint water. Dose, 2 to 6 grains in the usual case where iodine is employed.

4780. Solution of Chloride of Calcium. Dissolve 4 ounces fused (or 8 ounces crystallized) chloride of calcium, in 12 ounces water, and filter. Dose from 10 drops to 2 drachms, for scrofulous tumors, &c.; also used as a test for sulphuric acid, in contact with which it throws down a white precipitate insoluble in nitric acid.

4781. Solution of Sulphate of Morphia. Dissolve 16 grains sulphate of morphia in 4 drops dilute sulphuric acid, 1 fluid ounce water, and 1 fluid drachm rectified spirit. Dose, 5 to 10 drops.

4782. Solution of Nitrate of Baryta. Dissolve 4 grains nitrate of baryta in 80 grains water. This is used in the same manner as chloride of barium (*see No.* 4774) for testing sulphuric acid, with the same results.

4783. Solution of Nitrate of Silver. Dissolve 1 drachm crystals of nitrate of silver in 1 fluid ounce distilled water. It must be protected from the action of light. This is employed as a test for soluble chlorides, any of which, slightly acidulated with nitric acid, will give a white, curdy precipitate (chloride of silver) when brought in contact with diluted nitrate of silver.

4784. Liquor of Potassa; Solution of Potash; Soft-Soap Lye. Take 1 gallon boiling distilled water; use sufficient of this to slack 8 ounces recently burnt lime in an earthen vessel; in the remainder of the water dissolve 15 ounces carbonate of potassa, and add the slacked lime. Cork the mixture closely in a vessel, and shake it frequently until cold, then allow it to settle and decant the clear liquid into clean, well-stoppered green-glass bottles. Liquor of potassa is antacid, diuretic, and resolvent. In indigestion, acid eructations, heartburn, &c., it may be taken with great benefit. It neutralizes the acid, and counteracts the morbid tendency of the stomach to acid secretion. Dose, 10 drops, gradually increased to 40. It is powerfully poisonous, and should be greatly diluted in anything not acidulous. When pure, it does not effervesce with acids, nor give a precipitate with lime-water, or with a solution of oxalate of ammonia. (*See No.* 101).

4785. Liquor of Soda; Solution of Soda; Soda Lye; Hard-Soap Lye; &c.

The proportions are, crystallized carbonate of soda, 32 ounces (troy); recent quicklime, 9 ounces (troy); boiling water, 1 gallon; the lime being slacked with a little of the water. The product is stated to have specific gravity 1.061, and to contain about 5 per cent. of pure caustic soda. The process by which the above is made is similar to that noticed under "Liquor of Potassa." The test of its purity, and uses, are also the same. (*See Nos.* 4784 *and* 102.)

4786. Solution of Chloride of Lime. This solution, usually called *bleaching liquor*, is prepared of 1 part chloride of lime to 10 parts of distilled water (both by weight). That is, 2 ounces to the pint, or 1 pound to the gallon. This is the ordinary strength of that of the shops; but in that which is sold as Concentrated Solution of Chloride of Lime, the proportions are usually 3 parts of the chloride to 20 of water. That is, $1\frac{1}{2}$ pounds per gallon. The British Pharmacopœia directs the chloride to be triturated with the water in a wedgwood-ware or porcelain mortar, and having transferred the whole to a stoppered bottle, to be well shaken, several times, for the space of 3 hours; lastly, the solution is to be filtered through muslin, and preserved in a stoppered bottle. The specific gravity of that of the Pharmacopœia is 1.035. On the large scale, the ingredients are usually placed in a carboy, or a stone-ware bottle, which they will only $\frac{3}{4}$ or $\frac{4}{5}$ fill, and, after being corked or bunged close, agitated frequently for a day or two. A cork or bung of bees'-wax or gutta-percha should be used for the purpose, unless the vessel is a stoppered one. After repose for 2 or 3 days, the clear portion is decanted through a funnel choked with crushed glass into bottles. The last should be closely corked (preferably stoppered), and kept in a cool and dark place. Nothing metallic should be allowed to come in contact with it. (*See No.* 104.) A better plan of filtering the above is as follows: The neck of the funnel should be choked with some fragments of broken glass, over which a layer of smaller ones should be placed, and, over all, a thick layer of coarsely powdered glass. This is all the filtration necessary, and is much superior to that ordered in the Pharmacopœia, as the contact with the muslin, and the longer exposure, weaken the solution. The U. S. Pharmacopœia directs the solution of chloride of lime to be prepared by mixing 12 troy ounces muriatic acid with $\frac{1}{2}$ pint distilled water; gradually adding 6 troy ounces marble in small pieces. Towards the close of the effervesence, apply a gentle heat, and, when the action has ceased, pour off the clear liquid, and evaporate to dryness. Dissolve the residue in $1\frac{1}{2}$ times its weight of distilled water, and filter through paper.

4787. Solution of Chloride of Potash. This solution is also known as *Javelle's Bleaching Liquid; Eau de Javelle*, &c. This is best made by passing gaseous chlorine into a solution of 1 part of carbonate of potash in 10 parts of water, until the gas ceases to be absorbed. It may also be made by adding a solution of carbonate of potash to a solution of chloride of lime, with agitation, as long as a precipitate forms; the liquid being afterwards decanted or filtered. These processes are precisely similar to that for the soda solution, an equivalent portion of carbonate of potash being used. (*See Nos.* 4788, *&c.*)

4788. Solution of Chloride of Soda. Also variously called *Solution of Chlorinated Soda; Solution of Hypochlorite of Soda; Labarraque's Disinfecting Fluid; Eau de Labarraque.* Take of crystallized carbonate of soda, 12 ounces avoirdupois; distilled water, 1 Imperial quart; dissolve, and pass through the solution the chlorine evolved from a mixture of common salt, 4 ounces; binoxide of manganese, 3 ounces; sulphuric acid, $2\frac{1}{4}$ fluid ounces, previously diluted with 3 fluid ounces water, heated in a retort together, and the gas purified by passing through a wash bottle containing 5 ounces water, before it enters the soda solution.

4789. Solution of Chloride of Soda. To a solution of chloride of lime (formed of chloride of lime, $\frac{1}{2}$ pound; water, 3 pints), add a solution of carbonate of soda (formed of carbonate of soda, crystallized, 7 ounces; water, 1 pint), and, after agitation for about 10 minutes, decant or filter, and preserve the filtrate in a well-stoppered bottle, and in a cool and dark place. This is the formula of the Dublin Pharmacopœia, and often more convenient than the preceding one. A writer in Boettger's Notizblatt recommends that in preparing this solution from chloride of lime, bicarbonate of soda be used in place of salsoda. There is no question but that the precipitate will be much less bulky, and more of the liquid will be recovered in a concentrated form by decantation.

4790. Solution of Ammonio-Sulphate of Copper. Dissolve 1 drachm of the ammonio-sulphate in 1 pint water, and filter. This is stimulant and detergent. Applied as a lotion to indolent ulcers; and, largely diluted, to remove specks on the cornea. Also used as a test for arsenical compounds, with which it throws down a green precipitate.

4791. Solution of Indigo. Place a stone-ware vessel containing 8 parts oil of vitriol in a tub of very cold water; add 1 part fine powdered indigo very gradually, to prevent the mixture from heating. The mixture should be stirred occasionally with a glass rod; and, when the solution is complete, allow it to repose for 48 hours. Then dilute with twice its weight of soft water, adding this also very gradually, to prevent heating. This precaution is necessary to prevent partial decomposition of the indigo, which would result in the formation of sulphurous acid and indigo green. This is the sulphate of indigo or *liquid blue* of trade. This solution is preferably prepared by using 5 parts *fuming* sulphuric acid instead of the 8 parts oil of vitriol. (*See No.* 98.)

4792. Solution of Carbonate of Ammonia. This is prepared by dissolving 1 part sublimed carbonate of ammonia in 3 parts water, and adding 1 part ammonia-water. Used in chemical analyses, and as a very delicate test for the presence of lime, from a solution of which it forms a white precipitate soluble in nitric or hydrochloric acid.

4793. Solution of Sulphuretted Hydrogen. Pass sulphuretted hydrogen gas through cold distilled water, recently boiled, until it will absorb no more. Keep in small bottles securely stoppered.

4794. Solution of Santonin. The insolubility of santonin in water impairs its utility as a vermifuge. Water, cold or warm, takes up the merest trace. Chloroform, absolute alcohol, the strongest acetic acid, turpentine, hot olive oil, and hot glycerine, are the only simple fluids that dissolve any appreciable quantity. But it separates from the oil and glycerine on cooling; and water added to the other solutions produces the same result. By the use of the following formula, however, a useful and effective solution may be obtained. Put 20 grains bicarbonate of soda and 3 ounces distilled water into a flask; keep the liquid near the boiling point and add 12 grains santonin, finely powdered, about 2 grains at a time, until the whole has dissolved. Solution is effected in about half an hour, during which time the water is reduced to 2 ounces, or, if not, may be reduced to that bulk, when 1 ounce will contain a full dose—6 grains of santonin. The solution is bright and permanent, strongly alkaline, free from odor and taste, except that of carbonate of soda. Carefully neutralized with acetic acid, an equally bright and permanent solution is formed. Both may be diluted to any extent with hot or cold water without impairing the solution of santonin. The whole, or nearly the whole, of the santonin is precipitated in its original form of colorless rectangular plates, with bevelled edges, immediately by mineral acids, and after some hours by excess of acetic acid.

4795. Miscible Copaiba. Mix transparent balsam of copaiba with half its volume of strong liquid of potassa of double strength. Different samples often require slightly different quantities of the solution of potassa; it is therefore best to mix them gradually and cautiously together. Should the mixture be opaque, a little more of one or other of the ingredients, as the case may be, will render it clear. No heat should be used. This article is miscible with water, with which it forms a kind of milk; and, from containing all the volatile oil of the copaiba, is a very valuable preparation. Its activity is considered equal to the balsam itself, and is given in similar doses.

4796. Solution of Permanganate of Potassa. M. Leconte prepares this solution in the following manner: Caustic potassa, 6 drachms; chlorate of potassa, 5 drachms; binoxide of manganese, 5 drachms. Dissolve the caustic potassa and the chlorate in a small quantity of water, and add the manganese; get rid of the water by evaporation, stirring constantly, and calcine the dry mass to a dark red for an hour in an untinned iron cup; allow to cool, and add a quart of plain water. Then boil for 5 minutes in a china capsule, and you will obtain a fluid of a slightly purplish tint; decant the solution, and wash the residue with such a quantity of water as to make altogether 2 quarts. When filtering is thought necessary, the liquid should be passed, not through paper, but through very fine sand. For dressing foul wounds, or for injection, use 1 drachm of this solution to from 3 drachms to 5 of spring water.

4797. Reveil's Solution of Permanganate of Potassa. The officinal solution of the British pharmacopœia consists of 80 grains of the permanganate dissolved in 1 imperial pint distilled water. This is about 1 part by weight to 110 parts water. M. Reveil recommends a standard solution of 10 parts permanganate to 90 of water, so that the solution contains 10 per cent. of permanganate. This latter strength is endorsed by the U. S. Dispensatory, which also recommends extreme cleanliness in its preparation and use, and of the bottles containing it, as organic matter more or less neutralizes its disinfecting and cleansing powers. The same authority orders the pencil or brush used for its application to be made of amianthus, or asbestos, in order to ensure its fullest effects. (*See No.* 1701.)

4798. Directions for Using Permanganate of Potassa. Reveil's standard solution (*see No.* 4797) may be used at its full strength for dressing cancerous sores and ulcers, applied with a pencil made of asbestos, or sprinkled over a dressing of the same material. For simple wounds or for injections, $\frac{1}{2}$ fluid ounce of the solution may be diluted with 1 pint of water. For gangrenous wounds and scrofulous ulcers, or as a gargle in unhealthy ulcers of the mouth and throat, 1 fluid ounce to a pint of water. For a gargle in croup and diphtheria, or as a wash for the hands after dissecting, 2 fluid ounces to the pint. A dose administered internally may consist of 10 to 30 drops of the standard solution. (*U. S. Disp.*)

4799. Aceto-Carbolic Solution. Acetic acid (pyroligneous) 8°, 20 parts; pure carbolic acid, 5 parts; water, 75 parts. Mix the two acids and add the water. The acetic acid favors penetration through the epidermis. For tinea, apply the liquid once a day over the diseased parts by means of a brush. For scabies, sponge all the parts. The clothes, &c., of the affected individual should also be treated with the liquid. (*Lemaire.*)

4800. Solution of Carbolic Acid in Water. To obtain uniform solution, it is better to slack the carbolic acid with four times its bulk of hot water, and then to add a sufficiency of cold water; or the carbolic acid may be first mingled with alcohol, which causes more ready solubility, before the addition of cold water. Water will not dissolve more than one-twentieth of its bulk of carbolic acid.

4801. Frank's Specific Solution of Copaiba. Boil 2 parts balsam of copaiba, 3 parts liquor of potassa, and 7 parts water together for 2 or 3 minutes; put the mixture into a separator, and let it stand for 5 or 6 days; then draw it off from the bottom, avoiding the upper stratum of oil, and add to the clear liquid 1 part sweet spirits of nitre, perfectly free from acid; should it turn milky, a very little liquor of potassa will usually brighten it; but if it does not, place it in a clean separator, and let it stand, closely covered, for a few days, then draw it off from the bottom as before, and it will be perfectly transparent.

4802. Mackenzie's Solution of Nitrate of Silver. This is used for sponging the throat and fauces, for affections of those parts. Dissolve 20 grains nitrate of silver in 1 fluid ounce distilled water.

4803. Solution of Hydrosulphuret of Ammonia. Saturate strong water of ammonia with sulphuretted hydrogen gas, then add a second portion of water of ammonia, equal to that first used, and put into well-stoppered bottles.

4804. Fowler's Solution; Solution of Arsenite of Potassa. Boil 64 grains arsenious acid (in small pieces), and 64 grains bicarbonate of potassa, in 12 fluid ounces water, until the acid is entirely dissolved. When cold, add ½ fluid ounce compound spirit of lavender, and sufficient distilled water to make the whole mixture measure a pint. (*U. S. Ph.*)

4805. Solution of Citrate of Magnesia. Crystallized citric acid, 37 drachms; water, 268 drachms; carbonate of magnesia, 22 drachms. Dissolve the acid in the water, and mix the magnesia with it under constant stirring; filter, and add to the filtrate so much water as to bring the weight of the whole to 40 ounces. To prepare the lemonade, take of aromatized simple syrup, 4 ounces; pulverized citric acid, 48 grains; bicarbonate of soda, 64 grains. Fill into bottles of suitable size, add water and so much of the magnesia solution as is required, and cork and tie immediately. Keep in a cool place. This solution contains 80 grains of citrate of magnesia to the ounce of fluid.

4806. Parisel's Solution of Citrate of Magnesia. M. Parisel recommends the following method of preparing this article, which he has followed during two years, as being both simple and effectual: Take of powdered and well dried citric acid, 20 parts by weight; carbonate of magnesia, 12 parts; mix accurately, and enclose the powder in a slightly warmed and well-dried bottle, which must be kept well stopped. The mixture thus made is rapidly dissolved in three times its weight of water at the ordinary temperature; and, if the water be pure, the solution in a few minutes becomes perfectly transparent, without any precipitate. The salt preserves its solubility for a long time.

4807. Solution of Tartrate of Soda. Take of carbonate of soda, 1¾ pounds; tartaric acid, 1¼ pounds; crushed sugar, 2 pounds; hot water, 2 gallons. Dissolve the soda in 1½ gallons of the water; the sugar in 1 quart; and the acid in 1 quart. When all have dissolved and cooled down, add the acid slowly to the soda solution, and mix with the sugar. Filter into strong 12-ounce bottles, to each of which must be added a few drops of strong essence of lemon, and 35 grains of bicarbonate of soda. Cork immediately and tie or wire the bottles; will keep for any length of time. This is considered a good substitute for soluton of citrate of magnesia.

4808. Solution of Citrate of Potassa. Take of citric acid, ½ ounce, troy; bicarbonate of potassa, 330 grains; water, ½ pint. Dissolve the acid and bicarbonate in the water, and strain the solution through muslin. (*U. S. Ph.*)

4809. Effervescing Citrate of Magnesia. Take of citric acid, dried and powdered, 7 parts; heavy carbonate of magnesia, 5 parts; mix, and preserve in well-corked bottles.

4810. Effervescing Citrate of Magnesia. Take of powdered citric acid, 2½ ounces; powdered sugar, 8 ounces; mix and triturate to a fine powder, and drive off the water of crystallization by the heat of a water-bath. Add citrate of magnesia (prepared by fusion), 4 ounces; oil of lemon, 10 drops; and mix immediately; then add bicarbonate of soda, 3 ounces; and again triturate until the whole forms a fine powder, which must be preserved in well-stoppered bottles. From 1 to 3 table-spoonfuls, mixed in a tumbler of water, furnishes an effervescing draught, in which the undissolved portion is so nicely suspended that it can be taken without inconvenience.

4811. Effervescing Citrate of Magnesia. Take of crystallized citric acid, 20 grains; carbonate of magnesia, 14 grains; mix in a tumbler of cold water and drink the mixture whilst effervescing.

4812. Solution of Citrate of Bismuth. Put 2 ounces pure sub-nitrate of bismuth into a porcelain dish, add 1450 grains nitric acid of specific gravity 1.44; heat over a spirit lamp until the bismuth is dissolved; then add one fluid ounce water, and let stand until cold; then gradually add water, constantly stirring with a glass rod, until a further addition produces milkiness, or until the whole measures 1¼ pints. Filter and set aside.

Next, dissolve 3 ounces citric acid in 1½ pints water, and exactly neutralize the acid with carbonate of soda dissolved in water. It is important that there shall be no excess of soda, as the resulting citrate of bismuth would be contaminated with the oxide after decomposition. Put the bismuth solution in a suitable vessel, and add, stirring constantly with a glass rod, sufficient of the solution citrate soda exactly to decompose; the precise quantity is known to have been added, when, after placing the whole upon a cloth filter, the washings, after having been suffered to run awhile until clear, first, fail to precipitate bismuth when dropped into water, and, second, show no precipitate upon the addition of a few drops of ternitrate of bismuth, a small quantity of which should be reserved for this purpose. When the liquid portion has mostly passed, pour water upon the filter until thoroughly washed from nitrate of soda, or until the water passes tasteless; then, after draining, transfer to bibulous paper, and dry by gentle heat.

4813. Bartlett's Preparation of Citrate of Bismuth. Dissolve 1 troy ounce subcarbonate of bismuth in 720 grains nitric acid; after effervescence has ceased, gradually introduce 1½ fluid ounces distilled water; add to this slowly, and with constant stirring, a solution of 600 grains citrate of potassa in 2 pints distilled water. Nitrate of potassa and citrate of bismuth are formed; the latter, being insoluble, is precipitated, and, being thoroughly washed with distilled water, may be dried on bibulous paper with a gentle heat.

4814. Solution of Citrate of Bismuth and Ammonia. Rub some citrate of bismuth with sufficient distilled water to reduce it to a uniform pasty consistence, and add cautiously, with constant trituration, strong water of ammonia until a solution is obtained, observing to avoid an excess of ammonia. Filter the liquid through paper, returning the first portions that pass, should they be turbid.

4815. Solution of Citrate of Iron. Dilute 1 pint of solution of tersulphate of iron with 2 pints distilled water; precipitate with water of ammonia in slight excess, constantly stirring. Transfer the precipitate to a muslin strainer, and wash it with water until the washings are nearly tasteless. Drain it, and put half of it in a porcelain capsule on a water-bath heated to 150° Fahr., add 5¾ troy ounces citric acid in coarse powder, and stir until the precipitate is nearly dissolved; then add sufficient of the reserved precipitate to fully saturate the acid. Lastly, filter the liquid, evaporate it at a temperature not over 150° Fahr., until it measures a pint. (*U. S. Dis.*)

4816. Styptic Solution of Perchloride of Iron. Mix together 12 fluid ounces muriatic acid and 5 fluid ounces water; pour the mixture, a small portion at a time, on 2 ounces avoirdupois of iron wire; aiding the complete solution of the wire by a gentle heat. Add 6 fluid drachms nitric acid, previously mixed with 2 ounces water; and evaporate the whole to 5 fluid ounces. Lastly, add water sufficient to make the whole up to 10 fluid ounces. (*U. S. Dis.*)

Lotions.

Lotions. Solutions of medicinal substances in water, employed for external application. They may be made of any soluble medicaments that are capable of exerting their action by contact with the skin. Lotions have been divided into classes, as sedative, anodyne, stimulant, &c. *Sedative* and *refrigerant* lotions are commonly employed to allay inflammation; *anodyne* and *narcotic* lotions to relieve pain; *stimulant* lotions to assist the ripening of tumors, &c.; *detergent* lotions to clean foul ulcers, &c.; *repellant* and *resolvent* lotions to disperse tumors, remove eruptions, &c. Lotions are usually applied by wetting a piece of linen with them, and keeping it on the part affected, or by moistening the part with the fingers previously dipped into them. Lotions are more agreeable if made with rose water. A number of these preparations are here given, and others will be found by referring to the index, under their respective headings.

4818. Lotion of Nitric Acid. Mix together 2 drachms dilute nitric acid and 1 pint water. This lotion is stimulating and cleansing. It is very useful when applied to foul and fœtid ulcers; it is likewise of considerable value in ulceration of the bone and threatened inflammation. It was the favorite lotion of Sir Astley Cooper in cases of unhealthy ulcerations requiring the application of a stimulant.

4819. Anodyne Lotion. Crude opium, 2 drachms; warm water, 1 pint. Rub the opium for a few minutes in a mortar with a little of the warm water, then pour in the remainder of the water and mix them well. This is an excellent wash for painful and irritable ulcers and swellings.

4820. Astringent Lotion. Sulphate of zinc, 2 drachms; water, 1 pint; camphorated spirit of wine, 2 drachms; mixed together. This is an excellent lotion for piles, used night and morning.

4821. Compound Alum Lotion. A detergent and astringent lotion for old ulcers, chilblains, excoriations, &c., and, largely diluted, as an eye-wash and injection. Dissolve 1 ounce each of alum and sulphate of zinc in 3 pints boiling water; filter, if necessary.

4822. Camphorated Lotion. Diluted solution of diacetate of lead, 8 fluid ounces; spirit of camphor, 2 drachms; mix, and shake well. Refrigerant and anodyne. Employed in erysipelatous inflammations, burns, contusions, sprains, excoriations, &c.

4823. Spackman's Lotion for Inflamed Parts. Mix 1 drachm tincture of myrrh; 3 drachms tincture of camphor; 1 ounce rectified spirits of wine; 1 drachm Goulard's extract; 1 ounce solution of sulphate of morphia; 2 ounces tincture of arnica, and 4 ounces water.

4824. Lotion of Acetate of Lead. Dissolve sugar of lead, ½ ounce avoirdupois, in distilled or soft water, 1 Imperial pint. Sometimes a little vinegar is added, a like quantity of water being omitted. Used in excoriations, burns, sprains, contusions, &c.; also as an occasional cosmetic wash by persons troubled with eruptions.

4825. Preventive Lotions. These are washes intended to prevent infection from personal contact with those laboring under contagious diseases. Most of the nostrums of this character are mere weak solutions of chloride of lime, corrosive sublimate, potassa, or acetate or diacetate of lead. (*See No.* 4830.)

4826. Lotion of Muriate of Ammonia. Dissolve sal-ammoniac in coarse powder, 1 to 4 drachms (avoirdupois), in water, 1 Imperial pint. A useful wash in itch, old ulcers, tender feet, sweaty feet and hands, swelled joints, &c.

4827. Strong Lotion of Hydrochlorate of Ammonia. Dissolve sal-ammoniac, 1 to 2 avoirdupois ounces, in water, 1 Imperial pint. In bruises and contusions, extravasations, glandular swellings and indurations, chilblains, &c., when the skin is not broken. Vinegar is often substituted for the whole or a part of the water, and sometimes ⅙ or ⅛ part of rectified spirit, or some brandy or rum is added.

4828. Lotion of Muriatic Acid. Mix hydrochloric acid (specific gravity 1.16), 1 fluid ounce, with water, 19 fluid ounces. For unbroken chilblains. Diluted with an equal bulk of water, it forms a useful lotion in lepra and other scaly skin diseases.

4829. Lotion of Nitrate of Silver. Dissolve crystallized nitrate of silver, 1 to 2 drachms avoirdupois; concentrated nitric acid, 20 drops; in distilled water, 1 ounce. Used as a liquid caustic to destroy corns and warts.

4830. Lotion of Chloride of Lime. Take of chloride of lime (best, fresh), ½ ounce avoirdupois; pure water, 1 Imperial pint; mix in a bottle, and agitate, occasionally, for 2 or 3 hours; after repose, filter the clear portion through a piece of calico that has been previously moistened with water, and preserve the filtrate in a stoppered bottle.

4831. Lotion of Chloride of Soda. As the last, but substituting chloride of soda for chloride of lime. Or: Take of chloride of lime, ½ ounce avoirdupois; water, ¾ Imperial pint; mix, &c., as before; then add of crystallized carbonate of soda, 3½ drachms; previously dissolved in water, ¼ pint; agitate the whole for 12 or 15 minutes, and filter, &c., as before.

4832. Lotion of Chloride of Potassa. As the last, but substituting 3 drachms dry carbonate of potassa for the carbonate of soda.

4833. Lotion of Prussic Acid. Mix medicinal prussic acid, ½ fluid drachm, with rectified spirit, 1 fluid ounce, and distilled water, 2 fluid ounces; cover the bottle with thick purple paper, and keep it in the shade. Recommended by Dr. Elliotson as a lotion to moisten the face both before and after shaving, as being very soothing to an irritable skin. It is poisonous.

4834. Sulphuretted Lotion. Dissolve sulphuret of potassium, 1 drachm avoirdupois, in distilled water, 1 pint Imperial. Used to render the skin soft, white, and smooth, particularly when there is a tendency to slight eruptions of a pustular or vesicular character. The addition of ½ to 1 ounce of glycerine improves it for present use.

4835. Carbolic Acid Lotion. Dissolve 5 grains carbolic acid in crystals, in 1 ounce water. As a lotion for foul ulcers, carbuncles, scabies, and lepra.

4836. Carbolic Acid Lotion for Burns. Mix 1 drachm liquid carbolic acid with 3 ounces linseed oil and 3 ounces lime-water.

4837. Lotion of Arnica for Bruises, Sprains, Burns, &c. Take 1 ounce of arnica flowers dried, and put them in a wide-mouthed bottle; pour just enough scalding water over them to moisten them, and afterwards about 1 or 1½ pints spirits of wine. In case of a burn or bruise, &c., wet a cloth in the arnica and lay it on the part affected. Renew the application occasionally, and the pain will soon be removed.

4838. Balm of Gilead Lotion. Balm-gilead buds, bottled up in new rum, are very healing to fresh cuts or wounds. An excellent preparation to have in the house.

4839. Glycerine Lotion for Irritation of the Skin. Mix 1 ounce of glycerine with 1 pint water. It allays itching and removes dryness, &c., in various skin diseases. With the addition of 2 or 3 drachms of borax, it removes chaps from the lips, hands, and nipples.

4840. Startin's Glycerine Lotion to Allay Irritation. Take ½ drachm trisnitrate of bismuth; 1 fluid drachm tincture of foxglove; 1 fluid drachm dilute nitric acid; 4 drachms glycerine; and 8 fluid ounces rose-water. To allay the irritation in itch and some other skin diseases.

4841. Glycerine Lotion for Burns, Scalds, &c. Take 1 ounce glycerine, 2 ounces thick mucilage (gum-arabic dissolved in water), and 7 ounces lime water. For burns, scalds, chaps, excoriations, &c.

4842. Startin's Glycerine Lotion for Bruises, &c. Triturate together 1 ounce glycerine, 1 drachm extract of belladonna, and 3 ounces soap liniment. (*See No.* 4869.) For bruises, sprains, and swelled joints; also gouty, neuralgic, and rheumatic pains.

4843. Evaporating Lotions. These lotions are soothing and refrigerant if allowed to evaporate by free exposure; and stimulant, if the evaporation is prevented by covering the part with the hand, or a piece of oiled silk. They are useful applications in nervous headaches, restlessness, irritability of the skin, &c. Mix 1½ fluid ounces each of sulphuric ether, rectified spirit, and solution of acetate of ammonia, with 3½ fluid ounces rose-water. A simple evaporating lotion may be made with 1 part rectified spirit, and 4 to 6 parts water.

4844. Camphorated Evaporating Lotion. Dissolve ½ drachm camphor in 4 ounces rectified spirit and ½ ounce elder flowers; digest 24 hours and strain. This is a good calming lotion.

4845. Tar Lotion. Quicklime, 6 ounces; water, 48 ounces; slack, add tar 4 ounces, and boil to one half. This liquid consists of a solution of pyrolignite of lime and pyrogenous oil and resin. It may be advantageously employed in various chronic skin diseases, especially those affecting the heads of children.

4846. Lotion of Galls. Bruised galls, 2 drachms; boiling water, 1 pint; infuse an hour, and strain. Astringent. An excellent application to sore nipples, or to strengthen them before suckling; spirit of wine, 3 ounces, may be advantageously added, and a like portion of water omitted.

4847. Mercurial Lotion; or Black Wash. Calomel, 1 drachm; lime water, 1 pint; mix, and shake well. These are the usual proportions. The bottle should be well shaken before the lotion is applied. Black wash is a favorite application to all kinds of syphilitic sores.

4848. Yellow Lotion or Wash, Sometimes called Red Wash. Corrosive sublimate (in powder), ½ drachm; lime water, 1 pint; mix, and shake well. It should be well shaken before use. A common application to syphilitic and scrofulous sores.

4849. Lotion of Belladonna. Extract of deadly night-shade, 1 drachm; diluted solution of diacetate of lead, 1 pint; dissolve. Applied to tumors and glandular enlargements.

4850. Cazenave's Antipsoric Lotion. Sulphuret of potassium, 1 drachm; soft soap (pure), 2 drachms; water, 8 ounces; dissolve. An excellent remedy for the itch. It leaves but little smell behind, and does not soil the linen.

4851. Iodine Lotion. Tincture of iodine, ½ fluid ounce; iodide of iron, 12 grains; chloride of antimony, ½ ounce. Mix for a wash. It is a remedy for corns. Apply with a small brush. Or: Iodine, 1½ grains; spirits of wine, 3 tea-spoonfuls. Dissolve, and add 1 pint of water. A most excellent wash for scrofulous sores.

4852. Disinfecting Lotion. Liquor of common salt, 1 fluid ounce; water, ½ pint; Or: Chloride of lime, 3 drachms; water, 1 pint; dissolve. Both are good washes for foul ulcers, the itch, the teeth, to sweeten the breath and remove the smell of tobacco smoke, and for various similar purposes.

4853. Valuable Lotion for Wounds, &c. Camphor, 5 drachms, cut into small pieces, and dissolved in half a pint of spirits of wine in a closely corked bottle; when fully dissolved, add ½ pint of ox-gall and 60 drops of laudanum. Shake it well, and bottle for use. This has been a patent medicine, and is very efficacious in the cure of fresh wounds, cuts, bruises, swellings, sores, and inflamed and pained parts.

4854. Lotion for Mange. Corrosive sublimate, ¼ ounce; spirits of salt (muriatic acid), ½ ounce; water, 1 quart. Or: Corrosive sublimate, 1 drachm; sal-ammoniac, ½ ounce; water, 1 pint Or: To the last add strong decoction of white hellebore, ½ pint. Used for mange in horses, cattle, and dogs, when sulphur ointment fails.

4855. Lotion for Galls. Vinegar and spirit of wine, of each 4 ounces; sugar of lead, ½ ounce; water, ½ pint; mix. Or: Soap liniment and solution of acetate of ammonia, equal parts. Or: Sal-ammoniac, 1 ounce; muriatic acid, 3 drachms; water, 1 pint. Used by farriers for saddle-galls or warbles.

4856. Lotion of Chlorate of Potassa. —sometimes called *Cosmetic Solution of Potassa*—for bad breath. Dissolve powdered chlorate of potassa, ½ ounce, in distilled water, 12 ounces, and rose-water, 2½ ounces. Used as a wash in foul mouth, gums, &c., particularly where there is a scorbutic or syphilitic taint; also extensively by smokers, to deodorize the breath. Its daily use is said to give a rich healthy hue to the gums and lips.

Liniments.

A semi-fluid ointment or soapy application for painful joints, swellings, burns, &c. The term is also occasionally extended to various spirituous and stimulating external applications. When they are of a thinner consistency they are called *embrocations,* although this distinction is not always observed. Liniments are generally applied by friction with the hand or fingers, or with some substance, such as a piece of flannel, capable of producing some amount of irritation of the skin. Sometimes a piece of linen rag dipped in them is simply laid on the part. The greater number of cerates and ointments may be converted into liniments by reducing their substance with almond or olive oil, or oil of turpentine. Besides those here given, others will be found in the index under their proper heads.

4858. Good Samaritan, or Immediate Relief from Pain. Take 2 quarts of 95 per cent. alcohol, and add to it the following articles: Oils of sassafras, hemlock, spirits of turpentine, balsam of fir, chloroform, and tincture of catechu and guaiacum, of each 1 ounce; oil of origanum, 2 ounces; oil of wintergreen, ½ ounce, and gum camphor, ½ ounce. The above is a noble liniment, and may be successfully employed in rheumatism, bruises, neuralgia, sprains, headache, burns, and spinal affections.

4859. Hemlock Liniment. Oil of hemlock, ½ ounce; camphor, in gum, ¼ ounce; opium, ¼ ounce; spirits of wine, 1 pint. Mix. It is a first-rate rubefacient in inflammatory rheumatism, gout, quinsy, inflamed breast, white swellings, &c.

4860. Morphia Liniment. An excellent anodyne, which often allays pain when other means have failed. Put 3 grains pure morphia into a mortar; add gradually, during trituration, 1 fluid ounce warm oil of almonds; when the morphia is dissolved, add 1 ounce camphor liniment. (*See No.* 4880).

4861. Magic Liniment. Alcohol, 1 quart; gum camphor, 4 ounces; turpentine, 2 ounces; oil of origanum, 2 ounces; sweet oil, 1 ounce. For cuts or calks in horses or cattle in winter it has no equal; but it must be applied often. For human flesh use twice the amount of alcohol, and no liniment will be found superior to it.

4862. Spirits of Camphor. The gum resin camphor readily dissolves in alcohol, forming spirits of camphor. About 2 ounces camphor are generally dissolved in about 1 pint spirits. It is used as an external application for sprains, local pains, and stitches. It is applied by rubbing with the hand upon the painful part. To secure the full benefit of the application, the part should be afterwards covered with a piece of flannel of suitable size, more or less wetted with the spirits, and the whole covered with oil silk for the purpose of restraining evaporation.

4863. Camphorated Oil. This is a camphor liniment. The proportions are the same as in the preceding formula, substituting olive oil for the alcohol, and exposing the materials to a moderate heat. As an external stimulant application it is even more powerful than the spirits; and to obtain its full influence, the part treated should be also covered with flannel and oil silk. It forms a valuable liniment in chronic rheumatism, and other painful affections, and is specially valuable as a counter-irritant in sore or inflamed throats, and diseased bowels.

4864. Arnica Liniment. Add to 1 pint sweet oil, 2 table-spoonfuls tincture of arnica; or the leaves may be heated in the oil over a slow fire. Good for wounds, stiff joints, rheumatism, and all injuries.

4865. London Liniment. Take chloroform, olive oil, and aqua-ammonia, of each 1 ounce; acetate of morphia, 10 grains. Mix, and use as other liniments. Very valuable.

4866. Valuable Embrocation. Take ½ ounce camphor, cut it into small pieces, and dissolve it in ½ pint spirits of wine in a closely corked bottle; when completely dissolved, add 1 pint ox-gall (which can be had of any butcher), and about 40 or 50 drops laudanum; shake it well and bottle it for use. Apply lint dipped into it.

4867. Hungarian Counter-Irritant Liniment. Macerate for a week 1 drachm powdered cantharides, 1 drachm sliced garlic, 4 drachms each camphor, bruised mustard seed, and black pepper, in 6 fluid ounces strong vinegar and 12 fluid ounces rectified spirit; then filter.

4868. Liniment for Wounds. In 1 quart alcohol dissolve 1 ounce each saltpetre and gum camphor, and 1 table-spoonful of salt. When dissolved the liniment is ready for use, and is a magical remedy.

4869. Steer's Opodeldoc, or Soap Liniment. White castile soap, cut small, 2 pounds; camphor, 5 ounces; oil of rosemary, 1 ounce; oil of origanum, 2 ounces; rectified spirit, 1 gallon; dissolve in a corked bottle by the heat of a water-bath; and when considerably cool, strain, then add liquor of ammonia, 11 ounces; immediately put it in bottles, cork close, and tie over with bladder. It will be very fine, solid and transparent, when cold.

4870. Liquid Opodeldoc. Take 2 ounces castile soap shavings, and dissolve it in 1 quart alcohol, with gentle heat, then add 1 ounce camphor, ½ ounce oil rosemary, and 2 ounces spirits hartshorn.

4871. Belladonna Liniment for Skin Diseases. Take 4 drachms extract of belladonna, 1 ounce glycerine, and 6 ounces soap liniment. (*See No.* 4869.) For rheumatism, neuralgia, painful swellings, &c.

4872. Black Oils. Best alcohol, tincture of arnica, British oil, and oil of tar, of each 2 ounces; and slowly add sulphuric acid, ½ ounce. Extensively used as a liniment, particularly in cases where there is much inflammation.

4873. Factitious Oil of Spike. Oil of turpentine, 3 pints; oil of lavender, 1 pint; mix. Used by enamelers to mix their colors in. Or: Oil of turpentine, 1 gallon; Barbadoes tar, 4 ounces; alkanet root, 2 ounces; digest a week. Used as a liniment for horses.

4874. Liniment of Cantharides. Powdered Spanish flies, 1 drachm; oil of turpentine, 1 fluid ounce; digest 2 hours and filter. Or: Tincture of cantharides and soap liniment (*see No.* 4869), equal parts; mix. Both the above are irritant and stimulant, but should be used cautiously, lest they produce strangury.

4875. Hydrochloric Acid Liniment. Take of olive oil, ¼ Imperial pint; pure spermaceti and camphor, of each ½ ounce avoirdupois; balsam of Peru, ¼ ounce; mix by a gentle heat, add ½ fluid ounce hydrochloric acid, and stir until quite cold. An excellent friction for chilblains before they break. The balsam of Peru may be omitted if the cost be an object.

4876. Compound Chloroform Liniment. This is composed of 1 ounce each chloroform, ether, spirit of camphor, and laudanum, and ½ ounce tincture of cayenne pepper. For rheumatic pains.

4877. Petroleum Liniment. Mix together 1 ounce petroleum, ½ ounce camphor, and ¼ drachm alcohol.

4878. Opium Liniment. Mix 2 ounces laudanum with 6 ounces soap liniment. (*See No.* 4869.) It constitutes an excellent soothing application in rheumatism, sprains, and other painful affections.

4879. Belladonna Liniment for Lead Colic. Take 40 grains extract of belladonna, 1 drachm rectified ether, and 2 fluid ounces cherry-laurel water. As a friction to the abdomen in lead colic.

4880. Compound Camphor Liniment, or Essence for Headache. Take of camphor, 2½ ounces avoirdupois; oil of lavender, 1 fluid drachm; rectified spirit, 15 fluid ounces; dissolve, then add of liquor of ammonia (specific gravity .882–.880), 5 fluid ounces, and shake them until mixed. It is powerfully stimulant, rubefacient, and counter-irritant. A piece of folded linen wetted with it applied to the part, and then covered with a towel, and pressed with the hand, or covered with a piece of oiled silk, will generally relieve superficial pains.

4881. Liniment Volatile, or Magic Pain Killer. Spirit of hartshorn, 1 ounce; olive oil, 1½ ounces; cayenne pepper, 2 drachms; laudanum, 2 drachms; 1 table-spoonful of salt and 2 of brandy. Shake well in a bottle. Rub the affected part with it, apply afterwards a rag saturated with it. It removes pains and swellings. It is a magic remedy.

4882. Instantaneous Pain Killer. Another and even more instant cure of pain is made as follows: Take aqua-ammonia, sulphuric ether, and alcohol, equal parts, and apply over the pain.

4883. Chilblain Liniment. Take 1 ounce of camphorated spirit, ½ ounce of the liquor of subacetate of lead. Mix and apply 3 or 4 times a day. This is Sir Astley Cooper's prescription, and a very efficacious remedy for chilblains.

4884. Rheumatic Liniment. Tincture of cayenne, oil of turpentine, olive oil, hemlock oil, gum camphor, sassafras oil, tincture of prickly ash, of each 1 ounce; powdered capsicum, or cayenne, 1 ounce; spirit of wine, 2 quarts; vinegar, 1 quart; ammonia, 1 quart; add 2 ounces gum camphor. Mix, put in a vessel, and stir occasionally till mixed and dissolved. This is a magic liniment, soon giving ease in rheumatic pains, gout, neuralgia, sprains, &c., &c. It seldom or never fails. "Good Samaritan" is also an excellent remedy for rheumatism. (*See No.* 4858.) Bathe the parts affected freely, and wet a piece of flannel and bind on the parts.

4885. Good Liniment for Rheumatics. Take 1 gill each of alcohol, beef's gall, spirits of turpentine and sweet oil, and 4 ounces gum camphor. Put them all in a bottle and shake it up; use it 2 or 3 times a day, a tea-spoonful at a time. Apply it to the parts affected, before the fire. It is good, also, for frost-bites.

4886. Liniment for Old Rheumatic Pains. A powerful liniment for old rheumatic pains, especially when affecting the loins, is the following: Camphorated oil and spirits of turpentine, of each 2 parts; water of ammonia, 1 part; laudanum, 1 part; to be well shaken together.

4887. Gebhard's Liniment for Sprains and Bruises. Mix together 2 ounces each oil of spike and British oil; 1 pint tanner's oil; $\frac{1}{2}$ pint spirits of turpentine; put it into an iron or copper kettle placed over a fire, and carefully stir in $\frac{1}{2}$ ounce sulphuric acid. When the whole becomes quite hot, cool and bottle. This is an excellent liniment for all kinds of sprains and bruises, and for horses or cattle it cannot be surpassed.

4888. Stimulating Liniment. Cayenne, $1\frac{1}{2}$ ounces; salt, 1 table-spoonful; spirits of wine, 2 ounces; camphor, $\frac{1}{2}$ ounce; spirits of turpentine, $\frac{1}{4}$ pint. Bottle, and shake now and then during one day. Then add $\frac{1}{2}$ pint vinegar. It is excellent for sponging the body in cases of pain, debility, inflammation, rheumatism, gout, sore throat, numbness, neuralgia, &c.

4889. Embrocation for Bruises. Pour upon 2 ounces carbonate of ammonia (smelling salts) as much distilled vinegar as will dissolve it, then add $1\frac{1}{2}$ pints common rectified spirit, and shake the whole together in a bottle. It is a good remedy for sprains and bruises.

4890. Cajeput Liniment. Mix together 7 ounces soap liniment (*see No.* 4869), $\frac{1}{2}$ ounce camphor, and 1 ounce oil of cajeput.

4891. Cantharides Liniment for Chilblains. Mix together 2 ounces soap liniment and 1 ounce tincture of Spanish flies. Apply at intervals during the day.

4892. Compound Mustard Liniment. Take of oil of mustard, 1 fluid drachm; ethereal extract of mezereon, 40 grains; camphor, 120 grains; castor-oil, 5 fluid drachms; alcohol, 4 fluid ounces; dissolve the extract of mezereon and camphor in the alcohol, and add the oil of mustard and castor-oil.

4893. Nerve and Bone Liniment. Take 1 ounce spirits of turpentine, $\frac{1}{2}$ pint brandy, and 1 gill neat's-foot oil. Simmer over a fire till mixed; then put it into bottles for use.

4894. Mustard Oil Ointment. Crude mustard-seed oil, 16 fluid ounces; ethereal oil of mustard, 30 drops; water of ammonia, 4 fluid ounces, or a sufficient quantity to form into a soap. Mix and bottle in broad-mouthed phials containing about 2 ounces.

4895. Wonderful Ointment. The following liniment is good for all sprains, bruises, lameness, &c.: Mix together 2 ounces oil of spike; 2 ounces origanum; 2 ounces hemlock; 2 ounces wormwood; 4 ounces sweet oil; 2 ounces spirit of ammonia; 2 ounces gum camphor; 2 ounces spirits turpentine. Add 1 quart 95 per cent. alcohol, mix well together, and bottle tight. This is an unequaled horse liniment, and, by omitting the turpentine, it constitutes one of the best liniments ever made for human ailments, such as rheumatism, sprains, &c.

4896. Horse Embrocation. Take $\frac{1}{2}$ ounce each of oil of spike, oil of monarda (horsemint), and strong ammonia water; $\frac{1}{4}$ ounce acetate of opium, 1 ounce chloroform, 2 ounces tincture of camphor, 1 ounce oil of origanum, and 2 ounces oil of camphor. This is said to be an excellent preparation.

Pills.

This form of medicine is particularly adapted for administering nauseous substances, and such as operate in small doses. Extracts may be made into pills either alone or with the addition of any simple powder, as that of liquorice, to increase their consistence. Powders are usually beaten up with syrup, mucilage, conserve of roses, or extract of liquorice. Castile soap is frequently used for substances that are not decomposed by alkalies. When the mixed ingredients are made into a mass, it should be preserved in a bladder placed in a covered stone pot, and occasionally moistened with a little spirit, or spirit and water, to prevent it getting hard. In all cases the dry ingredients should be reduced to fine powder, and the whole beaten into a uniform mass of a proper consistence for rolling into pills. This is effected by rolling it on a slab into a convenient thickness, and dividing into pieces of the requisite weight, lastly rolling them between the thumb and finger, to give them a globular form. A *pill machine* is usually employed for dividing the roll and shaping the pills. In ordinary cases, rolling the pills in carbonate of magnesia or powdered starch is usually adopted, to prevent them sticking together while moist. For other pills not under this heading, see *Index.*

4898. To Sugar-coat Pills. To sugar-coat, place the pills dry and smooth in a round copper pan or porcelain dish. In another pan dissolve white sugar in water in the same proportion as for making simple syrup; and, when dissolved, slowly evaporate the syrup until it feathers; that is, when a small portion taken out with a ladle and drawn up between two fingers forms a thread. The pan with the pills is next suspended over a slow fire, a little fine flour is sprinkled over them, and immediately after a spoonful of the syrup is poured on, or enough to cover. The pan is now kept swinging or moving over the fire, care being taken not to burn the sugar by too much heat, until it is reduced to a fine dust. Then more sugar is added, and the swinging and drying continued until a coat of sufficient thickness is obtained.

4899. To Silver or Gild Pills. Pills are gilded and silvered by rolling them between the fingers slightly moistened with mucilage, and then shaking them up in a small gallipot covered with a piece of paper, along with a little gold or silver leaf, or a little powdered gold or silver.

4900. Aloes Pills. Make 1 ounce aloes and 1 ounce soap into a mass with water. Divide into 240 pills.

4901. Aloes and Assafœtida Pills. Take $\frac{1}{2}$ ounce each powdered aloes, assafœtida, and soap, made into a mass with water. Divide into 180 pills.

4902. Aloes and Myrrh Pills. Mix 1 ounce aloes, $\frac{1}{2}$ ounce myrrh, and $\frac{1}{4}$ ounce saffron, with sufficient syrup to make a mass. This is sufficient for 240 pills.

4903. Assafœtida Pills. Mix into a mass with water $\frac{3}{4}$ ounce assafœtida and $\frac{1}{4}$ ounce soap. Make into 120 pills.

4904. Sulphate of Quinine Pills. Mix $\frac{1}{4}$ ounce sulphate of quinine with 1 drachm powdered gum-arabic, and make into a mass with honey. To make 240 pills, each of which will contain 1 grain of quinine.

4905. Quinia Pills for Chronic Intermittent Fever. Mix 20 grains sulphate of quinia, 2 grains powdered opium, and 5 minims oleo-resin of pepper, with sufficient syrup of gum-arabic to make a mass. Make into 20 pills. Dose, 2 pills every hour in the morning of an expected chill.

4906. Alterative Pills. Take 24 grains blue mass, 3 grains pulverized opium, and 3 grains powdered ipecacuanha. Make into 24 pills.

4907. Vegetable Anti-bilious Pills. Take 54 grains pulverized compound extract of colocynth, and 6 grains podophyllin (extract of may-apple or mandrake root). Make into 24 pills.

4908. Anti-chill Pills. Take 20 grains chinoidine, 40 grains ferrocyanuret of iron, 20 grains oil of black pepper, and 1 grain arsenic. Make up into 20 pills.

4909. Aperient Pills. Take 8 grains nux-vomica, 12 grains extract of henbane, and 48 grains compound extract of colocynth. Make into 24 pills.

4910. Diuretic Pills. Take 40 grains powdered castile soap, 40 grains dry carbonate of soda, and 20 drops oil of juniper. Make into 20 pills.

4911. Gonorrhea Pills. Take 48 grains powdered cubebs, 24 grains solid balsam of copaiba (powdered), 12 grains sulphate of iron, and 36 grains Venice turpentine. Make into 24 pills.

4912. Mandrake Mercurial Pills. Take 6 grains *podophyllin* (extract of mandrake or may apple), and 48 grains blue pill. Make into 24 pills.

4913. Podophyllin, Aloes, and Iron Pills. Take 3 grains podophyllin, 15 grains socotrine aloes, 15 grains extract of nux-vomica, 45 grains dry sulphate of iron, 10 drops oil of cloves, and sufficient syrup of gum-arabic to make into a mass. Divide into 30 pills. Dose, 1 pill immediately before each meal. A good remedy for indigestion, with costiveness.

4914. Opium Pills. Mix 2 drachms opium and 24 grains soap with water, to make 120 pills.

4915. Iodide of Iron Pills. Mix 1 drachm sulphate of iron, 4 scruples iodide of potassium, 10 grains tragacanth, and $\frac{1}{2}$ drachm sugar with syrup. Make into 40 pills.

4916. Compound Iron Pills. Triturate together 2 drachms myrrh and 1 drachm carbonate of soda; then add 1 drachm sulphate of iron, and make up with syrup into 80 pills.

4917. Compound Cathartic Pills. Take $\frac{1}{2}$ ounce compound extract of colocynth, 3 drachms extract of jalap, 3 drachms mild chloride of mercury, and 2 scruples gamboge; mix with water to make 180 pills.

4918. Copaiba Pills. Mix 2 ounces copaiba with 1 drachm fresh magnesia; set it aside to dry, and, when the mass is of proper consistency, make into 200 pills.

4919. Mercurial Pills. These are commonly known as *blue* pills. Rub 1 ounce mercury with $1\frac{1}{4}$ ounces confection of roses; add $\frac{1}{2}$ ounce liquorice root, and divide into 480 pills.

4920. Calomel Pills. Mix $\frac{1}{2}$ ounce mild chloride of mercury with 1 drachm powdered gum-arabic. Make up with syrup, into 240 pills.

4921. Compound Galbanum Pills. 6 drachms myrrh, and 2 drachms assafœtida, mixed with sufficient syrup. Make 240 pills.

4922. Rhubarb Pills. Mix 3 drachms powdered rhubarb and 1 drachm soap with water to make 60 pills.

4923. Compound Rhubarb Pills.

Form into a mass with sufficient water, 1 ounce rhubarb, 6 drachms aloes, ½ ounce myrrh, and ½ fluid drachm oil of peppermint. Divide into 240 pills.

4924. Compound Pills of Squill. Mix 1 drachm powdered squill, 2 drachms ammoniac, and 2 drachms ginger, with 3 drachms soap. Make up with syrup into 120 pills.

4925. Compound Storax Pills. Take 6 drachms of storax, 2 drachms of powdered opium, and 2 drachms of saffron; work up to the proper consistency of a pill mass. Dose, from 5 to 10 grains.

4926. Sulphur Pills. The following formulæ furnish a convenient and neat method of administering sulphur when this useful medicine is required to be given as an alterative in chronic rheumatism and certain diseases of the skin: Take sulphur, 42 grains; castile soap, 18 grains. Mix and divide into 12 pills. 1 to 3 pills for a dose, morning and night. Or: Take sulphur and acetate of potassa, of each 24 grains. Make up with sufficient confection of roses into 12 pills. 1 or 2 twice a day in scorbutic and scrofulous cases, and when sulphur generally is indicated.

4927. Sulphite of Soda Pills. Dr. Polli, who introduced the sulphites to the notice of the medical profession in certain blood diseases, recommends the following formula: Take powdered sulphite of soda, 36 grains; powdered ginger, 12 grains. Make up with mucilage into 12 pills. Dose, 1 to 3 soon after eating. These are given when the stomach is foul, and the food ferments and becomes putrescent. The sulphite of magnesia, Dr. Polli says, is better for this purpose than sulphite of soda. Sulphur obtained by decomposing precipitated sulphide of copper, called *brown sulphur*, is stated by Dr. J. Hannon, an English Physician, to be a most powerful remedy against gout and rheumatism.

4928. Pepsine and Iron Pills. Mix together 2 drachms 34 grains starchy pepsine, and half that weight of iodide of iron in crystals, with sufficient syrup to make 100 pills. Cover them with 2½ drachms reduced iron, and finish with sugar-coating.

4929. Compound Taraxacum Pills. Take ½ drachm extract of taraxacum, and 10 grains blue pill. Make into 10 pills. Dose, 1 pill three times a day, in dropsy with disease of the liver.

4930. Pills of Iodide of Iron. Mix ½ troy ounce iodine with 1 fluid ounce water in a thin glass bottle; add 2 drachms iron wire in small pieces, and shake together until a clear green solution is formed. Mix 1 troy ounce sugar, ½ troy ounce marshmallow, 1 drachm gum-arabic, and 1 drachm reduced iron, all in fine powder, in a porcelain capsule. Filter upon them, through a small filter, first the green solution, heated, and afterwards 2 fluid drachms water. Evaporate over a water-bath with constant stirring, to a mass, and divide it into 300 pills. Dissolve 60 grains balsam of tolu in 1 fluid drachm ether, shake the pills in the solution until uniformly coated, and place them on a plate, occasionally stirring them until dry. Keep in a well stoppered bottle. (*U. S. Ph.*) The iodide of iron pills, as ordinarily prepared, crumble by time and exposure; but, made according to the above formula, they will undergo no change. This is the plan proposed by Prof. Procter in imitation of *Blancard's Pills.* (*U. S. Dis.*)

Ointments, Salves, and Cerates.

Ointments are unctuous preparations, that merely differ from cerates in consistence, being made and used in a similar manner. Their solidity should not exceed that of good butter, at the ordinary temperature of the atmosphere. When the active ingredients are pulverulent substances, nothing can be more suitable to form the mass of the ointment than good lard, free from salt; but when they are fluid, or semi-fluid, prepared suet, or a mixture of suet and lard, will be necessary to give a proper consistence to the compound; in some few instances wax is ordered for this purpose. Glycerine is now frequently prescribed in ointments, and is difficult to mix. Suppose it be ordered with zinc ointment, as is often the case, do not use ready-made zinc ointment, but weigh the proper quantity of oxide, rub the glycerine with it, and then add the lard. This makes a good smooth ointment which does not separate. Of course, the same plan can be adopted with any other powder. If there be no powder, melt the ointment, but do not let it get too hot, and beat the glycerine in and stir till cold; it then mixes much better; but still, if there be a large proportion of glycerine, it will separate after a time. (*See No.* 5009, *&c.*) Unctuous preparations may be prevented from getting rancid, by dissolving in the fat a little gum-benzoin or benzoic acid. The term *cerate* is applied to those unguents which contain wax. A number of these preparations are given here, and others will be found, by referring to the Index, under their respective headings.

4932. Simple Cerate. Melt together 8 ounces lard, and 4 ounces white wax, stirring constantly until cold. (*U. S. Ph.*)

4933. Spermaceti Cerate. Melt together 2 ounces spermaceti, 8 ounces white wax, and 1 pint warm olive oil, and stir assiduously until cold. This is used as a soft cooling dressing. As soon as the materials are melted, they should be moved from the fire, strained into a clean vessel, and stirred until cold. To facilitate the cooling, the vessel may be placed in cold water or a current of cold air. This will render the product both whiter and finer than when allowed to cool by itself. The operation of melting should be performed in a water-bath. On the large scale lard or suet is substituted for oil, by which means less wax is required. The following is a good form where a cheap article is wanted: Clarified mutton suet, 5½ pounds; white wax and spermaceti, of each ¾ pounds. As above.

4934. Chilblain Ointment. Take of gall-nuts, in very fine powder, 1 drachm avoirdupois; spermaceti cerate (*see No.* 4933), 7 drachms; mix, add pure glycerine, 2 drachms, and rub the whole to a uniform mass. An excellent application to obstinate broken chilblains, particularly when used as a dressing. When the parts are very painful, 1 ounce of compound ointment of galls may be advantageously substituted for the galls and cerate ordered above. (*See No.* 5006.)

4935. Family Salve. Take the root of yellow dock and dandelion, equal parts; add good proportion of celandine and plantain. Extract the juices by steeping or pressing. Strain carefully, and simmer the liquid with sweet cream, or fresh butter and mutton tallow, or sweet oil and mutton tallow. Simmer together until no appearance of the liquid remains. Before it is quite cold, put it into boxes. This is one of the most soothing and healing preparations for burns, scalds, cuts, and sores of every every description.

4936. Salve for All Wounds. Take 1 pound hog's lard, 3 ounces white lead, 3 ounces red lead, 3 ounces bees'-wax, 2 ounces black resin, and 4 ounces common turpentine; all these ingredients must be put together in a pan, and boil ¾ of an hour; the turpentine to be put in just before it is done enough, and give it a gentle boil afterwards. This is an excellent cure for burns, sores, or ulcers, as it first draws, then heals afterwards; it is excellent for all wounds.

4937. Lard Ointment. Melt 2 pounds pure lard, add 3 fluid ounces rose-water, and beat them well together while hot. When cold, separate the congealed fat from the water. This is simple lard ointment.

4938. Savine Ointment. Savine tops, dried and in fine powder, 1 drachm; ointment of white wax (simple ointment), 7 drachms; mix by trituration.

4939. Simple Ointment of White Wax. Olive oil, 5½ fluid ounces; white wax, 2 ounces; melted together and stirred while cooling.

4940. Spermaceti Ointment. Melt together 5 ounces spermaceti, 14 drachms white wax, and about 1 pint olive oil. The article commonly sold as spermaceti ointment is composed of 1 pound spermaceti, ½ pound white wax, and from 3 to 6 pounds pure lard.

4941. Camphor Ointment. Camphor, finely powdered, 1 ounce; lard, 2 ounces. Mix. It is designed to ripen indolent tumors.

4942. Compound Iodine Ointment. Mix 1 drachm iodide of potassium in very fine powder, with 2 ounces lard; then add ½ drachm iodine dissolved in 1 fluid drachm rectified spirit.

Fresh lard cannot always be got, and as long as simple cerate is directed to be made with white wax, an already rancid body, it happens very often that an ointment of iodide of potassium gets yellow, instead of being perfectly white. A few grains of hyposulphite of soda dissolved in a little water, added to such ointment, will have the effect of turning it snow-white.

4943. Compound Belladonna Ointment. Mix 1 drachm fresh extract of belladonna with 7 drachms of compound iodine ointment. (*See No.* 4942.) For dispersing glandular tumors, &c., which it is not desirable to mature.

4944. Ammoniacal Ointment. Melt 1 ounce each of suet and lard, in a strong wide-mouthed bottle; add 2 ounces liquor of ammonia of specific gravity .923, and close the bottle immediately. Then mix, by shaking the bottle, until the contents harden. The fat should not be heated any more than is sufficient to melt it, to prevent unnecessary loss of ammonia.

4945. Catechu Ointment for Tropical Climates. An astringent ointment may be prepared, which is not likely to become soon rancid, as is the case with ointments made with fat. Melt 4 ounces resin in ½ pint olive oil; add 1 ounce alum and 3 ounces catechu, both finely powdered.

4946. Stramonium Ointment. Mash ½ bushel of green stramonium, or jimson leaves, to a pulp (this is best done by mashing a few leaves at a time), put the pulp in an iron kettle over a slow fire. Add 2¼ pounds fresh lard, and simmer to a crisp. Strain and box for use. Or: Take extract of stramonium, 1 drachm; lard, 1 ounce, and mix by trituration. This ointment is excellent for strengthening broken limbs after the bones have healed. It is also good for skin diseases, painful piles, ulcers, burns and scalds. It is probably the best ointment that can be kept in a family for general use.

4947. Citrine Ointment. Dissolve 1½ ounces mercury in 3½ ounces nitric acid. Stir till effervescence ceases. Heat 16½ ounces lard to 200° Fahr., in an earthen vessel, and add the solution, stirring constantly until thoroughly amalgamated.—(*U. S. Ph.*)

4948. Mercurial Ointment. Triturate 24 ounces mercury with 12 ounces each lard and suet. *Mild Mercurial Ointment* is composed of 1 part lard added to 2 parts Mercurial Ointment.—(*U. S. Ph.*)

4949. Magnetic Adeps. This is a prepared fat used for making mercurial ointment, as it will reduce 30 to 40 times its weight of quicksilver to salve. It is made by pouring melted lard, in a small stream, into cold water, placing the thin fragments thus obtained in a sieve covered with paper, or other suitable apparatus, and exposing it to the air for 3 or 4 months.

4950. Ointment of Iodide of Sulphur. Reduce 30 grains iodide of sulphur to a fine powder, rub it with a small portion taken from 1 troy ounce lard, then add the remainder of the ounce of lard, and mix them thoroughly. (*U. S. Ph.*)

4951. Ointment of Borax. This is also called *Pomade de Toscanie.* Take of borax in very fine powder, 1 drachm avoirdupois; spermaceti ointment, 1 ounce; mix by trituration. In excoriations, chaps, &c. It also forms an excellent lip-salve. A drop of neroli, or ½ drop of otto of roses, renders it more agreeable.

4952. Glycerinated Ointment of Borax. To the borax ointment, as prepared in the foregoing receipt, add 1 drachm avoirdupois pure glycerine, using a slightly warmed mortar for the mixture. This is a very effective ointment.

4953. Ointment of Creosote, or Creosote Pomade. Take of creosote, 1 fluid drachm; spermaceti ointment (*see No.* 4940), 1 ounce avoirdupois; triturate them together in a slightly warmed mortar until perfectly united, and subsequently until nearly cold. It is used as a dressing for scalds and burns, chilblains, &c. It is very useful in ringworm and some other skin diseases; also as a friction in facial neuralgia or tic-douloureux.

4954. Ointment for the Itch. The usual treatment of itch has been noticed elsewhere, and various lotions, ointments and pomades, of more or less value in its treatment, will be found under the names of their leading ingredients. Here are two additional formulæ:

4955. French Hospital Itch Ointment. Take of chloride of lime, 1 drachm avoirdupois; rectified spirit, 2 fluid drachms; rub them together, add ½ fluid ounce sweet-oil; soft-soap, 2 ounces aviordupois; oil of lemon, ¼ fluid drachm; mix perfectly, and then further add common salt and sulphur, of each 1 ounce. Cheap, very effective, and much less offensive than sulphur ointment.

4956. Stavesacre Ointment. Melt together 1 ounce powdered stavesacre (staphisagria), and 3 ounces lard; digest for 3 or 4 hours, and strain. A cleanly remedy for itch, and for destroying body vermin.

4957. Ointment for Baker's Itch. Mix well together ¼ ounce ointment of nitrate of mercury (*see No.* 4947), and 1 ounce palm oil.

4958. Venice Turpentine Ointment. Venice turpentine, 2 ounces; tar, 1 ounce; butter, 4 ounces. Simmer until they are well mixed. This is very good for scald-head, ringworm, &c. First wash the head well with soap and water, and then apply the ointment.

4959. Brown Ointment. Extract of henbane, 1 drachm; yellow wax, ½ ounce; red precipitate, 2½ drachms; pure zinc, powdered, 1¼ drachms; fresh butter, 3 ounces. Melt and mix, and add 1½ drachms camphor dissolved in olive oil. This ointment is good for ringworm, all cutaneous eruptions, for ulcers, sore lips, itch, chronic ophthalmia, &c.

4960. Tar Ointment. Tar and mutton suet, equal parts; melt together, and stir till cold. This is an excellent remedy for scald-head and ringworm.

4961. Tobacco Ointment. Fresh tobacco leaves, chopped small, 1 ounce; lard, 1 pound; boil till crisp, and strain through lime. Used for ringworm, irritable ulcers, and other diseases of the skin. It should be used with caution.

4962. Salt Rheum Ointment. Mix in an earthen vessel, 1 ounce aqua-fortis, with 1 ounce quicksilver; when effervescence has ceased, incorporate with it 1 pound lard and 1 ounce dissolved hard soap; then work into the mixture 1 ounce prepared chalk and ½ table-spoonful spirits of turpentine.

4963. Magnetic Ointment. Lard, raisins cut in pieces, and fine-cut tobacco, equal weights; simmer well together, then strain and press out all from the dregs. This is an excellent ointment for salt-rheum and other skin diseases. It is also good for piles, bruises, and cuts.

4964. Basilicon Ointment. Take 10 ounces resin, 4 ounces yellow wax, and 16 ounces lard; melt them together, strain through muslin, and stir constantly until cool. This is the *resin ointment* of the U. S. Pharmacopœia. The British officinal preparation contains only 8 ounces resin, and substitutes simple ointment for the lard.

4965. Yellow Basilicon Ointment. Yellow wax, 8 ounces; burgundy pitch, 3 ounces; Venice turpentine, 4 ounces; linseed oil, 10 ounces. First melt the resin, to which add the wax and the burgundy pitch. When the whole is melted, remove from the fire, and slowly put in the oil, stirring well till it is cold. For healing cuts, abscesses, &c.

4966. Black Basilicon Ointment. Black basilicon, yellow wax, and yellow resin, 10 ounces; common pitch, 5 ounces. Melt as before, and add 10 ounces linseed oil when taken from the fire.

4967. Green Basilicon Ointment. Yellow wax and yellow resin, of each 3 ounces; Venice turpentine, 6 ounces; powdered verdigris, 1 ounce; lard, 6 ounces. Melt first the resin, &c., as before. Very efficacious in healing cuts, abscesses, and local affections of any kind.

4968. Saturnine Cerate. Powdered acetate of lead, 2 drachms; white wax, 2 ounces; olive oil, ½ pint. Melt the wax in the oil, and add gradually the acetate of lead, separately rubbed down with a portion of the oil reserved for that purpose.

4969. Hemlock Salve. Hemlock ointment, 12 ounces; spermaceti, 2 ounces; white wax, 3 ounces; melt the last two, then add them to the first, softened by a gentle heat. Used for inveterate cancerous, scrofulous, and other sores.

4970. Green Stick Salve. According to the American Dispensatory, this is prepared by taking white gum turpentine, bayberry wax, of each 2 ounces; melt together, strain, and stir till cold; adding olive oil will give it the consistence of an ointment.

4971. Black, or Healing Salve. Olive oil, 1 pint; common resin, ½ ounce; bees'-wax, ½ ounce; Venice turpentine, ¼ ounce. Melt, raising the oil nearly to the boiling point; then gradually add 2 or 3 ounces powdered red lead while on the fire; do not burn it; boil slowly till it becomes a dark brown; remove from the fire, and add 1 drachm powdered camphor when it is nearly cold. This is a first-rate healing salve, superior to most; is wonderful in burns, scalds, scrofulous, fistulous, and all other ulcers. Spread on linen, and renew daily.

4972. Red Salve. Red lead, 1 pound; bees'-wax and resin, of each 2 ounces; linseed and sweet oils, of each 3 table-spoonfuls; spirits of turpentine, 1 tea-spoonful; melt all, except the first and last, together, then stir in the lead and stir until cool, adding the turpentine. Good for all inflamed sores.

4973. Green Salve. White pine turpentine and lard, ½ pound each; honey and bees'-wax, ¼ pound each; melt all together and stir in ½ ounce of very finely pulverized verdigris. This ointment cannot be surpassed when used for deep wounds. It prevents proud flesh from forming, and keeps up a healthy discharge.

4974. Green Ointment. Take prepared subacetate of copper, ½ drachm; ointment of white wax (*see No.* 4939), 7½ drachms. Triturate the subacetate of copper with the ointment until they are intimately mixed. A mild caustic, applied to venereal ulcers of the mouth and tonsils, and to the ulcerated sore throat of scarletina.

4975. Cod-Liver Oil Ointment. Melt together 1 part white wax, 1 part spermaceti, and 7 parts pale cod-liver oil. Used for ophthalmia, scrofulous sores, rheumatism, stiff joints, and some skin diseases, including ringworm. Scented with oil of nutmeg and balsam of Peru it forms an excellent pomade for strengthening and restoring the hair.

4976. Ointment for Old Sores. Red precipitate, ½ ounce; sugar of lead, ½ ounce; burnt alum, 1 ounce; white vitriol, ¼ ounce or a little less; all to be very finely pulverized; have mutton tallow made warm, ½ pound; stir all in, and stir until cool. Good.

4977. Bitter-Sweet Ointment. Bark of bitter-sweet root, 2 ounces; cover with spirits of wine, and add, unsalted butter, 8 ounces. Simmer and strain. Excellent for swelled breasts, tumors, ulcers, &c. It may be applied twice a day.

4978. Astringent Ointment. Triturate 1½ drachms powdered catechu with 2 fluid drachms boiling water; add, gradually, 1½ ounces spermaceti ointment, continuing the trituration until the mass concretes. This is an excellent dressing for sores and ulcers, especially during hot weather.

4979. Neuralgia Ointment. Take 2 drachms each of cyanide of potassium, and chloroform, and make into a salve with 1 ounce lard, for external application.

4980. Ointment of Lead. Take of olive oil, ½ pint; white wax, 2 ounces; sugar of lead, 3 drachms. Let the sugar of lead, reduced to a fine powder, be rubbed with some of the oil, and added to the other ingredients, previously melted together, stirring them till quite cold. This cooling astringent ointment may be used in all cases where the intention is to dry and skin over the part, in scalding, &c.

4981. Zinc Ointment. Mix 1 ounce oxide of zinc and 6 ounces lard. This is astringent, desiccative, and stimulant; an excellent and useful application for burns, excoriations, and skin diseases attended by discharges.

4982. Chloroform Ointment for Neuralgic Pains. Mix 1 drachm chloroform with 1 ounce spermaceti ointment. (*See No.* 4933.) This should be kept in a wide-mouthed, stoppered phial.

4983. Belladonna Anodyne Ointment. Mix 3 drachms fresh and good extract of belladonna, ½ drachm powdered opium, and 3 drachms lard. For neuralgia, &c., apply with friction for 6 to 8 minutes.

4984. Aconitine Ointment. Aconitine, 16 grains; alcohol, 12 drops; olive oil, ½ drachm; lard, 1 ounce. Rub the aconitine with the spirit, then add the oil by drops, and, after it is thoroughly mixed, pour in the lard rendered nearly liquid by heat; stir well until cold. A small portion is applied by the tips of the fingers and gentle friction, in neuralgic and rheumatic affections, &c.

4985. Ointment for Sore Nipples. Glycerine, rose-water, and tannin, equal weights, rubbed together into an ointment, is very highly recommended for sore or cracked nipples.

4986. Tannin Ointment for Piles. Tannin, 2 drachms; water, 2 fluid drachms; triturate together, and add lard, 1½ drachms. An excellent application for piles.

4987. Spackman's Pile Ointment. Mix together 1½ ounces carbonate of lead; 6 grains sulphate of morphia; 1 ounce stramonium ointment (*see No.* 4946); and sufficient olive oil to make into a salve.

4988. Ointment for Piles. Triturate 8 grains morphia in 1 ounce melted spermaceti ointment (*see No.* 4940), until the morphia is dissolved; then add 1½ drachms of galls in impalpable powder, 12 to 15 drops essential oil of almonds, and stir until the mass is cool.

4989. Pile Salve. Take 1 scruple powdered opium, 2 scruples flour of sulphur, and 1 ounce of simple cerate. (*See No.* 4932.) Keep the affected parts well anointed. Be prudent in your diet.

4990. Salve for Sore Breasts. Take 1 pound tobacco, 1 pound spikenard, ½ pound of cumfrey, and boil them in 3 quarts chamber-lye till almost dry; squeeze out the juice, add to it pitch and bees'-wax, and simmer it over a moderate heat to the consistence of salve. Apply it to the part affected.

4991. Iodide of Lead Ointment. An ointment of iodide of lead composed of 4 parts iodide of lead, 4 parts chloride of ammonium, and 50 of lard, is either of a yellow or white color, according to the manner in which these ingredients are brought together. When rubbed together dry, the color of the mixture is yellow; but when the chloride of ammonium, in order to facilitate the mixing, is first liquefied in a small quantity of water before being added to the iodide of lead, the yellow color of the latter disappears, owing to the formation of two colorless salts, the chloride of lead and iodide of ammonium. It is well in cases like these to adhere strictly to the directions of the prescription. (*Eymael.*)

4992. Ingall's Iodoform Ointment. Dissolve ½ drachm iodoform in sufficient rectified alcohol, and make into an ointment with 7½ drachms lard. Iodoform is extensively and successfully used in the treatment of syphilitic ulcers and rupia. The above formula is the one adopted by Dr. Ingalls, attending surgeon of the Boston city hospital.

4993. Carbolic Cerate. Melt together 5 ounces lard, and 2½ ounces white wax; add ½ ounce balsam of fir, and when it begins to cool, stir in ½ ounce carbolic acid. The addition of balsam fir to this preparation corrects the disagreeable odor of the acid, and renders it slightly adhesive, which is quite desirable when used as a dressing for burns, old sores, &c. (*See No.* 4996.)

4994. Ointment of Tannate of Manganese. Mix 3 grains tannate of manganese with 1 troy ounce cold cream. (*See No.* 1125.) This is a good application for bad wounds.

4995. Tartar Emetic Ointment. Take 2 drachms potassio-tartrate of antimony, and rub it well into 1 ounce lard. This will produce an eruption on the skin very similar to small-pox in appearance.

4996. Carbolic Salve. There are different formulæ recommended for this salve, containing different amounts of carbolic acid; the character of the disease will determine which to use. The carbolic acid employed is the crystallized article, sold in bottles, and taken out by warming the latter in hot water, or the fluid resulting from the crystals, which are melted in warm weather, or are dissolved by absorbing a little water, when the bottles are not perfectly stoppered.

I. Take carbolic acid, ½ fluid drachm, and lard, 1 ounce. Triturate together in a porcelain mortar.

II. Take carbolic acid, 1 fluid drachm, and lard, 3 ounces. Melt the lard at a gentle heat, add the carbolic acid, and triturate until the mixture is cold.

III. Take carbolic acid, 1 fluid drachm, and ointment of white wax (*see No.* 4939), 7 drachms. Prepare as No. II. (*See No.* 4993.)

4997. Cerate of Savine. Moisten 3 troy ounces savine in fine powder with ether; pack it firmly in a cylindrical percolator, and displace with ether until the percolate passes nearly colorless. Evaporate spontaneously to the consistence of syrup, add it to 12 troy ounces resin cerate softened by a gentle heat, and mix thoroughly.

4998. Sulphur Ointment. Mix together 1 ounce sublimed sulphur and 2 ounces lard.

4999. Itch Ointment. Washed sulphur, 1½ ounces; chloride of lime, 2 drachms; hog's lard, 4 ounces. Mix and make into an ointment.

5000. Cucumber Ointment. Take of oil of sweet almonds, 7 fluid ounces; spermaceti, 18 drachms; white wax, 5 drachms; glycerine, 1 fluid ounce; green cucumbers, 4 pounds. Cut the cucumbers in small pieces, mash them in a wedgwood mortar, let them macerate in their own liquor for 12 hours, express and strain; melt the almond oil, spermaceti, and wax together, by means of a water-bath; add to it the strained liquor, stirring constantly so as to incorporate the whole together. Set aside in a cool place (an ice-chest preferred) till it becomes hard, then beat with a wooden spoon, so as to separate the watery portion of the cucumbers from the ointment; pour off the liquor thus obtained, and mix the glycerine with the ointment without the aid of heat, by working it with the hands until it becomes thoroughly incorporated. Put up in 4-ounce jars, cover with a layer of rose-water, and set aside in a cool place.

5001. Foot-Rot Ointment. Lard and Venice turpentine, 4 ounces of each; melt and add 1 ounce blue vitriol. Good for cows or sheep.

5002. Cracked Hoof Ointment. Tar and tallow, equal parts melted together.

5003. Compound Resin Cerate. Melt together 12 troy ounces each of resin, suet, and yellow wax; 6 troy ounces turpentine, and 7 troy ounces flax-seed oil. Strain through muslin, and stir constantly till cool. (*U. S. Ph.*) This preparation, also known as *Deshler's Salve*, should be kept well protected from the air, as it is liable to become tough by exposure. (*U. S. Dis.*)

5004. Egyptiacum Salve. Take 1½ ounces verdigris, 1½ ounces alum, ½ ounce sulphate of copper, ⅛ ounce corrosive sublimate, all in powder; boil over a slow fire with 2½ ounces vinegar and ½ pound honey until of a proper consistence. Stir up well before using.

5005. Egyptian Ointment. A detergent application for foul ulcers, &c. Mix by heat and agitation, 10 parts verdigris, 1 part calcined alum, 14 parts strong vinegar, and 32 parts thick purified honey.

5006. Compound Gall Ointment. Rub together 6 drachms very finely powdered gall-nuts, 1½ drachms powdered opium, and 6 ounces lard.

5007. German Black Salve. Lard, 24 parts; white oxide of zinc and Peruvian balsam, of each 3 parts; nitrate of silver, finely pulverized, 1 part. This formula is taken from the Hamburg Pharmacopœia.

5008. To Keep Ointment from Becoming Rancid. About 2 per cent. of finely powdered gum benzoin, or a less quantity of benzoic acid dissolved in the fatty matter by heat, will greatly retard, if not wholly prevent, the ointment from turning rancid.

5009. Schacht's Glycerine of Starch, or Plasma. The use of fatty matter as the vehicle for drugs in preparing ointments and cerates is sometimes open to objection. The remedies introduced are frequently insoluble in fat, which consequently acts to a certain extent in defending the skin from, instead of facilitating the perfect action of the remedy. Aqueous remedies are difficult to mix with fat without soap or some otherwise needless addition. Another strong objection is the tendency of fatty matter to become rancid in contact with the skin. Mr. G. F. Schacht proposes a substitute consisting of 1 fluid ounce pure glycerine and 70 grains starch powder. These are mixed while cold, and then gradually heated to about 240° Fahr., constantly stirring; he gives this preparation the name of *plasma*. This constitutes a basis whose consistence is good, and does not vary with changes of temperature; it is soluble in water, and may consequently be removed from tender surfaces with the greatest ease; it dissolves and thoroughly mingles with all materials that are soluble in water, and therefore presents such remedies in the condition most favorable for absorption; and, lastly, it is not liable to rancidity. With plasma substituted for fat, may be produced preparations corresponding to most of the cerates and ointments of the Pharmacopœia, but free from the special objections before alluded to. The plasma should be kept in a closely corked bottle. The following plasmas are proposed by Mr. Schacht as improvements on the corresponding ointments of the Pharmacopœia.

5010. Schacht's Cantharides Plasma. Evaporate the decoction of Spanish flies to an extract, and mix with the plasma, using the same proportions as laid down for cantharides ointment. (*See No.* 5017.)

5011. Schacht's Mercurial Plasma. Mix 14 drachms starch with 6 fluid ounces glycerine, gradually adding 12 ounces mercury, and stirring till the globules disappear. Then add 6 fluid ounces glycerine, and heat to 240° Fahr., constantly stirring.

5012. Schacht's Glycerinated Nitrate of Mercury. Take 1 drachm terbasic nitrate of mercury, and 1 ounce plasma.

5013. Schacht's Glycerinated Iodide of Potassium. Dissolve 2 drachms iodide of potassium in 2 fluid ounces glycerine; add 140 grains starch, and heat to 240° Fahr.

5014. Schacht's Glycerinated Petroleum. Rub 1 drachm petroleum with 70 grains starch until quite smooth, then add gradually 1 fluid ounce glycerine.

5015. Glycerinated Iodine. This is recommended for loss of voice, and is composed of 16 grains of iodine in 1 ounce inodorous glycerine. The addition of starch to this is not advisable, as it would convert the iodine into iodide of starch.

5016. Narcotic Glycerole, for external use, applied on lint. Take 1 part aqueous extract of opium, 4 parts extract of belladonna, and 60 parts glycerine.

5017. Cantharides Ointment. Infuse for 12 hours 1 ounce avoirdupois of cantharides in 6 imperial fluid ounces olive oil in a covered vessel. Place the vessel in boiling water for 15 minutes, press through muslin, and add 1 ounce melted yellow wax, stirring constantly till cool. (*Br. Ph.*)

Poultices.

External applications, used to promote suppuration, allay pain and inflammation, resolve tumors, &c. They are generally prepared with substances capable of absorbing much water, and assuming a pulpy consistence, so as to admit of their application to any surface, however irregular. Their curative action principally depends on the liquids with which they are moistened, and the heat retained by the mass. The addition of a little lard, olive oil, or, still better, glycerine, to a poultice, promotes emollient action and retards hardening. A fold or two of lint dipped in hot water, either simple or medicated, and covered with a thin sheet of gutta-percha, or India-rubber cloth, to prevent evaporation, may often be conveniently employed instead of a poultice. Spongio-piline (*see No.* 5039) is still better for this purpose than lint. The following are the principal poultices, but others may be found by referring to the Index.

5019. Bread Poultice. Take stale bread in crumbs, pour boiling water over it, and boil till soft, stirring it well; then take it from the fire, and gradually stir in a little glycerine or sweet oil, so as to render the poultice pliable when applied.

5020. Slippery Elm Poultice. Take a sufficient quantity of pulverized slippery elm bark; stir it in hot or warm milk and water, to the consistence of a poultice. This is a most efficacious poultice; is of almost universal application, and removes inflammation sooner than any other. If tincture of myrrh be added, it is valuable in boils, ulcers, carbuncles, &c.

5021. Mustard Poultice. Take equal parts of ground mustard and ground flax-seed, and mix them thoroughly together, with barely enough of water to make them of the thickness of common paste. To prevent sticking, a little glycerine or sweet oil is to be added. The addition of bread crumbs serves to diminish, that of a little vinegar to increase the irritating power of the mustard.

5022. Strong Mustard Poultice. Mix the best English ground mustard with strong vinegar; spread it on a piece of book or tarleton muslin, to prevent its adhesion to the skin. Wet the part first with vinegar, and apply the poultice.

5023. Linseed Poultice. Take of linseed, powdered, 4 ounces; hot water, ½ pint. Gradually sprinkle the powder into, and stir well with a spoon. This is good and convenient for many cases. It is preferable to the bread and milk poultice so much in use, as it is not so liable to become brittle and hard when dry. It is very useful in carbuncle, obstinate inflammation, &c.

5024. Carrot Poultice. Take of boiled carrots, bruised, 1 pound; flour, 1 ounce; butter, ½ ounce. Mix them with a sufficient quantity of hot water to form a pulp. This will be found a valuable application in ulcerated sores and swellings, scrofulous sores of an irritable kind, and many other inveterate ulcers.

5025. Poultice for Sprains and Bruises. Carbonate ammonia, 2 ounces; vinegar, 2 pints; proof spirits, 3 pints. Mix the ammonia and vinegar; when the effervescence ceases, add the spirit. For inflammation of the joints, of some standing, mix with aniseed meal, and use as a poultice twice a day. It is also valuable for sprains, bruises, and other injuries.

5026. Charcoal Poultice. Linseed meal, ½ pound; charcoal powder, 2 ounces; hot water, sufficient to give it the necessary consistence. Or: Soak 2 ounces bread in ½ pint boiling water; add to this, by degrees, 10 drachms linseed meal; and, afterwards, 2 drachms powdered fresh charcoal; then sprinkle 1 drachm powdered charcoal on the surface of the poultice. This poultice is highly antiseptic; that is to say, it has great power in cleansing ulcers and correcting a tendency to mortification. The power is derived from the charcoal, which is remarkable for its purifying energy. It should be frequently renewed. Dr. Bird, in his work on the medical uses of charcoal, gives numerous proofs of the efficacy of this application. Besides purifying and healing, it conteracts the offensive smell arising from putrid sores.

5027. Yeast Poultice. Take of milk, blood-warm, 1 pint; yeast, 1 gill. Stir in fine slippery elm bark, to form a poultice. This is a good antiseptic and refrigerant poultice. Applied to gangrenous ulcers, it is more efficacious than any others; it sooner arrests mortification, used with proper auxiliaries. It is also very serviceable in other species of inflammation.

5028. Indian Turnip Poultice. Take of the tops and roots of Indian turnip, if green; if dry, the roots only; simmer in water, and add slippery elm bark sufficient to form a poultice. This poultice is used in the treatment of scrofula with the best effect. It is superior to every other poultice in scrofula, in a state of swelling and inflammation.

5029. Potato Poultice. Boil the common potato, mash or bruise soft, and then stir in finely pulverized slippery elm bark. This poultice has been used with success in ophthalmia (inflammation of the eyes) of an acute character, when other means have failed.

5030. Goulard's Poultice. It is thus made: Take 1½ drachms extract of lead (solution of acetate of lead); rectified spirit of wine, 2 ounces; water, 12 ounces; bread-crumb, sufficient to make the whole into a proper consistence. This poultice is an excellent application to reduce swelling and inflammation, and to allay irritation.

5031. Lobelia Poultice. Linseed meal, ½ ounce; slippery elm, 1 ounce; powdered lobelia, 1½ ounces; ginger, 1 ounce; whiskey sufficient to make it. Good for all inflamed parts, as the side in pleurisy, liver complaints, rheumatism, lumbago.

5032. Poultice for a Fester. Boil bread in lees of strong beer; apply the poultice in the general manner. This has saved many a limb from amputation.

5033. Alum Poultice. Take of alum, in fine powder, 1 drachm avoirdupois, and the white of 2 eggs; shake them together until they coagulate. Formerly much used in broken chilblains, chaps, sore nipples, chronic inflammation of the eyes, &c., applied on linen, and covered with a piece of fine muslin.

5034. Hemlock Poultice. Make a poultice of 4½ ounces linseed meal in ½ pint boiling water; spread on its surface 1 ounce extract of hemlock softened with a little hot water. This is an anodyne application for irritable and painful cancerous, scrofulous, and syphilitic sores, tumors, &c.

5035. Gout Poultice. Dissolve 6 drachms balm of Mecca in 16 ounces rectified spirit; next digest for 48 hours, 1 ounce each of red cinchona bark, sarsaparilla, and sage, and ½ ounce saffron, in 32 ounces rectified spirits; filter this, mix it with the solution of balm of Mecca, and add twice their weight of lime-water. Sprinkle 2 fluid ounces on the surface of a hot linseed meal poultice, large enough to surround the affected part.

5036. Soap Poultice. Dissolve 1 ounce scraped or sliced white soap in ¼ pint boiling water, and mix with sufficient bread to make a poultice. This is good for scalds and burns.

5037. Vinegar Poultice. Soak bread in vinegar and apply cold; for bruises, extravasations, black-eyes, &c.

5038. Chlorinated Poultice. Mix gradually 4½ ounces linseed meal with 6 fluid ounces boiling water; add 2 fluid ounces of a solution of chlorinated soda (chloride of sodium), applied to foul ulcers, &c.

5039. Spongio-piline. This is the name of a very ingenious contrivance, recently introduced abroad, which may be used either as a poultice or as a means of fomentation. It consists of wool and small particles of sponge, apparently felted together, and attached to a skin of India-rubber. It is about half an inch in thickness. It will be found of great value and convenience for either of the purposes referred to. It retains heat for a considerable time, and vinegar, laudanum, camphor, hartshorn, etc., can be, by its means, placed on the skin, accompanied by heat and moisture, much more readily, and with greater cleanliness, than by means of ordinary poultices.

Plasters.

External applications that possess sufficient consistence not to adhere to the fingers when cold, but which become soft and adhesive at the temperature of the human body. Plasters are chiefly composed of unctuous substances united to metallic oxides, or to powders, wax, or resin. They are usually formed whilst warm, into ½ pound rolls about 8 or 9 inches long, and wrapped in paper. When required for use, a little is melted off the roll by means of a heated iron spatula, and spread upon leather, linen, or silk. The less adhesive plasters, when spread, are usually surrounded with a margin of resin plaster, to cause them to adhere. In the preparation of plasters, the heat of a water-bath, or steam, should be alone employed.

5041. To Spread Plasters. In spreading plasters convenience requires and neatness demands an uncoated marginal edge. This is usually secured by pasting strips of paper along the edges of the skin or other material used, and removing them after the spreading of the plaster is affected. It is just here that a practical difficulty frequently arises. The paper edges are liable, from drying of the paste, to adhere so strongly that either paper or skin will give way upon an attempt at their removal; the application of water will then be necessary to soften the attachment, and the final results may be expected to present a daubed and uncleanly aspect. This difficulty may be entirely avoided by applying to the paste brush a little glycerine before the adjustment of the marginal strips. (*Ebert*).

5042. To Prevent Plasters from Adhering to Paper. It is recommended to dust the latter over with powdered French chalk. If a piece of thin paper, moistened with olive oil and then wiped dry, be laid over a plaster, it will prevent adhesion to the wrapping paper.

5043. Litharge, Lead, or Diachylon Plaster. Take 5 pounds litharge in very fine powder, 1 gallon olive oil, and 1 quart water. Or: 5 ounces litharge, 12 fluid ounces olive oil, and 8 fluid ounces water. Unless the oil is fully 2½ times the weight of the litharge, the plaster soon gets hard and non-adhesive. Put the water and litharge into a perfectly clean and well polished tinned copper or copper pan, mix them together with a spatula, add the oil, and boil, stirring constantly until the plaster is sufficiently hard when thoroughly cold. This process usually occupies from 4 to 5 hours. The operation may be completed in from 20 to 30 minutes by adding to the litharge and water ¼ pint colorless vinegar, for each pound of litharge employed, previous to adding the oil.

5044. Mahy's White Lead Plaster. Boil together 1 pound pure carbonate of lead, 32 fluid ounces olive oil, and sufficient water, constantly stirring until perfectly incorporated; then add 4 ounces yellow wax, and 1½ pounds lead plaster; when these are melted, and the mass somewhat cooled, stir in 9 ounces powdered orris root. This is an application much used for inflamed and excoriated surfaces, bed-sores, burns, &c.

5045. Deschamp's Plaster. Fasten a piece of fine muslin, linen, or silk, to a flat board; give it a thin coating of smooth, strained flour paste. When dry, apply 2 coats of colorless gelatine, made into size with warm water. This is said to be superior to the ordinary court plaster.

5046. Adhesive Resin Plaster. Resin plaster, spread upon muslin, forms the well-known *Strapping* or adhesive plaster, so extensively used for protecting raw surfaces, supporting parts, dressing ulcers, retaining the lips of recent cuts and wounds in contact, &c. It is gently stimulant, and is thought to assist the healing process; it is also employed as a basis for other plasters. Mix by a moderate heat, 1 ounce resin with 5 ounces litharge plaster. (*See No.* 5043.) Or: 4 ounces resin, and 2 ounces powdered castile soap, with 2 pounds litharge plaster.

5047. Cancer Plaster. White oak-bark, 4 ounces; bruise it well, and add urine sufficient to cover it. Infuse four days, boil it till it becomes as thick as molasses. Add 2 ounces honey and 2 ounces strained turpentine gum. To make this plaster caustic, add 2 drachms white vitriol. Spread on soft leather or linen. It may be applied to all kinds of ulcers and white swellings. For cancers it is invaluable.

5048. Anodyne Plaster. Melt an ounce of adhesive plaster, or diachylon (*see No.* 5043), and, whilst cooling, add a drachm of powdered opium, and the same quantity of camphor, previously dissolved in a small quantity of olive oil. Spread on leather. This soon relieves an acute local pain. Or: Powdered opium, ½ ounce; resin of the spruce fir, powdered, 3 ounces; lead plaster, 1 pound. Melt the plaster and resin together, then add the opium and mix the whole. Useful for rheumatic pains.

5049. Strengthening Plaster. Litharge plaster, 24 parts; white resin, 6 parts; yellow wax and olive oil, of each 3 parts; red oxide of iron, 8 parts. Let the oxide be rubbed with the oil, the other ingredients added, melted, and mix the whole well together. This is an excellent plaster for relaxation of the muscles and weakness of the joints arising from sprains and bruises. The plaster spread over leather should be cut into strips 2 inches wide, and strapped firmly round the joints.

5050. Cough Plaster. Castile soap, 1 ounce; lead plaster, 2 drachms; sal-ammoniac, 1 drachm. Melt the soap and lead plaster together, and add the ammoniac when the mixture is nearly cold. This plaster must be applied to the chest immediately after it is spread, and must be renewed every 24 hours. It is often of great service in whooping-cough and coughs of an asthmatic character.

5051. Resolvent Plaster. Purified ammoniac, 1 pound; purified mercury, 3 ounces; sulphuretted oil, 1 fluid drachm. The mercury must be rubbed with the sulphuretted oil till the globules disappear, and the ammoniac, previously melted, added gradually, and the whole mixed together. This plaster has great efficacy in promoting the absorption of glandular swellings and indolent tumors. It is of much use also as an application to corns and bunions. It can be obtained from the apothecary, and is usually known as the plaster of ammoniac and mercury.

5052. Burgundy Pitch Plaster. Melt together 2 pounds strained burgundy pitch, 1 pound prepared frankincense, and 4 ounces each yellow resin and bees'-wax; add 2 fluid ounces each olive oil and water, and 1 ounce expressed oil of nutmeg; stir constantly until evaporated to a proper consistence.

5053. Blister or Cantharides Plaster. Melt together 7½ ounces each yellow wax and suet; 6 ounces lard, and 3 ounces resin; when mixed, remove from the fire, and, a little before they concrete, sprinkle in and mix thoroughly 1 pound very finely powdered cantharides.

5054. Strong Blistering, or Cantharides Plaster. Mix at a heat below 212° Fahr., 4½ ounces Venice turpentine, 3 ounces each of burgundy pitch and cantharides, 1 ounce bees'-wax, ½ ounce finely powdered verdigris, and 2 drachms each of powdered mustard and black pepper.

5055. Warm Plaster. For this plaster, take 1 part of blistering plaster, and of burgundy pitch 14 parts; mix them by means of a moderate heat. This plaster is stimulant, slightly irritating the skin, and is of use in ordinary coughs and whooping-cough, sciatica, and other local pains.

5056. Homœopathic Mustard Plaster. For chronic inflammation, colds, sore throats, inflammations of the lungs, liver, and bowels, sprains, &c. Take 1 part by measure of mustard; 5 parts flour; and 5 of Indian meal. Mix the mustard in a little hot water, and, when smooth, add about 2 parts boiling water, and when all is dissolved stir in the flour, and then the meal, thoroughly; adding more boiling water if necessary. Spread on a thick cloth double folded, to retain heat and moisture. Cover with mosquito netting, or lace, and nothing closer, sew around the edges, apply to the painful spot; fasten with bandages, and wear till dry, or for 24 hours, and then put on a fresh one. Continue to renew these for 1 or 2 weeks. When the skin becomes too tender, add 1 more spoonful of flour and meal each. When these plasters can no longer be borne, use powdered ginger instead of mustard, and then finish with plain Indian meal poultice alone. (*Leggett.*)

5057. The Best Mustard Plaster. Take a piece of waste linen, and, if crumpled, iron it smooth; or paper will do. Procure a small quantity of black mustard seed, and bruise it to a coarse powder, in a pestle and mortar or otherwise. Spread over the linen a thin solution of gum, and sprinkle the powder equally over it. Dry in a warm place. When wanted, plasters may be cut of any size or shape; and when applied should be momentarily dipped in tepid water, and tied over the affected part with a bandage. These plasters are more simple, cleanly, and effective than the ordinary mustard poultices. This preparation may be had at the drug stores, made in 3 different strengths, No. 1 being the most powerful.

5058. Court Plaster. This plaster is merely a kind of varnished silk, and its manufacture is very easy. Bruise a sufficient quantity of isinglass, and let it soak in a little warm water for 24 hours; expose it to heat over the fire till the greater part of the water is dissipated, and supply its place by proof spirits of wine, which will combine with the isinglass. Strain the whole through a piece of open linen, taking care that the consistence of the mixture shall be such that, when cool, it may form a trembling jelly. Extend a piece of black or flesh-colored silk on a wooden frame, and fix it in that position by means of tacks or twine. Then apply the isinglass (after it has been rendered liquid by a gentle heat) to the silk with a brush of fine hair (badgers' is the best). As soon as this first coating is dried, which will not be long, apply a second; and afterwards, if the article is to be very superior, a third. When the whole is dry, cover it with two or three coatings of the balsam of Peru. This is the genuine court plaster. It is pliable, and never breaks, which is far from being the case with spurious articles sold under that name.

5059. De Rheims' Healing Paper. Make a strong tincture of capsicum-pods by steeping them for several days, in a warm place, in twice their weight of rectified spirits of wine. Dissolve gum-arabic in water to about the consistency of molasses. Add to this an equal quantity of the tincture, stirring it together with a small brush or a large camel's-hair pencil, until they are well incorporated. The mixture will be cloudy and opaque. Take sheets of silk or tissue-paper; give them with the brush a coat of the mixture; let them dry, and then give another; let that dry, and, if the surface is shining, there is enough of the peppered gum; if not, give a third coat. This paper, applied in the same way as court plaster to chilblains that are not broken, and burns that are not blistered, speedily relieves the itching and the pain. It acts like a charm, and effects a rapid cure. The same with cuts and discolored bruises. It likewise allays rheumatic pains in the joints. Its great value is that, besides acting as ordinary sticking-plaster, it abates suffering and hastens the process of healing.

5060. Cooley's Corn Plaster. In a piece of card, cut a round hole the size of the central portion of the corn; lay the card on a piece of adhesive plaster, and warm the spot of plaster exposed by the hole in the card, by holding a hot iron near it for a second or two; then remove the card and sprinkle some finely powdered nitrate of silver on the warm spot of the plaster. When cold, shake off the loose powder, and apply to the corn. Two or three applications seldom fail to cure.

5061. Carbolic Plaster. Carbolic glycerine, 34 parts by weight; prepared chalk, 94 parts. Mix well by kneading, and enclose in closely-stoppered jars.

5062. Irritating Plaster. Boil together 1 pound tar, ½ ounce burgundy pitch, 1 ounce white pine turpentine, and 2 ounces

resin. Finely powder 1 ounce each mandrake root, blood root, poke root, and Indian turnip. Stir these into the melted tar, &c., before it cools. This plaster, spread on muslin and renewed daily, will raise a sore, which is to be wiped with a dry cloth, to remove matter, &c. The sore must not be wetted. This is a powerful counter-irritant, for removing internal pains, and in other cases where an irritating plaster is necessary.

Gargles

Gargles are simple remedies well adapted to domestic practice in sore throats of various kinds. According to the nature of the ingredients of which they are made, they allay irritation and inflammation, invigorate the membrane lining the mouth and throat, and promote suppuration. The particular purpose for which they are required ought to be kept in view in their preparation.

5064. Potassa Gargle for Sore Throat. Strong sage tea, 1 pint; strained honey, 2 table-spoonfuls; chlorate of potassa, 1 tea-spoonful; mix and use as often as necessary, being careful to shake before using. Also poultice the throat with hops and warm vinegar. Brewers' yeast substituted for the chlorate of potassa makes a very effectual gargle.

5065. Gargle for Sore Throat. Very strong sage tea, ½ pint; strained honey, common salt, and strong vinegar, of each 2 table-spoonfuls; cayenne (pulverized), 1 rounding tea-spoonful; steeping the cayenne with the sage, strain, mix, and bottle for use, gargling from four to a dozen times daily, according to the severity of the case.

5066. Carbolic Acid Gargle. Used as a gargle for sore throat, attended with foul breath. Take 2 grains of the crystals to 1 ounce of water.

5067. Gargle for Ulcerated Sore Throat. Water, ½ pint; decoction of Peruvian bark, ½ pint: sulphate of zinc, 1 drachm. Mix.

5068. Gargle for Inflammation of the Throat. Purified nitre, 2 drachms; barley water, 7 ounces; acetate of honey, 7 drachms; mix the ingredients. To be used frequently.

5069. Gargle for General Domestic Use in Sore Throat. Take 3 tea-spoonfuls vinegar, 2 tea-spoonfuls tincture of myrrh, 2 of honey, a glass of port wine, and 3 or 4 wine-glasses of warm water; mix all these ingredients, and the gargle is ready for use. A decoction of the leaves of the black currant may, with good effect, be added instead of the warm water. This makes both a pleasant and most useful gargle.

5070. Mucilaginous Gargle for Inflamed Throat. Tincture of myrrh, 3 drachms; mucilage of gum-arabic, 7 ounces. Mix. This gargle is of use in defending the parts when the saliva is of an acrid character.

5071. Gargle for Threatened Mortification of the Throat. Tincture of capsicum, 6 drachms; honey of roses, 3 drachms; infusion of roses, ½ pint. Mix. Or: Tincture of capsicum, 6 drachms; infusion of Peruvian bark, 5 ounces; port wine, 3 ounces. Mix.

5072. Gargle to Promote Suppuration. Barley water and infusion of linseed. This gargle is to be used warm. It must be kept in view that this mild gargle acts by softening the parts of the throat, and hastening the suppuration by its heat; and it is requisite, therefore, that the temperature of the gargle be kept up.

5073. Carbolized Gargle for Diphtheria, Tonsilitis, &c. Carbolic acid, 20 minims; acetic acid, ½ drachm; honey, 2 fluid ounces; tincture of myrrh, 2 fluid drachms; water, 6 fluid ounces. The carbolic and acetic acids to be well shaken together before the other ingredients are added. *(Charles Sedgwick.)*

Caustics

Caustics. Subtances that corrode or destroy the texture of the skin and organized bodies. Their action is commonly called burning. The principal caustics employed by surgeons are nitrate of silver, caustic potassa, sulphate of copper, red oxide of mercury, and the nitric and acetic acids.

5075. Vegetable Caustic. Burn oak or beech wood to ashes. Make a lye from them, and simmer it till it becomes rather thicker than cream; the evaporation may be continued in the sun. Spread on leather when used. It is valuable in cancers, fistulas, scrofulous and indolent ulcers, where there is proud flesh.

5076. Medicated Lint. Dissolve 20 to 30 grains nitrate of silver in 1 fluid ounce distilled water; saturate ½ ounce of dry lint with the solution, and expose it in a saucer to the light and air until it becomes black and dry.

5077. Iodine Paint; Iodine Caustic. Take of iodide of potassium, ½ ounce avoirdupois; iodine, ¼ ounce; proof-spirit, 3 ounces; dissolve by agitation. Used as a paint in cases in which it is desired to apply iodine, in a strong form, locally; also as a caustic for corns, warts, &c. *(Soubeiran.)* The tincture of iodine of the Pharmacopœia is, however, more generally employed; but it is only of about one-third the strength of the above.

5078. To Prevent Iodine from Staining. By adding a few drops of liquid carbolic acid to the iodine tincture, the latter will not stain. According to Dr. Bogs, of the Indian Service, carbolic acid also renders the efficacy of tincture of iodine more certain. He recommends the following formula, whenever injections of the latter are indicated: Alcoholic tincture of iodine, 45 drops; pure liquid carbolic acid, 6 drops; glycerine, 1 ounce; distilled water, 5 ounces. In blennorrhœa and leucorrhœa, this mixture is said to be superior to tar-water.

5079. Caustic for Corns. Take of liquid terchloride of antimony and tincture of iodine, of each 2 drachms avoirdupois; protiodide of iron, 7 grains; mix, and preserve it in a well-stoppered phial. Applied, with care. Two to four applications are said to effect a cure.

5080. Convenient Vehicle for the Application of Nitrate of Silver. At University College Hospital (London) they have adopted the plan of dissolving nitrate of silver in nitrous ether; it can then be spread with a camel's-hair brush over a surface, and the ether immediately evaporates.

Rubefacients

Rubefacients. Substances or agents, which, when applied for a certain time to the skin, occasion a redness and increase of heat without blistering. They act as counter-irritants. Mustard or powdered ginger, made into a paste with water, hartshorn and oil, and ether or alcohol (when their evaporation is prevented), are among this class of remedies.

5082. Counter-Irritants. Substances applied to the surface of the body to establish a secondary morbid action, with the view of relieving one already existing. Those best known are blisters, mustard poultices, hartshorn and oil, and liniment of ammonia.

5083. Blistering Tissue. These blistering compositions are superior to the common cantharides blisters, from their greater cleanliness, efficiency, and ease of application, and their being less liable to produce excessive irritation.

5084. Strong Blistering Tissue. Powdered cantharides is exhausted with sulphuric ether by percolation (*see No.* 41), and the resulting tincture reduced to the consistence of molasses by distillation; the extract is then mixed with twice its weight of yellow wax, melted by a very gentle heat, and spread on waxed cloth.

5085. Blistering Tissue. Digest 3 drachms powdered cantharides in 1 ounce ether for a day or two; decant and add 4 drachms sandarach, 2 drachms mastic, ½ drachm turpentine, and 10 or 12 drops oil of lavender; mix and spread as above.

5086. Blistering Tissue. Mix 2 parts acetic extract of cantharides, and 1 part each of resin cerate and bees'-wax; use as before.

5087. Blistering Plaster. Infuse 3 drachms powdered cantharides in 4 ounces acetic ether for 8 days; decant and evaporate as in No. 5084; then add 4 drachms resin, and spread on court plaster.

5088. Management of Blisters. Spread the plaster thinly on paper, or linen, and rub over it a few drops of olive oil. In this way the blister acts speedily, and with less irritation than usual.

5089. To Camphorate Blisters. M. Deschamps d'Avallon has suggested, when it is desirable to camphorate a blister, it may be readily accomplished by dropping on its surface a few drops of a saturated solution of camphor in chloroform, made by adding 2 parts of the latter to 4 of the former.

Balsams

Balsams. Balsams are semi-liquid resinous substances, having for the most part the consistence of honey. Some, however, are solid, and the greater number harden by exposure to the air and age. They are generally aromatic, soluble in alcohol, partly soluble in ether, and not at all so in water. Their usual constituents are resin and benzoic acid, mixed with a large portion of aromatic essential oil. Some of the substances falsely called balsams contain no benzoic acid, as the balsam of copaiba, &c.; and many preparations, from the presumption that they possess balsamic qualities, have also received this name.

5091. Friar's Balsam, or Jesuit's Drops. Take gum benzoin, 6 ounces; strained storax, 2 ounces; pulverized aloes and myrrh, each ½ ounce; balsam Peru, 1 ounce; balsam tolu, 2 ounces; extract of liquorice, 2 ounces; alcohol, 2 quarts. Let it stand for 2 weeks, with occasional agitation, and filter the whole through paper. A good application for wounds and cuts; and as such was very effectual in the hands of the old friars. Internally, it is stimulant, expectorant, and anti-spasmodic, and is useful in asthma, catarrh, consumption, and languid circulation. Dose, ½ a drachm on loaf sugar.

5092. Balsam of Horehound. Dissolve 2 ounces each extract of horehound and extract of liquorice, in ½ pint hot water; when cold, add ¾ pint paregoric, 6 ounces oxymel of squills, 2 ounces tincture of benzoin, and 10 ounces honey. Mix well and strain through flannel. Dose for an adult, ½ to 1½ tea-spoonfuls, accompanied by a dose or two of aperient medicine.

5093. Balsam of Honey. Balsam of tolu, 1 ounce; gum storax, 1 drachm; purified opium, 15 grains; best honey, 4 ounces; rectified spirits of wine, 1 pint. Digest them together for a week, and strain the liquor. This prescription is of great use in colds and habitual coughs, unaccompanied by feverish symptoms. The dose is from 1 to 3 teaspoonfuls occasionally.

5094. Balsam Riga. Young shoots of fir (collected in March), 2 pounds; rectified spirit and water, of each 5 pints. Bruise the fir-shoots and macerate in the spirit and water for 3 or 4 days, then distill 1 gallon. Or: Mix together rectified spirit, 8 ounces; oil of juniper and compound tincture of benzoin, of each 1 ounce; agitate well and filter. Stimulant and diuretic; also used for sprains and bruises.

5095. Glycerine Balsam. This is designed to whiten and soften the skin, remove roughness, chaps, chilblains, and irritations from common causes. Take pure white wax, 1 ounce; spermaceti, 2 ounces; oil of almonds, 9 ounces. Melt together by a moderate heat in a glazed earthenware vessel, and add pure glycerine, 3 ounces; balsam of Peru, ½ ounce. The mixture is to be stirred until nearly cold, and then poured into pots. Instead of balsam of Peru, 12 or 15 drops of attar of rose may be employed.

5096. Universal Wound Balsam. Gum benzoin, in powder, 6 ounces; balsam of tolu, in powder, 3 ounces; gum storax, 2 ounces; frankincense, in powder, 2 ounces; gum myrrh, in powder, 2 ounces; socotrine aloes, in powder, 3 ounces; alcohol, 1 gallon. Mix them all together and put them in a digester, and give them a gentle heat for 3 or 4 days; then strain. 30 or 40 drops on a lump of sugar may be taken at any time, for flatulency or pain at the stomach; and in old age, where nature requires stimulation. This valuable remedy should be kept in every family ready for use; it cannot be surpassed as an application for cuts and recent wounds, and is equally good for man or animals.

5097. Pectoral Balsam. Tincture of tolu and compound tincture of benzoin, of each 2 ounces; rectified spirit, 4 ounces; mix. As a pectoral in coughs and colds. Dose, 1 tea-spoonful.

5098. Anodyne Balsam. Take of white soap, 1 ounce; opium, unprepared, 2 drachms; rectified spirit of wine, 9 ounces; digest them together by a gentle heat for 3 days; then strain off the liquor, and add to it 3 drachms of camphor. This balsam is of service in violent sprains and rheumatic complaints, when not attended with inflammation. It must be rubbed with a warm hand on the part affected, or a linen rag moistened with it, and renewed every third hour till the pain abates.

5099. Balsam of Turpentine. Melt by a gentle heat black resin, 1 pound; remove the vessel from the fire and add oil of turpentine, 1 pint.

5100. Canada Balsam. This balsam is the product of the Canadian balsam fir, a tree of very common growth in Canada and the State of Maine. When fresh, it has the consistence of thin honey, an agreeable odor, an acid taste, and a pale yellow color, nearly white. It should be perfectly transparent, and soluble in rectified oil of turpentine, with which it forms a beautiful glassy and colorless varnish, which is much used for preparing a semi-transparent copying-paper. A factitious kind is sold, but is wholly deficient of some of the properties of the genuine balsam.

5101. Factitious Canada Balsam. Dissolve 3 pounds of clear yellow resin in 1 gallon of oil of turpentine; then add ½ pint of pale linseed oil, and ¼ ounce each of essence of lemon and oil of rosemary.

5102. Factitious Balsam of Tolu. Dissolve orange shellac and gum benzoin, of each 1 pound, in coarse powder; in rectified spirit, 5 pounds (in a close vessel); filter and distill off the spirit until the residuum has a proper consistence, then add a few drops of the oils of cassia and nutmeg, dissolved in a little essence of vanilla. Or: Take of balsam of tolu, 4 ounces; white resin, 16 ounces; sheep's suet, 1½ ounces, or sufficient to make it soft enough, according to climate or season.

5103. To Detect Factitious Balsam of Tolu. The genuine balsam is perfectly soluble in alcohol, forming a transparent solution. By exposure to the air it becomes hard and brittle. It is frequently adulterated, in which case it has a weaker smell, is less soluble in alcohol, and the tincture formed with that fluid is opaque.

5104. Factitious Balsam of Copaiba. Powdered gum benzoin, 4 ounces; castor oil, 1 gallon; yellow resin, 3 pounds; balsam of Canada, 2 pounds; oil of juniper, 2 ounces; oil of savine, 1 ounce; essences of orange and lemon, of each ½ ounce. Melt the resin, then add a little of the castor oil and the powdered benzoin, and withdraw the heat; when well mixed add the remainder of the castor oil, and, when nearly cold, the essences; mix well, and filter through a Canton flannel bag, adding a little coarsely powdered charcoal.

5105. Imitation Balsam of Copaiba. Balsam of Canada, 8 pounds; yellow resin, 2 pounds; castor oil, 3 pounds; oil of juniper, ¼ ounce; essential oil of almonds, 15 drops; oil of savine, 20 drops. As above.

5106. Reduced Balsam of Copaiba. Balsam of copaiba, 4 pounds; castor oil, 3 pounds; mix. Or: Balsam of copaiba, 7 pounds; castor oil, 4 pounds; yellow resin, 2 pounds. Or: Equal parts of balsam of copaiba and balsam of Canada mixed together. Or: To the last add 2 pounds of Venice turpentine. Or: Balsams of Canada and copaiba, and nut or castor oil, equal parts. Or: Copaiba, 7 pounds; nut oil, 3 pounds; yellow resin, 2 pounds; balsam of Canada, 1 pound. The above are the forms for the reduction of copaiba balsam, that have from time to time been circulated in the drug trade. For the mode of distinguishing such compounds from the pure balsam, see next receipt.

5107. To Detect Factitious or Reduced Balsam Copaiba. Chevallier recommends the following test: Place a drop of the balsam on a piece of unsized paper, and heat it until all the essential oil be expelled; it should then form a semi-transparent, well-defined spot; but if the balsam has been adulterated with a fat oil, it will be surrounded by an oily areola. According to Planche, the pure balsam, when shaken with liquid ammonia specific gravity .965, becomes clear and transparent in a few moments. Vigne says: 2½ parts pure balsam with 1 part liquor of ammonia, form a transparent mixture, which may be heated to 212° without becoming opaque. Boiled with 50 times its weight of water for 1 hour, it should lose at least half its weight.

Dr. Hager recommends the following simple mode as very reliable for detecting adulteration of copaiba balsam with turpentine oil: 5 or 6 drops of water and about 1 drachm of the balsam are mixed in a small porcelain dish with as much litharge as will make a thin ointment. This mass, at the common summer temperature, exhales the characteristic odor of oil of turpentine, even if the balsam is adulterated with only 10 per cent. of the oil.

5108. Factitious Balsam of Peru. Balsam of tolu, 1 pound; gum benzoin, 3 pounds; liquid storax, 1 ounce; sufficient rectified spirit. The gum benzoin in coarse powder is dissolved in a little of the spirit, and then mixed up with the balsam of tolu and storax, adding as much spirit as is necessary to reduce it to a proper consistence.

5109. Reduced Balsam of Peru. Balsam of Peru, 3 pounds; balsam of tolu, 2 pounds; rectified spirit enough to reduce it to a proper consistence. As above. Or: Balsam of Peru, 3 pounds; gum benzoin dissolved in the least quantity of spirit possible, 1 pound. As above.

5110. To Detect Factitious or Reduced Balsam of Peru. Genuine balsam of Peru should possess the following characteristics: It should have a consistence and appearance resembling molasses, and an aromatic odor between that of benzoin and vanilla. It should be entirely soluble in alcohol. It should undergo no diminution in volume when agitated with water. 1000 parts of the balsam should saturate exactly 75 grains of pure crystallized carbonate of soda. Its specific gravity should not be less than 1.150, nor more than 1.160.

5111. Factitious Balm of Gilead. Also called *Baume de la Mecque*. Gum benzoin, 1 pound; resin, 4 pounds; oil lemon, rosemary, caraway, of each 4 ounces; alcohol, sufficient quantity, till of proper consistence. Or: 4 ounces gum benzoin may be dissolved by heat in 1 pound Canada balsam, and to the mixture, when cold, ¼ ounce each of the oils of rosemary, lemon, and cassia, added.

5112. Hoffmann's Life Balsam, found in Continental Pharmacopœias under the name *Mistura oleoso-balsamica*, and other titles, is prepared as follows: Take 1 fluid ounce each of the oils of lavender, cloves, cinnamon, thyme, lemon peel, and mace; 3 fluid ounces each oil of bergamot and balsam Peru; and 5 pints alcohol. The oils and balsam are gradually added to the alcohol, the whole well shaken and allowed to rest for a few days in a cool place, when it is filtered and ready for use. Different European Pharmacopœias vary from each other somewhat in the proportion of the oils.

5113. Nervine Balsam or Baume Nerval. Expressed oil of mace, and prepared ox-marrow, of each 4 ounces melted together; oil of rosemary, 2 drachms; oil of cloves, 1 drachm; camphor, 1 drachm; balsam of tolu, 2 drachms; the last two dissolved in rectified spirit, 4 fluid drachms; and the whole stirred till cold.

5114. Balsam of Sulphur. Boil together in a vessel, tightly covered, 1 part flowers of sulphur and 4 parts olive oil, until they assume the consistence of a thick balsam.

5115. Balm of Rakasiri. Oil of rosemary dissolved in common gin.

5116. Balsam de Malta. Gum benzoin, 2 ounces; gum aloes, 1 ounce; alcohol, 2 pints. Mix.

Tonics. Medicines that increase the tone of the muscular fibre, and impart vigor to the system. The principal mineral tonics are iron, zinc, copper, silver, arsenic, bismuth, mercury, and the mineral acids. The principal vegetable tonics are cinchona or Peruvian bark, cinchonine, quinine, the vegetable bitters, and some of the aromatics. Of the above, iron, bark, and its preparations, and the aromatic bitters, are those generally employed, and which prove most genial to the constitution.

5118. Stomachic Elixir. Pare off the thin yellow rinds of 6 large oranges, and put them in a quart bottle with 1 ounce gentian root, scraped and sliced, and ½ drachm cochineal. Pour over these ingredients a pint of brandy; shake the bottle well several times during that and the following day; let it stand 2 days more to settle, and clear it off into bottles for use. Take 1 or 2 tea-spoonfuls morning and afternoon, in a glass of wine or in a cup of tea. This elegant preparation is a most valuable tonic.

5119. Stomachic Elixir. Gentian root, 2 ounces; bitter oranges, sliced, 1 ounce; Virginia snake-root, ½ ounce. Bruise, and infuse for 4 days in 1 pint of brandy; then add 1 pint of water. A wine-glassful to be taken occasionally. Good for flatulency, indigestion, want of appetite, &c.

5120. Tonic Infusion. Gentian root, sliced. ½ ounce; dried orange peel, bruised, coriander seeds, bruised, of each 1 drachm; boiling water, 12 ounces. Macerate for an hour in a lightly covered vessel, and strain the liquor. This infusion is often most beneficially employed in general debility, chronic gout, indigestion, and other ailments. The dose is from 1 to 2 ounces taken 3 or 4 times a day.

5121. Infusion of Calumba. Calumba root, 1 drachm; boiling water, ½ pint. Macerate for 4 hours and strain, adding afterwards ¼ ounce of spirit of cinnamon. The dose is 1½ or 2 ounces. It is an excellent tonic, and is held in high esteem by many eminent physicians, who employ it in the latter stage of diarrhœa, bilious intermittent fever, and puerperal fever. It is also a good preparation for allaying the nausea and vomiting which often accompany pregnancy.

5122. Orange Tonic. Orange peel, 1 ounce; chamomile flowers, 1¼ ounces, and a little ginger. Put in 1 pint of boiling water. Add ½ a wine-glassful of brandy. Take a wine-glassful at a time.

5123. Spackman's Tonic and Nervine Mixture. Take ½ drachm sulphate of quinine, 6 grains tannin, 1 ounce ginger syrup, 6 drachms fluid extract of valerian, and 2 drachms compound tincture of cardamoms. Dose, a tea-spoonful 4 times a day.

5124 Tonic Aromatic Mixture. Digest in a close vessel for 3 days, agitating frequently, 1 ounce powdered pale cinchona bark, 3 drachms powdered calumba root, 2 drachms bruised cloves, and ½ ounce iron filings in 16 fluid ounces peppermint water; strain, and add 3 fluid ounces compound tincture of cardamoms, and 3 fluid drachms tincture of orange peel. Dose, 1 or 2 table-spoonfuls or more, 3 or 4 times a day.

5125. Tonic Pills. Extract of gentian, 2 scruples; sulphate of iron, 16 grains; sulphate of quinine, 10 grains. Mix, and form into pills. Take 1 pill three times a day.

5126. Tonic Tincture. Peruvian bark, bruised, 1½ ounces; orange peel, bruised, 1 ounce; brandy, or proof spirit, 1 pint. Infuse 10 days; shake the bottle every day. Pour off the liquor, and strain. Take a tea-spoonful in a wine-glassful of water twice a day, when you feel languid.

5127. Decoction of Red or Peruvian Bark. Bruised red bark, 1 ounce; water, 1 pint. Boil for 10 minutes in a covered vessel, and strain the liquor while hot.

5128. Infusion of Red or Peruvian Bark. Red bark, bruised, 1 ounce; boiling water, 1 pint. Macerate for 2 hours in a covered vessel, and strain. This is of great use in convalescence from acute diseases. It contains a considerable amount of the febrifuge and strengthening qualities of the quinine.

5129. Dr. Thompson's Bitters. Balmony bark, 1 part; poplar bark, 5 parts. Boil in water sufficient to strain 2½ gallons of water from a pound of the bark, to which add sugar, 3½ pounds; nerve powder, 2½ ounces; while hot, strain, and add best Malaga wine, 3½ gallons; tincture of meadow-fern, 1 quart. A less quantity may be made by observing the proper proportions. Dose, from half to a wine-glassful twice a day. These bitters are excellent. They are sure to correct the bile, and create an appetite by giving tone to the digestive powers.

Anodynes.

Medicines which allay pain. Some act by actually assuaging pain; others by inducing sleep; a third class give ease by stupefying the senses, or lessening the susceptibility to pain. Among the principal anodynes are opium, morphia, camphor, ether, chloroform, nitrous oxide or laughing gas, &c.

5131. Anodyne Powder. Opium, ½ ounce; camphor, 3 drachms; valerian, 1 ounce; cayenne pepper, 1 ounce. Put the opium and camphor into a close bag; place it on the oven top to harden. Powder and mix. Take ¼ tea-spoonful at a time. Most valuable in colic, cramp, and severe pains.

5132. Anodyne Substitute for Opium. Take 2½ drachms each tincture of lupuline (hops), and tincture of henbane; 5 drachms camphor water. A tea-spoonful of the mixture may be given every 2 hours in cases where opium cannot be administered.

5133. Anodyne Cigars. The leaves of the belladonna (deadly nightshade), 4 parts, moistened with 1 part tincture of opium, dried and made into cigarettes of 1 drachm each; or the leaves alone, without the addition of opium, form an effective anodyne in troublesome coughs, tooth-ache, sore throat, &c.

Diaphoretics.

Medicines that increase the perspiration. Those that produce this effect in a powerful degree are generally called sudorifics. The principal diaphoretics are warm diluents, as gruel, tea, barley-water, &c.; salts of the alkalies, as the citrates of potassa and soda, acetate and carbonate of ammonia, sal-ammoniac, nitre, &c.; preparations of antimony, as tartar emetic, antimonial powder, &c.; also Dover's powder, opium, camphor, ipecacuanha, alcohol, wine, &c. The use of diaphoretics is indicated in most diseases accompanied by fever and a dry skin.

5135. Balm Tea. Balm leaves, 1 ounce; fine sugar, 1 spoonful; lemon juice, 1 ounce; infused in a pint of boiling water for 20 minutes. This forms a useful drink in colds or fevers. Or it may be made just like common tea, without the lemon. Let the patient drink it frequently, especially the last thing at night, and keep himself warm during the perspiration.

5136. Herb Drink for Fevers. Infuse 1 ounce each of balm, elder-flowers, marshmallow, spearmint, and arnica-flowers, with ½ ounce anise-seed, in boiling water.

5137. Fever Mixture. Mix 2 scruples nitrate of potash with 3 drachms sweet spirits of nitre, 3 ounces solution of acetate of ammonia, 4½ ounces camphor water, and 2 drachms lemon syrup. Dose for an adult, 2 table-spoonfuls every 4 hours. Children in proportion. This mixture is excellent where the fever affects the head.

5138. Infusion to Produce Sweating. Infuse 1 ounce pleurisy root for 30 minutes in 1½ pints water. A tea-spoonful taken warm as often as the stomach will bear it.

5139. Boneset Tea. Infuse 1 ounce boneset in 1 pint boiling water for 30 minutes. A wine-glassful as hot as possible every half hour will produce a profuse perspiration.

5140. Blessed Thistle Tea. The leaves of the blessed thistle prepared and administered in the same way as boneset (*see last receipt*), but not sufficient to produce nausea, will have a similar effect.

5141. Febrifuge Wine. The following mixture is highly recommended for fever and ague: quinine, 25 grains; water, 1 pint; Epsom salts, 2 ounces; brandy, 1 gill; sulphuric acid, 12 drops; loaf sugar, 2 ounces. Color with tincture of red saunders. Take a wine-glassful three times a day.

5142. Sweating Drops. Take of camphor, saffron, ipecacuanha, opium, and Virginia snake-root, ½ ounce each; Holland gin, 1½ pints; infuse 2 or 3 days. A wonderfully efficacious cure for fever and ague, after suitable evacuants. Dr. Beach says he finds this the best medicine for fever and ague of any with which he is acquainted. In two cases this tincture removed the paroxysms where other remedies failed.

5143. Spirit of Mindererus, or Solution of Acetate of Ammonia. Take of diluted acetic acid, 2 pints; carbonate of ammonia, in powder, a sufficient quantity. Add the carbonate of ammonia gradually to the acid, until it is saturated. This is a valuable diaphoretic, and is much employed in fevers and inflammatory diseases.

5144. Houseleek for Fevers. It is used as a cooling application to sores, ulcers, &c. The juice mixed with cream is good for inflammation of the eyes, and erysipelas. Taken inwardly it is good for fevers, cooling them down wonderfully. First give a purgative to cleanse the stomach and bowels; then bruise the houseleek; adding to the juice its weight in fine sugar to form a syrup. A table-spoonful every 2 hours. Drink balm or catnip tea. This receipt is worth gold.

5145. Sudorific, or Fever Powder. Crawley root, 1 ounce; lobelia herb, ½ ounce; pleurisy root, 1 ounce; skunk cabbage, ½ ounce. Powder, and mix them together. Dose, from ¼ to ½ tea-spoonful every one hour and a half till perspiration is produced. It may be given in balm or common tea. In fevers, inflammations, influenza, and colds, this powder is invaluable. It subdues irritation, corrects the pulse, improves respiration, and promotes sound natural sleep. It is sure, if properly administered, to arrest a fever. Keep it in a bottle, well corked.

Diuretics.

Medicines which promote the secretion of urine. The principal diuretics are aqueous fluids, which act by increasing the watery portion of the blood, and certain substances which promote the secretion of urine, by stimulating the kidneys. Among the former may be classed nearly all aqueous liquids, as most of them produce diuresis, if the skin be kept cool. Among the latter may be mentioned the nitrate, acetate, and bitartrate of potassa; oils of juniper, turpentine, cajeput, and copaiba; dilute spirit and sweet spirits of nitre; decoction of common broom, &c.

5147. Diuretic Drops. Tincture of kino, ½ ounce; balsam of copaiba, spirits of turpentine, of each 1 ounce; sweet spirits of nitre, 2 ounces; queen of the meadow, 1 ounce. Mix, and add 1 scruple of camphor. Take nearly a tea-spoonful in mucilage. Most valuable for scalding urine, inflammation of the kidneys, &c.

5148. Diuretic Infusion. Parsley seeds, ½ ounce; cleavers, ½ ounce; burdock seeds, ½ ounce; coolwort, ½ ounce; spearmint, ½ ounce; juniper berries, ½ ounce; linseed, ½ ounce; gum arabic, ½ ounce. Pour upon these 2 quarts boiling water; infuse 2 or 3 hours, covering the vessel. Strain, and add ½ pint of best gin, 4 ounces of honey, and 3 table-spoonfuls of slippery elm. This is a most valuable diuretic; it is cooling, allays all urinary affections, gravel, scalding of urine, and causes an easy and sufficient flow of the same.

5149. Diuretic Pills. Calcined magnesia, 1 drachm; solidified copaiba, 2 ounces; extract of cubebs, 1 ounce; oil of turpentine, 4 drops; oil of juniper, 6 drops; form into 3-grain pills. Take 1 or 2 a few times a day. A sovereign remedy for diseases of the kidneys, bladder, urethra, gravel, whites, and venereal complaints.

5150. Buchu Leaves. They are diuretic and tonic, and a most valuable remedy in rheumatism, irritable bladder, gravel, stricture, &c. They are given in infusion and tincture. Infuse ½ ounce of leaves in ½ pint of boiling water, for 3 or 4 hours. A wine-glassful for a dose 2 or 3 times a day; or from 1 drachm to ½ ounce of the tincture.

5151. Compound Spirit of Juniper. Stimulant and diuretic, administered in doses of 2 to 4 drachms. This spirit, when mixed with 2 or 3 times its weight of proof spirit, makes a fair imitation of Holland gin. Take 15 ounces bruised juniper berries, 2 ounces each of bruised caraway and fennel, 1 gallon proof spirit, and about 1 quart water. Distill 1 gallon. The wholesale preparation is a solution of 2 drachms oil of juniper, ½ drachm each of the oils of caraway and sweet fennel, in 5 quarts proof spirit. If not clear, filter through magnesia.

Electuaries.

These are chiefly mixtures of vegetable substances combined with syrup or honey, so as to be of a moderate consistence, neither liquid nor solid. The object of such preparations is to secure a vehicle by which medicines may be administered, so that their taste may be covered by the mixture with which they are combined.

5153. Aperient Electuary. Cream of tartar, 1 ounce; milk of sulphur, 1 ounce; sub-borate of soda, 2½ drachms; syrup of ginger, of sufficient quantity to give the required consistence. The dose is 1 or 2 tea-spoonfuls at bedtime. This will be found a mild and excellent laxative, and often is of great use in uterine obstructions.

5154. Lenitive Electuary. The mode of preparing this electuary is the following: Take of the best senna leaves reduced to a fine powder, 4 ounces; pulp of prunes, 1 pound; pulp of cassia, ½ pound; pulp of tamarinds, 3 ounces; molasses, 1½ pints; essential oil of caraway, 2 drachms. Boil the pulps with the molasses to the consistence of honey, add the senna, and when the mixture is nearly cold, add the oil of caraway, and, lastly, mix the compound thoroughly. This preparation is a mild aperient, suited to constipation from whatever cause. It is admirably suited to children and delicate persons. United with an equal quantity of flowers of sulphur, it is an admirable remedy for piles. Dose, from 1 to 3 tea-spoonfuls at bed-time.

Fomentations.

In domestic practice hot fomentations are, although a simple, yet a very useful remedy for allaying pain, relieving irritation, relaxing and removing spasms, and inducing not only local, but even general perspiration. Cloths dipped in very hot water, wrung out and instantly applied on the seat of the pain, will be frequently of very great service. But in some cases it adds to the efficacy of the application to employ substances possessing medical properties in addition to the mere application of heat. In every process of fomentation there should be two flannels, each (say) three yards long, with the ends sewed together, to admit of the boiling water being wrung out of them, and the one flannel should be got ready whilst the other is applied.

5156. Anodyne Fomentation. White poppy heads, 3 ounces; elder flowers, ½ ounce; water, 3 pints. Boil until the liquor is reduced to ⅔ of its original quantity, and strain it; 2 or 3 tea-spoonfuls tincture of opium or laudanum, and 30 drops tincture of cayenne, may in some cases be added to it. This fomentation relaxes spasm, and relieves acute pain.

5157. Fomentation for Ordinary Occasions. Dried mallows, 1 ounce; chamomile flowers, dried, ½ ounce; water, 1 pint. Boil for ¼ hour, and strain the liquor.

5158. Strengthening Fomentation. Decoction of oak bark, 2 pints; alum, 3 drachms. Mix. This is a powerful astringent, and often of great use when applied to weak parts.

5159. Arnica Fomentation. Flowers of arnica, 2 ounces; rue leaves, 1 ounce; boiling water sufficient to strain 6 fluid ounces of infusion after an hour's maceration at nearly boiling temperature. Used in contusions, especially as an application to black eyes.

5160. Stimulating Fomentation. Cayenne pepper, 3 ounces; mustard seed just bruised, 2 ounces; whiskey, 2 quarts. Simmer all together a few minutes. Excellent external application in cholera, paralysis, palsy, rheumatism, &c. A less quantity may be made.

Alteratives.

Medicines which effect some alteration in the nature or the quality of the vital action, and occasion a change in the habit or constitution, establishing the healthy functions of the body without producing any sensible evacuation by perspiration, vomiting, or purging. The preparations of mercury and iodine, when properly administered, are among the most useful and generally employed alteratives.

5162. Alterative Pills. Lobelia seeds, 2 drachms; mandrake, 2 drachms; blue flag, 2 drachms; blood root, 2 drachms; cayenne pepper, 1 drachm; gum guiacum, 2 drachms; extract of dandelion, 6 drachms; oil of peppermint, 3 or 4 drops; simple syrup to form into pills. Dose, 2 pills twice or thrice a day. These pills are of great service in bilious and liver complaints, diseased joints, boils, carbuncles, cutaneous eruptions, scrofula, syphilis, &c.

5163. Alterative Syrup. Tincture of cayenne, ½ ounce; tincture of lobelia and tincture of myrrh, of each 2 ounces; molasses, ½ pound. Mix. A tea-spoonful 2 or 3 times a day. Noted for its effectual cure of cutaneous sores, boils, indigestion, and some chronic complaints.

5164. Dandelion Alterative. A useful alterative medicine, especially in cases where the function of the liver is at fault. Dose, fluid extract of dandelion, a dessert-spoonful, twice daily, with or without a little water.

5165. Blood Maker and Purifier. Mix ½ ounce sulphate of manganese with 1 pint water. Dose, a wine-glassful 3 times a day. This can be used in the place of iron tonic, or in connection with it.

5166. Pancoast's Alterative and Tonic Pills. 1 scruple extract of Ignatia amara (the bean of St. Ignatius), 1½ drachms bromide of potassa, ½ drachm saccharine carbonate of iron, 1 scruple piperine, and 1 scruple extract of henbane. Make into 60 pills and take 2, fifteen minutes after each meal.

Emetics.

Medicines which induce vomiting. The principal emetics are ipecacuanha and tartarized antimony, and their preparations; and the sulphates of zinc and copper. Ipecacuanha is usually administered in substance or infused in wine. The use of tartar emetic and antimonial wine is generally followed by nausea, relaxation of muscular power and of the circulation. Sulphate of zinc acts promptly and energetically, and its effects cease as soon as ejected from the stomach; hence it is employed to eject poison. Sulphate of copper is more violent and disagreeable, and its intense metallic taste is a great objection to its use. The operation of emetics is powerfully promoted by drinking copiously of diluents, especially of warm or tepid water. This latter is itself an emetic when taken in quantity. Its use prevents, in a great degree, excessive straining accompanying vomiting.

5168. Emetic Mixture. Ipecacuanha wine, ½ ounce; water, 1 ounce; simple syrup, ½ ounce. Mix. For a child, 20 drops or more, every quarter of an hour until vomiting ensues. An adult may take from ½ to 1 ounce.

5169. Eclectic Emetic Powder. Ipecacuanha and lobelia, of each 2 ounces; blood root, 1 ounce. Powder, and mix well. Take half a tea-spoonful every 20 minutes till it operates.

5170. Simple Emetic. Half a glass of warm water, 1 heaping tea-spoonful of salt, and another of mustard. These materials are usually to be had at a moment's notice, and form a very efficient emetic.

Patent and Proprietary Medicines.

The following receipts embrace a variety of domestic, popular, and proprietary remedies, and include many compounds which, without being proprietary, are better known by the names of the practitioners who have brought them into prominent notice than by any other title. A variety of articles not included in this place are noticed along with other preparations of the class to which they belong, or under the names of their proprietors.

5172. Dalby's Carminative. Take oils of caraway, fennel, and peppermint, each 10 drops; rub them up with 10 ounces white sugar and 5 ounces carbonate or lump magnesia, then add 1½ drachms sal-tartar and 2 ounces laudanum. Mix with 3½ pints of water.

5173. Kitchener's Peristaltic Persuaders. Turkey rhubarb, in powder, 2 drachms; oil of caraway, 10 drops; simple syrup, 1 drachm by weight; mix, and divide into 40 pills. Dose, 2, 3, or more. From 2 to 4 will generally produce one additional motion within 12 hours. The best time to take them is early in the morning.

5174. Barclay's Antibilious Pills. Extract of colocynth, 2 drachms; extract of jalap, 1 drachm; almond soap, 1½ drachms; guiacum, 3 drachms; tartarized antimony, 8 grains; oil of juniper, 4 drops; oil of caraway, 4 drops; oil of rosemary, 4 drops.

5175. Lee's Antibilious Pills. Take pulverized jalap, aloes, and rhubarb, each ½ ounce; calomel, 3 drachms; pulverized gamboge, 1 drachm; form the whole into a mass with shavings of castile soap and syrup; then make into pills.

5176. Dover's Powder. Ipecacuanha, in powder, 1 drachm; powdered opium, 1 drachm; powdered saltpetre, 1 ounce. All well mixed. Dose, from 8 to 20 grains.

The U. S. Pharmacopœia directs 1 ounce sulphate of potassa instead of the saltpetre (nitrate of potassa); in other respects the formula is the same as the above.

5177. Thompson's "Number Six." Gum myrrh, bruised, 1 pound; powdered cayenne pepper, 1 ounce; macerate for 8 days in 1 gallon strong alcohol (95 per cent.) This is an excellent stimulant and tonic.

5178. Thompson's Composition Powder. Take bayberry, 8 ounces; ginger, 8 ounces; poplar bark, 4 ounces; cayenne pepper, 3¼ ounces; cloves, ½ ounce. Powder all these ingredients, and mix intimately. Dissolve a tea-spoonful in a cup of boiling water, sweetened. Valuable to remove colds, influenza, fever, relax, pain in the bowels, cold extremities. As a sudorific, or for removing morbific matter, the cause of disease, it is invaluable. When taken, the patient should go to bed, and make use of any of the various appliances for promoting perspiration.

5179. Thompson's Hot Drops. Gum myrrh, 2 ounces; cayenne pepper, 1½ drachms; spirit of wine, 1 pint. Put in a bottle, and shake several times a day for a week. Take a tea-spoonful or more in a little warm tea. It is a fine remedy for rheumatism. It will relieve the headache by taking a dose, bathing the head with it, and snuffing it up the nose. It is good for bruises, sprains, swollen joints and old sores, &c., &c.

5180. Anderson's Scott's Pills. Barbadoes aloes, 24 ounces; colocynth, 1 ounce; gamboge, 1 ounce; Spanish soap, 4 ounces; oil of anise, ½ ounce; water, a sufficient quantity. To be made into 3-grain pills.

5181. Marshall Hall's Dinner Pills. Take of powdered Barbadoes aloes, soap, and powdered extract of liquorice, of each equal parts. Make a mass with molasses, and form into pills of 4 grains each.

5182. White's Gout Pills. Take of calomel, powdered socotrine aloes, powdered ipecacuanha, and acetic extract of colchicum, of each 1 drachm. Make a mass with syrup, and form into 60 pills.

5183. Abernethy's Pills. Take of powdered socotrine aloes, 48 grains; powdered ipecacuanha, 20 grains; extract of henbane, 48 grains; blue pill mass, 24 grains. Make a mass with water, and form into 24 pills.

5184. Triplex Pills. Take of powdered socotrine aloes, 24 grains; blue-mass, 12 grains; podophyllin, 3 grains. Mix thoroughly, with sufficient syrup to form a proper consistency. The mixture is to be made up into 12 pills.

5185. Peter's Pills. Aloes, jalap, gamboge, and scammony, of each 2 drachms; calomel, 1 drachm.

5186. Walter's Indian Vegetable Pills. Socotrine aloes, 1 pound; powdered gamboge, 6 ounces; compound extract of colocynth, castile soap, and Aleppo scammony, of each 3 ounces; extract of butternut, 2 ounces; African cayenne, ½ ounce; oil of cloves, 1 drachm. Mix and make into 4-grain pills.

5187. Becquerel's Gout Pills. Mix together 106 grains sulphate of quinine, 15¼ grains extract of digitalis (fox-glove), and 38½ grains of colchicum seeds. Make into 50 pills. Dose, from 1 to 3 daily for several days in succession.

5188. Health Pills. *Pill salutis.* Take 2 drachms socotrine aloes, 1 drachm extract of henbane. 16 grains extract of nux-vomica, and 10 grains powdered ipecacuanha. Mix, and make into 60 pills.

5189. Leake's Pill of Health. *Pill salutaria.* Take 2 drachms calomel, 2 drachms precipitated sulphuret of antimony, ½ ounce powdered gum guiacum, and ½ ounce molasses. Mix, and make into 240 pills.

5190. Thomas' Colocynth and Mandrake Pills. Take ½ drachm compound extract of colocynth, and 3 grains resin of podophyllin. Mix, and make into 12 pills. Dose, 2 at bed-time.

5191. Parrish's Aloes and Mandrake Pills. Take 24 grains aloin, 12 grains resin of podophyllin, and 4 minims oleo-resin of ginger. Mix, and make into 24 pills. Dose, as a laxative, 1 pill; as a purgative, 2 or 3 pills.

5192. Chirayta Pills and Mixture. Dr. Reece's pills. Extract of chirayta (chiretta), 2 drachms; dried soda, 20 grains; ginger, 15 grains; mix, and divide into 36 pills. Two twice a day. Mixture: Infusion of chirayta, 8 ounces; subcarbonate of soda, 1 drachm; 2 table-spoonfuls 3 times a day.

5193. Bateman's Pectoral Drops. Compound spirit of aniseed, 16 fluid ounces; opium, 1 drachm; camphor, 1 drachm; oil of fennel, 20 drops; cochineal, 2 drachms.

Or: Proof spirit, 4 gallons; red saunders, 2 ounces; digest 24 hours, filter, and add powdered opium, 2 ounces; camphor, 2 ounces; catechu, 2 ounces; oil of aniseed, 4 fluid drachms; digest for 10 days. (*Philadelphia College of Pharmacy.*) The old wine gallon is here intended.

5194. Clutton's Febrifuge Spirit. The original formula is: oil of sulphur by the bell, oil of vitriol and sea salt, of each 1 ounce; rectified spirit, 3 ounces; mix, digest for a month, and distill to dryness.

5195. Clutton's Febrifuge Tincture. Febrifuge spirit, 8 fluid ounces; angelica root, serpentary, cardamom seed, of each 1½ drachms; digest and strain. Water acidulated with these, and sweetened to the taste, forms a cooling diuretic and diaphoretic julep. Though never admitted into the Pharmacopœia, these preparations are favorites with some practitioners.

5196. Lartigue's Gout Pills. Compound extract of colocynth, 20 grains; extract of colchicum, 60 grains; extract of opium, 1 grain; mix, and divide into 18 pills. Dose, one or more, according to their purgative effect.

5197. Baillie's Pills. Compound extract of colocynth, 1½ drachms; extract of aloes, 1½ drachms; castile soap, ½ drachm; oil of cloves, 15 drops. Make into 38 pills. 3 at bed-time occasionally.

5198. Marseilles Vinegar. Also called *vinaigre de quatre voleurs*, or *thieves' vinegar.* Dried tops of large and small wormwood, rosemary, sage, mint, rue, lavender-flowers, of each 2 ounces; calamus root, cinnamon, cloves, nutmeg, garlic, of each ¼ ounce; camphor, ½ ounce; concentrated acetic acid, 2 ounces; strong vinegar, 8 pounds. Macerate the herbs, &c., in the vinegar for 2 weeks, strain, press, and add the camphor dissolved in the acetic acid. It is said that this medicated vinegar was invented by four thieves of Marseilles, who successfully employed it as a disinfectant during a visitation of pestilence.

5199. Collier's Wine of Quinine. Take disulphate of quinine, 18 grains; citric acid. 15 grains; sound orange wine, 1 bottle, or 24 fluid ounces.

5200. Chlorodyne. The composition of this well known secret remedy has excited much attention among chemists; many formulæ have been published, but it is difficult to determine which of them approaches nearest to the chlorodyne of J. Collis Browne, its originator. There can be no doubt about the three important ingredients, chloroform, morphia, and hydrocyanic acid, nor can there be about oil of peppermint and molasses. The question is whether anything else exists in the compound. Hitherto, of the formulæ which have been published, two—one by Dr. Ogden, the other by Mr. Squire—have attracted most attention. The difference between these lay essentially in the presence of Indian hemp and capsicum as indicated by Ogden, their absence in the formula given by Squire. But besides this, the proportion of morphia, as given by the two authorities, differed greatly. Mr. Edward Smith has recently investigated the question, and published the result in the London Pharmaceutical Journal. He puts the composition of chlorodyne as follows: Mix together 4 fluid drachms chloroform, 20 grains muriate of morphia, 2 fluid drachms rectified ether, 8 minims oil of peppermint, 4 fluid drachms diluted hydrocyanic acid, 6 fluid drachms tincture of capsicum, 1 fluid ounce acacia mixture, and add 4 fluid ounces molasses. This does not give as dark a compound as the original, because the latter contains caramel; but as this has no medicinal or other value, he omits it, making up to the required volume with the molasses. Mr. Smith thinks there is no Indian hemp, because the alcoholic extract is soluble in water; but then there is capsicum, as, after the chloroform and ether (which also give pungency to the mixture) have been distilled off, the substance left behind has a hot, peppery taste. He seems to have taken much pains with the analysis.

5201. Ogden's Chlorodyne. The following receipt will furnish a preparation having the pharmaceutical properties of chlorodyne, according to Dr. Ogden: To 8 grains muriate of morphia and ½ fluid drachm water, add 20 drops perchloric acid of 25° Baumé, and heat until a clear solution is obtained; then add ½ fluid ounce molasses, previously warmed to render it fluid; heat the mixture and agitate well. When cold, add 1½ fluid drachms chloroform, 12 drops hydrocyanic acid, 1 fluid drachm tincture of Indian hemp, 2 drops oil of peppermint, and 1 drop oleo-resin of capsicum. Mix thoroughly.

5202. Groves' Chlorodyne. The following is an improvement by Mr. Groves, on the receipt of Dr. Ogden. Take chloroform, 4 drachms; ether, 1½ drachms; oil of peppermint, 8 drops; resin of Indian hemp, 16 grains; capsicum, 2 grains; macerate for 2 or 3 days, and filter. Then dissolve hydrochlorate of morphia, 16 grains, in 1 ounce of syrup; add perchloric acid and water, ½ drachm each, assisting the solution by a water-bath; then, when cold, add hydrocyanic acid (Scheele's), 96 drops. Mix the solutions.

5203. Squire's Chlorodyne. Dissolve 8 grains muriate of morphia, and 16 minims oil of peppermint, in 4 ounces rectified spirit; add 4 ounces chloroform and 1 ounce ether; next dissolve 2½ ounces extract of liquorice in 17½ ounces syrup, and add 4 ounces molasses. Mix, and add 2 ounces prussic acid (2 per cent.). (*Comp. to Br. Pharm.*, 11*th Ed.*)

5204. Chandler's Chlorodyne. Take 8 grains muriate of morphia, ½ drachm fluid extract of cannabis indica, 10 drops oil of peppermint, 15 drops tincture of capsicum, 2 drachms chloroform, and 1 ounce each of 98 per cent. alcohol and pure glycerine. Dose, 10 to 30 drops in a wine-glass of water every 3 hours. This preparation is of a clear greenish color.

5205. Horsley's Chlorodyne. The following formula is the result of an analysis made by Mr. Horsley. Burnt sugar, 1 drachm; muriate of morphia, ½ grain; distilled water, 2 drachms; oil of peppermint, 6 minims; dilute prussic acid, 5 minims; tincture of capsicum, 7 minims; and chloroform, 1 drachm. Mix. It must be observed that the water is perhaps an error, as it will not mix with the chloroform, which will be found on the bottom of the bottle.

5206. Chlorodyne. Mix together ½ fluid ounce chloroform, 90 minims sulphuric ether, 8 drops oil of peppermint, 8 drops resin of Indian hemp (cannabis Indica), and 2 drops capsicum; shake the mixture occasionally and allow it to stand for a few days. Dissolve 16 grains muriate of morphia, by heat, in 2 drachms water; when cold, add 65 minims Scheele's hydrocyanic acid, 1 fluid drachm perchloric acid, and 2 fluid ounces molasses. Add this gradually to the first mixture, and then add sufficient molasses to make the whole measure 4 fluid ounces. Dose, 30 minims.

5207. Chlorodyne. Mix together 6 fluid drachms chloroform, 1 fluid drachm chloric ether, ½ fluid drachm tincture of cayenne pepper, 2 drops oil of peppermint, 8 grains muriate of morphia, 24 drops dilute hydrocyanic acid, 20 drops perchloric acid, 1 fluid drachm tincture of Indian hemp, and 1 fluid drachm molasses. Dose, 20 drops, as a soporific; 30 drops to 1 fluid drachm, as an anodyne in cholera or violent paroxysms of pain. (*Cooley.*)

5208. Eau Médicinale d'Husson. It is prepared, according to Dr. Williams, from the juice of colchicum flower with half the quantity of brandy; mix, and, after standing a few days, decant into small bottles. But it was more probably made from the root, as prescribed in the followiog formulæ:

Dry colchicum, 60 parts; in sherry, 125 parts. 20 drops for a dose. (*Paris Codex.*)

4 ounces of the fresh root, sliced, macerated in ½ pint of proof spirit. (*Want.*)

5209. Bates' Anodyne Balsam. Soap liniment, 2 parts; tincture of opium, 1 part.

5210. Delamott's Golden Drops. Muriate of iron, 1 ounce; spirit of sulphuric ether, 7 ounces; dissolve and expose to sunshine in a closely-stopped bottle till it becomes divested of color.

5211. Gregory's Powder. Calcined magnesia, 2½ ounces; powdered Turkey rhubarb, 1 ounce; powdered ginger, ½ ounce. Mix. The above is Dr. Gregory's formula. Some receipts add powdered chamomile. Rhubarb, 1 ounce; ginger, ¼ ounce; powdered chamomile, ½ ounce; magnesia, 2 ounces. Mix. Some druggists prepare it with the heavy carbonate of magnesia, instead of the calcined. (*See No.* 5414.)

5212. Black Draught. Infusion of senna, 10 drachms; sulphate of magnesia, 3 drachms; syrup of ginger, 1 drachm; aromatic spirit of ammonia, 20 drops.

5213. Standert's Red Mixture. Carbonate of magnesia, 4 drachms; powdered rhubarb, 2 drachms; tincture of rhubarb, 1½ ounces; tincture of opium, 1 drachm; oil of aniseed, 24 drops; essence of peppermint, 30 drops; water, 1½ pints; mix. A popular remedy for bowel complaints in the west of England.

5214. Graves' Gout Preventive. Orange peel, 2 ounces; rhubarb, 1 ounce; hiera picra, 2 ounces; brandy, 1 quart. Digest for a week.

5215. Elixir of Bromide of Sodium. Prepare this like elixir of bromide of potassium, substituting bromide of sodium for bromide of potassium, and omitting the color.

5216. Bacher's Tonic Pills. Alkaline extract of black hellebore, 2 drachms; extract of myrrh, 2 drachms; powder of holy thistle, 1 drachm; mix, and divide into 4-grain pills.

5217. Daffy's Elixir. This is similar to the compound tincture of senna; but different makers have their peculiar formulæ. The following is one of them. Avoirdupois weight seems to be intended. Senna leaves, 3¾ pounds; jalap, aniseed, caraway seed, of each 20 ounces; rectified spirit, 18 pints; sugar, 5 pounds. Infuse the senna 2 or 3 times in sufficient boiling water to yield, when strained with pressure, 4 gallons in the whole. Add to this the tincture made with jalap and seeds digested with the spirit for a week. Pour off the clear liquor and add the sugar and brandy coloring if required.

5218. McLean's Neuralgic Liniment. Mix together 4 grains extract of belladonna, 6 fluid ounces ammonia water, ½ fluid ounce oil of turpentine, ½ fluid ounce olive oil, and 2 fluid ounces tincture of opium. Apply during the paroxysms.

5219. Hayes' Pile Liniment. Melt 1 pint lard to the consistence of honey; stir in briskly 1 ounce muriatic acid until thoroughly incorporated; and add 1 ounce tincture of opium, 2 ounces oil of turpentine, and 2 drachms camphor.

5220. Graham's Neuralgic Liniment. Mix together 1 fluid ounce chloroform, 2 fluid drachms oil of cajeput, 1½ ounces camphor, 12 grains veratrine, and 1½ fluid ounces tincture of aconite root.

5221. Mexican Mustang Liniment. Take 2 fluid ounces petroleum, 1 fluid ounce ammonia water, and 1 fluid drachm brandy. Mix.

5222. Heyle's Horse Embrocation. Mix together 1 ounce oil of spike, 1 ounce ammonia water, 2 ounces oil of camphor, ½ ounce oil of origanum, ½ ounce tincture of opium, 1 ounce spirits of turpentine, and 2 ounces olive oil.

5223. Barrell's Indian Liniment. Alcohol, 1 quart; tincture of capsicum, 1 ounce; oils of origanum, sassafras, pennyroyal, hemlock, of each ½ ounce, and mix.

5224. Allen's Nerve and Bone Liniment. Take oil of origanum, oil of rosemary, oil of amber, oil of hemlock, of each 4 ounces; spirits of turpentine, 2 gallons; linseed oil, 3 gallons. Mix, and color with anchusa root.

5225. Glycerine Jelly. Used as an application to chaps and roughened parts of the skin. It may be made of pure glycerine thickened with tragacanth powder and scented with otto of roses. An imitation may be prepared in the following manner: Mix ½ drachm good soft soap intimately with 2 drachms purified honey; gradually add 5 ounces pale olive oil, stirring without intermission until all is taken up. Care must be taken not to mix in the oil too fast. Finally perfume as desired.

5226. Glycerine Paste. A stiff glutinous compound, recommended by Dr. Tilt as a basis for plaster. It is made by boiling 100 or 150 grains common starch in 1 ounce of glycerine. This is similar to Schacht's plasma. (*See No.* 5009.)

5227. King's Cordial. Dissolve in ½ pint of proof spirits, 1½ drachms each of the oils of caraway and cinnamon; extract the stones from 3 pounds of black cherries, and mash the fruit in a pan; grate 1 nutmeg; take 2 quarts of Madeira wine, 2 quarts of brandy, and 1 gallon of syrup; mix all together, and color with red saunders wood.

5228. Squire's Elixir. Opium, 1 ounce; camphor, 1 ounce; spirit of aniseed (compound), 4 pints; tincture of serpentaria, 1 pint; water, 4 pints; tincture of ginger, ½ ounce. Some receipts add a little aurum musivum.

5229. Ward's Essence for the Headache. Spirit of wine, 2 pounds; roche alum in fine powder, 2 ounces; camphor, 4 ounces; essence of lemon, ½ ounce; strong water of ammonia, 4 ounces; stop the bottle close, and shake it daily for 3 or 4 days.

5230. Henry's Magnesia. A solution of Epsom salts is precipitated by one of carbonate of potash in the cold; the precipitate is well washed, rose water being used for the last washing; it is then made up while drying into large or small cubes.

5231. Hill's Balsam of Honey. Balsam of tolu, 2 ounces; styrax, 2 drachms; opium, ½ drachm; honey, 8 ounces; spirit of wine, 32 fluid ounces.

5232. Battley's Senna Powder. Senna leaves heated until they become light in color, reduced to powder, and mixed with some finely powdered charcoal.

5233. Munro's Cough Medicine. 4 drachms paregoric with 2 drachms sulphuric ether and 2 drachms of tincture of tolu. Dose, 1 tea-spoonful in some warm water.

5234. Griffin's Tincture for Coughs. Oil of caraway and anise, each 2 drachms; saffron, ½ ounce; benzoic acid, ¾ ounce; opium, 5 drachms; camphor, ½ ounce; spirit, 6 ounces; honey, 6 ounces. When mixed and dissolved, color with burnt sugar.

5235. Derbyshire's Patent Embrocation for Preventing Sea-Sickness. Boil 2 ounces opium, 2 drachms extract of henbane, 10 grains mace, and 2 ounces mottled soap, in 3 pints of water for ½ hour. When cold, add 1 quart of rectified spirit and 3 drachms spirit of ammonia.

5236. Papier Fayard et Blayn. This preparation is now made officinal in the Paris Codex, under the name of *Papier dit Chimique*. Heat 200 parts olive oil in a capacious dish over an open fire, until vapors begin to be given off. Then add gradually, with stirring, 100 parts finely powdered minium (red lead). As soon as the first effervescence is over, continue to stir and heat the mixture until it begins again to effervesce. Then remove from the fire and stir rapidly, to remove the white scum on the surface, and at once add 6 parts white wax. This is applied to paper or muslin with a sponge or brush.

Before spreading on the paper or muslin, it must have been prepared a week earlier with the following varnish, to make it impenetrable: olive oil, 100 parts, and garlic, 10 parts, are heated together over the open fire until the moisture of the latter is dispelled and they turn a brown color, after which they are strained. To this mixture are added 80 parts oil of turpentine, 40 parts subcarbonate of iron, and 15 parts carbonate of lead (white lead) in oil. It is also laid on with a brush or sponge.

5237. Papier Fayard. Gout paper. Euphorbium, 3 drachms; cantharides, 6 drachms; powdered and digested with 4 ounces alcohol; and 3 drachms Venice turpentine added to the strained tincture. Fine paper is dipped into it and dried in the air. Mohr directs 4 drachms cantharides and 1 drachm euphorbium to be digested in 5 ounces of highly rectified spirit; filter, and add 1½ ounces Venice turpentine previously liquefied with 2 ounces resin. To be spread on the paper while warm.

5238. Papier Epispastique de Vée. This is of three strengths, distinguished by the colors white, green, and red. The composition is made by boiling cantharides for an hour with water, and lard, green ointment, or lard colored with alkanet; adding white wax to the strained fats, and spreading on paper, silk, or linen. No. 1 is made with 10 ounces cantharides to 4 pounds of lard; No. 2 of 1 pound flies to 8 pounds of green ointment; and No. 3 of 1½ pounds to 8 pounds of colored lard; and to each are added 2 pounds of white wax.

5239. Bateman's Itch Ointment. Carbonate of potassa, ½ ounce; red sulphuret of mercury, 1 drachm; hog's lard and flowers of sulphur, each 11 ounces; bergamot, 30 drops; rose water, 1 ounce. Mix the potassa and powders with a little of the lard, and rub them well together; then add the remainder of the lard, previously softened by heat, afterwards add the rose water, gently warmed. Stir till cold.

5240. Smith's Itch Ointment. Flowers of sulphur, 2 ounces; sulphate of zinc, 2 drachms; powdered hellebore, 4 drachms; soft soap, 4 ounces; lard, 8 ounces. Mix.

5241. Wiegand's Tetter Ointment. Powder and mix 2 drachms submuriate of mercury (calomel) with 1 drachm acetate of lead, and ½ drachm red precipitate. Make 42 grains of the above powder into an ointment with 2 drachms of lard or simple cerate.

5242. Wiegand's Tetter Salve. Take 8 grains of the powder in the last receipt, mix with 20 drops glycerine, 5 grains powdered camphor, ½ ounce simple cerate, and 2 drops oil of lemon.

5243. Bailey's Itch Ointment. Sweet oil, 1 pound; suet, 1 pound; root alkanet, 2 ounces. Melt and macerate until sufficiently colored, then add powdered nitre, 3 ounces; powdered alum, 3 ounces; powdered sulphate of zinc, 3 ounces; powdered vermilion, to color; oil of aniseed, oil of spike, and oil of origanum to perfume.

5244. Beddoe's Pills, for gravel, &c. Carbonate of soda, dried without heat, 1 drachm; soap, 4 scruples; oil of juniper, 10 drops; syrup of ginger, sufficient quantity for 30 pills.

5245. Mathieu's Vermifuge. Tin filings, 1 ounce; fern root, ⅔ ounce; worm-seed, ½ ounce; resinous extract of jalap, 1 drachm; sulphate of potassa, 1 drachm; honey to form an electuary. A tea-spoonful every 3 hours for 2 days; then substitute the following: jalap, 2 scruples; sulphate of potassa, 2 scruples; scammony, 1 scruple; gamboge, 10 grains; made into an electuary with honey, and given in the same dose.

5246. Swaim's Vermifuge. Wormseed, 2 ounces; valerian, rhubarb, pink-root, white agaric, of each 1½ ounces; boil in sufficient water to yield 3 quarts of decoction, and add to it 30 drops oil of tansy, and 45 drops oil of cloves, dissolved in a quart of rectified spirits. Dose, 1 table-spoonful at night.

5247. Calvetti's Manna Lemonade. Dissolve 1 ounce pure mannite in 10 ounces boiling water, and add sufficient lemon juice to flavor. To be drunk cold or iced. *Mannite* is a peculiar saccharine principle obtained n crystalline form from manna.

5248. Bond's Compound Mixture of Iron. Take 1½ drachms gum myrrh in tears, 6 drops oil of wintergreen, 2 drops oil of nutmeg, 2 scruples carbonate of potash, 1 ounce loaf sugar, ½ drachm sulphate of iron, and 7 ounces distilled water. Rub down the myrrh with the oils, add gradually a portion of the water, making a milk of myrrh; then add the potash and sugar. Dissolve the iron in the remainder of the water, and mix the two mixtures by trituration. To be bottled and well corked directly.

5249. Mialhe's Syrup for Hoarseness. Take 15 parts syrup of gum-arabic, 5 parts syrup of tolu, 5 parts maiden-hair, 1 part nitrate of potassa, and 1 part cherry-laurel water. Dose, a table-spoonful in a cup of sweet balm tea, in short draughts.

5250. Dewees' Carminative. Take ½ drachm carbonate of magnesia, 1 drachm loaf sugar, 60 drops tincture of assafœtida, 20 drops tincture of opium, and 1 fluid ounce water. Dissolve the sugar in half the water; add this to the tinctures previously mixed in the bottle. Rub the magnesia with the remainder of the water; then mix together the two preparations. Direct the mixture to be shaken before used.

5251. Golden Tincture. Take 3 parts sulphuric ether, 2 parts acetated tincture of opium, and 1 part compound spirit of lavender.

5252. Golden Tincture. Sulphuric ether, 1 ounce; laudanum, 1 ounce; chloroform, ¼ ounce; alcohol, 1 ounce. Mix. This preparation is extensively used by the German physicians. Dose, from 3 to 30 drops, according to circumstances. It makes an excellent local application in neuralgia and other painful affections.

5253. Napoleon's Pectoral Pills. Ipecacuanha, 30 grains; powdered squills and ammoniac, of each 40 grains; mucilage to mix; divide into 24 pills. It is said that the above was a favorite remedy with the first Emperor of France for difficulty of breathing, bronchitis, and various affections of the organs of respiration. Dose, 2 pills night and morning.

5254. Gedding's Piles Ointment. Carbonate of lead, 4 drachms; sulphate of morphia, 15 grains; stramonium ointment, 1 ounce; olive oil, sufficient to make into an ointment.

5255. Ditchett's Remedy for Piles. Spermaceti ointment, 8 ounces; powdered galls, 1 ounce; powdered opium, 1 drachm; solution of diacetate of lead, 1½ ounces. Mix well.

5256. Brown's Bronchial Troches. Take 1 pound pulverized extract of liquorice, 1½ pounds pulverized sugar, 4 ounces pulverized cubebs, 4 ounces pulverized gum-arabic, and 1 ounce pulverized extract of conium (hemlock). Mix.

5257. Roche's Embrocation, or Whooping Cough Liniment. Olive oil, 8 ounces; oil of amber, 4 ounces; oil of cloves, a sufficient quantity to give it a strong scent. Mix. Rubbed on the chest it stimulates the skin; it is useful in general for the coughs of children; in whooping-cough, however, it ought not to be used for the first ten days of the disease. This liniment is understood to be the same as the celebrated embrocation of Roche.

5258. Dupuytren's Pills. Take 120 grains powdered guaiacum, 4 grains corrosive chloride of mercury (corrosive sublimate), and 5 grains powdered opium; make into 40 pills.

5259. Anodyne Necklaces. Beads formed of the root of henbane, and used as necklaces, to allay the pain of teething.

5260. Digestive, or Live-long Candy. Powdered rhubarb, 60 grains; heavy magnesia, 1 ounce; bicarbonate of soda, 1 drachm; finely-powdered ginger, 20 grains; cinnamon powder, 15 grains; powdered white sugar, 2 ounces; mucilage of tragacanth, sufficient quantity; beat together and divide into square, flat cakes of 20 grains each.

5261. Cholagogue. Quinine, 20 grains; Peruvian bark, 1 ounce; rhubarb, 1 ounce; sulphuric acid, 15 or 20 drops, or 1 scruple tartaric acid; brandy, 1 gill, and water to make 1 pint. Dose, 2 spoonfuls every 2 hours in absence of fever.

5262. Malone's Mixture for a Cough or Cold. Take 1 tea-cupful of flaxseed, soak all night. In the morning put in a kettle 2 quarts water, 1 handful of liquorice root (split up), ½ pound good raisins (cut in half). Boil them until the strength is thoroughly extracted, then add the flaxseed, which has been previously soaked. Let all boil about half an hour more, watching and stirring, that the mixture may not burn. Then strain and add lemon-juice and sugar to taste. Take any quantity, cold, through the day, and half a thimbleful, warm, at night. The above is a most excellent receipt.

5263. Chapman's Copaiba Mixture. Make a mixture of ½ ounce copaiba, ½ fluid ounce sweet spirits of nitre, 2 drachms powdered acacia, 1 drachm sugar, 4 fluid ounces distilled water, 2 fluid drachms compound spirit of lavender, and 1 fluid drachm tincture of opium. Dose, a table-spoonful 3 times a day. A specific remedy for gonorrhœa.

5264. Morton's Copaiba Mixture. Take ½ ounce each copaiba and powdered cubebs, 2 drachms each acacia and sugar, 7 fluid ounces water, and ½ fluid ounce camphorated tincture of opium. Make into a mixture. Dose, a table-spoonful every 3 hours. An efficacious remedy for obstinate gonorrhœa.

5265. Jackson's Pectoral Syrup. Macerate 1 drachm sassafras pith and 1 ounce acacia in 1 pint water for 12 hours; add 21 ounces sugar, dissolve the sugar in it without heat, filter, and then add 8 grains muriate of morphia. Dose, 1 tea-spoonful every 3 hours.

5266. Ayer's Wild Cherry Expectorant. Mix together 3 grains acetate of morphia, 2 fluid drachms tincture of blood-root, 3 fluid drachms each antimonial wine and ipecacuanha wine, and 3 fluid ounces syrup of wild cherry bark. Dose, 1 tea-spoonful in catarrh, bronchitis, and influenza.

5267. Ayer's Cherry Pectoral. The following receipt is said to be somewhat near to, if not exactly identical with the receipt after which this well known article is compounded: Take of syrup of wild cherry, 6 drachms; syrup of squills, 3 drachms; tincture of blood-root, 2 drachms; sweet spirits of nitre, 2 drachms; antimonial wine, 3 drachms; wine of ipecacuanha, 3 drachms; simple syrup, 1½ ounces; acetate of morphine, 2 grains. Mix, and add oil of bitter almonds,

2 drops; dissolved in alcohol, 1 drachm.

5268. **Donovan's Mixture of Cyanide of Potassium.** Mix together 1 grain cyanide of potassium, 3½ fluid ounces distilled water, and ½ fluid ounce lemon syrup. Dose, a table-spoonful every 2 hours. Useful to check vomiting, and allay cough; and, in much smaller doses, for whooping cough in children.

5269. **Regnault's Pectoral Paste.** Flowers of mallow, flowers of cudweed, flowers of coltsfoot, and flowers of red poppy, 1 ounce of each; boil in a quart of water, strain, then add 30 ounces of gum-arabic, 20 ounces of white sugar, and 2 drachms tincture of tolu; dissolve, strain, and evaporate to the proper consistence.

5270. **Dennis' Patent Anti-spasmodic Tincture.** Take 1 ounce each tincture of scullcap, valerian, myrrh, and capsicum; 2 ounces tincture of lobelia; a *little* soda; and sufficient water.

5271. **Goitre Jelly.** Better known, perhaps, under the French name *Gelée pour le Goitre.* Dissolve 1 ounce white soap in 2½ ounces of proof spirit by a gentle heat; and add to it, while still warm, a warm solution of 5 drachms iodide of potassium in 2¼ ounces proof spirit. A few drops of any fragrant and essential oil may be added.

5272. **Mettauer's Aperient Solution.** Take of socotrine aloes, 2½ ounces; super-carbonate of soda, 6 drachms; water, 4 pints; compound spirits of lavender, 2 ounces. After digesting 14 days, the clear liquor may be either decanted or allowed to remain. Age is said to improve both the powers and taste of the solution. The common dose is 1 drachm, which may be increased, if necessary, to an ounce. It is recommended as a valuable remedy in most forms of constipation, taken soon after meals.

5273. **Coxe's Hive Syrup.** Put 1 ounce each squills and Seneca snake-root into 1 pint water; boil down to one-half and strain. Then add ½ pound clarified honey containing 12 grains tartrate of antimony. Dose for a child, 10 drops to 1 tea-spoonful, according to age. An excellent remedy for croup.

5274. **Bateman's Sulphur Wash.** Break 1 ounce sulphur, and pour over it 1 quart of boiling water; allow it to infuse for 12 or 14 hours, and apply it to the face 2 or 3 times a day, for a few weeks. This application is equally useful in removing that roughness of the skin which generally succeeds pimples.

5275. **Allcock's Porous Plaster.** The only difference between this plaster and ordinary adhesive plasters is, that rubber is used in the place of lead plaster. It is a good addition, and very generally recognized by makers of adhesive plasters. Take rubber, 1 pound; pitch, ½ pound; thus, ½ pound; and capsicum, 30 grains. The plaster, as offered for sale, is spread upon muslin or linen, in which small holes have been punched out, allowing vent for perspiration, and affording increased flexibility. These plasters adhere very firmly, frequently requiring the application of heat (by means of a hot towel or warm flat-iron), for their removal. The skin may be cleansed after the removal of the plaster, by rubbing with sweet oil, until the remains of the plaster are dissolved; wiping it off, and washing with warm water and soap.

5276. **Poor Man's Plaster.** Take bees'-wax, 1 ounce; tar, 3 ounces; resin, 3 ounces. To be melted together and spread on paper or muslin.

5277. **Universal Plaster.** A plaster is officinal in several of the European Pharmacopœias, under different names, which appears to be identical with *Keyser's Universal Plaster,* which is sold extensively in this country as a nostrum. The following is the formula of the Prussian Pharmacopœia: Take of red-lead, in very fine powder, 8 ounces; olive oil, 16 ounces. Boil them in a proper vessel with constant agitation until the whole has assumed a blackish-brown color, then add yellow wax, 4 ounces; and after this has been melted and well mixed, add 2 drachms camphor, previously dissolved in a little olive oil. Pour it out into suitable boxes, or into paper capsules, to be cut into square cakes when cold.

5278. **Devil Plaster.** Cases of severe wounds are said to have healed without suppuration after 17 or more days by the use of this plaster. It has also been successfully applied to fractures and tumors. Take 15 drachms black pitch, 15 drachms dry resin, 2½ drachms dried earth-worms in powder, 8 drachms essential oil of turpentine, and 1 scruple crude alum. Mix well. This plaster was much used by an old surgeon of Morello, and by his sons, for the cure of wounds without the loss of substance. The composition, which they kept secret, is now published to the world by M. Escorihuela. He obtained the secret from one of the heirs.

5279. **Wallace's Pills.** Take socotrine aloes, scammony, and soap, all in powder, blue mass and compound extract of colocynth, 1 scruple each, to make 20 pills.

5280. **Canada Liniment.** Take water of ammonia, olive oil, oil of turpentine, and alcohol, of each 1 ounce; oil of peppermint, ½ ounce. Mix.

5281. **St. John Long's Liniment.** White and yolk of 1 egg; oil of turpentine, 6 ounces; acetic acid, 1 ounce; oil of lemon, 12 drops; and rose-water, 5 ounces. Mix.

5282. **Brodie's Liniment.** Take of sulphuric acid, 1 drachm; olive oil and oil of turpentine, of each 1 ounce. Add the acid gradually to the olive oil, stirring it in a mortar; when cool, add the oil of turpentine and mix.

5283. **Good Old Samaritan Liniment.** Mix together 2 gallons alcohol, 12 ounces oil origanum, 4 ounces oil hemlock, and 2 ounces each of oil of cedar, balsam of fir, spearmint, balsam of life (*see No.* 5112), oil of sassafras, oil of wintergreen, spirits of turpentine, and sulphuric ether. Mix.

5284. **Physic's Issue Ointment.** Powdered cantharides, ½ ounce; rose water, 2 fluid ounces; tartar emetic, 15 grains. Apply heat and evaporate the rose-water one-half; strain, and add olive oil, 3 ounces; white wax, 1½ ounces; spermaceti, 1 ounce. Mix, and apply a gentle heat until all the water has been driven off. When the manipulations have been conducted with care, the cerate is light in color.

5285. **Beach's Black Plaster or Healing Salve.** Take of olive oil, 3 quarts; common resin, 3 ounces; bees'-wax, 3 ounces. Melt these articles together, and raise the oil almost to boiling heat; then gradually add of pulverized red lead 2½ pounds, if in the summer; if in the winter, ¼ pound less. In a short time after the lead is taken up by the oil, and the mixture becomes brown or a shining black, remove from the fire, and, when nearly cold, add ½ ounce pulverized camphor.

5286. **M'Kenzie's Ointment.** Powdered sulphate of zinc, 4 ounces; liquid storax, 1 ounce; melted lard, 16 ounces. Mix by means of heat and triturate over a water-bath for about an hour. A useful application for tetter and scald-head. Apply night and morning, first washing the part with Castile soap and warm water.

5287. **Conklin's Salve.** Take resin, 12 ounces; bees'-wax, mutton suet, and tallow, of each 1 ounce. Melt together, strain the mixture through muslin, and work into rolls in a bath of cold water.

5288. **Newell's Compound Tar Ointment.** Lard and mutton suet, of each 12 ounces; tar, 6 ounces; bees'-wax, 3 ounces; powdered black hellebore, 4 drachms; melt and strain, then add flowers of sulphur, 4 ounces. Used for tetters, salt rheum, itch, &c.

5289. **Turner's Cerate.** Take of sweet oil, 2 pounds; yellow wax, carbonate of zinc, powdered, of each 1 pound. Mix at a low heat.

5290. **Allison's Tobacco Ointment for Gathered Breasts.** Tobacco leaves (fresh and sliced), 10 ounces; dilute acetic acid, 4 pints; basilicon ointment (*see No.* 4964), 13 ounces. Boil the tobacco in the acid, strain and evaporate the decoction over a warm bath to 4 fluid ounces; add this to the basilicon ointment, heated, and stir the whole together until cold. Apply spread upon linen or soft kid skin.

5291. **Allison's Acetated Ointment of Tobacco.** Tobacco leaves, sliced, 10 ounces; cider vinegar (or officinal dilute acetic acid), 4 pints; basilicon ointment (*see No.* 4964), 13 ounces. Boil the tobacco in vinegar to 1 pint, strain, reduce in a water-bath to 6 fluid ounces, and add this fluid extract to the melted ointment, stirring constantly till it is cool. A fine remedy for gathered breasts.

5292. **Parrish's Compound Ointment of Tobacco.** Basilicon ointment (*see No.* 4964), 13 ounces troy; powdered camphor, 29 drachms; extract of belladonna, 2 ounces; fluid extract of tobacco (made as in the above formula), 6 ounces. Dissolve the extract of belladonna in the fluid extract of tobacco and add to the melted ointment, in which the camphor should be previously dissolved. Stir constantly till cool. Dr. Parrish has stated, in the New Jersey Medical Reporter, that he uses this ointment in nearly every case of mammary abscess, with entire satisfaction.

5293. **Mege's Rheumatic Ointment.** Take 160 parts lard, 6 parts each of the extracts of opium, belladonna, and cinchona, 7 parts ammonia water.

5294. **Mitchell's Ointment of Three.** Mix together equal parts of tar ointment, sulphur ointment, and red oxide of mercury ointment.

5295. **Berthold's Chilblain Wash.** Boil for 15 minutes 1½ ounces bruised nut-galls in ½ pint water, and strain. Apply to the chilblains 2 or 3 times a day. Tannic acid dissolved in glycerine has a very similar effect, but in a neater form for application.

5296. **Lapis Divinus.** This preparation, called also *cuprum aluminatum,* is the *pierre divine* of the French codex. It is made by mixing in powder, 3 ounces each of sulphate of copper, nitrate of potassa, and alum; heating the mixture in a crucible so as to produce watery fusion; then mixing in 1 drachm powdered camphor; and finally pouring out the whole on an oiled stone to congeal. The mass, when cold, is broken into pieces, and kept in a well-stopped bottle. When this preparation is used as an eye lotion, a filtered solution is made, of the average strength of 30 grains to a pint of water.

5297. **Lapis Miraculosus.** Fuse together sulphate of copper, 3 parts; sulphate of iron, 6 parts; verdigris and alum, of each 1 part; sal-ammoniac, ½ part. It is used for ulcers only.

5298. **Biett's Solution.** This is a solution of 1 grain of arseniate (not arsenite) of ammonia in 1 troy ounce of water. It is not as safe a preparation as either Fowler's or Pearson's solution, owing to the ready decom position of the ammonia salt.

5299. Pearson's Arsenical Solution. This is an aqueous solution of arsenite of soda, containing 1 grain of the salt in a fluid ounce.

5300. Sampson's New York Pills. The 1¼ grain pills consist of powdered coca, 25; extract of coca, 30; powdered iron, 35 parts.

5301. Oil of Stone. Take crude American petroleum, and Barbadoes petroleum, of each 2 pints; oil of turpentine, 6 pints.

5302. Chelsea Pensioner. Take powdered rhubarb, 2 drachms; cream of tartar, 1 ounce; guaiacum, 1 drachm; sulphur, 2 ounces; 1 nutmeg grated fine; clarified honey, 16 ounces. Mix. Dose, 2 tea-spoonfuls night and morning. A very good remedy for chronic rheumatism.

5303. Indian Cathartic Pills. Reduce to a fine powder, 1 ounce each aloes and gamboge; ¼ ounce each mandrake, blood-root, and myrrh; 1¼ drachms camphor (*see No.* 4358) and cayenne; with 4 ounces ginger. Mix thoroughly and make into ordinary-sized pills with thick mucilage. Dose, 2 to 4 pills.

5304. Turlington's Balsam is much like the compound tincture of benzoin of the Pharmacopœia of the U. S., though it is somewhat more complicated. To make it, take benzoin, 12 ounces; liquid storax, 4 ounces; balsam of Peru, 2 ounces; myrrh and aloes, each 1 ounce; balsam of tolu and extract of liquorice, each 4 ounces; angelica root, ½ ounce; alcohol, 8 pints. Digest for 10 days, and strain.

5305. Thibault's Balsam. Myrrh, aloes, and dragon's blood, of each 1 drachm; flowers of Saint John's wort, 1 handful; spirit of wine, ½ pint; Canada balsam, ½ ounce. Digest the flowers in the spirit for 3 days, then express the liquor and dissolve the other ingredients therein. To heal cuts and wounds, and to stop bleeding. Internally diuretic, in doses of 1 to 2 tea-spoonfuls; given in gonorrhœa.

5306. Locatelle's Balsam. Yellow resin, olive oil, and Venice turpentine, of each 1 pound; shavings of red saunders wood, 1 ounce. Boil to the consistence of a thin ointment, and strain.

Or: Yellow wax, 4 ounces; olive oil and Venice turpentine, of each 1 pound; alkanet root, 2 ounces; as last. Used as a pectoral in coughs and colds. Dose, ½ to 1 tea-spoonful mixed with the same quantity of conserve of roses.

5307. Bell's Gargle. Take of pure borax, 2 drachms; yeast and honey, of each ½ ounce; boiling water, 7 ounces. Mix.

5308. Mrs. Wheeler's Nursing Syrup. Mix together 35 ounces sugar, 4 ounces lime-water, ¼ ounce aqueous extract of podophyllin, 4 ounces fluid extract of poppy, and 1 drachm oil of anise in 2 ounces rectified spirit. The aqueous extract of podophyllin is of the same strength as the ordinary fluid extracts, 16 troy ounces to the pint. The above syrup will be found to contain about 2 drops fluid extract of poppy in each tea-spoonful.

5309. Mrs. Wheeler's Worm Confection. Triturate to a fine powder, 1 drachm mild chloride of mercury and 10 drachms sugar; add 25 ounces sugar and 6 drachms santonin; mix all together and make into 360 tablets. Each tablet will therefore contain ⅙ grain of calomel and 1 grain santonin.

5310. Brodie's Decoction of Pareira Brava. Take ½ ounce bruised pareira root, and 3 pints boiling water; boil down gently to 1 pint, and filter. Dose, 1 wine-glassful every 2 hours. An excellent remedy for chronic inflammation of the bladder.

5311. Hufeland's Diuretic Drops. Take ¼ fluid drachm oil of juniper, and 3 fluid drachms each sweet spirits of nitre and tincture of digitalis. Dose, 30 drops every 3 hours.

5312. Stephens' Infusion of Cayenne Pepper and Salt. Macerate ½ ounce powdered cayenne pepper, and 1 drachm chloride of sodium (table salt) for 1 hour in 8 fluid ounces each boiling vinegar and boiling water. Filter. Dose, 1 table-spoonful every 2 hours. This has been administered with great success in malignant scarlet fever; used both internally and as a gargle.

5313. Magendie's Acid Solution of Veratria. Dissolve 1 grain veratria in 2 fluid ounces distilled water and 5 drops aromatic sulphuric acid. Dose, 1 tea-spoonful, in gouty affections.

5314. Ryan's Gleet Powder. Take 2 scruples powdered ergot, 1 ounce powdered cubebs, ½ drachm powdered cinnamon, and 1 drachm sugar. Make into 8 powders. Dose, 1 powder 3 times a day, for leucorrhœa and gleet.

5315. Channing's Mixture. Dissolve 3½ grains iodide of potassium in 1 fluid ounce distilled water; then add 4½ grains red iodide of mercury. Dose, from 2 to 5 drops, in cases of secondary symptoms, and obstinate skin diseases.

5316. Thomas's Cathartic Pills. Take ½ drachm compound extract of colocynth, and 3 grains resin of podophyllin. Make into 12 pills. Dose, 1 or 2 at bed-time. 1 pill acts as a laxative; 3 as a free purgative.

5317. Parrish's Cathartic Pills. Take 24 grains aloin, 12 grains resin of podophyllin, and 4 minims oleo-resin of ginger. Make into 24 pills. Dose, the same as directed in the last receipt.

5318. Becquerel's Anti-Gout Pills. Take 2 drachms sulphate of quinine, 15 grains alcoholic extract of digitalis, and 2 scruples acetic extract of colchicum. Make into 50 pills. Dose, 1 pill every 3 hours.

5319. Butternut Pills. Take ½ drachm extract of butternut, 1 scruple powdered jalap and 10 grains soap. Make into 15 pills. Dose, 3 pills, and, if these do not operate, administer 2 more. Butternut is highly recommended as a cathartic in fevers, dysentery, &c.

5320. Chapman's Peristaltic Persuaders. Take 1 drachm powdered rhubarb, 10 grains powdered ipecacuanha, and 10 drops oil of caraway. Make up with sufficient powdered acacia into 20 pills. Dose, 2 pills at bed-time, in obstinate constipation.

5321. Composition Powder. Finely pulverize 2 pounds bayberry bark, 1 pound hemlock bark, 1 pound ginger, 2 ounces cayenne pepper, and 2 ounces cloves. Mix them together. This is an excellent remedy for weak stomach, dyspepsia, &c. Put ½ tea-spoonful of the mixture with a tea-spoonful of sugar into a cup of boiling water. After standing for a few moments, drink the contents. (Some omit the hemlock bark.)

5322. Le Gros's Itch Ointment. Take of iodide of potassium, ½ drachm avoirdupois; lard, 1 ounce; mix. Cleanly harmless, and effective.

5323. Stokes' Liniment. The formula here given for this preparation is the one adopted by the Maryland College of Pharmacy, and is believed to be as originally prescribed by Dr. Stokes. Take 3 fluid ounces oil of turpentine, ½ fluid ounce strong acetic acid, the yolk of 1 egg, 3 fluid ounces rose-water, and 1 fluid drachm oil of lemon.

5324. Mother's Cordial. Take 4 ounces each of starwort (helonias dioica), high cranberry bark (viburnum opulus), and blue cohosh (caulophyllum thalictroides), and 1 pound of partridge-berry (mitchella repens). Bruise or grind the ingredients, and macerate for 3 days with enough strong alcohol to cover; then displace from them with more alcohol 3 pints of tincture, which are set aside, and the ingredients exhausted with hot water until it passes tasteless. Add 2 pounds sugar and evaporate with a gentle heat to 5 pints; then mix with the 3 pints of tincture and flavor with sassafras.

5325. Wyndham's Pills. Gamboge, 3 ounces; aloes, 2 ounces; Castile soap, 1 ounce; nitre, ½ ounce; extract of cow-parsnip, 1 ounce. In pills of 5 grains each. (*Lec.*)

5326. Anderson's Pills. Barbadoes aloes, 24 ounces; soap, 4 ounces; colocynth, 1 ounce; gamboge, 1 ounce; oil of aniseed, ½ fluid ounce. Mix, and divide into pills of 3 grains each.

5327. Morrison's Pills. No. 1 consists of equal parts of aloes and cream of tartar; No. 2 consists of 2 parts of gamboge, 3 of aloes, 1 of colocynth, and 4 of cream of tartar, made into pills with syrup.

5328. Ayer's Sarsaparilla. Take 3 fluid ounces each of alcohol, fluid extracts of sarsaparilla and of stillingia; 2 fluid ounces each fluid extracts of yellow-dock and of podophyllin; 1 ounce sugar, 90 grains iodide of potassium, and 10 grains iodide of iron. This is from a receipt given by Dr. Ayer himself.

5329. Henderson's Lotion for Corns. Take tincture of iodine, ½ ounce; iodide of iron, 12 grains; chloride of antimony, ½ ounce. Pare the corn, and apply with a camel's-hair pencil. This lotion has been much commended for destroying corns.

5330. Velpeau's Black Caustic. Triturate in a porcelain mortar 1 ounce powdered liquorice root, and add sulphuric acid in small quantities until a mass is obtained neither too hard nor too liquid. This preparation forms a well-marked hard black scab.

5331. Jarave Spanish. Pour 4 gallons of boiling water on 2 pounds Rio Negro sarsaparilla, 8 ounces powdered guaiacum bark, 4 ounces each of rasped guaiacum wood, anise seed, and liquorice root, 2 ounces of bark of mezereon root, 2 pounds of molasses, and 12 bruised cloves. Shake it thrice a day, and keep it in a warm place. When fermentation has set in, it is fit for use. Dose, a small tumblerful.

5332. Bouyer's Syrop de Lait Iodique. Take cow's milk 200 parts; cane sugar, 60 parts; iodide of potassium, ⅙ part; and a *little* soda. Mix, and evaporate to 100 parts.

5333. Cephalic Snuff. Dried asarabacca leaves, 3 parts; marjoram, 1 part; lavender flowers, 1 part; rub together to a powder.

5334. Boeli's Cephalic Snuff consists of 2 drachms valerian, 2 drachms snuff, 3 drops oil of lavender, 3 drops oil of marjoram; mix. This is said to relieve the eyes as well as the head.

5335. Radway's Ready Relief, according to Peckolt, is an ethereal tincture of capsicum, with alcohol and camphor.

5336. Radway's Renovating Resolvent. A vinous tincture of ginger and cardamom, sweetened with sugar. (*Hager* and *Jacobsen.*)

5337. Swedish Essence of Life is made in this country, under various names. As usually made by apothecaries, it is a tincture prepared from 4 parts aloes, 1 each of agaric, rhubarb, zedoary, gentian, myrrh, and theriac, with 100 to 120 parts dilute alcohol. The medicine manufacturers usually substitute cheaper articles for the high-priced saffron and rhubarb. (*See No.* 5365.)

5338. Walker's Jesuits' Drops. Balsam of copaiba, 6 ounces; gum guaiacum, 1 ounce; Chio turpentine, ½ ounce; subcarbonate of potash, ½ ounce; cochineal, 1 drachm; rectified spirit, 1 quart.

5339. Molinari's Remedy for Sea-Sickness. Digest for 12 hours in 1¾ Imperial pints of wine vinegar, ½ ounce each of rue, thyme, mint, rosemary, absinthe, turmeric, and green walnut rind; ⅛ ounce annatto; ¼ ounce pearlash; and 1 poppy-head. After digestion boil for half an hour; then strain through linen; in this decoction are moistened or dipped some 4 or 5 strips of filtering paper 7 or 8 inches long, and then dried; upon one side of these strips some light stuff is fastened by the corners and some loose wadding placed inside. Strings are next fastened to the bandage and it is then tied around the body so as to cover the region of the heart. This preventive of sea-sickness has been patented in England.

5340. Redwood's Nervine Balsam. Melt together 4 ounces oil of mace and 4 ounces beef marrow. Dissolve in 4 drachms alcohol, 2 drachms each oil of rosemary and balsam of tolu, and 1 drachm each of camphor and oil of cloves. Mix all together. A good liniment in rheumatism.

5341. Chaussier's Obstetric Ointment. Extract belladonna, 2 drachms; water and lard, each 2 drachms. Mix well.

5342. Dutch Drops, or Haerlem Drops. There is considerable difference in the ingredients and quality of these long-celebrated drops; but the most common preparation, perhaps, is made according to the following formula: Take balsam of turpentine, 2 ounces; oil of turpentine, 10 ounces. Mix. The following is also one of the imitations of it made in this country: Linseed oil, 1 quart; resin, 2 pounds; sulphur, 1 pound; boil together over a slow fire; when combined remove from the fire, and add 1 pint oil of turpentine, and 50 drops liquor of ammonia; stir well together and bottle. The genuine drops are the residuum of the rectification of oil of turpentine. Dutch drops are of course stimulant and diuretic in their therapeutical effects; but they have been regarded by the common people as possessed of many other virtues, and have been much applied to wounds and other external injuries of the surface.

5343. Russia Salve. Take equal parts of yellow wax and sweet oil, melt slowly, carefully stirring; when cooling, stir in a small quantity of glycerine. Good for all kinds of wounds, &c.

5344. James' Oil of Gladness. Take oil of hemlock, 1 ounce; linseed oil, 1 quart.

5345. Green Mountain Salve. Take 2 pounds resin, ¼ pound Burgundy pitch, ¼ pound bees'-wax, ¼ pound mutton tallow; melt them slowly. When not too warm, add 1 ounce oil hemlock, 1 ounce balsam fir, 1 ounce oil origanum, 1 ounce oil of red cedar, 1 ounce Venice turpentine, 1 ounce oil wormwood, ½ ounce verdigris. The verdigris must be very finely pulverized and mixed with the oils, then add as above and work all in cold water until cold enough to roll. This salve has no equal for rheumatic pains or weakness in the side, back, shoulders, or any place where pain may locate itself. Where the skin is broken, as in ulcers, bruises, &c., use without the verdigris.

5346. Keating's Cough Lozenges. These are said to be composed of lactucarium, 2 drachms; ipecacuanha, 1 drachm; squills, ¾ drachm; extract of liquorice, 2 drachms; sugar, 6 ounces. Made into a mass with mucilage of tragacanth, and divided into 20-grain lozenges.

5347. Milburn's Mixture. Precipitated prepared chalk, loaf sugar, and gum-arabic, of each 2 drachms; green mint water, 4½ ounces; laudanum, 10 minims; spirits of lavender, 2 drachms; simple syrup, 1½ ounces; tincture of kino, 1 ounce. Mix. Useful in loose bowels in children, and can be given to them after each evacuation, regardless of number. Dose, from ½ to 1 table-spoonful. Shake the mixture well each time before using it.

5348. Ricord's Aromatic Wine. Take rue, sage, hyssop, lavender, absinth, rose-leaves, thyme, and elder flowers, of each 4 ounces. Digest for 2 weeks in 9 pints claret. Then add tannic acid, alum, wine of opium, of each 9 ounces.

5349. Beyran's Wash. Dissolve chloride of zinc in 100 times its weight of pure water. This solution is used as a wash for chancres, and spontaneously or artificially opened buboes that are extending both in size and depth, and show no signs of cicatrization. It is applied twice a day by means of lint moistened with it. As soon as the vitality of the parts becomes favorably modified, Dr. Beyran replaces this wash by Ricord's wine of cinchona or aromatic wine. (*See No.* 5348.)

5350. Charta Epispastica. White wax, 4 parts; spermaceti, 1½ parts; olive oil, 2 parts; resin, ¾ parts; Canada balsam, ¼ part; cantharides in powder, 1 part; distilled water, 6 parts. Digest all the ingredients excepting the Canada balsam in a water-bath for 2 hours, stirring them constantly; then strain, and separate the plaster from the watery liquid. Mix the Canada balsam with the plaster melted in a shallow vessel, and pass slips of paper over the surface of the hot liquid, so that one surface of the paper shall receive a thin coating of plaster.

5351. Brodum's Nervous Cordial. Take equal parts of iron wine, compound sprits of lavender, tinctures of calumba, gentian, cinchona, and cardamoms.

5352. Atkinson's Infant Preservative. Carbonate of magnesia, 6 drachms; white sugar, 2 ounces; oil of aniseed, 20 drops; spirit of sal-volatile, 2½ drachms; laudanum, 1 drachm; syrup of saffron, 1 ounce; caraway water to make a pint.

5353. Boyle's Fuming Liquor. Take quicklime and sulphur, each 3 parts. Triturate together, adding water sufficient to form a paste, and incorporate 7 parts sulphate of ammonia dissolved in water; let the whole stand, then decant, wash the residuum, rubbing it with a small portion of water, unite the solutions, and filter. This is the *sulphuretted hydrosulphate of ammonia*, and is used in medicine as a powerful alterative in constitutional diseases.

5354. Hall's Solution of Strychnia. Take pure crystals of strychnia, 16 grains; water and alcohol, of each 7½ ounces; acetic acid and compound tincture of cardamoms, of each ½ ounce. Mix for solution. Dose, 20 to 30 drops, once or twice a day.

5355. Flemming's Solution of Strychnia. Take of strychnia, 2 grains; distilled water, 5 fluid drachms; muriatic acid, 1 drop, or sufficient to dissolve the strychnia. Dissolve by trituration, and add diluted alcohol enough to make 10 fluid drachms. Dose, in the beginning, 10 minims.

5356. Brandish's Alkaline Tincture of Rhubarb. Coarsely powdered rhubarb, 1 ounce; Brandish's alkaline solution, 32 fluid ounces. The original formula directs only ½ ounce rhubarb, but as smaller doses than were given by Dr. Brandish are now usually prescribed, the quantity of rhubarb is here increased. Or an alkaline infusion of rhubarb may be made by pouring boiling water, 8 parts, on rhubarb, 3 parts, and carbonate of potash, 1 part.

5357. Brandish's Alkaline Solution, or Caustic Alkali. American pearl-ashes, 6 pounds; quicklime, 2 pounds; wood ashes prepared by burning the branches of the ash, 2 pounds; boiling water, 6 gallons; slack the lime, add the rest of the water and the pearl-ashes, and lastly stir in the wood-ashes; let it stand in a covered vessel for 24 hours, and decant. To each pint add 1 drop of true oil of juniper berries. Keep it in stoppered bottles of green glass. The common liquor of potassa is usually sold for the above solution.

5358. Coating for Pills. Durden recommends collodion as a covering for pills; others, a solution of gutta percha in chloroform; but the ready solubility of these materials in the stomach may be questioned. Blanchard uses balsam of tolu dissolved in ether. Baildon recommends chloroform instead of ether for dissolving the balsam.

5359. Garrot's Covering for Pills. Soak 1 ounce purified gelatine in 2 or 3 drachms water; keep it liquefied in a salt-water bath. The pills are stuck on long pins, and dipped in the solution; when cold the pins are withdrawn, after being heated by a small flame, which melts the gelatine and closes the hole.

5360. Bochet's Syrup. Compound syrup of sarsaparilla, with senna, and 1 per cent. of iodide of potassium. Used for scrofulous affections.

5361. Betton's British Oil. Oil of turpentine, 8 ounces; Barbadoes tar, 4 ounces; oil of rosemary, 4 drachms; mix.

5362. British Oil, or Oil of Stone. Take oils of turpentine and linseed, each 8 ounces; oils of amber and juniper, each 4 ounces. Barbadoes tar, 3 ounces; seneca (petroleum) oil, 1 ounce. Mix. This is an excellent application to cuts and bruises, swellings and sores of almost any description whatever.

5363. Cochrane's Cough Medicine. This consists of an acidulated syrup of poppies.

5364. Godfrey's Cordial. The Philadelphia College of Pharmacy, to prevent the mischief arising from the different strength of this compound, directs it to be prepared as follows: Dissolve 2½ ounces carbonate of potash in 26 pints of water, add 16 pints molasses; heat together over a gentle fire till they simmer, remove the scum, and, when sufficiently cool, add ½ ounce oil of sassafras dissolved in 2 pints of rectified spirit, and 24 fluid ounces of tincture of opium, previously mixed. It contains about 16 minims of laudanum, or rather more than 1 grain of opium in each fluid ounce.

5365. Baume de Vie. Socotrine aloes, 2 drachms; rhubarb, 6 drachms; saffron, 2 drachms; liquorice root, 1 ounce; proof spirit, 8 ounces. Digest for 8 days and filter. The original Swedish form is this: Aloes, 9 drachms; rhubarb, gentian, zedoary, saffron, theriaca, agaric, of each 1 drachm; proof spirit, 2 pints. (*See No.* 5337.)

5366. Jozeau's Copahine-mege. The intention of M. Jozeau in devising this form of copaiba was to furnish an article that the stomach would be more able to digest than the crude article. To this end he proposed to himself to oxidize the copaiba, which he accomplishes by mixing nitric acid with it. The essential oil is acted on, and hyponitrous acid gas escapes into the atmosphere. The copaiba thus treated is then washed with water, until it no longer reddens litmus paper, and one-tenth part of cubebs in fine powder are added to it, the same proportion of carbonate of soda, and one-sixteenth part of calcined magnesia. The mixture is allowed to stand until it is quite solidified, and in that state it is made into small masses, which are then carefully covered with sugar.

5367. Ford's Balsam of Horehound is said to be prepared according to the following formula: horehound herb, 3½ pounds; liquorice root, 3½ pounds; water, 8 pints. Infuse for 12 hours, then strain off 6 pints, to which add camphor, 10 drachms; opium and benzoin, of each 1 ounce; dried squills, 2 ounces; oil of aniseed, 1 ounce; proof spirit, 12 pints. Macerate for 1 week, then add honey, 3½ pounds. Mix and strain.

5368. Holloway's Ointment. Take butter, 12 ounces; bees'-wax, 4 ounces; yellow resin, 3 ounces. Melt, and add vinegar of cantharides, 1 ounce. (*See No.* 1178.) Evaporate and add Canada balsam, 1 ounce; oil of mace, ½ drachm; balsam Peru, 15 drops.

5369. Holloway's Pills. Take aloes, 4 parts; myrrh, jalap, and ginger, of each 2 parts. Mucilage to mix.

5370. Sydenham's Laudanum. According to the Paris Codex this is prepared as follows: opium, 2 ounces; saffron, 1 ounce; bruised cinnamon and bruised cloves, each 1 drachm; sherry wine, 1 pint. Mix and macerate for 15 days and filter. Twenty drops are equal to one grain of opium.

5371. Riegler's Fever Tincture. Take of aloes, ½ ounce; camphor, 4 scruples; orange peel and elecampane root, of each 8 ounces. Bruise and digest with 10 pints alcohol (80 per cent.) for 8 days. Then express, add 12 ounces dilute sulphuric acid, 6 ounces sulphate of quinine, and 1½ ounces Sydenham's laudanum. (*See last receipt.*) After the use of a purgative or emetic if required, 2 drachms of this tincture are given 3 hours before the paroxysm is expected, with short diet. On the seventh, fourteenth, and eighteenth day, after the last attack, the same dose is given. This remedy fails only in very exceptional cases. It is in use in the Austrian military hospitals.

5372. Kitridge's Salve. Make a decoction in rain water of 1½ pounds each bittersweet root and sweet elder root; ½ pound each hop-vines, hop leaves, and garden plantain tops, with ½ pound of the root of the last named plant, and ½ ounce plug tobacco. Strain, and press through a thick cloth, and evaporate to ½ pint. Then mix with 1 pound sweet butter and 1 ounce each resin and bees'-wax. Heat gently until the water has all evaporated. This is a good curative salve for sores on the human body as well as on animals.

5373. Thirlault's Glycero-pomade of Iodide of Potassium. Melt glycerine (of 28° to 30° Baumé), 100 parts; powdered animal soap, 50 parts, powdered iodide of potassium, 130 parts; in a warm bath; then pour out into a warm porcelain mortar, and triturate well for ¼ hour. Then flavor with 2 parts oil of bitter almonds.

5374. Elixir of Bromide of Ammonium. Prepared from bromide of ammonium as in No. 5449, without the coloring.

5375. Patent Dysentery Cordial. Take of rhubarb, catechu, and camphor, 2 parts each; laudanum, 4 parts; and a little oil of anise. Dose, 15 to 60 drops after each operation.

5376. Whitwith's Red Drops. Take oil of thyme, 4 drachms; tincture of myrrh, 2 ounces; tincture of camphor, 2 drachms; compound spirits of lavender, 2 ounces; alcohol, 8 ounces. Mix. Dose, 25 drops in some suitable vehicle, two, three, or four times a day. This is the original receipt, but it has been varied in many ways.

5377. George's Myrrhine. Glycerine, 38 parts; myrrh, 7 parts; arrow-root, 5 parts; chalk, 54 parts; oil of cinnamon, 1 part. For the preservation of the teeth.

5378. Kirkland's Neutral Cerate. Mix together 4 ounces litharge plaster, 1½ drachms acetate of lead, and 2 ounces each olive oil, precipitated chalk, and acetic acid.

5379. Hufeland's Zinc Cerate. For sore nipples, ulcerations of the breast, &c. Mix 15 grains each oxide of zinc and lycopodium, with ½ ounce simple cerate and about ½ ounce of spermaceti cerate.

5380. Deschamps' Fuligokali Ointment. This ointment has been considerably used in obstinate chronic diseases of the skin as a detersive, resolvent, and stimulant application, and is made by taking of fuligokali, 16 to 30 parts (*see next receipt*); lard, 1 ounce. Rub together.

5381. To Obtain Fuligokali. Take of potassa, 20 parts; bright soot, 100 parts; water, sufficient; boil for an hour, cool, dilute with water, evaporate to dryness, and keep in well-stoppered bottles.

5382. Hooper's Female Pills. Take 1 drachm dry sulphate of iron, 15 grains powdered jalap, 1 drachm powdered aloes and cinnamon, and 8 grains myrrh. Mix with syrup, and make into 30 pills. Dose, 2 or 3 at bedtime for several nights in succession. They purge smartly, and act beneficially as an emmenagogue. According to a recent analysis, the iron is in a peroxidized state; probably the sulphate is partially calcined. The Philadelphia College of Pharmacy gives the following formula: Barbadoes aloes, 8 ounces; dried sulphate of iron, 9½ drachms; extract of black hellebore, 2 ounces; myrrh and soap, each 2 ounces; canella, 1 ounce; ginger, 1 ounce; water sufficient to form a mass. Divide into pills of 2½ grains each.

5383. Nuremberg Plaster. Mix 8 ounces red lead with 1 pound olive oil, and expose to a heat until the mixture becomes brown or blackish; add ½ ounce resin, 1½ ounces yellow wax, and 2 drachms camphor. The red lead should not be added to the oil until so far heated as to scorch a feather dipped into it.

5384. Green Coloring Powder. Mix together 1 part indigo and 10 parts curcuma root, and reduce to a fine powder. (*Hager.*)

5385. Green Oil. Digest for 2 days, with frequent agitation, 1 part green coloring powder (*see last receipt*) in 20 parts olive oil. Decant the clear, and filter. Keep in glass bottles carefully stopped. Or: Boil 1 part fresh plantain in 8 parts olive oil, until crisp; press and filter. (*Hager.*)

Either of these will produce an oil whose appearance is identical with the oil of henbane, and is probably sometimes sold for it.

5386. Plunket's Ointment for Cancer. White arsenic, sulphur, powdered flowers of lesser spearwort and stinking chamomile, levigated together, and formed into a paste with white of egg.

5387. Hope's Camphor Mixture. Take 4 ounces camphor water, 30 drops fuming nitric acid, and 20 to 40 drops tincture of opium. Dose, a table-spoonful every 2 hours.

5388. Murphy's Carminative. Take ½ pint tincture of valerian, 10 fluid drachms acetated tincture of opium, 128 grains pulverized camphor, 248 grains carbonate of potassa, 2 ounces carbonate of magnesia, 40 minims each oil of anise and oil of mint, and 1½ pints water. Dose for an infant, 20 to 25 drops. This is said to be an improvement on Dewees' carminative. (*See No.* 5435.)

5389. Eisenmann's Opiated Wine of Colchicum. This consists of a mixture of 6 parts wine of colchicum seed and 1 part wine of opium.

5390. Pierlot's Solution of Valerianate of Ammonia. Dissolve 3 scruples extract of valerian in 7 fluid ounces spring water; add 3 fluid drachms fluid extract of valerian, and filter; then add 2 drachms valerianate of ammonia, 6 fluid drachms orange-flower water, and 6 fluid drachms simple syrup. Dose, 1 tea-spoonful 3 or 4 times a day.

5391. Brandreth's Pills. According to Dr. Hager's analysis, these consist of 10 grains extract of may-apple, 30 grains poke berry juice, 10 grains saffron, 10 grains powdered may apple root, 15 grains powdered cloves, and 3 drops oil of peppermint. This is made into 30 pills with powdered liquorice root.

5392. Foucher's Dressing for Wounds. Dissolve 2 drachms chlorate of potassa in 4 fluid ounces glycerine, and add 2½ ounces alcohol. This forms a clear liquid which is readily absorbed by linen, and does not soil the clothing. It keeps the dressings moist for 24 hours, is easily washed off with lukewarm water, and is well adapted for soft granulations.

5393. Atler's Nipple Wash. Take ½ drachm powdered gum-arabic, 10 grains borate of soda, and 1 drachm tincture of myrrh.

5394. Beach's Neutralizing Cordial. Mix together 1 ounce coarsely powdered Turkey rhubarb, ½ ounce peppermint leaves, and 1 ounce bicarbonate of potash. Put the materials in a stone jar, and add 1 pint boiling water; let it stand till cold, and then add ½ pint best brandy and ½ pound loaf sugar. Digest for a day or two, and strain through flannel. Bottle for use.

5395. Hager's Vermin Ointment. Mix together 12 parts sulphate of quinine, 2 parts muriatic acid, and 200 parts lard.

5396. Mayes' Substitute for Osgood's Indian Cholagogue. Dr. Mayes, of Mayesville, S. C., gives the following receipt, which he declares to be very similar to, if not identically the same, in taste, smell and effects, as Osgood's Indian cholagogue. Take 2 drachms sulphate of quinine; 1 drachm Tildens' fluid extract of leptandra; 4 ounces saturated tincture of queens' root; 3 drachms Tilden's extract of podophyllin (may-apple); 10 drops each of oil of sassafras and oil of wintergreen; and sufficient best New Orleans molasses to make the whole up to 8 ounces. This mixture to be well shaken up before a dose is measured; as the quinine (not being dissolved) will settle to the bottom of the bottle. The dose for adults is from 1 to 3 tea-spoonfuls 3 times a day. The dose is, however, a matter dependent entirely upon the nature of the case; and may be less or more, according to circumstances. It usually requires at least one 8-ounce bottle of the mixture to insure a permanent cure. When Tilden's fluid extracts cannot be had, saturated tinctures may be used, but in increased quantities; say rather more than double the quantity given of the fluid extract. In order, then, to preserve the due balance, the mixture must be made to measure 10 ounces, and a corresponding increase of dose must be made.

5397. Norris's Soda Mint. Soda mint, so much employed as an antacid and carminative for over-fed infants and dyspeptics, was originally a favorite prescription of Dr. Geo. Norris. His formula was the following: **Mix** together ½ ounce bicarbonate of soda, 1 ounce aromatic spirits of ammonia, and 1 pint peppermint water. Dose, from a dessert-spoonful to a table-spoonful for adults; from ½ to 1 tea-spoonful for infants.

5398. Foy's Muriatic Acid Chilblain Lotion. Muriatic acid, 1 part; water, 16 parts. To be used occasionally as a wash.

5399. Foy's Sulphuric Acid Chilblain Liniment. Sulphuric acid, 2 drachms; olive oil, 2½ ounces; and oil of turpentine, 1 ounce. Mix. Applied with gentle friction where the skin is not broken.

5400. Balsam of Peru Liniment for Chilblains. Balsam of Peru, ½ drachm; muriatic ether, 2 drachms; and laudanum, 2 drachms. To be used as a friction.

5401. Gassicourt's Turpentine Chilblain Lotion. Oil of turpentine, 4 parts; sulphuric acid, 1 part; olive oil, 10 parts. To be applied to the affected part night and morning.

5402. Saunders' Petroleum Chilblain Embrocation. Mix together petroleum, ½ ounce; alcohol, ½ ounce.

5403. Radius' Camphor Chilblain Ointment. Lard, suet, oil of bayberries, and wax, of each ½ ounce. Melt together and add camphor, 1 drachm.

5404. Compound Creosote Ointment for Chilblains. Creosote, 10 drops; solution of subacetate of lead, 10 drops; extract of opium, 1½ grains; lard, 1 ounce.

5405. Deschamps' Pastils for Bad Breath. Take of dry hypochlorite of lime, 2 drachms; sugar, 8½ ounces; starch, 8 drachms; gum tragacanth, 1 drachm; and carmine, 2½ grains. The pastils should be made so as to weigh about 2½ grains; 5 or 6 may be taken in the space of 2 hours. By employing starch in the preparation of the lozenges, Deschamps wishes to prevent the yellow color which they would otherwise assume.

5406. Soubeiran's Lotion of Veratria. Take 15 grains veratria, dissolve it in sufficient dilute muriatic acid, and add 5 drachms glycerine.

5407. Noble's Tonic Elixir. Take 1 ounce each of rhubarb root, orange peel, and caraway (or fennel) seed; percolate with 1 pint brandy. Dose, a tea-spoonful 3 times a day, after each meal.

5408. Delioux's Wine for Rheumatism, Gout, and Neuralgia. Take 5 parts tincture of colchicum seed, 2 parts tincture of aconite leaves, 1 part tincture of fox-glove, and 200 parts white wine. Dose to commence with, ½ table-spoonful 3 times a day.

5409. Ludlam's Specific. Take 2 drachms extract of rhatany, 1 drachm alum, 1 ounce cubebs, all in powder; 1 fluid ounce balsam of capaiba, and sufficient carbonate of magnesia. Dose, a small piece every 3 or 4 hours.

5410. Davis' Pain Killer. This preparation is said to be prepared as follows: Take 20 pounds powdered guaiac, 2 pounds camphor, 6 pounds powdered cayenne pepper, 1 pound caustic liquor of ammonia, and ½ pound powdered opium; digest these ingredients in 32 gallons alcohol for 2 weeks, and filter.

5411. Hunter's Red Drop. Triturate in a glass mortar, 10 grains corrosive sublimate in 12 drops muriatic acid, and add gradually 1 fluid ounce compound spirit of lavender. Dose, 5 to 20 drops in wine. A powerful alterative in syphilitic diseases, and will not salivate.

5412. Battley's Sedative Solution of Opium. Take 6 ounces sliced opium, 1½ ounces bruised nutmegs, ½ ounce Spanish saffron, and 4 pounds verjuice. Boil together, and add 4 drachms yeast; let the whole ferment 6 weeks, in a warm place. Decant, filter, and bottle; add a little sugar to each bottle. One drop of this sedative is equivalent to 3 drops of black drop.

5413. Nimmo's Solution of Croton Oil. Mix together ½ drachm alcoholic solution of croton oil, 2 drachms each simple syrup and guaiac mucilage, and ½ ounce distilled water. This quantity constitutes a dose; a little milk to be swallowed before and after.

The alcoholic solution referred to is formed by adding 8 drops croton oil to 1 fluid ounce rectified spirit of wine (90 per cent.)

5414. Gregory's Powder. Mix together 6 drachms calcined magnesia, 3 drachms powdered rhubarb, and 1 drachm powdered ginger. (*See No.* 5211.)

5415. Remoussin's Anti-Syphilitic Gargle. Take 1 ounce of a decoction of black nightshade and hemlock, and 3 grains bichloride of mercury.

5416. Ricord and Favrot's Capsules of Copaiba. Take 270 grammes (4167 grains) balsam of copaiba, 60 grammes (926 grains) neutral pepsin, 12 grammes (185 grains) subnitrate of bismuth, and 18 grammes (277½ grains) calcined magnesia. This is sufficient for 600 gelatine capsules. Dose, from 15 to 18 capsules a day.

5417. Ricord and Favrot's Capsules of Copaiba and Tar. Take 220 grammes (3395⅕ grains) balsam of copaiba, 20 grammes (308⅔ grains) Norwegian tar, and 15 grammes (231½ grains) calcined magnesia. To make 400 gelatine capsules. Dose, 15 every day.

5418. Hamburg Tea. This formula for *Hamburger Thee* is given by Hager. Mix together 8 parts senna leaves, 4 parts manna, and 1 part coriander.

5419. Persian Balsam. This is also known under the names of *Traumatic Elixir*, *Balm of the Innocents*, and *Baume du Commandeur*. Digest 1 ounce angelica root and 2 ounces St. John's wort, for 8 days in 5 pints 80 per cent. alcohol. Strain, and digest with 1 ounce each gum myrrh and gum olibanum. Then add 6 ounces each balsam of tolu and gum benzoin; macerate for 2 weeks; then filter.

5420. Grahame's Elixir of Bismuth. Dissolve 10 minims oil of orange flowers, 1 drop oil of cinnamon, 1 drop oil of cloves, and 2 drops oil of anise, in 1½ fluid drachms deodorized alcohol; add 2 fluid drachms syrup, and shake the mixture well. Dissolve 136 grains ammonio-citrate of bismuth in 2 fluid ounces distilled water and 1½ fluid ounces rose-water, adding sufficient aqua ammonia to produce a perfect solution. Mix the two solutions, add 1¼ fluid ounces alcohol, and, after standing for a short time, filter until perfectly clear; if not bright, add about 2 fluid drachms more alcohol. This is a fine preparation, each tea-spoonful containing about 2 grains of bismuth salt.

5421. Lugol's Iodine Solution. This consists of 1 part iodine dissolved in 2 parts iodide of potassium and 20 parts water. The solution of this strength is the one generally understood as Lugol's solution.

5422. Iodine Solution for External Use. Lugol devised two other solutions of different degrees of strength from the one given in No. 5421. As follows:

Rubefacient solution, containing 1 part iodine to 2 parts iodide of potassium and 12 parts water.

Caustic solution, consisting of 1 part iodine, 1 part iodide of potassium, and 2 parts water.

5423. Camphorated Dover's Powder. Pulverize 5 drachms camphor with ether, add 5 drachms prepared chalk, 5 drachms pulverized liquorice, and 17 grains sulphate of morphine. Dose, from 1 to 10 grains, used in all kinds of fevers, and as an anodyne.

5424. Davis' Neutralizing Cordial. Take 8 ounces rhubarb, 2 ounces each saffron, cardamoms, nutmeg, and carbonate of soda; 2 pounds white sugar, and 2 ounces essence of peppermint, with sufficient brandy and water to make up to 2 pints. Dose, 1 to 2 tea-spoonfuls.

5425. German Tea for the Chest. The compound known as German *Brust-Thee* is composed of the following ingredients, cut up small and mixed together: Take 4 ounces marsh-mallow root, 1½ ounces liquorice-root, ½ ounce Florentine orris root, 2 ounces colt's foot leaves; 1 ounce each red poppy flowers, mullein flowers, and star anise seed.

5426. Frey's Vermifuge. Take 1 ounce castor oil, 1 ounce aromatic syrup of rhubarb, 30 drops oil of Baltimore wormseed, and 5 drops croton oil.

5427. Velpeau's Erysipelas Lotion. Dissolve 1 ounce sulphate of iron in 1 pint water. Apply to the part affected every 2 or 3 hours.

5428. Procter's Vermifuge. To expel stomach worms from young children. Mix 16 grains santonin with 2 fluid ounces fluid extract of pink-root and senna. Dose, for a child 2 years old, 1 tea-spoonful night and morning, until purging takes place.

5429. Laurence's Hemorrhage Solution. Dissolve 2 drachms chloride of iron in 1 fluid ounce water. Apply with a brush, to prevent gangrene and arrest homorrhage.

5430. Laurence's Styptic Solution. If the solid perchloride of iron be kept in a bottle, a small portion deliquesces after a time, forming a thick brown liquid. This, applied to a bleeding surface by means of a brush of spun glass, arrests the flow of blood almost immediately.

5431. Monsel's Styptic Solution. This consists of a solution of subsulphate of iron, and is applicable for the same purpose as Laurence's hemorrhage solution. (*See No.* 5429.) The preparation of the *solution of subsulphate of iron* is thus given in the U. S. Ph. Mix 510 grains sulphuric acid and 780 grains nitric acid with ½ pint distilled water; heat to the boiling point, and add, ¼ part at a time, 12 troy ounces sulphate of iron, in coarse powder, stirring after each addition until effervescence ceases. Boil the solution until nitrous vapors are no longer perceptible, and the color assumes a deep ruby tint. When nearly cold, add sufficient distilled water to make up to 12 fluid ounces.

5432. Patterson's Emulsion of Pumpkin-Seeds. This is a good preparation for expelling tape-worms. Take 2 ounces pumpkin seeds, peel and pound to a paste with 1 ounce sugar; then add by degrees 8 fluid ounces water. The whole to be taken in 2 or 3 draughts, at short intervals, fasting. Dr. H. S. Patterson has prescribed this repeatedly with success.

5433. Teft's Dental Anæsthetic. Mix 1 fluid ounce each tincture of aconite root, purified chloroform, and alcohol, with 6 grains morphia. Used to diminish the pain in extracting teeth, by applying two plugs of cotton, moistened with the liquid, to the gums around the tooth for 1 or 2 minutes. It must not be swallowed.

5434. Parrish's Quinine Pills. Place

20 grains sulphate of quinia on a slab, drop upon it 15 minims aromatic sulphuric acid, triturating it with a bone spatula until it becomes a thick paste. Then divide rapidly into the required number of pills. A 3-grain pill made in this manner will not be inconveniently large.

5435. Grimault's Matico Injection. The matico injection, used by Grimault, of Paris, for gonorrhœa, is prepared, according to Bjoerklund, by dissolving 4 grains sulphate of copper in 8 ounces infusion of matico. The 8 ounces of infusion are made from ½ ounce matico.

5436. Storm's Specific. Take 2 ounces sweet spirits of nitre, 1½ drachms oil of cubebs, 2 ounces balsam of copaiba, 1 drachm oil of turpentine, 20 drops oil of cinnamon, 3 ounces mucilage of gum-arabic, and sufficient tincture of cochineal to color. This preparation is preferred by many to the capsules.

5437. Milhau's Emulsion of Cod-Liver Oil. Take 1 fluid ounce syrup containing sufficient saccharate of lime to represent 6 grains of the hydrate of lime; 5 fluid ounces water, 9 fluid ounces cod-liver oil, and 6 drops essential oil of almonds. Make into an emulsion.

5438. Bumstead's Opium Injection for Gonorrhœa. An injection, composed of 1 scruple extract of opium, 1 fluid ounce glycerine and 3 fluid ounces water, passed into the urethra after every passage of urine, affords relief in local pain, and diminution of discharge.

5439. Ricord's Gonorrhœa Injection. Take 20 grains each sulphate of zinc and acetate of lead, and 4 fluid ounces rose-water. The bottle to be well shaken before using.

5440. Condy's Fluid. Dissolve ½ drachm permanganate of potassa in 1 pint water. This is an excellent lotion for burns, ulcers, and suppurating surfaces, relieving the pain and removing the fetid odor.

5441. Dewees' Tincture of Guaiacum. Digest for a few days 4 ounces powdered gum-guaiac, 1½ drachms carbonate of soda (or of potassa), and 1 ounce powdered allspice, in 1 pint dilute alcohol. Add 1 or 2 drachms volatile spirit of ammonia to every 4 ounces of the tincture. To be administered in doses of 1 tea-spoonful in a little sweetened milk, or in sherry wine, morning, noon and night, in cases of suppressed menses. This is an excellent and well-tried remedy.

5442. Powell's Cough Balsam. Mix together 2 drachms syrup of tolu, 1 ounce paregoric elixir, and 2 ounces liquorice juice.

5443. Steer's Opodeldoc. I. Rectified spirit, 1 quart; castile soap, 5 ounces; camphor, 2½ ounces; oil of rosemary, 2½ drachms; oil of origanum, 5 drachms; weaker ammonia, 4 ounces; digest till dissolved, and pour while warm into wide-mouthed bottles.

II. Rectified spirits, 8 pints; white soap, 20 ounces; camphor, 8 ounces; water of ammonia, 4 ounces; oil of rosemary, 1 ounce; oil of horsemint, 1 ounce; dissolve the soap in the spirit by a gentle heat, and add the other ingredients. Bottle whilst warm.

5444. Falk's Antacrid Tincture. This alterative has been found a valuable remedy in secondary syphilis and other disorders. Macerate for 7 days 1 ounce powdered guaiac, 1 ounce Canadian balsam, and 2 fluid drachms oil of sassafras, in 8 fluid ounces alcohol. Filter, and add 1 scruple corrosive sublimate. Dose, 20 drops in wine or sweetened water, night and morning.

5445. Hufeland's Pectoral Elixir. Take 3 parts saffron, 4 parts each benzoin, myrrh, gum-ammoniac, aniseseed, and purified liquorice-juice; 8 parts each sneezewort root, Florentine orris-root, and squill-bulbs; macerate for a week in 93 parts rectified spirit, stirring frequently, then filter.

5446. Hufeland's Aperient Elixir. Reduce to coarse powder 4 parts each of aloes, myrrh, and gum-guaiacum; add 4 parts bruised rhubarb-root, 2 parts bruised saffron, 8 parts carbonate of potassa, 8 parts muriate of ammonia, 48 parts spirit of horse-radish root, and 144 parts distilled water. Macerate the ingredients for a few days, frequently stirring, then filter. (*Hager.*)

5447. Hufeland's Anticatarrh Elixir. Take 60 parts extract of blessed-thistle, 20 parts extract of bitter-sweet, dissolve them in 480 parts fennel water and 60 parts bitter-almond water. Dose, 60 drops 4 times a day. (*Hager.*)

5448. Meyer's Water of Life. Take 18 parts fresh myrtle-berries, 12 parts orange-peel, 8 parts cinnamon, 2 parts galanga-root, 2 parts zedoary-root, and 1 part cardamoms. Reduce them by bruising and cutting, and digest them for 3 days with frequent agitation, in 600 parts rectified spirit and 680 parts water; then strain with pressure, and let it settle; decant the clear, filter it, and add 120 parts white sugar. (*Hager.*)

5449. Elixir of Bromide of Potassium. Dissolve 2 ounces and 16 scruples bromide of potassium in 1 pint simple elixir; add 20 minims solution of oil of orange and 10 minims of solution of oil of bitter almonds, and filter; color with cochineal color. The officinal formula for preparing bromide of potassium is given in No. 4198.

5450. Hufeland's Infant Powder. Take 2 ounces valerian root; 3 ounces orris root, 1 ounce aniseed; 2 drachms saffron, all in powder, and 2 ounces carbonate of magnesia.

5451. Granville's Counter-Irritant Lotions. These consist of three ingredients, viz.: *strong water of ammonia* (specific gravity .872) more than 3 times the strength of officinal liquor ammonia; of *spirit of rosemary*, made by infusing 2 pounds of the fresh tops of rosemary in 8 pints alcohol for 24 hours, and distilling 7 pints; and of *spirit of camphor*, composed of 4 ounces camphor dissolved in 2 pints alcohol.

The lotion is prepared of two different strengths; the *milder lotion* consists of 4 drachms of the ammonia, 3 drachms of the spirit of rosemary, and 1 drachm of the camphor spirit.

The *stronger lotion* contains 5 drachms of the ammonia, 2 of the spirit of rosemary, and 1 of the camphor. The milder is generally sufficient to produce full vesication in from 3 to 10 minutes. The stronger is seldom used except in apoplexy, and to produce cauterization.

5452. Startin's Borax and Glycerine Lotion. For sore lips, chapped hands, &c. Take ½ drachm borax, ½ fluid ounce glycerine, and 7½ fluid ounces rose-water.

5453. Brainard's Solution for External Use. Dissolve 16 grains lactate of iron in 2 fluid drachms distilled water.

5454. Birch's Pills for Habitual Constipation. Take ½ drachm alcoholic extract of rhubarb, 24 grains extract of taraxacum, and 2 grains sulphate of quinine. Mix together and make into 12 pills. One to be taken either on rising in the morning or at dinner time, or even at both periods when the constipation is very obstinate. This is a very gentle stomachic and tonic evacuant, particularly useful for the delicate.

5455. Da Costa's Chronic Constipation Pill. Take 1 grain podophyllin, 1 grain extract belladonna, 5 grains capsicum, and 20 grains powdered rhubarb; mix and divide into 20 pills. One pill to be taken 3 times a day.

5456. Birch's Constipation Pill. Take 12½ grains compound extract of colocynth and 40 grains extract of henbane. Mix and divide into 20 pills. This is an excellent pill for occasional use, especially for constipation in old age.

5457. Ricord's Copaiba and Pepsine Pills. Take 11½ drachms balsam of copaiba, 2½ drachms neutral pepsine, 31 grains nitrate of bismuth, and 46 grains calcined magnesia. Divide into 100 capsules. Administer 15 to 18 daily.

5458. Lime Juice and Glycerine. Lime (or lemon) juice, ½ pint. Heat in a porcelain mortar to near the boiling point, and add gradually rose water, elder-flower water, and rectified spirit, of each 2 ounces. Agitate the whole well together. After 24 hours' repose, decant or filter through calico or muslin, then add pure glycerine, 2½ ounces; oil of lemons, ½ drachm. Again agitate them together for some time, and by careful manipulation you will have a somewhat milky liquid; but it should be quite free from any coarse floating matter or sediment.

5459. Boudault's Pepsine Pills. Mix 2 drachms and 34 grains starchy pepsine, with sufficient powdered tragacanth to make 60 pills. Dose, 3 pills before and 3 after each meal, and sometimes 3 during the meal.

5460. Hogg's Pepsine Pills. Mix 2 drachms 34 grains starchy pepsine, 1 drachm 17 grains nitrate of bismuth, and 38½ grains lactic acid. Make into 100 pills, and coat with sugar and balsam of tolu. Dose, 4 to 12 pills 1 hour after meals.

5461. Angelot's Remedy for Ulcerated Gums. Take of hypochlorite of lime, from 10 to 25 grains; mucilage of gum-arabic, 1½ to 4 drachms; syrup of orange peel, 1½ to 2 drachms. Mix thoroughly. This mixture is employed as a lotion to the ulcerated gums.

5462. Angelot's Pastils for Bad Breath. These preparations are better adapted than liquids for carrying on the person. Take of hypochlorite of lime, 7 drachms; sugar flavored with vanilla, 3 drachms; gum-arabic, 5 drachms. The pastils are made so as to weigh from 10 to 11 grains. 2 or 3 of these pastils are sufficient to remove from the breath the disagreeable odor produced by tobacco smoke. The pastils thus prepared have a grey color and become quite hard.

If pastils of whiter color are required the following substances are employed: Take of dry hypochlorite of lime, 20 grains; pulverized sugar, 1 ounce; gum tragacanth, 16 grains. The hypochlorite of lime is triturated in a glass mortar, and a small quantity of water is poured upon it; it is then left to repose, decanted, and a second quantity of water added; the two liquids are filtered, and the gum and sugar added so as to form a paste. This is divided into pastils weighing from 12 to 16 grains. If it is desired to aromatize the paste, 1 or 2 drops of any essential oil may be added to the sugar and gum before the paste is formed.

5463. Santonin Lozenges. Take 5 troy ounces white sugar in powder, ¼ troy ounce fine starch, 10 grains finely powdered tragacanth; the whole well mixed with the white of 5 eggs previously beaten to a dense froth; place in a porcelain dish over the water-bath, and, with constant stirring, keep at a temperature not exceeding 100° Fahr. until a sample taken from the mixture no longer runs from the spatula. An intimate mixture of 50 grains powdered santonin and 100 grains powdered sugar is incorporated with the mass, and the whole, by means of a syringe, formed into 100 lozenges, each containing ½ grain of santonin. They are deposited on smooth or waxed paper, and when hard are to be placed between cotton-wadding and protected from the light.

5464. Quesneville's Ferruginous Powder. Bicarbonate of soda, 4 parts; tartaric acid, 7 parts; pure sulphate of iron, 4 parts; sugar, 8 parts. Powder each fine, then mix and keep the powder in a well-corked bottle. Dose, 1 spoonful in 6 or 7 ounces of sweetened water.

5465. Tronchin's Cough Syrup. Powdered gum-arabic, 8 ounces; precipitated sulphuret of antimony, 4 scruples; anise, 4 scruples; extract of liquorice, 2 ounces; extract of opium, 12 grains; white sugar, 2 pounds. Mix, and form lozenges of 6 grains, one of which is to be taken occasionally in catarrh and bronchial affections.

5466. Pierquin's Cough Syrup. Kermes mineral, 2 grains; gum-arabic, 1 drachm; syrup, 5 ounces. Mix. A spoonful occasionally when expectoration is difficult.

5467. Kermes Mineral. Dissolve 23 troy ounces carbonate of soda in 16 pints boiling water; add 1 troy ounce finely powdered sulphuret of antimony, and boil for an hour. Filter rapidly into a warm earthen vessel, cover closely and cool slowly. After 24 hours decant the fluid, drain the precipitate on a filter, wash it with cold water (previously boiled), and dry without heat. Keep in a well-stopped bottle, protected from the light. (*U. S. Ph.*) This is the *oxysulphuret of antimony.*

5468. Rousseau's Laudanum. Dissolve 12 ounces white honey in 3 pounds warm water, and set it aside in a warm place. When fermentation begins add to it a solution of 4 ounces selected opium in 12 ounces water. Let the mixture stand for a month at a temperature of 86° Fahr.; then strain, filter, and evaporate to 10 ounces; finally strain and add 4½ ounces proof alcohol. Seven drops of this preparation contain about 1 grain of opium.

5469. Bonnamy's Dentifrice. Take prepared chalk, 1 part; burned hartshorn, 1 part; hydrate of alumina, 1 part; perfume with oil of cinnamon. This is an excellent dentifrice.

5470. Extract of Milk. *Condensed milk* is thus prepared: Take 10,000 parts fresh cows' milk, 50 parts white sugar, and 2 parts pure carbonate of soda. Place them in a porcelain vessel, and, with constant stirring, evaporate to the consistence of a thick extract, either in a vacuum or by the heat of a vapor bath of 140° to 160° Fahr. One part of the extract will represent 10 parts of fresh milk. (*Hager.*)

5471. Milk Powder. Take 10,000 parts fresh cows' milk, 2 parts dry caustic potassa, and 2 parts borax. Evaporate these in a vacuum to about 2000 parts. Then mix in thoroughly 50 parts precipitated phosphate of lime, 15 parts table salt, 100 parts powdered gum-arabic, and 200 parts powdered sugar. Evaporate the whole to a dry powder at a heat of 95° to 110° Fahr. (*Hager.*)

5472. Schwarz's Liniment for Scalds and Burns. Take 16 parts linseed oil, 8 parts white of egg, and 1 part tincture of opium; mix them thoroughly by trituration with 2 parts acetate of lead. Spread upon lint and apply to the wound. (*Hager.*)

5473. Hungarian Liniment. Pulverize 5 parts cantharides, 20 parts each mustard seed, black pepper, and camphor; macerate for 2 days in 200 parts wine vinegar, then add 400 parts rectified spirits. Strain with pressure, and filter. (*Hager.*)

5474. Bland's Ferruginous Pills. Take equal weights of sulphate of iron and carbonate of potassa; make into a mass with mucilage of tragacanth and powdered liquorice root.

5475. Castillon's Powders. Take 1 drachm each sago, jalap, and tragacanth, all in powder; 1 scruple prepared oyster shell, and sufficient cochineal to color. Boil 1 drachm of this mixture in a pint of milk, and use the decoction as a diet in chronic bowel affections.

5476. Goulard's Cerate. This is the same preparation as the *cerate of subacetate of lead* of the U. S. Pharmacopœia. Mix 4 troy ounces melted white wax with 7 troy ounces olive oil. When it begins to thicken, gradually pour in 2½ fluid ounces solution of subacetate of lead, stirring constantly with a wooden spatula until cool. Then mix in 30 grains camphor dissolved in 1 fluid ounce olive oil.

5477. Gondret's Ammoniacal Ointment. Take 32 parts lard and 2 parts oil of sweet almonds. Melt together by a gentle heat, and pour the mixture into a wide-mouthed bottle. Add 17 parts of a solution of ammonia of 25° Baumé, and mix thoroughly until cold. Keep it in a cool place, and in a bottle with an accurately fitting stopper. It will vesicate, or raise a blister under the skin in 10 minutes if properly prepared.

Medical Receipts.

The scope of this work does not allow of the insertion of much beyond general and specific remedies for ailments of every-day occurrence; it being understood that, in all serious cases, the guidance of a physician is indispensable. Advice and directions are given for the treatment of some severe cases requiring prompt action, that may be followed with benefit until the arrival of the doctor. No particular school of medicine is adhered to, the efficacy of each receipt being the primary consideration in inserting it. The list includes many popular and domestic remedies, together with prescriptions of celebrated and leading physicians.

5479. To Cure Prairie or Seven Years' Itch. Use plenty of castile soap and water, and afterwards freely apply iodide of sulphur ointment; or take any given quantity of simple sulphur ointment and color it to a light-brown or chocolate color with the subcarbonate of iron, and perfume it. Apply this freely; and, if the case is severe, administer mild alteratives in conjunction with the outward application.

5480. Sulphur Bath. The bath may be prepared either by adding 1 ounce sulphuret of potassium for every 10 or 12 gallons of water used, or 1 ounce sulphuret of calcium for every 15 gallons of water. The sulphur bath is a powerful remedy in every description of skin disease. Leprosy (the most obstinate of all) has been completely cured by it; the common itch requires only 1 or 2 applications to eradicate it; all scurfy and moist skin affections, &c., speedily yield to its influence.

5481. Benzine for Itch. Benzine, it is said, will effect a cure for scabies in the course of half an hour, after which the patient should take a warm bath for 30 minutes. This has been highly recommended.

5482. To Cure Salt Rheum. Wash the part affected with castile soap and water, dry with a soft cloth; then wet with tincture of iodine, and let it dry; after which apply a little citrine ointment. (*See No.* 4947.) When the eruption is on an exposed part, a wash composed of 1 drachm corrosive sublimate, 2 scruples white vitriol (sulphate of zinc), 3 drachms sal-ammoniac, 2 drachms salt, and 3 ounces sugar of lead, mixed with 1 pint soft water, may be used alternately with the tincture of iodine.

5483. Salt Rheum from Photographic Chemicals. Make a salve by steeping queen of the meadow root over a slow heat in fresh hog's lard for from 2 to 6 hours—the longer the more powerful the salve. Apply this to the eruptions as often as convenient, and in a short time there will be a decided improvement and a cure will be effected in from 1 to 6 weeks. If the stomach or blood should seem out of order, take Winchester's hypophosphites of lime and soda. Use this medicine and no other, as it acts without fail and to the point, not being in any way injurious. Avoid using either iron or mercury, as they do no good and are very apt to do injury. Where the disease is not hereditary a cure will be effected in a short time; where it has become a chronic difficulty the cure will be slower. When buying the root, ask for queen of the meadow root. Be careful not to get queen's root, commonly called stillingia, many druggists not knowing the difference.

5484. Baker's Itch. This disease is of common occurrence on the hands of bakers; hence the vulgar name. The treatment is as follows: Frequent ablution in warm water, keeping the bowels open with saline purgatives, and the nightly use of the ointment given in No. 4957 will generally effect a cure. Salt food should be avoided as much as possible, as well as keeping the hands covered with dough and flour; the latter being the cause of the disease.

5485. Remedy for Tetter, Ring-worm, and Scald-Head. Take 1 pound simple cerate; sulphuric acid, ¼ pound; mix together, and it will be ready for use.

5486. Remedy for the Tetter. Dissolve 1 ounce sulphuret of potash in 1 quart of cold soft water; put it into a bottle and keep it tightly corked. Bathe the eruption 5 or 6 times a day, with a sponge dipped in a little of this solution. If the tetter reappear in cold weather, repeat the treatment. This is an excellent remedy.

5487. Remedy for Barber's Itch and Tetter. A simple and effectual cure. Moisten the part affected with saliva (spittle) and rub it over *thoroughly* three times a day with the ashes of a good Havana segar. Simple as this remedy may appear, it has cured the most obstinate cases.

5488. To Cure Ring-worm. To 1 part sulphuric acid, add 16 to 20 parts water. Use a brush or feather, and apply it to the parts night and morning. A few dressings will generally cure. If the solution is too strong, dilute it with more water; and if the irritation is excessive, rub on a little oil or other softening application; but always avoid soap.

5489. Cure for Ring-worm. Wash the head with soft-soap every morning, and apply the following lotion every night: 1 drachm sub-carbonate of soda, dissolved in ½ pint of vinegar.

5490. To Cure Pimples and other

Eruptions of the Skin. Never tamper with any breaking-out on the skin; even though it be a single red spot, do not apply to it so simple a thing as water, hot or cold, but let it alone, and omit a meal or two; if it does not abate, consult a physician. If one is not at hand, then live on half allowance until it disappears.

5491. Glyconine, or Glycerine Varnish for Cutaneous Affections. Take yolk of egg, 4 parts by weight; rub in a mortar with 5 parts glycerine. Applied to the skin it forms a varnish which effectually excludes the air, and prevents its irritating effects. It is unalterable (a specimen having laid exposed to the air for 3 years unchanged), and is quickly removed by water. These properties render it serviceable for erysipelas and cutaneous affections, of which it allays the action. It is also very valuable for soothing the irritation resulting from burns.

5492. Cure for Eruptions of the Skin. Take 2 ounces rasped sarsaparilla root, 1½ ounces solanum dulcamara (bitter-sweet, or woody nightshade), 1½ ounces mezereon bark, ½ ounce rasped guaiacum wood, and ½ ounce sassafras bark. Pour on these 1 quart boiling water, let it stand 24 hours, and then boil away slowly to 1½ pints; press, strain, and add 2 pounds sugar and 1 ounce diluted spirits of wine. Take a wine-glassful 3 times a day with 1 grain precipitated sulphuret of antimony.

5493. Treatment of Sprains. The great remedy is rest; when severe, rest for days, to save weeks; the best treatment is warm fomentations at the time of accident, to prevent or reduce the swelling and pain, and arnica, applied by means of rags, to prevent pain and give strength to the part. The tincture of arnica is the preparation used. If the skin is not broken, about 20 to 30 drops, or even, in severe cases, 60 drops may be added to a wine-glassful of water. If the skin is broken, or any abrasion is present, the strength of the tincture must be considerably reduced; from 5 to 10 drops will then be sufficient, and if any redness or inflammation occurs in or about the sprain, in consequence of using the lotion, leave it off at once and use only cold water. A firm bandage will be useful to support the part. Walking should, for a considerable time, be only sparingly indulged in after a severe sprain.

5494. Remedy for a Sprain or Bruise. Wormwood boiled in vinegar, and applied hot, with enough cloths wrapped around to keep the sprain moist. This is an invaluable remedy.

5495. Sprains of the Wrist and Ankle. As soon as possible after the accident, get a muslin bandage 1 or 2 yards long, and 2 or 3 inches wide; wet it in cold water, and roll it smoothly and firmly around the injured part. Keep the limb at rest, exposed to the air, and continually damp with cold water. The sooner after the accident the bandage is applied, the less pain and swelling there will be; but if pain becomes excessive, care must be taken to slightly loosen the bandage.

5496. Sprains of the Muscles of the Back. Take of Canada turpentine, ½ ounce; soap liniment, 6 ounces; and 1 drachm of laudanum. Mix, and rub well in before a hot fire.

5497. Sprain in the Back. The first thing is rest; take night and morning 15 or 20 drops of the balsam of copaiba. If the part is inflamed, apply cold water cloths. Let the bowels be kept gently open by aperients. When the inflammation is gone, rub the part with stimulating liniment. (*See No.* 4888.)

5498. Treatment of Scratches. Do not neglect them. Wash them in cold water; close them as much as you can, and cover with diachylon plaster. If there is inflammation, apply a bread poultice, or one of slippery elm.

5499. To Extract Splinters. Thorns and splinters finding their way under the skin frequently give considerable pain, and, unless extracted, the annoyance may be very great, as inflammation will in all probability ensue, which is the process nature adopts for getting rid of the cause of irritation. If the splinter or thorn cannot be immediately extracted, for which purpose a needle will be found in most cases a sufficient surgical instrument, linen dipped in hot water ought to be bound around the place, or the part may be bathed in hot water. In the event of inflammation, which may probably issue on the production of an ulcer, the steam of hot water should be applied, and afterwards a poultice of bread and milk.

5500. Treatment of Cuts. The divided parts should be drawn close together, and held so with small pieces of strapping or adhesive plaster stretched across the wound, or by the application of collodion. If the part be covered with blood, it should be first wiped off with a sponge. When the wound is large, and the parts much exposed, a good method is to sew it up. The application of a little creosote will generally stop local bleeding, provided it be applied to the clean extremities of the wounded vessels. A good way is to place a piece of lint, moistened with creosote, on the wound previously wiped clean, or to pour a drop or two of that liquid upon it. Friar's balsam, quick-drying copal varnish, tincture of galls, copperas water, black ink, &c., are popular remedies applied in the same way. A bit of the fur plucked from a black beaver hat is an excellent remedy to stop the bleeding from a cut produced by the razor in shaving. For light cuts with a knife, or any sharp instrument, the Riga balsam usually stops the bleeding immediately. (*See Lockjaw.*)

5501. Artificial Skin for Cuts, &c. A small quantity of collodion applied with a brush to a cut or wound will produce a perfect artificial covering which is more elastic than plaster, and sufficiently insoluble in cold water.

5502. Traumaticine, or Water-proof Covering for Wounds. This article is simply a solution of white and dry pure unmanufactured gutta-percha in bisulphuret of carbon. Dropped on a wound or raw surface, it almost instantly forms a pliable, water-proof, and air-tight defensive covering to the part, resembling, in appearance, gold-beater's skin. The fetid odor of the bisulphuret is lost in a few seconds. Chloroform, which has an agreeable odor, may be used as the solvent, but is very much more expensive than the bisulphuret of carbon.

5503. Treatment of Bed-Sores. Remove the excessive discharge by gently pressing the part with a bit of cotton wadding; then paint the sore over with prepared collodion (*see No.* 4744), using a soft camel-hair pencil. The application may be repeated daily, and when it has well dried place a bit of soft lint or cotton wadding over the part for protection.

5504. Detergents. Deterge means to cleanse. Detergents remove unwholesome matters adhering to and obstructing the vessels; usually applied to foul ulcers, &c., as tincture of myrrh, honey, alum, water, turpentine, &c.

5505. Treatment of Ulcers. An ulcer is an injury done to the flesh, from which issues matter, or some kind of discharge, with more or less pain and inflammation. The common ulcer should be kept clean and cool, and protected from the atmosphere, especially in frosty or cold weather. It should be washed now and then with warm soap-water. Put upon it a little lint, wet occasionally with salt and water, and put over it the black salve. (*See No.* 4971.) Perhaps the best dressing is the saturnine cerate. (*See No.* 4968.) Poultices made of the oak bark or sumach bark may be used alternately.

5506. Treatment of Severe Ulcers. Sometimes ulcers are very irritable, tender, and painful, and discharge a thin acrid fluid. They should be steamed every night with a bitter decoction, and occasionally washed with an infusion of chamomile flowers, or a strong decoction of wild cherry bark, with a little spirit. Apply a poultice of slippery elm, mixed with a strong decoction of poplar bark, and a trifle of salt. Repeat as required. If the ulcer or ulcers are indolent, steam as before, and apply the cancer plaster (*see No.* 5047,) with only a trifle of the white vitriol mixed with it; or, sprinkle the ulcer with powdered blood-root. Sometimes ulcers become very much inflamed, and assume a livid color; they are covered with small vesicles or blisters, as in mortification. Wash the ulcer with tincture of myrrh, and apply a poultice made of charcoal, yeast, slippery elm, ginger, and a minute portion of tincture of cayenne. Bear it as long as possible. Then apply the saturnine cerate. (*See No.* 4968.)

5507. Beach's Remedy for Ulcers. The following is recommended by Dr. Beach: Take sweet clover tops and stalks, burdock leaves, and parsley, a handful of each; get the strength out by boiling; strain, and add 1 pound of resin and ¼ pound of fresh butter; simmer until of a proper consistence. A cold water cloth constantly applied is a good remedy. Put a little cerate on the ulcer previously. Attend to the general health by cleansing the stomach and bowels, and then giving tonics.

5508. To Disinfect and Deodorize Foul Ulcers. Permanganate of potassa disinfects rapidly the most fetid ulcers, in the proportion of 2 scruples of the salt to 8 ounces of water as a lotion or injection. The most favorable method is to cover the wound with lint soaked with that substance, and to place above this a layer of raw cotton, the latter having the property of filtering the air, and to retain the germs which determine putrid fermentation. In cancers of the womb it is necessary to repeat the injections several times a day.

5509. Ulcers in the Mouth. If the ulcers are not of a syphilitic origin, a local wash of carbolic acid or permanganate of potassa will speedily cure them; say 1 part of acid or permanganate to 100 of water. If they are, however, syphilitic, the wash of carbolic acid, perhaps 2 or 3 times as strong, in combination with internal treatment, will be found beneficial; the wash may be used 3 times a day.

5510. Treatment of Running Sores on the Legs. Wash them in brandy, and apply elder leaves, changing twice a day. This will dry up all the sores, though the legs were like a honey-comb. Or, poultice them with rotten apples. But take also a purge once or twice every week.

5511. Fine Clay as a Dressing to Sores. Dr. Schreber, of Leipsic, recommends the use of clay as the most energetic, most innocent, simple, and economical of palliative applications to surfaces yielding foul and moist discharges. He moreover considers that it has a specific action in accelerating the cure. Clay softened down in water, and freed from all gritty particles, is laid, layer by layer, over the affected part. If it becomes dry and falls off, fresh layers are applied to the cleansed surface. The irritating secretion is rapidly absorbed by the clay, and the contact of air prevented. The cure thus goes on rapidly. This clay ointment has a decisive action in cases of fetid perspiration of the feet or armpits. A single layer applied in the morning will destroy all odor in the day. It remains a long time supple, and the pieces which fall off in fine powder produce no inconvenience. (*Brit. Med. Journ.*) We can corroborate Dr. Schreber's observations, having used fine clay poultices for several years, chiefly, however, in cases of local inflammation requiring the application of cold. Rags wet in water, or Goulard water, so rapidly become dry and hot that the benefit from the cold application is completely lost. There is no dirt when the clay is enveloped in a piece of fine linen, and is not too fluid in consistence. (*Braithwaite.*)

5512. Treatment of Burns. In regard to the treatment of burns there is a great diversity of opinion, scarcely any two surgeons agreeing as to the remedies. All of them are doubtless valuable, but there is one which has a great reputation (namely, carron oil, *see No.* 5513). The great objection to it is its offensive odor, rendering an entire hospital ward disagreeable. In all cases of burns and scalds, it is necessary to observe that, if fever should ensue, laxative medicines ought to be given; as castor oil, or salts and senna.

5513. Carron Oil. This is composed of equal parts linseed oil and limewater, and should be well shaken before using.

5514. Treatment of Recent Burns. When recently inflicted, nothing tends more decidedly to soothe or deaden the suffering than cold water; the burnt part should, therefore, be immediately placed in cold water, or thin cloths dipped in cold water should be applied and frequently renewed over the injured surface. After the lapse of a short time, when the cold fails to relieve, rags dipped in carron oil (*see No.* 5513) are to be substituted for the water, care being taken to keep the rags moist with the oily mixture until the burn heals; this is the main point in the treatment; the rag or linen must not be removed or changed. The carron oil may be applied from the first if it is at hand; but, cold water being nearly always to be had, will be found very grateful until assistance arrives. A large bottle of carron oil should be kept in every nursery cupboard, or in every house, in a place easy of access, a large label being affixed to it, with plain directions.

5515. Treatment of Superficial Burns. When the burn is very superficial, simply inflaming or vesicating the part, covering it up with flour, and then placing a layer of cotton over it, so as to exclude the air, makes a very comfortable dressing. Another method consists in applying cold water; and another, warm water covered with oiled silk and a bandage. Glyconine or glycerine varnish (*see No.* 5491) is also a valuable remedy. Lard, deprived of salt, and simple cerate, make pleasant applications.

5516. Gross' Treatment of Burns. The profession is indebted to Prof. Gross for the introduction of white lead and linseed oil in the treatment of burns. It is one of the very best applications which can be used, effectually excluding the air, and being always grateful to the patient. In all cases, no matter whether merely the skin or the deeper structures are involved, white lead, rubbed up with linseed oil to the consistence of paste or paint, and placed on with a brush, will be found productive of great relief. There does not appear to be any risk from the constitutional influence of the lead, though it has been suggested, to counteract any tendency of this kind, that the patient should take occasionally a little sulphate of magnesia.

5517. Burns and Scalds. Every family should have a preparation of flaxseed oil, chalk, and vinegar, about the consistency of thick paint, constantly on hand for burns and scalds. A noted retired physician states that he has used it in hospital and private practice for the past forty years, and believes that no application can compare with it, as regards relief of pain and curative results.

5518. Remedy for Scalding. Apply a poultice of slippery elm bark and milk, and, when the inflammation has left, apply black salve. (*See No.* 4971.) For very slight burns, the black salve alone will cure. The slippery elm poultice is a sovereign remedy, and has effected the greatest cures. Dr. Beach relates a case of severe scalding, in which a poultice of slippery elm bark and olive oil alone very soon arrested the inflammation and acute sufferings of the patient, to the astonishment of all who witnessed the cure.

5519. Remedy for Scalded Mouth. In cases of scalding the mouth with hot liquids, gargle with a solution of borax, and then hold in the mouth a mucilage of slippery elm, swallowing it slowly, if the throat also has been scalded; the slippery elm bark may be mixed with olive oil. Some recommend soap liniment, but the latter must not be swallowed.

5520. To Cure Slight Burns. When a burn is only trifling, and causes no blister, it is sufficient to apply a compress of several folds of soft linen upon it, dipped in cold water in which has been dissolved a little carbonate of soda; to be renewed every 15 minutes until the pain is removed. Dr. Tissot says, in cases of blisters, beat up an egg with 2 table-spoonfuls olive oil or linseed oil, spread it on soft linen, and apply it to the affected part. For very slight burns or scalds, the black salve alone is sufficient to remove the pain and inflammation. (*See No.* 4971.) If the skn is not broken, cover the part with a layer of flour or starch, place cotton wool over it, or a linen rag, and bind it over lightly. If a blister has been burst or cut, use a cerate.

5521. Carbolic Acid for Burns or Scalds. The best application in cases of burns or scalds is a mixture of 1 part of carbolic acid to 8 of olive oil. Lint or linen rags are to be saturated in the lotion, and spread smoothly over the burned part, which should then be covered with oiled silk or gutta-percha tissue, to exclude the air. The dressing may be left on from 2 to 3 days, and should then be reapplied, exposing the burn as short a time as possible to the air.

5522. Oil of Brown Paper. Dip a piece of thick brown paper into the best salad oil. Set the paper on fire upon a plate, and the oil that drops from it is a good remedy for burns.

5523. Treatment of Burns and Discolorations Caused by Gunpowder. Dr. Davies, in a recent number of the London Lancet, states that he has found the following treatment most successful: Smear the scorched surface with glycerine, by means of a feather, then apply cotton wadding; lastly, cover with oil silk. In one case the discoloration was very great, the patient looking more like a mummy than a living being. It entirely subsided in a month by the above treatment.

5524. Nature of Rheumatism. Rheumatism is a diseased condition of the fibrous and muscular tissues, chiefly affecting the larger joints; the heart and diaphragm are also liable to be affected by it. It is a promoting cause of heart disease. The principal forms of rheumatism are these: When the joints about the back and loins are affected the complaint is known as lumbago; pains in the hip joints are designated sciatica.

5525. Causes of Rheumatism. The causes of rheumatism are various. Vicissitudes of temperature are the most common; occupying a damp bed for a single night is sufficient to engender the disease. Such persons as blacksmiths, who are exposed to severe changes of temperature, are generally victims to the complaint. Miners and persons employed in smelting-furnaces are often severe sufferers. There is likewise a hereditary tendency to the malady, which a slight cold will develop. Rheumatism proceeds from a vitiated condition of the blood. A hereditary taint in the circulating fluid may be developed by a slight cold, but more commonly the blood becomes vitiated through mal-assimilation and a faulty digestive action. The precise principle of the poison engendered has not yet been fully ascertained. It is generally believed to be lactic acid.

5526. Premonitory Symptoms of Rheumatism. An attack of rheumatism is imminent when a stiffness is felt in the joints, combined with a dryness of the skin and a burning thirst. The variety of the complaint of which these signs are the precursors is termed acute. The other variety is chronic rheumatism. The latter may be described as an aggravated condition of the former, though some persons not only describe them as quite distinct, but introduce a variety between them.

5527. Treatment of Rheumatism. In the early stages, when there is much thirst, a refreshing saline drink will be beneficial; cold water may be freely allowed, but acid drinks must not be given without consulting the doctor, as they may not agree with his medicines. A correspondent of the Medical Circular vouches for the relief he has experienced in the liberal use of lime (or lemon) juice, while laboring under the paroxysms of rheumatism. By persistent use of the above simple acid for the space of 3 days, avoiding all stimulating liquids, the most confirmed rheumatism will, he says, relax, and the tone of the muscular and nervous system will be restored to its usual character.

5528. Local Remedies for Rheumatism. Unless anything else is ordered, cotton-wadding wrapped around the swollen joints, and covered with oil silk, will be found grateful; a kind of local vapor bath is produced by it. If this is not agreeable, rags may be dipped in a saturated solution of nitre in water, and applied, care being taken to keep them moist; oiled silk should be applied round these as well.

5529. Treatment of Chronic Rheumatism. When rheumatism becomes chronic, the general health, particularly the diet in connection with the digestive powers, must be attended to with great care. The attacks often arise from pure debility, and will then be best cured by tonics and good food.

5530. Simple Remedy for Rheumatism. Bathe the parts affected with water in which potatoes have been boiled, as hot as can be borne, just before going to bed; by the next morning the pain will be much relieved, if not removed. One application of this simple remedy has cured the most obstinate rheumatic pains.

5531. Dover's Rheumatic Powder. Ipecacuanha powder, and purified opium, of each 1 part; sulphate of potassa, 8 parts; triturate them together to a fine powder. Be very careful to reduce the opium, and intimately mix with the rest. This powder is recommended by Dr. Dover as an effectual remedy for rheumatism. The dose is from 2 to 5 grains, repeated. Avoid much drinking after taking it, or it might act as an emetic.

5532. Remedy for Rheumatism. Take ½ ounce each black cohosh root, golden seal, and nerve powder; 1 pint of rum. Mix. Dose, ½ table-spoonful 3 times a day. The most obstinate cases of rheumatism have yielded to the above simple remedy.

5533. Speedy Cure for Rheumatism. Dr. R. H. Boyd states that he cures inflammatory rheumatism in from 3 to 7 days by the following method: Give first a full emetic dose of tartar emetic (1½ to 2 grains), and when this has operated, 5 drops laudanum and 5 drops tincture of colchicum, every 3 or 4 hours, and a tea-spoonful of a half-pint mixture, containing 4 drachms acetate of potassa, every hour. When the patient becomes very hungry, and is quite free from pain, having fasted several days, he allows 2 table-spoonfuls of milk or 1 oyster 3 times a day, increasing the quantity gradually each day.

5534. Remedy for Inflammatory Rheumatism. Gelseminum, administered in doses of 5 to 30 drops, is a very serviceable remedy. The dose should be repeated at intervals until the pain and inflammation disappear.

5535. Rheumatic Alterative. Macerate for 3 or 4 days ½ ounce each colchicum seed and black cohosh root, both well bruised, in 1 pint best rye whiskey. A dessert-spoonful 3 times a day, before meals, has been found a valuable remedy in chronic rheumatism.

5536. Indian Remedy for Rheumatism. Macerate the following ingredients for a few days in 1 quart rye whiskey: 1 ounce bark of wahoo root, 1 ounce blood root, 2 ounces black cohosh root, ½ ounce swamp hellebore, 1 ounce prickly ash bark, and 1 ounce poke root cut fine. Dose, 1 tea-spoonful every 3 or 4 hours, increasing the dose as the stomach will bear it.

5537. Spanish Cure for Chronic and Syphilitic Rheumatism. Take 4 ounces sarsaparilla, 1 ounce rasped guaiacum wood, 2 ounces extract of sarsaparilla, ½ ounce crude antimony. Tie them in a linen rag with 10 drachms English walnut hulls (or black walnut), and boil in 3 pints water down to 2 pints; strain. Dose, a wine-glassful every hour.

5538. Jackson's Cure for Chronic Rheumatism. 1 drachm cajeput oil; ½ ounce syrup of myrrh; 3½ ounces syrup of gum-arabic. Dose, 1 tea-spoonful 3 times a day.

5539. Caution to Painters. Painters should seldom wash their hands in turpentine, as the practice, if persisted in, will lead to the most serious results, even to the loss of power in the wrist joints. It has a tendency to enlarge the finger joints, renders the hands more sensitive to cold in winter, and lays the foundation of rheumatism.

5540. Rheumatic Decoction. Virginia snake-root, 1 drachm; sarsaparilla in powder, 6 drachms; burdock seed, 2 drachms; poke root, 2 drachms; wine-pine bark, 2 drachms; cayenne pepper, ½ drachm. Powder them, and add 3 quarts of water. Boil down to 2 quarts. A cupful 2 or 3 times a day. It is most valuable in chronic rheumatism.

5541. Lumbago. It is a species of chronic rheumatism, which affects the muscles of the lower part of the back, causing great pain and stiffness. The patient can scarcely stir without having the most piercing pain. It may be confined to one side, or affect the loins generally. Its attacks are generally sudden, immediately after or in stooping, or rising from bed. Lumbago is connected with derangement of the stomach, bowels, and kidneys.

5542. Remedy for Lumbago. Rectified oil of turpentine, 25 drops; sulphuric ether, 1 scruple; mucilage of gum-arabic, 3 drachms; syrup of poppies, 1 drachm; rose-water, 1½ ounces; make into a draught; take at bed-time.

5543. Remedy for a Weak Back. Take a beef's gall, pour it into 1 pint alcohol, and bathe frequently.

5544. Remedy for Neuralgia. A remedy said to be efficacious consists in applying bruised horse-radish to the wrist on the side of the body where the pain is.

5545. Excellent Remedy for Neuralgia. A remedy, which is sometimes instantaneously successful, is mixing equal parts of sweet oil, spirits of hartshorn, and chloroform; shake it well, and before time is allowed for its particles to separate, wet a bit of rag or lint, place it on the painful spot for about a minute, or less if relieved sooner, but hold a handkerchief on the lint, so as to confine the volatile ingredients; if kept on too long, the skin may be taken off.

5546. Effective Cure for Neuralgia. Apply a blister of Spanish flies, and let it remain until it draws the skin *red* (not longer); then take it off, and apply a morphine powder. This is often very effectual.

5547. Jackson's Neuralgia Remedy. Mix 1½ drachms iodide of potassa, 15 grains sulphate of quinine, 1 ounce ginger syrup, and 2½ ounces water. Dose, a table-spoonful every 3 hours.

5548. Whitlow, or Felon. The severity of the inflammation in whitlow varies considerably; there is the mild form, which generally yields to fomentation with hot water cloths or poultices; and if matter forms, if relieved by the lancet, it speedily heals; but there is a much more formidable affection, in which the deep textures of the finger are involved, accompanied by severe pain, throbbing, and much redness, heat, and swelling. This form is only to be relieved by free and early incisions with the lancet; for if this be neglected, the bones will become affected, and will be destroyed. It would therefore be advisable to submit the finger to the inspection of a surgeon when it does not easily yield to fomentations or a poultice.

5549. Treatment of Whitlow. Steam the whole hand with bitter herbs for 30 or 40 minutes; bathe it frequently in strong hot lye water. The steaming must not be dispensed with. Or: Immerse the diseased finger in strong lye as long and as hot as can be borne several times a day. Apply a poultice of linseed and slippery elm, with a little salt and brandy. The formation of matter is indicated by a small white spot in the center of the swelling. When this appears, open it with the point of a large needle or probe, that the matter may escape. Repeat if necessary. If proud flesh appears, apply the vegetable caustic or chloride of potassium, diluted. A poultice of powdered hops is very effectual to relieve pain. Apply the *black salve* (*see No.* 4971), to heal it. Attend to the general health, by giving aperients, tonics, and nutritious cooling diet.

5550. Simple Cure for a Felon. As soon as the parts begin to swell get the tincture of lobelia, and wrap the part affected with cloth saturated thoroughly with the tincture, and the felon is dead. An old physician says that he has known it to cure in scores of cases, and it never fails if applied in season.

5551. Bone Felon. The following receipt for the cure of bone felon is given by that high authority, the London Lancet: As soon as the disease is felt, put directly over the spot a blister of Spanish fly, about the size of the thumb nail, and let it remain for 6 hours, at the expiration of which time, directly under the surface of the blister may be seen the felon, which can be instantly taken out with the point of a needle or a lancet.

5552. To Cure Felons. Stir ½ tea-spoonful water into 1 ounce Venice turpentine with a rough stick until the mixture appears like granulated honey. Wrap a good coating of it round the finger with a cloth. If the felon is only recent, the pain will be removed in 6 hours.

5553. Treatment of Boils. When these appear, suppuration should be promoted by poultices of bread and linseed meal, to which a little glycerine or fat or oil may be added, to prevent their getting hard. When poultices are inconvenient, exposure to the vapor of hot water, or the application of stimulating plasters, may be adopted instead. When sufficiently ripe, the boil should be opened with a lancet, the matter evacuated, and the wound dressed with a little simple ointment spread on a piece of clean lint or linen. The diet may be full and liberal until the maturation of the tumor and the discharge of the matter, when it should be lessened, and the bowels opened by some saline purgatives, as salts or cream of tartar. When there is a disposition in the constitution to the formation of boils, the bowels should be kept regular, and tonics, as bark or steel, taken, with the frequent use of sea-bathing, if possible.

5554. Carbuncle. A carbuncle is a species of boil, but larger, and much more painful. It shows debility in the constitution. Carbuncles are very dangerous, and medical advice should at once be obtained.

5555. Astringents. Substances that constrict the animal fibre, and coagulate albumen. When employed to check bleeding, they are called *styptics*. The principal vegetable astringents are catechu, kino, galls, and oak bark; the principal mineral astringents are sulphate of iron, nitrate of silver, chloride of zinc, sulphate of copper, acetate of lead, &c.

5556. To Stop Bleeding. If a man is wounded so that blood flows, that flow is either regular or by jets or spirts. If it flows regularly, a vein has been wounded, and a string should be bound tightly around below the wounded part, that is, beyond it from the heart. If the blood comes out by leaps or jets, an artery has been severed, and the person may bleed to death in a few minutes; to prevent which, apply the cord above the wound, that is, between the wound and the heart. In case a string or cord is not at hand, tie the two opposite corners of a handkerchief around the limb, put a stick between, and turn it round until the handkerchief is twisted sufficiently tight to stop the bleeding, and keep it so until a physician can be had. This appliance is called a *tourniquet*.

5557. To Stop the Bleeding from Leeches. Make a ball of cotton about the size of a pea; put this pellet of cotton or lint upon the wound; press it down firmly; keep up the pressure for a quarter of an hour. Remove the finger cautiously, taking care to let the pellet remain.

5558. Pancoast's Styptic. Take carbonate of potash, 1 drachm; castile soap, 2 drachms; alcohol, 4 ounces. Mix. This styptic has been found preferable to the persulphate of iron in many of the minor cases of hemorrhage, inasmuch as it leaves the surface of the stump in a healthy condition, and does not produce the thick incrustation so often objectionable after the application of the iron.

5559. Styptic Collodion. This is made by uniting equal parts of collodion and chloride of iron. It is recommended for erysipelas.

5560. Ehrle's New Preparation of Cotton for Stanching Hemorrhage. American cotton of the best quality should be cleansed by boiling it for an hour in a weak solution of soda (about 4 per cent.), then repeatedly washed in cold water, pressed out, and dried. By this process it will be perfectly cleansed and adapted to more ready absorption. After this it should be steeped once or twice, according to the degree of strength required, in liquid perchloride of iron, diluted with ⅓ water, pressed, and thoroughly dried in the air—neither in the sun nor by the fire—then lightly pulled out. The cotton so prepared will be of a yellowish-brown color. It must be kept very dry, as it is affected by the damp.

5561. Styptic Paper. A mode for carrying about chloride of iron as a ready styptic has been invented in Paris, which consists in dipping paper in a decoction of 1 pound benzoin and 1 pound alum in 4 gallons water, which has been kept boiling for 4 hours, with renewal and skimming. The paper is left in the filtered solution for some time until saturated; it is then dried, and painted over with a neutral solution of perchloride of iron; this is then dried, folded, and wrapped in an impervious cover.

5562. New Styptic Collodion. Collodion, 100 parts; carbolic acid, 10 parts; pure tannin, 5 parts; benzoic acid, 5 parts. Agitate until the mixture is complete. This preparation, which has a brown color, leaves on evaporation a pellicle exactly similar to that of ordinary collodion. It adheres strongly to the tissues, and effects the instantaneous coagulation of blood and albumen. Tannin effects a consistent coagulation of the blood, whilst benzoic acid has a cicatrizing action on the tissues.

5563. Spitting of Blood. In cases of spitting of blood, it is often difficult to determine whether it proceeds from the internal surface of the mouth, from the throat, from the stomach, or from the lungs. When the blood is of a florid or frothy appearance, and brought up with more or less coughing, preceded by a short tickling cough, a saltish taste, anxiety, and tightness across the chest, its source is the lungs. The blood proceeding from the lungs is usually of a florid color, and mixed with a little frothy mucous only. It may be distinguished from bleeding from the stomach, by its being raised by hacking or coughing, and by its florid and frothy appearance; that from the stomach is vomited in considerable quantities, and is of a dark color.

5564. Treatment for Spitting of Blood. Moderate the discharge of blood by avoiding whatever tends to irritate the body and increase the action of the heart. A low diet should be strictly observed, and external heat and bodily exercise avoided; the air of the room should be cool, and the drink (which should consist chiefly of barley-water, acidulated with lemon-juice), taken cold, and the patient not suffered to exert his voice. After the operation of a little gentle aperient medicine, as lenitive electuary, or an infusion of senna, with a little cream of tartar dissolved in it, take 10 drops of laudanum and 10 drops of elixir of vitriol in half a cupful of cold water. If there is no cough, the laudanum may be omitted. A little salt and water given will often check spitting of blood, when it comes on. Put the feet in warm water, and give as above, the elixir of vitriol, &c. Give also ipecacuanha powder in small doses of from 1 to 2 grains every 4 hours.

5565. Bleeding from the Nose. This may be caused by violence, or may arise from an impoverished state of the blood. When it occurs in persons of middle age it is more serious, as it is then often a symptom of some other disease. The bleeding can generally be stopped by making the patient raise both his arms above his head, and hold them there for some time. Sponging with cold or iced water to the forehead and face, or applying a towel wet with cold water between the shoulders, will, in most cases, succeed. The application of a strong solution of alum or iron-alum to the inside of the nostrils, or plugging the nostrils with lint or cotton wool soaked in the solution, may be necessary if the bleeding is profuse. The health of persons subject to these attacks should be improved by nutritious diet, animal food, with potatoes, water-cresses, and fruit. The following prescription may be relied on: Tincture of steel, 2 drachms; dilute muriatic acid, 1 drachm; syrup of orange peel, 1 ounce; infusion of calumba, 7 ounces. Mix. For a child, 1 table-spoonful in a wine-glass of water before meals; for an adult the dose may be increased.

5566. To Stop Bleeding at the Nose. Placing a small roll of paper or muslin above the front teeth, under the upper lip, and pressing hard on the same, will arrest bleeding from the nose, checking the passage of blood through the arteries leading to the nose.

5567. Astringent for Leech-Bites. Dissolve 1 part of crystallized perchloride of iron in 6 parts of collodion very gradually. A drop or two of the product forms an admirable styptic.

5568. Antispasmodics. Medicines that allay spasms and other pains. Bark, opium, camphor, ether, musk, castor, assafœtida, valerian, and chalybeates, are antispasmodics.

5569. Nervines — sometimes called neurotics—are substances or agents which relieve disorders of the nerves. Antispasmodics, chalybeates, and vegetable tonics belong to this class.

5570. Treatment of Nervousness. The cure of nervousness is best effected by restoring the healthy action of the stomach and bowels, and by the use of proper exercise, especially in the open air. The stomach should not be overloaded with indigestible food, and the bowels should be occasionally relieved by the use of some mild aperient. Abernethy's injunction to a nervous and dyspeptic lady, "Dismiss your servants, madam, and make your own beds," should be recollected by all as a proof of the importance that eminent surgeon attached to exercise. Valerian is a medicine of great use in nervous disorders, hysteria, lowness of spirits, restlessness, and diseases of the bladder, &c. The common dose is from a scruple to a drachm, in powder; and in infusion from 1 to 2 drachms. Its unpleasant flavor may be neutralized by the addition of mace. Assafœtida is also recommended. Take assafœtida, 1½ drachms; water, 6 fluid ounces. Dose, 1 to 3 table-spoonfuls thrice or oftener, daily. But there is no remedy for nervous disorders of every kind, comparable to the proper and constant use of magnetic electricity.

5571. Nerve Powder. Take 1 ounce each of scullcap, valerian and catnip; and cayenne, 1 drachm; coriander seeds, ½ ounce. Pulverize, and mix. Take 1 tea-spoonful in a cupful of boiling water, leaving room for milk and sugar. Repeat according to the symptoms. This powder tranquillizes the most irritable nerves without debilitating and deadening their sensibility. It greatly strengthens the nerves.

5572. Nervous Mixture. Liquid carbonate of ammonia, ½ drachm; compound tincture of cardamom, ½ ounce; oil of lavender, 8 drops; mint water, 3 ounces; mix, and take in two or three doses. It is invaluable.

5573. Nervous Pill. Assafœtida, extract of hops, carbonate of ammonia, of each 1 ounce; extract of valerian, 20 grains. Dissolve the first two ingredients over the fire, then take off, and add the others; mix well, and with a few drops of the oil of lavender, and a little powdered liquorice, form into pills. Dose, 1 or 2 once or twice a day. Valuable in all nervous and hysterical disorders.

5574. Nervous Tincture. Compound tincture of bark, 2 ounces; ammoniated tincture of valerian, 1½ ounces; compound tincture of aloes, ½ ounce. Mix. Good for general weakness, low spirits, and nervous irritability. Two tea-spoonfuls twice a day. (*See No.* 5570.)

5575. Mixture of Valerian and Carbonate of Ammonia. An excellent remedy for nervous headache and depression of spirits. Mix 3 drops oil of valerian and 10 grains carbonate of ammonia with 1½ fluid ounces cinnamon water and ½ fluid ounce syrup. One-half to be taken every 4 hours.

5576. Remedy for Spasms. Take of acetate of morphia, 1 grain; spirit of sal-volatile and sulphuric ether, of each 1 fluid ounce; camphor julep, 4 fluid ounces. Mix. It should be kept closely corked, in a cool place, and should be well shaken before use. Dose, 1 tea-spoonful in a glassful of cold water or wine, as required.

5577. Hypochondriasis, or Low Spirits. Hypochondriasis, low spirits, or "blues," is a peculiar state of the mind, accompanied with indigestion. The principal objects of treatment are, to remove the indigestion, to strengthen the body, and to enliven the spirits; and one of the best plans with which we are acquainted for this is constant exercise and change of place, with a warm bath about thrice a week; early hours, regular meals, and pleasant conversation; the bowels being at the same time carefully regulated by the occasional use of a mild pill, and the stomach strengthened by some appropriate tonic medicine.

5578. To Dissolve Quinine. Sulphate of quinine (sometimes called simply quinine) when forming a part of a fluid mixture, must be dissolved in sulphuric acid before compounding with the other ingredients. In most of the fluid receipts which contain quinine, a small quantity of the acid is prescribed solely for this purpose; it should be added to the quinine drop by drop, and only sufficient used to make a perfect solution.

5579. Remedy for Fever and Ague. Peruvian bark, 2 ounces; wild-cherry tree bark, 1 ounce; cinnamon, 1 drachm, all pulverized; capsicum, 1 tea-spoonful; sulphur, 1 ounce; port wine, 2 quarts. Let stand a day or two. Always buy the Peruvian bark and pulverize it, as most ready pulverized articles are adulterated. This is the reason why more cures are not performed by it. Dose, a wine-glassful every 2 or 3 hours in the day until broken; then 2 or 3 a day until all is used. This mixture will be found an infallible cure for intermittent fever and fever and ague. It removes the disease when all other means fail, and may be used by those who object to quinine.

5580. Cure for Ague. To 5 tea-spoonfuls water, add 50 drops tincture of gelseminum and 10 grains quinine. Shake well before using. Administer 1 tea-spoonful in a wine-glass of sugar water every 2 hours. This medicine has a tendency to affect the head and vision, and produce physical prostration. When these symptoms become developed, cease the doses, and the effects will pass off, leaving the patient completely restored. These directions must be adhered to, as gelseminum, administered after its effects have become apparent, may be attended with serious consequences. (*See No.* 5578.) This is an excellent remedy.

5581. Dr. Krieder's Ague Pills. Take 20 grains quinine, 10 grains Dover's powder, (*see No.* 5176), 10 grains sub-carbonate of iron; mix with molasses or mucilage of gum-arabic, and divide into 20 pills. Dose, 2 each hour, commencing 5 hours before the chill should set in. Then take one night and morning until all are taken. (*See No.* 5584.)

5582. Quinine Mixture for Children. For small children nothing is better than 5 or 6 grains dissolved (*see No.* 5578) quinine in a 2-ounce vial, 1 table-spoonful of white sugar, then fill with water. Dose, 1 table-spoonful as above.

5583. Caution in the Use of Quinine. In all cases where quinine is to be administered, first give a cathartic to cleanse the stomach and bowels.

5584. Ague Mixture. Dissolve 20 grains quinine, mix it with 1 pint diluted gin or port-wine, and add 10 grains Dover's powder (*see No.* 5176), and 10 grains sub-carbonate of iron. Dose, a wine-glass each hour until the ague is broken, and then 2 or 3 times a day till the whole has been used. This is receipt No. 5581, in a liquid form. It may be used when the pills are objectionable.

5585. Remedy for Cold in the Head. Pollion, of France, recommends the inhaling of hartshorn for curing colds in the head. The inhalation by the nose he recommends 7 or 8 times in 5 minutes. Spirits of camphor may be used in the same manner with beneficial results.

5586. Catarrh. There is perhaps no complaint so common as catarrh, or cold in the head; it occurs both in winter and summer; and it is generally said that a summer cold is more difficult to get rid of than a winter one. The attack sets in with pains in the limbs and back, lassitude, and a sense of tightness across the forehead, repeated sneezing, watery and inflamed eyes, and increased discharge from the nose; sometimes there is inflammation of the throat and tonsils, and an eruption of vesicles about the lips.

5587. To Cure Catarrh. Remedies without number have been recommended for catarrh, but few are better than the old-fashioned plan—putting the feet into hot water, giving 10 grains of Dover's powder (*see No.* 5176) a hot drink, and plenty of blankets.

5588. Brown Mixture. Take powdered extract of liquorice and powdered gum-arabic, of each 2 drachms; hot water, 4 fluid ounces; mix, and add spirit of nitrous ether, 1 fluid drachm; antimonial wine, 2 fluid drachms; and tincture of opium, 40 minims. A table-spoonful for a dose. This is an excellent remedy in the early stages of catarrh; it is the well-known *compound liquorice mixture* of the Pharmacopœia.

5589. Flaxseed Tea. Macerate 1 ounce flaxseed and ½ ounce bruised liquorice root in 1 pint boiling water for 2 hours, in a lightly closed vessel; filter, and add 1 fluid ounce lemon juice. This is a good drink in cases of catarrh.

5590. To Relieve a Cough. The troublesome cough caused by an accumulation of phlegm in the throat, especially in the morning, experienced mostly by persons affected with chronic catarrh, can be relieved instantly by taking a tea-spoonful of the following mixture, which has also the advantage of being harmless to the stomach, rather improving the appetite. Put into an 8-ounce phial, ½ ounce muriate of ammonia and ½ ounce pulverized gum-liquorice; fill the phial nearly full with hot water, and shake thoroughly, to prevent the liquorice from becoming solid; shake also before using.

5591. Hay Fever. This very peculiar disease appears generally as a severe attack of catarrh, with asthmatic symptoms superadded. The lining membrane of the eyes, nose, throat, and lungs is all more or less affected. The patient suffers from headache, sometimes severe, sneezing, irritation of the nose and throat, with a dry harassing cough. The asthmatic attacks come on generally towards evening, and last from 1 to 3 hours, causing great distress. Hay fever is not a very common complaint, and only attacks those persons who, from some peculiarity of constitution, are susceptible to the causes producing it. It is supposed to be caused by the inhalation of the pungent aroma of spring grass and hay, but the inhalation of the powder of ipecacuanha will also produce it in certain individuals. In places where the rose is largely cultivated, similar attacks sometimes occur; it is then called *rose fever or rose catarrh.*

5592. Treatment of Hay Fever. The best treatment for hay fever is change of air, to the sea-side if possible. During the attacks, antispasmodics, such as sal-volatile, ether, or an emetic, if the patient is able to bear it, inhalations of hot steam medicated with creosote, carbolic acid, or turpentine, will be found useful. When the attack passes off the general health should be improved by tonics, diet, &c.

5593. Asthma. This disease is well known. It manifests itself in temporary fits of difficult breathing, is accompanied with wheezing, cough, a sense of suffocation, and constriction of the chest. The causes are, hereditary predisposition; cold and moist atmosphere; sudden changes of temperature; intense study; suppression of long accustomed evacuations; certain fevers; irritation of the air cells of the lungs; irritation of the stomach, &c. When this disease is attended with expectoration, it is called humoral asthma; and when there is no discharge it is named dry asthma. It is remarkable that what will excite the disease in one patient will often prove a means of relieving it in another.

5594. To Alleviate Asthma. For moderating the asthmatic paroxysm, no agent is more valuable in many cases than tobacco. A pipe often acts as a charm, and enables the patient to sleep and forget his troubles. In others, the wearing of a gauze veil over the face quite prevents the effects of the evil. It is most important to see that the bowels be freely opened at the commencement of an attack.

5595. Expectorants. Medicines that promote the secretion of the tracheal and bronchial mucus. According to Dr. Good, true expectorants are those medicines which rather promote the separation of the viscid phlegm with which the bronchiæ are loaded, than simply soften and dilute it; though these are also treated as expectorants by many writers. Numerous articles of the materia medica have been denominated expectorants, of which the following are the principal: Tartarized antimony, ipecacuanha, squills, garlic, assafœtida, ammoniacum, the oily resins, the balsams of tolu and Peru, benzoin, styrax, benzoic acid, the fumes of vinegar, tar, and of many of the volatile oils, and the smoke of tobacco and stramonium. Chlorine and ammoniacal gases have also been called expectorants. Medicines of this class are commonly employed in pulmonary complaints and affections of the air-tubes, attended by a vitiated state of the mucus, or an imperfect performance of the natural functions of the secretory vessels. (*Cooley.*) Of all classes of the materia medica, none are more uncertain in their action than expectorants. (*Pereira.*) The act of ejecting matter from the chest is called expectoration.

5596. Bronchitis. An inflammation of the mucous lining of the bronchia, or smaller ramifications of the windpipe. In its milder form it is commonly called a cold on the chest. The usual symptoms are hoarseness, dry cough, a slight degree of fever, followed by expectoration of mucus, at first thin, and afterwards thick and copious. In the severer forms there is more fever, cough, and oppression at the chest, &c. The generality of cases of bronchitis yield to small and repeated doses of ipecacuanha and antimonial diaphoretics, at the same time adopting a light diet, and keeping the bowels open with mild purgatives.

5597. How to Cure a Cold. Dr. G. Johnson, Professor of Medicine in King's College, London, in a recent lecture gives the following cure for a cold: On the whole, the plan which combines the greatest degree of efficiency with universal applicability, consists in the use of a simple hot-air bath, which the patient can have in his own bed-room. All that is required is a tin spirit lamp, with a sufficiently large wick, and holding sufficient spirit to burn for half an hour. The patient sits undressed in a chair with a lamp between his feet, rather than under the chair, care being taken to avoid setting fire to the blankets, of which an attendant takes two or three, and folds them around the patient from his neck to the floor, so as to inclose him and the lamp, the hot air from which passes freely around the body. In from a quarter to half an hour there is usually a free perspiration, which may be kept up for a time by getting into bed between hot blankets. Headache, pain in the limbs, and other premonitory indications of a severe cold, may be entirely removed in the course of half an hour by the action of the *hot-air bath.*

Another simple and efficient mode of exciting the action of the skin consists in wrapping the undressed patient in a sheet wrung out of warm water, then over this folding two or three blankets. The patient may remain thus packed for an hour or two, until free perspiration has been excited.

5598. Cough Pill. Extract of hyoscyamus, balm of gilead buds, with pulverized ipecacuanha or lobelia, and balsam of fir, of each ¼ ounce; oil of anise a few drops, to form into common sized pills. Dose, 1 or 2 pills, 3 or 4 times daily. Dr. Beach says he endeavored

for more than 25 years to obtain a medicine to fulfill the indications which are effected in this cough pill, particularly for ordinary colds and coughs; and this admirably answers the intention, excelling all others. It allays the irritation of the mucus membrane, the bronchial tubes, and the lungs, and will be found exceedingly valuable in deep-seated coughs and all diseases of the chest.

5599. To Cure a Troublesome Cough. 2 or 3 table-spoonfuls of linseed, a small bunch of horehound; boil to a jelly, and strain. Add ½ pound sugar candy, ¼ pound honey, ½ pound loaf sugar. First boil the horehound in 1 quart water, then add the strained linseed and the other articles. Simmer for 2 hours. When cold, add of chlorodyne, 3 table-spoonfuls. Bottle it and cork tight. A small quantity of spirits of wine or brandy to keep it. When the cough is troublesome, take a table-spoonful. This is an excellent remedy.

5600. Pulmonary Syrup. Blood-root, boneset, slippery elm bark, coltsfoot, elecampane, of each 2 ounces; white root, spikenard root, of each 4 ounces; comfrey root, poplar bark, of each 1 ounce; lobelia, horehound, snake-root, of each ½ ounce. Pour upon them 2 quarts of boiling water; stir well, add 1 pound molasses, and, when cool, 1 quart Holland gin. It is one of the best remedies for asthma, coughs, hoarseness, &c. A table-spoonful every hour, or a wine-glassful 3 times a day.

5601. Pulmonary Balsam. Horehound plant, comfrey root, blood root, elecampane root, wild cherry bark, spikenard root, penny-royal plant, of each 4 ounces. Pour 3 quarts boiling water upon them; infuse for 3 hours; then heat the water again, and pour it upon the plants to infuse 5 or 6 hours. Sweeten with sugar candy. It is very serviceable in diseases of the lungs, chronic coughs; it removes constriction of the chest by promoting expectoration. Take half a small tea-cupful 3 or 4 times a day, or oftener if necessary.

5602. Blood-Root Syrup. Bruised blood-root, 2½ ounces; lobelia, ½ ounce; white sugar, 1½ ounces; water, 1½ pints; gently simmer half an hour, till it thickens; when cool, add a tea-spoonful of paregoric elixir. Take a table-spoonful occasionally; for a child, a tea-spoonful or less. This syrup is very valuable in chest complaints, bronchial affections, coughs, and difficult breathing.

5603. Cough Syrup. Tincture of lobelia, 1 ounce; Iceland moss, 2 ounces; white poppy capsules, bruised, 2 ounces; pearl barley, 2 table-spoonfuls; water, 2 quarts; molasses, 2 ounces. Boil down to 3 pints, and strain. Dissolve in it from 4 to 8 ounces of sugar candy. It effectually allays a tickling cough. A table-spoonful when the cough is troublesome. It does not constipate, like laudanum and paregoric.

5604. Cough Remedy. Take lobelia herb, horehound, boneset, of each 1 ounce; comfrey root, spikenard, St. Johns' wort, poppy capsules, of each ½ ounce. Infuse in 3 pints boiling water for 3 hours. Strain and add ½ pound loaf sugar boiled to a syrup. Add a wine-glassful of best rum. A table-spoonful is a dose. This is a valuable receipt for cough, hoarseness, &c.

5605. To Cure a Cold with a Cough. Make a decoction of the leaves of the pine tree, and sweeten with loaf sugar. Drink it freely, warm, before going to bed, and cold, through the day. It is a certain cure in a short time.

5606. Inhalation of Cubebs and Carbolic Acid. Mix together ½ fluid ounce tincture of cubebs and 20 drops liquid carbolic acid. Add the mixture to ½ pint hot water in an inhaler, and use every 3 or 4 hours, taking full respirations. A very efficient remedy in dry cough.

5607. Cough Mixture. Take muriate of morphia, ½ grain; glycerine, 2 fluid ounces. Mix. Dose, a tea-spoonful when the cough is troublesome.

5608. Treatment for Ulcerated Sore Throat. Chlorate of potassa, in cases of putrid ulcerated sore throat, has been used with the most decisive success. Its internal application more effectually allays thirst and abates fever than any other medicine; and, when applied as a gargle to inflamed or ulcerated sore throats, it has been found to disperse the inflammation and cleanse the ulcers more effectually than the infusion of rose-leaves with sulphuric acid, the gargle generally resorted to in those cases. The chlorate of potassa may be given in the dose of from 20 to 30 grains in a half glass of water, 3 or 4 times a day. For the purpose of gargling the throat, 4 drachms of the chlorate may be added to ½ pint of water. (*See No.* 5637.)

5609. Bell's Gargle for Syphilitic Sore Throat. Mix together 2 grains corrosive sublimate; 1 ounce rectified spirits of wine; 3 ounces tincture of Peruvian bark, and 1 ounce each honey of roses and tincture of myrrh.

5610. Atlee's Cough Mixture. 2 grains acetate of morphia; 1 drachm each tincture of belladonna and tincture of nux-vomica; 3 drachms each antimonial wine and syrup of ipecacuanha root; 1 ounce fluid extract of wild cherry bark, and 2 ounces syrup of balsam of tolu. A tea-spoonful 4 times a day relieves chronic or hacking cough.

5611. Hope's Cough Mixture. 2 ounces ammonia mixture; 5 ounces camphor mixture; 1 drachm tincture of digitalis (foxglove); ½ ounce each sweet spirits of nitre and syrup of poppies; 2 drachms solution of sulphate of morphia. A table-spoonful of this mixture is to be taken 4 times a day.

5612. Treatment of Consumption. It seems at first sight as superfluous to state that in a disease of debility like consumption, patients should breathe pure air, as that they should have good nourishing food, but it is not so. Theoretically, the value of pure air is accepted; but practically it is universally neglected. Healthful respiration has yet to be applied not only to every-day life, but in the treatment of disease. In ill health, and particularly diseases of the respiratory organs, the dictates of science and common sense are grossly outraged. If those persons who have consumption, or who have an inclination to it, would spend an hour every day in breathing pure air to the fullest extent to which their lungs are capable of taking it in, they would do more to prevent and cure this disease than it is possible to do by medication.

5613. Inhalation of Tar for Consumption. Mix together 16 ounces liquid tar and 1 fluid ounce liquor of potassa; boil them for a few minutes in the open air; then let it simmer gently in an iron vessel over a spirit or other lamp in the chamber of the patient. This may, at first, excite a disposition to cough, but in a short time allays it, and removes any tendency to it.

5614. Syrup of Blood-root. 1 ounce blood-root, ½ ounce aniseseed, and ½ ounce liquorice boiled in 2 pints water down to a pint, and then mixed with 4 ounces honey. This is highly recommended in consumptive cases attended with dyspeptic symptoms.

5615. Blood-root for Consumption. 25 to 40 drops saturated tincture of blood-root, taken 2 or 3 times a day, afford great relief.

5616. Cigars for Pulmonary Consumption. Dissolve 1 part arseniate of soda in 30 parts water. Dip white unsized paper into the solution and form into small rolls, 3 or 4 inches long. In pulmonary consumption inhale 4 or 5 whiffs as many times a day.

5617. Goddard's Cure for Loss of Voice. Wet bibulous paper with a solution of 1 part arsenite of potash in 25 parts water; dry and roll strips of 3 inches by 1 inch into cigarettes. The smoke to be inhaled, 8 or 10 inspirations, 3 times a day. In connection with this use $\frac{1}{16}$ grain ammoniated mercury mixed with 10 drachms powdered sugar, apply a little to the throat with the end of the finger. This is an excellent remedy.

5618. To Cure Hoarseness. Saturated solution of iodine, 20 drops; alcohol, 1 ounce; 5 drops of the above on a lump of loaf sugar every two hours will be found invaluable.

5619. Cigars for Hoarseness, Asthma, &c. Soak thick unsized paper in a solution of saltpetre, and dry. Then brush over with tincture of cascarilla; and, when nearly dry, with compound tincture of benzoin. In about half an hour, cut it into pieces 1½ by 4 inches, and roll into cigarettes. Excellent for hoarseness, loss of voice, and asthma.

5620. Remedy for a Sudden Hoarseness. Mix 1 tea-spoonful of sweet spirits of nitre in a wine-glassful of water. This may be taken 2 or 3 times a day.

5621. To Prevent Hoarseness. A celebrated singer states that the greatest benefit is derivable from taking, during 5 or 6 days, twice a day, 5 or 6 drops of nitric acid in a glass of sugared water. If from use the acid loses its efficacy, the dose may be increased to 10 or 12 drops.

5622. Snuffles. A troublesome complaint, to infants especially. The mucous membrane of the nose, through the taking of cold, being much swollen, the child is no longer able to breathe through its nose, as it was accustomed to do, but is compelled to breathe through the mouth. The difficult breathings are attended by a peculiar snuffling noise, which, in sleep, becomes a regular loud snore. It often interferes with its sucking at the breast; as soon as it seizes the nipple a threatening suffocation compels it to desist. While this complaint lasts the child may be partially fed with the spoon; give it a very mild purgative; bathe its legs frequently in warm water. Rub the nose with tallow, and apply a slippery elm poultice mixed with cream.

5623. Influenza Mixture. Mix ½ ounce paregoric elixir, 1 ounce syrup of squills, and 2 drachms antimonial wine, with 6 ounces water. A tea-spoonful every 15 minutes until relieved.

5624. Treatment of Asthma. Relief is often obtained by smoking a pipe of tobacco. To a person unaccustomed to smoking, a pipe of latakia, or other mild description of tobacco; this soon produces exhaustion, while, directly the feeling of nausea comes on, the attack ceases. This remedy is often very useful in preventing an attack when one is impending. Stronger tobacco should be used by inveterate smokers. The fumes of burning filtering or blotting-paper, which has been soaked in a saturated solution of nitre, and dried, afford much relief in some cases (*see No.* 5619); and, lastly, there are instances where palliation is soonest obtained from a stimulant, as a glass of whiskey or brandy toddy, or a cup of very strong coffee. A mustard poultice over the front of the chest is often effective. Sometimes an attack may be arrested by taking off the patient's coat and vest, and exposing his back to the heat of a good fire. (*See No.* 5764.)

5625. Croup. This is a dangerous disease. It is common to infancy, and rarely occurs to adults. It is an inflammation of the larynx, trachea, and contiguous tissues. It derives its name from the peculiar sound of the voice and breathing, being of a whistling or crowing character, owing to a contraction of the glottis. It generally commences with a common cold and catarrh, hoarseness, cough, and increased difficulty of breathing, and the crowing already spoken of. It demands prompt treatment.

5626. Treatment of Croup. The great object is to diminish the inflammation and irritation, and to relax the spasmodic state of the muscles in the parts diseased. The vessels in those parts are overcharged with blood, by an imperfect action of the exhalants. Place the feet in warm water, and give an emetic. (*See No.* 5169.) After bathing, rub the legs and feet well with flannel. Then give a vapor bath, if the patient can bear it. Repeat the process, if needful. The perspiration will be greater by applying to the feet and each side hot bricks, and wrapped in flannel saturated with vinegar and a little water. At the same time give an aperient, to produce a free action on the bowels. Apply this tincture to the throat, viz.: ½ tea-spoonful of cayenne pepper, nearly a cupful of vinegar; simmer 10 minutes, and strain. This tincture may be diluted with warm water, according to the strength of the patient. Rub it well on the throat for 5 or 10 minutes; and next saturate a flannel with it, and apply it to the throat. This application tends to relieve the internally congested blood-vessels. Repeat the application as necessary. Mustard plasters may be applied to the feet, the upper part of the chest, and between the shoulders alternately. Even a large sponge dipped in as hot water as the hand can bear, squeezed half dry, and renewed before it is cool, is of great advantage. It has been recommended to steep hops in hot vinegar, and the patient to inhale the vapor. Keep the atmosphere of the room at a regular temperature. Aid the perspiration by warm drinks, as balm tea, &c.

5627. Remedy for Croup. Turpentine is a sovereign remedy for croup. Saturate a piece of flannel with it, and place the flannel on the throat and chest, and in very severe cases 3 to 5 drops on a lump of sugar may be taken inwardly. Every family should have a bottle of turpentine on hand.

5628. To Prevent a Return of Croup. To prevent a return of this disorder, keep the child warm, avoid wet feet, cold, damp, easterly winds, &c. Children whose constitutions dispose them to croup ought to have their diet properly regulated, and be kept from all crude, raw, and trashy fruits.

5629. Mumps. This is a specific contagious inflammatory affection of the salivary glands, especially the largest, situated below the ear. It begins with slight feverish symptoms, with pain and swelling, extending from beneath the ear along the neck to the chin. The attack generally reaches its height in 4 days and then declines. The treatment is very simple—a mild diet, gentle laxatives, occasional hot fomentations, and wearing a piece of flannel around the throat.

5630. Quinsy. Inflammation of the tonsils, or common inflammatory sore throat, commences with a slight feverish attack, with considerable pain and swelling of the tonsils, causing some difficulty in swallowing; as the attack advances these symptoms become more intense, there is headache, thirst, a painful sense of tension, and acute darting pains in the ears. The attack is generally brought on by exposure to cold, and lasts from 5 to 7 days, when it subsides naturally, or an abscess may form in the tonsil and burst, or the tonsil may remain enlarged, the inflammation subsiding.

5631. Treatment of Quinsy. The patient should remain in a warm room, the diet chiefly milk and good broths, some cooling laxative and diaphoretic medicine may be given; but the greatest relief will be found in the frequent inhalation of the steam of hot water through an inhaler, or in the old-fashioned way, through the spout of a teapot. Relief will also be experienced from the following treatment: Roast 3 or 4 large onions. Peel them quickly, and beat them flat with a rolling-pin. Immediately place them in a thin muslin bag that will reach from ear to ear, and about 3 inches deep. Apply it speedily as warm as possible to the throat. Keep it on day and night, changing it when the strength of the onion appears to be exhausted, and substituting fresh ones. Flannel must be worn around the neck after the poultice is removed.

5632. Treatment of Whooping Cough. The attack generally begins as a common cold, with slight feverish symptoms. In 8 or 10 days the fever partially subsides, and the child gets attacks of convulsive coughing, accompanied by the peculiar whoop which gives the disease its name. The number of attacks varies from 1 or 2 to 10, or even 15 in the 24 hours, according to the severity of the disease. The child should be kept in a warm room. He ought to be clothed in flannel; his diet should be light and nourishing, such as fish, milk, light puddings, and new-laid eggs. The following prescription is strongly recommended by Dr. Valentine Mott: Hydrocyanic acid, 6 drops; extract of belladonna, 2 grains; paregoric elixir, 3 drachms; syrup of balsam of tolu, 1 ounce; and water, 3 ounces. Mix. 1 tea-spoonful 3 or 4 times daily. When the severity of the disease has passed off, change of air will be found most useful; and if the child has become debilitated, tonics, with nutritious diet, should be given. This disease being very infectious, great care should be taken to prevent communication of any kind with houses where there are children who have not already had whooping-cough.

5633. Syrup for Whooping-Cough. Onions and garlics, sliced, of each 1 gill; stew them in 1 gill sweet oil, in a covered dish, to obtain the juices; then strain and add honey, 1 gill; paregoric and spirits of camphor, of each ½ ounce; bottle and cork tight for use. Dose, for a child of 2 or 3 years, 1 tea-spoonful 3 or 4 times daily, or whenever the cough is troublesome, increasing or lessening, according to age.

5634. Atlee's Cure for Whooping-Cough. Take 1 drachm each powdered cochineal and strong aqua-ammonia; 1 ounce rectified spirits of wine. Mix. Dose for a child one year old, 10 drops in sweetened water 3 times a day.

5635. Cure for Whooping Cough. Pure carbonate of potassa, 1 scruple; cochineal, 1 grain; dissolve in 6 ounces of water sweetened with sugar. Dose for a child four years old, 1 tea-spoonful 3 times a day, to be taken before meals. This is an excellent remedy.

5636. Treatment of Diphtheria. Make two small bags to reach from ear to ear, and fill them with wood-ashes and salt; dip them in hot water, and wring them out so that they will not drip, and apply them to the throat; cover up the whole with a flannel cloth, and change them as often as they become cool, until the throat becomes irritated, near blistering. For children it is necessary to put flannel cloths between the ashes and the throat, to prevent blistering. When the ashes have been on a sufficient time, take a wet flannel cloth and rub it with castile soap until it is covered with a thick lather; dip it in hot water, and apply it to the throat, and change as they cool; at the same time use a gargle made of 1 tea-spoonful each of cayenne pepper, salt, and molasses, in a tea-cupful of hot water, and when cool, add ¼ as much cider vinegar, and gargle every 15 minutes, until the patient requires sleep. A gargle made of castile soap is good to be used part of the time.

5637. Remedy for Diphtheria. Permanganate of potassa has been administered with great success in cases of diphtheria. The proportions used for external use are 1 drachm of the permanganate to a pint of water; the dose for internal use, 1 tea-spoonful of a solution of 1 drachm in 1½ pints water. (*U. S. Dis.*)

5638. Remedy for Diphtheria. A gentleman who has administered the following remedy for diphtheria, says that it has always proved effectual: Take a tobacco pipe, place a live coal in the bowl, drop a little tar upon the coal, and let the patient draw smoke into the mouth, and discharge it through the nostrils. The remedy is safe and simple.

5639. Roche's Remedy for Diphtheria. M. Roche recommends the following mode of treatment. The false membranes are first freely cauterized with lunar caustic, and injections then made every hour against the fauces with a solution of common salt, the strength of the solution being such as not to create nausea. Chlorate of potassa may be also given internally; and tincture of iodine as a local application; but M. Roche considers that the irrigations with the solution of common salt are the chief agents in the case.

5640. Stiff Neck. Occasionally an attack is severe, and confinement to the house or bed, with wrapping up of the neck with cotton-wadding or flannel, together with attention to the state of the digestive powers, is necessary. The diet in these cases must be regulated, and an aperient, such as the lenitive electuary (*see No.* 5154), or castor oil, taken if required by the state of the bowels. If the stiffness be obstinate in its duration, it had better be rubbed with camphorated oil, or some other appropriate liniment.

5641. Anthelmintics. Medicines that destroy worms. Among the principal anthelmintics are santonin (worm-seed), calomel, tin powder, castor oil, oil of turpentine, cowhage, pinkroot, male-fern, and gamboge. A good plan for removing worms from children, is to give 3 to 5 grains of calomel in sugar, over-night, and a dose of castor oil the next morning. The motions should be observed, and if worms be found, the same treatment may be followed once a week, until they are wholly removed.

5642. Worms. The worms found in the human body are mostly the ascarides, the thread worm, infesting the lower intestine, causing much itching and irritation about the anus. The teres, or long round worms, are generally seated in the small intestines, and stomach. The symptoms denoting the existence of worms are common to the different species, viz.: indigestion, with a variable appetite; foul tongue; offensive breath; hard, full, and tense belly, with occasional gripings and pains about the navel; heat and itching sensation in the rectum and about the anus;

the eyes heavy and dull; itching of the nose; short dry cough; grinding of the teeth; and starting during sleep, attended often with a slow fever.

5643. Dr. Freeman's Vermifuge Oil. Oil of worm-seed, ½ ounce; oil of turpentine, 2 drachms; castor oil, 1½ ounces; pink root, ½ ounce; hydrastin, 10 grains; syrup of peppermint, ½ ounce. Dose, for a child 10 years old, a tea-spoonful 3 times a day, 1 hour before each meal; if it purges too freely, give it less often. This is an excellent vermifuge, and never fails to expel worms when administered for that purpose. Where no worms are present, it answers the purpose of a tonic, correcting the condition of the mucous membrane of the stomach and bowels, and operating as a mild cathartic.

5644. Spackman's Worm Syrup. Take ½ ounce pink root; 2 drachms rhubarb root; 1 drachm worm-seed; ½ drachm savine leaves; 2 drachms colombo root, and 1 drachm cardamom seeds. Infuse these ingredients in ½ pint boiling water in a covered vessel; when cool, add ½ pint molasses and a table-spoonful brandy. Dose for a child 1 year old, 2 tea-spoonfuls 3 times a day.

5645. Remedy for Worms. Take 6 grains santonin; 2 grains powdered gamboge; 3 grains calomel; and 12 grains powdered white sugar. Make into 6 powders. Give 1 powder 3 times a day for a child one year old, and a dose of castor oil the day after taking the powders.

5646. Oil of Worm-seed Mixture. Take 1½ fluid drachms oil of worm-seed, 3 ounces castor oil, and 10 drops oil of anise; mix them together, and add 1 fluid ounce aromatic syrup of rhubarb. Shake well before using. Dose for a child of 2 years, 1 tea-spoonful night and morning.

5647. A Simple and Safe Vermifuge. Powdered rust of iron is a good vermifuge. It expels the worms and strengthens the constitution. To a child 6 years old from 10 to 40 grains may be given. An adult may take ¼ ounce or more. It may be given in molasses or in beer. Dr. Rush says that he knows of no safer and more certain remedy than this simple preparation of iron. It should always be followed by an aperient.

5648. Worm Pills. Ethereal extract of male-fern, 30 drops; extract of dandelion, 1 drachm; powdered gum enough to make 30 pills. Dose, from 6 to 20; followed half an hour later by a strong dose of castor oil.

5649. Tape-Worm. The common male-fern root is a certain remedy for the tape-worm. 2 or 3 drachms of the powdered root to be taken in the morning, no supper having been taken the night before. It generally sickens a little. A brisk purgative is to be given a few hours after, which sometimes brings off the worm entire; if not, the same course must be followed at due intervals. For the success of this remedy, the root should be recently gathered; as, after being kept long in the stores, its activity is diminished or destroyed.

5650. Dowler's Treatment of Tape-Worm. Dr. Dowler expelled a tape-worm 135 feet long by prescribing the continued use of elm-bark. He ordered the bark to be chewed and swallowed in moderate quantities.

5651. Beach's Treatment of Tape-Worm. Dr. Beach effectually cured a patient who had been tormented with a tape-worm for 25 years. His treatment was as follows: Cowhage stripped from the pod, a small tea-spoonful 3 times a day; to be taken, fasting, in a little arrow-root jelly; then occasionally a purgative of mandrake. In connection with this, eat freely of garlic and fine common salt. This treatment is to be continued until the tape-worm is killed or so sickened that it will lose its hold on the bowels, when it will be expelled entire. When once the tape-worm begins to pass the bowels, care must be taken not to break it off, for it will then grow again; it has this peculiar property.

5652. Diarrhea. The following excellent remarks on this disease are extracted from Dr. Hall's Journal of Health: Cholera is nothing more than exaggerated diarrhea. It may be well for travelers to know that the first, the most important, and the most indispensable item in the arrest and cure of looseness of the bowels, is absolute quietude on a bed; nature herself always prompts this by disinclining us to locomotion. The next thing is, to eat nothing but common rice, parched like coffee, and then boiled, and taken with a little salt and butter. Drink little or no liquid of any kind. Bits of ice may be eaten and swallowed at will. Every step taken in diarrhea, every spoonful of liquid, only aggravates the disease. If locomotion is compulsory, the misfortune of the necessity may be lessened by having a stout piece of woolen flannel bound tightly round the abdomen, so as to be doubled in front, and kept well in its place. In the practice of many years, we have never failed to notice a gratifying result to follow these observances.

5653. Velpeau's Remedy for Diarrhea and Cholera Morbus. Take 1 ounce each tincture of opium, paregoric elixir, and tincture of rhubarb; 10 drachms essence of peppermint; and 6 drachms tincture of capsicum. This is the original receipt for this celebrated remedy. Dose for an adult, a tea-spoonful in ½ a wine-glass sweetened water; and, if required, half a dose after each loose evacuation.

5654. Diarrhea Tincture. Compound tincture of myrrh, 6 ounces; tincture of rhubarb, and spirits of lavender, of each 5 ounces; tincture of opium, 3 ounces; oils of anise and cinnamon, with gum camphor and tartaric acid, of each ½ ounce. Mix. Dose, 1 tea-spoonful in a little warm water sweetened with loaf sugar; repeat after each passage. This is a magic remedy.

5655. Chlorodyne Mixture. Shake together 2½ fluid drachms each chlorodyne and rectified spirit; add 1 fluid ounce syrup, and shake again well; then add a little at a time, with brisk agitation, 4 fluid ounces distilled water and 3 fluid drachms mucilage. Dose, 1 to 2 table-spoonfuls in diarrhea, cholera, &c. Shake well before using.

5656. Goddard's Diarrhea Remedy. Dr. Paul Goddard gives the following remedy: Take ½ ounce tincture of catechu, 2 drachms each tincture of opium and tincture of camphor, and 1 drachm aromatic spirits of ammonia. 40 drops every hour will afford speedy relief.

5657. Remedy for Diarrhea. Tincture of opium, spirits of camphor, essence of peppermint, ethereal tincture of capiscum, of each ½ ounce; syrup of kino, 1 ounce; neutralizing cordial, 2 ounces (*see No.* 5663); brandy, 2 ounces. Mix. Dose, one table-spoonful, may be given every twenty minutes if the case is urgent. In dysentery give 1 table-spoonful 3 times a day. This is an excellent remedy.

5658. Blackberry Cordial. To 1 quart blackberry juice, add 1 pound white sugar, 1 table-spoonful each cloves, allspice, cinnamon, and nutmeg. Boil all together 15 minutes, add a wine-glass of whiskey, brandy, or rum. Bottle while hot, cork tight and seal. This is almost a specific in diarrhea. Dose is 1 wine-glassful for an adult, half that quantity for a child; will often cure diarrhea. It can be taken 3 or 4 times a day if the case is severe.

5659. Remedy for Summer Complaint. A tea made of the seeds of the sunflower, roasted like coffee berries, is an admirable remedy for all species of summer complaint. ½ pint of the seed is sufficient. It should be remembered, however, that serious results often follow the too sudden stoppage of diarrhea by astringents, and with this, as all remedies of a similar nature, caution should be used.

5660. Remedy for Bilious Diarrhea. Infuse ½ ounce Angostura bark for 2 hours in 1 pint boiling water, and strain; is a remedy for bilious diarrhea, especially in southern latitudes.

5661. Treatment of Diarrhea in Infants. Dr. Smith recommends the following prescriptions, if the bowels are rather loose, with dark, slimy, offensive stools. Tincture of opium, 8 minims; castor oil, 1 drachm; syrup of ginger and mucilage of acacia, each 1 ounce. A tea-spoonful 3 times daily. In the screaming fits, accompanied by constipation, this combination of castor-oil with laudanum is very valuable. (*Med. News.*)

5662. Treatment of Cholera. The following excellent directions are given for the treatment of cholera by Dr. Pratt: For the *stage of diarrhea.* This may come on insidiously, painless, and hence not alarming, but should be met promptly. The remedy is the cholera mixture, so called, consisting of equal parts of laudanum, tincture of rhubarb, and spirits of camphor. Begin with 30 drops, taken clear and unmixed, with a little sugar placed in the mouth afterward. Repeat the dose after every evacuation, increasing it if the case becomes urgent to 60 drops (a tea-spoonful), or 90 drops if necessary. If the diarrhea is not controlled by this means, an injection of from 30 to 90 drops laudanum, in a table-spoonful of starch, will prove a valuable help. This may be often repeated. If the diarrhea ceases, do not entirely intermit the medicine, but give in gradually diminished doses, every 1 or 2 hours, for a period of 12 or even 24 hours.

5663. Treatment for the Vomiting Stage. Dr. Pratt's remedy is laudanum, tincture of capsicum, tincture of ginger, and tincture of cardamom seeds, equal parts; to be given from 40 to 60 drops undiluted, and followed by sugar, after every fit of vomiting; taking care to give it as soon as the fit ceases, when it will be more likely to be retained. An excellent assistant to this is a large mustard poultice applied to the abdomen.

5664. Treatment for the Stage of Malignancy. According to Dr. Pratt, the only remedy is stimulants, especially brandy, which must be given with great freedom, from 2 to 4 tea-spoonfuls every half or even quarter hour, till heat returns, and pulse and sensibility of extremities are restored. It is always to be given undiluted. Alcohol, or other spirits, will answer the purpose, if brandy is not to be had. It will be necessary to combine with this, artificial heat, bottles of hot water to the body and extremities, friction of the limbs (which no one need fear to apply), and mustard, perhaps, to the feet and hands, stomach and limbs. Remember that boldness, to the verge of rashness, is better than excess of caution, and that no danger is to be apprehended from any of these remedies so long as the symptoms for which they are given are uncontrolled. The use of cold water must be strictly forbidden, except merely to gargle the throat; a very small quantity, swallowed, will bring on the diarrhea after it has been stopped for hours. A little water of gum-arabic may be allowed, a tea-spoonful at a time; or, perhaps, lumps of ice might be taken with safety. For the typhoid fever, which often follows an attack, chamomile or sage tea, and diaphoretic (*see No.* 5134) treatment, will be all that is needed, beside a moderate use of stimulants, for convalescence.

5665. Cholera Preventive. A Burgundy-pitch plaster worn over the region of the stomach during the prevalence of the disease. It should be warmed a little before it is put on, the person standing erect when it is applied, so that the plaster shall not interfere with the motions of the body. It is asserted that a British regiment supplied with such plasters lost only five men during a severe visitation of cholera, and these had refused to wear then. The efficacy of this preventive is also corroborated by other well-authenticated evidence.

5666. Neutralizing Mixture. Powdered rhubarb, 3 scruples; saleratus, or crude bicarbonate of potash, 3 scruples; powdered peppermint plant, 3 scruples; boiling water, ½ pint; decoction of aniseed, ½ pint. Mix. Strain, sweeten with sugar, and add 3 table-spoonfuls of brandy. Take 1 or 2 table-spoonfuls as often as the symptoms require it. For children, a less dose. Very valuable in cholera, bowel complaints of children, laxity of the bowels, flux, &c.

5667. Spackman's Cholera Mixture. Take 1 ounce gum camphor; 2 ounces gum kino; ½ ounce gum catechu; 2 ounces ground cinnamon; 1 ounce ground cloves; 2 drachms African capsicums. Moisten these with brandy and digest for 48 hours. Displace (*see No.* 41) 18 ounces; then add 20 drachms tincture of opium and 1 ounce chloroform. Dose for an adult, 60 drops after every passage.

5668. Brown's Cholera Mixture. Mix together 1 ounce essence of Jamaica ginger; 2 ounces each camphorated tincture of opium and aromatic spirits of ammonia; and 1 ounce spirits of camphor. Dose, a teaspooonful every hour.

5669. Troth's Cholera Mixture. Digest for 10 days 1 ounce each opium, camphor, oil of cloves, and African capsicums, in 1 pint Hoffman's anodyne (*see No.* 4749); administer 20 to 40 drops every 2 hours.

5670. Austrian Cholera Specific. Take 20 grains sulphuric acid specific gravity 1.500; 15 grains each sugar and gum; distilled water sufficient to make the whole weigh exactly 1 ounce. 1 table-spoonful of the above mixture is to be taken in water on the first appearance of premonitory symptoms, followed by the free use of ice-cold water. A second dose ½ an hour after is generally sufficient to arrest the disease, but occasionally 4 or 5 doses are required. A table-spoonful in a pint of cold water may afterwards be drunk as often as desired. When collapse sets in, double doses are to be given, and repeated after every attack of vomiting, until the sickness and cramp abate. After which, the doses are to be repeated until 5 or 6 doses are retained by the stomach. Quiet sleep or drowsiness should not be interfered with. The free use of cold water or acidulated water is to be allowed until perspiration sets in and the warmth of the body returns. The use of warm drinks, wine, spirits, &c., are to be carefully avoided as so much poison. The above was adopted by the Austrian Government in 1849, after 18 years' successful trial.

5671. Homœopathic Cholera Preventive. Dissolve 1 drachm camphor in 6 drachms rectified spirit, and preserve it in a well-corked bottle. Dose, 2 drops on a lump of sugar 2 or 3 times a day.

5672. Homœopathic Cholera Remedy. Repeat the dose of the mixture in foregoing receipt every 10 or 15 minutes, followed by draughts of ice-cold water until the symptoms abate.

5673. Use of Calomel in Cholera. When cholera is prevailing, a single large, thin, painless, weakening action of the bowels may be cholera begun, and the business man should start for home in a vehicle instantly, calling on his physician on his way, and take him home with him; or, if he cannot be found immediately, get into bed as soon as possible, dress up warm, eat ice if thirsty, bind a thick warm flannel tightly around the abdomen, and wait for his doctor's arrival. A physician should be called always on the instant of an attack, but when it is impossible to procure his services within an hour, 10 or 20 grains of calomel should be taken in pill or powder, as a means of stopping the discharges, and of thus arresting the disease, until the physician arrives. Calomel is generally easy to be procured, will remain on the stomach, from its heaviness, when even cold water is ejected as soon as swallowed, and is the most certain of all medicines known to stimulate the liver to action, this want of action being the fundamental cause of the disease. (*Hall.*)

5674. Cholera Tincture. Tinctures of rhubarb, cayenne, opium, and spirits of camphor, with essence of peppermint, equal parts of each, and each as strong as can be made. Dose, from 5 to 30 drops, or even to 60, and repeat until relief is obtained, every 5 to 30 minutes. Many lives have been saved by the timely use of this valuable medicine.

5675. Treatment of Dysentery. A slight attack will often yield to the employment of a dose of castor oil; warm fomentations or mustard poultices being applied over the belly; the patient being confined to bed, and only allowed to partake of food the most simple in its nature, that is, farinaceous food, cream, or milk (with one-third of lime-water, if requisite), thin broths, &c. Perfect rest in the horizontal posture is almost essential. A warm bath for 20 minutes, or a shorter time if the patient feels faint, will often give great relief. Stimulants should be forbidden in mild cases; but where the patient is becoming weakened by the disease, port wine, as the best stimulant in these cases, may be given in beef-tea, or alone. And the rule of little and often may be strictly observed. Early treatment is most important in dysentery, and therefore the medical man should be sent for without loss of time, in case the simple means recommended are ineffectual.

5676. Indian Cure for Dysentery. In diseases of this kind, the Indians use the root and leaves of the blackberry bush, a decoction of which in hot water, well boiled down, is taken in doses of a gill before each meal, and before retiring to bed. It is an almost infallible cure.

5677. Simple Remedy for Dysentery. The following simple remedy has been known to cure the most obstinate and malignant forms of dysentery when all the ordinary methods were ineffectual: Take hot water, 1 gill; vinegar, ½ pint; mix; then continue to add common salt as long as it will be dissolved, stirring and irritating it freely and frequently. Give for an adult 1 table-spoonful every hour until the bloody discharges cease, or until it operates freely on the bowels. The patient must remain in bed.

5678. Antacids. Medicines that neutralize the acid of the stomach, and thus tend to remove heartburn, dyspepsia, and diarrhea. The principal antacids are the carbonates of potassa, soda, ammonia, lime, and magnesia. Ammonia is the most powerful, and when the acidity is conjoined with nausea and faintness, is the best; when great irritability of the coats of the stomach exist, potash is preferable; when accompanied with diarrhea, carbonate of lime (prepared chalk); and when with costiveness, magnesia. The dose of the carbonates of potassa and soda in powder is half a tea-spoonful; of chalk, a tea-spoonful; of magnesia, a dessert-spoonful; and of carbonate of ammonia, 10 grains, or a tea-spoonful of the solution. All these are taken in water.

5679. Dyspepsia. If a man wishes to get rid of dyspepsia, he must give his stomach and brain less to do. It will be of no service to follow any particular regimen—to live on chaff bread or any such stuff—to weigh his food, etc., so long as the brain is in a constant state of excitement. Let that have proper rest, and the stomach will perform its functions. But if he pass 10 or 12 hours a day in his office or counting-room, and take no exercise, his stomach will inevitably become paralyzed; and if he puts nothing into it but a cracker a day, it will not digest it. In many cases it is the brain that is the primary cause. Give that delicate organ some rest. Leave your business behind you when you go to your home. Do not sit down to your dinner with your brows knit, and your mind absorbed in casting up interest accounts. Never abridge the usual hours of sleep. Take more or less of exercise in the open air every day. Allow yourself some innocent recreation. Eat moderately, slowly, and of just what you please. If any particular dish disagrees with you, however, never touch it or look at it. Do not imagine that you must live on rye bread or oat meal porridge; a reasonable quantity of nutritious food is essential to the mind as well as the body. Above all, banish all thoughts of the subject. If you have any treatises on dyspepsia, domestic medicines, etc., put them directly out of your reach. If you are constantly talking and thinking about dyspepsia, you will surely have it. Endeavor to forget that you have a stomach. Keep a clear conscience; live temperately, regularly, cleanly; be industrious, too, but avoid excess in that, as in all other things.

5680. Artificial Digestion. A London physician, Dr. Marcet, has announced a process by which natural digestion is imitated by artificial means, and solid food may thereby be prepared for invalids. Dr. Marcet takes 58 grains muriatic acid having a specific gravity of 1.1496; 15 grains of pepsin—the organic principle procured from the stomach of a pig or other animal. Diluted in a pint of water and added to a pound of raw meat, the whole is allowed to simmer over a water-bath at about the temperature of the body, 98° Fahr. When the meat is by this means sufficiently broken up, it is strained, and the acid neutralized by 81 grains of bicarbonate of soda. The product is of a most agreeable character, easily digested and vastly more nutritious than beef tea. Where pepsin cannot be obtained, the doctor has found strips of calves' stomach answer very well.

5681. Dick's Cure for Dyspepsia. Mix together ½ ounce bicarbonate of soda; 2 drachms aromatic spirits of ammonia; 6 drachms compound tincture of gentian; 6 drachms tincture of henbane; 2 drachms tincture of ginger; 3 drops creosote; ½ ounce ginger syrup, and 3 ounces water. A table-spoonful taken after each meal will cause a speedy cure.

5682. Dick's Dyspepsia Pills. Make the following ingredients into 40 pills: 2 scruples each compound extract of colocynth, and compound rhubarb pill (*see No.* 4923); 1 scruple blue mass (*see No.* 4919); 55 grains soap; 1 drachm extract henbane; 3 drops oil of cloves. Take 2 pills at bed-time.

5683. Spackman's Anti-Dyspeptic Pills. Make into a mass, 6 drachms 24 grains powdered aloes; 3 drachms 20 grains each gamboge, scammony, and compound extract of colocynth; 96 grains soap; 15 drops each oil of caraway and oil of anise; with 1 drachm water. Divide the mass into 16 parts, and make each part into 24 pills; 384 pills altogether. A dose consists of 3 pills.

5684. Absorbents are medicines administered to counteract acidity in the stomach or intestinal canal. In most cases, emetics and aperients are given previous to their being taken; they are carbonate of ammonia, in doses of from 5 grains to 1 scruple; liquor of ammonia, 10 to 20 drops; aromatic spirit of ammonia, 20 to 30 drops; lime water, 2 ounces to ½ pint; magnesia, calcined, 20 to 40 grains; carbonate of magnesia, ½ to 2 drachms; carbonate of potassa, 10 grains to ½ drachm; carbonate of soda, 10 grains to ½ drachm; soda water, ½ pint. (*See No.* 5678.)

5685. To Correct Acidity of the Stomach. The neutralizing mixture (*see No.* 5666) is very effectual in curing this disorder. Or, 10 grains of calumba, powdered, and 10 grains of magnesia, well mixed. Magnesia and a little finely powdered chalk will be of great service.

5686. Remedy for Acidity of Stomach. This is a common symptom of weak or disordered digestion, and should be treated with small doses 3 or 4 times daily of the carbonate or bicarbonate of potassa, soda, or ammonia; or of sal-volatile or ammonia water, to which some tonic bitter may be added. Diet should be light and nutritious, with as much out-door exercise as possible. The bowels should be kept regular by the occasional use of some mild aperient.

5687. Carminatives. Medicines that allay flatulency and spasmodic pains. Among the principal carminatives are aniseed, caraway-seed, cardamoms, cassia, cinnamon, ginger, peppermint and the peppers; including ardent spirits and most aromatic essences and tinctures.

5688. Flatulency in Children. It often arises from a mother's impure milk; when it is so she must take the neutralizing mixture (*see No.* 5666); and if not effectual, administer it to the infant. Also foment the stomach with warm brandy and water, to which add a little salt. Give also the carminative drops. (*See No.* 5689.)

5689. Carminative Drops, for expelling wind. Angelica, 2 ounces; lady's slipper, 1 ounce; sweet flag, ½ ounce; anise, 1 ounce; fennel seed, ½ ounce; catnip flowers, 1 ounce; mother-wort, 1 ounce; pleurisy root, 2 ounces. Infuse in a pint of spirits of wine for 3 or 4 days, often shaking, keeping it in a warm place; then add a pint of water and a table-spoonful of tincture of cayenne. Excellent in flatulency, colic, nervous affections, promoting perspiration and refreshing sleep.

5690. Heartburn. Anxiety and pain about the region of the stomach, generally attended by a sense of gnawing and heat; hence called heartburn. Faintness, nausea, and eructation of a thin, acidulous, watery liquid, especially in the morning, are common symptoms of this complaint. The usual causes of heartburn are excess in eating or drinking, the use of improper food, and sedentary habits. A good remedy is a tea-spoonful of carbonate of magnesia, or carbonate of soda, in a glass of peppermint or cinnamon water, to which a little powdered ginger may be added with advantage. This dose may be taken 2 or 3 times daily until the disease is removed. Articles of food that easily undergo fermentation should at the same time be avoided, and a dry diet had recourse to as much as possible. Soda-water, toast and water, and weak spirits and water, are the most suitable beverages in this complaint.

5691. To Cure Water-Brash. When there is a tendency to confined bowels, some aperient must be administered occasionally until proper dieting, &c., renders it unnecessary. Fluid magnesia, or the lenitive electuary (*see No.* 5154), will probably be all that is necessary. The diet must be carefully attended to in all cases; and as the disorder often arises from the use of innutritious or unwholesome food, the adoption of a more varied and generous diet, including a sufficient proportion of meat, is essential to the permanent success of any remedy.

5692. Treatment of Colic. Let it be remembered that colic may occur as the prelude to an inflammatory attack; and that if neglected or unskillfully treated, such tendency is very considerably increased. In the treatment of colic, very great advantage results from the external application of warmth; hot fomentations, bags of hot salt or bran, or flannel wrung out of turpentine, or mustard poultices, should be diligently employed. While these means are being used, a dose of laxative medicine should be administered; for, as in the great majority of cases of colic the pain depends on some obstruction in the bowels—very likely on the presence in them of some deleterious and indigestible food, &c. —it is of essential importance that free passage should be obtained as speedily as possible. A full dose (1½ ounces) castor oil, is a safe and good medicine for the purpose; to be repeated in 2 or 3 hours if there has been no action of the bowels. If the medical man has arrived meanwhile, he will very likely order some stronger medicine, as, if the oil has not acted, steps must be taken to clear the bowels as soon as possible. If the pain is very severe, a tea-spoonful of powdered ginger, or a little cayenne pepper may be added to the oil or taken after it. When free action of the bowels is obtained, the pain soon ceases. After such attacks great caution is requisite in the matter of diet for some time; only the plainest and most digestible food being taken.

5693. Treatment of Lead or Painters' Colic. In cases of colic arising from poisoning by lead, called lead colic, so often experienced by plumbers, painters, workers in shot towers, &c., the great object is to obtain free action of the bowels, as in common colic; and medical assistance should be obtained at once. Of course every care should be taken to prevent any further entrance of lead into the system. In order to obviate the occurrence of lead-poisoning in those who are of necessity exposed in a greater or less degree to its influence, frequent ablutions of the hands and surface of the body should be practiced; while sulphuric acid lemonade should be used as a beverage.

5694. Fainting Fits. If a person faints, let him be placed on his back until he comes to. Do nothing else. He has fainted because the heart has stopped beating. It will come to of itself as soon as nature desires it, and it will be easier to propel the blood in a horizontal direction, when lying down, than perpendicularly to the head, chest, and arms, when sitting up. And yet the very first effort of bystanders when a person is observed to have fainted, is to place him on a chair, or lift up his head. (*Hall*). If the patient be a female, place her on her back, with the head low, loosen all clothes about the neck and chest, sprinkle cold water on the face, and apply smelling salts to the nostrils. When the patient can swallow, give some cold water, with 20 or 30 drops of sal-volatile, or a little brandy.

5695. Fits. If a person falls in a fit, let him remain on the ground, provided his *face be pale;* for should it be fainting or temporary suspension of the heart's action, you may cause death by raising him upright, or by bleeding; but if the *face be red or dark colored*, raise him on his seat, throw cold water on his head immediately, and send for a surgeon, and get a vein opened, or fatal pressure on the brain may ensue.

5696. Cure for Cramps. Mix 2 drachms chloroform, 1 drachm oil of camphor, 6 drachms mucilage of gum-arabic, and 1½ grains acetate of morphia. Dose, 40 drops every 2 hours.

5697. Remedy for Dropsy and Liver Affections. Mix 8 ounces infusion of dandelion (taraxacum); ½ ounce extract of dandelion; 2 drachms carbonate of soda; 6 drachms tartrate of potassa; 8 drachms tincture of rhubarb; 1½ ounces tincture of henbane. Dose, a table-spoonful every 2 hours.

5698. Cure for Liver Complaint. Take ½ ounce each extract of taraxacum (dandelion) and tartrate of potassa; 45 grains carbonate of soda; ½ ounce sweet tincture of rhubarb, and 6 ounces spring water. Dose, a tea-spoonful 3 times a day.

5699. Remedy for Liver Complaint. Mix ½ ounce each fluid extract of rhubarb and of senna with 4 ounces water. Then add ½ ounce extract of taraxacum; 3 drachms acetate of potassa; ½ ounce compound tincture of gentian; and 1 drachm muriatic ether. Dose, a table-spoonful 3 times a day.

5700. Dandelion Pills. Take 30 grains extract of dandelion, and 6 grains calomel; make into 10 pills. 2 taken 3 times a day are a useful remedy for dropsy in the belly arising from disorder of the liver. (*See No.* 5697.)

5701. Infusion of Dandelion. Steep 2 ounces bruised dandelion root in 1 pint boiling water. After 24 hours strain. 2 table-spoonfuls 4 times a day is a remedy for dropsy. (*See No.* 5697.)

5702. Sick Headache. This usually proceeds from acidity and overloading the stomach. When it is not from improper eating, all that is necessary is to soak the feet in hot water for 15 minutes, drink some warm herb tea, retire to bed, and take a good sweat for about an hour. This will give relief. If the trouble arises from over-eating, relief may be obtained by taking an emetic. (*See No.* 5169.)

5703. Periodical Sick Headache. Those who are afflicted periodically with sick headache, accompanied with nausea and sometimes with vomiting, may obtain relief by soaking the feet in hot water, and using the emetic directed in No. 5169. This treatment should be followed by taking the lenitive electuary. (*See No.* 5154.)

5704. Nervous Headache may be relieved by using one of the evaporating lotions. (*See No.* 4843.) An application of the "Good Samaritan" is also very effectual. (*See No.* 4858.) Any of the remedies under the head of neuralgia are also recommended for severe attacks. (*See Nos.* 5544, *&c.*)

5705. To Relieve Nervous Headache. From 10 to 20 drops sal-volatile (aqua ammonia) in ½ wine-glass of water will frequently give relief; a dose of 10 drops, and repeated at intervals of 10 minutes, seldom fails.

5706. Remedy for Sick Headache.

It is stated that 2 tea-spoonfuls of finely powdered charcoal, drank in half a tumbler of water, will give immediate relief to the sick headache, when caused, as in most cases it is, by too much acid on the stomach. This remedy has been highly recommended. (*See also Antacids, No.* 5678.)

5707. Bisulphide of Carbon a Remedy for Headache. Dr. Kennion thus describes the mode of application of this remedy: A small quantity of the solution (about 2 drachms) is poured upon cotton-wool, with which a small wide-mouthed glass-stoppered bottle is half filled. This, of course, absorbs the fluid; and, when the remedy has to be used, the mouth of the bottle is to be applied closely (so that none of the volatile vapor may escape) to the temple, or behind the ear, or as near as possible to the seat of pain, and so held for from 3 to 5 minutes. After it has been applied for a minute or two, a sensation is felt as if several leeches were biting the part; and, after a lapse of a few minutes more, the smarting and pain become rather severe, but subside almost immediately after the removal of the bottle. The effect of this application is generally immediate. (*British Med. Journ.*)

5708. Simple Remedy for Piles. Take fresh white pine pitch in pills, from 12 to 20 a day, and sit in a tub of cold water 4 or 5 times a day, 30 minutes each time, for a month. A very obstinate case of piles was cured by this treatment.

5709. Internal Remedy for Piles. Pulverize in a mortar and mix thoroughly, 1 ounce each of cream of tartar, jalap, senna, flowers of sulphur, and golden seal, and ½ ounce saltpetre. Dose, a tea-spoonful 3 times a day.

5710. External Remedy for Piles. Boil some of the inner bark of white oak in water, and strain; evaporate to a thick extract. To ½ pint of this extract, add ½ pint of oil rendered from old, strong bacon. Simmer together till mixed, and let it cool. Apply with the finger inside the rectum every night and until cured.

5711. Persulphate of Iron for Piles. An ointment made of ½ drachm persulphate of iron, and 1 ounce simple salve, has been found especially beneficial in cases of ulcerated hemorrhoid. Dr. Geo. S. Cartwright describes a case of hemorrhoid in which there was an external tumor of the size of a large pea, protruding, at certain times, to the size of a walnut. He applied lead water freely to the part, with an application of this salve before the patient retired at night, and the effect was almost immediate, relieving the pain and cauterizing the part. The effect of this salve is permanent. The same physician occasionally uses the ointment with double the above proportion of the persulphate.

5712. Treatment for Irregular Menstruation, or Monthly Flow. Where the flow is absent, or irregular. The treatment of cases of this kind should embrace every possible means of improving the general health, particularly the enjoyment of pure air, and the use of the shower or hip-bath; moderate exercise, especially on horseback; with a wholesome nutritious diet. The medical treatment must not be trifled with, as it requires considerable watching; it should therefore be carried out under the eye of a skillful physician. When the slightest appearance of menstruation takes place, the patient should be kept as quiet as possible; and, in order to encourage the flow, recourse should be had to the use of the warm hip-bath; indeed, very frequently it will be found that a hot hip-bath, containing a handful of the flowers of mustard, used every night for the week preceding the regular time for the flow to appear, and accompanied by a good rubbing with a rough towel of the hips and lower part of the front of the body, will greatly assist in bringing on the flow.

5713. Treatment for Interrupted or Suppressed Menstruation. The same suggestions in the way of treatment apply as in No. 5712. When interruption has taken place suddenly, recourse should be had to the warm hip-bath, bed, and some warm drink, such as sherry and water, or a little brandy, or hot ginger water. When cessation for one or more periods has occurred, then it is specially important to favor, as much as possible, its restoration by attention to those particulars of general treatment already adverted to.

5714. Treatment of Excessive Menstruation. Those who are liable to this form of irregular menstruation should be careful in their diet, choosing a plain and nutritious one. They should attend to the function of the bowels, and maintain a horizontal posture from the time when the discharge commences till its cessation. In addition, if the discharge, besides being copious, is continuous, recurring over and over again, it is necessary to have recourse to powerful remedies. When the discharge is so profuse as rapidly to reduce the patient's strength, still more, if by it, as has happened sometimes, life be brought into peril, local means of arresting bleeding must also be adopted; foremost among these is the application of cold—cold cloths placed over the lower part of the body, and to the groins. Injections of cold water may further be employed if the nurse or relatives are skilled in the use of the injecting instrument, but not otherwise.

5715. Difficult or Painful Menstruation. The most common form of this complaint is ranged under the head of neuralgia, for the violent pain with which it is accompanied bears a close resemblance to neuralgic pains experienced in other parts of the body. In such, if the affection is of long standing, the nervous system generally has probably sympathized, and headache, with hysteria and many other distressing symptoms, accompany the menstrual disorder. Many cases of this nature are connected with marked constitutional derangement, more particularly with gout and rheumatism. For the relief to the extreme pain which accompanies the complaint, soothing remedies are rendered indispensable, and the most suitable medical ones will be prescribed by the medical attendant. In his absence, or conjoined to the medicines, the warm hip-bath may be tried, followed by the application of mustard poultices, or flannel wrung out of hot water and sprinkled with turpentine, over the lower part of the back. In the general treatment, the greatest attention must be paid to diet and regimen.

5716. Remedy for Suppressed Menstruation. Make into 12 pills, 12 grains sulphate of iron, 6 grains powdered aloes, and 12 grains white turpentine. Dose, 1 at bedtime. (*See No.* 5441.)

5717. Ashwell's Injection for Obstructed Menstruation. Mix 1 to 2 fluid drachms liquor of ammonia with 1 pint of milk; use thrice daily, commencing with the least quantity of ammonia.

5718. Injection for Obstructed Menstruation. Take 1 fluid drachm liquor of ammonia, 1 ounce mucilage, and 9 fluid ounces water; use in the same way as the last receipt.

5719. Pills for Suppressed Menstruation. Take dried sulphate of iron, 1 scruple; powdered aloes, 2 scruples; powdered cloves, 5 grains; Venice turpentine, sufficient to make a mass, and divide into 20 pills. One pill 3 times a day.

5720. To Relieve Vomiting During Pregnancy. Mix 2 ounces sweet tincture of rhubarb, and 1 ounce compound tincture of gentian. Dose, a tea-spoonful 3 times a day.

5721. To Cure Vomiting in Pregnancy. Mix 1 drachm carbonate of magnesia, ½ ounce tincture of colombo, 5½ ounces peppermint water. Take a table-spoonful 3 times a day.

5722. Citric Acid in After-pains. Dr. J. B. Chagnon recommends citric acid for the pains following labor, and declares that it has never failed in his hands. He gives 5 grains in 2 or 3 ounces of water every 5 hours. It acts as a nervine, and as a preventive of inflammation.

5723. Pills to Remove Obstructions in Females. Aloes and lobelia, 1 drachm each; black cohosh, gum myrrh, tansy, unicorn root, 1 ounce each; cayenne, ½ ounce. Mix, and form into pills with solution of gum. These pills remove female obstructions, and are good for headaches, lowness of spirits, nervousness, and sallowness of the skin.

5724. Female Regulating Pills. Aloes, red oxide of iron, white turpentine, 1 ounce each. Melt the turpentine, and strain; mix well; form into pills with mucilage. Take 2 or 3 per day.

5725. Alum Injection for Leucorrhœa. Compound solution of alum, 6 drachms; water, 1 quart. Mix, and use it lukewarm.

5726. Lead Injection for Leucorrhœa. Sugar of lead, 60 grains; water, 1 quart. Mix.

5727. Catechu Injection for Leucorrhœa. Catechu, 1 drachm; myrrh, 1 drachm; lime-water, 12 ounces. Mix, and dilute with water.

5728. Caustic Injection for Leucorrhœa. Nitrate of silver, 35 grains; water, 1 quart. Mix.

5729. Zinc Injection for Leucorrhœa. Sulphate of zinc, 40 grains; water, 1 quart. Mix.

5730. To Cure Sore Nipples. This painful affection of the breast, especially so during the period of nursing, may be cured as follows: Arrest the bleeding by a slight application of compound tincture of benzoin, carefully dry the parts with a soft muslin handkerchief; apply a solution of gutta-percha, so as to completely surround the nipple and cover all abrasions, giving it three or four coatings, allowing each to dry thoroughly before repeating the application. During the act of suction, a boxwood shield, with calf's teat, should be used, and in the course of a few days all will be well. The solution of gutta-percha is prepared by dissolving 1 drachm gutta percha in a bottle containing 3 drachms chloroform. The film rapidly formed by the evaporation of the chloroform is firm, elastic, and harmless, and, should it rub off, is very easily replaced. The almost painless nature of the treatment, the effectual protection from the contact of the air and irritation of the infant's mouth recommend it strongly to general use.

5731. Harland's Gonorrhœa Cure. Mix together 1½ ounces powdered cubebs; ½ ounce balsam copaiba; ½ ounce powdered gum-arabic; and 3 ounces cinnamon water. A table-spoonful of the mixture to be taken at intervals 8 times a day.

5732. Harland's Gonorrhœa Injection. Mix 2 scruples Armenian bole, and 10 grains sulphate of zinc, with 4 ounces water. Inject 3 or 4 times a day.

5733. Goddard's Gonorrhœa Mixture. Take 2 drachms oil of cubebs; ½ ounce balsam of copaiba; 1 ounce each syrup of tolu and syrup of poppy; 2 drachms *strong* liquor of potassa; 1 drachm oil of juniper; and 2½ ounces peppermint water. A table-spoonful 3 times a day.

5734. Goddard's Gonorrhœa Injection. Mix 3 drachms solution of iodide of iron with 4 ounces spring water. Apply with a syringe 3 times a day.

5735. Spackman's Copaiba Mixture. Mix together 2 drachms syrup of gum-arabic; ½ ounce balsam of copaiba; 24 drops oil of cubebs; 1 ounce syrup of balsam of tolu; ½ ounce each sweet spirits of nitre and compound tincture of opium; 20 drops tincture of opium; 3 drops oil of lavender, and 3 drachms compound spirits of lavender. Dose, a table-spoonful 3 times a day.

5736. Permanganate of Potassa in Gonorrhœa. Dr. John G. Rich has employed this remedy with great success. He begins the treatment with a purgative, and then uses as an injection, 3 times a day, 6 grains of permanganate of potassa dissolved in 1 ounce water.

5737. To Apply Caustic to the Urethra. A weak solution of nitrate of silver (2 or 3 grains in 1 ounce rose-water), may be used as an injection twice a day. Some prefer a stronger solution of 10 grains to the ounce, injected every 2 or 3 days. It may be also administered as an ointment of 10 to 20 grains to the ounce, smeared on a bougie and introduced into the urethra. This is perhaps better for severer cases of gonorrhœa; the injections answering the purpose for milder cases, and gleet.

5738. Ricord's Gonorrhœa Injection. Mix 15 grains each sulphate of zinc and acetate of lead, with 6½ ounces rose-water. Inject 3 times a day.

5739. Cure for Nocturnal Emissions. Mix 50 grains bromide of potassa with 25 grains each aromatic powder and white sugar. Make up into 12 powders, 1 to be taken 2 or 3 times a day.

5740. Remedy for Difficulty in Urinating. Mix together 1 scruple each oil of turpentine, extract of henbane, and soap. Make it into 12 pills, and administer 1 pill 3 times a day.

5741. To Relieve Spasm of the Bladder. To relieve the spasm, place the patient in a hot bath immediately, and keep him there, supplying fresh hot water when required, until he is relieved, or he becomes at all faint or fatigued. Then put him into a bed which has previously been well warmed, and keep hot cloths, hot salt, hot bran, or hot tins applied, to prevent a return of the pain if possible; and as the medical treatment is of great consequence, lose no time in summoning the medical man.

5742. Remedy for Disease of the Kidneys. Boil 1 ounce pareira brava in 3 pints of water until it is only 1 pint. Dose, a wine-glassful 3 times a day.

5743. Remedy for Incontinence of Urine. Put 4 drops tincture of aconite root in a tumbler of water. Dose, a tea-spoonful every hour until relieved.

5744. Remedy for Nocturnal Incontinence of Urine. Nocturnal incontinence of urine has been treated successfully by administering from 15 to 20 minims of tincture of belladonna 3 times daily.

5745. Remedy for Incontinence of Urine of Old People. The continued use of 1 to 6 drops tincture of iodine daily has proved a successful remedy.

5746. Remedy for Spermatorrhœa. Gelseminum, ½ grain; lupulin, 3 grains. To be taken each night on retiring. Gradually diminish the dose as the patient shows signs of improvement.

5747. Belladonna as a Remedy for Typhoid Fever. Dr. B. Kelly, of Dublin, has met with great success in the use of belladonna in typhoid fever. Within 24 hours after the first dose, he found delirium, &c., vanish, succeeded by calm, natural sleep, clearness of intellect, and complete repose of the system, accompanied by regular evacuations. Dr. Lewis S. Pilcher, of the U. S. Navy, reports equally successful results from the use of this drug. The amount and frequency of the dose will probably be understood by every physician, as the authorities above quoted do not specify these points.

5748. Remedy for Festering Wounds and Cancers. Professor Boettger recommends gun cotton, saturated with a solution of permanganate of potassa, put up in the form of a poultice, and held over an open wound by a bandage, as the best disinfectant for bad odors that can be conveniently applied. The strength of the solution of permanganate, best adapted for the purpose, is 1 part, by weight, of the dry permanganate, in 100 parts water. Ordinary cotton cannot be taken, as it readily decomposes, but gun cotton is permanent, and not liable to explosion when in a moist state.

5749. Treatment for Measles. In the treatment of the ordinary cases of measles occurring in children otherwise than delicate, little is necessary beyond attention to the temperature of the room, the amount of the bed-clothes, preventing the access of too strong a light, which affects the eyes, &c. Great care should be taken that draughts of cold air are avoided, lest they might prove the cause of increase in the chest complaint, which generally attends the attack; and, while the room is not overheated, it must not, for the same reason, be allowed to be cool. It must be remembered that in measles, as in all fevers accompanied by an eruption, the patient will require a more abundant supply of blankets, &c., before the eruption appears, than after it. Indeed, afterwards, he generally desires light clothing. The room should be well ventilated; all excrements and dirty linen immediately removed. Disinfectants should be used. The sense of heat and dryness of the body, sometimes most distressing to the patient, can be much alleviated by washing the surface with soap and tepid water; too great exposure being avoided by one part of the body being cleansed, dried, and covered, before the rest is exposed. The feeling of tension of the hands and feet can be relieved by rubbing these parts with some greasy matter, such as lard or simple salve. All sources of annoyance or irritation, all noises, should be avoided, and thus sleep is promoted, a condition which most materially affects the welfare of the patient, sleep lessening the fever and increasing the appetite. Food, light and nutritious, such as arrowroot, gruel, good beef-tea, milk, chicken, or veal broth, plain wine, jellies, &c., should be given at the usual hours. The quantity should be moderate, great care being taken that the digestion be not impaired by too large a quantity being taken. Should the patient be very weak, the food must be administered in small quantities at frequently repeated intervals. There is no stimulant so important as food. The prospect of recovery in all fevers is very greatly if not mainly dependent on the power of digesting and assimilating food possessed by the patient. The bowels should be moved by a mild laxative, such as the lenitive electuary (*see No.* 5154), effervescing magnesia, or castor oil; and, so as to produce a little perspiration, a small dose (for a child, a tea-spoonful), of mindererus spirit (*see No.* 5143), in a little water, may be given at intervals of 2 or 3 hours. If the rash is long in appearing, or shows a disposition to disappear, the development of the eruption may be secured by placing the child in a warm bath; if the child appears sunk and the pulse be feeble, a little warm wine and water may be administered. In ordinary cases, the early appearance of the eruption will be favored by administering a dose of sulphur (a small tea-spoonful for a child, in milk); and if there be much hoarseness, and croupy character of the breathing and cough, it will be expedient to apply a hot sponge over the throat. (*See No.* 5626.) With the appearance of the eruption, these symptoms usually decline. Measles not unfrequently terminate in an attack of bowel complaint; this may be slight, and if so, will not require any medicinal treatment; indeed, it is salutary, but, on the other hand, when severe, and occurring in a delicate child, prompt means for arresting it must be adopted (such as are mentioned under the head of Diarrhea. (*See No.* 5652, *&c.*) If there be often-repeated sickness, food of the very blandest nature, pounded raw meat (the fat and gristle being removed before pounding), beef-tea, uncooked white of egg diluted with water, barley water, &c., should be given in small quantities, and be very frequently repeated. Thirst, and the consequent restlessness, must be allayed by drinks. Large draughts should be prohibited, as they tend to impair the digestion, and sometimes cause diarrhea; small quantities, swallowed slowly, or ice to suck, are sufficient to allay thirst, and also prove grateful to the patient. The patient, however, must be allowed to take larger quantities of fluids than in health, as an increased quantity is required by the system during the existence of fever. Acid, or acid and bitter drinks are generally found to lessen thirst to a greater degree than mere water, and are, moreover, grateful to the patent. Lemonade with very little sugar, or raspberry vinegar and water, will be found useful. Stimulants are administered to support the strength of the patient. This they do in a great measure by promoting digestion, and by also directly increasing the force of the heart's action. The administration and quantity of stimulants given cannot be regulated by the condition of the patient. Medical advice is particularly necessary here. Various complications are apt to take place, so that, if possible, advice should be had early in the day. If no advice is at hand, the symptoms must be treated according to the directions given under the particular heads.

5750. Scarlet Fever. The preliminary treatment for this disease is very similar to that for measles. Give the patient a gentle cathartic, and keep very warm in bed until the eruption appears. (*See No.* 5749.) The after-treatment consists of administering a gargle every 15 minutes, when the patient is awake. Make a gargle of 2 table-spoonfuls each brewer's yeast and strained honey, mixed with 1 pint strong sage tea, and alternate it with the potassa gargle. (*See No.* 5064.) Keep the skin of the patient moist by washing all over, at least 3 times a day, with a solution of saleratus and water as hot as it can be borne; after each washing grease the patient all over thoroughly, with a piece of fat bacon. Great care must be taken to prevent the patient from catching cold in every stage of the disease, and the same cautions about ventilation, warmth, diet, &c., given under the head of measles, must also be observed in the treatment of scarlet fever. The patient must not be exposed to any great or sudden changes of temperature, even 3 weeks after convalescence, as a relapse might be the consequence.

5751. Preliminary Treatment of Scarlatina and Measles. The preliminary treatment is simple: from ½ grain calomel, for children, to 5 grains for adults, should be placed on the tongue and swallowed. About an hour after, the first dose of the ammonia (*see next receipt*) is to be given, and repeated every 3 or 4 hours, as long as the disorder takes a favorable course. If the disorder increases in violence, the medicine must be given every 2 hours, or every hour, or sometimes even more frequently, till the graver symptoms are subdued. This medicine has been found to possess similar powers over diphtheria.

5752. Treatment of Scarlatina and Measles. Dr. Witt states that sesquicarbonate of ammonia is an antidote to scarlatina and measles. The dose in these complaints varies from 3 to 10 grains, according to the age of the patient, given at longer or shorter intervals, according to the mildness or severity of the attack. The suitable dose dissolved in as small a quantity of cold water as will admit of its being swallowed with as many grains of loaf sugar, merely to make it palatable, is all that is required. Any admixture with other medicines, as salines, bark, &c., and all acidulous drinks, are to be avoided.

5753. Preventive of Scarlet Fever. Belladonna has been found to render persons unsusceptible to the fever, in places where it is raging. It is to be given in extract, $\frac{1}{10}$ grain morning and evening.

5754. Remedy for Dropsy in Scarlatina. Mix together 1½ drachms acetate of potassa; 6 grains extract of foxglove; 2 drachms vinegar of squill; 6 drachms syrup of ginger; and 2 ounces water. Dose, 1 tea-spoonful every 3 hours.

5755. Atlee's Scarlet Fever Remedy. ½ ounce each chlorate of potassa and hydrochloric acid, and ½ ounce spring water. Dose, 10 drops in a wine-glassful of cold water every 2 hours.

5756. Intermittent Fever Pills. Take 10 to 12 grains white oxide of arsenic; 1 drachm muriate of ammonia, and 12 grains gum opium. Make into 64 pills. Dose, 1 to be taken morning, noon and night, with or without fever.

5757. Intermittent Fever Mixture. Take 5 grains tannin, 16 grains sulphate of quinine, 1 ounce syrup of ginger, and ½ ounce cinnamon water. Take 1 tea-spoonful every hour, in the absence of the fever.

5758. Treatment of Small-Pox. Advice should always be obtained as soon as the earliest symptoms appear; often the only symptom understood by the parents or friends is the eruption. In the absence of advice, the simpler cases of small-pox, unattended by much eruption, scarcely require any further treatment than confinement of the patient to bed, adminstering at the commencement a dose of aperient medicine, such as effervescing magnesia (*see No.* 4805, *&c.,*) or castor oil, &c., and, until the eruption appears, of a few doses of mindererus spirit (*see No.* 5143), to promote perspiration. In the more severe cases there are individual symptoms of an unfavorable nature not unlikely to be developed, and these must be met by appropriate treatment. The imperfect filling of the pustules is generally accompanied by a low form of fever, requiring the use of stimulants, wine or brandy; these must, of course, be administered with great caution. In all stages, if the patient present a sunken look, and the pulse be feeble, the necessity for stimulants is indicated. By giving them with caution is meant that only just sufficient to keep up the vital powers should be given.

5759. To Prevent Pitting in Small-Pox. The following has been found very effectual: The application consists of a solution of india-rubber in chloroform, which is painted with a camel-hair pencil over the surface of the skin, where exposed, when the eruption has become fully developed. When the chloroform has evaporated, which it very readily does, there is left a thin elastic film of india-rubber over the face. This the patient feels to be rather comfortable, as it removes itching and all irritation; and, what is more important, pitting, once so common, is thoroughly prevented by the application. In making the solution, the india-rubber must be cut into small pieces, and chloroform added till it is dissolved. Gutta-percha has been tried, but has not answered, on account of its non-elasticity. Should any of the solution, from some cause, be torn off, apply the solution as before.

5760. Dr. George's Treatment to Prevent Pitting in Small-Pox. Dr. George recommends the following treatment: Firstly, from the commencement of the disease cover the whole body, face and all, with calamine, shaken through a common pepper-box, taking care that the powder does not remain in masses. The inflammation on each pustule is by these applications much lessened, a point of great consequence. Secondly, sprinkle about 1 ounce powdered camphor every 2 or 3 nights between the under sheet and blanket, the whole length of the body, putting more about the shoulders and neck. The relief obtained by this, few would credit until they had had experience. Thirdly, in the advanced stage of the disease, should hardened incrustations have formed, they may be removed, and without much pain too; for in one case every portion of the cuticle was removed from the whole face, forehead, and even eyelids, the calamine applied, and in a few days the cuticle was formed again without a blemish.

5761. Calamine. Native carbonate of zinc. It is prepared and purified for medicinal purposes by heating to redness, and pulverizing it, afterwards reducing it to an impalpable powder in the same manner as directed for prepared chalk. (*See No.* 1292.)

5762. To Remove Pitting and Old Pock-Marks. To remove pitting and old pock-marks, simple oil, pomade, or ointment, medicated with croton-oil, and of a strength just sufficient to raise a very slight pustular eruption, is probably the safest and most effective and convenient of all the preparations that are employed for the purpose. It has for some years been successfully employed in France and has there received medical approval. Dr. Cooley says he has seen it succeed to admiration, when every other method has failed. It should be applied at intervals extending over several weeks, as the feelings, experience, and convenience of the party concerned may indicate, due care and caution being observed the whole time.

5763. Treatment of an Attack of Apoplexy. Loosen the clothes, especially those about the neck and throat, and send *at once* for a physician. Meanwhile, remove the patient into a cool, well-ventilated room, raise the head above the level of the body, and apply cold to the head, either by means of rags dipped in water, never allowing them to become warm, or by ice in a bladder, &c. The diet will require great care when the patient is reviving. Only very small quantities of milk, beef-tea, &c., must be given until he is able to digest more. Supposing the patient to recover from the fit, great care will be afterwards required to prevent a second attack. Strong medicines, great excitement, or much mental occupation are to be avoided. The diet ought to be light, but nutritious; milk is useful, taken to the extent of 1½ or 2 pints in the day; and, as a rule, no spirits or wine should be allowed.

5764. Remedy for Shortness of Breath. Take spirits of ether, 1 ounce, and camphor, 12 grains. Make a solution, of which take a tea-spoonful during the paroxysm. This is usually found to afford instantaneous relief in difficult breathing, depending on internal disease and other causes, where the patient, from a quick and very laborious breathing, is obliged to be in an erect posture.

5765. To Relieve Shortness of Breath. Take ½ ounce powder of elecampane root, ½ ounce powder of liquorice, as much flower of brimstone and powder of aniseed, and 2 ounces sugar-candy powdered. Make all into pills, with a sufficient quantity of tar; take 4 large pills when going to rest. This is an incomparable medicine for asthma.

5766. Palpitation of the Heart. Soda water, either the usual carbonated water, or prepared from effervescing soda powders, frequently gives instant relief in an attack of palpitation of the heart.

5767. To Relieve Palpitation of the Heart. Take 40 drops tincture of digitalis (fox-glove); 20 drops tincture of aconite; 2 drachms tincture of henbane; 6 drachms camphor-water. Dose, a tea-spoonful 3 times a day.

5768. Biliousness. Persons subject to bilious attacks should be particularly careful to guard against excess in eating and drinking, and should especially avoid those articles of food which, from experience, they find to disagree with them. A mutton chop undercooked is an excellent article for the breakfast or lunch of a bilious patient; and mutton or beef, either broiled or roasted, so that the gravy be retained, is better for dinner than many articles apparently more delicate. Beer and porter should be particularly avoided, as well as puddings and most articles of pastry, as they are very indigestible. Hard cheese, butter, unripe fruit, and especially beans, peas, and nuts, are also objectionable. An attack of bile may frequently be prevented by the use of a saline purgative, and it may generally be removed by a blue pill, followed with a mild purgative.

5769. To Remove Tumors. To remove tumors, Dr. Simpson, of Edinburgh, introduces a hollow acupuncture needle, or very fine *trocar* (a surgical instrument in the form of a fine hollow needle) into their tissue, and injects a few drops of some irritant liquid, such as a solution of chloride of zinc, perchloride of iron, or creosote. The effect has been to destroy the vitality of the tumors so treated, and they have been separated. A similar plan has been adopted in Paris by M. Maisonneuve. He had slender stylets made of a paste composed of flour, water, and chloride of zinc. These are baked. A puncture is made in the tumor, the caustic stylet is inserted, broken off, and left. Several malignant tumors have been successfully treated in this manner, and in some cases a healthy granulating surface was left, after the separation of tumors which had been destroyed in this manner.

5770. Treatment of Rupture. Rupture is generally caused by a strain or an accident, and should be attended to by a surgeon as soon as possible. Meanwhile the patient must be laid upon a sofa or bed with his hips and legs slightly raised, so as to give him ease and to place the rupture in the most favorable position for being restored to its proper place. If the patient is faint, support him by giving wine and water, or sal-volatile, or a little broth, but do not over-stimulate him. In other respects he must be kept perfectly quiet.

5771. To Relieve Lockjaw. Let any one who has an attack of lockjaw take a small quantity of spirits of turpentine, warm it, and pour it on the wound, no matter where the wound is, or what its nature is, and relief will follow in less than 1 minute. Nothing better can be applied to a severe cut or bruise than cold turpentine; it will give certain relief almost instantly.

5772. Cure for Cancer. The use of clover tea is said to effect speedy and effectual cures of cancer, even in its most malignant form, and of long standing. The red clover is used; the tops are boiled in water, and the tea is used externally and internally. About a quart a day should be administered internally, and the tea should be used as a wash twice every day.

5773. Remedy for Scrofula. Put 1 ounce aqua-fortis in a bowl or saucer; drop in it 2 copper cents; when the effervescence ceases, add 2 ounces strong vinegar. The fluid will be of a dark green color. It should and will smart. If too severe, dilute it with a little rain-water. Apply it to the sore, morning and evening, by a soft brush or a rag. Before applying it, wash the sore with water. This receipt comes well recommended for curing old sores and other scrofulous eruptions.

5774. Anti-Scrofulous Mixture. Mix 30 drops tincture of bichloride of gold; 40 drops tincture of iodine; 1 fluid drachm tincture of gentian; 7 fluid drachms simple syrup, and 5 fluid ounces rose-water. Dose, a dessert-spoonful 3 or 4 times daily, in a wine-glassful of water, observing to shake well before pouring out the liquid.

5775. White Swelling. This is a very painful disease; it more frequently affects the knee than any other joint; sometimes the hip, ankle, and elbow. At first a severe pain is felt penetrating the joint, or only one particular part of the joint. The least motion aggravates the pain. It soon begins to swell considerably, and suppuration takes place. Matter is discharged from several openings or ulcers, the bones are affected; and if the disease is not arrested the life of the patient is endangered.

5776. Treatment of White Swelling. Attend to the stomach and bowels, giving an emetic and an aperient, if needed; to be followed by bitter tonics occasionally, giving the alterative syrup (*see No.* 5163), diluted when first taken; or a decoction of sarsaparilla, sassafras, guaiacum, queen's delight, unicorn root, cleavers, and prickly ash berries, of each 1 ounce. Simmer in a covered pan with 2 quarts water down to 3 pints. Sweeten. A dessert-spoonful 3 or 4 times a day. Steam the part with bitter herbs, and now and then give a vapor bath to the whole body. After steaming the affected part, rub the limb with the rheumatic liquid. (*See No.* 4884.)

5777. Beach's Cure for White Swelling. Oil of hemlock, oil of sassafras, gum camphor, tincture of opium, ½ ounce each, and a pint of spirits of wine. When dissolved and properly mixed, bathe the part with it frequently. Then apply an oatmeal and bran poultice, mixed with a little finely powdered charcoal, salt, and cayenne pepper. If the pain is great, sprinkle on the poultice ¼ ounce laudanum. Keep it on as long as possible, and then steam.

5778. To Relieve Sea-Sickness. Take camphorated spirit, sal-volatile, and Hoffman's anodyne, a few drops of each, mixed in a small quantity of water, or upon a small lump of sugar. This often relieves when other prescriptions fail.

5779. To Prevent Sea-Sickness. The neutralizing mixture (*see No.* 5666) is a good preventive. So is a tea-spoonful of bicarbonate of soda in ½ pint of water. Take an aperient before a voyage. One of the best means of counteracting the tendency to sea-sickness, is to keep a horizontal position. A little chloroform has lately been suggested as a good remedy. 5 to 10 drops on a piece of lump sugar.

5780. Treatment of Debility. This arises from a diseased action of the stomach; the occasional use of mild aperients, followed by bitters and tonics, is the best treatment. When, from a general laxity of the solids, and there are no symptoms of fever, nor a tendency of the blood to the head, a course of iron tonics will prove advantageous. Either of the following may be adopted for this purpose: Pure sulphate of iron, 1 drachm; extract of gentian and powdered ginger, of each 1½ drachms; beat together into a mass, and divide into 120 pills, 1 to be taken morning, noon, and night. Or: Sulphate of iron and powdered myrrh, of each 1 drachm; sulphate of quinine, ½ drachm; conserve of roses, sufficient to form a pill mass. Divide into 120 pills, administered as the last.

5781. Remedy for Sick Stomach and Vomiting. Mix 24 drops creosote, 1 drachm each white sugar and gum-arabic, with 3 ounces water. Administer a tea-spoonful every 2 hours, until vomiting ceases.

5782. Sunstroke. This is a sudden prostration due to long exposure to great heat, especially when much fatigued or exhausted. It commonly happens from undue exposure to the sun's rays in summer, but the same effects have been produced in a baker from great heat of the bake-room. It begins with pain in the head, or dizziness, quickly followed by loss of consciousness and complete prostration. Sometimes, however, the attack is as sudden as a stroke of apoplexy. The head is often burning hot, the face dark and swollen, the breathing labored and snoring, and the extremities cold.

5783. Treatment of Sunstroke. Take the patient at once to a cool and shady place, but don't carry him far to a house or hospital. Loosen the clothes thoroughly about his neck and waist. Lay him down with the head a little raised. Apply wet cloths to the head, and mustard or turpentine to the calves of the legs and the soles of the feet. Give a little weak whiskey and water if he can swallow. Meanwhile let some one go for the doctor. You cannot safely do more without his advice.

5784. Precautions Against Nightmare. Avoid all exciting causes, as too much abstruse thinking, late and heavy suppers, food difficult of digestion, cold feet, costiveness, and flatulence.

5785. To Prevent the Nightmare. To prevent the nightmare, mix together 10 grains carbonate of soda; 3 drachms compound tincture of cardamoms; 1 drachm simple syrup, and 1 ounce peppermint water. Repeat for several nights in succession; afterwards use for a few weeks the tonic aromatic mixture. (*See No.* 5124.) Also a little cayenne in scullcap tea will prevent an attack. Those who are habitually subject to nightmare should not sleep in a room alone, but have some person near them, to arouse them when attacked with it. A person is most liable to nightmare when sleeping on his back; in fact, it rarely occurs in any other posture. Those subject to it should therefore avoid sleeping in a bed which is hollow in the centre, as this induces the sleeper to lay on his back. The bed should be level and not too soft, and the pillow moderate in thickness, so that the head is not raised too high.

5786. To Restrain Perspiration. Spring water, 2 ounces; diluted sulphuric acid, 40 drops; compound spirits of lavender, 2 drachms; take a table-spoonful twice a day.

5787. Remedy for Night Sweats of Consumption. M. Guyot recommends as particularly useful, in the sweats of consumption, the phosphate of lime in quantities of from ½ to 1½ drachms in the day. In a small proportion of cases it may be inert; but in the majority it will diminish or quite remove the trouble.

5788. Treatment for Night-Sweats in Consumption. Powdered borax, 5½ drachms; washed sulphur, 1 ounce; subnitrate of bismuth, 1½ drachms; divide into 40 powders, 1 to be given every 2 hours (12 a day). 4 to 5 days of treatment will suspend or diminish this troublesome and exhausting symptom, and give much relief to the patient.

5789. To Relieve Night-Sweats. Dissolve 15 grains sulphate of quinine in ½ ounce essence of tansy, ¼ ounce alcohol, ¼ ounce water, and 30 drops muriatic acid. A tea-spoonful taken 2 or 3 times during the day and at bed-time. In connection with this remedy, cold sage tea is recommended to be used freely as a drink.

5790. Squinting. It is well known that in infancy there is not unfrequently a tendency to squint; this often passes away as the child increases in age; but it sometimes becomes quite a fixed habit, requiring a surgical operation for its permanent cure. A means of rendering this operation unnecessary by curing the tendency in early life has been suggested, which is worthy of trial. A pair of spectacles is procured without any glasses in them. One of the orifices opposite the eye that squints is to be filled with thin horn or with ground glass, and in the centre of the horn or glass is to be made a small hole. It is obvious that to see with the squinting eye it is necessary for the child to look directly through the orifice in the centre. He will thus acquire the habit of looking forward towards an object, instead of looking to the right or left hand of it. It is not at all improbable that the slight squint, which in infancy is apparently only a habit, may be remedied by this means.

5791. Treatment of Styes. A stye is a small boil which projects from the eyelid, much inflamed, and very painful. The application of ice to the part will sometimes check it in the beginning. Apply a poultice of linseed meal, or bread and milk, and take at the same time an aperient. If the stye is ripe, puncture it, and then apply spermaceti ointment.

5792. To Treat a Black Eye. This is usually caused by a blow. If attended with inflammation and pain, wash the eye often with very warm water, in which is dissolved a little carbonate of soda; or with equal parts of tincture of opium and water. If the pain be acute, foment with a decotion of stramonium leaves, simmered in spirits. Wash the eye, and bind on the leaves; often repeat. Perhaps the best application is a poultice of slippery elm bark. Mix with milk and put it on warm.

5793. To Cure a Black Eye. To remove the discoloration of the eye, bind on a poultice made of the root of Solomon's seal. Culpepper says it is available for bruises, falls, or blows, to dispel the congealed blood, and to take away the pains, and the black and blue marks that remain after the hurt. The root may be washed, the dark-colored skin carefully cut off, then scraped like horseradish, and applied direct to the eye in the way of a poultice, cold. A tingling sensation is the consequence; when this sensation ceases, another fresh application should be made, and repeated until the whole discoloration is absorbed. It is often found sufficient to apply the scraped root at bed-time to the closed eye, when the blackness has disappeared by the morning. Or: Moisten with tepid water, and then with a piece of lint apply pure extract of lead; continue to keep the lint wet with the extract for a couple of hours. Leeches ought not to be used. A lotion often used by surgeons with advantage is prepared thus: Take nitrate of potassa and sal-ammonia, each 1 part; water, 48 parts; vinegar, 4 parts. The part bruised to be kept wet with this by means of a bandage.

5794. To Remove Dirt or Foreign Particles from the Eye. Take a hog's bristle, double so as to form a loop. Lift the eyelid and gently insert the loop up over the ball, which will occasion no disagreeable feeling. Now close the lid down upon the bristle, which may now be withdrawn. The dirt will surely be upon the bristle. M. Renard, in the case of small movable bodies which become entangled beneath the upper eyelid, recommends the following simple process: Take hold of the upper eyelid near its angles, with the forefinger and thumb of each hand, draw it gently forwards and as low down as possible over the lower eyelid, and retain it in this position for about a minute, taking care to prevent the tears from flowing out. When, at the end of this time, you allow the eyelid to resume its place, a flood of tears washes out the foreign body, which will be found adhering to, or near to, the lower eyelid. If lime gets into the eyes, a few drops of vinegar and water will dissolve and remove it. Almond or olive oil will do away with any hot fluid that may reach the eye.

5795. To Expel Insects, Dirt, &c., from the Eye. The first thing to be done when a mote or spark gets into your eye, is to pull down the lower part of the eyelid, and with a handkerchief in your hand blow your nose violently at the same moment. This will frequently expel the mote without further trouble. A mote will, in many cases, come out of itself, by immediately holding your eye wide open in a cup or glass filled to the brim with clear cold water.

5796. To Extract Particles of Iron or Steel from the Eye. A particle of iron or steel may be extracted from the eye by holding near it a powerful magnet.

5797. Eye-Waters. Eye-waters should be perfectly clear, and free from any floating matter, however trifling. To secure this, it is in general necessary either to filter them through bibulous paper, or a piece of clean, fine muslin, or to carefully decant them after sufficient repose to allow the impurities to subside. When pure distilled water is used in their preparation, only some of them will require filtering. In using eye-waters, a little of the liquid should be poured into a clean cup, gallipot, or glass, or into the clean palm of the left hand, when the eye should be thoroughly wetted with it, either by means of a small piece of clean sponge or soft white rag, or the clean tips of the fingers of the right hand. In all cases it is advisable to bathe or wash the eyes in tepid water, and to wipe them dry, before the application of the eye-water; and, in most cases, this is absolutely necessary to insure benefit from their use. In the preparation of eye-waters, substances of crystalline formation are better when used dry, that is, deprived of their water of crystallization. (*See No.* 2065.)

5798. Astringent Eye-water. Take of sulphate of zinc, 20 grains; distilled water, ½ pint; dissolve. An excellent astringent eye-water, in chronic as well as ordinary ophthalmia, as soon as the inflammatory symptoms subside; also in weak, lax, watery, irritable eyes, &c. If there be much pain and irritability, 5 or 6 grains of acetate of morphia, or 2 fluid drachms of wine of opium, may be added.

5799. Eye-water for Weak Eyes. Take ½ ounce rock salt and 1 ounce of *dry* sulphate of zinc; simmer in a perfectly clean covered porcelain vessel with 3 pints of water until all are dissolved; strain through thick muslin, add 1 ounce of rose-water; bottle and cork it tight. To use it, mix 1 tea-spoonful of rain-water, with 1 of eye-water, and bathe the eyes, if weak, frequently. If it smarts too much, add more water; if not enough, make it a little stronger by adding more eye-water. This is an admirable wash for weak eyes. It cannot be excelled.

5800. Wash for Inflamed Eyes. Take 10 drops extract of lead (the liquor of acetate of lead); distilled vinegar, 2 drachms; distilled water, 4 ounces. This is an excellent wash for inflamed eyes.

5801. Anodyne Eye-water. Solution of acetate of ammonia, 2 ounces; distilled water, hot, 6 ounces; soft extract of opium, 10 grains. Dissolve the opium in the hot water, strain through fine muslin, and add the solution of the acetate of ammonia. This application frequently affords great relief from the pain and irritation accompanying inflammation.

5802. Eye-water for Specks on the Eye. Oxymuriate of mercury, ½ grain; best rose-water, 4 ounces. This solution is of much use in removing the indolent inflammation and the white specks which an acute inflammation of the eyes frequently leaves after it.

5803. Bates' Eye-water. Dissolve in 4 fluid ounces boiling water, 15 grains *dry* sulphate of copper (*see No.* 5797), and 4 grains camphor. When cold, add water to make it 4 pints, and filter. Good in purulent ophthalmia.

5804. Goulard's Eye-water. Solution diacetate of lead, 10 drops; rose or elder-flower water, 6 fluid ounces. Mix. Good in inflammatory stage of ophthalmia.

5805. Wash for Removing Particles of Iron or Zinc from the Eye. Muriatic acid, 20 drops; mucilage, 1 drachm; mix with 2 fluid ounces rose-water.

5806. To Allay Temporary Irritation or Weakness in the Eye. Temporary inflammation, produced by cold or external causes, is rapidly allayed by frequently bathing the eye with lukewarm milk and water, or rose-water; applied either with a linen rag or by means of an eye-glass. A poultice of tea-leaves (the wet leaves left in the tea-pot) is also an excellent remedy. Probably the best remedy of all is to put a table-spoonful of salt in a basin of water (say ½ gallon), immerse the face in this twice a day, opening the eyes under the water, and using fresh salt and water every day. The eyes should under no circumstance be rubbed, as that will increase the irritation.

5807. Atropine Paper. Green tissue paper imbued with a solution of sulphate of atropia, so that a piece one-fifth of an inch square contains as much as a drop of a solution 2 grains to 1 ounce of water. The paper is hung up and turned about while drying. A piece of the size named will dilate the pupil if placed on the sclerotic, and the lids closed over it and tied with a handkerchief.

5808. Belladonna Mixture for Cataract. Triturate together 1 drachm each extract of belladonna and glycerine. Used for dilating the pupil of the eye in cataract, by anointing the eyebrow and temple.

5809. Taylor's Remedy for Deafness. Digest 2 ounces bruised garlic in 1 pound oil of almonds for a week, and strain. A drop poured into the ear is effective in temporary deafness.

5810. Treatment of Earache. M. Emile Duval says that he has, in person, found relief in severe earache, after other means had been tried in vain, from the use of a mixture of equal parts of chloroform and laudanum; a little being introduced on a piece of cotton. The first effect produced is a sensation of cold; then there is numbness, followed by scarcely perceptible pain and refreshing sleep.

5811. Cure for the Earache. Take a small piece of cotton batting or cotton wool, make a depression in the centre with the finger, and fill it up with as much ground pepper as will rest on a five-cent piece; gather it into a ball and tie it up; dip the ball into sweet oil and insert it in the ear, covering the latter with cotton wool, and use a bandage or cap to retain it in its place. Almost instant relief will be experienced, and the application is so gentle that an infant will not be injured by it, but experience relief, as well as adults. 1 part laudanum and 6 parts sweet oil dropped in the ear is also very effectual.

5812. Simple Cure for Earache. Take a common tobacco-pipe, place a wad of cotton in the bowl, drop upon it 8 or 10 drops of chloroform, and cover with another wad of cotton; place the stem to the affected ear, then blow into the bowl, and in many cases the pain will cease almost immediately.

5813. Remedy for Inflammation of the Ear. Swelling and redness, attended with throbbing, indicates it. If caused by accumulation of wax, syringe the ear forcibly with tepid water. If by cold, a poultice of warm hops, soaking the feet. If the pain is great, 1 drop laudanum and 2 drops sweet oil of almonds dropped into the ear 3 times a day, or juice of onions and laudanum. A slice of onion, toasted and tied on hot outside the ear, is a good remedy for earache in children, and often effective with adults. If very severe, a mustard poultice can be held behind the ear. If the stomach is out of order use an emetic. If no relief comes, call a physician.

5814. Remedy for Temporary Deafness. If deaf from hardened wax in the ear, a mixture of sassafras oil, 10 drops; glycerine, 1 fluid drachm; olive oil, ½ fluid ounce, mixed, may be dropped into the ear every day. If deaf from other causes, go to the physician.

5815. Cure for Temporary Deafness. Inject warm water into the ear by means of a proper syringe, the head being placed with that side upwards during the operation.

5816. To Destroy Insects in the Ear. Insects may be destroyed by pouring a spoonful of warm olive oil, or camphorated oil, into the ear over night, retaining it there until the next morning by means of a piece of cotton wool, when it may be washed out with a little mild soap and warm water.

5817. To Cure Habitual Drunkenness. The following singular means of curing habitual drunkenness is employed by Dr. Schreiber, a Russian physician: It consists in confining the drunkard in a room, and in furnishing him at discretion with his favorite spirit diluted with ⅔ of water; as much wine, beer, and coffee as he desires, but containing ⅓ of spirit; all the food—the bread, meat, and the vegetables steeped in spirit and water. The poor patient is continually drunk. On the fifth day of this treatment he has an extreme disgust for spirit; he earnestly requests other diet; but his desire must not be yielded to, until he no longer desires to eat or drink; he is then certainly cured of his love of drink. He acquires such a disgust for brandy, or other spirits, that he is ready to vomit at the very sight of it.

5818. Tonic After Drinking to Excess. Mix together 5 grains sulphate of quinine; 10 drops aromatic sulphuric acid; ½ ounce compound tincture of gentian; 2 drachms compound tincture of cardamoms; 1½ ounces ginger syrup; and 2 ounces water. A table-spoonful administered 3 times a day will remove the prostrating effects of drinking to excess.

5819. Remedy for Chafing. Stout persons suffer greatly, especially in warm weather, from chafing. We know of nothing better than a wash of alum dissolved in water, and applied with a linen or cotton rag.

5820. Lotion for Bed-Sores. To 1 table-spoonful of powdered alum put 1 teacupful of whiskey and bathe the sore part several times a day.

5821. To Relieve Irritation in Bed-Sores. Apply to the sores the white of an egg, well beaten, and mixed with spirits of wine.

5822. To Prevent and Cure Chapped Hands. Wash the hands with fine soap; and before removing the soap, scrub the hands with a table-spoonful of Indian meal, rinsing thoroughly with soft tepid water, using a little meal each time except the last; wipe the hands perfectly dry; then rinse them in a very little water containing a tea-spoonful of pure glycerine, rubbing the hands together until the water has evaporated. This is an excellent remedy, but the glycerine must be pure, or it will irritate instead of healing.

5823. Treatment of the Nails. The nails should be kept clean by the daily use of the nail-brush and soap and water. After wiping the hands, but while they are still soft from the action of the water, gently push back the skin which is apt to grow over the nails, which will not only keep them neatly rounded, but will prevent the skin cracking around their roots (hang-nails), and becoming sore. The points of the nails should be pared at least once a week; biting them should be avoided.

5824. To Remove Warts. A daily application of either of the three following remedies is effective in dispersing warts: Touch the wart with a little nitrate of silver (lunar caustic); or with nitric acid or aromatic vinegar. The lunar caustic produces a black, and the nitric acid a yellow stain, which passes off in a short time; the vinegar scarcely discolors the skin. Sparks of frictional electricity, repeated daily, by applying the warts to the conductor of an electrical machine, have been also successfully employed as a cure for these troublesome and unsightly excrescences.

5825. Wart or Corn Powder. Ivy-leaves dried and ground to fine powder. A popular and useful remedy for warts and soft corns. The part having been moistened with strong vinegar, a pinch of the powder is sprinkled on it, and then bound on with a strip of rag. This is sometimes called *cosmetic vegetable caustic.* A mixture of equal parts of savine and verdigris also make an efficacious wart powder.

5826. To Remove Moles. Croton oil, under the form of pomade or ointment, and potassio-tartrate of antimony (tartar emetic), under the form of paste or plaster, have each recently been successfully employed for the removal of ordinary moles and birth-marks. The following is the mode of using the latter adopted by an eminent French surgeon: Take tartar emetic in impalpable powder, 15 grains; soap plaster, 1 drachm; and beat them to a paste. Apply this paste to nearly a line in thickness (not more), and cover the whole with strips of gummed paper. In 4 or 5 days eruption or suppuration will set in, and, in a few days after, leave, in place of the birth-mark, only a very slight scar. Croton oil ointment effects the same, but less completely unless repeated, by producing a pustular eruption, which, however, does not permanently mark the skin. (*See No.* 5762.)

5827. Ingrowing Toe Nails. This most painful of the diseases of the nails is caused by the improper manner of cutting the nail (generally of the great toe), and then wearing a short, badly-made shoe. The nail beginning to grow too long, and rather wide at the corners, is trimmed around the corner, which gives temporary relief. But it then begins to grow wider in the side where it was cut off; and, as the shoe presses the flesh against the corner, the nail cuts more and more into the raw flesh, which becomes excessively tender and irritable. If this state continue long the toe becomes more and more painful and ulcerated, and proud-flesh sprouts up from the sorest points. Walking greatly increases the suffering, till positive rest becomes indispensable.

5828. Treatment of Ingrowing Toe Nails. Begin the effort at cure by simple application to the tender part of a small quantity of perchloride of iron. It is found in drug stores in a fluid form, though sometimes in powder. There is immediately a moderate sensation of pain, constriction or burning. In a few minutes the tender surface is felt to be dried up, tanned or mummified, and it ceases to be painful. The patient, who before could not put his foot to the floor, now finds that he can walk upon it without pain. By permitting the hardened, wood-like flesh to remain for 2 or 3 weeks, it can be easily removed by soaking the foot in warm water. A new and healthy structure is found firm and solid, below. If thereafter the nails be no more cut around the corners or sides, but always curved in across the front end, they will in future grow only forwards; and by wearing a shoe of reasonably good size and shape, all further trouble will be avoided.

5829. To Prevent the Nail Growing into the Toe. If the nail of your toe be hard, and apt to grow round, and into the corners of your toe, take a piece of broken glass and scrape the top very thin; do this whenever you cut your nails, and, by constant use, it makes the corners fly up and grow flat, so that it is impossible they should give you any pain. Do not fail to try this.

5830. Remedy for Blistered Feet from Long Walking. Rub the feet, at going to bed, with spirits, mixed with tallow dropped from a lighted candle into the palm of the hand.

5831. Method of Preventing Cold Feet at Bed-time. Draw off your stockings just before undressing, and rub your ankles and feet well with your hand, as hard as you can bear the pressure, for 5 or 10 minutes, and you will never have to complain of cold feet in bed. It is hardly conceivable what a pleasurable glow this diffuses. Frequent washing of the feet, and rubbing them thoroughly dry with a linen cloth or flannel, is very useful.

5832. Chilblain. This is an inflammatory swelling, of a purple or lead color, produced by the action of cold. Children, especially those of a scrofulous habit, and elderly persons, are generally most liable to chilblains. The common cause is holding the hands or feet to the fire, after exposure to cold. The sudden change of temperature partially destroys the vitality, and prevents the proper flow of blood through the part. As chilblain is only another name for a languid circulation in the part affected, indicated by a congested skin, or a low form of inflammation, the value of most of the following receipts will be apparent when it is noticed that they are all calculated to act as stimulants of the blood-vessels, and thus promote the motion of the partially stagnant blood which gives rise to the heat and itching that are so distressing. (*See No.* 4883.)

5833. Remedy for Broken Chilblains. Mix together 4 fluid ounces collodion, 1½ fluid ounces Venice turpentine, and 1 fluid ounce castor oil.

5834. Zinc Wash for Chilblains. Dissolve 1 ounce sulphate of zinc in 1 pint water. Apply several times a day.

5835. Chilblain Lotion. Dissolve 1 ounce muriate of ammonia in ½ pint cider vinegar, and apply frequently. ½ pint alcohol may be added to this lotion with good effects.

5836. Petroleum Liniment for Chilblains. Nothing appears of such uniform utility for allaying the inflammatory irritation, as the ordinary petroleum or kerosene oil.

5837. To Cure Chilblains. M. W. E. Schaller says that the fluid concentrated chloride of iron is an unfailing remedy for chilblains, its application to them for a single day effecting a cure. It may also be used with advantage for frost-bites.

5838. Remedy for Severe Chilblains. From 10 to 60 grains nitrate of silver dissolved in 1 fluid ounce water has been sometimes found useful after other applications had appeared of no benefit. Tincture of cantharides, to stimulate almost to blistering, has also been used in the more intractable forms of the disease. The tincture of capsicum has been presented as a specific in this disease.

5839. Chilblain Balm. Boil together 10 fluid ounces olive oil, 2 fluid ounces Venice turpentine, and 1 ounce yellow wax; strain, and while still warm add, constantly stirring, 2½ drachms balsam of Peru, and 9 grains camphor.

Another formula for making this balm adds ½ ounce alkanet root, but employs ½ drachm less of the balsam of Peru. This is applied by being spread on a soft cloth and laid on the part affected.

5840. Chilblain Liniment. Mix together 1 fluid ounce rectified oil of turpentine, 15 drops sulphuric acid, and 2 ounces olive oil. This, rubbed gently on the chilblains twice a day, is generally very effective.

5841. To Cure Chilblains. The following remedy was published by order of the Wirtemberg government. Mutton tallow and lard, of each ¾ pound avoirdupois; melt in an iron vessel and add hydrated oxide of iron, 2 ounces; stirring continually with an iron spoon, until the mass is of a uniform black color; then let it cool, and add Venice turpentine, 2 ounces; and Armenian bole, 1 ounce; oil of bergamot, 1 drachm; rub up the bole with a little olive oil before putting it in. Apply several times daily by putting it upon lint or linen. It heals the worst cases in a few days.

5842. Russian Remedy for Chilblains. Slices of the rind of fully-ripe cucumbers, dried with the soft parts attached. Previous to use they are softened by soaking them in warm water, and are then bound on the sore parts with the inner side next them, and left on all night. This treatment is said to be adopted for both broken and unbroken chilblains.

5843. Remedy for Itching Feet from Frost-bites. Take hydrochloric acid, 1 ounce; rain water, 7 ounces; wash the feet with it 2 or 3 times daily, or wet the socks with the preparation until relieved.

5844. To Cure Slight Frost-bites. The remedy for this is long-continued friction with the hands or cold flannel, avoiding the fire or even a heated apartment.

5845. To Correct an Offensive Smell in the Feet. Bathe them in a weak solution of permanganate of potassa; 1 scruple of he salt to 8 ounces of water. (*See No.* 1701.)

5846. Powder for Absorbing Excessive Perspiration of the Feet. Mix together 7 ounces carbonate of magnesia, 2 ounces powdered calcined alum, 7 ounces orris root, and ½ drachm powdered cloves.

5847. Corns. Corns are entirely owing to continued pressure, such as wearing small boots or shoes. At first they are the production of the outer skin only, but by gradually thickening they at length come to be connected with the true skin beneath, and even with the subjacent muscles. (*See Nos.* 5079 *and* 5060.)

5848. To Prevent Corns. Prevention is better than cure. Wear woolen stockings, and see that there is no local and permanent pressure on any part of the foot.

5849. To Cure Corns. If a cure be requisite, soak the corn for ½ hour in a solution of soda, and pare as close as possible; then apply a plaster of the following ingredients: Take of purified ammonia and yellow wax, of each 2 ounces; and acetate of copper, 6 drachms. Melt the first two ingredients together, and, after removing them from the fire, add the acetate of copper just before they grow cold. Spread this ointment on a piece of soft leather or on linen, and apply it to the corn, removing it in two weeks.

5850. To Cure Soft Corns. The soft corn occurs between the toes, and is produced in the same manner as the common corn; but in consequence of the moisture existing in this situation, the thickened scarf-skin becomes saturated, and remains permanently soft. The soft corn is best relieved by cutting away the thick skin with a pair of scissors, avoiding to wound the flesh; then touching it with a drop of Friar's balsam, and wearing habitually a piece of cotton wool between the toes, changing the cotton daily. Tincture of arnica, applied on a piece of cotton wool, is also said to be an excellent remedy.

5851. To Cure Soft Corns. Dip a piece of linen rag in turpentine and wrap round the toe on which the corn is situated, night and morning. The relief will be almost immediate, and in a few days the corn will disappear.

5852. To Relieve Hard Corns. Bind them up at night with arnica, to relieve the pain. During the day, occasionally moisten the stocking over the corn with arnica, if the shoe is not large enough to allow the corn being bound up with a piece of linen rag.

5853. Remedy for Corns. Soak the feet well in warm water, then with a sharp instrument pare off as much of the corn as can be done without pain, and bind up the part affected, with a piece of linen or muslin thoroughly saturated with sperm oil, or, what is better, the oil which floats upon the surface of the pickle of herring or mackerel. After 3 or 4 days the dressing may be removed, and the remaining dead cuticle removed by scraping, when the new skin will be found of a soft and healthy texture and less liable to the formation of a new corn than before.

5854. To Relieve Corns. Take a lemon, cut off a small piece, then nick it so as to let in the toe with the corn, tie this on at night, so that it cannot move, and in the morning you will find that, with a blunt knife, you may remove a considerable portion of the corn. Make two or three applications, and great relief will be the result.

5855. Remedy for Corns. The pain occasioned by corns may be greatly alleviated by the following preparation: Into a 1-ounce phial put 2 drachms of muriatic acid and 6 drachms of rose-water. With this mixture wet the corns night and morning for 3 days. Soak the feet every evening in warm water without soap. Put one-third of the acid into the water, and, with a little picking, the corn will be dissolved.

5856. Liquid Solvent for Corns; Corn Solvent. A saturated solution of salt of tartar or pearlash. It is commonly obtained by exposing the article, contained in a jar or wide-mouthed bottle, in a damp place, until it forms an oil-like liquid.

5857. To Cure Bunions. A bunion is a swelling on the ball of the great toe, and is the result of pressure and irritation by friction. The treatment for corns applies also to bunions; but, in consequence of the greater extension of the disease, the cure is more tedious. When a bunion is forming it may be stopped by poulticing and carefully opening it with a lancet.

5858. To Cure a Corn on the Sole of the Foot. A corn on the sole of the foot is usually difficult to cure, as the weight of the body causes a constant pressure on it. The application of an ordinary corn-plaster, with a hole in the centre, will relieve the pressure from the corn, but it causes an inequality under the foot, which is not only uncomfortable, but likely to produce other corns. The following method never fails: Cut a piece of stout cardboard (or thin binders' board) to fit inside the sole of the boot. This should be large enough in every way to prevent it shifting under the foot in walking. Next cut a round hole in this inner sole, exactly where the corn rests, the hole being rather larger than the corn. This arrangement relieves the corn from pressure and allows of its rapid cure, at the same time affording instant relief and freedom in walking.

5859. To Cure a Disagreeable Breath. This most disagreeable infliction may be alleviated or cured by one or other of the following remedies, provided that the teeth do not require a dentist's assistance. Chlorine water, as supplied by a good chemist, a tablespoonful to half a tumbler of water, to be used as a wash and gargle for the mouth; no harm will be done if a few drops are accidentally swallowed in so doing. Charcoal in tea-spoonful doses of the powder, or as charcoal biscuits, or the use of prepared chalk as a tooth-powder. A frequent cause of foul breath is a torpidity of some of the excretory organs, such as the skin, kidneys, bowels, liver, lungs. When these cease performing their functions one of the others will be called upon to perform an extra office. In this way, when the bowels or skin become affected, the lungs, being an excretory organ, will be called upon to throw off an additional waste from the system. If so, the breath becomes tainted. Should the foul breath be depending upon the stomach, it must be corrected by some skillful physician.

5860. Remedy for Bad Breath. Take of dry hypochlorite of lime, 3 drachms; distilled water, 2 ounces troy. Triturate the hypochlorite of lime in a glass pestle and mortar; when the hypochlorite has been thoroughly pulverized add a portion of the distilled water; allow the mixture to rest until the liquid has become transparent; then decant; add a second portion of water, triturate and allow to rest, again decant; this process is repeated a third time. The three liquids which have been decanted are then mixed, and 2 troy ounces of 85 per cent. alcohol, and 4 drops oil of roses or some other essential oil are added. The solution thus prepared may be employed to remove the fetid odor which is given off by the gums—an odor often due to the diseased condition of the tissues. To employ it, ½ tea-spoonful is poured into a tumblerful of water, and the gums are washed with the mixture, employing for the purpose a sponge-brush. The same preparation may be employed to remove the odor of tobacco, rinsing the mouth several times with water to which has been added a tea-spoonful of the liquid. Inasmuch as the odor of the essential oil is gradually diminished in time, said diminution taking place at the expense of the chlorine of the hypochlorite, it is suggested that this inconvenience may be obviated by preparing the solution with water and the hypochlorite of lime, and keeping it in one bottle, while the aromatic alcoholic solution (prepared of 2 ounces of 85 per cent. alcohol and 4 drops of essential oil) is preserved in another, both being well stoppered. When it is desired to use the liquids, a half tea-spoonful of each of the solutions is poured into a glass of water, which is then employed as described above.

5861. Remedy for Bad Breath. Take 5 to 10 drops hydrochloric acid in half a tumbler of spring water, a little lemon juice, and loaf sugar rubbed on lemon peel to flavor it to suit the palate. Let this mixture be taken 3 times a day for a month or six weeks, and, if useful, then continued occasionally. It is a pleasant refrigerant and tonic draught.

5862. Remedy for Bad Breath. Bad or foul breath will be removed by taking a tea-spoonful of the following mixture after each meal: 1 ounce liquor of potassa, 1 ounce chloride of soda, 1½ ounces phosphate of soda, and 3 ounces water.

5863. Bad Breath from Constipation. When the breath is affected by constipation of the bowels, the following mixture will be useful: Take 4 drachms Epsom salts, 8 drachms tincture of columba, 6 ounces infusion of roses; well shake the phial each time you take the draught, which should be every other morning for a month or six weeks, a wine-glassful each time.

5864. To Remove the Smell of Onions from the Breath. Parsley eaten with vinegar will remove the unpleasant effects of eating onions.

5865. To Correct the Odor of Decay-

ed Teeth. To correct the odor of decayed teeth, 2 drops of a concentrated solution of permanganate of potash may be used in a glass of water as a wash, or a few drops of a weak solution may be introduced in the cavity of the tooth on a small piece of cotton. (*See No.* 1701.)

5866. To Preserve the Teeth and Gums. The teeth should be washed night and morning, a moderately small and soft brush being used; after the morning ablution pour on a second tooth-brush, slightly damped, a little of the following lotion: carbolic acid, 20 drops; spirit of wine, 2 drachms; distilled water, 6 ounces. After using this lotion for a short time the gums become firmer and less tender, and impurity of the breath (which is most commonly caused by bad teeth) will be removed. It is a great mistake to use hard tooth-brushes, or to brush the teeth until the gums bleed. (*See Nos.* 1288, *&c.*

5867. Magnetic Pain-Killer for Acute Pain and Toothache. This is one of the very best receipts for relieving acute pain and toothache. Laudanum, 1 drachm; gum camphor, 4 drachms; oil of cloves, ½ drachm; oil of lavender, 1 drachm; add these to 1 ounce alcohol, 6 drachms sulphuric ether, and 5 fluid drachms chloroform. Apply with lint; or, for toothache, rub on the gums, and upon the face against the tooth.

5868. Blake's Cure for the Toothache. Take alum, reduced to an impalpable powder, 2 drachms; spirits of nitric ether, 7 drachms. Mix, and apply them to the tooth. This is said to be an infallible cure for all kinds of toothache unless the disease is connected with rheumatism.

5869. Chloral for Toothache. Dr. Page recommends chloral hydrate as a local application in cases of toothache. A few grains of the solid hydrate introduced into the cavity of the tooth upon the point of a quill speedily dissolves there; and in the course of a few minutes, during which a not unpleasant warm sensation is experienced, the pain is either deadened, or, more often, effectually allayed. A second or third application may be resorted to if necessary. (*Brit. Med. Journ.*)

5870. To Cure Toothache. To 1 drachm flexible collodion add 2 drachms carbolic acid. A gelatinous mass is precipitated, a small portion of which inserted into the cavity of an aching tooth invariably gives immediate relief.

5871. Chlorate of Potassa as a Cure for Toothache. According to the experience of eminent dentists, chlorate of potassa affords quick relief in toothache. If the hollow tooth is in the lower jaw, a small crystal of this salt may be put in the cavity; but perhaps it is more advisable to use a solution of 1 part of the potassa in 20 of water.

5872. Paste for Toothache. Take of root-bark of pellitory, 1 drachm; muriate of morphia, 5 grains; triturate until reduced to fine powder, then add, finest honey, 3 drachms; oil of cloves (or of cajeput), 20 drops; concentrated tincture of pellitory, a sufficient quantity to form the whole into a smooth paste. Very effective.

5873. Cure for Toothache. Take equal parts of burnt alum and salt. Saturate a piece of cotton, cover with the mixture, and put in the tooth. Or saturate a small bit of clean cotton wool with a strong solution of ammonia, and apply it immediately to the affected tooth. Immediate relief will be experienced.

5874. Perry's Essence for the Toothache. A concentrated tincture of pellitory made with about equal parts of ether and rectified spirit largely charged with camphor. Though a nostrum, it is an excellent preparation. (*See No.* 4532.)

5875. Pieste's Toothache Essence. This is laudanum mixed with about twice its volume of liquor of ammonia specific gravity .960. Applied on lint, like other toothache drops, it often rapidly relieves the pain.

5876. Cottereau's Odontalgic Essence. A nearly saturated ethereal solution of camphor, mixed with $\frac{1}{16}$ to $\frac{1}{12}$ its volume of liquor of ammonia (specific gravity .880 to 882). A very useful preparation.

5877. To Kill the Nerve of a Hollow Tooth. Take ½ drachm white oxide of arsenic; 1 drachm sulphate of morphia; mix with a little creosote, and apply to the cavity of the tooth, previously cleansed.

5878. Tooth Cements. These are preparations for filling up cavities, cracks, &c., in defective teeth, the object being either to restore or preserve them, or to cure or prevent toothache. (*See Nos.* 3549, *&c.*)

5879. Diamond Tooth Cement. Take of anhydrous phosphoric acid in fine powder, 12 grains; pure caustic lime, fresh burnt, and in fine powder, 13 grains; mix them rapidly, by trituration, in a porcelain or wedgwood-ware mortar, and apply the powder, in the dry state, as quickly as possible, as it soon becomes moist. The powder, after being well pressed in the crack or cavity of the tooth, is smoothed off with the finger moistened with a drop of water. It soon acquires great hardness, is white, very durable, and does not become discolored by age. The compound that results from the combination of the ingredients almost exactly resembles the natural earthy matter of the teeth, and is, therefore, unobjectionable. Its color closely resembles, and will soon become that of the teeth to which it is applied, provided they possess ordinary whiteness. To cause it at once to imitate the color of the teeth, the mixture may be rendered slightly grey by adding to it a mere trace of carbon. This may be done by holding the pestle, used to mix the powders, over the flame of a candle or lamp, for an instant. A faint yellowish shade may be given to it by a trace of sulphuret of cadmium or a little yellow ochre; and a faint shade of red or flesh-color by a trace of jeweler's rouge or peroxide of iron, or a very little light-red (burnt yellow-ochre). This stopping, from its composition and other qualities, is, perhaps, superior to all others; but, except in the case of hollow teeth, its use requires some degree of skill and expertness, which is, however, readily acquired.

5880. Gutta-Percha Stopping for Teeth. This is pure, uncolored, native gutta-percha. A small piece is softened in hot water, and at once applied. It answers well for filling hollow teeth with central cavities, and is efficient and durable.

5881. White Gutta-Percha. The Journal of Applied Chemistry gives the following method of preparing this, for dentists' use and for other purposes. 4 ounces of pure gutta-percha are digested with 5 pounds of methyl-chloroform until the solution is thin enough to pass through filtering paper. It is then filtered (an additional pound of chloroform will facilitate this), and should then be clear and nearly colorless. Alcohol is now added in sufficient quantity to precipitate the gutta-percha in a voluminous white mass, which is washed with alcohol, pressed in a cloth, and dried in the air. It must finally be boiled in water in a porcelain vessel for half an hour, and, while still hot, rolled into sticks. The chloroform can be separated from the alcohol by adding water, and the alcohol recovered by distillation. (*See No.* 1725.)

5882. How to Fill or Plug Teeth. One of the most important points to attend to in filling or stopping teeth, is that each tooth must be thoroughly cleaned out, and wiped perfectly dry, before inserting or applying the cement, of whatever kind it be. Without careful attention to this matter, the cement will not adhere, or will soon become loose, and drop out or off, and the operation prove a failure. When a defective tooth is conveniently situated it may often be stopped by the party himself, by the exercise of a little skill and care, particularly if it be a hollow one with a clearly defined central cavity. When the reverse is the case, it is generally necessary that the operator should be a second party. A hollow tooth with a central and nearly circular hole in it may, in general, be effectively filled with a plug of dry soft wood, or of bone or ivory. If the hole be not round, it may be made so. Such stopping will often last for years.

5883. To Remove Tattoo Marks from the Skin. Inquiry is frequently made for methods for the successful removal of tattoo marks in the skin. While these are generally asserted to be indelible, if produced by the insertion of some carbonaceous matter, a correspondent of the Chemical News says that the marks disappeared by being first well rubbed with a salve of pure acetic acid and lard, then with a solution of potash, and finally with hydrochloric acid.

5884. To Remove Freckles. If the exact cause of freckles were known, a remedy for them might be found. A chemist in Moravia, observing the bleaching effect of mercurial preparations, inferred that the growth of a local parasitical fungus was the cause of the discoloration of the skin, which extended and ripened its spores in the warmer season. Knowing that sulpho-carbolate of zinc is a deadly enemy to all parasitic vegetation (itself not being otherwise injurious), he applied this salt for the purpose of removing the freckles. The compound consists of 2 parts of sulpho-carbolate of zinc, 25 parts of distilled glycerine, 25 parts of rose-water, and 5 parts of scented alcohol, and is to be applied twice daily for from half an hour to an hour, then washed off with cold water. Protection against the sun by veiling and other means is recommended, and in addition, for persons of pale complexion, some mild preparation of iron.

5885. To Remove Liver-spots. These are well-defined, brownish blotches on the skin, and generally appear on the forehead. Notwithstanding their name, they do not always proceed from the liver alone, but usually from some derangement or unhealthy state of the internal organs. In the first place, the general health must be thoroughly cared for, in order to have a fair prospect of success in any external local application. A pomade composed of 20 grains of sulphate of zinc and 1 ounce elder-flower ointment should be applied over-night to the spot, entirely within its limits, and not on the surrounding skin. In the morning wash it off with white castile soap and water, and bathe it repeatedly during the day with a lotion composed of 30 grains citric acid and ¼ pint infusion of roses. The spots should yield to this treatment in about 2 weeks, and their recurrence may be prevented by a regular use of borax and glycerine lotion. (*See No.* 4839.)

5886. To Remove Birth-marks. Mix together, with frequent agitation, 1 part pure carbonate of potash, 4 parts rose-water, 2 parts Hoffmann's Life Balsam (*see No.* 5112), and 2 parts distilled water. Apply to the mark twice a day, shaking the bottle well before using. (*Hager.*) (*See No.* 5826.)

5887. To Disguise the Taste of Medicines. Instead of attempting to flavor the medicine, or to remove the disagreeable taste from the mouth *after* taking the medicine, it is far more efficacious to prepare the mouth beforehand with some strong aromatic flavor, such as orange or lemon peel, or cachou aromatise. (*See No.* 1336.) In preparing the mouth for bitters, liquorice is the only *sweet* that should be used, all others creating a peculiarly disagreeable compound taste.

5888. To Disguise the Taste of Castor Oil. Castor or cod-liver oil may be taken with porter by pouring a little in the bottom of the glass, and then a little on top of the oil, but the best method of covering the nauseous flavor is to put a table-spoonful of strained orange-juice in a wine-glass, pour the castor oil into the centre of the juice, and then squeeze a few drops of lemon-juice upon the top of the oil, and rub some of the juice on the edge of the glass.

5889. French Method of Administering Castor Oil to Children. Pour the oil into a pan over a moderate fire; break an egg into it, and stir up; when it is done, add a little salt or sugar, or some currant jelly. The sick child will eat it agreeably, and never discover the disguise.

5890. To Destroy the Taste of Castor Oil. A good way is to beat the castor oil with the white of an egg until both are thoroughly mixed.

5891. To Disguise the Taste of Epsom Salts. Peppermint water almost prevents the nauseous taste of Epsom salts; a strong solution of extract of liquorice covers the disagreeable taste of *aloes;* milk, that of Peruvian bark; and cloves, that of senna.

5892. Agreeable Mode of Taking Senna. Dr. Linthner says that senna leaves (1 or 2 drachms to 1 or 2 cups of water) should be allowed to infuse all night in cold water. With the strained infusion coffee is prepared next morning, as if with water; and an aperient which does not taste of senna, and does not cause griping, is thus produced.

5893. Restoration of Persons Apparently Dead from Drowning. The following rules for the restoration of persons apparently dead from drowning, are given by Professor Benjamin Howard, of this city, and sanctioned by the Metropolitan Board of Health of the City of New York.

I. Unless in danger of freezing, never move the patient from the spot where first rescued, nor allow bystanders to screen off the fresh air, but instantly wipe clean the mouth and nostrils, rip and remove all clothing to a little below the waist, rapidly rub and dry the exposed part, and give two quick, smarting slaps on the stomach with your open hand. If this does not succeed immediately, proceed according to the following rules to perform artificial breathing:

II. Turn the patient on his face, a large bundle of tightly rolled clothing being placed beneath his stomach, and press heavily over it upon the spine for half a minute.

III. Turn the patient quickly again on his back; the roll of clothing being so placed beneath it as to make the short ribs bulge prominently forward, and raise them a little higher than the level of the mouth. Let some bystander hold the tip of the tongue out of one corner of the mouth with a dry handkerchief, and hold both hands of the patient together, the arms being stretched forcibly back above the head.

IV. Kneel astride the patient's hips, and with your hands resting on his stomach, spread out your fingers so that you can grasp the waist about the short ribs. Now throw all your weight steadily forward upon your hands, while you at the same time squeeze the ribs deeply, as if you wished to force everything in the chest upwards out of the mouth. Continue this while you can slowly count—one—two—three; then suddenly let go, with a final push, which springs you back to your first kneeling position. Remain erect upon your knees while you can count—one—two; then throw your weight forward again as before, repeating the entire motions—at first about 4 or 5 times a minute, increasing the rate gradually to about 15 times a minute, and continuing with the same regularity of time and motion as is observed in the natural breathing which you are imitating.

V. Continue this treatment, though apparently unsuccessful, for 2 hours, until the patient begins to breathe; and for a while after this, help him by well-timed pressure to deepen his *first* gasps into full, deep breaths; while the friction of the limbs, which should, if possible, have been kept up during the entire process, is now further increased.

VI. As soon as the breathing has become perfectly natural, strip the patient rapidly and completely. Wrap him in blankets only. Put him in bed in a room comfortably warm, but with a free circulation of fresh air, and, except for the administration of internal treatment, let him have perfect rest. Give him a little hot brandy and water, or other stimulant at hand, every 10 or 15 minutes for the first hour, and as often thereafter as may seem expedient.

5894. Abstinence as a Cure for Disease. Disease may often be cured by abstinence from all food, especially if the disorders have been produced by luxurious living and repletion. The latter overtaxes nature, and it rebels against such treatment. Indigestion, giddiness, headache, mental depression, &c., are often the effects of greediness in meat and drink. Omitting one, two, or three meals, allows the system to rest, to regain strength, and allows the clogged organs to dispose of their burdens. The practice of drug-taking to cleanse the stomach, though it may give the needed relief, always weakens the system, while abstinence often secures the same result, and yet does no injury.

5895. Antidotes for Poison. It need hardly be said that medical assistance must be sent for at once; but, meanwhile, as it is of the greatest importance to administer some aid as soon as possible, the subjoined directions may be followed. When any poisonous or other hurtful thing has been swallowed, take instantly half a glass of water—cold, not hot—put into it a heaping tea-spoonful of salt, and another of ground mustard; stir it rapidly 3 or 4 times; if there is no salt at hand, use mustard alone; catch the patient by the nose and toss it down. The reason for using cold water is that, in the hurry, the water may be hotter than thought for, and may scald the throat, causing eventual, if not instant death. The salt and mustard make the speediest emetic known, and are almost everywhere to be had in a moment. It brings up the contents of the stomach more or less completely. And for fear that some remnant may be left, administer a cupful of strong coffee, and then the white of 2 or 3 raw eggs, either first, as may be the quickest had, because these are two domestic articles which are found in every house, and nullify the effects of a greater number of virulent poisons than perhaps any other articles known. (*Hall.*)

5896. Treatment in Cases of Poisoning. Dr. Hall says: Whatever is done must be done quickly. The instant a person is known to have swallowed poison by design or accident, give water to drink, cold or warm, as fast as possible, a gallon or more at a time, and, as fast as vomited, drink more; tepid water is best, as it opens the pores of the skin and promotes vomiting, and thus gives the speediest cure to the poisonous article. If pains begin to be felt in the bowels, it shows that part at least of the poison has passed downwards; then large and repeated injections of tepid water should be given, the object in both cases being to dilute the poison as quickly and as largely as possible. Do not wait for warm water—take that which is nearest at hand, cold or warm, for every second of time saved is of immense importance; at the same time send instantly for a physician, and as soon as he comes turn the case into his hands, telling him what you have done. This simple fact cannot be too widely published; it is not meant to say that drinking a gallon or two of simple water will cure every case of poisoning; but it will cure many, and benefits all by its rapidly diluting quality. (*Journal of Health.*) A short summary of the antidotes resorted to in reference to particular poisons is given below. They should, of course, be administered as speedily as possible.

5897. Antidotes for Acid Poisons. Hydrochloric acid; nitric acid; oxalic acid (often mistaken for Epsom salts); *acetic* acid. For this form of poison, give quickly large draughts of chalk, whiting, magnesia, or soap and water, about as thick as cream; followed by albuminous diluents, such as milk, and white of egg mixed with water. Or, if these cannot be procured at once, warm water; and promote vomiting by the emetic recommended in No. 5895.

5898. Antidotes for Arsenic. The first endeavor, in cases of poisoning by arsenic, should be to remove, if possible, the poison from the stomach; for this purpose strong emetics or the stomach-pump should be had recourse to, after which the hydrated peroxide of iron in a dose thirty times greater than that of the poison may be administered. (*See No.* 4155.)

5899. Antidotes for Baryta in all its Forms. Sulphate of magnesia (Epsom salts), sulphate of soda (Glauber's salts), or any alkaline or earthy sulphate.

5900. Antidotes for Antimony, or Tartar Emetic. Administer large doses of warm water to induce vomiting (*see No.* 5896); give the powder of Peruvian bark, and, as soon as it can be prepared, the infusion of bark, which decomposes the tartar emetic.

5901. Antidotes for Alkalies, Soda, Potash, Ammonia, &c. Vinegar and lemon-juice are the best antidotes for potash, and all other alkaline poisons. A glassful of water, mixed with a table-spoonful of vinegar or lemon-juice, should be given frequently; and in defect of these, simple water, in such quantities as to cause vomiting. Emetics, and other irritating means, are to be avoided. Olive oil may also be administered.

5902. Antidotes for Corrosive Sublimate, or Calomel. The white of eggs beaten up with cold water is the best antidote for these. If eggs are not at once to be had, milk may be used with great success. Warm water should be given afterwards, to induce vomiting, also free purging in most instances.

5903. Antidote for Corrosive Sublimate. In case of poisoning by corrosive sublimate, if a dose of the hydrated protosulphuret of iron (*see No.* 4149) be administered it instantly renders the poison innocuous. This antidote is almost useless unless taken within 15 or 20 minutes after swallowing the poison.

5904. Antidotes for Verdigris and Sulphate of Copper. The treatment is the same as for corrosive sublimate. (*See No.* 5902.)

5905. Antidotes for Nitrate of Silver. Same as for corrosive sublimate (*see No.* 5902), with copious draughts of warm water and salt. (*See No.* 5895.)

5906. Antidote for Phosphorus. Same as for corrosive sublimate. (*See No.* 5902.) Phosphorus is the principal ingredient used in the manufacture of matches.

5907. Antidote for Sulphate of Zinc. Solution of carbonate of soda; also cream, butter, and chalk, are good antidotes for sulphate of zinc (white vitriol). Give water after the antidotes.

5908. Antidotes for Lead. *Litharge, red lead, white lead, sugar of lead, and Goulard's extract.* In the first stage, or the irritant form of injury, administer sulphate of magnesia, potash, or soda. The phosphate of soda is a good antidote. When palsy supervenes, the regimen must be regulated carefully.

5909. Antidotes for Opium and its Preparations. Emetics of the sulphate of zinc, ½ drachm or 2 scruples; the stomach pump, or injections of tartar emetic, must be employed to bring away the poison. The patient should be constantly roused by dragging about the floor, throwing cold water in the face, and giving ammonia, assafœtida, and strong coffee.

5910. New Antidote for Opium. In a case of accidental poisoning by an overdose of morphia, the administration of 18 drops of Norwood's tincture of green hellebore was followed by a complete cure. The narcotic had obtained such mastery over the unfortunate patient that the pupils of the eyes had contracted, and the jaws had to be forced open to give the medicine, which was mixed with 2 ounces of brandy. All appearance of poisonous effects had vanished in an hour.

5911. Antidotes for Prussic Acid. Small quantities of ammonia water diluted with 10 or 12 parts of water; also the fumes inhaled. The joint administration of carbonate of potash and sulphate of iron. This has been lately very strongly recommended. Cold affusion should be adopted in all cases, and is almost of itself a certain cure, if employed before the convulsive stage is over; and it is often successful even during the stage of insensibility and paralysis. Artificial respiration should also be attempted. Unfortunately, the poisonous action of prussic acid is so rapid that life is usually extinct before antidotes can be applied. (*See No.* 5913.)

5912. Antidotes for Strychnia and Nux-vomica. Evacuate the stomach with the stomach pump or emetics. (*See No.* 5896.) No antidote is known.

5913. Antidotes for Carbonic Acid Gas. When asphyxia from the inhalation of carbonic acid gas occurs, the patient should be immediately removed into the open air, and placed upon his back with the head slightly raised. Cold water should be dashed over the body, hot water applied to the feet, and ammonia to the nostrils. Brandy and water, and other stimulants, may be administered. Friction on the surface of the body is also recommended. If the patient has ceased to breathe, artificial respiration should be attempted. This may be done by pressing down the ribs, forcing up the diaphragm, and then suddenly withdrawing the pressure. (*See No.* 5893, *Rule V.*)

5914. Antidotes for Poisonous Mushrooms. The best antidote to poisonous mushrooms is tannin, or an infusion or decoction of galls. A strong emetic should also be given to remove them from the stomach.

5915. Antidote for Carbolic Acid. Dr. Crace Calvert states that the best antidote after the stomach pump is large doses of olive or almond oil, with a little castor oil. Oil is a solvent, and consequently a diluent of carbolic acid, and may be used to stop the corrosive effect of the acid when the action on the skin is too violent. Dr. Husemann, of Gottingen, suggests, for counteracting its effects on the stomach, a new preparation which he calls calcaria saccharata (*saccharate of lime*), prepared by dissolving 16 parts refined sugar in 40 parts water, and adding 5 parts slacked lime. Digest the mixture for 3 days, stir occasionally, filter, and evaporate to dryness.

5916. Antidote for Poisoning by Chlorine. Chlorine gas is an irritative poison, and the best antidotes are said to be ammoniacal gas, or the vapor of warm water, of wine, or of ether. The effects of chlorine have been known to pass off in the open air; leaving, in a certain instance, a violent cough, which disappeared in a few hours.

5917. Hodgen's Simple Stomach Pump. Attach 4 feet of india-rubber tubing to a stomach tube, fill both with water by simply dipping it in the liquid, end first, then compressing the elastic tube between the thumb and finger to keep the fluid from running out, introduce the stomach tube down the throat of the patient, lower the outer end of the elastic tube, and the contents of the stomach pour out as readily as if from an open vessel, the rubber tube acting as a syphon. When the fluid ceases to flow, dip the outer end of the tube beneath the surface of the water, elevate the vessel containing it above the level of the patient's mouth, and the stomach is soon filled; lower again the outer end of the tube and the stomach is emptied. This can, of course, be repeated as often as is necessary. The advantages claimed for this simple contrivance are, that it is of speedy and easy application, has no valves to become obstructed or deranged, and is far less expensive than a stomach pump.

5918. Cure for Ulcers Caused by Cyanide of Potassium. This substance is used in electroplating and other arts, and sometimes occasions ulcers on the hands. Protosulphate of iron in fine powder, ground in raw linseed oil, is recommended by a practical man, as the most effectual application for relieving the pain and healing the sores.

5919. Treatment for Hydrophobia. First dose, 1 ounce elecampane root, boiled in 1 pint milk until reduced to ½ pint. Second dose (to be taken 2 days after the first), 1½ ounces elecampane root in 1 pint of milk, boiled as the first. Third dose, the same as the second (to be taken 2 days after); in all, 3 doses. Mr. J. W. Woolston, a respectable citizen of Philadelphia, vouches for the above receipt. He says: "I have known of its being tried in one case, and no inconvenience has been felt. A friend of mine, of whom I obtained the receipt, knew of 20 instances where it was successfully given." We give the above for what it is worth, but we have no great faith in it.

5920. Cure for Hydrophobia. Cut out completely the wounded part before the poison can be absorbed. It is recommended, in order to do this quickly and thoroughly, that a stick be whittled to a shape resembling a dog's tooth, and inserted in the wound. This supports the part, and renders the cutting more easy and certain. This should be followed by cauterization, either by the use of a hot iron, or some strong caustic substance.

5921. To Remove the Virus in Hydrophobia. Suck the bitten part well, spitting out the fluid obtained from the wound; then apply some strong nitric acid, or lunar caustic, and bind the part up as tightly as the patient can bear it. Only one cauterization is needed.

5922. Youatt's Cure for Hydrophobia. Youatt (the great horse doctor) says he has been bitten eight or ten times and always cured himself by rubbing nitrate of silver in the wound. It should be applied as soon after the accident as may be. In 6 weeks the virus is disseminated through the system and then hope is gone.

5923. Preventive of Hydrophobia. The production of profuse perspiration is sometimes of great use in preventing the bad effects of a bite, so it should be tried.

5924. Bibron's Antidote to the Poison of the Rattlesnake. Iodide of potassium, 4 grains; corrosive sublimate, 2 grains; bromine, 5 drachms. 10 drops of this mixture, diluted with 1 or 2 table-spoonfuls of brandy, wine, or whiskey, constitute a dose, to be repeated if necessary. It must be kept in glass-stoppered phials, well secured, as the air will affect it. The salts may, in case of emergency, be first dissolved in a little water, before adding the bromine, as this dissolves them very slowly. This is a valuable remedy. Dr. Hammond, in speaking of the remedy, says that during a recent expedition to the Rocky Mountains, he had frequent opportunities to test its efficiency. The results were satisfactory, and he thinks that, when taken in time, it may be entirely depended upon in the poisonous wounds of the rattlesnake.

5925. To Extract the Poison from a Rattlesnake Bite. The most direct and efficient means of counteracting the absorption of the poison is suction, and this is most effectually done by exhausting a cupping-glass over the wound. The cupping-glass must be applied as soon after the injury as possible, and kept exhausted until all danger has passed. It has been proved that the bites of vipers, both on man and animals, were rendered entirely harmless by the application of these glasses.

5926. Cure for Snake Bites. As many as 8000 persons die annually in British India and Burmah, from the effects of snake bites. The Inspector of Police to the Bengal Government reports that of 939 cases in which ammonia was freely administered, 702 victims have recovered, and in the cured instances, the remedy was not administered till about 3¼ hours after the attack, on the average. In the fatal cases, the corresponding duration of time was 4½ hours.

5927. To Cure the Stings of Hornets, Wasps, Bees, and Spiders. Swelling may instantly be arrested by an application of equal parts common salt and bicarbonate of soda, dissolved in warm water, and well rubbed in on the place bitten or stung. (*See also No.* 5929.)

5928. Cure for Stings of Wasps, &c. Rub the part affected with a mixture of 1 part spirits of hartshorn and 2 parts olive oil.

5929. To Cure the Bites of Insects. Dissolve 1 ounce borax in 1 pint water that has been boiled and allowed to cool. Instead of plain water, distilled rose-water, elder, or orange-flower water is more pleasant. The bites are to be dabbed with the solution as long as there is any irritation. For bees' or wasps' stings the borax solution may be made of twice the above strength.

**5930. To Cure Poisoning by Poison

Ivy, Oak, or Sumach. Bathe the poisoned part thoroughly with hot water, without soap. When dry, paint the place liberally, 2 to 4 times a day, with a feather dipped in strong tincture of lobelia. Avoid bringing the tincture in contact with any fresh wound or excoriation.

5931. Remedy for Poison Ivy, &c. In some cases, where lobelia (*see last receipt*) does not succeed quickly, an application, in a similar manner, of fluid extract of gelseminum sempervirens (yellow jessamine) will rarely fail to cure. Both of these are excellent remedies, generally acting like magic.

5932. Remarks on Poison Ivy, &c. Poison ivy, &c., act very differently upon different people. Some people are entirely proof against its effects, and can, with impunity, rub it on without any ill effect. Others are poisoned by simple contact with clothing that has touched it. This difference of susceptibility to the poison seems to apply equally to the remedies, as what will cure one person has little or no effect on another.

5933. Applications for Poison Ivy. Various applications have been used for the same purpose; bathing the parts with a decoction of hemlock boughs, or of oak leaves; or with a table-spoonful of copperas (sulphate of iron) in a small tea-cupful of boiling water; or painting over with fresh lime-water; or rubbing wet salt on the poisoned part; or bathing the parts affected freely with spirit of nitre. If the blisters be broken, so as to allow the nitre to penetrate, more than a single application is rarely necessary. It will scarcely be possible to fail in finding, in one or other of the remedies here given, a means of cure suited to the party affected.

Tables of Weights, Measures, &c.

The following tables have been compiled for the purpose of aiding the reader to determine with facility, the relative values of different weights and measures; and to furnish in a convenient group a mass of valuable information that would otherwise have to be sought for in a number of volumes not easy of access. Most of the tables have been made expressly for this work, and all of them have been carefully recalculated, revised, and corrected by a competent mathematician.

5935. Avoirdupois Weight is employed for weighing all goods, except those for which Troy or Apothecaries weight are used. The ton is subdivided into hundred-weights, quarters, pounds, ounces, and drachms. (*See No.* 6031.) Some goods are sold by the hundred-weight of 100 pounds, instead of the hundred-weight (cwt.) of 112 pounds; a ton composed of 20 hundreds would then contain only 2000 pounds. The pound avoirdupois consists of 7000 Troy grains. The drachm avoirdupois is therefore 27.34375 Troy grains. The standard avoirdupois pound of the United States is the weight of 27.7015 cubic inches of distilled water, at 39.83° Fahr., the barometer being at 30 inches.

Ton.		Cwt.		Qrs.		Lbs.		Oz.		Dr.
1	=	20	=	80	=	2,240	=	35,840	=	573,440
		1	=	4	=	112	=	1,792	=	28,672
				1	=	28	=	448	=	7,168
						1	=	16	=	256
								1	=	16

5936. Equivalents of Avoirdupois in Troy Weight.

Avoirdupois.		Lbs.	Oz.	Dwt.	Grains.
1 Ton	=	2922	2	13	8
1 Cwt.	=	146	1	6	16
1 Qr.	=	34	0	6	16
1 Lb.	=	1	2	11	16
1 Oz.	=			18	5½
1 Dr.	=			1	3$\frac{11}{32}$

5937. Value of Avoirdupois Weight in Apothecaries Measure.

		f℥	fʒ	♏
1 pound	=	15	2	53.3622
1 ounce	=		7	40.8351

5938. Value of Avoirdupois in Apothecaries Weight.

Avoirdupois.		Apothecaries. ℔	℥	ʒ	℈	Gr.
1 pound	=	1	2	4	2	0
1 ounce	=			7	0	17½
1 drachm	=				1	7$\frac{11}{32}$

In the new British Pharmacopœia the weights are expressed in pounds, ounces, and grains avoirdupois. (*See No.* 6031.)

5939. Decimal Equivalents of lbs., qrs., and cwt.

qrs.	lbs.	cwt.	qrs.	lbs.	cwt.	qrs.	lbs.	cwt.	qrs.	lbs.	cwt.
0	0½	=.0044	1	0	=.25	2	0	=.5	3	0	=.75
0	1	.0089	1	1	.2589	2	1	.5089	3	1	.7589
0	2	.0178	1	2	.2678	2	2	.5178	3	2	.7678
0	3	.0268	1	3	.2768	2	3	.5268	3	3	.7768
0	4	.0357	1	4	.2857	2	4	.5357	3	4	.7857
0	5	.0446	1	5	.2946	2	5	.5446	3	5	.7946
0	6	.0535	1	6	.3035	2	6	.5535	3	6	.8035
0	7	.0625	1	7	.3125	2	7	.5625	3	7	.8125
0	8	.0714	1	8	.3214	2	8	.5714	3	8	.8214
0	9	.0803	1	9	.3303	2	9	.5803	3	9	.8303
0	10	.0892	1	10	.3392	2	10	.5892	3	10	.8392
0	11	.0982	1	11	.3482	2	11	.5982	3	11	.8482
0	12	.1071	1	12	.3571	2	12	.6077	3	12	.8571
0	13	.1160	1	13	.3660	2	13	.6160	3	13	.8660
0	14	.125	1	14	.375	2	14	.625	3	14	.875
0	15	.1339	1	15	.3839	2	15	.6339	3	15	.8839
0	16	.1429	1	16	.3929	2	16	.6429	3	16	.8929
0	17	.1518	1	17	.4018	2	17	.6518	3	17	.9018
0	18	.1607	1	18	.4107	2	18	.6607	3	18	.9107
0	19	.1696	1	19	.4196	2	19	.6696	3	19	.9196
0	20	.1786	1	20	.4286	2	20	.6786	3	20	.9286
0	21	.1875	1	21	.4375	2	21	.6875	3	21	.9375
0	22	.1964	1	22	.4464	2	22	.6964	3	22	.9464
0	23	.2054	1	23	.4554	2	23	.7054	3	23	.9554
0	24	.2143	1	24	.4643	2	24	.7143	3	24	.9643
0	25	.2232	1	25	.4732	2	25	.7232	3	25	.9732
0	26	.2321	1	26	.4821	2	26	.7321	3	26	.9821
0	27	.2411	1	27	.4911	2	27	.7411	3	27	.9911

5940. Decimal Equivalents of Pounds and Ounces.

oz.	lb.	oz.	lb.	oz.	lb.	oz.	lb.	oz.	lb.
¼	.015625	3	.1875	6½	.40625	10	.625	13½	.84375
½	.03125	3½	.21875	7	.4375	10½	.65625	14	.875
¾	.046875	4	.25	7½	.46875	11	.6875	14½	.90625
1	.0625	4½	.28125	8	.5	11½	.71875	15	.9375
1½	.09375	5	.3125	8½	.53125	12	.75	15½	.96875
2	.125	5½	.34375	9	.5625	12½	.78125	16	1.
2½	.15625	6	.375	9½	.59375	13	.8125		

5941. Avoirdupois Weight Expressed in Grams.

Avoirdupois.		Grams.		
1 Ton	=	1,015,938.84	=	1.016 Milliers
1 Cwt.	=	50,796.94	=	5.080 Myriagrams
1 Quarter	=	12,699.23	=	1.270 Myriagrams
1 Pound	=	453.54	=	4.535 Hectograms
1 Ounce	=	28.34	=	2.834 Dekagrams
1 Drachm	=	1.77		

5942. Troy Weight is used by jewelers for weighing gold, silver, platina, and all precious stones except the diamond; and is the weight adopted by the mint. The pound Troy contains 5.760 grains.

Pound.		Ounces.		Pennyweights.		Grains.
1	=	12	=	240	=	5760
		1	=	20	=	480
				1	=	24
						1

5943. Diamond Weight. Diamonds are weighed by a separate method; the carat, equivalent to 3.2 grains Troy, is thus subdivided.

Carat.		Grains.		Parts.		Troy Grains.
1	=	4	=	16	=	3.2
		1	=	4	=	.8
				1	=	.2

5944. Troy Weight Compared with Avoirdupois.

Troy.		Avoirdupois. Oz.	Dr.
1 Pound	=	13	2.65
1 Ounce	=	1	1.55
1 Dwt.	=		0.877

5945. Equivalents of Troy in Apothecaries Weight.

Troy.		Apothecaries. ℔	℥	ʒ	℈	Gr.
1 Pound	=	1	0	0	0	0
1 Ounce	=		1	0	0	0
1 Dwt.	=				1	4
1 Grain	=					1

5946. Troy Weight Expressed in Grams.

Troy.		Grams.
1 Pound	=	373.202, or 3.732 Hectograms
1 Ounce	=	31.100, or 3.110 Dekagrams
1 Dwt.	=	1.555
1 Grain	=	.0648, or 6.48 Centigrams.

5947. Approximate Values of Troy in Metrical Weight.

Troy weight.		Weight.		Measure.
32 oz.	=	1 kilogramme,	=	1 litre.
16 oz.	=	½ kilog. = 500 grams,	=	.500 "
4 oz.	=	125 grams,	=	.125 "
1 oz.	=	32 grams,	=	.32 "
1 dr'm.	=	4 grams,	=	.4 "
15 grains	=	1 gram,	=	.1 cubic centimetre.
1½ gr'ns	=	1 decigram.		

5948. Assayer's Gold Weights. The richness or purity of gold is expressed in carats. Pure gold is spoken of as containing

24 carats, of 12 grains each; and any sample containing 12, 18, 22, or any other number of parts of pure gold, in 24 parts, is said to be of so many carats fine. In the process of assaying gold, the real quantity taken is very small, generally 6 or 12 grains; and this is termed the "assay pound." It is nominally subdivided into 24 carats, and each carat into 4 assay grains, and each grain into quarters. When the assay pound is only 6 grains, the quarter of the assay grain will only weigh the $\frac{1}{96}$ of a grain; hence the most accurate system of weighing must be adopted.

5949. Assayer's Silver Weights. The richness or purity of silver is either expressed in pennyweights or $\frac{1}{1000}$. In the first case, it is supposed that the mass of silver to be examined consists of 12 equal parts, called pennyweights; so that if an ingot weighs an ounce, each of the parts will be $\frac{1}{12}$ of an ounce. Hence, if the mass of silver be pure, it is called silver of 12 pennyweights; if it contain $\frac{1}{12}$ of its weight of alloy, it is called silver of 11 pennyweights; if $\frac{2}{12}$ of its weight be alloy, it is called silver of 10 pennyweights; and so on in proportion for other qualities. It must be observed here, that the assayers give the name pennyweight to a weight equal to 24 real grains, which must not be confounded with their ideal weights. The assayer's grains are called fine grains. An ingot of fine silver, or silver of 12 pennyweights, contains, then, 288 fine grains; if this ingot contain $\frac{1}{288}$ of alloy, it is said to be silver of 11 pennyweights and 23 grains; if it contain $\frac{4}{288}$ of alloy, it is said to be 11 pennyweights, 20 grains, &c. The purity of silver is now more frequently expressed in $\frac{1}{1000}$, which admits of greater accuracy.

5950. Table for Converting Troy into Avoirdupois Weight.

Troy Ounces.		Avoirdupois Ounces.	Avoirdupois Grains.	Troy Ounces.		Avoirdupois Ounces.	Avoirdupois Grains.
1	=	1	42½	7	=	7	297½
2	=	2	85	8	=	8	340
3	=	3	127½	9	=	9	382½
4	=	4	170	10	=	10	425
5	=	5	212½	11	=	12	30
6	=	6	255	12	=	13	72½

175 Troy ounces are equal to 192 avoirdupois.

Troy. ℔		Avoirdupois. ℔	Oz.	Gr.	Troy. ℔		Avoirdupois. ℔	Oz.	Gr.
1	=	0	13	72½	18	=	14	12	430
2	=	1	10	145	19	=	15	10	65
3	=	2	7	217½	20	=	16	7	137½
4	=	3	4	290	30	=	24	10	425
5	=	4	1	362½	40	=	32	14	275
6	=	4	14	435	50	=	41	2	125
7	=	5	12	70	60	=	49	5	412½
8	=	6	9	142½	70	=	57	9	262½
9	=	7	6	215	80	=	65	13	112½
10	=	8	3	287½	90	=	74	0	400
11	=	9	0	360	100	=	82	4	250
12	=	9	13	432½	175	=	144	0	0
13	=	10	11	67½	200	=	164	9	62½
14	=	11	8	140	300	=	246	13	312½
15	=	12	5	212½	400	=	329	2	125
16	=	13	2	285	500	=	411	6	375
17	=	13	15	357½	1000	=	822	13	312½

5951. Apothecaries Weight is a subdivision of the Troy pound into ounces, drachms, scruples, and grains. It is used in compounding medicines, and is the officinal standard of the U. S. Pharmacopœia.

℔		℥		ʒ		℈		Grs.
1	=	12	=	96	=	288	=	5760
		1	=	8	=	24	=	480
				1	=	3	=	60
						1	=	20
								1

5952. Apothecaries Weight Compared with Avoirdupois Weight.

Apothecaries.		Avoirdupois. Oz.	Dr.
1 Pound	=	13	2.65
1 Ounce	=	1	1.55
1 Drachm	=		2.19
1 Scruple	=		0.73

5953. Apothecaries Weight Compared with Troy Weight.

Apothecaries.		Troy. Lb.	Oz.	Dwt.	Gr.
1 Pound	=	1	0	0	0
1 Ounce	=		1	0	0
1 Drachm	=			2	12
1 Scruple	=				20

5954. Value of Apothecaries Weight in Apothecaries Measure.

Weight		f℥	fʒ	♏
1 Pound	=	12	5	7.2238
1 Ounce	=	1	0	25.6020
1 Drachm	=	0	1	3.2002
1 Scruple	=	0	0	21.0667
1 Grain	=	0	0	1.0533

5955. Apothecaries Weight Expressed in Grams.

1 Pound	=	3.732 Hectograms
1 Ounce	=	3.110 Dekagrams
1 Drachm	=	3.887 Grams
1 Scruple	=	1.296 "
1 Grain	=	6.4 Centigrams.

5956. Apothecaries, or Wine Measure, is the gallon of liquid measure divided into pints, fluid ounces, fluid drachms, and minims. The minim being equivalent to one drop of water. The symbols or abbreviations used in this table will be found explained in No. 5964. In all the tables of comparison between apothecaries measure and avoirdupois or other weights, the basis assumed is the weight of a cubic inch of water at a temperature of 39.83° Fahr., the barometer being at 30 inches, and is equivalent to 252.693 Troy grains. A *grain measure* is the capacity or bulk of a grain of water weighed at its maximum density; a grain measure of any fluid, therefore, weighs more or less than a grain, according as its specific gravity is greater or less than water at standard temperature.

Cong.		O.		f℥		fʒ		♏		Cubic Inches.
1	=	8	=	128	=	1024	=	61440	=	231
		1	=	16	=	128	=	7680	=	28.875
				1	=	8	=	480	=	1.8047
						1	=	60	=	.2256
								1	=	.0376

5957. Graduated Fluid Measures. Fluids are measured by means of glass vessels having a graduated scale engraved on their sides. These are of different capacities, to measure 8 ounces, 2 ounces, 1 ounce and 1 drachm respectively; the scale of each being graduated to represent the aliquot parts of their respective capacities.

1 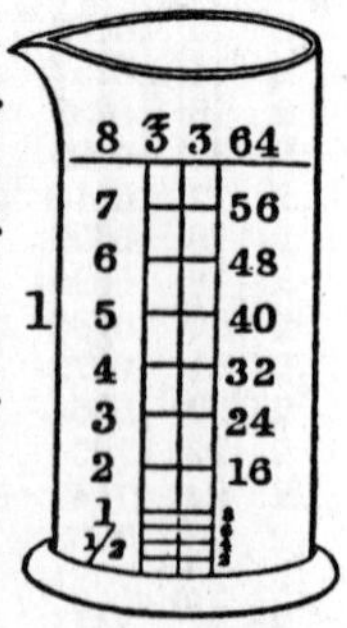

2 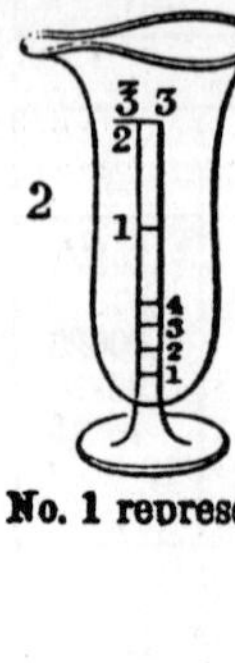

3 

4 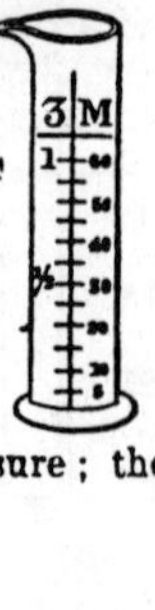

No. 1 represents an 8-ounce measure; the figures on the left of the graduated scale denote ounces, and those on the right, drachms; the first ounce being divided into quarters of 2 drachms each. No. 2 is a 2-ounce measure, the first half-ounce being divided into drachms. Nos. 3 and 4 are 1 ounce and 1 drachm measures respectively; the former is graduated in drachms, the first of which is divided into halves; the latter is marked in divisions of 5 minims each.

5958. Relative Value of U. S. Apothecaries and British Imperial Measure. (*See No.* 6031.)

U. S. Apothecaries Measure.		Imperial Measure.			Pints.	Fl. oz.	Fl. dr.	Minims.
1 Gallon	=	.83311	Imp.	Gallon,	or, 6	13	2	22.85
1 Pint	=	.83311	"	Pint,	or,	16	5	17.86
1 Fl. Oz.	=	1.04139	"	Fl. Oz.,	or,	1	0	19.87
1 Fl. Dr.	=	1.04139	"	Fl. Dr.,	or,		1	2.48
1 Minim	=	1.04139	"	Minim,	or,			1.04

5959. Apothecaries Measure Expressed in Litres.

1 Gallon	=	3.78515 Litres.
1 Pint	=	4.73143 Decilitres
1 Fluid ounce	=	2.95715 Centilitres
1 Fluid drachm	=	3.69644 Millilitres
1 Minim	=	.06160 "

5960. Value of Apothecaries Measure in Avoirdupois Weight.

1 Gallon	=	8.332698 Pounds
1 Pint	=	1.041587 Pounds
1 Fluid Ounce	=	1.041587 Ounces

5961. Value of Apothecaries Measure in Troy Weight.

Apothecaries Measure.		Troy Weight. Lbs.	Oz.	Dwt.	Grains.
1 Gallon	=	10	1	10	8.88
1 Pint	=	1	3	3	19.11
1 Fluid Ounce	=			18	23.69
1 Fluid Drachm	=			2	8.96
1 Minim	=				.95

5962. Value of Apothecaries Measure in Apothecaries Weight.

Measure.		℔	℥	ʒ	℈	Grains		Grains
1 Gallon	=	10	1	4	0	8.88	=	58328.886
1 Pint	=	1	3	1	1	11.11	=	7291.1107
1 Fluid ounce	=			7	1	15.69	=	455.6944
1 Fluid drachm	=				2	16.96	=	56.9618
1 Minim								.9493

5963. Miscellaneous Measures and their Equivalents.

Tea-spoonful.................about 1 fl. drachm.
Dessert " .... " 2 "
Table " .... " 4 "
Wine-glassful................ " 2 fl. ounces.
Tea-cupful................ " 4 "
Breakfast-cupful........... " 8 "
Tumblerful................ " 8 "
Thimbleful................ " ¼ fl. drachm.
Pinch (of leaves and flowers) " 1 dr. (Troy).
Handful " " 10 "

5964. Signs and Abbreviations Used in Medical Prescriptions.

℞.......Recipe...........................Take
āā.......Ana...........................Of each
℔.......Libra...........................Pound
℥.......Uncia...........................Ounce
ʒ.......Drachma...........................Drachm
℈.......Scrupulus...........................Scruple
Cong....Congius...........................Gallon
O.......Octarius...........................Pint
f℥.......Fluid Uncia.............Fluid Ounce
fʒ.......Fluid Drachma.............Fluid Drachm
♏.......Minimum...........................Minim
Chart....Chartula.............Small paper
Coch.....Cochlear...........................Spoonful
Collyr...Collyrium.............Eye-water
Decot...Decoctum.............Decoction
Ft.......Fiat...........................Make
Garg.....Gargarysma.............Gargle
Gr.......Granum...........................Grain
Gtt......Gutta...........................Drop
Haust...Haustus.............Draught
Infus...Infusum.............Infusion
M.......Misce...........................Mix

Mass....Massa........................Mass
Mist.....Mistura.....................Mixture
Pulv.....Pulvis......................Powder
Q. S.....Quantum Sufficit.Sufficient Quantity
S.........Signa.........................Write
S. S.....Semis..........................Half

5965. Strength of Doses at Different Ages. The following gradations for doses of medicines apportioned to the age of the patient were originally drawn up by Gaubius.

Under	½ year	$\frac{1}{16}$	of a full dose.
"	1 "	$\frac{1}{12}$	"
"	2 years	$\frac{1}{8}$	"
"	3 "	$\frac{1}{6}$	"
"	4 "	$\frac{1}{5}$	"
"	7 "	$\frac{1}{3}$	"
"	14 "	$\frac{1}{2}$	"
"	20 "	$\frac{2}{3}$	"
Above	21 "	the full dose.	
"	63 "	$\frac{11}{12}$	"
"	77 "	$\frac{5}{6}$	"
"	100 "	$\frac{2}{3}$	"

Dr. Young gives the following simple formula: For children under 12 years, the doses of most medicines must be diminished in the proportion of the age to the age increased by 12. Thus, at 2 years, the dose will be $\frac{1}{7}$ of that for an adult, viz:

$$\frac{2}{2+12}=\frac{1}{7}$$

Sex, temperament, constitutional strength, and the habits and idiosyncrasies of individuals, must be taken into account. Nor does the same rule apply to all medicines. Calomel, for instance, is generally borne better by children than by adults; while opium affects them more powerfully, and requires the dose to be diminished considerably below that indicated above.

5966. Liquid Measure. This is used for all liquids which are sold by measure. The United States Government standard gallon, adopted by the Treasury Department in 1832, has a capacity of 231 cubic inches, and contains 58,372.2 troy grains of distilled water, at 39.83° Fahr., the temperature of its maximum density.

Gal.		Quarts.		Pints.		Gills.		Cubic Inches.
1	=	4	=	8	=	32	=	231
		1	=	2	=	8	=	57.75
				1	=	4	=	28.875
						1	=	7.2175

A Barrel	contains		31½ gallons.
A Tierce	"		42 "
A Puncheon	"2 tierces,	or	84 "
A Hogshead	"2 barrels,	or	63 "
A Pipe	"2 hogsheads,	or	126 "
A Tun	"2 pipes,	or	252 "

5967. Liquid Measure Compared with Apothecaries Measure. The gallon and pint are the same in both measures. A liquid gill contains 4 fluid ounces, or 32 fluid drachms, or 1920 minims.

5968. Relative Value of U. S. Liquid Measure in English Imperial Measure.

U. States.		Imperial.	Quart.	Pint.	Gill.
1 Gallon	=	.83311 gal., or	3	0	2.66
1 Quart	=	.83311 qt., or		1	2.66
1 Pint	=	.83311 pt., or			3.33
1 Gill	=	.83311 gill, or			0.83

5969. Liquid Measure Expressed in Litres.

1 Gallon	=	3.785148 Litres
1 Quart	=	9.46287 Decilitres
1 Pint	=	4.73143 "
1 Gill	=	1.18286 "

5970. Dry Measure. The Winchester bushel, formerly used in England, contained 2150.42 cubic inches; this was superseded in 1826 by the Imperial bushel of 2218.192 inches, or 80 pounds of distilled water at 62° Fahr., and the barometer at 30 inches. In the United States, the Winchester bushel of 2150.42 inches has been generally adopted, which holds 77.627413 pounds of distilled water at 39.83° Fahr., the temperature of its maximum density, and 30 inches barometric pressure. In New York the bushel is declared to contain 80 pounds distilled water at its maximum density, under the mean pressure of the atmosphere at the level of the sea. This would make the New York bushel contain 2216.128 cubic inches, somewhat less than the Imperial bushel, owing to the different standard of temperature of the water. The "small measure" used in the markets should contain 2 quarts, or ¼ peck.

Quarter.		Bushels.		Pecks.		Quarts.		Pints.		Capacity in Cubic Inches.
1	=	8	=	32	=	256	=	512	=	17203.36
		1	=	4	=	32	=	64	=	2150.42
				1	=	8	=	16	=	537.605
						1	=	2	=	67.200
								1	=	33.600

5971. Dry Measure expressed in Litres.

1 Bushel	=	35.23661 Litres
1 Peck	=	8.80915 "
1 Quart	=	1.10114 "
1 Pint	=	.55057 "

5972. Relative Value of United States Dry Measure and Imperial Dry Measure.

United States.		Imperial.		Imperial Bush.	Pecks.	Gals.	Pints.
1 Quarter	=	.96945 quarter,	or	7	3	0	.36
1 Bushel	=	.96945 bushel,	or		3	1	6.04
1 Peck	=	.96945 peck,	or			1	7.51
1 Quart	=	.24236 gallon,	or				1.94
1 Pint	=	.96945 pint,	or				.97

5973. Weight of a Barrel of Various Articles. Some things which are sold by weight or measure are also sold by the *Barrel*, the quantity being different for different articles. The weights are here given. For rice, 600 pounds. Flour, 196 pounds. Powder, 25 pounds. Corn, as bought and sold in Kentucky, Tennessee, &c., 5 bushels of shelled corn. As bought and sold at New Orleans, a flour-barrel full of ears. Potatoes, as sold in New York, a barrel contains 2½ bushels. Pork, a barrel is 200 pounds, distinguished in quality by "clear," "mess," "prime." A barrel of beef is the same weight.

5974. Weight of a Bushel of Various Commodities. The term bushel is also applied to a certain arbitrary weight varying with different articles. Wheat, beans, potatoes, and clover seed, 60 pounds to the bushel. Corn, rye, flax-seed, and onions, 56 pounds. Corn on the cob, 70 pounds. Buckwheat, 52 pounds. Barley, 48 pounds. Hemp seed, 44 pounds. Timothy seed, 45 pounds. Castor beans, 46 pounds. Oats, 35 pounds. Bran, 20 pounds. Blue grass seed, 14 pounds.

5975. Lineal or Long Measure. The standard of linear measurements, by which all measures of capacity are also regulated, is derived from the length of a pendulum vibrating seconds in a vacuum. This, in the latitude of London, is equal to 39.1393 inches, and in the City Hall of New York, 39.1012 inches.

By scientific persons, parts of an inch are represented by a decimal fraction, but for mechanical purposes the inch is divided into a half, quarters and eighths.

Mile.		Furlongs.		Rods.		Yards.		Feet.		Inches.
1	=	8	=	320	=	1760	=	5280	=	63360
		1	=	40	=	220	=	660	=	7920
				1	=	5½	=	16½	=	198
						1	=	3	=	36
								1	=	12

5976. Long Measure Expressed in Metres.

		Metres.		
1 Mile	=	1609.30634	=	1.609 Kilometres
1 Furlong	=	201.16329	=	2.012 Hectometres
1 Rod	=	5.02908	=	5.029 Metres
1 Yard	=	.91438	=	9.144 Decimetres
1 Foot	=	.30479	=	3.048 Decimetres
1 Inch	=	.02539	=	2.539 Centimetres

5977. Comparative Scale of Inches in French Metres.

Inches. Millimetres.

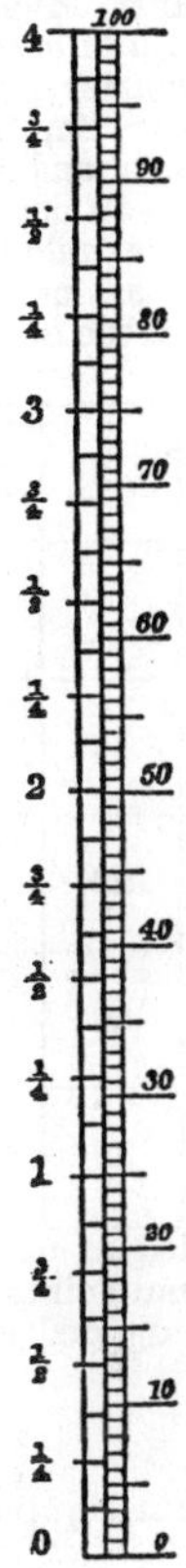

5978. Value of Inches and Feet in French Metres.

Inches.		
$\frac{1}{8}$	=	.00317
$\frac{3}{16}$	=	.00475
$\frac{1}{4}$	=	.00635
$\frac{5}{16}$	=	.00794
$\frac{3}{8}$	=	.00952
$\frac{7}{16}$	=	.01111
$\frac{1}{2}$	=	.01269
$\frac{9}{16}$	=	.01428
$\frac{5}{8}$	=	.01586
$\frac{11}{16}$	=	.01745
$\frac{3}{4}$	=	.01904
$\frac{13}{16}$	=	.02063
$\frac{7}{8}$	=	.02221
$\frac{15}{16}$	=	.02379
1	=	.02539
2	=	.05079
3	=	.07619
4	=	.10159
5	=	.12699
6	=	.15239
7	=	.17779
8	=	.20319
9	=	.22859
10	=	.25399
11	=	.27939
12	=	.30479
Feet.		
2	=	.60958
3	=	.91438
4	=	1.21916
5	=	1.52395
6	=	1.82874
7	=	2.13353
8	=	2.43832
9	=	2.74311
10	=	3.04791
11	=	3.35270
12	=	3.65750

5979. Decimal Equivalents of Fractional Parts of an Inch.

Decimals.		Parts of an Inch.	Decimals.		Parts of an Inch.
.03125	=	$\frac{1}{32}$	.53125	=	$\frac{17}{32}$
.06250	=	$\frac{1}{16}$	.56250	=	$\frac{9}{16}$
.09375	=	$\frac{3}{32}$	.59375	=	$\frac{19}{32}$
.12500	=	$\frac{1}{8}$	.62500	=	$\frac{5}{8}$
.15625	=	$\frac{5}{32}$	.65625	=	$\frac{21}{32}$
.18750	=	$\frac{3}{16}$	.68750	=	$\frac{11}{16}$
.21875	=	$\frac{7}{32}$	.71875	=	$\frac{23}{32}$
.25000	=	$\frac{1}{4}$	.75000	=	$\frac{3}{4}$
.28125	=	$\frac{9}{32}$	.78125	=	$\frac{25}{32}$
.31250	=	$\frac{5}{16}$	.81250	=	$\frac{13}{16}$
.34375	=	$\frac{11}{32}$	.84375	=	$\frac{27}{32}$
.37500	=	$\frac{3}{8}$	.87500	=	$\frac{7}{8}$
.40625	=	$\frac{13}{32}$	.90625	=	$\frac{29}{32}$
.43750	=	$\frac{7}{16}$	.93750	=	$\frac{15}{16}$
.46875	=	$\frac{15}{32}$	.96875	=	$\frac{31}{32}$
.50000	=	$\frac{1}{2}$			

5980. Pendulum Measure. 6 points = 1 line. 12 lines = 1 inch.

5981. Shoemakers' Measures. No. 1 is 4⅛ inches in length, and every succeeding number is ⅛ inch. There are 28 divisions, in two series of numbers, viz.: from 1 to 13 and 1 to 15.

5982. Square or Superficial Measure.

Acre.	Roods.	Poles.	Yards.	Feet.	Inches.
1 =	4 =	160 =	4,840 =	43,560 =	6,272,640
	1 =	40 =	1,210 =	10,890 =	1,568,160
		1 =	30¼ =	272¼ =	39,204
			1 =	9 =	1,296
				1 =	144

5983. Square Measure in Square Metres.

1 Acre =	4046.66700	sq. metres =	40.46667 Ares
1 Rood =	1011.66675	" =	10.11667 "
1 Pole =	25.29167	" =	25.29167 Centares
1 Yard =	.83609	" =	.83609 "
1 Foot =	.09289	" =	9.289 Milliares
1 Inch =	.000645	" =	.0645 "

5984. Government Land Measure. A Township—36 sections, each a mile square. A Section—640 acres. A Quarter Section, half a mile square—160 acres. An Eighth Section, half a mile long, north and south, and a quarter of a mile wide—80 acres. A Sixteenth Section, a quarter of a mile square—40 acres. The Sections are all numbered one to thirty-six, commencing at the northeast corner, thus:

6	5	4	3	2	NW NE SW SE
7	8	9	10	11	12
18	17	16*	15	14	13
19	20	21	22	23	24
30	29	28	27	26	25
31	32	33	34	35	36

The Sections are all divided in quarters, which are named by the cardinal points, as in section 1. The quarters are divided in the same way.

*School Section.

5987. To Find the Square Surface or Area of a Circle. Square the radius (half the diameter), and multiply that by 3.14159; for small calculations 3 1/7 is nearly the same as 3.14159. Thus, to find the area of a circle whose diameter is 8 feet: The radius is 4 feet, this squared is 16; then 16 times 3.14159 is 50.265 square feet. If the diameter is 8 inches, the area would be 50.265 square inches.

Breadth in Inches.	Area of a Lineal Foot.	Breadth in Inches.	Area of a Lineal Foot.
¼	.0208	6¼	.5208
½	.0417	6½	.5416
¾	.0625	6¾	.5625
1	.0834	7	.5833
1¼	.1042	7¼	.6042
1½	.125	7½	.625
1¾	.1459	7¾	.6458
2	.1667	8	.6667
2¼	.1875	8¼	.6875
2½	.2084	8½	.7084
2¾	.2292	8¾	.7292
3	.25	9	.75
3¼	.2708	9¼	.7708
3½	.2916	9½	.7917
3¾	.3125	9¾	.8125
4	.3334	10	.8334
4¼	.3542	10¼	.8542
4½	.375	10½	.875
4¾	.3958	10¾	.8959
5	.4167	11	.9167
5¼	.4375	11¼	.9375
5½	.4583	11½	.9583
5¾	.4792	11¾	.9792
6	.5		

Example. To find the square feet in a board 14½ feet long and 9¼ inches wide.

The decimal in the table opposite 9¼ inches is .7708

Multiply by 14½

30832
7708
3854

Answer 11.1766 feet,
Or about 11 1/6 feet.

5985. Decimal Equivalents of the Divisions of a Foot.

	0	1	2	3	4	5	6	7	8	9	10	11
		.08333	.16666	.25	.33333	.41666	.5	.58333	.66666	.75	.83333	.91666
1/16	.00521	.08854	.17187	.25521	.33854	.42187	.50521	.58854	.67187	.75521	.83854	.92187
⅛	.01041	.09374	.17707	.26041	.34374	.42707	.51041	.59374	.67707	.76041	.84374	.92708
3/16	.01562	.09895	.18228	.26562	.34895	.43228	.51562	.59895	.68228	.76562	.84895	.93229
¼	.02083	.10416	.18750	.27083	.35416	.43759	.52083	.60416	.68750	.77083	.85416	.93750
5/16	.02604	.10937	.19270	.27604	.35937	.44270	.52604	.60937	.69270	.77604	.85937	.94270
⅜	.03125	.11458	.19791	.28125	.36458	.44791	.53125	.61458	.69791	.78125	.86458	.94791
7/16	.03646	.11979	.20312	.28646	.36979	.45312	.53646	.61979	.70312	.78646	.86979	.95312
½	.04166	.12500	.20832	.29166	.37500	.45833	.54166	.62500	.70832	.79166	.87500	.95833
9/16	.04687	.13020	.21353	.29687	.38020	.46354	.54687	.63020	.71353	.79687	.88020	.96354
⅝	.05208	.13541	.21874	.30208	.38541	.46875	.55208	.63541	.71874	.80208	.88541	.96875
11/16	.05729	.14062	.22395	.30729	.39062	.47395	.55729	.64062	.72395	.80729	.89062	.97395
¾	.06250	.14583	.22916	.31250	.39583	.47916	.56250	.64583	.72916	.81250	.89583	.97916
13/16	.06771	.15104	.23437	.31771	.40104	.48437	.56771	.65104	.73437	.81771	.90104	.98437
⅞	.07292	.15625	.23958	.32292	.40625	.48958	.57292	.65625	.73958	.82292	.90625	.98958
15/16	.07813	.16146	.24479	.32813	.41146	.49479	.57813	.66146	.74479	.82813	.91146	.99479

To use the above table—suppose it is required to find what decimal of a foot is equivalent to 8 inches—look for the column headed 8, and the figures at the top of that column, .66666, is the decimal required. Again, to find the decimal of a foot equal to 5¾ inches, look in the column under figure 5, run the finger down that column until it is level with the ¾ (marked on the left side of the table); the figures .47916 give the decimal required.

5986. To Find the Square Feet in Boards. Multiply the decimal in the table, corresponding to the width of the board, by the length of the board in feet.

5988. Table Showing the Square Inches Contained in a Circle from Ten to Seventy-Three Inches in Diameter.

Diameter of Circle.	Square Inches.	Diameter of Circle.	Square Inches.
10	78.54	42	1388.59
11	95.03	43	1452.20
12	113.10	44	1520.53
13	132.73	45	1590.43
14	153.94	46	1661.91
15	176.71	47	1735.00
16	201.06	48	1809.56
17	226.98	49	1885.74
18	254.47	50	1963.50
19	283.54	51	2042.82
20	314.16	52	2123.72
21	346.36	53	2206.19
22	380.13	54	2290.23
23	415.47	55	2375.83
24	452.39	56	2463.00
25	490.88	57	2551.76
26	530.93	58	2642.00
27	572.56	59	2734.00
28	615.75	60	2827.44
29	660.20	61	2922.47
30	706.86	62	3019.00
31	754.77	63	3117.25
32	804.25	64	3217.00
33	855.30	65	3318.31
34	907.92	66	3421.20
35	962.00	67	3526.66
36	1017.88	68	3651.69
37	1075.20	69	3739.29
38	1134.00	70	3848.46
39	1194.60	71	3959.20
40	1256.64	72	4071.51
41	1320.26	73	4185.40

The area may also be obtained by multiplying the square of the diameter by .7854. This method is deduced from the first one, and is founded on the fact that the square of any number is always 4 times as much as the square of half the number. In the first method the radius or *half diameter* is to be squared, and multiplied by 3.14159; in the second, the *whole diameter* is squared, which will result in just 4 times as much as the square of the radius; the multiplier must be therefore the fourth part of 3.14159, or .7854.

5989. To Find the Area of a Parallelogram or Square. Multiply the length of one side by the perpendicular height.

5990. To Find the Area of a Triangle. Multiply the base by ½ the perpendicular height. Or, to find the area from three sides given, from the half sum of the three sides subtract each side separately; multiply the half sum and the three remainders together, and the square root of the product will be the area.

5991. To Find the Area of a Trapezoid. Multiply the sum of the two parallel sides by ½ the perpendicular height.

5992. To Find the Area of a Sector of a Circle. Multiply the radius of the circle by ½ the arc of the sector.

5993. To Find the Area of a Segment of a Circle. Find the area of a sector of a circle having the same arc, and deduct the triangle formed between the two radii and the chord of the arc.

5994. Cloth Measure, used for measuring dry goods.

Yard.		Quarters.		Nails.		Inches.
1	=	4	=	16	=	36
		1	=	4	=	9
				1	=	2¼

The height of horses is measured by the "hand" of 4 inches.

5995. Gunter's Chain. This is the

measure generally adopted in land surveying, is 22 yards in length, and contains 100 links, each link, consequently being 7.92 inches long. The length of the chain was fixed at 22 yards, because a square whose side is 22 yards (1 chain) contains exactly $\frac{1}{10}$ acre; in other words, a rectangular plot of ground 1 chain in width and 10 chains in length contains an acre. 80 chains make 1 mile in length; and, consequently, a square mile contains 640 acres. For surveying and laying out plots and building lots, a chain of 50 feet, or one of 25 feet (the usual frontage of a lot) is usually employed by surveyors.

5996. Cubic or Solid Measurement.

Yard.		Feet.		Inches.
1	=	27	=	46,656
		1	=	1,728

5997. American Cord-Wood Measure. Timber is measured by the ton of 50 cubic feet of round, or 40 cubic feet of hewn timber. Cord-wood is measured by the cord, which consists of a pile 8 lineal feet long and 4 feet high; and, as the wood is reckoned to be 4 feet in length, contains 128 cubic feet. A stick of cord-wood should measure 4 feet 4 inches from end to end, to compensate for the slope or bevil of the cut, and provide for an equivalent of 4 feet of solid wood. The contents of each lineal foot of the length of the pile is called a cord foot, and contains one-eighth part of a cord, or 16 cubic feet. A New York load of wood is one-third of a cord.

A shipping ton contains 42 cubic feet.

Also, the cubic foot being considered unity, or 1, a cylinder 1 foot in diameter and 1 foot in length = .7854.

A sphere 1 foot in diameter = .5236.

A cone 1 foot in diameter at the base and 1 foot in height = .2619.

5998. Cubic Measure in Cubic Metres.

1 Yard	=	.76450	Cubic Metres
1 Foot	=	28.31486	Cubic Decimetres
1 Inch	=	16.38591	Cubic Centimetres

5999. Table of Solid Feet reduced to Solid Inches.

Feet.	Inches.	Feet.	Inches.	Feet.	Inches.
2=	3456	35=	60480	68=	117504
3	5184	36	62208	69	119232
4	6912	37	63936	70	120960
5	8640	38	65664	71	122688
6	10368	39	67392	72	124416
7	12096	40	69120	73	126144
8	13824	41	70848	74	127872
9	15552	42	72576	75	129600
10	17280	43	74304	76	131328
11	19008	44	76032	77	133056
12	20736	45	77760	78	134784
13	22464	46	79488	79	136512
14	24192	47	81216	80	138240
15	25920	48	82944	81	139968
16	27648	49	84672	82	141696
17	29376	50	86400	83	143424
18	31104	51	88128	84	145152
19	32832	52	89956	85	146880
20	34560	53	91584	86	148608
21	36288	54	93312	87	150336
22	38016	55	95040	88	152064
23	39744	56	96768	89	153792
24	41472	57	98496	90	155520
25	43200	58	100224	91	157248
26	44928	59	101952	92	158976
27	46656	60	103680	93	160704
28	48384	61	105408	94	162432
29	50112	62	107136	95	164160
30	51840	63	108864	96	165888
31	53568	64	110592	97	167616
32	55296	65	112320	98	169344
33	57024	66	114048	99	171072
34	58752	67	115776	100	172800

6000. Measurement of Stone and Brick-Work.

1 *Perch, Masons' or Quarrymen's Measure.*

16½ feet long, 16 inches wide, 12 " high, } = { 22 cubic feet. To be measured in wall.

1 ½ feet long, 18 inches wide, 12 " high, } = { 24.75 cubic feet. To be measured in pile.

1 cubic yard = 3 feet×3 feet×3 feet = 27 cubic feet. The cubic yard has become the standard for all contract work of late years. Stone walls less than 16 inches thick count as if 16 inches thick to mason; over 16 inches thick, each inch additional is measured.

Number of Bricks required in Walls for each Square Foot of Face of Wall.

Thickness of Wall.		Thickness of Wall.	
4 inches	7½	24 inches	46
8 "	15	28 "	52½
12 "	22½	32 "	60
16 "	30	36 "	67½
20 "	37½	42 "	75

Cubic yard = 600 bricks in wall.

Perch (22 cubic feet) = 500 bricks in wall.

To pave 1 sq. yard on flat requires 41 bricks.

" 1 " edge " 68 "

6001. To Find the Cubical Contents of a Cylinder. Find the area of the circular end, as directed in No. 5987, and then multiply the area by the length of the cylinder; the product will be the cubical content. The same denomination of measurement must be adhered to throughout the calculation, as, if the diameter or area is in inches, the length must be in inches. Thus: to find the cubical content of a cylinder 8 inches in diameter and 3 feet long; we find in No. 5987 that the area of a circle 8 inches in diameter is 50.265 square inches; multiply this by 36 inches (3 feet reduced to inches, the same denomination as the given diameter), and the product is 1809.54 cubic inches, or 1 foot, 81.54 cubic inches.

6002. Table of Spherical Contents, &c. This table shows the relative proportions between the diameter, surface, and capacity (or cubical contents) of spheres.

Diameters.	Surfaces.	Capacities.
1	3.141	.523
2	12.567	4.188
3	28.274	14.137
4	50.265	33.51
5	78.540	65.45
10	314.159	523.6
15	706.9	1767.1
20	1256.6	4189.
25	1963.5	8181.
30	2827.	14137.
40	5026.	33510.

6003. To Find the Cubical Contents of Spars or Other Round Timber. If the spar or timber were the same thickness through its entire length, the diameter of all parts would be the same, and one measurement would suffice to obtain the correct diameter; its cubical contents could then be found in the same way as for a cylinder; but this is hardly ever the case, as the thickness or diameter is different in every part. If the spar tapers regularly from one end to the other, measure the diameter at each end, add the two measurements together, and divide their sum by 2; this will give the *average* diameter. A piece of timber of irregular thickness must be measured in portions, each portion extending as far as the tapering is regular, and the contents of the different portions added together to get the contents of the whole. Having obtained the correct diameter in inches, look for it in the next table, and opposite it, in the next column to the right, will be the contents in feet of 1 foot of timber in length; multiply this by the length of the timber in feet, and the result will be the contents of the whole.

Thus, to find the contents of a 16-foot log whose average diameter is found to be 13½ (that is, 13.5) inches, we find the figures on the next right hand column in the table are .99; this means that a log 1 foot long and 13½ inches in diameter contains .99 or $\frac{99}{100}$ of a cubic foot. Multiply this .99 by 16, the length of the log in feet, and we get 15.84, or about 15¾ cubic feet, which is the contents of the whole log.

About 10 per cent. should be deducted from the results given in the table when toll is charged on rafts of spars or logs, for the reason that many sticks of timber taper suddenly, and others are unequal in diameter when the average is taken.

Diameter Inches.	Contents. 1 foot long.	Diameter Inches.	Contents. 1 foot long.
4.	.0872	27.5	4.12
5.	.137	28.	4.28
6.	.196	28.5	4.43
7.	.267	29	4.59
7.5	.31	29.5	4.75
8.	.35	30.	4.91
8.5	.39	30.5	5.07
9.	.44	31.	5.24
9.5	.49	31.5	5.41
10.	.55	32.	5.58
10.5	.60	32.5	5.76
11.	.66	33.	5.94
11.5	.72	33.5	6.12
12.	.79	34.	6.31
12.5	.85	34.5	6.49
13.	.92	35.	6.68
13.5	.99	35.5	6.87
14.	1.07	36.	7.07
14.5	1.15	36.5	7.27
15.	1.23	37.	7.47
15.5	1.31	37.5	7.67
16.	1.40	38.	7.88
16.5	1.48	38.5	8.09
17.	1.58	39.	8.30
17.5	1.67	39.5	8.51
18.	1.77	40.	8.73
18.5	1.87	40.5	8.95
19.	1.97	41.	9.17
19.5	2.07	42.	9.61
20.	2.18	43.	10.08
20.5	2.29	44.	10.555
21.	2.40	45.	11.044
21.5	2.52	46.	11.541
22.	2.64	47.	12.049
22.5	2.76	48.	12.566
23.	2.89	49.	13.095
23.5	3.11	50.	13.635
24.	3.14	51.	14.186
24.5	3.27	52.	14.747
25.	3.41	53.	15.320
25.5	3.55	54.	15.904
26.	3.69	55.	16.499
26.5	3.83	56.	17.104
27.	3.98	57.	17.720

6004. Capacity of Cubical Boxes. A box 1 foot and 1 inch each way, *i. e.*, length, breadth, and depth, will contain 1 standard bushel.

Feet.	Inches.		Bushels.
1	1	=	1
1	4½	=	2
1	6¾	=	3
1	8½	=	4
1	10$\frac{1}{16}$	=	5
1	11½	=	6
2	¾	=	7
2	2	=	8
2	3	=	9
2	4	=	10

6005. Capacity of Boxes of Different Dimensions. A box 4 feet 7 inches long, and 2 feet 4 inches in width, and 2 feet 4 inches in depth, will contain 20 bushels. The dimensions of a cylinder containing 1 United States standard bushel are 18½ inches inside diameter, and 8 inches deep. A box 24 inche by 16 inches square, and 28 inches deep will contain a barrel, 5 bushels. A box 24 inches by 16 inches square, and 14 inches deep, will contain a half barrel. A box 24 inches by 11.2 inches square, and 8 inches deep, will contain 1 bushel. A box 12 inches by 11.2 inches square, and 8 inches deep, will contain ½ bushel. A box 8 inches by 8.4 inches square, and 8 inches deep, will contain 1 peck. A box 8 inches by 8 inches square, and 4.2 inches deep, will contain 1 gallon. A box 7 inches by 8 inches square, and 4.8 inches deep, will contain 1 gallon. A box 4 inches by 4 inches square, and 4.2 inches deep, will contain 1 quart.

6006. To Find the Amount of Lumber any Log will Make. Find the length of the log in the left-hand column of the next Table; then on the top of the page find the diameter, and under the same will be found the quantity of lumber the log will make; calculated for any length from 10 to 25 feet, and for any diameter from 12 to 44 inches.

Table Showing the Number of Feet of Inch-Board in a Log of Timber.

Length in Feet	Diameter in Inches.																
	12	13	14	15	16	17	18	19	20	21	22	23	24	25	26	27	28
10	49	61	72	89	99	116	133	150	175	190	209	235	252	287	313	342	363
11	54	67	79	98	109	127	147	165	192	209	230	259	278	315	344	377	400
12	59	73	86	107	119	139	160	180	210	228	251	283	303	344	375	411	436
13	64	79	93	116	129	150	173	195	227	247	272	306	328	373	408	445	473
14	69	85	100	125	139	162	187	210	245	266	292	330	353	401	439	479	509
15	74	91	107	134	149	173	200	225	262	285	313	353	379	430	469	514	545
16	79	97	114	142	159	185	213	240	280	304	334	377	404	459	500	548	582
17	84	103	122	151	168	196	227	255	297	323	355	400	429	487	531	582	618
18	89	109	129	160	178	208	240	270	315	342	376	424	454	516	562	616	654
19	93	116	136	169	188	219	253	285	332	361	397	447	480	545	594	650	692
20	98	122	143	178	198	232	267	300	350	380	418	470	505	573	625	684	728
21	103	128	150	187	208	243	280	315	368	399	439	495	530	603	656	719	764
22	108	134	157	196	218	255	293	330	385	418	460	518	555	631	688	753	800
23	113	140	164	205	228	266	307	345	403	437	480	542	571	659	719	787	837
24	118	146	172	214	238	278	320	360	420	456	501	566	606	688	750	821	873
25	123	152	179	223	248	289	333	375	438	475	522	589	631	717	781	856	910

Length in Feet	Diameter in Inches.															
	29	30	31	32	33	34	35	36	37	38	39	40	41	42	43	44
10	381	411	444	460	490	500	547	577	644	669	700	752	795	840	872	925
11	419	451	488	506	539	550	602	634	708	734	770	828	874	924	959	1017
12	457	493	532	552	588	600	657	692	772	801	840	903	954	1007	1046	1110
13	495	534	576	598	637	650	712	750	836	868	910	978	1033	1091	1135	1203
14	533	575	622	644	686	700	766	807	901	934	980	1053	1113	1175	1222	1295
15	571	616	666	690	735	750	821	865	965	1001	1050	1129	1192	1259	1309	1388
16	609	657	710	736	784	800	876	923	1029	1068	1120	1204	1272	1343	1396	1480
17	647	698	755	782	833	850	931	980	1094	1134	1190	1279	1351	1427	1484	1573
18	685	739	799	828	882	900	985	1038	1158	1201	1260	1354	1431	1511	1571	1665
19	723	780	843	874	931	950	1040	1096	1222	1268	1330	1430	1510	1595	1658	1758
20	761	821	888	920	980	1000	1095	1152	1287	1335	1400	1505	1590	1679	1745	1850
21	800	863	932	966	1029	1050	1150	1210	1351	1401	1470	1580	1669	1763	1833	1943
22	838	904	976	1012	1078	1100	1204	1268	1415	1468	1540	1655	1749	1847	1920	2035
23	876	945	1021	1058	1127	1150	1259	1322	1480	1535	1610	1730	1828	1931	2007	2128
24	914	986	1065	1104	1176	1200	1314	1380	1544	1601	1680	1806	1908	2015	2094	2220
25	952	1027	1109	1150	1225	1250	1369	1438	1608	1668	1750	1881	1987	2099	2182	2313

6007. Measure of Time.

Lunar Month.	Weeks.	Days.	Hours.	Minutes.	Seconds.
1 =	4 =	28 =	672 =	40,320 =	2419,200
	1 =	7 =	168 =	10,080 =	604,800
		1 =	24 =	1,440 =	86,400
			1 =	60 =	3,600
				1 =	60

The year of 365 days is divided into 12 calendar months, 7 of which have 31 days; 4 have 30 days; and 1, February, 28 days. The solar year consists of 365 days, 5 hours, 48 minutes, and 49 seconds; this excess over 365 days, nearly 6 hours, or ¼ day, is allowed to accumulate through each 4 years, and provided for every fourth, or leap year, by adding 1 day to February; but as this is adding a trifle too much, every 400 years one leap year is omitted, and this occurs when the year is divisible by 400 without remainder.

In the year 1582, the fact was observed by Pope Gregory XIII that, in consequence of this discrepancy not having been taken into account since the commencement of the Julian system (*see No.* 6064), the true time exceeded the time as then reckoned by 10 days; and therefore ordered the 11th of March to be accounted the 21st. The Pope's edict was generally observed by the nations subject to his authority, but the Protestant countries continued the use of the Julian reckoning. This gave rise to the two modes of computation still found in Europe, called the old style and new style. The latter was adopted in England in 1752, by making the 1st of September the 12th.

Whenever the date of the year is divisible by 4 without remainder, February has 29 days, and that year is called Bissextile.

6008. Table Showing the Number of Days from any Date in One Month to the Same Date in any Other Month.

From To	Jan.	Feb.	Mar.	April.	May.	June.	July.	Aug.	Sept.	Oct.	Nov.	Dec.
January.	365	31	59	90	120	151	181	212	243	273	304	334
Feby....	334	365	28	59	89	120	150	181	212	242	273	303
March...	306	337	365	31	61	92	122	153	184	214	245	275
April.....	275	306	334	365	30	61	91	122	153	183	214	244
May.....	245	276	304	335	365	31	61	92	123	153	184	214
June....	214	245	273	304	334	365	30	61	92	122	153	183
July.....	184	215	243	274	304	335	365	31	62	92	123	153
August..	153	184	212	243	273	304	334	365	31	61	92	122
Sept.....	122	153	181	212	242	273	303	334	365	30	61	91
October.	92	123	151	182	212	243	273	304	335	365	31	61
Nov.....	61	92	120	151	181	212	242	273	304	334	365	30
Dec......	31	62	90	121	151	182	212	243	274	304	335	365

Example: How many days from the 2d of February to the 2d of August? Look for February at the left hand, and August at the top, in the angle is 181. In leap year, add one day if February be included.

6009. Table Showing Difference of Time at 12 o'Clock (Noon) at New York.

New York......12.00	N.	Boston.........12.12	P. M.
Buffalo.........11.40	A. M.	Quebec.........12.12	"
Cincinnati......11.18	"	Portland........12.15	"
Chicago.........11.07	"	London......... 4.55	"
St. Louis.......10.55	"	Paris........... 5.05	"
San Francisco.. 8.45	"	Rome........... 5.45	"
New Orleans...10.56	"	Constantinople. 6.41	"
Washington....11.48	"	Vienna......... 6.00	"
Charleston......11.36	"	St. Petersburg.. 6.57	"
Havana.........11.25	"	Pekin, night...12.40	A. M.

6010. Geographical or Nautical Measure.

Great Circle.		Degrees.		Leagues.		Geo. Miles.
1	=	360	=	7200	=	21600
		1	=	20	=	60
				1	=	3

The geographical or nautical mile, according to Brande, is equivalent to 1.153 statute miles; this would give 2029.3 yards to the nautical mile, 69.18 statute miles to the degree, and about 24.905 miles for the earth's equatorial circumference. According to one of the very best authorities, Chambers' Encyclopædia, the nautical mile contains 2029 yards; on this basis, a degree would measure about 69.17 statute miles, and the earth's circumference about 24.901 statute miles. A great circle of the earth is an imaginary line or belt so drawn round the earth as to divide it into two equal parts or hemispheres; the equator and the ecliptic are great circles. In navigation, sailors measure depth of soundings and short distances by the *fathom* of 6 feet, and the *cable-length* of 120 fathoms.

6011. Nautical Time. The hour of the day or night is noted on board a ship by 1, 2, 3, &c., up to 8 bells. The 12 hours between midnight and noon, or noon and midnight, are divided into 3 portions of 8 bells each, the duration of time between bells being half an hour. During the course of each 12 hours, the same number of strokes of the bell will necessarily be used to denote three different hours or periods of time.

Bell.		Clock-Time.	Clock-Time.	Clock-Time.
1	denotes	12.30	4.30	8.30
2	"	1.	5.	9.
3	"	1.30	5.30	9.30
4	"	2.	6.	10.
5	"	2.30	6.30	10.30
6	"	3.	7.	11.
7	"	3.30	7.30	11.30
8	"	4.	8.	12.

6012. Capacity of Cisterns, &c.

Diameter in Feet and Inches.	Depth in Feet and Inches.	Number of Wine Gallons.	Number of Barrels.	No. of Hhds.	No. of Gallons in 10 Inches Depth.
2 ft.	2 ft.	45	$1\frac{9}{21}$	$\frac{45}{63}$	19
2 ft. 6 in.	2 ft. 6 in.	90	$2\frac{6}{7}$	$1\frac{6}{14}$	30
3 ft.	3 ft.	158	5	$2\frac{1}{2}$	44
3 ft. 6 in.	3 ft. 6 in.	252	8	4	60
4 ft.	4 ft.	374	$11\frac{55}{63}$	$5\frac{59}{63}$	78
4 ft. 6 in.	4 ft. 6 in.	524	$16\frac{40}{63}$	$8\frac{20}{63}$	97
5 ft.	5 ft.	732	$23\frac{15}{63}$	$11\frac{39}{63}$	122
5 ft. 6 in.	5 ft. 6 in.	976	31	$15\frac{1}{2}$	148
6 ft.	6 ft.	1267	$40\frac{2}{9}$	$20\frac{1}{9}$	176
6 ft. 6 in.	6 ft. 6 in.	1614	$51\frac{15}{63}$	$25\frac{39}{63}$	207
7 ft.	7 ft.	2016	64	32	240
8 ft.	8 ft.	3004	$95\frac{23}{63}$	$47\frac{43}{63}$	313
8 ft. 6 in.	8 ft. 6 in.	3600	$114\frac{18}{63}$	$57\frac{9}{63}$	353
9 ft.	9 ft.	4276	$135\frac{47}{63}$	$67\frac{55}{63}$	396
9 ft. 6 in.	9 ft. 6 in.	5027	$159\frac{37}{63}$	$79\frac{50}{63}$	441
10 ft.	10 ft.	5868	$186\frac{18}{63}$	$93\frac{9}{63}$	489
11 ft.	11 ft.	7814	$248\frac{4}{63}$	$124\frac{2}{63}$	592
12 ft.	12 ft.	10152	$322\frac{18}{63}$	$161\frac{9}{63}$	705
13 ft.	13 ft.	12901	$409\frac{5}{9}$	$204\frac{49}{63}$	827
14 ft.	14 ft.	16111	$511\frac{29}{63}$	$255\frac{45}{63}$	959
15 ft.	15 ft.	19818	$629\frac{8}{63}$	$314\frac{36}{63}$	1101
20 ft.	20 ft.	46992	$1491\frac{51}{63}$	$745\frac{57}{63}$	1958
25 ft.	25 ft.	91770	$2913\frac{1}{8}$	$1456\frac{2}{3}$	3059

Example: Suppose you desire to ascertain the capacity of a cistern 4 feet 6 inches in diameter and 4 feet 6 inches in depth. Find the diameter in the left hand column, and directly opposite you will see that the cistern will hold 524 gallons of 231 cubic inches each, equal to $16\frac{40}{63}$ barrels, or $8\frac{20}{63}$ hogsheads. The right hand column shows the number of gallons contained in 10 inches of depth. By this standard you may easily increase or diminish the capacity at pleasure. Thus, if you wish the above cistern to hold 97 gallons more, make it 10 inches deeper; or 194 gallons more, 20 inches deeper.

6013. Log Lines. 1 knot = 51.1625 feet, or 51 feet $1\frac{3}{4}$ + inches. 1 fathom = 5.11625 feet, or 5 feet $1\frac{1}{3}$ + inches, estimating a mile at $6139\frac{1}{2}$ feet, and using a 30″ glass. If a 28″ glass is used, and eight divisions, then 1 knot = 47 feet 9 + inches. 1 fathom = 5 feet $11\frac{3}{8}$ inches. The line should be about 150 fathoms long, having 10 fathoms between the chip and first knot for stray line. Miles × .87 = knots. Knots × 1.15 = miles. Feet per minute × .01 = knots per hour. 1 knot = 6082.66 feet; 1 statute mile = 5280 feet.

6014. The Decimal System of Weights and Measures. A permissive law has already been passed by the American and British governments, adopting the decimal system as applied to weights and measures. It is substantially the same as the French decimal system, and founded on units of the same value. The multiples and subdivisions of the different units are the same; Greek prefixes being used to denote the multiples, and Latin prefixes the fractional parts of the units.

The Greek prefix		DEKA	means	10	units
"	"	HECTO	"	100	"
"	"	KILO	"	1000	"
"	"	MYRIA	"	10000	"
The Latin prefix		DECI	"	$\frac{1}{10}$	of a unit
"	"	CENTI	"	$\frac{1}{100}$	"
"	"	MILLI	"	$\frac{1}{1000}$	"

The fundamental unit of all the decimal weights and measures is the METRE; the standard length of which is the $\frac{1}{10000000}$ of a quadrant of the earth's meridian, equivalent to 39.371 inches. The unit of dry and liquid measures of capacity is the LITRE, which is the $\frac{1}{1000}$ of a cubic metre, and contains 61.028 cubic inches. These figures are as exact as a calculation involving twelve places of decimals will bring it. The government standard, adopted as sufficiently correct for all practical purposes, is 61.022 cubic inches; this is based on a metre of 39.3685 inches, which would make the gram 15.432 grains. The GRAM or unit of weight is the weight of a cubic centimetre ($\frac{1}{100}$ of a metre) of water at 39.83° Fahr., and is equivalent to 15.434 grains. For post-office purposes, the $\frac{1}{2}$ ounce avoirdupois is declared equivalent to 15 grams. The ARE, or unit of surface measurement, is the $\frac{1}{100}$ of a square metre, or 119.6 square yards. This system of weights and measures has not as yet come into general use, either in America or England. Its advantages are indisputably great for facilitating calculation as well as establishing uniform international standards; but its adoption necessarily meets with much opposition, as it overthrows not only all the old, arbitrary units of measurement, but their multiples and subdivisions also. It seems so natural to halve and quarter, and count by the dozen, that even in our decimal currency we cannot dispense with the half and quarter dollar and eagle; in fact, the advantage of our decimal currency cannot be appreciated to its full extent until the custom of counting by the dozen is entirely superseded by the decade. The dozen, 12, is divisible by 2, 3, 4, and 6; the decade, 10, by 2 and 5 only; and, although this is a matter of little moment as far as regards calculation, it makes a great difference for practical subdivision. Old rooted customs are difficult to eradicate, but there is no doubt that the dozen, half, and quarter, those stumbling-blocks in the way of the decimal system, will eventually disappear as entirely as the now totally obsolete eighth and sixteenth of a dollar, the Mexican shilling and sixpence.

6015. Official Standard Metre. The following information was lately given by Mr. Hilgard, of the United States Coast Survey, to the Journal of the Franklin Institute: "There are, in the custody of the Treasury Department, at the Office of Weights and Measures, the following authentic copies of the standard metre and kilogramme of France, viz.: Metre of platinum, compared and certified by Arago; metre of steel, compared and certified by Silbermann; kilogramme of platinum, compared and certified by Arago; kilogramme of brass (gilt), compared and certified by Silbermann. The length of the metre is 39.3685 inches of the United States standard scale, and the kilogramme is 15432.2 grains, or 2 pounds, 3 ounces, 119.7 grains avoirdupois. There is also another metre, the property of the American Philosophical Society, which is one of the twelve original metres made by the French Government, and was brought to this country by Mr. Hassler, the originator of the United States Coast Survey. A comparison between this bar and the standard of France at the Conservatory of Arts and Trades was made by Dr. F. A. P. Barnard, with the result that, at the temperature of melting ice, there is no appreciable difference, by the most delicate means of comparison, between the platinum standard of the Conservatory and this iron metre."

The above standard metre of 39.3685 inches would make the equatorial circumference of the earth measure 24.854 statute miles. Bessel's calculations, given in Chambers' Encyclopædia, give the equatorial circumference at $24.901\frac{4}{11}$ miles. If this measurement be correct, the standard metre should be 39.371 inches. This difference, however, is so trifling that it would not be appreciable for all practical purposes.

6016. Decimal Measures of Length.

Myriametre	=	10,000	metres.
Kilometre	=	1,000	metres.
Hectometre	=	100	metres.
Dekametre	=	10	metres.
Metre	=	1	metre.
Decimetre	=	$\frac{1}{10}$	metre.
Centimetre	=	$\frac{1}{100}$	metre.
Millimetre	=	$\frac{1}{1000}$	metre.

6017. Value of Metric Measures of Length in Long Measure.

		Miles.	Yds.	Ft.	Inches.
Myriametre	=	6	376	1	2
Kilometre	=		1093	1	11
Hectometre	=		109	1	1.1
Dekametre	=		10	2	9.71
Metre	=		1	0	3.371
Decimetre	=				3.937
Centimetre	=				.394

For general purposes, or small calculations, the following equivalents will be found sufficiently accurate: 1 millimetre is equal to $\frac{1}{25}$ inch; 1 centimetre is equal to $\frac{2}{5}$ inch; 1 decimetre is equal to $3\frac{9}{10}$ inches; 1 metre is equal to $39\frac{1}{3}$ inches; $\frac{91}{100}$ metre is equal to 36 inches or 1 yard.

6018. Value of Metres in Inches.

Millimetre.		Metre.		Inches.
1	=	.001	=	.03937
2	=	.002	=	.07874
3	=	.003	=	.11811
4	=	.004	=	.15748
5	=	.005	=	.19685
6	=	.006	=	.23622
7	=	.007	=	.27560
8	=	.008	=	.31497
9	=	.009	=	.35434
Centimetre.				
1	=	.01	=	.3937
2	=	.02	=	.7874
3	=	.03	=	1.1811
4	=	.04	=	1.5748
5	=	.05	=	1.9685
6	=	.06	=	2.3622
7	=	.07	=	2.7559
8	=	.08	=	3.1497
9	=	.09	=	3.5434
Decimetre.				
1	=	.1	=	3.9371
2	=	.2	=	7.8742
3	=	.3	=	11.8113
4	=	.4	=	15.7484
5	=	.5	=	19.6855
6	=	.6	=	23.6226
7	=	.7	=	27.5597
8	=	.8	=	31.4968
9	=	.9	=	35.4339

6019. Value of Metres in Feet.

Decimetres.		Feet.
1	=	.328
2	=	.656
3	=	.984
4	=	1.312
5	=	1.640
6	=	1.968
7	=	2.297
8	=	2.625
9	=	2.953
Metres.		
1	=	3.281
2	=	6.562
3	=	9.843
4	=	13.124
5	=	16.405
6	=	19.686
7	=	22.967
8	=	26.248
9	=	29.529

Dekametre.		Metres.		
1	=	10	=	32.81
2	=	20	=	65.62
3	=	30	=	98.43
4	=	40	=	131.24
5	=	50	=	164.05
6	=	60	=	196.86
7	=	70	=	229.67
8	=	80	=	262.48
9	=	90	=	295.29
Hectometre.				
1	=	100	=	328.1
2	=	200	=	656.2
3	=	300	=	984.3
4	=	400	=	1312.4
5	=	500	=	1640.5
6	=	600	=	1968.6
7	=	700	=	2296.7
8	=	800	=	2624.8
9	=	900	=	2952.9

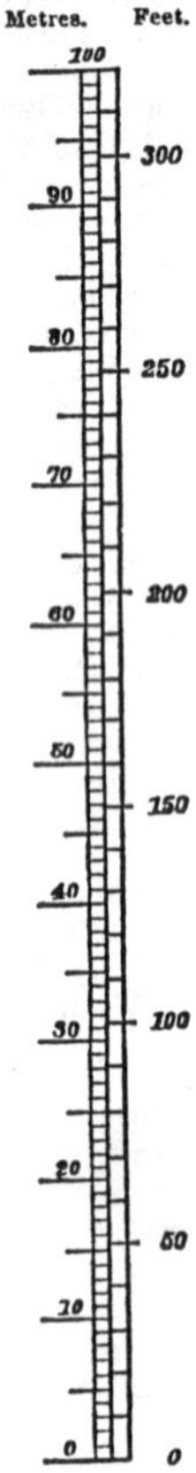

The foregoing scale may be used for any other portion of the metrical system; for instance, if millimetres be used instead of decimetres, the relative scale of feet will consist of the same figures, with the decimal point removed one place to the left, to divide by 10, the millimetre being $\frac{1}{10}$ decimetre.

6020. Decimal Measures of Capacity.

Names.	Number of Litres.	Cubic Measure.
Kilolitre, or stere	1,000	1 cubic metre
Hectolitre........	100	$\frac{1}{10}$ cubic metre
Dekalitre........	10	10 cu. decimetres
Litre............	1	1 cub. "
Decilitre........	$\frac{1}{10}$	$\frac{1}{10}$ cub. "
Centilitre.......	$\frac{1}{100}$	10 cu. centimetre
Millilitre........	$\frac{1}{1000}$	1 cu. centimetre

The following are approximate values, correct enough for rough calculations. One millilitre is equal to 15½ grain measures of water; one centilitre is equal to 154 grain measures, or 3 fluid drachms; one decilitre is equal to 1,540 grain measures, or 3½ fluid ounces; one litre is equal to 15,406 grain measures, or 2$\frac{1}{10}$ pints; one cubic centimetre of water at its maximum density weighs 15½ grains, and is ¼ fluid drachm.

6021. Value of Metric Measures of Capacity in U. S. Dry Measure.

		Bush.	Peck.	Quart.	Pint.
Kilolitre	=	28	1	4½	
Hectolitre	=	2	3	2	1.6
Dekalitre	=		1	0	1.6
Litre	=				1.816
Decilitre	=				.181
Centilitre	=				.018

6022. Value of Metric Measures of Capacity in U. S. Liquid Measure.

		Gals.	Quarts.	Pints.	Gills.
Kilolitre	=	264	0	1	1.6
Hectolitre	=	26	1	1	1.36
Dekalitre	=	2	2	1	0.136
Litre	=		1	0	0.413
Decilitre	=				.841
Centilitre	=				.084

6023. Equivalent of Metric Measures of Capacity in U. S. Apothecaries Measure.

		Gal.	Pint.	Fluid Ounce.	Fluid Drachm.	Minims.
Hectolitre	=	26	3	5	5	20
Dekalitre	=	2	5	2	1	20
Litre	=		2	1	6	32
Decilitre	=			3	3	3
Centilitre	=				2	42

6024. Value of Metric Measures of Capacity in Imperial Dry Measure.

		Bush.	Pecks.	Gals.	Pints.
Kilolitre	=	27	2	0	0.800
Hectolitre	=	2	3	0	0.080
Dekalitre	=			2	1.608
Litre	=				1.760
Decilitre	=				.176

6025. Value of Metric Measures of Capacity in Imperial Liquid Measure.

		Hhds.	Gals.	Qts.	Pts.	Gills.
Kilolitre	=	3	31	0	0	3.200
Hectolitre	=		22	0	0	0.320
Dekalitre	=		2	0	1	2.432
Litre	=				1	3.040
Decilitre	=					.704

6026. Decimal Measures of Surface.

		Equivalents in Square Measure. Acres.	Sq. yds.	Sq. ft.
Hectare..	10,000 square metres	2	2279	5.76
Are......	100 square metres		119	5.4
Centare...	1 square metre		1	1.76

6027. Decimal Weights.

Names	Number of Grams.	Weight of what quantity of Water at maximum density.
Millier, or Tonneau	1,000,000	1 cub. metre
Quintal............	100,000	1 hectolitre
Myriagram........	10,000	10 litres
Kilogram or kilo..	1,000	1 litre
Hectogram.......	100	1 decilitre
Dekagram........	10	10 cu. cent're
Gram............	1	1 cu. cent're
Decigram.........	$\frac{1}{10}$	$\frac{1}{10}$ cu. cent're
Centigram........	$\frac{1}{100}$	10 cu. milm's
Milligram.........	$\frac{1}{1000}$	1 cu. milm'e

6028. Equivalent of Metric Weights in Avoirdupois Weight.

		Lbs.	Oz.	Dr.
Millier	=	2204	9	1.6
Quintal	=	220	7	4.96
Myriagram	=	22	0	11.69
Kilogram	=	2	3	4.37
Hectogram	=		3	8.44
Dekagram	=			5.64
Gram	=			.56

6029. Equivalent of Metric Weights in Troy Weight.

		Lbs.	Oz.	Dwts.	Grains.
Millier	=	2677	1	19	20.
Quintal	=	267	8	11	23.6
Myriagram	=	26	9	5	4.77
Kilogram	=	2	8	2	12.48
Hectogram	=		3	4	6.05
Dekagram	=			6	10.21
Gram	=				15.43
Decigram	=				1.54
Centigram	=				.15

6030. Equivalent of Metric Weights in U. S. Apothecaries Weight.

		Lbs.	Oz.	Dr.	Scr.	Grs.
Millier	=	2677	1	7	2	16.
Quintal	=	267	8	4	2	7.6
Myriagram	=	26	9	2	0	4.77
Kilogram	=	2	8	1	0	0.48
Hectogram	=		3	1	2	2.05
Dekagram	=			2	1	14.21
Gram	=					15.43
Decigram	=					1.54

For general purposes the following values are sufficiently correct: 1 milligram is equal to $\frac{1}{64}$ grain; 1 centigram is equal to $\frac{1}{6}$ grain; 1 decigram is equal to 1½ grains; 1 gram is equal to 15½ grains; 1 dekagram is equal to 154 grains; 1 hectogram is equal to 1,543 grains; 1 kilogram is equal to 15,432 grains.

6031. English Weights and Measures. Avoirdupois and Troy weight are exactly the same as used in the United States, and the tables will be found in Nos. 5935, &c. In the new British Pharmacopœia, the weights are expressed in pounds, ounces, and grains, avoirdupois; thus superseding the Apothecaries weight as now in use in the United States. The old British avoirdupois drachm ($\frac{1}{16}$ ounce or 27.344 grains) is now obsolete, except in weighing silk. The new drachm is ⅛ ounce.

6032. Imperial Standard Measure.

Gal.		Quarts.		Pints.		F. Oz.		F. Dr.		Minims.
1	=	4	=	8	=	160	=	1280	=	76.800
		1	=	2	=	40	=	320	=	19.200
				1	=	20	=	160	=	9.600
						1	=	8	=	480
								1	=	60

The standard unit of this measure is the gallon which is declared by statute to contain 10 pounds avoirdupois (70,000 Troy grains) of distilled water at a temperature of 62° Fahr., the barometer being at 30 inches. The weight of a cubic inch of water, under the foregoing conditions, is 252.458 grains; the capacities of the measures are therefore as follows:

Imperial Gallon	=	277.274	Cubic Inches.
" Quart	=	69.3185	"
" Pint	=	34.65925	"
Fluid Ounce	=	1.73296	"
" Drachm	=	.21662	"

Thus it will be seen that there is a slight difference in weight between the English and United States unit of capacity, viz.: The cubic inch of water; the English being weighed at 62° Fahr., and the United States at 39.83°. (*See No.* 5935.)

6033. Imperial Measure Expressed in Litres.

1 Gallon	=	4.54339 Litres
1 Quart	=	1.13585 "
1 Pint	=	5.67925 Decilitres
1 Fluid Ounce	=	2.83962 Centilitres
1 " Drachm	=	3.54952 Millilitres
1 Minim	=	.05916 "

6034. Measure of Capacity for all Liquids.

Tun.		Pipes.		Hhds.		Bbls.		Gallons.		Quarts.		Pints.		Gills.
1	=	2	=	4	=	8	=	252	=	1008	=	2016	=	8064
		1	=	2	=	4	=	126	=	504	=	1008	=	4032
				1	=	2	=	63	=	252	=	504	=	2016
						1	=	31½	=	126	=	252	=	1008
								1	=	4	=	8	=	32
										1	=	2	=	8
												1	=	4

The gallon is the Imperial measure of 277.274 cubic inches; and the gill contains 5 ounces avoirdupois of water. In addition to the above measures, there is the Tierce of 42 gallons, and the Puncheon of 84 gallons.

6035. Comparative Value of Imperial Measure and U. S. Liquid Measure.

Imperial.		United States.		Gall.	Qt.	Pints.	Gills.
1 Gallon	=	1.20032 Gallons,	or	1	0	1	2.41
1 Quart	=	1.20032 Quarts,	or		1	0	1.60
1 Pint	=	1.20032 Pints,	or			1	0.80
1 Gill	=	1.20032 Gills,	or				1.20'

6036. Imperial Liquid Measure Expressed in Litres.

1 Hogshead	=	2.86234 Hectolitres
1 Barrel	=	1.43117 "
1 Gallon	=	4.54339 Litres
1 Quart	=	1.13585 "
1 Pint	=	5.67925 Decilitres
1 Gill	=	1.41981 "

6037. Dry or Corn Measure.

Quarter.	Bushels.	Pecks.	Gallons.	Pints.	Capacity in Cubic Inches.
1 =	8 =	32 =	64 =	512 =	17,745.536
	1 =	4 =	8 =	64 =	2,218.192
		1 =	2 =	16 =	554.548
			1 =	8 =	277.274
				1 =	34.659

The above capacities are for *struck* measure; the heaped measures contain nearly ⅓ more, the heaped bushel containing 2815½ cubic inches.

6038. Relative Value of Imperial Dry Measure and United States Dry Measure.

Imperial.	United States.	Qr.	Bush.	Pecks.	Qts.	Pints.
1 Quarter	= 1.03151 Quarters, or	1	0	1	0	0.133
1 Bushel	= 1.03151 Bushels, or		1	0	1	0.016
1 Peck	= 1.03151 Pecks, or			1	0	0.404
1 Gallon	= 4.12604 Quarts, or				4	0.252
1 Pint	= 1.03151 Pint, or					1.031

6039. Relative Value of Imperial Measure and United States Standard Apothecaries Measure.

			Gal.	Pint.	Fl. Oz.	Fl. Dr.	Minims.
1 Imp. Gallon	= 1.20032	U. S. Gallons, or	1	1	9	5	7.66
1 " Pint	= 1.20032	" Pints, or		1	3	1	38.45
1 " Fluid Ounce	= .960256	" Fluid Ounces, or				7	40.92
1 " Fluid Drachm	= .960256	" Fluid Drachms, or					57.62
1 " Minim	= .960256	" Minims, or					.96

6040. Imperial Dry Measure Expressed in Litres.

1 Quarter	=	2.90777 Hectolitres
1 Bushel	=	3.63471 Dekalitres
1 Peck	=	9.08677 Litres
1 Gallon	=	4.54338 "
1 Pint	=	5.67922 Decilitres

6041. The English Last is an English measure of various articles. A last of soap, ashes, herrings, and some other articles, is 2 barrels. A last of corn is 10 quarters. A last of gun-powder, 24 barrels. A last of flax or feathers, 1,700 pounds. A last of wool, 12 sacks.

6042. The Scotch Pint. A Scotch pint contains 105 cubic inches, and is equal to 4 English pints. 21¼ Scotch pints make a farlot of wheat.

6043. English Wood Measures. Wood is sold in England by the stack, skid, quintal, billet, and bundle. A stack is 108 solid feet, and usually piled 12 feet long, 3 feet high, and 3 feet wide. A quintal of wood is 100 lbs. A skid is a round bundle of sticks, 4 feet long. A one-notch skid girts 16 inches. A two-notch skid, 23 inches. A three-notch skid, 28 inches. A four-notch skid, 33 inches. A five-notch skid, 38 inches. A billet of wood is a bundle of sticks 3 feet long, and girts 7, 10, or 14 inches, and these bundles sell by the score or hundred. A score is 20, and comes from the count by tally, or marks. Faggots of wood are bundles of brush, 3 feet long, 2 feet round. A load of faggots is 50 bundles.

6044. English Coinage. English money is reckoned in pounds, shillings, pence and farthings thus symboled and relatively valued.

£.	s.	d.	q.
1 =	20 =	240 =	960
	1 =	12 =	48
		1 =	4

The farthing, or fourth part of a penny, is always written in the form of the fraction of a penny, one farthing being ¼ penny, 2 farthings ½ penny, &c. The standard sovereign is made of 22 carats pure gold and 2 carats copper alloy. The coin weighs 123.274 Troy grains; and the standard value of gold is £3, 17, 10½ ℔ Troy ounce, or £46, 14, 6 ℔ Troy pound. The half-sovereign, or 10 shilling gold coin is of the same standard, and half the weight and value of the sovereign. The standard shilling is composed of 222 parts pure silver alloyed with 18 parts copper. The coin weighs 87$\frac{3}{11}$ Troy grains; and the standard value is £3, 6, 0 ℔ pound troy; consequently 66 shillings weigh exactly 1 pound Troy. The crown, or 5 shilling piece, the half-crown, value 2s, 6d, and the six pence, are of the same standard and relative weights.

6045. French Weights and Measures. There are two systems of weights in use in France; the système usuel, or old Binary, and the more modern Decimal system. The former is still the most used in buying and selling, but the decimal system is already employed for all scientific purposes.

6046. French Binary Weights. These are more or less in common use in France, but are gradually being superseded by the decimal system.

Kilogram.	Livre.	Once.	Gros.	Denier.	French Grains.
1 =	2 =	32 =	256 =	768 =	18,432
	1 =	16 =	128 =	384 =	9,216
		1 =	8 =	24 =	576
			1 =	3 =	72
				1 =	24

6047. French Binary Weights Compared with Avoirdupois Weight. French Apothecaries weight is the same as the above, except that the livre contains 12 instead of 16 onces. The old French grain was equivalent to .820 of a Troy grain, but the new French grain (of 1812) is equal to .8365228 grains Troy. This would make the French Binary weight, as compared with Avoirdupois weight.

		Avoirdupois.
1 French Grain	=	.8365 Grains
1 " Denier	=	20.0765 "
1 " Gros	=	60.2296 "
1 " Once	=	1.1023 Ounces
1 " Livre	=	1.1023 Pounds
1 " Kilogramme	=	2.2046 "

6048. French Binary Weights Compared with U. S. Apothecaries Weight.

	Lbs.	Oz.	Drms.	Scruples.	Grains.
1 French Livre (16 oz) =	1	4	0	1	9.3941
1 " " (12 oz) =	1	0	0	1	2.0456
1 " Once =		1	0	0	1.8371
1 " Gros =			1	0	0.2296
1 " Denier =				1	0.0765
1 " Grain =					.8365

6049. Value of French Binary Weights in Troy Weight.

	Lb.	Oz.	Dwt.	Gr.
1 French Livre (16 oz.) =	1	4	1	5.184
1 " " (12 oz.) =	1	0	0	21.888
1 " Once =		1	0	1.824
1 " Gros =			2	12.228
1 " Denier =				20.076
1 " Grain =				.8365

6050. Value of French Binary Weights in Grams.

1 French Livre (16 oz.)	=	4.9957 Hectograms
1 " " (12 oz.)	=	3.7468 "
1 " Once	=	3.1223 Dekagrams
1 " Gros	=	3.9028 Grams.
1 " Denier	=	1.3009 "
1 " Grain	=	.0542 "

6051. Old French Linear Measure. The former measures of length in France were the

Toise	= 1.949	metres, or	6.3945 feet
Foot (pied)	= .32484	" =	12.785 inches
Inch (pouce)	= .02707	" =	1.0654 "
Line (ligne) or $\frac{1}{12}$ inch		=	.002256 metre

The metre is equal to 3 ft. 11 lines old French measure.

6052. French Decimal Weights and Measures. The French Gramme, litre, metre and are, are precisely the same as in the American Decimal system. They are founded on the same standard unit, the metre; and therefore represent respectively the same lengths, weights and capacity. The measures of capacity in France are multiples and divisions of the litre, which is the measure occupied by a kilogram (15434 Troy grains) of distilled water at its greatest density. It exceeds the old Paris pinte by $\frac{1}{16}$, and is equal to 35 fluid ounces and 103 minims, or 1.7608 Imperial pints, or 61.028 English cubic inches. 4½ litres make an Imperial gallon, within about 12 fʒ. The following table will show the relations between the Litre and the Imperial gallon of 277.2738 cubic inches.

Litres.		Cubic Inches.	Gals.	Pts.	Imperial Fl.℥	Imperial Fl.ʒ	Min.
$\frac{1}{1000}$	Millilitre	.061028					16.9
$\frac{1}{100}$	Centilitre	.61028				2	49
$\frac{1}{10}$	Decilitre	6.1028			3	4	10.36
1	Litre	61.028		1	15	1	43.69
10	Dekalitre	610.28	2	1	12	1	16.9
100	Hectolitre	6102.8	22	0	1	4	49
1000	Kilolitre	61028.	220	0	16	6	40
10000	Myrialitre	610280.	2201	(or 275¼ bushels.)			

6053. French Money. In France money is reckoned in *francs* and *centimes.* The centime is the $\frac{1}{100}$ part of a franc, 5 centimes being represented by a sou; so that 20 sous are equivalent to a franc. The same system of coinage is also at present in use in Belgium, Switzerland, and Italy.

6054. Foreign Medicinal Weights. The following are divided as our Apothecaries' weight: The pound of Austria weighs 6482.42 grains; Bavaria, 5556.24; Holland, 5787.75; Lubec, 5697.09; Nuremberg (German pound), 5522.96; Poland, 5533.25; Prussia, 5113.99; Sweden, 5498.01; Venice (sottile), 4649.17.

The division of the following differs in the scruple being divided into 24 grains: Bologna, 5026.32; Lucca, 5162.67; Modena, 5254.61; Parma, 5062.35; Portugal, 5312.23; Rome, 5233.25; Spain, 5325.84; Tuscany, 5240.49; Piedmont (Turin), 5123.49. The Naples pound contains 5490.63 Troy grains; the ounce contains 10 drachms; the scruple 20 grains.

The old Paris pound was divided into 16 ounces; the scruple into 24 grains. The pound by which drugs are weighed in Turkey is the Tchegy, equal to 4957 grains, and is divided into 100 drachms, each drachm into 16 killos, and each kilo into 4 grains.

The obolo is half a Spanish scruple; 3 silicua make 1 obolo, and 4 grains a silicua.

The commercial pound in several countries differs from the pharmaceutical. The civil pound of Bavaria and mark of Vienna are each about 19¾ avoirdupois ounces. That of Holland is the French kilogram, or 12 grains more than 2 pounds 3¼ ounces avoirdupois. The mark is half a kilogram. The Coburg commercial pound is nearly 18 ounces avoirdupois.

The unit of the British India system of weights is the tola, equal to 180 Troy grains. 32 tolas are equal to 1 pound Troy. The maund is equal to 100 Troy ounces.

6055. Foreign Money, Weights, and Measures, Compared with American.

	MONEY.		LENGTH.		LIQUID.		WEIGHT.	
	Name of Coin.	Value in American Dollars, Gold.	Name of Measure.	Length in Inches, English.	Name of Measure.	Contents in Cubic Inches.	Name of Weight.	Ounces Avoird.
England	Sovereign	4.80	Foot	12	Gallon	277¼	℔ Avoird.	16.
America	Dollar	1.00	Foot	12	Gallon	231	Pound	16.
Austria	Florin	.48½	Foot	12.45	Eimer	3452	Pound	19.76
Denmark	Dollar	.53	Foot	12.35	Anker	2355	Pound	17.65
France	Franc	.19	Metre	39.37	Litre	61.028	Kilogran	35.28
Holland	Florin	.40	Foot	11.14	Anker	2331	Pound	35.28
Portugal	Milreis	1.12	Foot	12.96	Almude	1040	Pound	16.19
Prussia	Dollar	.70	Foot	12.36	Eimer	4200	Pound	16.51
Russia	Rouble	.79½	Foot	12	Veddras	752	Pound	14.44
Spain	Dollar	1.00	Foot	11.03	Arroba	978	Pound	16.23
Sweden			Foot	12	Eimer	4794	Pound	15.

The rate of exchange varies, but the value of money is taken, reckoning silver at $1.20 per ounce.

6056. Foreign Measures. The kanna of Sweden = nearly 2.62 litres, or about 4 pints 12 ounces imperial.

The pott (half kanne) of Denmark = .9653 litre.

The arroba of Spain = 16.073 litres.

The almude of Portugal = 16.451 litres.

The barile of Naples = 43.6216 litres; of Rome, 58.5416 litres; of Tuscany, 45.584 litres.

The wedro of Russia (10 stof or 30 Russian pounds) = 12.29 litres, or 21 pints 12 ounces 12½ drachms imperial.

The mass of Wurtemburg = 1.537 litres, or about 3 pints 14¾ ounces imperial.

6057. Roman Money. The Romans, like other ancient nations, at first had no coined money, but either exchanged commodities with one another, or used a certain weight of uncoined brass, or other metal. Hence the names which indicated certain pieces of money, when coin came to be used, were the same as those which were used to indicate weights.

6058. Roman Brass Coins. The first brass coin that was used at Rome was called As, made in the reign of Servius Tullius; and being stamped with the heads of oxen, sheep, swine, &c., was called pecunia, from pecus. Hence Æs, brass, is often put for money; Ærarium, for treasury, &c. Some time afterwards the stamp was changed, and on one side it bore the figure of Janus; on the other the beak of a ship. The As originally weighed a pound, but was gradually reduced, and in the first Punic war, Asses were coined of only 2 ounces in weight; in the second Punic war, of only 1 ounce; and in the year of the city 563, of only half an ounce. The other brass coins were the Semissis, the Triens, the Quadrans or Teruncius, and the Sextans. The As, in value of our money, about 1½ cents; the Semissis, half an As; Triens, one-third; Quadrans, or Teruncius, one-fourth; Sextans, one-sixth.

6059. Roman Silver Coins. Silver was first coined in the year of the city 484, five years before the first Punic war; the impressions upon which were usually, on one side, carriages drawn by two or four beasts, and on the reverse, the head of Roma, with a helmet. On some were stamped the figure of Victory. The coins of silver were the Sestertius, Quinarius, Denarius, and Centussis. Sestertius, marked L.L.S. for libra libra semis, or by abbreviation H. S., worth 2½ Asses, or, in our money, 3¾ cents; Quinarius, marked V, worth 5 Asses, 7½ cents; Denarius, marked X, worth 10 Asses, 15½ cents; Centussis, worth 10 Denarii, nearly $1.60.

6060. Roman Gold Coins. Gold coin was first struck in the year of the city 546, in the second Punic war, and called Aureus. The stamps upon it were chiefly the images of the Emperors. The Aureus, at first, was equal in value to 25 Denarii, or 100 Sestertii; or, in our money, to $3.98. Soon afterwards it was debased, and under the later Emperors was worth only $3.70. Accounts were kept in Sestertii and Sestertia. The Sestertium was not a coin, but a shorter expression of 1000 Sestertii, or, in our money, about $40. We find also mentioned the Libra, containing 12 ounces of silver, worth $15, and the Talentum, worth about $965. Besides the ordinary coins, there were various medals struck to commemorate important events, properly called Medallions; for what we commonly term Roman medals were their current money.

6061. Roman Measures of Length. The Roman measures of length or distance were feet, cubits, paces, stadia, and miles.

	M.	Yds.	Ft.	In.
Foot	0	0	0	12
Cubit	0	0	1	6
Passus, or Pace	0	0	5	0
Stadium, or Furlong	0	208	3	0
8 Stadia, or 1000 Paces	1	0	0	0

The Roman Acre contained 240 feet in length, and 120 in breadth, that is, 28,800 square feet.

6062. Roman Weights. The chief weight among the Romans was the As, or Libra, a pound, equal in English Troy weight to 10 ounces 18 dwt. 13 grains; this Libra was divided into 12 parts, Unciæ (ounces), and these Unciæ into several weights of lower denominations.

6063. Roman Measures of Capacity. The most common measure of capacity was the Amphora, called also Quadrantal or Cadus, containing nearly 9 English gallons. They had also a measure called Congius, equal to ⅛ of an Amphora, or 1⅛ gallon English; and another called Sextarius, equal to ⅙ of the Congius, or about 1½ pints.

6064. Roman Division of Time. Romulus is said to have divided the year into 10 months, beginning with March; Numa added the other 2 months. When Julius Cæsar became master of the State, he adjusted the year according to the course of the sun, and assigned to each month the number of days which it still contains. This is the famous Julian Year, which continues in use to this day in all Christian countries, without any variation except that of the old and new style, occasioned by Pope Gregory, A. D. 1582. The Romans divided their months into three parts, by Calends, Nones, and Ides. The 1st day was called the Calends, the 5th day the Nones, and the 13th the Ides; except in March, May, July, and October, when the Nones fell on the 7th, and the Ides on the 15th. The custom of dividing time into weeks was introduced under the Emperors, being derived from the Egyptians; and the days of the week were named from the planets, viz.: Dies Solis, Sunday; Lunæ, Monday; Martis, Tuesday; Mercurii, Wednesday; Jovis, Thursday; Veneris, Friday; Saturni, Saturday. In marking the days, they counted backwards; thus they called the last day of December, Pridie Calendas Januarii, or the day before the Calends of January; the 30th day they called the third day before the Calends of January; and so on through the year. In leap-year the 24th and 25th days of February were both called the 6th day before the Calends of March, and hence this year is called Bissextilis. The day, as with us, was divided into 12 hours, and lasted from six o'clock in the morning till six in the evening. The night was divided into four watches, each consisting of three hours. The Romans had no clocks or watches, and the first dial is said to have been erected in Rome so late as 447 years after the building of the city.

6065. Scriptural Measure of Length.

	M.	Yds.	Ft.	In.	B.C.
A Finger	0	0	0	0	2¼
A Hand breadth	0	0	0	3	1¼
A Span	0	0	0	10	2¼
A Cubit	0	0	1	9	2¼
A Fathom	0	2	1	3	1¼
Ezekiel's reed	0	3	0	0	0
Do. according to others	0	3	1	11	0¾
The Measuring Line	0	48	1	11	0
A Stadium or Furlong.	0	243	0	6	0
A Sabbath-day's Journey		1216	0	0	0
The Eastern Mile	1	672	0	0	0
A Day's Journey	33	288	0	0	0

6066. Scriptural Liquid Measure.

	Gals.	Qts.	Pts.
The Log	0	0	0¾
The Firkin or Metretes	0	3	1¼
The Hin	1	1	0
The Bath	7	2	0¼
The Homer or Cor	75	2	1½

6067. Scriptural Dry Measure.

	Bush.	Pks.	Pts.
The Cab	0	0	2⅚
The Omer	0	0	5
The Seah	0	1	1
The Ephah	0	3	3½
The Lethech	4	0	0½
The Homer	8	0	1½

6068. Scriptural Weights.

	Lbs.	Oz.	Dwts.	Gr.
A Shekel	0	0	9	2½
A Maneh	2	3	6	10
A Talent	113	10	1	10

6069. Scriptural Money.

		Cts.
A Gerah		2
A Zuzah		12
A Bekah		25
A Shekel (Silver)		50
Golden Daric, or Dram	5	18
A Shekel of Gold	9	00
A Maneh or Mina	29	50
A Talent of Silver	1,707	00
A Talent of Gold	27,320	00

6070. Jewish Method of Reckoning Time. The day, reckoning from sunrise, and the night, reckoning from sunset, were each divided into 12 equal parts, called the 1st, 2nd, 3rd, 4th, &c., hours. The first watch was from sunset to the third hour of the night. The second, or middle watch, was from the third hour to the sixth. The third watch, or cock-crowing, was from the sixth hour to the ninth. The fourth, or morning watch, was from the ninth hour of the night to sunrise.

6071. Russian Money. In Russia, money is calculated in *Roubles* and *Kopeks*, the silver Rouble consisting of 100 Kopeks, and equivalent to about 79½ cents of our money.

6072. Russian Weights. The Russian pound is 6317½ grains, or the weight of 25.019 cubic inches of water. The *Pood*, about 36 pounds, 1½ ounces avoirdupois.

6073. Russian Lineal Measure. The Russian foot is the same as the American.

1 Werst		500 Sashens
1 Sashen	=	3 Arsheens
1 Arsheen	=	2⅓ Feet

6074. Russian Measures of Capacity. The *Chetwert* is equivalent to 5 bushels 6¼ gallons imperial. The *Tschetwerick*, 5¾ imperial gallons. 10 Tschetwericki make 1 *Kuhl* or *Sark*.

The *Wedro* consists of 3¼ wine gallons, and 40 Wedroja make 1 *Fass*.

6075. Austrian Money is reckoned in *Florins* and *Kreutzers;* the Florin being equivalent to about 48½ cents American.

20 Kreutzers	=	1 Zwanziger
60 "	=	1 Florin
2 Florins	=	1 Thaler
1 Ducat	=	4⅔ Florins

6076. Austrian Weights. The Austrian pound is rather less than 1¼ pounds avoirdupois.

1 Sanne	=	275 Pounds
1 Pound	=	4 Vindlinge
1 Vindlinge	=	4 Unzen
1 Unze	=	2 Loth

6077. Austrian Lineal Measure. The Austrian *foot* measures 1[illegible]½ inches; the *Nult* is equivalent to 4⅝ miles.

6078. Austrian Measures of Capacity. The *Muth* is 50⅗ imperial bushels.

1 Muth	=	30 Metz
1 Metz	=	64 Moasel

The liquid *Mass* or *Kanne* is about 2½ imperial pints, or 1.415 litres.

6079. Roman Money. This was reckoned in *Paoli* and *Bajochi*, the latter being about equal to 1 cent American.

1 Scudo	=	10 Paoli
1 Paolo	=	10 Bajochi

6080. Prussian Money. The Prussians count their money in *Thalers, Silbergroschen* and *Pfennings.*

1 Thaler	=	30 Silbergroschen
1 Silbergroschen	=	12 Pfennings.

The Friedrich d'or is equal to 5 Thalers 20 Silbergroschen.

6081. Prussian Weights. The Prussian *pound* is 16½ ounces avoirdupois.

1 Cwt.	=	110 Pounds
1 Shipping last	=	400 Pounds

6082. Prussian Lineal Measure. The Prussian *foot* is 12⅓ inches English.

1 Ruthe	=	12 Feet
1 Foot	=	12 Inches
1 Inch	=	12 Linien
1 Faden	=	6 Feet
1 Mile	=	4⅔ Miles English

6083. Prussian Measures of Capacity. The *Scheffel* is equal to 1½ bushels.

1 Wispel	=	24 Scheffel
1 Scheffel	=	16 Metz

The Prussian liquid quart is equivalent to 1.145 litres, or nearly 2⅖ pints American.

6084. Money of the Netherlands is reckoned in *Guilders* and *Cents*, the guilder (or silver florin) being about 41 cents of our money. The *Ducat* is equivalent to 5.55 guilders, and the *Stuiver* to 5 cents.

6085. Weights Used in the Netherlands. The *pound* is 1 pound 1⅓ ounces avoirdupois.

1 Pound	=	10 Lood
1 Lood	=	10 Wigtj
1 Wigtj	=	10 Korrels

6086. Lineal Measure of the Netherlands. The *ell* is the same as the metre of America.

1 Roede	=	10 Ells
1 Ell	=	$\frac{1}{10}$ Palm
1 Palm	=	10 Duim
1 Duim	=	10 Streep
1 Myl	=	1000 Ells or ⅝ mile English

6087. Dry Measure of the Netherlands. The *Mudde* contains a little more than 2¾ bushels imperial.

1 Last	=	30 Mudden
1 Mudde	=	10 Schepel
1 Schepel	=	10 Kop
1 Kop	=	10 Maajtes

6088. Liquid Measure of the Netherlands. The *Vat* contains 22$\frac{1}{10}$ imperial gallons.

1 Vat	=	100 Kann
1 Kann	=	10 Maajtes
1 Maajte	=	10 Vingerh

6089. Portuguese Money. In Portugal, money is reckoned in *Reis*. For the value of the coins see No. 6055

1 Vintem	=	20 Reis
1 Crusado	=	400 "
1 Milrei	=	1000 "
1 Conto de reis	=	1000 Milreis

6090. Dutch Weights and Measures. The following are the points in which Holland differs from the rest of the Netherlands.

Dutch.	English.
1 Foot	=11⅕ Inches
1 Ell	=27$\frac{1}{12}$ "
1 Corn last	=10 qrs. 5¼ Winchester Bushels
1 Aam	=41 Imperial gallons
1 Hoed	= 5 Chaldrons
1 Freight last	=4000 Pounds.
1 Ballast last	=2000 Pounds

6091. Spanish Money. The *Dollar* of Spain contains 20 *Reals*, and is about the same value as the American. The coins used in different parts of Spain are various; almost every Province having a different system of coinage.

6092. Spanish Weights. The Castilian *Marca* is 7 ounces 3.16 dwts. Troy.

1 Marca	=	8 Onzas
1 Onza	=	8 Ochaves
1 Ochavo	=	72 Granos

The quintal is equivalent to 101¼ pounds avoirdupois.

1 Quintal	=	4 Arrobas
1 Arroba	=	25 Libras
1 Quintal Macho	=	6 Arrobas

Precious stones are weighed by the *ounce* of 431½ Troy grains.

1 Ounce	=	140 Quilates
1 Quilate	=	4 Granos

6093. Spanish Lineal Measure. The *Pie* equals 11⅛ inches, and the *Legua* 4¼ English miles.

1 Estado	=	2 Varas
1 Vara	=	3 Pies

6094. Spanish Dry Measure. The *Fanega* is 12½ imperial gallons.

1 Cahiz	=	12 Fanegas
1 Fanega	=	12 Almudes
1 Almude	=	4 Cuartillos

6095. Spanish Liquid Measure. The *Cantaro* or *Arroba Mayor* contains 3 gallons 3¾ pints imperial; the *Arroba Menor* for oil is 2 gallons 5¼ pints imperial.

1 Cantaro	=	8 Azumbres
1 Azumbra	=	4 Cuartillos
1 Moyo	=	16 Cantaros
1 Pipa	=	27 "
1 Bota	=	30 "

6096. Swedish Money. The *Riksdaler banco* is worth about 40 cents of our money, and is divided into 48 skillings.

6097. Swedish Weights. The *Skal pound* is 15 ounces avoirdupois. The *Schip pound* is equivalent to 400 skal pounds. The *Mark*, used in weighing gold, consists of 6 oz. 16 dwt. Troy.

6098. Swedish Lineal Measure. The Swedish *Foot* is the same as ours.

1 Faam	=	3 Alnar
1 Alnar	=	2 Feet
1 Foot	=	2⅝ Verthum

6099. Swedish Dry Measure. The *Tonn* is equivalent to 4 imperial bushels.

1 Tonn	=	8 Quarts
1 Quart	=	4 Kappar
1 "	=	7 Cans
1 Can	=	8 Quarrtiers

6100. Swedish Liquid Measure. The *Fuder* contains 2 pipes.

1 Fuder	=	4 Oxhoofte
1 Oxhoofte	=	3 Eimer
1 Eimer	=	60 Stop

6101. Swiss Money is reckoned in *Francs*, the franc being subdivided into 10 *Batzen.* The value of the franc is about 27 cents. This is the old system. (*See No.* 6053.)

6102. Swiss Weights.

1 Hundred-weight	=	50 Kilogrammes
1 Kilogramme	=	2 Pounds.

The hundred-weight is equivalent to 110¼ pounds avoirdupois; the pound is therefore about 17⅔ ounces avoirdupois.

6103. Swiss Lineal Measure. The Helvetian foot is equal to 11$\frac{7}{10}$ inches English.

1 Stab or Staff	=	2 Ells
1 Ell	=	2 Feet
16,000 Feet	=	1 Hour or Mile

The Swiss mile is consequently a trifle over 3 English miles.

6104. Swiss Dry Measure. The *Malter* is about 4 bushels 1 gallon Imperial measure.

1 Malter	=	10 Viertel
1 Viertel	=	10 Immir

6105. Swiss Liquid Measure. The Swiss *Ohm* contains 33 Imperial gallons.

1 Ohm	=	100 Maas.

6106. Turkish Money. In Turkey,

money is reckond by the *Piaster*, 22 of which are equivalent to $1.00.

1 Sequin	=	100 Piasters
1 Piaster	=	40 Paras
1 Para	=	3 Aspers
1 Piaster (grush)	=	100 Aspers.

6107. Turkish Weights. The Turkish *Chequi* is $11\frac{1}{3}$ ounces avoirdupois.

1 Cantaro	=	44 Okas
1 Batman	=	6 Okas
1 Oka	=	4 Chequi
1 Chequi	=	100 Drachmas.

6108. Turkish Lineal Measure. The Turks use, for measuring length, the large *pik halebi*, or 27 $\frac{9}{10}$ inches; and the small *pik andassa* of 27$\frac{1}{16}$ inches.

6109. Turkish Measures of Capacity. The dry *Killow* contains $7\frac{1}{2}$ imperial gallons; the *Fortin*, 4 killows. A killow of rice should weigh 10 okas. The liquid *almud* contains $1\frac{3}{5}$ imperial gallons.

6110. Chinese Money. The Chinese *Tael* is $1.56.

1 Tael	=	10 Mace
1 Mace	=	10 Candarin
1 Candarin	=	10 Cash

6111. Chinese Weights. The *Catty* is $1\frac{1}{3}$ pounds avoirdupois.

1 Pecul	=	100 Cattys
1 Catty	=	16 Taels
1 Tael	=	10 Mazas
1 Maza	=	10 Candarins
1 Candarin	=	10 Cash.

6112. East Indian Money. In Hindostan, money is reckoned in *Rupees*, *Annas*, and *Pice*, the Rupee being about 45 cents of our money.

1 Rupee	=	16 Annas
1 Anna	=	12 Pice
10,000 Rupees	=	1 Lakh

6113. Mexican Money. The Mexican gold dollar is worth about 96 cents United States coin; the Mexican silver dollar is reckoned equal to the United States gold dollar.

1 Doubloon	=	16 Dollars
1 Dollar	=	8 Reals.

6114. Monte-Video Money. The *Dollar* or *Peso Corriente* is equal to 80 cents United States coin.

1 Dollar = 8 Reales = 100 Centesimos

6115. Brazilian Money. In Brazil, money is reckoned in *Reis*, 4000 of which are equal to £1 sterling, or $4.84 United States coin.

1 Milreis	=	1000 Reis

6116. Brazilian Lineal Measure. The Brazilian *Pe* or Foot is the same as the English foot; the *Palma* is equivalent to $9\frac{1}{2}$ English inches. 10 Palmas equal 1 Braça or $2\frac{3}{8}$ English yards. The Braça is also subdivided into 2 *Varas* and $3\frac{1}{2}$ *Covados*. The *Legoa* or mile is about $4\frac{4}{5}$ English miles.

6117. Brazilian Weights. The weights in ordinary use are as follows, the *Quintal* being equal to $91\frac{1}{2}$ pounds avoirdupois.

1 Quintal	=	4 Arrobas
1 Arroba	=	32 Arratels

Gold and silver are weighed by the *Marco* of 7 ounces $7\frac{3}{4}$ dwts Troy.

1 Marco	=	8 Onças
1 Onça	=	8 Oitavas
1 Oitava	=	72 Granos

Precious stones are sold by the *Quilate*, equal to $4\frac{13}{20}$ dwts. Troy.

1 Oitava	=	3 Escrupulos
1 Escrupulo	=	3 Quilates
1 Quilate	=	4 Granos

6118. Brazilian Dry Measure. The Brazilian *Mayo* is equivalent to $22\frac{1}{2}$ imperial bushels.

1 Mayo	=	15 Fanegas
1 Fanega	=	4 Alqueires

6119. Decimal Approximations for Facilitating Calculations.

Lineal feet multiplied by		.00019	=	miles.
" yards	"	.000568	=	"
Square inches	"	.007	=	square feet.
" feet	"	.111	=	square yards.
" yards	"	.0002067	=	acres.
Circular inches	"	.00546	=	square feet.
Cylindrical inches	"	.0004546	=	cubic feet.
" feet	"	.02909	=	cubic yards.
Cubic inches	"	.00058	=	cubic feet.
" feet	"	.03704	=	cubic yards.
" "	"	6.2321	=	imperial gallons.
" inches	"	.003607	=	" "
Bushels	"	.0476	=	cubic yards.
"	"	1.284	=	cubic feet.
"	"	2218.2	=	cubic inches.
Cubic feet	"	.779	=	bushels.
Cubic inches	"	.00045	=	bushels.
Pounds	"	.009	=	hundredweights.
Pounds	"	.00045	=	tons
Cylindrical feet	"	4.895	=	imperial gallons.
" inches	"	.002832	=	" "
Cubic inches	"	.263	=	pounds of cast iron.
" "	"	.281	=	" wrought do.
" "	"	.283	=	" steel.
" "	"	.3225	=	" copper.
" "	"	.3037	=	" brass.
" "	"	.26	=	" zinc.
" "	"	.4103	=	" lead.
" "	"	.2636	=	" tin.
" "	"	.4908	=	" mercury.
Cylindrical inches	"	.2065	=	" cast iron.
" "	"	.2168	=	" wrought iron.
" "	"	.2223	=	" steel.
" "	"	.2533	=	" copper.
" "	"	.2385	=	" brass.
" "	"	.2042	=	" zinc.
" "	"	.3223	=	" lead.
" "	"	.207	=	" tin.
" "	"	.3854	=	" mercury.

6120. Memoranda Connected with Water. 1 cubic foot of water = 62.4 pounds. 1 cubic inch = .036 pounds. 1 gallon imperial = 10 pounds; or = 0.16 cubic feet. 1 cubic foot of water = 6.2321 imperial gallons; or, approximately = $6\frac{1}{4}$ gallons. 1 cwt. of water = 1.8 cubic feet = 11.2 gallons. 1 ton of water = 35.9 cubic feet = 224 gallons. Cubic feet of water × .557 = cwt. approximately. Cubic feet of water × .028 = tons approximately. 1 cubic foot of sea water = 64.14 pounds. Weight of sea water = weight of fresh water × 1.028.

6121. Pressure of the Atmosphere. In engineering, the common pressure of the atmosphere, 14.6 pounds to the square inch, is taken as a standard of that exerted by other elastic fluids. Thus, steam, or air condensed so as to exert a pressure of 30 pounds to the square inch, is said, in round numbers, to be of 2 atmospheres; at 45 pounds to the inch, 3 atmospheres, &c.

6122. Memoranda Connected with Light. Velocity of light 192,000 miles per second, nearly. Decomposition of light: The seven prismatic colors of a ray of light are violet, indigo, blue, green, yellow, orange, red. Violet is the maximum chemical or actinic color; yellow the maximum illuminating color, and red the heat color.

6123. Force of the Wind.

Miles per Hour.	Feet per Minute.	Feet per Second.	Force in lbs. per Sq. Foot	Description.
1	88	1.47	.005	Hardly perceptible.
2	176	2.93	.020	Just perceptible.
3	264	4.4	.044	
4	352	5.87	.079	Gentle breeze.
5	440	7.33	.123	
10	880	14.67	.492	Pleasant breeze.
15	1320	22.	1.107	
20	1760	29.3	1.970	Brisk gale.
25	2200	36.6	3.067	
30	2640	44.	4.429	High wind.
35	3080	51.3	6.027	
40	3520	58.6	7.870	Very high wind.
45	3960	66.	9.900	
50	4400	73.3	12.304	Storm.
60	5280	88.	17.733	Great storm.
70	6160	102.7	24.153	
80	7040	117.3	31.490	Hurricane.
100	8800	146.6	49.200	

6124. Velocity of Sound. In air, 1.142 feet per second. In water, 4,900 feet. Through iron, 17,500 feet. Through copper, 10,378 feet. Through wood, 12,000 to 16,000 feet.

Distant sounds may be heard on a still day: Human voice, 150 yards. Rifle 5,300 yards. Military band, 5,200 yards. Cannon 35,000 yards.

6125. Heat-conducting Power of Building Materials. Conducting power of substances, slate being 1000.

Slate	1000	Chalk	564
Lead	5210	Asphaltum	451
Flagstone	1110	Oak	336
Portland stone	750	Lath and plaster	255
Brick	600 to 730	Cement	200
Fire-brick	620		

6126. Properties of the Circle. Diameter × 3.14159 = circumference. Diameter × .8862 = side of an equal square. Diameter × .7071 = side of an inscribed square. Radius squared, × 3.14159 = area of circle. Diameter squared, × .7854 = area of circle. Radius × 6.28318 = circumference. Circumference ÷ 3.14159 = diameter. Circumference = $3.54\sqrt{\text{area of circle}}$. Diameter = $1.128\sqrt{\text{area of circle}}$.

6127. To Determine the Weight of Live Cattle. Measure in inches the girth round the breast, just behind the shoulder-blade, and the length of the back from the tail to the forepart of the shoulder-blade. Multiply the girth by the length, and divide by 144. If the girth is less than 3 feet, multiply the quotient by 11; if between 3 feet and 5 feet, multiply by 16; if between 5 feet and 7 feet, multiply by 23; if between 7 feet and 9 feet, multiply by 31. If the animal is lean, deduct $\frac{1}{20}$ from the result. Or: Take the girth and length in feet, multiply the square of the girth by the length, and multiply the product by 3.36. The result will be the answer in pounds. The live weight, multiplied by .605, gives a near approximation to the net weight.

6128. To Measure Corn in the Crib. Corn is generally put up in cribs made of rails, but the rule will apply to a crib of any size or kind. Two cubic feet of good, sound, dry corn in the ear, will make a bushel of shelled corn. To get, then, the quantity of shelled corn in a crib of corn in the ear, measure the length, breadth, and height of the crib, inside of the rail; multiply the length by the breadth, and the product by the height; then divide the result by 2, and you have the number of bushels of shelled corn in the crib. In measuring the height, of course the height of the corn is intended. And there will be found to be a difference in measuring corn in this mode between fall and spring, because it shrinks very much in the winter and spring, and settles down.

6129. Percentage of Pork to Live Weight. The following table shows the proportion of pork to live weight of fat swine:

Live Weight in Stones of 14 pounds.	Per Cent. of Pork.
Above 40 stones	87 to 88
From 35 to 40 stones	84 to 86
" 30 to 35 "	83 to 84
" 25 to 30 "	81 to 82
" 20 to 25 "	80
" 15 to 20 "	77 to 78
Under 15 "	75 to 77

6130. Measures for Housekeepers.

Wheat flour	1 pound	is 1 quart.
Indian meal	1 " 2 oz.	" 1 "
Butter when soft	1 "	" 1 "
Loaf sugar, broken	1 "	" 1 "
White sugar, powd.	1 " 1 oz.	" 1 "
Best brown sugar	1 " 2 oz.	" 1 "
Eggs	10 eggs are 1 pound.	
Flour	8 quarts " 1 peck.	
Flour	4 pecks " 1 bushel.	
16 large table-spoonfuls are		½ pint.
8 large table-spoonfuls are		1 gill.
4 large table-spoonfuls are		½ gill.
2 gills are		½ pint.
2 pints are		1 quart.
4 quarts are		1 gallon.
A common sized tumbler holds		½ pint.
A common sized wine-glass		½ gill.
25 drops are equal to		1 tea-spoonful.

6131. Sizes of Drawing Paper.

Wove Antique	52 ×31 in.
Uncle Sam	48 ×120 in.
Double Elephant	40 ×26 in.
Emperor	40 ×60 in.
Atlas	32 ×26 in.
Colombier	33¾×23 in.
Elephant	27¾×23¼ in.
Imperial	29 ×21¼ in.
Super Royol	27 ×19 in.
Royal	24 ×19 in.
Medium	22 ×18 in.
Demy	19 ×15½ in.
Cap	13 ×16 in.

6132. Barometrical Rules for Prognosticating the Weather. I. After a continuance of dry weather, if the barometer begins to fall slowly and steadily, rain will certainly ensue; but if the fine weather has been of long duration, the mercury may fall for 2 or 3 days before any perceptible change takes place, and the longer the time that elapses before rain comes, the longer the wet weather is likely to last.

II. Conversely, if, after a great deal of wet weather, with the barometer below its mean height, the mercury begins to rise steadily and slowly, fine weather will come, though 2 or 3 wet days may first elapse; and the fine weather will be the more permanent, in proportion to the length of time that passes before the perceptible change takes place.

III. On either of the two foregoing suppositions, if the change immediately ensues on the motion of the mercury, the change will not be permanent.

IV. If the barometer rises slowly and steadily for two days together, or more, fine weather will come, though for those two days it may rain incessantly, and the reverse; but if the barometer rises for two days or more during rain, and then, on the appearance of fine weather, begins to fall again, the fine weather will be very transient, and *vice versa*.

V. A sudden fall of the barometer in spring or autumn indicates wind; in summer, during very hot weather, a thunder-storm may be expected; in winter, a sudden fall after frost of some continuance indicates a change of wind with thaw and rain; but in a continued frost a rise of the mercury indicates approaching snow.

VI. No rapid fluctuations of the barometer are to be interpreted as indicating either dry or wet weather of any continuance; it is only the slow, steady, and continued rise or fall, that is to be attended to in this respect.

VII. A rise of the mercury late in the autumn, after a long continuance of wet and windy weather, generally indicates a change of wind to the northern quarters, and the approach of frost.

6133. Melting or Boiling Point of Metals, Liquids, &c.

Degrees Fahr.	
3080°	Platinum melts.
2786	Cast iron melts; 2696° (*Morveau*).
2500	Steel melts.
2016	Gold melts (*Daniell*); 2200° (*Kane*).
1996	Copper melts (*Kane*); 2548° (*Daniell*).
1873	Silver melts (*Makins*); 2233° (*Daniell*).
1869	Brass melts (*Daniell*).
1000	Iron, bright cherry red (*Poillet*).
980	Iron, red heat (*Daniell*).
914	Zinc burns (*Daniell*).
810	Antimony melts.
773	Zinc melts (*Daniell*); 793° (*Gmelin*).
644	Mercury boils (*Daniell*); 662° (*Graham*).
630	Whale oil boils (*Graham*).
612	Lead melts (*Crighton*); 609° (*Daniell*).
600	Linseed oil boils.
560	Sulphur ignites.
545	Sulphuric acid boils (*Phillips*); 620° (*Graham*).
476	Bismuth melts (*Phillips*); 518° (*Gmelin*).
442	Tin melts.
380	Arsenious acid volatilizes.
372	Saturated solution of nitrate of ammonia boils.
356	Metallic arsenic sublimes.
336	Saturated solution of acetate of potassa boils.
320	Cane sugar melts, 320° to 400°, baking heat of an oven.
315	Oil of turpentine boils (*Kane*).
304	Saturated solution of nitrate of lime boils.
302	Etherification ends.
275	Saturated solution of carbonate of potash boils.
256	Saturated solution of acetate of soda boils.
248	Nitric acid, specific gravity 1.42, boils.
238	Saturated solution of nitre boils.
236°	Saturated solution of sal-ammoniac boils.
226	Sulphur melts (*Fownes*); 232° (*Turner*).
220	Saturated solution of alum, carbonate of soda, and sulphate of zinc boils.
218	Saturated solution of chloride of potassa boils.
216	Saturated solution of sulphate of iron, sulphate of copper, and nitrate of lead boils.
213	Water begins to boil in glass (or 213½°).
212	Water boils in metal, barometer at 30 inches.
199	Milk boils.
194	Sodium melts.
185	Nitric acid, specific gravity 1.52, boils.
180	Starch dissolves in water.
176	Rectified spirit boils. Benzole distills.
173	Alcohol, specific gravity 796 to 800, boils.
151	Bees'-wax melts (*Kane*); 142° (*Lepage*).
150	Scalding heat. Pyroxylic spirit boils (*Scanlan*).
145	Albumen coagulates.
140	Chloroform and ammonia, specific gravity .945, boils.
136	Potassium melts (*Daniell*).
132	Acetone (pyroacetic spirit) boils (*Kane*).
130	Butter melts (130° to 140°).
122	Mutton suet and styracine melts.
120	Phosphorus inflames. Friction matches ignite.
116	Bisulphuret of carbon boils (*Graham*).
112	Spermaceti and stearine melt.
111	Beef tallow melts.
110	Highest temperature of the human body (in lockjaw).
106	Mutton tallow melts.
99	Phosphorus melts (99° to 100°).
98	Ether, specific gravity .720, boils. Blood heat.
88	Acetous fermentation ceases. Water boils in a vacuum.
81	Mean temperature at the equator.
77	Vinous fermentation ends; acetous begins.
67	Lowest temperature of the human body (in cholera).
65	Best temperature of a room (65° to 68).
62	Oil of anise liquefies; congeals at 60°.
60	Mean temperature at Rome.
$50\frac{1}{10}$	Mean temperature at London.
42	Sulphuric acid, specific gravity 1.741, congeals (41° to 42°).
41	Mean temperature of Edinburgh.
36	Olive oil freezes.
32	Water freezes.
30	Milk freezes.
28	Vinegar freezes.
20	Strong wine freezes.
— 4	Mixture of snow and salt.
— 7	Brandy freezes.
—39	Mercury freezes (30° to 40°). (*See also Nos.* 7, 3353, 3459 *and* 1687, *&c*).

6134. Weight of Earth, Rocks, &c. A cubic yard of sand or ground weighs about 30 cwt. Mud, 25 cwt. Marl, 26 cwt. Clay, 31 cwt. Chalk, 36 cwt. Sandstone, 39 cwt. Shale, 40 cwt. Quartz, 41 cwt. Granite, 42 cwt. Trap, 42 cwt. Slate, 43 cwt.

To find the weight of a cubic foot of any of the above, divide the weight of a cubic yard by 27. Thus, a cubic foot of sand weighs $\frac{30}{27}$, or $1\frac{1}{9}$ cwt., equivalent to about 124 pounds.

6135. Weight of Various Minerals. One cubic foot of water weighs at a temperature of 60° Fahrenheit, $62\frac{1}{2}$ pounds avoirdupois. By ascertaining the specific gravity of a substance and multiplying with $62\frac{1}{2}$ pounds, the exact weight of one cubic foot is obtained.

	Sp. Gr.	Pounds Avoirdupois. Cubic foot Weighs.
Anthracite coal	1.5	94
Antimonial copper, tetrahedrite, or grey copper	5.0	300
Antimonial silver	9.5	600
Antimony ore, grey sulphuret	4.5	279
Antimony metal	6.5	400
Apatite, or phosphate of lime	3.0	186
Arsenical iron pyrites, mispickel	6.0	370
Asbestos	3.0	186
Asphaltum, mineral pitch	1.0	62
Baryta sulphate	4.5	310
Baryta carbonate, witherhite	4.0	248
Bismuth	9.7	600
Bituminous coal	1.5	90
Black lead, graphite	2.0	125
Black jack blende, sulphuret of zinc	4.0	250
Bog iron ore	4.0	250
Brown hæmatite	4.0	250
Building stones, comprising granite, gneiss, syenite, &c.	3.0	186
Calamine	3.3	190
Chromic iron	4.5	260
Copper pyrites	4.0	260
Derbyshire spar, fluor spar	3.0	186
Feldspar	3.0	190
Flint	2.5	110
Loose sand	—	95
Franklinite	5.0	310
Galena	7.5	465
Gold (20 carats)	15.7	1000
" (pure)	19.2	to 1200
Gypsum	2.3	130
Iron—cast iron	—	450
" magnetic ore	5.0	310
" spathic ore	3.0	200
" " pyrites	5.0	310
" pyrrhotine, or magnetic pyrites	4.5	280
" specular ore	4.5	290
" wrought	—	487
Limestone, hydraulic	2.7	150
" magnesian	2.5	130
Manganese, binoxide of	4.8	294
Malachite	4.0	248
Mica	2.8	160
Novaculite, or whetstone	3.0	186
Ochre	3.5	217
Platinum, metal and ores	16 to 19	1116
Porcelain clay	2.0	140
Pyrites, iron	4.5	280
Quartz, pure, compact	2.6	155
" loose, angular, and round sand	—	100
Trap	3.0	186
Vitreous copper, copper glance	5.5	341
Wood tin, stream tin	7.0	434
Zinc, sulphide or blende	4.0	250
Zincite, red zinc ore	5.5	331
Zinc carbonate	4.4	268
Zinc silicate	3.4	200

(*Feuchtwanger*).

6136. Table of the Relative Hardness and Weight of the Principal Precious Stones, &c.

Substances.	Hardness.	Specific Gravity.
Diamond from Ormus	20	3.7
" (pink)	19	3.4
" (bluish)	19	3.3
" (yellowish)	19	3.3
" (cubic)	18	3.2
Ruby	17	4.2
" (pale, from Brazil)	16	3.5
Sapphire	16	3.8
Topaz	15	4.2
" (whitish)	14	3.5
" (Bohemian)	11	2.8
Ruby (spinelle)	13	3.4
Emerald	12	2.8
Garnet	12	4.4
Agate	12	2.6
Onyx	12	2.6
Sardonyx	12	2.6
Amethyst (occidental)	11	2.7
Crystal	11	2.6
Cornelian	11	2.7
Jasper (green)	11	2.7
" (reddish yellow)	9	2.6
Schoerl	10	3.6
Tourmaline	10	3.0
Quartz	10	2.7
Opal	10	2.6
Chrysolite	10	3.7
Zeolite	8	2.1
Fluor	7	3.5
Calcareous spar	6	2.7
Gypsum	5	2.3
Chalk	3	2.7
Glass		2.3 : 3.62
" (plate)		2.5 : 2.6
" (crystal or flint)		3.0 : 3.616

6137. Weight of Hemp and Wire Rope.

Hemp.		Iron Wire.		Steel Wire.	
Circumference.	Lbs. Weight per Fathom	Circumference.	Lbs. Weight per Fathom.	Circumference.	Lbs. Weight per Fathom
2¾	2	1	1	—	—
—	—	1½	1½	1	1
3¾	4	1⅝	2	—	—
—	—	1¾	2½	1½	1½
4½	5	1⅞	3	—	—
—	—	2	3½	1⅝	2
5½	7	2⅛	4	1¾	2½
—	—	2¼	4½	—	—
6	9	2⅜	5	1⅞	3
—	—	2½	5½	—	—
6½	10	2⅝	6	2	3½
—	—	2¾	6½	2⅛	4
7	12	2⅞	7	2¼	4½
—	—	3	7½	—	—
7½	14	3⅛	8	2⅜	5
—	—	3¼	8½	—	—
8	16	3⅜	9	2½	5½
—	—	3½	10	2⅝	6
8½	18	3⅝	11	2¾	6½
—	—	3¾	12	—	—
9½	22	3⅞	13	3¼	8
10	26	4	14	—	—
—	—	4¼	15	3⅜	9
11	30	4⅜	16	—	—
—	—	4½	18	3½	10
12	34	4⅝	20	3¾	12

6138. Miscellaneous Statistics.

TIMBER.	Specific Gravity	Weight in lbs. per Cubic Foot.	Tenacity in lbs. per Square Inch.	Crushing Force in lbs. per Square Inch.
Ash	.8	50	17.200	9.000
Beech	.69	43	11.000	9.000
Birch	.71	44	15.000	5.500
Cedar	.48	30	11.000	5.600
Deal, Christiana	.7	44	12.000	6.000
Elm	.6	37	13.000	10.000
Hornbeam	.75	47	20.000	7.000
Larch	.55	34	9.000	5.500
Memel	.6	37		
Mahogany, Spanish	.8	50	16.000	8.000
Oak, English	.93	58	17.000	10.000
Oak, Canadian	.87	54	10.000	6.000
Pine, red	.65	41	12.000	5.800
Pine, yellow	.45	28	11.000	5.100
Teak, Moulmein	.65	41	15.000	12.000
Yew	.8	50	8.000	
MISCELLANEOUS.				
Asphaltum	.9	56		
Gutta-percha	.98	61		
India-rubber	.94	59		
Ivory	1.8	112		
FLUIDS.			Boiling Point.	Expansion.*
Alcohol	.8	50	173°	.11
Ether	.74	46	100	.07
Oil	.90	56		.08
Water, fresh	1.000	62.4	212	.047
Water, sea	1.028	64.1	213	

GASES.	Water 1.	Comparative Weight (Air being 1.)	Weight of Cubic Foot in Grains.
Air	.0012	1.000	527
Carbonic acid	.0018	1.524	800
Carburetted hydrogen	.0005	.420	220
Hydrogen	.00008	.069	43
Oxygen	.00125	1.103	627

* Expansion of fluids is calculated between 32° and 212° Fahrenheit.

6139. Weight of Copper and Lead.

Weight of a Square Foot of Copper and Lead in pounds, from 1/32 to 1/2 inch in thickness.

Thickness.	Copper.	Lead.
1/32	1.45	1.85
1/16	2.90	3.70
3/32	4.35	5.54
1/8	5.80	7.39
5/32	7.26	9.24
3/16	8.71	11.08
7/32	10.16	12.93
1/4	11.61	14.77
9/32	13.07	16.62
5/16	14.52	18.47
11/32	15.97	20.31
3/8	17.41	22.16
13/32	18.87	24.00
7/16	20.32	25.85
15/32	21.77	27.70
1/2	23.22	29.55

6140. Weight of Cast-Iron Plates.

Weight of Cast-Iron Plates, 12 inches square.

Thickness.	Weight.	Thickness.	Weight.
1/8 inch..	4 lbs. 13 3/8 oz.	5/8 inch.	24 lbs. 2 3/4 oz.
1/4 " ..	9 " 10 5/8 "	3/4 " .	29 " 0 "
3/8 " ..	14 " 8 "	7/8 " .	33 " 13 3/8 "
1/2 " ..	19 " 5 3/8 "	1 " .	38 " 10 3/4 "

6141. Weight of Sheet Iron.

Weight of a Square Foot of Sheet Iron in pounds avoirdupois, the thickness being the number on the wire gauge. No 1 is 5/16 of an inch; No. 4, 1/4; No. 11, 1/8, &c.

No. on Wire Gauge.	Pounds Avoir.	No. on Wire Gauge.	Pounds Avoir.
1	12.5	12	4.62
2	12.	13	4.31
3	11.	14	4.
4	10.	15	3.95
5	9.	16	3.
6	8.	17	2.5
7	7.5	18	2.18
8	7.	19	1.93
9	6.	20	1.62
10	5.68	21	1.5
11	5.	22	1.37

6142. Weight of Boiler Iron.

Weight of a Square Foot of Boiler Iron, from 1/8 to 1 inch thick, in pounds.

Thickness.	Weight.	Thickness.	Weight.
1/8 inch	5 pounds.	5/8 inch	25 pounds.
3/16 "	7.5 "	11/16 "	27.5 "
1/4 "	10 "	3/4 "	30 "
5/16 "	12.5 "	13/16 "	32.5 "
3/8 "	15 "	7/8 "	35 "
7/16 "	17.5 "	15/16 "	37.5 "
1/2 "	20 "	1 "	40 "
9/16 "	22.5 "		

6143. Properties of Metals.

METALS.	Weight of a Cubic Inch in Lbs.	Specific Gravity.	Weight of a Cubic Foot in Lbs.	Tenacity in Lbs. per Square Inch.	Crushing Force in Lbs. per sq. Inch.	Melting Point. Fahr.	Expansion between 32° & 212°	Conducting Power.	Specific Heat.
Aluminum	.092	2.56	160			*1800°			
Antimony, cast	.242	6.7	418	1,066		810°	.0011		.0507
Bismuth	.35	9.82	605	3,250		497°	.0014		.0288
Brass, cast	.3	8.4	525	17,978	10,300	1869°	.002		
" wire		8.5	531	49,000					
Copper, cast	.32	8.89	555	19,072	11,700	1996°	.0017		.0949
" sheet		8.95	559	33,000				898	
" wire		9.	562	61,000					
Gold	.7	19.25	1203	20,400		2016°	.0016	1000	.0298
Gun-metal	.3	8.4	525	36,000					
Iron, wrought bar	.28	7.7	481	60,000	38,000		.0012	347	.1100
" Swedish		7.6	475	70,000					
" wire				85,000					
" cast	.26	7.18	448	19,000	92,000	2786°	.0011		
Lead, cast	.41	11.35	709	1,824	7,000	612°	.0028	180	.0293
" sheet				3,328					
Mercury	.49	13.56	847			—39°	.016		.0330
Silver	.38	10.47	654	41,000		1873°	.0019	973	.0557
Steel	.282	7.8	487	120,000		2500°	.0011		
" puddled		7.78	485	80,000					
Tin	.263	7.29	455	5,000	15,000	442°	.0021	304	.0514
Zinc	.253	7.	437	8,000		773°	.0029	363	.0927

* Approximate; no well-authenticated experiments on Aluminum.

6144. Weight of Round and Square Shafts of Wrought Iron, 1 Foot Long.

Size in Inches.	Weight in Lbs. Round.	Weight in Lbs. Square.	Size in Inches.	Weight in Lbs. Round.	Weight in Lbs. Square.
1/8	.042	.053	4 3/4	59.7	76.0
1/4	.166	.211	5	66.2	84.3
3/8	.372	.474	5 1/4	72.9	92.9
1/2	.662	.843	5 1/2	80.1	102
5/8	1.03	1.32	5 3/4	87.5	111
3/4	1.49	1.90	6	95.3	121
7/8	2.03	2.58	6 1/4	103	132
1	2.65	3.37	6 1/2	112	142
1 1/8	3.35	4.27	6 3/4	121	154
1 1/4	4.14	5.27	7	130	165
1 3/8	5.00	6.37	7 1/4	139	177
1 1/2	5 97	7.58	7 1/2	149	190
1 5/8	7.00	8.90	7 3/4	159	203
1 3/4	8.11	10.3	8	169	216
1 7/8	9.31	11.8	8 1/4	180	229
2	10.6	13.5	8 1/2	191	244
2 1/8	11.9	15.2	8 3/4	203	258
2 1/4	13.4	17.1	9	214	273
2 3/8	14.9	19.0	9 1/4	227	288
2 1/2	16.5	21.1	9 1/2	239	304
2 5/8	18.2	23.2	9 3/4	252	320
2 3/4	20.0	25.5	10	265	337
2 7/8	21.9	27.9	10 1/2	292	372
3	23.8	30.3	11	320	408
3 1/4	28.0	35.6	11 1/2	350	448
3 1/2	32.4	41.3	12	381	486
3 3/4	37.2	47.4	12 1/2	414	527
4	42.4	54.0	13	447	570
4 1/4	47.8	60.9	13 1/2	483	614
4 1/2	53.6	68.2	14	519	661

6145. Weights of Wrought-Iron and Steel.

Round Iron.—Multiply the square of the diameter in inches, by the length in feet, and by 2.63, and the product will be the weight in pounds avoirdupois, nearly.

Square Iron.—Multiply the area of the end of the bar in inches, by the length in feet, and by 3.36; the product will be the weight in pounds avoirdupois, nearly.

Square, Angled, T, *Convex, or any figure of Beam Iron.*—Ascertain the area of the end of each figure of bar, in inches, then multiply the area by the length in feet, and that product by 10, and divide by three; the remainder will be the weight in pounds, nearly.

Square Cast Steel.—Multiply the area of the end of the bar in inches, by the length in feet, and that product by 3.4; the product will be the weight in pounds, nearly.

Round Cast Steel.—Multiply the square of the diameter in inches, by the length in feet, and that product by 2.67; the product will give the weight in pounds avoirdupois, nearly.

6146. Number of Nails per Pound.

The following table shows the length of the various sizes of nails and the number of each in a pound:

Size.	Length.	Number.
3-penny,	1 inch long,	557 per pound.
4 "	1 1/4 "	353 "
5 "	1 1/2 "	232 "
6 "	2 "	167 "
7 "	2 1/4 "	141 "
8 "	2 1/2 "	101 "
10 "	2 3/4 "	98 "
12 "	3 "	54 "
20 "	3 1/2 "	34 "
Spikes	4 "	16 "
"	4 1/2 "	12 "
"	5 "	10 "
"	6 "	7 "
"	7 "	5 "

The term "penny," designating the size of nails, appears to mean "pound." Ten-penny nails weighing 10 pounds per thousand, four-penny nails 4 pounds per thousand, &c. (*Webster.*) This is probably the weight the nails were originally made; according to the foregoing table they have since learned economy in the material.

6147. Calendar for Ascertaining on what Day of the Week any Given Day will Fall within the Present Century.

YEARS 1801 TO 1900.											31 Jan.	28 Feb.	31 Mar.	30 April.	31 May.	30 June.	31 July.	31 Aug.	30 Sept.	31 Oct.	30 Nov.	31 Dec.
1801	1807	1818	1829	1835	1846	1857	1863	1874	1885	1891	4	7	7	3	5	1	3	6	2	4	7	2
1802	1813	1819	1830	1841	1847	1858	1869	1875	1886	1897	5	1	1	4	6	2	4	7	3	5	1	3
1803	1814	1825	1831	1842	1853	1859	1870	1881	1887	1898	6	2	2	5	7	3	5	1	4	6	2	4
1805	1811	1822	1833	1839	1850	1861	1867	1878	1889	1895	2	5	5	1	3	6	1	4	7	2	5	7
1806	1817	1823	1834	1845	1851	1862	1873	1879	1890		3	6	6	2	4	7	2	5	1	3	6	1
1809	1815	1826	1837	1843	1854	1865	1871	1882	1893	1899	7	3	3	6	1	4	6	2	5	7	3	5
1810	1821	1827	1838	1849	1855	1866	1877	1883	1894	1900	1	4	4	7	2	5	7	3	6	1	4	6

To ascertain any day of the week in any year of the present century, first look in the table of years for the year required, and under the months are figures which refer to the corresponding figures at the head of the columns of days below.

For Example: To find what day of the week January 1 will be in the year 1873, look in the table of years for 1873, and in a parallel line under January is figure 3, which directs to column 3, in which it will be seen that January 1 will fall on Wednesday.

LEAP-YEARS.				31 Jan.	28 Feb.	31 Mar.	30 April.	31 May.	30 June.	31 July.	31 Aug.	30 Sept.	31 Oct.	30 Nov.	31 Dec.
1804	1832	1860	1888	7	3	4	7	2	5	7	3	6	1	4	6
1808	1836	1864	1892	5	1	2	5	7	3	5	1	4	6	2	4
1812	1840	1868	1896	3	6	7	3	5	1	3	6	2	4	7	2
1816	1844	1872		1	4	5	1	3	6	1	4	7	2	5	7
1820	1848	1876		6	2	3	6	1	4	6	2	5	7	3	5
1824	1852	1880		4	7	1	4	6	2	4	7	3	5	1	3
1828	1856	1884		2	5	6	2	4	7	2	5	1	3	6	1

1	2	3	4	5	6	7
Mon..... 1	Tues..... 1	Wed..... 1	Thur..... 1	Fri....... 1	Sat....... 1	Sun...... 1
Tues..... 2	Wed..... 2	Thur.... 2	Fri...... 2	Sat....... 2	Sun...... 2	Mon..... 2
Wed..... 3	Thur.... 3	Fri....... 3	Sat....... 3	Sun...... 3	Mon..... 3	Tues..... 3
Thur.... 4	Fri...... 4	Sat...... 4	Sun...... 4	Mon..... 4	Tues..... 4	Wed..... 4
Fri...... 5	Sat...... 5	Sun...... 5	Mon..... 5	Tues..... 5	Wed..... 5	Thur.... 5
Sat...... 6	Sun...... 6	Mon..... 6	Tues..... 6	Wed..... 6	Thur.... 6	Fri...... 6
Sun...... 7	Mon..... 7	Tues..... 7	Wed..... 7	Thur.... 7	Fri...... 7	Sat...... 7
Mon..... 8	Tues..... 8	Wed..... 8	Thur.... 8	Fri...... 8	Sat...... 8	Sun...... 8
Tues..... 9	Wed..... 9	Thur.... 9	Fri...... 9	Sat...... 9	Sun...... 9	Mon..... 9
Wed.....10	Thur....10	Fri......10	Sat......10	Sun......10	Mon.....10	Tues.....10
Thur.....11	Fri......11	Sat......11	Sun......11	Mon.....11	Tues.....11	Wed.....11
Fri......12	Sat......12	Sun......12	Mon.....12	Tues.....12	Wed.....12	Thur....12
Sat......13	Sun......13	Mon.....13	Tues.....13	Wed.....13	Thur....13	Fri......13
Sun......14	Mon.....14	Tues.....14	Wed.....14	Thur....14	Fri......14	Sat......14
Mon.....15	Tues.....15	Wed.....15	Thur.....15	Fri......15	Sat......15	Sun......15
Tues.....16	Wed.....16	Thur.....16	Fri......16	Sat......16	Sun......16	Mon.....16
Wed.....17	Thur....17	Fri......17	Sat......17	Sun......17	Mon.....17	Tues.....17
Thur....18	Fri......18	Sat...... 18	Sun......18	Mon.....18	Tues.....18	Wed.....18
Fri......19	Sat...... 19	Sun......19	Mon.....19	Tues.....19	Wed.....19	Thur.... 19
Sat......20	Sun......20	Mon.....20	Tues.....20	Wed.....20	Thur.... 20	Fri......20
Sun......21	Mon.....21	Tues.....21	Wed.....21	Thur.... 21	Fri......21	Sat......21
Mon.....22	Tues.....22	Wed.....22	Thur....22	Fri......22	Sat......22	Sun......22
Tues.....23	Wed.....23	Thur.....23	Fri......23	Sat......23	Sun......23	Mon.....23
Wed.....24	Thur....24	Fri......24	Sat......24	Sun......24	Mon.....24	Tues.....24
Thur.... 25	Fri......25	Sat...... 25	Sun......25	Mon.....25	Tues.....25	Wed.....25
Fri...... 26	Sat......26	Sun......26	Mon.....26	Tues.....26	Wed.....26	Thur....26
Sat......27	Sun......27	Mon.....27	Tues.....27	Wed.....27	Thur....27	Fri......27
Sun......28	Mon.....28	Tues.....28	Wed.....28	Thur....28	Fri......28	Sat......28
Mon.....29	Tues.....29	Wed.....29	Thur.....29	Fri......29	Sat......29	Sun......29
Tues.....30	Wed.....30	Thur.....30	Fri......30	Sat......30	Sun......30	Mon.....30
Wed.....31	Thur....31	Fri......31	Sat......31	Sun......31	Mon.....31	Tues.....31

6149. Loss Sustained by Different Substances in Drying.

Grains.		Dried at	Lose Grains.
100	Gallic Acid	212°	9.5
100	Sulphate of Quinine	212°	14.4
100	Arseniate of Soda	300°	40.38
100	Alum	400°	47.
100	Carbonate of Soda	Dull Redness	63.
100	Phosphate of Soda	"	63.
100	Sulphate of Soda	"	56.
100	Carbonate of Potassa	"	16.
Grains.		**Dried at**	**Leave Grains.**
29	Oxide of Silver	Redness	27 Metallic Silver
10	Oxalate of Cærium	"	4.8 Oxide with Peroxide
100	Oxalate of Iron	"	27 Peroxide of Iron
50	Tartrate of Iron	"	15 Sesquioxide of Iron
50	Carbonate of Magnesia	"	22 Magnesia

6148. Proportions of a Beautiful Body. The height should be exactly equal to the distance between the tips of the middle fingers of either hand when the arms are fully extended. Ten times the length of the hand, or seven and a half times the length of the foot, or five times the diameter of the chest from one armpit to the other, should also each give the height of the whole body. The distance from the junction of the thighs to the ground should be the same as from that point to the crown of the head. The knee should be precisely midway between the same point and the bottom of the heel. The distance from the elbow to the tip of the middle finger should be the same as from the elbow to the middle line of the breast. From the top of the head to the level of the chin should be the same as from the level of the chin to that of the armpits, and from the heel to the toe.

6150. Table of Symbols and Equivalents of Metallic Elements. The specific gravity of the following are given at water standard. The equivalents are multiples of hydrogen, which is adopted as the basis, or 1.

	Symbol.	Equivalent. U. S. Dis.	Equivalent. Ure.	Sp. Gr.
Aluminum	Al	13.70	13.67	2.56
Antimony (Stibium)	Sb	122.00	129.00	6 70
Arsenic	As	75.00	75.00	5.67
Barium	Ba	68.70	68.50	4.70
Bismuth	Bi	210.00	213.00	9.80
Boron	B	10.90	11.00	2.68
Cadmium	Cd	55.80	56.00	8.63
Calcium	Ca	20.00	20.00	1.58
Cerium	Ce	46.00	46.00	
Chromium	Cr	26.30	26.27	5.90
Cobalt	Co	29.50	29.50	8.53
Columbium (Tantalium)	Ta	185.00		
Cæsium	Cæ		123.00	
Copper (Cuprum)	Cu	31.70	32.00	8.72
Didymium	D	47.50	48.00	
Erbium	E	56.30		
Glucinium	G	7.00	6.97	
Gold (Aurum)	Au	199.00	98.33	19.4
Ilmenium	Il	60.20		
Indium	In	74.00		
Iridium	Ir	98.80	98.56	18.63
Iron (Ferrum)	Fe	28.00	28.00	7.84
Lantanium	La	44.30		
Lead (Plumbum)	Pb	103.60	104.00	11.30
Lithium	L	7.00	7.00	.59
Magnesium	Mg	12.00	12.00	1.75
Manganese	Mn	27.70	26.00	8.00
Mercury (Hydrargyrum)	Hg	200.00	200.00	13.50
Molybdenum	M	48.00	48.00	8.60
Nickel	Ni	29.50	29.50	8.63
Niobium	Nb	94.00		
Norium	No			
Osmium	Os	99.70	99.41	10.00
Palladium	Pd	53.30	53.24	11.50
Pelopium	Pe			
Platinum	Pt	98.90	99.00	21.50
Potassium (Kalium)	K	39.20	39.00	.86
Rhodium	Ro	52.20	52.16	11.20
Rubidium	Rb	85.40	85.00	
Ruthenium	Ru	52.20	52.11	8.60
Silicon	Si	21.30	21.00	
Silver (Argentum)	Ag	108.00	108.00	10.43
Sodium (Natrium)	Na	23.30	23.00	.97
Strontium	Sr	43.80	44.00	2.54
Tellurium	Te	64.00	64.08	6.30
Terbium	Tb			
Thallium	Tl	204.00		
Thorium	Th	59.60	59.50	
Tin (Stannum)	Sn	59.00	59.00	7.29
Titanium	Ti	25.00	24.12	5.28
Tungsten (Wolfram)	W	92.00	92.00	17.20
Uranium	U	60.00	60.00	10.15
Vanadium	V	51.20	68.46	
Yttrium	Y	30.85		
Zinc	Zn	32.30	32.52	6.91
Zirconium	Zr	33.60	33.58	

6151. Table of Symbols and Equivalents of Non-Metallic Elements. The specific gravity of these are given in their gaseous form, air being the standard or 1.000. The equivalents are multiples of hydrogen which is adopted as the basis or 1.

	Symbol.	Equivalent.		Specific Gravity.
		U. S. Dis.	Ure.	
Bromine	Br	78.4	80.0	5.4110
Carbon	C	6.0	6.0	.8290
Chlorine	Cl	35.5	35.5	2.4530
Fluorine	Fl	18.7	19.0	1.3270
Hydrogen	H	1.0	1.0	.0692
Iodine	I	126.3	127.0	8.7827
Nitrogen	N	14.0	14.0	.9713
Oxygen	O	8.0	8.0	1.1056
Phosphorus	P	32.0	32.0	4.2840
Selenium	Se	40.0	40.0	7.6960
Sulphur.	S	16.0	16.0	2.2140

6152. To Reduce Parts by Volume or Measure to Parts by Weight. Multiply the parts by volume or measure by the specific gravity of the different substances; the result will be parts by weight.

6153. To Find the Length of the Day or Night. To find the length of any day, double the time of sunset. Double the hour of sunrise will be the length of the night.

6154. To Reduce a Liquid to a Given Density. It has been already stated in No. 52 that the actual weight of any substance may be found by weighing an exactly equal bulk of water, and multiplying the weight found by the specific gravity of the substance; the product is the actual weight. To simplify this, suppose that a liquid has a specific gravity of 1.325; also that a certain bulk of water (say any 1 measure) weighs 100 grains; then a similar bulk (1 measure) of the substance would weigh $100 \times 1.325 = 132.5$ grains. Now, supposing we wish to reduce the weight of this liquid, so that 1 measure of it shall weigh only 115.5 grains (that is, shall have a specific gravity of 1.155), how much water, whose specific gravity is 1.000, must be added to it to produce this result?

From the nature of the proposition, it follows that the bulk of the substance (1) multiplied by its specific gravity (1.325), added to the bulk of added (unknown) water multiplied by its specific gravity (1.000), must be equal to the aggregate bulk of the substance and of the water combined, multiplied by its required specific gravity (1.155).

Putting the above words into shape, and assuming x to be the required bulk or quantity of water

$$(1 \times 1.325) + (x \times 1.000) = (1+x) \times 1.155$$

or $1.325 + 1.000x = 1.155 + 1.155x$

by subtracting 1.155 and 1.000 x from each side we have

$$.170 = .155x$$

in other words the required

bulk of water, ... $x = \frac{.170}{.155} = 1.097$

If, as supposed above, the measure assumed was such that it weighed 100 grains of water, we should have to add $109\frac{7}{10}$ grains of water to 1 measure of the substance to produce a mixture of specific gravity 1.155.

6155. Gay Lussac's Light Areometer Reduced to Specific Gravity. This instrument ranges from 0° to 50°, 0° corresponding with water at 59° Fahr.

Degree.	Sp. Gr.	Diff.	Degree.	Sp. Gr.	Diff.
0°	1.0000	.0095	30°	.7692	.0057
5	.9524	.0087	35	.7407	.0053
10	.9090	.0079	40	.7143	.0049
15	.8696	.0073	45	.6897	.0044
20	.8333	.0067	50	.6667	
25	.8000	.0062			

This table gives the specific gravity corresponding to every 5 degrees of the scale. To find the specific gravity of intermediate degrees, the average difference between each degree is given in the third column, each given difference referring to the four degrees following the degree opposite which the difference is placed. Thus: To find the specific gravity corresponding with 33 degrees of the scale, look in the table for the specific gravity of the nearest lower degree given, in this instance 30°; and we find .7692; 33° is 3° more than 30°, hence we must deduct 3 times the given difference (.0057), or .0171; this last deducted from .7692 = .7521, which is the approximate specific gravity corresponding to 33° of the scale.

The intermediate degrees of other areometers may be determined in a similar manner.

The corresponding degrees of different areometers may also be found by a comparison with their respective specific gravities; allowance being made for difference of temperature.

Information showing the practical use of some of the areometers will be found in Nos. 58 to 68.

6156. Gay Lussac's Heavy Areometer Reduced to Specific Gravity. This areometer ranges from 0° to 50°, 0° representing water at 59° Fahr.

Degree.	Sp. Gr.	Diff.	Degree.	Sp. Gr.	Diff.
0°	1.0000	.0105	30°	1.4286	.0220
5	1.0526	.0117	35	1.5385	.0256
10	1.1111	.0131	40	1.6667	.0303
15	1.1765	.0147	45	1.8182	.0363
20	1.2500	.0167	50	2.0000	
25	1.3333	.0191			

The specific gravity of the intermediate degrees is found in the same manner as in No. 6155, only that the differences must be added instead of subtracted.

6157. Gay Lussac's Alcoholmeter Reduced to Specific Gravity. This instrument exhibits the percentage of alcohol by volume in different alcoholic mixtures at 59° Fahr.

Per cent. of Alcohol by Volume.	Sp. Grav.	Diff.	Per cent. of Alcohol by Volume.	Sp. Grav.	Diff.
100	.7947	.0044	60	.9141	.0021
95	.8168	.0036	55	.9248	.0020
90	.8346	.0031	50	.9348	.0018
85	.8502	·0028	45	.9440	.0016
80	.8645	.0031	40	.9523	.0014
75	.8799	.0022	35	.9595	.0002
70	.8907	.0024	10	.9656	.0034
65	.9027	.0023	0	1.0000	

The specific gravity of the intermediate degrees is found as explained in No. 6155, only that the difference must be added instead of subtracted.

6158. Beck's Heavy Areometer Reduced to Specific Gravity. This ranges from 0° to 76°, 0° corresponding with water at 54½° Fahr.

Degree.	Sp. Gr.	Diff.	Degree.	Sp. Gr.	Diff.
0°	1.0000	.0061	45°	1.3600	.0113
5	1.0303	.0064	50	1.4167	.0123
10	1.0625	.0068	55	1.4782	.0134
15	1.0968	.0073	60	1.5454	.0147
20	1.1333	.0078	65	1.6190	.0162
25	1.1724	.0084	70	1.7000	.0179
30	1.2143	.0090	75	1.7895	
35	1.2592	.0097	76	1.8085	
40	1.3077	.0105			

The specific gravity of the intermediate degrees is obtained as shown in No. 6155, the differences being added instead of subtracted.

6159. Beck's Light Areometer Reduced to Specific Gravity. The scale on this areometer marks from 0° to 70°, 0° representing water at 54½° Fahr.

Deg.	Sp. Gr.	Diff.	Deg.	Sp. Gr.	Diff.
0°	1.0000	.0057	40°	.8095	.0038
5	.9714	.0054	45	.7907	.0036
10	.9444	.0051	50	.7727	.0034
15	.9189	.0048	55	.7555	.0033
20	.8947	.0046	60	.7391	.0031
25	.8718	.0043	65	.7234	.0030
30	.8500	.0041	70	.7083	
35	.8293	.0040			

The equivalents of the intermediate degrees may be found by the method given in No. 6155.

6160. Dutch Light Areometer Reduced to Specific Gravity. This areometer ranges from 0° to 60°, 0° denoting water.

Deg.	Sp. Gr.	Diff.	Deg.	Sp. Gr.	Diff.
0°	1.0000	.0067	35°	.8045	.0044
5	.9664	.0063	40	.7826	.0041
10	.9351	.0059	45	.7619	.0039
15	.9057	.0055	50	.7423	.0037
20	.8780	.0052	55	.7236	.0035
25	.8521	.0049	60	.7059	
30	.8276	.0046			

The specific gravity of the intermediate degrees may be found in the same manner as directed in No. 6155.

6161. The Heavy Areometer of Brix. This instrument is graduated from 0° to 200°, 0° denoting water at 60° Fahr.

Deg.	Sp. Gr.	Diff.	Deg.	Sp. Gr.	Diff.
0°	1.0000	.0025	105°	1.3559	.0047
5	1.0127	.0026	110	1.3793	.0048
10	1.0256	.0027	115	1.4035	.0050
15	1.0390	.0027	120	1.4286	.0052
20	1.0526	.0028	125	1.4545	.0054
25	1.0667	.0029	130	1.4815	.0056
30	1.0811	.0029	135	1.5094	.0058
35	1.0058	.0030	140	1.5385	.0060
40	1.1111	.0031	145	1.5686	.0063
45	1.1268	.0032	150	1.6000	.0065
50	1.1429	.0033	155	1.6326	.0068
55	1.1594	.0034	160	1.6667	.0071
60	1.1765	.0035	165	1.7021	.0074
65	1.1940	.0036	170	1.7391	.0077
70	1.2121	.0037	175	1.7777	.0081
75	1.2308	.0038	180	1.8182	.0085
80	1.2500	.0039	185	1.8605	.0089
85	1.2698	.0040	190	1.9047	.0093
90	1.2900	.0042	195	1.9512	.0098
95	1.3115	.0044	200	2.0000	
100	1.3333	.0045			

The specific gravity of the intermediate degrees is obtained as in No. 6155, by adding the differences instead of subtracting them.

6162. The Light Areometer of Brix. This areometer is graded from 0° to 200°, 0° corresponding with water at 60° Fahr.

Degree.	Sp. Gr.	Diff.	Degree.	Sp. Gr.	Diff.
0°	1.0000	.0025	105°	.7921	.0016
5	.9876	.0024	110	.7843	.0015
10	.9756	.0024	115	.7767	.0015
15	.9638	.0023	120	.7692	.0015
20	.9524	.0022	125	.7619	.0014
25	.9412	.0022	130	.7547	.0014
30	.9302	.0021	135	.7477	.0014
35	.9195	.0021	140	.7407	.0014
40	.9091	.0020	145	.7339	.0013
45	.8989	.0020	150	.7273	.0013
50	.8889	.0020	155	.7207	.0013
55	.8791	.0019	160	.7143	.0013
60	.8696	.0019	165	.7080	.0012
65	.8602	.0018	170	.7018	.0012
70	.8511	.0018	175	.6957	.0012
75	.8421	.0018	180	.6897	.0012
80	.8333	.0017	185	.6838	.0012
85	.8247	.0017	190	.6780	.0011
90	.8163	.0016	195	.6723	.0011
95	.8081	.0016	200	.6667	
100	.8000	.0016			

To obtain the specific gravity of the intermediate degrees see No. 6155.

6163. Dutch Heavy Areometer Reduced to Specific Gravity. The range of this instrument is from 0° to 75°, 0° corresponding with water.

Deg.	Sp. Gr.	Diff.	Deg.	Sp. Gr.	Diff.
0°	1.0000	.0072	40°	1.3846	.0140
5	1.0359	.0077	45	1.4545	.0155
10	1.0746	.0083	50	1.5319	.0172
15	1.1163	.0090	55	1.6180	.0193
20	1.1613	.0098	60	1.7143	.0217
25	1.2101	.0106	65	1.8228	.0246
30	1.2631	.0116	70	1.9459	.0282
35	1.3211	.0127	75	2.0869	

The specific gravity of the intermediate degrees is easily obtained by following the directions laid down in No. 6155, adding the difference instead of subtracting it.

6164. Twaddel's Areometer Reduced to Specific Gravity. The range of this areometer or saccharometer is from 0° to 200°, 0° corresponding with water.

Degrees.	Sp. Grav.	Degrees.	Sp. Grav.
0°	1.000	105°	1.525
5	1.025	110	1.550
10	1.050	115	1.575
15	1.075	120	1.600
20	1.100	125	1.625
25	1.125	130	1.650
30	1.150	135	1.675
35	1.175	140	1.700
40	1.200	145	1.725
45	1.225	150	1.750
50	1.250	155	1.775
55	1.275	160	1.800
60	1.300	165	1.825
65	1.325	170	1.850
70	1.350	175	1.875
75	1.375	180	1.900
80	1.400	185	1.925
85	1.425	190	1.950
90	1.450	195	1.975
95	1.475	200	2.000
100	1.500		

In the above table the difference between the degrees is .005, throughout; the specific gravity of the intermediate degrees can be found by following the method given in No. 6155, adding instead of deducting the difference. (*See No.* 68.)

6165. Baumé's Heavy Areometer. This instrument marks from 0° to 75°, 0° being water at 63½° Fahr.

Deg.	Sp. Gr.	Diff.	Deg.	Sp. Gr.	Diff.
0°	1.0000	.0071	40°	1.3746	.0135
5	1.0353	.0076	45	1.4421	.0149
10	1.0731	.0081	50	1.5166	.0165
15	1.1138	.0088	55	1.5992	.0184
20	1.1578	.0095	60	1.6914	.0207
25	1.2053	.0103	65	1.7948	.0234
30	1.2569	.0112	70	1.9117	.0266
35	1.3131	.0123	75	2.0448	

The specific gravity of the intermediate degrees can be obtained as directed in No. 6155, adding the difference instead of subtracting. A ready method of calculating the specific gravity corresponding to the degrees of this areometer, sufficiently correct for common purposes, will be found in No. 66; the table given in No. 65 is made on that principle, and based on 1000 as the unit representing water, instead of 1.

6166. Baumé's Light Areometer. This areometer ranges from 10° to 60°, 10° denoting water at 54½° Fahr.

Deg.	Sp. Gr.	Diff.	Deg.	Sp. Gr.	Diff.
10°	1.0000	.0066	40°	.8294	.0046
15	.9669	.0062	45	.8065	.0043
20	.9358	.0058	50	.7848	.0041
25	.9067	.0055	55	.7642	.0039
30	.8794	.0051	60	.7447	
35	.8537	.0049			

The specific gravity of the intermediate degrees is found by following the directions given in No. 6155. A simple method for converting the degrees of this areometer into specific gravity, applicable in cases where great accuracy is not required, is given in No. 66. A table, similar to the above, will be found in No. 62, sufficiently accurate for general practical purposes.

INDEX.

In the compilation of this Index, especial pains have been taken to economise space as much as possible, without impairing its usefulness for ready reference. With this end in view, classification of items has been largely resorted to; so that, in many cases, a single entry will embrace several receipts, varying in number from two or three to twenty or more.

Some discretion is, therefore, advisable in searching the Index for any desired receipt. If, for instance, it is required to find out "How to put out a fire in a Chimney," it will naturally be found under "Chimney," the object to be operated upon. Again: in searching for some preparation of a compound body, "Solution of Citrate of Magnesia," for instance, it would be found under "Citrate of Magnesia," the principal ingredient, and not under "Magnesia," which, although its base, is an entirely different substance.

Proprietary preparations and processes will be found only under the name of the inventors; thus, "Brandreth's Pills" are indexed under "Brandreth," and not under the head of "Pills;" this latter heading including only such as have no such distinctive designation. This is done to avoid needless repetition, and thereby save space.

The subjects covered here from Dick's formula encyclopedia fall under the general categories of chemistry and medicine. Seeming gaps are simply deletions from this volume of subjects falling in the category of products such as shoepolish, toothpaste, metallurgy, liquors, etc., which are in or will be in other Atlan Formularies publications

THE SCIENTIFIC AMERICAN CYCLOPEDIA

NEW YORK
SCIENTIFIC AMERICAN PUBLISHING CO.
1925

LABORATORY APPARATUS

Wire Apparatus for Laboratory Use.

For most of the apparatus shown, some oxidizable wire should be selected, such as brass or tinned iron, and the tools for forming these articles of wire consist of a pair of cutting pliers, a pair of flat and a pair of round-nosed pliers, a few cylindrical mandrels of wood or metal, made in different sizes, and a small bench vice. Any or all of the articles may be in different sizes, and of different sizes of wire for different purposes.

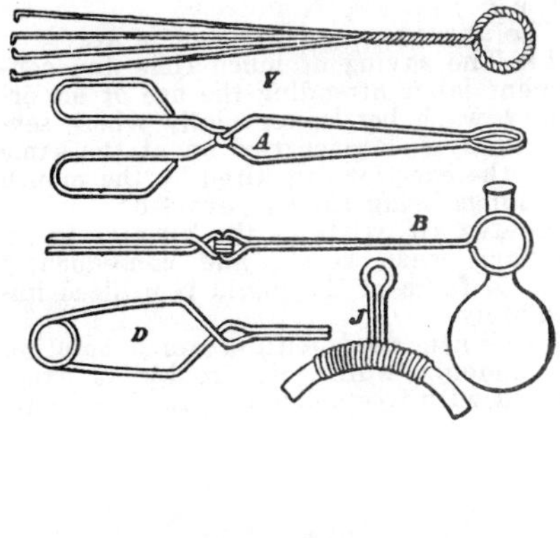

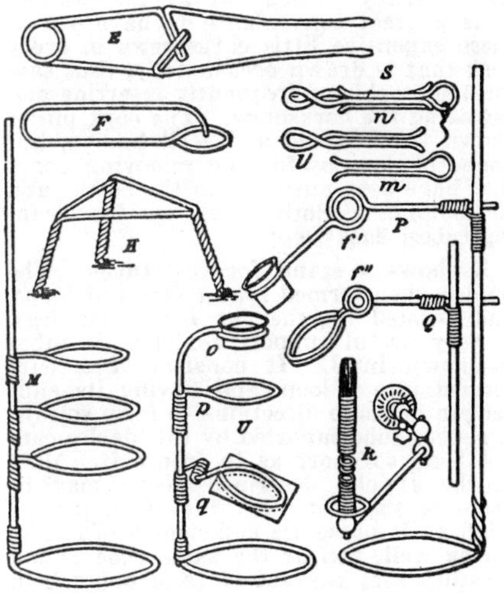

Wire Apparatus for Laboratory Use

A shows a pair of hinged tongs, which are useful for handling coals about the furnace, for holding a coal or piece of pumice for blowpipe work, and for holding large test tubes and flasks, when provided with 2 notched corks, as shown in B and O. These tongs are made by first winding the wire of one half around the the wire of the other half to form the joint, then bending each part at right angles, forming on one end of each a handle, and upon the other end a ring. By changing the form of the ring end the tongs are adapted to handling crucibles and cupels and other things in a muffle.

C shows a pair of spring tongs, the construction of which will be fully understood without explanation. It may be said, however, that the circular spring at the handle end is formed by wrapping the wire around any round object held in the vice; the rings at the opposite end are formed in the same way. The best way to form good curves in the wires is to bend them around some suitable mandrel or form.

D shows a spring clamp for holding work to be soldered or cemented. It may also be used as a pinch cock.

E represents a pair of tweezers, which should be made of good spring wire flat-

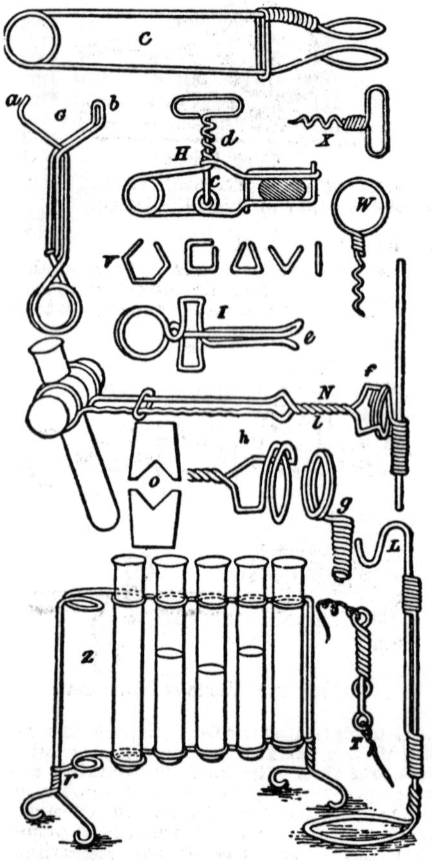

Wire Apparatus for Laboratory Use.

tened at the ends. F is the clamp for mounting microscope slides, and for holding small objects to be cemented or soldered. G is a pinch cock for rubber tubing; its normal position is closed, as in the engraving, but the end *a* is capable of engaging the loop *b*, so as to hold the pinch cock open. H shows a clamp or pinch cock having a wire *c* hooked into an eye in one side, and extending through an eye in the other. This wire is bent at right angles at its outer end to engage a spiral *d*, placed on it and acting as a screw. The open spiral is readily formed by wrapping 2 wires parallel to each other on the same mandrel, and then unscrewing one from the other. The handle will of course be formed by aid of pliers. I shows still another form of pinch cock. It is provided with 2 thumb-pieces, which are pressed when it is desired to open the jaws. K is a tripod stand, formed by twisting 3 wires together. This stand is used for supporting various articles, such as a sand bath or evaporating dish, over a gas flame. It is also useful in supporting charcoal in blowpipe work.

L shows a stand adjustable as to height for supporting the beak of a retort, or for holding glass conducting or condensing tubes in an inclined position. The retort or filter stand, represented in M, is shown clearly enough to require no explanation. Should the friction of the spiral on the standard ever become so slight as to permit the rings to slip down, the spirals may be bent laterally, so as to spring tightly against the standard. N shows an adjustable test tube holder, adapted to the standard shown in M, and capable of being turned on a peculiar joint, so as to place the tube in any desired angle. The holder consists of a pair of spring tongs, having eyes for receiving the notched cork, as shown in O. One arm of the tongs is corrugated to retain the clamping ring in any position along the length of the tongs. The construction of the joint by which the tongs are supported from the slide on the standard is clearly shown in O *a*. It consists of 2 spirals *g h*, the spiral *h* being made larger than the spiral *g*, and screwed over it, as shown in O. This holder is very light, strong and convenient.

P represents a holder for a magnifier, which has a point *f'*, similar to the one just described. The slide *k* is formed of a spiral bent at right angles and off-set to admit of the two straight wires passing each other. This holder may be used to advantage by engravers and draughtsmen. Q shows a holder for a microscope condenser, the difference between this and P being that the ring is made double to receive an unmounted lens.

R shows a Bunsen burner, formed of a common burner, having a surrounding tube made of wire wound in a spiral, and drawn apart near the top of the burner to admit the air, which mingles with the gas before it is consumed at the upper end of the spiral.

S represents a connector for electrical wires, which explains itself. The part with a double loop may be attached to a fixed object by means of a screw. Another electrical connector is shown in T, one part of which consists of a spiral having an eye formed at each end for receiving the screws which fasten it to its support, the other part is simply a straight wire having an eye at one end. The connection is made by inserting the straight end in the spiral. To increase the friction of the two parts, either of them may be curved more or less.

A microscope stand is shown in U. The magnifier is supported in the ring *o*. The ring *p* supports the slide, and the double ring *q* receives a piece of looking-glass or polished metal, which serves as a reflector.

V shows a set of aluminum grain weights in common use. The straight wire is a 1 gr. weight, the one with a single bend is a 2 gr. weight, the one having two bends and forming a triangle is a 3 gr. weight, and so on. W and X are articles now literally turned out by the million. It is a great convenience to have one of these expensive little corkscrews in every cork that is drawn occasionally, thus saving the trouble of frequently inserting and removing the corkscrew. The cork puller shown in Y is old and well known, but none the less useful for removing corks that have been pushed into the bottle, and for holding a cloth or sponge for cleaning tubes, flasks, etc.

Z shows a stand for test tubes. The wire is then formed into a series of loops, and twisted together at *r* to form legs. A very useful support for flexible tubes is shown in J. It consists of a wire formed into a loop, and having its ends bent in opposite directions to form spirals. A rubber tube suported by this device cannot bend so short as to injure it. Most of the articles described above may be made to the best advantage from tinned wire, as it possesses sufficient stiffnsss to spring well, and at the same time is not so stiff as to prevent it from being bent into almost any desired form. Besides this the tin coating protects the wire from corrosin, and gives it a good appearance. —*George M. Hopkins.*

Wash Bottle.

By this simple device the washing of precipitates and the cleansing of vessels used in the process of analysis, which before required the use of the ordinary wash bottle, can now be done with much more facility and in a shorter time. It consists essentially of a thin glass flask C, placed about 3 ft. above the level of the working desk, and closed by a 3-hole rubber stopper. Through one of the holes issues a rubber tube D (or glass with rubber connections), descending to the desk and ending in a glass nozzle. Connection is made by a second hole in the stopper with a reser-

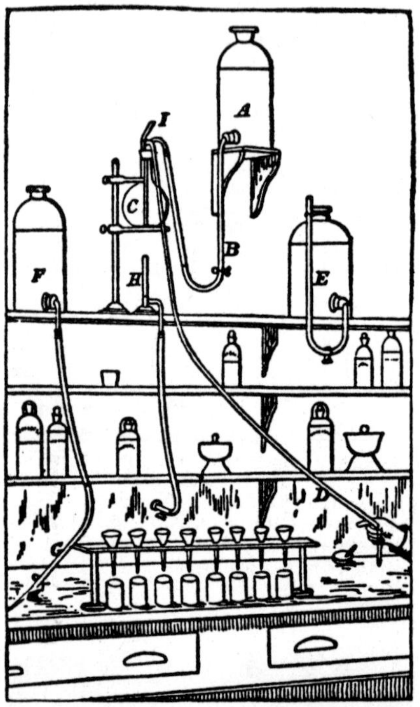

Laboratory Table Showing Wash Bottles.

voir bottle A, placed above the top of the wash bottle. In the third hole is placed a glass tube bent at an angle to keep out dust. On filling the flask from the reserby a pinch cock placed conveniently to the hand, the height of the water flask voir—the flow being stopped by a pinch cock—the water is started by suction from below, and the stream through the nozzle can be regulated or stopped at will furnishing the pressure, which is sustained by the syphon.

A Bunsen burner H is placed underneath the flask, and the water can be heated when it is so desired. Hot water as well as cold can thus be used in treating precipitates. Other solutions can be employed equally as well as water. (See bottle F.).

The advantages of the system are:

1.—The saving of much time and consequent labor attending the use of an ordinary wash bottle, especially where several analyses are carried on at the same time, the exertions required by the mouth and lungs being thereby avoided.

2.—No air exists in the tube, as in an ordinary wash bottle, and consequently the full force of the liquid is utilized immediately.

3.—When used with a wash solution of ammonia water, no trouble is experienced with free ammonia, which ordinarily is quite hurtful to the mouth and eyes.

The large bottle E with the accompanying tube shows a convenient arrangement for holding any solution and delivering the same.

The shelves of a laboratory should be widest at the bottom and should become of less depth at the top to accommodate smaller bottles. The large acid bottles should be put on the bottom shelves. Reagent bottles with the names and symbols blown in are very convenient.

A wash bottle is easily constructed with the aid of a couple of glass tubes and a flask or any bottle of convenient size. One of the glass tubes should be drawn out to the fine point, and the other should be inclined so that it is easily introduced into the mouth. Any desired quantity of water may be forced through the fine powder by moderate blowing. In some

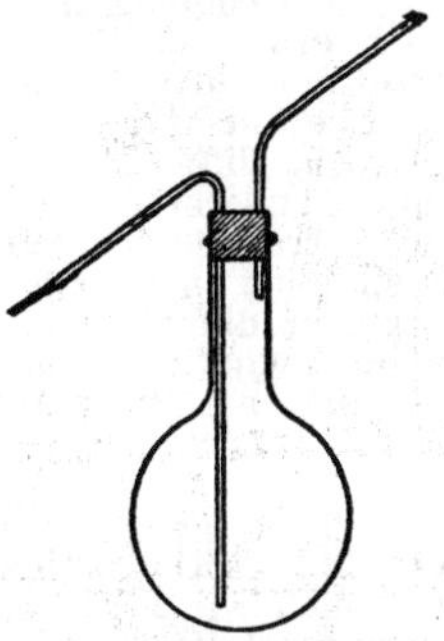

Wash Bottle.

cases the wash bottle is more efficacious when warm. For fine chemical work still water should preferably be used.

Syphons.—Our engravings show handy glass syphons adapted for small operation, the former being without, the latter with stop cock *c* for regulating the flow.

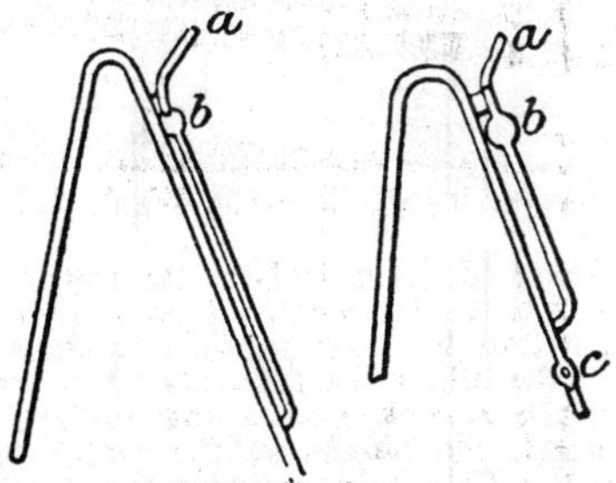

Glass Syphons.

The current is started in these by applying the mouth to the end *a* of the tube, and employing it as an air pump to exhaust the air till the fluid rises into the bulb *b*. With harmless liquids, a simple

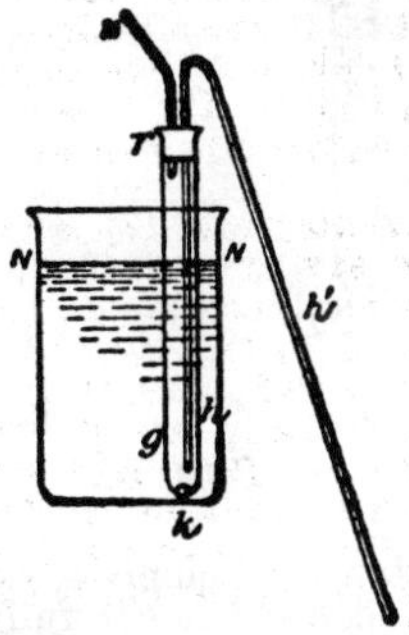

Improved Syphon.

bent glass tube may suffice as a syphon; but suction with the mouth at the end of the longer arm is somewhat inconvenient. The arrangement shown above is simple, and presents certain advantages: A glass tube *g*, ¾ in. wide, and 12-16 in. long, contracted at the lower end, has, at its upper end, a cork stopper, in which the mouthpiece M and the syphon *h h'* are fixed air-tight. The shorter arm *h* of the syphon reaches nearly to the bottom of the tube, and limits the play of a glass ball *k*, which acts as a valve. The diameter of the ball is about ½ in., that of the syphon ¼ in. The instrument thus arranged, being dipped into the vessel to be discharged, the tubes *g* and *h* become filled with liquid to the surface N N. Instead of now sucking, as with the common syphon, one blows into the mouth-piece M; and in consequence of the compression of air, the lower opening is shut by the ball *k*, while the liquid rises in *h*, and begins to flow through *h'* in the usual way. If the vessel to be emptied is not full, or the column of liquid is a small one, it is necessary before blowing into the mouthpiece, to suck it slightly, in order to obtain a larger volume of the liquid in *g*; as one condition for the right action of the instrument is that *h h'* should be filled before the column of liquid in *g* sinks to the mouth of the syphon at *k*, when one blows through M.

Cork Work.

Corks are of the greatest possible use in all laboratories. Boxes of corks may be had of all drug companies and a plentiful supply should be kept at all times. It would probably be necessary to buy larger corks separately. It is frequently necessary to perforate corks, and for this purpose a set of cork borers should be bought; they come in sets. An iron rod passes through the small holes, forming a handle. A rotary motion should be given to the hand at the same time pressure is applied. There is considerable knack in boring corks, but it is soon attained. After the glass tubes have been passed through the corks the corks can be swelled to insure a firm joint. Files and rasps are convenient for altering the

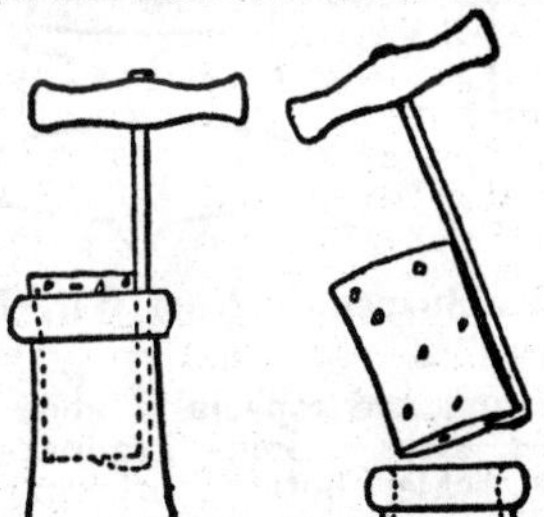

Cork Puller.

shape of corks. Rubber corks are very expensive, but are better for many purposes. They may be purchased already perforated. The ordinary cork borer may, however, be used, wet with dilute ammonia. Pieces of rubber tube of various sizes, and also pieces of hog's bladder for joints, and heavy linen thread for tying the same, should always be at hand.

A cork press will save its cost in a short time. The form shown in our engraving is very effective. Corks which have been compressed give better results than those which are used dried. In the type of press shown, the cork is revolved at the same time it is being compressed,

Cork Press.

thus giving a uniform compression. Corks having a taper should be selected.

Stands, Clamps, etc.

The amateur who has a shop at his disposal will have little difficulty in constructing all necessary supports, which

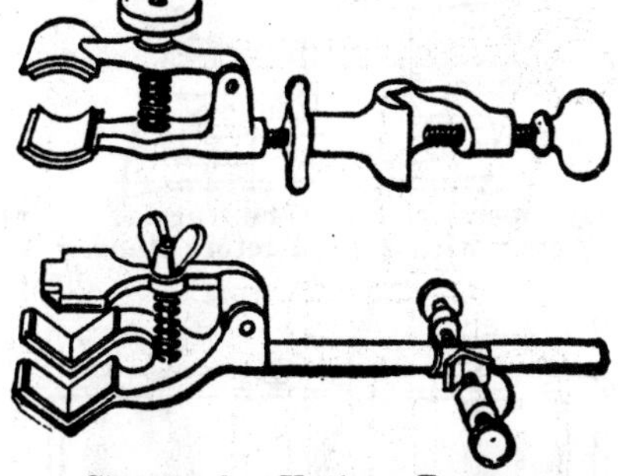

Clamps for Various Purposes.

will tend to materially assist his labors. To those who have no natural mechanical ability, or who have no facilities, are recommended to purchase such apparatus ready prepared of dealers in chemical supplies. A good retort stand is of prime importance, and one of our engravings shows how a retort stand may be used for several purposes at once. Iron retort stands are better than the wooden ones, and there should be at least 4 or 5 rings. The base should be of sufficient weight to make the stands firm at all times. If the base of the retort stand is too light it can be filled with lead. Our engravings also show a variety of clamps which are very useful for a great number of purposes; at least 2 or 3 such clamps should be provided. Nearly every dealer in chemical apparatus lists 15 or 20 different types at all prices. Where rubber tubes are used, pinch cocks will

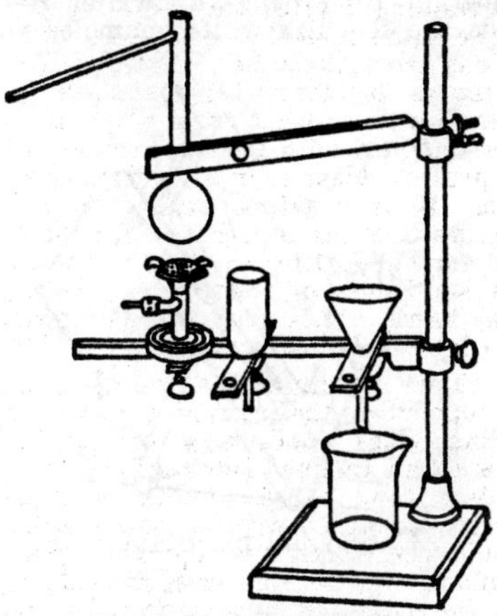

Simple Retort Stand.

be found of value in cutting off the supply of the gas. They can be readily made by the amateur according to the designs given under WIRE APPARATUS in this section.

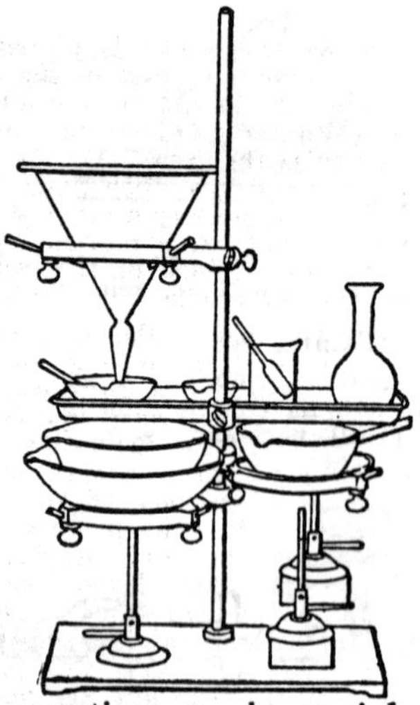

Many operations can be carried on at once with a good retort stand.

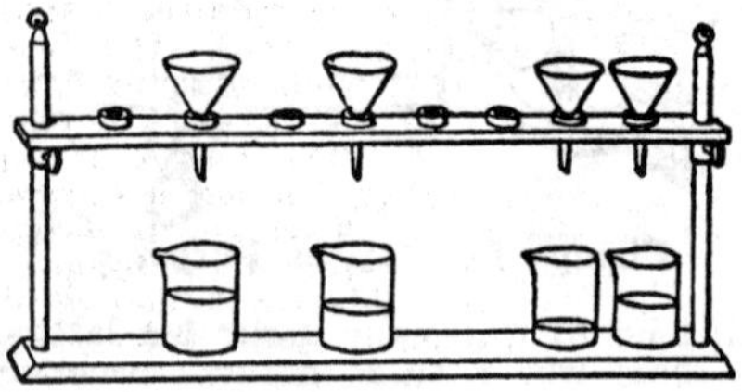

Simple Filter Holder.

A Triangular Holder.

Measuring Liquids.

Liquids may be measured in dishes or containers, of which there are a large number of patterns. The writer recommends the Swedish white enameled ware

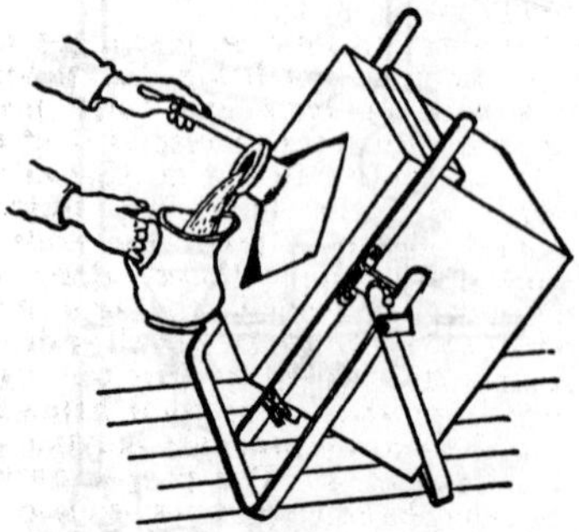

Carboy Tilting Stand.

as indicating at once if there is any dirt in the article. Almost any large dealer in household furnishings would be able to supply a large number of vessels for measuring liquids required by technologists and chemists. Copper measures last a long time, but are very hard to keep clean. They are good for alcoholic liquors. A porcelain measure with graduations inside is very useful. An article of this kind will save its cost in a short time for much work that is done in a laboratory.

Glass graduates form an essential part of the equipment of all laboratories, no matter how small or for what purpose.

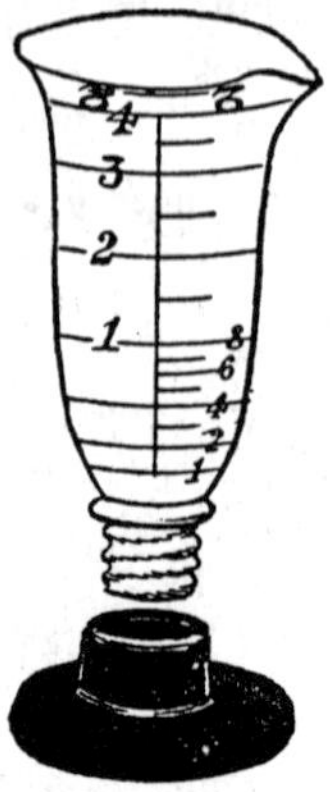

Graduate with Rubber Foot.

Glass graduates of 2, 4, 8, 16, and 32 oz. are recommended. The chemical graduates are easier to get clean than the cylindrical ones. Glass graduates having a beaker shape lessen the liability of

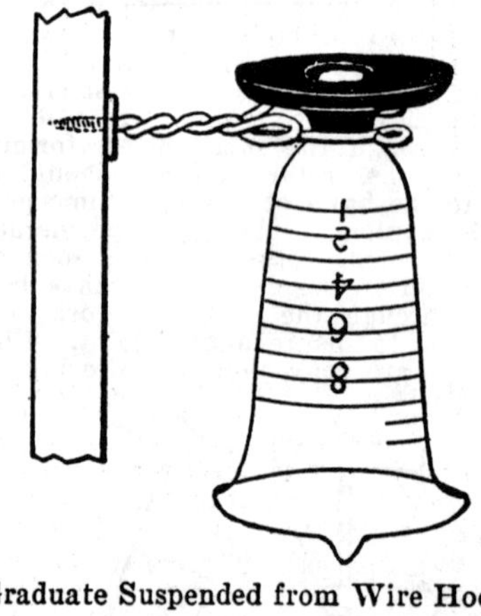

Graduate Suspended from Wire Hook.

breakage and are especially good for 16 and 32-oz. sizes. Some graduates have a double scale, both apothecary's and metric; these are specially recommended where mixed formulas are used calling for both systems. Their use will save much time and calculations, and are specially useful in photographic work where many of the formulas are now given exclusively in metric system. A graduate is "no stronger than its foot," and this is the most vulnerable part of the glass measures. Rubber feet with the screw socket into which the top of the graduate screws have come into quite general use, and are recommended as they tend to decrease the breakage to a considerable extent. When graduates are not in use they should be hung up by the foot, as illustrated in one of our engravings.

For beginning with small quantities of liquids the pipette is recommended, and the simplest form is like the well-known fountain pen filler. Small pipettes can be

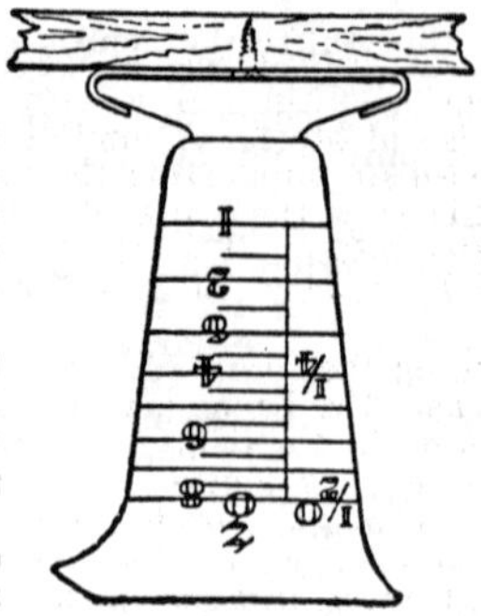

Graduate Slung under Shelf.

obtained shaped like a fork so that they can be used as such in small bottles. For volumetric work and for other accurate determinations, graduated pipettes are sold, but they are comparatively high in price. Small drops of liquid can be readily drawn out of a bottle and distributed with the aid of the pipette. The drop, however, is different from almost every substance, and the number of drops a minim varies from 60 to 250. An excellent table showing the number of drops in a fluid dram of different weights with the weights in grains and grams will be found in Remington's Practice of Pharmacy.

Scales.

A good ordinary scale costing from \$6 to \$10 is recommended. Scales should have a capacity of at least 10 lb. Any sensitive weighing such as required in analytical work, assaying, etc., should not be attempted with scales of this kind. Where corrosive substances which would corrode metal scale pans are in use, the glass tanks should be used, or the substance should be weighed in glass bottles or other containers.

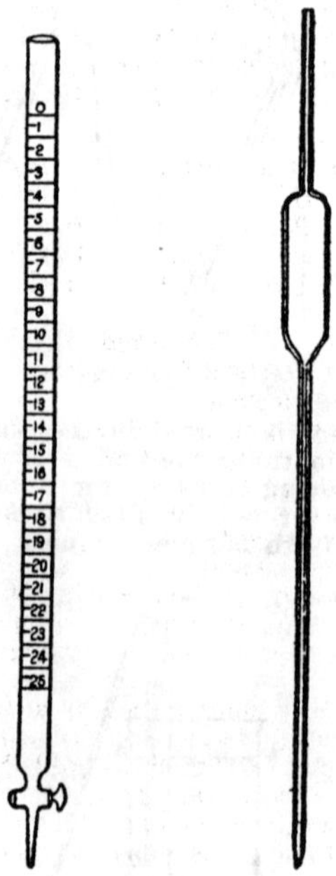

Pipettes.

The Balance is simply a pair of scales, made and adjusted so carefully as to show very small differences in weight of two substances.

The beam is supported in the middle by

a wedge of hard steel, or of agate—a "knife-edge"—resting in a very shallow groove, also of steel. A similar arrangement is used for supporting the scale pins, but in this case the knife-edge is on the end of the beam. The steel should be protected by a very thin coating of vaseline.

By turning the screw placed outside the balance case, the beam may be raised so as to allow it to swing, or lowered so as to prevent any motion. When not in use it should always be lowered.

A pointer is fixed to the middle of the beam, and when the beam is swinging, the end of this pointed moves over a white graduated scale. When the two pans balance, the pointer will move over the same number of divisions on each side of the zero position.

The weights to be used range from 50

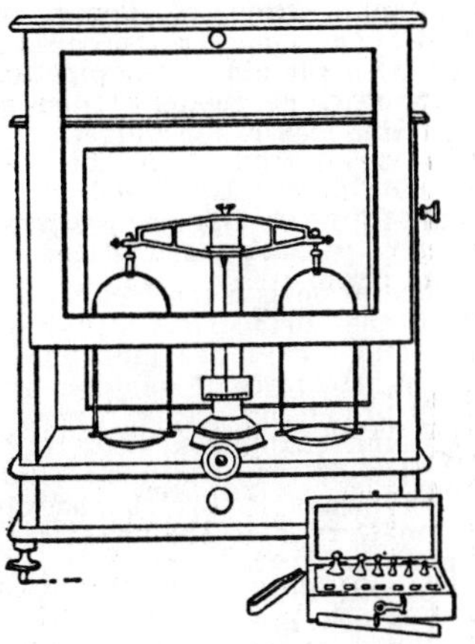

A Balance of Precision.

grams to 1 milligram. The weights below 1 cgrm. may be made of aluminum wire. Each weight should have a separate place in the box. The weights are arranged as follows:

grams.	grams.	grams.	grams.	grams.
50	5	0.5	0.05	0.005
20	2	0.2	0.02	0.002
10	2	0.1	0.01	0.001
10	1	0.1	0.01	0.001

Rules to be Observed in Weighing:

a.—Put the weights on the right-hand pan of the balance.

b.—Never put anything on the balance pans, or take anything off, while the balance is free to swing.

c.—Always use the forceps provided for lifting the weights.

d.—On commencing to weigh, find a weight which is too great, then, after removing this, try the succeeding weights in order. Never pick out weights at random.

e.—Do not put the small weights in a heap. Arrange them in order round the larger weights, which should be in the center of the balance pan.

f.—Place yourself opposite the center of the graduated scale while weighing.

g.—Do not remove any weight from the balance pan until the values of all have been written down, and check your result as the weights are replaced.

h.—Be careful to put the weights back in their proper place.

i.—Never attempt to weigh anything which is not quite cold. In addition to injuring the balance, the weighing will not be accurate.

This mode of pulverization, though particularly applicable to fibrous substances, is sometimes used for metals and hard materials. In the latter case the files may have finer and sharper teeth, and in both instances be particularly clean, and free from grease and dust.

To Make a Balance.—A balance suitable for weighing small articles can be made easily and cheaply. Such a balance can be made sensitive to the weight of one-quarter of a postage stamp, and capable of sustaining a weight of several ounces. It is made chiefly of wood. All the parts are common articles, and only ordinary tools are required. Only certain features require careful attention; in other respects, rough work is permissible, says "School Exercises in Plant Production," by D. J. Crosby, in Farmer's Bulletin No. 408. The essential parts of a balance (see cut) are the base (*a*), the pillar (*b*), the beam (*c*), and the trays or pans, as they are usually called (*d, d*). The beam is balanced by means of the balancing nuts (*e, e*). The pointer (*f*) indicates on the scale (*g*) the effect of weights on the trays. A screw-eye (*h*) encircling the pointer serves to hold the

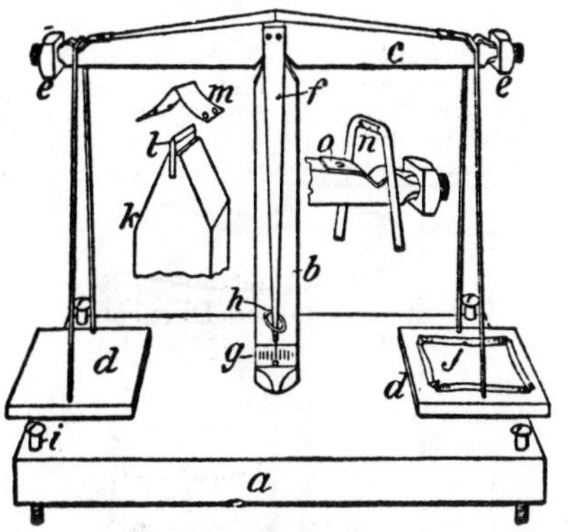

A Simple Balance

beam at rest, or permits it to swing, as desired, according as the screw-eye is turned. Four screws (*i*) at the corners of the base serve to level the balance.

In making the balance thoroughly dry, soft pine wood is preferable. Screws are preferable to nails. The base is 12 or 14 in. long by 7 in. wide and 1 in. thick. The pillar is 1 in. square and about 9 in. high. It can be set in an inch hole in the center of the base. Care should be taken to have it stand perpendicular to the base. The upper end of the pillar is beveled on the right and left sides, as shown at *k*. A slot is sawed in the end to receive a knife edge, as shown at *l*. The beam is made from a stick 1 in. square and about 10 in. long. Its lower face is left straight; the other faces are beveled from the center to the ends, which are left ⅜ or ½ in. square. A notch 1 in. wide and ½ in. deep is accurately cut in the center of the flat or bottom face. This receives the central bearing (*m*) of the beam. An inch from each end of the beam a notch ¼ in. deep is cut to receive the tray bearings. Each end is rounded to receive the balancing nuts. The nuts should cut well defined threads in the wood and move easily and smoothly. Applying a little soap to the threads helps this. A strong pointer (*f*) is firmly fastened to the beam by two or more screws. Its lower end is provided with a needle, colored black so as to be readily seen. The screw-eye (*h*) is placed near the end of the pointer and in the center of the pillar. It should turn easily and smoothly. When the balance is otherwise completed, turn the screw-eye so as to hold the pointer firmly, then paste to the pillar back of the pointer a strip of white paper (*g*) bearing scale marks, 1-16 in. apart, with the 0 mark of the scale directly back of the needle.

The three bearings of the beam are the most exacting features of the construction. Each consists of a knife edge, acting within a groove formed of bent tin. The knife edge (*l*) for the central bearing may be made of a pocket or case knife blade, or of a piece of hard brass filed to a straight, sharp edge. The knife edges for the end bearings are made by filing the lower side of the tray wires where they cross the beam, producing a straight, sharp edge (*n*) about ¾ in. long. The tins forming the grooves of the bearings are made of thin tin, such as is used in oyster and vegetable cans. Bright pieces are selected. The central bearing requires a strip 1 in. wide and 2 in. long (*m*). It is bent across at the middle, the bend being lightly hammered flat on a flatiron. The ends are then separated. The halves of the strip curve somewhat, leaving a narrow angle at the bend. This tin is firmly held in the central notch of the beam by four small screws. The tin strips for the end bearings are about ½ in. wide. They are bent in the same way as the other. One end of the strip is longer than the other, and is punched to receive a single screw holding it to the beam, as shown at *o*. The bending of the tin strips roughens the surface of the groove. It must be polished by rubbing the back of the point of a knife blade back and forth in the groove for some time. To insure success, the grooves must be very narrow to prevent side slipping, yet not so narrow as to bind on the knife edge. The highly polished groove and sharp knife edge produce the least friction, and increase the sensitiveness of the balance.

The trays are made of common No. 12 wire. The trays are 3 by 3 in. and ¼ in. thick. Two holes near opposite edges receive the wires, which are bent in opposite directions beneath the trays, thereby holding them firm and level. If the trays tend to swing from front to back of the balance, the tins of the bearings may be slightly twisted by inserting a knife blade under them.

The balance can now be tested for use. When in working condition the pointed will slowly swing back and forth many times, and finally come to rest at 0 of the scale. It probably will not do this at the first trial. Set the balancing nuts at about equal distances from the ends of the beam, then stand tacks along the lighter beam arm until the two arms nearly balance. The tacks are then driven in permanently. If tacks are too light, use brads or screws. The final balancing can then be done by properly moving one or both of the nuts. The proper adjustment of the balancing nuts should be tested each time the balance is used.

Weights, and objects to be weighed, can be held on the trays by cardboard dishes (*j*). A pair of forceps can be made from a strip of spring brass, or even of hickory wood, the points being properly sharpened.

A set of metric weights ranging from 20 grams to 1 centigram, and suitable for use with this balance, can be had for $1 or less.

Fuels.

The technologist has a wide choice of fuels at the present day. In certain localities wood is plentiful and is well adapted for various processes. It is, however, very sooty and cannot be used for many purposes. Charcoal is much in use and is not expensive. It can be used freely when a quick, strong heat is required. Coal is an excellent fuel for general purposes. Anthracite coal is better now for general use than bituminous coal, although the latter makes the hotter fire. The deposit of soot is often very objectionable. Coke may be had almost anywhere and affords a clean, hot fuel. It is easily kindled. Gas is perhaps the best all-round medium for the production of heat, except where manufacturing operations are to be carried on. A large number of devices calling for the use of gas

A Convenient Alcohol Lamp.

are illustrated in the present book. The Bunsen burner is perhaps the most generally used type of burner. The flame should be blue, and the air regulation is usually accomplished by a ring at the bottom. There are scores of types of

The Blowpipe Flame.

Bunsen burners. For very intense heat the multiple Bunsen burners are recommended. Radio burners using the Bunsen principle are largely used in all of the mechanical arts. Gas can also be used to drive a small hot-air engine for small power laboratories. There are many apparatus which give increase by stirring or agitating where a small caloric engine, or water or electric motor, can be used to advantage. All of the dealers in chemical apparatus furnish petroleum, gasoline and benzine burners as well, so that those who are away from large cities or towns will find their wants very well supplied.

Where considerable quantities of hot water are required, a hot water heater run preferably by gas should be provided. They are not so expensive, and produce

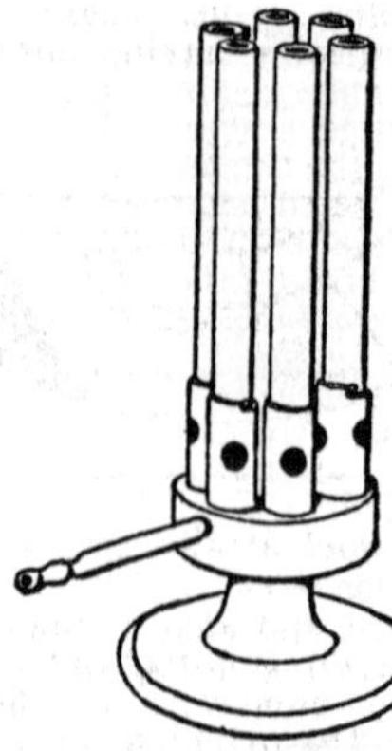

Multiple Bunsen Burner.

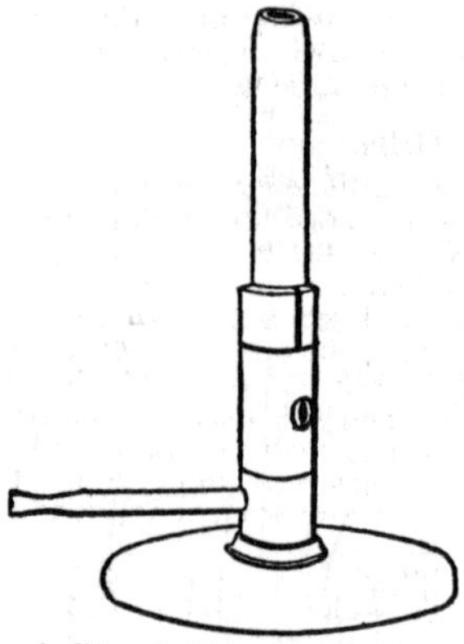

A Simple Bunsen Burner

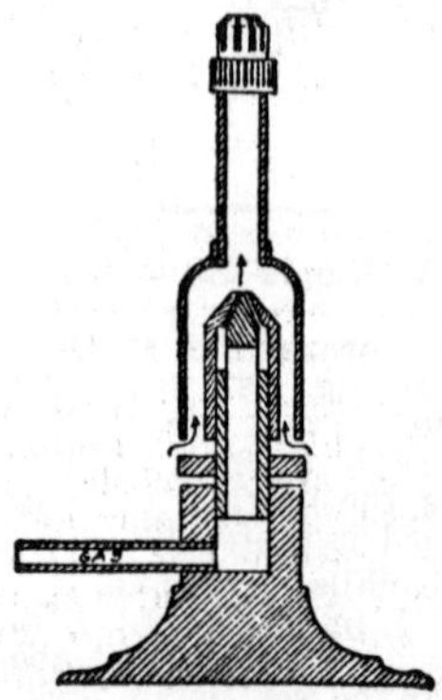

Improved Bunsen Burner.

large volumes of hot water at moderate cost. Perfect control and safety of gas has a great deal to recommend it.

Electricity, though well adapted for all classes of technical work, is very little used owing to the great expense of the initial apparatus and the cost of current, and the length of time which is also required to heat up the hot plate or other device militates against the use of elec-

Burner for Slow Heat.

tricity. The writer has used electrical stoves for heating purposes, and he cannot see that they are of any advantage over hot plates heated by gas. Should it be desired, however, to install electrical apparatus, great care should be taken when

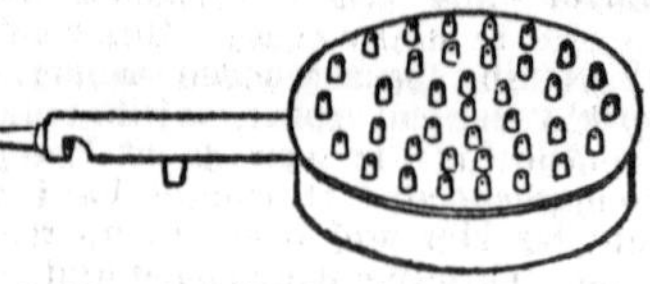

A Good Type of Burner for Evaporation

ordering the equipment that the voltage is the same as the feed mains, as otherwise the electrical apparatus will surely be destroyed.

The blowpipe and charcoal are very useful things to have about the laboratory in connection with the Bunsen burner. Numerous small operations can be conducted with their aid. Blowpipe analysis is a very valuable means of determining minerals and other substances.

I

COMMINUTION OR DIVISION OF SUBSTANCES

This operation is a mechanical process, by which the surface and points of contact of solid bodies are multiplied, thus diminishing the force of cohesion, and consequently promoting greater access to its particles, and enabling a more ready and rapid action of reagents upon solid matter. The means by which the division of solid matters is accomplished are manifold, and those who are using technical formulas will often have to resort to methods which are not in use even by pharmacists.

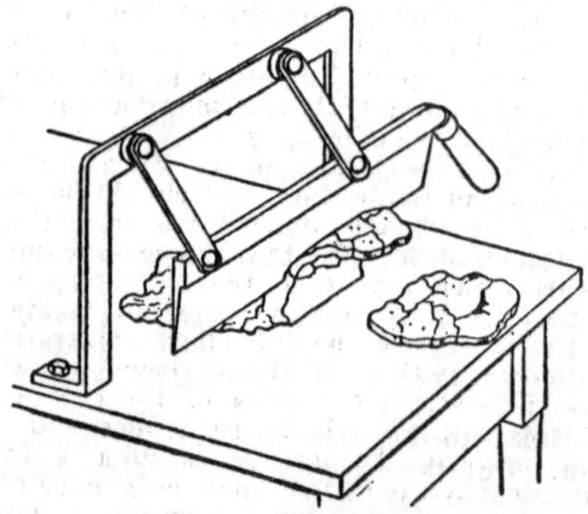

Draw Knife Slicer

Slicing.

This process applies to fibrous matters, and is largely practiced with a lever knife similar to that used by tobacconists for cutting tobacco. This slicing renders the substance in better form for maceration, and, moreover, admits of readier desiccation, a necessary process when it is required to be further reduced under the pestle or by being grated on a coarse rasp. On a large scale, rotary cutters are in use, but they are far beyond the reach of the amateur.

Contusion.

This is a bruising operation, which is very frequently resorted to to reduce a substance to particles, by striking a plurality of blows. A mortar and pestle is perhaps the most used apparatus for this purpose. Corrosive or caustic matter should never be pulverized in metallic mortars, and such substances as chlorate of potash should only be reduced with the greatest possible care. Mortars are made of various materials, such as glass, wedgewood ware, wood and marble. Marble mortars are only recommended where the manufacture of toilet preparations, etc., is to be conducted on a considerable scale. Wooden mortars are useful in many cases. Boxwood mortars are the best wooden mortars. A sheepskin conical cover, with a hole in the center for the passage of the pestle, is recommended. It should be fastened around its rim and over its mouth with a string. Circular pasteboard and wooden covers are often substituted for the sheepskin cover. All substances of an organic nature should be previously dried, so as to afford greater facility for pulverization. A previous reduction of ores and coarse, hard substances into lumps, by concussion with a hammer upon an anvil, and of roots and like substances into slices or bits with a lever knife, are preliminary processes which greatly facilitate their pulverization. The substance to be struck upon the anvil can be wrapped in strong brown paper before crushing.

Silicious stones are pulverized much more readily after having been heated to redness in a crucible, and in that state thrust into cold water. This increased friability is occasioned by the unequal cooling of the mass. •Metals, alloys, and the like, which are pulverized with difficulty while cold, may be readily crushed when heated to redness. When it is required to reduce the substance into small frgaments only, it can be broken down by a succession of blows with the pestle. If the substance is very hard, the force of the arm should be added to the descending weight of the pestle, so as to impart power to the blow. A subsequent circular, grinding motion of the pestle, continued for a length of time, will further reduce these fragments to fine powder, and consequently this movement must be avoided when only a comminution is desired. The mortar should always rest on a sound foundation, and should be occasionally shaken during the operation of pounding, in order that the coarser particles which mount to the sides may be forced back to the center of the mortar so as to receive the full effect of the descending pestle. It should never be allowed to strike the sides of the mortar. If the substance is to be reduced to a fine powder, the process is greatly facilitated by operating upon only a small portion at a time, as the pestle is less liable to become clogged.

Grinding and Pulverizing.

These terms refer to the reduction of substances, by mechanical means, to coarse particles, this being usually referred to as grinding, while the word "pulverizing" is used to distinguish the reduction to fine particles. These processes are of great technical importance, and grinding mills are modified for the various purposes for which they are used,

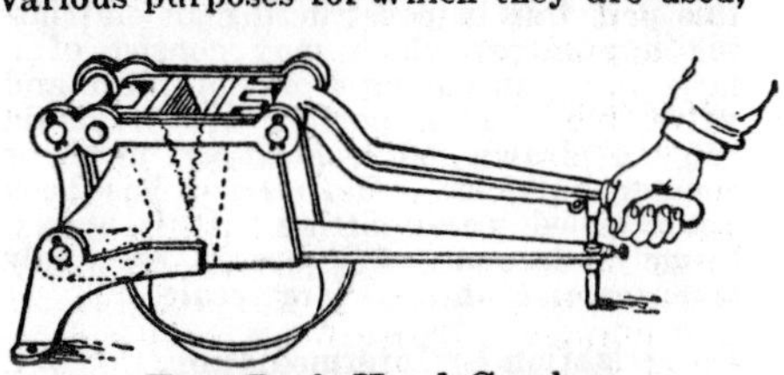

Fine Rock Hand Crusher

and are manufactured by many concerns. Burr stones, roller mills, chaser mills, pebble mills, and mills having antagoniz-

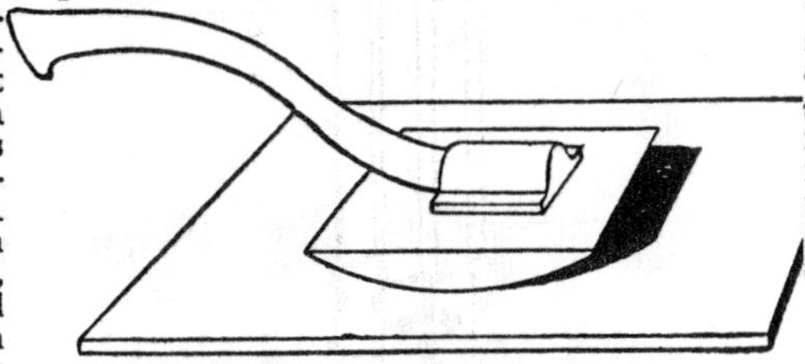

Bucking Board and Muller for Reducing Ores

ing grinder plates, and also various crushing and disintegrating mills, and machinery almost too numerous to mention. Hand mills, on the principle of the coffee mill, are of a great deal of use. The drug-mill type is recommended. For certain classes of grinding, the ordinary meat chopper will answer, such as for the cutting up of herbs.

Grinding Mills.

Grinding mills may be purchased for all purposes. It is impossible to recommend any one mill which will be of universal application. If work is to be carried on on a large scale, an appropriate mill will prove an economy, even at first. The pebble mill is particularly recommended for general use. It consists of a porcelain jar, made of imported porcelain; these jars are impervious to the action of heat and such materials as ink.

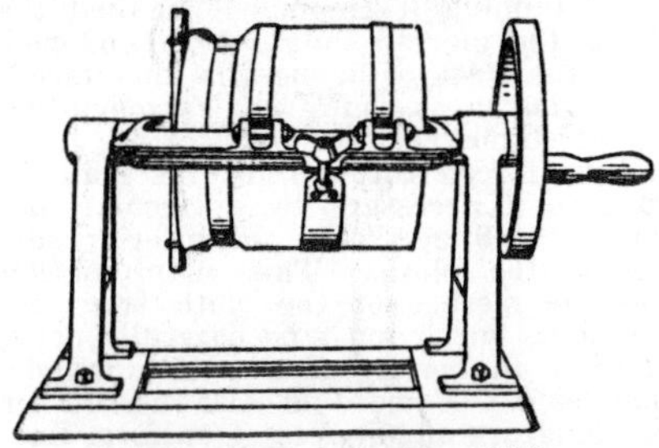

Abbe Porcelain Jar Mill

The effect is produced largely by friction: the sliding, tumbling and rolling inside of the mill of flinty pebbles or balls, which are mixed with the substances to be ground. The movement is caused by revolving the mill at a regulated speed. The type of mill which we illustrate will handle material up to 5 lb. in weight, and

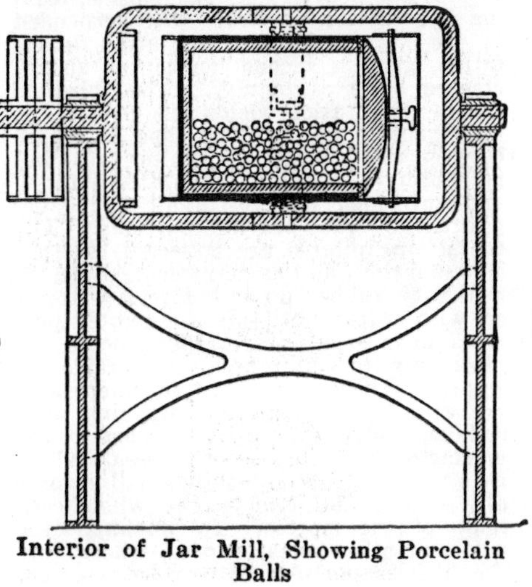

Interior of Jar Mill, Showing Porcelain Balls

is turned at about 60 revolutions per minute. It weighs about 120 lb. Those who are going to manufacture on a large scale will find a large variety of mills of this type. The action is very well shown by our section of the mill. The mills referred to are particularly adapted for hard substances. Articles of a vegetable origin may be ground in a drug mill, which may be had of any size. A spatula is absolutely essential; in fact,

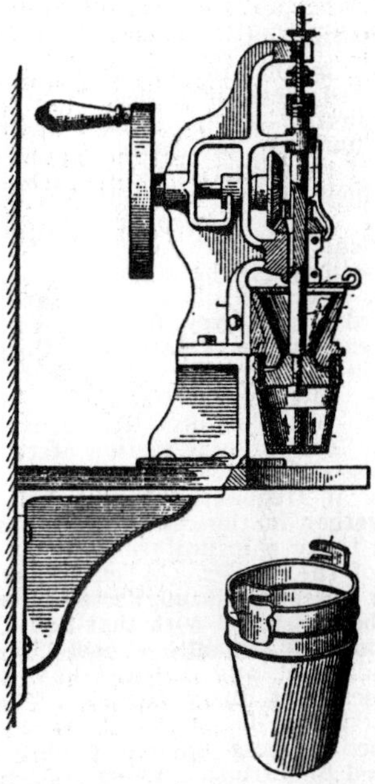

Hand Power Sample Grinder

two or three of them will not come amiss. A steel spatula, and one of horn or rubber should be provided. Strange to say, the spatula is one of the most convenient implements to have in the kitchen.

Braun Type of Pulverizing Mill

Trituration.

This mode of manipulating with the pestle is applicable to those substances which are friable and fall to powder by being merely rubbed up by a circular or grinding motion of the pestle, and which would soften and become obstinate by being pounded. Chalk and the like, and most of the salts, are in the first category, the rosins and gum rosins in the second. The pestle is given a circular or spiral motion, accompanied by downward pressure. The operation is continued until pulverization is effected. Sand is added to facilitate the reduction of the rosins and similar substances, which cake under the pestle, only when they are intended for maceration or solution. Under other circumstances the medium would be an adulterant, on account of the impossibility of separating it. The process of trituration is also often performed with the aid of spatulas or flexible steel blades attached to handles, and is useful in the kitchen as in the laboratory. It is possible to get spatulas made of hard rubber for making preparations which contain corrosive substances.

Porphyrization.

This means of pulverization is only employed when it is desired to give the comminuted substance the greatest possible fineness, and takes its name from that of the material of which the vessels in which it is practiced were formerly made. A small porphyry mortar, hemispherical interiorly, or preferably a slab and miller, is the apparatus employed. Flint, and even glass, which are equally as hard as porphyry, form economical substitutes for that material. Porphyrization is usually effected by rubbing the coarse powder between a flat slab and muller until reduced to an impalpable state. The circular motion of the muller disperses the powder over the slab, rendering it frequently necessary to collect it together in the center with a spatula, so as to keep it uniformly under the action of the muller. When the substance under operation is unaffected by water it may be moistened with that liquid, which, by converting it into a paste, facilitates its reduction, and prevents any waste by the escape of dusty particles. The powdered paste is easily dried by being dropped in dots upon a porcelain plate exposed to warmth. Those matters which are soluble in, or affected by, water, must be porphyrized in a dry state.

Sifting.

The impossibility of reducing the whole of a substance at once to a uniform state of fineness by any of the preceding processes renders necessary an occasional separation, during the progress of pulverization, of the more comminuted portions from the grosser particles. This is effected by means of a sieve, of which there should be several in the laboratory. A wooden cylinder of about 4 in. depth, with an accompanying ring of the same materials, constitutes the frame, over which can be stretched a cloth of any required fineness. For coarser articles, fine brass wire is the best material for the cloth, but when the powder is to be impalpable, bolting cloth (raw silk), or gauze, is requisite. Sieves are also covered with haircloth, buckram, book muslin, and iron wire of different sized meshes, each of which has its appropriate application. The metallic sieves should have their cloths permanently fitted to them. For all the rest, two frames, as above described, one of much larger dimensions than the other, will serve, as it is only necessary to remove the ring when it is desired to substitute one kind of covering for another. The sieve of cloth, of graduated fineness, can be kept in some secure place, and withdrawn as wanted, and thus we have the economical means of possessing a full suite of sieves, from the metallic wire, through all the grades of fineness, up to the closest wrought bolting cloth. After the separation of the finer portions by the sieve, the coarser particles are again subjected to grinding and sieving as often as is necessary to convert the whole into the requisite state of uniform fineness. Where a more ex-

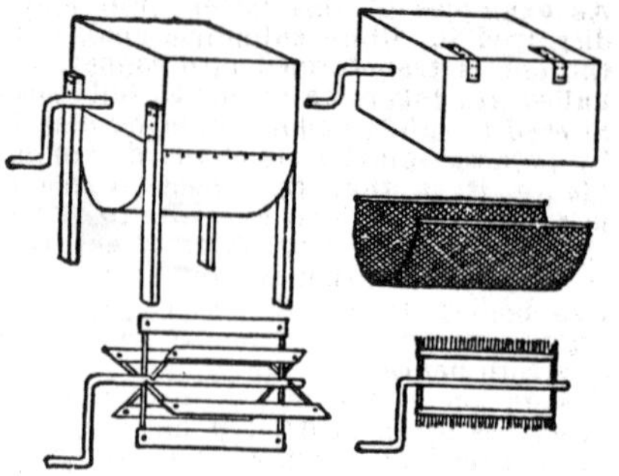

Home-made Sifter

tensive sifter is necessary, the one shown in our engraving can be used. Its construction will be readily seen by referring to the engraving. Horn scoops, or porcelain spoons or ladles, are the proper implements for transferring the contents of the mortar to the sieve. In some cases a stiff pasteboard card, being more pliable, is a convenient substitute. The use of the hand for this purpose should always be avoided, as a slovenly practice. A platinum, horn or bone, or—less preferably—steel spatula, may be used to detach the particles adherent to the sides of the mortar. A round jarring motion will force through some of the coarser particles, and thus destroy the uniformity of the powder, and hence the common practice of tapping it frequently against the side of the mortar should be abandoned, unless the state of fineness is immaterial. Some substances, however, as magnesia, etc., which obstruct the pores of the cloth, must be forced through in this manner, and even if necessary by a circular motion of the fingers over the interior surface of the cloth. This manipulation frees the meshes of the cloth from obstructions, but it must be carefully done, otherwise the safety of the cloth will be endangered. A sieve is also useful for the admixture of powders of uniform fineness.

Levigation.

Is that mode of mechanical reduction which is practiced by first rubbing the substance into a smooth paste, and then separating the finer from the coarser portions by agitating the bruised matters with water. After a sufficient repose the grosser and heavier portions subside, leaving the lighter particles still suspended in the water. This water, after decantation, gives a second deposit of an increased state of tenuity. The third or fourth decantation yields the powder of impalpable fineness. The time of repose between the decantations, unless great impalpability is required, should be limited, and only long enough to allow the deposition of the heavier portions. The coarse precipitates are collected together a second time, and as many more times as necessary, rubbed up as before, and treated with water until all the lighter portions have separated. This process applies only to substances unalterable by water. When uniformity of fineness is not at all important, one washing even suffices, and can be accomplished in the mortar without the use of glasses. Alternate poundings and washings will eventually reduce and remove the whole contents of the mortar. In washing over gold and other metallic ores, where only the heavier portions are to be reserved, the water may be allowed to flow directly into the mortar, which, being held in an inclined position, permits its exit, together with the fine dusty portions, which are kept in suspension by trituration with the pestle.

This process of levigation is founded upon the different specific gravities of the coarse and fine bruised matters, and is, therefore, not only applicable for the separation of the particles of homogeneous matters, but also of equally fine matters of unequal densities. In the latter case it takes the name of elutriation.

All minerals for analysis which have to undergo ignition with alkalies should be previously levigated, in order that decomposition may be complete; for if the powder is not uniform, the larger particles will escape decomposition.

Pulverization in this manner, by uniformly comminuting the particles, promotes their equal expansion and the escape of contained moisture, and thus prevents the decrepitation of substances when heated.

The deposited powder must always be dried, by exposure, previous to subjecting it to any other process.

Reduction by Granulation.

The reduction of metals to a pulverulent state is effected by fusing them in a crucible, and pouring the melted matter, from an elevation, in a thin stream, very gradually, into a bulk of cold water, which is, during the process, kept in constant agitation with a stirrer. The fineness of the resultant granules is proportional to the slowness with which the fused metal was poured into the water. It is more convenient to transfer the metal from the crucible into a ladle, and project it into the water from that more handy vessel, which enables a frequent change of the position of the descending stream, and thus prevents the formation of clots instead of smaller and more solid granules. The fusion of zinc for granulation must be in a covered crucible, otherwise it becomes oxidized while hot, and partially sublimes by exposure in an open vessel. Zinc may also be finely divided by being beaten, while hot, in a heated mortar. The process of fusing metals and then agitating the melted matter in a wooden box until cool, reduces them to a state of minute division, but at the same time promotes their oxidation. For general purposes, however, it is not objectionable, and the particles of charred wood with which it becomes mixed can be separated by elutriation. The sides of the box are generally well chalked, to prevent any adherence of the metal; this also is separable by elutriation.

Elutriation.

Elutriation is a process of obtaining substances in a very fine powder by the aid of water. The heavier particles fall to the bottom first, and the lighter particles follow. Advantage may be taken of this principle in constructing an elutriating apparatus, which may consist of a large iron pan having 4 or 5 openings and valves, so that a portion of the liquid can be drawn off containing finer or coarser particles. Elutriation has been aptly called water sifting. It is an extremely economical process, especially when carried on on a large scale.

Pulverization by Intermediation.

This mode is both mechanical and chemical, and applies particularly to the noble metals, in foil, which are difficult of pulverization. Honey, sugar, salts, etc., are the most usual media. By binding the particles together it assists their minute division, and prevents their escape from the mortar. The addition of boiling water solves out the medium without action upon the metallic powder, which then only requires to be thrown upon a filter and dried. Phosphorus may be finely divided by fusing it with alcohol over a water bath and shaking the contents of the flask until thoroughly cooled. The phosphorus subsides at the bottom in pulverulent form. Camphor, which is obstinate under the pestle, readily yields to its power when mixed with a few drops of alcohol or ether to destroy its elasticity.

II

SOLUTION AND EXTRACTION

Solution.

When a substance added to a liquid is wholly or partially taken up by that liquid it is said to be soluble therein. The liquid employed is termed the solvent, and its combination with the dissolved particles a solution; and if the liquid has exerted its solvent power to the fullest extent, then the solution which it forms is said to be saturated, because it can hold no more. The variable degree of solubility in different liquids serves as a distinctive characteristic of bodies, particularly those which are solid. Solution is either wholly mechanical, or else chemico-mechanical. In the first case it is a molecular division of a body, or, in other words, a diffusion of its particles in an appropriate liquid without any alteration of its original properties, save as to form and cohesion. Thus, for example, an aqueous solution of sugar or salt yields the whole of its charge by evaporation, and one of sulphate of lime by addition of alcohol, in which it is insoluble. Ethereal or spirituous solutions deposit their dissolved matter by distillation or crystallization; and some other kinds, that of gutta percha, in chloroform, for instance, by precipitation with ether or alcohol. When the dissolved particles are thus recoverable again in an unaltered state, chemically considered, their solution may be styled *simple*.

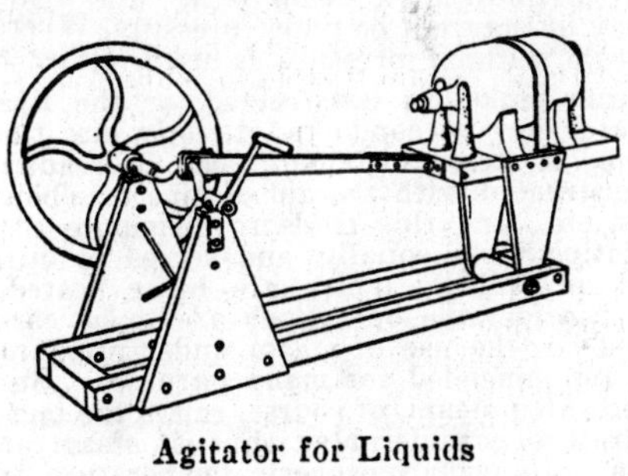

Agitator for Liquids

In the second case, chemico-mechanical solution, in contradistinction to that which is purely mechanical, is a process requiring the modification of a body by chemical action previous to its solution. Thus, for example, copper, iron, or any other base or acid, insoluble in the ordinary solvents, may be readily taken up by liquid acids or bases. But the liquid holds in solution a newly formed body entirely dissimilar to the original substance in properties, as appears when it is separated. In this, therefore, consists the difference between a simple, or mechanical, and a chemico-mechanical solution. As examples of this latter, iron may be dissolved in dilute sulphuric acid, but in the act is transformed into copperas; alkalies are taken up by acids, but become altered to salts; and oil, in being dissolved by potassa solution, is changed into soap. Hence it is that the chemical reaction is a preliminary step requisite to promote simple solution. The point of saturation in chemical solution is that at which the two bodies, invariably of opposite properties, have combined in proportions adequate to neutralization.

Solution is one of the most important processes in chemistry: it not only facilitates chemical reaction, but allows the separation of soluble from insoluble bodies, or parts of the same, and consequently the purification of the solution by subsequent filtration, evaporation and crystallization.

As regards the power of dissolving the greatest number of substances, water is the first in the rank of simple solvents, alcohol the next, and ether third. Then follow spirits of turpentine, pyroxylic spirit, the volatile and fixed oils, chloroform, and a host of other liquids suitable to particular substances. Of the alkalies, aqua ammonia, or potassa, are most used; the former preferably because of its volatility, and that of most of its salts. All of the common acids are employed, though some few only are of general application, such as the muriatic, nitric, sulphuric, acetic and tartaric.

A very convenient way of testing the solubility of a substance is by means of a test tube. If solid, a small portion, in powder, is to be introduced, and covered with distilled water, or the solvent to be used, and repeatedly agitated by the hand, the forefinger closing the mouth to prevent the escape of particles. If the matter is wholly soluble, there will be no deposit at the bottom of the tube; if partially soluble, the deposit will have decreased in bulk; if totally insoluble, it will occupy the same space as at first. To determine as to the two latter results, a minute portion of the supernatant liquid is decanted and evaporated in a small platinum spoon, or strip of window glass, over a spirit lamp; if a residue remains, it indicates that matter has been taken up. When heat is required, the lamp affords a convenient means of application. The procedure in such cases is the same as that above indicated.

1.—There are certain conditions which greatly facilitate the solution of substances: First, comminution, which increases the extent of surface; second, agitation, which promotes the frequent contact of all parts of the surface with fresh portions of solvents; third, the freedom from impurity of both the solvent and the body to be dissolved; fourth, it is also influenced by the quantity and state of dilution of the solvent; fifth, by the temperature; sixth, by the mode in which the process is conducted.

2.—Agitation is effected by stirring with glass rods when the containing vessel is open at the top. The rod should be rounded at the end over the blowpipe flame, and to prevent its rolling from the table or top of the vessel upon which it should be placed, may be square, instead of cylindrical, as usual. A very convenient and effective mode of bringing all portions of the liquid successively in contact with the substance to be dissolved is to place the latter in a colandered diaphragm suspended beneath the surface of the liquid. The first stratum of liquid, in becoming saturated, increases its density, and consequently descends, and displaces a lower and fresher portion, which, being in the same way surcharged in its turn, gives way to successive strata, and so the operation continues until the whole of the matter, or so much as can be, is taken up. This mode keeps the substance in constant contact with new portions of liquid, and is, in fact, a kind of *displacement* process. When flasks or bottles are used, the same effect may be produced by repeated shaking. Trituration in a mortar, and alternate decantation and fresh additions of the solvent, greatly facilitate the solution of solid substances.

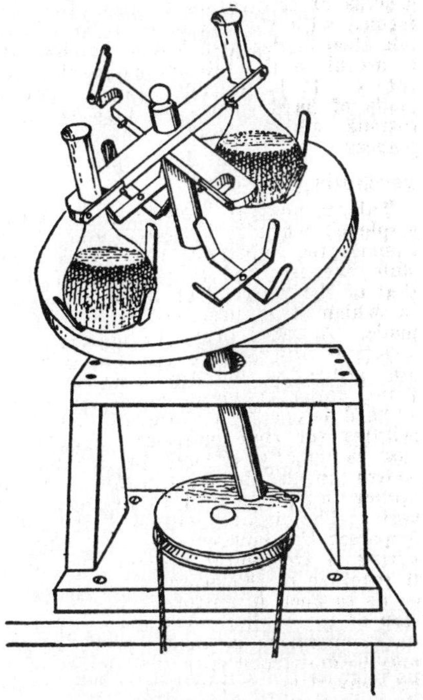

Power Mixer for Liquids

3.—The purity of the solvent is an important consideration, for if it contains foreign matters they may impart a dissolving power which is not inherent in the pure liquid, or diminish that already possessed by it.

4.—In regard to the quantity and state of dilution of a solvent, it must be remembered that some substances require more of it than others for their solution, and that it should be in a greater degree of dilution. Therefore, in examining the solubility of a body, always commence with small quantities, and increase both quantity and strength gradually as may be required.

5.—Temperature exerts a considerable influence in the solution of bodies, and though in a few instances, as in the solution of lime, magnesia and anhydrous sulphate of soda in water, its elevation impairs the power of the solvent, yet, as an almost universal rule, it facilitates its action. The temperature must be adapted to the nature of the solvent and the substance to be dissolved, and of the solution formed.

It may be as well to mention that the caloric rendered latent at the moment of the liquefaction of a solid, which is being dissolved in a liquid, causes a decrease of temperature. Solution in volatile liquids should be, in most cases, performed in the cold, and, when of small quantities, in narrow-necked flasks. If heat is required, especially when the vapors are inflammable, a retort or covered still must be used; and if the distillate is valuable, a recipient may be annexed to receive as much as comes over.

The mode of effecting solution varies with the substance under process: Maceration, decoction, infusion, digestion, boiling and displacement have each and all appropriate application.

In ordinary solution, the solid should be added in portions, and sufficient interval allowed for the solution of those in the liquid before fresh are added. In case of foaming or effervescence, an additional amount of fluid will produce a calm.

Some volatile substances which are insoluble in water under ordinary circumstances are taken up by it in the state of vapor. For this purpose both should be distilled together.

When solutions emitting corrosive or disagreeable fumes are being made in open vessels the operation should be conducted under a hood the barrel of which connects with the chimney flue, so as to insure their exit. The containing vessels should be those which resist the action of heat, acid, alkalies and corrosive liquids.

For making saturated solutions of most substances, ebullition is necessary. For this purpose the solid must be boiled with the solvent until the latter, on cooling, deposits some of its charge. The cooled solution is then to be filtered.

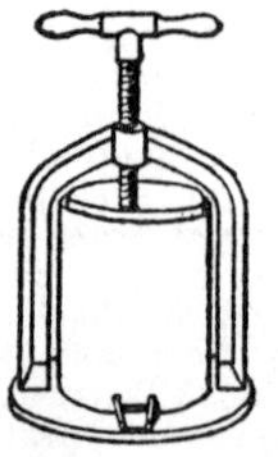

Hand Press

Expression.

By expression we are to understand the process of separating solids from liquids by means of force. Presses are usually used for expression, and are divided into screw presses, lever presses, hydraulic presses, etc. The ordinary screw press shown in our engraving is of great use. The ordinary meat chopper, with a knife in one piece, and costing $1.50, is a valuable aid to expression. Horizontal screw presses of the same general appearance express as well as cut.

Maceration.

The soaking or steeping of a substance in a liquid, at the ordinary temperature, is termed maceration. It is almost exclusively applicable to organic substances, being most frequently resorted to as a means of hastening and facilitating the after solution of the extractive parts of hard, compact or impervious wood, roots, stems and leaves, by the more active methods of *displacement* and *ebullition.* It is employed when the soluble principles are alterable by heat, and is also made use of to effect the solution of a substance containing several principles, the solubility of which varies with the temperature applied, as it leaves those which are not taken up in the cold to be acted upon by the aid of heat. Thus, for example, in the treatment of most vegetable substances, starch, which is generally present, and is only soluble at the boiling point of water, will remain untouched, while all other principles soluble without heat can be separated from it.

The mode of performing the process is merely to place the solvent and the substance to be dissolved together in a vessel, and allow them to remain a longer or shorter time, according to the nature of the substance. For ordinary purposes, a loosely covered pan of blue stoneware is very convenient. In delicate operations, a beaker glass, or solution jar, is more appropriate. When the solvent is volatile, a wide-mouthed, stoppered bottle may be used.

Infusion.

This process is likewise applicable almost solely to organic substances. Instead, however, of the solid remaining in contact for a length of time with the solvent, the latter is first heated to boiling and then poured upon the former.

This mode is used for the exhaustion of flowers, leaves, roots, seeds, and other substances of delicate texture, which are easily penetrable and readily yield their soluble matters; and especially for the purpose of extracting volatile ingredients. The heat applied to the solvent increases its energy; but as the material is only in contact for a limited time, the interval between the commencement and completion of the operation is not sufficient to affect the material or solution, even though one or more of its components are alterable by heat.

Decoction.

This mode of solution, which is so important to the pharmaceutist, is chiefly employed for the purpose of exhausting those vegetable substances the components of which will not readily yield to other means. It is merely an extension of the last process, and consists in that contact of the material to be dissolved with a hot solvent in a covered vessel, which is continued until all soluble matter is taken up. Most volatile matters are epelled by decoction, but those which are insoluble, save by prolonged action of heat, are dissolved or suspended, as it were, by favor of other principles present. Decoction is only used with liquid solvents which are not decomposable by heat.

In all of the preceding processes, as well also in others in which solid vegetable matter is subjected to the solvent action of liquids, the colandered ladle of tinned wire is most useful for transferring the residue to the press, for removal of any retained liquid.

Digestion.

This mode of solution differs from maceration in requiring the assistance of heat, and consists in exposing a body to the prolonged action of a liquid in a covered vessel, at any temperature between 90° F. and several degrees less than the boiling point of the solvent. The method of heating varies with circumstances, and can be by a gentle fire, or by the sand, steam, water or saline bath, as the nature of the operation requires.

In analysis, glass or platinum vessels are used, but in less important operations those of other materials are more convenient and economical.

A very important advantage of digestion is that it allows the perfect solution of all soluble portions of a substance without modifying the nature of the solvent. It is especially useful for the decomposition of ores, minerals, and other substances with difficulty acted upon by acids or other solvents, and also for effecting the synthesis of compounds requiring a long continued heat. Moreover, it is very available in preparing alcoholic and aqueous solutions, medicinal oils and other pharmaceutical products.

Evaporating Dishes.

Special evaporating dishes of porcelain, glass, or enameled steel, can be purchased of all dealers in supplies, and are specially recommended. Broad, shallow vessels should be usually selected. If glass evaporating dishes are to be used, they should be heated in a sand bath. The evaporation is aided by stirring; glass rods, or porcelain or wood stirrers, should be used. If the reader is going to use large quantities of the same materials, various means of stirring artificially will present themselves. Evaporation of many substances should be carried on under a hood, which may be of sheet iron or galvanized iron, like the hood over a blacksmith's forge, or the work may be carried on in an evaporating chamber, which may be likened to a closet with the lower portion boarded up so that the floor of the closet is of a convenient height to be reached with the hands. There should be a closed window in the closet, which should be well ventilated to the outside by galvanized iron or asphaltum painted ventilating tight. All the arrangements for gas, etc., should be at the front of the evaporating chamber, so that it will not be necessary to reach over hot plates, etc.

Steam Baths.

Steam is very largely used in the arts for maintaining a steam bath. The steam may or may not be under pressure. Where steam without pressure is used, either a steam jacket is constructed, or the live steam may be conducted directly into the top. A steam distributor can be readily constructed with the aid of pipe or elbow Ts, etc., and this tends to distribute the heating more equally, and serves to mix the ingredients which are being heated. If considerable operations are to be carried on, the use of steam under pressure is recommended for many purposes. Superheated steam, of course, raises the temperature considerably; thus, if steam at the ordinary atmospheric temperature is to be increased, a temperature of 240° may be obtained by a pressure of 40 lb. to the square inch, while with a pressure of 80 lb. to the square inch a temperature of 312° can be obtained. It is possible to build a water bath with a jacket in which steam at high pressure is gnerated directly in the water jacket.

Attemperating Baths.

There are many substances which have to be treated moderately to heat, so as to prevent the decomposition or destruction of the substance which is being treated. This is especially the case with medical preparations. Various attemperating baths have been devised, many of which are extremely ingenious, and are fully illustrated in the catalogues of dealers in chemical apparatus. The sand bath is one of the best-known means of producing an even heat without burning. It

can be readily made by putting sand in a pan over the naked fire and putting next in porcelain or other vessels as it becomes necessary. Oil and paraffine baths are used for certain purposes, as are also glycerin baths. The water bath is perhaps the most widely distributed and best-known means of regulating the heat which is applied to substances. The water bath may be extemporized, or the special baths furnished by dealers in chemicals may be used, which are more satisfactory, being specially adapted to the purpose. Salt-water baths are also largely used. The action of salt in the water is to raise the boiling point.

DRYING AND DESICCATING

Mechanical Methods.

Foremost among mechanical appliances for this purpose ranks the centrifugal machine, or hydro extractor. In principle, this apparatus consists of an upright drum, which can be made to revolve with great velocity on a vertical axle. The drum may have its sides constructed of sheet metal, perforated with a multitude of fine holes, of wire gauze properly supported, or of basket work, according to the nature of the substances to be treated. The drum, being charged with material, is set in quick rotation. The water present is thus expelled through the perforated sides, in the form of a fine shower. This

Hood For Chemical Work

process is exceedingly well adapted for removing the greater part of the moisture from cloth, yarn, unspun wool, etc.; also from crystalline and granular substances. It is not so well adapted for drying wet powders, pastes, etc., since in such cases a very considerable proportion of the solid matter is projected away along with the liquid, so the holes may get choked up. Thus it has not hitherto been found satisfactory for drying sewage mud. Its use requires, further, special modifications where the liquid to be got rid of is not pure water, but holds useful or hurtful matters in solution. A recent very simple improvement has considerably extended the use of the hydro extractor. The materials, instead of being put into the drum loose, are inclosed in bags of some suitable material, thus preventing the dispersion of the solids. This method has been very successfully adopted with butter. It must, however, be remembered that no substance, especially if of organic nature, can be rendered absolutely dry by the use of the hydro extractor.

Another mechanical agency for desiccation is the press, more especially that device known as the filter press, which has proved itself invaluable for separating solids from fluids when the latter largely predominate. This apparatus contains a number of cells, each consisting of a couple of cast-iron plates, lined, when in use, with suitable cloths. The inner surface of each plate shows a number of ridges. The liquid paste is forced by a pump or press into each cell, through an aperture, and the water escapes through the cloth, and trickles down between the grooves formed of the ridges to the pipe at the bottom.

The filter press, like the centrifugal machine, only expels a part of the water in mud, etc.; thus, if a sewage mud contains at the outset 90 to 95% of moisture, it may be reduced by the filter press down to 50 to 60%, according to the time during which the pressure is maintained. It is only in a few cases that hydraulic presses, screw presses, etc., can be employed for desiccation.

Small Hot-Air Baths or Closets for Laboratory and Other Purposes.

(*a*) The ordinary steam or hot-air chambers for laboratory use, although meeting the most of the requirements for which they are designed, have the disadvantage of being more adapted for experimental than manufacturing purposes. The want of a cheap and convenient apparatus induced Maben to bring under notice a design, due to Hyslop, one of his apprentices, who intended it for drying photographic gelatine plates; but, by slight modifications of the interior, it is perfectly adapted for the purposes of the laboratory.

The chamber consists of a strong wooden box, *a*, 18 in. high by 18 in. wide, and 14 in. deep. To the front a door is attached, hinged in this instance, but a vertical sliding movement would be more convenient. To two sides of the box are fixed wooden supports, which serve to receive teak spars for supporting drying trays or evaporating dishes. The bottom of the box has a perforation of 3 in. diameter, into which a zinc cylinder, *b*, is securely fitted, and to this is soldered the upper end of a copper cone, *c*, with a flat bottom, while into this latter a bent tube of 2½ in. diameter and 9 in. total

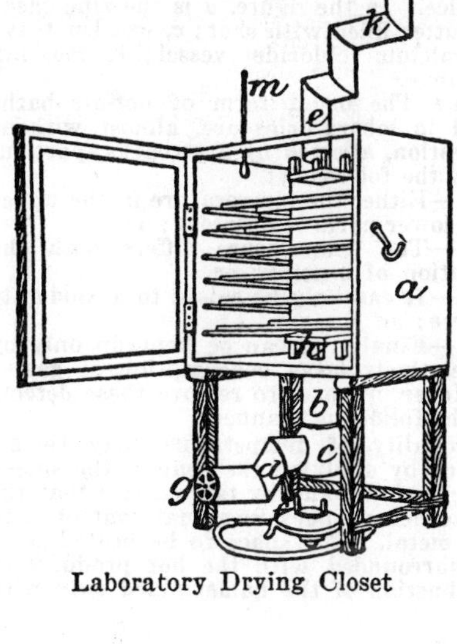

Laboratory Drying Closet

length is securely inserted in the manner shown. A corresponding perforation is made in the top for receiving a tube to answer the purposes of a chimney.

Using a Bunsen burner or a spirit lamp as the source of heat, the flame is directed to the bottom of the cone, *c*, with the result that the heated air ascends into the chamber, being diffused by means of a dispersion board, *h*, about 4 in. square, which is placed over the orifice. At the end of the tube, *d*, is fitted a "hit-and-miss" regulator, *g*, which consists of a series of triangle-shaped holes, with a revolving disc behind, so that the size of the apertures can be increased or diminished, thus enabling the amount of air entering to be under partial control. The highest temperature to which the air in the chamber has been raised is 180° F. (82° C.) which is sufficiently high for most operations. If a uniform temperature of say 100° F. (38° C.) be required, the admission of air must be regulated accordingly by means of the regulator, *g*, accuracy being insured by the insertion of a thermometer, *m*, into a perforated cork fitted into a ½-in. aperture on the top of the chamber. By this means there is no difficulty in keeping within 2½° less or more of the desired temperature.

If a rapid current of warm air is desired, this can be had by placing an angular tube, *k*, on the top of the chimney, *e*; by heating the angle of the tube a draught is quickly created.

It is desirable in some cases to filter the admitted air; this can be done by stretching a piece of lint or other suitable material between the regulator, *g*, and the tube, *d*, by which means dust particles are effectually excluded.

The metallic parts of the apparatus being made to screw off and on, they can be detached at will, so that we can thus have a series of wooden chambers suited to different purposes. In this instance, the chamber being intended for drying gelatine plates, it was of course constructed so that the light would effectually shut out, but it is obvious that a small glass window would add greatly to its value for most other purposes. The advantages of this chamber are its simplicity, its perfect security against overheating, and its small cost—it can be made for a few shillings. It is light and easily handled, and is always ready for work, a current of pure hot air being obtained in a very few minutes after the application of the Bunsen flame. It is specially adaptable in the preparation of granular and scale compounds, for drying precipitates, hardening pills previous to coating, and in other operations requiring a current of hot air.

(*b*) A writer describes his drying closet as being made of teak 1 in. thick, with light-tight door in front; the ends project beyond the bottom to form legs; the top and bottom are both double (4 in. apart), and the air enters through a slit 3 in. wide, and reaching right across the box. This slit is at one end, and the air has then to pass along the double bottom to the other end, where it gets into the box through a similar slit, thus keeping out the light; and it gets out at top in a similar way. Over the exit at top is fitted a tin or copper chimney 3 ft. high, in which burns a Silber lamp, giving a good draught, and drawing a large quantity of air through. Inside the box are brackets (each having a leveling screw through it, with the point upward), projecting from the ends, on which are laid plate-glass shelves cut the width of the box, but 3 in. shorter, so that when the shelves are in place, if one is pushed close to the right end of the box and the next to the left, and so on, the air has to pass backwards and forwards over the plates. His box has 3 shelves, 13 in. wide and 32 in. long, and will dry 6 photographic plates 15 in. by 12 in., or, of course, anything less that will lie in the same space. Some have an arrangement for drying and warming the air before it enters the box; but this sometimes induces blisters and frilling. Shelves should be far enough apart to get the hand in easily, say 6 in.

Our next engraving shows a sectional view of another form of photographic drying box. *a* are shelves on which to put plates. In the drawer, *b*, are placed some lumps of calcium chloride. This absorbs moisture very rapidly, and the air in passing through it is thoroughly dried. In the flue, *d*, is a small gas burner, and below is a light trap, *c*, made of tin. The gas jet is for the purpose of causing an extra current of air to pass over the plates. It is better to confine the plates as much as possible to the 2 middle shelves, as there they are sure to be safe. At *e* is a sketch showing how

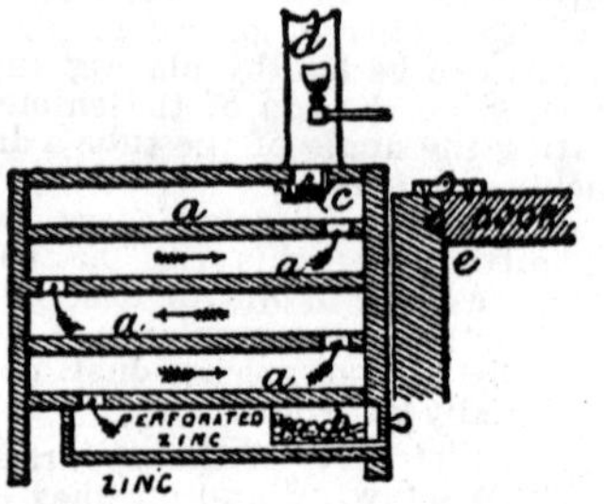

Photographic Drying Box.

the door of the box should be rebated into the side.

(*c*) England's drying closet is simply a light-proof box with wires stretched across the interior to suport the articles to be dried; *e.g.*, photographic plates. Through the center runs a 1-in. gas pipe, open at both ends, with a small gas jet burning inside at the lower end. At the top and bottom of the box 2 draught holes are cut, to which a tin tubing of about 3 in. diameter is attached. The gas tube gets warmed with a very small jet of gas burning in it, a mere pin-hole being sufficient exit for the gas. This warms the air in contact with the tin tube, and also slightly the air inside the cupboard. The consequence is, that a current of slightly warm air is set up, and circulates among the plates while supported on the wires, and the drying of the films takes place rapidly. Some 5 to 6 hours is a sufficient time in which to dry the plates, while without the gas jet it would take 24 hours or more. In the inside of the cupboard, and near the top and bottom, are placed 2 cardboard discs to stop the possibility of any stray light entering, and as the whole affair is placed in the dark room, the chances of any such access even without it would be small. Inside the cupboard door is a thermometer, and the jet is regulated so that a temperature of about 70° F. is indicated—80° would do no harm to the plates; beyond that temperature it might not be safe to go. The small gas jet used is the same as seen in tobacconists' shops; the hole in the end is plugged up, and a very small hole drilled at the side.

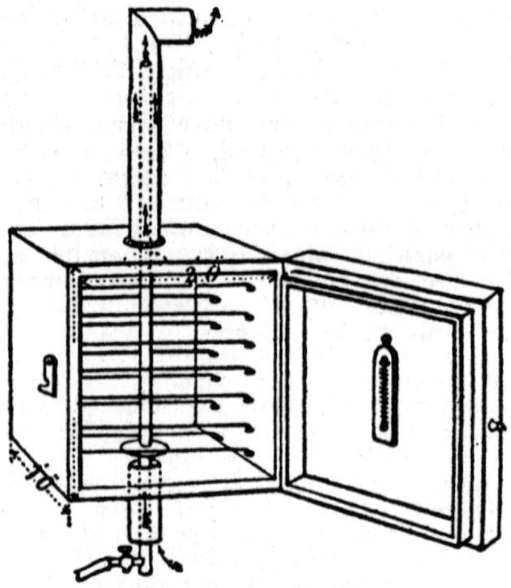

England's Drying Closet.

(*d*) A photographer adopted a large zinc case with a lid of the same material. He cut a long opening at one end of the bottom, and had another bottom soldered inside with an opening at the opposite end. He then had a Russian iron chimney fastened on one of the sides, and fitted this with a gas flame placed as shown, so that it might produce the necessary current of air. To make the cover fit air and light-tight was rather more difficult. This, however, he managed in the following manner. He had a rim soldered

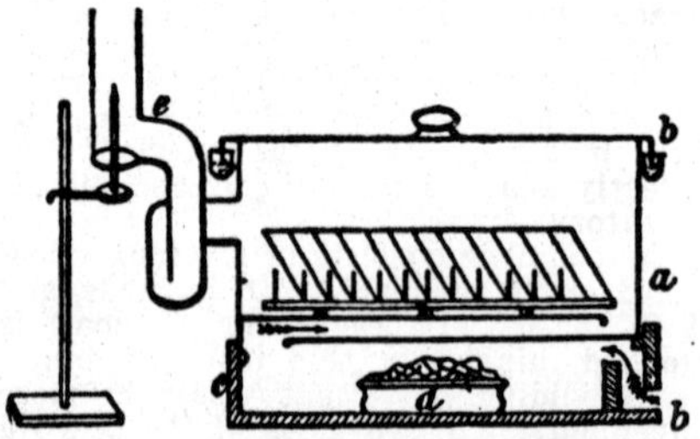

Calcium Chloride Drying Box.

all round in the shape of a gutter, the edge of the lid sinking into the bottom of the gutter, and then filled the latter with small shot, and thus obtained a most perfect closure. This box has been in use ever since, and, with the addition of a wooden tray, and of an iron vessel full of calcium chloride, has done very good service. In the figure, *a* is the zinc case; *b*, gutter filled with shot; *c*, wooden tray; *d*, calcium chloride vessel; *e*, Russian chimney.

(*e*) The usual form of hot-air baths used in laboratories are, almost without exception, affected by drawbacks, particularly the following:

1.—Either the temperature in the upper and lower parts is different; or

2.—The temperature differs with the duration of heating; or

3.—It can only be raised to a moderate degree; or

4.—Finally, it can be kept up only by a relatively large consumption of gas.

Meyer proposes to remove these defects in the following manner:

Equality of temperature may be attained by applying the heat at the side—never below—and by taking care that the flame never comes in actual contact with the metal. The space to be heated is to be surrounded with the hot products of combustion of the flame mixed only with the smallest possible excess of air, in such a manner that a triple layer of heated gases, proceeding from without in-

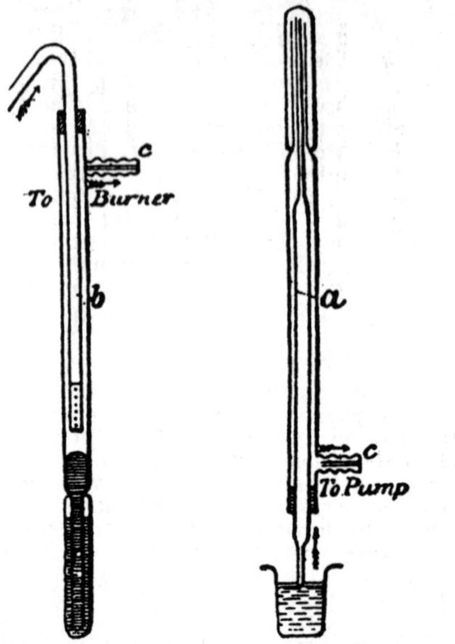

Fig. *a* Fig. *b*

ward, surrounds the inner mantle. Besides, the outer, or hottest layer, must be protected from too rapid cooling by applying a suitable coating of bad conductibility for heat.

Equality of temperature for any length of time may be best attained by a regulator constructed on the principle of Andrea's, which contains, in a small, confined space a small quantity of a liquid having a boiling point a trifle below the degree of temperature to be maintained. The author prefers the modified form suggested by Kemp, and improved by Bunsen, which is wholly constructed of glass except the lower end of the gas tube, this being made of perforated sheet platinum.

In order to fill it, the gas tube, *a*, Fig. *a*, is temporarily replaced by a tube, *b*, drawn out at both ends and reaching down into the reservoir of the regulator (top of Fig. *b*). The lateral branch, *c*, is now connected with the vacuum pump, the whole inverted (as in Fig. *b*), and contracted end dipped, first into the liquid to be used as regulator, and then into mercury, until the chamber is almost, but not quite, full. The apparatus is now turned over, a little more mercury poured in, and the gas tube, *c*, is inserted. When using the apparatus, the gas tube is first drawn upwards, and, when the proper temperature has been reached, pushed down into the mercury, until the supply of gas is reduced to a minimum. By cautious adjustment, it is easy to find the position at which the tension of the vapor developed in the tube raises the column of mercury sufficiently to just close the orifice of the tube, *c*, at the proper temperature. As the air bath cools off very slowly, but heats up rapidly, it is of advantage to adjust the regulator to a slightly lower temperature than actually required.

It is best to have a series of such regulators, charged with substances, the boiling points of which are about 30° C. apart, and to keep them in a proper receptacle for use. Suitable substances are, for water baths: ethyl chloride, ether, carbon disulphide, mixtures of ether and alcohol, benzole; for air baths: water, toluol, xylol or amylic alcohol, cymol or

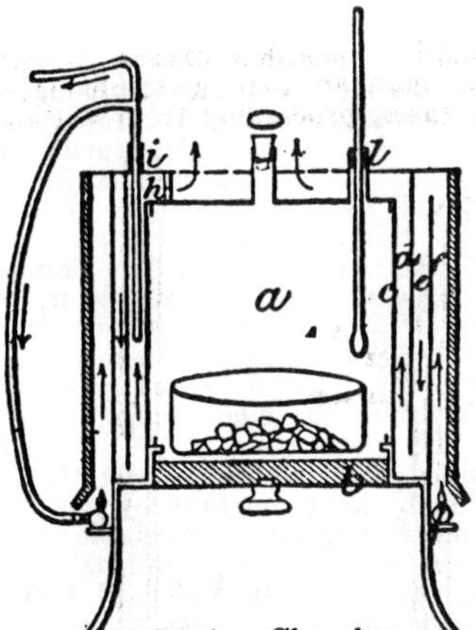

Drying Chamber.

oil of turpentine, aniline or phenol, naphthaline, diphenyle or diphenylmathane, diphenlyamine, and perhaps also anthracene. It is not at all necessary to use these in a pure state, particularly those which are solid at ordinary temperature, since they melt more easily when impure. Only very little of solid substances should be introduced, for the excess distils off, and may clog up the gas tube.

The annexed engraving shows an approved air bath.

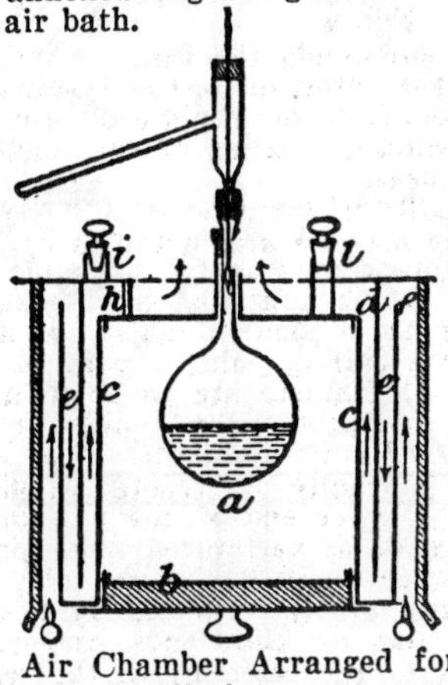

Drying Air Chamber Arranged for Distillation.

It consists of 4 concentric walls of sheet copper, 2 of which are attached to the upper plate, and the others to the bottom plate. It can be arranged for the dry distillation of substances which should not be heated beyond a certain point (for instance, citric acid in the preparation of aconitic acid, etc.).

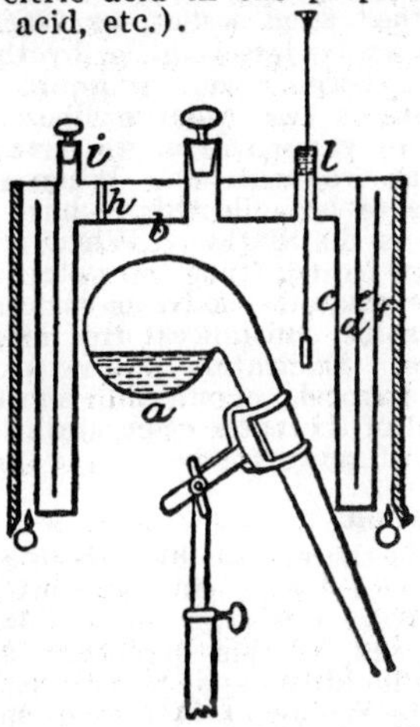

Drying Chamber Arranged for Dry Distillation.

*The air chambers illustrated above are not square, but round. The illustrations represent a vertical section through the center.

The innermost cylinder* surrounds the space, *a*, to be heated, which is closed from below by a double bottom, *b*, fastened by a bayonet-clamp. The upper cover also double (the 2 walls being kept parallel by inner supports, of which one is shown at *h*), has 2 tubulures, one, *l*, for the insertion of a thermometer, another, *i*, for the regulator, and another for the escape of the heated vapors. To this cover the 2 cylinders, *d* and *f*, are attached, while *e* and *c* are soldered to the bottom piece, which is also provided with 3 legs. The heating is done by a brass ring attached to the legs, with a supply of gas controlled by the regulator, *i*. The ring has holes of 2 to 3 mm. bore in intervals of 3 cm. The little flames thus produced burn quietly and may easily be regulated. With the same amount of gas which is furnished by a gas cock supplying an ordinary Bunsen's burner, the space in a (= about 5 l.) may readily be heated to 300° C. and over, even when it is not closed below. But in order to obtain this result, the intervals between the several cylinders, in which the products of combustion circulate, must not exceed 10 mm. Besides, the outer cylinder, *f*, must be protected with a non-radiating cover. The best, for this purpose, is a layer of asbestos (in sheet), to be applied so as to leave a little space between it and cylinder *f*, which space is to be filled out with silicious earth ("kieselguhr") or mineral wool.

If tubes are to be heated, the modification shown herewith may be used. It is also here of importance that the channels through which the warm air circulates are very narrow, scarcely 1 cm. apart. The 8 iron tubes pass through the narrow walls, which latter are not double but covered with little flaps hinging upwards (one corresponding to each tube), as closely as possible fitting to the surface of the outer cylinder, but remaining slightly distant from the ends of the tubes. In case a glass tube (inserted in one of

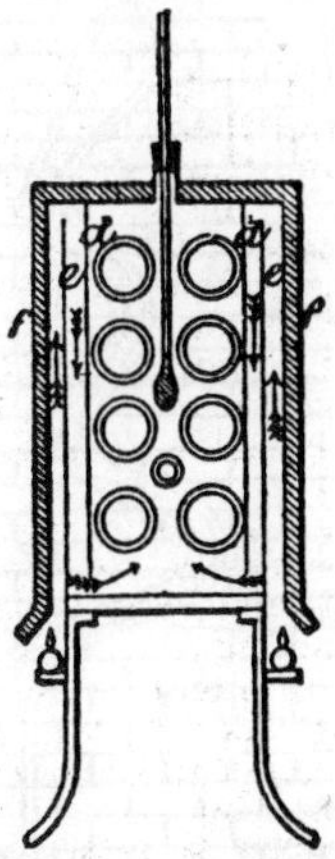

Drying Chamber Arranged for Tubes.

the iron tubes, for being heated) should explode, its fragments are caught by the loosely hanging flaps. Between the iron tubes, a Babo's regulator may be inserted.

For special uses the above forms of air baths may be still further modified. It is, however, of importance to remember that the heated gases should surround the space to be heated in a triple layer; that the hottest layer should be near the outside, and that the intervals between the walls should admit as little excess of air as possible. The gases escaping above must have the property of extinguishing a glowing splinter of wood.

(f) The air bath ordinarily used in chemical laboratories for drying precipitates, for making determinations of water by loss, and for similar purposes, is usually a rather expensive piece of apparatus. The iron or copper closet, with its door, tubulure for thermometer, shelves, stand,

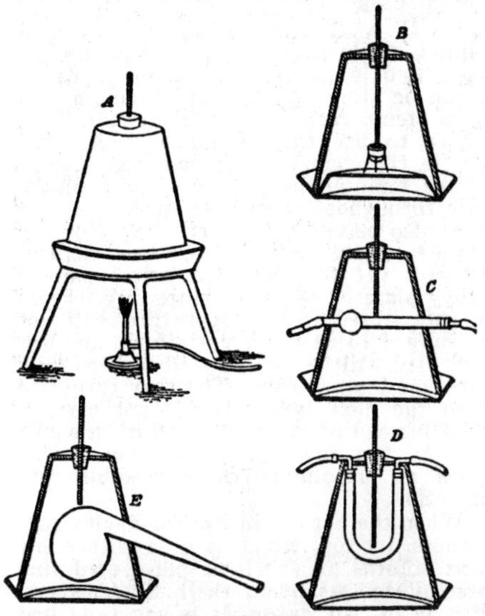

Air Baths.

etc., works no more satisfactorily because of its somewhat elaborate or difficult construction. In our engravings are shown a simple substitute for this apparatus, that as regards simplicity cannot well be excelled, while its other good features certainly operate to commend it. It consists of an inverted flower pot sustained upon an ordinary tin pan or sand bath, the whole being carried by a tripod or retort stand. The aperture at the top serves to receive a perforated cork through which a thermometer is passed. An ordinary Bunsen burner is used to heat it. As the sand bath directly over the burner becomes very hot it is advisable to invert a second smaller sand bath within the first as shown in B. This prevents too direct a radiation of heat from the hot metal. Upon this the little stand or bent triangle supporting the crucible or watch glass containing the substance to be heated may be placed. The thermometer should be thrust down through the cork until its bulb is near the substance to be dried, so as to obtain a correct indication of the temperature at that point. The entire arrangement is shown in external view in A.

To place the vessel in it or to remove one, the flower pot is lifted off the sand baths. It will be observed that its porous nature provides a species of ventilation, while its composition assures it against corrosion. It even protects the plates below to a considerable extent, as drops of water or other fluid cannot run down its sides as it cools.

But convenient as it is in the rôle of air bath for simple drying operations, it will be found more so where drying tubes or retorts have to be manipulated at constant temperature. The flower pot can be perforated at any place, and holes of any size or shape can be drilled and cut through it with an old knife, file, or other implement. Thus in C it is shown in use for drying a substance at constant temperature in a straight drying tube. The holes to receive this tube can be drilled in a few minutes. The arrangement as shown is of the simplest kind, but if the usual bath was used, it would require a special tubulation to be introduced or contrived for the tube to pass through. Flower pots cost so little that there need be no hesitation in preparing them for special uses.

In D a U tube is shown as being heated, while in E a retort occupies the bath, and is in use for fractional distillation or other operation requiring a constant temperature. In all cases it is better to use the second bath inverted within the chamber. It conduces greatly to the maintenance of an even temperature throughout the whole space. A hint may also be taken from the heavy drying plate formerly perhaps more used than at present. If for the light metal pans a heavy plate of ½ in. or more in thickness is substituted, the temperature will not be subject to as rapid variations, and less difficulty will be experienced in keeping a constant temperature. The tray furnished with the next large size of pot may be used instead of the sand bath upon which to rest the inverted flower pot. This gives an absolutely non-corrodible construction.

When the bath is in use for drying substances, its top, which is at a rather low heat, affords an excellent place of drying precipitates wrapt in their filter papers. It acts in two ways. It is generally just hot enough to dry them with reasonable quickness without danger of spurting, and it also acts by capillarity to absorb the water directly. It represents in the last respect the porous tile or blotting paper—appliances too little appreciated by chemists here. It must be remembered that the drying of a precipitate by evaporation leaves all the impurities of the wash water concentrated therein, while capillary absorption removes a great part of both wash water and its impurities, thus conducing to the accuracy of the work.

Water-heated Air Baths and Ovens.

(*a*) The accompanying sketch of a combined steam oven and distilled water apparatus, so arranged as to be left to itself for a long period of time without the risk of the boiler going dry, may perhaps be of interest to many, and a few words only are necessary to describe the working. The steam oven, *a*, is of the ordinary construction, but is fitted at the side with a tube connecting it with the condenser, *b*. Heat is applied to *a* by means of a radial burner, connected with the gas supply by metallic tubing; the steam generated circulates around the drying chamber, escapes through the copper tube, *c*, thence through block-tin worm, and falls as distilled water in the receiver, *d*. The cistern, *e*, fitted with a Mariotte's tube, holds cold water, which falls through the tube, *f*, enters the condenser, where it rises slowly, absorbing heat from the condensing worm, until it reaches the tube leading to the boiler at a high temperature. For a cistern, an 18-gal. ale cask, supported on a stool, has been found to answer admirably, having the advantage of holding sufficient water on the top to secure the 2 corks being airtight. By a suitable adjustment of the Mariotte's tube, *h*, the rate of flow of the water can be so regulated that the level of water in the condenser is constant, or, if desired, allowed to drop slowly into the waste pipe, while the water evaporated from *a* is renewed by water

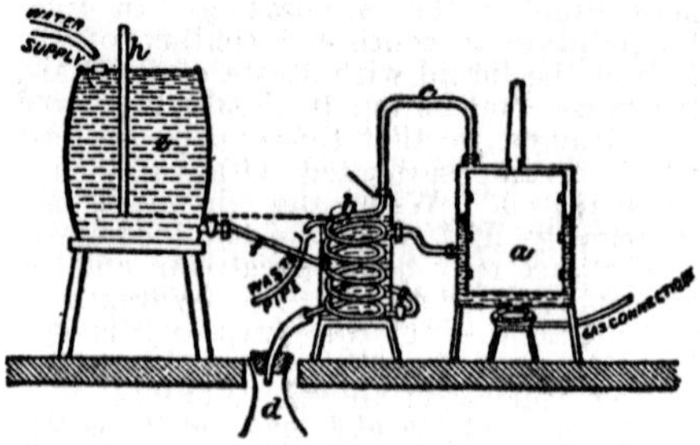

Steam Oven and Distilled Water Apparatus.

already near boiling. In practice it has been found necessary to allow the water to waste at the rate of about 2 drops per minute, the 18 gal. lasting for over 72 hours, during which time 10 to 11 gal. of distilled water are collected. When this apparatus was first fitted up in the laboratory, it was intended to have connected the condenser directly with the town water supply, but as the waterworks authorities would sanction no such connection, we had recourse to the cistern, with the satisfactory result that we are in this respect quite independent of the caprice of the waterworks turncock. The several connections are made by union joints, to allow the apparatus to be taken to pieces and the boiler freed from scale. The whole apparatus may be supported upon a strong shelf, which should be protected from the heat of the burner by means of slates or asbestos millboard. With this arrangement, bulky precipitates may be allowed to remain in the steam oven all night and found ready for further treatment next morning.

(*b*) In the annexed engraving is shown a constant water bath, consisting of a square box, A, supported over a Fletcher's solid flame burner. The top of the box, 15 x 15.5 in., is formed by a brass plate, ⅛ in. thick, which thus is stiff enough to support a considerable weight without yielding, the sides and bottom being sheet copper. From the point, B, projects a ½-in. brass tube, B C, which turns up at right angle. At E is a stop cock, which is connected by a thick rubber tube with the glass tube, D F, which is fastened against the adjoining wall. Connected with C by a rubber joint is a ½-in. block tin tube of 20 ft. length, which extends up the wall in the manner shown to the highest point, T, and thence returns and ends just over the slightly funnel-shaped top of the glass tube at D. The bath being filled with water to just the level, B *b*, may be kept constant by boiling for many days without appreciable loss of water, the steam being condensed in its passage up, or, if uncondensed before it reaches the point, T, in its passage down the block tin tube. In flat-bottomed platinum or porcelain capsules, evaporation goes on very rapidly when placed on top of this water bath. The whole surface of the bath is nickel plated.

Constant Water Bath.

Automatic Cut-off for Gas for Drying Chamber.

III
VAPORIZATION

By the term "vaporization" we are to understand certain mechanical operations by which volatile substances are separated from other fixed bodies, or from bodies which may be less volatile, by the action of heat. When a volatile liquid is separated from a less volatile liquid, by the process of vaporization, we have what is known as evaporization. When a volatile liquid is to be collected we have what is known as distillation. When a solid is to be separated from the volatile liquid, we have what is known as desiccation, in which solid substances are deprived of moisture. Excication is the process by which a solid, crystalline substance is deprived of its water of crystallization, by the aid of powerful heat.

Granulation.

This is the process by which a powder is produced by heating a solution until the moisture has evaporated. Many salts are treated in this manner. The heat which should be applied in this process should be strong at first, and then gradually reduced. The stirring should be constant. When vaporization is used to separate a volatile solid from another body, it is known as sublimination. It can also be called a process of distilling volatile solids. It is a process which is largely used in the manufacture of chemicals, and is not so largely used in the laboratory.

Evaporation.

When any liquid is heated for the purpose of expelling vaporizable matter, and the process is conducted solely with a view to saving its fixed portion, the operation is termed evaporation. It thus far differs from distillation, which has for its object the preservation of the volatilized portion, in most cases, regardless of the solid. By its aid we can decrease the volume of or concentrate solutions for crystallization and chemical reaction, expel valueless volatile ingredients from those which are more fixed, obtain dissolved matter in a dry state, and prepare extracts and other pharmaceutical products.

Liquids evaporate more or less at all temperatures, those having the lowest boiling point yielding the most readily; but there are certain conditions which greatly promote this tendency. It must be remembered, therefore:

1.—That evaporation is more rapid in dry atmospheres, and that consequently the transit of a constant stream of air over the surface af the heated liquid effects a continual removal of each stratum as it becomes saturated with vapor.

2.—That evaporation is confined to the surface, and consequently that the breadth of the evaporating vessel must be extended at the expense of its depth.

3.—That heat greatly facilitates evaporation by lessening the cohesive force of the particles of a liquid, and consequently that the evaporating vessel should present a broad surface to be heated.

4.—That a diminution of the atmospheric pressure also facilitates evaporation, for the more perfect the vacuum the lower the boiling point of a liquid.

For analytical purposes, capsules of Berlin porcelain are by far the best implements. The capsules should be very thin, with steep sides, spout for pouring, nearly flat bottomed, and glazed throughout. Watch glasses answer for small experiments, but require to be very cautiously heated, as they are readily fractured.

Beaker glasses are also used for evaporating solutions which would lose by being transferred. Broad-mouthed glass flasks are of but limited application for evaporating, and are only employed for slow processes with valuable liquids, which are liable to alteration by too much exposure when ebullition is necessary.

For the larger operations of the chemist or pharmaceutist, vessels of copper, tin, enamelled iron, tinned copper, and for some purposes very large porcelain capsules are more suitable.

Retorts are used when the vaporized particles are of sufficient value to be condensed, as in the process of distillation.

Spontaneous Evaporation.

Those liquids which are very volatile or which become altered by heat, are evaporated by mere exposure to the atmosphere at its ordinary temperature. To this end they are poured into broad shallow vessels, and placed aside until the dissipation of all vaporizable matters, or until crystallization; this mode of evaporation being also employed for procuring large crystals, which are better defined than those obtained by rapid evaporation. The more dry and hot the atmosphere the more rapid is the evaporation. In order to maintain a continued contact of the face of the liquid with strata of fresh air, the vessel containing it should be placed in a draught, so that those portions of air which become saturated with vapor may be displaced. When the air might act injuriously, and a vacuum is unnecessary, a substance may be evaporated in another atmosphere, for instance, of hydrogen or carbonic acid. For this purpose it is only necessary to adjust the disengagement leg of the apparatus to the tubulure of a retort, so that its end may reach nearly to the level of the liquid in the latter. The generated hydrogen passes into the retort heated to the required temperature, and promotes the discharge of the vapors into a recipient attached to the beak of the retort, and fitted with a small tube in its other tubulure for the disengagement of uncondensed portions.

For the evaporation of solutions of sulpho-bases, of sulpho-salts, and of all substances readily oxidizable by exposure, this process is better applicable than that with the air pump, which is apt to be attacked when the eliminated vapors are corrosive.

This process is much used in crystallization, for concentrating alterable solutions, and drying precipitates.

Evaporation in Vacuo.

We have already referred to the happy influence of diminished atmospheric pressure in facilitating evaporation, and shall now speak of the means by which it is accomplished, and the particular instances in which it is employed.

This mode is resorted to for hastening the evaporation of all liquids, but more especially of those which are alterable by exposure.

Evaporation by Heat in Open Air.

Having already noted the effects of heat in facilitating evaporation, we proceed to make known its modes of application. As the boiling points of solutions differ, so accordingly their evaporations are effected at varying temperatures. For example, aqueous or other solutions of unalterable matter may be evaporated over the fire; others which are destructible by heat require the intervention of baths. In whatever mode the operation is performed, the general principles are the same, and whether the vessel be a porcelain capsule or metallic pan, the greater its width in proportion to its depth the more rapid is the evaporation. Constant agitation with a stirrer is also promotive of the process.

Evaporation Over Water and Saline Baths.

When solutions are alterable at a temperature of 212° F., the capsule or containing vessel is heated over the water bath. If it requires a higher heat, but one not exceeding 300° F., then the water must be replaced by a saline bath.

Evaporation by Steam.

This mode has many advantages over all others, not among the least of which is that with the aid of the generator any number of vessels may be heated simultaneously, and in any part of the laboratory, it being only necessary to have conduits of sufficient length to convey the steam to them. Moreover, convenient stop cocks allow a regulation of the heat, and consequently all danger of injury to the evaporating solution is avoided. By increasing the pressure of the steam, the temperature of the solution is also elevated.

Steam is applied through metallic coils placed at the bottom of the containing vessels, and having an exit pipe leading into the neighboring flue, or else by means of metallic casings.

Evaporation Over Sand Baths.

This mode is much used in analyses and for careful evaporations, requiring temperatures greater than 212°, and yet not so high as those given by the naked fire. The position and arrangement of the vessels are as directed under the head *Sand Baths.*

Evaporation by Heated Air.

This mode is admirably adapted for the inspissation of the natural juices of plants or for preparing dry extracts. It is also applicable to the completion of evaporations which have been carried as far as is safe over the naked fire. Porcelain plates or panes of window glass are the vessels used, and a stove or apartment for their reception heated from 95 to 110°, with a free draught passing through are the means of obtaining the required temperature. The juice evaporates either to thin scales or else to a spongy mass, as in the case of tannin extracted by ether, and as soon as it reaches dryness, the plates or panes are to be withdrawn, and their contents removed with a spatula.

Evaporation Over the Naked Fire.

The tendency of many substances to decomposition over fire, especially organic, even when in solution, renders this mode inapplicable save when the solvent and substance dissolved are both inalterable below the boiling point of the former. It is resorted to for expediting evaporations, but otherwise is far more inconvenient than steam, because of its affording less facility for the regulation of the heat and requiring greater attention. The containing vessel should be placed over a furnace of small dimensions, and its contents continually stirred with a porcelain spatula —this precaution preventing decomposition or carbonization, provided the temperature is not allowed to exceed the boiling point of the solvent.

In analysis and other processes, the heating implement is generally the gas or spirit lamp. The capsule filled to about 2-3 its depth with liquid, being placed in position, the flame is applied gradually and maintained just low enough to prevent ebullition; and in order to facilitate the process, and at the same time to allay turbulence, it should be frequently stirred with a glass rod. The same directions apply when the operation is performed in a beaker glass, as is done in some analytic experiments. A cover of white paper prevents access of dust without retarding the process, but care must be taken that the contents of the vessel be not ejected against it, thus causing a loss. In evaporating to dryness, towards the end of the process the flame must be so managed as to impart a uniform heat to all parts of the thickened solution. The interposition of a very thin plate of sheet iron between the flame of the lamp and the bottom of the heating vessel is an additional means of preventing spirting. These precautions and constant stirring will prevent the loss of particles which is liable to occur upon disengagement of the last portions of liquid. If the liquid drops a powder during the operation, the vessel must be inclined, and in order to prevent spirting, heated above the deposit.

Distilling.

Small Apparatus for General Purposes.—(a) All ordinary distilling apparatus consists of 2 parts—one in which the heat is applied to the body to be distilled and vaporized (called the "still"), and the other into which the vapors that are

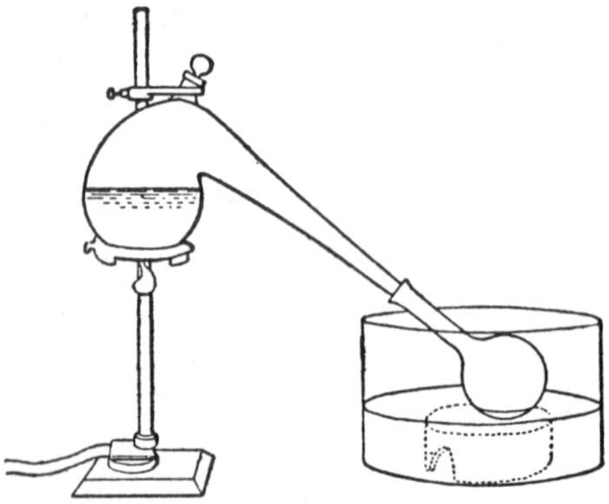

A Simple Distilling Apparatus.

formed enter in order to undergo the cooling that condenses them (termed the "condenser"). One of the simplest forms of distilling apparatus used in laboratories consists of a still into which is introduced the liquid to be distilled, and which is placed upon a furnace. The neck of this fits into that of a sphere whose opening must be wide enough to allow the orifice of the still to reach the spherical part of the receiver. Finally, the sphere dips into a vessel full of cold water, and is cooled on its external surface by a wet cloth. The heated mixture begins to boil, and its vapors, escaping from the retort, cool and condense upon the cold sides of the spherical receiver. This latter serves at once as a condenser and a vessel for receiving the distilled product.

In the beginning, the empty receiver weighs less than the volume of water that it displaces, and tends to float. This may be remedied by using a sufficiently heavy ring of lead into which the neck of the receiver may be introduced, and which may rest upon the latter's bulge. Upon fixing a similar ring under the receiver, the latter will be prevented from turning laterally and even from getting broken.

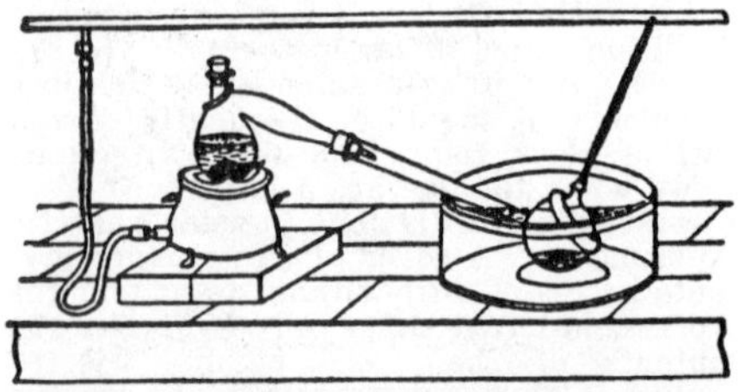

Small Apparatus for General Purposes.

The water in the external vessel is renewed so as to keep it cold.

A simple arrangement of this kind is not adapted for materials that have a low boiling point, since a large proportion of the vapor escapes, and makes its exit through the neck of a receiver, which is kept hot by the vapors coming from the still. The following, which is just about as simple, is a much more perfect arrangement.

The narrow part of the still is fixed into the neck of a long, tubular receiver by means of a cork which it traverses. This annular cork exactly closes the space between the neck of the still and that of the receiver. On the other side, in the tubulure of the receiver, there is fixed by means of a cork, perforated and arranged like the preceding, a long and narrow glass tube.

When the still has been filled with the substance to be distilled, and placed upon a furnace covered with wire gauze, the receiver is immersed, as above stated, in cold water. The vapors that are formed become cooled in traversing the elongated neck of the receiver, and are thoroughly condensed in the immersed part, provided the ebullition is not too rapid. In this latter case, the narrow tube, which presents the only open orifice, becomes heated, and indicates to the operator that the fire must be moderated.

The inconvenience of every apparatus of this kind is that the vapors which enter the receiver are not compelled to impinge against the sides, and may go directly to the exit-tube, or, in other words, the refrigeration is not methodical. Moreover, the refrigerating surface continues to diminish in measure as the receiver fills. Finally, if the receiver breaks, the entire distilled product comes in contact with the water. Despite these disadvantages, the rapidity with which such apparatus may be arranged, causes them to be frequently employed.

The use of refrigerators permits of a more exact and methodical condensation of the vapors. These are arranged as follows: The 2 orifices are placed in contact by means of a rubber tube, 3 to 4 cm. in length, into one end of which is introduced the neck of the retort, a, and into the other tube of the refrigerator. The latter being held in an inclined position by means of a clamp, a current of water traversing it from top to bottom, and a bent tube being adapted to its lower extremity, the free extremity of the bent one is fixed into the flask that is to collect the product. We may also suppress the central tube of the refrigerator in the flask, b, kept inclined. To facilitate this arrangement, the neck of the retort is cut at a point where it has the same external diameter as the tube of the refrigerator, and is then edged with a flame.

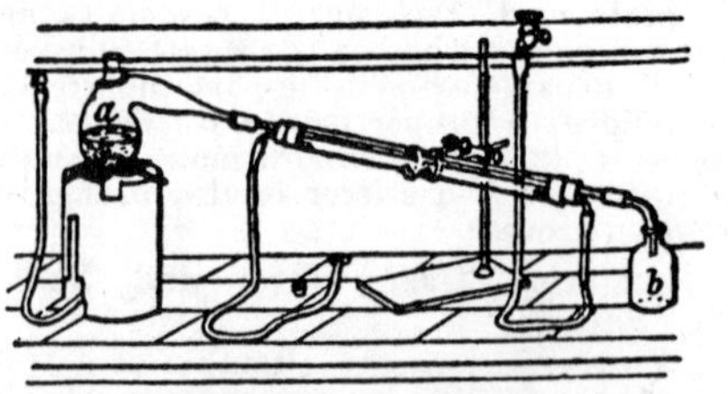

Type of Laboratory Condenser.

Again, if the difference between the diameters is considerable, we may, by means of a flame, draw out slightly the one of the two tubes that is the larger, and cut it at the proper point to obtain an equality in the diameters. Finally, we may solder to the extremity of the refrigerator a cylindrical tube, 2 or 3 cm. in diameter and 6 or 7 in. length, into which is fitted the neck of the retort previously provided with a cork. This latter contains an aperture running in the direction of its axis, and the whole is arranged so as to form a tight joint.

When the substance distilled attacks cork or rubber, the neck of the retort is drawn out to a sufficient length to allow the tube that terminates it to enter the refrigerator to some depth. The rubber with which the two parts of the apparatus are connected is thus nearly out of the range of the vapors.

Tin Can Still.

(b) One of the simplest forms of still consists of a tin can or bottle in which the water is boiled, and to this a tin tube is adapted by means of a cork, one end of this tin tube terminating in a coil passing through a tub or other vessel of cold water. A gas burner, as shown, is a convenient source of heat, and in order to insure a complete condensation of the vapor, the water in the cooling tub must be changed now and again.

(c) Sometimes the vapor is condensed by being allowed to play against the inside of a conical cover which is adapted to a saucepan, and is kept cool by the external application of cold water; and in this case the still takes the form represented by our next engravings; the condensed water trickles down on the inside of the cone, and flows out at the spout.

(d) An extemporized arrangement of a similar character may be made by passing a tobacco pipe through the side of a tin saucepan as shown in the engraving, and inverting the lid of the saucepan; if the

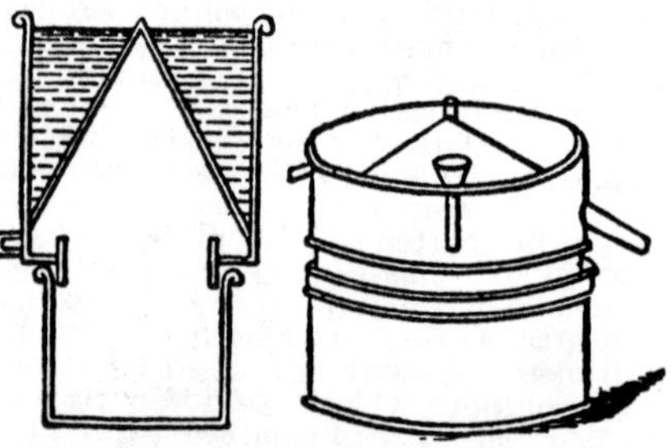

Simple Externally-Condensed Still.

lid is now kept cool by frequent changes of water inside it, and the pipe is properly adjusted, so as to catch the drippings from the convex side of the lid, a considerable quantity of distilled water may be collected in an hour or so.

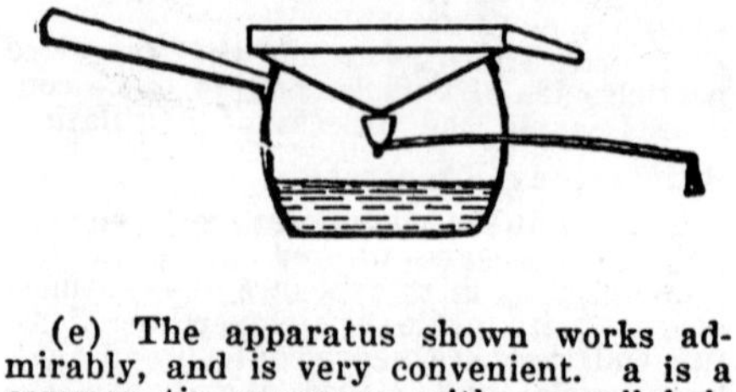

(e) The apparatus shown works admirably, and is very convenient. a is a common tin saucepan, with a small hole in the side, for a tobacco pipe; b, a "steamer," on top, with a bottom like an inverted cone, 1 in. of wire being soldered at the apex.

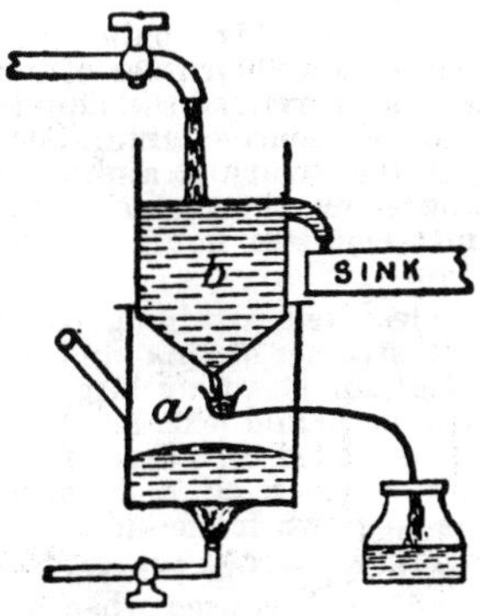

Tap-Cooled Still.

A gas jet (Bunsen's, if possible) boils the water in the saucepan; the ascending steam is condensed on the lower surface of the steamer, runs down to the point of the wire, down the pipe into the bottle. A small jet of cold water keeps b cool.

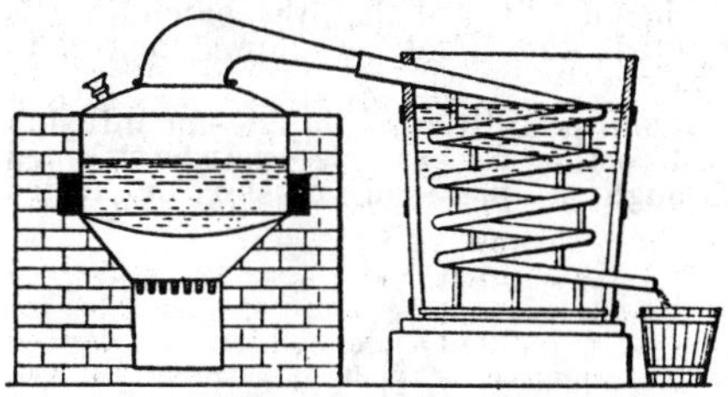

An Old Fashioned But Efficient Still.

(f) The arrangement shown is one that may readily be adapted to, and is specially suited for, the old fashioned stills which are in frequent use among pharmacists for the purpose of distilling water. The idea is extremely simple, but thoroughly efficient in actual practice. The still is thin copper, 2 gal. capacity, and the condenser is the usual worm surrounded with cold water.

Tinctures, Extracts, etc.

(a) A very convenient and complete still is shown herewith. The body holds

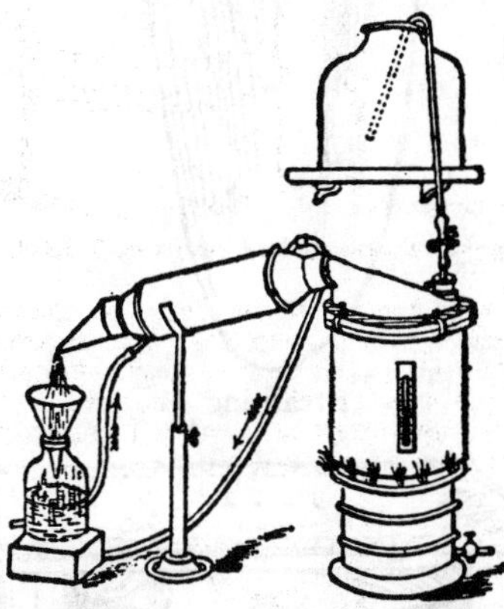

Tincture and Extract Still.

over 3 gal.; the condenser has 7 straight tubes surrounded with the cold water introduced by a rubber from a hydrant or bucket of water placed higher than the still, and carried off as it becomes warmed by another tube as indicated by the arrows. By the siphon arrangement shown in the cut, it is possible to feed the still from a reservoir while distillation is in progress, thus using a 3-gal. still where a much larger one would have been necessary. The still may be set into a kettle partly filled with water, and thus used as a water bath, or a shallow dish, with flat rim, which accompanies the still, may be placed between the two brass ring bands and clamped securely.

(b) Stevens arranged the apparatus as shown for continuous distillation. As soon as the water passes out of the boiler,

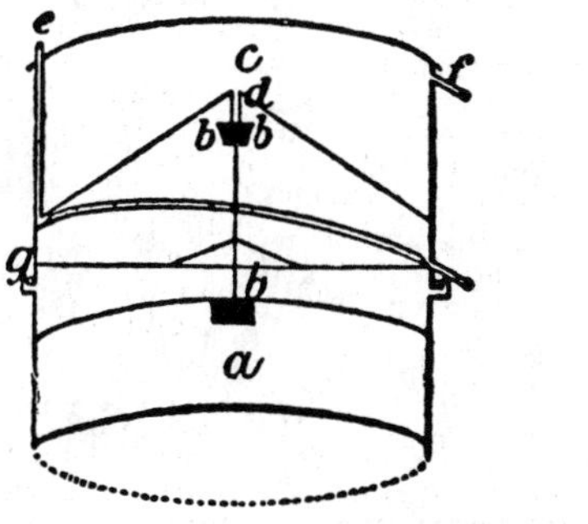

Apparatus for Continuous Distillation.

a, the float, b, lowers, letting a fresh supply of water from the condenser, c, through d, thereby keeping the water in the boiler at a constant level. This avoids the necessity of adding a large quantity of cold water at once, the effect of which would be to reduce the temperature of the water below the boiling point.

Cold water is supplied to the condenser through e, and as it becomes heated and rises to the top, it is carried off through f. The boiler and condenser are joined at g.

By leaving out the float and closing the inlet, d, with a cork, it can be used for distilling other liquids.

The apparatus is not patented, and should any pharmacist desire to make one for his own use, he can do so.

(c) The distilling apparatus represented herewith is intended primarily for the use of pharmaceutical chemists or druggists, but it possesses features which will recommend it to many who have need of a trustworthy and quick-acting still. The wide delivery tube is a useful feature, allowing as it does for the accumulation of vapor, and permitting the introduction of the hand. The body of the still is of wrought iron or copper, with a lid fitting on ground edges, and held together by screw clamps, as seen in the engraving. A gauge is fitted to show the quantity of liquid in the still. The cendenser consists of a number of glass tubes, which, if they are 1 in. diameter and 24 in. long, expose a surface of 264 in., while that of the surrounding cylinder is only 188½ in. The ends of the condenser tubes are drawn together and tapered, as shown in cut, to permit, if desired, the collection of the distillate in a narrow-mouthed bottle. The advantage gained by this apparatus, aside from the general one of convenience, is thus seen to be in the notable increase of condensing surface it exposes, which to that extent increases the effectiveness of the device, *i.e.* its rapidity of action. Compared with a Liebig condenser of similar dimensions, this apparatus exposes probably 3 times as much condensing surface. The idea of a tubular condenser, employed in the manner set forth, is, in the opinion of the *American Journal of Pharmacy*, an excellent one, that may find useful imitation in the chemical laboratory and elsewhere. The device illus-

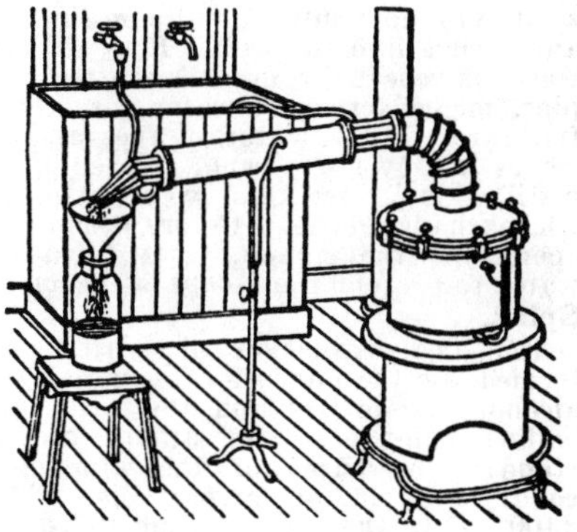

Remington's Still.

trated and described was invented by Joseph P. Remington, whose recommendation of its merits is based upon a continuous use of it for years.

(d) *Flowers, Plants or Seeds.*—To obtain the essential oils, from flowers, plants or seeds, the oleiferous material is placed in an iron, copper or glass still, of 1 to 1,000 gal. capacity, and is covered with water; superposed is a dome-shaped lid, terminating in a coil of pipe, placed in a vessel of cold water, and protruding therefrom with a tap at the end. On boiling the contents of the still, the essential oil passes over the steam, and is condensed with it in the receiver; the oil and water separate on standing. A great improvement, introduced by Drew, Heywood and Barron, is the use of a steam-jacketed still, as shown. Steam is supplied from a boiler by the pipe, a, into the jacket, b; within the head of the still is fixed a "rouser," c, a double-branched stirrer curved to the form of the pan, and having a chain attached and made to drag over the bottom, the whole being set in motion by means of the handle, d. The still is charged, and nearly filled with

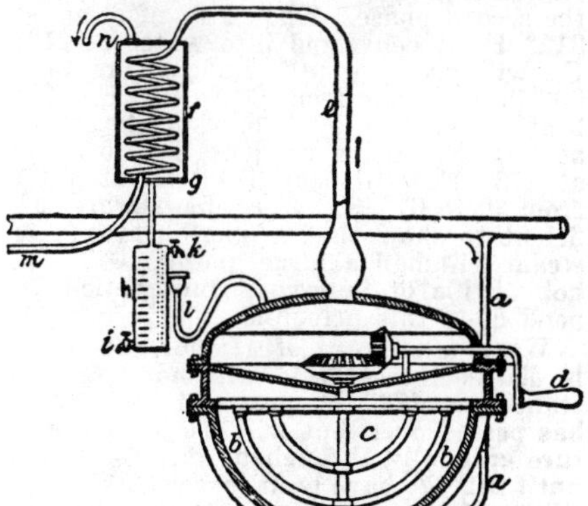

Steam Jacketed Still.

water; the head is then bolted on, steam is admitted into the jackets, the contents are well stirred, and soon the oil and steam are carried up the pipe, e, condensed in the refrigerator, f, and let out at g into the receiver, h. Here the oil and water separate, and escape by different taps. In the illustration it is supposed that the oil obtained is heavier than water; it will then sink, and be drawn out by the lower tap, i, and as son as the water reaches the level of the upper tap, k, it will flow into the siphon-funnel, l, and thence into the still. Thus the same water is repeatedly used in the still. The pipe, m, conveys cold water into the refrigerator f; the water escapes as it becomes hot by the pipe n. When the oil distilled is lighter than water, the taps, i k, exchange duties Before commencing operations the siphon, l, is filled with water to prevent the escape of vapor.

Spirit.

(a) The distillation of spirit is performed for the purpose of separating the alcohol more or less from the water. The boiling point of water at the ordinary standard pressures of the atmosphere, equal to 30 in. of mercury, is 212° F. (100° C.), that of alcohol 173.1° F. (78.5° C.). At the sea-level, the pressure of the atmosphere may frequently vary between 28.5 and 30.5 in.; the boiling points of water corresponding to these temperatures are 210° F. and 213° F. Indeed, changes in the weather may cause the boiling point of water to vary as much as 5° F. in our climate. These alterations in pressure would cause corresponding changes in the boiling point of alcohol. If we gradually raise the temperature of alcoholic fluids to a point when vapors are freely formed, it is observed that though there is a continuous absorption of heat, yet the liquid does not increase in temperature. The heat which is absorbed during the first period is doing work of a different character from that employed subsequently. There are two phases in the process, and two different kinds of work performed by the heat employed in boiling even a kettle of water.

The first phase is indicated by a rise of temperature from 60 to 212° F.; the second phase by a change of state, from that of a liquid at 212° F. to a vapor at the same temperature. The quantities of heat required by different liquids in these changes varies greatly, but the variation is greatest when they pass through the second phase. Thus 1 lb. of steam at 212° F., if converted into water at 212° F., will give up heat sufficient to raise 996 lb. of water from 60 to 61° F. The heat rendered up by 1 lb. of alcohol vapor at 173° F. during condensation to liquid at 173° F., will heat 374.9 lb. of water from 60 to 61° F. These figures are sufficient to show that a small quantity of steam will boil a large quantity of alcohol. Stills of improved construction depend upon this principle.

When a mixture of alcohol and water is distilled, the liquid will not boil constantly at 173° F. until all the alcohol has passed over, but will rise in temperature gradually throughout the distillation until 212° F. have been reached. The distillate, if separated into fractions boiling between fixed points, consists of a series of mixtures of alcohol and water in definite proportions. The mixtures richest in alcohol come over first; that is to say, at the lowest temperature.

The latent heat of the vapor of a liquid with a high boiling point can be made to boil a liquid with a lower boiling point. For instance, steam at 212° F. can boil alcohol at 173° F., and alcohol at 173° F. in turn can boil ether at 94.8° F. With a simple still, strong alcohol can be obtained from wash by repeated distillation only. Woulffe realized the fact that this wasteful and tedious process could be dispensed with by connecting together a number of rectifying chambers in such a manner that the vapor driven off from the chamber nearest the fire should be condensed in the second, and by the heat given out by its condensation cause the more volatile portions of the liquid of the second to distil into the third chamber, and those of the third into the fourth, and so on, until a sufficient degree of concentration is attained.

IV
PRECIPITATION AND SEPARATION

Edulcoration.

The affusion of water on any substance for the purpose of removing the portion soluble in that liquid. Edulcoration is usually performed by agitating or triturating the article with water, and removing the latter, after subsidence, by decantation or filtration. It is the method commonly adopted to purify precipitates and other powders which are insoluble in water. The washing bottle is a most useful instrument for the edulcoration of precipitates.

Precipitation.

By precipitation we are to understand a process of separating a solid substance from a solution by the action of chemicals, heat, or light. The precipitate easily drops to the bottom of the receptacle, although sometimes it may rise or be held in suspension. The solid substance is called the precipitate; the added agent which produces the effect is called the precipitant, while the liquid which remains in the vessel is called the supernatant liquid. Precipitation is one of the most valuable aids to the analytical chemist, and is constantly employed, but is also of great use in the arts. It is sometimes used to bring the substance into a powdered state; again, it is used for purification, or to separate substances which are insoluble in the liquid. It is sometimes necessary to heat the solution in order to obtain precipitation. Some preparations, such as silver salts, are precipitated by the action of light. A special precipitating jar is inexpensive, and is very convenient. The precipitated matter is usually collected with the aid of a filter and a filter paper.

Straining.

Straining is best accomplished through some textile fabric, as felt, muslin, Canton flannel, gauze, etc. Felt strainers are particularly recommended where chemical work is being done, but for the amateur's use they are apt to be expensive, as the felt takes up a great deal of the odor of the material. Canton flannel is cheap, and the bleached Canton flannel is recommended. One or two funnels or tunnels should be provided. The white enameled ones, which are imported from Sweden, are particularly recommended. Hard-rubber funnels are good for certain purposes; also copper funnels. Special funnels are provided for hot filtration, as shown in one of our engravings. This is particularly recommended when we deal with preparations containing wax, jellies, ointments, etc. The jacketed hot-water funnel is perhaps the most convenient means of obtaining heat. Steam may also be used, if available, and is both cheap and handy.

Colation.

Colation or straining is a process which does not differ from filtration in principle, but the term is applied to the removal of insoluble particles of a relatively large size by passing the liquid through a medium of coarser texture than filter paper. The ordinary straining media are felt, flannel, muslin and calico, through which materials the liquid will flow with considerable rapidity.

A seamless felt straining bag is illustrated. A strainer of this kind is particularly useful for straining large quantities of syrups or liquid extracts. When in use it is suspended by means of tapes over a suitable receiver, or is supported by a frame, as is shown in the figure.

Our next engraving illustrates a form of strainer which is used when bulky precipitates are required to be filtered, washed and drained. Ferric hydroxide is precipitated in large quantities for the manufacture of the scale preparations of iron, and it is conveniently separated and washed on a piece of strong calico stretched over, and fastened by means of nails, to a rectangular wooden frame supported on short wooden legs. In this case it should be noted that the precipitate is wanted; the filtrate is allowed to run to waste.

Small quantities of liquid—an infusion or decoction, for example—may be strained through a piece of muslin or calico

Straining

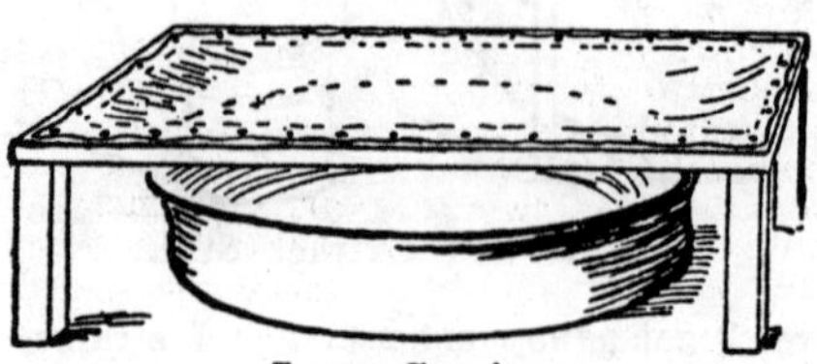

Large Strainer

stretched over the top of an ordinary funnel.

Clarification.

Clarification is the process of separating the suspended matter contained in a

liquid or semi-liquid substance without recourse to filtration. It may be effected in a variety of ways. The official method adopted for the clarification of honey, the viscid nature of which renders ordinary filtration somewhat impracticable, is the application of heat. The honey is heated on a water bath in an open, shallow dish, under which treatment it becomes much more fluid, and the suspended particles of solid matter rise to the surface, or sink, according to their specific gravity. By skimming, or by straining through flannel while the honey is still hot, the solid foreign particles can be easily separated out. In the same way, vegetable juices can be clarified by heat, albuminous material forming a coagulum which can be separated by filtration.

Certain liquids which are difficult to filter, and which do not yield a satisfactory filtrate, are sometimes clarified by the use of white of egg or of gelatine. In the former case a relatively small quantity of the white of egg is thoroughly mixed with the turbid liquid, and the whole is then heated to about 80° C., at which temperature white of egg coagulates. The particles which rendered the liquid turbid are enclosed in the coagulum formed, which is easily removed from the liquid by the ordinary process of straining. Gelatine is useful, particularly when the turbidity of a liquid is due to tannin bodies, with which the gelatine readily combines to form an insoluble gelatine tannate, which can be readily removed by filtration through paper or by straining through calico.

Centrifugation.

By centrifugal force is meant the force exerted by any whirling body. A solid

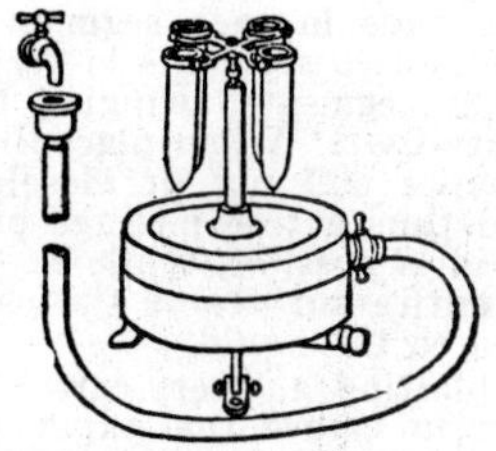

Water-Drive Centrifuge

body contained in suspension in a liquid can be readily separated by rapid rotation, the heavier particles of solid always tending to fly to the outer rim of the revolving ring of fluid. Centrifugation is thus another means of separating a solid from a liquid, and is a method especially useful when dealing with small quantities of liquid which contain in suspension minute quantities of a solid body which it is difficult to collect satisfactorily on a filter paper.

Centrifugal machines are constructed to various patterns, but the simple form illustrated will serve to show the principle of their construction. They consist essentially of two or four, or sometimes more, glass tubes (G) enclosed in metal tube holders (F), the tubes themselves being constructed with a somewhat conical-shaped bottom. The tubeholders are swung upon a horizontal axis (E), which can be rotated at a rate of from 2,000 to 3,000 revolutions a minute. The whole apparatus is clamped firmly to the laboratory bench, as shown in the figure. When in use, the tubes are filled with the liquid so that they are equally balanced, and the machine is turned rapidly for a

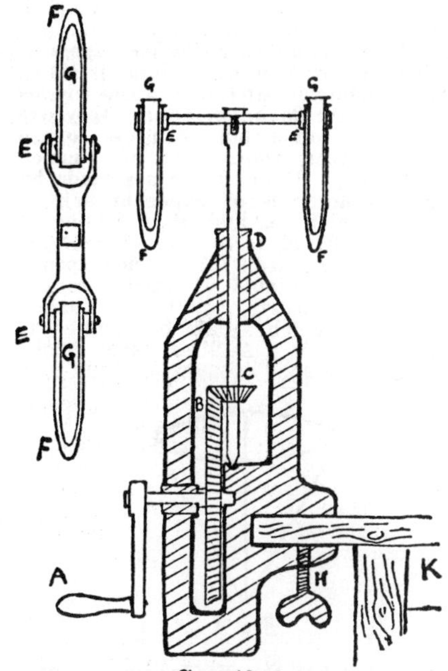

Centrifuge

few minutes, at the end of which time the solid particles will be found compacted together at the bottom of the glass tube, leaving a clear layer of supernatant liquid, which can be poured off.

A centrifuge is used in the laboratory for the rapid determination of fat in milk. A measured quantity of the milk is put into a graduated centrifuge tube and a little amylic alcohol, hydrochloric acid, and some concentrated sulphuric acid are added, in order to secure a better separation of the fat. A second tube, containing a similar quantity of liquid, is placed

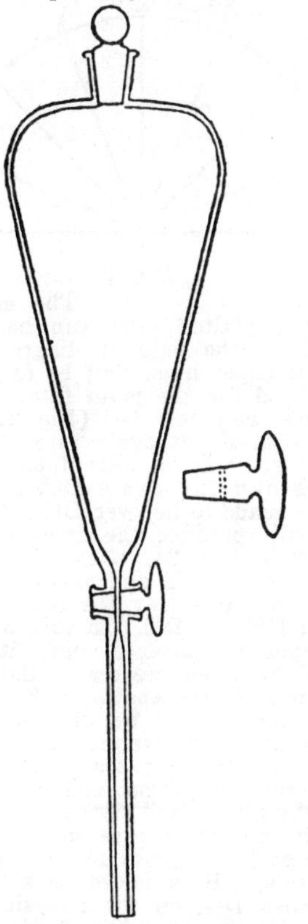

Separating Funnel

on the opposite side of the machine in order to secure a proper balance, and the apparatus is then rotated for one or two minutes, at the end of which time all the fat will have collected in the neck of the tube, and the percentage can be directly calculated. The centrifuge is also extremely useful for collecting for microscopical examination the deposit in a small quantity of liquid, the deposit in a sample of urine being best collected in this way.

The Separation of Immiscible Liquids.

The separation of two liquids which are more or less insoluble in one another is an operation important in many pharmaceutical and manufacturing processes. When relatively large quantities of immiscible liquids have to be separated, a tubulured jar or a siphon may be used, as has been already described under DECANTATION; but for quantities of a few ounces some other means must be adopted.

The alkaloidal assay of the galenical preparations frequently necessitates the separation of a layer of ether or chloroform or other organic liquid from a watery solution with which it is immiscible. In the assay of opium, for example, a layer of mixed alcohol and ether has to be separated from an aqueous layer, and in this case the Pharmacopœia directs the use of a pipette. A pipette, as shown, consists of an elongated bulbed glass tube, open at both ends, the lower end being drawn out into a narrow orifice. It is used by dipping the lower end under the surface of the top layer of liquid and applying suction with the mouth at the upper end of the tube. The bulb may be large enough to hold from 5 to 50 mils, and when as much as possible of the layer has been drawn into the bulb the moistened tip of the forefinger is placed firmly over the upper end of the tube, the liquid being thus kept from flowing out until the finger is removed. A glass syringe may be used for the same purpose as a pipette, but it is somewhat more clumsy.

Separating Funnels.

A more convenient means of separating layers of immiscible liquids is by the use of a glass separating funnel. An elongated pear-shaped separator, as illustrated, is a good form by means of which two liquids can be separated with greater accuracy than with a separator of a cylindrical shape.

For the separation of two liquids neither of which is particularly volatile, an ordinary glass funnel, the neck of which is provided with a stopcock, is sometimes used, but a separator of this pattern is quite unsuitable for assay processes, since it is impossible to shake the two layers together before they are set aside to separate.

Decolorization.

Decoloration is a process of rendering colored liquids colorless, and this is accomplished by the aid of animal charcoal or bone black. Decolorization may be accomplished in an ordinary filtering funnel or in a percolator.

Filtration and Other Processes of Separation.

Filtration is a process of separating a liquid from solid matter mechanically suspended in it, by passing it through some porous medium which does not allow the solid particles to pass through. In some cases it has for its object the collection of the suspended matter; in others it is used for obtaining the liquid in a clear state. Filtration is a simple process in principle, but in manufacturing, as well as in processes on a smaller scale, where liquids are employed, there is perhaps no operation of wider application, hence it is of great importance that the process shall be carried out in an economical and expeditious manner. Among the substances which are used as filtering media are various kinds of cloth, flannel, unglazed porous paper, engineer's waste, absorbent cotton wool, glass wool, asbestos, sand and charcoal. For small quantities of a liquid which filters easily, and in which the suspended matter is in coarse particles, a pledget of absorbent cotton wool placed in the throat of a funnel is often sufficient to produce a satisfactory filtrate. For extensive laboratory processes, however, the latter simple device is seldom of much service, for the small extent of filtering surface will soon lead to imperfect filtration, or possibly to complete blocking of the filter. The form of filter used, and the character of the filtering medium, depends not only upon the nature of the liquid to be treated, but also upon the amount of liquid that is required to be filtered.

Filtering Media.—Of the filtering media in common use, fine porous unglazed paper is the most universal for small operations, a piece of paper of suitable size being folded into a cone and fitted into a funnel. The funnels used for supporting filter papers are made of glass, glazed earthenware, or of metal, and those which are intended for rapid filtration are usually deeply ribbed or fluted on the inside, the space between the filter paper and the glass permitting a free passage of the filtered liquid. The same end is sometimes attained by placing thin glass rods or quills between the filter paper and the sides of the funnel. Filtering paper may be obtained in many qualities, the best quality consisting of practically pure cellulose. For the majority of purposes, white filter paper should be used, and this is made from pure flax fiber. The gray paper, on the other hand, contains a varying amount of wool, and although on account of its low cost it is used for the filtration of some galenical preparations, it is liable to color certain solutions, particularly alkaline ones, yellow. Such paper frequently contains also a considerable amount of chlorides, calcium carbonate, and iron salts, all of which are liable to pass into solution. For analytical work, particularly in ignition processes, a Swedish filter paper of very fine quality is necessary; such filter papers, in the course of preparation, are washed with hydrofluoric and hydrochloric acids, and by this means are rendered practically free from mineral impurities, and yield, on ignition, a very minute quantity of ash.

The suitability of filter paper for ordinary pharmaceutical purposes may be determined by the application of a few simple tests. Distilled water which has been passed through the paper should leave no residue on evaporation, showing that the paper contains no soluble mineral substances. Similarly diluted hydrochloric acid, after passing through the filter paper, should give none of the reactions of the alkaline earths, while the paper should not blacken with ammonium sulphide, proving the absence of many of the metals; nor should it be colored by a solution of salicylic acid, which would indicate the presence of iron.

Methods of Folding Filtering Papers.—Filtering paper is sold cut into circles of varying diameter, and since these circles merely require doubling for use, they are much more convenient than the square sheets of paper, which must be trimmed after folding. Plain filters are made by doubling the circle of paper in half to form a semicircle, and then folding it again in half, so as to form a triangle, with a convex base. This, when opened out (Fig. 1), should fit exactly to the sides of a properly constructed funnel, the sides of which should be inclined at an angle of 60°. A filter paper folded in this way is good enough for many purposes, but it has the disadvantage of presenting three thicknesses of paper to one-half of the funnel and only one thickness to the other half.

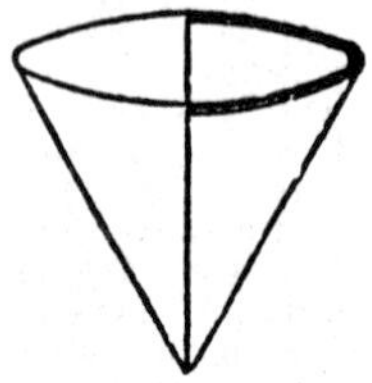

Fig. 1

The "plaited filter" affords a means of furthering rapid filtration, and at the same time it overcomes the objection of the unequal distribution of the paper on the sides of the funnel.

While, assuming that the funnel used has plain and not fluted sides, the filtration will not proceed with as much rapidity, since the sides of the paper will fit closely to the glass.

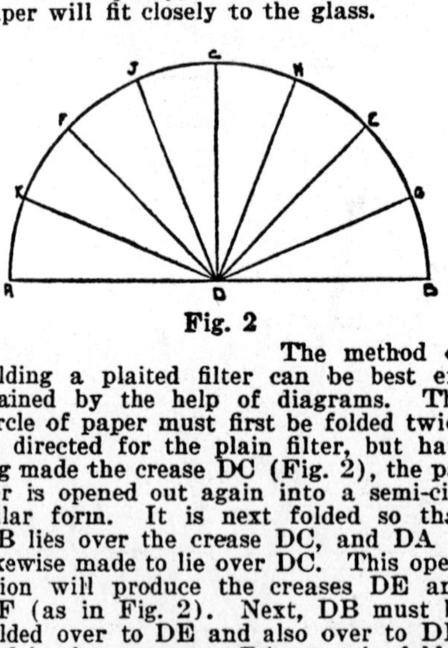

Fig. 2

The method of folding a plaited filter can be best explained by the help of diagrams. The circle of paper must first be folded twice as directed for the plain filter, but having made the crease DC (Fig. 2), the paper is opened out again into a semi-circular form. It is next folded so that DB lies over the crease DC, and DA is likewise made to lie over DC. This operation will produce the creases DE and DF (as in Fig. 2). Next, DB must be folded over to DE and also over to DF, and in the same way DA must be folded over to DF and DE. In this way, when the paper is flattened out, it will be marked by seven creases, radiating from the center, D (as shown in Fig. 2), and the semicircle will be divided by these creases into eight segments. Up to the present all these creases have been made in the same direction, and now, to complete the filter, each segment must be divided by another crease *made in a direction opposite to those already made.* To effect this, DB is folded back so that it lies under DG, on the opposite face of the semicircle; in other words, the new crease DL (Fig. 3) is in an opposite direction to any of the other creases previously made. In a similar fashion, DG is folded back so that it lies under DE, producing a new crease, DM (Fig. 3), which has the same direction as the crease DL, but is in an opposite direction to DG or DE. This process is repeated until the semicircle is divided into sixteen segments by fifteen creases, the eight new creases (illustrated by dotted lines) all being in an opposite direction to the first seven creases. The paper can now be opened out, as shown in Fig. 4, and it will be found divided into thirty-two segments, two of which, situated opposite to one another, have both edges in the same direction, and in order to prevent these two segments from lying flat against the glass when the paper is placed in a funnel a new crease, pointing inward, should be made in each segment so that each of these two segments is divided into two smaller segments, bringing the total up to thirty-four. When placed in a funnel the paper will not fit closely to the glass, and thus a free passage of the filtered liquid is possible, while at the same time the entire surface of the paper will be exposed to the liquid.

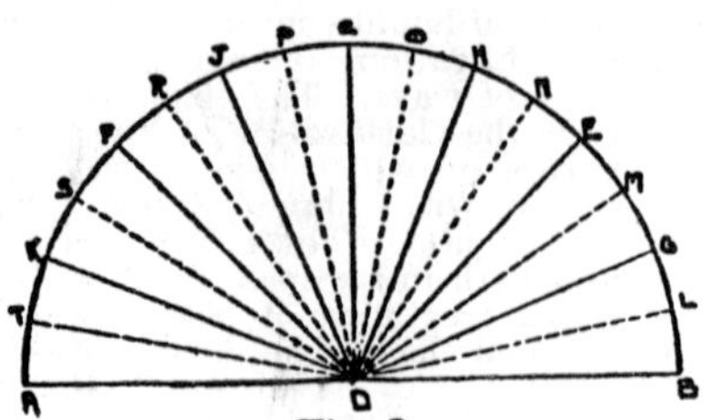

Fig. 3

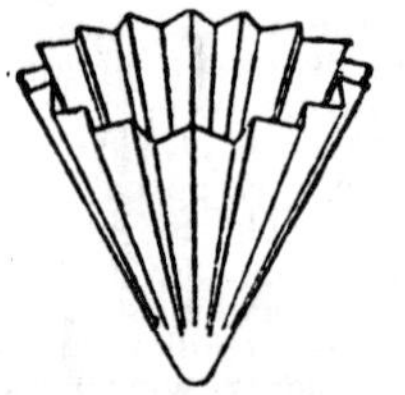

Fig. 4

When plaiting a filter, care should be taken not to crease the paper down to the extreme center of the circle (D), otherwise the apex of the filter may be so weakened as to break with the weight of the liquid poured upon it. The weakest part of a filter paper, whether plain or plaited, is always the extreme apex, and various suggestions have been made with a view to overcoming this weakness. One method is to dip the apex into strong nitric or sulphuric acid; the latter acid converts the paper into parchment paper, and thus renders it impervious to the passage of fluids, but the former treatment merely toughens the fiber of the paper. In either case care must be taken to wash the filter free from all traces of acid. The apex of a filter may also be supported by a small cone made of platinum foil, or more simply by means of a smaller filter paper folded and placed in the funnel first, or a pledget of cotton wool may be used for the same purpose. When filtering large quantities of liquid the paper is sometimes supported with calico to avoid breakage, the cloth is usually folded up with the paper, the double filter being placed in the funnel in the usual way. The fact that the apex of a filter paper is always a source of weakness has led to the adoption of another method of folding filter papers. The circle of paper is, as usual, first folded into a semicircle.

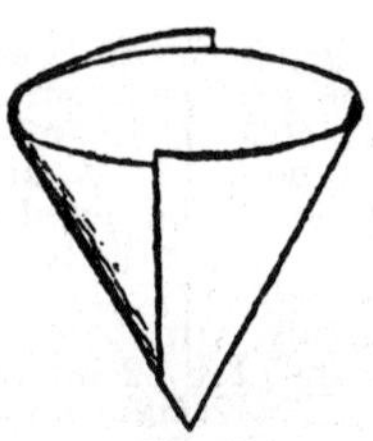

Fig. 5

Next, EB (Fig. 6) is folded over, with the crease in the position marked by the line EH; the point E, it will be noted, is not the center of the circle of filter paper. The paper is now turned completely over, and DA is folded over in the position marked by the line, DF, the crease,

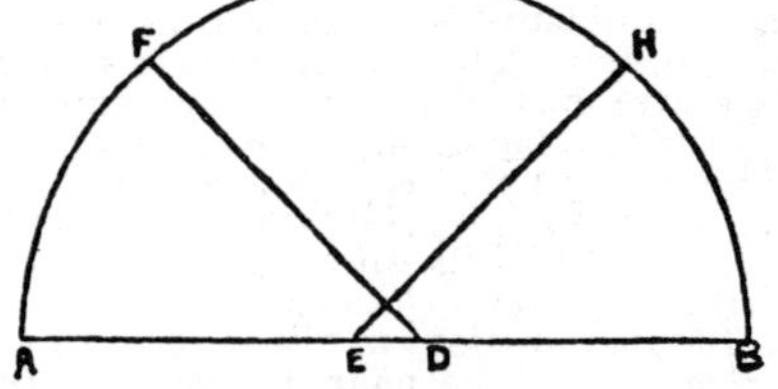

Fig. 6

DF, being, of course, in the opposite direction to the first crease, EH. When the paper is opened out (Fig. 5), it will fit into a funnel having the proper angle of 60°, while the apex will be strengthened by the presence of a double thickness of paper.

A liquid should never be poured in a sudden stream on to the apex of a filter paper, but should always be poured gently against the side of the filter, where, if dealing with small quantities, it may be conveniently directed by means of a glass rod (as shown in Fig. 7). In this figure the student should note the small strip of paper (A) inserted between the neck

Fig. 7

of the flask and the funnel tube. This precaution is necessary if the end of the funnel fits closely into the receiver, in order that there may be a free escape of air as the filtered liquid enters the receiver. A filter paper placed in a funnel should never reach above the rim of the funnel, for, if such be the case, the liquid will be sucked by capillary attraction into the projecting edges, and there will be considerable loss by evaporation from the exposed edges. Even when the filter paper does not protrude over the rim of the funnel there is always some loss by evaporation, especially when the liquid is a particularly volatile one, and the room temperature is high. In order to lessen the loss by evaporation during a slow filtration, a piece of plate glass may be placed on the top of the funnel.

Continuous Filtration.—It is frequently inconvenient for an operator to give constant attention to a filtration process, hence a "self-feeding" filter is of great service. On a small scale, the following simple method, illustrated in Fig. 8, works well. An inverted Winchester quart, containing the unfiltered liquid, is arranged at such a height that the mouth of the bottle is in the liquid at the level at which it is desired to keep the funnel filled. The liquid in the funnel acts as a valve, and until air enters the bottle none of the liquid will flow out, since the atmospheric pressure is sufficient to support a column of water 32 ft. in height. As, however, the liquid in the funnel passes through the filter, it sinks in due course below the

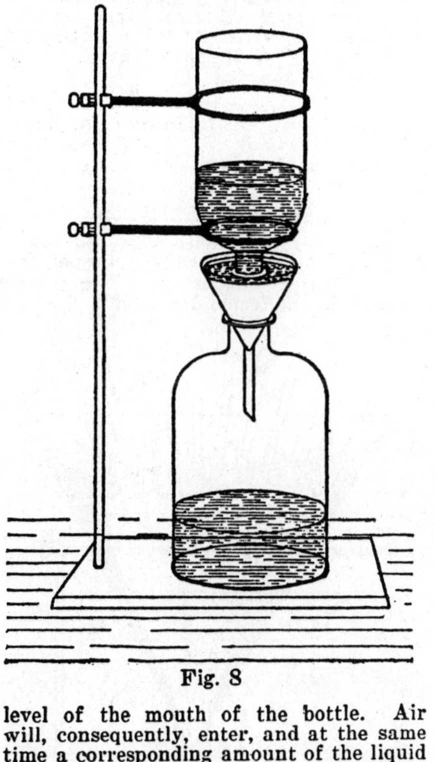

Fig. 8

level of the mouth of the bottle. Air will, consequently, enter, and at the same time a corresponding amount of the liquid will flow from the bottle into the funnel. This process will go on automatically until the bottle is empty. The method is similar to that adopted for obtaining a continuous supply of menstrum for percolation, a process which has been already described. An arrangement which is similar in principle to the above has been adopted for the continuous washing of a precipitate. In Fig. 9 is shown a specially constructed tube fitted into the neck of an inverted flask by means of an india-rubber cork. As in the case of the inverted Winchester, water will flow out of the flask at E as soon as the level of the liquid in the funnel falls below the level of where the side tube joins the main tube (C), air entering the flask through the open side tube (D). The process is continuous so long as any liquid remains in the inverted flask.

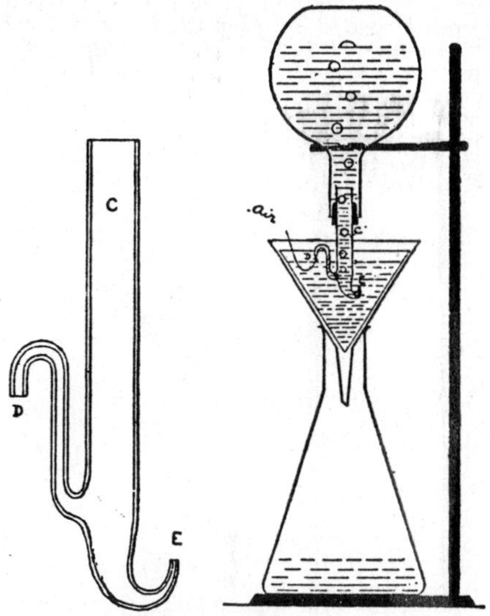

Fig. 9

Asbestos Filters.—In some cases, the turbidity of a liquid is due to the suspension in it of particles of matter so minute that their removal is not easily effected by the ordinary method of filtration through paper. In such cases, a clear and bright filtrate can often be obtained by shaking up with the turbid liquid some substance by means of which the minute particles are entangled, and can no longer pass through the pores of the filtering medium. For this purpose, paper pulp, prepared from waste scraps of filter paper, calcium phosphate, kieselguhr, kaolin, French chalk, magnesia, and finely shredded asbestos, have all been recommended. Whichever one of these substances is chosen, a small quantity of it is well shaken up with the liquid to be filtered, or the filter itself is first coated by shaking up a little of the filtering agent with water, pouring the mixture over the filter and allowing the latter to drain. Usually, with either method, the first few drops of the filtrate are not very clear, hence the first runnings should be returned to the filter until the filtrate is obtained bright.

For rapidly filtering turbid liquids, especially those which are cloudy from the presence of minute globules of essential oil, the "Seitz" asbestos filter has proved successful. The apparatus consists of a conical filter of fine brass-wire gauze, suitably supported. The turbid liquid is thoroughly shaken with a small quantity of finely shredded asbestos fiber, and is then transferred directly to the gauze filter. With most liquids, a rapid flow of bright, transparent filtrate is obtained.

Hot Filtration.—It is sometimes necessary to filter through paper substances, such as fats and waxes, which are not liquid at ordinary laboratory temperature. In such a case, a rough and ready plan is to arrange the funnel over a circular low-power gas burner (Fig. 10), but a better plan is to use a hot-water jacket for the funnel. In Fig. 11 a funnel suitable for hot filtration on a small scale is illustrated. The jacket is usually constructed of copper; at some point around the top rim there is an opening (A) through which water is introduced, and this water is kept at the desired temperature by means of a Bunsen gas burner or a spirit lamp placed under the projecting arm. In practice, the substance to be filtered is first melted, and is then poured into the funnel, which has previously been allowed to become properly heated in the copper jacket. As the heating is continued, some of the water in the jacket will be lost by evaporation, since the opening, A, must not be closed

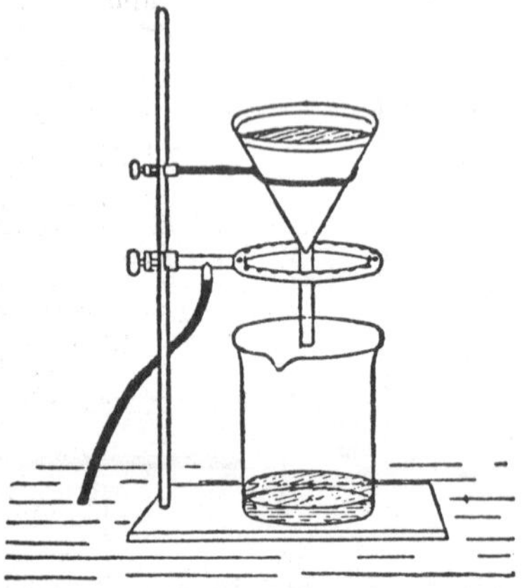

Fig. 10

on account of the pressure which the steam would produce if this were done; hence from time to time a little more water must be poured into the jacket. Fig. 12 shows an improved type.

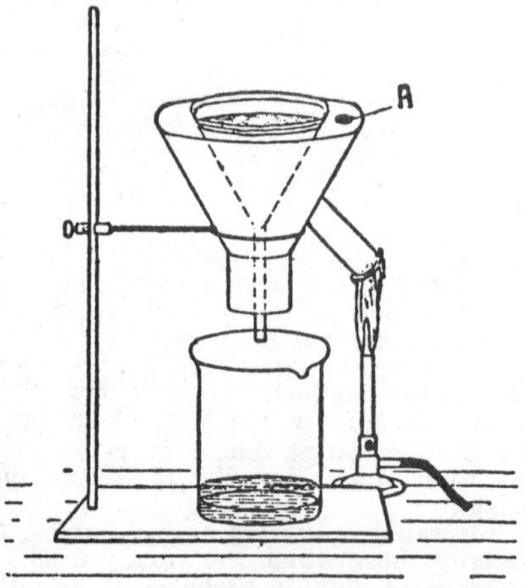

Fig. 11

Accelerated Filtration.—The rapidity at which filtration is effected depends upon several factors, the chief of which are: The extent of the filtering surface, the viscosity of the liquid, the porosity of the filtering medium, and the pressure or force by which the liquid is impelled through the pores of the filter.

In filtration as ordinarily carried out, the only pressure exerted is that due to the liquid itself resting on the filtering medium; but by increasing the height of this column of liquid the pressure is increased, and filtration is consequently accelerated. One of the principles of hydrostatics is that the thrust exerted by a liquid of given depth on the base of the containing vessel is independent of the shape of the remaining portion of the vessel, hence the column of liquid need not be of equal diameter throughout in order to produce uniform pressure.

Acting on this principle, a simple means of filtering oils or other liquids has been suggested. A filter bag is firmly attached to the lower end of a long tube, while to the upper end of the tube is fixed a funnel, into which is poured the liquid that is required to be filtered. Under such conditions the pressure exerted is that due to the weight corresponding to the total height of the column of liquid, and the filtrate is forced through the filter bag and collected. Instead of a filter bag an ordinary inverted funnel may be used; the filtering medium is tied securely over the broad mouth of the funnel, it being necessary always to support filter paper between layers of calico.

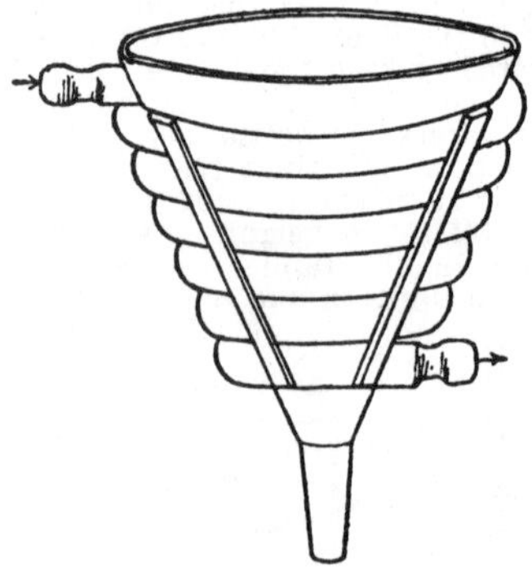

Fig. 12

A Device for Rapid Filtration.

Glass filter rods with a hooked end set over the edge of the ordinary funnel, form a corrugated support for filter paper, which is unaffected by liquids likely to

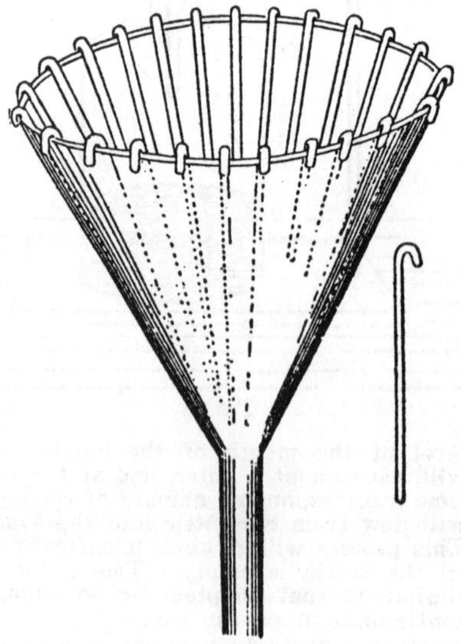

Glass Filter Rack

be filtered through the glass funnel, and can be effectually cleaned with a minimum of labor.

Percolation.

This is a kind of filtration, commonly called "by displacement," employed for extracting the essence from roots, herbs, seeds, barks, etc. It is effected in the following manner: It is first necessary that the articles to be acted upon should be ground in a drug mill to the condition of a coarse powder; then moisten the mass thoroughly with alcohol, allowing it to "macerate" for 12 hours in a vessel well covered. Next is required a hollow instrument of cylindrical form, having one end shaped like a funnel, so that it can be inserted in the neck of a glass bottle, and having inside, near the lower end, a partition pierced with numerous small holes, like the strainer of a French coffee pot, which is a simple coffee percolator; in the absence of such a partition, soft cotton, or any insoluble substance, may be substituted, and being placed in the inside at the lower end of the instrument, will answer as well as the strainer. This instrument is called a percolator. Boullay's filter or percolator is usually employed. Macerate the ingredients to be acted upon, for the time named, introduce them into the percolator, and slightly press them upon the partition. Any portion of the liquid used in the maceration not absorbed by the powder should be poured upon the mass in the instrument, and allowed to percolate. Now gradually pour into the percolator sufficient of the alcohol, or other liquid to be filtered, to drive before it, or "displace," the liquid contained in the mass; the portion introduced must, in like manner, be "displaced" by another portion, and so on till the required quantity of filtered liquor is obtained. This extract is called a tincture. In case the liquor which first passes through should be thick and turbid, again introduce it into the instrument, being very careful not to have the powder too coarse or loosely pressed, or it will permit the liquid to pass too quickly; and, on the other hand, it should not be too fine or compact, or it may offer an unnecessary resistance. Should the liquor flow too rapidly, return it to the instrument, and close it beneath for a time, and thus permit the finer parts of the powder to subside, and cause a slower percolation.

The first portion of liquid obtained by the method of displacement is always in a state of high concentration. In general, it is a simple solution of the soluble ingredients of the crude drug in the fluid employed. But sometimes the solvent, if compound, is resolved into its compound parts, and the fluid which passes through it at any given time is only one of these, holding in solution only the most soluble parts of the drug.

Thus, if diluted alcohol be poured over the powder of myrrh, in the cylinder of the percolator, the fluid which first drops into the receiver is a solution of an oily consistency, chiefly composed of rosin and volatile oil dissolved in alcohol. In like manner, when the powder of gallnuts is treated in the same way by hydrated sulphuric ether, two layers of fluid are obtained, one of which is a highly concentrated solution of tannin in the water of the ether, and the other a weak solution of the same principle in pure ether. In all cases, therefore, in which it is not otherwise directed, it is absolutely necessary to agitate the several portions of the liquid obtained by percolation together, in order to insure a product of uniform strength or activity.

To illustrate the operation of displacement, and describe an excellent percolator for making perfume tinctures, we will suppose that benzoin is under treatment. The apparatus, made wholly of glass, having been arranged, as shown, and a plug

Percolator for Perfume

of raw cotton dropped loosely at a, the benzoin, in coarse powder, is then poured into the portion, b, until it reaches the line, c. Alcohol, 95%, is next added until it rises to the line, d. As soon as the first portion sinks into the benzoin a fresh addition must be made; and thus the succeeding relays go on displacing those which preceded them without mingling with them. Each stratum becomes more and more charged with soluble matter as it descends; and when it reaches the bottom of the mass, under the pressure of the superincumbent liquor, it runs out saturated. When, by successive additions of fresh alcohol, the benzoin under treatment has become exhausted, the liquid passes through the mass and falls into the receiver, e, as tasteless and colorless as when first poured in. This indicates the completion of the process.

As atmospheric pressure is an important element in the operation, it will not answer to shut it off by closing the top of the displacer without making some compensation; and, therefore, a communication between the upper and lower vessels is established by means of a latent tube arrangement, f. In this manner the apparatus is kept close, and the evaporation of alcohol prevented, while the pressure produced is distributed throughout the apparatus, and rendered uniform. As the runnings are clear, filtration is rarely necessary. The quantity of alcohol thus consumed need not be more than sufficient to exhaust the material; and the resulting tincture must therefore be diluted to the proper strength. For perfumes, deodorized alcohol must always be used.

The method of displacement has the advantage of expedition, economy, and yielding products possessing uniformity of strength, but it requires considerable experience to adapt it to all substances. The art rests in properly packing the ingredients in the cylinder, some substances requiring considerable pressure to be used, while others, when even lightly packed, scarcely permit the fluid to pass through them. An excellent plan, applicable to all substances, but especially those of a glutinous or mucilaginous nature, is to mix the powder with an equal bulk of well washed sand before rubbing it up with the menstruum. The coarseness of the powder must also be attended to. Substances that readily become soft and pappy when wetted by the menstruum should not be used so fine as those that are more woody and fibrous. The method of displacement answers well for the preparation of all tinctures that are not of a resinous nature, and for most infusions of woody and fibrous substances, as roots, woods, barks, leaves, seeds, insects, etc. It is especially adapted for the preparation of concentrated infusions and essences, as they may thus be obtained of any required strength, without loss, or requiring concentration by heat, which is so destructive to their virtues.

When ordinary tinctures are made in large quantities, displacement is never likely to supersede maceration on account of any practical advantages it may possess. If the prescribed directions be duly attended to, the process of maceration is unexceptionable. The process is more simple than the other; the mode of operation more uniform; it is, in fact, always the same; it requires less of skill and dexterity in conducting it; it requires less constant attention during its progress, which, in operating on large quantities, is a consideration; and finally, the apparatus required is less complicated. When, however, only small quantities are to be made at a time, and kept in stock, the adoption of the process of displacement will often be found convenient and advantageous. It offers the means of making a tincture in two or three hours, which, by the other process, would require as many weeks.

Dialysis.

This is a process of separating substances which do not crystallize from those which do, by means of a porous diaphragm which sets in water. The apparatus which is used is called a dialyzer, which consists of a cylinder over whose bottom is secured a sheet of parchment paper. This sets in a dish of water. The liquid which is to be treated is placed in the upper dish, and the whole is put away for a time, when the separation will be found complete. This process is more useful in pharmacy than in the arts.

Crystallization.

When a body, in the act of passing from a liquid or gaseous to a solid state, arranges itself in symmetrical forms, the process is termed crystallization, and the parts of the body so aggregated are called crystals.

By this process we can separate crystallizable from amorphous substances dissolved in the same menstrua; purify crystals from foreign and coloring matters, and in qualitative examinations be enabled to determine the composition of bodies by a reference to the characteristics of figure.

The modes of crystallization are by *fusion, sublimation, solution* and *chemical reaction.*

Crystallization by Fusion.—Sulphur, lead, bismuth, tin, antimony, silver, numerous alloys, anhydrous salts, and other fusible substances which are unalterable by heat, are crystallizable by *fusion.* To this end they are melted at the lowest possible temperature, and allowed to cool very gradually. As soon as a crust forms upon the top, which may be readily seen by the surface becoming furrowed, it must be pierced with a rod, and the still fluid portion decanted with sufficient dexterity to prevent it from cooling during the process, and at the same time from injuring the crystals coating the interior of the vessel. The liquid matter should be placed so as to be free from all vibration. The greater the mass of the material, and the more slowly it is cooled, the more voluminous and better defined will be the crystallization.

Crystallization by Sublimation.—Volatile solids, as iodine, camphor, several metallic chlorides and mercurial compounds, arsenic, benzoic acid, iodide of lead, etc., when heated as directed in *sublimation,* yield vapors which, in cooling, take the form of crystals.

Crystallization from Solution.—When it is desired to obtain a substance in crystals it must first be liquefied, or made into a *solution* with an appropriate liquid. If, after making the solution, there be any insoluble residue, it must be separated by *filtration;* and subsequently, if the solution is capable of decolorization by such means, it should be boiled with a small portion of clean bone or ivory black, and again filtered. As it is the almost universal law that heat increases the solvent power of bodies, the solution should generally be made and clarified at the boiling point, so that the excess of matter taken up at the high temperature may separate, on cooling, in the form of crystals. So long as a solution is dilute it yields no crystals; these latter are only formed when the containing liquid is supersaturated; or, in other words, holds more than it can retain; and consequently, in diminishing the quantity of the liquid by *evaporation,* we increase the density of that which remains, and hence, upon cooling, it deposits that excess of the dissolved substance which it only held by virtue of its high temperature. Some instances are so easily soluble, and to such an unlimited extent, that their solutions form crystals immediately upon cooling; others, again, are taken up with such difficulty, even at high heats, unless in large bulks of liquid, that although exposed to prolonged ebullition they require to be evaporated in order to separate what has been dissolved. As the mode of evaporating has an important influence upon the form and size of crystals, we give some hints as to the proper manner of performing it.

If large and well defined crystals are required, the solution should be subjected to spontaneous evaporation, for the more slow and uniform the concentration the more regular and gradual will be the superposition of material required to make distinct and large crystals. A slight addition of solution of gelatine will, in some instances, it is said, give the crystals the form of plates, as in the case of boracic acid. The solution should be removed from the fire as soon as drops, withdrawn by a glass rod, and deposited upon a watch glass or clean spatula, give small crystals upon cooling. If, however, a very dense crystallization is required, the concentration may be continued until a pellicle forms upon the top, but then the solidified masses are confused and less brilliant. These essays indicate that the liquid is evaporated to a point at which it cannot retain all of its soluble matter. The vessels are then placed aside to cool gradually and uniformly, that the excess may crystallize out of the liquid. The temperature should be regular, for slight variations may alter the form of the crystals.

Bodies equally soluble in cold and hot water, as well as those which are deliquescent, require a prolonged evaporation, as they only crystallize from very dense solutions.

When the liquid is to be converted *wholly* into solid, then the process is termed *granulation,* and is practiced by concentrating it to a syrupy consistency, removing the vessel from the fire and stirring its contents *constantly* until the mass has cooled into granules. This mode is adapted for purifying pearlash and converting it into *sal tartar*, and also for graining brown sugars.

Emulsions and Emulsifying.

To emulsify an oil consists in rendering it capable of mixing with water to form a uniform milky fluid, by the aid of an intervening medium, generally saccharine or mucilaginous.

Milk being the most perfect emulsion obtainable, the mixture of fat which stimulates this compound most closely must likewise be regarded as superior in the degree that these qualities are intensified. To be sure, an artificial emulsion always represents a greater percentage of fat than milk, and its preservation is, therefore, relatively easier than in that obtained from nature; but this fact merely modifies the result, and does not involve the principle. The greater proportion of water in milk also favors decomposition, but on the other hand, the minute, perhaps even molecular, division of the fat globules renders it possible to withstand decomposition longer than an equally dilute artificial emulsion, wherein the oil globules are not so thoroughly disseminated.

We, of course, recognize the fact that milk contains different animal bodies not present in ordinary artificial emulsions, which are prone to decomposition, so that the similarity drawn between the two is based more upon physical characteristics than their presenting any features in common chemically.

But it is this attempt at compromising its principal physical feature—fluidity—with permanency, which makes the preparation of an emulsion so difficult. To so change a fat as to render it miscible with water is a matter of easy execution, but when we attempt to embody the desirable feature of fluidity then we are thwarted by physical laws, and resort to chemical means as a compromise.

Condensed milk is a striking illustration wherein by a change of its physical condition. complete preservation has been attained much more satisfactorily than milk in its natural form could be preserved, even with chemical means. It is for this reason that *consistency* is the most desirable feature to insure the permanence and preservation of any emulsion, natural or artificial.

It is well known that a perfect and permanent emulsion can be made with cod-liver oil and malt extract, owing to the consistency of the preparation solely, as we have attempted to use the same agents represented in malt extract, namely, dextrine and glucose, and discovered that as soon as the consistency was abandoned these agents did not possess any advantage over those usually employed for emulsifying fats. To the albumen in milk has been ascribed the high degree of and most permanent emulsification, and therefore gelatine is employed in artificial emulsions, with not much better success, however, than other agents, when semi-fluid consistency is abandoned.

We will now consider what should be used as emulsifying agents, and also such as, while largely used, are not desirable, for obvious reasons.

Unfortunately, the well-worn maxim, so justly applied to most classes of pharmaceutical preparations, "The sacrifice of medicinal value for elegance," has not been lost sight of in the preparation of emulsions. Periodically, different substances from all the different kingdoms of nature have been proposed, enjoyed a short, fashionable stay, and then been relegated to their well merited oblivion.

The vegetable gums, acacia and tragacanth, have been the longest in use, and the first mentioned of these has probably answered the purpose of a reliable, convenient, and at least innocuous emulsifying agent better than the majority of latter-day substitutes.

The late Prof. Wm. Procter announced the proportion to be used of gum acacia to produce a perfect temporary emulsion. His directions were as follows: "Mix

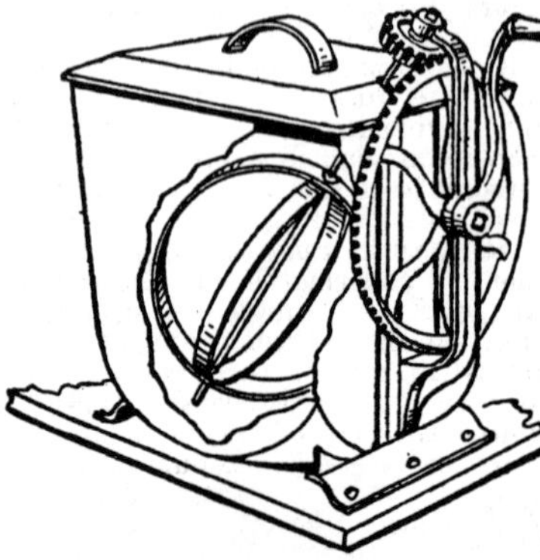

Emulsifier

intimately, in a perfectly dry mortar, the oil with one-half its weight of powdered acacia; to this add at once one-half as much water as the combined weight of oil and gum, and triturate briskly until the mixture has assumed the color and consistency of a thick cream, which produces a crackling noise when the pestle is moved rapidly around the sides of the mortar." This is the emulsion proper, and to this can be added any amount more of water or other desirable vehicle or medicament to bring the finished preparation up to the quantity prescribed.

If perfectly made, this emulsion will stand any degree of dilution with watery mixtures; in fact, its quality is proved when, by a large addition of water, the oil globules will not separate or aggregate at the top of the liquid.

Practice has demonstrated that the proportion of gum can be varied according to the nature of the oil employed, but the constant relation between the water used for the emulsion proper, and the mixture of oil and gum, must be scrupulously adhered to as insuring infallible results.

Fixed oils rich in gum, *per se*, as copaiba, castor oil, etc., do not require as large an amount of gum as cod-liver oil, while in the case of ethereal oils, for instance, oil of turpentine, an equal amount of gum, or weight for weight, is necessary. To prepare an emulsion from turpentine not unfrequently presents difficulties, and so much the more is this to be guarded against, as it is a powerful remedy, and if presented in a merely mechanical mixture will prove irritating, and perhaps engender serious consequences.

But then, if by careful observance of this method we can obtain a perfect emulsion, what more is desired? Although this emulsion is perfect, it is not permanent, and to circumvent this negative feature is the problem for solution.

While we have not discovered any means or process whereby this problem can be solved, yet we have found agents capable of preventing this separation in a great degree, being guided in their selection by a knowledge of the constituents which are most favorable to this separation and those that are not.

An emulsion should be palpable, and for this reason it is always sought to make it sweet by the introduction of cane sugar or glycerine. These two agents are the cause of the most dissatisfaction with emulsions. Sugar, owing to its affinity for water, and density, favors separation very rapidly, precipitating while the emulsified oil forms a compact, creamy and gradually diminishing stratum at the top of the vessel. Glycerine, probably from the same causes, and its incompatibility with fixed oils, behaves in a similar manner, and for these reasons these otherwise desirable vehicles cannot be represented in an emulsion when permanence is to be obtained.

As no other agents present themselves for fulfilling the sweet object in view, we have been in the habit of preparing emulsions without attempting to make them sweet, and, we believe, without detracting from their palatability, while enhancing their appearance.

Now, then, let us consider what agent will favor the homogeneity of the emulsion; that is, prevent separation or precipitation, bearing in mind that the preparation must not be changed physically or chemically.

Gelatine has been used with some satisfaction, as it retards the separation for a considerable length of time; in fact, it answers the purpose so well that for the extemporaneous preparing of emulsions it leaves nothing to be desired. But in common with other agents used for this purpose, it gradually loses its power of preserving the homogeneity of an emulsion, and eventually the separation and decomposition, so called, alluded to above, take place.

The proportion of gelatine employed is about 40 gr. to 1 pt. of the emulsion; it should be dissolved in the water, and added at any time of the operation. By increasing this amount so that a jelly is formed of the emulsion, a perfectly permanent and stable preparation is obtained. But this result is obtained because the physical character of the emulsion has been changed—fluidity abandoned for consistency. Unhappily, we cannot take advantage of this condition, and therefore "consistency is not a jewel" pharmaceutically.

Chemical agents such as change the character of an emulsion by saponifying the oil, have been largely advocated, and to the employment of this class of substances is principally due the elegance and permanence of ready-made emulsions. That this is attained at the sacrifice of medicinal value of the preparation we have no doubt, but medical authorities have also demonstrated it to be a questionable procedure to chemically change the constitution of a fat intended for internal administration by what should be a simple pharmaceutical process—emulsification—and now condemn the use of alkalies with balsams and rosins. Copaiba is no more exhibited with solution of potash, and alkalies are generally conceded as operating to break up the sensitive electronegative principles of rosins, upon which their medicinal value chiefly depends. Animal fat, and especially cod-liver oil, when rendered alkaline, undoubtedly suffers decomposition in those very constituents to which its superior digestibility is due, and thus what has been gained on one hand is more than lost on the other. The saponification which has been produced by the use of the alkali renders the preparation very prone to rancidity if exposed to the air, and even when freshly made it possesses inferior palatability; but then this has been of secondary importance to homogeneity or elegant appearance.

V
IGNITION

Substances frequently require to be ignited to redness, either as the sole process of their preparation, or as a preliminary step to subsequent operations.

Ignition of Filters.

In analyses, the filters containing the insoluble or precipitated substances which are to be estimated are ignited or "burned off," to expel carbonaceous and volatile matters, before being weighed. The im-

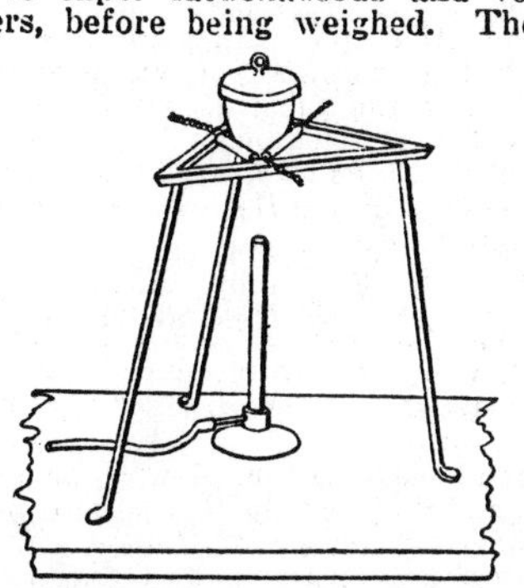

Heating Porcelain Crucible

plements for this purpose are porcelain or platinum crucibles, either having their appropriate application.

As it is necessary that the filter should be wholly or partially dry, it must be carefully removed from the funnel, so as not to lose a particle of its contents, compressed between the folds of bibulous paper, and, further, dried in a capsule over a sand or water bath, or in a drying stove (desiccation), at a temperature of about 200° F., or less. The dried filter is then to be transferred to the crucible, which has been previously weighed. The transfer must be made without the loss of the least particle, and for this purpose the crucible may be placed upon a sheet of glazed white paper, so that any particles that accidentally fall may be preserved. The filter should be placed in the crucible with its apex upwards, after having been freed as much as possible from the adherent precipitate by gently rubbing the sides together between the thumb and forefinger. The force used for this purpose must not be sufficient to abrade the paper, otherwise the matter will reach the fingers, and a loss thus be occasioned by adherence.

When substances are to be ignited for the determination of their hygroscopic, volatile, or organic matter, the heat of the lamp should be gradually applied without the blast, and, for the former purpose, only to the production of a dull red heat. In these instances, the crucible should be weighed first, so that the loss sustained by a given weight of its contents during ignition, may be ascertained in one weighing merely by subtracting the weight of the crucible and contents after ignition from the combined weight of the two before the same process. The loss gives the amount of the volatile matter.

In analyses of coals, the moisture can be determined by heating the crucible in a hot sand bath, or very gently over a low flame. After the loss thus occasioned is determined by weighing, the amount of carbon may be ascertained by subjecting the crucible and contents to a much higher heat.

When the substances are to be exposed to heat, the crucible and contents must

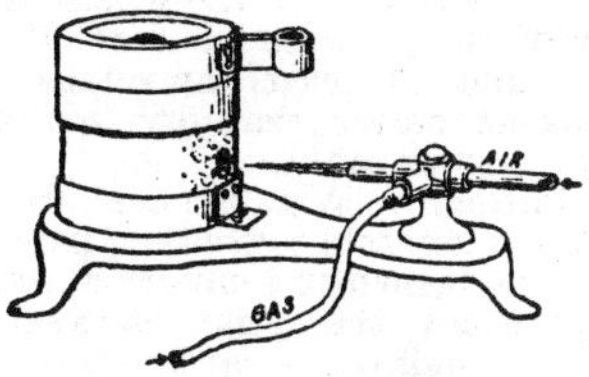

Gas Crucible Furnace with Air Blast.

likewise be weighed separately before ignition. The loss of weight gives the amount of volatile matter driven off. The ignited matter can then be removed from the crucible by hot water alone or acidulated.

Scoriae may be removed from platinum crucibles by covering them with a paste of borax and carbonate of soda, heating them to redness, and when cold, dissolving out the saline matter with boiling water. A repetition of the process is necessary to brighten the crucible perfectly if it had been very dirty. One of our engravings represets an assaying plant of gas furnaces as arranged by Walter Lu Brour.

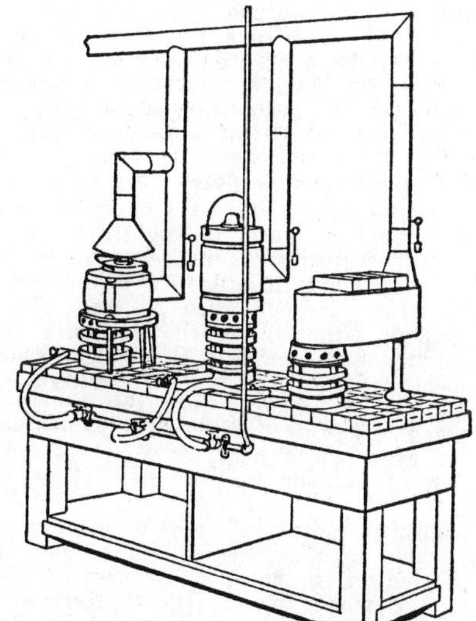

Assayer's Plant of Gas Furnaces.

The furnace to the right is for roasting, the middle is for crucible fusions, and to the left is one for scorification and cupellation.

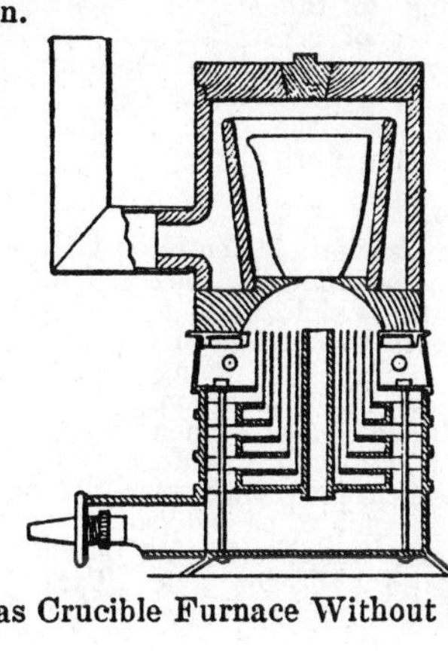

Gas Crucible Furnace Without Blast.

Fusion.

Fusion is a process of liquefying solid bodies by heat without a solvent, such as wax melting. Gas melting arrangements as shown are recommended. With this apparatus a sound 2-oz. ingot of gold or silver can be molded in 2 min. A crucible of molded carbon is supported by a sheet-iron slide or plate which is clamped to an ingot mold by a clamp which swivels in the U-shaped cast-iron stand. The metal to be melted is placed in the crucible, and the flame of the blowpipe directed on it until it is perfectly fused. The whole is then tilted over by means of the upright handle at the back of the mold. The waste heat serves to make the ingot mould hot. No flux should be used with the carbon crucibles.

The plate mold will cast an ingot 1⅜ x 1⅜ x 3-16 in. thick; wire mold, 3-16 x 3-16 x 2¾ in. long.

For melting up to 2 oz. of gold or silver rapidly, without the use of a furnace. In this arrangement the two parts of the ingot mold slide on each other, to enable ingots of any width to be cast, and the blowpipe is part of the rocking stand.

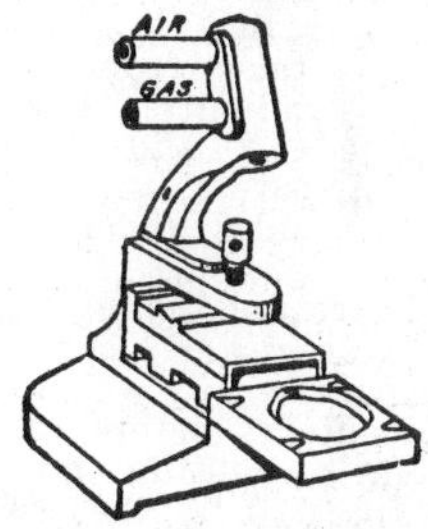

Ingot Casting Arrangement.

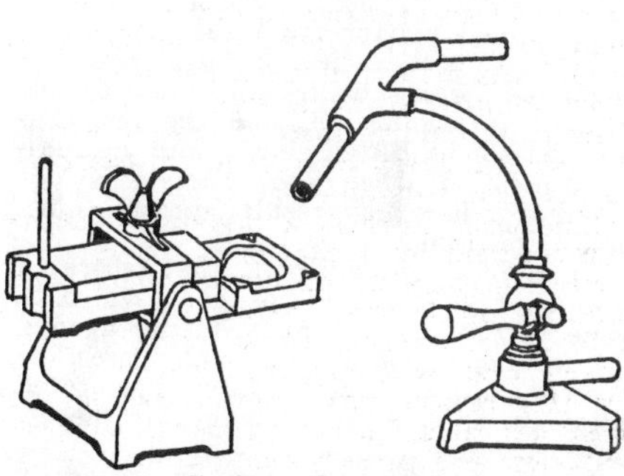

Carbon Crucible.

When the metal is melted in the shallow crucible of molded carbon, till the whole apparatus over so as to fill the ingot mold.

Calcination.

The separation (in a dry way) of volatile from fixed matter, by heat, is termed calcination. The process is applicable:

To the expulsion of water from salts, minerals, coals and other substances.

To the expulsion of carbonic acid from certain carbonates.

To the expulsion of arsenic and sulphur from cobalt, nickel and other sulphuretted compounds.

To the expulsion of bituminous matter from coals, and certain minerals and ores.

To the ignition of quartz and silicious minerals to promote their disintegration.

For the purpose of expelling the combined water of argillaceous minerals, and of thus rendering them more obstinate to the solvent action of acids and reagents.

If the substance under process is organic, its calcination in a close vessel by a medium heat usually effects only partial decomposition, the gaseous matter generated escaping through interstices and the fixed components remaining with a portion of unaltered carbon. Performed in this manner, the process takes the name of coking, familiar instances of which are the formation of coke by distilling coal in closed retorts, the manufacture of charcoal from wood, and of bone black from bones.

By increasing the temperature and admitting the air, the whole of the alterable and volatile matter is expelled, the fixed matter remaining as ashes. The process is then styled incineration, and in this way the coke, charcoal and ivory black, obtained as above directed, may be entirely reduced to their incombustible portions or ashes.

Calcination is effected in platinum spoons or crucibles, in delicate experiments, over a spirit lamp; but in large operations a furnace is required, and the containing vessels are crucibles of either metal or earthenware, according to the nature of the substance to be heated, though the latter are often unsuitable for temperatures above a red heat.

When the operation is finished, the crucible should be taken from the fire and allowed to cool gradually. The cover is then to be lifted off and the contents taken out with a spatula, and the portions adhering to the sides removed with a feather.

If the substance undergoing calcination is fusible, it is necessary when quantities are to be ascertained, to weigh both the crucible and contents before ignition, so that the amount of volatile matter driven off may be expressed by the weight lost in heating. Water alone or acidulated, with the aid of heat, generally removes the calcined matter from the crucible.

A body decripitating by heat should be powdered before being subjected to the process of calcination, and the temperature should be raised slowly and gradually, otherwise when the crucible is not covered, a loss may result from the ejection of particles.

To avoid contact with the generated vapors or with the atmosphere, which to some substances act as reducing agents, the crucible should in such cases be covered, and if tightly luted perforated with one or more small holes for the escape of vapor.

Roasting (as the term is generally used) is a kind of calcination to which many ores are submitted before their final reduction to the metallic state, for the purpose of expelling ingredients which would either delay that process or be injurious to the metal when extracted. In this way water, carbonic acid, sulphur, selenium, arsenic, and sometimes other substances, are driven off from the ores containing them. The term is also applied to other processes, among the most important of which is that of the exposure to heat and air by which metals become altered in composition. Thus, copper becomes oxidized, and antimony and arsenic acidified by union with oxygen.

Roasting is always effected in broad, shallow open vessels, so that the air may have free access; and in order to promote the absorption of oxygen or the escape of the volatile substances, the surface of the body to be heated should be increased by previous pulverization, and it should be constantly stirred during the operation so as to present as many points of contact as possible. The most suitable vessel is a baked earthenware saucer or capsule placed in a muffle or upon the bars of a calcining furnace. Sometimes a crucible is used, and then the position of the vessel in the furnace should be slightly inclined on one side. In either case the vessels should be heated to dull redness previous to receiving their charge.

Deflagration.

That species of roasting termed deflagration is effected by rapidly heating the substance to be oxidized, together with some additional body as an oxidizing agent, as a nitrate or chlorate for instance. The powdered mixture is added portionwise to the crucible previously heated, and maintained at redness during the operation. The vivid and sudden combustion which ensues modifies the composition of the original substance and increases its amount of oxygen at the expense of the addendum. Thus, for instance, sulphuret of arsenic is deflagrated with niter to produce arseniate of potassa, titanium and certain other metals to be transformed into oxides.

Deflagration is also used as a means of detecting the presence of nitric or chloric acids. For this purpose the suspected substance is to be heated with cyanide of potassium, in a small platinum spoon. If deflagration ensues it is a test of the presence of one of them, or a compound of one of them.

The crucibles may be of clay or metal, according to the nature of the substances to be heated. The roasting of substances for the expulsion of organic matter may be effected in platinum vessels, provided the heat is not caried sufficiently high to produce fusion of the substance being roasted.

The heat must, at first, be very gradually applied, and at no time be made great enough to fuse or agglutinate the material, otherwise the process will have to be suspended in order to repulverize the matter. Proper care at the commencement will obviate the necessity of this additional trouble. When the heat has been cautiously raised to redness and all liability of fusion is over, the fire may be urged to the production of a yeilowish red or even white heat, so that the expulsion of volatile matter may be complete.

Roasting operations which disengage deleterious or disagreeable fumes should be carried on in the open air or under a hood, and when the volatile matters are valuable they may be condensed as directed in *Distillation* and *Sublimation*.

Decrepitation.

This frequently occurs and occasions loss by ejections of particles of the mixture, owing to the sudden vaporization of the water of crystallization, which in finding vent scatters the confining substances with a crackling noise. To prevent this loss, the crucible should be loosely covered until decrepitation ceases.

Reduction.

This operation is employed for the separation of metallic bases from any bodies with which they are combined; but is generaly confined to the extraction from an oxide—that being the kind of combination most commonly met with. The combined action of heat and certain reagents is required to effect this result, the temperature varying with the nature of the substance to be reduced.

The most usual reducing agents are charcoal and hydrogen gas. Tallow, oil and rosin are sometimes used, but being easily decomposed they are dissipated before entire reduction has occurred. Sugar and starch are also occasionally employed. We shall, however, confine our remarks to the two principal articles.

Reduction by Charcoal.

Charcoal is used for this purpose in two ways, either in powder and directly mixed with the substance, or as a lining coat to the crucible in which the reduction is accomplished. The first mode is objectionable, because the excess of coal which is required to be used interferes with the agglomeration of the particles of reduced metal. Whenever it is adopted, the quantity of coal dust to be added, which must be sufficient to transform all the oxygen of the oxide into carbonic acid, can be determined by calculation. This amount is then mixed thoroughly with the oxide previously powdered, and is transferred to a crucible, taking care to place the charge in the center and to cover the contents with a layer of the dust. The whole is then to be subjected to the heat of a furnace, assisted if necessary by a blast. The reduction in this way, the most convenient for large quantities, is rapid and complete, but the metallic residue is often mixed with coal dust.

Incineration.

This is a process of heating organic substances with air until all the carbon is consumed, the product sought being the ash.

Carbonization.

This is a process calling for the heating of organic substances without exposure to the air until all the volatile products are given off and the residue remains as a kind of charcoal. Bone black is a good example.

Sublimation.

When simple compound bodies which are either wholly or in part capable of assuming the aeriform state are subjected to heat, they or their most volatile constituents, upon reaching the required temperature, rise in the form of vapor. If these vapors, in their transit, are intercepted by a surface of a lower temperature, they condense and take a solid or liquid form, according to their nature. If the product is a solid, it is termed *sublimate*, and the process by which it is obtained is *sublimation*. If it is liquid or gas, it takes the name of *distillate*, and the operation which yields it that of *distillation*.

Both of these processes are indispensably useful in chemistry, for they afford the facility of taking advantage of the unequal volatility of bodies for their separation.

As instances of sublimation, we have calomel and corrosive sublimate made by heating equivalent proportions of sulphate of mercury and comon salt; benzoic acid evolved from the gum; pure indigo from the commercial article, and camphor from the crude material. Iodine is sublimed to free it from impurities; biniodide of mercury to convert it into crystals; naphthaline to free it from empyreumatic matter, and succinic acid to separate water.

Specific Gravity.

The specific weight of a substance is its weight in comparison with weights of similar bulks of other substances. This comparative heaviness of solids and liquids is conventionally expressed in relation to water; they are considered as being lighter or heavier than water. Thus, water being regarded as unity = 1, the relative weight, or specific weight, of ether is represented by the figures .720 (it is nearly three-fourths, .750, the weight of water), oil of vitriol by 1.843 (it is nearly twice, 2.000, as heavy as water). The specific weight of substances is, moreover, by generally accepted agreement, the weight of similar volumes at 15° C. (59° F.), except in the case of alcohol and wine, which are at present taken at 15.6° C. (60° F.), to maintain consistency with the United States laws and regulations; for the weight of a definite volume of any substance will vary according to temperature, becoming heavier when cooled and lighter when heated, different bodies (gases excepted) differing in their rate of contraction and expansion. While, then, specific weight—or, conventionally, specific gravity—is truly the comparative weight of equal bulks, the numbers which in America commonly represent specific gravities are the comparative weights of equal bulks at 15° (59° F.), water being taken as unity.

The true weight of the body is its weight in air plus the weight of an equal bulk of air, and minus the weight of a bulk of air equal to the bulk of brass or other weights employed; or, in other words, its weight *in vacuo* uninfluenced by the buoyancy of the air; but such a correction of the weight of a body is seldom necessary, or, indeed, desirable. Density is sometimes improperly regarded as synonymous with special gravity. It is true that the density of a body is in exact proportion to its specific gravity, but the former is more correctly the comparative bulk of equal weights, while specific gravity is the comparative weight of equal bulks.

The standard of comparison for gases was formerly air, but is now usually hydrogen.

Specific Gravity of Solids Lighter than Water.—This is obtained in a manner similar to that for solids heavier than water; but the light body is sunk by help of a piece of heavy metal, the bulk of the water which the latter displaces being deducted from the bulk displaced by both; the remainder is the weight of a bulk of water equal to the bulk of the light body. For instance, a piece of wood weighing 12 grams (or grains) is tied to a piece of metal weighing 22 grams, the loss of weight of the metal in water having been previously found to be 3 grams. The two, weighing 34 grams, are now immersed, and the loss in weight found to be 26 grams. But of this loss 3 grams have been proved to be due to the buoyant action of the water on the lead; the remaining 23, therefore, represent the same effect on the wood; 23 and 12, therefore, represent the weights of equal bulks of water and wood. As 23 are to 12, so is 1 to .5217. Or, shortly, as before, divide the weight in air by the weight of an equal bulk of water; .5217 is the specific gravity of the wood. Another specimen of wood may be found to be three-fourths (.750) the weight of water, and others heavier. Cork varies from .100 to .300.

The specific gravity of a very minute quantity of a heavy or light substance may be ascertained by noting the specific gravitly of a fluid in which it, being insoluble, neither sinks nor swims, or by immersing it in a weighed piece of paraffine whose specific gravity is known, noting the specific gravity of the whole, and deducting the influence of the paraffine.

Specific Gravity of Solids in Powder or Small Fragments.—Weigh the particles; place them in a counterpoised specific-gravity bottle of known capacity, and fill up wtih water, taking care that the substance is thoroughly wetted; again weigh. From the combined weights of water and substance subtract amount due to the substance; the residue is the weight of water. Subtract this weight of water from the quantity which the bottle normally contains; the residue is the amount of water displaced by the substance. Having thus obtained the weights of equal bulks of water and substance, a rule-of-three sum shows the relation of the weight of the substance to 1 part of water—the specific gravity.

Or suspend a cup, a short tube, or bucket from a shortened balance-pan; immerse in water; counterpoise; place the weighed powder in the cup, and proceed as directed for taking the specific gravity of a solid in a mass.

Specific Gravity of Solids Soluble in Water.—Weigh a piece of sugar, or other substance soluble in water; suspend it from a balance in the usual manner, and weigh it in turpentine, benzol or petroleum, the specific gravity of which is known or has been previously determined; the loss in weight is the weight of an equal bulk of the turpentine. Ascertain the weight of an equal bulk of water by calculation:

As is the specific gravity of turpentine to the specific gravity of water, so is the observed bulk of turpentine to an equal bulk of water.

The exact weights of equal bulks of

SPECIFIC GRAVITY.

Tables showing a comparison of the degrees of Baumé, Cartier, and Beck's Areometers, with specific gravity degrees.

For Liquids Lighter than Water.

Degrees of Baumé, Cartier, Beck.	Baumé.	Cartier.	Beck.
	Sp Gr.	Sp. Gr.	Sp. Gr.
0	....	....	1.0000
1	....	....	0.9941
2	....	....	0.9883
3	....	....	0.9826
4	....	....	0.9770
5	....	....	0.9714
6	....	....	0.9659
7	....	....	0.9604
8	....	....	0.9550
9	....	....	0.9497
10	1.000	....	0.9444
11	0.993	1.000	0.9392
12	0.986	0.992	0.9340
13	0.979	0.985	0.9289
14	0.973	0.977	0.9239
15	0.967	0.969	0.9189
16	0.960	0.962	0.9139
17	0.954	0.955	0.9090
18	0.948	0.948	0.9042
19	0.942	0.941	0.8994
20	0.935	0.934	0.8947
21	0.929	0.927	0.8900
22	0.924	0.920	0.8854
23	0.918	0.914	0.8808
24	0.912	0.908	0.8762
25	0.906	0.901	0.8717
26	0.901	0.895	0.8673
27	0.895	0.889	0.8629
28	0.889	0.883	0.8585
29	0.884	0.877	0.8542
30	0.879	0.871	0.8500
31	0.873	0.865	0.8457
32	0.868	0.859	0.8415
33	0.863	0.853	0.8374
34	0.858	0.848	0.8333
35	0.853	0.842	0.8292
36	0.848	0.837	0.8252
37	0.843	0.831	0.8212
38	0.838	0.826	0.8173
39	0.833	0.820	0.8133
40	0.829	0.815	0.8095
41	0.824	0.810	0.8061
42	0.819	0.805	0.8018
43	0.815	0.800	0.7981
44	0.810	....	0.7944
45	0.806	....	0.7907
46	0.801	....	0.7871
47	0.797	....	0.7834
48	0.792	....	0.7799
49	0.788	....	0.7763
50	0.784	....	0.7727
51	0.781	....	0.7692
52	0.776	....	0.7658
53	0.771	....	0.7623
54	0.769	....	0.7589
55	0.763	....	0.7556
56	0.759	....	0.7522
57	0.755	....	0.7489
58	0.751	....	0.7456
59	0.748	....	0.7423
60	0.744	....	0.7391
61	0.740	....	0.7359
62	0.736	....	0.7328

For Liquids Heavier than Water.

Degrees of Baumé, Beck.	Baumé.	Beck.
	Sp. Gr.	Sp. Gr.
0	1.000	1.0000
1	1.007	1.0059
2	1.014	1.0119
3	1.020	1.0180
4	1.028	1.0241
5	1.034	1.0303
6	1.041	1.0366
7	1.049	1.0429
8	1.057	1.0494
9	1.064	1.0559
10	1.072	1.0625
11	1.080	1.0692
12	1.088	1.0759
13	1.096	1.0828
14	1.104	1.0897
15	1.113	1.0968
16	1.121	1.1039
17	1.130	1.1111
18	1.138	1.1184
19	1.147	1.1258
20	1.157	1.1333
21	1.166	1.1409
22	1.176	1.1486
23	1.185	1.1565
24	1.195	1.1644
25	1.205	1.1724
26	1.215	1.1806
27	1.225	1.1888
28	1.235	1.1972
29	1.245	1.2057
30	1.256	1.2143
31	1.267	1.2230
32	1.278	1.2319
33	1.289	1.2409
34	1.300	1.2500
35	1.312	1.2593
36	1.324	1.2680
37	1.337	1.2782
38	1.349	1.2879
39	1.361	1.2977
40	1.375	1.3077
41	1.388	1.3178
42	1.401	1.3281
43	1.414	1.3386
44	1.428	1.3492
45	1.442	1.3600
46	1.456	1.3710
47	1.470	1.3821
48	1.485	1.3934
49	1.500	1.4050
50	1.515	1.4167
51	1.531	1.4286
52	1.546	1.4407
53	1.562	1.4530
54	1.578	1.4655
55	1.596	1.4783
56	1.615	1.4912
57	1.634	1.5044
58	1.653	1.5179
59	1.671	1.5315
60	1.690	1.5454
61	1.709	1.5596
62	1.729	1.5741
63	1.750	1.5888
64	1.771	1.6038

sugar and water being obtained, the weight of a bulk of sugar corresponding to 1.000 of water is shown by a rule-of-three sum; in other words, divide the weight of sugar by that of the equal bulk of water; the quotient is the specific gravity of sugar. The stated specific gravity of the sugar ranges from 1.590 to 1.607.

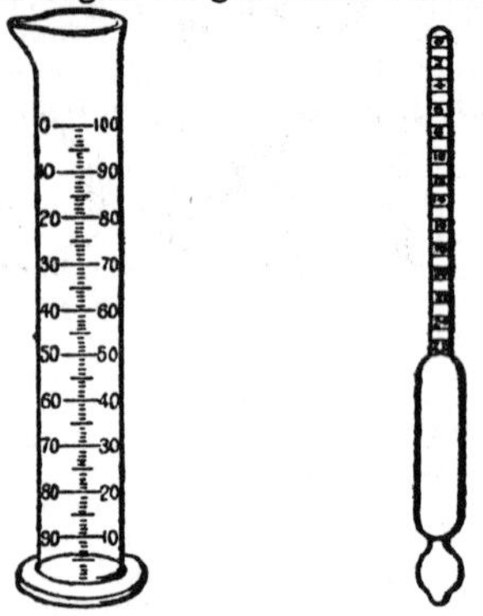

Hydrometers and Jar

Hydrometers.—The specific gravity of liquids may be ascertained without scales and weights, by means of a hydrometer, an instrument usually of glass, having a graduated stem, and bulb or bulbs at the lower part. The specific gravity of a liquid is indicated by the depth to which the hydrometer sinks in the liquid, the zero of the scale marking the depth to which it sinks in pure water. Hydrometers require a considerable quantity of liquid to fairly float them, and specific gravities observed with them are less delicate and trustworthy than those obtained by the balance; nevertheless, they are exceedingly useful for many practical purposes where the employment of a delicate balance would be inadmissible.

Hydrometers are of two kinds: First, those which are always immersed in the same depth in still water and the liquid to be tried, small weights being used for the purpose, as in Fahrenheit's and Nicholson's hydrometers; and second, those which are suffered to rise or sink freely in the liquid, as in Syke's and Baumé's. In both cases a correction must be made for any variation in temperature.

In conducting technical experiments, the hydrometer will often be found of great use, even to those who are not chemists. The Baumé instrument seems to be falling into disuse, a hydrometer having a graduated scale in which the graduations represent the specific gravity, taking its place. A hydrometer jar and two specific gravity scale hydrometers should be used, one for liquids heavier than water, and one for liquids lighter than water. For special purposes, or if the equipment of the laboratory is large, a considerable number of hydrometers may be provided. When constructed for special purposes they often have special names. In the catalogue of a prominent manufacturer of chemical apparatus and materials we find the following special hydrometers for special purposes. The prices run from 75 cents to $2.00, although some special types cost more, and some are only sold in sets. These special hydrometers are for testing the following substances: Alcohol, alkali, ammonia, bark (tannometer), battery fluid, beer, beer and wort, benzine, blood, chlorine, cider, coal oil, ether, gasoline, glycerine, milk (lactometer), naphtha, oil, salt solution' (salimeter), silver solution, sugar, vinegar, wine and must. If the liquid is too warm, the hydrometer jar containing it should be cooled to the proper temperature; if the temperature has fallen too low, the hydrometer jar can be slightly warmed with the hand. Many of the hydrometers found in the older books have either dropped out of use, or are rarely used in this country by chemists. The Pralles hydrometer is largely used by distillers in this country, and by the Government for making alcoholic determinations. Twaddell's hydrometer is very often employed in tanneries and other technical works, especially in England. If work in specific gravity is to be performed, a specific gravity balance is recommended. The tables of specific gravity will be found in the chapter on WEIGHTS AND MEASURES. Tables of specific gravity, and the method of using the same, are presented herewith.

Thermometer Scales.

Much annoyance is caused by the great difference of thermometer scales in use in the different civilized countries. The scale of Reaumur prevails in Germany. As is well known, he divides the space between the freezing and boiling points into 80°. France uses that of Celsius, who graduated his scale on the decimal system. The most peculiar scale of all, however, is that of Fahrenheit, a renowned German physicist, who in 1714 or 1715, composed his scale, having ascertained that water can be cooled under the freezing point without congealing. He therefore did not take the congealing point of water, but composed a mixture of equal parts of snow and sal ammoniac, about — 14° R. The conversion of any one of these scales to another is very simple, and easily made. To change a temperature, as given by Fahrenheit's scale, into the same as given by the centigrade scale, subtract 32° from Fahrenheit's degrees, and multiply the remainder by 5-9. The product will be the temperature in centigrade degrees.

To change from Fahrenheit's to Reaumur's scale, subtract 32° from Fahrenheit's degrees, and multiply the remainder by 4-9. The product will be the temperature in Reaumur's degrees.

To change the temperature, as given by the centigrade scale, into the same as given by Fahrenheit, multiply the centigrade degrees by 9-5 and add 32° to the product. The sum will be the temperature by Fahrenheit's scale.

To change from Reaumur's to Fahrenheit's scale, multiply the degrees on Reaumur's scale by 9-4 and add 32° to the product. The sum will be the temperature by Fahrenheit's scale.

See No. 92, page 234

INDEX

ELEMENTS OF GLASS-BLOWING

BY

H. P. WARAN

M.A., PH.D.(CANTAB.), F.INST.P.

LONDON

G. BELL AND SONS LTD

1923

INDEX PAGE 434

PREFACE

In writing this book my aim has been to present the junior student as well as the inexperienced research worker in the laboratory with a concise and connected account of the systematic methods to be followed for success in glass-blowing. In preference to dealing with the details of manipulation for a number of individual pieces of apparatus, I have confined myself to the various processes classified under distinct heads, with typical examples illustrating them. With the help of the directions given under these headings, one ought to be able to devise a combination of some of them suited to one's own individual needs and skill. In cases where professional methods are unsuited to the limited skill of the average laboratory experimenter, alternative methods are suggested. In order to ensure that the diagrams shall be explanatory of the text, they have all been specially prepared by myself. A few abbreviations have been used in them. Where a piece is to be handled in a particular orientation by the hands, the left and right are indicated by the letters L and R, and generally in all the diagrams the left and right hand ends correspond to the left and right hand side of the page. Where a tube is to be cut, this is indicated by a full line (|), while a dotted line (.....) indicates a joint. Since the way in which the flame is directed against the glass controls the shape of the zone of glass softened, it is indicated wherever necessary by short arrows (↗). Skill in glass-blowing is necessarily a result of long practice on the right lines, and, above all, the one thing to be avoided is undue haste in the manipulations.

I must thank Prof. A. W. Porter, D.Sc., F.R.S., for much valuable criticism in the preparation of this work for the press.

H. P. W.

UNIVERSITY COLLEGE,
LONDON, *January* 1923.

CONTENTS

ELEMENTS OF GLASS-BLOWING

CHAPTER I

PRELIMINARY CONSIDERATIONS

Introductory.

A moderate knowledge and experience of the operations of glass-blowing are of invaluable help in modern experimental work, and as its importance cannot be over-emphasised, it is to be regretted that it has not yet found a permanent place in the curriculum. This knowledge is very useful to the average student, as it enables him to set up his own apparatus and effect minor repairs. To the research student it becomes an absolute necessity. How many times has a promising piece of research had to be postponed for want of some training in glass-blowing, and how often have inefficient substitutes and dodges had to be resorted to! This training is still more important (1) to workers in out-of-the-way stations and laboratories; (2) in times of political trouble, when supplies of ready-made apparatus are not available; and (3) in modern high vacuum work where practically the whole of the apparatus is in glass.

As in any other art, skill in glass-blowing is the outcome of long practice combined with the knowledge of many important practical rules to be adhered to if time

is to be saved, and a maximum of personal convenience and comfort for work are to be obtained.

The standpoint adopted here not being that of the professional glass-blower, but of the laboratory student, a short description of the necessary implements will not be amiss.

A separate steady table about 3 feet high should be set apart for this use, and its top surface preferably covered with some non-combustible material, such as asbestos board. It should be located in a comparatively dark corner of the laboratory, free from draughts and glaring illumination, so that the worker may always be able to make a proper judgment of the portion of glass heated, and the temperature it has attained from its colour. Since the usual professional practice is always to handle the longer and heavier part with the left hand, there should be some clear space beyond the table on the left-hand side of the worker.

Tools.

Beyond the usual bellows and blowpipe, the tools of the glass-blower are few and simple, since he relies more on his skill than on the tools to produce the desired results. Yet a few words about them may be of some advantage. The blowpipe largely used is of the Herapath type, and the one illustrated in Fig. 1, and obtainable of any dealer in scientific requisites, is provided with a multiplicity of adjustments, which is greatly in its favour. In addition to the adjustable cocks for the air and gas supply, it has also the different sizes of jets and the flexibility of a universal joint to produce the particular types of flame required in any direction according to the needs of the work in hand. In the use of this type of blowpipe care must be exercised to keep the air jet central and the external sleeve A drawn out a few milli-

metres beyond the air nozzle to produce a quiet and hot blue flame.

A very simple and easily made blowpipe largely used by glass-blowers for professional work is the one shown

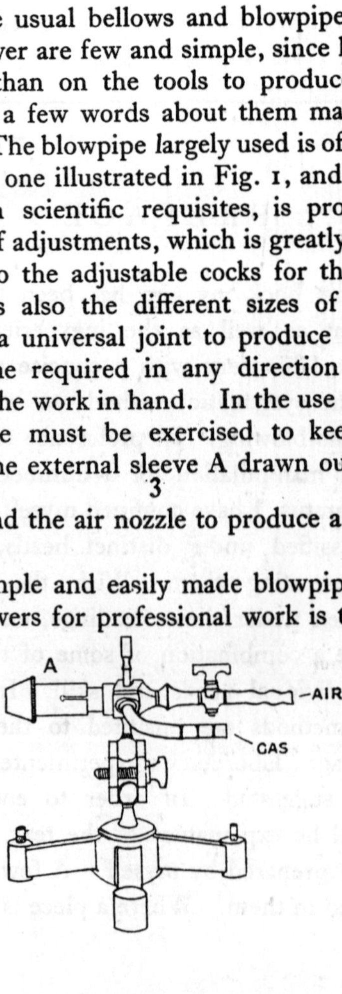

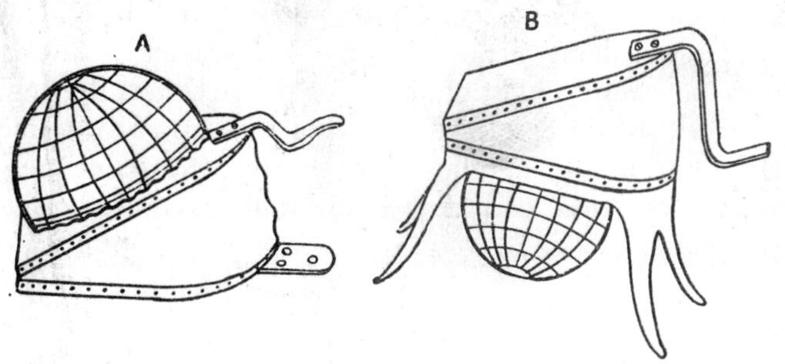

Fig. 1.—In bellows of the type A the rubber pressure chamber, being on the top, is readily accessible for repairs, though it has the drawback of exposing this sensitive portion to pieces of hot glass falling off the table. In the type B, though the pressure chamber is well protected from above, it is likely to be punctured by tips of glass projecting from the floor.

in Fig. 2. It is, in fact, a double blowpipe, composed of a small one mounted directly on top of a large one, the air jets for both being ordinary glass tubes fitted through a cork at one end. Centering screws are provided for the air-jet tube; and though between the two such a

4

compound blowpipe is capable of dealing with any job that demands a quick change from a large flame to a small one, additional variation is also provided in the form of a set of air-jet tubes of different sizes, fitted with corks ready at hand.

The Fletcher type of leather foot-bellows, as shown in Fig. 1, with air chamber of rubber protected by a net,

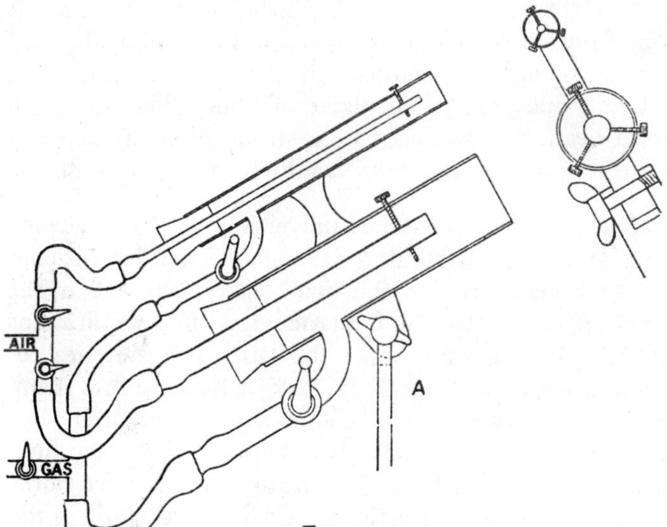

Fig. 2.

ensures a fairly steady blast of air when one has got accustomed to its use, and hence is widely popular. Professionals, however, prefer the larger and more elaborate semi-automatic types, with leather chambers that give a longer and steadier blast, without necessitating a too frequent kicking with the foot, with consequent shake, that may result in the deformation of the work in the flame. In addition to the extra control it allows over the flame, the small pair of bellows has also the

5

great advantage of easy portability, a necessary quality when an elaborate piece of apparatus is being assembled *in situ* on a research table. The many forms of motor-driven automatic blowers combine the advantages of both types, and are recommended if available.

The rest of the tools in frequent use are a small new triangular file, or, better, a hardened steel (glass) knife for cutting purposes, and a piece of arc-light carbon with

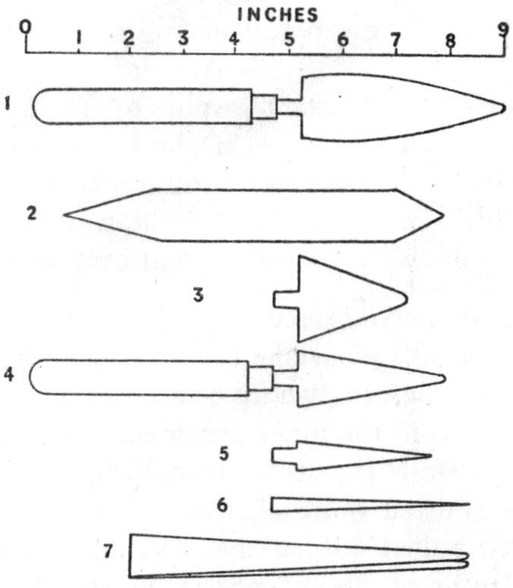

Fig. 3.

the end filed to form a cone ending in a blunt point, for use in enlarging the ends of narrow tubes. Professionals and experienced amateurs can spin glass in the flame, and they find very effective use for another tool made of a strip of stiff tapering metal sheet fixed on to a wooden handle as shown in Fig. 3. This type of tool is generally useful in three sizes as shown, and with them, the carbon cone can be entirely dispensed with. A few minor auxiliaries are also required occasionally, such as small

6

corks, and short pieces of rubber tube with one end closed by a bit of glass rod or tube. Professionals accustomed to quantitative work are able to arrange their procedure in such a way as to dispense with these accessories. It may be mentioned in this connection that it is better to avoid their use, and follow the procedure suggested later in the chapter, since thereby many chances of accidents leading to failures are eliminated.

A metre length of small-bore rubber tubing forms a very useful auxiliary for use in blowing through when working a large or complicated piece in the flame, that cannot be conveniently taken up to the mouth for blowing. In the choice of this tube, as well as in those used for air and gas connections, care must be exercised to select tubing with thicker walls than that commonly used in the laboratory for gas connections, and the harder and springy variety of rubber that is not too flexible is to be preferred.

This is to guard against the consequences, possibly damaging to the work, if accidentally the tubes get choked or squeezed at a critical stage of the operation. However, should such an accident happen, its consequences would be nullified if the worker takes care to have always at hand a small idle flame such as a burning candle, so that the blowpipe may be relit without delay, thus preventing the consequent cooling and cracking of the work. Some of these minor precautions are very profitable in the long run, and if systematically adopted would eliminate a great deal of failure and disappointment.

The management of the blowpipe to obtain the proper flames for the various stages of the work is a matter of some importance, and the beginner will do well to practise a little before starting on the work. The inner nozzle in the blowpipe is for air, and care must be taken

7

to see that it is connected properly to the air supply (bellows) and not by mistake to the gas supply, in which case the gas on lighting would be seen to burn as a small conical flame from the inner nozzle and not the outer as it should. Before the flame is lighted, the air nozzle should be adjusted quite central to the outer, and the outer sleeve pulled out so that its end is a few millimetres in front of the air nozzle, as shown in Fig. 2. When the gas is turned full on and lighted, a large smoky flame results, whose size can be adjusted by the amount of gas let in. If a little air is now turned on it will just project this flame forward and yet keep it fairly large and bright, and such a flame is very useful for warming up tubes before applying the hotter flame to them, and for gradual cooling or annealing after the work has been done in a hot flame.

When the air supply is increased still further, the flame becomes smaller, and turns bluish-grey in colour, getting very hot, since complete oxidation of the gas is taking place towards the tip of the flame. If the air nozzle is fine (1 mm. diameter), a pointed flame, Fig. 4 (B), of great heating power just inside the very tip is the result, and such a flame is very useful for local heating of small and narrow patches. If a wider nozzle and stronger blast are employed, a brush-shaped flame, as shown in Fig. 4 (D), of great heating power is the result, and is generally the one in use for softening larger masses of glass for blowing bulbs and working large joints. For average work an intermediate type of flame Fig. 4 (C) is obtained by so adjusting the air supply as to get a slightly noisy blue-grey flame with a small flicker of bright flame just at the base. The best-sized flame for average work is about 10 to 15 cm. long, with its largest diameter near the tip, equal to the diameter of the tube worked in the flame. By the relative adjustments of gas, air, and size

8

of nozzles, the size, shape, and character of the flame can be varied to a great extent; and in general the aim should be to use a flame of not too noisy a character, resulting from too strong a blast of air. For the strongest heating, the work should be placed just inside the tip of the flame and right in the middle of it, so that the flame may envelop the tube and heat it all round symmetrically.

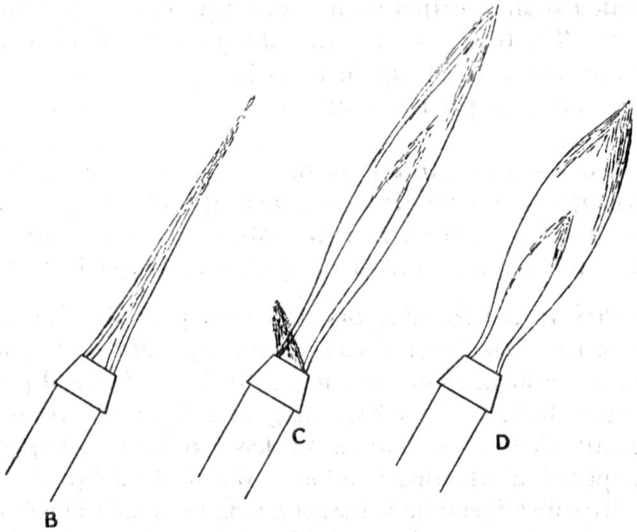

FIG. 4.

A very little practice will teach one the relative adjustments of gas and air necessary to secure the proper type and size of flame, as well as the regions of it that produce the greatest and most uniform heating of the glass. A big flame of low heating power is in general obtained by an excess of gas and less air through a wide nozzle, while a fine pointed hot flame of blue colour for small local heating is produced by a strong blast of excess of air through a fine nozzle with a diminished supply of gas.

9

Soft soda glass is almost universally used for blowing purposes in the laboratory, though the harder varieties of lead glass and combustion glasses, as well as the modern pyrex glass and quartz glass, find applications in high temperature work—the latter two varieties requiring the higher heat of the oxygas flame to fuse them. For general vacuum work, apparatus are generally improvised out of soft soda glass, because of its easy fusibility, though it is far more sensitive to imperfect workmanship. Since this variety of glass softens easily in the flame, it is advantageous to use, wherever possible, tubes with comparatively thick walls and wide bore, especially if the worker has not sufficient skill to manipulate the softened glass without serious deformation. On this ground, tubes 5 to 10 mm. bore with walls about a millimetre thick are the safest to use.

To be free from the contaminating influence of unknown impurities it is quite essential, especially for high vacuum work, to clean and dry the glass thoroughly before starting work on it; and in many cases it is an easier plan to clean the tubes themselves first rather than the finished apparatus afterwards. A handful of crystals of potassium dichromate shaken with some concentrated sulphuric acid forms a very active and effective cleansing reagent, the acid being entirely washed out afterwards by a liberal use of tap water and alternate brushings with a swab of cotton wool to dislodge sticking dust and dirt. The tube is washed free of tap water with distilled water, and drained dry in a vertical position, the open ends being plugged by cotton wool to exclude dust.

We will now proceed to discuss briefly some of the principles of operation in actual glass-blowing most suitable for the student in the laboratory, paying attention at the same time to the processes adopted by the

10

expert to secure reliability and beauty of the finished product. The expert secures this symmetry of good work by the adoption of a systematic and calculated procedure from the very beginning, and these are essential points that do not seem to have received adequate emphasis or attention in other works on the subject.

The first operation is that of taking the required length of tubing out of the usual 6 or 8 feet lengths supplied by the makers, which we have already submitted to a thorough cleaning and drying process. The beginner proceeds straightaway to cut the tube with a file or knife, but the expert employs a method which, though seemingly complicated, is undeniably the best. We will discuss it at some length, since it is of such fundamental importance.

Uniformity of thickness all round worked glass is essential, since in its absence the unequal strains set up on cooling invariably lead to fracture. With a stationary flame, this uniformity can be secured only by a uniform rotation of the tube in the flame. This cannot be properly done unless there is an axis round which to rotate the tube. Hence the expert starts his glass work by drawing this axis at the two ends of the piece he is going to work. The forming of this axis or spindle (about 6 inches in length) is by no means as easy as it looks. In addition to being stiff enough to bear the weight of the tube, when held by its stem in the left hand, as shown in Fig. 5, it has also to be quite axial to the tube to permit a rapid rotation without undue wobbling.

Though the rotation of the glass in the flame is effected by the twirling motion of the spindle between the thumb and forefinger of the left hand, the other end of the tube has to be suitably supported by the upturned forefinger and thumb of the right hand, and for this a spindle drawn out at this end of the tube is a great advantage.

11

Two such good spindles drawn out axially to the tube are useful also for closing or opening either end for blowing purposes, without having resource to corks, tubes, plugs, etc., which are liable to slip out or get burnt at the most awkward moment.

As a general rule, except when alternate blowing at either end is required in special pieces, the blowing is always done through the open end at the right hand side, while the left end is kept closed by a momentary application of the tip in the flame; it can be opened instantaneously, when required, by breaking off the extreme tip by a scratch. To be long enough for these purposes and to give a comfortable hold, the spindle should be not less than 6 inches in length. In addition to enabling a rapid rotation of the glass by a small rubbing motion of the fingers, the narrowness of these spindles serves also to make the rise of pressure inside gradual while blowing, and at the same time accidental blowing off of glass that is too soft by a momentary excess pressure is less likely to occur.

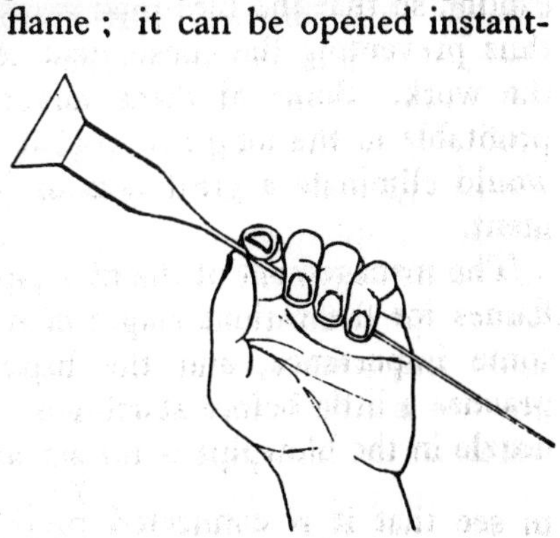

FIG. 5.—View from below.

Let us consider now the expert's way of drawing this all-important spindle, an operation he effects simultaneously while taking out the required length of tubing for his work. Let us suppose this is about a foot. He holds the tube in his left hand, allowing its weight to rest on the table in a V notch cut on its edge; and allows the flame to soften the glass at the very end of the

12

tube, rotation being imparted to the tube by the left hand fingers. When sufficiently softened, a waste piece of glass is stuck into this softened mass and drawn out to a 6-inch length, and the blob at the end melted off as indicated in Fig. 6 (1). This is a preliminary operation done simply to close this end of the tube. Incidentally it may be of interest to note that, though this spindle is fairly thick walled, it is comparatively flexible, because of the concavity of the shape at A in Fig. 6 (1), a thing to be avoided in drawing out the required length with the correct spindle from the other end of the tube as indicated below.

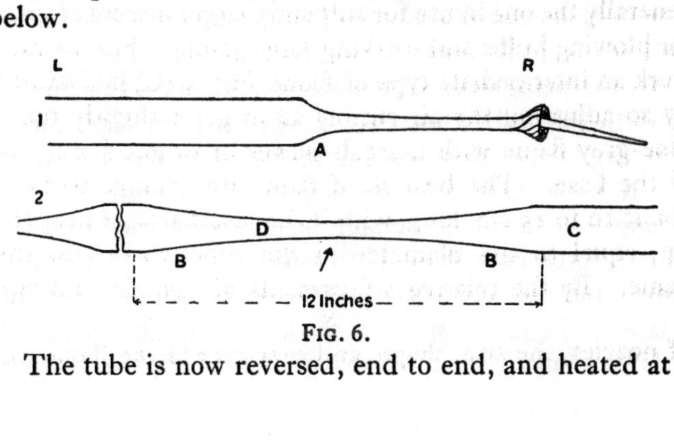

FIG. 6.

The tube is now reversed, end to end, and heated at a

point a foot from the open upper end, until about 1 cm. length of the tube (in general a length equal to the diameter of the tube) is softened, particular care being taken not to get the softened portion into a screw in this operation. It is taken out of the flame and a puff of air blown in quickly through the open upper end in order to enlarge this softened portion to a bulb in diameter about one and a quarter times that of the tube. The two pieces are then immediately drawn apart, elongating this bulb to a length of about 12 inches ; meanwhile, the rotation of the tube is steadily kept up until the drawn-out stem has cooled down to rigidity. The shape obtained

13

by this procedure is conical, as shown at B in Fig. 6 (2), and though the walls are comparatively thin, the stem or spindle is rigid enough to take the weight of the glass without flexure. It is now melted off in the middle, and we get the required piece C with a 6-inch spindle attached to it at one end. While the required length has been cut off from the main tube by this process, we have another good spindle D formed already at the upper end of the longer tube ready for use in its turn. It will take some preliminary practice at this work before the spindle can be got quite axial, which is the primary desideratum. Any small residual eccentricity can easily be rectified by heating the spindle at B and doing a little of trial-and-error centering as with a lathe. Since the portions finally discarded as useless are only these spindles, it is easy to see how economical this procedure is, a factor of paramount importance in commercial operations.

The principal operations of glass-blowing may roughly be classified under five distinct heads, and any complicated apparatus can be made by a judicious combination of a few of these operations. We will discuss the details of these operations in the next few chapters, and endeavour to give under each the various aspects in which the job may present itself.

14

CHAPTER II

CUTTING AND BENDING OF TUBES

THOUGH it is advisable to dispense with the cutting of glass (by knife or file) and apply instead the above-mentioned drawing-off process, sometimes a cut has to be made, and this may be accomplished as follows :

At the place where the cut is to be made—this applies only to thin tubes less than a centimetre in diameter—a fine single scratch a few millimetres long (not all round the tube) is made with the help of a glass knife or a new triangular file. This scratch being kept turned upwards and a force exerted as if to bend the ends of the tube down, the two halves are pulled apart. The tube readily snaps at the mark, and if a single scratch has been made at right angles to the axes of the tube a sharp square cut is the result. If the tube does not snap easily, as is the case sometimes, the heated tip of a waste glass spindle of diameter less than a tenth of an inch rubbed gently along the scratch for a moment starts the crack, and the two halves may be pulled apart quite easily. In making a cut like this between two heavy halves, it is a wise precaution to have both the sides well supported, since the crack sometimes goes right round the tube instantaneously, permitting the two halves to fall apart.

For cutting tubes of a larger bore a simple procedure is to make a scratch, start a crack as above, and lead it right round carefully by the repeated applications of the hot point just in front of the crack. Though the small

15

irregularities of a cut made by this process disappear in the fire polishing, or by rubbing with carborundum paper, the following is the process professionally adopted to ensure clean square cuts :

A stout piece of iron wire, not less than an eighth of an inch in diameter and 15 inches in length, has one end bent round to form a smooth semicircle of a diameter equal to the external diameter of the tube to be cut, so that when this end is made red hot and applied to the tube it will be in contact with the glass as shown in Fig. 7. When the tube, after the preliminary small scratch, is laid horizontally on this hot wire and rapidly turned on its axis by the spindle, a crack flies right round the tube, giving a perfectly clean cut such as is required for joining the ends internally in the making of a liquid air Dewar tube of large bore.

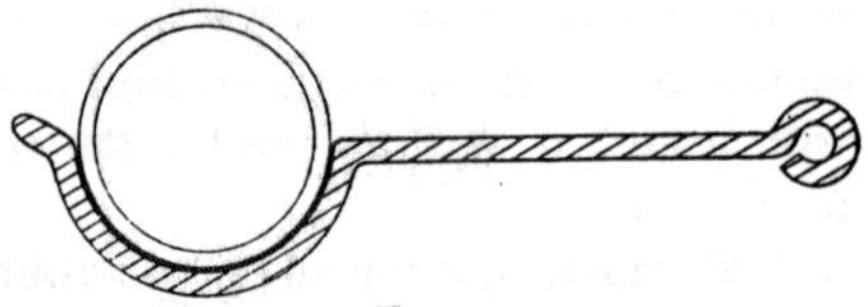

FIG. 7.

This method of cutting large tubes is only a particular aspect of the general procedure adopted in glass works for similar purposes, and it may be of some use to describe this procedure, as it is very convenient for cutting large bottles or other ware in the laboratory. A piece of diamond is employed to make the initial scratch, and the work is mounted centrally on a turn-table which is kept in slow rotation. A very small gas flame, not more than a tenth of an inch in size, suitably mounted on a stand, is adjusted to play against the rotating glass cylinder at the level of the scratch. In a few seconds a crack starting at the scratch flies right round the hot zone, giving a perfectly even cut.

There are occasions when a quick cut has to be made, as, for example, beyond a joint that has to be reheated when joining on a fresh tube to the cut end. The above-

16

mentioned processes are too slow for this purpose. In such a case an efficient method is simply to draw off the excess length of the tube by fusing it, and then, after the removal of the excess glass, to blow out a hole at this end. The edges of this hole, when fire polished after the removal of the films of glass adhering to them, give a fairly good square-cut end. Such an end may often be constricted a little, and hence may require to be enlarged to the full bore by the use of a charcoal cone or one of the spinning tools as described in a later chapter. If the end to be cut off is rather short and not more than a centimetre in diameter, an equally good—though somewhat daring—method is to make a sharp scratch and, resting the *diametrically opposite* part of the tube on the edge of the table, to give a sharp forward knock on the bit beyond the scratch with the wooden handle of one of the tools. It generally knocks the bit off, giving a sharp square cut, in the minimum of time.

It may happen that the cut is irregular, leaving a piece or two projecting out of the otherwise regular and square-cut end. This is specially annoying to the student, whose aim should be to make the end perfectly uniform before attempting a joint on to it. A rugged end of this nature can be easily made uniform by any one of the following methods. The projecting pips may be ground off on a carborundum paper or wheel, but a much less complicated way is to chip away the corners by filing with the flat of the triangular file, the filing being done towards the axis of the tube, in a slanting direction. If the end of the tube is given a preliminary warming before the filing is commenced, the glass chips off readily in small fragments, and the chances of breaking away a huge chip and thus inceasing the irregularity is minimised to a great extent. If there is only a single large projecting pip, it may as readily be softened in the flame and judiciously drawn off at the

17

end of a waste spindle of glass. Similarly, a small hollow in the edge may as easily be filled by glass from the softened end of a spindle before a tube is joined on to this end.

Bending of Tubes.

The guiding principle in this operation to ensure a smooth round bend is to soften lightly a length of tube equal to at least three times its own diameter for a right-angle bend. It is then an easy enough task to be accomplished in a flame wide enough for the purpose, whether it be a special fish-tail burner or only the blowpipe flame adjusted to give a large smoky flame. In the use of the latter, care must be taken to soften the glass uniformly and slowly. There are, however, two distinct ways of effecting this operation. The first one is practised with small-bore tubes of comparatively thick walls that do not collapse easily, when a length of tubing is *lightly* softened, enough to be gently and gradually forced to the required shape. This is evidently the method of the inexperienced. The second one is that of the professionals, and requires considerable skill and experience when attempted on tubes of thin wall and large bore. For a nice U-bend, a length of about five times the diameter of the tube has to be melted down and gathered to form an elongated thick bulb in the middle of the tube. After a strong heating the operator pulls this apart gently, at the same time bending the tube to the required shape and blowing in all the while to expand it to the diameter of the rest of the tube, all these operations being effected simultaneously before the glass loses its heat. Even then it may require local correction.

Spirals.

The making of spirals of glass tubing is a fascinating part of glass work. It demands but a very limited skill

18

on the part of the operator, since the tube employed is usually thick walled and hence possesses considerable rigidity, even when softened enough to permit of easy bending into a spiral shape in small sections. There are two methods, and in both of them a large length of the tubing must be softened by a large bright flame. This must be played along the length of the tube, to ensure a uniform softening for the graceful bending to the curve required.

The first method is a free-hand process. Requiring no extra mechanical assistance, it is largely used when slight irregularities of shape in the finished spiral are not of any consequence. A metre length of tube about 3 mm. in bore, with walls not less than 1 mm. thick, is the average size of tube employed for the purpose. One end is closed by drawing a small spindle. About 15 cm. from this end the tube is bent sharply at right angles (Fig. 8 (1)) and the short limb AB serves as a handle in line with the axes of the spiral to be formed. The longer piece BC is then bent towards the operator at a point D by another right angle, the distance BD being the radius of the spiral to be formed. We are now ready to bend the tube from D to C as the required spiral, holding AB slantingly in the left hand and resting C lightly on the palm (upturned) of the right hand, the flame playing along DC, as shown in Fig. 8 (2). During this operation, a backwards and forwards motion of DC along its axis ensures the heating of a longer length than that covered by the large flame, and a to-and-fro semi-rotary motion about DC, exerted by the left hand holding and swinging the arm AB, ensures a fairly uniform softening of the length of tube all round. When the length D*b* shows signs of yielding because of the softening, it is taken off the flame and (AB being kept vertical in the left hand and DC horizontal) a clock-wise twisting force

19

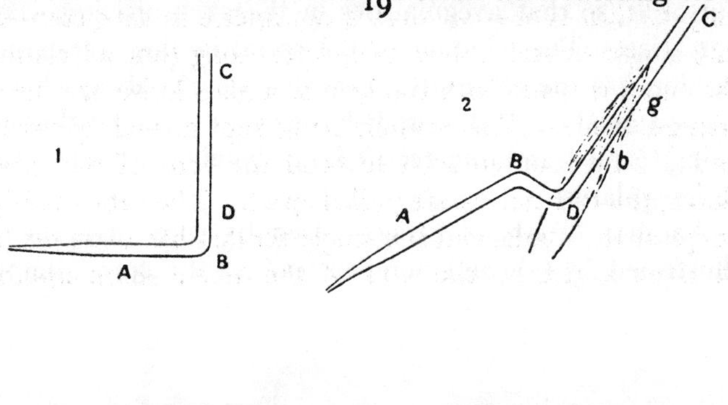

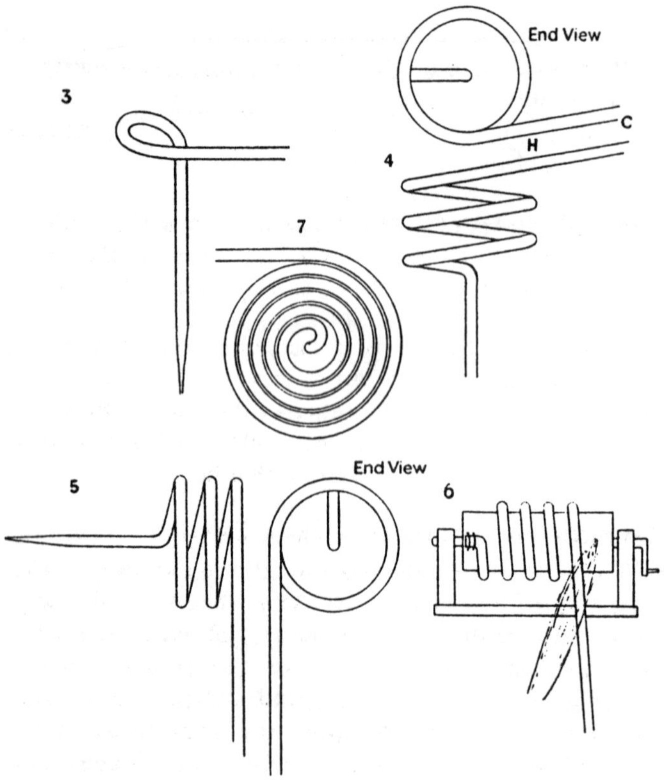

FIG. 8.

20

is exerted about the axes of AB. The bending of the tube D*b* under this strain is aided by a counter-clock-wise push of DC by the right hand. This forms a short sector of the circular bend for the first turn of the spiral, as indicated in Fig. 8 (3). It is presented to the flame again in position (2) and a further length *bg* softened and formed into a continuation of the arc of the circle previously formed. This operation is repeated until the required number of turns of the spiral have been formed. If the original length is found short, a fresh length may be joined on now and the work completed. After the first turn has been formed rather carefully, and made quite circular and axial to AB, the further turns are much easier, since we have only to see that the softened section of the tube is always bent into coincidence with the outline of the turn below it, when looked at axially, keeping of course the axial spacing between the successive turns by a normal or radial view. The straight portion HC of tube (4) must always be adjusted horizontal and in plane with the turn of the spiral just formed out of it, so that irregularities in the axial spacing may not arise. The winding of the spiral in this free-hand fashion has the great advantage that we can always have a good axial as well as radial view of the shape that is being formed, and thus rectify immediately any tendencies to irregularity.

Another free-hand process used by some experts is illustrated in (5), and here, instead of the spiral being wound up vertically, it is wound horizontally in one continuous operation, the softened region steadily winding up into a spiral and thus drawing a fresh portion of the tube into the flame. Since it is difficult to have a control over the radius of the turns everywhere from only a side view of it, the process is risky in practice and demands a higher degree of manipulative skill and eye judgment.

21

The second method requires mechanical aid and is practised when a true cylindrical shape and definite size are required with tubes of larger bore for the spiral. The tube is wound over a hollow drum of sheet iron or copper covered with sheet asbestos and horizontally pivoted with a winch handle at one end, as in Fig. 8 (6). If the edges of the metallic sheet forming the mandrel are overlapped, it can be reduced in diameter by squeezing it round, thus facilitating the withdrawal of the mandrel from the finished spiral. A bent end of the tube is tied to the mandrel by wire. The tube, as well as the region of the mandrel in the neighbourhood, is then heated by a large flame. As the tube softens, the winch is turned round slowly to wind the tube on and draw fresh portions into the flame.

Another useful type is the flat spiral of Fig. 8 (7). This is easily wound by the first free-hand process, the axial view permitting each turn to be adjusted in plane with the previous one and with a uniform separation from it.

Though not coming directly under the heading of this chapter, we may at this stage describe the procedures adopted to round or flatten the closed end of a tube. In either case the main principle is to remove the excess glass that accumulated while the tube was being closed by softening and drawing off the spindle. To round the right end of a tube A, for example, in Fig. 9 (1), heat it at *h* to draw off the spindle, when the glass will thicken and collapse down (2). This extra glass must be removed by repeated softening of this end and sticking the end of the spindle on to it and drawing it off; the drawn glass being wound on to itself while it is soft. Finally, shape (3) is secured for the end, where the wall is comparatively thin and nearly flat. After the small projecting pip *p* has been melted down, this end is softened by a flame (*r*) directed as indicated, and a gentle puff blown

22

in; the end then gets blown round like (4) and not into the shape (5), where it is larger than the diameter of the tube. If the end is to be flattened and not rounded, the heating is done largely on the end face alone of the tube by a flame (*f*) directed as shown, and after a gentle puff just enough to give the end a very slight bulge outwards (6), it is rubbed down flat while still soft by a few strokes with the flat blade of tool (4) of Fig. 3.

23

CHAPTER III

JOINING TUBES

GLASS work is largely a question of joints, and the operations of joining can be roughly classified under eight different heads, whose salient points, including the precautions to be observed under each, are briefly considered below. The principal aim ought to be to work the glass at the joint so uniform in thickness with the walls of the tube on either side as to render the joint almost invisible. To achieve this end, the external diameter must also be made equal to that of the tube on either side by blowing out or drawing down. Half the trouble is over if the ends are cut quite square and of equal bore; and the two ends must be heated quite soft and uniform all round before they are put together. This latter operation must be done carefully and without undue haste, care being taken that they are in proper alignment, and that an all-round sealing contact of glass (without any holes) has been established. This is readily seen when a puff of air is blown in immediately to expand the accumulation of glass at the joint.

Since trouble is experienced at times in making a square cut by the usual methods, an alternative method is to blow out a hole after closing and rounding the end of the tube. Though this method is largely used, to get the hole of the correct size and shape (Fig. 9 (11)), is a matter requiring some experience. To do this, the closed end of the tube is rounded and then strongly

24

heated with a pointed flame so as to soften a circular region (at the end) equal to the diameter of the required

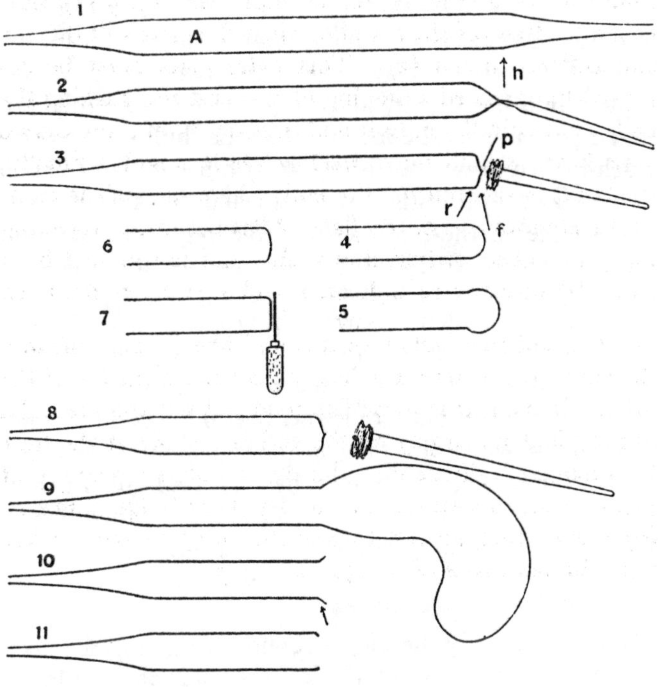

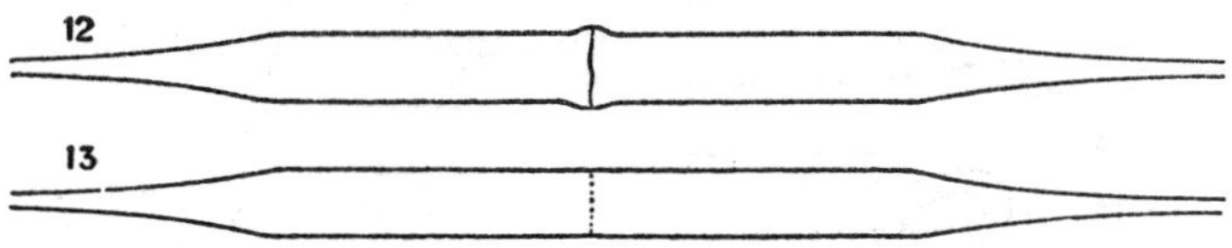

FIG. 9.

hole. A strong puff blows this off into a thin kidney-shaped film which is readily swept away by the knife, leaving the open end as shown in Fig. 9 (10). The minor

25

irregularities of the edges of this cut disappear when the end is softened, and by the concentration of the greater heating on the one or the other of the ends to be joined together, the openings can be made equal and thus to fit each other when they are put together.

Straight-line Joints—(1) *Equal Tubes.*

Three cases of this type usually occur in practice. The first is the joining of two short tubes of equal or very nearly equal bore, a simple enough job considering that it permits of easy rotation about the axial spindles at the ends of the tubes. The joint is softened down symmetrically all round, and after any small twist of the softened region has been carefully and immediately neutralised by a judicious turn of the one half relative to the other, it is gently blown out to form a small bulb as shown in Fig. 9 (12), and pulled apart (13) as the glass cools, so as to make the bore and thickness of glass in this region equal that of the tube. This operation is repeated over and over again until uniformity is secured at the joint. In cases where inconvenient lengths or shapes of glass or the lack of sufficient skill makes this plan impracticable, the beginner's course of going round the joint fusing patches by patches and blowing them out to the level of the tubes is the obvious one to follow. A point to be remembered in the adoption of the latter procedure is to employ only a small heating flame, while going round the joint rather slowly, thus allowing sufficient interval for cooling down before proceeding to heat the diametrically opposite region. In this way the joint does not get out of shape in inexperienced hands. The joining together of thick-walled tubes of capillary bore requires preliminary preparation of the ends, and is for convenience described under the third heading.

26

(2) *Unequal Tubes.*

The second case is when tubes to be joined are of unequal bore, as in a pipette; hence this may be called a pipette joint. The right end of the wider tube is drawn down to form a thick-walled short cone, and cut squarely off at a place where the thickness of wall and bore equal that of the smaller tube to be joined on to it. If the line of joint is arranged to lie exactly at the angle where the two tubes meet, the joint is rendered practically

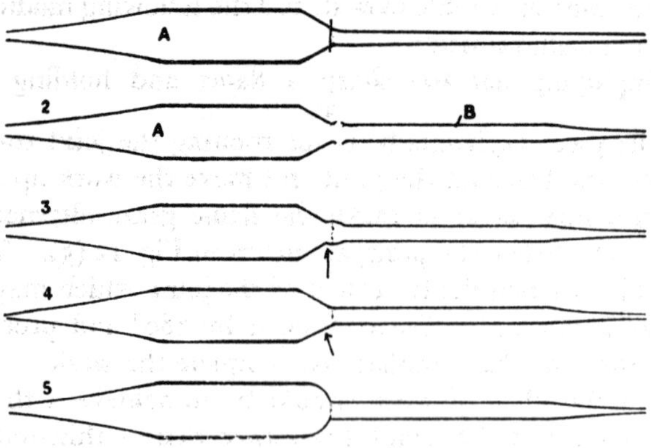

FIG. 10.

invisible; and since this is of fundamental importance in many of the subsequent operations, the procedure to be followed is illustrated in detail in Fig. 10, where A and B are the two pieces to be joined together. After putting the two heated ends together, the joint is worked out quite uniform as explained above, the heating being more on the side of the wide tube to ensure less risk of deformation at the joint. The narrower tube should be maintained to be of perfectly uniform bore and thickness right up to the joint beyond which the wider tube starts as a sharp cone. After this shape has been
27
secured (3) to the well-made joint, the conical portion of the wider tube is softened right up to the joint (4) and blown out slowly into a rounded hemispherical shape, a slight pressing together of the two halves during the final stage of this operation effectively counteracting any tendency to draw the two halves apart (5).

(3) *Capillary Tubes.*

The third case consists in joining tubes of great inequality in the thickness of the walls as well as in bore. A common example of this type is met with in making a Plucker discharge tube, where a millimetre bore capillary with thick walls has to be joined on to wide tubes of anything up to a centimetre or more in diameter. In such cases the end of the capillary tube has to be first fused and closed; then by blowing through the other end a small thick-walled bulb is formed as shown in Fig. 11. The outer semicircle of this is strongly heated and blown off, leaving at this end a short conical end of large bore with thick walls to which a glass of any size could be sealed on easily by the operations explained above. The wider tube side of the joint should be finally blown out hemispherical as indicated. A simpler case of this kind is the joining together of two pieces of capillary tubing. Heating the two square-cut ends and putting them together is too risky to be adopted generally, because of the liability of the narrow ends getting choked on softening or when put together improperly aligned. The safest course is to form the widened ends as described above and then join them together as a case of simple straight-line joint. The slightly wider bore which this operation is likely to leave at the joint can easily be eliminated, if required, by gathering up the glass on either side of the joint and blowing it into an elongated bulb of very thick walls, and then drawing it down
28
carefully to the diameter and bore of the rest of the tube.

Greater obstacles are encountered in dealing with thermometer tubes of very fine bore, since it is very difficult to blow through them, and inadvisable to do so, because of the difficulty of removing the moisture condensed inside the fine bore. The tendency of the fine

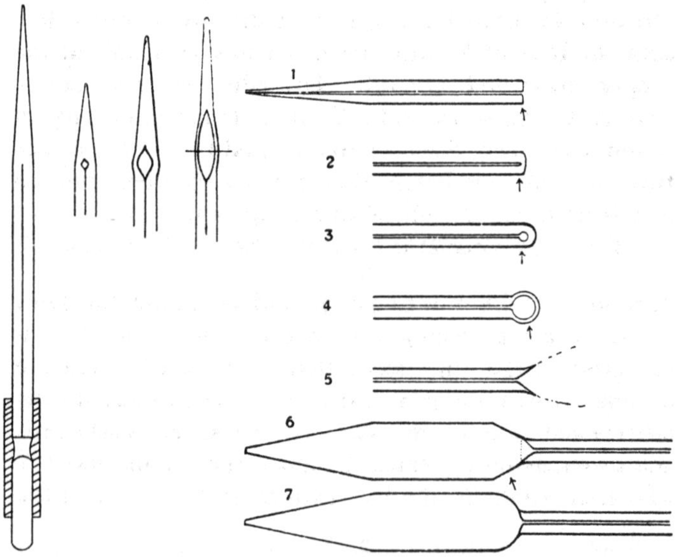

FIG. 11.

bore to get closed on the slightest softening can be effectively overcome by maintaining an air pressure inside while heating it, and by a careful manipulation this air pressure can be made to blow the bulb at the end, which is then drawn out into a short tube and cut to give the required end with large bore and thick walls. A very simple method of exerting the requisite pressure is to slip on the end of the thermometer tube a 2-inch
29
length of stout-walled pressure tubing as shown in Fig. 11, after closing the far end in the operation of drawing a spindle. The outer end of the rubber tube is plugged tightly by a bit of glass rod, and the compression of the enclosed air when the plug is pushed home exerts the requisite pressure, which may be increased still further by pinching the rubber tube. Professional practice differs from this, and is discussed at a later stage.

(4) *Angle Joints and T-pieces.*

Having got ready the pieces X and Y (Fig. 12), the first thing to do is to blow out a hole on the side of the tube X to which the piece Y is to be joined on. This hole should be of a slightly conical shape projecting out from the side of the tube and of a bore smaller than that of Y. This is because on heating, while the former tends to enlarge, the latter tends to contract under the capillary forces of the softened glass. If the side of the tube is simply heated to a bright red spot and blown out, there is risk of the hole being too large and elliptical, in addition to being formed flush with the tube as shown in Fig. 12 (1). This arises from our inability to heat a round spot of the correct size directly on the side of the tube. The right method of procedure is, first, to heat the spot to a dull red by a pointed flame, and then by sticking on to it the heated end of a waste spindle, to draw it out to a conical projection as in (2). The end region of this cone can now be heated properly to a disc of diameter less than the bore of Y, and blown out to give a hole of the proper shape and size (3). The edges of this hole as well as the tip of the tube Y are softened uniformly all round, care being taken to attain equality of bore of the two softened ends before they are put together, pressed in a little to ensure all-round sealing contact, and blown out slightly to expand the thick glass here. The beginner's[30] method is to blow through Y, and go round the joint patch by patch, taking care not to bulge out or thin the glass at the joint more than is required to ensure uniformity with the rest of the tube. The expert, on the other hand, rotates the joint in the flame backwards and forwards by a semicircle about the middle piece Y held horizontally in the left hand as an axis. When the

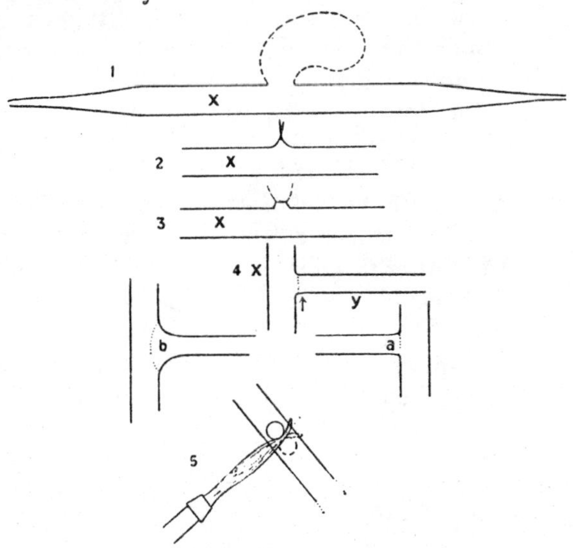

FIG. 12.

joint has thus become softened uniformly all round, by the use of a wider flame applied mainly on the side of the tube joined on, it is blown out a little, the side piece being kept vertical, with a gentle outward pull exerted by the right hand. Though this is the ideal procedure to obtain a good joint at a single heating, beginners are apt to come to trouble over it, and the following modification is recommended:

Employing not too sharp a flame and holding the[31] middle piece horizontally as for rotating the joint round it, play the flame on the joint and move the work up and down a little so as to make the flame graze alternately above and below the joint, as shown in Fig. 12 (5). The effect is to soften the front half of the joint, which may be gently blown out. Then rotate it by 180° and proceed with the other half similarly to complete the work. The aim in this class of work should be to achieve a thick-walled joint of the shape (*a*) rather than a thin-walled rounded type (*b*), and to secure this the joint should be heated more on the side of the tube joined on, and any tendency to pull the pieces apart carefully guarded against. Before annealing and cooling the work, the deformations due to the tube having got bent or out of plane during the work must be rectified by a judicious general heating.

(5) *Internal Joint.*

The internal joint, so called because it is located at the place where one tube passes through the walls of another, is of great application in high vacuum work, since by its liberal use leaking and troublesome joints of other types can be entirely dispensed with. Two types of this work arise in practice according as the smaller tube passes through the end or side of the wider tube, and the job can be satisfactorily done by either of the processes described under each.

In the first case, illustrated in Fig. 13, the right spindle of the wider tube A is melted off, and after the removal of the excess glass at the tip, the end is fused and blown to give a thick-walled hemispherical end. The central spot of this end is heated red and a cone pulled out as described above (2), and the whole of this conical projection softened and blown out to give a hole wider than the tube B to be slipped in. A narrow zone in the middle of tube B is softened all round and blown out a little, and[32] the two halves pressed together so as to form in B an enlargement as shown at C (4). The shape and size of this are of prime importance, and its angular edge should be quite circular and normal to the axes of the tube and of such a size as just to cover the hole at the end of A when slipped in (5). These two preparatory operations must be done in quick succession. Before the glass has cooled down, the pieces A and B are put together as in (5)

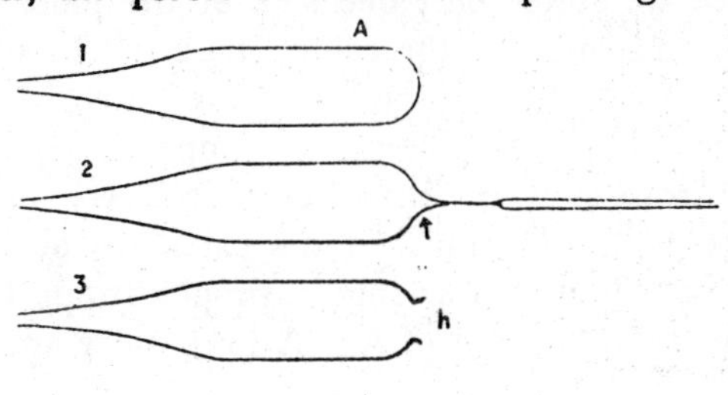

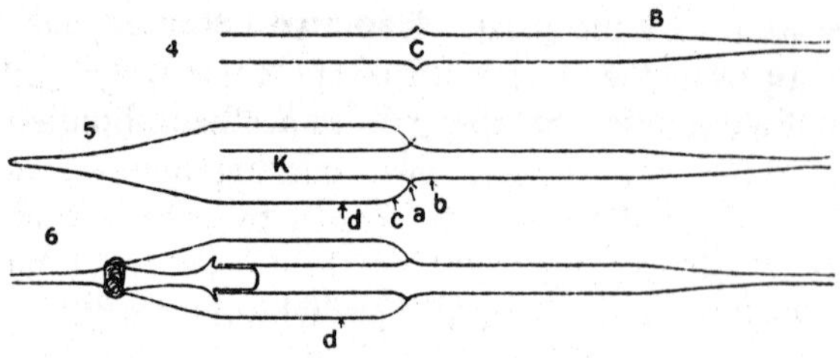

FIG. 13.

(after warming the edges), and a pointed flame applied as shown (*a*) to fuse together the line of contact. Then the pieces are pressed together a little to ensure that no holes are left at the joint. To work the joint uniform a larger flame is applied to the right side of the joint (*b*) and the work proceeded with as in the case of a line joint, only the left edge of the flame playing against the actual joint. The attempt should be to make the line of contact narrow, uniform, and circular, and the walls on either side

33

thick and slightly and gracefully curved, as shown. To get the line of joint narrow, which is essential the following procedure is found to be very useful in practice: The hole *h* at the end of the tube A is first blown larger than the bulb C. Then by the softening down of its edges which get thickened in the process, the hole is made small enough to fit C. If the joint is now proceeded with as before, probably because of the thickened edges of A coming into contact with the surface of C only along a very narrow line, the resulting line of joint is found to come out much narrower and hence better than usual. When the joint has been worked fine and smooth a broader flame is applied at *c* and the deformations as well as the thick glass walls here are blown out to a uniform hemisphere, an action assisted by pressing the two halves lightly together. If the glass has been kept in uniform rotation during the heating as well as blowing, the inner portion K of B keeps fairly central and axial, and by a slight softening of the tube at *d* any small defect in the centering of the inner tube is readily effected at this stage without deforming the joint in any way, and then B can be brought into axial alignment with K by softening it at *b*. A joint of this nature requires careful and prolonged annealing before being allowed to cool.

A modification to this procedure has to be adopted if the inner piece K is heavy and will not keep fairly central while the joint is being worked up. But in general, since we soften only one side of the joint at a time, *i.e.* (*b*) and then (*c*), the cooler glass of the other side provides a retaining support strong enough for the purpose, when the inner piece K is comparatively light and short. In cases where the piece K is heavy, or for any reason requires to be finished in a particular orientation, it is easy enough to arrange a proper support for this through the other end of A. A long waste spindle

34

stuck in through the left end and fixed in by a blob of sealing-wax as shown (6) is recommended for beginners.

The second case of internal joint is (1) shown in Fig. 14, where the downward bend of the smaller inner tube, in addition to its passage through the walls of the outer tube, adds extra complexity to the work. In such cases the general procedure is to have the inner and outer pieces (A and B) separate, and join them on either side of the walls of the bigger tube, blowing out the wall of

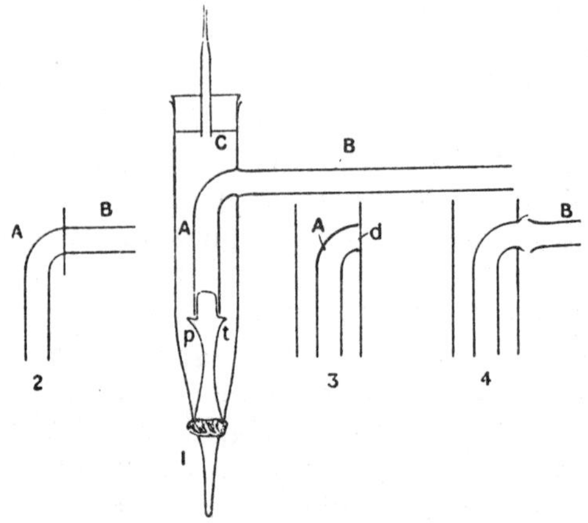

FIG. 14.

glass between them during the operations, and thus obtaining through communication. The inner tube is first bent (2) and cut to give the two pieces A and B. A good central support for the inner piece A is provided as described in the last section, with retaining prongs *p* and *t* for accurate work, and the piece A slipped into position through the wide end of C, which is then either drawn down to a spindle or closed by a cork with a central tube for blowing. The bent end of A is arranged to be almost in contact with the walls of C (3), and a

35

large spot *d* opposite the end of A is softened to a flat disc and permitted to establish all-round sealing contact with the end of A. After this joint is well worked by repeated heating and blowing, the central region *d* is blown off to give an opening on to which the piece B is sealed on as in making a T-piece (4). Since the heating and working are practically on one side of the tube alone, dangerous strains are likely to be set up by the opposite side unless the tube is periodically heated all round (as in the case of a large T-joint) by a large flame, and finally annealed and cooled with great care.

We may now proceed to discuss the characteristics of three further types of joints peculiar to general vacuum technology.

(6) *Joints* in situ.

The first type is the one required when assembling together the parts of an elaborate scientific apparatus;

these parts being small enough to be handled and worked directly in the flame. Sometimes a number of these pieces are available ready-made, and the operation consists only in making this type of joint between the pieces, using tactics peculiar to this work. To begin with, every opening except the two ends (preferably of thick walls and equal bore) to be joined and a third to blow through must be closed by corks or pieces of rubber tube plugged at one end. The opening to blow through must be fitted with a sufficiently long rubber tube ending in a short bit of narrow-bore glass tube serving as a mouth-piece and preferably arranged to open downwards and located beyond an intervening clamp or support, as shown in Fig. 15. In this way the weight of the tube or any accidental pull on it will neither cause a strain on the ends joined, nor a sharp bend choking the passage. Since it is not advisable to contaminate the inner walls

of the work by the condensation of moisture from the mouth it is better to interpose in the middle of this blowing tube a short length of glass tube containing a

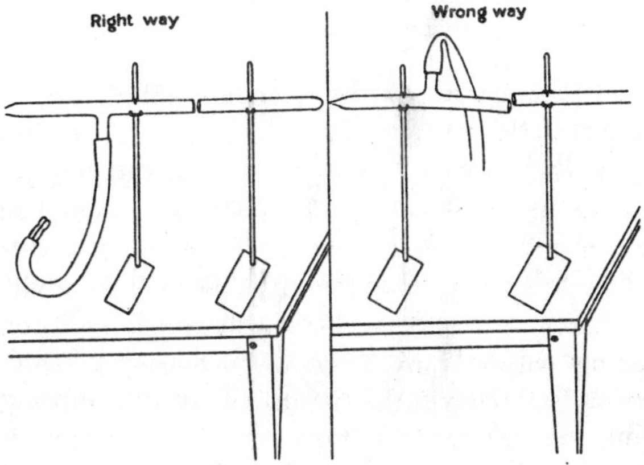

Fig. 15.

few lumps of fused calcium chloride or other drying agent. Such a drying tube shown in Fig. 16 is easily improvised, and preserved when out of use by connecting the two ends by the rubber tube used for blowing, the

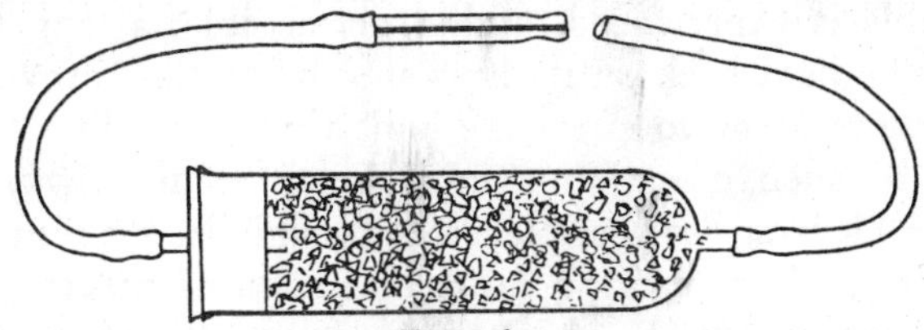

Fig. 16.

corked end of this design permitting of easy refilling with fresh drying agent when required. Since the blow-pipe (and bellows) have now to go to the joint instead of the joint being taken to them, it is screwed out of its base for use as a hand blowpipe and fitted with extra lengths

of rubber tubing to permit of the free movement of the flame all round the joint. The two common accidents resulting from the movement of the connected tubes are either their slipping off through an accidental pull or the flame going out. These two contingencies must be carefully guarded against, the former by secure fastenings of the tubes, and the latter by the provision of an idle flame near by to relight the flame without delay.

The component parts are assembled on the table and clamped lightly so that the weight of the apparatus is properly supported without strain at any joint. While the weights of both the ends to be joined are fully taken by the supports on either side, at least one of the ends must have a freedom of motion along the axes of the tubes to the extent of about 1 cm., and this motion must be effected without bringing into operation any unbalanced forces. It is a wise plan to design the apparatus and its component parts so that this condition is easily fulfilled, irrespective of the ends to be joined being arranged horizontally or vertically. The ends to be joined having been properly aligned, the right end is clamped rigid and the left end kept half a centimetre away from the right. They are then softened uniformly all round by a wide flame waved about by the right hand, the mouthpiece being already in position for blowing. When the ends have been sufficiently softened, the free end is pushed (by the left hand) into sealing contact with the right end and the joint worked uniformly all round as a case of line joint, by the patch by patch method, a smaller flame being used.

The beginner will find that there are many things apt to go wrong in this operation. While he is heating the ends the portion of the tube facing the flame gets hotter and softer, and shrinks away, as in Fig. 17 (1), from the capillary action of the fluid glass. If one side

gets shrunk more than the other, the shape is distorted, and when they are put together sealing contact all round the joint cannot be secured. This is likely to happen especially when the ends have not been well softened all round, and hence would not permit of being pushed together close enough to seal up all openings. To guard against this, the tubes should be carefully and uniformly softened all round by waving the flame as much as possible, and at least a quarter of a centimetre of the glass on both ends well softened before the tubes are put together. If a hole develops, notwithstanding all these precautions, there are two possible ways of closing it. The first is to heat the whole of the joint with a large flame, push the ends together by the reserve freedom of half a centimetre of motion, and see whether the hole gets closed. If it does, the amount of local distortion this operation may cause at the joint may be set right by judicious heating and blowing. The other remedy is to have (always ready at hand) a piece of waste spindle drawn out of the same kind of tube, and then, having plugged up the hole by the sphere of glass formed on heating the end of this piece, to soften the joint and blow out the irregularity caused by this blob of glass. Whenever a hole like this is inadvertently caused, the main thing to be avoided is to soften the glass in its neighbourhood, since this tends only to make the hole larger and larger.

In such remedial operations, since the glass gets considerably distorted at times, it cannot be worked uniform again without an adequate knowledge of the characteristics of heated glass. If, as in Fig. 17 (2) adjacent regions of thick and thin glass are formed, a general softening of the whole, followed by a blowing out, would only tend to make the thin region thinner still, leaving the thicker region scarcely affected. On the other hand, if

the region is slowly softened down to shape (3) the thicker region has time to get softened and can then be blown out after a short delay to let the thinner region cool; the effect is to blow out the thicker region alone, and thus tend towards uniformity (4). This is a consideration of prime importance regulating the type of heating and blowing to be adopted in general glass work, and we shall have to refer to it again when we come to blowing bulbs. Another point to be remembered in the use of the mouthpiece is to keep the lips open during the heating of a joint, as otherwise the expanding air

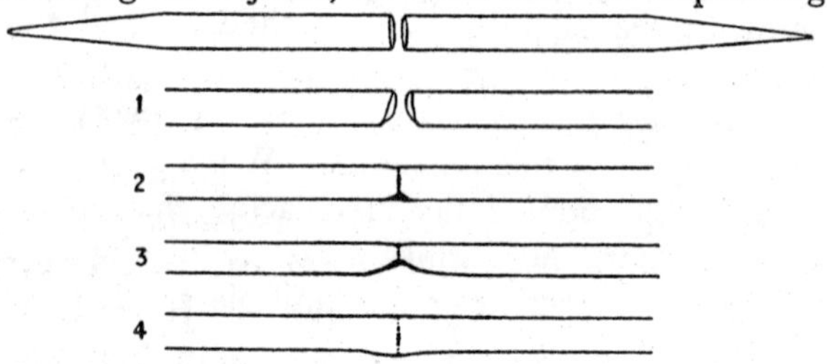

FIG. 17.

inside is likely to burst through the softened region at the joint. In awkward situations the pointed high pressure gas flame burning at the tip of a glass nozzle a millimetre in bore may be used instead of the blowpipe and bellows for effecting joints in tubes up to about 5 or 6 mm. in bore.

(7) *Cross Joints.*

This type of joint, rarely required and better avoided whenever possible, is illustrated in Fig. 18 (1), (4), and success in doing it is more a matter of satisfactory design than of difference in the working of the actual joint. Two tubes, A and B, having been joined on to the main

tube C′ as two T-pieces, their ends have to be bent and joined together, a problem difficult to do according to the design (2), but easier by (3). The trouble with (2) arises because of the short leverage about BC, which gives no play to permit of the two softened ends being

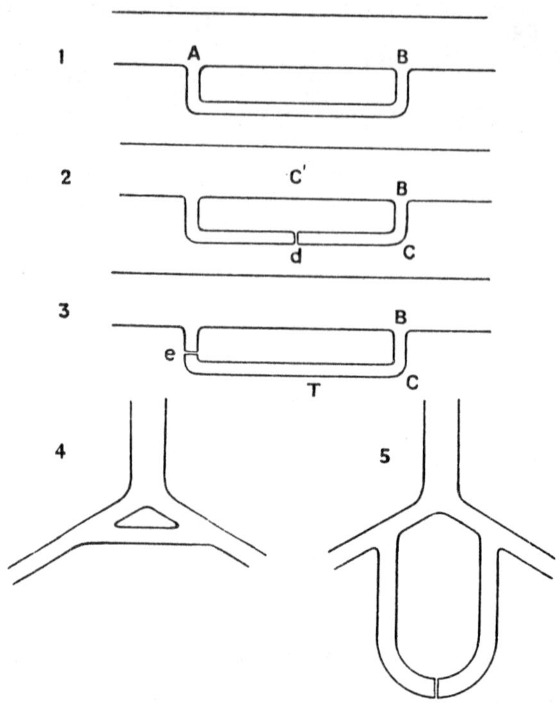

FIG. 18.

pushed together into sealing contact. With (3), on the other hand, the leverage about C*e* is larger, and the ends when pulled apart a millimetre or two can be softened and sprung together into sealing contact without undue strain on the joint at B. The joint is finished as usual, though at the end it must be subjected to a general

softening without any force on the arm T, and well annealed to relieve the joints of strains during the cooling and resultant contraction. Considerations based on this thermal contraction are also in favour of (3) and against (2), since the cooling of a joint at *d* is likely to strain the joints at A and B far more than one located at *e*. By such a procedure any cross-joint can be managed if the distance is sufficient to provide the requisite play. But where a short cross-joint (4) is required, trouble is experienced, and if permissible, a long tube may be bent back on itself as shown (5), thus providing a long leverage to effect the joint without strains. The aim in such cross-joints must be to secure always as large a leverage as possible, so as to minimise the resultant strains; thus ensuring not only an easier process, but also a greater factor of safety.

(8) *Joint on to Thin Glass or Different Kind of Glass.*

This is a special type of joint often required in general vacuum work, where a lead has to be taken out of a thin-walled vessel of the nature of a bulb to be exhausted and sealed. The usual process is entirely out of the question, since the reheating of a hole blown out on such a thin wall tends to enlarge it considerably. In such cases a blob of very hot glass is formed at the end of a tube or rod, and applied to the warmed surface at the spot where the joint is to be made as indicated in Fig. 19 (1). After a momentary interval to allow the extra heat of this blob to melt the thin wall in contact with it as well as cool round its edges, air is blown into the bulb, which operation tends to blow the blob of glass into a bulb on the side of the main bulb (2). If the glass is pulled out gradually during the blowing, this bulb is formed out into a conical thick-walled tube (3), which may be cut at the thicker place and a tube joined

42

on (4) without any difficulty, a comparatively fine flame being used if the cut is made near the bulb. The joint round A is perfectly uniform and thin, and hence likely to stand even if the glasses on either side of it differ slightly. Owing to this valuable property, two

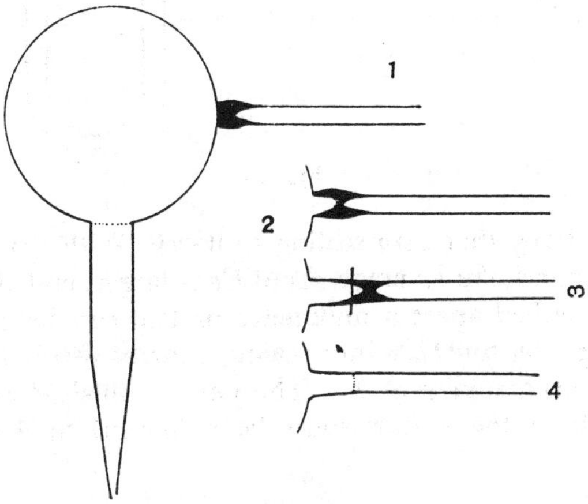

Fig. 19.

glasses differing considerably in their composition and coefficients of expansion may be sealed together by the interposition of a graded series of one or more joints of this type, blown thinner than usual out of glasses having intermediate expansion.

43

CHAPTER IV

BLOWING BULBS

In general, the blowing of bulbs may be classified under three principal heads.

(1) *At the End of a Tube.*

The first and simplest is the blowing of bulbs at the end of a tube. When the required bulb is small and with comparatively thin walls, the spindle on one end is melted off and the end rounded as for a test-tube. A length of tube at this end not exceeding twice the bore of the tube is softened down in a fairly large flame, a slow and uniform rotation being maintained to ensure equal heating and softening all round. When the tube has collapsed down to shape (2), it may be blown out gradually to the desired diameter (generally less than four times the bore of the tube if its walls are fairly thick), the glass being kept in steady rotation all the while. The successive stages of the operation are illustrated in Fig. 20. Any tendency to irregularity in the bulb (4) proves inequality of heating round it, and it must be carefully collapsed down again by means of a large brush flame and blown out again when the softened glass shows uniform colour indicative of uniform heating.

This operation is one of the simplest in glass-blowing, and even a beginner can manage to make a small bulb without much practice; he may then work on a larger bore tube to produce a larger size bulb. The point to

44

be specially careful about is not to get a waist of thin glass just above the softened blob of glass (6), and as a safeguard against this it is better to blow out a little during the heating and form shape (5) rather than (2) before proceeding to the final heating and blowing out to full size (3).

A good position to adopt in heating is to hold the tube almost vertically in a flame going up at about 45°, and to dip the tube gradually into the flame so that the extreme end is hottest and there is a minimum of heating at the neck. A convenient position for blowing when, as here, the mass of glass is small, is to have the tube pointing downwards and blow gradually with an increasing pressure as the glass cools.

We may discuss at this stage the shape the bulb takes according to the position in which it is held during the blowing. When the softened mass of glass is taken out of the flame for blowing, the air surrounding it, getting heated, rushes past upwards giving rise to a stream of cold air blowing upon it from below. If we hold the hot end down during the blowing, as in Fig. 20, this rush of cold air cools the bottom and hardens it while the sides and top regions are yet soft and continue to expand under the pressure of blowing. The effect is to produce a rather flattened bulb; thicker at the bottom. Though this thickness at the bottom is by no means a disadvantage when we have to join on a tube here, the tendency may be counteracted by giving an extra heating to the end regions of the softened blob before starting to blow the bulb. Thus quite a good spherical shape can be produced by this process. On the other hand, if the bulb is blown with the softened glass upwards the tendency will be to get a rather elongated bulb thin at the bottom and thicker at the neck. If the tube be held horizontally and carefully rotated while being blown to

45

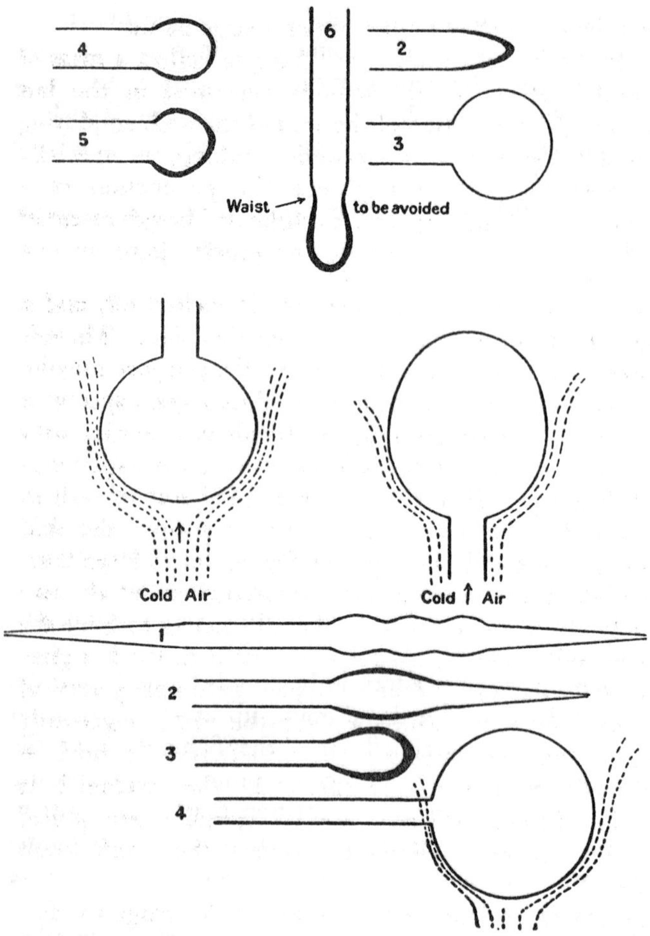

FIG. 20.

46

counteract the sagging tendency of the soft glass under its own weight, the cooling effect of the air current is equal all round and the result is a better approximation to a spherical shape. But evidently the use of this position demands greater skill, though any slight want of alignment between the bulb and the tube can be rectified by a judicious heating of the neck and centering by trial afterwards. But all these positions for blowing have their special advantages and uses as helps towards rectifying a wrong shape or thickness of glass, and by experience an operator may employ all these and other intermediate positions to remedy a faulty shape. For example, if at one blowing a flat bulb with a rather thick bottom is obtained, to correct it, after collapsing the bulb, the blowing must be done with the tube pointing upwards, so that the flat bulb tending to get elongated blows out the extra thickness on top, and produces a spherical shape of uniform thickness.

When the bulb is to be rather large, the procedure demands some skill in collecting and handling the larger mass of glass required to give a reasonable thickness of wall when blown out. A much larger flame and even a bigger blowpipe may be necessitated to ensure uniform heating of the large mass. A device used with larger work is to have a heat deflector as well, so that a large zone of heat between it and the flame may be secured to soften the glass uniformly and quickly all round. This heat deflector is generally a slab of brick, fireclay, or asbestos fixed just beyond the flame.

There are two ways of securing the large mass of glass for blowing. The first method is to collect it out of the tube itself gradually by blowing a few small bulbs adjacent to each other, as in Fig. 20 (1), and collecting them together little by little by the combined operations of blowing out a little and pressing together (2). For

47

tubes of average thick walls an approximate rule is to collect together a length of tube equal to the diameter of the bulb required, and during this stage the glass should not be blown out larger than is required to collect the glass. Finally, the upper spindle that served as a handle is removed, the conical glass left at this end blown round in continuation of the thick walls of the small bulb of glass, and after a strong final heating, the bulb is blown out gradually in successive puffs to the full size. During the collecting of the glass, it is likely to become distorted a little here and there, and the hot blob before blowing may have differences of temperature as well as thickness all round it, which may lead to distortion in the finished bulb. In such a case it must be collapsed back again slowly by means of a brush flame large enough to envelop the bulb, with occasional puffs to retain the spherical shape, and after a final heating blown out again with the necessary correction for the distortion observed.

At this stage, where we may have to correct for inequalities of thickness round the equatorial parts of the bulb, we may consider factors of importance other than the position of blowing, that influence the result. Since we endeavour to produce uniformity by thinning down the thicker regions rather than the reverse, a useful guide is to have always the thicker and hence hotter and brighter region facing upwards, and thus sheltered from the cold draft from below. Thus in the rotation of the work during the blowing, if a momentary stop is made as the brighter spots come up, it tends to make the bulb more uniform. For the same reason, the blowing must be gradual and not too sudden, so that the thinner regions are cooled earlier and stop expanding, while the thicker and hotter portions continue to expand, making the glass uniform in thickness and symmetrical in shape.

48

In the second method, a large mass of uniformly collected glass, in the form of a short tube of wide bore with thick walls, is used for the bulb, and the tendency to distort the glass during the collection as well as the time required for it are both considerably diminished. As in Fig. 21, a short length of the thick-walled tube is drawn (1), and a narrow region in its middle softened

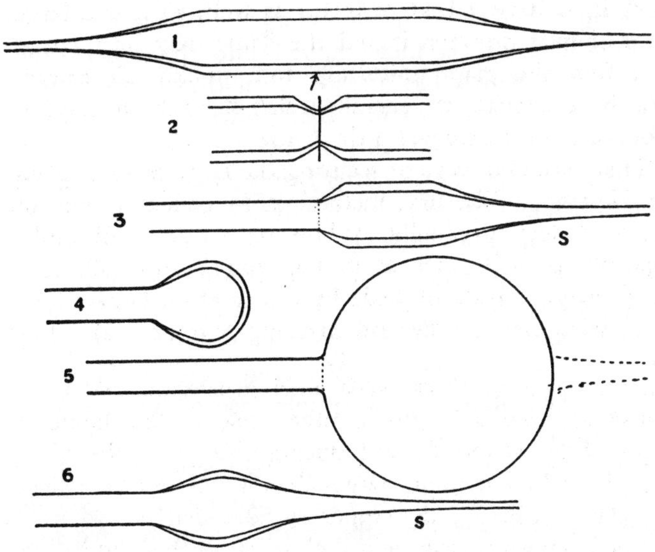

FIG. 21.

and pulled apart to give the thinner constricted waist (2). The two halves are severed by a cut at this waist, and the thick glass for two bulbs obtained at one operation, a procedure, adopted in the case of a pipette as well, that results in great economy of glass. One of these halves is joined on to the tube on which the bulb is required by a well-made pipette joint (Ref. p. 26) (3). The spindle S is drawn off and the end rounded and the

49

mass of glass slowly heated and formed into shape (4), and blown out as before. If the softening of the glass is carried right up to the pipette joint, and not beyond it, the bulb when blown will have the joint exactly at the neck, a position in which it is visible only on very close scrutiny.

(2) *Bulbs in the Middle of a Tube.*

The second type of work occurs when a bulb has to be blown in the middle of a tube. This is a specially troublesome job for the beginner, since the softening of the glass in the middle, combined with his lack of dexterity, makes continuous rotation and uniform heating impossible, without the softened glass in the middle becoming screwed in a short time. Experience in rotating by both hands at the same rate has to be gained by practice. It is easier to do it very slowly, so that the eye can follow the motion and check readily any tendency of one half of the tube to gain on the other. During the heating an inclination to pull the two halves apart should be strongly checked, and any tendency to push them together slightly encouraged. When the glass has become fairly soft, but not much collapsed, it is taken off the flame and blown, the hands keeping their distance apart and maintaining the rotation without distorting the glass in any way. Small-size bulbs can easily be blown like this by a single heating if a sufficient length of the tube is softened by the use of a large flame.

For the larger sizes we shall have to collect a mass of glass by either of the methods described in the last section. The adoption of the second method, employing a middle section of thick-walled tubing, is specially advisable for such work, since the production of a symmetrical shape is somewhat difficult, though essential in this work. After effecting the pipette joint on cne

50

side, the spindle on the other side is melted off, and a hole blown out to make the joint on this side. Though a presentable piece of work to serve the purpose may be managed without much skill, to produce a good specimen with a large-sized bulb, symmetrical and coaxial with the tubes on either side, is a matter of great experience and dexterity. In fact, it forms a good test of skill in glass-blowing, and a good exercise to acquire the skill is to practise blowing shape 1 of Fig. 22 into a large thin-walled bulb (2), centred and symmetrical about the two spindles. A useful guide in this practice is to have the two spindles well centred and coaxial with the hot glass before starting to blow with the glass held nearly vertical in the left hand, which rotates the work vigorously without any restraint by the upper spindle held in the mouth with the right hand. By the residual heat of the thicker corners the two spindles are pulled out into straight alignment before the work cools down.

It may be of interest to mention at this stage the fact that, with a little experience, it is found easier to produce a symmetrical large bulb between a pair of supporting spindles rather than at the very end of a tube which is more suitable for use at the glass furnace. Hence large bulbs at the end of a tube are generally blown through the upper spindle used for the collecting together of the glass. After the thick piece has been attached to the tube forming the neck, as shown in Fig. 21 (3), the glass is collected into shape (6), and blown through S to form the bulb (5), the neck, held in the left hand, forming a substantial spindle to rotate the work in this operation. If it is too wide, then the stout spindle at its end is used for the purpose. The upper spindle S is then very carefully melted off, leaving only a very small pip at the bottom of the bulb indicative of its

51

existence during the blowing of the bulb. The pip itself, as well as the region of the bulb surrounding it, are softened a little for a final blowing of this region into uniformity with the rest of the bulb.

Such processes are in general suitable only for moderate size bulbs with comparatively very thin walls. The blowing of such articles as a big flask or the bulbs for an X-ray or positive-ray tube is in general beyond the scope of table blowing, since it would be inconvenient and difficult to gather and work economically the large

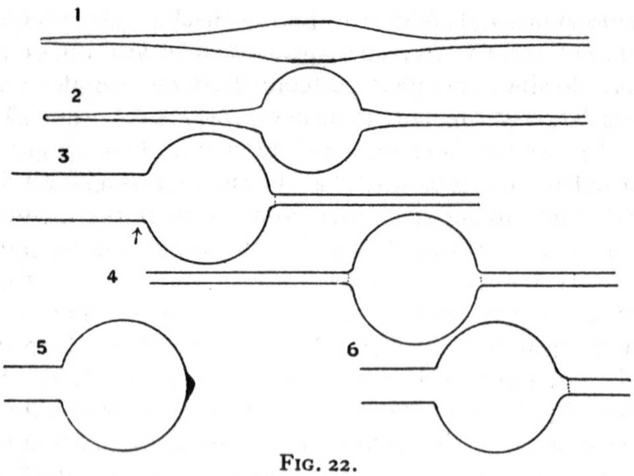

FIG. 22.

mass of glass required to give the walls of the large bulb a thickness of the order of about 1 mm., to enable further joints to be made on to them with comparative ease. Such bulbs are blown directly at the glass works, where the requisite mass of glass can be readily gathered directly from the pot and blown to any size without difficulty. It is advisable to make the tubes (for the purposes of joining on to the bulb at the table) preferably out of the same melting at the same time, so that trouble may not

52

be experienced in joining the tubes to the bulb at a later stage in the laboratory.

In concluding this section we may also mention a beginner's device. He can easily improvise a good-sized bulb in the middle of a small tube by first blowing a large stout-walled bulb at the end of a large tube, see Fig. 22 (3), and then joining on to its bottom a small-size tube. This operation is repeated on the other side if necessary after the larger tube has been melted away close to the bulb, and the glass in the neighbourhood formed into continuation with the rest of the bulb by a careful softening of this region round the tube, and blowing (4).

Joining on to the thin walls of a bulb is rather risky at times, and if the special procedure suggested for it earlier is not adopted, a useful method is to form a thicker region of glass locally by adding a blob of glass (5) where required, and spreading it out by blowing. Now, if a hole is blown out in the middle of this thickened patch, a joint can be worked here satisfactorily without introducing any serious deformation.

53

CHAPTER V

ENCLOSED WORK

A TYPICAL work of this nature is the making of a double-jacketed condenser. It serves very well to illustrate the special feature of this class of work, namely, the process of working the second internal seal. The inner piece A, whether tubular or spiral, is got ready first, and its open ends on either side slightly enlarged to form thin protruding lips. It is then slipped into the outer tube, one end of which has already been drawn down to a spindle, and then this end is also drawn out as shown in Fig. 23 (1), care being taken that the outer tube is rather too long than too short. After being softened down at ↗, it is pulled out a little to form the thick-walled enlargement C (2), the object of which we shall see later. The narrow waist beyond it is now softened down to a smaller bore than the enlargement at the end of A (1), which is then slid into contact and the junction fused down to establish all-round sealing contact. It is then blown out and the joint worked uniform as a simple case of internal joint (Ref. p. 31). The bulb C is now softened and blown out and drawn apart to form the connecting tube *d* (3). If there is not enough glass in C to form the tube *d* to the required dimensions, the short tube so formed may be cut off a centimetre beyond the joint as indicated, and a regular tube joined on, the internal joint being carefully kept hot during this operation by the use of a large flame, and the joint partly grazed as well at times.

54

Before the joint is allowed to cool, the side tube *s* has also to be joined on, and to prevent the cooling of the internal joint in this operation it is safer to do it quite close to this internal seal in the position shown rather than in the dotted position indicated. Further, unless the whole of the joint has been kept hot during the working by the frequent use of a large flame, it is inadvisable to try any annealing over a smoky flame. It is safer to let the work cool by itself, and rely on the annealing the joint has received during the working of the other joints. The joint on the other side is started in the same way, and when sealing contact has been established all round, the blowing has to be done through two openings on either side of the joint, since there is no through communication from one side to the other. The first puff is blown through S, the side less heated, and the rest through 4, the glass outside the joint being hotter and thus retaining the heat during this slight delay. The same effect can also be achieved by joining T and S by a rubber tubing and blowing through 4 alone, but is not recommended here for the reason that it will seriously interfere with the free rotation of the joint in the flame. The joining of a longer outer tube, as well as a side tube on the other side for the outflow, are also effected immediately as before and in close proximity to the internal seal. This joint should be kept hot all the while, and cooled off after a carefully prolonged annealing, to ease the strains resulting from the fixity of the other end. The wider tube D for corks may also be joined on if necessary before the spindles are all cut off, and the edges polished in the flame to give the finished specimen.

A spherical condenser is only a special case of this kind of work employing a sufficiently wide and short length of outer tube to enclose the inner bulb. By a judicious blowing of the outer tube beyond the joints into a regular

55

hemispherical shape, with no cylindrical connecting tube between them, the tube form of the outer envelope dis-

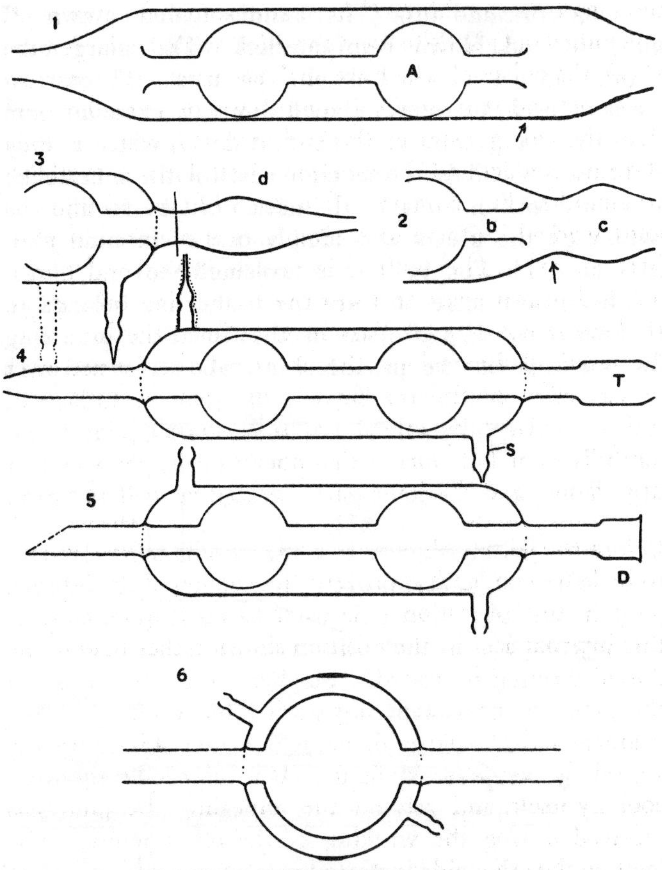

FIG. 23.

appears, giving place to a nearly spherical shape, as shown in Fig. 23.

An alternative method of doing jobs of this kind is

56

to divide the outer envelope in two by a sharp and square cut, taking care not to chip off any portion of the edges if the method of leading the crack round by a hot point is used. After the inner piece is introduced, the two halves are put back in their original position, and because of the perfect and almost air-tight fit all round the cut, the joint can be fused together again without leaving any marked trace. The internal seals on either side are made as usual, and in case there is no lead from this outer envelope, a temporary lead for blowing purposes must be formed near one of the internal joints, though it may be sealed off after effecting the second seal.

Bulb inside a Bulb and Vacuum Vessels.

It is often a case of wonder to those who do not know the process, how a large bulb (as in a Dewar flask) got inside another through the narrow neck. The process does not seem to have been described in any books on glass-blowing, though it is comparatively simple, and finds large application in general vacuum work. The usual procedure is to blow the outer bulb first (or get it blown from the glass works) at the end of a comparatively large bore tubing, and then join on to it the smaller piece T at the opposite end for blowing and other purposes, as in Fig. 24. At the end of another tube B that will readily slip through the neck of the wider tube with ample clearance, enough glass for the inner bulb is collected, and when softened down to a size that will easily slip through the wide neck of A, it is carefully introduced into the bulb through the neck and blown in this position (1), due care being taken that the hot glass does not come into contact with the outer bulb anywhere during the rotation of the work in blowing the inner bulb. Because of the outer envelope of glass

57

the inner glass remains soft much longer, and hence the blowing must be carefully regulated to prevent its being blown out into contact with the walls of the outer bulb. It is better to keep the outer bulb clamped for convenience in doing small-size work, while for the bigger-size bulbs an assistant's aid is also required to warm up the outer bulb while the glass for the inner is being softened, and then to bring it up and hold it in position for the blowing.

Dewar Flasks and Tubes.

Having described the process of blowing the inner bulb, we may at this stage discuss the further procedure by which the ends of the necks are joined together to form a vacuum flask. The neck of the outer tube having been already cut square by the bent-wire method (see p. 15), the neck of the inner tube has to be cut longer by nearly double the clearance between the two bulbs, so that there may be enough glass to push the line of joint on to the outside of the neck and thus away from the hot or cold fluid introduced into the finished vacuum flask. By a wooden bar pushed through the back tube the neck of the inner bulb is kept pushed out as in Fig. 24 (3), and the edge softened and enlarged to a size large enough to cover the neck of the outer bulb.[1] The two ends are now fused together into contact without holes, and the joint blown out uniform through the back tube T. During this blowing the tendency of the neck of the inner bulb to be pushed outwards as illustrated (Fig. 24 (5)) must be counteracted by pushing

[1] The risk of softening the neck of the outer flask in this operation, and causing it to stick to the inner, can be avoided by inserting a piece of asbestos cord round the neck of the inner flask, to keep it well separated from the outer flask. After effecting the enlargement, this separation piece can be taken out with the help of a bent wire before sealing the lips together.

58

in a wooden rod as shown by the right hand (4). Now the irregularities of the joint will get blown uniform with-

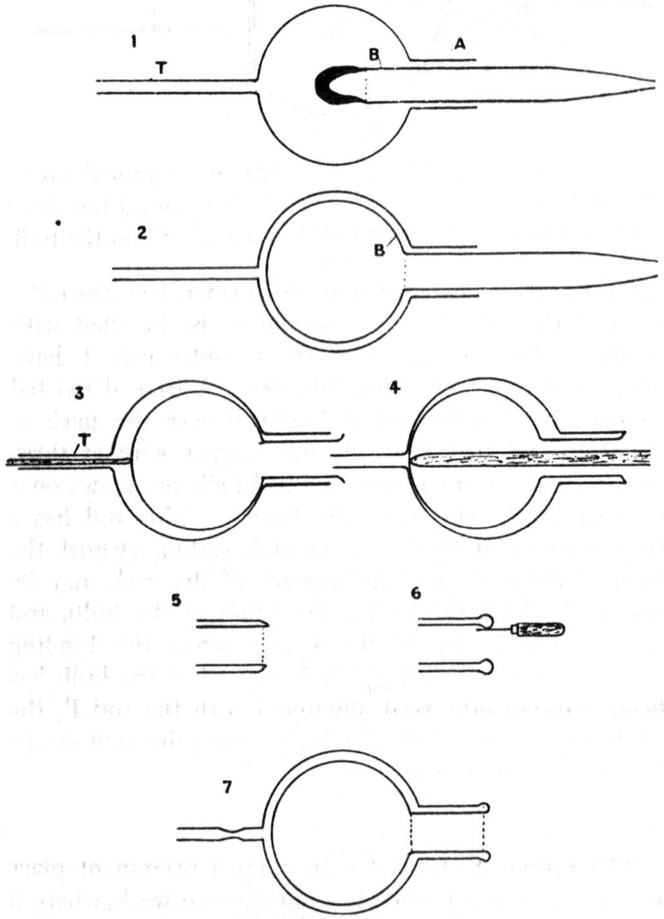

FIG. 24.

out the line of joint being pushed out beyond the small extent permissible and necessary to bring the inner bulb

59

concentric with the outer. Another tendency of the joint noted at this stage will be for it to get blown too round and bulging, and thus constrict the aperture at the neck as shown (6). This must be rectified by turning this softened enlargement pressed against the spinning tool as shown, without exerting any undue force to push back the bulging glass. Before annealing and cooling the joint, the inner bulb must be adjusted concentric with the outer. The finished bulb, with the joint located well outside the tip of the mouth, is illustrated in Fig. 24 (7).

For a cylindrical Dewar tube, two wide tubes, one sliding inside the other, with an all-round clearance of at least 2 mm., have to be drawn off with stout spindles at one end and the other end cut square at the right length by the bent-wire method. Corks provided with central tube have also to be got ready to close these open ends when required, as shown in Fig. 25 (1). The spindle of the outer tube can be drawn off, the end rounded to perfect hemispherical shape, and the end tube 2 joined on to a hole pierced at this end (2). The inner tube being held by its spindle, its square-cut end is enlarged to cover the bore of the outer, and then closed with its cork and tube; its spindle is then drawn off and the end blown hemispherical. The inner tube when cold can be slipped into the outer, the ends in contact fused all round, and the joint worked uniform and smooth as above. The heating should be only right on the end face, and tendencies of the inner tube to project out or bulge at the neck must be rectified as before by the use of the stick and the spinning tool, as indicated in Fig. 25 (4), and explained above.

The manufacture of Thermos flasks also belongs to this class of work, and is done on the same lines, excepting for a change in the procedure owing to the narrower

60

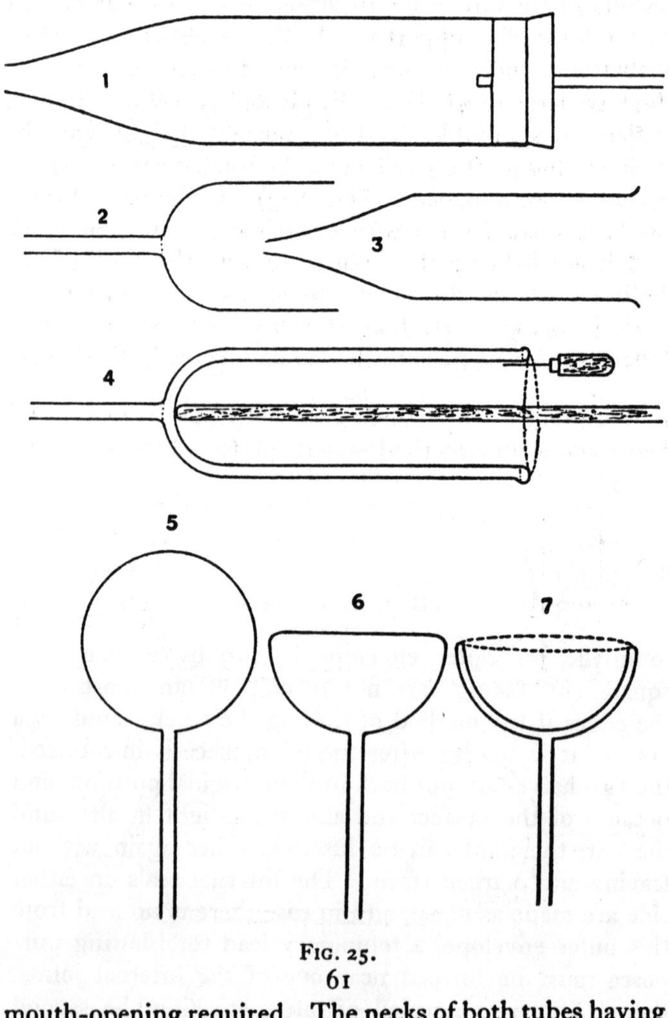

FIG. 25.

61

mouth-opening required. The necks of both tubes having been drawn out to the narrow bore and cut at the right length, the inner tube has its bottom rounded, and is slipped into the outer tube through the wider bore at the back as shown in Fig. 26 (1). The excess length of the outer tube is drawn off behind into a stout spindle, and the projecting lip of the inner tube enlarged and the joint at the neck effected and worked uniform as before. The glass at the head of the spindle is blown to a hemispherical shape and the glass of the spindle itself collected and drawn to form the tail tube for exhaustion and sealing purposes (2).

An interesting and simple work of this class is the vacuum bowl illustrated in Fig. 25 (7). For this a bulb is blown at the end of a tube, and the outer hemisphere of it slowly softened down in a large brush flame to shape (6) and then gently sucked in, when the shape (7) results without further ceremony.

In all these and like cases, where a comparatively heavy mass of glass has to be supported and handled in uniform rotation by one hand, the supporting tube T, or the spindle drawn for the purpose, must be substantial and extra strong to take up the load without danger of fracture at a critical moment. Mechanical aid in the form of a support to take the weight is resorted to sometimes, and in general it takes two shapes. In the first, we have two horizontally pivoted cylindrical wooden rollers on a suitable stand of adjustable height, and the bulb resting on these rollers can be rotated easily without any exertion, as shown in Fig. 26 (3). In the second type, we have a wooden bar with a notch at one end, the other end being held by the right hand with the neck of the bulb rested in the notch, when a region beyond the neck is being worked as shown. But evidently this latter process keeps both the hands occupied, in addition

62

to providing a very fatiguing and unsteady support, and hence the former method is to be preferred in every way.

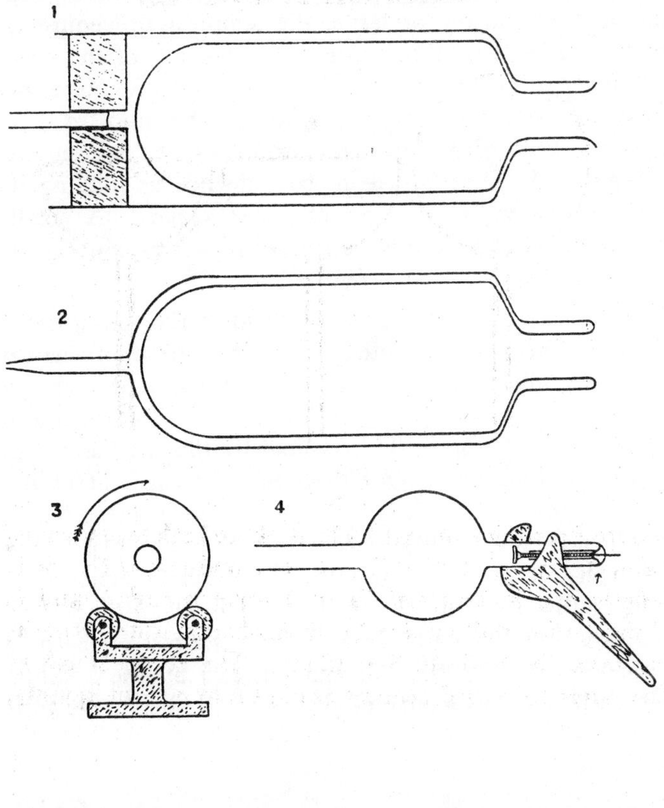

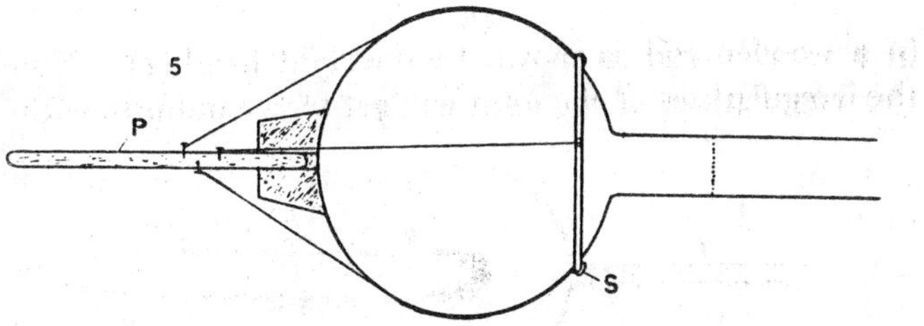

FIG. 26.

Occasions arise when a tube has to be joined on to the neck of a ready-made bulb, which is found too short for the purpose. It is difficult to keep a hold on the bulb

63

by the left hand fingers even in the smaller sizes, since the walls of the bulb become too hot to be handled with comfort. Fig. 26 (5) illustrates a contrivance I have often used to get over this difficulty. A ring of twisted copper wire S is formed and slipped over the neck of the flask and tied up with fine copper wire at three or more equidistant places, to small nails or screws on a wooden rod P serving as the handle. This rod has a large cork fitted tightly at its end, resting against the back of the bulb, and the surface of the cork may be hollowed a little to suit the curvature of the bulb, and thus ensure better frictional grip when the binding wires are twisted and tightened up. After the bulb has been adjusted into axial alignment with the rod P, the work may be rotated steadily for effecting the joint on the neck or anywhere beyond it.

Electrodes.

The fitting of electrodes occupies a prominent place in glass work and peculiarly so in vacuum work, where it is frequently required, and that sometimes on a massive scale. Electrodes form a constant source of trouble if not fitted properly, especially if the work begins to reveal its leaking or cracking propensities only at a later stage when it is difficult and often dangerous to reheat it again for repair.

In the simplest cases if a piece of fine platinum wire is put through a hole in the walls of the glass tube, and the glass well melted down on to the platinum and then blown out uniform, a satisfactory air-tight joint is produced, establishing electrical connection sufficient to serve the purpose. But any want of equality between the coefficients of expansion of the glass and the platinum leads to cracks or leaks sooner or later, especially if the wire employed is thick, and hence it is the general practice

64

to introduce an intermediate material between the two in the form of an enamel, of which there are a lot of varieties in various colours. Ordinary colourless soft lead glass or the variety of blue or white enamel glass serves the purpose admirably, and they are available in the form of small sticks about a tenth of an inch in diameter. The resulting procedure for a simple electrode at the end or side of a tube is illustrated in Fig. 27. The enamel glass is drawn into a fine rod less than a millimetre in diameter, and the platinum wire being held in the flame by a pair of pincers (very necessary in this work), an oval bead of the enamel glass is melted on to its middle and well fused in position. A small patch of glass on the tube is heated by a pointed flame and pulled out by sticking a waste spindle on to it. The tip of this conical projection is broken by a scratch from the knife or file and a hole smaller than the bead on the platinum wire is the result. The electrode is then introduced through this hole and held in position by the pincers, while the bead of enamel is fused on to the outer edges of this hole by a sharp flame. The whole of this electrode region is well fused together without softening an unduly large area, and care is taken that the platinum wire inside does not sag down and make a contact with the glass elsewhere. It is then blown out into an imperceptible hump on the tube, and the operation repeated to secure a symmetrical and graceful shape before annealing and cooling the work.

In general it is not safe to leave the platinum wire sticking out and make electrical connections straight on to it, since it breaks easily at the joint if the wire gets bent this way and that a few times. Hence the usual practice is to loop the wire outside and embed this end also in the enamel at the joint, thus providing greater mechanical strength for the platinum loop. But this

65

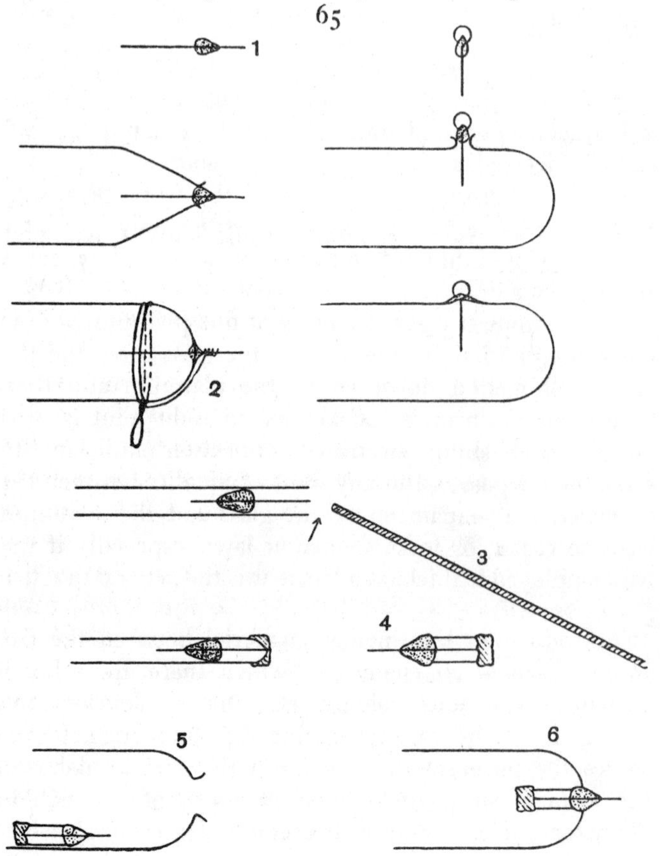

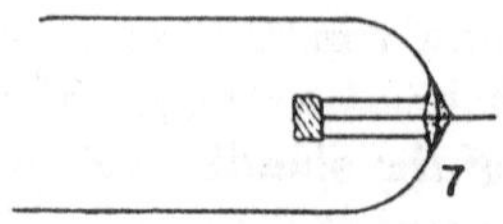

FIG. 27.

66

necessitates a greater length of platinum wire, which is rather costly, and hence to be very sparingly used. A common and simpler plan is to leave a short length of platinum wire projecting out straight, the electrical connection to it being made through a piece of fine copper wire twisted a few times round the glass tube, and then the end twisted round the platinum wire as shown (2). Thus all mechanical strains are taken off the platinum permanently, the connections being made always to the outer end of the copper wire twisted into a loop.

For general vacuum work these simple electrodes are not satisfactory enough, since the unprotected state of the platinum wire inside limits its current-carrying capacity. In discharge tubes, for want of sufficient surface, the wire will get too hot if the discharge starts directly from it, and hence it must be sheathed by a small tube of glass and thus used only as a conductor to the electrode proper, having a greater mass and surface. Further, the marked sputtering platinum exhibits makes it unsuitable for an electrode. In this respect, aluminium is far superior to platinum, and its lightness is also greatly in its favour wherever large electrodes to carry heavy currents are required. In such cases the electrode surface of aluminium is fitted on to the sheathed platinum in either of two ways.

With small discharge tubes, where the electrode need not be extra heavy or large, the simplest procedure is to melt a bead of aluminium on to the top of the projecting platinum wire itself as follows: A piece of aluminium wire about 30 cms. long and 2 mm. in diameter is kept ready for such purposes (the long length being simply for the sake of convenience in holding it comfortably when one end of it is heated). A small length of narrow bore-glass tubing (a waste spindle) of

67

about 2 mm. external diameter and 1 mm. bore, has a small length of platinum wire with a bead of enamel glass already formed on it, placed at its one end as shown in Fig. 27 (3), with a millimetre or two of the platinum wire projecting out. This is held by the left hand, and the aluminium wire by the right, and the end of the aluminium wire is then melted in a rather pointed flame, and coated on to the projecting platinum and glass. An irregular bead of aluminium in metallic contact with the platinum is thus formed at one end of the sheath of glass. It is now broken off from the rest of the tube just opposite the bead of enamel, and by the addition of a further lot of enamel all round the bead, the tube is sealed to the enamel at this end as shown (4). The bead of enamel should be now larger in diameter than the glass tube forming the sheath, and of the conical shape indicated in the figure.

When a heavier electrode is desired, the usual practice is to employ a short length of stout aluminium rod or wire for the purpose, and in this case the platinum wire may be fused to one end of the aluminium wire filed conical, as shown in Fig. 28 (1) and (2). Another method of effecting the joint between the two is to drill a small hole at the end of the aluminium rod, and after introducing one end of the platinum wire into it, to roll the aluminium down tight into gripping contact on the wire. The bead of enamel may now be formed round the platinum and a close-fitting protecting tube of sufficient length slipped over the rod of aluminium. The bead of enamel is now fused into contact with the end of this tube to give the finished electrode element, as shown in Fig. 28 (1), (2), (3).

The procedure of fusing this electrode in position has to be somewhat different, in so far as the electrodes have to be introduced into the tube before the end is narrowed

68

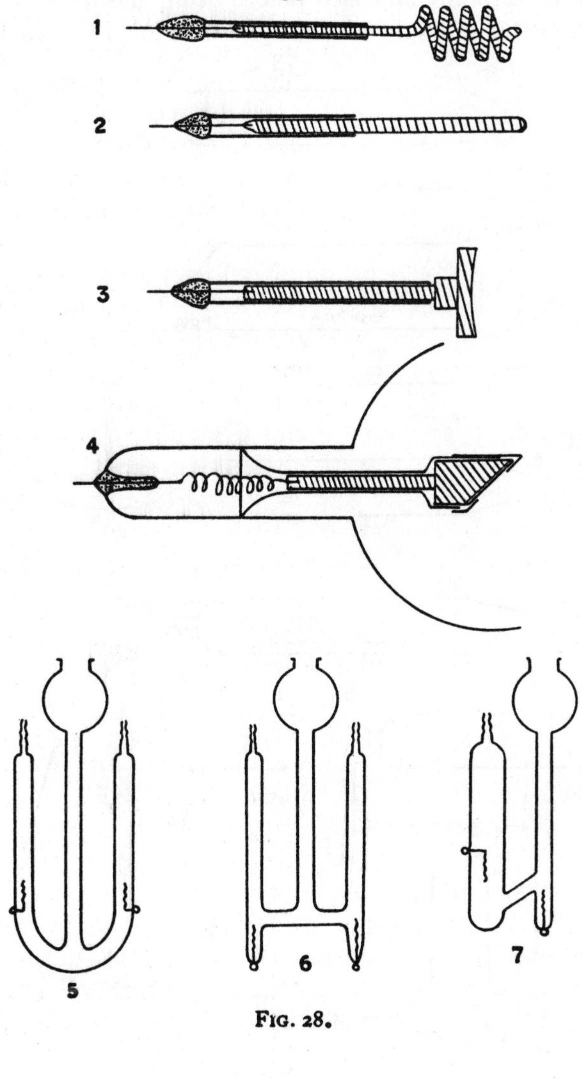

Fig. 28.

69

down to effect the seal. The best way is to pierce at the end of the tube where the electrode is desired, a hole larger than the largest diameter of the electrode. It is then slipped in in the right direction, and the end of the tube rotated horizontally and heated ; as shown, the hole is thus contracted down until it is just too small to let the end bead of enamel slip out (Fig. 27 (6)). The electrode is now jerked into position and kept horizontal by holding the projecting end of the platinum wire by a pair of pincers ; in this position a sharp flame directed to the joint fuses it on to the outer tube. The joint is worked uniform as usual with alternate softening and blowing, care being taken in the heating to see that the electrode does not get bent too much out of the axial position. When the joint has been worked uniform and symmetrical, the whole of the conical end of the tube is softened and blown out to a hemispherical shape as shown in Fig. 27 (7), and annealed. If the initial bead of enamel has been made conical and elongated, the inner glass sheath would not have any contact with the outer walls except through the short intermediate column of enamel, and at the very root of the projecting platinum wire there will be a tiny conical lump of enamel projecting out of the hemispherical end, as shown in Fig. 27 (7). This latter is an advantage rather than a disadvantage, since it provides additional mechanical support to the projecting platinum.

When still heavier electrodes are required, as with some types of X-ray bulbs and rectifying valves, the large mass of metal forming the electrode has to be given extra mechanical support. The tube forming the tight-fitting sheath round the stem of the electrode has its outer end enlarged and joined on to the walls of the outer tube by a well-worked internal seal, as shown in Fig. 28 (4). The metallic electrode is also suitably clamped

70

to the glass sheath, and the connection to the short length of platinum, sealed in as usual, is through a spiral of thin copper wire.

Another class of electrode work occurs in improvising an electrolysis apparatus, of which the Hoffman's apparatus of Fig. 28 (5) forms a well-known example. Here two electrodes of platinum with leading-in wires have to be sealed to the side or end of two tubes. The fitting of the electrodes offers no difficulty, and for more than one reason it is better to have them at the end as shown in (6) rather than at the side. They are often found fitted on the side and projecting vertically up as shown (5), though there is great risk in this method—not so much in the making, but in actual use. By the generation of gas for some time, when the level of water sinks below one of the electrodes the current becomes interrupted and sparks take place between the wire and the liquid ; this often occurs right at the joint, with the unpleasant result of a crack at the joint. If the electrode is bent down instead of being turned up, the interruption of current and spark takes place at the metal tip far away from the glass or joint without any serious consequences. The form of electrolysis apparatus shown in Fig. 28 (7) is specially suitable for vacuum work, where often only one of the gaseous products is required to be fed steadily into an apparatus. By the use of this modified form for the electrode the battery may be permanently connected to the apparatus, and the current is switched on or off automatically as the level rises or falls by the escape or generation of gas in excess of the outflow.

CHAPTER VI

SPINNING GLASS

THE spinning of glass consists in shaping the ends and other portions of glass tubes into certain conical and cylindrical shapes by turning the softened glass in contact with an angular blade of metal. This is an operation largely used by experts in table glass-blowing work, though rarely described in any book on laboratory glass-blowing. The significant omission is probably due to the fact that the spinning of glass is more a professional than an amateur's type of work. But it is erroneous to think that it requires very great skill, and with a little practice a lot of work done by the amateur can be finished much more satisfactorily than at present. Further, the absence of any adequate description of the process leaves the laboratory student entirely at a loss to know how certain types of work, such as stop-cocks and conical ground joints, were ever made by a process of glass-blowing at the table. For this reason I propose to treat the subject in detail.

The few simple tools Nos. 3, 4, 5, and 6 of Fig. 3 are essential for this work, and all of them may easily be improvised in the laboratory, the blades of the right size and shape being cut out of $\frac{1}{16}$-inch brass or copper sheet and fitted on to cylindrical wooden handles. When the blade is brought into contact with the softened glass it gets too hot and is liable to stick to the glass, unless it is periodically rubbed over a lump of beeswax or paraffin

wax, that acts also as a lubricant. Further, since the glass has to be rotated backwards and forwards, while pressed lightly against the blade or tool, a truly centred strong axial spindle is an absolute necessity to produce circular symmetry in the worked end.

We can best illustrate the general procedure by describing how a trumpet enlargement is to be effected at the end of a glass tube. The piece of tube with the stout axial spindle is held in the left hand as usual, and its outer end softened in the flame uniformly all round, and most of all at the extreme tip. Tool No. 4 of Fig. 3, after a preliminary warming and rubbing over the wax, is applied horizontally into the end of the tube by the right hand, and with a gentle pressure against the side as the glass is rotated, the tool is gradually swung round towards the left hand so that the enlargement is first formed and then gradually bent back like the mouth of a trumpet, as shown in Fig. 29 (1) and (2). In fact, this is the way in which all enlargements at the end of a tube, big or small, are produced, since the application of a conical charcoal tends to compress, thicken, and distort the glass.

Three points have to be carefully attended to if success in the operation is required. The first is to have a gradation in the heating from the outer end inwards. The extreme end should be softened enough to look quite orange-yellow, gradually merging into the faintest red of the unsoftened portion of the tube. The second point is not to exert undue pressure in forcing the glass into shape. The glass must be soft enough to be gently pushed into shape on the merest contact with the tool. Thirdly, a vigorous to-and-fro rotation of the work against the tool must be maintained throughout the spinning operation until the glass sets hard. It should also be carefully annealed before being allowed to cool.

A better example of this type of work is given by a

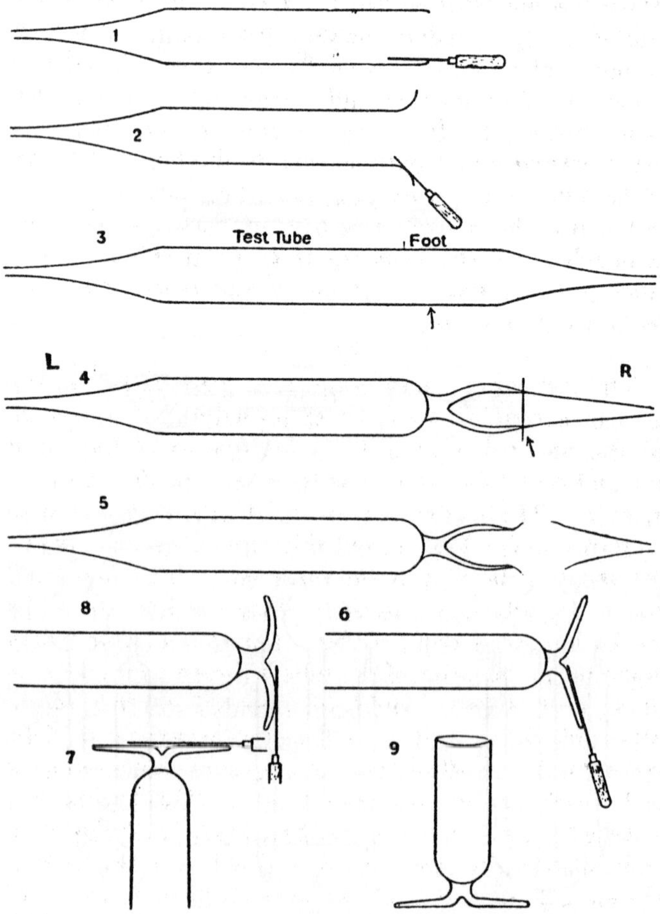

FIG. 29.

74

test-tube on foot. A tube of the form (3), Fig. 29, with spindles on either side, is drawn out, having an extra length equal to about twice the diameter of the tube with which to form the foot. The ends of the spindle are open on both sides, and if the tube beyond the portion required for the foot is softened down, it may be contracted and collapsed into contact as shown (4), forming a cylindrical short neck between the test-tube portion on the left and the foot on the right. By repeated heating at this junction and blowing in first at the left and then at the right end, the bottom of the tube is rounded and the glass for the foot collected together as for a bulb, care being taken to see that the cylindrical neck of solid glass is axial to the tube throughout the operation. The right spindle may now be cut off as indicated (4), or, still better, be blown off from it in a novel way as follows (5): The neck of the spindle is thoroughly softened and quickly blown out through R into a thin bulb or film which may easily be brushed aside to give the opening at this end. The edges of this opening are then softened and enlarged to make the glass for the foot form a cone. Then, after a strong heating and softening of the whole of this glass, it is enlarged out by spinning against the tool whose point conveniently rests in the central hollow of this cone which has not been softened. The result is the shape indicated (6), and before the glass loses its heat it is held upright and stroked horizontally with a flat-bladed tool to make the foot quite flat and horizontal but for a dimple in its centre (7). If during the spinning operation the tool is swung round quite normal to the horizontal axes of the tube the edges are likely to curl up as indicated (8), and hence the above procedure of flattening out immediately after the spinning is to be preferred. The other end may now be cut square and polished to give the test-tube on foot.

75

The making of a measuring-jar or glass on foot is also similar, except in the order of procedure. The mouth of the measuring-jar with lip has first to be formed at the end of a tube which has its other end drawn into a spindle. The mouth is then fitted with cork and tube as shown in (1), Fig. 30, and this tube forms the spindle for shaping the foot at the other end. The foot itself has to be attached separately, since the bottom of the measuring-jar is required flat. For this a short length equal to the diameter of the tube is drawn with a spindle at one end, the other end being rounded and then blown off to give the piece A (2). The spindle at the end of the piece B is drawn off and the excess glass carefully removed as for making a flat end for the tube. This end is then softened and blown flat against a flat-bladed tool, Fig. 3 (1). Immediately after, the piece A is joined on to the back of the piece B as shown (3), the joint being fused very well together and blown first into B and then into A. The right spindle may now be blown off as before, and the projecting short tube softened and spun out to form the foot as shown (4).

Another common laboratory apparatus made by the process of spinning is the common thistle funnel. The usual method given is to blow a bulb at the end of a tube and then soften and blow out the outer hemisphere of it. This leaves the rim quite rugged even after the best of trimming and polishing, and the finished article of commerce with its thick uniform rim is a source of mystery to the student who has tried making it. The real procedure is as follows: A length of wide-bore tubing is joined on to the tube forming the stem (5), and about three-quarters of the length of the wide thick-walled tube blown into a stout-walled bulb of diameter equal to that of the funnel desired (6). The right spindle beyond the short thick neck is blown off, and the cylindrical neck as

76

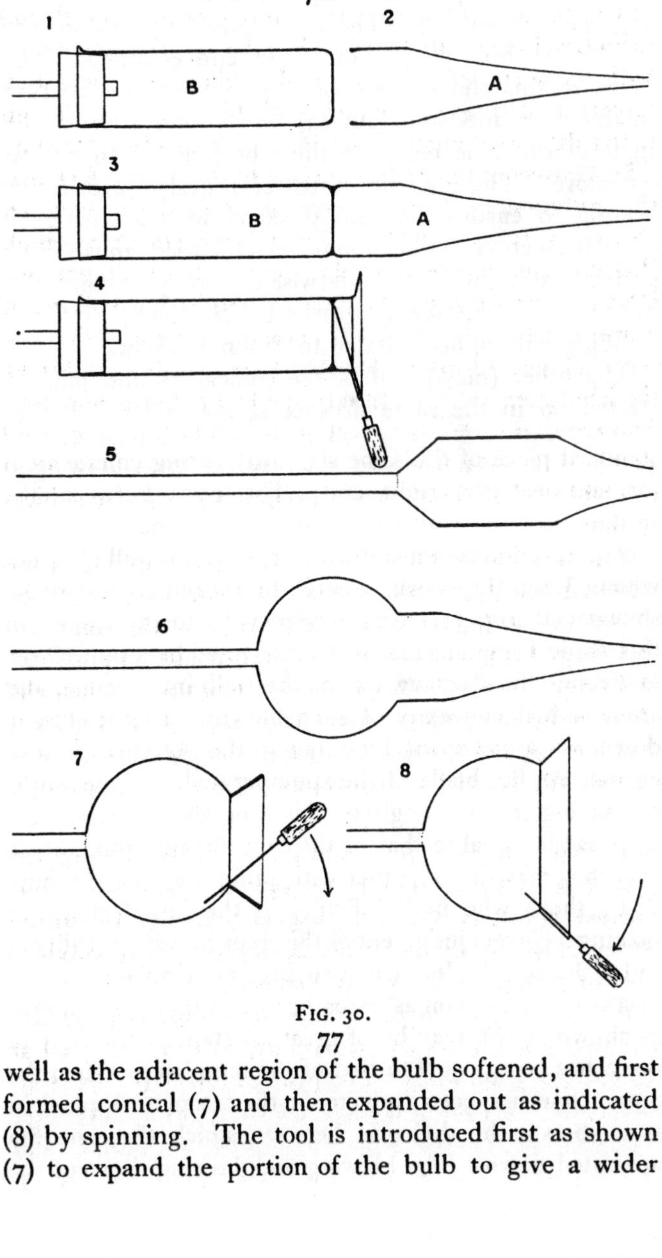

Fig. 30.

77

well as the adjacent region of the bulb softened, and first formed conical (7) and then expanded out as indicated (8) by spinning. The tool is introduced first as shown (7) to expand the portion of the bulb to give a wider neck, and then gradually swung round into position to spread it out as shown (8).

After this preliminary consideration of the process of spinning, we may now describe the two principal operations in vacuum technology that require the aid of the spinning tool.

The first and simplest one is the making of ground joints between tubes, since a large number of these are used in modern work. The ends of the tubes have to be shaped into cones of approximately equal angles and then ground together with an abrasive into a tight-fitting contact over the whole conical surface. The first piece to be made is generally the outer or (female) cone, the procedure for which is as follows: The standard piece having been drawn out (1), Fig. 31, the spindles are recentred with extra care, and at a distance from the right end equal to the diameter, the tube is softened round by a sharp flame and pressed together to give a ring enlargement (2). The right spindle is now melted off and the hole blown out at this end enlarged. The whole of this end is then softened right up to the ring enlargement and spun out (3) slowly into a cone, the angle of the cone being anything between 2° to 5° and not more. The operation must be repeated with lesser heating to ensure the conical shape with a true axial circular section. The spinning tool used must have perfectly straight edges, otherwise concavity or convexity in the conical surface may easily be produced, and extra grinding will be necessitated to eliminate them.

The other (male) half of the conical ground joint is started on in the same manner at the end of another

standard piece of the same size, with a ring enlargement formed near the right end. The glass between the spindle and the ring enlargement is softened, collected together, and blown into a small thick-walled bulb, which, when drawn out a little, forms the conical shape shown (5), Fig. 31. Any tendency to form a waist in this conical region must be counteracted by a puff blown in during the drawing out of the bulb into a cone, and any residual convexity of such blowing is either drawn down or spun down by rotating the softened surface against the flat blade of the spinning tool. It goes without saying that the angle of this cone should be as near as possible equal to that of the cone already spun out, so that they may fit each other with a minimum of grinding.

To those who find difficulty in the initial stages in making a correct judgment of this approximate equality at sight, the simple device of two angular templates cut out of a scale marked on asbestos board, or thin sheet metal, as shown at (6), may be of great assistance, the angle of the cones being equal to that of the template. When the outer cone has been spun to fit the template applied in it as shown, the divisions between which the cone fits are noted down. The blowing and drawing down of the male cone is now repeated until it also makes an approximate fit between the corresponding divisions of the V template. During the working it is dangerous to test the fit of the glass cones themselves, since they invariably stick and crack at the first contact. Another important point to be remembered in blowing and drawing the male cone is to have its largest and smallest external diameters CD and XY smaller than the corresponding internal dimensions of the female cone, so that the inner cone can advance freely into the outer to take up the wear of the grinding. Otherwise shoulders will be cut at the ends where the glasses bear against each other,

as shown in (8), and any further grinding tends only to widen and distort the gap between the two. In such cases, if the end responsible for this action is ground down a millimetre or two against a flat sheet of iron with carborundum powder, the grinding together of the cones may be proceeded with, and a satisfactory joint obtained. It is best to adopt the first process, where this trouble is entirely eliminated.[1] The excess length of the inner cone is cut off after the grinding by the hot joint, and the outer sharp edge of the cut rounded by rubbing over a piece of emery or carborundum paper.

[1] To keep the male cone axial during the initial grinding, it is useful to have a cork of suitable size fitted on to the spindle. The spindle is cut off after the grinding.

An alternative method of forming these cones is by spinning them in a lathe, though it is generally practised only in the case of large cones with thick walls, where greater accuracy of initial finish is essential. The tube is then chucked in a lathe, and held firmly and well centred by a sheet of rubber wrapped round it. A quarter-inch brass or copper rod has a cone of the required angle turned for some length at one end, and forms the spinning tool when held horizontally in the slide rest, with its axes parallel to its travel along the lathe bed. The end of the tube is heated by a blowpipe flame in the left hand, as the tube is rotated, and when it is beginning to soften the tool is gradually advanced inwards by the right hand operating the slide rest. The end thus becomes expanded readily into an exceedingly uniform cone of angle equal to that of the tool. The lathe should be run slow to reduce the centrifugal action on the softened glass, and the rate of advance of the tool inwards adjusted to the degree of softening of the glass. After it has been carefully annealed to release the strains due to any unconscious forcing of the

imperfectly softened glass, it may be cooled off and removed.

The male cone also should be spun in immediately after without the tool being unclamped in any way, so that the two cones may be of exactly the same angle to fit each other with the minimum of grinding. The corresponding tube having been chucked and centred on the lathe, the end is softened as before, and the tool advanced on to the work, with this difference, that it is now made to bear on the external, and not on the internal, surface of the tube, as shown in (10), Fig. 31. To secure this, the tool will have to be shifted back horizontally by a few millimetres before starting on the work of the inner cone. An exact counterpart to fit the cone already made is easily secured if the tool is advanced on to the work to the same extent so that the length of the conical surface is nearly the same.

The grinding together of the cones is most quickly accomplished in a lathe, though the smaller sizes can quite easily be finished by hand work. The best abrasive to use is fine carborundum, since it cuts much faster than emery. In the use of the former abrasive, it must be borne in mind that the best grinding takes place when the cones are held against each other without undue pressure. In the process of grinding the two halves must be lifted apart frequently and put together in a different orientation. When a good fit has been obtained without the slightest wobble, indicating that the two conical surfaces are in contact throughout, the grinding may be finished with a finer abrasive to give a smoothing finish to the surfaces. Any slight misfit between the two cones resulting from the slightest differences in their angles may be rectified quicker in the initial stages if the abrasive is judiciously applied only to that end that is coming into premature contact. If hand grinding

81

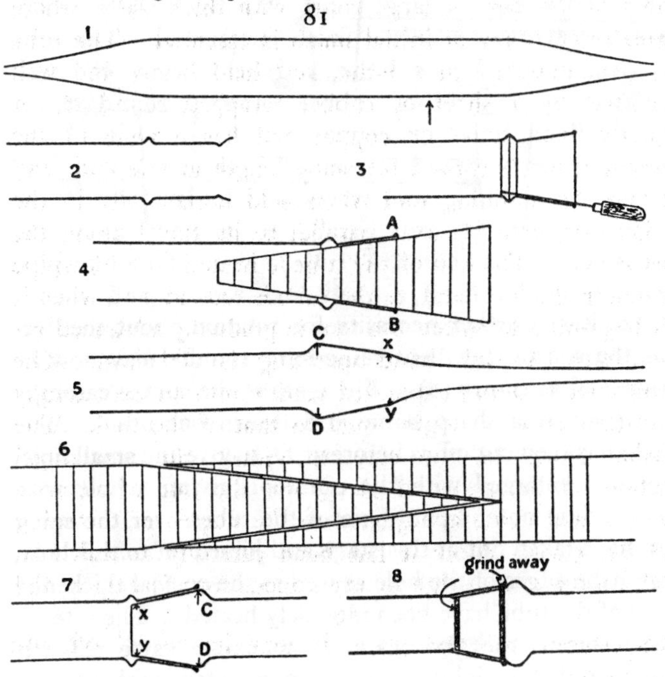

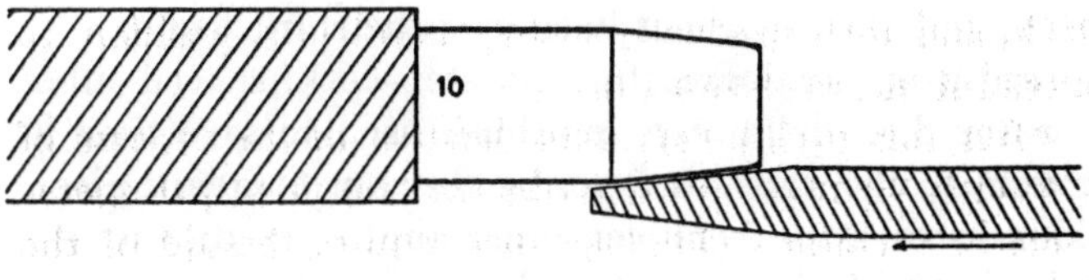

FIG. 31.

82

is resorted to, a further modifying influence can be exerted by holding the work vertical in such a way that the larger particles of the abrasive fall down under gravity towards that end that is giving the tight fit too early. During the grinding the abrasive should also be periodically moistened with a drop of water, just enough to prevent the cones working dry. After the grinding is finished, the tubes should be thoroughly cleaned and every particle of even the finest abrasive removed by scrubbing and then washed clean before drying. This last precaution is particularly essential if the joint is to be greased and used for vacuum work. A great many of the leakages observed with such ground joints are due to gritty particles remaining lodged between them, and often they are only the remnants of the abrasive that have not been washed away in the cleaning.

The second and more complicated work pertaining to vacuum technology is the making of glass stop-cocks. If any good commercial stop-cock is closely examined, traces of faint circular marks can be readily seen round the barrel and the thickened ends, indicating that they were made by spinning and not by a process of cutting out the length from the conically drawn end of a thick-walled tube. The actual process is to form the centre barrel or socket portion by spinning (for which operation ingenious ways of handling are necessary), and then to join on the side tubes to holes formed on the side of the socket. The side tubes with conically shaped thick-walled ends have first to be made, since they must be ready at hand to be joined on as soon as the socket has been spun out.

To make them, a foot-length of thick-walled quill-tubing is softened in the middle, and the glass pressed together and thickened so as to be of about the same external diameter, though the bore inside is reduced to

83

about a third of what it was, as shown in Fig. 32 (1). Before it has cooled down, a sharp scratch in the middle of this thickened region, followed by a brisk knock, severs the two halves, giving a square cut at this thickened region. Since this thickened end has to be heated later on, it should not be allowed to cool in the interval, and is therefore rested on a tripod stand with a spirit lamp flame a few centimetres below it. Such a practice is always used whenever thickened or worked regions have first to be got ready for use later on in the course of the work. One after another the thickened ends are softened at the extremity and shaped into a cone by spinning as shown (2), and then put back over the spirit flame.

To make the socket for the stop-cock, special tubes with thick walls (2 to 3 mm.) are used to save time in collecting the glass, and out of this our standard piece, with extra stout, well-centred spindles, is drawn out, as shown in Fig. 32 (3), the tubular portion in the middle being about 3 to 4 cm., according to the length of the socket required. The heating and softening of such thick-walled tubes should be done very gradually, and the whole of this thick-walled region is first blown out a little and drawn apart to form a conical shape. The right spindle on the wider side is then melted off, the excess glass removed, and the end blown away. This extreme end having been softened sharply and turned back a little by an act of compression against the spinning tool, the thickened end is formed into the triangular section shape by repeated spinning against the edge and flat of the tool. The whole of the conical barrel portion is now lightly softened and spun against a long tapering tool to finish a true conical surface inside. By an alternate application of a flat tool horizontally against the external surface, the latter also is worked quite conical without any convexity or concavity (6).

84

To form the other end of the socket the work has to be supported in a novel way on a bent temporary spindle which enables that end to be spun out into regular shape with equal ease. For this purpose, a spot on the side of the conical barrel where the side tube is to be joined on later, as well as the end of a foot-length of small size quill tubing, are intensely heated and stuck together to make a temporary joint. This tube is then bent round as shown (7), and adjusted to be in true axial alignment with the spindle on the other side. Now and hereafter, irrespective of the region actually softened for work, the whole of the socket must be kept hot by a periodical application of a broad brush flame, care being taken not to soften the joint of the bent temporary spindle in this operation. The work being rotated by the provisional spindle, the spindle on the right side may be fused and drawn off, and the conical thin projecting glass D left at this end is removed by a sharp scratch at its base followed by a gentle sideways knock. The small hole thus formed here is enlarged by spinning in the flame, first with a pointed pin tool, and subsequently with one of the tapering blade tools. The end ring is formed as before by compressing the softened glass and spinning it into shape. Any deformity in the conical shape of the socket at this end is also rectified by softening this end region and spinning it without enlarging it in any way. In fact, if the edge of the tool is applied flush with the conical portion already formed at the other end, the risk of such a deformation does not arise at all, and the end region easily makes a regular continuation of the cone already formed.

The socket being thus shaped, it remains only to join on the two side tubes. For this another temporary spindle, in the shape of a straight quill tube, is joined on to the diametrically opposite spot on the socket, and since

85

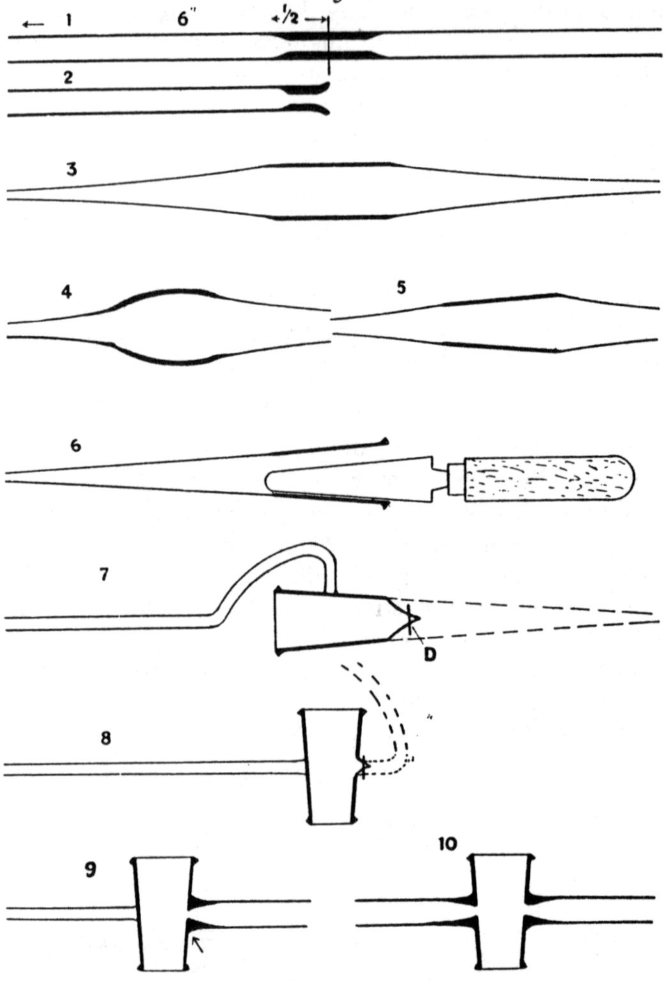

Fig. 32.

86

there is no further necessity for rotation about the axes of the socket, the bent temporary spindle may now be fused off as shown (8). The glass at the spot is carefully softened by a sharp flame and drawn out into a cone, which is then cut off as before at D, to give the small hole. This is enlarged with the pin tool, giving a hole with protruding outer lips. One of the tubes over the spirit flame is taken out at this stage and joined on to this hole, after the edges of the hole as well as the conical thickened end of the tube have been intensely heated. The intense heat softens the glass to such an extent as to make it flow, and the joint becomes completed without any further work. A little more finishing may be attempted with advantage by the application of a fine pointed flame all round the joint as shown (9), so as to make the glass at the surface fuse into continuity all round and flow from the thickened end of the tube on to the barrel. After this joint has cooled enough to be rigid, it is reversed, end for end, and the other temporary spindle melted off, the hole formed, and the second tube over the spirit flame joined on as before (10). The whole work must be thoroughly warmed up in a large flame and annealed very carefully before being allowed to cool, a separate annealing chamber being used in case a large number of stop-cocks are being made.

When the side tubes joined on are small-bore capillary tubes, it is often advisable to have at hand the assistance of a suitably mounted pin with which to open out the hole at the joint, as this is likely to contract or close altogether during the operation of joining the tube on to the socket. A long thin steel wire (knitting needle) may be mounted vertically on a block of wood, and if the hole has been choked up, the joint, after being strongly heated and softened, may be plunged straight on over the wire as shown, Fig. 33 (1), and momentarily pressed down

87

to re-form the hole. If only a widening of the hole is required, the short bent conical end of a steel wire on a handle may be applied momentarily, as shown in Fig. 33 (2), from the inside through the wide end of the socket, the joint having been thoroughly softened before and after to relieve all strains introduced in this process.

The inner piece or plug of the stop-cock may be solid or hollow, and the process of making them differs accordingly.

The solid plug is the easier to make, especially on a commercial scale, when it may simply be squeezed to the

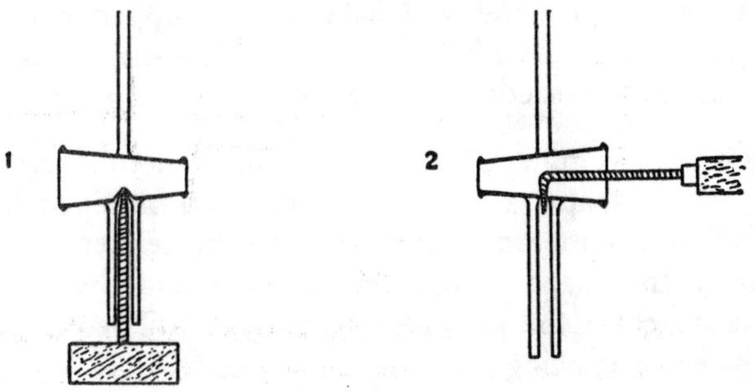

FIG. 33.

approximate size and shape in a mould by the glass-maker himself. It is also easily made in the laboratory out of glass rod worked into the approximate size in the flame, though the impossibility of enlarging a constricted portion by blowing necessitates a certain degree of skill and judgment in getting the piece into shape by a process of pressing together and pulling apart of the softened glass.

For the cross-bar or handle a small-size glass rod is drawn off with spindles as in Fig. 34 (1). It is then softened and constricted in the middle by pulling apart (2), and by a judicious series of softenings and drawings apart

88

shaped into (3), and then bent to the shape of the left half of (6), the horizontal spindle P being centred about the centre C of the cross-arm. For the barrel or plug portion enough glass out of a thicker glass rod is first

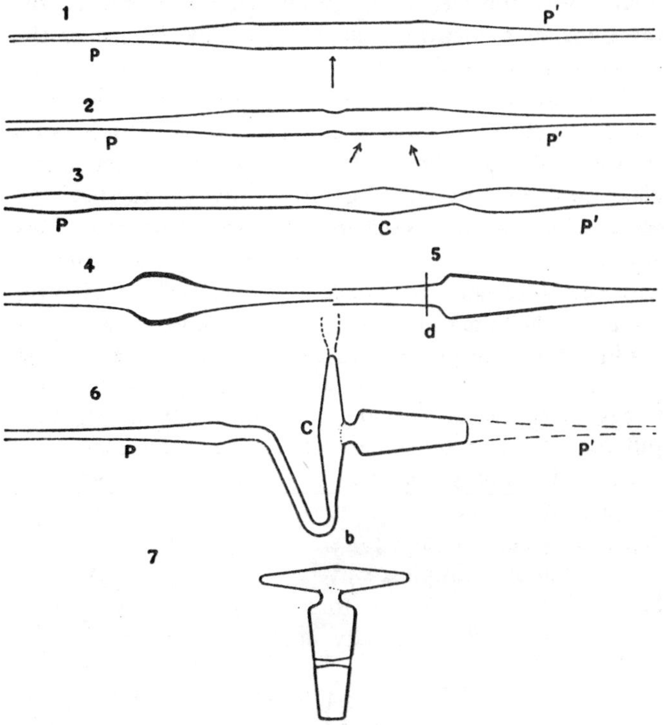

FIG. 34.

collected into shape (4) and then carefully drawn out to shape (5), a template being used as suggested before to get it of the approximate size and angle to fit the socket. The left spindle of this piece having been fused off beyond a short neck, the cross-arm and the barrel are fused together as shown at (6), the glass being heated and

68

softened almost white before being put together to effect the joint. Since the joint permits of easy rotation about the spindles P and P′, the joint is thoroughly fused into contact all round in a pointed flame. The spindle P′ on the right may now be melted off at the right length, and the end of the plug fused and compressed flat. The whole of the plug must be heated to redness and annealed to release all strains. When it has cooled, the bent supporting spindle P may be fused off at *b*, giving the finished blank for the solid stopper.

The stopper can now be ground in the socket with carborundum powder as before, until a good fit without any wobble is obtained. The position for the hole is now marked on opposite sides and the hole itself drilled halfway through from each side until they meet in the centre (7). The drilling is most conveniently done in a vertical drill, where the very gentle pressure with which the drill should bear on the glass is fully under control by the hand. The drill itself is a conical square piece of hardened steel with its ruggedly broken end as well as the sharp ground-down edges acting as the cutting points. The ground-down end of a small triangular file has the requisite hardness, and out of this the drill may easily be improvised. A freshly prepared solution of camphor in turpentine acts as a good lubricant, giving the drill a grip on the glass.

The hollow stopper is made largely (in the larger sizes) to lighten the stop-cock, and it is blown at the table with comparative ease, though it is difficult for the student of glass-blowing to guess the process by which the inner communication tube is sealed in. The initial procedure is exactly identical with that already indicated in Fig. 34, with this difference, that instead of glass rods, glass tubes are used, and they are worked to the same shapes. For joining the cross-piece to the barrel a hole is blown out

90

at O, as well as at *d*, and the heated ends having been put together the joint is worked uniform by rotation in a pointed flame as before. The short inner communicating tube is introduced at this stage by either of the two following processes:

The first and simplest way is to soften and enlarge the barrel of the plug and then form two small holes as shown in Fig. 35 (1) diametrically opposite each other. The

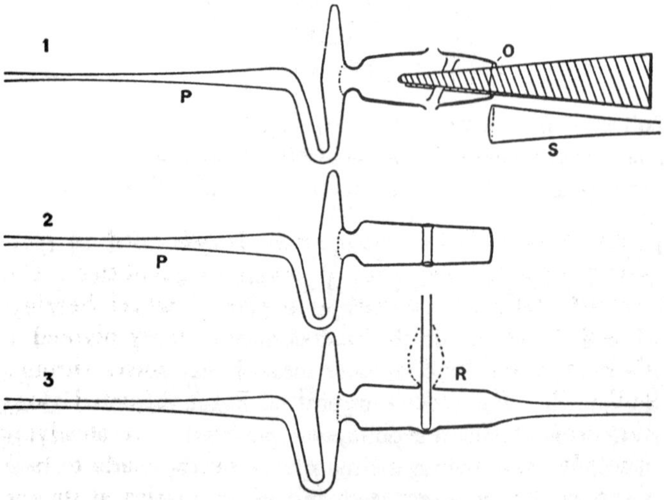

FIG. 35.

right spindle is now cut off temporarily, giving a fairly wide opening O. A short length of thick-walled tubing cut out of a waste spindle is now introduced through O by a pair of pincers as shown, and one end fused on over one of the holes. The cross-tube is gently pushed in vertical and the opposite side of the plug softened and collapsed on to the other end, care being taken that the hole comes into sealing contact all round the rim of the inner tube. Keeping this region hot by using a large

91

flame, the operator again fuses on the right spindle S, and, blowing through it, softens, blows out, and pulls apart the whole of the plug, forming it into conical shape by spinning. The internal pressure of the blowing tends to close the holes, and hence they require to be periodically enlarged with a pin tool. The circular section for the stopper is best effected by softening and spinning against a flat tool applied horizontally until it has come down to the size of the template. The spindle S may now be drawn off and the end shaped down flat (2) by blowing through P.

The second method is to blow out a hole on the side of the plug, and through this opening introduce the closed end of a piece of narrow-bore tubing to form the cross-piece inside. After the closed end has been thoroughly fused into contact, the projecting cross-tube is softened round at R (3), Fig. 35, and blown out thin against the plug so as to come into sealing contact all round the edges of this hole. It is worked thereafter as before, and usually it is finished off with the ends of the cross-tube closed, so that they have to be drilled out after or during the grinding.

The grinding of the stopper to fit the socket offers no special difficulty if care has been taken in the beginning to give them a true conical shape and approximate equality by the use of templates as suggested before. The stopper may be ground inside the socket itself straightaway, especially if the stopper is solid, the hole across being drilled after a good fit has been obtained. But with a hollow stopper the definite position of the cross-tube sets a limit to the grinding, which is reached when the cross-tube comes opposite the two side tubes, and further grinding will have to be stopped whether a good fit has been obtained or not. Further, if the inequalities of section between the two are great, or if

92

there are prominent departures from a circular section or conical shape, a proper fit cannot be obtained by a grinding together of the two, since the tendency is to deform the shape of both the stopper and the socket. In such cases the commercial method will have to be resorted to as being the best.

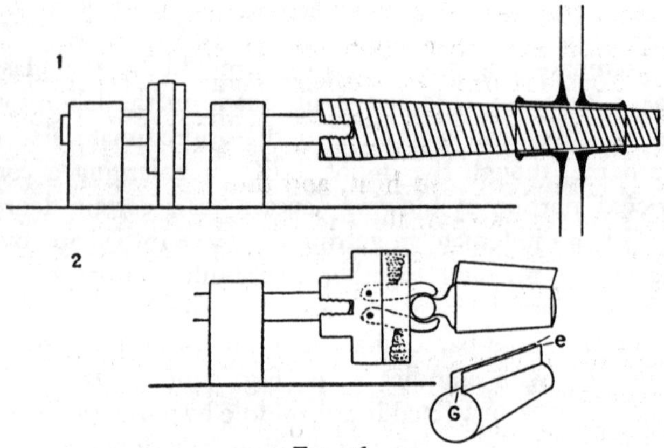

FIG. 36.

Here the procedure is to grind the two parts separately in corresponding metal cones until equality in size is established between the two. Then the stopper is given a final grinding in its socket by the use of very fine abrasive to give it a smoothing finish and perfect fit.

Since the taper of stop-cocks made in one works is very nearly the same irrespective of the size, a single conical rod of iron chucked in a crude lathe (Fig. 36) serves to grind out all the sockets to the same taper or angle, the size varying according to the distance to which they enter along the conical rod. A few strips of sheet iron, $\frac{1}{16}$ inch to $\frac{1}{32}$ inch thick, and about 2 inches wide, are bent to fit this grinding cone at various distances from the end, and they form the corresponding tools

93

to grind out the stoppers to the size and taper to fit the sockets. A little gap G of a millimetre left between the upturned ends *e* of the iron strip gives the stopper-grinding tool a small range of variation from one size to the next by the mere squeezing together of the two ends as the tool is pressed round the stopper, which runs in a lathe, its cross-arm being held between a pair of rubber cheeks on the chuck. By this method the grinding together of a stopper and socket to give perfect fit is only a matter of a few minutes if fine carborundum is used as the abrasive.

There is another very useful little device practised by some experts for giving a temporary support to a small piece that has to be worked or spun out in the flame. We have already seen how the barrel of the socket for the stop-cock was supported by a bent spindle that permitted

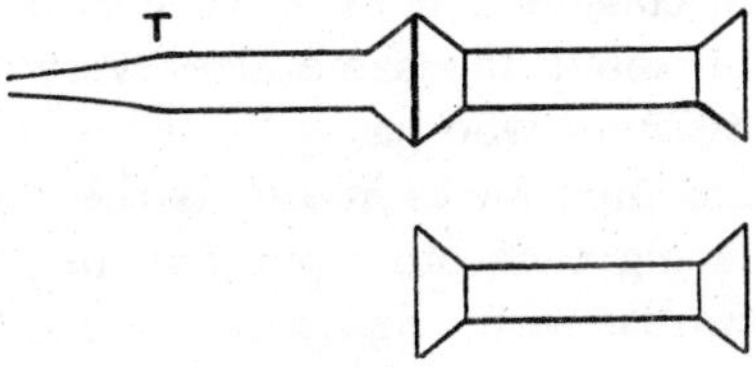

Fig. 37.

of rotary motion for spinning the second end. Suppose a glass spool of small dimensions is the object desired. Evidently a mark will be left if a bent spindle is used to support it by the middle, so this is out of the question for the present purpose. In such a case, the end of a small supporting tube T is first thickened and then spun out as shown in Fig. 37, and kept hot over an auxiliary flame. After one end of the spool has been spun out at the end of a normal spindle, the end of the supporting tube is lightly stuck on to it at a dull red heat, and then the spindle on the right side drawn off, the end enlarged and spun out to form the second flange for the spool. At the close of the operation a gentle knock on the spool severs it off

94

from the supporting tube, a slight warming of the joint momentarily in the flame being useful sometimes to make the spool come off the holder.

At the close of this section we may describe a very

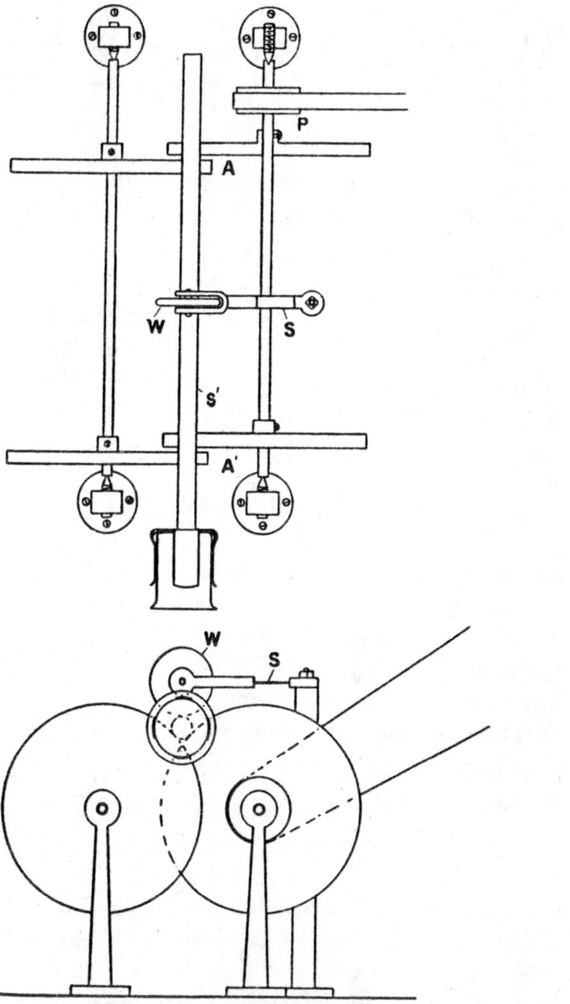

Fig. 38.

95

useful type of lathe arrangement largely used in glass works for producing spun glassware in quantities. Described simply, it consists in a pair of wheel bearings A and A′ on a pair of parallel axes suitably pivoted at their ends and kept in slow rotation by power through pulley P. The wheels permit of being adjusted along the axes and clamped in any position. A steadying wheel W on a springy swing arm S can be made to bear down on the rotating stem S′, which carries at its end in a suitable clamp the glassware, such as a beaker that requires its mouth to be enlarged by spinning. In the spinning of small work like conical ground joints or sockets for stop-cocks, the glass itself forms the stem S′, and the work is directly spun out at its end.

96

CHAPTER VII

COMMERCIAL METHODS FOR LABORATORY GLASSWARE

It is often of great interest to know the processes adopted for the manufacture of laboratory glassware on a commercial scale, and, moreover, such knowledge enables students in the laboratory to devise ways and means of overcoming special difficulties experienced by them in their glass work. It would be difficult, even if the information were available, to enter into the details of commercial practices which differ widely from place to place. All that we can hope to do is to indicate on broad lines the ways in which certain pieces of laboratory glassware are made.

Since many of these are bulky articles, requiring too much glass and of too unwieldly a size to be comfortably handled in a blowpipe flame, they are made directly at the glass works, where the necessary facilities exist. In general the arrangements for this kind of work are as follows:

A high degree of purity in the glass is essential, and hence the ingredients of the necessary purity are gradually introduced in small quantities into a covered fireclay crucible or melting-pot kept at a high temperature in the furnace. If this temperature is sufficiently high, the ingredients readily melt and mix thoroughly during the frothing that occurs on fusion. When this high temperature and consequent fluidity of the glass has been main-

tained for some time, the process of "fining" takes place, and the glass becomes comparatively free from bubbles of any large size. In this stage the glass is too mobile and hot to be gathered at the end of the iron tubes for work. It is then allowed to cool down to the working consistency at which quantities of it may readily be gathered at the end of the tubes. These are clean iron tubes about 4 feet in length and about half an inch in diameter, and have a slight trumpet-shaped enlargement at the end to give a better support for the glass, though at high temperatures the glass itself sticks quite fast on to the clean iron of the tube, so that large masses of glass can be gathered at the end of the pipe and handled without fear of its dropping off during the operations of blowing, spinning, and shaping of the glass.

It may be of interest to mention here a few of the natural forces on which the glass-blower depends to alter the shape of the blob of glass he has gathered out of the pot. It may be too hot and soft, and then he can cool it a little in water before starting work on it. It may not be symmetrical in shape, and he can make it so by rolling it gently over a flat sheet of iron. He can distend it to form a spherical bulb by blowing into the tube, and he can elongate this bulb into a cylinder by merely swinging the tube like a pendulum, when the combined actions of gravity and centrifugal force produce the desired effect. By merely rotating the tube rapidly about its axis he can bulge out the sides of the bulb by means of the centrifugal force acting on the plastic glass. He can reheat and soften the mass of glass periodically at the mouth of the pot, where a fierce heat exists. For spinning operations on the glass he sits on a sort of arm-chair provided with a pair of knife edges or rails over the arm-rests. On these he can roll the blowing tube with the left

hand and give the glass the rapid rotation it requires while the shaping or spinning tool is held against it by the right hand.

After this brief survey of the mode of work of the glass-blower at the factory, we may proceed to discuss in brief the procedure adopted for some particular specimens of laboratory glassware.

The primary article is the glass tube out of which so many other things are made in the laboratory as well as in the finishing shop of the glass works. To make a full length of glass tube at one stretch, a large quantity of glass has first to be gathered at the end of the blowing-tube. This can be done only in three or four successive stages, since the glass in the pot is too fluid to adhere to the tube in large masses at a time. After gathering an initial blob of as large a size as possible, it is blown out a little, made pear-shaped, and allowed to cool and gain in rigidity. It is then reintroduced into the pot and a fresh layer gathered all round this initial blob. When the requisite mass of glass has been collected by a repetition of this process, it is blown out into an elongated bulb whose size and thickness of walls are governed by the size of tube required to be formed out of it. It is a matter of considerable experience to be able to reproduce tubes of about the same size every time. At this stage an assistant who has collected a blob of glass at the end of his pipe sticks it on to the bottom of this bulb, and facing each other the two workers begin to walk backwards with the mass of glass stretched horizontally between their blowing-tubes. The rate at which the men recede from each other is again a matter of considerable experience, governing the ultimate size of the finished tube, both as regards its bore and thickness of walls. In general, for large tubes the men move back only slowly, just taking the sag of the glass between them,

while for narrow quill tubes they move away from each other at a good walking pace.

During this operation of drawing the tube horizontally it cools on the under side, owing to the current of cold air rising past it. To make the effects of this cooling uniform all round the tube, the glass is periodically rotated through the same angle simultaneously at both ends at a prearranged call signal. Boys with fans are also frequently employed to cool down quickly any portion of the glass that tends to get drawn out thinner or smaller than the rest of the tube. The tube as drawn out is laid on a wooden rack like a ladder, and the masses of glass at the ends having been cut off, the central region of uniform tubing is cut up into 6 or 8-foot lengths by the application of a pair of cold tongs, and the resulting bundle is taken straightway to be annealed.

In modern works it is usual to find glass tubes of all sizes drawn by semi-automatic and automatic processes. In the former the glass is gathered and shaped to the initial bulb as usual, and then drawn by machinery in a vertical tower while held between a fixed and a moving platform. In the latter, a cone of fireclay rotating at high speed draws a continuous ribbon of glass out of a tank furnace, and as the glass moves towards the point of the cone, a central jet of air blows it out to form a bulb or cylinder, which is drawn off between rollers at a steady rate.

The formation of a triangular section with white inlaid backing, as shown in Fig. 39 (1), for certain classes of tubing used largely in thermometry, is of some interest. The white backing is readily laid on between the successive gatherings of glass, and in the case of a clinical thermometer tube the gathering finally approximates to a blob about 4 inches in diameter, with a central hollow nearly an inch in diameter. Blowing in to form

100

this hollow in such a thick mass is done with the help of a rubber bulb stuck on to the end of the blowing-pipe, and squeezed by an assistant. By swinging, the mass is then elongated to form a cylinder about 3 inches in diameter, and after the outer layers of glass have been re-softened, it is pressed against an iron flat and the required shape of section given to the glass. The walls being thick and the glass a little cool and rigid, the procedure does not materially affect the section of the cylindrical central hollow. It is then drawn out as usual by hand, great care being taken to see that no twist is introduced in the glass.

Knowledge of these operations is useful as a guide for work in the laboratory, where short lengths of tubes for special purposes may be easily made out of any other size of tubing available. In fact, very often a short length of tube wider or narrower than the tube worked on may easily be formed out of the tube itself, and thus an unsightly joint saved.

The next article of interest is the test-tube of all sizes. In one works the lengths for the wider test-tubes are first cut out of tubing, a diamond being used to give a circular initial scratch inside. One edge of these tubes is then expanded out to form the mouth, and then the other end sealed and rounded. A special blowpipe, giving a very hot, pointed flame, is used for rapid work, and differs from the ordinary blowpipe in that the gas and air supply tubes are heated by a row of fine jets of burning gas.

Flasks, beakers, funnels, and bottles of all sizes are all blown into moulds, of special design, whose halves fit together so perfectly as to leave no trace of a wing or line of joint in the finished article. The interior of the moulds (nowadays made of metal) are faced with suitable greasy dressings that do not stick to hot glass,

101

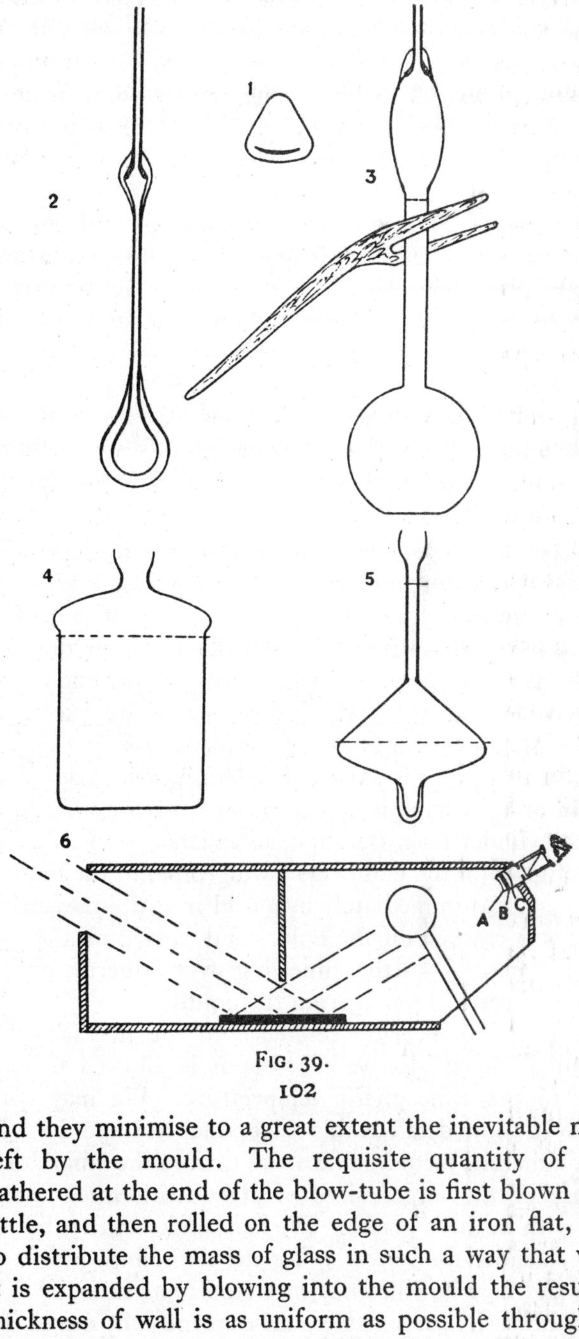

FIG. 39.

102

and they minimise to a great extent the inevitable marks left by the mould. The requisite quantity of glass gathered at the end of the blow-tube is first blown out a little, and then rolled on the edge of an iron flat, so as to distribute the mass of glass in such a way that when it is expanded by blowing into the mould the resulting thickness of wall is as uniform as possible throughout. For a flask an extra mass of glass has to be provided at the end, and then by a swing as the neck gets elongated (2) it is introduced into the mould (whose halves are closed around it by pressing a pedal) and blown, the work being rotated inside the mould until it is rigid. The blob of glass outside the mould is now blown out thin, and suspended by it the flask is taken out of the mould (3). At this stage an assistant applies the forked end of a wooden stick across the neck of the flask and by a twist breaks it off the iron blow-tube, and takes it straight to the annealing oven. The beakers and funnels present the appearances shown in Fig. 39 (4), (5) as they come out of the mould, and because of their dome covering they present but little resemblance to the finished article. These idle domes are cut off in the finishing shop, the method being to mount them on a rotating table and lead a crack, started by a diamond scratch, right round by a fine pointed flame as described on p. 15. The beaker pieces are then chucked on to the horizontal spinning lathe (p. 95), and their lips, heated by a row of about nine radial blowpipes, are opened out, and the spout also formed immediately by pressing down a softened region of the lip, the lip on either side of it being supported by two stationary rods. They are immediately sent to the reannealing oven to ease the strains resulting from these final finishing operations.

A large measuring-jar is another interesting laboratory article. Its cylindrical body is first blown in a mould

103

of the proper shape. It is then transferred to the shaper for chair work or spinning, by which the foot is attached and worked to shape. The glass for the foot is held in readiness by an assistant who sticks it on to the end of the cylinder as soon as it has cooled down enough to rotate rigidly at the end of the pipe. The excess of glass is pinched off, and the blob of soft glass left on the end of the cylinder is flattened and worked to shape with the aid of a few simple spinning tools. Before breaking off the cylinder from the pipe, an assistant sticks a small blob of glass (at the end of a blowing-pipe) to the bottom of the jar, and immediately by a cold iron the measuring cylinder is cracked off the holder and carried on the end of this "pontil" to the annealing oven, where a gentle knock suffices to sever it from the pontil.

We have often made reference to the necessity for annealing blown glassware before it is allowed to cool down to the atmospheric temperature. We may consider it in some detail at this stage, since its importance in the commercial factory is far more than in the laboratory, because of the larger sizes of glass, with thick walls, dealt with in these places. We may understand it better if we consider the necessity for the annealing, how it is done, and how its efficiency may readily be tested.

The necessity for annealing arises principally out of the peculiar thermal and physical properties of glass. It is a poor conductor of heat, and hence equalisation of temperature through a thickness of glass can take place only very slowly. It has an appreciable coefficient of expansion which is not of negligible magnitude. It cannot bear great strains of tension or compression without fracturing readily. Now, in the finishing off, say, of a cylindrical piece of glassware with thick walls, either at the laboratory or at the factory, the external surface begins to cool at once by radiation and by conduction

104

through the surrounding air, and tends to contract, while the inner surface as well as the mass of glass in the interior is shielded from such action. No harm is done so long as the mass of glass is plastic enough to yield to the forces brought about by the shrinkage of the outer envelope. But, as the glass cools down and becomes too hard to yield, a critical stage sets in and strains begin to develop. In the case considered they are readily seen to be of the nature of a tension on the inner surface and compression on the outer. The strains are thus set up mostly by the contraction of the inner layers during its cooling from the critical stage to that of the atmospheric temperature, and are thus proportional to this range of temperature. The more rapid the cooling and the thicker the glass, the bigger this range and hence the larger the strain. But if we keep the glass at or near the critical temperature for a time and allow the interior masses to cool down to this temperature, the whole mass is quite solid before the final cooling begins ; and if this latter cooling also is conducted slowly, the resulting strains are of negligible magnitude. This process of maintaining the article at the critical temperature and then allowing it to cool slowly is called annealing.

An annealing oven in a glass works is a large brickwork chamber kept at a temperature a few degrees below the critical temperature at which glass begins to soften and deform under its own weight. All articles as soon as they are blown and before they have cooled down far below the critical temperature are placed in this oven, and the strains that were developing are eased away gradually during the initial rise of temperature. The oven when full of articles is closed up and allowed to cool by itself very gradually, a process which may occupy a week or more in the case of massive articles.

Such annealed articles often require further working

105

before they are made into the final form of laboratory apparatus. Thus the rough ends of flasks, measuring-jars, etc., and the domes of open articles such as funnels, beakers, etc., have all to be cut away along the dotted lines indicated in Fig. 39, and the ends smoothed out or slightly expanded by spinning. Such operations introduce strains of a minor order which are removed by rapidly passing the article through a reannealing oven. This is generally a long tunnel over 50 feet in length, in which iron trays carrying the articles can move along a pair of rails. About a quarter of the way down the tunnel a set of gas heaters is arranged to maintain this part of the tunnel at the annealing temperature. As each tray of glassware is pushed in, those in front of it are first gradually pushed into the hot zone and then gradually cooled as they move towards the other end, the time of passage of an article through the tunnel occupying generally two or three hours.

We may emphasise at this stage another point of importance in annealing. To ensure a finished product having no residual strains it is not essential that the very slow rate of cooling should be maintained from the critical temperature right down to that of the atmosphere. When the whole of the glass has set hard after its passage through the critical temperature, thin-walled laboratory articles at any rate may be cooled down fairly rapidly, the consequent residual strains disappearing as soon as the article attains the final uniform temperature of the atmosphere.

Table-blown glassware of the laboratory may also be given an annealing with advantage in a small annealing oven heated by a gas flame. A convenient brickwork chamber for the laboratory or small workshop is one about 2 feet square in area and about 1 foot deep. It should be provided with a small chimney outlet with a sliding

shutter, by which it may be closed down tight. The door of entry must also have a similar arrangement for closing the chamber when necessary. The articles to be annealed must necessarily be well finished, as otherwise they would fly to pieces during the heating up of the oven. They are assembled on an iron tray supported at a height of 6 inches from the floor. The heating gas-jets are most conveniently formed by a dozen millimetre holes in a ½-inch iron pipe connected to the gas mains through a control cock outside. After the articles have been assembled in the oven, the chamber is heated up slowly, and then the gas cut off and the oven shut down to cool.

The efficiency of these annealing operations in producing articles free from strain can easily be tested in polarised light, since such strains induce in glass double refraction of a type similar to that of crystalline media. When light polarised by its passage through a Nicol prism or otherwise is examined by another Nicol prism, a position can be found by rotating the latter at which the field of view is quite dark, and then the nicols are said to be crossed. If a double refracting medium like strained glass is now interposed between the two nicols, patches of light appear through the glass, and often these are separated from each other by broad dark crosses which may even be coloured in the case of seriously strained pieces of glass. Such glassware should never be used for any purpose, since it is likely to fly to pieces spontaneously at any time.

A very convenient strain viewer for the laboratory can easily be improvised by using a closed box fitted with nicols at either end, and an opening on the side through which any piece of glassware can be introduced and viewed between the nicols after the analysing nicol has been adjusted to give the darkest field of view. A strong illumination through the polarising nicol is

essential for the detection of the faint traces of strain noticeable in much glassware of daily use in the laboratory. For simplicity the polarising nicol may often be dispensed with, and the polarised illumination of the object obtained by reflecting the light of the sky on a sheet of black glass at a suitable inclination, as indicated in Fig. 39 (6). An ordinary piece of plate glass may also be used for polarising the light by reflection, but, since the light reflected by its bottom surface passes through its thickness of glass, it must be a piece free from strains within itself.

An interesting type of commercial instrument largely used in glass works for testing the efficiency of annealing is the Hilger strain viewer. A closed wooden box shutting off all extraneous illumination of the object is used, and the initial polarisation of the diffused light from an incandescent lamp is effected by reflection at a black glass plate as usual. In front of the analysing nicol a compound lens is interposed, which has a half-wave plate of mica (B) cemented between the concave (A) and convex (C) elements of the lens. The object of this complication is to give the field of view a uniform magenta tint, as it is then easier to locate the strained regions by their change in tint.

Apart from the bulky and large-sized articles made at the glass works directly from the furnace, there is a whole family of laboratory glassware made out of tubes. These are made in commercial quantities either at the finishing shop of the glass works, or in the glass-blowing department of the instrument makers. Considering the variety of designs into which these are made, and that the processes and procedures differ from place to place and from worker to worker, it is not possible to enter into their details here. The broad principles on which such table-blown glassware is made have already been discussed in detail with examples in the foregoing chapters,

and with their aid any one can easily set out the procedure suitable for any particular design of apparatus.

Simple pieces like pipettes and burettes are easily made by processes of straight-line joints, though some care and judgment are required to get them of the right dimensions for the particular capacity. Often in the case of a pipette, after the delivery tube has been joined to the short length of wider tube to form the middle reservoir, it is filled with alcohol or methylated spirit drawn out of a standard pipette, so that some idea may be formed of where the wider tube should be drawn off for joining on the narrower tube on the other side. These liquids are used in preference to water, as they dry easily and so cause no delay in proceeding with the work. Great attention is paid to such matters in devising the scheme of procedure for the production of articles in quantities. Thus one cannot afford to wait idle for the joint to cool before pouring in the liquid. Hence the glass-blower uses the interval in making a number of other pipettes, and does the approximate gauging for each later on as they cool down in turn. In the case of a burette, a stop-cock is joined on to the end of a suitable size of tubing, and in all such articles involving line joints (Fig. 10) the joint, being carefully located exactly at the angle of the shoulder, remains practically invisible.

We may emphasise at this stage the fact that in making all complicated glass apparatus the correct order of procedure previously settled upon facilitates the attainment of success without accidents and failures. This is a matter to which great attention must be paid before work is commenced on the piece, and the possibilities of failure well thought out and amply provided against, especially when parts that have to be made to dimensions are concerned. Any complicated or unwieldy apparatus

109

is usually made in parts, which are united together after being mounted on a temporary stand in their respective positions. This is essential to secure the correct alignment of the parts which adds so greatly to the elegance of the finished apparatus. It is then transferred over to its regular mount. Even the best finished of such articles leave enough clues in it to enable one to devise a procedure suitable for making up a similar article.

But there are some small lamp-blown articles, such as hollow stoppers for flasks, where very little of such traces are to be seen, and often when one has to replace one of these broken articles it is useful to know the procedure adopted in making it. So we will take a hollow stopper as a typical example, the various stages in the making being indicated in Fig. 40. A short standard piece (1), drawn out of thick-walled tubing, has a thick-walled constriction formed in it at one end, as shown at *x* (2). The glass between it and the spindle is then collected to form an elongated bulb, which is immediately pressed down flat between a pair of metal pincers. The neck *y* beyond the constriction is softened and blown out to round it, and then pressed together to give the flattened shape indicated in (3). The tubular glass beyond this neck is then collected together and drawn out a little to form the conical body for the stopper (4). The excess length is then melted off and the end blown flat against a plate, as in the case of giving a flat end to a tube (5). When the piece has cooled, or after its return from the annealing chamber, the spindle by which it was handled may be melted off, and before the pip cools down, if the flattened bulb is partly warmed, the expansion of the air inside blows out the pip into a nice rounded shape as indicated (6).

We have already touched upon the laboratory processes for conical ground joints, stop-cocks, and other articles

110

produced by spinning or shaping. Since considerable personal skill is required for this work it is costly in practice, and we may describe here the somewhat simpler commercial process, in which much less skill is required. The spinning lathe of Fig. 38 is used extensively for the

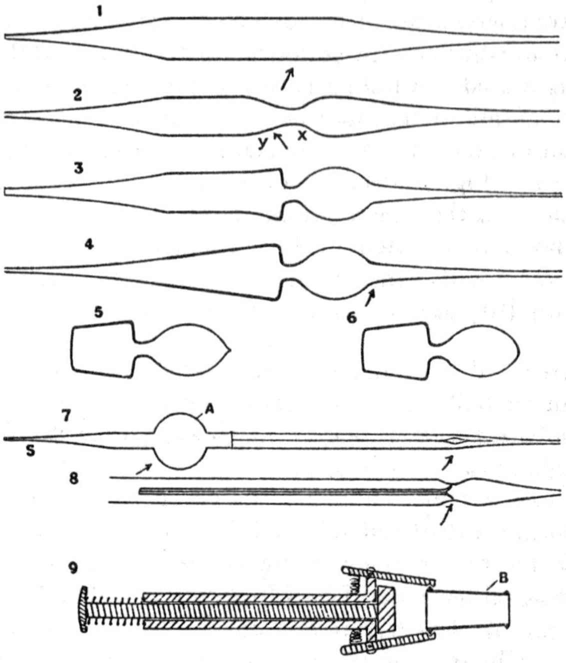

FIG. 40.

purpose, and the spinning out of conical ends for ground joints or other similar articles calls for no special comment. For the sockets of stop-cocks, the conical barrel with the thickened end is first formed at the end of a tube and spun to shape in the lathe, and the piece is then sent to be annealed. On its return from the oven the conical portion is cut away from the tube, and the piece being

111

held at the thickened end with a special form of spring holder sketched in Fig. 40 (9),[1] the other end is also spun out with the thickened edge, and the piece sent off again to be annealed. At this stage it presents the barrel appearance B indicated in Fig 40 (9). On its return from the annealing oven to the blowing table, the piece is handled as before by the spring holder, and the side tubes put on as described before without any difficulty. The piece is then sent away again to be annealed. This simpler process of successive stages is rendered possible by the thorough annealing that takes place between the successive stages. The stopper, nowadays always solid, offers no difficulty, since it is readily pressed in moulds directly at the furnace with or without the cross handle. The hollow cross handle when required is fitted on at the blowing table, as described before.

[1] In the actual holder there are three or more jaws, though only two are indicated in the figure.

The thermometer is an important piece of laboratory glassware produced in commercial quantities, and the commercial and laboratory processes do not differ very much in practice. The process of blowing the bulb at the end of the fine capillary tube offers certain difficulties, and we have already discussed on p. 28 a simple way of forming an enlargement at the end of such a tube. In commercial practice a different procedure is used to exert an adjustable air pressure inside the tube while it is being heated. A bulb A is blown at the middle of a short piece of quill tube, and if one end of this tube is softened it can be stuck on to the warmed end of the capillary as shown in Fig. 40 (7), and a good temporary joint is the result. In this operation, since the end of the capillary has not been softened, the bore does not get choked at all. The outer end of the spindle S as well as the other end of the capillary are now sealed off, and air pressure can be

112

exerted and maintained in the capillary by warming the air in the bulb, and any portion of the capillary may now be softened without fear of the passage being closed. If a region near the end is carefully softened it gets blown out into a bulb which may immediately be drawn out to form a cylindrical thin-walled tube. The cylindrical bulb for the thermometer may be formed of this by melting off the excess length and carefully blowing round the pip of glass left at this end. When comparatively large bulbs are required, the usual process is to cut off the cylindrical tube at its neck and join on to this enlarged end of the capillary a fresh tube of suitable thermometer glass. This is then blown out into a thin-walled bulb and drawn down to form the cylindrical bulb with specially thin walls. In the case of internally sealed thermometers, the fine thermometer stem is first prepared by this process to have an enlargement at one end, which is then internally sealed in the constriction drawn in a tube which forms the casing on the one side and bulb on the other, as indicated in Fig. 40 (8). For the production of spherical bulbs at the end of comparatively wide bore thermometer tubing, regular mouth-blowing is sometimes resorted to. It is somewhat hard work to exert the requisite air-pressure by the lungs, and special care has to be taken to keep the mouth dry and the tongue in a peculiar position so that the amount of moisture forced into the tube is a minimum. Templates of wood with holes in it are frequently used to gauge the size of the bulbs.

Another very interesting commercial process is machine glass-blowing, by which bottles and other laboratory articles are made nowadays in very large quantities. The bulbs for incandescent electric lights are also made by this process, and a single machine is capable of producing as many as 20,000 bulbs per day without human

113

assistance at any stage of the work. In the actual manufacture of these bulbs into lamps, processes of automatic machine-blowing are again used, though assisted at certain stages by the operators in charge of the machine. These are some of the modern processes by which scientific glassware is being manufactured in quantities, and though they are all very interesting the mechanical appliances involved are rather complicated, and space would not permit us to enter into their details here.

INDEX

When this book was compiled a whole page of Dick's was left out. This is that page. Rather than reworking a third of the whole book it was considered acceptable to let you refer to this page for the missing items.

4251. Chloride of Cobalt. Dissolve carbonate of cobalt in muriatic acid; the solution deposits rose-colored crystals on standing, which contain water. By evaporating the solution, anhydrous blue crystals of the chloride are obtained. (*Cooley.*)

4252. Carbonate of Cobalt. This is precipitated from a solution of nitrate of cobalt, by carbonate of potassa, producing a pale peach-colored powder, soluble in acids. (*Cooley.*)

4253. Acetate of Cobalt. The acetate is obtained by dissolving carbonate of cobalt in acetic acid. Acetate of cobalt forms a sympathetic ink. (*See No.* 2540.)

4254. Manganese. A hard, brittle, greyish-white metal, very easily oxidized, fuses with difficulty, unaffected by cold water, but dissolving freely in dilute sulphuric acid, evolving hydrogen gas. It has a specific gravity of 8.013. It is obtained by calcination in a crucible, at a strong heat, of 10 parts by weight of an oxide of manganese, made into a paste with oil, and combined with 1 part calcined borax.

4255. Peroxide of Manganese. The *black oxide* is the only oxide of manganese that is directly employed in the arts. It is a plentiful mineral production in a crude state; and is purified by grinding the native mineral or *pyrolusite* in mills, and removing the earthy matter by washing. The blackest samples are esteemed the best.

4256. Alum. The alum of commerce is a sulphate of alumina and potassa, obtained by lixiviation (*see No.* 23) from crude alum ore, or *schist.* It is obtained in large crystals, slightly efflorescent. It is applied in the arts to a great variety of purposes. When deprived of its water of crystallization by heat, it becomes *burnt* or *dried alum.* Pure *red* or *roche alum* was originally imported from Italy, where it is found in a native state. This has a reddish tinge, which extends more or less through the crystals.

4257. Hydrate of Alumina. Dissolve alum in 6 times its weight of boiling water, add a solution of carbonate of potassa, in slight excess, agitate the mixture for a few minutes, and then allow it to repose. After a time, pour the clear supernatant liquor from the precipitate or sediment, and wash the latter three or four times with tepid distilled or soft water. Next collect the precipitate on a fine calico filter, and again wash it with tepid water. When it has drained, press it between bibulous paper, and, lastly, dry it either without heat, or at a temperature not higher than 120° to 130° Fahr. The product is a soft white powder. (*Cooley.*)

4258. Acetate of Alumina. Add a solution of acetate of baryta to another of sulphate of alumina, and filter. Or, add 5 parts alum to 6 parts sugar of lead, each being first dissolved separately in hot water, and allowed to cool before mixing; decant the clear liquor. The pure acetate is made from pure hydrate of alumina, by digesting it in cold, strong acetic acid, until the latter is saturated. By spontaneous evaporation long transparent crystals form.

4259. Sulphate of Alumina. Saturate dilute sulphuric acid with hydrate of alumina; evaporate and crystallize.

4260. Butyrine. An oily fluid obtained from butter. Keep clarified butter in a porcelain vessel, at a heat of 66°, for some days; carefully collect the oily portion which separates, and agitate it with an equal weight of absolute alcohol for 24 hours, then pour off the clear and evaporate, treat the oily residuum with a little carbonate of magnesia to remove any free acid, and wash off the *butyrate* of *magnesia* thus formed with water; next heat the remaining fatty matter in alcohol, filter, and evaporate, to obtain the butyrine.

4261. Bromine. A dark reddish-colored liquid, having an odor resembling chlorine. It freezes at —4°, boils at about 135° Fahr., is very soluble in ether, less so in alcohol, and only slightly so in water. With hydrogen it forms *hydrobromic acid*, and, with the bases, compounds called BROMIDES or HYDROBROMATES. It is obtained as follows: A current of chlorine is passed through the uncrystállizable residuum of sea-water, called *bittern*, which then assumes an orange tint, in consequence of bromine being set free from its combinations; sulphuric ether is then agitated with it, and the mixture allowed to stand until the ethereal portion, holding the bromine in solution, floats upon the surface. By decanting, and evaporating the ether, a crude bromine may be obtained at once. To get it pure, the ethereal solution is carefully decanted, and agitated with a solution of potassa, by which means bromide of potassium and bromate of potash are formed. The whole is next evaporated to dryness, and submitted to a dull red heat; the residuum is then powdered, mixed with pure peroxide of manganese, and placed in a retort; sulphuric acid, diluted with half its weight of water, is now poured in. Red vapors immediately arise, and condense into drops of bromine, and are collected by plunging the neck of the retort to the bottom of a small receiver containing cold water. The bromine forms a stratum beneath the water, and may be collected and further purified by distillation from dry chloride of calcium. (*Cooley.*)

4262. Iodide of Cadmuim. This is prepared by mixing iodine and cadmuim filings in a moist state. This is freely soluble in water or alcohol, and may be crystallized by evaporation from ether solution, in large white transparent crystals. (*U. S. Disp.*)

4263. Bromide of Cadmium. This is made from cadmium filings and bromine, in the same manner as the iodide of cadmium from iodine. It consists of long, white, efflorescent, crystalline needles.

4264. Hydriodate of Quinine. To a concentrated solution of neutral sulphate of quinine, add, drop by drop, a concentrated solution of iodide of potassium; dry the precipitate in the shade; or, heat the liquid nearly to the boiling point, and allow it to crystallize.

4265. Sulphate of Quinine. This is the disulphate of quinia. Boil 48 troy ounces coarsely powdered yellow cinchona, in 13 pints of water containing $1\frac{1}{8}$ troy ounces muriatic acid, and strain through muslin. Boil the residue twice successively with the same quantity of water and acid as before, and strain. Mix the decoctions, and, while the liquid is hot, gradually add 5 troy ounces finely powdered lime, previously mixed with